Malaysia Technical Universities Conference on Engineering and Technology (MUCET 2017)

MATEC Web of Conferences Volume 150 (2018)

Penang, Malaysia
6 – 7 December 2017

Part 1 of 2

Editors:

S.A. Aljunid
M.A.A. Mohd Salleh
C.B.M. Rashidi
P.J. Soh
K.N.F. Ku Azir

ISBN: 978-1-5108-5872-5

Printed from e-media with permission by:

Curran Associates, Inc.
57 Morehouse Lane
Red Hook, NY 12571

Some format issues inherent in the e-media version may also appear in this print version.

Printed by Curran Associates, Inc. (2018)

For additional information, please contact EDP Sciences – Web of Conferences
at the address below.

EDP Sciences – Web of Conferences
17, Avenue du Hoggar
Parc d'Activité de Courtabœuf
BP 112
F-91944 Les Ulis Cedex A
France

Phone: +33 (0) 1 69 18 75 75
Fax: +33 (0) 1 69 28 84 91

contact-edps@webofconferences.org

Additional copies of this publication are available from:

Curran Associates, Inc.
57 Morehouse Lane
Red Hook, NY 12571 USA
Phone: 845-758-0400
Fax: 845-758-2633
Email: curran@proceedings.com
Web: www.proceedings.com

TABLE OF CONTENTS

PART 1

PART 2

 https://doi.org/10.1051/matecconf/201815000001

Malaysian Technical Universities Conference on Engineering and Technology (MUCET 2017)
6th – 7th December 2017, Park Royal Hotel, Penang, Malaysia

Preface

MUCET 2017 aims to provide a platform for academics and researchers to share knowledge and ideas, widen networking, present their research findings and explore any other potential benefits that might be contribute to the better future.

This conference highlights the contributions of innovative research in science, engineering and technology. Topics presented includes of Electronic and Electrical Engineering, Civil Engineering, Chemical Engineering & Natural Resources, Mechanical & Manufacturing Engineering, Education, Social Science & Technology Management, Science, and Information & Communication Technology.

Research fields of the presented papers including:

- Electronical and Electronic Engineering (EE)
- Civil Engineering (CE)
- Chemical Engineering & Natural Resources (CHEM)
- Mechanical & Manufacturing Engineering (MME)
- Education, Social Sciences & Technology Management (ESTM)
- Information & Communication Technology (ICT), Science (SCI) & Mathematics (SM)

Malaysian Technical Universities Conference on Engineering and Technology
(MUCET 2017)
6th – 7th December 2017, Park Royal Hotel, Penang, Malaysia

Welcoming Note

CHAIRMAN OF MALAYSIAN TECHNICAL UNIVERSITIES CONFERENCE ON ENGINEERING AND TECHNOLOGY (MUCET 2017)

Assalamualaikum w.b.t and Salam 1Malaysia.

It is great pleasure to welcome all delegates to the Malaysian Technical Universities Conference on Engineering and Technology (MUCET 2017).

Competitive advantage in today's globalized economy lies principally in fostering the skills, knowledge, creativity and competencies of people and these cuts across nations, regions, sectors and organizations. This is very significant as it is explicit recognition that individual human development is fundamental to our collective development as societies. As the success in the 21st century is driven by excellence in engineering and technology innovation, by pushing the boundaries of knowledge and braving through discoveries of new, innovative and improved product and processes, we, as technical universities must play key role in fostering these skills and provides the best effort to be the pillar for these success. While we face significant competition for developing and instilling the engineering and technology skills within our region, other nations are also putting in place measures to boost and strengthen their capacity. Hence we must seize every opportunity to stand at par with other countries, while making sure sustainability and dynamic growth of Malaysia economy.

The theme of this year's MUCET, "Leading Innovation in Engineering, Science and Technology" is indeed in tandem with the government's strategic initiatives at advancing, in a more effective manner on innovative human capital towards economic growth by utilizing every aspects of engineering, science and technology. Innovation, in deed is the key to global competitiveness that be able to provide new and better jobs, a resilient economy and strategic key in gaining the national goals. As a developing nation, Malaysia relies on its human capital and universities have a pivotal role in nurturing and producing this much-needed pool of competent and highly skilled labour force at every levels in our endeavour towards national transformation 2050 (TN50).

I sincerely wish that this conference will provide a useful platform for participants, especially between technical universities academia, to interact, brain storming, and come up with concrete resolutions towards more effective collaboration in the area of science, engineering and technology. It is hoped that from this conference, participants can grab the opportunities to emerge stronger and more dynamic in research to stimulate the development of the new products, processes and so that they will match with the latest global industrial trends.

On behalf of the Research, Management and Innovation Centre, UniMAP, I would like to record my deepest gratitude and thanks to all MUCET 2017 committee, Universiti Malaysia Perlis (UniMAP), other MTUN research and management offices and as well as to all parties, for your dedication in ensuring the success of this conference.

Thank you.

PROFESSOR DR SYED ALWEE ALJUNID SYED JUNID
Chairman
Malaysian Technical Universities Conference on Engineering and Technology (MUCET 2017)

edp sciences

MATEC Web of Conferences

Statement of Peer review

In submitting conference proceedings to MATEC Web of Conferences, I certify to the Publisher that I adhere to the **Policy on Publishing Integrity** of the journal in order to safeguard good scientific practice in publishing.

1. All articles have been subjected to peer review administered by the proceedings editors.
2. Reviews have been conducted by expert referees, who have been requested to provide unbiased and constructive comments aimed, whenever possible, at improving the work.
3. Proceedings editors have taken all reasonable steps to ensure the quality of the materials they publish and their decision to accept or reject a paper for publication has been based only on the merits of the work and the relevance to the journal.

Title, date and place of the conference

MALAYSIA TECHNICAL UNIVERSITIES CONFERENCE ON ENGINEERING AND TECHNOLOGY (MUCET) 2017
6 - 7 DECEMBER 2017
PARK ROYAL HOTEL, PENANG MALAYSIA

Proceedings editor(s):

PROFESSOR DR. SYED ALWEE ALJUNID SYED JUNID

DR. MOHD ARIF ANUAR MOHD SALLEH

DR. MOHD RASHIDI CHE BESON

DR. SOH PING JACK

DR. KU NURUL FAZIRA KU AZIR

Date and editor's signature

30TH DECEMBER 2017

Partial Least Square with Savitzky Golay Derivative in Predicting Blood Hemoglobin Using Near Infrared Spectrum

Mohd Nazrul Effendy Mohd Idrus[1], and *Kim Seng* Chia[1,*]

[1]Faculty of Electrical and Electronic Engineering, Universiti Tun Hussein Onn Malaysia, Parit Raja, Batu Pahat, Malaysia

Abstract. Near infrared spectroscopy (NIRS) is a reliable technique that widely used in medical fields. Partial least square was developed to predict blood hemoglobin concentration using NIRS. The aims of this paper are (i) to develop predictive model for near infrared spectroscopic analysis in blood hemoglobin prediction, (ii) to establish relationship between blood hemoglobin and near infrared spectrum using a predictive model, (iii) to evaluate the predictive accuracy of a predictive model based on root mean squared error (RMSE) and coefficient of determination r_p^2. Partial least square with first order Savitzky Golay (SG) derivative preprocessing (PLS-SGd1) showed the higher performance of predictions with RMSE = 0.7965 and r_p^2 = 0.9206 in K-fold cross validation. Optimum number of latent variable (LV) and frame length (*f*) were 32 and 27 nm, respectively. These findings suggest that the relationship between blood hemoglobin and near infrared spectrum is strong, and the partial least square with first order SG derivative is able to predict the blood hemoglobin using near infrared spectral data.

1. Introduction

Hemoglobin concentration can be used to diagnose anemia [1]. Hemoglobin (Hb) is a protein molecule in red blood cells that contains an iron molecule to carries oxygen from the lungs to the rest of body [2]. Generally, people with hemoglobin level 12.0 g/dL or higher was defined as non-anemia, 10 to 11.9 g/dL was mild anemia, 9.9 to 7.0 g/dL was moderate anemia and lower than 7.0 g/dL was severe anemia [1]. Normally, blood hemoglobin was measured using Cyanmethemoglobin method with some blood drawn from patient to be mixed with reagent chemicals for analysis [3]. However, this method has limitations in term of time consuming, require reagent chemical in analysis and invasive method. In this way, NIRS method is a promising fast response, noninvasive and prominent technique to measure blood hemoglobin.

NIRS is a simple and reliable technique widely used in various field such as medical [4–10], food [11–15], agrochemical [12, 13, 15], and fuel [16]. NIRS measures overtones and combination tones of the fundamental molecular vibrations in especially the asymmetric vibrations and these properties make NIR useful for analyzing in biological system [17]. However, different reviews have demonstrated that fundamental reasons which limit the use of NIRS based on several factors; interference resulting in poor S/N ratio, calibration issues, baseline drift, thermal noise and proper selection of wave- length [18]. These problems of NIRS can degrade accuracy performance of the predictions. Due to these issues and nonlinearity of NIRS spectral data, multivariate calibration modelling method need to be developed for quantitative analysis of the target component in complex samples. Preprocessing, calibration and validations is a common process rely in developing multivariate calibration modelling [19].

Different types of multivariate calibration methods have been applied into NIRS spectral data to extract the relevant part of information for a large dataset to predict concentration from samples [16, 20]. The major concern for these multivariate calibration methods with spectral data in data nonlinearity [20, 21]. An appropriate calibration modelling need to be investigated to give an optimum performance predictions from the spectral sample data. Partial least squares (PLS) was famous predictive modelling used in NIRS due to its advantages of rapidity, simplicity and practicability [21, 22]. With a linear method combination of principal multi linear regression (MLR) and component analysis (PCA), PLS be able to handle data with strong co-linearity and noise, as well as in situations with the number of variables more than the number of samples [22]. PLS model show superior model in number of component in terms of effectiveness compared to multi linear regression (MLR,) principal component regression (PCR) [20]. Moreover, PLS has shown much better performance compared to artificial neural network (ANN) in term of RMSEP [24]. However, conventional PLS model need to use prior preprocessing step result to confront with the change in interferent structure in the test set and

* Corresponding author: kschia@uthm.edu.my

reducing the prediction error [8].

SG preprocessing method has been used successfully to remove unwanted signal from spectral data and overcome most common issues in raw spectral data from NIR [5, 10, 14, 15] . However, little studies have been conducted to investigate the effect of the preprocessing to the predictive accuracy of predictive models [25]. Thus, PLS combined with different SG preprocessing techniques (i.e. smoothing, first and second order derivative) is proposed to be investigated to predict the blood hemoglobin concentration in this study.

2. Material and methods

2.1 Research methodology

Figure 1 describe the general idea of the research methodology in this research. Raw spectral data from near infrared spectroscopy (NIRS) will be preprocessed with SG derivative. After that, PLS multivariate calibration will be used for modelling the spectral data and generate prediction value of blood hemoglobin.

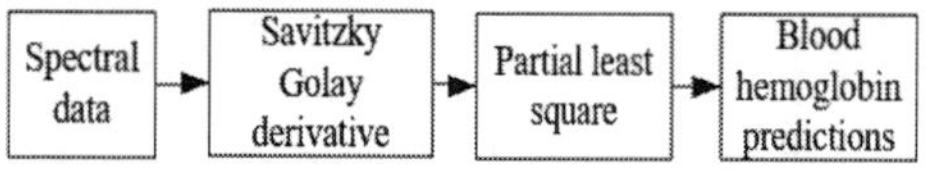

Fig. 1. The flow of blood hemoglobin predictions.

2.2 Spectral data

The spectral dataset were adopted from IDRC ShootOut 2010 that provided by Karl Norris [26]. Blood samples were analyzed during the period from 1990 to 1992 with an NIRSystems 6500 spectrometer with a transmission amplifier mounted in the sample transport. All spectra have 700 variables, from 1100 to 2498 nm, with a 2 nm interval as shown in **Fig. 2**.

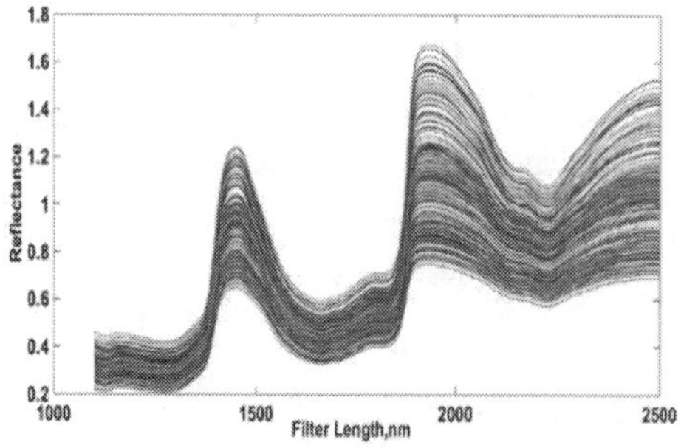

Fig. 2. Blood hemoglobin reflectance raw spectral data.

The data set contain n=231 sets for calibration and n=194 testing sets for blind test used to measure concentration of blood hemoglobin from blood constituent. The blood hemoglobin reference method was measured by Coulter STKS monitor, which is made by the Coulter Corporation of Hialeah, FL [27]. Descriptive statistics of the samples and reference blood hemoglobin showing, number of samples (n), minimum (Min), maximum (Max), mean and standard deviation (Std) as shown in Table 1.

Table 1. Descriptive statistics of the blood hemoglobin

	n	Min	Max	Mean	Std
Calibration	231	10.30	17.30	13.78	1.66
Testing	194	6.50	18.20	12.20	2.83

2.3 Savitzky Golay preprocessing

Preprocessing works to performed data loading, preprocessing zero order SG, first order SG derivative and second order SG derivatives to remove unwanted signal before the spectral data going to modelling process. Sets of 231 sample data $(X = 1{:}231)$ was processed with three level of SG preprocessing method produced smoothing SG (SG0), first order SG derivative (SG1) and second order SG derivative (SG2) data. The coefficients of SG (C0, C1 and C2) were generated by using built in matrix routine function from MATLAB simulation software (MATLAB® Version8.4 (R2014b)). Middle value from desired order derivative can be estimated by dot product of each value of C0, C1 and C2 represented coefficient differentiation filter with spectral data using following equation:

$$Y_j = (C \otimes y)_j = \sum_{i=-\frac{m-1}{2}}^{\frac{m-1}{2}} C_i y_{i+j} \quad (1)$$

Where C_i is set of SG coefficient; y_{i+j} is related set of data before treatment; Y_j is observed value after treatment. The range spectral data was treated between m set data $\frac{m-1}{2}$ and $n - \frac{m-1}{2}$. Where m and n is measured number and total number of frame length.

2.4 Partial least square

Partial least square was carried out using MATLAB simulation software (MATLAB® Version8.4 (R2014b)). General concept idea behind model of PLS modelling is to decompose both the design matrix predictor X and matrix of response Y as following equations:

$$X = TP^T \quad (2)$$

$$Y = UQ^T \quad (3)$$

Where X is an $n \times m$ matrix of predictors, Y is an $n \times p$ matrix of response. T and U is $n \times l$ matrix that are projections of X score and Y score respectively. P and Q are $m \times l$ and $p \times l$ orthogonal loading matrices respectively. The algorithm will yield the PLS regression estimates B_{ij} and B_o after estimating the factor and loading matrices T, U, P and Q for the linear regression as following equations:

$$Y_p = \sum X_{ij} B_{ij} + B_o \quad (4)$$

Where B_{ij} and B_o is PLS regression coefficient. Y_p is predicted value of blood hemoglobin. In this research, the coefficients of PLS regression were generated by using the MATLAB built-in matrix routines function from MATLAB. PLS with SG preprocessing have four stage process start with preprocessing, training, validation and testing process.

2.5 Validations

K-fold cross validation has been used to evaluate performance of PLS model [27, 28]. There are three steps in K-fold cross validation. First, the data set was randomly divided into 5 disjoint folds with approximately equal size. Second, each fold turn to be test the model induced from the other k -1 folds with certain arrangement. After that, root mean square error of cross validation (RMSECV), root mean square error of prediction (RMSEP) and coefficient of determination of prediction r_p^2 can be determined to characterize prediction accuracy capacity of created model. The RMSECV, RMSEP and r_p^2 were calculated as follows:

$$\mathrm{RMSECV} = \sqrt{\frac{\sum_{i=1}^{n}(\hat{y}-y)^2}{n}} \qquad (5)$$

where n is the total number of samples, $\hat{y}$ and y denote the predicted blood Hb and reference Blood Hb from calibration data set, respectively. While root mean squared error of prediction (RMSEP) is used to measures the accuracy of the predictions of the calibration model with new unseen of data set can be computed as

$$\mathrm{RMSEP} = \sqrt{\frac{\sum_{i=1}^{n}(\hat{y}_p-y_p)^2}{n}} \qquad (6)$$

where n is the total number of samples, $\hat{y}$ and y denote the predicted blood Hb and reference Blood Hb from new unseen data set, respectively. The coefficient of determination of prediction r_p^2 used interpreted proportion of the variance in the predicted from reference value output of regression analysis is defined as

$$r_p^2 = 1 - \frac{\sum(y-\hat{y})^2}{\sum(y-\bar{y})^2} \qquad (7)$$

Where $\bar{y}$ is mean of reference data, $\hat{y}$ and y denote the predicted blood Hb and reference Blood Hb from new unseen data set, respectively. SSE and SST denoted as residual sum of squares and total sum of squares respectively.

3. Results and discussion

3.1 Savitzky Golay preprocessing

Fig. 3 shows the output of smoothing SG after raw spectra data from near infrared spectroscopy has been applied smoothing SG preprocessing. The result indicates that the spectra has been smooth and signal to noise ratio (SNR) has been increase without greatly distorting the original signal.

The raw spectra data has been treated with a set of 37nm frame length to produce convolution coefficients relatively. Thus, 38nm frame length from starting point (1100 to 1176nm) and end point (2422 to 2498nm) has been remove because convolution coefficients that has been applied to all data sub-sets to give estimates of the smoothed signal at the central point of each subset for 37nm frame length. It means 76 from 700 variable information of spectra data has been lost during preprocessing. Therefore, the use of number of frame length should be optimized to avoid more elimination information during preprocessing. Output spectra data after first order SG derivative preprocessing (SGd1) as shown at Fig. 4. The result indicate that baseline shift effect has been remove after the SGd1 process. The useful information of original spectra still available for modelling is between range (1306 to 1630 nm, 1824 to 2150 nm and 2232 to 2460).

In spite of that, all information spectral data from 1136 to 2460 nm can be used for modelling process. While Fig. 5 indicate more information from spectral data has been neglected after second order SG derivative preprocessing has been applied.

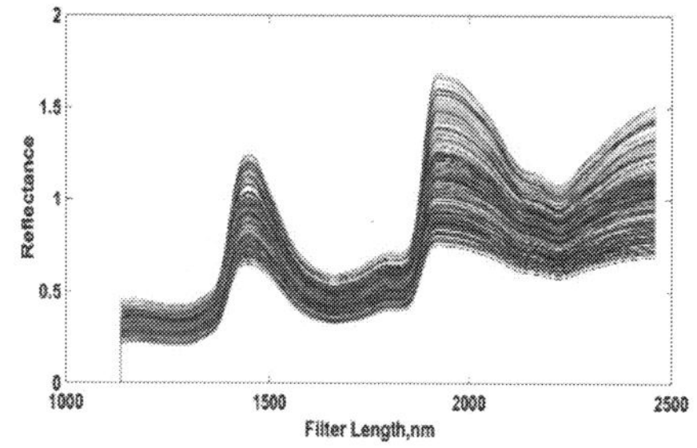

Fig. 3. The preprocessed spectral data after SG smoothing

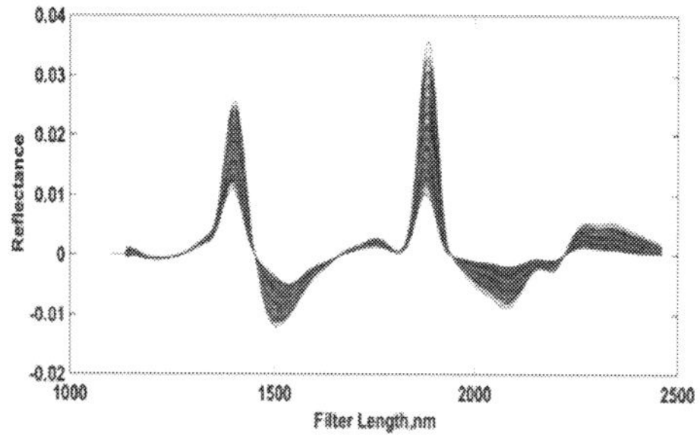

Fig. 4. The preprocessed spectral data after 1st order SG derivative

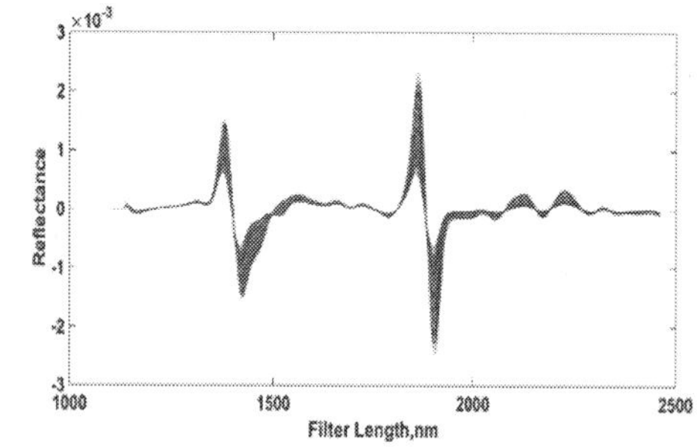

Fig. 5. The preprocessed spectral data after 2nd order SG derivative

3.2 Performance calibration of SG-PLS

Table 2 shows the RMSECV of PLS with smoothing, first order and second order SG derivative pre-processed spectral data with optimal filter and different LVs. As can be seen, best five number RMSECV from each type of preprocessing were presented. It indicated best of 25 models have been developed.

Table 2. RMSECV of PLS with Smoothing, 1st and 2nd Order SG Derivative pre-processed spectral data with optimal filter and different LVs.

Pre-processed method	*f* (nm)	LV	RMSECV (gd/L)
PLS Smoothing Savitzky-Golay (PLS-SGd0)	77	35	0.2194
	95	32	0.223
	97	34	0.2231
	93	35	0.2235
	95	33	0.2238
PLS First Order Savitzky-Golay Derivative (PLS-SGd1)	27	32	0.2178
	21	30	0.2182
	27	33	0.2184
	37	35	0.2185
	67	34	0.2201
PLS Second Order Savitzky-Golay Derivative (PLS-SGd2)	79	31	0.2163
	79	32	0.2185
	97	34	0.2189
	97	32	0.2190
	97	33	0.2195

Minimum value of RMSECV is 0.2163 gd/L follow by 0.2178 gd/L while highest is 0.2238 gd/L. Generally, value of RMSECV of 25 selected models is not much different with 2.18% different. Optimization on the use of an appropriate of frame length after PLS-SGd0 pretreatment resulted in high scale between 93nm to 99nm. This may happen because of the data still have baseline shift and slope effects after smoothing pretreatment process.

RMSECV value for PLS-SGd1 having a minimum number of frame length 37nm and range number of LVs is between 21 and 37. This slight improvement performance might have resulted from removing the slope effect. When higher filter length used to optimize performance of the model, the more original signal will be neglected because new value after preprocessing is a central value from dot product of convolution coefficient with spectral data. Optimal frame lengths were used for PLS-SGd2 are between 79nm to 99nm and LVs are between 31 to 33. It indicated that much information has been lost due the preprocessing process. Different order of SG preprocessing influencing number of frame length and latent variable used to perform optimum RMSECV.

3.3 Performance prediction of SG-PLS

192 new unseen samples from calibration has been used to measured performance prediction of PLS. From the result, indicates that optimization of frame length and number of LVs can improve performance of modelling for example PLS-SGd1 showed the higher performance of predictions (f = 27, LV = 32, r^2_p = 0.9206) followed by PLS-SGd0 (f = 77, LV = 35, r^2_p = 0.9190) as shown in Table 3. This could be due to the minimum number of frame length and lower number of LVs used. The results indicate that PLS-SGd1 have more robustness in testing new unseen test sets samples where the samples were frozen for storage and thawed before used for analysis.

Table 3. The Performance of PLS with different SG pre-processed spectral data with optimal filter and latent variable (LV) using Cross Validation

Pre-processed method	*f* (nm)	LV	RMSECV (gd/L)	RMSEP (gd/L)	r^2_p
PLS-SGd0	77	35	0.2194	0.8046	0.9190
PLS-SGd1	27	32	0.2178	0.7965	0.9206
PLS-SGd2	79	31	0.2163	0.8200	0.9159

4. Conclusions

As a conclusion, partial least square modelling is promising to predict the blood hemoglobin from near infrared spectral data. With the optimal frame length and latent variables, partial least square with first order Savitzky Golay derivative preprocessing (f = 27, LV = 32) showed the highest performance of prediction, i.e. RMSEP = 0.7965 gd/L and r^2_p = 0.9206 in K-fold cross validation. Thus, optimization in frame length and latent variable is crucial to improve the prediction performance. Next, findings also show that the smaller size the frame length, and the lower the number of LVs give a better prediction.

The author would like to acknowledge Research and Innovation Fund provided by the Office for Research, Innovation, Commercialization and Consultancy Management (ORICC), RMC, Universiti Tun Hussein Onn Malaysia (UTHM) for providing financial support, and Faculty of Electrical and Electronic Engineering, UTHM for providing facilities for this study.

5. References

1. M. Chan, "Haemoglobin concentrations for the diagnosis of anaemia and assessment of severity," *Geneva, Switz. World Heal. Organ.*, pp. 1–6, 2011.
2. F. B. Jensen, A. Fago, and R. E. Weber, "Hemoglobin Structure and Function," *Fish Physiol.*, vol. 17, no. March 2016, pp. 1–40, 1998.

3. J. Rosenblit *et al.*, "Evaluation of three methods for hemoglobin measurement in a blood donor setting.," *Sao Paulo Med. J.*, vol. 117, no. 3, pp. 108–112, 1999.
4. P. Venkatesan, C. Dharuman, and and S. Gunasekaran, "squares s regression with A comparative study of principal component regression and partial least square application to FTIR diabetes data," *Indian J. Sci. Technol.*, vol. 4, no. 7, pp. 740–746, 2011.
5. N. N. M. Yatim *et al.*, "Noninvasive glucose level determination using diffuse reflectance near infrared spectroscopy and chemometrics analysis based on in vitro sample and human skin," *Syst. Process Control (ICSPC), 2014 IEEE Conf.*, no. March 2016, pp. 30–35, 2014.
6. S. Ramasahayam, K. S. Haindavi, B. Kavala, and S. R. Chowdhury, "Non invasive estimation of blood glucose using near infra red spectroscopy and double regression analysis," *Proc. Int. Conf. Sens. Technol. ICST*, pp. 627–631, 2013.
7. A. Al-Mbaideen and M. Benaissa, "Determination of glucose concentration from NIR spectra using independent component regression," *Chemom. Intell. Lab. Syst.*, vol. 105, no. 1, pp. 131–135, 2011.
8. M. Goodarzi, S. Sharma, H. Ramon, and W. Saeys, "Multivariate calibration of NIR spectroscopic sensors for continuous glucose monitoring," *TrAC - Trends Anal. Chem.*, vol. 67, pp. 147–158, 2015.
9. J. Hennrich, C. Herff, D. Heger, and T. Schultz, "Investigating deep learning for fNIRS based BCI," *Proc. Annu. Int. Conf. IEEE Eng. Med. Biol. Soc. EMBS*, vol. 2015–Novem, pp. 2844–2847, 2015.
10. C. Wood *et al.*, "Near infra red spectroscopy as a multivariate process analytical tool for predicting pharmaceutical co-crystal concentration," *J. Pharm. Biomed. Anal.*, vol. 129, pp. 172–181, 2016.
11. K. S. Chia, H. A. Rahim, and R. A. Rahim, "Artificial Neural Network Coupled with Robust Principal Components in Near Infrared Spectroscopic Analysis," *IEEE 8th Int. Colloq. Signal Process. its Appl. Artif.*, no. 36, pp. 19–22, 2012.
12. S. Manager and T. Chemometric, "Artificial Neural Networks and Near Infrared Spectroscopy - A case study on protein content in whole wheat grain," no. 1, pp. 1–6, 2013.
13. P. Lin, Y. M. Chen, Y. He, G. W. Hu, X. L. Fu, and C. L. Gu, "Study on Nonlinear Multivariate Methods Combined with the Visible Near-Infrared Spectroscopy (Vis/NIRS) Technique for Detecting the Protein Content of Cheese," *Food Bioprocess Technol.*, vol. 7, no. 12, pp. 3359–3369, 2014.
14. N. C. T. Mariani, G. H. De Almeida Teixeira, K. M. G. De Lima, T. B. Morgenstern, V. Nardini, and L. C. C. Júnior, "NIRS and iSPA-PLS for predicting total anthocyanin content in jaboticaba fruit," *Food Chem.*, vol. 174, pp. 643–648, 2015.
15. T. R. Viegas, A. L. M. L. Mata, M. M. L. Duarte, and K. M. G. Lima, "Determination of quality attributes in wax jambu fruit using NIRS and PLS," *Food Chem.*, vol. 190, pp. 1–4, 2016.
16. H. Abdul Rahim, C. Kim Seng, and R. Abdul Rahim, "Prediction of soluble solid content in pineapple using adaptive linear neuron," *Sensors and Transducers*, vol. 168, no. 4, pp. 243–248, 2014.
17. L. Norgaard, B. Rasmus, and S. B. Engelsen, "Principal Component Analysis and Near Infrared Spectroscopy," *A white Pap. from FOSS*, pp. 2–7, 2014.
18. J. Yadav, A. Rani, V. Singh, and B. M. Murari, "Prospects and limitations of non-invasive blood glucose monitoring using near-infrared spectroscopy," *Biomed. Signal Process. Control*, vol. 18, pp. 214–227, 2015.
19. M. Blanco and I. Villarroya, "NIR spectroscopy: A rapid-response analytical tool," *TrAC - Trends Anal. Chem.*, vol. 21, no. 4, pp. 240–250, 2002.
20. R. M. Balabin, R. Z. Safieva, and E. I. Lomakina, "Comparison of linear and nonlinear calibration models based on near infrared (NIR) spectroscopy data for gasoline properties prediction," *Chemom. Intell. Lab. Syst.*, vol. 88, no. 2, pp. 183–188, 2007.
21. H. Yang, P. R. Griffiths, and J. D. Tate, "Comparison of partial least squares regression and multi-layer neural networks for quantification of nonlinear systems and application to gas phase Fourier transform infrared spectra," *Anal. Chim. Acta*, vol. 489, no. 2, pp. 125–136, 2003.
22. X. Shao, X. Bian, J. Liu, M. Zhang, and W. Cai, "Multivariate calibration methods in near infrared spectroscopic analysis," *Anal. Methods*, vol. 2, no. 11, p. 1662, 2010.
23. S. Wold, M. Sjöström, and L. Eriksson, "PLS-regression: A basic tool of chemometrics," *Chemom. Intell. Lab. Syst.*, vol. 58, no. 2, pp. 109–130, 2001.
24. X. Jintao, Y. Liming, L. Yufei, L. Chunyan, and C. Han, "Noninvasive and fast measurement of blood glucose in vivo by near infrared (NIR) spectroscopy," *Spectrochim. Acta - Part A Mol. Biomol. Spectrosc.*, vol. 179, pp. 250–254, 2017.
25. K. S. Chia, H. Abdul Rahim, and R. Abdul Rahim, "Evaluation of common pre-processing approaches for visible (VIS) and shortwave near infrared (SWNIR) spectroscopy in soluble solids content (SSC) assessment," *Biosyst. Eng.*, vol. 115, no. 1, pp. 82–88, 2013.
26. C. S. Rules, "Chemometrics ShootOut Rules," *Int. Diffus. Reflectance Conf. 2012*, no. June, 2012.

27. J. T. Kuenstner, K. H. Norris, and W. F. McCarthy, "Measurement of hemoglobin in unlysed blood by near-infrared spectroscopy," *Appl. Spectrosc.*, vol. 48, no. 4, pp. 484–488, 1994.
28. G. Jiang and W. Wang, "Error estimation based on variance analysis of k -fold cross-validation," *Pattern Recognit.*, vol. 69, pp. 94–106, 2017.
29. T.-T. Wong, "Performance evaluation of classification algorithms by k-fold and leave-one-out cross validation," *Pattern Recognit.*, vol. 48, no. 9, pp. 2839–2846, 2015.

RetComm 1.0: Real Time Condition Monitoring of Rotating Machinery Failure

Lee Chun Hong[1], Abd Kadir Mahamad[1, 2, *], and Sharifah Saon[1, 2]

[1]Faculty of Electrical and Electronic Engineering, Universiti Tun Hussein Onn Malaysia
[2]Internet of Things Focus Group, Faculty of Electrical and Electronic Engineering, Universiti Tun Hussein Onn Malaysia

Abstract. The breakdown of motor proves to be very expensive as it increases downtime on the machines. Development of cost-effective and reliable condition monitoring system for the protection of motors to avoid unexpected breakdowns is necessary. Therefore, RetComm 1.0 is developed as assistant tool for bearing condition diagnosis system. The smartphone accelerometer is used to collect the vibration signal data and send it to computer by using the Android application named Matlab Mobile. The Matlab software is used to implement a program which is the RetComm 1.0 system to analyse the vibration signal and monitor the condition of the bearing. The algorithm used to observe the condition of bearing is trained by using Artificial Neural Network (ANN). In this project, the ANN is trained by using Matlab software. This proposed method is implemented for early diagnosis purposes. The diagnosis process can be done by just attached the smartphone onto the bearing for data collection. In conclusion, the bearing condition can be identified with this system. The bearing condition are shown in text to let the user know the bearing conditions. The raw data and power spectrum graph plotting are to let the user more further to understand the health condition of the bearing.

1 Introduction

The breakdown of motor proves to be very expensive as it increases downtime on the machines. So it becomes necessary to develop some cost-effective and reliable condition monitoring system for the protection of motors to avoid unexpected breakdowns [1]. Condition monitoring techniques which available in the market are obvious choice for the industries. Early prediction provides engineers information scheduling the maintenance [2].

Based on the previous researchers, the condition monitoring methods of the bearing faults are included temperature, acoustic emission, grindings, vibration and oil film resistance [3-6]. Vibration diagnosis is the most effective method and widely used at this present time.

The rolling element bearing is one of the most important components in the rotating machinery which contain of inner race, outer race, ball and cage. Figure 1 shows the basic structure of the rolling bearing.

The defect location of bearing can be determined by calculating the defect frequencies using these 4 equations [7].

$$BPFI = \frac{N}{2} F \left(1 + \frac{B}{P} \cos\theta \right) \quad (1)$$

$$BPFO = \frac{N}{2} F \left(1 - \frac{B}{P} \cos\theta \right) \quad (2)$$

$$FTF = \frac{F}{2} \left(1 - \frac{B}{P} \cos\theta \right) \quad (3)$$

$$BSF = \frac{P}{2B} F \left(1 - \left(\frac{B}{P} \cos\theta \right)^2 \right) \quad (4)$$

where BPFI, BPFO, FTF and BSF are the inner race defect frequency (Hz), the outer race defect frequency (Hz), the cage defect frequency (Hz), and the ball spin frequency (Hz), respectively. While, N is number of balls bearing, F is shaft frequency (Hz), B is ball diameter (mm), P is pitch diameter (mm) and θ is contact angle.

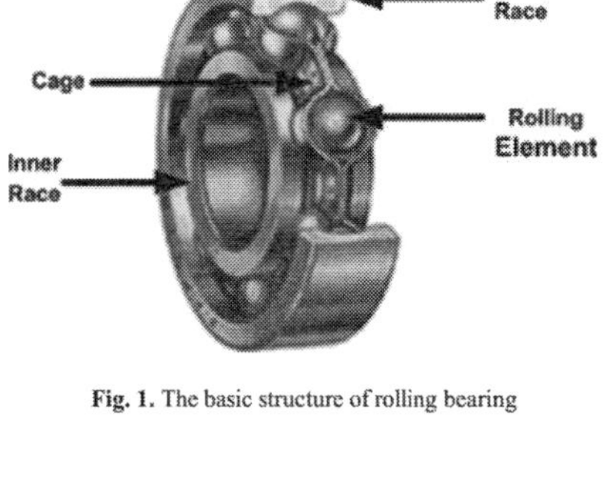

Fig. 1. The basic structure of rolling bearing

* Corresponding author: kadir@uthm.edu.my

2 Methodology

Figure 2 illustrates the block diagram of the overall system and the figure 3 shows the flowchart of system architecture for RetComm 1.0 system. Initially, the system will measured the vibration signal of the machine by using the accelerometer application named Matlab Mobile and the raw data is save in .mat format file. Sugru is a self-setting rubber and almost hold everything. The Sugru is used to hold the smartphone on the machine instead of other holding items because it not absorb the vibration and affect the result. It stickiness is strong enough and easy to take off. The raw data is send to PC from smartphone via Wi-Fi Direct.

Matlab software is used as the back-end of the system and also to analyse vibration signal from the accelerometer. An algorithm used to calculate the condition of bearing is trained by using Artificial Neural Network. A Graphic User Interface (GUI) standalone program is implemented and used as the front-end system for users to review.

2.1 The implementation of backend system

The implementation of back-end system can be divided into 2 phases, the data collection phase and data analysis phase.

For data collection phase, the Matlab Mobile is installed to the smartphone for the measuring of the vibration signal. Matlab Mobile is an all-in-one diagnostic tools that record the status of the sensors inside the smartphone and accelerometer is one of it. The Matlab Mobile can direct send the accelerometer data into computer and save the data in Matlab format. Figure 4 shows the Matlab Mobile that collecting the status of the accelerometer inside the smartphone.

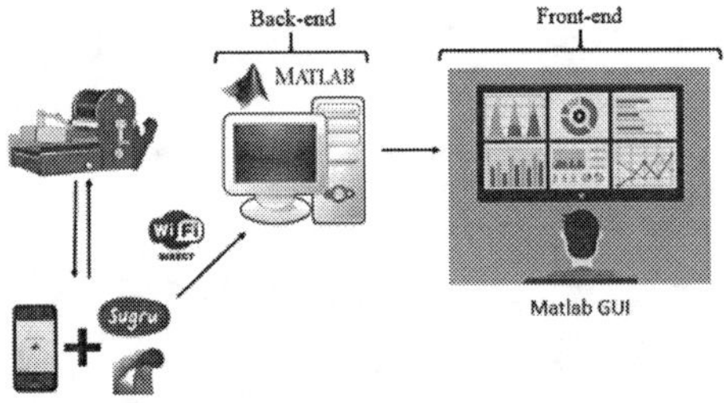

Fig. 2. Block diagram of the system

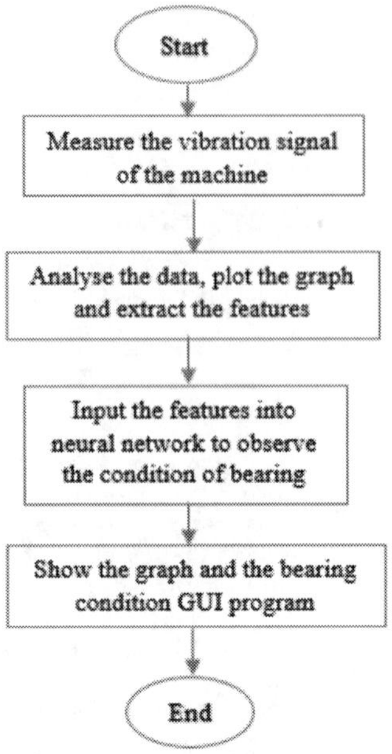

Fig. 3. The flowchart of the system

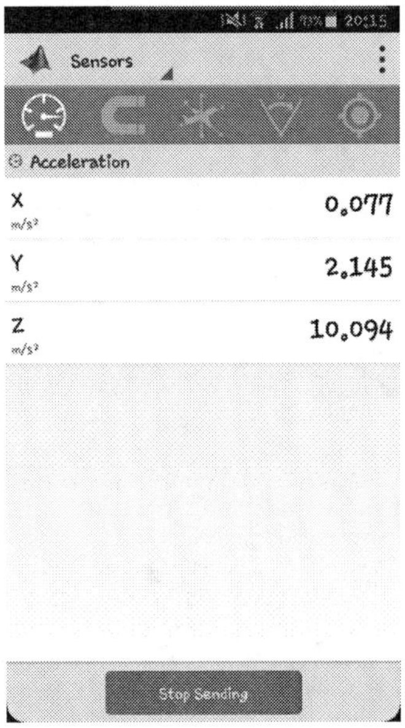

Fig. 4. The Matlab Mobile Apps that measuring the vibration signal

To enable the connection between the Matlab software from computer and Matlab Mobile Apps from smartphone, the Matlab software in computer must have the support packages for Android Sensor.

For data analysis phase, the analysis part is divided into 2 parts, the vibration signal analysis and the bearing condition analysis. After applied vibration signal analysis, the raw data and power spectrum graph of the vibration signal are plotted and the results are generated. The generated results are the eight characteristics of the vibration signal which will be used in bearing condition analysis. These nine characteristics of the vibration signal are:

a) Maximum value
b) Minimum value
c) Kurtosis
d) RMS
e) Median
f) Variance
g) Standard deviation
h) Maximum envelope frequency
i) Maximum power spectrum

In previous studies that have been done by other researchers, the ANN is used to monitor the bearing condition and detect the bearing faults [8-11]. The inputs for the ANN are the extraction features from the data collected. By collecting the features from different conditions of the bearing, the neural network is trained with these features. Therefore, the evaluation result by the ANN is very accurate and trustable. In this project, neural network is used to train an algorithm for bearing condition analysis by input the eight features of the vibration signals.

The condition of bearing that can be monitored by Retcomn 1.0 are divided into four conditions, which are normal condition as class 1, ball defect as class 2, inner race defect as class 3 and outer race defect as class 4. After finished the ANN network training, the best network will be selected as an optimum network, to generate it as standalone Matlab function for the implementation of GUI program.

2.2 The implementation of frontend system

The main role of the front-end of this system is to visualize the data and give to the end user for review. The standalone GUI program is used to visualize the data and act as the front-end for this system. The graphs which visualized are the raw data and the frequency spectrum of the vibration signal. The figure 5 shows the overview of the front-end system.

There are two push buttons in this GUI program which are the analysis button and clear button. The analysis button is to analyse the collected vibration signal data and be used to plot raw data and power spectrum graphs. The raw data graph is plotted into axes1 box meanwhile, the power spectrum graph is plotted into axes2 box. Besides, it also extract the features of the vibration signal data and put into neural_function. The function is loaded the features and check the bearing condition. The condition of the bearing is showed out in the static test box. The neural_function which is trained from the ANN network training is coded under analysis button. The codes for data analysis is also inside analysis button. The clear button is to reset the GUI layout into default layout and ready for next analysis.

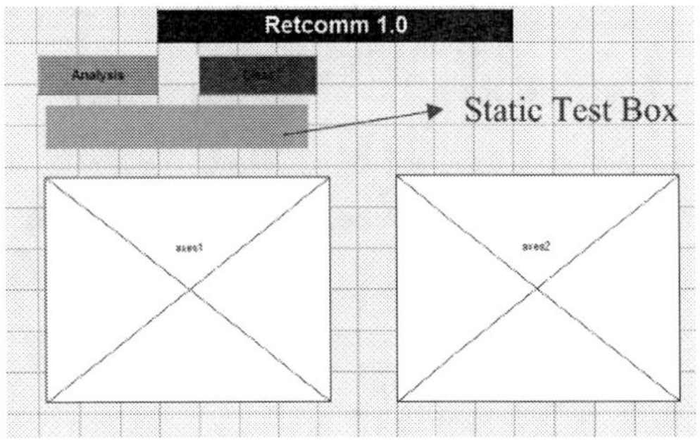

Fig. 5. The overview of the RetComm 1.0 GUI program

3 Results

3.1 Accelerometer data analysis

Figure 6 shows the plotting graph of raw data in Matlab. While, figure 7 shows the plotting graph of power spectrum in Matlab. The vibration signal used is the example of the normal condition of bearing.

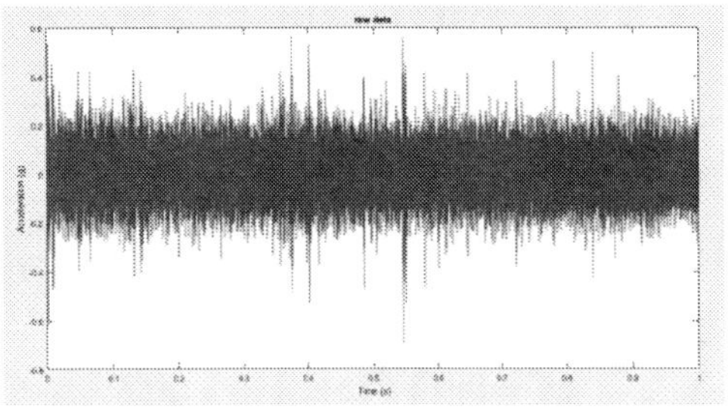

Fig. 6. The raw data graph

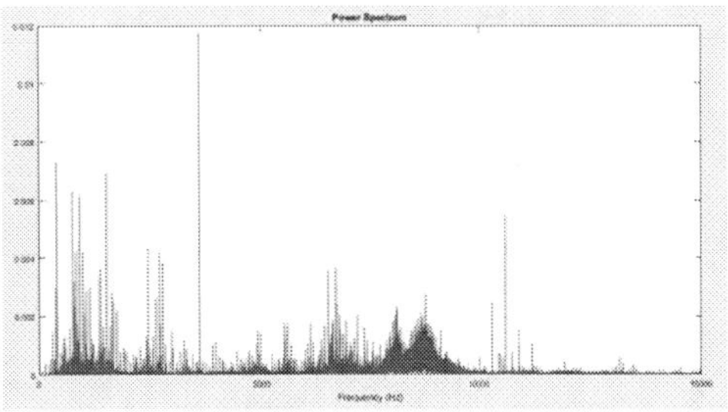

Fig. 7. The power spectrum graph

3.2 ANN network training

Figure 8 shows the combination plot of the ANN network training results. The dashed line represents the perfect result, which are outputs = targets. The solid line represents the best fit linear regression line between outputs and targets. The R-value is an indication of the relationship between the outputs and targets. From the figure 8, the R is equals to 1, which indicates that there is an exact linear relationship between outputs and targets.

3.3 RetComm 1.0 frontend system

Figure 9 shows the RetComm 1.0 GUI program layout. The raw data graph and power spectrum graph are plotted out and the bearing condition are shown up in static test box after pressed the analysis button. The clear button also function very well.

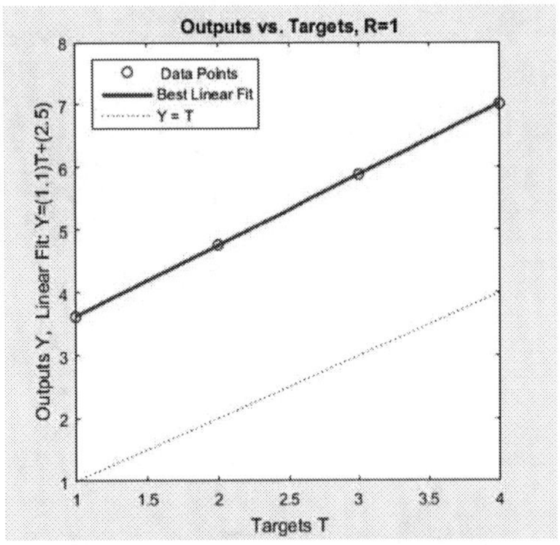

Fig. 8. The combination plot of the ANN network training results

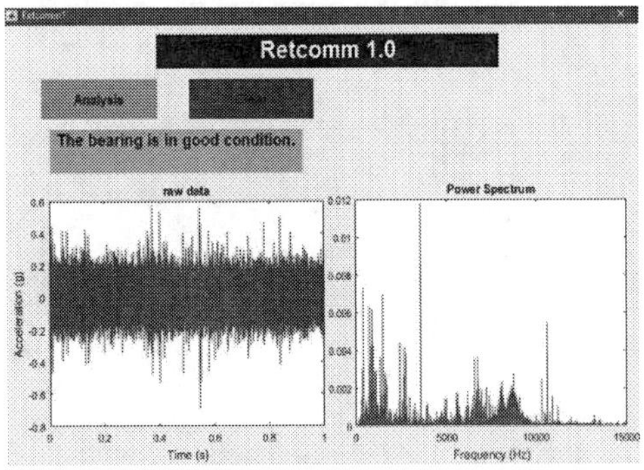

Fig. 9. The RetComm 1.0 GUI program layout

4 Discussion

For the raw data graph, the sampling frequency is 8000 samples/second and the length of time data sampled is 1 second. It mean the raw data is sampled 1-second frame with 8000 samples. The values of raw data are the acceleration of the motor vibration. The power spectrum is calculated by applying Fast Fourier Transform (FFT) into the vibration signal data.

The neural network is trained with nine features of the vibration signal as input and give the result of the class as the bearing condition. The hidden layer of the network used tansig (tangent-sigmoid) transfer function and the output layer used purelin (linear) transfer function. This is to configure the network can approximate to any function. For the ANN network training technique for optimization, the Levenberg Marquardt (trainlm) algorithm is used because it gives the better performance among the other network training algorithms. By train it with different number of hidden nodes from 2 to 20, the least validation error among them implies that it is the best neural network, which are with 4 hidden nodes and 2.10E-2 validation error, as shown in table 1.

Table 1. The number of hidden nodes trained with respective of training error, validation error and testing error

No. of Nodes	Training	Validation	Testing
2	2.15E-01	2.57E-01	3.55E-01
3	7.89E-02	1.19E-01	2.73E-01
4	3.75E-10	2.10E-02	8.50E-02
5	4.80E-03	1.02E-01	3.98E-01
6	1.75E-04	2.30E-02	9.87E-02
7	3.41E-04	2.35E-01	1.25E+00
⋮	⋮	⋮	⋮
19	4.29E-02	9.94E-02	4.61E-01
20	2.05E-02	8.51E-02	2.86E-01

5 Conclusion

Through the completion of RetComm 1.0 system, the objective of this project which is to design RetComm 1.0 system that helps to monitor the condition of bearing has been achieved. In addition, the objective which is to analyze the performance of this system has been achieved as well in this project.

With RetComm 1.0 system, the user can monitor the condition of the bearing by just attach the smartphone onto the machine for a few minutes. It can help to save a lot of time to measure all of the machine in a certain area in the industry which are in a huge number of rotating machinery. By using accelerometer, the affected machine is no need to be shut down in order to carry out the diagnosis process. In this case for further investigation, industries required expert supervisor/technician to predict the condition of that bearing through vibration signal or power spectrum analysis and other diagnostic tools such as acoustic emission and tomography. It may cause the company lost in terms of profit and time. By using RetComm 1.0, it can diagnosis the machine in simple way without spent a lots of time for investigation and not required an expert to decide the machine condition.

For the analysis of this system performance, the bearing condition can be identified after insert the features of the vibration data of the machines. The bearing condition are shown in text to let the user know. The raw data and power spectrum graph plotting are to let the user more further to understand the health condition of the bearing. To use this system, the user just need the install the software application into the computer.

The financial support received from Office for Research, Innovation, Commercialization and Consultancy Management (ORICC), Universiti Tun Hussein Onn Malaysia is gratefully acknowledged.

References

1. M. Irfan, N. Saad, R. Ibrahim, V.S. Asirvadam, An intelligent diagnostic system for the condition monitoring of AC motors, *8th IEEE Conference on Industrial Electronics and Applications (ICIEA)*, pp. 1248-1253 (2013)
2. H. Wasif, A. Aboutalebi, D. Brown, L. Axel-Berg, Condition monitoring system for process industries a business approach, *IEEE Symposium on Industrial* Electronics and Applications (ISIEA), pp. 251-256 (2012)
3. M.A.A. Elmaleeh, N. Saad. Acoustic emission techniques for early detection of bearing faults using LabVIEW, *5th International Symposium on Mechatronics and Its Applications*, pp. 1-5 (2008)
4. Ling Xiang, Lanlan Hou, Feature analysis of interaction on rub-impact and oil-film faults for a rotor-bearing system, *12th International Conference on Ubiquitous Robots and Ambient Intelligence (URAI)*, pp. 246-254 (2015)
5. G.A. Skrimpas et al. Detection of generator bearing inner race creep by means of vibration and temperature analysis, *IEEE 10th International Symposium on Diagnostics for Electrical Machines, Power Electronics and Drives (SDEMPED)*, pp. 303-309 (2015)
6. Wei Feng et al., Modeling and Simulation of Process-Machine Interaction in Grinding of Cemented Carbide Indexable Inserts, *Hindawi Publishing Corporation, Shock and Vibration*, Article ID 508181 (2014)
7. B.P. Graney, K. Starry, Rolling Element Bearing Analysis, *The American Society for Nondestructive Testing Inc., Materials Evaluation*, **70**(1), pp. 78-85 (2012)
8. L.S. Dhamande, M.B. Chaudhari, Detection of Combined Gear-Bearing Fault in Single Stage Spur Gear Box Using Artificial Neural Network, *Procedia Engineering, Elsevier*, **144**, pp. 759-766 (2016)
9. Z.X. Li, J.J. Zhu, X.F. Shen, C. Zhang, J.W. Guo, Fault Diagnosis of Motor Bearing Based on the Bayesian Network, *Procedia Engineering, Elsevier*. **16**, pp. 18-26 (2011)
10. H.S. Kumar, P.P. Srinivasa, N.S. Sriram, G.S. Vijay, ANN-based evaluation of performance of wavelet transform for condition monitoring of rolling element bearing, *Procedia Engineering, Elsevier*, **64**, pp. 805- 814 (2013)
11. A.K Mahamad, T. Hiyama, Diagnosis and Prognosis of Bearing Failure in Rotating Machinery using Acoustic Emission and Artificial Neural Network, *IEEJ Transactions on Industry Applications*, **130**, pp. 443-449 (2009)

E-VPMS 1.0: Enhanced Vehicle Parking Monitoring System

Lim Yeu Shian[1], Sharifah Saon[1, 2, *], and Abd Kadir Mahamad[1, 2]

[1]Faculty of Electrical and Electronic Engineering, Universiti Tun Hussein Onn Malaysia
[2]Internet of Things Focus Group, Faculty of Electrical and Electronic Engineering, Universiti Tun Hussein Onn Malaysia

Abstract. Due to lack of vehicle parking slot in front of Block Q, Faculty of Electrical and Electronic Engineering (FKEE), UTHM, the enhanced vehicle parking monitoring system (E-VPMS 1.0) is developed to identify parking slots available in the parking area as well as to detect double parking problem. Once this system is implemented, time consume for searching available parking slots as well as double parking problem that may cause unwanted accident can be reduced. In the development of E-VPMS 1.0 system, whole system are designed and developed in three different phases; which are video processing phase, login and registration phase, and webview development phase. In video processing phase, video sources are processed through few video processing function. In login and registration phase, series of webpage function have been created that accessible in both smartphone and laptop. While, in webview development phase, webview function have been developed and through this function, content of E-VPMS 1.0 system can be displayed directly via web browser for both smartphone and laptop. Through E-VPMS 1.0, availability parking slots can be checked by registered user through input recorded video before entering the specific parking area. Other than that, performance analysis in terms of numbers of registered user and parking slot availability at different situations such as during peak event or normal event can be managed and controlled by admin through visualisation data method.

1 Introduction

The number of cars on the road is increases day by day, due to country development and economic growth. Therefore parking can be a major problem, especially in urban areas. Therefore, the number of parking slots is increasingly limited and charges are gradually high [1]. The parking monitoring system is part of smart parking system for the smart city which is integrated with the concept of Internet of Things (IoT). The smart city is a potential city that has an important role to form a global economy and society [2].

Fully occupied parking slots are a major problem especially during activities or programs scheduled at Block Q, FKEE. Therefore, double parking issues may increase, as car owners have difficulty finding an alternative parking slot or in a hurry. On the other hands, it's may lead to unwanted crashes or car accident. Using the Internet of Things (IoT) concept, E-VPMS 1.0 has been designed and developed. Through this developed system, the time taken to find the existing parking slot and the scenario for double parking problems can be reduced.

2 Vehicle Parking Monitoring

There are several parking systems that implemented all around the world based on several technique, such as vision-based [3,4], sensor-based [5,6,7] as well as IoT-based [8]. All of this parking system are getting advance through research and development in the industry and implemented using various detection technology or algorithm.

In vision-based, availability of parking space will be detected through surveillance camera, where this surveillance camera will be placed at higher position to ensure wide-angle vision of the whole parking space [3,4]. For sensor-based, availability of parking space will be detected through various type of sensors, such as IR sensor [5], ultrasonic sensor [6] and ZigBee [7], that made up by either wired-sensor circuit or wireless-sensor circuit. For wired-sensor circuit, the sensors are connected together to a centralized controller that stored all information regarding a parking space. For wireless-sensor circuit, it is a type of special ad hoc networks which consists of few wireless nodes that can communicated directly through wireless channel. For IoT-based, availability of parking space will be detected through interconnection between computing device and between machine to machine. Data collected by embedded sensor will be processed and transferred via internet [8]. Vision-based monitoring system is chosen due to the developed system can be integrated directly to the existing CCTV system, thus installation cost can be reduced.

* Corresponding author: sharifa@uthm.edu.my

In E-VPMS 1.0 system, video will be recorded through camera or surveillance camera as the input for vision-based parking system. Large area of parking space can be covered through single camera. In addition, double parking problem can be easily detected through algorithm and solution proposed. Besides that, with Fusionex GIANT, data will be analysed into graphical analysis and the trend can be easily detected.

3 Methodology

To design E-VPMS, there are 3 phases required to achieve the main objective which are Phase 1: video processing phase, Phase 2: login and registration phase and Phase 3: webview development phase.

3.1 Video processing phase

In video processing phase, the recorded video processed with background subtraction method and blob analysis which the process flow as shown in Figure 1. In this phase, Microsoft Visual Studio Community 2015, OpenCV 3.1 and Open Broadcast Software (OBS) are used as the simulation tools. Started with pumping the recorded video to Microsoft Visual Studio. Followed by detected and tracked the vehicle using blob analysis as well as background subtraction method. From the video input, when the vehicle passed through the threshold line which red in colour, the total number of vehicle counted will be increased one by one. Then, the processed video will be recorded and be proceed for live stream via YouTube using Open Broadcast Software.

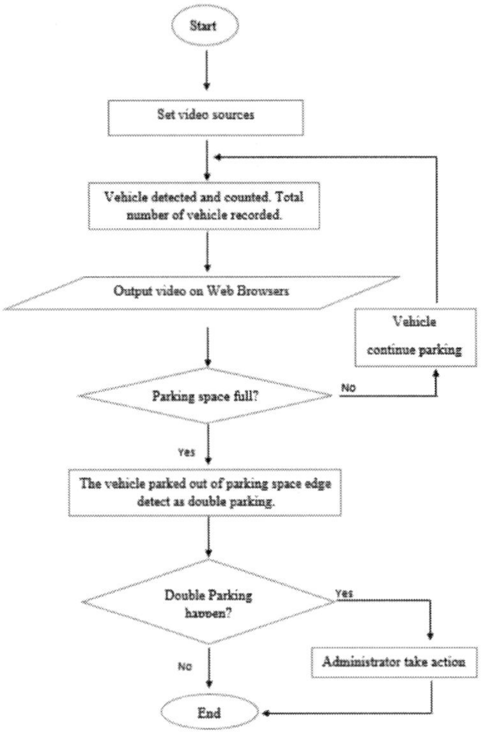

Fig. 1. Video Processing Phase

3.2 Login and registration phase

All the functions in this phase are shown in figure 2 that created using Hypertext Preprocessor (PHP) and HyperText Markup Language (HTML), which are the scripting language for web development.

At first, several information on the register page have been filled upon first registration by users and admin. Once registration successful and at login page, users will be directed into home page after logged in. In home page, button name 'video' and 'logout' can be accessed by users whereas all function page with addvideo, database, graph, and logout page can be accessed by admin.

In addvideo page, users will be directed to the YouTube live stream video and total car counted in the parking area will be shown in the data page and able to view by users. For the purpose of searching an available parking at particular area, users need to enter into the parking slot selection page. In this page, car plate number and parking slot selection will be inserted by users and it was visible to all user. In case of accidentally same parking slot number selected by two different users, then one of them need to change the parking slot number via edit page. This system based on first come first serve basis. The registered user can have the parking slot with no time limit.

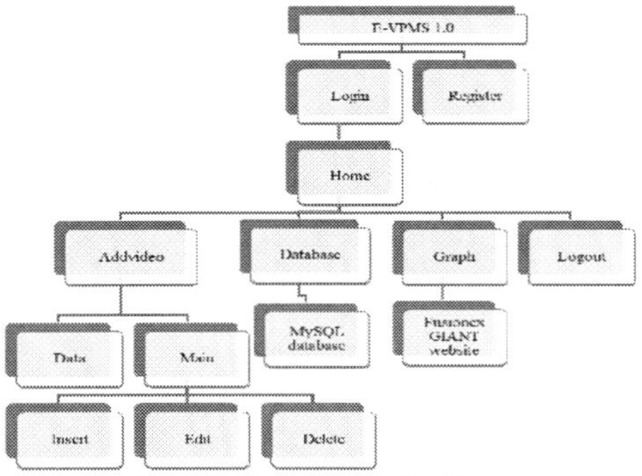

Fig. 2. Login and Registration Phase

The registered number need to be deleted by users when they leave the parking slot, and it can be done at the delete page. If unregistered user park their car at the parking slot selection that have been registered, admin have an authority to request the driver to moved their car to other parking slot which is not yet registered.

As for the admin, they can enter MySQL database page with the fixed username and password, which is all recorded data for the login and registration as well as the parking slot selection were shown. In addition, admin can enter graph page which will be directed to the website of Fusionex GIANT for data visualisation.

To visualise data using Fusionex GIANT, data from admin MySQL database will be exported in CSV format to Fusionex GIANT for further analysis process.

3.3 Webview development phase

In webview development phase, function provided in E-VPMS 1.0 accessible by user and admin through web browser via smartphone or laptop. E-VPMS 1.0 on the smartphone was developed using Android Studio. The android application name E-VPMS 1.0 need to be downloaded by users and admin in order to access all the function.

4 Results and discussion

Figure 3 shows the video insertion for video processing using Microsoft Visual Studio Community 2015. This video sources recorded in front of Block Q, FKEE. When the moving object which is vehicle detected as blob, it will be tracked and bounded with red colour rectangle as shown in figure 4. While, figure 5 shows the output of blob after extracted from the background in sequence frame. White colour box showed the blobs that extracted out from the background, using background subtraction method. Red line as shown in figure 6 act as a threshold in this video, when the blob passed through the threshold, the vehicle will have counted as one and will keep on increasing. The counted number will be display in green at the top right corner of the viewer.

Fig. 3. Video insertion.

Fig. 4. Vehicle detection.

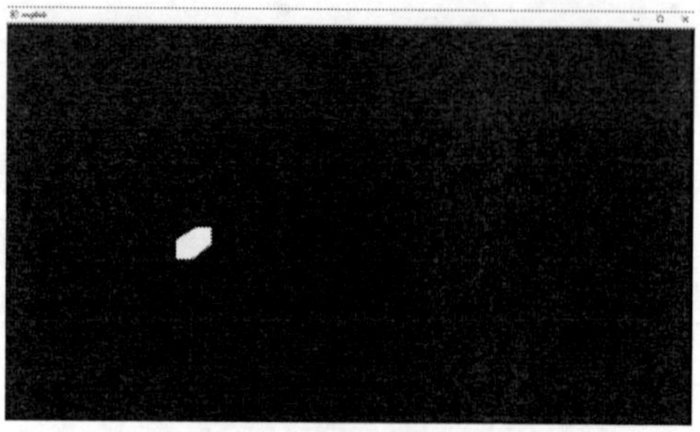

Fig. 5. Blob analysis.

Fig. 6. Vehicle counting.

Figure 7 shows the login page of E-VPMS 1.0 system. To login into E-VPMS 1.0, registered email and password need to be enter by users or admin at the email column and password column respectively. Figure 8 shows the register page of E-VPMS 1.0 system. Before login into E-VPMS 1.0, important information, which are full name, email, password, and car plate number need to be registered by first-timer registration for users or admin at specific column.

MySQL database login which can access only in laptop web browser platform as shown in figure 9. Only admin are allowed to access this page. This page created for admin to export and stored data from MySQL database for system analysis. Figure 10 shows the Fusionex GIANT Application Process Interface (GIANT API) login which can accessed by admin via web browser platform from laptop. While, figure 11 and 12 shows the table of login registrations and parking selection in MySQL database respectively

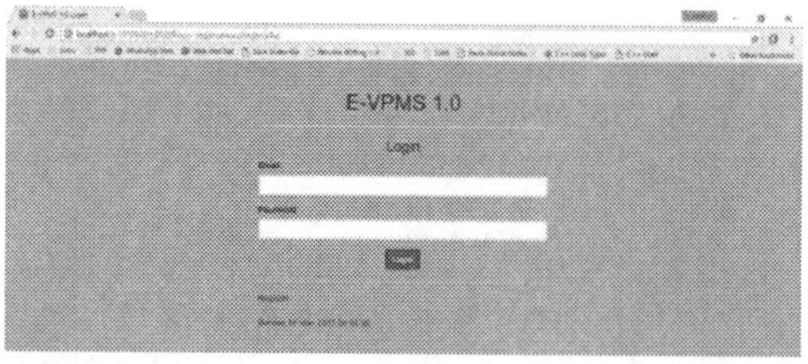

Fig. 7. Login page

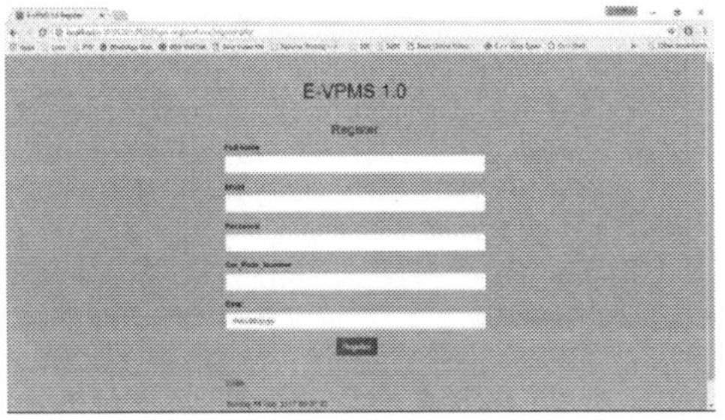

Fig. 8. Registration page

Fig. 9. MySQL database login

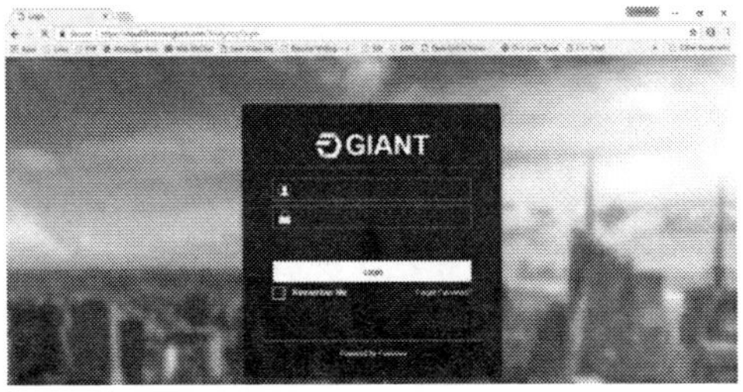

Fig. 10. Fusionex GIANT login

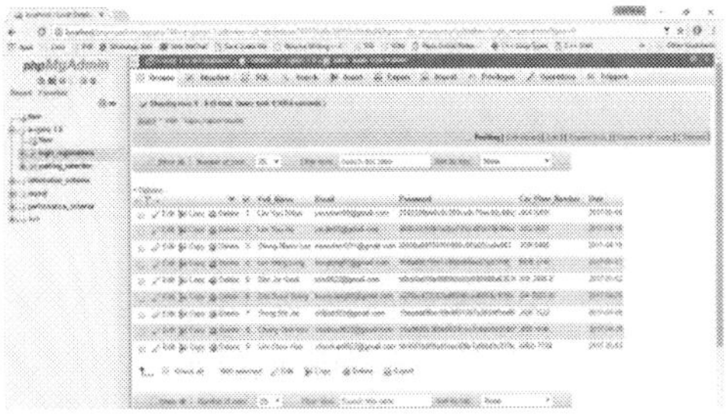

Fig. 11. Table of login registrations in MySQL database

Figure 13 shows graph of id against date of total registered vehicle in specific parking slot generated from the total count of id for each date. Besides that, total registered vehicle can be compared by admin in specific parking slot when the date of peak event or the date of normal daily event.

Fig. 12. Table of parking selection in MySQL database

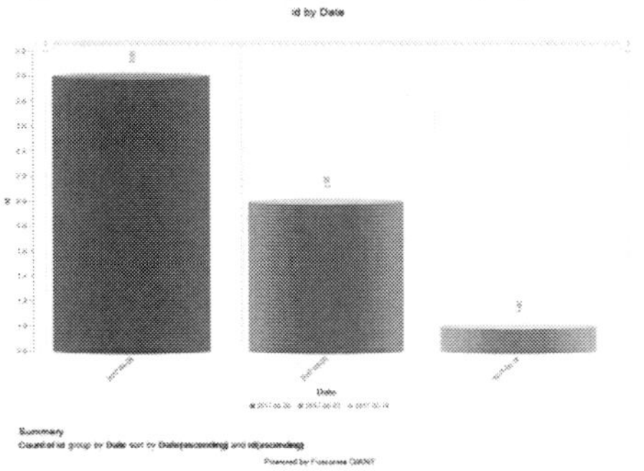

Fig. 13. Total registration vehicle

Figure 14 shows the pie chart of id against date in terms of month for total registered vehicle in specific parking slot generate from the total count of id in terms of month. Data label for orange colour legend represent of 3 registered vehicles in specific parking slot over total of 6 registered vehicles, which can be calculated in percentage to be 50% in April 2017. Data label for white colour legend represent of 3 registered vehicles in specific parking slot over total of 6 registered vehicles, which can be calculated in percentage to be 50% in May 2017.

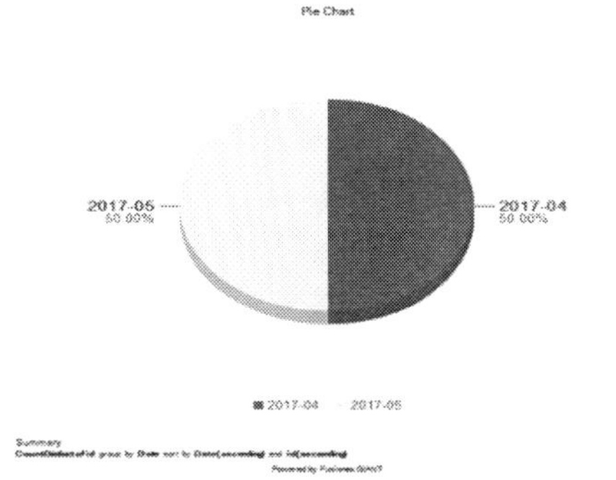

Fig. 14. Total vehicle registration in term of month

5 Conclusion

E-VPMS 1.0 expected able to help in identify parking slots availability. In addition, to analysed performance of the system.

From E-VPMS 1.0, user can identify parking slots availability through the live stream video as well as the table for parking slot selection. This can help user to secure their parking slot as well as reduce time for finding parking before reach the parking area. In other context, admin can manage the parking slots in the parking area as well as control and reduce the double parking problem through the vehicle detection and vehicle counting from both live stream video as well as table for parking slot selection.

For performance analysis of the system, admin can export collected data from MySQL database and visualise it in a graph through Fusionex GIANT tools. This system worked in any duration no matter daily, weekly, monthly or annually, depends on the system setting.

The financial support received from Office for Research, Innovation, Commercialization and Consultancy Management (ORICC), Universiti Tun Hussein Onn Malaysia is gratefully acknowledged.

References

1. G. Yan, W. Yang, D. B. Rawat, S. Olariu, Smart Parking, A secure and intelligent parking system: Observations, *IEEE Intell. Transp. Syst. Mag.*, **3**(1), pp. 18–30 (2011)
2. V. Albino, U. Berardi, R. M. Dangelico, Smart Cities, Definitions, Dimensions, Performance, and Initiatives, *J. Urban Technology*, **22**(1), pp. 3–21 (2015)
3. C. Wah, Parking Space Vacancy Monitoring, Retrievedfrom http://charlotte.ucsd.edu/classes/wi09/cse190-a/reports/cwah.pdf
4. E. Màrmol, X. Sevillano, QuickSpot: a video analytics solution for on-street vacant parking spot detection, *Multimedia Tools Applications*, **75**(24), pp. 17711-17743 (2016)
5. M. A. Al Mamun, S. A. M. M. Rahman, N. U. Ahamed, N. Ahmed, L. A. Hassnawi, Z. B. M. Yusof, Automatic Car Parking and Controlling System Using Programmable Logic Controller (PLC), *International Journal of Applied Engineering Research*, **10**(1), pp. 69-75 (2015)
6. A. Kianpisheh, N. Mustaffa, P. Limtrairut, P. Keikhosrokiani Smart Parking System (SPS) Architecture Using Ultrasonic Detector, *International Journal of Software Engineering and its Applications*, **6**(3), pp. 51-58 (2012)
7. H. A. Sulaiman, M. F. M. Afif, M. A. Othman, M. H. Misran, M. A. M. Said, Wireless based Smart Parking system using Zigbee, *International Journal of Engineering and Technology*, **5**(4), pp. 3282-3300 (2013)
8. R. Grodi, D. B. Rawat, F. Rios-Gutierrez, Smart parking: Parking Occupancy Monitoring and Visualisation System for Smart Cities. *SoutheastCon*, pp. 1–5 (2016)

Power Factor Improvement Using Automatic Power Factor Compensation (APFC) Device for Medical Industries in Malaysia

Maryam Nabihah Zaidi[1,*], *Adlan* Ali[1]

[1]Department of Electrical Engineering Technology, Universiti Teknikal Malaysia Melaka, Melaka, Malaysia

Abstract. This paper present the project designed to correcting power factor for medical industries in Malaysia automatically. Which with hope to make the cost and energy usage efficient, because the energy source are depleting due to increase in population. Power factor is the ratio of real power and apparent power. This definition is mathematically represented as kW/kVA where kW is active power and kVA is apparent power (active + reactive). Reactive power is the non-working power generated by the magnetic and inductive load to generate magnetic flux. The increase in reactive power increase the apparent power so the power factor will decrease. Low pF will cause the industry to meet high demand thus making it less efficient. The main aim of this project is to increasing the current power factor of medical industries from 0.85 to 0.90. Power factor compensation contribute to reduction in current-dependent losses and increase energy efficiency while expanding the reliability of planning for future energy network. As technology develops, the gradual cost and efficiency penalty should reduce. Therefore, automatic power factor compensation device should become cost-effective and smaller device over time. That is the reason this project is using programmable device as it is a miniature architecture device.

1 Introduction

Electrical power has been proven to be one of the most important resource in Malaysia and due to its high demand and widely used, it has become a very expensive resource. The cause of its high cost is mainly because of the generation cost, transmission cost and distribution cost. Plus, the demand of commercial and industrial customers varies greatly throughout the day and for them (the industrial consumers) the maximum demand plays an important role in their overall electricity bills. Therefore, they need to ensure that their max demand is low as possible for their substantial savings. This is the main purpose of this project where it is to overcome the unwanted problem, power factor need to be improved by installing Automatic Power Factor Compensation (by adding capacitor load to offset the inductive load present in the power system). The capacitors can be installed at the service entrance of the plant or on the load side of the metering equipment. These capacitors may supply part, or all of the reactive power required by the plant. There are many benefits by having Automatic Power Factor Compensation device. For the industrial usage, the equipment will have a longer life span and the maintenance costs remain low. According to Sandesh, Singh and Phulambrikar (2012), power factor improvement leads to a huge drop of apparent power drawn from the ac source which in turn protects energy and minimizes the transmission losses.

This project will make the consumers actually see the effect of high and low power system in their consumption of energy. When the power factor is very low, the may be able to see the effect and reach out to make the operation system change as needed. They can increase the power factor by using the automatic power factor compensation which use capacitor and microcontroller as the main components. Thus, this will help in improving the power factor of a system. Plus, the current demand on energy is increasing day by day and the industries growth are inclining. Plus, according to Ministry of Health Malaysia (MoH), medical industries had been tremendously grow for the past decades, 8% to 10% annually. But on the other hand, the energy sources are depleting due to increase in population.

1.1 Power Factor

Ware, J. (2006) described that power factor is the ratio between the useful (true) power (kW) to the total (apparent) power (kVA) consumed by an item of alternating current (A.C) electrical equipment or a complete electrical installation. It is a measure of how efficiently electrical power is converted into useful work output. The ideal power factor is unity, or one. Anything less than one means that extra power is required to achieve the actual task at hand. Power factor is expressed as a percentage (%).

$$Power\ factor = \frac{Real\ power\ (kW)}{Total\ power\ (kVA)} \times 100 \qquad (1)$$

Real power or true power, P is the power that is used to do work on the load. Active power is measured in *watts (W)* and in the electricity billing, it is stately in kilowatt-hours (kWH). It is the power drawn by the electrical

* Corresponding author: zmaryamnabihah@yahoo.com

https://doi.org/10.1051/matecconf/201815001004

resistance of a system doing useful work. Reactive power is the power not used to do work on the load.

Kanchi et al. (2013) have described that generally, power factor is called as the cosine of angle between the voltage and current. In an a.c circuits there is generally a phase difference between voltage and current. If the circuit is inductive, the current is lags behind the voltage and the power factor is called lagging power factor and if the circuit is capacitive then current leads to voltage and power factor is said to be leading power factor.

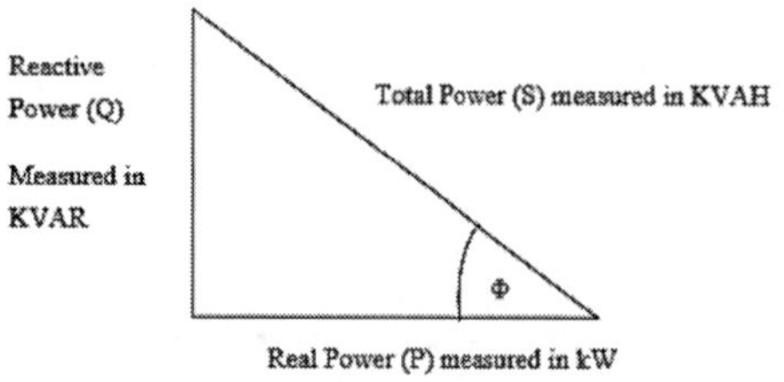

Fig. 1. Power triangle (source: Tagwira, 2014)

The power factor of an alternating-current device or circuit is the ratio of the active power P to the apparent power S. Power factor is expressed as a simple number, or as a percentage. It is given by the equation;

$$Power\ factor = P/S \qquad (2)$$

where; P = active power delivered or absorbed by the circuit or device [W]
S = apparent power of the circuit or device [VA]

The power factor never exceed 100 percent or unity because the value of active power can never be greater than apparent power. Power factor is also a measure of the phase angle between voltage and current. Where this leads to another equation,

$$Power\ factor = \frac{P}{S} = \frac{EI_p}{EI} = I_p/I = \cos\theta$$

Neha Shrivastava (2015) stated that the AC circuits are made of resistive, inductive and capacitive components. For purely resistive load like incandescent lighting, the current and voltage are in phase where the current follows the voltage. For inductive loads case, the current is out of phase with voltage where it lags behind the voltage except for a few purely resistive loads and synchronous. However, in a capacitive load the current leads the voltage thus means the current and voltage are out of phase. The most common capacitive loads are the capacitors used for the correction of power factor of the load.

2 Power Factor Correction

Power factor correction can be applied by the electrical utility to improve the stability and efficiency of the transmission network or, it can be installed by the electrical consumer to reduce costs charged by the electrical provider.

Wildi, T. (2006) mentioned that power factor correction (or improvement) is economically practicable whenever the decrease in the annual cost of electricity exceed the pay back cost of installing the required capacitors. In some cases, the customer has no choice but must obey with the minimum power factor specified by the utility company. The power factor may be amended by installing capacitors at the commercial enterprise or at the service entrance of the factory. In other cases, if the power factor is particularly low, it may be necessary to correct the power factor of the device or machine individually.

Power factor compensation can be defined as a process of correcting the lagging current by producing leading current so that the angle between voltage and current reduces. There will be no power loss if the current and voltage are in phase, thus it can improve the power factor value to nearly unity or even unity. The process is done by connecting few capacitors at the service entrance where the power factor is attuned by an adequate value of capacitance.

The electrical loads that work on alternating current need apparent power which is the product of real power and reactive power. The reason why the electrical load has power factor less than one is, the existence of reactive power makes the apparent power higher than real power. The reactive power increases the current flow from the power sources and the load. When this happens, power losses increases in both distribution and transmission lines. Thus, this results in financial losses for the power provider like Tenaga Nasional Berhad (TNB) which made them require their consumers and customers, especially those who use large loads to maintain their power factor above 0.85 or higher.

Besides the increased in operation costs, reactive power requires the use of higher current capacitance components like wires, switches, transformer and circuit breakers. Power factor can be improved by few methods such as installed capacitor banks in parallel, synchronous condenser and phase advancer.

2.1. Passive power factor compensation

Passive PFC needs lager inductor than active PFC but it still cost less. This is an easy way to correct the nonlinearity in a load by using capacitor bank but it is not as effective as active PFC.

2.2 Active power factor compensation

Active PFC is a power electronic system that modified the wave shape of current drawn by a load to correct the power factor. The reason is to make the power factor corrected appear purely resistive so that the voltage and current are in phase and the reactive power consumption is zero. This allows the most efficient distribution of electrical power from the power company to the consumer. There are few types of active PFC which are Boost, Buck and Buck-boost.

2.3 TNB power factor surcharge

According to Tenaga Nasional Berhad (TNB), for customers taking supply at 33 kV or below, the value of the power factor to be retained is more or equal to 0.85. For customers taking supply at 132 kV or above, the value of the power factor to be retained is more or equal to 0.90. If the consumers cannot oblige the value needed, they will be penalised by the power provider in power factor surcharge. A high Power Factor index (e.g. above 0.85 or 0.90) indicates an efficient level of electricity usage. On the other hand, a low Power Factor index (e.g. under 0.85 or 0.90) shows an inefficient use of electricity, indicating electricity wastage.

Plus, according to Ministry of Energy, Green Technology and Water (KeTTHA) regulates the electricity tariffs charged by the utilities to final consumers in Peninsular Malaysia and Sabah in pursuance of the Electricity Supply Act 1990. Among the principles applied in determining tariff rate is:

- The tariff should reflect the cost supply.
- Provide adequate revenue for the development of the power sector.
- Competitiveness among the industries and services.
- Affordability of the consumers and social economic objectives of the government.

TNB imposes on all industrial and low power consumer to maintain their power factor to no less than 0.85. Failing to maintain the required PF will result on a penalty being imposed on them. The penalty is incorporated into the tariff structure. The consumer should use their best endeavours to obtain highest possible in PF in the operation of any of their electrical installation. Based on TNB, power factor below 0.85 and up to 0.75 $(0.85 > PF \geq 0.75)$ lagging, a supplementary charge 1.5% of the bill for the month for each 0.01 unit below 0.85 and up to 0.75 lagging PF will be added to the bill for that month. On the other hand, power factor below 0.75 lagging, in addition to the charge payable under the above condition, a supplementary charge of 3% of the bill for that month for each 0.01 unit below 0.75 lagging PF will be added to the bill of the month.

Effectively, on 1st June 2011, the electricity tariff was reviewed. It is structured to balance between protecting low income group and sustaining the nation's competitiveness. Prior to that, the last review was carried out in 1st March 2009. The electricity tariff rate are based on the following guideline:

- The amount of energy consumed (kWh)
- The maximum demand at which energy consumed (kW)
- Peak or off peak period
- The power factor of the load
- The connected load

2.4 The uses of capacitors in APFC

An automatic power factor correction unit have a number of capacitors that are switched by means of contactors. These contactors are controlled by a regulator that measures power factor in an electrical network or individual motors. This assures that only the needed capacitor is energized depending on the load and power factor of the network to ensure the power factor remains above a selected value. For this type of application, usually a fixed capacitor is used because it is the simplest and cheapest way of power factor correction, based on Tagwira (2014) research.

A capacitive load bank is like a reactive load bank in rating and purpose, except leading power factor loads are created. These loads simulate certain electronic or non-linear loads typical of telecommunications, computer or UPS industries. An automatic capacitor system should be used for multiple loads in order to reassure a proper value of power factor capacitor is always connected to the system. Therefore, a microcontroller is used to form the automatic power factor compensation.

2.5 Location for APFC

Failing to correct the power factor value will make the source carry excess current and collapsing the magnetized field. This can limit the capacity of components in transmission line. Because the capacitor delivers the magnetizing current the cable from the power source only need to supply the load current. The cable and other components where the capacitor links in, supply load current only to the motor. Therefore, they do not need to be sized to handle the magnetizing current of the motor. Load current has been accentuated because the current flow to the load has only been corrected upstream of the point in the system where the capacitor is link in. There is no change to the current flow downstream of the tie-in point, which is still carrying the magnetizing current. Thus, finding the PF correction location is significant in achieving the required power factor.

2.5.1 PFC at motor

This way is the most advantageous power factor correction location but it is expensive if it is not prepared at early construction. One good thing about this is that, when the motor is turned off, the capacitor that is not required removes from the circuit.

2.5.2 PFC at motor control centre

This can dismiss the equipment upstream of power factor correction location from having to supply the magnetizing current. Which can cause less amperage on the components that carry power to the motor control centre. But, the drawback is the capacitors may deliver too much reactive load when some of the loads are turned off.

 https://doi.org/10.1051/matecconf/201815001004

2.5.3 PFC at utility meter

This method is to simply cut down the need to pay large reactive demand charges (TNB power factor surcharge) where the capacitor bank is installed at the service entrance. This will profit the utility line because they no longer need to deliver magnetizing current and they have more current carrying capacity to be sold to others.

2.5.4 PFC at substation transformer

Placing the power factor compensating device at the substation transformer can help to re-feed the loads from another source. The position of the capacitors need to be on the correct side of the transformer, so it is better to link it on the secondary side. By this way, the equipment upstream of the link in point will be benefited.

3 Flow of the system

The automatic power factor compensation device is to collect and analyse the voltage and current in the electrical system to make a correction in the power factor. Firstly, the two interrupt pins of the microcontroller are provide with time lag between the zero voltage pulse and zero current pulse duly created by suitable operational amplifier circuits in comparator mode. It shows the time lag between the current and voltage on an LCD. The program takes over to actuate suitable number of relays from its output to bring shunt capacitors into the load circuit to get the power factor till it reaches near unity. The microcontroller used in the project belongs to 16PF877A. The construction of the automatic power factor correction circuit are as shown in Figure 2.

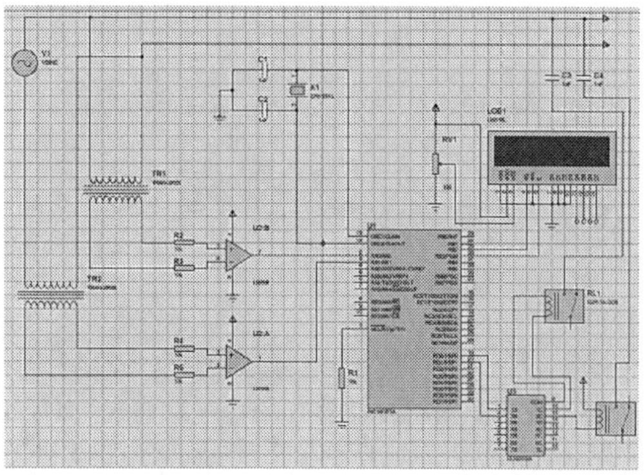

Fig. 2. Automatic Power Factor Correction circuit

This project consist of five main circuit which can be explained in simple block diagram. Those circuits are:

i. Voltage sensing circuit
ii. Current sensing circuit
iii. Data calculation circuit
iv. Display circuit
v. Power factor correcting circuit

3.1 Voltage sensing

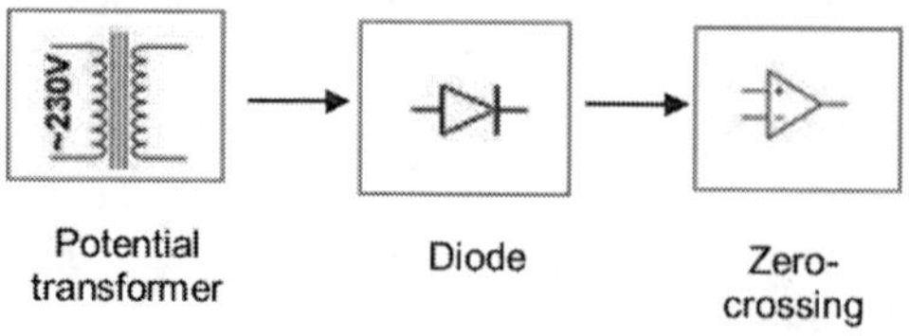

Fig. 3. Block diagram of voltage sensing

The transformer used for this project is a step-down transformer. The potential transformer is fed with alternating voltage with a value of 220V and the value is step down to less than 5V. The output of the transformer is connected to diode. The diode act as rectifier which converted the voltage wave into pulsating D.C. wave to be detected by the microcontroller. The output of the diode is then connected to the zero-crossing detector where zero crossing detection is used to sense sine wave zero crossing from positive half cycle to negative half cycle or vice-versa

3.2 Current sensing

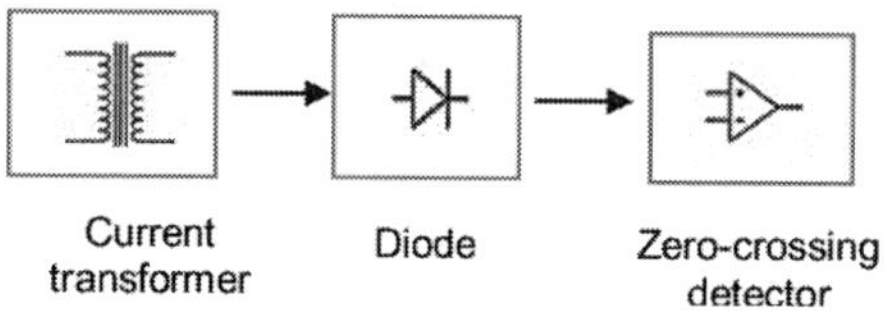

Fig. 4. Block diagram of current sensing

The current sensing circuit works the same as voltage sensing circuit but this circuit detects current. The input signal is connected to the step down current transformer where it converts high current in to low current output for the use of different electrical circuits. The reason current transformer is used because, Op-amp requires current less than 50mA. Output of current transformer is connected to diode. Diode function as a rectifier in the circuit and it is connected to zero-crossing detector. This zero-crossing detector detects sine wave zero crossing.

3.3 Data calculation circuit

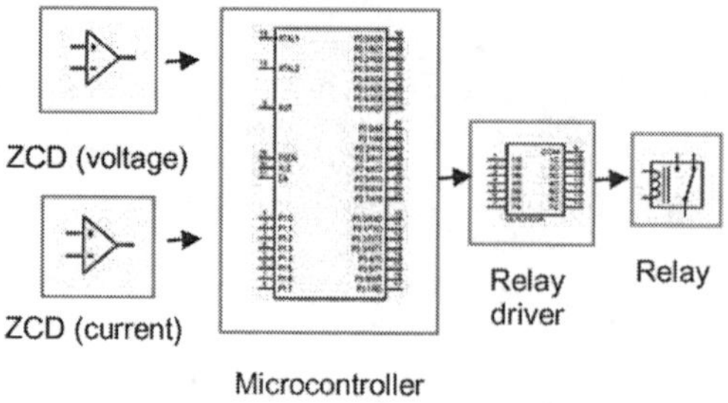

Fig. 5. Block diagram of data calculation process

The data calculation process is the most important circuit on this project. Zero crossing detector converts sinusoidal wave to square wave and it uses op-amp as the comparator. When the sinusoidal signal has value greater than 0 reference, op-amp output has positive value. But, output of op-amp drop to 0 when the sinusoidal signal drops below 0. This way, the zero crossing detector can convert sinusoidal signal to square signal or PWM. The op-amp proposed to be used is the LM385 because it can operate at a very low voltage (3V) and very high voltage (32V). The output from the zero crossing detectors are now fed to the pins of PIC of microcontroller and the data collected are send to the microcontroller. Then, microcontroller uses its capabilities and program to calculate the phase angle and phase difference between two waveforms.

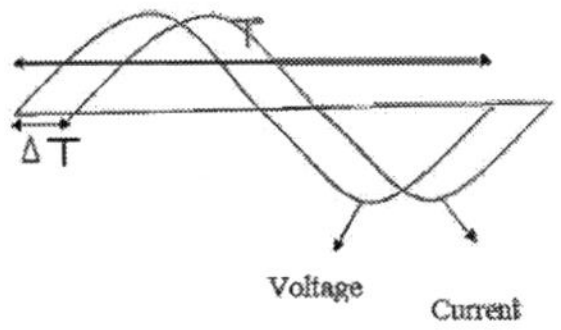

Fig. 6: Voltage and Current wave with time difference

3.4 Result display

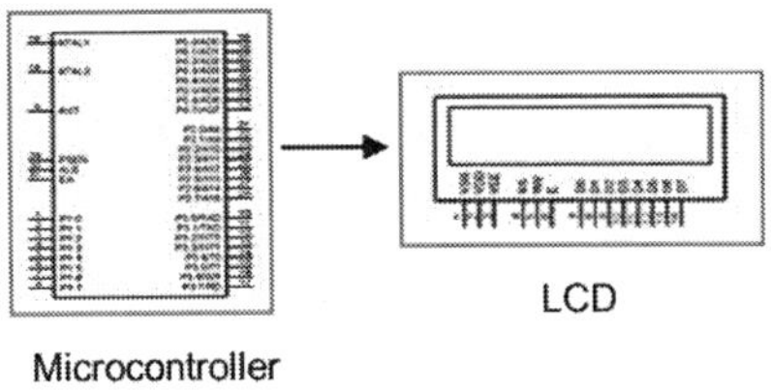

Fig. 7. Block diagram of result display

The pins of microcontroller are connected to the LCD pins. Once the microcontroller program calculates the phase difference between two sinusoid waves and power factor, the information is then displayed on the LCD screen. Liquid Crystal Display screen is very simple module and is used very frequently in various circuits for display purpose. It has a wide range of applications.

3.5 Power factor correcting

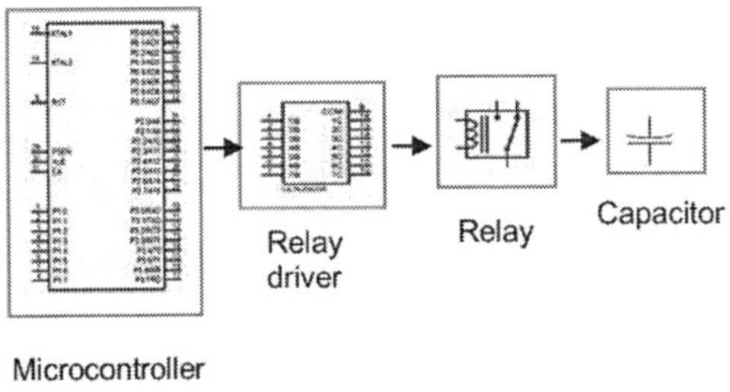

Fig. 8. Block diagram of power factor correcting process

If the power factor is less than the approved value then microcontroller produce command to turn on the relay. Turning on relay will add capacitor into the circuit which can help to improve power factor. Capacitors add reactive load in the circuit which will help to increase power factor. Number of capacitors need to be added depends on the power factor of the circuit. As soon as the power factor drops from a specific value, the relay will add capacitor to the circuit. If power factor's value drops a little from the wanted value then one capacitor is added. In case of power factor falls more than prescribed value then second capacitor is added and the system will keep on working this way. The circuit will continue to add capacitor in parallel to the load until a good value of power factor is attained.

4 Effectiveness and Efficiency

The main purpose of automatic power factor compensation is to help correcting the excess reactive power generate by inductive loads in the industry. With an efficient use of this device, the industry can improves efficiency of the system by reducing losses and reduced apparent power demand charges. Thus, eliminates low power factor penalties levied by TNB. The aim power factor proposed is 0.90 which is convincing as it gives the power factor compensation system a sufficient control capability.

According to European Community, 94% of reactive power demand comes from sectors of industry and services and when they maintained their power factor to 0.95, the sum of active energy and reactive energy falls from 3173TVAh (without power factor compensation) to 2730TVAh. That is 15% drops and leads to reduction in network losses and reduces the loads on the network. For instance;

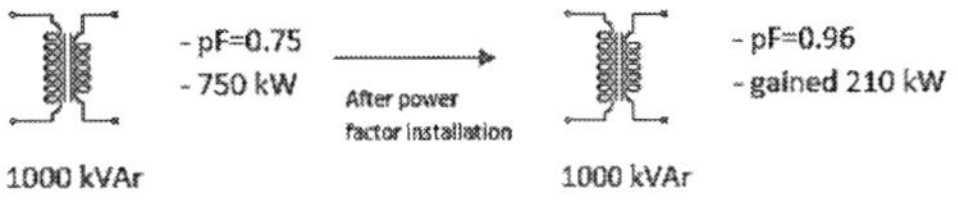

Fig. 9. Effect of installing power factor compensation device on transformer

Based on Fig. 9, after installing the power factor device, the transformer gained 210kW which is 28% from the initial value. This device can limit energy losses in the cable by the Joule effect (limit voltage drop) given the reduction in the current carried in the installation.

According to Ecos, an electronic device with a power factor of 0.40 draws 2.5 times the current compared to an ideal load with a power factor of 1.00. By installing power factor compensation device, it can reduce building wiring losses up to 6.25 times less than the ideal case.

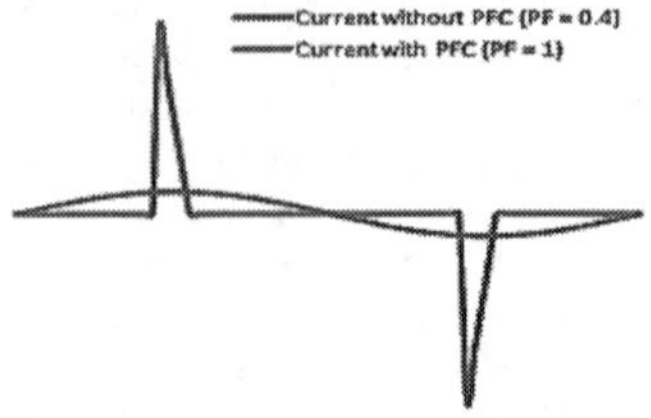

Current draw over one 60 Hz cycle for a 100 W device with and without PFC.

Fig. 10. Effect of installing power factor compensation device on current wave

One of the manufacturer for power factor correction panel in France, Alpes Technologies stated that when the power factor of the industry had been corrected to certain values, it will provide an additional power for the transformer. For example, if the power factor value is corrected to 0.8, it can give up to 7% more power to the transformer. And if the power factor value reached the ideal value, 1.0, the additional power to the transformer is around 33%. But the contrast of the current device is, the size is too big and the cost for installing those device is a bit expensive. While, the automatic power factor compensation that been proposed is a miniature architecture device by using a programmable integrated circuit which can cut the cost and space.

M.N. Zaidi and A.Ali gratefully thanks to Universiti Teknikal Malaysia Melaka (UTeM) for providing support for this paper via special funding Paper ID 1570380395. The authors would also want to express their thank you to all parties involved for their effortless support towards the completion of this paper.

References

1. Babu, I. S., Raju, P.V.R.K.B.A.N., Rao, G. V. S. K., International Journal of Science, Engineering and Technology Research (IJSETR). *Simulation of Active Power Factor Correction Using Boost Type Converter*, 3 (10), pp. 2755-2759, (2014)
2. Kumar, P., Sharma, P. R., Kumar, A., European Scientific Journal. *Simulation & Design Of Power Factor Correction Prototype For Bldc Motor Control*, 9 (12), pp. 141-153, (2013)
3. Mishra, A., Khan, U., Kazi, F., International Journal for Scientific Research and Development. *Power Factor Compensation (Reducing Electricity Bill)*, 3 (12), pp. 845–849, (2016)
4. Nazarkar, S., Shelar, S., International Journal Of Innovations In Engineering Research and Technology. *Design & Simulation Of Active Power Factor,* 3 (3) , pp. 1–8, (2016)
5. Sullivan, K. R. (no date). *Understanding relays*, pp. 1–20.
6. Tade, P. U., Garat, S., Pathilkede, D., Patil, S., Yadav, S., *Automatic Power Factor Correction using PIC*, (2017)
7. Tagwira, M. P., Design Of An Automatic Power Factor Correction System, (2014)
8. R. Saidur, M. Hasanuzzaman, S. Yogeswaran , H.A. Mohammedc, M.S. Hossain., *An End-use Energy Analysis in a Malaysian Public Hospital,* (2010)
9. FRAKO, *Improving Energy Efficiency by Power Factor Correction,* (2008)
10. Ecos, Power Factor Correction: *An Energy Effieciency Perspective,* (2011)

A High Torque Segmented Outer Rotor Permanent Magnet Flux Switching Motor for Motorcycle Propulsion

Enwelum Mbadiwe I [*], *Erwan* Sulaiman, and *Ahmad* Md Zarafi

Research Centre for Applied Electromagnetics, Universiti Tun Hussein Onn Malaysia, 86400 Batu Pahat, Johor Malaysia

Abstract. Electric scooters also known as electric motorcycle are viable and personal means of road transportation have been making their ways into the world markets now because in them, combustion engine with the use of fuel oil for propulsion have been completely eliminated for economic and environmental imperatives. Electric motor which converts electrical energy into mechanical energy is used to overcome the complication of combustion engine. As it is, everyone is opting for combustion engine free and fuel-less type of vehicle. For this reason, manufacturers have exhibited interest, making research on electric motor very attractive. Meanwhile, surface permanent magnet synchronous motor (SPMSM) has been successfully developed having output torque of 110 Nm, the assembly of motor lacked mechanical strength between the rotor yoke and the mounted permanent magnet (PM) which heats up during speed operation, resulting to poor performance. To overcome the challenges laced with SPMSM, this paper presents a novel design of 24 stator 14 pole outer rotor-permanent magnet flux switching motor (SOR-PMFSM) capable of high torque and high performance. It employs an unconventional segmented rotor which has short flux path flow. It also embraces alternate stator tooth windings to reduce material cost. Design specifications and restriction with input DC current are the same with SPMSM. The 2D-FEA by JMAG, version 14 is used to examine the performance of the proposed motor in terms of cogging torque, back-emf, average torque, power and efficiency. Preliminary results showed that torque, power output and efficiency of the proposed motor are 1.9Nm times, 5.8kW times more than SPMSM and efficiency of 84% thus, can sustain acceleration for long distance travel.

1 Introduction

For more than a century now, combustion engine (CE) vehicles which include conventional motorcycles or scooters have used fuel oil to create energy for propulsion for road transportation [1-2]. As it is, cost of maintaining these complex engines with the gloomy fuel crises has been a major concern [3-4]. To reduce fuel consumption in conventional scooters, hybrid electric scooters (HES), also known as hybrid electric scooters have combined combustion engine and electric motor in inner rotor structure, to greatly minimize the intake of oil [5-6].

The four electric motor candidates viable for HES are permanent magnet direct current motor (PMDC), induction motor (IM), switched reluctance motor (SRM) and permanent magnet synchronous motor (PMSM) respectively. Meanwhile, there are different types of HES with various speed operating ranges from 40 to 65 kmph [7]. Obviously, these speed ranges are not suitable for long distance travels.

PMDC was used because of the advantages of simple control which was developed with the tendency of high torque but due to the problems of commutator and regular maintenance of brushes renders it not reliable for driving operation [8-9]. Induction motor has been employed in electric scooter propulsion due to its robustness, low maintenance and ability to operate in unfriendly environments. Unfortunately, it has fundamental limitation which lies on low efficiency at low speed-light load region, low starting torque and speed which cannot be easily changed [10].

Obviously, SRM does not contain permanent magnets, the operation of stators is the same as brushless DC motors. SRM is reliable, rugged and low cost with advantage of low maintenance [11-12]. SRMs have lower torque capability and are generally noisy. They are suitable for low cost, medium to high speed applications where controllability and noise/shaft torque ripple are not critical. The torque is produced by the predisposition of the rotor to adjust with excited stator pole. The path of torque created is an element of the rotor position with respect to the energized phase, and is independent of the current flow direction through the phase winding. Constant torque can be produced by aligning stator pole and phase's excitation with the rotor rotor position [13].

For PMSM, especially IPMSM, has the advantage of high torque density and power density, its design tends is very difficult as permanent magnet (PM) is inserted inside the rotor with assembly lacking mechanical strength thus, degrading the performance of motor for driving system [14-15].

[*] Corresponding author: mb.fkee.uthm@gmail.com

https://doi.org/10.1051/matecconf/201815001005

While each of these electric motors has assisted in reducing fuel oil intake, emission of hazardous gas remains worrisome [16-17]. As a viable solution for complete elimination of the complex combustion engine with fuel oil together, high torque electric motor is crucially necessary due to economic imperatives. Meanwhile, surface mounted permanent magnet synchronous motor (SPMSM) [18-19] has been developed that has completely eliminated CE and the use of fuel oil from HES, thus pure electric scooter vehicle (ESV) in the following scooters: Kingday electric scooters, Aima electric scooters, Mobility electric scooter and Terra motors with operating ranges between 35-60 km/hr and power output of 3kW-5kW respectively. However, the torque capability of the motor operates at low speed range thus, the continuous search for a high torque motors for high performance suitable for long distance travels. Figure 1 illustrates the growth demand of ESVs from 2014 – 2024 [20].

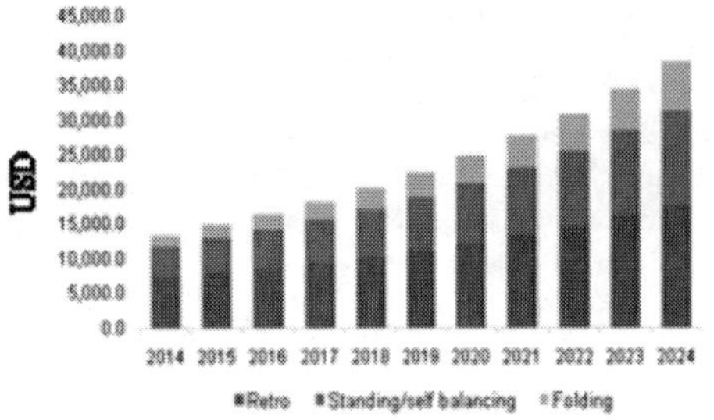

Fig. 1. Demands/brands of electric scooters

For this reason, research for a new motor with high torque capability is of importance. Meanwhile, surface mounted permanent magnet synchronous have been successfully developed with torque output of 110 Nm and speed range of 100km/hr with 6kW [21]. It consisted of 24 stator slots with single tooth armature windings but embracing surface mounted permanent magnet rotor. While SPMSM has exhibited longer speed range of travel, motor's torque of 110 Nm cannot sustain acceleration for long distance travels as presented in figure 2.

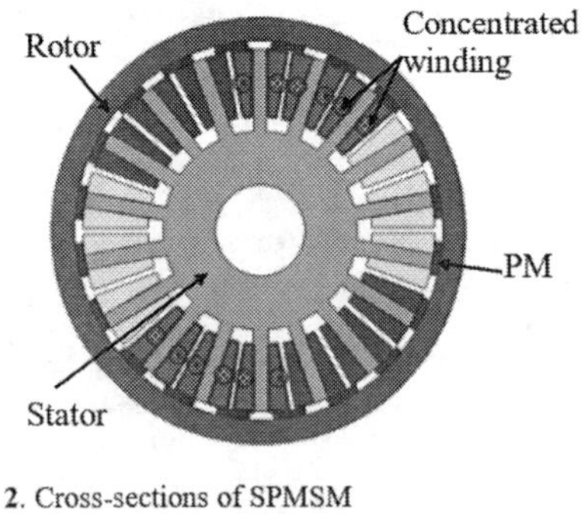

Fig. 2. Cross-sections of SPMSM

Meanwhile, the development of flux switching motor (FSM) [22-23], has turned around motor performance. FSM is a combination of the structures of SRM and induction machine (IM) but locates all active parts on the stationary stator only. It encompasses a new class of brushless motor that does not locate any active part on the rotating rotor, thus immaterial rotor. Obviously, FSM is the electric motor for high torque, high speed, high performance, constant power, simple cooling option, light-weight, high efficiency and cost competiveness.

In the main time, FSM has been applied in an aerospace, domestic appliances and automotive [24-25]. Furthermore, FSM is categorized into three sources of flux excitations namely, permanent magnet FSM, field excitation FSM and hybrid excitation FSM respectively.

To overcome the problems of SPMSM and its limitations, segmented outer rotor permanent magnet flux switching motor (SOR-PMFSM) for high torque electric scooter propulsion is presented.

2 Research method

The design restriction and parameter specifications of the proposed motor for electric scooter propulsion are the same with SPMSM as listed in Table 1. In the same vein, the electrical restrictions related with the inverter are set the same. Furthermore, the limit of the current density is maximum of 30 A/mm2 for the armature coil. The outer radius of motor, stack length, and shaft radius of the target motor are the same with PMSM.

The motor design employed the commercial 2D-FEA package, JMAG-Software version 14.1 released by Japan Research Institute (JNI). At first, the JMAG Geometry Editor to draw the rotor, stator, armature coil and permanent magnet respectively. Flow chart for design process is shown in figure 3.

At the completion of the entire drawing, the motor is uploaded on the JMAG-Designer where materials, conditions, circuits and properties are set. The PM material used is NEOMAX-35AH having the residual flux density and coercive force at 20ºC are 1.2T and 932kA/m, the electrical steel 35H210 is used for rotor and stator parts. Additionally, the external rotor shaft is a non-ferromagnetic material of aluminium employed to secure the isolated segmented rotor members. Coil arrangement test was carried to affirm the operating principle of SOR-PMFSM. Furthermore, flux lines, flux distribution, cogging torque, induced back-emf voltage and finally, torque characteristics and efficiency were all examined.

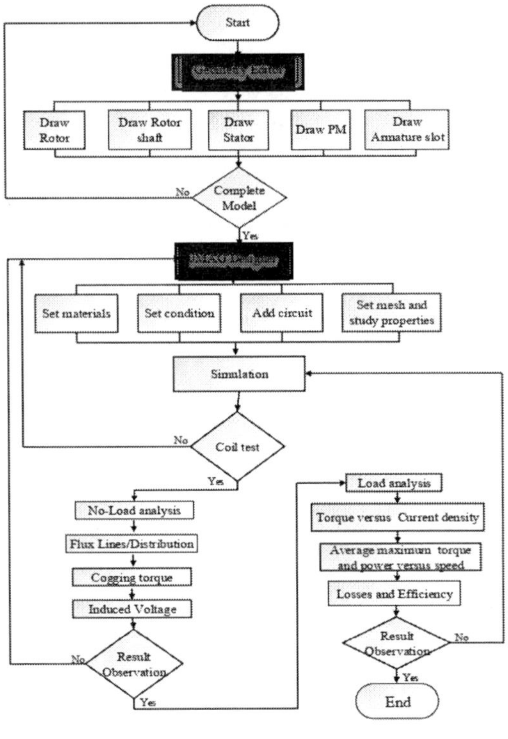

Fig. 3. Flow chart for design process

2.1 Design motor parameters/rotor segment

Each of the motor components such as rotor thus dimensioned as D1-D3, stator, PM D4-D5 and armature coil D6-D7 respectively were sketched in the JMAG Geometry Editor followed the process which was arranged as region mirror and region redial respectively as shown in figure 4. Stator topology with 12 slots limited to 7, 8 and 10 segmented rotor pole and 24 stator slots limited to 10, 14 and 22 segmented rotor pole respectively were investigated for three-phase to embrace two or four set of armature windings for the purpose of selecting the topology with optimal torque capability.

Moreover, the choice of rotor segment is significant to allow an appropriate permanence for the flow of flux. Already, some segmented rotor geometries have been considered as illustrated in figure 5. The shape in (a) shows an earliest circumferential extending pole applied in inner rotor synchronous reluctance machine [26]. Though circumferential extending pole is ideal in terms of performance in inner rotor configuration, it needs notable modifications for in-wheel/outer rotor application. However, to retain these segments for speed operation, design provision is necessary so that non-ferromagnetic material is used to secure them without adding excessive load that will affect motor's performance. The modification led to dovetail structure in part (b) and the transerving design -

Table 1. Design dimension, electric supply and motor performance

Descriptions	SPMSM	SOR-PMFSM
No of phases	3	3
No of rotor poles	50	14
No of stator poles	24	24
Motor radius (mm)	139.7	139.7
Air gap length (mm)	0.5	0.5
PM mass (kg)	2	1
Stack length (mm)	100	100
Stator shaft (mm)	30	30
DC-voltage inver. (V)	415	415
Inverter current (A_{rms})	360	360
Number of conductors	18	18
Slot area (mm^2)	432	432
Avg. max. torque (Nm]	110	>>110
Power output (kW)	6	>>6
Rated speed (rev/min)	1900	1900

part (c) [27]. While the dovetail segment is applied in the FSM as compromise, it lacks the mechanical strength to be used for speed application. In order to overcome it, this proposed segment geometry provides a feature so that the segmented rotors will be properly secured using an external stator. Therefore, achieving this geometry follows the relationship that exists between length of an arc and segment of a circle using the convention given in Eqn. (1). Segment span, θ is calculated as:

$$\theta = \frac{180 \text{ x rotor tooth width radius (mm)}}{\pi \text{ x inner rotor radius (mm)}} \qquad (1)$$

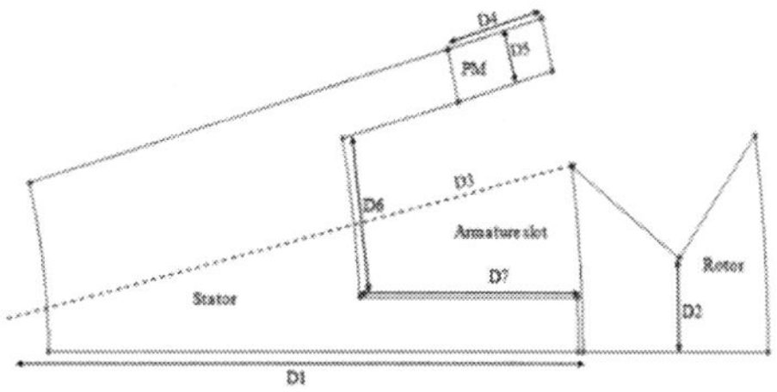

Fig. 4. Design motor parameters

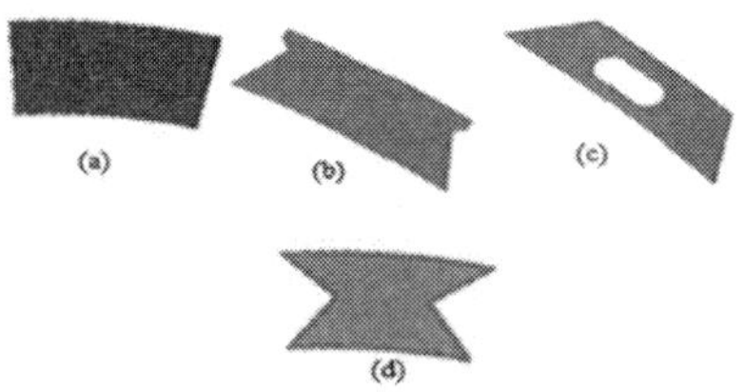

Fig. 5. Segmented rotors under consideration

Furthermore, completed motors are shown in figure 6.

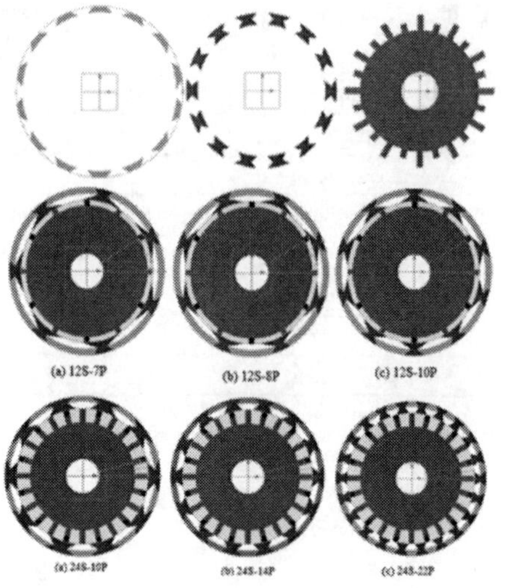

Fig. 6. Completed motors designed in Geometry Editor

3 Results and discussions

This section explained the results of research and at the same time, given the comprehensive discussion. Results are presented in figures, graphs that make the reader to follow through the process. The discussions are made in the following sub-topics.

3.1 Coil test and operating principle

Coil test was conducted on the proposed motor to properly set the correct position of each armature coil phase so as to confirm the operating principle of SOR-PMFSM. To accomplish this, a coil arrangement test examination was conducted on each armature coil separately to witness whether they were wound in the same direction or in opposite direction so that each coil phase would be correctly set in the same way. After this, the zero rotor position was clarified.

By comparing the flux linkages of different coils, the armature coil phases were defined according to conventional three-phase system. Figure 7 defines the U-phase flux linkage for each of the motors. Zero rotor position is the point in the x-axis where the magnitude of U coil phase cuts it at 90° and 270° respectively. This confirms the operating principle.

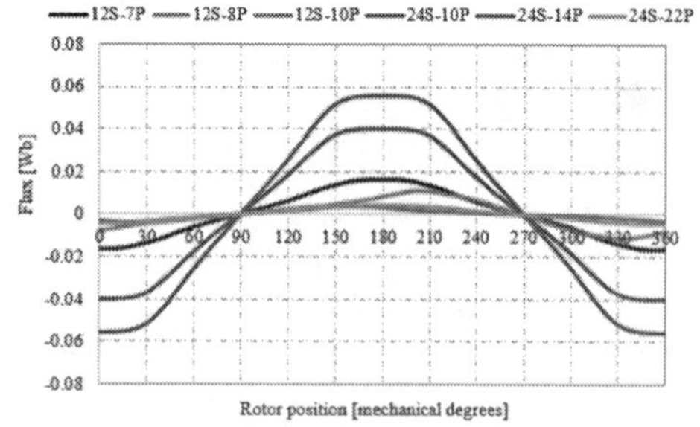

Fig. 7. U-phase of motors at various rotor poles with PM flux

3.2 Cogging torque

Cogging torque is the undesirable component that results to jerking of motor when value is high. For the proposed designs, the cogging torques with the PM flux source in open circuit are shown in figure 8. Taking a cursory look them all, 12S-10P has the highest peak-to-peak cogging torque with 48 Nm and followed by 12S-8P with 24 Nm. Other topologies have low cogging torque but 24S-14P has favorable sinusoidal waveform which is considered more acceptable.

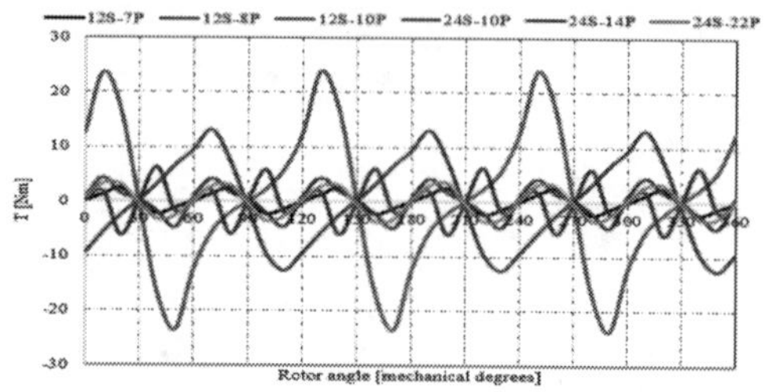

Fig. 8. Cogging of SOR- PMFSM

3.3 Induced back-emf

The motors induced emf presents the voltage generated due to PM only at the speed of 1900 rev/min are depicted in figure 9. From the plot, all the motors have induced back-emf which appeared distorted. Meanwhile, 24S-10P has the highest induced emf of 148V followed by 24S-14P of 110V. While 12S-7P has emf amplitude of 48V, 12S-8P and 24S-22P could not achieve sinusoidal waveform.

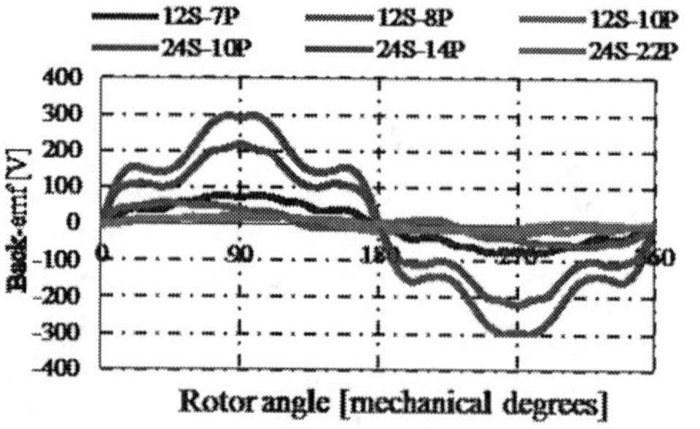

Fig. 9. Comparison of induced back-emf of SOR- PMFSM

3.4 Load analysis

To establish the best configuration, load analysis is conducted to examine torque performances of all the motors. At first, the maximum armature current density of 30 Arms/mm^2 is injected into the circuit and the torque is investigated and analyzed. The results obtained for torque at various armature current density are plotted in figure 10. It is obvious that torque production of 24S-14P is the highest with 209 Nm at the maximum

 https://doi.org/10.1051/matecconf/201815001005

armature current density. Meanwhile comparison of various torques obtained for the six topologies of SOR-PMFSM are provided in Table 2.

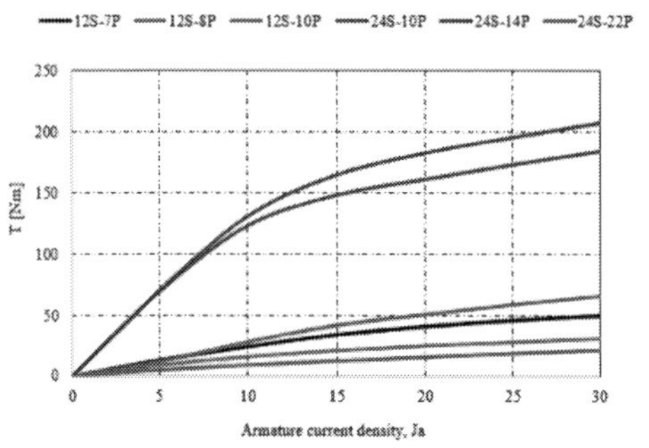

Fig. 10. Torque versus armature current density (A/mm^2)

Table 2. Comparison of torque capability of SOR-PMFSM topologies

S.No.	SOR-PMFSM topologies	Ave. output torque (Nm)
1	12S-7P	50
2	12S-8P	31
3	12S-10P	21
4	24S-10P	184
5	24S-14P	209 (Highest)
6	24S-22P	66

3.5 Torque and power versus speed characteristics

24S-14P topology of SOR-PMFSM was selected to examine the torque and power versus speed characteristics for securing the highest torque among the motors. Figure 11 shows the torque and power versus speed characteristics of the motor in which the power has increased from to almost 30kW and remained constant throughout the speed operation at the base speed of 1374 rev/min.

As shown in the plot, electric motors generate maximum torque at zero speed to pull away from a stop, for it uses a magnetic field to rotate. If applied a set current, a set torque is generated, thus maximum torque at zero speed. Furthermore, the power output is seen to be constant throughout the operating speed region.

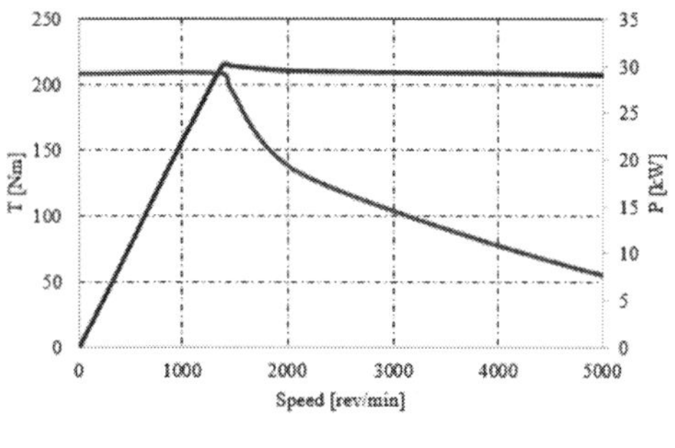

Fig. 11. Torque and power versus speed characteristics.

3.6. Losses analysis and efficiency of 24S-14P

Motor losses such as iron loss in all the laminated cores and copper loss in armature coil and the efficiency of 24S-14P have been conducted and calculated on the basis of 2D-FEA. The specific operating points such as the maximum torque and highest speed at the operating points are chosen for motor's losses and efficiency analysis. The points are outlined in Figure 12. While points 1 and 2 are selected at maximum torque and highest speed, other operating points of 3 to 8 are selected at the speeds of 1000 rev/min, 2000 rev/min and 3000 rev/min respectively.

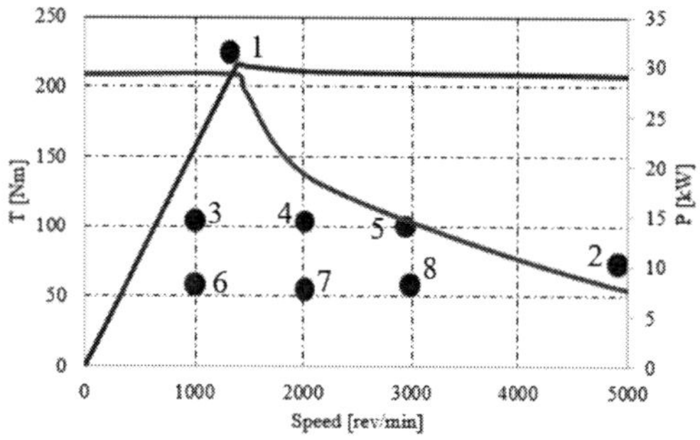

Fig. 12. Specific operating points for loss analysis

Table 3. Detailed loss analysis and motor efficiency of 24S-14P SOR-PMFSM

Point	Pout (W)		Piron (W)	Pcopper (W)	Ptotal (W)	Eff. (%)
1	332328		414.114	16482.66	349224.8	83.16
2	196910		3969.002	457.8516	201336.9	84.35
3	100000		72.53145	2770.735	102843.3	84.42
4	200000		210.1237	2770.735	202980.9	85.18
5	300000		396.203	2770.735	303166.9	85.0
6	5000		54.87562	1354.508	51409.38	84.09
7	10000		158.4833	1354.508	101513	85.28
8	150000		293.2833	1354.508	151647.8	85.31

Average Efficiency- 84.6%

4 Conclusion

This paper has presented the performance characteristics of segmented outer rotor permanent magnet flux switching motor (SOR-PMFSM) for electric scooter capable of high torque. Six topologies were designed for comparison to select the topology with optimum performance. The 2D-FEA was by JMAG version 14, to examine its characteristics. Topology 24S-14P gauged the lowest cogging torque, low induced back emf and highest torque. While the conventional SPMSM achieved maximum torque of 110Nm, the proposed SOR-PMFSM outshines it by securing 209 Nm which is 1.9 times higher. This preliminary result reveals that there are more gains using segmented rotor in terms of manufacturing cost, torque capability, stable power and high performance. This result is expected to improve during design optimization which will be reported soon. Therefore, SOR-PMFSM, serves a better motor for high torque and high performance for electric scooter propulsion.

The authors would like to thank Research, Innovation, Commercialization and Consultancy Office (ORICC) Universiti Tun Hussein Onn Malaysia (UTHM) for sponsoring this work.

References

1. N. Ravi, S. Ekram, D. Mahajam, *International Conference on Power Electronic, Drives and Energy Systems* (ICPEDE), 1-4. (2006)
2. F. Zhan, L. Jin-hee Lee, C. You-young, K. Byung-il, *International Conference on Electrical Machines and Systems (ICEMS),* 949-953. (2013)
3. A. Boglietti, A. Cavagnino, A. Feraris, L.M. Lazzari, *IEEE Industrial Electronics*, **2** (4): 32-37. (2008)
4. W. Fei, P.C.K. Luk, J. Shen, Y. Wang, *International Conference on Power Electronics Systems and Applications, (ICPES).* (2009)
5. Y. Amamra, L. Vido, M. Gabsi, E. Hoang, F. Chabot, *IEEE Vehicle Power Propulsion Conference (IEEE VPP)*, **58**, (5) 1-6. (2006)
6. M. Z. Ahmad, E. Sulaiman, T. Kosaka, *IEEE International Conference on Electrical Machines, (ICEM)*, pp. 691-696. (2015)
7. M. Alexander, *Electric Power Research Institute, (EPRI),* **2,** 152. (2015)
8. K.T Chau, D. Zhang, J. Z. Jiang, C. Liu, Y. Zhang, *IEEE Transactions on Magnetics (TOM)*, **43** (6), 2504-2506. (2007)
9. S.R. Macminn, T.M. Jahns, *IEEE Transactions on Industry Application (TIA)*, **27**, 997-1004. (1991)
10. M. Ehsanim, K.M. Rahman, H.A Toyilat, *IEEE Transactions on Industrial Electronics, (TIE)*, **44**,19-27, (1997)
11. R. E. Araujo, H. Teixeira, J. Barbosa, V. Leite, *Proceedings of the IEEE International Symposium on Industrial Electronics (ISIE)*, 1499-1504, (2005)
12. D.P. Kothari, I. J. Nagrath, *Electric Machines* (Third Edition, Tata, McGraw-Hill Publishing Company, Copyright 2004, 508-509, (2004)
13. N. Bianchi, S. Bolognani, M. Zigliotto, *IEEE Transactions on Industry Applications (AIE)*, **37**, (5), 1348-1355, (2001)
14. C. T. Liu, K. C. Chuang, *Conference on IEEE Transactions on Industry Application (IAS)*, 377-383. (2002)
15. B. Kou, L. Chunyan, S. Cheng, *IEEE Transactions on Plasma Science, (TPS)*, **39** (1), 511-511, (2005)
16. T. Raminosoa, M. Hamiti, M. Galea, C. Gerada, *In Energy Conversion Congress and Exposition (ECCE)*, 298-2804. (2011)
17. S. I. Kim, G. H. Lee. J. P Hoang, T. U. Jung, *IEEE Transaction on Magnetics (AIE)*, **44**, (6), 1590-1593. (2008)
18. R. E. Araujo, H. Teixeira, J. Barbosa, V. Leite, *Proceedings of the IEEE International Symposium on Industrial Electronics (ISIE)*, 1499-1504. (2005)
19. M. Risticevic, D. Iles, A. Moeckel, *in International Symposium on Power Electronics, Electrical Drives, Automation and Motion (ISPEESAM)*, 221–225, (2016)
20. Electric Scooters Market By Product, Available *http://www.grandviewresearch.com/industry-analysis/electric-scooters-market,* (2014)
21. Eclimo-EB25 (n.d) Retrieved from *http://techologicvehicles.com/en/details/920/eclimo-eb25-prix-et-fiche-techinique*, (2016)
22. Evans, A. Stevens, *IEEE Conference on Electrical Machines (ICEM),* (2010)
23. B. M. Song, K. C. Chang, J. Y. Choi, *International Power Electronics Conference (ECCE),* 2763-2742. (2010)
24. A. Affanni, A. Bellini, G. Franceschini, P. Guglielmi, C. Tassoni, *IEEE Transactions on Industry Electronics (TIE)*, **52** (5), 1343–1349, (2005).
25. Fodorean, Daniel, L. Szabo, *In IEEE Power Electronics, Electrical Drives, Automation and Motion, (PEEDAM).* (2012)
26. A. Zulu, B.C. Mecrow, M. Armstrong, *IEEE Trans. Industry Application (AIP)*, **48**, (6), 2259–2267, (2012)
27. M. Galea M, Gerada C, Hamiti T, *In International Conference on Electrical Machines (ICEM).* (2010)

https://doi.org/10.1051/matecconf/201815001006

Cyclist Monitoring System using NI myRIO-1900

Wong Siong Kwong[1], *Aizan* Ubin[1] , *Mohd Zarar* Mohd Jenu[2]

[1]Department of Communication Engineering, Faculty of Electrical and Electronic Engineering, Universiti Tun Hussein Onn Malaysia, 86400 Parit Raja, Johor, Malaysia
[2]Department of Electrical Engineering Technology , Faculty of Engineering Technology, Universiti Tun Hussein Onn Malaysia, Higher Education Hub Pagoh, 84600 Panchor , Johor, Malaysia

Abstract. This paper presents cyclist monitoring system that tracks cyclist path and shows the cyclist training condition. Cyclist monitoring system is using NI myRIO-1900 that integrates the sensors and modules and then monitors the cyclist data using Ubidots web interface. In this system, heart rate sensor, magnetic sensor, GPS module and GSM module are connected to NI myRIO-1900. Heart rate sensor was used to calculate the heart rate of cyclist from time to time. Magnetic sensor was attached to the frame of bicycle to calculate the speed of bicycle. GPS module was used to track location of cyclist. GSM modules was used to send SOS message if there has an accident happened on cyclist. At the first, NI myRIO-1900 initialized the state of the GPS signal. After initializing, NI myRIO-1900 received the signal from magnetic sensor and heart rate sensor. After receiving all of the signals from those sensors, the data was uploaded to the GUI in LabVIEW and the web interface that run by Ubidots. The parameters such as speed, heart rate, distance travel and power were displayed on the GUI and Ubidots for cyclist to monitor their condition of body during cycling. The S.O.S button was installed on the bicycle. If the cyclist is in dangerous situation, cyclist can press this button to send the emergency signal to police or the person that is important to them.

1 Introduction

Cycling is one of five sports that have been in every modern Olympics. It has two different judging elements which are speed and endurance. In 2016, Azizulhasni Awang, a Malaysian cyclist, had created the path for Malaysian in cycling sport while he won the first Olympic bronze medal in Olympics cycling competition.

In fact, the cycling technology is not well developed in Malaysia. They have no complete bicycle monitoring system to train the cyclists. The monitoring system should include cyclists' heart rate, cycling speed, distance travelled, and power consumption during training section in order to enhance cyclists' performance without putting the cyclists in risk of injured [1]. In addition, Global Positioning System (GPS) is another useful system that enables cyclists to send their location when there is an emergency happened.

This paper presents a cyclist monitoring system using NI myRIO embeddes device. It enables the athletes to monitor and record their state and condition of their bicycle training. The data can be transferred to the database of website for monitoring purposes. The integration of GPS module, GSM module, heart rate sensors and magnetic sensors with the wireless platform are essential to monitor the heart rate, speed, distance travelled, and location of the cyclist from embedded device, NI myRIO by using the IoT technology. This monitoring system is useful in planning optimal training intensities for the athlete [2]. The development of this project is very significant because it encourages the development of cycling technology in sport industry of Malaysia.

The main objectives of this project are:

- To develop training monitoring system for cyclist using wireless technology.
- To integrate the sensors and wireless platform by using website.
- To produce comprehensive performance report for cyclist.
- To ensure the safety of the cyclist and send the emergency signal by using wireless platform.

* Corresponding author: aubin@uthm.edu.my

https://doi.org/10.1051/matecconf/201815001006

1.1 Previous Studies

Recently, the cyclist monitoring system was developed in the sports field and it was same in cycling activities, which enable the user to monitor their health, safety and even exercise performance by using Smart phone application.

In 2006, student from Lappeenranta University of Technology generates a cyclist monitoring system that made out of two entities: bicycle sensor module and mobile phone application that enable the exercise and health monitoring. The bicycle sensor module consists of heart rate, pedal and tire sensors that installed onto the bicycle with power sourcing from bicycle dynamo. This collected data which was transmitted to the mobile phone via bluetooth wireless technology. The application in mobile phone enables it to calculate the distance travel, speed, time interval, pulse values and calories burn.

Jose have introduced the control of cycling effort using electric bicycles and mobile device [3]. This system enable the automatic control of pedaling resistance by using smartphone in varies external environment such as oil, water, dust and vibrations from the terrain. The variation of the effort or torque is calculated using exponentially weighted moving average (EWMA) technique. The data was analysed using a smart phone application via bluetooth network connection.

Chris and Frauke proposed Smart e-bike monitoring system (SEMS) which is a safety monitoring system to locate the bicycle position. In this system, a GPS module was used to provide the information of bike's location in each trip. Information such as longitude, latitude, altitude and time can be viewed on the online map. The location of the bicycle was being recorded from time to time is for the safety purposes to avoid bicycle lost and the recorded data can be further monitored for health care purposes.

Antonio and Jose had promoted wireless cycling posture monitoring system that used a sensor module and smartphone via Bluetooth Low Energy (BLE) communication. The sensor module consists of the accelerometers, magnetometers and gyroscopes that enable posture calculation. The sensors module was attached to the cyclist and it provides the trunk and knee angles of the cyclist that transfer to mobile phone via Wi-Fi to calculate the posture in pitch, roll and yaw.

Raluca, Paul and Marimuthu present a wireless monitoring system and performance analysis for professional cycling [1]. In this system wireless motion sensor nodes is placed onto the body of the cyclist and provide the cyclist pedaling movement from time to time. It had been used to improve performance and decreases the injured risk of the cyclist via this monitoring system. When cycling, the characteristics of the cyclist is emphasized in cleat, saddle height and tilt, stem height and length and handlebar height and this data was collected and used to monitor the knee and ankle joints movement to avoid injury happened.

Yoshiaki, Kodai and Hirosyuki generate a road surface condition monitoring system for ensuring the safety cyclist. This system used the ultrasonic distance sensor by emitting and receiving ultrasonic signal that enable the detection of obstacles on the road surface. This monitoring system is divided into two parts which are the data collection using sensor and microcontroller Arduino, and a bluetooth module chip and the application in mobile phone. The transmission and reception of the data take place between the data collecting enable the mobile phone to store and process the data from the sensors and alert the cyclist when an obstacle is detected.

To conclude, the previous project had been made focusing on the wireless data transmission from sensing module to a mobile phone either using bluetooth or Wi-Fi. This data enable the monitoring process in the prospects of health and safety of the cyclist. However there are challenges during the implementation of the system in actual circumference in the environment and this will be discussed in next section.

1.1.1 Existing Challenges

The existing monitoring system is facing problems in the data transmitting itself, power lasting and external environmental conditions.

The communication between the sensors module and the processing module are always in wireless bluetooth and Wi-Fi internet. The signal transmission may lost due to the targeted object are moving. The signal is a in the form of frequency, during the transmission process, the signal may interrupt or degraded by the external factor in the environment like noise (similar frequency profile) signal and long distance data transmission and this situation is considered as signal lost. This problem will bring dangerous to the cyclist when dealing with the safety monitoring [4].

An addition problem is the energy used to power up the sensor module in the bicycle. These sensors module are needed an external power source to operate except the Bluetooth low energy module or using dynamo power generation [3]. This problem occurs because of the sensor unit is not an active unit that can power up themselves. It is similar to the smart phone. The inconsistent of data collection in the monitoring system made this problem to become worst. It is because the sensor need more power to work and generate signals.

Other than that, the external factor especially the weather conditions affect the monitoring system due to the external force applied to the sensor [5]. This problem become significant when dealing with motion detecting sensing. The weather condition like wind and speed direction will either oppose or impose direction which will degrade the quality and reliability of the data collection. It is because this weather condition is able to deteriorate the data recorded due to the present of external force. This problem made the system not giving a precise and correct reading.

In short, the existing bicyclist monitoring system still face with some unsolved problems and it can be improved further in one day. However, the sensors used are also important to ensure the reliability and quality of the data collected.

2 Methods

Cyclist monitoring system involve the design of hardware and software. It integrates the sensors and wireless platform in order to provide acquire specified data from cyclist. The embedded device National Instrument (NI) myRIO was applied to integrate the sensors and to transmit the specified data to the web application through wireless platform. The performances of cyclist is recorded through Internet of things (IoT) where the meaningful data such as heart beat, cyclist speed, training path and location can be viewed through IoT platform. The chosen IoT platform is Ubidots. The diagram of the overall system is shown in Figure 1.

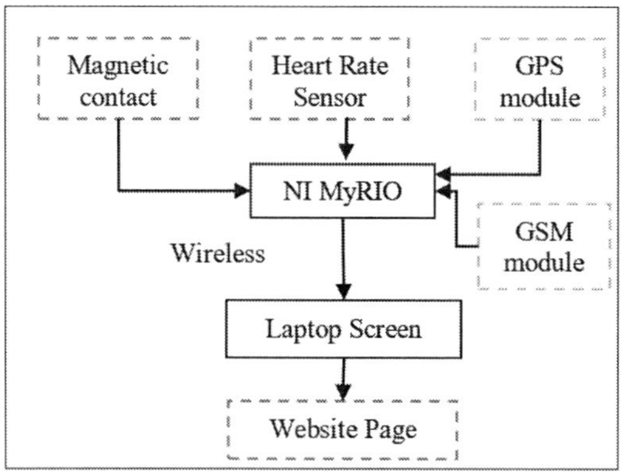

Fig. 1. Overall System Design and Flow

The main parts of the project are heart beat monitoring, speed monitoring, GPS navigation module, GSM module and IoT user interface through Ubidots.

In heart beat monitoring system, an analog heart rate sensor was applied to acquire data. Two heart rate sensors are installed at both of the hand grip of bicycle.

The system enables the speed detection by using magnetic contact switch. This sensor marked every cycle of the wheel rotates. The magnetic contact switch or reed switch act as speedometer of the bicycle to measure the speed of bicycle.

In addition, there was a GPS module use for navigation purposes. It enable the coach to track training path and current location of the cyclist. GPS module was connected to the microcontroller using UART connection. The connection is in a serial communication.

GSM module was attached to the microcontroller for allowing cyclist to send emergency signal to required person.

After the sensors finish data collection, the logging process is started in order to send the data to the website through internet in a specified period. This enable the data are displayed on the computer.

In short, this system enables the users to record down their performance and plan for their training in order to enhance and improve their performance through practices. In this project, the project development is divided into three phases. The phases are circuit design, hardware setup and programming design.

2.1. Circuit Design

The circuit design for this system was drawn using Proteus software. The main controller, NI myRIO acts as a brain to the sensors like GPS module, Analog Pulse Sensor, Accelerometer, and Magnetic Contact Sensor. Figure 2 shows the schematic circuit of the system.

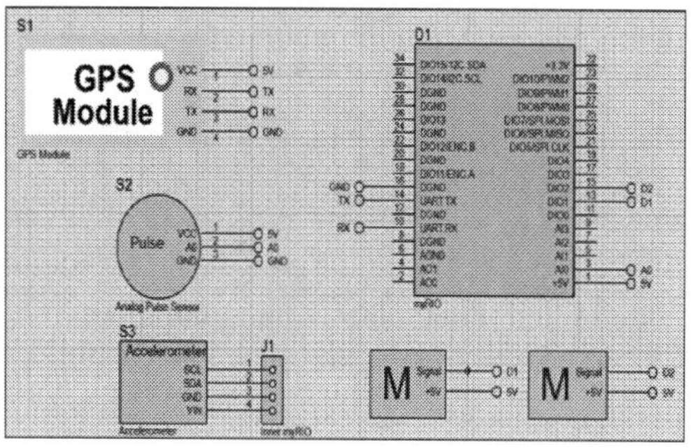

Fig. 2. Schematic of Cyclist Monitoring System Design

It shows the circuit design of the sensors with MyRIO. The transmit (TX) and Receive (RX) pins from GPS Module are connected to the TX (pin 14) and RX (pin 10) of MyRIO respectively. The A0 pin from pulse sensor or heart rate sensor is connected to the pin 3 of MyRIO for detecting the changes in analog changes of pulse sensor. The signal pin of magnetic sensor is connected to the D1 (digital input) of MyRIO. SOS button is connected to D0 (digital input) of MyRIO to send SMS alert to a certain number by triggering the GSM module.

2.2 Hardware Setup

Once the hardware are installed into a bicycle, the system was ready to record and analysis the data from time to time. All the hardware components like sensors

https://doi.org/10.1051/matecconf/201815001006

and microcontroller were ready to be configured and drivers are installed. The sensors were connected to expansion port (MXP) connectors of MyRIO. The driver were installed by using compact dish that comes together with the LabVIEW. Figure 3 shows the hardware layout installed in the bicycle.

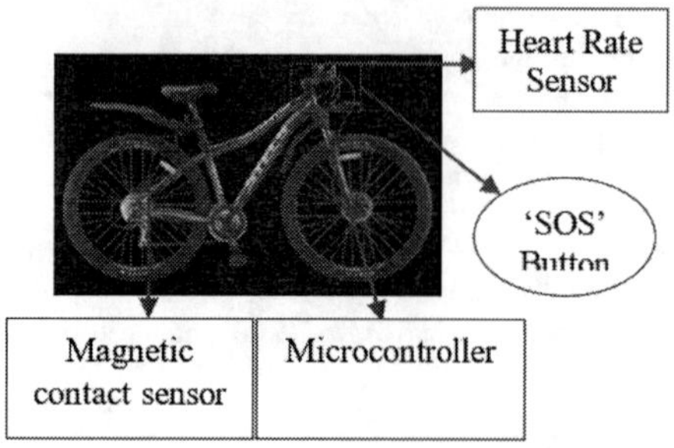

Fig. 3. Hardware layout of Cyclist Monitoring System

2.3 Programming Flow Chart and Design

The proposed system was designed and constructed using LabVIEW graphical Programming Language which is feasible and user-friendly.

LabVIEW is an integrated development environment designed specifically for engineers and scientists. Native to LabVIEW is a graphical programming language (G) that uses a dataflow model instead of sequential lines of text code, empowering you to write functional code using a visual layout that resembles your thought process. This means you spend less time worrying about semicolons and syntax and more time solving the problems that matter. The system programming was constructed based on two different parts which are data collecting and data display. The basic programming flowchart is shown in Figure 4.

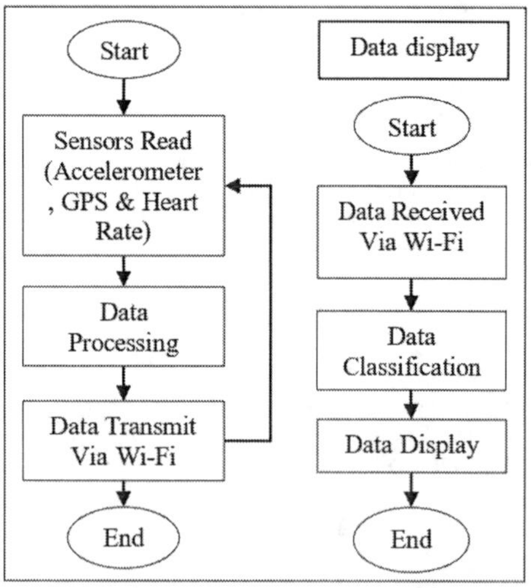

Fig. 4. Programming Flowchart of Cyclist Monitoring System

3 Results

Cyclist monitoring system was installed in a bicycle and tested according to a standard procedure. During this process the data recorded was ready and the results is shown on the laptop screen.

This system was fully connected and all the data transmission was sent using WiFi access. The graph was plotted according to the data collected, which was shown in Figure 5. The data collected was based on the sensors range and with its SI units.

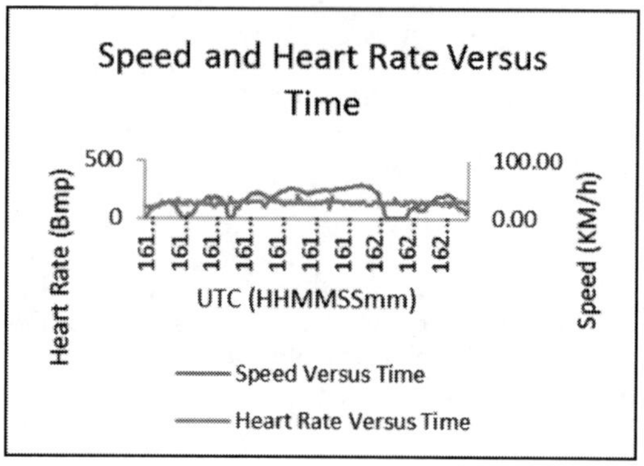

Fig. 5. Speed and Heart Rate Versus Time

The result obtained via the cyclist monitoring system by using LabVIEW and Internet of things (IoT). The system performances can be analysed using LabVIEW.

The GPS data shows the actual location of the experiment, it could easily verified by using google map software. With this data, the bicyclists are safe when there is an emergency, they can get help easily by just pressing the 'SOS' button. However, the GPS performance degraded while it is use in indoor or shielded location and it is shown in Figure 6.

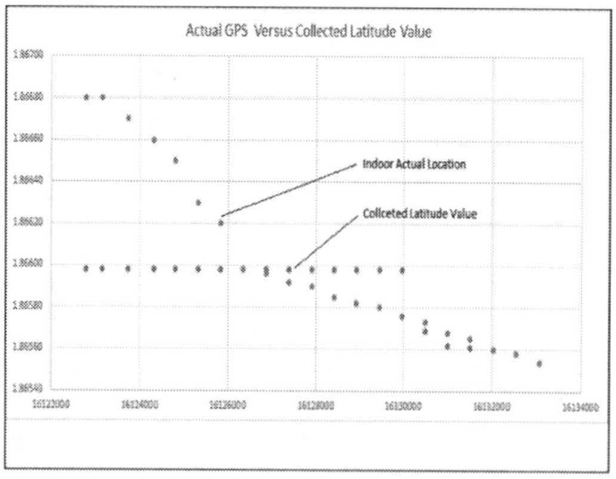

Fig. 6. The invalid GPS data in shielded area.

According to the speed data collected, there is a delay of around 0.5s to 1s while compare to the phone GPS tracking Application. These shows the system is not a real time embedded system yet. However the speed data is valid despite the delay of its performance. This problem is clearly showed in Figure 7.

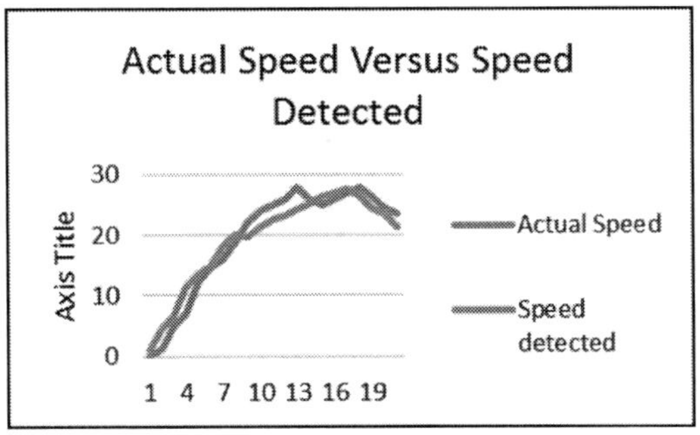

Fig. 7. Actual Speed Versus Speed Detected

Heart rate sensing is one of the most difficult elements, while cyclist perform the exercise, the heart rate sensor is unable to function well due to the inconsistency of analog signal. The signal collected is needed to undergo a long formula in order to get the heart rate in bmp. Therefore, the heart rate data are not precise and inconsistence.

The myRIO controller performed in the expected requirement. It interfaces all the sensors by receiving the data in a couple of seconds which is around 200ms every circle. Then, it directly transmit the data using WiFi to the required port.

The signal strength of the WiFi system is important in this project. The cycling activities can be done in either indoor or outdoor. However the data for cyclist location for indoor is not valid since the unavailability of GPS signal.

The results from the device are transferred to the web interface that run by Ubidots. The results was obtained from the GPS device, magnetic sensor and heart rate sensor. GPS data logger is a real time application that can track the location of cyclist went and the green path is the road that cyclist went through as shown in Figure 8. Figure 9 shows the dashboard of the Ubidots after login by user. It shows several variables such as GPS data logger, maximum current speed of the day, average current speed of the day, the sum of distance travelled and the sum calories burn of the day.

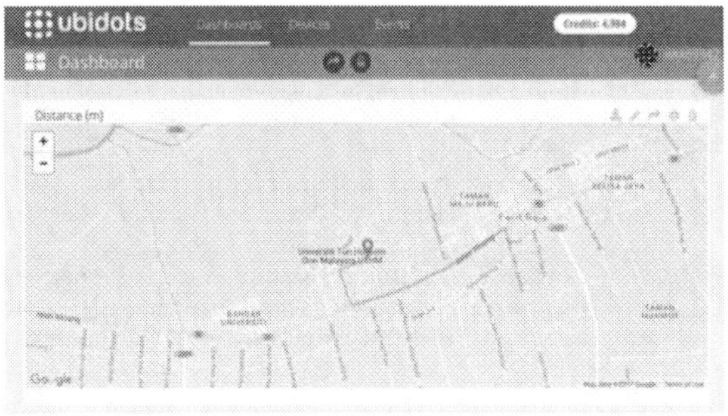

Fig. 8. Tracking of Cyclist in Google Maps

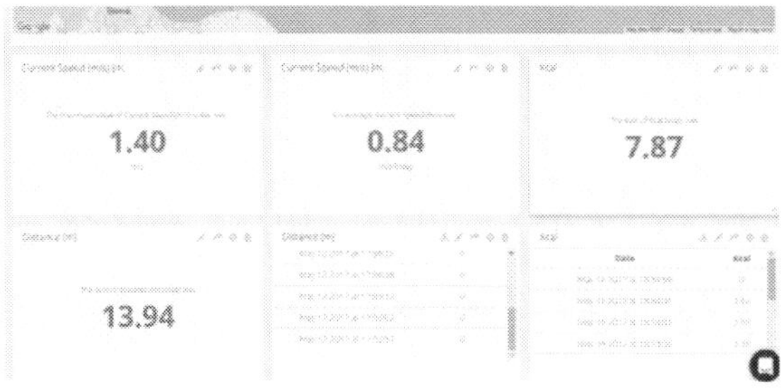

Fig. 9. Dashboard on Ubidots

4 Conclusions

The system was developed to monitor the cyclist performance in cycle sport especially in Malaysia. There are many previous projects that have the similar with this system. The projects include the smart bike system and bicycle renting system. The previous systems was mainly focus on business purpose. However, this paper presents a system which is designed for monitoring the performance of the cyclist in cycle sport. The contribution of this project is to facilitate the coach and management in cycle sport to supervise their athletes. The data such as heart rate, cyclist speed, training path and location are important data which helps coach to optimise the training load and to record personal best performances. This project exploits the WiFi technology using embedded device platform myRIO by National Instrument. The implementation of Internet of Things (IoT) in this project will reshape the sport industry in Malaysia which emphasize on sporting performance and safety and security of athletes.

The accomplishment of this undertaking could not have been possible without the participation and encouragement of so many people whose names may not all be enumerated but their involvements and contributions are sincerely appreciated and gratefully acknowledged. At the first I would to express my gratitude to Mr. Aizan Bin Ubin for his endless support. I also would like to thank my friend Jackson Tan Jia Soon for giving me a helping hand to finish this report and project. I would like to thank to National Instruments (MyRIO) and UTHM for giving me support to finish this project.

References

1. R. Marin-Perianu *et al.*, "A performance analysis of a wireless body-area network monitoring system for professional cycling," *Pers. Ubiquitous Comput.*, vol. 17, no. 1, pp. 197–209, (2013).
2. A. E. Aubert, B. Seps, and F. Beckers, "Heart Rate Variability in Athletes," *Sport. Med.*, vol. 33, no. 12, pp. 889–919, (2003).
3. J. A. Afonso, F. J. Rodrigues, D. Pedrosa, and J. L. Afonso, "Automatic Control of Cycling Effort Using Electric Bicycles and Mobile Devices," vol. I,(2015).
4. Y. Taniguchi, K. Nishii, and H. Hisamatsu, "Evaluation of a bicycle-mounted ultrasonic distance sensor for monitoring obstacles and holes on road," *Int. J. Simul. Syst. Sci. Technol.*, vol. 16, no. 6, p. 1.1-1.6, (2015).
5. C. Kiefer and F. Behrendt, "Smart e-bike monitoring system: real-time open source and open hardware GPS assistance and sensor data for electrically-assisted bicycles," *IET Intell. Transp. Syst.*, vol. 10, no. 2, pp. 79–88, (2016).

Solar Energy Measurement Using Arduino

Siti Amely Jumaat[1], *Mohamad Hilmi* Othman[1]

[1]Green and Sustainable Energy Focus Group, Faculty of Electrical and Electronic Engineering, Universiti Tun Hussein Onn Malaysia. 86400 Parit Raja, Batu Pahat, Johor, Malaysia.

Abstract. This project aims to develop a measurement of solar energy using Arduino Board technology. In this research, four parameters that been measured are temperature, light intensity, voltage and current. The temperature was measured using temperature sensor. The light intensity was measured using light dependent resistor (LDR) sensor. The voltage was measured using the voltage divider because the voltage generated by the solar panel are large for the Arduino as receiver. Lastly for the current was measured using the current sensor module that can sense the current generated by the solar panel. These parameters as the input value for the Arduino and the output was display at the Liquid Crystal Display (LCD) screen. The LCD screen display output of the temperature, the light intensity, the voltage and the current value. The purpose of Arduino to convert the analog input of parameter to the digital output and display via LCD screen. Other than that, this project also involve with a design to ensure that device case are easy to be carry around.

1 Introduction

Rising fossil fuel and burning fuel such as coal, global warming and severe weather conditions have compelled many nations to look for alternative sources to reduce reliance on fossil based fuels. Solar energy is one of the most promising renewable sources that is currently being used worldwide to contribute for meeting rising demands of electric power. [1] Solar power is a conversion of sunlight into electricity, sunlight was collect either directly by using photovoltaics or indirectly using concentrated of solar energy. [2]

Photovoltaics was initially use as a power source for a small and medium-size applications from the calculator powered by a single solar cell to a remote homes powered by an off-grid rooftop photovoltaics system. As the cost of solar electricity has fallen, the number of grid-connected solar photovoltaics systems has grown into the millions and utility scale solar power stations with hundreds of megawatts are being built. Solar photovoltaic is becoming inexpensive, low-carbon technology to harness renewable energy from the sun. [3] This paper presented by Arindam Bose *et. al* [4] describe a potential a solar system using two set of stepper motor, the light sensor and the concave mirror. The purpose of this project to improve the power collection efficiency 65% with developing the track of solar panel perpendicular. This paper presented by Mohsen Taherbaneh *et. al* [5] proposed the method based on simulation of two fuzzy controllers in order to maximize the generate the output power of solar panel in a photovoltaic system. The output I-V curve from this project are the maximum current is 1.56A nad the maximum voltage are 20V with solar irradiance = 500W/m^2 and temperature = 34.5°C. The results of the fuzzy-based maximum power point tracking with 23W with 51% and 11W with 24.5% from the nominal output power. On the other hand, the combination of fuzzy-based maximum power point tracking and fuzzy-based sun tracking is 35W with 78% of the nominal output power. Yi-Hua Liu *et.al* in [6] presented the maximum power point tracking (MPPT) method for low power photovoltaic system, 87W PV system. The MPPT circuit using the low power operational amplifier (OP-AMP) with high and low irradation line and an analog switch. The simple DC-DC converter using such as the interface the voltage from the PV system. The type of multi-crystalline solar cell KC85T with the nominal open-circuit voltage of 21.7V and the nominal voltage at the maximum power point of 17.4V. The simulation results shows with 1000W/m^2 irradiance, 25°C and the maximum power is 87.46W. From experimental results with 1000W/m^2, the maximum power is 87.39W. The advantages of analog MPPT are very simple, the low-power analog component, the fast speed and can using to all types of power convertes. This paper [7] propose the method the Fibonnaci search technique for narrow down possible locations. The results shows that 1000W/m^2 is 112.4W with 13% efficiency of panel solar. Mohammad H. Moradi presented in [8] to improve the performance of the maximum power point tracking in solar panel. The PV panel with Aria Solar are 60W, with 25°C,1000W/m^2

* Corresponding author: sitiamely1979@gmail.com

https://doi.org/10.1051/matecconf/201815001007

and the current is 2.5A and the voltage is 23.1V. The proposed algorithm are the set point calculation and the fine turning loops. The method was simulated with Matlab/Simulink and also development of the prototype.

Concentration of solar energy are concentration of the light from the sun to a point by using lens or mirror and tracking system. The light from the sun also produce heat and if we concentrate at a point. The point will experience the heat produce from the sun [9-10].

This project are focus on measure the solar power using Arduino [10]. This design project are to measured parameters: light intensity, voltage and current and temperature using multiple sensor. The main part in this project are the solar panel, the light sensor, the temperature sensor, a voltage divider, the current sensor and the LCD screen to display.

2 Specification of Components

This part are discusses the main components that used on this research.

2.1. Solar Cell

In this research, the panel solar is polycrystalline type are be used with 12V, 250mA, 3W as a source [9]. The size of the panel is 145mm X 145mm as shown in **Figure 1.**

Fig. 1. The Solar Panel Polycrystalline Type

2.2. Arduino Board

The Arduino Uno is microcontroller board based on the ATmega328 datasheet as shown in Figure 2. It has 14 digital input and output pins: 6 pins used PWM outputs and 6 pins is analog input such as the clock speed is 16MHz, the ceramic resonator, the USB connection, the power jack, the ICSP header and the reset button [10].

Fig. 2. The Arduino Board

2.3. Light Sensor

A Light Dependent Resistor (LDR) or photo resistor is device whose resistivity is a function of the incident electromagnetic radiation which means it is light sensitive devices. These components also called a photo conductors, photo conductive cells or simply photocells [11]. The LDR function on the principle based on photo conductivity which an optical phenomena. Figure 3 shows the resistance vs illumination graph of LDR.

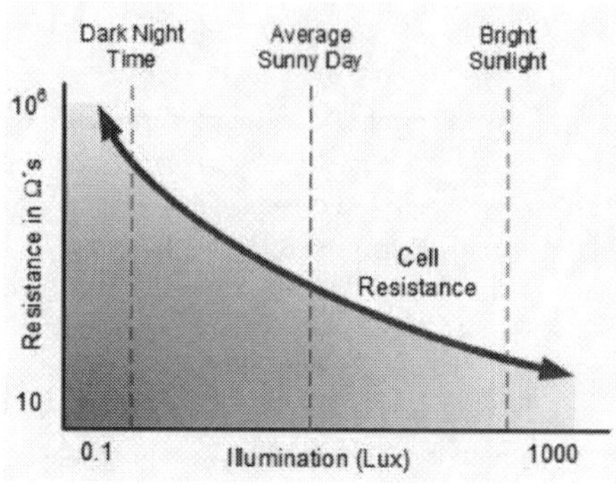

Fig. 3. The resistance VS illumination Graph of LDR

2.3. Circuit Simulation

Figure 4(a) and (b) shown the block diagram of the research and the complete of simulation circuit for this project that have been develop by using Proteus 8 Professiona [12-16]. In this simulation circuit, it consist three main parts : the voltage divider, the LED light indicator, the temperature sensor, LDR sensor and the LCD screen display. The power generated from the solar panel is 12V approximately. Inside the circuit have LDR sensor for detect the light intensity. Next, the temperature sensor have detect the temperature changing [17-18, 20]. In this project, the main controller are use the Arduino Uno and its need the power supply [16, 19, 21]. The power supply for this controller is 5V. Then, the Arduino Uno must have coding for it to function as desired. Lastly, the LCD is to display the output that have written in the coding inside the Arduino Uno [22-23].

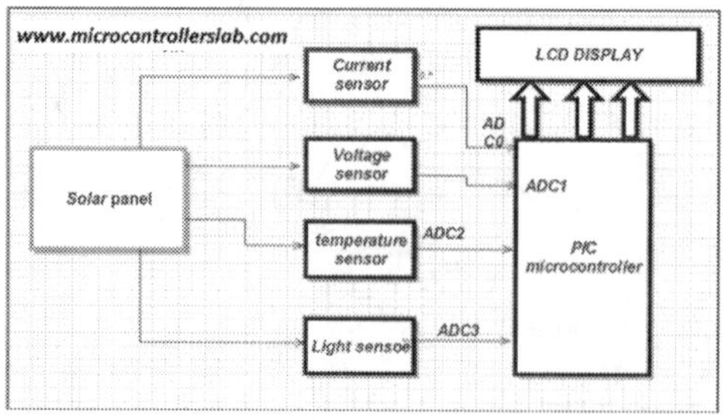

Fig. 4(a). The block diagram of the research

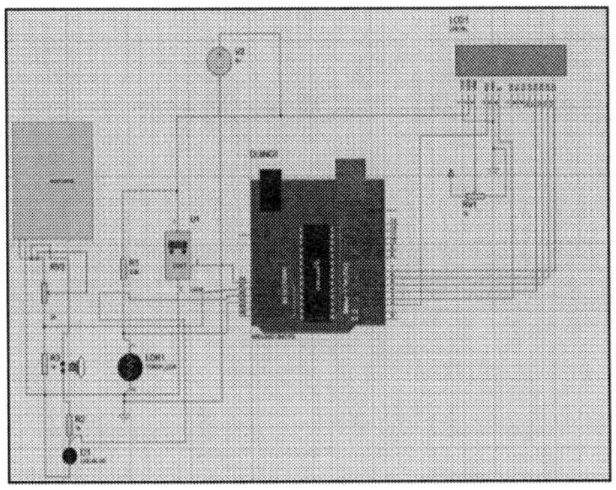

Fig. 4(b). The Simulation Circuit

3 Result and Analysis

This section are discusses the results based on the simulation results and the measurement result.

3.1. Simulation Results

Figure 5 shown the results of the temperature . It varies depend on the temperature sensor setting during the simulation. The Table 1 shown the simulation result of light intensity and the output voltage. From the simulation results shown that the circuit are function is very well and can be continue to the hardware development.

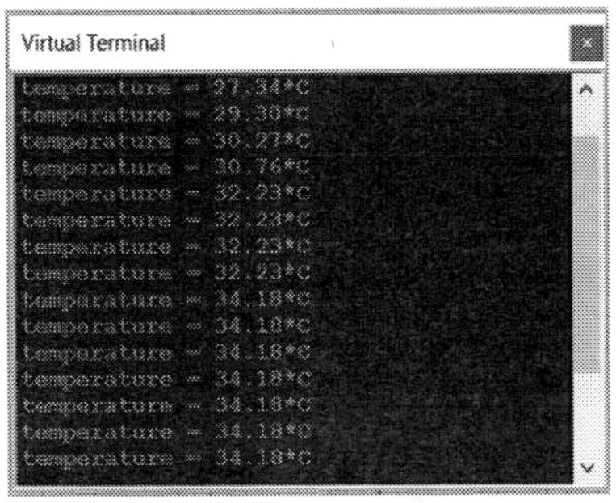

Fig. 5. The temperature virtual terminal output

Table 1. LDR simulation result

LDR (Lux)	LDR intensity	Output Voltage (V)
0.1	11	0.0538
10.1	372	1.8174
20.1	520	2.5379
30.1	607	2.9655
40.1	667	3.2545

3.2. Measurement Results

These sections are divided to three main parts: the light intensity, the voltage versus light intensity and the output power.

3.2.1 Results of Light Intensity

The light intensity have been recorded in the three days with the solar panel in the sunrise position as shown in Figure 6. Based on the result, the highest light intensity was 980 Lux at 2.00pm, while the lowest light intensity was 700 Lux at 5.00pm. The light intensity have been recorded in the three days with the solar panel in the sunrise position as shown in Figure 7. Based on the result, the highest light intensity was 970 Lux at 11.00am, while the lowest light intensity was 350 Lux at 5.00pm.The light intensity have been recorded in the three days with the solar panel in the sunrise position as shown in Figure 7. Based on the result, the highest light intensity was 970 Lux at 11.00am, while the lowest light intensity was 350 Lux at 5.00pm. The light intensity have been recorded in the three days with the solar panel in the sunrise position as shown in Figure 8. Based on the result, the highest light intensity was 950 Lux at 12.00pm, while the lowest light intensity was 830 Lux at 5.00pm.

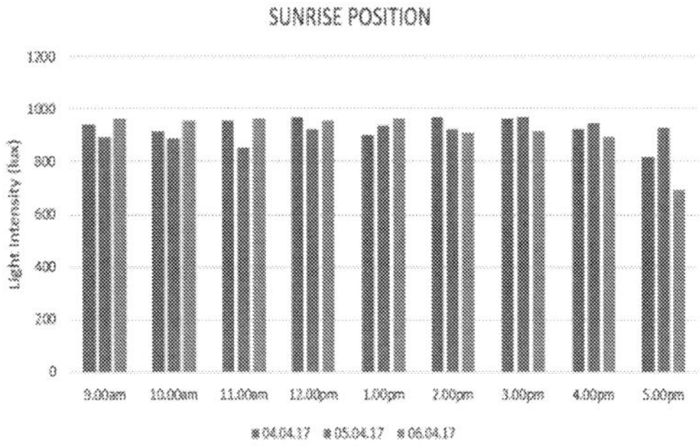

Fig. 6. Result of light intensity for sunrise position

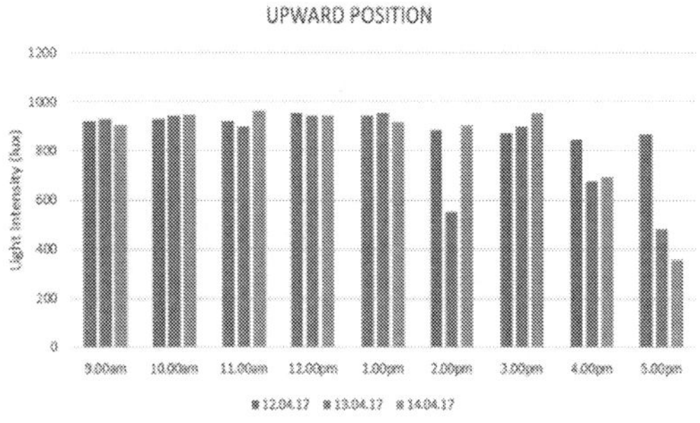

Fig. 7. Result for light intensity for upward position

https://doi.org/10.1051/matecconf/201815001007

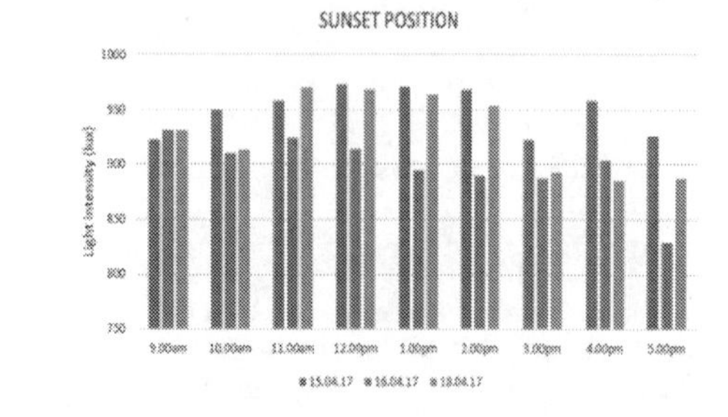

Fig. 8. Result for light intensity for sunset position

Table 2 shown the maximum and the minimum of the light intensity that have been recorded according to their solar panel position.

Table 2: Maximum and minimum light intensity.

Position	Sunrise	Upward	Sunset
Maximum (Lux)	980 Lux at 2.00pm	970 Lux at 11.00am	950 Lux at 12.00pm
Minimum (Lux)	700 Lux at 5.00pm	350 Lux at 5.00pm	830 Lux at 5.00pm

3.2.2 Results of Voltage versus Light Intensity

The result voltage versus light intensity for the solar panel in sunrise position as shown in Figure 9. The highest voltage recorded was 14.75V at 11.00am with light intensity was 945 Lux and the lowest voltage recorded was 9.17V at 5.00pm with light intensity was 695 Lux. The result voltage versus light intensity for the solar panel in upward position as shown in Figure 10. The highest voltage recorded was 13.11V at 10.00am with light intensity was 929 Lux and the lowest voltage recorded was 6.3V at 5.00pm with light intensity was 357 Lux. The result voltage versus light intensity for the solar panel in sunset position as shown in Figure 11. The highest voltage recorded was 12.57V at 1.00pm with light intensity was 964 Lux and the lowest voltage recorded was 11.19V at 9.00am with light intensity was 931 Lux.

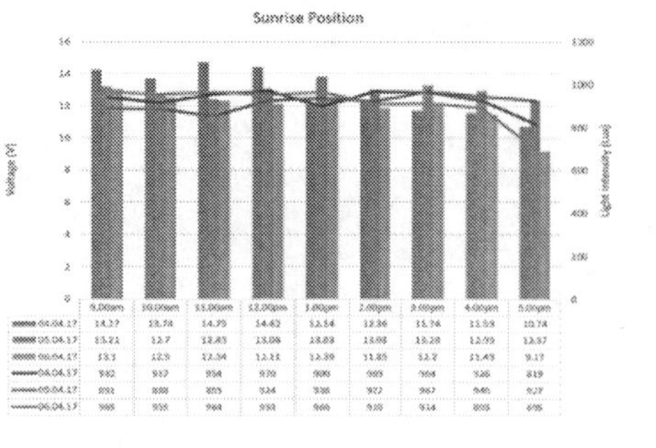

Fig. 9. Voltage versus light intensity for sunrise position.

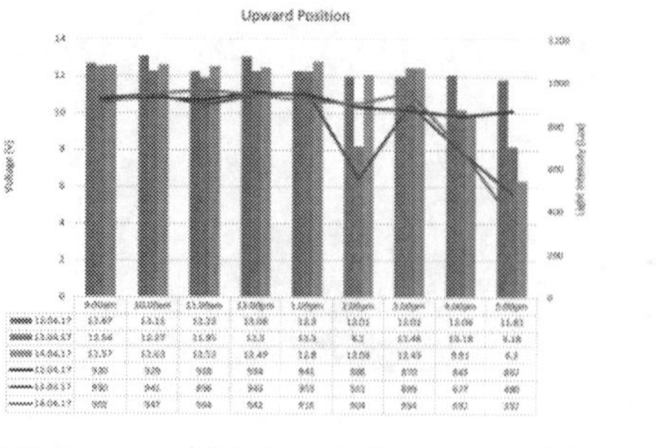

Fig. 10. Voltage versus light intensity for upward position.

Table 3 shown the maximum and minimum value of voltage and light intensity for each of the position. From the shows that, the higher voltage value, the light intensity also have higher value. The present of light intensity will affect the voltage produce by the solar panel.

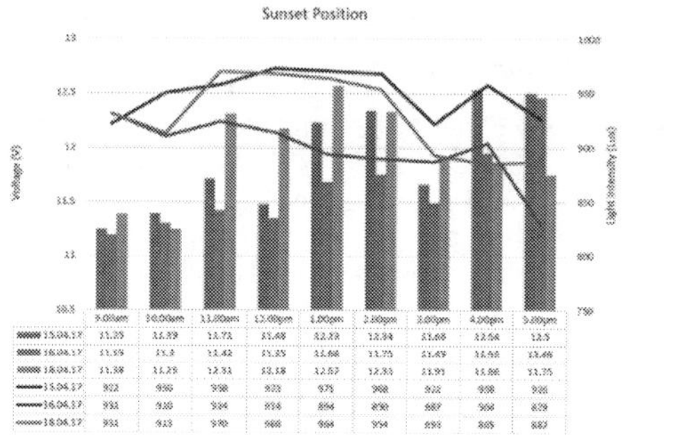

Fig. 11. Voltage versus light intensity for sunset position.

Table 3. The maximum and minimum value for each position.

Position	Sunrise	Upward	Sunset
Maximum	14.75V (11.00am) with 954 Lux	13.11V (10.00am) with 929 Lux	12.57V (1.00pm) with 964 Lux
Minimum	9.17V (5.00pm) with 695 Lux	6.3V (5.00pm) with 357 Lux	11.19V (5.00pm) with 931 Lux

3.2.3 Results of Output Power

The result of output power with the solar panel in the sunrise position was shown in Figure 12. The highest power produce was 2.4W at 4.00 pm, while the lowest power produce 0.3W at 5.00 pm. The result of output power with the solar panel in the upward position was shown in Figure 13. The highest power produce was 1.7W at 12.00pm, while the lowest power produce 0.38W at 5.00pm. The result of output power with the solar panel in the sunset position was shown in Figure

https://doi.org/10.1051/matecconf/201815001007

14. The highest power produce was 1.63W at 5.00pm, while the lowest power produce 1.21W at 9.00am.

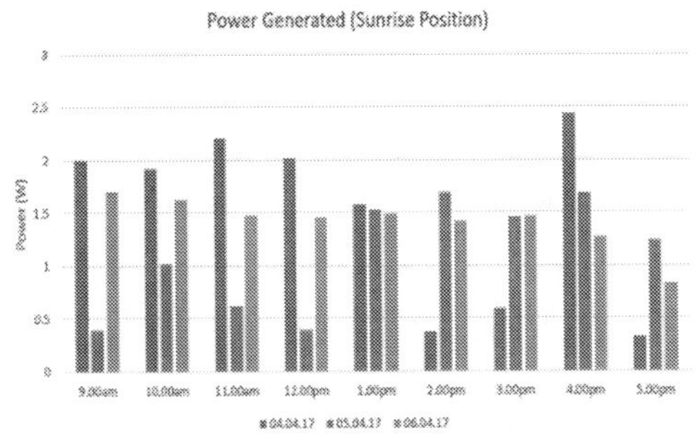

Fig. 12. Power from the sunrise position.

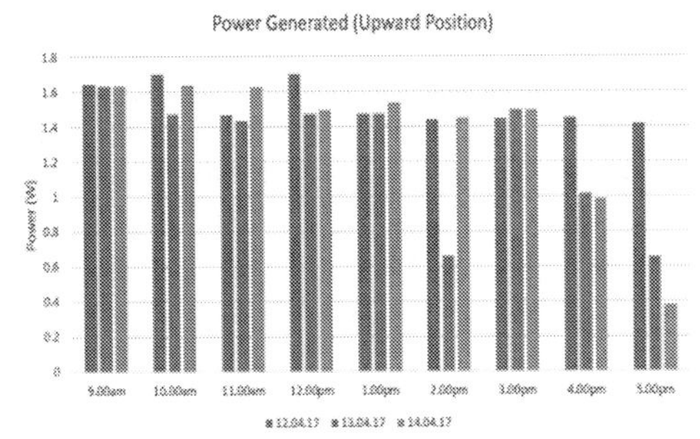

Fig. 13. Power from the upward position.

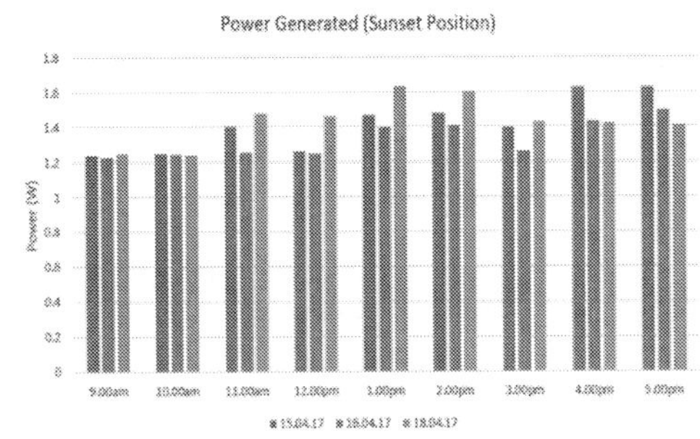

Fig. 14. Power from the sunset position.

Table 4 shown the maximum and minimum power generated by the solar according to its type of position. For the sunrise position, the maximum power generated is 2.4W, while the minimum power generated is 0.4W. For the upward position, the maximum power generated is 1.7W and the minimum power generated is 0.38W. For the sunset position, the maximum power generated is 1.63W and the minimum power generated is 1.21W.

3.2.4 Prototype of Project

Figure 15 shows the prototype of solat energy measurement using arduino uno. The best position of the solar panel energize was the sunrise position with the highest voltage value which is 14.75V at time 11.00am have been recorded. At this time the light intensity was

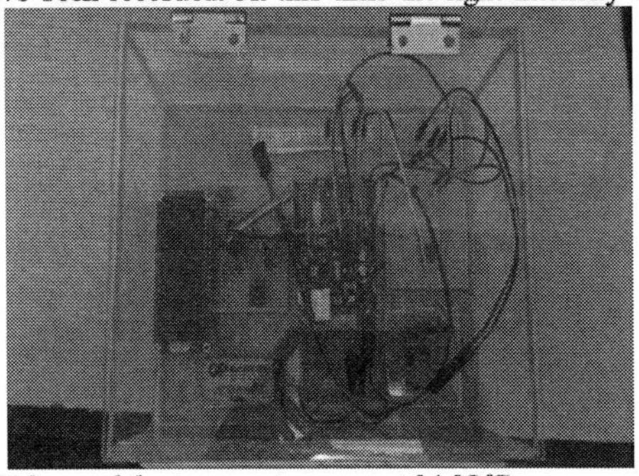

954 lux and the temperature was at 34.32℃.

Fig. 15. The Prototype of Solar Energy Measurement

Table 4. Maximum and minimum output power.

Position	Sunrise	Upward	Sunset
Maximum (W)	2.4W at 4.00pm	1.7W at 12.00pm	1.63W at 5.00pm
Minimum (W)	0.4W at 5.00pm	0.38W at 5.00pm	1.21 at 9.00am

4 Conclusion

In the conclusion, the project is achieve all of the objective are : to measure solar panel parameter such as the temperature, light intensity, voltage and current. Using the temperature sensor that sense the changes in surrounding temperature, for the light intensity parameter was by using the LDR sensor, for the voltage parameter was by using the voltage divider method in order to reduce the maximum value of the solar panel to the voltage value suitable for the Arduino of power supply and lastly the current parameter was by using the current sensor module. Next, to find the best position and time for the solar power effectively energize the electricity. The data from measurement part shows that the best position of the solar panel effectively energize was the sunrise position with the highest voltage value which is14.75V at time 11.00am have been recorded. At this time the light intensity was 954 lux and the temperature was at 34.32℃. Lastly, to develop a portable device for measuring the solar energy can be achieve with developing the light in weight of the casing of the device and the neat arrangement of the electrical component inside the casing.

The authors would like to acknowledge the Research Management Center (RMC), Universiti Tun Hussein Onn Malysia (UTHM), Batu Pahat, Johor, Malaysia for the financial support of this search. This research is partly by RMC under the U861 (Tier 1) Grant.

References

1. M.R. Al Rashidi, M.F. Al Hajri, K.M. El-Naggar, A.K. Al-Othman, "*A new estimation approach for determining the I–V characteristics of solar cells*", Electrical Engineering Department, College of Technological Studies (PAAET), Shuwaikh, Kuwait, (2011).
2. V. Ryan, *what is solar energy?* (2016, September 20) Retrieved from http://www.technologystudent.com/energy1/solar1.htm
3. John Balfour, "*Introduction to Photovoltaic*", United States of America, (2013).
4. Arindam Bose, Sounak Sarkar, Sayan Das, "*Helianthus-a Low Cost High Efficient Solar Tracking System Using AVR Microcontroller*", International Journal of Scientific & Engineering Research, **Volume 3**, Issue 10, October (2012).
5. Mohsen Taherbaneh, A. H. Rezaie, H. Ghafoorifard, K. Rahimi and M. B. Menhaj, "*Maximizing Output Power of a Solar Panelvia Combination of Sun Tracking and Maximum Power Point Tracking by Fuzzy Controllers*", Hindawi Publishing Corporation, International Journal of Photoenergy, **Volume 2010**, (2010).
6. Yi-Hua Liu, Jia-Wei Huang, " *A fast and low cost analog maximum power point tracking method for a low power photovoltaic systems*", Department of Electrical Engineering, National Taiwan University of Science and Technology, 13 September (2011).
7. R. Ramaprabha, M. Balaji, B. L. Mathur. " *Maximum power point tracking of partially shaded solar PV system using modified Fibonacci search method with fuzzy controller*", Department of EEE, SSN Collage of Engineering Chenmai, India, 10 July (2012).
8. Mohammad H. Moradi, Ali Reza Reisi, "*A hybrid maximum power point tracking method for photovoltaic systems*", Department of Electrical Engineering, Faculty of Engineering, Bu Ali Sina University, Hamedan, Iran, 14 September (2011).
9. Mybotic (2010) *Solar cell 12V 250mA* (2016, November 20), retrieve from http://www.mybotic.com.my/webshaper/store/viewProd.asp?pkProductItem=831
10. Chua Hock-Chuan (2015) *Getting Started with Arduino* (2016, October 30), retrieve from https://www.ntu.edu.sg/home/ehchua/programming/arduino/Arduino.html
11. Electrical 4U.com (2011) *Light Dependent Resistor / LDR and working Principle of LDR* (2016, November 2) retrieve from http://www.electrical4u.com/light-dependent-resistor-ldr-working-principle-of-ldr/
12. Taxes Instrument (2016) *LM35 precision centigrade temperature sensors* (2016, November 4) retrieve from http://www.ti.com/lit/ds/symlink/lm35.pdf
13. SparkFun (2015) *Voltage Divider* (2016, November 6) retrieve from https://learn.sparkfun.com/tutorials/voltage-dividers
14. Engineer Garage (2012) *LCD* (2016, November 7) retrieve from http://www.engineersgarage.com/electronic-components/16x2-lcd-module-datasheet
15. Wikipedia (2016) *Proteus Design Suite* (2016, November 21) retrieve from https://en.wikipedia.org/wiki/Proteus_Design_Suite
16. Arduino (2016) *Overview of the Arduino*, (2016, November 21) retrieve from https://www.arduino.cc/en/Main/ArduinoBoardUno
17. Mohd Rashid B. Mohd Kamlon, "*Solar Power Measurement System*", Universiti Teknikal Melaka Malaysia, (2007).
18. Ganiyu R. A., Arulogun O. T. and Okediran O. O, (2014), *Development of A Microcontroller-Based Traffic Light System For Road Intersection Control, Journal of Scientific and Technology Research*, **vol. 3**, p.1-4.
19. Vulnerable space blogspot (2016) *Arduino Welcome to World*, (2016, December 14) retrieve from http://vulnerablespace.blogspot.my/2016/01/arduino-welcome-to-world-of.html.
20. Wikipedia (2016) *Current sensor module 5A*, (2016, December 22) retrieve from https://www.elecrow.com/wiki/index.php?title=ACS712_Current_Sensor-_5A.
21. Siti Amely Jumaat, Firdaus Mohammad, Shamsul Aizam Zulkifli, *Development of Portable Case Solar Battery Charger*, Electricial and Electronics Engineering 2016, 6(4) : 55-61, DOI : 1-.59923/j.eee.20160604.01 (2016).
22. Idris. I, Robian. M.S, Mahamad. A.K, Saon. S, '*Arduino based maximum power point tracking for photovoltaic system*', APRN Journal of Engineering and Applied Sciences, **Volume 11**, Issues 14, (2016), pp 8805-8809.
23. Zulkifli. S.A, Hussin M.N, Saad A.S, '*MATLAB-Arduino as a low cost microcontroller for 3-phase inverter*', 2014 IEEE Student Conference on Research and Development, SCOReD 2014.

Size-dependent separation of polydisperse samples in a tapered microfluidic device

Ida Laila Ahmad[1,*] and *Mohd Ridzuan* Ahmad[2]

[1]Department of Electronic Engineering, Faculty of Electrical and Electronic Engineering, Universiti Tun Hussein Onn Malaysia, 86400 Batu Pahat, Johor, Malaysia
[2]Department of Control and Mechatronics Engineering, Faculty of Electrical Engineering, Universiti Teknologi Malaysia, 81310 Skudai, Johor, Malaysia

Abstract. In this paper we introduce the use of tapered microfluidic device for size-based separation of polydisperse samples using passive platform. The separation mechanism works based on hydrodynamic principle which is further amplified by sedimentation effect. The tapered structure allows for generation of gradient pressure as a result from velocity distribution. From this device, we achieved 98% purity of the samples collected from the outlets with low complexity design. Furthermore, the simulation and experimental results are in agreement whereby suggesting that different sized particles were collected at different outlets. The presented microfluidic system is useful and versatile due to its simplicity and ease of operation.

1 Introduction

Separation of micro-particles is important in various industries for example food processing, biological analysis, mineral synthesis and environmental [1]. For example, separation of harmful bacteria and unwanted microbes during fermentation process can eliminate the chances of spoiled processed food [2]. While fractionation of human blood to its components such as red blood cell (RBC) and platelets may help to provide early disease diagnosis related to blood especially blood hereditary problems, blood clotting, anaemia and it can also be used for circulating tumour cells (CTC) detection [3]. The advancements in microfluidic technology facilitate progress in various developments of separation devices. Microfluidic based separation device offers vast advantages for example it uses less reagents, reduced costs, operates at low power, uses smaller sample volumes, small and portable, flexible and allow for integration with other modules and with minimum operator handling [4], [5].

Passive separation techniques work using fluid manipulation or particle interaction for example field flow fractionation (FFF) [6], [7], filtration [8], hydrodynamic [9]–[11] and sedimentation [12]. It serves as the most fundamental approach and broadly investigated. Microfluidic based hydrodynamic separation manipulates fluid flow and suitable for separation based on size and density. It offers simple design and straightforward operation. In addition, the devices have been tested for separation of both non-biological and biological particles such as microbeads, polymer, colloids, cancer cells and RBC [13], [14].

Although the techniques have been tested and proven to be efficient, it still suffers from being slow, need for sheath fluid and need for longer channel dimensions [7], [15], [16]. In this study, we describe a simple passive separation device for separation of polydisperse samples based on size. The device performs the separation by utilizing hydrodynamic properties of fluid and improves the samples purity through sedimentation effect resultant from the tapered microfluidic design.

This paper is organized as follows: Section 2 describes the methodology used which includes device design, working principles, finite element simulation, device fabrication and experimental procedures. In Section 3, the results obtained from both simulation and experiments were discussed thoroughly. A fair comparisons have been carried out and the device performance have been evaluated.

2 Material and method

2.1 Device design

The use of tapered design microfluidic were widely adopted by other researchers for various purpose; stiffness measurements, deformability detection and retention time [17]–[19]. However, the applicability of the design for separation applications has yet to be reported. The design of tapered microfluidic device consists of a single inlet without any sheath inlet and three separation outlets as shown in Figure 1. Length of tapered area is 4 mm, diameter of inlet and outlets are 2 mm and the height is 10 μm. Four devices with different taper angles were designed with 6°, 12°, 20° and 25°.

* Corresponding author: ida@uthm.edu.my

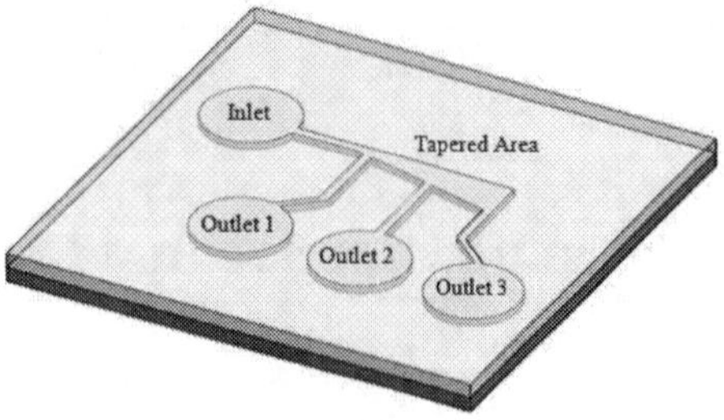

Fig. 1. Tapered microfluidic device design

The design was implemented by using an equivalent resistive circuit model to represent the resultant circuit during steady state operation as given by Figure 2.

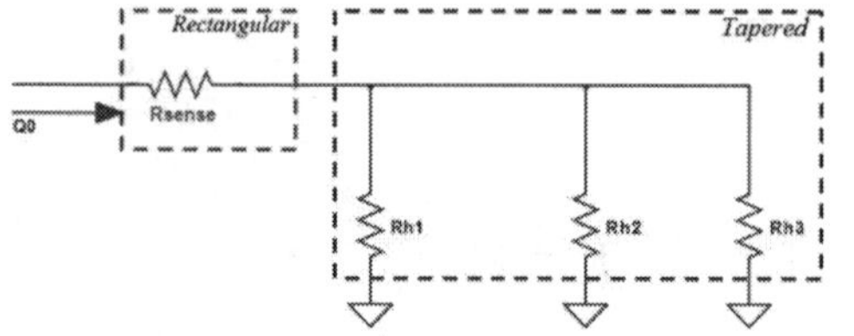

Fig. 2. Equivalent resistive circuit model

2.2 Working principle

Figure 3 illustrates the working principle of the proposed device.

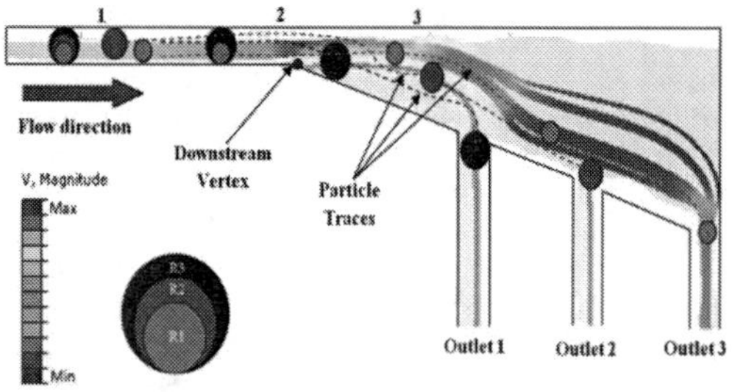

Fig. 3. Working principle of tapered microfluidic device

Considering samples of different sizes were flown into the inlet, these multi-particles will be aligned along the narrow focussing length towards the tapered area. This eliminates the need for sheath flow during particle focussing. Upon reaching at the downstream vertex, each particle starts to travel using its own trajectory which is dictated by the particle radius. The generated flow profile inside tapered microfluidic has a lower mean velocity at the central while higher velocity at the inlet and outlets created a gradient streamlines [12]. At point 3 where mean velocity is very small, gravity takes effect and sedimentation on larger particle is more significant. Larger particle tend to deviate from the original fluid path and try to escape using the nearest outlet (Outlet 1). On the contrary, smaller particle will follow the fluid path and will travel towards the lowest hydrodynamic resistance outlet (Outlet 3) [15].

Therefore, successful separation can be observed with largest particles collected at Outlet 1 and smallest particles collected at Outlet 2. Hypothetically, the intermediate sized particles will be collected at the Outlet 2. To understand this phenomena better, it is basically controlled by Reynolds number and Stokes number. These dimensionless parameters determine the laminarity profile of the fluid flow inside tapered microfluidic and the behaviour of particles moving inside the fluidic environment [20]. Larger particles will have higher Stoke number causing them to sediment faster as compared to its smaller counterparts. Stokes number can be described as the following equation:

$$St = \frac{\tau_r}{\tau_f} = \frac{\rho_p a^2/18\eta}{D_h/V} \quad (1)$$

Where ρ_p is the particle density, a is the diameter, η is the kinematic viscosity of fluid, V is velocity and D_h is the hydraulic diameter [21].

2.3 Finite element simulation

In this research, Abaqus 6.12 developed by Dassaults Systemes Simulia Corp, Providence, RI, USA was used. For simulation purpose, a tapered microchannel was designed using Eulerian material and was assigned with water properties. Hexahedral meshing was used and different boundary conditions were assigned at the respected inlet (inlet velocity) and outlets (free flow) as shown in Figure 4. Finite element models were developed for both polydisperse samples containing polystyrene microbeads and human cervical epithelial carcinoma (HeLa) cells. Deformable parts were meshed with tetrahedral elements and materials assignment were defined according to polystyrene microbeads and HeLa properties comes from previous literature [22]. Figure 5 shows the developed model. Finally, Figure 6 shows the assembly setting for simulation condition.

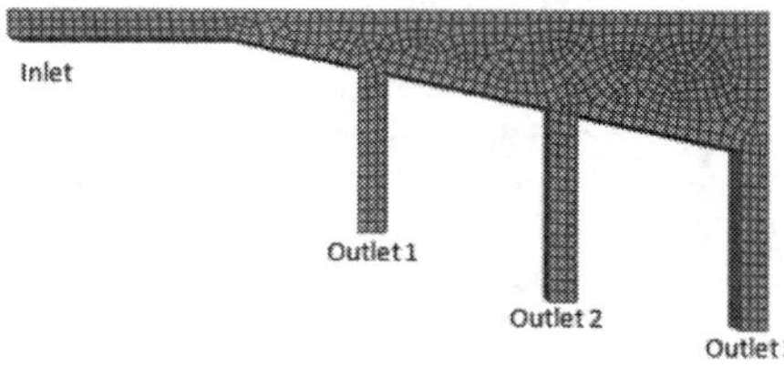

Fig. 4. Hexahedral meshing for tapered microfluidic device

Fig. 5. Tetrahedral meshing for polystyrene microbeads and HeLa cell

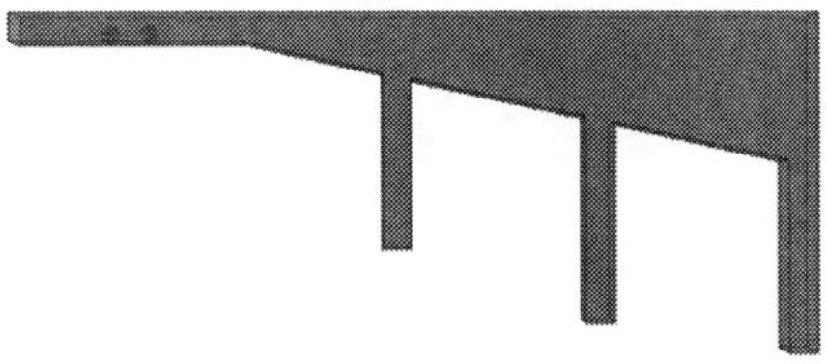

Fig. 6. Assembly setting for finite element simulation consists of two particles representing polydisperse samples

2.4 Device fabrication

Tapered design for the microchannel was drawn using LayoutEditor software. A 2.5 inch glass chrome mask was used to transfer the design by using a direct laser writing technique on µPG machine (Heidelberg Instruments, Germany). Master mould was prepared using a silicone wafer as the substrate. Negative photoresist SU8-3005 (Kayaku Microchem, Japan) was poured onto the substrate and spun using a spin coater for even coating. The spinning speed were programmed at 500 rpm for 10 seconds and followed with 4000 rpm for 30 seconds.

Then, the silicone substrate was put onto a hot plate for soft baked at 95°C for 3 minutes. Following this, a mask aligner was used to align the chrome mask and the substrate for design pattern transfer via photolithography technique whereby ultraviolet (UV) exposure was set to intensity level of 200 mJ/cm for 11.8 seconds. Two stages post-baked procedures were conducted by setting the temperature at 65°C for 1 minute and 95°C for 3 minutes to harden the substrate. Etchant was used to etch away the unexposed SU-8 layer and further developed using SU-8 developer (Microchem, Newton, MA) leaving a positive relief containing microchannel pattern. Isopropyl alcohol (IPA) was used to further clean the master mould.

The usage of liquid polydimethylsiloxane (PDMS) is widely adopted for various microfluidic applications. PDMS generally is inexpensive, flexible and optically transparent [23]. It is also biocompatible and highly permeable to carbon dioxide (CO_2) and oxygen (O_2) allowing for biological assays and cell culturing experiments. Oxidation of PDMS channels can create high surface energy allowing polar liquids like water to easily fill the channel. In order to fabricate the PDMS device, the liquid PDMS pre-polymer and curing agent (SILPOT 184, SILPOT CAT Dow Corning, USA) were mixed in the ratio of 9:1. After that, air bubble removal and degassed procedures were carried out and the mixture was poured over the master mould. For a good cross-linking effect the mixture was cured in a convection oven at 65°C for 2 hours. The cured PDMS was peeled from the master leaving the negative cast of the microchannel pattern. Inlets and outlets were punched manually using 2 mm puncher (Harris, Redding, CA). Plasma oxidation treatment was done at 20 mA for 3 minutes to seal the PDMS with glass slide. Tygon tubings were connected to the outlets and glued to plastic vials secured at adjustable X-Y manipulator for output collection. Figure 7 shows the completed PDMS chip.

Fig. 7. Completed PDMS chip for tapered microfluidic device

2.5 Experimental setup

To demonstrate the operation of fabricated device non fluorescent polystyrene (PS) microbeads with diameter of 3 µm (Polysciences. Inc.) and HeLa cells were used as samples. Next, HeLa cells were prepared following cell harvesting procedures. The cell tube contained $5x10^5$ of cells/mL was thawed using water bath and cell culturing was prepared inside clean bench. After culturing process was done, the remaining cell was kept in an incubator at 37°C with CO_2 level 5. Cell viability was checked under confocal microscope before usage. A mixture containing PS microbeads and HeLa cells were prepared and diluted further then centrifuged before re-suspended in culture medium.

During experiment, KDS Scientific syringe pump was connected to the inlet and was given a constant flow rate throughout the experiment. Phosphate Buffer Saline (PBS) or culture medium solutions were introduced into the microchannel first to eliminate unwanted bubbles. Then, the samples were introduced to the inlet. The separation behaviour was recorded using high speed charge-coupled-device (CCD) camera (Hamamatsu ORCA ER) and inverted microscope (IX-71 Olympus Corp., Japan).

3 Results and discussions

3.1 Finite element simulation results

Polydisperse separation simulation results are shown on Table 1 below.

Table 1. Separation simulation results

Device angle (°)	Monodisperse samples	Polydisperse samples
6	No separation	No separation
12	No separation	No separation
20	No separation	Outlet 1: HeLa cells Outlet 2: None Outlet 3: PS Microbeads
25	No separation	Outlet 1: HeLa cells Outlet 2: None Outlet 3: PS Microbeads

 https://doi.org/10.1051/matecconf/201815001008

Based on the simulation results, successful separation can be seen with the use of larger taper angles devices (20° and 25°). Larger angles provide more widening and significantly enhanced the sedimentation effect experienced by larger particles [12]. As a result, the larger particle (HeLa) will be collected at the Outlet 1. This phenomena also is supported by the Stoke number theory. From the calculation, St for HeLa cell is higher than the microbeads, indicating that it has the preference to travel on its own path. In comparison, hydrodynamic resistance values are the highest at the furthest outlet (Outlet 3). Smaller particle which have smaller St number will flow according to fluid streamlines moving towards Outlet 3 [15].

3.2 Experimental results

Prior to the separation experiment, the tapered microfluidic was tested for leaking. It was observed that any flow rate higher than 3.0 μl/min will cause device leakage. Therefore flow rate was set to be in between 0.5 – 3.0 μl/min was used during the experiment. Table 2 shows the separation results. It can be seen that the simulation results are in agreement with the experimentation results although different flow rates were used.

Table 2. Separation experiment results

<table>
<tr><th>Device angle (°)</th><th>Flow rate (μl/min)</th><th>Polydisperse sample</th></tr>
<tr><td>6</td><td rowspan="4">0.5
1.0
1.5
2.0
2.5
3.0</td><td>No separation</td></tr>
<tr><td>12</td><td>No separation</td></tr>
<tr><td>20</td><td>Outlet 1: HeLa cells
Outlet 2: None
Outlet 3: PS Microbeads</td></tr>
<tr><td>25</td><td>Outlet 1: HeLa cells
Outlet 2: None
Outlet 3:PS Microbeads</td></tr>
</table>

The following Figure 8(a)-(c) shows the collected samples from the three outlets. It can be seen that, HeLa cells were collected from Outlet 1, microbeads were collected from Outlet 3 and Outlet 2 was empty.

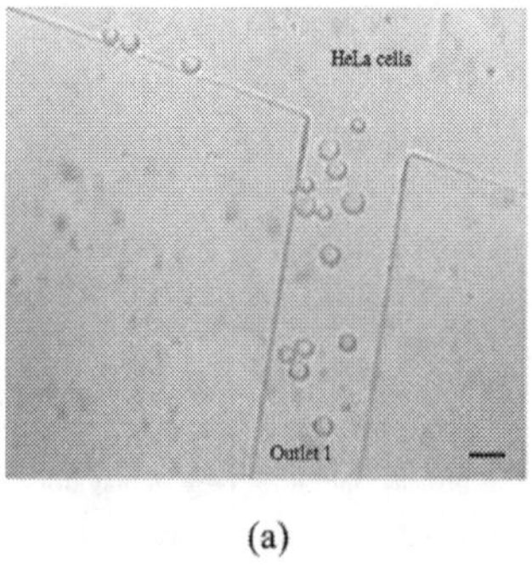

(a)

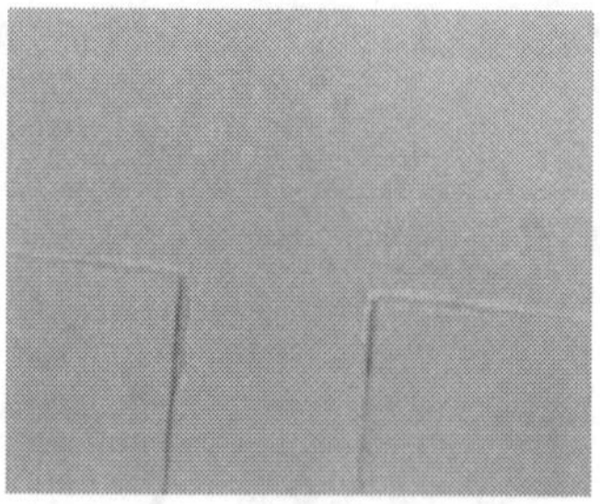

(b)

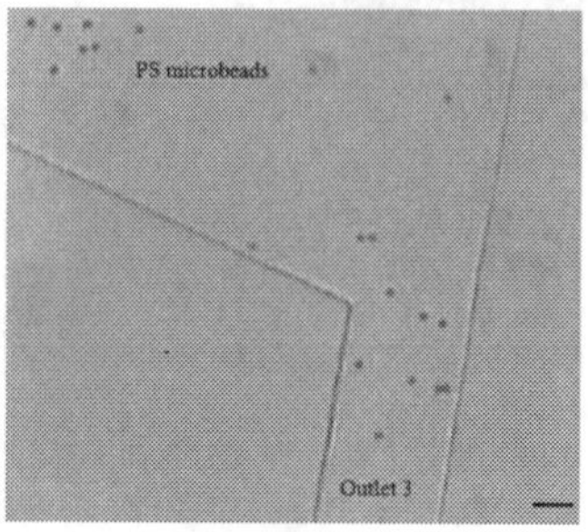

(c)

Fig. 8. Separation experiment results. (a) HeLa cells collected from Outlet 1 (b) Nothing is collected from Outlet 2 (c) Microbeads collected from Outlet 3. Scale used is 10μm

Overall, the performance of the proposed microfluidic separator device is in agreement with the simulations. It further proves the co-existence between the hydrodynamic effect and sedimentation due to uniquely designed tapered microchannel. The separation behavior can be understood better by comparing how rigid (PS microbeads) and deformable particle (HeLa cells) travel inside a microfluidic channel. The comparison is listed by Table 3 below.

Table 3. Traveling behavior inside microchannel

Parameter	Rigid	Deformable	Ref.
Speed	High	Slow	[24]
Deformation	Low	High	[25]
Density	Light	Dense	[26]
Nature	Hard	Soft (Fluidic)	[27]
Travel pattern	Travel along centerline	Travel near channel wall	[28]

These behaviors imply that, both rigid and deformable particle possess different characteristics during traveling inside fluidic condition. Therefore, good device design and parameter tuning can be manipulated to achieve successful separation. After experiment, samples collections from the outlets were evaluated to test the device performance. A good microfluidic separator should meet with these three basic requirements; high purity samples, acceptable flow rate (speed) and low complexity design. Purity of samples can be translated into the following formula:

$$\text{Purity} = \frac{(N_{Target})_{Outlets}}{(N_{Target+Waste})_{Outlets}} \times 100\% \quad (2)$$

Where N_{Target} was the number of target particles and N_{Waste} was the number of waste particles [14]. The percentage of purity was calculated for all flow rates tested and the results can be seen from Figure 9 below.

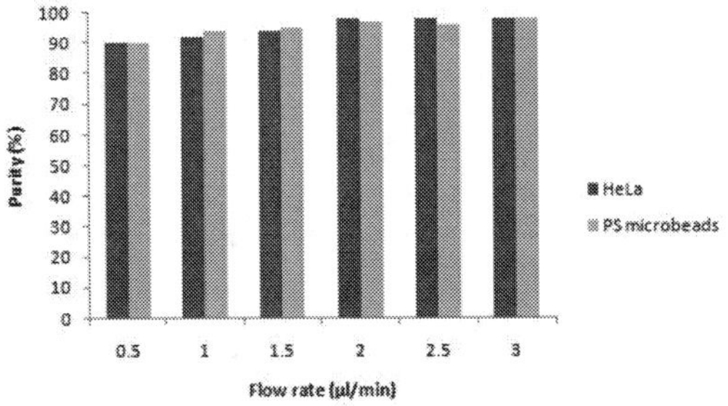

Fig. 9. Purity of samples collection for all tested flow rates

Highest purity of 98% can be achieved from the proposed device as illustrated by Figure 9. Therefore, the use of tapered microfluidic was proven to be effective in improving samples purity.

4 Conclusion

We have demonstrated a novel, passive tapered microfluidic device for separation of polydisperse sample containing PS microbeads and HeLa cells. The present study demonstrated high purity of samples collection at 98% for most of flow rates tested. The proposed device mechanism works by using coupling between hydrodynamic principle and sedimentation effect. Even though both techniques are well established, but the capability for polydisperse separation is demonstrated by this study. Since the proposed device dimensions are easily tunable and the operation is straight-forward, specific application development is possible. Above all, the capability to realize a label free sorting with high purity can be useful for various applications such as samples preparation and medical diagnostic

The authors wish to extend their gratitude to the financial aid sponsored by Ministry of Higher Education of Malaysia (Grant Nos. 4L640), and Universiti Teknologi Malaysia (Grant Nos. 03E11, 03G47, 4J148, 02G46 and 03H80). Special thanks for research colleagues at Micro-Nano System Engineering Department, Nagoya University, Nagoya, Aichi, Japan (Dr. Masaru Takeuchi and AP. Dr. Masahiro Nakajima) for beneficial discussions, guidance and laboratory facilities.

References

1. P. Sajeesh and A. K. Sen, "Particle separation and sorting in microfluidic devices: A review," *Microfluid. Nanofluidics*, vol. 17, no. 1, pp. 1–52, 2014.
2. D. R. Gossett, W. M. Weaver, A. J. Mach, S. C. Hur, H. T. K. Tse, W. Lee, H. Amini, and D. Di Carlo, "Label-free cell separation and sorting in microfluidic systems," *Anal. Bioanal. Chem.*, vol. 397, no. 8, pp. 3249–3267, 2010.
3. H. W. Hou, L. Wu, D. Amador-Munoz, M. P. Vera, B. D. Levy, R. M. Baron, and J. Han, "Microfluidic blood margination (µBM): A 'cell-based' blood purification platform for broad spectrum immunomodulation in murine model of sepsis," *18th Int. Conf. Miniaturized Syst. Chem. Life Sci. MicroTAS 2014*, vol. 6, no. 2, pp. 333–335, 2014.
4. C. Rivet, H. Lee, A. Hirsch, S. Hamilton, and H. Lu, "Microfluidics for medical diagnostics and biosensors," *Chem. Eng. Sci.*, vol. 66, no. 7, pp. 1490–1507, Apr. 2011.
5. A. M. Streets and Y. Huang, "Chip in a lab: Microfluidics for next generation life science research," *Biomicrofluidics*, vol. 7, no. 1, p. 11302, Jan. 2013.
6. M. Seki, M. Yamada, M. Yasuda, and Y. Sai, "Continuous separation of particles using a microfluidic device equipped with flow rate control valves," *J. Chromatogr. A*, vol. 1127, no. 1–2, pp. 214–220, 2006.
7. J. Takagi, M. Yamada, M. Yasuda, and M. Seki, "Continuous particle separation in a microchannel having asymmetrically arranged multiple branches.," *Lab Chip*, vol. 5, no. 7, pp. 778–784, 2005.
8. Y. Park, C. a Best, K. Badizadegan, R. R. Dasari, M. S. Feld, T. Kuriabova, M. L. Henle, A. J. Levine, and G. Popescu, "Measurement of red blood cell mechanics during morphological changes.," *Proc. Natl. Acad. Sci. U. S. A.*, vol. 107, no. 15, pp. 6731–6, Apr. 2010.
9. M. Yamada, K. Kano, Y. Tsuda, J. Kobayashi, M. Yamato, M. Seki, and T. Okano, "Microfluidic devices for size-dependent separation of liver cells," *Biomed. Microdevices*, vol. 9, no. 5, pp. 637–645, 2007.
10. D. W. Inglis, J. a Davis, R. H. Austin, and J. C. Sturm, "Critical particle size for fractionation by deterministic lateral displacement.," *Lab Chip*, vol. 6, no. 5, pp. 655–658, 2006.
11. J. P. Beech, S. H. Holm, K. Adolfsson, and J. O. Tegenfeldt, "Sorting cells by size, shape and deformability," *Lab Chip*, vol. 12, no. 6, p. 1048, 2012.

12. D. Huh, J. H. Bahng, Y. Ling, H. H. Wei, O. D. Kripfgans, J. B. Fowlkes, J. B. Grotberg, and S. Takayama, "Gravity-driven microfluidic particle sorting device with hydrodynamic separation amplification," *Anal. Chem.*, vol. 79, no. 4, pp. 1369–1376, 2007.
13. J. Alvankarian, A. Bahadorimehr, and B. Yeop Majlis, "A pillar-based microfilter for isolation of white blood cells on elastomeric substrate," *Biomicrofluidics*, vol. 7, no. 1, p. 014102(1)-014102(16), 2013.
14. Y. Y. Chiu, C. K. Huang, and Y. W. Lu, "Enhancement of microfluidic particle separation using cross-flow filters with hydrodynamic focusing," *Biomicrofluidics*, vol. 10, no. 1, p. 011906(1)-011906(13), 2016.
15. Z. Wu and K. Hjort, "Microfluidic Hydrodynamic Cell Separation: A Review," *Micro Nanosyst.*, vol. 1, no. 3, pp. 181–192, 2009.
16. D. Di Carlo, J. F. Edd, D. Irimia, R. G. Tompkins, and M. Toner, "Equilibrium separation and filtration of particles using differential inertial focusing," *Anal. Chem.*, vol. 80, no. 6, pp. 2204–2211, 2008.
17. C.-H. D. Tsai, M. Kaneko, S. Sakuma, and F. Arai, "Observability of cell stiffness in micro-channel method," *2013 IEEE Int. Conf. Robot. Autom.*, pp. 2807–2813, May 2013.
18. G. Guan, L. Wu, A. A. Bhagat, Z. Li, P. C. Y. Chen, S. Chao, C. J. Ong, and J. Han, "Spiral microchannel with rectangular and trapezoidal cross-sections for size based particle separation.," *Sci. Rep.*, vol. 3, no. C, p. 1475, 2013.
19. Y. N. Luo, D. Y. Chen, Y. Zhao, C. Wei, X. T. Zhao, W. T. Yue, R. Long, J. B. Wang, and J. Chen, "A constriction channel based microfluidic system enabling continuous characterization of cellular instantaneous Young's modulus," *Sensors Actuators B Chem.*, vol. 202, pp. 1183–1189, Oct. 2014.
20. J. Happel and H. Brenner, *Low Reynolds number hydrodynamics: with special applications to particulate media*. Springer Netherlands, 2012.
21. E. Ortega-Rivas, *Unit Operations of Particulate Solids: Theory and Practice*. CRC Press, 2016.
22. K. Tomankova and P. Kolar, *Mechanical characterisation of HeLa cells using atomic force microscopy*. 2012.
23. S. K. Sia and G. M. Whitesides, "Microfluidic devices fabricated in poly(dimethylsiloxane) for biological studies," *Electrophoresis*, vol. 24, no. 21. pp. 3563–3576, 2003.
24. S. Huang, H. Bow, M. Diez-silva, and J. Han, "Applying A Microfluidic 'Deformability Cytometry' To Measure Stiffness of Malaria Infected Red Blood Cells at Body and Febrile Temperatures," in *14th International Conference on Miniaturized Systems for Chemistry and Life Sciences*, 2010, no. October, pp. 259–261.
25. E. M. Nascimento, N. Nogueira, T. Silva, T. Braschler, N. Demierre, P. Renaud, and A. G. Oliva, "Dielectrophoretic sorting on a microfabricated flow cytometer: Label free separation of Babesia bovis infected erythrocytes," *Bioelectrochemistry*, vol. 73, no. 2, pp. 123–128, 2008.
26. J. Rousselet, G. H. Markx, and R. Pethig, "Separation of erythrocytes and latex beads by dielectrophoretic levitation and hyperlayer field-flow fractionation," in *Colloids and Surfaces A: Physicochemical and Engineering Aspects*, 1998, vol. 140, no. 1–3, pp. 209–216.
27. A. A. S. Bhagat, S. S. Kuntaegowdanahalli, and I. Papautsky, "Enhanced particle filtration in straight microchannels using shear-modulated inertial migration," *Phys. Fluids*, vol. 20, no. 10, p. 101702, 2008.
28. S. Kuntaegowdanahalli and A. Bhagat, "Inertial microfluidics for continuous particle separation in spiral microchannels," *Lab Chip*, vol. 9, no. 20, pp. 2973–80, 2009.

Experimental and Numerical Comparison Q-Switched Fiber Laser Generation using Graphene as Saturable Absorber

Noor Azura Awang*, *Nor Syuhada* Aziz, *Atiqah Nabieha* Azmi, *Fatin Shaqira* Hadi and *Zahariah* Zakaria

Science Department, Faculty Science, Technology and Human Development, University Tun Hussein Onn Malaysia, Parit Raja, 86400, Batu Pahat, Johor

Abstract. We demonstrated the comparison experimentally and numerically a compact Q-switched erbium-doped fiber (EDF) laser based on graphene as a saturable absorber (SA). By optically driven deposition of graphene on a fiber core, the SA is constructed and inserted into a diode-pumped EDF laser cavity. Lasing in CW region starts at 10 *mW*, whereas stable self-starting Q-switching with a central wavelength of 1530 *nm* begins at 18 mW. In this paper, at 35 *mW*, the maximum pulse energy reaches at 2 μJ with pulse repetition rate of 1 *MHz* and the narrowest pulse width is around 10 μs is obtained. The stability of the pulse is verified from the radio-frequency (RF) spectrum with a measured signal-to-noise ratio (SNR) of 48 *dB*. In this study, the design is compared with the simulation using the Optisystem software. The output power of the experimental study is also compared with the simulation to examine the performance.

1 Introduction

Q-switched fiber laser based on saturable absorber has been useful in light source and widely potential for use in application such as telecommunication, medicine, fiber optic sensing, remote sensing, range finding and industrial processing [1]. The generation of short pulses is also called a Q-switching operation. A laser ring cavity requires a gain medium such as erbium-doped fiber (EDF) and also a 'Q-switcher' component in order to obtain stable short pulses. The most widely utilized Q-switching technologies employ a nonlinear effect, such as nonlinear polarization rotation (NPR) as reported in [2] or a nonlinear optical loop mirror (NOLM) as proposed in [3, 4], along with the use of a fast saturable absorber (SA) element such as semiconductor saturable absorber mirrors (SESAMs) [5], carbon nanotubes (CNT) [6] or graphene [7]. Chronology, first SESAM usually has drawbacks of cost-ineffective and time-consuming fabrication process lastly has limited pulse [8] compare to CNT which is simple and cost effective alternative [9]. Graphene said by Pi Li Huang et al good in optically visualized, high transperancy and linear absorption that can be characterized in optical properties. Graphene as a novel nanomaterial can be used as a fast SA with wide spectrum range [7] few-layered graphene is needed for the implementation as SA.

There are few complicated fabrication methods of few-layered graphene, such as chemical-vapor deposition (CVD) [10], the laser-induced surface decomposition method, pulsed-laser deposition (PLD), chemical synthesis and the mechanical exfoliation [11] however we filter out the easier alternatives with comparable saturable-absorption performance. Graphene is still the most favourite selection in most experiment recently in mode-locked and Q-switched fiber laser.

Fiber that doped with rare earth element mostly used in research as active gain medium [12][13] such as Ytterbium (Yb) doped (YDFL), Erbium Doped (EDFL), Er and Yb co-doped (EYDFL), Bismuth (Bi) doped (BDFL), Praseodymium (Pr) doped (PDFL) ,and Thulium (Th) doped (TDFL) [14].

In this work, we determined successful operation spectrum at 1530 nm of a Q-switched fiber laser. It only required a simple and cheap graphene SA. The graphene layer was fabricated using mechanical exfoliation method. This fiber laser is cooperation with two meters of erbium doped fiber (EDF) as a gain medium to amplified light when energy is pumped into active laser medium. The schematic setup between Q-switched and mode locked exactly the same. Q-switched fiber lasers ring are area of most interest because they have clear technical advantages. In a single ring, we use graphene as the SA because the features are almost ideal SA as fast recovery times, low saturation intensity, low cost and easy fabrication. The Q-switched fiber laser generation is compared with Optisystem software.

* Corresponding author: norazura@uthm.edu.my

2 Deposition and characterization grapheme on ferrule

The first stage in setting up Q-switched fiber laser is deposited graphene on fiber ferrule as SA which plays role as a passive Q-switched. The process deposition of graphene was done using mechanical exfoliation technique. First, flakes of graphite were pasted onto a strip of scotch tape and pressed and peeled off repeatedly, in order to enhance the graphene thickness. The result of graphene strip was then pressed against the end face of the optical fiber ferrule. The scotch tape was slowly peeled off so that the graphene layer remained on the end facet of the ferrule and was connected to another fiber pigtail via a connector (FC/PC), to form a simple SA device for the fiber laser as shown in Fig. 1.

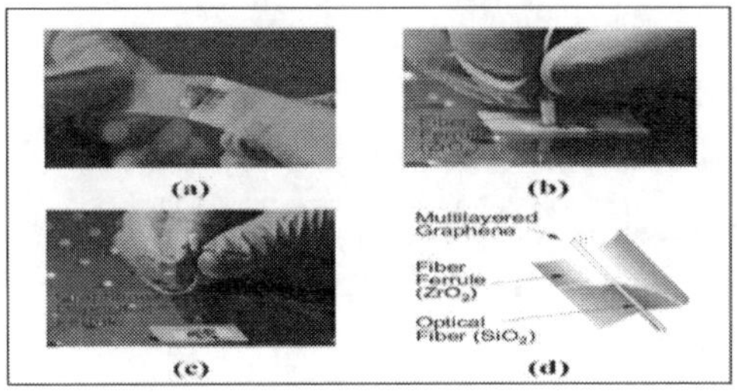

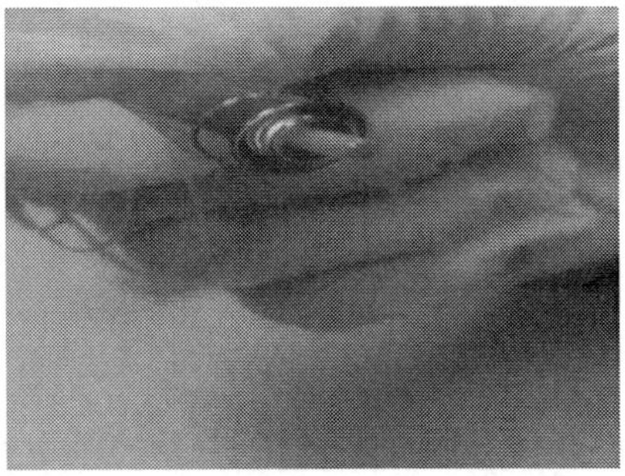

Fig. 1. Graphene deposited on ferrule by mechanical exfoliation

Fig. 2 shows (a) Single Natural Graphite flakes are opaque and have a metallic look under Meiji Microscope. After several times peels off, graphite flakes are identified under Optical Microscope as shown in Fig. 2(b). In Fig. 2(c) mechanical exfoliation processtransferred the graphite layers to the surface of the fiber ferrule identifying by fiberscope. Fig. 2(d) Scanning electron microscopy (SEM) of the optical fibers with graphite flakes optical deposited at core region.

(a)

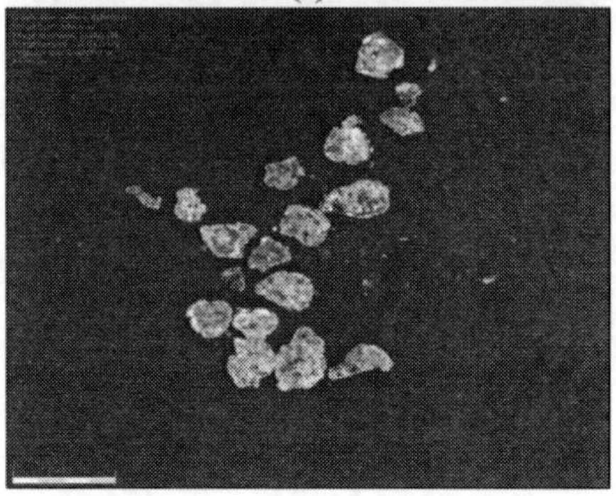

(b)

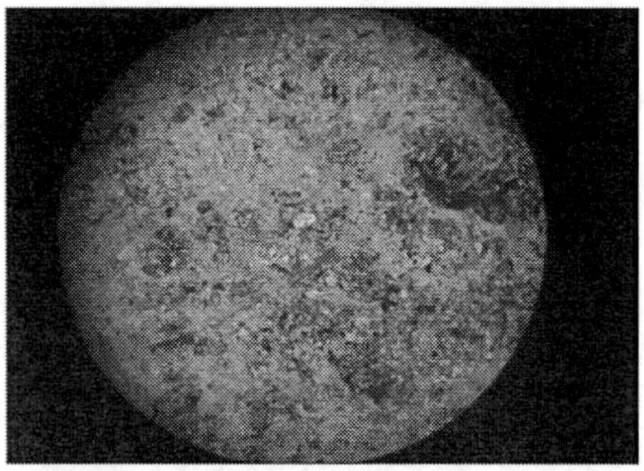

(c)

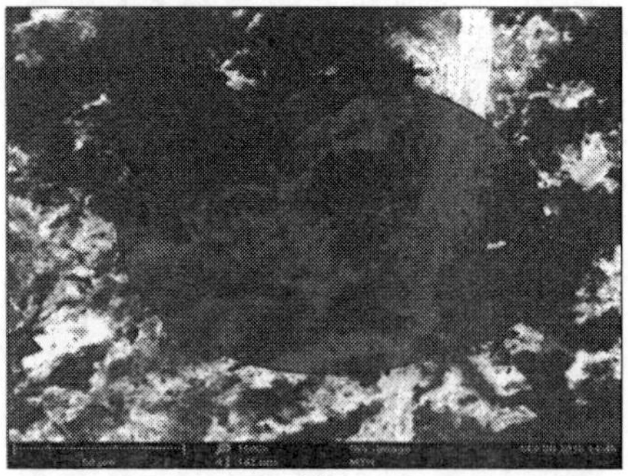

(d)

Fig. 2. Graphene under a) Meiji Microscope, b) Optical Microscope, c) Fiberscope, d) Scanning electron microscopy (SEM)

The graphene layers were transferred onto the fiber ferrule, a series of measurements including scanning electron microscopy (SEM) and energy dispersive X-ray spectroscopy (EDS) were conducted so that entirely cover of fiber core clearly seen. Fig. 3 shows the images of graphene on ferrule by mechanical exfoliation method under Semiconductor Electron Microscope (SEM). The image was zoomed on the core surface of the ferrule. Graphite flakes on ferrule close up using scanning electron microscope. SEM image micrograph of graphene ferrule on the core arrange loosely around the core. Dark grey spot on fiber core consist of high content carbon while light grey spot content low of carbon.

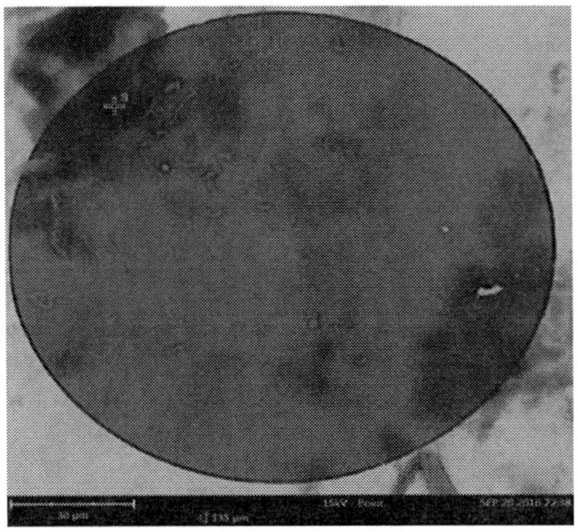

Fig. 3. Fiber-end with Graphene optically-deposited in the core area

Fig. 4 shows the EDS analysis on the ferrule deposited with graphene. The analysis of the percentage of element was placed on Table 1. The EDS have traced three elements on fiber ferrule such as silicon, carbon and zirconium. The EDS observed the silicon existed because the fiber ferrule is made of silicon which represented by the surface part of the ferrule.

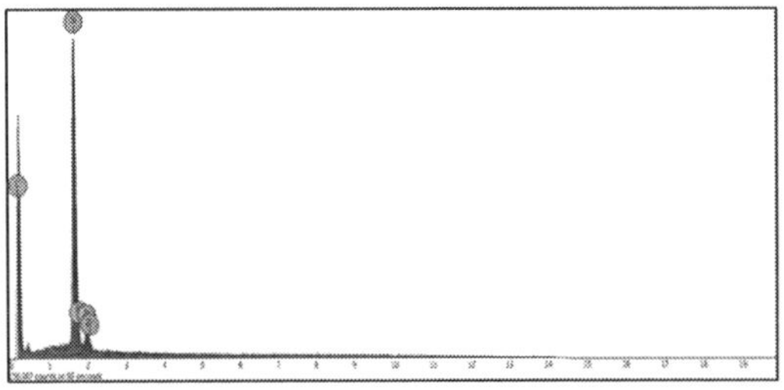

Fig. 4. Energy Dispersive Spectrum (EDS) of thegraphite flakes deposited onfiber ferrule.

Table 1. Percentage of elements on ferrule deposited with graphite.

Element	Weight (%)	Atomic (%)
Silicon	16.44	8.10
Carbon	79.14	91.22
Zirconium	4.42	0.67

The composition of the transferred graphene was confirmed by the energy dispersive spectroscopy (EDS). EDS is a technique that analyzes X-rays generated by the bombardment of the sample by an electron beam.The analysis of the spectroscopy data presented in Fig. 4 above confirms that the transferred material have a strong intensity peak of Carbon, along with small portions of Silicon and Zirconium that existed because of the fiber ferrules that were used. The silicon originate from the substrate ferrule which is the optical fiber made from high purity silicone dioxide (SiO2). Graphene has natural form of hexagonal crystal structure and consist of pure carbon. Based on study, chemicalsymbol of graphite is C referring to most well-known allotropes of carbon,which purely consists of carbon. It is proved under EDS the carbon existed on ferrule.

2 Experimental setup

Fig. 5 shows a design of Q-switched fiber laser. It includes a 980 nm laser diode as a pump source, a wavelength division multiplexer (WDM), a two meter long EDF as a gain medium, an optical isolator, polarization controller (PC), graphene SA prepared above, 90:10 dB output coupler, 50:50 output coupler, oscilloscope (OSC) and optical spectrum analyser (OSA). It was pumped by a 980 nm diode laser via a wavelength division multiplexer (WDM). An isolator was employed to ensure the unidirectional operation, a polarization controller employed to fine tune the linear cavity birefringence and utilized to adjust the polarization for Q-switched optimization. The light will transmit to the 90:10 directional optical coupler which 10% maintain as an outputand the another 90% will be continuously transmitted to the fiber ring. The 10% optical coupler connected to the 50:50 optical coupler which split the 10% of the output power to OSC and OSA by 50% respectively. The propagation processes would occur inside the coupler, so the output spectrum spectral resolution 20.0 nm/div and repetition rate were shown by OSA and OSC.

Experiments of Q-switched fiber laser can be compare between optical system software. Optisystem is the optical communication system design software that enables users to plan, test, and simulates optical links in the transmission layer of modern optical networks. This software can create a model of an optical system, execute the model and view measures of the system's performance. The optical system demonstrated a fiber laser, based on the passive Q-switched using SA as shown in Fig. 6. It is a unidirectional ring cavity and every component in the software application acted as a real component in the fiber laser.

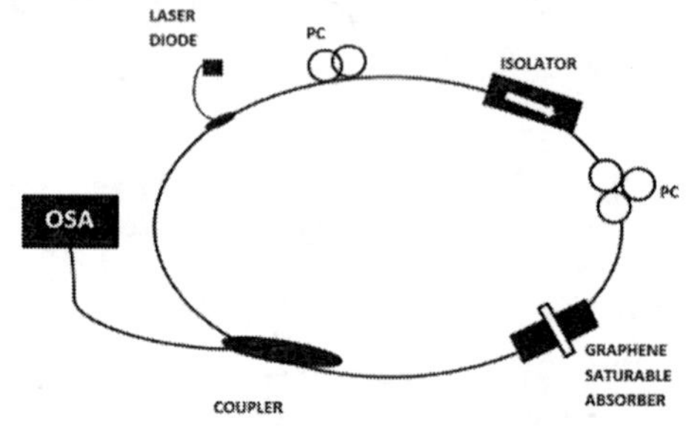

Fig. 5. Schematic configuration of single Q-switch fiber laser.

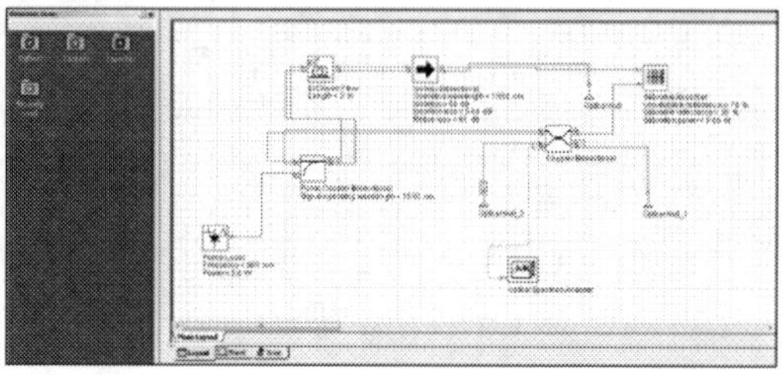

Fig. 6. Schematic circuit simulation for single ring in optisystem

3 Results and Discussion

In experimental, broad wavelength can be generated in the range of 1500 nm until 1600 nm in C-band and Q-switched is available this spectrum range which trends in this research. The Fig. 7 the optical spectrum taken from the OSA at central wavelength of 1550 nm. Lowest pump current 50 mA did not appear any single peak wavelength. Q-switched start to appear at current 300 mA and consistency increasing the peak with increasing the current value. The Q-switching operation starts at pump current 300 mA and continues until 600 mA.

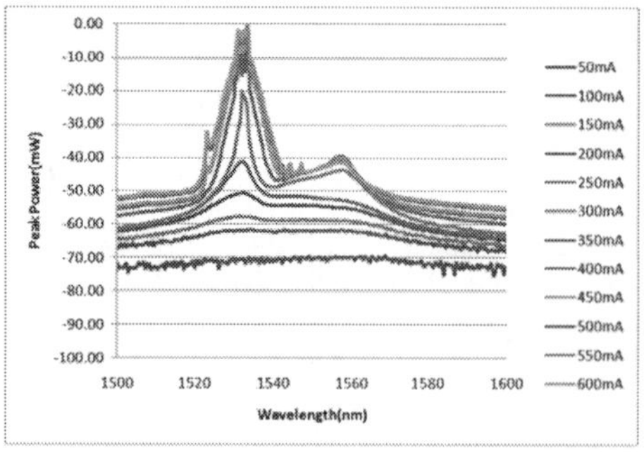

Fig. 7. Output spectra of the Q-Switch at different current pump power.

The spectrum seemed to have little change when increased the pump current and show a good Q-switched stability as shown in Fig. 7. The spectral bandwidthis 13.6 nm at 600 mA was centered at 1530 nm, measured by an optical spectral analyzer with resolution of 20nm/div. The maximum peak power was 2.53×10^{-6} mW at 600 mA pump current.

In experimental, stable Q-switched starts when pump power reach by slightly tuning the PC. The fiber laser can operate in a stable Q-switched state at optical spectrum as shown in Fig. 8. This output spectrum was obtained by using optical spectrum analyser (MS9740A). Q-switched works at the wavelength of 1530 nm with output power is generate spectral bandwidth 13.6 nm. For comparison purpose with simulation, the spectrum was combined in a single graph as shown in Fig. 4. Both of the experimental result and simulation result of Q-switched fiber laser was compared.

The optical system which using simulation demonstrates a fiber laser, based on passive Q-switched a SA, which efficient in producing short pulse, however due to limitation of study we study the bandwidth between experiment and simulation as shown in Fig. 8. By comparing, the spectral bandwidth of the simulation is 9×10^{-9} nm. A slight different value between this two bandwidth, can be said due to loss in fiber during progress experiment. The experimental result is more efficient compare to simulation because experimental exceed high power, which is needed the most to carry information for long haul – transmission.

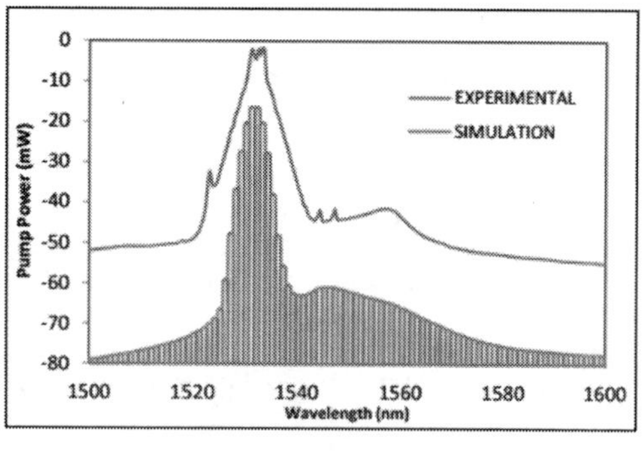

Fig. 8. Result comparison between experimental and simulation.

The definition of a threshold is the entrance or start of something. In Fig. 9 threshold cut level wavelength is 3.00 dB. This is to identify spectrum width ($\Delta\lambda$) for each pump power after q-switched started to appear. The spectrum width minor fluctuated between 2.00 - 3.00 nm started from pump current 350 mA which pump power 17.25 mW, threshold product 1.179 nm. Stable threshold can be seen during pump current 350 mA until 550 mA.

https://doi.org/10.1051/matecconf/201815001009

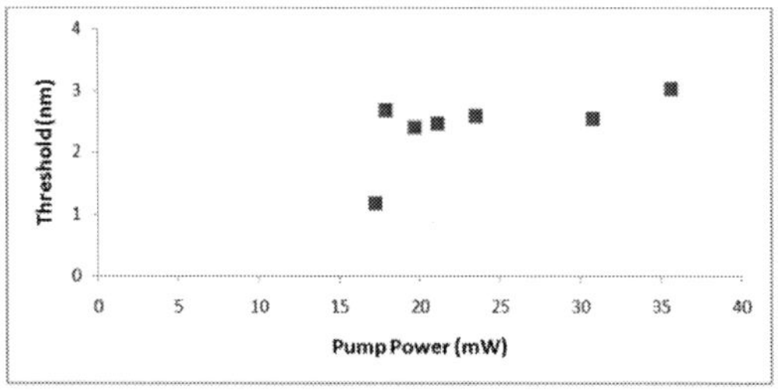

Fig. 9. Threshold against Pump Power

A stable Q-switched is self-started when pump power is fixed at pump power. The measured OSA were recorded every 5 min for 1 h as shown in Fig. 10. A stable operation is clearly observed for Q-Switched. It can be seen that the overall shape Q-switched spectra remain unchanged despite a slight time shifts and optical power drift due to environmental temperature changes.

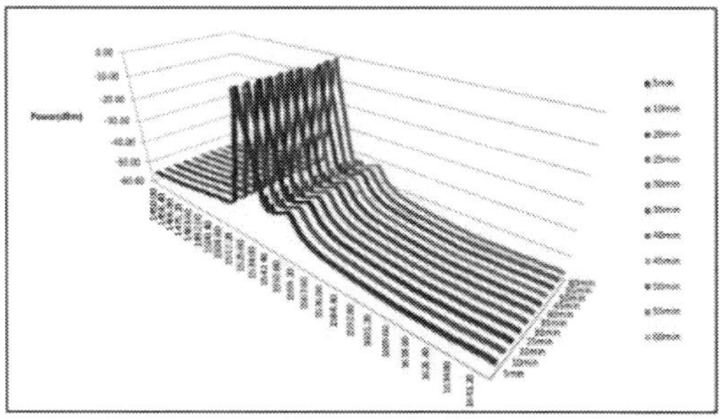

Fig. 10. The measured of OSA traces every 5 min for 1h for Q-Switching.

As expected, the increment the pump power lead to increment of the pulse train or repetition rate and decrement the pulse width. As the input pump was raised from 200 mW to 600 mW, the repetition rate expanded from 26.18 kHz to 52.23 kHz while the pulse width diminished starting from 0.2 µs to 0.1 µs. Fig. 11 show the pulse train from the oscilloscope for 600 mA. Then, toward each pump energy and pulse repetition rate, no amplitude modulations in these pulse trains were observed and the Q-switched pulse output was stable. This demonstrates that there might have been no self-mode locking impact throughout the Q-switching operation. Lastly, the value of pulse spacing is 0.2 µs.

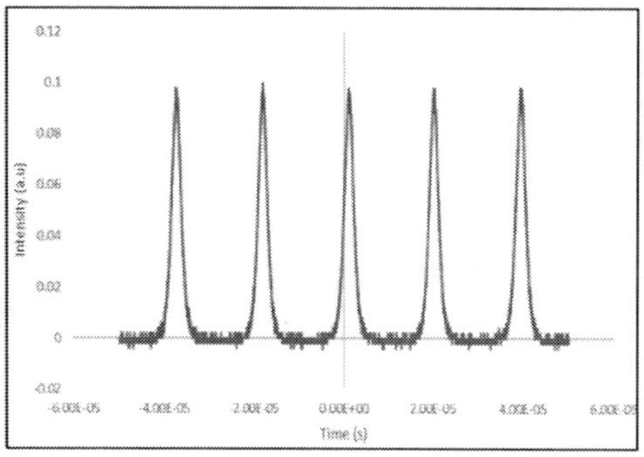

Fig. 11. Pulse train from the oscilloscope for 600 mA

4 Conclusions

In summary, Q-switching of erbium-doped fiber laser (EDFL) is reported by using graphene layer as a saturable absorber. The graphene layer SA was sandwiched between two fiber ferrules to act as a passive Q-switch device. CW operation is achieved at 10 mW, whereas stable self-starting Q-switching with a central wavelength of 1530 nm begins at 18 mW. In this paper, at 35 mW, the maximum pulse energy reaches at 2µJ with pulse repetition rate of 1 MHz and the narrowest pulse width is around 10µs is obtained. The stability of the pulse is verified from the radio-frequency (RF) spectrum with a measured signal-to-noise ratio (SNR) of 48 dB. This work highlights the feasibility and simplicity of graphene layer as SA for femtosecond pulse generation that may have great potential in many applications such as supercontinuum generation.

We would like to thanks to Universiti Tun Hussein Onn Malaysia (UTHM) for providing GPPS Grant 0f (U753), (U757), RAGS Grant (R046) for sponsoring this work.

References

1. H. Ahmad, N. E. Ruslan, M. A. Ismail, S. A. Reduan, C. S. J. Lee, S. Sathiyan, S. Sivabalan, and S. W. Harun, Appl. Opt., vol. **55** (2016)
2. Zhao L M, Bartnik A C, Tai Q Q and Wise F W 2013 *Opt. Lett.* **38** 1942
3. Kharenko D, Podivilov E, Apolonski A and Babin S 2012 *Opt. Lett.* **37** 4104
4. Aguergaray C, Broderick N G R, Erkintalo M, Chen J S Y and Kruglov V 2012 *Opt. Express* **20** 10545
5. Okhotnikov O G, Jouhti T, Konttinen J, Karirinne S and Pessa M 2003 *Opt. Lett.* **28** 364
6. Yamashita S, Inoue Y, Maruyama S, Murakami Y, Yaguchi H, Jablonski M and Set S Y 2004 *Opt. Lett.* **29** 1581

7. H Ahmad, Julian M Semangun, Saaidal R Azzuhri, M Z Zulkifli, N A Awang and S W Harun, Laser Phys. **25** (2015)
8. H. H. Kuo and S. F. Hong, J Nanomater., vol. **14**, (2014)
9. B. Xu, A. Martinez, S. Y. Set, C. S. Goh, and S. Yamashita, IEEE Photonics Technol. Lett., vol. **26** 2 (2014)
10. J. Sotor, I. Pasternak, A. Krajewska, W. Strupinski, and G. Sobon, Optics Exp. vol. **23**, no. 21, (2015).
11. J. Lee, J. Lee, J. Koo, and J. H. Lee, Appl. Opt., vol. **55**, no. 2, (2016).
12. H. Ahmad, J.M. Semangun, S.R. Azzuhri, M.Z., Zulkifli, N.A. Awang and S.W. Harun, Laser Physics, 25 (4) (2015)
13. A. Hamzah, M.C. Paul, N. A. Awang, H. Ahmad, M. Pal, S. Das, M. A. Ismail, and S. W. Harun Optics and Laser Technology, vol. **47** (2013)
14. Z. Sun, T. Hasan, and A. C. Ferrari, Phys. E Low-dimensional Syst. Nanostructures, vol. **44**, no. 6 (2012)

The effect of pump parameters on dual-pump fiber optical parametric amplifier

N. Othman[1], *N.S.* Mohd Shah[2], *K.G.* Tay[1,*], *N.A.* Cholan[1] and *R.* Talib[1]
[1]Faculty of Electrical and Electronics Engineering, Universiti Tun Hussein Onn Malaysia, 86400 Batu Pahat, Malaysia.
[2]Faculty of Engineering Technology, Universiti Tun Hussein Onn Malaysia, 86400 Batu Pahat, Malaysia.

Abstract. The impact of pump parameters on dual-pump (2-P) fiber optical parametric amplifier (FOPA) is investigated. The four-wave model of coupled amplitude equations with fiber losses and pump depletion are solved numerically for the calculation of the parametric gain. Simulation results indicate that an increase in pump powers does not only enhances the parametric gain but also flattening the amplification bandwidth. Moreover, separating the pumps wavelengths further from each other can also improve the flatness of the bandwidth. Besides that, the parametric gain flatness can be improved when the separation between the pump central wavelength and zero-dispersion wavelength is small. In essence, the optimal performance of 2-P FOPA can be achieved when the pump parameters are carefully tailored.

1 Introduction

Four-wave mixing (FWM) is a nonlinear effect which involves the parametric process, where a signal is amplified with the pump waves and consequently the new wave name idler is generated. This effect is detrimental in some of the applications. Nevertheless, FWM is still favorable for a few optical applications such as FWM-assisted lasers [1], optical regenerators [2] and parametric oscillators [3]. Fiber optical parametric amplifier (FOPA) is one of the devices that also exploits the effect of FWM [4]. There are two types of FOPA, i.e. one-pump (1-P) FOPA and dual-pump (2-P) FOPA. Both types of FOPA are capable in providing adjustable gain spectra and center frequency, wavelength conversion, phase conjugation, pulse operation for signal processing and 0-dB noise figure [5]. These advantages of FOPA have surpassed the limitation of conventional amplifiers i.e. Raman amplifier (RA) and Erbium-doped fiber amplifier (EDFA), and consequently become the interest of researchers to explore the potential of FOPA in exceeding the current limit of optical communication systems.

Practically, a FOPA is required to demonstrate good performance i.e. high parametric gain and large amplification bandwidth. The parameters such as pump wavelength and pump power do contribute to the FOPA performance [6]. As for 2-P FOPA, the pumps wavelengths separation will also affect the performance of FOPA amplification bandwidth [7]. In addition, a wide amplification bandwidth of 2-P FOPA can be obtained by adjusting the separation of pump central frequency, ω_c from zero-dispersion frequency. Besides that, a research conducted in [8] successfully verified that the amount of power for both pumps is crucial in determining the parametric gain of 2-P FOPA.

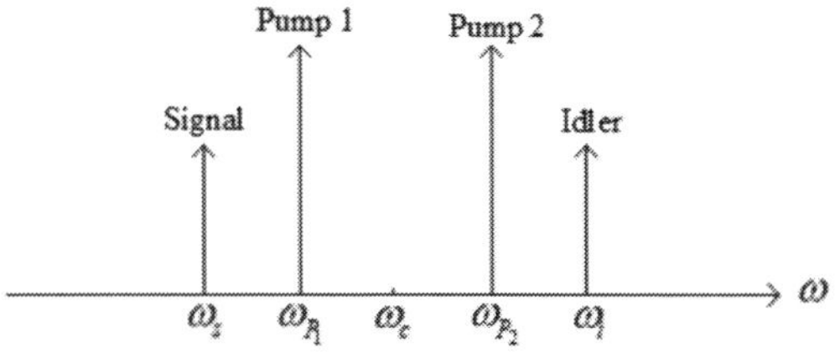

Fig. 1. Schematic of non-degenerate FWM in frequency domain.

This simulation work concentrate on four-wave model of FWM which is known as non-degenerate FWM. The schematic of the model is illustrated in Fig.1. The non-degenerate FWM process include the interaction of three waves which are two pump waves and a signal wave. The beat process of these three waves give rises to a new wave, idler. The aim of this paper is to numerically investigate the effect of pump parameters such as wavelength and power, on the four-wave model of 2-P FOPA. The performance of the parametric gain and amplification bandwidth are critically observed by varying the mentioned pump parameters. The highly-nonlinear dispersion shifted fiber (HNL-DSF) was used as a gain medium and pump depletion as well as fiber losses were taken into account. All simulations were carried on using Matlab 2014.

2 Mathematical model

The non-degenerate FWM process in Fig.1 can be represented by the amplitude equations for pump 1 (P_1), pump 2 (P_2), signal (s) and idler (i) along the fiber-length z. The equations are given by [5]

* Corresponding author: tay@uthm.edu.my

$$\frac{dA_{P_1}}{dz} = i\gamma\left[\left(|A_{P_1}|^2 + 2|A_{P_2}|^2 + 2|A_s|^2 + 2|A_i|^2\right)A_{P_1}\right] + 2i\gamma A_{P_2}^* A_s A_i e^{i\Delta\beta z} - \frac{\alpha}{2}A_{P_1}, \quad (1)$$

$$\frac{dA_{P_2}}{dz} = i\gamma\left[\left(|A_{P_2}|^2 + 2|A_{P_1}|^2 + 2|A_s|^2 + 2|A_i|^2\right)A_{P_2}\right] + 2i\gamma A_{P_1}^* A_s A_i e^{i\Delta\beta z} - \frac{\alpha}{2}A_{P_2}, \quad (2)$$

$$\frac{dA_s}{dz} = i\gamma\left[\left(|A_s|^2 + 2|A_{P_1}|^2 + 2|A_{P_2}|^2 + 2|A_i|^2\right)A_s\right] + 2i\gamma A_i^* A_{P_1} A_{P_2} e^{-i\Delta\beta z} - \frac{\alpha}{2}A_s, \quad (3)$$

$$\frac{dA_i}{dz} = i\gamma\left[\left(|A_i|^2 + 2|A_{P_1}|^2 + 2|A_{P_2}|^2 + 2|A_s|^2\right)A_i\right] + 2i\gamma A_s^* A_{P_1} A_{P_2} e^{-i\Delta\beta z} - \frac{\alpha}{2}A_i, \quad (4)$$

where γ is the fiber nonlinearity, α denotes the fiber loss and A_j^* for $j \in \{P_1, P_2, s, i\}$ represent the complex conjugate of A_j. The linear phase-mismatch, $\Delta\beta$ can be expressed as [9]

$$\Delta\beta = 2\sum_{m=1}^{\infty}\frac{\beta_{2m}}{(2m)!}\left[(\Delta\omega_s)^{2m} - (\Delta\omega_P)^{2m}\right], \quad (5)$$

where β_{2m} is higher-order dispersion coefficient given by the $2m^{th}$ derivative of mode-propagation constant $\beta(\omega)$, at the central frequency $\omega_c = (\omega_{P_1} + \omega_{P_2})/2 = (\omega_s + \omega_i)/2$. Meanwhile, $\Delta\omega_s = \omega_s - \omega_c$ and $\Delta\omega_P = \omega_{P_1} - \omega_c = (\omega_{P_1} + \omega_{P_2})/2$.

Generally, Eqs. (1)-(4) are solved numerically by using the fourth-order Runge-Kutta method. Then, the parametric gain (in dB) at the respective wavelength are calculated by the ratio of the output signal power, $P_{s,out}$ to the input signal power, $P_{s,in}$ such as

$$G = 10\log\left(\frac{P_{s,out}}{P_{s,in}}\right), \quad (6)$$

where $P_s = |A_s|^2$, for both input and output signal powers.

3 Result and Discussions

In order to investigate the performance of 2-P FOPA effectively, the parametric gains which have been obtained by using (6) were plotted against the signal wavelengths. The parametric gains were computed by varying the parameters such as the power of P_1 and P_2, the separation between two pumps wavelength, $\Delta\lambda_P = \lambda_{P_2} - \lambda_{P_1}$ and the wavelength distance of the central wavelength from zero-dispersion wavelength (ZDW), $\lambda_c - \lambda_0$. A 500 m-length HNL-DSF of OFS company was used in this simulation work and the fiber has ZDW at $\lambda_0 = 1556.5$ nm, $\alpha = 0.82$ dB/km and $\gamma = 11.5$ W^{-1}km^{-1}.

Firstly, the parametric gain was computed by manipulating the power of both pumps. The other parameters were fixed as $P_{s,in} = -40$ dBm, $\lambda_{P_1} = 1548.1$ nm, $\lambda_{P_2} = 1568.03$ nm and $\lambda_c = 1558$ nm. The second-order and fourth-order dispersion coefficients at $\lambda_c = 1558$ nm are $\beta_2 = -1.97 \times 10^{-2}$ ps^2/km and $\beta_4 = 6.231 \times 10^{-5}$ ps^4/km, respectively. The resulted parametric gains were plotted in Fig. 2. As predicted, the parametric gain increases as the pumps power increased. This is similar with 1-P FOPA behaviour when the pump power increases [10]. However, unlike in the 1-P FOPA, the increase of pump power in the 2-P FOPA does not only affects the amplification bandwidth but also its flatness. As is seen, when the power of $P_1 = 1$ W, the bandwidth of 2-P FOPA is flatter if compared with the bandwidth of the lowest pump power, $P_1 = 0.25$ W. This implies that rather than just increasing the parametric gain and amplification bandwidth the increasing pump power in 2-P FOPA also resulted in flatter gain spectrum.

Next, in order to attain optimum performance of 2-P FOPA, the selection of both pumps wavelengths are crucial. The selected pump wavelengths will determine the separation between two pumps wavelengths. Hence, this time the parametric gains were calculated while manipulating the distance between P_1 and P_2 wavelengths, $\Delta\lambda_P$. In the meantime, the other parameters value were assigned as $P_1 = P_2 = 0.5$ W and $\lambda_c \approx 1558$ nm, thus $\lambda_c - \lambda_0 \approx 1.5$ nm. The value of β_2 and β_4 at λ_c are similar as before. The obtained parametric gain are

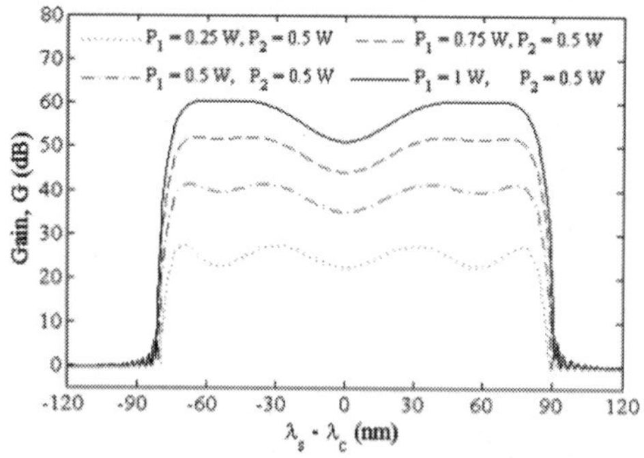

Fig. 2. Gain spectrum for different amount of pump power.

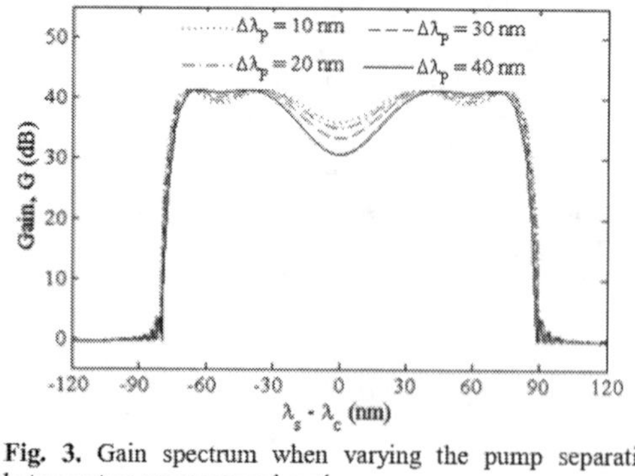

Fig. 3. Gain spectrum when varying the pump separation between two pumps wavelengths.

exhibited in Fig. 3. It shows that the flatness of amplification bandwidth changed when 2-P FOPA experience different values of $\Delta\lambda_p$. When the pumps wavelengths are positioned further from each other, it shows that the bandwidth flatness at the signal wavelength far from the central wavelength, λ_c is improved. However, the parametric gain corresponds to the signal wavelength near λ_c shows significant reduction as the $\Delta\lambda_p$ increases.

Research in [8] proved that the flatness near the λ_c can be engineered by adjusting the λ_c position with reference to the ZDW of HNL-DSF. Thus, to overcome the reduction of the parametric gain near λ_c, the parametric gain was computed while varying the distance of λ_c from λ_0. The results were plotted in Fig. 4. The other parameters are fixed as in Fig. 3 except for the λ_{p_1}, λ_{p_2} and the corresponding λ_c as well as β_2 and β_4. Although the λ_{p_1} and λ_{p_2} are different in each calculation which depend on the desired λ_c, the $\Delta\lambda_p$ is fixed to 20 nm. There are four different values of λ_c that are used in calculating the parametric gain, and the corresponding higher-order dispersion coefficients given as $\beta_2 = 2.239 \times 10^{-2}$ ps^2/km and $\beta_4 = 6.022 \times 10^{-5}$ ps^4/km for $\lambda_c = 1555.3$ nm, $\beta_2 = 0$ ps^2/km and $\beta_4 = 6.009 \times 10^{-5}$ ps^4/km for $\lambda_c = \lambda_0 = 1556.5$ nm, $\beta_2 = -3.169 \times 10^{-2}$ ps^2/km and $\beta_4 = 6.292 \times 10^{-5}$ ps^4/km for $\lambda_c = 1558.13$ nm, and as for $\lambda_c = 1558$ nm the value of β_2 and β_4 are similar as previous. As is seen, by bringing λ_c close to λ_0 i.e. $\lambda_c - \lambda_0 = 1.5$ nm, it is possible to attain broader and flatter amplification bandwidth far from λ_c as well as higher gain. Noteworthy, the parametric gains near λ_c are slightly increase. Conversely, by increasing the distance of $\lambda_c - \lambda_0$, the 2-P FOPA gain spectrum totally loses its uniformity. However, the assignment of λ_c in the normal regime (less than or exactly at λ_0) i.e. $\lambda_c - \lambda_0 = 0$ nm and -1.2 nm, need to be avoided. This is because in the normal regime the resulted phase-mismatch is large, and as a consequence, the poor gain spectrum is obtained. Hence, $\lambda_c - \lambda_0$ need to be positive and small if a wide and flat 2-P FOPA spectrum is desired.

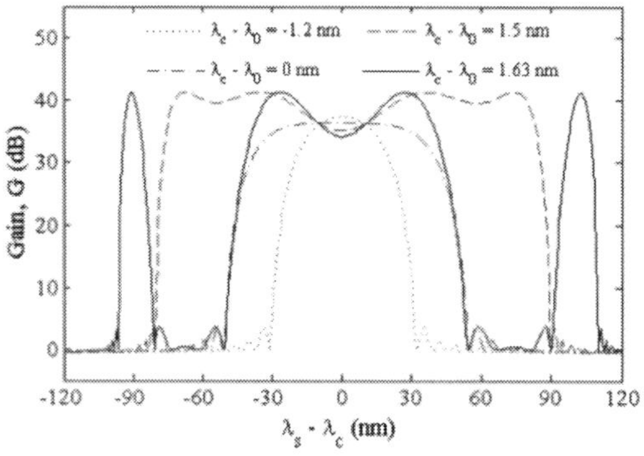

Fig. 4. Gain spectrum when varying the wavelength distance between the central wavelength and ZDW.

4 Conclusion

This paper has numerically investigated the impact of pump parameters on 2-P FOPA. The optimum parametric gain require high pump powers. The high pump powers also lead to a flatter amplification bandwidth. The flatness of the amplification bandwidth can also be enhanced by positioning the pumps wavelengths further from each other. However, the parametric gains near the central wavelength experience a reduction as the distance between two pumps wavelengths become larger, hence reduced the bandwidth flatness near the central wavelength. This problem can be overcome by adjusting the central wavelength close to ZDW, then a flatter 2-P FOPA amplification bandwidth can be attained.

This work was supported by Fundamental Research Grant Scheme (FRGS) Vot 1537.

References

1. N. A. Cholan, M. H. Al-Mansoori, A. S. M. Noor, A. Ismail, M. A. Mahdi, Appl. Phys. B Lasers Opt. **115**, 251–256 (2014)
2. N. S. Mohd Shah, M. Matsumoto, Opt. Commun. **284**, 4687–4694 (2011)
3. B. Sun, K. Hu, D. Chen, Y. Wei, S. Gao, S. He, J. Light. Technol. **30**, 1937–1942 (2012)
4. M. E. Marhic, P. A. Andrekson, P. Petropoulos, S. Radic, C. Peucheret, M. Jazayerifar, Laser Photon. Rev. **9**, 50–74 (2015)
5. M. E. Marhic, *Fiber Optical Parametric Amplifiers, Oscillators and Related Devices* (Cambridge University Press, United Kingdom, 2008)
6. P. S. Maji, P. R. Chaudhuri, Appl. Opt. **54**, 3263–3272 (2015)
7. N. Wong, K. K. Y. Wong, Opt. Commun. **272**, 514–520 (2007)
8. P. Parolari, L. Marazzi, E. Rognoni, M. Martinelli, J. Light. Technol. **23**, 2524–2530 (2005)
9. M. E. Marhic, F. S. Yang, L. G. Kazovsky, Y. Park, Opt. Lett. **21**, 1354–1356 (1996)
10. N. Othman, N. S. Mohd Shah, K. G. Tay, N. A. Cholan, *International Conference on Advances in Electrical, Electronic and System Engineering 2016*, 326–330 (2016)

Design a Fuzzy Logic Controller for a Rotary Flexible Joint Robotic Arm

Jamaludin Jalani*, and *Suman* Jayaraman

Department of Electrical Engineering Technology, Facuty of Engineering, University Tun Hussein Onn Malaysia, Beg Berkunci 101, 86400, Parit Raja, Batu Pahat, Johor Darul Takzim

Abstract. The purpose of this research is to design a fuzzy logic feedback controller (FLC) in order to control a desired tip angle position a rotary flexible joint robotic arm. The FLC is also employed to dampen the vibration emanated from a rotary flexible joint robotic arm when reaching a desired tip angle position. The performance of FLC is tested in simulation and experiment. It is found that the FLC is successfully designed, applied and tested. The results show that fuzzy logic controller performed satisfactorily control a desired tip angle position and reduce the oscillations.

1 Introduction

Fuzzy Logic Control has been renowned control approach over the centuries. Fuzzy set theory was first introduced in a seminar paper by Lotfi Zadeh. It was published in the rather obscure journal of *Information and Control* [1]. Zadeh an electrical engineer by training was one of the leading authorities in control theory in the 1950s and early 1960s. However, during the process of writing a book on linear systems with Charles Desoer in 1963, he noticed that in spite of the richness of the existing mathematical theory of control, we have been able to deal with a very special case of systems that are linear and time invariant, or nonlinear but with a specific property [2]. He traced this problem back to the Aristotelian notion of absolute truth and falsehood, and generalized such notion to the case of partial truth and partial membership in a set.

By the late 1980's, with the advent of fuzzy chips, fuzzy logic was established successfully in Japan's industry. In contrast, the US contributed less application of fuzzy logic for their industry. This mainly due to the fact that, fuzzy logic lacked basic theoretical verifications concepts such as stability and robustness (i.e. mathematical model not available) which normally used by theoreticians for classical control. When a mathematical model is not available, it makes little sense to talk about stability or any other structural properties. Several researchers have tried to come up with stability conditions for fuzzy systems and have reported some success [3]. However, the primary goal of fuzzy logic is to develop an alternative to mathematical modeling for systems which either lack a proper mathematical model because it is either too ill-defined, or the model is so complicated that it is of no practical use [4].

The application of fuzzy logic began in 1974, when Mamdani and his associates used a fuzzy logic controller to control the temperature in a rotary cement kiln [5]. After that, more applications of fuzzy logic control were presented by researchers all over the world in various fields [6]–[12]. For example, the FLC has been used in pattern recognition, classification, and clustering and presented in [13]. Several applications for robotic have been reported in [4] and [5]. Moreover, [16] had applied the FLC to suppress the swing of gantry crane system. Meanwhile, application of fuzzy logic for renewable energy system has been presented by [17]. More application of FLC can be found in [18]–[20].

The application of FLC to control a rotary flexible joint module has been reported by several researchers. [21] had developed a composite collocated proportional-derivative (PD) and fuzzy logic control for tracking and vibration control. [22] had proposed a cascade fuzzy logic controller (FLC) to remove link vibrations and to obtain fast trajectory tracking performance. [23] had modelled and controlled the flexible joint for vertical orientation. Other control strategies for controlling a rotary flexible joint module can be found in [24], [25] and [26]. Although a fuzzy control approach has been used in their work, most of the control technique is complex and not straight forward. Hence a direct fuzzy control strategy is proposed for a rotary flexible joint in this paper. In particular, a fuzzy Mamdani is employed to overcome oscillations emanated from a rotary flexible joint robotic arm.

2 Mathematical Equation

In brief, Figure 1 depicts the flexible joint module coupled to the SRV02 plant. The Module is attached to the SRV02 load gear by two thumbscrews. The Main Arm is attached to the module body by two identical springs thus resulting in the flexible joint. In this paper, only the state space representation of the complete

* Corresponding author: jamalj@uthm.edu.my

system is given. This allows us to investigate the fuzzy logic control performance in simulation. The detailed derivation of the mathematical equation of flexible joint module can be found in [15].

Fig. 1. Flexible Joint Module

$$\begin{bmatrix} \dot{\theta} \\ \dot{\alpha} \\ \ddot{\theta} \\ \ddot{\alpha} \end{bmatrix} = \begin{bmatrix} 0 & 0 & 1 & 0 \\ 0 & 0 & 0 & 1 \\ 0 & \dfrac{K_{stiff}}{J_{eq}} & \dfrac{-\eta_m \eta_g K_t K_m K_g^2 + B_{eq} R_m}{J_{eq} R_m} & 0 \\ 0 & \dfrac{-K_{stiff}(J_{eq} + J_{ARM})}{J_{eq} J_{ARM}} & \dfrac{\eta_m \eta_g K_t K_m K_g^2 + B_{eq} R_m}{J_{eq} R_m} & 0 \end{bmatrix} \begin{bmatrix} \theta \\ \alpha \\ \dot{\theta} \\ \dot{\alpha} \end{bmatrix} + \begin{bmatrix} 0 \\ 0 \\ \dfrac{\eta_m \eta_g K_t K_g}{J_{eq} R_m} \\ \dfrac{-\eta_m \eta_g K_t K_g}{J_{eq} R_m} \end{bmatrix} V_1 \qquad (1)$$

3 Experimental Set Up

Similar experimental set up as in [24] is used for controlling the rotary flexible joint. In brief, WinCon™ is a real-time Windows 98/NT/2000/XP application. It allows running code generated from a Simulink diagram in real-time on the same PC (also known as local PC) or on a remote PC. Data from the real-time running code may be plotted on-line in WinCon Scopes and model parameters may be changed on the fly through WinCon Control Panels as well as Simulink. The automatically generated real-time code constitutes a stand-alone controller (i.e. independent from Simulink) and can be saved in WinCon Projects together with its corresponding user-configured scopes and control panels.

4 Fuzzy Logic Controller

The goal of this project is to dampen oscillations of a rotary flexible joint robotic arm with fuzzy systems. The fuzzy logic controller is used because it is an innovative technology to design solutions for multi-parameter and non-linear control problems. In addition, it uses human experience and experimental results rather than a mathematical model for the definition of a control strategy. Mathematical models and difference equations generate crisp descriptions of systems. This is fine if the mathematical properties and physical laws of the system are known or can be calculated. However, for non-linear processes, the underlying dynamics of the system can be too difficult, or indeed, impossible to model. In these situations, it is more useful to describe the system as a series of if-then rules. This is essentially what a fuzzy model is. It is a mapping of input space to output space by means of a rule base. Hence, the model requires no strict mathematical equations and the range of uses for the fuzzy model is vast. Figure 2 shows how fuzzy logic is used to control the position and to dampen the oscillation.

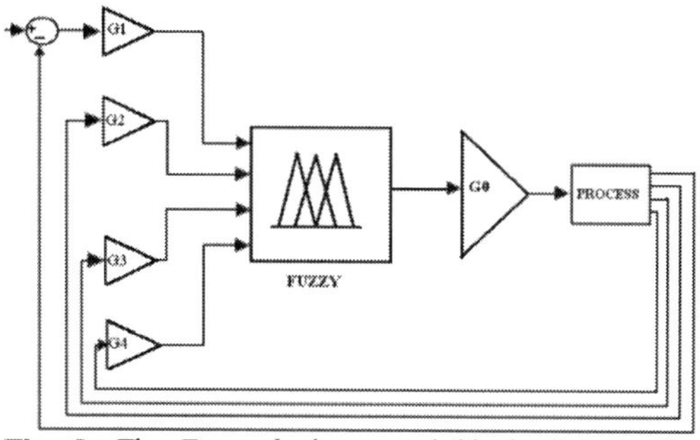

Fig. 2. The Fuzzy logic control block diagram for controlling flexible joint module

In general, the fuzzy controller has four components namely rule-base, interface mechanism, fuzzification interface and defuzzification interface as shown in Figure 3. A brief description of each component is as follows:

- ***The rule-base***
  Here, the knowledge for the control of the system is held as a set of IF – THEN statements.
- ***The inference mechanism***
  The inference mechanism evaluates which control rules are relevant at the current time and then decides what the input to the plant should be.
- ***Fuzzification interface***
  The fuzzification interface can be regarded as the input interface. It has the function of modifying/scaling the inputs so that they may compare to the rules in the rule base.
- ***Defuzzification interface***
  The defuzzification interface acts as an output interface. The output from the inference mechanism is converted so that it may be fed into the input for the process.

Figure 4 depicts the Rotary Flexible Joint controller with the fuzzy logic controller used for the experiment. Notice that both the actual system and an exact simulation are running in parallel thus to allow us comparing the actual and simulated results. This model has the I/O connection blocks linking to the physical plant as well as a simulated block to compare experimental and simulated results.

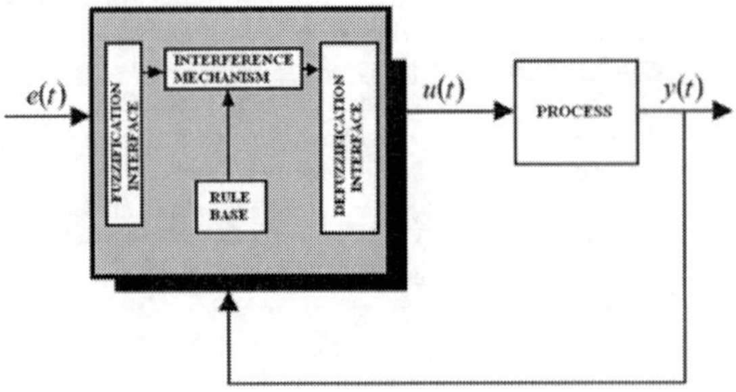

Fig. 3. Control system with fuzzy controller

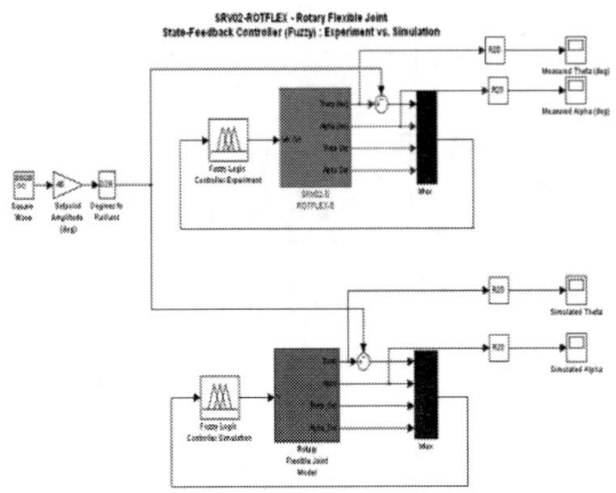

Fig. 4. Rotary Flexible Joint controller with the Fuzzy Logic controller.

4.1 Fuzzy Mamdani Controller

4.1.1 Input membership function

Theta Error (rad)	→	neg-high, neg-low, zero, pos-low, pos-high.
Alpha (rad)	→	neg-high, neg-low, zero, pos-low, pos-high.
Theta Dot	→	neg, zero, pos.
Alpha Dot	→	neg, zero, pos.

4.1.2 Output membership function

Voltage (V) → neg, zero, pos

4.1.3 Fuzzy Logic Rule-Base

Fuzzy Logic Rule-Base applied for the rotary Flexible Joint Robotic Arm are as follows:

R1 : If (thetaerror is zero) and (alpha is zero) and (thetadot is zero) and (alphadot is zero) then (Voltage is zero)

R2 : If (thetaerror is neg-high) and (alpha is neg-high) and (thetadot is pos) and (alphadot is negative) then (Voltage is zero)

R3 : If (thetaerror is pos-high) and (alpha is pos-high) and (thetadot is neg) and (alphadot is positive) then (Voltage is zero)

R4 : If (thetaerror is neg-low) and (alpha is pos-high) and (thetadot is pos) and (alphadot is zero) then (Voltage is zero)

R5 : If (thetaerror is pos-low) and (alpha is neg-high) and (thetadot is neg) and (alphadot is zero) then (Voltage is zero)

R6 : If (thetaerror is neg-low) and (alpha is pos-low) and (thetadot is zero) and (alphadot is positive) then (Voltage is pos)

R7 : If (thetaerror is pos-low) and (alpha is neg-low) and (thetadot is zero) and (alphadot is negative) then (Voltage is neg)

R8 : If (thetaerror is neg-high) and (alpha is neg-high) and (thetadot is pos) and (alphadot is negative) then (Voltage is pos)

R9 : If (thetaerror is pos-high) and (alpha is pos-high) and (thetadot is neg) and (alphadot is positive) then (Voltage is neg)

R10 : If (thetaerror is zero) and (alpha is zero) and (thetadot is zero) and (alphadot is positive) then (Voltage is zero)

R11 : If (thetaerror is zero) and (alpha is zero) and (thetadot is zero) and (alphadot is negative) then (Voltage is zero)

R12 : If (thetaerror is zero) and (alpha is pos-high) and (thetadot is zero) and (alphadot is zero) then (Voltage is zero)

R13 : If (thetaerror is zero) and (alpha is neg-high) and (thetadot is zero) and (alphadot is zero) then (Voltage is zero)

R14 : If (thetaerror is zero) and (alpha is neg-low) and (thetadot is zero) and (alphadot is zero) then (Voltage is zero)

R15 : If (thetaerror is zero) and (alpha is pos-low) and (thetadot is zero) and (alphadot is zero) then (Voltage is zero)

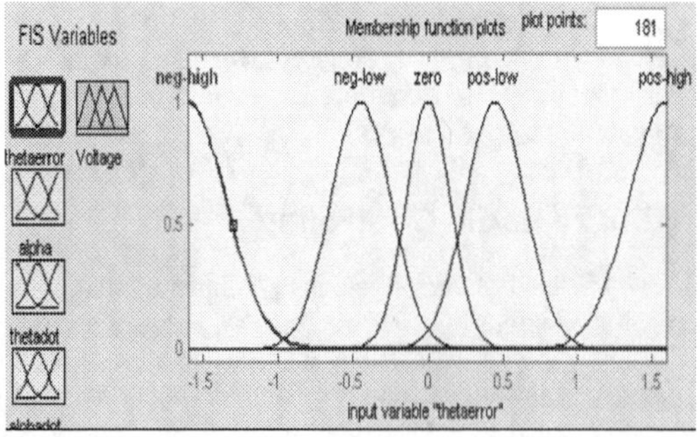

Fig. 5. FSMF for Theta Error.

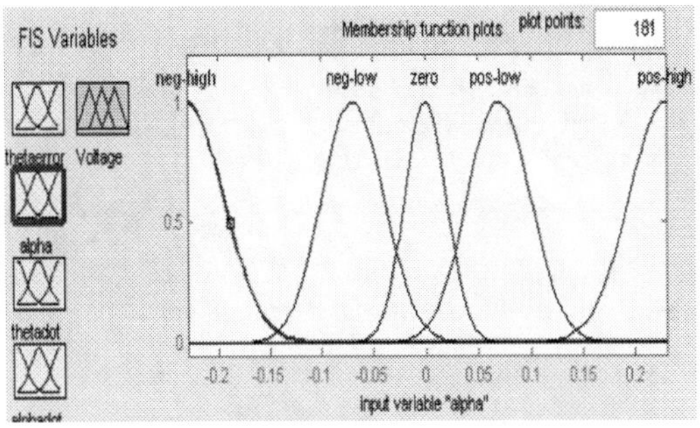

Fig. 6. FSMF for Alpha

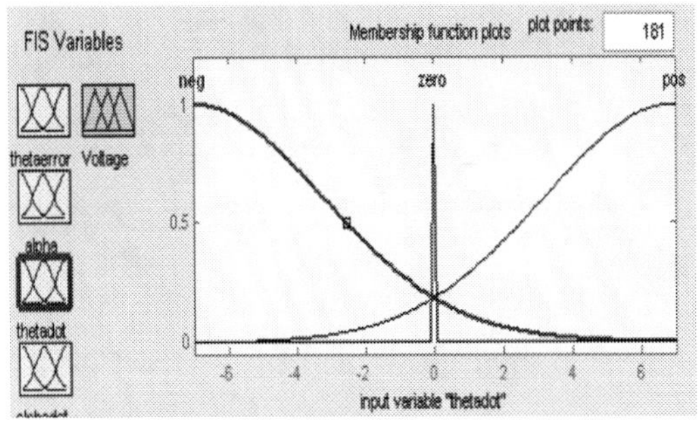

Fig. 7. FSMF for Theta Dot

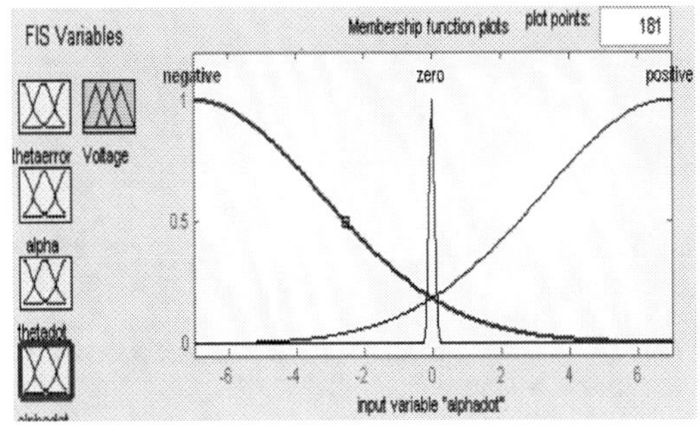

Fig. 8. FSMF for Alpha Dot

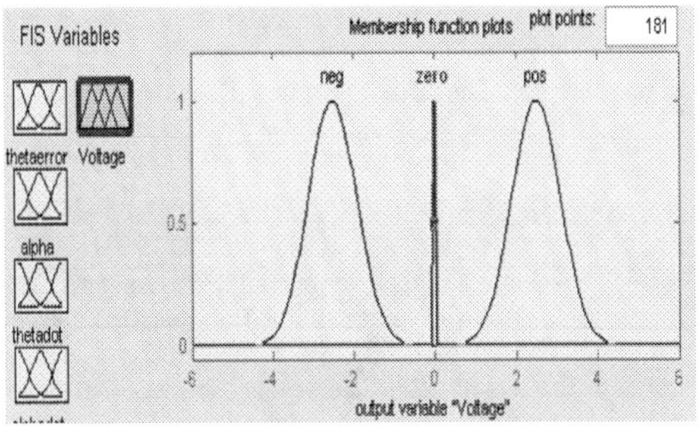

Fig. 9. FSMF for Voltage

The fuzzy set membership functions (FSMF) for the inputs are illustrated in Figure 5 to Figure 8. Meanwhile, the FSMF for the output illustrated in Figure 9.

5 Results

System performances by using Mamdani fuzzy controller is compared with the system without a controller in simulation and experiment. Simulated results are presented in Figure 10 and Figure 11. Meanwhile, experimental results are depicted in Figure 12 and Figure 13. The results show that the servo load angle did not achieve 45°. The maximum arm deflection angle is about 8°.

5.1 Without Controller

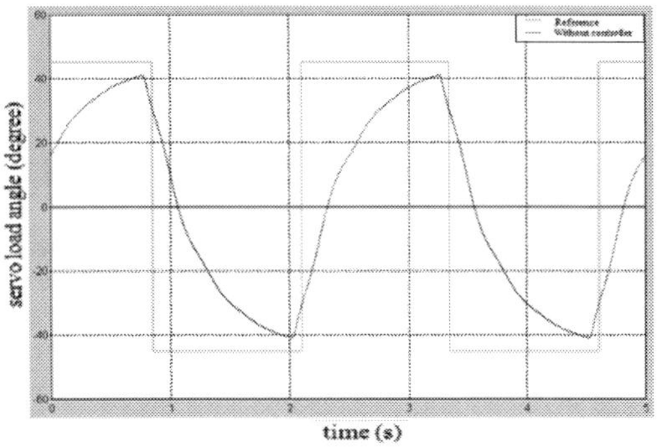

Fig. 10: Simulated plot for servo load angle without a controller.

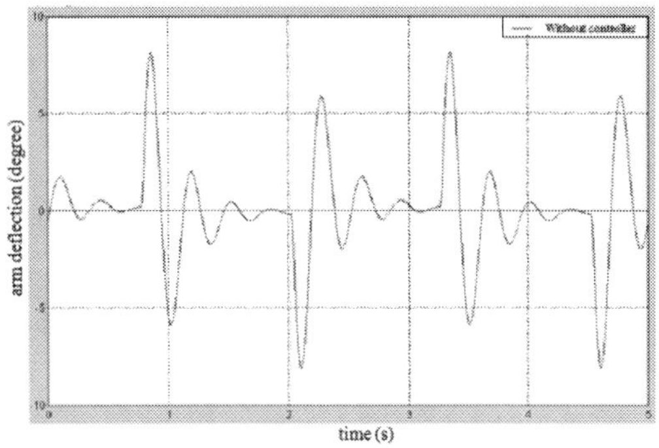

Fig. 11: Simulated plot for arm deflection angle without controller

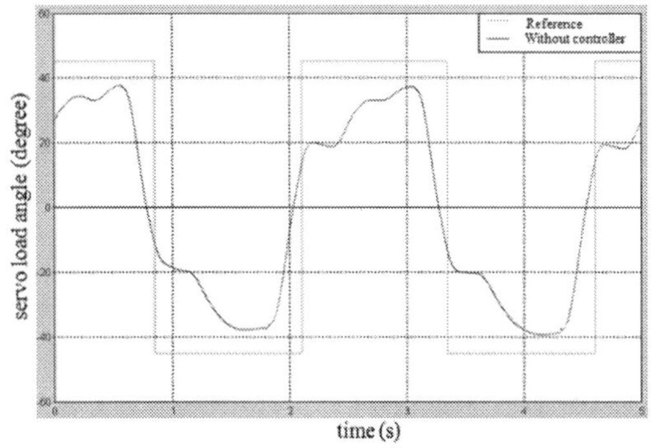

Fig. 12. Experimental plot for servo load angle without controller

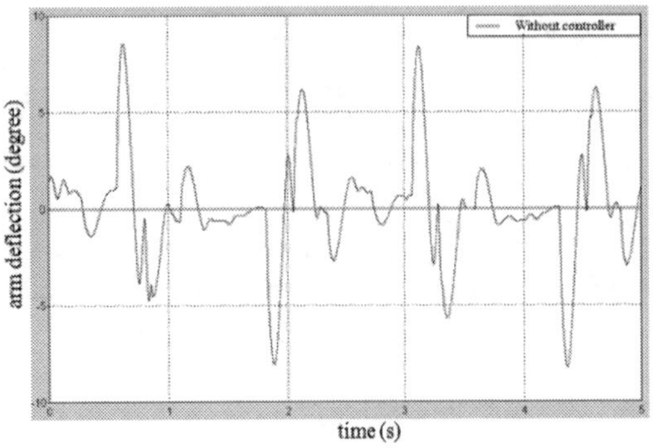

Fig. 13. Experimental plot for arm deflection angle without controller

5.2 Simulation Results

Simulated results for servo load angle and arm deflection angle with Mamdani Fuzzy controller is presented in Figure 14 and Figure 15.

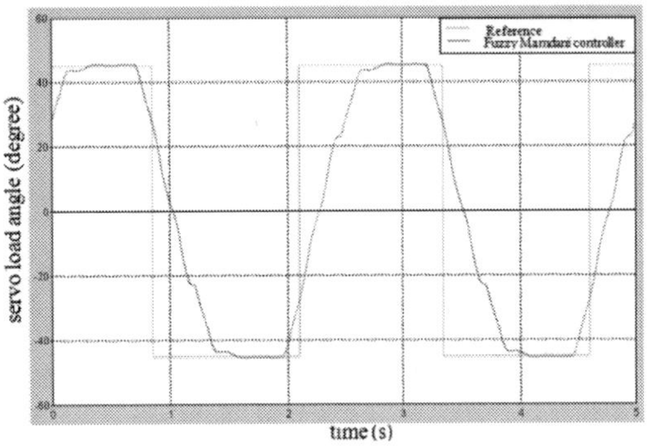

Fig. 14. Simulated plot for servo load angle with Fuzzy Mamdani controller

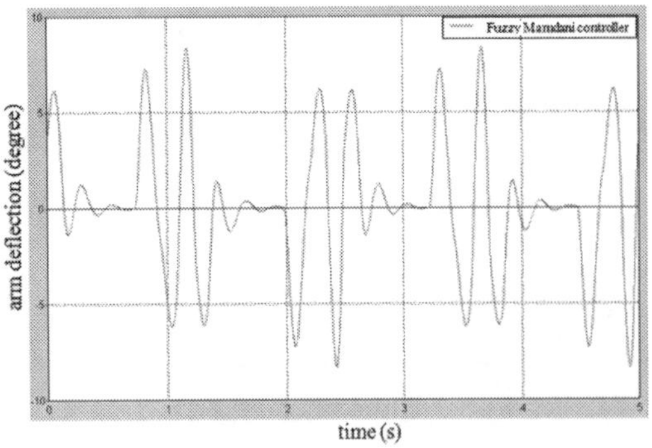

Fig. 15. The simulated plot for arm deflection angle with Fuzzy Mamdani controller.

5.3 Experimental Results

Experimental results for servo load angle and arm deflection angle with Fuzzy Mamdani controller is presented in Figure 16 and Figure 17.

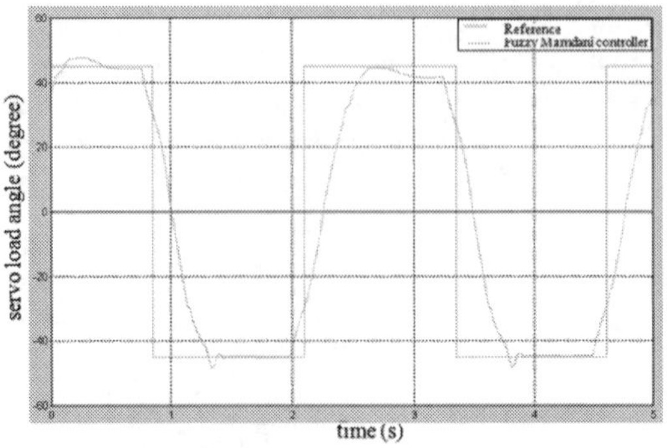

Fig. 16. Experimental plot for servo load angle with Fuzzy Mamdani controller

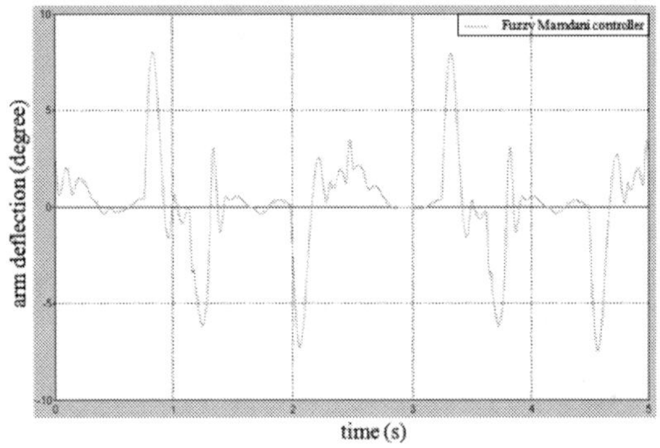

Fig. 17. The experimental plot for arm deflection angle with Fuzzy Mamdani controller.

A slight overshoot occurs in the experimental plot for servo load angle compared to the simulated plot. Meanwhile, the result of arm deflection angle results showed that the experimental plot is better than simulated plot. The arm deflection angle is about 8°.

6 Conclusion

The fuzzy logic controller was applied, designed and tested in the Rotary Flexible Joint Robotic Arm system. The Gaussian membership function was considered to develop the fuzzy rule. The results showed that fuzzy controller performed satisfactorily to suppress the oscillations. In future, the FLC should be further tuned so that the performances for servo load angle and arm deflection angle become better. The design may consider different membership function such as triangle and trapezoidal. The Fuzzy Sugeno may also be considered for future control design.

The authors wish to thank the Faculty of Engineering Technology, Universiti Tun Hussein Onn Malaysia for providing a platform to carry out the research activities.

References

1 Zadeh and L. A, "Fuzzy Sets," *Information and Control*, vol. 8, no. 3, pp. 338–353, 1965.
2 L. A. Zadeh and C. A. Desoer, Linear System Theory: The State Space Approach. McGraw-Hill, 1963.
3 S. Thongchai and K. Kawamura, "Application of Fuzzy Control to a Sonar-Based Obstacle Avoidance Mobile Robot," Proceeding of the IEEE International Conference on Control Applications, pp. 425–430, 2000.
4 Pedro Albertos and Antonio Sala, "Fuzzy logic controllers Advantages and drawbacks," in *IEEE Transactions On Control System*, pp. 833–844.
5 E. H. Mamdani, "Application of Fuzzy Logic to Approximate Reasoning Using Linguistic Synthesis," *IEEE Transaction Computer*, vol. C–26, no. 12, pp. 1–7, 1977.
6 D. D. B. D. Rubin, G. Baselli, G. F. Inbar, and S. Cerutti, "An Adaptive Neuro-Fuzzy Method (ANFIS) for Estimating Single-Trial Movement-Related Potentials," *Biological Cybernetics*, vol. 75, pp. 63–75, 2004.
7 L. Birglen and C. M. Gosselin, "Fuzzy Enhanced Control of an Underactuated Finger Using Tactile and Position Sensors," *Proceeding IEEE International Conference on Robotics and Automation*, no. April, pp. 2320–2325, 2005.
8 J. Jantzen, "Design Of Fuzzy Controllers," *Report Department of Automation, Technical University of Denmark*, vol. 864, no. 98, pp. 1–27, 1998.
9 S. M. Hoseini, A. J. Koshkouei, and M. Farrokhi, "Chattering Free Fuzzy Sliding Mode Controllers For Robot Manipulators," *Control*, 2006.
10 F. Nagata, Y. Kusumoto, K. Hasebe, K. Saito, M. Fukumoto, and K. Watanabe, "Post Processor Using a Fuzzy Feed Rate Generator for Multi-Axis NC Machine Tools with a Rotary Unit," *Proceeding of International Conference on Control, Automation and System*, pp. 438–443, 2005.
11 J. Jalani, "Hardware-In-The-Loop Simulation (HILS) Based Design and Evaluation of an Intelligent Gantry Crane System," *International Journal of Automation Technology*, Vol. 1, No. 1, Pp. 11-19, 2007.
12 A. Bicchi, "Hands for Dexterous Manipulation And Robust Grasping: A Difficult Road Toward Simplicity," *IEEE Transactions on Robotics and Automation*, vol. 16, no. 6, pp. 652–662, 2000.
13 P. Melin and O. Castillo, "A Review On The Applications of Type-2 Fuzzy Logic In Classification And Pattern Recognition," *Expert System Application*, vol. 40, no. 13, pp. 5413–5423, 2013.
14 T. Doersam and O. Hammerschmidt, "An Adaptive Fuzzy Control for a Three Fingered Robot Gripper," *Proceeding of Third Australian and New Zealand Conference on Intelligent Information Systems. ANZIIS*, pp. 158–163, 1995.
15 L. C. Hun, J. Djuanda, W. E. Kiong, L. C. Ying, and T. C. Looi, "DSP-Based Fuzzy Logic Controller Using Memory Based Position Recognition For Robotic Arm Manipulator," *Proceeding of the 2nd International Conference on Mechatronics*, pp. 323–330, 2005.
16 J. Jalani, "Design and Implementation of Fuzzy Logic Controller for Intelligent Gantry Crane System," *Proceeding of the 2nd International Conference on Mechatronics*, pp. 345–351, 2005.
17 L. Suganthi, S. Iniyan, and A. A. Samuel, "Applications of Fuzzy Logic In Renewable Energy Systems – A Review," *Renewable and Sustainable Energy Reviews*, vol. 48, pp. 585–607, 2015.
18 S. Lee, D. Kim, and G. Park, "Identification of Nonlinear Dynamic Systems via the Neuro-Fuzzy Computing and Genetic Algorithms," *Proceeding of International Conference on Control, Automation and System*, pp. 1892–1896, 2005.
19 V. V. K. Reddy and M. Sydulu, "Fuzzy and Genetic Approach for Optimal Capacitor Placement in Distribution Systems," 2006.
20 P. P. Bonissone, "Adaptive Neural Fuzzy Inference Systems (ANFIS): Analysis and Applications," pp. 1–41, 2002.
21 M. A. Ahmad, R. Mohd, T. Raja, M. S. Ramli, M. A. Zawawi, and M. H. Suid, "Composite PD and Fuzzy Logic Approach for Vibration of Elastic Joint Manipulator," International Conference on Control, Automation and Systems, 2010.
22 I. H. Akyuz, Z. Bingul, and S. Kizir, "Cascade Fuzzy Logic Control of A Single-Link Flexible-Joint Manipulator," *Turkish Journal of Electrical Engineering and Computer Sicence*, vol. 20, no. 5, pp. 713–726, 2012.
23 K. Groves and A. Serrani, "Modeling and Nonlinear Control of a Single-Link Flexible Joint Manipulator," no. 3, p. 13, 2010.
24 D. Sendrescu, "Nonlinear Identification of a Rotary Flexible Joint Experiment."
25 M. Rostami Kandroodi, M. Mansouri, M. Aliyari Shoorehdeli, and M. Teshnehlab, "Control of Flexible Joint Manipulator via Reduced Rule-Based Fuzzy Control with Experimental Validation," *ISRN Artificial Intelligence*, vol. 2012, pp. 1–8, 2012.
26 R. A. Koshy and S. Thomas, "A Two-surface Discrete Sliding Mode Control based on Approach Angle Reaching Law," *Proceeding of the International Multiconference of Engineers and Computer Scientists*, vol. 2, 2012.
27 S. Yurkovich, S. Yurkovich, K. M. Passino, and K. M. Passino, "A Laboratory Course On Fuzzy Control," *Educatiom IEEE Transaction*, vol. 42, no. 1, pp. 15–21, 1999.
28 S. Yurkovich and K. M. Passino, *Fuzzy Control*. Addison Wesley Longman, 1998.

MATEC Web of Conferences **150**, 01012 (2018) https://doi.org/10.1051/matecconf/201815001012
MUCET 2017

Development of Op-Amp Based Piezoelectric Rectifier for Low Power Energy Harvesting Applications

Muhamad Syazmie Bin Sepeeh[1], *Farahiyah* Binti Mustafa[1*], *Anis Maisarah* Binti Mohd Asry[1], *Sy Yi* Sim[1], *Mastura* Shafinaz Binti Zainal Abidin[2]

[1]Department of Electrical Engineering Technology, Faculty of Engineering Technology, Universiti Tun Hussein Onn Malaysia, 86400 Parit Raja, Batu Pahat, Johor, Malaysia
[2] Department of Electronic and Computer Engineering, Faculty of Electrical Engineering, Universiti Teknologi Malaysia, 81310 Skudai, Johor, Malaysia

Abstract. In this study, the development of operational amplifier (op-amp) based rectifier for piezoelectric energy harvesting applications was studied. The two stage op-amp full wave rectifier was used to convert the AC signal to DC signal voltage received by piezoelectric devices. The inverted half wave rectifier integrated with full wave rectifier were designed and simulated using MultiSIM software. The circuit was then fabricated onto a printed circuit board (PCB), using standard fabrication process. The achievement of this rectifier was able to boost up the maximum voltage of 5 V for input voltage of 800 mV. The output of the rectifier was in DC signal after the rectification by the op-amp. In term of power, the power dissipation was reduced consequently the waste power decreases. Future work includes optimization of the rectifying circuit to operate more efficiently can be made to increase the efficiency of the devices.

1 Introduction

Energy harvesting is a process that captures small amounts of energy that would otherwise be lost as heat, light, sound, vibration and movement. Piezoelectric energy harvesting, has been investigated since the late 1990s [1]. Energy harvesting technology has ability to create autonomous, self-powered electronic systems that do not rely on battery power for their operation [2]. There are available energy from the environment that allow self-sufficient energy to supply to low power consumption load such as sensor or radio transmitters. This available energy needs to convert before being used. The kinetic energy converts to electrical energy by using the piezoelectric device (PD) as a transducer.

In order to convert kinetic energy to electrical energy, some vibration or shocks need to apply on the surface of piezoelectric devices, then it will mechanically deform the piezoelectric crystal with tension or pressure to generate electrical charges. The electrical charges that being generated can measure as voltage. The circuitry of the energy harvesting connected to the piezoelectric devices by using the operational amplifier (op-amp) based rectifier to amplify the output and give the better desired output to the system.

The power dissipation will reduce as the waste power decreases. The rectifier used to convert the alternating current signal voltage to direct current signal voltage. It connected with the piezoelectric devices as a source energy, then flow through the rectifier. Based on the piezoelectric device, it is capabl e to generate AC signal from the vibration or movement that being apply on it. The AC signal from piezoelectric device converts to DC signal using rectifier.

In order to maximize the power transfer of the transducer to the load, rectifier will be the most important part in the piezoelectric energy harvesting. A recent method of rectifier used comparator-based active diode, but it faces oscillation problem and dc offset. This problem cause the lower power efficiency of the rectifier [3]. In this paper the rectifier used the active diodes based on op-amp. The reason for using the op-amp rectifier compared to other rectifier because it reduces the power dissipation and increase the power efficiency.

In this study, the design and development of op-amp based rectifier for low power energy harvesting applications reported. The rectifier simulated using MultiSIM software and the characterization carried out in order to check capability of the circuit to work as energy harvesting circuit.

2 Research Method

In this study, two stage op-amps full wave rectifier circuit has been proposed to convert the AC signal to DC signal. Figure 1 shows the schematic circuit and printed circuit board (PCB) of Op-amp based rectifier.

For negative input voltage, the output first Op-amp should be positive, then the D_1 ON and D_2 OFF. Since, action of the first Op-amp is forcing the input

* Corresponding author: farahiyah@uthm.edu.my

voltage of the Op-amp to the same level, the output is hold at the ground potential when the D_1 is ON. For positive input signal V_i, the output goes negative, the D_2 is ON and D_1 is OFF. The output of first Op-amp then acts as inverting amplifier. The gain set to 1 by R_2/R_1 which means $R_2 = R_1$. The output of first Op-amp is 0V follow the negative half cycle of the input for positive half cycle. This output becomes one of the input for the second Op-amp which summed with input waveform. At the first stage it has 2 relative gains to the input which means $R_3 = 2R_4$

For second stage, at negative half cycle, the input wave summed with potential ground result the output of second Op-amp is a positive gain. The inverted wave of the half wave rectifier summed with input waveform and the gain of the half wave rectifier of first Op-amp for the positive half cycle results the positive output. The summer performs as an inverted again giving $V_o = |V_i|$ as shown in Figure 2 and Figure 5. The result is a full wave rectifier. Typically the gain of the half wave rectifier is set to 1 which are $R_2 = R_1$ and ratio of R_3 and R_4 should be 1:2 so the summing of the input with the output of the half wave rectifier output will be sum up correctly. The light emitting diode (LED) used as a load in this rectifier circuit.

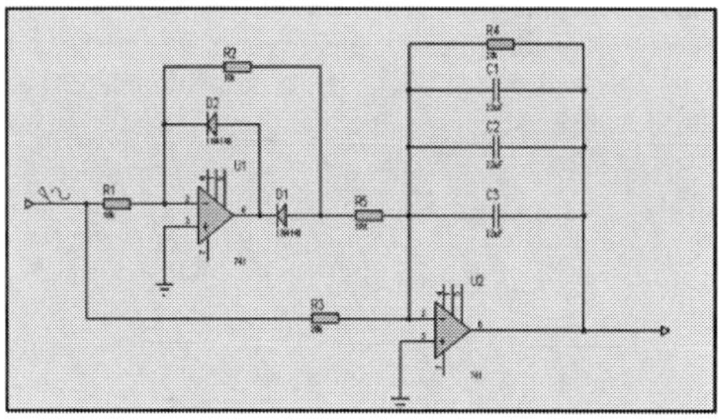

(a)

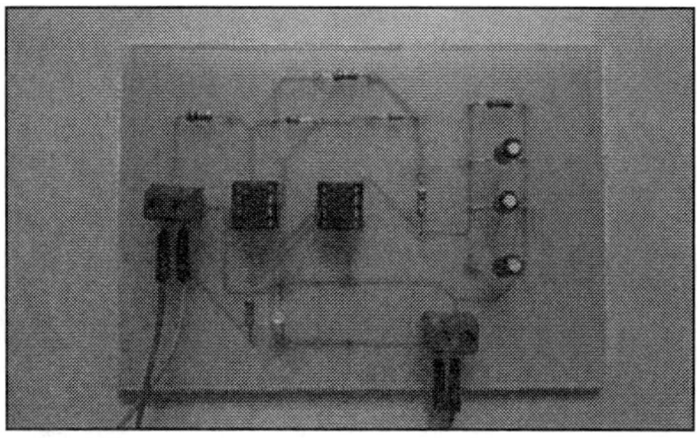

(b)

Fig 1. (a) Schematic and (b) PCB of op-amp based rectifier circuit.

3 Result and Discussions

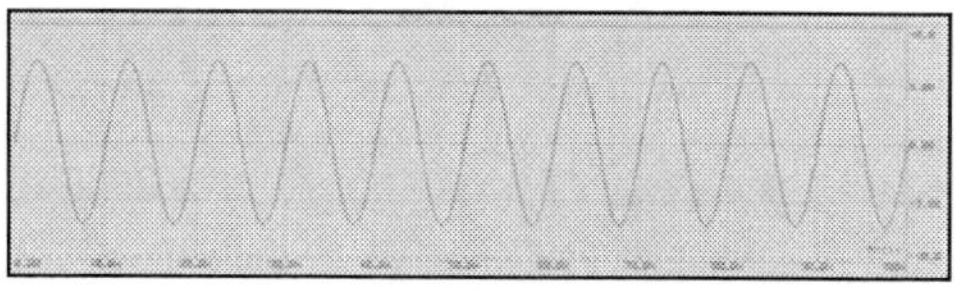

(a)

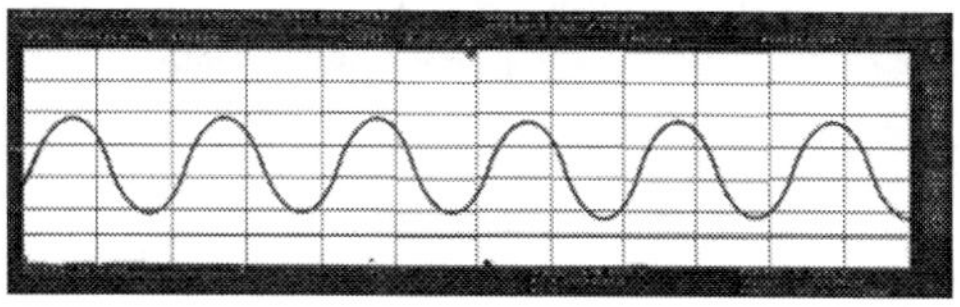

(b)

Fig 2. a) The simulation and b) experimental input waveform of the Op-amp based rectifier.

Based on the arrangement of the circuit, the input V_{rms} set at 5V at the source of the circuit. The gain of the inverting op-amp is 1. As shown in Figure 2 the amplitude of V_{rms} output is 5V. There are no different between the simulation and experimental.

$$A_V = -\frac{R_2}{R_1} = \frac{10k\Omega}{10k\Omega} = -1 \qquad (1)$$

By calculation the value of V_{out} :

$$A_V = \frac{V_{out}}{V_{in}} \qquad (2)$$

$$1 = \frac{V_{out}}{V_{in}} \qquad (3)$$

$$\therefore V_{out} = 5V$$

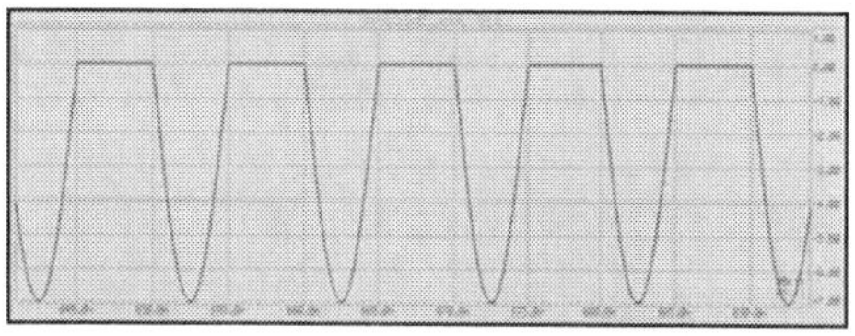

(a)

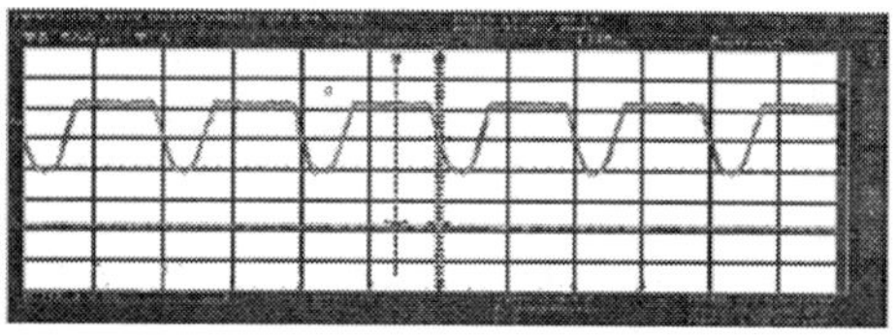

(b)

Fig 3. a) The simulation and b) experimental inverted half-wave rectified waveform of output by first Op-amp.

Since inverting Op-amp used, the output waveform inverted as the output of the first stage of Op-amp is half wave. During the negative half-cycle of a sinewave input, the output should positive. During the positive half-cycle, the output should zero. The circuit that accomplishes this amazing feats looks like the inverting amplifier with a couple of diodes added because only one actually does the rectifying action.

The other simply keeps the Op-amp in control while the signal output holds at zero. For the positive half-cycle input, the Op-amp output goes negative forcing D_2 to turn ON and D_1 to shut OFF. The output sits at 0V. This is because the Op-amp still does its job of holding *V-* at 0V and because there is no current through R_2 and get V_o = *V-* = 0V. The only thing left to reckon with is the current running through R_1:

$$I = \frac{V_i}{R_2} = \frac{1}{10k} = 0.5mA \tag{4}$$

During this half-cycle, the Op-amp's output swings negative to turn on D_1 enough to pull 0.5 mA through the diode.

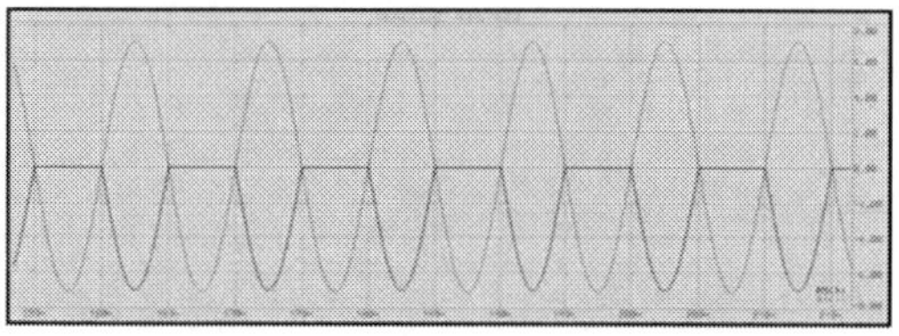

(a)

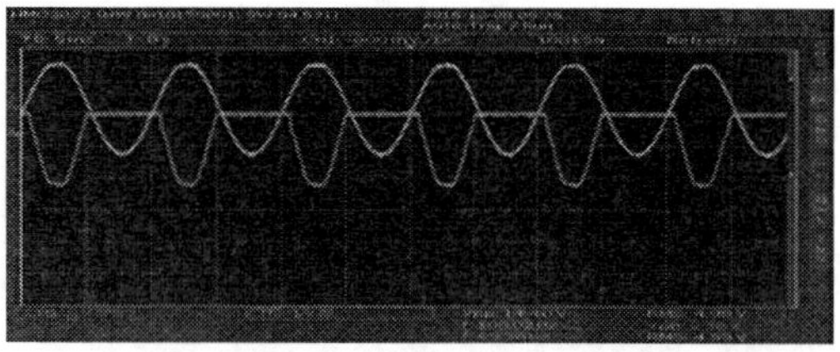

(b)

Fig 4. a) The simulation and b) experimental inverted half-wave with the input waveform.

As shown in Figure 4. The amplitude of the V_{rms} is 5V that equal to the $V_{p\text{-}p}$ = 7.07V due to the gain of the operational amplifier.

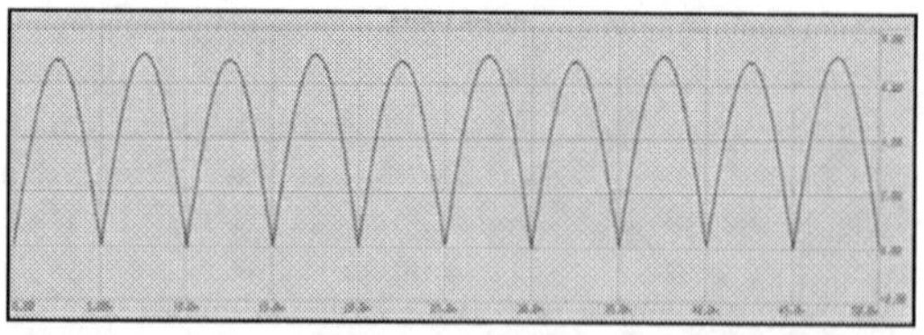

(a)

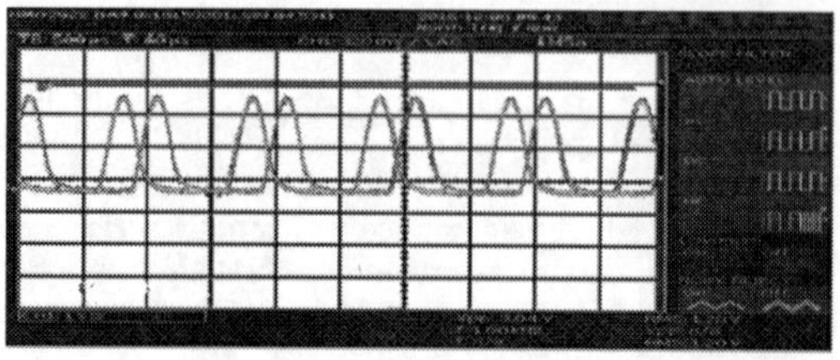

(b)

Fig 5. a) The simulation and b) experimental Full Wave rectified waveform of output by second Op-amp.

Two separate circuits used for each half of the full AC cycle and combined signal used to get the full wave rectified output [4]. From Figure 5, the output waveform generated by the second Op-amp. The output voltage V_{out} form the second Op-amp are 7.07$V_{p\text{-}p}$ that equivalent to 5V of V_{rms}. There are different in phase shift when compared between the simulation and experimental.

$$V_{out} = -R_f\left[\left(\frac{V_{in1}}{R_5}\right) + \left(\frac{V_{in2}}{R_3}\right)\right] \tag{5}$$

$$V_{out} = -20k\Omega\left[\left(\frac{1.05mV}{10k\Omega}\right) + \left(\frac{7.07V}{20k\Omega}\right)\right] - 7.07V \tag{6}$$

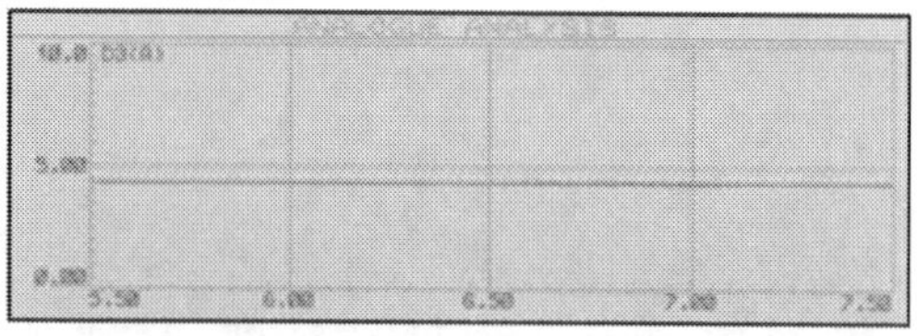

(a)

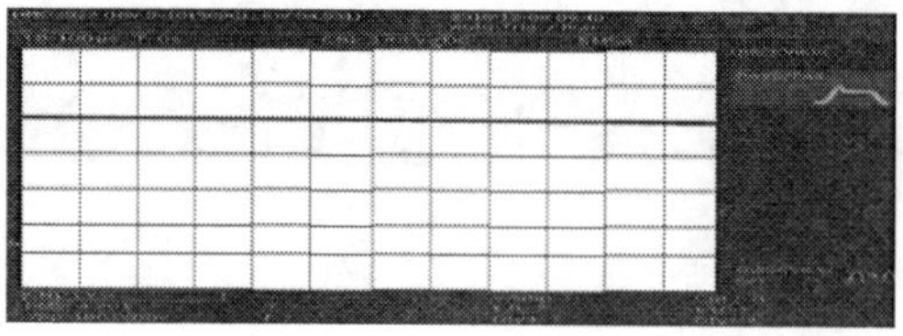

(b)

Fig 6. a) The simulation and b) experimental output of Op-amp based rectifier.

The output of Op-amp in Figure 6 rectified by 22uF capacitor filter. 3 units of capacitor used to rectify the peak of the full wave. As the input and current was 1.05mA and 1.0mA respectively, the V_{p-p} of the rectified waveform is:

$$V_{P+} + V_{P-} = V_{P-P} \quad (7)$$

$$4.90 + 4.79 = 9.69V \quad (8)$$

The V_p of the waveform:

$$\frac{V_{P+} + V_{P-}}{2} = 4.845V \quad (9)$$

The V_{DC} of the waveform:

$$\frac{4.90V + 4.79V}{2} = 4.845V \quad (10)$$

Therefore, the efficiency of rectifier:

$$Efficiency, \eta = \frac{P_{out\,DC}}{P_{in\,AC}} \times 100\% \quad (11)$$

$$P_{out\,DC} = V_{out\,DC} \times I_{out\,DC} \quad (12)$$

$$P_{out\,DC} = (4.845V \times 1mA) = 4.845mV \quad (13)$$

$$P_{in\,AC} = V_{rms} \times I_{rms} \quad (14)$$

$$P_{in\,AC} = (5V \times 1.05mA) = 5.25mW \quad (15)$$

$$Efficiency, \eta = \frac{4.845mV}{5.25\,mV} \times 100\% \quad (16)$$

$$\therefore 92.28\%$$

The output DC voltage of the Op-amp based rectifier is 4.845V. Compare to the simulation output is 4.35V. The voltage dropped through the circuit since LED used as the load of the circuit design.

Table 1 shows the comparison between this proposed method and previous proposed method. The input voltage for proposed design is 5V in AC and the voltage output is 4.845 V in DC measured. The proposed design was using the full wave rectifier based on op-amp with the capacitor filter and achieved the highest power efficient of 92.28%. Although the proposed method also used the operational amplifier as a rectifier but with some modification made, it achieved the highest value of voltage output in DC, power output in DC and power efficiency.

Table 1. Comparison with previous rectifier method

Parameter	**Lam Rectifier**	**Synchronized Switching Rectifier**	**A Self-Powered High-Efficiency Rectifier**	**Proposed Method**
$Vout_{DC}$ (V)	2.78	1.8	2.9	4.845
$Pout_{DC}$ (mW)	0.081	0.143	0.090	4.854
η (%)	90.00	77.00	91.20	92.28

4 Conclusion

In this paper, the development of Op-amp based rectifier circuit has been present. The design circuit utilizes two of Op-amps with the capacitor filter through precise in rectification of the waveform. The configuration gives the advantages of precise rectification over a wide range of operating frequencies, a high input impedance and the ability to give the less power dissipation. An improved rectifier circuit employing the principles similar from the previous research.

References

1. S. Bhalla, S. Moharana, N. Kaur, V. Talakokula, *Piezoelectric Materials,* (John Wiley & Sons, London, 2016)
2. M.N. Gupta, S.K. Yadav, Suman, AEEE, **4**, 313 (2014)

3. T. Gunasheela, E.K .Sharmitha, K. Senthilprakash, *An Efficient Rectifier Design for Piezo Electric Energy Harvesting System* (ICSEIE, Erode, 2015)
4. R. Dixit, C. K. Dwivedi, IJETAE, **4**, 388 (2014)
5. Y. Sun, N. H. Hieu, C.J. Jeong, S.G. Lee, IEEE Trans. Power Electron. **27**, 623 (2012)
6. X. D. Do, H. H. Nguyen, S.K. Han, D. S. Ha, S.G. Lee, JSTS, **15**, 319 (2015)
7. X. D. Do, H. H. Nguyen, S.K. Han, D. S. Ha, S.G. Lee, IEEE Trans. Very Large Scale Integr. (VLSI) Syst. **23**, 444 (2015)

Development of Electrocardiograph Monitoring System

Khairul Affendi Rosli[1*], *Mohd. Hafizi* Omar[1], *Ahmad Fariz* Hasan[1], *Khairil Syahmi* Musa[1], *Mohd Fairuz* Muhamad Fadzil[1], and *Shu Hwei* Neu[1]

[1]Department of Electronic Engineering Technology, Faculty of Engineering Technology, Universiti Malaysia Perlis, Malaysia.

Abstract. Electrocardiograph (ECG) monitoring system is one of the diagnostic tools which can help in reduce the risk of heart attack. A cardiologist may be able to determine heart condition from the ECG signal that recorded from subject. The purpose was to design an ECG monitoring system which consists of ECG circuit and digital signal processing system to deny the unwanted signal. In general, the ECG signal is nature weak and only around 1mV amplitude. Therefore filter and amplifier circuits were designed into 3 stages with a total gain of 1000 to bring the signal to around 1V. Circuit designed included of instrumentation amplifier, bandpass filter and notch filter. The frequency bandwidth of ECG is between 0.05Hz until 100Hz. Schematic circuit was tested by software simulation before proceeding to hardware implementation. Simulation analysis was done by using Software Proteus 8 Professional while the further signal processing was done in MATLAB software environment. A PQRST ECG waveform can be seen clearly after digital filtering stage in MATLAB environment. Digital signal processing in MATLAB software included of pre-filtering, Fast Fourier transform and peak detection. As conclusion, the time interval between peaks can be determined automatically which can provide useful information in clinical aspect.

1 Introduction

Cardiovascular disease is still the first killer in the world although there was a rapid development in medicine and medical technology. ECG system is one of the diagnostic tools used for the heart patient monitoring. ECG is a graphical recording of the cardiac cycle which the electric current generated by the heart muscle during a heartbeat. Electrode is used as a sensor like an antenna to detect and receive the electrical signal generated during the cardiac cycle. In 1901, a high sensitivity of new string galvanometer was developed successfully and implemented in Dr Willem Einthoven's electrocardiogram with weighed 600 pounds [1]. Three electrodes were positioned according to Einthoven's Triangle on the body parts: right arm, left arm and left leg of the human body [2]. Refer Figure 1.

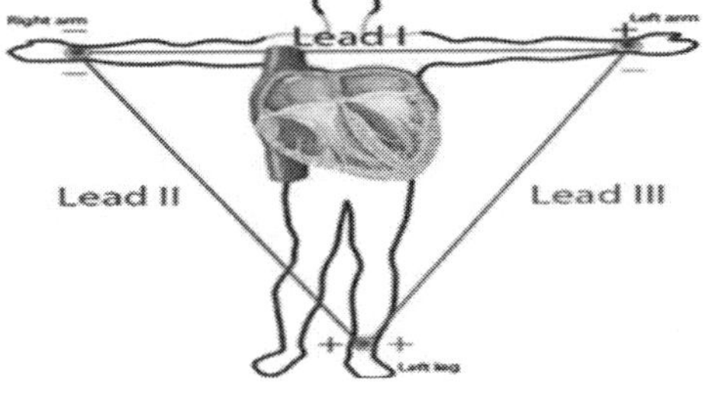

Fig. 1. Diagram of Einthoven Triangle [3]

A typical electrocardiogram shows the deflections that express the alternate contraction of the parts of human heart. In general, heart of human consists of four chambers which are two upper atria and two lower ventricles. Figure 2 shows the ECG signal. The first wave is called "P wave", is generated by upper chamber. After a "P wave" there is a flat line indicated that the impulse transfers to lower chamber. The next wave is called "QRS" complex. QRS wave is making by the right and left bottom ventricles. The last wave is called "T wave" which represented the ventricles return to its resting state. It provided information of the heart condition and performance to the medical worker to have a good clinical diagnosis of cardiac disease.

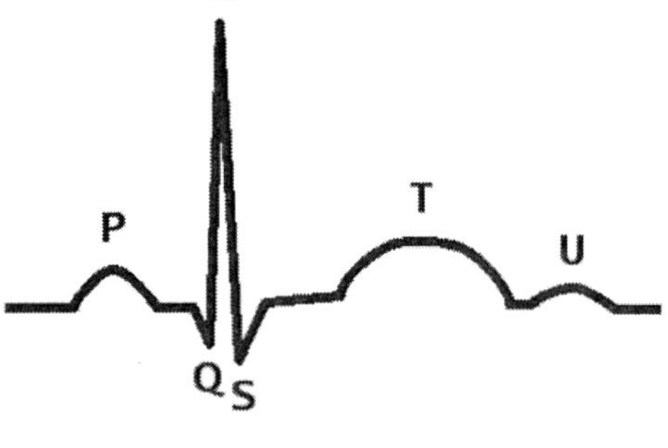

Fig. 2. ECG Signal (PQRSTU Wave) [4]

* Corresponding author: khairulaffendi@unimap.edu.my

 https://doi.org/10.1051/matecconf/201815001013

1.1 ECG Signal Characteristics

For diagnosis purpose, the recording bandwidth between 0.01 to 300 Hz was used by using the standard 12 lead ECG. For monitoring purpose such as the ambulatory patient, the bandwidth was limited from 0.05 to 100 Hz. Thus the restricted bandwidth attenuates the higher frequency noise caused by contractions of muscle and the lower frequency noise caused by motion of the electrodes [5]. The third bandwidth used to maximize the signal to noise ratio and represented the QRS wave. Therefore a filter that can eliminate the non-QRS complex and pass the QRS complex is needed. The non-QRS complex that needs to reject including the P wave and T wave. This filter helps to detect the QRS complexes but distorts ECG so much that the appearance of the filtered signal is not clinically acceptable. These are some small higher-frequency that consists in the QRS complex. An ECG amplifier normally has a gain of about 1,000 in order to amplify the weak peak signal from 1 mV to a range of about 1 V. The table below shows the amplitude and the duration time of the ECG waveform interval.

Table 1. ECG Waveform [5]

Amplitude	Durations
P-wave-0.25 mV	P-R interval: 0.12 to 0.20
R-wave-1.60	Q-T interval: 0.35 to 0.44 s
Q-wave- 25%R wave	S-T interval:0.05 to 0.15 s
T-wave-0.1 to 0.5 mV	P-wave interval: 0.11s
	QRS interval: 0.09s

1.2 Noise

ECG technology still need very much research phase to identify and investigate the characteristic of the noise sources and their features in order to obtain a desired real time analysis of ECG signal. Electrode placed on the human body act as an antenna received the signal from heart and also the unwanted signal called noise. Some of the noises and interference signal are difficult to filter away. There were several types of disturbance occurred during the ECG signal collection. The primary noise sources which are electromyography (EMG) noise, power line interference, source impedance unbalance noise and motion artifacts and instrumentation noise.

2 Design Step

2.1 Circuit Development

The ECG circuit consisted of two stages which are Stage 1: Instrumentation amplifier stage and Stage 2: Filter and amplifier stage. Instrumentation amplifier was the first stage and used to differentiate the voltage difference between two electrodes that act as the input signal of the system. The signal needs to be amplified to a significant gain to be proceeding to the next stage.

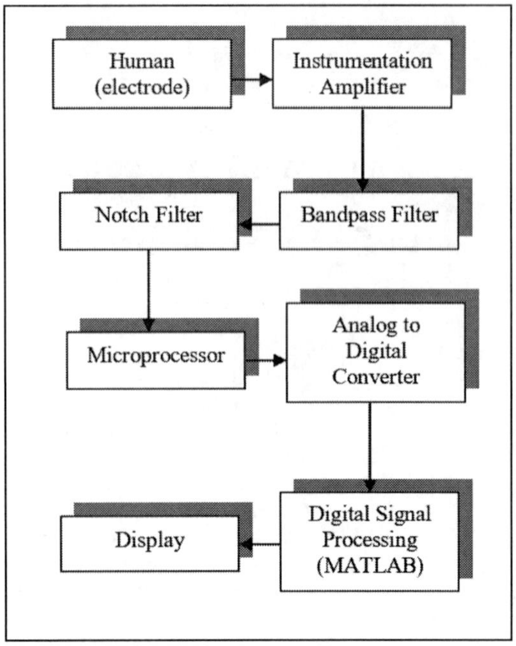

Fig. 3. Basic block diagram for ECG processing

Different type of filter circuit was constructed to remove the unwanted noise and allowed the desired signal to pass through. An active bandpass filter which combined both low pass and high pass characteristic was designed to pass a board range of frequencies. It matched the requirement bandwidth for ECG signal (0.05Hz to 100Hz) and attenuated the signals which are out of the pass band. At the same time, the signal was amplified by a filter to a gain of 11. For power line interference 50Hz, notch filter was used to remove this particular frequency. For improvement, the twin-T notch filter with variable Q consisted of low pass circuit and high pass circuit with a variable Q was utilized.

2.1.1 Instrumentation Amplifier

First step for ECG monitoring where a very small voltages passed out from the probes and require certain amplification before go to next stages [6]. The differences between two input potentials were amplified by the instrumentation amplifier while any signals that are common to both inputs rejected. The AD620 which has a low current noise was chose for ECG monitoring. It amplified the ECG signal to around 50 times by an external resistor, R_1 connection. The value of capacitor C_1 was chose to maintain the stability of the right leg drive loop. The gain values are as follows.

$$Gain = \frac{49.4k\Omega}{R_1} + 1 \quad (1)$$

Where $R_1 = 1\ k\Omega$, therefore

$$Gain = \frac{49.4k\Omega}{1k\Omega} + 1$$
$$= 50.4$$

The active grounding is connected to produce a higher gain [7].

2.1.2 Bandpass

The active bandpass filter was designed to passed a range of ECG frequencies and attenuated the unwanted frequencies signal. The lower cut-off frequency (f_L) is 0.05Hz and higher cut-off frequency (f_H) was designed to 100Hz. This bandpass filter was designed with a gain of 11. The lower cut-off frequency (f_L) is determined by :

$$fL = 1 / 2\pi R_1 C_1 \quad (2)$$

The higher cut-off frequency (f_H) is determined by :

$$fH = 1 / 2\pi R_4 C_2 \quad (3)$$

The gain of the filter is:

$$Gain = (R_3 / R_2) + 1 \quad (4)$$

Based on the calculated values, fL = 0.048 Hz, fH = 106.1 Hz and Gain = 11.

2.1.3 Notch Filter

For an ideal notch filter, it can completely flat response to the frequency range and remove the particular notch frequency. In reality the perfection is not achievable. It can be improved when using an active notch filter which employs negative feedback around the operational amplifier [8].Twin-T notch filter with variable Q was designed to remove the 50Hz power line interference [9]. The higher the Q, the narrow the notch can be achieved and higher the level of attenuation needed to remove the unwanted signal. Q is designed to 100 so degree of filter's performance increased [10].Gain of the filter was designed to 2.

Notch frequency is calculated by :

$$f_{notch} = 1 / 2\pi RC \quad (5)$$

Gain of the notch filter:

$$Gain = (R_5 / R_4) + 1 \quad (6)$$

Q value:

$$Q\ ratio = (R_7 / R_6) \quad (7)$$

2.1.4 Proposed ECG Circuit Design

The proposed ECG circuit design consisted of an instrumentation amplifier with gain of 50.4, a bandpass filter with a pass band of 0.05Hz-100Hz with gain of 11 and a notch filter of 50 Hz with gain of 2. This QRS detector circuit amplify the weak ECG signal from around 1mV to approximately 1V.

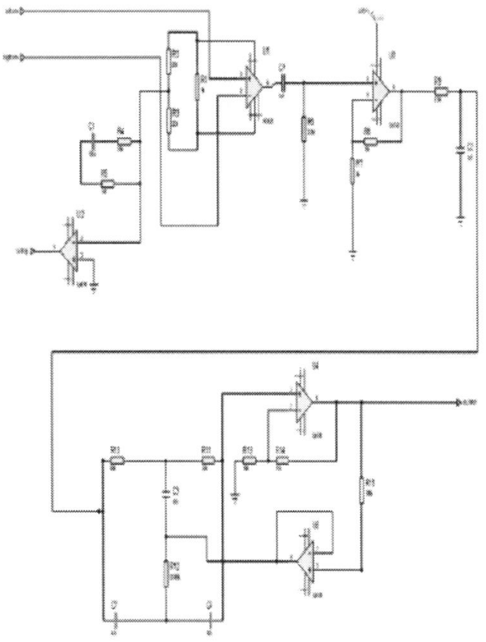

Fig. 4. Proposed ECG circuit design schematic

2.2 Signal Processing Method

The signal processing method can divided into 3 steps which are pre-filtering, squaring the signal and rate calculation. ECG signal from circuit still contain the noise which cannot be removed or fail to remove. The unwanted signal was still exist because it shares the same frequency contents with ECG signal bandwidth. In pre-filtering step, MATLAB were used to perform Fast Fourier Transform (FFT) in order to remove the low frequency signal in frequency domain. After the low frequency signal was removed, the Inverse Fast Fourier Transform (IFFT) applied to restore the data to time domain. A notch filter was designed to remove the 50 Hz again to the signal. After that, the signal was squared to make the peak signal become more significant and last is peak detection to count the heart beat rate.

3 Results and Discussion

The proposed circuit design was simulated in the Proteus 8 Professional. The simulation result was recorded as to compare it to the hardware simulation result. The percentage of error was calculated to determine the performance of the circuit.

$$\textit{percentage of error} = \frac{(\textit{hardware or software value}) - \textit{theoritical value}}{\textit{theoritical value}} \times 100\% \quad (8)$$

3.1 Bandpass

The digital oscilloscope failed to display the result as the lower cut-off frequency for bandpass is too small (0.05Hz), in software simulation. The graph in Figure 5 showed that the higher cutoff frequency of the bandpass filter is 101.8Hz based on the software simulation. Hardware simulation was done by using oscilloscope and signal function generator in lab. The input voltage of signal function generator was adjusted to different value of frequency and the changes of output voltage can be seen in the oscilloscope. Figure 6 showed that the cutoff frequency of the bandpass filter by hardware simulation is100.5Hz. The percentage error of software and hardware simulation compare to theoretical value is -4.053% and -5.278% respectively.

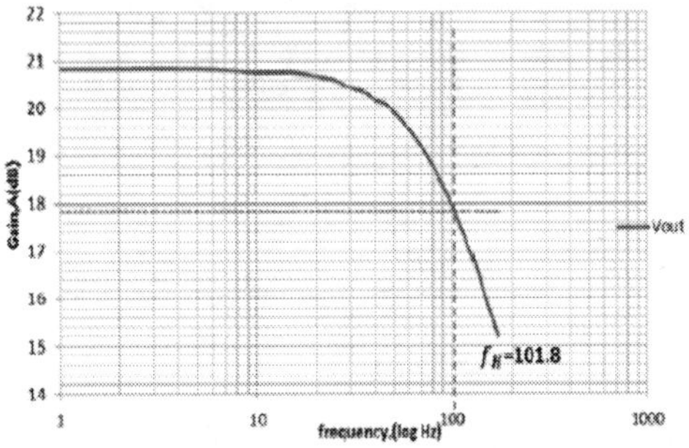

Fig. 5. Bandpass Filter (Software)

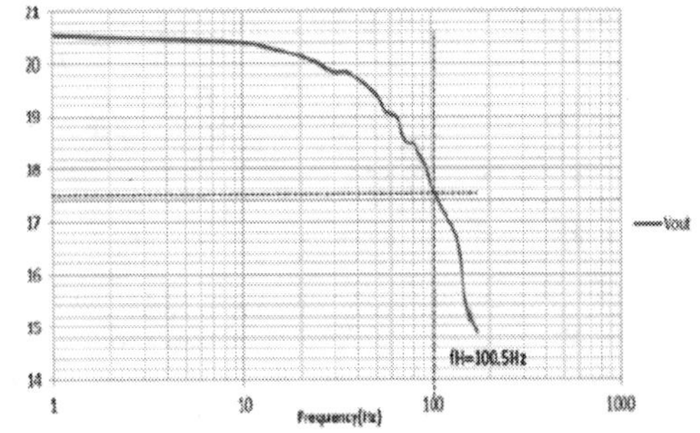

Fig. 6. Bandpass Filter (Hardware)

Table 2. Percentage error of Bandpass Filter

Simulation	Cutoff frequency(Hz)	Percentage of error (%)
Calculation	106.1	-as reference
Software	101.8	-4.053
Hardware	100.5	-5.278

3.2 Notch Filter

The frequency was adjusted at the input voltage while the output voltage was recorded. The graph in Figure 7 showed that the notch frequency is 50Hz based on the software simulation. Figure 8 showed that the cutoff frequency of the notch filter by hardware simulation is 48Hz The error percentage of software and hardware simulation compare to theoretical value is 0.4218% an - 3.595% respectively.

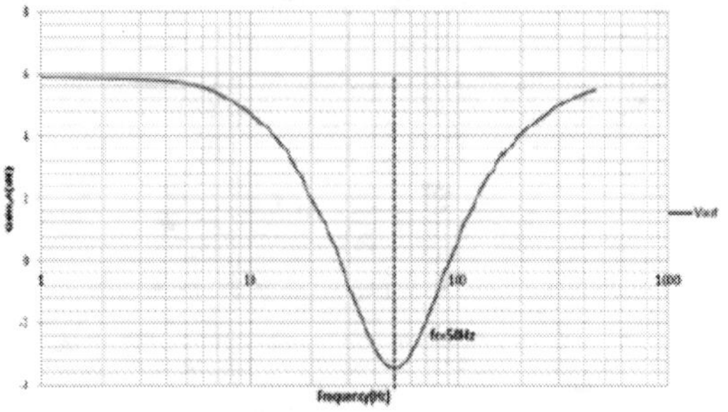

Fig. 7. Notch Filter (Software)

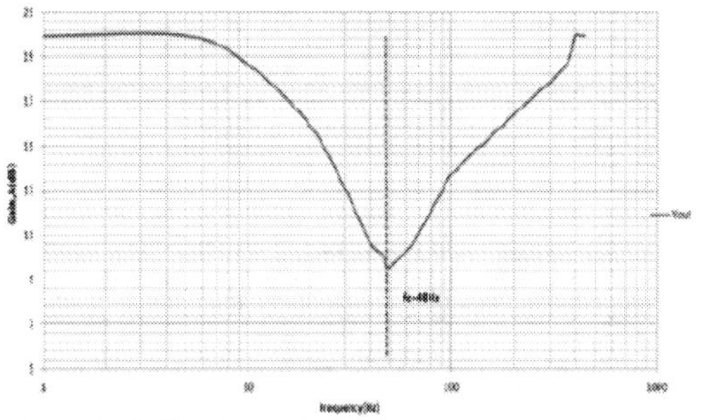

Fig. 8. Notch Filter (Hardware)

Table 3. Percentage error of Notch Filter

Simulation	Cutoff frequency(Hz)	Percentage of error (%)
Calculation	49.79	As reference
Software	50	0.4218
Hardware	48	-3.595

3.3 ECG Circuit

Electrode sensors were placed to human skin to detect the electrical impulse generate from the human heart by Einthoven Triangle method (refer Figure 1) and the waveform is read by using the oscilloscope. A QRS complex with noise was detected by the circuit. The R peak was amplified to average of 1.46V as the figure shown. This signal was collected for further signal analysis in MATLAB.

3.4 Analog signal interference to computer

The output signal from the circuit was in analog form, by using an analogue to digital converter (ADC), it can convert the analog signal from circuit to digital signal so that, the digital data can be read by computer and saved for the further signal analysis. The data was exported to excel and converted to text form and also into MATLAB.

3.5 Signal Processing Method

The signal which collected in circuit was imported into MATLAB. Figure 9 showed the ECG signal collected from the hardware circuit without any filtering process in MATLAB. On the other hand, Figure 10 showed the ECG signal which after filtering stage in MATLAB. The noise was removed and a clear PQRST ECG signal can be seen clearly. The beat rate can be calculated by detecting the peak of the signal.

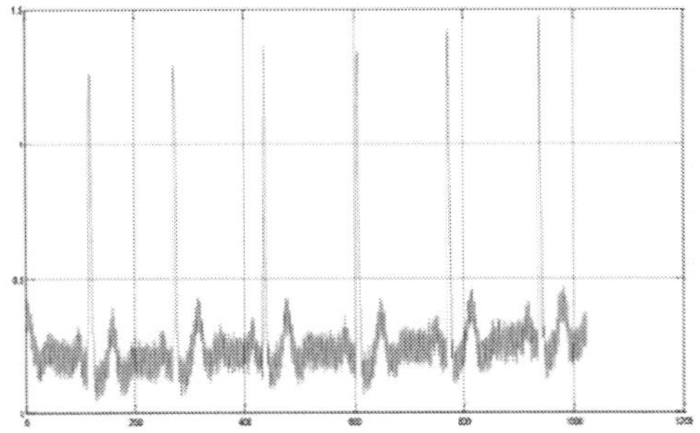

Fig. 9. Without filtering process

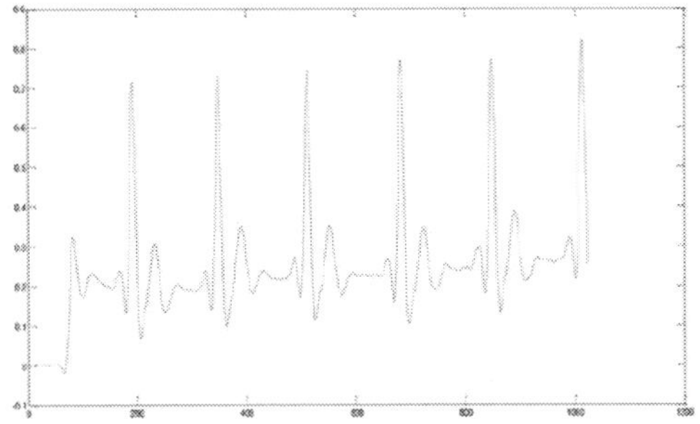

Fig. 10. With filtering process

4 Conclusion

The ECG monitoring system was developed successfully. The ECG analog signal was collected and digital signal processing was done in the MATLAB. The signal display was low noise disturbance and acceptable SNR ECG waveform. The data of the signal can be used to study the heart condition by analyses the peak voltage and the time interval between each signal such as coronary heart disease, heart muscles disease, arrhythmia, and coronary thrombosis. For further improvement on the project, ECG signal can be improve to real time analysis. The signal of ECG can be recorded in a real time and displayed on the computer screen. A real time analysis on signal received can be done and enable it to diagnose the heart condition immediately.

References

1. M. Al Ghatrif, J. Lindsay, "A brief review: history to understand fundamentals of electrocardiography", J. of Community Hospital Internal Medicine *Perspectives* **2**, 1 (2012).
2. "ECG timeline - History of the electrocardiogram", Ecglibrary.com, 2016. [Online]. Available: http://www.ecglibrary.com/ecghist.html. [Accessed: 10- May- 2016].
3. "Cardiovascular Lab: Electrocardiogram: Setup", Medicine.mcgill.ca, 2016. [Online]. Available: http://www.medicine.mcgill.ca/physio/vlab/cardio/setup.htm. [Accessed: 10-May- 2016].
4. "Adaptive Noise Removal of ECG Signal Based On Ensemble Empirical Mode Decomposition", 2016. [Online].Available:applications/adaptive-noise-removal-of-ecg-signal-based-on-ensemble-empirical-mode-decomposition. [Accessed: 8- June- 2016].
5. PCbasedpatientmonitoringsystem,DevelopmentofECGamplifier",Library.utem.edu.my, 2016. [Online]. Available:http://library.utem.edu.my/index2.php?option=com_docman&task=doc_view&gid=817&Itemid=208.[Accessed: 10- Apr- 2016].
6. M. G. Naazneen., Sumaya Fathima, Syeda Husna Mohammadi, Sarah Iram L. Indikar, Abdul Saleem, Mohamed Jebran, "Design and Implementation of ECG Monitoring and Heart Rate Measurement System", Int'l. J. Eng. Sci. and Inno. Tech. (IJESIT)**2**, 3, (2013).
7. M. J. Burke, D. T. Gleeson, "A micropower dry electrode ECG preamplifier," IEEE Trans. Biomed. Eng., **47**, 2, pp. 155–162, (2000).
8. A. S. Alkhader, A.A Alomar, A.S Althonaibat, B.S Hiyari, M.A Alshira'ah, "ECG Interface Circuit Design for Improving The Quality of ECG Signal," Int'l J. Res. and Rev. Phar. and App. Sci., **5**, 2, (2015).
9. C.M. Chang, et al., "Voltage-mode notch, lowpass and bandpass filter using current-feedback amplifiers," Electronics Letters, **30**, (1994).
10. F. Moulahcene, N. Bouguechal, I. Benacer, S. Hanfoug, "Design of CMOS Two-stage Operational Amplifier for ECG Monitoring System Using 90nm Technology," Int'l J. Bio-Sci. and Bio-Tech. **6**, 5 (2014).

Effect of Load Model Using Ranking Identification Technique for Multi Type DG Incorporating Embedded Meta EP-Firefly Algorithm

Siti Rafidah Abdul Rahim*[1], *Ismail* Musirin[2], *Muhammad Murtadha* Othman[2], and *Muhamad Hatta* Hussain[3]

[1]Center of Excellence for Renewable Energy (CERE), UniMAP, Universiti Malaysia Perlis, 02600, Arau, Perlis, Malaysia.
[2]Center for Electrical Power Engineering Studies, Faculty of Electrical Engineering, Universiti Teknologi MARA, 40450 Shah Alam, Malaysia
[3]School of Electrical System Engineering, Universiti Malaysia Perlis, 02600, Arau, Perlis, Malaysia

Abstract. This paper presents the effect of load model prior to the distributed generation (DG) planning in distribution system. In achieving optimal allocation and placement of DG, a ranking identification technique was proposed in order to study the DG planning using pre-developed Embedded Meta Evolutionary Programming–Firefly Algorithm. The aim of this study is to analyze the effect of different type of DG in order to reduce the total losses considering load factor. To realize the effectiveness of the proposed technique, the IEEE 33 bus test systems was utilized as the test specimen. In this study, the proposed techniques were used to determine the DG sizing and the suitable location for DG planning. The results produced are utilized for the optimization process of DG for the benefit of power system operators and planners in the utility. The power system planner can choose the suitable size and location from the result obtained in this study with the appropriate company's budget. The modeling of voltage dependent loads has been presented and the results show the voltage dependent load models have a significant effect on total losses of a distribution system for different DG type.

1 Introduction

The terms of distributed generation (DG) is small generating units installed at strategic points of the distribution system. The introduction of DG on the distribution network can significantly impact the flow of power, voltage conditions and power quality at customers and utility equipment [1]. The ranking identification technique is developed in order to determine the suitability of DG planning in the distribution system. This technique was conducted using the developed automatic algorithm for single DG by incorporating the Embedded Meta EP-Firefly Algorithm (EMEFA) technique as optimization technique[2]. In this study, the different loads are modeled prior to the different DG type installation for single objective implementation considering loss minimization.

The load in the distribution system is divided into three categories i.e. residential, commercial and industrial. These three load types are voltage dependent, and active and reactive power components respond differently to variations in voltage. The voltage dependent load has a main impact on distribution system planning studies. The load models termed as residential, commercial, industrial and mixed loads will give impact to active and reactive intake power and system losses [3][4].

The research to date has tended to focus on the loss minimization scheme of DG capacities where the locations are dependent on the load model. However, in this study, the proposed technique was also performed for different type of DG. The analysis was performed in order to investigate the effect of the different load model during DG installation. Detailed explanations on each procedure are described in the sub-sections. In this study, the convergence is set to 100 iterations in order to minimize the burdensome on the optimization. The iteration is set to that value in order to limit the computational time due to improper location. This situation may refer as the location is not suitable to locate any DG or not recommended to install the DG due to the system did not converge.

2 Methodology

Generally, in most distribution system studies, it is assumed that active and reactive power demands are constant values (i.e., constant power model). However, in actual distribution systems, different categories of load models or even mix of load might be present for the system [5]. The nature of these types of loads is such that their active and reactive powers are voltage dependent. Moreover, load characteristics have considerable effects on power flow solutions and system power losses.

* Corresponding author: rafidah@unimap.edu.my

The loss minimization is required in determining the size of DG. First, A_{ij} and B_{ij} need to be obtained using equation (2) and (3) and then these values are substituted into equation (1) to calculate the total line losses in the system:

$$P_{loss} = \sum_{i=1}^{n}\sum_{j=1}^{n} A_{ij}\left(P_iP_j + Q_iQ_j\right) + B_{ij}\left(Q_iP_j - P_iQ_j\right) \quad (1)$$

$$A_{ij} = \frac{R_{ij}\cos(\delta_i - \delta_j)}{V_iV_j} \quad (2)$$

$$B_{ij} = \frac{R_{ij}\sin(\delta_i - \delta_j)}{V_iV_j} \quad (3)$$

P_i and Q_i are real and reactive power of bus i respectively, P_j and Q_j are the real and reactive power of bus j respectively, R_{ij} is the line resistance between bus i and bus j, V_i and V_j are the voltage magnitude at bus i and bus j respectively, and δ_i and δ_j are the voltage angle of bus i and bus j respectively. Due to the main objective to study the effect of load model prior to the DG installation for single objective implementation considering loss minimization, the optimization technique is to minimize the total real power loss, so the P_{loss} is subjected into the objective function, *Of* which can be described in (4).

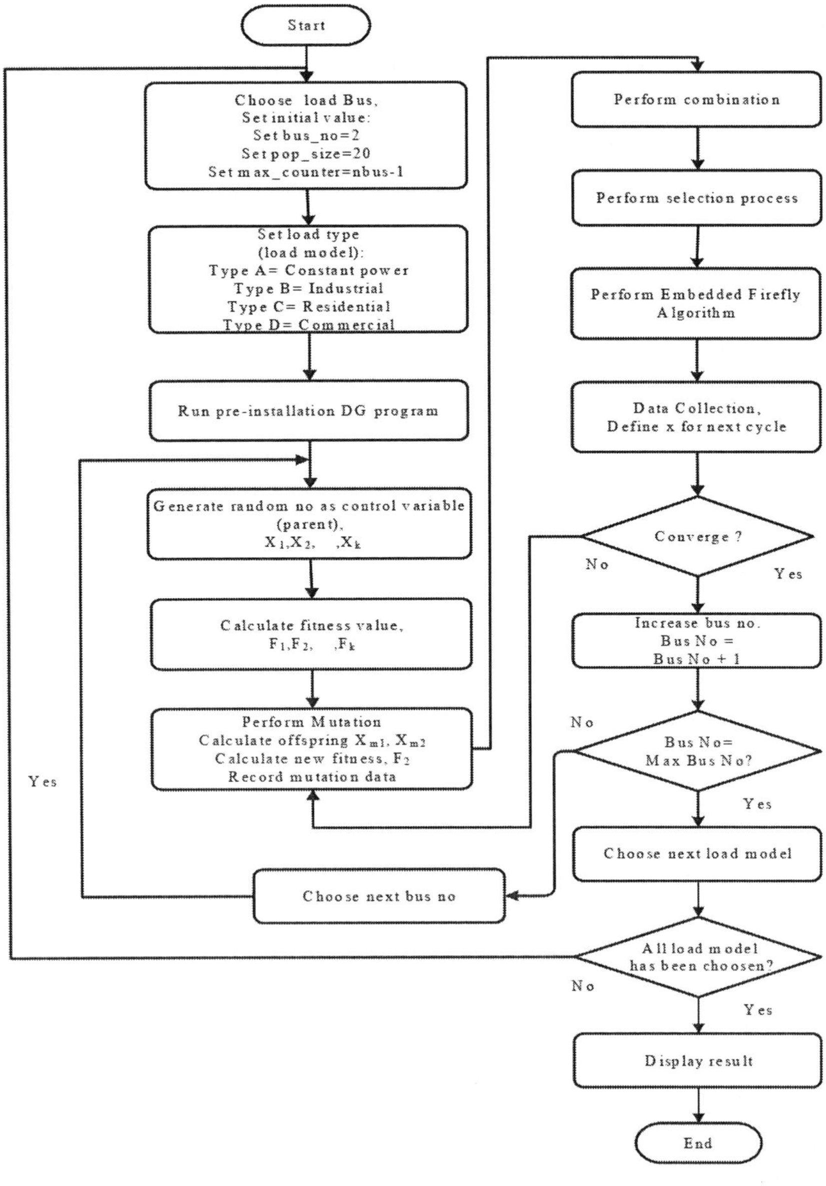

Fig. 1. Flowchart for complete Ranking Identification using the EMEFA technique of DG planning

Further study also presented the development of the Loss Reduction Index (LRI) as shown in equation (5). This index is derived from the total losses obtained from the based case that will be compared to the total losses with the DG installation in the system. The highest percentage of LRI indicates the more loss reduction in the system. The result obtained from this index can contribute to the DG planner to allocate the suitable DG location in the system based on the percentage of LRI.

$$\text{Of} = \min\left(P_{loss} = \sum_{i=1}^{n}\sum_{j=1}^{n} A_{ij}\left(P_iP_j + Q_iQ_j\right) + B_{ij}\left(Q_iP_j - P_iQ_j\right)\right) \quad (4)$$

$$LRI = \frac{\left(\text{Total Loss}_0 - \text{Total Loss}_{DG}\right)}{\text{Total Loss}_0} \times 100\% \quad (5)$$

Where TotalLoss_0=Total loss without DG in MW and TotalLoss_{DG}= Total Loss with DG in MW.

The load models are very much influenced by the nature of DG types [6]. Each tested scenario will consider all the 4 load models. Integration between techniques, scenarios and load model is called study cases. The general scenario depends on DG type as shown in Table 1. In this table, the lists of possible scenarios for DG installation dependent on load model is tabulated for different DG type.

Table 1. Scenarios of case study considering DG installation dependent on load model.

Case Study	Scenario	Load Model			
		Model 1	Model 2	Model 3	Model 4
A (DG Type I)	I-1000	1	0	0	0
	I-0100	0	1	0	0
	I-0010	0	0	1	0
	I-0001	0	0	0	1
B (DG Type II)	II-1000	1	0	0	0
	II-0100	0	1	0	0
	II-0010	0	0	1	0
	II-0001	0	0	0	1
C (DG Type III)	III-1000	1	0	0	0
	III-0100	0	1	0	0
	III-0010	0	0	1	0
	III-0001	0	0	0	1
D (DG Type IV)	IV-1000	1	0	0	0
	IV-0100	0	1	0	0
	IV-0010	0	0	1	0
	IV-0001	0	0	0	1

The ranking identification technique was conducted in order to identify the suitable location for DG planning. The DG was located at all bus individually except slack bus. The simulation is identified automatically based on the optimal DG sizing for every single unit of DG at all busses individually to determine the total loss in the system. Then, the total loss will be ranked in descending order to identify the minimum total loss for each DG location. The pre-developed EMEFA technique was performed with an objective function is to minimize the total losses in the system [7]. The complete algorithm of ranking identification using the EMEFA technique for DG planning is demonstrated in Fig. 1.

3 Results and discussion

This section describes the results of the Ranking Identification technique in order to determine the suitability of DG planning in the distribution system. The simulation is identified automatically based on the optimal DG sizing for every single unit of DG at all busses individually to determine the total loss in the system. Then, the total loss will be ranked in ascending order to identify the minimum total loss for each DG location for the possible scenarios of case study for Ranking Identification technique considering DG installation dependent on load model. To address the effectiveness of the proposed technique for DG planning, several scenarios have been considered. The scenario will not be consistent for both test systems.

The proposed technique was performed for DG location and planning on IEEE 33-Bus Distribution System. The simulation is identified automatically based on the optimal DG sizing for every single unit of DG at all buses individually to determine the total loss in the system. In this study, the DG was tested into 32 load buses individually. Then, the results for total loss will be recorded and ranked in descending order to identify the minimum total loss for each DG location. In this analysis, the results of the DG location, DG sizing and total loss are recorded and the results are tabulated the top five locations that are chosen as the suitable location.

a) DG Type 1 (Case A)

In this scenario, the analysis was done for different DG type and load model. The DG is located as dependent on the load model using Ranking Identification technique. Table 2 indicates the results for Scenario I-1000 until I-0001. For the first scenario I-1000, the DG is modelled as type I and load is modelled as constant load. The Ranking Identification technique tabulated the top five of suitable location at bus 25, 26, 27, 28 and 29 with the minimum total loss is located at bus 25 with 0.097 MW. From the table, the value for DG size is the optimized DG values. For bus 25, the optimal DG size is 2.2076 MW. The total loss is evaluated based on the objective function to minimize total loss in the system. The minimum and maximum voltage was recorded in order to show the improvement of the voltage profile with the existing of DG unit.

Comparable study was performed for scenario I-0100. The result ranked the suitable location at bus 30, 25, 26, 29 and 27 based on minimum total loss as objective function. The minimum total loss is located at bus 30 with 0.0767 MW and the optimal DG size is 1.3690 MW. For scenario I-0010, the result is ranked the location of DG at bus 30, 25, 26, 27 and 29 with the minimum total loss is 0.0816 MW at bus 30. Lastly, the determination of DG location is located for Scenario I-0001. The result is ranked the location of DG at bus 30, 25, 26, 27 and 29. The minimum total loss at bus 30 with the optimal DG size is 1.3440 MW and the total loss is 0.0837 MW.

b) DG Type II (Case B)

In this scenario, the DG is modelled as Type II, which is the DG is capable of injecting reactive power. A comparative study was performed to consider the impact on the different load model. Table 3 indicates the results for Scenario II-1000 until II-0001. For the first scenario II-1000, the DG is modelled as Type I and load is modelled as constant load. The Ranking Identification technique tabulated the suitable location at bus 30, 29, 31, 28 and 32 with the minimum total loss is located at bus 30 with 0.1190 MW. From the table, the optimal DG size for bus 30 is 1.1862 MVAR.

The similar study was performed for scenario II-0100. The result ranked the suitable location at bus 30, 29, 31, 32 and 33 based on minimum total loss as objective function. The minimum total loss is located at bus 30 with 0.1185 MW and the optimal DG size is 0.8513 MVAR. The next scenario II-0010 also indicates the suitable location at bus 30, 29, 31,32 and 33 with the minimum total loss is located at bus 30 with 0.9136 MVAR. The similar study was performed for scenario II-0001. The minimum total loss is located at bus 30 with 0.1150 MW and the optimal DG size is 0.9328 MVAR.

c) DG Type III (Case C33)

To further investigate the effect of different DG type and load model, the DG is modelled as type III, which is the DG is capable of injecting reactive power. A similar study was performed for Scenario III-1000 until III-0001 and the results is presented in the Table 4. In scenario III-1000, the load is modelled as constant load. This Ranking Identification technique tabulated the suitable location at bus 30, 29, 28, 27 and 26 with the minimum total loss is located at bus 30 with 0.0537 MW and the optimal DG size is 1.3386 MW and 0.6780 MVAR. The total loss is evaluated based on the objective function to minimize total loss in the system. The minimum and maximum voltage was recorded in order to show the improvement of the voltage profile with the existing of DG unit.

Similar study was performed for scenario III-0100. The result ranked the suitable location at bus 30, 29, 28, 27 and 26 based on minimum total loss as objective function. The minimum total loss is located at bus 30 with 0.0537 MW and the optimal DG size is 1.3353 MW and 0.6764 MVAR. The next scenario I-0010 also indicates the suitable location at bus 30, 29, 28, 27 and 26 with the minimum total loss is located at bus 30 with 0.0537 MW. From the simulation, the optimal DG sizing is 1.3236 MW and 0.6704 MVAR. The similar study was performed for scenario I-0001. The minimum total loss is located at bus 30 with 0.0537 MW and the optimal DG size is 1.3025 MW and 0.658 MVAR.

Table 2 Result for Case A

Scenario I-1000				
DG	DG Size		Total Loss	
Location	P_{DG}(MW)	Q_{DG}(MVAR)	Ploss (MW)	Qloss (MVAR)
25	2.2076	0	0.097	0.0704
26	2.0514	0	0.0977	0.0704
27	1.9862	0	0.0985	0.0707
28	1.6448	0	0.1001	0.0711
29	1.4795	0	0.1005	0.0706
Scenario I-0100				
DG	DG Size		Total Loss	
Location	P_{DG}(MW)	Q_{DG}(MVAR)	Ploss (MW)	Qloss (MVAR)
30	1.369	0	0.0767	0.0543
25	2.1738	0	0.0933	0.0681
26	2.1118	0	0.0971	0.0702
29	1.473	0	0.0975	0.0685
27	1.987	0	0.0979	0.0703
Scenario I-0010				
DG	DG Size		Total Loss	
Location	P_{DG}(MW)	Q_{DG}(MVAR)	Ploss (MW)	Qloss (MVAR)
30	1.3586	0	0.0816	0.0575
25	2.1856	0	0.0943	0.0688
26	2.1163	0	0.0972	0.0703
27	1.9704	0	0.098	0.0704
29	1.4429	0	0.0982	0.0689
Scenario I-0001				
DG	DG Size		Total Loss	
Location	P_{DG}(MW)	Q_{DG}(MVAR)	Ploss (MW)	Qloss (MVAR)
30	1.344	0	0.0837	0.0589
25	2.1642	0	0.0946	0.069
26	2.1383	0	0.0973	0.0704
27	1.9722	0	0.0981	0.0704
29	1.4542	0	0.0985	0.0692

Table 3 Result for case study B33

Scenario II-1000				
DG	DG Size		Total Loss	
Location	P_{DG}(MW)	Q_{DG}(MVAR)	Ploss (MW)	Qloss (MVAR)
30	0	1.1862	0.119	0.0783
29	0	1.2278	0.1209	0.0795
31	0	1.028	0.1244	0.0829
28	0	1.3216	0.1248	0.083
32	0	0.9809	0.1261	0.0845
Scenario II-0100				
DG	DG Size		Total Loss	
Location	P_{DG}(MW)	Q_{DG}(MVAR)	Ploss (MW)	Qloss (MVAR)
30	0	0.8513	0.1185	0.0781
29	0	1.1844	0.1206	0.0793
31	0	0.9838	0.1239	0.0826
28	0	1.2911	0.1247	0.0829
32	0	0.9097	0.1254	0.084
Scenario II-0010				
DG	DG Size		Total Loss	
Location	P_{DG}(MW)	Q_{DG}(MVAR)	Ploss (MW)	Qloss (MVAR)
30	0	0.9136	0.1164	0.0767
29	0	1.1991	0.1195	0.0786
31	0	0.9957	0.1223	0.0815
32	0	0.9419	0.123	0.0825
28	0	1.3239	0.1243	0.0827
Scenario II-1000				
DG	DG Size		Total Loss	
Location	P_{DG}(MW)	Q_{DG}(MVAR)	Ploss (MW)	Qloss (MVAR)
30	0	0.9328	0.115	0.0758
29	0	1.195	0.1187	0.0781
31	0	0.9921	0.1211	0.0807
32	0	0.9224	0.1213	0.0812
28	0	1.3045	0.1239	0.0825

Table 4 Result for case study C33

Scenario III-1000				
DG	DG Size		Total Loss	
Location	P_{DG} (MW)	Q_{DG} (MVAR)	Ploss (MW)	Qloss (MVAR)
30	1.3386	0.678	0.0537	0.0402
29	1.4252	0.7219	0.0546	0.0412
28	1.5774	0.799	0.0576	0.0448
27	1.9062	0.9655	0.0604	0.0487
26	2.0233	1.0248	0.0608	0.0492
Scenario III-0100				
DG	DG Size		Total Loss	
Location	P_{DG} (MW)	Q_{DG} (MVAR)	Ploss (MW)	Qloss (MVAR)
30	1.3353	0.6764	0.0537	0.0402
29	1.4114	0.7149	0.0546	0.0411
28	1.573	0.7967	0.0576	0.0448
27	1.905	0.9649	0.0604	0.0487
26	2.0217	1.024	0.0608	0.0492
Scenario III-0010				
DG	DG Size		Total Loss	
Location	P_{DG} (MW)	Q_{DG} (MVAR)	Ploss (MW)	Qloss (MVAR)
30	1.3236	0.6704	0.0537	0.0402
29	1.3993	0.7088	0.05461	0.0411
28	1.5652	0.7928	0.0576	0.0448
27	1.8727	0.9486	0.0605	0.0488
26	2.008	1.0171	0.0608	0.0492
Scenario III-0001				
DG	DG Size		Total Loss	
Location	P_{DG} (MW)	Q_{DG} (MVAR)	Ploss (MW)	Qloss (MVAR)
30	1.3025	0.6597	0.0537	0.0402
29	1.3954	0.7068	0.0546	0.0411
28	1.5654	0.7929	0.0576	0.0448
27	1.8978	0.9613	0.0604	0.0487
26	2.014877	1.0206	0.0608	0.0492

Table 5 Result for case study D33

Scenario IV-1000				
DG	DG Size		Total Loss	
Location	P_{DG} (MW)	Q_{DG} (MVAR)	Ploss (MW)	Qloss (MVAR)
30	1.3447	-1.9489	0.0538	0.0403
29	1.4169	-0.4095	0.0546	0.0412
28	1.5772	-2.7462	0.0576	0.0449
27	1.9082	-2.0788	0.0605	0.0488
26	2.0105	-1.4771	0.0609	0.0493
Scenario IV-0100				
DG	DG Size		Total Loss	
Location	P_{DG} (MW)	Q_{DG} (MVAR)	Ploss (MW)	Qloss (MVAR)
30	1.3347	-2.1756	0.0537	0.0403
29	1.4159	-1.3634	0.0546	0.0412
28	1.5761	-2.8988	0.0576	0.0449
27	1.9156	-2.1981	0.0605	0.0488
26	2.0228	-2.213	0.0608	0.0493
Scenario IV-0010				
DG	DG Size		Total Loss	
Location	P_{DG} (MW)	Q_{DG} (MVAR)	Ploss (MW)	Qloss (MVAR)
30	1.3218	-1.5812	0.0538	0.0403
29	1.4044	-3.8557	0.0546	0.0412
28	1.5583	-1.2274	0.0576	0.0449
27	1.9096	-0.3885	0.0605	0.0488
26	2.0184	-4.9472	0.0608	0.0493
Scenario IV-0001				
DG	DG Size		Total Loss	
Location	P_{DG} (MW)	Q_{DG} (MVAR)	Ploss (MW)	Qloss (MVAR)
30	1.2998	-2.1713	0.0537	0.0403
29	1.3889	-2.5971	0.0546	0.0412
28	1.5511	-2.5921	0.0576	0.0449
27	1.8984	-1.367	0.0605	0.0488
26	2.0152	-5.0073	0.0608	0.0493

d) DG Type IV (Case D33)

To compare the effect of different DG type, similar study was performed for case D33. The DG is modelled as type IV, which is the DG is capable of injecting active power while consuming the reactive power.

The results are obtained from Ranking Identification technique was tabulated in Table 5. It indicates the results for Scenario IV-1000 until IV-0001. For the first scenario I-1000, the load is modelled as constant load. The results are tabulated the suitable location at bus 30, 29, 28, 27 and 26. The minimum total loss is located at bus 30 with 0.0537 MW. The value for DG size is 1.3405 MW and consumes 0.679 MVAR. The results for scenario IV-0010 till IV-0001 the also indicates the top five locations at bus 30, 29, 31, 32 and 33 with the minimum total loss is located at bus 30. Based on the analysis, the different load model does not affect the DG sizing and the voltage profile for the system.

Further analysis was done by determining the LRI for different scenario and the result is illustrated in Fig. 2. For Case A, the results at bus 25 indicate the highest LRI at case I-0100 with 55.26 %, followed with the Case I-0010 at 52.37 % as illustrated on the graph in Fig. 2(a). The percentage of LRI for Case I-0001 is 51.14 % and the minimum LRI at Case I-1000 with 43.38 %. The result for DG located at bus 26, 27, 28 and 29 shows. Based on the results, the load model for industrial load gives the highest LRI followed by the residential, commercial and constant load.

For Case B, the results at bus 30 indicate the highest LRI compared to others cases. The result shows the similar trends for each case for DG located at bus 29, 31, 32 and 28. Based on the results, the load model for industrial gives the highest LRI followed by the commercial, residential and constant load.

Similar study was performed for LRI for Case C. From the graph in Fig 2 (c), the results at bus 30 show the highest LRI at all cases with the percentage is 68.64 %. This result shows different trends compared to Case A and B. In Case C, when the DG is installed as type III, it is found that the different load model does not affect the total loss. In the meantime, the result shows the LRI remains the same for all scenarios. This result shows similar trends in Case D, when the DG is installed for active and consumes reactive power. It is found that the different load model does not affect the total loss. In the meantime, the result shows the LRI remains the same for all scenarios.

4 Conclusion

The analysis of load model which considers the loss minimization and the voltage performance was successfully implemented and tested on IEEE 33 and IEEE 69 bus test system. The Ranking Identification technique using EMEFA is used to determine the optimal output of DG with different DG type and load models. The ranking identification technique is developed in order to determine the suitability of DG planning in the distribution system. This technique was conducted using the developed automatic algorithm for single DG by incorporating the EMEFA technique as optimization technique. The simulation is identified automatically based on the optimal DG sizing for every single unit of DG at all busses individually to determine the total loss in the system. Then, the total loss will be ranked in descending order to identify the minimum total loss for each DG location. The modeling of voltage dependent loads, i.e. residential, commercial and industrial loads have been presented and the results show the voltage dependent load models have a significant effect on total losses of a distribution system for different DG type consequently would affect the loss of the system.

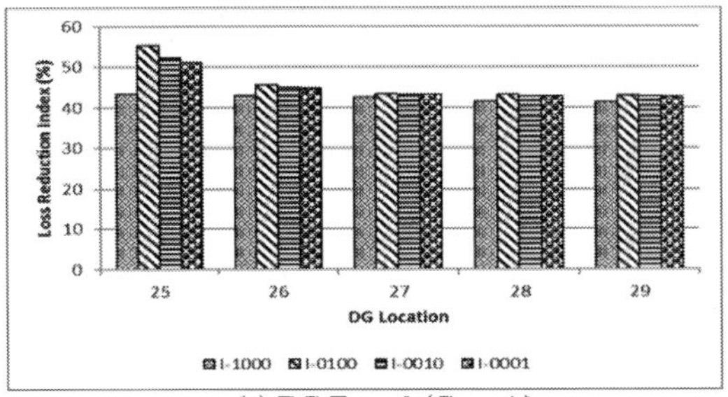

(a) DG Type 1 (Case A)

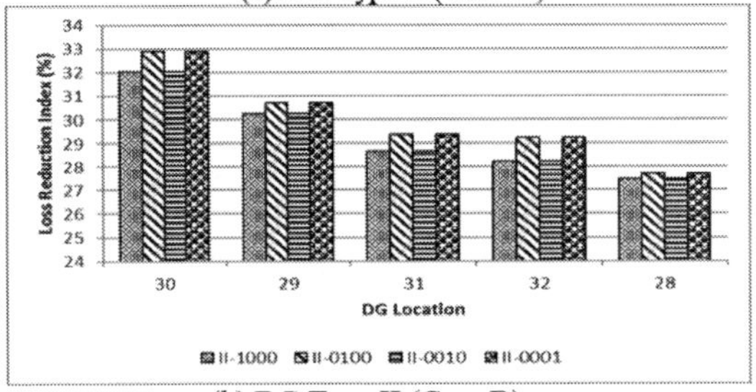

(b) DG Type II (Case B)

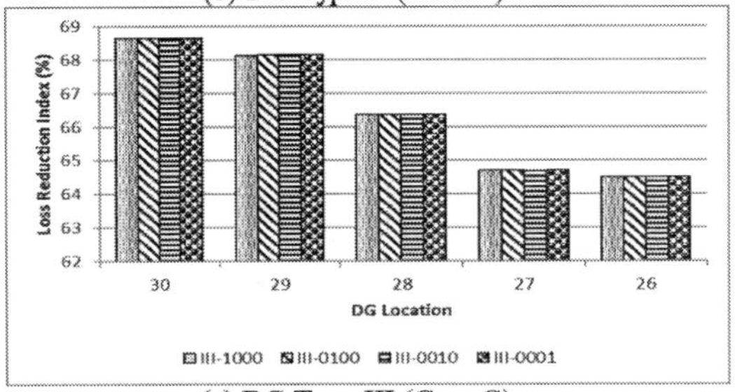

(c) DG Type III (Case C)

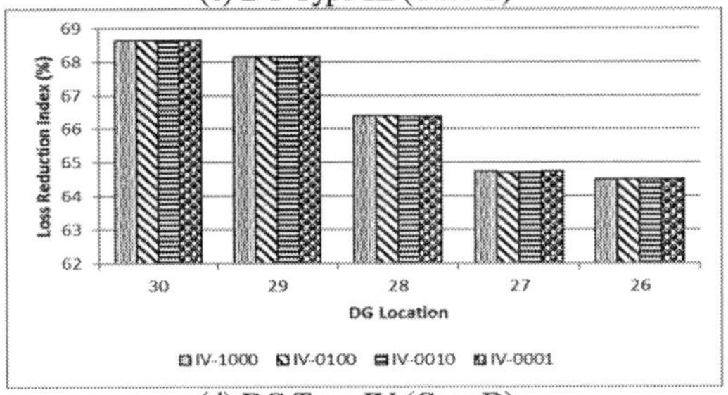

(d) DG Type IV (Case D)

Fig. 2. Comparative study on LRI for different scenario

The authors would like to acknowledge Centre of Excellence for Renewable Energy (CERE), UniMAP, School of Electrical System Engineering, UniMAP, Ministry of Higher Education Malaysia (MOHE) and Universiti Malaysia Perlis for the financial support of this research. This research is supported by MOHE under the Fundamental Research Grant Scheme with project code: (File No: FRGS/1/2016/TK04/UNIMAP/02/9).

References

1. Z. M. Yasin, T. K. A. Rahman, I. Musirin, and S. R. A. Rahim, "Optimal sizing of distributed generation by using quantum-inspired evolutionary programming," *4th Int. Power Eng. Optim. Conf.*, 2010.
2. S. R. A. Rahim, I. Musirin, M. H. H. Hussain, and M. M. M. Othman, "EMEFA Approach for DG Installation in Distribution System," *Int. Rev. Model. Simulations*, vol. 5, no. 6, pp. 2546–2553, 2012.
3. A. K. Bohre, G. Agnihotri, M. Dubey, and S. Kalambe, "Impacts of the Load Models on Optimal Planning of Distributed Generation in Distribution System," *Electr. Eng. Dep. Azad Natl. Inst. Technol. (MANIT), Bhopal 462003, India*, vol. 2015, pp. 1–10, 2015.
4. H. Nasiraghdam and S. Jadid, "Load model effect assessment on optimal distributed generation (DG) sizing and allocation using improved harmony search algorithm," *Conference on Smart Electric Grids Technology (SEGT)*, 2013, pp. 210–218.
5. A. Zidan, R. S. Al-Abri, and E. F. El-Saadany, "Load model effect on distributed generation allocation and feeders' reconfiguration in unbalanced distribution systems," *2015 IEEE 8th GCC Conf. Exhib. GCCCE 2015*, pp. 1–4, 2015.
6. A. M. El-Zonkoly, "Optimal placement of multi-distributed generation units including different load models using particle swarm optimisation," *IET Gener. Transm. Distrib.*, vol. 5, no. 7, p. 760, 2011.
7. S. R. A. Rahim, Muhammad Murthadha Othman, and I. Musirin, "Clustering Technique for DG Planning Considering Load Models," *Appl. Mech. Mater.*, vol. 793, no. August 2016, pp. 457–461, 2015.

A Study of Accuracy and Time Delay for Bilateral Master-Slave Industrial Robotic Arm Manipulator System

Nuratiqa Natrah Mansor[1,*], *Muhammad Herman* Jamaluddin[2], *Ahmad Zaki* Shukor[3] and *Chee Cong* Lok[4]

[1,2,3,4] Centre for Robotics and Industrial Automation (CERIA), Faculty of Electrical Engineering, Universiti Teknikal Malaysia Melaka, Malaysia

Abstract. Bilateral master-slave industrial robotic arm manipulator system is an advanced technology used to help human to interact with environments that are unreachable to human, due to its remoteness or perilous. The system has been used in different areas such as tele-surgery, autonomous tele-operation for sea and space operation and handling explosive or high radiation operation fields. It is beneficial both for science and society. Remarkably, the system is not common and generally used in Malaysia. Likewise, the number of research conducted that focused about this technology in our country manufacturing industry are not yet discovered and existent. The implementation of this bilateral manipulator system in an industrial robot could be useful for industrial imminent and development over our country and people, specifically for production yield size and human operative. Hence, the study of bilateral robotic arm manipulator system in an industrial robot and analyzation of its performance and time delay in 3 differ controllers will be discussed to attest the efficiency and its effectiveness on the said design system. The experiment conducted was on KUKA youBot arm in V-Rep simulation with three different controllers (P, PD,PID).

1 Introduction

In robotic technology, haptic system is known to be a medium that can give force sensation to user from real environment. Basically, haptics system can be in synchronous and asynchronous multilateral control for walking motion [1]. Haptic information can also be transmitted into two directions which is unilateral and bilateral. Unilateral is a transmission of information in one direction only without giving feedback whereas bilateral direction is transmission of information that consist of action and reaction from environment contact [2]. While, bilateral master-slave telerobotic is a potential useful tool to substitute human for carry out difficult or unreachable load and task. Position or velocity and force information are bidirectionally transmitted in this type of control system. These physical variables are controlled locally [3]. The efficiency of the master-slave robot in term of synchronization of position between master and slave arm and time delay is rest on the accuracy from the tasks. In the complexity of communication network between computer and master-slave telerobotic system, time-delay is unavoidable during the data packet exchanged. This is because the subsystems, which is master and slave robot may have located far from each other [4]. The delay give rise to latency and interference in terms of the system performance during transmitting the tactile sensation to other pair of its subsystem. It deteriorates overall performances and thereby may destabilize the system works [5]. In this century, this kind of telerobotic arms have become an emerging technology in this century and has many potential applications in numerous research fields and works including robotic rehabilitation [6]. The main redundancy of this technology is that it helps human to interact with environments that are not accessible and also risky. There are five basic parts involved in master-slave telerobotic system which include human operator, master robot, slave robot, communication channel and the environment.

Therefore, the motivation of this research is to develop a bilateral master-slave telerobotic arm manipulator system by simulation work of V-Rep and Python (x, y) software. Besides, the project also proposed to analyse the time-delay and the performance of robotic arm system in term of accuracy as well as its efficiency when different type's controllers are applied and tested on the control system.

2 Research methodology

In this section, there are three steps that will be implemented and discussed to develop the proposed system. In order to achieve the listed objectives as stated in previous introduction, every following course of procedure have been conducted step by step in this project. To visualize the research flow, every major step of the research is presented briefly in flow chart below as in Fig. 1.

* Corresponding author: m011620002@student.utem.edu.my

https://doi.org/10.1051/matecconf/201815001015

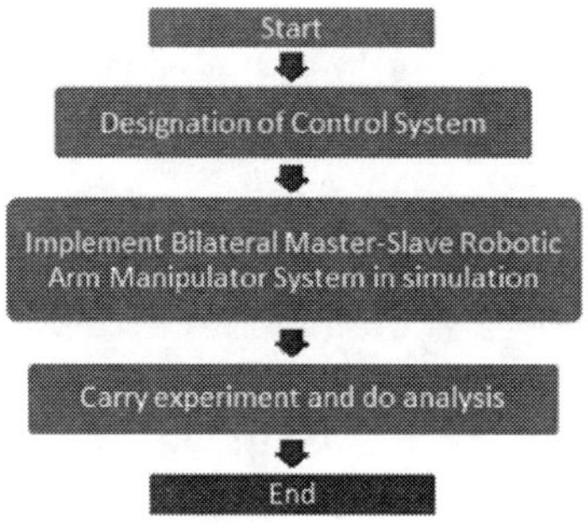

Fig. 1. Flow Chart of the Research

2.1 Designation of control system

A reliable and ideal design of control system is utmost important for a precise motion control. In bilateral control, the accuracy of motion between master and slave robot is emphasized, hence a good control system is necessary in this project. To make it clear, several types of controller in control systems have been tested for verification, comparison and analysation of respective performances. It is for selecting the controller which has the best performance for an ideal bilateral master-slave manipulator system. Eventually, there are 3 types of control system which are P control, PD control and PID control. A block diagram of a single link motor control system without any controller is indicated as in Fig. 2. After that, P controller which carry an important variable is added as shown in Fig. 3.

P control also known as proportional term, K_p. The input of K_p is the difference between input angle and output angle of the system. K_p is used to reduce the rise time of the system which give system a quick respond and reduce the steady state error. The shorter the rise time and settling time, the better the performance of a system. Therefore, the system is more stable when overshoot is smaller. K_p purpose in a control system is for achieving desired response of critical damping. This controller also is much simple in tuning the gain.

Afterwards, a derivative term known as K_d is added into the P control system as shown in Fig. 4 to become second design of control system which is PD control system. In PD control system, K_d is multiplied by difference between the rate of change of output angle and the rate of change of input angle. K_d is used to help reducing the settling time, eliminate the overshoot problem and improve the stability of the system.

The third design control system is further improved with adding another term which is integral term, K_i into the current PD control system. This results in forming a new design of control system as shown in Fig. 5. K_i is used to reduce and eliminate steady- state error. At most cases, there are some small error exists due to the external disturbance from environment although K_i able to lead the system into zero steady-state error, which is an ideal situation for a built control system.

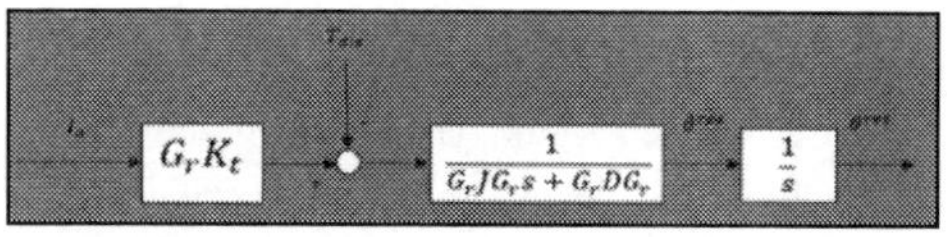

Fig. 2. Block diagram of motor

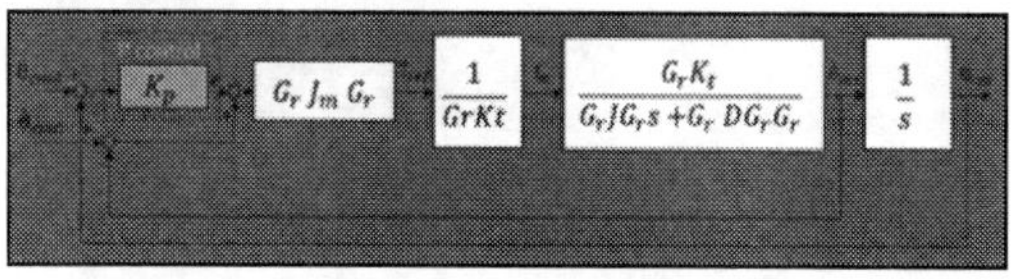

Fig. 3. Block diagram adding P control to the manipulator system

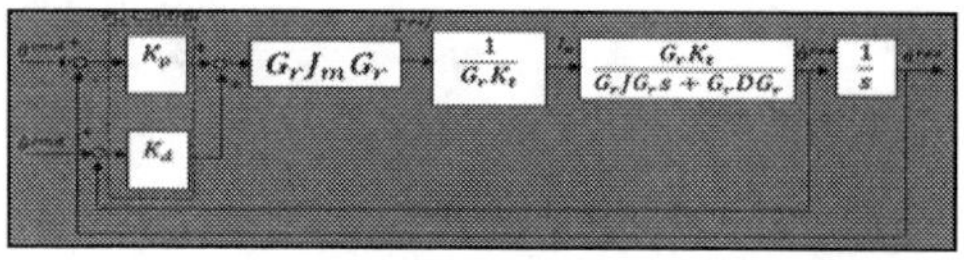

Fig. 4. Block diagram adding PD control to the manipulator system

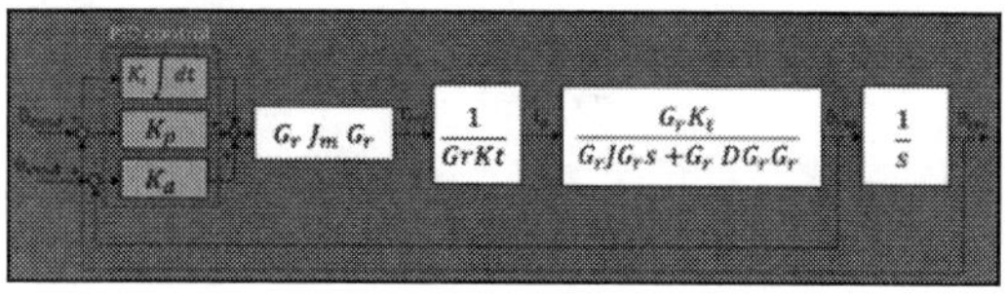

Fig. 5. Block diagram adding PID control to the manipulator system

where;
θ_{res}[rad]: Respond angle of motor rotation
θ_{cmd}[rad]: Command angle of motor rotation
I_a[A]: Torque electric current
T_{dis} / T_{ref} [Nm]: Motor torque
G_r: Gear ratio
K_t[Nm/A]: Torque constant
J_m[kgm^2]: Inertia of the motor shaft
D[Nmrad/s]: Viscous friction factor
K_p: Proportional gain in P controller
K_d: Proportional gain in PD controller
K_i: Proportional gain in PID controller

2.2 Implementation of bilateral master-slave telerobotic arm manipulator system in simulation

To verify the designed control systems, a simulation need to be done to prove the designed control system is robust. In this project, V-Rep and Python (x, y) are used. V-Rep is used due to it already has the built in KUKA youBot and much easier to design the scene and object properties while python is used for programming language as it has more functions and easy to be connected with V-Rep through remote API address for communication channel.

When simulation starts, external arm will swing and exerted force on master youBot. Expectedly, slave robot will track the movement of master robot accurately. An obstacle is placed on the slave robot to block it from moving while rotating. At the same time, master robot arm will also stop at the same position even though there is no obstacle comes in contact with the master of youBot arm. This situation can be referred in Fig. 6.

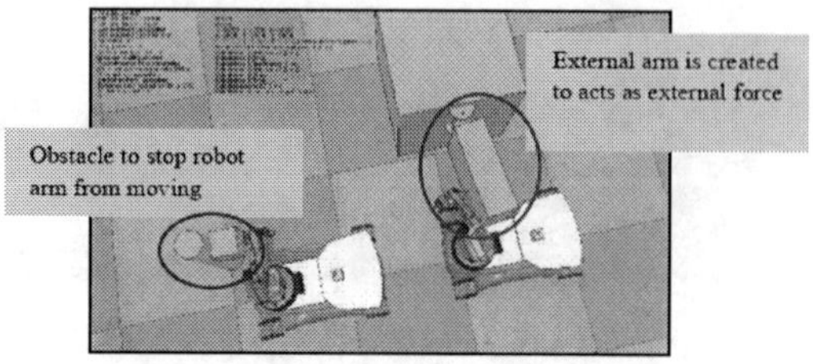

Fig. 6. Simulation of two bilateral master-slave KUKA youBot (master on the right and slave on the left)

In bilateral concept, every command signal applied on the operator at master manipulator can be executed by the end effector at slave manipulator [7]. Moreover, operator can measure the stiffness of executing subject or object at the end-effector of slave manipulator. To visualize the result, the angle of the master youBot arm and slave youBot arm are extracted and plotted in graph. The experiments have been carried out 6 times with different values of K_p, K_i and K_d respectively. The accuracy of movement between the master and slave robot, time delay of the movement between the master and slave robot and precision of final angle to the reference angle could be calculated by using the formula in Equation 1, 2 and 3.

$$Accuracy = \left[1 - \frac{(y_m - x_m) - (y_s - x_s)}{(y_m - x_m)}\right] x\ 100\% \qquad (1)$$

where;
x_m = Initial position for master robot
x_s = Final position for slave robot
y_m = Initial position for master robot
y_s = Final position for slave robot

$$Time\ delay = (t_{fm} - t_{im}) - (t_{fs} - t_{is}) \qquad (2)$$

where;
t_{im} = Initial time taken for master robot
t_{is} = Initial time taken for slave robot
t_{fm} = Final time taken for master robot
t_{fs} = Final time taken for slave robot

$$Precision = \left[1 - \frac{(\theta_{ref} - \theta_f)}{\theta_{ref}}\right] x\ 100\% \qquad (3)$$

where;
θ_{ref} = Reference angle of the robot
θ_f = Final angle of the robot

To transfer the information in haptic communication between the bilateral telerobotic system, position tracking and law of action and reaction between master and slave system are necessary. Based on the block diagram in Fig. 7, the control system are similar as the previous PID control system as in Fig. 5 but, the changed is there are two similar block diagram applied regard to robotic arm system (master and slave). From the block diagram, T_m represent the torque of master manipulator provided by human operator or external force while T_s represents the torque of slave manipulator applied by environment. General equation of bilateral master-slave telerobotic system is as shown in equation 4.

$$\ddot{\theta} = K_p(\theta^{cmd} - \theta^{res}) + K_d(\dot{\theta}^{cmd} - \dot{\theta}^{res}) \qquad (4)$$

From the general equation, the equation of motion in master and slave system are in equation 5 and 6 correspondingly.

$$\ddot{\theta}m = K_p(s - \theta m) + K_d(\dot{\theta}s - \dot{\theta}m) \qquad (5)$$

$$\ddot{\theta}s = K_p(\theta m - \theta s) + K_d(\dot{\theta}m - \dot{\theta}s) \qquad (6)$$

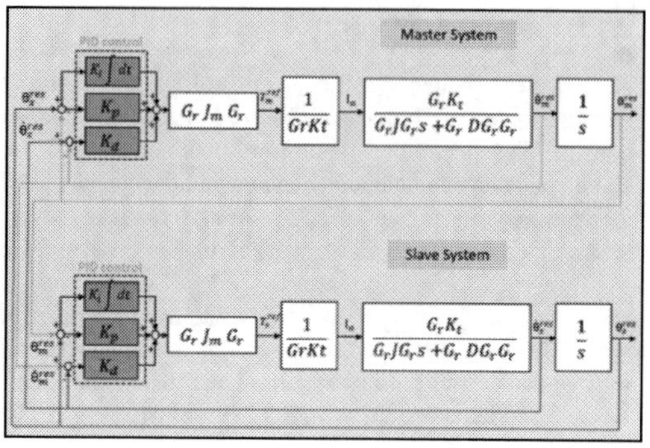

Fig. 7. Block Diagram of bilateral telerobotic arm manipulator

3 Result and Discussion

To prove the common mode and differential mode of bilateral master-slave telerobotic arm manipulator system, two separate experiment has been conducted. In common mode, the summation of external torque from master and slave system must always be zero. Hence, for this part, different maximum torque of KUKA youBot Joint 1 (lower part of the arm robot as shown in Fig. 8) is set to show different results in order to prove the common mode is achieved.

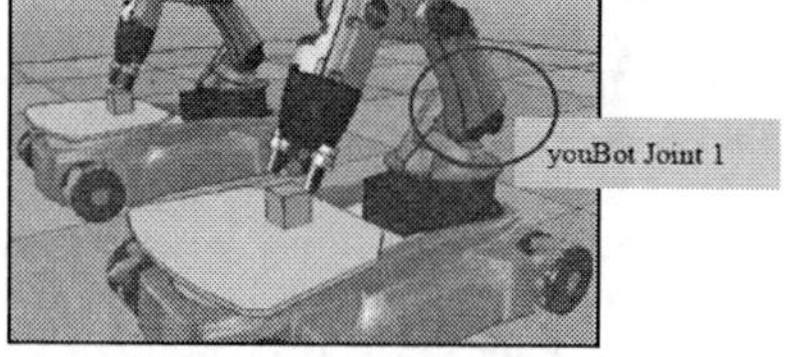

Fig. 8. Joint 1 of KUKA youBot

Meanwhile, the maximum torque of created external arm is adjusted to a suitable torque that will makes the robot to exert its maximum torque while in the same time the master robot arm would not be moved when slave youBot is being blocked. This is because when the torque of extra arm is too large and move the master robot arm, the master robot arm unable to reach its maximum torque when the robot is moving and hence, did not give the desired and expected result. K_p is set to

1. After the simulation starts, the external arm will swing its arm and hit master KUKA youBot arm and accordingly, slave youBot arm will following the motion of master until it is being blocked. This will cause slave youBot arm having maximum reaction force against the obstacle. After the slave is being blocked by the corresponding obstacle, the master will also follow the motion from the slave. This experiment is repeated for three times with different maximum torque of external arm and KUKA youBot joint 1 arm as shown in Table 1.

Forces reading for both arms are recorded and plotted as shown in graphs in Table 2.

Table 1. Maximum torque of external arm and KUKA youBot arm.

External arm maximum torque (Nm)	youBot maximum torque (Nm)
10	5
35	15
55	25

Table 2. Graph of forces for master and slave arm for common mode

a) Graph of forces for master and slave arm vs time for 5N	b) Graph of forces for master and slave arm vs time for 15N	c) Graph of forces for master and slave arm vs time for 25N
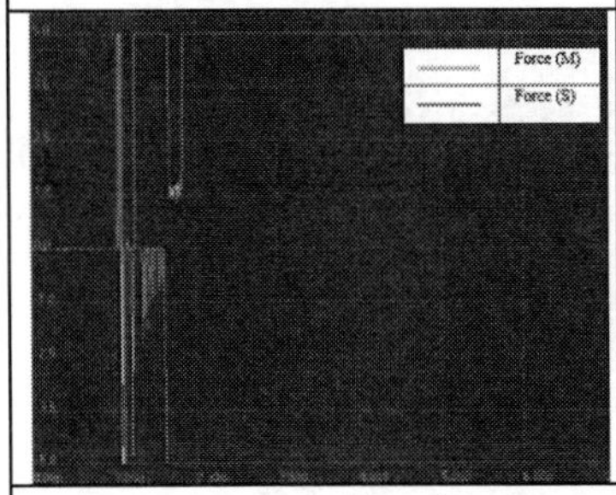	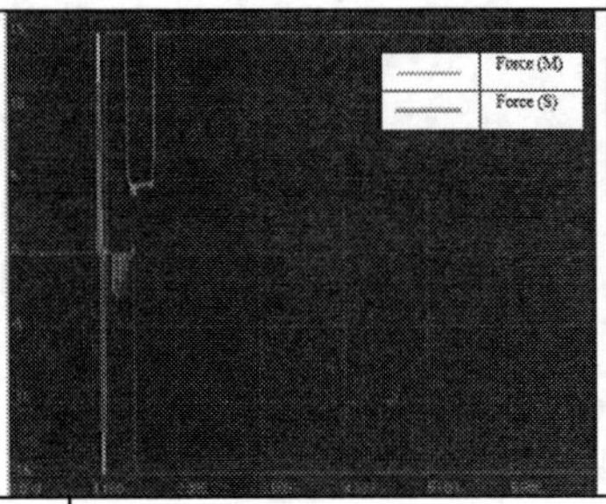	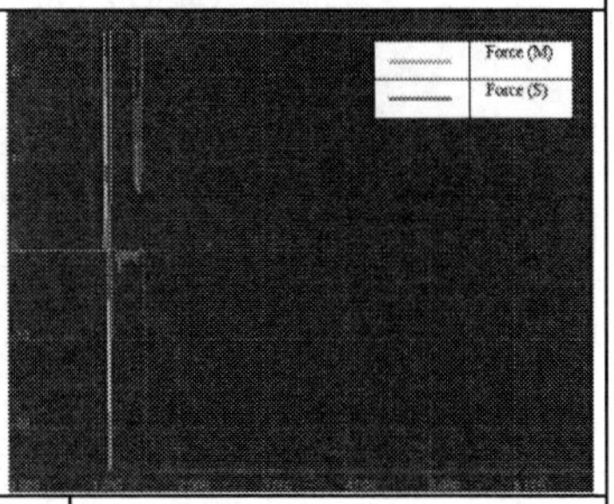
Remarks	—— Force for master	—— Force for slave

Referring to figure a) in Table 2, it shows that time before 1.5s, the force of slave youBot arm is zero, this is because the slave arm is in free contact form any obstacle. However, once it is being blocked by an obstacle, it immediately has a reaction force against the obstacle. The graph shows that master youBot arm has giving its maximum torque of 5N while slave youBot arm has giving its maximum torque of -5N in the experiment. This proved that this bilateral master-slave telerobotic manipulator arm system has achieved the common mode and obey the equation which state that the summation of external torque from master and slave system must always be zero. To prove the validity of the result, experiment is repeated with different maximum torque set to both master and slave youBot arm and the result is shown as in figure b) and c) in Table 2. The results are almost same between these 3 figures, but differ in force values and responded force pattern. In short, the first requirement on designing bilateral system is success.

In differential mode, the subtraction of position between the master youBot arm and slave youBot arm is emphasised. In this experiment, the fixed variable will be maximum torque of KUKA youBot arm, which will fix the maximum torque youBot arm to 5Nm and the external arm to 35Nm and the manipulated variable will be the parameter value of K_p, K_d and K_i. In expected result, slave youBot should follow the motion of master youBot after the simulation is started.

Table 3. Parameter value of K_p, K_d and K_i

Value of K_p	Value of K_i	Value of K_d
1	0.80	0.0100
5	0.10	0.0010
8	0.50	0.0050
10	0.20	0.0005
20	1.60	0.0050
50	4.65	0.0500

Table 4. Graph of Force and Angle of master and slave arm for each controller for differential mode between different controllers

P Controller	PD Controller	PID Controller
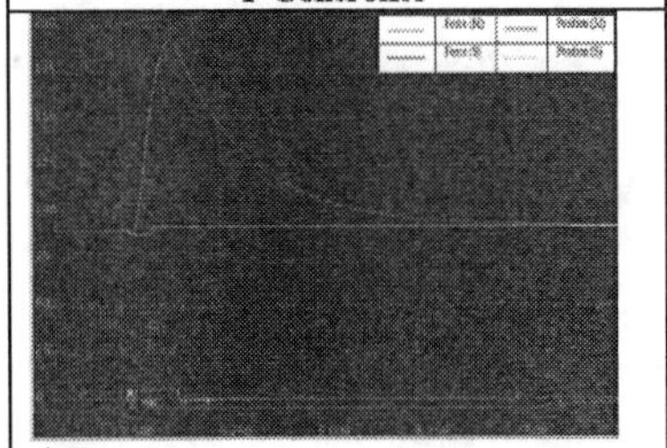	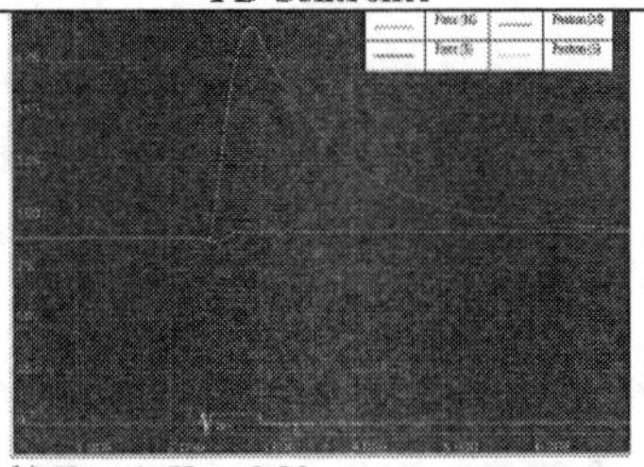	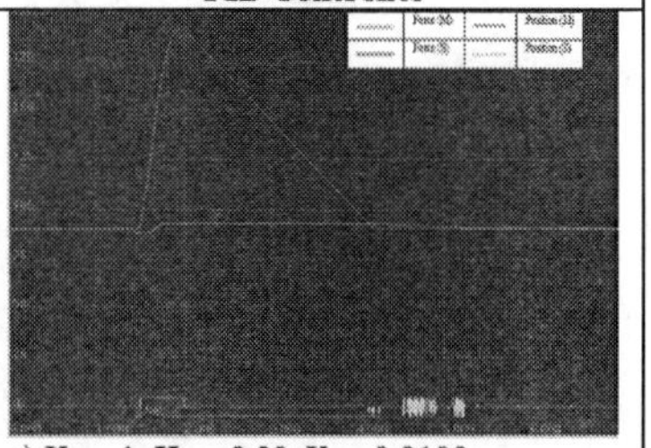
a) $K_p = 1$	b) $K_p = 1$, $K_d = 0.80$	c) $K_p = 1$, $K_d = 0.80$, $K_i = 0.0100$

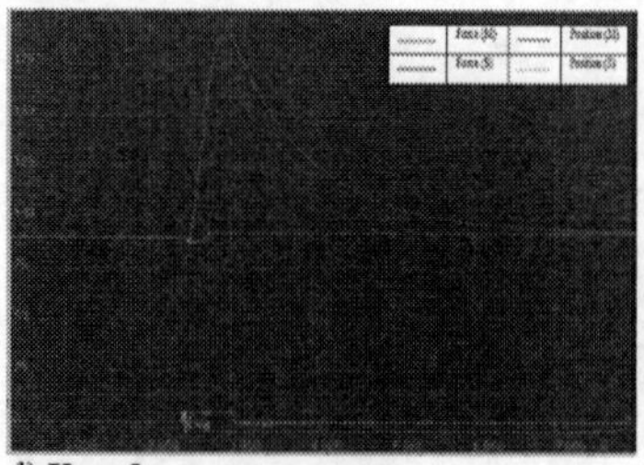
d) $K_p = 5$

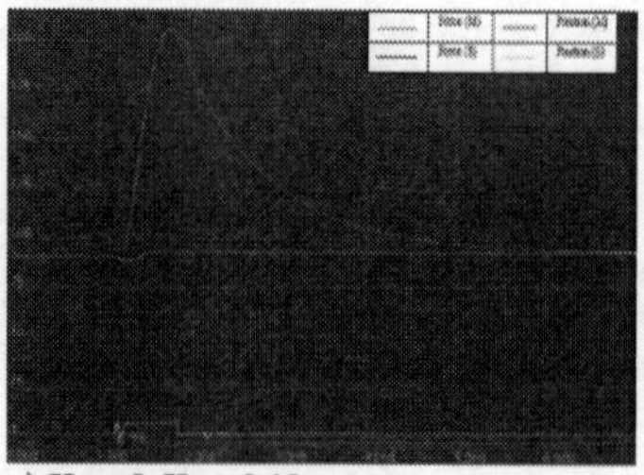
e) $K_p = 5$, $K_d = 0.10$

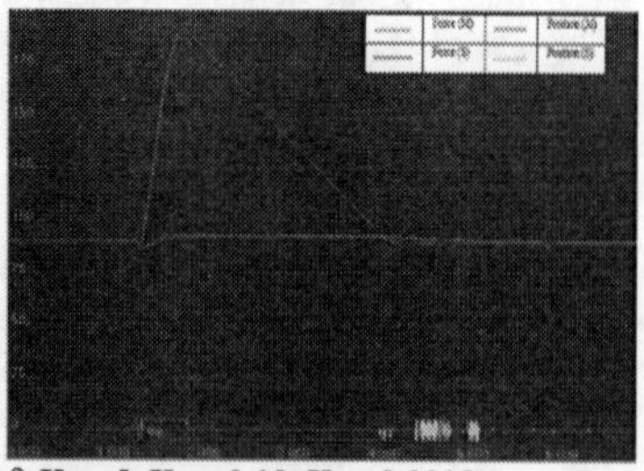
f) $K_p = 5$, $K_d = 0.10$, $K_i = 0.0010$

g) $K_p = 8$

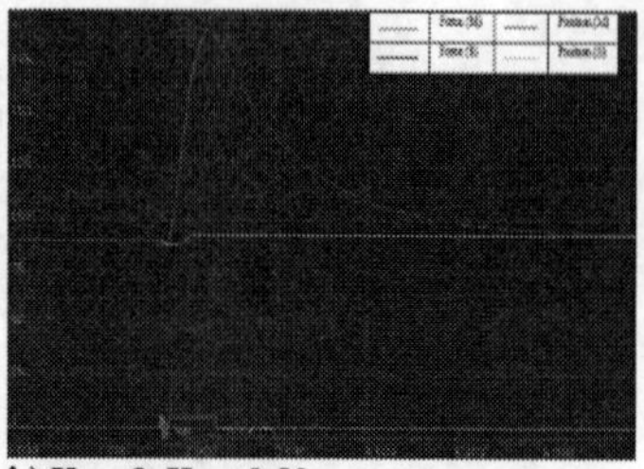
h) $K_p = 8$, $K_d = 0.50$

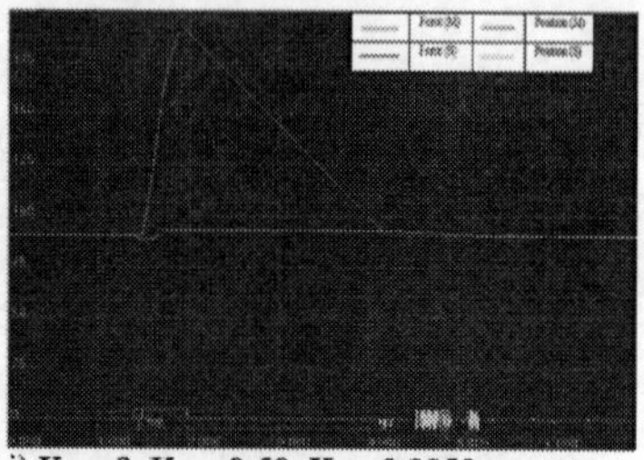
i) $K_p = 8$, $K_d = 0.50$, $K_i = 0.0050$

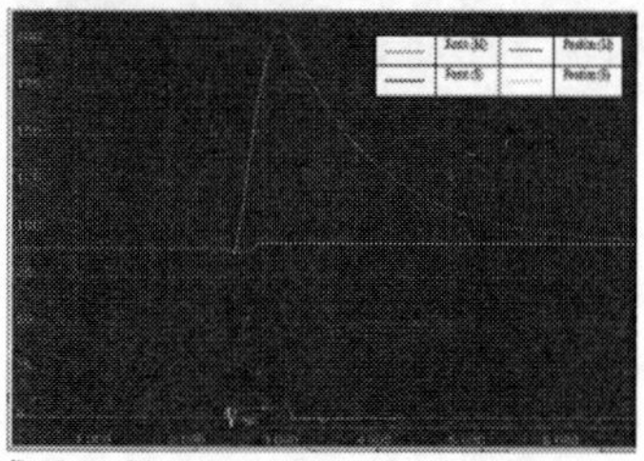
j) $K_p = 10$

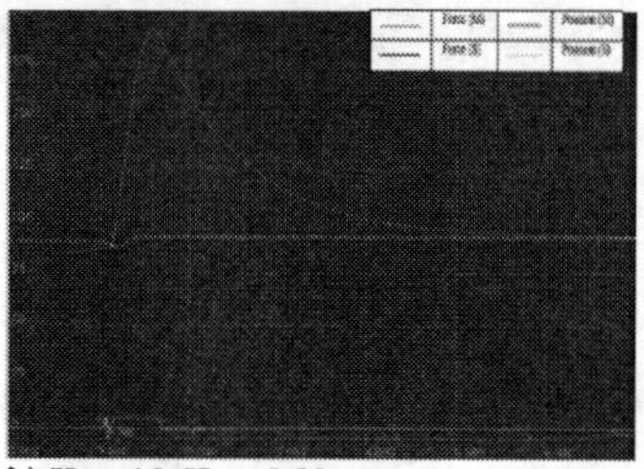
k) $K_p = 10$, $K_d = 0.20$

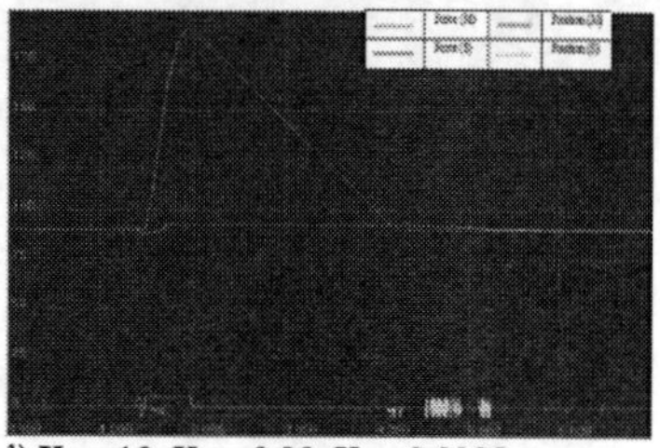
l) $K_p = 10$, $K_d = 0.20$, $K_i = 0.0005$

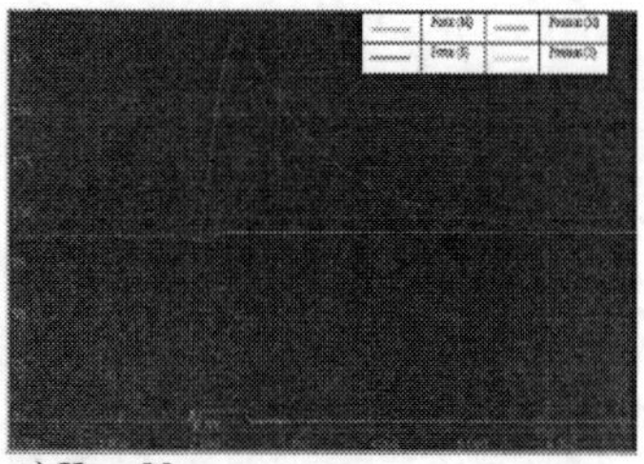
m) $K_p = 20$

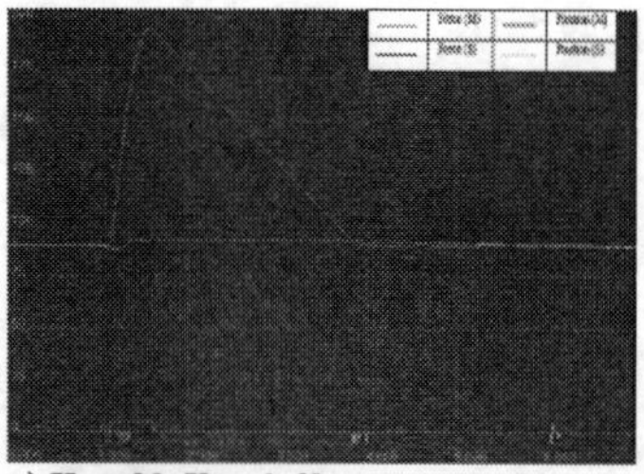
n) $K_p = 20$, $K_d = 1.60$

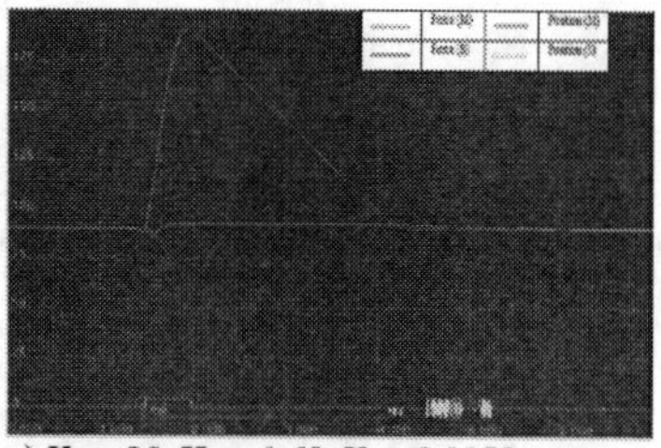
o) $K_p = 20$, $K_d = 1.60$, $K_i = 0.0050$

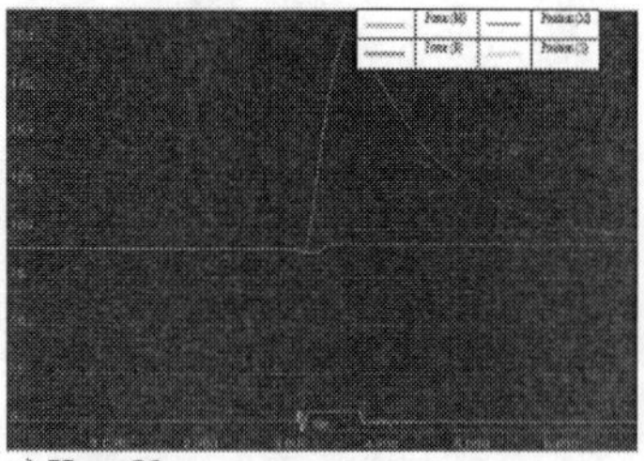
p) $K_p = 50$

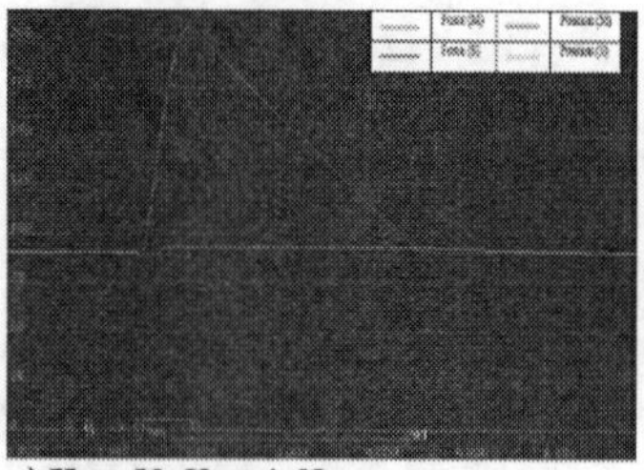
q) $K_p = 50$, $K_d = 4.65$

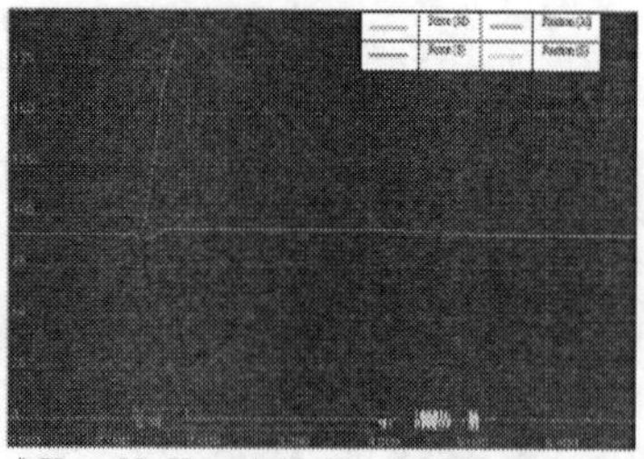
r) $K_p = 50$, $K_d = 4.65$, $K_i = 0.0500$

Remarks	——	Force for master	——	Position for master
	——	Force for slave	——	Position for slave

Table 5. Accuracy result for P control system

Kp	Accuracy (%)	Time-delay (s)	Precision (%)
1	99.44	0.5344	99.45
5	99.75	0.2273	99.24
8	99.74	0.2130	99.35
10	99.62	0.2050	97.85
20	94.11	0.1855	96.68
50	55.77	0.1767	94.54

Table 6. Accuracy result for PD control system

Kp	Kd	Accuracy (%)	Time-delay (s)	Precision (%)
1	0.80	98.06	0.1909	98.12
5	0.10	99.92	0.2075	99.85
8	0.50	99.24	0.1850	99.53
10	0.20	99.99	0.1866	99.24
20	1.60	97.85	0.1699	99.98
50	4.65	86.41	0.1780	99.98

Table 7. Accuracy result for PID control system

Kp	Kd	Ki	Accuracy (%)	Time-delay (s)	Precision (%)
1	0.80	0.0100	97.13	0.1921	99.92
5	0.10	0.0010	99.63	0.2072	99.98
8	0.50	0.0050	97.81	0.1876	99.99
10	0.20	0.0005	99.89	0.1873	99.99
20	1.60	0.0050	99.18	0.1690	99.97
50	4.65	0.0500	87.52	0.2380	99.92

According to the six-plotted graph in figure a), d), g), j), m) and p) in Table 4 when the system used P controller, the angle of master and slave youBot will reach almost same final angle exceptional for Figure p). This means that the system achieved the differential mode which obey the equation of subtraction of position from master and slave must be zero. The system also reached its steady state without fluctuated. Accuracy and time-delay between master and slave arm are calculated using Equation (1) and (2) as stated in Section 3. The result is plotted in Table 5. By referring to Table 5, the most accurate for system performance is when K_p=5, whereby it reached 99.7526% while the least accurate system is when value of K_p is 50 which merely 55.7752%.

Eventhough it is least accurate, the delay is the shortest which is only 0.1767s while K_p=1 results 0.5344s, making it the longest delay. Therefore, it can say that the bigger the value of K_p, the lowest the value for time-delay. Thus, for P controller, the most suitable proportional value is when K_p=5 as it obeys both common and differential mode, has critical damping, near to zero steady state error, high accuracy and short time-delay between both bilateral manipulator arms.

For PD Controller, based on the six-plotted graph in figure b), e), h), k), n) and q) in Table 4 when the system used PD controller, the angle of master and slave youBot will reach almost same final angle. This means that the system achieved the differential mode which obey the equation of subtraction of position from master and slave must be zero. The system also stable and reached its steady state without fluctuated. By referring to Table 5, The highest accuracy of system is when K_p=10, and K_d=0.2 whereby 99.9996% and shortest in terms of time-delay with 0.1699s. However, the least accurate system is when K_p=50 and K_d=4.65. The accuracy is just 86.4099% which improves about 30% of the performance comparing when the system use P controller. In short, by using PD controller, the most suitable proportional and derivative value is when K_p=10 and K_d=0.2 as it obeys both common and differential mode, has critical damping, zero steady state error, high accuracy and acceptable time-delay between both bilateral manipulator arms.

On the same time, PID controller also responded that it achieved differential mode and obey the equation whereby the subtraction for each position resulting to zero. Based on Figure c), f), i), l), o) and r), each system reaches slight steady state from the reference angle. From the observation, there are three system that has reached over 99% but the system with K_p=10, K_d=0.2, K_i=0.05 has outperformed others by reaching 99.8894%. Whereas, the shortest time delay in the system is when K_p=20, K_d=1.6, K_i=0.005 when it responded in just 0.169s. Hence, the most suitable value for proportional, derivative and integral term are K_p=10, K_d=0.2, and K_i=0.0005. This is because the system achieved differential mode, has least steady state error, high accuracy and acceptable time-delay.

4 Conclusion and Recommendations

In a nutshell, all objectives have been achieved in this project. From the findings of results, the performance and feedback of said system in differential mode is proven. The suggested bilateral master-slave control system has been proved by carrying experiments with certain set of variables. Responded data, which are force and also angle (position) of the bilateral control system are used for plotting graphs on the system response. As the basic requirement for common mode and differential mode has been validated, the development of bilateral master-slave telerobotic arm manipulator system is realized and successfully proven.

Overall, by carrying out experiments by using different value from different kind of controllers (K_p, K_i and K_d), most of the system achieved very high accuracy between master and slave youBot arm simulation which is over than 95%. The time delay is also short whereby around 0.2s. However, the precision of youBot rotated angle to the reference angle and the settling time has showed in huge difference with different control system and different value of proportional term, K_p, derivative term, K_d, and integral term K_i. It can be concluded that most suitable controllers are when using K_p =5 for P controller, the PD controller with K_p = 10 and K_d = 0.2 for the most precise control with zero steady state error, or PID controller with K_p =10, K_d=0.2, and K_i=0.0005 with little steady-state error but with shorter settling time which is 1s.

For future research, the system designed is expected to be carried out with the real KUKA youBot so that the differences between real time and simulation performances can be observed, analysed and understand.

Furthermore, future works can be developed by integrated the manipulator system by other control system such as disturbance observer (DOB) and reaction force observer (RFOB) as these two well-known controllers are potentially to gain the efficiency and smooth feedback for precise control system.

We wish to express our gratitude to Universiti Teknikal Malaysia Melaka (UTeM) especially for Robotics and Industrial Automation (RIA) research group in the Centre of Excellence in Robotis and Industrial Automation (CeRIA), Centre of Research and Innovation Management (CRIM) and to the Faculty of Electrical Engineering from UTeM by giving us financial (University Short Term Research Grant) support as well as moral support for completing this project successfully.

References

1. Muis and K. Ohnishi, "Cooperative Mobile Manipulator with Dual Compliance Controllers based on Estimated Torque and Visual Force," *2006 IEEE International Conference on Mechatronics*, Budapest, pp. 619-624 (2006)
2. L. J. Wei, A.Z. Hj. Shukor, M.H. Jamaluddin, "Investigation on Force Scaling for Multi Degree of Freedom Bilateral Teleoperation Control System," IJMME, vol.**16**, no. 16, pp. 1-11 (2016)
3. H. Morimitsu, S. Katsura and K. Ohnishi, "Performance enhancement of bilateral control under time delay using nonlinear filter," *2014 IEEE 13th International Workshop AMC*, Yokohama, pp. 735-740 (2014)
4. S. Nishimura and S. Katsura, "Multilateral control under time delay for decoupling force and velocity controllers," *2016 IEEE 25th ISIE*, Santa Clara, CA, pp. 1258-1263 (2016)
5. K. Cetin, A. Bayrak and E. Tatlicioglu, "Online time delay estimation in networked control systems with application to bilateral teleoperation," *2016 ECC*, Aalborg, pp. 1007-1012 (2016)
6. C. Morito et al., "Development of a haptic bilateral interface for arm self-rehabilitation," 2013 IEEE/ASME International Conference AIM, Wollongong, NSW, pp. 804-809 (2013)
7. M. H. Jamaluddin, T. Shimono and N. Motoi, "Motion navigation in haptic bilateral system based on vision-based force compliance controller taking object coordinate into account," 2014 IEEE 23rd ISIE, Istanbul, pp. 2238-2243 (2014)

An Electro-Hydraulic Servo with Intelligent Control Strategy

Saeed Mohammed[1], *Chong* Chee Soon[1], *Rozaimi* Ghazali[1,*], *Ahmad Anas* Yusof[2], *Yahaya* Md Sam[3], and *Chai* Mau Shern[1]

[1]Centre for Robotics and Industrial Automation, Faculty of Electrical Engineering, Universiti Teknikal Malaysia Melaka, Hang Tuah Jaya, 76100 Durian Tunggal, Melaka, Malaysia.
[2]Faculty of Mechanical Engineering, Universiti Teknikal Malaysia Melaka, Hang Tuah Jaya, 76100 Durian Tunggal, Melaka, Malaysia.
[3]Department of Control and Mechatronic Engineering, Faculty of Electrical Engineering, Universiti Teknologi Malaysia, 81310 Skudai, Johor, Malaysia

Abstract. Versatile engineering applications have been developed to assist, reduce, and avoid human being from any heavy or harmful manufacturing processes. The gradually increased demand in force and position controls have simultaneously increased the usage of Electro-Hydraulic Servo (EHS) system. However, the time varying characteristics such as high-speed, outburst starting and stopping dynamic have led the EHS system to suffer from uncertainties and nonlinearities effects. Therefore, in order to enhance the performance of an EHS to surmount the uncertain and nonlinear effects, a hybrid Fuzzy-PID control strategy is developed which particularly improve the accuracy of the system by enhancing the control performance during the positioning tracking. By measuring the performance of the proposed control approach, the transient response and steady-state analysis will be performed which taking linear and intelligent control strategies as the references in the assessment process. The finding indicates the capability of a hybrid Fuzzy-PID controller in reducing the control effort applied to the EHS system.

1 Introduction

An actuator system is a device that produces mechanical movement by changing different type of energies into mechanical energy. The sources of the force are whether a supply from electrical motor, pneumatic pressure or hydraulic pressure. Hence, an actuator is classified according to the source of supply. The hydraulic actuator is an actuator system that utilizes pressurized hydraulic fluid, which is functioned as a drive or a transmission system to generating a dynamic [1]. A common structure of an EHS system composed of hydraulic fluid, control valve and cylinder [2].

The hydraulic actuator is widely used in industry due to its ability in generating large torque, capable of generating high power, and produce good positioning with fast motion [3]. Due to its capabilities, the hydraulic system has been used in many applications such as power system [4], braking systems [5], manufacturing system [6], flight simulation [7, 8], and cranes [9-11]. The low input power is converted into a high torque movement that used to control hydraulic applications.

However, the nonlinear electro-hydraulic system is suffering from nonlinearities and time varying characteristics such as high speed, outburst starting and stopping dynamic that produced by the flow and the pressure in the hydraulic system. The nonlinear properties causing a backlash in the control valve, actuator friction, distinction in fluid volume that make the system models and controller designs more complex [12].

The nonlinear properties that are produced through pressure and flow rate of the hydraulic system required a suitable controller to achieve better performances. In the previous works, there are many type of control techniques have been reported, which can be utilized to control the tracking capability of a nonlinear electro-hydraulic actuator system. Each of the control techniques required a proper tuning technique and some of the advanced tuning techniques have been reported recently such as Particle Swarm Optimization (PSO) [13-18], Genetic Algorithm (GA) [19-21], and Differential Evolution (DE) [22, 23].

All the controllers may be successfully controlled the nonlinear electro-hydraulic actuator. However, the output performance will be different. The numerical analysis such as system overshoot, settling time and steady-state error will be different for each type of control techniques. This paper proposed a hybrid Fuzzy-PID control approach to dealing with the nonlinear characteristic of the EHS system. It is observed from the literatures, there are not much comparison studies for this type of controller in term of transient response and steady state error evaluation. Therefore, an attempt to compare these numerical analyses for the proposed control strategies have been implemented in this paper.

* Corresponding author: rozaimi.ghazali@utem.edu.my

By using a hybrid Fuzzy-PID controller, the performance of the system can be adjusted to having less overshoot, steady state error, faster settling and rising time.

This paper is organized as, the common introduction regarding the EHS system, the basic mathematical modelling of the EHS system, and the theoretical explanation of the Fuzzy-PID controller are presented in Section 2. Section 3 presents the performance of the proposed control approach implemented in the EHS system. Finally, the conclusion will be drawn in Section 5.

2 Basic Structure of the EHS System

The development of the physical model for the EHS system will be done by using MATLAB/Simulink 2016 software. In this study, the EHS system will be first modelled according to the mathematical modelling of the EHS system based on the first principles of the physical law as discussed in [24]. After the formation of the EHS system, the proposed Fuzzy-PID controller will be applied, which is particularly applied to control the displacement of the EHS system. Fig. 1 [25] indicates the basic structure of the EHS system in this study.

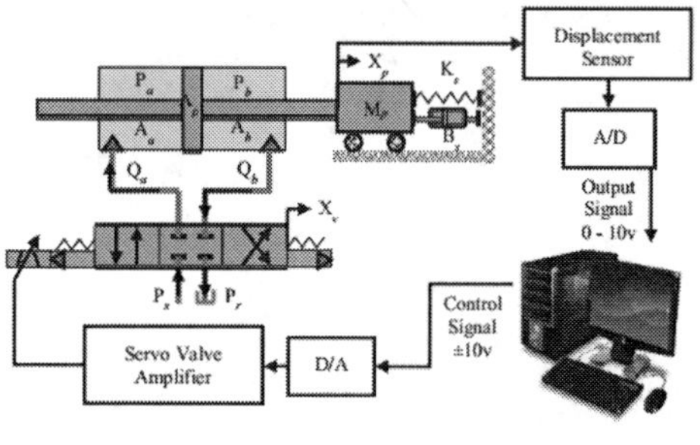

Fig. 1. Basic structure of the EHA system

The error value is identified by the sensor that measure the output value, which will be subtracted with the desired input value. This error value is then flows to the controller followed with some calculation of the controller parameters to obtain new output parameters.

Theoretically, the dynamic of the hydraulic servovalve can be illustrated in general second order equation as:

$$\frac{d^2x_v}{dt^2}+2\xi\omega_n\frac{dx_v}{dt}+\omega_n{}^2=\frac{I}{I_{sat}}\omega_n{}^2 \quad (1)$$

where x_v is the displacement of the servovalve spool, w_n is the natural frequency, ξ is the damping ratio, I/I_{sat} is the normalized input current, and I is the input current.

The flow rate through the servovalve of the hydraulic system is directly proportional to the spool motion and the square root under constant load or varying load condition. An ideal servovalve has a perfect geometry and written as:

$$Q=K\cdot x_v\cdot\sqrt{\Delta P_v} \quad (2)$$

The cylinder of the hydraulic consists of two chambers A and B. Each chamber has volume and pressure as shown in the equations (3)-(6).

$$V_a=V_{line}+A_a(x_s+x_p) \quad (3)$$

$$P_a=\frac{\beta}{V_{line}+A_p(x_s+x_p)}\int\left(Q_a-q_{ab}-q_a-\frac{dV_a}{dt}\right)dt \quad (4)$$

$$V_b=V_{line}+A_b(x_s-x_p) \quad (5)$$

$$P_b=\frac{\beta}{V_{line}+A_p(x_s-x_p)}\int\left(\frac{dV_b}{dt}-Q_b-q_{ba}-q_b\right)dt \quad (6)$$

From the total dynamic equations, the force exerted by the hydraulic cylinder against the load is shown in equation (7).

$$F_p=M_p\frac{d^2x_p}{dt^2}+B_s\frac{dx_p}{dt}+K_sx_p+F_f \quad (7)$$

2.1 Integration of the Fuzzy-PID Controller

The development of the EHS system which consists of controller and sensor are demonstrated in the block diagram in Fig. 2.

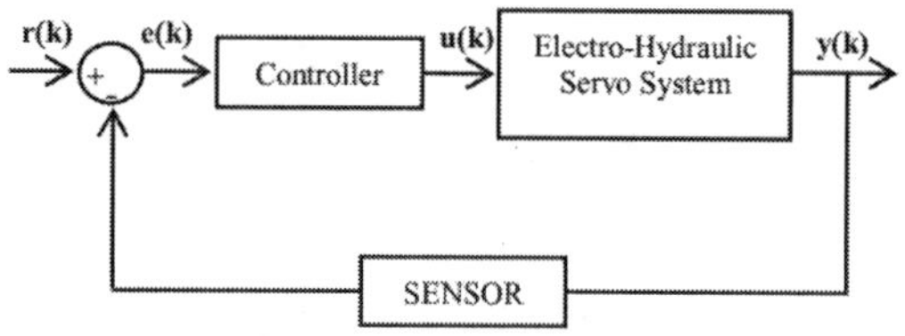

Fig. 2. Basic block diagram of the EHA system

where $r(k)$ denotes the required performance, $u(k)$ is the input signal fed to the hydraulic plant, $y(k)$ is the output performance of the positioning tracking by using the proposed hybrid controller, and $e(k)$ represents the error occurred during the execution process.

By designing the Fuzzy-PID control strategy which is developed based on Fuzzy control theory, the basic concept of the Fuzzy control approach is illustrated is block-diagram in Fig. 3.

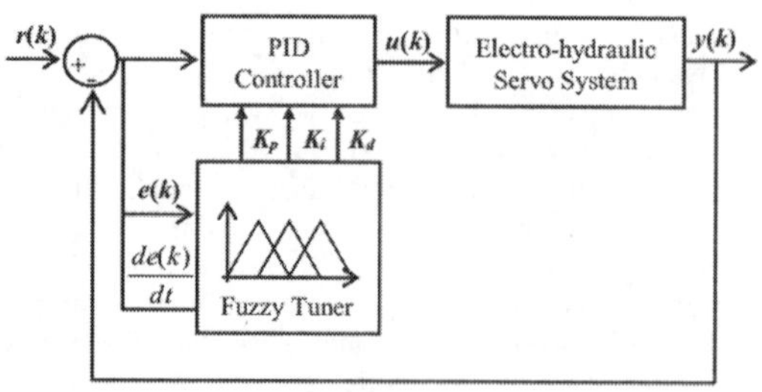

Fig. 3. Integration of the hybrid Fuzzy-PID controller

Since the assistant of the Fuzzy Logic control approach applied to the PID controller is proven to be performed better [7], therefore, the important properties of the Fuzzy Logic will be used to tuned the PID controller to obtain an optimal result for the EHS system that dealing with large amount of parameter variation [26].

Firstly, the conventional PID parameters obtained through the auto tune and the Ziegler-Nichols tuning techniques will be used as the initial value. Then, the Fuzzy control theory will be applied to tune the PID controller for the purpose of improving the positioning tracking capability of the controller. By adopting the gain value that effect on the transient response and the steady-state error, the inputs and outputs of the Fuzzy Logic will be first defined. The inputs consist of the feedback error *e*(*t*) and the derivative of the error *de*(*t*)/*dt*. While the outputs are the PID gains, including K_p, K_i and K_d as shown in Fig. 4.

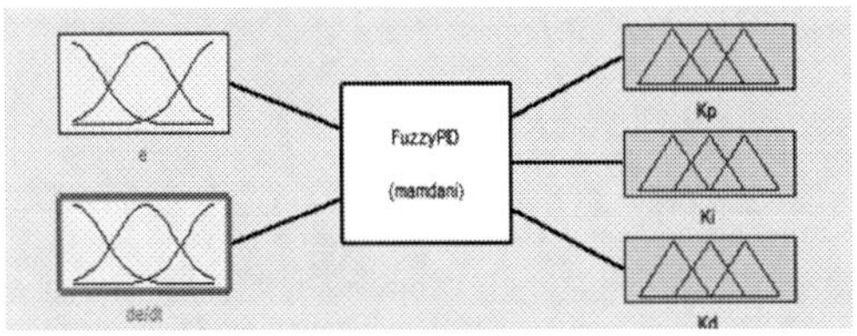

Fig. 4. Inputs and outputs of the Fuzzy-PID controller

The design rules for Fuzzy-PID related to the characteristic of the processes will be set according to the membership functions of the input error, which are composed of three linguistic variables for the change of error, and four rules are applied to design the Fuzzy-PID controller for example, if the error is negative and the change of error is zero, then the outputs must be K_p negative, K_i positive medium, and K_d positive medium as shown in Fig. 5.

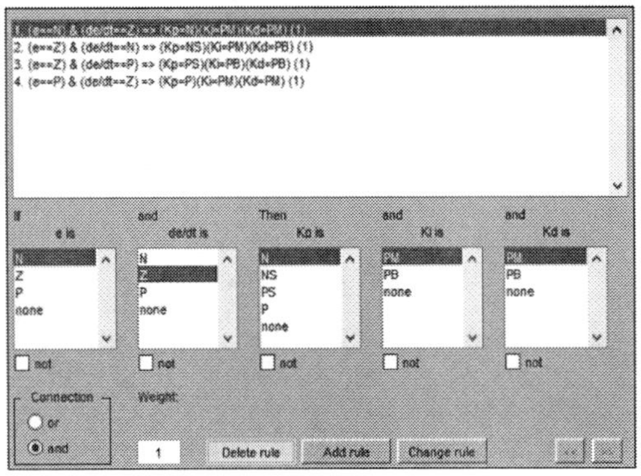

Fig. 5. The design rules for the Fuzzy-PID controller

3 Results and Discussion

In order to control the positioning tracking of the nonlinear EHS system, the PID controller has been first applied. Then, the Fuzzy controller has been utilized to improve the performance of the EHS system. Followed by the integration of the Fuzzy-PID controller that will significantly enhance the performance of the EHS system. Fig. 6 depicts the Fuzzy-PID development in the Simulink environment. The parameter values of the PID controller are tuned by using Fuzzy Logic according to the membership function. The scaling factors obtained through the Fuzzy Logic tuning approach including K_1, K_2, and K_3 are 20.5, 0.458 and 1.5 respectively.

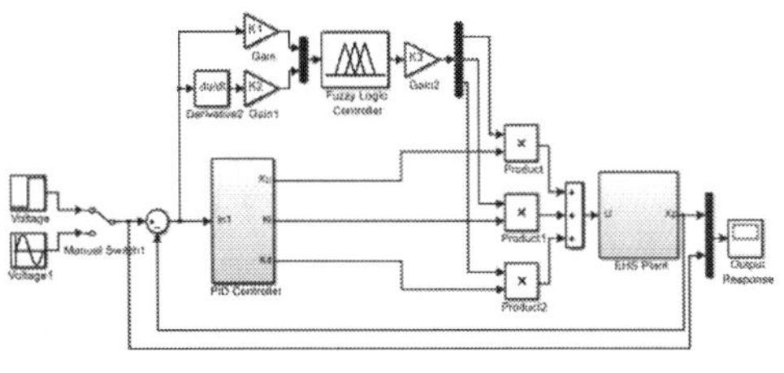

Fig. 6. Development of Fuzzy-PID controller in the Simulink environment.

Fig. 7 demonstrates the Simulink model of the EHS system that composes of the PID controller, Fuzzy logic controller, and Fuzzy-PID controller.

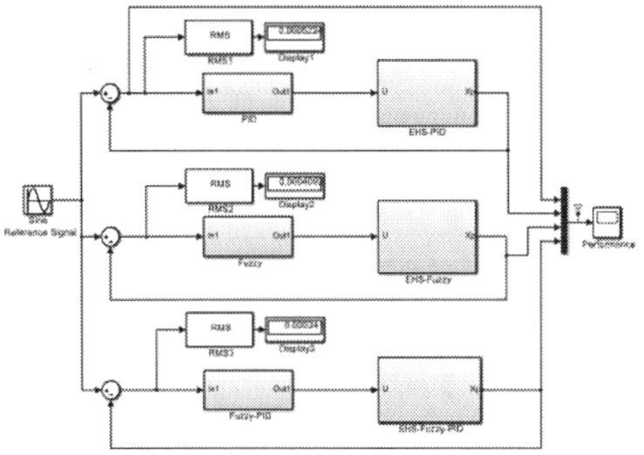

Fig. 7. Three different controllers applied to the EHS system

For the evaluation purposes, the EHS system has been executed with the step and sinusoidal references signal as depicted in Fig. 8 and 9, respectively.

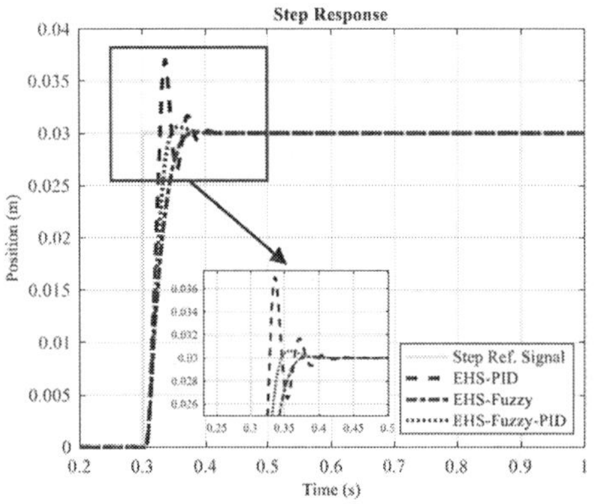

Fig. 8. The response of the controllers to a step reference signal

For the step reference signal with the results of PID, Fuzzy and Fuzzy-PID, the performance is assessed in term of its steady state error (e_{ss}) and transient response performance, including overshoot (OS), settling time (T_s), and rise time (T_r). The numerical analysis is tabulated in Table 1. As indicates in Table 1, the overshoot of the PID, Fuzzy, and Fuzzy-PID controllers are 23.33%, 0.33%, and 2% respectively. The results indicate the Fuzzy Logic controller is able to eliminate the overshoot better than the others two controllers. However, for the settling time and rise time, the Fuzzy-PID controller is outperformed PID and Fuzzy logic controller. The fast response is essential for the EHS system in real-time applications since EHS system is dealing with the time delay in nature. Although the overshoot of the Fuzzy-PID is higher, the limit is in acceptable boundary.

Table 1. The transient response and steady state error analysis

Controller	OS%	T_s (s)	T_r (s)	e_{ss}
PID	23.33	0.447	0.326	0
Fuzzy Logic	0.333	0.420	0.345	0
Fuzzy-PID	2.000	0.411	0.340	0

For the evaluation of the positioning tracking capability, sinusoidal reference signal has been fed to the EHS system as demonstrated in Fig. 9. It is clearly depicted in Fig. 9, the nearest output signal produced by these controllers responded to the required sinusoidal reference signal is Fuzzy-PID, followed by Fuzzy logic and PID controller.

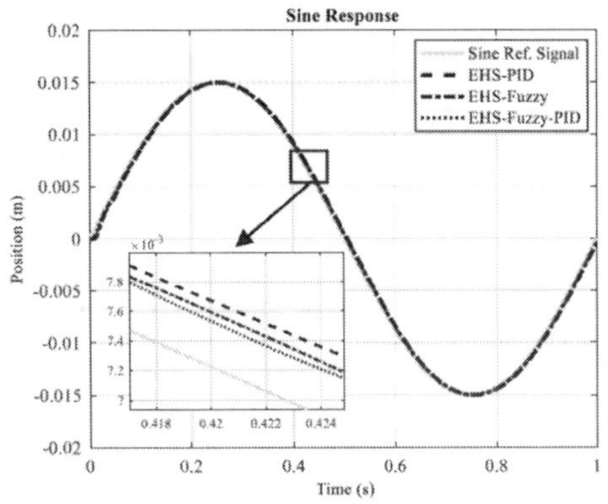

Fig. 9. The response of the controllers to a sinusoidal reference signal

For the more specific analysis, the Root Means Square (RMS) evaluation has been performed as indicated in Table 2.

Table 2. The RMS analysis for the tracking error

Controller	RMS (step)	RMS (sinusoidal)
PID	$7.248x10^{-9}$	$5.224x10^{-4}$
Fuzzy Logic	$2.099x10^{-11}$	$4.092x10^{-4}$
Fuzzy-PID	$9.950x10^{-13}$	$3.410x10^{-4}$

The RMS analyses clearly indicate the lower error has been achieved by the Fuzzy-PID controller whether is in step or sinusoidal reference signal. Hence, it can be concluded that by using smallest amount of control effort, the Fuzzy-PID controller is capable to perform better than the conventional Fuzzy and PID controllers.

4 Conclusion

In this paper, three different type of controllers, which are PID, Fuzzy logic and Fuzzy-PID controllers have been successfully developed and applied to the nonlinear EHS system. The performances of these controllers have been evaluated and compared to the transient response and the steady-state error analysis. The positioning tracking capability also evaluated via different type of trajectories. As a result, the numerical analyses indicate the hybrid Fuzzy-PID is able to perform better compared to PID and Fuzzy controllers. Approximately 20% of improvement in the tracking error has been obtained by using the Fuzzy-PID controller.

Acknowledgement

The support of Centre for Research and Innovation Management (CRIM), Universiti Teknikal Malaysia Melaka (UTeM) and Ministry of Education (MOE) are greatly acknowledged. The research was funded by Short Term Grant No. (PJP/2017/FKE/HI11/S01534) and Fundamental Research Grant Scheme (FRGS) Grant No. FRGS/1/2017/TK04/FKE-CERIA/F00333.

References

1. R. Ghazali, C. C. Soon, H. I. Jaafar, Y. M. Sam, and M. F. Rahmat, "System Identification of Electro-hydraulic Actuator System with Pressure and Load Effects," in *IEEE International Conference on Control System, Computing and Engineering*, 2014, pp. 256–260.
2. C. C. Soon, R. Ghazali, H. I. Jaafar, S. Y. S. Hussien, Y. M. Sam, and M. F. Rahmat, "Controller Parameter Optimization for an Electro-hydraulic Actuator System based on Particle Swarm Optimization," J. Teknol., vol. **78**, no. 6–13, pp. 101–108, 2016.
3. C. C. Soon, R. Ghazali, H. I. Jaafar, S. Y. S. Hussien, Y. M. Sam, and M. F. Rahmat, "The Effects of Parameter Variation in Open-Loop and Closed-Loop Control Scheme for an Electro-hydraulic Actuator System," Int. J. Control Autom., vol. **9**, no. 11, pp. 283–294, 2016.
4. M. Heikkilä and M. Linjama, "Displacement control of a mobile crane using a digital hydraulic power management system," Mechatronics, vol. **23**, no. 4, pp. 452–461, 2013.
5. R. He, X. Liu, and C. Liu, "Brake Performance Analysis of ABS for Eddy Current and Electrohydraulic Hybrid Brake System," Math. Probl. Eng., vol. **2013**, 2013.
6. R. Ghazali, Y. M. Sam, M. F. Rahmat, D. Hanafi, R.

Ngadengon, and Zulfatman, "Point-to-point trajectory tracking with discrete sliding mode control of an electro-hydraulic actuator system," in *Proceedings - IEEE Student Conference on Research and Development, SCOReD*, 2011, pp. 148–153.

7. H. Songshan, J. Zongxia, W. Chengwen, and S. Yaoxing, "Fuzzy Robust Nonlinear Control Approach for Electro-Hydraulic Flight Motion Simulator," Chinese J. Aeronaut., vol. **28**, no. 1, pp. 294–304, 2015.
8. J. Zhao, G. Shen, W. Zhu, C. Yang, and S. K. Agrawal, "Force Tracking Control of an Electro-Hydraulic Control Loading System on a Flight Simulator Using Inverse Model Control and a Damping Compensator," Trans. Inst. Meas. Control, pp. 1–13, 2016.
9. H. C. Pedersen, T. O. Andersen, and B. K. Nielsen, "Comparison of Methods for Modeling a Hydraulic Loader Crane With Flexible Translational Links," J. Dyn. Syst. Meas. Control, vol. **137**, no. 10, pp. 1–15, 2015.
10. S. Chakraborty and R. Meena, "Saving Time by Tip Control & Automation of Knuckle and Boom Hydraulic Crane," Int. J. Hybrid Inf. Technol., vol. **9**, no. 4, pp. 235–254, 2016.
11. Y. Chu, F. Sanfilippo, V. Asoy, and H. Zhang, "An Effective Heave Compensation and Anti-Sway Control Approach for Offshore Hydraulic Crane Operations," in *IEEE International Conference on Mechatronics and Automation, (ICMA)*, 2014, pp. 1282–1287.
12. N. H. Izzuddin, A. 'Athif M. Faudzi, M. R. Johary, and K. Osman, "System Identification and Predictive Functional Control for Electro-hydraulic Actuator System," in *IEEE International Symposium on Robotics and Intelligent Sensors (IRIS)*, 2015, pp. 138–143.
13. J. Yao, G. Jiang, S. Gao, H. Yan, and D. Di, "Particle Swarm Optimization-Based Neural Network Control for an Electro-Hydraulic Servo System," J. Vib. Control, vol. **20**, no. 9, pp. 1369–1377, 2014.
14. C. C. Soon, R. Ghazali, H. I. Zaafar, S. M. Rozali, Y. M. Sam, and M. F. Rahmat, "The Effects of Pressure Variation in Sliding Mode Controller with Optimized PID Sliding Surface," in *Asian Simulation Conference*, 2017, pp. 104–115.
15. J. Yao, H. Yu, M. Dietz, R. Xiao, S. Chen, T. Wang, and Q. Niu, "Acceleration Harmonic Estimation for a Hydraulic Shaking Table by Using Particle Swarm Optimization," Trans. Inst. Meas. Control, vol. **39**, no. 5, pp. 738–747, 2017.
16. C. C. Soon, R. Ghazali, H. I. Jaafar, S. Y. S. Hussien, S. M. Rozali, and M. Z. A. Rashid, "Optimization of Sliding Mode Control using Particle Swarm Algorithm for an Electro-Hydraulic Actuator System," J. Telecommun. Electron. Comput. Eng., vol. **8**, no. 7, pp. 71–76, 2016.
17. S. M. Rozali, N. S. Farhana, M. N. Kamarudin, A. F. Z. Abidin, M. F. Rahmat, A. R. Husain, and C. C. Soon, "Robust Control Design of Nonlinear System via Backstepping-PSO with Sliding Mode Techniques," in *Asian Simulation Conference*, 2017, pp. 27–37.
18. C. C. Soon, R. Ghazali, H. I. Jaafar, and S. Y. S. Hussien, "Sliding Mode Controller Design with Optimized PID Sliding Surface Using Particle Swarm Algorithm," Procedia Comput. Sci., vol. **105**, pp. 235–239, 2017.
19. Z. Chen, X. Yuan, B. Ji, P. Wang, and H. Tian, "Design of a Fractional Order PID Controller for Hydraulic Turbine Regulating System Using Chaotic Non-Dominated Sorting Genetic Algorithm II," Energy Convers. Manag., vol. **84**, pp. 390–404, 2014.
20. A. Montazeri, C. West, S. D. Monk, and C. J. Taylor, "Dynamic Modelling and Parameter Estimation of a Hydraulic Robot Manipulator Using a Multi-objective Genetic Algorithm," Int. J. Control, vol. **90**, no. 4, pp. 661–683, 2016.
21. T. Samakwong and W. Assawinchaichote, "PID Controller Design for Electro-hydraulic Servo Valve System with Genetic Algorithm," Procedia Comput. Sci., vol. **86**, pp. 91–94, 2016.
22. M. Dangor, O. A. Dahunsi, J. O. Pedro, and M. M. Ali, "Evolutionary Algorithm-Based PID Controller Tuning for Nonlinear Quarter-Car Electrohydraulic Vehicle Suspensions," Nonlinear Dyn., vol. **78**, no. 4, pp. 2795–2810, 2014.
23. Y. C. Hung, F. J. Lin, J. C. Hwang, J. K. Chang, and K. C. Ruan, "Wavelet Fuzzy Neural Network with Asymmetric Membership Function Controller for Electric Power Steering System via Improved Differential Evolution," IEEE Trans. Power Electron., vol. **30**, no. 4, pp. 2350–2362, 2015.
24. C. C. Soon, R. Ghazali, H. I. Jaafar, and S. Y. S. Hussien, "PID Controller Tuning Optimization using Gradient Descent Technique for an Electro-hydraulic Servo System," J. Teknol. Sci. Eng., vol. **77**, no. 21, pp. 33–39, 2015.
25. C. C. Soon, R. Ghazali, H. I. Jaafar, S. Y. S. Hussien, S. M. Rozali, and M. Z. A. Rashid, "Position Tracking Optimization for an Electro-hydraulic Actuator System," J. Telecommun. Electron. Comput. Eng., vol. **8**, no. 7, pp. 1–6, 2016.
26. P. Pratumsuwan and A. Junchangpood, "Force and Position Control in the Electro-Hydraulic System by using a MIMO Fuzzy Controller," in *IEEE Conference on Industrial Electronics and Applications, (ICIEA)*, 2013, pp. 1462–1467.

Kalman filter estimation of RLC parameters for UMP transmission line

Siti Nur Aishah Mohd Amin, *Hamzah* Ahmad *and Mohd Rusllim* Mohamed

Faculty of Electrical & Electronics, Universiti Malaysia Pahang, 26600 Pekan, Pahang

Abstract. This paper present the development of Kalman filter that allows evaluation in the estimation of resistance (R), inductance (L), and capacitance (C) values for Universiti Malaysia Pahang (UMP) short transmission line. To overcome the weaknesses of existing system such as power losses in the transmission line, Kalman Filter can be a better solution to estimate the parameters. The aim of this paper is to estimate RLC values by using Kalman filter that in the end can increase the system efficiency in UMP. In this research, matlab simulink model is developed to analyse the UMP short transmission line by considering different noise conditions to reprint certain unknown parameters which are difficult to predict. The data is then used for comparison purposes between calculated and estimated values. The results have illustrated that the Kalman Filter estimate accurately the RLC parameters with less error. The comparison of accuracy between Kalman Filter and Least Square method is also presented to evaluate their performances.

1 Introduction

Electrical energy is generated at power stations which are typically located far away from the load centres [1]. Thus, a network of conductors between the power stations and the consumers is required in order to utilize the power generated. This network of conductors is divided into two main components, which are the transmission system and the distribution system. At a generating plant, electric power is "stepped up" to several thousand volts by a transformer and delivered to the transmission line. At numerous substations on the transmission system, transformers step down the power to a lower voltage and deliver it to the different load centres in the distribution lines [2]. This is because high voltage improves efficiency by reducing heat loss in the power line. But high voltages are too dangerous, so step down transformers are used to reduce the voltage to a required and safe levels.

The power transmission line is one of the major components in electrical power system. All transmission lines in a power system posses the electrical properties of resistance, inductance and capacitance [3]. These passive elements are usually referred to investigate and assess the transmission line characteristics and efficiency which makes them irresistible to be analysed as part of system performance. The series resistance relies basically on the physically composition of the conductor at a given temperature. While the series inductance and capacitance is produced by the presence of magnetic field and electric field around the conductors. Besides, the series inductance and capacitance depends on the geometrical arrangement in a transmission line [4]. The important consideration in the design and operation of transmission line are the determination of voltage drop, line losses and efficiency of transmission line. These values are greatly influenced by the constant resistance (R), inductance (L) and capacitance (C) in the transmission line [5]. According to the previous paper, resistance (R), reactance (X) and susceptance (B) also affected the transmission line performance in a system [3]. The frequency involve in this three constant parameter indicates the performance of a system. When considering high frequency, the magnetic field generated drive the currents to the outer edge of the conductor that carries them. As a result, the higher the frequency, the thinner the layer of metal available to carry the current and the higher the effective resistance of the cable [6]. In fact, the voltage drop in the line depends on the values of above three lines constant [7]. In addition, the resistance of transmission line is the most important cause of power losses in the line and it determines the transmission line efficiency. Since there are many research has been done on R, X and B values in medium transmission line, so this paper will focuses on R, L and C values in UMP short transmission line. This is because of calculation of power factor relies on these parameters.

Power system state estimation is a main tool of energy management system. State estimation was first applied to power systems by Schweppe and Wildes in the late 1960s [8][9]. Since then, several applications and techniques of Kalman filter have been one of the famous methods in power system estimation. It has been used for damping and frequency [7], modal changes [10], state estimation [11], power factor detection and etc. However all of these researches did not highlighted important indicators that give great effects in the system which are RLC parameters. Most of these papers are focused solely

* Corresponding author: aishahamin_92@yahoo.com

on IEEE system and are not tested on a real system. One of the systems that have been tested is long transmission line [12]. Theoretically, values of R, L and C in long transmission line are supposedly higher than short transmission line. But due to some condition, the values of R, L and C can be opposed the actual properties such as type of cable, size of cable, length of cable, number of winding and etc. All of these properties will indicate the values of R, L and C in the transmission line.

Since the real situation is different from the theoretical analysis due to unexpected condition such as the presence of noise and system dynamics, practical analysis in a real system also one of the essential aspects to be investigated to understand how the system behaves. In fact, by using a real system for analysis purposes, certain parameters can be determine such as the power losses, fault, and voltage drop. Therefore, this will create an opportunity for researcher to provide a suitable corrective action according to the circumstances. As a result, the determination for the future condition in transmission line can be identified. If the theoretical analysis and the real system have produced similar results then the system physical properties such as type, size and length of cable can be determined more precisely by the load applied in the system.

An accurate understanding of transmission line parameters is necessary for accurate power flow modelling, fault location, transient stability and etc. Previously, calculations based in conductor dimensions, sag and line length were used to estimates these parameters [13]. However, due to the assumptions and approximation involved, the estimation was not reliable for real - time application [13]. Therefore, this situation has inspired this research to examine UMP transmission line as shown in Figure 1. It is also worth to mention that, currently UMP do not have any information about the parameters of the transmission line. This situation has produced difficulties to reduce the operating cost and maintenance operations. To overcome this issue, Kalman filter method is used for estimation of R, L and C values for short transmission line in UMP. The analysis mainly investigates the R, L, and C parameters. Mathematical model for UMP short transmission line is designed by using standard parameters. Our analysis attempts to provide information for Department of Property Management and Development in UMP about the transmission line to understand the system performance as well as to execute a corrective action accordingly based on the given results. The Kalman filter model for estimation of UMP transmission line will be examined in the next section.

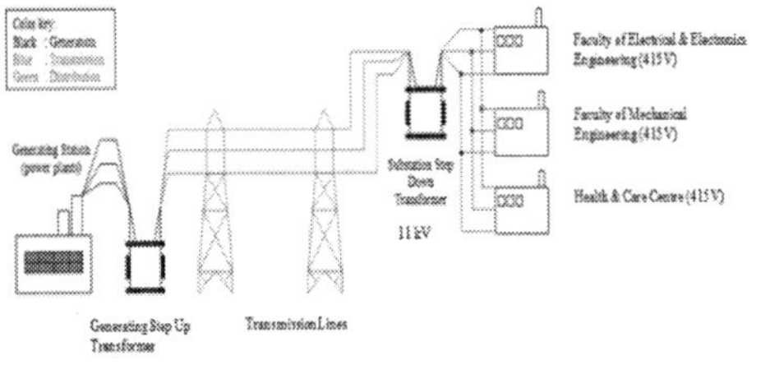

Fig. 1. Power distribution system in UMP

2 Kalman filter method based on RLC estimation

A transmission line can be defined as a short length line if its length is less than 50 km and the voltage is comparatively lower than 20kV. In this case, the shunt capacitance (C) effect is negligible and only the resistance and inductive reactance are taken into account. Balanced conditions of the line can be represented by the equivalent circuit of a single phase with resistance (R), and inductance (L) in series (series impedance) [14], as shown in Figure 2.

If the transmission line has a length between 50 km and 150 km and the voltage is between 20 kV – 100 kV, the line is considered as a medium-length line and the equivalent circuit with resistance (R), inductance (L) and capacitance (C). Both short and medium length transmission lines use approximated lumped-parameter models [14] [15]. The lumped element model simplifies the description of the behaviour of spatially distributed physical systems into a topology consisting of discrete entities that approximate the behaviour of the distributed system under certain assumptions [5] [15].

However, if the transmission line is longer than 150 km, and the voltage is higher than 100 kV, the model will be considered as a long-length line. The appropriate series impedance and capacitance are found by solving the corresponding differential equations, where voltages and currents are described as a function of distance and time [14]. As for our case, the short transmission line is considered for UMP.

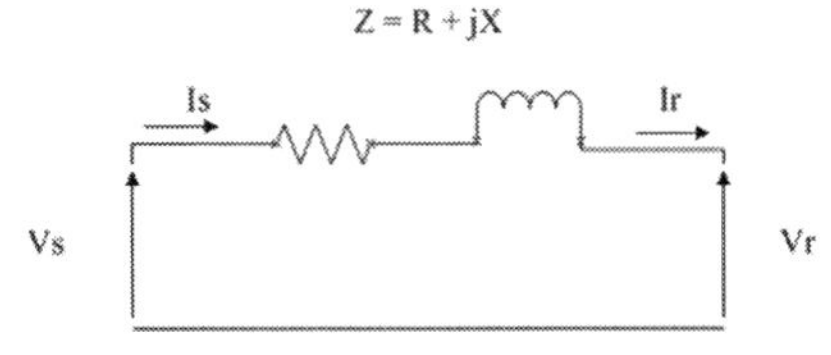

Fig. 2. Equivalent circuit for short transmission line

A. Kalman Filter

For a linear system, the process model of Kalman filter is described as

$$X_k = AX_{k-1} + BU_k + W_k \quad (1)$$

$$Z_k = HX_k + V_k \quad (2)$$

where X_k is the estimation value of X, X_{k-1} is the estimate value on the previous state, U_k is the control input, W_k and V_k is noise covariance matrix, A, B and C will be numerical constant.

The estimation value of X_k is represented by R, L and C as

$$X_k = \begin{bmatrix} R \\ L \\ C \end{bmatrix} \quad (3)$$

where R is resistance, L is inductance and C is capacitance. In this study, current, A is used as a control input, U_k.

$$U_k = \begin{bmatrix} A & 0 & 0 \\ 0 & A & 0 \\ 0 & 0 & A \end{bmatrix} \quad (4)$$

The estimated value on the previous state, X_{k-1} is represented by

$$X_{k-1} = \begin{bmatrix} R_{k-1} \\ L_{k-1} \\ C_{k-1} \end{bmatrix} \quad (5)$$

B. *Prediction and update*

Kalman filter recursively computes estimation for a state X_k which includes R, L and C parameters according to the process and observation model [17]. The stages of Kalman filter algorithm are as follows:

- *Prediction* (time update) to estimate priori estimation of state and its error covariance matrix:

$$X_k = AX_{k-1} + BU_k \quad (6)$$

$$P_k = AP_{k-1}A^T + Q \quad (7)$$

where P_k is the prior error covariance. This prior value is used in the measurement update equations.

- *Correction* (measurement update) to provide correction based on the measurement Z_k to yield the posteriori state estimate and its error covariance:

$$K_k = P_k H^T (HP_k H^T + R)^{-1} \quad (8)$$

$$X_k = X_k + K_k (Z_k - HX_k) \quad (9)$$

$$P_k = (1 - K_k H) P_k \quad (10)$$

where K_k is the Kalman gain.

Figure 5 shows the the Kalman filter model for the UMP transmission line. The analysis mainly observes the RLC parameters based on different noise a condition which describes various unexpected issues during transmission in UMP. It is expected that from the proposed technique, UMP management is able to plan and execute suitable actions to gain the desired system performance.

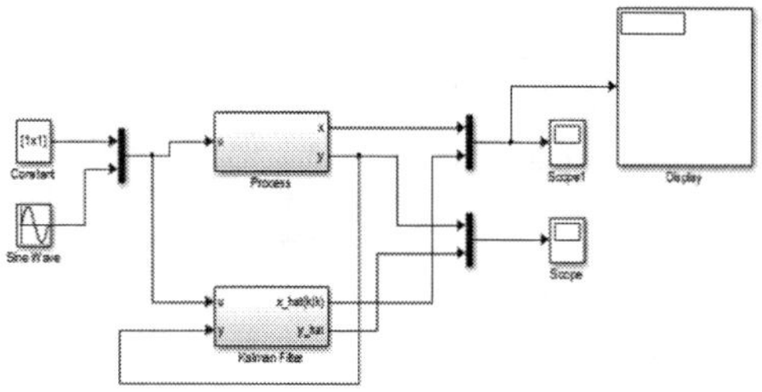

Fig. 5. Kalman filter model

3 Theoretical analysis

A 11 kV, 5 km long, 60 Hz transmission line is considered in this research to model the UMP short transmission line. This section analyses the calculated R, L and C parameters of the transmission line based on Figure 2 as this is the main reference for estimation purposes. Other quantities such as active power and impedance are calculated by using Ohms Law shown in below equations.

$$P = \sqrt{3}\, VI \cos\theta \quad (11)$$

$$\cos\theta = 0.93 \quad (12)$$

$$V = IZ \quad (13)$$

$$Z = 49.84\ \Omega$$

Figure 6 shows the power triangle characteristics which defined ϕ as the cosine of the angle between voltage and current phasor [18]. The angle of ϕ will be used to derive the value of R, L and C of the short transmission line.

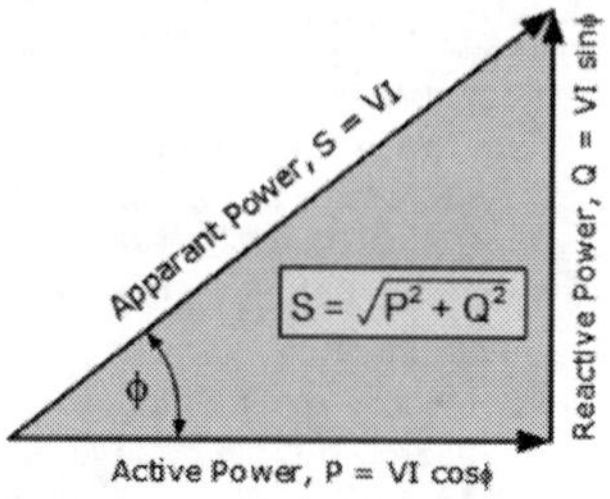

Fig. 6. Power triangle

$$pf = \cos\theta \quad (14)$$

$$pf = \frac{R}{Z} = \frac{P}{S} \quad (15)$$

$$pf = 0.93$$

$$R = 46.35\ \Omega$$

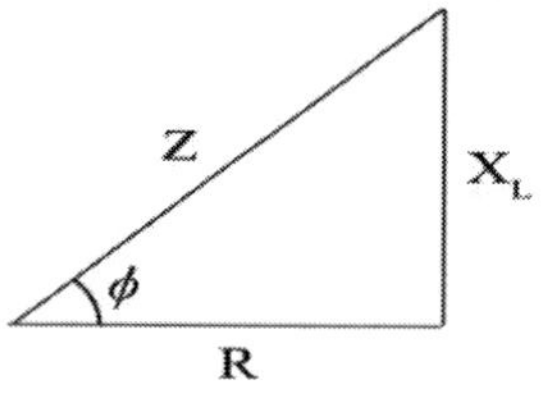

Fig. 7. Phasor diagram

From the phasor diagram above[18],

$$\theta = \tan^{-1}\left(\frac{\omega L}{R}\right) \quad (16)$$

It is known that,

$$\omega = 2\pi f \quad (17)$$

Substituting all the related value into (17), the equation becomes,

$$\omega = 376.99$$

By referring to equation (14), the value of pf is substitute to get the value of θ,

$$\cos\theta = 0.93$$

$$\theta = 21.5652$$

In order to get the value of L, all the values are substitute into equation (16), and exhibit

$$L = 48.58\ mH$$

As demonstrated above, the R, L and C values have been calculated. The phase per metre line parameters R, L, C are 46.35 Ω, 48.58 mH and 0 F respectively. The line is operated in a power of 3911 kW at receiving end. These values will be assessed for estimation analysis covered in the next section.

4 System model and analysis

This subsection investigates the results of having different noise conditions to the estimation of UMP short transmission line. This paper takes into account system noises of 0.1 and 10 to simulate the unknown parameters which can effects the estimation results such as size, length and type of cables. Previous section has described how the value of R, L and C are calculated. This section analyse how Kalman Filter works to estimate those value efficiently. The results obtained are described as follows.

Simulation was conducted for the equivalent circuit of short transmission line to simulate the UMP condition and tuning process also take place in order to achieve a good estimation results. Figure 8, Figure 9 and Figure 10 below shows the calculated and estimation values of R, L and C over a period. The values of calculated R, L and C simulated by using Kalman filter are 46.35 Ω, 48.58 mH and 0 F while the average values of estimated are 46.34 Ω, 48.57 mH and 0 F respectively with noise of 0.1. There are small differences between the previous and new estimated values of R, L and C. Both of the calculated and estimated parameters are seen overlapping with each other due to the approximate parameter values. The result shows the higher consistency involved in R, L and C value due to the approximate results between calculated and estimated values. According to Figure 10, the value of capacitance is 0F as it is assumed negligible due to the properties of short transmission line. It must be noted that the initial values of estimation error covariance matrix affects the estimation of parameter values.

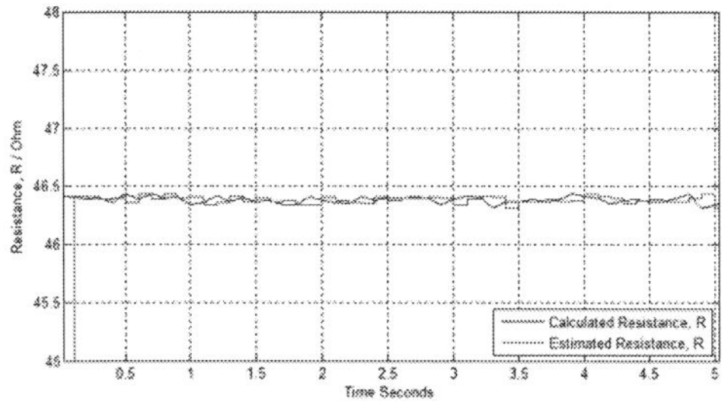

Fig. 8. Graph of calculated and estimated value of resistance, R

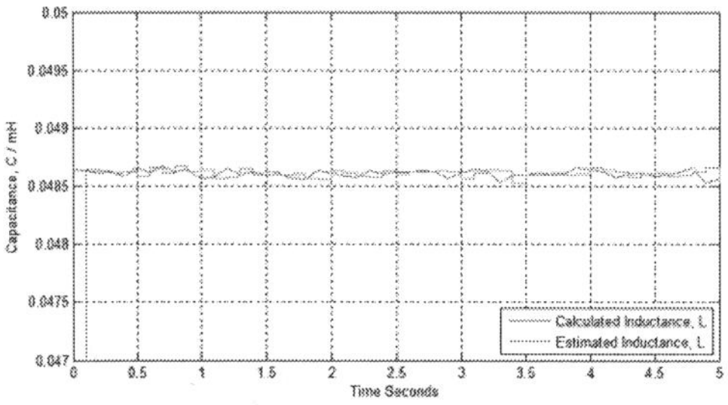

Fig. 9. Graph of calculated and estimated value of inductance, L

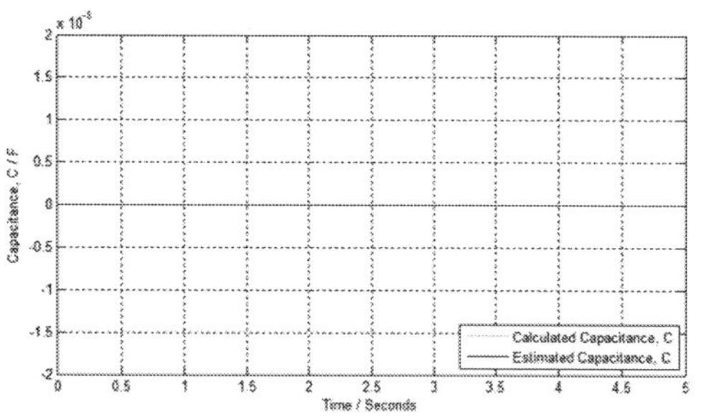

Fig. 10. Graph of calculated and estimated value of capacitance, C

Figure 11, Figure 12 and Figure 13 below also shows the calculated and estimation values of R, L and C in the system. The calculated values of R, L and C are 46.35 Ω, 48.58 mH and 0 F while the estimated average values of R, L and C are 46.22 Ω, 48.43 mH and 0 F respectively with higher system noise of 10. Both of the calculated and estimated parameters are seen less consistency and the results are slightly different due to the greater noise in the system. Figure 13 shows the value of capacitance which is 0 F.

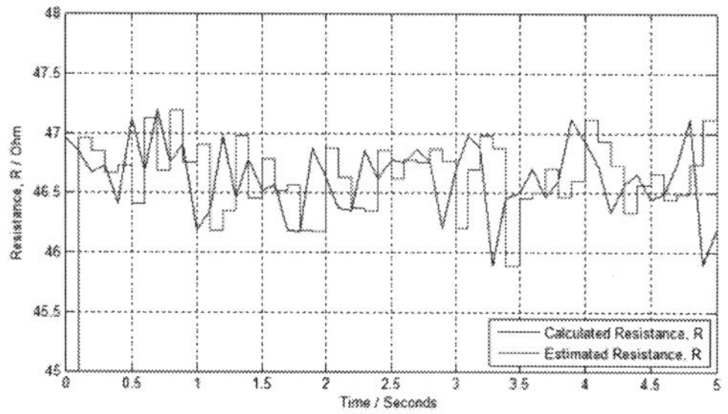

Fig. 11. Graph of calculated and estimated value of resistance, R

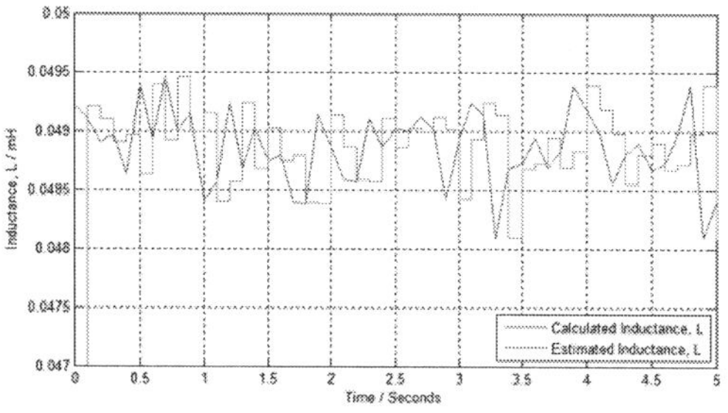

Fig. 12. Graph of calculated and estimated value of inductance, L

https://doi.org/10.1051/matecconf/201815001017

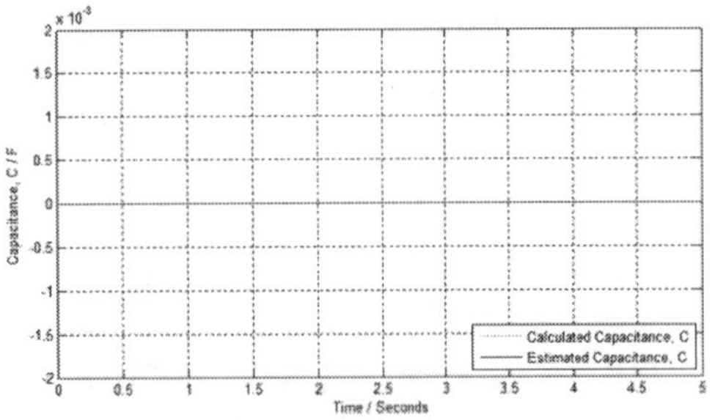

Fig. 13. Graph of calculated and estimated value of capacitance, C

Table 1 defines the system parameters of the simulink model with the small value of noise which is 0.1 and 10. It shows the accuracy of resistance and inductance in the system more precise with the small value of noise which is 99.98 % and 99.98 % respectively. Thus, it is shown that higher estimation accuracy will be achieved with the decreasing values of the noise. This table also defines the system parameters of the simulink model with the greater value of noise which is 10. It shows the accuracy of resistance and inductance in the system is inaccurate with the greater value of noise which is 99.72 % and 99.69 % respectively. Thus, it is proved that the accuracy in the system will decrease with the increasing noise.

Table 1. Parameters value with noise

Noise = 0.1			
Parameter	Calculated value (average)	Estimation value (average)	Accuracy (%)
R	46.35 Ω	46.34 Ω	99.98
L	48.58 mH	48.57 mH	99.98
C	0 F	0 F	-.
Noise = 10			
R	46.35 Ω	46.22 Ω	99.72
L	48.58 mH	48.43 mH	99.69
C	0 F	0 F	-.

According to this table and figures presented previously, the value of noise affected the accuracy of R, L and C estimation in the system. The higher the value of noises or disturbance in the system, the performance degrades. These results also indicates the possibility to estimate the RLC parameters accurately even in different places especially when the same cable is being used. Higher voltage increased the noise problem in a system. These situations occur due to the electron movement within the electrical circuits.

Comparison of accuracy between least square method and Kalman filter also has been done [3]. From Table 2 below, the results presented that Kalman filter give a higher accuracy compared to the Least Square method. Note that, Least Square method is used to estimate the values of R, X and B in long transmission line located in East Midlands, England [3] while Kalman filter is used to indicates the values of R, L and C in UMP. Eventhough the research is conducted in different places but the value of accuracy can still be taken into account. A better accuracy will come from short transmission line compared to long transmission line. This is due to the unknown noises which bigger system will have more unknown noises compared to the small system. From these two results of accuracy, it can be proven that Kalman filter is a better method that can be used for estimation.

Table 2. Parameters value

Least Square method		Kalman filter	
Parameter	Accuracy (%)	Parameter	Accuracy (%)
R	96.28	R	99.98
X	99.97	L	99.98
B	94.53	C	-.

Remark that, it is important to consider of R, L, and C and R, X and B values as they affected the determination of power transfer, voltage drop and efficiency in the transmission line. The voltage drop in the line depends upon the values of above three lines constant. Similarly, the resistance of transmission line is the most important cause of power losses in the line and determine the transmission efficiency. Same as R, L and C, values of R, X and B also affected the determination of efficiency in a system.

5 Conclusions

This paper presented the analysis of R, L and C estimation in a system by using Kalman filter. It was shown that although there is a little data from UMP short transmission line, the estimation of R, L and C parameters is still possible. From this research, it has been revealed that noise in the system indicates the performance of UMP short transmission line and the parameter estimation can help the engineers in estimation for future growth even in a different area. In addition, with the different in accuracy according to the method used, this can be proven that Kalman filter is one of the optimal method that can be used for estimation for real - time line parameter estimator.

Author would like to thank University Malaysia Pahang and Ministry of Higher Education for the supports in conducting the research under RDU160379 and RDU160145.

References

1. P. Leader and M. V Venkatasubramanian, "Coordination of Transmission Line Transfer Capabilities," 1996.
2. H. B. Harrison, "Transmission Line," no. May, pp. 111–113, 1982.
3. D. Ritzmann, P. S. Wright, W. Holderbaum, and B. Potter, "A Method for Accurate Transmission Line Impedance Parameter Estimation," *IEEE Trans. Instrum. Meas.*, vol. 65, no. 10, pp. 2204–2213, 2016.
4. G. Chen and E. G. Friedman, "Effective Capacitance of RLC Loads for Estimating Short-Circuit Power," pp. 2065–2068, 2006.
5. S. Ma, Z. Bo, and A. Klimek, "The research on lumped parameter equivalent circuit of transmission line," pp. 2–6, 2009.
6. B. Curran, I. Ndip, and K. Lang, "A Comparison of Typical Surface Finishes on the High Frequency Performances of Transmission Lines in PCBs," pp. 1–3, 2017.
7. M. Larsson and C. Rehtanz, "Detection of Oscillations in Power Systems using Kalman Filtering Techniques," pp. 183–188, 2013.
8. A. Saikia and R. K. Mehta, "Power system static state estimation using Kalman filter algorithm," 2016.
9. G. Gede, "Introduction to Kalman Filtering," *Robotica*, no. June, 2011.
10. R. A. Wiltshire, G. Ledwich, S. Member, and P. O. Shea, "A Kalman Filtering Approach to Rapidly Detecting Modal Changes in Power Systems," vol. 22, no. 4, pp. 1698–1706, 2007.
11. K. Vadirajacharya and B. A. T. U. Lonere, "Kalman filter analysis in dynamic state of power system," pp. 1170–1173, 2015.
12. E. C. M. Costa and S. Kurokawa, "Estimation of transmission line parameters using multiple methods," vol. 9, pp. 2617–2624, 2015.
13. C. Mishra and V. A. Centeno, "Kalman-Filter Based Recursive Regression for Three- Phase Line Parameter Estimation using Synchrophasor Measurements," 2015.
14. S. Gajare, A. K. Pradhan, and V. Terzija, "A Method for Accurate Parameter Estimation of Series Compensated Transmission Lines Using Synchronized Data," *IEEE Trans. Power Syst.*, vol. 8950, no. c, pp. 1–1, 2017.
15. M. A. Shicong, G. U. O. Jianbo, S. U. N. Huadong, and Y. U. Yongjin, "The comparison and analysis of lumped parameter equivalent circuits of transmission line," 2011.
16. A. R. J. Araújo, R. C. Silva, and S. Kurokawa, "Comparing Lumped and Distributed Parameters Models in," pp. 1–5, 2014.
17. G. Bishop and N. Carolina, "An Introduction to the Kalman Filter," 2001.
18. Ravish R. Singh, "Electrical Networks," vol. 352, no. August, pp. 35–42, 2003.

Adsorption of Benzene by "Green" functionalization of Montmorillonite

Hirra Anjum[1], *Mansoor* Abdullah Obaid[1], *Muhammad* Yasir Shamim[1], *Thanabalan* Murugesan[1*]

[1]*Department of Chemical Engineering, Universiti Teknologi PETRONAS, Bandar Seri Iskandar 32610, Perak, Malaysia.*

Abstract. The capacity of organically modified Montmorillonite (MMT) clay to adsorb nonpolar organic compound (benzene) in an aqueous solution, was investigated under the batch process. MMT was pretreated (centrifuged) and then functionalized with green intercalating agent i.e. 1-hexyl-3-methyl imadazolium chloride [HMim][Cl]. The characterization through Fourir Transoform Infrared Spectroscopy (FTIR), Differential Scanning Calorimeter (DSC) and Field Emission Scanning Electron Microscope (FE-SEM) confirmed the presence of the oxygen containing functional groups, changes in melting point and variation in the morphological properties. The governing parameters for the sorption of benzene such as the effect of contact time, pH, adsorbent dose and rotation were studied. The kinetic data conformed to pseudodo 2nd order kinetic model and the isotherm experimental data were a better fit to Langmuir model with maximum adsorption capacity of 588.23mg/g under experimental conditions. Overall, MMT intercalated with 1-hexyl-3-methyl imadazolium chloride is a promising environmental friendly adsorbent for the abatement of benzene in an aqueous solution.

1 Introduction

In the last few years, the application of energy and power sources has been increased in the world. Nowadays, the main sources of environmental pollutants are industrialization, civilization, agricultural activities and other environmental changes. Large quantities of dangerous pollutants are emitted by these sources. Benzene has been consigned as one of the dangerous hazardous material. A severe exposure by inhalation or ingestion can be lethal. The maximum allowable concentration of benzene in drinking water is 0.01mg/L. Therefore, many conventional desalination technologies have been employed for wastewater treatment. But, excessive chemical usage and sludge accumulation prove them to be ineffective methodologies. However, adsorption is found to be superior in performance in terms of simple design, ease of operation and nontoxic adsorbent utilization. Furthermore, availability, cost-effectiveness and environmental friendly nature are the major factors in the selection of green adsorbent [1].

The crystalline clay minerals are the fundamental constituents of the phyllosilicates. Their network structure encapsulates two, three or four individual sheets. These sheets are usually comprised of tetrahedral $(SiO_4)^{4-}$ and octahedral $(AlO_3(OH)_3)^{6-}$ sheets. Furthermore, the silicate layers are capable of arranging themselves in the form of stacks having a regular van der Waals gap (interlayer) between them. These layers assimilate small cationic structures and some of the peaks are connected to protons (OH) because they are occupied by oxygen [2]. However, isomorphous substitution within the sheets is the only factor responsible for the presence of exchangeable cations. These ionic substitutions are managed by the environmental chemistry and the reaction kinetics during the generation of clay minerals [3-5]. Thus, the characteristics that make it a suitable adsorbent for several contaminants abatement from polluted water are the surface electric charge, internal porosity, surface functional groups and predominantly, exchangeable cations inside crystal lattice. The pre-eminent consumers of these moieties are petroleum industries, steel mills, paint and varnish industries etc. [6].

However, the hydrophilic clays are proved to be ineffective adsorbent for abatement of aromatic compounds because of very low sorption capacity and selectivity. Probably, the water cluster formation is held responsible for minimization of adsorption efficiency. Taking into account these considerations, various approaches i.e., the introduction of exchangeable cations, pillarization [7], thermal treatment[8] and acid activation [9] have been employed to improve the configuration, conformation and framework of clay for uptaking the organic pollutants.

On the other hand, the uptake of organic/inorganic pollutant via clay adsorbent is a complex phenomenon involving more than one mechanisms. These techniques inculcate surface adsorption [10], ion exchange [11], ligand exchange [12], surface partition [13], surface precipitation [14] and structural incorporation [15].

* Corresponding author: murugesan@utp.edu.my

Furthermore, structure, molecular size, partition coefficient, hydrophobicity, polarity [16] are major responsible factors for the adsorption affinity interactions of benzene for clay adsorbent [17].

Organoclay minerals prepared with small groups tend to sorb targetted pollutants by surface adsorption because small molecules occupy outer energetic sites immediately and very few adsorb inside the interpenetrating network structure. However, clay adsorbent with large electron donor/acceptor group tends to remove the hazardous material by partitioning as the big structures occupy a large number of active sites of the interlamellar spaces of the clay [18]. It was approved by Jaynes et al. (1999) in which they concluded that SWy-TMA and SWy-Adam (small groups) removed benzene through partitioning process while SWy-HDTMA (bulky group) followed the surface adsorption path [19].Since the ordered structure with large molecular weight lead towards higher sorption capacity. While working with bentonite clays, Bertagnolli et al. (2012) deduced that organophilization altered the hydrophilic character of clay structure into hydrophobic channels by replacing inorganic cations with organic cations that in turn, caused basal spacing to increase 15.2 to 23.9 Å. This factor was solely, responsible for the strong interactions between BTX and organoclay framework [20]. Additionally, Koh et al. (2001) proved that the bulky groups (BDTDA) of surfactants attached to the clay minerals, cause the multiple interlayer adsorption as compared to the smaller groups (BTMA) that show the reluctance for second interlayer sorption for benzene and toluene [21]. More importantly, steric constrains of cations attached to the clay framework also affects the overall uptake of organic compounds. The study revealed that the cations used in organoclays may accommodate BTX either in parallel or tilted orientation with clay surface. The probability of BTX to be arranged in either manner depends upon the size distribution of free area. Thus, the smaller active sites favor, the tilted orientation while, as the size increases from benzene to xylene, cationic molecules are forced to arrange in parallel manner relative to clay layers [22].

However, Carvalho et al. (2012) endorsed that polarity effect of BTEX compounds is mainly due to the inductive effect (presence of delocalized π-electrons in ring structure) stimulated by methyl and ethyl groups. Moreover, adsorption isotherm data suggested that the heterogeneous sorption sites exist on the surface of semectite clay[23]. Additionally, the interaction between water molecules and sorbent with oxygen bearing functionl groups create some competition in sorption of organic compounds by establishiing hydrogen bond interactions. This effect is observed more in the large nonpolar molecules. While, Qu et al. (2009) summarized that cross-sectional area, critical volume, polarizability and enthalpy of vaporization are the influencing parameters for adsorption capacity. they elucidated that the dispersion interactions of BTEX compounds represent π-π electron interactions (weak hydrogen bond) while acetone comes up with dipole induced dipole forces (medium hydrogen bond).thus, because of stronger intermolecular interaction, acetone was considered to posess high adsorption affinity than that of BTEX [24]. Nevertheless, pH of the solution and humidity are not the controlling factor in the adsorption of aromatic compounds through organo-clays. Moreover, the controlling reagents to adjust the pH, cause the structural changes from porous bed to mud like structure in organo-clay. Similarly, Nourmoradi et al. (2012) reported the same results for removal of BTEX by modified MMT. They found that the increment in pH values from 4 to 12 does not have the potential influence on the removal capacity. This depicts that within solution range, modified MMT has high solubility [25].

In the present study, the sorption efficiency of economically viable and green (eco-friendly) MMT was evaluated for the removal of benzene in an aqeous solution. The equilibrium isotherms and kinetics adsorption models were also investigated.

2 Materials & Methods

2.1 Materials

Montmorillonite (MMT-K10, 250 m^2/g) and 1-hexyl-3-methylimadazolium chloride (≥97%, MW : 202.72) were purchased from Sigma Aldrich.

2.2 Methodology

10g of MMT was dissolved in 1 Litre deionized water. The suspension was stirred with a mechanical stirrer at 600 rpm for 24 h. The solution centrifuges at 4000 rpm for 15 minutes, consequently the impurities such as quarts were at the bottom of the centrifuge tube. Then, MMT dried at 110°C for 24 h and crushed to 124 μm particles. After the pre-treatment step, 4g MMT and 40 ml of 100 mmol/L 1-hexyl-3-methyl imadazolium chloride were mixed. The mixture was agitated at 150 rpm for one day at room temperature and centrifuged for half an hour. It was then dried in an oven at 40°C for 8 h. The schematic diagram of the surface modification of MMT is represented in Figure1.

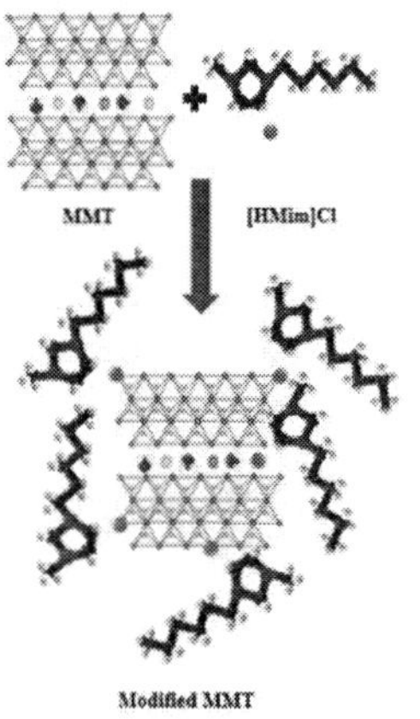

Fig.1. The schematic diagram of the surface modification of MMT.

3 Results and discussion

3.1 Infrared analysis

FTIR analysis of MMT before and after the modification was performed to identify the functional groups over the surface of clay in Figure 2. Pure MMT has only significant peak confirming functional groups to abate benzene is at 3490 cm^{-1}. On the other hand, modified MMT confirms the presence of oxygen containing functional groups at 3200 cm^{-1}, 3000 cm^{-1}, C-H stretch at 2892 cm^{-1}, C=C stretching vibration at 1550 cm^{-1}, 1440cm^{-1} that can not be found in Pure MMT.

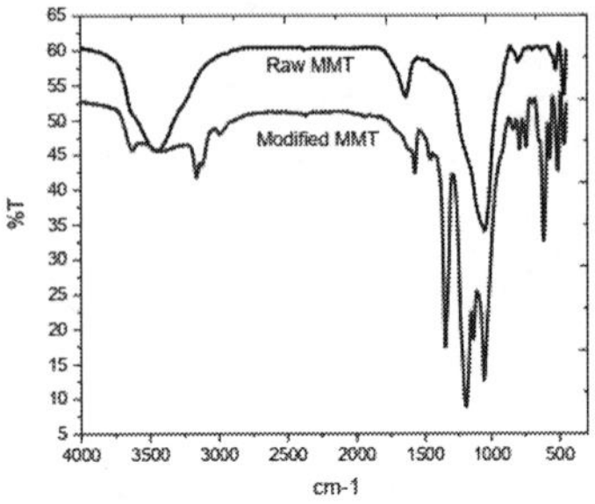

Fig. 2. IR spectra of raw and modified MMT.

3.2 DSC analysis

The thermal behavior of MMT before and after the surface modification is measured by DSC Q2000 V24.11 build 124, with a heating and cooling ramp of 10 K/min from 0°C to 300°C under nitrogen flow and the values were recorded during the second heating scan. The endothermic peak which corresponds to the melting temperature is presented at 120°C for pure MMT whereas it goes down to 99.6 °C for

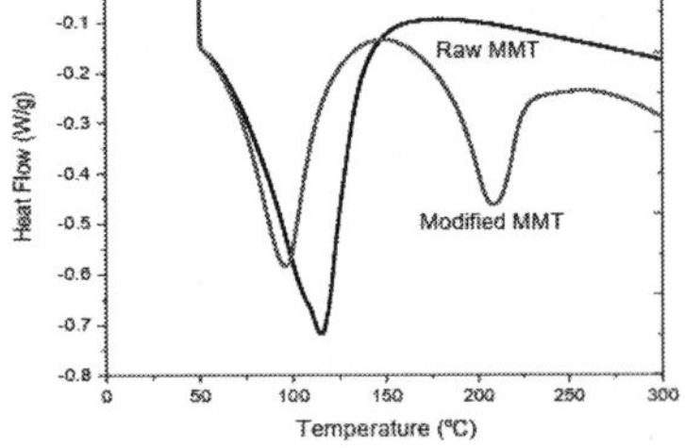

Fig. 3. DSC results of raw and modified MMT.

functionalized MMT. The crystallization temperature goes down from 175°C to 150°C which confirms the presence of functional group over the surface of pure MMT.

3.3 Scanning electron microscope analysis

Figure 4 shows the SEM images of structure of the MMT before and after the surface modification. Figure 3a depicts the surface roughness before the modification step. However, after modification, the surface exfoliation takes place as well as agglomeration is reduced due to the presence of functional groups that have an affinity for benzene as shown in Figure 4b.

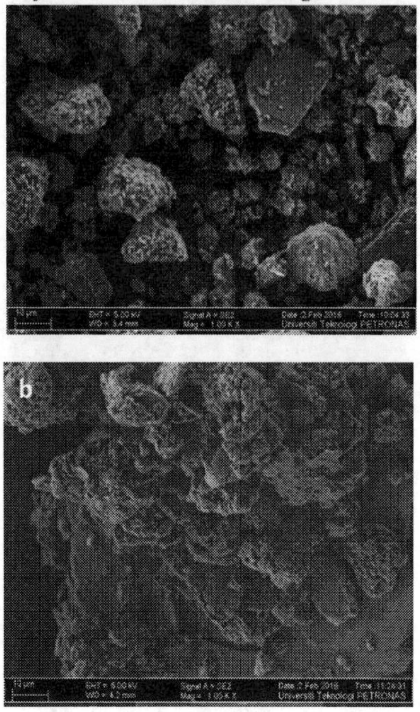

Fig. 4. FE-SEM results of a) raw and b) modified MMT.

However, the EDS spectra shows that the oxygen content has been increased after surface modification which consequently improve the sorption capacity of MMT.

3.4 Kinetic experiments

3.4.1 Effect of contact time

In order to design the batch experiments the determination of sorption rate is the mandatory requirement. The effect of contact time on the adsorption of benzene was investigated. Figure 5 shows the amount of benzene removed by the modified MMT with respect to time. It can be clearly seen that adsorption rate initially increased rapidly, and the optimal removal efficiency was reached within about 70min. the adsorption capacity is recorded to be 179.63mg/g. Further increase in contact time did not manifest significant change in equilibrium concentration.

https://doi.org/10.1051/matecconf/201815002001

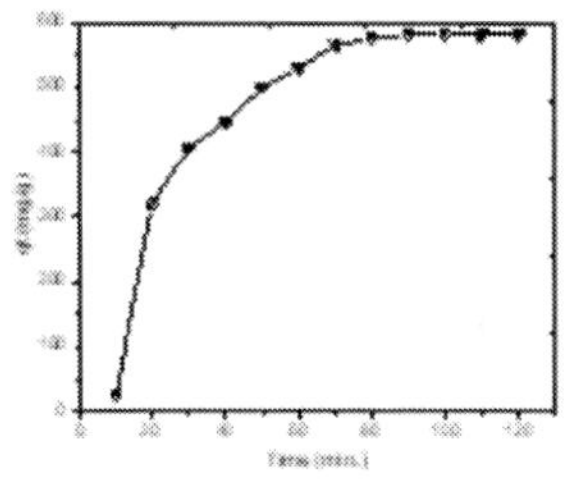

Fig.5. Effect of contact time on the adsorption of benzene.

The absorption rate decreases after equilibrium, which may be due to the fact that all the high energy sites are occupied and ions face difficulty in finding active vacant pores. This decreases the sorption capacity, which is because of the process of Intraparticle diffusion. Since, after the transfer of the ions to the surface this process occurs in the second stage of adsorption which is very slow and ultimately create repulsive forces between bulk phase and solute [26].

3.4.2 Effect of pH

The pH of a solution is an important factor in the adsorption process because it causes protonantion or deprotonation effect. Commonly, at extremely acidic and basic conditions the adsorption capacity comes out to be very low. It is because of layophobic behavior of adsorbent. Furthermore, at extreme acidic pH, the surface functional groups of adsorbent tend to compete with H^+ ions in background solution. Similarly, H_3O^+ ions compete with the surface charges at extreme basic conditions [27]. During this study, results revealed that the removal of benzene was strongly dependent on the pH of the solution. During the experimentation we varied the pH from 3 to 12, and found that maximum amount of benzene is adsorbed at pH 8.

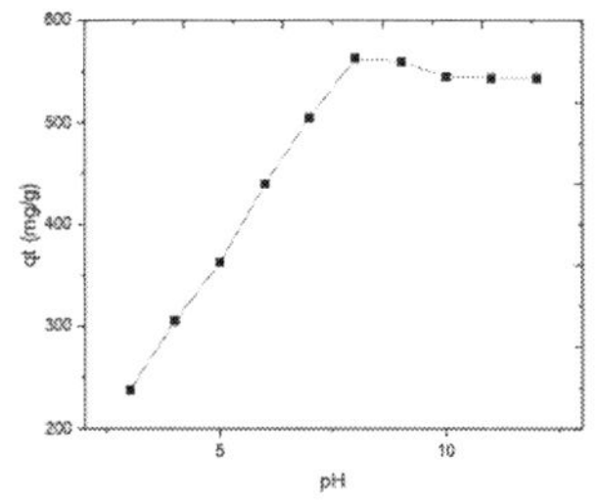

Fig. 6. Effect of pH on the absorption of benzene.

The amount increased rapidly with the increase in pH of the solution initially, and the optimal pH was observed at pH 6.5. Further increase in pH caused a decrease in the amount adsorbed. Thus, it can be deduced that the weak electrostatic forces of attraction exist between functionalized MMT and benzene. However, these interactions operate simultaneously. Therefore, the role of hydrogen bonding cannot be ignored as an interaction mechanism between adsorbate and adsorbent.

3.4.3 Effect of adsorbent dose

As adsorbent dosage increases the amount of benzene adsorbed, decreases. As adsorbent dosage increases benzene uptake decreases as shown in the Figure 7. At lower adsorbent concentration number of active sites is higher. With the increase in adsorbent dosage aggregation of particles take place, as a result benzene uptake decreases [28].

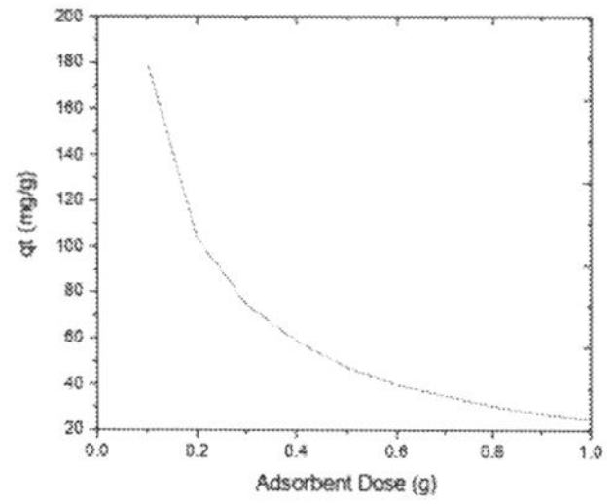

Figure 7: Effect of adsorbent dose on the adsorption of benzene.

3.4.4 Effect of rotation

As the rotation increases the amount of benzene removed is not much affected as presented in Figure 8. This describes that the solution RPM does not have a direct impact on the process of adsorption.

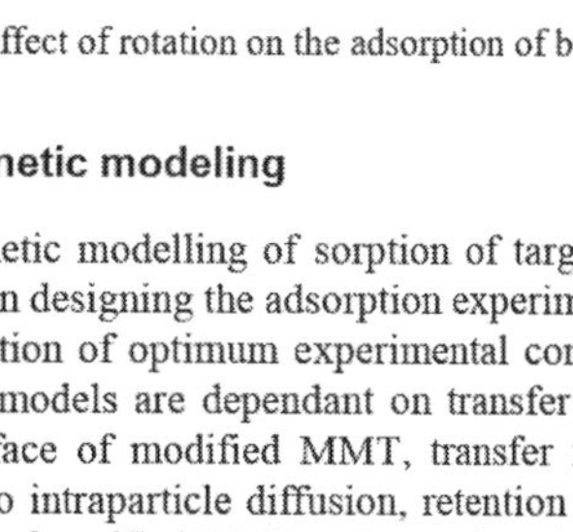

Fig. 8. Effect of rotation on the adsorption of benzene.

3.5 Kinetic modeling

The kinetic modelling of sorption of targetted pollutant guides in designing the adsorption experiments and helps in selection of optimum experimental conditions. These kinetic models are dependant on transfer of benzene to the surface of modified MMT, transfer from modified MMT to intraparticle diffusion, retention of ions on the surface of modfied MMT and complexation phenomena. In order to investigate the mechanism of removal of benzene from the surface of modified MMT, pseudo 1st order, pseudo 2nd order and intra-partile diffusion models were employed to fit the experimental data [29]. The kinetic study reveals that the sorption of benzene follows the pseudo 2nd order kinetics. All three kinetic models are summerized in Table 1.

Table 1. Kinetic parameters for the adsorption of benzene by modified MMT.

Model	Parameters	R^2
Pseudo 1st order	q_e = 977.5mgg^{-1} K_1 = 0.056min^{-1}	0.945
Pseudo 2nd order	q_e = 588.23mgg^{-1} K_2 = 0.0001gmg^{-1}min^{-1}	0.995
Intraparticle diffusion	K_{ID}= 106.06mgg^{-1} min$^{-1/2}$ C = 49.34gmg^{-1}min^{-1}	0.881

3.6 Adsorption Isotherms

The absorption isotherms are employed to fit the experimental data in order to gain more insight of adsorption experiments. Thus, to understand the interaction between adsobent and adsorbate, isotherm study is very important. In this regard, Langmuir isotherm assumes that the surface of modified MMT is homogenous in nature and the active sites for the sorption of benzene are identical. However, Freundlich isotherm is the empiral representation and explins the heterogeneity of the sorpbent and is not restricted by the monolayer formation[30].

Table 2. Parameters of isotherm models for the adsorption of benzene by modified MMT.

Model	Parameters	R^2
Langmuir	Q_o = 588.23 K_L = 2.06	0.989
Freundlich	n = 0.707 K_F = 1.138	0.979

The isotherm model study summerized in Table 2, implies that Langmuir isotherm model best fits the calculated amount adsobed benzene. However, heterogeneity of the modified MMT suggests the favorable adsorption process. Hence, the values of R^2 suggest that the isotherm model is adequate to describe the adsorption process well.

3.7 Comparative analysis of sorbents

The sorption capacities of several different types of clays (with or without modification) for the removal of benzene are summarized in Table 3. The comparative evaluation is difficult because the sorption capacities available in literature were determined at different pH, temperatures, adsorbent dosage and pollutant initial concentration. By keeping these caveats under consideration, selected adsorbents are compared. Although modified semectite clay removed benzene efficiently, but the introduction of ionic liquid treated MMT came up with much higher efficiency alongwith improved physicochemical properties.

Table 3. Comparative analysis for the adsorption of benzene by various types of clays.

Adsorbent	Pollutant	Modifier	Removal Efficiency (mg/g)
MMT (This study)	Benzène	[HMim]Cl	588.23
Bentonite [20]	Benzène	Cetyl pyridinium chloride salt	0.22
MMT [21]	Benzene	BDTDA/ BTMA/ Hyamine	28
Sericite mica [21]	Benzene	BDTDA/ BTMA/ Hyamine	21
Smectite [22]	Benzene	TMA	160
Semectite organoclay [23]	Benzene	HDTMA chloride	2.04
Na-MMT [25]	Benzene	PEG	3.58
K-10 MMT [31]	Benzene	Raw	-
Na-MMT [31]	Benzene	MPPyr	1.1
Wyoming montmorillonite [31]	Benzene	HDTMA TMA	-
Bentonite [32]	Benzene	Raw	0.28

Some of the reported modified clays have higher surface area after modification step but the sorption efficiency for the removal of benzene turned out to be very low. This may be because of the crowded surface of MMT with oxygen containing functional groups that may cause water cluster formation. But, surface area, alone cannot be the sole governing factor in determining the sorption efficiency.

4 Conclusion

In the present study, montmorillonite (MMT) was modified by 1-hexyl-3-methyleimadzolum chloride. The functionalization was approved by analytical tools i.e., FTIR, DSC and FE-SEM. However, intensive investigation of adsorption capacity through batch adsorption experiments revealed that the sorption capacity of benzene gradually increased to 581.25mg/g, which is the indication of the affinity of surface functional groups for aromatic ring structure. The adsorption equilibrium data was in good agreement with Langmuir isotherm with maximum adsorption capacity of 588.23mg/g whereas it follows pseudo second order model for adsorption kinetic data calculations. Hence, the modified MMT is considered highly efficient for wastewater treatment.

References

1. J. i. Konta. Clay and man: clay raw materials in the service of man. Appl. Clay Sci., **10(4)**, 275-335 (1995).
2. M. Brigatti, E. Galan, & B. Theng. Structure and mineralogy of clay minerals. *Handbook of Clay Science, Developments in Clay Science,* Elsevier, **5(2)**, 21-81 (2013).
3. S. Ismadji, F.Soetaredjo, A. Ayucitra. Clay Materials for Environmental Remediation, Springer, **1**, (2015).

4. S. Xu, G. Sheng, & S. A. Boyd. Use of organoclays in pollution abatement. *Advances in agronomy (USA)* (1997).
5. D. M. Moore, & R. C. Reynolds. *X-ray Diffraction and the Identification and Analysis of Clay Minerals*: Oxford University Press (1997).
6. M. Chaves, & M. Oliveira. Mechanisms underlying plant resilience to water deficits: prospects for water-saving agriculture. J. exp. Bot., ***55*(407)**, 2365-2384 (2004).
7. S. Guerra, L. Merat, R. San Gil & L. Dieguez. Alkylation of benzene with olefins in the presence of zirconium-pillared clays. Catal. Today., ***133***, 223-230 (2008).
8. M. Toor, B. Jin, S. Dai, & V. Vimonses. Activating natural bentonite as a cost-effective adsorbent for removal of Congo-red in wastewater. *J. Ind. Eng. Chem.*, ***21***, 653-661 (2015).
9. P. Komadel, & J. Madejová. Acid activation of clany minerals. *Handbook of Clay Science*, 385-408 (2013).
10. P. A. O'Day, & D. Vlassopoulos. *Mineral-based amendments for remediation. Elements (Quebec, Quebec)*, **6(6)**, 375 (2010).
11. A. A. Zagorodni. *Ion exchange materials: properties and applications: properties and applications*: Elsevier, **1**, (2006).
12. G. Yuan, B. Theng, G. Churchman, & W. Gates. Clays and clay minerals for pollution control. *Handbook of Clay Science*, **1**, 587-644 (2013).
13. C. T. Chiou. *Partition and adsorption of organic contaminants in environmental systems*: John Wiley & Sons (2003).
14. M. Gräfe, M. Nachtegaal, & D. L. Sparks. Formation of metal-arsenate precipitates at the goethite-water interface. Environ. Sci. tech., ***38*(24)** , 6561-6570 (2004).
15. G. E. Brown, & G. Calas. Environmental mineralogy–understanding element behavior in ecosystems. Comptes Rendus Geoscience, ***343*(2)**, 90-11 (2011).
16. T. Anirudhan, & M. Ramachandran. Adsorptive removal of tannin from aqueous solutions by cationic surfactant-modified bentonite clay. J. Colloid Interface Sci., ***299*(1)**, 116-124 (2006).
17. S. Bedin, M. Oliveira, M. Vieira, M. Silva. & O. Santos. Adsorption of toluene in batch system in natural clay and organoclay. Chem. Eng. Trans, ***32***, 313-318 2013).
18. J. A. Smith, P. R. Jaffe, & C. T. Chiou. Effect of ten quaternary ammonium cations on tetrachloromethane sorption to clay from water. Environ.l sci. tech., ***24*(8)**, 1167-117 (1990).
19. W. Jaynes, & G. Vance. Sorption of benzene, toluene, ethylbenzene, and xylene (BTEX) compounds by hectorite clays exchanged with aromatic organic cations. Clays Clay Miner., ***47*(3)**, 358-365 (1999).
20. C. Bertagnolli, & M. Silva. Characterization of Brazilian Bentonite Organoclays as sorbents of petroleum-derived fuels. *Mater. Res.*, ***15*(2)**, 253-259 (2012).
21. Koh, S.-M., & Dixon, J. B. Preparation and application of organo-minerals as sorbents of phenol, benzene and toluene. Appl. Clay Sci., ***18*(3-4)**, 111-122 (2001).
22. W. Jaynes, & S. Boyd. Hydrophobicity of siloxane surfaces in smectites as revealed by aromatic hydrocarbon adsorption from water. Clay Miner., ***39*(4)**, 428-436 (1991).
23. M. Carvalho, M. Da Motta, M. Benachour, D. Sales, & C. Abreu. Evaluation of BTEX and phenol removal from aqueous solution by multi-solute adsorption onto smectite organoclay. J. Hazard. Mater., ***239***, 95-101 (2012).
24. F. Qu, L. Zhu, & K. Yang. Adsorption behaviors of volatile organic compounds (VOCs) on porous clay heterostructures (PCH). J. Hazard. Mater., ***170*(1)**, 7-12 (2009).
25. H. Nourmoradi, M. Nikaeen, & Khiadani M. Removal of benzene, toluene, ethylbenzene and xylene (BTEX) from aqueous solutions by montmorillonite modified with nonionic surfactant: Equilibrium, kinetic and thermodynamic study. Chem. Eng. J., ***191***, 341-348 (2012).
26. S. Noreen, H. Bhatti, N. Nousheen, Sadaf. H. Batch and fixed bed adsorption study for removal of Drimarine Black CL-B dye from aqeous solution using a lignocellulosic waste: a cost effective adsorbent. Ind. Corps Produc, ***50***, 568-579.
27. K. Shroff, V. Vaidya. Kinetics and equilibrium studies on biosorption of nickel from aqeous solution by dead fungal biomass of Mucar heimalis. Chem. Eng. J. ***171***, 1234-1245 (2011).
28. I. Ullah, R. Nadeem, M. Iqbal, Q. Manzoor. Biosorption of chromimum onto native and immobilized sugarcane baggase waste biomass. Ecol. Eng. ***60***, 99-107 (2013).
29. M. Iqbal, M. Abbas, M. Arshad, T. Hussain, A. U. Khan, N. Masood, M.A.Tahir. Gamma radiation treatment for reducing cytotoxicity and mutagenicity in industrial wastewater. Polish J. Environ. Stud. *24*, 2745–2750 (2015).
30. P.D Saha, A. Dey, P. Marik. Batch removal of chromium(VI) from aqueous solutions using wheat shell as adsorbent: process optimization using response surface methodology. Desalin. Water Treat. *39*, 95–102 (2015).
31. G. Morozov, V. Breus, , S. Nekludov & I. Breus. Sorption of volatile organic compounds and their mixtures on montmorillonite at different humidity. Colloids Surf. A. Physicochem. Eng. Asp., ***454***, 159-171 (2014).
32. G. Alther. Using organoclays to enhance carbon filtration. Waste Management, ***22*(5)**, 507-513 (2002).

Chemical Resistance of Glass Composite Materials Made From Incinerated Scheduled Waste Slag and SLS Waste Glass

Jariah Mohamad Juoi[1,*], *Dilip* Arudra[1], *Zulkifli Mohd* Rosli[1], *Khairu Ilwani* Karim[1] *and Farah Hanum* Suzain[1]

[1]Advanced Manufacturing Centre, Faculty of Manufacturing Engineering, Universiti Teknikal Malaysia Melaka, 76100 Durian Tunggal, Melaka, Malaysia

Abstract. Incineration of scheduled waste and landfilling of the incineration residue (Bottom Slag) is extensively practised in Malaysia as a treatment method for scheduled waste. Land site disposal of Bottom Slag (BS) may lead to environmental health issues and reduces the availability of land to sustain the nation's development. This research aims in producing Glass Composite Material (GCM) incorporating BS and Soda Lime Silicate (SLS) waste glass as an alternative method for land site disposal and as an effort for recycling SLS waste glass. SLS waste glass originates from the urban waste has been a waste stream in most of the nation whereby the necessity for recycling is in high priority. Batches of powder mixture is formulated with 30 wt. % to 70 wt. % of BS powder and SLS waste glass powder for GCM sintering. The powder mixtures of BS and SLS waste glass is compacted by uniaxial pressing and sintered at 800^{0}C with heating rate of 2^{0}C/min and 1 hour soaking time into tiles of 18mm×18mm. The GCM porosity and water absorption increases as the BS waste loading increases. Meanwhile, its bulk density increases as the BS waste loading decreases. The GCM tiles made from BS 30 wt. % and 70 wt. % SLS waste glass are determined to have the lowest water absorption of 1.17 % and porosity percentage of 2.2 % with the highest bulk density of 1.88 g/cm^3. It was also found is found that the chemical resistance of these GCM tiles is classified as ULA (No visible Effect) and UHA (No visible Effect) after 5 day immersions in low and high concentration of acid and alkali solution; respectively (determined using MS ISO10545-13:2001(Ceramic Tile: Determination of chemical resistance) test. However, the chemical resistance is weak upon increased duration of 12 immersion days where severe corrosion effects on both surface tiles in low and high concentration chemical solutions. The penetration of chemical in attacking the samples are related to the presence of pores. Hence, further work should have aimed to reduce the amount of porosity presence in the produced tiles.

1 Introduction

Scheduled waste is defined by The Environmental Quality (Scheduled Waste) Regulations 2005 (amendment 1989) under the Environmental Quality Act 1974 [Act 127] as any waste falls within the 77 categories of hazardous waste listed in the First Schedule [1]. At present, incineration is one of the treatment method for scheduled waste in Malaysia. (Source: Kualiti Alam Sdn.Bhd). Incineration of waste is defined as the controlled burning of a substance whereby controlled refers to clearly defined parameters such as temperature range, oxygen input, turbulence, atmospheric pressure and other aspects of combustion environment. [2]. A waste incinerator typically, consists of a Rotary Kiln (Primary Combustion Chamber) which can incinerate up to 1100^0 C and a secondary combustion chamber which operates more than 1000^{0}C, meeting the EU incineration requirement (Source: Kualiti Alam Sdn.Bhd). About 13,300 tonnes of bottom slag and tonnes of fly ash are produced annually by incineration [3]. According to the classification of scheduled waste by the Department of Environment Malaysia, incinerated scheduled waste bottom slag falls under the SW501 code whereby the code is defined as any residue from treatment or recovery of scheduled waste [4]. At present, the disposal of bottom slag is done by landfilling.

* Corresponding author: jariah@utem.edu.my

The practice of landfilling creates issue of contaminated land whereby land containing substance when present in sufficient concentration, may cause harm to human, animals and environment [5]. Moreover, landfilling does not provide a sustainable solution whereby more land area is required due to increasing waste volume, hence decreases the land availability for nation's development and reducing the land capacity [6]. Utilization of incinerated scheduled waste bottom slag into producing a beneficial product or into a glass composite material would be an effective method in overcoming the inefficiency of current scheduled waste treatment. It reflects advantages in reducing disposal cost, preserving environment, and economic impact [7]. The bottom slag can be classified as a silicate waste due to high content of SiO_2. Studies on silicate wastes originated from metallurgic waste incineration, fly ash from coal production, incinerator fly ash and slag from industrial waste, fly ash and slag from municipal solid waste incinerator, incinerated scheduled waste bottom ash has been carried out extensively into producing glass ceramic or glass composite material. Based on this review, there is lack of sufficient work being carried out on incinerated Malaysian scheduled waste bottom slag and this research work highlights on producing glass composite material utilizing this waste [7, 8].

Thus, in this work, Malaysian incinerated scheduled waste BS (bottom slag) is incorporated into glass matrix to produce GCM. SLS (soda lime silicate) glass is obtained from urban waste (used glass bottle and household glass containers). Through recycling SLS waste glass, the raw material consumption is reduced yielding economic and environmental benefits [9]. Thus, by incorporating two waste streams (SLS glass and BS), the initiative to promote recycling can be achieved. Glass composite material produced by incorporating industrial waste is a low investment approach and has shown good ability to accommodate desirable properties for the application of floor and roof in industrial and public building, wall covering and road surfacing [10]. The properties of GCM can be tailored based on selection of suitable glass powder, volume fractions, powder processing and firing schedule giving higher possibility to achieve the objective of this research work [11]. It should be highlighted that chemical resistance is an important property for floor and wall covering as tiles are usually subject to chemical aggression as a result of cleaning operations, environmental pollution outdoors and product spill on flooring. Nevertheless, reported work on the chemical resistance of a tile made from waste is limited. Thus, in this paper reports on the evaluation of the chemical resistance of GCM tiles made from BS and SLS waste glass.

2 Methodology

Incinerated scheduled waste BS (bottom slag), is obtained from Kualiti Alam Sdn.Bhd situated in Bukit Nanas, Negeri Sembilan. The collected BS was ready to be landfilled. The BS is than pulverized and sieved for a particle size of <75 micron. Transparent SLS waste glass (bottles) was obtained from urban waste. The bottles are milled with alumina balls in a milling machine and sieved for particle size of <75 micron. The sieved bottom slag (BS) and SLS waste glass powder is then mixed to obtain a mixture of BS and SLS waste glass powder according to weight ratios as shown in Table 1. Maximizing the BS waste utilization will be an alternative route to reduce the current practice of landfilling BS waste. The waste loading based on weight ratios is selected from 30 wt % to 70 wt % BS waste in reference of research work on production of glass ceramic from industrial waste and immobilization of incineration ashes into GCM [12, 13].

The prepared batch mixture are mixed for 30 minutes using mixing machine. The samples are then pressed at 90MPa in a mould of rectangular shape of 18mm x 18mm. The pressed samples were then sintered in a furnace with 2°C/min heating rate at 800°C.

Table 1. Ratios of BS to waste glass powder (wt %)

BS Powder [wt %]	Waste Glass Powder [wt %]
30	70
40	60
50	50
60	40
70	30

The physical analyses of water absorption, bulk density and apparent porosity were determined according to ASTM C3739[14]. While, the determination of chemical resistance was carried out only on sintered samples from batch formulation of BS 30 wt. % and 70 wt. % SLS waste glass. This is due to the optimum performance of low porosity, water absorption and high density during physical analyses. Chemical resistance determination were carried out according to MS ISO 10545-13:2001 Ceramic tiles-Determination of chemical resistance (ISO 10545-13:1995(E)[15]. In this test, samples were immersed in a low concentration test solution (Hydrochloric acid solution, 3% (V/V), Citric acid solution, 100g/l, Hydrochloric acid solution, 18% (V/V) and high concentration test solutions (Hydrochloric acid solution, 18% (V/V), Lactic acid solution, 5% (V/V), Potassium hydroxide solution, 100 g/l) for 12 days. Later, the test specimens were examined after day 5 and day 12 immersion at a distance of 25 to 30 cm, with the naked eye of with spectacles if usually worn, for changes on the proper surface and non-cut edge to determine the class of chemical resistance.

In order to evaluate the effect of the chemical attack on the microstructure of the GCM tiles, further observation on the surface and cross section area of samples that were immersed in potassium hydroxide solution (the most reactive solution) were conducted using scanning electron microscopy SEM EVO 50 Carl Zeiss SMT model operated at 20 kV.

3 Results and Discussion

3.1 Porosity, Water Absorption and Bulk density

Figure 1, 2 and 3 shows the apparent porosity, bulk density and water absorption of sintered glass composite with increased BS waste loading ranging from 30 wt. % to 70 wt. %. It is observed that the porosity [%] increases as the BS waste loading [%] increases. A gradual increase in porosity (Fig. 1) is observed from BS waste loading of 30 wt. % up to 60 wt. % followed by a drastic increase from 60 wt. % to 70 wt. %. The lowest porosity value is given by 30 wt. % BS waste loading with 70 wt. % of SLS waste glass sintered samples which is 2.20 %. The water absorption (Fig. 2) shows similar trend as the porosity. The minimum water absorption value is given by 30 wt. % BS waste loading with 70 wt. % of SLS waste glass sintered samples which is 1.17 %. Bulk density (Fig. 3) shows a decreasing trend with increasing waste percentage. 30 wt. % BS waste loading with 70 wt. % SLS waste glass sintered samples shows the highest bulk density of 1.88 g/cm^3. Based on these physical properties, sintered samples from batch formulation of BS 30 wt. % and 70 wt. % SLS waste glass has shown optimum performance. Porosity contributes to the bulk density and water absorption characteristics of the sample.

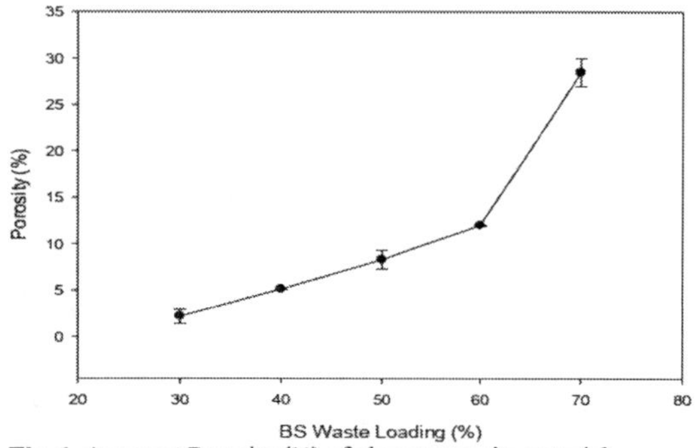

Fig. 1. Apparent Porosity (%) of glass composite material according to BS waste loading (wt %).

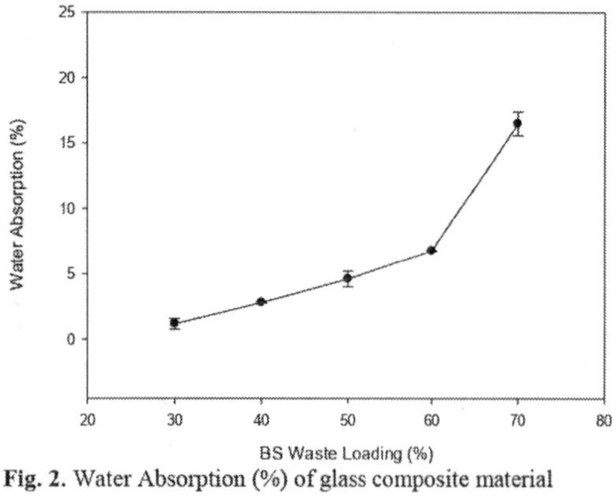

Fig. 2. Water Absorption (%) of glass composite material according to BS waste loading (wt %)

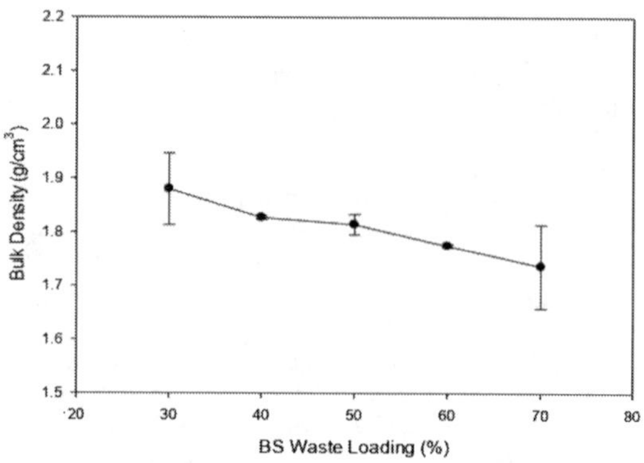

Fig. 3. Bulk density of glass composite material according to BS waste loading (wt %)

3.2. Chemical Resistance Classification

Table 2 and Table 3 shows the chemical resistance class of GCM tiles made from batch formulation of BS 30 wt. % and 70 wt. % SLS waste glass after the 5 days and 12 day immersion in the six different types of chemical solutions. Using naked eye in the daylight, the samples were observed using the correct procedure to determine the chemical effects on the samples. The results show that there are no visible effects on the GCM tiles after day 5 immersion. In contrast, severe effects on the GCM tiles were observed after 12 day immersion in low and high concentration chemical solution with a classification of ULC and UHC. Moreover, it was also observed that the GCM tiles are more prone to the acid solution compared to the alkali solution. This could be attributed to the properties of the SLS used in the GCM tiles making.

Table 2. Chemical Resistance Class of GCM tiles made from 30 wt. % BS and 70 wt. % SLS waste glass after day 5.

Types of Low Concentration Solution	Chemical Resistance Class
Hydrochloric acid solution, 3% (V/V)	ULA
Citric acid solution, 100g/l	ULA
Potassium hydroxide solution, 30 g/l	ULA
Hydrochloric acid solution, 18% (V/V)	UHA
Lactic acid solution, 5% (V/V)	UHA
Potassium hydroxide solution, 100 g/l	UHA

For low concentration test solution :
*Class ULA : No visible effect
*Class ULB : Visible effect on cut sides
*Class ULC : Visible effect on cut sides, non-cut sides and on the proper surface

For high concentration test solution :
*Class UHA : No visible effect
*Class UHB : Visible effect on cut sides
*Class UHC : Visible effect on cut sides, non-cut sides and on the proper surface

Table 3. Chemical Resistance Class of GCM tiles made from BS and SLS Waste after day 12

Types of High Concentration Solution	Chemical Resistance Class
Hydrochloric acid solution, 3% (V/V)	ULC
Citric acid solution, 100g/l	ULC
Potassium hydroxide solution, 30 g/l	ULC
Hydrochloric acid solution, 18% (V/V)	UHC
Lactic acid solution, 5% (V/V)	UHC
Potassium hydroxide solution, 100 g/l	UHC

3.3. Microstructural Analyses

Samples from batch formulation of BS 30 wt. % and 70 wt. % SLS waste glass and that were immersed in the most reactive chemical solution (potassium hydroxide solution of 100g/l) were chosen to be further analysed using SEM. Figure 4 shows the SEM micrographs on the surface of the GCM tiles before the chemical resistance test. It is observed that the microstructure consist of 4 main features which is even dense surface, uneven dense surface, coarse feature, closed pores and open pores (figure 4 (a)). Chemical attack were observed after the samples immersed in high concentration test solutions. The chemical attack had affect the dense surface microstructure where a rough and crippled surface is observed at this region after the 12thd day immersion in the potassium hydroxide solution (Figure 4(b)). It is also observed that the open pores are larger with the edge of the pores is clearly visible.

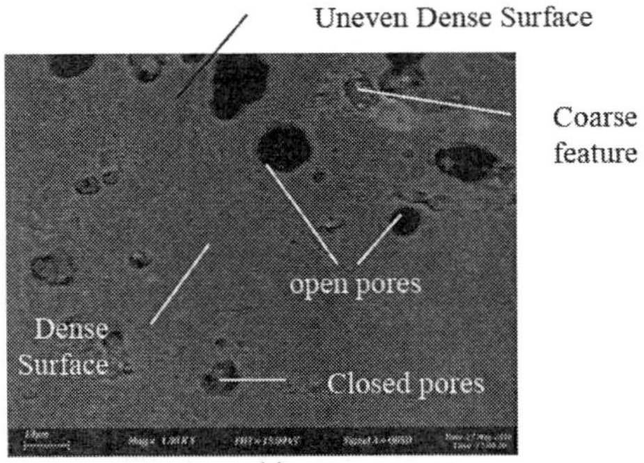

(a)

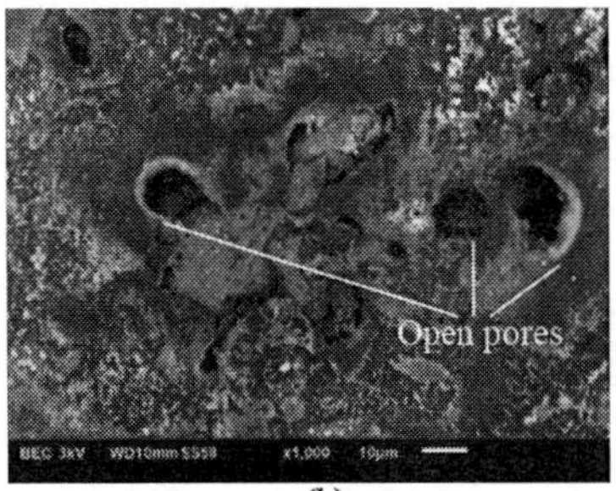

(b)

Fig. 4. SEM surface micrographs of GCM tiles made from 30 wt. % BS and 70 wt. % SLS waste glass a) before and b) after the chemical resistance tests.

Figure 5 shows the SEM micrographs on the cross sectional of the GCM tiles after the chemical resistance test. Chemical attack on dense surface, edge of closed pores were observed during SEM analyses as shown in figure 5(b). The dense surface that had been observed before the chemical test had become a rough surface due to the reaction of the high concentration test solution during the 12 immersion days. This is a similar observation as on the surface of the GCM tiles.

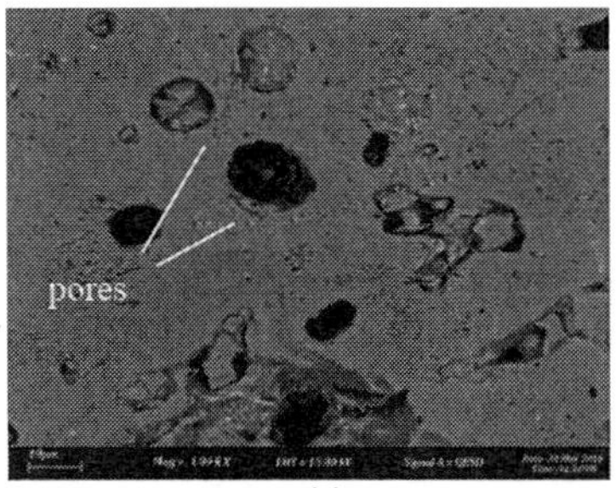

(a)

(b)

Fig. 5. SEM cross section micrographs of GCM tiles made from 30 wt. % BS and 70 wt. % SLS waste glass before and after the chemical resistance tests.

5 Conclusion

Chemical resistance of sintered GCM tiles are related to its physical properties. GCM tiles made from 30wt. % BS and 70 wt. % transparent SLSG (with the lowest water absorption of 1.17 % and porosity percentage of 2.2 % with the highest bulk density of 1.88 g/cm^3) has shown good chemical resistance at early chemical immersion (5 days) to the test solution. However, the chemical resistance is weak upon increasing duration of immersion. The penetration of chemical in attacking the samples are related to the presence of pores. This is evident during the SEM analyses where the pores is observed to become more visible and larger due to the chemical reaction that take place. In consequence, further work on producing GCM tiles from BS and SLS waste glass should aimed to reduce the amount of porosity presence in the sintered GCM tiles.

We would like to acknowledge Ministry of Higher Education Malaysia for awarding the PRGS grant (Green Innovative Glass Ceramic: Towards Resources Efficiency In Tiles Industry) and support from members of Sustainable Materials for Green Technology (SM4GT) Research group of Advanced Manufacturing center, Faculty of Manufacturing Engineering, UTeM.

References

1. http://www.doe.gov.my/v2/files/legislation/pua0294y 2005.pdf (Enviromental Quality Scheduled Wastes Regulations 2005,Accessed on 18/10/2011
2. http://www.epa.gov/osw/inforesources/pubs/training/ incin.txt (Accessed on 2/8/2011)
3. S. Naganathan, H. AbdulRazak, S. Nadzriah, Preliminary Investigation of Incinerator Bottom Slag as Controlled Low-Strength Material, Int. Con. On Construction And Building Technology, 2008
4. www.doe.gov.my (Accessed on January 2012)
5. Z. A. Rashid, A.B. Alias, Aris, M.J, El-Harbawi, M, Rahman, N.A. and Som, A.M, Int. J. Environ. Eng., **2**,139–158 (2010)
6. M. Shahnor Bani, Z. Abdul Rashid, K. Halim Ku Hamid, J. App. Sci.,**11**, 2358-2363, (2011)
7. H. Isa, Int. J. Phy. Sci., **6**, 30, 6781 – 679, (2011)
8. S. Naganathan, H. Abdul Razak, S. Nadzriah Abdul Hamid, J. Mat. and Design, **33**, 56–63, (2012)
9. T. Tarvornpanich, G.P. Souza, Use Of Soda-Lime-Silica Waste Glass As An Alternative Flux In Traditional Ceramics, Int.l Con. on Geo. of Thai: Towards Sustainable Development and Sufficiency Economy,2007
10. M. Erol, S. Ku¨c¸u¨kbayrak, A. Ersoy-Meric¸boyu, J. Cer. Int., **35**, 2609–2617(2009)
11. M. Eberstein, S. Reinsch, R. Müller, J. Deubener, W. A. Schiller, J. Eur.Cera. Soc., **29**, 2469–2479(2009)
12. Bianka V. Mangutova, Emilija M. Fidancevska, Milosav I. Milosevski, Joerg H. Bossert, APTEFF, **35**,1-280 (2004)
13. M. I. Ojovan, J. M. Juoi, W. E. Lee, J Pak. Mater. Soc., **2**, 72-76(2008)
14. American Standard Testing Material (2006). Standard Test Method for Water Absorption, Bulk Density, Apparent porosity, and Apparent Specific Gravity of Fired Whiteware Products, Ceramic Tiles, and Glass Tiles, West Conshohocken, U.S, ASTM C373-88.N.
15. Malaysian International Standard. Ceramic Tiles-Part 13: Determination of Chemical Resistance, Malaysia, MS ISO 10545-13:2001 (ISO 10545-13:1995 (E)).

Preparation and Characterization Silver Nanoparticle Embedded Polyamide Nanofiltration (NF) Membrane

Tengku Anisa Tengku Sallehudin[1], *Mazrul Nizam* Abu Seman[1,*], and *Syed Mohd Saufi* Tuan Chik[1]

[1]Faculty of Chemical and Natural Resources Engineering, Universiti Malaysia Pahang, 26300 Kuantan, Pahang, Malaysia

Abstract. In this study, silver nanoparticles (AgNPs) were synthesized from tea leaves extract and its antimicrobial properties was tested on *Escherichia coli* (*E. coli*) using agar well method. The synthesized nanoparticles were characterized by using UV-vis spectroscopy, transmission electron microscopy (TEM) and X-ray diffraction (XRD). The result from XRD analysis shows that the synthesized AgNPs are a face-centered cubic (fcc) structure with an average particle size of 28±15 nm AgNPs which confirmed by TEM. The synthesized AgNPs were then used in the preparation of thin film composite NF membrane via interfacial polymerization method. Separation performance of the produced membrane was evaluated in term of membrane permeability and solute rejection (vitamin B12, NaCl and Na_2SO_4 solutions). Based on the structural parameters (pore size, r_p and Ax/A_k) values obtained from vitamin B12 test, all membranes can be considered as tight NF membranes. From the salts rejection test, membranes with the addition of AgNPs exhibited higher salt rejection compared to the neat membranes.

1 Introduction

Nowadays, applications of membrane separation have many advantages but their disadvantages also become a huge issue. Despite the widely uses in separation processes, membrane process faces a common problem namely fouling. Fouling could affect the membrane performance including flux and selectivity. In normal practice, to restore the membrane performance, physical and chemical cleaning approaches are applied. Besides that, membrane properties also not less important to be taken into consideration since it also may help minimizing the fouling problem.

One of the popular methods is by embedded the hydrophilic nanoparticles in membrane matrix in order to change the membrane surface properties, hence, reduce fouling. The common nanoparticles used to reduce fouling (especially biofouling against bacteria) are silver (Ag) and titanium oxide (TiO_2). TiO_2 self-assembled thin film composite membrane could only work effectively by an assistant of UV illumination where the additional energy required for UV light system (increase operational cost) [1]. As an alternative, Ag self-assembled membrane could be produced for antibacterial properties without extra cost (energy for UV light). Few studies have reported that AgNPs were used to synthesize low biofouling UF membranes by phase inversion method [2-5]. Ag self-assembled UF membrane synthesized by phase inversion was very efficient against bacteria. However, it was found that this type of membrane facing a problem with depletion of silver from membrane surface [2]. In the previous report, Lee et al. prepared polyamide NF membrane consisting commercial AgNPs by *in situ* interfacial polymerization method between aqueous M-Phenylenediamine (MPD) and organic Trimesoyl Chloride (TMC)-Ag solutions and tested its anti-biofouling effect on bacteria *Pseudomonas* [6]. Chemical reaction method was employed to synthesize AgNPs in the presence of stabilizing agent to prevent the unwanted colloids agglomeration. However, AgNPs from chemical synthesis are considered hazardous to the environment, expensive and consume high energy [7-10]. In order to overcome this problem, AgNPs was synthesized by the green method as it was simple and cost effective technique [11]. Tea leaf extract (*Camellia sinensis*) has been used to synthesize AgNP since it contains polyphenols and terpenoids, such as β-cariophyllene, linalool, cis-jasmone, α-terpineol, δ-cadiene, indole, geraniol, among the major bio-components, which have bactericidal and antioxidant activity, and several other useful properties [12,13]. Here, we report on the preparation and characterization of low biofouling thin film composite polyamide NF membrane by interfacial polymerization method between organic TMC and aqueous MPD solution consisting AgNPs synthesized by greener method.

2 Methodology

2.1 Materials

Silver nitrate was obtained from Sigma Aldrich. Nutrient broth and Mueller Hinton agar powder, a product of Oxoid were supplied by Thermo Scientific. Dry tea leaves were bought from Malaysia tea plantations. The strain employed throughout this study *Escherichia coli*

* Corresponding author: mazrul@ump.edu.my

(0157:H7) (ATCC 43888) was obtained from Central Laboratory University Malaysia Pahang (UMP). It was cultured on nutrient agar slants and kept at 4° C. The commercially available Polyethersulfone Ultrafiltration (UFPES50) flat-sheet membrane (Nominal MWCO: 50kDa) was supplied by Amfor Inc (China) and used as a substrate for thin film composite NF membrane. Both M-Phenylenediamine (MPD) and Trimesoyl Chloride (TMC) were supplied by Acros. N-hexane used as the organic phase solvent was obtained from Hamburg Industries Inc.

2.2 Production of Silver Nanoparticles (AgNPs)

2.2.1 Preparation of tea extract

To prepare the tea extract, 5 g of ground tea leaves was boiled in 500 ml distilled water, and the beaker was thoroughly agitated for 1 hour using a heated stirrer and left to settle. The extract solution is then filtered through a 0.2 μm membrane filter and used for further experiments.

2.2.2 Synthesis of AgNPs

Firstly, a dilute solution of silver nitrate (0.1 M) was prepared by dissolving 1.6980 g $AgNO_3$ in 100 ml distilled water. To carry out the reaction, 2 ml of 0.1 M $AgNO_3$ was mixed with 10 ml of tea extract solution and stirred at a constant speed at room temperature (25°C), to ensure thorough mixing for about 24 hours. Any colour changes were observed.

2.2.3 Washing process of AgNPs

A desktop centrifuge was used in the washing process of AgNPs synthesized from the tea extract. The washing process could guarantee that the unreacted polyphenol content is removed from the AgNPs for purification. The AgNPs were separated and concentrated by repetition of 4 to 5 times mixture centrifugation at 2633×g for 20 minutes. The supernatant was washed by distilled water each time.

2.3 Characterizations of AgNPs

2.3.1 UV-Vis

AgNPs in tea extract solution was first characterized by using UV-Vis spectrophotometer (Hitachi U-1800). UV-Vis spectrophotometer with a resolution of 1 nm between 300-800 nm possessing a scanning speed of 300nm/min was used. The reduction of pure Ag^+ ions was monitored by measuring the UV–Vis spectrum of the reaction medium after diluting a small aliquot of the sample with deionized water. One millilitre (1 ml) of the sample was pipetted into a test tube and diluted with 4 ml of distilled water and subsequently analysed at room temperature.

2.3.2 TEM

Investigation of the size and morphology of the AgNPs were carried out under a transmission electron microscope, TEM (Hitachi H7100) with an accelerating voltage of 120 kV. A drop of the silver nanoparticles solution was deposited onto a carbon-coated copper grid and was allowed to evaporate at room temperature. The size distribution of the AgNPs was based on diameter measurement of >200 particles on TEM micrographs using image analysis software I-Solution (IMT).

2.3.3 XRD

The x-ray diffraction (XRD) analysis was carried out using Rigaku Miniflex II software θ - 2θ diffractometer with Cu, K_α radiation at 30 kV, 15 mA and wavelength (λ)=1.54056 Å. Scans were typically in the range 10–70° 2θ, with 0.02° step sizes that were held for 2 seconds each.

2.4 Thin Film Composite (TFC) Membrane

2.4.1 Preparation of PA and PA/Ag Membranes

The polyamide (PA) membrane was prepared by interfacial polymerization on the surface of commercial UFPES membrane support layer between 2% w/v MPD and 0.15% w/v TMC as described in our previous report [14]. For silver nanoparticles polyamide (PA/Ag) membrane preparation, 20 ml of 3g/L (3000μg/ml) neat silver nanoparticles colloids (produced as described in section 2.2) then added to 80ml MPD aqueous solution making the final concentration of AgNPs in the 100 ml of mixture aqueous solution (MPD + AgNPs) was 0.6mg/L. Overall, six different types of NF membranes were prepared at different reaction time and summarized in Table 1.

Table 1. Preparation of NF membranes

Types of Membrane	#AgNPs loading (ml)	Reaction Time (s)
PA-10	-	10
PA-30	-	30
PA-60	-	60
PA/Ag-10	20	10
PA/Ag-30	20	30
PA/Ag-60	20	60

20 ml is an optimum AgNPs loading as reported in [14].

2.4.2 Antibacterial assay of PA/Ag membrane

Antibacterial activity of the prepared membranes was investigated by agar diffusion method against *E.coli.* Kirby Bauer technique was applied in the preparation of media-culture plate containing *E.coli.* The media agar used was purchased from Merck. The membranes were cut into circular-disks, autoclaved and put on the bacteria media-culture at 37 °C incubation temperature overnight.

The antibacterial activity of the membrane was determined by inhibition ring that formed after 24 h incubation. The growth inhibition was visually detected on the next day after incubation.

2.4.3 Permeation Experiment

Separation performance of the produced membrane was evaluated in term of pure water flux, membrane permeability and solute rejection (vitamin B12, NaCl and Na_2SO_4 solutions). All the experiments were conducted by using Amicon stirred cell (Model 8200) at room temperature. For uncharged solute (100 g/L of vitamin B12) and charged solute (0.001M NaCl and 0.001M Na_2SO_4), the stirring speed was held constant at 300 rpm to minimize concentration polarization effect.

2.4.4 Approximate Characterization Method using Uncharged Solute (Vitamin B12)

For TFC nanofiltration membranes, effective pore radius (r_p), and the ratio of effective membrane thickness over porosity ($\Delta x/A_k$) are best characterized using approximate characterization method. Through this method, the flux and rejection data were fitted using the Donnan-Steric-Pore Model (DSPM) and Hagen-Poiseuille equation as described in previous studies [15,16]. For this study, r_p was determined from the following equation by using vitamin B12 rejection data.

$$R_{real} = 1 - \frac{C_p}{C_w} = 1 - \frac{K_{i,c}\Phi}{1 - \exp(-Pe_m)[1 - \Phi K_{i,c}]} \quad (1)$$

where the Peclet number, Pe_m is defined as:

$$Pe_m = \frac{K_{i,c}}{K_{i,d}} \frac{V\Delta x}{D_{i,\infty} A_k} \quad (2)$$

$\Delta x/A_k$ is then determined based on following Hagen-Poiseeuille equation:

$$J_w = \frac{r_p^2 \Delta P}{8\mu(\Delta x / A_k)} \quad (3)$$

3 Results and Discussion

3.1 UV-Vis Spectroscopy

The formation of AgNPs was confirmed by using UV-visible spectroscopy (UV-vis). As shown in Figure 1, the peak observed the broad plasma resonance peak around at (a) 353 nm for the solution endures non-washing step while (b) 453.5 nm for the samples through the washing step. Similar result also found by other researcher [17].

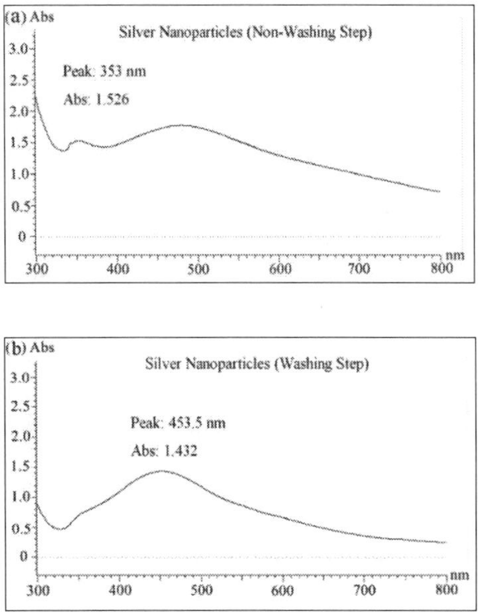

Fig. 1. UV/Vis spectrum shows the formation of AgNPs after 24 hours of reaction which the supernatant endured. (a) Non-washing step (b) Washing step.

3.2 Transmission Electron Microscopy (TEM)

The size, shape and morphology of synthesized AgNPs were characterized by using TEM, and the image of AgNPs is shown in Figure 2. As can be seen in Figure 2, TEM image clearly shows that the morphology of AgNPs is nearly spherical in shape. Figure 3 displays the histogram of size distribution of AgNPs with sizes ranged from 10-80nm. About 64 particles were measured from the TEM image and the average particle size is 28±15 nm. The difference in the particle size by TEM may be because of the nanoparticles were formed at different times.

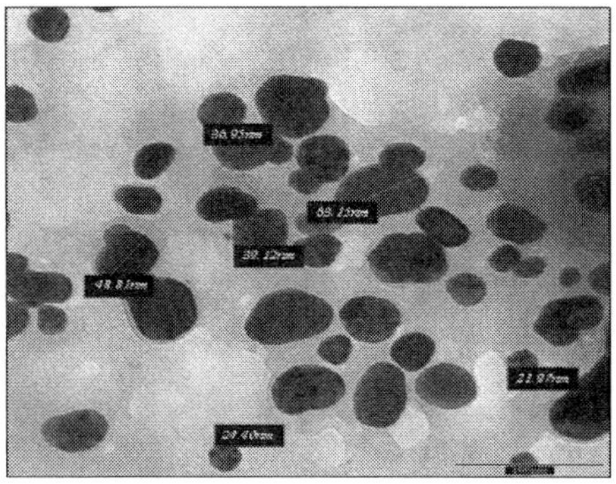

Fig. 2. TEM images of AgNPs

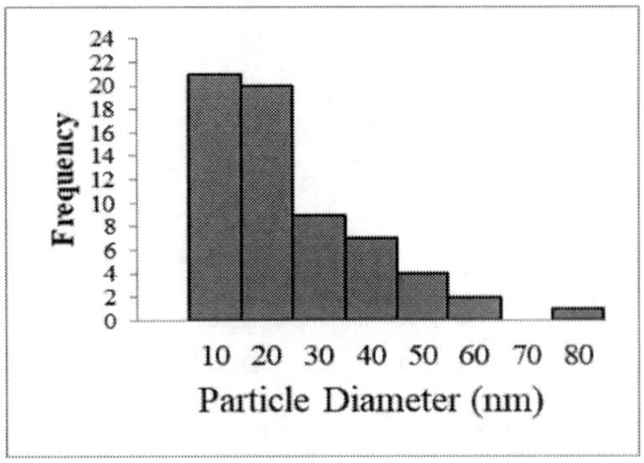

Fig. 3. The corresponding size distribution histogram of AgNPs

3.3 X-ray Diffraction (XRD) analysis

Figure 4 shows the XRD pattern of the Ag nanopowders obtained from the experiment. The pattern consists of three distinct peaks at corresponding 2θ values of 38.1°, 44.3° and 64.5°. These values reveal that it is a face-centered cubic (fcc) structure. The discernible peaks can be indexed to (111), (200) and (220) planes of the cubic unit cell, which corresponds to the cubic structure of silver comparable to International Centre for Diffraction Data (ICDD), silver file No. 04–0783.

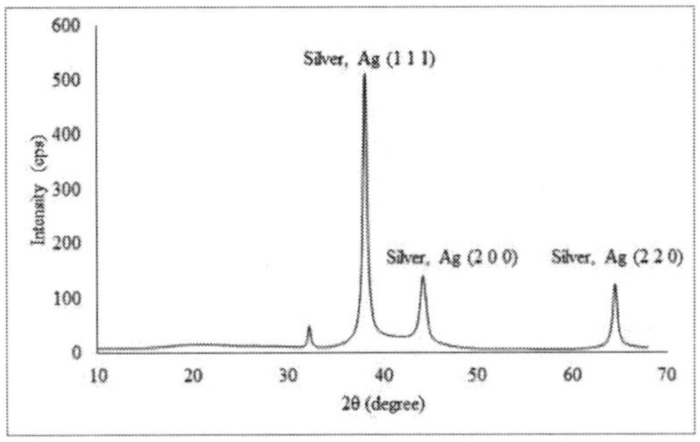

Fig. 4. XRD patterns of AgNPs

Debye-Scherrer equation was used to determine average crystallite diameter from half width of the diffraction peaks: Eq.4,

$$D = K\lambda/(\beta \cos \theta) \qquad (4)$$

where D is the size of the particles (Å), K is the shape dependent Scherrer's constant (1.05), β (rad) is the full peak width, λ is the wavelength of radiation, (θ rad) is the diffraction angle. The average crystallite sizes of synthesized AgNPs are listed in Table 2.

Table 2. Crystal sizes of AgNPs using Scherrer formula

Peak position	Average size (nm)
38.094 ° (1 1 1)	21
44.230 ° (2 0 0)	13
64.500 ° (2 2 0)	19

The crystal sizes of AgNPs as presented in Table 2 are very close or even smaller than the average value (28 nm) obtained by TEM analysis. XRD size is usually equal or smaller than that obtained by TEM and similar result also observed by Akbari et al., when characterized the nanoparticles size [18]. The (1 1 1) plane was chosen as the average crystalline size of AgNPs since it showed an intense peak from XRD analysis. Therefore, the size of the AgNPs estimated from Debye-Scherrer formula is 21 nm. The contradiction regarding the difference in the diameter between the value determined from the XRD patterns (21 nm) and the one from TEM observations (28 nm) is unavoidable, as it was correlated with the limitations of the use of Debye-Seherrer formula, which is only applicable to near-spherical shape particles [19]. However, the difference in size between TEM and XRD analysis is too small. On top of that, it seems that the size of AgNPs crystal (in the range of 10-20 nm) as tabulated in Table 2 are represent the majority size of AgNPs analysed by TEM (10 nm and 20 nm with the highest frequency; see Figure 3).

3.4 Membrane Characterization

3.4.1 Antibacterial Assays of PA/Ag Membrane

Disc diffusion method was used for susceptibility testing of *E. coli* bacteria. Literally, the antibacterial capability of produced PA/Ag NF membranes could be seen from the inhibition zone area.The results shown in Figure 5 are the antibacterial test on *E. coli* from this study and the antibacterial test results from Basri et al. [2]. Membrane (c) and (d) were produced through phase inversion technique with 0.54 wt% and 0.9 wt% of silver nanoparticles as deduced by EDX, respectively. Meanwhile, membrane (b) represent PA/Ag-60 was produced by IP with 0.7wt% of silver nanoparticles.

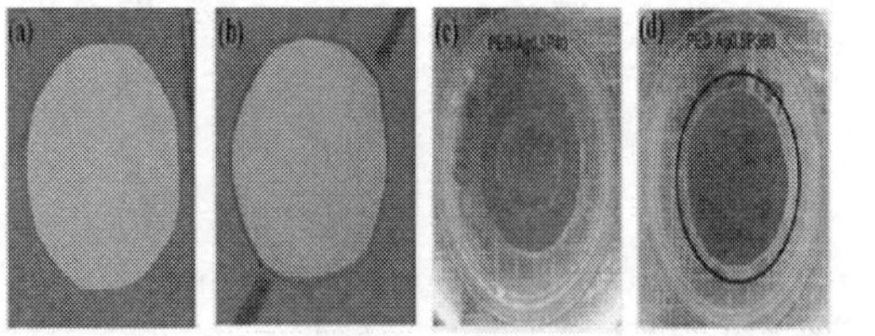

Fig. 5. Result of the Antibacterial Assays; (a) control PES (without AgNPs) [14] (b) PA/Ag-60 with 0.7wt% AgNPs by EDX [14](c) PES-Ag0.5P40 with 0.54wt% by EDX [2] (d) PES-Ag0.5P360 with 0.9wt% by EDX [2].

Based on Figure 5, membrane (b) showed clearer inhibition zone compared to membrane (c) and (d) even the amounts of silver nanoparticles were approximately similar. Therefore, it makes IP as a promising technique to incorporate silver nanoparticles on the membrane surface.

3.4.2 Characterization using Uncharged Solutes (r_p and $\Delta x/A_k$ parameter)

Data of pure water permeability, vitamin B12 rejection, pore size (r_p) and $\Delta x/A_k$ of six different types of membranes are listed in Table 3. The water permeation results show that the pure water permeability (PWP) for all membranes are in the range of reverse osmosis membrane [20]. However, pure water permeability alone is not enough to classify the type of membrane. Therefore, vitamin B12 rejection data was used to calculate the membrane pore size. The obtained pore size (r_p) revealed that the values were in the range of 29 commercial available NF membranes studied by Bowen and Mohammad [15]. The lower pure water permeability may due to the fact that the thin layer form on the top surface is denser/compact and thicker than the commercial membrane. It can be clearly seen from the higher values of $\Delta x/A_k$ of the prepared membranes compared to commercial one (maximum $\Delta x/A_k$ =16.9 μm [15]). However, the values of $\Delta x/A_k$ obtained in the current study are almost similar to the self- fabricated NF membranes as reported by Lau and Ismail [21]. Based on these two structural parameters, the prepared membranes in this study could be considered as tight NF membranes. In general, there is no specific correlation between reaction time and membrane structural parameters and water permeability. In addition, the presence of AgNPs also not significantly changes membrane performance.

Table 3. Permeability, Vitamin B12 rejection and Structural Parameters For Different Types of Membrane

Types of Membrane	Permeability (L/m^2.h.bar)	$^+$Vitamin B12 Rejection, R_{real}, %	$^*r_p$ nm	$^{**}\Delta x/A_k$ μm
PA-10	0.435	75.77	1.3	198
PA-30	0.619	65.00	1.4	176
PA-60	0.434	70.02	1.3	229
PA/Ag-10	0.294	85.54	1.1	223
PA/Ag-30	0.429	73.49	1.3	212
PA/Ag-60	0.475	56.53	1.6	292

$^+$concentration of 100 mg/L tested at 4 bar
*calculated based on Eq (1)
**calculated based on Eq (3)

3.4.3 Salt Rejection

For salt rejection determination, the experiments were carried out with 0.001M NaCl and 0.001M Na_2SO_4 solutions at 400 kPa. The percentage rejection of both NaCl and Na_2SO_4 solutions for all types of the membrane were summarized in Figure 6.

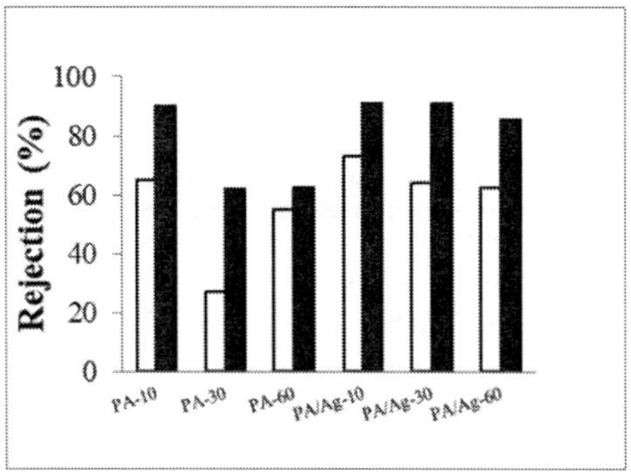

Fig. 6. Rejection of 0.001M NaCl (□) and 0.001M Na_2SO_4 (■) salt solutions at 400 kPa

PA/Ag membranes exhibited better salts removal than PA membrane. It can be clearly seen from Figure 6, higher rejection for both salts were obtained for all PA/Ag membranes. In other words, the presence of AgNPs gives a positive impact on salt removal. Among all prepared membranes, PA/Ag-10 shows the highest rejection value of 73% and 91% for monovalent (NaCl) and divalent salt (Na_2SO_4), respectively. In general, for all available membranes, higher rejection was observed for divalent ions than monovalent ions, and this represents the characteristic of NF membrane [22,23]. This result strengthens our previous claim that the membranes prepared in this study are in the range of NF membranes.

4 Conclusions

AgNPs have been successfully synthesized by a green approach using tea leaves extract. The method used in this study is an efficient, simple and clearly a green method without using any harmful reducing and capping agents. The spectroscopic characterization from UV-visible, XRD and TEM supports the stability of the biosynthesized nanoparticles. PA and PA/Ag membranes were successfully prepared by interfacial polymerization of aqueous MPD and TMC organic solution. The membranes were characterized in term of water permeability, and structural parameters (pore size, r_p and $\Delta x/A_k$) and results show that the membranes prepared were in the range of NF membranes. From salt rejection test, membranes with the addition of AgNPs exhibited higher salt rejection when compared to membranes without AgNPs. Among all six membranes, PA/Ag-10s membrane showed the highest NaCl and Na_2SO_4 salt rejection which is 73% and 91%, respectively.

The authors wish to express thanks to Ministry of Education Malaysia through the financial aid from ERGS grant (RDU 130602) and MyBrain15. Appreciation is also expressed to all colleagues and technical staffs of FKKSA Lab, Universiti Malaysia Pahang for providing necessary facilities and guidance throughout this study.

References

1. H.S. Lee, S.J. Im, J.H. Kim, H.J. Kim, J.P. Kim, B.R. Min, Desal. **219**, 48 (2008)
2. H. Basri, A.F. Ismail, M. Aziz, Desal. **273**, 72 (2011)
3. D.Y. Koseoglu-Imer, B. Kose, M. Altinbas, I. Koyuncu, J. Memb. Sci.**428,** 620 (2013)
4. J. Huang, G. Arthanareeswaran, K. Zhang, Desal. **285,** 100 (2012)
5. A. Mollahosseini, A. Rahimpour, M. Jahamshahi, M. Peyravi, M. Khavarpour, Desal. **306,** 41 (2012)
6. S.Y. Lee, H.J. Kim, R. Patel, S.J. Im, J.H.Kim, B.R. Min, Polym. Adv. Technol. **18**, 562 (2007)
7. P. Kshirsagar, S.S. Sangaru, M.A. Malvindi, L. Martiradonna, R. Cingolani, P.P. Pompa, Colloids Surfaces A Physicochem. Eng. Asp. **392,** 264 (2011)
8. P. Mohanpuria, N.K. Rana, S.K. Yadav, J. Nanoparticle Res. **10**, 507 (2008)
9. J.-H. Li, X.-S. Shao, Q. Zhou, M.-Z. Li, Q.-Q. Zhang, Appl. Surf. Sci. **265,** 663 (2013)
10. L. Mulfinger, S.D. Solomon, M. Bahadory, A.V. Jeyarajasingam, S.A. Rutkowsky, C. Boritz, J. Chem. Educ. **84,** 322 (2007)
11. Q.H. Tran, V.Q. Nguyen, A.-T. Le, Adv. Nat. Sci. Nanosci. Nanotechnol. **4,** 033001 (2013)
12. I. Kubo, M. Himejima, S. Chemical, J. Agric. Food Chem. **40,** 245 (1992)
13. Y.Y. Loo, B.W. Chieng, M. Nishibuchi, S. Radu, Int. J. Nanomedicine. **7,** 4263 (2012)
14. T.N.A.T. Sallehuddin, M.N. Abu Seman, Journal of Membr. Sci. and Res. **3**, 29 (2017)
15. W.R. Bowen, A.W. Mohammad, Trans IChemE. **76,** 885 (1998)
16. W.R. Bowen, A.W. Mohammad, N. Hilal, J. Memb. Sci. **126**, 91 (1997)
17. N. Ahmad, S. Sharma, V.N. Singh, S.F. Shamsi, A. Fatma, B.R. Mehta, Biotechnol. Res. Int. **2011,** 1 (2011)
18. B. Akbari, M.P. Tavandashti, M. Zandrahimi, Iran. J. Mater. Sci. Eng. **8,** 48 (2011)
19. Z.H. Dhoondia, H. Chakraborty, Nanomater. Nanotechnol. **2,** 15 (2012)
20. B.-H. Jeong, E.M.V. Hoek, Y. Yan, A. Subramani, X. Huang, G. Hurwitz, J. Memb. Sci. **294,** 1 (2007)
21. W.J. Lau, A.F. Ismail, Journal of Memb. Sci. **334**, 30 (2009)
22. J. Schaep, B.V. der Bruggen, S. Uytterhoeven, R. Croux, C. Vandecasteele, D. Wilms, E.V. Houtte, F. Vanlerberghe, Desal, **119**, 295 (1998)
23. N. Hilal, H. Al-Zoubi, N.A. Darwish, A.W. Mohammad, Abu Arabi M., Desal. **170,** 281 (2004)

Effects of Solvents on Polymorphism and Shape of Mefenamic Acid Crystals

Siti Kholijah Abdul Mudalip[1*], *Mohd Rushdi* Abu Bakar[2], *Parveen* Jamal[3], *Fatmawati* Adam[1], *Rohaida* Che Man[1], *Siti Zubaidah* Sulaiman[1], *Zatul Iffah* Mohd Arshad[1], *Shalyda* Md. Shaarani[1]

[1]Faculty of Chemical & Natural Resources Engineering, University Malaysia Pahang, Lebuh Raya Tun Razak, 26300 Gambang, Pahang, Malaysia.
[2]Department of Pharmaceutical Technology, Kulliyyah of Pharmacy, International Islamic University Malaysia, Bandar Indera Mahkota, 25200 Kuantan, Pahang, Malaysia.
[3]Department of Biotechnology Engineering, Kulliyyah of Engineering, International Islamic University Malaysia, 50728 Kuala Lumpur, Malaysia.

Abstract. Mefenamic acid [2-(2, 3-dimethylphenyl) amino benzoic acid] is an active pharmaceutical compound that exist in different polymorphic form and shape. In this work the effect of solvents on polymorphism and shape of mefenamic acid crystals were examined. The solvents used were ethanol, isopropanol, ethyl acetate, dimethyl acetamide, dimethyl formamide, and acetone. Natural cooling was employed during the crystallisation process. The crystals produced were dried and analysed using optical microscopy, differential scanning calorimetry, thermal gravimetric analysis, x-ray powder diffraction (XRPD) and fourier transform infrared spectroscopy (FTIR). The analysis confirmed that the crystals obtained using ethyl acetate, ethanol, isopropanol, and acetone are pure Form I with a needle-like flat shape. Meanwhile, the crystallisation using DMF produced polymorphic Form II in cubic shape.

1 Introduction

Solvent demonstrate an important role in the crystallisation process as it can contribute to the polymorphic selectivity of targeted pharmaceutical compound [1, 2]. This is because, at the molecular level, the solvent can selectively adsorb onto the crystals faces or interact with the solute clusters. The solvent-solute interactions, i.e. hydrogen bonding have been proven to affect the birth of different polymorphic forms [3]. A solvent molecule that has a stronger ability to donate or accept hydrogen bonding than the solute molecule may establish hydrogen bonding network with the solute molecule, resulting in selective nucleation [4]. For example, crystallisation of sulfathiazole using water produces polymorphic Form II and III. The polymorphic Forms I and IV are obtained from acetone, while n-propanol gives only Form I [5].

The types of solvent used during the crystallisation process also influence the shape of the crystals. Garti and co-worker reported that the crystallisation of cholesterol using ethanol and acetonitrile produced needles and plate-like crystal, respectively [6]. Ibuprofen crystals that crystallized from polar solvents, such as ethanol and acetone and non-polar solvents, such as diethyl ether and hexane, produced needle-like crystals. In contrast, crystallisation of ibuprofen using dichloromethane and acetonitrile produced cubic and spherical agglomerates, respectively [7]. The external structure has profound effect on orientation of particle, and thus may cause differences in packing, powder flow, compaction, syringability, stability of suspension, and dissolution characteristics [2].

Mefenamic acid (2-[(2,3-dimethylphenyl) amino] benzoic acid, $C_{15}H_{15}NO_2$) is a widely used nonsteroidal anti-inflamatory and analgesic agent for treatment of pain cause by menstrual disorders was reported to exist in three polymorphic forms: Form I, Form II and Form III [8-10]. The polymorphs are enantiotropically related and show different physicochemical properties which in turn affect the process acceptability, bioavailability and performance of the final product. The habit of mefenamic acid crystals can be either plate or needle shaped depending on the crystallisation conditions. Numerous research works on mefenamic acid crystals have resulted in the availability of different methods using various solvents for producing a particular polymorph [8, 9]. However, most of the literatures are difficult to follow as they do not state the initial solute concentration.

This paper presents the work on investigation of various types of solvents, which classified into polar

* Corresponding author: kholijah@ump.edu.my

protic (ethanol and isopropanol) and dipolar aprotic (ethyl acetate, dimethyl acetamide, dimethyl formamide, and acetone) on polymorphism and habit of mefenamic acid crystals. Apolar aprotic solvents such as hexane, cyclohexane and heptane were not considered as crystallisation solvent due to its poor solubility characteristics [11]. In order to understand and differentiate the characteristics of crystals produced, comprehensive characterizations were performed using various analytical techniques, i.e. thermal, microscopy, infrared spectroscopy and x-ray diffraction. The experimental results were compared with those of previous workers whenever possible.

2 Methodology

2.1. Material

Mefenamic acid powder with purity of 98 wt% was obtained from Baoji Tianxin Pharmaceutical Co. Ltd., China. The solvents used namely ethanol, ethyl acetate (EA), isopropanol (IPA), dimethyl acetamide (DMA), dimethyl formamide (DMF) and acetone were analytical grade with >99 wt % purity and supplied by Fisher Scientific.

2.2 Crystallisation method

Saturated solution of mefenamic acid was prepared by dissolving known amount of mefenamic acid in 50 mL of solvents studied. The amount of mefenamic acid used as tabulated in Table 1 is based on solubility data reported in our previous work [11]. The solution prepared was heated on a hot plate at 60°C except for DMF and acetone, which was at 50°C until dissolution. Then, the solution was allow to cool to room temperature through natural cooling. The crystals produced were filtered and dried in an oven at 60°C. The crystals were periodically dried and weighed until constant weight was achieved. The dried crystals were stored in a screw cap glass vials for further characterization works.

Table 1. Amount of mefenamic acid used in different solvents.

Method	Solvent	Amount of Mefenamic Acid (g)
1	Ethyl actetate	1.59
2	Ethanol	1.14
3	Isopropanol	1.23
4	Dimethyl formamide	8.60
5	Dimethyl acetamide	33.87
6	Acetone	2.70

2.3 Characterisation methods

The images of the crystals were captured using Leica microscope DM750 with a total magnification of 200x4x/0.10 and processed using Leica Application Suite Software version 3.6. The FTIR spectra of crystals from 500 to 4000 cm^{-1} were obtained using Perkin Elmer's ATR-FTIR Spectrometer (Frontier). The spectra were analyzed using OMNIC software with an average scan of 16 [12].

The melting properties analysis was performed using a Mettler Toledo DSC-1 from temperature of 25 to 300°C at a heating rate of 10 °C/min under constant purging of nitrogen [9]. About 2-10 mg of crystals was weighed into a 40 μL standards aluminum pan with lid and sealed hermetically prior to analysis. An empty aluminum pan was used as a reference in all the runs. The results were analyzed using Mettler Toledo Stare SW 9.10. The TA Instruments (Q500/50) was also used to study the changes of weight derivatives with temperature. Sample with a weight range of 4-6 mg was placed on a platinum pan and heated from 25 to 300°C at a constant heating rate of 10 °C / min.

The x-ray powder diffraction patterns were recorded using Shimadzu XRD 6000 equipped vertical x-ray goniometer and Cu Kα radiation with angle reproducibility of ±0.001° (2θ). Prior to analysis, the crystals were initially placed into an aluminum holder and gently grinded, pressed and flatten softly using spatula and glass plate. The measurement conditions used were: 40 kV of voltage; 30 mA of current; 5-50° (2θ) of scan range; 0.05° of step size and 3°/min of scan mode in a continuous mode [9].

3 Results and Discussion

3.3.1 Shape of crystals

Shape or habit of crystals of may influence the bulk density, flowability and mechanical strength of the pharmaceutical compound [13]. Fig. 1 shows the shapes of crystals produced using different solvents. As seen in Fig. 1 (a), (b), (c) and (f), mefenamic acid which crystallised using EA, ethanol, IPA, and acetone, respectively are elongated needle-like flat shaped crystals and consistent with those reported in the literature [8, 14, 15]. Fig. 1 (d) shows that the crystals crystallised from DMF are cubic-shaped. Meanwhile, the crystallisation using DMA produced a plate-like-shaped crystals (Fig. 1 (e)). A comparison, however, cannot be made for crystals obtained using DMA as it is yet to be reported in the literature.

The difference in crystal shapes produced in this work are probably due to the variation of solvents' polarity which results in different solute-solvent interactions at the crystal's facets. The crystallographic structure of mefenamic acid obtained from Lee and co-workers shows the presence of –OH, -C=O and $–C_6H_5$ which were exposed on the crystal facet (0 0 1) [16]. The –OH and -C=O have significant hydrogen bonding ability and thus, may permit strong hydrogen bonding interactions between facet (0 0 1) and solvents with high polarity. Based on the polarity index, DMF is expected to form stronger interactions with the facet (0 0 1)

https://doi.org/10.1051/matecconf/201815002004

followed by DMA, ethanol, propanone IPA, and EA. The strong hydrogen bonding interactions between the facet and solvent will consequently reduce the interfacial tensions and enhance the growth of the crystal facet. The relative contribution of the facet, however, decreases with the increase of solvent's polarity [15]. This phenomenon explained the reduction of crystal length produced using DMF and DMA solvents as shown in Fig. 1 (d) and (e), respectively.

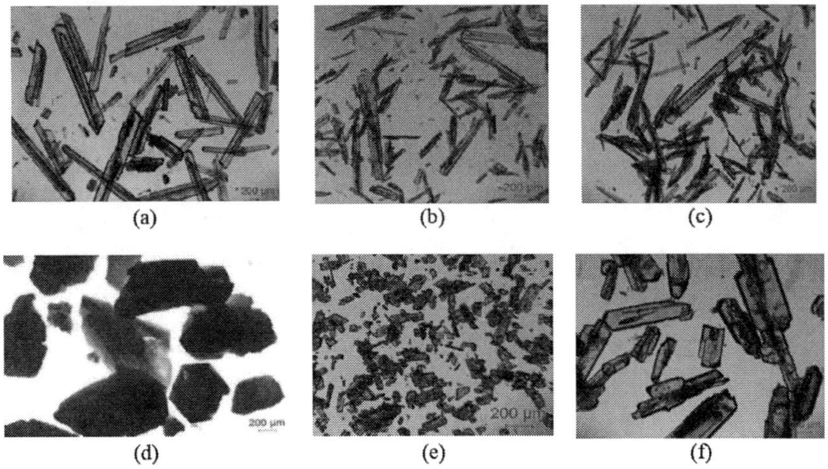

Fig. 1. Optical microscopy images of crystals obtained from crystallisation using different solvents: (a) ethyl acetate; (b) ethanol; (c) IPA; (d) DMF; (e) DMA; and (f) acetone.

3.3.2 Infrared spectroscopy

Fig. 2 shows partial infrared spectra for mefenamic acid crystals as received from the supplier and after crystallisation using different solvents. As shown by the dashed vertical line in Fig. 2, the crystals obtained after crystallised using EA, ethanol, IPA, DMA, and acetone have N-H stretching band at approximately 3313 cm^{-1}, which corresponds to Form I of mefenamic acid. The crystals obtained using DMF as solvents show N-H stretching at 3347 cm^{-1}, which correspond to Form II. The higher wavelength observed for the N-H stretching of mefenamic acid Form II suggests a weaker hydrogen interaction between carboxylic groups in comparison with mefenamic acid Form I [14]. Previous works also highlighted that the N-H stretching band that occurs between 3300 and 3350 cm^{-1} is an important spectrum that can be used to distinguish either Form I or Form II of mefenamic acid [17, 21].

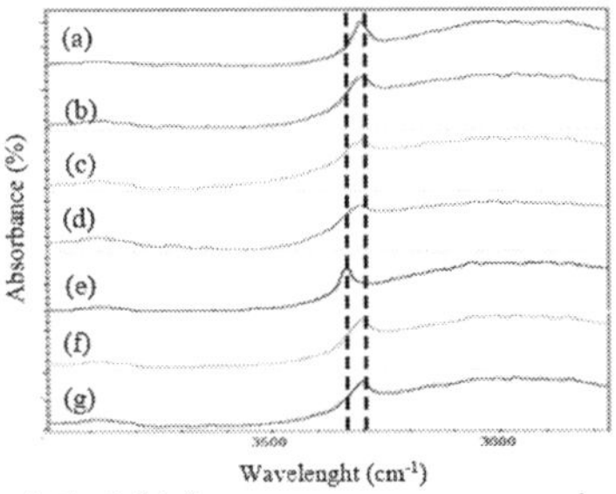

Fig. 2. Partial infrared spectra of of mefenamic acid samples: a) as received, and after crystallised using (b) ethyl acetate, (c) ethanol, (d) IPA, (e) DMF, (f) DMA, and (g) acetone.

3.3.3 Thermal behaviour

The differential scanning calorimetry (DSC) and thermal gravimetric (TG) curves of mefenamic acid after crystallisation using different solvents are shown in Fig. 3 (a) until (f). The DSC curve in Fig. 3 (a), (b), (c), (e), and (f) have two endothermic peaks. The onset temperature of the first endothermic peak of mefenamic acid shown in those figures varied from 168.82 to 190.16°C. Meanwhile, for the second endothermic peak, the values are from 226.75 to 232.08°C. These values concur with those values of mefenamic acid Form I reported in the literature [9, 12, 14, 17]. The temperature of the first and second endothermic peaks represent the transition temperature, T_T of Form I to Form II and the melting or fusion, T_f of Form II, respectively. The presence of an endothermic peak before the melting point shows that the mefenamic acid polymorphs are enantiotropic, where it can reversibly transform into one another by cooling or heating process [14].

The DSC curve of mefenamic acid crystallised using DMF as shown in Fig. 3 (d) on the other hand, only has one endothermic peak with molar enthalpy of fusion of 38.31±5.51 $kJmol^{-1}$. This value is comparable with those reported by literature, which is 38.243, 38.7 and 40.39 $kJmol^{-1}$ [12, 18]. The endothermic peak is associated with the molecule decomposition [14]. The DSC curve with one endothermic peak is consistent with mefenamic acid Form II characteristic [14]. The derivative TG curves shown in Fig. 3 (a) until (f) are presented to provide correlation between the weight loss and the calorimetric profile. The curves show a significant derivative weight change between 165 and 245°C, which is approximately at the second endothermic peak of the DSC curve. The change in the derivative weight is believed due to the decomposition of mefenamic acid crystals during the melting process. In addition, no significant derivative weight was detected in the region of 80 to 100°C, whereas in many cases, it may cause the decomposition of the substrate material or volatilisation of residual solvents [19].

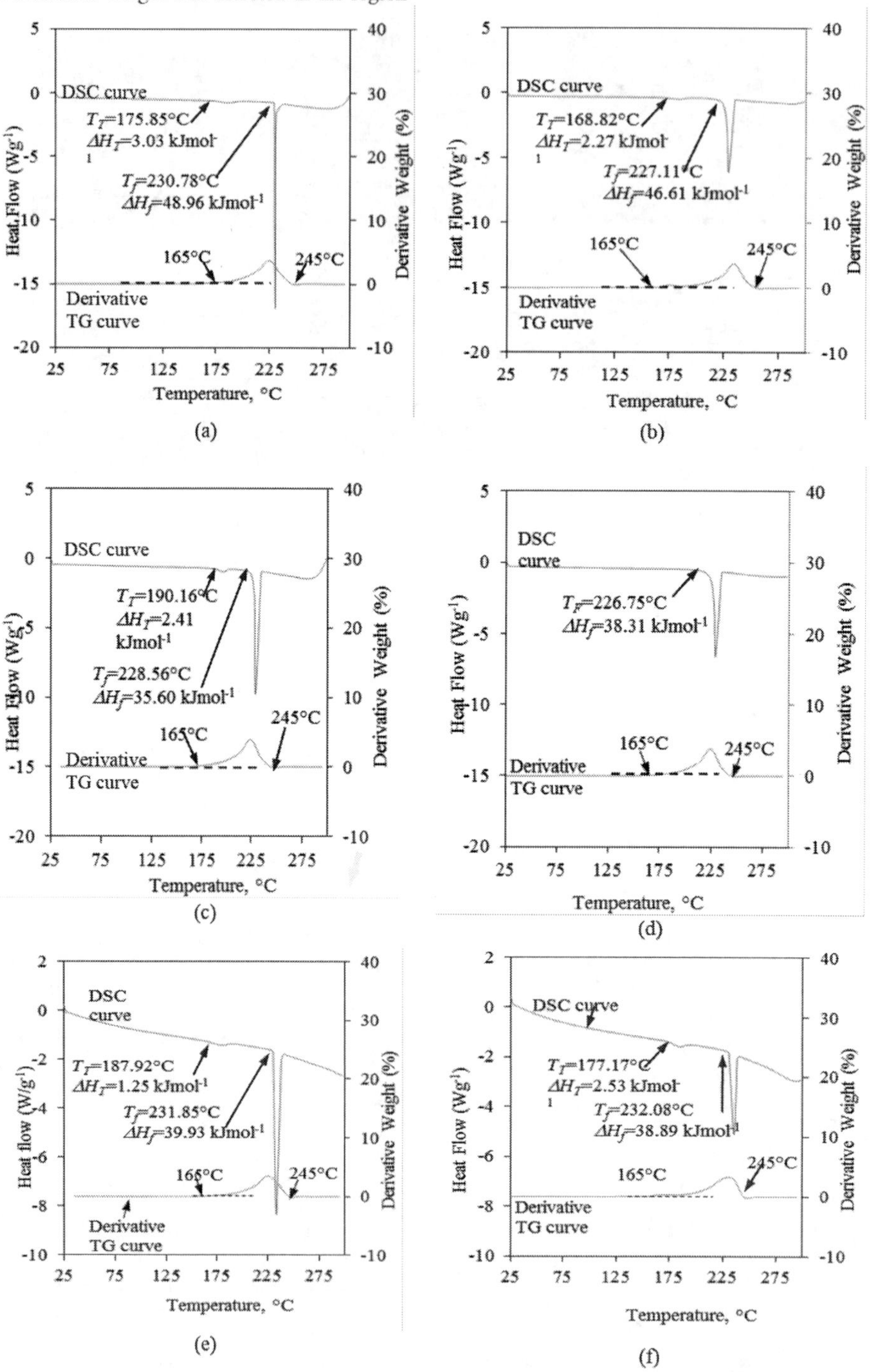

Fig. 3. DSC and derivative TGA curves of mefenamic acid crystals obtained by (a) ethyl acetate; (b) ethanol; (c) IPA; (d) DMF; (e) DMA; and (f) acetone.

3.3.4 Phase analysis

The XRPD pattern of mefenamic acid crystals obtained from EA, ethanol, IPA, DMA and acetone in comparison with some major reflections in the reference pattern of Form I reported literature [8, 20] is shown in Fig. 4 (a), (b), (c), (d) and (e), respectively. The pattern concurs with the literature. In addition, any significant peaks at 11.8°, 17.9°, 23.8° and 25.6°, as well as 9-12° (2θ) which represent Form II of mefenamic acid were not observed. This confirms that pure Form I crystals were successfully produced using ethyl acetate, ethanol, IPA, DMA and acetone.

The comparison of XRPD pattern of the crystals obtained from DMF with some major reflection of Form II reported by Kato et al. [20] and Cesur and Gokbel [8] are illustrated in Fig. 4 (f). The crystallisation of mefenamic acid using N, N-dimethyl formamide as a solvent produced an XRPD pattern that concurred very well with XRPD pattern of Form II reported by those literature [8, 20]. This finding demonstrates that crystallisation from DMF able to produced mefenamic acid crystals Form II.

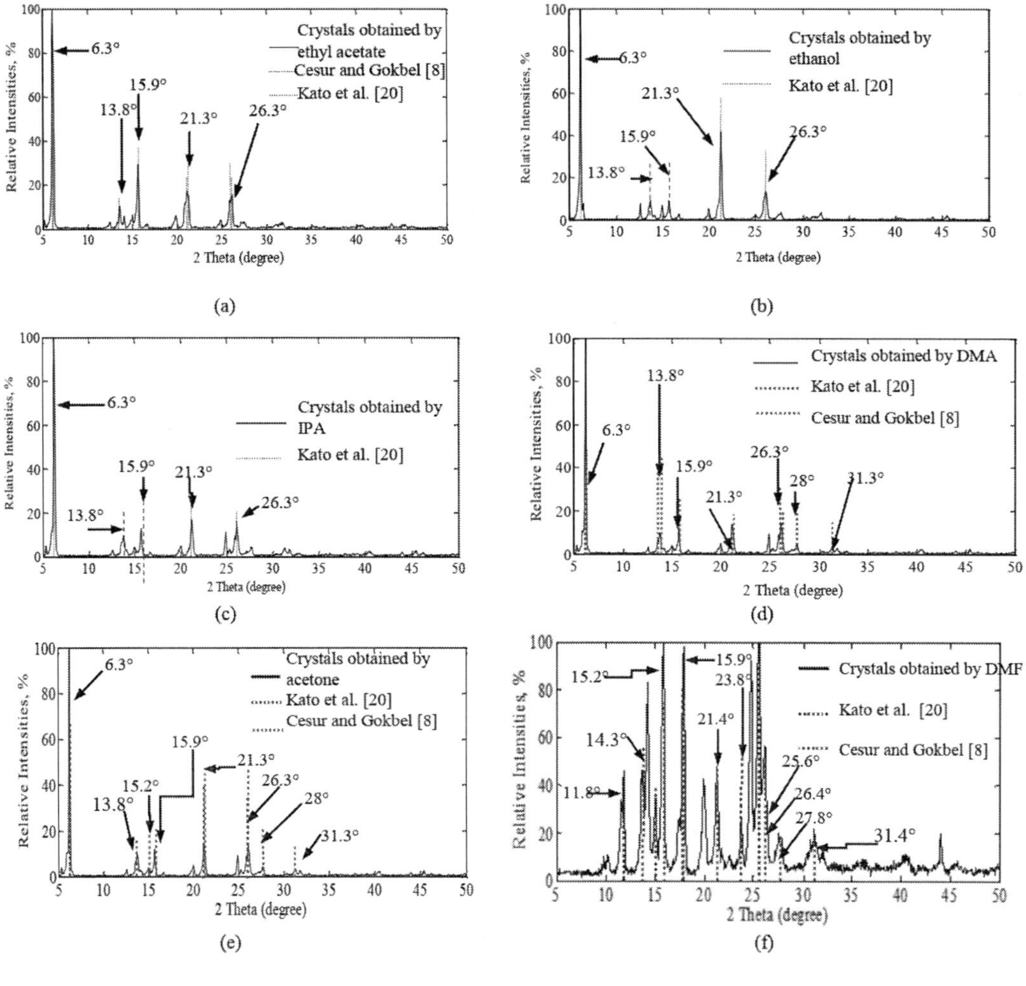

Fig. 4. XRPD pattern of crystals obtained from (a) ethyl acetate; (b) ethanol; (c) IPA, (d) DMA, (e) acetone and (f) DMF in comparison with some major reflections with the reference pattern of Form I or Form II obtained from literature.

Conclusion

The crystallisation of mefenamic acid using different solvents with natural cooling mode produced crystals with different polymorphic forms and habits. The results of the thermal analysis (DSC), FTIR spectroscopy and XRPD have shown that the crystals obtained from EA, ethanol, IPA, DMA and acetone are pure Form I, while DMF is pure Form II. No solvent entrapment in the mefenamic acid crystals were observed during thermal studies and the crystals were completely decompose during melting. The crystals produced were needle-like and cubic shaped crystals, depending on the solvents used.

* Corresponding author: kholijah@ump.edu.my

Special thanks are dedicated to University Malaysia Pahang for financial support under Internal Grant Research Scheme RDU150359.

References

1. C. Stoica, P. Verwer, H. Meekes, V.H.P.J.C. M., F.M. Kaspersen, E. Vlieg, *Understanding the effect of a solvent on the crystal habit*, Cryst.Growth & Des., **4**, 765-768, (2004).
2. I. Benmessaoud, O. Koutchoukali, M. Bouhelassa, A. Nouar, S. Veesler, *Solvent screening and crystal habit of metformin hydrochloride*, J. Cryst. Growth, 451, 42-51, (2016).
3. I. Weissbuch, M. Lahav, L. Leiserowitz, *Toward stereochemical control, monitoring, and understanding of crystal nucleation*, Cryst. Growth & Des., **3**, 125-150, (2002).
4. R. Davey, J. Garside, From molecules to crystallizers: *An introduction to crystallisation*, Oxford University Press Inc., United States, (2002).
5. M.R. Abu Bakar, Z.K. Nagy, S.E. Dann, C.D. Rielly, *Investigation of the riddle of sulfathiazole polymorphism*, Int. J. Pharm., **414**, 86-103, (2011).
6. N. Garti, L. Karpuj, S. Sarig, *The effect of solvents and crystallisation conditions on crystal habit of cholesterol*, Cryst. Res. Technol., **16**, 1111-1115, (1981).
7. N. Rasenack, B.W. Muller, *Properties of ibuprofen crystallized under various conditions: a comparative study*, Drug Dev. Ind. Pharm., **28**, 1077-1089, (2002).
8. S. Cesur, S. Gokbel, *Crystallisation of mefenamic acid and polymorphs*, Cryst. Res. Technol., **43** 720-728, (2008).
9. R. Panchagnula, R. Sundaramurthy, O. Pillai, S. Agrawal, *Solid-state characterization of mefenamic acid*, J. Pharm. Sci., **93**, 1019-1029, (2004).
10. S. SeethaLekshmi, T.N. Guru Row, *Conformational polymorphism in a non-steroidal anti-inflamatory drug mefenamic acid*, Cryst. Growth & Des., **12**, 4283-4289, (2012).
11. S.K. Abdul Mudalip, M.R. Abu Bakar, P. Jamal, F. Adam, *Solubility and dissolution thermodynamic data of mefenamic acid crystals in different classes of organic solvents*, J. Chem. Eng. Data, **58**, 3447-3453, (2013).
12. S. Romero, P. Bustamante, B. Escalera, M. Cirri, P. Mura, *Characterization of the solid phases of paracetamol and fenamates at equilibrium in saturated solutions*, J. Therm. Analy. and Calorim., **77**, 541-554, (2004).
13. R. Ho, M. Naderi, J.Y. Heng, D.R. Williams, F. Thielmann, P. Bouza, D.J. Burnett, *Effect of milling on particle shape and surface energy heterogeneity of needle-shaped crystals*, Pharm. Res., **29**, 2806-2816, (2012).
14. V.R. Cunha, C.M. Izumi, P.A. Petersen, A. Magalhaes, M.L. Temperini, H.M. Petrilli, V.R. Constantino, *Mefenamic Acid Anti-Inflammatory Drug: Probing Its Polymorphs by Vibrational (IR and Raman) and Solid-State NMR Spectroscopies*, The Journal of Physical Chemistry B, **118**, 4333-4344, (2014).
15. U.V. Shah, D. Olusanmi, A.S. Narang, M.A. Hussain, J.F. Gamble, M.J. Tobyn, J.Y. Heng, *Effect of crystal habits on the surface energy and cohesion of crystalline powders*, Int. J. Pharm., **472**, 140-147, (2014).
16. C.-W. Lee, S.-J. Kim, Y.-S. Youn, E. Wiidjojokusumo, Y.-H. Lee, J. Kim, Y.-W. Lee, R.R. Tjandrawinata, *Preparation of bitter taste masked cetirizinedihydrochloride/β-cyclodextrin inclusion complex by supercritical antisolvent (SAS) process*, J. Supercrit. Fluids, **55**, 348-357, (2010).
17. S. Romero, B. Escalera, P. Bustamante, *Solubility behavior of polymorphs I and II of mefenamic acid in solvent mixtures*, Int. J. Pharm. **178**, 193-202, (1999).
18. J.T. Su, P.B. Duncan, A. Momaya, A. Jutila, D. Needham, *The effect of hydrogen bonding on the diffusion of water in n-alkanes and n-alcohols measured with a novel single microdroplet method*, J. Chem. Physics, 132, 044506, (2010).
19. D.Q.M. Craig, G.A. K., *Thermal Analysis of Pharmaceuticals*, CRC Press, Taylor & Francis Group, Boca Raton, 193-220, (2007).
20. F. Kato, M. Otsuka, Y. Matsuda, *Kinetic study of the transformation of mefenamic acid polymorphs in various solvents and under high humidity conditions*, Int. J. Pharm., **321**, 18-26, (2006).
21. R.K. Gilpin, W. Zhou, *Infrared studies of the thermal conversion of mefenamic acid between polymorphic states*, Vib. Spectrosc., **37**, 53-59, (2005).

Surface Modification of Nanoclay for the Synthesis of Polycaprolactone (PCL) – Clay Nanocomposite

Kamal Yusoh[1,*], *Shamini Vesaya* Kumaran[1], and *Fadwa Sameeha* Ismail[1]

[1]Faculty of Chemical and Natural Resources Engineering, Universiti Malaysia Pahang, 26300 Gambang, Pahang Malaysia

Abstract. This paper presents a new modification method to modify the surface of nanoclay (Na-MMT) to increase its *d*-spacing using Aminopropylisooctyl Polyhedral Oligomeric Silsesquioxane (AP-POSS) and the fabrication of Polycaprolactone (PCL) nanocomposite through solution intercalation technique. The structure and morphology of pure nanoclay, modified nanoclay (POSS-MMT) and the PCL nanocomposite were characterized by X-ray Diffraction (XRD), Fourier Transform Infrared Spectroscopy (FTIR) and Field Emission Scanning Electron Microscopy (FESEM). XRD revealed that the *d*-spacing of the POSS-MMT is increased by 0.64 nm as compared to pure nanoclay. FTIR and FESEM results also showed that AP-POSS were well dispersed and intercalated throughout the interlayer space of Na-MMT. An exfoliated structure was also observed for PCL/POSS-MMT nanocomposite. Thermal properties of the nanocomposite were investigated using Thermal Gravimetry Analysis (TGA) which recorded highest degradation temperature for PCL/POSS-MMT 1% nanocomposite which is 394.1°C at 50% weight loss ($T_{50\%}$) but a decrease in degradation temperature when POSS-MMT content is increased and Differential Scanning Calorimetry (DSC) analysis which showed highest melting and crystallization temperature for PCL/POSS-MMT 5% nanocomposite which is 56.6°C and 32.7°C respectively whereas a decrease in degree of crystallinity for PCL/POSS-MMT nanocomposite as compared to PCL/Na-MMT nanocomposite. This study affords an efficient modification method to obtain organoclay with larger interlayer *d*-spacing to enhance the properties of polymer nanocomposite.

1 Introduction

Plastics are ideal for many applications such as in packaging, building materials and commodities but it can lead to waste disposal problems. Accumulation of plastic at the end of its life cycle had increased drastically on the earth causing serious pollution problems. Therefore, this motivated many researchers to conduct studies to produce a biodegradable and environmental friendly polymer nanocomposite to produce plastics. There has been a strong emphasis in the development of polymer nanocomposites for the last 20 years. Polymer nanocomposites consist of inorganic nanofiller and organic polymers represent a new class of materials that exhibit improved performance compared to their microcomposite counterparts [1].

Polycaprolactone (PCL) is considered to be a good biodegradable polymer because of the low production cost and easy processibility in large scale production. PCL have good commercial potential for plastics but their low thermal and mechanical properties for further processing restrict their use in a wide range of applications [2]. Therefore, these properties can be improved through the preparation of PCL nanocomposite by incorporating nanofiller such as nanoclay.

Sodium montmorillonite (Na-MMT) is the most abundant nanoclay mineral that is easily available with a good swelling capacity, high cation exchange capacity and high surface area [3]. However, the hydrophilic nature of nanoclay hinders homogenous dispersion of nanoclay in the hydrophobic polymer matrix [4]. Modifying the nanoclay surface through organic treatment will give a hydrophobic environment to the galleries of the Na-MMT and enhance its compatibility with the polymer matrix. Conventional modification method is the ion-exchange reaction using cationic salts such as alkyl ammonium or phosphonium salts. There are some drawbacks in this modification process such as low thermal stability of the cationic salts and the compounds are less readily intercalated in the polymer melts. Alkyl ammonium salt will decompose at temperature above 170-180°C which shows that it is not suitable for the high temperature melt processing techniques [5].

Thus, a new modification method will be the main aim of this study. The surfactant method by using Aminopropylisooctyl Polyhedral Oligomeric Silsesquioxane (AP-POSS) is developed to give a better surface modification for the Na-MMT. The main

* Corresponding author: kamal@ump.edu.my

purpose of the surface modification of Na-MMT is to expand the interlayer *d*-spacing of the nanoclay gallery so that polymer molecules can penetrate as well as to enhance its miscibility with polymer to achieve a good dispersion. AP-POSS has a high thermal stability which is up to 300°C. Therefore, an enhanced PCL nanocomposite can be produced with modified nanoclay.

Polymer nanocomposites usually can be prepared by using three different techniques such as in-situ polymerisation, melt blending and solution intercalation. Solution intercalation method is considered in this study because only a small amount of polymer nanocomposite is to be fabricated in a laboratory scale whereas the other two techniques is preferred for industrial scale. Thus, the main aim of this study is to develop a new modification method for dispersing nanofillers in PCL nanocomposite and to fabricate PCL nanocomposite through solution intercalation technique.

2 Experimental

2.1. Materials

Sodium montmorillonite (Na-MMT) was purchased from Southern Clays, Texas, US. The organic modifier, Aminopropylisooctyl Polyhedral Oligomeric Silsesquioxane (AP-POSS) was purchased from Hybrid Plastic Co., USA. Polycaprolactone (PCL), ethanol, acetic acid and chloroform were purchased from Fisher Scientific.

2.2 Modification of Na-MMT using Surfactant Method

Five (5) g of Na-MMT was suspended in 250 ml of deionised water at 25°C with stirring for 2 hours. 2.332 g of AP-POSS was dissolved in 6 ml of ethanol and acidified with 4 ml of acetic acid. It is then added dropwise to the Na-MMT suspension over a period of 20 min [6]. The mixed suspension was then continuously stirred at 70°C for 24 hours to obtain the organoclay (POSS-MMT). Then the organoclay was harvested by centrifuge, dried in oven for 48 hours and ground to pass mesh sieve.

2.3 Fabrication of PCL Nanocomposite

The nanoclay (1wt%, 3wt% and 5wt%) was stirred in chloroform for $1\frac{1}{2}$ hours. PCL (1g) was also stirred with chloroform (20ml) to get a solution of PCL/Chloroform. That solution was then dropped into nanoclay/chloroform suspension and stirred for $4\frac{1}{2}$ hours. The suspension was poured into a clean petri dish and left covered for 2 days. After 2 days, the film was dried in a vacuum oven for 24 hours at 40°C to remove the residual traces of chloroform.

2.3 Fabrication of PCL Nanocomposite

XRD data was recorded on a MiniFlex Rigaku using graphite-filtered Cu-Kα radiation (1.25, 0.154 nm).The diffractometer was controlled using Diffrac Plus XRD Commander and the raw data was manipulated using EVA software [6]. All the data were collected between the ranges of 3° to 80° 2θ with a resolution of 0.02°. The *d*-spacing of the nanoclay were calculated by using the Bragg's law.

$$Sin\ \theta = n\lambda /2d \qquad (1)$$

Where, *d* is the spacing between the layers of the nanoclay, λ is the wavelength of the X-Ray which is 0.154 nm, θ is the angle at the maximum point of the first peak in the spectra, n is the order of diffraction which 1.

FTIR was recorded by using Nicolet Omnic 3 FTIR spectrophotometer. The samples were used in the solid form. The background of the samples was in KBr standard. After the background is completed, the reading for the samples was collected.

FESEM analysis was done by using the JEOL FESEM JSM-7100F. 'Smart FESEM User Interface' program was run. The samples were mounted on the universal holder with stub forceps. The focus knob was turned to sharp the scanned images. FESEM images then were saved.

2.4 Thermal Analysis of PCL nanocomposite

TGA analysis was carried out by using Perkin Elmer TGA Q500. The samples were heated from 25°C to 600°C at a heating rate of 5°C/min under nitrogen atmosphere. A mass change versus temperature curve was obtained.

DSC analysis was done using Perkin Elmer DSC Q1000 under nitrogen atmosphere. Measurement was performed from room temperature to 160°C at heating rate of 10°C/min. The samples were stand stilled for 5 minutes to erase any previous thermal history. Subsequent cooling and heating cycles was recorded respectively. The degree of crystallinity (X_{cr}) of the sample was calculated by using equation shown below.

$$X_{cr}\ (\%) = (\Delta H_f)/(W_{PCL}\ x\ \Delta H_{100})\ x\ 100\% \qquad (2)$$

ΔH_{100} represents heat of fusion of 100% crystalline PCL [136.1 J/g [7]], ΔH_f represents experimental heat of fusion and W_{PCL} represents weight fraction of PCL.

3 Results and Discussion

3.1 Characterization of Nanoclay

X-ray Diffraction (XRD) analysis was used to characterize the interlayer *d*-spacing of the nanoclay galleries and the dispersion of nanoclay in the polymer nanocomposite structure. The pure nanoclay, Na-MMT shows the first characteristic peak at 2θ = 14.9° and the modified nanoclay, POSS-MMT shows at 2θ = 7.16° as indicated in the **Figure 1**. It can be clearly observed that there is a shift of characteristic reflections to lower angle in POSS-MMT compared to Na-MMT. The *d*-spacing of the nanoclays were calculated by using **EQN 1**. The *d*-spacing of the nanoclays increased from 0.593 nm for Na-MMT to 1.233 nm for POSS-MMT. This proved that the AP-POSS surfactant was successfully intercalated into the nanoclay galleries consequently pushing the nanoclay layers and confirms that modification process has taken place. Thus, this result corresponds to the study conducted by Zhao *et al.* (2009) [8] where the *d*-spacing increased from 0.96 nm for Na-MMT to 1.25 nm for POSS-MMT. This intercalation is an efficient way to increase the material surface hydrophobicity of Na-MMT which is a fundamental requirement for good compatibility between the polymer matrix and the Na-MMT surface [9]. Surfactant method gives a better increase in the *d*-spacing of the nanoclay compared to ion exchange reaction by alkyl ammonium salts where in the studies conducted by Uhl *et al.* (2004) [10], cetyltrimethylammonium bromide (CTMA) increases the nanoclay galleries only by 0.5 nm while the surfactant method used in this study could increase the nanoclay layer by 0.64 nm.

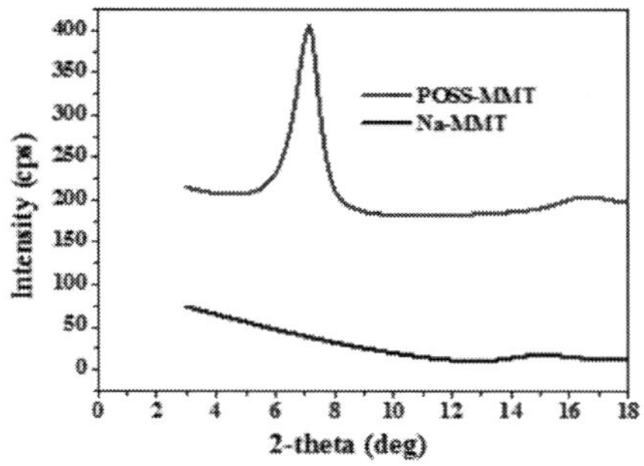

Fig. 1. XRD Spectra for Na-MMT and POSS-MMT

Fourier Transform Infrared Spectroscopy (FTIR) was used to analyse the presence of certain functional group in a molecule. In **Figure 2**, Na-MMT shows characteristic peak at 3630.97 cm^{-1} which represents the hydroxyl stretching vibration of OH group and at 3460.63 cm^{-1} which represents the broad band of stretching vibration of the interlayer water. Na-MMT also exhibits strong peaks at 1045.79 cm^{-1} for the Si-O stretching vibration, 523.87 cm^{-1} for the Al-O-Si deformation and 466.81 cm^{-1} for Si-O-Si deformation bonds [11]. The H-O-H bending vibration band can be observed at 1639.14 cm^{-1}. All these bands are also present in POSS-MMT. The AP-POSS surfactant exhibits characteristics absorption peak for its aliphatic C-H stretching vibration from CH_2 groups at 3000-2800 cm^{-1}. The stretching vibration of N-H bond also might be present in the same band. AP-POSS also shows a peak at 1226.37 cm^{-1} for Si-C bond [12]. The symmetrical Si-O-Si bond in the silsesquioxane cage is at stretching band of 1092.76 cm^{-1} [12]. All these bands also can be observed in the POSS-MMT which indicates the intercalation of AP-POSS surfactant into the nanoclay galleries. After the intercalation of AP-POSS surfactant, POSS-MMT not only had the characteristic bands of Na-MMT but also exhibited the characteristic bands of AP-POSS surfactant.

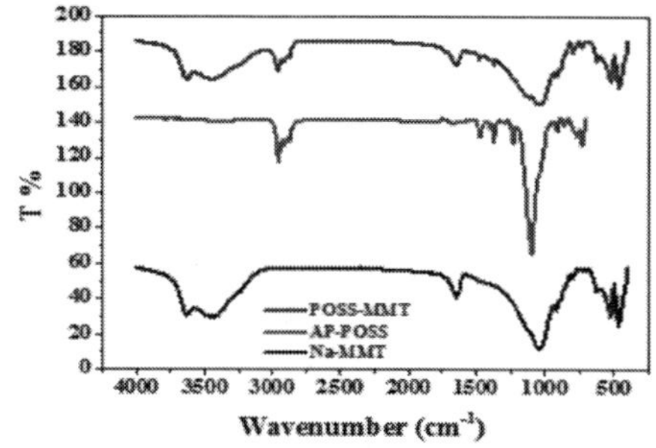

Fig. 2. FTIR Spectra for Na-MMT, AP POSS and POSS-MMT

The morphology of Na-MMT and POSS-MMT were analysed by means of FESEM. **Figure 3** shows the FESEM images for both the nanoclays. In **Figure 3 (a)**, the Na-MMT shows aggregated and curved plates. It also has a number of flakes with severely crumpled structure. In **Figure 3 (b)**, POSS-MMT shows a finer and uniform morphology. The aggregated, curved and crumpled structure of the Na-MMT disappears and well dispersed particles obtained after modification. Finer and more uniform morphology is obtained in the modified nanoclay. This result is speculated might be because of the well intercalated nanoclay galleries with increased *d*-spacing which are due to the modification by using AP-POSS.

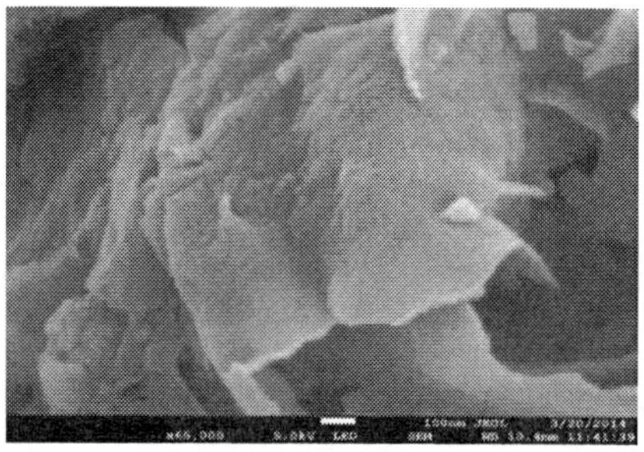

3(a)

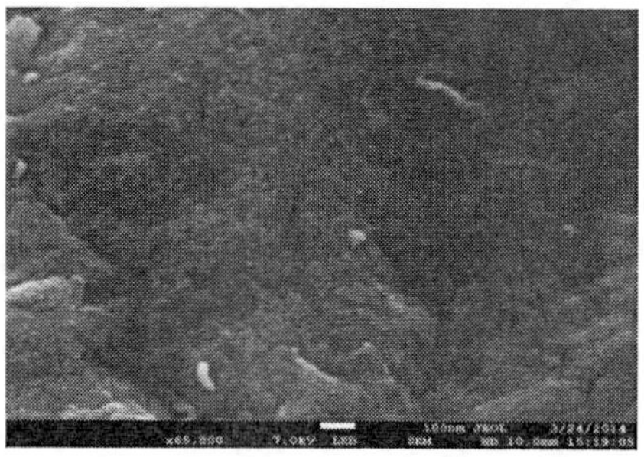

3(b)

Fig. 3. FESEM images of nanoclay (a) Na-MMT and (b) POSS-MMT

3.1 Characterization of PCL Nanocomposites

XRD traces for the POSS-MMT and PCL/POSS-MMT nanocomposite is shown in **Figure 4**. The POSS-MMT has a characteristic peak at 2θ of 7.16°, corresponding to interlayer *d*-spacing of 1.233 nm. There are no visible peaks for PCL/POSS-MMT nanocomposite which validates the formation of an exfoliated structure in the nanocomposite. The octyl chains in AP-POSS will be in straight alignment position by pushing the nanoclay interlayer galleries after the modification and PCL chains enter into the interlayer galleries of the nanoclay. This occurrence might have caused the destruction of the ordered geometry of nanoclay and resulting in complete delamination of the nanoclay layer on PCL matrix and forms an exfoliated structure. Similar observation has been reported by Liu *et al.* (2005) [13] where epoxy nanocomposite fabricated with POSS-modifed MMT does not exhibit any diffraction peaks in the XRD pattern representing exfoliated structure.

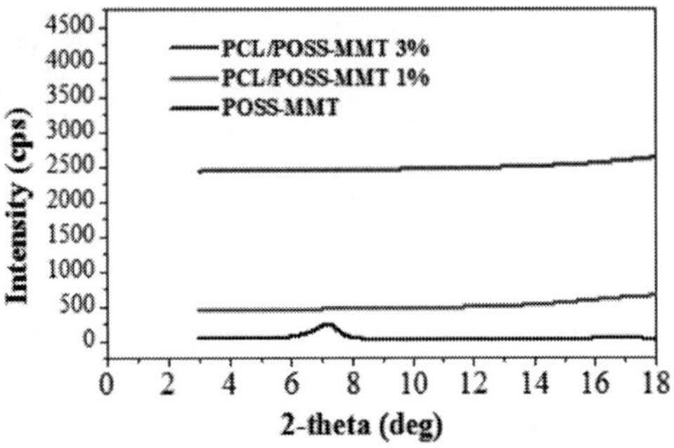

Fig. 4. XRD Pattern for POSS-MMT and PCL/POSS-MMT Nanocomposite

The **FTIR** spectra for pure PCL and PCL nanocomposite samples are shown in **Figure 5**. The pure PCL have peaks located at 2930-2945 cm^{-1} representing CH_2 bonds stretching vibration. The vibration of C=O bonds present at 1724 cm^{-1} band. The region of 1275 cm^{-1} to 1050 cm^{-1} indicates the C-O-C aliphatic ether stretching vibrations. All these important characteristic bands are situated in each PCL nanocomposite structure and show the presence of PCL. A new characteristic peak can be observed at the region of 3470-3485 cm^{-1} in the entire PCL nanocomposites. These peaks represent the broad band of stretching vibration of the interlayer water in the Na-MMT silicate. This band shows a slight increase in its intensity correlated with the increasing nanoclay content in each nanocomposite samples [14]. The most intense characteristic bands for Na-MMT are located in the region 1050-990 cm^{-1} attests the Si-O stretching vibrations. Unfortunately, those characteristic bands cannot be distinguished because PCL has multiple bands in the same IR spectral range [14]. The presence of important principal characteristic bands of POSS-MMT falls in the region between 3000 cm^{-1} to 2800 cm^{-1}. These bands which represent the CH_2 stretching vibration and the N-H bonds are also present in the pure PCL structure. It also can be seen a small increase in the intensity of that particular peaks with increasing POSS-MMT content. This result clearly verifies that intercalation of PCL with both unmodified and modified nanoclay has occurred.

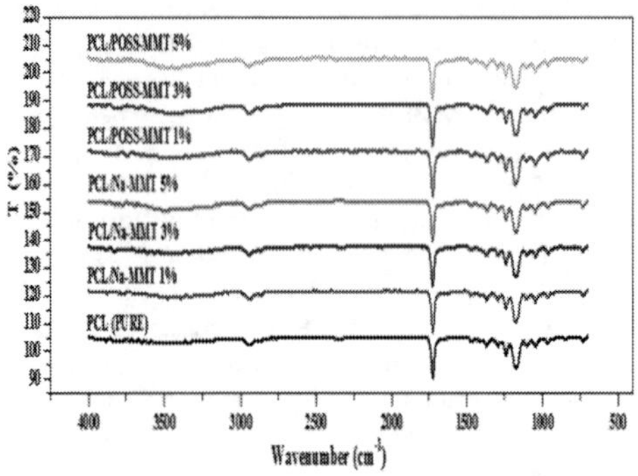

Fig. 5. FTIR Spectra of Pure PCL and PCL Nanocomposite

The **FESEM** analysis for PCL nanocomposite shows that POSS-treated nanoclay nanocomposite gives a matrix with well distributed particles compared to pure nanoclay nanocomposite as shown in **Figure 6**. In PCL/Na-MMT nanocomposite, nanoclay tends to stack together and do not disperse uniformly in the PCL matrices as shown in **Figure 6(a)**. The Na-MMT tends to be agglomerated and stacked on the PCL matrix. This is because the immiscibility between Na-MMT and PCL is a major issue due to poor interfacial adhesion [15]. The distribution of modified nanoclay layers in PCL/POSS-MMT nanocomposite appeared more uniform throughout the polymer matrix. The difference in morphology of PCL/Na-MMT 5% and PCL/POSS-MMT 5% nanocomposite is very obvious where the agglomerated particles disappear completely in PCL/POSS-MMT 5% nanocomposite as shown in **Figure 6(b).** This occurrence of finely dispersed modified nanoclay throughout the entire PCL matrix is because the compatibility of Na-MMT and PCL has been increased by modifying the Na-MMT.

6(a)

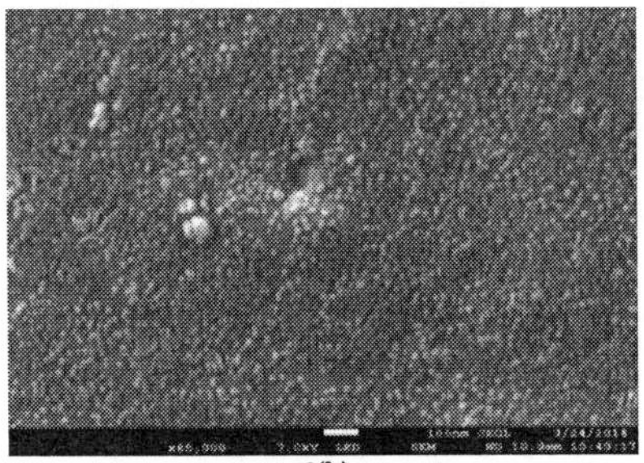

6(b)

Fig. 6. FESEM images for PCL nanocomposite (a) PCL/Na-MMT (5%) and PCL/POSS-MMT (5%)

Thermal degradation of pure PCL and PCL nanocomposite was monitored through TGA. PCL materials degrade with a large peak in single step as observed for each sample in **Figure 7**. The degradation temperatures for 50% weight loss ($\mathbf{T_{50\%}}$) decreases gradually from 392.9°C for pure PCL to 391.4°C for PCL/Na-MMT 5% nanocomposite and this is due to the incompatibility of Na-MMT with PCL matrix together with its hydrophilic character which can be related to the decomposition of physically adsorbed water and water molecules surrounding the exchangeable sites in Na-MMT. A noticeable increase in degradation temperature at $T_{50\%}$ can be observed for PCL/POSS-MMT 1% nanocomposite which is 394.1°C. The shift to higher degradation temperature is because POSS-MMT may act as a mass transport barrier to the volatile product that is generated during decomposition or diffusion hindrance of the decomposited volatiles through nanocomposite [16]. In a study conducted previously, the degradation temperature was higher for PCL/MMT-C_8H_{17} nanocomposite and lower for PCL/Na-MMT nanocomposites compared to neat PCL at $T_{50\%}$ [16]. PCL/POSS-MMT 3% and 5% nanocomposite have the lowest degradation temperature among other samples which is 382.3°C and 364.8°C respectively. This can be associated with two major effects as the content of POSS-MMT increased. Firstly, the POSS-MMT could have functioned as an inorganic catalyst for polymer degradation [17]. Secondly, the AP-POSS surfactant can be decomposed at lower temperatures following the Hoffman elimination where these resulting degradation products of AP-POSS could have possibly contributed in catalyzing the degradation of PCL matrix [9]. The resistance to thermal degradation was improved when the organically modified nanoclay content is 1 wt% in PCL matrix but the effect levels off independently of the modified nanoclay beyond that content [16].Thus, the results obtained becomes obvious that the modified nanoclay may possess two opposing functions in thermal stability of polymer nanocomposites where it improves the thermal stability by barrier effect and decreases the thermal stability by catalytic effect on the degradation of the polymer matrix.

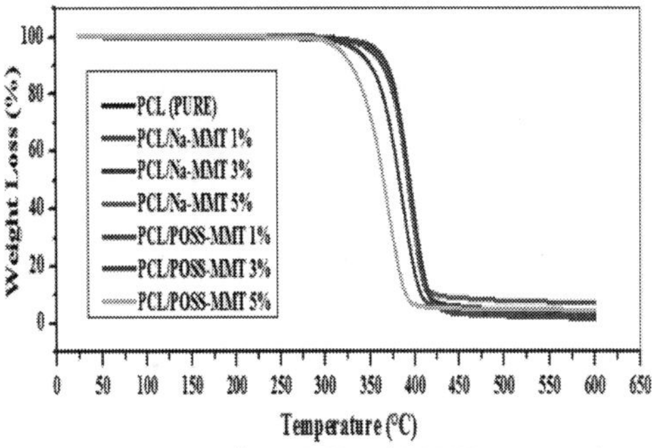

Fig. 7. TGA curves of Pure PCL and PCL Nanocomposite

The effect of nanoclay on the melting temperature ($\mathbf{T_m}$), crystallization temperature ($\mathbf{T_c}$) and degree of crystallinity ($\mathbf{X_{cr}}$) of PCL was analysed by DSC. Based on **Table 1**, it can be clearly seen that PCL/POSS-MMT nanocomposites possess a higher T_m compared to PCL/Na-MMT nanocomposite loaded with equal amount of nanoclay and pure PCL where it shows 56.6°C for PCL/POSS-MMT 5% nanocomposite. In the case of T_c, the PCL/Na-MMT nanocomposite experiences a decrease in the T_c from pure PCL to PCL/Na-MMT 5% nanocomposite which is from 31.9°C to 31.0°C then increases when POSS-MMT is incorporated into the PCL which is up to 32.7°C for PCL/POSS-MMT 5% nanocomposite. Similar results were reported by Di *et al.* (2003) [18], where T_c of the polymer matrix increased with the addition of a small amount of C30B, an organoclay to the PCL matrix. The increase in both these temperatures are because of increased order of lamellar crystals of PCL or the molecular mobility of PCL segments was suppressed by spatial confinement due to the presence of organoclay particles [19]. There also could be nano-reinforcement effect of Na-MMT layers together with the intercalating AP-POSS surfactant which were well dispersed in the PCL matrix. The T_m and T_c of nanocomposite depend on the degree of dispersion of nanoclay and its dispersed sizes in the polymer matrix. The PCL/Na-MMT nanocomposites have a higher X_{cr} (calculated by equation 2) compared to the PCL/POSS-MMT nanocomposite with the same amount of nanoclay loading. The presence of surfactant at the interface can prevent the nucleation process and leads to decrease in the crystallinity [20]. The decrease in X_{cr} is also due to the reduced mobility of the polymer chains in presence of strongly interacting nanoclay [18]. Thus, incorporation of modified nanoclay into PCL matrix improves the T_m and T_c compared to unmodified nanoclay but reduces the X_{cr}.

Table 1. DSC Data for Pure PCL and PCL Nanocomposite

Sample	Melting Temperature (°C)	Crystallization Temperature (°C)	Degree of Crystallinity (%)
PCL (Pure)	55.9	31.9	33.6
PCL/Na-MMT 1%	56.4	31.5	40.9
PCL/Na-MMT 3%	56.3	30.6	43.8
PCL/Na-MMT 5%	56.3	31.0	40.2
PCL/POSS-MMT 1%	56.5	31.8	37.3
PCL/POSS-MMT 3%	56.4	32.2	37.2
PCL/POSS-MMT 5%	56.6	32.7	35.8

4 Conclusion

The new AP-POSS based surfactant method for modification of nanoclay was done successfully where a significant increase in the interlayer *d*-spacing from 0.593 nm in Na-MMT to 1.233 nm in POSS-MMT has been achieved. The AP-POSS surfactant was successfully intercalated into the interlayer galleries of Na-MMT. The disappearance of diffraction peak in XRD for PCL/POSS-MMT nanocomposite and FESEM images verifies a good dispersion of POSS-MMT layers in the PCL matrix. TGA results showed highest degradation temperature of 394.1°C at $T_{50\%}$ for PCL/POSS-MMT 1% nanocomposite. DSC data showed highest T_m of 56.6°C and T_c of 32.7°C for PCL/POSS-MMT 5% nanocomposite with a lower X_{cr} of 35.8% as compared to PCL/Na-MMT nanocomposite. PCL/POSS-MMT nanocomposite has enhanced thermal properties compared to pure PCL. Organic modification of nanoclay with new surfactant method gives good potential for incorporation with various biodegradable polymers such as Polybutylene Succinate and Polylactic Acid to give enhanced properties.

The authors gratefully acknowledge the financial supported by UMP under research funding RDU150398 and Ministry of Higher Education Malaysia under FRGS scheme RDU160149

References

1. Kango, S., Kalia, S., Celli, A., Njuguna, J., Habibi, Y., & Kumar, R. (2013). Progress in Polymer Science, 38, 1232– 1261.
2. Sinha, R.S.S., & Bousima, M. (2005). Prog. Mater. Sci., Vol. 50, 962–1079
3. Hossain, M. D., Kim, W. S., Hwang, H. S., & Lim, K. T. (2009). Journal of Colloid and Interface Science, 336, 443–448.
4. Gorrasi, G., Tortora, M., Vittoria, V., Pollet, E., Lepoittevin, B. D., Alexandre, M., & Dubois, P. (2003). Polymer, 44, 2271–2279.
5. Sarier, N., Onderb, E., & Ersoyb, S. (2010). Colloids and Surfaces A: Physicochem. Eng. Aspects, 371, 40–49.
6. McLauchlin, A., Bao, X., & Zhao, F. (2011). Applied Clay Science, 53, 749-753.
7. Yam, W. Y., Ismail, J., Kammer, H. W., Schmidt, H., & Kummerlowe, C. (1999). Polymer 40, 5545–5552.
8. Zhao, F., Wan, C., Bao, X., & Kandasubramanian, B. (2009). Journal of Colloid and Interface Science, 333, 164–170.
9. Achaby, M. E., Ennajih, H., Arrakhiz, F.Z., Kadib, A. E., Bouhfid, R., Essassi, E., & Qaiss, A. (2013). Composites: Part B, 51, 310–317.
10. Uhl, F. M., Davuluri, S. P., Wong, S. C., & Webster, D. C. (2004). Polymer, 45, 6175–6187.
11. Zhou, Q., Pramoda, K. P., Lee, J. M., Wanga, K., & Loo, L. S. (2011). Journal of Colloid and Interface Science, 355, 222–230.
12. Fu, H. K., Kuo, S. W., Yeh, D. R., & Chang, F. C. (2008). Journal of Nanomaterials, Article ID 739613, 7 pages.
13. Liu, H. Z., Zhang, W., & Zheng, S. (2005). Polymer, 46, 157–65.
14. Viville, P., & Lazzaroni, R. (2003). Langmuir, 19, 9425-9433.
15. Ahmed, J., Auras, R., Kijchavengkul, T., & Varshney, S. K. (2012). Journal of Food Engineering, 111, 580–589.
16. Lepoittevin, B., Pantoustier, N., Alexander, M., Calberg, C., Jerome, R., & Dubois, P. (2002). J Mater Chem, 12, 3528–32.
17. Qin, H., Zhang, S., Zhao, C., Feng, M., Yang, M., Shu, Z., et al. (2004). Polym Degrad Stabil, 85, 807–13.
18. Di, Y., Iannace, S., Maio, E. D., & Nicolais, L. (2003). J Polym Sci Part B: Polym Phys, 41, 670–8.
19. Wu, T., Xie, T., & Yang, G. (2009. Applied Clay Science, 45, 105–110.
20. Eleonora, E., Fernanda, M., Antonia, & T.M., Hugo, D. (2011). Journals of Material Science and Engineering A, 778-789.

Electrochemical Evaluation of Fluorinated MnO_2 for Supercapacitor Application

Siti Nur Najwa Mohd Yusof[1], *Nurul Khairiyyah* Mohd Zain[1], *Izan Izwan* Misnon[1,*] and *Rajan* Jose[1]

[1]Nanostructured Renewable Energy Materials Laboratory, Faculty of Industrial Sciences & Technology, Universiti Malaysia Pahang, 26300 Gambang, Kuantan, Pahang, Malaysia

Abstract. Supercapacitors (SCs) functioning as alternative energy storage is useful in most electronic devices, renewable energy system and hybrid vehicles that have high demand in these days. Excellent electrochemical performance, environment friendliness and low cost material are needed to fulfil the energy demand by most developed country. In this study, fluorination treatment on manganese oxide (MnO_2) is considered as an effective way to develop better energy storage due to fluorine electronegativity and reactivity when correlate with other element. Hydrothermal method is used to synthesis MnO_2 (α–MnO_2 and δ–MnO_2) and the effect of fluorination (F–α–MnO_2 and F–δ–MnO_2) on MnO_2 surfaces is investigated on the charge storage ability. The crystallinity and functional groups of the samples was confirmed by the X-ray diffractogram and fourier transforms infrared spectroscopy (FTIR). The cyclic voltammetry (CV) and galvanostatic charging–discharging (CDC) analysis in 0.5 M K_2SO_4 electrolyte shows that F-δ-MnO_2 gives the highest *Cs* value of 184 F g^{-1} at scan rate of 5 mV s^{-1} and 66 F g^{-1} at current density of 0.3 A g^{-1}. The electrochemical impedance spectroscopy shows that the F–δ–MnO_2 has the lowest electrode resistances and charge transfer resistance which contributes to high *Cs* and the high conductivity of electrode.

1 Introduction

Electrochemical capacitors which known as supercapacitors (SCs), are compatible for application such as hybrid vehicles, portable electronic devices and renewable energy system that require high charge–discharge rates and high power density [1]. This is owing to their excellent power densities than conventional batteries (eg: lithium ion batteries), high charge and discharge rate also have long life cycles which is up to 10^4 cycles [1]. In accordance to charge storage mechanisms, SCs could be categorized into two types which are electrochemical double layer capacitors (EDLCs) and pseudocapacitors (PCs). The EDLCs operate on surface absorption at the electrode materials and electrolyte interfaces. Numerous carbon materials such as carbon aerogels, graphene and mesoporous carbons with high surface area have been used as electrode materials for SCs [2]. The PCs are based on faradaic redox reactions which occur at the electrode surfaces and its bulk fragment. Transition-metal oxides (TMOs), hydroxides, chalcogenides (eg: RuO_2, MnO_2, Co_3O_4, $Ni(OH)_2$ and MoS) and conducting polymers (eg:polypyrrole and polythiophenes) basically shows pseudocapacitive behavior. Basically, the PCs have much greater capacitance compared to the EDLCs.

Tremendous effort have been done by researchers in order to improve energy and power densities of the device using TMOs due to their availability along with multiple electron transfer and variable oxidation states during faradaic reactions [2,3]. Among TMOs, MnO_2-based electrodes are considered as most preferable for SCs fabrication due to their environmental friendly, natural abundance, low cost, wide potential window, and good electrochemical performance compared to RuO_2 electrode which is limited by its high cost and low abundance [2,4–7]. Manganese oxides can be existed in several allotropes including α, β, γ, δ, and λ due to the different arrangement of linkage of the basic [MnO_6] octahedral unit. The α, β, and γ-type MnO_2 exhibit 1D tunnel structure, whereas δ-type MnO_2 shows a 2D layered structure [8].

Several methods was used to synthesis MnO_2 including precipitation, self-reacting microemulsion, hydrothermal methods, sonochemical and room-temperature solid reaction. Among this, hydrothermal method is considered as the most popular synthesis approaches to synthesize numerous forms of MnO_2 and gives many advantages includes a wide range of initial precursors and mild synthesis conditions such as temperature and pH can be used. A solvothermal or hydrothermal method has been engaged to prepare one-dimensional nanoscaled materials, for instance, α-MnO_2, without using catalysts or templates. This method is remarkable to traditional methods since the equipment required is not particular and expensive for the preparation of nanostructured materials at a low

* Corresponding author: iezwan@ump.edu.my.

temperature. Preparation of MnO_2 by hydrothermal method is generally involve the phase transformation of granular MnO_2 precursors or redox reactions of MnO_4^- and/or Mn^{2+}[9]. Layered MnO_2 structure can be obtained by chemical treatments of Mn ions in a basic solution [10].

Zhang et al reported that, a layered structure of mesoporous birnessite-type manganese oxide (δ-MnO_2) with large specific surface area (SSA) consists of pores on nanometer scale and bicontinuous networks of solid become an attractive allotrope for active electrode material [5]. Commonly, specific capacitance (C_S) of the cathode materials is associated to ionic transport within the pores, electrical conductivity in the solid phase and its specific surface area. Therefore, α-MnO_2 and δ-MnO_2 is acceptably to be chosen electrodes for high-performance SCs due to their crystallographic structure and accessible nanostructures where MnO_2 with high C_S, low fabrication cost and good cyclic stability are expected [8].

Fluorination treatment is the best method to improve electrode performances because fluorine is the most reactive and most electronegative correlate with other elements. As a result, when fluorine is combined with the other element, it will forms stronger bonds by attracting more ions towards it. Kim et al. reported on fluorinated activated carbon (ACs) which shows improving capacitance performance compared to raw ACs due to its semi-ionic bonding character between AC and fluorine [1,9]. The potential of fluorination on TMOs is still lacking in the energy storage field. Hence in this research, the main target is the study of fluorination treatment on α-MnO_2 and δ-MnO_2 surface as well as to examine the effect to its capacitance.

2 Experimental

2.1 Material synthesis

The MnO_2 structures were synthesized using hydrothermal method. In this experiment, two samples were prepared using different precursor to oxidant ratio, 1:1 for α-MnO_2 and 1:6 for δ-MnO_2. The α-MnO_2 is prepared by dissolving $KMnO_4$ (0.05 mol) in 100 ml of deionized water. Then, the mixture is added with (0.05 mol) ammonium persulfate ($NH_4S_2O_8$). Next, the solution is stir for about 30 minutes and place in the Teflon-lined stainless steel reactor. The sample was heated in an oven for 24 hours at the temperature of 90 °C. After that, the reactor is cooled down to room temperature and then the brownish precipitate that produced is filtered. The sample is washed using deionized water followed by ethanol. The brownish precipitate is then dried for 24 hours at 50 °C. The fluorination of MnO_2 series was carried out by immersion into 50 mL of ammonium hexafluorosilicate [$(NH_4)_2SiF_6$] solution for 30 minutes. Next, the slurry was separated by filtration, washed clean using distilled water and then dried it at 110 °C for 24 hours to produce F-α-MnO_2 and F-δ-MnO_2.

2.2 Material characterization

The manganese dioxide structures were characterized by a range of characterization techniques such as XRD, FTIR, FESEM, EDAX and BET to find their crystals structure, molecular structures, topographical and elemental information, surface area and electrical properties.

2.3 Electrode fabrication and electrochemical evaluation

The MnO_2 sample is mixed with conducting carbon and polyvinylidene fluoride (PVDF) in the mass ratio of 80:10:10 in glass vial. Few drops of N-methyl-2-pyrrolidinone (NMP) was added into the mixture and stirred for 8 h to produce homogeneous slurry. The obtained slurry was coated onto nickel (Ni) foam current collector with an effective area of ~1 cm^2 by using brush technique and dried at 70 °C in the oven for 24 h. After that, the electrode was pressed at 5 MPa to produce better active material in contact with current collector.

Cyclic voltammetry (CV) were carried out in the potential range of 0.0–1.0 V at different scan rates of 2–100 mV s^{-1} for all electrodes with respect to Ag/AgCl as a reference electrode in 0.5 M K_2SO_4.The C_S of the electrodes were calculated from the CV curve using Equation 1.

$$C_S = \frac{1}{2mv(E_2-E_1)}\int_{E_1}^{E_2} i(E)dE \qquad (1)$$

where m is the mass of the active material, v is the scan rate, E_2–E_1 gives the potential window, and $i(E)$ is the current at each potential.

Galvanostatic charge discharge (CDC) was obtained in the potential range 0.0–1.0 V at different current density (0.3, to 10.0 A g^{-1}) in 0.5 M K_2SO_4 electrolyte. The Cs from the discharge curves can be calculated using the Equation 2.

$$Cs = \frac{I \times t}{m \times \Delta v} \qquad (2)$$

where I, t, m and Δv are applied current, discharge time, active mass and potential difference, respectively.

The electrochemical impedance spectroscopy (EIS) which illustrate the relationship between imaginary and real resistance is conducted in the frequency range of 100 kHz – 10 mHz determines the bulk resistance (R_S), charge transfer resistance (R_{ct}) and ion diffusion resistance (R_d).

3 Results and Discussion

3.1 Material analysis

Figure 1 shows the XRD patterns of all synthesized MnO_2 powders which are α-MnO_2, δ-MnO_2, F-α-MnO_2 and F-δ-MnO_2. The main characteristic peak of α-MnO_2 (Fig. 1a) at 2θ = 12.9, 18.2, 28.8, 37.7, 42.1, 49.8, 55.9,

60.2, 65.5, 69.5 and 73.0°which indexed to formation of pure tetragonal phase of α-MnO_2 with space group of *I4/m* (JCPDS 44-0141) [11]. The unit cell parameter of α-MnO2 are a= 9.73 Å, b= 1.3062 Å and c= 2.847Å. The prepared α-MnO_2 has a good crystallinity as it clearly illustrates sharp and clear diffraction peak [13]. The δ-MnO_2 is clearly demonstrated to be a layered structure consisting distinct peaks which correspond to long range of crystalline order in the compound. However, the peaks in Fig. 1b showing that the pattern was similar with previous MnO_2 sample as both pyrolusite and α-MnO_2 phase were present at each peaks. This could be due to excessive heat and reaction during δ-MnO_2 sample preparation. Nayak et al. suggest that it might be the H_2O was expelled from the interlayer gap of δ-MnO_2 which changed the structure into tunneled α-MnO_2 structure [14]. The peak present in Fig. 1c is highly crystalline tetragonal phase of F-α-MnO_2 which confirmed by the formation of peaks at 2θ = 12.63, 17.82, 26.13, 28.54, 36.51, 37.444 and 68.97° with space group P42/mnm (JCPDS card no. 041-1445). The phase present at each peak is α-MnO_2 and manganese fluoride phase. Similarly in Fig. 1d, manganese fluoride present at the peak with addition of pyrolusite phase with same space group of P42/mnm.

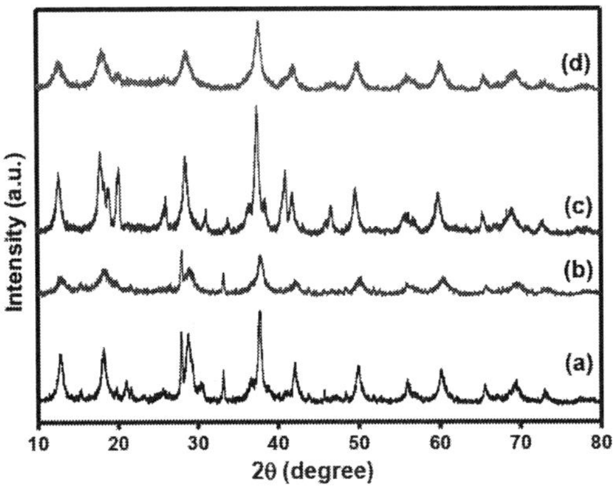

Fig. 1. X-ray diffraction pattern of (a) α-MnO_2, (b) δ-MnO_2, (c) F-α-MnO_2 and (d) F-δ-MnO_2.

The infrared spectra in Fig. 2 show all synthesized MnO_2 have the same absorbance peaks pattern. In the wavelength range of 3700-3000 cm^{-1} regions, the broad peak corresponds to stretching vibrations were recorded and attributed to the hydroxyl group on the MnO_2 surface. The peak detected at range of 1600 cm^{-1} indicate the bending vibration of water molecules whereas at wavelength of 1400 cm^{-1} are correlated with hydration of water and bending of Mn-OH in MnO_2 sample [15]. The band centered at ~1100 cm^{-1} correspond to bending and absorption of the Mn-OH functional groupand peaks at ~700 cm^{-1} attribute to the characteristics peak of manganese oxide. The IR result proposed the presence of water bonded in all of the MnO_2 structure, which responsive to the electrochemical activity of the material [16]. The IR result indicates the presence of some bound water in the sample. The fluorine vibration is not detectable using this technique due to low element loading.

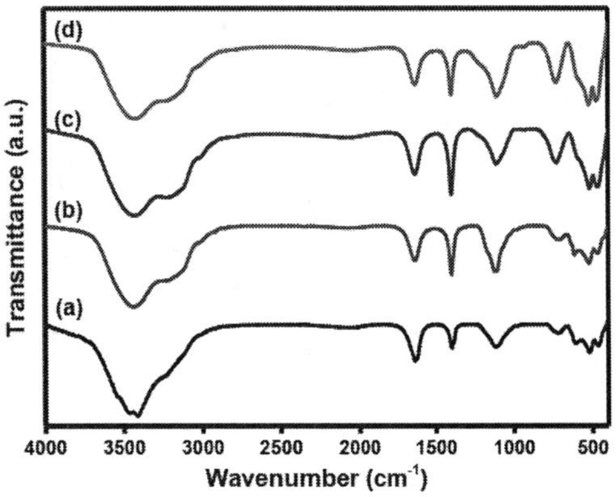

Fig. 2. FTIR spectra of (a) α-MnO_2, (b) δ-MnO_2, (c) F-α-MnO_2 and (d) F-δ-MnO_2 in the region of 4000-400 cm^{-1}.

The morphology of MnO_2 sample was observed using FESEM at low and high magnification as shown in Fig. 3. The image at low magnification (Fig. 3a) illustrates wool-like morphology while at high magnification the sample of α-MnO_2 showing a bundle of very fine, randomly-oriented needles that are about 80 nm in diameter. Similar morphology is observed for the sample of F-α-MnO_2 (Fig. 3c) that modified with fluorine which is approximately 50 nm in diameter of the needle-like structures. Fluorination will not change the structure of MnO_2 but its electrochemical properties. It can be further explain from the result of the energy-dispersive X-ray spectroscopy (EDAX) (not shown) that shows small proportion of ~11.68 wt% of fluorine followed by 54.00 and 34.32 wt% of oxygen and Mn in the sample. Fluorine was attached to the surface of MnO_2 because of reduction by fluoric acid. The decrease in oxygen content of F-MnO_2 from the pure MnO_2 may be correlated with the substitution of fluorine for oxygen, which largely covered the surface dangling bonds of MnO_2. The morphology of the δ-MnO_2 and F-δ-MnO_2 samples (Fig. 3b&d) showing the combination of spherical and wool morphology with growing needle-like structure onto the surface. This can be correlated with XRD result where there is two phases present in the sample which are pyrolusite and α-MnO_2 for prepared δ-MnO_2 also manganese fluoride and pyrolusite for F-δ-MnO_2. The well grown needle-like structures are ~60 nm in diameter in F-δ-MnO_2 whereas about 50–100 nm in diameter are observed in δ-MnO_2. The weight percentage (wt %) of fluorine is about 9.58 % followed by 26.14 and 64.28 % of Mn and O phase resulting from EDAX analysis.

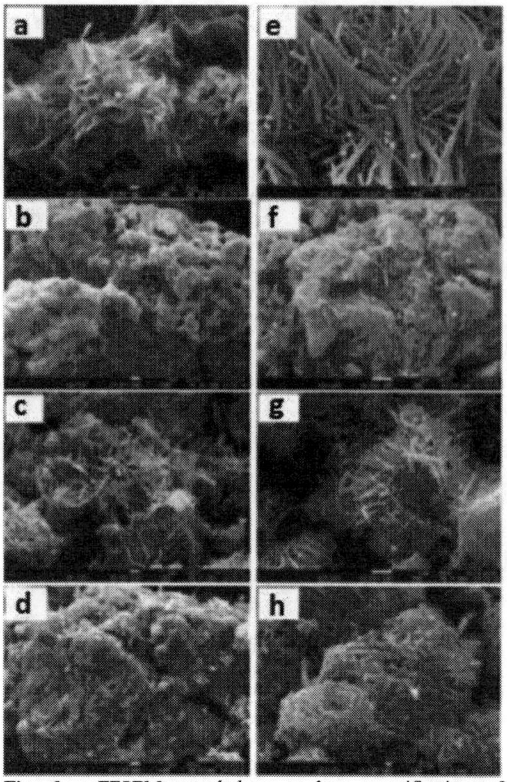

Fig. 3. FESEM morphology at low magnification of (a) α-MnO_2, (b) δ-MnO_2, (c) F-α-MnO_2 and (d) F-δ-MnO_2 and high magnification of (e) α-MnO_2, (f) δ-MnO_2, (g) F-α-MnO_2 and (h) F-δ-MnO_2.

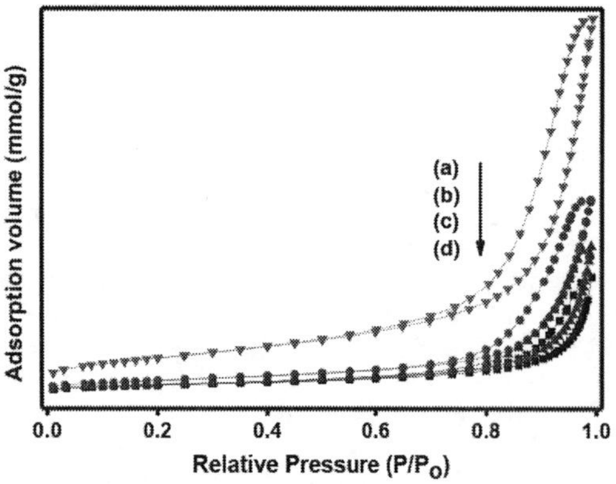

Fig. 4. Nitrogen adsorption-desorption analysis for (a) α-MnO_2, (b) δ-MnO_2, (c) F-α-MnO_2 and (d) F-δ-MnO_2.

Nitrogen adsorption of MnO_2 was measured to determine the specific surface area and pore volume between pure and fluorinated MnO_2 sample. Figure 4 shows the isotherms of α-MnO_2, δ-MnO_2, F-α-MnO_2 and F-δ-MnO_2 which belongs to type IV, showing hysteresis loop that indicates the mesoporous nature of the samples. At higher pressures, the absorption of nitrogen increase as pores was filled. The pore volumes were calculated correspond to Barrett–Joyner–Halenda (BJH) methods. The BET surface area, total pore volume, and average pore diameter for all samples are listed in Table 1. The F-δ-MnO_2 have the highest BET surface area and pore volume which are 121.7574 m^2 g^{-1} and 0.485473 cm^3 g^{-1} compared to the other sample. This indicates fluorinated MnO_2 is more porous than the pure MnO_2 and proved the assumption that fluorinated MnO_2 can exhibit high electrochemical activity. The highly porous structure can shorten the diffusion path for charge-carrier ions, while the large liquid-solid interface can facilitates ion exchange between the electrode and electrolyte.

Table 1. BET analysis of MnO_2 sample.

Sample	BET surface area (m^2 g^{-1})	Total pore volume (cm^3/g)	Average pore size (Å)
α-MnO_2	34.0539	0.133775	181.215
δ-MnO_2	50.2462	0.256872	197.544
F-α-MnO_2	37.1871	0.174289	208.740
F-δ-MnO_2	121.7574	0.485473	160.087

3.2 Electrochemical analysis

Figure 5 shows the CV curves of typical MnO_2 sample at various scan rate. The CV curves illustrate near rectangular-like shape of an ideal pseudocapacitive behavior.

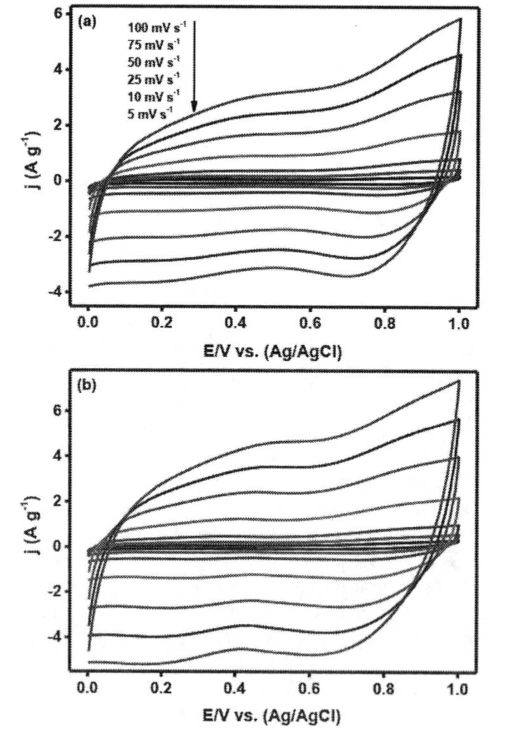

Fig. 5. Typical CV curves of (a) δ-MnO_2, and (b) F-δ-MnO_2 at scan rate of 5 – 100 mV s^{-1}.

The C_S obtained from the CV curves are 63, 158, 76 and 184 F g^{-1} for α-MnO_2, δ-MnO_2, F-α-MnO_2 and F-δ-MnO_2 electrodes respectively at the scan rate of 5 mV s^{-1}. The C_S decreases when the scan rate is increases. This is because electron cannot attach at the surface of the electrode because of the fast reaction of scan rate. Basically the utility of cyclic voltammetry is highly dependent on analyte used which has to be redox. If higher scan rate is used, it creates peak with large current and resistance which result in distortions.The result clearly shows that the C_S value is increased in following order: α-MnO_2 < F-α-MnO_2 < δ-MnO_2 < F-δ-MnO_2 which prove that the fluorination treatment on MnO_2 surface can improve the capacitance of the electrode.

Figure 6 represent the galvanostatic charging–discharging (CDC) curves of all samples with potential range of 0.0 – 1.0 V. The C_S at discharge current density of 0.3 A g^{-1} are 25, 50, 23 and 66 F g^{-1} calculated using formula in Eq. 2. In CDC, the higher the current densities cause the lower the C_S value. This result was affected by the value of discharging time. The higher the current density, the shorter the discharging time and causes the C_S to be lowered.

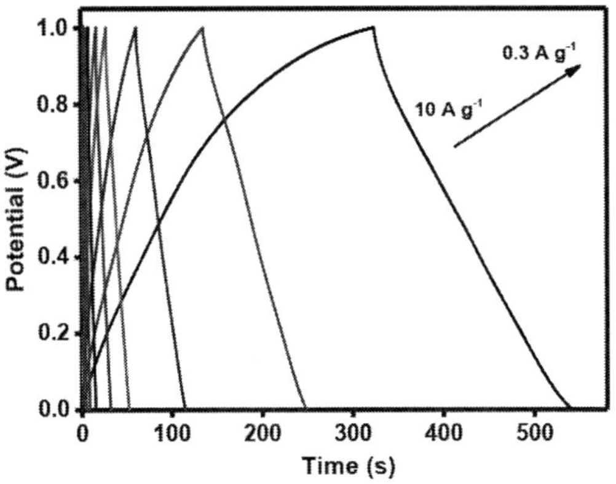

Fig. 6. Typical CDC curves of F-δ-MnO_2 at scan rate range of 0.3 – 10 A g^{-1}.

Figure 7 clearly shows that the C_S value is increase in following order: F-α-MnO_2 < α-MnO_2 < δ-MnO_2 < F-δ-MnO_2. The fluorination treatment on MnO_2 surface should improve the capacitance of the electrode. However the specific capacitance value in fluorinated α-MnO_2 is much smaller than the pure sample. This may be due to incomplete transfer of ion during discharging process.

The Nyquist plot in Fig. 8 illustrate of all sample immersed in 0.5M K_2SO_4 electrolyte which examined at the range of 0.01 to 10 000 Hz. At high frequency region, the resulting plot showed a half semi-circle whereas at low frequency region, a straight line is formed. The bulk resistance (R_S) is determined from the high frequency off-set in the real part of complex impedance of the Nyquist plot. The estimated R_S values are 1.07, 1.08 1.09 and 1.05 Ω for α-MnO_2, δ-MnO_2, F-α-MnO_2 and F-δ-MnO_2, respectively.

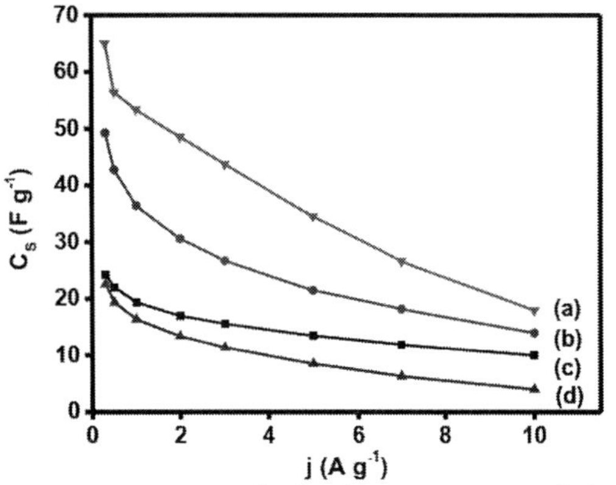

Fig. 7. C_S at current density from 0.3 – 10 A g^{-1} for (a) F-δ-MnO_2, (b) δ-MnO_2, (c) α-MnO_2 and (d) F-α-MnO_2 electrode.

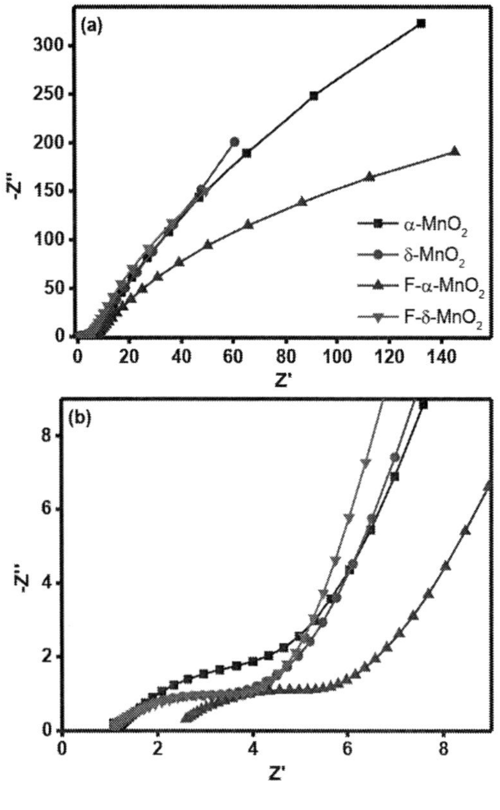

Fig. 8. (a) Nyquist plots of α-MnO_2, δ-MnO_2, F-α-MnO_2 and F-δ-MnO_2 and (b) magnified portion showing the semicircle at the high-mid frequency region.

The magnitude of the charge transfer resistance (R_{ct}) can be derived from diameter of the semi-circle. The estimated *Rct* values are 2.63, 2.51, 2.67 and 2.43 Ω for

α-MnO_2, δ-MnO_2, F-α-MnO_2 and F-δ-MnO_2. The ion diffusion resistance (R_d) is due to combination of resistance to ion diffusion extracted at the first segment of the straight line of Nyquist plot. The estimated R_d values are 4.64, 4.07, 4.66 and 3.97 Ω for α-MnO_2, δ-MnO_2, F-α-MnO_2 and F-δ-MnO_2 respectively. The R_S, R_{ct} and R_d values for F-δ-MnO_2 are the smallest which demonstrate a lower electron transfer resistance, resistance of electrochemical reaction, and ion diffusion resistance compare to α-MnO_2 δ-MnO_2, and F-α-MnO_2.

4 Conclusions

Four different types of electrode materials for supercapacitors were synthesized and their electrochemical performances were evaluated in three-electrode configuration using 0.5 M K_2SO_4. Cyclic voltammetry and galvanostatic charging–discharging (CDC) analysis shows that F-δ-MnO_2 gives the highest C_S value of 183.012 F g^{-1} at scan rate of 5 mV s^{-1} and 65.04 F g^{-1} at current density of 0.3 A g^{-1}. The EIS shows that the F-δ-MnO_2 has the lowest electrode resistances and charge transfer resistance which contribute high C_S and the high conductivity of electrode. Hence, F-δ-MnO_2 is considered as the best electrode material for supercapacitors compared to α-MnO_2, δ-MnO_2 and F-α-MnO_2.

The authors would like to acknowledge Universiti Malaysia Pahang for the UMP Internal Grant (RDU150354).

References

[1] M.H. Kim, J.H. Yang, Y.M. Kang, S.M. Park, J.T. Han, K.B. Kim, K.C. Roh, Colloids Surf. Physicochem. Eng. Asp. **443**, 535 (2014)

[2] Y.X. Zhang, M. Kuang, X.D. Hao, Y. Liu, M. Huang, X.L. Guo, J. Yan, G.Q. Han, J. Li, J. Power Sources. **270**, 675 (2014)

[3] S.G. Krishnan, M. Harilal, I.I. Misnon, M.V. Reddy, S. Adams, R. Jose, Ceram. Int. **43**, 12270 (2017)

[4] V. Subramanian, H. Zhu, B. Wei, Electrochem. Commun. **8**, 827 (2006)

[5] I.I. Misnon, R. Jose, New J. Chem. **41**, 6574 (2017)

[6] B.L. Vijayan, S.G. Krishnan, N.K.M. Zain, M. Harilal, A. Yar, I.I. Misnon, J.O. Dennis, M.M. Yusoff, R. Jose, Chem. Eng. J. **327**, 962 (2017)

[7] M. Harilal, S. G. Krishnan, A. Yar, I.I. Misnon, M.V. Reddy, M.M. Yusoff, J. Ojur Dennis, R. Jose, J. Phys. Chem. C. (2017)

[8] X. Zhang, X. Chang, N. Chen, K. Wang, L. Kang, Z. Liu, J. Mater. Sci. **47**, 999 (2012)

[9] S.C. Pang, S.F. Chin, C.Y. Ling, J. Nanomater. **2012**, 1 (2012)

[10] H.Y. Lin, Y.P. Sun, B.J. Weng, C.T. Yang, N.T. Suen, K.H. Liao, Y.C. Huang, J.Y. Ho, N.S. Chong, H.Y. Tang, Electrochimica Acta. **52**, 6548 (2007)

[11] Z. Li, Z. Liu, D. Li, B. Li, Q. Li, Y. Huang, H. Wang, J. Mater. Sci. Mater. Electron. **26**, 353 (2015)

[12] Y.S. Lee, *Adanced Fluoride-Based Materials for Energy Conversion* (Elsevier, Amsterdam,2015)

[13] J. Zang, X. Li, J. Mater. Chem. **21**, 10965 (2011)

[14] P.K. Nayak, T.R. Penki, N. Munichandraiah, J. Electroanal. Chem. **703**, 126 (2013)

[15] I.I. Misnon, R.A. Aziz, N.K.M. Zain, B. Vidhyadharan, S.G. Krishnan, R. Jose, Mater. Res. Bull. **57**, 221 (2014)

[16] X. Wang, A. Yuan, Y. Wang, J. Power Sources. **172**, 1007 (2007)

Mercury Removal from Wastewater Using Palm Oil Fuel Ash

Imla Syafiqah Mohd Salleh[1], *Nor Akmalina Mustazar*[1], and *Hafizuddin Wan Yussof*[1*]

[1]Faculty of Chemical Engineering & Natural Resources, Universiti Malaysia Pahang, 26300 Gambang, Pahang, Malaysia

Abstract. Palm oil fuel ashes (POFA) were used as the adsorbents for the removal of Hg II) ions from wastewater. From the preliminary experimental results, it shows that the POFA had good adsorption capability for Hg (II) ions. Initially, five factors were screened using 2^{5-1} factorial analysis in batch mode. From these factors, only two of them gave the significant effect which are contact time and agitation speed. Both also have obtained a maximum point which is likely to be the optimum point and possible for the optimization process. This study aims to optimize these two variables (contact time and agitation speed) using central composite design (CCD) which is a subset of response surface methodology (RSM). Quadratic model was developed for Hg (II) percentage removals. A 2^2 factorial CCD was used to get the mutual interaction between variables. The optimum adsorption conditions were obtained at contact time of 5 hr and agitation speed of 150 rpm with desirability of 0.95. At these optimum points, the mercury removal efficiency was calculated at 98.93 %. Based on the predicted and experimental results presented, the experimental values were in good agreement with the predicted values proposed by the model with an error less than 5 % and proved to be an adequate model. The results indicated that POFA has the potential to be used as an adsorbent for the removal of Hg (II) from aqueous environments due to its significant adsorption capacity and naturally abundance at low cost value.

1 Introduction

The presence of heavy metals in waste stream and ground water pose a very serious environmental concern since these metal ions are reported to be toxic to human beings as well as other living organisms. Heavy metals cause significant pollution especially for aqueous systems thus, in recent years, there has been increased research interest concerning the removal of heavy metal contaminants both in relation to improving existing methods and the development of new treatment approaches [1]. Mercury are toxic metals found in various industrial wastewaters. Several industrial activities, such as electroplating, leather tanning, metal finishing and petroleum refining, can cause water pollution by these ions. Mercury, known as a kind of remarkably toxic and nonbiodegradable metal and derives from both natural sources and industrial activities [2]. Therefore, an urgent need to effectively treat wastewater containing mercury(II) ions.

Most studies in recent years for heavy metal removal focused on the development of cheap and effective adsorbent. Researchers also looked into the availability of waste products especially on the biomass waste from agricultural to be used as an adsorbent [1] Adsorption is one the conventional technology to eliminate toxic metals from wastewater [3]. Adsorption process through utilization of adsorbents in the removal of heavy metal ions from aqueous solution is one of the most promising alternatives in substituting conventional methods like precipitation, membrane filtration, electrolyte or liquid extraction, electrodialysis and reversed osmosis [4].

Palm oil industries produce a significant amount of empty fruit bunches (EFB), oil palm fibres, and oil palm shells as wastes. POFA is a waste material obtained from burning of palm oil husk and palm kernel shell as fuel in palm oil mill boilers. The conventional method to discard these wastes is by incineration in the boilers as fuels for steam production in the oil palm mill. However, this method raises some environmental concerns due to the production of excessive amount of black soot during combustion and the production of large amounts of ash [5]. Previous studies have shown the possibility of using POFA as an adsorbent for dye removal [6], generation pf electricity [7] and as cement replacement material [8].

Adsorption method is considered flexible and easy to operate with much less sludge disposal problems. Various adsorbents such as graphene oxide [9]; coal fly ash [10]; nanoclay [11] and coconut husk [12] have been reported in the literature for the removal of mercury from wastewater. Among several chemical and physical methods, adsorption process is one of the effective techniques that have been successfully employed for mercury removal from wastewater. The main advantages of this technique are the low operating cost, improved selectivity for specific metal and short operation time. However, for industrial application, the selection of adsorbent material is mostly done by availability of

* Corresponding author: hafizuddin@ump.edu.my

waste material and applicability of the adsorption method taking into account on space, cost, and the amount of wastewater (Mondal et al. 2013). However, new adsorbents with local availability as well as economic suitability are still needed.

The aim of this research was to study the effects of contact time and agitation speed in influencing the mercury removal efficiency using POFA as adsorbent. Respond Surface Methodology (RSM) with five replicates at the center point was performed to evaluate the contribution of the factors. The optimum conditions for the mercury removal using POFA were also determined.

2 Materials and methods

2.1. Preparation of adsorbent

POFA was collected from from palm oil mill boiler at Kilang Sawit Felda Lepar, Kuantan. Samples were washed with distilled water several times to remove impurities and then dried in an oven at 110 °C for 24 h. 50 g of POFA were activated with 0.5 M H_2SO_4 for 24 hr in a round-bottom flask. It was sieved through 100 µm and stored in airtight container.

2.2. Experimental set up

A mercury stock solution was prepared at 1000 mg/L. This stock solution was then diluted to 1 mg/L and 5 mg/L using deionized water. All the adsorption experiment was carried out at room temperature. The required quantity of Hg^{2+} solution was dissolved in ultrapure water. Then, a small amount of HNO_3 was added to preserve the solution. The pH of the solution was adjusted using 0.1 M of HCl and 0.1 M NaOH. The mixture of the sample is shaken by using incubator shaker at the constant room temperature. The suspension is filtered using a vacuum filter and the filtrates are analysed using the direct mercury analyser RA-3310 (Nippon Instrument Corporation, Japan). The removal efficiency of the mercury was calculated according to the following formula (equation 1):

$$R\,(\%) = ((C_o - C_e))/C_o \times 100 \qquad (1)$$

where R is the percent removal of mercury, Co and Ce are the initial and residual concentration (mg/L) of mercury ion, respectively.

2.3 Experimental design and optimization

To understand the effect of the interaction between the adsorption-determining parameters, a design of experiment (DOE) was created with 95% confidence limit. The optimum experimental conditions were found under response surface methodology (RSM) and performed through central composite design (CCD) using Design Expert Software, version 7.1.6 (State-Ease, Inc.). Optimization was performed using two independent variables which were identified earlier in screening process as having significant effect on mercury removal efficiency. The independent variables considered were contact time and agitation speed rpm as presented in Table 1.

The experimental data were fitted by regression to a quadratic model as mentioned in equation 2:

$$Y = \beta_0 + \sum_{i=1}^{n} \beta_i X_i + \sum_{i=1}^{n} \beta_{ii} X_i^2 + \sum_{i=1}^{n} \cdot \sum_{j=1}^{n} \beta_{ij} X_i X_j + \varepsilon \qquad (2)$$

where Y represents the value of the predicted response, β_0 is a constant, βi, βj and βij are the linear, quadratic and interaction coefficients, respectively, and Xi and Xj are the experimental variables which levels are being optimized.

Table 1. Levels of the mercury removal variables tested in the CCD.

Factor	-α	-1	0	+1	+α
A: Contact Time	2	4.5	5	5.5	6
B: Agitation speed	100	125	150	175	200

3 Results and discussion

3.1. Central composite design (CCD)

Table 2 shows the results obtained from the 13 experiments performed through RSM approach. High mercury removal efficiency was achieved at the center point level (standard order 9-13). The second order polynomial model was utilized to express the mercury removal efficiency as a function of independent factors as shown in equation 3:

$$\textit{Mercury removal efficiency} = +\,94.65 + 2.13\,A - 1.36\,B + 0.52\,AB - 4.52\,A^2 - 6.14\,B^2 \qquad (3)$$

where the mercury removal efficiency represents the response, and A and B are the coded values of contact time and agitation speed, respectively. The term of A and B are denoted as the main effects, while AB is the interaction involves in the mercury removal efficiency. Quadratic effects are presented through A^2 and B^2 to imply the presence of curvature in the model. This model is found to be statistically significant because its p-value (<0.0001) is less than 0.05 [13].

Table 3 presents the analysis of variance (ANOVA) and regression coefficients. A model is considered significant if the p-values is less than 0.05 indicates that only a 5 % chance of noise can occur in the model. It is apparent from the table that the model obtained was significant with p-value is 0.0005, while the p–values for contact time and agitation speed are 0.0632 and 0.2010, respectively. The p-values was lower than 0.05, indicating that the model may be considered to be statistically significant [14]. Meanwhile, lack of fit is a vital expression which is express the pure error. As shown in table 3, the p-value of lack of fit (0.6332) is not significant, it means suitable and feasible model for predicting and describing the determination of Hg (II) ions as well as an appropriate fit. A coefficient of determination (R^2) for this study is 0.9354. It demonstrates that the predicted model perfectly satisfies the experimental data. According to [15] values for predicted and adjusted R^2 strongly proof high applicability and efficiency of model.

Table 2. Experimental design and response for optimization.

Std.	Contact time (hr)	Agitation speed (rpm)	Mercury removal efficiency (%)
1	4.5	125	84.67
2	5.5	125	85.07
3	4.5	175	79.33
4	5.5	175	81.80
5	4	150	71.53
6	6	150	82.87
7	5	100	72.67
8	5	200	68.80
9	5	150	93.20
10	5	150	95.07
11	5	150	90.20
12	5	150	98.93
13	5	150	98.40

Table 3. Formatting sections, subsections and subsubsections.

Source	SS[A]	DF[B]	MS[C]	F-Value	P-Value	
Model	1133.51	5	226.70	20.28	0.0005	Significant
A-Contact Time	54.40	1	54.40	4.87	0.0632	
B-Agitation Speed	22.28	1	22.28	1.99	0.2010	
Ab	1.07	1	1.07	0.096	0.7659	
A^2	468.51	1	468.51	41.90	0.0003	
B^2	863.29	1	863.29	77.21	< 0.0001	
Residual	78.27	7	11.18			
Lack of Fit	25.10	3	8.37	0.63	0.6332	Not significant
Pure Error	53.16	4	13.29			
Cor Total	1211.77	12				
V. =3.94%; R2=0.9354; Adjusted R2=0.8893; Adeq. Precision=12.008.						

[a]Sum of squares.
[b]Degree of freedom.
[c]Mean Square

3.2 Response surface plot

The response surface methodology (RSM) was then applied for improving the optimization and evaluation of the relative significance and all significance interaction of variables on adsorption processes. Figure 1 and figure 2 shows the mutual interactions of the combination of self-determining variables on mercury removal efficiency in the style of 3D surface and contour plots.

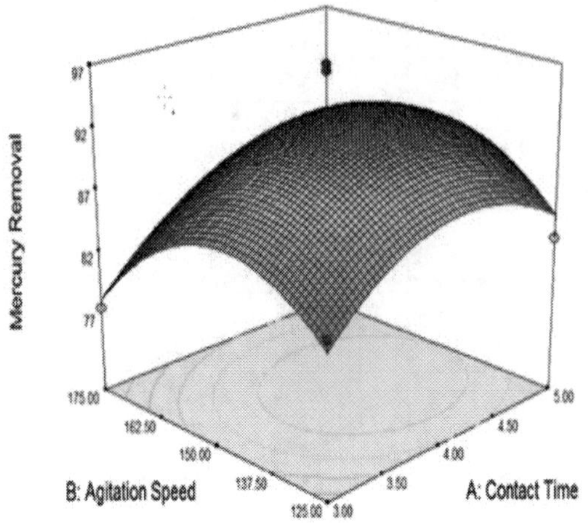

Fig. 1. 3D surface for interactive effect variables

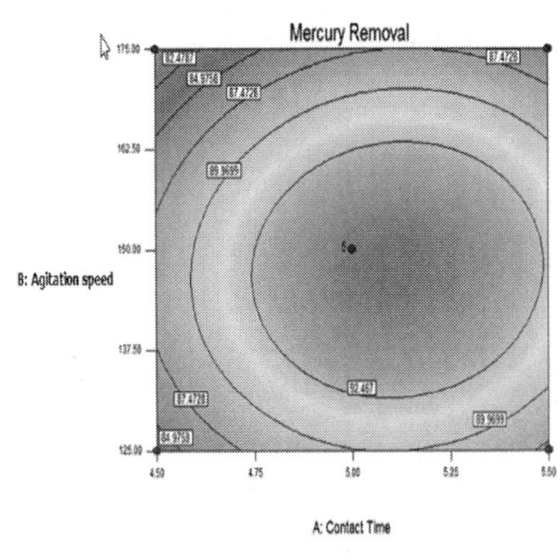

Fig. 2. Contour plots for interactive effect variables

The 3D dome-shaped curve and the 2D contour plot demonstrate the main relationship between contact time and agitation speed and the highest mercury removal efficiency. It is noticed that the percentage of mercury removal increased and reached the maximum level at contact time at 5 hr and 150 rpm of volume of agitation speed, and decreased beyond that. Adequacy of the model was reconfirmed by carrying out three replications at the estimated optimum point. This 3D plot illustrates the surface where the maximum point is placed inside the experimental region. The red zone is the optimal condition for mercury removal efficiency. At different agitation speed, the mercury removal efficiency reaches maximum value at approximate 5 hr and slightly decreases above the value.

4 Conclusions

In this study, the effect of contact time and volume of agitation speed was well described by RSM. The optimum condition at 150 rpm of agitation and 5 hr of contact time had successfully increased the percentage of mercury removal to 98.93%. A high correlation second order polynomial equation between observed and predicted responses was developed by RSM for the mercury removal efficiency. It presents the relationship between each factor and response efficiently. Therefore, it was verified that palm oil fuel ash has high potential for removing mercury in industrial wastewater.

This research was funded by Universiti Malaysia Pahang under the internal grant RDU170350. The authors would like to thank the Faculty of Chemical and Natural Resources Engineering, Universiti Malaysia Pahang for providing the infrastructure facilities for this study.

References

1. H. Xiyili, S. Çetintaş, and D. Bingöl, Process Safety and Environmental Protection, **109** 288-300 (2017)
2. M. Xu, P. Yin, X. Liu, X. Dong, Y. Yang, Z. Wang and R. Qu, Desalination and Water Treatment, **51**(22–24) 4546–4555 (2013)
3. B.S. Kruthi and M.R. Rajani, International Journal of Technical Research and Applications, **3(4)** 169–173 (2015)
4. Z.N. Garba, I. Bello, A. Galadima, and A.Y. Lawal, Karbala, International Journal of Modern Science, **2(1)** 20–28 (2016)
5. C. Acquah, L. Sie Yon, Z. Tuah, N. Ling Ngee, M.K. Danquah, Journal of Cleaner Production, **139** 1098-1104 (2016)
6. W.A. Khanday, F. Marrakchi, M. Asif, B.H. Hameed, Journal of the Taiwan Institute of Chemical Engineers, **70** 32-41(2017).
7. B.S. Thomas, S. Kumar, H.S. Arel, Renewable and Sustainable Energy Reviews, **80** 550-561 (2017).
8. S. Pourakbar, A. Asadi, B.B.K. Huat, M.H. Fasihnikoutalab, Transportation Geotechnics, **3** 24-35 (2015)
9. T. Esfandiyari, N. Nasirizadeh, M. Dehghani and M.H. Ehrampoosh, Chinese Journal of Chemical Engineering, **25(9)** 1170-1175 (2017)

* Corresponding author: hafizuddin@ump.edu.my

10. M. Attari, S.S. Bukhari, H. Kazemian, and S. Rohani, Journal of Environmental Chemical Engineering, **5(1)** 391–399 (2017)
11. M. Soleimani, and Z.H. Siahpoosh, Chinese Journal of Chemical Engineering, **23(11)** 1819–1833 (2015).
12. K. Johari, N. Saman, S.T. Song, C.S. Chin, H. Kong, and H. Mat, International Biodeterioration & Biodegradation, **109** 45–52 (2016).
13. D.O. Aksoy and E. Sagol, Fuel, **183** 609–616 (2016)
14. J. Aravind, C. Lenin, C. Nancyflavia, P. Rashika and S. Saravanan, International Journal of Environmental Science and Technology, **12(1),** 105–114 (2015)
15. K. Dashtian and R. Zare-Dorabei, Journal of Colloid and Interface Science, **494** 114–123 (2017)

Heavy Metals (Fe, Cu, and Cr) Removal from Wastewater by *Moringa Oleifera Press Cake*

Eman N. Ali[1], and *Hong* Tien Seng[1]

[1]Facutly of Chemical & Natural Resources Engineering, Universiti Malaysia Pahang, Pahang Darul Makmur, Malaysia.

Abstract. The consumption of water nowadays is increasing and the wastewater from variety of industry and domestic are increasing as well. If this wastewater is being untreated, it will be wasted and the chemicals inside the water especially heavy metal are harmful to the living organism. There are few conventional method used to treat the wastewater but most of them are costly, hence, natural coagulant is introduced in this study. One of the multipurpose tree *Moringa oleifera*, (MO) which is environmental friendly and low cost become a choice for removal of heavy metal from wastewater. The *Moringa oleifera* press cake (MOPC) are able to remove the heavy metal but the removal efficiency for different metals need to be investigated. Therefore, experiment on heavy metal removal by *Moringa oleifera* press cake was conducted. The wastewater samples and MOPC were collected from Balok River, Kuantan and Mitomasa Sdn. Bhd, Kuala Lumpur respectively. MOPC was immersed in distilled water to remove the oil and different concentration of MOPC solution were made. Jar test were carried out with different concentration of MOPC solution. The initial and final heavy metal (Fe, Cu, and Cr) concentration were measured by Atomic Absorption Spectrometer and the removal percentage was calculated. The MOPC removed 69.99% Fe, 88.86 % Cu and 93.73% Cr at optimum concentration of 10000 ppm, 5000 ppm and 15000 ppm, respectively. The MOPC also reduced 34.94% of COD and 81.6 % of turbidity. However, BOD of wastewater increased from 9.0 ppm to 18.3 ppm. The pH values was not changed with the addition of MOPC. The removal of heavy metal was increased proportionally with MOPC concentration until optimum removal was obtained.

1 INTRODUCTION

Water is very important and essential for human to survive and for all the living organism. Recently, the consumption of water have increased due to the increase of population worldwide. Meanwhile, large amount of wastewater containing contaminant such as suspended solid, pathogens, nutrients and heavy metals which may be harmful and dangerous are generated from the field of domestic and industrial agriculture. If they are kept untreated, safely disposed back to the sea or rivers, many pollution problems will be introduced (Paula et al., 2014). From World Health Organisation 2004, drinking water supplies around the world have been identified to have thousands of chemicals considered potentially hazardous to human health at relatively high concentrations (Vikashni et al., 2012). Besides, pollution of water has received special attention as an environmental issue and causes shortage of water supply especially in the developing and third world countries which have inadequate financial resources (Ali et al., 2009).

One of the contaminant in wastewater is the heavy metal. This wastewater with heavy metals are mainly contributed by anthropogenic sources like electronics assembly and fabrication, textiles, usage of fertilizers and pesticides, and mining activities (Kalavathy and Miranda, 2010). There are mainly 59 elements of heavy metal where cadmium (Cd), chromium (Cr), copper (Cu), lead (Pb) and zinc (Zn) are considered highly toxic (Vikashni et al., 2012). Even at low concentrations, these heavy metals can be toxic to organisms, including humans (Obuseng et al., 2012). Conventional methods used to remove heavy metal ions from wastewaters includes: ultra-filtration, reverse osmosis, ion exchange, solvent extraction, sedimentation and chemical precipitation (Obuseng et al., 2012). However, most of these methods has some disadvantages such as incomplete metal removal and Miranda, 2010). For instance, excess concentration of toxic sludge disposal or treatment problem (Kalavathy and aluminium sulphate

* Corresponding author: eman@ump.edu.my

(alum) which as a chemical coagulant in water treatment has been reported to cause Alzheimer's or other neuro-degenerative diseases (Paula et al., 2014). Cost of the chemical used in conventional method are too expensive for most of the developing country. Besides, for country like Malaysia which has frequent land development and rainy season, the cost of water treatment may be increasing due to unstable quality of water and high turbidity (Ali et al., 2009).

To solve this problem, natural coagulants which are more economical and environmentally friendly are studied. Natural coagulants are biodegradable and can be locally grown. Moreover, they produce less sludge and are safer for human compared to chemical coagulants. Three examples of the natural coagulants that are being tested by Kazi and Virupakshi (2013) are *Cicerarietinum, Seed Moringa,* and *Cactus Opuntiaficus*. It was found that the removal of turbidity by *Cicerarietinum* and *Seed Moringa oleifera* are 90% and 82.02%, respectively. Other research work by Patil et al., (2013) showed that both leave and latex of *Calotropisprocera* can be used in bioremediation of heavy metals. In addition, Sago (*Metroxylon*spp) is found to be one type of coagulant to remove turbidity and heavy metal which grown in Malaysia (Aziz et al., 2000). Since *Moringa oleifera* is one of the best natural coagulant discovered yet, it can be used to replace the conventional methods widely all around the world. It is a small tree which is fast growing and can reach up to 12 meters in height (Araújo et al., 2013). *Moringa oleifera* is one type of tropical multipurpose tree which is adaptable to semi-arid climates. Leaves and seeds of *Moringa oleifera* are also widely used in water treatment since they have no significant side effect and are non-toxic and biodegradable (Ali et al., 2014). Research done by Bodlund, (2013) on MO showed that coagulation activity is similar to the alum. The MO contain cationic polyelectrolytes which have proved to be effective in water treatment as a substitute for aluminium sulfate (Araújo et al., 2013). The flocculation activities of MO adsorb the metal cations based on the electrostatic charge mechanism (Muyibi et al. 2002). The use of *Moringa oleifera* in water treatment also do not have significant effect on pH (Sajidu et al., 2005). In this research, the removal of iron (Fe), copper (Cu) and Chromium (Cr) from wastewater by using *Moringa oleifera* press cake was performed and the efficiency was calculated. The properties of treated water were tested as well such as: pH, Turbidity, BOD, COD.

2 MATERIALS AND METHODS

The wastewater used for the experiment was obtained from the Balok River, Gebeng, Kuantan. While the *Moringa oleifera* Press cake (MOPC) used in the experiment was obtained from MitoMasa Sdn. Bhd., Kuala Lumpur. In order to use the MOPC, it was immersed in distilled water for 24 hours to remove any remaining oil after mechanical press. Then the MOPC was then filtered and stored in the chiller at 4°C.

2.1 Metal Standard Solution Preparation

Different concentration of 0 mg/L, 1 mg/L, 2 mg/L and 4 mg/L of iron, copper, and chromium solution were prepared for calibration curve. A drop of nitric acid was added to the solution for preservation (Vikashni et al. 2012).

2.2 *Moringa oleifera* Solution Preparation

The MO solution was prepared by adding 10 g (dry basis mass) of MO into 500 mL of distilled water to make 20,000 ppm MO solution. The solution was stirred so the MO can be dissolved in the distilled water. Then the solution was further diluted to 10 mL to 1,000 ppm, 5,000 ppm, 10,000 ppm, 15,000 ppm and 20,000 ppm.

2.3 Wastewater Treatment

Each beaker in jar test was filled with 500 mL of wastewater sample. 10mL of MOPC with different concentration prepared were added into each beaker. The stirring speed was set as 200 rpm for 4 minutes followed by 40 rpm for 30 minutes (Lea, 2010).

2.4 Heavy Metal Measurement

The heavy metal removal test was done by using Atomic Absorption Spectrometer (AAnalyst 400. Perkin Elmer). Calibration curve for each type of metal was obtained from the stock solution prepared from each standard at different concentration. The wastewaters obtained after the jar test were filtered by using vacuum pump through a 0.45μm nylon membrane filter before tested by AAS (Ali et al. 2014).

2.5 Turbidity and pH Measurement

The turbidity and pH of wastewater sample were measured by turbidity meter and pH meter respectively.

2.6 Biochemical Oxygen Demand (BOD) Measurement

BOD measurement was done by dissolved oxygen meter. Dilution water was prepared by mixing 3 mL of phosphate buffer, magnesium sulfate, calcium chloride, ferric chloride solution into 3 L volumetric flask. 10 mL of samples after jar test were mixed with 300 mL of distilled water into incubation bottle. Water was added to the BOD test bottle and covered with an aluminium foil. All the bottles were then put into BOD incubator for 5 days at 20°C. The final DO was measured and BOD of each sample was calculated according to Suhartini et al. (2013).

2.7 Chemical Oxygen Demand (COD) Measurement

Spectrophotometer was used to measure the COD. Around 2 mL of de-ionized water (control) and each sample from jar test were added into COD digestion reagent vials. Then, the samples were inserted into COD reactor and heated for 2 hours at 150°C with strong oxidizing agent (potassium dichromate solution). The solutions were then cooled down to room temperature (Kasmawati and Lee, 2007).

3 RESULTS AND DISCUSSION

3.1 Heavy Metal Removal Percentage

The concentration of heavy metal of Fe and Cr metal in the wastewater of the Balok River, Gebeng, Kuantan were approximately to 1 mg/L while Cu concentration was approximately 0.4 mg/L. To determine the optimum dosage or concentration of *Moringa oleifera* solution added into the wastewater, a jar test was done by adding 10 mL of 1000 ppm, 5000 ppm, 10000 ppm, 15000 ppm and 20000 ppm into 500 mL of wastewater. Figure 1 shows the results of the heavy metal concentration of the wastewater from Balok River after the treatment with different concentration of *Moringa oleifera* solution.

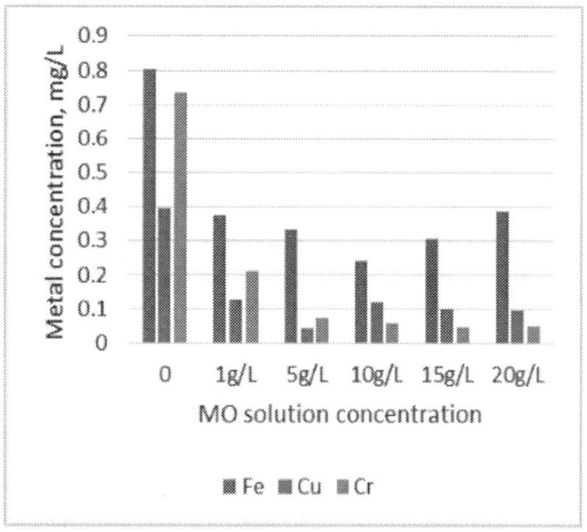

Fig.1. Metal Concentration after Jar test

From the results of the experiment, the heavy metals were successfully reduced after the treatment with the MOPC. The removal percentage of Iron, copper and chromium reached 69.99%, 88.86% and 93.73%, respectively. Generally, the metal removal percentage increase proportionally to the concentration of MOPC solution. Besides, the optimum MOPC concentration for iron, copper and chromium removal were at 10000 mg/L, 5000 mg/L and 15000 mg/L, respectively. The removal of iron in this experiment were lower than the research done by Sajidu et al. (2005), which had a removal of 92.14 %. The removal percentage of chromium in this research was higher compared with the research done by Vikashni et al. (2012), which was only 60% and almost similar to Ravikumar and Sheeja (2013), which was 73%. The copper removal efficiency in this research is approximately similar to the literature statement of Vikashni et al. (2012) which was approximately to 90%.The trend of the removal percentage of copper and chromium were found to be alike as the trend in the research done by (Ali, 2016; Mataka et al., 2010).

The results of the removal percentage of each metal is shown in Figure 2.

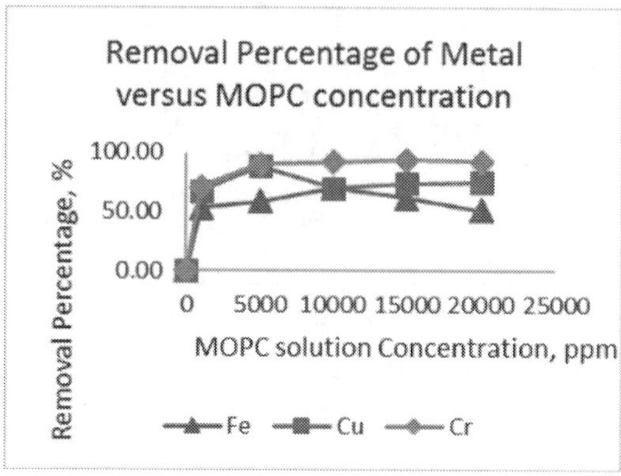

Fig.2. Metal Removal Percentages

3.2 Turbidity

The turbidity of the wastewater was determined by using the turbidity meter and results are shown in the Figure 3

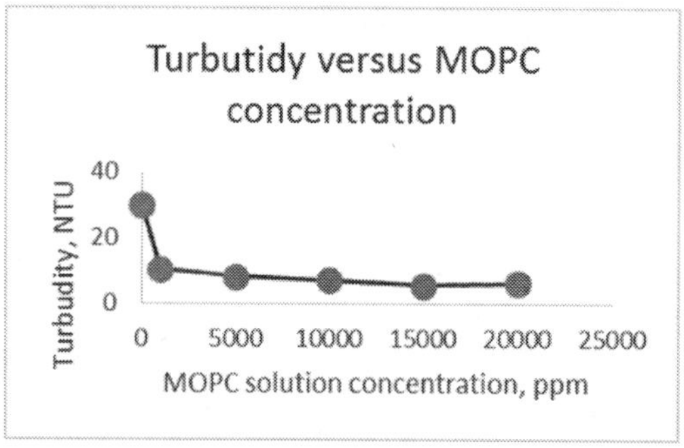

Fig.3. Turbidity of treated wastewater

From the results obtained in the experiment, the wastewater was having turbidity of 30.1 NTU. By adding MOPC, the turbidity was decreased to a minimum value of 5.54 NTU at the optimum concentration of 15000 ppm MOPC solution. Figure 3, shows the trend of decreasing in turbidity as the MOPC solution concentration increase. However, there are a limitation for the turbidity removal and the continue increase of the MOPC will lead to increase residual turbidity. The highest removal percentage of the turbidity was 81.6%.

3.3 pH Value

The pH value was measured by pH meter and the results are shown in the Figure 4.

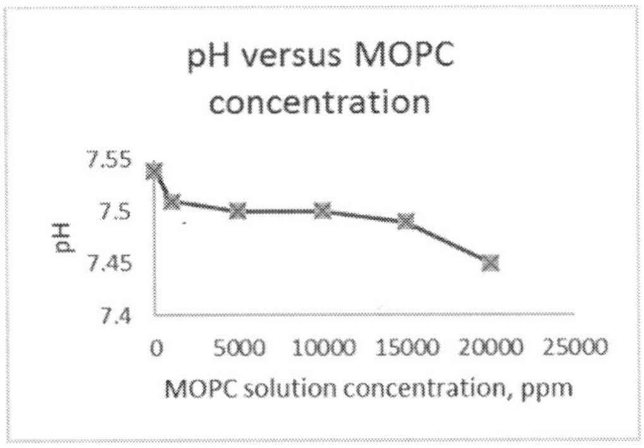

Fig.4. pH value of the wastewater

There was no significant difference in the pH value which is between 7.45 and 7.54 after treatment with different concentration of MOPC as shown in Figure 4, which is in the range accepted by the World Health Organization with permissible limits of 6.5 to 8.5 for drinking water.

3.4 Chemical Oxygen Demand

The COD of the wastewater were determined by using the spectrometer (HACH DR2800) after heating using (HACH DRB200) for 2 hours at 150°C with strong oxidizing agent (potassium dichromate solution) in COD digestion reagent vials. The results of COD are shown in the Figure 5.

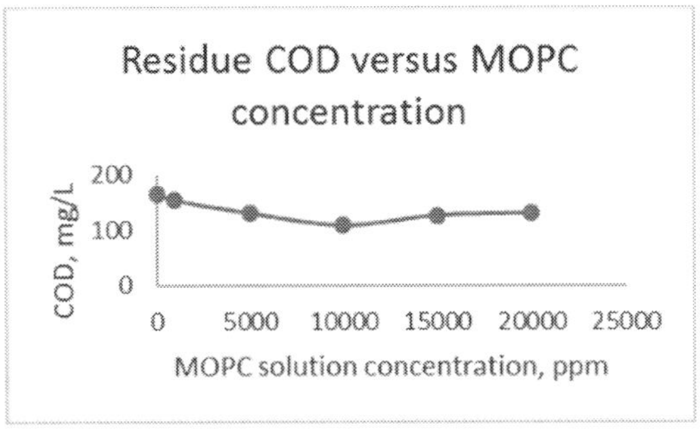

Fig.5. COD of the wastewater

Figure 5, shows that the COD was decreasing by adding MOPC. The COD of the wastewater decrease from 166 mg/L to 108 mg/L at the optimum concentration of 10000 ppm. Then the COD increase slightly with the increase of MOPC solution concentration after it reached the optimum concentration. The maximum COD removal percentage was 34.94 %.

3.5 Biochemical Oxygen Demand

The DO of the wastewater initially and after 5 days were determined by using the dissolved oxygen. The results are shown in Figure 6.

According to the results shown in Figure 6, the DO reduced from 1.31 mg/L to 0.68 mg/L and the BOD value increase as the MOPC solution concentration increase. The BOD had increase from 9 mg/L to 18.3 mg/L when 25000 ppm of the MOPC solution was added to the wastewater. The result are different from the previous research which had shown the removal of the BOD by the *Moringa oleifera*. The increase in BOD indicated the increase of the oxygen demand when higher concentration of the MOPC solution was added to the sample. This is due to the presence of organic compounds in the MOPC, which enable the growth of organism in the water. The decrease in DO value might be caused by the increase in oxygen needed by MOPC to oxidize the organic substance inside the wastewater (Yuliastri et al., 2016).

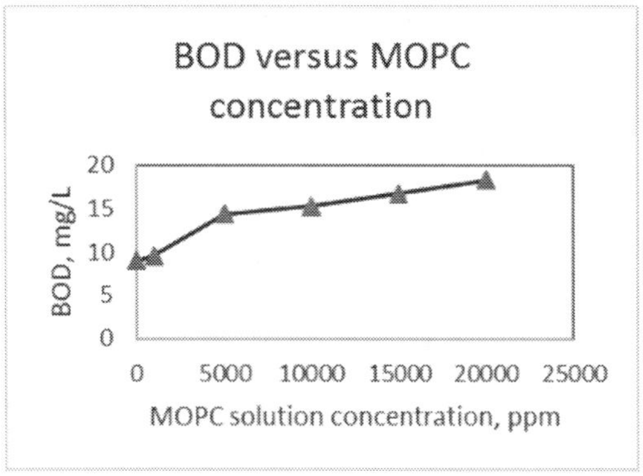

Fig.6. BOD of the wastewater

4 CONCLUSION

The MOPC is a good heavy metal removal. As the MOPC concentration increases the heavy metal removal percentage was increased until the optimum concentration was used. Moreover, the turbidity was reduced as the pH value was not affected, which can be considered as an added value for removal of turbidity and heavy metal where no pH adjustment is needed. MOPC can reduce the COD but the BOD was increased due to presence of some oil in MOPC, and it can suggested to make sure that the oil is totally removed from press cake to reduce BOD. The *Moringa oleifera* press cake can be recommended to be a good coagulant and heavy metal removal.

Acknowledgment

The authors would like to thank the Research and Innovation Department at Universiti Malaysia Pahang, Pahang, Malaysia, for financial support under grant # **RDU 1703172**.

References

Ali, E. N. (2016). Removal of Copper from Water System using *Moringa oleifera* Press Cake.*International Conference on Agricultural, Civil and Environmental Engineering 16,* 15-18.

Ali, E.N., Muyibi, S.A., Salleh, H.M., Salleh, M.R.M. and Alam, M.Z. (2009). *Moringa oleifera* Seeds as Natural Coagulant for Water Treatment. *Thirteenth International Water Technology Conference of Hurghada Egypt*,163-168.

Ali, E.N., Tan, C. S. and Makky, E. A. (2014). Impact of *Moringa oleifera* Cake Residue Application on Waste Water Treatment. *Journal of Water Resource and Protection*, 6, 677-687.

Araújo, C.S.T., Carvalho, D.C., Rezende, H.C., Almeida, I.L.S., Coelho, L.M., Coelho, N.M.M., Marques, T.L. and Alves, V.N. (2013). Bioremediation of Waters Contaminated with Heavy Metals Using *Moringa oleifera* Seeds as Biosorbent. *INTECH*, 227-255.

Aziz, H.A., Adlan, M.N., Mohamed, A.M.D., Raghavan, S., Mohd Isa, M.K. and Abdullah, M.H. (2000). Study on The Anionic Natural Coagulant Aid for Heavy Metals and Turbidity Removal in Water at pH 7.5 and Alum Concentration 25 mg/L- Laboratory Scale. *Indian Journal of Engineering & Materials Sciences*, 7, 195-199.

Bodlund, I. (2013). Coagulant Protein from plant materials: Potential Water Treatment Agent. *Royal Institute of Technology Stockholm*, 1-37.

Kasmawati, M. and Lee, K. K. (2007). Methods of Analysis for Water and Wastewater. UiTM. Shah Alam.

Kalavathy, M.H. and Miranda, L.R. (2010). *Moringa oleifera*—A Solid Phase Extractant for The Removal of Copper, Nickel and Zinc from Aqueous Solutions. *Chemical Engineering Journal*, 158, 188-199.

Kazi, T. and Virupakshi, A. (2013). Treatment ofTannery Wastewater Using Natural Coagulants. *International Journal of Innovative Research inScience, Engineering and Technology*, 2(8),43-47.

Lea, M. (2010) Bioremediation of Turbid Surface Water Using Seed Extract from *Moringa oleifera* Lam. (Drumstick) Tree. *Current Protocols in Microbiology*, 16, 1G.2.1-1G.2.14.

Mataka, L. M., Sajidu, S. M. I., Masamba, W. R. L., &Mwatseteza, J. F. (2010). Cadmium sorption by *Moringa stemopetala* and *Moringa oleifera* seed powders: Batch, time, temperature, pH and adsorption isotherm studies.

Muyibi, S.A., E.S.M. Ameen, M.M.J.M. Noor, and F.R. Ahmadum, (2002). Bench Scale Studies for Pre-treatment of Sanitary Landfill Leachate with *Moringa oleifera* seed Extract. *International Journal of Environmental Studies*, 59(5), 513-535.

Obuseng, V., Nareetsile, F. and Kwaambwa, H.M. (2012). A Study of The Removal of Heavy Metals from Aqueous Solutions by *Moringa oleifera* Seeds and Amine-Based Ligand 1, 4-bis [N, N-bis (2-picoyl) amino] butane. *Analytica Chimica Acta*, 730, 87– 92.

Patil, B.R., Kezia, B., and Mugeraya, G. (2013). Bioremediation of Heavy Metals and Waste Water Treatment Using Leaves and Latex of *Calotropisprocera. International Journal of Engineering Research and Science & Technology*, 2(4), 190-197.

Paula, H.M., Oliveira Ilha, M.S. and Andrade, L.S. (2014). Concrete Plant Wastewater Treatment Process by Coagulation Combining Aluminum Sulfate and *Moringa oleifera* Powder. *Journal of Cleaner Production,* 76, 125-130.

Ravikumar, K., &Sheeja, A. K. (2013). Heavy metal removal from water using *Moringa oleifera* seed coagulant and double filtration. International Journal of Scientific & Engineering Research, 4(5), 10-13.

Sajidu, S.M., Henry, E.M.T., Kwamdera, G., Mataka, L. (2005). Removal of Lead, Iron and Cadmium Ions by Means of Polyelectrolytes of The *Moringa oleifera* Whole Seed Kernel. *WIT Transactions on Ecology and the Environment*, 80, 251-258.

Suhartini, S., Hidayat, N. and Rosaliana, E. (2013) Influence of Powdered *Moringa oleifera* Seeds and Natural Filter Media on the Characteristics of Tapioca Starch Wastewater. *International Journal of Recycling of Organic Waste in Agriculture*, 2, 1-11.

Vikashni, N., Matakite, M., Kanayathu, K. and Subramanium, S. (2012) Water Purification Using *Moringa oleifera* and Other Locally Available Seeds in Fiji for Heavy Metal Removal. *International Journal of Applied Science and Technology*, 2, 125-129.

Yuliastri, I. R., Rohaeti, E., Effendi, H., & Darusman, L. K. (2016). The use of *Moringa oleifera* Seed Powder as Coagulant to Improve the Quality of Wastewater and Ground Water. In IOP Conference Series: Earth and Environmental Science (Vol. 31, No. 1, p. 012033). IOP Publishing.

The Significant Effect of Interface Shear Strength Between Soil Liner and Geotextile with Different Percentages of Bentonite and Sodium Bentonite with Geosynthetic

Roslizayati Razali[1,*], *Nor Azizah* Che Azmi[2], *Diana* Che Lat[3], *Mazidah* Mukri[4] and *Farah Hafifie* Ahamad[5]

[1,2,3,5] Lecturer, Department of Geotechnical and Transport, Faculty of Civil Engineering, Mara University of Technology, Malaysia
[4]Senior Lecturer, Department of Geotechnical and Transport, Faculty of Civil Engineering, Mara University of Technology,41450 Shah Alam, Malaysia

Abstract. This research focuses on the interface shear performance between geotextile and soil with various percentages of bentonite and sodium bentonite for landfill stability by using Direct Shear Test. Admixtures are mixed with clay to improve its performance as a liner material due to their ability to fill the voids between soil particles and also the most effective sealants. The aim of this study is focused on evaluating the effectiveness of bentonite and sodium bentonite on improving interface shear strength of fine soil. In this study, direct shear box was used to measure the interface shear strength of soil with different percentages of bentonite and sodium bentonite at optimum moisture content. Analysis shows that the most suitable percentage is of 5 % of bentonite and 2.5 % of sodium bentonite due to the highest interface shear strength of the mixture with geosynthetic. It is clearly shown that 2.5 % sodium bentonite is the most suitable percentage to be used as admixture for landfill stability as it gives higher interface shear strength.

1 Introduction

Landfills have been the most common methods of organizing waste disposal. In most cases, open dumping is being practiced and takes place at about 50% of the total landfills. In addition, many landfills have been opted to be closed rather than setting up the new ones while the quantity of waste generated yearly is much faster than the natural degradation process [1]. The long term problems caused by landfill are related to the generation of leachate which can cause considerable environmental problems. Leachate is a highstrength wastewater formed as a result of percolation of rain-water and moisture through waste in landfills [2]. Leachate from solid – waste has been found to contain a number of heavy metals. During the formation of leachate, organic and inorganic compounds are transferred

* Corresponding author: roslizayati440@johor.uitm.edu.my

from waste to the liquid medium [3] and pose hazardous effects to the receiving water bodies and thereby pose a potential threat to human health. In order to accomplish this, it is necessary to restrict the flow of water through the upper and lower boundaries of the landfill, referred to as the cover and the base liner of the landfill (McCartney et al., 2002). Recently, geosynthetic clay liner (GCLs) provides an alternative to compacted clay liners [6]. It has been used to replace and reduce the thickness of compacted clay liners (CCLs) required in composite liner or cover systems for waste containment (Eid and Stark, 1998). Due to the importance of the interface shear strength between geotextile and residual soil mixed with adhesive bentonite, this study is conducted to determine the difference of interface shear strength between geotextile and clay with different percentage of bentonite and sodium bentonite. A unique characteristic of sodium bentonite is that it can draw water from adjacent soils, possibly reaching water contents in excess of 100 percent [6].

2 Scope of Study

To achieve the objectives of this study, a stage of laboratory work has been performed so that the result can be presented systematically. All the standard tests were conducted based on British Standard as references. The basic laboratory tests that have been performed are Particle Size Distribution, Particle Density Test and Atterberg Limit which includes Plastic Limit (*PL*) and Liquid Limid (*LL*) in order to determine the Plasticity Index (*PI*). Inaddition, specific gravity test, shrinkage limit test and pH test were also included in this study. All of the basic properties tests excluding Particle Size Distribution were performed repeatedly with different percentage of bentonite and sodium bentonite. After all of the test have been conducted, soil samples were prepared and mixed together using various percentages of bentonite and sodium bentonite (2.5%, 5%, 7.5% and 10%) to get the best value of optimum moisture content of the admixtures using Standard Proctor Test. The value was used for the preparation of remoulded samples for direct shear test. The shear box is normally the most preferred method for soil shear testing because it is a fast and economical method of measuring shear strength. A specimen is placed in a shear box which has two stacked halves to hold the sample; the contact between the two halves is at approximately the mid-height of the sample. Soil without admixture at the top, bound by geotextile at the middle and soil with the admixture (bentonite or sodium bentonite) is placed at the bottom half of the apparatus. Geotextile and soil with admixture were placed exactly at the failure plane. A normal load was applied to the specimen and the specimen was sheared across the pre-determined horizontal plane between the two halves of the shear box. The size of the shear box used is 60 mm × 60 mm, which can give a more efficient result for soil shear strength as stated by Ingold (1982).

3 Theoretical background

Over the past few years, different geosynthetics manufacturers have proposed several unique GCL products that attempt to minimize hydraulic conductivity while maximizing shear strength [7]. Throughout the years, geosynthetic clay liners(GCLs) has been increasingly selected to replace compacted clay liners (CCLs) in composite liner and cover systems for waste containment facilities (Stark, 1998). GCLs are prefabricated geocomposite materials used in hydraulic barriers as an alternative to compacted clay liners due to their affordability and excellent barrier performance against contaminants. During the landfill construction, there are some critical element parts that need to be considered. One of the critical parts in this stage is when designing the landfill slope. Hence, engineers are required in designing slope which is safe in terms of not exceeding the safe slope angle for liner components, internal properties and their respectiveinterface parameters within the system (Saravanan et al, 2004). In this study, various admixtures of bentonite and sodium bentonite were used in order to determine the significant effect of interface shear strength between soil liner and geotextile with different percentages of admixtures.

4 Test Result and Discussion

Each sample has been processed at the laboratory accordance with British Standard Method – BS1377: Part 2. Table 1 and 2 show the list of basic properites of the soil.

Table 1: List of the basic properties tests of soil with different percentages of bentonite.

Sample	SG (Mg/m3)	LL (%)	PL(%)	PI (%)	PH	SL (mm)
SL-B	2.4	44.26	37.59	7	3.96	1.23
SL-2.5B	2.25	64.24	42.31	22	4.22	3.13
SL-5B	2.42	66.03	43.99	22	4.63	5.73
SL-7.5B	2.47	68.16	46.88	21	4.83	7.33
SL-10B	2.49	72.35	48.64	24	5.4	7.93

Table 2: List of the basic properties tests of soil with different percentages of sodium bentonite

Sample	SG (Mg/m3)	LL (%)	PL(%)	PI (%)	PH	SL (mm)
SL-SB	2.4	44	38	6	3.96	11.67
SL-2.5SB	2.46	63	49	14	4.37	13.33
SL-5SB	2.44	54	37	17	4.76	14
SL-7.5SB	2.6	61	48	12	5.15	15.67
SL-10SB	2.62	65	45	20	5.28	16

The results show that the increasing percentages of the bentonite and sodium bentonite give higher plasticity of the soil. The presence of bentonite in the soil gives the chances for more water to retain at the condition of suction. It was proved by Osinubi and Amadi in 2010 where in their previous studies, it was shown that liquid limit is increased in increasing of bentonite. When bentonite is added to the soil, it displays a clayey nature at relatively high water contents. Bentonite which contains swelling clay minerals such as montmorillonite is inferred to expand and fill the voids in the buffer and backfill materials (Komine and Ogata, 1996). In addition, it is proved that the presence of bentonite reduces the acidity of the natural soil and this is proven by Zanzinger et. al in 2002 [11], where bentonite typically have the value of pH 9 or more.

Compaction test

Proctor standard compaction test, according to BS 1377-1990: Part 4 was applied to determine the maximum dry density (MDD) and the optimum moisture content (OMC)
for all samples added with bentonite and sodium bentonite. It is observed from Fig. 1 that the percentage of optimum water content is increased relatively with the increased percentage of bentonite whereas Maximum Dry Density (MDD) is decreased slightly with the increased percentage of bentonite. Chalermyanont [12] stated that with more bentonite added, the optimum water content increases and maximum dry density decreases. Fig. 2

 https://doi.org/10.1051/matecconf/201815002009

shows the combination of compaction curve added with sodium bentonite and it could be clearly seen that by adding sodium bentonite to the soil samples, the Maximum Dry Density (MDD) increases as compared to blank sample.

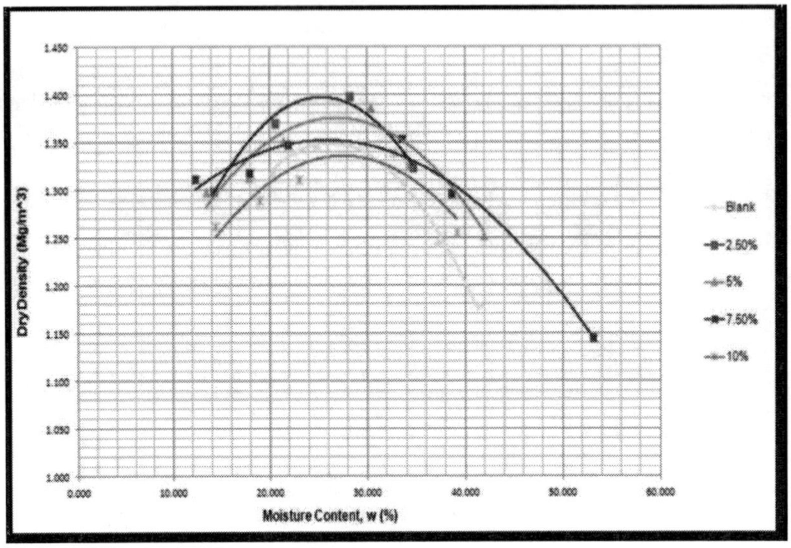

Fig.1: The combination of Maximum Dry Density against Optimum Moisture Content

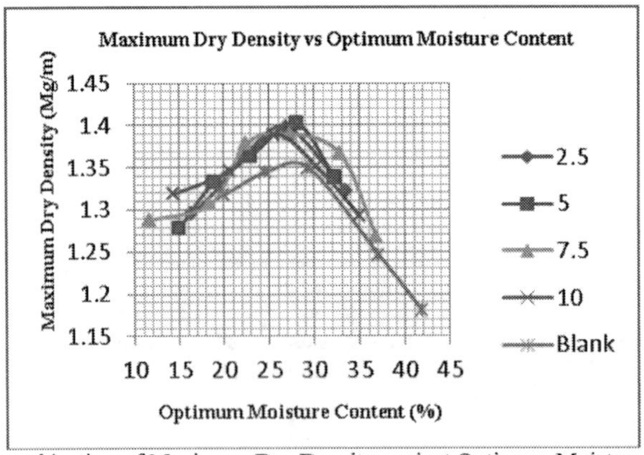

Fig. 2: The combination of Maximum Dry Density against Optimum Moisture Content

Shear box test

The size of shear box use in this study is 60 mm x 60 mm. Soil at top is to bound by geotextile at the middle and clay with the admixture (bentonite or sodium bentonite) is placed at the bottom part of the apparatus (Fig. 2.0). Geotextile and soil with admixture are placed exactly at the failure plane. The geotextile is wrapped on the surrounding upper area of the shear box to ensure that there is no folding of the geotextile during the shearing stage. Direct shear Test was performed with applied normal stress of 100 kPa, 200 kPa and 300 kPa for both samples of bentonite and sodium bentonite. Fig. 3 until Fig. 7 show five (5) samples prepared using the optimum moisture content that was obtained from the compaction tests. The graphs below present the relationship between Shear Stress and Horizontal Displacement of bentonite for SL-B, SL-2.5B, SL-5B, SL-7.5B and SL-10B.

Figure 2.0 sample preparation

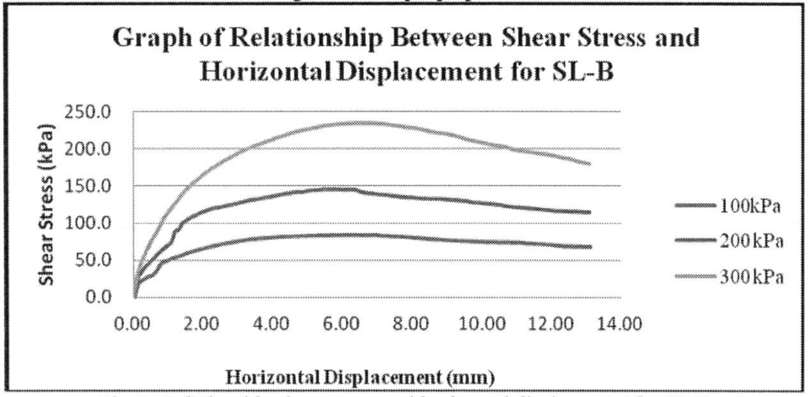

Fig. 3: Relationship shear stress and horizontal displacement for SL-B

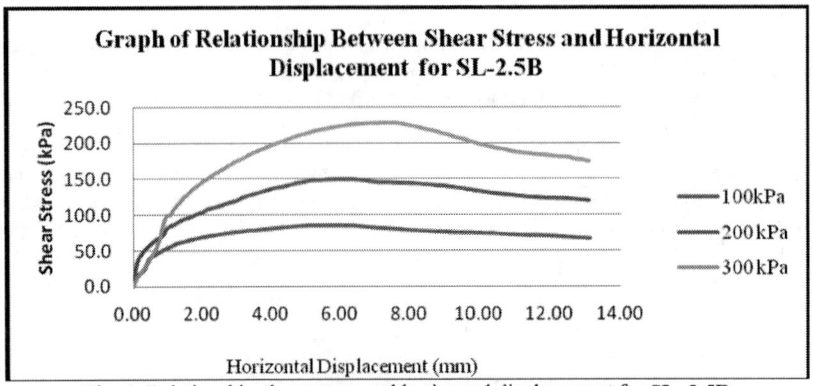

Fig. 4: Relationship shear stress and horizontal displacement for SL- 2.5B

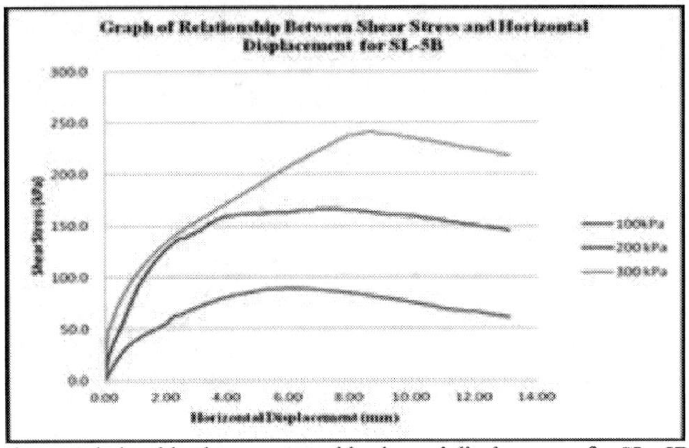

Fig. 5: Relationship shear stress and horizontal displacement for SL- 5B

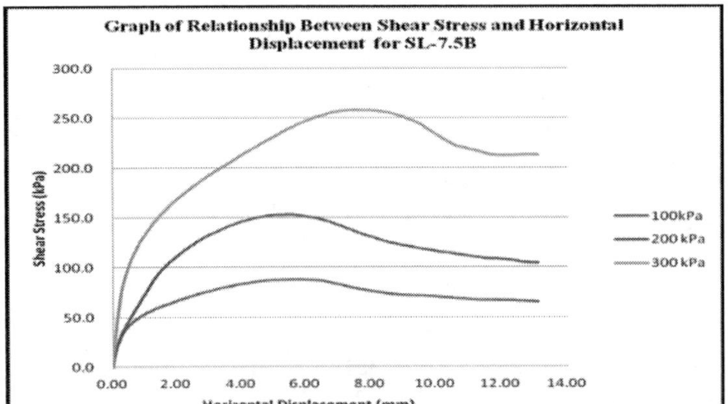

Fig. 6: Relationship shear stress and horizontal displacement for SL-7.5 B

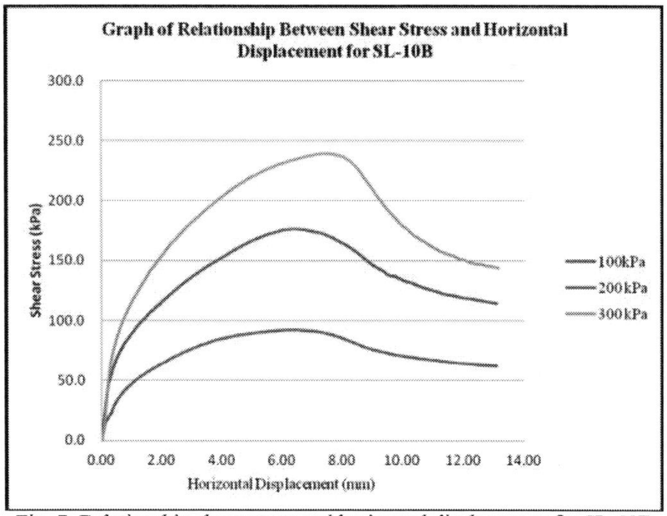

Fig. 7: Relationship shear stress and horizontal displacement for SL-10B

Based on the maximum value of shear stress with applied normal stress, graph of cohesion and internal angle can be determined. The summary of cohesion and friction angle of the soil is presented in Table I. From the result, the presence of bentonite in the soil did not give a serious impact on the cohesion value and the angle of interface friction between samples and the geotextile. Based on the result, cohesion angle without bentonite gives the minimum value of cohesion and friction angle which are 10.85 kPa and 35.16°. The increasing of bentonite showed the increment until the additional of 5% of the additive. However, there is a limitation for bentonite content beyond which is indicated by the drop of parameters after 7.5% of the bentonite content due to the reaching of optimum limit of bentonite added to the soil. The increase in the percentage of bentonite can help in increasing the cohesion and friction angle for the tested samples of SL-2.5B and SL- 5B up to 37.24° for friction angle and 13.29 kPa.

Table I: Cohesion and Friction value for Bentonite

Sample	Ø°(degree)	C(kPa)
SL-B|	35.16	10.85
SL-2.5B	35.45	12.45
SL-5B	37.24	13.29
SL-7.5B	36.80	12.54
SL-10B	36.60	12

Fig. 9 until Fig. 13 below present the relationship of Shear Stress and Horizontal Displacement of sodium bentonite for SL- 0SB, SL-2.5SB, SL-5SB, SL-7.5SB and SL 10SB. This test was performed with applied normal stress 100 kPa, 200 kPa and 300 kPa. The normal stress was kept constant during test and the maximum value of shear stress can be measured during the direct test.

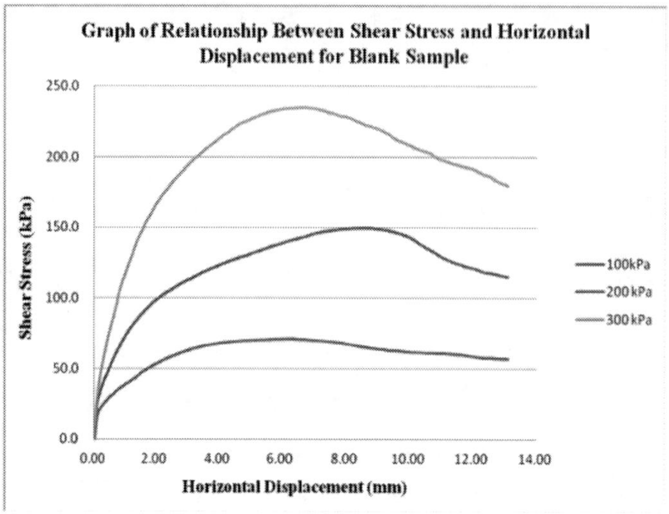

Fig. 9: Relationship shear stress and horizontal displacement for SL-0SB

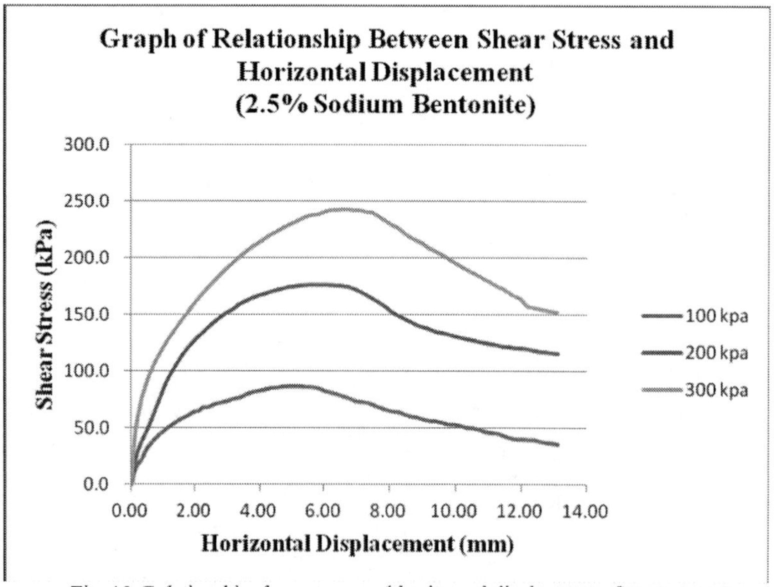

Fig. 10: Relationship shear stress and horizontal displacement for SL-2.5S B

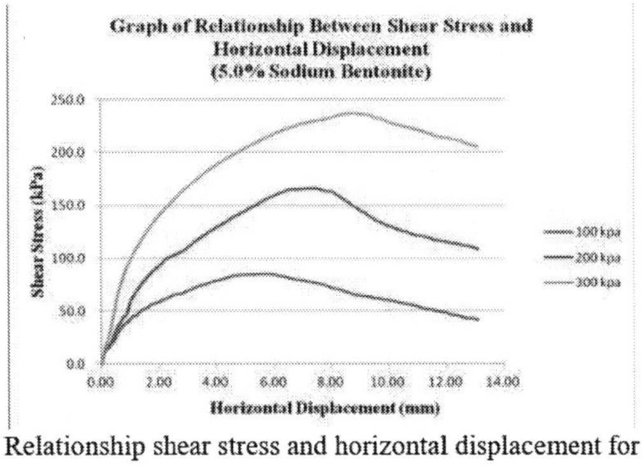

Fig. 11: Relationship shear stress and horizontal displacement for SL-5SB

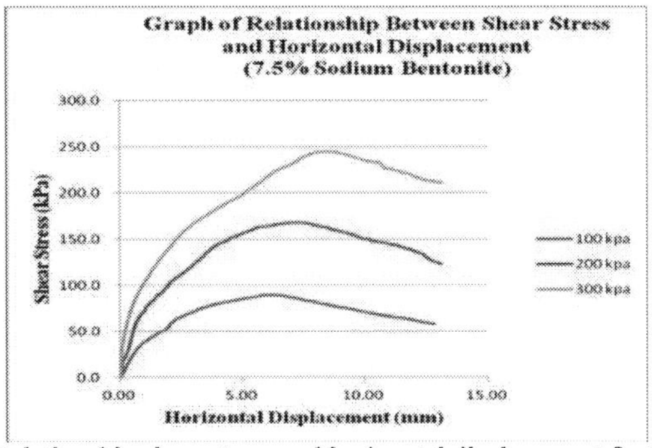

Fig. 12: Relationship shear stress and horizontal displacement for SL-7.5SB

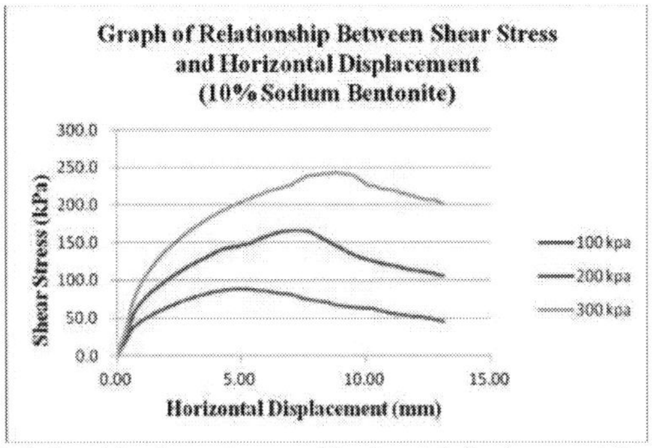

Fig. 13: Relationship shear stress and horizontal displacement for SL-10SB

Based on Table 2, it shows that the higher value for c is at soil added with 2.5% of sodium bentonite that gives the value of 12.35kPa and friction of 37.95%. As for the rest, the soil that was added with 10% of sodium bentonite is found to produce the lowest value for c with 11.27 kPa and friction of 37.70%. It can be said that the more sodium bentonite were added to the soil, the lower the shear strength tend to be, due to the swelling of the soil. It can be presumed that the cohesion value from the result is insignificant as the shear between soil and geotextile will affect the movement of the soil particle. Only natural soil sample gives lower value of cohesion but the other samples give quite similar value for cohesion. Apart from that, the value for the friction angles did not show any apparent change with the increasing percentage of sodium bentonite, and the result is quite similar to each other.

Table II: Cohesion and Friction value for sodium Bentonite

Sample	°(degree)	c(kPa)
SL-0SB	10.63	36.3
SL-2.5SB	12.35	37.95
SL-5SB	11.43	37.07
SL7.5SB	12.3	37.76
SL-10SB	11.27	37.70

From the parameters, the shearing resistance of the soil is calculated for applied normal stress of 100kPa, 200kPa and 300kPa. Shearing resistance at failure for the tested samples shows in Table III.

Table III(a): Value of shear strength for Bentonite

Normal stress (kPa)	Shear strength (kPa)				
	SL-B	SL-2.5B	SL-5B	SL-7.5B	SL-10B
100	81.29	83.64	89.3	87.35	86.27
200	151.73	154.85	165.32	162.16	148.53
300	222.16	226.04	241.33	236.97	234.8

Table III(b): Value of shear strength for sodium bentonite

Norm al stress (kPa)	Shear strength (kPa)				
	SL-0B	SL-2.5SB	SL-5SB	SL-7.5SB	SL-10SB
100	84.09	90.34	86.98	89.76	88.56
200	157.54	168.33	162.52	166.10	165.85
300	231	246.32	238.07	244.67	243.14

In comparison to experimental results, it can be stated that the increment percentage of additives added give different value of shear strength. Maximum value of shear strength tested at 100 kPa, 200kPa and 300kPa applied normal stress when added with 5% bentonite (SL-5B) and shear strength start to drop after that. However, based on the result that has been conducted for sodium bentonite, the best value of shear strength reach at 2.5% of sodium bentonite added (SL – 2.5SB) and start to decrease after more sodium bentonite is added.

5 Conclusion

Based on the findings and data of physical properties and engineering properties, the increase of the percentages of bentonite and sodium bentonite tends to increase the liquidity Index of fine soil. For compaction test, the percentage of optimum water content is increased relatively with increasing bentonite but given vice versa result for the value of the Maximum Dry Density (MDD). In case of sodium bentonite, the result for compaction test did not give good relationship to the soil as the optimum moisture content and Maximum Dry Density (MDD) are not consistent. In Direct Shear Test, it shows that even the presence of bentonite and sodium bentonite in the sample is proved to give higher interface shear strength to the tested soil, additional bentonite and sodium bentonite tend to decrease its shear strength. This may occur when increasing content of bentonite and sodium bentonite are added to the soil, increase in water content thus results in the decrease in density and shear strength. It shows that sodium bentonite is the most effective sealants and suitable for landfill stability as it gives higher interface shear strength compared to bentonite

6 Acknowledgment

The authors are thankful to Research Management Institute (RMI) University Technology MARA (UiTM0, for funding this paper and providing a place to carry out the experimental work. Also, special thanks to Dr Mazidah Mukri who has given numerous advices, information, encouragement, criticism and support during this research. In addition, special thanks to the laboratory staffs of Universiti Teknologi MARA (UiTM) for their valuable assistances in handling the laboratory equipment.

References

1. Mohd Masirin, Mohd Idrus and Ridzuan, Mohd Baharudin and Mustapha, Shahabudin and Adon@Mohd Don, Rashidah, *An overview of landfill management and technologies: a Malaysian case study at Ampar Tenang*. In: 1st National Seminar on Environment, Development and Sustainability (PSISenviro2008), 28-29 July 2008, Selangor, Malaysia, (2008)
2. Hasar, H., Unsal, S. A., Ipek, U., Karatas, S., C nar, O., Yaman,C. and K nac, C. , Stripping/ flocculation/ membrane bioreactor/reverse osmosis treatment of municipal landfill leachate. *Journal of Hazardous Materials*, 171, 309-317, (2009)
3. Bohdziewicz, J., Neczaj, E. and Kwarciak, A., Landfill leachate treatment by means of anaerobic mem brane bioreactor, Desalination, 221, 559–565, (2008)
4. John Scott McMcartney, Jorge G. Zomberg and Robert H. Swan, Internal and Interface Shear Strength Of Geosynthetic Clay Liners (Gcls), (2002)
5. Stark, H. Eid. I, Shear Behaviour of an Unreinforced Geosynthetic Clay Liner. *Geosynthetic International*, 645- 658, (1998)
6. Daniel, D.E. Shan, H,Y,. And Anderson, J.D., Effects Of Partial Wetting on the Performance of the Bentonite Component Of Geosynthetic Clay Liners, *Proceedings Of Geosynthetics'93, IFAI,* Vol, 3, March 1993, Pp. 1483- 1496, (1993)
7. Bouazza, A., Geosynthetic Clay Liners, Geotextile and Geomembrane, 20, 3-17, (2002)
8. Stark, T.D., Williamson, T.A. and Eid, H.T., HDPE Geomembrane/Geotextile Interface Shear Strength, Journal of Geotechnical Engineering. ASCE. 122(3): 197-203, (1996)
9. Osinubi K.J & Amadi A.A, Variations in Soil Water Characteristic Curves of Lateritic Soil Treated with Bentonite. *Experimental and Applied Modellingog Unsaturated Soils, ASCE,* (2010)
10. Komine, H., and Ogata, N. ''Observation of swelling behaviour of bentonite by new electron microscope.'' *Proc., 2nd Int. Congress on Environmental Geotechnics,* 1, 563–568, (1996b)
11. Zanzinger, H., Koerner, M. & Gartung, E., Clay Geosynthetic Barriers. *Proceedings of International Symposium is [i.e. In] Nuremberg 2002, Nuremberg/Germany*, 12-15, (2002)
12. Chalermyanont, T., & Arrykul, S., Compacted Sand-Bentonite Mixtures for Hydraulic Containment Liners. *Songklanakarin Journal Science and Technology* , 315-322, (2005)
13. British standard institution, Method of test for soils for civil engineering purposes, BS 1377 – 1990 : Part 2 and 4, (1990)

Comparison of Pigtail with J Anchor Bolt in Normal Concrete

Mohamad Hairi Osman[1,*], *Mohamad Nur Mustaqim* Abd Shukor[1], *Suraya Hani* Adnan[1], *Mohamad Luthfi* Ahmad Jeni[1], *Mohd Sufyan* Abdullah[1], *Salman* Salim[2], *Hannifah* Tami[3], and *Nor Azira* Abdul Rahman[4]

[1]Faculty of Engineering Technology, Universiti Tun Hussein Onn Malaysia, Parit Raja, Batu Pahat, Johor, Malaysia
[2]Centre For Diploma Studies, Universiti Tun Hussein Onn Malaysia, Parit Raja, Batu Pahat, Johor, Malaysia
[3]Faculty of Civil and Environmental Engineering, Universiti Tun Hussein Onn Malaysia, Parit Raja, Batu Pahat, Johor, Malaysia
[4]Politeknik Sultan Abdul Halim Mu'adzam Shah, Bandar Darulaman, Jitra, Kedah Darul Aman, Malaysia

Abstract. Anchor bolts have been used to attach the steel structure of concrete and transfer load into the concrete. Anchor bolts in concrete are to withstand the shear force as they connect steel beams to the reinforced concrete foundations. The research was carried out to increase the understanding and investigating the performance of anchor bolts Pigtail since there is a lack of research on the bolt and to compare it with anchor bolts J which have been widely used by the industry. According to the methodology, testing the tensile load was used in this research to get a maximum load capacity of the anchor bolt in concrete. The sample was embedded in concrete cylinders of 75mm radius and 300mm height. Depths of embedment were 200mm, 230mm, and 260mm. The Universal Testing Machine, UTM was used to test the strength of tensile. 12 samples were used, each type and depth used two samples to get the average value using concrete grade 30. The samples underwent the process of curing for 28 days. The anchor bolts J used a 40mm bending radius and the length of hook was 100mm. Anchor bolts pigtail also used the 10 bending radius with depth of 8mm that have been compressed using 16mm radius rigs with a load of 500kN. Results of the research showed the depth of 260mm for Pigtail almost the same with J 60.529kN and 53.628kN and anchor bolts J were 75.557kN and 76.332kN. Difference of the values was not too far vary when compared with the 200mm and 230mm depths. Each comparison showed the ability of each bolt and anchor bolt failure occurred. Performance of the anchor bolts pigtail can be used on a structure or a higher load at an embedment of 260mm. Embedment of 200mm and 230mm can be used on a lighter load if steel material saving is a priority in every usage.

1 Introduction

J anchor bolt has been widely used for construction industry today. However, Pigtail anchor bolts had less exposure and use in the construction industry. Normally Pigtails anchor bolts were used as a connection between the structures. The problem which always occurred is that J anchor bolt is difficult to install in the reinforcement bar due to its shape. Furthermore, lack of research has been done to prove that Pigtail anchor bolts may be used as the same function as J anchor bolts.

The objective of this research is to investigate the performance of Pigtail and J anchor bolts embedded in normal concrete under tensile load test. Next objective is to find out the optimum of embedment and investigate the type of failure that occurs until maximum load is achieved.

1.1 Scope of study

In this study, all the tests were using Universal Testing Machine (UTM). This research used two types of bolt systems which were Pigtail and J anchor bolts. The tests were limited to tensile load only for Pigtail and J as the anchor bolt in concrete. The embedment length of anchor bolts samples J1A, J2A, P1A and P2A in the concrete was 200mm. Next samples, J1B, J2B, P1B and P2B were 230mm embedded in the concrete. Next samples, J1C, J2C, P1C and P2C were embedded 260mm in the concrete. All samples were using the same dimensions of concrete cylinders which were 75mm in radius and 300mm in height. These concrete were going through the curing process of 28 days. Concrete plate was supported by a platform with dimensions of 150mm x 150mm x 300mm when tested using the UTM. The plate design platform and anchor bolt tensile capacity were based on BS 5950-1: 2000.

1.2 J anchor bolt, L anchor bolt and pigtail anchor bolt

Bolt anchors J and L are always used in the constructions. 180° and 90° bending by the end of each of the components aim to transfer load from upper structure to the lower structure. The anchor positions

* Corresponding author: mhairi@uthm.edu.my

https://doi.org/10.1051/matecconf/201815003001

must be coordinated with the reinforcement due to large constructions area and making it easy for installation at the construction site so that errors can be reduced [15]. A common problem during site installation there is. This is a critical job and is often done by skilled labour [6]. After that, it is not guaranteed that we can use after the installation is done because in the process of hardening the concrete, there is a possibility of movement or vibration in the anchor bolts in which to anchor the resulting movement. If bolt size or design does not follow the specifications after the concrete is harden, it needs to be re-issued in order to replace with the new anchor bolt [14]. Several types of anchors Cast-in-place are hex headed, L-bolt, J-bolt and bolt Welded headed (see Fig.1) [9]

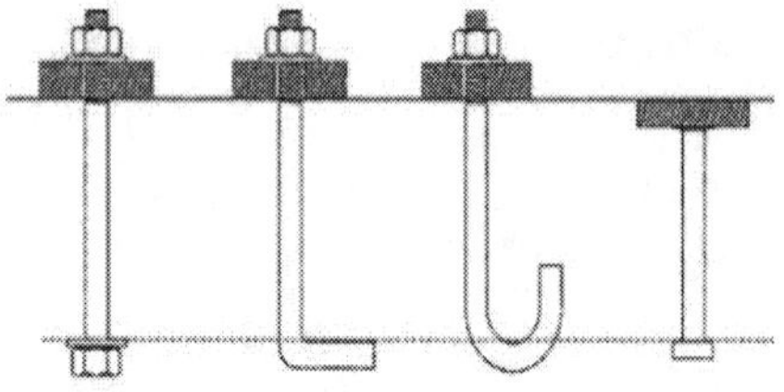

Fig. 1. a) Hex headed bolt b) L-bolt c) J-bolt d) Welded headed bolt

Pigtail anchor is with 10 bends in the concrete. Pigtail anchor bolt has a unique design, which gradually increases the depth of each groove to increase the length of the flexible anchor. The front of the anchor line allows transferring the pressure on the entire length. The gradient is positive that anchors are in high friction with the bend on the anchor (see Fig 2) [21]

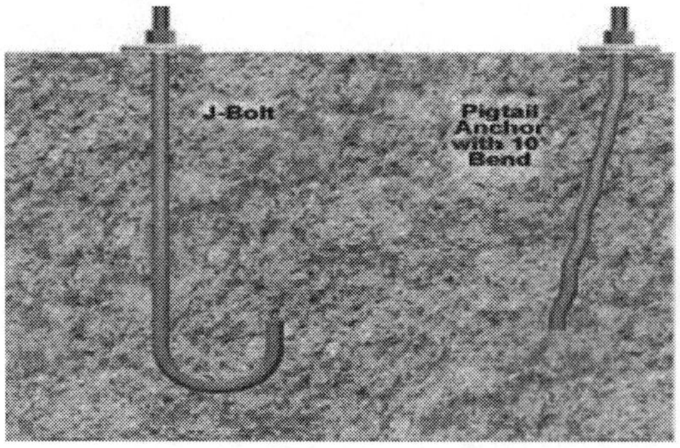

Fig. 2. Anchor bolt of J and pigtail

2 Experimental Work

Tensile strength test has been used in this experiment to obtain the capacity of load for the anchor bolts.

2.1. Samples Preparation

Pigtail Anchor bolts were made of reinforcement bar Y12. The diameters of the bolts were 12mm or 16mm [3, 11]. Preparation of the pigtail anchor bolts using a compression machine to form bending was as shown in Fig.3. The reinforcement bar Y12 was placed between the reinforcement rigs to be compressed with 500KN. The rig was made of 16mm diameter bars which were 16mm bars between each bending. Bonding connections on the rig had been ensured to have no movement or failure during compression. Fig.4 shows the rig and pigtail anchor bolt formed after the compression was completed.

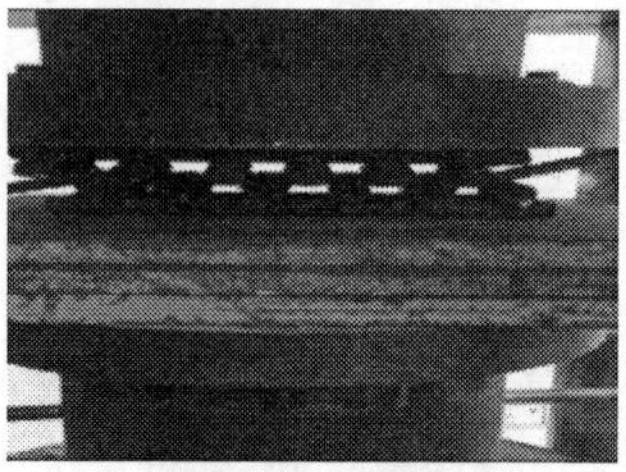

Fig. 3. Compression on bar

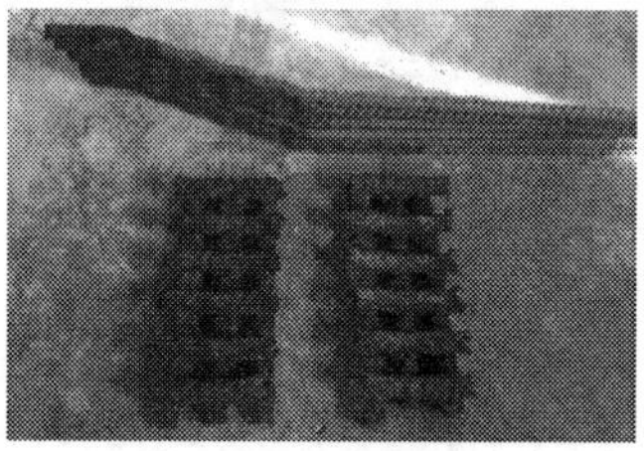

Fig. 4. Anchor bolt Pigtail and rig

Samples of the pigtail anchor bolts which have 10 bending with 16mm of diameter and also the distance at the end of pigtail anchor bolts from the original position was 50mm. All samples were following the same distance to avoid the negative impact on the results. The radius of the J hook or curve of each sample was 40mm and the length of the hook to the J anchor bolt all designed with 100mm in each anchor bolt and was set to the same length [3]. Table 1 shows the embedment of anchor bolts which have been used in this research. The symbol of J is for J-Bolt and P is for Pigtail. Symbol A, B and C is for depth 200, 230 and 260mm. Symbol 1 and 2 to show sample 1 and 2.

Table 1. Embedment J and Pigtail.

Sample	Embedment Depth (mm)			Sample	Embedment Depth(mm)		
	200	230	260		200	230	260
J1A	√			P1A	√		
J2A	√			P2A	√		
J1B		√		P1B		√	
J2B		√		P2B		√	
J1C			√	P1C			√
J2C			√	P2C			√

2.2 Cast-in-place

The concrete formwork was using polyvinyl chloride (PVC) with a diameter of 150mm and pre-cut the mould to make it easier in the process of removing the concrete from the PVC. Fig.5 shows the harden concrete in a cylindrical mould. DOE mix design methods were used to measure the exact amount of cement, fine aggregate, coarse aggregate and water in this experiment. Concrete mix design was to achieve a compressive strength of 30N /mm2 in 28 days and the design of slump was between 30mm to 60mm. The volume of concrete cubes was 0.1 mm3 each and concrete anchor bolts for the sample was 5.3 mm³ each. The curing process has been carried out and the water level was ensured at the same level as the concrete cylinder to prevent corrosion of the anchor bolt as it will affect the performance of the bolt [4].

Fig. 5. Bolt embedded in Concrete Cylinder

2.3 Pull-out test

Universal Testing Machine has been designed to perform various types of tests or experiments on tensile or compressive experiments or tests. Tests were conducted at Jamilus Research Centre. The sample was placed between the top and bottom and it was attached to the holder at the bottom and the upper parts. First test started with sample J200 and ensured that the jaws must grip tightly for a perfect performance [7]. The machine began to draw a graph of the tensile load until it reached the maximum capacity of the sample. The machine was stopped and the results were recorded in the computer.

3 Results and Discussions

The average results for the compression test for 7 days and 28 days were 28.5Mpa and 30.8Mpa which has achieved grade 30 as the scope of this research. Pull-out tests were carried out by using the Universal Testing Machine (UTM) to measure the strength of the bond between J and Pigtail in normal concrete. Results data is shown in Table 2 and fig. 6 for pigtail anchor bolt while the bolt anchors J in table 3 and figure 7. Load was increased gradually to achieve maximum value until failure occurred. Each depth has two samples to get an average value of the maximum tensile load. Based on the average result, 200mm embedment reached the lowest maximum value at 37.89kN. The highest average reading was 56.94kN with a depth of 260mm. The results showed that the higher the embedment, the higher maximum capacity achieved until failure occurred on the samples. This was because the bond between the anchor bolt and the concrete depended on the surface area or the length of the anchor. The elongation length was also one of the factors in order to achieve maximum value [5]. If the elongation was increased but no load was imposed as seen on bolt anchors P230B, jaws of the universal testing machine (UTM) and rig were loosen during start up. Next, the test was continued with a 4mm elongation until maximum tensile load was reached. The value of elongation of 200A and 200B were the highest because of the extension of the original position between the concrete and the anchor has a weak bonding with a depth of 200mm and length of anchor bolts the lowest length of 220mm [10]. The factors affecting the strength of tensile load were the adhesion concrete to anchor bolts and the friction force between concrete and anchor bolts[20].

Table 2. Result Load Tensile on Pigtail Anchor Bolt.

Sample	Embedment (mm)	Length of anchor bolt (mm)	Maximum load (kN)	Average Maximum load (kN)	Elongation (mm)
P1A	200	220	28.134	37.89	14.276
P2A	200	220	47.642		14.269
P1B	230	250	45.05		5.397
P2B	230	250	37.746	41.40	12.875
P1C	260	280	60.529		6.242
P2C	260	280	53.628	56.94	8.239

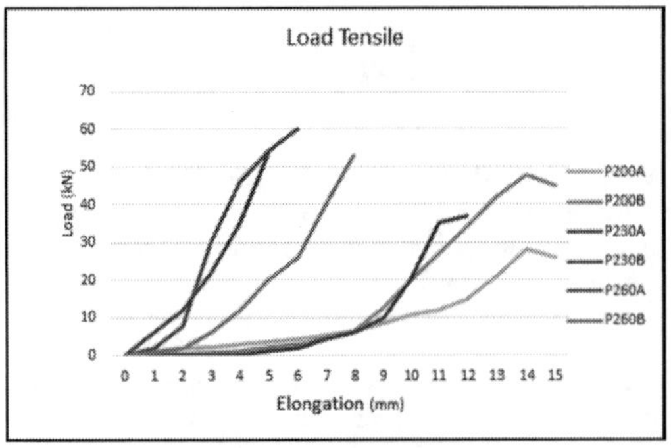

Fig. 6. Pullout test for Pigtail

Sample J2B with an embedment of 230mm with 360mm length of anchor has the highest maximum tensile load which was 80.642kN and the lowest maximum value for sample J2A embedment of 200mm was 68.12kN. Based on the expected results of embedment depth of 260mm, which will achieve the highest result, otherwise, the embedment of 230mm is the highest maximum tensile load. This was due to the 260mm steel failure which occurred and more load was distributed to the steel. Based on Fig.7 it shows that the range of the maximum tensile load does not differ too much. Furthermore, based on the average of the maximum load for each depth of 200mm, 230mm and 260mm the difference was insignificant and the adhesion and friction forces were almost the same [7]. In conclusion, the average value of the maximum load, the depth of 230mm is the optimal depth between 200mm and 260mm with the use of reinforcement bar Y12 in the concrete cylinder.

Table 3. Result Load Tensile on J Anchor Bolt.

Sample	Embedment (mm)	Length (mm)	Maximum load (kN)	Average maximum load	Elongation (mm)
J1A	200	360	74.237	71.18	14.53
J2A	200	360	68.124		15.63
J1B	230	390	71.761		13.65
J2B	230	390	80.462	76.11	13.50
J1C	260	420	74.557		21.56
J2C	260	420	76.332	75.45	27.00

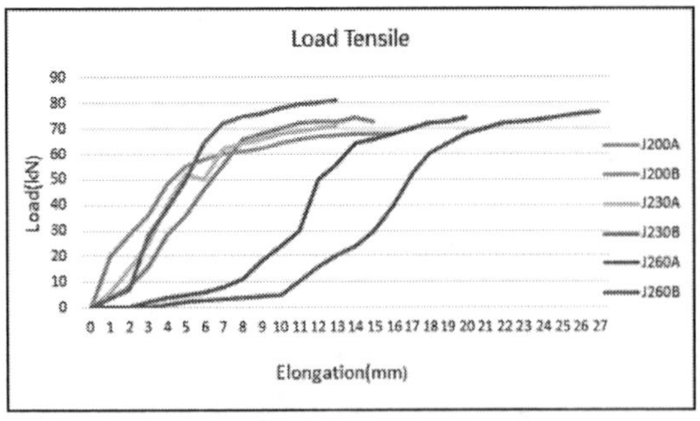

Fig. 7. Pullout test for J

Table 4. Comparison Pigtail with J Anchor Bolt

Sample	Embedment (mm)	Length (mm)	Maximum load (kN)	Sample	Embedment (mm)	Length (mm)	Maximum load (kN)
P1A	200	220	28.134	J1A	200	360	74.237
P2A	200	220	47.642	J2A	200	360	68.124
P1B	230	250	45.05	J1B	230	390	71.761
P2B	230	250	37.746	J2B	230	390	80.462
P1C	260	280	60.529	J1C	260	420	74.557
P2C	260	280	53.628	J2C	260	420	76.332

Table 4 shows the comparison between Pigtail and J anchor bolts. The difference of the maximum tensile load for an embedment of 200mm and 230mm for each type of anchor bolt that is very significant to 200mm pigtail anchor bolt are 28.134 kN and 47.642 kN, while the J are 74.237kN and 68.124 kN. Embedment of 230mm anchor bolts for pigtail are 45.05 kN and 37.746 kN, while the J about 71.761 kN and 80.462 kN. These results indicated that anchor bolt J is better than the Pigtail because the design of the hook of J anchor bolt increases the ability of cling in the concrete[19]. Furthermore, the friction force is also increased because the length of the anchor bolt embedment in J is more than in the Pigtail.

Embedment of 260mm for Pigtail was almost the same with J, maximum load for Pigtail were 60.529 kN and 53.628 kN. Anchor bolts J maximum tensile load were 75.557 kN and 76.332 kN. This was because the highest friction and the highest bonding of the anchor bolt with J260 was 420mm length of the anchor [2]. The difference was not too far vary from the embedment of 200mm and 230mm. Each performance indicated the ability of each anchor bolt according to the type and embedment.

Results of this study can be divided into three types. Fig 8 shows the failures that occurred. Steel failure happened to anchor bolt J2A, J1B, J2B, J1C, J2C. It shows all the samples of type J, a strong bond in the concrete due to cracks or failure did not occur in this sample. This was because the load distribution was focused on the steel until failure occurred. Failure blowout occurred on sample P1A, J1A, P1C, P2C. The failure occurred because the dimensions were limited and load distribution were also focused on the tip of the anchor bolt interiors or the end of the anchor bolt cracks begin to occur on the concrete samples. Finally, splitting failure occurred in sample P2A, P1B and P2B also because the end of anchor bolt was positioned close to the edge of the concrete and showed the load spread evenly until failure occurred in the axis indicating the failure was torn in two. Fig 8 shows the common failure in this research.

https://doi.org/10.1051/matecconf/201815003001

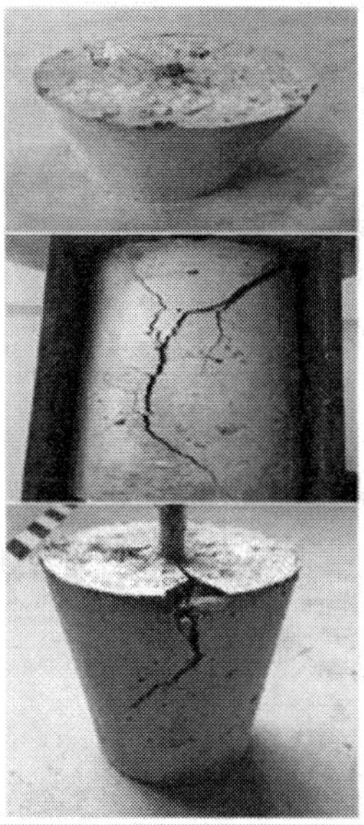

Fig. 8. Types of failure

4 Conclusions

J anchor bolt is widely used in the industry for connecting the structures as precast structure or industrial building system. It is very difficult to be installed in a confined space and needed so much of a skilled labour. Pigtail is a new design but lack of research for this type of anchor bolt design. This research has been done to compare the maximum tensile load, optimum embedment and types of failure for both type of bolts.

It can be concluded, the larger the surface area, the higher maximum tensile load achieved. Friction force was increased between concrete and anchor. Effect from the result for sample P230 was the holder or grip of UTM was loosen and load did not work until elongation reached 4mm. Every test for the next research is to ensure the grip is tight and replace with a new grip to show perfect result.

Recommendation from this research is to increase the bending of Pigtail to get hook ability to increase maximum tensile load. Besides, increasing the area or volume of samples to get cone failure and adding new design to compare both of these performances.

The author would like to give sincere appreciation to Office for Research, Innovation, Commercialization and Consultancy Management (ORICC) for the supports especially on giving the budget for fee and guidelines given throughout the duration for this project.

References

1. American Building Institution, Building Code Requirement for Structural Concrete (ACI 318-11). Farmington Hills: ACI 318-11 (2011)
2. Cao Chen , Jan Nemcik, Ren Ting, & Naj Aziz, International Journal of Mining Science and Technology,**23**(1) pp. 79-88 (2013)
3. Chanakya A, *Design of Structural Elements*: Concrete, Steelwork, Masonry, and Timber Designs to British Standards and Eurocodes. Third Edition. New York: Taylot & Francis,**87**(2009), pp. 1069 – 1064. .(2009)
4. Chuka, Chinwko Emmanuel, Odio.B.O Chukwuneke J. L Sinebe J. E, *Investigation Of The Effect Of Corrosion On Mild Steel In Five Different Environments*, International of journal scientific and technology research vol.**3** (issue 7.)pp. 306-310 (2014)
5. D. A. Winful, K.A. Cashell, A. M. Barnes & R. J. Pargeter, *High Strength Steel in Fire* (2015)
6. Eligehausen, R., Mallee, R. and Silva, **J Anchorage in Concrete Construction**, Ernst & Sohn: A Wiley Company (2006)
7. Glenn P.Peterson, Pullout testing of cast-in-place.Brigham Young University:Master,thesis. (1998)
8. John G. Shipp And Edward R. Haninger, *Design of Headed Anchor Bolts*. Engineering Journal, **20**(2), pp. 58-69 (1983)
9. Konstantinos Daniel Tsavdaridis, Mohamed A. Shaheen b, Charalampos Baniotopoulos c & EmadSalem, *Analytical approach of anchor rod stiffness and steel base plate calculation under tension*, **6**, pp. 207-218. (2016)
10. Maria Teresa Gomes Barbosa & Souza Sánchez Filho. *Investigation of Bond Stress in Pull Out Specimens with High Strength Concrete*..Journal of Researches in Engineering Civil And Structural Engineering, 13(3), pp. 55-64. (2013)
11. Marko Pavlović , Zlatko Marković , Milan Veljković & Dragan Buđevac Paredis, *J. Bolted shear connectors vs. headed studs behaviour in push-out test* Constructional Steel Research, **88**,pp. 134-149 (2013)
12. Muhammad Saleem & Tatsuya Tsubaki, *Two-Layer Model For Pull-Out Behav- Ior Of Post-Installed Anchor*, Journal of Japan and Society,**8**(1) pp. 49-62 (2012)
13. Razaqpur A.G., S. Foo, A. Mostafa and M. Saatcioglu, *Sesimic Respose Of Three Types Of Post-Installed Anchors In*, World Conference on Earthquake Engineering (2008).
14. Robert H. Lesage (Ed.) *Cast-in-place and drop-in anchor*. Washington D.C: The Aberdeen Group (1971)

15. R.W.Cannon , D.A. Godfrey, F. L Moreadith, *Guide To The Design Of Anchor Bolt and Other Steel Embedments*. Concrete International. (1981)
16. S. A. Al- Ta'an & A. A. Mohammed, *Tensile Strength of Short Headed Anchors Embedded in Steel Fibrous Concrete*. Civil engineering, Iraq. pp 35-49 (2009)
17. Sato Hiroshi, Kazunori fujikake & Sidney mindess, *Study On Dynamic Pullout Strength Of Anchors Based On Failure Modes*.13th World Conference on Earthquake Engineering(854) (2004)
18. Seira Owa, Yasutoshi Yamamoto, Tatsuya Kondo & Christian Fogstad, *Study on Strength and Ductility of Post-installed Adhesive Anchoring System*. 15WCEE / 15th World Conference on Earthquake Engineering, Lisbon (2012)
19. Sheikh Abdul Muhaimin, Anchor Bolt Position In Base Plate In Term Of "T" and "J" Anchor Bolt.University Tun Hussien Onn Malaysia:Degree.Thesis (2016)
20. X. Fu and D.D.L. Chung, *Sensitivity Of The Bond Strength To The Structure Of The Interface Between Reinforcement And Cement, And The Variability Of This Structure.* Cement and Concrete Research, **28**(6), pp. 787–793, (1998)
21. Williams Form Engineering, Cast-in-Place Concrete Anchors. Portland: Pigtail anchor bolt (2011)
22. Wooyoung Jung , Minho Kwon, Kim Jinsup, and Buseog Ju. Performance E valuation of the Post-Installed Anchor for Sign Structure, Construction and Building Materials, 40 (12), pp. 496-506. (2013)

Rating of Sound Absorption for EFBMF Acoustic Panels according to ISO 11654:1997

Emedya Murniwaty Samsudin[1*], *Lokman Hakim* Ismail[1], *Aeslina* Abd Kadir[1], *Ida Norfaslia* Nasidi[1] and *Noor Sahidah* Samsudin[2]

[1]Faculty of Civil and Environmental Engineering, UTHM, Johor
[2]Faculty of Architecture, Planning and Surveying UiTM, Seri Iskandar, Perak

Abstract. Empty fruit bunch fibre (EFB) and mesocarp fibre (MF) have been used in the fabrication of a new acoustic panel as a sound absorber for building. Measurements were carried out following ISO 354 in the mini reverberation chamber and the sound absorption performance of EFBMF acoustic panels were rated based on ISO 11654. Measurements of the new EFBMF acoustic panel involves five panel designs of 100 EFB dust panels, 80:20 dust panels, 100MF coir panels, 90:10 coir panels and 50:50 coir panels with 5 cm of initial thickness. Results showed that 100MF coir panel achieved α_w of 0.90 coefficient and was rated as Class A absorber followed by 90:10 coir panels with α_w of 0.85 coefficient and 100 EFB dust, 80:20 dust and 50:50 coir panels having α_w of 0.80 coefficients and been rated as Class B absorber. This research has successfully defined that EFB and MF are viable to be used as raw fibre for acoustic absorber for building.

1 Introduction

In the acoustic measurement for sound absorption of a building material, reverberation room method or ISO 354 [1] is often used. Even though an absorption coefficient of any material could be obtained from the normal wave incidence testing, but in many practical applications the random incidence testing is desired [2]. Random incidence test method requires materials to be tested in a reverberation chamber to specify the sound absorption performance.

Reverberation room method utilized Sabine's formula as shown by Equation 1 to define the sound absorption coefficient value (α_s) of a material at specific one-third-octave band of frequency. The range of frequencies for reverberation room testing was in between 100 Hz to 5000 Hz as been specified in the ISO 266: 1997 [3].

$$\alpha = \frac{0.16V}{S}\left(\frac{1}{Ts} - \frac{1}{Te}\right) \quad (1)$$

In the Sabine's formula, V represent the closed volume of test chamber, S is the total absorbing area and Ts and Te represent the reverberation time inside an occupied chamber and empty chamber respectively. Theoretically, reverberation time is defined as the time required by the sound pressure level to decay by 60 dB after the sound source has stopped with RT_{60} or T_{60} symbol normally been used to represent the value.

Rating of sound absorption for building material is done reffering to α_s value at mid-band frequencies of 250 Hz, 500 Hz, 1000 Hz, 2000 Hz and 4000 Hz. According to ISO 11654: 2007 [4], sound absorption material are rated to be either Class A, B, C, D, E or not classified based on the single-number weighted sound absorption coefficient (α_w) value. Materials having absorption coefficient value more than 0.5 are considered as a good absorber while materials having α_s= 0.2 or less are a reflective for acoustic application [5].

1.1 Utilization of palm oil wastes as sound absorber

1.1.1 Palm oil wastes in Malaysia

Oil palm planted in Malaysia is from *Elaeis guineensis* species and was first found in West Africa. Resulting from British Industrial Revolution, oil palm plantation was later expanded and commercially planted in South East Asia especially in Malaysia and Indonesia. Oil palm was first introduced to Malaysia in 1870 by British colony and the first commercial oil palm plantation in Malaysia was set up in year 1917 in Tennamaran Estate, Selangor [6]. The first Malaya company actively planted palm oil in Malaysia is Guthrie and Co. during 1924 in Kluang, Johor. Oil palm industry in Malaysia keeps on growing each year with the growth of world population and vast economic development by most countries importing oil for food. Palm is edible oil product that becomes one of the most important dietary components and source of fats worldwide. The demand for oil palm is also increase tremendously with the increasing of world population each year [7] and [8].

* Corresponding author: emedya@uth.m.edu.my

Oil palm production produced plenty of biomass which includes wastes from oil palm trunks (OPT), oil palm frond (OPF), empty fruit bunches (EFB), palm pressed fibres (PPF), palm kernel shells (PKS) and palm oil mill effluent (POME) [6] and [9]. The presence of huge amount of accumulated wastes resulting from the oil palm processing activites has created major waste disposal problems to the country. Among those wastes, EFB and MF are the larger lignocellulose biomass been produced [6] followed by MF and PKS [9]. Hence, government needs to allocate more money and assign more hectares of land to dispose those wastes. Moreover, accumulated palm oil wastes in the landfill could also cause another environmental issues if not been reused [10].

1.1.2 Sound absorption performance of palm oil fibres

The most recent local research utilizing palm oil fibers as sound absorption materials were done by [11],[12], [13] and [14].

Research by [11] used oil palm empty fruit bunches in different masses and thicknesses to create samples with different densities value. This research has found that higher density samples able to improve the sound absorption performance at high frequency region while air gap is necessary to improved absorption performance at low frequency region. Moreover, for 50mm thick OPEFB sample having 292 kg/m^3 of density could achieve SAC value of 0.90 in average at frequency above than 1000 Hz.

In [12], different weight percentage of oil palm mesocarp fiber was mixed with different percentage of polyurethane (PU). Impedance tube measurement following ASTM E1050-09 was used as the guidance to obtain the sound absorption coefficient for 50mm thick samples at low and high frequency regions. This research has concludes that sample containing 90% of mesocarp and 10% of PU achieved the highest NRC value of 0.66 compared to the others. Besides, this research also evidences that when PU content increased, density of sample will increase but lowering its porosity. This will result in lower sound absorption capability.

Earlier, [13] also utilizing mesocarp fiber and PU in their research for samples having different thicknesses. In this research the mix ratio is 70%MF: 30% PU with thickness ranging from 10mm to 40mm thick. However, 5mm and 10mm of air gap were incorporated at the back of the sample during conducting the impedance tube testing. Findings show that thick samples having 30mm and 40mm thickness are good at medium and low frequency region. It was also found that by incorporating an air gap to the back of the test samples able to improve absorption performance of thin samples at most of the frequencies but showing less effect for thicker samples.

Finally research by [14] compares on the absorption performance of 30 mm and 50 mm thick oil palm EFB (OPEFB) with date palm fiber (DPF) using impedance tube following ISO 10534 and ASTM E1050 :98. Results shows that 50mm OPEFB achieve its peak absorption of almost 90% absorption at higher frequency with considerably good performance at frequency 2000 Hz. However, when results were compared to DPF, OPEFB performed slightly lower at the most of the testing frequency.

Most of earlier research involving palm oil fibers as a new sound absorption material were done following either ISO 10534 or ASTM E 1050. Both methods used impedance tube as the measurement equipment with smaller sample size which is very suitable for material development besides low in cost. However, this method is only utilizing standing wave/direct wave incident which does not reflecting the actual performance of the samples in the real application. Therefore, it is necessary to conduct the same research using reverberation room method to acknowledge the samples performance as sound absorber in the real application.

2 Methodology

This research used raw EFB and MF collected from local oil palm plantation (Kulim Plantation (M) Sdn. Bhd) located in Kluang, Johor. EFBMF panels used in this classification were fabricated in five different panel mix designs consisted of 100 EFB dust panels, 80:20 dust panels, 100MF coir panels, 90:10 coir panels and 50:50 coir panels having 5 cm of initial thickness. Hot pressed compression method was used in the fabrication with urea formaldehyde (UF) as the wood binder for the panel.

Sound absorption testing was conducted in a mini reverberation chamber following ISO 354:2003 [1]. This testing measures the reverberation time (RT_{60}) using sound level meter type 01dB connected to a symphonie analyzer and analyzed the sound absorption coefficient/Sabine's absorption coefficient (α_s) values using dBBati32 software analyzer. Measurements of RT were done in four different microphone positions with turning position by 90° in each set of measurement (refer Figure 1).

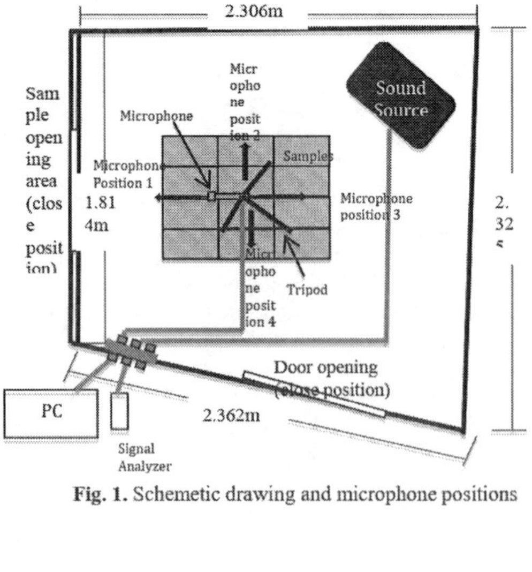

Fig. 1. Schemetic drawing and microphone positions

2.1. Pre-treatment and Panels fabrication

Earlier before fabrication, the raw fibers were treated with 2% of NaOH to eliminates any impurities and excessive oil attached on the fibers. Research by [15] revealed that pretreatment using 2% NaOH solutions is more effective than hot water treatment in removing oil and the fiber will also remain in shape with less surface damages. Fibers were first been soaked with cold water for 24 hours and drained out to removed all dirt and unwanted particles. Afterward, the drained fibers were soaked again into water containing 2% of NaOH within 30 minutes and rinse off using clean-running water to eliminate the residue oil and excessive sodium content. Finally, the washed fibers were sun-dried for 2 to 3 days depending on weather conditions to reduce its moisture. The purpose of sun-drying the fiber is to reduce energy usage from electrical oven, which thus, could contributes to less carbon footprint from the heating process. However, to ensure that all fiber containing less than 10% of moisture before been used for fabrication, approximately 15 to 30 minutes of oven-dried using 110°C ± 5°C is needed.

Next, EFB and MF were shredded and grinded using specific machine in the laboratory before sprayed with the UF. This research used 15% of UF liquid resin added with 1% of NH_4Cl as the hardener to fasten the hardening process of the liquid binder. Hot pressing work was done using hot press machine in the Timber Fabrication Laboratory, UTHM. This machine needed to be heated up to 180 °C before loading the wet mix fibers into. The temperature were maintained at 180 °C throughout the pressing time of 15 minutes using 1000 psi of pressing load. Finally, samples were removed from the machine and left to cold under room temperature before removing the mould.

2.2 Reverberation room testing

Reverberation room testing for this research was done in a downscale chamber (RT Mini Chamber) located at Faculty of Architecture, Planning and Surveying, University Technology MARA, Shah Alam. This chamber consisted of two interconnected chambers and was build to following BS EN ISO 140-1:1998. Mini chamber were designed to be use for measurement of small-scale sample sized $\pm 1m^2$. It is a quadrangular chamber having total surface area of $\pm 30m^2$ and total volume of $11.8m^3$.

Figure 2 shows the arrangement for test samples, sound source and microphone in the mini chamber during testing. In this testing, 15 test panels sized 0.30 meter x 0.22 meter x 0.05 meter thick of each design mix were laid in array resting on the floor. Samples were placed at the centre of the test chamber with the perimeter surface area was completely sealed with a reflective material, leaving only the top surface of the panels exposed to recieve the sound incidence.

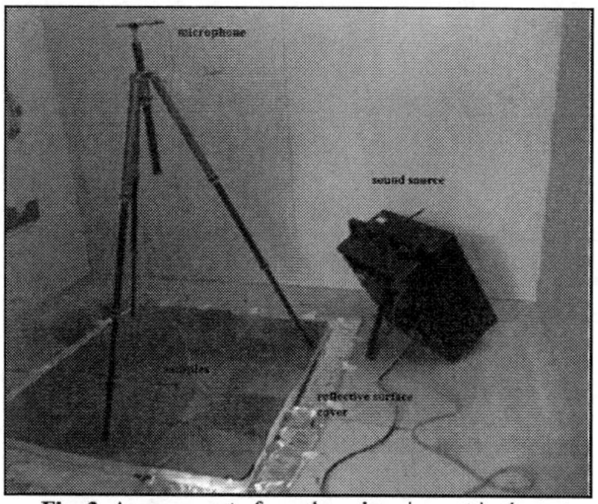

Fig. 2. Arrangement of panels and equipment in the test chamber.

Measurements of reverberation time were done in seconds (s), starting from measurement of an empty chamber followed by an occupied chamber containing EFBMF acoustic panels by mix design. Random pink noise was used in a full blast sound power of 100 dB inside the closed chamber. Since most of the surfaces inside the chamber are reflective, therefore, it will results in higher RT values when the sound signal stopped. When an absorbent placed into the chamber, the shorter RT will be obtained due to dissipation of sound energy that happen when the sound wave hits onto the absorptive surface.

3 Results and discussion

Theoretically, the shorter the reverberation time, means the more sound were absorbed and the gap between curve-lines representing the empty chamber with the occupied chamber will be distance. In this measurements, it was found that different panel design does not giving large effect to the reverberation time since the total absorbing surface area of those panels are almost the same. The longest reverberation time in the empty chamber was found at 2.08 s at frequency 1600 Hz. On average, reverberation time in the occupied chamber achieved less than 1.2 s in between frequency 400 Hz to 5000 Hz. Since results for all panels designs shows alignment for all curve-lines laid very close to each other, therefore, it was predicted that performances of all panel mixes are almost similar to each other. Results for reverberation time of an empty chamber and occupied chamber is presented in Figure 3.

 https://doi.org/10.1051/matecconf/201815003002

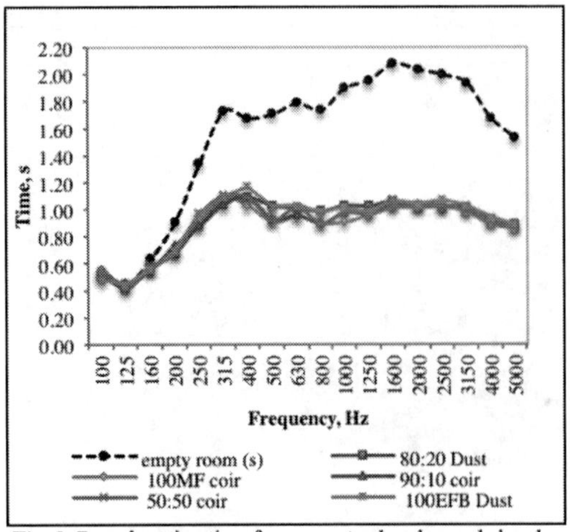

Fig. 3. Reverberation time for an empty chamber and chamber occupied with EFBMF acoustic panels

Sabine's absorption coefficient or represented as α_s, relates between the ratio of room volume to the total absorbing surface area in a room. The decrease in RT_{60} inside the test chamber containing test samples indicates that the test material successfully absorbed the sound incident during testing. Figure 4 show the Sabine's absorption performance of EFBMF acoustic panels in the reverberation chamber. At below than frequency 300 Hz, absorption performance of all samples are slightly unstable with all coir panels seems unable to give positive α_s value especially at frequency 100 Hz and 125 Hz. This made the results for below than 200 Hz unreliable for this testing. However, results for below than 250 Hz did not influence in the sound absorber rating according to ISO 11654 [4]. Therefore, the results could be ignored.

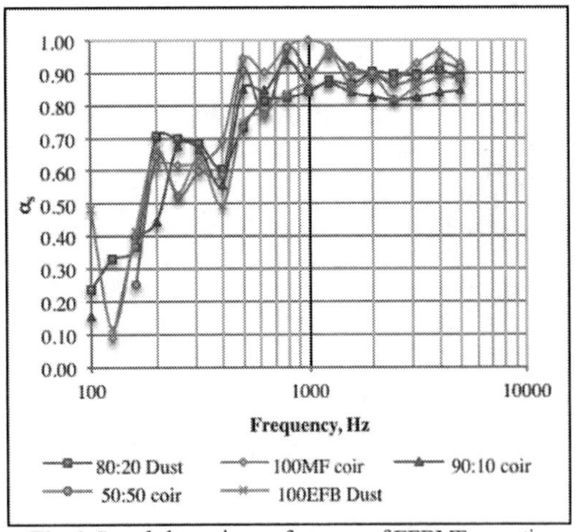

Fig. 4. Sound absorption performance of EFBMF acoustig panels in the reverberation chamber

In Figure 4, the highest α_s value achieved by 100 Mf coir panel at frequency 900 Hz to 1000 Hz which is near to 1.0 coefficient. Overall, 100 MF coir panel also exhibits higher absorption coefficient value compared to the other panels throughout the testing frequencies. This provides an indicator that 100 MF coir panel will have the highest rangking absorber according to the standard. Meanwhile, Figure 5 illustrates the rating curves for EFBMF panels according to ISO 11654:1997 [4], while Table 1 summarized the final results for all tested panels.

As been predicted earlier, 100 MF coir panel owned the highest rank followed by 90:10 coir panel and the three other panels (50:50 coir, 100 EFB dust and 80:20 dust). The α_{pi} values for 100 MF coir panel are 0.70, 0.90, 0.90, 0.90 and 0.80 at frequency 250 Hz, 500 Hz, 1000 Hz, 2000 Hz and 4000 Hz respectively. The weight of sound absorption value for this panel achieved 0.90 coefficients, which is the highest among others.

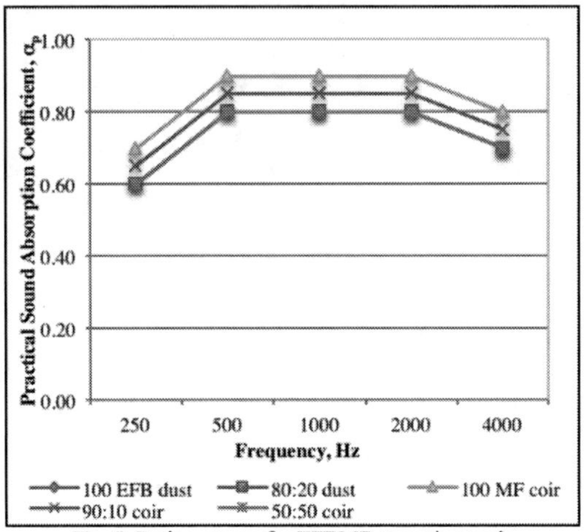

Fig. 5. Rating curves for EFBMF acoustic panels

As been summarized in the Figure 5, the α_w for EFBMF acoustic panel represented by all five designs demonstrates that 80% or more of sound energy are successfully absorbed in the test chamber. Based on the α_w value obtained, 100 MF coir panels has the highest absorption coefficient of 0.90, followed by 90:10 coir panels with 0.85 coefficient and the other three panels (50:50 coir, 100 EFB dust and 80:20 dust) with 0.80 coefficients. With reference to this result, 100 MF coir panel is rated as a class A absorber while the other panels were categorized as the class B absorber.

Class A absorber is material that able to absorb more than 90% of sound incident while class B absorber is capable to absorb in between 80 % to 85% of sound incident. Materials that able to absorb more than 80% of sound incident is considered as very absorptive sound absorption material for building [16]. This research has found that coir panels dominates as a good absorber compared to dust panel. Of it, mesocarp coir panel is the most absorbing panel compare to the others.

Table 1. Rating of sound absorption for EFBMF acoustic panels according to ISO 11654 :1997

Frequency (Hz)	250	500	1k	2k	4k	α_{pi}	α_w (with shape indicator)	Rating Class
100 MF coir	0.70	0.90	0.90	0.90	0.80	0.85	0.90	A
90:10 coir	0.65	0.85	0.85	0.85	0.75	0.80	0.85	B
100 EFB dust	0.60	0.80	0.80	0.80	0.70	0.75	0.80	B
80:20 dust	0.60	0.80	0.80	0.80	0.70	0.75	0.80	B
50:50 coir	0.60	0.80	0.80	0.80	0.70	0.75	0.80	B

4 Conclusions

This research has successfully utilized EFB and MF as a sound absorber for building acoustic use. Findings from this research evidance that EFB, MF and mixture of both fibers are comparable to be use as the raw fiber replacing the conventional synthetic sound absorber in the market. Utilizing natural fibers as a new material for building acoustic enable to promotes sustainablity approach in material development for construction.

Authors would like to express special acknowledgement to the Ministry of Higher Education Malaysia (MoHE) for providing the scholarship for this study and to the Faculty of Civil and Environmental Engineering, the Faculty of Architecture, Planning and Surveying, Universiti Teknologi MARA, Shah Alam for providing the technical support and facilities, without which this study would have not been possible. This research represent a part of the main objective of the 1st author's doctorate study

References

1. BS EN ISO 354 *Acoustic Measurement of sound absorption in a reverberation room*. (2003)
2. Hughes, W. O., Mcnelis, A. M., Nottoli, C., and Wolfram, E. *29th Aerospace Testing Seminar* (2015).
3. BS EN ISO 266 *Acoustic-Preferred frequencies*. (1997)
4. BS EN ISO 11654 *Acoustic –Sound absorbers for use in building-Rating of sound absorption*. (1997)
5. Egan, M. D. *Architectural Acoustics*. p 43, (1988).
6. Abdullah, N., and Sulaiman, F. *Biomass Now – Sustainable Growth and Use*, 75–100, (2013).
7. Murphy, D. J. *Journal of Oil Palm Research*, *26*(MAR), 1–24, (2014).
8. Basri Wahid, M., Abdullah, S. N. A., and Henson, I. E. *Plant Production Science*, 8(3), 288-297, (2005).
9. Jahaan, F., Aanifah, M., Yee, P. L. A. I., and Wasoh, H. *Journal of Oil Palm Research*, *26*(4), 321–331, (2014).
10. Kadir, A. A., Mohd Zahari, N. A., and Azizi Mardi, N. *Advances in Environmental Biology*, *7*(12), 3826–3834, (2013).
11. Hee, O. K., Putra, A., Nor, M. J. M., Selamat, M. Z. and Ying, L. Z. *23rd International Congress on Sound and Vibration*, 1-8, (2016).
12. Latif, H. A., Yahya, M. N., Zaman, I., Sambu, M., Ghazali, M.I and Hatta, N. M. *ARPN Journal of Engineering and Applied Sciences*, 11, 2462-2466, (2016)
13. Latif, H. A., Yahya, M. N., Rafiq, M. N., Sambu, M., Ghazali, M. I. and Hatta, M. N. M. *Applied Mechanics and Materials* 773-774, 247-252, (2015).
14. Al Rahman, L., Raja, R. I., Rahman, R. A. and Ibrahim, Z. *Research Journal of Applied Sciences, Engineering and Technology*, 7(8), 1656-1661 (2014).
15. Ramli, R., Shaler, S., and Jamaludin, M. A. *Journal of Oil Palm Research*, *14*(2), 34–40, (2002).
16. Cavanaugh, W. J. and Wikes, J. A. *Architectural Acoustics: Principles and Practice*, p59, 1999.

https://doi.org/10.1051/matecconf/201815003003

The Effect of Different Fibre Length and Different Urea Formaldehyde (UF) content on Sound Absorption Performance of Empty Fruit Bunch (EFB)

Ida Norfaslia Nasidi[1*], *Lokman Hakim* Ismail[2], *Emedya Murniwaty* Samsudin[1], *Muhamad Firdaus* Abdul Khodir[1] and *Muhammad Aizat* Kamarozaman[1]

[1]Faculty of Civil and Environmental Engineering, Universiti Tun Hussein Onn Malaysia
[2]Department of Design Engineering and Architecture, Universiti Tun Hussein Onn Malaysia

Abstract. Noise control is part of the major requirements to improve the living environment. One of the best methods to reduce noise is by employing sound absorber material into a space. Since traditional sound absorber in the market was believed could cause health problems to human, the need for alternative material is desired. This research intended to utilize waste materials from palm oil empty fruit bunch (EFB) in the production of natural sound absorber for noise control in building. Two parameters were investigated; the fibre length and binder content to determine their effect on sound absorption. Samples were tested using impedance tube using low and high frequencies according to ISO 10534-2:2001. Consequently, both parameters show that different fibre length and binder content affected the ability of the fibre to absorb sound. Results show that the optimum fibre size is in between 2mm to 5mm length while the optimum amount of binder is between 0% to 5% and 15% from fibre weight. This research has found that the highest Noise Reduction Coefficient (NRC) using 2mm to 5mm fibre size achieved 0.70 coefficients while the use of optimum amount of binder (particularly Urea Formaldehyde) could obtain 0.75 NRC. The newly develop panels are considered as a good sound absorbent and suitable to use as an alternative material replacing the synthetic absorber.

1 Introduction

Recently, noise has become one major source of human disturbance. Excessive noise also may deteriorate human physiologically and psychologically [1]. Many people has demanded the solution to this problem. Usually, using sound absorber materials are among the prominent way to solve this problem. However, traditional sound absorber that is extensively used in building compartment made up from synthetic materials including glass and rock wool. These materials are believed to emit carbon dioxide (CO_2), methane and nitrous oxide in their production [2]. Plus, this usage of materials also gives negative impact to human health and contributes to global warming issues [3].

Therefore, to counter this problem, many researchers have driven their interest by using alternative materials from natural fibre in purpose of replacing synthetic materials as sound absorber. Natural fibres are low in density, have good mechanical properties, easy to be processed, high in stability, health friendly, cheap and reduce the impact to environment during their productions [4]. Previous works have been done by employing natural fibre as sound absorber including sugarcane [5], coconut coir fibre [6], palm oil male flower spikes [7] and mesocarp fibre [8].

Mahzan et al. [9] produce rice-husk waste as reinforced composite and testing using two-microphone transfer function method. 5%, 10%, 15%, 20%, 25% and 30% of rice-husk added with polyurethane foam (PU). They found that most sample having high SAC value at 250Hz with highest SAC 0.899 for 25% rice-husk sample. Meanwhile, lowest SAC obtained by 15% rice-husk sample. By calculating the NRC value of the samples, 30% sample of the rice husk obtained the highest value. They conclude that the rice husk waste has a potential to become a sound absorber, especially at low frequency range.

Ismail et al. [10] utilized arenga pinnata as sound absorber at four distinct thickness, 10mm, 20mm, 30mm and 40mm. Based on the result, the thickness significantly influence the SAC value where thicker sample having high SAC at high frequency region with maximum SAC for 40mm sample thickness is 0.88. Arenga Pinnata fibre was then compared to coir fibre and palm oil fibre. The result indicate that SAC value of Arenga Pinnata was higher compared to coir fibre, around 0.7 to 0.9 but slightly lower than palm oil fibre.

Corresponding author: idalia_nasidi@yahoo.com.my

Another study by Putra et al. [5] aimed to utilize sugarcane waste as acoustic absorber with different percentages of fibre to binder. Up to 40% binder, absorption coefficient were not too affected. 3g of fibre with composition 70% of fibre with 30% of binder show good SAC among the other compositions. The SAC value gained over 0.5 at frequency 1000Hz to 4500Hz. This may be due to denser sample which has greater flow resistivity to create more paths for sound to be absorbed on the sample. By maintaining binder composition, with different amount of fibre 1g and 3g, shows an increasing amount of fibre to enhance the SAC value. Sample was then compared with three layer of woven cloth absorber. They found that, sugarcane sample were comparable with woven cloth absorber.

Research made by Samsudin et al. [11] differentiated the absorption performance between coir form and dust form of an empty fruit bunch (EFB) at three thickness 6mm, 12mm and 18mm. Thicker sample (18mm) for both dust and coir form fibre enhanced the SAC performances as the frequency increased. 18mm thick dust form sample had maximum SAC 0.9 at 3250Hz. While coir form 18mm sample gained maximum SAC lower than dust form, 0.62 at 3500Hz to 5000Hz. Dust form showed great performances of absorption compared to coir form caused by the small particle creating high-density sound absorption material and can easily turn sound energy to heat energy.

Therefore, this paper is discussing the potential of natural fibre waste in Malaysia to be an alternative as acoustic absorber which is the empty fruit bunch (EFB). Even this fibre have been discussed and studied by other researchers, but this research is focusing on the other parameters that are yet to be explored using this kind of natural fibre. The main parameters in this study are the fibre length and percentage of urea formaldehyde (UF) in the sample.

1.1 Empty Fruit Bunch (EFB)

Palm oil or known as *Elaeis guineensis Jacq.* was first planted int Selangor before it became a major plantations in Malaysia. Stated by Malaysian Palm Oil Board (MPOB), most of the states in Malaysia have palm oil plantation area as an agriculture by products with total planted area of 5,642,943 hectares. Subsequently, this oil production generates waste products including empty fruit bunches (EFB), palm kernal shells, mesocarp fruit fibers (MF) oil palm trunks (OPT), oil palm frond (OPF) and palm oil mill efflent (POME).

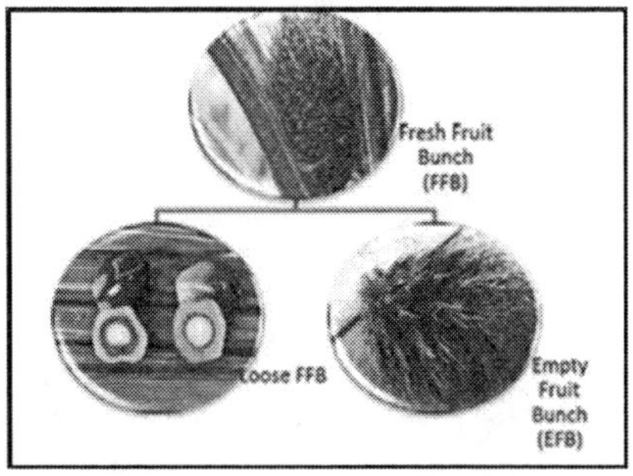

Fig. 1. Breakdown of EFB from FFB

EFB are the residue in solid form that are mostly produced from fresh fruit bunches [12]. Out of 100% of FFB, 22% of it is EFB which is the second largest product after sterilization and stripping process where 67% out of it is POME [13]. EFB that resulted from the extraction process of oil have the potential to be utilized as raw materials for new products in order to reduce the amount of disposal. Traditional way to dispose EFB are by turning it into fuel for production of steam in palm oil mills but faced environmental issues afterwards [14]. Figure 1 shows the breakdowns of EFB from the FFB.

2 Methodology

2.1 Pre-treatment Stage

EFB fibre supplied by Kulim Plantation, Ladang Tereh Mill located at Kluang Johor. First, fibre was shredded to get into loose form hair like materials. Fibre then soaked in water for 24 hours to remove unwanted dirt or particle retained on fibre. After 24 hours, fibre was soaked with 2% of Sodium Hydroxide (NaOH) solution for 30 minutes [15]. Finally, fibre was rinsed using clean water and sun dried 2 to 3 days depending on weather to reduce heating rate thus reduce carbon footprint. Controlled moisture content of 10% [16] was obtained by oven dry fibre for 30 minutes at 110°C.

2.2 Fabrication Stage

Dry fibres were weighted to require proportion. Urea Formaldehyde (UF) used as the binder obtained from S.A. Wood Chemicals Sdn Bhd was sprayed on the fibre using spray air gun to ensure fibres are evenly mixed. Then fibre is fitted into round shape mould, 100mm and 28mm diameter for Impedance Tube testing with 50mm thickness. Hot compression machine was used to compress the panel at 150°C for 20 minutes. Figure 2 shows the sample of sound absorber after taken out from the mould.

Fig. 2. 100mm and 28mm diameter sample

Two parameters were taken throughout these study, fibre length and UF content. For UF content parameter, 0%, 5%, 10%, 15% and 20% UF used with 0.4g/cm^3 target density. While for fibre length parameter, fibre was sieve to get the required length as presented in Table 1 with 65% solid content used in fabrication stage [17].

Table 1. Length of fibre

Number	Fibre Length
1	Retain sieve 15mm
2	Retain sieve 12mm
3	Retain sieve 5mm
4	Retain sieve 2mm
5	Below size 2mm

3 Result and Discussion

The results of impedance tube measurement at high and low frequency from 250 Hz to 5000 Hz for empty fruit bunch (EFB) with two distinct parameter, fibre length and UF content are shown in Figure 4 and Figure 5.

Figure 4 shows the SAC result of five different lengths of fibre. Although there were inconsistencies in values all samples demonstrated a uniform pattern of absorption. It can be observed that the sound absorption coefficient increased between frequency 1600 Hz and 2000 Hz by shifting the peak from low frequency range

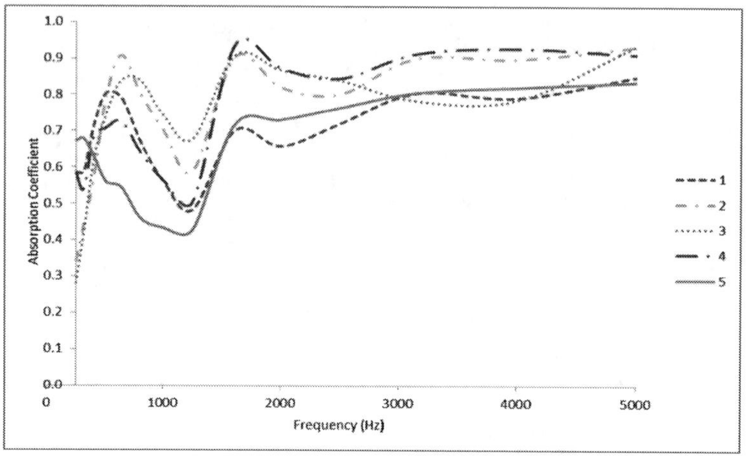

Fig. 4. The sound absorption coefficient with different fibre length

2.3 Sound Absorption Measurement

Sound absorption coefficient of the sample was measured using impedance tube according to ISO10534-2:2001, i.e. the transfer-function method [18]. Impedance tube able to measure absorption coefficient of materials using small sample size that was easy to assemble and disassemble [4]. Loudspeaker was placed at the one end of impedance tube and generates broadband, stationary random sound waves while sample placed at the other end. This study used Impedance Tube with AFD1001 software at Building Services Laboratory, Faculty of Civil and Environmental Engineering, Universiti Tun Hussein Onn Malaysia (UTHM). Figure 3 shows Impedance Tube set up.

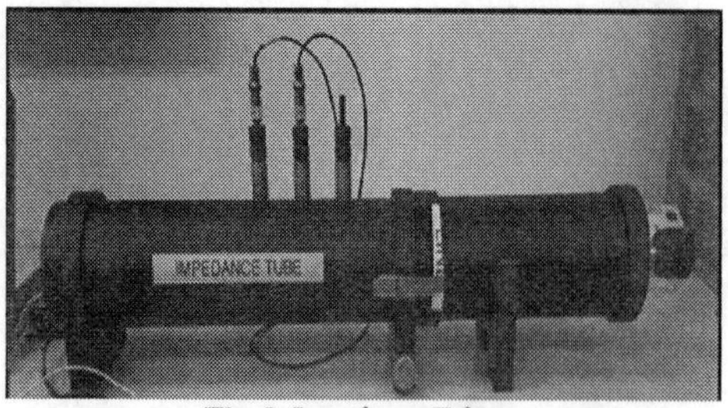

Fig. 3. Impedance Tube

to middle frequency range. Sample 2 showed superior performance of SAC at low range frequency while Sample 4 showed a good performance at high range frequency. Highest SAC recorded by Sample 4 with $\alpha = 0.937$ at frequency 1600 Hz. Conversely, Sample 5 had least SAC compared to the other sample. Theoretically, these were due to the fibre content on the sample which contained dust form fibre and filled the empty space between the fibre and decreasing the porosity value of the sample that affected the SAC performances. It is found that the fibre length effects the acoustical performances of the material.

Table 2. Density and NRC value for fibre length parameter

Sample	Density (g/cm^3)	NRC
1	0.39	0.65
2	0.39	0.65
3	0.41	0.65
4	0.44	0.70
5	0.42	0.60

Table 2 tabulated different density and NRC performances for each sample for fibre length parameter. NRC is a term that describes the absorption properties of a materials, by averaging value of SAC of materials at frequencies of 250 Hz, 500 Hz, 1000 Hz and 2000 Hz

 https://doi.org/10.1051/matecconf/201815003003

and rounded off to the nearest multiple 0.05 [10]. Therefore, NRC value is a number between 0 to 1 which indicated no absorption for zero and full absorption for value one [19]. Generally, density affected the NRC value of sample. Denser structure gave better NRC value from mid to high frequency range but tends to decrease at the low frequency [8].

Urea Formaldehyde (UF) as the binder used in this study with different percentage, 0%, 5%, 10%, 15% and 20%. From Figure 5, the SAC value for all sample showed a uniform pattern. SAC performance is good above 500 Hz. Sample with highest UF content (15%) exhibited a higher SAC with $\alpha = 0.957$. All samples had high absorption at the low frequency region, declined at the middle frequency region but then increased at the high frequency region. High content of UF (20%) showed the least absorption than other samples. Theoretically, high amount of binder will create less porosity on the sample and reduce the absorption performances of the materials. Jayamani et al. [20] claimed high porosity in sample will cause an increasing friction on the surface where resulted in the dissipation of the sound.

percentage of UF increase, the density of samples also increase but significantly decrease the porosity value. According to Bratu et al. [21], high porosity value on sample results in good absorption of the sound wave.

4 Conclusion

Empty fruit bunch were studied as an acoustic absorber which have a potential to replace the common synthetic sound absorber materials such as glass wool, rock wool and asbestos. The results indicated that fibre length and binder content plays a role in the improvement of sound absorption performances. High amount of binder and too much small particles of fibre will create less porosity value, thus creating less SAC caused by the sound transmitted on it cannot be changed into heat energy. 15% of UF percentage as optimum binder with high absorption and resulting good sample without any depreciation on the sample. Therefore, for fibre length parameter, the optimum SAC of EFB is obtained when the fibre length is in between 2mm to 5mm.

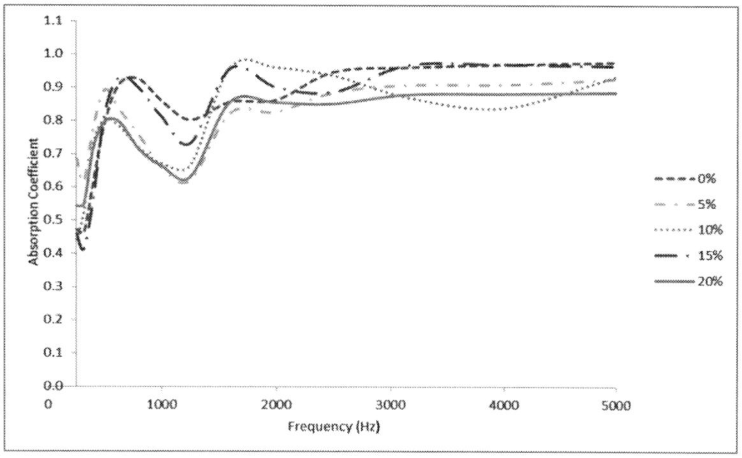

Fig. 5. The sound absorption coefficient with different percentage of UF

UF content parameter having a further analysis by calculating the noise reduction coefficient (NRC) and going through porosity testing.

Table 3. Density, Porosity and NRC value for different UF percentage

Sample	Density (g/cm^3)	Porosity (%)	NRC
0%	0.24	79.95	0.75
5%	0.29	79.69	0.75
10%	0.31	76.62	0.70
15%	0.32	71.30	0.75
20%	0.33	74.00	0.70

Table 3 summarizes the result of density, porosity and NRC value for all samples. All samples show about 70% of sound is absorbed. It is also indicated that as the

Absorption properties of natural fibres can be increased by controlling the porosity value of newly developed panels. As a green technology product, these innovative absorption panels may have a bright future to replace available synthetic absorber in the markets. However, further research is required to investigate the properties of the natural sound absorber to ensure their safety and strengthness before becoming a commercial acoustic material.

The authors thank Centre for Graduate Studies Universiti Tun Hussein Onn Malaysia (CGS UTHM) for sponsor and financial support in publishing this paper and Office for Research, Innovation, Commercialization and Consultancy Management (ORICC) Universiti Tun Hussein Onn Malaysia, vote U791.

References

[1] Fouladi, M. H., Nassir, M. H., Ghassem, M., Shamel, M., Peng, S.Y., Wen, S. Y., Xin, P. Z. & Nor, M. J. M. (2013). Utilizing Malaysian Natural Fibers as Sound Absorber, *Modeling and Measurement Methods for Acoustic Waves and for Acoustic Microdevices*, ISBN: 978-953-51-1189-4

[2] Arenas, J.P. and Crocker, M.J. (2010) Recent Trends in Porous Sound-Absorbing Materials. *Sound & Vibration*, 44, 12-17

[3] Asdrubali, F. (2007). Green and sustainable materials for noise control in buildings. *19th International Congress on Acoustic Madrid*, (September 2007)

[4] Berardi, U., Iannace, G., & Di, M. (2016). Challenges and Solutions in Acoustical Measurements and Design : Paper ICA2016-238 Characterization of sheep wool panels for room acoustic applications Characterization of sheep wool panels for room acoustic applications

[5] Putra, A., Abdullah, Y., Efendy, H., Farid, W.M., Ayob, M.R., & Py, M. S. (2013). Utilizing Sugarcane Wasted Fibers As A Sustainable Acoustic Absorber, *Procedia Engineering*, 53, 632 – 638

[6] Nor, M. J. M., Ayub, M., Zulkifli, R., Amin, N. & Fouladi, M. H. (2010). Effect of Different Factors on the Acoustic Absorption of Coir Fiber, *Journal of Applied Sciences*, 10(22), 2887-2892

[7] Masrol, S. R., Rosdin, M. K. R. & Ibrahim, M. N. (2013). Sound Absorption Characteristics of Palm Oil Male Flower Spikes Fiber Reinforced Composite. *International Confrence on Mechanical Engineering Research (ICMER2013)*, 1-3

[8] Latif, H. A., Yahya, M. N., Zaman, I., Sambu, M., & Ghazali, M. I. (2016). The Influence of Physical Properties and Different Percentage of The Oil Palm Mesocarp Natural Fiber, *11*(4), 2462–2466

[9] Mahzan, S., Zaidi, A. M. A., Yahya, M. N., & Ismail, M. (2009). Investigation on Sound Absorption of Rice-Husk Reinforced Composite. *In: Proceedings of MUCEET2009 Malaysian Technical Universities Conference on Engineering and Technology*

[10] Ismail, L., Ghazali, M. I., Mahzan, S., Mujahid, A., & Zaidi, A. (2010). Sound Absorption of Arenga Pinnata Natural Fiber, 4(7), 804–806

[11] Samsudin, E. M., Ismail, L. H., & Kadir, A. A. (2016). Comparison On Acoustic Performance Between Dust And Coir Form Empty Fruit Bunches (Efb) As Sound Absorption Material, *Jurnal Teknologi*, 78 (5), 191-196

[12] Piarpuzán, D., Quintero, J. A., & Cardona, C. A. (2011). Empty fruit bunches from oil palm as a potential raw material for fuel ethanol production. *Biomass and Bioenergy*.

[13] Abdullah, N., & Sulaiman, F. (2013). The Oil Palm Wastes in Malaysia. *Biomass Now – Sustainable Growth and Use*, 75–100

[14] Kadir, A. A., Azian, N., Zahari, M., & Mard, N. A. (2013). Utilization of Palm Oil Waste into Fired Clay Brick, *7*, 3826–3834

[15] Izani, M. A. N., Paridah, M. T., Astimar A. A., Nor, M. M. Y. & Anwar, U. M. K. (2012). Mechanical and dimensional stability of medium density fibreboard produced from treated oil empty fruit bunch, *Journal of Applied Science*, 12(6), 561-567

[16] Karlinasari, L., Hermawan, D., Maddu, A., Martianto, B., Lucky, I. K., Nugroho, N. & Hadi, Y. S. (2012). Acoustical Properties of Particleboards Made from Betung Bamboo (Dendrocalamus asper) As Building Construction Material, *7(4)*, 5700–5709

[17] Seddeq, H. S., Aly, N. M., A, A. M., & Elshakanekery, M. (2013). Investigation on sound absorption properties for recycled fibrous materials. *Journal of Industrial Textiles*, *43*(1), 56–73. doi:10.1177/1528083712446956

[18] British Standards Institution (2001). *Acoustics — Determination of sound absorption coefficient and impedance impedance tubes—Part 2:Transfer-function method*. United Kingdom: BS EN ISO 10534-2:2001 Incorporating Amendment No. 1 to BS ISO 10534-2:1998 (renumbers the BS ISO as BS EN ISO 10534-2:2001)

[19] Lechner, N. (2012). *Plumbing, Electricity, Acoustics Sustainable Design Methods for Architecture*. United States of America: John wiley & Sons, Inc

[20] Jayamani, E., Hamdan, S., Rahman, M., Soon, K., and Bakri, M. (2014). Processing and Characterization of Epoxy/Luffa Composites: Investigation on Chemical Treatment of Fibers on Mechanical and Acoustical Properties. *BioResources*, 9(3), 5542-5556

[21] Bratu, M., Dumitrescu, O., Vasile, O., Cristea, A. C., & Muntean, M. (2014). Research On The Sound-Absorbing Properties Of New Composite Materials With Some Wastes, *Romania Journal of Material*, 44(2), 160–168.

 https://doi.org/10.1051/matecconf/201815003004

Assessment of the Potential Occurrence of Dry Period in the Long Term for Pahang State, Malaysia

Ahmad Saifuddin Othman[1, 2,*], and *Nurul Nadrah Aqilah* Tukimat[1]

[1]Universiti Malaysia Pahang, Faculty of Civil Engineering and Earth Resources, 26300 Gambang, Kuantan Malaysia
[2]Universiti Malaysia Pahang, Centre for Earth Resources Research and Management, 26300 Gambang, Kuantan Malaysia

Abstract. The interference of climate circulation and continuous rising of surface temperature every year has caused the atmosphere composition change which gives serious impact to water resource management. Pahang is among of the affected states by El Nino that hit Malaysia in recent years which led to water depletion at several water plants. Based on the current situation, this study focuses on 1) simulate the average rain pattern using statistical downscaling; 2) identify the severity index and dry duration occurrence in the catchment area. Predicting potential changes in the climate events is important to evaluate the level of climate change in the critical region. Therefore, the integration of Statistical Downscaling Model (SDSM) and Standard Precipitation Index (SPI) have been conducted to study the potential occurrence of the dry period due to climate change for year 2020s and year 2050s. The results reveal that the dry condition is high during the mid-year. The lowest SPI value is estimated to reach -2.2 which can be classified as extreme. The potential dry period is expected to increase 2.5% and 3.3% in 2020 and 2050, respectively.

1 Introduction

Nowadays, climate change has become the important issue due to its impact on the ecosystem, water resources, environment and human life [1]. The probability of climate change of an area also refers to the historical condition or the projection of future climates such as monsoons, droughts and floods. Climate change refers to the continuous change in natural climate variability properties in long term due to changes in the global climate as a result of human activities [2]. Industrial advancement is one of the main factors that led to the uncontrolled greenhouse gas emissions especially carbon dioxide (CO_2) thereby raising the world temperature. The enhancement of CO_2 was measured by computing the hydrogen ion of sea water concentration which showed a reduction in pH value to 0.1 which equivalent to an increase of 26% of acid [3]. The dramatic increase in temperature due to climate change will cause high evapotranspiration which affects the water resources [4, 5,6]. In the 20th century, the temperature is expected to increase more than 2°C in the tropical and temperate regions and more than 4°C globally. This will threaten the global security [3]. The projected mean surface temperature in Malaysia is expected to be higher with an average increase of 0.6°C–1.2°C [7]. Drought is a natural phenomenon that occurs directly but the impact to the region and the occurrence frequency are the features that distinguished from other phenomena. Dry basis is the long term dry conditions with low moisture content which the studies indicate that the information of rainfall and temperature changes.

General Circulation Models (GCMs) is a method developed in the present, for the purpose of projections of climate change scenarios studied by the scale set. The uses of GCMs need to be identified in order to focus and narrow down the grids to get the right hydrological cycle information at site studied. Statistical downscaling model, the National Centers for Environmental Prediction (NCEP) data have been used in designing the predictor-predictant relationship and GCM data provided by Canadian Centre for Climate Modeling and Analysis (CCCma) used to project the climatic changes due to greenhouse gases (GHGs) contribution in the future year [8, 9]. Two main approaches for a downscaling model are dynamical downscaling (DD) and statistical downscaling (SD). In this study, statistical downscaling techniques have been chosen and the universally multiple linear regression models called Statistical Down-Scaling Model (SDSM).

There are many future climate change projection studies conducted globally such as Sylhet and Moulvibazar districts in North-eastern region of Bangladesh [10], Marsyangdi River Basin, Nepal [11] and Lisbon [12]. While in Malaysia, the studies in the Muda Irrigation Scheme in north of peninsular Malaysia by N. N. A. Tukimat et al. [13] and the Kurau River by Hasan et al. [14] are the preliminaries studies available in Malaysia at this time. The study is initiated by Hassan et al. [14] in Kurau River are conducted to assess the impact

*Corresponding author: othman.ibnu@yahoo.com

of future climate change that may affect a dam which located at the downstream of the river. The dam is main water supply to meet domestic and industrial demands at Kerian District as well as Larut Matang District. The impact are concerned may also disturb the irrigation water for double cropping planting intensity to the Kerian-Sg. Manik irrigation scheme as rice cultivation is the main economic activity in that area. The study also conducted at several cropping areas in Malaysia including Alor Setar (Kedah), Subang (Selangor) and Senai (Johor) in order to investigate the irrigation water demand changes in an intensive irrigated area of Malaysia under climate change scenario [15]. Pahang experienced the latest dry event in 2016 which affects four plants around Pahang which are Lubuk Kawah Water Plant in Temerloh, Jelai and Batu 9 Water Plant in Lipis and Chini Water Plant in Pekan, Pahang. This study focuses on 1) simulate the average rain pattern using statistical downscaling; 2) identify the severity index and dry duration occurrence in the catchment area. Hence, the projection study on severity index and dry occurrence pattern are conducted in Pahang state to provide significant information for water resources management in facing the uncertain climate changes.

2 Background Theories

This steps using statistical downscaling model (SDSM) described by Z.Hassan et al. [14] which defined the rainfall occurs on each day or not as shown below:

$$W_t = \propto_0 + \sum_{j=1}^{n} \propto_{t-1} \hat{u}_t^{(j)} + \propto_{t-1} w_{t-1} \qquad (1)$$

Where t is time [days], W_t is the conditional possibility of rain occurrence on day t, $\hat{u}_t^{(j)}$ is the normalized predictor, α_j is the regression parameter deduced by an ordinary least square method, and W_{t-1} and α_{t-1} are the conditional probabilities of rain occurrence on day t-1 and lag-1 day regression parameter, respectively based on the studied region and predictand. The uniformly distributed random number r_t $(0 \leq r_t \leq 1)$ was used to determine the rain occurrence and the possibility that the rain would happen if $w_t \leq rt$. The estimated value of rainfall on each rainy day is determined. This can be represented with a z score:

$$Z_t = \beta_0 + \sum_{j=1}^{n} \beta_j \hat{u}_t^{(j)} + \beta_{t-1} + \varepsilon \qquad (2)$$

where, Z_t is the z-score on day t, β_j is the calculated regression parameter, and β_{t-1} and are Z_{t-1} are the regression parameter and the z-score on day t-1, respectively. Thus, rainfall y_t on day t can be written as:

$$y_t = F^{-1}[\emptyset(Z_t)] \qquad (3)$$

where, $\emptyset$ is the normal cumulative distribution function and F is the empirical function of y_t. A transformation of the fourth root was applied, to take account of the skewed nature of the rainfall distribution. During the model development (calibration) of SDSM, some parameters such as event threshold and variance inflation were adjusted in order to obtain the good agreement between observed and simulated climate variables.

3 Study Area

Pahang is one of the state which located at 4° 11' 10"N and 104° 03' 45''E on the east coast of Peninsular Malaysia. Pahang State with area 35,965 km^2 as shown in Figure 1. The area experienced wet and dry seasons throughout the year due to its geographical position which located in equator lines. The climatic cycle in the region is influenced by four seasonal changes, known as the monsoon season. The northeast monsoon and the southwest monsoon are two major monsoon that occurs every year with two inter monsoons [16].

The prediction of highest and lowest seasonal average temperature for Peninsular Malaysia until this 20th century is 3.7°C and 3.3°C, respectively [17]. The average global surface temperature is projected will increase by 1.4°C to 5.8°C over the period 1990 to 2100 [2]. Total rainfall for three consecutive months from July to September recorded at 151.30 mm, 141.10 mm and 142.39 mm, respectively.

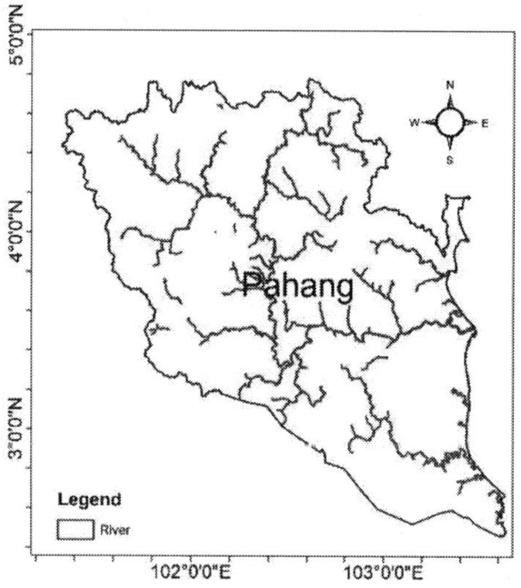

Fig. 1. Locations of Pahang state

4 Methodology

4.1 Data and Sources

12 rainfall stations are selected in this study after the screening process that only considered below 1% missing data. The rainfall stations were selected due to the completion of history data for 30 years. Table 1 shows the list of rainfall stations used in this study which cover the whole of Pahang State. All stations are monitored by the Department of Irrigation and Drainage (DID) of Malaysia. The prediction of rainfall for year 2010 to 2069 (60 years) are using the historical data from the year 1979 to 2008

(30 years). The historical data then divided by two sets of a time period which are (1979-1993) and (1994-2008). The first 15 years (1979-1993) was used in the calibration process, while the other 15 years (1994-2008) for validation purpose. The historical data and NCEP parameters are used to obtain the relationship between local climate pattern and information of the atmospheric circulation by mathematical model simulation. The SDSM model generates the synthetic precipitation for 12 rainfall stations. Multiple regression techniques were used in the SDSM model to construct the predictor-predictand relationship. The hydrological data such as rainfall, inflow, spill, seepage, water level and water release were purchased from Department of Irrigation and Drainage (DID) of Malaysia.

Table 1. Details rainfall stations at Pahang State

NO	Station ID	Coordinate	
		Latitude (N)	Longitude (E)
1	3931014	3°54'	103°08'
2	4122067	4°07'	102°12'
3	4320066	4°21'	102°04'
4	3325086	3°23'	102°32'
5	3431099	3°29'	103°08'
6	3429096	3°29'	102°56'
7	3532101	3°32'	103°14'
8	3122142	3°10'	102°14'
9	3318127	3°19'	101°51'
10	3221001	3°12'	102°09'
11	4514032	4°31'	101°25'
12	4414038	4°26'	101°27

4.2 Statistical downscaling tool

A spatial data of GCMs were used in SDSM to downscale the daily scenarios from the predictor-predictand relationships. The predictor variables such as Climate Model Inter-comparison Project Phase 5 (CMIP 5) provide daily information concerning the large-scale state of the atmosphere, while the predictand such as rainfall describes the condition at the regional scale. Daily local rainfall data are required at least 30 years historical data to generate the future climate trend based on the emission level in the region. This model implies the statistical relationships to downscale the large-scale resolutions of GCMs denoted as predictors into the local climate variables known as predictand. It allowed the raw data to transform into standard predictor variables to produce nonlinear regression models before calibration and validation steps were applied. The data series can also be shifted forward or backward by any number of time steps to produce lagged predictor variables.

The major challenge in climate downscaling especially in extreme rainfall indices is a selection of appropriate predictors. The predictors are expected to be highly correlated with extreme rainfall indices. Furthermore, the predictors should be accurately projected by available GCMs for the future projection of climate. There are no general guidelines for the selection of predictors in different parts of the world and therefore, a comprehensive search of predictors is necessary. Rainfall is set to be as conditioning process because there is an intermediate process between the regional forcing and local weather, which the local rainfall amount is correlated with the occurrence of wet days. As the distribution of rainfall is skewed, a fourth root transformation is applied to the original series to convert it to the normal distribution, before applying the regression analysis. Table 2 shows the predictors variables in the downscaling experiments.

Table 2. List of predictors variables

No	Predictors	Code
1	Surface Meridional Velocity	p_v
2	Surface Vorticity	p_z
3	Surface Wind Direction	p_th
4	500hPa Zonal Velocity	p5_u
5	500hPa Meridional Velocity	p5_v
6	500hPa Vorticity	p5_z
7	500hPa Wind Direction	p5thgl
8	850hPa Meridional Velocity	p8_v
9	850hPa Vorticity	p5_z
10	850hPa Geopotential Height	p850
11	850hPa Wind Direction	p8th
12	Specific Humidity at 850hPa Height	s850
13	Surface Specific Humidity	shum

4.3 SPI Method

Standardized Precipitation Index was created by T.B. McKee, N.J. Doesken, and J. Kleist [18]. This method is very flexible to be able to monitor the state of a variety of time scales. SPI has been used to study the effects of rainfall which is fall to be as underground water, reservoir storage, soil moisture, and water discharge. The index which looks forward to the probability of rainfall for varies time scale. Thus, SPI considers the time scale in the calculation the drought index such as 1, 3, 6, 9, 12, 24, and 48 time scales that represents the number of the month. The spi-1 is the percentage of normal rainfall for the period of 30 days. The value of 1-month obtained is more accurate because monthly distribution is a regular cycle. This calculation is short-term closely related to soil moisture. SPI-1 month calculation is suitable with the catchment area condition has to decrease monthly rainfall value. This SPI variation value to negative or positive more uniform because reducing monthly rainfall difference percentage.

https://doi.org/10.1051/matecconf/201815003004

5 Results and Discussions

5.1 Calibration and Validation Analysis

Figure 2 and 3 show the simulated results produced for calibration (1979-1993) and validation (1994-2008) processes using NCEP predictor variables compared to observed data. There are 5 parameters used in calibration process include 1) specific humidity at 850hPa height; 2) surface specific humidity; 3) 500hPa zonal velocity; 4) surface vorticity and 5) surface meridional velocity. The graphs result of calibration and validation indicate that the observed rainfalls are very close to the simulated results. For calibration, the annual calibrated shows the slight decrease of rainfall than the observed rainfall recorded with 28.58mm. While, the validation part shows the increment of rainfall with 17.8mm. The difference between the simulated and observed data are affected by the climate change. The percentage of calibrated and validated error are 1.34% and 0.81%, respectively. The amount of rainfall in the calibration period used was higher than the amount of rainfall in the validation period. However, this model is applicable to use as the validation error are reduced to 0.81% from 1.34% in the calibration process. These results show that the projection analysis results are acceptable at this stage.

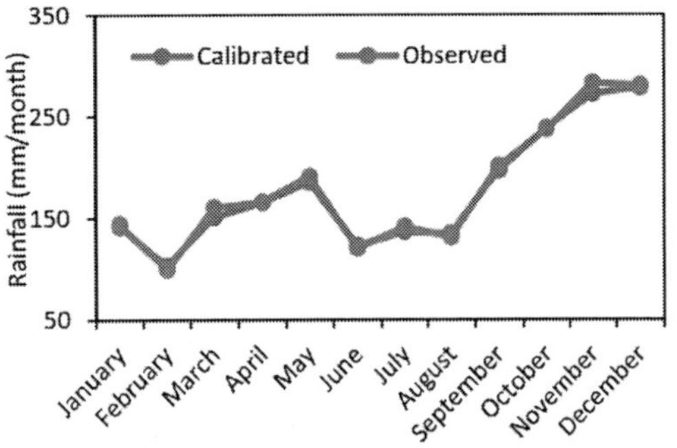

Fig. 2. Results of calibrated rainfall (1979 to 1993)

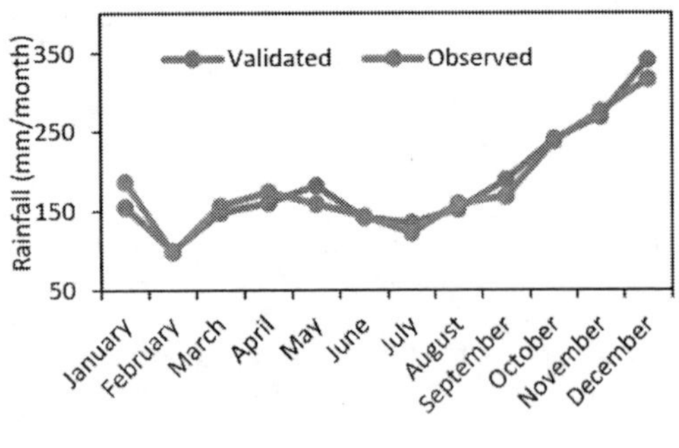

Fig. 3. Results of validated rainfall (1994 to 2008)

5.2 Performances of Projected Rainfall by AR5 Scenarios

Figure 4 present the constant predictors used to project the future rainfall trend in the same grid box provided by AR5 type 26, 45 and 85 scenarios. The scenarios are divided based on radiative forcing quantities concentrations in a region of up to the year 2100. The diversity of scenarios in the simulation resulting in diverse analysis results in each study area. This graph compares the annual historical rainfall data in the year 2010 with the annual prediction of total rainfall in the year 2010 at Pahang State. After data screening process, 2010 was identified as the year with the complete rainfall history data for all rain stations. The graph shows that the projection and historical pattern are quite similar. It shows that the decreasing amount of rainfall until the end except in month February, October and November. The simulated rainfall for month February, October and November were estimated to increase more than the historical rainfall recorded in February, October and November with 47.3mm, 96.8mm and 33.8mm, respectively. The percentage error difference of this 3 months are 48.3%, 39.0% and 11.6%. The historical total rainfall in other month shows the difference result with the slightly percentage error compared to projected rainfall with January (2.4%), March (11.8%), April (23.6%), May (5.9%), June (35.5%), July (23.7%), August (13.2%), September (20.1%), and December (5.6%). However, the annual projection rainfall percentage decrease with 185.28mm and 7.98% error differences. SDSM Model are specialized in analysing the future pattern based on the normal cycle of hydrology historical pattern. Thus, the disaster event will not be analysed. Moreover, disaster event such as flood, flash flood, landslide, thunderstorm and so on are rare and unexpected conditions. The projection analysis the model is reliable to project the future rainfall.

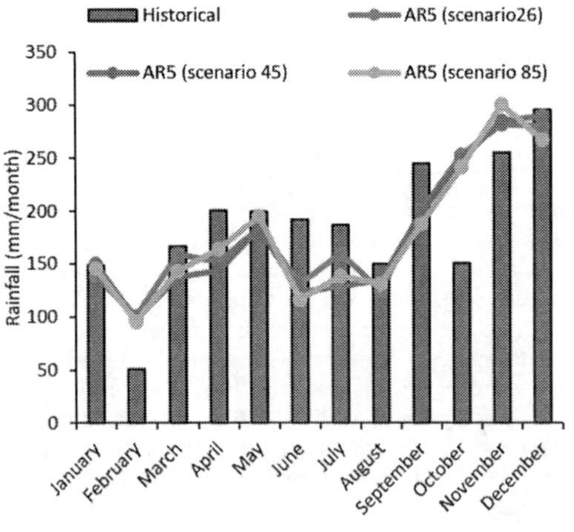

Fig. 4. Performance of annual monthly total rainfall (mm) against historical data in year 2010

5.3 Climate Simulation and Projection

The prediction of rainfall in the Pahang State for 60 years are covered from year 2010 to 2069 which divided into two periods of time with 15 years each. The mid year of two periods which are 2020s (2010–2039) and 2050s (2040–2069) are used for future pattern analysis. The result of future emission downscaling are shown in Figure 5 and Figure 6. The 100 ensembles of synthetic daily time series are produced for AR5 (26, 45 and 85 scenarios). The figures also show the average history rainfall data for 30 years (1979-2008) by monthly basis. The figures reveal that the rainfall increasing thread in all future time horizons in overall months except February and June to August. The rainfall intensity achieved 47.8% in February and 13.4% to 56.5% in June to August in year 2020s. Meanwhile, in the year 2050s rainfall intensity achieved 48.0% in February and 13.8% to 56.3% in June to August. The heaviest rainfall is expected to occur in December during Northeast season. However, the rainfall intensity is predicted to become lesser in year 2020s and 2050s achieve 8.1% and 8.6% respectively compare to the historical rainfall.

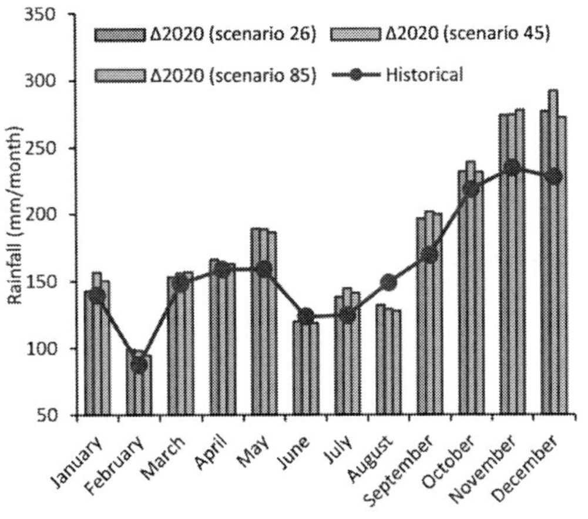

Fig. 5. Average annual rainfall at year Δ2020s

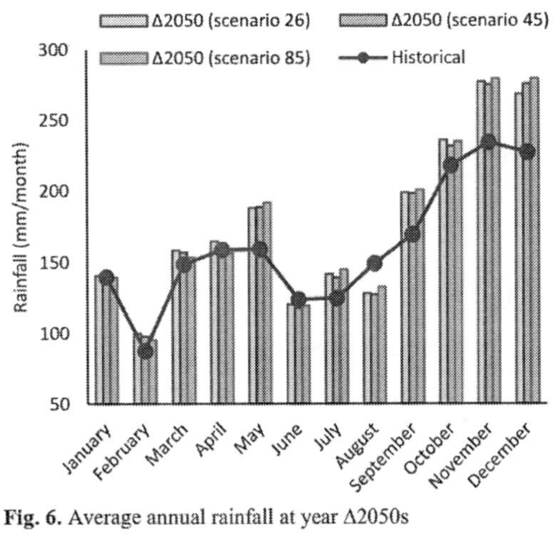

Fig. 6. Average annual rainfall at year Δ2050s

5.4 SPI Projection and Simulation

The analysis used monthly rainfall simulation data for a minimum period of 30 years. Monthly rainfall simulation data for each selected station has been analysed before the SPI analysis can be carried out. Graph 7 and 8 are result of analysis SPI-1 value suitable with study carried out. SPI-1 values were plotted with the yearly time scale for the observed annual dry pattern. The analysis shows there is dry condition can be detected in the future year. Dry pattern that become increase in the middle of year. The SPI values in May and September are expected low and potentially to reach -2.2 and -1.5, respectively. These conditions can be classified as extreme and severe dry pattern which might cause drought event. In the early year phase, the SPI value is expected to be normal between -1.1 to 1.2. The percentage in the first period 2018 to 2039 year expect 2.5% dry pattern will potential. On the next period 2040 to 2069 years expectation condition increase of 3.3% dry pattern will potential in every 30 years. In generally, Southwest Monsoon from late May to September normally signifies relatively drier weather.

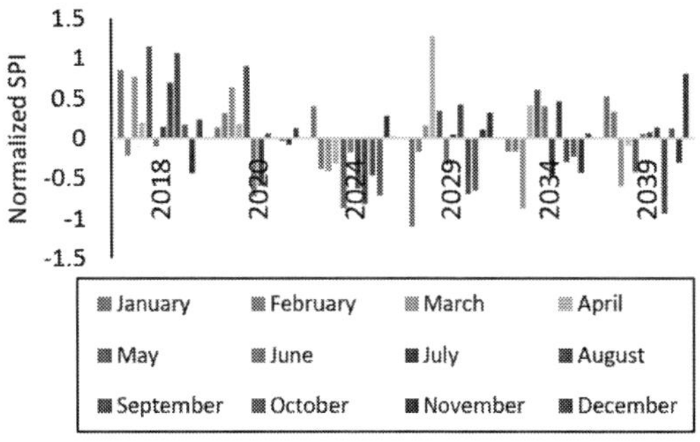

Fig. 7. SPI values for 1-month time scale at year 2018 to 2039

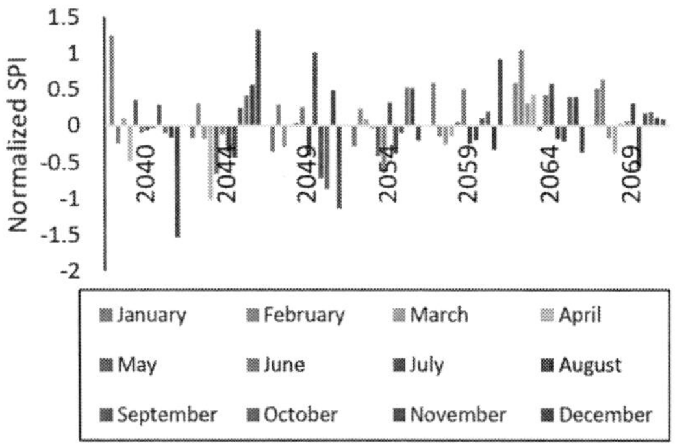

Fig. 8. SPI values for 1-month time scale at year 2040 to 2069

Rainfall data obtained are consistent data for analysis, although there are few missing data, but it can still provide a satisfactory and consistent results. Through this analysis, dry conditions have been detected in the catchment areas based on monthly depleted rainfall data. This analysis show the relation between dry conditions in the catchment to the rainfall distribution cycle. The SPI index is a simple method to apply because it only required rainfall data. Rainfall data will be analysed by using gamma distribution parameters to obtain the distribution of flows well. This calculation obtained the probability of cumulative flows for each station studied. However, the analysis of the drought level should be detailed by requiring other calculations such as the severity level to make the results more accurate and effective.

6 Conclusions

This study contributes towards the ability of the SDSM to project the future rainfall pattern and the dry future pattern estimation by using SPI. The correlation relationship method is proposed to manage the complexity of predictor selections in SDSM model. The General Circulation Model (GCMs) parameters were employed to project the climate trend which considered the estimated emission level projection in the future year. In general, the dry climates result in significant changes to monthly rainfall decrement which affect the future dry scenario in the catchment area. Through SPI analysis, a dry scenario was detected in a state on monthly rainfall data basis within 30 years. The dry pattern analysis in all station by using a consistent data provided has demonstrated the suitability in detecting a dry pattern occurred in this study area. Therefore, the effective water management planning is needed by the local authorities to face the climate changes challenges in the future.

The authors would like to thanks, Universiti Malaysia Pahang (UMP) for the funding support through grant number RDU150113. A lot of thanks to Department of Irrigation and Drainage (DID) and Malaysian Meteorological Department (MMD) for the technical support and data provided.

References

1. T. Jiang, Y.D Chen, C. Xu, X. Chen, V.P. Sigh, Comparison of hydrological impacts of climate change simulated by six hydrological models in the Dongjiang Basin, South China, *J Hydrol* **336**, 316-333 (2007)
2. Intergovernmental Panel on Climate Change, Climate change 2007: synthesis report (2007)
3. Intergovernmental Panel on Climate Change, Climate change 2014: synthesis report (2014)
4. L. Zhang, Z.M. Ma, S.Z. Kang, Analysis of impacts of climate variability and human activity on stream flow for a river basin in arid region of northwest China, J. Hydrol **352(3–4)**, 239–249 (2008)
5. N. N. A. Tukimat, S. Harun, S.Shahid, Comparison of different methods in estimating potential evapotranspiration at Muda irrigation scheme of Malaysia, *J. Agric. Rural Dev. Tropics and Subtropics* **113(1)**, 77–85 (2012)
6. X.-J. Wang et al, Climate change and water resources management in Tuwei River basin of northwest China, *Mitigation Adaptation Strategies Global Change* **19(1)**, 107–120 (2014)
7. National Hydraulic Research Institute of Malaysia, Final Report: Study of the Impact of Climate Change on the Hydrologic Regime and Water Resources of peninsular Malaysia, *Ministry of Natural Resources and Environment, Kuala Lumpur, Malaysia* (2006)
8. K.V. Salzen, J.F. Scinocca, N.A. McFarlane, J. Li, J.N. Cole, D. Plummer, The Canadian fourth generation atmospheric global climate model (CanAM4) part I: representation of physical processes, Atmosphere–ocean (2013)
9. V.K. Arora, J.F. Scinocca, G.J. Boer, J.R. Christian, K.L. Denman, G.M. Flato, V.V. Kharin, W.G. Lee, W.J. Merryfield, Carbon emission limits required to satisfy future representative concentration pathways of greenhouse gases, *Geophysical Research Letters* **38(5)**, 3–8 (2011)
10. A. H. Nury, M. J. B. Alam, Performance Study of Global Circulation Model HADCM3 Using SDSM for Temperature and Rainfall in North-Eastern Bangladesh, J. Sci. Res. **6(1)**, 87-96 (2014)
11. Dibesh Khadka, Dinesh Pathak, Climate change projection for the marsyangdi river basin, Nepal using statistical downscaling of GCM and its implications in geodisasters, *Geoenvironmental Disasters* 3, 15 (2016)
12. Pedro Lopes, Assessment of statistical downscaling methods – application and comparison of two statistical methods to a single site in Lisbon, *Earth and Environmental Science*, **6** (2009)
13. N. N. A. Tukimat, S.Harun, S. Shahid, Modeling Irrigation Water Demand in a Tropical Paddy Cultivated Area in the Context of Climate Change, *Journal of Water Resources Planning and Management* **143(7)**, (2017)
14. Zulkarnain Hasan, Sobri Harun, Application of statictical downscaling model for long lead rainfall prediction in Kurau River Catchment of Malaysia, Malaysian Journal of Civil Engineering **24(1)**, 1-12 (2012)
15. Z. Hassan, S. Harun, S. Shamsudin, Application of SDSM and LARS-WG for simulating and downscaling of rainfall and temperature, *Theoretical and Applied Climatology* **116**, 243-257 (2014)
16. J. Suhaila, S.M. Deni, W Zin, A.A. Jemain, Spatial patterns and trends of daily rainfall regime in Peninsular Malaysia during the southwest and northeast monsoons : 1975 – 2004, *Meteorology Atmospheric Physics* **110**, 1 – 18 (2010)
17. Malaysian Metrological Deparment (MMD), Malaysia Scientific report: Climate Change Scenarios for Malaysia 2001–2009, *MetMalaysia Office Department, Malaysia* (2009)
18. T.B. McKee, N.J. Doesken, J. Kleist, The Relationship of Drought Frequency and Duration to Time Scales, *Conference on Applied Climatology 8th*,179–184 (1993)

Investigation of Salt Intrusion Condition in the Belat Estuary

Jacqueline Isabella Gisen[1, 2], *Siti Syuhaida* Adnan[1, 2], *Ahmad Amirul* Ahmad Tajudin[2], and *Tian Xian* Chan[1]

[1] Universiti Malaysia Pahang, Faculty of Civil Engineering and Earth Resources, 26300 Gambang, Kuantan, Malaysia
[2] Universiti Malaysia Pahang, Centre for Earth Resources Research and Management, 26300 Gambang, Kuantan, Malaysia

Abstract. Awareness on salt intrusion problem is still lacking in Malaysia due to high precipitation in the region. However, the El-Nino phenomenon that occurred recently has caused extremely low fresh water discharge in the Kuantan River which allowed the sea water to intrude further into its water intake region. Consequently, the Belat River may become potential water resources alternative to build new water intake station for the water supply in the Kuantan River Basin. The aims of this study are to: 1) investigate the salinity distribution in the Belat Estuary; 2) evaluate the applicability of the 1-D analytical salt intrusion solution; 3) determine the calibration parameters in the salt intrusion model. Salt intrusion measurements was conducted during the dry season at spring tide when the fresh water discharge is the minimum. Collection of data such as hydrological data, river cross section and salinity were collected to be used in the salt intrusion analysis. The results obtained show good agreement between the simulated and observed salinity distribution in the estuary with low RMSE and high NSE values. This indicates that the model is reliable and can become an important tool for water manager in conducting salt intrusion study for this area.

1 Introduction

Nowadays, salt water intrusion problems have become critical and serious issues that need to be addressed, especially in arid and semi-arid region. The concentration of salinity level is vital especially when the water in an estuary is used for human consumption. High salt concentration in water can bring harm to human health. This problem also can affect the natural cycle of the estuarine ecosystem especially the biodiversity in the region. Uncertainties of climate change, rapid population growth and industrial advancement nowadays also affect the water quality in rivers. Apart from that, human activities such as dredging and construction of dams potentially causing salt intrusion problems that affect the quality of water resources and give a threat to marine species. Therefore, the study of salt intrusion necessary to ensure the quality of the intake water can be consumed for human use [1].

There are many research studies have been conducted to investigate the saline intrusion in the estuaries worldwide such as along the Bay of Bengal in Bangladesh [2], the Mekong Estuary [3], Chao Phraya Estuary [4], and the Red River Delta [5]. In Malaysia, several studies have been done by different researchers such as H.R Wallingford [6] for Kuantan Estuary, SMHB et al. [1] for the Rompin and Ulu Sedili Besar Estuary, Van Breemen [7] for the Selangor Estuary, and Gisen et al. [8] for six estuaries in the region. The six estuaries covered includes the Kurau, the Perak, the Bernam, the Selangor, the Muar and the Endau Estuary. HR Wallingford [6] conducted the salt intrusion study in the Kuantan Estuary following the case where the Kobat water intake station was affected by saline water. Thus, a barrage was built to prevent the salt water intrusion. For the Rompin and Ulu Sedili Besar Estuary, SMHB et al. [1] carried out the salt intrusion investigation to fulfil the requirements stated in the National Water Resources Study in which potential water intake stations are to be identified for water supply purposes. Van Breemen [7] study shows that the changes of the salinity levels in the Selangor Estuary may contribute to deterioration of the mangroves and Berembang Trees which consequently causing the fireflies population moving towards extinction. The latest studies by Gisen et al. [8] in the six estuaries in Malaysia have shown that the salt intrusion problem in those areas are still under control without any major issues.

There are several salt intrusion model available in the market including one-dimensional (1-D), two-dimensional (2-D) and three-dimensional (3-D) models. These models have been widely applied in simulating salinity distribution in estuaries worldwide. 2-D and 3-D models are preferably use in cases where high accuracy is required for the modelling results. Meanwhile, 1-D analytical solution is sufficient for cases in which

* Corresponding author: isabella@ump.edu.my

moderate accuracy is required. The benefits of using a 1-D model compared to 2-D and 3-D models are its simplicity and ability to be applied in data-poor environments. These type of model only requires only requires directly measurable parameters such as geometry, fresh water flow and tide, and most importantly it is costs effective [9]. 1-D analytical salt intrusion model developed by Savenije [9-11] has been tested in many estuaries globally and proven reliable particularly in minimally gauged estuaries [5, 8, 12]. The researchers confirmed that the 1-D Salt intrusion model can be used not only for single reach estuaries but also multi-branch estuaries.

In March 2016, Malaysia was hit by extreme El-Nino phenomenon. Due to this extremely dry weather, the water supply system in the Kuantan River Basin have been affected. Several complaints were received by the water authority from the Kuantan citizen claiming that the tap water has a taste of saltiness. The authority also confirmed that this is due to the pumping of saline water into the treatment plant. Although the Kobat Barrage that has been built in 1980's to prevent the salt water intrusion, but the problem still occur after 30 years. Hence, salt intrusion investigation in the Belat Estuary was conducted in this study to introduce the Belat River as potential water resources alternative where new water intake stations can be built for the future water supply in the Kuantan River Basin. The aims of this study are to: 1) investigate the salinity distribution in the Belat Estuary; 2) evaluate the applicability of the 1-D analytical salt intrusion solution; 3) determine the calibration parameters in the salt intrusion model. It is believed that the results obtained from this study can provide useful information for water manager in managing the saline intrusion condition in the estuary.

2 Background theories

There are two components in the 1-D analytical salt intrusion model of Savenije [9-11] which are the geometry and longitudinal salinity distribution analyses. The geometry analysis involved several parameters namely the channel cross sectional area, width and depth. According to Savenije [10], the geometry of an estuary can be presented in exponential functions as shown below:

$$A = A_0 e^{-\frac{x}{a_1}} \quad \text{for } 0 < x \leq x_1 \tag{1}$$

$$A = A_1 e^{-\frac{(x-x_1)}{a_2}} \quad \text{for } x > x_1 \tag{2}$$

$$B = B_0 e^{-\frac{x}{b_1}} \quad \text{for } 0 < x \leq x_1 \tag{3}$$

$$B = B_1 e^{-\frac{(x-x_1)}{b_2}} \quad \text{for } x > x_1 \tag{4}$$

$$h = h_0 e^{\frac{x(a_1-b_1)}{a_1 b_1}} \quad \text{for } 0 < x \leq x_1 \tag{5}$$

$$h = h_1 e^{\frac{(x-x_1)(a_2-b_2)}{a_2 b_2}} \quad \text{for } x > x_1 \tag{6}$$

where parameters A [m^2], B [m] and h [m] are the cross sectional area, width and average depth at any distance x [m]. Subscription (0) in the equations represents the location of $x = 0$ which in this case the mouth of the estuary. The longitudinal distance, x_1 [m] is the inflection point where the convergence length changed. For the inflection point, all the symbols for the geometry characteristics are accompany by a subscription (1). Meanwhile, for the cross-sectional and width convergence lengths, the downstream part is given the symbols a_1 [m] and b_1 [m], respectively. As for the upstream reach, they are given the subscription (2). In this geometry analysis, the average depth is computed by dividing the area over the width. Generally, in a natural alluvial estuary, the ideal convergence lengths a and b are equal lading to a constant depth [10].

Substituting the geometry Equations (1) to (6) and integrating the Van der Burgh (1972) equation into the salt balance equation, the longitudinal salinity distribution for a steady state condition can be written as:

$$\frac{S-S_f}{S_0-S_f} = \left(\frac{D}{D_0}\right)^{\frac{1}{K}} \quad \text{for } 0 < x \leq x_1 \tag{7}$$

$$\frac{S-S_f}{S_0-S_f} = \left(\frac{D}{D_1}\right)^{\frac{1}{K}} \quad \text{for } x > x_1 \tag{8}$$

where S [kgm^{-3}] and D [m^2 s^{-1}] are the salinity and dispersion as a function of any distance. The freshwater salinity is expressed by S_f [kgm^{-3}], which is normally close to 0. The salinity and dispersion equations are expressed as:

$$\frac{D}{D_0} = 1 - \frac{Ka_1}{\alpha_0 A_0}\left(\exp\left(\frac{x}{a_1}\right) - 1\right) \quad \text{for } 0 < x \leq x_1 \tag{9}$$

$$\frac{D}{D_1} = 1 - \frac{Ka_2}{\alpha_1 A_1}\left(\exp\left(\frac{x-x_1}{a_2}\right) - 1\right) \quad \text{for } x > x_1 \tag{10}$$

With:

$$\alpha_0 = \frac{D_0}{Q_f} \quad \text{for } 0 < x \leq x_1 \tag{11}$$

$$\alpha_1 = \frac{D_1}{Q_f} \quad \text{for } x > x_1 \tag{12}$$

where Q_f [m^3 s^{-1}] is the freshwater discharge and K is the Van der Burgh coefficient, which ranges between 0 and 1; Since the data for freshwater discharge is always not available in the tidal region and the dispersion is not measurable, their ratio is then expressed as the mixing number, α_0 [m^{-1}] and α_1 [m^{-1}] in the calibration process. The ultimate objective in the salt intrusion study is to determine the furthest distance the salt water intrudes into the river system. Thus, from Equations (9) and (10), the salt intrusion length L are computed for $D=0$ at tidal average condition by the following equations:

$$L^{TA} = \alpha_1 ln\left(\frac{1}{\beta_0}+1\right) \text{ for } 0 < x \leq x_1 \quad (13)$$

$$L^{TA} = x_1 + \alpha_2 ln\left(\frac{1}{\beta_1}+1\right) \text{ for } x > x_1 \quad (14)$$

In order to convert the intrusion length from tidal average to high water slack (HWS) condition where the salinity length is the most, Equations (7) and (8) are shifted for $x = -E/2$ are shown as below:

$$S^{HWS}(x) = S^{TA}\left(x - \frac{E}{2}\right) \quad (15)$$

Hence, the salt intrusion length equation becomes:

$$L^{HWS} = \alpha_1 ln\left(\frac{1}{\beta_0{}^{HWS}}+1\right) \quad (16)$$

3 Study Area

The Kuantan River Basin is situated within the East Coast Range next to the Pahang-Terengganu border with the catchment area of 1630km². The Belat Estuary is one of the tributaries for the Kuantan River and is situated near Kampung Belukar Baharu, Kampung Baharu Batu Empat, and Perumahan Sungai Isap Satu residential area. Meanwhile, Sungai Air Hitam and Sungai Soi are the tributaries of the Belat River. The exact location of the Belat Estuary is at the coordinate of 4° 08' 00" N and 103° 24' 00" E on the east coast of Peninsular Malaysia. Belat Estuary meets the Kuantan River at the confluence of about 6.5km from the Kuantan River mouth as displayed in Figure 1.The main economic activities surround the mouth of the Belat Estuary is the large scale commercial fishing activities. In terms of flora and fauna, mangrove plants can be found along the river bank of the Belat Estuary. The topography of the Belat River is shallow and flood tends to occur at the residential area around the region during heavy rainy season.

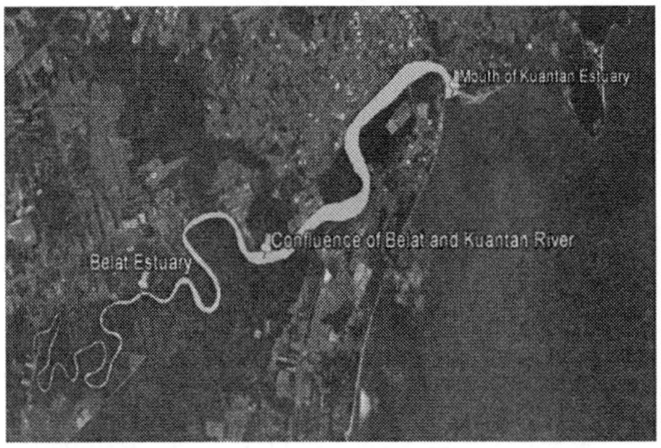

Figure 1. Aerial view of the Belat Estuary

In normal condition, freshwater discharge from the Belat River is drained into the Kuantan River. However, during the dry and hot season which occur from April to September, the freshwater discharge is reduced to a low level causing the saltwater from the tide to intrude into the Belat Estuary. Since the Belat River is a tributary of the Kuantan River, the tidal mixing pattern at the confluence is rather complex. During the ebb and flood event, part of the saltwater from the Kuantan Estuary may flow into the Belat Estuary causing instability in the salinity distribution.

The tidal information of the Kuantan Estuary are used as a reference since the location of the Belat Estuary mouth is not open to the ocean. Figure 2 shows the tidal oscillation of the Kuantan Estuary in April 2017. From the tidal oscillation pattern, it can be seen that the tidal mechanism in the studied area is mixed-diurnal with 12 hours tidal cycle.

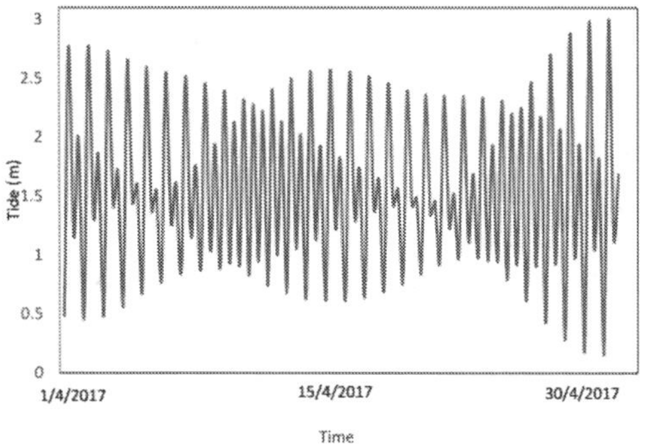

Figure 2. Tidal oscillations for Kuantan Estuary in April 2017

4 Methodology

4.1 Data Collection and Observation

A salinity field survey was conducted on 28 April 2017 when the dry season began in the region. The correct timing and tidal condition are crucial for the survey because salt intrusion is most severe during the dry season in conjunction with the spring tide (maximum high water). Field survey were carried out for both high water slack (HWS) and low water slack (LWS) when the water is stagnant at zero velocity. The salinity measurements were conducted by adopting a moving boat technique starting from the estuary mouth moving towards the inlands with the speed of the tidal wave celerity. Measurements stopped when the salinity level reads 0.1ppt. A Global Positioning System (GPS) was used to record the waypoint of each measurement while the portable handheld depth sounder was used to capture the water depth at every salinity measurement points. For the salinity measurement, a Horiba Multi-Parameter Water Quality Meter attached with 30 m cable was used. Water temperature, conductivity, turbidity, salinity and pH values of the river water were measured simultaneously with this instrument. A weight was attached to the cable of the water quality meter so that the measuring probe can submerge in the water at all time during the measurement.

4.2 Geometry analysis

For the geometry analysis, measurement at the Kuantan Estuary was not carried out on the first and second survey as existing data was available. The geometry analysis for the Belat Estuary was carried out based on Savenije's [9] theory for alluvial estuaries, where the change in the geometry of estuary varies exponentially over the distance. In this study, the width, depth and cross section of the Belat estuary was determined and plotted in a semi-log graph according to Equations (1) to (6). Then from the analysis, the value for A_0, B_0, and h_0 were obtained by fitting the simulated geometry characteristics to the observed data. This geometry information was used as boundary input data for the salinity analysis.

4.3 Salt intrusion modelling

For the salinity analysis, the 1-D analytical salt intrusion solution developed by Savenije [9-11] was applied for the Belat Estuary. In this model, the salinity distribution was simulated at Tidal Average (TA) condition and later converted to HWS and LWS situation to determine the maximum and minimum salt intrusion length. These boundary condition are important to determine the salt water intrusion limit. The governing equations applied to generate the simulated salinity profile are Equations (7) to (12). From the TA simulated curve, the salinity profile for HWS and LWS was generated by shifting the TA salinity seaward and landward by half of the tidal excursion (E). Salinity curves at HWS and LWS represent the envelopes of the salinity variation in a tidal cycle.

4.4 Model Calibration

The parameters that were calibrated in the salt intrusion analysis are the Van der Burgh's coefficient K, Dispersion coefficient D, Tidal Excursion E, and the Sea salinity S_0. K is the 'shape factor' which influence the shape of the salt intrusion curve at the toe. It is dependent on the changes in channel topography. Dispersion is a mathematical artefact that cannot be directly measured [13]. Since D is highly influenced by the discharge in which the data is always unavailable, the dispersion is calibrated in term of the mixing number α representing the ratio between dispersion D and discharge Q. In this study, tidal excursion has to be calibrated because the tidal velocity amplitude was not measured.

4.5 Error Analysis

RMSE is the one of the error analysis tools used to validate the reliability and performance of the salt intrusion model. The equation of RMSE is as below:

$$RMSE = \sqrt{\frac{\sum_{i=1}^{n}(P_i - O_i)^2}{n}} \tag{20}$$

Where P_i is the predicted value and O_i is the observed value.

NSE coefficient was also used to access the predictive power of the salt intrusion model. The equation of NSE is as below:

$$NSE = 1 - \frac{\sum_{i=1}^{n}(P_i - O_i)^2}{\sum_{i=1}^{n}(O_i - \bar{O})^2} \tag{21}$$

Where P_i is the predicted value and O_i is the observed value and $\bar{O}$ is the average of the observed value.

5 Result and Discussion

5.1 Geometry of the estuary

The geometry analysis for the Belat River Estuary was performed by utilizing the Equations (1) to (6). Data of the observed and simulated geometry characteristics of the estuary were plotted in a logarithmic scale which is shown in Figure 3. For the simulated geometry, the computed results were analysed and presented in a regression form. The geometry of the Belat Estuary consists of two reaches, Kuantan Estuary at the downstream and the Belat Estuary at the upstream (starting from distance about 6.5km from the mouth). From the plots, it can be seen that the shape of the estuary is well expressed by exponential function, confirming the theory of Savenije [11]. Table 1 shows the geometry characteristics results obtained from the best fit regression analyses. The inflection point was identified at distance 6km where the cross sectional area and width convergence length changed.

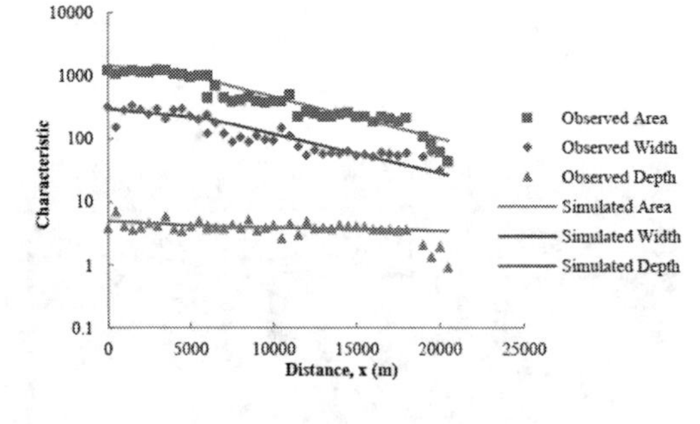

Figure 3. Geometry of Belat Estuary

Table 1 Geometry characteristics

Characteristic	Mouth, x_0	Inflection point, x_1
Area, A (m²)	1500	870
Width, B (m)	300	210
Area convergence, a (m)	11000	6500
Width convergence, b (m)	17000	7000
Average depth, h (m)	3.82	3.82

5.2 Salinity Distribution

Figures 4 and 5 show the vertical salinity distribution of the Belat Estuary during both High Water Slack (HWS) and Low Water Slack (LWS). From the patterns displayed, the mixing mechanism for the Belat Estuary can be classified as well-mixed because the distribution of saline water is almost evenly mixed from the bottom to the surface. Well-mixed mechanism in estuaries occurs when the discharge of fresh water is small compared to the tidal flows (tidal dominant). In Figure 5, there is a significant stratification occurs at distance 6.5km. The stratification is identified at the confluence of the Kuantan-Belat Estuary. When there is a split in the river system, the confluence receives fresh water discharge from both rivers where the amount of total discharge increase drastically. From the plot, there is no significant deviation of salinity level in the Belat Estuary during the LWS. This means that the confluence does not affect much on the vertical salinity distribution during ebb tide and the results are satisfying.

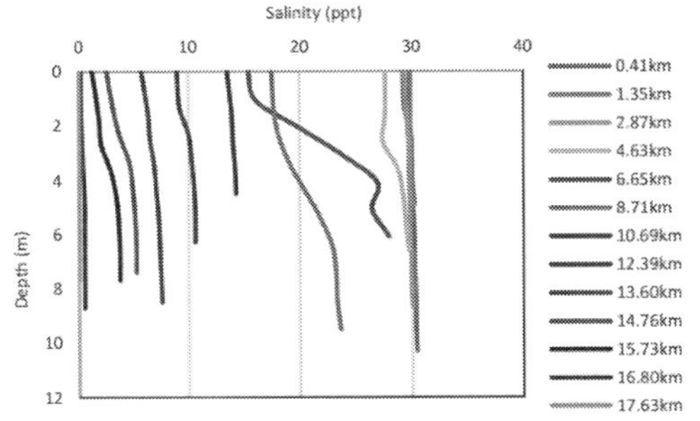

Figure 4. Vertical salinity distribution of the Belat Estuary during HWS

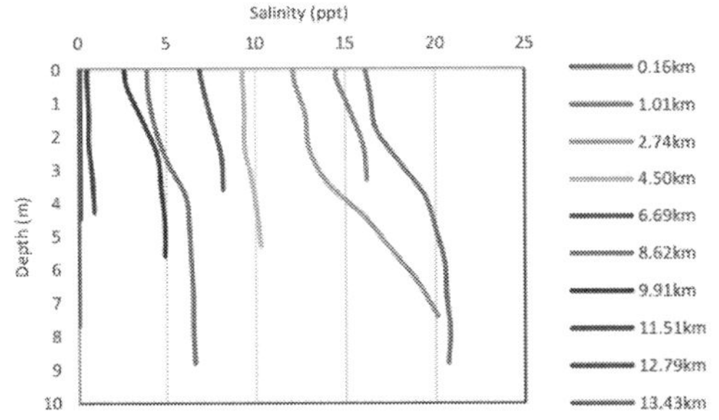

Figure 5. Vertical salinity distribution of the Belat Estuary during LWS

Figure 6 displays the measured and simulated longitudinal salinity profile of the Belat Estuary. As can be seen from the plot, the 1-D salt intrusion model is able to simulate the salinity profile in the Belat Estuary very well when compared to the observed data. From the simulation, the calibration parameters K and α obtained to fit the curve against the observed data are 0.45 and 13m^{-1}, respectively. Since the velocity amplitude was not measured in this study, the tidal excursion E was calibrated based on the HWS and LWS tidal envelop. The values obtained for the tidal excursion are 11000m at the mouth and 7800m at the inflection point with a convergence length e of 20000m.The salinity at the mouth for the HWS is 34ppt while for the LWS is about 22ppt indicating a difference of 12ppt.

In overall, the simulated longitudinal profile has a good fit with the observed data. However, at distance 6.5km, there are some outliers. These is due to the split in the Kuantan River (Belat – Kuantan). For the downstream part near the mouth, it can be seen the salinity distributions show some inconsistency, this is because the downstream part of the Belat Estuary belongs to the Kuantan River and this river is no longer in an equilibrium condition due the construction of the Kobat Barrage.

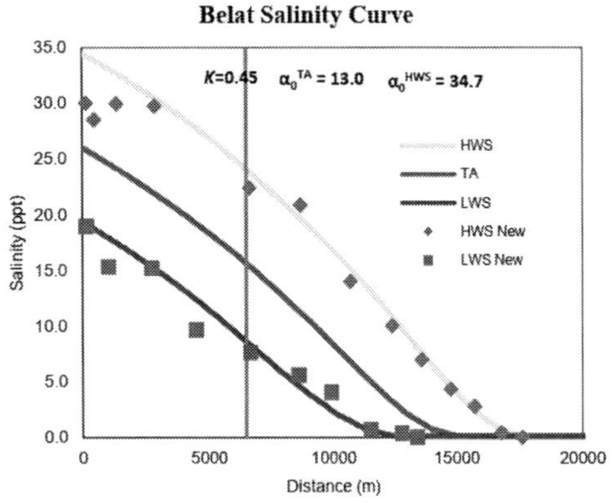

Figure 6. Measured and simulated longitudinal salinity profile of the Belat Estuary

5.3 Error Analysis

For the model performance, the RMSE obtained for the Belat River Estuary during HWS and LWS are 2.18ppt and 3.58ppt, respectively. At LWS, the RMSE value is higher because the difference between the observed and simulated values is bigger for the outliers. On the other hand, analysis shows high NS efficiency for both HWS and LWS conditions with the factor of 0.998 and 0.993, respectively which is very near to unity. This means that the salt intrusion model is very reliable and has very high efficiency.

6 Conclusions

In a nutshell, the investigation of salt intrusion in the Belat River Estuary has been successfully carried out. The geometry profile of the Belat Estuary has been well analysed adopting an exponential function and is proven valid in this case study. Salinity intrusion investigation have been successfully conducted for the Belat Estuary during both High Water Slack (HWS) and Low Water

Slack (LWS). The vertical salinity distribution results show that the mixing mechanism in the Belat Estuary can be classified as well-mixed. The calibration parameters, Van der Burgh coefficient K, mixing coefficient α, and tidal excursion E, give the values of 0.45, 13m^{-1} and 11000m, respectively. The performance of the 1-D analytical salt intrusion model has been evaluated and the results show accuracy. The average RMSE and NSE values obtained for both the HWS and LWS are 2.88ppt and 99%, respectively. The calibrated values had successfully fit the simulated data and the observed data. As a conclusion, the salt intrusion model is very reliable for the salt intrusion investigation in the Belat River Estuary. Hence, it is believed the 1-D salt intrusion model can be a useful tool to the water manager in deciding the appropriate location to install any water intake station according to the salinity distribution of the Belat River Estuary.

Acknowledgement

The authors would like to express their gratitude to: Universiti Malaysia Pahang (UMP) for the funding support through grant number RDU160326, Centre for Earth Resources Research Management (CERRM) and other technical members for their invaluable support and assistance in completing the fieldworks in Kuantan and Belat River; and the Department of Irrigation and Drainage (DID) Malaysia for providing the hydrological data.

References

1. SMHB, Ranhill, and Perunding Zaaba, Salinity studies for Sg. Rompin and Sg. Sedili Besar, *Malaysia National Water Resources Study, DID Malaysia* **5** (2000)
2. S.A. Haque, Salinity problems and crop production in coastal regions of Bangladesh, *Pak. J. Bot.*, **38(5)**, 1359–1365 (2006)
3. A.D. Nguyen , H.H.G. Savenije , D.N. Pham , D.T. Tang, Using salt intrusion measurements to determine the freshwater discharge distribution over the branches of a multi-channel estuary: The Mekong Delta case, *Estuar. Coast. Shelf Sci*, **77(3)**, 433–445 (2008)
4. S. Wongsa, Impact of Climate Change on Water Resources Management in the Lower Chao Phraya Basin, *Journal of Geoscience and Environment Protection*, **3**, 53–58 (2015)
5. D.H. Nguyen, M. Umeyama, T. Shintani, Importance of geometric characteristics for salinity distribution in convergent estuaries, *J Hydrol*, **448–449**, 1–13 (2012)
6. H.R. Wallingford, Kuantan River, Malaysia : Prediction of salinity intrusion, *Hydraulic Research Station* (1978)
7. M.T.J. van Breemen, Salt intrusion in the Selangor Estuary in Malaysia Model study with Delft3D estuary in Malaysia, (2008)
8. J.I.A. Gisen, H.H.G. Savenije, R.C. Nijzink and A.K. Abd. Wahab, Testing a 1-D analytical salt intrusion model and its predictive equations in Malaysian estuaries, *Hydrolog Sci J*, **60:1**, 156-172 (2015)
9. H.H.G. Savenije, A one-dimensional model for salinity intrusion in alluvial estuaries, *J Hydrol*, **85(1–2)**, 87–109 (1986)
10. H.H.G. Savenije, Salinity and tides in alluvial estuaries. *Amsterdam: Elsevier* (2005)
11. H.H.G. Savenije, Salinity and tides in alluvial estuaries, *Delft University of Technology* (2012)
12. E. F. Zhang, H.H.G. Savenije, S. L. Chen, and X. H. Mao, An analytical solution for tidal propagation in the Yangtze Estuary, China, *Estuar. Coast. Shelf Sci*, **91(4)**, 492–501 (2012)
13. H.H.G. Savenije, Predictive model for salt instrusion in estuaries, *J Hydrol*, **148(203–218)**, 203–218 (1993)

MATEC Web of Conferences 150, 03006 (2018) https://doi.org/10.1051/matecconf/201815003006
MUCET 2017

Overview Of Application Of Traffic Simulation Model

Nurul Nasuha Nor Azlan[1], *Munzilah* Md Rohani[2,*]

[1]Faculty of Civil and Environmental Engineering, Universiti Tun Hussein Onn Malaysia, Parit Raja, 86400, Johor, Malaysia
[2]Faculty of Civil and Environmental Engineering, Universiti Tun Hussein Onn Malaysia, Parit Raja, 86400, Johor, Malaysia

Abstract. Traffic simulation is a widely used method applied in the research on traffic modelling, planning and development of traffic networks and systems. From the literature study, a variety traffic simulation models were found in experiments and applications with aims to imaginary real traffic operations. The traffic simulation models can be categorised into three namely, microscopic modelling, macroscopic modelling and mesoscopic modelling. This report is aimed to overview these traffic simulation models, in term of its function, limitation and application.

1 Introduction

Traffic modelling is aimed to accurately recreate traffic as observed and measured on street. Traffic modelling assumed the appearance of a traffic system without replicating. It was developed based on modeller's experience that integrate the mathematical models into traffic system [1]. Traffic modelling plays an important part in traffic engineering. It can be applied to plan and manage the traffic within certain road network [2]. For example, making a smooth traffic at an intersection and etc [3].

Principally, simulation models focus on three output values to solve traffic problems [4]. First is the traffic flows. In traffic flows, alternative routes can be identified based on the number of vehicles. By using the simulation model, modeller can devise on how to reduce the levels of congestion of certain roads. Second output is the network element. Network element in traffic simulation consists of link, merge, link cross and other elements of the road [5]. This is related to the geometric layout of the road. Using appropriate simulation software, the road geometric design can be changed to see how it can influence the current traffic situation. Third output is the skim category. Simulations model can help to estimate the time and cost of travel. This is especially used when the assessment of traffic improvement is needed to be measured. The transport planner can easily make a performance comparison without any extra cost of money and time.

2 Classification of Traffic Simulation Models

The models were grouped according to their areas of application. The three classification of traffic simulation models, are microscopic modelling, macroscopic modelling and mesoscopic modelling.

* Corresponding author: munzilah@uthm.edu.my

2.1 Microscopic Modelling

Microscopic modelling based on the characteristics of various vehicle movements such as cars, buses, motorcycles and so on in the traffic flow. Microscopic modelling aimed to collect data parameters, such as, flow, density, speed, travel and delay time, long queues, stops, pollution, fuel consumption and shock waves. The characteristics of microscopic modelling methods were based on car-following model, lane-changing models and gaps of the individual drivers.

2.1.1 Car Following Models

These algorithms introduced a concept that a driver recognize and follow a lead vehicle in a lower speed [6]. This can be described as a situation such as in a car platoon without being able to change the lane. Generally, the following car algorithm determined the relationship with the lead vehicle at times. It was the function of spacing, speed and acceleration at times. The closer the following vehicle to the lead vehicle the more sensitive the reaction of the following vehicle to the lead vehicle. This sensitivity also increases with speed. This is because if the main vehicle was driven in a lower speed, the following vehicles will then reduce the speed. This resulted the occurrence of car platoon and traffic congestion.

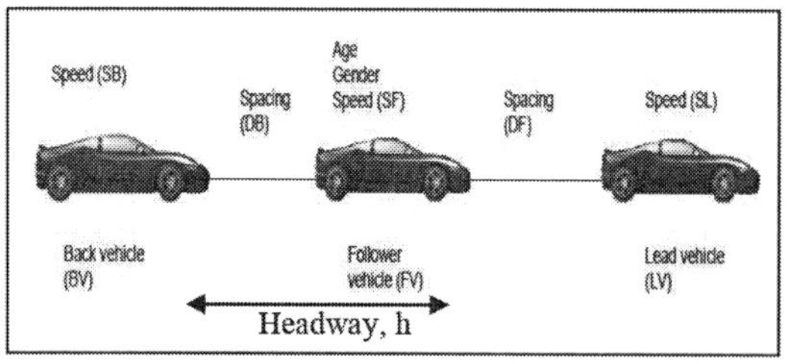

Fig. 1. Car Following Model

https://doi.org/10.1051/matecconf/201815003006

In the figure above, headway (h) is the horizontal distance between two vehicles. Spacing (s) is the distance measured from the rear-bumper of the following vehicle to rear-bumper of the lead vehicles pass at a certain point within time interval. The time headway is defined as the time difference of successive vehicles crossings at the point. Microscopic modelling is based on the movement of individual vehicles and their relative time and space.

2.1.2 Lane Changing Models

Gipps has proposed a development of the lane-changing model [7]. The lane-changing model is a decision process to estimate the driver's behaviour in making a lane change within the time given. Lane changing can be categorised into two types. Mandatory lane change, which occurs when a driver change lane to a specified lane. For instance, a driver changed to the right lane when he wanted to make a right turn at the next intersection and discretionary lane change, which the driver made to the next lane to avoid following trucks and increase speed.

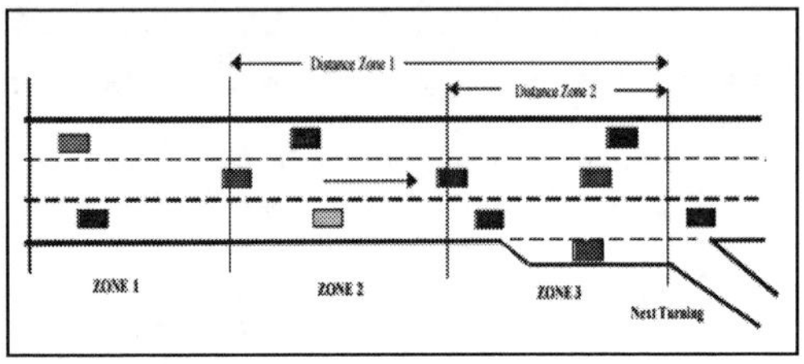

Fig. 2. Lane Changing Zones

- Zone 1 : The furthest distance from the next turning to measure how the driver:

  i. Change vehicles lane and speed of driver
  ii. Speed and distance of lead vehicles
  iii. Speed and distance upcoming lead vehicles at the other lane

- Zone 2 : The intermediate zone. The vehicles is looking for a gap to turn into the turning lane which affects the lane-changing decision.

- Zone 3 : This is the shortest distance to the next turning point. The vehicle has started to change lane at the merging lane to the next turning. It is necessary to reduce the speed to provide a big gap for vehicle to change the lane.

2.1.3 Gap Acceptance Models

Gap acceptance are used to determine the number of vehicles of traffic that can pass through a conflict point. A gap is defined as the time headway in between the lead and lag vehicles in the target lane. The lead gap is the gap to the lead vehicle in the target lane. The lag gap is the gap to lag vehicle in the target.

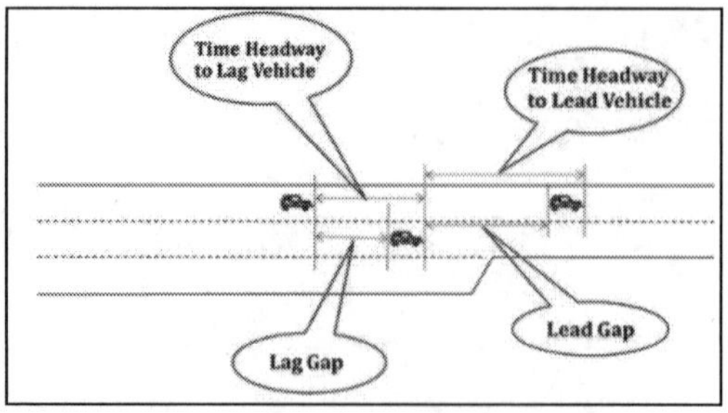

Fig. 3. Lane Changing with Consideration of Lead and Lag gaps

Gap acceptance models were used to determine the size of gap that will be accepted or rejected by a driver who aimed to merge or cross the intersection. Critical gap is defined as the number of accepted gap shorter than its equal to the number of rejected gaps. The driver needs time to clear the intersection and decide. The parameters of the gap acceptance model are the acceleration rate, desired speed and speed acceptance. Among these, the acceleration rate, the maximum give-way time and visibility distance at the intersection are the most important. The acceleration-rate is the ability of the vehicle to accelerate with required safety gap. The maximum give-way time is to determine when a driver starts to get intolerant if they cannot identify the gap.

2.2 Macroscopic Modelling

Macroscopic modelling describes the intersections at a low level of detail [8]. In macroscopic model, traffic stream is represented in an aggregate measured in terms of characteristics like speed, flow and density. Flow, speed and density are the three main characteristics of traffic. The researchers have studied speed-flow-density relationships and have attempted to develop mathematical descriptions for these curves, which is Greenshield's Model and Greenberg Model. Greenshield's Model is used to develop a model of uninterrupted traffic flow [9] It is a fairly accurate and simple modelling. There are three characteristics of the graph.

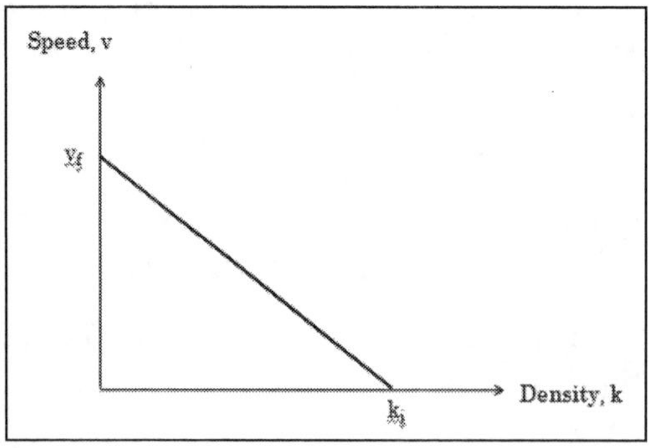

Fig. 4. Speed, v versus Density, k

Greenshied's Model shows the relationship between speed and density is linear. When density increases, flow decreases to an optimum when more vehicles were on the road. The speed also decreases due to the interaction of vehicles.

$$v = v_f - \frac{v_f}{k_j} k \quad (1)$$

Where,

v = speed (km/h)
k = density (veh/km)
v_f = free mean speed (veh/km)
k_j = jam density (veh/km)

Eq. 1. Speed, *v*

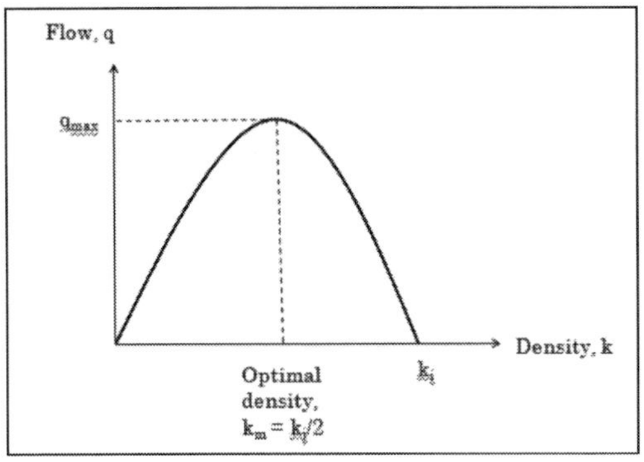

Fig. 5. Flow, *q* versus Density, *k*

Greenshield's model shows that the relationship between flow and density is a parabolic curve.

$$q = \left(v_f - \frac{v_f}{k_j} k\right) k \quad (2)$$

Where,

q = flow

Eq. 2. Flow, *q*

When flow is very low, speed is higher. The drivers are able to travel at a desired speed. As the flow increases, speed gradually decreases. The highest flow shows the transition of non-congested to congested condition.

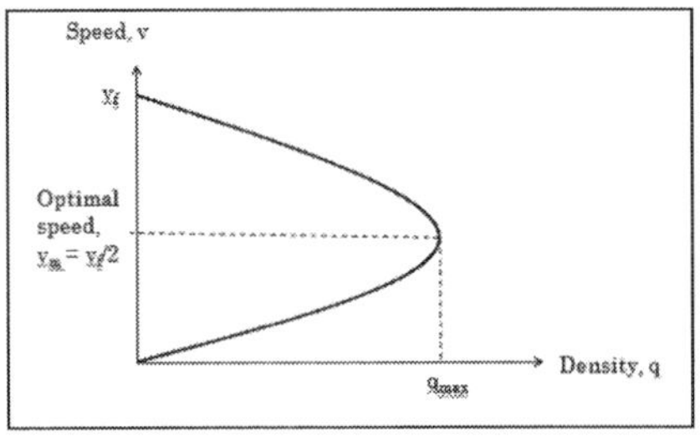

Fig. 6. Speed, *v* versus Density, *k*

Greenshield's model shows the relationship between speed and density.

$$q_m = v_m \, k_m \ or \ q_m = \frac{v_f k_f}{4} \quad (3)$$

Where,

q_{max} = maximum flow
v_{max} = optimal speed
k_{max} = optimal density

Eq. 3. Maximum flow

When the capacity (q_{max}) is in the maximum flow, the density is increasing and the speed will decrease due to the maximum number of vehicles passing to a certain point. The characteristics of congested unstable flow is high density and low speed. There is no gap for vehicles to enter. The characteristics of uncongested stable flow is slow density and high speed, when there are gaps for merging lane.

2.3 Mesoscopic Modelling

Mesoscopic modelling describes the analysed transportation elements in small groups. Mesoscopic models are a combination of microscopic modelling and macroscopic modelling. In this model, platoon dispersion is stimulated. There are two methods of mesoscopic modelling which is platoon dispersion and vehicle platoon behaviour.

The first method is platoon dispersion. As a platoon moves downstream from an upstream intersection, the distance between the vehicle which may be due to the differences in the vehicle speeds, vehicle interactions (lane changing and other interferences), (parking, pedestrians and others). This phenomenon is called as platoon dispersion. The second method is vehicle platoon behaviour is a group of vehicles, travel at a short headway and moving at the same speed. Vehicle platoon behaviour is to predict the arrival of vehicle platoon with time, the total arrival time.

The important considerations in modelling of platoon behaviour are the number of vehicles in a platoon, speed of a platoon and the distribution of speed [10]. Platoon are formed when the traffic signal turns to red and the vehicles start to queue. For an isolated traffic signal, the lenght of queue is the key performance measures that is afftected by the vehicle arrival pattern [11]. The platoon will disperse when the traffic signal starts to change from yellow to green, so it starts with incerasing the speed, then the vehicles start to move. Once the vehicles are released from the stop line, they will start to move in a tight platoon position with short time headways and will start to disperse travel further downstream [12].

Table 1 presents the various software and modelling applied by previous research on traffic simulations. From the figure, it shows that the simulation modelling can solved a range of problems related to transportation particularly in public transport.

https://doi.org/10.1051/matecconf/201815003006

Table 1. Traffic Simulation Software

Modelling	Software	Project Research	Aim	Parameter	Developed By (Referrence)
Microscopic	Vissim	Alignment of public transport lines. Examples Light Rail Transit (LRT) [13]	1) To develop the model urban traffic transit operations. 2)To provide the simulation results the animation of vehicle movements, traffic signal and travel time [14].	1)Travel 2)Delay 3)Signal timing 4)Detector actuations 5)Environmental indicators 6)Queue length	PTV System Software and Consulting GMBH, France (Integrated) [15]
Microscopic	AIMSUN [16]	Taharoto Road area, located in Takapuna, North Shore City, New Zealand - proposed Bus Rapid Transit project, including changes to bus routes [17]	1)To reproduce real traffic conditions in an urban network, such as expressways and arterial routes. 2)To stimulate any kind of measureable traffic detector, such as counts, occupancy and speed.	1)Flow 2)Speed 3)Travel time	J.Barcelo and J.L.Ferrer, Polytechnic University of Catalunya, Barcelona (Integrated) [18]
Microscopic	SUMO [19]	1)Simulation of traffic based on the daily activities of population of the city of Cologne [20].	1)To compare and evaluate different models and algorithms related to traffic and traffic management. 2)To evaluate traffic light system.	1)Signal timing (green time and cycle length) 2)Turning direction 3)Intersection type 4)Connection of lane	D.Krajzwwicsz, Institut Fur Verkehrssystem Technik, Berlin, Germany (Integrated) [21]
Microscopic	FRESIM (FREway [22]	1)Stimulate expressway traffic flow in Singapore [23]	1)To enhance improvements to the geometrics and operational capabilities and employs car-following and lane changing models 2)To stimulate geometric conditions	1)Ramp metering 2)Driver behaviour 3)Warning sign 4)Off ramp	Federal Highway Administration,1996 (Freeway) [24]
Microscopic	CORSIM [25]	Design alternatives in Seattle, WA [26]	1)To stimulate traffic and traffic control system	1)Stop/yield sign 2)Traffic signal 3)Ramp metering	Federal Highway Administration (FHWA), USA (Integrated and Freeway) [27]
Microscopic	Paramics [28]	Modelling ITS in Southern California [29]	1)Simulation of congested traffic networks at the level of individual vehicles	1)Type of intersection (Signal, stop sign, roundabout)	Edinburgh Parallel Computing Centre in Scotland (Integrated and freeways) [30]
Macroscopic	Visum [31]	1)Planning for feeder bus services using VISUM : A case study of Mumbai, India. [32]	1)To analyse traffic flows, forecast and geographic information system (GIS). 2)To plan public transport services. 3)To develop advanced transport strategies	1)Mapping (GIS) 2)Environmental indicator 3)Travel time 4)Traffic flow 5) Delay time 6)Queue length	PTV VISUM [33]
Macroscopic	Saturn [34]	1)Off-line calculation of traffic signal timing in coordinated traffic signal networks in the UK, USA and Australia [35]	1)To test the impacts of one way streets, traffic control measures and bus-only streets. 2)To analyse and evaluate of traffic management systems	1)gap acceptance 2)junction coordinates 3)speed 4) saturation flow 5)signal data1	Institute of Transport Studies, University of Leeds, 1979 [36]

3 Conclusion

The objective of simulation model is to presents a real traffic situation in to dynamic model. There are three simulation model available to be used in traffic engineering namely macroscopic, microscopic and mesoscopic. The application on each model is depending on the aim of study to be conducted. Selecting the right model according to study aims is an important step towards traffic problem resolution.

The authors acknowledge the contribution of the Ministry of Malaysian of Higher Education (MOHE) and the University Tun Hussein Onn Malaysia (UTHM). This research is funded under FRGS Grant scheme (vot 1451) and GPPS Grant scheme (vot U778).

References

1. J.G. Shanthikumar, R.G. Sargent. A Unifying View Of Hybrid Simulation/Analytic Models And Modeling. *Operations research*, **31**(6), pp.1030-1052 (1983)
2. A. D. May, *Traffic Flow Fundamentals,* Prentice Hall Englewood Cliffs, New Jersey (1990).
3. S. Mahajan, A. Umadekar, K.Jethwa. New Concept Of Traffic Rotary Design At Road Intersections. Procedia-Social and Behavioral Sciences, **96**, pp.2791-2799 (2013)
4. M. Friedrich. *Multimodal Transport planning*. Stuttgart University, Germany (2015). (Lecture Notes).
5. J. Barcello. *Fundamentals of Traffic Simulation*. International Series in Operations Research and Management Science. Barcelona, Spain, pp. 68-69 (2010)
6. R.M. Michaels. *Perceptual Factors In Car Following*. In Proceedings of The 2nd International Symposium on The Theory of Road Traffic Flow, London, England, OECD (1963)
7. P.G. Gipps. A Model For The Structure Of Lane-Changing Decisions. Transportation Research Part B: Methodological, **20**(5), pp. 403-414 (1986)
8. A.K. Rathi, Z.A. Nemeth. *Freesim: A Microscopic Simulation Model Of Freeway Lane Closures (abridgment)*, **91**, pp. 21-24 (1996)
9. B.D. Greenshields, W. Channing, H.A. Miller. *A Study Of Traffic Capacity*. In Highway research board proceedings. National Research Council (USA), Highway Research Board, **14**, pp. 448-477 (1935)
10. H. Qing, L. Head, J. Ding. PAMSCOD: Platoon-Based Arterial Multi-Modal Signal Control With Online Data. Transportation Research Part C: Emerging Technologies, **20**(1), pp. 164-184 (2012)
11. *Highway capacity manual*. Washington, DC (2000)
12. L. Yu. Calibration Of Platoon Dispersion Parameters On The Basis Of Link Travel Time Statistics. Transportation Research Record: Journal of the Transportation Research Board, **1727**, pp. 89-94 (2000).
13. R. Cervero. Journal Report: Light Rail Transit And Urban Development. Journal of the American Planning Association, **50**(2), pp.133-147 (1984)
14. J. Barceló, E. Bernauer, L. Breheret, G. Canepari, C.D. Taranto, J. Ferrer, R. Liu. Simulation Modelling Applied to Road Transport European Scheme Tests (SMARTEST)-Review of MicroSimulation Models. In Institute for Transport Studies, University of Leeds (1998)
15. *PTV VISSIM 5.30-05 User Manual*. Karlsruhe: Planning Transport, Verkehr AG (2011)
16. J. Barceló, J. Casas. *Dynamic Network Simulation With AIMSUN*. Springer US. pp. 57-98 (2005)
17. C.P. Haas. Assessing Developments Using AIMSUN. In *Institution of Professional Engineers New Zealand. Annual conference (*2001)
18. J. Barcelo, L. Ferrer, R. Martin. Simulation Assisted Design and Assesment of Vehicle Guidance Systems. Department d'Estadistica I Investigacio Operativa, Polytechnic University of Catalunya, (1997)
19. M. Behrisch, L. Bieker, J. Erdmann, D. Krajzewicz. *SUMO–Simulation Of Urban Mobility: An Overview*. In Proceedings of SIMUL 2011, The Third International Conference on Advances in System Simulation. ThinkMind (2011)
20. Hertkorn, G, Wagner, P. *The Application Of Microscopic Activity Based Travel Demand Modelling In Large Scale Simulations* (2004)
21. D. Krajzewicz, G. Hertkorn, C. Rössel, P. Wagner. *SUMO (Simulation of Urban Mobility)-An Open-Source Traffic Simulation*. In Proceedings of the 4th middle East Symposium on Simulation and Modelling, pp. 183-187 (2002)
22. H. Rakha, B. Crowther. *Comparison And Calibration Of FRESIM And INTEGRATION Steady-Statecar-Following Behavior*. Transportation Research Part A: Policy and Practice, **37**(1), pp. 1-27 (2003)
23. R.L. Cheu, X. Jin, K.C. Ng, Y.L. Ng, D. Srinivasan. Calibration Of FRESIM For Singapore Expressway Using Genetic Algorithm. *Journal Of Transportation Engineering*, **124**(6), pp. 526-535 (1998)
24. *Federal Highway Administration*. US Department of Transportation (1981)
25. A. Halati, H. Lieu, S. Walker. *CORSIM-Corridor Traffic Simulation Model*. In Traffic Congestion and Traffic Safety in the 21st Century: Challenges, Innovations And Opportunities (1997)
26. L. Bloomberg, J. Dale. Comparison of VISSIM and CORSIM traffic simulation models on a congested network. Transportation Research Record: Journal of the Transportation Research Board, **1727**, pp. 52-60 (2000)
27. *Federal Highway Administration*. US Department of Transportation (1981)
28. G.D.B. Cameron, G.I.D. Duncan. PARAMICS — Parallel Microscopic Simulation Of Road Traffic. The Journal of Supercomputing, **10**(1), pp. 25-53 (1996)

* Corresponding author: munzilah@uthm.edu.my

29. B. Abdulhai, J.B. Sheu, W. Recker. Simulation Of ITS On The Irvine FOT Area Using" Paramics 1.5" Scalable Microscopic Traffic Simulator: Phase I: Model Calibration and Validation. California Partners for Advanced Transit and Highways (PATH) (1999)
30. M. Smith, G. Duncan, S. Druitt. PARAMICS: Microscopic Traffic Simulation For Congestion Management (1995)
31. C. Hildebrand, S.A. Hörtin. Comparative Study Between Emme And Visum With Respect To Public Transport Assignment. (2014)
32. B.R. Muley, U. Chande, C. S. R. K. Prasad. *Planning For Feeder Bus Services Using Visum: A Case Study Of Mumbai, India* (2015)
33. *PTV VISUM User Manual* (2014)
34. J. Aron. *Using 'SATURN' To Determine The Effect Of Data On Model Accuracy And Operation Efficiency On A Network Of A Small CBD* (Doctoral dissertation) (2014)
35. L. He, C. Fu, L. Yang, S. Tong, Q. Luo. Coordinated Real-Time Control Algorithm For Multi-Crossing Traffic Lights. In Natural Computation (ICNC), 2014 10th International Conference. Pp. 128-133. IEEE (2014)
36. T. Worsley, P. Mackie. *Institute for Transport Studies, University of Leeds* (2015)

Evaluation of Weather Research and Forecasting (WRF) Microphysics single moment class- 3 and class- 6 in Precipitation Forecast

Syeda Maria Zaidi[1,*], *Jacqueline Isabella* Anak Gisen[1, 2]

[1]Faculty of Civil Engineering & Earth Resources, University Malaysia Pahang, Gambang, Malaysia
[2]Centre for Earth Resources Research & Management (CERRM), University Malaysia Pahang, Gambang, Malaysia.

Abstract. In this study, the performance of two different Microphysics Scheme options in Weather Research and Forecasting (WRF) model were evaluated for the estimating the precipitation forecast. The schemes WRF single moment class-3 (WSM-3) and single moment class-6 (WSM-6) were employed to produce the minimum, medium and maximum precipitation for the selected events over the Kuantan River Basin (KRB). The obtained simulated results were compared with the observed data from eight different rainfall gauging stations. The results comparison indicate that WRF model provides better forecasting at some rainfall stations for minimum and medium rainfall events but did not produce good result during maximum rainfall overall. The WSM-6 scheme is found to produce better result compared to WSM-3. The study also found that to acquire accurate precipitation results, it is also required to test some other physics scheme parameterization to enhance the model performance.

1 Introduction

The new generation Weather Research and Forecasting (WRF) model is the collaborative product of National Central of Atmospheric Research (NCAR) the mesoscale division, the national environmental prediction and forecast system lab (FSL) of Nation Oceanic and Atmospheric Administration (NOAA), the Naval Research Laboratory (NRL), the Department of Defence's Airforce Weather Agency (AFWA) and the Centre for Analysis and Prediction of Storms (CAPS) from the University of Oklahoma and the contribution of number of research scientists along with Federal Aviation Administration (FAA). The model is designed for both operational forecasting and weather research [1].

The WRF model is suitable for broad spectrum forecasting applications ranging from meters to thousands of kilometre. WRF contains many configured physics schemes to be easily implemented by the model user which include the advantages such as economical computation, existing default parameters, and relatively large spatial resolution grid. The offered numerous physics schemes in WRF are varies from simple to more sophisticate and computationally costly [2]. The physical schemes in WRF play a crucial role in WRF modelling. The parameterization of the schemes typically depends on the objective of the experiment design and the features of study area. The definition of parameterization refers to the process of selecting required parameters for a specific model. Besides that, parameterization in WRF also describes the phenomena occurs at sub gird scale. A significant amount of research has been conducted to evaluate the influence of physics parameterization and sensitivity of WRF in meteorological scenarios [3-8]. The parameterization of schemes is an important aspect related to the model forecasting accuracy. The blending of schemes depends on the domain of the research selection.

Since the last decades, a broad spectrum of physical parameterization schemes have been executed in weather models. Every scheme was developed based on different criteria which serve as a key source to identify the model's constraint and inaccuracies [9]. However, to define the optimum parameterization of every physics schemes is a highly difficult task as the number of these schemes have been increasing. Hence, this paper aims to to evaluate the performance two available Microphysics scheme options; single moment class -3 (WSM-3) and single moment class-6 (WSM-6) to determine the physical characteristic of precipitation with in the clouds.

1.1 Microphysics Scheme WSM class-3 and WSM class-6

Microphysics scheme plays a vital role in clouds, climate, and climatic models. It is used to regulate the process of water vapour and the formation of raindrops, hail and snow inside or outside the clouds [10]. It is also used to determine the type, size, and distribution of precipitation. The scheme is essentially required to attain accurate rainfall simulation output. Currently, the WRF system is supporting 13 microphysics options which range from simple to more complex physical options depending upon the criteria of a model designed and area

* Corresponding author: mariazaidi26@gmail.com

topography. In order to determine the physical characteristics of precipitation WRF Microphysics parameterization can be applied by simple scheme single moment class or advance scheme double moment [11].

The single moment scheme uses as a single prediction equation for each moment of precipitation to determine the particle size distribution. Whereas, the double moment is used to regulate the number and concentration of double moment of precipitation with an additional process; to determine the influence of aerosol on clouds [12].The single moment WSM-3 is a basic simple option which has clouds formation by considering three states of water namely cloud water, ice water and rainwater [13].This scheme is computationally efficient and simple suitable for meso-scale grid. On the other hand, The WSM-6 scheme is suitable for high resolution gird. The scheme class-6 has ice, snow, and rain with the extended association of graupel processes. The scheme is also suitable to use for cloud resolving grids [14].

2. Material &Method

2.1. Description of Study Area

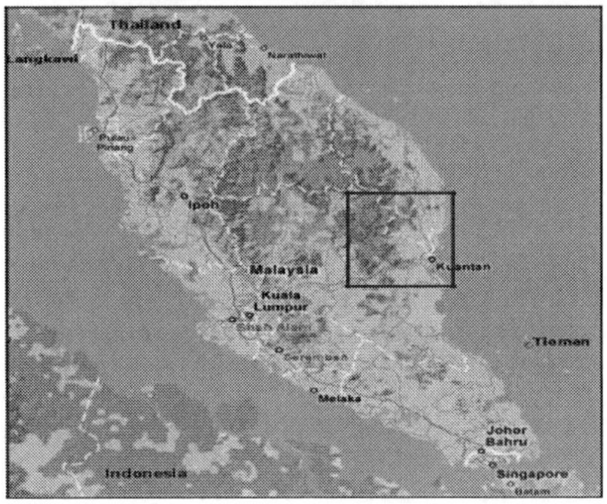

Fig. 1. Location of Kuantan River Basin

Kuantan River Basin (KRB) is the one of most important river basin located at the north east end of Pahang state in the peninsular Malaysia. KRB with the catchment area of 1630 km2 starts from Sg. Lembing passing through Kuantan city and ends at the South China Sea. KRB has tropical climatic condition. The area has experienced massive precipitation during the period of north east monsoon season from October to March which caused spilling over of river and produce floods. Additionally the Kuantan River is the main source of water supply in that region.

2.2. Data and Design of Experiment

In this study, the observed daily precipitation data records were obtained from the Department of Drainage and Irrigation (DID) Kuantan. The dataset was used to validate the accuracy of WRF model precipitation forecast at eight different rainfall gauging stations. The performance of the selected two microphysics options in the model was evaluated by comparing the simulated precipitation with gauged rainfall data for three different events. The selection of events and categorization of precipitation was based on minimum, medium and maximum recorded rainfall. The study has chosen 10th Jan 2012 as the minimum received rainfall, whereas, the rainfall events on 11th and 12th Jan 2012 set as medium and maximum precipitation respectively. WRF model version 3.01 was setup the with horizontal outer domain of grid resolution at 36km (d01), inner domain (d02) at 12km resolution and domain (d03) at 4km resolution grid, as shown in Fig.2. The Peninsular Malaysia is covered at 12km gird resolution. The boundary condition for all the simulation used in this study was 1o x 1o re-analysis data from the National Centre for Environmental Prediction (NCEP) Global Final Analysis (FNL). In total 6 simulations were conducted for both microphysics schemes (WSM-3 and WSM-6) with combination of other selected physics options for the study as shows in the given table (1).

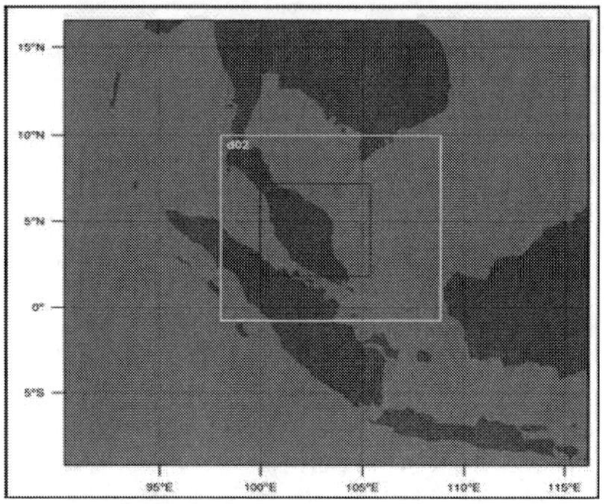

Fig. 2. WRF Model Grid Domains

Table 1. Selected WRF Model Physics

Physics Scheme	WRF Option
Cumulus	Betts-Miller_Jinjic (BMJ)
Longwave Radiation	Rapid radiative transfer model (RRTM)
Shortwave Radiation	Dhudiha Scheme MM5 short wave
Surface Layer	Monin–Obukhov similarity theory
Planetary boundary Layer	PBL scheme (MRF)

https://doi.org/10.1051/matecconf/201815003007

3. Results

The impact of two-available microphysics schemes in WRF model were tested for three different rainfall scenarios. Fig.3. shows the simulated precipitation which was conducted for minimum rainfall event on 10th Jan 2012 at different gauging stations. The result indicates that WRF model estimate uncertain forecasting values at different gauge locations. In comparison to the WSM-6, the scheme WSM-3 overestimated the rainfall at station Paya Besar, Kg. Sg Soi and Pulau Manis and lowest rainfall produced at Rumah Pam. However, both the scheme produced significantly rain accuracy at Ladang Kuala Raman compared to the observed rain.

In the in case of medium rainfall event on 11th Jan 2012 as shown in Fig.4. The simulated rainfall from WSM-3 produced slightly higher rain rate at stations Paya Besar and JPS Negeri Pahang. However, for the other stations both the WSM-3 and WSM-6 schemes have underestimated rainfall compared to the observed.

Fig 5. Shows the maximum predicted precipitation values from the model, simulated for the date 12th Jan 2012. The performance of the WRF model seems to be not satisfactory in producing heavy precipitation forecast. Although the schemes WSM-3 and WSM-6 shows the same pattern of forecasting but both the schemes did not produce compatible precipitation result to the observed for all the rainfall stations. Despite the fact, the microphysics scheme WSM-6 was found to produce the slightly substantial level of performance in all simulations output compared to WSM-3.

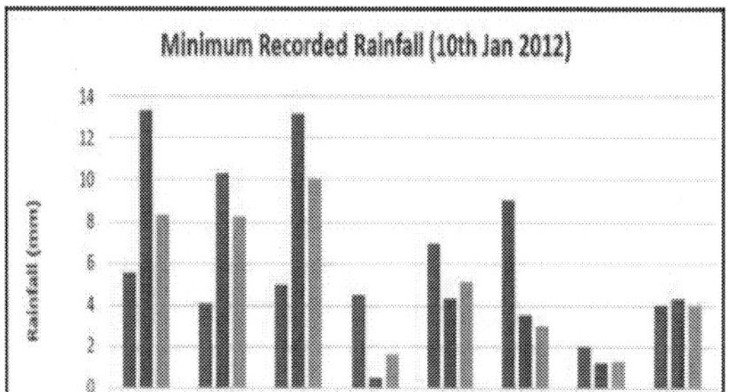

	Paya Basar	Pulao Mannis	Kg Sg Soi	Ruma Pam	JPS Negeri Pahang	PCCL Sg Lembing	LDN Nada	LDN Kuala Raman
Obseved	5.5	4.1	5.0	4.5	7.0	9	2	4.0
WSM-3	13.31	10.34	13.20	0.52	4.3	3.5	1.18	4.3
WSM-6	8.3	8.2	10.1	1.6	5.1	3	1.3	4.0

Fig 3. Minimum Simulated and Recorded Rainfall on 10th Jan 2012

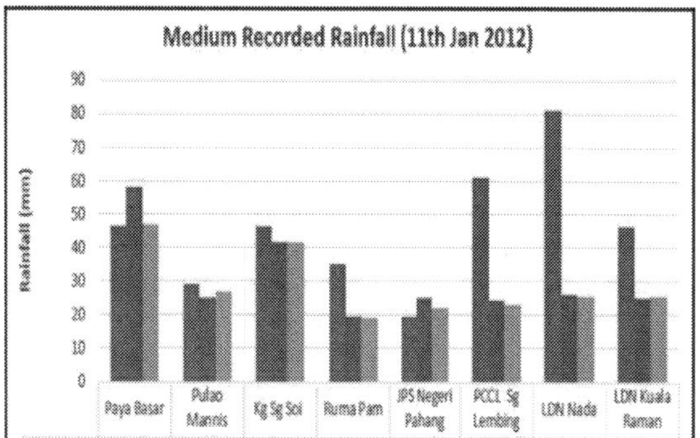

	Paya Basar	Pulao Mannis	Kg Sg Soi	Ruma Pam	JPS Negeri Pahang	PCCL Sg Lembing	LDN Nada	LDN Kuala Raman
Observed	46.5	29.0	46.5	35.0	19.5	61.0	81.0	46.5
WSM-3	58.0	25.3	41.6	19.7	25.3	24.2	25.9	25.3
WSM-6	46.77	26.82	41.55	19.2	22.1	22.89	25.7	25.7

Fig. 4. Medium Simulated and Recorded Rainfall on 11th Jan 2012

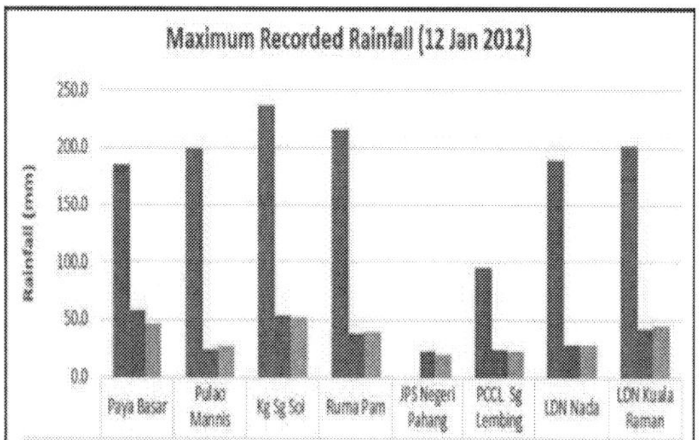

	Paya Basar	Pulao Mannis	Kg Sg Soi	Ruma Pam	JPS Negeri Pahang	PCCL Sg Lembing	LDN Nada	LDN Kuala Raman
Observed	185.0	199.5	237.5	216.5	0	96.0	190.0	201.5
WSM-3	58.0	24.3	53.6	38.6	22.0	24.5	28.9	42.2
WSM-6	46.77	26.82	51.55	38.8	20.2	22.89	28.7	45.69

Fig. 5. Maximum Simulated and Recorded Rainfall on 12th Jan 2012

4. Discussion

Considering all three rainfall scenarios, It found the microphysics scheme single moment class 6 (WSM-6) parameterized with other given physics option in WRF is able to produce better result in simulating minimum and medium precipitation events. Nevertheless, this scheme parameterization cannot be employed for all rainfall station as shown in the results. This may cause by the variation in topographic characteristics and atmospheric properties at certain stations. Therefore, it is necessary to test other model physics options to find the best suitable combination of parameterized scheme for the area. With regards to the model underestimation in simulating heavy rainfall, there has been numerous influenced factors which play key role to the WRF model accuracy. For instance, the selection of domain size, down scaling of lateral boundary conditions as well as the parameterization of appropriate physics scheme options, topographic features and climatic condition of the region.

This study tested the effects of two microphysics options WSM-3 and WSM-6 over the WRF model performance. For that reason, model simulations results found to be uncertain and inappropriate to produce precipitation forecast. Therefore, to obtain better WRF performance, it is also important to consider other influential model factors as discussed.

5. Conclusion

Microphysics scheme plays a vital role in atmospheric science. It has the significant role to deal with process of vapour, clouds and precipitation. In other words, microphysics scheme is one the responsible factor to produce rainfall. WRF model has available 13 microphysics options which are used to determine the formation, distribution and concentration of vapours, rain/snow, ice and clouds. This study was aimed to determine the precipitation forecast pattern of WRF model at Kuantan River employed by two simple options of microphysics single class (WSM-3 and WSM-6). The model was designed to simulate three different events to produced minimum, medium and maximum rate of rainfall. The event 10th Jan2012 was selected for minimum precipitation whereas, 11th Jan 2012 and 12th Jan 2012 events were selected for medium and maximum precipitation respectively. The result was validated from the observed rainfall at eight rainfall stations. From the study outcomes, it can be conclude that although microphysics Scheme WSM-6 showed better performance compared to the WSM-3 but it still produced uncertainties in results at different rainfall stations. However for the maximum precipitation forecast, both the schemes provided underestimate precipitation forecast. Considering the outcome of the study, it is cautious to state the exact performance of both schemes in forecasting heavy precipitation. This is because the study was focused on very specific objective. Therefore, more experiments are required to conduct considering other model factors including different physics option parameterizations, appropriate boundary condition and gird resolution to enhance the WRF model accuracy.

The authors are grateful to the University Malaysia Pahang (UMP) for the financial support of this research through the grant numbers RDU150127 and FRGS/1/2015/WAB08/UMP/02/1.

References

1. Michalakes, J., et al. The weather research and forecast model: software architecture and performance. in Proceedings of the Eleventh ECMWF Workshop on the Use of High Performance Computing in Meteorology. 2005. World Scientific: Singapore.
2. Xue, Y., et al., A review on regional dynamical downscaling in intraseasonal to seasonal simulation/prediction and major factors that affect downscaling ability. Atmospheric research, 2014. 147: p. 68-85.
3. Bukovsky, M.S. and D.J. Karoly, Precipitation simulations using WRF as a nested regional climate model. Journal of applied Meteorology and Climatology, 2009. 48(10): p. 2152-2159.
4. De Silva, G., et al. Application of WRF with different cumulus parameterization schemes for precipitation forecasting in a tropical river basin. in Proceedings of the 13th Asian Congress of fluid Mechanics. 2010.
5. Maussion, F., et al., WRF simulation of a precipitation event over the Tibetan Plateau, China-an assessment using remote sensing and ground observations. Hydrology and Earth System Sciences, 2011. 15(6): p. 1795.
6. Sohrabinia, M., W. Rack, and P. Zawar-Reza, Analysis of MODIS LST compared with WRF model and in situ data over the Waimakariri River basin, Canterbury, New Zealand. Remote Sensing, 2012. 4(11): p. 3501-3527.
7. Im, E.-S., S.-R. In, and S.-O. Han, Numerical simulation of the heavy rainfall caused by a convection band over Korea: a case study on the comparison of WRF and CReSS. Natural Hazards, 2013. 69(3): p. 1681-1695.
8. Mooney, P., F. Mulligan, and R. Fealy, Evaluation of the sensitivity of the weather research and forecasting model to parameterization schemes for regional climates of Europe over the period 1990–95. Journal of Climate, 2013. 26(3): p. 1002-1017.
9. Awan, N.K., H. Truhetz, and A. Gobiet, Parameterization-induced error characteristics of MM5 and WRF operated in climate mode over the Alpine region: an ensemble-based analysis. Journal of Climate, 2011. 24(12): p. 3107-3123.
10. Hong, S.-Y., J. Dudhia, and S.-H. Chen, A revised approach to ice microphysical processes for the bulk parameterization of clouds and precipitation. Monthly Weather Review, 2004. 132(1): p. 103-120.
11. Dudhia, J., WRF modeling system overview. www2.mmm.ucar.edu/wrf/users/tutorial/201201/WRF_Overview_ Dudhia. ppt. pdf, 2014.
12. Skamarock, W., Coauthors, 2008: A description of the advanced research WRF version 3. NCAR Tech. Note. NCAR/TN-475+ STR.
13. Dudhia, J., Numerical study of convection observed during the winter monsoon experiment using a mesoscale two-dimensional model. Journal of the Atmospheric Sciences, 1989. 46(20): p. 3077-3107.
14. Hong, S.-Y. and J.-O.J. Lim, The WRF single-moment 6-class microphysics scheme (WSM6). J. Korean Meteor. Soc, 2006. 42(2): p. 129-151.

Effect of Polypropylene Fibres on the Thermal Conductivity of Lightweight Foamed Concrete

Ashfaque Ahmed Jhatial[1], *Wan Inn* Goh[1*], *Noridah* Mohamad[1], *U. Johnson* Alengaram[2] and *Kim Hung* Mo[2]

[1] Jamilus Research Center, Faculty of Civil and Environmental Engineering, Universiti Tun Hussein Onn Malaysia, Batu Pahat 86400, Johor, Malaysia.
[2] Department of Civil Engineering, Faculty of Engineering, University of Malaya, 50603 Kuala Lumpur, Malaysia.

Abstract. With the reduction in the permanent load on the structure and excellent insulation properties, the lightweight foamed concrete is a potential thermal insulating building material to counter the urban heat island effect, which increases the temperature of urban areas due to the concentration of infrastructures constructed using conventional concrete that absorbs the solar radiation. The lightweight foamed concrete whose dry density ranges from 400 kg/m^3 to 1600 kg/m^3, has lower thermal conductivity compared to conventional concrete. But reduced density attributes to reduced compressive strength. In this study, to enhance the mechanical properties, the foamed concrete of 1600 kg/m^3 density is reinforced with polypropylene fibres (PP). Four percentages of PP fibres, 0% (controlled), 0.2%, 0.25% and 0.30% were added into the foamed concrete. The compressive strength as well as the thermal conductivity of foamed concrete reinforced with PP fibres were determined. Based upon the findings, the optimum percentage of PP was determined to be 0.20% which gave higher compressive strength while thermal conductivity of foamed concrete was observed to decrease upon addition of PP fibres. Thus, addition of PP fibres improves the thermal resistance in the foamed concrete along with enhancing the mechanical properties.

1 Background

Metropolitan and urban areas where conventional concrete is widely being used, faces urban heat island (UHI) effect. UHI is the phenomenon in which higher temperatures are developed in urban areas compared to the temperatures of surrounding suburban and rural areas [1]. Large surface materials such as concrete have influence on the UHI [2, 3]. Solar radiation is absorbed by conventional concrete, which stores the heat and releases it later, that can either be during the daytime or night-time. Urban area of 1 million and above dwellers can be up to 1.3°C warmer than the adjacent rural areas, while in night-time this difference can be up to 12°C [4, 5]. Foamed concrete is a solution to the raising UHI effect in urban areas.

Air bubbles, introduced by the mixing of foaming agent in foamed concrete, enables it to not only be lightweight but also achieve thermal insulating ability. Foamed concrete, being lightweight, helps in reducing the self-weight which the conventional concrete has, causing additional permanent load to the structure. The foamed concrete can be obtained in wide range of densities ranging from 400 to 1600 kg/m^3 [6]. The reduction in density however has a negative impact on the compressive strength, it is a fact that by reducing the density of concrete, the compressive also decreases.

Therefore, to improve the mechanical properties of foamed concrete, polypropylene (PP) fibres is used to reinforce the foamed concrete in this study. The addition of PP fibres in foamed concrete helps increase the tensile strength [7], controlling of post cracking which leads to micro-cracks, toughness and improves the deformation characteristics of the composite [8].

2 Literature Review

The conventional concrete is made up of mainly cement, sand, coarse aggregates and water, and the dry density ranges from 2200 kg/m^3 to 2600 kg/m^3 [9]. Although conventional concrete is widely used in applications ranging from buildings to pavements, bridges to dams, the use of coarse aggregates in concrete, tends to increase the self-weight of the concrete, ultimately putting unnecessary permanent load on the structure, while the surface of concrete absorbs more solar radiation causing the UHI. The thermal conductivity of 2400 kg/m^3 density concrete is 1.3 W/mK [10], which allows the concrete to transfer heat absorbed to the interior of the building and significantly raising the energy consumption of the building.

Due to the UHI effect, people living in the cities are more exposed to thermal stress compared to surrounding rural areas. The intensity of UHI is maximum during night-time when the temperature difference between urban areas and that of rural can reach up to 10°C [11]. Recent study [12] revealed that not only large cities but also villages and small built-up urban areas can have

* Corresponding author: wigoh@uthm.edu.my

https://doi.org/10.1051/matecconf/201815003008

higher temperatures than their surrounding areas, though UHI is more common in where concentration of infrastructure is. The concentration of construction and human activities in urban areas are causes to UHI along with global warming. Increase in UHI effect has significant health risk for urban dwellers [13], due to thermal stress an increase in mortality rate has been seen [14]. Therefore, thermal insulating buildings are being focused due to their resistance to heat flow which reduces the energy consumption of the building, ultimately reducing the UHI effect.

Foamed concrete has been known to construction industry for its excellent thermal properties and reduced self-weight apart from being fire-resistant and self-levelling concrete. The use of coarse aggregates is eliminated in the foamed concrete making it lightweight concrete and the reduced density helps in reducing the self-weight. Air bubbles are introduced in the form of foaming agent, which make it a porous structure and have air voids in its matrix, while producing micro-porous which reduce the interfacial bonding and strong plasticizing effect on the foamed concrete [15, 16]. The thermal conductivity value for foamed concrete is lower than conventional concrete, ranging from 0.1 W/mK for 300 kg/m^3 density to 0.7 W/mK for 1600 kg/m^3 density [17, 18]. The lower conductivity of heat through the foamed concrete is due to the air voids in the matrix which slows down the flow of heat by restraining the air from passing. Thermal resistance is created by the trapped air which gives the foamed concrete its excellent thermal insulating properties [17]. Although, reduction in density of foamed concrete can lower the thermal conductivity but it also reduces the compressive strength. Micro-cracks develop in concrete with curing and rapidly propagate upon applied stress resulting in low tensile strength [8]. Reduction in density also increases the micro-cracks in the foamed concrete. It has been found that 500 kg/m^3 density foamed concrete develops more micro-cracks than 1000 kg/m^3 [19, 20]. The mechanical behaviour is affected by the development of micro-cracks in foamed concrete.

To over-come cracks, the concrete is reinforced with steel bars. Though these reinforcements help in reduction in cracks but also increase the self-weight. Over the years different fibres have been used to reinforce the concrete, in replacement of steel bars. These reinforcing fibres range from synthetically organic and inorganic fibres to natural organic and inorganic fibres.

2.1 Characteristics of Polypropylene Fibres

Polypropylene fibres, which are thermoplastic polymers, are synthetically organic fibres which are produced when a gaseous by-product of petroleum refining, known as propylene, is polymerized. The propylene monomers are converted into long chains of polymers during polymerization. PP fibres are chemically inert, hydrophobic and stable in alkaline environment of concrete [21].

PP fibres have interestingly been used in concrete since 1960's with addition in concrete for construction of blast resistant buildings for the US Corps of Engineers [8]. Both conventional concrete and foamed concrete has seen the addition of PP fibres. A study [22] was carried out on conventional normal concrete in which PP fibres were added at 0.15%, 0.30% and 0.45% dosages and steel fibres at 1% and 2%, based on the results it was observed that 0.15% PP fibres achieved highest compressive strength while 0.30% achieved maximum flexural strength compared to controlled sample but when steel fibres along with PP fibres were added the compressive and flexural both increased. PP fibres content of 0%, 0.05%, 0.10% and 0.15% were also added as additives in foamed concrete of 1600 kg/m^3 and 1800 kg/m^3, it resulted in 0.15% achieving higher tensile strength and interfacial adhesion between the foamed concrete and PP fibres was observed [7]. 0% to 0.5% with an increment of 0.1%, PP fibres by volume have been used in lightweight concrete, which showed that 0.2% gained the highest compressive strength while lightweight foamed concrete achieved the most the tensile strength when 0.5% of PP fibres were added [23]. By adding PP fibres, it can prevent the formation of microcracks or reduce its scale owing to the crack arresting, thinning and crack bridging effect of the fibres [24].

3 Experimental Method

3.1 Materials

Ordinary Portland cement (OPC), sand which was passed through 5 mm, water, foaming agent and PP fibres were used. The foaming agent was used to create foam in a ratio of 1:20 (1 part of foaming agent and 20 parts of water). The water-cement ratio was taken as 0.55 while the sand-cement ratio was 2:1. In this study the density of foamed concrete was maintained 1600 kg/m^3. Four different percentages of PP fibres are utilized in this study, 0% (controlled sample with no PP content), 0.20%, 0.25% and 0.30% of PP fibres.

3.2 Procedure

The work procedure began with prepared materials for every single set. OPC, sand and percentage of PP fibres were mixed together in concrete mixer. Water was added into the mixer, along with pre-foam generated by using foamed generator and mixed together with cement mortar in concrete mixer. The addition of foam was done according to BS EN 12350-6:2009 [25] which continued gradually until target density of 1600 kg/m^3 was achieved. Foamed concrete was formed and was poured into the mould and kept for 24 hours.

3.3 Experimental Test

Compressive strength test conducted complied with BS EN 12390-3:2009 [26]. The sample size for the test was 100 mm x 100 mm x 100 mm. Before conducting the test, the sample was under air curing for 28 days. 12 samples have been prepared for the test. During the test, an axial load was applied to the cube samples and increased continuously without shock at a nominal rate within the range from 0.2 N/(mm^{2}s) to 0.4 N/(mm^{2}s) until the cube samples could not sustain further load. The average compressive strength of each set of samples was then recorded.

According to BS EN 12664:2001 [27], the panel samples were tested for thermal conductivity using the thermal conductivity apparatus shown in fig. 1. The panel samples for the test were of 300 mm x 300 mm x 50 mm size. Before conducting the test, the samples underwent air curing for 28 days. 8 samples had been prepared for the thermal conductivity test. Each sample had four holes drilled into its surface with depths of 10, 15, 20, and 30 mm, as shown in fig. 2, to insert the thermocouples. The samples were then placed between hot and cold plates. The temperature for cold plate has been set at 18 ^{0}C and the hot plate has been kept at 40 ^{0}C to simulate the exterior temperature and interior temperature. The hot and cold plates have been covering with one layer of barrier cushion while all sides of the sample were insulated with cork sheet [28]. The temperature of the hot and cold plate was recorded every 30 minutes for 24 hours.

Fig. 1. Thermal Conductivity Apparatus

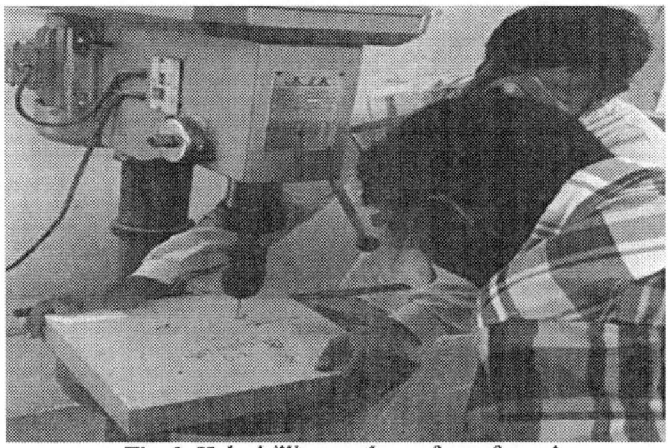

Fig. 2. Hole drilling on the surface of panels

4 Results

4.1 Compressive Strength

The results of the compressive strength test on the cubes are shown in table 1 represented graphically in fig. 3.

Table 1. Result of Compressive Strength

Target Density Kg/m^3	% of PP Fibres	Compressive Strength (MPa)
1600	0%	7.459
	0.20%	10.783
	0.25%	9.918
	0.30%	7.775

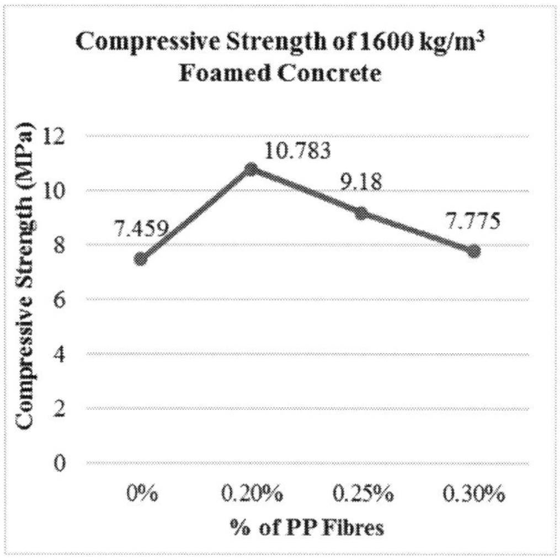

Fig. 3. Compressive Strength vs. % of PP Fibres

From the results, it can be noted that with the addition of 0.20% of PP fibres in the foamed concrete, the compressive strength increases 44.6% compared to the controlled sample. However, with the further addition of PP fibres started to lose the compressive strength but even at 0.30% PP fibres the compressive strength was 4.23% higher than controlled sample. Foamed concrete reinforced with PP fibres achieved higher compressive strength compared to the controlled sample. This gain in strength is due to PP fibres fill up the voids which are created by air. Hence, bonding mechanism between PP fibres and the foamed concrete is achieved. However, it can also be noted that the increase in PP fibres content beyond the determined optimum dosage also reduced the compressive strength. This may be due to PP fibres being hydrophobic material, can retain water and as a result voids are created. Moreover, the amount of fibre will obstruct the void which will cause weak bonding between the matrixes. Hence, the compressive strength will decrease.

4.2 Thermal Conductivity

The thermal conductivity was conducted on the panels and Table 2 and fig. 4 shows the results of thermal conductivity of foamed concrete with different percentage of PP fibres.

Table 2. Result of Thermal Conductivity

Target Density Kg/m^3	% of PP Fibres	Thermal Conductivity (W/mK)
1600	0%	0.68
	0.20%	0.67
	0.25%	0.71
	0.30%	0.66

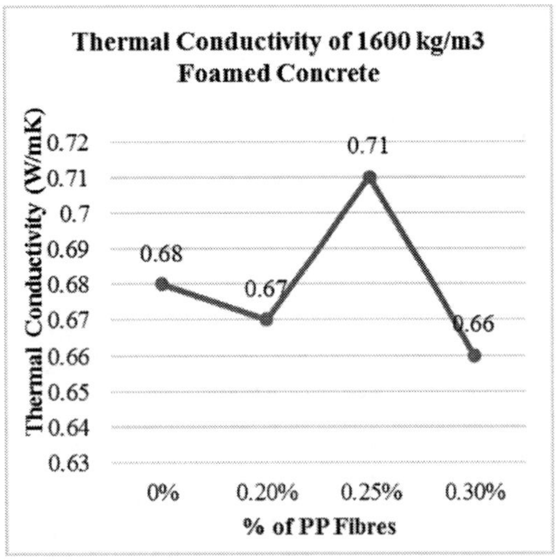

Fig. 4. Effect of PP Fibres Content on Thermal Conductivity

The addition of PP fibres in foamed concrete also had an impact on the thermal conductivity. The controlled foamed concrete sample achieved 0.68 W/mK while when the foamed concrete was reinforced with PP fibres, it was observed that the conductivity value dropped slightly. Foamed concrete reinforced with 0.30% PP fibres exhibited 2.94% reduction in the heat flow. This is due to the characteristics of PP fibres, being, hydrophobic; hence it retains water in foamed concrete. So, when the water dried out, air pores (voids) are be formed. The more air voids in the formed concrete, the lower the thermal conductivity. So, when more PP fibres is added into foamed concrete, it will result higher percentage of porosity. Hence, the thermal conductivity will decrease when more PP fibres are added. However, the sample with 0.25% PP fibres showed higher thermal conductivity, this may be due to non-uniform distribution of PP fibres in the concrete.

5 Conclusion

Based on the study, it can be concluded that:

1. The optimum percentage of polypropylene fibres needed to be added in 1600 kg/m^3 density foamed concrete to achieve the highest compressive was 0.20%.

2. The compressive strength of foamed concrete which contained polypropylene fibres achieved higher strength compared to controlled sample due to PP fibres filling in the air voids and reinforcing the concrete.

3. The thermal conductivity of foamed concrete decreased with the addition of polypropylene fibres, this is due to polypropylene fibres being hydrophobic in nature, retains water. When the water is dried out the air occupies the voids created. The more air voids the slower the transfer of heat thus lower thermal conductivity was obtained.

4. Thus, foamed concrete with the addition of polypropylene fibres can act as thermal resistance building material as with the addition of polypropylene fibres the thermal conductivity reduces as compared to the controlled sample. The addition also improves the mechanical properties of foamed concrete

The authors would like to thank the Universiti Tun Hussein Onn Malaysia (STG-528) and Research Management Centre (RMC) UTHM for the financial support for this project.

References

1. M. Santamouris, Using cool pavements as a mitigation strategy to fight urban heat island—A review of the actual developments. Renewable and Sustainable Energy Reviews, **26**, pp. 224-240 (2013).

2. Y. Wang, U. Berardi and H. Akbari, Comparing the effects of Urban Heat Island Mitigation Strategies for Toronto, Canada. Energy and Buildings, (2015).

3. M. A. Hart and D. J. Sailor, Quantifying the influence of land-use and surface characteristics on spatial variability in the urban heat island. Theoretical and Applied Climatology, **95**: pp. 397–406, (2009).

4. M. Ozyavuz, B. C. Bilgili and G. Elkıran, Temperature distribution and environmental impact of Tekirdag Ataturk Forest Nature Park, Int. J. Global Warming **8(1)** pp. 102-113, (2015).

5. Environmental Protection Agency (2017) Heat Island Impacts. Retrieved on June 20, 2017, from https://www.epa.gov/heat-islands

6. K. Ramamurthy, E. K. K. Nambiar and G. Indu Siva Ranjani, A classification of studies on properties of foam concrete, Cement and Concrete Composites **31(6)** pp. 388-396, (2009).

7. A. R. Hazlin, Iman Afif, N. Mohamad, W. I. Goh, L. M. Sia, A. A. A. Samad and Noorwirdawati Ali, Microstructure and Tensile Strength of Foamed Concrete with Added Polypropylene Fibres. MATEC Web of Conferences **103**, 01013 ISCEE 2016 (2017)

8. L. Madhavi, Swamy Raju and Deepak Mathur, Polypropylene Fibre Reinforced Concrete-A Review, International Journal of Emerging Technology and Advanced Engineering **4(4)**, pages 114-118 (2014).

9. A. M. Neville, *Properties of Concrete*, 4th ed., England: Longman (2000).

10. A. W. Hago, K. S. Al-Jabri, A. S. Al-Nuaimi and A. H. Al-Saidy, Concrete blocks for thermal insulation in hot climate. Cement and Concrete Research 35, Pg. 1472-1479, (2005).

11. D. Lauwaet, H. Hooyberghs, B. Maiheu, W. Lefebvre, G. Driesen, S. Van Looy and K. De Ridder, Detailed Urban Heat Island projections for cities worldwide: dynamical downscaling CMIP5 global climate models. Climate, **3**, pp. 391-415, (2015).

12. J. Lindén, C. S. B. Grimmond and J. Esper, Urban warming in villages, Adv. Sci. Res., **12**, pp. 157-162, (2015).

13. A. Lehoczky, J. A. Sobrino, D. Skoković and E. Aguilar, The Urban Heat Island Effect in the City of Valencia: A Case Study for Hot Summer Days. Urban Science, **1(1)**, (2017).

14. P. T. Nastos and A. Matzarakis, The effect of air temperature and human thermal indices on mortality in Athens, Greece. Theor. Appl. Climatol, **108**, 591–599, (2012).

15. W. I. Goh, N. Mohamad, R. Abdullah and A. A. A. Samad, Compression test and finite element analysis of foamed concrete cube. Journal of Engineering and Technology **5(1)**: pp. 1-9, (2014).

16. N. Mohamad and M. H. Mahdi, Testing of Precast Lightweight Foamed Concrete Sandwich Panel With Single and Double Symmetrical Shear Truss Connectors Under Eccentric Loading. Advanced Materials Research **335-336**: pp. 1107-1116, (2011)

17. N. M. Zahari, I. A. Rahman and A. M. A. Zaidi, Foamed Concrete: Potential Application in Thermal Insulation Heat in Heat out Air- voids Foam Concrete, *Proceedings of MUCEET 2009*, pp. 47–52, (2009).

18. M. R. Jones and A. McCarthy, Behaviour and assessment of foamed concrete for construction application. In Use of Foamed Concrete in Construction. Dhir, R.K., Newlands, M.D., McCarthy, A., Eds.; Thomas Telford: London, UK, 61– 88, (2005).

19. S. Wei, C. Yiqiang, Z. Yunsheng and M. R. Jones, Characterization and simulation of microstructure and thermal properties of foamed concrete, Construction and Building Materials, **47(10)** : pp. 1278–1291, (2013).

20. D. Aldridge, Introduction to foamed concrete: what, why, how? Use of Foamed Concrete in Construction, 1-14, (2005).

21. G. Martínez-barrera, F. Ureña-nuñez, O. Gencel, and W. Brostow, Mechanical properties of polypropylene-fiber reinforced concrete after gamma irradiation, Composites Part A: Applied Science and Manufacturing, **42(5)**, 567–572, (2011).

22. A. K. Singh, A. Jain and D. Singh, Evaluation of Mechanical Properties for Polypropylene and Steel Fibre Reinforced Concrete. International Journal of Engineering Research & Technology (IJERT), **2(4)**, pp. 1507 – 1517, (2013).

23. A. J. Hamad, Lightweight Concrete Reinforced with Polypropylene Fibres. International Journal of Advances in Applied Sciences (IJAAS) **4(2)**, pp. 45 – 49, (2015).

24. A. O. Mydin and S. Soleimanzadeh, Effect of polypropylene fiber content on flexural strength of lightweight foamed concrete at ambient and elevated temperatures, Advances in Applied Science Research **3(5)**, pp. 2837-2846. (2012).

25. BS EN 12350-6:2009, *Testing fresh concrete. Density*, British Standard Institution.

26. BS EN 12390-3:2009, *Testing hardened concrete. Compressive strength of test specimens*, British Standard Institution.

27. BS EN 12664:2001, *Thermal performance of building materials and products. Determination of thermal resistance by means of guarded hot plate and heat flow meter methods. Dry and moist products of medium and low thermal resistance*, British Standard Institution.

28. U. J. Alengaram, B. A. Al Muhit, M. Z. Jumaat and M. Y. J. Liu, A comparison of the thermal conductivity of oil palm shell foamed concrete with conventional materials. Materials and Design, **51**, 522–529, (2013).

A STUDY OF NUTRIENT DISTRIBUTION IN SEDIMENT LAYER AT SEMBRONG DAM

Abu Talib SH[1,*], *Syed Hashim SIN*[2], and *Abustan MS*[1]

[1]Senior Lecturer, Faculty of Civil and Environmental Engineering, UTHM, Johor
[2]Postgraduate Student, Faculty of Civil and Environmental Engineering, UTHM, Johor

Abstract. Reservoir is one of the key sources of water supply as it provides hydroelectric power, domestic usage, agriculture, farming, recreation and provides flood protection. Sedimentation in the reservoir makes the storage of water loss and the sediment sink contain a source of nutrients. Sediment deposition creates habitats for aquatic life but if too much sediment, it can destroy their habitats and even physically alter a waterway. The use of nutrients such as Total Phosphorus (TP) is key to growing plants and animals thus to feed growing populations. An excess of nutrients in the reservoir such as from agricultural activity makes algae growth rapidly and it can affected the water quality in the reservoir. This research was conducted at the Sembrong dam as this dam is a major water source for people in Kluang and parts Batu Pahat and the water quality become decreased because of algae bloom. The main objective of this study is to determine the distribution of nutrient in different location and to determine the nutrient distribution in vertical sediment layer. The study were conducted by taking sediments samples throughout the reservoir by using sediment corer and the sediment have been sliced every 5 cm. The samples were digested using USEPA Method 3050B which is acid digestion for sediments, sludge and soil for TP. The laboratory testing on these nutrients was conducted by following Standard Method for the Examination of Water and Wastewater (APHA, 1999) using spectrophotometer. The highest value of TP obtain is at outlet of the reservoir which is 30.72 mg/l. The results obtained shows that the top of sediment has high number of TP and it decreasing by depth. However, continuous and periodic monitoring should be done to avoid the increasing of the concentration of TP in the sediment to ensure the life of Sembrong dam can be extending for future.

1 Introduction

The Sembrong dam is a downstream flood protection up to 100-years and the storage capacity about 18 million m3 and other function is for water supply. It is big enough to supply drinking water for thousands of people surrounding it. Major water resource for some 120,000 people in the districts of Kluang and parts of Batu Pahat was "slowly dying" due to an algae bloom that cause to decreased the water production and the existence of all marine life in it. Besides, the sediments in the reservoir are also having been shown to exert an important influence on this process.

The sediment is an important part of a lake's ecosystem providing both a sink and a source for nutrients. Sembrong dam surrounding mostly covered by farming, agriculture activities and planting of oil palm trees which means the TP are widely used in fertilizers. Phosphorus and nitrogen are important nutrients that farmer used for the growth of plants in water bodies. With these two nutrients, phosphorus is most often considered to be the nutrients that control the production of algae in water bodies [1]. Furthermore, algae and algal blooms should not be considered as harmful to water quality in a waterway. In fact, algae are an important part of stream and river ecosystems. Algae oxygenate water and convert inorganic matter into organic matter and act as waterway's food chain. With slow moving rivers, the impacts of algae blooms grow rapidly than high stream flow. In fact, nutrients loads from sediment are the primary determinant of algae growth in rivers and the nutrient levels especially TP need to be found. High levels of algae can contribute to eutrophication in rivers and it affected the water supply for people in Batu Pahat and Kluang.

2 Literature Review

2.1 Chemical and Physical Aspects of Lakes

The most important chemicals in a lake and reservoir are nitrogen and phosphorus. Both chemicals allow the nutrient-rich plants and algae to grow. Besides, other organisms feed off these plants and algae, it creating a complex and healthy ecosystem. The balance nutrients in water may be altered by biological phenomena such as

* Corresponding author: hidayaht@uthm.edu.my

"algal bloom" and algae reproduces so rapidly and it prevents any nutrients from reaching below the lake's surface [2]. Sediment is delivered from water erosion that causes the water quality degradation and gets more waterway damages. Most sediment in surface waters derives from the surface erosion and a mineral component involved, arising from the erosion of bedrock and also an organic component arising during soil-forming processes [3]. Sediments carried into a reservoir may deposit throughout its full length, thus it can form rising of the bed elevation and may cause of aggradations. The combination of sediment trapping and flow regulation can gives big impacts on the ecology, water transparency, sediment balance and nutrient budget. Besides that, sedimentation can reduce the storage capacity for flow regulation and reduce environmental benefits that depend on releases from storage. Since sedimentation reduces the storage capacity, it also disturbs the flood control benefits, water supply, recreation and hydropower. Reservoir sedimentation is a complex process with varies of the watershed sediment production, rate of transportation, and mode of deposition. Reservoir sediment is depends on the river regime, flood frequencies, reservoir geometry and operation, sediment consolidation, density currents and possible land use changes over the life expectancy of the reservoir. The sedimentation makes the storage of water loss and many types of sediment related problems can also occur to both upstream and downstream of dams [4].

2.2 Nutrient in Sediment

Nutrient is an element or compound that important for animal and plant growth. The common nutrients are nitrogen, phosphorus and potassium. Without enough nutrients, they cannot sustain life and if they get in high concentrations, nutrients can be harmful [5]. According to [6], the loss of nitrogen, phosphorus and potassium through application of fertilisers use gives a negative nutrient balance in most soils. There are many sources of nutrients causing eutrophication of lakes and reservoirs. For example, all the activities in the entire drainage area of lake or reservoir that are reflected directly or indirectly in the water quality of water bodies [7]. Direct effect towards water bodies is has high of phytoplankton biomass which means microalgae in suspension in fresh water or marine waters and indirect effect is lack of oxygen in water bodies. These happened when the changes of flora and fauna in the bottom layer [8]. Nutrients enter the aquatic ecosystems through surface area, air or groundwater [9]. Drainage water from agricultural land is contains the phosphorus and nitrogen, usually the amount of nitrogen is higher than phosphorus because phosphorus is usually bound to soil components. The examples of nutrients enter from air is rain water. Rain water contains phosphorus and nitrogen from air pollution, usually nitrogen is over 20 times more concentrated than phosphorus because nitrogen is more freely in the atmosphere than phosphorus [10].

3 Materials and Methods

3.1 Sampling Location

In this study, sampling locations more focused at 3 parts of the reservoir which is at the inlet of the reservoir, middle of the reservoir and the outlet of the reservoir. Source of water in the Sembrong dam catchment cause of inflows stemming from the major river water flowing which is Sembrong river that flows directly into the dam. However, the presence of this river affects the quality of water in the dam because Sembrong river is covered by agriculture activities, so that the substances or chemical that human used will flow into the water through particles of soil that deposited as sediment into the Sembrong dam. The reason of the sampling taken at three locations mentioned is to achieve the objective which is to understand the nutrient distribution of sediments involved in Sembrong reservoir. The sampling location is divided by three zones and the samples were taken at different checkpoint for each zone in the Sembrong dam.

3.2 Sampling Method

The samples were taken carefully and keep safely, ensuring that the sediment will not be affected such as decreasing the volume of sediment besides to avoid the sediment sample mix together. The sediment samples were taken by using gravity corer as shown in Figure 1 and the sample were slicing for every 5cm. To take the sample of sediment, procedures of the sediment corer need to be following cautiously to ensure the sample is still in good condition for laboratory test. The sample of the sediment with 5cm is needed to be taken to the experimental work in the laboratory testing such as TP. The samples need to follow the digestion method before the further testing can be done.

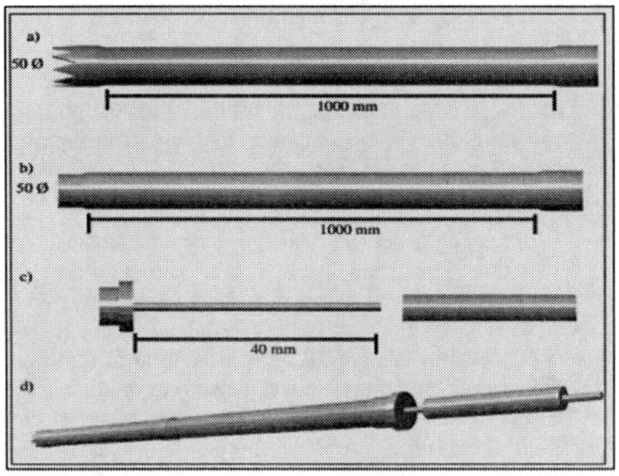

Fig, 1, Sketch diagram of Gravity Corer [11]

3.3 Laboratory Testing

The experiment to the sediment samples was conducted in the laboratory of wastewater laboratory and micropollutant research centre (MPRC), Faculty of Civil and Environmental Engineering, UTHM. The laboratory testing was conducted for digestion and determining the TP. In addition, before doing the test, the sediment samples need to change from solid to liquid by using digestion method. All the experiments will be carried out according to the procedures manual spectrophotometer DR6000 and USEPA method 3050B. Digestion method is using a very strong acid digestion that will dissolve almost all elements that could become environmentally available. For samples of digestion, it is representative 0.5 gram sample is digested with repeated additions of nitric acid (HNO3) and hydrochloric acid (HCL) (USEPA, 1996). For this method of digestion, the experiment will be following the USEPA method 3050B which is acid digestion for sediments, sludge and soil for TP. For determine the amount of TP, the testing that have been conducted is using a reagent tube which is total and acid hydrolysis test vial, 1.54 N Sodium Hydroxide Standard Solution and PhosVer 3 Reagent. Manual procedures of spectrophotometer DR6000 is a tool that will going to use for determining the TP.

4 Results and Discussion

4.1 Nutrient Distribution in Sembrong Reservoir

The analysis that have been done in laboratory is TP testing. The laboratory testing was conducted for 6 different locations and each sample that have been sliced for 5cm. The results of TP was presented in graph and been compared to the previous study which is TP in water. The graph of the comparison is shown in Figure 2 below.

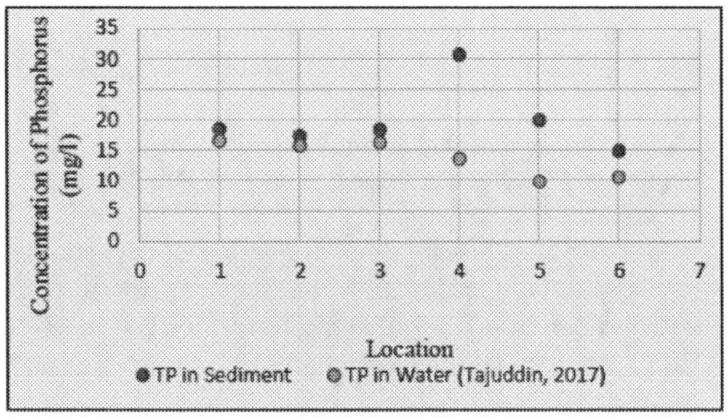

Fig, 2, Total Phosphorus (TP) distributions in water and sediment

From the Figure 2 above, the results of the TP in sediment is higher than the TP in water. The highest value of the concentration of phosphorus is at location 4 which is TP in sediment is 30.72 mg/l compared to the TP in water at location 4 is just only 13.6 mg/l. It is quiet big of number of the difference compared to other location. For TP in water, the highest value is at location 1 which is 16.6 mg/l and the lower TP is at location 5 which is 9.81 mg/l. At Zone I which is location 1 and 2, the TP in water and sediment is almost same amount. This is because at this location, it is nearest to the inlet area which means the nutrient and sediment introduce from the catchment and the flow is starting to reduce and accumulated all the nutrient and sediment. At Zone II, the number of TP in sediment is high which is 30.72 mg/l at location 4. According to the previous study from [11], the location is nearest to the spillway which is outlet of water in Sembrong and at this location it has a highest sediment accumulated at this area. Phosphorus release from sediments can contribute up to 99% of the TP input in some shallow lakes even there are at the surface sediment [12]. At Zone III which is location 5 and 6, the TP in sediment at location 5 has higher than TP in location 6. As in figure above, the value of TP in sediment and water at location 5 has high difference than location 6. At this zone, the flow almost stagnant in the reservoir. That is the reason of TP in sediment and water at location 6 tend to have low difference than location 5. The importance of sediments as a potential source of phosphorus has been highlighted in shallow lakes [13].

4.2 Vertical Distribution of Total Phosphorus Forms in Sediment

Figure 3 below shows the graph of vertical distribution of Total Phosphorus at Zone I, Zone II and Zone III with different location for each zone. For graph a) which means at Zone I for location 1 (CP1) and location 2 (CP2) shows the concentration of phosphorus versus of sediment depth in Sembrong dam. For location 1, the highest value of phosphorus concentration is 18.51 mg/l at top of the surface and the value of concentration of phosphorus is decreased by the depth. The lower of the concentration for location 1 is 12.21 mg/l at depth 30cm. For location 1, the sediment depth is 30 cm compared to location 2 with only 25 cm depth. For location 2, the highest value of the concentration of phosphorus is 17.37 mg/l at top of the sediment and the lower of the concentration of phosphorus is 6.63 mg/l at depth 25cm. The graph shows the value for both location is decreasing by sediment depth. According to the value from both location, the value TP concentration is highest at top of the sediment as Zone I is near from the inlet of the Sembrong dam. The nutrient content from water flow which is phosphorus comes from the landuse activity at the catchment area. The TP concentration is decreasing by depth as the water flow at the inlet is high and will flushed away the sediment particle to the middle part of the reservoir. As this process repeated, the sediment and it pollutant including nutrient will not retain at the same position.

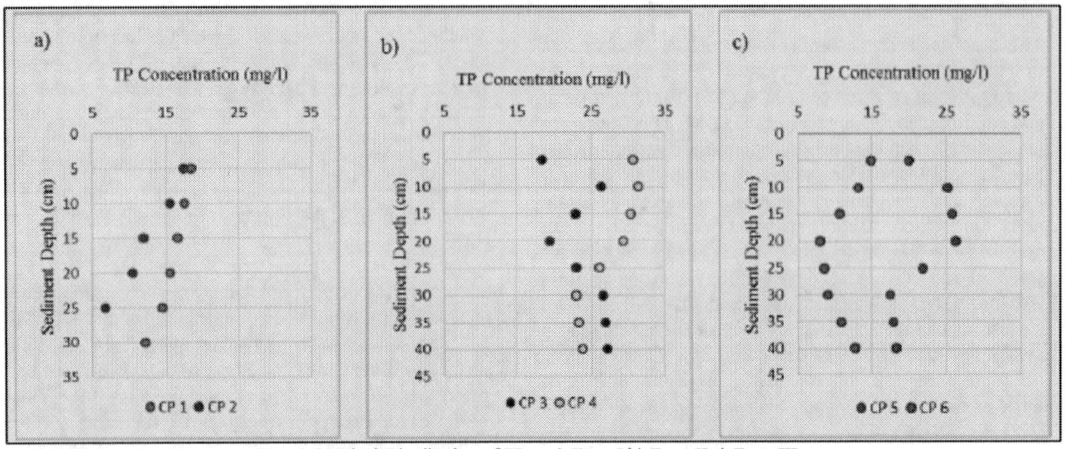

Fig, 3, Vertical Distribution of TP at a) Zone I b) Zone II c) Zone III

From Figure 4.2 b) above which mean for Zone II, the location 3 (CP3) and location 4 (CP4) were conducted in this zone. For location 3, the highest value of concentration of phosphorus is 27.15 mg/l at bottom of the sediment which is at 40 cm and the lowest value at location 3 is 18.36 mg/l at surface or top of the sediment. The value of the concentration of phosphorus at location 3 is lower to higher according to the depth. For location 4, the highest value of concentration of phosphorus is 31.41 mg/l at depth 10 cm from the surface of the sediment in Sembrong dam and the lowest value is at depth 30 cm which is 22.98 mg/l. The concentration of phosphorus at location 4 according to the depth is higher to lower compared to the location 3. In addition, location 3 and location 4 have a high value of concentration of phosphorus compared to other location because the location is near to the spillway or outlet in the Sembrong dam. The nutrient from activity surrounding especially phosphorus was flow throughout the reservoir and seep into the sediment due to the low of flow of water because of the distance from inlet and outlet quite distant. Besides that, from Figure 4.2 c) which mean for Zone III, the location 5 (CP5) and location 6 (CP6) were conducted in this zone. For location 5, the highest value of the concentration of phosphorus is 26.28 mg/l at depth 20 cm while the lowest value of concentration of phosphorus is 18.3 mg/l at depth 40 cm. For this trend of graph as shown in figure above, the concentration value of phosphorus is the lowest and higher at the middle of the sediment depth and lowered at bottom of the sediment. For location 6, the highest value of concentration of phosphorus is 14.94 mg/l at depth 5 cm which mean at top of the sediment and the lowest value of concentration of phosphorus is 7.98 mg/l at depth 20 cm. For this type of graph, the trend of graph of concentration of phosphorus is higher in top, lower at middle and increasing at bottom of the sediment. From trend of graph for location 5 and 6, it is opposite result for both value because at location 5, at middle sediment depth tend to high to low by sediment depth and location 6 tend to low and higher by sediment depth. For location 6, the nutrient flow from water and sediment especially phosphorus is from Sungai Merpoh and at this zone, it surrounded by oil palm include dead trees and stump. Overall review about graph of vertical distribution of Total Phosphorus at Zone I, Zone II and Zone III is the value of TP is mostly higher at top of sediment and decreasing by the sediment depth. Zone II has a high number of TP than Zone I and Zone III because of the location is near to the spillway of the reservoir.

4.2 Sedimentation and Phosphorus Content at Surface of Sembrong Dam

Based on the Figure 4 below, the Total Phosphorus (TP) is plotted on the map according to the location that have been decided and devided followed by the zone. The different colour of the plotted on the map is according to the results taken from the experiment. The value that has been taken is from concentration of phosphorus at surface or top of the sediment. Based on the map, it shows that the lower of water depth which means the depth of sediment higher has contain a lot of concentration of phosphorus in the sediment. One of the point which is CP4 has a higher number of concentration of phosphorus in sediment at surface and the value is in between 30.0 mg/l and 31.0 mg/l. Besides that, CP1 and CP3 has a value of concentration of phosphorus in range between 18.0 mg/l and 19.0 mg/l.

* Corresponding author: hidayaht@uthm.edu.my

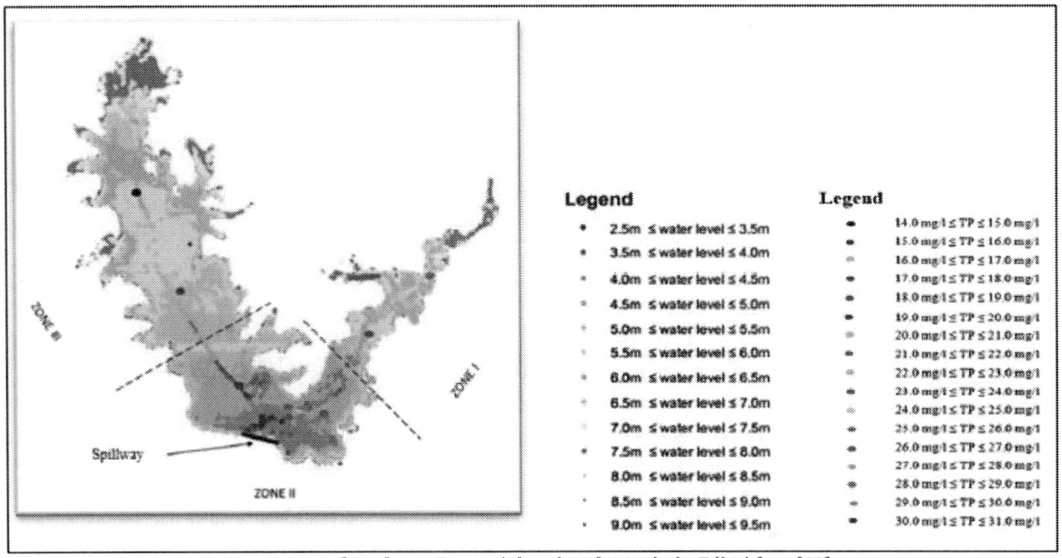

Fig, 4, Map of Sembrong Reservoir based on the Analysis. Edited from [11]

Furthermore, the lowest number of concentration of phosphorus according to the figure above is at Zone III which is CP6 and the the value is in between 14.0 mg/l to 15.0 mg/l. The reason of the lowest number of concentration of phosphorus at this location because the surrounding area covered by oil palm, stump and dead trees. The phosphorus of this matter is not as high as zone that covered by farming and agriculture activites. Overall review towards map of Sembrong reservoir based on the analysis is at Zone I which is the inlet of the reservoir, it contain lower of sediment depth other than Zone II and Zone III. The sediment depth taken from the Zone I is only 30 cm and 25 cm respectively. Other location the sediment taken is more than 30 cm which is up to 40 cm. Based on the graph and map plotted, as Zone 1 is the inlet of the reservoir, so that the nutrient content from water flow especially phosphorus that comes from the landuse activity such as farming and agriculture activity was flow into the reservoir. The TP in water and sediment at Zone I is almost same amount because of the sediment and water flow into the reservoir is always change for new particles and new amount of nutrient. The sediment and nutrient content will not retain at the same position as the water flow from inlet and the flow of water is start to reduce and accumulate all the nutrient and sediment into the middle part of the reservoir. As the reservoir of external phosphorus and the source of internal phosphorus, lake sediments play an important role in TP cycling [14].

At the middle part of the Sembrong dam which is in Zone II, it contain the highest value of TP at top of the sediment other than Zone I and Zone III because at this area, the sediment accumulate is higher than other zone. Based on the map, the shallowest part of the reservoir occurs at outlet of reservoir because the sediment and nutrient flow from Zone I has accumulated at this zone. According to [13], TP has been highlighted in shallow lakes. This can proof the result of this study that the shallowest part of water level has high content of TP in sediment. Furthermore, location 5 at Zone III has a high amount of TP in sediment and lower at location 6. This is because at this zone the flow of water almost stagnant. Location 5 has a high number of TP in sediment because of the sediment at location 4 has a highest number of TP and it flow to the location 5. For location 6, the value of TP in sediment is the lowest TP in sediment among other location because the sediment that accumulates at this location is not change because of the flow is almost stagnant.

5 Conclusion

Based on the results, the highest amount of TP in sediment is at Zone II which is at location 4 and the value is 30.72 mg/l. The TP were significantly enriched in the surface sediments and tend to higher in upper layers [13]. According to the map, water level at this zone is the shallowest part other than zone because this area is near to the outlet. As the water flow towards the spillway or outlet of the reservoir, the water velocity has decrease and causing suspended solid deposited and accumulates at this area [4]. For Zone I, the amount of TP in sediment obviously decreased with depth because the zone is nearest to the inlet of reservoir. As the water and sediment flow, the nutrient is not retaining at the same position and always gets new sediment and nutrient depends on the landuse activity. The TP content was clearly related to the regional distribution. The distribution characteristics of TP forms to the entrance of reservoir were very similar [13]. For Zone III, the

* Corresponding author: hidayaht@uthm.edu.my

amount of TP in sediment is decreasing because of the flow of water is almost stagnant and at location 6, the amount of TP in sediment is the lowest among other location which is 14.94 mg/l. According to [4], the sediment deposition at this area with no activity along the Merpo river does not affect huge changes of the sediment in the reservoir. As the flow almost stagnant in this area, so that the nutrient content is not flush out or in as much as in Zone II. Determining the nutrient distribution in vertical sediment is also been achieved as the results that has been plotted in Figure 2. The TP in the Sembrong river sediment has been proven by the presence of high value of TP in sediment layer at upper layer and according to [13], the TP is significantly enriched in the surface of the sediments. Mostly the top of the sediment in Sembrong dam which is at surface has high number of TP and it decreasing by depth.

Acknowledgement

We thank the Batu Pahat Department of Irrigation and Drainage for their hospitality, Universiti Tun Hussein onn for funding this project under the Research University Grant (Geran Penyelidikan Kontrak Khas 2015) vot no. U422 and Research, Innovation, Commercialization, Consultancy Office UTHM (ORICC) for Postgraduate Research Grant (GPPS).

References

1. D. W. Schindler, *Evolution of phosphorus limitation in lakes*. Science, **195(4275)**, 260-262 (1997).
2. K. Rutledge, M. McDaniel, D. Boudreau, T. Ramroop, S. Teng, E. Sprout, H. Costa, H. Hall, H and J. Hunt, *The Encyclopedic Entry Lake*. National Geographic (2011).
3. N. Junakova & M. Balintova. *Assessment of nutrient concentration in reservoir bottom sediments*. Procedia Engineering, **42**, 165-170 (2012).
4. T. Sumi, & T. Hirose. *Accumulation of sediment in reservoirs*. Water storage, transport and distribution. UNESCO-IHE and EOLSS Publishers Co. Ltd., Paris, France, 224-252 (2009).
5. L. Gardiner. "Traveling Nitrogen." Windows to the Universe. National Earth Science Teachers Association (2005).
6. R. Gensch. *Urine as Liquid Fertilizer in Agricultural Production in the Philippines: A Practical Field Guide*. Xavier University Press (2011).
7. W. Rast, & M. Straš kraba. *Lakes and Reservoirs: Similarities, Differences and Importance*, **Vol 1**.United Nations Environment Programme-International Environmental Technology Centre; International Lake Environment Committe Foundation (2000)..
8. S. H. A. Talib. *Sedimentation Modelling and Nutrient Profile in Bukit Merah Reservoir. Universiti Sains Malaysia* : Doctor of Philosophy Report (2014).
9. M. Selman, S. Greenhalgh, R. Diaz, & Z. Sugg. *Eutrophication and hypoxia in coastal areas: a global assessment of the state of knowledge*. World Resources Institute, 284, **1-6** (2008).
10. W. Rast, & M. Straš kraba. *Lakes and Reservoirs: Similarities, Differences and Importance* **(Vol. 1).** United Nations Environment Programme-International Environmental Technology Centre; International Lake Environment Committe Foundation (2000).
11. M. W. Z. Nasru. *Sedimentation assessment in Sembrong Dam, Batu Pahat, Johor. Universiti Tun Hussein Onn Malaysia* : Bachelor Degree's Project Report (2017).
12. L. Volterra, M. Boualam, J. P. Duguet, X Bonnefoy, A Ménesguen, and J Duchemin. *Eutrophication and Health. Report from European Union and World Health Organization* (2002).
13. D. A. Shilla, T. Asaeda, & M. Kalibbala. *Phosphorus Speciation in Myall Lake Sediment, NSW, Australia*. Wetlands Ecol Manage, DOI **10.1007/s 11273-008-9087-5** (2008).
14. S. L. Xiang, & W. B. Zhou. *Phosphorus forms and distribution in the sediments of Poyang Lake, China*. International Journal of Sediment Research, **26(2),** 230-238 (2011).
15. L. Q. Xie, P. Xie, & H. J. Tang. *Enhancement of Dissolved Phosphorus Release from Sediment in Hyper-Eutrophic, Subtropical Chinese Lake*. Environmental Pollution, **No. 122**, pp. 391 – 399 (2003).

The Effect of Water Cement Ratio on Cement Brick Containing High Density Polyethylene (HDPE) as Sand Replacement

Noorwirdawati Ali[1,*], *Nor Fazlin* Mohd Yusup[2], *Faisal* Sheikh Khalid[1] *Shahiron* Shahidan[1] and *Siti Radziah* Abdullah[1]

[1]Repair, Strengthening and Rehabilitation of Structure Focus Group, Jamilus Research Centre, Faculty of Civil and Environmental Engineering, Universiti Tun Hussein Onn Malaysia, 86400 Parit Raja, Batu Pahat, Johor, Malaysia
[2]Abd Manas Bin Abd Wahab, No. 3A, MARA Building, Iskandar Street, 33100 Pengkalan Hulu, Perak, Malaysia

Abstract. Waste disposal can contribute to the problem of environmental pollution. Most of the waste material is plastic based, because the nature of difficult of plastic degradable by itself. In order to overcome the problem, many study has been conducted on the reuse of plastic material into various field such as civil engineering and construction. In this study, municipal solid waste (MSW) in the form of High Density Polyethylene (HDPE) plastic was used to replace sand in cement sand brick production. The HDPE used in this study was obtained from a recycle factory at Nilai, Negeri Sembilan. 3% of HDPE replacement was applied in this study, with the cement-sand mix design of 1:6 and water-cement ratio 0.35, 0.40, 0.45 and 0.50 respectively. All specimens were tested for compressive strength and water absorption at 7 and 28 days. The density of the bricks was also recorded. The finding show that brick with 3% HDPE content and 0.45 of water-cement ratio at 28 days of age curing show the highest compressive strength, which is 19.5N/mm^2 compared to the control specimen of 14.4 N/mm^2.

1 Introduction

Increased waste from day to day cause the management of waste disposal a major problem to the world. Inefficient management of waste disposal can contribute to the problem of environmental pollution. Most of the waste material is plastic based, because the nature of difficult of plastic degradable by itself.

Plastics are divided into two categories, namely thermo-plastic and thermosetting plastics. Thermo-plastic such as polyethylene (PE), polystyrene (PS), polypropylene (PP), polyethylene terephthalate (PET) and high density polyethylene (HDPE) which can be melted by heating and hardened by cooling. Whereas, thermosetting plastics cannot be melted or softened by heating.

In effort to an increase demand of brick production, the brick industry continues to search and explore the alternative markets to improve their brick's quality and productivity. On top of that, many research have been conducted on the use of recycle materials into construction or building material [1 - 7]. Plastic is one of the materials that receive great attention among researcher to be used to substitute sand in the production of brick or concrete. Since plastic is non-biodegradable, study on the use of this recycle material become crucial as it can helps in reducing plastic wastes to be dumped to landfills.

In this study, high density polyethylene (HDPE) is used as a sand replacement in the production of cement-sand bricks. In the experimental work conducted, 3% HDPE was used as sand replacement. 3% replacement was chosen because from literature review conducted, 3% of plastic material to substitute sand in brick or concrete was found to be the optimal value for replacement. Meanwhile, the ratio of water to cement used is varied by 0.35, 0.40 (as a control sample), 0.45 and 0.5. These various value chosen also from the literature review conducted. This study aimed to investigate the effect of water-cement ratio on the compressive strength and water absorption of cement-sand bricks containing HDPE.

2 Material and methods

The materials used in the preparation of specimens were Ordinary Portland cement, river sand and high density polyethylene (HDPE) obtained from a recycle factory at Nilai, Negeri sembilan. The HDPE used is in the shape of plastic seed that is used as materials to produce plastic-based products as shown in Figure 1. The HDPE average size was 3mm in shape. In this study, 3% HDPE was used as sand replacement in the cement-sand production. The mix design of HDPE cement-sand brick are shown in Table 1 while Table 2 shows the density for

* Corresponding author: wirda@uthm.edu.my

 https://doi.org/10.1051/matecconf/201815003010

each raw materials used in this study. For each mix design, a total of 12 specimens were prepared where 6 specimen for water absorption test and another 6 specimen for compressive strength test. For each test, 3 specimens were tested at 7 day of curing while another 3 specimen were tested at 28 day of curing. The specimen for compressive strength test was also used to determine the brick density. Before specimens were placed for compression test, it will be weighed to determine the weight of the brick. Figure 2 and 3 shows the process of specimen production.

Fig. 1. HDPE used in this study

Table 1. Mix design

HDPE (%)	Water : cement (w/c)
3	0.35
0 (Control)	0.4
3	0.4
3	0.45
3	0.5

Table 2. Density of raw materials

Density	Value
Cement	1506kg/m^3
Sand	1600kg/m^3
Water	1000 kg/m^3
High Density Polyethylene (HDPE)	625 kg/m^3

Fig. 2. Brick specimen production

Fig. 3. Compaction of the brick

Specimen were left for air curing. They were placed at indoor environment and left for 7 dan 28 days of curing as shown in Figure 4. Specimens were then tested to determine the density, water absorption and compressive strength. Figure 5 shows water absorption test on specimen. The results were analyzed and compared with control specimen. Conclusions were drawn based on the test result.

Fig. 4. Curing of specimen

Fig. 5. Water absorption test on specimen

3 Results and discussions

3.1 Density

The density of cement bricks affected by the raw material and the manufacturing process. Cement bricks must meet the standards of BS EN 771-3: 2003 [8] in order to meet the needs of a satisfactory compressive strength and high durability. This is because the mechanical properties of cement brick containing HDPE affected by its density. The results of the density of the brick are displayed in Table 3 and Figure 6.

Table 3. Results of density of bricks

Water : cement (w/c)	Average density at 7 day (kg/m³)	Average density at 28 day (kg/m³)
0.35	1730	1690
0.4 (control)	1900	1840
0.4	1550	1840
0.45	1870	1930
0.5	1940	1850

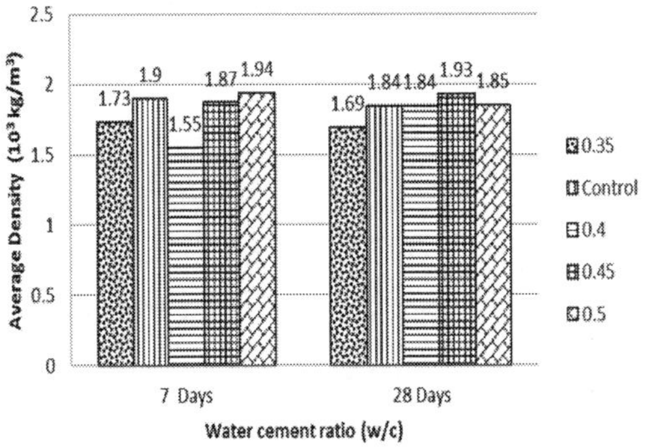

Fig. 6. Results of density of bricks

Based on the results obtained, the density of brick containing 3% of HDPE with various water cement ratio showed no significant difference with the control sample. From the graph of the results obtained, 0.50 water-cement ratio recorded the highest density at 7 days of curing with the value of 1940 kg/m³, while the water-cement ratio of 0.45 recorded the highest average density at 28 days of curing with the value of 1930 kg/m³. The lowest density was recorded for brick with water-cement ratio of 0.4 for 7 day at 1550 kg/m³ while brick with water-cement ratio of 0.35 recorded the lowest density at 28 day with 1690 kg/m³.

Water-cement ratio is the key factor in determining the strength and workability to be achieved in the production of bricks. Workability is an important element for determining the density of a brick. Low water-cement ratio will increase the strength and hardness, but it will reduce the workability while high water-cement ratio will increase the workability of the mix during the production process of bricks. A brick with good workability will encourage the production of high-density brick. These factors explain the existence of differences in density that occurs between all the samples tested.

3.2 Water absorption

Water absorption is important to determine the quality of bricks produced. Water absorption rate of a unit brick is closely related to the porosity. High water permeability by a brick will cause swelling of the soil particles in the bricks that have been stabilized and will cause a decrease in strength.

Table 4 and Figure 7 shows the average water permeability for cement bricks sample produced. Specimens with water cement ratio of 0.5 showed the lowest reading at the age of 7 and 28 days of curing which are 2.17% and 1.89% respectively. Meanwhile, the water-cement ratio of 0.35 recorded the highest value at the age of 7 and 28 days of curing, at 6.85% and 10.33% respectively.

Table 4. Results of water absorption test

Water : cement (w/c)	Average water absorption at 7 day (%)	Average water absorption at 28 day (%)
0.35	6.85	10.33
0.4 (control)	2.94	5.32
0.4	6.79	4.18
0.45	3.36	2.17
0.5	2.17	1.89

 https://doi.org/10.1051/matecconf/201815003010

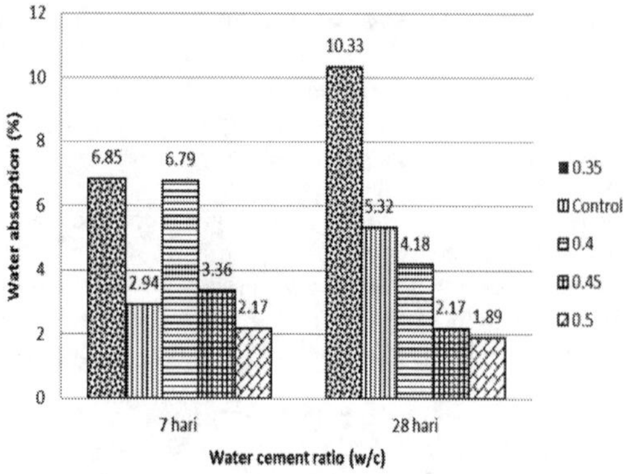

Fig. 7. Results of water absorption test

Based on the results obtained, it can be concluded that the use of water-cement ratio in the production of bricks have an impact on increasing the percentage rate of water absorption in the brick. This is because water is the agent of chemical reactions in cement to bind all the raw materials in mixing a brick.

The use of 3% HDPE in cement bricks did not leave much impact on improving water permeability. This situation can be evidenced by the rate of water absorption of cement brick with 3% HDPE which was lower than the control specimen.

3.3 Compressive strength

Compressive strength of a brick is much influenced by the cement content in the mix, the type of raw material and the water content of a cement brick. The compressive strength is an important aspect in assessing the load bearing capability of a brick. Based on the analyzed data, it is found that compressive strength of bricks containing HDPE meets the standards set in MS 76: 1972 [9-10].

From Table 5 and Figure 8, it was found that the average compressive strength of brick at age 7 and 28 days with an addition of 3% HDPE and 0.45 water-cement ratio recorded the highest compressive strength of 15.5 N/mm^2 and 19.5 N/mm^2 respectively. This mix also recorded the highest density at the age of 28 day.

Table 5. Results of compressive strength test

Water : cement (w/c)	Compressive strength at 7 day (%)	Compressive strength at 28 day (%)
0.35	7.4	9.8
0.4 (control)	11.7	14.4
0.4	11.1	15.9
0.45	15.5	19.5
0.5	12.5	14.4

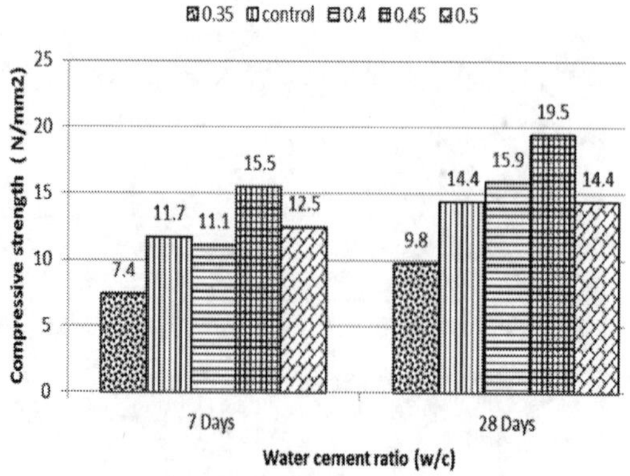

Fig. 8. Results of compressive strength test

The content of water-cement ratio in the brick plays an important role in determining the compressive strength of the brick. This is because it affects the air voids in the brick and affect the strength. Based on the results obtained, the compressive strength at 28 days of age showed and increased of compressive strength with the increased of water cement ratio up to 0.4 w/c. Beyond water cement ratio 0.4, the compressive strength of the brick decreased as shown in Figure 4.

Meanwhile, the use of 3% of HDPE in the cement brick does affect the compressive strength of bricks produced. Figure 8 above shows the reading of compressive strength using 3% of HDPE with a 0.40 ratio of cement was higher than the control specimen at 28 days of age the curing with the value of 15.9 N/mm^2. It shows that with the same water-cement ratio, the value of compressive strength recorded almost similar value with the added of 3% of HDPE into the mix.

From the experimental results, it is also noted that the lowest compressive strength was recorded by brick with water-cement ratio of 0.35 which is the lowest water-cement ratio used. It can be deduced here that low water-cement ratio not only reduce the workability but also reduce the compressive strength of the brick. This results can also be supported by the results from water absorption test where the specimen with the lowest water-cement ratio which is 0.35, recorded the highest water absorption and therefore it also can be deduced that with high water absorption resulted in decreased of compressive strength of the brick. This mix also recorded the lowest density at the age of 28 day and the second lowest of density at the age of 7 day.

4 Conclusions and Recommendations

From the results presented above, some conclusion can be deduced as follows:

1) Water-cement ratio of 0.45 is the optimal ratio for cement bricks containing 3% HDPE because it recorded the highest compressive strength and the second lowest of water absorption at 28 day.

2) This study also shows that HDPE can be use as sand replacement in brick production where the utilization of HDPE can reduce the waste dump at the landfill.
3) The ratio of water used in the production of cement bricks affect the percentage of water absorption. From the results of experimental work that have been carried out, the percentage of water absorption decreases with increasing water cement ratio in the production of the bricks.
4) Results also shows that compressive strength increased with the increased of water cement ratio up to water cement ration of 0.45. This shows that the use of excessive water-cement ratio can reduce the compressive strength of bricks produced.

In order to enhance the research of using HDPE in cement brick, there are several recommendations that is suggested to be investigated. The recommendation includes:

1) Percentage of replacement can be varied to get more range of data.
2) Study the effect of different mix design ratio should also be considered.
3) Other type of testing can be conducted for example thermal testing.

This research is funded by Short Term Grants (STG) U340, Universiti Tun Hussein Onn Malaysia and Fundamental Research Grant Scheme (FRGS) 1576.

References

[1] J. Hadipramana, A.A.A. Samad, A.M.A. Zaidi, N. Mohammad, A. Wirdawati. Influence of Polypropylene Fiber in strength of foamed concrete. *Advanced Materials Research*, **488-489**, pp. 253-257 (2012)

[2] N. Ali, K.Y. Yaacob, M.K. Burhanudin, S. Shahidan, S.R. Abdullah. Investigation of compressed earth brick containing ceramic waste. *ARPN Journal of Engineering and Applied Sciences*, **11** (8), pp. 5459-5462 (2016)

[3] F. S. Khalid, M. I. Juki, N. Othman, M. H. Wan Ibrahim. Pull-out strength of polyethylene terephthalate bottle fibre in concrete matrix. *Malaysian Construction Research Journal*, **21**(1), 2017, pp. 75-85 (2017)

[4] A.S. Leman, S. Shahidan, M.Y. Yusuf, S.S. Mohd Zuki, and N.A. Misnon. Workability and compressive strength for concrete with coconut shell aggregate behaviour of concrete with coconut, MATEC Web of Conf., **87**, no. 1017, pp. 1–6 (2017)

[5] M. Frigione. Recycling of PET bottles as fine aggregate in concrete. *Waste management*, **30**(6), 1101-1106 (2010)

[6] N. Saikia and J. de Brito. Mechanical properties and abrasion behaviour of concrete containing shredded PET bottle waste as a partial substitution of natural aggregate. *Construction and building materials*, **52**, 236-244 (2014)

[7] F. S. Khalid, H. S. Herman, N. B. Azmi, and M. I. Juki. Sand cement brick containing recycled concrete aggregate as fine-aggregate replacement, MATEC Web of Conferences. **103**, 01016 (2017)

[8] BS EN 771. Part3: Aggregate concrete masonry unit (dense and light-weight aggregates) (2003)

[9] MS76:1972.Malaysian Standard. Specification for bricks and blocks of fired brick earth, clay or shale, (1972)

[10] J.C. Morel, A. Pkla, and P. Walker. Compressive strength testing of compressed earth blocks. *Construction and Building Materials*, **21**(2), pp.303–309 (2007)

Printing of Wearable Antenna on Textile

Rd. Khairilhijra Khirotdin[1], *Muhammad Adzeem* Mahadzir[2], *Muhammad Musawwir* M Nor Nazli[3], *Fauziahanim* Che Seman[4], *Samsul Haimi* Dahlan[5], *Che Ku Afifah* Che Ku Alam[6] and *Nurhaffizah* Hassan[7]

[1, 2, 3]Faculty of Mechanical and Manufacturing Engineering, University of Tun Hussein Onn Malaysia, 86400 Parit Raja, Batu Pahat, Johor, Malaysia
[4, 5, 6]Faculty of Electrical and Electronics Engineering, University of Tun Hussein Onn Malaysia, 86400 Parit Raja, Batu Pahat, Johor, Malaysia
[7]Faculty of Engineering Technology, University of Tun Hussein Onn Malaysia, 84600 Pagoh, Muar, Johor, Malaysia

Abstract. A wearable antenna which is meant to be a part of the clothing used for communication purposes includes tracking, navigation and mobile computing has been seen in demand due to the recent miniaturization of wireless devices. Printing of conductive ink provides flexibility properties on electronics thus allowing it to be used on conformal surfaces. However, the current printing techniques mostly suffer from ink incompatibility and limited of substrates to be printed with. Hence, this paper intend to discloses the printing of wearable antenna using alternative technique via syringe-based deposition system with conductive ink on textile. A validation between simulation and measurement of return loss, (S_{11}) and radiation pattern of the antenna printed is also performed. It was found that a functional antenna is successfully printed on textile since the performances obtained are as expected. The antenna resonated at a minimum resonant frequency of 1.82 GHz which the S_{11} gathered at -18.90 dB. The radiation pattern for both simulation and measurement is as predicted since both have a larger magnitude of the main lobe than the side lobe. The magnitude of the main lobe from measurement was observed to be 8.83 dB higher than the magnitude of the main lobe of the simulation which is only 3.77 dB. It is proven that the syringe-based deposition system is capable of printing functional antenna on textile.

1 Introduction

Wearable electronic devices are becoming very popular in personal communications, computer systems and wireless monitoring of vital functions [1]. A main advantage of wearable antennas is that they are designed as elements of clothing able to transmit or receive wireless signals [2]. Integrating antennas into clothing can improve the desirability of such systems, and novel methods of fabricating conducting structures on textiles will be of benefit to diverse applications. Commonly, wearable antenna requirements for all modern application require light weight, low cost, almost maintenance-free and no installation [3]. Silver and copper conductive ink are frequently adapted by researchers in experimenting printed electronics due to its availability in market. Silver ink is used in this experiment because the conductivity properties are slightly better than copper ink. There are a lot of reports regarding the usage of conductive ink such as glass [4], printed circuit board [5], polyethylene terephthalate (PET) [6] and paper [7] via inkjet, flexography, aerosol and gravure respectively. However, most of these printing techniques only capable of printing on planar substrates and some of these techniques suffer from clogging as well as limited type of materials to be printed with [8] as reported in several previous study which is gas sensor via screen printing [9] and organic thin film transistors via inkjet printing [10, 11]. Unlike the syringe deposition system, it is capable of printing electronics structure via strain gauge on both planar and non-planar substrates with variety of materials [12, 13]. The performance of printing the strain gauge on non-planar substrates were also evaluated respectively [14]. Furthermore, the syringe deposition system provides larger building volume due to its nozzle movement that utilize three-dimensional coordinate axis system as well as can deposit wide range of viscosity level of ink, flexible and easy to handle. With the potential of this technique provides, an expansion of the usage particularly in printing antenna is worthwhile since there are no reports regarding the capability of syringe deposition system in fabricating a functional printed antenna on textiles. Besides, by realizing this, it will potentially open up a new scale of target market especially in the field of personal communications, computer systems and wireless monitoring of vital functions [1] and structural and health monitoring system.

* Corresponding author: adzeemmahadzir@gmail.com

 https://doi.org/10.1051/matecconf/201815004001

2 Antenna Design

2.1. Specification of antenna design

A type of micostrip patch antenna was first identified and selected and its detail dimension and shape was determined accordingly and modeled using AutoCAD software. Figure 1 shows the design of the antenna with its detail dimension respectively. The antenna is designed to be operated at resonance frequency of 1.575 GHz to fulfil the requirement set by Multimedia Communication and Malaysian Commission (MCMC) for rescue signal. There were four essential parameters to be specified including the operating frequency (f_o), dielectric constant of substrate (ε_r), the height of the dielectric substrate (h) and the thickness of the conductor patch (t) and Table 1 below shows the specification for each of the parameters considered.

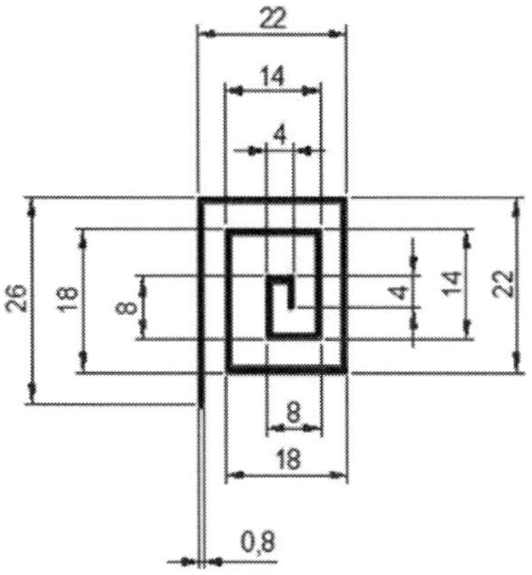

Fig. 1. Microstrip patch antenna design with its detail dimension (units in mm).

Table 1. Specification of Antenna designed.

Specification	Width (mm)	Length (mm)	Thickness (mm)
Silver conductive ink track	0.80	158.00	0.22
Substrate (polyester fabric) with dielectric constant, $\varepsilon_r = 1.6$	30.00	30.00	0.15
Ground Plane	30.00	30.00	0.11

The total length of the antenna designed is 158 mm long with a 0.80 mm width and 0.22 mm thickness of the track.

2.2 Simulation of Antenna Designed

The antenna designed was initially simulated using Computer Simulation Technology (CST) Microwave Studio software to measure its resonance frequency as designed. The simulation process also is employed to determine the antenna return loss, S_{11} and to measure whether the antenna is radiated according to the pattern intended; omni-direction with bigger main lobe and smaller side lobe as required.

3 Experiment

3.1 Deposition of ink and curing process

An automatic syringe-based deposition system (Model: FISNAR 3-axis, F4200N.1) equipped with a teach pendant and industry's leading dispensing software is used to deposit the ink on textile substrate via polyester fabric as shown in Figure 2. The system has been widely used in electronic industries especially in solder paste machine and surface mount technology. A series of pilot tests were conducted to determine the range of suitable printing parameters and these printing exercises were performed with several range of printing parameters to determine the optimum printing parameter in order to obtain the desired dimension set previously. The dimensions are measured by taking an average from several points on the sample. The optimum printing parameter used is illustrated in Table 2. The other two parameters including a height deposition and nozzle diameter were set to be constant.

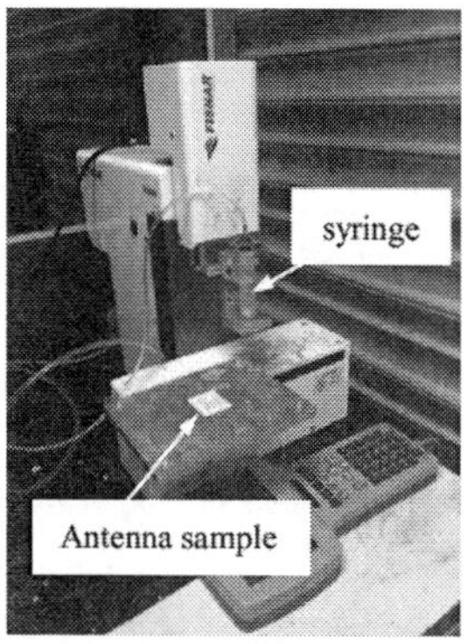

Fig. 2. Automated Syringe Deposition System.

Table 2. Optimum Printing and Curing Parameter selected.

Printing Parameter				
Pressure (kPa)	Speed (mm/s)	Height Deposition (mm)	Nozzle Diameter (mm)	Viscosity of ink (Pa.s)
175	25.00	1.20	0.51	2.40
Curing Parameter				
Temperature (°C)		**Time (minutes)**		
160		45.00		

https://doi.org/10.1051/matecconf/201815004001

A curing process is then employed using an oven (Memmert model 100-800) to reduce the percentage of unwanted material in conductive ink since conductive ink; either oil-based or contain water or solvents is mundanely comes in liquid and needs to be cured usually by heating process to expose its metallic contents [15]. The curing process played an important role to provide a good conductivity of the ink tracks. Oven is used as a curing process since it provides a consistent and stable temperature and the samples could be cured at a higher temperature if needed. In addition, oven is easy to set up by simply adjust the timer and the temperature gauge. Several pilot tests as well were previously performed to determine the optimum curing parameters that result in the lowest resistance. The optimum curing parameter selected is depicted in Table 2 previously shown. A total of ten samples were printed from the simplification method made using Taguchi method. Figure 3 shows the antenna that had been successfully printed using the syringe deposition system and cured with oven.

Fig. 3. Microstrip patch antenna printed using syringe deposition system and cured with oven

3.2 Material and substrate

A conductive ink used throughout the whole study is silver epoxy-based ink material (Model : AG806) as shown in Figure 4 and direct usage of the ink is not possible due to high solid fraction thus an adjustment of viscosity via dilution with Toluene solvent is necessary. The fabric material used is made of polyester and since the aim is to embed antenna on a life jacket, polyester material is much suitable because it mimics the actual fabrics made for life jackets.

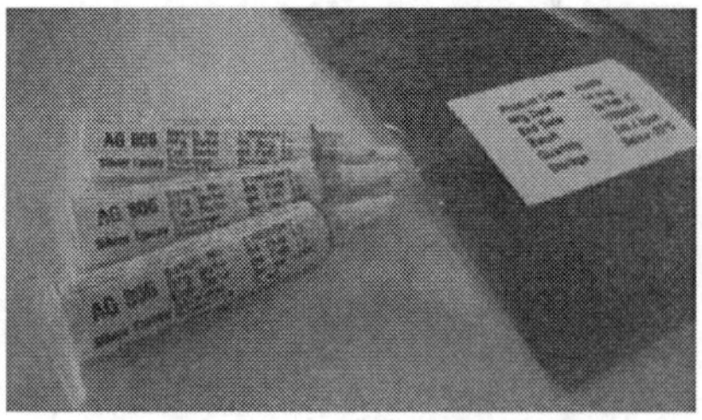

Fig. 4. Silver-epoxy based ink (Model: AG806)

3.3 Measurement

Analysis of the antenna printed is consisted of the measurement of the line width and thickness of the ink track using an optical microscope and its electrical properties via inductance (L), capacitance (C) and resistance (R) (LCR) meter. A cross-sectional area of the ink track and its total length were determined to calculate the resistivity. A measurement of resonance frequency and return loss (S_{11}) of the antenna were measured using Vector Network Analyzer (VNA) (Model: Rohde & Schwarz ZVB14). The samples were then tested in semi-anechoic chamber to measure its radiation pattern.

4 Results

4.1 Simulation

An electromagnetic performance of the antenna is performed numerically in time domain solver using CST Studio software. The simulated return loss (S_{11} parameter) of the antenna is shown in Figure 5. The figure implies that the antenna radiates best at 1.575 GHz as designed (indicated by dash line). The magnitude of the antenna is found to be -14 dB. In the meantime, the simulated radiation pattern of the antenna designed is depicted in Figure 6 where the farfield directivity of the antenna is observed to be 3.77 dB at 1.575 GHz with the main lobe direction is 5.0 degree. Furthermore, according to the radiation pattern, the side level of the antenna is -2 dB. Both results prove fulfilled the requirement for antenna designed previously.

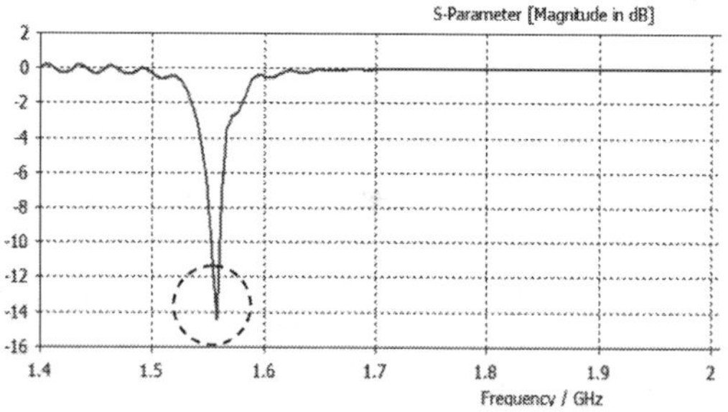

Fig. 5. Simulated return loss (S_{11}) of the antenna

https://doi.org/10.1051/matecconf/201815004001

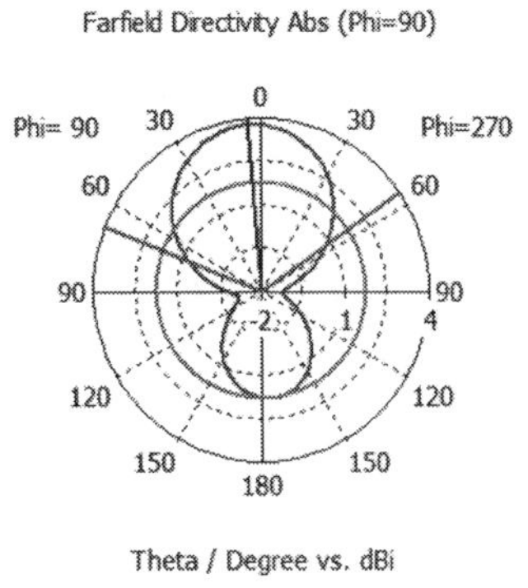

Fig. 6. Simulated radiation pattern of the antenna

4.2 Measurement of resonance frequency, return loss (S_{11}) and radiation pattern

The measurement of the antenna was conducted to validate the simulated result made earlier. The resonance frequency of the fabricated antenna was determined using VNA as shown in Figure 7 below. The return loss (S_{11} parameter) of the antenna is also measured accordingly and the result is illustrated in Figure 8. The result indicates that the antenna radiates best at 1.824 GHz with a magnitude of -18.896 dB.

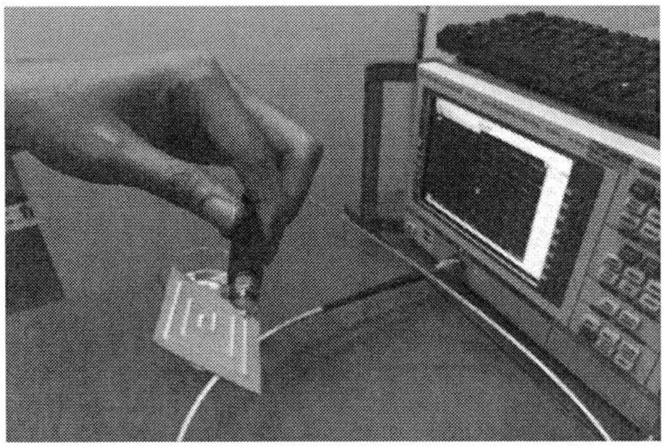

Fig. 7. Measurement of return loss of the antenna using VNA.

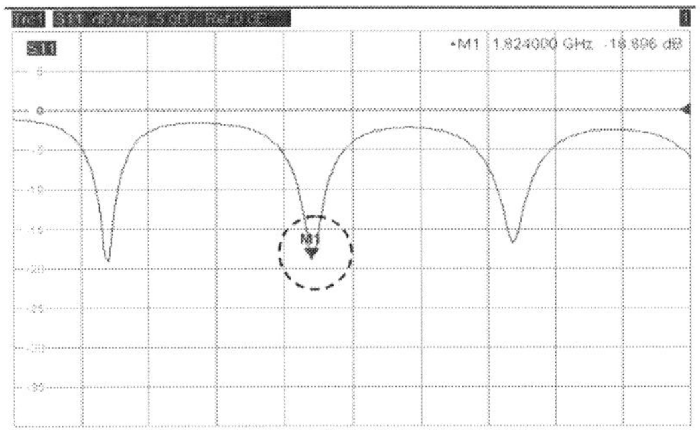

Fig. 8. Measurement result of the return loss of the antenna

A far-field radiation patterns result of the antenna was measured in the semi-anechoic chamber. The antenna was aligned first to a horn antenna with adjustable polarization. Figure 9 (a) and (b) show the position of the antenna and the horn antenna during measurement of radiation pattern performed in the semi-anechoic chamber. The result of the radiation pattern measurement is depicted in Figure 10 where the condition indicates that the antenna designed has an omnidirectional directivity at 1.824 GHz. The farfield directivity is observed to be of 8.83 dB magnitude. The main lobe direction is 6.3 degree and the side level magnitude of the antenna is 5.78 dB.

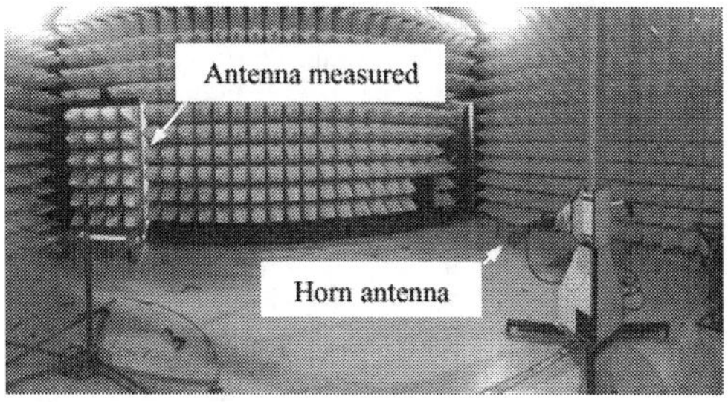

(a)

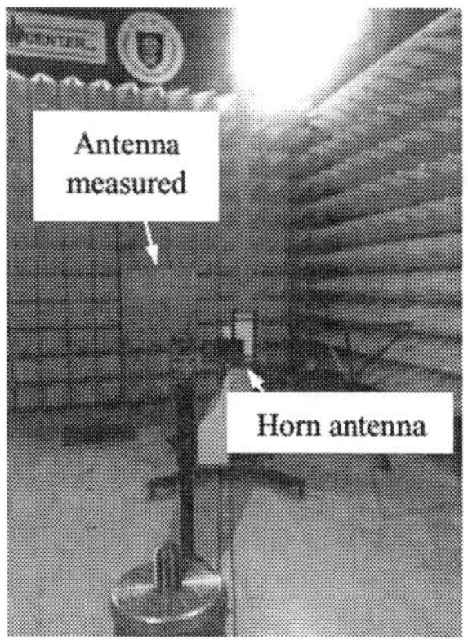

(b)

Fig. 9. Position of antenna and horn antenna during measurement (a) side view (b) antenna facing horn antenna

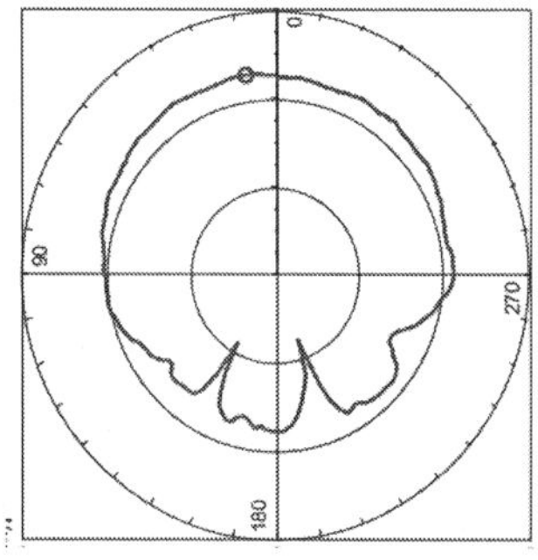

Fig. 10. Measurement result of the radiation pattern of the antenna

5 Discussion

From the simulation, the frequency of the designated wearable antenna operated at 1.575 GHz but in measurement, the resonance frequency measured is shifted to 1.824 GHz. The slight difference observed is might due to the imperfect soldering process that makes the unmatching condition on the port. The return loss (S_{11} parameter) from the simulation resulted in a magnitude of -14 dB while in measurement, the return loss is -18.896 dB. The return loss of both simulation and measurement was passed below -10 dB and it means that more than 10% of the incident power is reflected back towards the source from the antenna as required. The performance of the printed antenna is considered better in performance since the return loss of the measurement gives a higher return loss. In addition, the radiation pattern for both simulation and measurement gave a good result since both have a larger magnitude of the main lobe than the side lobe. The magnitude of the main lobe from measurement was observed to be 8.83 dB higher than the magnitude of the main lobe radiation pattern of the simulation which is only 3.77 dB. The difference between the simulation and the measurement may due to the manufacturing tolerance especially on soldering the SMA connector to the track which is result in imperfect connection in transmitting the signal. Besides, a cross-sectional shape of the antenna track from simulation is drawn in rectangular shape to replicate the conventionally fabricated antenna while in measurement the resultant cross-sectional shape of the ink track deposited by the syringe deposition system is always in semi-circular shape. There is a potential of this semi-circular shape in producing a better result in antenna performance as observed from the aforementioned results. A more thorough investigation on the effect on these two different cross-sectional shapes of the ink track should be performed to validate the produced results.

6 Conclusion

In summary, the wearable antenna printed is able to operate near its designated frequency. It was also observed that several factors caused a slight difference in simulation and measurement results due to imperfect soldering that makes unmatching condition on the port. Another difference perceived is the cross-sectional shape of the ink track designed where the actual cross-sectional shape of the fabricated antenna is not the same with the designed ink track of the antenna in simulation. These two factors were found to be the main reasons of the slight difference on the results between simulation and measurement. However, the results from measurement offers a better performance in antenna performance since the return loss, S_{11} parameter and the magnitude of the main lobe of radiation pattern is higher than the simulation. Besides, from the results gathered, it had proven that the syringe-based deposition system is capable of fabricating a functional electronics structure particularly antenna since the wearable antenna printed showing a good results. The antenna could operate and radiate at its designated frequency thus, it will potentially open up another dimension in communication and might become a life saver in the near future. A further investigation on the antenna performance especially in various conditions including bending and wet condition is highly recommended.

References

[1] H. Adel, R. Wansch, C. Schmidt, *Antennas and Propagation Society International Symposium, 2003.IEEE*, **1**, 471, ST-Antennas for a body area network (2003)

[2] E. G. Lim, Z. Wang, J. C. Wang, M. Leach, R. Zhou, C. U. Lei, K. L. Man, *Eng. Lett.*, **22**, 2, 94–101 (2014)

[3] N. H. M. Rais, P. J. Soh, F. Malek, S. Ahmad, N. B. M. Hashim, P. S. Hall, *Loughborough Antennas and Propagation Conference, LAPC 2009 - Conference Proceedings*, 225–228 (2009)

[4] M. Inoue, Y. Tada, H. Muta, Y. Hayashi, T. Tokumaru, *14th International Conference on Electronic Materials and Packaging, EMAP 2012* (2012)

[5] K. V. Abhinav, R. Rao, V. K. Karthik, S. P. Singh, *RSC Adv.*, **5**, 79, 63985–64030 (2015)

[6] E. Halonen, T. Viiru, K. Ostman, A. L. Cabezas, M. Mantysalo, *IEEE Trans. Components, Packag. Manuf. Technol.*, **3**, 2, 350–356 (2013)

[7] R. Lakhmi, H. Debeda, I. Dufour, C. Lucat, *IEEE Sens. J.*, **10**, 6, 1133–1137 (2010)

[8] S. Laschi, I. Palchetti, M. Mascini, *Sensors Actuators, B Chem.*, **114**, 1, 460–465 (2006)

[9] A. S. G. Reddy, B. B. Narakathu, M. Z. Atashbar, M. Rebros, E. Hrehorova, B. J. Bazuin, M. K. Joyce, P. D. Fleming, A. Pekarovicova, *Sens. Lett.*, **9**, 2, 869–871 (2011)

[10] J. Li, B. An, J. Qin, Y. Wu, *Proceedings - International Symposium on Advanced Packaging Materials*, 91–93 (2011)

[11] A. S. G. Reddy, B. B. Narakathu, M. Z. Atashbar, M. Rebros, E. Rebrosova, M. K. Joyce, *Procedia Engineering*, **25**, 956–959 (2011)

[12] R. K. Khirotdin, N. Hassan, H. H. Siang, and M. H. Zawahid, Engineering Letters, **25**, 3 (2017)

[13] R. K. Khirotdin, M. I. S. M. Zainuri, A. I. Ishak, N. Hassan, K. Kamarudin, M. R. Ibrahim, R. A. Haq, M. Ibrahim, N. Sau'de, O. M. F. Marwah, 8th. International Conference of Mechanical and Manufacturing Engineering, Langkawi, Malaysia (2017) (to be published)

[14] K. Khirotdin, M. F. Ngadiron, M. A. Mahadzir, N. Hassan, Material Science & Engineering Conference Series, **226**, 1, 012019 (2017)

[15] S. Ahmed, Polymer and Science: Coatings and Adhesives (2007)

An Investigation of Finite Element Analysis (FEA) on Piezoelectric Compliance in Ultrasonic Vibration Assisted Milling (UVAM)

Rasidi Ibrahim[1] *Haris* Rachmat[2], *Damayanti* Dida[2], *Anis* Radzi[1,] and *Tatang* Mulyana[2]

[1]Faculty of Mechanical Engineering and Manufacturing, University Tun Hussien Onn Malaysia, Batu Pahat, Johor

[2]Fakulti Rekayasa Industrial, Telkom Universitas, Bandung Indonesia

Abstract. Finite element analysis for piezoelectric actuator has been developed in Ansys Software which are a program that can analyses and simulate the dynamic behaviour of piezoelectric. The Ultrasonic Vibration assisted Milling (UVAM) experimental having a difficulty to investigate the effect of vibration mechanism where existence of error in material, mechanism and attachment of piezoelectric thus affect the amplitude and frequency of mechanical compliance during the machining of UVAM. This paper will investigate the modelling of piezoelectric compliance and follow the procedures of FEA to accurately predict the dynamic behaviour of compliance. The parameters for simulation of piezoelectric are voltage, electromechanical coupling and frequency. The compliance mechanism is model by using SolidWorks 2014 and imported to Ansys Mechanical APDL Software were the piezoelectric are embedded on the mechanism. Modal analysis and harmonic analysis has been used in order to obtain the mode shape and displacement. The displacement of the compliance mechanism will be compare between simulation and experimental. The dynamic behaviour was discussed in simulation to study the reliability of the compliance mechanism before it safely used in UVAM.

1 Introduction

Ultrasonic Vibration Assisted Milling (UVAM) has been developed rapidly in the last period of twenty years ago. UVAM process consist a combination of milling operation and vibration system on the compliance mechanism. UVAM characterized by separate type cutting is a cutting method in which vibration at regular frequency within an ultrasonic range is imposed on the tool or the work piece to achieve better cutting effects [1,2,3]. Inside of the compliance mechanism contain of piezoelectric actuator that give a vibration to the compliance mechanism. Piezoelectric Actuator is widely use in vibration system because it has ability to generate an electric charge in response to applied mechanical stress. The piezoelectric can expansion and contraction when source of voltage applied to it. Based on the previous study the piezoelectric is a transducer that transforms part of the mechanical vibration energy into electrical energy that can be conveniently dissipated in circuit [4].

Piezoelectric is popular in recent years, due to the superiority of the piezoelectric material that have simple mechanical properties, light weight and small volume. It has been use in various application such as vibration control of rotating machine. Finite element analysis (FEA) is a method of simulating or analyzing the behaviour of engineering structures and components under a variety of conditions. In previous research, Finite element analysis is used to study the response of plate structure on which piezoelectric active devices are mounted [5]. It is an advance engineering tool to predict how a product react to real world forces, vibration, heat, fluid flow and other physical effects. Finite element analysis shows whether a product will break or work after it was designed. It is called analysis, but in the product development process, it used to predict what is going to happen when the product is used [6].

The compliance mechanism for UVAM processes has been evaluate and study in order to define the mechanical behaviour of the compliance. Nowadays, design world computer-aided engineering (CAE) tools, especially finite element analysis (FEA) plays an important role in manufacturing engineers because it can solve complex engineering problems for wide range which by simulating the design in order to improve the product designs. The mechanical equilibrium equations is apply to the complex model by integrating contributions from small and simple element shapes. Thus, mechanical and thermal responses of complex objects can be approximate through computer simulation.

*Corresponding author: rasidi@uthm.edu.my

 https://doi.org/10.1051/matecconf/201815004002

2 Simulation Setup

2.1. Material and methods

Piezoelectric is used to expand and contract the compliance after it being applied voltage and frequency on the piezoelectric. Type material of piezoelectric that has been used in this paper is PIC 151 which are modified lead zirconate – lead titanate material with high permittivity, high coupling factor and high piezoelectric charge constant. It is suitable for low power ultrasonic transducer and low frequency sound transducers [5]. The compliance mechanism is a structural steel while for piezoelectric is a discrete piezoelectric stack with two flat end plates with the material properties from PIC 151 Ceramics. The dimension of the piezoelectric are 18mm width, 5mm height and 5mm depth. For compliance mechanism of piezoelectric was modelled in SolidWorks 2014 as shown in Figure 1 and save in SAT file and then imported in the Ansys Mechanical APDL 16.0.

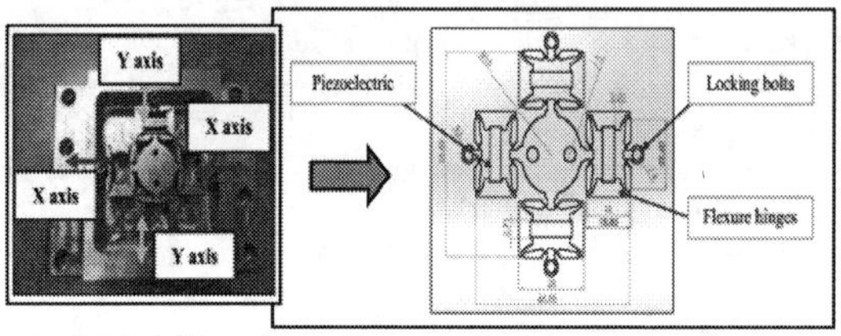

Fig. 1. Modelling of compliance mechanism and piezoelectric

The piezoelectric has been embedded on compliance in Ansys software with the voltage and electromechanical coupling as their loading condition. Element for piezoelectric are solid226 which is analysis electroelastic/piezoelectric. This model particularly has generates 69423 numbers of nodes and 37360 element. The meshing are using a smart size which are 5 fine and has been mesh at entire of the model by tetrahedral shape of meshing.

Fig. 2. Constraints (Fixed support), voltage and coupling

The constraint are zero displacement at X, Y and Z-axis as a locking bolt on the hole of compliance. Voltage and coupling are apply with 20V at the top and 0V at the bellow of the piezoelectric as shown in Figure 2. The frequency range are 0 to 18000Hz with Damping of stiffness matrix multiplier piezoelectric also has been apply with 3 x 10^{e}-9

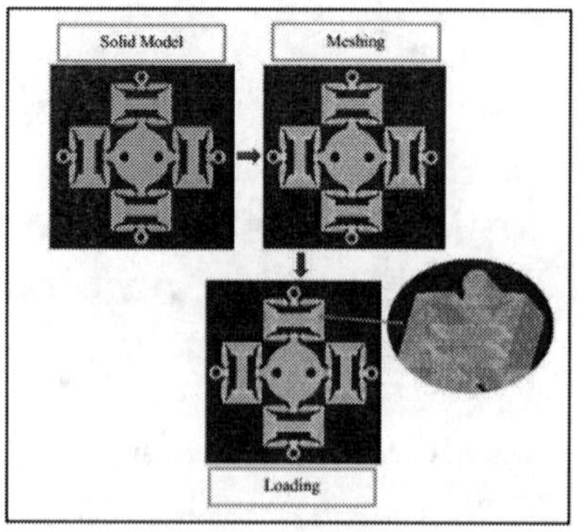

Fig. 3. Finite element analysis of compliance mechanism and piezoelectric

Figure 3 represent the compliance mechanism and piezoelectric after has been applied the meshing and loading. The solid model was transform into generation of meshing in order to create a partition space into element and nodes. The structure of the compliance are changes after it being applied the loading which are fixed constraint, voltage and electromechanical coupling.

2.2 Experimental of UVAM

Ultrasonic Vibration Assisted Milling (UVAM) is a machining process that has ultrasonic vibration on compliance mechanism by piezoelectric source. This machining process are suitable for cutting difficult material that has hardened surface. Figure 4 represent the illustration of vibrating worktable.

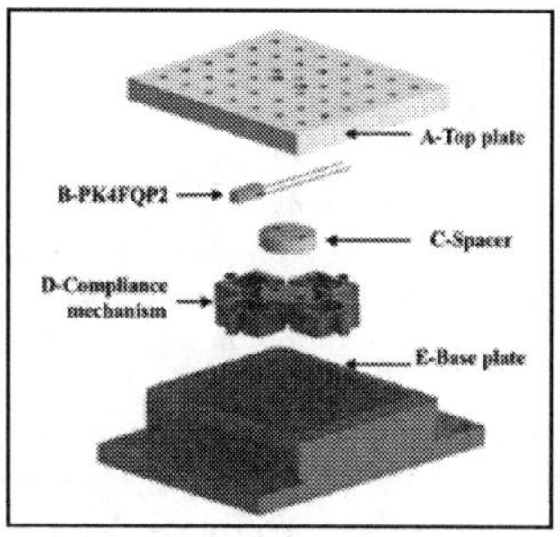

Fig. 4. Illustration of vibrating worktable.

The experiment was done in isolated room, controlled room temperature and on the granite table. The granite table is being used because it is an anti-vibration or vibration isolator. UVAM worktable consist of compliance mechanism and piezoelectric. Piezoelectric was embedded on the compliance mechanism as a source of the vibration to the compliance mechanism. The laser displacement was mounted on the tripod that provide stability at 90° angle as shown in Figure 5 which is important for the reading accuracy. The vibrometer port was connect to PDV-100 and laser displacement was pointed to

UVAM worktable. The function of PDV-100 is to detect the vibration at the worktable.

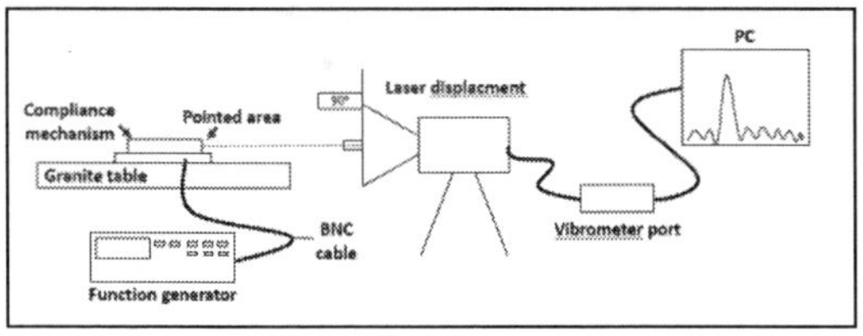

Fig. 5. Schematic diagram of instrumentation setup

Figure 5 shows the schematic diagram of instrumentation setup. Function generator are connected to compliance mechanism to produce electrical waveforms over wide of frequencies up to 1 MHz. Typical waves that produce are square, sine and triangular. In this experiment the frequency range are 0 to 18000Hz to define the frequency response of the structure. Vibration amplitude of the workpiece was measured by this high-accuracy instrument delivered by the function generator.

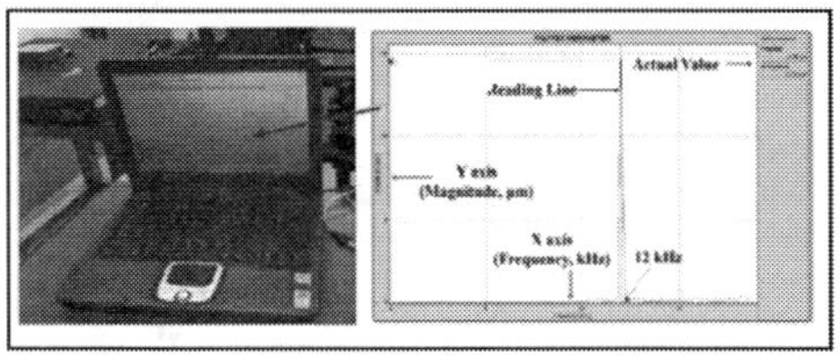

Fig. 6. Polytec Software

Polytec Vibrometer software in the PVD-100 was use to interpret the graph of displacement against frequency. The graph was set as X-axis for frequency while the Y-axis set as displacement or amplitude. The reading for the frequency at X-axis limited at 18.75 kHz. The actual value of displacement on compliance will take the peak of the graph. Figure 6 shows graph obtained when measurement is being taken.

3 Result and Discussion

3.1 The mode shapes of piezoelectric compliance

Modal analysis is a technique to specify the characteristic of vibration which are mode shapes and natural frequencies of the structure compliance mechanism. This analysis is very important because the mode shapes and natural frequencies are important parameters in the design of a structure for dynamic loading conditions and for this analysis, the mode shapes and natural frequency was obtained.

Table 1. Natural frequency obtained in FEA

Mode	Natural Frequency FEA (Hz)
1	317.0
2	555.9
3	2040.1
4	15002.0

Table 1 show the result of modal analysis of natural frequency that was determine in order to avoid resonance effects during vibration assisted machining. Natural frequency happen because of the wavelength or the speed of the piezoelectric. It is important to predict the possible failure modes in order to define the best performance.

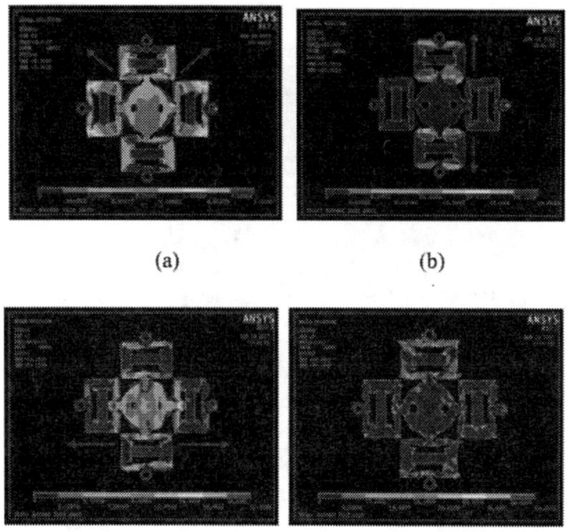

(a) (b) (c) (d)

Fig. 7. Mode shapes (a) F1 = 317.05Hz, (b) F2 = 555.95Hz, (c) F3 = 2040.1Hz, (d) F4 = 15002.0Hz

The response of this compliance mechanism is different at each of the natural frequencies. For the first mode shape which are F1 = 317.05Hz (Figure 7a), the deformation pattern is horizontal bending mode at 150° angle of the compliance mechanism. For the second mode shape F2 = 555.95Hz (Figure 7b), shows the mode shape are vertical bending (up and down) corresponding to Y-axis of the compliance mechanism. In the Figure 7(c) indicate the third mode shape where the deformation pattern are horizontal bending (left and right).

The shape for the fourth mode shape are rupture because the movement of the compliance mechanism has a little deformation that it just move at their position only. Based on second and third mode shape, it is predicted that the maximum displacement occurs at compliance direction Y-axis and X-axis where the work piece is attached on the compliance. The red arrow at all Figure of the mode shape represents the direction of the mode shape action on the natural frequency. All the

https://doi.org/10.1051/matecconf/201815004002

mode shape are important for the future analysis as it can predict the surface roughness by including the frequency response function. From the modal shape, it is possible to differentiate in which part the maximum value of error or deflection occurs and what range of natural frequency.

3.2 Harmonic analysis – Displacement

Harmonic analysis is a technique to determine the steady-state response of linear structure to loads that vary sinusoidal with time. From this analysis we can obtain the response of the piezoelectric compliance in Ultrasonic Vibrate Assisted Milling at several voltage and frequency. In additional, harmonic analysis also can determine the displacement, stress, strain, and amplitude of the piezoelectric compliance. Through this analysis, the problem such as breaking or fatigue of this structure can be overcome.

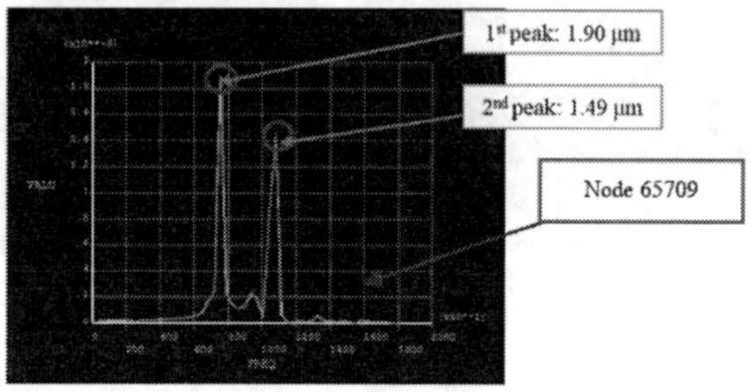

Fig. 8. Graph displacement vs frequency of compliance (Y-axis)

Based on Figure 8, the graph displacement against frequency for overall frequency from 0 to 18000Hz. This graph are taken at node 65709 where it located on the compliance. The node has been choosen at area above of the compliance because in experimental the workpiece are attached on the compliance to obtain the response of the compliance. The maximum displacement of compliance are 1.90µm at frequency 7560Hz. For the second peak of graph displacement is 1.49µm at frequency 10900Hz. From the graph 4.10 show that the maximum displacement on compliance occoured between 7,000Hz to 110,00Hz.

3.3 Comparison displacement of compliance piezoelectric between simulation and experimental

In finding the displacement between experimental and simulation is important to correlate the result between them. This is because during machining, the displacement will affect the accuracy of the vibration of the work piece. The displacement in simulation was taken at node 65709 where it located on top of the compliance.

Table 2. Comparison of displacement compliance mechanism

Readings	Displacement (Simulation), µm	Displacement (Experimental), µm	Percentage Error, %
1	1.89	1.72	9.88
2	1.48	1.23	20.33
3	0.85	1.12	24.11
Average	1.41	1.36	18.11

Table 2 represent the comparison of displacement compliance mechanism between simulation and experimental. The readings was selected based on the 3 highest reading in the data displacement of FEA and experimental. Based on the Table 2, the value of the percentage error is slightly higher at reading 2 and 3 which is 20.33% and 24.11%. However by comparing the result for the reading 1, the comparison error are acceptable. The data displacement from simulation and experimental was turn into comparison graph. Refer Figure 11 to see the correlation between simulation and experimental. All the readings has different frequency response where the compliance are displace. According to reading 1, 2 and 3, the simulation frequency are 7560Hz, 10800Hz and 10440Hz. While for the experimental are 18000Hz, 15000Hz and 17640Hz. The frequency are different because in experimental source of voltage are not directly pointing to the piezoelectric compare to simulation the source of voltage are directly through the piezoelectric. Besides that, in experimental the attachment of piezoelectric on the compliance is loose and has a gap inside the flexure hinges that might affect the frequency of the compliance. However the peak of the displacement between simulations and experimental is near similarly

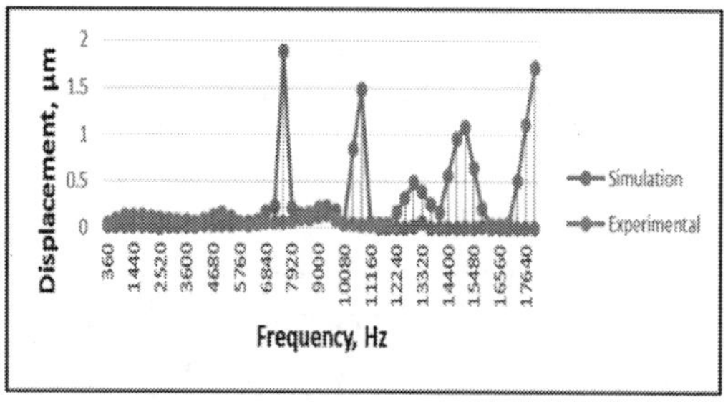

Fig. 9. Comparison graph displacement vs frequency between simulation and experimental

4 Conclusion

Overall of this paper was presented about a finite element analysis of compliance mechanism and piezoelectric that based on the simulation and experimental to achieve the aims and objectives. The data was presented through the Figure, Table and graph to show the trend of the data in order to correlated the results between the simulation and experimental. All the data has been analysed and discussed based on previous review. The displacement is important for compliance mechanism because it will affect the accuracy of the vibration of the work piece during machining. Natural frequency mode shape is a modal parameter that we can analyse and prevent the vibration problem causes by resonance.

The authors would like to express their appreciation to the Office for Research, Innovation, Commercialization and Consultancy Management (ORICC), Universiti Tun Hussein Onn Malaysia for the financial support. This work also a collaboration with FRI Telkom Univesitas Bandung, Indonesia.

References

1. Ibrahim, M. R, Rahim, Z., Rahim, E., Tobi, L, Cheng, K. and Ding, H, " An Experimental Investigation of Cutting Temperature and Tool Wear in 2 Dimensional Ultrasonic Vibration Assisted Micro-Milling,". 3rd International Conference on Mechatronic and Mechaical Manufacturing (ICMME16)., Vol. 95, (2017).

2. Ibrahim, R., Rafai N.H., Rahim, E.A., Cheng, K.and Ding, H,"A Performance of 2 Dimensional Ultrasonic Vibration Assisted Milling in Cutting Force Reduction on Aluminium AL6061". Vol. 11. Issue 18, (2016) Pp. 11124-11128.

3. J. Kim, V. V Varadan, and V. K. Varadan, "Finite Element Modelling of Structures Including Piezoelectric Active Devices," Int. J. Numer. Methods Eng., vol. 40, no. 5, pp. 817–832, (1997).

4. Ibrahim, M. R, Rahim, E.A, ghazali, M. I, Chai, M. H and Goh, Z. O, "Use of finite element structural models in analyzing machine tool chatter," Finite Elem. Anal. Des., vol. 38, no. 11, pp. 1029–1046, (2002).

5. P. Technology, "Piezoelectric Ceramic Products," pp. 1–44, (2014).

6. Rachmat, H. Ibrahim, M. R and Hasan, S. B. " Design Selection of and Innovative Tool Holder for Ultrasonic Vibration Assisted Turning (IN-UVAT) using Finite Element Analysia Simulation" 7th International Conference on Mechanical & Manufacturing Engineering, Sustainable Energy Towards Global Synergy, (ICME2016), Vol. 1831,

7. X. H. Shen, J. H. Zhang, H. Li, J. J. Wang, and X. C. Wang, "Ultrasonic vibration-assisted milling of aluminum alloy," Int. J. Adv. Manuf. Technol., vol. 63, no. 1–4, pp. 41–49, 2012

8. J. R. Baker and K. E. Rouch, "Use of finite element structural models in analyzing machine tool chatter," Finite Element Analysis and Design, vol. 38, no. 11, pp. 1029–1046, 2002

9. J. Kim, V. V Varadan, and V. K. Varadan, "Finite Element Modelling of Structures Including Piezoelectric Active Devices," Int. J. Numer. Methods Eng., vol. 40, no. 5, pp. 817–832, 1997

10. A. J. Moheimani, S.O.R. , and Fleming, Piezoelectric Transducers for Vibration Control and Damping Piezoelectric Shunt Damping. 2006

Behaviour of Condensation for High Steam Energy and Water in Vertical Annular Channel

W.M.Faizal Wan Abd Rahim[1,2], *M.Zamri* Zainon[2,3], and *N.Nazri* Nik Ghazali[2]

[1] Mechanical Department of Engineering Technology, Faculty of Engineering Technology, University Malaysia Perlis,02100 Padang Besar, Perlis, Malaysia
[2]Department of Engineering, Faculty of Engineering, University of Malaya, 50603 Kuala Lumpur, Malaysia
[3]Center of Energy Science, University of Malaya, 50603 Kuala Lumpur, Malaysia

Abstract. In this paper, condensation for high steam energy in vertical annular channel was investigated. Simulations were carried out to investigate the behaviour of condensation using FLUENT 14.5 in static water and steam velocity 0.2ms^{-1}. The results show that, the steam outer surface show rapid condensation activity effect from the steam velocity. Due to this effect, further investigate may require to show condensation behaviour with the various steam velocity.

1 Introduction

Nuclear Power Plant has multiple safety system to ensure three basic system secure function such as controlling the reactor, controlling the fuel and containing the radiation[1].

Consequences of this awareness, proper study need to be carrying out. Referring to the latest accident in the nuclear power plant, such as Fukushima in 2011 will give bad impression to the world wide scale. The main concern for the public on the both either long term and short term to the public health, environment and energy policy and security of the neighbour country.

Safety factor is the most important factor need to focus related to the nuclear reactor. Several factors need to be main focus activity for example rector system and structure and components. Reactor system is the main agenda need to be focus for this study. That is strategies to enhance safety for this RTP upgrading, refurbishment of RTP primary cooling system, refurbishment of RTP instrumentation and control, Neutronics & Thermal Hydraulic Analysis, Ageing Management Program and Quality Assurance.

Refurbishment of primary cooling system is the main focus for this thesis. Cooling system is the most important system in the nuclear reactor. The most famous phenomena that related to the cooling system failure are the Loss of Coolant Accident (LOCA), where dried liquid on the heating element cannot manage the heat produce by the fuel rod. Radiation contamination not only for safety issue but also affect to the environmental regulation [2-4].

To overcome this issue, condensation in annular channel is the correct prototype to simulate this safety issue. To understand condensation phenomena in the annular channel, all parameter that contribute to the condensation need to analyse such as condensation heat transfer coefficient, surface contact area and also need to understand flow pattern involve in the particular condition. Condensation also involve in the two-phase phenomena. By understand properties in two-phase flow, condensation can be determined more accurate.

The appreciative of heat transfer knowledge in convection and conduction are compulsory to investigate the nuclear reactor safety system. By looking the basic knowledge of convection, Newton's Law of Cooling,

$$\dot{Q}=hA(T\_s-T\_w) \quad (1)$$

where h = convection heat transfer ceoffcient, W/m2K , A = heat transfer surface area, m^2, T_s = temperature of the surface, °C, T_w = temperature of the fluid sufficiently far from the surface,°C.

2 Physical Model

The condensation model has been investigate experimentally and computationally by many researchers as pervious mentioned. Condensation is a complex field in the heat transfer area. This area involve by two phase, high steam energy (vapor) and liquid (water). The early stages of the condensation phenomena were introduce by Nusselt and the proposed the first theoretical solution to solve heat transfer coefficient by assuming linear temperature profile cross the laminar film thickness liquid.

Condensation for the purpose in nuclear safety, these phenomena mostly encounter inside the containment. The behaviour inside containment, condensation phenomena will change and directly affect the pressure, temperature and local mixture concentration.

Corresponding author: wanmohd@unimap.edu.my

Condensation in annular channel can be classified into 2 main areas. Firstly, Direct Contact Condensation (DCC) or Bubble Condensation, these phenomena usually used to studies in the Nuclear Safety. Secondly, condensation occurs on the surface (plate) usually used to studies for heat exchanger. These two phenomena occur in the same condensation situation.

Based on the statement above, the important of the condensation need to be focus. Numerical method or computational approaches currently become famous nowadays to analyse phase change phenomena[5]. Many researcher studies the condensation in high speed velocity as steam plume condensation [6-9]. Steam Injector (SI) is the passive system in nuclear reactor as a secondary during emergency accident[10]. This article will show the importance of the velocity that affected the condensation in selected value.

3 Mathematical Models

In this research, Volume of Fraction (VOF) method in Mathematical Model will be used to calculate the void fraction of each fluid along the domain was tracked to understand condensation behaviour [11]. Interface mass transfer was tracked by using the thermal equilibrium model. The governing equation of continuity, momentum, energy and the interface model used are presented in the following sections.

2.1 Mass Conservation Equation

The mass conservation equation describes the mass flux ration into and out of a control volume and the accumulation in a corresponding control volume. For the qth phase, this equation has the following form:

$$\frac{1}{\rho_q}\left[\frac{\partial}{\partial t}(\alpha_q\rho_q) + \nabla\cdot(\alpha_q\rho_q v_q) = S_{\alpha q} + \sum_{p=1}^{n}(\dot{m}_{pq} - \dot{m}_{qp})\right] \quad (2)$$

The volume fraction equation is not solving for the primary phase and the primary-phase volume fraction is computed based on the following constraints:

$$\sum_{q=1}^{n}\alpha_q = 1 \quad (3)$$

where v_q is the velocity of phase q and α_q is the volume fraction of the qth phase. $\dot{m}_{qp}$ is the mass transfer from the phase p to the phase q. This term is very useful to calculate condensation model in annular channel in understanding void fraction of the high steam energy (vapour).

2.1 Momentum Transfer Equation

A single momentum equation is solved throughout the domain, and the resulting velocity field is shared among the phases. The momentum equation, shown below, is dependent on the volume fractions of all phases through the properties ρ and μ.

$$\frac{\partial}{\partial t}(\rho v) + \nabla\cdot(\rho v v) = -\nabla p + \nabla[\mu(\nabla v + \nabla v^T)] + \rho g + F \quad (4)$$

where ρ, μ and v are the density, viscosity and velocity of phase in the control volume, respectively.

The term on the right-hand side describe the force acting on the element in the control volume. These are the overall pressure gradient, the stresses, the gravitational force and effect of an external body. Generally speaking, the external body term consists of interfacial forces such as lift, virtual mass forces, surface tension, and etc.

2.2 Energy Conservation

The energy equation also share among the phases, is shown below

$$\frac{\partial}{\partial t}(\rho E) + \nabla\cdot(v(\rho E - p)) = \nabla\cdot(k_{eff}\nabla T) + S_h \quad (5)$$

The VOF models treat energy, and temperature, T, as mass average variable.

2.3 Condensation Model

In this work, the phase change model, which is describe the phase change induced by interphase heat transfer of intrinsic flow, was simulated the condensation model. The models only applicable to the phase change in pure substance and consider heat transfer process on each side of the phase model [12].

For the interfacial area is require to specified volumetric heat transfer coefficient, and has the form

$$A_{fg} = \frac{6\alpha_g}{d_g} \quad (6)$$

where d_g(m) and $\alpha_g$are the mean diameter and volume fraction of the high steam energy injected to the liquid region, respectively.

Interfacial heat transfer coefficients are requiring them to calculate the heat and mass transfer occur between steam and water phases.

3 Simulation Set-ups

Using the finite volume formulation, two-dimensional simulation was carried out to model the condensation behaviour of vapour in annular vertical channel. The numerical approach need to be done to analyse and understand the important of velocity towards condensation in vertical orientation for upwards flow. Even the simulation only carried out single situation, this analyse only to understand the basic affect the velocity towards the condensation.

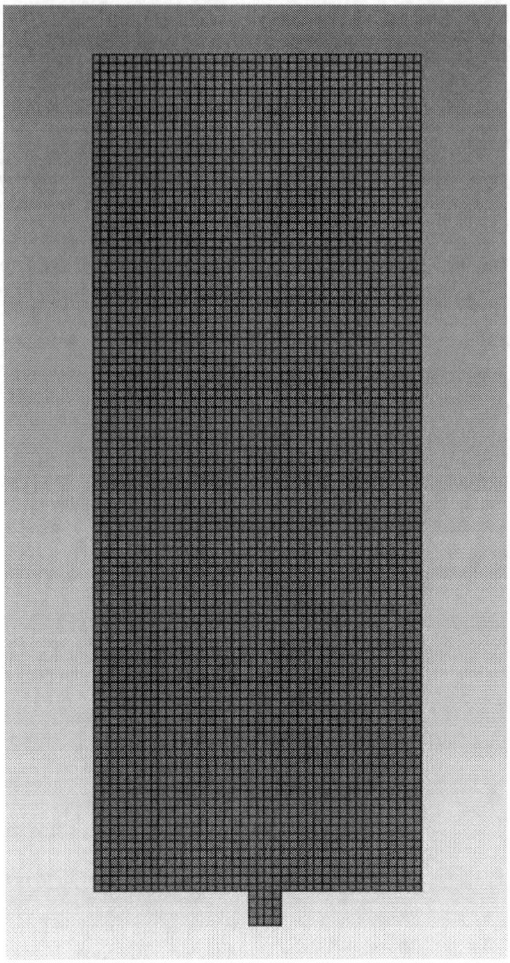

Figure 1: The schematic diagram of annular channel

The SIMPLE algorithm was employed for pressure-velocity coupling and the implicit scheme was used in the time discretization.

All the solution was considered to be fully converged. Here the setting parameter for this simulation shown in Table 1.

Table 1
Parameter used in FLUENT simulation

	Inlet Velocity (ms^{-1})	Temperature Inlet (°C)	Tank Temperature (°C)
Vapor	20	400	-
Water	-	-	323.15

The Figure 1 showed the annular channel as shown in with fully dimensional.

The steam inlet was set for fully turbulent with the velocity 20 ms^{-1} to show the competency the real situation in the nuclear power plant. By using ANSYS FLUENT 14.5, this solver supports the simulation of multiphase flow. The solver also supports the incompressible two phase flow with phase change using VOF model. In order to support the condensation phenomena, energy equation need to be used during this simulation.

This simulation will be run for the 6 sec with the time step size 0.01s. For

4 Results and Discussion

To understand the behaviour of the condensation, the temperature distribution will be used as indication to understand these phenomena.

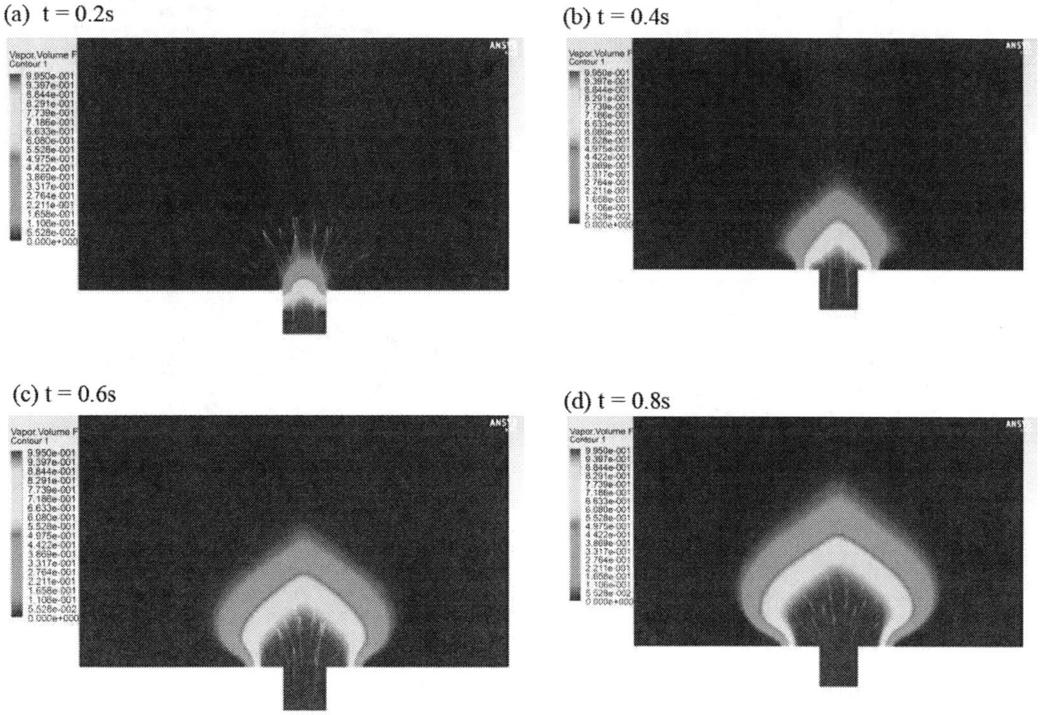

Figure 2: Contour of the void fraction for t = 0.2s, 0.4s, 0.6s, 0.8s

https://doi.org/10.1051/matecconf/201815004003

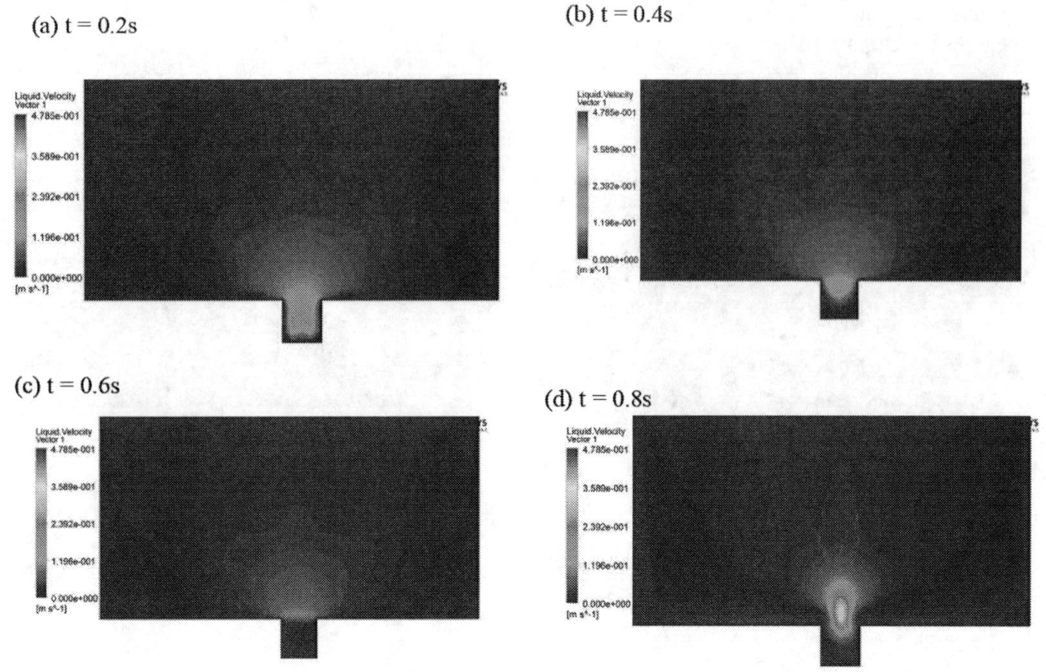

Figure 3: Contour of the liquid velocity for t = 0.2s, 0.4s, 0.6s, 0.8s

Figure 2 show the contour of the void fraction for the time 0.2s, 0.4s, 0.6s and 0.8s. These figures show the distribution of the vapour in the liquid region. The boundary area in the figure is the region occur the condensation in the annular channel. This is the important part in this simulation for this analysis.

Figure 3 show the liquid velocity with the same interval time 0.2s, 0.4s, 0.6s and 0.8s. The important of this figure to show vapour trajectory in liquid. For this case of moving liquid and vapour, slip velocity is important to determine.

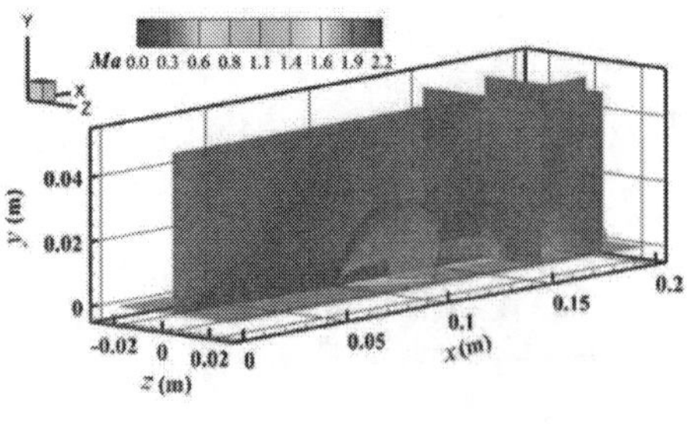

Figure 4: Distribution of Mach number[13]

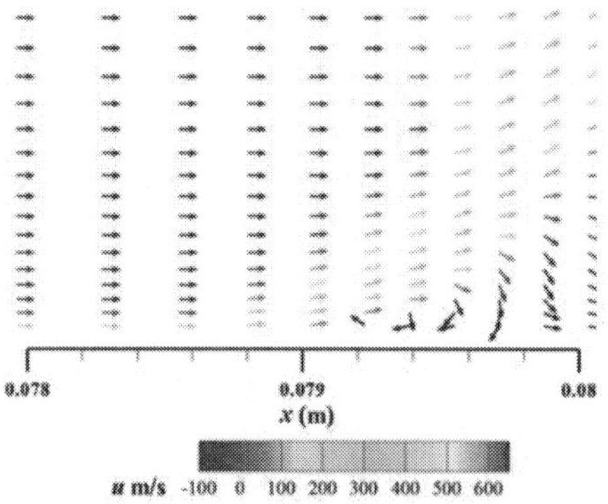

Figure 5: Recirculation Zone[13]

The importance of the velocity been done before by Liu et. al [13]. This author shown that, the velocity or Mach number have main contribution affected the condensation in two-phase flow. Phase change phenomena depend on many parameters such as Nusselt Number, Prandlt Number, viscosity and phase of phase. This is main reason, this article to the importance of velocity that need to focus in analysis heat transfer coefficient or phase change phenomena as shown in figure 4 and 5.

This work is supported by the Short-Term Grant Unimap (STG 9001-00559) and Postgraduate Research Grant (PPP) Universiti Malaya (PG206-2015B)

References

1. Zong, X., et al., *Experimental study on the direct contact condensation of steam jet in subcooled water flow in a rectangular mix chamber.* International Journal of Heat and Mass Transfer, 2015. **80**: p. 448-457.
2. Yeoh, G.H. and J.Y. Tu, *Two-fluid and population balance models for subcooled boiling flow.* Applied Mathematical Modelling, 2006. **30**(11): p. 1370-1391.
3. Ozar, B., et al., *Interfacial area transport of vertical upward steam-water two-phase flow in an annular channel at elevated pressures.* International Journal of Heat and Mass Transfer, 2013. **57**(2): p. 504-518.
4. Evgenidis, S.P. and T.D. Karapantsios, *Effect of bubble size on void fraction fluctuations in dispersed bubble flows.* International Journal of Multiphase Flow, 2015. **75**: p. 163-173.
5. Kharangate, C.R. and I. Mudawar, *Review of computational studies on boiling and condensation.* International Journal of Heat and Mass Transfer, 2017. **108**: p. 1164-1196.
6. Tejedor Sastre, M.T. and C. Vanhille, *A numerical model for the study of the difference frequency generated from nonlinear mixing of standing ultrasonic waves in bubbly liquids.* Ultrason Sonochem, 2017. **34**: p. 881-888.
7. Li, S.Q., et al., *Experiment study on steam-water direct contact condensation in water flow in a Tee junction.* Applied Thermal Engineering, 2017. **120**: p. 99-106.
8. Joshi, J.B., et al., *Bubble generated turbulence and direct numerical simulations.* Chemical Engineering Science, 2017. **157**: p. 26-75.
9. Qu, X.-h., H. Sui, and M.-c. Tian, *CFD simulation of steam–air jet condensation.* Nuclear Engineering and Design, 2016. **297**: p. 44-53.
10. Qu, X.-h. and M.-c. Tian, *Acoustic and visual study on condensation of steam–air mixture jet plume in subcooled water.* Chemical Engineering Science, 2016. **144**: p. 216-223.
11. Xu, Q.G., Liejin., *Direct contact condensation of steam jet in crossflow of waterin a vertical pipe. Experimental investigation on condensationregime diagram and jet penetration length.* International Journal of Heat and Mass Transfer, 2016. **94**.
12. Schlegel, J.P. and T. Hibiki, *A correlation for interfacial area concentration in high void fraction flows in large diameter channels.* Chemical Engineering Science, 2015. **131**: p. 172-186.
13. Liu, H., Y. Guo, and W. Lin, *Numerical simulations of spray jet in supersonic crossflows using an Eulerian approach with an SMD model.* International Journal of Multiphase Flow, 2016. **82**: p. 49-64.

Model Updating of Friction Stir Welding for Aluminium and Magnesium Plate Structure

Nazrotul Afina Nazri[1], *Mohd Shahrir Mohd* Sani[1,2*], *Muhammad Nasiruddin* Mansor [2] and *Siti Norazila* Zahari[1]

[1]Advance Structural Integrity and Vibration Research (ASIVR), Faculty of Mechanical Engineering, Universiti Malaysia Pahang, 26600 Pekan, Pahang, Malaysia,
[2]Automotive Engineering Centre, Universiti Malaysia Pahang, 26600 Pekan, Pahang, Malaysia

Abstract. Friction stir welding (FSW) of aluminium and magnesium alloys face high demands in automotive and aerospace application due to its advanced and lightweight properties. FSW is an emerging solid state joining process in which the material that is being welded does not melt and recast. The main objectives of this project are to perform model updating based on finite element analysis (FEA) and experimental modal analysis (EMA) of dissimilar material of aluminium alloy AL 7075 and magnesium alloy AZ 31B. Modal properties such as natural frequencies, mode shapes are obtained and compared between FEA and EMA. The discrepancies of first five modes natural frequencies are below than 10% and the model updating have been conducted to minimize the error between two methods. This model updating are based on sensitivity analysis in order to make sure which parameters are given more influence in this structural dynamic analysis. Young's modulus and Poisson's ratio both materials are selected in the model updating process. After perform model updating, total average error of the natural frequencies of dissimilar friction stir welding plate is improved significantly.

1 Introduction

Many industrial fields give much attention which focusing on aluminium and magnesium alloys due to escalating appeal for further fuel-efficient vehicles to deduct the devouring of energy and more environmental friendly [1-2]. For their practical applications in market production, bonding and welding technologies should be given a colossal attention in addition to consider issues such as alloy design, microstructure control, plastic forming, casting, and surface treatment [3]. Both of these alloys are attractive to be applied in automotive structure for improving energy efficiency, which also leads to lessening the emission of greenhouse affecting gases. In order to ensure these alloys take place in automotive structures, dissimilar joint between magnesium and aluminum alloys is required [4]. Dissimilar metal joining also escalates in demand due to the good mechanical properties through the combination and also gives good corrosion resistance [5]. Joining method chosen is friction stir welding (FSW) which can be describes as a solid-state welding process, relatively novel joining technology, which has caught the interest of many industrial sectors, not only in automotive, but also aeronautic and transportation due to variation of advantages and vivid industrial potential. The process adds new possibilities within component design and allows more economical and environmentally efficient use of materials [6]. It was notified that using Al alloys and Mg alloys can provide distinct weight savings instead of using steel and cast iron [7].

For this paper, finite element analysis (FEA) method was picked for the purpose of theoretical approach by collective data of natural frequencies and mode shape in order to diagnose basic dynamic structure of plate. FEA involved the utilization of CAD Simulation software to illustrate the plate. Meanwhile for experimental modal analysis (EMA), impact hammer test is conducted to gain the modal properties. In this paper, modal properties are rectified using both methods, FEA and EMA which normal analysis mode in FEA order to select the most reliable model. From result obtained, model updating will be conducted to modify the correlation between experiment and numerical counterparts.

2 Welding

In industry, application of welding joining method experiences a high demand and become one of the most frequently method chosen for a mass production [8]. Welding is a fabrication process that is used to join materials together such as metals or thermoplastics in which the materials placed together exactly at the point to be bonded. FSW is being introduced approximately around last decade [9], is a solid state welding process

* Corresponding author: mshahrir@ump.edu.my

that normally used for welding soft materials like aluminium and magnesium alloys and in addition, to dissimilar metals that frequently facing a problem due to the hard in joining by conventional fusion techniques [10-11]. FSW reveals many advantages such as low distortion and wide chemical tolerance due to the heat input in the process is below the melting point [12]. Besides, the mechanical strength is high due to grain refinement in the heat affected zone (HAZ), low pollutions and low costs [11,13].

In order to acquire an excellent result in joining of FSW and good weld quality, tool design if one of the vital criteria that needed to be consider [14-15]. Tool materials should have good thermal fatigue strength to resist repeated heating and cooling cycles, good fracture toughness to resist the damage during plunging and dwelling, good machinability to ease manufacture of complex features on the shoulder and probe, and have an affordable cost [16]. AISI H13 is a type of steel chosen for fabrication of welding tool in this study due to ease in machinability. Image and dimension of FSW tool can be seen in Figure 1 and Table 1 respectively below.

Fig. 1. FSW tool design

Table 1. FSW tool dimension

Tool Parts	Dimensions
Holder diameter, a (mm)	10.0
Holder length, b (mm)	19.0
Shoulder diameter, D (mm)	9.0
Shoulder length, c (mm)	5.0
Pin diameter, d (mm)	3.0
Pin length, l (mm)	1.7
D/d ratio of tool	3.0

FSW experiment process was conducted by using vertical milling machine. Dissimilar FSW process is carried out by placing the high strength AL 7075 at the retreating side (RS) and by placing the AZ 31B at the advancing side (AS) since if the weaker alloy is located at retreating side (RS), the fabricated weld will become weaker than the weaker alloy is at retreating side (RS) [17]. Figure 2 shows the setup of specimen for FSW. The process parameters which have the greater influence on the tensile strength of dissimilar FSW joints are identified as rotational speed (RS), transverse speed (TS) and tilt angle. It is important in order to produce a high quality joint for FSW process. Welding parameter set up for this study can be referring in Table 2.

Fig. 2. FSW specimen set up

Table 2. Welding parameter for AL7075 and AZ31B

Rotational Speed (rpm)	Transverse Speed (mm/min)	Tilt Angle (°)
700	170	1

3 Finite Element Modeling and Analysis

Finite element analysis (FEA) is a computational system generally used to predict the behavior of structures. It is mostly valuable to analyze those conditions which cannot be directly studied directly by experimentation, and thus, able to provides valuable information that cannot be obtained in more straight ways [18]. In FEA, normal mode analysis is performed on the finite element model of the dissimilar plate structure in order to obtain the dynamic properties of the structure, which are the natural frequencies, damping ratios and the mode shapes. Elements of dissimilar plates are model in MSC Nastran Patran as shown in Figure 3. Nominal properties of both structure are tabulated in Table 3.

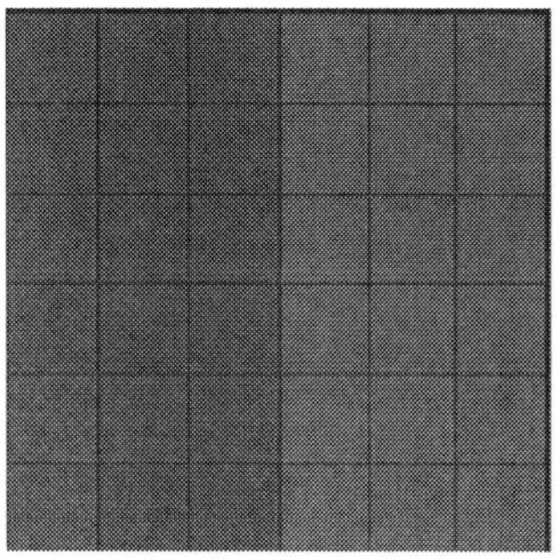

Fig. 3. Meshing element of dissimilar plates in FE

Table 3. Material properties of AL 7075 and AZ 31B

Properties	AL 7075	AZ 31B
Density (kg/m^3)	2820	1770
Modulus of elasticity (GPa)	72	45
Poisson Ratio	0.33	0.35

 https://doi.org/10.1051/matecconf/201815004004

From the numerical result, the values of natural frequencies and mode shapes of the dissimilar plates are obtained and tabulated in Table 4 below.

Table 4. Natural frequencies and mode shapes from FEA

Mode	Natural Frequency (Hz)	Mode Shape
1	180.85	
2	236.51	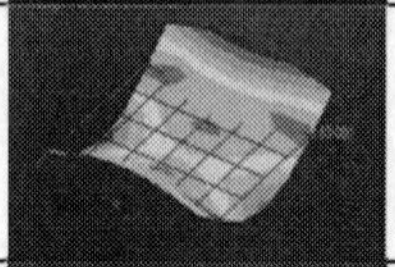
3	364.70	
4	429.27	
5	478.90	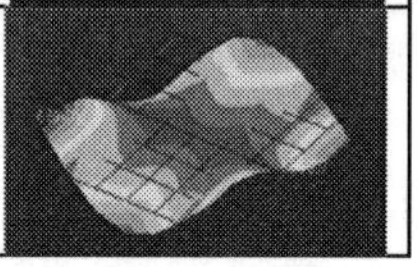

4 Experimental Modal Analysis

Experimental modal analysis (EMA) or sometimes called as modal testing, is the process of extracting dynamic characteristics of a system, machinery or structure experimentally [19]. Confirmation for the accuracy of a result obtained in finite element analysis is usually validated by EMA [20-22]. The most regularly used technique of conducting the experiment is by using impact hammer as an actuator, an accelerometer as the sensor and a Fast Fourier Transform (FFT) analyzer to obtain the system Frequency Response Function (FRFs) [23].

Impact hammer test is a method of testing that allows user to acquire natural frequency, damping ratio and mode shapes of test structure. Indeed, this test is simple to implement but it is tough to obtain a consistent results. In this project, free-free boundary condition is set up for the flat plates by placing the specimen on sponge. Roving accelerometer method is chosen, and 36 node points are picked for knocking point. Figure 4 shows how the specimen is placed on the sponge and the accelerometer is placed on 1st nodes on the specimen. The experiment is set up as shown in Figure 4.

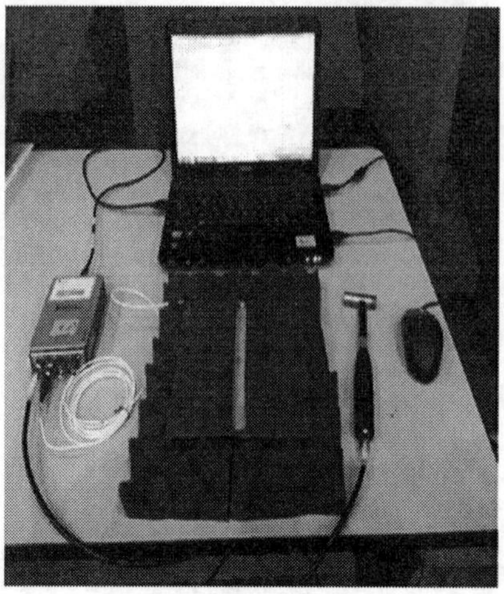

Fig. 4. Set up for EMA

From the experimental modal testing, results of natural frequencies and mode shapes are tabulated in Table 5 below.

Table 5. Natural frequencies and mode shapes from EMA

Mode	Natural Frequency (Hz)	Mode Shape
1	178	
2	228	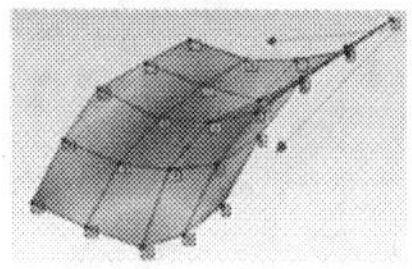
3	318	
4	384	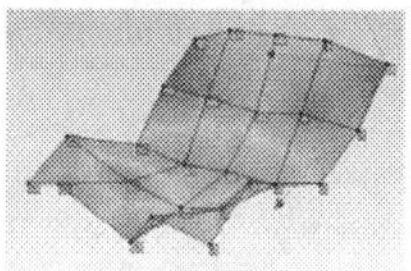

https://doi.org/10.1051/matecconf/201815004004

5	426	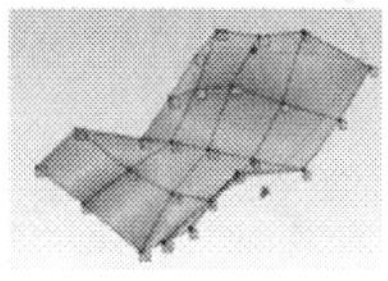

5 Correlations and Finite Element Model Updating

Correlation of data that was gained through finite element analysis and modal testing are showed in order to analyze the discrepancies existed between those two sets of data. In addition, correlation of data is essential in order to have accurateness estimation on the existing plate model. Table 6 exhibits the correlation of natural frequencies of dissimilar plates structure that was gathered via experimental and numerical (finite element) analysis. Value of discrepancies between those two sets of data was calculated by accepting the value obtained through experiment as the actual value. As illustrated, the percentage of error that exists is significantly high in second and third modes.

Table 6. Correlation of natural frequencies between EMA and FEA.

Mode	Numerical Natural Frequency (Hz)	Experimental Natural Frequency (Hz)	Percentage of Error (%)
1	186.85	178	4.87
2	236.51	228	3.73
3	364.70	318	14.69
4	429.27	384	11.79
5	478.90	426	12.42
		Total average error	9.52

In order to reduce the existing discrepancies, model updating procedure was implemented on the finite element model of dissimilar plate structure, hence improves the model to have better correlation with the actual structure. There are dual approaches of model updating which are direct method and sensitivity method [19]. Model updating will be performed as an optimization technique and being presented by the structural optimization capability [24-27]. Several updating parameters which are Young's modulus and Poisson ratio for both materials were considered to be included in model updating procedure. The number of updating parameters was kept to be less than number of modes to be updated in order to avoid ill-conditioning problem in updating procedure. EMA result becomes as the benchmark for the comparison as the results are more reliable based on the FRF and coherence graph while taking results. The objective function for the prediction error in model updating optimization is defined as

$$g(x)=\sum_{i=1}^{n} W\left(\frac{w_i^e}{w_i^c}-1\right)^2 \qquad (1)$$

where w_i^e and w_i^c are the experimental and computational natural frequencies respectively, with W as the real positive weighing factor. The prediction of the modal data is bestowed for detraction in the updating operation. Structural optimization is proven to be an effective algorithm for solving the sensitivity of parameters.

Table 7. Comparison of discrepancies between initial and updated results.

Mode	Natural Frequency (Hz)				
	EMA	Initial FEA	Initial (%) of Error	Updated FEA	After Updating % of Error
1	178	186.85	4.97	178.46	0.26
2	228	236.51	3.73	226.93	0.47
3	318	364.70	14.69	357.93	12.56
4	384	429.27	11.79	413.50	7.68
5	426	478.90	12.42	460.14	8.01
		Total average error	9.52	Total average error	5.61

Table 8. Changes of updating parameters from the initial values.

Parameter	I	II	III
	Initial Value	Updated Value (Upper)	Updated Value (Lower)
Young's Modulus of AL7075 (GPa)	72	64	80
Young's Modulus of AZ31B (GPa)	45	46	44
Poisson Ratio of AL7075	0.33	0.36	0.30
Poisson Ratio of AZ31B	0.35	0.38	0.32

6 Conclusions

This study was carried out to correlate the experimental modal data to the data gathered from finite element analysis of a dissimilar plate structure. A finite element model of the plate structure was produced and the percentage of errors between those two sets of data was obtained and the study was furthered into performing model updating procedure in order to reduce the existing percentage of errors. The updating procedure is considered as parameter identification and strikes to improve numerical prediction to be as closely as possible to the actual structure.

Four parameters were chosen as the updating parameters and correlation was done based on the first five measured natural frequencies. The results of this investigation show that discrepancy is unavoidable when constructing model for a structure because of inaccuracies in parameters assumption and simplification in process of modeling. However, this inconsistency between the prediction model and the actual structure can be improved by carrying out model updating procedure.

It is recommended that further research be undertaken by including the joint element during modeling process. Therefore, procedure of model updating can be carried out at the joint element and joint properties can be chosen as one of the updating parameters. At the end, the updated finite element model structure can possibly exhibit more accurate and reliable numerical prediction model for future research.

Acknowledgements

This work was supported and funded by Universiti Malaysia Pahang (UMP) under RDU 150335.

References

[1] M. K. Kulekci, "Magnesium and its alloys applications in automotive industry," *The International Journal of Advanced Manufacturing Technology,* vol. 39, pp. 851-865, 2008.

[2] W. S. Miller, L. Zhuang, J. Bottema, A. J. Wittebrood, P. De Smet, A. Haszler, *et al.*, "Recent development in aluminium alloys for the automotive industry," *Materials Science and Engineering: A,* vol. 280, pp. 37-49, 3/15/ 2000.

[3] Y. J. Kwon, I. Shigematsu, and N. Saito, "Dissimilar friction stir welding between magnesium and aluminum alloys," *Materials Letters,* vol. 62, pp. 3827-3829, 8/31/ 2008.

[4] R. Borrisutthekul, Y. Miyashita, and Y. Mutoh, "Dissimilar material laser welding between magnesium alloy AZ31B and aluminum alloy A5052-O," *Science and Technology of Advanced Materials,* vol. 6, pp. 199-204, 3// 2005.

[5] E. Taban, J. E. Gould, and J. C. Lippold, "Dissimilar friction welding of 6061-T6 aluminum and AISI 1018 steel: Properties and microstructural characterization," *Materials & Design (1980-2015),* vol. 31, pp. 2305-2311, 5// 2010.

[6] D. Muruganandam and D. S. lal Das, "Friction Stir Welding Process Parameters for Joining Dissimilar Aluminum Alloys," *International Journal of Mechanical Engineering & Technology (IJMET),* vol. 2, pp. 25-38, 2011.

[7] L. Liu, D. Ren, and F. Liu, "A review of dissimilar welding techniques for magnesium alloys to aluminum alloys," *Materials,* vol. 7, pp. 3735-3757, 2014.

[8] Y. Gori and R. P. Verma, "Experimental Fatigue Life Estimation of AA5083 Aluminium Alloys Welded by Two Welding Processes-Gas Metal Arc (GMA) Welding and Friction Stir Welding (FSW)," *Journal of Graphic Era University,* vol. 5, pp. 10-15, 2017.

[9] J. P. Bergmann, M. Grätzel, R. Schürer, A. Regensburg, and M. Weigl, "Advances and Potentials in Friction Stir Welding of Aluminum Alloys," in *Key Engineering Materials,* 2016, pp. 137-142.

[10] S. D. Meshram, A. G. Paradkar, G. M. Reddy, and S. Pandey, "Friction stir welding: An alternative to fusion welding for better stress corrosion cracking resistance of maraging steel," *Journal of Manufacturing Processes,* vol. 25, pp. 94-103, 1// 2017.

[11] A. F. Hasan, C. J. Bennett, P. H. Shipway, S. Cater, and J. Martin, "A numerical methodology for predicting tool wear in Friction Stir Welding," *Journal of Materials Processing Technology,* vol. 241, pp. 129-140, 3// 2017.

[12] R. Thakur and P. S. Bajwa, "Friction Stir Welding of 5xxx series Aluminium Alloys A Literature Survey," 2016.

[13] Z. Zhang, Q. Wu, and H. Zhang, "Prediction of fatigue life of welding tool in friction stir welding of AA6061-T6," *The International Journal of Advanced Manufacturing Technology,* vol. 86, pp. 3407-3415, 2016.

[14] A. Fall, M. H. Fesharaki, A. R. Khodabandeh, and M. Jahazi, "Tool wear characteristics and effect on microstructure in Ti-6Al-4V friction stir welded joints," *Metals,* vol. 6, p. 275, 2016.

[15] K. Kumar and S. V. Kailas, "The role of friction stir welding tool on material flow and weld formation," *Materials Science and Engineering: A,* vol. 485, pp. 367-374, 6/25/ 2008.

[16] Y. Zhang, X. Cao, S. Larose, and P. Wanjara, "Review of tools for friction stir welding and processing," *Canadian Metallurgical Quarterly,* vol. 51, pp. 250-261, 2012.

[17] S. R. Kumar, V. S. Rao, and R. Pranesh, "Effect of welding parameters on macro and microstructure of friction stir welded dissimilar butt joints between AA7075-T651 and AA6061-T651 alloys," *Procedia Materials Science,* vol. 5, pp. 1726-1735, 2014.

[18] D. Burr, "The use of finite element analysis to estimate the changing strength of bone following treatment for osteoporosis," ed: Springer, 2016.

[19] S. N. Zahari, A. A. R. Zakaria, and M. S. M. Sani, "A review on model updating of joint structure for dynamic analysis purpose," 2015.

[20] N. A. Z. Abdullah, M. S. M. Sani, M. M. Rahman, and I. Zaman, "Correlation of Numerical and Experimental Analysis for Dynamic Behavior of a Body-In-White (BIW) Structure," 2016.

[21] M.H.N. Izham, N.A.Z. Abdullah, S. N. Zahari, and M.S.M. Sani, "Structural dynamic investigation of frame structure with bolted joints," in *Journal MATEC Web of Conference*, 2017, pp. 1-9.

[22] M.S.M. Sani, N.A. Nazri, S.N. Zahari, N.A.Z Abdullah, and G. Priyandoko, "Dynamic Study of Bicycle Frame Structure," in *IOP Conf. Series: Materials Science and Engineering 160 (2016) 012009 doi:10.1088/1757-899X/160/1/012009*, 2016, pp. 1-8.

[23] N. A. Z. Abdullah, M. S. M. Sani, and N. A. Husain, "Model Updating of Go-Kart Structure in Structural Dynamics," 2015.

[24] S. N. Zahari, M. S. M. Sani, N. A. Husain, M. Ishak, and I. Zaman, "Dynamic analysis of friction stir welding joints in dissimilar material plate structure," *Jurnal Teknologi,* vol. 78(6-9), pp. 57-65, 2016.

[25] N.A.Z. Abdullah, M.S.M. Sani, M.M. Rahman, and I. Zaman, "A review on model updating in structural dynamics," in *IOP Conference Series: Material Science and Engineering*, 2015, pp. 1-6.

[26] M.N. Abdul Rani, D. Stancioiu, M.A. Yunus, H. Ouyang, H. Deng and S. James, "Model updating for a welded structure made from thin steel sheets," *Appl. Mech. Material, Vol 70,* Aug 2011, pp. 117-122.

[27] M.N. Abdul Rani, S. Kasolang and M.H. Othman, "Finite element modelling and modal based updating for the dynamic behaviour of a laser spot welded structure," *23rd International Congress on Sound and Vibration (ICSV2016).*

Conceptual Design for an Anti-Thief Donation Box

Febrian Idral[1,*], *Ahmad Fuad* Ab Ghani[1], *Nur Amanina* Syazwani[2], *Tan* Chee Teck[2], *Nur Aliyah* Abd Mutalib[2], *Mohd* Aiman Mohd Fadhil[2], *Muhamad* Dzulhisham Idris[2]

[1] Centre for Advance Research on Energy, Universiti Teknikal Malaysia Melaka, Hang Tuah Jaya, 76100, Durian Tunggal, Melaka, Malaysia
[2]Faculty of Mechanical Engineering. Universiti Teknikal Malaysia Melaka., Hang Tuah Jaya, 76100, Durian Tunggal, Melaka, Malaysia

Abstract. The donation box is commonly used to collect money to from the society for charity purpose. Due to the increase case of the stealing involving donation box, it indicates that the current design of the box is not safe. A research survey are being conducted and the newly design Anti-Thief Donation Box is is introduced in this paper. The material chosen for prototype is mild steel upon its strength and durability. The manufacturing method involved are welding, grinding and drilling. Special specification on this donation box includes the dual manual locking mechanism and the button controlled alarm mechanism.

1 Introduction

Donation boxes are often being seen in public places that receive a lot of visitor such as mosque or mall as a means of generating additional revenue in small increments. In churches, it is common to find a Mite box or Alms Box, and in Shinto shrines a Saisen-bako.

The term of mite box or generally known as donation box (also alms box or poor box) refers to a box that is used to collect coins and money bills for charitable purposes. The mite boxes are collected by the church, mosque or temple, and the donations are given to the poor.

Donation box is firstly used to collect money for the worshipping place funding rather than giving some alms [1]. It is also helps the people to do charity more conveniently as they only need to insert the money or clothes inside the box. A man has been arrested by the public member as he was attempting to steal from a donation box from the local Hindu temple [2]. Mohd [3] also wrote news about the caught of a band of thieves that are targeting the donation box placed inside the mosque. These bands of thieves are believed to be involved also in the other several cases of the stolen donation box around Malaysia.

Based from these two statements, donation box which is being left unattended has become the target of thieves. This happen because the current design of the donation is lack of safety feature, making it easier for the thieves to stole the money inside the donation box. The safety feature of the current donation boxes is solely depending on the locking mechanism which normally includes a lock only. Therefore, it is crucial to design an anti-thief donation box which include extra safety features in order to reduce the risk where the donation are being stolen.

2 Anti-Thief Donation Box Design Material and Method

The aim of this project is to design an anti-thief donation box that can overcome or reduce loss of donation to the public. The objective is to design a locking mechanism that will act as the safety feature of the donation box and designing a donation box that made up from material that can withstand high force and is very durable.

2.1 Proposed Design

Four concept of the purpose design of the Anti-thief donation box are shown in Figure.1. The characteristic of safety measurement, material strength and durability, product durability, and the design are considered during the design stage for this anti-thief donation box[4]. The detailed explanations for each design are also included.

*Corresponding author: febrian@utem.edu.my

https://doi.org/10.1051/matecconf/201815004005

Concept 1

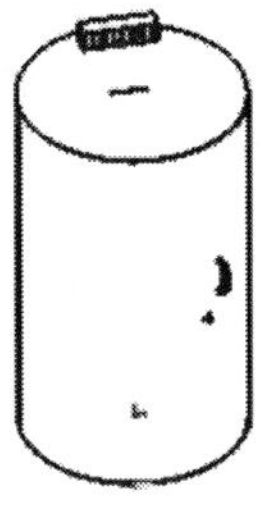

No	Characteristic	Type
1	Shape	Cylinder
2	Height	100 cm
3	Material	Iron
4	Alarm System	Siren
5	Power System	Alternating current
6	Locking System	Sensor Locking System
7	Lock Type	Mechanical Cupboard Lock
8	Bill and coin slot	Normal Slot
9	Compartment of Donation Box	Single
10	Door Handle	Rectangular Shape

Concept 2

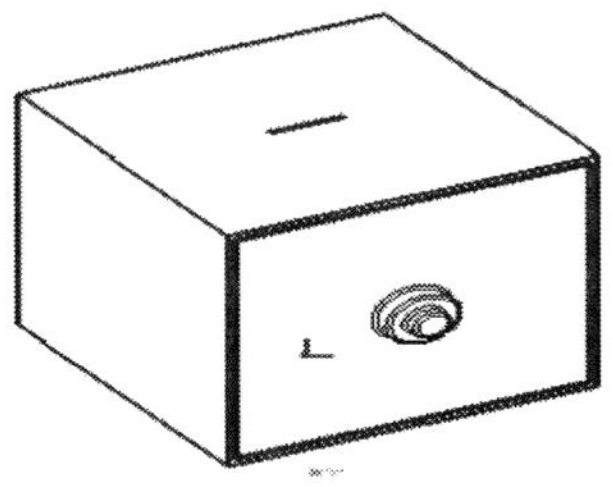

No	Characteristic	Type
1	Shape	Rectangular
2	Height	50 cm
3	Material	Cement
4	Alarm System	Magnetic
5	Power System	Direct current
6	Locking System	Simple Dua-Locking System
7	Lock Type	Combination lock and mechanical cupboard lock
8	Bill and coin slot	Rectractable Slot
9	Compartment of Donation Box	Single
10	Door Handle	Semicircle Shape

Concept 3

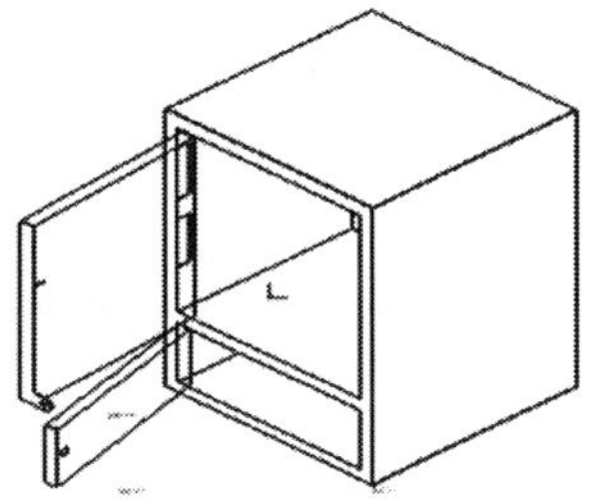

No	Characteristic	Type
1	Shape	Rectangular
2	Height	100 cm
3	Material	Two steel plates
4	Alarm System	Speaker
5	Power System	Direct current
6	Locking System	Dual- Locking System involving mechanical mechanism
7	Lock Type	Mechanical cupboard lock
8	Bill and coin slot	Normal slot
9	Compartment of Donation Box	Double
10	Door Handle	L-shaped

Concept 4

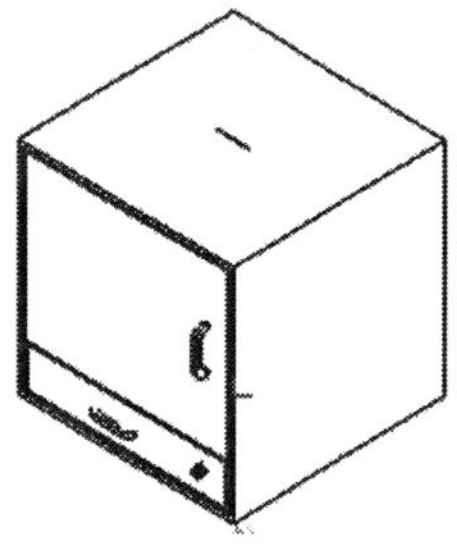

No	Characteristic	Type
1	Shape	Cube
2	Height	50 cm
3	Material	Steel plates
4	Alarm System	Speaker
5	Power System	Direct current
6	Locking System	Simple Dual-locking System
7	Lock Type	Mechanical cupboard lock
8	Bill and coin slot	Normal slot
9	Compartment of Donation Box	Double
10	Door Handle	Semicircle shape

Fig 1. Concept 1,2,3,4 for the Anti-Thief Donation Box

2.2 Conceptual Design of Anti-Thief Donation Box

Concept design 3 is selected as the best design to be fabricated because the design fulfils all the criteria listed by the customers. This design has the best locking mechanism and alarm system that can improve the security aspect of this box. The material for this box that is mild steel has the perfect strength and durability to make the box capable to withstand big forces.

A block diagram of product component decomposition is drawn as shown in Table 1. The donation box is designed to prevent thief therefore it is created to have dual locking system, two compartment that is upper and lower, an alarm system and finally the outer part and inner part. This ensures the security, durability and strength of the product. Every single part has its own function in order for the device to function well.

Table 1: Decomposition of the Anti-thief donation box

Anti-Thief Donation Box	
Strength and Durability Aspect	**Safety Measure**
• Consist of outer and inner part • Have upper and lower part of compartment • Fully consist of mild steel	• Dual Locking System • Digital lock • Combination Lock • Alarm System

The Anti-Thief donation box is functioning by activating the alarm in the upper compartment. The user can control the alarm by pushing the button that is installed in the lower compartment. To open the upper compartment, the user needs to open the lower compartment door first. Then, after unlocking the mechanical lock, the user needs to insert the saved pass code into the digital lock.

Table 2: The Functional Decomposition Of The Anti-Thief Donation Box.

Transfer the patient from wheelchair to bed			
Lower Compartment	**Mechanical Lock**	**Digital Lock**	**Alarm System**
Equipped with a combination lock The mechanical lock is installed	Control the upper compartment. Steps requirement Increase the security	pass code is inserted mechanical lock to increase the safety	Safety measure. Gives warning alert Button Controlled

3 Results

Pugh concept section method is used to evaluate the concept alternatives. Criteria choose are high efficiency, high reliability, low maintenance, low cost and light weight. Concept 1 was selected as Datum which is a benchmark comparison of a product.

Table 3: Pugh Method Selection

Criterion	Importance Weightage (%)	Concept			
		Concept 1 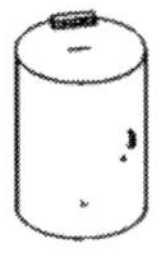	Concept 2 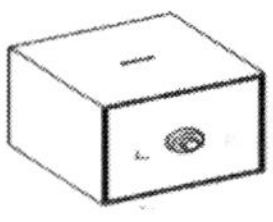	Concept 3 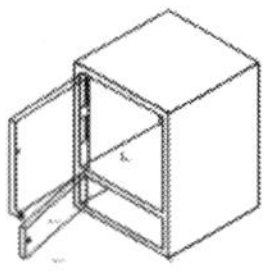	Concept 4 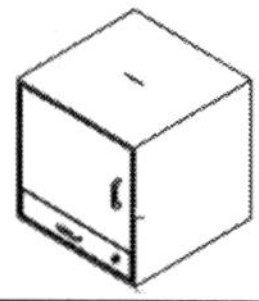
High strength of the product	20	DATUM	+	+	+
Long life span	10		+	+	+
High security of the product	25		-	+	-
High endurance of heat	10		+	+	S
Manufacturing cost (RM/cost)	10		S	-	S
Easy access for maintenance	5		-	-	S
Complicated locking system	10		S	+	-
High durability	10		+	+	+
Total +			50	85	40
Total -			30	15	35
Total S			20	0	25
Weighted total = (Total+) – (Total -)			20	70	5

The concept 1 is set as datum column. +, -, = sign is used against datum to evaluate and compared the other concepts. Finally, select the best alternative configuration as the final configuration where were concept 3.

3.1 Fabrication of Product

For this product, the fabrication process involved is cutting and welding. First of all, the mild steel plate is being measured and cut into desired size. Then the parts will be combining by welding the steel plate together to form a box [5-6]. The electronic component of the product is being installed using hot glue gun and soldering. Figure 2 below shows the final product after the fabrication process.

Figure 2. Anti-Thief Donation Box

3.2 Mechanical Analysis Aspect

From the analysis aspect, the box is designed to be able to withstand strong forces. That is why, for this product, mild steel is selected as it base material. Mild steel has the perfect ultimate strength of 400 MPa. Based on the Finite Element Analysis and hardness testing, mild steel has the most suitable strength and durability.

The dual-locking mechanism is designed to be fully mechanically functioning. This because, we want to create locking system that is easy to operate and maintain. The dual locking mechanism does no required outer power source as it is design to function only by retracting it forward and backward.

3.3 Health, Safety, Environment and Sustainability

The Anti-thief donation box was designed to have a perfect dimension where it suitable for people to insert money without having to bend themselves down. This design was different from the current design of donation box which comes in small sizes that need people to bend when they want to insert the money into the donation box money slot. Other than that, the upper compartment that keeps the money storage is designed to ease the owner to pull the box out and installed it back into position. The material use for the product is user friendly and did not cause any harm to the human health.

This donation is designed to fulfil the anti-thief theme. However, the safety aspect is considerate which is harmless to the people surrounding. Therefore, the donation box did not have any sharp edges. The donation box is also designed to have two compartments which have separate function. These designs make sure the product is compatible and safe to the user and people surrounding. The processes of creating the Anti-Thief donation box are also being done under the supervision of the specialist.

Sustainable design or also known as environmental design an environmentally sustainable design, is the philosophy of designing physical objects, the built environment, and services to comply with the principles of social, economic, and ecological sustainability. Eliminating negative environmental impact completely through skilful and sensitive design has been the main intention of sustainability design. Sustainability design manifestations require renewable resources, minimally impact the environment and at the same time connecting the user with the natural environment [7]. Other than the "elimination of negative environmental impact", sustainable design must also create projects that can provide meaningful innovations that can shift behaviour. A dynamic balance between economy and society, intended to generate long-term relationships between user and object or service and finally to respect and considered the Mother Nature and social differences.

Therefore, by following the twelve-sustainability design principle [8-9], the product was followed as an environment-friendly product. The material used to make the thief-nightmare donation box is did not contain any harmful substance and chemical. The mild steel that is being used for the product can be recycled when the product is not functioning anymore.

Not only are the materials, the process of creating the product also following the principle. The processes that involved when creating the product are being done in specifically place. For example, the process of combining the box required the mild steel plate to weld together. Welding process itself taking place in the welding laboratories which are specifically designed for the process. During the process, we did not do anything that can cause harm to the environment.

4 Conclusion

Anti-Thief Donation Box has been studied, designed and fabricated at Faculty of Mechanical Engineering (FKM), Universiti Teknikal Malaysia Melaka (UTeM). A few methods such as Product design specification (PDS), concept generation, Pugh selection method for criteria decision making process during design has been performed in order to decide and select the best design from safety measure, material strength and durability, and product design perspective. The donation box concept design 3 has been selected due to matching the best criteria set in Pugh selection method.

The authors would like to acknowledge Centre for Research Innovation Management (CRIM), UTeM especially faculty Centre for Advanced Research on Energy, (CARe) and Innovation and Sustainability in Machine Technologies (i-SMAT) as a research group for providing facilities in carrying out this study.

6 References

[1] V.J.B. Carrera (1828). Encyclopaedia Dictionary History, Imp. Roca. Pp.457- Retrieve 6 April 2013.

[2] B. Mohd (2017). Three friends jailed for stealing from mosque donation box. Retrieved May 21, 2017, from http://www.malaysiakini.com/news/370265

[3] X. Liu, Z. Kuang, L. Yin, L. Hu,(2017) Structural reliability analysis based on probability and probability bos hybrid model. Jounal of Structural safety, Volume 68, pp 73-84.

[4] Idral. F, (2010), Conceptual Design of a Single Seater Formula Racing Car Final Year Project Report, (Universiti Teknikal Malaysia Melaka (UTeM), Melaka, Malaysia.

[5] Gerard, Barbara (2015)."What is ASTM International? Retrieved 1 February 201 from http://info.craftechind.com/blog/what-is-astm-international

[6] Steel Construction Manual, 8th Edition, second revised edition, American Institute of Steel Construction, 1986, Ch. 1 pp. 1–5.

[7] J.F. McLennan (2004). The Philosophy of Sustainable Design. Ecotone Publishing. Seattle, USA.

[8] Anastas, P., & Zimmerman, J. (2007). Design Through the 12 Principles of Green Engineering. *IEEE Engineering Management Review,35*(3), 16-16. doi:10.1109/emr.2007.4296421
[9] F. Ceschin, (2016), Evolution of Design for Sustainability; From Product Design to Design for System Innovation and Transitions, Design Studies, Volume 47, pp 118-163.

A study on the use of renewable energy for coconut dehusking

Shamsul Shamsudin[1], *Bonny* Dominic Merikan[2,*], *Safarudin* Herawan[1],and *Juhari* Ab Razak[1]

[1] Centre for Advanced Research on Energy (CARe), Universiti Teknikal Malaysia Melaka, 75450 Ayer Keroh, Melaka, Malaysia
[2] Institut Kemahiran MARA Bintulu, km 20 Jalan Bintulu-Sibu, 97000 Bintulu, Sarawak, Malaysia

Abstract. A solar photovoltaic (PV) coconut dehusking machine may become one important application of renewable energy especially in rural areas like in Sarawak. It will not only enhance the socio-economy in such areas but also indirectly reduce the risk of injuries from the process of dehusking coconuts manually. Thus the purpose of the study is to simulate the performance of the PV coconut dehusking machine by circuit modeling in Simulink and PVsyst software. Simulation results showed that it could match the rpm and torque required in the dry coconut shear test. Furthermore, when implemented, such a PV-powered machine is expected to offer better cost of ownership in the long term when compared to a dehusking machine that runs on fossil fuel based on the PVsyst software analysis.

1 Introduction

Coconut dehusking machines are capable to improve production and efficiently eliminate injuries to operators while performing the coconut dehusking task. These difficulties might change people's mind set to perform this job fulltime as their core business. While the government implement continuous drive to improve the rural economy, coconut production must not be neglected as it has high potential to generate consistent income. Thus, coconut dehusking machine is a critical element for successful production of coconut products, yield better productivity and save time. The coconut husk waste is also be used for both commercial by-products and environmental protection.

The recent increase in coconut milk demand for the past few years has contributed to a rapid growth in coconut dehusking activities. In 2009, coconut crops were planted in 12.23 million hectare throughout various places in the world. Malaysia, among other coconut producers, yields about 400 million coconuts which is still short of the 543 million coconuts for domestic demand [1]. Thus, this is quite an impressive opportunity and handsome pay for some of the rural communities especially in Malaysia.

The safety of the operation is taken into consideration at this stage. Thus, many versions of the machine has been developed to meet the challenge. In 1998, Kwangwaropas used a single-phase 1.5 kW electric motor to design an electric dehusking machine [1]. It comprises of two revolving cylindrical rotors with six fins on each rotor. Like previous designs, the operator needs to force the coconut by pushing it toward the rotors, but this time it is done by using a cover. Another successful version of the dehusking machine uses double spike rollers with a hydraulic-actuated mechanism. It incorporates a 4.1 kW petrol-powered motor to generate the hydraulic pressure which is used to drive the double spike rollers. The coconut is fed between the two rollers to remove its husk.

All the technologies and mechanisms of the dehusking machines available might be functional and efficient but they might not suit everyone's needs especially where the area lack basic facilities such as road access, electricity and petrol stations. Thus, this paves the way for a new innovation to develop a green energy dehusking machine for rural people of Bintulu division to equip them well and boost their economic capabilities. Renewable energy alternatives are also necessary for sustaining the environment, affording the basis for shrewd long-term rural production system. This would also minimize direct costs to producers for preserving good environmental quality and indirect costs caused by consumers when environmental quality declines. Furthermore, users may benefit greatly from the high environmental quality in air, soil and water.

A coconut is a well-known favourite fruit and can be found easily in the tropics. Young coconuts are harvested for its juice and flesh which bring sweet taste. The old coconut is a source for coconut milk which also an ingredient in consumer food. This basic crop is greatly in demand particularly in Bintulu, Sarawak. There, the coconut trees can be found widely in rural area such as *Sebemban, Setulan, Serupai* and other areas along the seaside. It becomes one of the main economic

* Corresponding author: bonny@ikmbintulu.edu.my

 https://doi.org/10.1051/matecconf/201815004006

sources of income besides the palm oil for people living in those areas.

Dealing with coconuts can be considered lucrative for the farmers. The current market price for husked coconut varies from RM 1.00 to RM 1.80 each. Normally, the farmer can easily sell up to 100 kernels daily as they do this by a conventional dehusking method. Most of the farmers are old as the young ones work in the cities. Thus, doing this daily could drain their energy, time and jeopardize their health. Unfortunately, they cannot afford to stop doing their routine, as they also need some income to support their own livelihood.

A new techniques or a mechanical dehusking machine is what they might be craving for to support their production. Currently, they need something that is portable and not powered by electricity, at least not from the grid. Some areas are on-grid while some are not. Even in the areas with electricity, their farms can be far from power socket points. Thus, electric-powered dehusking machines cannot be used directly. Another solution is probably to introduce a petrol or diesel engine to power the dehusking machine.

In short term, this might be the best solution but the nearest gas filling station might be around 80 km away in many areas. Another weakness is the regulation, which discourages one from simply filling their empty drum or fuel canisters. Thus, an alternative approach is to use solar (photovoltaics) PV coconut dehusking machine in order to meet their needs. It can be made portable, without the need for fossil fuel or the electrical grid. Hence, this would enable the users to produce at least the same amount of work.

This paper consists of five sections. The background of study is already presented in Section 1. In Section 2, the review of related works like PV systems, energy storage technologies, and food production technologies are given in brief. Section 3, the discussion is on the methodology to achieve the objectives. In addition, Section 4 reports the results and discussions. Lastly, Section 5 highlights the conclusions of this work and suggestions that can be done to improve this study.

2 State of the art

2.1 Understanding coconuts

Coconut dimension is a critical element in designing the dehusking machine. The dimensions of a mature coconut vary depending on the area it is planted and its variant. An efficient machine has to meet a range of coconut sizes and able to increase productivity. Venkataramanan et al. in [2] conducted a research to measure the dimensions of the coconut by using external callipers. Most of coconuts appeared to have almost identical dimensions in term of the Y and Z directions. However, the most similarity had been found in the X dimension [2]. The dimensions of X, Y and Z are shown in Figure 1.

The data was collected from various places in order to get the average measurement of the coconut. Table 1 shows the dimension range of the coconut with husk. Meanwhile, Table 2 lists the dimension range of the husked coconut.

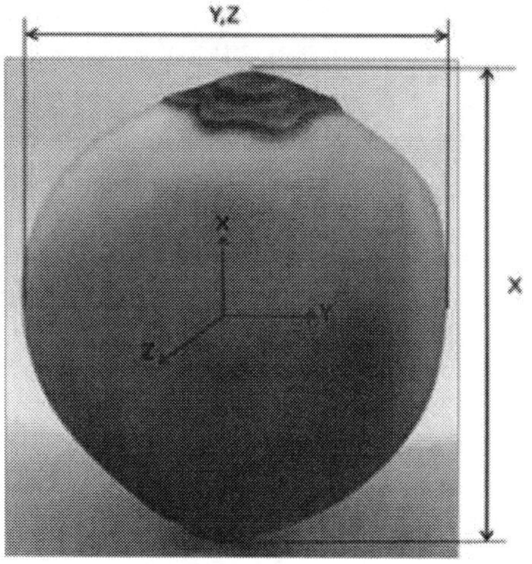

Fig. 1. Mature coconut dimensions [2].

Table 1. Dimension of the coconut with husk [2]

No	X-axis (mm)	Y-axis (mm)	Z-axis (mm)
1	121	113	109
2	157	146	142
3	182	161	165
4	246	218	214
5	276	230	227

Table 2. Dimension of the dehusked coconut [2].

No	X-axis (mm)	Y-axis (mm)	Z-axis (mm)
1	97	86	86
2	114	98	96
3	125	115	109
4	147	129	125
5	162	135	134

Another criterion required to synthesise the coconut dehusking machine is the force required to remove the coconut crown. To find the force needed, or the shear load, the coconut is tested with the Universal Testing Machine (UTM). Dry and mature coconuts in various sizes as mentioned in Table 2 were tested and the results shown in Table 3 and Table 4.

Table 3. Shear load required to dehusk mature coconuts [2].

No.	X-axis (mm)	Shear Load (kN)
1	128	0.32
2	145	0.35
3	178	0.39
4	212	0.45
5	248	0.52

Table 4. Shear load required to dehusk dry coconuts [2].

No.	X-axis (mm)	Shear Load (kN)
1	133	0.24
2	154	0.26
3	174	0.29
4	208	0.36
5	252	0.42

2.2 Cutting torque

Based on the experimental data, the torque required in dehusking a coconut can be calculated. From this torque value, the drive reduction between the motor and the rollers can be designed.

$$T_s = n_s F r \quad (1)$$

where $n_s = 1.25$ is the service factor. F is the cutting force from the shear load whereas r in the average radius of the coconut. If, for instance, the average radius considered is 80 mm while the force is taken to be 450 N, the resulting cutting toque is 45 Nm. Next, torque at the motor can be determined as

$$P_m = T_m \omega \quad (2)$$

Here, some parameters are decided for the electric motor that include ω = 1445 rpm and the power set to be 5.5 kW. Hence, the torque is found to be $T_m \approx 36$ Nm.

One contemporary machine includes rotating rollers with sharp pin blades designed to remove the husk. Although it consumes less time, it still depends on the capability of an operator. The operator still needs to apply a downward force on the coconut using a special clamping tool or even their bare hands to avoid the coconut from being thrown out of the blades. Figure 2 displays the horizontal roller dehusking machine.

Fig. 2. A horizontal roller dehusking machine [2].

Venkataraman et al. [2] noted the current machines or tools are very useful for removing the husk and crown before the coconuts can be processed. The crown of the coconut is also a crucial part of the coconut that is removed by the newly proposed machine. The crown is the fibrous portion that is protruding from the top of a coconut. It is usually seen covering the eyes of the coconut. As mentioned before, it can be very difficult to manually separate the crown from the nut.

2.3 Energy source

Energy source is one of important selection criteria for designing a mechanical. Energy source can be from fossil fuel like gasoline or kerosene, wind, solar and biomass to name a few. Next, each will be elaborated further.

Wind turns out to be mechanically helpful when convective streams made by the sun empower air spread over the surface of the Earth. When sun radiation reaches the Earth and warms its surface, impression of the warmth heats up the ambient air. Warm air, which is less dense than cool air, rises and chill air slips off. In this procedure, the cool air gets the chance to be warmed when it accomplishes the Earth surface and rises up to the higher height [3]. The cycle continues when it brings warm air up and in the long

Wind speed changes directly with stature over the ground [4]. Wind moving over the Earth surface overcomes grinding brought on by turbulence over and

https://doi.org/10.1051/matecconf/201815004006

around structures, trees, mountains, and different deterrents. This rubbing process diminishes with expanding height over the surface until unhampered wind stream is accomplished.

The most confounding thought in wind power estimation is the mean force in the wind. Contingent upon the wind speed scattering (or the recurrence of event of various wind speeds), two indistinguishable wind machines situated at various destinations won't as a matter of course deliver the same force yield at the same mean wind speed [3]. The reason is that the mean force in the wind identifies with the mean of the blocks of the wind speeds over that time and this is distinctive to the shape of the mean wind speed.

The recurrence of frequency of various wind velocities can be assessed by a factual bend utilizing the Weibull dissemination [5].

Another source of energy is the sun, which is commonly used in crop drying and sun oriented warmth gathering. Sunlight based schemes can be specifically changed to produce power by utilizing either thermoelectricity or PV cells [6].

Despite the efforts of many researchers, the highest general efficiency could not exceed 18% at present with the best conductivity materials and innovation [7]. Since PV cells for direct generation of power from the sun are as of now the most encouraging and broadly accessible innovation, PV is underlined in this work.

The measure of sun based radiation impinging on the surface of the PV array originates from the edge at which the sun's beams strike. As the PV surface is slanted far from the sun, the force of the radiation on the array becomes less intense. The measure of sun oriented radiation caught by the surface differs as the cosine of the occurrence point between the sun's beams and the normal of the surface [8].

In light of the mind boggling geometry of the position of the sun in connection to the Earth amid different seasons, a framework is needed to track the sun for PV arrays [9]. The following framework can be introduced for better focused PV modules. However, frameworks that take into account the concentrated daylight can by and large acknowledge just beams crossing a constrained scope of edges.

The configuration of PV frameworks depends basically on the accessibility of exact sunlight based radiation information set. This depends not only on gross topographical traits like scope, elevation and atmosphere classes, but also on land highlights [10].

Sadly, exact sun based radiation information are not really available from secluded rural places where numerous PV frameworks are to be introduced. In any case, a few methodologies for anticipating sun orientation stations have been developed taking into account regularly accessible daylight hours and satellite overcast spread information set or on direct nearby estimations [3].

3 Methodology of Research

This section lays out the methods of study employed in this research. Some methods were applied in order to accomplish the objectives which included simulation using Simulink, calculations and comparison based on the data obtained through the simulation.

3.1 Simulink

Simulink is used create direct models to investigate more practical nonlinear models, considering in contact, air resistance, gear slippage, hard stops, and alternate events that portray real system dynamics. Simulink tries to simulate in the PC what should be done in a physical laboratory for displaying and breaking down frameworks that would not be conceivable in a real experiment [11].

After a model is characterised, it can be represented in a schematic fashion to see the dynamic response using scientific methods, either from the Simulink pull-down menus or by entering orders in the MATLAB Command Window. The re-enactment results from Simulink can be placed in the MATLAB workspace for extra post preparing and representation [11].

In this work, the modeling system design processes started by modeling the PV system with storage system. The design is based on the basic designing for a PV system. Then, the total load and rpm will be simulated to ensure the machine able to reach the target. The modeling of the PV system is based on the basic design that contains loads, energy storage technology (batteries), charger, and PV module. The model is then constructed in Simulink as shown in Figure 3.

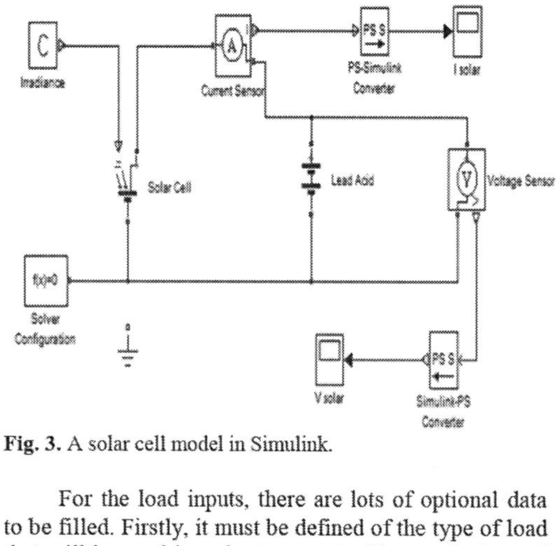

Fig. 3. A solar cell model in Simulink.

For the load inputs, there are lots of optional data to be filled. Firstly, it must be defined of the type of load that will be used in order to operate. There only type of current used was DC. The next step is entering daily profile in the data source. It is easier than choosing import time series data file because the annual average electricity use in kW per day is collected. Furthermore, the data file for electric usage per day for each hour could be imported as a series data file.

Lastly, the data were keyed in to determine the annual average (kWh/d) scale. In the random variability, the data is automatically simulated by the software itself.

For PV inputs, a few things such as the size, capital, replacement, operation and maintenance, lifetime, de-rating factor and output current produce by the PV system have to be considered into an account. In this example, the cost for 230W PV was RM 1300. The cost of replacement was assumed to be the same as the initial cost. Operating and maintenance cost was assumed to be zero since it is negligibly small. In the properties, output current must be selected DC current and the average lifetime of the silicon type solar system is 20 years and the de-rating factor for the PV is 80%. For slope and azimuth, just set it to zero in the ground and 20% according to the default setting of the software.

For battery inputs, there are many types of battery that can be found in the market. All the data that needed can be found from the manufacturer in its recommendation sheet. A serial set of batteries is needed because it is an important storage for the solar system. In this project, the batteries were used is a Lead Acid Energy Storage.

The design is based on electric motor specification that able to dehusk the coconut. All the data that need are being provided by the manufacturer recommendation sheet. In this project, the DC motor was used and the model of it in Simulink is shown in Figure 4.

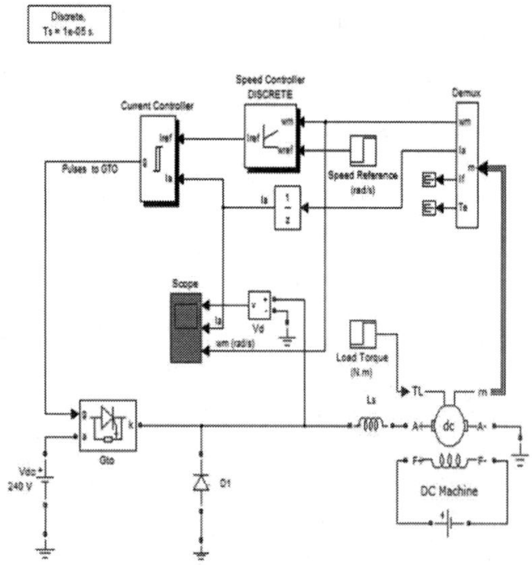

Fig. 4. A DC motor model in Simulink.

3.2 PVsyst Analysis of investment

PV framework size is composed of energy required to run the framework and what number of PV modules are expected to create it. A PV framework needs to create enough energy to cover the energy utilization of the losses (DC Motor) and energy utilized by the framework itself. The exact configuration of a PV framework is completed by utilising PVsyst, which figures the energy yield of the PV framework for a specific area.

The size and arrangement of sun based array is then streamlined keeping in mind the end goal to coordinate the energy yield of the framework to the energy utilization of the framework. The energy yield of a PV framework relies on upon the kind of PV modules, the attributes of a PV inverter, the introduction of the modules, and meteorological conditions.

Here, the solar PV is designed based on the power require to operate the 5.5 kW DC motor in order to achieve the torque for the dehusking the coconut. The system is optimised by using PVsyst simulation and the best power achieved is 6.9 kW.

Table 5. PVsyst parameters for coconut dehusking machine.

Specification	Description	Total Unit
PV module – Si poly	230 Wp / Unit	30
Battery – Lead Acid	24 V/ Unit	20
Inverter	3.5 kW	2
Operation Time	hour/ day	3 hours

From Table 5, 230 Wp Si poly PV module is chosen based on availability of supplier in Bintulu area. Batteries and inverters are acted as an auxiliary device to meet the 6.9 kW design power. Figure 5 shows one of the many window pages in the PVsyst setup before executing the project.

4 Results

The simulation shows motor voltage (V_a), current (I_a) and speed (ω_m) on the scope in Figure 5. The reference rate is expanded from 120 rad/s to 160 rad/s. The pace controller directs the velocity in roughly 0.25 s, and the normal current balances out at 6.6 A. Amid the transient period, the current is still restricted to be under 30 A. The load torque increased abruptly from 5 Nm to 25 Nm.

The PV Solar coconut dehusking machine is driven by the Direct Current (DC) motor and the previous designed which is developed by previous researcher is driven by alternate current (AC) motor. The overall performance is summarised in Table 6. The simulation conduct for the PV solar dehusking machine was able to turn the ABB 5.5 kW DC motor to produce a torque of 36 Nm at 1445 rpm. This would be enough torque if ns = 1 is used in equation 1. However, if the required torque is still expected to be 45 Nm, then a belt or gear system would be able to step up the torque.

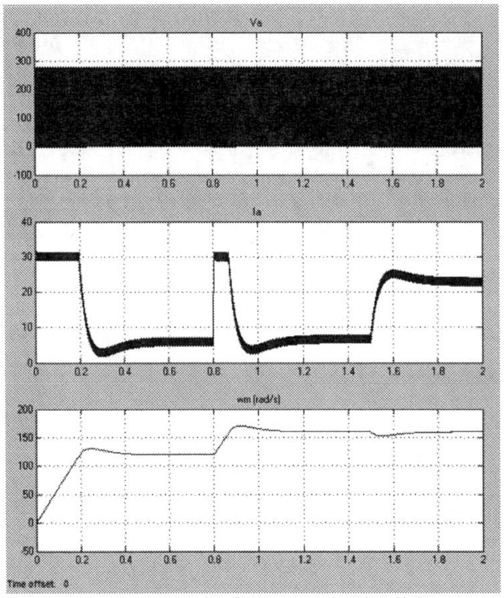

Fig. 5. : Simulation result of DC motor in Simulink.

Table 6. PV dehusking machines performance

Specification	New PV Drive
Energy source	DC
Power	5.5 kW
Torque	36.33 Nm
Speed of the Motor	1451 rpm

The PV Solar coconut dehusking machine is driven by the Direct Current (DC) motor and the previous designed which is developed by previous researcher is driven by alternate current (AC) motor. The overall performance is summarised in Table 6. The simulation conduct for the PV solar dehusking machine was able to achieve about 5.5 kW power at 1445 rpm. Still from Figure 5, the calculations can be shown as

$$P = IV = (24\,A)(230\,V) = 5.52\,kW \quad (3)$$

and

$$\omega = 152\,rps(\frac{30}{\pi}) = 1451\,rpm \quad (4)$$

Meanwhile, PVsyst was able to report the peak electricity produced each month throughout the year for the designated place, Bintulu. The total electricity produced was predicted to be 6867.6 kWh and at its peak was in December that recorded a power of 648.4 kW. The system simulation produced 6868 kWh annually with RM 51,516 as the initial capital or RM0.60/kWh. Here, a diesel engine coconut dehusking machine becomes the main reference for this simulation.

From Table 7, the initial investment is estimated at RM 51,516 for the PV Solar Dehusking Machine. However, as shown in Figures 6 and 7, the cost to produce energy is at RM 0.82 /kW and this means it could save 69.398 tons of CO_2 emission into the atmosphere over the period of 20 years. In a long term, the PV solar system is considered cheaper as its operational and maintenance cost is at RM 5,634 annually [12].

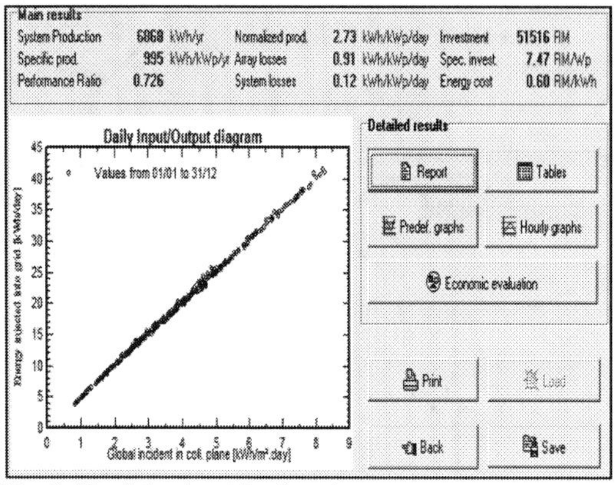

Fig. 6: Simulation input details and parameters.

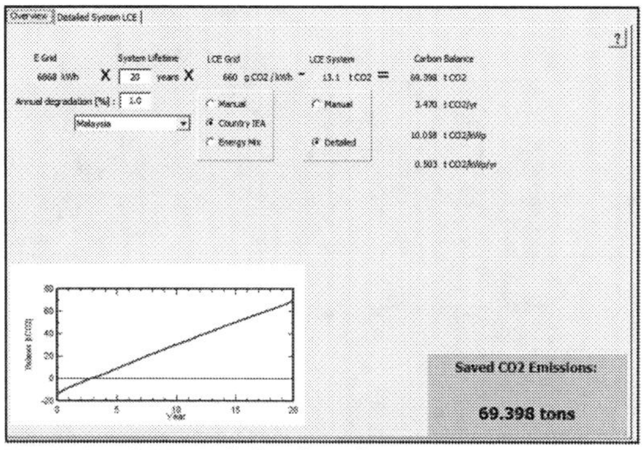

Fig. 7: Saved CO_2 Emissions by Solar PV System.

Table 7. Overall system cost.

Specification	Quantity	Unit Price (Rm)	Total Price (Rm)
Pv Module – Si Poly	30	1,300.00	39,000.00
Battery – Lead Acid	20	300.00	6,000.00
Inverter	2	800.00	1,600.00
Installation and Wiring	Lump Sum	2,000.00	2,000.00
GST	6%	2,916.00	2,916.00
TOTAL			**51,516.00**

The general scheme of this new PV dehusking machine can be seen in Figure 8. The main power should come from the PV module that generates DC current. The voltage can be regulated by the Zener diode system and the power can be supplied directly to the machine. The batteries can also be used to store energy. Alternatively, the machine may also get power from the grid through the use of an adaptor.

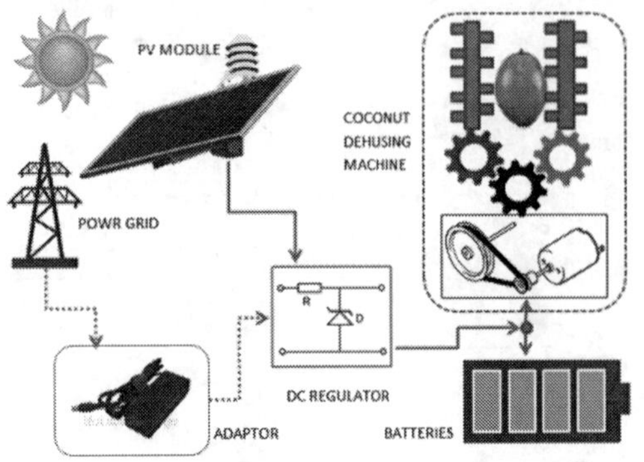

Fig. 8: The general scheme for the new dehusking machine.

Bonny in [12] suggested the machine that resembles much of the good aspects in Venkataraman et al. in [2]. This is evident in Figure 9.

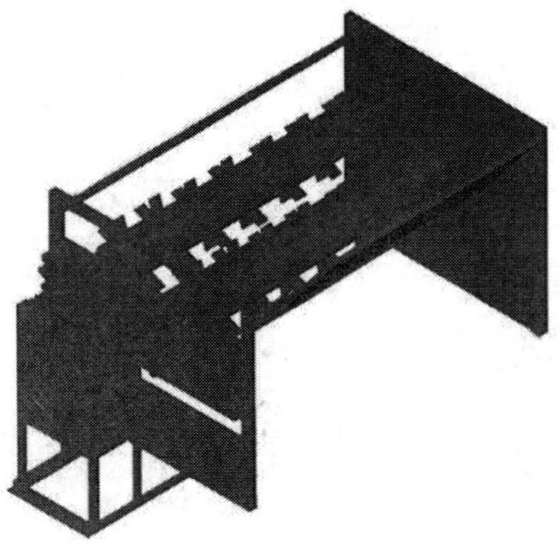

Fig. 9: Solar PV coconut dehusking machine model.

5 Conclusions

The computer simulation results showed that the PV Solar Dehusking Machine was capable to achieve at least 36 Nm of torque at about 1450 rpm which is required to dehusk some coconuts. Furthermore, it offers better cost of ownership in long term when compared to the on-grid coconut dehusking machine. It could give big advantages for the owner not to take care of the fluctuating fossil price which could yield higher operation cost and harm to environment. While this work has not been proven by experiments, the results serve as a guide for future work should a physical machine is to be assembled.

The authors would like to thank the Majlis Amanah Rakyat (MARA) and the Universiti Teknikal Malaysia Melaka (UTeM) for their support in this work.

References

1. A. Taufik, H. Akhir, J. Trop. Agric. and Fd. Sc, **42**, 2 183–190 (2014)
2. S. Venkataramanan, R. Abhinav, R. Rahul, Int. J. of Sciences: Basic and Appl. Res. **13**, 2 183–219 (2014)
3. N. Argaw, R. Foster, A. Ellis, Nat. Ren. Ener. Lab. NREL/SR-500-30361(2003)
4. T. Ackermann, L. Söder, Ren. and Sust. Ener. Rev. **6**, 1 67-127 (2002)
5. M.R.S Siti, M. Norizah, M. Syafrudin, Int. J. Chem. and Environ. Eng. **2**, 4 284 (2011)
6. S. Kamaruzzaman, H. Abdul-Rahman, C. Wang, S. Karim, T. Lee, Maejo Int. J. Sci. Technol. **6**, 2 196-215 (2012)
7. M. Albadia , R. Al Abria , M. Masoud, et al. Int. J. of Smart Grid and Clean Ener. **3**, 4 401-409 (2014)
8. B. Elhab, K. Sopian, S. Mat, et al. Sci. Res. Essays **7**, 42 3758-3765 (2012)
9. O. Idowu, O. Olarenwaju, O. Ifedayo, Int. J. of Ener. and Environ. Eng. **4**, 29 (2013)
10. U. Fumichika, H. Yousay, Y. Hitosi, *The seventh International Conference on Urban Climate* (2009)
11. A. Tyagi, *MATLAB and SIMULINK for Engineers* (2012)
12. B. Dominic Merikan, Master's report, Universiti Teknikal Malaysia Melaka (2016)

Raman Spectroscopy and XRD investigation on TiO_2 sol-gel dip coating thin films synthesizes with and without solvents

Shuhadah A. Yazid[1], *Zulkifli Mohd. Rosli*[1, *], *and Jariah Mohamad Juoi*[1], *Nur Dalilah Johari*[1]

[1]Faculty of Engineering Technology, Universiti Teknikal Malaysia Melaka, 76100 Durian Tunggal, Melaka, Malaysia.

Abstract. TiO_2 coating films were prepared by sol-gel dip coating method using TTiP precursors with and without ethanol as a medium to hydrolyze the HCl catalyst. The prepared samples of titanium dioxide (TiO_2) coating were dried at 110°C for 30 minutes then heated at 500°C and 600°C for 1h and 3h respectively to study effect of annealing temperature and soaking time. The crystallinity of the TiO_2 coating were identified by using X-ray Diffraction technique and crystallite size was calculated. Raman Spectroscopy also used as the confirmation of phases formation of TiO_2 coating films. Crystallite size of TiO_2 films were obtained at 12.35 nm, 17.29 nm for anatase ,21.71 nm, 28.95 nm for rutile and ~2.19 nm for brookite. Thus, the establishment of the desired phases, crystallinity and grain size of TiO_2 thin films sol-gel can be controlled and encouraging to explore as an effort toward producing a sustainable photocatalytic..

1 Introduction

TiO_2 is an appropriate material for industrial use at present and also perhaps in the future. Besides, TiO_2 materials are play an important role in helping solve many serious environmental and pollution challenges because of has the most efficient photocatalytic activity [1,2] the highest stability [3]and the lowest cost. There are commonly known have three polymorphs of TiO_2 found in nature such as rutile (tetragonal), anatase (tetragonal) and brookite (orthorhombic). TiO_2 can be prepared in the form of powder, crystals, or thin films. Liquid-phase processing or sol-gel technique is one of the many examples and the most conventional and utilized methods in synthesizing TiO_2 because it offers many advantages such as low temperature processing, easy coating of large surface, provides high surface homogeneity and most importantly has a low cost [4]

In the sol gel, there are four parameters are considered can give their effect on the structural and morphology of TiO_2 such as TiO_2 precursor, solvent, acid and water as a medium to hydrolyze of metal alkoxide as a TiO_2 precursor. In this study, solvent content in sol gel preparation is focused. Yu et al. [5] was studied about the hydrolysis of TTiP in pure water or the EtOH/H_2O mixed solution on the photocatalytic activity of nano-sized TiO_2 powders. The nature of solvent was reported can affect the rate of hydrolysis and condensation. Higher reaction rates of hydrolysis and condensation will lead to the formation of larger particles with rough surface films whereas lower reaction rate tend to form columnar particles with a smooth surface [6]. In the work of Edusi et. al [7], they used aerosol-assisted CVD technique to deposit TiO_2 phases and found that solvents like ethanol or propanol will lead to anatase phases while rutile phase favored when methanol was used. They concluded that, the use of different solvents can have a direct effect in controlling the TiO_2 phases when deposited as a film [7].

In the other hand, mixed anatase/rutile/brookite give an attractive in photocatalytic compare to pure anatase phase. For instance, thermal treatment at ~500 - 600°C of powder precipitated with a significant excess of water will formed the mixed anatase/brookite/rutile TiO_2 powders, with potentially attractive photocatalytic has been the subject of study Dorian et. al [8]. Therefore, the phase transformation was important to gain the best photocatalytic activity with the correlation ratio of EtOH/TTIP solution. Besides, annealing temperature also play an important role toward film structure, change in some properties and chemical composition. Ahmad et al. [9] stated that annealing temperature will affected the grain size. As the annealing temperature increase the grain size also improve the electron migration [9].

However, to our knowledge, the effect of solvent content can pose serious threat to the environment. Therefore, the element of solvent and annealing temperature becomes the studied subject in promoting TiO_2 phases and crystal size in supporting green process.

2 Methodology

2.1 Preparation of TiO_2 Sol

Titanium tetra isopropoxide ($C_{12}H_{28}O_4Ti$), Hydrochloric acid (HCl), Deionised water (H_2O) and Ethanol

* Corresponding author: zmr@utem.edu.my

(C_2H_5OH) were used to prepare the coating solution. All chemicals were of analytical grade and were used without further purification. Firstly, the deionised water was added with the half volume of solvent (ethanol) and stirred about 30 minutes as a first solution. Then, the HCl was added dropwise into the first solution and continue stir about 30 minutes again. Meanwhile, the second solution was prepared the precursor (TTIP) was added into another half volume of solvent (ethanol) and stirred about 30 minutes. After that, the second solution was added into the first solution as a TiO_2 sol. Stirring process was performed for about three hours. Different combinations of TiO_2 sol were prepared by altering the volume of ethanol in the TiO_2 solution for preparing TiO_2 thin films, chemical composition in volume TTIP: C_2H_5OH: H_2O: HCl shown in Table 1.

Table 1. Chemical composition used for prepared TiO_2 thin films

TTIP (ml)	EtOH (ml)	H_2O (ml)	HCl (ml)
2.0	0.0	32.0	0.4
2.0	16.0	32.0	0.4

2.2 Preparation of thin films coating

The TiO_2 thin films were immobilized on glass substrates. The cleaned glass substrate was dipped into the TiO_2 solution with the withdrawing rate was constant at 0.5mm/s and immersing time in 5s. After each coating TiO_2 films were dried at 110°C for 30 min. The dip-coating and drying process were repeated for ten times. The prepare film were heated at 500°C and 600°C for 1 h and 3 h at 5°C/ min respectively. The process flow of TiO_2 thin films preparation given in Figure 1.

2.3 Characterization of TiO_2 thin films

Phases of TiO_2 thin film were analyzed using X-ray diffraction technique of an X'Pert Pro model with Cu K-Alpha of 1.54060 A° and generator settings at 30 mA and 40 kV in the range of 10-90°. The grain size of TiO_2 films was calculated from XRD line broadening using Debye-Scherrer's formula given by equation:

$$D = \frac{0.9\lambda}{\beta_{\frac{1}{2}}\cos\theta} \quad (1)$$

Where D is the crystal size; λ is the wavelength of the X-ray radiation (λ=0.15406 nm) for CuKα; and β is the line width at half-maximum height. Affirmation of the TiO_2 phases were confirm with UniRAM-3500 Raman spectroscopy with laser excitation wavelength, λ= 532 nm (Nd:YAG).

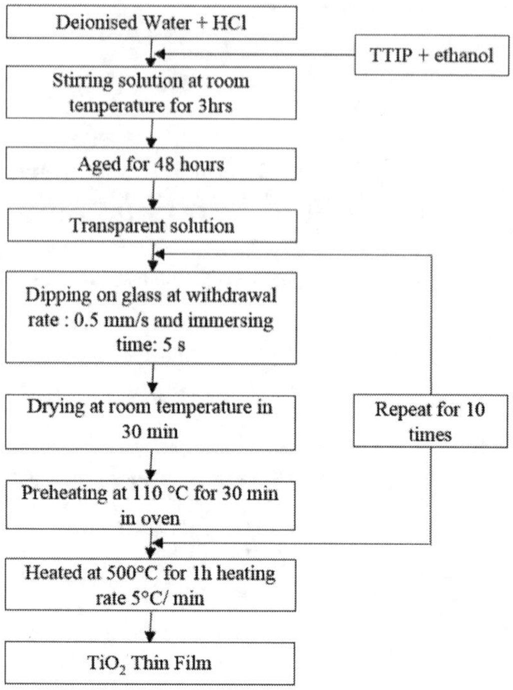

Fig. 1. Process flow for Preparing TiO_2 thin film.

3 Results and discussion

3.1 XRD analysis

Figure 2 show the XRD results of TiO_2 thin films from different ethanol content in TiO_2 solution after through heat treatment 500°C and 600°C for 1h and 3h at the final stage. All the films show anatase peak at θ=25° as main peak has (1 0 1) orientation except in Figure 2(b) at 600°C. However, the films of TiO_2 solution with ethanol (16 ml) produced mixed phase of anatase and brookite peak at θ=25° has (1 0 1, 2 1 0) orientation. At 600°C pure brookite will be observed at (2 1 0) for the 1h and 3h soaking temperature shown figure 2(b). Besides, rutile phase was growth as the brookite transformation at 600°C without the presence of ethanol. This is probably due to the fact that ethanol solvent suppresses the hydrolysis of Titanium alkoxide and rapid crystallization of the TiO_2 particles by adsorbing on the surfaces of TiO_2 thin film [10]. In the other hand, Dorian et al [8] state that presence of brookite likely to have enhanced the formation of rutile through a higher density of phase interfaces and consequently higher interfacial energy and also act as a nucleation site for rutile formation in anatase.

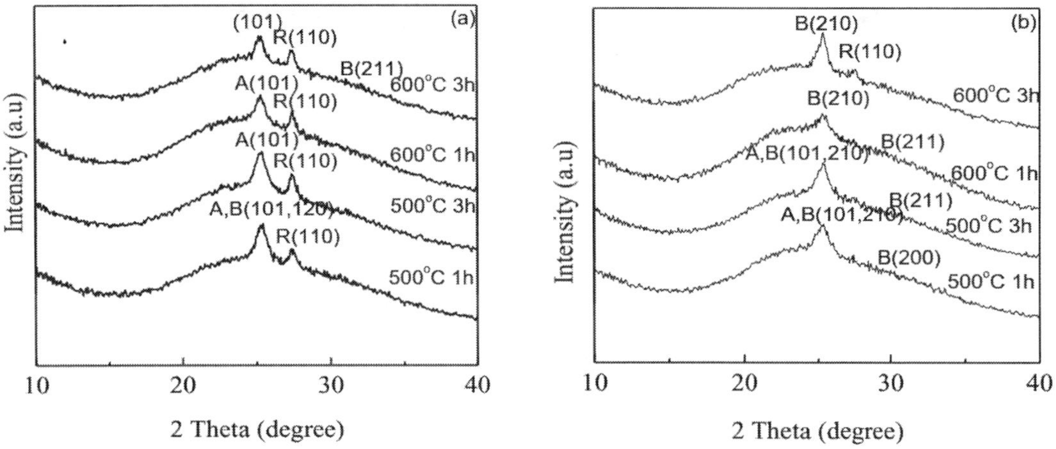

Fig. 2. XRD pattern of the deposited TiO_2 thin films versus ethanol concentrations (a) 0 ml and (b) 16 ml. A=anatase, R=rutile and B=brookite.

3.2 Crystallite Size

Figure 3 and Table 2 shown the crystallite size of the of the deposited TiO_2 thin films produced against ethanol content calculated using Debye-Scherrer's equation. At, 16 ml of ethanol brookite phase will be dominant at peak $\theta = 25°$ and 31° while anatase peak will be dominant without the presence of an ethanol at peak $\theta = 25°$. Rutile peak at $\theta = 27°$ will apppeared as the annealing temperature and soaking temperature increased. In the other hand, the pattern of the phase formation was also depend on the presence of the ethanol as a solvent. As the annealling temperature increased, the crystallite size of the deposited TiO_2 film also increased. Therefore, the effect of annealing temperatures on crystallites size was significant[12]. At 500 °C, mixed anatase-brookite and anatase at peak 25° form the crystallite size at 12.35 nm for both soaking time. Moreover, without presenting the ethanol, crystallite size was growth at 17. 29 nm. The same pattern was observed at peak 27° at 1 hour (21.71 nm) and 3 hour (24.81 nm) respectively. At 600°C, crystallite size of rutile increased at 28.95 nm then decreased to 24.80 nm due to the nucleation of brookite phase at $\theta = 31°$ with 3.66 nm. Besides, at 16 ml of ethanol mixed phase anatase-brookite were presence at peak 25° with 17.29 nm and decreased to 10.81 nm as increased the annealing temperature. Furthermore, soaking time also play an important role to increase the crystallite size and retarded the brookite phase at peak 31°. Then,rutile peak at 27° presence due to anatase to rutile phase transformation with the offset slower transformation kinetics in the absence of brookite [8]

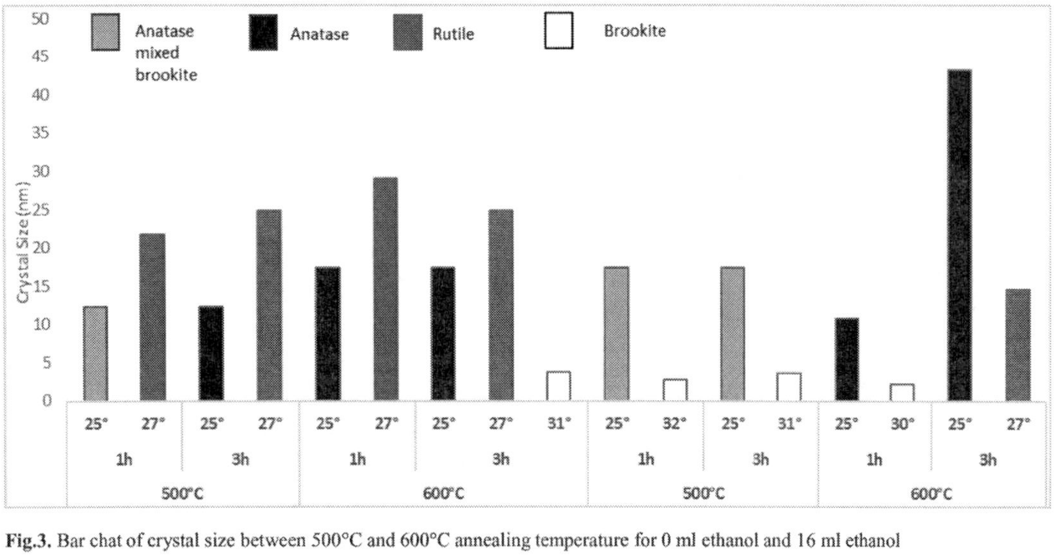

Fig.3. Bar chat of crystal size between 500°C and 600°C annealing temperature for 0 ml ethanol and 16 ml ethanol

Table 2. Crystal size of the deposited TiO_2 film (Anatase=JCPDS No: 01-071-1167, Rutile= JCPDS No: 01-072-4813 and Brookite= JCPDS No: 00-029-1360)

Solvent (ml)	Temp. (°C)	Soaking Time (h)	Postion (2 Theta)	d-spacing (Å)	Crystal size (nm)
0	500	1	25.3629	3.51175	12.35
			27.4091	3.25406	21.71
		3	25.2973	3.52071	12.35
			27.4605	3.24809	24.81
	600	1	25.2597	3.52587	17.29
			27.4344	3.25111	28.95
		3	25.2845	3.52247	17.29
			27.2661	3.27081	24.8
			31.8558	2.80925	3.66
16	500	1	25.3097	3.51901	17.29
			32.153	2.78397	2.74
		3	25.3821	3.50915	17.29
			31.0923	2.87648	3.65
	600	1	25.3741	3.51024	10.81
			30.5765	2.92382	2.19
		3	25.3497	3.51356	43.23
			27.5302	3.24002	14.48

3.3 Raman spectrum and analysis

Figure 4 shows the Raman spectrum of the deposited TiO_2 thin films. Overall, there are six characteristics of active bands for the Raman spectrum of anatase which are (A_{1g} + $2B_{1g}$ + $3E_g$). Brookite has thirty-six active bands of ($9A_{1g} + 9B_{1g} + 9B_{2g} + 9B_{3g}$) and rutile have four Raman actives bands of (A_{1g}+B_{1g} +B_{2g} +E_g)[11]. Thus, for all deposited TiO_2 samples, Raman bands appeared at 144, 396, 514, and 636 cm^{-1} and 142, 394, 514 and 638 $cm^{-1}$are very distinctive and can be assigned to anatase phase shown in figure 4(a) and 4(b). For the 0 ml of ethanol, the Raman bands appeared at 442 cm^{-1} can be appointed to rutile phase but then slowly disappeared as the ethanol content was added 16 ml shown in figure 4(a) and (b) respectively. There are also others Raman bands noticeable at 248 and 319 cm^{-1} for the 0 ml while 16 ml of ethanol Raman bands was noticeable at 240 and 322 cm^{-1} which can be assigned to the brookite phases. As an increased the annealling temperature and soaking time rutile phase was clearly observed due to the thermal treatment [9].

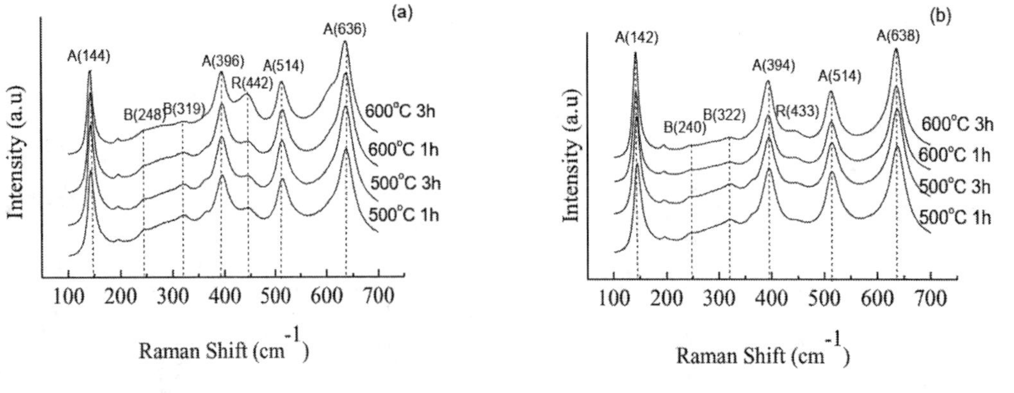

Fig. 4. Raman spectrum of the deposited TiO_2 thin films versus ethanol concentrations (a) 0 ml and (b) 16 ml. A=anatase, R=rutile and B=brookite.

4 Conclusions

TiO_2 thin films were successfully deposited on glass slide substrate from the ethanol content of 0 ml and 16 ml via sol gel dip coating technique. Anatase was the main phases appeared throughout all TiO_2 thin films without the presence of ethanol while pure brookite and mixed phase anatase-brookite will appearred in the presence of 16 ml of ethanol. Rutile phases were found to presence with the annealling temperature. The pattern of the crystal size of all phases increased as the annealling temperature increased from 500°C to 600°C. These observations lead to a conclusion that without the used of ethanol as a function of annealing temperature, desired TiO_2 thin film can be controlled and encouraging to explore as an effort toward producing a sustainable photocatalytic activity.

The authors would like to thank the financial support given by the Ministry of Higher Education Malaysia and Universiti Teknikal Malaysia Melaka (UTeM) through Grant FRGS/1/2016/TK05/FKP-AMC/F00319 and all SM4GT group members under Advanced Manufacturing Centre for the support given throughout this research

References

1. N.Mok, *Diss.*UMP(2009)
2. M.Jose, J. Kumar, and S. Shukla. Nano Bull., **2**.1 (2013)
3. D.L Marchisio, F. Omegna, A.A. Barresi, and P. Bowen, Ind. & Eng. Chem. Research, **47**,19(2008)
4. N. Fagnern, R. Leotphayakkarat, C. Chawengkijwanich, M. P. Gleeson, N. Koonsaeng & S. Sanguanruang, J. Phys. Chem. Solids, **73**,12 (2012)
5. J.C Yu, L. Zhang,H.Wo, Jour.of Photochem. & Photobiology,**148**,1 (2002)
6. L. Hu, T.Yoko, H. Kozuka & S. Sakka, Thin Solid Films (1992).
7. C. Edusi, G. Sankar & I. P. Parkin,.Chem. Vap. Depos. **18** (2012)
8. H.N. Dorian, G.Tirani, C.C. Sorell, Adv. in App.Cer. **111**,3 (2012)
9. M. K. Ahmad, N. A. Rasheida, A. Zain Ahmedc, S. Abdullah, & M. Rusop,Jour. Phys. Mas. , **29**,3(2008)
10. S. Ito,S. Inoue, H. Kawada, Jour.of coll. Int. Sci., **216**,1(1999)
11. M. Rezaee, S. M. Mousavi Khoie & K. H. Liu, Cryst.Eng.Comm. **13** (2011)
12. F. Sayilkan, Y. Önal, M. Akarsu, Turk J Chem **29** (2005)

Effect of starch sizes particle as binder on short pineapple leaf fiber composite mechanical properties

Mohd Zulkefli Selamat[1,*], *Muhammad Syazwan Zhafri* Tahir[1], *Ayu Natasya* Kasim[1], *Sivakumar* Dharmalingam[1], *Azma* Putra[1], *Mohd Yuhazri* Yaakob[1] and *Mohd Ahadlin* Mohd Daud[1]

[1]Centre of Advanced Research on Energy (CARe), Faculty of Mechanical Engineering, Universiti Teknikal Malaysia Melaka, Hang Tuah Jaya, 76100 Durian Tunggal, Melaka, Malaysia.

Abstract. Pineapple leaf fiber (PLF) is one of the natural fibers that abundantly can be found in Malaysia, but the usage of the pineapple plant is limited only on their fruit and the other parts to be a waste. In this study, PLF is used as the reinforcement material and starch (SH) used as the matrix or binder. Both materials were combined with several compositions ratio (weight percentage) of PLF/SH composites which are 50PLF/50SH, 60PLF/40SH and 70PLF/30SH. Before undergo the fabrication process, the fiber has gone through an alkaline treatment to increase the strength of the fiber and chopped with an approximate size range from 0.5 mm to 5 mm. Besides that, SH powder is sieved to gain several particulate sizes which are 75 µm, 100 µm and 250 µm. The related tests such as flexural, hardness, density tests and macrostructure analysis have been done to determine their mechanical properties of composite. Based on the results, the sample with composition of 70PL/30SH with 75 µm has shown the highest result for flexural stress which is 14.49 MPa. While, the composite with the same composition of 70PLF/30SH with particulate size SH of 250 µm has shown the highest result in the hardness of 67 Shore-D and density of 1.36 g/cm3 respectively.

1 Introduction

A composite is a material that combined from two or more materials to produce a new material that has a new or improved ability from its original individual components [1-2]. Basically most of the materials that exist or we see is made of a composite materials. For examples, bones, wood, and stone are natural composite items that develop by a natural process. The leaf itself consist of natural fibre and usually use in making a new composite materials because of their unique characteristics which have good mechanical properties, stronger, lighter, biodegradable and less expensive compared to the synthetic fibre. Because of that, natural fibres have huge potential to be an alternative to synthetic fibres such as the glass fibre and carbon fibre [3-4].

There are a few examples of natural fibres that can be extracted from a plant such as pineapple leaf, palm leaf, banana leaf, hemp, kenaf, bamboo and coconut shell fibre [4]. From all natural fibres, pineapple leaf fibre (PLF) seems to have the highest cellulose content which makes the fibres can produce good mechanical properties. In order to give the unique ability for the natural fibres, binder/matrix such as the starch, epoxy and polypropylene are added to enhance the existing mechanical properties or called as a reinforcement of a material. In other words, reinforcement is strengthening the structure or material itself. For an example, back in the ancient years, clay was reinforced by straw to build walls. In this case, clay will become the binder holding the straw together thus; make the construction become stronger [5-6].

As the previous studies shown that fibre-reinforced plastic (FRP) is a very well-known composite that being used in structure engineering like mostly in the field of aerospace, buildings and also offshore platforms as they are considered to have more strength, non-corrosive, light in weight and are easily moulded or constructed. But the materials or fibres are usually from the glass and carbon combining with the plastic polymer as the binder [7]. While in this study, PLF will be used as the reinforce material and starch as the binder or the matrix which may potentially give a good result in mechanical properties besides it characteristics which is an environmental friendly, renewable, recyclable and biodegradable.

Nowadays, the implementations of natural fibres are still in research phase. Some of the problems arise is that synthetic fibres is widely use are hardly to decompose and are not sufficiently eco-friendly. Using natural fibres will allow many environmental problems to be solved. The combination of natural fibres such as pineapple leaf, kenaf, hemp and jute fibres with certain type of binder/matrix to produce a composite materials seems can be compete with existing synthetic composites which they have a good mechanical properties [8-10]. Recently, natural fibres have been

*Corresponding author: zulkeflis@utem.edu.my

awakening the industry to substitute their products by using the natural composite has been widely use in the automotive industry. Thus, because by using the natural fibres they can produce a higher strength of automotive interior components but cheaper in price [11-14]. Biodegradable composite can be defined as composite material or the combination between matrix and reinforcement of natural fibres. These kinds of materials have their own function which keep strengthening the properties of the matrix that was used, but always providing biodegradable compatibility. The constituents inside the bio-degradable composite consist of reinforcement from natural fibre and matrix from natural polymer. Bio-degradable composite totally comply with the green composite concept where all the material are from natural source [15-17]. Bio-degradable composite become prominent product since it is eco-friendly and safe too human application. It is also known as composite material made from natural fibre and polymer matrix derived from petroleum such as polylactic acid (PLA), cellulose and thermoplastic starch [18-20] (Sahari & Sapuan, 2011). Thus, the matrix part of the biodegradable composite can be classified into two parts which are fully degradable and partly degradable as shown in Figure 1.

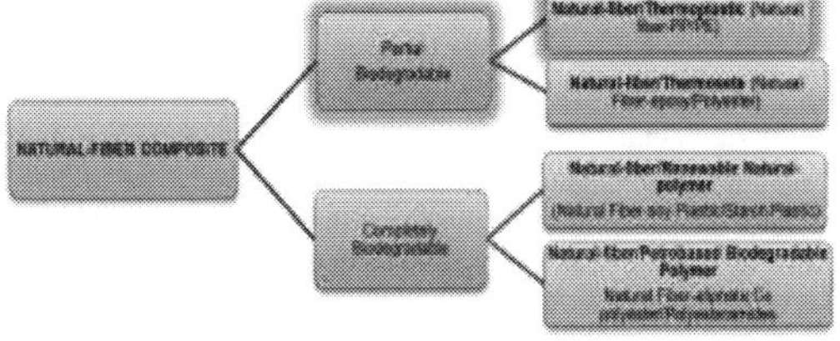

Fig.1. The classification of biodegradable composite [21]

In this project, the pineapple leaf fibre (PLF) is chopped with the range of size approximately from 0.5 up to 5 mm was used as the reinforce material as the fiber orientation as if the composite material will be the random-discontinuous, while starch as the binder and the difference particulate size of starch range from 75, 100 and 250 µm are used to produce the PLF/SH composite. The composite will be tested to determine the mechanical properties of the material. The related tests done will be based on the American Standard Testing Method (ASTM) and by referring the previous studies. The tests that will be covered or done are the hardness test, density test, fractured strength test and macrostructure analysis. The composite seems to have a good potential that can be widely use in the industry like for example for the plastic industries product that can provide more benefit to the environment.

2 Methodology

This research is studies the effects PLF loading on the mechanical properties of PLF/SH composite. The various ratio of PLF/SH composite were been selected and the ratio of composition were fixed as shown in Table 1. The materials used in the fabrication of FRP composites are PLF and SH. Figure 2 (a) and (b) shows the image of PLF and ST. Meanwhile, Table-2 shows the specific mechanical properties of various PLF types. An alkaline treatment has be conducted to extract the pineapple leaf bundles and enhance the pineapple leaf properties. After that the PLF is chopped into smaller size with the approximate in the range size of 0.5 up to 5 mm and starch is sieved to gain several sizes of particulate from which are 75, 100 and 250 µm before the formation process of PLF/SH composite by using hot press. The mechanical properties of pineapple leaf-starch composite will be determined by using flexure test, hardness test, density measurement and macrostructure analysis.

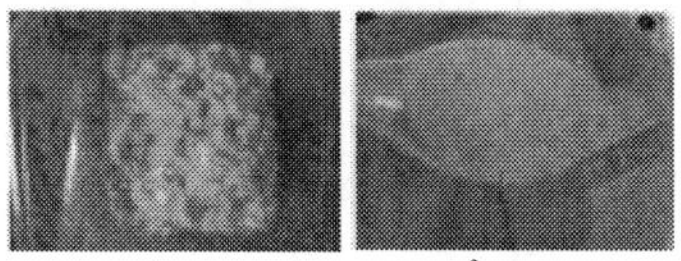

c b

Fig.2. (a) Chopped PLF and (b) SH powder

Table 1. Composition of PLF/SH composite

PLF (wt%)	SH size particle (µm)	SH (wt%)
50	75	50
	100	
	250	
60	75	40
	100	
	250	
70	75	30
	100	
	250	

Table 2. Types of pineapple cultivars and physical properties [13]

Properties	Cultivar		
	Moris Gajah	Josapine	Sarawak
Average no. of leaves	50	65-70	65-70
Length of leaves (mm)	63	61	70
Width of leaves (mm)	61	46	65
No. Of fiber bundles per leaf	90	80	> 100
Diameter (µm)	120-440	105-300	170-340

3 Results And Discussion

3.1 Effect of particulate size of SH in PLF/SH composite with different composition on Flexure Stress

The flexural properties of the composites material are shown in Figure 3. The results shows based on the different composition from 50PLF/50SH, 60PLF/40SH to 70PLF/30SH with each composition contain of different sizes of particulate size of SH which are from 75 µm, 100 µm and 250 µm. Based on previous study, the higher the fiber content in a composite material, the stronger the mechanical properties will be [6].

Figure 3 shows the comparison for all nine samples of the composite materials. Based on the loading of PLF, the flexural stress is decreased with the increment of SH particle size for all composition. Meanwhile in general, the trend was found that the flexural stresses are gradually increased as the composition of the PLF content was increased. This can be compared by same particulate size of the starch, SH that 50% content of PLF in the 75μm is only gained 8.17 MPa. The result increasing as the PLF content rose to 60% and the flexural stress shows result of 11.69 MPa. The results substantially increased to 14.49 MPa as the PLF content keep rising until 70%. The results obtained from this experiment are compatible with the results that had been found by Selamat (2016) which used short PLF with the length of 2 cm, 4 cm and 6 cm with a matrix of SH. The results of the findings show that the higher the fiber content also will gradually increase the value of the flexural stress.

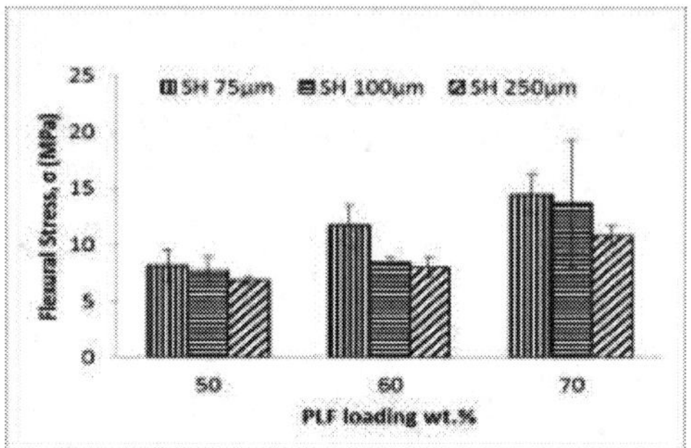

Fig. 3. The Flexural Stress with various PLF loading and size of particle

Instate of composition of the fiber and binder/matrix, this research also highlight about the different sizes of the binder in the composite material whether it will gives a significance results on the mechanical properties or not. Based on Figure 3, it shows that the finest size of binder will gives the higher value in flexural stress of the composites material itself. This is due to the ability of the binder as a thermoplastic to react and melted with heat faster as it being pressed makes it hold to the fiber firmly and thus make the mechanical properties went stronger [22]. This experiment is repeated a few times and the average results is then been calculated. This is to obtain more accurate results as the sample might have some discontinuities during to the fabrication process as it shows random error bars in the results. This is due to the smaller particle SH sizes that will lead to be more even dispersion, thus make the flexure stress to be stronger.

3.2 Effect of particulate size of SH in PLF/SH composite with different composition on Hardness

Based on Figure 4, the graph shows the results of the hardness (Shore-D) with the various percentage of PLF content and the sizes of particulate SH respectively. The results show that the hardness increases with the increment of the fiber loading and the particulate size of the binder materials for all the samples. The experiment had been repeated at several points to more accurate results.

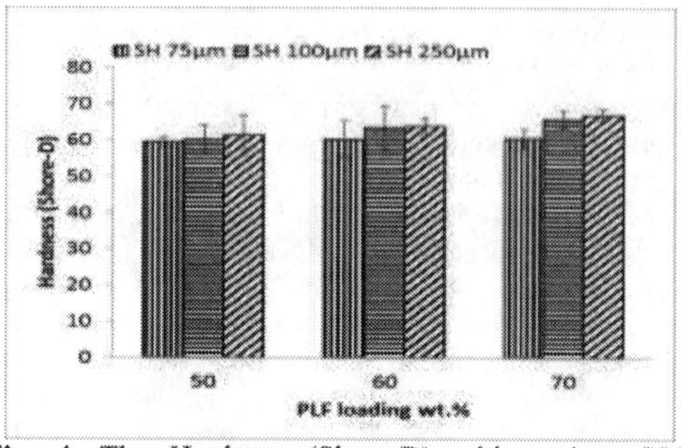

Fig. 4. The Hardness (Shore-D) with various PLF loading and size of particles

The result obtained for the 75 μm shows a small increment of 1.34% in results for the composition of 50PLF/50SH to 60PLF/40SH and increasing 5.79% for the 70PLF/30SH composition of the composite material. Since the reinforcement material is in the random discontinuous orientation, it leads to uneven dispersion of the matrix. In other words, the higher the matrix content it makes it harder for the PLF to mix well with the SH. Hence, make the sample with the larger content of PLF give the highest value of hardness compare to the lower amount of PLF composition of the composite material [23].

Meanwhile, at the composition of 60PLF/40SH with the particulate size of 100 μm and 250 μm show a very small difference of hardness which is 63.6 and 63.8 in values. For fiber loading, it can be seen clearly that, with the highest content of fiber loading 70%, has shown the highest value of 67 of hardness. The declination of the results may due to the unsmooth surface or the matrix and the fiber are not properly mix during the fabrication process and make the material does not melt properly. Thus, the sample cannot withstand the wear resistance of the composites. This is the reason why the experiment has to be repeated for a several times as the error bars showed on the graph indicates the ability of the maximum and minimum hardness value for the composite material to withstand any wear resistance. In a research by Selamat and Mashitah (2015) the materials used are PLF for the reinforcement material and SH as the binder. The results shows a abreast result with the results obtained from this experiment which proved that the higher PLF content will slightly rose the results of the hardness test.

3.3 Effect of particulate size of SH in PLF/SH composite with different composition on density

Based on the data of Density (g/cm3) vs. PLF Loading (%) as shown in Figure 4, the lowest density is in 50% fiber loading with 250 μm size of SH. The bar chart on the 50% fiber loading shows a same value of density at 75 μm and 100 μm which is 1.292 g/cm3 but fall dramatically when the size of SH increased to 250 μm with a density value of 1.166 g/cm3.

Figure 5 also shows that the increasing in fiber loading and the size of matrix will increase the density measurement of the samples. This can be seen clearly at the composition of 60% and 70% which shown the data collected to be slightly grown as the size of SH increased. This can be concluded that the density of the composite material is directly proportional to the fiber loading and size of the matrix materials, in this case is starch. In this research it is proved that the percentage of fiber loading and the size of the matrix will affect the density measurement of the composite material.

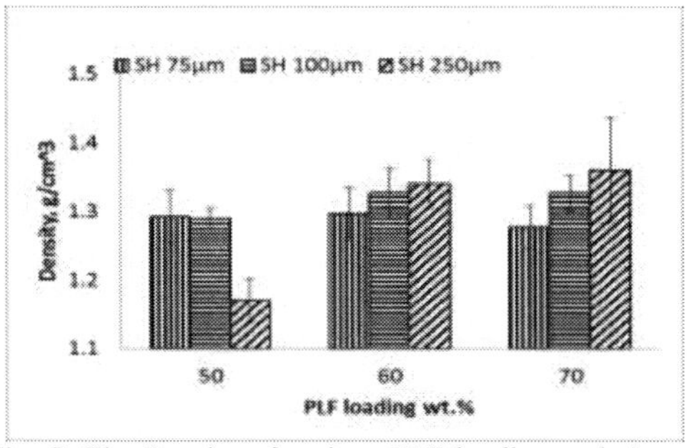

Fig. 5. The Density of various PLF loading and size of particles

In this case, the highest density is found at the 70PLF/30SH composition is because of the ability of the starch to fully melt as it is in fewer composition. This make the SH is fully covered the PLF, thus make it less water absorption through the samples. To make sure the results is acceptable, the test is repeated with three different specimens for each samples and the average data is taken out as the comparison among all other sample with different composition and particulate size of SH.

The error bar data shows that each of samples with different content of fiber loading can be rises and drop up to the indicated values of the error. If the value is above or less that the indicated standard deviations, it means that the results is not acceptable and the experiment is failed. To avoid this, the experiment can be repeated with more specimen and make sure all the dimensions of the specimens is the same as it may affect the result and the absorption of water if they have a different dimensions.

3.4 Macrostructure Analysis

Based on Figure 6 it shows all the macrostructure analysis for all the composition based on the percentage of fiber content and also the size of the matrix materials. The sample that with higher composition of SH tends to have voids compare to the one who has less SH content. This is may cause from the higher level of matrix make them hardly to be mix properly and the probability for the SH to accumulate at a certain spot is much higher. Thus, make the SH more difficult to melt and this make the bonding is weaker between the fiber and matrix. This result is contradicted with the composition of 70PLF/30SH which previews a very good surface of composite materials with no voids at all.

Fig. 6. Microstructure of PLF/SH composites

Composition	Microstructure		
	Size of particles (μm)		
	75	100	250
50PLF/50SH			
60PLF/40SH			
70PLF/30SH			

○ PLF mixed with PLF
○ SH accumulates at certain sport
○ Voids
○ Impurity

For the size of particulate SH, it seems that the finer size will contribute a much better surface compare to those with larger size of particulate SH. The sample that fabricated from 75 μm of SH shows a good adhesion between chopped PLF and SH with a clean surface and minor presence of impurities compared to 100 μm and 250 μm. This shows that the finest size of SH is easier to be mix with the chopped PLF and will be melt evenly making a stronger bonding between the fiber and also the matrix material [8, 24]. Thus, this contributed a big part in affecting the mechanical properties of the composite material.

3. Conclusion

This research is basically the studies on how to produce a natural composite which is formed by a combination of natural fiber and natural polymer. This green composite material basically can be used in the plastic industries. Pineapple leaf fiber is used as the reinforcement material while starch is used as the matrix or binder. The main objective of this research is to study the effect of pineapple leaf fiber with the difference particulate sizes of starch on the composite material mechanical properties. The expected result that may be obtained from this project is the finer size of the particulate matrix may have a greater value in the mechanical properties compared to the larger size of the particle. From the results obtained in this research, the mechanical properties have showed that these biodegradable materials can be useful in the production of composite material. Based on the flexural test, it can be concluded that the higher percentage of fiber content which is 70% will give the highest value of 14.49 MPa of flexural stress with the finest size of SH. This result is in line with the expected result of this research. In the hardness test also shows an interesting results, where a sample with composition of 70PLF/30SH give the greatest hardness with value of 67 Shore-D. Based on these findings, it can be concluded that the fiber loading will affect the mechanical properties of the composites material itself. Besides that, it can be conclude that the finest size of the matrix will give an advantage base on the ability to be mixed properly with the fiber and also easier to be melt in the fabrication process. Thus, this also gives a hint that the finer size of matrix will be a good potential in producing a higher performance of the composite material of PLF/SH.

Acknowledgement

The authors would like to thank the Malaysia Ministry of Higher Education, Malaysia Ministry of Science, Technology and Innovation for sponsoring this work under Grant FRGS/2/2014/SG06/FKM/02/F00237 and Advanced Material Research Group (A-MAT), Faculty Mechanical Engineering, University Teknikal Malaysia Melaka (UTeM) for financially sponsoring, facilities and gratefully knowledge during this research.

References

1. Siva N, Paul J. J, Fabrication and Experimental Analysis of Pineapple Leaf Fiber Based Composite Material, 243-246. (2015)
2. 5980 Floor Model Systems for High-Capacity Universal Testing. (n.d.). Retrieved December 11,(2015)
3. Liu W, Misra M, Askeland P, Drzal L. T, Mohanty A. K,.'Green' Composites from Soy Based Plastic and Pineapple Leaf Fiber: Fabrication and Properties Evaluation, 1-3 (2005)
4. R.Jumaidin, S.M. Sapuan, M. Jawaid, M.R. Ishak and J.Sahari. Characteristic of thermoplastic suger plam Starch/Agra blend : Thermai, tensile and physical properties. International Journal of Biologcal Macromolecules, 89, 575-581 (2016)
5. Drzal L. T, Mohanty A. K, Misra M,. Bio-Composite Materials as Alternatives to Petroleum-Based Composites for Automotive Applications, 1-7 (2009)
6. Mohd Zulkefli Selamat, Mashita Razi, Ayu Natasya Kasim, Sivakumar Dharmalingam, Azma Putra, Mohd Yuhazri Yaakob and Mohd Ahadlin Mohd Daud. Mechanical Properties of Starch Composite Reinforced by Pineapple Leaf Fibre (PLF) from Josapine Cultivar, ARPN Journal of Engineering and Applied Sciences, 9783-9788 (2016)
7. Wambua, P., Ivens, J. and Verpoest, I.. "Natural fibers: Can they replace glass in fiber reinforced plastics?", Composites Science and Technology **23**,1259-1264. (2003)
8. Masuelli M. A, (2013). "Introduction of Fibre-Reinforced Polymers – Polymers and Composites: Concepts, Properties and Processes", 14 Oct, (2015)
9. Arib, R., Sapuan, S., Ahmad, M., Paridah, M., & Zaman, H.. "Mechanical properties of pineapple leaf fibre reinforced polypropylene composites". Materials & Design, 391-396 (2004)
10. Selamat MZ, Sahari J, Muhamad N, Muchtar A. THE EFFECTS OF THICKNESS REDUCTION AND PARTICLE SIZES ON THE PROPERTIES GRAPHITE – POLYPROPYLENE COMPOSITE. International Journal of Mechanical and Materials Engineering (IJMME), Vol.**6** (2011), No.2, 194-200
11. Bledzki AK, Gassan J. "Composites reinforced with cellulose based fibers". *Prog Polym Sci;* **24(2)**, 221–74. (1999)
12. Dhal, J., & Mishra, S. "Processing and Properties of Natural Fiber-Reinforced Polymer Composite". Journal of Materials, 1-6 (2012)
13. Dagade P.C. Shaikh S. M, "Fabrication of Composite Materials by using Short Pineapple Leaf Fiber (PALF)", IJIERT Journal, **2** (2015)
14. Nabilah A.. "Effect of Pineapple Leaf Fiber Loading on the Properties of Pineapple Leaf Fiber-Polypropylene Composite", Thesis UTeM, 44-51(2014)
15. Mohamed A.R., Sapuan S.M, Shahjahan M, Khalina A, "Characterization of pineapple leaf fibers from selected Malaysian cultivars". (2008)
16. Kasim, A.N, Selamat, M.Z, Aznan, N., Sahadan, S., Daud, M.A.M, Salleh, S. & Jumaidin, R. Effect of pineapple leaf fiber loading on the properties of pineapple leaf fiber–polypropylene composite. Jurnal Teknoloy. **21**, 117-123 (2015)
17. Kasim, A.N., Selamat, M.Z., Daud, M.A.M., Yaakob, M.Y., Putra, A. & Sivakumar, D.. Mechanical properties of polypropylene composites reinforced with alkaline treated pineapple leaf fibre from Josapine cultivar. International Journal of Automotive and Mechanical Engineering. **13**, 3157-3167 (2016)
18. Selamat, M.Z., Kasim, A.N., Shamsudin, S.A., Mohd Daud, M.A. & Dhar, S. Effect of Bamboo Fibre Length on The Mechanical Properties of Bamboo Fibre/Polypropylene Composite. 8th MUCET (2014)
19. Selamat, M.Z., Razi, M., Kasim, A.N., Sivakumar, D.M., Azma, P., Daud, M.AM. & Yuhazri, Y. Mechanical properties of starch composite reinforced by pineapple leaf fiber (PLF) from josapine cultivar. ARPN Journal of Engineering and Applied Sciences. 11(**16**) 9783-9788 (2016)
20. Mohamed, A.R., Sapuan, S.M., & Khalina, A. Selected properties of hand-laid and compression molded vinyl ester and pineapple leaf fiber (PALF) reinforced vinyl ester composites. International Journal of Mechanical and Materials Engineering. 5(**1**), 68-73 (2010)
21. Abdul Khalil, H.P.S., Bhat, I.U.H., Jawaid, M., Zaidon, D.Hermawan, A., Hadi. Y.S. 2012. Bamboo Fiber Reinforced Biocomposites: A review. Materials and Design. **42**, 353-368 (2012)
22. Kalapakdee, A. & Amornsakchai, T. 2014. Mechanical properties of preferentially aligned shortpineapple leaf fiber reinforced thermoplastic elastomer: Effects of fiber content and matrix orientation. Polymer Testing **37**, 36-44.
23. George, J., Bhagawan, S. S., Prabhakaran, N. S. & Thomas, S. Short pineapple-leaf-fiber-reinforced low-density polyethylene composites. Journal of Applied Polymer Science. **57**, 843-854 (1995)
24. Luo, S. & Netravali, N. Mechanical and thermal properties of environment-friendly ''green'' composites made from pineapple leaf fibres and poly (hydroxybutyrate-co-valerate) resin. Polym Compos. **20**, 367–78 (1999)

Fabrication and Machining Performance of Powder Compacted Alumina Based Cutting Tool

Hadzley Abu Bakar[1*], *Naim* Fahmi[1], *Faiz* Mokhtar[1], *Norfauzi* Tamin[2], *Umar* Azlan[2], *Abdul Aziz* Adam[1], *Raja* Izamshah[1] and *Shahir* Kasim[1]

[1]Faculty of Manufacturing Engineering, Universiti Teknikal Malaysia Melaka (UTeM), Hang Tuah Jaya, 76100 Durian Tunggal, Melaka, Malaysia
[2]Faculty of Engineering Technology, Universiti Teknikal Malaysia Melaka (UTeM), Hang Tuah Jaya, 76100 Durian Tunggal, Melaka, Malaysia

Abstract. This study focuses on the fabrication of alumina based cutting tool for machining application. Specific weight of alumina powders were pressed in a mold using hydraulic press at 8 tons before compressed through Cold Isostatic Press (CIP) at 30000 psi. Green body then was dried and sintered at 1700 °C at 4 hours sintering time to form solid cutting tools. These cutting tools were then tested in turning operation to machine AISI 1045 with different cutting speeds. The results shows that CIP pressed alumina cutting tool obtained the hardness of 83.2 HRA which is considered adequate to machine AISI 1045 steel. The shrinkage of alumina powders recorded about 7% form green compact to sintered body. In terms of flank wear, the alumina cutting tool demonstrated decreased wear rate as the cutting speed increased from 150 m/min to 225 m/min. Wear area focused at the edge of cutting tool due to small nose radius with the minimum wear rate recorded at 0.0025 mm/s for 225 m/min cutting speed. The newly fabricated cutting tools can be improved if finer or secondary reinforced particles were used.

1.0 Introduction

Ceramic cutting tool is a material produced by the combination of ceramic powders that are compacted and applied at machining operations. There are many example of ceramic cutting tool such as silicon nitride, silicon carbide, sialon and titanium carbide [1]. Ceramic cutting tools have been widely used in machining hard material due to its excellent properties especially in high temperature and high speed machining. This cutting tool tool can be fabricated by the powder metallurgy processes. Specific composition of ceramic powders can be compacted with mixture and milling with other reinforced materials to create refractory body for high performance in tribology applications [2].

There are many technique to process ceramic powders such as slip casting, ball milllling and slurry mixing. Another secondary processes that involve are pressing, sintering grinding or polishing. The efficiency of ceramic cutting tools depended on many factors. Among dominant one are characteristics of powders that used as raw materials, additive materials, sintering temperature and the geometry of cutting tools. In order to evaluate the performance of ceramic cutting tool, several criteria can be employed such as hardness, density, wear performance and surface roughness of the machined surface [3].

Wear performance of ceramic compact is very important in order to run prolong machining operation. The major causes of tool wear are mechanical, thermal, chemical and abrasion [4]. Most of the cutting processes in machining are done by the nose radius of the inserts. The stability of nose radius during contact action with the workpiece material rely on the structure and bonding strength between composition of particles that integrated within cutting tools. Strong bonding between particles would lead to higher tribological wear resistance. In addition, the integrated particles should also have high hot hardness to retain its stability at elevated temperature [5].

Among many ceramic cutting tools that available in industry, alumina based materials are among the most popular choices to be used as cutting tools. Alumina based cutting tool reportedly has high resistance to heat, abrasion and chemical inertness against the enviroment and the workpiece [6]. In the application of alumina as a cutting tool, important criteria such as microstucture, grain size, grain uniform and reinforced particle are very important. Fine and uniform grain size would provide high resistance to wear when this powder compact engages with other materials [7].

Due to recent development and advancement in ceramic technology, some better quality of alumina powders have been discovered and reportedly have excellent performance in machining [8]. Such ceramic powders are prepared from various type of processes in order to enhance the properties so that the powders can be manufactured into excellent products. Therefore, it is a very good oppurtunity to fabricate a cutting tools from these ceramic powders and study their potential in

* Corresponding author: hadzley@utem.edu.my

machining. This paper presents the fabrication of ceramic cutting tool based on the specific composition of alumina powder. As the ceramic powders were compacted and sintered, the performance of ceramic cutting tool was evaluated based on wear performance and surface roughness of the machined surface. Selected cutting tools were used to machine AISI 1045 by the CNC machine at various cutting parameters. The evaluation of this research was examined using stereo microscope. This project also presented the failure mode of the newly fabricated cutting tool to highlight the possible weaknesses from the manufacturing processes. In the end, this result will be used to propose some improvement or refinement for the cutting tool development in the future.

2.0 Methodology

2.1. Fabrication of ceramic insert

Specific amount of 4g alumina was prepared and shaping process was done in a mould by pressing using hydraulic hand press with pressure of 8 tons. The pressure was increased by 2 tonns and rested for 1 minutes during the compaction process. After the pressure reached 8 tonns, the green body was ready to be ejected from the mould. Then, the green body was pressed using cold isostatic press (CIP) with 30000 psi pressure. After the process completed, the compacted green body was taken out from the chamber. Series of sintering were implemented at 1700 °C with 4 hours soaking time. Figure 1 shows the sequence of cutting tool manufacturing produced in this study. Shrinkage of compacted alumina were measure based on length and thickness and hardness were evaluated using Rockwell hardness tester.

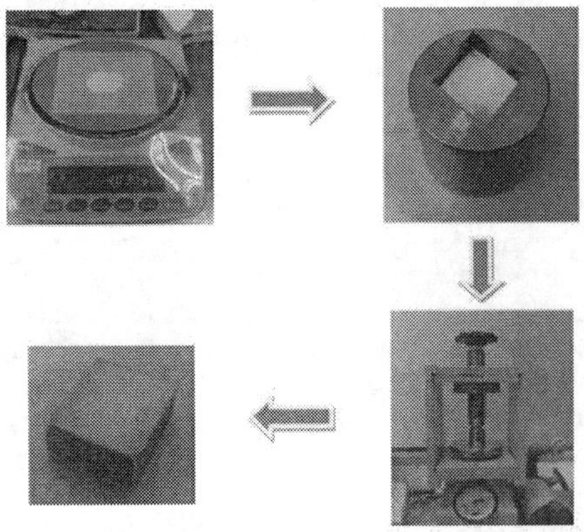

Fig. 1. Pre-compaction process of alumina powder into compacted green body

2.2 Machining tests

The second stage of this experiment was the machining performance test on the newly fabricated alumina cutting tools. The machining test was performed using CNC lathe machine as shown in Figure 2. The workpiece materials selected was AISI 1045 medium carbon steel with straight turning process as shown in Figure 3 (a-d). The dimension of the workpiece was150 mm length and 50 mm diameter.

Fig. 2. CNC lathe machine DMG Mori Seiki CTX 310 Ecoline

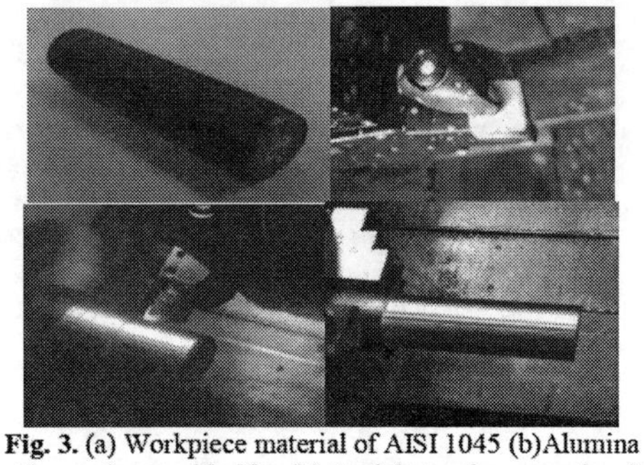

Fig. 3. (a) Workpiece material of AISI 1045 (b)Alumina cutting tool on tool holder (b) Straight turning operation (c) machined surface of AISI 1045

The machining trials was designed by varying the cutting speeds as shown in Table 1. Only one straight layer of workpiece was machined within the cutting length around 150 mm. Tool wear on cutting tool was measured using a stereo microscope based on average values. The wear rate of the cutting tool was calculated by dividing the average flank wear on both sides with the cutting time.

Table 1. Machining parameters

Cutting Tool	Cutting Speed (m/min)	Feed Rate (mm/rev)	Depth of Cut (mm)
Tool 1	150	0.05	1
Tool 2	175	0.05	1
Tool 3	200	0.05	1
Tool 4	225	0.05	1

3.0 Results and Discussion

3.1 Properties of fabricated cutting tool

Figure 4(a) shows the image of the compacted green body after being pressed with hydraulic hand press and cold isostatic press at 30000 psi, before sintering process. The sample appeared in brownish white with scar marks at the side of cutting tool. The size of this green compact was around maximum 13.00 mm length and 5.9 mm thickness. Figure 4(b) shows the fabricated alumina cutting tool after sintering. The colour of alumina compact changed to milky white. The size of sintered cutting tool was reduced up to 12.7 mm length and 5.3 mm thickness. Overall, the total shrinkage of alumina compact before and after sintering recorded maximum 7.01%. Table 2 shows the details of shrinkage dimension for all samples measured.

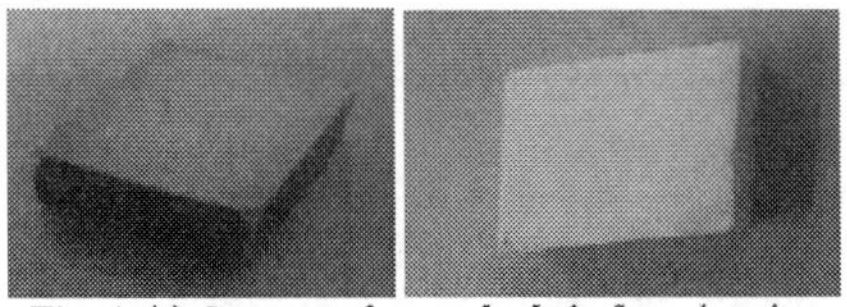

Fig. 4. (a) Compacted green body before sintering (b) Alumina cutting tool after sintering

Table 2. Percentage of shrinkage of the alumina cutting tool after sintering

Cutting Tool	Before Sintering		After Sintering		Shrinkage (%)
	Length (mm)	Thickness (mm)	Length (mm)	Thickness (mm)	
Tool 1	12.9	5.9	12.8	5.5	6.78
Tool 2	12.8	5.7	12.7	5.3	7.01
Tool 3	13.0	5.9	12.9	5.4	8.47
Tool 4	12.9	5.8	12.8	5.3	8.62

During sintering, the high temperature that absorbed by the ceramic compacts resulting isolated grain growth of the alumina particles. This resulting nucleation between particles, which in the end reduced the porosity between particles compact. As the grain particles bind closely to each other, the diffusion among particles resembles a close packed mass of powder particles. As a result, the structure of the alumina cutting tool became denser, translated into shrinkage of solid material [9].

The densification of alumina compact after sintering enable these inserts to increse their hardness and mechanical properties. As shown in Table 3, maximum hardness of 85.3 HRA was recorded and this value considered 4 times higher as compared to AISI 1045. Such adequate hardness benefitted the cutting tool to shear the steel effectively ecspecially when the dense structure finely integrated at the tool nose radius. This highlighted their potential to be a promising cutting tool, even with simple manufacturing procedure.

Table 3. Hardness value of alumina cutting tools

Cutting Tool	Hardness Value (HRA)			
	1	2	3	Average
Tool 1	83.4	85.1	75.9	81.5
Tool 2	83.8	84.6	82.7	83.7
Tool 3	85.3	84.7	82.1	84.0
Tool 4	82.8	84.3	83.0	83.4
Average hardness of all tools				83.2

3.2 Wear performance of fabricated cutting tool

After the fabrication process, the fabricated cutting tools were tested by turning process at various cutting speeds. The performance of the cutting tools were analyzed based on wear performance and failure modes. Table 4 shows the average wear and wear rate measured at a cutting distance around 150 mm cutting length. Figure 5 shows the trend in flank wear rate of the alumina cutting tool when machining AISI 1045 with different cutting speeds at constant 0.05 mm/rev feed rate and 1 mm depth of cut. Based on the graph, the wear rate dropped slightly from cutting speed 150 m/min to 175 m/min. However, the wear rate decreased drastically from cutting speed 175 m/min to 200 m/min. The wear rate mantained similar value on the last run. From this graph, it shows that the highest flank wear rate occurred at cutting speed 150 m/min with the value of 0.0041 mm/s. The lowest flank wear rate recorded at 0.0025 mm/s which occurred at cutting speeds of 200 m/min and 225 m/min.

Table 4. Wear rates of each cutting tool at different cutting speeds.

Tool	Cutting Speed (m/min)	Average Wear (mm)	Wear Rate (mm/s)
1	150	0.62	0.0041
2	175	0.43	0.0040
3	200	0.18	0.0025
4	225	0.15	0.0025

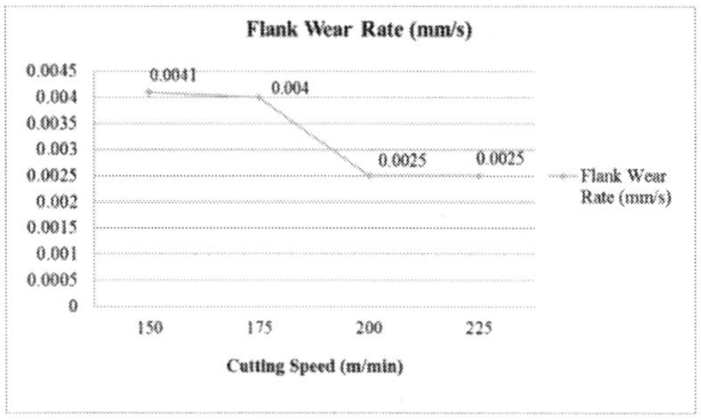

Fig. 5. Flank wear rate of alumina cutting tools at different cutting speeds

3.3 Failure modes of fabricated cutting tool

Figure 6 shows the failure modes of cutting tools tested. It shows that the wear of the fabricated cutting tools concetrated at the edge of cutting tool, notched far to the flank face. Obviously, wear occured at the flank face, combined with the abrasive and notch wears. Small nose radius could be the major reason of poor wear performance. Flank wear happen when the relief face of a cutting tool and the workpiece rub each other, resulting material loss due to abrasion mechanisms. During contact shearing between cutting tool and workpiece, high friction occurred at the tool-workpiece interfaces caused the temperature generated. At that particular point, high temperature at concentrated region weakening the edge of tool nose radius. Particles of the cutting tool gradually slide away until certain limit where the cutting tool could not retain its structural integrity, facilitating major damaged to the shape of nose radius. As shape of nose radius damaged, cutting force would not be uniform anymore and caused more friction to occur. Thus, machining action will be no more efficient. Further cutting action will destroy the edge of cutting tool [10].

Another observation of failure modes that occurred at the cutting edge is crater wear. This is shows in Figure 7 where the crater region apparently appeared on the rake face of the cutting tool. Crater wear occurred due to heavy sliding from the back side of chip that gradually abrade the contact surfaces of cutting tool. With fast chip movement and high temperature, some portion of particles on the rake face were ploughed as a result of disintegrate structure of cutting edge. There are evidence of flaking on the rake face probably due to higher pressure to the tool edge. At some extent, the flaking phenomena would expose the intermediate structure of cutting tool to the sliding chip. With high cutting temperature, some portion of chip could diffuse into the rake face resulting diffusion wear to occur. Diffusion wear would enhanced structure disintegration at the tool edge resulting heavy damage to the cutting tool [11].

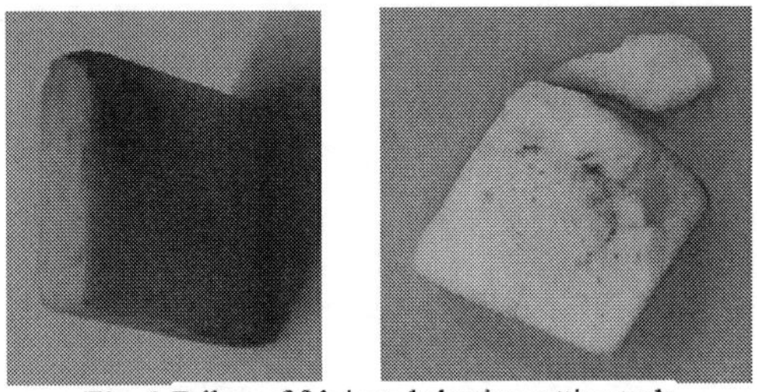

Fig. 6. Failure of fabricated alumina cutting tools

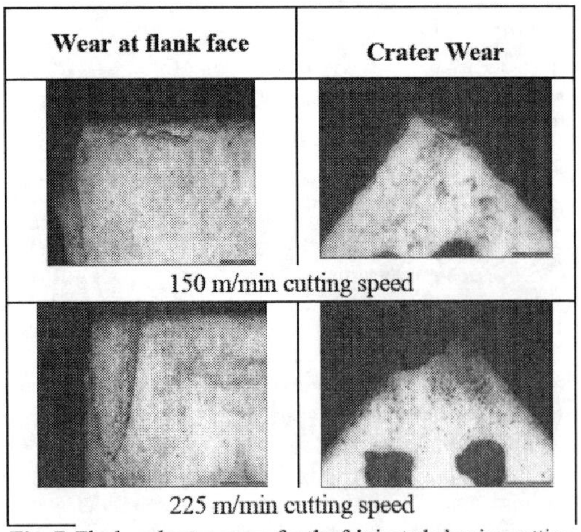

Fig. 7. Flank and cater wears for the fabricated alumina cutting tool at lowest and highest cutting speeds

These findings are differ from the research done by Rejab et. al, 2011 [12] where the flank wear of the alumina cutting tools increased when the cutting speed used is increased. There are several possible factors that can influence the results of this experiment. The first one is, there are some inconsistency in the shape and structure of all four cutting tools that were used for the machining test. The alumina cutting tools were not having perfect shapes for the machining. There were tiny chippings on the structure of the cutting tools which could lead to bigger wear during machining. Therefore, the results of the wear rate were inconsistent among all four cutting tools.

Another possible factor that could lead to this issue is the difference between the grain sizes of the alumina cutting tools. Milak et al. (2015) [13] stated that the flank wear decreases with the reductions in alumina grain sizes. This proves that when a cutting tool has smaller grain sizes as compared to the others, the wear of the tool will be smaller. The grain sizes of tool 1 used at 150 m/min might be larger or not consistent than the tool used at 225 m/min, hence caused higher wear rate.

Other than that, it is expected that the temperature generated during machining AISI 1045 at 150 m/min is lower as compared to the cutting speed 225 m/min. Such low temperature probably not sufficient to soften the steel during machining process and therefore provided difficult for the cutting tool to shear the workpiece material [14]. Due to this problem, the cutting tool was subjected to a higher cutting force which led to higher wear on the cutting edge. On the other hand, when cutting speed of 225 m/min applied, higher temperature is expected to be generated during the machining in this facilitate easy cutting process due to softening condition of the workpiece.

Overall, fabrication of alumina-based cutting tools is still in its infancy and has yet to be done extensively. It should be proposed that, even though the wear performance of this newly fabricated cutting tool still far behind to compete the commercial cutting tools, some promising aspects can be demonstrated. The alumina powders selected in this study still capable to be compacted without fracture, even with simple manufacturing procedure. There are still many flaws occurred during the fabrication process that need some improvements so that these newly fabricated cutting tools can match the performance of the other commercialized cutting tools. Therefore for the next experiment, there are some aspects that can be done to improve the properties of the cutting tools. The first approach should be replacing the alumina powder used now with a more refined alumina powder. This is because particle size of the alumina powder used are not refined enough to produce a compact structure. Usage of finer powder will fix some of the weaknesses in terms of density, agglomerates and reduction of porosity. Addition of several additive powders such as zirconia, chromia or silicon carbide could also improve the mechanical properties of the cutting tool.

3.0 Conclusion

This study involved fabrication of alumina cutting tool and evaluation of the performance of the cutting tool by machining test on AISI 1045 medium carbon steel. Based on the experimental finding the following conclusions can be drawn:

1) The alumina powder used in this study able to be compacted in the mold to form the required shape of cutting tool insert.
2) Sintering of alumina compact resulting denser and stronger structure. All cutting tools had shrinkage at average of 7% from the initial size and achieved an average of 83.2 HRA hardness.
3) The flank wear rate decreases as the cutting speed increases. The lowest wear rate produced at 0.0025 mm/s when cutting speed 225 m/min was used.
4) Failure modes of fabricated cutting tool appeared to be flank and crater wears. Small edge of tool nose radius suspected the be main concern for the tool failure.

The authors would like to thank Faculty of Manufacturing Engineering, Faculty of Engineering Technology and Universiti Teknikal Malaysia Melaka (UTeM) for their support that enabled this work to be carried out through the grant of FRGS/1/2017/TK03/UTEM/02/11.

References

1. S. Kalpakjian, S.R. Schmid. 4th ed. New Jersey: Prentice-Hall (2001)
2. M. Cheng, H. Liu, B. Zhao, C. Huang, B. Wang, Cer. Int. 43-16, 13869-13874, (2017)
3. A.B. Hadzley, S. Sarah, R. Izamshah, M. Amran, M. Shahir, M. Amri, N. Fatin, App. Mech. and Mat 465-466, 1049-1053 (2014).
4. Z., Yin, C., Huang, J., Yuan, B., Zou, H., Liu, & H. Zhu. Cer. Int., *41*(5), 7059-7065. (2015)
5. Xiabon X. Cui, F. Jiao, P. Ming, J. Guo, Cer Int 43-13, 10109-10122 (2017)
6. A. A. Zahirani, L. C. Choong, H. Mohamed, M.M. Ratnam, Z. A. Ahmad, J. Ally. Comp. 513, 91-96, (2012)
7. A. A. Zahirani, H. Mohamad, M.M. Ratnam, Z.A. Ahmad., J Ally. Comp. 497, 1-2, 316-320 (2010)
8. A. A Sanusi, A.A Aziz, M.A. Amran, M.S. Kasim, M. Zaimi, R. Izamshah, iDECON 2015
9. N.J. Loh, L. Simao, J. Jiusti, A. De Noni Jr, O.r.K Montedo, Cer. Int. 43-11, 8269-8275 (2017)
10. M. R. Fatin, A.B. Hadzley, R. Izamshah, M.A. Amran, Appl. Mech. Mat. 761, 257-261 (2015)
11. M.S. Kasim, C.H. Che Haron, J.A. Ghani, M.A. Sulaiman, M.Z.A. Yazid, Wear 302, 1-2, 1171-1179 (2013)
12. N. A., Rejab, A. Z. A., Azhar, K. S., Kian, M. M., Ratnam Z. A. Materials Science and Engineering: A, 595, 18-24. (2014)
13. P. C. Milak, F. D. Minattom A. De Noni Jr., O. R. K. Montedo, Cerâmica, 61, 88-103 (2015)
14. M.A. Kamely, M.Y. Noordin, A.Y. Bani Hashim, S.H. Yahya, M.M.P. Dan, J. Adv. Man. Tech. 6-1 (2012).

Development of Low-Cost Solar Water Heater Using Recycled Solid Waste for Domestic Hot Water Supply

Abdul Talib Din[1], Khairul Azri Azlan[2]

[1,2] Faculty of Mechanical Engineering, Universiti Teknikal Malaysia Melaka, Malaysia

Abstract. This research is focused on the development of a low-cost solar water heater (SWH) system by utilizing solid waste material as part of system elements. Available technologies of the solar water heater systems, heat collectors and its components were reviewed and the best system combinations for low cost design were chosen. The passive-thermosiphon system have been chosen due to its simplicity and independency on external power as well as conventional pump. For the heat collector, flat plate type was identified as the most suitable collector for low cost design and suits with Malaysia climate. Detail study on the flat plate collector components found that the heat absorber is the main component that can significantly reduce the solar collector price if it is replaced with recycled solid waste material. Review on common solid wastes concluded that crushed glass is a non-metal material that has potential to either enhance or become the main heat absorber in solar collector. A collector prototype were then designed and fabricated based on crashed glass heat collector media. Thermal performance test were conducted for three configurations where configuration A (black painted aluminum absorber) used as benchmark, configuration B (crushed glass added partially) that use glass for improvement, and lastly configuration C (black colored crushed glass) that use colored glass as main absorber. Result for configuration B have shown a negative effect where the maximum collector efficiency is 26.8% lower than configuration A. Nevertheless, configuration C which use black crushed glass as main heat absorber shown a comparable maximum efficiency which is at 82.5% of the maximum efficiency for configuration A and furthermore have shown quite impressive increment of efficiency at the end of the experiment. Hence, black colored crushed glass is said to have quite a good potential as the heat absorber material and therefore turn out to be a new contender to other non-metal heat absorber such as plastic and rubber.

Keywords: Low cost heater, domestic hot water, recycled solid waste, crashed glass

1 Introduction

Currently, most of houses in Malaysia is using the "instantaneous" or "tankless" electric water heater type. As the name implies, this type of water heater does not require storage tank thus eliminates energy losses at the tank. Theoretically, this type of system is also enables an unlimited hot water supply. In addition to that, by detecting the water flow through the valve, it will ensure that this system is only works "on demand" as the heating device of the instant water heater is only operates when the hot water is in used [1]. This mechanism delivers another advantage of saving electricity through the elimination of unnecessary heating as it doesn't require for the system to maintain the hot water temperature inside a storage tank. Basic construction of the electric tankless water heater can be seen in Figure 1.

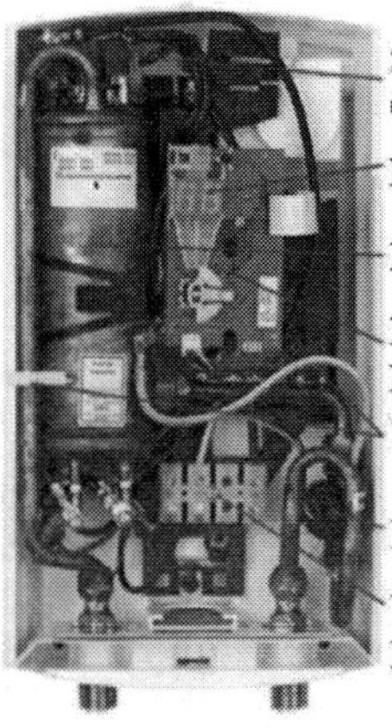

Figure 1: Example of Electric Water Heater System (Air & Water, 2015).

Nevertheless, despite of these advantageous, "tankless" water heater consume quite a vast amount of electrical energy. Typical types of electric water heater that is used in Malaysia consume power of 3 kW and above, hence a simple calculation of constant use of an hour per day will contribute a minimum 90 kWh of monthly electrical consumption.

Electric water heater draws more instantaneous power at a time in order for the water to be heated quickly, hence the operational cost will be much expensive if the demand charge is introduce by the electric utility company [2]. In Malaysia currently, luckily the maximum demand charge set by Tenaga Nasional Berhad (TNB) is only applicable for commercial customer with a supply of 6.6 kV and above and does not include the residential area [3]. Though, there is no guarantee that this tariff will remain as it is, especially when the trend of the energy pricing is increasing. Hence, for a long term benefits, a more sustainable and more economic technology such as solar water heater need to be promoted.

Principally, in determining the value of energy generated by RE system, there are two most critical factors that come into consideration which are the initial cost of the installation and the amount of energy that can be produced by the system annually [4]. Therefore, in the direction of increasing the awareness and promoting the use of green technology, it is very important to ensure that the technology is affordable for everyone.

Although the technology of the SWH system is readily available in Malaysia for a long time, the awareness on its existence and benefits is still very low mainly due to the high cost of this technology. Despite the promising annual solar irradiance in Malaysia which resulting an impressive annual saving up to RM 704 of the energy cost with a payback period of 7.4 to 10.7 years, the high initial installation cost of the solar water have made most of people in Malaysia prefer to choose electric water heater as the installation cost is very much lower in comparison with SWH [5]. Therefore, the idea for this project is to initiate the development of an affordable solar water heater by selecting a system types with a good potential of cost reduction. This will be done through the utilization of recycled solid waste as a material for its heat absorbing component.

In general, the municipal solid waste (MSW) comes from six typical sources which are residential, institutional, industrial, commercial, agricultural, and municipal services [6]. The most common method in organizing solid waste in Malaysia is through the landfill. Current numbers of landfills in Malaysia are listed in Table 1.

In Malaysia, it is reported that the generation of municipal solid waste is at approximately 30,000 tonnes for each day, or more than 11 million tonnes per year which has increased for about 60% for the last 10 years and 170% from the last 30 years [7]. As a consequence, vast amount of solid waste can cause shortage of the landfill, as the rate of the solid waste generation is much faster than the natural degradation process.

Furthermore, there are also environmental issues that will come up such as surface and ground water contamination, health hazard, bad odours and so on as most of the landfills in Malaysia are crude dumping grounds [8].

Table 1: Number of Operating and Terminated Solid Waste Landfill by State until 31 March 2015

State	Operating Landfill	Terminate d Landfill	Total
Johor	14	23	37
Kedah	8	7	15
Kelantan	13	6	19
Melaka	2	5	7
Negeri Sembilan	7	11	18
Pahang	16	16	32
Perak	17	12	29
Perlis	1	1	2
P. Pinang	2	1	3
Sabah	19	2	21
Serawak	49	14	63
Selangor	8	14	22
Tergganu	8	12	20
Kuala Lumpur	0	7	7
Labuan	1	0	1
Total	**165**	**131**	**296**

Therefore, in order to reduce the environmental impact and creating a more sustainable society, good waste management practices need to take place and one of the ways is by promoting the 3R which stands for reduce, reuse, and recycle [9]. Hence, the aim of this project is to reutilize the solid waste from a worthless item to become a valuable and cost effective product which is the solar water heater, hence converting "waste" to become "wealth".

The objectives of this research are:

i. To conduct a review on recent technologies available for SWH systems and select a system that have potential for low-cost production.
ii. To identify which components in the selected SWH system that have potential for significant cost reduction.
iii. To identify the best solid waste material that can be used to build the selected component by performing experimental analysis to see its performance.

2 Theoretical Analysis

Fundamentally, in order to successfully design an efficient solar water heater, it is essential to understand the basic concept of the heat transfer. In thermodynamic knowledge, the interest is on the amount of heat that is transferred. On the other hand, in

https://doi.org/10.1051/matecconf/201815004010

heat transfer study the concern is more towards the rate of the heat transfer. There are three basic ways on how heat can be transferred which are conduction, convection and radiation. Conduction is a way of energy transfer from particles of a material that are more energetic to the adjacent less energetic particles by interacting each other's, and it can occur in solids, liquids or gasses. Convection on the other hand is a mean of heat transfer through the interaction that occurs between solid surface and moving fluid or gas which also involves conduction and fluid motion. Heat can also be transferred by radiation where the energy can be emitted through the electromagnetic waves form due to the alteration occurred to the atoms or molecules electronic configurations [10]. The basic equation for each of these mechanisms is as in mathematical models below.

2.1 Mathematical Model of the Solar Collector.

By taking the glazing transmittance (τ) and absorbtivity (α) of the absorber plate into consideration, the rate of heat gain by the receiver $\dot{Q}_r$ is given as follows (Robert Foster, Majid Ghassemi, 2009):

$$\dot{Q}_r = \dot{q}_r A_c = (\tau\alpha)_{eff} I_T A_c \text{ ....(1)}$$

Where,
$\dot{q}_r$ = heat flux gain by the receiver,
A_c = collector aperture area,
$(\tau\alpha)_{eff}$ = effective optical fraction of energy absorbed,
I_T = incident radiation fof tilted solar collector.

Not all of the heat energy can be transferred to the working fluid. There are some heat loss occurred mainly through convection and radiation. The amount of heat loss by convection $\dot{Q}_{conv}$ is determined as below:

$$\dot{Q}_{conv} = \dot{q}_{conv} A_r = UA_r(T_r -) \text{ ....(2)}$$

Where,
$\dot{q}_{conv}$ = heat flux loss by convection,
A_r = receiver's area,
T_r = receiver's temperature,
T_a = ambient's temperature,
U = overall heat coefficient.

The heat loss by radiation $\dot{Q}_{rad}$ is determined as below:

$$\dot{Q}_{rad} = \dot{q}_{rad} A_r = \varepsilon_{eff} \sigma A_r (T_r^4 - T_a^4) \text{ ....(3)}$$

Where,
$\dot{q}_{rad}$ = heat flux loss by radiation,
A_r = receiver's area,
ε_{eff} = effective emissivity of the collector,
σ = Steffan Boltzmann constant = 5.67×10^{-8} ($W/m^2.K^{-4}$).

By subtracting the rate of heat loss from the heat gain, we can get the useful energy collected, $\dot{Q}_u$ as follows:

$$\dot{Q}_u(t) = \dot{q}_u A_c = (\tau\alpha)_{eff} I_T A_c - UA_r(T_r - T_a) - \varepsilon_{eff} \sigma A_r (T_r^4 - T_a^4) \text{ ....(4)}$$

The thermal efficiency, η with respect to the useful heat collected and the incoming solar radiation at a particular moment are calculated as follows:

$$\eta = \frac{\dot{Q}_u}{I_T A_c} = (\tau\alpha)_{eff} - \frac{UA_r}{I_T A_c}(T_r - T_a) - \frac{\varepsilon_{eff}\sigma A_r}{I_T A_c}(T_r^4 - T_a^4) \text{ ....(5)}$$

Hence, the overall efficiency within a time frame are determined as:

$$\eta = \frac{\int_{t_1}^{t_2} \dot{Q}_u dt}{A_c \int_{t_1}^{t_2} I_T dt} \text{ ....(6)}$$

The heat loss due to radiation are relatively smaller than convection especially for a low temperature collector such as flat plate type hence it can be neglected:

$$\eta = \frac{\dot{Q}_u}{I_T A_c} = (\tau\alpha)_{eff} - \frac{UA_r}{I_T A_c}(T_r - T_a) \text{ ....(7)}$$

As the temperature of the receiver, T_r is hard to be measured as it is located inside the collector casing, a constant named as effectiveness removal factor, F_R is use to allow equation to be in the terms of inlet temperature, T_{in} of the working fluid:

$$\eta = F_R\left[(\tau\alpha)_{eff} - \frac{UA_r}{I_T A_c}(T_{in} - T_a)\right] \text{ ....(8)}$$

The useful energy collected therefore can be written as below:

$$\dot{Q}_u = \eta I_T A_c = I_T A_c F_R\left[(\tau\alpha)_{eff} - \frac{UA_r(T_{in} - T_a)}{I_T A_c}\right] \text{ ....(9)}$$

It can also be represented as below:

$$\dot{Q}_u = \dot{m} C_p (T_{out} - T_{in}) \text{ ....(10)}$$

Where,
$\dot{m}$ = fluid mass flow rate,
C_p = heat capacity at constant pressure ,
T_{out} = outlet temperature of the working fluid.
T_{in} = intlet temperature of the working fluid.

The last equation can be used to perform thermal testing for the solar collector in order to measure it performance.

2.2 Crushed Glass as Heat Absorber.

The findings from past research indicate [12], among all three solid waste materials of metal, plastic and glass that have potential to be used as material for heat absorbing component in a solar collector, it is found that glass waste is having the least scrap value among others. Glass is also listed among the highest solid wastes composition in Malaysia with 3.3 % as shown in Table 2.

Table 2: Solid Waste Composition In Malaysia for 2012 (KPKT, 2015) [11].

Waste	Percentage
Food waste	44.5
Plastic	13.2
Paper	8.5
Disposable diapers	12.1
Garden waste	5.8
Glass	3.3
Metal	2.7
Textiles	3.1
Tetra Pak	1.6
Rubber	1.8
Leather Goods	0.4
Wood	1.4
Household Hazardous Waste	1.3
Others	0.5
Total	**100**

It is also found in Table 2 that food waste, plastic, paper, disposal diapers, and garden waste are the top five of the highest solid waste materials in Malaysia.
Nevertheless, these materials are either not suit for design or heat insulator material hence they are not suitable to be used as heat absorbing component in solar collector. Plastic material for instance is generally is an insulator, and only few types of plastic can be used as heat absorber but for low temperature collector and it might involve recycling process with advance material technology. Glass is listed as the sixth highest solid waste composition in Malaysia with 3.3 %. Therefore, glass have been chosen in this research due to the following reasons:

1) Glass holding quite a great portion of total solid waste composition in Malaysia.
2) Crushed glass is a free or a very cheap material.
3) Glass recycling facing a lot of challenges and is less worthy. Hence, a more creative solution is needed for glass to be reused for different purpose especially for non-container or mixed crush glass which cannot be recycled.
4) To avoid or reduce glass dumping in the landfill as it is non-biodegradable material which will create more environmental issue.
5) As suggested by the research conducted by Din et al.(2015) which shown that glass is among the materials that have shown good heat transfer performance after aluminum, iron and granite.
6) Although the thermal conductivity for glass (k_{glass}=1.4 W/m-K) is relatively very much lower than metals (e.g. $k_{alumimum}$ >100 W/m-K), it has a higher thermal conductivity than plastic ($k_{plastic}$ < 0.5 W/m-K). Glass also have a lower thermal expansion (CTE_{glass} < 9.0x10^{-6} (°C)$^{-1}$) which is better than plastic ($CTE_{plastics}$ > 70x10^{-6} (°C)$^{-1}$) (Callister, 1994). Therefore, these thermal advantage shown that glass can compete with plastic in non-metal heat absorber category.
7) There is still no research that have been conducted to see the performance of a solar collector that use glass as it heat absorber or enhancer found in previous literature. Therefore, the prototype that using crashed glass was fabricated and tested in this project to see the performance and efficiency of crushed glass for the development of low cost solar collector.

3 Methodology

In order to answer all of the research questions, a proper arrangement of activities is required. The research process flow was started with the identification of the existing technologies of the solar hot water system that are available in the market. The low cost system with low operation was selected as a "guide system" for this research. It is called the "guide system" as it will only be used as guidance in developing the collector since the scope of this research is to focus on the collector and not the overall system. Next, numbers of collector types were studied and a low cost solar water heater that is suitable for Malaysia climate was selected. Each of the components of the selected collector was studied and the component that can significantly reduce the cost of the collector and have prospect to make use of recycle solid waste was selected. Subsequently, the selected component was studied in detail to know the basic requirement on its material properties and characteristic in order to be able to perform its function. Few solid wastes materials were evaluated and the one that fulfill the basic requirement and have the potential to replace or enhance the efficiency of the selected component was chosen. The following are summary of the selected "guide system", collector type, component, and solid waste in this research:

1) Guide system : Passive, Thermosiphon.
2) Collector type : Flat plate
3) Component : Heat absorber.
4) Solid waste : Crushed glass.

The test rig for the above elements with 3 type of configuration A,B and C were brought to the test lab to proceed for the official experiment and the experimental data were recorded. The data were then analyzed and will then be repeated for different configurations A, B and C which will be explained in details later in this chapter. The experimental setup is as per shown in figure below.

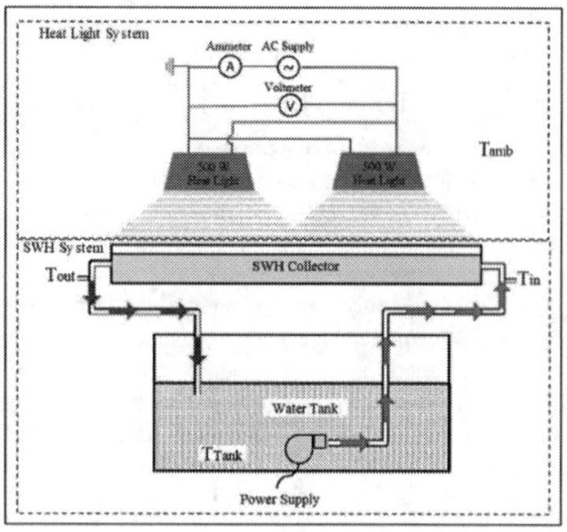

Figure 2: Schematic Diagram for Thermal Performance Test Rig.

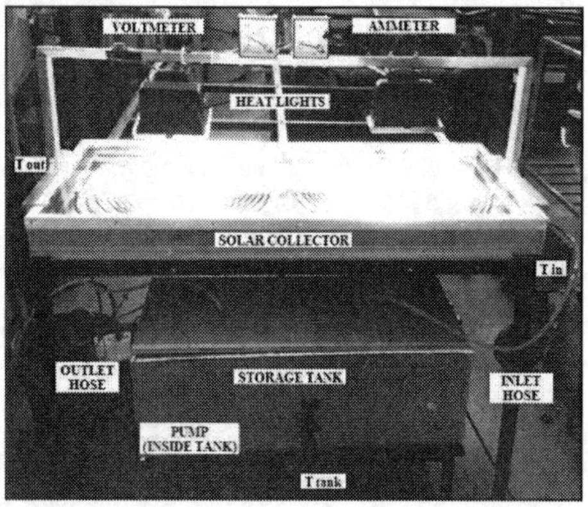

Figure 3: Actual Setup of Thermal Performance Test Rig.

Configuration A, B and C are similar set up as the above diagram in Figure 2 with difference in heat absorber materials. Configuration A having Non-Selective Black Painted Aluminum Absorber Plate. Configuration B having Non-Selective Black Aluminum Absorber Plate with crushed Glass as the filler medium. While configuration C having Non-selective Black painted Aluminum Absorber Plate with black crushed glass.

4 Results and Discussion

The results that were obtained from the thermal performance test are presented and discussed. Results from each configuration are delivered separately, then compared to each other, and finally discussed at the end of the research.

By following the measurement procedures that were described in the previous section, the result that have been obtained for the water mass flow rate by manual measurement are as follows:

Time, $t = 46.50\ s$

Flow Rate, $\dot{V} = \frac{V}{t} = \frac{250\ ml}{46.5\ s} = 0.0054\ ls^{-1}$

$= 0.0054\ \times 10^{-3} m^3 s^{-1}$

Mass Flow Rate, $\dot{m} = 0.0054\ \times 10^{-3} m^3 s^{-1} \times 1000\ kgm^{-3}$

$= 0.0054\ kgs^{-1}$ @ $19.44\ kgh^{-1}$

The flowrate which is measured at $19.44\ kgh^{-1}$, is assumed to be constant for all configuration A,B and C. The water volume inside the tank was set to 20 liters to allow the water to circulate in 2 cycle within 2 hours so that the effect of the temperature rise inside the tank can be seen clearly.

The thermal performance test were then proceeded for each configuration and the data that were measured and recorded every 10 minutes are shown in Figure 4, 5 and 6 in the following section.

4.1. Configuration A: Non-Selective Black Aluminum Absorber Plate

The test data that were measured and recorded for configuration A are summarized and plotted in the following graph in Figure 4.

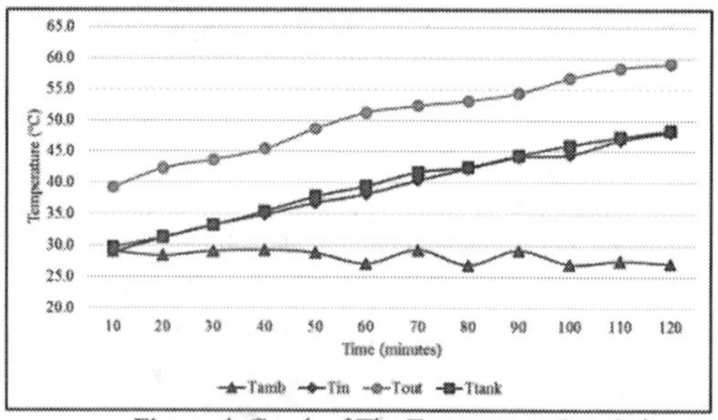

Figure 4: Graph of The Temperature Result for Configuration A.

Thermal performance test for configuration A were conducted at the noon. The drizzle environment that occurred during that time have caused the ambient temperature a bit fluctuating with a minimum temperature of 26.9 °C (minutes of 80) and a maximum temperature of 29.3 °C (minutes of 40 and 70). Whereas, for the collector inlet temperature, the minimum temperature occurred at the first 10 minutes with a value of 28.9 °C and gradually increase its temperature with time in a nearly linear pattern until it reach the maximum temperature of 48.1 °C before the test ended (minutes of 120). The outlet temperature on the other hand have the highest temperature among all with a minimum temperature of 39.3 °C (minutes of 10) and gradually increase until a maximum temperature of 59.1 °C (minutes of 120). The water temperature inside the tank were slightly higher than the inlet temperature with a minimum of 29.7 °C (minutes of 10) and a maximum of 48.4 °C (minutes of 120). The trend of temperature increase for inlet, outlet and water inside tank temperature is quite parallel. The above configuration A only serve as a bench mark to observe the performance of the crushed glass in absorbing heat.

4.2 Configuration B: Non-Selective Black Aluminum Absorber Plate with Crushed Glass.

The test data for configuration B are summarized in Figure 5 below

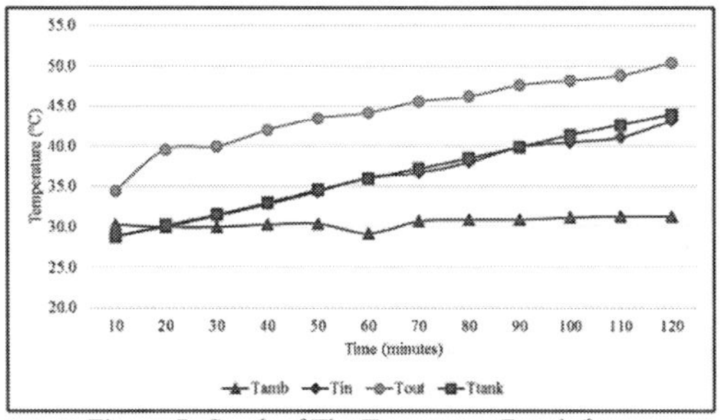

Figure 5: Graph of The Temprature Result for Configuration B.

For configuration B, the thermal performance test were conducted at 30 minutes pass noon. The ambient temperature is quite steady during the test as it is a sunny day outside with a minimum temperature of 29.2 °C (minutes of 60) and a maximum temperature of 31.3 °C (minutes of 110 and 120). The collector inlet in this configuration shown a steady temperature increase of its temperature with minimum temperature occurred at the first 10 minutes with a value of 28.9 °C and until it reach the maximum temperature of 43.2 °C at the end of the test (minutes of 120). Whereas for the outlet temperature, the temperature rise is quite slow with the minimum value is only at 34.5 °C (minutes of 10), ramp up for the next 10 minutes but almost flat after that. Then after 30 minutes onwards, the temperature was gradually increase until it reach a maximum temperature of 50.4 °C (minutes of 120). As what have been shown in previous test, the water temperature inside the tank temperature were also slightly higher than the inlet temperature with a minimum of 28.8 °C (minutes of 10) and a maximum of 44.0 °C (minutes of 120).

4.3 Configuration C: Non-Selective Black Aluminum Absorber Plate with Black Crushed Glass.

The data for ambient temperature together with water temperature at collector inlet, collector outlet and water tank for configuration C were plotted in Figure 6.

The last thermal performance test is for configuration C which is conducted earlier (morning) than configuration A and B. Hence the ambient temperature a bit fluctuating during the first hour with a minimum temperature of 26.6 °C (minutes of 40) and become more stable after that and reach it maximum temperature of 29.6 °C at the end of experiment (minutes of 120).

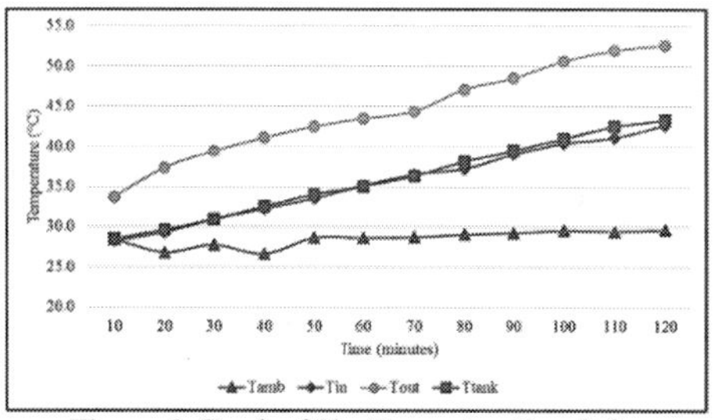

Figure 6: Graph of The Temprature Result for Configuration C.

For the collector inlet temperature, the temperature gradually increase where the minimum temperature occurred at the first 10 minutes with a value of 28.2 °C and reach the maximum temperature of 42.7 °C before the test ended (minutes of 120). The outlet temperature have its minimum temperature of 33.7 °C (minutes of 10) and gradually increase until a maximum temperature of 52.6 °C (minutes of 120) which is higher than configuration B. The water temperature inside the tank have a minimum of 28.5 °C (minutes of 10) and have a maximum temperature of 43.3 °C (minutes of 120). The temperature growth in this configuration is quite different from previous configuration especially for outlet temperature.

Based on the equation described in section 3.8, the efficiency data for each configurations were plotted against time in Figure 7.

From the graph in Figure 7, we can see that the highest efficiency is configuration A which is the benchmark in this test.

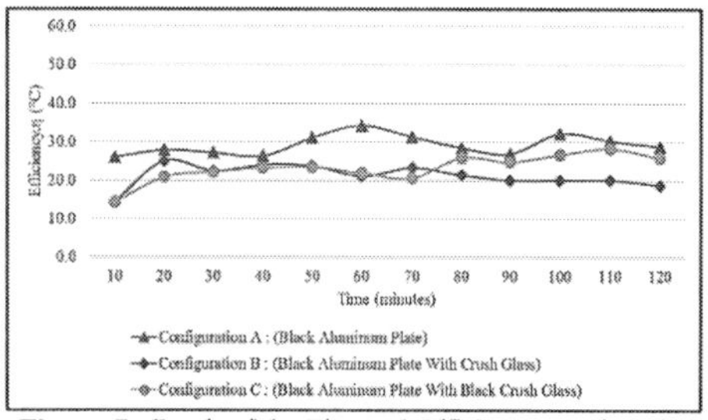

Figure 7: Graph of the Thermal Efficiency vs Time for Configuration A, B and C.

The efficiency for configuration A is a bit fluctuating with time with the minimum value of 26.1 % at the first 10 minutes of the experiment with no significant rise until after 40 minutes, it start to increase until it reach its highest value at minutes of 60 with 34.3 % efficiency. However, the following value of the efficiencies shown a wavy trend until the experiment ended. For configuration B, the efficiency is very low at the first 10 minutes but ramp up after that until it reach it highest value at minutes of 20 with 25.1 % efficiency. However the trend was not last long where it start to show fall and rise pattern and become worst after 70 minutes where it gradually decrease until the experiment ended. Lastly, configuration C shown almost same efficiency with configuration B for the first 10 minutes but have shown a steady increase after that. Although shown a decrease pattern after 50 minutes, it start to rise again and have shown significant increase after that where the efficiency have reached the value of 28.3% which is much better than configuration B and almost reach the same efficiency of configuration A in the first place.

5 Conclusion

Based on the study on the state-of-the-art technologies of the solar water heater, the passive-thermosiphon system have been identified as the most suitable solar water heater due to its simplicity and no need of external power or conventional pump. This advantage contribute to the reduction of the manufacturing, operating, and maintenance costs respectively. Next, it is also found that flat plate type solar collector is the best collector for low cost design and suit with Malaysia climate. This study on solid waste concludes that crushed glass have been selected due to it has low scrap value, abundant, have good potential as heat absorber [12] and have not been explored by previous researchers. From the test result of the thermal performance test, it is also found that black colored crushed glass have a good heat absorber properties which is good for the low-cost solar water heater.

Reference

1. Kibert, C.J., 2012. *Sustainable Construction: Green Building Design and Delivery,* 3rd ed. New Jersey: John Wiley & Sons.
2. Consumer Energy Center of California Energy Comission, 2015, *Water Heaters.* [Online] Available at:http://www.consumerenergycenter.org/residential/appliances/waterheaters .html [Accessed on 21 August 2015].
3. Tenaga Nasional Berhad, 2015, *Maximum Demand.* [Online] Available at: http://www.tnb.com.my/business/for-commercial/maximum-demand.html [Accessed on 21 August 2015].
4. Foster, R., Ghassemi, M. and Cota, A., 2009. *Solar Energy: Renewable Energy and the Environment* Ghassemi, A., (ed.). Florida: CRC Press.
5. Jing, O.L., Bashir, M.J.K. and Kao, J.-J., 2015. Solar radiation based benefit and cost evaluation for solar water heater expansion in Malaysia. *Renewable and Sustainable Energy Reviews*, 48 (C), pp.328–335.
6. Uriarte and Filemon A., J., 2008. *Solid Waste Management: Principles and Practices : an Introduction to the Basic Functional Elements of Solid Waste Management, with Special Emphasis on the Needs of Developing Countries*. Quezon: UP Press.
7. Fauziah, S. and Agamuthu, P., 2012. Trends in sustainable landfilling in Malaysia, a developing country. *Waste Management & Research*, 30 (7), pp.656–663.
8. Masirin, M.I.M., Ridzuan, M.B., Mustapha, S. and Don, R.A.@ M., 2008. An Overview of Landfill Management and Technologies : A Malaysian Case Study at Ampar Tenang 1. *1st National Seminar on Environment, Development and Sustainability (PSISenviro2008)*. 2008 Selangor, Malaysia, 28-29 July 2008: Universiti Tun Hussein Onn (UTHM).
9. Pariatamby, A. and Tanaka, M. eds., 2013. *Municipal Solid Waste Management in Asia and the Pacific Islands: Challenges and Strategic Solutions*. Singapore: Springer.
10. Cengel, Y.A., 2007. *Heat & Mass Transfer: A Practical Approach,* 3rd ed. Noida: McGraw-Hill Education (India).
11. Kementerian Kesejahteraan Bandar, Perumahan, dan Kerajaan Tempatan, 2015. *KPKT Selected Statistics Until 31 March 2015*. Putrajaya.
12. Din, A.T., Hafiz, N.M. and Rus, A.Z.M., 2015. Exploration to Find Heat Absorbent Materials from Recycled Solid Wastes for the Design and Development of Low Operational Cost and Affordable Water Heating System for Domestic Application. *International Journal for Innovation Education and Research*, Vol. 3-4, p15

E-Learning Tutoring System for Sijil Pelajaran Malaysia (SPM) English

Munirah Mohd Yusof[1], *Ng* Lee Wah[1], *Rozlini* Mohamed[1] and *Muhaini* Othman[1]

[1]Faculty of Computer Science & Information Technology, Universiti Tun Hussein Onn Malaysia, Johor, Malaysia

Abstract. An E-Learning tutoring system for English (SPM) is an e-learning platform for Sijil Pelajaran Malaysia English subject. E-learning is a web based application that supports the delivery of learning, skills and knowledge related to the English subject in SPM. E-learning use the technology to enable people to learn anytime and anywhere. Besides, it is developed with the purpose to evaluate student's performance through an online quiz. Nowadays, tuition center have a lot of students, hence insufficient time to analyze the individual performance in learning English. In addition, e-learning is able to reduce the printing cost and administrative cost at the same time reducing usage of paper and printed materials associate in learning. The system developed by using Javascript and PHP language based on system prototyping methodology. The result will produce student's performance in monthly quiz and teacher are able to monitor the performance of each students. Although the development or this e-learning is based on requirement gathered via tuition center, however it can be proposed to be by any learning center or to self-study. It is hoped that this e-learning will be able to improve student knowledge in English language and subsequently help student to obtain the best result in SPM by providing more guided references and practices.

1 Introduction

An E-Learning tutoring system for SPM English is an e-learning platform for Sijil Pelajaran Malaysia (SPM) English. It is a web based application that supports the delivery of learning, skills and knowledge related to the English subject in SPM where it enable people to learn anytime and anywhere. E-learning tutoring system is developed with the purpose to evaluate student's performance through an online quiz. Quizzes are carried out monthly to evaluate student's English knowledge. Nowadays, tuition center have a lot of students, hence insufficient time to analyze the individual performance in learning English. In addition, e-learning is able to reduce the printing cost and administrative cost at the same time reducing usage of paper and printed materials associate in learning. The system developed by using Java Script and PHP language based on system prototyping methodology. The result will produce student's performance in monthly quiz and teacher are able to monitor the performance of each students. This system could be proposed to be used by education field such as tuition center. An e-learning tutoring system, user does not need to install any software to support and run it. End user only need to have a web browser to access the web application via internet.

The rest of this paper is organized to provide a brief explanation of e-learning and related work in Related Works section. The following section is Methodology while, in the next section will discuss on System Development. Section 5 focus on Implementation and Testing and Section 6 discuss on Evaluation. Discussion and Conclusion will be discussed in the last section.

2 Related Works

In recent year, the knowledge based economy has exhibited a pervasive and ever increasing ways of delivering education, which has led to dramatic changes in learning technology. As the new economy requires more people learn new knowledge and skills in a timely and effective manner, the advancement of computer and networking technologies are providing a diverse means to support learning in a more personalized, flexible, portable and on-demand manner. There radical changes in learning needs and technology are fueling a transition in modern learning in the era of the Internet, referred as e-learning [1].

*Corresponding author: munirah@uthm.edu.my, leewah92@hotmail.com, rozlini@uthm.edu.my, muhaini@uthm.edu.my

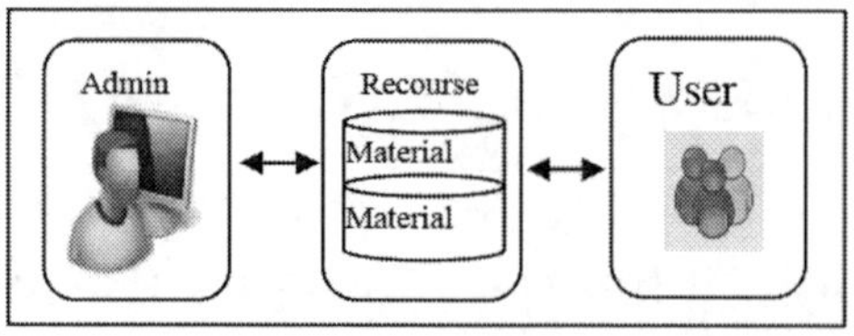

Figure 1: E-learning architecture

E-learning is the use of electronic media, educational technology and information and communication technologies (ICT) in education. E-learning includes numerous types of media that deliver text, audio, images, animation, and streaming video, and includes technology applications and processes such as audio or video tape, satellite TV, CD-ROM, and computer-based learning, as well as a local intranet or extranet and web-based learning **[2]**. Basically, the e-learning architecture as shown in Figure 1 are divided into three part; resources, user and administrator. Resources consist of teaching materials such as note, assessment and others. Users are instructors and learners; where instructors are the person that responsible to provide the courses and place on the internet while learners are the person that uses the courses. The administrator is an important person that responsible to manage the e-learning in a whole.

With the rapid growth in internet technology e-learning implementation becomes more advanced. Internet accessibility is one of the major factors that contribute to the successful implementation of the online e-learning **[3]**. With the growth of Internet accessibility in various places and growth of Internet subscribers among Malaysians, it is believed that it will bring a positive element in digital lifestyle towards the usage of e-learning in Malaysia.

The very first reference to e-Learning was made way back in 1999 in the government Multimedia Super Corridor (MSC Malaysia) blueprint for Smart School Flagship Application **[4]** It comprised with browser-based teaching-learning materials (and related print materials) for Bahasa Melayu, English language, Science and Mathematics. In 2007, K-Perak e-Learning Cluster (KPEC) was launched to provide professional development for teachers and build e-learning capability in a cluster of five selected schools in the state of Perak. Teaching and learning content is made available to teachers through the project website, which features an interactive forum. As an initiative to support e-Learning activities Malaysian Grid for Leaning (MyGfl) was introduced to ensure conformance and best practice in e-learning content and systems. One of the projects under MyGfl was Cikgu.net that launched on 2000. The portal used Malay language as a teaching medium and Cikgu.net attracted thousands of registered users. Besides that, private sector play an important role in fostering e-Learning in Malaysians schools. For instances, Smart Utusan Education Portal has been set up by Utusan Melayu Berhad contains teaching and learning material for all levels of school education. Most of the main subjects in school for instances mathematics, science and English have been covered. This initial implementation shows that e-learning been established in Malaysia and it progress not only in primary school but also in secondary school and higher institution learning.

Thus this paper based on one of the subject that studied from primary school until higher institution learning which is English. However, this study scope is on English for SPM which is for secondary school. English has known as an important second language and is widely spoken especially in the countries once colonized by United Kingdom or the United States of America **[5]**. It has been the aims of *Kurikulum Baru Sekolah Menengah* (KBSM) to extend learners English language proficiency in order to meet their needs to use English in certain situations in everyday life. Traditional English subject were conducted in class where teacher will delivered their lesson and handling exercise throughout the class session **[5]**. The good of this traditional method is that, students can have lesson and ask any questions face to face with their teachers, however time frame is insufficient for each lesson. Students need to study by themselves to grasp more knowledge and understanding in English subject alone. However, improving proficiency in English language through a subject called English in the school curriculum is insufficient as educational planners at the Education Ministry and teachers in schools have realized over the years **[5]**. Hence, e-learning can be one of the platform to help students improve their English. With e-learning, students can enrolled in virtual English class, download learning material, answer exercise and quiz to evaluate their understanding level and compare their results easily. Besides, some e-learning platform placed forum as one way for students to communicate and discuss with others students or teachers.

3 Methodology

The methodology used for developing an e-learning tutoring system for SPM English is system prototyping. System prototyping able to develop a simplified version of proposed system and give it to the user evaluation and feedback. After collecting the feedback from the users, the developer will reanalyze, redesign and re-implement a second prototype that correct deficiencies and add more features in order to develop a quality of system that meet user requirement [4]. In addition, Prototype Model able to identify all the system error test after received user's feedback. The model performed the analysis, design and implementation phases repeatedly in a cycle until the system is completed. The prototype model include five phases which is planning, analysis, design, system prototype and implementation. However, the prototype model may increase the complexity of the system may expand beyond original plans. Nevertheless, prototyping ensures that the end users constantly work with the system and provide a feedback which is incorporated in the prototype to result in a useable system. They are excellent for designing good human computer interface systems. The process model of the Prototype Model approach is shown in Figure 2. Section 3.1 until 3.5 will discussed on the phases in details.

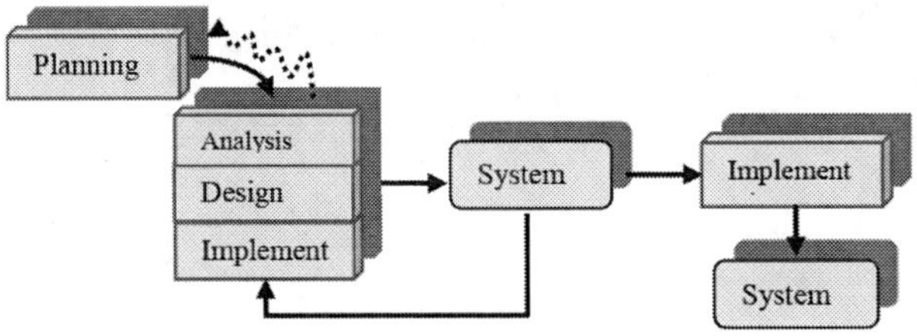

Figure 2: Prototype Model [6]

3.1. Planning Phase

In planning phase is to discuss the development of a project based on the given time period. The problem statement, objective, scope of system are well defined. In addition, interview with related field has been carried out to identify the problem statement and scope of system. It works as a guideline to ensure this project able to complete on time.

3.2. Analysis Phase

This phase will design the context diagram which showing the entities that interact with it, entity relationship diagram, data flow diagram and flow chart for every entities. The diagrams able describe the interaction among the user, proposed system and the database. In addition, interview with the owner of tuition center Pusat Tuisyen Minda Proaktif, Puan Nurul Husna Zakaria and Miss Erni asn English teacher had been done to get more information about the process of conducting the class and studies material she used during English class. Hence, the information gained from the interviewer and determine the user requirement.

3.3 Design Phase

In design phase, user interface and database will be designed based on the gathered information which is based on the user requirement identified on the previous phase. Input and output of the proposed system are designed as well as context diagram, entity relationship diagram and flow chart. The user interface consists of user login page for admin, teacher and students. Designing the database are using MySQL to store information of teacher, students and studies materials.

3.4 Implementation Phase

In implementation phase, the development of proposed system is constructed according the planning, analysis and design phase. After constructed the propose system, testing and debugging will be carried out. The testing will be conducted by allowing users to access to the system and then distributing questionnaires to gain feedback. Those feedback will be analyze to determine whether the objectives have been achieved.

3.4 Implementation Phase

User feedback is obtained and analyze and new action will be carried out in order to improve the system. Each prototype and design of interface may redesign and make improvement. This activity the system is believed to be able fulfill the requirements and provide the best quality to the users.

As a conclusion, by using Prototype model in the development of web based tutoring system for English SPM it will help development process more systematic.

4 System Development

E-Learning Tutoring System for SPM English is divided into four users; admin, teacher, student and parent. Each of this user has their own modules. User modules are built on online platform consists of six modules which are login module, tutoring module, assessment module, teacher module, parent module and result module. Tutoring and assessment module is the main part of this system used by students. In tutoring module, various notes were uploaded in order to help students learn with more leisure while assessment module help in identifying student understanding on what they learn through the answer given for several questions prepared. In teacher module learning material can be uploaded and managed so the notes will be changed in some period of time. Teacher will manage exercise and quiz materials and they can access and view student assessment result. For parents, they can access their children details and assessment result. Meanwhile, system administrator are able to register new students information and manage students details by add, delete, and update their information.

This E-Learning Tutoring System for SPM English is an alternative way for the students in for their learning purposes. It have a user friendly interface design and provides good interaction between systems and user. The system will start with login module where the user need to insert username and password to access the system after registration. After the user was authenticated, main page will be display. New student need to be registered in order to use this system. As for teacher, one of the tasks is to added quiz or exercise questions as in Figure 3. Figure 4 shows interface that allowed teacher in managing students' assessment report. If `VIEW` icon been clicked, it will show students results for each of the answered questions and their total assessment results. Figure 5 shows the interface of assessment results for students. Each student can only view their own results and the system stored all their previous results in database. This will help students in comparing their performance after each quiz is taken. As for Figure 6, the interface shows that parents can view all of their children' assessment report. With this, parents can observe their children performance and taken extra measures in making sure that their children will perform well in this subject.

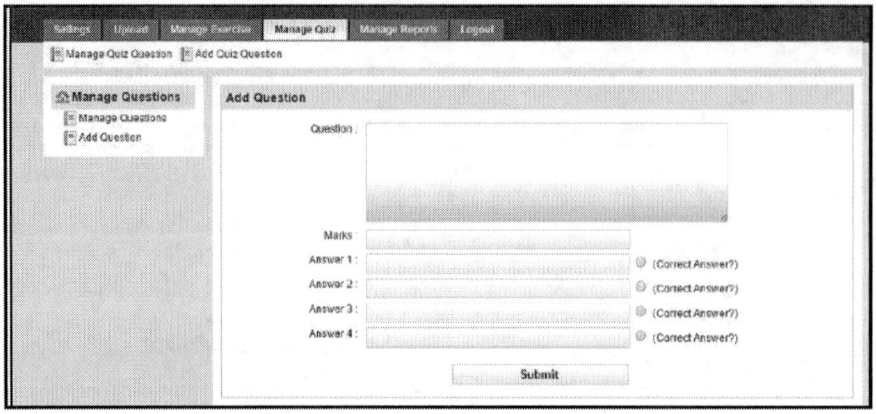

Figure 3: Adding new question for assessment module

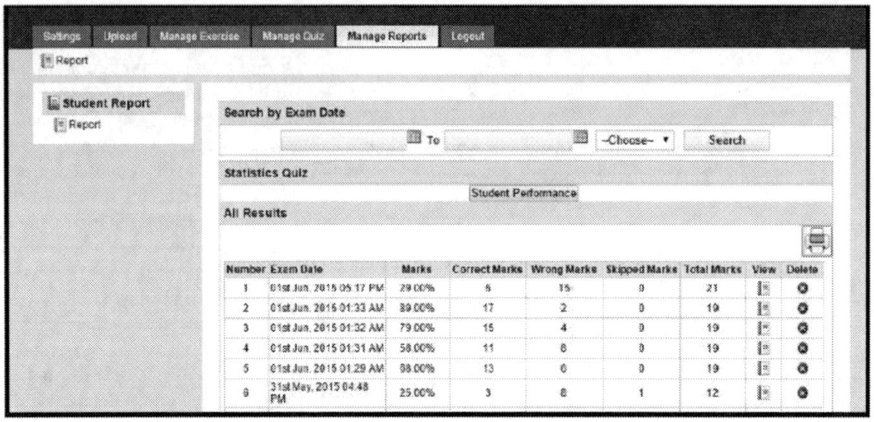

Figure 4: Teacher managing students' assessment report

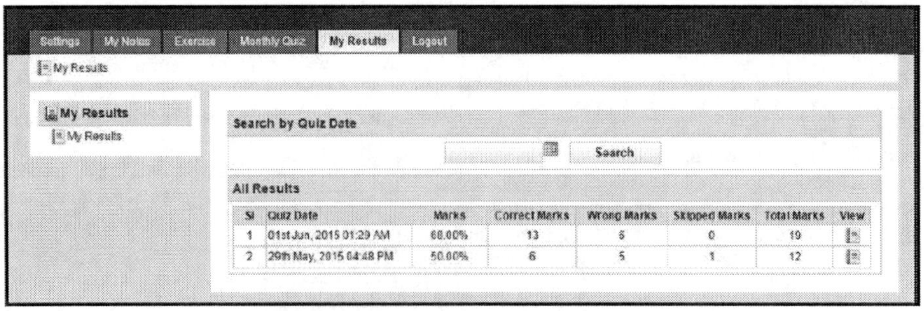

Figure 5: Interface for students viewing their own assessments result

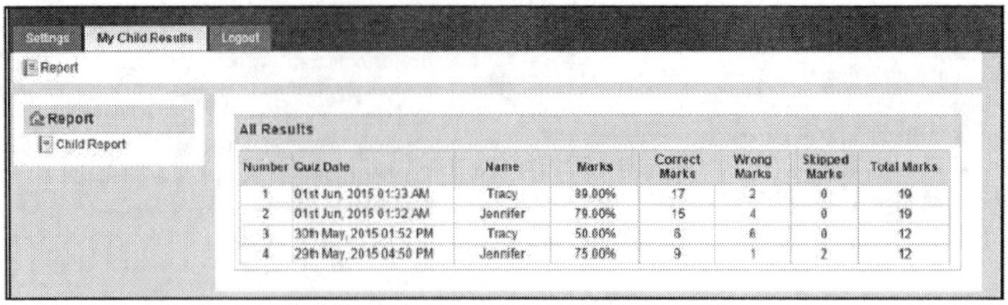

Figure 6: Interface for parents viewing their children assessment results

5 Implementation and Testing

Implementation and testing phases play a role to test the functionality of the system in order to ensure that the systems developed, meet the needs of the design mentioned previously. Implementation and testing phases also ensure that all the processes involved in the system are functioning properly. The implementation phase consists of design interface, writing system coding and documentation.

A. System Implementation

The Implementation Phase described how the information system is deployed, installed and transitioned into an operational system. At the end of the implementation phase, the result is evaluated according to the list of requirements created earlier. The system also will be evaluated according to the design. This phase is complete when all of the requirements have been met and when the result corresponds to the design.

B. Functional Testing

Functional testing was conducted to examine the functionality of all the module created in the system. This testing involved user input and output where during this process all the error will be reviewed and fixed. In order to ensure the system functionality testing was correctly done, a test plan was divided into three user level which are admin, teacher and student. Table 1 shows the validation for the administrator site. The admin module consists of managing user account, learning module material, test bank and student class groups. Table 2 shows the test plan for parent module where parent should be able to change their password, view children details and view children assessment report. Table 3 shows the test plan for student module and Table 4 shows test plan for teacher module. Student test plan are based on whether students are able to update profiles, download learning materials, submits assessment answer and view their results. Teacher unit testing plan are based on where users are able to upload, create, edit and delete questions, setting the exercise time range and view students' performance.

Table 1. Unit testing of test plan for admin module

Test cases	Expected output	Actual output
Click manage user	Register teacher and student, able to edit user	The manage user is initiated
Click view feedback	View public feedback	The feedback is initiated

Table 2. Unit testing of test plan for parents module

Test cases	Expected output	Actual output
Change password	Password changed	Password change
View children details	Children details viewed	Children details viewed
View children assessment report	Assessment report viewed	Assessment report viewed

Table 3. Unit testing of test plan for student module

Test cases	Expected output	Actual output
Student profile updated.	'Update the new information.	Student profile updated.
Download learning material files.	Files downloaded.	Files download.
Submitted assessment answer.	Assessment answer save into databases system.	Assessment answer added.
Download button.	Files downloaded.	Files downloaded.
Question button to do all the exercises.	Exercise displayed and time set.	Exercise displayed and time set.
View the result.	Result review.	Result review.

Table 4. Unit testing of test plan for teacher module

Test cases	Expected output	Actual output
Teacher profile updated.	Updated the new information.	Teacher profile updated.
Search student by group and session.	Display student by class group and session.	The student list by group and session is initiated.
View student list profile.	Display registered student list and profile.	The student list and profile is initiated.
Upload learning material files.	Files uploaded.	Files uploaded.
Manage student assessment answer.	Marking student assessment answer.	Assessment answer marked.
Click generate mark.	Student assessment mark.	Assessment mark generated.
Save student mark.	Added student mark into database.	Students mark recorded.
Add, delete and update learning material files.	Learning material added, deleted and updated	Learning material added, deleted and updated

6 Evaluation

In order to validate the performance of the proposed system, user acceptance testing (UAT) been carried out through questionnaire distributed among the user. UAT is important phase in this study as to validate the capability of the software in handling required tasks in real world scenarios. The testing was conducted by 1 admin and 1 teacher in the Minda Proactive Tuition Center and 20 students. The implementation of system testing is successfully conducted. The actual outputs are similar to the expected results. Comments and suggestions from the user acceptance testing are gathered and adopted to improve the system in the future. Figure 6 shows the number of satisfied user with reference to system interface design.

Figure 7 showed result for six interface design questions answered by 22 user with scale 1 (least agree) to 5 (strongly agree). The question are based on homepage design, text and button style, text color, background color, interface design and level of user friendliness. From Figure 7, it shown that most user agree and strongly agree with the system design interface and none of the user disagree and strongly disagree with the design interface.

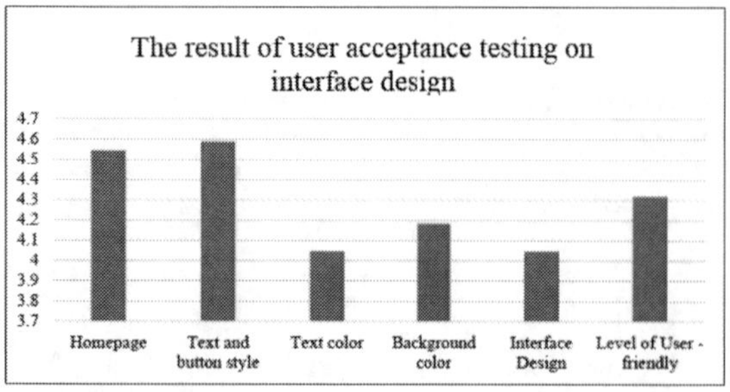

Figure 7: Result of user acceptance test on interface design

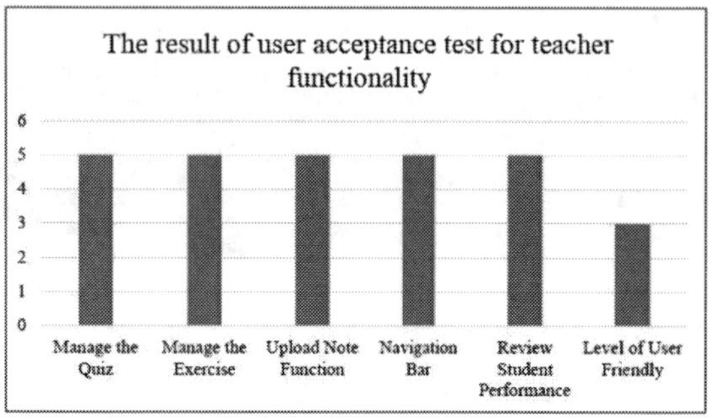

Figure 8: Result of user acceptance test for teacher functionality

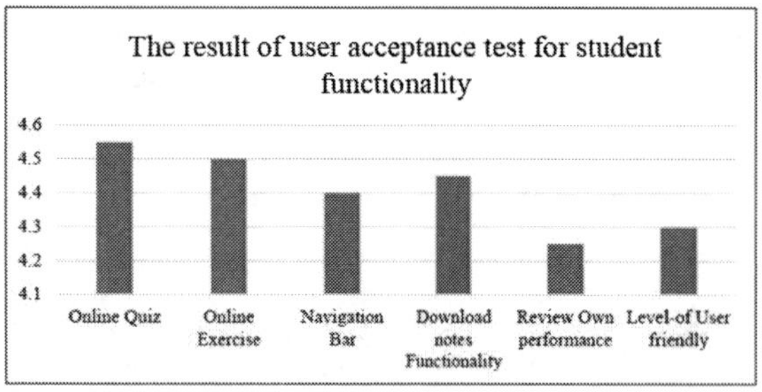

Figure 9: Result of user acceptance test for student functionality

From Figure 8, it shows the number of satisfied user according to the teacher functionality. It showed result for six function with scale 1 (weak) to 5 (very good). The tested aspects are; manage quiz, manage exercise, upload learning material, navigation bar aspect, review student performance and level of user friendliness.Besides, Figure 8 show that most user agree and strongly agree with the system functionality and none of the user disagree and strongly disagree with the functionality. Figure 9 shows the number of satisfied user according to the student functionality. It showed result for six function answered by 20 user with scale 1 (weak) to 5 (very good). The tested aspects are on online quiz, online exercise, on navigation bar, downloading learning material functionality, review their own performance and level of user friendliness. From Figure 9, it shown that most user agree and strongly agree with the system functionality and none of the user disagree and strongly disagree with the functionality.

7 Conclusion

E-Learning Tutoring System for SPM English was developed in assisting secondary school student in learning English using alternatives ways through web platform. It can be used by any students who are interested in English subject. Furthermore it helped in student in tracing and comparing their result while doing quiz through this e-learning system. Although this system has been successfully developed, it still can be enhanced for instances adding more interactive components such as videos as learning material. SMS notification for parents should be one of the improvements that will make the information of their children delivered faster rather than email notification. This improvement can also be applied in the future to produce a more functional and useful system, yet easy to use and more compatible. In conclusion, it is hope that this research will contribute to education domain particularly in SPM English subjects and it will reap the benefit from e-learning approach.

Acknowledgement

This paper has been supported by, Short Term Grant, Universiti Tun Hussein Onn Malaysia (UTHM) (vot U539) for the financial support. This research also supported by GATES IT Solution Sdn. Bhd under its publication scheme.

References

1. D. Zhang, J. L. Zhao, L. Zhou and J. F., Nunamaker Jr, "Can e-learning replace classroom learning?", Communication of the ACM, **47**(5), 75-79, (2004).
2. D. Tavangarian, M. E., Leypold, K. Nolting, M. Roser and D. Voigt, "Is e-learning the solution for individual learning", Electronic Journal of E-Learning, **2**(2), 273-280, (2004).
3. C. L. Goi and P. Y. Ng, "E-learning in Malaysia: Success factors in implementing e-learning program", Journal of Int J Teach Learn Higher Education **20**(2) 237-46, (2009).
4. http://digitallearning.eletsonline.com/2008/11/tracing-e-learning-initiatives-in-malaysian-schools-2/ access on 27/03/17.
5. J. Thirunsaku and M. Y. Melur, "The Many Faces of Malaysian English", International Scholarly Research Network, 2012.English language issues in Malaysia revisited,

   https://www.malaysiakini.com/letters/306589#ixzz4cniS7Rd1, access on 27/3/2017
6. Dennis, A. Roth, R. M. and Wixom, B. H. "System Analysis and Design", 5th Edition, Wiley, (2013)

Exploring the Elements of Integrity in Peer Assessment

Noor Atikah Zainal Abidin[1,*], *Alias* Masek[2], and *Nur Sofurah* Mohd Faiz [3], and *Shafizza* Sahdan[4]

[1,2, 3& 4] Faculty of Technical and Vocational Education, Universiti Tun Hussein Onn Malaysia, 86330 Batu Pahat, Johor, Malaysia.

Abstract. Peer assessment is one of the best methods to evaluate group work of students. However, integrity of students as assessors is still questionable. Hence, the purpose of this study is to explore the elements of students' integrity in carrying peer assessment. The qualitative approach using the document analysis and interviews is used to explore the elements of integrity in peer assessment. In an effort to increase the acceptance of this evaluation approach, the results show that the student's integrity in conducting peer assessment consists of three elements, which are self-integrity, integrity in social interaction and work commitment.

1 Introduction

Periodic assessments carried out during the teaching and learning process will reveal students' progress over time. Tracking students' academic performance on a regular basis provides feedback to the teachers about the problems faced by students [1]. Do students have difficulties in learning and mastering the knowledge and skills that have been taught? If the answer is in the affirmative, the teachers will investigate the root cause. The teachers can then use the information obtained from the investigation to change their teaching approach or provide special training; this will help students overcome the problems they face. Without this special assistance, students' problems will remain unsolved, and they will continue to fall behind in learning. Peer assessment is one of the alternatives to monitor students' learning progress, through which the teachers can carry out corrective measures to improve students' performance.

According to [2], peer assessment refers to a process in which students think critically and suggest the best grade for their friends. In peer assessment, students use a checklist or rubric, which helps them in various aspects of learning, whether they are assessed or assess others. Sharing the criteria of assessment empowers students to evaluate their own performance and their friends. The ability to work effectively with others in a group is one of the critical ways to achieve learning success. [3] states that peer assessment is a process in which students assess the quality of work of their fellow students and provide each other with feedback.

According to [1], peer assessment is a rule for students to consider and determine the extent, value, or quality of products or the performance of other students. Students provide marks for the work or show their friends the standard of learning using pre-defined criteria [4,5]. In order for students to determine the fair value of marks, the lecturer must create an environment that gives them the opportunity to participate [5]. Previous studies often discuss issues related to the implementation of peer assessment. This is due to the social influences that may affect the reliability and validity of peer assessment. The relationship between partners is the main factor that affects the reliability and validity of peer assessment and it includes these aspects: the bonds of friendship, enmity, individual popularity, negative perceptions and discrimination between genders [1].

Therefore, the aim of this study is to explore the elements of student's integrity in conducting peer assessment and to raise the acceptance level of this assessment approach among them.

2 Methodology

This study was conducted using the qualitative approach to explore the integrity of students in carrying out peer assessment. Two methods were used to obtain the results of the study: analysing documents and conducting interviews. Documents from several sources were analysed, including previous studies, model of integrity and assessment policy, principles and guidelines. The document analyses were carried out on scientific papers published from the years 1996 to 2017. Search engines were utilised to gather papers from various sources, including Google Scholar and selected journal databases relevant to assessment, i.e. ELT journal, journal computers in human behaviour, radiography journal, assessment in education, assessment and evaluation in education, and Malaysian online journal of educational technology. These are the search terms deployed for data collection: peer assessment, bias of assessment, factor affect peer assessment process, marking influence by demography, gender bias in peer assessment, friendship bias in peer assessment, policy marking in assessment,

* Corresponding author: ika.unimas@gmail.com/hb14079@siswa.uthm.edu.my

model of integrity, conceptual of integrity and disadvantage of peer assessment.

This research study involved six experts in the fields of Professional Education and Technical and Vocational Education (TVET); they are familiar with and experienced in teaching and learning, especially in matters related to peer assessment. Interviews were conducted to collect these data: information of peer assessment, bias of assessment, factors affecting peer assessment process, marking influence by demography, gender bias in peer assessment, and friendship bias in peer assessment. The data were then analysed along thematic lines. The thematic analyses were conducted in order to organise the data collected into clear groups of topical information needed to answer the research questions.

The qualitative data analysis was conducted based on the six steps recommended by [6]. These are the six steps of conducting thematic analysis:

(i) A researcher needs to be familiar with the data; he or she should know the data that are important. The researcher needs to read the documents carefully and take note of the initial ideas.

(ii) The researcher generates the initial codes based on the data or transcripts. He or she should collate the data related to each code.

(iii) The researcher searches the themes of the data and classifies them into groups.

(iv) The researcher needs to review the data and ensure the suitability of the themes.

(v) The researcher must define and name themes that interpret the groups of data. He or she will then generate a definition for each theme.

(vi) The researcher writes the report based on the findings.

3 Findings

A total of three main themes have emerged from the document analysis and thematic analysis of the interview data, representing the group elements of students' integrity in peer assessment discussed in detail as follows:

Element	Sub	1	7	8	9	10	11	12	13	14	15	16	17	18	19	20	21
Self-Integrity	Self-		√	√													√
	Courage or				√	√	√										√
	Self-Discipline							√	√								√
	Transparency									√							
Integrity in social interaction	Honesty	√									√					√	√
	Fairness											√				√	√
	Consistency												√	√			√
	Trustworthiness			√												√	
	Unity														√		
Work Commitment	Diligence													√			√
	Responsibility							√									√

3.1 Self-Integrity

Self-integrity is one of the elements of a student's integrity during peer assessment. Self-integrity is defined as something that happens inside an individual or in his/her mind. In peer assessment, self-integrity of a student influences the integrity of conducting the assessments that have several sub-elements. There are three sub-elements of self-integrity, namely self-motivation, courage or assertiveness, and self-discipline.

 https://doi.org/10.1051/matecconf/201815005002

3.1.1 Self- Motivation

Self-motivation is defined as an inner drive of a person to achieve the goals set. Students who have self-motivation are more likely to motivate themselves to evaluate the work of others, and at the same time, rated by other students; peer assessment based on such positive attitudes and practices will produce results of high validity and reliability. Helpful and patient teachers are a good example who will motivate students to become effective evaluators. However, there are students who are not self-motivated to conduct an assessment. When students are not self-motivated, they feel peer assessment is a burden to them. Hence, self-motivation plays a crucial role for students to conduct peer assessments.

"*..He/she does not have knowledge..*" (Participant 1)

"*..What I am expecting from students...maybe they cannot conduct a good assessment due to the fact that he/she does not have knowledge, commitment, or he/she does not trust, dishonest among their friends..*" (Participant 2)

"*..Peer assessment can encourage students to exchange opinions and accept others' opinions... indirectly it can increase students' motivation in groups, to give full commitment in group activities..*" (Participant 3)

"*..At the same time, the absence of transparency causes a student who is assessed with no transparency to have less motivation in learning. This is because they have done as well as possible, but peer assessments have destroyed their brilliant performance..*" (Participant 4)

"*..I think the best thing is to see the view from others, others' angle, other than from the lecturer's view... assessment toward, from the lecturer's view, maybe from the soft skills that students assess themselves, through peers...how they conduct their field work, communication with client...with the absence of lecturer, students in that group know he/she is good in communication, they can rank them..*" (Participant 5)

3.1.2 Courage or Assertiveness

Courage or assertiveness is the quality of a person who acts upon and maintains the principles they believed in. Past studies found that some students are stressed out in conducting peer assessment. This proves that the students do not have the courage to make a decision in giving marks to their friends. Students think that they are not qualified to give marks to their friends as their knowledge is at about the same level as their friends. Courage within students is needed to overcome the integrity issues in peer assessment.

"*..we want to train students to become brave, confident in assessing their friends..*" (Participant 2)

"*..A positive learning environment, for example, students should be brave and firmly give an opinion during peer assessment..*" (Participant 3)

"*..Other than that, students who assess other friends should be credible, not easily influenced and reliable. In this case, when they assess, they must strongly stand on their judgment and cannot be influenced by others..*" (Participant 4)

3.1.3. Self-Discipline

Self-discipline refers to the quality of someone who can follow the guidelines given or regulations that have been set. Students who have a high degree of self-discipline are able to follow the rules and guidelines of evaluation while conducting peer assessment. Results from the thematic analysis found that students need to evaluate the work of their colleagues using guidelines and the criteria that have been set, so that the validity and reliability of an assessment are unquestionable. When students follow the guidelines, they will focus on the job of assessment rather than other things such as their personal problems.

"*..If students want to evaluate their friend, he/she will not take it seriously..*" (Participant 2)

"*..Evaluating according to rubric that had been provided, discipline should be taken into consideration in order to avoid distraction on integrity..*" (Participant 3)

3.1.4 Transparency

Transparency means students must be honest with the evaluators when a peer assessment is being conducted. Transparency is inward quality or honest behaviour of the students. It is an important ingredient in creating integrity of students. When students have a good trait of transparency in conducting an assessment, they will exhibit the same behaviour in other matters.

"*..If he/she likes, he/she will give high marks to that person. But if we ask a female student, for sure she will give high marks..*" (Participant 1)

"*..When a friend assesses other peers' cooking styles, he/she should judge based on the rubric and give criticism for improvement, and praise if it is brilliant, so that this marking is transparent ... In this case, a student must be open and accept the views and criticism of others to avoid mistakes, improve any weaknesses..*"(Participant 4)

3.2 Integrity in social interaction

Social interaction is the act of an individual associated with or has contact with other individuals. Integrity of social interactions occurs outside of student's life. Integrity of social interactions can be related to external factors that contribute to a student's criteria. There are

https://doi.org/10.1051/matecconf/201815005002

five sub-elements under the integrity of social interactions: honesty, fairness, consistency, trustworthiness and unity.

3.2.1 Honesty

Honesty refers to intention and ability to do something in a transparent manner, be open to yourself and others. In conducting an honest appraisal, a student rates transparently and openly without being influenced by feelings of bias. Analyses found that students tend to give high marks because of their compassion, relationship with the student being assessed and the status of that student. Honesty is hard to measure, but when people around behave in an open and transparent manner, they create an environment of honesty; hence, it is possible to train and develop a good and honest student. People around with negative attitudes can influence the honesty of students as they tend to listen to their friends' opinions and do not care about the integrity when they assume the role of an assessor.

"..We know that in the class he is a friend of that person, so, we need to ask him to evaluate others.... From the teacher's aspect, this student may not be able to conduct the assessment well because he does not have knowledge, is not committed or because he is not trustworthy, and is dishonest with her friends.." (Participant 2)

"..Students' perceptions are varied; hence, their implications will expect something different, will ultimately lead to unequal and illogical scoring systems ... There are those who are good friends and the impact is high marks and vice versa .." (Participant 4)

"..Being dishonest when they are assessors in peer assessment...depicts low performance but high scores .." (Participant 6)

3.2.2 Fairness

Justice is a quality of someone who can make fair decisions, without bias and prejudice, especially when the decisions will affect the welfare of other people. A peer assessment will be useful when the process is carried out in a fair and equitable manner. Students who are fair in the assessment will produce high-quality assessment grades that reflect the real performance of those being evaluated. The study findings reveal that the grades of a fair and equitable assessment conducted by a friend are not different from the scores assessed by the teacher. The practice of fairness among students could create an environment that will generate good and quality marks in peer assessment; the students know how to be objective and handle the process of giving marks with a cool head. Sometimes, it is hard for students to be as fair as they can; teachers, as the facilitators, can play their roles by reminding and encouraging students to maintain the stance of impartiality when carrying out peer assessments.

"..Students must be transparent to evaluate a friend fairly...do it fairly, without being influenced by the feeling of giving more scores to certain individuals.." (Participant 4)

3.2.3 Consistency

Consistency refers to the quality of someone who organises something repeatedly based on the same principles. Students must be able to carry out evaluations with integrity consistently without being affected by other factors. The findings of the study show that when students assume the role of an assessor, they should adhere to the criteria specified strictly in evaluating the work of their friends; the marks given will then be fair and unbiased. Consistent marks given during peer assessments are of paramount importance; such practice gives the assurance, both to students and teachers, that peer assessment is a workable and reliable method of evaluation. If students give marks according to their moods or swayed by their assignment burden, the marks given will not be valid and cannot be used as an instrument to gauge the performance of a student.

"..What the teacher expects is that, the student may not be able to do well because he has no knowledge, no commitment or because he is not trustworthy, dishonest among his friends.." (Participant 2)

"..Their assessment must be based on a rubric set by the teacher rather than on emotional pursuing.." (Participant 4)

"... if it's an individual, it's possible to have integrity, for example we are helping each other ... in a group of 5 students ... we ask each of the students to assess their friends ... but if it is in the group, they need to discuss first ... for example in a group of 5 students, the first group will present and the score for the first group will be discussed among the 4 groups ... "(Participant 5)

3.2.4 Trustworthiness

Trust is the feeling of confidence in someone that he or she has the ability to do something. Students believe that the results given by their friends as peer assessors are valid and reliable. The data from thematic analysis found that students trust their friends who play the role of evaluators while conducting peer assessment. However, there are researchers who said that students are worried with the marks given by their friends. This is because students think that their friends cannot be trusted and it is hard even to trust themselves. This normally happens when students themselves do not have the confidence, and they also have no confidence in their friends; hence, in their mind, no one around them is trustworthy. Good assessors should exhibit the quality of trustworthiness during peer assessment, so that the marks given are valid and can be used as positive feedback to improve their friends' performance in the future.

"..Incurred losses in assessment due to a miscalculation, resulting in a student getting low marks, who was supposed to get a high score.." (Participant 4)

3.2.5 Unity

Unity refers to the consensus of fellow students in the class in giving assessment feedback without being influenced by any factors. Unity of students in the class is important as the value and quality of scores in peer assessment depend on them. When all the students are united and cooperate with one another in conducting peer assessment, the process will be smoother and acceptable to all people.

"..They will know if their friends do the work well, as they work in a group.." (Participant 2)

"..Students are educated with collaborative skills, which will create unity among students. If the students are united, there will be no student with no integrity... If one person does something wrong, all the group members will be involved later.."(Participant 3)

3.3 Work Commitment

Work commitment refers to the strong belief of students to do a job well while conducting peer assessment as an evaluator. A student needs to view work commitment as a serious matter and incorporate it into their lifestyle, as this can train them to be more responsible in the future. Work commitments are important in the process of peer assessment; each score given by a committed evaluator is valid and can be used to measure students' performance. Students who do not have work commitment will think that peer assessment is just another normal evaluation exercise; they do not take the assessment seriously and therefore the results are not useful. The sub-elements of work commitment are commitment, diligence, responsibility and ethics.

Commitment is the quality of someone who has the passion in carrying out an activity or assignment. Students who have a high level of integrity are committed to do their assignments. Previous studies found that students are comfortable when they are evaluated by their friends, and this method can be used to improve their communication skills [1]. Students must be encouraged to embrace commitment when they are in the school; when they are working in the future, the quality of commitment will show up in the workplace, leading to excellent work performance. A student who does not have commitment in conducting a peer assessment will not do a good job and produce results that are not useful to anyone.

3.3.1 Diligence

Diligence refers to the quality of someone who is hard-working in carrying out the tasks given and does something earnestly. A diligent person will endeavour to carry out a task to the best of his ability. Peer assessments are an on going process that needs an appraiser who can diligently assess and give marks to the work of a peer, commensurate with the quality.

"..Not focus. He/she does not have the commitment, does not know what to assess ...... but to me if the administrator observes, it will be much easier because we've prepared. We have tried. But peer assessment may not be carried out .. "(Participant 1)

"..They compete, which group will be able to collect the most stars, and they will get a prize." In conclusion, they would get the star for the sake of what I have commented.."(Participant 2)

3.3.2 Responsibility

Responsibility in this study is the quality of someone who is willing to be accountable for a duty given in order to achieve the goals and objectives set. Responsible evaluators are students who accept the job of peer assessments in accordance with the procedures laid down by the lecturer without being influenced by other factors. According to the searched results of research literature, responsibility is an element of the student integrity, and students are aware of the job responsibilities of an assessor when they conduct a peer assessment.

"..first is that students should know that they should be sincere to take up the responsibility as someone who needs to assess other friends...Sometimes, students assume the task of evaluating, which is the duty of the teacher, and they think they do not need to be involved with the assessment task .." (Participant 4)

3.3.3 Ethics

The element of ethics refers to the quality of a student who sticks to good moral principles when conducting a peer assessment. Ethical students can distinguish between good and bad actions while conducting the assessment. With the practice of good ethics among students, the running of the assessment process will be smooth, which in turn will produce useful feedback for the assessors and their peers. Ethics are an important part in the social development of students' life; when students do not have any ethics, they will live their lives carelessly without rules and principles. This will affect the employability of students in the future.

"..Students must understand what needs to be evaluated and what the values are according of what the lecturer wants him to value... Do not evaluate just like that..." (Participant 3)

"..If you have some kind of secrecy, it is about the marks.." (Participant 5)

4 Discussions

In an effort to increase the acceptance of peer assessment, three elements of student's integrity need to be considered in conducting peer assessment, which are self-integrity, integrity in social interaction and work-oriented commitment. The element of self-integrity has four sub-elements, and they are self-motivation, courage or assertiveness, self-discipline and transparency. The statement is supported by a previous study that states students are not prepared to conduct peer assessment [7]. In addition, they are worried that their opinions expressed during peer assessment might be used by other students, and some students might not be serious in conducting peer assessment [8]. Some researchers found that there is a problem in controlling the cost of conducting peer assessment, and there is a lack of trust in peer assessment [9]. This is due to the reason that students might give higher marks to their teammates [9,10]. There are several factors that can influence the self-discipline of students such as nature of a subject, number of products or performance assessed, gender difference, racial and cultural background [12,13]. From the findings of a previous study, the gender-based differences may influence the transparency of a student whilst doing the assessment [14].

Integrity in social interaction has five sub-elements: honesty, fairness, consistency, trustworthiness and unity. Honesty of a student can be influenced by friendship, reference, purpose (development or grading), feedback (effects of negative feedback on future performance), collusive behaviours (lack of differentiation) bias, enmity or other power processes and the popularity of individuals [15, 1]. Hence, friendship bias does not affect the outcome of peer assessment as marks given by teachers, peers and non-peers are the same and according to the quality and performance [16]. In addition, students do not understand the criteria of peer assessment, which will not lead to quality assessment [17]. Teachers should provide some guidelines by giving practical examples so that students can use the right procedure to conduct peer assessments [18]. In peer assessment, students think that they should be assessed by higher-level students in order to receive feedback from persons who are more experienced [8]. In the same way, marks from several assessors should be taken into consideration because the average of the scores could neutralise the effects of bias on a student's marks [19].

The element of work commitment has three sub-elements: diligence, responsibility and ethics. Training and briefings on evaluation criteria should be conducted to assist students in understanding the role of the assessor and the process of assessing [18]. Past studies do not discuss the sub-elements of responsibility and ethics in peer assessment. However, responsibility [21] and ethics are important sub-elements and should be emphasised in assessment policies and models. This effort will help to improve the quality of peer assessment.

These elements and sub-elements measure the integrity of students. Each element or sub-element is supported by evidence and therefore is suitable to measure the integrity in peer assessment. Each element or sub-element has been validated by a few experts, and hence all the elements and sub-elements are suitable instruments with the right construct. Each sub-element has been improved based on the interviews of experts in Profession Education and Technical and Vocational Education. The sub-element of ethics has been improved and added. No specific framework of integrity in peer assessment has been developed in previous studies. Hence, the framework of student integrity in conducting peer assessment has been developed in this study. New elements as contained in Figure 1 must be taken into account in order to measure the overall integrity of students in peer assessment.

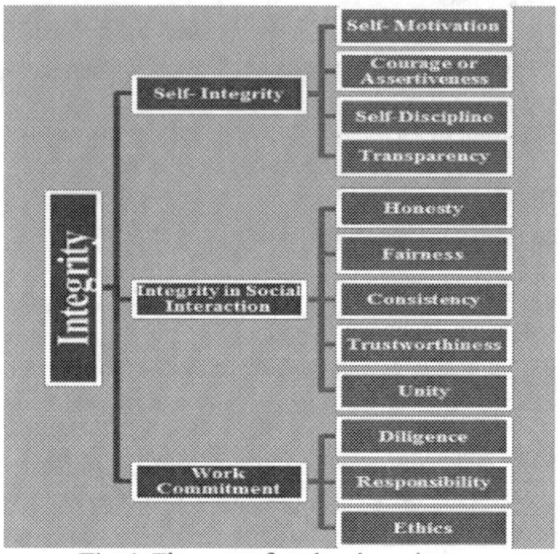

Fig. 1. Elements of student integrity

5 Conclusion

In conclusion, students need to have integrity in order to conduct peer assessments of high validity and reliability [22]. Highly valid and reliable peer assessments are a crucial factor that will convince all lecturers and education experts to accept and implement this alternative evaluation method. A good student who does his or her job with integrity will perform well in the workplace in the future. A student who has a high level of integrity in their learning process will attain better results; they are able to pursue self-regulated learning more consistently and achieve excellent academic performance [23]. Quality students are good assessors and will do a good job during peer assessments [22]. Integrity of students is important and indeed necessary to improve the validity of peer assessments. When peer assessment is perceived as a valid and reliable method of evaluation, many people, particularly those in the field of education will implement this alternative assessment tool in their teaching process. There is no doubt about it; peer assessment can be an avenue to churn out more skilled students in the future. The personality and reputation of a student can be a factor affecting the outcome of a peer assessment [23]. Students who have experienced the benefits of peer assessments think that this is a good practice in the learning process, but some students think negatively of this assessment method [24]. This will adversely affect the integrity of elementary students in the process of peer evaluation. Thus, the elements of the student integrity should be explored further to increase the validity of this assessment, and to motivate teachers to use the peer assessment as a method of teaching and learning. It is suggested that future researchers look into the development of a model for the study of the relationship between the elements of integrity. Lecturers can examine the integrity levels of students while conducting peer assessments.

The author would like to thank the Ministry of Higher Education (Mybrain 15) and Universiti Tun Hussein Onn Malaysia (UTHM) for the funding of this research project.

References

1. K. Topping. Peer Assessment. Theory Into Practice. **48 (1)**, 20 – 27 (2009).
2. A.S. Roberts, S.I. Benrimoj, T.F. Chen, K.A. Williams, P. Aslani. Implementing cognitive services in community pharmacy: A review of facilitators used in practice change. International Journal of Pharmacy Practice, **14**, 163 – 170 (2006)
3. V.D.I. Berg. Peer assessment in universiti teaching : Evaluating seven course designs. Assessment and Evaluation in Higher Education, **31 (1)**, 19 – 36 (2006)
4. N. Falchikov. *Learning together : Peer tutoring in higher education.* London : Routledge Falmer, (2001)
5. F. Tiew. Business student's of peer assessment on class participation. International Education Studies, **3 (3)**, 126 – 131 (2010)
6. V. Braun, V. Clarke. Using thematic analysis in psychology. Qualitative Research in Psychology , 3 (2), 77 – 101 (2006)
7. J. Mok. A case study of students' perceptions of peer assessment in Hong Kong. ELT Journal, 65(3), 230 -239 (2011)
8. W. Elshami, M. E. Abdalla. Diagnostic radiography students' perceptions of formative peer assessment within a radiographic technique module. Radiography, 23(1), 9 – 13 (2017)
9. C. Domínguez, A. Jaime, A. Sánchez, J. M. Blanco, J. Heras. A comparative analysis of consistency and difference among online self-, peer-, external- and instructor- assessment. Journal Computers in Human Behavior, 60, 112 - 120 (2016)
10. G. Xianwie, M. Samuel, A. Asmawi. A model of critical peer feedback to facilitate business english writing using Qzone weblogs among Chinese undergraduate. Malaysian Online Journal of Educational Technology, 4(4), 1 - 17 (2016)
11. A. Konak, S. Magluilo, S. Kulturel-Konak. Behaviorally anchored rating scales for teamwork peer assessment. 2016 IEEE Integrated STEM Education Conference, 168 – 172 (2016)
12. K. Topping. Self and peer assessment in school and university : Reliability, Validity and Utility, 55 -8 7 (2003)
13. H. Meihami, S. A. Razmjoo. An emic perspective toward challenges and solutions of self- and peer-assessment in writing courses. Asian Pacific Journal of Foreign Language Education, 1 - 20 (2016)
14. N. K. L. Pope. The impact of stress in self and peer assessment. Assessment & Evaluation in Higher Education, 30(1), 51 – 63 (2005)
15. H. Saito, T. Fujita. Charactesristics and user acceptance of peer rating in EFL writing classrooms. Language Testing, 25(4), 553 - 581 (2004)
16. M. Azarnoosh. Peer assessment in an EFL context : attitudes and friendship bias. Language Testing in Asia, 3(11), 1 – 10 (2013)
17. P. Orsmond, S. Merry, K. Reiling. A study in self-assessment :tutor and students' perceptions of performance crateria. Assessment & Evaluation in Higher Education, 22(4), 357 - 368 (1996)
18. Sabri & Yunus (2014)
19. Xiao (2010)
20. M. Monga. Intergrity and its antecedent : A unified conceptual freamwork of integrity. Proceeding of the Asia Pacific Conference on Business and Social Sciences 2015, Kuala Lumpur, 171 – 180.
21. A. Barnard. The competencies of integrity. Journal of Psychology in Africa, 21 (2), 267 - 273
22. N.A.B.Z. Abidin, S.B. Sahdan. Peer assessment : A challenge in the new generation. Advanced Science Letters, **23 (3)**, 908 – 911 (2017)
23. S.B. Sahdan, N.A.B.Z. Abidin. Self-regulated learning : A literature review for 21st century learning technology. Advanced Science Letters, **23 (3)**, 912 – 915 (2017)
24. R.L. Hulsman, J.F. Peters, M. Fabriek. Peer assessment of medical communication skill : The impact of students' personality, academic and social reputation on behavioural assessment. Patient Education and Counseling, 92, 246 -354 (2013)
25. H. Asikainen, V. Virtanen, L. Postareff, P. Heino. Peer assessment in a large introductory class of gene technology. Studies in Continuing Education, 43, 197 – 205 (2014)

Effective ways to test changes of practical intelligence in order to assess unintentional learning in laboratory classes

Mohd Hisam Daud [1], *Zol Bahri* Razali [1*], *Mohamed Mydin* M. Abdul Kader[1]

[1]Robotics and Automation Technology, Faculty of Engineering Technology, Universiti Malaysia Perlis, 02100 Padang Besar, PERLIS, MALAYSIA

Abstract. Previous research shows that experience or academically called 'practical intelligence' in the field of engineering can be measured through comparing experience to the experts in the field. The expert teaches and shows the students the ways to complete some practical tasks based on their own practical intelligence, which they developed for years and becomes their behaviors. Thus after some or repeated hands-on exercises, the level of students' practical intelligence continuously developed and close to the experiences possesses by the experts. By this way, the level of practical intelligence can be assessed if it is referring to the experts score. A measuring instrument, an 'Automated Engineers Testing Kits' consisted of a partially completed circuit in which a battery provides power for a flash light. This is a semi-completed circuit which requires students to diagnose why the light does not work and complete the necessary connections. The authors hypothesize that practical intelligence measured in the context of constructing simple circuits used for laboratory experiments will be correlated with performance in real constructing tasks on similar complicated electrical circuits. The results show the correlations between the level of practical intelligence and the ability to construct the circuits. The methodology is described in the paper.

1 Laboratory Classes

Engineering is a practicing profession; a profession devoted to harnessing and modifying the three fundamental resources those humankinds has available for the creation of all technology: energy, materials and information. The overall goal of engineering education is to prepare students to practice engineering and in particular to deal with the forces and materials of nature. Thus, from the earliest days of engineering education, instructional laboratory classes have been an essential part of undergraduate and, graduates programs. Indeed, prior to the emphasis of engineering science, it could be said that the most engineering instruction took place in the laboratory.

The emphasis on laboratories has varied over the years. While much attention has been paid to curriculum and teaching methods, relatively little has been written about laboratory classes. One reason for the limited research on practicing laboratory classes may be a lack of consensus on the basic objectives of the laboratory experience. While there seem to be general that laboratories are necessary, little has been said about what they are expected to gain. In most papers about the practicing of laboratories, no expected outcomes or implicit knowledge are listed, even though it is not unusual for the author to state in the conclusion that the objectives of the course were met. An accepted set of fundamental objectives for laboratories would help engineering educators focus their efforts and evaluate the effectiveness of laboratory experiences or 'practical intelligence' [1].

Conventional approaches of assessment in engineering education focus mostly on explicit cognitive learning outcomes. Typically laboratory classes have been evaluated by assessing explicit knowledge and student perceptions of their laboratory experience. The authors have not been able to find any research undertaken to measure tacit knowledge or 'practical intelligence' acquired during laboratory work. Of course, workshop skills have been traditionally assessed by observing students performing their work and the quality of the artifacts created in the process. Practical intelligence is a critical part of these skills. Developing ways to include effective assessment of practical intelligence could be one way to overcome this difficulty.

1.1 Practical intelligence in laboratory classes

As outlined in [2], practical skills include the ability to acquire useful knowledge from experience, including "tacit knowledge" and unwritten know-how that is not explicitly taught and is often difficult to articulate, and to apply this knowledge to solving complex everyday problems. Complex everyday problems are distinguished

*Corresponding author : mhisam.daud@gmail.com

from academic problems in that they are practical, must be solved with incomplete information, and often do not have a single correct answer...

Research on engineering practice confirms the importance of unwritten know-how or practical skill in everyday laboratory problems. Careful studies of engineering practice (e.g. [3]) have revealed that extensive technical knowledge is needed. Most of this unwritten knowledge is acquired after completing university courses and much of it is surprisingly basic. For example, engineers need to know the components and materials used in their discipline as practiced within a given firm, at least to the extent that they can recognize components and understand what they are used for. Much of this knowledge is so relatively simple on the one hand, and so specific to a particular firm or industry sector on the other hand, that it would not be appropriate to attempt to teach it in university engineering courses [3].

However, students need to appreciate the significance of this 'implicit' knowledge or 'practical intelligence' in engineering practice. However, since engineering courses restrict most assessment to explicit knowledge (the students have to write or speak to convey their knowledge), it is possible that the perceived relative value of practical intelligence and tacit knowledge may be reduced in the view of students. This might help to explain why employers often criticize the quality of the practical skills of engineering graduates [4].

Through their laboratory experience, the authors expect that students may acquire practical intelligence. It is possible they may learn enough for troubleshooting: to be able to detect and solve problems or diagnose faults in the equipment. This experience develops either intentionally or unintentionally and the authors hypothesize that informal learning is an important aspect of laboratory work.

1.2 Assessing unintentional learning through diagnosing circuits' fault

In current situation, complicated engineering process and systems need a systematic method of diagnosis. Thus, in recent years research efforts have shown that the technical problem due to faults can be detected in an early state by using diagnostic system. The diagnostic system must be equipped with the programming knowledge of how to relate faults and their effect to the operating state [5], but the programmers through their explicit knowledge develop that programming knowledge.

In term of human, the diagnostic engineer or technical person must be well self-enhancing knowledge of how to relate faults and the implications, which to learn from experience or gained by unintentional learning. This self-enhancing knowledge is developed through their working experience, and either explicit or tacit, but is expected mostly unintentional. By utilizing this knowledge, they will be expected to provide information of diagnostics for failure localization, planned preventive maintenance and service staff [6].

In order to increase critical thinking and awareness in terms of the equipment's fault, the need for training in fault diagnosis has been identified [5-8]. Thus, to become an engineer who will be able to diagnose the equipment faults, the engineering students have the opportunity to practice these qualities in their laboratory work. Hence the exercise on diagnosis of equipment fault is good practice for them. However, to achieve the performance in the exercise, pedagogical support was needed. Regarding to [9] in his experiential learning theory, four stages in experimental i.e. active experimentation, concrete experience, reflective observation and abstract conceptualization, should be incorporated in the practical exercise. When this practical exercise placed in this sequence, the stages form the experimental learning cycle, i.e. learning-by-doing [10-11].

Johnson [12] studied novice and expert trouble-shooters extensively in order to understand their cognitive processes and skills. This and many other similar studies demonstrate that trouble-shooters make extensive use of tacit and implicit knowledge which has to be developed through experience. This is a powerful argument in support of the need for engineering students to practice and value the acquisition of practical intelligence.

The question is, do the students who gain experience during their laboratory classes possess a high level of practical intelligence acquired through informal learning which might allow them to diagnose the faults of equipment. Therefore, in this study, the authors examine informal learning through experience of laboratory work and the subsequent ability to diagnose equipment faults. Before the authors can achieve this goal, and given the well-known influence of assessment practices on student learning, the authors need to develop reliable ways to measure and assess the acquisition of practical intelligence. Psychologists, as shown above, have provided the required methods. All that remains is to develop specific testing instruments in the context of fault diagnosis.

2 Research Methodologies

2.1 Purpose of research

The aim of this research is to develop ways to measure changes in practical intelligence in order to assess unintentional learning or tacit knowledge, in engineering laboratory classes. In other words, the authors wish to develop ways to measure the experiential and "hands-on" component of learning laboratory classes.

The authors would also like to test the relationship between practical intelligence acquired in laboratory classes with the ability to complete circuit faults in laboratory equipment arrangements.

2.2 Research Design

The study uses a quasi-experimental design of "Non-equivalent control group design" [13] with 2 x 2 factorial designs. The experiment design is as below:

G ----------- O1 ------------ X ----------- O2

G - Experimental and control sample groups
O1 - pre-test (using Automated Engineers Testing Kit)
X - Treatment (lab work for group experiments)
O2 - post-test (using Automated Engineers Testing Kit)

The testing instrument was used to test a number of students (n=39) before and after they performed the relevant laboratory experiment tasks (named as *experiment group*). The pre-test and post-test contained the same circuits faults. A control group (n=20) was recruited from a similar population students who were due to enroll in the same subject in the following semester. The control group completed the pre-test and post-test diagnosing circuits faults twice with a similar elapsed time between exposures, but without experiencing laboratory task. Seven domain experts such as laboratory demonstrators and electronics technicians provided reference scores as mentioned above

.2.3 Circuits' fault diagnosis instrument

An instrument of circuits fault diagnosis is developed and known as 'Automated Engineers Testing Kits' is consisted of a partially completed circuit in which a battery provides power for a flash light. Participants were presented with the instrument and they were asked to diagnose why the light does not work and complete the necessary connections. Although the task seems very simple, almost trivial, it was necessary to design a task for which the students' scores would provide sufficient variation to provide statistically meaningful results. A substantially more challenging task may have resulted in performance being more related to random chance than acquired practical intelligence. **Figure 1** shows the test kit for diagnosing the circuits' faults.

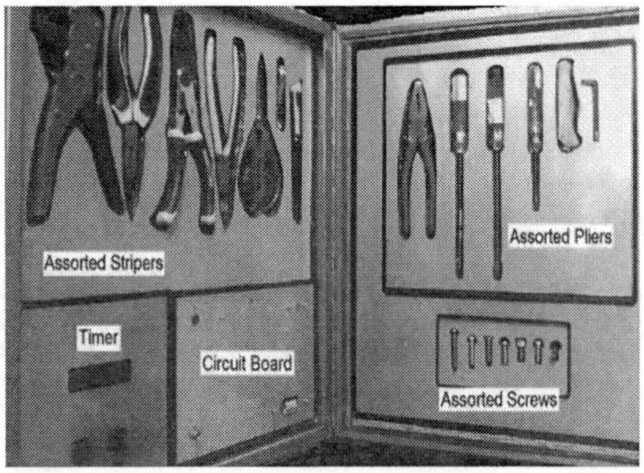

Figure 1: Automated Engineers Testing Kits

.2.4 Research Hypothesis

The authors proposed a null hypothesis:
Hypothesis: *The correlation between practical intelligence and the ability to diagnosis/repair basic circuits' faults is statistically insignificant.*

If this hypothesis is proved to be false, the researcher can conclude that the acquisition of practical intelligence during the laboratory sessions significantly helps the participants to complete and diagnose faults in the circuit equipment used for the experiment.

This study is important to explore the relationship, if any, between Practical Intelligence score and ability to compete and repair the circuit's faults. The authors also need to explore the relationship, if any, between these measures and conventional assessment scores from lab reports, tests and examinations. Knowing that engineers need to be able to diagnose equipment faults and decide on repairs or modifications in their work, it would be useful to understand the kinds of assessment that could predict performance in this kind of work.

2.5 Circuits' faults testing items

There were 5 faults items (**Table 1**) in the circuit where the participants had to diagnose and repair it.

Table 1. The faults and diagnosis/repair tasks

Location	Faults	Diagnosis/Repair task
Point 2	Spring terminal gripped the insulation instead of a wire metal.	Detect the fault and let the spring terminal grip the metal wire
Point 3	There was no screw to connect wires from point 2 and to point 3	Choose an appropriate screw from assorted screws provided.
Point 3	Tighten the screw ensuring that the wires are secured together	Choose an appropriate tool from various tools provided. Prevent the wires from becoming detached from the screw terminal.
Point 3 and 4	Two unstripped wires are too short to connect two terminals required.	Choose an appropriate stripping tool from various tools provided and stripping the two unstripped wires.
Point 3 and 4	Connect the two short wires	Choose an appropriate way to connect the wires
Entire circuit		Measure the time to complete repairs to the circuit and for the light to come ON.

3 Analyses

The main aim of this study is to find the value of practical intelligence acquired by correlating the score of diagnosing and repairing circuit faults for students and experts. In this analysis, the authors are interested in calculating the Pearson product-moments correlation between the pair of variables. For this, the authors made an analysis to test the possibility of correlation between the variables (students' scores and expert's scores).

3.1 Analysis across samples

The results of this investigation shows in **Table 2** demonstrated that the original null hypothesis was false. These results demonstrated that the changes of practical intelligence (PI) can be measured by calculating the difference between students' ratings and the experts' ratings.

Table 2: Results of practical intelligence tests

No	Analyses	Mean (close to experts' mean = 0)	Std. deviation	Sig. (2 tailed)
1	Pre-test (Treatment vs. control)	113.3 128.7	35.34 36.15	p = 0.078
2	Treatment group (pre-test vs. post-test)	113.3 68.3	35.34 18.95	p = 0.000**
3	Control group (Pre-test vs. post-test)	128.7 119.3	36.15 33.80	p = 0.076
4	Post-test (Treatment vs. control)	68.3 119.3	18.95 33.80	p = 0.000**

** Significant at the 0.01 level (2-tailed).

The finding of the primary study showed that there is a statistically significant difference in Practical Intelligence scores (PI) for the treatment group measured before and after exposure to the laboratory class experience, compared to the control group where they did not participate in the laboratory. State differently, the treatment group participants acquired Practical Intelligence during the "hands-on" exposure while engaging with the laboratory experiments.

Based on Table 2

1. Both groups had the same level of initial PI as indicated by the pre-test scores.
2. There is a significant difference for treatment group, with an increment in the post-test close to experts' mean score. Data of standard deviation also shows that the spread of data point tends to be closed to the experts' score. The results suggest that, the treatment group is expected to acquire practical intelligence by performing laboratory tasks. Thus they were able to perform better in the post-test.
3. In contrast, for the control group, there is no significance difference between the pre-test and the post-test scores. Even though, there was an increment in the post-test score, the difference is not statistically significant. The results suggest that the intervening course work on other unrelated studies does not contribute toward PI improvement.

3.2 Analysis between subgroup

The comparison between testing conducted for this study by referring to experts mean score is summarized in **Table 3** and illustrated in **Figure 2.** The participants in each group (categorized by the practical intelligence score) performed the diagnosis/repair equipment faults and were awarded a diagnosis score.

Table 3: Summary of test conducted

Group	Difference in PI means score (%)	Treatment	Difference in FD means score (%)
Higher-post-test (treatment)	19.69136	diagnose/ repair equipment faults test	13.84615
Lower-post-test (treatment)	34.93827		25.38462
Post-test (control)	44.25926		44.61538
Experts (for comparison)	0		0

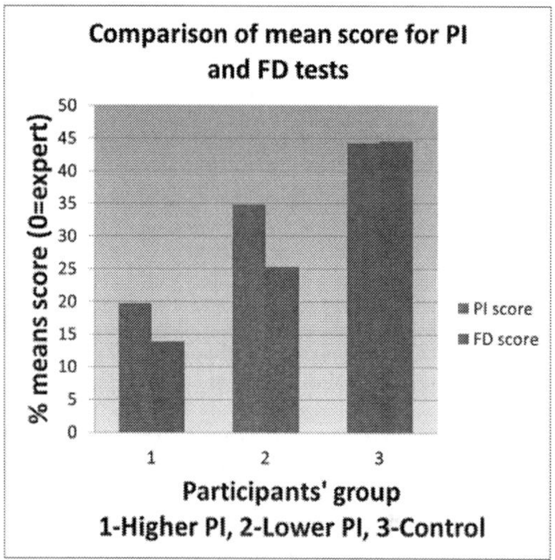

Figure 2. Comparison of the mean for each group with experts

Participants in the *higher post-test (treatment)* group have the lowest difference in practical intelligence score (mean 19.7%) and awarded 13.8% diagnosis/repair mean score. The results revealed that the participants in this group performed well in the practical intelligence test even though are not correlated with the fault diagnosis/repair test. Practical intelligence that they acquire during the laboratory experiments helped them to diagnose and repair the equipment faults, able to use an appropriate tool and complete the exercise promptly.

For the other two groups, the means score for practical intelligence test and diagnosis/repair test are higher than the first group. Even though there are no significant correlation, the authors found that the participants were able to do the diagnose/repair tasks and use the tools although some of them exceed the time given.

3.3 Analysis entire samples

The results of the testing for the entire samples are summarized in the section 6.5.2. There is a correlation between the combination of the post-test scores and diagnosis/repair score, with the value of correlation is 0.752, while the 2-tailed p-value is 0.001. The results show that there is a correlation between the practical intelligence scores and diagnosis/repair scores. The results of the diagnosis/repair test provide further clarification of the relationship between Practical Intelligence and the ability to diagnose/repair equipment faults. **Figure 5** deficits the relationship between the Practical Intelligence score for the 3 groups of participants in the fault diagnosis study with their ability to repair and diagnose equipment faults.

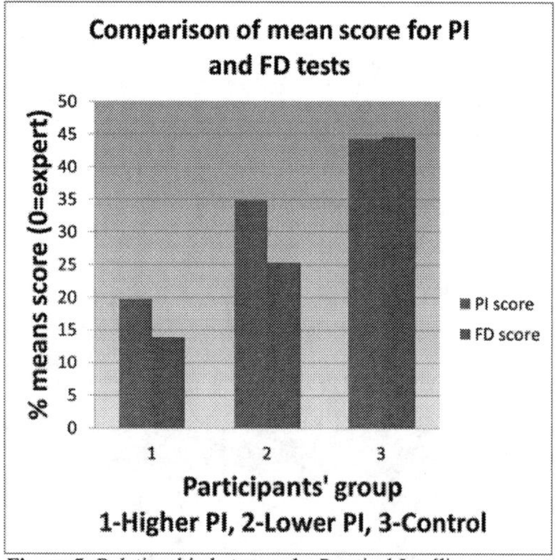

Figure 5: Relationship between the Practical Intelligence score for the 3 groups of participants

The results show that the higher PI group gained lowest difference score in the diagnosis/repair test (slightly parallel in practical intelligence). The post control group gained lesser, but their score is slightly proportional to the practical intelligence. For lower PI group, their score is scattered. The result for this separate analysis shows there is no obvious correlation.

Further research on the relationship between Practical Intelligence vs. ability of diagnosing equipment faults showed a novel relationship (**Figure 3**). The score of the fault diagnosis test is proportional to the practical intelligence score, the higher the practical intelligence score, the higher the fault diagnosis score. Therefore the results suggest that PI scores predict ability to diagnose equipment faults in similar laboratory equipment.

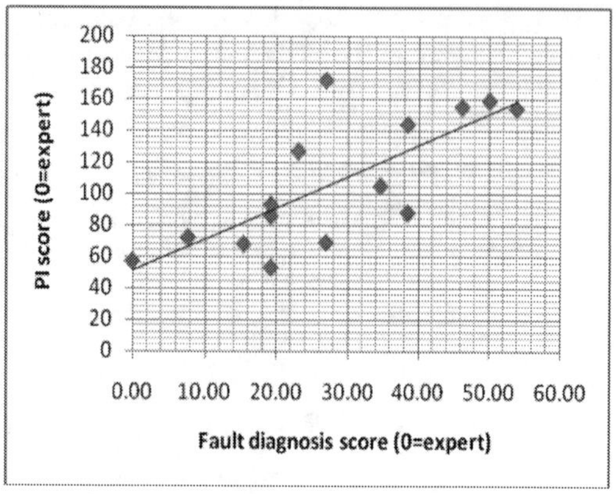

Figure 3: Results of practical intelligence (PI) proportional to fault diagnosis

4 Discussions

Constructing a survey instrument was not an easy exercise in the first instance. The authors were surprised by the relative lack of practical knowledge demonstrated by the students and it was not easy to construct a test which would result in meaningful scores. The authors learned through this research that a close examination and analysis of actual student behaviour in the laboratory is essential, over an extended period of time [14]. This provides significant insights into the learning by students. Through conversations with the students, one can begin to appreciate how learners are constructing new understandings based on their pre-existing knowledge [15]. An understanding of Zygotsky's notion of zones of proximal development was also useful [16] : students can only learn something if they have sufficient pre-existing knowledge on which to construct the required new knowledge.

It is possible that the author may be able to alter student learning behavior by including practical intelligence tests in assessment processes. It is well known that assessment practice drives student learning behavior [10]. The testing may motivate students to acquire the ability to learn practical intelligence which could ultimately make them more effective as practicing engineers. It is possible that they will learn to value the practical intelligence and possibly relate better to tradespeople and technicians on whom engineers need to rely to achieve practical results from their work.

Of course, the actual PI measurement instruments used in this research can only be applied in the context of similar laboratory classes with the same equipment. While the same problem situations might be applicable in other institutions, it is essential to use photographs and diagrams of the actual equipment used by the students.

However, before we can apply this research with confidence, we need many more similar investigations with large numbers of student participants. This paper has outlined the approach we followed. Once the concepts are understood and the necessary detailed observations of student behaviour have been completed, it is possible to readily create an appropriate survey instrument in order to collect data and, hopefully, later, to use the same instrument as a student assessment tool. The same approach can be applied in the context of our remote laboratories. In our situation, we allow the students to have access to the physical equipment. We introduce them to the equipment before they start, both in lectures and tutorials.

Creating a practical intelligence instrument, as described, would require extensive detailed observation of students, both in laboratory working alongside the actual equipment, and remotely operating the equipment somewhere else. However, such an instrument would provide a useful alternate means to assess tacit and implicit learning, and might be a better predictor of troubleshooting performance than conventional written assessments that currently dominate education practice.

5 Conclusions

For this research, the authors hypothesized that practical intelligence measured in the context of constructing simple circuits used for laboratory experiments will be correlated with performance in real constructing tasks on similar complicated electrical circuits. The results show the correlations between the level of practical intelligence and the ability to construct the circuits. The results proved the previous research that experience or academically called 'practical intelligence' in the field of engineering (diagnosing circuit faults) can be measured through comparing experience to the experts. Through some repeated hands-on exercises, the level of students' practical intelligence continuously developed. This is tested and by using 'Automated Engineers Testing Kits' where level of practical intelligence of students' is proved can be assessed.

The results demonstrate that the author can devise effective ways to measure changes on practical intelligence acquired in order to assess unintentional learning from laboratory experiences. This would provide a third means to evaluate engineering laboratory class experiences, beyond the established methods of comparing student performance in explicit assessment tasks (e.g. reports, tests) and measurement of student perceptions of their laboratory experience. The study on fault diagnosis provided a clear relationship demonstrating the possibility that practical intelligence predicts fault diagnosis ability.

The author would like to acknowledge the support from the Fundamental Research Grant Scheme (FRGS) under a grant number of **FRGS/1/2014/ICT06/UNIMAP/** 9003-00442 from the **Ministry of Higher Education Malaysia**.

References

1. L.D. Feisal, A.J. Rosa. Journal of Engineering Education, 94, 1 (2005)
2. R.J. Sternberg. *College and University,* 82, *1* (2006)
3. J.P. Trevelyan. Journal of Engineering Education, 96, 3 (2007)
4. F.V. Christiansen, C. Rump *European Journal of Engineering Education,, 32*, 4 (2007)
5. R. Isermann, *Automatica*, 29, 4 (1993)
6. L. Monostori, *Robotics and Computer-Integrated Manufacturing*. 4, 3 (1988).
7. S.D. Johnson. *Journal of Industrial Teacher Education, 26*, 3 (1989)
8. Z.B. Razali, J.P.Trevelyan, *Annual Conference of the Australasian Association for Engineering Education*, Queensland (2008).
9. D.A. Kolb *Experiential Learning: Experience as the source of learning and development...* New Jersey: Englewood Cliffs, NJ: Prentice Hall (1984)
10. G. Gibbs. *Research, Teaching and Learning in Higher Education S. B and B. S* (1988)
11. K. Garmer, L. Sperling. *Applied Ergonomics*. 33, 1 (2002).
12. D. Halpern. *Thought and knowledge: An introduction to critical thinking*. Hillsdale, NJ: Lawrence Erlbaum Associates (1989).
13. S. Issac, M. Michael, *Research Design in Education*. New Jersey: Englewood Cliffs, NJ: Prentice Hall (1963)
14. Z.B. Razali, J.P.Trevelyan. *Jurnal Teknologi (Social Science) UTM, 64(1) (2013) p11-16.*
15. J.D. Bransford, A.L. Brown, R.R. Cocking, M.S. Donovan & J.W. Pellegrino (Eds.). (2000). *How People Learn: Brain, Mind, Experience and School*. Washington DC, USA: National Academy Press.
16. A.L. Brown, D. Ash, M. Rutherford, K. Nakagawa, A. Gordon & J.C. Campione (1993). Distributed expertise in the classroom. In G. Salomon (Ed.), *Distributed cognitions: psychological and educational considerations* (pp. 188-228): Cambridge University Press.

The Readiness Level of Using e-Learning Among Teacher Graduate Programme (TGP)

[1]*Noorazman* Abd Samad, *Wan Mohd Rashid* Wan Ahmad, *Hairuddin* Harun, *Siti Noor Fazelah* Mohd. Noor, *Affero* Ismail, *Mohd Wafi* Abd. Rahman, Randy Justin.

[1]Faculty of Technical and Vocational Education, Universiti Tun Hussein Onn Malaysia,

Abstract. Using e-Learning among Teacher Graduate Programme (TGP) to students who are newly exposed in Teaching and Learning (T&L) proses. The major problem among student's TGP is the readiness level in knowledge, attitude and skills using e-Learning during Teaching and Learning (T&L) . This study is to identify any significant differences of the knowledge level between male and female TGP students. The design of this study is to survey research using instruments of questionnaire which are distributed to respondents among TGP students at the Faculty of Technical and Vocational, UTHM. The sample is given to 248 respondents and analysis using version SPSS 21.0. Data presented using descriptive method such as, Mean, Standard Deviation(SD), inferens statistic using Mann-Whitney U to identify the knowledge difference between male and female TGP students. The result showed that the level of application of e-Learning among TGP students in terms of knowledge, attitude and skill at moderate level. The study also showed that there are no significant knowledge difference among male and female TGP students because the result of $p= 0.044$ value is less <0.05 can be stated that Ha failed rejected. The results of this study is important to provide awareness especially to the TGP students, lecturers and also UTHM in providing new T&L using e-Learning.

Keywords: e-Learning, attitudes, skills, knowledge

1 Introduction

Globalization leads to cultural change, social and political. Nowadays, the concept of a borderless world makes learning globally, liberalization and global change information exists is because of the globalization. According [6] advocated that Information and Communications Technology (ICT) have a huge impact on the shift change and the management of education systems in developing countries such as Malaysia. ICT is also a need in everyday life due to the use in the increasingly sophisticated today. Life on this day is much dominated by ICT, while the future development of a country is dominated by the development of science and technology.

The development of science and technology is closely linked to the education system in Malaysia. [20], stated that the learning system is one of the factors that affect the competence of approval among students. Changes in the education system had n developed quite rapidly in line with the development of ICT that has brought a huge change for the advancement of education. The development of the learning method also had undergone many changes, both for individual learning, instructional media, and the learning process. According to [5] and [17], the change is based on the development of ICT. However, students need to have satisfaction with tradisional learning system before e-Learning can be implemented.

The disadvantage of tradisional learning needs to be improved in terms of delivery of instruction for teachers to overcome. The statement is in line with [13] where it is believed that tradisional learning system is ineffective due to the rapid technological developments and will cause institutions that still use traditional learning systems be left behind. Technological transition will leave tradisional learning and in turn to e-Learning.

Therefore the development of education in Malaysia started with the traditional teaching which adopted the teacher-centered approach and widely used in 21st century education. [15] stated that some teachers use methods of

*Corresponding author:noorazman@uthm.edu.my

speaking and writing (chalk and talk) and one-way communication only. This method had proved that fast learners get bored easily and had caused students to be more passive. The learning process should be dependent on the teaching content in the classroom. According to [2], the development of internet usage leads to on-line learning, better known as e-Learning. E-Learning is one of the alternative methods used today for traditional learning methods against the weakness in terms of one-way communication between students and teachers. According [16[, also argues that, communication between teachers and students, students and students and students with the global web of material occurs in the shortest time through this network and electronic learning makes students not only active but also responsible for learning.

This statement was also supported by [2], where students can make plans and find information with its own efforts and initiatives. Information search is not limited because the materials can be selected by the student, whether perceived or not. Students do not shy to repeat the same title to achieve the level of satisfaction of their understanding before moving on to other topics. In addition, its more flexible and less time pressure because it can be done anywhere as long as there is a computer and internet facilities.

2 Research Problem

The use of e-Learning in Malaysia is still new compared to the developed countries, where e-Learning is used as teaching and learning process But according to the study [14], the lack of e-.Learning is limited only to download and upload by lecturers and students involved in a class or lecture only. Therefore, there is constraint information transmitted from several parties such as lecturers and students using e-learning. This statement is in line with reality [17], who stated that students who possess positive attitude towards computers are also more willing to use it compared to the students who have a negative attitude toward computers

The obstacle that led to the implementation of e-Learning is the difficulty to achieve information and a lack of students' skills in using e-Learning. Based on studies by [1], found that meaningful e-Learning can be achieved through the use of the resources available from the internet by students. However, this situation will not happen if the teacher lack the knowledge that can help facilitate the use of ICT among students. According [8], stated that in the process of implementing a program of computer-based information technology proficiency among teachers is required. It is important not only to improve the effectiveness of teaching and learning but also from the administration and management of schools.

Among other obstacles identified are lack of knowledge. Knowledge refers to certain levels of knowledge belonging to the individual [22]. Therefore, it is closely linked to experience and expertise. According to [6], knowledge and attitudes towards computers is affected by the frequency of computer use, the more often a person uses a computer, knowledge level will also rise. The level of knowledge of information technology teachers can be improved through information technology courses. The study [8], shows the role of teachers who are less skilled in producing electronic teaching materials of high quality and difficult to form a teaching based on the level of knowledge and skills in a variety of e-Learning delivery failed to achieve effectiveness.

Based on the issues described as traditional learning weaknesses and failures of the use of e-Learning, the study of the use of e-Learning in learning among the students of TGP can be very effective since they are the main focus. Therefore, e-Learning is an innovative method to improve the effectiveness of teaching and learning process. But awareness and readiness for change among students and lecturers is also an important factor in determining the success of e-Learning.

3 Research Questions

The study was conducted to find answers based on the objective in this study. The research questions is :

i) What is the level of readiness of TGP students in the application of e-learning from aspects of knowledge to the teaching and learning process in UTHM?

ii) What is the level of readiness of TGP students in the application of e-learning from aspects of attitudes to teaching and learning processes in UTHM?

iii) What is the level of readiness of TGP students in the application of e-learning in terms of skills for the teaching and learning process at UTHM?

iv) Is there a significant difference in the level of readiness of TGP students in the aspect of knowledge in the application of e-learning between male and female students?

4 Methodology

The design of this study is using survey design. According [10], the survey methods should be used to obtain samples of a large size because the process using other methods is difficult and complex study design is used as a guide to find a clear answer to the question.

4.1 Population and Sample

Sample is a subset of the study population or part of the population. The study involved population 725, TGP students Cohort 1 at UTHM. The number of samples to be used are a total of 248. This representing a large population, sampling was used to take part set of respondents from this population [3]. Rational use of samples is to facilitate the review to be carried out. Use the sample is large enough to represent the entire population regardless of population studies using proper sampling method [10].

4.2 Research Instrument

Questionnaire as a instrument tool consist Part A- Demographic, Part B- Knowledge, Part C- Attitude and Part C- Skill that have been developed with items Likert scale. Data is analyzing using SPSS Version 21.0. Data presented using a descriptive analysis such as Mean, Standard Deviation (SD) and inferens statistic such as Mann-Whitney U.

5 Result

Demographic analysis shows the number of female participants noted that the number of respondents was higher, namely 130 respondents (54%) than male respondents were 114 respondents (46%). The difference in number of female respondents and male respondents in total of 20 respondents (8%). The total number of male and female respondents is 248 people (100%). Analysis of demographic (gender) is required to analyze if there are significant differences in the level of readiness of TGP students in the application of e-Learning between male and female using Man-Whitney.

From the data analysis, the data showed that the overall level of readiness students' knowledge, skill and attitude among TGP to use e-Learning is still moderate level. The overall result average is the mean value of knowledge mean= 3.26, attitude mean=2.94 and skills mean= 2.97. Mann-Whitney U test shows the result of the score no difference between male and female TGP students knowledge in the application of e-Learning. This is from the results obtained the value of $P < 0.05$. This means the value 0.044 and value is less $p<0.05$ can be stated that Ha failed rejected.

6 Finding And Discussion

Students' TGP knowledge, skills and attitude is importance refers to the level of their readiness in e-Learning because its provides control over the computer, knowledge and perception of e-Learning. This results showed that the level of students' knowledge, skills and attitude among TGP student's using e-Learning is still at moderate level. This findings also revealed that , most respondents have knowledge of e-Learning at a moderate level. This assertion is supported by a study of the [19], who stated at the beginning of the use of information technology requires knowledge in the handling of equipment and software. He also said if someone trainers are not knowledgeable in technology it is difficult for them to fully express the meaning clearly. The statement was in line with reality [11] that an educator should enhance knowledge in the field of technology to build a deeper knowledge in the field. Apart from attending the course, teachers also need to explore all science-related technologies that are widespread on the Internet [12]. TGP students' attitudes rather get learning materials directly than do discussions and also active as leaving comments on social services author system. In conclusion, the student TGP attitude towards the application of e-Learning is still at a moderate level. Student TGP should be a change of attitude towards e-Learning and also there should be awareness of the use of computers to improve their ability to use technology. Skills level of TGP student's used e-Learning is still a moderate level. From the research conducted, most of the respondents have moderate level of skills using e-Learning, this also shows their readiness is also at a moderate level.

Consequently skills are very important in the application of e-Learning in which skills students need to be in line with TGP knowledge so that it can be applied perfectly. For example in the use of power point slides, without knowledge, creativity and skills, chances main objective of attracting students to focus on the course will not be obtained [9]. This is because the interface is creative and draw attention to prevent students from getting bored and can help students to better understand the information to be communicated clearly. The statement in line with the statement by the [18], where the skills and knowledge must be reconciled. Lack of knowledge will cause it to be a constraint in the use of information technology as well as the teaching and learning process. Therefore, the skills and the use of technology will not be obtained without knowledge.

Mann-Whitney U test was conducted to see whether there are significant differences between the level of readiness of students TGP in knowledge application of e-Learning between male and female. The findings show that there are differences between the level of readiness of students TGP in knowledge in the application of e-Learning between male and female. Therefore, it is concluded that there are significant differences between male and female students against the use of e-Learning knowledge TGP students at the UTHM.

7 Conclusion

Overall findings of this study can have implications for UTHM especially lecturers on the importance of e-Learning in teaching and learning process. In addition, this study is also expected to bring awareness to students in UTHM in order to make use of e-Learning to improve the quality of teaching and learning more effective. In addition, the administration can also take advantage of in-campus study as an opening to find the best way to produce lecturers who are able to perform their teaching and learning effectively. Based on the findings showed that the application of e-Learning is still at a moderate level.

Therefore, this study is very important to our students to build on the existing system especially using author system. UTHM needs to constantly improve the quality of e-Learning systems and appropriate to the needs of lecturers and students so that they can enjoy optimal use of e-Learning. With the studies like this then it will make it easier for our students to know the weaknesses in systems in particular, author. Therefore, UTHM can design the best strategy to implement the use of e-Learning graduates in their learning process. The issue of this research, the researchers want to overcome the system of e-Learning undertaken by students TGP to improve the quality Teaching & Learning (T&L) but the results showed e- Learning is still at an average level. The main factor researchers realize that online learning or e-Learning is still new for to students to use of e-Learning among students TGP should be upgraded among TGL students in T&L at UTHM

Acknowledgement to U226 - RSGS (Research Supporting Grant Scheme), ORICC Universiti Tun Hussein Onn Malaysia (UTHM).

References

1. Abu Bakar, R.. *Kajian Terhadap Penggunaan Perisian Dan Perkakasan Multimedia Di Kalangan Guru Dalam P & P Di Sekolah Menengah Zon Skudai.,* Tesis. *Universiti Teknologi Malaysia*,(2009)
2. Abdul Rahman M.A. & Shamsudin, F. *Penggunaan E-Pembelajaran Di Kalangan Pelajar Jabatan Pendidikan Teknik Dan Kejuruteraan, Universiti Teknologi Malaysia. Penggunaan E-Pembelajaran Di Kalangan Pelajar Jabatan Pendidikan Teknik dan Kejuruteraan,.* Tesis (Unpublished). Universiti Teknologi Malaysia (2010).
3. Chua, Y.P. *Kaedah dan Statistik Penyelidikan: Kaedah Penyelidikan.* Kuala Lumpur: McGraw Hill (Malaysia) Sdn. Bhd. (2006).
4. Daud, S. M., Hashim, Y., & Saleh, M. N. Faktor-Faktor Yang Mempengaruhi Keberkesanan Pengajaran Dan Pembelajaran Melalui Sidang Video Dalam Pendidikan Jarak Jauh. *Malaysian Journal of Distance Education, 2(1), 67–90)(2000).*
5. Graham,C.R.,Allen, S.. Designing Blended learning Environments, 562-570.In encyclopedia Of Distance Learning Vol.2. (2009).
6. Hassan, J., & Kamisan, S. N.. Halangan terhadap penggunaan komputer dan ICT di dalam pengajaran dan pembelajaran (P&P) di kalangan guru di sekolah menengah kebangsaan luar bandar di daerah Kulai Jaya, Johor. N/A, 1–10. (2010).
7. Hassan, J., Rahman, A., & Rashida, F. Penggunaan ICT dalam proses pengajaran dan pembelajaran di kalangan pendidik Fakulti Pendidikan Universiti Teknologi Malaysia, Skudai, Johor. Unspecified, 1–9. (2011)
8. Ibrahim, N., Wong, S. L., & Ayub, A. F. M. Sikap terhadap Komputer di kalangan Pelajar ICT Tingkatan Empat. Jurnal Teknologi Pendidikan Malaysia, 1(1).(2011).
9. Kassim, Z. & Ahmad, A.R.. E-Pembelajaran: Evolusi Internet Dalam Pembelajaran Sepanjang Hayat .Proceedings of Regional Conference on Knowledge Integration in ICT 2010.209-218. (2010).
10. Konting,M.N. Kaedah Penyelidikan Pendidikan. Kuala Lumpur. Dewan Bahasa dan Pustaka.(2004).

11. Krishnan, J., Said, M., Haruzuan, M. N., Atan, N. A., & Hassan, J. Faktor-Faktor Yang Mempengaruhi Penggunaan e-Pembelajaran Di Kalangan Pelajar Tahun Akhir Fakulti Pendidikan, Universiti Teknologi Malaysia.(2007).
12. Mohamed, W., Azlinda, W., & Hassan, H. Hubungan faktor demografi dan faktor literasi komputer dalam menentukan kesediaan pekerja industri terhadap e-Latihan.(2010).
13. Mohd Balwi, M.K.Perkembangan Pembangunan dan Penerimaan e-Pembelajaran Di Institusi Pengajian Malaysia. *Jurnal Teknologi*, 41(E), 55-72.(2014).
14. Omar,R, & Ahmad,J. Kesedaran,Penilaian dan Penerimaan e-Pembelajaran dalam Kalangan Ahli Akademik. *Jurnal Pendidikan Malaysia* 34(1) (2009):155-172. (2009).
15. Ramli, M.E.J..Hubungan antara tahap kesediaan Pembelajaran Arahan Kendiri Dengan Tahap Penggunaan E-Pembelajaran Di Kalangan Pelajar Di Universiti Tun Hussein Onn Malaysia Batu Pahat, Johor. Tesis Sarjana, .Universiti Teknologi Mara.(2009).
16. Rahman, A., Anuar, M., & Ibrahim, E. Y. Penggunaan E-Pembelajaran Dan Hubungannya Dengan Pencapaian Akademik Pelajar Di Fakulti Pendidikan, Universiti Teknologi Malaysia, Skudai, Johor., 1–7.(2010).
17. Singh,H..Building effective blended learning program, Educational Technology, 43(6),pp 51-54.(2003).
18. Suhaini, M. Kompetensi Pelajar Dalam Aplikasi Komputer Bagi Pembelajaran di Kolej Vokasional Pahang, Tesis Sarjana, Universiti Tun Hussein Onn Malaysia (2014).
19. Su, L. T. Kemahiran teknologi maklumat di kalangan pelajar Fakulti Pendidikan, UTM dan hubungannya dengan sikap terhadap e-Pembelajaran. Universiti Teknologi Malaysia, Faculty of Education.(2007).
20. Tahar, I. Hubungan kemandirian belajar dan hasil belajar pada pendidikan jarak jauh. Jurnal Pendidikan Terbuka Dan Jarak Jauh, 7, 91–101.(2006).
21. Yaacob, M., Zainab, A. N., Edzan, N. N., & Abdullah, A. Teknologi Maklumat dan Komunikasi (TMK) pengupaya masyarakat Melayu berasaskan pengetahuan. (2002).

Elements of Technology Enabled/Enhanced Active Learning (TEAL) to Enhance Quality and Employability of Bachelor's Students

Nur Farha Bte Hassan[1,*], *Saifullizam Bin Puteh*[1], and *Amanina Binti Muhamad Sanusi*[1]
[1]Faculty of Technical and Vocational Education, Universiti Tun Hussein Onn Malaysia, 86400 Batu Pahat, Johor, Malaysia.

Abstract. The application of technology innovation is rapidly increasing in industries and educational institutions. This phenomenon has led to the emergence of Technology Enabled/Enhanced Active Learning (TEAL) which emphasizes the use of various techniques and technologies. TEAL is a new learning format that combines educational content from a lecturer, simulation, and student's experiences using technological tools to provide a rich collaborative learning experience for students. This approach is used to provide academic professional development that brings innovation to the learning content, practically by using pedagogy, technology and classroom design. TEAL ensures the enhanced development of student's knowledge and skills in order to produce quality skilful workers with adequate employability skills. Technology is an effective tool used to facilitate the teaching and learning process, which can, in turn, create an active environment for students to build their knowledge, skill and experience. This paper determines the elements of TEAL based on interview sessions with expert academicians and from a systematic literature review. The selection of TEAL elements for this study was carried out using thematic analysis approach. Findings show that these TEAL elements would help institutions to promote students in involving themselves in active learning in order to enhance the quality of graduates in improving their technical knowledge, thereby enhancing their employability skills.

Keywords: Technology Enabled/Enhanced Active Learning (TEAL), Technologies, Pedagogies and Content Knowledge (TPACK)

1 Introduction

Nowadays, engineering and technology industries have a high demand for bachelor degree graduates from Technical and Vocational Education and Training (TVET) programs. TVET programs prepare individuals for work in technical areas and provide these individuals with skills, knowledge, experience and marketability in applied innovation technology [1]. The unemployment rate of bachelor graduates is high, and this has been closely linked to the problem of inadequate knowledge, skills and attitudes of graduates in applied technology and technical work for labour market requirements in the industrial sector [2,3]. Figure 1.1 shows that the highest unemployment rates in Malaysia were bachelor degree graduates with a 27.9% unemployment rate compared respective groups as reported in the data source obtained from the Ministry of Higher Education and Bank Negara Malaysia estimates [4].

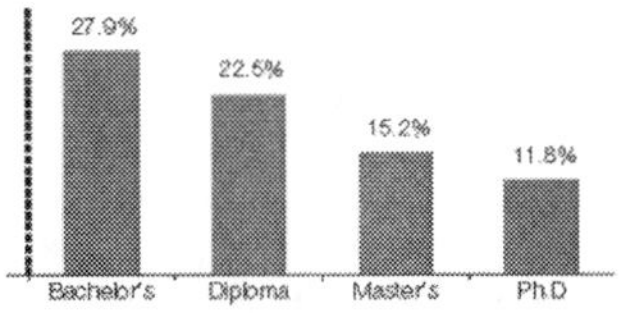

Fig. 1.1: Unemployment Rates of Graduates by Qualification (Ibrahim & Mahyuddin, 2017)

*Corresponding author: hb140185@siswa.uthm.edu.my

This indicates that TVET institution needs to prepare bachelor degree graduates with communication skills, technology skills, employability skills and marketability skills to improve their performance gradually [5]. More specifically in providing opportunities for students to apply technology tools practically while engaging in an active learning environment [6].

Thus, Technology Enabled/Enhanced Active Learning (TEAL) can be applied in the learning process. It will help students to meet the demand of industry needs. That is graduates who are skilful, talented, knowledgeable, hand's on practices, with a right attitude and willingness to share ideas to improve the quality of jobs in industrial organizations [6,7]. Therefore, an innovative learning format in applying digital technology, Information and Communications Technology (ICT) equipment, electronic media, application tools, multimedia tools and machine tools are needed. This innovative learning format for these crucially selected elements is required in order to improve the quality of learning in simulations, hands-on and short lectures from instructors [8].

In this paper, to produce technology active learning in education, there is need to refer to the learning pyramid by. [9,10] to help students learn in productive ways so that student can master knowledge related to employability skills. Similarly, in order to produce active learning model, students should be involved in the learning activities.

Students will be given a chance to practice with group discussions, collaborative activities; practical hands-on and active learning during the class session [7]. From figure 1.2 shows the learning pyramid. Effective learning was divided into two parts; namely passive learning such as lectures, reading, audiovisuals and demonstrations; and active learning such as group discussion, practical and collaborative activities [9, 11].

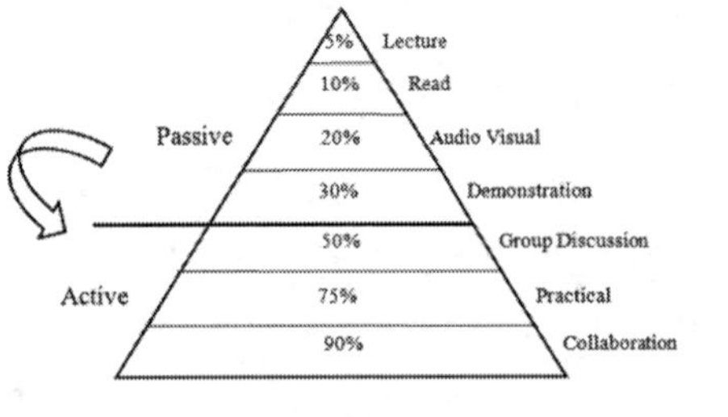

Figure1.2: Learning Pyramid (Lalley, James, Miller & Robert,2007)

The learning pyramid shows that 50% of students need to participate in group discussion, 75% for practical hands-on and 90% for collaborative activities to create active learning environment [9]. In addition, to create more efficient learning, technology will be applied to active learning activities to ensure that students will obtain a better understanding of the learned concept during class sessions [11]. This will help students to become high-quality workers when they employed in various job roles in industries.

Therefore, the theory of Practice Technologies, Pedagogies and Content Knowledge (TPACK) needs to be used as a platform to build active learning environment [7, 12]. This study utilized the TPACK theory to establish the TEAL concept that requires a combination of active pedagogical learning with technology-rich classrooms in producing an effective learning environment. The application of technology is used to provide activities which constitute the implementation of active learning strategy, as it is more focused on learning outcomes and not on technology tools only and involves students in the learning process [13]. It can help instructors to create effective learning by increasing their learning content and pedagogical knowledge in developing a global mindset to become flexible workers in industries after graduation [6,12].

According to the study [14], TPACK can help institutions to increase knowledge about applied technology in curriculum activities; it will prepare students in practical work, knowledge, and skills in the use of technological equipment and machine tools during classroom sessions. To create an active learning environment, there is a need to start from the basic knowledge of technologies, pedagogies and content to incorporate effective learning [12, 14].

Active learning is fundamental to the educational system, which is in line with global technology needs in learning activities needed to produce skilful graduates who will become manpower for industries, who are knowledgeable, productive, competent in work, excellent talented, have high potentials, are tech-savvy and innovation-minded in applied machine tools especially in technical areas that meet industry demands and institution needs [5,15,16].

Moreover, TEAL needs to be applied in teaching and learning to enhance the quality of graduates by using a variety of methods. That is to assist students to be more interested in the learning process, especially in some elements of online learning, blended learning, collaborative learning, practical learning and interactive learning [17, 18, 19].

This paper determines the elements of TEAL in enhancing quality and the employability of TVET graduates based on the TPACK theory for preparing graduates to be more active and excellent in applied technology in their studies. The paper is organized under the following headings: Section 2 methodology of the study, section 3 is about the elements of TEAL; Section 4 discusses the outcomes of the study, summaries and conclusion of this paper.

2 Methodology

The purpose of this study was to determine the elements of TEAL in solving the problems of unemployed graduates based on the theory of practice (TPACK). This study used thematic analysis approach to identify the element of TEAL. Thematic analysis was performed on data obtained from interview sessions with expert academicians and review of the literature. This paper is necessitated as a response to the need in helping graduate become better-prepared workers after graduation from studies.

3 Elements of Technology Enabled/Enhanced Active Learning

The selection of crucial elements of TEAL is based on a summary of interview data and documents from previous research. From the literature and from the perspectives of expert academicians, elements of online learning, blended learning, collaborative learning, practical learning and interactive learning is an approach that is often discussed as being related to learning technology to improve quality and employability of TVET students. TEAL is a technology innovation learning that provides opportunities for TVET students to participate in active lessons and to be involved in quality environment studies as stated in the Malaysia Education Blueprint (Higher Education) 2015-2025 [20,21,22].

This is to prepare students with a variety of approach to learning such as online learning, blended learning, collaborative learning, laboratory learning and interactive learning to build their knowledge and skills related to employability [23]. According to experts, these five elements should be applied in higher education learning and should be embraced by all graduates who want to find a job in industries because they will encounter these elements as a prerequisite to measuring the abilities and skills of graduates while working.

Based on the feedback from experts, it is more about the graduates needing to prepare themselves with the various skills and knowledge expected by prospective employers. The education system and skills according to [24] have also been highlighted as efforts needed to develop more effective learning in terms of combining aspects of online learning, blended learning, collaborative learning, laboratory learning and interactive learning to enhance quality and producing world-class employable graduates that can achieve goals. Summary of these key elements of TEAL is shown in the following table. Table 3.1 shows key elements of technology-enabled active learning

Table 3.1: Key Element of Technology Enabled Active Learning

Key Element of Technology Enabled Active Learning		
No	Elements	Descriptions
1.	Online learning	Students can build social and interpersonal skills to find information, learn through the internet and share information in online conferences.
2.	Blended learning	Students are able to build organizational and management skills in face-to-face learning by using online technology.
3.	Collaborative learning	Students are able to build communication skills by sharing opinions and working together in teams to build motivation and control their emotions.
4.	Laboratory learning	Students can build technical skills in applied technology equipment/machinery to create a hands-on practice.
5.	Interactive learning	Students are able to build cognitive, creative and critical thinking skills in using multimedia software to produce a project or product.

3.1 Elements of online learning

Based on interview session and previous journal findings, researchers list the elements of online learning that support the TEAL approach. All these elements are based on the practice theory (TPACK), in producing quality graduates. It was suggested that technology should be used in class lectures. Technology facilitated learning such as using the internet to build knowledge and skills for finding/sharing information in learning. According to the expert's opinion, all these elements need to be applied in a learning session to help students participate in active learning environment using technology tools for improving their knowledge. Research literature [25,26,27], also discusses the effectiveness of applying computer technology in active learning environments and this is dependent on how the tools were used. For example, how accessible are information via the internet or Wi-Fi to facilitate active learning environment, how can a telecommunication media user facilitate online learning sessions with experts in improving student's technical knowledge, how can web application be used to share ideas and how can Internet of thing (IoT) be used to find information to support learning activities?

Additionally, pedagogical knowledge can also be used to help students adapt virtual and web learning to build social interactions. Interpersonal skills can be developed through the use of the internet to find and share information in online conferences [28, 29]. Thus, students are able to build content knowledge by engaging in extensive exploration activities and searching support learning materials in support of creating critical, creative and innovative thinking to solve problems through e-learning and educational web resources. This provides flexibility in the way and manner by which information is accessed at any time and anywhere [30,31,32]. Summary of the elements of online learning are shown in Table 3.2.

Table 3.2: Elements of Online Learning

Elements of Online Learning		
No	Elements	Description
1.	Computer Technology	Students can learn how to handle the physical hardware of computer technology to assist in the learning process
2.	Internet and Wi-Fi	Students can learn in an environment of learning by using the internet, intranet and satellite Wi-Fi in accessing information
3.	Telecommunication Media /Mobile /IPad	Students can more readily use ICT equipment and tool in building knowledge of meaningful learning
4.	Web Application (Web 2.0)	Students can share ideas and collaborate in innovative ways with expert and instructor in social networks.
5.	IoT (Internet of Things)	Students can find necessary support materials from the internet prior to classes to be better prepared before attending such classes or virtual learning
6.	Web and Virtual Learning	Students can easily get reference from e-learning, e-training and web learning
7.	Social Interaction	Students can interact with colleagues and experts to obtain or share information
8.	Experience Learning	Students use technology through online learning, such as simulations, quizzes, tests and debate discussions to build personal experience
9.	Critical, creative and innovative in Problems Solving	Students are able to improve the thinking of the mind to be more positive and vast in building knowledge
10.	Flexibility in Access Information	Students can search information online at any time and anywhere

3.2 Elements of Blended Learning

The elements of blended learning are summarised based on the interviews from experts and previous research that had been discussed about the use of a combination of lectures and online learning approach to ensure that the process of teaching and learning is more interesting by using telecommunication media and computer technology applications in class. Feedback from the experts reveals that this can help to build organizational and management skills of students primarily on developing the experience to conduct their own learning with technology. Blended learning is an essential element for enhancing student's competence in the use of digital and electronic technology in class through online conferences with experts to enhance the technical knowledge of students, that technologies application is parallel with [33,34,35,36] research.

Other than that, pedagogical knowledge in blended learning enables students to be more confident in social interaction either with friends or teachers to helps minimize ineffective lecture in the learning process that is supported by virtual learning and to create an active environment using technology applications [37,38]. In fact, content knowledge enables students to be able to gain experience in practical work and presentation skills in flexible conditions and explore the content that had been delivered to the audience [39]. It will help students to develop cognitive thinking and social innovation for preparing them with an occupation in technology equipment, software application and machine tools [40]. Summary of the elements of blended learning are shown in Table 3.3.

Table 3.3: Elements of Blended Learning

Elements of Blended Learning		
No	Element	Description
1.	Online Conference	Students can conduct online discussions to ask opinions from experts and partners in order to learn easily
2.	Telecommunication Media and computer	Students know how to apply the ICT tools in build technology knowledge
3.	Digital and Electronic Technology / Clicker Personal Response System	Students will be able to learn quickly through digital equipment using technology systems for sharing answers by using clicker
4.	Social Interaction	Students can connect between friend and teacher in class to ask questions and share information
5.	Mini-Lecture	Students listen to a short description from instructor in order to know what needs to be done in the learning process
6.	Virtual Learning	Students can review the teaching and learning notes via online platforms.
7.	Flexibility in the presentation and exploration of learning	Students will be able to explore the contents independently to aid their learning and share this knowledge during class lessons
8.	Cognitive and Social Innovation	Students can train to be more open-minded in innovation thinking for developing projects, products and work

3.3 Elements of Collaborative Learning

Based on interviews conducted and findings in previous journals, the elements of collaborative learning to identify technology knowledge, pedagogical knowledge and content knowledge of students in enhancing quality and employability skills in the future when they work in industries. According to the data collected and analysed, from the interview sessions, all experts highlighted that collaborative learning is critical to be applied in the class, especially in terms of technologies such as using desktop workspace to produce group projects, using web-based software in designing product, developing projects through internet networks and using digital tools to search information needed like those used in researching [41,42,43].

Moreover, group learning can be used to teach students to become confident in social interaction and communication with other people, to create a discussion group, a forum for sharing information, views and knowledge to build teamwork spirit to control emotions and self-motivation when receiving instructions from tutors and peers [44,45,46]. In order to create content knowledge, students must be given the chance to build accountability of individual and group to identify their own responsibilities in decision making and taking actions to solve problems, and also to improve critical thinking of students in class discussions [47, 48, 49]. These will help students to create effective learning and become active members of the class while enjoying their lessons at the same time. Summary of elements of collaborative learning are shown in Table 3.4.

Table 3.4: Elements of Collaborative Learning

Elements of Collaborative Learning		
No	Element	Description
1.	Desktop technology workspace	Students will be able to learn in workspaces, and this should be available for each group
2.	Web-based software	Students know how to use a variety of software to share information through web-based systems such as Dropbox, emails, Edmodo and so on.
3.	Internet Network and Digital Tools	Students can refer to the internet to get information and operate digital equipment in group work
4.	Social Interaction and communication	Students are able to interact with group members with tolerance and prudence.
5.	Discussion Group Forum	Students are more open to working as a group to share ideas and opinions.
6.	Instructions Tutor and Friend	Students will be able to receive instructions and adhere to the leader of a group discussion
7.	Accountability of	Students are aware of the

	Individual and Group	responsibilities that should be shouldered by each individual in the group
8.	Decision Making and Action	Students are able to make decisions in learning with group discussions
9.	Critical Thinking and Problem Solving	Students will be able to think outside the box and share ideas with fellow members of the group

3.4 Elements of Practical Learning

The summary of discussion with expert and literature form some previous studies, reveal that elements of practical learning are related to the use of scientific tools and machinery in the laboratory with ICT and computer equipment to provide opportunity for students to master a variety of skills through web-based software applications in order to build their technological knowledge, as reported in previous studies [50,51,52]. Based on the feedback from the experts, this practical learning is very important as it enables students to have the opportunity to learn from practical projects, as well as learning from products and solutions based on procedural learning to build their own experiences in a conducive learning environment while engaging in group work activities and improving pedagogical knowledge of graduates.

This learning increases the content knowledge of students to build hands-on and experimental skills in handling and controlling devices during practical session in the laboratory [53,54]. Through actual experiment, it enables students to perform testing by using hand tools and at the same time building technical experience for students with scientific skills in analysing data with a variety of research's techniques that can be applied in the industries [55, 56]. The summaries of elements of practical learning are shown in Table 3.5.

Table 3.5: Elements of Practical Learning

Elements of Practical Learning		
No	Element	Description
1.	Use of Scientific Tools and Machinery	Students are able to use scientific equipment to build work skills.
2.	ICT Equipment and Computer Equipment	Students know how to apply technology equipment and computers in doing practical work
3.	Web-Based Software	Student can use software in work simulation
4.	Practical Projects, Products and Solutions Based on Procedure Learning	Students will be able to exert more effort in trying to produce products or perform better in practical work
5.	Conducive /Experience Learning	Students can build work skills when using laboratory equipment correctly.
6.	Group Work Activities	Students are able to work in teams to perform practical work.
7.	Hands-On Experimental / Controlling Devices	Students are able to enhance expertise in building work skills
8.	Actual Experiment	Students have the opportunity to perform testing in the laboratory with the use of hand tools
9.	Scientific Skills and Technical analysis	Students will be skilled in analysing data with a variety of research techniques

3.5 Elements of Interactive Learning

Based on the interview conducted with expert academicians and findings from previous studies, the researcher identifies the elements of interactive learning to include knowledge of students in applied technology in the learning process, learning through technology media such as multimedia tools via computer networks and software programming to improve technical skills and the use of digital equipment to produce products [57, 58]. According to the experts, this interactive learning will help students in developing pedagogical knowledge. For instance, in demonstrating their work, animations and graphic designs with audio sound systems can be used to illustrate, while video can be used to display simulation and visualization in game applications can help students to be more interested in the learning process.

Students must be more creative and have high imagination in illustrating so that they can understand what is being learned by visualizing them in their mind [59,60]. This learning area is based on the level of student's thinking on how they can become with high productivity in their work, with a high level of cognitive thinking for increasing their learning environment. Thus, making it more flexibility in creating a high performance to produce graduates who are adequate in content knowledge [61,62]. Summary of elements of interactive learning are shown in Table 3.6.

Table 3.6: Elements of Interactive Learning

Element of Interactive Learning		
No	Element	Description
1.	Use of Scientific Tools and Machinery	Students are able to use scientific equipment to build work skills.
2.	ICT Equipment and Computer Equipment	Students know about applying technology equipment and computers in doing practical work
3.	Web-Based Software	Student be able to use software for help work simulation
4.	Practical Projects, Products and Solutions Based on Procedural Learning	Students will give more effort in trying to produce products or perform better in practical work
5.	Conducive /Experience Learning	Students can build work skills when using laboratory equipment properly
6.	Group Work Activities	Students are able to work in teams to perform practical work
7.	Hands-On	Students are able to enhance

	Experimental / Controlling Devices	expertise in building work skills
8.	Actual Experiment	Students have the opportunity to perform testing in the laboratory with the use of hand tools
9.	Scientific Skills And Analysis Technical	Students will be skilled in analysing data using a variety of research techniques.

4 Discussion and Conclusion

In this study, expert shared their opinions regarding the effectiveness of student's knowledge and skills in order to become skilful workers. This five elements of TEAL approaches needed in TVET institutions, especially for enhancing quality and employability skills of bachelor degree graduates were established from the study. These elements are essential to be applied in classroom settings to ensure that students are able to participate in an active environment to build experience in learning. The combination of all these elements will help graduates to improve their knowledge, skills and experience in technical work by using technology innovation tools.

The five elements of TEAL have their own benefit; online learning help students to find information for their studies; blended learning gives student the opportunity to become flexible in their work and to improve their communication skills; collaborative learning helps to build teamwork spirit in students to enable them to share their ideas and opinions; practical learning helps to ensure that students will be able to handle tools to enhance their hands-on skill; interactive learning helps to ensure that students are able to run multimedia or programming of software to create cognitive thinking.

Each student needs to activate their knowledge as a worker-in-training with various skills for helping them find jobs after graduation. Graduates must be prepared for all the challenges and industrial demand of skilled workers in technologies knowledge that is related to employability skills. This is in line with the theory of practice (TPACK) of pyramid learning for developing an active learning environment. The TEAL approach will guide students in achieving the goals of their studies and help them build knowledge and skills to gain experience.

In conclusion, this study aimed at identifying the elements of TEAL approach, to guide students and lecturers in the learning process in order to create active environments by using technology applications, machinery equipment, digital device, and ICT tools in the class. There are various methods that can be applied to teaching and learning that will attract student's interest to enjoy their studies more. At the same time, lectures will become facilitators to guide students to understand their lesson by applying technology in class either through online or face-to-face method. Lastly, students will become active persons by preparing themselves for industry requirement in job placement. For the next stage, all these elements will be validated by experts and a final checklist will be developed to specify the elements of TEA. Thereafter, it will be analyzed using Fleiss' Kappa agreements.

The author would like to thank to Research Management Centre (RMC), Universiti Tun Hussein Onn Malaysia for supporting this research.

References

[1] Pang Chau Leong. Key reforms in revitalizing technical and vocational education and training (TVET) in Malaysia. *Regional Conference HRD through TVET as a Development Strategy in Asia Colombo, Sri Lanka* (2011).

[2] Husain, M. Y. & Rasul, M. S. & Mustapha, R. & Malik, S. A. & Abd Rauf, R. A. (2013). Tahap kemahiran employability pelajar kejuruteraan dari perspektif majikan. Jurnal Teknologi. **62(1)**. 31-39.

[3] Yusof, N., Jamaluddin, Z., & Lazim, N. M. (2013). Persepsi Pelajar Prasiswazah Terhadap Kebolehpasaran Graduan dan Persaingan dalam Pasaran Pekerjaan (The Perception of Undergraduates' Student Towards the Marketability of Graduate and Competition in the Job Market). International Journal of Environment, Society and Space, **1(1)**. 43-61. Jurnal Personalia Pelajar, 16, 77-92.

[4] Ibrahim, D. H. M., & Mahyuddin, M. Z. (2017). Youth Unemployment in Malaysia: Developments and Policy Considerations.

[5] Zulkhairi,W., Ismail, R., & Amirmudin. Critic and different of philosophy in technical and vocational education (TVE). *Journal of Edupres*, **2(1)**, 287-300(2011).

[6] Manitoba Education. Technical vocational education overview. T*he Government of Manitoba, represented by the Minister of Education*. ISBN: 978-0-7711-5391-4 (2013).

[7] Hassan, N. F., Puteh, S., & Buhari, R. Student understanding through the application of technology enabled Active Learning in practical training. *Procedia-Social and Behavioral Sciences*, **204**, 318-325 (2015).

[8] Belcher, J. Technology enabled active learning, *Retrieved on May 5, 2007* (2005).

[9] Lalley, James P., Miller & Robert H., The Learning Pyramid: Does It Point Teachers in the Right Direction?. *Education Journal*. ISSN-0013-1172, **128 (1)**, 64-79 (2007).

[10] Baer, J. Lectures may be more effective than you think: The learning pyramid unmasked. *The International Journal of Creativity & Problem Solving*, **20(2)**, 7-21(2010).

[11] Dam, N.V. 25 Best practices in learning & talent development. *Second Edition USA: Lulu Publishers*. 136 (2008).

[12] Koehler, M., & Mishra, P). What is technological pedagogical content knowledge?. *Contemporary Issues in Technology and Teacher Education*, **9(1)**, 60-70 (2009).

[13] Hamilton Smith TEAL Classrooms, (2017). Technology Enabled Active Learning (TEAL) Classroom. www.unh.edu.

[14] Guthrie, H., Harris, R., Simons, M., & Karmel, T. Teaching for technical and vocational education and training (TVET). *International Handbook of Research on Teachers and Teaching*, **21**, 851-865 (2009).

[15] Hassan, N. F., & Puteh, S. A Survey of Technology Enabled Active Learning in Teaching and Learning Practices to Enhance the Quality of Engineering Students. *Advanced Science Letters*, **23(2)**, 1104-1108 (2017).

[16] Sanusi, A. M., & Puteh, S. An approach of excellence talent in engineering education programmed of enhancing the quality of students. *Advanced Science Letters*, **23(2)**, 1109-1112 (2017).

[17] Lee, Y., Boatman, E., Jowett, S., & Guenther, B. REAL: The technology-enabled, engaged, and active learning classroom. *International Journal of Designs for Learning*, **5(1)** (2014).

[18] Shieh, R.S., Chang, W., & Tang, J. The impact of implementing technology enabled active learning (TEAL) in university physics in Taiwan. *The Asia-Pacific Education Researcher*, **19 (3)**. 401–415 (2010).

[19] Dori, Y.J., & Belcher.W.J. How does Technology Enabled Active Learning affect undergraduate students' understanding of

electromagnetism concepts?. *The Journal of the Learning Sciences*, **14 (2)**, 243-279(2005).

[20] Cinganotto, L., Panzavolta, S., Garista, P., Guasti, L., & Dourmashkin, P. TEAL as an innovative teaching model. Insights from "educational avant-garde" movement in Italy. *Je-LKS Teaching Model.* **12**, 115–126 (2016).

[21] Van Hof, J. C. The effects of active learning technology on instructors 'practices and students 'engagement and grades: A Mixed Methods Study, (2016).

[22] Chiu, P. H. P. A Technology Enriched Active Learning Space for a New Gateway Education Programme in Hong Kong: A Platform for Nurturing Student Innovations. *Journal of Learning Spaces*, **5(1)**, 52–60 (2016).

[23] Ministry of education. Malaysia Education Blueprint 2015-2025 (Higher Education). ISBN 978-967-0334-99-8, 1-240 (2015).

[24] OECD. Education at a Glance 2016: OECD Indicators. *OECD Publishing, Paris*, (2016).

[25] Yee, R. Perceptions of online learning in an Australian university: an international students' (Asian region) perspective–quality of learning. *International Journal of e-Education, e-Business, e-Management and e-Learning*, **3(2)** (2013).

[26] Colorado, J. T., & Eberle, J. Student demographics and success in online learning environments. *Emporia State Research Studies*, **46(1)**, 4-10 (2010).

[27] Keengwe, J., & Kidd, T. T. Towards best practices in online learning and teaching in higher education learning, **6(2)**, 533–541 (2010).

[28] McBrien, J. L., Jones, P., & Cheng, R. Virtual spaces: employing a synchronous online classroom to facilitate student engagement in online learning. *International Review of Research in Open and Distance Learning*, **10(3)**, 1–17 (2009).

[29] Wise, B. The online learning imperative: a solution to three looming crises in education. *Education Digest: Essential Readings Condensed for Quick Review*, **76(3)**, 52-58 (2010).

[30] Butler Kaler, C. A model of successful adaptation to online learning for college-bound Native American high school students. *Multicultural Education & Technology Journal*, **6(2)**, 60-76 (2012).

[31] Means, B., Toyama, Y., Murphy, R., Bakia, M., & Jones, K. Evaluation of evidence-based practices in online learning: *A meta-analysis and review of online learning studies. U.S. Department of Education*, 94 (2009).

[32] Du, J., Liu, Y., & Brown, R. L. The key elements of online learning communities. In *Web-Based Education: Concepts, Methodologies, Tools and Applications*, 119–132 (2010).

[33] Vaughan, N. D., Cleveland. I.M., & Garrison, D. R. Teaching in Blended Learning Environments: Creating and Sustaining Communities of Inquiry. *Athabasca, AB: Athabasca University Press*, (2013).

[34] Norm, F. Report: Defining Blended Learning. *Learning spaces papers*, (2012).

[35] Bath, D. & Bourke, J. Getting Started with Blended Learning. Griith University. *Griith Institute for Higher Education*, ISBN: 978-1-921760-24-2, (2010).

[36] Yen, J. C., & Lee, C. Y. Exploring problem solving patterns and their impact on learning achievement in a blended learning environment. *Computers and Education*, **56(1)**, 138–145(2011).

[37] Edginton, A., & Holbrook, J. A blended learning approach to teaching basic pharmacokinetics and the significance of face-to-face interaction. *American Journal of Pharmaceutical Education*, **74(5)**, 88 (2010).

[38] Holley, D., & Oliver, M. Student engagement and blended learning: Portraits of risk. *Computers & Education,* **54(3)**, 693-700 (2010).

[39] Azizan, F. Blended Learning in Higher Education Institution in Malaysia. *In Proceedings of Regional Conference on Knowledge Integration in ICT*, 454–466 (2010).

[40] Wu, J.H., Tennyson, R. D., & Hsia, T.-L. A study of student satisfaction in a blended e-learning system environment. *Computers & Education,* **55(1)**, 155–164 (2010).

[41] Fernandez Luna, J. M., Huete, J. F., Perez Vazquez, R., & Rodriguez Cano, J. C. CIRLab: A groupware framework for collaborative information retrieval research. *Information Processing and Management*, **46(6)**, 749–761 (2010).

[42] Nam, N. H. Utilizing the active and collaborative learning model in the introductory physics course. *Journal of Education and Learning*, **3(3)**, 108–124 (2014).

[43] Verginadis, Y., Papageorgiou, N., Apostolou, D., & Mentzas, G. A review of patterns in collaborative work. *In Proceedings of the 16th ACM international conference on Supporting group work*, 283 (2010).

[44] Bell, T., Urhahne, D., Schanze, S., & Ploetzner, R. Collaborative inquiry learning: models, tools, and challenges. *International Journal of Science Education*, **32(3)**, 349-377 (2010).

[45] Kim, H.-N., Ji, A.-T., Ha, I., & Jo, G.-S. Collaborative filtering based on collaborative tagging for enhancing the quality of recommendation. *Electronic Commerce Research and Applications*, **9(1)**, 73–83 (2010).

[46] Shukor, N. A., Tasir, Z., Van der Meijden, H., & Harun, J. Exploring students' knowledge construction strategies in computer-supported collaborative learning discussions using sequential analysis. *Educational Technology & Society*, **17(4)**, 216-228 (2014).

[47] Filigree. Instructional Technology and Collaborative Learning Best Practices: Global Report and Recommendations. *Sponsor SMART Technologies* (2012).

[48] Nguyen, N. H. Utilizing the Active and Collaborative Learning Model in the Introductory Physics Course. *SSRN 2481765* (2014).

[49] Laal, M. Collaborative Learning; Elements. *Procedia Social and Behavioral Sciences*, **83**, 814–818 (2013).

[50] Ahmad, C. N. C., Osman, K., & Halim, L. Physical and psychosocial aspects of science laboratory learning environment. *Procedia Social and Behavioral Sciences*, **9**, 87-91(2010).

[51] Abdulwahed, M., & Nagy, Z. K. Applying Kolb ' s experiential learning cycle for laboratory education. *Journal of Engineering Education*, **98**, 283–294 (2009).

[52] Alam, F., Hadgraft, R. G., & Subic, A. Technology enhanced laboratory experiments in learning and teaching. *Using Technology Tools to Innovate Assessment, Reporting, and Teaching Prectices in Engineering Education, Hershey, United States*, 289-302 (2014).

[53] Razali, Z. B., & Daud, M. H. Approach of Assessing 'Hands-on'in Laboratory Exercises in Engineering Technology Courses. *In Malaysia University Conference Engineering Technology* (2014).

[54] Yao, J., Limberis, L., Warren, S., & Asee. Enhancing Laboratory Experiences with Portable Electronics Experiment Kits, *Asee Annual Conference* (2012).

[55] Korchnoy, E., & Verner, I. M. Characteristics of learning computer-controlled mechanisms by teachers and students in a common laboratory environment. *International Journal of Technology and Design Education*, **20(2)**, 217–237 (2010).

[56] Schuurbiers, D. What happens in the Lab: Applying Midstream Modulation to Enhance Critical Reflection in the Laboratory. Science and Engineering Ethics, **17(4)**, 769–788 (2011).

[57] Domagk, S., Schwartz, R. N., & Plass, J. L. Interactivity in multimedia learning: An integrated model. *Computers in Human Behavior*, **26(5)**, 1024–1033(2010).

[58] Leow, F. T., & Neo, M. Interactive multimedia learning: Innovating classroom education in a Malaysian university. *Turkish Online Journal of Educational Technology*, **13(2)**, 99–110 (2014).

[59] Zulkifli, A. N., Noor, N. M., Bakar, J. A. A., Mat, R. C., & Ahmad, M. A conceptual model of interactive persuasive learning system for elderly to encourage computer-based learning process. *In Proceedings-International Conference on Informatics and Creative Multimedia*, 7–12 (2013).

[60] Violante, M. G., & Vezzetti, E. Virtual interactive E-learning application: An evaluation of the student satisfaction. *Computer Applications in Engineering Education*, **23(1)**, 72–91(2015).

[61] Beauchamp, G., & Kennewell, S. Interactivity in the classroom and its impact on learning. *Computers & Education*, **54(3)**, 759-766 (2010).

[62] Sabry, K., & Barker, J. Dynamic Interactive Learning Systems. *Innovations in Education and Teaching International*, **46(2)**, 185–197 (2009).

Ergonomic Design of Public Bus in The Philippines with Provision for Senior Citizens and Persons with Disability

Ma. Janice J. Gumasing[1,*] and *Charm Hanalae C. Dela Cruz* [1]

[1] School of Industrial Engineering and Service Engineering Management, Mapúa University 658 Muralla St., Intramuros, Manila 1002, Philippines

Abstract. Comfortability in buses for commuters particularly those with special needs such as senior citizen and person with disability is a concern, which requires attention. Public buses are highly used mode of transportation in the Philippines, both provincial and city. In this work, the researchers aim to determine the design factors that affect the comfortability of commuters and eventually come up with an ergonomic design of public bus to be used in the Philippines with provision for senior citizen and persons with disability. The study covers and mainly focused on designing bus doors and its interior for the accessibility of the senior citizens and persons with mobility aid such as wheelchairs, crutches and canes ergonomically. The researchers have conducted review of related literature, direct observation, surveys, interviews and actual measurements of bus dimensions in order to describe factually and accurately the current design of buses in the Philippines. The researchers have improved the current design of provincial public buses having trips from Cavite to Manila (vice versa) with consideration of the issues and needs of the commuters in riding a bus.

1 Introduction

Transportation is the movement of people or goods from one place to another that supports human's daily activities. According to Dr. Romulo A. Virola, Secretary General of National Statistical Coordination Board (NSCB), transportation plays a vital role in the development of a country's economy. The most accessible and commonly used mode of transportation in the Philippines is by land and is known to contribute more than three times the combined Gross Value Added of water and air transportation with a 26.63 percent share to the total GVA in transport in year 2013 [1]. GDP defined by Webster as a measure in economics of the value of goods and services produced in a country while GVA is a measure in economics of the value of goods and services of an area, industry or sector.

It is one of the basic needs in life and is part of our normal routine. And with the vast increase in population, the NSCB predicts that people aging 60+ represented 5.97% is expected to grow to 13.8% of the population by 2040 while persons with disability is said to be 1.57 percent of 141.7 million Filipinos according to National Statistics Office (NSO) [2]. This will have a massive impact on public transport. Thus, there are things that unfairly prevent many incapable commuters from using public transports to fully participate in society.

On the study regarding mobility restriction of the elderly conducted by Matthias Fiedler [3], he stated, as the age increases, the likeliness of impairments that affects mobility also increases. In addition, the work done by [4] stated that recommended that in order to fulfill the needs of people with disabilities, the transport infrastructure and mobility factor for the disable has to be assessed and improved. A study done by [5] has stated that bus design and service provision has relative impact on the usage of bus by older people. The study by [6] has described that travel experiences of older and people with disability has to be taken into account to the progress of quality mode of transport for all. A study by [7] has discussed that for the people with disabilities, the current conventional urban public transport system could not deliver an efficient service and limitation still exist. A study by [8] concluded that restricted mobility caused by the existing transport system and design is still a major concern for older people using the bus service.

Therefore, this evidently shows the need to improve the quality of transportation for persons with disabilities and for senior citizens [9,10]. Prior researches focused on the evaluation of the measurements of curb, bus entrance/exits steps, safety handrails and the required distance between them, for the incapacitated passengers to board/alight [9,10]. The design devoted for passengers with mobility impairments and the older people seems to be an issue for other commuters. The problem on integration of the designs fit to use to various commuters continue to exist. Also, currently there is no existing studies/researches applicable here in the Philippines.

That being so, the researchers aimed to ergonomically design a bus including the entrance/exits doors and its

* Corresponding author: mjjgumasing@mapua.edu.ph

interior that is fit to use for Filipino commuters; the bus is equipped with area guards to ensure the security of the passengers during the loading and unloading process as well as the whole duration of the trip. This study also aimed to minimize the effort and assistance from all the concerned commuters and operator. The design is dedicated to all types of passengers but most especially for those who have mobility aids and are incapable of riding a bus with ease of use, safety and accessibility.

The study covers and mainly focused on designing bus doors and its interior for the accessibility of the senior citizens and persons with mobility aid such as wheelchairs, crutches and canes ergonomically. The researchers have improved the current design of provincial public buses having trips from Cavite to Manila (vice versa) with consideration of the issues and needs of the commuters in riding a bus. The researchers used the population within the area as respondents in this study. Having said that, it does not cover the engine specifics, aesthetics and other designing aspects of the bus. Engine specifics include the mechanical features to be used and as per aesthetics, it is a guiding principle in matters of artistic beauty and taste.

2 Methodology

In pursuing and coming up with a result, the study followed a framework that is shown in Figure 1 which provided this paper a structure and a guideline to come up and to have a clear idea on what and how the output design was constructed.

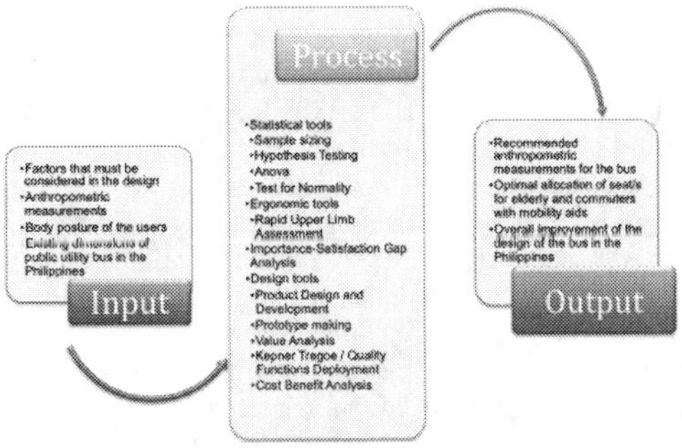

Fig. 1. Research Framework

2.1. Data Collection

The researchers have conducted review of related literature, direct observation, surveys, interviews and actual measurements of bus dimensions in order to describe factually and accurately the current design of buses in the Philippines. The researchers were able to identify factors that will be considered in the design of ergonomic bus in the Philippines such as anthropometric measurements of users, body posture of users, and actual dimensions of public buses in the Philippines. Data were collected by manually measuring the dimensions of the bus and afterwards incorporated for the proposed design. Along with the measurement of safety devices and anthropometrics, normal passengers and elderly were treated separately due to independent demand and measurements. • Body posture of the users assessed are boarding and alighting postures; this was done to eliminate tension, stress and musculoskeletal disorder in riding a public bus.

The researcher also conducted an assessment and the current state of the bus safety devices, measurements and its effectiveness and efficiency along with the anthropometric measurements that an average Filipino bus passenger would require from such specifics like height, width, and the like. The gathered results and measurements had shown the gap from the current measurements to the ideal and user-demanded specifics and had become the basis for formulation and design of the proposed improvement prototype intended for utilization for the senior citizens, persons with disabilities, and lastly average Filipino commuter. But actual measurements were gathered and measured on the buses routing Manila-Cavite and vice versa.

2.2. Data Analysis

Factors obtained from data gathering procedure were recorded, described, interpreted and analyzed. Statistical analyses and treatments such as sample sizing, hypothesis testing, analysis of variance and test for normality were also conducted in order to systematically and statistically investigate and explain the nature of relationship between factors considered in the study. The researchers also evaluated the current posture of bus users in the Philippines including the senior citizens and persons with disability through ergonomic tool such as Rapid Upper Limb Assessment (RULA). This tool estimates the risk of upper-limb disorders based on the posture of the subject. In addition to this, the researchers also performed Importance-Satisfaction Gap Analysis to determine the importance of the design attributes of the bus to the users. And finally, in order to come up with an ergonomic design of bus with provision for senior citizen and persons with disability, the researchers made use of design tools such as product design and development, prototype making, value analysis, Kepner-Tregoe technique, Quality Functions Deployment technique and cost-benefit analysis.

As per the output phase, once all data has been gathered, computed and processed, the researcher has come up into a conclusion of designing an effective and efficient bus layout specifically crafted with respect to the recommended anthropometric measurements that will cater the needs of an average elder and/or persons with disabilities and for them to take advantage of the transportation service that is widely provided by the bus lines within the coverage of the study. This research has delivered an overall improvement design of buses in the Philippines including the optimal number of seats for elderly and commuters with mobility aids that does not compromise the earnings and service rate of bus companies.

3 Results and Discussion

Subsequent to computation of sample size, bus lines are measured and tested using hypothesis testing and analysis of variance (ANOVA) for further assessment. Door clearance including step height, step width, step depth, vertical and horizontal clearance for the mobility aids that will enter the bus and the seat depth, seat width, seat height, leg clearance, backrest height, overhead handrails height, stanchion height, stanchion length, stanchion width, compartment level height and aisle width that depends on the measurements of people with disabilities (PWD) and elderly with consideration on their physical limitations and anthropometrics.

The researchers have managed to measure each specification on each bus line at Philippines including San Agustin, Erjohn and Almark, Don Aldrin, Starliner, Dionets, Lorna Express and St. Anthony of Padua. Each liner's newest bus units are chosen knowing that it was improved as compared to their old units.

As for validation of the bus design measurements, hypothesis testing and ANOVA was deployed using Statplus as shown in Table 1, where μ_0 are the bus lines.

Table 1. Hypothesis Testing

Hypothesis Testing	
a	Null Hypothesis $H_0 : \mu_1 = \mu_2 = \mu_3 = \mu_4 = \mu_5 = \mu_6 = \mu_7$
b	Alternative Hypothesis $H_a : \mu_1 \neq \mu_2 \neq \mu_3 \neq \mu_4 \neq \mu_5 \neq \mu_6 \neq \mu_7$
c	Confidence Level :95%
d	One - Way ANOVA

Based on the result of ANOVA, F value .08 is less than F critical, which is 2.19, the claim is said to be true, do not reject null hypothesis. Therefore, bus liners' design has no significant difference. This study could assess only one bus line as comparison for the proposed design with appropriate measurements and improvements. Table 2 shows the ANOVA results.

Table 2. Anova Results

Analysis of Variance (One-Way)				
Summary				
Groups	*Sample size*	*Sum*	*Mean*	*Var.*
San Agustin	15	457	30.47	545
Erjohn	15	457	30.47	545
Don Aldrin	15	434.5	28.97	490
Starliner	15	383.5	25.57	547
Dionets	15	452	30.13	531
Lorna Express	15	441.1	29.41	500
St. Anthony	15	436.	29.07	554
ANOVA				
Source of Variation	*SS*	*df*	*MS*	*F*
Between Groups	260.71	6	43.45	0.08
Within Groups	51,999.73	98	530.61	
Total	52,260	104		

On account of gathering anthropometric measurements, the researchers verified the data set of 100 senior citizen respondents if was well modeled by a normal distribution. Each measurement (stature height, tibial height etc.) were tested and concluded based on the result individually using NumXL which computed the following results as shown in Table 2.

Table 2. Anthropometric Measurement

Anthropometric Measurement	P-Value	Pass?
Stature	7.3%	TRUE
Tibial Height	5.7%	TRUE
Popliteal Height	24.7%	TRUE
Buttock to Knee	5.7%	TRUE
Buttock to Popliteal	5.7%	TRUE
Forward Reach	7.1%	TRUE
Sitting Height	5.7%	TRUE
Elbow to Wrist	5.7%	TRUE
Sitting elbow height	5.7%	TRUE
Hip Breadth	5.6%	TRUE

Null Hypothesis was accepted because the P-value of each anthropometric data satisfied the condition that it should exceed alpha level of 5%, and as computed by NumXL, the result above pass the Normality test. Therefore, the data collected of senior citizen anthropometric measurements are normally distributed and was verified to be reliable for use.

Behaviors of users at boarding and alighting were observed at bus stops. For the assessment of the body postures of commuters on different cases, the researchers took photos of persons with disabilities with mobility aids and elderly with normal condition and elderly using canes. The critical postures examined were the act of boarding and alighting because these are the positions that occur stress on the body of the commuters in accessing a bus and were also used for the Rapid Upper Limb assessment. The results of the RULA on different types of commuters has an average of 3-5 which dictates that the current bus design needs further investigation for improvement and implementation of the improved design as soon as possible to avoid risk on the users especially on PWD with mobility aids that requires more detailed and hazard free design considering their physical limitation and capabilities because it causes awkward position to its users that may lead to musculoskeletal disorder or even accidents.

Importance – Satisfaction Gap Analysis. The IS rating is a unique tool that allows the researchers to better understand both of the highly important decision making criteria of each of the service the bus company is providing. This method is based on the concept that companies will maximize commuter's satisfaction by improving the bus specifications where the level of satisfaction is relatively low and the perceived importance of the service is relatively high.

Common problem among senior citizens is the height issue that is inevitable because they are having difficulty in moving or by just boarding themselves in high-elevated flat-forms or stairs due to their physical capabilities, seniors also want their luggage near them or within sight. For the crutch users, aside from the height issues obviously because of their disability, leg clearance, space for maneuvering, and difficulty in reaching door safety handrails were the top concerns. Wheelchair users on the other hand has more demand as compared to other passengers since their mobility aid requires much space for securement and the mechanism of the doors so they board and alight safely. Last but not the least are the crutch users, they have the same assessment of the current bus as the others with addition on the importance of overhead bars and handrail heights, according to some PWD interviewed, since there is no strict policy of the reserved seats dedicated for them, they experience standing the whole trip so they needed something to hold on for stability and balance.

Now, House of Quality (HOQ) is where the design creation starts and it comes after all the assessment tools because customer issues and demands were already identified. The primary goal of Quality Function Deployment is to distinguish the problems that are vital to the planning process. By deploying HOQ, it helps the researcher to identify the technical attributes that has to be prioritized in the design aligned with the customer demands. The improved design was compared to competitor's design that allows the researcher to strategize on how to better satisfy the customer and identify where the design is weak that needs more attention. The researcher has translated customer demands to product specifications.

The proposed bus design measurements were established using anthropometrics, design and assessment tools. Table 3 shows the proposed bus door specifications. Table 4 shows the proposed bus seats specifications. Table 5 shows the proposed bus interior measurement

Table 3. Proposed Bus Door Specifications

Part Specification	Anthropometric Measurement	Measurement (inches)
Door Vertical Clearance	Standard	83
Door Clearance Width	Maximum Wheelchair Width + allowance	26.5 +4 =30.5
Ramp Step Height	Curb Height and is adjustable to ground	8
Ramp Length	Standard	30
Ramp Height	Standard	3
Stanchion Height	Sitting shoulder height (50th male and female) + seat height	21.07 +14 =35.07 =35
Stanchion Length	Seat Width x 2	24.6 33.4
Handrail Height	Standing Shoulder Height 5th female – ramp height	46.5 – 6 = 40.5
Handrail Length	Step Width up to platform	30

Table 4. Proposed Bus Seats Specifications

Bus Specification	Anthropometric Measurement	Percentile	Inches
Seat Height	Popliteal Height	5th female	Normal :14 Senior:13.41
Seat Depth	Buttock to Popliteal	5th femla	Normal:15.7 Senior:12.49
Seat Width	Hip Breadth	95th female	Normal:16.70 Senior:12.29
Seat Cushion Slope	Standard		8° - 15°
Arm rest Height	Sitting Elbow Height	5th Female	Normal:6.69 Senior:6.63
Arm rest Lenght	Elbow to Wrist or length of lower arm	5th Female	Normal 7.87 Senior:7.35
Leg clearance Normal: PWD:	(Buttock to Knee – Buttock to Popliteal) * 2 +allowance Buttock to Knee + tibial height – Buttock to popliteal	95th male	(24.37-20.47)*2 = 8 + 2 = 10 24.37+21.65-20.47 = 25.55
Backrest Height	Sitting Height	50th male	Normal: 33.47 Senior: 34.09
Backrest Width	Hip Breadth	95th female	Normal: 16.70 Senior: 12.29
Backrest Slope	Standard		8° - 17°

Table 5. Proposed bus interior measurement

Bus Specification	Anthropometric Measurement	Percentile	Inches
Roof Height	Standing Height	95th male	72
Grab bar Height	Roof Height – Luggage Space Height		60
Luggage Level Height	Shoulder ht. + length of upper arm - allowance	50th female	60
Luggage Space Height	-	-	12
Aisle Width			25.2
Hand Strips	Shoulder breath	50th	17

 https://doi.org/10.1051/matecconf/201815005006

(wheelchair area)	male		

Figures 2 ,Figure 3 and Figure 4 shows the existing bus layout and design. Figure 5, Figure 6 and Figure 7 shows the proposed bus layout and design.

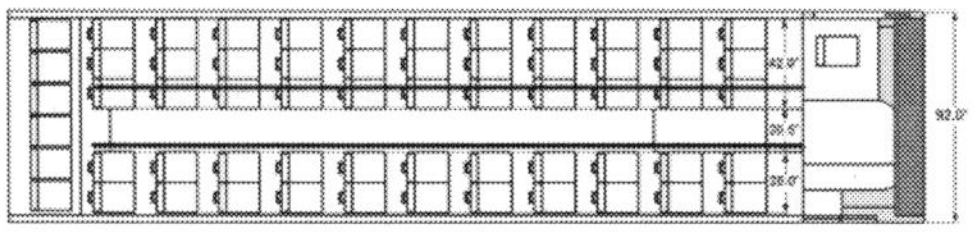

Figure 2. Bus layout top view (existing)

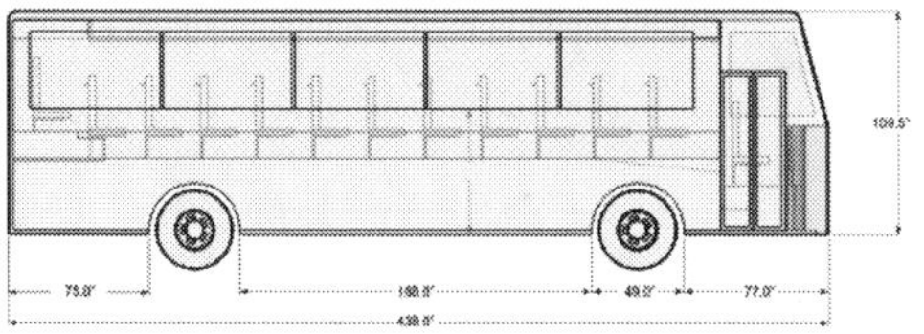

Figure 3. Bus layout side view (existing)

Figure 4. Bus 3D View (existing)

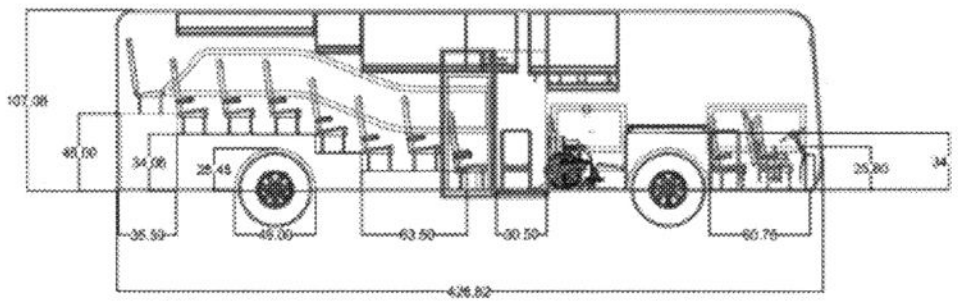

Figure 5.Bus layout top view (proposed)

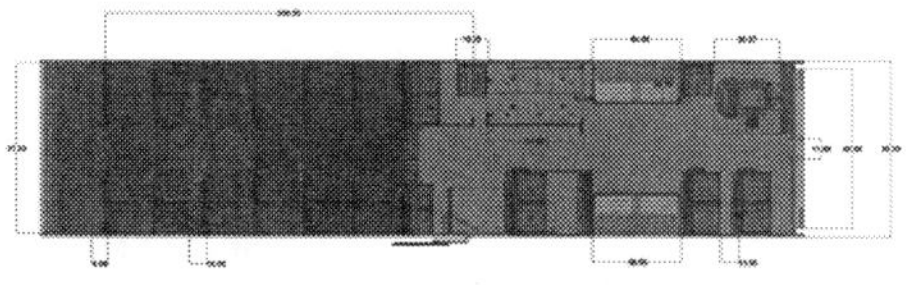

Figure 6.Bus layout side view (proposed)

Figure7. Bus 3D View (proposed)

In addition, the cost analysis was also done. The customization price for the proposed designed is estimated at Php 6,134,008. Thus by comparing with the cost of the existing bus design, the following additional cost has been calculated.

Total Cost = Proposed Bus – Current Bus
= Php 6,134,008 - Php 5,000,000
= Php 1,134,008

The total cost of improving the current bus design is the cost of the proposed design with inclusion of the customization and the all the improved and added parts of the bus minus the cost of the current bus design. The value computed is the amount that the company is willing to pay or disburse in exchange of the accessibility of PWD and comfortability of its users
Primarily, cost-benefit analysis has itemized all the improvements done in the creation of the design and discussed the benefits it will cause to the passengers as well as the things compromised to make way for the design prioritization to elderly and PWD. One of the highest cost contributor done to fulfill the objective of this study was decreasing the bus capacity, originally, a bus could accommodate up to 56-61 passengers on a single trip. Proposed bus design could now accommodate 43 passengers only including senior citizens and PWDs. If all passengers onboard are normal, 6 standing passengers could fill and utilize the wheelchair space. All in all, bus optimal capacity has reduced to 50, not much difference with the existing but with great improvement that provides ease, accessibility and comfort to the passengers.

4 Conclusion

The findings of this study have shown that major users of public buses who participated in this study were less satisfied and are having difficulty in the use of the current design of public buses in the Philippines. The dissatisfaction was evident on the scores generated by the rapid upper limb assessment computed from their body postures while boarding or alighting a bus that led to analysis and conclusion that they have physical limitations and issues. In addition, the current design measurements manifested to be not ergonomically designed or not fit for use for Filipino commuters. This claim is supported with statistical analysis, thus, the actual anthropometric measurements of Filipinos were used to ergonomically fit the design of the bus to the commuters. Furthermore, the equipment especially crafted for the passengers with mobility aids were added to suffice their necessity. The limitations and accessibility issues of the design has caused the passengers being deprived in utilizing the public transportation determined thru interviews and surveys of passengers aging 60 years of age above, passengers with disabilities and/or mobility aids. The customer demands or specifications were identified on the assessment conducted by the researchers using Importance-Satisfaction Gap and as a result, the design attributes found to be important for the passengers but has less satisfaction given to them were considered to be the target specifications in product design and development.

In other words, the concerns of the passengers and their requirement on the design were translated as final specifications of the proposed bus design as a guarantee of the fulfillment of the demand. However, the improvements of the current bus scenario cannot be possibly evaluated because this study does not require (and is not possible) to create a prototype.

The study has achieved its goal and made an overall improvement on the design of the current public bus in the Philippines with respect to Filipino anthropometric measurement favorable to the customer and the company.

References

1. Virola, R., Reyes, R., Polistico, F., and Oliveros, A. (2011) "Things Statisticians wanted to know about the Tourism Satellite Account but were afraid to ask," *Retrieved March*, vol. 3, p. 2014,
2. Virola, R. A., and Encarnacion, J. O. (2016) "The Washington Group on the Improvement of Disability Statistics Globally: Perspectives from the Philippines," in *International Measurement of Disability*, Springer, pp. 233–253.
3. Fiedler, M. (2007). Older people and public transport. *Challenges and changes of an ageing society. Final report. Retried from: http://www. emta. com/IMG/pdf/Final_Report_Older_People_protec. pdf.*
4. Cavinato, J. L., & Cuckovich, M. L. (1992). Transportation and tourism for the disabled: An assessment. *Transportation Journal*, 46-53.
5. Stern, S. (1993). A disaggregate discrete choice model of transportation demand by elderly and disabled people in rural Virginia. *Transportation Research Part A: Policy and Practice*, *27*(4), 315-327.
6. Ipingbemi, O. (2015). Mobility challenges and transport safety of people with disabilities (PWD) in Ibadan, Nigeria. *AFRICAN JOURNAL FOR THE PSYCHOLOGICAL STUDIES OF SOCIAL ISSUES*, *18*(3), 15-27
7. Broome, K., McKenna, K., Fleming, J., & Worrall, L. (2009). Bus use and older people: A literature review applying the Person–Environment–Occupation model in macro practice. *Scandinavian journal of occupational therapy*, *16*(1), 3-12.
8. Hine, J., & Mitchell, F. (2001). Better for everyone? Travel experiences and transport exclusion. *Urban studies*, *38*(2), 319-332.
9. Zhou, H., Hou, K. M., Zuo, D., & Li, J. (2012). Intelligent urban public transportation for accessibility dedicated to people with disabilities. *Sensors*, *12*(8), 10678-10692.
10. Wretstrand, A., Svensson, H., Fristedt, S., & Falkmer, T. (2009). Older people and local public transit: Mobility effects of accessibility improvements in Sweden. *Journal of Transport and Land Use*, *2*(2).

The Impact Of Information Quality And Ergonomics On Service Quality In The Use Of Automatic Teller Machines

Jeremy C. Elias [1,*], and *Rene D. Estember* [1],

[1]School of Industrial Engineering and Service Engineering Management, Mapúa University 658 Muralla St., Intramuros, Manila 1002, Philippines

Abstract. The introduction of automatic teller machines (henceforth ATMs) brought dramatic changes in the way people do banking. ATMs have become a daily necessity and they are widely used for their convenience and accessibility. However, there have been several major challenges and issues faced in delivering quality service in ATM banking. Since the use of ATMs involves an exchange of information between a machine and a customer, this study investigated the impact of information quality (i.e., quality of information the machine provides) and ergonomics (i.e., design quality of the machine) to the service quality of ATMs. To this end, the study geared towards providing an improvement framework for the ATM design to improve service quality. A random survey among ATM users in Metro Manila was conducted to gather data with regard to their usage and perceived service quality of ATMs. Through statistical analyses, it was found that ATM usage varied according to age, education and socio-economic status of customers. ATM users were typically young, more literate and have higher economic status. More importantly, it was found that components of information and ergonomic qualities had fair influence to service quality, with just a few components having significant impact.

1 Introduction

Superiority in quality has developed as an essential for administrative sustainability in this competitive world [1]. On top of that, the changing customers' perception of quality poses unique challenges. Administration of client coordination and service quality is the utmost efficient ways of making trust in the competitive organization industries such as banks [2].

The expansion of information and telecommunication technologies has allowed establishments to deliver excellent amenities to fulfill customers requirements [3]. Banks are amassed their services options based on technology to meet the demand of this competitive world and ATM facilities are one of such services. ATM is an automatic telecommunication device that provides instant financial transaction facility for customers without depending on the banking personal for manual transaction [4].

Banks keep developing the range of the ATM system to meet the needs of consumers with lower transaction cost for both bank and consumers. However, ATMs are public technology devices [5]. This means that people who may use ATMs vary from age, gender, literacy, behavior, social and economic status and other factors, thus raising major challenges and issues in delivering quality service in ATM banking.

Issues on the use of ATMs have been attended from a number of perceptions such as age [6], control modality including voice [7], language [8], attitudes to innovativeness and computers [9], interface design [10] literacy [9], and attributes of ATM users [11,12].

Most of the previous studies about ATMs were able to identify prospects and problems concerning the use of ATMs. Also, they have provided suggestions and recommendations to address these concerns such as design to accommodate the older population, providing training and additional different alternative ATM interface. However, none of these studies have provided an in-depth analysis on the service quality of ATMs. Service quality is the evaluation of the service level delivered to the customer based on the customers expectation [13]. It is the customer's verdict about an unit's overall quality or preeminence. This means that if the client recognizes facility at a certain level but estimated something different, then the consumer will be discontented. In manufacturing, consumers judge the quality by evaluating the product manufactured. In contrast, service customers judge service quality by experiencing the service delivery processes as well as the outcome. In other words, customer satisfaction is the outcome of service quality.

On the other hand, a study made by Najjar [14] investigated the impact of information quality and ergonomics on service quality in selected banks in Nebraska, United States. His study involved the traditional over-the-counter banking setup. A model postulating that process quality predicts product/service quality was proposed. Bank employees were asked to assess through a questionnaire how information quality

* Corresponding author: jeremyelias22@gmail.com

 https://doi.org/10.1051/matecconf/201815005007

and ergonomics (workstation, overall comfort, and environment) affect how they deliver the service. However, in the case of ATMs, in which service is delivered only by a machine, is the service quality predicted by the quality of information provided by the machine and the ergonomics of the machine itself?. With this, the researcher wanted to extend Najjar's [14] principle to the concept of ATMs. Nowadays, organizations increasingly rely on their information systems to integrate and support their business processes [15]. These information systems and the quality of the data they contain affect customer's perceptions of the quality of purchased products and services [16]. With that being said, since the ATM interface requires an exchange of information between the user and the machine, this study determined the impact of information quality (i.e., the quality of information the machine provides) and ergonomics (i.e., the machine design) in delivering quality service in the use of ATMs. In improving service quality, this study geared towards providing an improvement framework for the ATM design.

2 Methodology

In the course of the study, the researcher followed the methodological framework shown in Figure 1.

Fig. 1. Research Framework

2.1. Data Gathering

The sample for this study was composed of 250 individuals coming from Metro Manila who were actively using ATM services. ATM users from the three largest banks as to total assets reported by the Bangko Sentral ng Pilipinas (BSP) as of the end of December 2013 were selected. Coincidentally, these were also the banks with the largest number of ATM units as of the 2nd quarter of 2011.

A two-part random survey was conducted using questionnaires. The first part was the preliminary survey. The purpose of the preliminary survey was to gather information about customers' ATM usage and to identify customers' needs and demands as well. The second part of the survey asked the respondents to rate their respective banks in terms of their perceived quality of its service, information, and ergonomic design. The survey used a 10 point rating scale with 1 being the lowest and 10 being the highest.

Firstly, the survey helped draw a profile of customers' ATM usage based on their age, gender, education, and socio-economic status. Secondly, this helped determine the problems experienced by customers with ATMs. And third, it helped determine what customers demand for quality service, information and ergonomics and how they perceive the quality of service that ATMs of their respective banks provide.

2.2. Data Analysis

The data gathered from the survey was statistically analyzed. Descriptive statistics was used to draw a basic profile of ATM usage in Metro Manila and the problems customers experienced with its use. General Linear Model and T-test were used to find out significant variations in customer's ATM usage based on their age, gender, education, and socio-economic status. General Linear Model was also used to find out significant difference in the customer's perceived service, information, and ergonomic qualities of the ATMs of their respective banks.

The impact of information quality and ergonomics on service quality was tested using Simple Linear Regression. Each independent variable (i.e., information and ergonomic components) was used to predict the value or its impact to the dependent variable (i.e., service quality components).

Tools such as Pareto chart and Fishbone diagrams were used for cause and effect analysis. The Pareto chart was used to determine which of the many problems were significant and needed to be prioritized. The Fishbone diagram was then used to determine the causes of these problems, as well as what causes the causes or the so-called root cause. The findings in the cause and effect analysis served as minimum basic guidelines in creating an improvement framework for ATM design.

The primary output of this study is an improvement framework for the ATM design using House of Quality (HOQ). It was used to highlight the voice of the customers (i.e., customer wants). It is a implement to explain what the client desires into products or services that meet the consumer requirements in terms of engineering design standards by way of generating a relationship matrix. Basically, HOQ is comprises of six main mechanisms. They are technical requirements, customer requirements, an interrelationship matrix, competitor evaluation, a technical priorities and technical correlation matrix, benchmarks and targets section. Figure 2 shows project design framework.

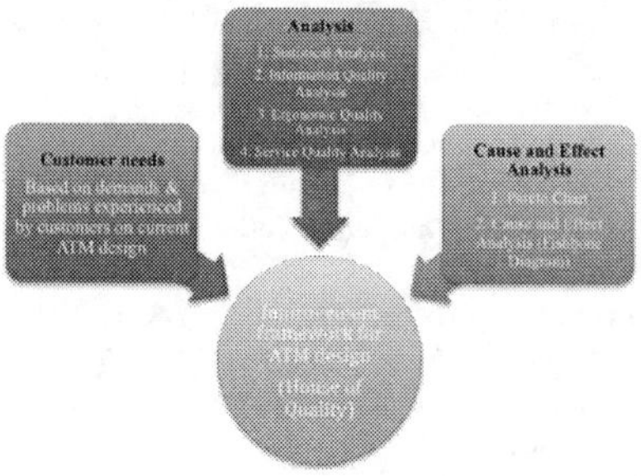

Fig. 2. Project Design Framework

3 Results and Discussion

Generally, the survey showed that ATM users were typically young, being more educated and having higher economic status. 65% of the respondents fell under the age of 39 years old and below. Around 97% have finished at least secondary education and 72% were earning from average to high income. Around 45 percent of the users have been using ATM for only 5 years or less as shown in Figure 3. In addition, the survey result showed that almost 70 percent owned one (1) ATM card only. As for the atm usage, it was found that around 46 percent were using ATMs twice (2x) within a week as shown in Figure 4. Typical ATM usage time among respondents ranged from 30 seconds to 1 minute. Convenience was the topmost reason for using ATMs and cash withdrawal had 100% utilization.

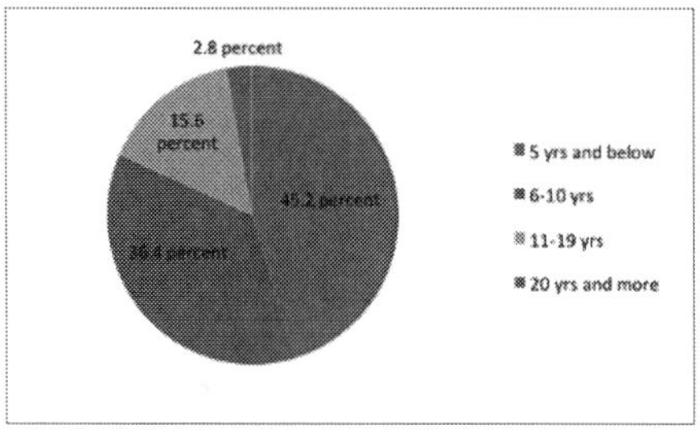

Fig. 3. Length of ATM Use

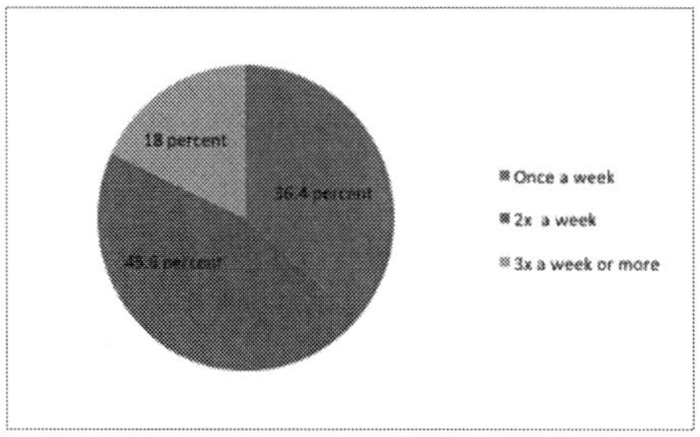

Fig. 4. Frequency of ATM Use

Respondents were asked for their reasons for using an ATM. From Figure 5, at approximately 97 percent, customers found convenience as their topmost reason for using an ATM. Easier accessibility of ATMs as compared to branch offices was also a major reason why 68 percent of the respondents use ATM. 47 percent of the respondents also identified 24-hour service as a major reason for using an ATM.

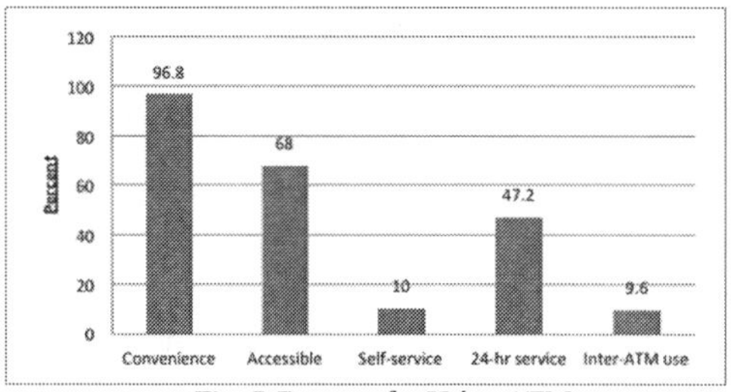

Fig. 5. Reasons for Using ATM

Table 1. Problem Experienced by Customers

Problem	frequency	Percentage
ATM is offline or out of service	234	93.6
Long queue	232	92.8
ATM not printing receipt	154	61.6
ATM running out of money	115	46
Difficulty seeing the keypad (due to cover)	82	32.8
Security problems	71	28.4
ATM not dispensing cash, but debited on account	40	16
Poor visibility of screen	39	15.6
ATM working too slowly	38	15.2
Inserting card incorrectly	34	13.6
ATM not releasing card (card captured)	30	12
Don't know w/c keys and buttons to press	23	9.2
Forgetting the PIN	22	8.8
Insufficient lighting at ATM facility	20	8
Not knowing if ATM is operating	19	7.6
Inconvenient location	15	6
ATM charge high fees	13	5.2
Unclear instructions	11	4.4
Difficulty reaching the slot (height of the machine)	2	0.8

Using, General Linear Model and T-test, ATM usage was found to be significantly different with customers' age, education, civil status, income and occupation with most of the P-values equal to <0.0001 ($\alpha = 0.05$). Basically, ATM usage pertains to the period of ATM use (i.e., how long the customer has been using ATM banking), card ownership (i.e., how many cards the customer own), frequency of ATM use (i.e., how often the customer use an ATM within a week), and ATM usage time (i.e., how much time it takes the customer in using ATM). On the other hand, ATM usage was found to be the same with both men and women except that men were faster in using ATMs.

In addition, comparison was done between local banks in Philippines namely BDO, Metrobank and BPI bank. Overall, the three banks were likely the same with their service quality components except for communication. With that, the three banks were found to be significantly different in terms of communication alone with P value equals to less than 0.0001. Descriptively, using the equivalent rating scale, the respondents were "extremely satisfied" with the quality of information their banks' Overall, the three banks were not significantly different in terms of ergonomic quality components. Overall service quality (SQ) rating was 8.199, 8.531 for information quality (IQ), and 7.810 for ergonomic quality (EQ).

Additionally, the banks were significantly different with the Communication component of service quality with P-value of <0.0001. This difference was mainly due to the language options incorporated in the Furthermore, the Simple Linear Regression test revealed that information and ergonomic quality components generally had fair impact or effect to service quality components, with only few components having significant impact. Consistency of information had significant impact to Competence of service quality with P-value of 0.0038. It was found out that for every 1 point increase in the Consistency rating there's approximately 0.15 increase in the Competence value of service quality. That is,

$$Competence = 6.67045 + 0.14719(Consistency) \quad (1)$$

Moreover, the ergonomic components of Lighting and Keypad had significant impact to the Tangibles of service quality.

$$Tangibles = 11.75384 - 0.27566(Lighting)$$
$$0.25396(Keypad) \quad (2)$$

It means that for every 1 point increase in the rating for Lighting there's around 0.28 decrease in the value of Tangibles. Also, for every 1 point increase in the rating for the Keypad there's around 0.25 decrease in Tangibles. However, since the parameter estimate was 11.75384 which exceeds the highest score which is 10, improvement in the ratings of both components would bring the Tangibles score in the acceptable range.

Furthermore, the House of Quality was utilized as a tool to provide an improvement framework for the ATM design in order to improve service quality. Table 2 shows the customer requirements which were derived from the survey and the technical requirements which were drawn from cause and effect analysis.

Table 2. Customer requirements and Technical Requirements

Customer Requirements (Whats)		Technical Requirements
SQ	Security of service(Safety)	Improve security measures
	Accessibility of service	Convenient locations
	Reliability	ATM always online
IQ	Security (Conficentiality of information)	Security biometrics
	Accuracy	Accurate transaction account details
	Accessibility of information	Avaiability of needed information
	Understandability	Language options
	Completeness	Issuance of receipt
EQ	Keypad	Improve keypad cover
	Keys & buttons on the screen	Mapping of keys & buttons
	Screen	Change screen glass material

In addition as illustrated in Figure 6, based on the framework, the ATM design improvement framework suggested that:

The technical requirements should be prioritized in the following order:

1. Improve security measures (including security biometrics)
2. Accessible and convenient locations
3. Keys and buttons on the screen
4. Improve keypad cover (trade-off between security biometrics)
5. ATM always online
6. Issuance of receipt
7. Availability of needed information
8. Accurate transaction/account details
9. Reduce glare on the screen
10. Incorporate language options

The achievement of all the technical requirements will have the most impact to the following customer requirements:

1. Security of service (safety)
2. Accuracy of information
3. Security of information (confidentiality)
4. Accessibility of service
5. Reliability of service
6. Keys and buttons on the screen
7. Accessibility of information
8. Screen
9. Understandability of information
10. Keypad
11. Completeness of information

Five of the targets are found only in the improved ATM design such as:

1. Panic button
2. Monitoring system of ATM traffic
3. Fingerprint or eye recognition technology
4. Transaction history option
5. Non-glare glass

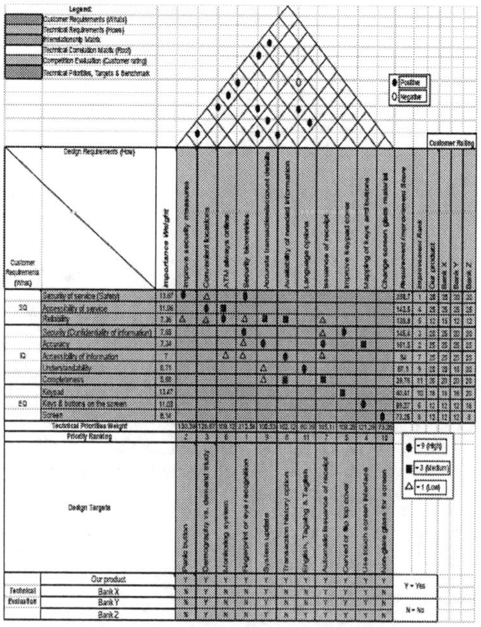

Fig. 6. Improvement Framework for ATM Design

4 Conclusion

In this study, the researcher found out that ATM users were typically young, being more educated, and having higher economic status. Common problems with the use of ATM were also identified in this study. ATMs being out of service and customers having to wait in long queues were the most commonly experienced problems. Additionally, ATMs not printing receipt and running out of money were also frequently experienced. Categorically, system-related problems were likely to occur the most. This means that most ATM related problems had the ATM's system itself as its source. Furthermore, ATM users were significantly different with their ATM usage in terms of their age, education, income, civil status, and occupation. Older, educated, married, employed and higher earning groups tend to have higher or greater ATM usage as compared to their counterparts. Moreover, as for this study, information quality (IQ) and ergonomic quality (EQ) components in general do not have significant impact or effect to service quality (SQ) components. Information, ergonomic, and service qualities are independent entities which people perceived separately. However, there were certain components of information and ergonomic qualities which were found to impact certain components of service quality. Particularly, consistency of information was found to impact competence of service, and the lighting and keypad were found to significantly affect the quality of design (tangibles) of ATMs. More importantly, this study was able to provide an improvement framework for the ATM design that could serve as a basis for future researches and projects in designing an automatic teller machine.

References

1. Lewis, B., Orledge, J., & Mitchell, V. (1994). Service quality: students' assessments of banks and societies. *International Journal of Bank Marketing 2(4)*, 3-12.
2. Athanassopoulos, A. (2000). Customer satisfaction cues to support market segmentation and explain switching behavior. *Journal of Business Research 47(3)*, 191-207.
3. Surjadjaja, H., Ghosh, S., & Anthony, J. (2003). Determining and assessing the determinants of e-service operations. *Managing Serving Quality 13(1)*, 39-53.
4. Dhungel, A., Acharya, B., & Dhungel, K. (2002). Perception of bank customers about automated teller machine (ATM). *Banking Journal 2(2)*, 23-38.
5. Baber, C., Stanton, N., & Johnson, G. (1998). Editorial. From public technology to ubiquitous computing: Implications for ergonomics. *Ergonomics 41*, 921-926.
6. Chan, C., Wong, A., Lee, T., & Chi, I. (2009). Modified automatic teller machine prototype for older adults: A case study of participative approach to inclusive design. *Applied Ergonomics 40*, 151-160.
7. Hone, K., Graham, R., Maquire, M., Baber, C., & Johnson, G. (1998). Speech technology for automatic teller machines: An investigation of user attitude and performance. *Ergonomics 41*, 962-981.
8. Dix, A., Finlayson, J., Abowd, G., & Beale, R. (1998). *Human-Computer Interaction, second edition*. London: Prentice-Hall.
9. Thatcher, A., Shaikh, F., & Zimmermann, C. (2005). Attitudes of semi-literate and literate bank account holders to the use of automatic teller machines (ATMs). *International Journal of Industrial Ergonomics 35(2)*, 115-130.
10. Mahmood, T., Shaikh, G.M. (2013). Adaptive automated teller machines. *Expert Systems with Application 40*, 1152-1169.
11. Khawaja, K., & Manarvi, I. (2009). Evaluating customer perceptions towards atm services in financial institutions; a case study of pakistani banks. *International conference on computers and industrial engineering* (pp. 1440-1445). CIE.
12. M. B. Calayag, D. M. M. Regala, C. J. R. Rivera, and D. M. N. Tancio, "A Comparative Study of Different Automated Teller Machine and Its Effectiveness Based on Customer's Satisfaction," *J. Ind. Eng. Manag. Sci.*, vol. 2017, no. 1, pp. 49–102, 2017.
13. Lewis, R., & Booms, B. (1983). The marketing aspects of service quality. In L. Berry, G. Shostack, & G. (. Upah, *Emerging Perspectives on Services Marketing* (pp. 99-107). Chicago, IL: American Marketing.
14. Najjar, L. (2002). *The impact of information quality and ergonomics on service quality in the banking industry*. Michigan: ProQuest Information and Learning.
15. Wang, R., & Kon, H. (1993). Toward total data quality management. In R. (. Wang, *Information technology in action: Trends and perspectives* (pp. 179-197). Englewood Cliffs, NJ: Prentice Hall.
16. Wang, R., & Strong, D. (1996). Beyond Accuracy: What data quality means to data consumers. *Journal of Information Management Systems 12(4)*, 5-34.

https://doi.org/10.1051/matecconf/201815005008

Preliminary Study: Self-Regulated Learning Procedure

Shafizza Sahdan[1*], Alias *Masek*[2], *Noor Atikah Zainal Abidin*[3], *Juliati Jusoh*[4]

[1,2,3] Faculty of Technical and Vocational Education, Universiti Tun Hussein Onn Malaysia, 86400 Batu Pahat, Johor.
[4] Department of Commerce, Polytechnic Sultan Haji Ahmad Shah, Semambu, 25350 Kuantan, Pahang.

Abstract. Self-regulated learning (SRL) is widely adopted by various educational institutions all over the world. A simple guideline is important and essential for the successful implementation of SRL that yields good results. Hence, a procedure and lesson plan of SRL for the subject of Principles of Management are developed and tested in a preliminary study. The objective of the preliminary study is to determine the reliability of the instruments so that the implementation procedure and lesson plan for the real study can be executed smoothly. This study was conducted for a cycle of SRL designed procedure. ANCOVA was used to analyse the effects of the procedure and lesson plan in relation to students' motivation, metacognitive awareness and academic achievement. Based on the results of the preliminary study, the items reliability index using alpha cronbach was 0.92, indicated that the questionnaire was good and effective. Each variable was significantly different, indicated that the procedure and lesson plan have positive effects on motivation, metacognition, and academic achievement. In conclusion, the procedure and lesson plan of Principles of Management developed are ready for an actual study.

1 Introduction

Education is a key to success in life. Education is started with a good learning strategy. A good strategy employed in learning can influence the academic success of students. Self-regulated learning (SRL) is one of the best strategies available thus far that focuses on developing students to become independent learners. With a systematic procedure of SRL and a combination of a variety of SRL activities in the lesson plan, there is a high probability that a student may succeed in their learning. SRL can be defined as an effective rule and strategy implemented in students' learning process in order to achieve their own goals [1].

SRL strategies have a very good and positive effect on individuals, especially for students who are undergoing the learning process. This statement is supported by the findings of [2], which state that students' achievement increases when they apply SRL in the classroom. The author [3] also said that students can increase their skills in the future when they receive feedback and comments from the lecturers to improve their learning strategies. The mastery learning, cooperative learning and computer-based learning for each student nowadays can be implemented using the SRL strategies.

Although the literature suggests that the implementation of SRL promises several advantages, most students and lecturers are sceptical about adopting SRL in teaching and learning. SRL is a more suitable learning strategy for students who are eager to search for advice and information on their own initiative [4]. This type of learning attitude really helps them to acquire knowledge more quickly and effectively. The students tend to look out for a more positive environment conducive to learning instead of staying in a negative environment [5].

According to [6], SRL is strongly related to motivation of a student, as motivation is one of the important factors in SRL. Hence, when students implement SRL as a tool, they will automatically be motivated to embrace learning. As the level of motivation increases, the level of academic achievement of the students also increases. Metacognition of students increases when they apply SRL. According to [7], students who apply SRL can think more creatively and generate more ideas based on their learning outcomes. In a sentence, SRL has a very good impact on students, and teachers should implement the strategy using the right procedure so that students can produce good outcomes of learning.

SRL undergoes three phases of activities, which are forethought, performance and self-reflection [8]. These are the main phases a student will go through when the SRL strategy is implemented as a learning approach. In this study, a combination of activities based on the SRL strategy is used to create a new procedure with a complete range of activities in each phase. The development of the procedure and lesson plan is based on three sources, namely the literature review, model and theories; course learning outcomes and expert panel from SRL; and Principles of Management. The development of the procedure and lesson plan of SRL is based on Kemp Model; it is used as a base for the instructional design.

* Corresponding author:hb140176@siswa.uthm.edu.my

The procedure implemented in the lesson plan will be tested in the teaching and learning process for the Principles of Management subject. Before the actual experiment is conducted, a preliminary study was carried out. Hence, the aim of this paper is to discuss the results of the preliminary study that was conducted before the real experiment is carried out. The first section outlines the objective of the preliminary study, the second section touches on the methodology used, the third section describes the overall results and the fourth section contains the discussions. Lastly, the conclusion of this paper is presented.

2 Objectives

The preliminary study was conducted in order to ascertain the effects of the procedure and the lesson plan on the students who are studying the subject of Principles of Management, before the real experiment is conducted. This preliminary study is necessary to ensure that the procedure in the real experiment will run smoothly according to the plan. It is also important to test the validity of the instruments used in this study, and find out if they are suitable for the study. The reliability of the instruments was measured before they are used in the real experiment. The results from the preliminary study were used to upgrade and improve the procedure and lesson plan that had been developed.

3 Methodologies

A one-week cycle of SRL procedure was conducted to run the preliminary study. The procedure of this study was the same as that for the real experiment. This is to ensure that the environment for the real experiment is properly developed. An analysis of covariance (ANCOVA) was used in order to find out the effects of SRL on motivation, metacognitive awareness, and the academic achievement of the students. Two main instruments were involved in this study. The first is a questionnaire that comprises two sections, namely motivation and metacognitive awareness. The second is a short test used to measure the students' academic achievement before and after the application of the procedure and lesson plan of Principles of Management.

3.1 Experimental procedure

The preliminary study involved two groups of students, namely the experiment group and the control group. A carefully designed SRL procedure and a lesson plan were used as the treatment for the experiment group within the context of Principles of Management subject. In this case, the selected class of this study was not informed of the research conducted to avoid bias. Similarly, the control group was set up to undergo the usual teaching and learning process using the traditional approach. Before the class started, a pre-test questionnaire and pre-test quiz were distributed to measure the students' existing knowledge. After one cycle of SRL procedure that consisted of two classes, a post-test questionnaire and post-test quiz were distributed to measure and monitor the difference that occurred, for both the experiment and control groups. The roles of lecturers are important in this study as the lecturers act as facilitators who guide students in following the procedure and lesson plan developed.

4 Results

Each variable was tested, and the results are explained. Before the analysis of the preliminary study was conducted, reliability of instrument was checked and then the ANCOVA analysis was conducted. The ANCOVA analysis shows that there is a difference in the results of the experiment and control groups for the score of each variable in relation to academic achievement, motivation and metacognitive awareness. The analysis results are used to interpret whether the procedure and lesson plan of Principles of Management have any effect on the students' academic achievement, motivation and metacognition.

4.1 Pilot Study

A pilot test was conducted on the instrument to ensure suitability of the questionnaire developed and modified. Before the questionnaire can be used, the questionnaire was checked by the supervisor and a few experts to ensure items/construct validity. Before the preliminary study was conducted, the questionnaire underwent back-to-back translation as the original questionnaire is in English. The translation was done in two processes: from English to Malay, and then Malay to English; through this procedure, it is quite certain the translations are accurate and the instrument can be used to achieve the objectives. The test used to measure the academic achievement of students also underwent a few processes. First, the test was developed based on the findings of the past year papers. Then discussions were held with a panel of experts who teach the subject of Principles of Management. These discussions are important to know the students' basic knowledge and mind-set so that they can understand the test developed. The roles of the expert panel are to verify the validity of the test if it suits the students.

The reliability of the questionnaire was measured using the alpha value. When the alpha value is larger than 0.60, the reliability of the instrument is acceptable, and it can be used in the real-world environment [9]. The results of the pilot test show that the reliability for the overall constructs is $0.921(\alpha > 0.60)$, which indicates that the instrument is very good and effective to be used. The alpha values of the constructs are 0.828 and 0.929 for metacognitive awareness and motivation respectively. The summary of each construct is shown in Table 1.

Table 1. Alpha Cronbach

Construct	Alpha cronbach
Metacognitive Awareness	0.828
Motivation	0.929
Overall	0.921

4.2 Effects of SRL Lesson Plan on Students' Academic Achievement

An analysis of covariance (ANCOVA) was used to investigate the effects of the variables; the post-test results reflect that the experiment group attained a higher level of academic achievement than the control group. The differences in the means and standard deviations for the both groups are shown in Table 2. The homogeneity of variances is not violated. This is due to the cell sizes of this study that are not similar, which are 21 and 34. The results show that there are significant differences between the experiment group and control group in the post-tests, F (1,51)=. 5.85, p=.019, partial eta^2 = .103; the pre-tests were controlled. Table 3 shows the means and standard deviations of the post-test achievements for both groups in relation to the subject of Principles of Management; the pre-tests of Principles of Management were controlled.

Table 2. Adjusted and unadjusted group mean (experiment and control) and variability for post-test using pre-test as a covariate.

Group	N	Unadjusted		Adjusted	
		M	SD	M	SE
Experiment	21	46.19	11.61	45.92	3.1
Control	34	44.41	17.44	43.52	2.4

Table 3. Analysis of covariance for post-test as a function of group (experiment and control), using pre-test as a covariate

Source	*df*	*SS*	*MS*	*F*	*p*
Pre test	1	530.83	530.83	2.74	.104
Group* Pre-test	1	1133.30	1133.30	5.85	.019
Error	51	9882.35	193.77		
Total	55	124600.00			
Corrected total	54	12774.55			

4.3 Effects of SRL Lesson Plan on Students' Metacognition

The results indicate that there are differences in the means for both experiment and control groups. The score differences of metacognitive awareness are shown in Table 4. The assumption of homogeneity of variance is violated due to the cell sizes that are dissimilar (21 and 34). Results indicate that after the controlling for the pre-test metacognition, there are significant differences between the groups (experiment and control) in the post-test metacognition, F(1,51)=. 15.30, p=.000, partial eta^2 = .231. Table 5 presents the post-test means and standard deviations of both groups for metacognitive awareness; the pre-test metacognitive awareness was controlled.

Table 4. Adjusted and unadjusted group mean (experiment and control) and variability for post-metacognitive using pre-metacognitive as a covariate.

Group	N	Unadjusted		Adjusted	
		M	SD	M	SE
Experiment	21	42.38	6.18	42.47	1.03
Control	34	43.06	5.83	43.42	.80

Table 5. Analysis of covariance for post- metacognitive as a function of group (experiment and control), using pre-metacognitive as a covariate

Source	*df*	*SS*	*MS*	*F*	*p*
Post test	1	261.24	261.24	11.95	.001
Group* Metacognitive	1	334.57	334.57	15.30	.000
Error	51	1115.02	21.86		
Total	55	102642.00			
Corrected Total	54	1890.80			

4.4 Effects of SRL Lesson Plan on Students' Motivation

An analysis of covariance (ANCOVA) was used to investigate the effects of the variables; the post-test results reflect that the experiment group attain a higher level of academic achievement than the control group. The differences in means and standard deviations for both groups are shown in Table 6. The assumption of homogeneity of variance is violated. It is because the cell sizes are dissimilar (21 and 34); this violation does not present an issue. After the pre-test for motivation was controlled, there are significant differences between the groups (experiment and control) in the post-test results, F(1,51)=4.49, p=.039, partial eta^2 = .08. Table 7 presents the means and standard deviations of the post-test motivation for the groups (experiment and control); the pre-test motivation was controlled.

Table 6. Adjusted and unadjusted group mean (experiment and control) and variability for post-test motivation using pre-test motivation as a covariate.

Group	N	Unadjusted		Adjusted	
		M	SD	M	SE
Experiment	21	5.40	.78	5.33	.14
Control	34	5.16	.76	5.26	.10

Table 7. Analysis of covariance for post- motivation as a function of group (experiment and control), using pre-motivation as a covariate

Source	*df*	*SS*	*MS*	*F*	*p*
Post test	1	7.28	7.28	20.36	.000
Group*Motivation	1	1.61	1.61	4.49	.039
Error	51	18.234	0.36		
Total	55	1547.19			
Corrected Total	54	32.01			

4 Discussions

In this study, several student-centred learning activities are involved in the SRL strategy. Those activities help students to acquire higher-quality knowledge from their learning process. In this study, the SRL procedure and lesson plan guide students to learn more effectively in the theoretical subjects, although they are boring and hard to learn. Hence, a procedure and lesson plan of SRL are developed and implemented to improve the quality of students in the future. The effects of the SRL procedure on motivation, metacognition, and academic achievement of students are discussed.

In this study, the results indicate that there is a significant difference between the academic achievement of the experiment group and that of the control group. Although the treatment was carried out within a very short period, the higher resultant test scores are a sign that students' achievement was improving. These findings are in line with the results from [10], which indicate that student receive a higher score if they were exposed to the adoption of SRL in important assignments. The findings from [11] state that SRL have a positive relationship with the academic performance of a student. Students who can improve their academic achievement can train their cognitive faculty to think more creatively. Students need a good and effective strategy to use advanced technology nowadays in order to achieve and improve their own performance in the future [12]. A student who has a good strategy will produce academic performance; on the other hand, a student may not show much improvement if he or she depends fully on an advanced technology but does not know how to use it effectively in the learning process. Based on the cognitive theory, when students think creatively, they can easily apply the theoretical principles to solve practical problems in the real life, and this can improve the quality of students [13].

In the context of metacognitive awareness, the results show that the experimental group has outperformed the control group in terms of scores. Metacognition indicates how students think, create and manoeuvre the information they have acquired from the learning process; the various forms of information and important facts can be used to answer questions in the examination. This thinking style is in line with the constructivism theory: students learn from their past, for example learning in class, and create new knowledge based on their own understanding [14]. This learning theory focuses on the thinking ability of students, rather than knowledge that is totally from the lecturer. This theory is in line with the objectives of the SRL strategy; it focuses on the student-centred learning. The students are the main character in the learning process and lecturers play the role of facilitator.

The findings are supported by the statement from [15]; when student are exposed to the SRL strategies, they will be trained to think more creatively in their learning process. The results from [7] show that the metacognition of students improve when they apply the SRL strategies, and they can use different types of strategies that suit them best. Peer assessment is one of the strategies that can be employed to evaluate the metacognition of a new-generation student; a more sophisticated and effective strategy is needed to evaluate a new breed of students today [16]. In peer assessment, students can train their metacognition as they act as an evaluator to assess the performance of their friends. Hence, they will automatically think broadly and effectively as they need to comment on their friends' work. This will help them to produce more knowledge for their friends and for themselves. Their metacognitive faculty will be used frequently, and their thinking skills will be sharpened; this will result in an increase in performance in the future.

The post-test results show that there was a statistically significant difference on students' motivation between the two groups (experiment and control). Students who have a high degree of motivation tend to take initiative to search for a better environment. This attitude is in line with the Gagne learning theory, whereby students try to search for new strategies in order to acquire the needed knowledge [17]. The findings of this study were also in line with the result from Yoon [18], which state that self-confidence of a student increases when he or she applies SRL. When self-confidence increases, students tend to be more motivated and have a greater desire to learn more new things. Lastly, they will be willing to learn harder subjects in order to gain more knowledge. These findings are supported by [5]: students are more energetic and motivated when they adopt and apply the SRL strategies. In contrast, there are previous studies that do not support the findings; the study results from [19] say that when students apply SRL, they feel that they do not receive the full attention from the lecturers as some of them need a lot of guidance from the lecturers. Hence, they tend to lose control of themselves and are unable to focus in the classroom. Although there are some shortcomings in the application of SRL, generally the adoption of SRL has implied on the positive changes on students' academic performance.

In summary, SRL has positive impacts on students' academic achievement, metacognition and motivation. Hence, for future research, the results obtained from this study can be used to upgrade and modify the procedure and lesson plan of Principles of Management. This will facilitate the smooth running of actual studies without any major difficulties and help future researchers to replicate the similar study to other context of research.

5 Conclusions

The SRL procedure and lesson plan within the Principles of Management subject have been carefully designed. The study has resulted an encouraging improvement on students' achievement, metacognition, and motivation. These preliminary study findings provide a proper framework for carrying out the actual study, which has shed light on the benefits of SRL. In addition, this study has yielded very encouraging findings that are applicable to students' learning process to attain a higher level of academic performance.

This research is supported by a scholarship awarded by the Malaysian Ministry of Higher Education (Mybrain15) and Universiti Tun Hussein Onn Malaysia.

References

1. J. L. Nietfeld, L. R. Shores, K. F. Hoffmann, Self-regulation and gender within a game-based learning environment. *Journal of Educational Psychology*, **106**, 961–973 (2014)
2. M. Leidinger, F. Perels, Training Self-Regulated Learning in the Classroom: Development and Evaluation of Learning Materials to Train Self-Regulated Learning during RegularMathematics Lessons at Primary School. *Education Research International*. Retrieved on Jun 12, 2014 from doi:10.1155/2012/735790 (2012)
3. J. Hattie, *Visible learning: A synthesis of over 800 meta-analyses relating to achievement.* New York: Routledge. (2009)
4. G. Clarebout, H. Horz, W. Schnotz, The relations between self-regulation and the embedding of support in learning environments. *Educational Technology Research and Development*, **58(5)**, 573-587(2010)
5. A.S. Labuhn, B.J. Zimmerman, M. Hasselhorn, Enhancing students' self-regulation and mathematics performance: The influence of feedback and self-evaluative standards. *Metacognition and Learning*, **5 (2)**, 173-194(2010)
6. M.H. Mahmoodi, B. Kalantari, R. Ghaslani, Self-regulated learning (SRL), motivation and academic achievement of Iranian EFL learners. *Procedia – Social and Behavioral Sciences*, 1062-1068 (2014)
7. K.D.Y. Campbell, *The Effects Of Self-Regulated Learning On Community College Students ' Metacognition And Achievement In Developmental Math Courses*. Old Dominion University: Tesis PhD (2013)
8. B.J. Zimmerman, *Attaining self-regulation: a social cognitive perspective*. In M. Boekaerts, P. R. Pintrich, & M. Zeidner (Eds.), Handbook of self- regulation.San Diego: CA: Academic Press (2004)
9. Y.P. Chua, *Kaedah penyelidikan (Kaedah dan Statistik Penyelidikan)*. 2nd.ed. Malaysia: McGraw-Hill Education. (2011)
10. J.S. Rolf, L. Scharff, T. Hodge, *Does "thinking about thinking" impact completion rates of pre-class assignments?* Paper presented at the Joint Mathematics Meetings, Boston, January 4.(2012)
11. N.A. Bakar, A. Shuaibu, R.A. Bakar, Correlation of Self-Regulated Learning and Academic Achievement among Universiti Sultan Zainal Abidin (UniSZA) Undergraduate Students. *International Journal of Academic Research in Business and Social Sciences*, **7(4)**, 254-268 (2017)
12. S.B. Sahdan, N.A.B.Z. Abidin. Self-regulated learning : A literature review for 21st century learning technology. *Advanced Science Letters*, **23 (3)**, 912 – 915 (2017)
13. J. Piaget, *Equilibration of cognitive structure*. New York: Viking Press. (1970)
14. M. Briner, *What is Constructivism?*. University of Colorado : Denver School of Education (1999)
15. J.A. Greene, C.M. Bolick, J. Robertson, Fostering historical knowledge and thinking skills using hypermedia learning environments: The role of self- regulated learning. *Computers & Education*. **54**. 230–243 (2010)
16. N.A.B.Z. Abidin, S.B. Sahdan. Peer assessment : A challenge in the new generation. Advanced Science Letters, **23 (3)**, 908 – 911 (2017)
17. R.M. Gagne, *The Conditions of Learning and Theory of Instruction*. 4th ed. New York : Holt, Rinehart & Winston. (1985)
18. T. Yoon, Are you digitized? Ways to provide motivation for ELLs using digital storytelling. International. *Journal of Research Studies in Educational Technology*, **2(1)**, 25-34. (2012)
19. A. Barrus, *Does Self-Regulated Learning-Skills Training Improve High- School Students' Self-Regulation, Math Achievement, and Motivation While Using an Intelligent Tutor?*.Arizona State University:Tesis Ph.D (2013)

It's Not Just About the Game: A Study on Attitudes and Motivation of Malaysian Secondary Sports School Students in East Malaysia towards Learning English as a Second Language

Nursyaheedah Muhammad Isa[1,*], *Noor Azlinda* Zainal Abidin[1], *Nabila* Abdul Malek[1], *Mohamad Hazeem* Mohmad Sidik[2] and Safirul Azli Abu Bakar[3]

[1]Centre for Modern Languages and Human Sciences, Universiti Malaysia Pahang, 26300 Gambang, Pahang, Malaysia
[2]Faculty of Industrial Management, Universiti Malaysia Pahang, 26300 Gambang, Pahang, Malaysia
[3]National Football Development Programme, National Sports Council of Malaysia, 57700 Kuala Lumpur, Malaysia

Abstract. In the past few years, Malaysian athletes have gained international recognition with their impressive, world-record breaking accomplishments. As a result of this, Malaysian athletes are finding themselves being interviewed and giving speeches at local and international events. For professional athletes, there is a need and a duty to be able to communicate their stories and experiences as well as giving real insight and opinions about their performances, confidently, to both local and global audiences. In addressing these audiences, it is important for these professional athletes to be confident and are able to converse well using the English language, especially so when addressing a global audience. However, some Malaysian athletes are facing difficulties in communicating confidently and coherently, especially in English. It is a recognised and acknowledged skill gap in which is believed should be tackled during the formative school years. It is possible that the skill gap stems from a lack of motivation and poor attitude beginning from secondary school. This paper discusses the findings of a study undertaken to investigate these issues further by carrying out a questionnaire survey on the attitudes and motivation of Malaysian secondary sports school students in East Malaysia towards learning English as a second language. The study reveals an overall positive attitude towards learning English by the sports students. In addition, the findings reveal that the sports school students are extrinsically and intrinsically motivated in learning English. The study concludes with recommendations for future research and is expected to make a contribution to the literature of attitudes and motivation especially for but not limited to secondary sports school students.

1 Introduction

Malaysian athletes are famous and looked up by other Malaysians due to their outstanding performances internationally and nationally. Due to their accomplishment, their popularity has made them public figures for some. Thus, they need to present themselves well not only in sports but also in other matters including the way they dress and even their communication skills especially in English. In order to master and be communicative competence in English, they should be equipped with skills related which help them in communicating. The need of communicating is not only meant for interacting when travelling abroad but it also helps them to present themselves better when being interviewed. As they are also known as the 'ambassador' of Malaysia, they have to represent the best for Malaysia not only in sports but also in communication.

However, due to limited practice and proper skills learnt, some athletes are not confident and ready to be interviewed by people especially in public. The lack of confidence and negative attitude towards language learning has made them poor in presenting themselves especially when being interviewed by the international journalists. Having low drive of motivation and negative perception towards language learning can impede the language learning process and results poor performance in communication. Therefore, this study investigates these issues on the attitudes and motivation of Malaysian secondary sports school students in East Malaysia towards learning English as a second language.

2 Literature Review

Learning or acquiring second language is often seen as difficult by most people especially students. In Malaysia, English language is learned and used as a second language but nevertheless the students are having difficulties in learning the language. English language is perceived as not important to the students' immediate needs thus decrease the motivation to learn the language (Oroujlou & Vahedi, 2011). Attitude and motivation towards learning the second language has been the major

* Corresponding author: syaheedah@ump.edu.my

concerns among researchers. According to Oroujlou & Vahedi (2011), motivation varies among students and it is difficult to keep students motivation at a high level all the time. Thus, it is important to further explore and understand the literature on motivation and attitude.

2.1 Motivation and Language Learning

Motivation in present context of language learning is seen according to Gardner (1985) as "referring to the combination of effort plus desire to achieve the goal of learning the language plus favourable attitudes towards learning the language." The desire to learn the language combined with the effort put in the process and the satisfaction experienced at the end of the process is what motivation is all about. It helps the students to strive for the better. There are two types of motivation that is relevant to this study; intrinsic and extrinsic motivation.

Deci and Ryan (2008) identify intrinsic motivation as self-satisfaction feeling resulted from the ability to complete a task. Extrinsic motivation, on the other hand focuses on involving in an activity because of different reason or consequence. For example earning a reward or to avoid a punishment.

Dornyei and Csizer (1998) indicate that motivation is responsible for achievement in language learning, and without adequate motivation, learners with distinctive talents will not be able to fulfil such objectives, even if they have a good curriculum or teachers.

In Malaysian context, Samsiah et al.'s (2009) suggested that Malaysian students are generally extrinsically rather than intrinsically motivated. It is believed that students do not perceived the usefulness of learning English language for self-satisfaction but more of the reward they obtain when learn the language especially for future careers and development. Samsiah et al.'s findings also revealed that motivation either extrinsic or intrinsic do not directly influence students' English language performance.

2.2 Attitudes and Language Learning

As cited by Elyıldırım & Ashton (2006), according to Brown (2000), attitudes are related to thoughts as well as to feelings and emotions. Attitudes govern how one approaches learning which in the case of language requires exposure to a different culture and also to the difficult task of mastering a second language.

Spolsky (2000) states that the attitudes towards the language hint at the learners' fears, feelings, or preconception about learning English as a second language. It is believed that students' decision to participate in language learning process are mostly depending on their experiences towards the process of learning and acquiring the language. The students' attitude are influenced by many factors. If the students experience success, the attitudes are reinforced, in contrast, if the students experience failure, the attitudes become more negative. This was supported by Samsiah et al. (2009), revealed that high achievers students have better attitude in learning English compared to the low achievers. In addition, female students have slightly better attitude in learning English compared to the male students. As such, attitudes may play an important role in language learning process as it is able to influence students' success or failure in their learning.

3 Methodology

3.1. Research Design and Context

This quantitative study employed a questionnaire survey to collect data. The procedure was carried out in a secondary sport's school in Gambang, Pahang. The school was established in 2012 and is the third Sport's School formed after Sekolah Sukan Bukit Jalil and Sekolah Sukan Tunku Mahkota Ismail. The school uses Bahasa Malaysia (BM) as medium of instruction.

3.2 Sample Population

47 male students participated in the survey irrespective of their English proficiencies. The sport's school accepted only male students into their football programme which resulted with the sample population being all male students. The population was made up of all Malay football students as there were only Malay students present during the data collection. 34 questionnaires were distributed and collected. In order to gauge their English proficiencies, the students' English Language result for the Ujian Penilaian Sekolah Rendah (UPSR) or also known as Primary School Achievement Test were used. Students were categorised into High Proficiency (HP) students, those who obtained Grade A for English, Average Proficiency (AP) students, those who obtained Grade B and Low Proficiency (LP) students, those who obtained Grades C, D and E.

3.3 Research Tool

A survey questionnaire adopted from Thang et al. (2011) was employed to investigate students' attitudes and motivation. The questionnaire adopted was originally translated into Bahasa Malaysia as translations into the mother tongue provides accuracy, clarity and flexibility for students in understanding the questionnaires (Ross, 2000). The survey questionnaire was previously adapted by Thang et al. (2011) from the Gardner's Attitude and Motivation Test Battery (AMTB) (Gardner, 1985). The questionnaire consists of 48 questions and has two sections. The first section consists of eight items. The first section was designed to elicit students' background information, the result of their USPR English paper, their perceptions of their own levels of motivation and their reading habits. The second section comprises of 40 items designed to extract information pertaining to the students' attitudes and motivation towards the learning of English. The Likert scale items were used to gauge students' responses and it ranged from 'strongly disagree' to 'strongly agree'. 20 items were aimed at eliciting information pertaining to students' attitudes while another 20 measured their motivation in learning English. For motivation, there were 10 items regarding intrinsic motivation and another 10 on extrinsic motivation.

3.4 Research Procedures

The questionnaire was distributed to the students during their English class after the students were briefed thoroughly on the purpose of the research. Students took about 15 minutes to complete the questionnaire. Scores were tabulated using SPSS Windows 23.0. The statistical measures used to analyse data were ranking of mean scores, one-way analysis of variance (ANOVA) and correlation coefficient. The purpose of ANOVA in this study was to compare the mean differences between attitudes and motivation of students of different English proficiency levels. The p-value for significance was set at 0.05. For differences in English proficiency levels, One Way ANOVA was used to determine significance since only one group of subjects was involved. In addition, the correlation coefficient was used to determine the correlation between the two variables; attitudes and motivation.

4 Data Analysis

The data were analysed using both descriptive and inferential tools. The descriptive tools employed were ranking and correlation coefficient and ANOVA was the inferential tool utilised.

4.1 Students' Background Information

The questionnaires were distributed to one class of form one students, 4 (11.8%) students were HP students. 8 (23.5%) were AP students and the remaining 22 (64.7%) were LP students.

4.2 Attitudes and Motivation towards Learning English: Ranking of Positive Attitudes

Table 1 shows the mean score ranking of 10 items for positive attitudes.

Table 1. Rank of mean score for positive attitudes

Rank	Items	Mean	Std. Deviation
1	English is an important part of the school programme	3.35	.646
2	Learning English is very great	3.24	.496
3	I have desire to learn good English	3.24	.654
4	I find learning English is very interesting	3.21	.538
5	I plan to learn as much English as possible	3.03	.627
6	I put a lot of effort in learning English	3.03	.627
7	I enjoy my English classes	3.00	.603
8	I really enjoy learning English	2.97	.577
9	When I have English exercises, I will finish it immediately	2.85	.702
10	I would read English newspapers and magazines as often as I could	2.74	.666
	Average mean	3.066	

Key:
Item 1 – Awareness of the importance of English
Item 3, 5, 6 – Expression of desire to improve English
Item 2, 4, 8 – Expression of interest in learning English
Item 7, 9, 10 – Actual effort towards learning and improving English

Item 1 which gauges the students on their awareness of the importance of English ranks the highest. This goes to show that students are very much aware of the importance of English either in their education, in their daily lives and for their futures. Subsequent items 2, 3, 4, 5, 6 & 8 depict interest in learning English and expression of desire to learn English. Since these items generally rank higher than 2.5, it shows that there is a tendency towards this direction and when translated into action it shows that the students actually put in the effort to as indicated by the high mean for items 7, 9 & 10 that show actual interest and effort in learning and improving English in and out of the classroom. The students, despite majority being of low English proficiency, are still very much interested in learning English and improving their English proficiencies. Regardless, teachers will still need to make continuous effort to keep making their English classes more interesting and interactive.

4.3 Attitudes and Motivation towards Learning English: Ranking of Negative Attitudes

Table 2 displays the mean score ranking of 10 items for negative attitudes.

Table 2. Rank of mean score for negative attitudes

Rank	Items	Mean	Std. Deviation
1	It is hard to learn English	2.62	.853
2	I prefer to read material in other languages rather than English	2.56	.660
3	I find doing English exercises difficult	2.44	.660
4	I force myself to listen to the teacher during English classes	2.41	.925
5	I feel stress in learning English	2.26	.567
6	I never think of improving my English languages	2.03	.717
7	During English classes, I become completely boring	1.88	.478
8	When I leave school, I shall give up the study of English entirely because I am not interested in it	1.82	.576
9	I think that learning English is dull	1.68	.589
10	Learning English is a waste of time	1.50	.615
	Average mean	2.361	

Key:
Item 1, 2, 3, 5 – Expression of difficulty in learning English
Item 4, 7, 9 – Perception of English as being uninteresting
Item 6, 8, 10 – Indication of no interest in learning or improving English

The mean scores for negative attitudes towards English are lower than positive attitudes. All of the mean scores are below the level of 3 (i.e. the agree level) which is a positive sign. The first four items that rank highest (with mean scores above or approaching 2.50) are items relating to the difficulty involved in learning English. It can be observed that the students have problems learning English and prefer learning another language. Items 7 & 9 investigate whether students find learning English uninteresting and the low mean scores (of below 2.30) suggests that they do not really find learning English boring with the exception of item 4. Item 4 has a mean score of 2.41 (approaching mean score of 2.50) which shows that the students find English to be uninteresting that they have to force themselves to listen to their English teachers. Finally, the low mean scores for items 6, 8 & 10 show that the students do not think that learning English is a waste of time and are still interested in improving their English when they leave school.

4.4 Motivation towards Learning English: Ranking of Intrinsic Motivation

Table 3 displays the mean score ranking of 10 items for intrinsic motivation.

Table 3. Mean scores ranking of intrinsic motivation towards learning English

Rank	Items	Mean	Std. Deviation
1	Learning English will allow me to converse with more varied people	3.62	.652
2	Learning English is good for self-development	3.56	.561
3	Learning English will enable me to better understand and appreciate English literature	3.41	.500
4	I learn English in order to improve my English language skills	3.38	.604
5	Learning English allows me to participate more freely in the activities of other cultural groups	3.35	.597
6	Learning English will allow me to be more at ease with English speakers	3.35	.597
7	I feel more confident in learning English	3.29	.629
8	I make better use of my time to learn English	3.26	.666
9	I study English because I enjoy learning it	3.21	.687
10	I learn English because it is something that I always want to do	3.00	.550
Average mean		3.343	

Key:
Item 2, 4 – Learning English to improve
Item 1, 3, 5, 6 – Learning English for social reasons
Item 9, 10 – Learning English to satisfy interest and enjoyment

Items 2 and 4 have mean scores of above 3.00 which depicts that students have interest to improve themselves for self-development, and to develop their language skills. The mean scores of above 3.00 for items 1, 3, 5, 6 further reveal that students seem to be willing to learn English for social purposes and also to understand the literature and cultures of other people. Similar for items 9 & 10, with the mean score of 3.21 and 3.00 respectively, suggests that students are willing to learn English for interest and enjoyment. With a mean score of 3.29 and 3.26 respectively, students are also confident in learning English (Item 7) and actually make better use of their time to learn the language (Item 8). Considering that all items display a mean score of more than 3.00 it shows that students possess positive intrinsic motivation towards learning English.

4.5 Motivation towards Learning English: Ranking of Extrinsic Motivation

Table 4 shows the mean score ranking of 10 items for extrinsic motivation.

Table 4. Mean score ranking of extrinsic motivation towards learning English

Rank	Item	Mean	Std. Deviation
1	Learning English is important for my future career	3.65	.544
2	I study English because I want to do well in my examination	3.62	.551
3	Learning English is useful in getting a good job	3.62	.604
4	I learn English because it makes me a more knowledgeable person	3.56	.561
5	I learn English because I need it to further my studies overseas	3.41	.701
6	Learning English is important for me because other people will respect me more if I am good in English	3.29	.629
7	I study English in order to please my family	3.15	.821
8	I learn English because I can contribute more to the society	3.15	.500
9	I need to learn English in order to finish high school	2.79	.946
10	I feel that no one is really educated unless he is fluent in English language	2.74	.751
Average mean		3.298	

Key:
Item 1, 3 – Learning English for future career
Item 2, 4, 5, 9 – Learning English for education
Item 6, 7, 8, 10 – Learning English due to external influences

The mean score of above 3.00 for items 1, 2, 3, 4, & 5 suggest that students have a strong desire to learn English for future career and for the purpose of education. However, there is an exception for item 9 whereby the mean score is 2.79 (below mean score of 3.00). This findings suggest that students slightly agree that they need to learn English to complete their high school. For items 6, 7 & 8 with a mean score of 3.29, 3.15 and 3.15 respectively shows that students are influenced by external factors such as earning the respect of other people, pleasing family and contributing to society, for wanting to learn English. However, for item 10 with a mean score of 2.74, contrary to popular beliefs, some students does not feel that intelligence is measured by one's fluency or proficiency in the English language.

4.6 Reliability Analysis

According to Thang et al. (2011), prior to comparing the group mean scores it is necessary to determine whether the grouping of items is reliable. For that purpose, the alpha reliability test was performed. The Cronbach Alpha reliability value for positive attitudes is 0.701, for negative attitudes is 0.709, for intrinsic motivation is 0.756, and for extrinsic motivation is 0.649. The Cronbach Alpha reliability values for positive attitude, negative attitude and intrinsic motivation are all above 0.700, which indicate that the grouping of the three groups of items is reliable. However, the The Cronbach Alpha reliability value for extrinsic motivation is 0.649 (less than 0.700). Nevertheless, according to Hair et al. (2014), the values between 0.690 – 0.600 is considered to be acceptable.

4.7 Overall Attitudes towards Learning English

The results of the students' attitudes towards the learning of English were divided into sub-categories. The mean scores of positive and negative attitudes were divided into two sub-categories. Mean scores of 1.00-2.49 was classified as disagree and mean scores of 2.50-4.00 was classified as agree.

Table 5. Students' overall attitudes towards the learning of English

Subcategories for positive attitudes	n	%	Subcategories for negative attitudes	n	%
Disagree (mean 1.00-2.49)	2	5.8%	Disagree (mean 1.00-2.49)	29	85.3 %
Agree (mean 2.50-4.00)	32	94.2%	Agree (mean 2.50-4.00)	5	14.7 %
	34	100.0		34	100.0

As shown in Table 5, 94.2% of the students agreed that they have positive attitudes in learning English and only 5.8% disagreed. The data portrays that students possess promising attitudes towards learning English. In addition, 85.3% of the students disagreed that they have negative attitudes and only 14.7% admitted that they have negative attitudes towards the learning of English, which is further supported the fact that the students in general, have encouraging, favourable and promising attitudes towards the learning of the English language.

4.8 Relationship between Attitudes and Motivation

To investigate the relationship between attitudes and motivation in learning English, Pearson correlation was carried out. The result is presented in Table 6.

Table 6. The result of correlation analysis performed on attitudes and motivation

	Positive attitude	Negative attitudes	Intrinsic Motivation	Extrinsic motivation
Positive attitude	1			
Negative attitudes	-0.065	1		
Intrinsic Motivation	0.699**	-0.055	1	
Extrinsic motivation	0.147	0.122	0.395*	1

**Correlation is significant at the 0.01 level (2-tailed)

* Correlation is significant at the 0.05 level (2-tailed)

As can be seen from Table 6, positive attitudes are positively correlated with intrinsic motivation (r= -0.699) at 0.01 significant level. Intrinsic motivation is significantly correlated to extrinsic motivation (r= 0.395) at 0.05 significant level. On the other hand, extrinsic motivation is not correlated to positive or negative attitudes. The findings illustrates that students who are intrinsically motivated have more positive attitudes as opposed to negative attitude in learning English, as compared to extrinsically motivated students. Additionally, these intrinsically motivated students according to Thang et al. (2011), are also more likely to be extrinsically motivated as well. Thus, the findings suggests that students whom are intrinsically motivated have a high probability to be a more successful English learners as having positive attitudes can and will drive them to success in learning English.

4.9 Comparing the Attitudes and Motivation of Students of Different Proficiency

Table 7 displays the differences in mean score of learning attitudes and motivation among students of different proficiency levels.

Table 7. The results of one way ANOVA performed on learning attitudes and motivation among proficiency

	Proficiency level of student	Mean	Std. dev	df	F	Sig (P-value)
Positive attitudes	HP	3.150	0.1291			
	AP	3.0125	0.1807	2	0.237	0.790
	LP	3.0682	0.3822			
Negative attitudes	HP	2.1250	0.2630			
	AP	2.1625	0.3068	2	0.074	0.929
	LP	2.104	0.3958			
Extrinsic motivation	HP	3.600	0.3652			
	AP	3.288	0.2696	2	1.338	0.277
	LP	3.318	0.3514			
Intrinsic motivation	HP	3.3250	0.320			
	AP	3.2750	0.3105	2	0.031	0.970
	LP	3.30	0.3532			

*The mean difference is significant at the 0.05 level

The results of one way ANOVA showed that in positive attitudes variables there are no significant difference between all of the proficiency levels of student as the p-value is not significant at 0.790. But, it can be seen that the means score for all of the Proficiency levels of student exceed 3.0 where they had positive attitude in learning English. In contrary, the ANOVA result in negative attitudes showed that students are in disagreement with negative attitudes regardless of any of the proficiency levels. In addition, result also indicated that there is no significant different between all of the proficiency levels of student as the p-value is not significant at 0.929. This indicated that students do have neither negative perception nor negative expression towards learning English.

Next, in the extrinsic motivation variable it showed that HP students have high mean value with 3.600 and AP students have the lowest mean value by on 3.288. But, in the ANOVA analysis it was found that there is no significant difference between all of the proficiency levels towards extrinsic motivation. Other than that, Students are also unanimously agree with intrinsic motivation towards learning English regardless of any proficiency levels. With the significant value of 0.970, it depicts that there is no significant different between the students for each of the proficiency levels towards intrinsic motivation.

5 Discussion and Conclusion

From the study, Malaysian secondary sports school students have shown positive attitude towards learning English language. Positive language attitudes let learners have positive bearings towards learning English and this is especially important as it will significantly contribute to the success of second language learning (Samsiah et al.,2009; Mohamad Jafre, 2012).

Positive relationship between higher proficiency level and positive attitudes and motivation to learn English among the students of Sport School are also seen in the result of the study. These findings correlate with the findings of Samsiah et al (2009) that Malaysian students are generally more extrinsically as opposed to intrinsically motivated in learning English and the similar results can be seen from studies carried out in other parts of the world (Gayton, 2010). Attitudes and motivation goes hand in hand as both elements need to coexist to achieve a successful language learning experience (Gardner, 1985). Students with positive attitudes and high level of motivation are generally more successful than those with negative attitudes and low motivation (Samsiah et al., 2009).

The findings also show that students' extrinsic (average mean = 3.343) and intrinsic (average mean = 3.298) motivation in learning English complement one another. Both of the motivation are fairly seen contributing to students' motivation. The intrinsic motivation which comes from learning to improve oneself and learning for social reasons will help to motivate students in learning English Language. Meanwhile, the external motivation they have is due to the external rewards which they get if they are able to be proficient in English. The students feel the importance of the English language as it benefits them for future career and education.

From the results given, the skill gap which found missing during the study in secondary school can be accomodated with proper trainings on English communication as part of their preparation to represent Malaysia in sports. These communication trainings should be incorporated just to keep them motivated in learning English, gain confidence and be competence in the language. The ability to communicate in English might be seen as small part to some, but as a whole it helps to make them confident in presenting themselves in public especially when interviewed. With input given in trainings and sufficient time of communication practices, it is hoped that they will make them better communicators. It is strongly recommended that contribution to the literature of attitudes and motivation especially for but not limited to secondary sports school students should be explored further.

References

Brown, H. D. (2000) *Principles of language learning and teaching*. 4th ed. White Plains, NY: Addison Wesley Longman.

Cook, V. (1991). Second language learning and language teaching. London: Edward Arnold.

Deci, E. L., & Ryan, R. M. (2008). Facilitating optimal motivation and psychological well-being across life's domains. Canadian Psychology, 49, 14-23. doi: 10.1037/0708-5591.49.1.14

Dornyei, Z. & Csizer, K. (1998). Ten Commandments for motivating language learners: results of an empirical study. 2(3). 205-229.

Elyıldırım, S. & Ashton, S. (2006). Creating Positive Attitudes towards English as a Foreign Language. 44(4). 2-21.

Gardner, R. C. (1985) .Social psychology and second language learning: The role of attitudes and motivation. London: Edward Arnold.

Gayton, A. (2010). Socioeconomic Status and Language - Learning Motivation: To What Extent does the Former Influence the Latter? Scottish Languages Review. 22, 17-28.

Hair, J., Hult, G. T. M., Ringle, C., and Sarstedt, M. (2014). *A Primer on Partial Least Squares Structural Equation Modeling (PLS-SEM)*. Los Angeles: SAGE Publications, Incorporated.

Karahan, F. (2007). Language attitudes of Turkish students towards the English language and its use in Turkish context. Journal of Arts and Sciences, 7, 73-87.

Mohamad Jafre Zainol Abidin. (2012). EFL Students Attitudes Towards Learning English Language: The Case of Libyan Secondary School Students. Asian Social Science, 8: 119-134

Oroujlou, N., & Vahedi, M. (2011). Motivation, attitude, and language learning. Procedia - Social and Behavioral Sciences, 29, 994-1000. doi: 10.1016/j.sbspro.2011.11.333

Ross, N. J. (2000). Interference and Intervention: Using Translation in the EFL Classroom.
Modern English Teacher, 9 (3), 61-66.

Samsiah Bidin, Kamaruzaman Jusoff, Nurazila Abdul Aziz, Musdiana Mohammad Salleh & Taniza Tajudin. (2009). Motivation and attitude in learning English among UiTM students in the northern region of Malaysia. English Language Teaching. Vol. 2(2), 16-20.

Spolsky, B. (2000). Language motivation revisited: Anniversary article. Applied Linguistics, 20 (2), 157-169. http://dx.doi.org/10.1093/applin/21.2.157

Thang, S. M., Ting, S. L., & Jaafar, N. M. (2011). Attitudes and motivation of Malaysian secondary students towards learning English as a second language: A case study. 3L; Language, Linguistics and Literature, The Southeast Asian Journal of English Language Studies, 17(1), 40-54.

Determining firm characteristics and the level of voluntary corporate governance disclosures among Malaysian listed property companies

Shabana Talpur[1*], *Mohd Lizam*[1], and *Nazia Keerio*[1]

[1]Faculty of Technology Management and Business, UTHM, Batu Pahat, Johor, Malaysia

Abstract. This study examined the level of voluntary corporate governance disclosures and the influence of firm characteristics (i.e., firm size, firm age, and firm market listing) on the level of these disclosures among Malaysian property listed companies. The check-list to measure the voluntary corporate governance disclosures was adopted from Malaysian corporate governance index 2011 by Minority Shareholder Watchdog Group (MSWG). The voluntary corporate governance disclosure practices and firm specific characteristics were obtained from annual reports of property listed companies on Bursa Malaysia for the period of 2012 to 2015. The findings suggested an improving voluntary corporate governance reforms in Malaysia. However, the firm size was found as an inflicting factor in determining the level and quality of voluntary corporate governance disclosure practices. On the contrary, the results found were contradicting the hypothesis related to firm age and firm market listing, as no relation of voluntary corporate governance disclosures and firm age and firm market listing. The study has made an interesting contribution toward the disclosure and corporate governance by contributing in understanding the importance of quality disclosure and good governance practices.

1 Introduction

Foreign investment is deemed to be vital source of foreign currency inflow through stock market investment. As in the case of Thailand in 1997-1998 crises, that occurred due to high withdrawal of foreign investment as the investor lost their confidence in the country's stock market [1, 2]. [3], contend that foreign investors face high level of risk due to the chances of expropriate of investment by insiders (managers and block-shareholders) of the company. The effective and efficient stock market, with high level of transparency through financial statements, boosts investors confident in making financial decisions [4](Jaffar, Jamaludin, Rahman, & Riduan, 2007). Investors gain confidence when they have sufficient information about the company to make investment decision. According to IASB (2010), "the two primary qualitative characteristics of information in financial statements are relevance and faithful representation". Relevance and faithful representation are basic characteristics for improving quality. The information is considered relevant and faithfully represented when it helps an investor to make sound investment decisions [5]. The disclosures made in annual report are the main source of periodical information for shareholders. The key objective of disclosure is to update analysts and the investors about the amount, timing, and uncertainty of future cash flows [6,7]. Boards and management must maintain quality disclosures and transparency as they are essential for informed decision-making as mentioned in Malaysian Code of Corporate Governance.

The significance of quality disclosures and usefulness of accounting information is a major concern by many empirical studies. In previous studies, the attributes and characteristics of firm and stock market are analysed to get the factor which influences the quality of corporate disclosures. [8] Salter (1998), in their study about emerging stock markets and quality of corporate disclosures, explained deficiency in good disclosure is the reason to limit the quantity of company's stock on foreign and local investor's portfolio. A good disclosure, may be mandatory or voluntary, are communication of economic information, portraying company's financial position and performance throughout financial year [9,10]. [9], contended the adoption and implementation of high quality disclosure practices is important to invite foreign and national investors, which creates reserves for developing countries.

This study evaluated the voluntary disclosure practices of corporate governance, which were developed and motivated to adopt by regularity bodies in Malaysia after surviving the late 1990's financial crisis. Along with that, these reforms have also motivated on global level to moderate the financial problem among developing economies. Malaysian code of corporate governance was introduced in 2001 for implementing the good governance reforms among both public and private companies. However, eventually the code was implemented through Bursa Malaysia listing requirements for listed companies. The Bursa Malaysia listing requirements mandate each listing company to disclose their corporate governance practices in annual reports – under the label of corporate governance statement. Along with that, the regulatory bodies also encourage listed companies to disclose additional information beyond statutory requirement to enhance the level and quality of information in annual report. Many studies have evaluated the corporate governance practices of emerging countries including Malaysia. However, this study is distinguished from previous literature. Firstly, it focuses on voluntary corporate governance practices of property sector of Bursa Malaysia. Secondly, Malaysian listed companies have majority of highly concentrated ownership companies as most of the companies are family owned. Lastly, the panel data set is rarely used in disclosure studies to evaluate the changes in disclosure practices in recent years.

This study is divided into four parts. First, the detailed previous literature is reviewed in the subsequent section, followed by section about research methodology adopted in this research. In the next section, results and analysis are presented. Lastly, this study is duly concluded in the end.

1.1 Types of disclosures in financial statements

Previous studies have considered quality of disclosures as the presentation of minimum level of mandatory as well as voluntary disclosures made by the company in annual financial statements. The mandatory disclosures are the information which is compulsory to disclose according to the rules and regulations imposed by financial regulatory bodies. [11], in their study about the interaction of mandatory and voluntary disclosure, differentiate the usage of both disclosure according to the information needed. They argue that while mandatory disclosure provides the information on past performance of company, voluntary disclosure proposed the information which helps to forecast about future earning of company.

1.1.1 Mandatory disclosures

Mandatory disclosures are amount of information in company's financial statements required by the regulatory bodies of a country to disclose up to a set standard [10]. Even though, mandatory disclosures are regulated by the laws and enforcement bodies, but many disclosure laps are observed in previous studies. [10], assessed the status of mandatory disclosure level among Zimbabwean companies listed on Zimbabwe Stock Exchange (ZSE). By taking all the listed companies as a sample size, the study found only 76.4 (mean disclosure of companies) out of 214 mandatory items were disclosed. The mandatory items were collected from three regulatory resources of the country (i.e., the adopted IASs standards, the Companies Act, and the listing rules of the ZSE).

1.1.2 Voluntary disclosures

Voluntary disclosures are additional information provided by company, beyond the statutory and legal requirements by regulatory bodies, like MASB etc., to enhance the quality of information disclosed, sufficient for making decision and assessing risk factors of company's securities by the user of annual reports. The research related to voluntary disclosures is an important determined identified since 1970's [2] and recently there has been increased focus on voluntary disclosures. [12], contended that voluntary disclosures are the way to communicate the shareholders of the company about the transparency and highlight company's good value.

Many previous studies discussed the reasons of why company discloses information in form of voluntary disclosure. Voluntary disclosures are the information which is beyond mandatory requirements [13], helps reduces information asymmetry [14, 15], building investor trust [16] and reduce cost of equity [17]. Furthermore, information asymmetry creates a negative image about company's stock which ultimately resulted in increased in the cost of capital because investor cover the risk associated with investment. [18], in their study examined the linkage between voluntary disclosure and stock liquidity. They suggested that companies should pursue high quality of disclosure which will reduce the information asymmetry and leads towards enhancement in stock market liquidity.

1.2 Corporate governance

The emergence of term "Corporate Governance" can be new, but the issue addressed in the system is discussed by the pioneer of economics. The study identified that the separation of ownership and control in the corporation can lead toward the coincidence of conflicts of interest. This discussion about conflicts of interest continues and in the early 1930s, [19] in their publications called "The Modern Corporation and Private Property", illustrated that with the separation of ownership and management of company may push the manager to not act in the best interests of stockholders. Additionally, [20] further developed the idea and proposed the fully-fledged concept of agency theory conflict and suggested corporate governance as mechanism to monitor and control management activities to avoid agency problem in company. The research and discussion on corporate governance were only restricted to the United States and Britain from 1970's to 1980's. However, in the early 1990's the research begins to expand and cover the major economies in the world such as Germany, Japan and United Kingdom. In the recent years, the explosion of research regarding corporate governance has been observed across both developed and emerging markets in the world. The academicians with multidiscipline have recognized the importance of the corporate governance either they are from economics and finance or from law, management and accounting fields. In the new era of globalization, along with the academicians, interest in corporate governance has been growing rapidly among the corporate world [21].

Notwithstanding, an optimum governance structure for a specific country is difficult to estimate, but the performance and transparency showed by a firm may indicate the most suitable corporate governance structure adapted by it. As stated by [22], the way corporate governance controls the operations, it guarantees companies involvement in value added activities for investor which is reflected by high quality disclosure provided in financial statements. [23], also maintained the quality of disclosures as an indicator for optimum corporate governance mechanism. The author defines it as "a mechanism for accountability, emphasizing the need to raise reporting standards in order to ward-off the threat of regulation. Higher disclosure results in improved transparency, which is one of the most essential elements of healthy corporate governance practices". Thus, transparency achieved through quality voluntary disclosures, reflects efficient and optimum corporate governance practices are being followed by a company [24]. Therefore, the disclosures regarding corporate governance information have appealed the interest of researchers in last decade. Numerous CG indices have been developed previously, but generally originated on developed economies. However, much less work has been performed on developing and emerging countries [25]. Therefore, this study is aimed to identify the disclosure practices regarding corporate governance among emerging countries as Malaysia.

1.3 Firm characteristics

As mentioned before, the voluntary corporate governance disclosures and quality financial reporting influence in reducing agency conflicts. The agency cost bear by firm can be reduced by increasing the level of information disclosed by firm [20]. Different factors have been examined in previous studies which play an important role in enhancing the level of voluntary disclosure by firm. In this context, firm characteristics are also considered influencing [26, 27, 16]. Following these researches, this study has also examined the voluntary corporate governance disclosure in the context of three firm characteristics: firm size, firm age and firm's market listing.

Firm size: Large firms are more evident as compared to smaller firms, and their public interest is also more exposed [28]. Therefore, firm size is a key factor while predicting the type of association a firm maintains with inside and outside environment. It is generally assumed that the firm with larger size has more influence on their stakeholders. As [29] claimed that the size plays an important role in companies' corporate disclosures. Larger firms have higher agency costs, compelling management to disclose more information to ease agency conflicts. Thus, firm size is considered as an important feature towards good corporate governance disclosure practices [25, 30]. It also improves the voluntary corporate governance disclosures [20]. In this context, firm size is considered as an essential variable while explaining the disclosure level by previous studies. Therefore, by following previous studies such as [31], this study also evaluated voluntary corporate governance disclosures based on company size.

H1: Firm size influences the level of voluntary corporate governance disclosure among property listed companies of Malaysia.

Firm age: The firm age can influence the level of voluntary disclosure because the age considered as the stage of growth and development of a firm. [10], argues that as younger firms have lack of attractive track record, along with the cost and complexities related to gathering information, can contribute toward the phenomenon that younger firms have less motivation to disclose additional information in their annual reports. On the contrary older firms have a reputation, brand name and sufficient capital which gave them advantage to disclose more information and become additionally compliant as compared to their counterparts. Therefore, the variable of firm age has been considered important by previous studies [32, 33] while determining the voluntary disclosure practices of firms. Thus, this paper examined the influence of firm age on the level of voluntary corporate governance disclosure among property listed companies of Malaysia.

H2: Firm age influences the level of voluntary corporate governance disclosure among property listed companies of Malaysia.

Firm market listing: When the company is listed on Bursa Malaysia, they must follow all listing prerequisites; for example, to follow Bursa Malaysia listing requirements, to get successfully listed. Similar condition is also applied in the case of code of corporate governance. Listed companies on Bursa Malaysia are required to make a statement in their annual reports about the compliance of corporate governance principles, as per stated in the Code (Malaysian Code of Corporate Governance 2012). Therefore, companies listing status, for how many years a company has been listed, affect the level of voluntary corporate governance disclosures practices of listed companies. Thus, this study analysed the level of voluntary corporate governance disclosures based on the number of years the sampled companies have listed.

H3: Firm market listing influence the level of voluntary corporate governance disclosure among property listed companies of Malaysia

2 Methodology and research design

2.1 Data Collection

In this paper, we investigated the annual reports of 85 listed companies under property sector of Bursa Malaysia for the period of 2012 to 2015. The voluntary corporate governance disclosure and firm characteristics data was collected through content analysis. All the annual reports were downloaded from Bursa Malaysia website and each company's website was also visited as per requirement. The voluntary corporate governance disclosure (VCGDS) checklist was adopted from Malaysian corporate governance index 2011 by Minority Shareholder Watchdog Group (MSWG), which evaluates all listed companies of Bursa Malaysia, against those criteria on yearly basis. The voluntary corporate governance disclosure score (VCGDS) is the score obtained by each company based on their compliances to each disclosure on the checklist.

2.2 Variable measurement

For examining the relationship between voluntary corporate governance disclosures and firm characteristics VCGDS has been calculated. The measurement of VCGDS and firm characteristics is discussed below:

2.2.1 Voluntary corporate governance disclosures score

The checklist consists of 71 observations related to corporate governance information. Each observation was scored 1 if the company is disclosing that information, otherwise it was scored 0 for non-disclosure. The maximum disclosure was 71 if a company discloses all the items on the checklist. So, the VCGDS was calculated as follows:

$$VCGDS = \frac{TD}{MD} \qquad (1)$$

where
VCGDS = Voluntary corporate governance disclosure score;
TD = Total Disclosures;
MD = Maximum Disclosures.

3.2.2 Independent variable

Firm size: A firm is considered as large if it owns a large number of assets. So, the variable firm size will be measured as natural log of total assets own by firm.
Firm age: Firm age is measured by the number of year passed a company has been founded.
Firm market listing: A firm market listing is measured as number of years a company has been listed on Bursa Malaysia.

2.3 Research model

For the second part, the study examined the relationship between voluntary corporate governance disclosure and firm characteristics for the sample period of 2012 to 2015. The following pooled least square regression model is used because of the panel data set:

$$VCGDSit=\beta0+ \beta1FIRMSIZEit + \beta2FIRMAGEit+\beta3FIRMMARLISTit+\varepsilon it \qquad (2)$$

where,
VCGDS = voluntary corporate governance disclosure score for each sample firm;
FIRMSIZE = Natural log of total asset of each firm;
FIRMAGE = Number of year a firm has been founded;
FIRMMARLIST = Number of year a firm has been listed on Bursa Malaysia;
$\beta1$ to $\beta3$= constants; and ε= error term;
i = N firm, and t = T time periods

3 Results and discussion

3.1 Descriptive statistics

The VCGDI consists of 71 voluntary disclosures which were derived from Minority Shareholder watchdog criteria for assessing corporate governance practices in Malaysia. As shown in Figure 5.1, the means voluntary corporate governance has improved in four years. The mean of disclosure in four years has shown improved disclosure practices among sampled companies from 19% in 2012 to 26% in 2015 (35% improvement). The average disclosure score in four years was 22%. The average score was reported below from other studies conducted in Malaysia, for example [32, 7, 34] and [26] reported 31.35%, 91.5% and 53.20%, respectively. The results of average voluntary disclosures were higher than many disclosure studies, for example [35, 36] and [37] from Italy, Australia and China, respectively. However, the average voluntary disclosure was reported below as compared to studies in other emerging economies. Such as [38, 39] and [40] reports 46% in Mexico, 61% in South African and 40% in Venezuela, respectively.

Descriptive statistics of company size was measured by natural log of company's total assets. The results show a range of 9.92 (2.03 ml) to 16.33 (12.26 bn), with means size of 13.72 (91.29 ml). The mean has evidenced an insignificant growth over the sampled period. Furthermore, Panel L presents the descriptive regarding company age, which range from 3 to 108 year during the sampled period, with mean age of 35.46 years among property listed companies in Malaysia. Lastly, the number of years the company has been listed on Bursa Malaysia is presented in Panel M. The descriptive reports a range of 0 to 54 years the companies have been listed, with the mean number of listing years was 21 years. It indicates that sample companies have included newly listed to very older listed companies.

3.2 Panel data regression analysis

As shown in Table 2, the results suggest a positive and significant coefficient of firm size. However, the results were insignificant of firm age and market listing. The result was consistent with previous studies, likewise [41, 39, 25] and [42]. These studies also found a positive impact of company's size on voluntary disclosures. [43], also contended that large size companies have higher voluntary disclosure level as they have more agency problems due to their complex capital structure.

Table 1. Descriptive statistics of voluntary corporate governance disclosure and explanatory variables.

Variables	All	2012	2013	2014	2015
Panel A: COMZIE					
Mean	13.7	13.67	13.73	13.67	13.8
Min	9.9	9.9	9.9	10.4	10.52
Max	16.3	16.3	16.33	15.66	15.82
STD	14.3	14.33	14.34	14.04	14.19
Panel A: COMMARLIST					
Mean	35.4	33.94	34.93	35.92	36.98
Min	3.0	3.0	4.0	5.0	6.0
Max	108	105	106	107	108
STD	17	17.13	17.15	17.17	17.18
Panel A: COMAGE					
Mean	21.1	19.61	20.62	21.62	22.62
Min	0.0	0.0	1.00	2.00	3.00
Max	54.0	51.00	52.00	53.00	54.00
STD	12.8	12.83	12.85	12.85	12.85

Table 2. Voluntary corporate governance disclosure Panel data regression analysis summary.

Variable	Co-EFF	STD	T-Statis	Prob
COMSIZE	0.008	0.035	0.2465	0.805
COMMARLIST	-0.011	0.035	-0.312	0.754
COMAGE	1.11E	4.10E	2.7122	0.0070
C	-17.59	5.1523	-3.4157	0.0007
R-squared	0.822	F-statistic	7.10947	
Adjusted R-squared	0.7904	Prob (F-statistic	0.000	

As far as firm age is concern, the findings are again consistent with [44], which also found no significant relationship between voluntary disclosures and company age. However, some studies, for example Adrian [45], found positive relationship between voluntary corporate governance disclosures and company's age. However, these results were inconsistent with previous studies, for example [9, 46] which found a positive and significant relationship between voluntary disclosures and company market listing.

4 Conclusion

The relationship between voluntary corporate governance disclosures and corporate governance mechanism has been assessed by many studies in the last decade. However, rare studies investigated the impact of firm characteristic's influence on level of voluntary corporate governance disclosures. This study has contributed in the vast literature regarding disclosures and corporate governance by empirical results.

The study found that the voluntary corporate governance disclosure practices among Malaysian property listed companies have been improved by significant percentage. It was interesting to found that, although the checklist consists of voluntary disclosures, which are not required by regularity bodies. But the property listed companies are motivated to disclose this information for the benefit and interest of shareholders. On the other hand, the larger the company, the higher the disclosure level was found in this study, which evidence that large size companies have more interest in disclosing additional information as compared to small size companies.

The study made an exempt to provide overview of the voluntary corporate governance disclosure practices of Malaysian property listed companies; however more detailed study can be done for all the listed companies on Bursa Malaysia.

The authors would like to thank Universiti Tun Hussein Onn Malaysia (UTHM) for supporting this research under Postgraduate Research Grant Vote No. U615.

References

1. A. Bin-Zulkafli, M. Adul Samad, & M. Ismail, Corporate Governance in Malaysia. *Malaysian Institute of Corporate Governance.* **1**, 18 (1999)
2. S. S. Ho, & K. S. Wong, J. Int. Account. Audit Taxat., **10(2)**, 139-156 (2001)
3. K. Cremers, & V. B. Nair, J. Finance, **60(6)**, 2859-2894 (2005)
4. R. Jaffar, S. Jamaludin, C. A. Rahman, & M. Riduan, *Malaysian Accounting Review* 6(2) (2007) (2007)
5. F. Van Beest, G. Braam, & S. Boelens, Nijmegen Center for Economics (NiCE). Working Paper, 09-108 (2009)
6. J. Banghøj, & T. Plenborg, Account. Finance, **48(2)**, 159-180 (2008)

7. L. L. Eng, & Y. T. Mak. J. Account Publ. Pol., **22(4)**, 325-345 (2003)
8. S. B. Salter, Int. J. Account, **33(2)**, 211-234 (1998)
9. M. Al-Akra, I. A. Eddie, & M. J Ali, BAR, **42(3)**, 170-186 (2010)
10. S. Owusu-Ansah, Int. J. Account, **33(5)**, 605-631 (1998)
11. E. Einhorn, J. Account. Res., **43(4)**, 593-621 (2005)
12. A. Alhazaimeh, R. Palaniappan,& M. Almsafir,. Procedia Soc. Behav. Sci, **129**, 341-348 (2014)
13. Y. Lan, L. Wang, & X. Zhang, China J. Account. Res., **6(4)**, 265-285 (2013)
14. Cheng, & Lo. J. Account Res., **44(5)**, 815-848 (2006)
15. O. Kim, & R. E Verrecchia, J. Account. Econ., **17(1-2)**, 41-67 (1994)
16. D. G Barako, Afr. J. Bus. Manag, **1(5)** (2007)
17. J. Piotroski,. Unpublished manuscript, University of Chicago (1999)
18. C. Leuz, & R. E. Verrecchia, J. Account. Res., **38**, 91-124 (2000)
19. A. A. Berle, & Gardiner, C. Means. 1932. *The modern corporation and private property*, 204-205 (1968)
20. M. C. Jensen, & W. H. Meckling, J. Financ. Econ., **3(4)**, 305-360 (1976)
21. L. A. Bebchuk, & M. S. Weisbach, Rev. Financ Stud., **23(3)**, 939-961 (2010)
22. A. Shleifer, & R. W. Vishny, J. Finance, **52(2)**, 737-783 (1997)
23. A. Cadbury, Cadbury report: *Tech reprt*, HMG, London (1992)
24. H. Li, & A. Qi, Corpor. Owner. Control, **5(2)**, 360-366 (2008)
25. K. Samaha, K. Dahawy, K. Hussainey, & P. Stapleton, Adv Account, **28(1)**, 168-178 (2012)
26. M. Akhtaruddin, M. A. Hossain, M. Hossain, & L. Yao, JAMAR, **7(1)**, 1-19 (2009)
27. K. Aljifri, Adv Account, **24(1)**, 93-100 (2008)
28. Z. Watts, *Positive accounting theory*. Edgewood Cliff, NJ (1986)
29. S. J. Bhayani, S. East Asian J. Manag, **17(4)**, 6 (2010)
30. Y. B. Tariq, & Z. Abbas, Econ Model, **35**, 565-575 (2013)
31. R. Bauer, B. Frijns, R. Otten, & A. Tourani-Rad, Pac. Basin. Finance J., **16(3)**, 236-251 (2008)
32. M. Akhtaruddin, & H. Haron. ARA, **18(1)**, 68-82 (2010)
33. M. Hossain, & M. Reaz, Corp. Soc. Resp. Env. Manag., **14(5)**, 274-288 (2007)
34. R. M. Haniffa, & T. E. Cooke, Abacus, **38(3)**, 317-349 (2002)
35. S. Lim, Z. Matolcsy, & D. Chow, Eur. Account Rev., **16(3)**, 555-583 (2007)
36. L. Patelli, & A. Prencipe, Eur Account Rev, **16(1)**, 5-33 (2007)
37. T. Vo, Brooklyn Law Review, **76**, 2011-2006 (2010)
38. R. Price, F. J. Román, & B. Rountree, J. Financ. Econ., **99(1)**, 76-96 (2011)
39. K. Jallow, C. G. Ntim, K. K. Opong, J. Danbolt, & D. A. Thomas, J. Appl. Account Res., **13(2)**, 122-144 (2012)
40. U. Garay, & M. González, Corp. Govern. Int. Rev., **16(3)**, 194-209 (2008)
41. H. Elzahar, & K. Hussainey, J. Risk Finance, **13(2)**, 133-147 (2012)
42. M. Allegrini, & G. Greco, J. Manag. Govern, **17(1)**, 187-216 (2013)
43. K. H. Chung, & H. Zhang, J. Financ. Quant. Anal., **46(01)**, 247-273 (2011)
44. N. H. Wan Abd Rahman, M. Mohamed Zain, & N. H. Y. Yahaya Al-Haj, Soc. Respon. J., **7(2)**, 181-201 (2011)
45. A. Cadbury, *Long Range Planning*, 32, 12-19 (1999)
46. A. Uyar, M. Kilic, & N. Bayyurt, Intangible capital, **9(4)**, 1080-1112 (2013)

Examining The Relationships between Attitude Towards Behaviour, Subjective Norms and Entrepreneurial Intention among Engineering Students

Ummi Naiemah Saraih[1,*], *Ain Zuraini* Zin Aris[1], *Suhana* Abdul Mutalib[2], *Tunku Salha* Tunku Ahmad[1], and *Mohd Harith* Amlus[1]

[1]School of Business Innovation and Technopreneurship, Universiti Malaysia Perlis (UniMAP), Malaysia
[2]School of Human Development and Technocommunication, Universiti Malaysia Perlis (UniMAP), Malaysia.

Abstract. The purpose of this research is to investigate the relationships between attitude towards behaviour, subjective norm and entrepreneurial intention among engineering students from Public Higher Educational Institution (PHEI) in Malaysia. This research is carried out by using the quantitative method (questionnaire). Data are gathered from 345 respondents which consisted of the final year students from one PHEI in Malaysia. Results presented that entrepreneurial intention are positively associated with attitude towards behaviour (β=.62, $p<.01$) and subjective norm (β=.25, $p<.01$). Thus, it is confirmed that both factors of Theory of Planned Behaviour (TPB), namely attitude towards behaviour and subjective norm are significantly related to entrepreneurial intention among the engineering students in this institution. Elevating the degree of attitude towards behaviour and subjective norm are the best strategies to enhance the level of entrepreneurial intention among the engineering students in this institution. Theoretical and practical implications of the results are discussed. In this line, recommendations for the institution management are provided

1 Introduction

Recent research claimed that most of countries faced with the uncertain and complex of unemployment problems. Thus individuals with creative, complex and diversified entrepreneurial knowledge and skills may assist the nation to boost the economies and reduce the unemployment rate [1]. As stated by [2] limited opportunities to grab jobs happened to the competitive environment. In line with this statement, [3] mentioned that graduates from university and college found some difficulties to secure their employment in both public and private sectors due to the current volatile economic environment. Therefore, there is a need to switch and focus on entrepreneurship as a gateway to the employment creation in order to stabilize the economies.

To this point, [4] presented that students from the background of non-economic and business programs are significantly related to students' intentions to become the entrepreneurs. Moreover, the importance of entrepreneurship development could solve the unemployment problems [5]. Therefore, entrepreneurship education is important key point to promote entrepreneurship development to the countries' economy. In the Malaysian context, Ministry of Higher Education Malaysia (MoHE) has taken initiatives to make entrepreneurship as a compulsory subject to all of educational programs under the public universities. Hence, entrepreneurial element has been focused among the most important element to produce students with the entrepreneurial minds (National Education Blueprint, Higher Education).

Self-employment as an entrepreneur is a career that must be given attention as an alternative for graduates besides those employed in facing the volatile limited job market and high retrenchment rate [6]. Hence, an investigation is warranted to reveal the level of entrepreneurial intention among students who are not in the non-economic and business programs (eg. engineering students). This is because entrepreneurship can provides a source of income when an economy cannot yet supply enough jobs [7].

Moreover, MoHE reported that the unemployment percentage for the first degree graduates from the sub field of technical and engineering were 33.7% in 2015 [8]; and 33.2% in 2014 [9]. It showed that the percentages of unemployment among graduates in these sub fields seemed high. Thus, it is assumed that the current engineering students in the Malaysian public universities have a low to moderate level of intention to become the future entrepreneurs. Thus, the researchers attempt to investigate the level of entrepreneurial intention among students from the engineering programs in one of public university in Malaysia. Based on the findings of this study, the appropriate initiatives can be well-planned to decrease the unemployment percentage among the future engineering graduates in Malaysia.

* Corresponding author: ummi@unimap.edu.my

2 Literature Review

2.1 Theory of Planned Behaviour (TPB)

Theory of Planned Behaviour, or TPB was a theory that linked between human belief and human behaviour. The concept of TPB is proposed by Icek Ajzen to improve on the predictive power of the Theory of Reasoned Action (TRA). Attitude towards the behaviour, social norm with respect to the behaviour, and perceived control over the behaviour are usually found to predict behavioural intentions with a high degree of accuracy. Behavioural intention is referred toan indication of an individual's readiness to perform a given behaviour. It is assumed to be an immediate antecedent of behaviour [10]. Behavioural intention is identified as the best predictor of behaviour [11]. In this study, behavioural intention is referred to the entrepreneurial intention.

2.2 Entrepreneurial Intention

Entrepreneurial intentions is defined by [12] as a commitment to start a new business. This study aimed to find out on students entrepreneurial intention in Saudi University based on several entrepreneurial activities (eg. market challenges, risk taking). In addition, [13] pointed that TPB was an appropriate tool to model the development of entrepreneurial intention through pedagogical processes and learning in among students in Portugal. Meanwhile, [14] defined entrepreneurial intention as a mental orientation such as desire, wish and hope which influence their choice of entrepreneurship. Other authors [15] investigated the impact of education on students' intentions to start-up businesses in Lithuania. In this study, entrepreneurial intention is measured based on a statement of "*Yes, I am thinking about own business in the future, when I have completed my studies*". With reference to the research results, it is confirmed that beside of the chosen educational program, students in Kaunas University of Technology, Lithuania are inclined to be the entrepreneurs once they completed their studies.

Although TPB factors have been proven as predictors to individuals' entrepreneurial intention, lack evidence is known regarding the influences of attitude towards behaviour and subjective norm on entrepreneurial intention among the engineering students in Malaysia. Therefore, this study aimed to investigate the relationship between two factors of TPB (eg. attitude towards behaviour, subjective norm) and entrepreneurial intention among the engineering students in the Malaysian context.

2.3 Attitude towards Behaviour, Subjective Norms and Entrepreneurial Intention

Attitude towards behaviouris the first predictor of entrepreneurial behaviour in this study. In general, the more favourable the attitude towards the behaviour, the stronger individuals intent to perform the behaviour [16]. For example, [18] presented that students with positive attitude towards entrepreneurship education inclined to become entrepreneurs once they completed their studies. Similarly, [15] confirmed that attitude towards entrepreneurship was the main factor of students' entrepreneurial intention in Lithuania. Besides, [19] found that personal attitude affected entrepreneurial intention among the secondary students in Portugal. Also, it is confirmed that entrepreneurial attitude is significantly related to entrepreneurial intention among the university students in Xi'an, China [14].

Subjective norm is the second predictor of entrepreneurial intention in this study. It is defined as the perceived social pressure to perform or not to perform the behaviour [16]. Perceived social pressure to perform or not to perform the entrepreneurial behaviour may be triggered from family, friends, teachers and other possible role models. For example, a study of [14] found that students' subjective norm in China is significantly related to entrepreneurial intention. Similarly, [17] found that subjective norm played major role and created influence on entrepreneurial intention among the business students in Bangalore. Moreover, findings of [18] research presented that entrepreneurship education stimulated students' subjective norm and their intentions towards the entrepreneurship. Based on the existing literature,the specific hypotheses are developed as follows:

H_1: There is a moderate level of entrepreneurial intention among the engineering students in this institution.

H_2: There is a significant relationship between attitude towards behaviour and entrepreneurial intention in this institution.

H_3: There is a significant relationship between subjective norm and entrepreneurial intention in this institution.

3 Methodology

This study involved a sample of the engineering students from one of PHEI located in the Northern Region of Malaysia. The adapted instruments used for this study has been presented in Table 1.

* Corresponding author: ummi@unimap.edu.my

Table 1. Instrumentation for Variables

Variables	No of Items	Authors
Entrepreneurial Intention	9	[20]
Attitude towards Behaviour	5	[21]; [22]; [13]; [23]
Subjective Norm	8	[24]; [25]; [26]

Out of 500 distributed questionnaires, 345 returned questionnaires have been used for further analysis.

4 Findings Analysis

4.1 Profile of the Respondents

Results from the descriptive analyses showed that 60.6% (209) in this survey were females and the remaining 39.4% (136) were males. Most of the respondents (83.5%) were in the age between 21 to 23 years old. This figure is followed by respondents from age of 24 to 26 years old (15.9%), age of 27 to 29 years old (.3%) and above than 30 years old (.3%). The majority of respondents were Malays which contributed of 64.6% (223 participants). The race categories are followed by 91 Chinese (26.4%), 21 others (6.1%) and 10 Indians (2.9%).

With regards the schools categories, the respondents came from the various types of schools/faculties. The majority of respondents came from School of Engineering Technology which contributed about 42.0%. This figure is followed by respondents from School of Computer and Communication Engineering (35.4%), School of Manufacturing Engineering (21.2%) and School of Mechatronics Engineering (1.2%). The small percentage of the respondents which contributed of .3% came from School of Microelectronics Engineering. Meanwhile, more than half of the respondents (65.8%) came from families which have the business background.

4.2 Reliability Analysis

As presented in Table 2, the Cronbach's Alpha values are reported at .68 for entrepreneurial intention, .79 for attitude towards behaviour and .80 for subjective norm.

Therefore, the reliabilities of variables used in this research are indicated as acceptable and good [27].

Table 2. Reliability Analysis

Variables	N	Cronbach Alpha	Remarks
Entrepreneurial Intention	345	.68	Acceptable
Attitude towards Behaviour	345	.79	Good
Subjective Norm	345	.80	Good

4.3 Mean and SD Analyses

As presented in Table 3, the means for all variables were relatively high, with the means for all variables exceeding the scale mid-point of three. In this study, the interpretation of mean scores have been interpreted in the ranged of low (1.00-1.66), moderate (1.67-3.33) and high (3.34-5.00). Based on the result, it is noted that the mean for entrepreneurial intention is also high at 3.67. Therefore, H_1 is not supported in this study. In the perspective of predictors, the means ranged from 3.43 for subjective norm to 3.64 for attitude towards behaviour. Hence, it can be confirmed that not only for entrepreneurial intention, respondents in this study also have high levels of attitude towards behaviour as well as their subjective norm.

Table 3. Mean and SD Values

Variables	N	Mean	SD
Entrepreneurial Intention	345	3.67	.54
Attitude towards Behaviour	345	3.64	.62
Subjective Norm	345	3.43	.63

The SD values for entrepreneurial intention, attitude towards behaviour and subjective norm are reported at .54, .62 and .63 respectively. These low levels of all SD values showed that most respondents were close to the mean for all variables in this study.

4.4 Correlation Analysis

As shown in Table 4, entrepreneurial intention are positively and significantly correlated with attitudes towards (r=.75, p<.01) and subjective norm (r=.59, p<.01).

Table 4. Correlation Analysis

Variables	N	1	2	3
1. Entrepreneurial Intention	345	-		
2. Attitude towards Behaviour	345	.75**	-	
3. Subjective Norm	345	.59**	.56**	-

***Correlation is significant at the .01 level (2-tailed)*

Also, analysis presented a significant positive correlation between attitude towards behaviour and subjective norm (r=.56, p<.01). Therefore, the correlations for all variables are reported as significantly and positively between each other.

4.5 Regression Analysis

As shown in Table 5, 61.0% of the variance in entrepreneurial intention is explained by attitude towards behaviour and subjective norm (R^2=.61).

Table 5. Regression Analysis

Variables	Entrepreneurial Intention	
	Beta	**Sig.**
Attitude towards Behaviour	.62**	.00
Subjective Norm	.25**	.00
$R^2 = 0.61$ Adjusted $R^2 = 0.61$ F = 266.853		

***Correlation is significant at the .01 level (2-tailed)*

To further clarify, the remaining of 39.0% of variance in entrepreneurial intention among these engineering students is explained by the other variables. In addition, the adjusted R^2 value also taken into account as .61. Based on the results, entrepreneurial intention is positively associated with attitude towards behaviour (β=.62, p<.01).

Thus, this implied that attitude towards behaviour is significantly related with entrepreneurial intention among the engineering students in this study. Also, it is noted that entrepreneurial intention is positively associated with subjective norm (β=.25, p<.01). Thus, this implied that subjective norm is significantly related with entrepreneurial intention among the engineering students in this study. Therefore, H_2 and H_3 are supported respectively.

5 Conclusion

This study presented that the level of entrepreneurial intention among the engineering students in one of the Malaysian PHEI is high with the mean value is reported at 3.67. Therefore, H_1: *There is a moderate level of entrepreneurial intention among the engineering students in this institution* is not supported in this study. In addition, this study provide some new knowledge pertaining the relationships between attitude towards behaviour, subjective norm and entrepreneurial intention among engineering students in this institution.

As indicated in Table 5, entrepreneurial intention are positively associated with attitude towards behaviour (β =.62, p<.01) and subjective norm (β=.25, p<.01). Therefore, H_2: *There is a significant relationship between attitude towards behaviour and entrepreneurial intention in this institution*; and H_3: *There is a significant relationship between subjective norm and entrepreneurial intention in this institution* are supported in this study.

Based on the results, it can be concluded that by increasing the level of attitude towards behaviour and subjective norm among the engineering students in this institution, the level of entrepreneurial intention among them can be enhanced. In the theoretical perspective, it can be seen that the influences of both factors of TPB namely attitude towards behaviour and subjective norm are significantly related to entrepreneurial intention among the engineering students in Malaysia. These findings were consistent with the previous findings as reported by several authors. For example, [13] presented that attitude towards behaviouris significantly related with entrepreneurial intention among the secondary students in Portugal. Similarly, [15] found that attitude towards entrepreneurship is one of the main factors of entrepreneurial intention among students to start-up business in Lithuania. This result implies that the more positive attitudes the student have towards entrepreneur behaviour, the more intention they will have on entrepreneurship.

In the perspective of the relationship between subjective norm and entrepreneurial intention, the finding is consistent with a study of [17]. These authors found that subjective norm played major role to influence theentrepreneurial intention among students in Bangalore. Similarly, [18] presented that entrepreneurship education stimulated students' subjective norm and intentions towards entrepreneurship in Botswana. This result implies that the higher level of subjective norm which students have in themselves, the higher level of intention they will have on entrepreneurship. One possible explanation for this result might be because students are influenced by real life examples which provided by their peers (eg. families, friends, etc.).

Practically, detail suggestions to increase the appearance of attitude towards behaviour and subjective norm on entrepreneurial intention may be channelled to the top management in this institution as well as to the students' families and friends. In order to increase attitude towards behaviour among students in the institution, these following actions can be taken by the top management: 1) To convince students that having their own business is the other option for their career; 2) To convince students that career as entrepreneur is also attractive; 3) To convince students to start a firm if they have the opportunity; 4) To convince students that being an entrepreneur would entail great satisfactions; and 5) To convince students that if they start their businesses, they will certainly be successful.

Meanwhile, to increase the level of subjective norm among students in the institution, some actions which can be taken by families, friends, and the close or important persons around the students: 1) To convince students that their families, friends and people who are close and important to them wants them to pursue a career as an entrepreneur; 2) To convince students that their parents are positively oriented towards their future career as an entrepreneur; 3) To convince students that their friends see entrepreneurship as a logical choice for them; 4) To convince students that in the universities' environment, students are actively encouraged to pursue their own business; and 5) To convince students that there is a well-functioning infrastructure in the universities to support the start-up of new firms.

References

1. Iqbal, A., Melhem, Y. & Kokash, H. Readiness of the University Students towards Entrepreneurship in Saudi Private University: An exploratory Study. European Scientific Journal, **8 (15)**, 109-131 (2012)
2. Keat, O.Y., Selvarajah, C. & Meyer, D. Inclination towards entrepreneurship among university students: An empirical study of Malaysian university students. International Journal of Business and Social Science, **2(4)**, 206-220 (2011)
3. Teshome, T. Attitudes of private higher education students towards entrepreneurship: A case of distance learners in Wolaita Sodo and Hossana Towns. Middle-East Journal of Scientific Research, **19**(2), 277-285 (2014)
5. Dilanchiev, A. Relationship between Entrepreneurship and Unemployment: The Case of Georgia. Journal of Social Sciences, **3(2)**, 5-9, (2014)
6. Mazdan, A. A. Involvement of graduates in the field of entrepreneurship in Kuala Lumpur, Malaysia. International Journal of Liberal Arts and Social Science, **3**(9), 46-53 (2015)
7. Kelly, D.J., Bosma, N. & Amoros, J.E. Global Entrepreneurship Monitor (GEM) (2010)
8. Ministry of Higher Education (MoHE). "*Laporan Kajian Pengesanan Graduan 2015*" Retrieved from: http://graduan.mohe.gov.my/skpg-report/Penerbitan/SKPG%202015/6)%20Bab%203%20-%20Status%20Pekerjaan%20Graduan%202015.pdf (2015)
9. Ministry of Higher Education (MoHE). "*Laporan Kajian Pengesanan Graduan 2014*" Retrieved from: http://graduan.mohe.gov.my/skpg-report/Penerbitan/SKPG%202014/6)%20Bab%203%20-%20Status%20Pekerjaan%20Graduan%202014.pdf (2014)
10. Ajzen, I. Perceived Behavioral Control, Self-Efficacy, Locus of Control, and the Theory of Planned Behavior. Journal of Applied Social Psychology, **32**, 665-683 (2002)
11. Fishbein, M. & Ajzen, I. *Belief, attitude, intention and behavior: An introduction to theory and research*. Reading, MA: Addison Wesley (1975)
12. Krueger, N.F. The Impact of Prior Entrepreneurial Exposure on Perceptions and New Venture Feasibility and Desirability. Entrepreneurship Theory and Practice, **18**, 5-21 (1993)
13. do Paco, A.M.F., Ferreira, J.M., Raposo, M., Rodrigues, G.G. & Dinis, A. Behaviours and entrepreneurial intention: Empirical findings about secondary students. Journal of International Entrepreneurship, **9(1)**, 20-38 (2011)
14. Peng, Z., Lu, G. & Kang, H. Entrepreneurial Intentions and Its Influencing Factors: A Survey of the University Students in Xi'an China. Creative Education, **3**, 95-100 (2012)
15. Remeikiene, R., Startiene, G. & Dumciuviene, D. *Explaining Entrepreneurial Intention of University Students: The Role of Entrepreneurial Education*. Croatia International Conference 2013, 299-307. (2013)
16. Ajzen, I. Theory of planned behavior. Organizational Behavior and Human Decision Processes. **50**, 179-211 (1991)
17. Krithika, J. & Venkatachalam, B. A Study on Impact of Subjective Norms on Entrepreneurial Intention among the Business Students in Bangalore. IOSR Journal of Business and Management, **16(5)**, 48-50 (2014)
18. Rudhumbu, N., Svotwa, D., Munyanyiwa, T & Mutsau, M. Attitudes of Students towards Entrepreneurship Education at Two Selected Higher Education Institutions in Botswana: A Critical Analysis and Reflection. Academic Journal of Interdisciplinary Studies, **5(2)**, 83-94 (2016)
19. Ferreira, J.J., Raposo, M.L., Rodrigues, R.G., Dinis, A. & Paço, A.D. A model of entrepreneurial intention: An application of the psychological and behavioral approaches.

Journal of Small Business and Enterprise Development, **19(3)**, 424-440 (2012)
20. Indari & Kristiansen. Empirical Studies of Self-Employment. Journal of Economic Survey, **13**, 381-415 (2003)
21. Nishantha, B. *Influence of Personality Traits and Socio-demographic Background of Undergraduate Students on Motivation for Entrepreneurial Career: The Case of Sri Lanka. Euro Asia Management Studies Association (EAMSA)*. Conference held on 5th December 2008 at Doshisha Business School, Kyoto, Japan. (2009)
22. Sagiri, S. & Appolloni, A. Identifying the Effect of Psychological Variables on Entrepreneurial Intentions. DSM Business Review, **1(2)**, 61-86 (2009)
23. Liñán, F. & Chen, Y.W. Development and Cross-Cultural Application of a Specific Instrument to Measure Entrepreneurial Intentions. Entrepreneurship Theory and Practice, **33**(3), 593-617 (2009)
24. Leroy, H., Maes, J., Sels, L. & Debrulle, J. *Gender effects on entrepreneurial intentions: A Tab Multi Group Analysis at factor and indicator level*. Paper presented at the Academy of Management Annual Meeting, Chicago (USA) (2009)
25. Leong, C.K. *Entrepreneurial Intention: An Empirical Study among Open University Malaysia Students*. Master Thesis, Open University Malaysia (2008)
26. Gurbuz, G. & Aykol, S. Entrepreneurial intentions of young educated public in Turkey. Journal of Global Strategic Management, **4(1)**, 47-56 (2008)
27. George, D. & Mallery, P. *SPSS for Windows step by step: A simple guide and reference 11.0 update (4^{th}ed.)*. Boston: Allyn & Bacon (2003)

The Mediating Role of Conflict Management Styles Between Organizational Justice and Affective Commitment Among Academic Staffs in Malaysian Public Universities

Muhammad Asyraf Mohd Kassim[1], Muhammad Safizal Abdullah[1], and Mohd Fitri Mansor[1]

[1]School of Business Innovation and Technoprenuership, (UniMAP)

Abstract. This study analyzes in deep the effects of two major dimensions in organizational justice such as procedural and distributive justice on affective commitment through three conflict management styles such as integrating, compromising, and avoiding styles. These relationships are analyzed in advance on the extent of academic staff in Malaysian Public Universities. Partial Least Squares of Structural equation modelling (SEM) and Statistical Package Social Science (SPSS) are utilized to determine the effect of the two variables and the mediating effect of the conflict management styles. The results exhibit that procedural and distributive justice is significantly related with integrating, and compromising styles while not significantly related to avoiding style. It also revealed that integrating and compromising styles were significant with affective commitment while avoiding style does not relate with affective commitment. In conclusion, the results also showed only integrating and compromising styles mediate the relationship between procedural and distributive justice and affective commitment.
Keywords: Organizational Justice, Affective Commitment, Conflict Management Styles, Structural Equation Modelling

1 Introduction

Rapid developments in Malaysian Higher Education (MHE) proved that this sector is decisive in transforming Malaysia to become high income nation by the year 2020. Academic staffs in public and private higher education institutions in Malaysia mostly are involved in teaching and research and also providing professional consultancy to people for high-level job requirements in industry as this significantly contributes to the development of social economics especially in human capital development (Choong, Keh, Tan & Tan, 2013). Therefore, academic staffs play an important role in sustaining the quality of education as they served as the backbone of any success achievements in higher education institutions.

Unfortunately, recently, MHE sector is facing with increasing turnover rate among the academic staffs as this is one of the critical issues in public and private universities in Malaysia. Choong et al. (2013) indicated that the rising turnover rate among academic staffs in universities will result to increasing recruiting cost, decreasing service quality, waste of management efforts and demoralized of other employees. Aizat, Nizam, Zaifudin and Turiman (2014) discovered that one of the factors influencing the academic staffs in Malaysian public universities leaving their organization is because lacking of affective commitment towards the organization. Academic staffs tend to leave their organization when they feel that they are not committed towards the organization (Alzahrani, 2013). Therefore, it is hard to ignore the issue of turnover among academic staffs' public universities in Malaysia.

The issue high turnover among academic staffs in Malaysian public universities has been identified as a major problem for the organizations' as it affects the organizations' performance and goals. Bahramzadeh & Yadegari, (2010) discovered that one of the critical factors that influence high turnover is lacking of affective commitment among the staffs which is resulted from procedural and distributive justice. Employee's perception on distributive and procedural justices can significantly affect the attitudes, behavior and performance of employees working in any organizations including academic sector (Seyed, Faraahi and Taheri, 2009).

Conversely, Alzahrani (2013) revealed that conflict management styles also impacts affective commitment. Rahim (2002) revealed that applying the dual concern model of five conflict management styles on affective commitment, the conflict management styles of avoiding and dominating is negatively associated with affective commitment, but is positively associated with compromising, obliging and integrating.

Previous researches about distributive justice, procedural justice and conflict management styles have been focusing on investigating the relationship of both justices and conflict management styles on

* Corresponding author: muhammadasyraf@unimap.edu.my

organizational commitment each of the variables individually and separately (Tatum & Eberlin, 2008). Hence, this has led the present study to integrate these variables in a single research framework which, will be focusing on the significance of distributive and procedural justice, conflict management styles and affective commitment among academic staffs. This study is embarked in an effort to bridge the gap in the study of affective commitment by integrating both distributive and procedural justice together with conflict management styles as the mediator. Specifically, the present study will examine the relationships between distributive and procedural justice and three dimensions of conflict management styles namely integrating, compromising and avoiding. In addition, this study also will examine the relationship between these three conflict management styles and affective commitment and the mediating role of conflict management styles between distributive justice, procedural justice and affective commitment.

1.1 Problem Statement

Affective commitment studies among academic staffs in Malaysia are very rare as not many researchers have studied the phenomenon in the higher education sector in Malaysia. The severity of the high turnover rate in Malaysia's higher education sector is becoming worst each year and this problem has been linked to the issue of affective commitment among the academic staffs. This was evidenced from a statistics showing the average turnover rate among academic staffs is increasing yearly. The turnover rate in MHE sector was about 30% in year 2015 as it increase about 7% from year 2014 (Sofiah, Zabid & Lionel, 2016). A study conducted by Azman, Amy, Elizabeth, Kong and Ju (2015) demonstrated that the high turnover trend is mostly happen among the academic staffs where they perceived they are treated unfairly by their organization in terms of payment and workload. Thus proving that there is a dire need to examine the effects of organizational justice and conflict management styles on organizational commitment as an effective method to reduce or prevent the increasing turnover rate.

2 Literature Review

2.1 Affective Commitment

Meyer, Allen and Smith (1993) defined affective commitment as positive feelings of identification with, attachment to and involvement in the work organization. Seto (2013) indicated those employees that have higher level of affective commitment towards their organization, the more effort they will exert in performing daily tasks. Additionally, the issue of affective commitment has received attention for the past several years as organizations look for management tools and methods to enhance their employees' involvement to boost their commitment and indirectly improving organizational performance.

Zayas-Ortiz, Rosario, Marquez, and Gruneiro (2015) discovered that it is difficult for the organizations to inculcate high affective commitment among their employees as it has been established to be a significant predictor of performance and once the organisational objectives has been fulfilled, this will indicate that the affective commitment has been fully optimised by the employees. For instance, employees who recognize justice through the distribution of tasks and work processes are more likely to develop a high degree of affective commitment. Therefore, affective commitment among academic staffs is very important in order to boost their satisfaction so that they will stay with their organization.

2.2 Distributive Justice

Greenberg and Baron (2003) defined distributive justice as employees' perceptions of fairness in the distribution of resources between people. Elamin (2012) indicated that in determining fairness, employees will compare the value of their work inputs such as hard work, commitment and passion to the outcomes or rewards received from their organizations. The outcomes or rewards gained are in terms of increase in salary, promotions and recognition. If the employees believe the outcomes of a decision is unfair, they may engage in a conflict with their organization. Conflicts occur because of perception of injustice by employees lead to dissatisfaction, poor performance, higher absenteeism and turnover intention. Kahn and Rashid (2012) proposed that how employees' perceive they are treated in the organization can influence the work behavior and productivity of the employees.

2.3 Procedural Justice

Procedural justice refers to the perceived fairness present in decision making process with regard to the distribution of rewards (Lemons & Jones, 2001). Procedural justice is a major focus of study in the field of organizational behaviour due to its impact on organizations. Colquitt (2001) indicated that distributive justice and procedural justice are moderately to highly correlated. However, distributive justice is more strongly correlated with reactions with regard to specific outcomes such as job satisfaction, whereas procedural justice is strongly related to attitudes with regard to the organizational system, institution, or authorities perceived to be responsible for the process and procedural decision (Tyler & Lind, 1992).

2.4 Conflict Management Styles

A conflict management styles model was developed by Rahim (2002) which consisted of two dimensions namely intended for self and concern for each other with each axis measured as high and low. The first dimension clarifies the level to which an individual pursues to fulfil his needs and the second one expresses one's concern about the needs of others (Rahim, 2002). Integrating, compromising and avoiding styles are some of the conflict management styles in Rahim's model.

Integrating style is considered as one the more effective problem solving approach. The resolution of conflict requires openness, information sharing and a clear expression of the problem among the parties involved in the conflict (Rahim, 2010). Meanwhile, compromising style is generally characterized as dividing resources in some equitable fashion without resorting to alternative solutions that may seemed unfair to each party's interests. On the other hand, avoiders normally ignore the problem and do not even make the effort to resolve the conflict at all.

3 Hypotheses

3.1 Relationship Between Distributive, Procedural Justice And Conflict Management Styles

Tatum & Eberlin (2008) presented two circumstances to demonstrate the influence of organizational justice and conflict management styles in an organizational setting. First scenario depicts the operation of high levels of justice (both distributive and procedural) within a potentially high conflict situation. The other scenario portrays low levels of justice within a potentially high conflict situation. Table 1 illustrates how sensitivity to organizational justice issues affects different uses of conflict styles when employees are embroiled in a conflict.

An employee, who is highly perceptive to justice related issues, does not steer clear from conflicts by avoiding himself unnecessarily during the resolution. Additionally, the most effective approach to addressing conflict for highly-sensitive employees is integrating and compromising styles. These conflict resolution methods afford the employees to reach an outcome that encourages information sharing, employee involvement, and a genuine demonstration of care and concern for each other.

Table 1. High and low justice and the relationship to conflict styles

Scenario	Avoiding (Style 1)	Compromising (Style 2)	Integrating (Style 3)
Employee is sensitive to distributive and procedural justice issues (High justice)	The employee does not avoid conflict because the employee would not be brought into the process and would not have access to information	The employee's willingness to compromise, however, would depend on what is best for both the employee and the organization as a whole	Collaboration (integration) promotes information sharing, involvement, fairness in the distribution of outcomes, and an expression of caring and concern
Employee is insensitive to distributive and procedural justice issues (Low justice)	An employee who does not care about fairness, openness, or trust may opt for avoiding conflict and thereby saving time and effort	If sharing information and trying to reach a compromise are viewed as the most direct way to complete the task, then the employee may adopt this style	Employee may select this approach, but not because of justice. The employee may adopt collaboration as the best way to complete the review with minimal employee resistance

Note. Source: Tatum and Eberlin (2008)

In contrast, employees who are less perceptive with regard to organisational justice tend to be more concerned with established procedure rather than the fairness or openness of processes (see Table 1). An employee who ignores matters relating to organisational justice is more likely to avoid conflict altogether to save time and effort. Oddly, the employee for whom organizational justice is of little concern might actually adopt the conflict resolution approach of integration. In this case, the employee's purpose is not to establish fairness in the exchange; but rather, he or she was attempting to conclude the resolution process with as little resistance as possible. Therefore, the hypotheses formulated will be:

H1: There is a significant relationship between organizational justice and conflict management styles

3.2 Relationship Between Conflict Management Styles And Affective Commitment

Ahmad & Marinah (2013) stated that integrating style in conflict management plays a key role in the formation of developing affective commitment toward the organization. Employees that utilise this style confront conflicts directly and try to find new and creative solutions to the problems by balancing between their own needs as well as that of the others.

In contrast, Griffin and Steen (2011) in a research on secondary schools found that the conflict management style of compromising attributed to principals were positively associated with secondary level teachers' affective commitment. Conversely, the management style of withdrawing of the principals was related to lower levels of teachers' commitment.

London and Howat (1978) found that the conflict management styles of avoiding were negatively associated with affective commitment, but compromising and integrating styles were positively related to affective commitment. Therefore, it appears that those conflict management styles which allow a subordinate's opinions to be expressed and utilised as part of decision-making process in the achievement of organizational goals are the ones which will be the more fruitful in producing employees' affective commitment. In sum, the hypotheses formulated are:

H2: There is a significant relationship between conflict management styles and affective commitment

3.3 Mediating Effect of Conflict Management Styles

Distributive and procedural justice has been found to be positively related to affective commitment (Suliman & Kathairi, 2013). Employees who perceive distributive and procedural justice to be high are more competent in negotiation and problem solving. The effectiveness of conflict management is the first step in enhancing employees' affective commitment towards their organization. Additionally, employees with higher levels of distributive and procedural justice have a better understanding of each other as they will work at resolving issues together without resorting to negative behaviour. For example, when faced with conflict in the workplace, these types of employees have the abilities to think positively and behave accordingly to pacify the conflicting parties (Ahmad & Marinah, 2013). Effective and appropriate conflict management behaviours in turn are likely to improve employees' perception of justice and affective commitment in their organization. Furthermore, employees with high distributive and procedural justice tend to share their ideas with others, obtain suggestions and help from colleagues and know how to maintain a long-term and cooperative relationship with their co-workers, which will result in enhancing affective commitment towards their organization (Ndubisi, 2011). Finally, employees that are high in both justices know how to express their own ideas appropriately even if there are conflicting opinions and are able to make their detractive novel ideas more agreeable. Therefore, the following hypotheses are developed:

H3: Conflict management styles mediate between distributive and procedural justice and affective commitment.

4 Method

Self-administered questionnaires were developed and distributed to academic staffs in public universities in Malaysia. Respondents were academic staff from public universities located at Perlis, Kedah, and Penang. The questionnaires were distributed using a simple random sampling method. In total, 1200 questionnaires containing justice-oriented assessments, conflict management styles scales and affective commitment-related scales were distributed by hand to the respective academic staffs. Due to response rate of 25% in public universities, 1200 questionnaires were distributed in order to obtain a sample size of 300 academic staffs. The scales were measured using a five-point Likert scale ranging from (1) "strongly disagree" to (5) "strongly agree". 330 questionnaires were returned and the overall response was 27.5 per cent. From the 330 responses, 303 were usable for analyses, representing an effective rate of 25.2 per cent. Descriptive statistics of the final sample are shown in Table 2. These variables (gender, race, educational background, and working length) are considered as control variables in the following analysis.

Table 2. Descriptive statistics of sample

Profile	Description	Frequency
Age	21-29	58
	30-39	104
	40-49	94
	> 50 years	47
Gender	Male	112
	Female	191
Length of Service in Current University	0-10	132
	11-20	94
	21-30	61
	> 30 years	16
Length of Service in Academic Profession	0-10	122
	11-20	95
	21-30	68
	> 30 years	18
Highest Academic Qualification	Masters	129
	PhD	174

PLS-SEM were used to assess the goodness of measures and to test the hypotheses for this study. The mediation was tested using the bootstrapping approach as suggested by Preacher and Hayes (2008).

4.1 Measures

Distributive justice (four items) and procedural justice (seven items) were assessed based on the measures developed by Rahim, Magner and Shapiro (2000). The Cronbach's alphas were both 0.93 for distributive justice and procedural justice. The scale was measured using a five-point Likert scale ranging from (1) "strongly disagree" to (5) "strongly agree".

The Rahim Organizational Conflict Inventory (ROCI-II) Form B (Rahim, 1983) was used in this study to measure the three styles of managing interpersonal conflict, consisting of 16 items which include integrating (seven items), compromising (four items), and avoiding (five items). The items were modified to measure subordinates' perception of their managers' styles in handling disagreements with them. Employees were instructed to indicate the extent of fairness to which they perceive the way their manager handled interpersonal conflicts within the organisation on a five- point Likert type scale, ranging from strongly disagree (1) to strongly agree (5). The reliability coefficients for integrating, compromising and avoiding styles were 0.89, 0.86, and 0.73 respectively. Affective commitment was measured based on the scales developed by Meyer, Allen, and Smith (1993) containing six items. The Cronbach's alpha for the scale is 0.82. The measures were measured using a five-point Likert scale ranging from (1) "strongly disagree" to (5) "strongly agree".

5 Findings

5.1 Goodness of Measures

The goodness of measures for the variables in the study was assessed using construct validity, convergent validity, discriminant validity, and reliability analysis. For the purpose of construct validity, Hair et al. (2013) indicated that every loading that are greater than 0.50 on a particular construct are considered significant. Therefore, loadings below 0.50 were removed and there were no reported cross loadings. As a result, the constructs were valid for this study. Table 3 shows the factor loadings for each construct.

Table 3. Loadings and cross loadings

	AC	AV	CO	DJ	IN	PJ
AC1	0.8					
AC3	0.8					
AC4	0.9					
AC5	0.8					
AC6	0.8					
AV1		0.6				
AV2		0.6				
AV3		0.6				
AV4		0.5				
AV5		0.6				
CO1			0.7			
CO2			0.6			
CO3			0.8			
CO4			0.7			
DJ1				0.7		
DJ2				0.9		
DJ3				0.7		
DJ4				0.9		
IN1					0.8	
IN2					0.7	
IN3					0.8	
IN4					0.8	
IN5					0.9	
IN6					0.8	
IN7					0.9	
PJ1						0.7
PJ2						0.7
PJ3						0.8
PJ4						0.8
PJ5						0.8
PJ6						0.7

Next, to determine the convergent validity, factor loadings, composite reliability (CR), and average variance extracted (AVE) were analysed. Table 4 exhibits that all factor loadings are above 0.50, the composite reliabilities are above 0.70, and the AVEs are above 0.50. Thus, the convergent validity of the constructs was accepted.

Table 4. Results of measurement model

Model Construct	**Items**	**Loadings**	**AVE**	**CR**
Distributive Justice	DJ1	0.7	0.8	0.8
	DJ2	0.9		
	DJ3	0.7		
	DJ4	0.9		
Procedural Justice	PJ1	0.7	0.7	0.8
	PJ2	0.7		
	PJ3	0.8		
	PJ4	0.8		
	PJ5	0.8		
	PJ6	0.7		
Integrating Style	INT1	0.8	0.8	0.8
	INT2	0.7		
	INT3	0.8		
	INT4	0.8		
	INT5	0.9		
	INT6	0.8		
	INT7	0.9		
Compromising Style	COM1	0.7	0.8	0.9
	COM2	0.6		
	COM3	0.8		
	COM4	0.7		
Avoiding Style	AVO1	0.6	0.7	0.7
	AVO2	0.6		
	AVO3	0.6		
	AVO4	0.5		
	AVO5	0.6		
Affective Commitment	AC1	0.8	0.7	0.9
	AC3	0.8		
	AC4	0.9		
	AC5	0.8		
	AC6	0.8		

Then, the correlations between the measures are compared with the square root of the AVEs in order to assess the discriminant validity. Table 5 exhibits all of the correlations between the measures were smaller than the square root of thee shown on the diagonals.

Therefore, the items measuring the constructs discriminant validity for this study are satisfactory. Finally, reliability was also accessed via Cronbach's alpha coefficient, which should have a value higher than the recommended value of 0.60 as indicated by Nunally and Berstein (1994). In addition, all of the constructs have alpha values of above the recommended value and therefore, the measures are reliable (see Table 5).

Table 5. Discriminant Validity of Construct

	AC	AV	CO	DJ	IN	PJ
AC	**0.8**					
AV	-0.3	**0.8**				
CO	0.5	-0.3	**0.9**			
DJ	0.6	-0.2	0.5	**0.9**		
IN	0.5	-0.3	0.6	0.4	**0.9**	
PJ	0.5	-0.2	0.5	0.7	0.6	**0.8**

Note: The bolded diagonals represent square root of the AVE while the off-diagonals represent the correlation.

5.2 Hypotheses Testing

The hypotheses in this study were tested using path analysis. The R^2 value for integrating, compromising, and avoiding style were 0.463, 0.419, 0.092 meaning that 46.3%, 41.9%, and 9.2% of the variance in conflict management styles are explained by distributive and procedural justice respectively. Looking at each predictors individually, distributive justice was a positive significant predictor for integrating style ($\beta = 0.432$, $p < 0.01$) and compromising style ($\beta = 0.178$, $p < 0.01$) but negative significant predictor for avoiding style ($\beta = -0.263$, $p < 0.01$). Additionally, procedural justice was also found to be a positive significant predictor for integrating style ($\beta = 0.372$, $p < 0.01$), compromising style ($\beta = 0.271$, $p < 0.01$) but negative significant predictor for avoiding style ($\beta = -0.262$, $p < 0.01$). Therefore, H1 was supported (see Table 6).

With affective commitment as the dependent variable, the $R2$ value revealed that 45%, of the variance for affective commitment was explained by integrating, obliging, compromising, dominating and avoiding style. Integrating style ($\beta = 0.293$, $p < 0.01$), obliging style ($\beta = 0.217$, $p < 0.01$) and compromising style ($\beta = 0.174$, $p < 0.01$) were found positively significant with affective commitment while avoiding style ($\beta = -0.197$, $p < 0.01$) was negatively significant with affective commitment. Therefore, H2 was supported (see Table 7).

Table 6. Coefficient beta and r^2 for conflict management styles

Hypotheses	Relationship	Coefficient (β)	R^2	Comment
1.1	Distributive Justice →Integrating Style	0.4***	0.4	Supported
1.2	Procedural Justice →Integrating Style	0.3***		Supported
1.3	Distributive Justice →Compromising Style	0.1**	0.4	Supported
1.4	Procedural Justice →Compromising Style	0.2***		Supported
1.5	Distributive Justice →Avoiding Style	-0.2***	0.09	Supported
1.6	Procedural Justice →Avoiding Style	-0.2***		Supported

Note. *** $p < 0.01$, ** $p < 0.05$, * $p < 0.10$

Table 7. Coefficient Beta and R^2 for Affective Commitment

Hypotheses	Relationship	Coefficient (β)	R^2	Comment
2.1	Integrating Style →Affective Commitment	0.293***	0.45	Supported
2.2	Compromising Style →Affective Commitment	0.174**		Supported
2.3	Avoiding Style →Affective Commitment	-0.197**		Rejected

Note. *** $p < 0.01$, ** $p < 0.05$, * $p < 0.10$

Next, to determine the mediating effect of conflict management styles between distributive and procedural justice and affective commitment, bootstrapping of 303 samples was conducted. After bootstrapping, both paths from distributive and procedural justice have significant impacts on affective commitment via integrating, compromising and avoiding style (see Table 8). Therefore, H3 also was supported.

Table 8. Mediating effects

H	Relationship	Path A (β)	Path B (β)	t-value (After Bootstrap)	Mediating Effect
H3.1	DJ → IN → AC	0.4***	0.2***	5.7***	Yes
H3.2	PJ → IN → AC	0.3***	0.2***	3.5***	Yes
H3.3	DJ → CO → AC	0.1**	0.1**	2.3***	Yes
H3.4	PJ → CO → AC	0.2***	0.1**	2.5***	Yes
H3.5	DJ → AV → AC	-0.2***	-0.1**	1.7	Yes
H3.6	PJ → AV → AC	-0.2***	-0.1**	1.7	Yes

6.0 Discussion

The present study discovered that both distributive and procedural justice have significant positive relationship with integrating, and compromising; and negative significant relationship with avoiding style consistent with prior studies such as Rahim et al. (2001) and Tatum and Eberlin (2008). This means that when employees observe that their organization distributes rewards fairly in terms of monetary benefits, career promotions, performance evaluation and they were treated with respect, they respond to their subordinates by supporting them in order to help the organization achieve its objective, and vice versa for avoiding style.

The result of this study also proved that integrating and compromising styles were positively significant with affective commitment while avoiding style was discovered to be related negatively and significantly to affective commitment. These findings reflected the results of previous empirical studies (e.g. Ahmad & Marinah, 2013; Ndubisi, 2011; Griffin et al., 2010; London & Howat, 1978). Thus proving that when employees perceive that the use of integrating,obliging or compromising styles is apparent in the organisation, it will reduce the likelihood of negative attitude and behaviours and generate a stronger sense of commitment to the organization. On the other hand, employees who utilize the avoiding style tend to exhibit discontentment, discourtesy, intention to quit and extremely dissatisfied with organisation.

The result of this study discovered that the integrating, compromising and avoiding style mediated between distributive and procedural justice and affective commitment. The results are consistent with the findings from Fisher et al. (2005) where conflict management styles (integrating style, compromising style and avoiding style) are the combinations that contribute to the prevention of conflict among employees in the organization.

7.0 Theoretical Implications

One of the contributions of the present study is in proving the mediating effect of conflict management styles between organizational justice and organizational commitment. As indicated by Ahmad et al. (2013), the research on conflict management as a mediator is scarce but it is an important management tool in stabilising the organization as it is a highly visible phenomenon to the employees. Likewise, the examination of the impact of distributive and procedural justice on integrating, compromising and avoiding styles and its indirect effect on affective commitment discovered that conflict management styles is critical issue in influencing organizational outcomes.

8.0 Practical Implications

A major contribution of this present study proves that integrating, compromising and avoiding styles are an effective management tool which can be utilized in the academic organization. It could act as a guideline in resolving organizational justice issues among the academic staffs. Understanding conflict management styles can help the academic staffs to recognize and evaluate situations that could lead to conflict. By having the ability to monitor and assess difficult situations, the academic staffs will be able to prepare coping strategy in anticipating such conflicts in organizations.

9.0 CONCLUSION

Overall, it could be concluded that the findings had achieved the objectives of this study. The academic staffs perceived integrating, compromising and avoiding styles to be present in their respective organization, thus, motivating them to be more committed to their organization. In addition, academic staffs who perceived integrating and compromising styles being practiced in their organizations have higher value-effort of affective commitment towards their organization. To maintain and ensure the professionalism and quality of conflict management within the organizations and among the staffs, it is recommended that the staffs' interpersonal skill and knowledge must be continuously developed in managing conflict. It can be implemented by improving the staffs' understanding on the strength and weaknesses of the styles of managing conflict and how to implement it at the right place and the right time. As a result, turnover which is the problem among the academic staffs can be overcome.

REFERENCES

1. Ahmad, A. R., & Marinah, A. (2013). Learning organization and organizational commitment in primary school. *International Proceedings of Economics Development and Research, 60* (12).
2. Ahmad, Z.M.S., Jamilah, O. & Jeffrey, L.D. 2013. The Effects of Perceived Leadership Styles and Organizational Citizenship Behavior

* Corresponding author: muhammadasyraf@unimap.edu.my

on Employee Engagement: The Mediating Role of Conflict Management. *International Journal of Business and Management, 8*(8), 91-99.
3. Aizat, A. R., Nizam, S. S., Zaifudin, S. Z., & Turiman, S. (2014). Turnover intention among academics: A case study of private higher learning education in Klang Valley. *Pertanika J. Soc. Sci. & Hum. 22 (1)*, 321 - 334.
4. Alzahrani, M. (2013). *A comparative study of the relationships between conflict management styles and job satisfaction, organizational commitment, and propensity to leave the job among Saudi and American Universities' faculty members*. Published PhD Diss., Florida Atlantic University.
5. Azman, I., Amy, Y., Elizabeth, Y., Kong, L.-K., & Ju, S. Y. (2010). Occupational stress features, emotional intelligence and job satisfaction: An empirical study in private institutions of higher learning. *Negotium, 16*(5), 5-33.
6. Bahramzadeh, H., & Yadegari, M. (2010). Analysis of the relationship between organizational justice and organizational commitment of employees of girls' schools in the different regions of Mashhad Education Departments. *Tommorow's Management Quarterly, 21*, 55-65.
7. Choong, Y. O., Keh, C. G., Tan, Y. T., & Tan, C. E. (2013). Impacts of demographic antecedents toward turnover intention amongst academic staff in Malaysian private universities. *Australian Journal of Basic and Applied Sciences, 7*(6), 46-54.
8. Colquitt, J. A. (2001). On the dimensionality of organizational justice: a construct validation of a measure. *Journal of Applied Psychology, 86*(3), 386 - 400.
9. Elamin, A. M. (2012). Perceived organizational justice and work-related attitudes: A study of Saudi employees. *World Journal of Entrepreneurship, Management and Sustainable Development, 8*(1), 71-88.
10. Fisher, S. G., Macrosson, W. D., & Semple, J. H. (2005). Control and belbin's team roles. *Personnel Review*, 30, 578-588.
11. Greenberg, J. & Baron, R. A. (2003). *Behavior in organization: Understanding and managing the human side of work.* 8th edition. New Jersey: Prentice Hall.
12. Griffin, D., & Steen, S. (2011). A social justice approach to school counselling. *Journal for Social Action in Counseling and Psychology, 3*(1), 74-85.
13. Hair, J. F., Ringle, C. M., & Sarstedt, M. (2013). Partial least squares structural equation modeling: Rigorous applications, better results and higher acceptance. *Long Range Planning, 46* (1), 1-12.
14. Khan, S. K., & Rashid, M. Z. A. (2012). The mediating effect of organizational commitment in the organizational culture, leadership and organizational justice relationship with organizational citizenship behavior: A study of academician in private higher learning institutions in Malaysia. *International Journal of Business and Social Science, 3*, 83-91.
15. Lemons, M. A., & Jones, C. A. (2001). Procedural justice in promotion decisions: Using perceptions of fairness to build employee commitment. *Journal of Managerial Psychology, 16*(4), 268 – 281.
16. London, M., & Howat, G. (1978). Relationship between employee commitment and conflict resolution behavior. *Journal of Vocational Behavior, 13*, 1-14.
17. Meyer J. P., Allen, N. J., & Smith, C. (1993). Commitment to organizations and occupations: Extension and test of a three-component conceptualization. *Journal of Applied Psychology, 78*, 538-551.
18. Ndubisi, N. O. (2011). Conflict handling, trust and commitment in outsourcing relationship: A Chinese and Indian study. *Science Direct Journal, 40*(1), 109- 117.
19. Nunnally, J. C., & Bernstein, I. H. (1994). *Psychometric theory.* 3rd edition. New York: McGraw-Hill.
20. Preacher, K. J., & Hayes A. F. (2008). Asymptotic and resampling strategies for assessing and comparing indirect effects in multiple mediator models. *Behavior Research Methods*, 40(3), 879-891.
21. Rahim, M. A. (1983). A measure of styles of handling interpersonal conflict. *Academy of Management Journal, 26*, 368- 376.
22. Rahim, M. A. (2002). Toward a theory of managing organizational conflict. *International Journal of Conflict Management, 13*(3), 206–235.
23. Rahim, M. A. (2010). *Managing Conflict in Organizations* (4th Ed.). Transaction Publishers. New York.
24. Rahim, M. A., Magner, N. R., & Shapiro, D. L. (2000). Do justice perception influence styles of handling conflict with supervisors? What justice perceptions, precisely? *International Journal of Conflict Management, 11*(1), 9-31.
25. Schlaerth, A., Ensari, N., & Christian, J. (2013). A meta-analytical review of the relationship between emotional intelligence and leaders' constructive conflict management. *Group Processes & Intergroup Relations, 16*(1), 126–136.
26. Seto, D. M. (2013). *Breach of psychological contract and workplace outcomes: The mediational role of job satisfaction, organizational commitment, and employee engagement.* Published PhD Diss., Alliant International University.
27. Seyed, J. S. R., Faraahi, M. M., & Taheri, A. G. H. (2009). How different dimensions of organizational justice affect the different dimensions of job and organizational

satisfaction. *Journal of Business Management, 1*, 55-70.

28. Sofiah, K. K., Zabid, M. A. R., & Lionel, K. V. (2016). The role of organisation commitment in enhancing organisation citizenship behaviour: A study of academics in Malaysian private universities. *Int. Journal of Economics and Management, 10*(2), 221 – 239.
29. Suliman, A., & Kathairi, M. (2013). Organizational justice, commitment and performance in developing countries: The case of the UAE. *Employee Relations, 35*(1), 98-115.
30. Tatum, B. C., & Eberlin, R. J. (2008). The relationship between organizational justice and conflict style. *Business Strategy Series, 9*(6): 297-305.
31. Tyler, T. R., & Lind, E. A. (1992). A relational model of authority in groups. *Advances in Experimental Social Psychology, 25*, 115 - 191.
32. Varman, R., & Bhatnagar, D. (1999). Power and politics in grievance resolution: Managing meaning of due process in an organization. *Human Relations, 52* (3), 349-380.
33. Zayas-Ortiz, M., Ernesto Rosario, E., Marquez, E., & Gruñeiro, P. C. (2015). Relationship between organizational commitments and organizational citizenship behaviour in a sample of private banking employees. *International Journal of Sociology and Social Policy*, *35*(1/2), 91-106.
34. Zulkiflee, D., Faizal, M. M. I., Shakizah, W. W. M. N. (2012). *The relationship between procedural justice and grievance handling styles.* International Conference on Management, Economics and Finance (ICMEF 2012) Proceeding.

Knowledge Construction Models: A view of Various Models for an Impeccable Learning Environment

Suhaizal Hashim, Mimi Mohaffyza Mohamad, Wan Hanim Nadrah Wan Muda

Faculty of Technical and Vocational Education, University Tun Hussein Onn Malaysia, Parit Raja, Johor, Malaysia.

Abstract. Knowledge construction plays an important role in determining the effectiveness of learning by focusing on how humans process information and transform it into knowledge. This paper tries to identify the models of knowledge construction process that is suitable for the development of learning environment. The process of constructing student knowledge through a learning system involves an analysis of several types of knowledge construction models used as analytical tools for online learning. The methodology that been used in this paper is thematic analysis. From the analysis, it seems that it is suitable for us to use the Alavi Knowledge Management Process Model to develop a learning environment. Besides, the analysis from Veerman and Veldhuis-Diermanse model also appears to be a better choice as the process in this model able to help construct a social constructivist principle through focusing on non-task related activities to determine the process of knowledge construction. The implementation of online collaborative discussions is said to have great potential to provide opportunities for students to engage intellectually in the process of knowledge construction and in the tradition of cognitive and constructivism learning. Therefore, teachers must be prepared to implement suitable developments and cultures in teaching strategies by providing basic information, involving students to practice and apply relevant skills and implement it in fitting time

1 Introduction

Knowledge construction is the heart of a learning process that determines the success of any learning effort. In the tradition of cognitive and constructivism learning, knowledge construction plays an important role in determining the effectiveness of learning by focusing on how humans process information and transform it into knowledge [1]. Therefore, regardless of the nature of any learning environment, the issue of knowledge construction is still relevant.

Rahman et al. [2] stated in their study that focusing on the process of knowledge construction in online discussions designed for use with mobile technology found that the majority of the respondents agreed about online discussions helped the process of knowledge construction through critical examinations of other people's ideas during the discussion. The way to cope with psychological barriers is one of an important aspects for online environment that facilitates student engagement to construct knowledge rather than using face-to-face discussions. Further analysis of the learning process reveals the pattern of steps used by the students to build meaning through the discussion process.

A case study report on the discovery of the level of knowledge construction among postgraduate students for six asynchronous computer conferences in blended learning environment Universiti Malaysia Sarawak showed that students were active in knowledge construction but limited to finding and giving opinions about knowledge among friends and explain about the knowledge. Students can benefit from their training in degree programs that incorporate a blended learning environment as a way for them to stay in touch with other students outside the classroom at any time and place [3].

In addition, Zhu [4] also conducted a study on identifying satisfaction, performance and knowledge construction through collaborative online discussion among students in two different cultural contexts. Students from both universities were first year students specializing in educational science from Flemish University and China University. The study found that Flemish students liked to study themselves according to their ability, while Chinese students were very satisfied when they could work with friends to complete their assignment. However, the problem of Flemish students is that working on an online task is very time consuming, while the biggest problem of Chinese students reported is lack of interaction between students and teachers.

Noroozi et al. [5] conducted a study to identify the construction of student knowledge in a computer

*Corresponding author: suhaizal@uthm.edu.my

learning environment. The combination of computer supported collaborative learning (CSCL) and face to face learning in a new learning environment where students can discuss in front of a shared computer has been used. Their findings show that both students are able to contribute to the better quality of knowledge construction for the second test than the first test.

In a study to evaluate the effectiveness of using video streaming and synchronous forums to provide feedback to presenters in oral skills tutorials by Thanasingam and Soong [6] which emphasizes the quality of knowledge in the form of feedback constructed by students. The findings show that discussion forums encourage understanding and erudition among students as they jointly process their comment through various phases of knowledge construction. This allows them to improve the quality and accuracy of their comment. Thus, this paper tries to identify the models of knowledge construction process that is suitable for the development of learning environment. The existence of an online learning environment based on internet technology enables the knowledge construction process for collaborative discussions held.

2 Different Models of Knowledge Construction Process

In order to identify the models of knowledge construction process that is suitable for the development of learning environment, a review process has been carried out as in Table 1 which shows the analysis of 3 models of knowledge construction process. These three models were analyzed by Paavola, Lipponen, and Hakkarainen [7] according to the type of model. Each of these models has the advantages and disadvantages in helping the process of students' knowledge construction.

Table I: Analysis Of The 3 Models Of Knowledge Construction Process By Paavola, Lipponen And Hakkarainen

Content	Model Types		
	Nonaka and Takeuchi	Engeström	Bereiter
The role of individual expertise	Black box method where individuals create knowledge	Social application	Expertise theory
The main focus	Implicit Knowledge (Overview)	Knowledge found in practice (Acts)	Knowledge object (Concept)
The type of process that is focused	Emphasize the process through personal deeds and experiences	Emphasize object-oriented material activity	Emphasize solving knowledge problems
The source of innovation	The transformation of knowledge is implied with clear knowledge	Overcoming tension, chaos, and ambiguity through progressive learning	Work deliberately to expand and create new knowledge objects
Scope of framework	Different levels of ontology (innovative teams, organizations and between individual level organizations)	Activity system and network activity system	Knowledge construction community
Applications in education	School to create Knowledge	School as a learning progress	School as knowledge construction society

One of the theories found in the process of knowledge construction is Knowledge-Construction Community. In this learning situation, students will develop knowledge-oriented issues that are more complex and at the same time the students will improve their learning progress. This approach developed by Carl Bereiter [8] who considers that in the process of learning, the approach is to put students as teachers, where in the learning situation, students will do something similar with teacher behaviour, even though they do not have the same knowledge or experience owned by the teachers.

Bereiter [8] has criticized the tendency of some mentalistic over-theory of learning and views the learning process as the amount of information in a person's mind and that thought is regarded as a record of knowledge. The process of knowledge construction that involves work in a particular competence can improve one's achievement as well as to achieve knowledge progress, not only on individual learning but also on group learning. The type of learning by Bereiter [8] is designed activities to create mutual knowledge, which requires the collaboration to create, develop, understand and criticize. Its main purpose is not only learning something but also to collaborate in developing new methods, theories, models and so forth that can be used in the future.

The basic point raised in the construction of this community knowledge is more towards the method of presentation that is learning with this method of presentation. It can be seen as one of the best methods to determine the ability to build knowledge for students and if it developed in schools, it can help them to build knowledge and train them to communicate and display themself. If we view it further and linked to the theories by Nonaka and Takeuchi [9], the main points raised in the theory of knowledge construction of this community are more on the method of delivery. According to Engeström [10], the main problem with Nonaka and Takeuchi model [9] is that it does not take into account the first two phases of the expansive model, which are questioning and analyzing the situation. Their model is based on too many ideas to share the implicit knowledge in the phase of socialization and does not take into account the importance of controversy and conflict in the creation of knowledge.

There are some drawbacks from the analysis of all these models and it seems that it is suitable to use the Alavi Knowledge Management Process Model [11] to established a suitable learning environment. The model in knowledge management involves the rotation of the activity for example knowledge needs to be developed, stored, channeled and need to be combined as well as refer to a systematic and manageable specific process for obtaining, handling and delivering both the tasit and explicit knowledge acquired from a student so they can use it to improve the efficiency and productivity of their assignments [12][13]. Alavi and Leidner [13] also suggest that knowledge management involves four key processes, namely knowledge creation, knowledge storage, knowledge transfer and knowledge application.

Knowledge Creation is the process of claiming that knowledge will be created when two different people share their knowledge and explicit knowledge with one another. The process of creating knowledge and content development can be achieve by organizing experiences and learning from lessons learned by collecting, synthesizing and interpreting information. Then, knowledge storage is a process whereby the knowledge received will be processed and stored. The following three phases that include in knowledge storage are indexing, filtering and linking references to resource management activities that include checking, classifying, cataloging, integrating and linking content from both knowledge sources both internal and external.

Furthermore, knowledge transfer is a process once the knowledge is acquired and stored in order to enable it to be used, this knowledge should be disseminated or transferred to the party requiring it to be used. Finally, knowledge application is a phase that determines of whether an organization can apply the knowledge they have to become more competitive. This phase uses knowledge that has been collected, stored and transmitted or otherwise, it is a level of knowledge received in the form of tacit and explicit that will be used in the same situation by other individuals. Figure 1 shows the Alavi Knowledge Management Process Model [11] which states that knowledge management refers to a systematic and manageable specific process for obtaining, handling and delivering both the tacit and explicit knowledge acquired from a student so that other students can also use it to increase the level of efficiency and productivity of their projects.

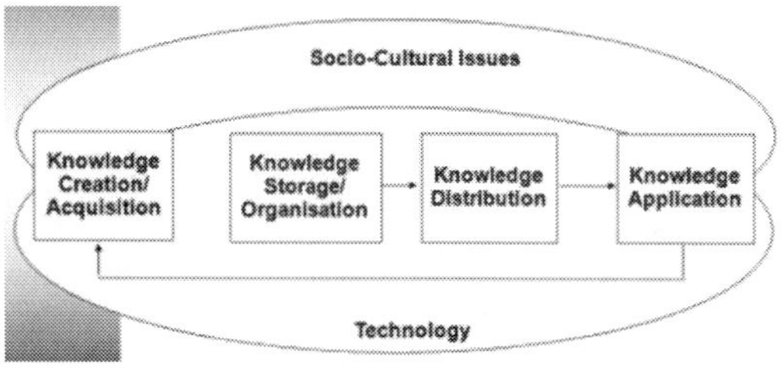

Figure 1: Alavi Knowledge Management Process Model

After a thematic analysis had been done, there are several factors that encourage the selection of this model amongst others as the involvement of Alavi itself who involved in several studies on the process of collaborative assessment and learning. One of the studies itself is to test the effectiveness of the use of the Group Decision Support System in the collaborative learning process to enhance Student learning experience and classroom assessment. According to Alavi, Yoo and Vogel [14], an innovation in management education is to use information technology to create value-added partnerships between universities to leverage student learning experiences through a diverse range of resources and technologies where students have the opportunity to interact with their friends and teachers simultaneously via video conference facilities. This study also shows how this information technology can enable collaborative learning and teaching to take place in different groups of students, diversity of teachers and integration of outside expertise.

Alavi and Dufner [15] stated that the technology from internet network open more space for collaboration and through their research also showed that compared to individualized learning, collaborative learning shows better psychological relationship and academic achievement, add self-confidence and it improve psychological level of a person. In order to increase the knowledge of teachers and students, the Internet technology is seen as one of the channels for the process of knowledge construction. The implementation of online collaborative discussions is said to have great potential to provide opportunities for students to engage intellectually in the process of knowledge construction and in the tradition of cognitive and constructivism learning. Knowledge construction plays an important role in determining the effectiveness of the learning process [16].

3 Analysis Of Knowledge Construction Process Based On Support Type

The analysis of the process of constructing student knowledge through a learning system involves an analysis of several types of knowledge construction models used as analytical tools for online learning. Online learning refers to a learning environment that uses internet technology to enable virtual learning sessions to take place. There are three interrelated elements used by Oliver [17] in describing web-based teaching designs which are learning resources, learning support and learning assignments. Table 2 is the analysis of the process of knowledge construction based on the type of support is the hypothesis that students use different knowledge construction models in the online learning environment compared to learning in a traditional setting.

Table II: Analysis Of Knowledge Construction Based On Support Type

Researcher	Research Objectives	Support Type	Knowledge Construction Analysis
Kanuka, H. and Anderson, T. [18].	This study presents the results of exploratory studies of various evaluation and analysis methods for online forum transcripts.	Online forums.	An analysis of five phase content model that explains the steps in knowledge construction (Gunawardena, Lowe and Anderson).
Schellens, T. dan Valcke, M. [19].	This study to identify collaborative learning in an asynchronous discussion group resulting in an increase in academic performance and knowledge construction.	Face-to-face lectures with options for participation in online discussion groups. Group discussions focus on the same learning objectives as a face-to-face lecture session.	Each complete message analyses using coding process from the Veerman and Veldhuis-Diermanse models.
Ortegano-Layne, Gunawardena [20].	This study to determine whether a concept map can be used to synthesize the construction of knowledge in online discussions and to determine whether a moderator carrying an online discussion can use a concept map as a tool that capable of assisting them.	Online courses designed using WebCT learning management systems.	An analysis of five phase content model that explains the steps in knowledge construction (Gunawardena, Lowe and Anderson).
Kumar, S. [21].	This study aims to determine whether face-to-face course discussions using asynchronous communication tools facilitate the construction of knowledge in the same way as online discussions in online courses.	Web-based course management system, CourseInfo.	An analysis of five phase content model that explains the steps in knowledge construction (Gunawardena, Lowe and Anderson).
Thanasingam, S. and Soong, S. K. A. [6].	This study assesses the effectiveness of video streaming and discussion of synchronous forums to provide feedback to speakers in oral skills tutorials	Online tutorial prototype	Knowledge construction phase of Gunawardena et.al.
García, B., Márquez, L., Bustos, A., Miranda, G. A., & Espíndola S. M. [23].	This study provides suggestions that consider the contextual factors that make up the input and interaction scenarios, the interaction process: type of interaction and its contents and the learning	Computer-mediated communication (CMC).	An analysis of five phase content model that explains the steps in knowledge construction (Gunawardena, Lowe and Anderson).

	strategies and decisions that involve the quality of knowledge constructed by participants.		
Kian-Sam, H., and Lee, J. A. C. [3].	This study to identify the level of knowledge construction among postgraduate students in six asynchronous computer sessions in a blended learning environment	Quickplace, an elearning system at Universiti Malaysia Sarawak.	Each complete message analyses using coding process from the Veerman and Veldhuis-Diermanse models.
Rahman, S., Yasin, R. M., Jusoff, K., Yassin, S. F. M., Nordin, N. M., & Yusof, M. M. [6].	The process of knowledge construction in online discussions designed for the use with mobile technology.	Students were asked to participate in online discussions about what they understand on the subject they have learned.	Not mention
Zhu, C. [4].	This study to examine the satisfaction, performance and build of knowledge through student online discussions group with two different cultural contexts.	Elearning using open source platforms based on Dokeos	The coding schemes of Veerman and Veldhuis-Diermanse) were used to analyze the distribution of communication types and coding schemes of Gunawardena, Lowe, and Anderson to analyze the social level of knowledge construction.

Noroozi, O., Biemans, H., Mulder, M. dan Chizari, M. [5].	Identify of student knowledge construction in a computer supported learning environment in the nutrition research on domain nutritionology methodology.	Combination of Computer Supported Collaborative Learning (CSCL) and face to face learning in a learning environment where students can discuss in front of shared computers.	Veldhuis-Diermanse encoding scheme was used to analyze the construction of student knowledge.
Hashim, S., & Harun, J. [23]	Examine the impact of CPA learning system as a medium for the process of knowledge construction in helping the learning and teaching process.	CPA learning system	Veldhuis-Diermanse encoding scheme was used to analyze the construction of student knowledge.

From the analysis, It seems that the analysis model of Veerman and Veldhuis-Diermanse [24] appears to be a better choice as the process in this model was able to construct a social constructivist principle through focusing on non-task related activities to determine the process of knowledge construction. The model of Veerman and Veldhuis-Diermanse distinguishes between task-oriented and non-task-oriented messages based on encoded behavioral features. These differences help to control interpersonal monitoring activities in the process of students' knowledge construction. This topology is more common than the Gunawardena et al. [25] as it is more specific by suggesting alternatives to assessing the construction of knowledge through social negotiations concluded with interaction analysis model that encompasses five phases and reflects the process of negotiations that occur in a social constructivist learning environment. Gunawardena et al. was found to be more suitable to determine the level of knowledge construction process.

The analysis model of Veerman and Veldhuis-Diermanse has also been used as a content analysis for several studies in online collaborative learning environments, among them Schellens and Valcke [19] who have conducted studies to identify collaborative learning in an asynchronous discussion group resulting in increased academic performance and knowledge construction. The results of the study confirmed that discussions of the students in the group were highly task-oriented and the ratio of high knowledge construction phase was also observed. There were also significant improvements in cognitive interaction, task, orientation and higher knowledge construction phase detected.

Walker [26] supports this statement by expressing written assignments such as answering and responding to other student discussion questions can increase knowledge construction. Besides, analyzing case studies, report writing, project summaries, research papers and collaborative written assignments such as preparing team reports on specific topics, group discussion questions, debates and criticisms also help students to increase knowledge construction. Parker [27] in his research demonstrates that in order to influence the students' knowledge, skills and behavior, teachers must be prepared to implement suitable developments and cultures in teaching strategies by providing basic information, involving students to practice and apply relevant skills and implement it in fitting time.

4 Conclusion

The development of multimedia technology demonstrates the great potential of transforming student's method of learning, how to obtain information and how to adapt information. Teachers' explanations about things can be easy to understand when assisted by a variety of materials and tools that capable to interpret a clear meaning about the matters. The use of attractive and concrete visual materials not only enhances students' understanding of the topic, it also has an insightful effect on the student's personality.

In building their own understanding, students may accept this new idea and throw away their old ideas because they feel that the idea does not fit their judgment. In every situation, they are actually building understanding on new ideas and constructing knowledge based on the experiences they have had or their existing ideas. Students who come to the classroom are individuals who carry out a variety of experiences and knowledge that will be the basis of new knowledge to be built.

This new knowledge is built on itself as this student aligns with the past experiences and the existing ideas they bring to the classroom. Opinion in constructivism will put students into centers in the teaching and learning process. Collaborative activity through discussion seems to be an important aspect of teaching and learning in constructivism. The discussion method is widely used for knowledge construction process as this method can provide a collaborative learning environment and facilitate students to participate in online communication.

The authors would like to thank the Universiti Tun Hussein Onn Malaysia (UTHM) for their support in making this project possible. Part of this work was supported by the Research University Grant initiated by UTHM under PPG VOT V003.

References

[1] Woolfolk, A. (2008). *Educational Psychology*.(10th ed.) New Jersey: Allyn and Bacon.

[2] Rahman, S., Yasin, R. M., Jusoff, K., Yassin, S. F. M., Nordin, N. M., & Yusof, M. M. (2011). Knowledge construction process in online learning. *Middle East Journal of Scientific Research*, *8*(2), 488-492. Kian-Sam, H., dan Lee, J. A. C. (2008). Postgraduate students' during asynchronous computer conferences in a blended learning environment: A Malaysian experience. *Australasian Journal of Educational Technology*, 24(1), 91-107.

[3] Kian-Sam, H. & Lee, J. A. C. (2008). Postgraduate students' during asynchronous computer conferences in a blended learning environment: A Malaysian experience. *Australasian Journal of Educational Technology*, 24(1), 91-107.

[4] Zhu, C. (2012). Student Satisfaction, Performance, and Knowledge Construction in Online Collaborative Learning. *Educational Technology & Society*, 15 (1), 127–136.

[5] Noroozi, O., Biemans, H., Mulder, M. & Chizari, M. (2010). Students' Knowledge Construction in Computer-Supported Learning Environments: A Comparative Study in the Domain of Nutritional Research Methodology Education. *In Society for Information Technology & Teacher Education International Conference*. March 2010. Vol. 2010, No. 1, pp. 2506-2513.

[6] Thanasingam, S. & Soong, S. K. A. (2007). Interaction patterns and knowledge construction using synchronous discussion forums and video to develop oral skills. In ICT: Providing choices for learners and learning. *Proceedings ascilite Singapore 2007*.

[7] Paavola S., Lipponen, L. & Hakkarainen K. (2002). *Epistemological Foundations for CSCL: A comparison of three modes of innovative knowledge communities*. In G. Stahl, (Ed), *4th CSCL: Foundations for a CSCL Community*, (CSCL-2002), Boulder, Colorado, January 2002, pp.24-32, LEA, NJ. USA.

[8] Bereiter, C. (2002). *Education and mind in the knowledge age*. Mahwah, NJ: Lawrence Erlbaum Associates.

[9] Nonaka, I. & Takeuchi, H. (1995). *The Knowledge-Creating Company: How Japanese Companies Create the Dynamics of Innovation*, Oxford University Press, New York.

[10] Engeström, Y. (2000). Activity theory and the social construction of knowledge: A story of four umpires. *Organization*, 7(2), 301-310.

[11] Alavi, M. (1997). *KPMG Peat Marwick U.S.: One Giant Brain*, Case 9-397- 108 Boston, MA: Harvard Business School.

[12] Alavi, M. & Leidner, D. E. (1999). Knowledge management systems: issues, challenges, and benefits. *Communications of the AIS*, 1(2es).

[13] Alavi, M. & Leidner, D. E. (2001). Review: Knowledge management and knowledge management systems: Conceptual foundations and research issues. *MIS quarterly*, 107-136.

[14] Alavi, M., Yoo, Y. & Vogel, E. R. (1997). Using information technology to add value to management education.*Academy of Management Journal*, 40(6), 1310–1333.

[15] Alavi, M. & Dufner, D. (2005). *Technology-mediated collaborative learning: a research perspective*. In S.R. Hiltz and R. Goldman (Eds.).*Learning Together Online: Research on Asynchronous Learning Networks*,. (pp. 191–213). Mahwah, NJ: Lawrence Erlbaum Associates.

[16] Rahman, S., Yasin, R. M., Jusoff, K., Yassin, S. F. M., Nordin, N. M., & Yusof, M. M. (2011). Knowledge Construction Process in Online Learning, *Middle-East Journal of Scientific Research* 8 (2): 488-492, 2011.

[17] Oliver, R. (2001). *Developing e-learning environments that support knowledge construction in higher 10.education*. In: S. Stoney and J. Burn, (Eds). *Working for excellence in the e-conomy*. Churchlands: Australia, pp: 407-416.

[18] Kanuka, H. & Anderson, T. (2007). Online social interchange, discord, and knowledge construction. *The Journal of Distance Education* / Revue de l'Éducation à Distance, 13(1), 57-74.

[19] Schellens, T. & Valcke, M. (2006). Fostering knowledge construction in university students through asynchronous discussion groups. *Computers & Education*, 46(4), 349-370.

[20] Cañas, A. J., Novak, J. D., González, F. M., Ortegano-layne, L. & Gunawardena, C. N. (2008). Synthesizing social construction of knowledge in online conferences using concept maps.

[21] Kumar, S. (2005). Student Knowledge Construction during Asynchronous Online Discussions in a Face-to-Face Course, *World Conference on Educational Multimedia, Hypermedia and Telecommunications*. Vol. 2005.No. 1. 2005.

[22] García, B., Márquez, L., Bustos, A., Miranda, G. A. & Espíndola, S. M. (2008). Analysis of Patterns of Interaction and Knowledge Construction in Online Learning Environments: A Methodological Strategy. *Revista Electrónica de Investigación Educativa*, 10(1), 2.

[23] Hashim, S. & Harun, J. (2014, April). Effects of CSCL Learning Approach for Students' Knowledge Construction Process in School Based Assessment (SBA) Environment. In *Teaching and Learning in Computing and Engineering (LaTiCE), 2014 International Conference on* (pp. 293-298). IEEE.

[24] Veerman, A. & Veldhuis-Diermanse, E. (2001). Collaborative learning through computer-mediated communication in academic education.

[25] Gunawardena, C. N., Lowe, C. A. & Anderson, T. (1997). Analysis of a global online debate and the development of an interaction analysis model for examining social construction of knowledge in computer conferencing. *Journal of educational computing research*, 17(4), 397-431.

[26] Walker, G. (2005). Critical thinking in asynchronous discussions.International Journal of Instructional Technology and Distance Learning, 6(2).

[27] Parker, J. K. (2010). *Understanding youth and digital media*. In. J.K. Parker (Ed.) *Teaching Tech-Savvy Kids: Bringing Digital Media Into the Classroom, Grades 5-12*. Thousand Oaks, California: Corwin.

Motivational Factors in Privacy Protection Behaviour Model for Social Networking

Muliati Sedek[1,*], *Rabiah* Ahmad[2], and *Nur Fadzilah* Othman[2]

[1]Center for Teaching and Learning, Universiti Teknikal Melaka, Universiti Teknikal Melaka, 76100 Durian Tunggal, Melaka
[2]Information Security and Networking Research Group (InForSnet), Center for Advanced Computing Technology, Faculty of Information Technology and Communication, Universiti Teknikal Melaka, 76100 Durian Tunggal, Melaka

Abstract. This study aims to investigate the determinants of the privacy protection behaviour strategies that been employed by users while utilising SNSs. By understanding the determinants of privacy protection will be able to generate awareness that can protect users and allow them to confidently impose their self-control through the execution of privacy protection behaviour strategies. The finding has shown that there was a significant relationship of perceived severity, perceived vulnerability, response efficacy and self-efficacy towards information privacy concern as well as a significant relationship of information privacy concern and privacy protection behaviour strategies. This research is crucial as it serves as a guide that provides instructions and guidelines that help users of SNSs to keep their privacy intact.

1 Introduction

Social Networking Sites (SNSs) have become a phenomenon amongst Malaysians. Statistics from the Malaysian Communication and Multimedia Commission (MCMC) has reported that 45.5% of the population or 13.3 million users are registered Facebook users (MCMC, 2014). SNSs such as Facebook and Instagram allow individuals to stay in touch with their friends, reconnect with old friends and create new relationships with other people through the plethora of activities provided, such as sharing photos and videos, archiving events, updating others on activities, sending messages privately and posting public testimonials (Boyd, 2008; Vithessonthi, 2010). Therefore, the nature of SNSs that offer an attractive way of online interaction and communications encourage users to use it to its zenith. Unfortunately, the excessive information sharing and activities performed while accessing SNSs in an uncontrolled manner, can lead to a privacy breach on the user's behalf.

Recently, privacy issues related to personal information has been widely discussed and deliberated by various researchers (Nemec Zlatolas, Welzer, Heričko, & Hölbl, 2015). Due to the rapid development and utilisation of communication technology, privacy has become a serious concern. Users are willingly to share their private information subconsciously without a clear idea of who is allowed to access to their personal information and what portion of it is really accessed. Even though SNSs themselves have been equipped with systematic safety features, there is no guarantee that one's privacy is fully protected (Salleh et al., 2012). Hence, there is an urgent need for an assessment mechanism that can detect threats from engaging in risky situations so that users can determine how much and what type of personal information should be shared and disclosed.

Therefore, this study aims to investigate the determinants of the privacy protection behaviour strategies that been employed by users while utilising SNSs. By understanding the determinants of privacy protection will be able to generate awareness that can protect users and allow them to confidently impose their self-control through the execution of privacy protection behaviour strategies.

2 Privacy Protection Behaviour Strategies

Literature has given various definitions of privacy. The concept of privacy ranges from a "right to be alone" as from the perspective of law (Warren & Brandeis, 1890), "state of limited access" in the aspect of philosophy (Schoeman, 1984), and to the "control over information about one's self" as given from the views of social science (Kokolakis, 2017).

Privacy protection behaviour can be defined as specific computer-based actions that individuals take to keep their information safe. Rogers (1983) has discussed that individuals are motivated to rely on protection behaviour in order to cope and adopt behaviour to control risk, threat and danger. Coping strategies can be divided into two dimensions, which are approach and avoidance (Chen, Beaudoin & Hong, 2017). Approach strategies include fabricating personal information and seeking

* Corresponding author: muliati@utem.edu.my

social support whereas avoidance strategies include withholding personal information. In the context of SNS, fabricating information as based on approach strategies refers to a user covering up or disguising their identity by using fake or false information in SNSs. Seeking social support means that users ask for advice and read privacy statements from SNS providers in order to gain knowledge that would help them adopt privacy protection behaviour strategies alongside increasing their privacy. As for avoidance strategies, refraining information means that users will refuse to provide their personal information to SNSs and instead even begin to patronize them over it.

Information privacy concern is the "extent to which an individual is concerned about organizational practices related to the collection and use of his or her personal information (Martin, Borah & Palmatier, 2017). Previous research has shown that information privacy concern has an impact on privacy protection behaviour strategies (Adhikari & Panda, 2017). Within the protection motivation theory, information privacy concern is considered to be a mediating variable that explains the relationship between the factors involved and privacy protecting behaviour strategies.

3 Theoretical Framework and Hypotheses

3.1. Protection Motivation Theory

Protection Motivation Theory (PMT) as introduced by Rogers (1975) postulates that an individual's motivation to protect from risk and threat, which comes from (1) perceived severity, (2) perceived vulnerability and (3) response efficacy. The PMT model was modified to explain failures concerned in protection behaviour by including (4) self-efficacy, (5) response cost and (6) rewards associated with risky behaviour (Rogers, 1983; Rogers, 1975). PMT has been principally used in the health industry (Grindley, Zizzi & Nasypany, 2008) this theory has also been used in more than 20 different health- related areas in order to study about intentions and behaviours. In the field of Information System (IS), PMT has also been widely used to examine protection behaviour in online transactions, awareness of employees in organizational information security policies and individual use of security software (Johnston & Warkentin, 2010).

3.1.1 Perceived Severity

Perceived severity refers to an individual's belief that the judgement of severity significance results from a threatening event (Palladino, BMenesini et al., 2017). Perceived severity evaluated how severe an individual believes that threat will interrupt their life. Individuals will adopt recommended action when they seriously perceive the negative consequence. Moreover, Adhikari and Panda (2017) found that an individual's motivation for engaging in risk-reducing behaviour is increased by perceived severity. For the purpose of this study, users will develop a perceived severity after losing information privacy to SNSs. They will significantly associate this loss with information privacy concern and in this form, indirectly motivate them to adopt privacy protection strategies in SNSs.

3.1.2 Perceived Vulnerability

Perceived vulnerability explains an individual's perception in experiencing possible negative effects that stem from performing risky behaviour (Kim & Kim, 2016). Based on the findings, it can be argued that perceived severity is found to have increased students' intention to perform malware avoidance behaviour. Subsequently, Schoeman (1984) agreed that one of the factors that contribute to users increasing information privacy concerns is perceived vulnerability. Conversely, perceived vulnerability has an insignificant impact on employees' intention to comply with IS security policies (Kokolakis, 2017). Thus, for this study, it is suggested that individuals who perceive the risk and threats of losing information privacy through SNSs will increase their information privacy concerns which will later motivate them to use privacy protection behaviour strategies.

3.1.3 Self-efficacy

Self-efficacy is defined as an individual's belief and capability to implement protective behavior while using SNSs (Compeau, Higgins, & Huff, 1999). Several studies provide evidence that self-efficacy plays an important role in a user's choice to perform risky online behaviour. While Lee et al. (2008), proves that self-efficacy should be the influential factors that stimulates protection behaviour. Hence, this study suggests that individuals who are self-efficacious in using SNSs are more likely concerned with their information privacy and are totally motivated to use privacy protection behaviour strategies.

3.1.4 Response efficacy

Response efficacy is the belief in coping with responses in avoiding a threat (Grindley et al., 2008). Research identifies that response efficacy is a significant predictor behaviour that determines the decision whether or not to implement security features on their networks, increase intentions to use anti-spyware software as protective technology, predicts backing up data on personal computers and use of malware devices (Martin et al., 2017). Therefore, the study posits that by having a good response efficacy could help users in decreasing the loss of data.

 https://doi.org/10.1051/matecconf/201815005014

3.1.5 Rewards

Reward refers to an individual's expectation in getting benefits when keeping with selective behaviour (Lee et al., 2008). Previous study has shown that individuals who find great enjoyment and satisfaction from sharing personal information are less inclined to make adaptive change for protection (Marett, Harris, & Mcnab, 2011). Additionally, individuals believe that when they are willing to disclose their information, they may experience a sense of being close to their friends and family and at the same time get satisfaction from the feeling of togetherness.

3.2 Information Privacy Concern

Information privacy concern is the "extent to which an individual is concerned about organizational practices related to the collection and use of his or her personal information. Previous research has shown that information privacy concern had an impact on privacy protection behaviour strategies. Within the protection motivation theory, information privacy concern is considered to be a mediating variable that explains the relationship between the factors involved and privacy protecting behaviour strategies (Stern & Kumar, 2017).

In this study, six aspects of PMT and information privacy concern were hypothesized. The following hypotheses are as follows, H1: Information privacy concern is significantly associated with privacy protection behaviour strategies, H2: Perceived severity is significantly associated with information privacy concern, H3: Perceived vulnerability is significantly associated with information privacy concern, H4: Self-efficacy is significantly associated with information privacy concern, H5: Response efficacy is significantly associated with information privacy concern and H6: Reward is significantly associated with information privacy concern. The proposed research model is presented in Figure 1.

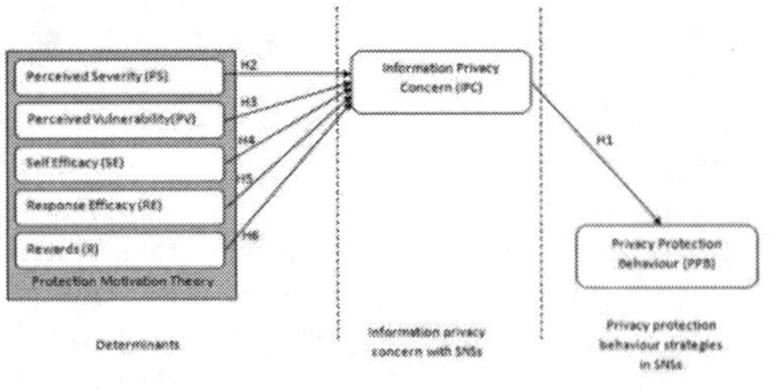

Figure 1: Proposed Research Model

4 Research Methodology

For the purpose of this study, a total of six hypotheses were tested and a quantitative approach was employed to them. Quantitative approach is the best method to use in order to test any existing theory as it involves a collection and statistical analysis of numerical data. (Ary, Jacobs, Razavieh, & Sorensen, 2010).

The instrument used in this study was a questionnaire that consisted of 44 items in total. All the items used a five point Likert scale, where 5 represented strongly agree and 1 represented strongly disagree.

4.1 Sample Selection and Data Collection

For the sampling process, stratified random sampling was used in this study. From the data given by the universities' administration on the number of active undergraduates as of February 26, 2015, there were approximately 9,205 undergraduates in total. As according to Sedek, Mahmud & Jalil (2012), they recommended that the ideal number for sample size suitable for analysis using SEM should be approximately between 300 to 800 samples. Of the 550 distributed, 499 were returned. 485 were usable for the purpose of this study with a response rate of 88%. Table 1 shows the profile of the respondents.

Variable	Type	Frequency	Percent
Gender	Male	2 5 4	5 2
	Female	2 3 1	4 8
Age	15-20	-	-
	21-25	449	93
	26-30	36	7
	31-35	-	-

5 Results

The first step conducted in SEM analysis was Confirmatory Factor Analysis (CFA). CFA was meant to identify the individual construct and was employed for three major purposes, which are (i) model fit, (ii) convergent validity and (iii) construct validity.

Maximum likelihood estimate (MLE) was used to estimate the structural model. Table 2 presents the test of overall model fit. All the fit indices were above recommended values.

Construct	Convergent Validity	
	(CR) (Above 0.6)	(AVE) (Above 0.5)
Privacy protection behaviour (PPB)	0.874	0.587
Information privacy concern (IPC)	0.917	0.650
Perceived severity (PS)	0.833	0.555
Perceived vulnerability (PV)	0.845	0.579

Self-efficacy (SE)	0.882	0.882
Response efficacy (RE)	0.867	0.623
Rewards (R)	0.903	0.702

Figure 2 presents the detailed result of the structural model. The R2 values for information privacy concern and privacy protection behaviour is 0.55 and 0.08 respectively. The root mean square error of approximation (RMSEA), which measures the discrepancy per degree of freedom, was 0.065. The goodness-of-fit index (GFI) was 0.851, comparative fit index (CFI) was 0.882 and the discrepancy Chi Square (Chisq/df) was 2.964.

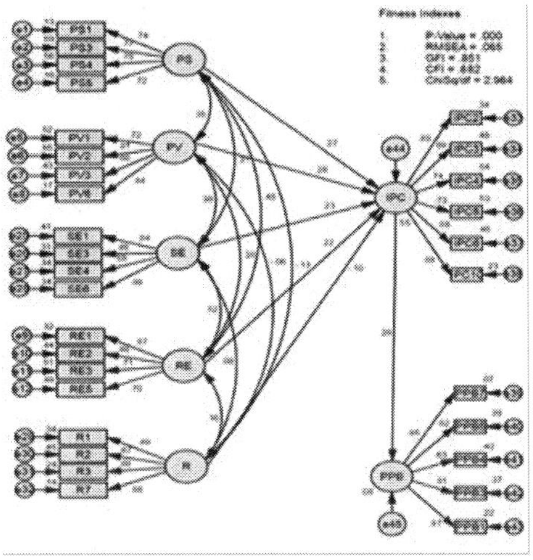

Figure 2: The Structural Model

Table 5: The regression path coefficients, significance values and hypothesis statement for every path and its conclusion

Source	Destination	Hypothesis Statement of Path Analysis	Estimates	P-value	Results on Hypothesis
IPC -->	PPB	H1.Higher information privacy concern will increase privacy protection behaviour.	0.28	0.003	Supported
PS -->	IPC	H2.Higher perceived severity will increase information privacy concern.	0.27	0.043	Supported
PV -->	IPC	H3.Higher perceived vulnerability will increase information privacy concern.	0.28	0.033	Supported
SE -->	IPC	H4.Higher self-efficacy will increase information privacy concern.	0.23	0.028	Supported
RE-->	IPC	H5.Higher response efficacy will increase information privacy concern.	0.22	0.001	Supported
R -->	IPC	H6. Higher rewards will reduce information privacy concern.	-0.10	0.605	Not Supported

6 Conclusion

The data analysis as executed from SEM reveals that there was a significant relationship of perceived severity, perceived vulnerability, response efficacy and self-efficacy towards information privacy concern as well as a significant relationship of information privacy concern and privacy protection behaviour strategies.

Similar with the results of prior research, individuals who are concerned with their information privacy in SNSs were found to use and adopt privacy protection behaviour strategies (Kim & Kim, 2016).). In order to materialise such concern and awareness, several determinants have been found. Perceived severity was found to be one such determinant. Users who feel that they will be seriously affected by the loss of information privacy will be more concerned with their information privacy, whereas those who think otherwise will not be as concerned. The next determinant that is significantly associated with information privacy concern is perceived vulnerability. The finding suggests that individuals who have been or are exposed to the loss of information privacy are more concerned with information privacy, whereas those who have not been or are not exposed to such a loss are less concerned. The next determinant proven in this study that contributes to information privacy concern is self-efficacy. Individuals who believe that they have the ability to use the protective strategies in SNSs will be more concerned with their information privacy. Finally, the last determinant that contributes to information privacy concern is response efficacy. One determinant however, was found does not support the hypotheses as stated in this study. It was found that reward is not significantly associated with information privacy concern.

References

1. Adhikari, K., & Panda, R. K. (2017). Users' Information Privacy Concerns and Privacy Protection Behaviors in Social Networks: Evidence from India.
2. Ary, D., & Jacobs, L. C., Sorensen, C., & Razavieh, A.(2010). Introduction to research in education.
3. Boyd, D. (2008). Facebook's Privacy Trainwreck: Exposure, Invasion, and Social Convergence. Convergence: The International Journal of Research into New Media Technologies, 14(1):13–20.
4. Chen, H., Beaudoin, C. E., & Hong, T. (2017). Securing online privacy: An empirical test on Internet scam victimization, online privacy concerns, and privacy protection behaviors. Computers in Human Behavior, 70, 291-302.
5. Compeau, D., Higgins, C. A., and Huff, S. (1999). Social cognitive theory and individual reactions to computing technology: A longitudinal study. MIS

Quarterly, 23(2):145–159.

6. Grindley, E. J., Zizzi, S. J., and Nasypany, A. M. (2008). Use of protectiomotivation theory, affect, and barriers to understand and predict adherence to outpatient rehabilitation. Physical Therapy. Jounal of American Physical Theraphy Association, 88(12):1529–1540.
7. Johnston, B. A. C. and Warkentin, M. (2010). Fear Appeals and Information security Behaviors: An Empirical Study. Ministry of Education Official Website, 34(3):549–566.
8. Kim, A. Y., & Kim, T. S. (2016). Factors Influencing the Intention to Adopt Identity Theft Protection Services: severity vs Vulnerability. In PACIS (p. 68).
9. Kokolakis, S. (2017). Privacy attitudes and privacy behaviour: A review of current research on the privacy paradox phenomenon. Computers & Security, 64, 122-134.
10. MCMC (2014). Communications & Multimedia Pocket Book of Statistic.
11. Marett, K., McNab, A. L., & Harris, R. B. (2011). Social networking websites and posting personal information: An evaluation of protection motivation theory. AIS Transactions on Human-Computer Interaction, 3(3), 170-188.
12. Martin, K. D., Borah, A., & Palmatier, R. W. (2017). Data privacy: Effects on customer and firm performance. Journal of Marketing, 81(1), 36-58.
13. Nemec Zlatolas, L., Welzer, T., Heriˇcko, M., and H¨olbl, M. (2015). Privacy antecedents for SNS self-disclosure: The case of Facebook. Computers in Human Behavior, 45:158–167.
14. Palladino, B. E., Menesini, E., Nocentini, A., Luik, P., Naruskov, K., Ucanok, Z., ... & Scheithauer, H. (2017). Perceived Severity of Cyberbullying: Differences and Similarities across Four Countries. Frontiers in Psychology, 8, 1524.
15. Salleh, N., Hussein, R., Mohamed, N., Abdul, N. S., Ahlan, A. R., and Aditiawarman, U. (2012). Examining Information Disclosure Behavior on Social Network Sites Using Protection Motivation Theory, Trust and Risk. Journal of Internet Social Networking & Virtual Communities, 2012.
16. Schoeman, F. (1984). Philosophical Dimensions of Privacy: An Anthology.
17. Sedek, M., Mahmud, R., Jalil, H. A., & Daud, S. M. (2012). Types and levels of ubiquitous technology use among ICT undergraduates. Procedia-Social and Behavioral Sciences, 64, 255-264.
18. Stern, T., & Kumar, N. (2017). Examining privacy settings on online social networks: a protection motivation perspective. International Journal of Electronic Business, 13(2-3), 244-272.
19. Vithessonthi, C. (2010). Knowledge sharing, social networks and organizational transformation. The Business Review, Cambridge, 15(2):99–109
20. Warren, S. D. and Brandeis, L. D. (1890). The Right to Privacy. Harvard Law Review, 4(5):193–220.

Tracking Outfield Employees using GPS in Web Applications

Vinothini Kasinathan[1,*], *Aida* Mustapha[2], and *Roshein* Chandran[1]

[1] *Faculty of Computing, Technology and Engineering, Asia Pacific University of Innovation and Technology, Technology Park Malaysia, Bukil Jalil, Kuala Lumpur 57000, Malaysia*
[2] *Faculty of Computer Science and Information Technology, Universiti Tun Hussein Onn Malaysia, 86400 Parit Raja, Batu Pahat, Johor, Malaysia*

Abstract. This paper presents e-Track, a web-based tracking system for outfield employees in order to cater for various business activities as demanded by the business owners. Such demands may range from a simple task assignment, to employee location tracking and remote observation of the employees' task progress. The objective of the proposed system is two-fold. First, the employees to access the application and clocks-in work. Second, a standalone web system for the employers to determine the approximate location of the staff assigned with outfield duties. The IP address recognition will ensure no buddy punching takes place. e-Track is hoped to increase efficiency among employees by saving time travelling between branches during outfield duties. In the future, e-Track will be integrated with claim and payment modules to support arrangement for outfield duties.

1 Introduction

Business owners are now looking for more ways to incorporate technology to their benefit. Each business has to have its own set of employees for a company to maintain sustainability. It is worth noting that running a business is a risky venture to which a survey attests to proving 49% of businesses fail within their first five years and approximately 30% of businesses don't even make it through the first two years [1]. One of the reasons as to why businesses fail, especially small businesses are due to a lack of employee productivity [2].

There are several problems employers and employees face which reduces their productivity in a working environment. In most organizational setting, there are no standard methods to track employee with outfield works that actually meets a business's needs. The way the working environment functions currently would not be optimal to nurture an individual's full potential. Employees have to check in to work using punch cards and would have to stay on in their office cabin throughout their working hours [3]. Even if an employee has outfield duties, it is a complicated procedure for an employee to travel to a destination and return to office to sign out before office hour ends. The scenario worsens if the company has multiple offices and employees have to be travelling constantly.

Another issue is buddy punching. According to Synerion[4], buddy punching is described as when one individual checks in for another colleague of his without them being there. This is a wastage of company resources as each employee are being paid accordingly. This issue may lead to employee dissatisfaction, as some employees who give full effort and those who slack off get paid the same amount. In the long run, there is a high chance of employee productivity decreasing. It is clear that the traditional method used to track employee activities based on card swipes or manual entry using a book keeping method is no longer effective.

To address the issues, a Web-based Outfield Employee Tracking System named e-Track is proposed. e-Track architecture is proposed to be on a client server basis whereby there will be a mobile application for employees to access and a standalone web application for the employers to determine the approximate location of the staff assigned with outfield duties. Any organization that sends their employees for outfield duties will find this application extremely useful as they would be able to assign duties to employees based on their locations. In addition, this application also allows companies with many branches to take advantage of this system by allowing checking in to work through the mobile application. The IP address recognition will ensure no buddy punching takes place. The advantages are mutually beneficial as employees will also feel secure that their whereabouts are known by the employee in case of any mishaps taking place.

The remainder of this paper is as follows. Section 2 presents the related works, Section 3 presents the architecture and prototype of the proposed system e-Track, and finally Section 4 presents discussions and conclusions with plans for future work.

2 Related Work

In this modern era, tracking devices have becoming a common household gadget in human daily lives for var-

*e-mail: vinothini@apu.edu.my

ious reasons, ranging from tracking missing pets to missing phones. The devices are small and convenient to use. Nonetheless, tracking devices have not been completely utilized in an organizational setting. In this section, three online time tracking applications are reviewed, which are TSheets, ActivTrak, and Veriato 360.

2.1 TSheets

TSheets is the mobile applications name wherein it allows users to track time spent from any location, using any device in real time [5] (Refer Figure 1). This application would be useful for employees who work in remote locations or switch jobs regularly. Employees would be able to track time easily as there is clock in and clock out reminders, employee breaks and overtime alerts as well to ensure all employees are aware and up to date of the current situation ongoing in the company. The main functionality of TSheets is the use of Global Positioning System (GPS) for online time tracking. The location at which the employee has clocked-in is visible with the aerial view.

Figure 1. User interface for TSheets Time Tracking Source.

Secondly, the time entry can be manual, punch or custom. GPS location points are immediately attached to the employee's timesheet when they log in and log out or even change job modes. Employees are allowed to log in and log out at real time. On extenuating circumstances, employees can enter their time manually and allocate time to custom projects and tasks. Finally, in the case of any absenteeism such as sick days, personal days or even a family vacation, TSheets tracks employee time off to manage employees paid time off. This function enables employees to request time offs directly from the mobile application and employers would be able to approve the request in a similar manner.

2.2 ActivTrak

ActivTrak is a free employee monitoring software and it is developed on the basis of a free cloud-based monitoring service [6] (Refer Figure 2). It allows employers to know who is doing what and for how long by tracking applications and web usage based on each workstations. ActivTrak has three main features, which are real-time monitoring, alarm, and website blocking.

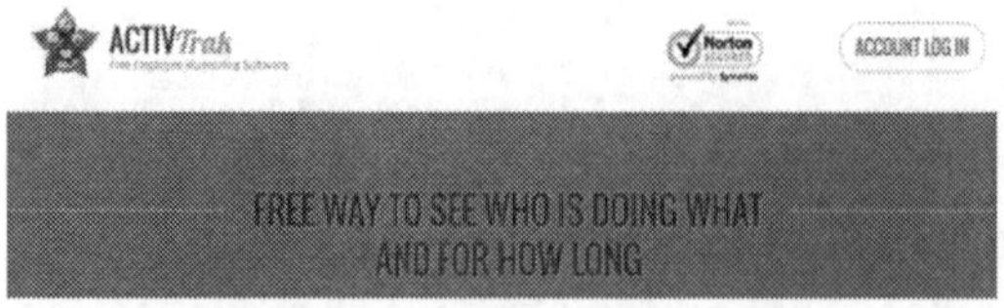

Figure 2. ActivTrak homepage.

For real-time monitoring, ActivTrak allows administrators to view a live data stream of the active window on any of the monitored devices screen. The administrators are able to see the time, device, user, title and even its public IP address. With real time monitoring, ActivTrak provide details on exactly the events happening in the monitored devices at any point from anywhere.

ActivTrack also has alarms that can be set to trigger at any actions of employees. For example, if an employee visits any restricted websites for longer than 20 seconds, an alarm will ring indicating them to close the tab. Each time an alarm is triggered, there is an option to take multiple screenshots of the complete screen to have a proof of evidence. The admin may also opt to display a popup message which would appear on the screen of the user. If all else fails and employees are still on the restricted website, there is an option to terminate the application which triggered the alarm.

Finally, websites which employers feel takes the employees productivity away can be blocked. A website can be blocked by entering the domain name and clicking the 'blocked column'. Websites can be blocked for individual users, groups or even everyone at once. Thus, there will be no complications if separate groups of people must be blocked from specific websites.

2.3 Veriato 360

Veriato 360 is a record-keeping system wherein it presents detailed, accurate and actionable data used in incident response together with high-risk insider monitoring and productivity reporting [7] (Refer Figure 3).

Figure 3. Veriato 360 homepage.

Veriato 360 has a different set of functionalities as compared to TSheets and ActivTrak. First, Veriato 360 has file and document tracking. This feature allows employers to view new files created and when existing files are edited, renamed or deleted. It is also useful as a supporting evidence in cases of leaks, breaches or theft.

Second is email recording, whereby Veriato 360 records emails between clients and employees. It supports reading traditional email clients such as Outlook and also recent popular webmail such as Yahoo and Gmail. Veriato 360 allows searching across all employee emails simultaneously and sorting them according to subject, CC, BCC and Web Mail Host columns. Finally is the keystroke logging feature. This option allows the recording of each and every thing typed on every keystroke. Applying this feature to highly positioned individuals in the company will ensure complete transparency among people with high level of access in the organization. Keystrokes can be determined on which application they are used at as well. For example, employee typing in Outlook will display in 'Outlook' and not just vaguely 'user'.

3 Architecture and Prototype of e-Track

The proposed project will make use of a mobile and web application architecture. Usually, a client server architecture will be applied for any form of mobile applications. However, with the development of technology and different architectures, the opportunity to take into consideration specific aspects related to mobile devices and their connectivity with servers is now available. Figure 4 shows the architecture for the proposed web-based tracking system for outfield employees, which is e-Track.

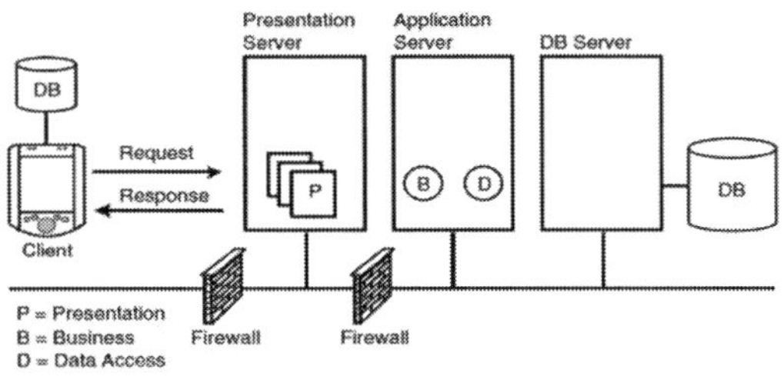

Figure 4. e-Track architecture.

Clients are divided into two categories on how they operate which are thin clients and fat clients [8]. Examples of clients are mobile device types ranging from cellular telephones, tablets and RIM devices. Thin clients do not have the ability to run custom application code therefore has to completely rely on the servers to obtain its functionality. However, the advantage is that it does not rely solely on the mobile device's operating system [9]. Thin clients uses widely available web and wireless application protocol to display html, xml and wml types of application content pages.

On the other hand, fat client usually has one to three layers of application code and can function without a server for a certain amount of time [8]. Practically, fat clients would be the most suitable client type to choose for businesses as even when communication between client and server is not established, the application will still be useful. For example, a fat client application will be able to accept the users input and store the date in a local database until connectivity the server is strong and the data can be transferred to the server. This feature would be essential to businesses as the user can still utilize the application even when not in contact with the server.

However, fat client does have its disadvantages. It relies heavily on the device's operating system and mobile device type. Thus, making the code tough to release and distribute. This also means the device has to support multiple code version on different devices for it to have a proper resolution (Aung, 2016). Fat client does have its variety, as it can be implemented in different ways ranging from one layer to three-layer application code. Although single layer may be useful on devices in a small scale project, it would be extremely hard to isolate individual functionalities and reuse the code in a large project. Table 1 shows the comparison between the two application code.

Nowadays, web page hosting has become a necessity in every mobile application development. This is to be able to display and service web pages using the mobile device even when the mobile client is only connected to a network and back-end system occasionally. Thus, to achieve that, the equivalent of a mini web server must be established on the mobile device.

3.1 Prototype

Figure 5 shows the user interface for login page. Users are required to input their own login details for their account which are email address and password. The login page has simple design and layout for users.

Figure 5. User interface for login.

Figure 6 shows the interface for tracking the users. Users can view the location of the employees using the google API implemented in the coding. Real-time location tracking is available when employees are logged into their accounts.

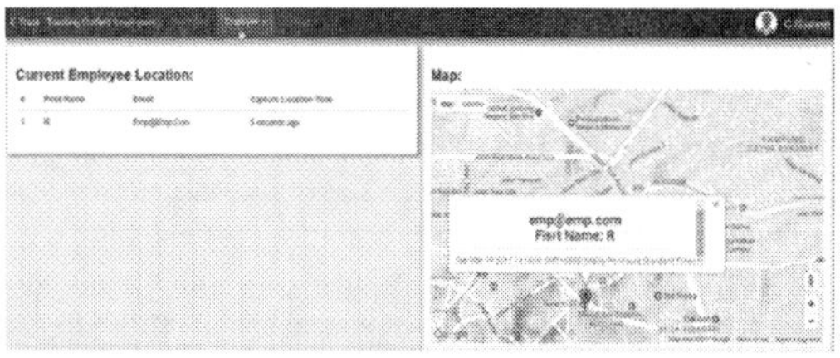

Figure 6. User interface for location.

Table 1. Comparison between one layer to three-layer application code.

Advantages	**Disadvantages**
Good scalability due to the distributed deployment of application servers.	Communication complexity increases due to increased distance between communication points.
Better data integrity as data corruption through client applications can be eliminated.	Additional effort is required due to increase of performance whereas the 2-tier model can handle the particular function using an automated tool.
Enhanced security through the implementation of several layers.	

Figure 7 shows the task assignment page. Administrators can assign tasks by allocating specific time slots to employees registered in the database.

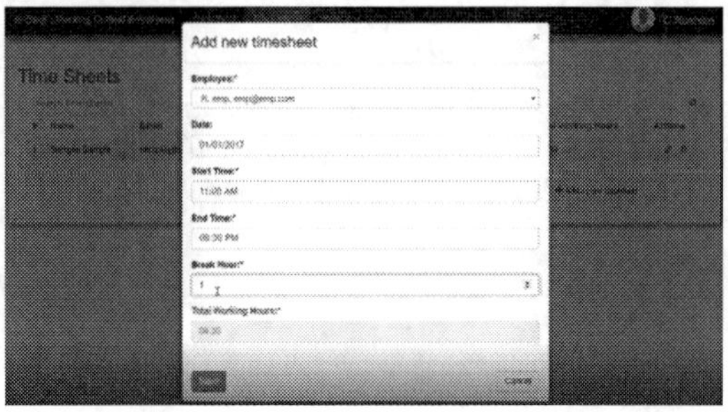

Figure 7. User interface fortask assignment.

Finally, e-Track generates list as shown in Figure 8. The users can search for specific employees and retrieve their details. The administrators will also be given power to edit information and delete employees when necessary.

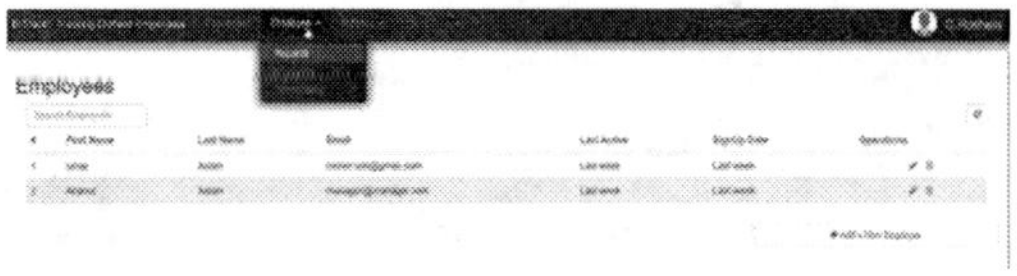

Figure 8. User interface to generate list.

3.2 Evaluation

e-Track was evaluated among people who work at an organization as the respondents. Questionnaires were distributed to three categories of people who are office employees, outfield employees and employees who juggle both office tasks and outfield duties. These three groups of people would be the primary users who would benefit from the usage of the system. The questionnaire was distributed through physical paper to the respondents and in order to gain a bigger pool of answers, both open-ended and close ended questions will be used. Figure 9 to Figure 13 show the evaluation findings.

Question 1: What do you think is the reason for employee productivity to decrease?

This question was done to determine the core reasons for declining employee productivity and its correlation with need of technological gadgets. The results prove that employees feel the need for the inclusion of technology to their working routing while simultaneously help address an organizations weakness.

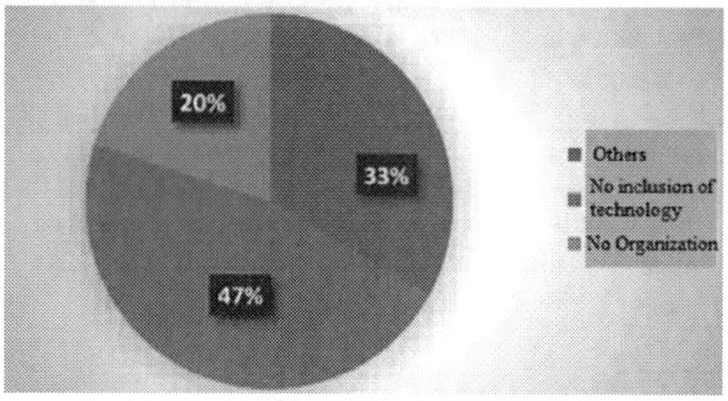

Figure 9. Findings for Question 1

Question 2: Do you do check in behalf of your colleague? If yes, why do you do it and how often?

This questions objective was to analyse if buddy punching occurs at the office place. The results show most percentage of the employees check in for their colleagues in behalf of them. This would possibly lead to a tremendous loss of revenue for any businesses.

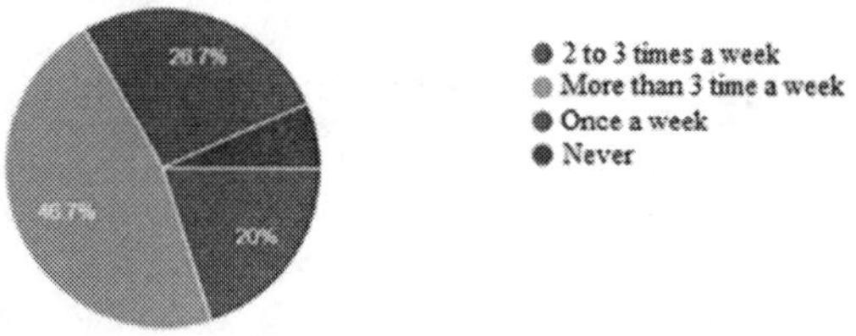

Figure 10. Findings for Question 2.

Question 3: How many hours would you save in a week if you had an agile work scope?

Time equals to money. Wasted time can't be recovered especially in a business setting. This leads to the motivation behind this question. The aim was to determine how many hours can be saved or maximized by a business owner by providing their employees a flexible and agile work scope. The highest percentage of answers was 2 to 4 hours. These many hours per week would means 10 to 20 wasted hours in a month.

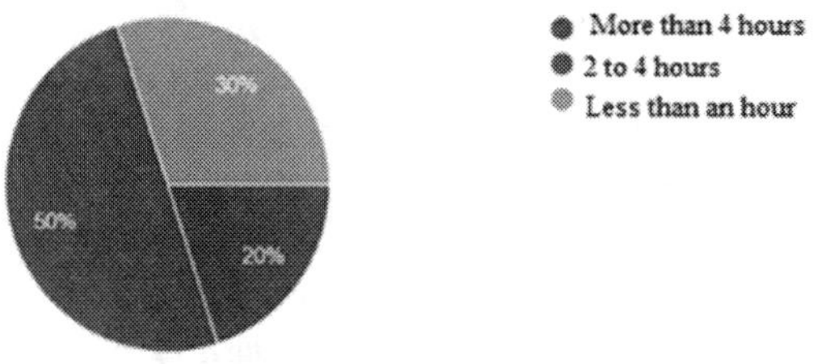

Figure 11. Findings for Question 3.

Question 4: Rate the level of satisfaction of the current attendance system which is being used?

The ranges were purposefully given to over a 4 rating as the evaluation is designed so the respondents unable a neutral answer as that would not be reflective of how adaptive they are towards the current system. The results show that equal percentage of users are happy with the current manual way of handling employees. The developer believes this would be due to a phenomenon of resistance to change.

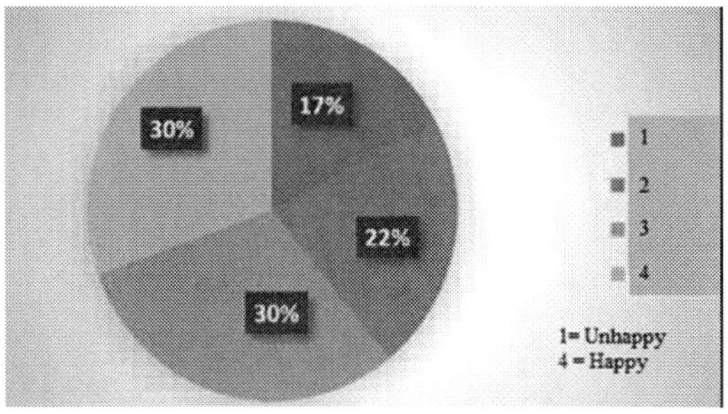

Figure 12. Findings for Question 4.

Question 5: Would you prefer if you had an agile work scope during your outfield duties where by your distances travelled will determine your payroll?

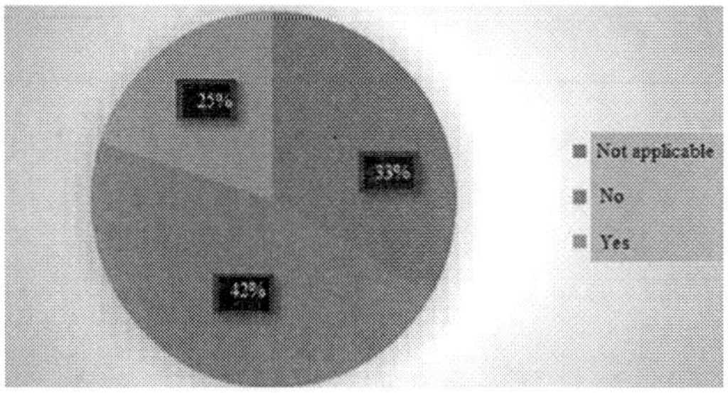

Figure 13. Findings for Question 5.

This analysis was done to decide the users requirement on if a payroll needs to be incorporated to the system for employees who travel for outfield duties. The results were mostly not receptive towards that idea. The author believes the respondents need to be further enlightened regarding the benefits of this functionality for it to be accepted wholly.

4 Discussions and Conclusions

The major aim of the proposed research is to design and implement a web-based tracking system for outfield employees in order to cater for various business activities as demanded by the business owners. Such demands may range from a simple task assignment, to employee location tracking and remote observation of the employees' task progress. The aim of the project is to provide both employees and business owners a convenient method to monitor business activities, employee attendance while simultaneously producing and implementing a more convenient, revolutionary method that can widen employees' work scope and increase the productivity.

The proposed e-Track System benefits both employees and business owners in a number of ways. One of the core benefit is the increased efficiency among employees. A more efficient working environment allows each employee to focus on developing their individual potential and talent in a more agile working environment. Next, the tracking system helps to reduce time wastages because employees do not have to waste unnecessary time travelling between branches during outfield duties just to sign out from their workspace. e-Track gives a time-out mechanism via a simple logout button from the mobile application.

From the perspective of business owners, the automation of the monitoring job will lessen the burden for HR department to manage their employees' whereabouts. This will enable the Human Resource (HR) department to focus on different tasks such as recruiting staffs and managing timely payrolls. A tracking mechanism based on Global Positioning System (GPS) may also reduce absenteeism cases especially when dealing with common excuses such as the traffic jam. This is because the system allow users to login from the nearest geolocation according to a specific radius required by the employer. Finally, the tracking system can also be used to locate employees during their outfield duties in case of mishappenings and emergencies.

There are also intangible benefits of the web-based outfield employee tracking system. First and foremost, employees will be able to enjoy complete freedom of workspace and improve their individual creativity in contributing to the organization, instead of being inhibited by the traditional system of having to be in the office from 9am-5pm. Secondly, a more secure attendance check-in feature provides a more secure way to enable employees to check in to work while simultaneously reducing buddy punching. Thirdly, absenteeism can be well managed by avoiding absenteeism due to minor reasons. The system will even alert users who have not checked in to the system 5 minutes before commencement of work. The system also allows employees to inform beforehand failure to attend work in cases of extenuating circumstances. Finally, the application that is based on GPS location provider is more reliable and available 24/7.

Future enhancement of e-Track is to make the system more reliable in terms of calculation of employee payment based on travelled location. This will give more precise payment to employee to plan their budget before making decision. Finally, this project could be developed into smart phone application for IPhone, Android, Symbian, and Blackberry platform to use e-Track application. This application will have basic features such as trip advisor and photo gallery for front camera image capturing.

Acknowledgements

This research is sponsored by Asia Pacific University of Innovation and Technology.

References

[1] S.W. Solovic, E.R. Kadin, *It's Your Biz: The Complete Guide to Becoming Your Own Boss* (AMACOM Div American Mgmt Assn, 2012)
[2] D. Preece, P. Iles, Personnel Review **38**, 286 (2009)
[3] *Punch card attendance system*, `http://www.bioenabletech.com/punch-card-attendance-system.html`
[4] T. Synerion, Retrieved May **1**, 2016 (2015)
[5] *Time tracking - #1 employee time tracking software*, `https://www.tsheets.com/`
[6] *Activtrak - free employee monitoring & tracking software*, `https://activtrak.com/`
[7] *Veriato 360*, `http://www.veriato.com/`
[8] W.W. Aung, *Preserving green computer labs in developing countries with thin technology*, in *MOOCs, Innovation and Technology in Education (MITE), 2015 IEEE 3rd International Conference on* (IEEE, 2015), pp. 21–25
[9] Q.F. Hassan, *Innovative Research and Applications in Next-Generation High Performance Computing* (IGI Global, 2016)

Towards the Development of Mobile App Design Model for Dyscalculia Children in Malaysia

Fiqa Azureen Abd Halim[1,*], *Mazeyanti* Mohd Ariffin[2], & *Savita* K. Sugathan[3]

[1, 2, 3] Computer & Information Sciences Department, Universiti Teknologi Petronas, Malaysia

Abstract. Learning disabilities (LD) are problem that influence the brain capacity to receive, process, analyse and store information. These processing issues can interfere with learning fundamental abilities, for example, reading (Dyslexia), writing (Dysgraphia) and math (Dyscalculia). Fortunately, in today's world technology, we have numerous option techniques for educating and supporting fundamental skills in reading, writing and math such as mobile learning by using mobile app. However, the research in mobile app design model for Dyscalculia children is still scarce: very few studies have focus on how to design and develop appropriate mobile app for Dyscalculia children from the perspective of mobile app designers and developers. The purpose of this paper is to identify the appropriate components that should be included in the mobile app model. Interview sessions were conducted with Dyscalculia practitioners that involved a teacher, paediatric and an educational psychology. This has resulted in identifying 16 components that can be included in the mobile app design model.

1 Introduction

According to Malaysian Ministry of Education (2016), 29 810 primary school children are known to enrol in special education integration program including learning disability (LD). In Malaysia, LD is included Dyslexia, Dyscalculia and Dysgraphia (Malaysian Social Welfare Department, 2016). LD is neurologically-based processing issues (Butterworth, 2013). These processing issues can meddle with learning essential aptitudes, for example, reading, writing and math. LD is a wide term depicting various others, more particular learning disability such as Dyslexia which refers to difficulty in reading, Dysgraphia refers to difficulty in writing and Dyscalculia refers to difficulty in math.

Nowadays, a lot of technologies have been developed and some of them can help the children with LD such as mobile app. According to Skiada, Soroniati, Gardeli, & Zissis, (2014), the advantages of utilizing mobile learning are to create interactive encounters and idealistic learning surroundings that can motivate and help children, along these lines offering them some assistance with addressing their disabilities in early age and potentially alleviate its negative impacts' differences. However, the research on the mobile app design model is still limited (Ok, Kim, Kang, & Bryant, 2016). Thus, the mobile app designer and developer is lacking of guidance on how to design appropriate mobile app for LD children (Draper Rodriguez, Strnadova, & Cumming, 2014). Hence, the objective of this research is to identify the potential components that can be included in the mobile app design model called Calculic Model.

As mentioned before, the designers and developers of mobile app are lacking of guidance to design and develop the effective mobile app particularly for teaching and learning for LD children (Draper Rodriguez et al., 2014). As a result, most of the mobile app for LD children are poorly designed (Papadakis, Kalogiannakis, & Zaranis, 2017). To address this issue, the purpose of this study is to develop a Calculic Model that can guide mobile app designers and developers in designing and developing mobile app for Dyscalculia children. In order to design and develop an effective mobile app for LD children, the developer and designer of mobile app should follow the model and guideline from the researchers and teachers (Cumming, Strnadov, & Singh, 2014). Hence, the interview session with Dyscalculia practitioners have been conducted to identify the appropriate components in Calculic Model for Dyscalculia children in teaching and learning.

2 Literature Review

2.1 Learning Disability

Dyslexia is a type of LD particularly in reading. According to Nayton, Hettrich, Samar, & Wilkinson (2017), Dyslexia alludes to a group of symptoms', which result in individuals experiencing issues with particular language abilities, especially reading. Children with Dyslexia usually encounter troubles with other language aptitudes, for example, spelling, writing, and professing words. British Dyslexia Association claims that Dyslexia

* Corresponding author: fiqazureen@gmail.com

frequently happens in association with the other formative disorder, such as Dyscalculia and Dysgraphia.

According to Drigas, Pappas, & Lytras (2016), children with Dyscalculia often have difficulties in number processing and basic math operation such as addition, subtraction, multiplication and division. Dysgraphia is a learning disability that related to writing difficulties. Dysgraphia was initially incorporated into the depiction of Dyslexia however is presently separated as a component inside the wide range of writing issue, alluding particularly to spelling, and indecipherable handwriting (Fletcher-Flinn, 2016).

Cortiella & Horowitz (2014) estimated that 2.4 million of American public schools' students were identified with LD. 66% of students identified with LD are male. Meanwhile, the percentage of female students with LD is about 51% (Cortiella & Horowitz, 2014). The male students are more likely to be diagnosed with the LD as compared to the female students.

2.2 Dyscalculia

Dyscalculia is induced at first as a persevering trouble in the learning or comprehension of number such as knowing fundamental number names, segregating the bigger or smaller number, counting number sequence, and troubles in basic math operation (Plerou, 2014). Research by Renwick (2014) found that possible indicators of Dyscalculia are numeric issues, spatial issues, memory issues and measure issue. Table 1 shows the early signs of Dyscalculia by Plerou (2014).

Table 1. Signs of Dyscalculia.

Signs of Dyscalculia	
Difficulties in basic number counting	Unable to memorize the number sequence
Difficulties in determine which objects are bigger or smaller	Difficulties in objects counting
Difficulties in connecting the number concept to the symbolic form (example: 5 = five)	Difficulties in number comparison which numbers are bigger or smaller
Poor in sense direction such as right or left.	Difficulties in clock reading.

According to Morin (2014) the children with Dyscalculia show different signs at different age and Dyscalculia tend to be more visible as children get older. However, it can be recognized as early as preschool. Dyscalculia can be detected at an early age by difficulties noticed in calculating the quantity of objects, in counting and in the recognition of the mathematical operations symbol (Plerou, 2014). British Dyslexia Association claims that Dyscalculia frequently happens in association with the other formative disorder, such as Dyslexia and Dysgraphia.

2.3 Mobile Learning & Dyscalculia

Mobile learning might be favoured by children, as it can possibly provide an extensive variety of function and segments which offer an exceptionally intelligent learning background. (Skiada et al.,2014). Dyscalculia children should utilize the mobile learning to make learning easier and easy to adapt this method of learning (Mulligan, 2012). Mobile apps have potential in education due to the graphic and sound that can help them to stimulate their brain to learn math effectively and grab the concept quickly (Nagavalli & Juliet, 2015).

A study conducted by Mohd Ariffin, Abd Halim & Abd Aziz (2017) reported that learning through mobile app show positive improvement for Dyscalculia children in math learning.

2.4 Initial Calculic Model

A study conducted by Abd Halim, K. Sugathan & Mohd Ariffin (in press), illustrated the initial Calculic Model that comes from the literature review. They have extended the components in the instructional factor by adding one new component which is language. Vukovic & Lesaux (2013) supported that language has a relationship in learning and teaching Maths and language also relate to Dyscalculia children mathematical performance. Language is important for the development of mathematical proficiency (Vukovic & Lesaux, 2013). Figure 1 shows initial Calculic Model.

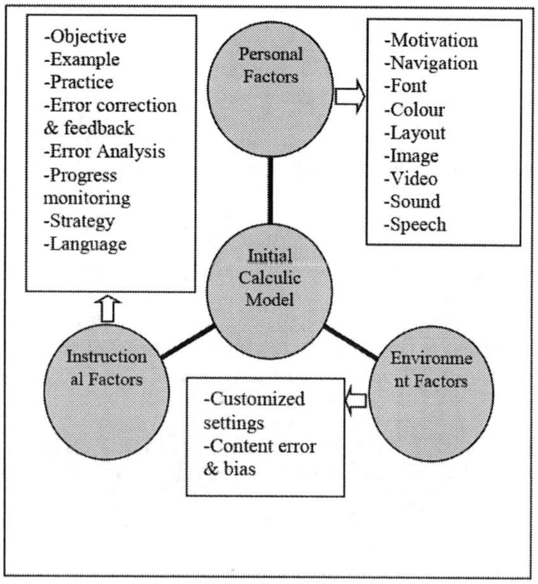

Figure 1. Initial Calculic Model.

The initial Calculic model has three 3 factors which are instructional factor, personal factor and environment factor. There are 8 components under the instructional strategic factor which are *'objective'*, *'strategy'*, *'example'*, *'practice'*, *'error correction and feedback'*, *'error analysis'*, *'progress monitoring'*, and *'language'*. Meanwhile, there are 9 components that fall in personal factor which are *'motivation'*, *'navigation'*, *'font'*, *'color'*, *'layout'*, *'image'*, *'video'*, *'sound'* and *'speech'*. There are 2 components under the environment factor which are *'customized settings'* and *'content error and bias'*.

Besides that, the presentation style is very important because inappropriate presentation layout can be confusing particularly when the text is written too close (Chinn & Ashcroft, 2017). According to Chinn & Ashcroft (2017), most of the mobile apps use a crowded layout and it is very difficult for Dyscalculia children to read the information in the correct order. Thus, this kind of presentation can distract Dyscalculia children. Drigas & Kokkalia (2016); Weng & Taber-Doughty (2015), proposed to use clear, simple and consistent layout design so that Dyscalculia children can navigate the app easier. The layout design should avoid cramping material, use short text and numbering or bullet point (Aziz, Husni & Jamaludin, 2013). Hence, Calculic Model have included the *'layout'* component to support learning for Dyscalculia children.

3. Methodology

As a preliminary study, an interview session was conducted with Dyscalculia practitioners that involved a teacher, paediatrician and educational psychology in order to support the components that come from literature review. The interview session was aimed to review the initial Calculic model that come from literature review. The interview session was conducted to gather information about Dyscalculia children, the appropriate method to teach them, and how to design an effective mobile app for Dyscalculia children.

3.1 Participants

Three Dyscalculia practitioners were recruited. The number of Dyscalculia practitioner in Malaysia is very limited. All the participants have more than 5 years' of experiences in handling Dyscalculia children. The details of the participants are shown in Table 2.

Table 2. Details of Participants.

Participant ID	Profession	Years of Experience
P1	Educational psychology	17 years
P2	Pediatrician	15 years
P3	Teacher	7 years

Hence, the role of P1 is to assess the difficulty of Dyscalculia children in learning and suggested the methods or strategies that can help Dyscalculia children in learning based on their cognitive style and behavioural. The main role of P2 is to do physical examinations to the children and diagnose the children with Dyscalculia. Thus, the main role of P3 is to teach Dyscalculia children particularly in maths. All the participants have different role in their profession.

3.2 Instruments

The author conducted face to face semi structured interview with P2 and P3. Meanwhile, the interview session with P1 was conducted via telephone. Each interview took between 45 minutes to 80 minutes. Participants were asked about: (1) their experience in handling Dyscalculia children; (2) to describe how they teach the Dyscalculia children and (3) what components or features that suitable for Dyscalculia children in learning. The main questions in the interview guide comprise:

1. How was your experience in handling Dyscalculia children?
2. How to teach Dyscalculia children in math learning?
3. What are the potential components that should have if you want to design mobile app for Dyscalculia children for math learning?

All of the components were probed with more specific questions and in depth discussion during interviews to understand which components are important for Dyscalculia children during learning through mobile app and how to design for each component.

3.3 Data Analysis

The data from the interview session were transcribed verbatim with consent. The data were analyzed to find which components are most important and which components are least important.

4 Results & Discussion

Dyscalculia children are different from normal children. As mentioned earlier, they have troubles in understanding numerical concept. Thus, they have different style of learning and it is crucial to identify the appropriate teaching strategy for Dyscalculia children.

4.1 Participants Perspective on Dyscalculia

Based on the interview session with Dyscalculia practitioners that involved educational psychology, paediatrician and a teacher, they found out that Dyscalculia children are special need children that have different learning style compared to the normal children. From the interview session, all the participants talked about multisensory approach. They agreed that multisensory approach is one of the teaching strategies for Dyscalculia children. The multisensory learning involved auditory, visual, tactile and kinaesthetic. The multisensory learning strategy can be applied in mobile app. Moreover, the multisensory approach is proven to be an effective teaching approach for Dyscalculia children (Duijzer, Shayan, Bakker, Van der Schaaf, & Abrahamson, 2017).

Based on the interview, Dyscalculia children can be cured if the teaching strategy is suitable for them in math learning. According to P3, the teaching material should be appropriate for Dyscalculia children because they require an appropriate learning material that can empower their motivation in learning. For example, the use of graphics and sounds in the mobile app can motivate them in math learning.

According to P1, the cognition style and the behavioural of Dyscalculia children are the important aspects that need to be identified before developing any learning material or tools. Usually the Dyscalculia children have major difficulties in memory and numerical processing.

According to P2, parents should be aware of the signs shown by their children at the early age because LD can be diagnosed as early as 6 years old. However, in Malaysia, most of the children are diagnosed at 7 years old. Hence, the intervention steps at the early age are crucial because the intervention is likely to be more effective.

4.2 Analysis of the Interview

From the interview session, preliminary analyses were performed to the collected data. Table 3 shows the analysis of the interview.

Table 3. Analysis from Interview

No	Components	P1	P2	P3
1	Motivation	/	x	/
2	Navigation	/	/	/
3	Font	/	/	/
4	Colour	/	/	/
5	Layout	/	/	/
6	Images	/	/	/
7	Video	/	/	/
8	Sound	/	/	/
9	Speech	/	/	/
10	Objective	/	/	/
11	Examples	/	/	/
12	Practice	/	/	/
13	Error correction & feedback	/	/	/
14	Error analysis	/	x	x
15	Progress monitoring	/	/	/
16	Strategy	x	x	x
17	Language	/	/	/
18	Customized setting	x	/	x
19	Content error & bias	/	/	/

Based on the initial analysis shown in Table 3, there are 3 components that need to be eliminated from the proposed model which are 'error analysis', 'strategy' and 'customized setting'. The components were eliminated based on the majority from the participants.

4.2.1 Error Analysis

Two participants (P2 and P3), both agreed that *'error analysis'* is not necessary to include in the model.
P2 said:
"It is not necessary to include error analysis in the mobile app. I think it is not important to keep record of the children error because this will make them demotivated"

Meanwhile, P3 said:
"It is not necessary because if they get incorrect answer they can repeat the same question until they get correct answer. They can do the same question in order to make them more understand"

4.2.2 Strategy

All the participants gave opinion that 'strategy' component can be eliminated from the initial Calculic Model. P1 said:
"The strategy can be used in examples such as show step by step example, break down small part, show step by step"

P2 said:
"Not necessary to include strategy because mobile app should know the appropriate strategy to be used to teach them. I think you can use a multisensory strategy in example and practice"

Meanwhile P3 said:
"The strategy should be include in the practice component such as give them hint for correct answer"

4.2.3 Customized Settings

Two participants (P1 and P3), gave opinion that 'customized setting' component is not necessary to include in the model. P1 said:
"It is not necessary. The target user is Dyscalculia children so just design specific to them"

Meanwhile P3 said:
"It is not necessary because the design of mobile app should be appropriate so that they can learn effectively without need to change anything"

4.3 Proposed Calculic Model

The Figure 2 shows the proposed Calculic Model after the elimination of 3 components.

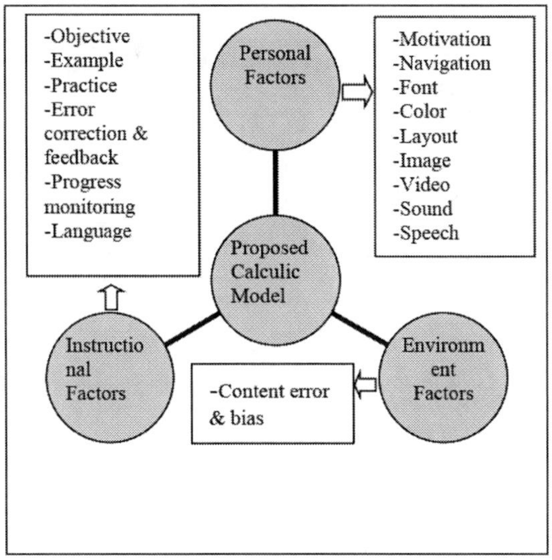

Figure 2. Proposed Calculic Model

Three components have been eliminated from the initial Calculic model. The proposed Calculic model has three 3 factors which are instructional factor, personal factor and environment factor. There are 6 components under the instructional strategic factor which are *'objective', 'example', 'practice', 'error correction and feedback', 'progress monitoring'*, and *'language'*. Meanwhile, there are 9 components that fall under personal factor which are *'motivation', 'navigation', 'font', 'color', 'layout', 'image', 'video', 'sound'* and *'speech'*. There are 1 components under the environment factor which is *'content error and bias'*.

Based on Figure 2, there are 16 components that are important for Dyscalculia children in learning. For example, 'font' component in personal factor is important because the font can influent their reading performance and word understanding (Rello & Baeza-Yates, 2013). Moreover, from the interview session, all the participants agreed that the font component is important. P1 said:
"The font is important and the font chosen should be suitable with Dyscalculia children and also readable"

While P2 said:
"The font is important component that should be include in the mobile app. I think the suitable font is Arial type font"

P3 said:
"The font is very important components that mobile app should include because the font can affect their reading. The suitable font should be non-serif font"

Moreover, Rello & Baeza-Yates (2013), reported that the san serif font and roman style lead to better readability to LD children. Research by Brittany & Ross, (2015), supported that the san serif category font is suitable for Dyscalculia children because the font is readable for Dyscalculia children. Thus, this research aimed to include font as one of the components to support learning for Dyscalculia children.

Besides, the 'practice' component in instructional factor is also important. Thus, practice should provide clear and consistent guidance for Dyscalculia children because they need sufficient practice to enhance deeper learning (OK & Bryant, 2015). From the interview session, all the participants agreed that practice is important that should be included in the model. P1 said:
"Practice is important in learning process. Practice should have a level of difficulty"

P2 said:
"The practice should be include in the mobile app so that they can try the practice after they have learn through example"

Meanwhile P3 said:
"The practice is important because it is the part of learning process. Provide them a guidance in doing the practice"

In addition, a study conducted by Mohd Ariffin, Abd Halim & Abd Aziz (2017) reported that guidance elements such as providing hint through font colour in the practice can encourage them to choose the correct answer. Hence, this research aimed to include 'practice' as one of the components to support learning for Dyscalculia children.

Next is the *'content error and bias'* component in environment factor. According to Bryant, Bryant, & Pfannenstiel (2015) leaning material is prohibited for content error because incorrect example or misstated formula can lead to confusion. The effective content of mobile app for Dyscalculia children is free from error, up-to-date and free from bias (Ok, Kim, et al., 2016). From the interview session, P1 said:
"Any learning material should free from bias and error. That's the good feature of any learning material"

P2 said:
"Content in the mobile app should be free from error and bias because they need an appropriate learning material in learning"

Meanwhile, P3 said:
"It is important that the mobile app should be free from error and bias to avoid the confusion"

The effective content of mobile app for Dyscalculia children is free from error, up-to-date and free from bias (Ok, Kim, et al., 2016). Thus, this research aimed to include content error and bias as one of the components to support learning for Dyscalculia children.

5. Conclusion & Future Work

This paper presented the development of the mobile app design model called Calculic Model. As mentioned earlier, the Calculic Model is for mobile app designers and developers to design and develop an appropriate mobile app for Dyscalculia children in Malaysia. An interview session was conducted with Dyscalculia practitioners that involved a teacher, paediatrician and an educational psychology to review the components of initial Calculic Model. Initially there are 19 components in Calculic Model. As a result, there were 16 components included in the proposed Calculic Model. Hence, 3 components were eliminated which are *'error analysis'*, *'strategy'* and *'customized setting'*. For the future work, the proposed Calculic Model will be verified by mobile app designers and developers.

This research work is currently undertaken by Fiqa Azureen Abd Halim, Msc student in Information Technology at Universiti Teknologi Petronas.

References

1. Abd Halim, F. A., K Sugathan, S., & Mohd Ariffin, M. (in press). Towards a Mobile App Design Model for Dyscalculia Children in Malaysia. In e-Learning, e-Management and e-Services (IC3e), 2017 IEEE Conference. IEEE.
2. Aziz, F. A., Husni, H., & Jamaludin, Z. (2013). Translating interaction design guidelines for dyslexic children's reading application. In Proceedings of the World Congress on Engineering (Vol. 2).
3. Butterworth, B. (2013). Handbook of Mathematical Cognition: Chapter 26: Developmental Dyscalculia. Mathematical Cognition, 455-468.
4. Bryant, D. P., Bryant, B. R., & Pfannenstiel, K. H. (2015). MathematicsInterventions : Translating Research Into Practice Mathematics Interventions : Translating Research Into Practice, (November). https://doi.org/10.1177/1053451214560893
5. Chinn, S. J., & Ashcroft, R. E. (2017). Mathematics for dyslexics and dyscalculics: a teaching handbook. Chichester, West Sussex: Wiley Blackwell.
6. Cortiella C. & Horowitz, S. H. The State of Learning Disabilities: Facts, Trends and Emerging Issues. New York: National Center for Learning Disabilities, 2014.
7. Cumming, T. M., Strnadová, I., & Singh, S. (2014). IPads as instructional tools to enhance learning opportunities for students with developmental disabilities: An action research project. Action Research, 12(2), 151-176. doi:10.1177/1476750314525480
8. Duijzer, A. C. G. C., Shayan, S., Bakker, A., Van der Schaaf, M. F., & Abrahamson, D. (2017). Touchscreen Tablets: Coordinating Action and Perception for Mathematical Cognition, 8(February), 1–19. https://doi.org/10.3389/fpsyg.2017.00144
9. Draper Rodriguez, C., Strnadova, I., & Cumming, T. (2014). Using iPads With Students With Disabilities: Lessons Learned from Students, Teachers, and Parents. Intervention in School and Clinic, 49(4), 244–250. https://doi.org/10.1177/1053451213509488
10. Drigas, A., & Kokkalia, G. (2016). Mobile Learning for Special Preschool Education. International Journal of Interactive Mobile Technologies (iJIM), 10(1), 67. doi:10.3991/ijim.v10i1.5288
11. Drigas, A., Pappas, M. ., & Lytras, M. (2016). Emerging Technologies for ICT based Education for Dyscalculia: Implications for Computer Engineering Education. International Journal of Engineering Education, 32(4), 1604–1610. Retrieved from http://search.ebscohost.com/login.aspx?direct=true&AuthType=ip,shib&db=bri&AN=117610015&site=ehost-live&scope=site
12. Fletcher-Flinn , C. M. (2016). Developmental Dysgraphia as a Reading System and Transfer Problem: A Case Study. Front. Psycho, 7(149). doi:10.3389/fpsyg.2016.00149
13. Mazeyanti Mohd Ariffin, Fiqa Azureen Abd Halim, & Norshakirah Abd Aziz. (2017). Mobile application for dyscalculia children in Malaysia in Zulikha, J. & N. H. Zakaria (Eds.), Proceedings of the 6th International Conference on Computing & Informatics (pp 467-472). Sintok: School of Computing.
14. Morin, A. (2014, March 10). Understanding Dyscalculia. Retrieved February 21, 2016, from https://www.understood.org/en/learning -attention-issues/child-learning - disabilities/dyscalculia/understanding-dyscalculia
15. Nagavalli, T., & Juliet, P. (2015). Technology For Dyscalculic Children. SALEM, 16, 1-10. Retrieved January 16, 2017
16. Nayton, M., Hettrich, E. L., Samar, S., & Wilkinson, C. (2017). Evidence-Based Assessment and Interventions for Problems with Reading in School Psychology. In Handbook of Australian School Psychology (pp. 173-195). Springer International Publishing
17. OK, M. W., & Bryant, D. P. (2015). Effects of a Strategic Intervention With iPad Practice on the Multiplication Fact Performance of ... Effects of a Strategic Intervention With iPad Practice on the Multiplication Fact Performance of Fifth-Grade Students With Learning Disabilities, (August). https://doi.org/10.1177/0731948715598285
18. Ok, M. W., Kim, M. K., Kang, E. Y., & Bryant, B. R. (2016). How to Find Good Apps: An Evaluation Rubric for Instructional Apps for Teaching Students With Learning Disabilities. Intervention in School and Clinic, 1053451215589179-. https://doi.org/10.1177/1053451215589179
19. Papadakis, S., Kalogiannakis, M., & Zaranis, N. (2017). Designing and creating an educational app rubric for preschool teachers. Education and Information Technologies. doi:10.1007/s10639-017-9579-0
20. Plerou, A. (2014). Dealing With Dyscalculia Over Time. Icicte, (2008), 1–12. https://doi.org/10.13140/2.1.4229.5681
21. Rello, L., & Baeza-Yates, R. (2013). Good fonts for dyslexia. ASSETS 2013: Proceedings of the 15th International ACM SIGACCESS Conference on Computers and Accessibility, 1–8. https://doi.org/10.1145/2513383.2513447
22. Skiada, R., Soroniati, E., Gardeli, A., & Zissis, D. (2014). EasyLexia: A Mobile Application for Children with Learning Difficulties. Procedia Computer Science, 27(Dsai 2013), 218–228. https://doi.org/10.1016/j.procs.2014.02.025
23. Vukovic, R. K., & Lesaux, N. K. (2013). The language of mathematics: Investigating the ways language counts for children's mathematical development. Journal of Experimental Child Psychology, 115(2), 227-244. doi:10.1016/j.jecp.2013.02.002
24. Weng, P., & Taber-Doughty, T. (2015). Developing an App Evaluation Rubric for Practitioners in Special Education. Journal of Special Education Technology, 30(1), 43-58. doi:10.1177/016264341503000104

A Review of Continuing Professional Development (CPD) of Training Competencies for Malaysian Mechanical Industries

Zeti Kasman[1,*], *Affero* Ismail[2], *Nadzri* Siron[1] and *Noorazman* Abd Samad[2],

[1]Manpower Department, Ministry of Human Resources, 62530 W.P Putrajaya, Malaysia
[2]Faculty of Technical and Vocational Education, Universiti Tun Hussein Onn Malaysia, 86400 Parit Raja Batu Pahat Johor, Malaysia

Abstract. The aim of this paper is to review issues in continuing professional development of vocational training to the mechanical industries. The spectacle of individuals entering the labour market without relevant qualifications is common in Malaysia. There are many who choose to work instead of pursuing further education after secondary school. In the labour market, these individuals are considered to be low-skilled workers because they had no training prior to employment. The role of employers in providing training and education to employees is vital in establishing career development of employees. Employers who contributed to their employees' training funds through Human Resources Development Council would provide opportunities to increase the skills of workers. Based on the Malaysia's Development Plan of Occupational Skill, the issues and challenges that have been identified in producing skilled workers in interpersonal and technical skills. This paper provided an opportunity to examine an enterprise-based approach to skill formation for workers with basic academic qualifications. It presents an alternative scenario to institution-based Technical and Vocational Education and Training (TVET), which many Malaysians are familiar with. A structured curriculum development of human resources learning needs according to the job profile of the typical individual and group work which will provide a clearer perspective on knowledge, competence and skill levels of employee behavior in performing tasks will be discussed. The biggest impact on this study is to produce high skill employees concerning customer satisfaction and increased organizational productivity towards high income nations.

1 Human Resources Development and Education

The development of skilled manpower is extremely needed and it is parallel to the industries demand. The review of the contemporary issues in Higher Education (HE) and Vocational Education and Training (VET) from a global perspective and identify principles and practices. Universities have evolved and expanded their identity by venturing into various concepts and style of administration, student attachment, funding and delivery. Considering the fact that this research explores the perspective of academics, this research will also consider the cultural environment of academics within the technical institute framework and industries.

The paper will also explore the position of mechanical technicians within the technical vocational training system and how they have evolved in relation to institutions. It is argued that the diversification of technical institution experienced by other countries would allow Malaysian higher education administrators to venture into a wider educational horizon mostly collaboration between industrial worker and technical instructor.

The discipline of HRD and adult education both consider the process of adult learning as being central to their theory and practice. Even so, the purpose of HRD and Adult Education differ, and their perspective on adult learning differs too [1]. The core differences is related to control of goals and purposes for which adult learning is employed – organizational versus individual control [2]. HRD professionals are in general agreement as to their goals. Most take the position that HRD should focus on increasing the performance requirements of its host organizations through the development of the organisation's work force [1, 3, 4]. Others believe HRD should focus on individual development and personal fulfilment without using organizational performance as the measure of worth [5]. It is the increase in performance resulting from HRD that justifies its existence. Consequently, it is the responsibility of HRD to focus on those goals as well as the goal of the individual employee.

If a nation can be perceived as a large organization that has a clear goal and objective, then it is possible to consider manpower planning that is designed to achieve the nation's goal and objective, then the whole adult learning process can be regarded as HRD. In this respect,

* Corresponding author: zeti.kasman@gmail.com

the Malaysian government has established a clear goal in its Vision 2020 announced by the Prime Minister in 1991. Thus the overall education and training can be categorised within the context of human resources development. As such, the Prime Minister's department outlined the concerted effort of the various organizations and ministries. This is explained in Figure 1:

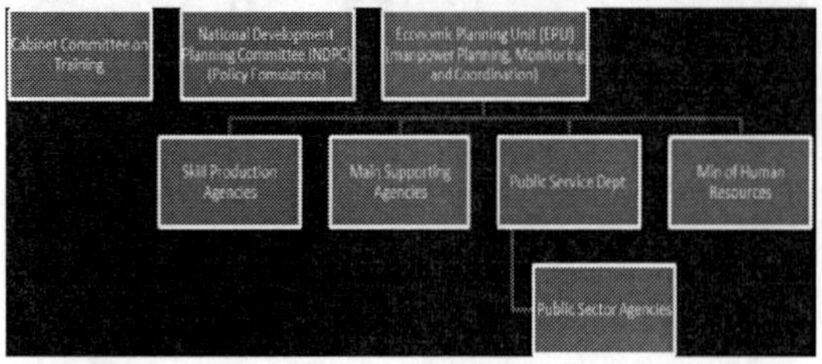

Figure 1: Human Resources Planning Machinery

There are two major organizations that share responsibilities in the development of human resources in the country. The Ministry of Education has full responsibilities on the education issues and encompasses the whole spectrum of education matters that include adult education particularly at universities and polytechnics. The Ministry of Human Resources (MoHR), together with few other ministries, is also actively involved in providing education and training but with more focus for adult VET. Therefore, this research wishes to explore the concern of 'power struggle' in educating and training, and try to examine the perception of academics about the matter.

1.1 Work-based learning in TVET

The term "work-based learning" is widely used in the literature. A general definition of work-based learning is "learning in the workplace, derived from work undertaken for or by an employer (i.e. in paid or unpaid work). It involves the gaining of competencies and knowledge in the workplace [6]. Other definitions of work-based learning are based on the context in which the term is used. For example, in higher education literature, especially from the UK, work-based learning refers to an approach for gaining academic qualifications through university–industry collaboration in which the university develops an academic program that is designed to meet the needs of a specific industry or organization [7- 9].

Work-based learning may also refer to the blending of workplace experiences with traditional classroom-based educational programs. In this context, work-based learning may exist in various forms, such as practicums, internships, and service training [10]. At the university level, work-based learning is compulsory for certain degree programs, especially those leading to professions in medicine, law, and architecture. It is less common at the school level, except in the US, where work-based learning is an intervention for low-performing high schools [11].

In organizational learning literature, work-based learning is seen as important for bridging knowledge and action in the workplace [12]. Members of organizations need to learn so they can keep up with the pace of change in the modern world. As skills can quickly become obsolete, new skills need to be learned quickly, and the natural way to do so is to learn at work. The whole notion of skills as a set of technical abilities to perform a job has itself become obsolete. Replacing skills is the new idea of learning, which will be commingled with the notion of work itself. Those who support learning at work argue that it is important for enhancing productivity, competitiveness, and innovation [13, 14].

Integrating work-based learning into traditional classroom-based educational programs helps bridge the gap between universities and workplaces [15]. Formal education is seen as too theoretical; thus, gaining practical experience would help students in many ways. Supporters of work-based learning argue that it has a positive effect on learners' attitudes, aspirations, and performance [16]. There is evidence that work-based learning increases the number of employment opportunities for graduates [17, 18] and improves the starting salaries for high performing interns [19]. In addition to improving employability, work-based learning improves academic performance, leading to, for example, better grade point averages and honours [20, 21].

In TVET, work-based learning takes many forms, and its implementation varies from country to country [22]. At one end of the spectrum, there are informal apprenticeships, in which young individuals learn trades from more experienced craftsmen. This type of learning is common in less developed countries. As Sweet pointed out, these apprenticeships are unregulated and are not linked in any way to a formal TVET qualification system, but they are still an important source of skills and provide opportunities for young people to develop specialized skills.

At the other end of the spectrum are regulated frameworks, such as the German dual-training system. Here, work-based learning is linked to a TVET qualification system; apprentices who successfully go through the process will gain a recognized qualification.

Between these two extremes are several variations of work-based learning within the TVET system. In the Malaysian TVET system, for example, which is largely institution-based, there is a compulsory portion of training in which students are required to learn within an actual workplace environment. For example, programs providing a Diploma Kemahiran Malaysia (Malaysian Skill Diploma) require that students undergo three months of industrial training in addition to institution-based training.

Formal apprenticeships are also available. In Malaysia, two agencies run national-level apprenticeship schemes: Pembangunan Sumber Manusia Berhad (PSMB) and the Department of Skill Development (DSD). Both are agencies within the Ministry of Human Resources. PSMB's apprenticeship scheme has been in action for years, but its scope is limited because it covers only mechatronics, the hotel industry, woodworking, plastics injection molding, tool and die making, industrial sewing, and multimodal transport. The number of apprentices is also small. In 2014, PSMB approved 865 apprentices, which cost the government about RM 8 million, or roughly RM 9,250 per apprentice (Ministry of Human Resources, 2015). The DSD runs a much larger apprenticeship program, called the National Dual Training System (NDTS), which is modelled after the German dual system. In 2014, there were 11,271 apprentices at 1,083 companies throughout the country.

2 Vocational Education Training at the Work Place

Liberal education, conducting technical and vocational education is normally capital intensive and more expensive to maintain. VET institutions need to have machinery and equipment to begin with, a continuously require a constant supply of materials to be used in training. It was quite difficult to encourage their private investors to establish technical and vocational institutions.

In order to implement the strategy, Human Resources Development Council (HRDC) was established and they were tasked to manage of Human Resources Development Funds (HRDF). Briefly, an Act of the Malaysian Parliament has entrusted HRDC to collect a levy from Malaysian employers depending on the number of employees they employed. This levy was to be used by the contributing companies for the purpose of training [23]. Basically the contributing companies are not losing their money, but it is kept in the fund for their own benefit.

Since then, the number of companies mounting training programs has increased. Big corporation such as Proton Car Manufacturer, Motorola, Sony, Hitachi, Malaysia Airline and many other companies have embarked on establishing their own training divisions. Therefore, the work place has been actively involved in providing education and training, at least for their own employees.

2.1. Improvement of Vocational Training System to Keep Meets with the Needs of Industries

Malaysia has achieved very rapid development centred on manufacturing industries. Therefore, manpower needs have been on the increase and the quality of manpower has been changed as well. The government of Malaysia has focused on developing human resources of industries, because advancement of industries being capable of meeting international competition and development human resource of industries are prioritized in the Ninth Malaysia Plan (2006-2010), Tenth Malaysia Plan (2010-2015), and Third Industrial Master Plan (2006-2020). The government of Japan has supported CIAST (Center for Instructor and Advanced Skill Training) and JMTI (Japan-Malaysia Technical Institute), but human resources policy officials felt that ILJTM (Vocational training institutions under JTM) and others vocational training providers have been left behind the needs of industry and some policy intervention should be made for this matter. In order to cope with this issue, JICA (Japan International Cooperation Agency) dispatched a Japanese short-term expert of Labor Policy Advisor to MoHR (Ministry of Human Resource Development) in 2006. The expert conducted survey for one month in November 2006, for 2 months in January and April 2007 respectively. [24-26]

Through the discussion with officers and training in Japan, officials of MoHR understood the following recognition how Japanese vocational training institutes catching up the needs of industries.

i. High priority on the curriculum which enables graduates to obtain knowledge, skills and attitude expected by employees.
ii. The collaboration between education/ training institutes and industries is very strong.
iii. The collaboration between education/ training institutes and community is very strong.
iv. Involvement of ex-graduates in student's career guidance.
v. Strong collaboration with secondary schools where their intakes come from.

MoHR officials drew a concept image showing the relation between vocational training institutes and industry, parents, teachers and ex-graduates as shown in the Figure 1 above. They named the concept of these collaborations as Community and Employment Support Services (CESS). The CESS concept was utilized to analyse the various activities.

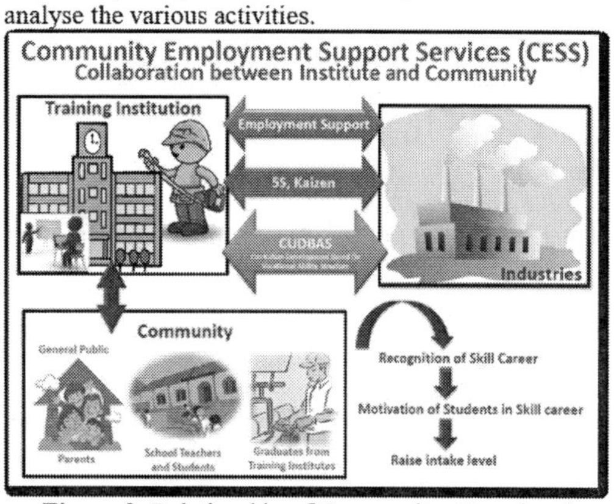

Figure 2: Relationship of Community, Vocational Training Institutes and Industries

Varieties of activities were identified such as employment support services, industry support, and management support as 5S/KAIZEN. 5S/KAIZEN has another role to improve the management of Vocational Training Institutes and at the same time, students are expected to learn practical implementation of 5S activities to develop their awareness on human capital improvement. Employment supports consist of technique on produce a marketable student by designing employment activities such in terms on package the student before sent them to the industries [27].

2.2 The Skill Training Fund Division

The Skills Development Fund Division was established within the structure of the Human Resources Ministry in the year 2000. The setting up of the SDF was approved by Parliament on 14 December 2000 as a Trust Fund under Section 10 of the Financial Procedure Act 1957. The fund is specifically used for the purpose of providing loans for skills training. This division was entrusted to disburse loans from the Skill Development Funds (SDF) to student pursuing technical and vocational courses in public and private training institutions. This fund took effect from 2nd January 2001.
The loans will be given to Malaysian full time and part time students in public and private vocational and technical institutions. The maximum loan available is RM5000.00 per annum and will be for the duration of the accredited course. An administrative charge of 4% interest per annum will be levied on the loan which is payable over maximum period of fifteen years. Repayment of the fund has to commence within six month after the completion of the course or when the borrower starts working, whichever comes first. The Skills Development Fund was modelled along the conventional student loan in many other developed nations.

2.3 Human Resources Development Council (HRDC)

The Human Resources Development Council (HRDC) was established to encourage greater participation of industry to implement staff training. HRDC was given the task of collecting levy from companies, and the contributing companies used this fund for staff training. The HRDC was established under the Human Resources Development Act 1992. In 2001, the HRDC was corporatized and change its identity to the Human Resources Development Limited (Pembangunan Sumber Manusia Berhad). While continuing the same task, the PSMB is now focusing on skills upgrading of the workforce, to equip them with the latest knowledge and specializing in up-to date skills in manufacturing and service sector. The functions carried out by PSMB include assessing and determining the type and extent of employee training. With regard to financial assistance, PSMB also serves to determine the terms and conditions under which financial assistance or grant is to be given. [23]

3 Continuing Professional Development (CPD)

Continuing professional development (CPD), when provided formally, is something that is easy to recognize but perhaps rather more difficult to define. Theoretical and empirical argument surrounds the scope and understanding of the concept. Definition is made more subtle by the different conceptualisations in use but the common denominator is that CPD concerns practices aimed at employees' development beyond that derived from their initial training.

Employees are the prime movers or the main vehicles in producing the organization's objectives whom are not only skilful and competence in hands-on, but also academically. In facing the globalization challenges, it is the responsibilities of the employer to give a proper and effective input so that the outputs will be productive an in parallel to the industry's needs.

The effort to enrich the employee CPD in their skill, knowledge and attitude should be monitored by the employee. The achievement to higher level of performance will strengthen the level of confident for the employee to undergo the skill training.

Wider opportunities must be given to all employees to exhibit their performance and competencies so that they will be proofed to all kinds' of situations they might encounter and working comfortably. In addition, current policies and practices in CPD are frequently based on assumptions about learning and practice that are in urgent need of more empirical research [28].

Effective participation in contemporary, technology-based, knowledge society implies an increasing importance for voluntary learning and development by employees [29, 30].

Traditionally, further professional development was focused on education and training, in more or less formal classroom-based settings. Although problems regarding the transfer to the workplace of skills learned during training have been recognized [31], training is still assumed to be highly important for organizations [32].

Partly in reaction to the problem of transfer, the range of possible learning activities has been broadened over time and currently includes informal learning activities in the workplace as well as formal ones [33-35]. The workplace is a typical place where informal learning can occur, as the workplace context entails a wide range of more or less structured environments, which are only rarely organized with learning in mind [36]. The existence of informal learning opens up new avenues in research into learning [34].

Moreover, life-long learning or CPD is only possible where employees have systematic and valid information about their capabilities, that is to say, if they are able to form accurate self-perceptions, to carefully identify the qualities they need for future career success, and if they are able to adapt their behaviour accordingly [37, 38].

Recent literature on workplace learning emphasizes that both individual learning needs and collective prerequisites for learning should be taken into account simultaneously if the learning is to be successful [39-41]. Consequently, individual motivation and the will to learn or develop oneself alone do not guarantee CPD. Or the other way around, opportunities for development offered by the employer and one's working organization do not necessarily lead to sought outcomes [42].

In addition to formal training activities, work organizations can facilitate employees' CPD by providing a climate that encourages individual development and change and by providing ample opportunities for informal learning [43]. In a similar vein, an organizational structure which allows experience-based learning at work [44].

Defining CPD broadly is not only a question of how to organize training and education for professionals to develop their professional skills and competencies, but also how to identify competent and skilled professionals as part of their working community [45]. Once we have a better understanding of CPD, quantitative approaches using psychometrically sound (reliable and valid) measures can be used to determine CPD practices, determinants and outcomes, in terms of both individual career success and organizational performance.

CPD could be defined as learning which takes place in workplaces and organizations but we enter a caveat here because so much of the research appears to be in the health services and teaching and these professions may not necessarily be an accurate paradigm for professional work more generally. They are often characterized by large, hierarchical organizations in the public sector operating within quite detailed government policy guidelines. Therefore, CPD needs to be more strongly linked with organizational HRD practices and organizational learning as a whole as well as with developments at the level of the profession. Thus, more studies and practical innovations are needed to build on workplace learning pedagogies or even a workplace curriculum as part of organizations' HRD practices and to examine how these relate to developments at the level of the profession.

3.2 The Capability Approach

The economic growth is an insufficient indicator for individuals' quality of life. This is because information on economic growth alone will not help us understand the barriers in our society to achieving equity for all members of society; a better indicator would be "human capabilities." The word "capabilities" refers to what people are able to be and do, rather than how much income or how many material goods they possess.

In the capability-based assessment of justice, individual claims are not to be assessed in terms of the resources or primary goods the persons respectively hold, but the freedoms they actually enjoy to choose the lives that they have reasons to value [46].

The core idea of the Capability Approach is that individuals should have the freedom to do what they value, or in other words, to live a life that they have reasons to value. There are three important concepts in the Capability Approach: capability, functioning, and agency. Capability is one's potential capacity, functioning is one's actual or the realized capacity, and agency is one's personal freedom to decide whether or not to use a particular capability. Individuals need resources in order to convert capability into functioning. These resources include the goods they possess and the services they have access to. Examples of resources are academic qualifications, paid employment, and education and training.

However, the extent to which individuals can generate capabilities from resources is influenced by personal, social, and environmental factors, called conversion factors [47]. Personal conversion factors include one's physical condition, literacy, and intelligence. Social conversion factors include public policies, social norms, discriminating practices, gender roles, societal hierarchies, and power relations. Environmental conversion factors include climate, geographical location, infrastructure, transportation systems, and communication systems.

Given its emphasis on expanding people's freedoms and capabilities to live a life of their own choosing, the capabilities approach offers a promising basis upon which worker wellbeing can be enhanced. The capabilities approach would ask how workers' freedom to achieve wellbeing can be increased, which transcends simply asking how resources may be distributed or how happy or content workers feel [48].

The Capability Approach was only recently suggested as an alternative philosophical or theoretical underpinning for TVET. The view that TVET is linked to productivity and economic development, writing that the current approach to TVET is "grounded in an outmoded model of development". He goes on to suggest several alternative philosophies to TVET, including the Capability Approach, the human rights approach, and the integrated human development approach [49].

Table 2. Valued capabilities at the workplace

Capabilities*	Work capability list**
The capability to be alive, to have physical security, and to be healthy	▪ Work that is safe and healthy in the short and long term ▪ Workplace that is free from harassment and unfair ▪ discrimination
The capability to be knowledgeable, to understand and reason, and to have the skills needed to participate in society	▪ Work that lets you develop your skills and abilities ▪ Work where you receive the training you need to do the job effectively ▪ Work with good communication among the people with whom you work ▪ Work that uses your skills, knowledge and experience ▪ Work hours that let you participate in the community
The capability to enjoy a comfortable standard of living, with independence and security	▪ Work that pays well ▪ Work with good job security ▪ Work where the people you work for treat you with respect

Capabilities*	Work capability list**
The capability to engage in productive and valued activities	▪ Work that is interesting ▪ Work where the people you work with are friendly and helpful ▪ Work that gives you a sense of accomplishment ▪ Work for which you receive recognition ▪ Work where your chances of career advancement are good
The capability to enjoy personal, family and social life	▪ Work that allows you to balance your work, family and personal life
The capability to participate in decision making and, have an influence	▪ Work that allows you freedom to do your job • Work where you can choose your own schedule within established limits ▪ Work that allows you to participate in decision-making ▪ Work that allows you to form and join civil organizations and solidarity groups, including trade unions ▪ Work that allows you to participate in the local community
The capability to be and express yourself, have self-respect, and know you will be protected and treated fairly by the law	▪ Work where the people you work for treat you with respect

Note: * From Vizard and Burchardt (2007)
** From Bryson (2013)

4 Past studies on work-based learning and ccompetencies

At the time this study was carried out, there were no studies utilizing Sen's Capability Approach to examine work-based learning among Malaysian workers. There are, however, several studies on training for industry workers. When discussing training for workers (whether vocational training, on-the-job training, formal training, informal training, etc.), the few articles on work-based training that are available usually emphasized the perspective of management [52, 53]. Studies that present ordinary workers' training stories are lacking.

There are some studies that try to present workers' views on certain issues. For example, Abdul Aziz et al. (2008) studied the views of ex-trainees from an electrical wiring course regarding whether the vocational training they received meets industrial needs [54]. Additionally, Muhamad and Idris studied learners' perspective on workplace learning, covering workers from various sectors in Malaysia. So far, the researcher has found no studies that discuss vocational training from the perspective of capability development for Malaysian workers [55]. It was estimated that about 64% of the Malaysian workforce has only an SPM qualification or lower [56]. This implies that many young Malaysians enter the job market without any post-secondary education or training. They had to acquire vocational knowledge and skills at the workplace. Although the acquisition of vocational skills at the workplace is common, the researcher found no studies on how workers' capabilities are developed at the workplace through vocational training. This particular gap will be addressed through this research.

Internationally, some studies have looked at TVET from a capability perspective of South African students at a vocational college. According to Powell, the education received by students at this vocational college gave them the "capability to aspire," or to dream and make plans about something that was previously unimaginable.

5 Methodology

The methodology which will be adopted in this study is to identify the competencies of the mechanical industries employees'. This stage begins with identifying problems followed by setting objectives of the study. Having all the objectives in place, related literatures were reviewed to build up knowledge and ideas on the topic. Further, the topic was reviewed in view of the objectives set to ensure the data collection study is in line with the objectives set and within the scope of the topic.

Reviews of previous studies were used in supporting to specify the improvement plan according to the issues faced. This paper used a document analysis technique of the important documents from Ministry of Human Resources called Curriculum Development Based on Ability Structure (CUDBAS) methodology. The competency chart, competency map and training development plan will be created according to the specific job position to the industries. The competency checklist will be used during pre and post survey as a data collection. Response from the competency map will be analysing based on the study objectives. A review of job competencies and the landscape of TVET in Malaysia mainly aimed that contribute to the improvement of mechanical industries employees'. In this study, data analysis planned to be performed concurrently with data collection to examine the data.

5 Conclusions

This study is concerned with the development of TVET in Malaysia for it has significant contribution in nation building. It is a fact, that TVET is within the subset of education and its development has to be in relation to its surrounding within educational sector. Though these study, it began with hypothesis that HRD to participate in VET development. The shortage of quality workforce experienced by Malaysian economy during the economic booming in late 80s and early 90s clearly provide lesson to policy makers that VET is obviously in need for better attention for development. Collaboration and means has to be explored, to capitalise these strengths and improve the weaknesses.

Trends that are relevant to the issues addressed by this study in particular regarding the issue of CPD of training competencies for Malaysian Mechanical Industries, human resources development, vocational education training at workplace and capability approach. All these issues were closely related to develop the career path in mechanical industries. In this respect, there has been clear diversification and changes in up to date industries and technologies in term of their concept of establishment in order to maintain a continuous response to the external environment. The recent trend on the competencies development frameworks is considered as effective mechanism to bring closer relationship or connection between the resources and industries.

References

1. Swanson, R.A., Human Resource Development: Performance Is The Key. 1995.
2. Knowles, M., Holton, E. F., & Swanson, R. A., The adult learner: The definitive classic in adult education and human resource development. 1998.
3. Knowles, M., The adult learner. A neglected species. 4th ed. 1990.
4. McLagan, P., Models for HRD practice. 1989.
5. Dirkx, J.M., Transformative Learning Theory in the Practice of Adult Education: An Overview. 1996.
6. Connor, H. (2005). Work-based learning: a consultation. London: Council for Industry and Higher Education.
7. Foster, E., & Stephenson, J. (1998). Work-based learning and universities in the U.K.: a review of current practice and trends. Higher Education Research & Development, 17(2), 155-170.
8. Garnett, J. (2001). Work based learning and the intellectual capital of universities and employers. The Learning Organization, 8(2), 78-82.
9. Courtney, J., Long, J., & Hall, D. (2013). Work-based learning partnerships: a case of learning and talent management intervention for developing future leaders and managers. HR Bulletin: Research and Practice, 8 (2). 13-18.
10. Sattler, P., & Peters, J. (2013). Work-integrated learning in Ontario's postsecondary sector: The experience of Ontario graduates. Toronto: Higher Education Quality Council of Ontario.
11. Herlihy, C.M., & Quint, J. (2006). Emerging evidence on improving high school student achievement and graduation rates: the effects of four popular improvement programs. Washington: National High School Center. Retrieved December 28, 2015 from http://files.eric.ed.gov/fulltext/ED501076.pdf
12. Raelin, J.A. (2008). Work-based learning: Bridging knowledge and action in the workplace. San Francisco, CA: Jossey-Bass.
13. Fenwick, T. (2003). Innovation: examining workplace learning in new enterprises. Journal of Workplace Learning, 15(3), 123-132.
14. Høyrup, S. (2010). Employee-driven innovation and workplace learning: basic concepts, approaches and themes. Transfer: European Review of Labour and Research, 16(2), 143-154.
15. Hills, J.M., Robertson, G., Walker, R., Adey, M.A., & Nixon, I. (2003). Bridging the gap between degree programme curricula and employability through implementation of work-related learning. Teaching in Higher Education, 8(2), 211-231.
16. Taylor, J.C. (1997). Learning at work in a work-based welfare system: opportunities and obstacles: Lessons from the school-to-work experience. Boston, MA: Jobs for the Future. Retrieved December 25, 2015 from http://www.jff.org/sites/default/files/publications/LrngatWork.pdf
17. Callanan, G., & Benzing, C. (2004). Assessing the role of internships in the career-oriented employment of graduating college students. Education + Training, 46(2), 82-89.
18. Huq, A., & Gilbert, D.H. (2013). Enhancing graduate employability through work-based learning in social entrepreneurship: A case study. Education + Training, 55(6), 550-572.
19. Gault, J., Leach, E., & Duey, M. (2010). Effects of business internships on job marketability: the employers' perspective. Education + Training, 52(1), 76-88.
20. Blair, B.F., & Millea, M. (2004). Student academic performance and compensation: The impact of cooperative education. College Student Journal, 38(4), 643-652.
21. Rawlings, P., White, P., & Stephens, R. (2005). Practice-based learning in information systems: the advantages for students. Journal of Information Systems Education, 16(4), 455-464.
22. Sweet, R. (2013). Work-based learning: Why? How? In K Ananiadou, (Ed.), Revisiting global trends in TVET: reflections on theory and practice (pp. 164-203) Bonn, Germany: UNESCO-UNEVOC. Retrieved June 20, 2014 from http://unevoc.unesco.org/fileadmin/up/2013_epub_revisiting_global_tre nds_in_tvet_book.pdf
23. (HRDC), H.R.D.C. Akta Pembangunan Sumber Manusia 1992. 2013.
24. Plan, T.M. 2010-2015, Percetakan Negara Malaysia.
25. Plan, T.I.M. 2006-2020, Percetakan Negara Malaysia.
26. Human Resource Development Policy. 2007.
27. Manpower Department, M.o.H.R., Community Employment & Support Services – 'Improvement of Vocational Training System to Keep Meet with the Needs of Industries'. 2010.
28. Kilminster, S., Bradburg, H., Frost, N. and Zukas, M. (2012), 'Critical perspectives on professional lifelong learning', Studies in Continuing Education, 34, 1, 1–4.
29. Evers, A. T., Van der Heijden, B. I. J. M., Kreijns, K. and Gerrichhauzen, J. T. G. (2011b), 'Organisational factors and Teachers' Professional

Development in Dutch secondary Schools', Journal of European Industrial Training, 35, 1, 24–44.

30. Maurer, T. J. (2002), 'Employee learning and development orientation: toward an integrative model of involvement in continuous learning', Human Resource Development Review, 1, 1, 9–44.
31. Baldwin, T. T. and Ford, J. K. (1988), 'Transfer of training: a review and directions for future research', Personnel Psychology, 41, 1, 63–105.
32. Smith, E., Smith, A., Pickersgill, R. and Rushbrook, P. (2006), 'Qualifying the workforce. The use of nationally-recognised training in Australian companies', Journal of European Industrial Training, 30, 8, 592–607.
33. Cheetham, G. and Chivers, G. (2001), 'How professionals learn in practice: an investigation of informal learning amongst people working in professions', Journal of European Industrial Training, 25, 5, 248–92.
34. Eraut, M. (2004), 'Informal learning in the workplace', Studies in Continuing Education, 26, 2, 247–73.
35. Marsick, V. J. and Watkins, K. E. (2001), 'Informal and incidental learning', New Directions for Adult and Continuing Education, 89, 25–34.
36. Evers, A. T., Kreijns, K., Van der Heijden, B. I. J. M. and Gerrichhauzen, J. T. G. (2011a), 'An organizational and task perspective model aimed at enhancing Teachers' Professional Development and occupational expertise', Human Resource Development Review, 10, 2, 151–79.
37. Van der Heijde, C. M. and Van der Heijden, B. I. J. M. (2006), 'A competence-based and multidimensional operationalization and measurement of employability', Human Resource Management, 45, 3, 449–76.
38. Van der Heijden, B. I. J. M., De Lange, A. H., Demerouti, E. and Van der Heijde, C. M. (2009a), 'Employability and career success across the life-span. Age effects on the employability-career success relationship', Journal of Vocational Behavior, 74, 156–64.
39. Collin, K. (2006), 'Connecting work and learning – design engineers' learning at work', Journal of Workplace Learning, 18, 7/8, 403–13.
40. Fuller, A. and Unwin, L. (2004), 'Expansive Learning Environments: Integrating Organizational and Personal Development', in H. Rainbird, A. Fuller and A. Munro (eds), Workplace Learning in Context (London: Routledge), pp. 126–44.
41. Hodkinson, P., Biesta, G. and James, D. (2008), 'Understanding learning culturally: overcoming the dualism between social and individual views of learning', Vocations and Learning, 1, 1, 27–47.
42. Collin, K. (2009), 'Work-related identity in individual and social learning at work', Journal of Workplace Learning, 21, 1, 23–35.
43. Bartram, D., Foster, J., Lindley, P. A., Brown, A. J. and Nixon, S. (1993), Learning Climate Questionnaire (LCQ): Background and Technical Information (Oxford: Employment Service and Newland Park).
44. Wexlberger, L. P. (1993), 'Lernen im Arbeitsprozeß [Learning in the Labour Process]', in C. Heidack (ed.), Lernen der Zukunft. Kooperative Selbstqualifikation – die effektivste Form der Aus- und Weiterbildung im Betrieb (München: Lexika), pp. 299–310.
45. Billett, S., Harteis, C. and Eteläpelto, A. (eds) (2008), Emerging Perspectives of Workplace Learning (Rotterdam: Sense Publishers).
46. Sen, A. (1992). Inequality re-examined. Oxford, UK: Oxford University Press.
47. Anand, P., Hunter, G., & Smith, R. (2005). Capabilities and well-being: evidence based on the Sen-Nussbaum approach to welfare. Social Indicators Research, 74(1), 9-55.
48. Anand, P., Hunter, G., Carter, I., Dowding, K., Guala. F., & Van Hees, M. (2009). The development of capability indicators. Journal of Human Development and Capabilities, 10(1), 125-152.
49. Vizard, P., & Burchardt, T. (2007). Developing a capability list: Final recommendation of the Equalities Review Steering Group on Measurement. Case Paper 121, Centre for Analysis of Social Exclusion, London School of Economics.
50. Bryson, J. (2013). Putting skill in its place. Journal of Education and Work, doi:10.1080/13639080.2013.835794
51. Lambert, M., Vero, J., & Zimmermann, B. (2012). Vocational training and professional development: a capability perspective. International Journal of Training and Development, 16(3), 164-182.
52. Hooi, L.W. (2010). Technical training in the MNCs in Malaysia: a case study analysis of the petrochemical industry. Journal of European Industrial Training, 34(4), 317-343.
53. Hamzah, A., & Ho, S. (1994). TQM training for small and medium industries in Malaysia. Training for Quality, 2(2), 27-35.
54. Abdul-Aziz, A.R., Jaafar, M., Enshassi, A., & Mohamed-Salleh, M.N. (2008). Does vocational training meet the construction industry needs in Malaysia? Feedback from ex-trainees of the basic electrical wiring course from one training institution. Journal of Engineering, Design and Technology, 6(3), 258268.
55. Muhamad, M., & Idris, K. (2005). Workplace learning in Malaysia: the learner's perspective. International Journal of Training and Development, 9(1), 62-78.
56. Department of Statistics (2010). Labour Force Statistics, Malaysia, 2010.

Empowering Self-Management through M-Health Applications

Muhaini Othman[1,*], *Norhafizah Mohd* Halil[1,**], *M. Mohd* Yusof[1], *R.* Mohamed[1], and *Mohd Hafizul Afifi* Abdullah[1]

[1]Faculty of Computer Science and Information Technology, Universiti Tun Hussein Onn Malaysia, 86400 Parit Raja, Batu Pahat, Johor, Malaysia

Abstract. The advancement in mobile technology has led towards a new frontier of medical intervention that never been thought possible before. Through the development of MedsBox Reminder (MBR) application for Android as a pilot project of M-Health, health care information system for patient self-management is made possible. The application acts as an assistant to remind users for their timely medicine intake by notifying them through their mobile phone. MedsBox Reminder application aims to facilitate in the self-management of patient's health where they can monitor and schedule their own medicine intake more efficiently. Development of the application is performed using Android Studio 1.4, Android SDK, MySQL database, SQLite, Java language and Netbeans IDE 8.1. Object-Oriented System Development (OOSD) methodology has been adapted to facilitate the development of the application.

1 Introduction

Non-communicable disease such as diabetic, cardiovascular diseases, stroke, and hypertension continues to increase in the communities worldwide [1]. Medical intervention and self-management are critical to ensure the well-being of patients suffer from the non-communicable disease since they need extra attention for a longer period of time. Family and close relatives could also act as a mediator for patients in monitoring their health. Most patients are elderly people, who tend to ignore their schedule medicine intake and could result in a severe health relapse.

Currently, there is abundance of advertisement of health awareness and mobile applications has been developed that improve the living condition, nevertheless, without a proper health management system to monitor and intervene patients the quality of life is decreased [2]. Toward empowering people in self-management of their health and manage their long-term illnesses conditions systematically, patients must have access to a timely information, advice, assessment, and treatment from the medical practitioners [3]. Thus, the progressive growth of mobile technology, social networking and medical applications made self-management of health and awareness possible.

As a pilot study, a prototype Android application called MedsBox Reminder (MBR) is developed to empower self-management of health, acting as a mobile-enhanced preventive care, wellness management system to monitor patient's health and their medicine intake. The application is capable of encouraging self-health management and connects medical practitioners with patients on a personal level for guidance, monitoring and medical intervention.

2 Related Works

In this section, the essential background knowledge of mobile health applications and its related works which represents the essence of our work are introduced.

M-Health can be defined as "mobile computing, medical sensor, and communications technologies for health-care" [4]. WHO [5] has similarly emphasizes the application of devices such as mobile phones, patient monitoring devices, personal digital assistants, and other wireless devices for supporting medical and public health practice.

Health care industry has benefited significantly through the advancement of M-Health technology. People are going mobile where 52% of smartphones users gather health-related information on their phones while 80% of physicians use smartphones and medical applications [6]. The success of M-Health due to its usability and cost saving, especially in-home health care of the elderly where it increases patient's mobility and accessibility to a better wellness program anytime and anywhere [7]. To enrich the future of M-Health programs, M-Health systems should be inter-operable with the existing e-health standards [8]. Through proper execution, M-Health system may assist in providing compliance patient treatment, health data collection and disease monitoring, health information systems, support tools for medical practitioners, promotes individual health care, disease prevention, emergency medical response systems, economical treatments, appointment reminders, referral of patients, remote health services, super specialist services, emergency medicine, mobile medical data, mobile robotic system, post-hospital care,

* Corresponding author: muhaini@uthm.edu.my

telecommunication collaboration, medical information, pre-operative, and post-operative treatments [9].

Medical practitioners play an important role in empowering self-health management by giving guidance, monitoring adverse events, and identify areas for improvement while giving patients independently self-management of their health [2]. Through computer-mediated technologies, systematic intervention from medical practitioners and community is feasible, thus improving quality of life for individuals [10]. A research was carried out to measure user acceptance of M-Health remote monitoring system using a racially diverse sample of stroke patients to enhance medication adherence and blood pressure control [11]. The research has received positive feedback where 85% of the patients were comfortable with a nurse or doctor monitoring their health information using M-Health systems, while 78.3% of them believed it could assist medical adherence. 83.3% were confident this technology is an effective way for them to communicate with their respective health providers [11].

Aside from intervention from the medical practitioners, community or group intervention is also a good practice to promote a healthy lifestyle. There a mobile-based application called NutriWalking (NW), which recommends personalised daily exercise goals and promotes healthy nutritional habits among small groups specifically for patients with type-2 diabetes mellitus and depression. The application gathers peer support from people who suffer from the same disease, and the group members can assist each other through this application, acting as a self-management support tool [12]. Much more application like NW is being developed and can be downloaded via Google Play Store for Android phone users. Some of the M-Health system are for monitoring patient's health or self-monitoring [13-15], reminder [16], diagnosing diseases [17], non-communicable disease prevention [18], and clinical decision making [19].

One research has proposed a low-cost mobile health care monitoring system for monitoring ECG, blood pressure, SaO_2 level, and body temperature to facilitate treatments of chronic diseases [13]. The collected data is visualised in Android devices to assist patients in monitoring their health. Nguyen *et al.* [14] have identified few mobile applications to enhance self-management of patients with gout. These applications allow the users to monitor the symptoms of their gout disease including serum uric acid level, acute attacks, and educate patients about gout. The other application is for self-management and education of cardiac patients fighting against chronic diseases, especially heart diseases, whereby medical and users' activities are recorded including medication time reminder [15]. There are also applications developed to assist in monitoring and prevention of non-communicable diseases such as Stroke Riskometer which provides estimates of the absolute risk of stroke within the next 5 to 10 years for individuals aged 20 years and above, providing user a baseline risk comparison which allows them to compare their stroke risk with someone of the same age and gender who has no risk factors [18]. Since 5 to 10 years prediction of stroke risk is still inadequate, there are several other researches related to personalized modelling using spiking neural networks methods, aimed to predict stroke risk event several days earlier before stroke occurrences [20-21].

Besides monitoring and self-management applications, few works have studied the use of medical applications as a reminder for immunization [16]. Through a survey, the usage of Android applications for immunization reminder has received positive feedbacks from physicians, parents, and guardians. The advancement in mobile technology and authentication leads towards improved home health care where a person who suffers from chronic disease requires a reminder application to signal them in taking medicine promptly [22]. Therefore, Internet of Things (IoT) has become an essential part of data transaction between source points to destination for the reminder applications. Finally, a mobile phone reminder application using text messages was developed for an anti-psychotic patient [23]. This tool is a practical and ideal approach towards aiding different type of mental illness whereby patients can select preferred text messages as a reminder for medication intake, treatment appointments, and support.

In conclusion, monitoring and reminder applications have been helpful for users in assisting them in monitoring and keep track of the patient's health care records as well as their medication intake. This motivates the development of MedsBox Reminder application with the intention to help the medical practitioners to provide patients with the access to the prescribed medication and acts as a reminder for the patients' daily medication intake.

3 Methodology

The methodology used for the development of MedsBox Reminder application is Object-Oriented System Development (OOSD) which consists of four phases: i) object-oriented requirement analysis, ii) object-oriented design, iii) object-oriented implementation, and iv) object-oriented testing.

During object-oriented requirement analysis phase, data collection techniques such as observations and questionnaire were utilized to study the pattern of medicine intake among various community groups involving the elderly and youth. Survey and questionnaires were distributed among patients and clinicians, interviews were conducted with registered clinicians and physicians, and observation in clinics and pharmacies were also done. Information gathered includes medicine intake rules and the preferred method to keep in touch with the patients. The respondents were invited to participate in the evaluation of the prototype.

In the object-oriented design phase, information for development of the proposed application is translated into a working prototype. The purpose of the phase is to ensure the developed application meets the end-user's requirements before the prototype is translated to the production application. Unified Modelling Languages (UML) are used as a method of designing activities involved in MedsBox Reminder which consists of use cases, sequence diagrams and class activity diagrams.

During the object-oriented implementation phase, the user interfaces for each module in the prototype will be converted into program codes and connected to the database to fulfill specified functionality for the final product. Figure 1 visualises the system architecture where the main entities described as meducal practitioners (clinicians/pharmacists) and patients, will have direct access to the system. Clinicians/pharmacists access the system via administrator website while patients need to download the MedsBox Reminder application and install it on their mobile phone or tablet, to gain access to their record.

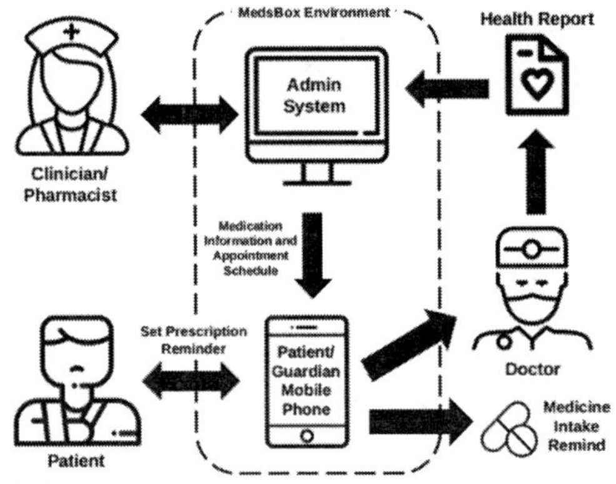

Fig. 1. MedsBox Reminder system architecture.

4 User Interface and Functionality

MedsBox Reminder application consists of 5 modules: login module, member registration module, patient's medicine module, appointment module, and medicine module. Patients have access to 3 modules: login module, appointment module, and medicine intake module (with notification reminder).

The administrator modules which are accessible by medical practitioners allows insert/update patient's data such as personal data (Figure 2), medication information for patients (Figure 3), and update a scheduled appointment for patient's check-up (Figure 4).

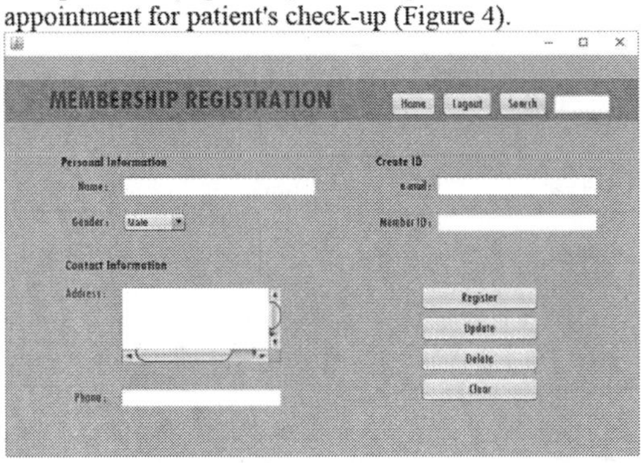

Fig. 2. Patient registration module (Admin).

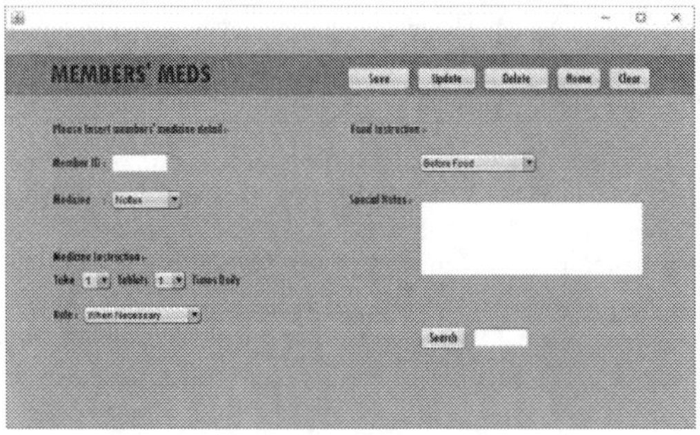

Fig. 3. Patient medicine record module (Admin).

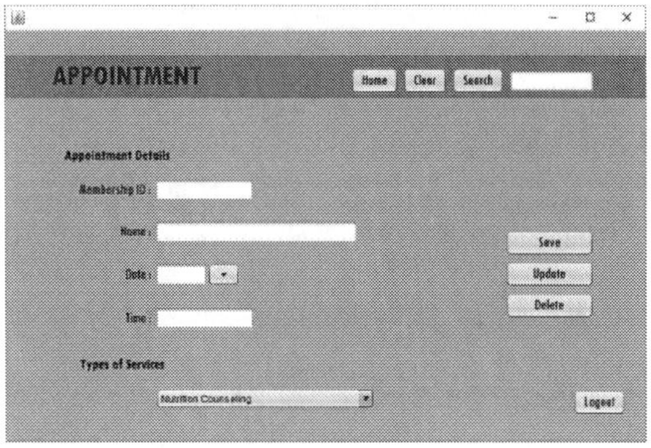

Fig. 4. Patient appointment module (Admin).

The administrator also can insert and update new medicine record into the database through medicine record page (Figure 5). The information about the medicine prescription can be accessed by patients via the MedsBox Reminder application.

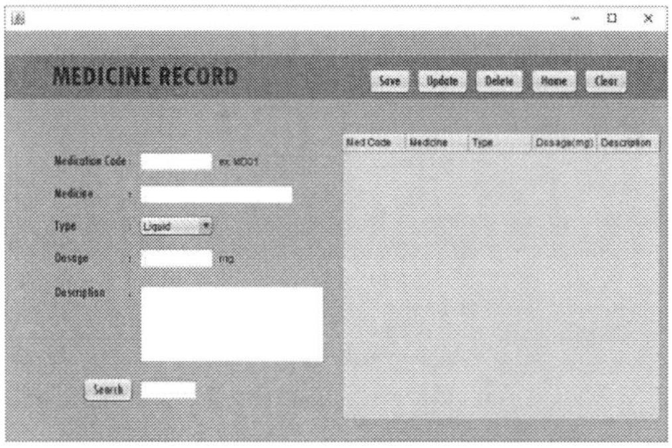

Fig. 5. Medicine record page (Admin).

MedsBox Reminder applications allow patients access to all information regarding medication prescribed for them. The application also provides medicine prescription reminder functionality to schedule personal medicinal intake according to the physicians' direction. Patients are able to view and reschedule their medical appointment via the MedsBox Reminder application. The patients have to log in into the application to verify that they are registered to the clinic

 https://doi.org/10.1051/matecconf/201815005018

or pharmacy. After logging into the mobile application, a menu as in Figure 6 will be displayed.

Figure 7 (a) shows list of medicine intake reminders set, where patients can delete and update previous reminders. The medicine reminder setting page is depicted in Figure 7 (b). Through this module, the user can schedule their medicine intake according to the instruction from clinicians/pharmacists.

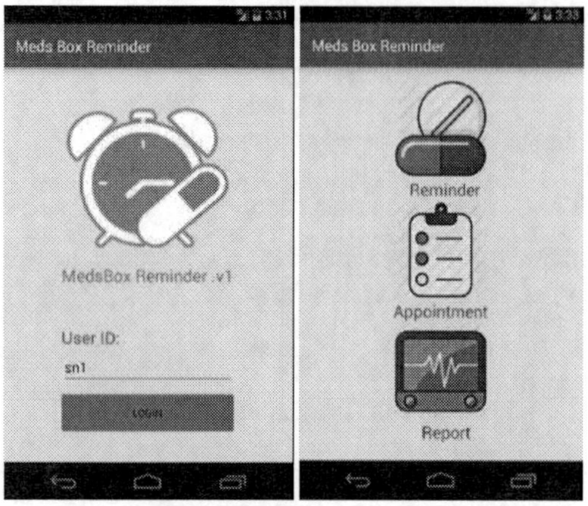

Fig. 6. Login module and the menu (Patient).

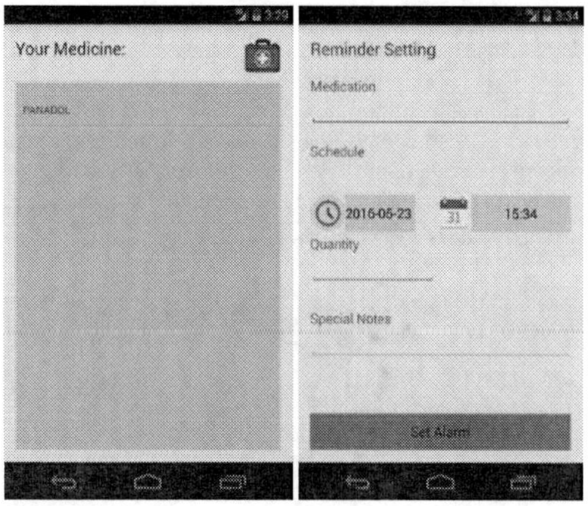

Fig. 7. (a) Reminder list. **(b)** medicine intake reminder setup (Patient).

After a reminder have been set, a notification will pop-up to remind patients for medicine intake. Figure 8 shows the notification reminder for medicine intake.

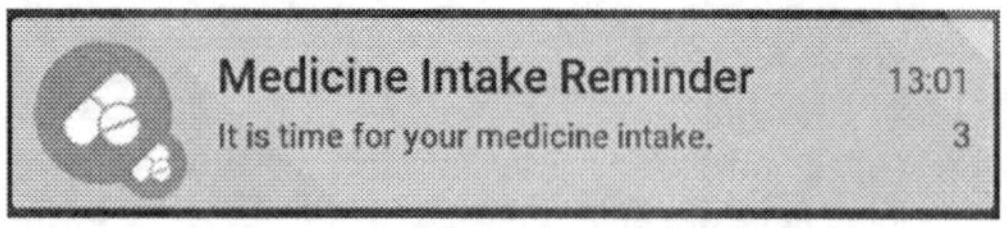

Fig. 8. Notification reminder for medicine intake (Patient).

Pseudocode 1 represents the workflow for medication intake reminder set by the patients. Pseudocode 2 represents the workflow for appointment reminder with the physician set by the clinician.

Algorithm 1 Medicine reminder

Input: medName, medNotes, medQty, and medDateTime
if *reminderId NOT exists* **then**
  Create new reminderId
  Store values from **Input**
  Save reminder
else
  Recall existing reminder
  Update value to existing reminder
  Add reminder to calendar
end

Algorithm 2 Appointment reminder

Input: aptName and aptDateTime
if *appointmentId NOT exists* **then**
  Create new appointmentId
  Store values from **Input**
  Save appointment
else
  Recall existing appointment
  Update value to existing appointment
  Add appointment to calendar
end

The above pseudocodes are vital to the success of this application where empowering self-management initiatives and medical intervention.

5 Result and Discussion

The developed application was deployed and installed into an Android system phones for system testing. The tests ensure the functionality of the modules perform properly and error-free. These tests involve the users; administrator and patients, covering all modules available according to the user access level.

After the system functionality testing is completed, users are invited to participate in user acceptance testing (UAT) to get feedback on whether the system and applications meet the user requirement. UAT is considered as the last phase of any software development process that helps in making sure that the software or applications can handle its required tasks in real world. UAT has provided users with the chance to interact with MedsBox Reminder and provide feedback to the development team for enhancement and bug-fixes before the final release of application.

UAT was conducted by selecting patients that are accustomed to the service received from a clinic. The questionnaire form was distributed to the respondents while testing the prototype. The UAT form consists of three sections: tester's demographic, application UI/UX UAT, and application functionality UAT. Table 1 shows the questions and criteria for scoring the UAT.

Table 1. UAT questions and criteria of scoring.

Section 1: Tester's Demographic	**Male**	**Female**
Gender	84%	16%
Age (in years)	18-55	24-36
Section 2: Interface and Experience	**Agree**	**Neutral**
Overall application design	76%	24%
Suitable background and colour	92%	8%
Suitable font family and font size	78%	22%
Contents are clear and readable	90%	10%
Smooth system flow and easily used	78%	22%
All buttons and links are working	88%	22%
Section 3: Application Functionality	**Agree**	**Neutral**
Suitable for monitoring medicine intake	60%	40%
Reminder notification works well	60%	40%
Overall application functionality	80%	20%

The implementation of system testing is successfully conducted with the clinicians and the obtained results is compared to the expected results. Comments and suggestions were gathered for future enhancement. Table 2 summarizes the UAT result from respondents.

Table 2. Overall UAT results from respondents.

Criterion of Score	**Agree**	**Neutral**	**Disagree**
Intuitive user interface	82%	18%	0%
Functions well as intended	67%	33%	0%
Overall score (average)	78%	22%	0%

Overall, MedsBox Reminder application has achieved the objective to be a well-functioning application which helps user in managing their daily medication intake. However, there is still room for improvements in making MedsBox Reminder better in the future.

6 Conclusion

The enhancement of mobile technologies initiates opportunities in the M-Health field especially for self-management of individual health, immediate medical interventions and assistance from medical practitioners, and also health awareness at an individual level. MedsBox Reminder application was developed to prove that M-Health is not only practical for improvement of the overall quality of life, but also much feasible for development and brings benefits to both health care industries and the community.

This work is supported by the Fundamental Research Grant (Vot 1612) from Ministry of Higher Education Malaysia, Universiti Tun Hussein Onn Malaysia (UTHM), and GATES IT Solution Sdn. Bhd.

References

1. World Health Organization Media Centre, Diabetes, http://www.who.int/mediacentre/factsheets/fs312/en/. Retrieved on October 27, 2016 (2015)
2. L. Whitehead, P. Seaton, J. Med. Internet Res. **18**, 5 (2016)
3. V. Zoffmann, Å. Hörnsten, S. Storbække n, M. Graue, B. Rasmussen, A. Wahl, M. Kirkevold, Patient Educ. Couns. **99**, 400-407 (2016)
4. R.S. Istepanian, E. Jovanov, Y.T. Zhang, IEEE Trans. Inf, Technol. Biomed. **8**, 405-414 (2004)
5. WHO Global Observatory for eHealth. New horizons for health through mobile technologies. Geneva: World Health Organization; 2011:112.
6. Govette, J. 30 Amazing Mobile Health Technology Statistics for Today's Physician, https://getreferralmd.com/2015/08/mobile-healthcare-technology-statistics/. Retrieved on October 27, 2016 (2015)
7. C.E. Smith, U. Piamjariyakul, M. Werkowitch, D.M. Yadrich, N. Thompson, D. Hooper, E.L. Nelson, Int. J. Sens. Netw. Data Commun. **5**, 1 (2016)
8. A.V. Heerden, M. Tomlinson, L. Swartz, Bull. World Health Organ. **90**, 393-394 (2012)
9. A. Rehalia, S. Prasad, mHealth **2**, 3 (2016)
10. K. Lorig, P.L. Ritter, R.M. Turner, K. English, D.D. Laurent, J. Greenberg, J. Med. Internet Res. **18**, 6 (2016)
11. C. Jenkins, N.S. Burkett, B. Ovbiagele, M. Mueller, S. Patel, B. Brunner-Jackson, R. Saulson, F. Treiber, mHealth **2**, 5 (2016)
12. A.L. Hartzler, A. Venkatakrishnan, S. Mohan, M. Silva, P. Lozano, J.D. Ralston, E. Ludman, D. Rosenberg, K.M. Newton, L. Nelson, P. Pirolli, IEEE Trans. Inf. Technol. Biomed. **2016**, 3277-3281 (2016)

13. Z.R. Nandi, A.C. Olivera, R.E.C. Valencia, A.A. Paredes, G.A. Alonso, Panam. Healt. Care **2015**, 1-4 (2015)
14. A.D. Nguyen, M.T. Baysari, D.R. Kannangara, A. Tariq, A.Y. Lau, J.I. Westbrook, R.O. Day, Int. J. Med. Inform. **94**, 67-74 (2016)
15. I. de la Torre-Díez, B. Martínez-Pérez, M. López-Coronado, J.J. Rodrigues, J. Arambarri, Iber. Conf. Inf. Syst. **2016**, 1-5 (2016)
16. J.L. Peck, M. Stanton, G.E. Reynolds, J. Pediatr. Health Car. **28**, 35.42 (2014)
17. O. Oluwagbemi,F. Oluwagbemi, C. Ughamadu, Info. Med. Unlocked **2**, 38-69 (2016)
18. P. Parmar, R. Krishnamurthi, M.A. Ikram, A. Hofman, S.S. Mirza, Y. Varakin, M. Kravchenko, M. Piradov, A.G. Thrift, B. Norrving, W. Wang, Int. J. Stroke **10**, 231-244 (2015)
19. P. Divall, J. Camosso-Stefinovic, R. Baker, Health Inform. J. **19**, 16-28 (2013)
20. M. Othman, N. Kasabov, E. Tu, V. Feigin, R. Krishnamurthi, Z. Hou, Y. Chen, J. Hu, IEEE IJCNN **2014**, 3197-3204 (2014)
21. N. Kasabov, N.M. Scott, E. Tu, S. Marks, N. Sengupta, E. Capecci, M. Othman, M.G. Doborjeh, N. Murli, R. Hartono, J.I. Espinosa-Ramos, L. Zhou, F.B. Alvi, G. Wang, D. Taylor, V. Feigin, S. Gulyaev, M. Mahmoud, Z.G. Hou, J. Yang, Neural Networks **78**, 1-14 (2016)
22. S.V. Zanjal, G.R. Talmale, Procedia Compt. Sci. **78**, 471-476 (2016)
23. K. Kauppi, K.A. Kannisto, H. Hätönen, M. Anttila, E. Löyttyniemi, C.E. Adams, M. Välimäki, Schizophr. Res. **168**, 514-522 (2015)

Factor Of Vendor Selection And Employees' Morale Towards Human Resource Outsourcing Decision In Organization

Tan Kok Hiean[1], *Irza Hanie Samah*[1*], Aidanazima Abashah[1], *S. Nurshahrizleen*[1] *and Nazmizan Muhammad*[2]

[1]School of Business Innovation and Technopreneurship, UniMAP.
[2]Faculty of engineering Technology, UniMAP.

Abstract. The purpose of this study is to investigate factor of vendor selection and employees' morale towards human resource outsourcing decision in organization. Study was conducted in Melaka within private organizations. Questionnaire was distributed and only 60 respondents were collected back. Data analysis was perform using SPSS version 21. Findings shows that employee's morale show high significant relationship *($r=0.761$, $p<0.05$)* with human resource outsourcing decision. Limitation and conclusion were discussed in this study.

1 Introduction

Nowadays, outsourcing is getting impetus in business corporations and organisations globally [1]. The majority of the companies tend to free from globalisation, their HR needs are becoming more complex and exacting, which is causing increased costs and more time and effort are required to take care of the employees. Therefore, cost is always persist a challenge to all the business organisations or enterprises to compete against their competitors by offering the best lower cost in total to end customers and to secure the market share in order to add value to the shareholders, especially in the competitive business environment and concurrently, facing the economic downturn [2]. They are always search for solutions and alternatives in order to stay competitive by reducing their costs and increasing their productivity [3].

Human Resource Outsourcing is the procedure of assigning and subcontracting human resource functions to an external vendor [4]. A remarkable current trend has been the detachment and transient labours, or the downright outsourcing of Human Resource activities were increasingly employed by the business organisations[5]. The multinational companies, business corporations from western countries, and the business process outsourcing (BPO) units in nations such as Malaysia, Russia, China and India even are expecting to outsource [4]. Reasons people to outsource their human resource because to focus on the core competencies of the business itself [6], reducing the cost of organization's operation, reducing headache of handling the issues of compliance, improving selection and recruitment, utilizing the resource efficiently [5].

Although human resource outsourcing activities have good reputation among organization, however there are some factors that organization need to consider when outsourcing their human resource especially when it comes to vendor. Every vendor providing unique outsourcing services depends on their culture and employees. Agent that have a similarity with the organization, outsourcing is likely to be success. Agent with different aim and goals may cause conflict and inefficiency. If your goals for outsourcing one or more functions aren't clear from the start, you can end up with a failed partnership. Next, employee's morale. Vendor that selected by the organization for outsourcing is consider as the third party beside the organization and employees. They are authorised on the human resource activities. Therefore, it will increase employee's insecurity, such as employee retention, performance appraisal, welfare, training which then will give an impact towards employee career development. Acording to [1] one third of human resource professionals are afraid of losing their job to work under vendors. At the same time, it will damage morale and motivation as jobs appear to be lost.

2 Literature Review

Human resource (HR) outsourcing is defined as the companies are transferring the administrative management of one of their most significant assets to third-party service vendor, which the third-party can

Corresponding Author: Irzahanie@unimap.edu.my

take on the administration of hiring new personnel and at the same time, improve the running of employee benefits management [7]. Human resource (HR) outsourcing is also defined as the action of partially or wholly contracting human resource function out to external vendors and give them the permission to execute human resource activities (such as training, payroll, talent recruitment, etc.) for an organisation, in proportion to internal procurement seeing that the particular function is executed by the employees of the organisation [8].

HR activities that the organisations itself would otherwise commonly Provide [9]. In HR outsourcing, the human resource intermediaries (HRIs) will play the role which is focusing on delivering human resource (HR) services to both public sector organisations and private companies or firms (Kock et al, 2012). Due to globalisation, the competitive pressures is increased, the organisations are being surpassed by competitors in the business areas where they are not expert and concurrently, rapid organisational change [7][10]. When the organisations decide to engage in outsourcing, they decide to ally an external vendor to obtain their competencies to meet in a need for certain services of which the organisations think they cannot perform best [11]

The objective of outsourcing should be free up management and infuse confidence among them to take on more risk in the core areas of the business which are most gaining value [7]. processes can be redesigned to reach higher levels of productivity and effectiveness, which convert into increased profitability [7]. Outsourcing allows the management to focus more time, attention and resources on building the organisation's core business and competencies [7].

If the organisations require professionals and they are unaffordable to recruit or train them, HR outsourcing may be a solution, especially outsource functions where it can be hard to hire high-skilled personnel in the labour market [1], [7]Specialized vendors are more efficient because they can extend the costs of training personnel and carrying out research and development across more users [1]. There are five types of human resource functions outsourced to the external vendors which are recruiting process outsourcing (RPO), training and development, compensation and salary surveys, organisational design and development, human resource information system/information technology, and payroll [12].

Vendor management involves both vendor selection and client–vendor relationship control. Many researchers have addressed the latter issue by the employment of transaction cost theory [13]or the analysis of incomplete contracts [14]. The objective is for the client to induce the optimal performance from the vendors. It is important to select the right vendor for the right task. Failure of human resource outsourcing is when the vendor has lack of knowledge about outsourcing especially when it comes to selection process [15]. This issues have been addresses by many scholars but focusing on the performance data, whereby such approaches guarantee promising future results. Such approaches imply that vendors' past performance guarantees future results, an assumption does not necessarily apply because of ever changing technology, a high degree of heterogeneity of outsourced projects, intrinsic variation, and low predictability in vendor performance [16].

By recognizing the approaches of human resource outsourcing, can help organization to improve the decision to choose vendor selection. Two stage approach of vendor selection; whereby the first stage involves trial phase; which clients choose the right vendor for the right task, at this stage there is an existence of client-vendor interaction and this is where game theory is explained. In the second stage, this is where client chosen the vendor and employ for project implementation. [16]

There has been an argument about vendor selection is a way to reduce the supply risk that arise because of unreliable outsourcing. Despite, the multiple vendor also has been proposed as a means to reduce such risks [17] [18]. However multiple vendor which having different characteristic could have the potential to supply risk because the chances of either one of them to dominates is there[19]. Therefore, it depends on the organization to make decision on choosing vendor. Which the organization can choose more than one vendor for its organization but the vendors must understand the job and the task given, and must follow the nature of business of the organization.

Moreover, the selection of vendors and the determination of the task to be placed with those vendors are important decisions. Vendor selection decisions have a significant impact on the competitiveness of a firm, as purchases from vendors account for a large percentage of total costs for many firms. Selection of vendors has long been regarded as one of the most important functions to be performed by organizations because it is one of the organization's expenses. Vendor selection decisions are complicated by the fact that various criteria must be considered in decision making process. Most research on vendor selection has been descriptive or empirical in nature which involve in quantitative analysis with base on the performance [20] this research highlighted the human factors whereby the decision making of vendor selection is up to the top management.

Previous study had mention that mostly company decision on vendor selections is based on the quantitative which something that they can see, however, none of them have addressed the issue of the quality [20] especially in subjective. Human resource outsourcing is considered as subjective matter, vendor need to connect and have the interaction with the clients so that they can understand what is the objectives that they need to deliver.

Employee's morale is defined as the "description of the emotions, attitude, satisfaction, and overall outlook of employees during their time in a workplace environment." It is implied that employee morale is thought to be directly related to part of effective productivity, which meant that the employees who are pleased and positive at work are considered to have high/positive morale while in contrast if employees who are unhappy and dissatisfied about their working environment are considered to have low/negative employee morale.

Zukin, & DiMaggio (2012)[24] mentioned that human capital is the pivot around which organisational effectiveness can be resulted in comparison to financial capital and it is found to be tightly associated with the employee morale in the organisations in productivity and employee satisfaction. The essential of positive receptions in way of psychologically being well in working environment do make sense in altering scenario of different business proposition. Employee morale is frequently connected with the work setting, which is the psychological bonding of personnel with the firm which self-motivates them to behave and perform respectively.

There are actually consists of six reasons of the importance of employee morale which are improved productivity (the higher the morale, the higher the level of productivity); improved performance and creativity (the higher the morale, the better the performance, which is lead to more creativity of solutions found); reduced number of leave days (the higher the morale, the less the absenteeism, which high absenteeism will causes the corresponding loss in productivity; .higher attention to detail (when the morale is low, the employees will more likely to become distracted and unable to truly focus on work which might lead to the occurrence of mistakes and slower job schedule); a safer workplace (when the morale is higher, the employee will able to pay more attention to safety rules which could reduce the accidents happen in workplaces and increased quality of work (the higher the morale, the higher the effectiveness of performance and production, the better the quality produced.) [21]

The effect on employee morale and performance is considered as one of the dominant risk in human resource outsourcing [23] mentioned, outsourcing is impacting employee morale. Employees resist the outsourcing because of grudged to change (Franceschini et al., 2003). According to Gilley, Greer, & Rasheed (2004). the research is studied about the issue about the effect of outsourcing movement on the employee's morale. Even though outsourcing has its benefits, the problems exist and can arise due to the dashed implementation of outsourcing arrangement (Power, 2006). Hence, the sudden restructuring in company and conflicting employee interests are a common factor turning up during outsourcing movement (Gilley, Greer, & Rasheed, 2004).The result shows that, once the employee morale is lowered, less effort is contributed to the company and job satisfaction eventually suffers, and this have directly been linked to an increase effect on the absenteeism and ill-health (Gilley, Greer, & Rasheed, 2004; Tekleab & Taylor, 2003). This situation definitely creates a direct impact and correlated to the decreased in job productivity. Furthermore, employee openness to other jobs will start when the employee morale is hurt and decreased, which will causes the increased in labour turnover, which may lead to a large cost factor for organisations (e.g. future separation costs, new personnel learning costs and new personnel acquisition costs)[22].

According to Belcourt (2006), organisations do provide the employees with a sense of identification and feelings of security and belonging while outsourcing is a form of restructuring or reforming that always results in dismissed employees. Most of the professionals resist outsourcing because they may face the risk of losing their jobs, or may be forced to work for a vendor and worry that the administration and management think that the "outsiders" are more capable than them [1].

Outsourcing arrangement such as employees transferred to outsourcing, internal transfer (e.g. other functions), outplaced, offering voluntary retirements, or even the occurrence of redundancies and layoffs to staff, may causes the employees are acting against these arrangements, with their implication of "serfdom" in which the "bondmen" are sold as capital equipment). Outsourcing can bring about the disintegration of an organisation's culture which is alienating and "deskilling" employees instead of empowering and appreciating employees, at the moment, the transferred employees will experience emotional loss and culture changing [1]. All the employees start to suffer anxiety that may be caused by scare of lose their job in the next phase of outsourcing as well as the feeling of security is disrupted, which resulting in lost production once the rumours of outsourcing arrangements are spread [1] [11]. Hence, the managers will have to cope with the responses and reactions of dismissed employees [1]

Based on the extensive of literature review, this research formulates the specific hypotheses are developed as follows:

H_1: There is a significant relationship between selection of vendors and effectiveness of HR outsourcing decision.

H_2: There is a significant relationship between employee's morale and effectiveness of HR outsourcing decision.

3 Methodology

150 Respondents were selected in this research where the respondents are the representative of the company which listed in private company in Melaka. Simple random sampling was adopted in this research, 60 respondents were collected back for data gathering. All questionnaires were collected over a period of 3 weeks. The response rate was 40%. Questionnaire was designed using likert scale ranging from 1 to 5 from strongly disagree to strongly agree.

4 Results and Discussion

Quantitative research design was used in this study. Analysis and findings in this research was perform using SPSS version 21.

Respondents Background. The results present that are 75.0% are male respondents while the remaining 25.0% are female respondents. Other than that, the majority of the respondents were aged between 31-40 years old with 45.0% or 27 respondents while the least group is aged 18-30 years old with 8.3% or 5 respondents only. There is also 15 respondents at the age between 41-50 years old which consist of 25.0%, 7 respondents at the age between 51-60 years old which is 11.7% and 6 respondents which is 10.0% are at the age above 60 years old.

Mean and SD Analyses.
Mean and standard deviation values are presented in table 1. The highest mean score was vendor selection (M=4.204, SD=0.468), followed by employee's morale (M=4.113, SD=0.483), and lastly human resource outsourcing (M=4.154, SD=0.516).

Table 1.: Mean and SD Values

Variables	N	Mean	SD
Vendor selection	60	4.204	0.468
Employee's morale	60	4.113	0.48369
Human Resource Outsourcing Decision	60	4.154	0.51686

Correlation Analysis. Pearson correlation coefficient (r) was perform to test the relationship of independent variables (vendor selection and employee's morale) and dependent variables (Human resource outsourcing).

Table 2.: Correlation Analysis

Variables	1	2	3
Vendor selection	-	-	-
Employee's morale	0.761**	-	-
Human Resource outsourcing decision	0.595**	0.647**	-

***Correlation is significant at the 0.01 level (2-tailed)*

The results shown are indicated that all of the independent variables involved in the analysis have strong relationships with the dependent variable, which all the values of correlation coefficient obtained for respective study variables are within the range between 0.50-0.69. Apparently, it can be concluded that all the hypotheses are considered accepted since all the study independent variables are having distinct positive correlation with the dependent variable (human resource outsourcing decision).

Regression Analysis.
result displays that the selection of vendors was found to be significant which the value is at a significance level of 1% (sig. = 0.249) with beta value 0.182. It can be considered that the selection of vendors has no significant relationship with the dependent variable, so the Hypothesis 1 (H1) is considered rejected. Simultaneously, the remaining study variable, which is employee morale is also determined to be significant at a significance level of 1% (sig. = 0.013) with a beta value 0.354. This result proved that a significant relationship subsisting between employee morale and effectiveness of human resources outsourcing decision, consequently, it means that the Hypothesis 2 (H2) is validated.

Based on the overall result of the multiple regression analysis, the value of R2 is 0.618 which indicates that the employee morale affects 61.8% of the effectiveness of human resource outsourcing decision. This shows that the result obtained from this research can provides an average explanation towards effectiveness of human resource outsourcing decision and the remaining percentage of 38.2% is influenced by other factors. Concurrently, the value of F is 22.24 and was found to be significant at significant at a significance level of 1%.

Table 3. Regression Analysis

Variables	Human Resource Outsourcing Desicion			
	Beta	Sig.	R^2	Adjusted R^2
Vendor selection	0.182**	.249	0.618	. 0.590
Emplooyee 's morale	0.354**	.00		

**Correlation is significant at the 0.05 level (2-tailed)

5 Conclusion

The result and finding of the research shows that it is essential for a business organization to consider about the factors that may influence the effectiveness of human resource decision. This is because factor such as employee's morale may cause ineffectiveness of human resource decision and bring negative impact to the business operation. Business may face crisis, loss or failure. Hence, when a business organization decided to outsource its human resources function, they should ensure that the vendor are able to understand all employees and the organization have to be smart to achieve effective human resource outsourcing.

This research consists of its limitation because it is conducted by using limited questionnaire survey as an instrument to collect data. The respondents may not be a true representation as a whole as the number of respondent are limited as 60. Besides, the questionnaires are only conducted in single language, which is English language. It is possible to cause inaccurate in the result due to respondents may not truly understand all the questions and statements of the questionnaire especially the technical terms. This may cause misinterpretation of the statement, as a result causing confusion and inaccuracy.

References

[1] M. Belcourt, "Outsourcing - The benefits and the risks," *Hum. Resour. Manag. Rev.*, vol. 16, no. 2, pp. 269–279, 2006.

[2] A. Islam, "ARTICLE DETERMINANTS OF OUTSOURCING DECISION IN THE MANUFACTURING INDUSTRY in BANGLADESH," *Rev. Lit. Arts Am.*, vol. 5, no. January 2008, pp. 127–149, 2010.

[3] D. L. Caruth, S. S. Pane Haden, and G. D. Caruth, "Critical Factors in Human Resource Outsourcing," *J. Manag. Res.*, vol. 12, no. 1, pp. 187–195, 2012.

[4] K. Mahmud, M. M. Billah, and S. M. R. Chowdhury, "Human Resource Outsourcing: A Study on Telecommunication Sector in Bangladesh," *Int. J. Bus. Manag.*, vol. 7, no. 10, pp. 74–84, 2012.

[5] J. Woodall, W. Scott-Jackson, T. Newham, and M. Gurney, "Making the decision to outsource human resources.," *Pers. Rev.*, vol. 38, no. 3, pp. 236–252, 2009.

[6] C. R. Greer, S. a. Youngblood, and D. a. Gray, "Human resource management outsourcing: The make or buy decision.," *Acad. Manag. Exec.*, vol. 13, no. 3, pp. 85–96, 1999.

[7] F. Franceschini, M. Galetto, a. Pignatelli, and M. Varetto, "Outsourcing: guidelines for a structured approach," *Benchmarking An Int. J.*, vol. 10, no. 3, pp. 246–260, 2003.

[8] M. Tremblay, M. Patry, and P. Lanoie, "Human resources outsourcing in Canadian organizations: An empirical analysis of the role of organizational characteristics, transaction costs and risks," *Int. J. Hum. Resour. Manag.*, vol. 19, no. 4, pp. 683–715, 2008.

[9] M. Seth and D. Sethi, "Human Resource Outsourcing : Analysis Based On Literature Review," *Int. J. Innov. Manag. Technol.*, vol. 2, no. 2, pp. 127–135, 2011.

[10] H. Kock, A. Wallo, B. Nilsson, and C. Höglund, "Outsourcing HR services: the role of human resource intermediaries," *Eur. J. Train. Dev.*, vol. 36, no. 8, pp. 772–790, 2012.

[11] P. Arroyo-López, E. Holmen, and L. De Boer, "How do supplier development programs affect suppliers?: Insights for suppliers, buyers and governments from an empirical study in Mexico," *Bus. Process Manag. J.*, vol. 18, no. 4, pp. 680–707, 2012.

[12] E. Aktaş, I. Çiçek, and M. Kiyak, "The effect of organizational culture on organizational efficiency: The moderating role of organizational environment and CEO values," in *Procedia - Social and Behavioral Sciences*, 2011, vol. 24, pp. 1560–1573.

[13] L. . Willcocks, M. . Lacity, and T. Kern, "Risk mitigation in IT outsourcing strategy revisited: longitudinal case research at LISA," *J. Strateg. Inf. Syst.*, vol. 8, no. 3, pp. 285–314, 1999.

[14] K. Nam, S. Rajagopalan, H. R. Rao, and A. Chaudhury, "A two-level investigation of information systems outsourcing," *Communications of the ACM*, vol. 39, no. 7, pp. 36–44, 1996.

[15] M. Bragg, *Outsourcing: A Guide to selecting the correct business unit, Negotiating the contract. maintaining control of the process*, vol. second edi. 2006.

[16] E. M. Snir and L. M. Hitt, "Vendor Screening in Information Technology Contracting With a Pilot Project," *J. Organ.*

Comput. Electron. Commer., vol. 14, no. 1, pp. 61–88, 2004.

[17] R. Ernst, B. Kamrad, and K. Ord, "Delivery performance in vendor selection decisions," *Eur. J. Oper. Res.*, vol. 176, no. 1, pp. 534–541, 2007.

[18] C. a. Weber, J. R. Current, and W. C. Benton, "Vendor selection criteria and methods," *Eur. J. Oper. Res.*, vol. 50, no. 1, pp. 2–18, 1991.

[19] R. V. Ramasesh, J. K. Ord, and J. C. Hayya, "Note: Dual sourcing with nonidentical suppliers," *Nav. Res. Logist.*, vol. 40, no. 2, pp. 279–288, 1993.

[20] R. G. Kasilingam and C. P. Lee, "Selection of vendors — A mixed-integer programming approach," *Comput. Ind. Eng.*, vol. 31, no. 1–2, pp. 347–350, 1996.

[21] D. L. Weakliem, "Morale and Workplace Performance," *Work and Occupations*, vol. 33, no. 3. pp. 335–361, 2006.

[22] M. Model, C. E. Michaels, and P. E. Spector, "Causes of Employee Turnover : A Test of the Mobley, Griffeth, Hand, and Meglino Model," *J. Appl. Psychol.*, vol. 67, no. 1, pp. 53–59, 1982.

[23] Taylor, P. (2012). Renegotiating relationships with employees in outsourcing operations. People and Strategy, 35(1), 32.

[24] Taylor, P. (2012). Renegotiating relationships with employees in outsourcing operations. People and Strategy, 35(1), 32.

Exploring the moderating effect of social intelligence on the relationship between entrepreneurial decision-making strategy and SME sustainable performance

Dayang Hasliza Muhd Yusuf[1*], *Mohd Salleh* Hj. Din[2], and *Muhamad Shahar* Jusoh[3]

[123]University Malaysia Perlis (UniMAP), School of Technopreneurship and Business Innovation, 01000 Kangar, Perlis, Malaysia

Abstract. The study reveals that causation, rather than effectuation, decision-making strategy is a more significant predictor of sustainable performance of SMEs. However, social intelligence was not found to be a significant moderator of entrepreneurial decision-making-sustainable performance relationship. The study uses data from a survey among 91 technology-based SMEs (TBS) in Malaysia and employs structural equation modelling techniques for data analysis. A new instrument to measure all three variables of entrepreneurial decision-making strategy, social intelligence, and venture performance is proposed based on adoption and adaptation of existing validated scales available in literature.

Key words: Effectuation, causation, entrepreneurial decision-making, social intelligence, SME, sustainable performance

1 Introduction

The study examined the moderating influence of social intelligence (SI) over the entrepreneurial decision-making strategy (EDM) and sustainable performance (SP) relationship of technology-based SMEs (TBS) in Malaysia. SP is defined as the extent to which an SME's achievement of the socio-efficiency objectives over its competitors [1]. While effectuation, rather than causation, has been argued as the more prominent thought process in uncertain environments [2]. Prior studies have established that due to the high uncertainty in starting a new venture, entrepreneurs need to be responsive to changes. Therefore, effectuation or non-predictive thinking is argued to make a company nimble and adaptable to the unpredictable shifts in the marketplace. At the same time, the importance of SI as the ability to understand, managed people and behave correctly in any social setting and has been associated with individual achievement in leadership, sales, and marketing [3]. However, whether SI can further improve the EDM-SP relationship need to be further explored.

1.2 Entrepreneurial Decision-Making Strategy

EDM has been argued to be within the realm of causative (or predictive) thinking and effectuation. Causation is planning-oriented based on entrepreneur's assessment of opportunities, whereas effectuation is along the stream of arguments that opportunities can be created rather than discovered [4, 2]. On the other hand, causation consists of strategies that combine available means with unanticipated contingencies to construct a series of stakeholder commitment [2].

1.3 Social Intelligence

Unlike general or cognitive intelligence, SI focuses on the ability to get along well with others and interacting successfully in any social setting [5]. Various authors look at SI from various perspectives of cognitive, behavioural, and psychometric [6, 7]. Authors offer varied definitions and measurement of the construct such as "the individual's fund of knowledge about the social world" [8], and how people perceive, react, and interact with the social world in which they live [6], and suggested the working definition of SI is suggested as "the ability to understand other people and how they will react to different social situations" [7]. SI has been suggested to be important for performance of selling and marketing activities [8, 9], and has influence over individual achievement. As among the challenge faced by TBS in Malaysia is marketing performance, there is a need to explore the effect of SI on the EDM-SP relationship. Therefore, this study aims to examine the EDM approach of entrepreneurs, and SI level of the entrepreneurs on the performance of TBS. The resource-based view (RBV) is adopted as the underpinning theory in our investigation.

1.4 Sustainable Performance

There is limited convergence among researchers on the definition and the measurement of organizational performance [10, 11]. For TBS, performance definition and measurement becomes more complex, and large firms, generic performance measures cannot be used for TBS [12]. Authors have also argued that the range of

*Corresponding author: dayanghasliza@unimap.edu.my

venture performance in entrepreneurship research need to be broadened to include the achievement of economic, environmental and social value creation objectives [1].

2 Study Framework and Development of Hypotheses

Prior studies have established the positive and significant relationship between the decision-making strategy and venture performance. However, the findings on the relationship between effectuation and venture performance has also not been consistent [2]. Research has also pursued the idea that social ability is relatively important for entrepreneurs, however little empirical evidence has supported the idea [9]. Thus, the researcher proposes that a new moderating variable of SI be introduced to explain the inconsistent findings. Figure1 depicts the study framework and research hypothesis developed.

2.2 Population and Sample

The population of the proposed study are owners of TBS registered with several agencies responsible for developing SMEs. As of 31st December 2013, there are a total 5,301,378 of businesses registered with SSM [13]. In terms of representation, SMEs percentage to total business establishments dropped from 99.2% in 2003 to 97.3% in 2010 [14].

Time, resources, and non-availability of data and contact information of the whole population understudy are among the factors which makes it impossible to collect data from the whole population. Using sample which is representative of the population would be more logical. The sample of the study are incubatees companies of several incubators and technology parks in Malaysia.

2.3 The Research Instrument

Figure 1 *Research framework depicting research hypothesis*

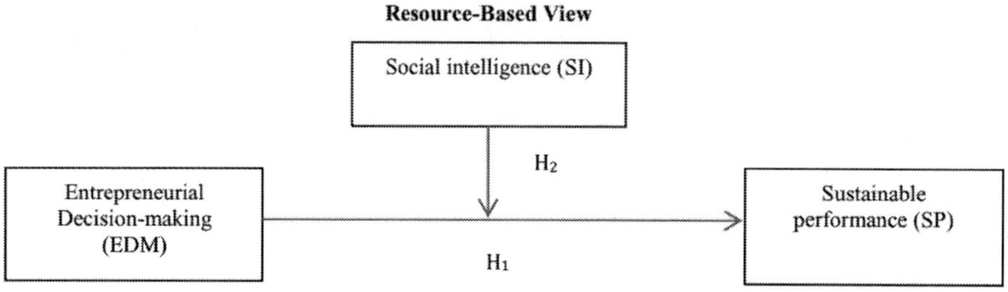

The expected relationship between all the variables, namely EDM, SI, and SP, used in developing the hypothesis for this study are as follows:

H_1: EDM is positively related to SP

H_2: Upon introduction of SI, the strength and direction of the formerly positive significant relationship between EDM and SP relationship will change.

2.1 Research Design

Data collection was carried out via a survey using questionnaires. The questionnaire was constructed with adaptation and adoption of existing instruments to measure the level EDM, SI and SP in decision making involving creation / discovery and exploitation of opportunities. Prior to data collection, a pre-test of the instrument developed was conducted to measure the validity and reliability test of the instrument. Data was analysed using the multivariate statistical techniques. The individual entrepreneur is the unit of analysis.

The questionnaire consists of four sections, namely, 1) demographic, 2) entrepreneurial decision-making, 3) social intelligence, and 4) sustainable performance.

In the beginning, 14 items represent the EDM construct (seven items measure causation while the other seven items measure effectuation). However, four items measuring effectuation were dropped due to low scores for reliability. While the scale for SI was adapted from the 3-factor, 21-item Tromso Social Intelligence Scale (TSIS) [7]. However, upon pilot test of the instrument, only 11 items were found to have the reliability criteria in representing the SI construct. On the other hand, SP is measured by 11 items measuring economic, social, and environmental performance adapted from existing studies. Total remaining items in the questionnaire is 32.

3 Findings and Discussion

The analysis involves looking at the effect on the direct relationship between EDM and SP (DM→SP), when a third variable, SI, is introduced in the model. The model tested is as per Figure 1. Discussion for the indirect effect of SI on EDM and SP (DM→SP) relationship was conducted starting with measurement model assessment, followed by goodness of fit assessment, and lastly with structural model assessment in subsequent topics.

3.1 Measurement Model Assessment

Based on the PLS algorithm report generated, the values for the assessment criteria can be derived. Table 1, Table 2, and Table 3 provide the values to be used for measurement model assessment of the study. Upon running the PLS-Algorithm function, three items measuring effectuation, and four items measuring SI were further dropped due to the low loadings of the items (<0.6). The remaining items were 25, of which seven items measure causation decision-making strategy, seven items for SI, and 11 items measures SP. Based on Table 1, the values of Cronbach's alpha for all variables are more than the recommended threshold of 0.8 [15], while the composite reliability (CR) values lie between 0.6 and 0.95. Thus, the all the items assigned to the EDM, SI, and SP constructs in the study, are found to have high internal consistency reliability [16].

Table 1. Internal consistency reliability assessment results

Constructs	Items	Composite Reliability (CR)	Cronbach's Alpha	Internal consistency reliability α > 0.8 and 0.60 < CR < 0.95
EDM	7	0.95	0.94	YES
SI	7	0.92	0.90	YES
SP	11	0.95	0.94	YES

EDM = Entrepreneurial decision-making strategy; SI=social intelligence; SP = sustainable performance

Next is the convergent validity test using examination of indicator reliability, and the average variance extracted (AVE) of all the constructs. Table 2 summarise the results of convergent validity tests based on AVE and factor loadings of each items. Items with loading values below 0.4 were deleted while remaining items with loading values below 0.708 were examined to see their effect on the construct's AVE score [17]. The values of factor loadings for the remaining 25 items lies above the threshold value of 0.708 as recommended by [18] indicating that all the items have much in common with other items within the same construct.

Table 2 *Convergent validity based on item loadings and AVE for SI→DM→SP*

Constructs	Item	Loadings	AVE	AVE >0.5	Loadings >0.708
Entrepreneurial decision making strategy (EDM)	CA01	0.851	0.719	YES	YES
	CA02	0.912			
	CA03	0.833			
	CA04	0.826			
	CA05	0.826			
	CA06	0.815			
	CA07	0.816			
Social intelligence (SI)	SI01	0.832	0.633	YES	YES
	SI02	0.798			
	SI03	0.812			
	SI04	0.803			
	SI05	0.771			
	SI06	0.785			
	SI07	0.768			
Sustainable performance (SP)	SP01	0.785	0.642	YES	YES
	SP02	0.768			
	SP03	0.758			
	SP04	0.813			
	SP05	0.822			
	SP06	0.744			
	SP07	0.856			
	SP08	0.838			
	SP09	0.826			
	SP10	0.811			
	SP11	0.815			

EDM = Entrepreneurial decision-making strategy; SI=social intelligence; SP = sustainable performance

While for the convergent validity assessment based on the AVE criterion, based on Table 2, the values of AVE for DM, SI and SP constructs are more than the recommended value of 0.5 [17]. Thus, EDM shared 71.9%, while SI shared 63.3%, and SP shared 64.2% of variance with the assigned items. In summary, the convergent validity tests were fulfilled by the study model. The next assessment is the discriminant validity which is conducted by comparing the cross loadings of item associated with the EDM, SI, and SP against the associated values of loadings of the same items on other constructs. All the item loadings are found to be highest on each construct the items are supposed to measure. Thus, the convergent validity requirement on cross loadings examination is met. While for the evaluation based on the Fornell-Larcker criterion, Table 3 indicates that the square root values of AVE for all constructs are higher than the associated cross-loading figures, thus, the discriminant validity requirements are satisfied at both the item and construct level. The difference between loadings of items assigned to the DM construct are also more than 0.1 of the loadings on SP and vice versa [19, 20].

Table 3 Fornell-Larcker criterion evaluation for SI*DM→SP relationship

	AVE	DM	SI	SP
DM	0.719	**0.848**		
SI	0.633	0.420	**0.796**	
SP	0.642	0.379	0.509	**0.801**

* *Diagonal elements and in bold are square roots of AVE.*

These results indicate that the indicators represent the assigned construct, and the extent to which all the constructs of DM, SI, and SP are truly distinct from each other. Lastly is the HTMT criterion which is the ratio of correlations within the constructs to correlations between the constructs [21, 22]. Based on Table 4, the HTMT ratio values are 0.654, and 0.536, which are below 0.85 [23], indicating that the discriminant validity is ascertained for the moderation effect of SI on DM→SP relationship.

Table 4. HTMT criterion for SI*DM→SP

	DM	SI	SP
SP	0.654 CI 0.90 (0.055, 0.334)	0.536 CI 0.90 (0.241, 0.563)	

Thus, for the indirect (moderating) effect of SI on EDM, and SP (DM→SP) relationship, all the measurement model assessment criteria of internal consistency reliability, convergent validity, and discriminant validity were met by the items and constructs of EDM, SI, and SP. Table 5 summarise all the results of the measurement model assessment conducted on the constructs and items of the study.

Table 5 Summary of results for measurement model assessment for SI*DM→SP

Type of test	Evaluation Criteria	Requirements	Results
Internal consistency reliability	Cronbach's alpha (α)	α > 0.8	Met for all items
	Composite Reliability (CR)	0.60 < CR < 0.95	Met for all constructs
Convergent validity	Indicator reliability	Outer loadings > 0.5	Met for all items.
	Average Variance Extracted (AVE)	AVE> 0.50	Met by all constructs
Discrimi-nant validity	Cross loadings of indicators	Indicator should have the highest loadings on the assigned construct compared to its loadings on other constructs.	Fully met by all items
	Fornell-Larcker Criterion (1981)	Square root of AVE > latent variable correlations	Fully met for all constructs
	HTMT criterion	HTMT < 0.85 Confidence interval does not have value of 1.	Met for all constructs

3.2 Goodness of Fit (GoF) for Moderated ET→SP relationship

This study will adopt the two assessment criteria, namely, SRMR and NFI assessment methods for the study model fit examination. Based on Table 6, the SRMR value is less than 0.10 which is the recommended threshold [24]. While the NFI value is 0.748 which is lower than the recommended value of at least 0.90 [25, 22]. Thus, the DM→SP model meets the model fit criteria for SRMR but not the NFI criteria. The next step is structural model assessment of the SI on DM→SP relationship, which is discussed in the following sub-topic.

Table 6. GoF assessment for SI*DM→SP

	Saturated Model	Estimated Model
SRMR	0.074	0.073
NFI	0.748	0.749

3.3 Structural Model Assessment

The structural model assessment of the study involves estimating the relationship among the exogenous (predictor), in this model DM, and endogenous (criterion) variables. The first step of the structural model assessment procedure is the model assessment for collinearity issues. The collinearity assessment was executed using the IBM SPSS software. The results of the collinearity assessment, which is the variance inflation factor or (VIF) estimates are shown in Table 7. All the VIF have values between 1.521 and 2.693, which are below 3.3, which is the threshold for collinearity for variance-based SEM, indicating that there is no problem of collinearity [26]. As collinearity is not an issue at this stage of analysis, the discussion proceeds to structural model assessment.

Table 7 VIF for DM, SI, and SP

DM	SI	SP
2.693	1.521	1.691

Figure 2 shows the moderation effect of SI on the EDM-SP relationship. While Table 8 lists the summary of results for the structural model assessment for the moderated DM→SP (SI*DM→SP) relationship.

The strength of DM→SP relationship is given as 0.504 which can be considered as strong as it is higher than the recommended value of 0.20 [15]. The empirical *t* statistics for this relationship is given as 6.447, which is higher than the critical *t* value at 1% level of confidence. Thus, H_1 which says that EDM is positively related to SP, is supported.

Whereas the strength of SI*DM→SP relationship is given as -0.047 which says that the interaction effect of SI on DM→SP relationship is negative. A one standard deviation point increase in SI will lead to 0.047 reduction in SP. The effect of SI*DM→SP can be considered as small [15]. The empirical *t* statistics for the SI*DM→SP is given at 0.406. This value is lower than the critical *t* value at 10% level of confidence. Thus, the study does not find that the moderated relationship between EDM and SP as significantly different from 0 at any level of confidence.

The second analysis is the coefficient of determination R^2 for the predictive accuracy of the model. In the case of the moderated effect of SI on the EDM and SP relationship (SI*DM→SP), the R^2 value is 0.433, which indicates that together, EDM, and SI can explain 43.3% of the variation in SP. The explanatory power of the model is slightly increased from 42.9% to 43.3% with introduction of the moderator variable of SI.

Figure 2 Moderation effect of SI on DM→SP relationship

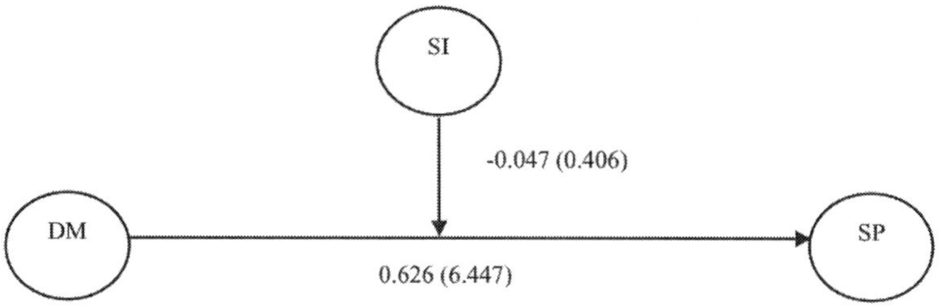

Table 8 Summary of structural model assessment for SI*DM→SP relationship

Hypothesis	Std β	SE	t-value	R^2	f^2	Q^2	Decision
H_1: Direct effect of DM→SP	0.504	0.099	6.447***	0.429	0.245	0.159	Supported
H_2: Moderation effect of SI on DM→SP	-0.047	0.008	0.406	0.433	0.006	0.250	Not Supported

Note: ***p<0.01

Additionally, the f^2 and Q^2 analyses were also adopted in evaluating the model's predictive ability. The effect size for the R^2 of DM is 0.245, indicating that DM has a moderate effect in producing the R^2 for SP. While the f^2 value for SI*DM is 0.006 indicating that SI*DM has a very small effect in producing the R^2 of SP, which indicates that the moderation effect of SI*DM does have much effect in producing the R^2 for SP. While the Q^2 analysis is for the predictive relevance of the model, the Q^2 value for the model is 0.250 (more than zero), which indicates that the overall model has a medium predictive relevance and validity for SP construct [15].

However, authors have argued that in the case of conditional effect analysis (moderation effect), the region of significance for the moderation effect is of more importance [27]. Thus, the analysis continues with examination of the region of significance for the moderation effect. From Figure 3, level of SI has an influence on SP but not on the strength of EDM influence on SP. The higher the level of SI, the higher is the level of SP, which is illustrated by the difference between the dotted line and the continuous line.

Figure 3 Region of Significance for Conditional Effect of SI on EDM – SP

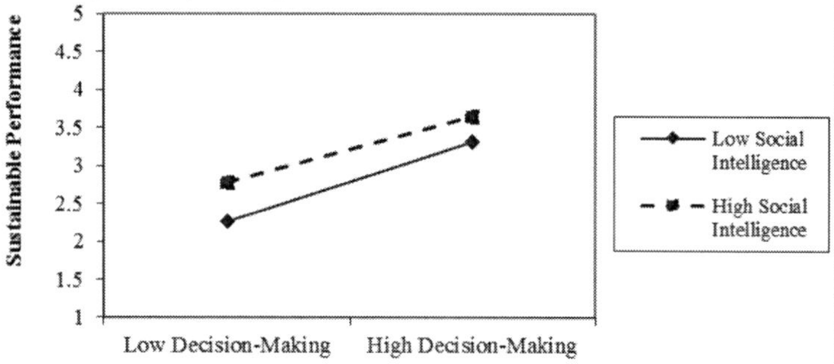

4 Conclusion and Recommendations

The study extends the understanding of the effects of EDM, and SI level of the entrepreneurs on the SP of SMEs. The resource-based view (RBV) is adopted as the underlying theory of our hypothesis development. The results reveals that although EDM has a positive influence on SP, it is causation, rather than effectuation that is the significant predictor, which contradicts prior studies [28, 2]. However, the moderated influence of SI on the EDM-SP relationship was not found to be significant. The exploratory study has proposed operationalization and measurement for the constructs of SP based on adoption and adaptation of previously validated measures available in entrepreneurship literature.

Reference

1. B. Cohen, B. Smith, R. Mitchell. Bus. Strtg. & Envr. **19(2)**, 107-119 (2008)
2. S, Read, M. Song, W, Smit. J. Bus. Vent. **24(6),** 573-587 ((2009)
3. D, Goleman, HBR. **86(9),** 74-81 (2008)

4. V. Parida, N.M. George, T. Lahti. J. Bus. Rsch. **69(11)**, 4815-4819 (2016)
5. J.F. Kihlstrom, N. Cantor. *Hndbk. Intel.* 359-379 (2000)
6. E.D. Heggestad. J. Pers. Assm. **90(1),** 102-104 (2008)
7. D. Silvera, M. Martinussen, T.I. Dahl. Scnd. J. Psy. **42(4),** 313-319 (2001)
8. R.E. Boyatzis, D. Good, R. Massa. J. Ledrsh. & Org. Std. **19(2),** 191-201 (2012)
9. J. Hartog, M. van Praag, J. van der Sluis. J. Eco. & Mgt. Strtg. **19(4),** 947-989 (2010)
10. P.J. Richard, T.M. Devinney, G.S. Yip, G. Johnson. J. Mgmt. **35(3),** 718-804 (2009)
11. J. van der Sluis, M. van Praag, W. Vijverberg. WBER. **19(2),** 225-261 (2005)
12. P. Garengo, S. Biazzo, U.S. Bititci. Int. J. Mgmt. Rev. **7(1),** 25-47 (2005)
13. NSDC. *SME Annual Report 2014/15* (2014)
14. NSDC. *SME Annual Report 2011/12* (2011)
15. J.F. Hair, G.T.M. Hult, M. Sarstedt. *A primer on partial least squares structural equation modeling (PLS-SEM)* (2013)
16. J. Henseler, C.M. Ringle, R.R. Sinkovics. Adv. Int. Mktg. **20(1),** 277-319 (2009)
17. J. Hulland. Strtg. Mgt. J. **20(2),** 195-204 (1999)
18. B.M. Byrne. *Structural equation modeling with AMOS: Basic concepts, applications, and programming.* (2016)
19. S.A. Snell, J.W. Dean Jr. T. Acad. Mgt. J. 467-504 (1992)
20. V.E. Vinzi, W.W. Chin, J. Henseler, H. Wang. *Handbook of partial least squares: Concepts, methods and applications.* (2010)
21. J. Henseler, C.M. Ringle, M. Sarstedt. J. Acad. Mktg. Sci. **43(1),** 115-135 (2015)
22. T. Ramayah, J. Cheah, F. Chuah, H. Ting, M.A. Memon. *Partial Least-Squares Structural Equation Modeling (PLS-SEM) using SmartPLS 3.0.* (2017)
23. R.B. Kline. *Convergence of structural equation modeling and multilevel modeling.* (2011)
24. L. Hu, P.M. Bentler. SEM: A Multdcpl. J. **691),** 1-55 (1999)
25. J. Lohmoller. *Latent variable path modeling with partial least squares.* (2013)
26. N. Kock, G.S. Lynn. J. Assoc. Info. Sys. **13(7),** 546-580 (2012)
27. A.F. Hayes. *Introduction to mediation, moderation, and conditional process analysis: A regression-based approach.* (2008)
28. S. Read, S.D. Sarasvathy. J. Priv. Eqt. **9(1),** 45-62 (2005)

The Effectiveness of Cognitive and Psychomotor Domain of Culinary Art Students' Performance after Internship in Private Colleges

[1]*Hairuddin* Harun, [2]*Abdul Wafi* Abdul Rahman, [1]*Noorazman* A. S.,[1]*Siti Noor Fazelah* Mohd Noor, and [2]*Adibah Aishah* Md Sahak,

[1]Universiti Tun Hussein Onn Malaysia,
[2]Kolej Yayasan Pelajaran Johor;

Abstract. With the demand of culinary arts graduates in hospitality industry, more higher learning institutions especially private colleges offer the programs. The course syllabus of culinary arts is specifically designed to provide a strong foundation for students who aspire to be chefs in the local and international fields. Students are equipped with a basic education in the culinary skills and knowledge associated with the cognitive and psychomotor domain. This study investigates the influence of the cognitive and psychomotor domain effect to private college student's performance after internship. The internship program is gradually enhancing the students' knowledge; confidence level and psychomotor performance which enable them to at least gain confidence when performing their practical assessment after coming back from internship. This is a positive indication in the beginning of the students' life before expose into a real life work situation. Thus, this research can be a guidance for the private institutional lecturers to look into the effectiveness of cognitive and psychomotor domain of culinary art students' performance in their internship programs.

Keywords: Private College, Cognitive and Psychomotor Domain, Performance, Internship

1 Introduction

Private High Education aimed to prepare the well-balanced students with theoretical knowledge and practical skill. The industry demands more on the practical skills and knowledge for students to embark the performance of students after internship. Furthermore, according to Arokiasamy (2011), the private colleges started to escalate in Malaysia in earlier in 1996 during enforcement on the Private Universities in Malaysia.

There are 600 private higher learning institutions college that complement the work of public institutions in providing high education. Depending on the syllabus from the institutions, the comprehensive knowledge and skills can affect the performance of the student after internship. It also improves student's prospect to perform well after internship to move to another stage and level of their study. Based on their syllabus, the Bloom taxonomy is the most important guideline to the educator in leading their students' learning objective outcome after internship.

Cognitive and psychomotor domain are among being measured by educational psychologist especially on the level of skills that requires coordination of movement of their physical, knowledge and the technique to assess the students after internship (Huit, 2009). From there, the educator know the level of confident and competencies of students after internship through the assessments given after internship. According to Neuman and Banghart (2001), the relationship between college and industry is the strength to achieve the college's goals to provide the high skills and knowledge on students comparable to industry's needs

By working in the industry for the internship for a few months, students will have the opportunity to expose themselves to apply knowledge learned in the theory class and gain the practical perspectives toward work and society. Chen (2012) agreed that the theoretical knowledge and practical experiences gained during internship programs maintain a well-trained workforce. Internship program has been

*Corresponding author: hairuddin@uthm.edu.my

practiced for long time to develop the skills and knowledge which is beneficial to the students

Through the internship program, the institutions create the goal and objectives to ensure the students acquire the specific outcomes from the internship programmed. With certain objectives before they go for the internship program, the adequate supply of knowledge and skills must be provided. Mani (2010), stated that Internship is the vital factor in the every firm to build the right ability and capability. Cognitive and psychomotor performance after internship is important to know to what extent their practical skills and knowledge can be improved after internship and the competency of the students after internship through the assessment.

Before going to the practical places, the basic knowledge and skills during prior semester before internship have already been taught by the lecturer based on their syllabus. Yusof (2011) to examine the content of those programs as well as the strength and weakness are important.

The serious and prolonged issues of the performance of the students after internship are currently focused on skills and knowledge, the problem detected from this situation in area internship program is the lacking of students theory skills. From the observation of this situation culinary students' who came back from their internship could not answer several basic questions. For instance some subjects were learned in the Basic Western Kitchen class. Fundamental question such like "What are five basic mother sauces", "French food terminologies" could not be answered well by majority of students who just came back from their internship is became an enquiry.

This situation is aligned with the problem faced by the hotel industry, where the problems were felt profoundly in the cooking tasks (Young and Corsun, 2009). Cooking tasks requires students to possess sound background skills from culinary. From the observation, cognitive and psychomotor domain were examined toward students' performance after internship to know the level of skills and knowledge of students based on their performance. The guideline before they do the internship gives impact toward performance of students with the adequate supply of basic skills and knowledge to sustain the flexibility of students involved. Mani (2010) supported that the internship is the vital factor in the every firm to build the right ability and capability depend on the basic skills and knowledge that students learned in the syllabus.

According to Zehr (2016), internship is the process of learning for student that involved the apprenticeship, cooperative education, experimental learning, field experience, industrial and workplace learning.

The cognitive domain related with the knowledge, comprehension and application where the psychomotor domain is involving with adaptation, origination and perception. The current study is to enclose this gap and add volume to the body of knowledge through understanding the dimensions of cognitive and psychomotor domain that significantly influence the performance of students after internship.

2 Method

The research were conducted in three private colleges around Johor Bahru. A total of 190 of students from the Culinary Arts, based on the information given by human resource private colleges in Johor Bahru took part in the study. Krejcie and Morgans (1970) stated that in determining sample size of 127 students with total population 190 students is sufficient to collect the data. The instrumentation of this study used five-point Likert scale. Other than that, the scale in the Likert scale refers to the total sum of all Likert items in the questions. This research also used interval scale. Value of 1 refer to (strongly disagree) the most disagree attitude in the scale, 2 (disagree), 3 (neutral), 4 (agree) and 5 (strongly agree) the most strongly attitude in the scale. The data was then coded and keyed for analysis using a Statistical Package of Social Science (SPSS), Version 20.Analysis of variance (ANOVA) used to explain the interaction effect of domain toward performance's student after internship. Hair et al., (2007) and Pallant (2011) suggested that the idea Cronbach's alpha coefficient of a scale should be above .70 to indicate high internal consistency. Hinton (2004) suggested that an alpha score in the range of .50 to .70 is commonly accepted and considered to be criterion for demonstrating the internal consistency of reliable scale. In certain cases, a minimum reliability level of .50 was also accepted and this is consistent (Hinton, 2004). Based on Hair et al., (2007), the reliability test can also be interpreted according to the strength using the rules of thumb.

3 Results and Discussion

This section discussed the research objectives and research questions. The research objectives are:

a) To identify the level of the cognitive domain of Culinary Arts students' performance after coming back from the internship.
b) To identify the level of psychomotor domain of Culinary Arts students' performance after coming back from the internship.
c) To examine the predictor of the cognitive domain and psychomotor domain.

Research Objective 1: **To identify the level of the cognitive domain of Culinary Arts students' performance after coming back from the internship.** This examines the first research question as well as the first hypothesis along with the three sub-hypothesis in this study:

As previously mentioned, this study exploits the standard linear regression. Cognitive domain that is a predictor for performance students after coming back from internship with the value of (β=.707, R^2=.499). Thus, answering the first research question. This results is significant and same goes with the sub-variables, whereby it is supported with a significant unique contribution to knowledge (β=.708, R^2=.502), comprehension (β=.561, R^2=.315) and application (β=.503, R^2=.253).

These results proved that knowledge contains strong relationship toward performance of students after internship, following by the comprehension and application there also have a strong relationship toward the performance of students after coming back from internship.

Research Objective 2: **To identify the level of the psychomotor domain of Culinary Arts students 'performance after coming back from the internship.** These examine the second research question as well as the second hypothesis along with the three sub-hypothesis in this study:

As previously mentioned, these studies exploit the standard linear regression. Psychomotor domain that is a predictor for performance students after coming back from internship with the value of (β=.525, R^2=.275). There have a strong relationship between independent variables (psychomotor domain) and dependent variables (performance of student after coming back from internship). For the sub-variables, all the sub-variables also have the strong relationship with the contribution to adaptation (β=.526, R^2=.277), origination (β=.359, R^2=.129) and perception (β=.459, R^2=.211).

The dimension of psychomotor domain measured in the research were involved between adaptation, origination and perception toward performance. The highest degree relationship is between adaptation and performance(r = 0.34**, <0.00) compared to perception and performance (r = 0.21*, P = 0.19) The very weak correlation is between origination and performance(r = 0.21*, p <0.03) For the relationship between cognitive domain toward performance of students after coming back from internship, the beta value recorded at .707 (p<0.001), which is higher and stronger compared to the beta value of psychomotor domain at .525 (p<0.001) respectively were recorded. It can be concluded that performance of students after coming back from internship have stronger contribution level of cognitive domain with sub-variables knowledge, comprehension and application.

From the second variables, it also showed that psychomotor domain significantly influence the performance of students after coming back from internship. The sub-hypothesis, adaptation is highest rather than perception and origination. It is conclude that, student know how to recognize task given by lecturer and chefs based of the knowledge they have before going to internship, in effect, after coming back from internship the level of adaptation, perception and origination became stronger.

For the lecturers, the basic knowledge of theory and practical must be parallel to expose the students on the experience how to use the knowledge. Doubtlessly, knowledge is cognitive domain which one of the most important components in all three domains and in facts as the tools for students to succeed their studies. Moreover, it is the most challenging piece of academic work which determines their rate of internship.

4 Conclusion

From this study, it is revealed that cognitive domain which comprises of knowledge, comprehension and application are significantly influenced the performance of Culinary Arts students after coming back from internship. The findings also showed during classroom/practical class students did not fully utilized their cognitive domain. Even though skill is an important part in the culinary programs but if the students are lack at cognitive level, it will lower their performance. In fact, the ability to think, understand and apply should be practiced in order to improve their performance. It is equally important in the psychomotor, students must have all elements of psychomotor domain (adaptation, origination and perception) with the existence of cognitive domain in order to perform well. In other words if students need to have a strong skills performance, he or she need to have strong knowledge similarly to students who want to be good in a theory part, he or she must have a good skills too. Indeed, students need to have both knowledge and skills in order to be a successful and good chef in this industry.

Reference

1. Arokiasamy.A.R.A. (2011). *An Analysis of Globalization and Higher Education in Malaysia*. Australian Journal of Business and management research.
2. Chen. T.L & Shen.C.C. (2012). *Today's intern,tomorrow's practitioners?* The influence of internship program's on student career development in the Hospitality Industry.

https://doi.org/10.1051/matecconf/201815005021

3. Huit, W. Bloom et al.'s (2009) taxonomy of the cognitive domain. Educational
4. Mani, V. (2010). *Evaluating effectiveness of Executive training*. International Buletin of Business Administration.
5. Neuman. R.B., & Banghart.S, (2001) *Industry-university "consulternship": an implementation guide*. International journal of education management.
6. Zehr, S. M. (2016). *Student internship experience and learning opportunities*: A mixed method study. University of Illinois at Urbana-Champaign,
7. Pallant, J. SPSS Survival Manual: *A Step by step guide to data analysis using SPSS* (4th Edition). Crows Nest, N.S.W: Allen & Unwin.
8. Hair, J. F., Money, A. H., Samouel, P., & Page, M. (2007). *Research Methods for Business*. West Sussex: John Wiley and Sons Limited
9. Yusof, M. F. M & Ahmad, G (2011).*Tourism and Hospitality education in Universiti Malaysia Kelantan*: An Entrepreneurial Approach
10. Young, C. A &Corsun, D. L Burned! (2009). *The impact of work aspect, injury, and job satisfaction on unionized cooks' intention to leave the cooking occupation*. Vol 34, Issue 1, 2010

Risky Riding Behaviour Among Malaysian Young Motorcyclists

Muhammad Safizal Abdullah[1], *Muhammad Asyraf* Mohamed Kassim[1], *and M. F.* Mansor[1]

[1] School of Business Innovation and Technopreneurship, (UniMAP)

Abstract. The main objective of this study was to explore the relationship between riding attitude, sensation seeking, anger and riding confidence with risky riding behaviour among motorcyclist at University Malaysia Perlis (UniMAP). Partial Least Squares of Structural equation modelling (SEM) and Statistical Package Social Science (SPSS) are utilized to determine the direct effect of the variables. The results discovered that riding attitude, sensation seeking, and anger had shown positive significant relationship towards risky riding behaviour while riding confidence has shown negative significant relationship towards risky riding behaviour.

Keywords: Riding Attitude, Sensation Seeking, Anger, Riding Confidence, Risky Driving Behaviour

1 Introduction

Road accident seems to be a never-ending issue in Malaysia. Statistics shows that road accidents record an increase each year as shown in Figure 1.

Figure 1. Road Accidents in Malaysia 2010 - 2015

Statistics provided by the Department of Road Transport has shown an increase in the number of accidents from year 2010–2015. In 2015, the number of accidents were 490,000 cases in which it has been an increase about 13.41% compared to year 2014, and the number of also increased from 6706 to 6674 in the respective time periods.

Motorcyclists are more prone to accident when they are on the road compared to drivers of other vehicles. According to the Road Safety Department of Malaysia, of the 131 deaths from road accidents, 85 cases involving motorcycles in 2015 from January to July. While, according to Royal Malaysia Police (RMP) report, among 6,706 deaths in 2015, 3,816 were motorcyclists and 387 were with the passengers. Individuals aged between 16 to 20 years old had the highest number of deaths in motorcycle road accident which amounted to 934 cases, followed by the age between 21 to 25 years (828 cases), 26 to 30 years (641), 31 to 35 years (579), and the remaining 36 years above. Therefore, this study will examine the risky riding behaviour among the young motorcyclists, and factors influencing risky riding behaviour will be investigated.

2 Underpinning Theory

Theory of Planned Behaviour (TPB) is widely used in various studies regarding human behaviour. According to TPB, there are three main predictors affecting individual intention towards behaviour, namely attitude toward the behaviour, subjective norm, and perceived behavioural control (Ajzen, 2005). Ajzen (1991) argued that individual's intention was a determinant of their behaviour. Intentions were cause by attitude, subjective norms, and perceived behaviouralcontrol. Therefore, this study utilised TPB theory. According to Ajzen (1991), TPB helps to understand how we change the behaviour of people. The TPB is a theory, which predicts deliberate behaviour, because behaviour can be deliberative and planned. Therefore, the theory TPB is a reference that can be used as an underpinning theory in this study.

3 Methodology

In this study, four predictors will be examined in relation to risky riding behaviour such as riding attitude, sensation seeking, anger, and riding confidence as the theoretical framework in figure 2.

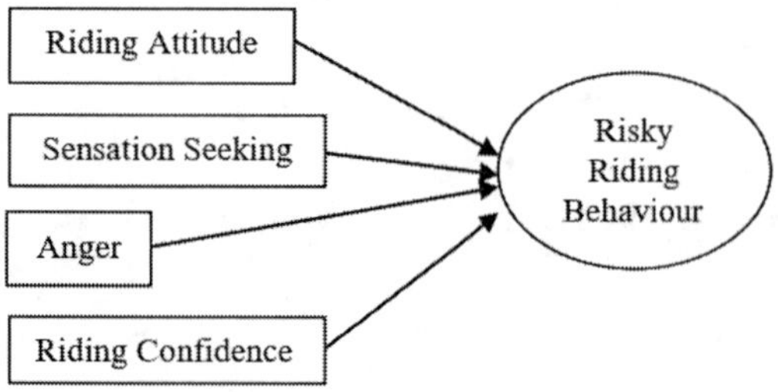

Figure 2. *Risky Riding Behaviour Theoretical Framework*

This framework shows the main factors that will have significant effect on risky riding behaviour. Therefore, the hypotheses of the study will be:

H1: Riding attitudes has positive significant effect on risky riding behaviour
H2: Sensation seeking has positive significant effect on risky riding behaviour
H3: Anger has positive significant effect on risky riding behaviour
H4: Riding confidence has negative significant effect on risky riding behaviour

Self-administered questionnaires were developed and distributed to young undergraduate students in UniMAP. Respondents were students aged 19 - 21 years old. The questionnaires were distributed using a simple random sampling method. In total, 600 questionnaires containing riding attitude, sensation seeking, anger and riding confidence scales and risky riding behaviour assessment were distributed by hand to the respective students. The scales were measured using a five-point Likert scale ranging from (1) "strongly disagree" to (5) "strongly agree". 285 questionnaires were returned and the overall response was 47.5 per cent. PLS-SEM were used to assess the goodness of measures and to test the hypotheses for this study.

4 Findings

4.1 Respondent's Profile

Descriptive statistics of the final sample are shown in Table 1. These variables are considered as control variables in the following analysis.

Table 1. Profile of respondents

Profile	Description	Frequency
Gender	Female	110
	Male	175
Marital Status	Single	280
	Married	5
Ethnic Group	Malay	134
	Chinese	107
	Indian	40
	Other	4
Types of Licenses	B2	140
	D	55
	B2 & D	72
	None	18
Preferred Vehicle	Motorcycle	200
	Motorcar	85
Driving Experience	Below 5 Month	55
	5 Month – 1 Year	103
	1 Year – 5 Year	98
	5 Years Above	29

4.2 Goodness of Measures

The goodness of measures for the variables in the study was assessed using construct validity, convergent validity, discriminant validity, and reliability analysis. For the purpose of construct validity, Hair et al. (2013) indicated that every loading that are greater than 0.50 on a particular construct are considered significant. Therefore, loadings below 0.50 were removed and there were no reported cross loadings. As a result, the constructs were valid for this study. Table 2 shows the factor loadings for each construct.

Table 2. Results of measurement model

Model Construct	Items	Loadings	AVE	CR
Riding Attitudes	RA1	0.778	0.761	0.863
	RA2	0.739		
	RA3	0.762		
	RA4	0.766		
	RA5	0.779		
	RA7	0.720		
	RA8	0.668		
	RA9	0.709		
Sensation Seeking	SS1	0.858	0.749	0.907
	SS2	0.830		
	SS3	0.832		
	SS4	0.820		
	SS5	0.812		
	SS6	0.771		
	SS7	0.795		
	SS8	0.887		
Anger	A1	0.826	0.756	0.929
	A2	0.682		
	A3	0.816		
	A4	0.593		
	A5	0.801		
Riding Confidence	RC1	0.742	0.746	0.914
	RC2	0.869		
	RC3	0.722		
	RC4	0.869		
	RC5	0.824		
	RC6	0.845		
	RC7	0.783		
Risky Riding Behaviour	RB1	0.784	0.883	0.945
	RB2	0.676		
	RB3	0.748		

	RB4	0.814		
	RB6	0.660		
	RB7	0.709		
	RB8	0.594		
	RB9	0.839		
	RB10	0.788		

Then, the correlations between the measures are compared with the square root of the AVEs in order to assess the discriminant validity. Table 5 exhibits all of the correlations between the measures were smaller than the square root of thee shown on the diagonals. Therefore, the items measuring the constructs discriminant validity for this study are satisfactory. Finally, reliability was also accessed via Cronbach's alpha coefficient, which should have a value higher than the recommended value of 0.60 as indicated by Nunally and Berstein (1994). In addition, all of the constructs have alpha values of above the recommended value and therefore, the measures are reliable.

Table 3. Results of discriminant validity

	A	RA	RC	RB	SS
A	**0.8**				
RA	0.5	**0.7**			
RC	0.6	0.6	**0.7**		
RB	0.5	0.6	0.5	**0.8**	
SS	0.6	0.6	0.5	0.7	**0.9**

Note: The bolded diagonals represent square root of the AVE while the off-diagonals represent the correlation.

4.3 Structural Model

The purpose of structural model is to determine the relationship between variables that were hypothesized in the framework. Hair et al., (2013) indicated that the structural model is a set of one or more dependence relationships linking the hypothesized model's constructs, such as the structural theory, which refers to the conceptual representation of the relationships between constructs. In addition, the structural model is more helpful in representing the interrelationships between constructs (Hair et al., 2013). On top of that, because of the main objective of PLS is to predict the structural model was assessed in terms of the coefficient of the variance explained (R2) and the significance of path coefficients. (Hair et al., 2013).

4.3.1 Direct Effects

As shown in Figure 3, risky riding behaviour had an R2 value of 0.817, which was on the high level of variance explained and also indicated that 81.7% of the variance in risky riding behaviour could be explained by the dimensions of riding attitude, sensation seeking, anger and riding confidence. The results from bootstrapping showed that riding attitude ($\beta = 0.41$, t = 6.00, p < 0.01 (1 tailed), sensation making ($\beta = 0.513$, $t = 10.015$, $p < 0.01$ (1 tailed), anger ($\beta = 0.823$, $t = 2.112$, $p < 0.01$ (1 tailed), and riding confidence ($\beta = -0.061$, $t = 1.280$, $p < 0.01$ (1 tailed) had positive significant effects on risky riding behaviour.

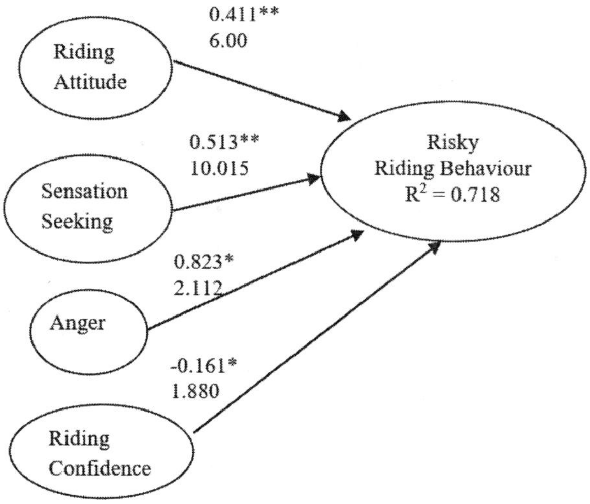

Figure 3. Results of structural model direct effects between riding attitude, sensation seeking, anger and riding confidence and risky riding behaviour.
Note. Value within parenthesis is t-value. $^{**}p < 0.01$ ($t > 2.326$), $^{*}p < 0.05$ ($t > 1.645$) (based on one tailed test).

5 Discussion

After assessing the validity and reliability of the measurement model, structural model is then examined to test the hypothesis of the study.

The first hypothesis (H1) assume that the attitudes of riding is significantly affects the risky riding behaviour. Results found that the assumptions are correct where (riding attitude) has a significant positive correlation with risky riding behaviour. In conclusion, H1 is supported or accepted.

The second hypothesis (H2) concluded that sensation seeking is significantly affects the risky riding behaviour. The results obtained showed that sensation seeking and risky riding behaviour has a significant positive relationship. Thus, H2 also was supported and accepted hypothesis.

The third hypothesis (H3) also assume that anger is significantly affects the risky riding behaviour. Results revealed that anger has a significant positive correlation with risky riding behaviour and the results have been made H3 also supported and accepted in the study.

Finally (H4) stated that confidence riding is significantly affects the risky riding behaviour. The results of these experiments, it has been shown that riding confidence provide results significantly negative correlation with risky riding behaviour. Thus, H4 is also a hypothesis that is accepted and supported. The summary of hypotheses is shown in 4.

Table 4. Summary of hypotheses

Hypotheses Statement	Results
H1: Riding Attitude has significance positive relationship with RB.	**Accepted**
H2: Sensation Seeking has significance positive relationship with RB.	
H3: Anger has significance positive relationship with RB.	
H4: Riding Confidence has significance negative relationship with RB.	

5.1 Riding Attitude, Sensation Seeking, Anger, Riding Confidence and Risky Riding Behaviour

The present study revealed that riding attitude, sensation seeking, anger and riding confidence have a positive significant relationship with risky riding behaviour.

This result can be supported by previous study on riding attitude such as by Jinn-Tsai Wong, Yi-Shih Chung, and Shih-Hsuan Huang (2009) and Deborah, Barry, Katherine, Melissa, Cynthid, and Darren (2011). This means that the riding attitude can lead to a safety risk to motorists. An attitude such as ignoring traffic rules must be ignored to escape from heavy traffic is one of the risky driving behaviour. Most young drivers age 17-21 were discovered to be more prone to ignore the traffic rules in order to escape from heavy traffic.

Next, the significance of sensation seeking and risky driving behaviour has been supported by previous studies such as Ching-Fu Chen (2009). Sensation seeking has been discovered to be one of the crucial factors of risky riding behaviour.

On the other hand, the findings of significance relationship of anger and risky riding behaviour is proven valid as motorcyclist will trigger a tremendous attitude like bringing speed, vehicle interjected, and trying to keep up with other vehicles when they feel angry because being honk by other vehicles as it will lead to the risk of other riders or other road users.

Lastly, findings revealed that riding confidence has negative significance relationship with risky driving behaviour as more experienced driver will have more confidence to drive a vehicle and this will reduce the risk to own and other on the road.

6 Conclusion

Study on risky riding behaviour is not a new investigation. This is because other countries are also facing the same problem on the issue of accidents is increasing in their respective countries. With this type of research, one of the goals of this study is to provide a deeper awareness to motorcyclists will attitudes that tend to risky riding behaviour, especially to motorcyclists in UniMAP, Perlis, Malaysia. To achieve this, the current study was to propose and validate a model consisting of four independent variables, riding attitude, sensation seeking, anger and confidence riding. The proposed model successfully provide significant results such as riding attitude, sensation seeking and anger found to be significantly positive with risky riding behaviour, while riding confidence has a negative significant with risky riding behaviour.

Last but not least, the issue of accidents involving attitudes rider or driver has received less attention in research to date. Therefore, the study of this must be improved in order to help bring awareness to the community about the importance of personal safety on the road.

References

Ajzen, I., 1988. Attitudes, Personality and Behaviour. Open Press University, Buckingham.

Malaysia, J. K. J. R. (2015). *Indeks Kematian Jalan Raya 2004-2014*: Jabatan Keselamatan Jalan Raya Malaysia.

Malaysia, J. K. J. R. (2015). *Indeks Kematian Jalan Raya 2004-2014*: Jabatan Keselamatan Jalan Raya Malaysia

Sadali, I. (2016). PDRM rekod 6,706 kematian akibat kemalangan jalan raya pada tahun lalu – purata 18 maut sehari [Electronic Version].

Langen, A. D. (2009). CHAPTER 3 Research Design and Methodology. *Example 3 Reseacrh Design and Methodology*.

Chen, C.-F. (2009). Personality, safety attitudes and risky driving behaviors—Evidence from young Taiwanese motorcyclists. *Accident Analysis and Prevention*.

Deborah J. Tunnicliff a, B. C. W., Katherine M. Whiteb, Melissa K. Hydeb,, & Cynthia C. Schonfelda, D. E. W. (2011). Understanding the factors influencing safe and unsafe motorcycle rider intentions. *Accident Analysis and Prevention*.

Jinn-Tsai Wong, Y.-S. C., Shih-Hsuan Huang. (2009). Determinants behind young motorcyclists' risky riding behavior. *Accident Analysis and Prevention*.

Mehri Ali, M. M. S. S., Morowatisharifabad Mohammad Ali,,&Haidar, N. (2010). Determinants of helmet use behaviour among employed motorcycle riders in Yazd, Iran based on theory of planned behaviour. *Injury*.

Nur Sabahiah Abdul Sukor, A. K. M. T., Satoshi Fujii. (2016). Analysis of correlations between psychological factors and self-reported behavior of motorcyclists in Malaysia, depending on self-reported usage of different types of motorcycle facility. *Transportation Research Part F*.

TürkerÖzkan, T. L., BurakDo˘gruyol, ZümrütYıldırım, Ahmet Coymak. (2011). Motorcycle accidents, rider behaviour, and psychological models. *Accident Analysis and Prevention*.

Wong, Y.-S. C. a. J.-T. (2011). Beyond general behavioral theories: Structural discrepancy in young motorcyclist's risky driving behavior and its policy implications. *Accident Analysis and Prevention.*

The Influences of Effectiveness, Competitive Advantages and Market Accessibility on SME Performance in Malaysia.

Razleena Razali[1,*], *Ummi Naiemah* Saraih[1], *Mohd Shahidan* Shaari[1], *Mohd Juraij* Abd Rani[1]
Aidanazima Abashah[1]

[1]School of Business Innovation and Technopreneurship, Universiti Malaysia Perlis (UniMAP), Malaysia

Abstract. The advent of online business has changed the pattern of doing business recently. This circumstance puts more pressure on SMEs to sustain their position in the fiercely competitive market. Unlike multinational companies, SMEs have several inevitable weaknesses in terms of planning strategy, technology exposure as well as resources that serve as a stumbling block to their better and competitive performance. Thus, online business appears to be a new medium to enhance the performance. Therefore, this study aims to investigate the influences of effectiveness, competitive advantage and market accessibility on SME performance in Malaysia. This study uses cross sectional study which focuses on SMEs that are engaged in online businesses. 200 questionnaires were distributed particularly to SMEs in food and beverages (F&B); apparel and textile (A&T); and health and cosmetic (H&C) industries in Malaysia. Results indicate that market accessibility is the only variable which has an influence on entrepreneurs in online businesses within SMEs performance (β=.48, p<.01). This study can shed light on SME performance to the government. For example policy makers can enhance programs and initiatives in the SMEs Master Plan for the year 2012-2020. This study also helps the SMEs avail themselves of the online business opportunity in enhancing their performance. Therefore, the implication of market accessibility in influencing SME performance is discussed in this study.

Keywords: Effectiveness; competitive advantage; market accessibility; online business; SME performance

1 Introduction

SMEs are one of the country's backbones in driving Malaysian economic growth close to other developed countries. Therefore, SMEs is encouraged to find ways such as engage in online business in ensuring their business remains competitive in the market. Online business helps firms become cost effective and at the same time improve productivity and quality. Thus, the online business appears to be a new medium to enhance business, especially in achieving better performance ad sustainability.

The idea of online business starts takes place and gradually replacing the traditional business system and some of business uses it as the channel of communication for their business, especially in creating a better and effective relationship with the customer and other stakeholders. In addition, the online business can be done faster and cost effective to the customer as well as for the businesses. Hence, that is why business starts chooses to go online, especially for SMEs in a certain business field such as food and beverages textile and apparel, as well as beauty and cosmetic.

According to [1], although 90% SMEs in Malaysia are using personal computers in their businesses, the IT implementations are still at the fundamental stage. The result is parallel with a research finding by [2] where the SMEs usage of online business among Malaysian SMEs is still low. To enhance the usage of online business, [3] has proposed online business or e- commerce adoption and implementation in SMEs business. This is because; the increasing of world population and internet user can be one of the opportunities for Malaysian SMEs to gain and improve their business performance. Besides, online business gives opportunity to improve SME's performance in domestic and international market. Unlike multinational companies, most of SMEs especially in Malaysia has weaknesses in term planning strategy, technology exposure as well as resources such as capital that become barriers for their expansion of business and operation abroad. Thus, the easiest and fastest strategy to reach the market is through online business strategies [4].

According to [5], online business will change Asia more than it changes the U.S. This shows that online business plays an important role to help the company change the way of doing business with local and overseas client and agents more cost effectively. This

* Corresponding author: razleena@unimap.edu.my

statement supports this study where SMEs in Malaysia is part of Asia and these SMEs will deal with the advancement of technology such as online business even though they will face a lot of challenges difficulties in implementing the online business. Hence, with the current scenario and justification made earlier, the primary aim of this study is to reveal the influences variables to the SME performance by the SMEs that already implemented the online business in their business.

2 Literature Review

2.1 SME Performance

The performance and growth of SMEs is a major driver and indices for the level of industrialization; modernization; urbanization, gainful and meaningful employment for all those who are able and willing to work, equitable distribution of income, the welfare, income per capital and quality of life enjoyed by the citizenry [6].

Performance is a widely used concept in many areas. Usually, performance is a measure of how well a mechanism or process achieves its purpose. In enterprise management, [7] defines an organization's performance as how well the organization is managed and "the value the organization delivers for customers and other stakeholders." For the purposes of this research, SME performance is related to sales based performance and organizational based performance, which is also focused on the financial part such as profitability, growth, cost reduction, market share and Return on Investment (ROI) and non-financial part which is customer satisfaction, employee satisfaction and productivity as well as innovation.

Studies have found that there is a significant relationship between competitive advantage and the sales-based performance of SMEs, when sales-based performance was measured by the level of sales revenue, profitability, return on investments, productivity, product added value, market share and product growth [8]. Additionally, other studies also further illustrate a significant relationship between effectiveness and the organizational-based performance of organizations, when organizational-based performance was measured by the emphasis on efficient organizational internal processes, customer satisfaction, employee development and job satisfaction [4].

Competitive advantage and business performance are two different constructs with an apparently complex relationship [9]. Since the 1980s, many scholars began to study online business on marketing activities of SMEs. Most scholars believe that with the development of internet technology, the performance of online business applications in SMEs has a significant promotion effect; in addition, some scholars also found some problems in the implementation of online business enterprises, and put forward the path and strategy of enterprises to optimize the mode of online business [10, 11]. A study of [12] investigated the impact of information technology on the competitive environment, and how to use internet technology to improve the competitive advantage of enterprises. The finding was the internet technology promotes the enterprises to develop new business, and bring more business opportunities for the enterprise.

As analyzed by [13], more than 40 SMEs from Canada, China, Japan and Mexico developed the online business in the international market. The result shows that online business can reduce product and service cost in a unique way, and combine the information and capital at the same time, then enhance enterprise effectiveness and competitive advantage. This is aligned with most previous studies on SMEs which is online business give a lot of impact towards SMEs' performance and it's just the SMEs do not take full charge of the opportunity of the online business. As discussed, the online businesses create effective business, competitive advantage and business have more market accessibility in domestically and internationally which directly associated with business performance. Therefore, studying SME performance would provide insights into the existing literature in the domain of effectiveness, competitive advantage and market accessibility.

2.2 Effectiveness

The key of online business effectiveness is its ability to target the right customers, cost efficiency and analytics. Adoption of online business is not only strengthening growth possibilities, but also creating a network with all other business in the world, can have cooperation, and can improve quality and knowledge [14].

According to [15], online business is important to SMEs because it will help to develop the efficiency of the organization. Several studies finding shows that use of e- business increase sales and can reduce business costs and improve productivity of the business [16]. The use of online business can improve business competitiveness with the internet providing numerous opportunities for SMEs to compete equally with large corporations [17]. The study conducted by [18] argued that the flow of information in an organization is the blood life of any business operating unit irrespective of its size. It is commonly accepted that online business provides many potential benefits to organizations so as to make them more efficient, effective and competitive [19].

Meanwhile, [3] suggests that SMEs should implement online business since it's really effective for business productivity and SME performance especially in online services. Additionally, online business is the kind of technology that enables flexibility and cost-effectiveness is now considered a requirement for SMEs in the current competitive global market. By the implementation, it adds value to the company in its day-to-day business transactions. Realizing the effectiveness of online business to business, SMEs are encouraged to involve more with technology. This is helping SMEs to become more effective in dealing with the challenging world [3].

2.3 Competitive Advantage

Competitive advantage is a base for a good strategy, and a good one creates competitive advantage [20]. Recently, [21] claim that IT and the Internet in particular can create a competitive advantage and improve performance and competitiveness and also it has been argued that achieving a position of competitive advantage is a necessary precursor to a SME significant performance. One of the ways for companies to be ahead of its competitor is to stay updated and in-trend is via engaging in online business. Online business is one of the strategies of the company and SMEs implement in their business nowadays.

Having online presence can be beneficial for many SMEs with adequate preparation prior to its implementation. They gain competitive advantage through online business. The online business can be the resourceful medium or platform for SMEs to gain competitive advantage against its competitor. This can be supported by the theory of Resource- Based View (RBV). RBV stipulates that the fundamental sources and drivers of competitive advantage and superior performance are chiefly associated with the attributes of resources and capabilities, which are valuable and costly-to-copy [22].

In literature on the other industry such as tourism and hospitality, result indicates the primary role of IT in the hotel industry is to improve its productivity and lead to a better SME's performance. Many studies on tourism and hospitality industry claim that ICT, especially online business is one of a competitive advantage factors. This can be the proof that online business gives competitive advantage to the companies. The existence of website, application for mobile and advertising via social media by this industry shows the importance of online business.

Online business has received much attention in the last two decades due to its effects and potential impact on organizational success especially for SMEs. However, according to [23], the degree to which a small-to SMEs may experience the internet as a threat or opportunity in developing business strategies will depend on industry and firm factors such as: The small firm's technical and internet knowledge, rate at which the market is growing, pace of innovation and change in the industry, technical and internet strengths of the larger competitors, sources of competitive advantage for the small business, strategic intent of the larger competitors, structure of the industry in which the firm competes.

2.4 Market Accessibility

Market access can be defined as the process to ensure that all appropriate patients who would benefit get rapid and maintained access to the brand, at the right price. The goal of business is to gain wider market accessibility domestically, internationally and globally so that their business can be a legacy or sustainability in the world besides making a profit and achieve business growth. Thus, one of the ways to make these goals become reality is through e- business or online business.

For many SMEs, the unpredictability of doing business and the complexity of procedures and regulations in many countries are perceived as major barriers. Furthermore, access to finance, new technology and good, reliable infrastructure, especially for communications and logistics, are all special problems for growth-oriented SMEs [24]. According to [25], the internet can be a critical factor in enhancing a firm's market reach and operational efficiency. It has been agreed by many authors that invest in, and proper utilization of online business is a major way of improving productivity among the vital SME sector.

Application of online business is found to have a significant positive relationship with increased performance of SMEs in Malaysia. This result is consistent with the prior research findings by [26] where an application of information technology will enhance the overall performance of SMEs in Malaysia. The adoption of online business will assist an organization in storing information as well as communicating with their international customer, suppliers and business partner who will facilitate business transaction oversea. As a result, it will lead to a better performance in reducing the operating expenses as a whole. Online businesses also help to facilitate SMEs to penetrate foreign markets. Basically, they have the potential market in their vision and online business helps to execute this vision into reality and an achievement. This is because, when market size increases, the benefits of internationalization will increase [27].

Given all the literature, the specific hypotheses are developed as follows:

H_1: There is a significant relationship between effectiveness and SME performance.
H_2: There is a significant relationship between competitive advantage and SME performance.
H_3: There is a significant relationship between market accessibility and SME performance.

3 Methodology

The population of this research included SMEs that already implemented online business in their business from three (3) types of industries which are food and beverages textile and apparel, as well as beauty and cosmetic. This study utilized the quantitative method (eg. questionnaire). Questionnaires are distributed randomly for 200 respondents. The SPSS software (version 22) has been used for the statistical analysis. The instruments used for all variables (performance, effectiveness, competitive advantage and market accessibility) have been adapted from previous measurements. SMEs need to respond their opinion towards four variables of performance (12 items) effectiveness (9 items), competitive advantage (8 items) and market accessibility (13 items).

4 Results and Discussions

Table 1 presented the response rate based on the total number of distributed questionnaires. Out of 200 distributed questionnaires, only 101 (50.5%) of the returned questionnaires have been used for further analysis.

Table 1. Response Rate

	Figures	Percentage
Number of Distributed Questionnaire	200	100%
Number of Returned Questionnaires	109	54.5%
Number of Usable Questionnaire	101	50.5%

4.1 Respondents' Background

Most of the respondents who participated in this study were female (55.4%), it meant that and the remaining of the repondents were male (43.6%). In term of business experiences, 49.5% of respondents have 1 to 5 years of business experience. This percentage was followed by 33.7% of respondents have 6 to 10 years of business experience. About 9.9% of respondents have 11 to 15 years of business experience and 5.9% of respondents have 16 to 20 years of business experience. The lowest percentages of responses (1.0%) are obtained from respondents who have more than 20 years of business experience.

In term of industry's type, 48.5% of respondents came from food and beverage (F&B) industry; 32.7% of respondents from the apparel and textile (A&T) industry; 10.9% of respondents from health and cosmetic (H&C) industry and the remaining 7.39% of respondents were from other industries.

In the aspect of respondents' perception towards their abilities to sell the products in the foreign market, most of respondents (67.3%) ticked 'yes' for the answer. This percentage was followed by 25.7% of the respondents ticked 'in the process'. A small percentage ofv 6.9% ticked 'no' for the answer.

4.2 Reliability Analysis

The Cronbach's alpha values are reported at .85 for effectiveness, .84 for competitive advantage and .92 for market accessibility. As for the SME performance, Cronbach's alpha value is reported at .94. According to [28] the Cronbach's alpha values showed that all of the variables are good to excellent in this study.

4.3 Mean and SD Analyses

In this study, the interpretation of mean scores has been referred to the previous authors [29]. According to these authors, the mean scores ranged from low (1.00-2.00), moderately low (2.01-3.00), moderately high (3.01-4.00) to high (4.01-5.00). As presented in Table 1, the mean value for SME performance was 4.01. Therefore, high acceptance is observed for this variable. Almost similar means score is recorded for effectiveness and market accessibility; 4.02, and 4.05. Competitive advantage has been recorded to have the lowest mean score at 3.73. However, moderately high acceptance is found for this variable.

Table 2. Mean and SD Values

Variables	N	Mean	SD
SME Performance	101	4.01	.48
Effectiveness	101	4.02	.44
Competitive Advantage	101	3.73	.54
Market Accessibility	101	4.05	.55

As for the SD analysis, it is noted that the SD values for SME performance (.48), effectiveness (.44), competitive advantage (.54) and market accessibility (.55) were also high.

4.4 Correlation Analysis

Pearson correlation coefficient (r) is a technique to determine the relationship between two quantitative and continuous variables. It measured the strength of the association between the variables. In this study, the correlation analysis had been run to determine the interdependency between SME performance, effectiveness, competitive advantage and market accessibility.

As presented in Table 3, low correlation is observed between SME performance and effectiveness ($r=.37$, $p<.01$). Next, moderate correlations are found between SME performance and competitive advantage ($r=.43$, $p<.01$) and between SME performance and market accessibility ($r=.58$, $p<.01$).

Also, moderate correlations are remarked for relationships between effectiveness and competitive advantage ($r=.67$, $p<.01$), effectiveness and market accessibility ($r=.63$, $p<.01$) and between competitive advantage and market accessibility ($r=.66$, $p<.01$). In sum, all correlations are reported as significantly and positively related between each other.

Table 3. Correlation Analysis

Variables	N	1	2	3	4
1. SME Performance	101	-			
2. Effectiveness	101	.37**	-		
3. Competitive Advantage	101	.43**	.67**	-	
4. Market Accessibility	101	.58**	.63**	.66**	-

***Correlation is significant at the 0.01 level (2-tailed)*

4.5 Regression Analysis

As presented in Table 4, 35.0% of the variance in SME performance is explained by effectiveness, competitive advantage and market accessibility (R^2=.35). To further clarify, the remaining of 65.0% of variance in entrepreneurial intention among these respondents is explained by the other variables. In addition, the adjusted R^2 value also taken into account as .33.

Based on the results, only market accessibility is positively associated with SME performance (β=.48, p<.01). Thus, this implied that only market accessibility is significantly related to SME performance among respondents in this study. Therefore, H_3: *There is a significant relationship between market accessibility and SME performance* is supported in this study.

Table 4. Regression Analysis

Variables	SME Performance	
	Beta	Sig.
Effectiveness	-.05	.72
Competitive Advantage	.09	.39
Market Accessibility	.48	.00
R^2 = 0.35 Adjusted R^2 = 0.33 F = 17.121		

***Correlation is significant at the .01 level (2-tailed*

Conversely, this study found that effectiveness (β=-.05, p>.01) and competitive advantage (β=.09, p>.01) are insignificantly related with SME performance among the respondents in this study. Thus, H_1: *There is a significant relationship between effectiveness and SME performance*; and H_2: *There is a significant relationship between competitive advantage and SME performance* are not supported respectively in this study.

5 Conclusions

The results of this study presented that the level of SME performance in this study was in the high indications with the mean value reported at 4.01. Not only that, the level of effectiveness and market analysis was also high with the mean values reported at 4.02 and 4.05 each. A moderately high level of competitive advantage is reported at 3.73.

As presented in Table 3, market accessibility is positively associated with SME performance (β=.48, p<.01). Therefore, only H_3: *There is a significant relationship between market accessibility and SME performance* is supported in this study.

Based on the results, it can be concluded that by increasing the level of market accessibility within the online business entrepreneurs, the future level of SME performance can be enhanced. In the theoretical perspective, these findings had been consistent with the previous findings as reported by several authors [25, 26, 27, and 28]. For example, [26] pointed out that internationalization is important for greater market access by SMEs for their sustainability strategy. In addition, the Internet which included of the online business can be a critical factor in enhancing SME market reach and its operational more efficiency. It has been agreed by many authors that invest in, and proper utilization of ICT is a major way of improving market accessibility among the vital SME sector. As a conclusion, this study presented that the online business created the entry mode by the market accessibility and it was positively and significantly related to SME performance.

In the practical perspective, the appropriate bodies such as policy maker or government possibly initiate and enhance appropriate policies and programs to help SME to utilize the online business presence and advantages. It can be enhanced by taking some initiatives to ensure both entrepreneurs and customers have access to the new technology by providing the strong and reliable infrastructure, particularly as a medium for entrepreneurs-customers interaction. These initiatives could ensure the existence of market accessibility for the current online business entrepreneurs in Malaysia and assist the online entrepreneurs to go further in the global market accessibility.

Also, this study provides awareness to the SME regarding the opportunities created by online business towards their business performance. The performance could be in term of satisfied customer and supplier relations, personnel, quality, flexibility, and planning.

Having respondents from SMEs industry in this study was not an easy process since not all online business entrepreneurs in this industry wanted to cooperate in providing their responses. Furthermore, it took about 2 to 3 months for the respondents to give their responses. Other than the large sample size, the researchers would like to propose for the future researchers to look on the other type of SME industry which focused on the services.

References

1. Hor, R. The Sun. SMI association launch E-SMI program (2001) AvailableWWW:http://asia.internet.com/Sun,%20SMI%20Association/Launch/ SMI/Program.
2. N. Muhd. NA Rahman et.al. The implementation of e-commerce application in bumiputera Small and Medium Enterprises (SMEs) in Malaysia, **2(2)**, 101-110 (2013) Available online at www.managementjournal.info34
3. SME Corp. Malaysia. (2013). [Online] Available:http://www.smecorp.gov.my/vn2/node/37.
4. Chelliah, S., Sulaiman, M., & Yusoff, Y. M. Internationalization and performance: Small and Medium Enterprises (SMEs) in Malaysia. *International Journal of Business and Management*, **5(6)**, 27–37 (2010)

5. Rao, M. "E-business will change Asia more than it changes the U.S." (2006) http://www.indiainfoline.com/nevi/inwi/mm52.html
6. Eniola, A. A. The role of SME firm performance in Nigeria. Arabian Journal of Business and Management Review, **3(12)**, 33-47 (2014)
7. Moullin, M. "Defining performance measurement." Perspectives on performance **2(2)**, 3 (2003)
8. Wang, Y., & Lo, H. Customer-focused performance and the dynamic model for competence building and leveraging: A Resource-Based View. *Journal of Management Development*, **22(6)**, 483–526 (2003)
9. Ma, J.X., Buhalis, D. & Song, H. ICTs and internet adoption in China's tourism industry. *International Journal of Information Management*, **23(6)**, 451-467 (2007)
10. Kim, D., Cavusgil, S.T., Calantone, R.J. Information system innovations and supply chain management: Channel relationships and firm performance, **34**, 40–54 (2006)
11. Tornatzky, L.G.; Fleischer, M. The processes of technological innovation; Lexington Books: Lexington, MA, USA, 145–156 (1990)
12. Kim, D.; Cavusgil, S.T.; Calantone, R.J. The role of information technology in supply-chain relationships: Does partner criticality matter? J. Bus. Ind. Mark, **20**, 169–178 (2005)
13. David, A.J., Lorna, W. The e-business capability of small and medium sized firms in international supply chains. inf. syst. e-bus. manage., **2**, 223–240 (2004)
14. Barba S.V, María D.P, Martínez R., Ana I.J.Z. Drivers, benefits and challenges of ICT adoption by Small and Medium Sized Enterprises (SMEs): A Literature Review, **5**, 103-114 (2007)
15. Wen Z., King J. and Jaska P. ICT and SMES in developing countries The IABPAD Conference Proceedings Dallas, Texas, April 24-27 (2008)
16. Ashrafi, R., & Murtaza, M. Use and impact of ICT on SMEs in Oman. Electronic Journal Information Systems Evaluation, **11(3)**, 125-138 (2008)
17. Alberto, B.M. and Fernando, L.L. A firm-level analysis of determinants of ICT adoption in Spain. technovation, **27**, 352-366 (2007)
18. Sharma, M.K. and Bhagwat, R. Practice of informati on systems, an evidence from select Indian SMEs. Journal of Manufacturing Technology, **17**: 199-223 (2006)
19. Fink, D. and Disterer, G. International case studies: To what extent is ICT infused into the operations of SMEs. Journal Enterprise Information, **19**: 608-624 (2006)
20. Analoui, F. & Karami, A. Strategic management. in Small and Medium Enterprises. Thomson Learning 2003 (2003)
21. Namasivayam, K., Enz, C. A., & Siguaw, J. A. How wired are we? Selection and use of new technology in U.S. hotels. *The Cornell Hotel and Restaurant Administration Quarterly*, **41**, 40–48 (2000)
22. Barney, J. Firm resources and sustained competitive advantage. *Journal of Management*, **17**, 99–120 (1991)
23. Drew, S. (2003). Strategic uses of e-commerce by SMEs in the East of England. *European Management Journal*, **21(1)**, 79–88.
24. OECD. Organizational change and firm performance. DSTI/DOC, 14 (Paris). (2002)
25. Porter, M. E., & Kramer, M. R. The competitive advantage of corporate philanthropy. *Harvard Business Review*, **80(12)**, 56–68 (2002)
26. Levy M., Powell P., & Yetton, P. SMEs: Aligning IS and the strategic context. *Journal of Information Technology*, **16**, 133-144 (2001)
27. Chen, Y., and B. Hu. Internationalization of the RMB: An evaluation framework. Economic and Political Studies **1(1)**, 5–20 (2013)
28. George, D. & Mallery, P. SPSS for Windows Step by Step: A simple guide and reference (4thed.). Boston: Allyn & Bacon (2003)
29. Norasmah, H.O. & Salmah, I. Kecenderungan terhadap pemilihan kerjaya keusahawanan mengikut persepsi peserta skim usahawan siswa. *Jurnal Teknologi*, **56**, 47-63 (2011)

The Optimization Concept of Service Fulfillment Measurement: A Research on Surgery Care

Norshahrizan Nordin[1,2*], *Razli* Che Razak[3], *Sharmini* Abdullah[4], *Ummi Naiemah* Saraih[1], Marniati[5] & Wan *Norsyafawati* W. Muhamad Radzi[1]

[1]School of Business Innovation and Technopreneurship (PPIPT), Universiti Malaysia Perlis (UniMAP), Kangar, Perlis, Malaysia
[2]Center of Excellence Geopolymer & Green Technology (CeGeoGTech), School of Material Engineering, Universiti Malaysia Perlis (UniMAP), Perlis, Malaysia
[3]Faculty of Business and Entrepreneurship, Universiti Malaysia Kelantan (UMK), PengkalanChepa, Kelantan, Malaysia
[4]Centre of International Languages, Universiti Malaysia Perlis (UniMAP), Pauh Putra, Perlis, Malaysia
[5]Universitas Ubudiyah Indonesia, Jln Alue Naga Desa Tibang, Banda Aceh, Indonesia

Abstract. This paper presents an optimization concept of service fulfillment measurement. Service delivery is assumed free from constraint, while service fulfillment is expected to achieve by maximizing the customer preferences. As the constraints are considered, the customer preferences are developed in a form of new model to set the optimum level of variables. The concepts will be further improved for medium and large scale optimization, which capable to evaluate the customer requirement in Kano-QFD integration. The results were also validated based on Kano Model and Quality Function Deployment (QFD) is integrated for patient attribute and service attribute prioritization. The new non-linear Kano-QFD service satisfaction model has been developed, tested and validated with Kano model to facilitate the analysis and decision making for better service delivery improvement.

1 Introduction

Healthcare organization is struggling to provide customer driven quality service. The key success of healthcare service is the ability to develop a comprehensive healthcare service and deliver the best service direct to customer as required. Besides, hospitals are classified into primary, secondary, and tertiary hospitals based on bed size. Tertiary hospital is the highest class distributed in Malaysia [6]. There were 98 hospitals without specialist, 80 single specialist hospitals, 94 secondary hospitals, and 83 tertiary hospitals [5]. In terms of functionality, MOH hospitals were classified into five types: State Hospitals, Major Specialist Hospitals, Non-Specialist Hospitals, and Special Medical Institutions, which were based on workload, number of inpatient beds, and scope of serviced rendered [3].

Excellent healthcare services delivered to patients would not necessarily result in excellent patient service fulfillment, meet the patient expectation, and gain high satisfaction level. Therefore, patient satisfaction is multifactorial and difficult to measure [4];[5]). Inversely, poor healthcare services might not result in high dissatisfaction [6]. All are due to the nature of uncertainty feedback, which can be translated as non-linear behavior [7] and possesses complex quadratic relationship [8].

Based on Health Facts 2014 by MOH, Planning Division, Health Informatics Centre, from 2013-2014, out of 355 hospitals, 141 public hospitals and 214 private hospitals had 14,033 beds [9]. However, in terms of bed size, out of total 43,437 beds, 39,728 beds were found in the public sector, which covered 91.46% of bed density in Malaysia. The higher percentage reflected the public hospital domination on acute care service in Malaysia. This had been in line with 75.07% of 28,949 total doctors in Malaysia placed in public hospitals and 11,697 doctors at private hospitals. Doctors and specialists are the backbone entities in national healthcare service delivery. As of 31st December 2013, the total number of doctors in Malaysia was 46,916 where 74% were in public hospitals and 26% were in private hospitals. This above doctor ratio to population shows the levels of stress and pressure among doctors that may affect the consistency of professionalism in daily medical practice. This is also the main reason for complaints on service by doctors had been the highest (22.16%) in public hospitals [10];[9].

Healthcare service providers need to reassess their strategies to cope with more challenging task with respond to continuation of customer demand. The healthcare challenging tasks are discussed in details by [11]. As a healthcare service provider, systematic reprogrammed and renewed assessments are the most significant step forward to cope with customer demand uncertainties. This is the only way to repositioning themselves in future. This scenario is true elsewhere. For example, an empirical study by [12] have observed that in 1999, 40% of respondents have rated the Singapore hospitals' service quality as poor or very poor, definitely below patients' expectations. These findings have taken seriously by hospitals and 80% of them have absorbed TQM philosophy and develop customer oriented strategy. To be competitive, an alternative customer prioritizing approach mainly incorporates the Quality Function Deployment (QFD) and Kano's Model is proposed conceptually. The new mechanism concept is hope to effectively address a complaint which not only satisfies the customer but also an opportunity to create positive experience with customers, building a healthier foundation, stronger brand value and avoiding legal penalties. It also provides fair balance information for decision making while facing constraints such as operational, legal, human resource and market pressures.

* Corresponding author: norshahrizan@unimap.edu.my

2 Kano-QFD Integration Model: A Surgery Care

The fast solution is needed to initiate the positive improvement of complaint handling. The solution should be incorporated with detail mechanism to quantify the complaints prioritization and analyze the complaints in regards to complainant and healthcare constraint perspective. To author's knowledge and support by literatures, there are no single model to incorporated with. A new optimization decision model needs to be developed comprehensively to incorporate and unite the prioritizing and analyzing element in a model. QFD can be mainly utilized effectively in healthcare services in two ways: First, to analyze customer expectations and characteristics of competitive services, and second, to define the prioritization of technical design characteristic for a new service design. Besides, QFD has been proven to be the prominent technique to resolve the marginal of uncertain customer requirement in a more effective way [13]. The ability to optimize and to analyze customer requirements in deciding the best service to be offered in advance is absolutely great in QFD. QFD and Kano model is found the best in its categories. Unfortunately, both are well success in product development but not in service sector as well as healthcare services [14]. In this paper, the new optimization decision model using QFD and Kano model is proposed and the related formulation in product development is replicated and modified to adapt with healthcare services.

The customer satisfaction coefficient indicates whether satisfaction can be increased by meeting requirement, or whether fulfilling this product requirement merely prevents the customer from being dissatisfaction [15]. The customer dissatisfaction coefficient indicates the other side [16].

Both can be expressed as:

Extent of satisfaction

$$CS = \frac{f_A + f_O}{f_A + f_O + f_M + f_I} \quad (1)$$

Extent of dissatisfaction

$$DS = -\frac{f_O + f_M}{f_A + f_O + f_M + f_I} \quad (2)$$

where f_A, f_O, f_M, f_I represents the frequency of A,O,M,I respectively. The minus sign in Eq.2 means that it is dissatisfaction. Eq.1 and Eq.2, CS and DS for each customer requirement can be calculated.

3 Methodological on Service Compliment (SCi) and Service Complaint Indexes (SCa)

The index of SCa and SCi were inserted into QFD Step 2 according to defined PAs, orderly. For the case of Surgery Care, Fig.1 shows the information of PAs, SCi and SCa that supply to QFD Phase 2 and 3.

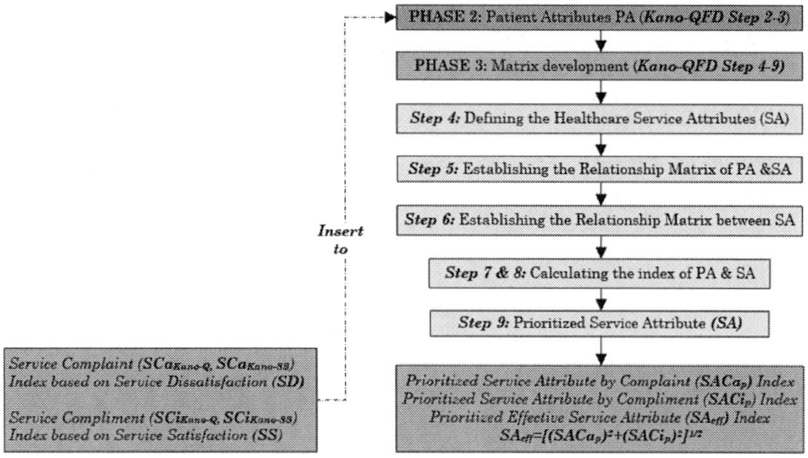

Fig.1. The flow of Kano-QFD execution for Phase 2 and Phase 3

4 Results and Discussion

Table 1. KQA Kano-SS statistical significant for Surgery Care

Variables		KQA	a	b	$\lvert a - b \rvert$		$1.65\sqrt{\frac{(a+b)(2N-a-b)}{2N}}$	KQA significant?
Surgery care	SCQ25	M	113	72	41	>	18.66	YES
	SCQ26	M	127	69	58	>	18.96	YES
	SCQ27	M	116	82	34	>	19.00	YES
	SCQ28	M	128	84	44	>	19.32	YES
	SCQ29	M	138	83	55	>	19.50	YES
	SCQ30	M	125	82	43	>	19.21	YES
	SCQ31	M	129	81	48	>	19.28	YES
	SCQ32	M	128	84	44	>	19.32	YES
	SCQ33	M	128	81	47	>	19.26	YES

Table 1 shows the *SCa* and *SCi* index for Surgery Care. It can be seen that index of "SCQ27: Flexible working hours" has the highest score of *SCi* = 0.38. That means, in surgery, the time punctuality and management is experienced with the highest satisfaction. From complaints, the highest index is referred to "SQ26: Equipped room" where *SCa* = 0.68. It can be concluded, the most compliment service element is not necessarily the most complaint element, particularly for Surgery Care.

Table 2. Surgery Care Attributes Index

Surgery Care	KQA	Compliment (SS)		*SCi* Index	Complaint (SD)		*SCa* Index	S_{eff}
		Kano-Q	Kano-SS		Kano-Q	Kano-SS		
SCQ25	M	0.43	0.35	0.39	-0.69	-0.61	-0.65	0.29
SCQ26	M	0.43	0.24	0.335	-0.69	-0.56	-0.63	0.25
SCQ27	I	0.43	0.18	0.305	-0.69	-0.39	-0.54	0.19
SCQ28	I	0.43	0.16	0.295	-0.69	-0.38	-0.54	0.19
SCQ29	M	0.43	0.13	0.28	-0.69	-0.55	-0.62	0.23
SCQ30	M	0.43	0.17	0.3	-0.69	-0.52	-0.61	0.23
SCQ31	M	0.43	0.16	0.295	-0.69	-0.48	-0.59	0.21
SCQ32	I	0.43	0.12	0.275	-0.69	-0.5	-0.60	0.21
SCQ33	M	0.43	0.17	0.3	-0.69	-0.54	-0.62	0.23

Table 2 tabulates the list of Surgery Care elements (SCQ25-SCQ33) to represent the patient attributes (PA) that contains KQA frequency and Berger's coefficient. Most of the KQA frequency ranking trend follows of M > I > A> O > Q and only one SC element demonstrated M > O> I> A > Q but quality attribute M still dominated the KQA frequency. It can be observed that "SCQ26: Equipped room" encountered the highest SDss= 0.66 and the highest SSss = 0.39, that means patients have responded that SC element with high satisfaction impact even experienced the high dissatisfaction level. This scenario is against the scenario of Doctor Care and Nurse Care.

Table 3. Kano-SS KQA for Surgery Care

Surgery Care	Kano Quality Attributes (frequency)						Σ	Berger's coefficient		KQA
	A	M	O	I	R	Q		SS_{ss}	SD_{ss}	
SCQ25	50	113	65	72	0	0	300	0.38	-0.59	M
SCQ26	47	127	69	56	0	1	300	0.39	-0.66	M
SCQ27	55	116	43	82	0	4	300	0.33	-0.54	M
SCQ28	47	128	37	84	0	4	300	0.28	-0.56	M
SCQ29	35	138	35	83	0	9	300	0.24	-0.59	M
SCQ30	47	125	44	82	0	2	300	0.31	-0.57	M
SCQ31	45	129	41	81	0	4	300	0.29	-0.57	M
SCQ32	50	128	35	84	0	3	300	0.29	-0.55	M
SCQ33	45	128	42	81	0	4	300	0.29	-0.57	M

Logically, this particular condition happened due to the high frequency of quality attribute of One-dimensional (O). Table 3 shows the frequency of O is higher than I and A. The satisfaction impact of O affects the Berger's coefficient value, significantly. Moreover, the lowest SDss is 0.54 which corresponds to "SCQ27: Flexible working hours". It refers the doctor and nurse that able and always ready for surgery and operation. The lowest SSss = 0.24 refers to "SCQ29: Easy appointment", that means the process of setting the surgery time and date is encountered some managerial problems.

Table 4. Kano-SS KQA and Berger's coefficient for Surgery Care

Service elements		Service gap			
		Pilot survey		Mass survey	
Surgery Care	SCQ25	-0.26	M	-0.25	M
	SCQ26	-0.32	M	-0.29	M
	SCQ27	-0.21	I	-0.34	M
	SCQ28	-0.22	I	-0.21	M
	SCQ29	-0.42	M	-0.25	M
	SCQ30	-0.35	M	-0.27	M
	SCQ31	-0.32	M	-0.25	M
	SCQ32	-0.38	M	-0.23	M
	SCQ33	-0.37	M	-0.29	M

Based on Table 4 & Table 5, the transition of KSS for Surgery Care is similar to Doctor Care and Nurse Care. In pilot survey, out of nine SCQ, only three SCQ are found significant (YES), but in mass survey, al the SCQ elements have resulted with significant (YES). SCQ33 has resulted with "YES" KQA statistical significant. This condition translates that all the SCQ is in normal distribution and the variation between first (a) and second (b) highest frequency is large enough to define the KQA for attribute M. The frequency difference (a - b) between attribute M and I is vary from 41 – 58 which is higher than KSS coefficient (18.66 – 19.50).

Table 5. Kano Statistical Significant for Surgery Care

SV	Codes	Mass Survey						
		KQA	KSS?	KQA				KSS?
Surgery Care	SCQ25	M	YES	M	41	>	18.66	YES
	SCQ26	M	NO	M	58	>	18.96	YES
	SCQ27	I	NO	M	34	>	19.00	YES
	SCQ28	I	YES	M	44	>	19.32	YES
	SCQ29	M	NO	M	55	>	19.50	YES
	SCQ30	M	NO	M	43	>	19.21	YES
	SCQ31	M	NO	M	48	>	19.28	YES
	SCQ32	M	NO	M	44	>	19.32	YES
	SCQ33	M	YES	M	47	>	19.26	YES

Table 6 shows the prioritized patient attributes (PACip) by compliments and (PACip) by complaints for Surgery Care. The "Hygiene rules and procedure" is found to be the highest complaints at w = 78.72 and compliments at w = 50.43 for Surgery Care. Similarly, the "Avoid unnecessary expenses" is shared for PACip and PACip as the lowest ranking. In terms of ranking order, both PACip and PACip listed in similar ranking order from rank 1 - 9.

Table 6. Prioritized patient attributes (PACip) by compliments for Surgery Care

Rank	%	w	Patient attributes (PA)	Codes
1	16.3	50.43	Hygiene rules and procedure	SCQ25
2	12.6	38.95	Equipped room	SCQ26
3	12.5	38.76	Flexible working hours	SCQ27
4	12.5	38.52	Comfortable waiting room	SCQ33
5	10.4	32.01	Easy appointment	SCQ30
6	10.3	31.68	Very helpful staffs	SCQ31
7	9.3	28.86	Relaxing environment	SCQ30
8	9.3	28.80	Easy access location	SCQ32
9	6.8	20.88	Avoid unnecessary expenses	SCQ28

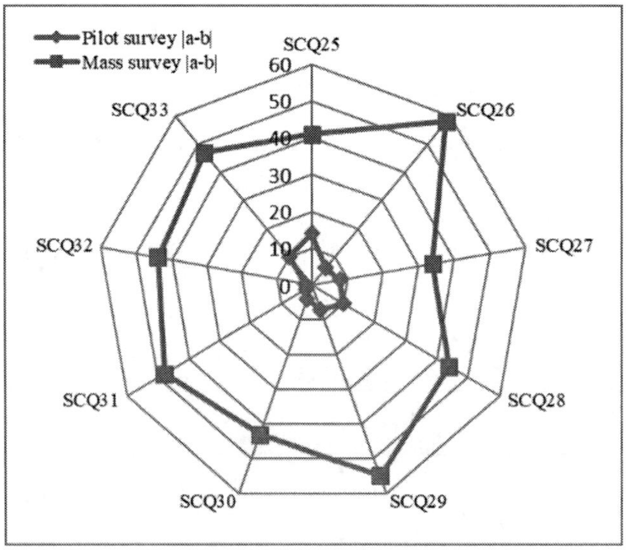

Fig.2. KSS spider chart map different |a - b| for pilot and mass survey comparison based on first and second highest Surgery Care KQA closed frequency

https://doi.org/10.1051/matecconf/201815005024

Fig.2 shows the KSS spider chart map different for pilot and mass survey comparison based on first a and second b highest frequency of Surgery Care KQA. It can be seen that most of insignificant KSS for pilot survey are laid below the value of 10. To be significant, the value of |a - b| must be over than 10. For mass survey, in order to be significant, the |a - b| must over 18.66 at average but the figure shows the value of |a - b| for all SCQ has lied over the value of 30. This result indicated the strong relationship along the SCQ elements based on KSS analysis.

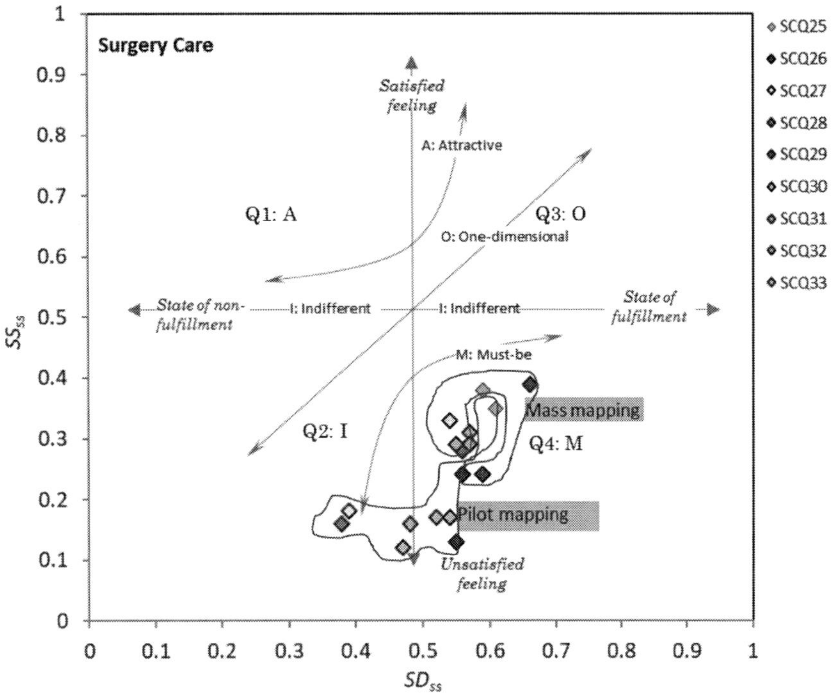

Fig.3. SS_{ss}-SD_{ss} grid mapping for Kano-SS Surgery Care

Again, the composed and concentrated Must-be M attribute has theoretically validated the present developed Kano-SS and Kano-Q in Kano-QFD integration by original Kano model. The similar scenario is happened to Surgery Care. Fig.3 shows SSss-SDss grid mapping for Kano-SS Surgery Care where the distribution of pilot mapping is scattered along the down side of unsatisfied feeling region. Most of the SCQ are placed in the Q2: I and near to the Unsatisfied feeling axis. Those attributes are marked as insignificant by KSS analysis. However, those attributes as placed well in upper region of Q1: M for mass mapping, where they have turned as significant and nearer to the state of fulfillment and lead towards satisfied feeling. That means, the proposed Kano-QFD satisfaction model suggests the HSP to improve the service attributes which were listed under "Medical Service Organization" and "Medical service delivery and technical care" in increasing the satisfaction level and reducing the level of dissatisfaction among patients.

5 Conclusions

In service design aspect, the patient's responds can be accounted based on satisfaction levels from the context of patient feelings and fulfilment. The complaint and compliments mechanism is proved to reflect the dissatisfaction and satisfaction of patients [17]. The non-linear relationship between patient attributes and service attributes has been successfully defined by Kano Quality Attributes (KQA) through Attractive (A), Must-be (M), One-dimensional (O), Indifferent (I) and Questionable (Q). The indexes of KQA translated the fulfilment of patients based on satisfaction and dissatisfaction [18]. The Kano-QFD integration provides the weightage between patient's attributes and service attributes and finally prioritised in ranking basis.

Acknowledgments

The authors are grateful for the research support of Ministry of Health (MOH) Malaysia and Universiti Malaysia Perlis (UniMAP) through awarded supportive facilities.

References

1. Onn, L. T., Sivasampu, S., & Ariza, Z., National healthcare establishments and workforce statistics 2008-2009, Kuala Lumpur: Clinical Research Centre, Ministry of Health Malaysia, **1**, 109 (2011)

2. Baba, N., Chong, T. S., Ibrahim, L., Salleh, A. R., Mokhtar, L. M., Ali, A.,Clinical Research Centre, Ministry of Health Malaysia 1 ed., **1**, 119-130 (2011)

3. Merican, M. I., Ministry of Health Malaysia, ISSN 1511-1512 (2011)

4. Quaschning, K., Korner, M., & Wirtz, M., Patient Education and Counseling, **91**, 167-175 (2013)

5. Rowell, R. M., & Polipnick, J., Journal of Manipulative and Physiological Therapeutics, vol. 31(8), pp. 602-610 (2008)

6. Driscoll, A., Currey, J., Marcia George OAM, R., & Patricia M. Davidson RN, B., Australian Critical Care, **26**, 55-57 (2013)

7. Lofgren, M., Witell, L., & Gustafsson, A., The TQM Journal, **23(2)**, 235-246 (2011)

8. Meng, Q., & Jiang, X., 2011 International Conference on Service System and Service Management (ICSSSM) (2011)

9. MOH, 2014 Health Indicators 2014, Planning Division, Health Informatics Centre (2014)

10. Baba, N., 2013 Retrieved Mac2013. http://www.pcb.gov.my/STATISTIK/Laman%20Web%20Statistik %20AduanTahun%202010%20BM.pdf, 2011.

11. Lim, P.C., & Tang, N.K.H., Managing Service Quality, **10(2)**, 103-111 (2000)

12. Lim, P.C., & Tang, N.K.H., International Journal of Healthcare Quality Assurance, **13(7)**, 290-299 (2000)

13. Chan, L. K. & Wu, M. L., European Journal of Operational Research, **143**, 463-497 (2002)

14. Chiou, C. C. & Cheng, Y. S., IEEE International Conference on Industrial Engineering and Engineering Management (2008)

15. Nordin, N., & Razak, R.C., Advanced Review on Scientific Research, **4(1)**, 2289-7887 (2014)

16. Lai, X., Xie, M. & Tan, K. C., IEEE International Engineering Management Conference (2004)

17. Nordin, N., Abdullah., M.M.A.B, Mat, F., Abdullah, M., Razak, R.C., & R., Adom, A.H., Assessment of Functional and Dysfunctional on Implant Stability Measurement for Quality of Life, MATEC Web of Conferences 97, 01116, (2017)

18. Nordin, N., Abdullah, M.M.A.B., Razak, R.C., Non-Linear Assessment and Deficiency of Linear Relationship for Healthcare Industry, AIP Conference Proceedings 1885, 020154 (2017)

Comparison on the Analysis on PM_{10} Data based on Average and Extreme Series

Nor Azrita Mohd Amin[1,*] and *Siti Aisyah* Zakaria[1]

[1]Institute of Engineering Mathematics, Universiti Malaysia Perlis, Kampus Pauh Putra, 02600 Arau, Perlis

Abstract. The main concern in environmental issue is on extreme phenomena (catastrophic) instead of common events. However, most statistical approaches are concerned primarily with the centre of a distribution or on the average value rather than the tail of the distribution which contains the extreme observations. The concept of extreme value theory affords attention to the tails of distribution where standard models are proved unreliable to analyse extreme series. High level of particulate matter (PM10) is a common environmental problem which causes various impacts to human health and material damages. If the main concern is on extreme events, then extreme value analysis provides the best result with significant evidence. The monthly average and monthly maxima PM_{10} data for Perlis from 2003 to 2014 were analysed. Forecasting for average data is made by Holt-Winters method while return level determine the predicted value of extreme events that occur on average once in a certain period. The forecasting from January 2015 to December 2016 for average data found that the highest forecasted value is 58.18 (standard deviation 18.45) on February 2016 while return level achieved 253.76 units for 24 months (2015-2016) return periods.

1 Introduction

Environmental quality management is more concerned about extreme situations due to its various hazardous impacts. However, most statistical methods are concerned primarily with what goes on in the centre of a statistical distribution, and do not pay particular attention to the tails of the distribution. Statistical modelling of extreme air quality has a very practical motivation since these events have major effects on human and ecosystem. Degradation of ambient air quality standards reduces visibility, impairs air, land and water transportation and seriously affects the economy.

The region's drier weather conditions has led to escalation in hotspot activities that are caused mainly by land clearing and "slash and burn" agricultural practices [1]. Improper management of open burning for commercial plantation sectors, heavy industries and business activities has made the situation worse. High PM10 levels have been a common problem in Malaysia especially in the dry season. During the haze periods, PM10 was found as the main pollutant while the other air quality parameters remained within the permissible healthy standards [2]. Study conducted by Dominick et al. [3] found that air pollution in eight selected air monitoring stations in Malaysia based on year 2008 to 2009 are predominantly influenced by PM_{10}. Yusof et al. [4] claimed that the hourly average PM_{10} data for Seberang Perai area (industrialized area) fit to lognormal distribution for 2000, 2001, and 2002 while 2003 and 2004 data fit to Weibull distribution model.

A time series is a collection of past values of the variable being predicted. Holt-Winters method was first proposed in the early 1960s. It uses a process known as exponential smoothing. Exponential smoothing in its simplest form should only be used for non-seasonal time series exhibiting a constant trend or also known as a stationary time series. The smoothed series depends on all previous values, with the most weight given to the most recent values. The Holt-Winters model uses a modified form of exponential smoothing. It applies three exponential smoothing formulae to the series which represent by α, β and γ hyper parameters. Firstly, the level (or mean) is smoothed to give a local average value for the series. Secondly, the trend is smoothed and lastly each seasonal sub-series (ie all the January values, all the February values and so on, for monthly data) is smoothed separately to give a seasonal estimate for each of the seasons. The Holt-Winters method set the hyper parameters γ to 0 which apply the exponential smoothing with trend and without seasonal component [5].

Extreme value (EV) theory is unlike other statistical approaches since its focus is on the tail of distribution either on maxima or minima values. The scope of EV theory has been widely explored in various fields. Recently, it has become a vigorously research area due to its implication in many applications. Literatures on EV theory among others are by Coles [6] and Haan [7] that provide from basic EV theory to the application of EVT in various fields. Basically, there are two main models for extreme series which is generalized extreme value distribution (GEV) and generalized Pareto distribution (GPD). This study focuses on extreme PM_{10}

* Corresponding author: norazrita@unimap.edu.my

concentrations based on GEV model. This paper aim to approximate the possible extreme PM_{10} levels in the future.

2 Methodology

There are two main approaches on the analysis of PM_{10} data in this research. The first approach is using the monthly average data where we use the time series approach to make a proper forecasting based on Holt's linear trend method. Second approach is based on EVT where the extreme data are extracted from the original data based on monthly maxima data.

2.1 Holt-Winters Method

Holt-Winters method consists in a simple yet effective forecasting procedure, based on exponential moving averages, covering both trend and seasonal models [5]. As mentioned earlier the Holt-Winters model assumes that the seasonal pattern is relatively constant over the time period. The exponential smoothing formulae applied to o a series with a trend and constant seasonal component using the Holt-Winters additive technique are:

$$\hat{a}_t = \alpha(Y_t - \hat{s}_{t-p}) + (1-\alpha)(\hat{a}_{t-1} + \hat{b}_{t-1})$$
$$\hat{b}_t = \beta(\hat{a}_t - \hat{a}_{t-1}) + (1-\beta)\hat{b}_{t-1}$$
$$\hat{s}_t = \gamma(Y_t - \hat{a}_t) + (1-\gamma)\hat{s}_{t-p}$$

where:

α, β and γ are the smoothing parameters
$\hat{a}_t$ is the smoothed level at time t
$\hat{b}_t$ is the change in the trend at time t
$\hat{s}_t$ is the seasonal smooth at time t
p is the number of seasons per year

2.2 Extreme Value Theory

EVT provides a concrete theoretical groundwork on which statistical models for describing extreme events are properly set up. General discussions about EVT are about their conventional forms which are Gumbel, Frechet and Weibull that is unified into its general form generalized extreme value distribution (GEV) and the recent approach is based on generalized pareto distribution GPD which depends on the threshold value of the data. This paper analyse the data based on GEV model where the series of extreme data were extracted based on monthly maxima (block).

2.1.1 Generalized Extreme Value Distribution

The cumulative distribution function (cdf) and probability density function (pdf) of the GEV distribution is given by (1) and (2) respectively. The GEV model has three parameters, $-\infty < \mu < \infty$, $\sigma > 0$ and $-\infty < \xi < \infty$ which refers to location, scale and shape parameter respectively. The ξ value determines the type of GEV distribution. $\xi > 0$ correspond to Frechet distribution, $\xi < 0$ correspond to Weibull distribution and $\xi = 0$ correspond to Gumbel distribution.

$$G(z) = \exp\left\{-\left[1+\xi\left(\frac{z-\mu}{\sigma}\right)\right]^{-1/\xi}\right\} \quad (1)$$

$$g(z) = \frac{1}{\sigma}\exp\left\{-\exp\left[-\left(\frac{z-\mu}{\sigma}\right)\right]-\left(\frac{z-\mu}{\sigma}\right)\right\} \quad (2)$$

Earlier works on EVT take on one of the three distributions and subsequently estimate the corresponding parameters. According to Coles [6], there are two weaknesses regarding this issue. First, a technique is required to choose the most appropriate distribution for the data analyzed and second, the inferences are made with the assumption that the choice is correct. Therefore, a better analysis could be done using GEV where the value of shape parameter, ξ itself will determine the most suitable tail behavior of the data.

2.1.2 Return Level

Application of EVT in air quality studies are concerned about how well the mathematical theory can be applied to further answer questions relating to the probability that pollutant concentration will exceed a certain level in a given period which refers to the return level. Awareness of the return levels of extreme air pollution events could benefit the development of air pollution risk management practices. This motivates the need to estimate the most terrible PM_{10} level that will be occur over a certain period in the future.

Return levels in extremes explain the value of extreme events that occur on average once in a given period. For example, what is the PM_{10} level that will be exceeded on average once in the next 100 years? It is convenient to interpret extreme value models in terms of quantiles or return levels rather than individual parameter values [8]. The return level for GEV with return period 1/p is defined by z_p.

$$z_p = \begin{cases} \mu - \dfrac{\sigma}{\xi}\left\{1-\left[-\log(1-p)\right]^{-\xi}\right\} & \xi \neq 0 \\ \mu - \sigma\log\left[-\log(1-p)\right], & \xi = 0 \end{cases} \quad (3)$$

3 Data Description

The air quality level in Malaysia is described in terms of Air Pollutant Index (API). API is an indicator of the air quality and is developed based on scientific assessment to indicate in a manner that can be easily understood, the presence of pollutants and its impact on health. The API scale and terms used in describing air quality levels are categorized as in **Table 1**. In extreme value analysis, most of the data considered range between moderate to very unhealthy level for which necessary actions need to be taken. The CAQM measures concentrations of five major pollutants in the ambient air, namely, PM_{10}, sulphur dioxide (SO_2), nitrogen dioxide (NO_2), carbon monoxide (CO), and ozone (O_3). PM_{10} is used to describe aerosol particles with diameter of less than $10\mu m$ for solids or liquids found suspended in the atmosphere.

The API is calculated based on the average daily concentrations of SO_2, NO_2, CO, O_3 and PM_{10}. Dominant air pollutant with the highest concentration considered as pollutants that will determine the IPU. Typically, concentration of PM_{10} is the highest compared to the pollutant others, and this determines the API readings. PM_{10} concentration is related to gases and particulates which are expected to originate mostly from industrial and vehicle emissions and also from some transboundary pollutions involving Malaysia. The three major sources of air pollutions especially in urban areas are mobile (motor vehicles), stationary (power stations, industrial fuel burning process and domestic fuel burning) and the burning of municipal and industrial waste.

Table 1. API status indicator

API scale	Air quality status
0-50	Good
51-100	Moderate
101-200	Unhealthy
201-300	Very unhealthy
301 and above	Hazardous

During haze periods, PM_{10} was found as the main pollutant while the other air quality parameters remained within permissible healthy standards. The extreme level of PM_{10} data is particularly due to the haze and biomass burning as well as industrial and vehicle emissions which usually contribute to high PM_{10} levels.

3.1 Sampling Site

The data collected is for Perlis region. Perlis is located in the north peninsular Malaysia and bounded with Thailand in the north and Kedah in the south with total area around 795 km2. Perlis is distributed to three parliament which is Kangar, Arau and Padang Besar and categorized as sub-urban area. Figure 1 shows the continuous air quality monitoring station (CAQMS) used to measure the air pollution located at Institut Latihan Perindustrian, Kangar. The device must be at 30 degree elevation from nearby building, which means that any taller building near it will interrupt the measurement to be recorded

In Perlis however, source of pollution is probably mainly from open burning of paddy fields, in most part of Perlis. Other than that is due to the haze from neighbouring country, Indonesia.

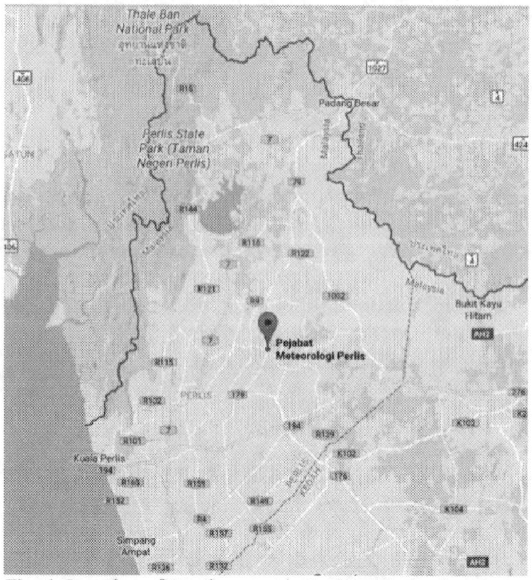

Fig. 1. Location of Continuous Air Quality Monitoring System Perlis.

3.2 PM_{10} Data

PM_{10} notation is used to describe aerosol particles with diameter less than $10\mu m$ in the form of solids or liquids found suspended in the atmosphere. Malaysia guideline for PM_{10} concentrations for 24-hours average and 12-months average are $150\mu g/m3$ and $50\mu g/m3$ respectively. In air quality control, PM_{10} is recognized as the most influencing atmospheric pollutant for air quality index in a majority of cities in Malaysia. The adverse effects of PM_{10} to human health and material damages are the main reasons for extensive explorations on the behaviour of this pollutant. Prolonged exposure to high concentrations of PM_{10} can be harmful to health especially on eye and throat irritations and respiratory problems among sensitive groups.

4 Results and Discussions

This section analyse the data based on the monthly average and monthly maxima PM_{10} data for year 2003 to 2014. Results were compare for both series using descriptive statistics and forecast value for year 2015 and 2016 based on Holt-Winters method with hyper parameters γ to 0 which apply the exponential smoothing with trend and without seasonal component

4.1 Data Analysis

Fig. 2. and **Fig. 3.** show the plot of monthly average temperature data with corresponding summary statistics in **Table 2.** In general both figures doesn't show any seasonal trend. The average data show obvious different on the level between January 2004 to January 2006 to the rest of the period. Transboundary haze is expected contributed to the higher PM_{10} recorded intermittently due to the land and forest fires in the Riau Province of

Central Sumatera, Indonesia. Additionally, quite persistent pattern from July 2006 to July 2013. Most of the data are lied between 30 and 50 units. The means for average data is 42.12 while maxima data is 144.53 units. From **Fig. 3.**, the obvious peaks are mostly happen in July especially for years 2005, 2008 and 2013. This is probably due to very hazy weather condition from forest fires in Sumatra. Haze is an annual problem during the monsoon season from May to September as winds blow the smoke across the Malacca Strait to Malaysia.

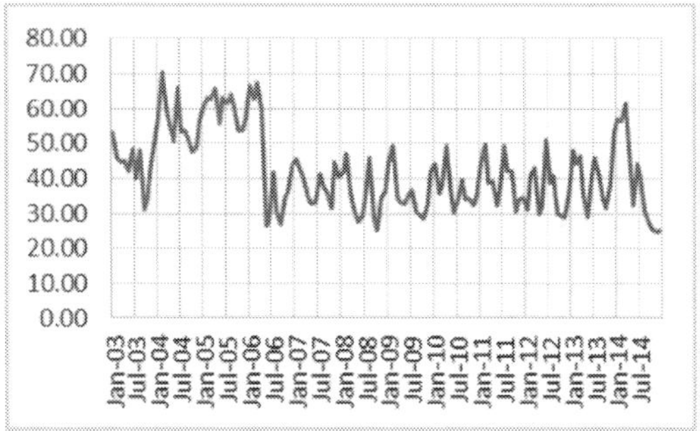

Fig. 2. Monthly Average of Temperature Data

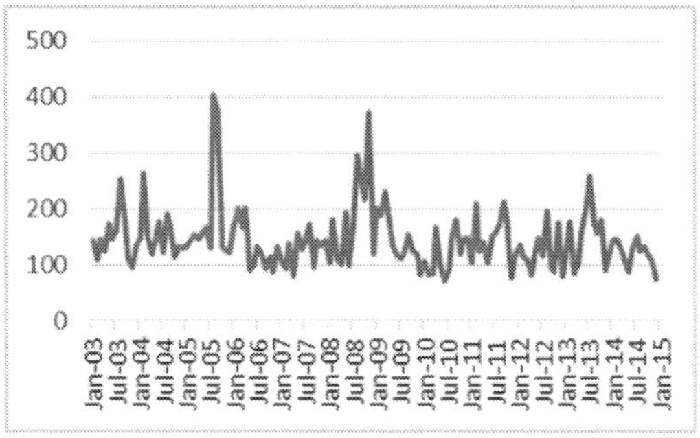

Fig. 3. Monthly Maxima of Temperature Data

Table 2. Descriptive Statistics for Monthly Average and Monthly Maxima Temperature Data for year 2003 to 2014

	Min	Max	Average	Std.
Monthly average	24.63	70.43	42.12	11.15
Monthly maxima	72.00	404.00	144.53	54.67

Inferences on the extremes of environmental events are essential as guidelines in designing structures in order to survive under the utmost extreme conditions. Extreme air pollutants caused various effects associated to human health and material damages. In many cases, the pollutants are responsible for huge impacts on economic performances. The EV theory is applied to model the extreme PM_{10} pollutant for three air monitoring stations in Johor. This study analyse the extreme PM_{10} data based on maximum likelihood estimation technique and hence estimate the return level. The estimated parameter for GEV model given in **Table 3.**

Table 3. Maximum likelihood estimates (with standard error) for GEV model

Parameter	μ	σ	ξ
MLE	119.67 (3.21)	34.01 (2.46)	0.14 (0.06)

4.2 Forecasting

The monthly average data is forecast by using Holt-Winters method. The observed, fitted and predicted monthly average PM_{10} values are shown in **Fig. 4**. The confidence intervals for the forecast values are plotted in the same plot. The forecast values found that the highest forecast value from January 2015 to December 2016 is 58.18 unit which is on Feb 2016. The details forecasting values are given in **Table 4**. Compare to GEV model, the return level for 24 months return period gives the return level value is 253.76 unit. This means that the highest value predicted using GEV model will reach very unhealthy level in two years period.

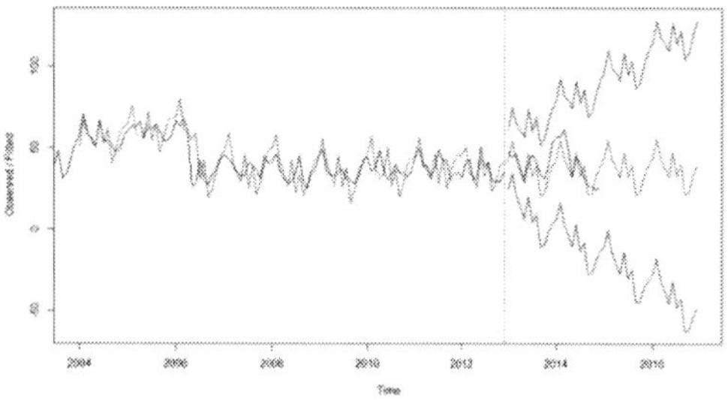

Fig. 4. Holt-Winters method of monthly average with 95% confidence limits in dashed blue lines. Original data are plotted in gray, fitted and predicted values in red.

Table 4. Forecast values of monthly average PM_{10} for year 2013 to 2016 with observed values for year 2013 and 2014 (bold)

Months	Jan	Feb	Mac	Apr	May	Jun	Jul	Aug	Sept	Oct	Nov	Dec
2013	**47.81**	**44.38**	**46.26**	**35.53**	**28.97**	**40.54**	**45.75**	**41.38**	**34.00**	**31.32**	**38.17**	**53.59**
	42.66	54.53	42.47	37.46	31.49	46.96	31.93	38.83	20.48	22.33	32.16	38.65
2014	**56.72**	**56.45**	**61.39**	**48.99**	**32.26**	**44.12**	**40.09**	**30.77**	**27.31**	**25.76**	**24.63**	**25.27**
	42.88	54.75	42.69	37.67	31.71	47.17	32.14	39.05	20.70	22.55	32.37	38.87
2015	43.09	54.96	42.90	37.89	31.92	47.39	32.36	39.26	20.91	22.76	32.59	39.08
2016	43.31	55.18	43.12	38.11	32.14	47.60	32.58	39.48	21.13	22.98	32.81	39.30

5 Conclusions

From the monthly average data, the PM_{10} data is fairly good except during periods of haze as indicated by its API. However, the average for monthly maxima series indicates that the pollutant in very unhealthy level. This obvious different is that the monthly average values explain the ordinary situation while the monthly maxima refer to the extreme cases only. Thus, the forecasting value for average data reflect the predicted values for the next two years period while the return level explain the maximum level that can be estimate in average for the next two years. Therefore, if our interest is on examining the extreme occurrence or catastrophic events, extreme value theory is the best methods that should be used.

The authors would like to thank Universiti Malaysia Perlis for the short term research grant (STG 9001-00532).

References

1. Amin N. A. M, Adam M. B and Aris A. Z. J. Teknologi **76(1)** 171–179 (2015).Payus C, Abdullah N, and Sulaiman N. Int. J. Env. Sci. Dev. **4** (4): 398-402 (2013).
2. Dominick D, Juahir H, Latif MT, Zain SM and Aris AZ. Atm. Env. 172-181 (2012).
3. Yusof NFFM, Ramli NA and Yahaya AS. Int. J. Env. Protect. **1**(4): 28-36 (2011).
4. Meyer D. The Holt-Winter Method in Package ts. The Newsletter of the R Project. **2(2)** (2002)
5. Coles S. G. An Introduction to Statistical Modeling of Extreme Values, 3rd ed. London: Springer-Verlag. (2001)
6. Haan D. and Ferreira L. Extreme Value Theory: An Introduction. Springer, New York (2006).
7. Gilli M. and Kellezi E. Computational Economics. 207-228 (2006).
8. Juneng L, Latif MT, Tangang F T and Mansor H. Atmospheric Environment **43** 4584-4594 (2009).

Digital Board Game: Is there a need for it in language learning among tertiary level students?

Zuraina Ali[1,*], *Mohd Amir Izzudin Mohamad Ghazali*[1], *Rosnani Ismail*[1], *Nurul Nadia Muhammad*[1], *Noor Azlinda Zainal Abidin and Nabila Abdul Malek*[1]

[1]Centre for Modern Languages & Human Sciences Department, Universiti Malaysia Pahang, 26600 Pekan, Pahang, Malaysia

Abstract. The concept of learning while playing to learn language is a pedagogy that can enhance students' learning interest. In this digital era, the use of board games, for instance, has provided a solution to learning a language. A considerable literature has grown up around the theme of digital board in language learning. Therefore, this paper aims at finding the significance of developing a digital board game for the purpose of vocabulary learning. Fifty-seven (57) students from Universiti Malaysia Pahang (UMP) were selected as samples using purposive random sampling to collect data for the current study. A questionnaire was employed to explore their needs of a digital board game to learn vocabulary. The findings revealed that digital board game has significant needs among students to learn vocabulary, and further support the idea of using digital board game in language classroom. Majority of them believed that digital board game may provide a significant impact on vocabulary learning. Overall, the current study proves to be particularly valuable to language teachers to incorporate digital board game into their instructional activity. Also, these findings suggest a role for interactive learning in motivating language learner to enrich their vocabulary size.

1 Introduction

It is a challenge for educators to keep the students' learning interest. Yet, learning while playing is one way that helps students to become passionate in their learning. It also helps to stimulate their interest in learning. The use of conventional board games is seen much beneficial to students as it is engaging them in playing while learning. However, with the emergence of technology, digital board games are more suitable to the interest of students nowadays [1]. Learning a language can be at ease with the existing games which are digitized. From learning the language in various skills such as reading, speaking, listening or writing, digital board games can also be used to learn vocabularies of a language. This study, therefore, explores whether or not there is a need of using digital board game in learning English, specifically vocabulary among tertiary level students.

It is important to understand that computer and games are enormously powerful learning tools, even though they are most often thought of as pure entertainment. Technology will not only help us to design better games but will also allow us to create effective new learning opportunities and tools. In designing digital game-based learning, two key premises need to be adhered to. First, learners demonstrate changes in some fundamentally important ways and second, computer games can provide a new way to motivate students to learn [2]. Moreover, successful learning mostly occurs when learning is more relevant and reflective to learners. This is true for board game as it is a game that is made of a text description, illustration cards, a game board and some simple properties. In fact, a good game design can maintain players' motivation and engagement [3]. In relation to the use of conventional board game in education, a good educational game design can keep learners engaged in learning. Yet, board game that is digitalized may amuse the participants or players to play the game. In addition, the teachers may provide more time paying attention to students' performance, rather than busy placing associated conversational cards in front of the learners.

2 Methodology

2.1 Research Sample

A total of fifty-seven (57) diploma and undergraduate students in Universiti Malaysia Pahang were selected as the samples of the current study. All of them were engineering students who, during the collection of the data were in their first, second and third year of study. A purposive random sampling was employed in selecting the qualified participants. This type of sampling is the most popular and rigorous form of probability sampling from a population under study and the advantage of it is

Corresponding author: zuraina@ump.edu.my

any bias in the population will be equally distributed among the people chosen [4].

2.2 Research Instrument

As for the instrument used, an online questionnaire was developed by using Google form and it consisted of three main sections. Section 1 concerns with the participants' demographic information such as gender, level of education, ethnicity, faculty and the semester of studying in UMP. On the other hand, Section 2 consists of two questions asking the participants about their familiarity of playing digital/online games and a list of options were provided for each question. The last part; Section 3 contains nine items – eight items are of 5-point Likert scale where the samples are asked to rate their needs for a digital board game and the last item is an open ended question in which they need to explain one significant way that digital board game may help them in their learning.

2.3 Research Procedures

In collecting the data for the current study, the researchers first set up the online questionnaire by using Google form. The reasons for using online questionnaire, particularly Google form were mainly because of its administration speed in which it can reach the participants faster and lower cost incurred as no budget spent on the photocopy purposes. After the set up was completed, the link to the questionnaire was shared to the participants, either through Edmodo or email. Prior to the administration of the questionnaire, the participating students were informed about the purpose of the study and the confidentiality of the results. Generally, each participant spent less than twenty minutes to complete the questionnaire. The obtained data was then analyzed using simple statistical analysis. Meanwhile, data that was gathered qualitative; more specifically interviews, were coded. Hence in reporting the results of the study 'P11' stands for 'Participant 11', 'P14' refers to 'Participant 14'; respectively. Figure 1 shows the research procedures of collecting the data for the study.

Basically, there are four (4) basic steps that were involved in the procedures and these are preparing an online questionnaire, identifying research sample, distributing questionnaire and analyzing data. These stages of collecting data took about a month in which researchers brainstormed to develop the items for the questionnaire. Several meetings were conducted to finalize the set of questionnaire. Later, a google form was created where all items were placed according to their suitable sections. Moreover, in identifying the participants that would be involved in the current study, it was decided that they would be the students that registered in the researchers' classes themselves. In distributing the questionnaire, the researchers had to ensure that students were not disturbed in conducting task. Therefore, they were informed of the link to click the via electronic communications.

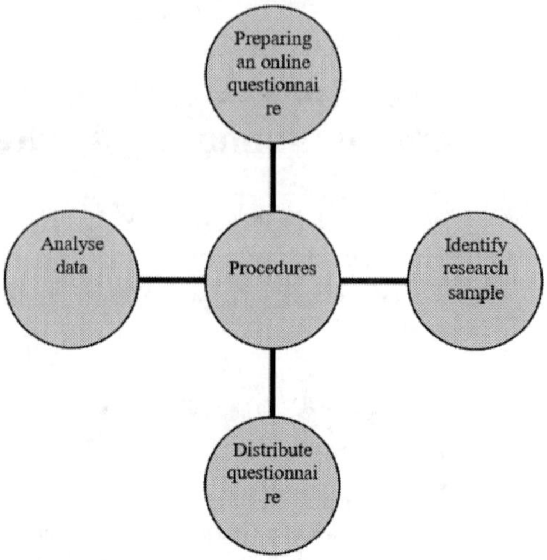

Fgure 1: Procedures of Collecting Data for the Study

3 Results and Discussion

The study aims to identify the need for a digital board game among tertiary students in learning English language. A simple statistical analysis was employed for the study. Also, the study presents qualitative findings to further elaborate the quantitative findings.

In general, these results indicate a significant need for a digital board game in learning English language. The participants provided positive feedback when asked on their views on the use of a digital board game in their language learning. Table 1 illustrates the data obtained from the questionnaire counting the percentage of students' responses in answering their views on the needs for a digital board game. In the table, it is apparent that majority of the participants (84.2%) agreed with Item 3 'Vocabulary digital game enrich a player's vocabulary size'. One of the participants commented that *"it [digital board game] allows players to widen their vocabulary choices"*. The view was echoed by P11, P14, and P21 in which they felt that *"while playing [digital board game], I can learn English vocabulary"*. These results are in accord with a previous study as they found that the incorporation of digital content into language learning can increase the students' vocabulary size [5]. The learning of vocabulary through digital learning environment helps to cater to different and unique individual's learning style. Similarly, another finding further supports the idea of incorporating digital learning into vocabulary learning in which the researcher found that the participants' vocabulary size increased significantly after learning from Quizlet [6]. This is very much true since learning is fruitful when it matches with students' learning aptitude [7]. The study reported that three (3) methods of vocabulary learning; namely contextual clues, dictionary strategy and Computer Assisted Language Learning (CALL) – in particular Tell Me More from Auralog, were employed to identify tertiary level students' ability in learning vocabulary. A post hoc test using Gabriel test (Anova) showed that students showed more positive attitudes when computers were employed in the students' learning of vocabulary compared to two other methods. The software that is interactive provides its learners with speech recognition technology and that it assists them to learn various languages (i.e. English, Spanish, Mandarin and others) in general, and vocabulary in particular with 37 different types of engaging activities. Moreover, the result is in the lines of earlier literature which found that students were able to widen their vocabulary as well as improve their language skills when they were able to play with other community of players [8]. The state-of-the-art for the kind of learning is described as gameplay-as-learning practices. Over a hundred tertiary level students took part in the study. They were reported to take active roles as gamers in building their community of gamers. Five (5) gamers whom their communication in forums were recorded, demonstrated that they communicated for game information and gameplay strategies.

Meanwhile, in response to Item 2, 82.5% agreed that 'Language digital game can improve a player's understanding of language'. According to the participants, digital board game can function as a tool to gain deeper understanding of English language. P32 noted that *"it helps me to improve my [English] language and I can know my weakness in that language"* and eventually, P1 noted that *"I can improve my academic results"*. As for P2 and P19, they believed that the use of digital board game can assist them in learning grammar and other language skills. Meanwhile, [9] examined teachers' perception on the use of digital games. The outcomes were positive as the teachers noted the students were actively engaged throughout the lesson and motivated them to learn the language. For Item 1, equal number of participants (N = 37) agreed to the statement in which participants responded that a digital board game can be played at leisure hours as they perceived the use of digital board game as fun and interactive language learning tool. P34 alluded that *"with interesting educational games, I will be able to make learning a fun process"* while P20 further elaborated that *"it [digital board game] is memorable due to its unique and simple"* approach of language learning. Furthermore, they can learn language in a stress-free environment as mentioned by P24, *"when playing, we do not [feel] stress so it is easier for us to learn..."* It is, in fact, in line with a previous study in which the respondents believed that learning digitally can make the experience more interesting and enjoyable [10]. Nevertheless, playing digital board games may not only be for learning as it may provide a popular pastime for all ages of life and genders. Playing the game enables its player to test his or her skill as in playing Chess and Poker or test someone's luck when he or she plays games; for instance Yahtzee and Pass the Pigs. The results of the study further support the idea that a digital board game provides students with a stress-free environment [11]. Such was demonstrated when forty participants expressed that there had more freedom to explore and move about when they play Tisch; a digital board game that provides interactive and interesting platform for its players to engage with. This is due to the fact that its players are able to monitor their potential as they may play various activities in the digital board game according to their preferences. Consistent with a series of article reviews pertaining to integrating games in language learning, mobile games are shown to be stress-free tools [12] that enable players to learn vocabulary in more contextualized manner. In fact, it is one approach that teachers may consider to ensure that classes would be interesting and effective. This mobile learning vocabulary games are able to help players/learners to improve their domain of vocabulary knowledge.

Overall, these results suggest that there is a significant need for a digital board game to be used by tertiary students to learn English language. It can be summarized that it is fundamental to develop a digital board game since the tool enables its players to play at their time and space, improve players' understanding of a language and enrich players' vocabulary size.

Table 1. Students' Perceptions of Need for Digital Board Game

No.	Items	Frequency (N)	Percentage (%)
1.	A digital board game can be played at leisure hours	37	64.9
2.	Language digital game can improve a player's understanding of language	47	82.5
3.	Vocabulary digital game can enrich a player's vocabulary size	48	84.2

Having stated the results of the current study, it may require that in the actual study, potential participants may need to be assigned to play the game outside the class hours. Such is considered due to ample time is forecast to be the factor that would engage players to play the game. Moreover, they may choose their co-players that are online at that particular time to play the game. Yet, before gauging their experiences of playing the board game, preparing story board is a requirement on part of the researchers. Due to the researchers' exposures of having to develop an online vocabulary game namely OnVac and a vocabulary app - VocBlast, it is hoped that a digital board game can be materialized through the concepts provided in the current study.

4 Conclusion

In this paper, the researchers aim at identifying the need for a digital board game among tertiary students in learning English language specifically in vocabulary learning. From the findings, it seems clear that majority of the participants believed that digital board game help to increase their vocabulary size. Learning vocabulary can be challenging sometimes and if conventional method is used throughout, the learning would result in years. Digital board game may be the alternative to let the students experimenting new vocabulary instead of memorizing it.

The participants also displayed positive attitude toward the use of language digital game in enhancing their understanding of the language which at the same time enhance their motivation to learn the language. Upgrading the board game digitally is an idea that is celebrated by majority of the participants in the current study as it helped them to comprehend the language use in a vivid presentation. Moreover, by upgrading the board game digitally, the game is enhanced with variety of auditory and visual stimuli that makes it a perfect educational tool that provides enjoyment, pleasure and motivation to the students [13]. Moreover, it is believed that students would gain higher self-confidence when they manage to complete the targeted level. This succession of emotions tends to keep the players immersed.

This study also found that students perceived digital board game as fun and interactive language learning tool because it can be played at leisure hours. Indeed, digital board game provides stress free environment for the students, and thus, makes learning a fun process whereby they can collaborate with other students through sharing information and learn from each other without the presence of the teacher. The students can develop critical thinking, increase self-esteem and also improve their skill and knowledge through this collaborative approach.

With pedagogical considerations in mind, we believed digital board game is an exceptional educational tool as it helps to reinforce important knowledge in an easy to understand format, reduce the time needed to learn, memorize and apply new knowledge plus at the same time promote discussion, collaboration and build communication. In short, digital board game has many benefits when it designed appropriately for learning.

Although the current study employed a small number of sampling size, we believed this study has added to the body of knowledge pertaining to the effectiveness of using digital board game in order to achieve better results in vocabulary learning and language development. Digital board game also provides genuine, authentic, hands-on and heads-on skill and knowledge as well as engaging atmosphere yet at the same time competitive environment for the students to learn the content through game elements such as discussion, problem solving and making decision. That being said, it is about time the tertiary education shifts forward in teaching language by incorporating digital board game in language classroom.

4 Recommendation

Further work needs to be done to establish students' perceived usefulness of the actual digital board game in learning English in tertiary education based on the results obtained in the current study. Particularly, future researchers may be able to examine issues in the use of digital board game in terms of the game content. Further investigation also may be carried out looking at students' progress in using digital board game to learn English language.

This research was supported by Universiti Malaysia Pahang (UMP) under research grant no. RDU1603144.

References

1. Agca, R. K., & Ozdemir, S. Foreign Language Vocabulary Learning with Mobile Technologies. *Procedia - Social and Behavioral Sciences* **83**, 781-785 (2013)
2. Prensky, M. *Computer Games and Learning: Digital Game-Based Learning*. (2007).
3. Pea, R.D. et al. Methods and apparatus for interactive map-based analysis of digital video content. Retrieved from https://www.google.com/patents/US7082572
4. Cresswell, J. W. Educational research: Planning, conducting, and evaluating quantitative and qualitative research (3rd ed.). New Jersey: Pearson Education (2007)
5. Agca, R. K., & Ozdemir, S. Foreign Language Vocabulary Learning with Mobile Technologies. *Procedia - Social and Behavioral Sciences* **83**, 781-785 (2013)
6. Dizon, G. Quizlet in the EFL Classroom: Enhancing Academic Vocabulary Acquisition of Japanese University Students. *Teaching English with Technology,* **16** *(2)*, 40-56 (2016)
7. Ali, Z., Mukundan, J., Baki, R., & Ayub, A. F. M. (2012). Second language learners' attitudes towards the methods of learning vocabulary. *English Language Teaching*, *5*(4), 24.
8. Chik, A. Digital gaming and language learning: Autonomy and community **18** *(2)*, 85-100. (2014)
9. Huizenga, J., Dam, G. t., Voogt, J., & Admiraal, W. Teacher Perceptions of the Value of Game-Based Learning in Secondary Education. *Computers & Education* **110**, 105-115 (2017)
10. Mora, S., Di Loreto, I., & Divitini, M. (2016). From interactive surfaces to interactive game pieces in hybrid board games. *Journal of Ambient Intelligence and Smart Environments*, *8*(5), 531-548.
11. Bergström, K., & Björk, S. (2014). The case for computer-augmented games. *Transactions of the Digital Games Research Association*, *1*(3).
12. Derakhshan, A., & Elham, D. K.. The effects of using games on English vocabulary learning. *Journal of Applied Linguistics and Language Research,* **2** *(3)*, 39-47. (2015)

13. Muhammad Sabri Sahrir, Nurkhamimi Zainuddin, & Mohd Shahrizal Nasir. Learning Preference among Arabic Language Learners via Mobile Learning Management System Platform (Mobile LMS) using I-Taleem. *International Journal of Current Research in Life Sciences* **5** *(1)*, 509-514 (2016)

Sport Tournament Automated Scheduling System

Raof R. A. A[1,*], *Sudin S.* [1], *Mahrom N.* [1], and *Rosli A. N. C* [2]

[1]School of Computer & Communication Engineering, Universiti Malaysia Perlis, 02600 Pauh Putra, Malaysia
[2]Faculty of Engineering Technology, Universiti Malaysia Perlis, 02600 Pauh Putra, Malaysia

Abstract. The organizer of sport events often facing problems such as wrong calculations of marks and scores, as well as difficult to create a good and reliable schedule. Most of the time, the issues about the level of integrity of committee members and also issues about errors made by human came into the picture. Therefore, the development of sport tournament automated scheduling system is proposed. The system will be able to automatically generate the tournament schedule as well as automatically calculating the scores of each tournament. The problem of scheduling the matches of a round robin and knock-out phase in a sport league are given focus. The problem is defined formally and the computational complexity is being noted. A solution algorithm is presented using a two-step approach. The first step is the creation of a tournament pattern and is based on known graph-theoretic method. The second one is an assignment problem and it is solved using a constraint based depth-first branch and bound procedure that assigns actual teams to numbers in the pattern. As a result, the scheduling process and knock down phase become easy for the tournament organizer and at the same time increasing the level of reliability.

1 Introduction

Over the last decade, interest in sports has increased tremendously. This is because sports helps to improve and maintain physical fitness, as well as provides entertainment to participants. There are many sports activities that are organized in schools, colleges, universities and also professional levels. Most of the time, sports start becoming very popular during college or university level since this is considered as the amateur phase. A schedule for sports tournament is a must when organizing any sports event at any level. Moreover, the tournament needs to be completed within a specific timeframe. Hence, the problem of sports tournament scheduling is of interest and the focus of this research proposal. A sports tournament scheduling is defined as the process of assigning essential sport activities in sequence with the time needed to complete each activity [1]. It is a challenging task due to the wide variety of different needs and requirements that must be taken into consideration, either at professional or amateur level [2-10]. The inefficiency of the solution, which is the tournament's schedule, normally exists when these requirements are not met and time consumed in constructing the schedules is relatively long. Hence, the search for the best and ideal solution approach is never ending.

Most of the time, the organizer of sports tournament faced problems on managing the schedule of the games involved in the round robin phase which will then lead to knock out phase [5][8]. The organizers will be responsible for coming out with a schedule before the tournament takes place, and then the organizers also need to calculate the mark and provide the standing result. All this while, the process of creating the schedule, providing the score sheet as well as standing result is being done manually by the organizer using manual spreadsheet. This will lead to problems such as wrong calculations of scores are being made, and the level of integrity of organizing committee is also being questioned.

Therefore, the development of web-based intelligent sports scheduling system is proposed. The system is expected to be able to automatically generate schedules for sport tournament and also auto calculate the marks and scores for these tournaments. In most of the sport tournaments, there are two parts of scheduling; round robin and knock out phase. The problem of scheduling the matches for the round robin and knock out phase in a sport league will be tackled. The intelligent scheduling system can be used to auto generate sport scheduling for any sport tournament. In most of the sport tournaments, there are two parts of scheduling; round robin and knock out phase. A round-robin (or all-play-all tournament) is a competition in which each contestant meets all other contestants in turn [5][8]. In a single round-robin schedule, each participant plays every other participant at least once. A knockout tournament or elimination tournament is divided into successive rounds; whereby each competitor plays in at least one fixture per round. The top-ranked competitors in each fixture progress to the next round. As the rounds progresses, the number of

* Corresponding author: rafikha@unimap.edu.my

competitors and fixtures decreases. The final round, which is usually known as the final or cup final, consists of only one fixture; the winner of which is the overall champion. To the best of knowledge, there is no such system which caters both processes (auto scheduling for round-robin and knock-out). Teams will be placed automatically in the schedule. Based on the auto scheduling system, scores will be keyed-in and the system will automatically calculate the points based on rules that have been pre-identified to determine the qualified team for next round.

2 Methodology

The development of the Sport Tournament Automated Scheduling System involves four phases, which are requirement identification phase, initial prototype development phase, prototype review phase, as well as revision and enhancement phase.

In the requirement identification phase, the user of the system will need to provide the name of the teams that will be involved in the tournament. Then, the system will automatically check for the duplicate teams' name (if any). There should not be any duplication of the teams' name. If the duplication occurs, the system will require the user to insert the team's name again before allowing proceeding to the next step.

It is assumed that altogether; there will be eight teams involved. The system will automatically generate groups if all eight teams' name have been entered correctly and saved to the system. Group generation process should be done by randomly choosing the names from the participating teams. The teams will be divided into two groups with each group having four teams.

Next process is generating the match schedule within the group. This process use round robin algorithm whereas it is assumed that there are enough fields or courts so that all the games in a round can be played simultaneously. If N is the number of competitors, then a round robin tournament requires $(N-1)$ rounds of games. Therefore, the time needed for the whole round robin tournament to complete is depending on the number of team or player. If the number is large, therefore it needs more time to complete the whole tournament with the condition that the number of courts or fields are enough. The number of teams (N) in round robin algorithm is fixed to four teams since the number of teams per group is also four. Team array is rotated every time after matches for each round is done. The algorithm resulting each team will be playing the match within three round with each round is with different team in a group.

The score for each team need to be keyed into the system in the results' table after the completion of each match. The user only need to fill in the section of the table noting the scores and the rest of section consisting the total and respective calculation will be automatically filled up by the system with respect to the inserted score.

The next process is calculating the total score of the teams. There are three rules applied in order to calculate the score. The team will get 2 points when winning the match, 0 point if losing it and 1 point if there is a draw. Then the points will be combined together to get the total marks. Once the total mark has been calculated, the standing of the team within the group will be shown according to the total mark.

The first and second runner-up from each of the group will be taken to the next process which is known as knock-down round. In this process, the first runner-ups of the group will be match with second runner-ups of the other group and vice versa. Their score will be inserted after this round and the schedule of the next match will be generated based on single elimination algorithm. This algorithm allows only one match to be held. If the team has higher score, then the team will be proceeding to the next match. On the other hand, the losing team will be eliminated. The match will be considered finished when the winner has been determined. The overall flow of the complete process is shown in Figure 1.

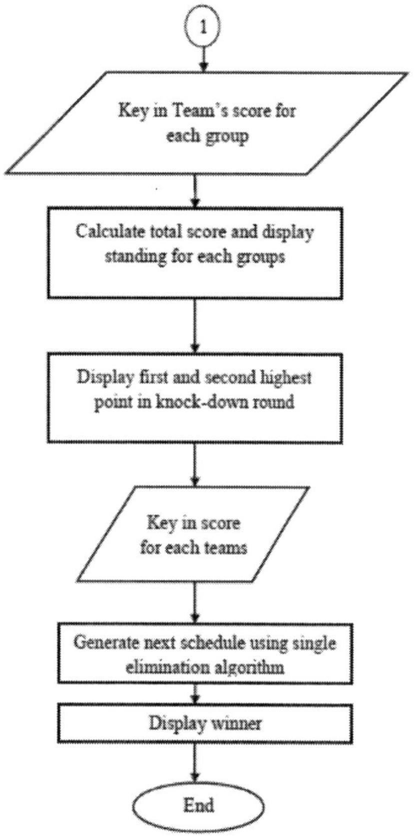

Fig. 1. Flowchart of methodology

3 Results

The system is developed using Microsoft Visual Studio 2015. This software is one of web developer platforms. Microsoft Visual Studio is an integrated development environment (IDE) from Microsoft. It is use in development of console and graphical user interface applications along with Windows Forms or WPF applications, web sites, web application and web services.

Figure 2 shows the interface for providing the name of the teams that will be involved in the tournament. If there is any duplication of names, the system will give alert as shown in Figure 3. There are 5 tabs on this form which are teams, groups, schedule, result and knock-down round.

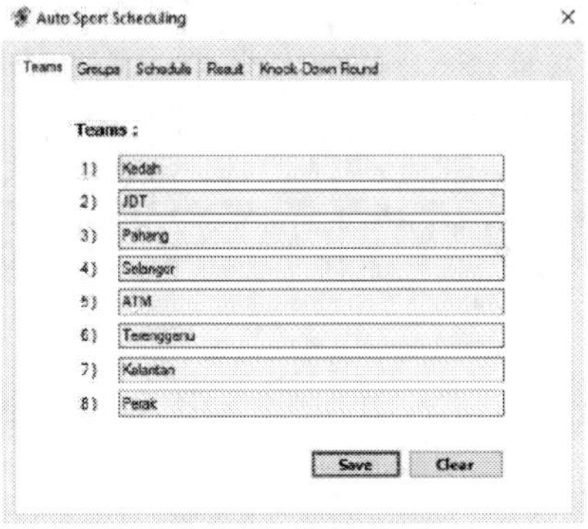

Fig. 2. Interface to provide teams' name

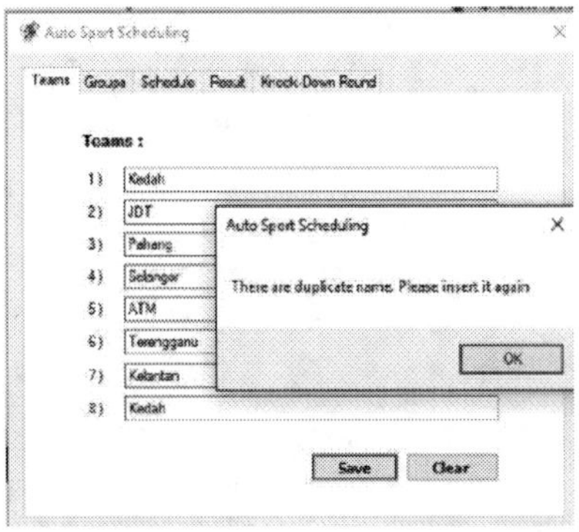

Fig. 3. Alert given on duplicate name of teams

Figure 4 shows the interface for generating groups for the participating teams. In this example, there are eight participating teams, so two groups are created with each group consisting of four randomly selected teams.

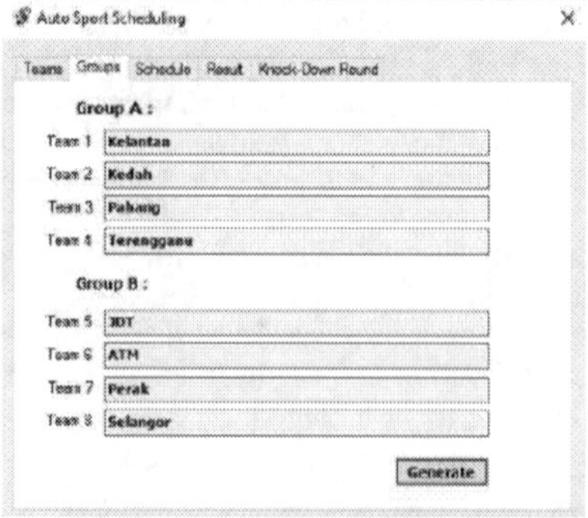

Fig. 4. Random group generation

The system will then proceed to generate the schedule for the participating teams within each group by applying the round-robin algorithm. Figure 6 shows the example of schedule that has been generated for all eight teams.

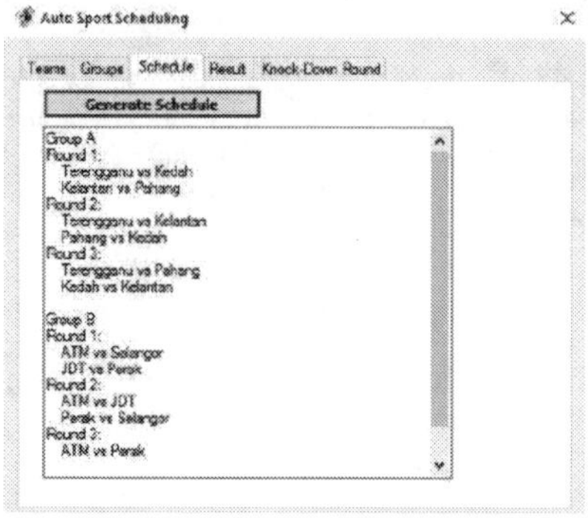

Fig. 5. Schedule generated from participating teams

Along the way of the tournament, the organizer will fill in the scores for each of the match, as shown in Figure 6. The total score and team standing will be automatically generated by the system. Each of the group will have their own standing separately.

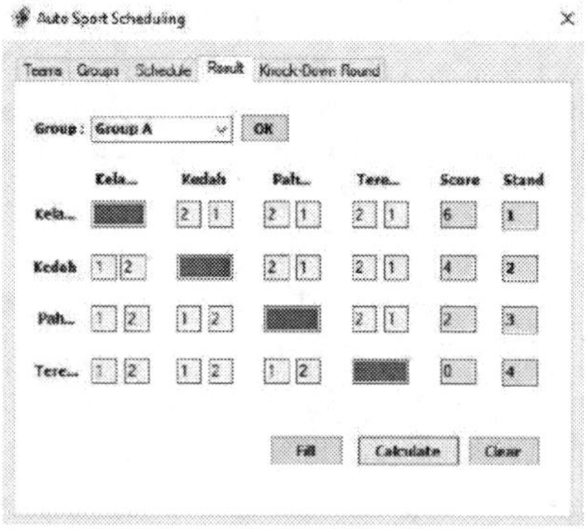

Fig. 6. Scores filling and standing generation interface

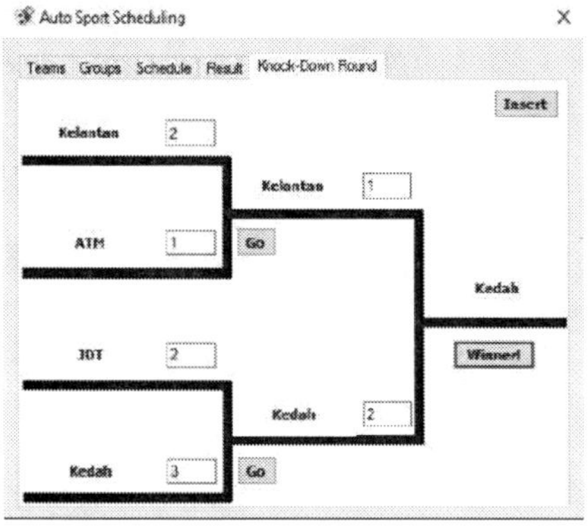

Fig. 7. Interface for knock-down round

Figure 7 shows the process during knock-down round. The first and second stander from each group previously will be displayed here automatically. The organizer then can insert the score after each match along the way of going to the final match.

4 Conclusion

In this paper, a sport scheduling system has been developed which is suitable for almost all type of sport tournament. This system has been designed based on Round Robin Scheduling Algorithm. In this algorithm every team plays against every other teams during the tournament and has equal chances against all other opponents. In Round Robin Algorithm, it is assumed that there are enough fields or courts so that all the games in a round can be played simultaneously. If N is the number of competitors, then in round robin tournament requires $(N - 1)$ rounds games. Therefore, the time needed for the whole round robin tournament to complete is depending on the number of team or player. If the number is large, therefore it needs more time to complete the whole tournament with the condition that the courts or fields are enough.

Another well-known type of tournament formats that sport schedulers used is the Single Elimination Tournaments. In this tournament, each player or team can defeat and eliminate a player or team from the tournament in just one match. A weak player can easily be defeated by a stronger player and eliminated from the tournament without given chances to play against other players. Therefore, this type of tournament is mainly depended on luck and not giving equal chances for all the players. However, time needed to complete a single elimination tournament is much shorter than a round robin tournament because one match is enough to determine the winner in the single elimination compared to the round robin which needed multiple matches to determine the winner.

In the future, the system will be expanded to include more types of tournament, with many more ways of score calculation. There are many variations in the way of sports handling and it is aimed that this system will be able to address as many way possible to accommodate the need of organizing sports tournament easily and effectively when it comes to scores calculation. The algorithms of how the scheduling is done also will be detailed out further. The authors also would look into constraints that are currently existing for creating this scheduling for some types of sports, if any, so that the problem can be overcome.

The authors wish to thanks Universiti Malaysia Perlis for funding this research projects under the Short Term Grant Scheme.

References

1. R. Ramli, C. J. Soong, H. Ibrahim, "A sports tournament scheduling problem: Exploiting constraint-based algorithm and neighborhood search for the solution", *Proceedings of the World Congress on Engineering 2012*, Vol I WCE 2012, July 4 - 6, 2012, London, U.K, (2012).

2. J. C. Hung, M. K. Chien and N. Y. Yen, "Intelligent Optimization Scheduling Algorithm for Professional Sports Games," Ubi-Media Computing (U-Media), *2011 4th International Conference on, Sao Paulo*, 2011, pp. 285-289 (2011).

3. T. Januario, S. Urrutia, D. de Werra, Sports scheduling search space connectivity: A riffle shuffle driven approach, Discrete Applied Mathematics, Volume **211**, 1 October 2016, Pages 113-120 (2016)

4. D. Goossens and F. C. R. Spieksma, "Sports scheduling with generalized breaks," *Computational Intelligence in Scheduling (SCIS), 2011 IEEE Symposium on*, Paris, 2011, pp. 54-57 (2011)

5. T. Januario, S. Urrutia, A new neighborhood structure for round robin scheduling problems, Computers & Operations Research, Volume **70**, June 2016, Pages 127-139 (2016)

6. L. H. Su, Y. Chiu, T. C. E. Cheng, Sports tournament scheduling to determine the required number of venues subject to the minimum timeslots under given formats, Computers & Industrial Engineering, Volume **65**, Issue 2, June 2013, Pages 226-232 (2013)

7. S. Duran, O. Ö. Özener, E. Yakıcı, League scheduling and game bundling in sports industry, Computers & Industrial Engineering, Volume **74**, August 2014, Pages 92-101 (2014)

8. W. Suksompong, Scheduling asynchronous round-robin tournaments, Operations Research Letters, Volume **44**, Issue 1, January 2016, Pages 96-100 (2016)

9. R. Lewis, J. Thompson, On the application of graph colouring techniques in round-robin sports scheduling, Computers & Operations Research, Volume **38**, Issue 1, January 2011, Pages 190-204 (2011)

10. D. Gupta, L. Goel and V. Aggarwal, "A hybrid biogeography based heuristic for the mirrored traveling tournament problem," *Contemporary Computing (IC3), 2013 Sixth International Conference on, Noida*, 2013, pp. 325-330 (2013)

Figurative Language in Malay to English Translation: An Analysis of the 2015 UniMAP VC's Keynote Speech

Sharmini A.[1], *Muhammad Bazli* Mahmood[1], *Khairul Hisham* Jamalludin [1], *Ahmad Hifzurrahaman* Ridzuan[1], *Mohamad Zaki* Abdul Halim[1], *Siti Nurul Jannah* Fital[1]

[1] Centre of International Language, Universiti Malaysia Perlis (UniMAP), Malaysia.

Abstract. Translating figurative language involves more than just replacing the figurative language with its equivalent in the target language. Therefore, it is not surprising for the translation of figurative language to have its own set of challenges. Problems the translator faces in translating the Malay Figurative Language into English include complexities in understanding, interpreting and recreating the Figurative language that are unique in the Source Language (SL) culture; which have to be explained and described in Target Language (TL) where such practices and customs are non - existent. Secondly, the Source Text (ST) figurative language may appear in a variety of types and have a distinct denotative and connotative meaning and reference; most often, it is difficult to find an equivalent which totally matches the original meaning or concept. This particular paper analyses the translation of figurative language extracted from UniMAP's Vice Chancellor Keynote Speech in 2015. Findings reveal that the three categories of figurative language identified were namely idioms, metaphors and similes. Translation strategies used are either not translated, paraphrased or translated with a similar meaning but in different form.

1 Introduction

Figurative Language according to Montgomery et al (2007) refers to the state of words or phrases that implies a non-literal meaning which does make sense or that could [also] be true. This notion is strongly supported by Nida (1975) who explains figurative word as a word which is used in place of another meaning or expression which is not its synonym but with which it has an association of ideas often mediated through a supplementary component such as idioms, simile and metaphors ». In short, figurative language occurs whenever an individual describes something by comparing it with something else.

Within the translation context, we can infer that figurative language may play an important role in the explanation of cultural concepts or can be used as the substitution of one idea or object with another, and to assist expression or understanding. However, translating figurative language is more than just handling terminology. In agreement with Abdul Wahid (2017, p. 17), "translating figurative language not only requires the knowledge of the language and cultures of both source and target language, but also choosing the appropriate word and translation strategies". Thus, the main challenge of translating figurative language lies in grasping the ST author's cultural meaning and transmitting that cultural meaning which is expressed via a figurative word into another language as accurately as possible in the way in which the SL author would most probably have expressed the content himself or herself, had he or she been a native speaker of the target language (Abdullah, 2014). In other words, the target language is presented in an accurate possible form, complete and that the information can be used correctly and effectively.

Bell (1991) stresses this challenging process of translating when he mentions that it is "very clear that the ideal of total equivalence is impossible in translation because languages are different from each other, they have different codes and rules regulating the construction of grammatical forms which have different meanings". (p. 6). Both Jakobson (1976) and Bassnett (1980) also emphasize this point. Jakobson (1976) believes there are "no exact synonyms between languages" (p. 26). Bassnett (1980) echoes this notion when she states that there cannot be complete equivalence in translation as "each unit contains within itself a set of non-transferable associations and connotations" (p. 15). She explains further that "sameness cannot even exist between two target language versions of the same text" (p. 29).

As such, translating figurative involves more than just replacing the figurative language with its equivalent in another language. Translating figurative language from Malay (source language, or SL) to English (target language, or TL) will require more than just reformulating the figurative word in one language (i.e. Malay) as a figurative word in a different language (i.e. English). This paper attempts to discuss the translation of figurative language namely idioms, metaphors and similes when translated from Malay to English, factors that influence the translation procedures used by the

*Corresponding author: sharmini@unimap.edu.my

translator and bring to light the possibilities and challenges of translating figurative language from Malay into English using examples extracted from the 2015 UniMAP VC"s Keynote Speech.

2 Literature Review

2.1 Translation

Newmark (1981) defines translation as a "craft consisting of the attempt to replace a written message and/or statement in one language by the same message and / or statement in another language" (p. 7). Bell (1991) further amplifies this definition in his interpretation of translation - "to produce as accurately as possible all grammatical and lexical features of the source language (SL) original by finding the equivalences in the target language". At the same time all factual information contained in the original text must be retained in the translation" (p. 13). These prescriptive definitions above highlight features of translation as either a product (the text that has been translated) or the process (the act of producing the translation).Translation as a process based on Hatim and Munday"s (2004) definition refers to "the role of the translator in taking the original or ST and turning it into a text in another language, the target text TT" (p. 3). Translation as a product on the other hand, "centres on the concrete translation product produced by the translator" (Hatim & Munday, 2004, p. 6).

Shuttleworth and Cowie (1997/2014) provides a more comprehensive definition as their definition incorporates a wider range of areas in translation – "an incredibly broad notion which can be understood in many different ways. For example, one may talk of translation as a process or product, and identify such sub types as literary translation, technical translation, subtitling and machine translation; moreover, while more typically it just refers to the transfer of written texts, the term sometimes also includes interpreting" (p. 183). For the purposes of this paper, the term translation will be used to refer to "a text containing instances of substitution, addition, omission, expansion, or modification and produced on the basis of a source text in the target language in terms of words, meaning, or sentence structure with the main purpose of making the target text suitable for a particular purpose, genre and audience" (Abdullah & Shuttleworth, 2013, 3(6), p. 608).

2.2 Equivalence in Translation

Identifying, selecting and creating an equivalent in the TL is not always an easy task. As such, translators would have to either alter, expand or omit items in their translations. The process of translating an SL is always to identify the suitable equivalents in the TL first. Shuttleworth & Cowie (1997) define equivalence (or Translation Equivalence) as "the nature and the extent of the relationships which exist between SL and TL texts or smaller linguistic units" (p. 49). These "smaller linguistic units" include levels of equivalents which range from the least significant level of language – that of morpheme - to the more complicated levels like a sentence between the SL and TL. It is not surprising that equivalence in translation is described as a "troubled notion" (Hermans, 1995, p. 217) because most translation scholars like Jakobson (1959/1966), Nida (1964/1969), Catford (1965), House (1977), Baker (1992), Vinay and Darbelnet (1995) express their understanding and interpretation of translation equivalence differently. However, regardless of the difference in views and interpretations, all unanimously suggest the fundamental role of equivalence - culturally and linguistically in the translation process. They view equivalence as a technique which reproduces the same situation in the TL as in the original; only with a completely different wording. They also mentioned that equivalence is the ideal method in dealing with figurative languages like idioms, clichés and proverbs.

It must be noted however, that this perception when applied now is somewhat old fashioned (Abdullah, 2014). "Translation is no longer determined by the principle of equivalence, but based on the adequacy in aacordance with the directions of translation. However, the production of the translation must be logical or coherent" (Abdul Wahid, 2017, p.3). Abdul Wahid's statement above mirrors Nord"s (2001) functionalist approach in translation. According to Nord (2001) each translation process does not only translate the language, but also the concept of a term in the source text.

Similarly, Vermeer (cited in Nord, 1997) implies that translation is a purposeful activity. "Any form of translational action, may be conceived as an action. Any action has an aim, a purpose" (p.12). Newmark (1988a) and Nord (1997) agree with Vermeer and further explain that if the aim of the commissioner is to convey the cultural aspects of the ST, then the translator will place more emphasize on the precise wording that it contains. If the purpose is to ensure that the translation conveys the same emotional and persuasive tones as the ST, the translator will then apply another set of strategies to help the target readers to achieve a better understanding of that textual level.

2.3 Translatability of Figurative Language

The translatability of figurative language has always been questioned as pointed out by Dagut (1976) and Newmark (1988a) below:

"The crucial question that arises is thus whether a metaphor can, strictly speaking, be translated as such, or whether it can only be "reproduced" in some way" Dagut, M. B. (1976, p. 24).

"whilst the central problem of translation is the overall choice of a translation method for a text, the most important particular problem is the translation of the metaphor" (Newmark, 1988a, p. 104).

The lack of clarity in figurative language makes translating and interpreting its meaning not always easy and straightforward. In translating figurative language from one language and culture to another, one may be hampered by linguistic and cultural differences between the two (or more) languages concerned. As such, translating figurative language can give rise to various challenges and this is predominantly so if the figurative in the source language are culture-bound as in the case of Malay to English Translation of figurative language. This means, that the translation of the figurative language may largely depend on the cultural associations (Dagut, 1976, p. 32; Bassnett , 1980 ; Snell Hornby, 1988/1995, p. 58, Abdullah, 2014, Abdul Wahid, 2017). Figurative language in the Malay language is aimed at expressing something implicitly and to create an impact on its target audience. As discussed above, challenges in translating figurative language arises due to cultural differences between the Malay and English texts.

Baker (1992) highlights four challenges in translating figurative language – i) a figurative language has no equivalent in the target language; ii) a figure of speech might have the same equivalent in the target language, but may vary with the context; iii) a figure of speech in the source language may be translated literally and idiomatic in the target language; iv) the use of figurative language in writing discourse, context and frequency of usage may differ with the source language and the target language.

In overcoming these four challenges, Baker (1992) suggests two translation procedures for translating the figurative language – i) using a figure of speech that is similar in meaning and form; ii) the use of figurative language is similar in meaning, but differ in form and the last is translated by the paraphrase. Baker (1992) further adds translation by paraphrasing is used when an equivalent match cannot be found in the TL (target language).

2.4 Previous Studies

Studies which discuss the issues of translating figurative language from Malay to English are a few. Available studies relating to figurative language dealt with translation of metaphors from English to Malay concern Malay poetry (Shunmugam, 2007), Malay figures of speech (Wan Ramli, 2014), classical Malay (Subagiharti, H.et al, 2015).

Shunmugam (2007) examined Malay into English pantun (poetry) translations of two women writers - Katherine Sim and Martha Blanche from the nineteenth century to the present times. Shunmugam investigated translation styles adopted by the translators and the extent to which skopos or translation purpose influenced the translation style. She conducted a comparative analysis of Lewis's and Sim's poetry translations with Wilkinson"s, Winstedt"s and Hamilton"s renditions. Findings indicate that Lewis"s and Sim"s translational styles are not distinctly different from the male British translators of their time.

Wan Ramli (2014), conducted a comparative study of English Similes and their translations of The Hunger Games. Strategies employed by the translator in translating the simile was analysed using strategies proposed by Pierini (2007). Findings revealed that out of the four identified strategies (i.e. literal translation, reduction of the simile, if idiomatic, to its sense, retention of the same vehicle plus explicitation of similarity feature(s) and omission of the simile) Literal translation seems to be the most used strategy by the translator. Wan Ramli concluded this is so probably because some of the similes could be understood and interpreted the same way in Malay. She adds translators did encounter problems in determining the Malay interpretation which conveyed the same meaning as in the SL. Wan Ramli (2014) highlighted the importance of conveying the SL meaning to the target readers rather than providing and accurate translation of the ST. Translators had a choice either to preserve the form or render the meaning by changing the form as long as the meaning is conveyed to the target readers.

On a slightly different note, Subagiharti, H. et al. (2015)'s research on Malay to English metaphor translation focused on the use of anthropomorphemic metaphors in Hikajat Abdullah (Abdullah Tales). Analysis revealed that the anthropomorphemic metaphors in classical Malay displayed specific characteristics as listed below and the form of metaphor in Hikajat Abdullah is determined by the target element:

1. The sentences are long, repetitive, and convoluted.
2. There are many passive sentences with specific vocabularies such as Ratna Mutu manikam, Masyghul (sad).
2. Many sentences are preceded by certain conjunctions, for example, sebermula „in the beginning," Once, Hatta, while, then, said sahibul saga, when, and while.
4. The sentences are rich of suffixes.

The relatively low number of existing studies related to figurative language in Malay to English Translation as discussed above justifies the need to conduct such studies in this area. Findings from this study will fill the gap in the literature, provide an insight on how and if the figurative language are translated and disclose current figurative language translation practices of the Malay to English Translators.

3 Methodology

The main aim of this study is to analyse the translation and translation procedures adopted by the TL translators to translate figurative language using examples extracted from the 2015 UniMAP VC"s Keynote Speech. The source text (ST) comprises of "Ucaptama Tahunan Naib Canselor 2015: "Mencarik Adat, Melarik Hasrat" while the target text (TT) consist of the slide presentation projected during the event. Translation Text : Slide presentation projected during the event. The methodology adopted is textual analysis whereby Hartmann"s (1980) Parallel Texts Technique and Lindquist"s (1989) Parallel Reading were applied to extract the figurative words from the texts. Bakers (1992) proposed two translation procedures as described in section 2.3 were employed to identify the translation procedures used by the TL translator.

4 Findings and Discussions

4.1 Types of Figurative Language

Analysis revealed the use figurative language in the UniMAP 2015 VC"s Keynote Speech and its English Translations. Table 1 below displays the types of figurative words identified and highlights the statistics for each type.

Table 1. Types of Figurative Language

Types of Figurative Language	Number
Simile	1
Idiom	2
Metaphor	10

The number and presence of different types of figurative words in the ST as presented in Table 1 above indicate that the academic context of the UniMAP 2015 VC"s Keynote Speech is certainly characterized by the use of figurative language – in this case idioms, metaphors and similes to describe, explain or express something implicitly or to create a far more effective impact on the readers.

4.2 Translation Procedures

In translating simile, the translator maintained the image or element of the ST as illustrated in Table 2. The translator used a simile of similar meaning and form. The phrase „bagai adat tiada bertukar, rahsia tidak berubah" has a similar meaning with "peas in a pod" and "hand and glove". As demonstrated, both the Malay and English simile refer to the same meaning with a similar use of lexical item.

Table 2. Simile in ST and TT

Source text (ST) - Malay	Target Text (TT) - English
Bekerja dengan semuanya bagai adat tiada bertukar, rahsia tidak berubah (maknanya perhubungan kita yang sangat akrab dan tidak dapat dipisahkan	Working together as peas in a pod, hand and glove, and (inseparable).

rather selected non- figurative English words which had a similar meaning. This finding corroborates Larson"s (1998:48-49) view that "a literal word-for-word translation of the idioms into another language will not make sense. The form cannot be kept, but the receptor language word or phrase which has the equivalent meaning will be the correct one to use for translation". As illustrated by the examples below in Table 3, the translator did not perform a "literal word-for-word translation of the idioms rather retained the Malay idiom in the TT and provided an explanation of the idiom in the TT. The wording „helping one another; the affluent helps those in need, the lettered assists those who are unlettered, and the influential protects those who are powerless" is equivalent to the meaning of „Adat Pulau, Limburan Pasang". For the ST idiom "bulat air kerana pembentung, bulat kata", the translator provides an equivalent idiom in the TT that has a similar meaning to the idiom in the ST - a pipe shapes flowing water, consensus shapes flowing words) denoting that a community creates its own destiny.

Table 3. Idiom in ST and TT

Source text (ST) - Malay	Target Text (TT) - English
1. Peribahasa Melayu ada menyebut „Adat Pulau, Limburan Pasang" (maknanya adat hidup bantu membantu; yang berharta membantu yang melata, yang berilmu membantu yang keliru, dan yang berkuasa melindungi yang derita).	There is a Malay proverb that says 'Adat Pulau, Limburan Pisang' (which means helping one another; the affluent helps those in need, the lettered assists those who are unlettered, and the influential protects those who are powerless)
2. Peribahasa Melayu mengingatkan kita bahawa bulat air kerana pembentung, bulat kata kerana muafakat. Begitulah adat kita.	A Malay proverb reminds us that bulat air kerana pembentung, bulat kata kerana muafakat (a pipe shapes flowing water, consensus shapes flowing words) denoting that a community creates its own destiny). Such is our tradition.

Table 3 displays the translation of idioms. The translator did not translate the three idioms as a literal translation

Larson, (1998, p.279) states that if an expression is properly understood by the target audience, the metaphor can be translated directly into the target language metaphor. In other words, where it concerns translating metaphors, translators can maintain metaphorical expression with the same image or substituting a different metaphor in the target language carries the same meaning as the metaphor in the SL as illustrated in Table 4 below.

Table 4. Metaphor in ST and TT

Source text (ST) - Malay	Target Text (TT) - English
1. darah penyalur tenaga; nadi penggerak usaha;	the lifeblood, the driving Pulse
2. dibajai	nourished.
3. subur dengan idea-idea baru, sentiasa subur dengan usaha-usaha penambahbaikan yang berterusan	fertile with new ideas, fruitful with constant Improvements

As seen in the examples above, the image of the SL metaphor resembles the same image in the TT. The translated image of "darah penyalur tenaga" to "the life- blood"; "nadi penggerak usaha" to the "driving pulse" ; "subur dengan idea-idea baru" to "fertile with new ideas"; "subur dengan usaha-usaha penambahbaikan yang berterusan" to "fruitful with constant improvements" resembles the same image in the TT. The translated image correlates with the image of the target culture and it can be understood by the TT audience.

5. Limitations of the Study

This study analysed the translation of figurative language extracted from UniMAP's Vice Chancellor Keynote Speech in 2015. The corpus selected presents as one of the limitations of the study. Firstly, the corpus size is relatively small mainly due to the specific genre, topic of the corpus and problems in identifying the suitable Malay (TT) translation for possible English Source Texts. Secondly, the time which was allocated to the TL translator to perform the translations serves as another limitation. According to the TL translator, the submission deadline was less than twelve hours. As a result, there was not enough time to edit nor consult or refer to more reliable sources to attain a more accurate translation for the problematic figurative words. This to a certain extent may have affected the quality of the translated text.

6. Conclusion

Translating figurative language can be trying especially when translating figurative words from one language into another. The challenge arises from the differences between languages in using figurative words and to ensure that the message is conveyed as accurately as possible In other words, figurative language comprise of unique linguistic characteristics and cultural elements that belong to the SL. Therefore, they are difficult to transfer into the TL. The more characteristic they are in the ST, the more complexities they pose for the translators who attempt to re-create the message they convey, their purpose and effect in the TT. The SL figurative words appear in a variety of types and have distinct denotative and connotative meaning and reference. These factors make it difficult for the translator to find a standard equivalent which totally complements the SL figurative meaning or concept. Therefore it is vital for the translator to not only possess a clear understanding and competence in the cultural, social and linguistic features of both the SL and TL figurative language; but also to ensure that the translation procedure adopted does not cause major changes to the structure of the ST.

References

[1] Montgomery, Mar; Durant, Alan; Fabb, Nigel; Tom Furniss; Sara Mills *Ways of Reading: Advanced Reading Skills for Students of English Literature*. Taylor & Francis. pp. 117–. ISBN 9780415346337. (2007)

[2] E. A. Nida, *Language Structure And Translation*. California, United States: Stanford University Press. (1975)

[3, 25, 36] P.R. Abdul Wahid,. *Translation Of Figurative Language From English To Malay: An Analysis Of The Translation Of The Harry Potter Series*. Proceedings of Research World International Conference, Phuket, Thailand, 9th-10th February, 2017, ISBN: 978-93-86083-34-0 (2017)

[4, 24, 35] S. Abdullah, *Translating Technical Metaphors from English into Malay: Possibilities (and Challenges)*. Proceedings of Symposium of International Languages and Knowledge. Penang, Malaysia, 14th -16th February 2017 (2014)

[5, 12] R. T. Bell, *Translation And Translating, Theory And Practice* (pp. 6).London and New York: Longman. (1991)

[6, 8] R. Jakobson, *Pëtr Bogatyrëv: Expert in Transfiguration*. Michigan Slavic Publications. (1976)

[7, 9, 10, 33] S. Bassnett, *Translation Studies* (pp. 15 - 29). London: Routeledge (1980)

[11] P. Newmark, *Approaches To Translation* (pp. 7). Oxford: Prentice Hall. (1981)

[13, 14] B. Hatim, & J. Munday, *Translation: An Advanced Resource Book* (pp. 3-6). USA & Canada Routledge. (2004)

[15, 17] M. Shuttleworth, & M. Cowie, *Dictionary Of Translation Studies*. Manchester: St Jerome Publishing. (1997)

[15] M. Shuttleworth, & M. Cowie, *Dictionary of Translation Studies*. (Reprint Ed). London & New York:Routledge, Taylor & Frances Group. (2014)

[16] S. Abdullah, & M. Shuttleworth, *Metaphors In The Translation Of English Technical Texts In to Malay: A Preliminary Study*. Journal of Asian Scientific Research, 3(6), 608-629. (2013)

[18] T. Hermans, "*Toury's Empiricism Version One: Review of Gideon Toury's In Search of a Theory of Translation*". In: The Translator, 1:2, pp.215-223. (1995)

[19] R. Jakobson, *On Linguistics Aspects Of Translation* (pp. 26). Selected Writings. Mouton: The Hague, 260-266. (1959)

[19] R. Jakobson, *Selected Writings: Slavic Epic Studies*. The Hague, Paris: Mouton & Co. (1966)

[20] E. A. Nida, *Toward A Science Of Translating* (pp. 159). Leiden, Netherlands: E.J. Brill. (1964)

[20] E. A. Nida, & C. R. Taber (1969/1982). *The Theory And Practice Of Translation* (pp. 200). Leiden: E. J. Brill.

[21] J. C. Catford, *A Linguistic Theory On Translation* (pp. 27 - 94). (London: Oxford University Press. (1965)

[22, 37-38, 44] M. Baker, *A Coursebook On Translation* (pp. 217). London: Routledge. (1992)

[23] J.P. Vinay, & J. Darbelent, *A Methodology For Translation*. Amsterdam/Philadelphia: John Benjamins. (1995)

[26] C. Nord, (2001). *Translating As A Purposeful Activity: Functionalist Approaches Explained*. Manchester, U.K.: St Jerome Publishers.

[27, 29] C. Nord, *Translating as a Purposeful Activity*. Manchester: St Jerome Publishing. (1997)

[28, 31] P. Newmark, *A Textbook Of Translation* (pp. 48-49, 109). New York: Prentice Hall. (1988a)

[30, 32] M. Dagut, *Can "Metaphor" Be Translated"?*, in Babel 22:1. 21-33. (1976)

[34] M. Snell-Hornby, *Translation Studies: An Integrated Approach* (pp. 19). (Revised Ed). Amsterdam & Philadelphia: John Benjamins Publishing Company (first published 1988). (1995)

[39] K. Shunmugam, *The Translation of Metaphors in Malay Pantuns into English*, in Proceedings, In Between Wor(l)ds: Transformation and Translation, The University of Melbourne. http://hdl.handle.net/11343/34691 (2007)

[40, 42] W.N.H. Wan Ramli, *The Translation Of Simile In The Hunger Games Novel: Translation Strategies*. (pp. 373-378). *E-proceedings of the Conference on Management and Muamalah (CoMM 2014), 26-27 May 2014. Synergizing Knowledge on Management and Muamalah (E-ISBN: 978-983-3048-92-2)* (2014)

[41] P. Pierini, *Simile in English: From Description to Translation*. CÍRCULO de Lingüística Aplicada a la Comunicación (clac) 29, 21-43. (2007)

[43] H. Subagiharti, et.al. *Classical Malay's Anthropomorphemic Metaphors in Essay of Hikajat Abdullah*. IOSR Journal of Humanities And Social Science (IOSR-JHSS).Volume 20, Issue 2, Ver. V (Feb. 2015), PP 01-09. e-ISSN: 2279-0837, p-ISSN: 2279-0845. www.iosrjournals.org. (2015)

[45-46] M. L. Larson, *Meaning-based translation: A guide to cross language equivalence*. Lanham, United States: University Press of America. (1998)

Level Readiness Participant Of Lifelong Learning (LLL) Program Venture Into Entrepreneurship

Mohd Shahzuan Ghazalan[1], *Noorazman* Abd. Samad[1], *Fazlinda* Ab Halim [1], *Nur Izeanty* Hamidon [1] *and Tun Ili Ayuni* Ahmad Hariri[1]

[1] Faculty of Technical and Vocational Education, University Tun Hussein Onn Malaysia (UTHM), Malaysia

Abstract. This study aimed to investigate the perception of Lifelong Learning (LLL) program participants on their readiness to venture into entrepreneurship. Aspects of the level of readiness measured in this study includes; motivation, abilities knowledge and interests of the LLL program participants. The research sample consists of 235 respondents from LLL program participants. The research design was a survey using a quantitative approach that was analyzed using descriptive statistics-specifically frequency counts, percentages and mean. SPSS version 25.0 was used to analyze the data. The instrument used was a questionnaire which had a reliability coefficient of α=0.9, indicating a very high instrument reliability. Findings reveal that the level of readiness of LLL program participants in entrepreneurial knowledge was at moderate level with mean value is 3.6. Meanwhile, readiness level of LLL program participants in terms of entrepreneurial interest was at high level with mean value is 4.1. Readiness in terms of entrepreneurial motivation was at a high level with mean value is 4.2. While readiness in terms of entrepreneurial abilities was moderately high level with mean value is 3.8. Thus, the researchers hope that the results obtained can be exploited by certain parties in order to improve the readiness of lifelong learning program participants in entrepreneurship, thereby producing quality entrepreneurs who can create more jobs to reduce the problem of unemployment in Malaysia.

1 Introduction

The Ministry of Higher Education in 2010 in the Executive Report Tracer reports that about 43.252 graduates are unemployed. Meanwhile, according to the National Statistics Department (2017) Report on Labor Force Statistics 2017, the percentage of unemployed people increased in Malaysia until July 2016 at a 3.5% rate[1-3]. Unemployment will continue to remain if responsible party do not try to find new alternatives to build their own career or get involved in economic activities likes as entrepreneurship that can help to lead the economy as well as provide employment opportunities for others. There is much room to be exploited in the community for people to venture into various aspect of livelihood such as self-employment, business and entrepreneurship. The areas that can be pursued however requires initiative and persistence. Generally, among the causes of unemployment are lack of experience, low quality certificate / diploma / degree that are not recognized and do not meet stipulated standards, the lack of soft skills, too many immigrants and attitudes of local employees.

However, through self-development, aspects of the causes of unemployment triggered by the lack of readiness within an individual begins with the readiness measured against knowledge, interest, motivation and abilities in the field of employment. In this context, it is suggested that the labor force expand the scope of their careers and make entrepreneurship as a career path. Entrepreneurship will help graduates develop their own careers and will also solve the issue of unemployment by creating more job market[4,6]. Entrepreneurship has been recognized by most researchers as a solution to the problem of unemployment[4,5].Development in the field of entrepreneurship has a significant impact on the business world at the moment. The Government always pay attention to the development of future entrepreneurs who are resilient and competitive in the global market. This can be seen in a variety of programs and activities in every country's development agenda, including

* Mohd Shahzuan Ghazalan: *gb150092@siswa.uthm.edu.my*

providing infrastructure to increase the number of entrepreneurs, especially among the indigenous people.

This effect has somewhat impacted participants Lifelong Learning Program at a community college to get a job after completing their program. As a solution, the ministry decided to utilize entrepreneurship as a means to solve the problem of unemployment in order to enable the Lifelong Learning program in community colleges to produce be more competent and resourceful entrepreneurs. Therefore, entrepreneurship is one of the areas of focus in community colleges due to the existence of the Lifelong Learning Program that has been provided. This is reflected in the government's efforts to produce entrepreneurs in various fields as an important agenda in ensuring that community colleges do not just produce graduates who are job seekers but also those who are knowledgeable to create jobs after graduation. This commitment can be seen with the establishment of Community College Entrepreneurship Center (PUSKOM) at the Department of Community Colleges and E-Tech Center in all Community Colleges throughout the country[7].

Thus, in 2011 community colleges have begun to see entrepreneurship as one of the branches of an alternative career choice for Lifelong Learning program graduates. Until 2013 the community college has produced nearly 400 entrepreneurs since the inception of the program. That effort continued throughout 2013 with community colleges organizing entrepreneurship challenges and competitions to develop entrepreneurship across community colleges. In this challenge, all community colleges were required to send at least one team. This effort is to nurture and cultivate entrepreneurship among participants in community colleges[7]. In addition, there were several approaches used to achieve the goal of creating a community of Malaysian entrepreneurial culture to attract Lifelong Learning program participants to entrepreneurship through entrepreneurship education in higher education. The goal was to provide knowledge, encouragement and entrepreneurial skills to these LLL participants[7,8]. The effectiveness of the program was viewed through the business performance in the third quarter of 2014 to the number of employees that showed a positive net balance of 5.4%[1,2].

Thus it was concluded that the program was able and adequate to nurture students to become entrepreneurs and change their mentality towards becoming self-employed[9,11,13,14].

Whereas in previous years, community colleges only promoted entrepreneurship education among participants in an attempt to produce more entrepreneurs or to make entrepreneurship as an alternate career path for students. In addition, entrepreneurs are still being educated in community colleges to ensure that they continue to be successful and be able to share the benefits with their friends and other participants in terms of experience, career opportunities or business opportunities. This is also seen as fostering a culture of social entrepreneurship in the colleges indirectly[10,12,14]. However, the question becomes to what extent is the readiness of participants in the Lifelong Learning program towards entrepreneurship in order to reduce the percentage of unemployment in the country. Thus, this study was conducted to determine the level of LLL program participant's readiness in terms of knowledge, interest, motivation and abilities to become entrepreneurs.

2 Research Objective

The objective is:

i. Investigate the perception of Lifelong Learning program (LLL) participants on their readiness to venture into entrepreneurship.

ii. Aspects of the level of readiness measured in this study includes; knowledge, interests, motivation and abilities of the LLL program participants

3 Methodology

This study utilized a quantitative approach, specifically the survey research design to assess the readiness of Lifelong Learning program participants in the field of pastry towards entrepreneurship at community colleges Johor.

3.1 Population and sample

The population in this research is composed of Lifelong Learning program participants in the field of pastry at Batu Pahat Community College (KKBP), Tanjung Piai Community College (KKTP) and Ledang Community College (KKL). The population consists of participants in the Lifelong Learning program from Batu Pahat Community College, Tanjung Piai Community College and Ledang Community College who completed short courses in Pastries in 2015 according to statistics of participation in 2014.

The sample for the study from Batu Pahat Community College (KKBP), Tanjung Piai Community College (KKTP) and Ledang Community College (MPA) consisted of 236 participants, randomly selected using Krejcie and Morgan's table for sample size determination as a reference[15].

3.2 Instrument

The instrument used for this study was a questionnaire adapted by the researcher form previously developed instruments [4,9,10]. The survey method is often used in educational research, especially in research on entrepreneurship education in higher education institutions. The questionnaire includes two parts, part A and part B. A total of 4 items covering sections A was designed to solicit demographic data from respondents. Meanwhile, part B contains a total of 60 items related to the perception of Lifelong Learning (LLL) program participants about their level of readiness towards entrepreneurship venture in the field of pastry. Part B which measures readiness also consist of 4 constructs namely knowledge, interest, motivation and ability towards entrepreneurship venture.

In this study, face and content validity was carried out by three experts in the field of entrepreneurship to verify the quality and clarity of the questionnaire. A pilot study was also conducted in this research, using a sample of 32 Lifelong Learning (LLL) program participants in Segamat Community. Thus, 32 questionnaires were distributed to the participants of short courses in field pastry, and all questionnaires were successfully retrieved. However, during the process of entering data into SPSS, two questionnaires were not useable as the responses do not meet the criteria for usability and as a result were disposed. Making the total useable questionnaire for the pilot study to be 30.

Data from the pilot study were prepared, cleaned and filtered for analysis using the software (SPSS V25.0). From the data analyzed, the reliability coefficient obtained for the instrument using Cronbach's Alpha reliability test was α = 0.9 which is more than the required alpha value for a highly reliable instrument, indicating that the questionnaire is highly reliable and capable of measuring the characteristics the researcher sought to measure (see Table 1).

Table .1. Results for Cronbach's Alpha

Score Cronbach's Alpha (α)	Level Cronbach's Alpha (Level of Reliability)
$\alpha = 0.9$	Highly Reliable Instrument

3.3 Data analysis

To analyze the data, the researcher used the software Statistical Package Social Science version 25.0 (SPSS V25.0) to answer four questions of the study. Descriptive statistics was used to analyze the data. More specifically, frequency counts, the mean and standard deviation were used to summarize and describe the findings of the study. Statistical analysis was carried out to investigate the mean level readiness of LLL program participants towards entrepreneurship venture in terms of knowledge, interest, motivation and ability.

4 Findings

The findings of the study which aimed at determining the level of readiness of LLL program participants from three community colleges in Johor towards entrepreneurship is presented in Table 1.

In overall, LLL program participant's level of readiness in terms of knowledge and abilities towards entrepreneurship venture is at medium-high level. While, in terms of their readiness towards entrepreneurial interest and motivation is at a high level (see Table 2).

4.1 Level of Readiness

In Table 2, the level of readiness for LLL program participants in terms of knowledge in entrepreneurship is at moderate level high at a mean value of 3.6. Meanwhile, their readiness level in terms of interest towards entrepreneurship was at a high level with mean values of 4.1, while their level of readiness in terms of their motivation towards entrepreneurial venture was at a high level with mean values of 4.2 and their level of readiness in terms of their ability to venture into entrepreneurship was at moderate-high level with mean values of 3.9 (see Table 2).

Table 2. Results for construct levels readiness

Constructs	Mean score	Level
Knowledge	3.6	Moderate High
Interest	4.1	High
Motivation	4.2	High
Ability	3.9	Moderate High

5 Conclusion

In conclusion, the Lifelong Learning (LLL) program participants showed a high level of entrepreneurship. It shows they have a tendency of becoming entrepreneurs. However, their level of readiness in terms of their knowledge of entrepreneurship needs to be improved. Concerned Community College stakeholders should play a role in designing and making continuous improvements in the curriculum to help the participants to be able to become more successful entrepreneurs in the future.

Thus, the researchers hope that the results obtained can be leveraged by these stakeholders in order to improve the implementation of Lifelong Learning (LLL) program participant's readiness for entrepreneurship, thereby producing quality entrepreneurs to contribute towards creating more jobs to reduce the problem of unemployment in Malaysia.

This paper was partly sponsored by the Centre for Graduate Studies UTHM.
Also, Acknowledgement to the MyBRAIN15 (MyPHD) because supporting researcher to finish this paper research.

Reference

1. Jabatan Perangkaan Negara (2017), Perangkaan Tenaga Buruh, Malaysia, Julai 2017. Dicapai pada September, 10, 2017 di https://www.dosm.gov.my/v1/index.php?r=column/pdf Prev&id=VzBiQlBKSVc0S0hRN0x1RDBqR09BZz09
2. Jabatan Perangkaan Negara (2017), Statistik Kecenderungan Perniagaan, Suku Pertama 2017. February, 2017. Dicapai pada September, 28 2017 di https://www.dosm.gov.my/v1/index.php?r=column/pdf Prev&id=eUd6T0YzWFU2aDRqeldQb2hMN1RGdz09
3. Laporan Siasatan Tenaga Buruh Usahawan Di Malaysia (2009), Perangkaan Tenaga Buruh, Malaysia. Siri 4 bil.1/2009 Dicapai pada November, 30 2014 di http://www.statistics.gov.my/portal/download_Labour/files/BPTMS/SIARAN_KHAS_(USAHAWAN_DI_MALAYSIA).pdf
4. Othman (2009). Pelaksanaan Kaedah Pembelajaran Berasaskan Konsultasi dalam Pembudayaan Keusahawanan. Prosiding Kogress Pengajaran dan Pembelajaran di Awana Porto Malai Langkawi. 13-17
5. Othman & Ishak (2009). Attitude Towards Choosing a Career in Entrepreneurship Amongst Graduates. Faculty of Education, Universiti Kebangsaan Malaysia,UKM. European Journal of Social Sciences, Vol 10, Num 3, pp. 1-16.
6. Osman (2007). Faktor-Faktor Yang Telah Mendorong Graduan Dan Alumni Universiti Teknologi Malaysia Menceburi Bidang Keusahawanan. 13-17
7. Pelan Stategik Pengajian Tinggi Negara 2007-2010 Fasa 1 (PSPTN), (2007) Membudayakan Pembelajaran Sepanjang Hayat. Dicapai pada Julai 23, 2014 dari http://www.mohe.gov.my/portal/images/utama/penerbitan/psptn.pdf
8. Buang & Yusof (2006). Motivating Factors that Influence Class F Contractors to Become Entrepreneurs .Faculty of Education, Universiti Kebangsaan Malaysia, UKM. Jurnal Pendidikan, pp s 107 – 121.
9. Hashim, Othman & Buang (2009). Konsep Kesediaan Keusahawanan Berdasarkan Kajian Kes Usahawan Industri Kecil dan Sederhana (IKS) di Malaysia. Jurnal Pendidikan Malaysia. pp 187 – 203.
10. Adam & Samsudin (2014). Pembangunan Keusahawanan dari Perspektif Kolej Komuniti. Prosiding Seminar Antarabangsa Kelestarian Insan 2014 (INSAN2014). pp 8 – 20.
11. Hanapi, Nordinb & Che Rus (2014). Unemployment Problem among Graduates of Technical Field: Competencies of the Graduates and Quality of the Education. Penerbit UTM Press. pp 1 -5
12. Halid (2012). Kesediaan Pelajar Politeknik Kejurteraan Elektrik Dalam Menceburi Bidang Keusahawanan. Jurnal Pendidikan Malaysia. pp 87 – 97

13 Nadzri, Shamsudin, & M. Sabri (2010). Faktor-Faktor Penyumbang Kepada Kejayaan Dan KegagalanPerusahaan Kecil Dan Sederhana (PKS) Bumiputera Di Malaysia. Jurnal teknologi Sains Sosial.Fakulti Pengurusan dan Muamalah. pp 1- 10

14 Yusof & Shafie (2012). Konsep Kesediaan Keusahawanan Di Kalangan Pelajar Kolej Komuniti Bandar Penawar. JurnalPendidikan Malaysiapp 1- 17

15 Krejcie, R. V. & Morgan, D. W. (1970). Determining Sample Size for Research Activities. Educational and psychology. 30: 607-610.

The Influence of Students' Experience on Alumni Giving in Malaysian Public Educational Institution

U. Naeimah Saraeh[1],*, *Nor Irwani* Abdul Rahman[1], *Norshahrizan* Noordin[1], *Sayang Nurshahrizleen* Ramlan[1], *Razli* Ahmad[2] and *Mohd Fo'ad* Sakdan[2]

[1]School of Business Innovation and Technopreneurship, Universiti Malaysia Perlis (UniMAP), Malaysia
[2]School of Human Development and Technocommunication, Universiti Malaysia Perlis (UniMAP), Malaysia.

Abstract. The purpose of this study is to examine the influence of students' experience on alumni giving in one of Malaysian Public Higher Educational Institution (PHEI). Utilizing the questionnaire method, data are obtained from 347 respondents among final year students in this institution. Results presented indicate students in this institution have low level of students' experience and alumni giving with the mean score are reported at 1.25 each. Results presented also indicate that students' experience is significantly related to alumni giving (β=.54, p<.01). This study contributed some new knowledge to the existing literature in the domain of alumni giving behaviour. Several recommendations can be suggested to Malaysian PHEIs to improve their level of students' experience, thus the level of alumni giving can be increased. Several recommendations are proposed to be enhanced so as to increase the satisfaction on students' experience during their undergraduates' life based on this study findings.

1 Introduction

Alumni giving is a crucial issue. Most of the institutions wonder how they can increase the level of alumni giving in both financial and non-financial aspects. However, it may become more important to understand the factors which may create the influence on alumni giving, particularly to support their undergraduate alma mater. Even there are many factors after graduation which contributes to alumni giving (eg. salary, alumni loyalty, alumni involvement); but the effect on how students' experience during their universities' life toward their future alumni giving behaviour is one of the factors which is salient to be examined. This is supported by [1] who presented that willingness to contribute for alma matter is affected by the undergraduates' experiences (students' experiences). Hence, it shows that factors on how student views of the institution during their life in campus are deemed important to be examined.

This answer why it is become crucial for the institution to identify in what aspects students' experience may create an influence on their future alumni giving while the students are still in the campus. Thus, aiming at the final year students to reveal the experience towards their future alumni giving could potentially contribute some information to the institutions. Consecutively, the institutions can develop their capacity to strengthen their future strategies.

Not only that, the Ministry of Higher Education Malaysia (MoHE) also has taken several initiatives by making alumni as one of the external parties to realize the goals of internationalization. Therefore, to realize the objective for universities' internationalization, it is expected that the factor of alumni giving can contribute to increase the institution's performance.

Moreover, all HEI including the public institution need such private (external) support such as from the alumnus as a revenue stream which can attribute to the continual decrease in the level of state appropriations [2]. In an effort to compensate for the decrease in state appropriations and increase in educational costs, institutions are placing more emphasis upon alternative on the private support (eg. alumni) to sustain a competitive advantage amongst the universities. As [2] highlighted that while there are several alternatives non-tuition revenue sources that could be explored (eg. alumni talk series, research grants); alumni giving continues to be a promising possibility. Moreover, in this current economic situation, alumni giving are something which all institutions cannot ignore.

Given that alumni giving is formed while they are still students [1]; thus, forming an understanding of the experiences which students have with their alma maters would provide insights into the future alumni giving behaviours. (eg. alumni financial and non-financial contribution). In other word, students' experience in the university is the best platform to reveal the intention of their future alumni giving.

In addition, lack of evidence of alumni survey is found in the local studies. According to [3] the three conceptual approaches of alumni survey are consisted of outcomes; students' engagement and competencies; and alumni giving. Outcomes are the assumption of the institutional quality and effectiveness which can be appraised on what alumni have accomplished. Meanwhile, students' engagement and competencies are

* Corresponding author: ummi@unimap.edu.my

the assumption of the institutional quality which can be appraised to what the students did at institution, learned and applied; and alumni giving are the assumption of a function of positive experiences with the institution and ability to contribute. Hence, it becomes a priority for the researchers to explore and examine one of the alumni survey especially on the issue of alumni giving in the context of Malaysia.

Moreover, in the local context, limited evidence of the local research is found on the relationship between students' experience and alumni giving. Thus, few developments can dramatically impact on the students' experience if this factor is found to influence the alumni giving level in the Malaysian HEI.

2 Literature Review

2.1 Students' Experience

To date, students' experience has a major impact on feelings and opinions towards the institution when they become the alumni [4, 5]. Previous authors claimed that students' experience became a popular variable among the higher educational managers but is still theoretically under-developed [6], thus the influence of students' experience should be examined particularly on the contribution of alumni giving.

In addition, [7] stated that alumni who enjoyed their students' experience gave positive feedback to people in their networks, acted as role models for the current students, shared their skills and experiences, and advised the program of the alma mater. Fostering a lifelong connection with current students, and later the alumni, is a potential source of support for institutions [8]. Thus, it shows that alumni are a significant part of any university community.

Other authors stated that undergraduate students who have a positive and satisfying experience while enrolled tend to develop lifelong relationships as alumni with the institution. For example, [9] claimed that positive experience includes enjoying classes, developing friendships and obtaining a useful degree. Hence, it can be assumed that being content with the undergraduate experience could mean obtaining a job upon graduation or getting connected to worthwhile efforts that offer challenge for the graduate. Therefore, students' experience is become relevant in shaping the future trend of alumni giving.

In a local context, it has been reported that in year 2016 students have moderate level of satisfaction with the function of one department in this institution. Based on this report, it becomes the opportunities for the researchers to identify the consistency of the students' satisfaction towards their experience in this institution by conducting another research on students' experience in 2017. Therefore, students experience when dealing with several departments in this institution has been investigated.

According to [10], empirical works stated that the strongest giving predictor (eg. alumni giving) is the satisfaction a student had with his or her students' experience. Other authors [11] added that graduates who are satisfied with their college experience are more likely to contribute to the college or university. Similarly, [4, 5] confirmed that students' experience while they in campus has a major impact on feelings and opinions toward the institution as alumni.

Thus, in this study the researchers attempt to investigate the satisfaction of students' experience in one of the PHEI in Malaysia. Based on report as presented by the Quality Department in this PHEI, the level of satisfaction among students towards this institution is still moderate. Therefore, studying students' experiences would provide insights into the existing literature in the domain of alumni survey particularly on alumni giving behaviour.

2.2 The Influence of Students' Experience on Alumni Giving

The research clearly establishes that among attitudinal variables, satisfaction with the student experience is by far the strongest predictor of alumni giving behaviour (2, 12, 13]. For example, [12] presented that the undergraduate experience became the most significant determinant of alumni giving in several investigation involving with several institutions. This result was supported by other authors [14] who found that donations are highly correlated to satisfaction with the college experience.

However, [15] who looked at the alumni donor participation rates in Pepperdine University, found that undergraduate experiences were not predictors of future alumni giving. The results of [15] study indicated that alumni donor status was better predicted by examining only the involvement of alumni post-graduation than the impressions that an institution left on graduates as their experiences.

According to [16], each study of alumni giving and alumni donor participation has attempted to address the overwhelming question of how to increase alumni donor participation. Early research on alumni giving was conducted largely by economists and sociologists and published rarely in peer-reviewed journals. For decades and until nowadays, most of this research was done as part of dissertation work and as a result, it remains difficult to access. Dissertations have been frequently cited in alumni giving research because of the disparity of articles in peer-reviewed journals on the topic [17].

As stated by [18] research on alumni giving focuses on four areas which consisted of individual donor characteristics, fund raising practices, the external environment, and institutional characteristics. However, in this study alumni giving will be explored in the areas of willingness of the current students to contribute in financial aspect (eg. donation) and the non-financial aspect (eg. alumni talk series such as experience sharing session with the current students).

As cited by [19] in [16], alumni giving in America began at Harvard College with a donation of land in

1648 by four Harvard alumni. The gift was followed by gifts from other alumni for the remainder of the seventeenth century, and contributed to the establishment and steadfastness of Harvard and other higher education institutions in America Meanwhile, the Yale System, an organized alumnus giving initiative for all alumni of Yale University, sought to increase smaller donations from many instead of relying on only large donations from the few. Literature on alumni development suggests colleges and universities that attain significant alumni giving rates tend to have well-developed and implemented strategies, with many focusing on developing current students into giving alumni. HEIs use a variety of traditional and creative methods to cultivate alumni and secure their support [20].

To date, alumni giving is can be assumed as a function of positive experiences with the institution and ability to contribute [3]. It is about the ability and inclination of alumni to support their alma mater through philanthropy, service, and advocacy. It can be measured by the perceptions of the quality (satisfaction) of college experience. Although [21] claimed that alumni giving rates do not adequately measure graduates' satisfaction with educational experience, however research clearly demonstrates that satisfaction with the college experience plays a significant role in alumni giving.

Therefore, this study will look at alumni giving behaviour and determine if students' experience can be identified for explaining the alumni contribution (financial, non-financial) from the Malaysian context. More specifically, the purpose of this research is to explore the level of alumnus future alumni giving in this population. Hence, this study will look at the influence of students' experience on alumni giving behaviour in one of PHEI in Malaysia. It is hoped that this research will stimulate the Malaysian HEIs to re-examine their types of services towards the current students to create and implement creative initiatives to increase alumni giving and, in turn, to strengthen their institutions' strategy.

It is the hypothesis of this research study that a students' satisfaction towards their experience with the institution may generate a greater emotional bond or affinity towards the institution after their graduation and thereby positively influence the alumni giving behaviour to their alma mater. Given the above literature, the specific hypotheses of this research are stated below:

H_1: There is a moderate level of students' experience in this institution.

H_2: There is a moderate level of alumni giving in this institution.

H_3: There is a significant relationship between students' experience and alumni giving in this institution.

3 Methodology

The population of this research involved the final year students from all faculties in the institution. This research used questionnaire to obtain data from all participants to the estimation of 500 respondents. The SPSS software (version 22) has been used for the statistical analysis. The instrument used for alumni giving has been adapted from the previous measurement; while the instrument used for students' experience has been developed based on the environment of the institution.

In this research, the final year students need to provide their satisfaction towards the four dimensions of experience consist of teaching and learning activities (3 items), dealing with several departments in the institution (6 items), co-curricular activities (2 items), and their intention as the future alumni (2 items).

As for the alumni future attitudes, 4 items of alumni giving have been developed based on the previous authors [2, 3, 8, 18]. Those items are (1) *As a future alumni, I believe that financial assistance from alumni is very important in the success of an activity involving alumni with the university*; (2) *As a future alumni, I believe that financial assistance from the alumni is very important to the development of interactions between alumni and universities*; (3) *As a future alumni, I would like to make a financial contribution to the university if I am capable*; and (4) *As a future alumni, I intend to contribute non-financial assistance such as sharing the knowledge, experience and skills that I have with prospective alumni who are still studying at university if I have spare time.*

Based on the pilot test, the reliability values of students' experience and alumni giving are reported at .82 and .76 each. In this study, the researchers used the criteria of Cronbach's alpha as established by [22]. Therefore, the reliabilities of variables used in this research are ranged from acceptable to good which mean that all variables can be used for further analysis.

4 Results and Discussions

A quantitative method has been utilized in this research. The response rate for this study was at 69.4% when 347 out of 500 of the distributed questionnaires have been used for this research. Majority of the respondents in this research were female, which contributed of 58.8% (204 participants). Hence, the remaining of the respondents was male which contributed of 41.2% (143 participants).

The highest percentage of responses (20.5%) was obtained from School of Business Innovation and Technopreneurship; and the lowest percentage of responses (7.6%) was obtained from School of Material Engineering. Other participants were from School of Electrical Systems Engineering (14.1%), School of Microelectronics Engineering (12.1%), School of Computer and Communication Engineering (9.5%), School of Human Development and Techno-Communication (9.8%), School of Bioprocess Engineering (8.9%), School of Environmental Engineering (8.9%) and School of Mechatronics Engineering (8.6%).

 https://doi.org/10.1051/matecconf/201815005030

4.1 Mean and SD Analyses

In this research, the interpretation of mean scores has been referred to [23] with the mean scores ranged from low (1.00-2.00), moderately low (2.01-3.00), moderately high (3.01-4.00) to high (4.01-5.00).

As presented in Table 1, the mean value for students' experience was low at 1.25. Therefore, low acceptance is observed for this variable. Similarly, the mean score for alumni giving was also low at 1.25. Thus, H_1: *There is a moderate level of students' experience in this institution;* and H_2: *There is a moderate level of alumni giving in this institution* are not supported in this study.

Table 1. Mean and SD Values

Variables	N	Mean	SD
Students' Experience	347	1.25	.21
Alumni Giving	347	1.25	.22

As for the SD analysis, it is noted that the SD values for students' experience (.21) and for alumni giving (.22) were also low. The low level of SD showed that most respondents were close to the mean for variables in this study.

4.2 Correlation Analysis

Pearson correlation coefficient (r) is computed to test the correlations and directions between students' experience and alumni giving. As shown in Table 2, a moderate relationship is observed between students' experience and alumni giving ($r=.53$, $p<.01$). Therefore, the correlation is reported as significantly and positively between each other.

Table 2. Correlation Analysis

Variables	N	1	2
Students'Experience	347	-	
Alumni Giving	347	.39**	-

***Correlation is significant at the 0.01 level (2-tailed)*

4.3 Regression Analysis

Regression analysis is used to investigate the relationship between independent variable and dependent variable as posited in the research hypotheses. By using SPSS, the data is performed for Coefficient for the data.

According to [24], regression is a technique that can be used to investigate the effect of one predictor variable (independent variable) on an outcome variable (dependent variable).

As shown in Table 3, regression analysis result presented that students' experience explained 29% of variance in alumni loyalty ($R_2=.29$). This clarified that 29% of variance in alumni giving is explained by students' experiences, while the rest of 71% of variance in alumni giving is explained by the other variables. Meanwhile, the adjusted R_2 value is reported at .30. This confirmed that students' experiences had strong influences on alumni giving ($\beta=.54$, $p=.00$).

Therefore, H_3: *There is a significant relationship between students' experience and alumni giving in this institution* is supported in this study.

Table 3. Regression Analysis

Variables	Students'Experience			
	Beta	Sig.	R^2	Adjusted R^2
Alumni Giving	.54**	.00	.29	.28

***Correlation is significant at the 0.01 level (2-tailed)*

5 Conclusion

Based on the research findings, it can be concluded that the level of students' experience as well as alumni giving among the final year students in this institution were in the low indications with the mean value reported at 1.25 each. Therefore, the first two hypotheses about the appearance of moderate levels of both students' experience and alumni giving behaviour in this institution are not supported this study.

In addition, this study provides some new knowledge on the influence of students' experiences on alumni giving. As presented in Table 3, students' experience is positively associated with alumni giving ($\beta=.54$, $p<.01$). Thus, H_3: *There is a significant relationship between students' experience and alumni giving in this institution*, is supported in this study.

Based on the results, it can be concluded that by increasing the level of students' experiences among the current students in this institution, the future level of alumni giving among them can be enhanced. In the theoretical perspective, these findings had been consistent with the previous findings as reported by several authors [2, 25, 26].

For example, [2] presented that undergraduates who were challenged by their alma mater, interacted with their respective faculty members, and were provided with the institutional support they needed to accomplish their academic goals, were more likely to giving (contribute) back to their alma mater. This is because the overall levels of satisfaction with undergraduate academic experience were variables that had the highest correlation for alumni giving [25]. Similar findings by [26] presented that satisfaction was greater with the student experience, and giving was also impacted, when students believe they had educational gains.

In order to increase the level of alumni giving in the financial and non-financial aspects, the level of students' experience need to be enforced first. By increasing the level of student satisfaction towards their experience, the future alumni giving could be inculcated. Thus, in the practical perspective, with the advent of contemporary information and technological innovations, the institution should increase the quality level of services as to improve the level of students' experience such as: 1)

To create a platform for further interaction between students and the academic staff even after teaching and learning session; 2) To create a platform for further interaction between students and the non-academic staffs even after office hours; 3) To enhance the implementation of the buddy system; and 4) To provide a more effective and interactive medium in attending students inquiries and complaints on university's facilities e.g. Library, Bursary, Information Communication and Technology (ICT), Residential Colleges, Students Affair and Alumni Department, Transportation, Cafeterias etc.

Furthermore, there is also a need for universities to show a compelling reason of importance of alumni giving as a cause worthy of donations. It is important to remember that successful alumni giving initiatives may not only take months but many decades. However, there is always opportunity there. Besides, this may continue to drive success between universities with NGOs (eg. Alumni Association Committee) that effectively raise most of their funding from alumni donors.

Today's millennial alumni are much more likely to give when introduced to a specific person who needs their help. What may and usually work with universities today is highlighting a specific student or faculty member and telling potential donors how their gift will make a difference through scholarships, new equipment, new facilities, program funding, etc. This may help when universities are able to show the need, tell how the alumni can alleviate the need and position themselves as a cause worthy of contributions. However, the strategy will not work unless there is a well-established relationship between the university and the alum. This relationship started in their early year in campus. In fact, in previous study, it is found that faculty relationships were hugely important to maintaining these university-alumni connections after graduation.

Regardless of how it is first introduced, the immediate goal for a university is to quickly get graduates into the habit of giving back to the alma mater. The ultimate goal is to sustain that generosity for a lifetime. A true indicator of the love for and loyalty to one's alma mater is the extent to which one is willing to support it financially. It also happens to be a reliable indicator of excellence. As it turns out, graduate giving is a very good barometer of how alumni view their university. Most of world Class University has long-standing traditions of alumni support based on their historic reputation for excellence and a campus life that binds students to the institution and to one another. Alumni financial contributions are one of the best indicators of a graduate's esteem for their university. As is often the case, money follows the heart.

Finally, government efforts to train and educate graduates in alumni giving should be maintained and enhanced. In recommending programs and policies to deal with these key issues, one size does not fit all. Recommendations with respect to alumni giving need to be tailored to the specific context in each PHEIs.

References

1. Dugan, K., Mullin, C.H. & Siegfried, J.J. Undergraduate financial aid and subsequent alumni giving behavior. *Williams Project on the Economics of Higher Education Discussion Papers, DP-57.* (2000)
2. Ropp, C.T. The relationship between student academic engagement and alumni giving at a public, state flagship university, Doctoral dissertation, University of Kansas (2014)
3. Cabrera, A.F., Weerts, D.J. & Zulick, B.J. Alumni Survey: Three conceptualizations to alumni research, *Paper presented before: Métodos de Análisis de la inserción laboral de los universitarios, Universidad de León, España* (2003)
4. Sun, X. A multivariate causal model of alumni giving at a Midwest, public university. Dissertation Abstracts International, 66 (12), (2005)
5. Thomas, J.A. The relationship between personal and social growth and involvement in college and subsequent alumni giving. Dissertation Abstracts International, 66 (12), 4323, (2005)
6. Koenig-Lewis, N., Asaad, Y., Palmer, A. & Petersone, E. The effects of passage of time on alumni recall of student experience. *Higher Education Quarterly*, **70(1)**, 59-80 (2016)
7. Hansen, L.L. Once students, forever alumni: A study analyzing drivers of alumni loyalty, Master Thesis, Carthage College (2015)
8. Pulley, J. The journey is the reward. *Currents*, **39(1)**, 16-21 (2013)
9. Berquam, L.M. Affinity development in undergraduate students at a large research institution, Doctoral Dissertation, Colorado State University (2013)
10. Stephenson, A.L. & Bell, N. Motivation for alumni donations: A social identity perspective on the role of branding in higher education. *International Journal of Nonprofit and Voluntary Sector Marketing,* **19**, 176-186 (2014)
11. Sun, X., Hoffman, S. & Grady, M. A multivariate causal model of alumni giving: Implications for alumni fundraisers. *International Journal of Educational Advancement,* **7(4)**, 307-332 (2007)
12. Monks, J. Patterns of giving to one's alma mater among young graduates from selective institutions. *Economics of Education Review,* **22(2)**, 121-130 (2003)
13. Radcliffe, S. A Study of Alumni Engagement and Its Relationship to Giving Behaviors, Master Thesis, Bucknell University (2011)
14. McDearmon, J.T. & Shirley, K. Characteristics and institutional factors related to young alumni donors and non-donors. *International Journal of Educational Advancement,* **9(2)**, 83-95 (2009).
15. Young, P.S. & Fischer, N.M. Identifying undergraduate and post-college characteristics that may affect alumni giving. *AIR 1996 Annual Forum* (1996)

16. Truitt, J. The relationship between student engagement and recent alumni donors at Carnegie baccalaureate colleges located in the South Eastern United States (2013)
17. Hoyt, J.E. Understanding alumni giving: Theory and predictors of donor status. *Paper presented at the meeting of the Association of Institutional Research* (2004)
18. Weerts, D.J. Toward an engagement model of institutional advancement at public colleges and universities. *International Journal of Educational Advancement,* **7(2)**, 79 -103 (2007)
19. Markoff, R.M. An analysis of the relationship of alumni giving and level of participation in voluntary organizations: A case study. Toledo, OH: University of Toledo (1978).
20. Ayers & Associates. Alumni giving in the new millennium: A guide to securing support. Washington, DC (2002)
21. Brant, K.E. & Regan, P.J. The spectrum of alumni involvement. *CASE Currents,* **28(2)**, 22-28 (2002)
22. George, D. & Mallery, P. SPSS for Windows step by step: A simple guide and reference (4thed.). Boston: Allyn & Bacon (2003)
23. Norasmah, H.O. & Salmah, I. Kecenderungan terhadap pemilihan kerjaya keusahawanan mengikut persepsi peserta skim usahawan siswa. *Jurnal Teknologi,* **56**, 47-63 (2011)
24. Sekaran, U. Research Methods for Business, a Skill Building Approach 4th edition. Wiley, New York (2003)
25. Conner, D.K. Factors that affect alumni giving at a southeastern comprehensive university. Doctoral Dissertation, Clemson University (2005)
26. Gaier, S.E. The impact of alumni satisfaction with their undergraduate academic experience on alumni giving and alumni participation, Doctoral dissertation, Purdue University (2003)

Integration of Augmented Reality into College Yearbook

Faharol Zubir[1,*], *Ina* Suryani[1], and *Norlizawati* Ghazali[2]

[1]Centre for International Languages, Universiti Malaysia Perlis.
[2]Academy of Language Studies, Universiti Teknologi MARA.

Abstract. Augmented reality (AR) has been used in many schools and colleges in developed countries. Apart from enhancing teaching to be more interactive, inventive and resourceful, the use of AR also has been reported to boost learning by fostering creativity and imagination, enhancing collaboration and team work, and making learning more engaging and stimulating. Despite the potentials of AR in enhancing school materials, this technology often has teachers and learners as the end users rather than developers. Given the benefits of AR in education, this study experimented on having the teachers and learners as developers in incorporating the use of AR to enhance the college yearbook. This paper describes the processes taken by the teachers and learners in integrating AR into a conventional reading material; specifically a college yearbook. In this study, an open source AR application is used and 32 markers are created. The method involves five major development processes which starts with planning, content, AR integration development, printing and finally AR integration construction. The 32 markers can be divided into a few sections which can be classified according to the types of content. At the end of this paper, the significance and challenges of this project are presented.

1 Introduction

College yearbook thus far, comes in hardback, paperback and lately in PDF formats. It is considered a conventional book since it comes with text and illustrations only. Often, publishing a yearbook has a lot of constraints particularly the on the finance and readership issues. Some colleges have stopped publishing the physical yearbook altogether due to the financial factor. One of the main problems is the high cost in making the yearbook interactive and attractive to the teenage readers. In order to attract the college students the yearbook needs to have many illustrations such as pictures and colorful pages of articles hence increase the printing cost further. Due to these issues, developing a conventional yearbook to students in the 21st century can be a daunting task for any yearbook advisor.

Opportunely, the integration of AR can enhance the yearbook to be more interactive and promote better readership without escalating the printing cost. Moreover, the yearbook is often associated with momentous occasions and it is constructed collectively by teachers and students as such, presents potentials for AR experimentation. Such meaningful, fun medium ease the learning of the new technology which otherwise is daunting to both teachers and students. While being meaningful as it records important events, yearbook construction involves limited number of editorial members. This research involves a project that accommodates participations from students and teachers. AR enhances experience of the real world by the use of virtual content which is overlaid onto the reality. AR presents new features that can be mixed with our perception. The new features in the form of 2D, 3D objects, audio, video files, textual information alters the perceptions and can be incorporated into the users' perceptions of the real world [1].

Recognizing these potentials, this study attempts to develop a yearbook by enhancing its quality and sustainability through the use of AR. In the process of the development, this study also examines the development of the instructional design for AR integration in teaching and learning. This paper reports on the development process of the yearbook, the incorporation of AR elements into the yearbook enhancement and the roles played by teachers and learners in completing this new yearbook.

2 Theory and Method

This section has two parts whereby the first part discusses on the previous researches on AR in education and the second part describes the methods used to integrate AR elements into the yearbook. The second part also illustrates the processes undertaken by the teachers and learners in achieving the expected yearbook.

2.1 Theory

Augmented reality (AR) can be defined as a technology which overlays virtual objects or augmented components into the real world. These virtual objects then display to coexist in the same space as objects in the real world [2]. AR is currently a popular technology that is widely used in marketing and entertainment settings. In recent years, AR has become the focus of many researches [3, 4]. The

* Corresponding author: faharol@unimap.edu.my

fact that AR no longer requires expensive hardware and sophisticated equipment, such as head-mounted displays (HMD) contributes to its popularity. The technology now can be accessed through computers or mobile devices. The rapid development of free AR application contributes to its vast accessibility towards users. Therefore, using AR technology is not as challenging as it was in the past.

Today, in the developed countries, AR is often used in every level of schooling [5, 6]; however this technology is hardly ever utilized or employed by the academics and students in Malaysia. One of the emerging application of AR in education is in the form of augmented books. The potential of digitally augmented books usage appeals to researchers and educators. Augmented books provide a means for enhancing regular 2D books with interactive visualizations, animations, 3D graphics, audios and videos. The purpose of integrating such features is to promote engaging user experiences and to enhance the reading experience.

To date, on the global scale, the number of book readers has declined rapidly. The development and production of conventional reading materials often limit readers' involvement particularly in the inclusion of additional reading materials. Therefore, the integration of multimedia using AR in reading materials, particularly yearbook, appeals to the readers' interest and may sustain the readership. Several research such have been done in Malaysia in terms of enhancing conventional reading material using AR integration such as [7, 8, and 9].

Despite the growing interest amongst publishers to incorporate AR into books, teachers and students are seldom involved or participated in the development of such book. Moreover, the technology is relatively new; consequently, the teachers and students only transpire as end users. Not wanting to miss out on the advantages, benefits and ease of incorporating AR in instructional materials, this study looks at the processes of transferring AR technology into teaching and learning materials. The yearbook is chosen to be the medium of transfer as the production of a yearbook is most likely to be a memorable project as it is associated with momentous occasions and constructed collectively. Such meaningful and fun medium ease the learning of the new AR technology which otherwise can appear daunting and overwhelming for the teachers and students. The knowledge can later be applied into other educational applications such as enhancements for textbooks, modules and students' projects.

2.2 Methodology

This study is conducted using a yearbook project which runs for 11 weeks. This yearbook project is a collaborative work by the International English Course (IEC) students of UniMAP and is supervised by the language teachers. This section explains briefly about the IEC background and the previous conventional yearbook. The development of the AR integration particularly on the process is described at the end of this section.

The International English Course aims to prepare international students for UniMAP undergraduate programs and to improve the international students' level of English so that they are able to gauge the learning at the undergraduate programs in UniMAP which are conducted in English. The students are taught according to the four components of the English language skills which are Reading, Writing, Listening and Speaking. The learners are grouped into three levels according to their proficiency.

Previously, the yearbook is developed by IEC learners from the Intermediate group only with the other levels as the content contributors. It is a project aimed to showcase students' ability to write in order to reflect on their language lessons. Students are given the freedom to suggest the article they want to write about and the instructors guide them through the process of writing the articles. There are now 3 volumes of the yearbook. Volume 1 consists of essays written by students from the Upper-intermediate class and focuses on academic topics. Volume 2 on the other hand has a mixture of contributors from Intermediate, Pre-intermediate and Elementary class. The focus of the written materials is more towards short paragraphs on experiences and simple topics. Both volumes do not stress on the format of the write up since the main purpose of having the yearbook is to promote students to write.

In this project, the yearbook which is integrated with AR is developed by the October 2016 Intake. Unlike the previous volume, Volume 3 is a collection of academic essays written using a process-based approach. These essays are enhanced through the integration of AR in the yearbook. The yearbook consists of 46 coloured pages and is printed in A4 size. The yearbook consists of six sections namely the front matters, academic essays, additional articles, interviews, IEC students' profile and credit page.

The front matters include table of content, note from IEC Program Chairperson and the Head Editor, information on the team of authors and instruction on utilizing AR integration in the yearbook. The front matters consist of 6 pages (page 1-6). The next section is the academic essay section consisting of 22 pages (page 8-29). This section displays a collection of academic essays written by the students. The topics of the essays are Tourism in Malaysia, Vacation, The Importance of Sports, How to Achieve Your Goals, The Amazing Ancient Egypt, Essential Skills for University Students and The Benefits of Internet.

The additional articles section consists of two articles, Our Life in IEC and Augmented Reality & Aurasma covering 3 pages (33-35). The next section is the Interview section which consists of 2 pages (37-38) about two IEC instructors. The last section is the IEC students section which displays a picture of each student in IEC October 2016 intake together with their names, country, offered undergraduate program and life motto. This consists of 7 pages (39-45). The last page consists of the credit page listing the name of the project advisors, head editor and compiler, team of writers and picture resources.

The AR integrated sections are the academic essays, additional articles and IEC students' profile. The selection of these sections is based on the purpose to showcase students' work. This is apparent in the academic essay and additional article section whereby the essays and articles are AR integrated with video overlays to provide additional information on each essay and article. The IEC students' profile on the other hand was chosen to maximize students' involvement in the yearbook. Each student records a 10 second video which are overlaid on their profile picture in the section. This ensures that more students can participate in the construction of the yearbook compared to the conventional way where only limited number of students can participate.

The process of integrating AR into the yearbook consists of 5 steps. The process starts with Step 1 which is the Planning step. In this step, the instructor briefs the students about the project that they are going to embark. The next process is to introduce students to AR and the chosen application which is Aurasma. This is then followed by assigning the person in charged for each section of the yearbook and also for the integration of AR in it.

Step 2 involves the process in developing the content of the essay which is the academic essays. Each student is assigned to write an academic essay for the yearbook. The first process in Step 2 is that each student is instructed to choose a topic of their interest and consult the instructor regarding their selection. Once both parties have agreed on the chosen topics, students can proceed to gather suitable materials needed to write the essay. The next process is to brainstorm for ideas and then construct an essay outline. Students are then required to submit the draft of their essay to the instructor. The instructor checks the drafts and suggests improvements for the drafts. The process is repeated until the drafts meet the required standards aimed by the International English Course. Once both parties are satisfied with the draft, the students proceed to submit the academic essay.

Step 3 includes the AR integration development process. The type, size and location of the markers are determined in this step. Next, the type, size, duration and resolution of the overlays are determined. Students are given the freedom to suggest and decide on these criteria for the markers and overlays after consulting the instructor. Step 4 involves the printing process of the yearbook. The formatting process is done and the yearbook is drafted. Once both parties, students and the instructor are satisfied with the draft, the printing process begins. The end product of this step is the hardcopy of the yearbook.

The next step is the AR integration construction. This step includes the process involved in integrating AR with the yearbook. Firstly, the selected open source AR application which is Aurasma is downloaded. Next, the technical editor registers a free Aurasma account and creates a channel for the yearbook. After that the technical editor tags the overlays to the markers available in the yearbook. Then the markers are tested to make sure that the overlays are retrievable. Once all the overlays for each marker works, the yearbook administrator, in this case the instructor, secures the tagging of the yearbook by controlling the access of the Aurasma account and channel for the yearbook. This is to ensure the safety of the content of the overlays triggered by the markers.

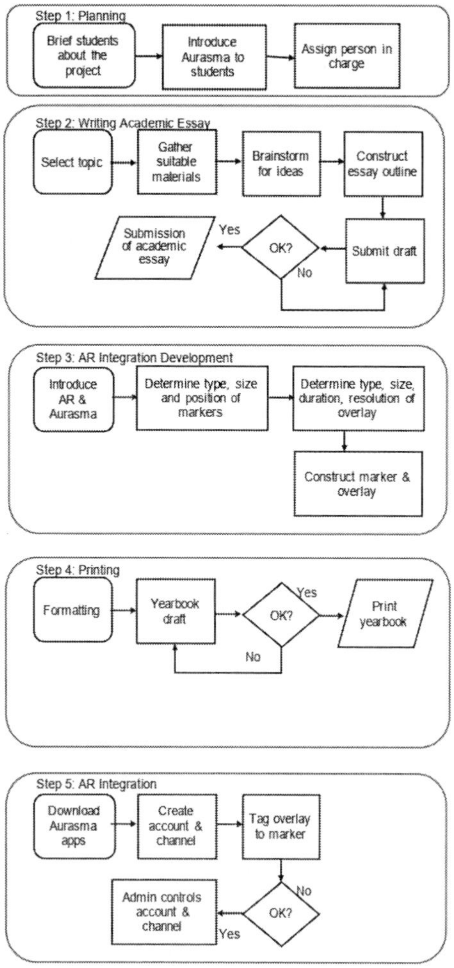

Fig. 1. The processes of AR Integration into yearbook

A survey was done to evaluate if the AR integration enhance the readers' interest and to compare the users' experience when reading the book with and without AR integration. The survey was adopted and adapted from [10] and added with five items for publication rating.

The survey was conducted among 77 college students taking English language course as part of their program. The survey was conducted in two stages. In the first stage the subjects were given 15 minutes to read the book. Then, they were asked to answer the readership survey question. Next, the integration of AR was explained and the use of it is briefed to the subjects. They were then given another 15 minutes to re-read the book by using the AR integration function. After that, they were asked to answer the readership survey question again. This is to compare their rating on the content, design, quality, user friendliness and reader appeal aspects of the yearbook. A 5 point Likert scale (1. very poor, 2. poor, 3. average, 4. good, and 5. excellent) were used in the readership survey questions.

3 Result

The outcomes of the project suggest three main considerations which are sections, markers and instrumentation. The first consideration on the section is related to the suitability of AR integration with the content. The yearbook consists of six sections namely the front matters, academic essays, additional articles, interviews, students' profile and credit page. AR is integrated throughout the yearbook except for two sections; front matters and interviews.

The second consideration is on the use of markers. Markers refer to the image in the yearbook that can be scanned in order to retrieve the overlays. Overlays on the other hand are materials; images, videos, links etc. that are projected on top of the marker once it is scanned. The markers need to be carefully planned and placed since the markers are visible in the reading material and once the printing process is completed, no further amendments can be done onto them. However, this is in contrast with the overlay which can be further enhanced or replaced even after the completion of the printing process. Markers utilize image recognition process and they work best if both pictures and texts are included in the same marker. The more complex and the more unique the markers are, the better they function. In this yearbook, the markers are combinations of images and texts. Each marker is bordered by a box with green background and an Aurasma logo. This is to ensure the visibility and uniformity of the markers for better recognition. The markers are relatively quite large, with some even covering a whole page. This is to ease users when they scan it. For the academic essays and additional articles, the markers are placed at the last page of each article. By doing so, readers are able to read the article without distractions from the markers. Since the overlays function as a summary for the articles, the articles need to be read first.

There are altogether 32 markers in the yearbook, which are put into three sections of the yearbook. Seven markers are included in the academic essay and additional articles section. Among the seven, four markers are triggered for a self-recorded video that are overlaid on top of the academic essay. These are the oral summary of the essays recorded by the respective authors of each essay. The remaining three markers are linked to short YouTube videos related to the essay. The students' profile section on the other hand consists the largest number of markers amounting to 24. These markers are linked to a 10 second self-recorded video of each student introducing themselves. The credit section which is the last page of the yearbook consists of one marker. The researcher labels this as a free marker since during the construction of the yearbook, even though the marker was positioned, there were no specific overlay for it. It is created as a backup marker should there be any need for additional overlay after printing. This comes handy as during the students' course completion ceremony, the students have come up with a farewell montage. This montage is not in the original plan. The idea for the montage crops up after the yearbook is printed. Such new idea is able to be accommodated into the yearbook by utilizing this free marker.

The third consideration is on the instrumentation. The integration of AR into reading materials requires both a software and a hardware. The software chosen for AR integration in the yearbook is the Aurasma application. It is an open source application and is among the popular applications used for AR in the education field. Aurasma is used by both the developer and the users of the yearbook. It can only be used online therefore internet connectivity is very important.

The hardware needed to access the integration of AR in the yearbook is the scanner. The scanner is important for both the developer and users of the yearbook. For the developer, the scanner is used in the process of tagging the overlays onto the markers while for the yearbook users, it is used to retrieve the overlays from the markers. The scanner must be equipped with a camera that is able to capture still images, record audios and videos, and equipped with a speaker, a microphone, a display screen and internet connection. Given these descriptions, smartphone is the most suitable scanner to be used for this yearbook. It does not only fit all the descriptions but it is also easily accessible since almost everyone nowadays owns one.

An instruction page on how to fully utilize the AR function of the yearbook is placed on page 6 of the yearbook. It is placed right after the front matters section and before the academic essay section since the integration of AR into the yearbook only starts after the academic essay section. It is important for readers to read the instruction before reading the academic essays to ensure a successful AR integration output. The instruction page includes a statement declaring the integration of AR in the year book. This is followed by a step by step instruction on how to acquire and utilize the tools needed to access AR features integrated in the yearbook, in this case the Aurasma application. Four illustrations consisting of pictures of the processes such as the selected tool (Aurasma), sample of the marker, scanning process, and result of the scan were also included in the instruction page to better assist readers in understanding the written instructions.

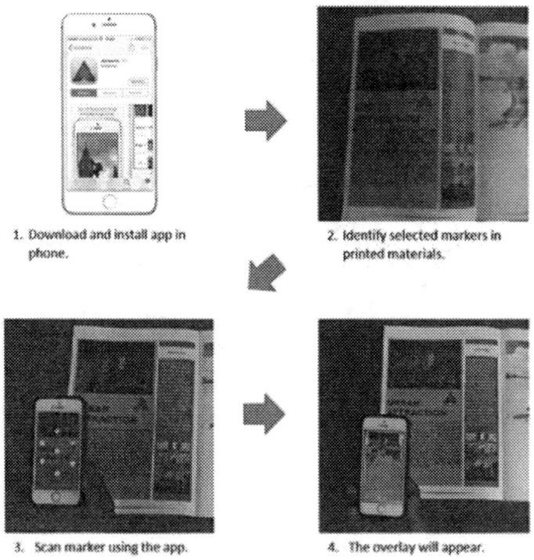

Fig. 2. Processes involved in retrieving AR content from the yearbook

Table 1. Reader appeal Mean and Standard Deviation for first and second reading

Item	First reading		Second reading	
	Mean	SD	Mean	SD
Reader appeal	4.1818	0.73855	4.4156	0.69492

With regard to the students' readership towards reading the AR yearbook, Table 1 presents the differences between reader appeal aspect for the first and second reading. The mean score for the first reading is 4.1818 (SD=0.73855) while the mean score for the second reading is 4.4156 (SD=0.69492). Hence, the students in this study consider the AR integrated yearbook having more reader appeal rather than conventional reading material.

4 Discussion

This study suggests a few merits on integrating AR elements into the yearbook particularly on extending the period for more material inclusion, boosting the appeal of the book and increasing teamwork and collaboration in the project. On a more important note, this study has encouraged knowledge transfer in relation of AR to the students for further application in various learning contexts.

In addition, the integration of AR in the yearbook extends the period for more material inclusion. The development of the yearbook often ends upon printing; whereas AR accommodates to further development even years after the printing is completed. More content can be added even after the printing process, for example more videos, clips, audio, images and montage can be layered onto the finished product. In other words, the content of the yearbook can be updated, for example the student leader can add a new video of him graduating, travelling or getting married. This addition can be done even years after the book is published and given out to the graduating students. With this added value on being more interactive, dynamic and stimulating, the yearbook is revolutionized into a more meaningful reading material. AR gives room for further development even after the printing process has ended.

Besides, the use of AR also enhance the appeal of the yearbook. Colourful and visually rich books are often more attracting to be read. However, the printing cost may be an issue to the project developers. The integration of AR technology provide ways to make a yearbook more attractive and interactive without inducing too much cost. By using AR, a college yearbook can be integrated with digital and multimedia content. The hardcopy of the conventional yearbook can be linked to additional digital content such as pictures, videos, clips, montages and others using AR. AR enhances a yearbook by making the pictures and articles on its page come to 'life'. Linking the articles to the multimedia and digital resources adds value to the publication by making it more interactive, explanatory and stimulating. The yearbook now can be enhanced and has value added since the interactive additional content can be integrated with the use of AR.

Furthermore, the use of AR in the yearbook also promotes teamwork and collaboration among students which otherwise are exclusive to the college editorial boards. The construction of the conventional yearbook involves only a small number of students who are the members of the editorial group. Often, the general readers are not involved in the development and only get to the see the end product. Therefore the yearbook is less interactive and may appear bland to some readers. The integration of AR into the yearbook is able to accommodate more involvement from a bigger number of students; teachers and even parents thus encourage more communal and collective effort. Moreover, with AR, the yearbook is more interactive thus sustain more readerships.

To boot, the AR integration process described in this study can be further applied in various teaching and learning context. The prospect of the application in teaching and learning is vast. The teachers can apply the AR learned to enhance textbooks and teaching modules for example by linking videos for language and History subject, 3D simulations for biology subject and latest

 https://doi.org/10.1051/matecconf/201815005031

updates for exam answering techniques. The students can incorporate AR into their projects and reports. The use of yearbook in this study is just a medium to assist acceptance on the integration of AR technology towards the education setting which otherwise would be daunting. The process described is not only limited in revolutionizing yearbook but can also be utilized to enhance other conventional printed materials thus embracing the 21st century teaching.

Acknowledgments

This study is supported by the Short Term Grant in part by Universiti Malaysia Perlis, under grant number 9001-00497.

References:

1. S.C. Yuen, G. Yaoyuneyong, & E. Johnson, J.Edu. Tech.(JETDE) **4** ,1 (2011)
2. R. Azuma, Y. Baillot, R. Behringer, S. Feiner, S. Julier, & B. MacIntyre, Comp. Grph. & App IEEE, **21**, 7 (2001)
3. H. Kaufmann, & D. Schmalstieg, J. Comp. & Grph, **27**, 3 (2003)
4. M. Dunleavy, C. Dede, & R. Mitchell, J. Sc. Edu. & Tech **18**, 1(2009)
5. M. Akcayir, Edu. Rsh & Rev, **20**, 1 (2017)
6. A. Dünser, L. Walker, H. Horner, & D. Bentall, *Proceedings of the 24th Australian Computer-Human Interaction Conference on - OzCHI '12*, 107–114. (2012)
7. Mahadzir, N. N., & Phung, L. F. (2013). J. Rsch. & Mthd. in Edu, **1**, (2013)
8. Matcha, W., & Rambli, D. R. A., *Computer & Information Science (ICCIS), 2012 International Conference on*, 1123-1126. (2012)
9. Valarmathie, G., Zulkifli, A. N., Faisal Mohamed, N. F., Alwi, A., Saidin, A. Z., Mat, R. C., & Abu Bakar, J. A. *Proceedings of the 1st International Conference on Educational Studies (ICES2015)*. (2015)
10. Cheng, K. H. Aus. J. of Edu. Tech., **33**,4, (2017)

Impacts of Psychological Capital on the Ethical Behavior of Teachers: A Case of Educational Sector in Pakistan

Fazal ur Rehman[1,*], *Rosman* Bin Md Yusoff[2], *Shafie* Bin Mohamed Zabri[3], and *Fadillah* Binti Ismail[4]

[1,3,4] FPTP, Universiti Tun Hussein Onn Malaysia, Parit Raja, Johor, Malaysia
[2] PPUK, Universiti Tun Hussein Onn Malaysia, Parit Raja, Johor, Malaysia

Abstract. This study aims at investigating the effect of psychological capital on the ethical behavior of teachers in the higher educational sector of Pakistan. This study applied quantitative approach to collect data through questionnaire based survey from the teaching staff of Rawalpindi and Islamabad region universities in Pakistan. This study used descriptive statistics and step wise regression to analyze the collected data. Results of the study indicate that psychological capital has positive and statistically significant effect on the ethical behavior of teachers in the higher educational sector of Pakistan. This study addresses the need to investigate the relationship of psychological capital with the ethical behavior of teachers in the higher educational sector of Pakistan, which can help in understanding the nature of these variables in the higher educational sector of Pakistan.

Keywords: Psychological Capital, Ethical Behavior, Teachers, Pakistan

1 Introduction

Teachers are the key assets of every educational organization and may influence the goals and objectives of organizations in positive or negative ways. The professional attitude and ethical behavior of teachers may influence on the careers and learning level of students. Their behavior with students and as a whole in organization may lead to gain competitive advantage and may enhance the level of learning and universities standard as well. However, psychological capital is one of the factors that may cause of enhancing the ethical behavior of teachers in universities. According to [1] psychological capital is "a core psychological factor of positivity in general, and positive organizational behavior (POB) meeting states in particular, that go beyond human and social capital to gain a competitive advantage through investment / development of 'who you are". However, previous literature has mentioned some key factors of psychological capital that are; self-efficacy, hope, optimism and life orientation [2] .These factors may affect the performance of teachers and help in behaving ethically in universities. As, [3] have assumed that psychological capital is helpful in developing positive attitude and professional skills of behaving in working environment.

On the other hand, ethical behavior may one of the interesting techniques to successfully accomplish the objectives of organization. Ethics are the code of conduct or a set of principles in a working environment to assess right or wrong imagination [4]. These codes of conduct may help in developing a culture of behaving in organizations. The predefined ethical principles may help in executing working environment in efficient ways. Therefore, this study has interest to investigate whether psychological capital has any relationship with the ethical behavior of teachers in Rawalpindi and Islamabad region universities in Pakistan or not? For the aim, a study was conducted in the higher educational sector of Pakistan to assess the relationship between psychological capital and ethical behavior of teachers in Rawalpindi and Islamabad region of Pakistan.

2 Literature Review

2.1 Psychological Capital

The term psychological capital originated in the field of economics and sociology to stare the performance at workplace and may cause of influence in behaving. The presumed behavior may cause confidence and self-esteem in employees [5]. However, this conception led towards development of "Positive Psychological Capital" in economics literature. Simply, it is known as psychological capital along with some key factors that are; "(a) Paradigm of positive psychology that highlights the

*Corresponding author: fazal_marwatpk@yahoo.com

significance of positivity and strength, (b) Positive organizational behavior, (c) Human and social capital i.e. what you know, who you know, who you are, (d) Investment and development for getting improvement in performance that may result in competitive advantage" [1]. Therefore, psychological capital is an optimistic approach of individual along with some features like; self-efficacy, hope, optimism and resiliency. These characteristics are described as;

2.1.1 Self-Efficacy

Self-efficacy is the comparison of thinking's, feelings and actions of individual in different situations to handle problematic situations [6]. Likewise, [7] have defined self-efficacy as the "individual's conviction... about his or her abilities to mobilize the motivation, cognitive resources, and courses of action needed to successfully execute a specific task within a given context". Therefore, low self-efficacious individual may face problems in challengeable tasks as compared to higher self-efficacious individual [8]. However, literature has mentioned some interesting features of self-efficacious individuals to distinguish them from others like, they often set demandable tasks for themselves, welcome challenges, possess higher level of self-motivation, have energy for completing goals and insists to problematic situations [9]. These features may provide sagacity to teachers to perform in efficient ways. Individuals with these features may not delay in demanding tasks and are called as "discrepancy reduction" as they develop discrepancies in unbreakable loops in attaining the tasks. They may have high level of confidence and repetition of failures or negative feedback has no influence on their minds [10].

2.1.2 Hope

Hope is the well-known term in daily life and is "the belief that one can find pathways to desired goals and become motivated to use those pathways" [11]. According to [12], hope is "a positive motivational state that is based on an interactively derived sense of successful (a) agency (goal-oriented energy) and (b) pathways (planning to meet goals)". However, [9] have noted that a lack of research in the term hope and its influence in working environment. Management that posses' high level hope is often perceived to perform well and will provide satisfactory working environment to employees [13]. However, [14] have linked it with job satisfaction and commitment towards organization. Likewise, [15] have differentiated hope from the other elements of psychological capital and assumed that it is theoretically a convergent element. Whereas, [16] have verified hope as the element with discriminate validity as compared to other elements of psychological capital.

2.1.3 Optimism

Tiger [17] has highlighted the conception of optimism and proposed a definition as "a mood or attitude associated with an interpretation about the social or material- one which the elevator regards as socially desirable to his or her advantage, or for his or her pleasure". Precisely, optimists are "those people who make internal, stable, attributions regarding positive events (e.g. task accomplishment) and those who attribute external, unstable, and specific reasons for negative events (e.g., a missed deadline)". Optimism is a vigorous and unchangeable open process and comprises of three steps that are; "leniency for the past, gratitude for the present and opportunity for future" [18, 19]. Therefore, optimism is a key element of psychological capital and cause of motivation and emotions [20]. Hence, optimism may have influence on the decisions and behavior of teachers to keep them involved in getting work done.

2.1.4 Resilience

Resilience is "the positive psychological capacity to rebound, or 'bounce back' from adversity, uncertainty, conflict, failure, or even positive change, progress and increased responsibility". According to [1] resilient people having abilities to "(a) a staunch acceptance of reality, (b) a deep belief, often buttressed by strongly held values, that life is meaningful, and (c) an uncanny ability to improvise and adapt to significant change". However, [21] have expressed their opinions that resilient people are efficient in getting work done and may be able to adjust themselves in threaten situations. [22] have noted lacks of knowledge in the term resilience but researchers are busy at various level to explore it, like; positive use of emotions, change in personal status [23] , improvement in efficiency [24], building sense of honesty and reality [25]. For example, [22] have assumed that positive emotions are the sources to get improvement in resilience.

2.2 Ethical Behavior

Ethical behavior may be the moral and professional set of principles in an organization for developing its culture and a set of executing regulation. These predefined principles may help to understand right and wrong perception of behaving in organization. Likewise, [4] has expressed his opinions that ethics are the set of principles to estimate the right and wrong indication in an organization and helps in developing acceptable culture of behaving. Therefore, ethical behavior may be used as a tool to assess the performance of employees and may affect the operation of a business. Sometime unethical behavior may be used as a tool to assess the ethical behavior of employees in an organization [26]. The study has heighted the conception the sometime unethical behavior can create confusion for management due to lack of predefined principles. Therefore, companies are defining ethical standard and strategies to smoothly execute the business operations, but the implementation may quit rare due to shortage of time.

2.3 Theoretical Framework

Considerable attention has been given towards behavioral aspects of employees, psychological capital and its role in organizations [26]. [9] have shown their intentions to explore various aspects of organizational behavior and psychological capital in organizational context. However, [27] have noted lack of researchers' interest in psychological capital as compared to organizational behavior. The study assumed that psychological capital can enhance the perceptions of organizational behavior. [28] has observed that positive emotions can lead to behave positively in organization. [29] have assessed the relationship of psychological capital with organizational citizenship behavior and organizational commitment in Indian context. Later on, [30] have assessed the relationship of psychological capital with organizations ethical climate components in Iranian context. At the same time, [31] have assessed the association of mediation in psychological capital with organizational climate and organizational citizenship behavior in Pakistani context. Meanwhile, [32] have conducted an empirical study to assess the relationship of organizational support with the ethical behavior. [26] have assessed the relationship of psychological capital with the ethical behavior of sales force in the Telecom Sector. The study suggested investigating the relationship of psychological capital with the ethical behavior in service as well as manufacturing industries. However, previous literature has mentioned psychological capital and its relationship with several other factors like job satisfaction, organizational citizenship behavior and ethical behavior in sales department. Hence, this study has interest to assess the relationship of psychological capital with the ethical behavior of teachers in Pakistani universities. Therefore, the current study is using the theoretical model of [26] to assess the relationship between variables in the higher educational sector of Pakistan.

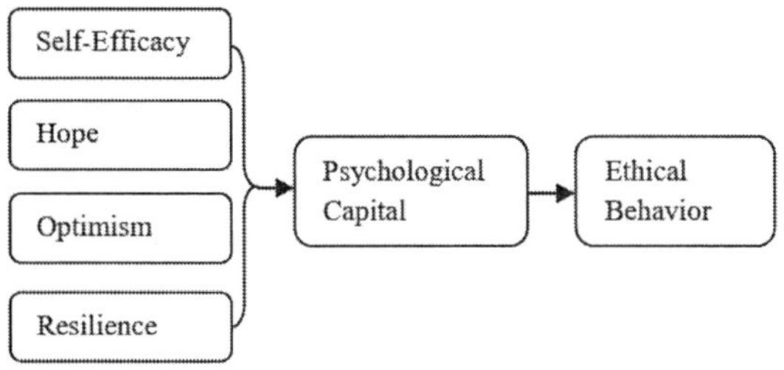

Figure 1 Rehman et al (2014) conceptual model

2.4 Hypothesis of the Study

Hypotheses are the predictions of the study that may be true or false and based on the theoretical concept of previous literature. Results will verify the developed hypothesis in this study as analyzed results decides to accept or reject the developed hypothesis [33]. This study has the following hypothesis;

H1: Self-efficacy has positive relationship with the ethical behavior of teachers.

H2: Hope has positive relationship with the ethical behavior of teachers.

H3: Optimism has positive relationship with the ethical behavior of teachers.

H4: Resilience has positive relationship with the ethical behavior of teachers.

3 Methodology

This study applied quantitative approach to collect data through questionnaires based survey form the faculty members of public and private sector universities in Rawalpindi and Islamabad region of Pakistan. Most of the respondents were below the age of 40 years. The measuring instrument was comprised of demographic features of respondents and information about other variables of the study. It was made clear to the respondents that the collected data will be used for research purposes only and will be kept as confidential. The instrument of the study was adopted from previous studies [26, 34, 35]. A pilot study was conducted to assess the reliability of adopted instrument. For the purpose, the study collected data from 60 respondents which included both gender male and female. The following table shows the result of pilot study;

Table 1. Value of Cronbach Alpha of Pilot Study

Variables	No of Items	Value of Cronbach Alpha
SE	10	0.721
HP	6	0.730
OP	8	0.746
RS	10	0.791
EB	11	0.732
*SE – Self Efficacy, * HP – Hope, *OP – Optimism, *RS – Resilience, *EB – Ethical Behavior		

After getting the aim of reliability, the study collected data for finding the relationship between main variables.

The collected data was analyzed by using correlation and regression statistic to find statistical results of the study.

4 Results

After data collection, the study applied correlation statistics to determine the association between variables. The results of correlation statistics are presented in the following table;

Table 2. Means, Standard Deviation, and Correlation between Variables

	Means	St. D	SE	HP	OP	RS	EB
SE	1.4315	.42144	1				
HP	2.0219	.33208	.041	1			
OP	1.6401	.32914	.047	.051*	1		
RS	1.5231	.48412	.031*	.032*	.049	1	
EB	2.0041	.29415	.030*	.019	.051*	.062	1

*Correlation is significant at the 0.05 level (2-tailed), **. Correlation is significant at the 0.01 level (2-tailed).
*SE- Self-efficacy, *HO-Hope, *OT-Optimism, *RS- Resilience, *EB-Ethical Behavior

Results of correlation statistics in the above table indicates that self-efficacy has positive relationship with hope, optimism, resilience and the ethical behavior of teachers in the higher educational sector of Pakistan. Hope is positively correlated with resilience, optimism and ethical behavior of teaching staff. Similarly, optimism is also positively correlated with resilience and the ethical behavior of teachers in the higher educational sector of Pakistan. Likewise, resilience is positively correlated with the ethical behavior of teachers in the educational sector in Rawalpindi and Islamabad region of Pakistan.

The study applied step wise regression to determine the statistical results of the study. This is the interesting part of study to identify the effect of psychological capital on the ethical behavior of teachers in the higher educational sector of Pakistan. The following table indicates the statistical results of regression analysis;

Table 3. Regression Analysis between Psychological Capital and Ethical behavior

	Beta Value	Standard Error	P Value
SE	.360	.035	.004
HP	.341	.031	.004
OP	.437	.024	.001
RS	.591	.040	.003

** $P < 0.01$, *$P < 0.05$

The above table indicates the result of regression and the magnitude of psychological capital (Self-efficacy, hope, optimism and resilience) on the ethical behavior of teachers in the higher educational sector of Pakistan. The table indicates that all the sub dimensions of psychological capital have positive influence on the ethical behavior of teachers. The values of beta coefficient of self-efficacy ($\beta = 0.360$), the value of beta coefficient for hope is ($\beta = 0.341$), for optimism ($\beta = 0.437$) and for resilience is ($\beta = 0.591$). Therefore, the results lead to the conclusion that psychological capital has positive and significant effect on the ethical behavior of teachers. It can be inferred that teachers in the higher educational sector of Pakistan may be able to handle the problems due to sufficient knowledge and skills. It can be inferred from the results that self-efficacy has positive influence on the ethical behavior of teachers as they may be able to handle the problematic situations independently. Similarly, it can be said that hope has positive effects on the ethical behavior of teachers as they may able to effectively accomplish their life objectives. Likewise, it can be inferred that optimism has also positive influence on the ethical behavior of teachers as they may be able to show positive attitude in different situations. It can also be said that resilience has positive effect on the ethical behavior of teachers as they may be able to adjust themselves in threaten situations. Therefore, all the dimensions of psychological capital have positive effects on the ethical behavior of teachers and leads to the conclusion to accept all the hypothesis of study.

5 Discussion

This study has investigated the impact of psychological capital on the ethical behavior of teachers in the higher educational sector of Pakistan. This study has applied quantitative approach to collect data through questionnaires based survey from the faculty members of both public and private sector universities in Rawalpindi and Islamabad region of Pakistan. The results of the study indicate that all the dimensions of psychological capital have positive and significant effects on the ethical behavior of teachers in the higher educational sector of Pakistan. The results of this study are supported by [26]. The results of this study are in line to with some previous studies which are mentioned as follow.

Golparvar and Azarmonabadi [30] have conducted a study to investigate the relationship of psychological capital with the components of ethical climate in Iranian context. The study distributed questionnaires among the railway employees in Isfahan city to collect data. The results of the study have shown a positive and statistically significant relationship between psychological capital and the components of ethical climate. This study is in line in the term of psychological capital. Likewise, [31] have conducted a study to assess the relationship between the mediation of psychological capital and organizational climate, organizational citizenship behavior in Pakistani context. The study has collected data from the 108 employees of multinational companies through questionnaire based survey. The results of the study have shown that psychological capital has significant effect on organizational climate and organizational citizenship behavior. In the same manner,

[36] has conducted a study to investigate the relationship of emotional intelligence, organizational commitment and job satisfaction with the ethical behavior of employees in Chinese context. The study has found a positive and significant relationship among main variables. It has noted that age has significant but negative influence on the ethical behavior of Chinese employees. This study is in line in the sense of ethical behavior. However, the managerial aspects of the study were to assess whether psychological capital has any relationship with the ethical behavior of teachers in the higher educational sector of Pakistan. The results of this study have contributed in the existing knowledge that psychological capital can be applied to influence the ethical behavior of teachers in the higher educational sector of Pakistan. The results of this study imply that educational institute can focus their efforts on the teacher's individual factors "psychological capital" to enhance the standard of ethical behavior in educational environment. The educational institutes can provide special training and practices to enhance the level of psychological capital for building smooth ethical environment. The educational institutes can establish special counters to handle the personal issues of teachers to monitor the display of unethical behavior educational environment.

6 Conclusion

This study aims at investigating the effect of psychological capital on the ethical behavior of teachers in the higher educational sector of Pakistan. The study collected data through questionnaire based survey from the faculty members in both public and private sector universities in Rawalpindi and Islamabad region of Pakistan. The collected data were analyzed via correlation and step wise regression to determine statistical results of the study. The results indicate that psychological capital has positive and significant effect on the ethical behavior of teachers in the higher educational sector of Pakistan. Therefore, while interpreting the results of the study it should be kept in mind that the data was collected from a small region of Pakistan and readers need to be cautious in generalizing the results of this study. This study is only limited to the higher educational sector of Pakistan within two cities only and future study should include manufacturing sector and addition of moderating variables like gender as well. This study has also a deficiency as it excluded the primary and secondary level of education and the future studies can assess psychological capital of the teachers in lower educational level.

References

1. Fred, L., Carolyn, M. Y. Human, Social, and now positive psychological capital management, investing in people for competitive advantage, Organizational Dynamics **33**. 2, 143-160 (2004)
2. Fred, O. W., Fred, L., James, B. A., Adegoke, O. Authentically leading groups: The mediating role ofcollective psychological capital and trust, Journal of Organizational Behavior **32**. 1, 4-24 (2011)
3. Fred, L., Steven, M. N., Bruce, J. A., James, B. A. The mediating role of psychological capital in the supportive organizational climate-employee performance relationship, Journal of Organizational Behavior **29**. 2, 219-238 (2008)
4. Hamza, A. Ethical Perceptions of Public-Sector Employees and Citizens and their Impact upon Attitudes against Unethical Behavior, Turkish Studies **13**.1, 27-44 (2012)
5. Arthur, H. G., Jonathan, R. V., William, D.J.R. The Impact of Psychological and Human Capital on Wages, Economic Inquiry **35**, 813-821 (1997)
6. Albert, B. Health promotion from the perspective of social cognitive theory, Psychology and Health **13**, 623-649 (1998)
7. Alexander, D. S., Fred, L. Self-Efficacy and Work-Related Performance: A Meta-Analysis, Psychological Bulletin 124.2, 240-261 (1998)
8. Albert, B. An agentic perspective on positive psychology In S. J. Lopez (Ed.) *Positive psychology: Exploring the best in people* (167–196). Westport, CT: Greenwood Publishing 1, (2008)
9. Fred, L., Bruce, J. A., James, B.A., Steven, M. N. Positive Psychological Capital: Measurement and Relationship with Performance and Satisfaction, Personnel Psychology **60**, 541–572 (2007)
10. Albert, B., Edwin, A. L. Bandura, A., Locke, E. A. Negative self-efficacy and goal effects revisited, Journal of Applied Psychology **88**. 1, 87–99 (2003)
11. Snyder, C. R., Lopez, S. J. *Handbook of Positive Psychology*, Oxford, UK: Oxford University Press. 257- 276 (2002)
12. Charles, R. S., Donlson, R.F (Eds.), Handbook of social and clinical psychology: The health perspective (285-305). Elmsford, NY: Pergamon (1991)

13. Suzanne, J. P., Fred, L. The positive impact and development of hopeful leaders, Leadership and Organizational Development Journal **24**, 26–31 (2003)
14. Fred, L., Susan, M. J. Hope: A new positive strength for human resource development, Human Resource Development Review **1**, 304–22 (2002)
15. Charles, R. S. Hope theory: Rainbows in the mind". Psychological Enquiry **13**. 4, 249-276 (2002)
16. James, C., Lauren, R. Construct validities and the empirical relationships between optimism, hope, self-efficacy, and locus of control, Work. **19**.2, 125-136 (2002)
17. Tiger, L. Optimism: The Biology of hope, New York: Simon & Schuster (1971)
18. Peterson, C. The future of optimism, American Psychologist **55**. 1, 44-55 (2000)
19. Schneider, S.I. In search of realistic optimism, *American Psychologist*, **56**. 3, 250-263 (2001)
20. Luthans, F. The need for and meaning of positive organizational behavior, Journal of Organizational Behavior **23**. 6, 695-706 (2002a)
21. Block. J., Kremen. A.M. IQ and ego-resiliency: Conceptual and empirical connections and Separateness, Journal of Personality and Social Psychology **70**.2, 349-361 (1996)
22. Tugade, M. M., Frederickson, B. L. Resilient individuals use positive emotions to bounce back from negative emotional experiences, Journal of Personality and Social Psychology **86**. 2 ,320-333 (2004)
23. Masten, A.S. Ordinary Magic: Resilience processes in development, American Psychologist **56**. 3, 227-239 (2001)
24. Luthans, F., Avey, J. B., Avolio, B. J., Norman, S. M., Combs, G. M. Psychological capital development: toward a micro intervention, Journal of Organizational Behavior. **27**. 3, 387-393 (2006)
25. Couty, D.L. How resilience works, Harvard Business Review **80**.3, 46-55 (2002)
26. Rehman, F.U., Javed, F., Rehman, S., Nawaz, T. Determining the Influence of Psychological Capital on the Ethical Behavior of Sales Force: A Case of Telecom Sector, European Journal of Business and Management, 6.37, 290-296 (2014)
27. Luthans, F., Avolio, B. J., Avey, J. B., Norman, S. M. Positive psychological capital: Measurement and relationship with performance and satisfaction, Personnel Psychology **60**. 3, 541-572 (2007)
28. Fredrickson, B.L., Joiner, T. Positive emotions trigger upward spirals toward emotional wellbeing, Psychological Science **13**, 172-175 (2002)
29. Shahnawaz. M.G., Jafri. H. Psychological Capital as Predictors of Organizational Commitment and Organizational Citizenship Behavior, Journal of the Indian Academy of Applied Psychology **35**, 78-84 (2009)
30. Golparvar. M., Azarmonabadi. A. R. Explaining Psychological Capital Components Through Organization's Ethical Climate Components, International Journal of Business, Economics and Management **1** .8, 216-228 (2014)
31. Faisal, Q., Hina, J. Mediation of Psychological Capital between Organizational Climate and Organizational Citizenship Behavior, Pakistan Journal of Commerce and Social Sciences **8**.2, 453- 470 (2014)
32. Qingshan, H., Xuansheng, C., Le, L., Hua, H. Empirical Research on the Influence of Organizational Support on Psychological Capital, American Journal of Industrial and Business Management **4**, 182-189 (2014)
33. Geoffrey, R.M., David, D., David, F. *Essentials of Research Design and Methodology*, Hoboken, NJ: John Wiley & Sons. 1-306 (2005)
34. Maria, H. L. The Socio-Demographic Correlations of Psychological Capital, European Scientific Journal **9**.29, 26-42 (2013)
35. Kathie, L. P., Michelle, C. B. Rebounding from Corruption: Perceptions of Ethics Program Effectiveness in a Public Sector Organization, Journal of Business Ethics **10**, 551-566 (2006)
36. Weihui, Fu. The Impact of Emotional Intelligence, Organizational Commitment, and Job Satisfaction on Ethical Behavior of Chinese Employees, J Bus Ethics. **122**.1, 137-144 (2013)

MATEC Web of Conferences 150, 05033 (2018) https://doi.org/10.1051/matecconf/201815005033
MUCET 2017

Cost Reduction and Business Strategy Matters to Human Resource Outsourcing? A Validation by HR Experts from Government Link Companies (GLC's).

Mohd Fitri Mansor[1,*], *Noor Hidayah* Abu[2], *Aida Nazima* Abashah[1,] *Muhammad Asyraf* Mohd Kassim[1]

[1]School of Business Innovation & Technopreneurship, Universiti Malaysia Perlis, Jalan Kangar- Alor Setar, 01000 Kangar Perlis, Malaysia.
[2]School of Technology Management & Logistic, College of Business, Universiti Utara Malaysia, 06010 Sintok, Kedah Malaysia.

Abstract. The phenomenon of outsourcing has spawned a rich body scholarly reseach in sveral decades. However, the answer to one of the pertinent question has remained elusive: Does cost reduction and business strategy really matters to the human resource outsourcing impacts. Looking at this question it is important for the organization to embark on the practice of human resource outsourcing to save operating cost and remain competitive. Therefore, the objective of this study is to investigate the effects of cost reduction and business starategy towards human resource outsourcing impacts. Both quantitative and qualitative methods were employed and the top management in Malaysian Government Link Companies (GLC') is the unit of analysis. The findings revelaved that, both variables are significant statistically and non statically i.e from the experts opinion. Finally, the study also provides useful directions for future research, HR Practitioners and policy maker particularly in managing and organizing human reseouce matters.

1 Introduction

Outsourcing is one of the most powerful trends in Human Resource Management (HRM). The growth in this, Human Resource Outsourcing (HRO), has promoted research into the optimal strategic conditions for externalization of HRM activities [20, 21, 35]. It refers to a contractual relationship for the provision of business service by an external provider or in simple words a company pays another company to do some work for it [4]. Again, Nordigarden et al (2014) [29] define outsourcing as the transference of internal governance activity to external control. It occurs when the organization contracts with another organization to provide service or products whether core or non-core activities. Works are traditionally done internally is shifted to an external provider and the employees of the said organization are sometime be transferred to the service provider.

The survey conducted by Gurchiek, (2005) [12] shown that nearly all organizations have outsourced parts of their HR functions. Intriguing facts some see outsourcing as a way of maintaining or increasing company competitiveness. However, many other views outsourcing in more negative light, focusing on job lost [38]. Failure to manage human resource outsourcing subject to risk in the organization such as reduces the organization capability for cross functional synergies and creativity. This risks and limitations, it is estimated between 1% to 20% of outsourced HR functions have been brought back in house [12, 31]. Organizations that outsource due to trends or "everybody is doing it" possibly faced unexpected costs and complications. Based on previous studies by Lutchen, (2004) [23] about one-half of the outsourcing arrangements entered into end up being terminated for various reasons. Some new vendors encounter financial difficulties or acquired by other firms with different procedures and priorities. For example Indian operations in business processing such as call centers and office handling payroll, accounting and human resource functions often loss 15-20 % of their workforces each year but the software programming skills are plenty, managerial experience is in very short supply (McKinsey Global Institute, 2003).

1.1 Problem Statement

HRO has grown popularly since early 1990s, particularly in the USA and Europe. It is up to 90% of companies outsource some HRM activities (William, 2000; Banham, 2003; Cook, 1999). Based on previous study by Galanaki & Papalexandris (2005), the positive implication of the company implementing the HR outsourcing is more efficient, economy of scale benefit, provide knowledge and competencies that is not available in house and able to reduce cost even more as compared to hiring staff to handle these HR activities. Again, it gains specialized HR expertise, achieving flexibility, reducing risks and enhancing access to advanced technology (Belcourt 2006; Lever, 1997; Scott et al., 2005). In addition, the benefits of HRO is also supported by Davidson (2005) where every organization entered in these activities reported that they significantly reduced HR costs through greater economies of scale, lower technology-investment costs, efficient HR processes and headcount reductions.

However, there are some issue and debate on HR outsourcing where it is seen as a threat that synonymous with downsizing or reductions in manpower that lead to generate little value to the core competence of the organization [8]. It is a danger in outsourcing some HR tasks while keeping others in-house. Major organizational

* Corresponding author: fitrimansor@unimap.edu.my

change become more difficult to achieve if part or whole of HR functions are outsourced, create rigidity, low morale and greater frustration among the employees that could leads to higher staff turnover [34]. Furthermore, it is still lack of study in the field of human resource outsourcing in the local context. The trend of outsourcing practice is widely practiced in manufacturing operations rather than government sector particularly in Government Link Companies [37] and indeed in some areas of business, outsourcing has been going on for quite some time. Despite an increase in HR outsourcing activities by organizations, research on this subject is not extensive [19] and is still at the seminal stage. Even though, the growth in HR outsourcing over the last decades has been remarkable, its effects are still vague and inexplicable [16].

Based on the above scenarios it shows that the gaps still exists between what is expected and the actual scenario particularly in human resource outsourcing practicing. In spite of the emerging trend in human resource outsourcing, few empirical investigations have been conducted into human resource outsourcing, particularly in Malaysia [14]. In fact, many studies emphasize outsourcing of other activities such as information technology [18], engineering [32], and manufacturing [9].

To some companies, it is still rarely applied the outsourcing to human resource function and more than that, it is a debatable issue whether this human resource outsourcing yield a good result or detrimental to the whole institution. In fact, results from the past studies are inconsistent [8]. Therefore, this study is pertinent for further investigate the impacts of cost reduction and business strategy towards HRO in Malaysia Government Link Companies (GLC's). The said findings then will be validated by HR Experts. Finally, the discussion and conclusion will be addressed.

2 Literature Reviews

2.1 Cost Reduction

Several studies are focusing on cost saving following outsourcing several activities to external experts. As for [16, 1], found evidence specifically linking outsourcing activity to cost saving in a way that firms with outsourcing implementation have an obvious significant advantage in term of cost efficiency compared to their counterpart with no single outsourcing approach at all.

In a study carried out among 51 publicly traded firm in US from 1990 to 2002 with applied outsourcing in their operations, authors have carried out the finding on which outsourcing can improve a firm's cost-efficiency [5]. Nevertheless the existing literature show that outsourcing improved their productivity and profitability. Previous researches has shown outsource activities to third party are proven to generate more productivity in businesses [24].

The objectives of outsourcing are to cut cost, enhancing productivity and time saving. According to Gilley (2000), [11] authors highlighted the need to outsourcing that can provide cost reduction to firm. This happen when the cost of upgrading maintenance in house is higher compare to outsourcing activities of hiring the third party expertise. For example, Malhotra (1995), [25] found that factors affected by outsourcing decisions are sharing risk on technology investment, operating cost reduction, access to specialized expertise, fixed cost due to the contract, and perception of efficiency. In addition, Bardhan et al (2006), [3] examined that the reason of why firm apply outsourcing is because it can increased capacity, higher quality of product and service and reduced cost.

In other setting, a study in UK workplace provides a detail acknowledgment by informant regarding the driving factors of HR outsourcing. Other than cost control and risk management gained, it is a relief for the HR specialist following the pressures they faced that drove them to consider outsourcing. Other outsourcing motivations besides reasons of cost and quality is a political and historical reason based on experience elsewhere in the business specifically influence of parent companies [40].

Dickmann and Tyson (2005), [7] conclude that the cost reduction was achieved through lesser investment made in IT software and future maintenance specifically made available for payroll function in HR. This research outlined of such advantage to a medium-sized organizations in payroll administration. Similarly, every company with adoption of outsourcing HR functions provide a significant reduction in HR costs through greater economies of scale efficient process, lower technology-investment costs and headcount reduction [7]. Based on the above reviews, the cost reduction can be hypothesized as following below:

H1: The cost reduction benefits has gained by adapting human resource outsourcing.

2.2 Business Startegy

Business strategy is a set of strategic statement that summarize on how an individual company will achieve its goals through meeting customer expectation and competitive advantage [17]. Business strategy can be defined in either several paragraphs or be written as a set of strategic statements. It is important to any organization since it is the determinant to the direction of business and what it will look like in the future. The elements in business strategy are core business activity, sharing and reducing risk, establishing strategic partnership, gaining recognition, raising innovation and streamline internal function. By defining business strategy clearly, organization can develop business or growth plan to achieve business and personal goals. Once it is defined, business strategy sets priorities for the company and management team and helps attract and retain the talented workers the organization need. Although each department in company may focus on different priorities to accomplish specific tasks, these priorities should have no conflict with the overall strategic direction of the company.

The strategy of business somehow indirectly related to human resource function. A study by Galanaki and Papalexandris (2005), [10] showed less developed country like Greece specifically in view of small size companies may benefit from service of HRO with the aim to improve human resource practices in order to increase competitiveness through people. The appropriate use of customization and sophistication of services from vendors could act as key players in the development of human resource function that eventually covered up for the lack of in-house resources.

HRO activity is associated with increasing partnership quality in a way that trust existed as an important component in establishing partnership. The trust leads to positive relationship building and establish the partnership quality which in turn yield a business strategy in partnership [13]. The effect can be seen in firms profit which depending on the ability of outsourcing partner to support the operation.

In addition, business strategy will help developing human resource management. It is a summary of how a company will achieve its goals, meet the expectations of its customers, and sustain a competitive advantage in the marketplace. Business strategy highlight why the company is in business, what is their competitive advantage, which customer should be focused, with what product or service should they continue to offer and why certain strategic direction is decided.

There is a relationship between business strategy and human resource outsourcing which is based on 4 distinctive business strategy integrated from [27, 28 33], which include quality-based, proactive, breadth and reactive strategy. For instance, organizations with a quality-based strategy will find it beneficial to outsource HR transactional function that include payroll and benefits while keeping recruitment function in the company. This is to enable company to enact policies that necessary to attract good candidates.

As for organization with reactive business strategy will prefer not to outsource all HR functions which are both transactional and traditional since their aim is to reduce overall cost. Therefore, Halim and Che-ha (2009), [13] have found a positive significant association between proactive business strategy and HR outsourcing of both functions. The HR outsourcing helps business to be more focused on creating innovative and differentiate products. Therefore, the business strategy can be hypothesized as:

H2: The business strategy will be achieved by adapting human resource outsourcing

3 Findings and Discussion

3.1 Quantitaive Findings

This study employed both quantitative and qualitative approaches by surveying employees working in Malaysia Government Link Companies (GLC). A total of 400 sets of questionnaire were distributed and 301 sets are usable. The data analyses were employed for quantitative are descriptive analysis, correlation analysis, reliability analysis and regression analysis.

3.1.1 Reliability Analysis

The reliability test was used for each variable to provide constant results and it is confirmed that there are no errors in the information provided. Table 1 showed the results of the reliability analysis test using the Cronbach Alpha for all variables. The Cronbach Alpha for value for all variables are .967, .971 and .969 for cost reduction, business strategy and HRO respectively. The reliability test shows that the variables are reliable and excellent. All of values are above .900 for the analysis conducted represent excellent. Hinton et al., (2004) have suggested four cut-off points for reliability, which includes excellent reliability for 0.90 and above, high reliability between 0.70-0.90, moderate reliability the range between 0.50-0.70 and low reliability between 0.50 and below. The higher Cronbach's Alpha values for all constructs represents they are internally consistent.

Table 1. Reliability Analysis

Constructs	Sample Size (n)	No. of items	Cronbach's Alpha (α)	Indication
Cost Red.	301	10	.967	Excellent
Buss. Strg	301	10	.971	Excellent
HRO	301	8	.969	Excellent

3.1.2 Regression Analysis

Regression analysis is a family of technique to explore the relationship between one continuous dependent variable and multiple independent variables [30]. Therefore, this analysis allows more sophisticated exploration of the interrelationship among variables. Refering to the Table 2 below, it contains several results such as beta, significant, R square and few more others. Each serves a different interpretation for data analysis. The beta with the largest value means that it makes the strongest unique contribution to explain the dependent variable. For the significant, the value with less than 0.05 is the variable that makes a significant unique contribution to the prediction of the dependent variable while value that is greater than 0.05 can be conclude with no significant unique contribution to the prediction of dependent variable. As for adjusted R square, it corrects the R square to provide a better estimate of the true population value in case a small sample is involved in the study.

The regression analysis result is computed in the Table 2 which was used to examine the independent variable and dependent variables. The results of regression analysis, it showed that hypothesis 1 (H1) has a positive and significant value .000 and with standardized coefficients (beta) 2.12 and t value is 3.702. The significant value of .00 is smaller than 0.05 which showed that the hypothesis is supported. Hence, the alternative hypothesis (*Ha*) shows significant impact on cost reduction towards HRO and is accepted.

Hypothesis 2 (H2) has a positive and significant value .000 and with standardized coefficients (beta) .534 and t value is 10.234, the significant value of .000 is smaller than 0.05 which showed that the hypothesis is supported. Hence, the alternative hypothesis (*Ha*) shows significant impact of business stategy towards HR outsourcing and is accepted. In addition, the R-square value of 0.693 indicates that 69.3% of the variance in the study can be explained by cost reduction and business strategy.

Table 2. Regression Analysis

	Unstandardized coefficients		Standardized coefficients		
Model	B	Std. error	Beta	t	sig
Cost Reduction	.202	055	.212	3.702	.000
Business Strategy	0551	.054	.534	10.234	.000

R Square = .693
Adjusted R square = .756
Durbin-Watson = 2.058

3.1.3 Correlation Analysis

Correlation analysis is used to describe the strength and direction of the linear relationship between variables. The strength of relationship is shown in r value while negative or positive sign of r value is used to determine the direction of the relationship. From the Table 3 below, the relationship between two independent variables (cost reduction and business strategy) and dependent variable (human resources outsourcing) are shown. First correlation is between cost reduction and human resources outsourcing the r-value are 0.770 or 77 percent at 99 percent confident interval. These values represent a very strong relationship. Meanwhile the correlation continues to correlate between business strategy and human resource outsourcing. The r-value is 0.844 or 84.4% confident interval and it is representing very strong relationship.

Table 3. Correlation Analysis

	Cost Reduction	Business Strategy	HRO
Cost Red.	1		
Buss Strg.	.756**	1	
HRO	.770**	.844**	1

***Correlation is significant at the .01 level (2-tailed)*
**Correlation is significant at 0.5 levels (2-tailed)*

3.2 Qualitative Findings

The study also employed qualitative method apart from questionnaire distribution. This method conducted done after quantitative data were analyzed and it is used for validation and comparison. Several HR practitioners representing each of GLCs were interviewed for this purpose and the results are summarized in the tables below.

Table 4. Content Analysis GLC A

GLC A		
Question	**Feedback**	
Respondent	Position : Human Resource Manager Gender : Male Experience : 20 years Education : Degree Job Scope : System Administrator (HRIS) Sector : Utility	
HR Outsourcing Activities	**Fully Outsource** • Ground work / technical • Sanitary worker • Health care	**Partially Outsource** • Training and development
Cost Reduction	• The HR expert agrees the statement that outsourcing some HR functions help reduced firm's operational cost. This can be seen from the point of monitoring employees in implementing their work at the workplace. The job being outsourced is high risks in which the employees are exposed to an accident or injury. By delegating the tasks of non-core activities to vendors, it eventually leads to time saving in HR management and operational cost. • HR manager recognized some cost reduction in seeking new employees. The company is recruiting new employees for the purpose of its core activities while workers for non-core activities are employed by the vendor in outsourcing. • In-house workers are given works that has nothing to do with functions being outsourced. It does not lead to time saving since jobs for in-house employees are not reduced.	
Business Strategy	• Although vendors are in charge of the labors, yet the risks are on the company to be accounted for any possible accidents occur. This is stated in Energy Commission Act as a new regulator for the energy industry in Malaysia which stipulates that the organization should accountable for any accident occurred. • Vendors play a crucial role in reaching the GLC's main objective of serving Malaysia users. It is important that the vendors are available whenever they are needed to support the organization. • Outsourcing assist in HR coordination in a long term even though it is quite difficult to be dealt particularly at the early stage of outsourcing. • Since outsourcing is crucial for this GLC, it does help making up for the lack of in-house resources.	

Table 5. Content Analysis GLC B

GLC B		
Question	**Feedback**	
Respondent	Position :General Manager (Human Resource) Gender :Male Experience :11 Years and above Education :Degree Job Scope :Human Resource Sector :Finance and banking	
HR Outsourcing Activities	**Fully Outsource**	**Partially Outsource**
	• Contact center • Healthcare benefits administration (Medical Care Organization) • Facility management	• Recruitment and selection (supporting staff) • Talent management (Training and development) • Executive development and Coaching • Compensation and Benefit (C&B) • Employees assistance and counseling • Employees' communication plans and strategies. • Retirement planning and pension administration • HR policy development and implementation • HR strategic planning
Cost Reduction	• At the initial implementation stage of outsourcing there are considerable amount of money involved. Nevertheless, the advantages of cost reduction are apparent to the GLC in a long term. For instance, depending on the market condition which alters the cost, the delegation of some functions to Medical Care Organization (MCO) is overly optimistic. This is happening due to the extensive knowledge as well as experiences of MCO in offering the best recovery plan for employees according to their injury. • It also benefits to the GLC in term of reducing the cost of employee recruitment since vendor directly involve in recruiting its own employees. • Timing is crucial and this factor is taken into account when considering the outsourcing of HR function. Thus, the time consumed for any function such as healthcare and benefit are outsourced to vendors. • Sometimes works can be reduced by using a third-party. In this company, a healthcare division has a responsible party for managing health-related problems that can be reduced workload and time.	
Business Strategy	• Employees' healthcare, without doubt, is a great relief to this GLC since this activity is being outsourced to the Medical Care Organization (MCO), no further grievances and complaints received compared than handled by in-house HR experts. It has reduced the risk of mishandling towards the cases. • There are many ways in which outsourcing helps this GLC in managing employees. Many improvement and adjustment is made to cope with the environment and the practice of HR activities. • Outsourcing recognized by HR manager to facilitate various activities starting from recruitment to the benefit administration. • The outsourcing implementation has covered up several deficiencies in training, recruitment, paper work and employee job-related assessment and helps the organization in in-house resources.	

4 Conclusion

In the nutshell, the current study is about the impacts of human resource outsourcing in Malaysian Government Link Companies (GLC's). Two variables carried out which are cost reduction and business startegy. A study employed both quantitative & qualitative methods in order to answer the research objectives. In addition, the results gained from the quantitative then validated from several HR experts. The study revealed that, the cost reduction and business strategy are positively impact HRO in Malaysian Government Link Companies from the statically and non statically evidences. The findings of the current study may help the government in updating and reviewing manpower policy in the future, as a basis of human resource outsourcing decision as well as a foundation for future research in investigating any new variables in human resource outsourcing filed.

References

1. Banerjee, A. and Williams, A. S. International service outsourcing: Using offshore analytics to identify determinants of value-added outsourcing. *Strategic Outsourcing: An International Journal,* **2**, 1 (2013)

2. Banham, R., HR Outsourcing Leads the Way. *Outsourcing Essentials,* **1**, 30 (2003)

3. Bardhan, I., Whitaker, J. W., & and Mithas, S. Information Technology, Production Process Outsourcing, and Manufacturing Plant Performance. *Journal of Management Information Systems,* **23**, 2 (2006)

4. Belcourt, M. Outsourcing; the benefits and the risk. *Human Resource Management Review*, **16** 2 (2006)

5. Bin Jiang, G. V. Outsourcing effects on firms' operational performance: An empirical study. *International Journal of Operations & Production Management,* **26** 12,(2006)

6. Cook, M. Outsourcing Human Resources Functions. Startegies for Providing Enhanced HR Services at Lower Cost, AMACOM, New York, NY. (1999)

7. Davidson, G. Why HR outsourcing continues to expand. *Human Resource Management International Digest,* **13** 3 (2005)

8. Delmotte, J. and Luc Sels. HR Outsourcing: Threat or opportunity. *Personnel Review*, **37** 5 (2005)

9. Dekkers, R., Decision models for outsourcing and core competences in manufacturing. *International Journal of Production Research*, **38** 7 (2000)

10. Galanaki, E. and Papalexandris, N. Outsourcing of human resource management services in Greece. *International Journal of Manpower*, **26** 4 (2005)

11. Gilley, K. M. Making more by doing less: An analysis of outsourcing andits effects on firm performance. *Journal of Management,* **26** 4 (2000)

12. Gurchiek, K. Record Growth: Outsourcing of HR Functions. *HR Magazine*, **50** 6 (2005)

13. Halim, A. H. & Che-Ha A. G. The influence of business strategy on the decision to outsource human resource activities. *Journal of Human Resource Costing & Accounting*, **13** 4 (2009)

14. Hasliza A.H., N. C. (2011). Embarking on HR Outsourcing- Do Organizational Size and Maturity Level Matter? *Int. Journal of Economics and Management,* **5** 1 (2011

15. Hinton, P. R., Brownlow, C., McMurray, I & Cozens, B. SPSS explained. Routledge Inc., East Sussex, England. (2004).

16. Jiang, B., Frazier, V. G. and Prater, L. E. Outsourcing effects on firms' operational performance. *International Journal of Operations & Production Management,* **26** 12 (2006)

17. Kauffman, F. Defining your business strategy. retrieved form http://www.entrepreneurship.org/resource-center/defining-your-business-strategy.aspx (2015)

18. Lacity, M.C. and Willcocks, L.P. An empirical investigation of information technology sourcing practices: Lessons from experience. *MIS Quarterly*, **22** (1998)

19. Leiblein, M.J., Reuer, J.J. and Dalsace, F. Do 'make or buy' decisions matter? The influence of organizational governance on technological performance. *Strategic Management Journal*, **23** (2002)

20. Lawler, E.E., Ulrich, D., Fitz-enz, J. and Madden, J.C. Human Resource Business Process Outsourcing; Transformiong How HR Gets its Work Done. Jossey-Bass, San Francisco, CA. (2004)

21. Lepak, D.P. and Snell, S.A. Virtual HR: Strategic Human Resource Management in the 21st century. *Human Resource Management Review*, **8** 3 (1998)

22. Lever, S. An analysis of managerial motivations behind outsourcing practices in human resources. *Human Resource Planning*, **20** 2 (1997)

23. Lutchen, M. Outsourcing IT headaches is no answer. *Chief Executive*, (2004)

24. M.Lindsay, J. R. (2001). *The Management and Control of Quality,Fifth Edition.*

25. Malhotra, Y. IS productivity and outsourcing policy: A conceptual framework and empirical analysis. Proceedings of Inaugural America's Conference on Information Systems (Managerial Papers). Pittsburgh, PA: *Association for Information Systems.*

26. McKinsey Global Institute. Offshoring: Is it a win-win game? San Francisco, Author. (2003)

27. Miles, R. E. and Snow, C. C. Organizational strategy, structure and process. McGrraw-Hill, New York, NY. (1978)

28. Miller, P. Strategic industrial relations and human resouce management: Distinction, definition and recognition. *Journal of Management Studies*, **24** 4 (1987)

29. Nordigarden, D., Rehme, J., Brege, S., Chicksand, D. and Walker, H. Outsourcing decisions. The case of parallel production. *International Journal of Operations & Production Management*, **34** 8 (2014)

30. Pallant, J. SPSS Survival Manual: A step by step guide to data analysis using IBM SPSS. 5th Edition. (2013)

31. Pollitt, D. Outsourcing HR: The contrasting experiences of Amex and DuPont, *Human Resource Management*, **12** 6 (2004)

32. Quin, J. B. Outsourcing innovation - the new engine of growth, *Sloan Management Review,* **41** 4 (2000)

33. Schuler, R. S. and Jackson, S. E. Linking competitive strategy with human resource management practices. *Academy of Management Executive*, **3** (1987).

34. Sim, M. The intangible costs of human resource outsourcing, *Human Resource Management International Digest*, **18** 6 (2010)

35. Stroh, L.K. and Treehuboff, D. Outsourcing HR Functions: when-and when not-to go outside, *Journal of Leadership and Organizational Studies*, **10** 1, (2003)

36. Scott-Jackson, W., Newham, T. and Gurney, M. HR Outsourcing: The key decisions, *Charted Institute of Personnel and Development*, London. (2005)

37. Wahrenburg, M., Hackethal, A., Friedrich, L. and Gellrich, T. Strategic decisions regarding the vertical integration of human resource organizations: evidence for an integrated HR model for the financial services and non-financial services in Germany, Austria and Switzerland. *International Journal of Human Resource Management,* **17** 10 (2006)

38. Weidenbaum M. Outsourcing : Prons and Cons", *Business Harizons*, **48** 1 (2005),

39. William M. Mercer and Cranfield School of Management. European Trends in HR Outsourcing, William M.Mercer and Cranfield School of Management, research report (2000)

40. Woodall, J., Jackson, S. W., Newham, T. and Gurney, M. (2009). Making the decision to outsource human resource. *Personnel, Review*, Vol. 38 Iss pp. 236-252

 https://doi.org/10.1051/matecconf/201815005034

Implementation of 5S in Manufacturing Industry: A Case of Foreign Workers in Melaka

San Chee Houa[1], *M.* Haslinda[1], *Sedek* Muliati[2], *Abdullah* Mariam Miri[1], *A.* F. Rahim[1]

[1]Faculty of Technology Management and Technopreneurship, Universiti Teknikal Malaysia Melaka, 76100 Durian Tunggal Melaka, Malaysia
[2]Centre of Technopreneurship Development (C-TeD), Universiti Teknikal Malaysia Melaka, 76100 Durian Tunggal Melaka, Malaysia

Abstract. Lean manufacturing system has been infiltrated in manufacturing sectors across the world. In fact, Lean manufacturing system is a practice which regards the use of the resources, creation of value for the end customers, and as the ways to eliminate the waste. There are several tools that can be used to eliminate the waste within the industry. This research is a study of the implementation of 5S in manufacturing industry. Despite this, the research study focused on manufacturing industry, which has been implemented 5S system in Melaka State. Although there are number of tools and technique available to help in improving the manufacturing process, however, there is only few industries could implement the tools successfully. In this research, foreign workers play a main role in implement the 5S systems as the manufacturing industry in Malaysia adopt large amount of foreign workers to work as employees. Therefore, it is important to ensure the foreign workers truly understand the concept of 5S system and adopt the best ways to implement it in order to have better performance. This research study has been proposed by the research model of the barriers to implementation of 5S in manufacturing industry among foreign workers. A several research method has been adopted to do the research, such as descriptive research design with quantitative methods, survey questionnaire and cross-sectional studies.

1 Introduction

According to N. Nordin et. al. [1] , the ultimate goal of a lean system is to create a smooth and high quality manufacturing that is able to produce finished products based on the customers demand without producing any waste during the operations. Lean thinking represents a set of principles and techniques for the identification and elimination of waste in manufacturing and administrative processes [2]. However, many industries in reality are not able to transform themselves to a lean manufacturing organization towards creating the world-class companies [3]–[5]. Due to that, Lack of knowledge is one of the most important problems of managers about familiarity with a convenient method to successfully improve the performance of the organization [6]. Furthermore, in this competitive environment, organizations have to make the best judgment and choose the best methods to achieve their target and not to lose very finite opportunities [7]. This is due to less effectiveness of implementing the 5S systems into the workplace.

Through 5S methodology, the management can create an environment where quality work is comfortable, clean and safe in the organization and it can ensure the compliance to standards and will further foster continuous improvement [8], [9]. 5S is a system that used to reduce unwanted waste and optimize productivity through continuously maintaining the system in the workplace in order to consistent the operational results. The 5S method is a tool for continuous improving lean management processes, where the task is to create a highly efficient, clean, and ergonomic working environment [10]. However, effort and participation from top management is a key factor that determines the success of the 5S practice [11]. Moreover, most important barrier for implementation of 5S effectively is poor communication and existing barrier between managerial level and shop floor employees and the poor training and awareness of 5S [12].

5S systems is similar as the quote "*A place for everything, and everything in its place*" by Benjamin Franklin which state out that the best solution is keep things in their correct positions to avoid from the waste or unwanted actions.

Before the industry starts to implement new manufacturing process, the top management should identify the true meaning of 5S, why and how to implement 5S system so that it will give beneficial for the industry. This report is concerned on the implementation of 5S in manufacturing industries with the performance of foreign workers which located in

* Corresponding author: haslindamusa@utem.edu.my

Melaka. All members within the industry, from the top management to the operator, should work together to have a better performance in implementing 5S system and create the situation as one of the competitive advantages for the industry.

2 5S System

Seiri (*Sort*) refers to the practice of going through all materials, tools, machines, etc., in the work area, identifying needs and unneeded items from them and keeping only the essential/needed items. When Sort is implemented in the workplace, the work area will be more apparent, and leading to a highly efficient workflow, at the same time the workers will be more satisfied with their environment Employee morale would increase as junk items or materials that are potential health hazards are identified and removed [2]. In 5S, it means removing from the shop floor where the items that the workers do not use routinely [13].

Seiton (*Set-in-Order*) deals with organizing the items that are needed in a way that best supports the employees doing the work. Is it the second step after implemented "Sort" [14]. Place things in such a way that they can be easily reached whenever they are needed. Set is the way to make the items look neat and reduce the time for searching the tools [15]. Set in Order focus on setting all the items in the industry into a certain place where the workers can simply find them. Manufacturing industry usually has many different tools and items, and that is the reason to set in order plays an important step to help the personnel save time to gather the items they need [16].

Seiso (*Shine*) step includes activities like cleaning workplace, maintaining its appearance and using preventive steps to keep the workplace clean [14]. Cleaning must be done not just after working shift, but apply on a regular schedule to remove dirt and dust from the workplace [17]. In other word, it is a way to notice problems early and to keep work areas and equipment in good operating condition even more to extend the duration. According to K. Marley and P. Ward [10], the idea for this stage is to have production operators clean their own workplaces at shift end, so that they notice details like spills, frayed cables, or broken lamps instead of just making the items become shine and bright.

Seiketsu (*Standardization*) is to create best practices and to get each team member to establish best practices the same way. According to S.S. Naqvi [18], this step focuses on identifying the best process in order to sustain the improvements and also continue to bring further improvements. Keeping standardization in industry workplace will helps in increasing the safety and meanwhile, reduction of the industrial pollution [2]. Standardization is the way to keep the first three S, and make it as a standard method. Standardization also brings three S into regular work duties.

Shutsuke (*Sustain/Self Discipline*) for every worker and manager has to follow the procedures in the work place and workshop with utmost sense of discipline and sincerity in following the principles and procedures. Sustain/Self Discipline is the final step in the entire 5S system, but also is the hardest step for 5S. The Sustain step is the most difficult because it requires continued diligence. M. Suarez and J. Ramis [19] explained that if 5S implementation failed in the middle way, it was due to the company never completed 5S implementation. On the other hand, if the organization implements 5S completely, a 5S program will have longevity [14].

The Framework of Employees' Motivation

N. Lukman et al., [20] have developed a theoretical model of employees' motivation I implementation of 5S System. The proposed model will help to understand the factors that contribute to the success of 5S implementation. In this framework, there are four independent variables that influence employees' motivation in implementing 5S System, organizational culture, management support, employee involvement and training. The results of the study revealed that organizational culture for 5S, management role for 5S and training for 5S have a significant relationship with employee motivation but insignificant for the employee involvement for 5S.

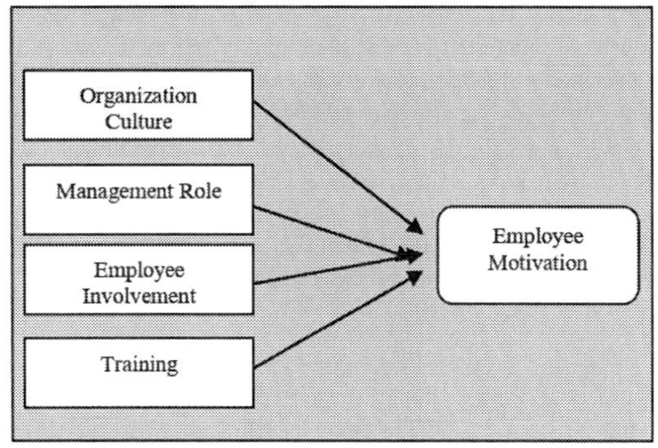

Fig. 1. The relationship of implementation of 5S and employee motivation

3 Methodologies

Hence, this 5S system will be used for improving the performance in manufacturing. There are several factors that will affect the successful implementation of 5S among foreign workers as shown in Figure 2.

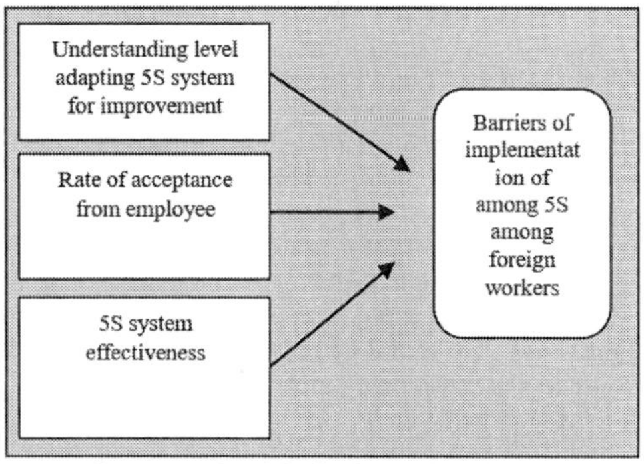

Figure 2. The conceptual model of implementation of 5S system

Pilot studies will be needed to refine and improve in order to allow the respondents understand fully when answering to the question before distribute the questionnaire respondents in manufacturing industry. According to J. Check and R. Schutt [21], as before the final form of the questionnaire is conducted, it is recommended that conduct a pilot study in order to determine if the items are yielding the kind of information that is needed so as to give advance warning regarding weaknesses in a proposed study.

According to Saunders et al. [22], it is advisable to distribute the pilot test to the friends or family instead of unknown persons. The reason to conduct the pilot test is it provides the researcher the ideas for refine the questions and these friends and family are more likely to explain more the weakness of the questions that conducted. It is recommended that pilot test will be conducted to pre-test the questionnaire about 20 respondents. Hence, the response collected will be beneficial to improve the questionnaire before conduct the questionnaire to the respondents.

Once the respondents answered the questionnaire, a questionnaire will be collected and will analyze by using Statistical Package for the Social Science (SPSS) software. This software is used to conduct the analysis of descriptive statistics for describing and summarizing the data includes the measure of central tendency (average), ANOVA (analysis of variance) and dispersion (the spread of data on how close each other is to measure of central tendency). The results will be shown by using SPSS, which in the form of descriptive statistics, reliability analysis, regression and Pearson correlation.

3 Results and Discussion

Most of the respondents were coming from Nepal, which 80 respondents out of 150 respondents. This was followed by 14% (21 respondents) are coming from Bangladesh, 11.3% (17 respondents) from Indonesia, and 10.7% (16 respondents) from India. There were minority respondents which came from the Philippines and Pakistan, with 6.7% (10 respondents) and 2.7% (4 respondents), respectively. Last but not least, 1.3% (2 respondents) are coming from others countries.

Inferential analysis is concerned with making prediction or inferences from the sample data about the population might think, and it is used to make judgments about the probability that an observed difference between groups is a dependable one or one that might have happened by chance in the research.

Table 1 showed a correlation between independent variables and a dependent variable. These independent variables have a positive relationship with successful implementation of 5S among foreign workers. This means that if there is an increase in understanding level adapting 5S systems for improvement, rate of acceptance from employees or 5S system effectiveness, it will also improve the successful implementation of 5S among foreign workers.

Table 1. Correlation

	Successful implementation of 5S among foreign workers	Understanding level adapting 5S system for improvement	Rate of acceptance from employees	5S system effectiveness
Successful implementation of 5S among foreign workers	1			
Understanding level adapting 5S system for improvement	.557**	1		
Rate of acceptance from employees	.516**	.503**	1	
5S system effectiveness	.525**	.559**	.589**	1

**.Correlation is significant at the 0.01 level (2-tailed)

Table 2 showed a model summary of multiple regression analysis. The R values between independent variables were in positive number, 0.639, which meant it was a good relationship between the independent variable and dependent variable. R square showed that the three independent variables influenced the successful implementation of 5S among foreign workers was equal to 0.409 which meant that 40.9%of the variables influenced the dependent variables. However, 59.1% (100% - 40.9%) was influenced by other factors which were not discussed in this study.

Table 2. Regression Analysis of the relation between independent variable and dependent variable

Model	R	R^2	Adjusted R^2	Std. Error of the Estimate	Sig.
1	0.658	0.409	0.397	0.231	0.00

Table 3 showed the highest significant value for independent variable is the 5S system effectiveness, which shown as 0.016 significant value. The unstandardized coefficient (B) is the partial regression coefficients as the values were taken into account the other predictor variables in the model and they will perform the predicted change in the dependent variable for every unit increase in the predictor. Understanding level has a significant positive influence on the successful 5S implementation among foreign workers, which has the strongest impact (t= 4.102, p= 0.000, β= 0.323). Rate of acceptance from employees has a significant positive influence on the successful 5S implementation among foreign workers where (t= 2.815, p= 0.006, β= 0.184). 5S system effectiveness has a significant positive influence on the successful 5S implementation among foreign workers where (t= 2.438, p= 0.016, β= 0.190).

Table 3. Regression Analysis on Coefficients

Model	Unstandardized Coefficients			t	Sig.
	B	Std. Error	Beta		
Understanding level adapting 5S system for ...	0.323	0.079	0.329	4.102	0.000
Rate of acceptance from employees	0.184	0.065	0.230	2.845	0.000
5S system effectiveness	0.190	0.078	0.207	2.815	0.006
Constant	**1.349**	**0.313**		**4.315**	**0.00**

The coefficient for the understanding level was 0.323. It meant that the understanding level was increasing 0.323 was predicted while other variables were set as constant. It also happen the same with another two variables which were rate of acceptance from employees and 5S system effectiveness. The coefficient for the rate of acceptance from employees was 0.184. It meant that the rate of acceptance from employees was increased to 0.184 was predicted while other variables were set as constant. Lastly, the coefficient for the 5S system effectiveness was 0.190. It meant that the 5S system effectiveness was increased to 0.190 was predicted while other variables were set as constant.

4 Conclusion

In conclusion, this study had been discussed on concluding the data analysis of the previous chapter, providing suggestion to overcome and improve the lean implementation and proposing recommendation for future study. The research question has been constructed and it had been determined and concluded by the result of data analysis, and finally answered the research objectives of this study.

References

1. N. Nordin, B. Deros, and D. Wahab, "A survey on lean manufacturing implementation in Malaysian automotive industry," *Int. J. Innov. Manag. Technol.*, vol. 1, no. 4, 2010.
2. V. Patel and H. Thakkar, "A Case Study: 5s Implementation in Ceramics Manufacturing Company," *Bonfring Int. J. Ind. Eng. Manag. Sci.*, vol. 4, no. 3, pp. 132–139, 2014.
3. I. Rajiani, H. Musa, and B. Hardjono, "Ability, Motivation and Opportunity as Determinants of Green Human Resources Management Innovation," *Res. J. Bus. Manag.*, vol. 10, no. 1, pp. 51–57, 2016.
4. H. Musa, N. A. Rahim, F. R. Azmi, A. S. Shibghatullah, and N. A. Othman, "Social Media Marketing and Online Small and Medium Enterprises Performance: Perspective of Malaysian Small and Medium Enterprises," *Int. Rev. Manag. Mark.*, vol. 6, no. 7S, pp. 1–5, 2016.
5. N. Nordin, B. M. Deros, D. A. Wahab, and M. N. A. Rahman, "A framework for organisational change management in lean manufacturing implementation," *Int. J. Serv. Oper. Manag.*, vol. 12, no. 1, p. 101, 2012.
6. P. Rai, "Effectiveness of 5S Implementation on Organizations Performance," *Abhinav Int. Mon. Ref. J. Res. Manag. Technol.*, vol. 5, no. 1, pp. 1–10, 2016.
7. S. R. Bin Ashraf, M. M. Rashid, and A. R. M. H. Rashid, "Implementation of 5S Methodology in the Small Scale Industry: a Case Study," *Int. J. Sci. Technol. Res.*, vol. 4, no. 3, pp. 1791–1796, 2017.
8. S. Sorooshian, M. Salimi, S. Bavani, and H. Aminattaheri, "Case report: Experience of 5S implementation," *J. Appl. Sci. Res.*, vol. 8, no. 7, pp. 3855–3859, 2012.
9. S. Kanamori, S. Sow, M. C. Castro, R. Matsuno, A. Tsuru, and M. Jimba, "Implementation of 5S management method for lean healthcare at a health center in Senegal: A qualitative study of staff perception," *Glob. Health Action*, vol. 8, no. 1, pp. 1–9, 2015.
10. K. A. Marley and P. T. Ward, "Lean management as a countermeasure for 'Normal' disruptions," *Oper. Manag. Res.*, vol. 6, no. 1–2, pp. 44–52, 2013.
11. M. N. A. Rahman, N. K. Khamis, R. M. Zain, B. M. Deros, and W. H. W. Mahmood, "Implementation of 5S practices in the manufacturing companies: A case study," *Am. J.*

Appl. Sci., vol. 7, no. 8, pp. 1182–1189, 2010.

12. A. Ghodrati and N. Zulkifli, "A Review on 5S Implementation in Industrial and Business Organizations," *J. Bus. Manag.*, vol. 5, no. 3, pp. 11–13, 2012.
13. M. Malmbrandt and P. Åhlström, "An instrument for assessing lean service adoption," *Int. J. Oper. Prod. Manag.*, vol. 33, no. 9, pp. 1131–1165, Aug. 2013.
14. R. S. Agrahari, P. A. Dangle, and K. V Chandratre, "Implementation of 5S Methodology in the Small Scale Industry: a Case Study," *Int. J. Adv. Res. Innov.*, vol. 3, no. 1, pp. 130–137, 2015.
15. A. Jain, R. Bhatti, and H. Singh, "Productivity Improvement Through 5S Implementation in Indian Manufacturing Industries," in *Productivity Improvement Through 5S Implementation in Indian Manufacturing Industries*, Springer, New Delhi, 2014, pp. 535–545.
16. C. HungLin, "5S implementation in Wan Cheng Industry Manufacturing Factory in Taiwan," University of Wisconsin-Stout, 2011.
17. S. Gupta and S. K. Jain, "An application of 5S concept to organize the workplace at a scientific instruments manufacturing company," *Int. J. Lean Six Sigma*, vol. 6, no. 1, pp. 73–88, Mar. 2015.
18. S. S. Naqvi, "Towards Successful Implementation of 5Ss in a U.S. Manufacturing Company with Indian SubContinent Workers," Eastern Michigan University , 2013.
19. M. F. Suárez-Barraza and J. Ramis-Pujol, "An exploratory study of 5S: a multiple case study of multinational organizations in Mexico," *Asian J. Qual.*, vol. 13, no. 1, pp. 77–99, Jun. 2012.
20. N. Lukman, W. A. Aziz, and A. H. Zakaria, "The relationship of implementation of 5S and employee motivation," in *conferences or management of higher education*.
21. J. Check and R. Schutt, *Research methods in education*. Sage Publications, 2011.
22. M. Saunders, *Research methods for business students, 5/e*, 5th ed. Pearson Education India, 2011.

The Autoregressive Distributed Lag Model to Analyze Soybean Prices in Indonesia

Mahjus Ekananda[1,*] and *T.* Suryanto [2]

[1]Faculty of Economy and Bussines, University of Indonesia, 16424, Indonesia
[2]IAIN, Indonesia

Abstract. The main objective of this study was to observe factors that affecting domestic soybean prices, including government intervention through BULOG. By using Bound Testing Cointegration method with ARDL approach. In the short term the world soybean price variables in the t-period and exchange rate affect the domestic soybean prices positively and significantly. The variable volume of soybean imports, GDP, and the role of BULOG as sole importer in the t-period does not affect the domestic soybean price significantly. In the long run, the t-period import tariff has a negative and significant effect.

1 Background

Food has a very important role because it is one of the primary human needs. The importance of food can be seen from the concept of food security. The concept of food security has three interrelated dimensions consisting of food availability, community accessibility to food, and stability of food prices [1]. The concept is in accordance with the World Health Organization (WHO) which defines three main components of food security consisting of availability, accessibility, and food utilization which then the Food and Agriculture Organization (FAO) added a fourth component, namely the stability of the three components in a long period. Thus, it can be concluded, this food security is not only seen from the affordability physically but also in terms of price and stability of both components.

The stability of food prices is definitely one of the focus of every country. However, in reality, world food prices experienced a fairly high fluctuation, especially in 2007 to 2008 that caused by the international food crisis [19]. Some of the food commodities affected are sugar, wheat, milk, meat, corn, and soybeans [8]. While in some Asia Pacific countries there is an increase in the price of rice (UN-ESCAP, 2011). This condition can be seen in Fig. 1 which shows a dramatic increase in world food prices since 2007 and started to decline in the second half of 2008 and then increased again in 2009 and showed a significant decline in 2010.

The fluctuation of commodity price which often become public attention is rice, corn, soybean, wheat flour, sugar, cooking oil, onion, chilli, egg, and milk [17]. Therefore, the

* Corresponding author: m.ekananda@gmail.com

government must control the price of these commodities to be affordable by all levels of society.

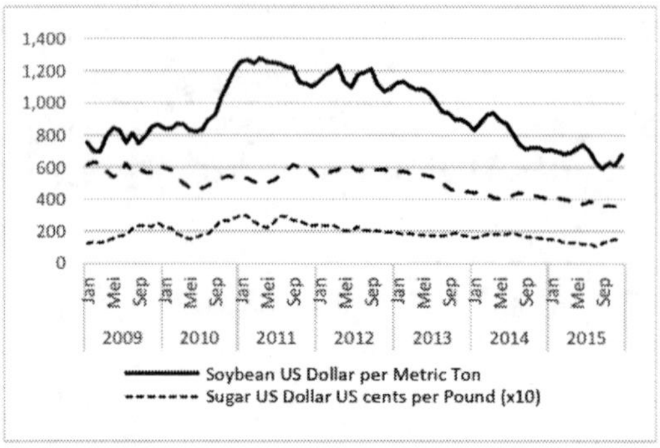

Fig. 1. Sugar World Price, Rice and Soybean Period January 2009 – December 2015.

Sources: IFS CD-ROM

Soybean is the most important food commodity after rice and corn which have strategic position in all national food policy because it is the raw material of daily staple food of most people of Indonesia that is Tofu and Tempe [18]. Fig. 2 shows the price of food in the country shows an increasing trend, although fluctuations in domestic food prices are not as sharp as world food prices, but quite unsettling the public.

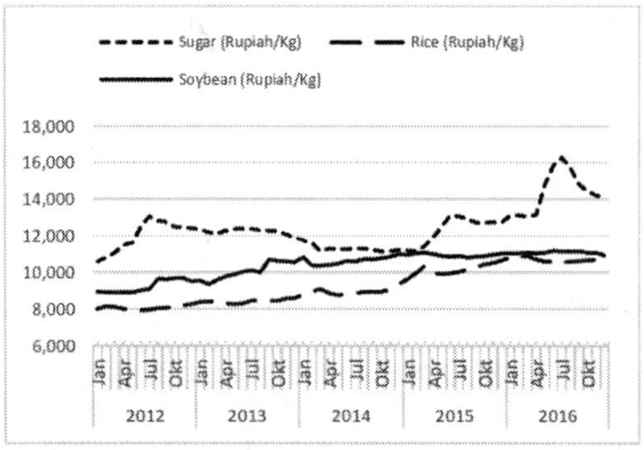

Fig. 2. Average Retail Price of Sugar, Rice and Soy Commodities, January 2012– Dec 2016.

Source: BPS and Ministry of Trade Sources: IFS CD-ROM

Fig. 3 shows a high increase in soybean prices since the end of 1997 and began to fall in the second half of 1999. Then there was an increase in early 2000 and a relatively large

 https://doi.org/10.1051/matecconf/201815005035

decline in early 2003. However, there was a sharp increase in 2008 and has continued to increase since then until now.

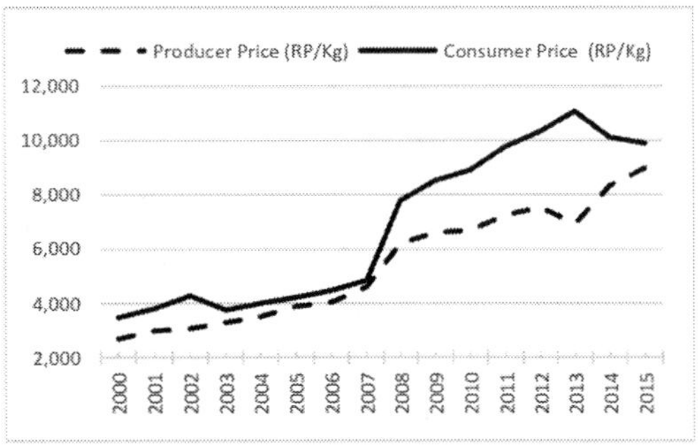

Fig. 3. Average Producer Price and Consumer Price of Soybean in Indonesia.

Source: Central Bureau of Statistics, 2015.

Table 1 shows the deficit of domestic soybean supply that must be met from import. There is a considerable increase in consumption in 2008 - 2012, placing Indonesia in the top ten soybean importers in the world. Nevertheless, the average import of Indonesia is only 2% of world imports while China is 49%. In addition, more than 95% of Indonesia's soybean imports are from the United States and only about 3% are obtained from Brazil. Thus, it can be said that the price of Indonesian soybean will be greatly influenced by US soybean prices.

Table 1. Production, Consumption, Bid Deficit, and Import Soybean in Indonesia
Source: BPS and Ministry of Trade (2016). Note: *) is a proxy number

Year	Production (Tons)	Consumption * (Tons)	Deficit (Tons)	Import (Tons)
2008	775,710	2,147,782	1,372,072	1,169,016
2009	974,512	2,288,686	1,314,174	1,314,620
2010	907,031	2,647,151	1,740,120	1,740,505
2011	851,286	2,781,533	1,930,247	2,088,616
2012	843,153	2,846,461	2,003,308	1,921,207
2013	791,334	2,990,961	2,199,627	1,774,252
2014	783,774	3,060,778	2,277,004	1,922,777
2015	776,286	3,132,224	2,355,938	1,639,104

Fulfillment of food can be done with domestic production and import. However, data show that the overall increase in soybean production in the period 2004 to 2013 is below a hundred thousand tons (Table 2). With the increasing demand for soybean commodities while domestic production is insufficient, then to cover the shortage the government takes import policy. Table 2 shows a considerable increase in soybean imports in the last ten years.

Table 2. Comparison of Productio Domestic and Soybean Imports in Indonesia
Source: BPS and Ministry of Trade

 https://doi.org/10.1051/matecconf/201815005035

Year	Domestic production (Tons)	Import Volume (Tons)
2004	723,483	1,115,793
2005	808,353	1,086,178
2006	747,611	1,132,144
2007	592,534	1,411,589
2008	775,710	1,169,016
2009	974,512	1,314,620
2010	907,031	1,740,505
2011	851,286	2,088,616
2012	843,153	1,921,207
2013	803733	1,774,252
2014	772183	1,922,777
2015	747586	1,639,104

The import policy can be accepted as one of the short-term solutions to reduce the effects of demand pressure on inflation. However, in the long run the import policy is known to have dragged Indonesia into various issues such as the potential of food cartel, import dependency, and inflation which is quite high due to the impact of rising international food prices. Fig. 4 shows that before 1999, the average volume of Indonesian soybean imports was only about 50,000 tons while after 1999 the average increased to about 150,000 tons. This indicates that when soybean imports are monopolized by BULOG, import volume is relatively lower compared to when soybean imports are handed over to private importers. The same is true for domestic soybean prices, which show a relatively more stable graph when the soybean import is monopolized by BULOG.

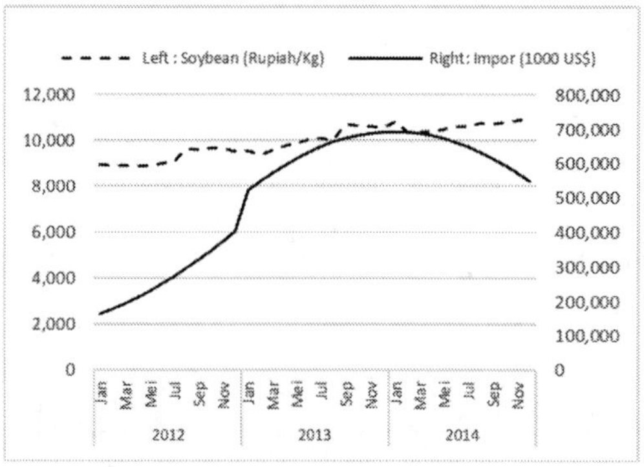

Fig. 4. Retail Price Trend and Import Volume of Indonesian Soybean, January 2012 – December 2014. Source: BPS and Ministry of Trade (processed)

The existence of dependence on food imports in Indonesia cannot be denied anymore. This condition can be shown from the Import Dependency Ratio (IDR) of food products especially for rice, beef, soybeans and sugar in Indonesia presented in Table 3.

Table 3. Import Dependency Ratio (IDR) of Food Products Indonesia Year 2008-2012. Source: DG of Livestock, Ministry and BPS, processed BP2KP, Ministry of Trade. Note: *) until October 2012

Commodity	IDR (%)				
	2008	2009	2010	2011	2012
Rice	1.2	1.0	2.3	7.9	5.8*
Beef	10.4	14.2	17.2	12.2	18.45
Soybean	60.1	57.4	65.8	71.1	71.46
Sugar	34.4	51.3	52.3	60.3	65.6*

From the table, it can be seen that IDR in the four commodities tends to increase from 2008-2012 and the highest ratio is found in soybean and sugar commodities which reach more than 60%. However, the ratio of import dependence is relatively larger on soybean commodity than sugar. The high ratio indicates that domestic soybean availability from domestic production is relatively smaller compared to sugar and on the contrary the fulfillment of national soybean consumption from import is relatively higher in soybean compared to sugar. It also implies that soybean commodity prices are relatively more vulnerable to the effects of international price fluctuations.

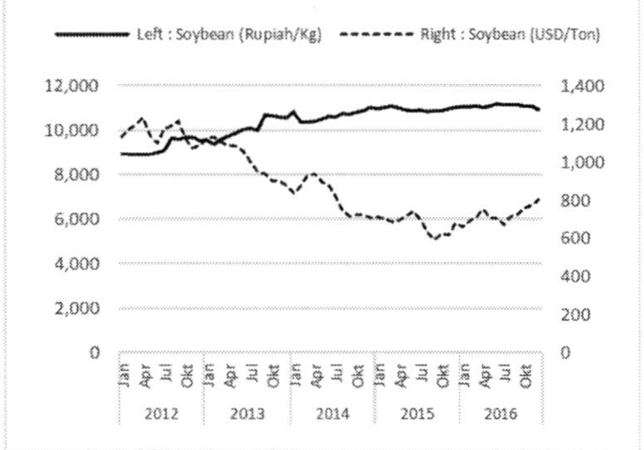

Fig. 5. Indonesia Soybean Retail Price Trend and Soybean Prices of the World Period January 2012-December 2016. Source: BPS and IFS CD-ROM

Fig. 5 shows that when BULOG is still a single importer, soybean prices tend to be more stable and unaffected by international soybean prices. However, after soybean imports were handed over to private importers, soybean prices tend to be more fluctuative and show price trends that are rising faster than before and tend to follow the world soybean price movements especially during world soybean price spikes.

This research is expected to provide benefits in the academic world as well as input for policy makers to develop more targeted policies in order to stabilize soybean prices. In this research, dummy variable of BULOG role is used to stabilize soybean price when becoming single importer and after not becoming a single importer. The stable definition meant in this research is BULOG able to intervene soybean price when price increase by increasing stock, for example through market operation. Thus, BULOG's role as a single importer of domestic soybean price is expected to be negative. Means that, at the time BULOG still be a single importer, namely in the period December 1993 - September 1998 then BULOG able to intervene in increased prices and lower it.

2 Literature Review

Fig. 6 shows that if there is no international trade, the equilibrium condition in Indonesian domestic market will occur at point A with the soybean price of PA and the amount of soybean demanded will be equal to the quantity of goods offered ie QA. Meanwhile, in foreign markets or referred to as 'rest of the world' (ROW), equilibrium conditions are at point B with the price of PB soybeans and the amount of soybean demanded will also be equal to the number of soybeans offered on QB. A country will import a commodity if its domestic price is above world price. If the condition is met then domestic economic actors have an incentive to buy imported goods (Pugel, 2004). If economic agents import soybeans, the total domestic soybean supply will rise and lower soybean prices in the domestic market resulting in excess demand. Suppose that domestic soybean price drops to PF, local production will decline to QSF1, the number of domestic demand rises to QDF1, and there are imports of QDF1 – QSF1. Meanwhile, in the ROW market, if the price is above the PB, say in PF, then the ROW producer will produce more that becomes QSF2, the ROW demand will decrease to QDF2, so there is an excess supply of QSF2-QDF2. Excess supply is what will be exported. Thus, international equilibrium will form at point E at the price level of PF and the quantity of goods demanded is equal to the quantity of goods offered by QF.

Fig. 6. Relation of Domestic and Domestic Soybean Prices. Source [16]

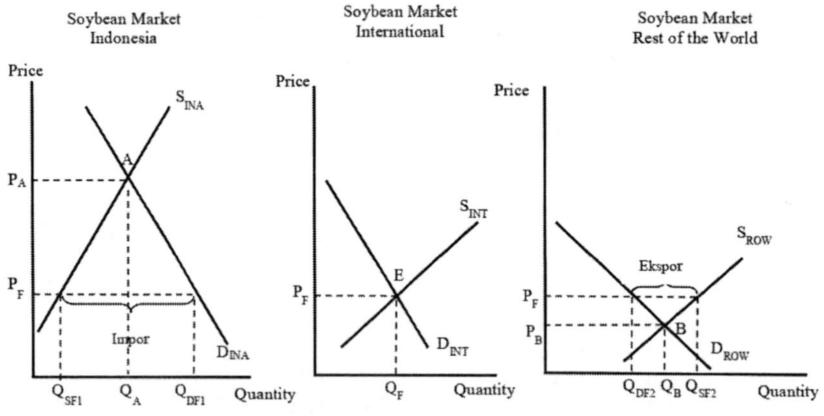

In order to control soybean prices in Indonesia, the government uses tariff instruments. Tariffs are taxes / duties / liabilities imposed on trade commodities that cross the borders of a country. Import tariff is the tax payable on imported commodities. There are three types of tariffs namely ad valorem, specific, and mixed. The ad valorem rate is the rate charged based on a certain fixed percentage of the value of the traded commodity. The specific tariff is the tariff charged at a fixed fixed value per unit of traded goods. A mixed or compound rate is the rate charged against a commodity by combining ad valorem and specific tariffs [16]. Tariffs on soybean commodity imports in Indonesia are subject to ad valorem method in accordance with the price stabilization policy set forth in the Minister of Finance Regulation (PMK).

https://doi.org/10.1051/matecconf/201815005035

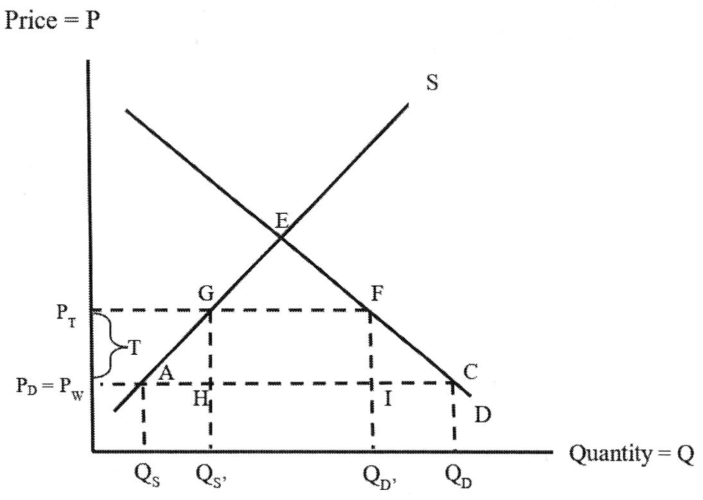

Fig. 7. Tariff Effect on Soybean Prices in Indonesia. Source: Pindyck & Rubinfeld (1996)

It is assumed that domestic soybean price is equal to world soybean price. If the government imposes a soybean import tariff of T% then domestic prices will rise to PT which is the sum of world soybean prices and tariffs, domestic production rises to QS', and domestic consumption decreases to QD'. Based on Fig. 7 it can also be seen that there is consumer loss of PTPWCF and producer surplus of PTPWAG, there are government revenue as much as GHIF, and deadweight loss of GAH and FIC. Based on the description above, we can conclude that there is a positive relationship between domestic prices and tariffs.

The impact of international price volatility on the spot market and the soybean futures market in China [24], which is divided into two periods before and after the global financial crisis to determine whether high levels of internationalization affect domestic price volatility in China. The method used is the Vector Error Correction Model (VECM).

The factors that affect the domestic price of cocoa beans in Ghana [2]. Based on the VECM estimation result, it is concluded that the domestic price of cocoa beans is influenced positively and significantly by government policy, domestic price of cocoa beans of the previous year and world price while consumer price index does not significantly affect the domestic price of cocoa beans.

Some researcher used Johansen Multivariate Cointegration and ECM approach with quarterly data from 1990-2006 to analyze factors influencing fluctuation of soybean price in Indonesia [20, 23]. He found that in the short term the domestic soybean price was positively affected by the domestic price of soybeans a quarter earlier, the nominal exchange rate of the previous quarter, and the GDP of Indonesia. While international soybean prices and import tariffs affect domestic prices positively but not significantly.

The determinants of domestic cocoa beans' domestic prices on the islands of Sumatra and Sulawesi using Engle Granger cointegration and ECM methods [23]. The estimation results show that: (i) in the long term the domestic price of fermented cocoa beans on Sumatra Island is significantly affected by the world price of cocoa beans and the nominal exchange rate. On the island of Sulawesi, the result is significantly influenced by world prices of cocoa beans and coffee prices; (ii) in the long term, the domestic price of non-fermented cocoa beans in

Sumatera Island is significantly influenced by the world price of cocoa beans. The price of coffee, the nominal exchange rate, and the export volume, while in the central Sulawesi Island is significantly affected by the world price of cocoa beans and Nominal exchange rate [13].

Table 4. Import Dependency Ratio (IDR) of Food Products

Independent Variables	Relationship
World Soybean Prices (HKD)	+
Soy Import Volume (VIK)	-
Nominal Exchange Rate (NER)	+
Gross Domestic Product (GDP)	+ / -
Import Rate of Soybean (TI)	+
DUMMY BULOG (DB)	-

Based on the above description and referring to research conducted by several researchers [7, 9, 11, 16], limited by data availability. The variables used in this research are world soybean price, import volume of Indonesian soybean, Nominal exchange rate, gross domestic product (GDP), import tariff of soybean, and dummy variable there is no influence of BULOG as a single importer. The hypothesis of the relationship of variables are presented in Table 3.

3 Methods of Econometrics and Data Sources

The general purpose of this study is to know how the influence of determinants on domestic soybean prices. To achieve the above objectives, the stages of research conducted are as follows. The models to be tested are as follows [7, 9, 11, 16] :

$$\mathrm{LHKI}_t = \propto_0 + \propto_1 LHKD_{t-1} + \propto_2 LNER_t + \propto_3 LVIK_t + \propto_4 LGDP_t + \propto_5 TI_t + \propto_6 DB_t + u_t \qquad (1)$$

Where LHKI_t: the logarithm of soybean price in Indonesia in the t-period, LHKD_t: the logarithm of world soybean price in the t-period, LER_t: nominal exchange rate logarithm in t-period, LVIK_t: logarithm of soybean import volume of Indonesia in the period t, LGDPt: logarithm of real GDP of Indonesia in period t, TI_t: variable of tariff of Indonesian soybean import in t-period, DB_t: dummy variable whether or no influence of BULOG as importer.

In research using time series data, often found data that is not stationary. If this unreformed data is reorganized, it is possible to produce false regression or spurious regression [6]. False regression can occur because the dependent variable and the independent variable used are not stationary and or the formation of the variable is not substantially related [12].

Some cointegration methods, such as those proposed by Engle-Granger and Johansen, require that the variables analyzed are integrated in the same order. In contrast to Pesaran, it allows the variables used to be integrated in I (0), I (1) or mixtures of I (0) and I (1) [15]. Since the initial identification of the stationarity level of the time series data used indicates that the possibility of non-stationary data at the same level, this research uses Bound Testing Cointegration method with Autoregressive Distributed Lag (ARDL) approach by Pesaran [15].

The ARDL contains the dynamic effects of a specific model by entering useful lag values to capture data degradation into the model[10]. With a simple transformation, the dynamics of Error Correction Model (ECM) can be derived from ARDL [3, 4, 14]. In addition, the coefficients of long-term estimation and ECM coefficients show the speed of adjustment of a variable in long-term equilibrium.

Before estimating the ARDL model, the optimum lag used should be set first. This lag selection criteria can use Schwarz Bayesian Criterion (SBC) or Akaike Information Criterion (AIC) with minimum value. Selection of the Best ARDL Model and Cointegration Test using Microfit 4.1 software. The ARDL estimation model is as follows:

$$\Delta LHKI_t = \propto_0 + \sum_{i=1}^{q} \propto_{1i}\Delta LHKI_{t-i} + \sum_{i=1}^{q} \propto_{2i}\Delta LHKD_{t-i} + \sum_{i=1}^{q} \propto_{3i}\Delta LNER_{t-i} + \sum_{i=1}^{q} \propto_{4i}\Delta LVIK_{t-i} + \sum_{i=1}^{q} \propto_{5i}\Delta LGDP_{t-i} + \sum_{i=1}^{q} \propto_{6i}\Delta TI_{t-i} + \propto_7 LHKI_{t-1} + \propto_8 LHKD_{t-1} + \propto_9 LNER_{t-1} + \propto_{10} LVIK_{t-1} + \propto_{11} LGDP_{t-1} + TI_{t-1} + \propto_{13} DB_t + u_t \quad (2)$$

The determination of the existence of cointegration is based on the F-count value compared to the critical value at a certain level of confidence in the bound table. The conclusion of cointegration is based on F-statistic bound table with the following criteria [15]:

- If the F-count is smaller than the lower bound (LB) then it can be concluded that there is no cointegration
- If the F-count is between the lower bound and the upper bound (UB) then the conclusion of cointegration can not be deduced.
- If F-count is above upper bound then it is concluded there is cointegration between variables [15].

To ensure the goodness of fit of the ARDL model a diagnostic test was performed to test for serial correlation, functional, normality, and heteroskedastic specification error in the model and stability test using cumulative sum of recursive (CUSUM) and cumulative sum of squares of recursive residuals (CUSUMQ). The ECM equation of the ARDL model for this study is as follows:

$$\Delta LHKI_t = \theta_0 + \sum_{i=1}^{q} \theta_{1i}\Delta LHKI_{t-i} + \sum_{i=0}^{q} \theta_{2i}\Delta LHKD_{t-i} + \sum_{i=0}^{q} \theta_{3i}\Delta LNER_{t-i} + \sum_{i=0}^{q} \theta_{4i}\Delta LVIK_{t-i} + \sum_{i=0}^{q} \theta_{5i}\Delta LGDP_{t-i} + \sum_{i=0}^{q} \theta_{6i}\Delta TI_{t-i} + \theta_7 DB_t + \theta_8 ECT_{t-1} + \mu_t \quad (3)$$

Where Δ is the first difference operator or the difference between two values of a variable in a consecutive time period, q is the optimum lag and μt is the term term equation 3

Table 5. Data and Data Sources

Variable	Description	Source
LHKI	Consumer prices obtained through surveys to soy retailers in markets across rural areas in Indonesia include DKI Jakarta in units of Rp / kg converted into natural logarithms.	BPS

Variable	Description	Source
LHKD	International soybean prices with HS HS 120100 in bushel / USD / million tons are converted into natural logarithms.	International Monetary Fund (*Retrieved from* IFS CD-ROM)
LNER	The nominal exchange rate in Rupiah / USD is converted into a natural logarithm.	Bank Indonesia
LVIK	Domestic Imported Soybean Volume in Rp / kg converted into natural logarithm.	Ministry of Trade
LGDP	GDP based on 2000 constant prices or quarterly real GDP in Billion Rupiah is interpolated to monthly series using Quadratic Match Sum method using E-VIEWS 7.0 software and converted into natural logarithm.	BPS
TI	Tariff on soybean import based on Regulation of the Minister of Finance in percentage	Ministry of Finance
DB	Dummy role variable of BULOG as a single importer. This BULOG dummy variable uses data 1 before September 1998 and 0 thereafter.	IMF

The data used in this study are secondary data derived from the publication of Central Bureau of Statistics (BPS), Bank Indonesia, Ministry of Finance, Ministry of Trade, and IFS CD-ROM. This study uses monthly time series data from 1993 to 2013. Selection of this time range is based on data availability and also to cover research objectives prior to the signing of Letter of Intent by Indonesia and IMF.

4 Empirical Results

The stationarity test is performed with the Phillips Peron (PP) Unit Root Test and the results are presented in Table 6. This test is performed to ensure that no variables are integrated in I (2). In the table, it can be seen that in the stationarity test at the level, all data is not stationary at the level of 95% confidence level except on the import data of soybeans. Then, the test results of the stationarity at the *first difference* level indicate that all data has been stationary.

Table 6. Stationary Test Results

Variable	Level			First Difference		
	PP Statistics		Conclusion	PP Statistics		Conclusion
	t-test	t-stat (α = 5%)		t-test	t-stat (α = 5%)	
LHKI	-0.12	-2.87	Not Stationary	-12.3	-2.87	Stationary
LHKPI	-1.26	-2.87	Not Stationary	-11.0	-2.87	Stationary
LHKD	-0.19	-2.87	Not Stationary	-11.2	-2.87	Stationary
LVIK	-9.18	-2.87	Stationary	-	-2.87	-
LER	-1.82	-2.87	Not Stationary	-13.6	-2.87	Stationary
LGDP	0.06	-2.87	Not Stationary	-8.09	-2.87	Stationary
TI	-2.2	-2.87	Not Stationary	-15.9	-2.87	Stationary

4.1 Estimation Results and Analysis of ARDL Model

Before estimating the ARDL model, the optimum lag used needs to be set first (Enders, 2007). The lag selection criteria use Schwarz Bayesian Criterion (SBC). In the cointegration

https://doi.org/10.1051/matecconf/201815005035

test using Bound Testing Cointegration obtained the F-statistic value of 11,833 while the upper and lower limit of the critical value at 95% confidence level respectively of 2.39 and 3.38. Since the F-statistic value is greater than the upper limit of the critical value, it can be concluded that there is a long-term relationship on soybean price variables of Indonesia, world soybean price, soybean import volume, exchange rate, GDP, and import tariff. Table 6 shows the selected ARDL model is ARDL(1, 0, 0, 0, 1, 0). The best model estimation results are presented as follows:

Table 7. Model ARDL

Variable	Coefficient	t-Statistic
LHKI(-1)	0.919***	43.902
LHKD	0.079***	3.257
LNER	0.089***	2.930
LVIK	-0.0006	-0.350
LGDP	-0.374*	-1.701
LGDP(-1)	0.352	1.553
TI	-0.001**	-2.167
DB	0.007	0.227
C	-0.281	-1.018
R-squared	0.997	
F-statistic	10771.59	
Prob(F-statistic)	0.000	

Information: ***) Significant on α = 1%, **) Significant on α = 5%, *) Significant on α = 10%

Diagnostic tests and stability tests are performed to avoid errors in interpretation and conclusions. The diagnostic test was performed by serial correlation and heteroschedasticity test. Table 8 below shows that the ARDL model in this study did not have serial and heteroscedastic correlation problems. The following test results are presented:

Table 8. Diagnostic Test Results

Test	P-Value	P-Value
Breusch-Godfrey Serial Correlation LM Test	3.043	0.049
Heteroschedasticity, Breusch-Pagan-Godfrey	2.217	0.027

Next is the stability test using CUSUM and CUSUMQ. The stability test below has shown that the model under test has been stable as long as the observation period. This can be seen from the CUSUM and CUSUMQ plots in *interval critical bound* 5%.

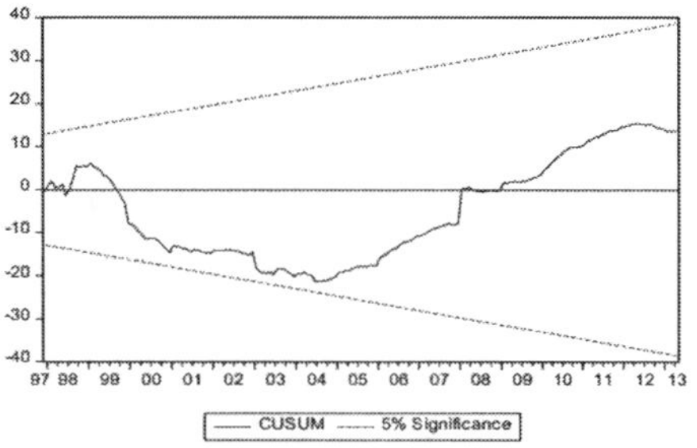

Stability test using Ramsey show the value F-statistic 12.691 and probability value 0.004, rejected hypothesis no stability. The ECM model contained in Table 9 is the result of the decline of the selected ARDL model. The expected ECM model is a model with significant (negative) and negative ECM coefficients (-1). The significant coefficient of ECM (-1) indicates that there is cointegration on the model and the negative sign ascertain whether there is a convergent correction mechanism against the deviation from the long-term balance. Based on ECM estimation results, it can be seen that there is negative and significant ECM coefficient (-1).

Table 9. ECM Estimated Results

Variable	Coefficient	t-Statistic
D(LHKD)	0.086**	2.139
D(LNER)	0.088***	2.897
D(LVIK)	-0.002	-1.267
D(LGDP)	-0.376**	-2.159
D(TI)	0.0001	0.061
D(DB)	-0.012	-0.330
CointEq(-1)	-0.081***	-9.053

The above results show that in the short term the world soybean price (LHKD), exchange rate (LNER) and GDP variables in the t-period affect the domestic soybean prices significantly with each coefficient of 0.086, 0.088. and -0.376 respectively. While the volume of imports (LVIK) in the t-period did not significantly affect the domestic soybean prices. However, import tariffs have a negative and significant influence that is inconsistent with the theory. Then, despite having a positive influence, but the role of BULOG as a single importer was not significant on the price of soybeans in the short term. The ECM variable coefficient (-1) shows the number -0.081. This negative sign indicates a convergent correction mechanism for deviations from the long-term equilibrium in the model.

4.2 Results and Analysis Long Term Model

The long-term coefficients are calculated from the short-run coefficients of the ARDL Model. The coefficients are presented in Table 10 below.

Table 10. Long Term Coefficient

Variable	Coefficient	t-Statistic
LHKD	0.985***	4.868
LNER	1.097***	3.332
LVIK	-0.008	-0.336
LGDP	-0.266	-0.660
TI	-0.016**	-2.356
DB	0.092	0.224
C	-3.468	-1.092

In the long term, the world soybean price and exchange rate variables have a positive and significant influence with each coefficient of 0.985 and 1.0297. However, soybean import volume and GDP variables did not have a significant effect. Just as in the short term, import tariffs have a negative and significant effect in the short term. This is not as expected. While in the long run, the role of BULOG as a single importer empirically does not have a significant effect on the decline in soybean prices.

The results confirm that world soybean prices have a positive and significant effect both in the short and long term. From the short-term results obtained if there is 1% change in world soybean prices will increase domestic soybean prices by 0.086%. Meanwhile, in the long term, if there is an increase in world soybean price of 1% it will cause the increase in domestic soybean prices by 0.985%. Thus, it can be said that there is a world soybean price transmission on Indonesian soybean prices. This discovery is in line with the results of research [23].

The results of the study must have been predictable because as previously described, more than 60% of Indonesia's soybean consumption is obtained through import. Thus, this dependence on imports causes domestic soybean prices to be severely affected by fluctuations in soybean prices abroad, especially US soybean prices. The increase in US soybean prices, for example, occurred in 2007-2008 caused by the decline in soybean production in the United States because farmers prefer to grow maize as raw material for biodiesel [18]. In addition, the increase in soybean prices in the United States also occurred in 2012 caused by bad weather and cause crop failure [22].

The nominal exchange rate has a positive and significant influence on the long term and the short term. Empirical results in the short-term show that if there is 1% depreciation of the rupiah against the US dollar then domestic soybean prices will rise by 0.088%. With soybean import policy, every fluctuation of rupiah exchange rate to US dollar, soybean price in domestic also will fluctuate.

Positive impacts are seen in empirical results that indicate incompatibility with the theory. However, long-term and short-term regression results indicate that the impact of import volume on soybean prices is not significant either in the short or long term. As mentioned earlier, soybean import is one of the government's policies in order to fulfill public consumption as well as one of the government's efforts to stabilize domestic soybean prices.

Increased real incomes will increase purchasing power so as to increase public demand for these goods and services which ultimately increase the prices of these goods and services. However, despite having a positive influence, the GDP variable has no significant effect on soybean prices.

Based on the results of this study, in the long term and short term, soybean import tariff has no effect. Meanwhile, in the long term, if there is a decrease of soybean import tariff by 1% it will decrease domestic soybean price by 0.016 %. This empirical study suitable to the theory. Based on this research, BULOG's role as a sole importer has positive and insignificant influence in long term and short term. This indicates that government intervention through BULOG does not have a significant impact in lowering soybean prices.

5 Conclusions and Policy Implication

Fulfilling national consumption of soybeans depends largely on soybean imports. Therefore, external turmoil has a greater effect on soybean prices in the country. The results of the analysis in this study, confirm the statement. By using Bound Testing Cointegration method with ARDL approach, it can be seen that in long term the world soybean price and exchange rate variable have positive and significant influence. Then, the volume variables of soybean import, GDP, and BULOG role as a single importer have no significant effect. However, import tariffs in the t-period have a negative and significant impact. Meanwhile, in the short term the world soybean price variables in the t-period and exchange rate affect the domestic soybean prices positively and significantly. Furthermore, the variable volume of soybean imports, GDP, and the role of BULOG as a sole importer in the t-period does not affect the domestic soybean price significantly. Just as in the long run, the t-period import tariff has a negative and significant effect. The results obtained on the relationship between domestic soybean prices and tariffs do not show conformity with the theory. However, if we look at the best ARDL model results, it can be seen that the import tariff of soybeans of a previous period has a positive and significant influence on domestic soybean prices. This indicates a lag in the implementation of the policy.

The ECM variable coefficient (-1) denotes the number -0.081. This negative sign indicates a convergent correction mechanism for deviations from the long-term equilibrium in the model. From the analysis results can be concluded that the intervention that has been done by the government in domestic soybean price control at this time, that is the import policy has not been effective in controlling domestic soybean price. While the soybean tariff, takes a period to affect the price of soybeans in the country. Then, the authority of BULOG has no significant effect. In other words, the role of BULOG as a single importer of soybean commodities has not been effective in intervening soybean prices. In the meantime, it is expected that there will be a quick response by the government on various information of soybean commodity as an effort to control soybean price and it is expected that the effectiveness of soybean market operation will be improved in the future. In addition, of course, the role of BULOG is certainly can be improved through the improvement of human resources capabilities.

References

1. B. Arifin, *Diagnosis Ekonomi Politik Pangan dan Pertanian*. Jakarta: Raja Grafindo Persada (2007).
2. Amendofu, M. *Determination of Cocoa Producer Price in Ghana*. Africa: Kwame Nkrumah University of Science and Technology (2009)
3. Banerjee, S, Recovery and Growth in Indonesia Industry: Element of A Future Policy Framework. UNSFIR Working Paper. Jakarta (2002)
4. Baulch, B. Testing for Food Market Integration Revisited. *The Journal of Development Studies, 33*(4), 512. (1997).
5. Enders, W., *Applied Econometric Time Series* (3rd ed.). New York: N.Y. John Wiley & Sons (2007).
6. Gujarati, D. N., & Porter, D. C.,*Basics Econometrics* (5th ed.). New York: The McGraw-Hill Companies, inc. (2009)
7. Hutabarat, B., Analisis Saling Pengaruh Harga Kopi Indonesia dan Dunia. *Jurnal Agro Ekonomi. , 24*(1), 21-40 (2006)
8. Jayasuriya S, Mudbhary P, Broca SS., *Food Price Spikes, Increasing Volatility and Global Economic Shocks: Coping with Challenges to Food Security in Asia: A Comparative Regional Study of the Experiences of Ten Asian Economies*. Bangkok: Food and Agricultural Organization (FAO) of the United Nations Regional Office for Asia and the Pacific (2012)
9. Joiya, S. A., & Shahzad, A. A, Determinants of High Food Prices – The Case of Pakistan. *Pakistan Economic and Social Review, 51*(1), 93-107 (2013).
10. Laurenceson, J., & Chai, J.,. *Financial Reform and Economic Development in China*. Inggris: Edward Elgar Publishing (2003)
11. Luo, F. Influence of External Shocks on Price Fluctuation of Agricultural Products of China: An Empirical Analysis Based on the VAR Model. (Q. Zhou, Ed.) *Applied Economics, Bussines, and Development*, 374-381 (2011).

12. Ekananda, M., Time Series Analysis for Research in Economy and Business, 2nd Edition, Publisher : Mitra Wacana Media, Jakarta (2016)
13. Ekananda, M., and Parlinggoman, D.J., The Role Of High-Tech Exports And Fdi On Economic Growth, European Research Studies Journal Volume XX, Issue 4A, (2017)
14. Nkang, N. M., Ndifon, H. M., & Odok, G. N., Price Transmission and Integration of Cocoa and Palm Oil Market in Cross River State, Nigeria: Implication of Rural Development. *Agriculture Journal, 2*(4), 457-463 (2007)
15. Pesaran, M. H., Shin, Y., & Smith, R. J., Bound Testing Approaches to the Analysis of Level Relationship. *Journal of Applied Econometrics, 16*, 289-326 (2001).
16. Pugel, T. A., *International Economics* (12th ed.). New York: The McGraw-Hill Companies (2004).
17. Sumaryanto. Retail Price Volatility of Some Food Commodities Using ARCH/GARCH Model. *Jurnal Agro Ekonomi, 27*(2), 135-163 (2009).
18. Supadi, Dampak Impor Kedelai Berkelanjutan terhadap Ketahanan Pangan. *Analisis Kebijakan Pertanian, 7*(1), 87-102 (2007).
19. Timmer CP., Managing Price Volatility: Approaches at the Global, National, and Household Levels. *Stanford Symposium Series on Global Food Policy and Food Security in the 21st Century*. California: Standford University (2011).
20. Suryanto, T., Implementation of Fair Value Accounting on Agency Problem Contract Mudharaba in Islamic Finance, International Journal of Economic Perspectives, 2015, Volume 9, Issue 4, 94-102 (2015)
21. Turner, P., & Dack, J. V., *Measuring International Price and Cost Competitivenes.* Basle: Bank for International Settlements (1993).
22. Westcott, Paul C., & Jewison, Michael, *Weather Effects on Expected Corn and Soybean Yields*. Florida: USDA (2013).
23. Widodo, S., Tesis: The Analysis of Long Term and Short Term Factors Affecting the Domestic Price of Soybean. Jakarta: *Universitas Indonesia* (2010).
24. Zhao, Y., Yang, M., Zhang, Y., & Qi, C. Impact on the Chinese Soybean Markets from International Prices Volatility: Empirical Study Based on VEC Model. *African Journal of Agriculture Research, 05*(15), 1943-1950 (2010)

The Roles of Product Attributes towards Generation-Y Shopping Interest

Abd Majid Nurul Fazura[1], *Musa* Haslinda[1], *Hasan* Hazmilah[2], *Othman* Norfaridatul Akmaliah[1], *Azmi* Fadhlur Rahim[1]

[1]Faculty of Technology Management and Technopreneurship, Universiti Teknikal Malaysia Melaka, 76100 Durian Tunggal Melaka, Malaysia

[2]Centre of Technopreneurship Development (C-TeD), Universiti Teknikal Malaysia Melaka, 76100 Durian Tunggal Melaka, Malaysia

Abstract. Generation Y is the second largest generational cohort growing up in a world infused with brands that has been socialized in a materialistic society with purchasing power and extraordinary impact on the economy. They often involve with the shopping behavior and purchase involvement that makes their preferences towards one product would most likely depending on the attributes of the product itself. Thus, this study has developed a framework that explored the generation-Y shopping interest towards product attributes. This study applied four (4) concepts; which are the functions of the product, features of the product, benefits of the product, and generation-Y shopping interest. In this study, the researcher used empirical study by employing the questionnaire survey to investigate its managerial implications to 150 respondents of generation-Y. The data collected is analyzed by using Statistical Package for Social Science (SPSS) that showed all features of the product, functions of the product, and benefits of the product are positively affected the generation-Y shopping interest.

1 Introduction

Generation Y has to be those who born between 1978 and 1994 that born straightaway after Generation X, and it is referred to the dot.com generation, eco boomers and millionaires [1, 2]. Generation Y often involve with the shopping behaviour and purchase involvement for books, apparel, and electronics such as clothing, computers, CDs, and electronics [3]. It has been socialized that the purchasing power of Generation Y has an extraordinary impact on the economy and this generational cohort is highly active in the marketplace [4] which is also held true in the Malaysian context and makes up the largest segment of Malaysia's population [1, 5]. Generational cohorts have different values, preferences and shopping behaviours but they have emerged as a big force with growing spending power which will unsurprisingly competing to other cohorts [5–8]. Furthermore, Generation Y which is so called Millennial, is a specific cohort who usually have similar characteristics and features that can be found in many fields of interest, e.g. lifestyle, choice of education or job and also in shopping behaviour [4]. They see themselves as special and enjoy activities such as listening to music, watching television, going to the movies, and eating out.

However, their preference towards one product is not fixing depending on the attributes of the product itself. For example, if there is slight of the Generation Y does not like to read and comprehend dull textual information through books because the features of the text is not interesting ,they will go for the electronic platforms as they found that the functions of the electronic gadgets is more stimulating and dynamic [10]. This assumption can be completely applied especially for one group of Generation Y which consists of mostly visual learners. Holistically, the product is evaluated for its fit with the desired benefits that not relied on a feature-by-feature comparison of alternative brands, but by the consumer choice that make consumers selectively attend to product features based upon the benefits sought [11].

2 Methodologies

For this research, the researcher used quantitative data analysis to examine relationships between variables which will be measured numerically and analysed using a range of statistical techniques. On this research, the data is measured on a numerical that has been specified by using statistical techniques and the result can be shown using charts, histograms, tables and graphs [12].

The researcher used primary data in this research. The primary data for this research is collected by using questionnaires. One set of questions is designed and distributed to 150 respondents and focus on generation Y in Melaka area. The questionnaires are based on this research topic. The questionnaire has been constructed carefully so that the information collected is valid. Secondary data sources such as books and journals are the subsequent publication of primary data sources. These publications are aimed at a wider audience. They

* Corresponding author: haslindamusa@utem.edu.my

are easier to locate than primary data sources as they are better covered by the tertiary data sources.

3 Results and Discussion

3.1.Respondents' Demographic Profile

The results obtained from 150 respondents are presented in Table 1. It shows summary of the respondent's background and characteristics. From 150 respondents, 34.7% are male, and 65.3% are female. In terms of percentage of qualification, 56% of the respondents own bachelor degree, followed by 30% of Diploma and 9.3% own Master & above. Lastly, 4.70% own SPM. Respondent within the age of 22-25 years old with value 49% as a majority of the respondents which is 74 people over 150 respondents in this result. This is followed by 28.0 % of respondents (42 people) among 26-30 years old. Furthermore, 18.0% of respondents (27 people) are 31-34 years old. Last but not least, only 4.7 % of respondents (7 people) in range of age 35-39 years old. Percentage of races between respondents who participated in this survey. 83% of the respondents are Malay, 10% are Chinese, and 7% are Indian. Malays has dominated this survey questionnaire for the research purpose. Employment status shows the highest percentage is students that are 49.30% which represent 74 numbers of respondents. The second highest of respondents' employment status is 21.30% which represents 32 number of respondents from private sector. This followed by 15.30% from government sector that indicates 32 respondents. The lowest is 14.00% of respondents who are self-employed which represents 21 numbers of respondents. The researcher can induce that most of the participants of the survey are students followed by private sector, government sector and self-employed. Based on the table 1, it shows the monthly income or allowance of the 150 respondents where 47.30% of them have the income that less than RM1000. The second is 27.30% for the income range of RM1001-RM2000 with 41 respondents and 26 respondents that is 17.30% from 150 respondents have the income of RM2001-RM3000. As most of the respondents are students, then most of the respondents' monthly income or allowance is below than RM1000.

Table 1. Respondents' demographic profile

	Frequency	**Percent**
Gender	**150**	**100.0**
Male	52	34.7
Female	98	65.3
Age	**150**	**100.0**
22-25	74	49.3
26-30	42	28.0
31-34	27	18.0
35-39	7	4.7
Qualifications	**150**	**100.0**
SPM	7	4.7
Diploma	45	30.0
Degree	84	56.0
Master and others	14	9.3
Race	**150**	**100.0**
Malay	125	83.3
Chinese	15	10.0
Indian	10	6.7
Occupations	**120**	**100.0**
Government sectors	23	15.3
Private sectors	32	21.3
Self-employed	21	14.0
Student	74	49.3
Monthly Income or Allowance (RM)	**150**	**100.0**
Less than 1000	71	47.3
1001-2000	41	27.3
2001-3000	26	17.3
3001 and above	12	8.0

3.2.Pearson's Correlation Analysis

Table 2. Pearson Correlation between the variable

	Gen-Y Shopping Interest	Function of the Product	Features of the Product	Benefits of the Product
Gen-Y Shopping Interest	1			

Function of the Product	.622 **	1		
Features of the Product	.476 **	.851 **	1	
Benefits of the Product	.605 **	.870 **	.848 **	1

**.Correlation is significant at the 0.01 level (2-tailed

Table 2 shows correlation between independent variables (functions of the product, features of the product, and benefits of the product) with dependent variable (gen-y shopping interest). There is moderate relationship between functions of the product and gen-y shopping interest. The value of the coefficient correlation is 0.622, falls under the coefficient range of "±0.41 to ±0.70".There is moderate relationship between features of the products and gen-y shopping interest. The value of the coefficient correlation is 0.476, falls under the coefficient range of "± 0.41 to ± 0.70". Last but not least, there is moderate relationship between benefits of the products and gen-y shopping interest. The value of the coefficient correlation is 0.605, falls under the coefficient range of "±0.41 to ± 0.70".

Table 3 shows Multiple Regression Analysis (MRA), R is 0.658 which is more than 0.5 that indicates positive relationship. This means that the respondent is good on shopping and product attributes. Furthermore, the positive sign of R implies a positive relationship. Coefficient of determination, R^2, is the proportion of variance in one variable associated with the variability in a second variable. From the result, it shows that 43.30% of the variables affected the gen-y shopping interest and that means 56.70% indirect factor other than variable that has been used for this study influenced the gen-y shopping interest.

Table 3. Multi Regression Analysis

Model	R	R^2	Adjusted R^2	Std. Error of the Estimate	Sig.
1	0.658	0.433	0.421	0.596	0.00

Based on Table 3, it shows the percentage of relationship between variables and the significant of each of variables that affect gen-y shopping interest. The first factor, functions of the product is significant to the gen-y shopping interest as its value amounted 0.000 and it is below than 0.005. The second factor which is features of the product is significant with the value of 0.007 and below than 0.05. Lastly, the last factor which is benefits of the product is also significant to the gen-y shopping interest as its value is 0.003 which is below than 0.05. The relationship can be denoted as the following equation from the analysis from the Table 4 below:

Y Gen-Y Shopping Interest= 1.272 + 0.575 Functions of the Product - 0.379 Features of the Product + 0.430 Benefits of the Product

According to Table 4, functions of the product have positive influence on gen-y shopping interest (t= 4.002, p= 0.000, β= -0.575). When other variables are held constant, every one unit increase in functions of the product will lead to increase in gen-y shopping interest by 0.575 units. From the analysis in table 4, it shows that features of the product have negative influence on the gen-y shopping interest (t= -2.737, p= 0.007, β= 0.379). It can be explained by every one unit increase in features of the product will lead to 0.379 units decreased in gen-y shopping interest. Table 4.19 also shows that benefits of the product was found to exert a significant positive influence on gen-y shopping interest (t= 3.017, p= 0.003, β= 0.430). It can be explained by every one unit increase in benefits of the product will lead to 0.430 units increased in gen-y shopping interest. It can be concluded the functions of the product is the factor that contributes the most to the gen-y shopping interest. The value of functions of the product that influences gen-y shopping interest is 0.575 which is 57.50% and it is the highest value compared to the other two variables.

Table 4. Multi Regression Analysis

Model	Unstandardized Coefficients			t	Sig.
	B	Std. Error	Beta		
Functions of the Product	0.575	0.144	0.561	4.002	0.000
Features of the Product	-0.379	0.139	-0.357	-2.737	0.007
Benefits of the Product	0.430	0.142	0.420	3.017	0.003
Constant	**1.272**	**0.281**		**4.527**	**0.00**

4 Conclusion

Research objectives are achieved and proved that the product attributes; features, functions, and benefits of the product are factors that significantly enhanced generation Y shopping interest.

References

1. M. Muda, R. Mohd, and S. Hassan, "Online Purchase Behavior of Generation Y in Malaysia," *Procedia Econ. Financ.*, vol. 37, pp. 292–298, 2016.
2. P. Kotler and G. Armstrong, *Principles of marketing*. Pearson education, 2010.
3. T. R. Kinley, B. M. Josiam, and F. Lockett, "Shopping behavior and the involvement construct," *J. Fash. Mark. Manag. An Int. J.*, vol. 14, no. 4, pp. 562–575, Sep. 2010.
4. P. Krbová, "Generation Y attitudes towards shopping: a comparison of the Czech Republic and Slovakia," *J. Compet.*, vol. 8, no. 1, 2016.
5. H. Musa, M. S. M. Taib, S. C. H. Li, J. Jabar, and F. A. Khalid, "Drop-sshipping supply chain: the characteristcs of SMEs towards adopting it," *Soc. Sci.*, vol. 11, no. 11, pp. 2856–2863, 2016.
6. H. Musa, S. C. H. Li, Z. A. Abas, and N. Mohamad, "Adoption Factor of Mobile Marketing: The Case of Small Medium Enterprises (SMEs) in Malaysia," *Int. Rev. Manag. Mark.*, vol. 6, no. 7S, pp. 112–115, 2016.
7. H. Musa, N. A. Rahim, F. R. Azmi, A. S. Shibghatullah, and N. A. Othman, "Social Media Marketing and Online Small and Medium Enterprises Performance: Perspective of Malaysian Small and Medium Enterprises," *Int. Rev. Manag. Mark.*, vol. 6, no. 7S, pp. 1–5, 2016.
8. T. J. Fogarty, A. Reinstein, and R. S. Heath, "Are Today's Young Accountants Different? An Intergenerational Comparison of Three Psychological Attributes," *Account. Horizons*, vol. 31, no. 2, pp. 83–104, Jun. 2017.
9. A. Bilgihan, "Gen Y customer loyalty in online shopping: An integrated model of trust, user experience and branding," *Comput. Human Behav.*, vol. 61, pp. 103–113, Aug. 2016.
10. H. Musa, "Analyzing factors towards adopting ICT within supply network in the UK," *J. Technol. Manag. Technopreneursh.*, vol. 2, no. 1, pp. 19–35, 2014.
11. A. Ponnam and J. Dawra, "Discerning product benefits through visual thematic analysis," *J. Prod. Brand Manag.*, vol. 22, no. 1, pp. 30–39, Feb. 2013.
12. M. Saunders, *Research methods for business students, 5/e*, 5th ed. Pearson Education India, 2011.

Overall Equipment Efficiency (OEE) Enhancement in Manufacture of Electronic Components & Boards Industry through Total Productive Maintenance Practices

Soo-Fen Fam[1,*], *Ser Lee* Loh[2], *M. Haslinda[1]*, Heri *Yanto[3]*, *Linda Mei Sui Khoo[4]*, *and Diana Hwa Yieng Yong[1]*

[1]Faculty of Technology Management and Technopreneurship, Universiti Teknikal Malaysia Melaka, 76100 Melaka, Malaysia
[2]Faculty of Electrical Engineering, Universiti Teknikal Malaysia Melaka, 76100 Melaka, Malaysia
[3]Fakultas Ekonomi, Universitas Negeri Semarng, 50229, Indonesia.
[4]Centre for Languages and Human Development, Universiti Teknikal Malaysia Melaka, 76100, Melaka, Malaysia

Abstract. In an environment of intense global competition, both creative and proven strategies need to be considered in order to bring about the effectiveness and efficiency in manufacturing operation. Total Productive Maintenance (TPM) is one of the effective maintenance strategy in enhancing the equipment effectiveness and to achieve a significant competitive advantage. This research paper addresses the impact of three TPM pillars namely planned maintenance (PM), autonomous maintenance (AM) and focused maintenance (FM) on overall equipment effectiveness (OEE) of die attach equipment in the production line of semiconductor industry. The effect of TPM on the OEE is also investigated depending on the equipment types, in where die attach process consist of two models- CANON and ESEC. The primary data was collected from an organization's database and was analysed by SPSS V23. The preliminary results of the analysis showed that the performance of OEE in ESEC is better than the CANON after the implementation of TPM. The analysis also showed that out of the three TPM practices deployed, planned maintenance of equipment by production and maintenance team played the biggest role in increasing the equipment effectiveness. In conclusion, this study provides insights the importance of implementing TPM in order to succeed in a highly demanding market arena.

1 Introduction

Malaysia government had launched the Economic Transformation Programme (EPT) in order to sustain and elevate country to developed-nation status by 2020. In various sectors in EPT, the Electrical and Electronics (E&E) NKEA sector is one of the key contributor to offset the hit taken from low oil price and the growth of Malaysia economy [1]. Hence, measure has been implemented by the E&E NKEA to enhance the capability and capacity of electronic manufacturing companies. In today's rapidly change environment, more and more companies are replacing their reactive, fire-fighting strategies for maintenance to proactive strategies, the TPM is an aggressive and promising strategy for improving maintenance performance in order to succeed in a highly demanding market arena.

In order to sustain and stay competitive, Malaysian semiconductor companies is urged to upgrade their facility and embrace automation and smart manufacturing. However, recent trends indicate that most of the system in use are not performing as intended, so far as cost effectiveness in term of their operation and support is concerned, in manufacturing system, some of the equipment often not fully utilized, with low productivity and thus the costs of producing products are high, reports show that 15% to 40% of total production cost is attributed to maintenance activities in the factory. Another study of Wireman [2] mentioned that the cost estimation for maintenance in selected companies increased from $200 billion in 1979 to $600 billion in 1989. Each company has their own focus on TPM implementation or the focus part of maintenance department. This study on semiconductor companies also led to interest on determine the effect of TPM on the OEE, in which OEE recognized as a metric to determine the effectiveness of TPM. It is also crucial to determine the effectiveness of TPM to improve OEE.

With the increased of global competition, the semiconductor companies is forcing to implement world class maintenance techniques will improve equipment utilization and thus reducing capital expenditure. This research is to understand the different elements of TPM that influence the effectiveness and efficiency in a manufacturing company.

2 Literature review

To begin with summary of the empirical view of the relationship between TPM and OEE, an understanding of basic conception of maintenance function is a need to be developed. Maintenance is the execution of activities that ensure the physical assets continue to do what their users

*Corresponding author: famsoofen@utem.edu.my

want them to do. The goal of any well run maintenance organization is to have the lowest cost of the sum of maintenance program, it is responsible for controlling the cost of manpower, materials, tools and overhead. According to Cross [3], the maintenance cost in the UK manufacturing industry is around 12-23% of the total factory operating cost. In last three decades, maintenance function has undergone serious change through many phases. The implementation of TPM is important in a manufacturing company and has the significant impact and relationship on manufacturing performance, in term OEE [4]. Hence, the main goal of TPM is to maximize equipment effectiveness and OEE is used as a measure.

TPM is an innovative strategy that developed at Nippondenso, Japan in 1971. Ahuja et al. [5] stated the TPM contains of three words, start with the word TOTAL which is emphasized to involve overall workforces from top to bottom in every aspect of equipment maintenance. Next word is PRODUCTIVE which signifies to ensure the operation is efficient maintenance problems/costs. While the last word MAINTENANCE is to ensure the equipment is keeping in a good condition.

Basically, the practices of TPM often known as the pillars or elements of TPM. Through its unique eight-pillar methodology, TPM is used for excellent planning, organizing, monitoring and controlling practices [4]. Pillar 1- autonomous maintenance: It is based on the concept that the equipment operators accept and share responsibility with the maintenance for the performance and health of the equipment. Pillar 2- Focused Maintenance: This concept is aim for small improvement by carried out continuously and involve entire workforce. Pillar 3- Planned maintenance: Is a system where the maintenance jobs and equipment stoppages are scheduled based on predicted or measured failure rates. Pillar 4- Quality Maintenance: Focus on achieving the zero defects and address the equipment problem and root causes. Pillar 5- Education and Training: Provide the knowledge of what daily maintenance is require to the operators to ensure the optimal operating condition of equipment. Pillar 6- Safety, Health and environment: Is used to create a safe workplace and an environment that will not be damaged by the process and procedure. Pillar 7- Office TPM: Focus on the improvement of the productivity and efficiency of administrative functions. Pillar 8- Development Management: Encourage the continuously developing ideas and procedure that can lead to the creating of maintenance improvement initiatives.

While OEE methodology incorporates metrics from all equipment manufacturing states guidelines in a measurement system that helps the company to enhance the equipment performance and thus, reduce the equipment cost of ownership [4]. The OEE concept is becoming well-known and his widely used as a quantitative tool that important for productivity measurement especially in semiconductor manufacture operations. Firstly, OEE used as a "benchmark" by comparing the initial OEE measures with future OEE values, thus quantifying the level of improvement require. This led to the improvement of maintenance policy and affect continuous improvement in the manufacturing system. Secondly, an OEE values of one manufacturing line can use to compare with the other line performance across the operation, which can help to identify the poor line performance and improvement needed [6]. Thirdly, OEE measurement is an effective way to analyzing and identifying the efficiency of a single machine, and then help in indicating where to focus for TPM resources. OEE tool is designed to identify the disturbances to the manufacturing process as the disturbances often lead to serious problem that can reduce the equipment effectiveness. The losses are measure by OEE, in where it is calculated by obtaining the product of availability of the equipment, performance efficiency of the process and rate of quality products. Equation 1 shows the calculation of OEE

$$\text{OEE} = \text{Availability (A)} \times \text{Performance Efficiency (P)} \times \text{Rate of Quality (Q)} \quad (1)$$

3 Research methodology

This study used primary data. Real time data was recorded by software to provide valuable equipment data for improvement, which known as Global Operator User Interface (GOUI) system. It is a middleware system between Equipment Tracking Interface (ETI) and different company's system. The main purpose of GOUI is to control equipment, process monitoring, manage user, verify material, data logging, efficiency losses measurement and so on without refer back to the machine vendor. Besides, SECS/GEM standard is also use as the interface protocol for equipment to communicate with the host and vice versa. It defines messages, equipment state and scenarios to enable factory software to control and monitor production equipment. With these software, the system can be measure and evaluate in all areas of production and offers users a comprehensive overview of the efficiency of production equipment at any time. Besides that, the main applications for the semiconductor industry – electronic manufacturing and micro-systems technology standard (SEMI E10 and SEMI E58 standards) is also applied. It is an ideal measurement in where it can assist manufacturing to monitor the machines and automatically determine availability indicators (SEMI E10, 1992) [7].

The data was collected and monitored weekly and from two different model of machines in die attach process- CANON and ESEC. Since Canon was a new machine model that started to run production on September 2016, hence the data started collected from the September 2016 and end on March 2017 due to time limitation of this study.

Figure 1 shows the TPM's variables used in this study are maintenance (FM), autonomous maintenance (AM), and planned maintenance (PM).

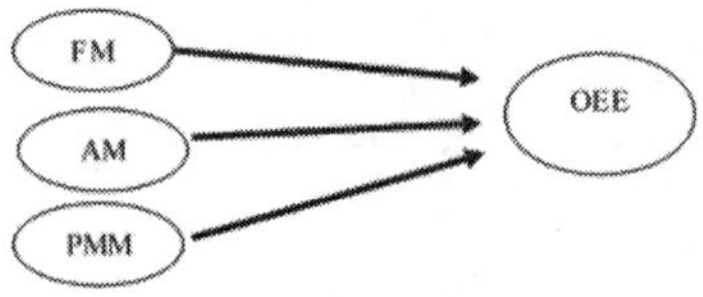

Fig. 1. Theoretical Framework

4 Result3s and discussion

In this part, the data for both model types of equipment is combined and two sample dependent or pair mean analysis was performed to determine the effect of the implementation of TPM pillars and OEE.

The independent t-test was used to compare the means between two unrelated group (CANON and ESEC) on the same continuous, dependent variables (OEE). The hypothesis for this study can be express as:

H0: μCANON = μESEC (Means of CANON and ESEC are equal.)

H1: μCANON ≠ μESEC (Means of CANON and ESEC are not equal.)

In the sample data, two variables had been used: Model and OEE. The Model variable consist of CANON and ESEC, which function as the independent variable in this T test, while the OEE is function as dependent variables and is a numeric duration variable (%). The result of analysis are summarized in Table 1 and 2.

Table 1. Group statistic of independent T-test

Model OEE	N	Mean	Std. Deviation	Std. Error Mean
Canon	31	72.49	8.42	1.51
ESEC	31	81.92	3.52	0.63

The first section, Group Statistics, provides basic information about the group comparisons, including the sample size (n), mean, standard deviation, and standard error for OEE by group. In this study, 31 weeks data had been collected for CANON and ESEC, respectively. The mean of OEE for CANON is 72%, and the mean OEE for ESEC is 82%.

Table 2. Independent T-test output

	Levene's Test for Equality of Variances		t-test for Equality of Means						
	F	Sig.	t	df	Sig. (2 tailed)	Mean Difference	Std. Error Difference	95% Confidence Interval of the Difference	
								Lower	Upper
Equal variances assumed	9.648	0.003	-5.97	60	.000	-9.337	1.639	-12.62	-6.06
Equal variances not assumed			-5.69	40.202	.000	-9.337	1.639	-12.65	-6.02

Second section, Independent Samples Test, displays the results most relevant to the Independent Samples t Test. There are two parts that provide different pieces of information: Levene's Test for Equality of Variances and t-test for Equality of Means. This study showed that the P= .003 is less than the significance level of 0.05. Hence, the null of Levena's test is rejected and concluded the variance in OEE of CANON is significant different than the ESEC. This also mean that the data need to be referred was the "Equal variance not assumed" row for the t-test results. The negative sign of T-values indicated that the mean of OEE for the first group, CANON is significantly lower than the mean of OEE for the second group, ESEC. Based on the result, below statements can be concluded:-

There was a significant difference in mean of OEE between CANON and ESEC

T (60) = -5.695, P=.003, which is less than the established significant level of 0.05.

Following is the results for Multiple regression analysis for the pair model as shown in Table 3.

Table 3. ANOVA of Multi Regression Analysis for CANON and ESEC.

ANOVA[a]

Model 1	Sum of Squares	df	Mean Square	F	Sig.
Regression	3434.604	3	1144.88	159.525	0.000[b]
Residual	416.250	58	7.177		
Total	3850.854	61			

a. Dependent Variables: OEE
b. Predictors: (Constant), FM, PM, AM

F- Ratio in the ANOVA Table 3 is used to determine the overall regression model is a good fit for the data. The table showed that independent variables statistically significantly predict the dependent variable, $F(3, 58) = 159.525$, $p < .05$. It indicated that the regression model is a good fit of the data.

Coefficients is a statistical measure of the degree to which changes to the value of one variable predict change to the value of another [8]. Table 4 shows the impact of independent variables- PM, AM and FM on the dependent variables of OEE.. The results yielded from the multiple regression showed that the significance level of planned maintenance and autonomous are less than the established significant level of 0.05, so they had significant impact on OEE, in where planned maintenance: t(61) = 14.398, P= .000 and autonomous maintenance: t(61) = 6.539, P= .000. For the focused maintenance, there is no significant impact on OEE with t (61) = .538, P= .593, in where the P value >.05.

Table 4. Coefficients of Multi Regression analysis for CANON and ESEC

Model 1	Unstandardized Coefficients		Standardized Coefficients	t	Sig.	95% Confidence Interval for B		Collinearity Statistics	
	B	Std. Error	Beta			Lower Bound	Upper Bound	Tolerance	VIF
(Constant)	11.586	12.865		.90	.372	-14.167	37.340		
PM	.501	.035	.775	14.39	.000	.431	.571	.643	1.555
AM	.218	.033	.429	6.54	.000	.151	.284	.433	2.309
FM	.076	.142	.037	.538	.593	.218	.361	.393	2.548

Multicollinearity in regression is viewed as disadvantage, because it practically inflates unnecessarily the standard errors of confidents in regression. The tolerance and VIF is used to determine the multicollinearity. There are various recommendation for the acceptable of VIF have been published in the literature. Perhaps more commonly, a value of 10 has been recommended as the maximum level of VIF [9-10]. However, a recommended maximum VIF value of 5 [11] and even 4 [12] can be found in the literature. In this study, the VIF of three independent variables (PM: T= .643, VIF= 1.555; AM: T= .433, VIF= 2.309; FM: T=.393, VIF= 2.548) are less than the established value of 10, so this indicated that there were no inter-correlation between the predictor of interest.

The general form of the equation to predict the relationship between the OEE and TPM variables (PM, AM and FM) was showed as below:

$$Y = 11.586 + 0.501\ X1 + 0.218\ X2$$

Where Y was the overall equipment effectiveness (OEE) and X1, X2 are the TPM practices of planned maintenance and autonomous maintenance respectively. Both the Y and X are measured in percentage, %.

The regression equation established by taking all factors into account constant at zero, the OEE rate will be at 11.586. It also indicate that for every unit increase in planned maintenance practices would actually result in a .501 increase of OEE, while every unit increase in the implementation of autonomous maintenance would also result in .218 increase of OEE.

5 Conclusion

In the nutshell, this study show that there is a significant and positive relationship between the implementation of total productive maintenance (TPM) practices at the semiconductor company from the September 2016 to March 2017 and the improvement in the overall equipment effectiveness (OEE) rates that was witnessed in the organization during that time period. The combined effect of the implementation of the three TPM practices of planned maintenance, autonomous maintenance and focused maintenance was an increase in all of the indicators of equipment effectiveness.

The finding of the study is supported by Paropate and Sambhe [13] who also performed a study on the implementation of TPM in a midsized cotton spinning plant in India. Their study proved that with established of TPM, the overall equipment effectiveness improved from 68.9866% respectively before TPM implementation to 71.465% respectively after TPM implementation.

The study of Sharma et al.[14] also support the result, in where the study is proved that TPM leads to increase in efficiency and effectiveness of manufacturing system in term of OEE index. In addition, the finding of Ateka [15] also highlighted that the increased quality and improved productivity as the key benefits resulting from TPM implementation. The study was examined the adoption of TPM practices in large manufacturing firms located in Mombasa County and also stated that the most important critical success factor of TPM is co-operation and involvement of both the operators and the maintenance workers in equipment maintenance.

This research is a cross-sectional study as the study will be carried out on a particular phenomenon and population at a particular time. Due to the time limitation and complexity of the data, the study could not be able to carry on for whole production line in the company. Therefore, the study was only focused on the equipment models of bottleneck process- die attach. Furthermore, the study used primary data in the company. Due to the sensitivity of the information, and the obligations placed on the researcher and the custodians of the information, the study was not able to establish more data points, and to obtain data for a longer time period, thereby limiting the scope of the period under the study.

The authors are grateful to Centre of Research and Innovation Management (CRIM), Universiti Teknikal Malaysia Melaka (UTeM) for the financial and technical support for this research

References

1. Economic Transformation Programme, Annual Report (2015). Available at https://www.pemandu.gov.my/assets/publications/annual-reports/ENG_PEMANDU_NTP_AR2015260416.pdf. Accessed on the 07th of September, 2017
2. T. Wireman, 1990. World Class Maintenance Management. Industrial Press, New York.
3. J. Cross, (1988), "Raising the value of maintenance in the corporate environment", Management Research News, Vol. 11 No. 3, pp. 8-11.
4. I.P.S., Ahuja, J.S. Khamba, and R. Choudhary, (2008), "Total productive maintenance: literature review and directions", International Journal of Quality & Reliability Management, Vol. 25 Iss 7 pp. 709 – 756
5. I.P.S., Ahuja, J.S. Khamba, and R. Choudhary, (2006), "Improved organizational behavior through strategic total productive maintenance implementation", Paper No. IMECE2006-15783,

2006 ASME International Mechanical Engineering Congress and Exposition (IMECE), Chicago, IL, November 5-10, pp. 1-8.
6. S.F. Fam, B.P. Lau, N. Ismail, 2016. Lean Manufacturing and Overall Equipment Efficiency in Paper Manufacturing and Paper Products Industry. International Conference on Design and Concurrent Engineering, IDECON 2016, ID146. Langkawi, Malaysia.
7. SEMI E10, "Semiconductor Equipment and Materials International (1992), SEMI Standards: Equipment Automation/Hardware Volume, Semiconductor Equipment and Materials International, Mountain View,CA," 1992.
8. R. Margaret (2013) Correlation Coefficient http://whatis.techtarget.com/definition/correlation-coefficient. Accessed on the 07th September, 2017.
9. P. Kennedy, (1992), "A Guide to Econometrics", Oxford: Blackwell.
10. D. W. Marquardt, (1970), "Generalized inverses, ridge regression, biased linear estimation, and nonlinear estimation", Technometrics, 12, 591–256.
11. P. A. Rogerson, (2001), "Statistical methods for geography", London: Sage.
12. Y. Pan and R.T. Jackson (2008). Ethnic difference in the relationship between acute inflammation and serum ferritin in US adult males. Epidemiology and Infection. 136. Pp. 421-431.
13. R. V. Paropate, & R. U. Sambhe, (2013). The Implementation and Evaluation of Total Productive Maintenance – A Case Study of mid-sized Indian Enterprise. International Journal of Application or Innovation in Engineering & Management (IJAIEM), 2(10). ISSN 2319 – 4847.
14. R. K. Sharma, D. Kumar, & ,P. Kumar (2006). Manufacturing excellence through TPM implementation: a practical analysis. Industrial Management & Data Systems, 106(2), 256-280.
15. T. N. Ateka, (2013). Adoption of total productive maintenance practices among large manufacturing firms in Mombasa County, Kenya (Unpublished MBA Thesis). University of Nairobi, Kenya

Implementation of Green New Product Development Among SMEs: Barriers and Critical Success Factors

Noor Hidayah Abu[1*], *Khoo* Kean Huat[1], and *M. F.* Mansor[3]

[1,] School of Technology Management and Logistics, University Utara Malaysia, Sintok, 06010 Universiti Utara Malaysia, Kedah
[3] School of Business Innovation & Technopreneurship, Universiti Malaysia Perlis, 01000 Kangar Perlis, Malaysia.

Abstract. Green New Product Development (GNPD) is very important due to environmental issues increasing. GNPD is defined as creating, designing and develop ecological product which could help in protecting the environment. Small and medium enterprise (SME) firms is one of the contributor to the environmental issues but they could not afford to implement GNPD because there are several obstacles which prevent the successful of GNPD implementation. The objective of this paper are to identify the critical success factors and to determine the obstacles of GNPD implementation among SMEs. The method used in this study is conceptual review of previous literature in green and NPD area of research. The result finding of this paper consist of 9 critical success factors and 12 obstacles. All the critical success factors and obstacles will be grouped. The finding of this study could be a source and fundamental guideline for future study in developing framework for GNPD implementation especially for manufacturing SMEs in Malaysia. Besides that, the result of this study could help organization especially SMEs's owner and manager in assessing their initiative to implement GNPD in organization. **Key words:** *Small and Medium Enterprise, Green New Product Development, Environment, Eco-design, Critical Success Factors*

1 Introduction

Today the environmental issues are increasingly integrated into international trade and market. Consumers worldwide are becoming concern for environmentally friendly products. One way to achieve and produce environmentally friendly products is through implementing Green New Product Development (GNPD) in organization. GNPD is a principle that was proposed in 1987 which is designing, developing, and producing "ecological" products by manufacture [1]. Effective implementation of Green New Product Development (GNPD) in organization will improve organization competitiveness, reducing operation cost, conserving environment, reduced energy and water consumption. In addition ecological product is an important key to reduce the negative impact on the environment as well as contribute to the improvement of product sustainability [2].

SMEs are considered as the fundamental to the development and growth of many countries. In Malaysia, SMEs from manufacturing sectors play a significant role in economic development [2]. Majority of businesses unit in Malaysia are consists of SMEs [3]. International Energy Agency stated that total emission produced by SMEs are expected to increase from 74% to 91% from year 2007 to year 2050.

Increasing customer awareness on eco-friendly product has led to a need for all manufacturers especially SMEs to implement GNPD. However, achieving successful implementation of GNPD is not easy for most SMEs especially for those who want to transform from conventional approach to green manufacturing approach. The process of GNPD implementation is involving numerous challenge and issues. Firstly, the creation, design and function of the activities must attract, satisfy and retain customer [4]. Secondly, [5] pointed out that GNPD and sustainability practises involve a huge cost for a company to implement although it will bring a lot of benefits. Thirdly, the process of GNPD is highly complex and it requires certain high skills and knowledge to implement [5]. Besides, there are many SMEs which still confuse about "what is a GNPD?" or "what is consider as a sustainable product" [4].

SMEs requires a systematic guideline in order to implement GNPD. However most of previous studies on GNPD were concentrated more on large organization [6,7,8,9] rather than SMEs. Theoretically, implementation of GNPD in SMEs must not adopt the "little big business" approach where GNPD in SMEs is viewed as a linear scalar of GNDP implementation within large organization. Hence, there is a need to fundamentally understand the process of GNPD implementation among SMEs, rather than fitting large organizational understanding [10].

This study is act as a beginning stage to assist SMEs in Malaysia to implement GNPD. Before executing GNPD, there are several critical factors and obstacles need to be identified. Therefore, the objective of this paper is to identify the critical success factors and to determine the obstacles of GNPD implementation among SMEs.

2 Literature Review

2.1 Concept and definition of NDP

Green NPD is evolved from NPD concept. The concept of NPD was introduced and popularized in the 1950s when manufacturers minimized the production costs through mass production [1]. As competition intensified in the 1990s, the increased awareness of green practices has triggered firms to act in an ethically and socially responsible manner in green NPD. New Product Development is the term described the complete process of new product or services that will launch in the market [11].

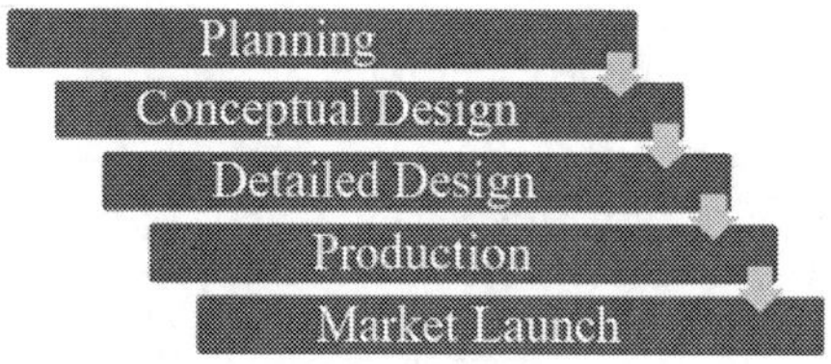

Figure 1.1: Process of New Product Development

Source: [1]

New product development stage is divided into 5 stages as shown in Figure1.1, which are planning, conceptual design, detailed design, production and market launch. Second stage and third stage is very important and must implemented properly in order to improve the product characteristics and the environmental issues [1].

2.2 Concept and definition of GNDP

The concept of GNPD is to minimize or eliminate wastages including hazardous chemical, emissions, energy and solid waste along designing, developing and producing product. The purpose of this concept is to improve the process and product performance in manufacturing industry based on requisite in environmental regulations. GNPD is an innovative and competitive tool for organizational sustainability to reduce environmental risk and to achieve both financial and environmental benefit simultaneously. GNPD is a general term which emphasizes a range of issues from the existing product, redesign the product and creating a new product based on green and eco-design concept [12]. GNPD is clarify as the development of new product which consider the environmental as a very importance issues and explicitly integrated in order to create as least product of environmental harmful as industries can [4].

2.3 Malaysian Manufacturing SMEs

An organization or manufacturing are classified as SMEs when they employed fulltime employees between 5-150 peoples or achieving annual sales less than RM25 millions [3]. According to Malaysian Statistical Department census 2014, there are total number of 645,136 SMEs in Malaysia. Table 1.1 shows the number of SMEs by sector in Malaysia.

Table 1.1 Types of SMEs in Malaysia
Source: [3]

Sector	Total Establishments	Total SMEs	Percentage of SMEs over total establish *100	Total Employment by SMEs
Overall Total	662,939	645,136	97.3	3,669,259
Services	591,883	580,985	98.1	2,610,373
Manufacturing	39,669	37,861	95.4	698,713
Agriculture	8,829	6,708	76.0	78,777
Construction	22,140	19,283	87.1	275,631
Mining & Quarrying	418	299	71.5	5,765

Within these 6 sectors of SMEs in Malaysia, each sectors will at least contributed to some extent of pollution and generate environmental issues. According to [1] the percentage of environment issues that contributed by SMEs is over 80% in waste. Table 1.2 shows the percentage of firm's activities which impact on the environment by considering 148 SMEs as respondent.

 https://doi.org/10.1051/matecconf/201815005038

Table 1.2 percentage of SME's activities which impact on the environment

Firm's activities	No. of Firm	Percentage of Firm (%)
Processes	51	34.5
Waste	127	85.8
Emissions	40	27.0
Purchasing Policies	34	23.0
Transport	74	50.0
Visual/Noise disturbance	20	13.5
Others	8	5.4
None of the above	9	6.1

3 Research Methodology

The research method used is conceptual based. Secondary sources were identified from literatures obtained from research journals, reports, articles, books, newspapers as well as empirical analysis on SMEs manufacturing. Most data is from journal or article written and published between years 2005 until 2016. Those article are searched and collected by using the search strings "Green Management", "Green Innovation", "Green Supply Chain Management", "Eco-Design Products", "Sustainable Practises", "Green New Product Development", "SMEs Practises in Green Management" and "Environment issues by SMEs". This paper will collecting, summarizing and analysing data collected from article resources. These analysed factors from most importance to least importance can provide SMEs a guideline to evaluate their performance when implementing GNPD.

4 Finding of paper

The finding have shown that there are nine Critical Success Factors (CSFs) that able to support effective implementation GNPD in SMEs. The CSFs are: commitment of leader, efficiency in energy and resources consumption, effective in product innovation, good relationship with supplier, strategic location, effective communication, established an environmental risk management system, skills & experience manpower, and sufficient manpower. The nine CSFs can be grouped into 7 core values categories, which shown in table 1.3

Table 1.3 Critical Success Factors for GNPD Implementation

	Critical Success Factors	Relevant Literature
Organization Resources Management	Efficiency in Energy and Resources Consumption	[13] Tochukwu et al (2016), [14] Lee, (2009), [15]Papadopoulos et al (2014)
Innovation initiative	Good in Product Innovation	[16] Stanley (2012), [17] Yi-Chun et al (2010), [18] Allen (2010)
Mentoring Management	Commitment of leader	[19] Sang et al., (2012), [13] Tochukwu et al., (2016)
Supplier Management	Good Relationship with Supplier	[19] Sang M. et al (2012), [18] Allen et al (2010), [20] Lee, (2008)
	Strategic Location	[13] Tochukwu et al (2016)
Risk Management	Established an Environmental Risk Management System	[15]Papadopoulos et al (2014), [18] Allen et al (2010)
Labour management	Sufficiency of manpower	[14] Lee, (2009), [18] Allen et al (2010)
	Require Skills and Work Experience	[14] Lee, (2009), [15]Papadopoulos et al (2014), [13] Tochukwu et al (2016)
	Effective Communication	[15]Papadopoulos et al (2014), [13] Tochukwu et al (2016)

From the literature, there are at list twelve obstacles that have been identified for GNPD implementation in SMEs. The barriers are: cost implication, financial constraint, negative perspective on sustainability, culture diversity, constraint of resources, resistance to adopt new technology, lack of technical expertise, lack of manpower, market competition and uncertainty, fear of failure, lack of top management commitment, and lack of customer awareness toward GNPD. Same as CSFs, these factors will be grouped into seven core values categories. Table 1.4 shows the obstacles of GNPD implementation.

Table 1.4 Obstacles of GNPD implementation

	Obstacles	Relevant Literature
Financial problem	Cost implications	[21] Sreejith (2012), [22] Hosseini (2007), Stanley (2012)
	Financial constraint	[15] Ioannis et al., (2014)
Organization culture	Negative perspective on sustainability	[23]Setthasakko, (2009)
	Culture Diversity	[23]Setthasakko, (2009)
	Resistance to adopt new technology	[24] Wu and Hang (2009),
	Fear of Failure	[25] Rao and Holt (2005),
Resources Problem	Constraints of resources	[16] Stanley (2012), [26] Jiju et al (2005),
Lack of expert and skills worker	Lack of technical expertise to implement GNPD	[27] Sharma et al. (2012), [14] Lee (2009),
	Lack of Manpower	[14] Lee, (2009)
Competition	Market competition and uncertainty	[28] Mudgal et al. (2010), [22] Hosseini (2007)
Mentoring Problem	Lack of Top Management Commitment	[23]Setthasakko (2009), [27] Singh et al. (2012),
Support from external organization	Lack of government support to adopt GNPD	[29] Daine et al., (2009)
	Lack of customer awareness toward GNPD	[21] Sreejith (2012), [28] Mudgal et al. (2009),

5 Conclusion and Discussion

From the critical review of previous literature, it can be conclude that there are nine Critical Success Factors under seven core values of GNPD implementation had been identified. In addition twelve obstacles had been identified from rigorous review of some article in GNPD. The finding could be a source and fundamental guideline for future research to develop GNPD implementation framework for manufacturing SMEs as illustrate in the figure 1.3 below:

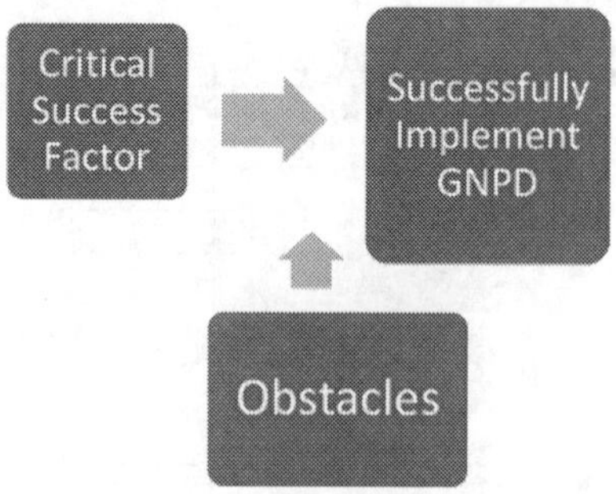

Figure 1.3: Proposal of GNPD implementation framework for manufacturing SMEs

For the future phase, the factors that has been identified in this study such as CSFs, obstacles and successful GNPD implementation will be validate via questionnaire. A set of questionnaire will construct and developed based from the finding. The questionnaire will be distribute among manufacturing SMEs in Malaysia. All Malaysian manufacturing SMEs that are certified in MS ISO 14001 will choose and involve in future study. Based on Standards and Industrial Research Institute of Malaysia (SIRIM) and Federation of Malaysian Manufacturers (FMM) directory in August 2014, there were 722 ISO 14001 certified manufacturing companies in Malaysia.

Environmental issues is critical nowadays and everyone have responsibilities to protect it especially SME industry. The purpose of this paper was to review current CSF and obstacle for GNPD implementation. The finding can be guideline for SMEs in their initiative on GNPD implementation activities. SMEs play a major role in the economic growth and provide most of new job for Malaysian. However at the same time SMEs pose serious environmental problems due to their high number and cumulative effect. It is important for SME to implement GNPD in order to protect the environment. Although Implementation of GNPD is importance for SMEs, SMEs cannot afford to make sure GNPD can be successful because there are several limitation of SME that prevent GNPD from Successful such as lack of knowledge, expertise, skills, finance and human resources. The finding of this paper have shown several critical success factors and obstacles for GNPD implementation. This finding can provide SMEs Malaysia a guideline to evaluate the performance of GNPD implementation.

Reference

1. Pochat S. L., Gwenola Bertoluci, Daniel Froelich - Integrating ecodesign by conducting changes in SMEs - Journal of Cleaner Production - Vol. 15, p.671-680 – 2006
2. Paul Knight/James O. Jenkins (2006). Adopting and Applying Eco-Design Techniques: A Practitioners erspective. Research and Development Department.
3. SME Corporation Malaysia (2016) Online:smecorp.gov.my
4. Dangelico R. M., Devashish Pujari (2010). Mainstreaming Green Product Innovation: Why and How Companies Integrate Environmental Sustainability. Journal of Business Ethics, 471–486.
5. Hart, S. (2007), "Beyond greening: strategies for a sustainable world", in Harvard Business Review on Green Business Strategy, Harvard Business School Publishing Corporation, Boston, MA.
6. Keith Goffin Colin New, (2001),"Customer support and new product development - An exploratory study", International Journal of Operations & Production Management, Vol. 21 Iss 3 pp. 275 – 301
7. Woodcock D.J. Mosey S.P. Wood,T.B.W. (2000),"New product development in British SMEs", European Journal of Innovation Management, Vol. 3 Iss 4 pp. 212 – 222
8. Bjarne JensenHanne Harmsen, (2001),"Implementation of success factors in new product development ?", European Journal of Innovation Management, Vol. 4 Iss 1 pp. 37 – 52
9. Naga Vamsi Krishna Jasti Aditya Sharma Shashikantha Karinka , (2015),"Development of a framework for green product development", Benchmarking: An International Journal, Vol. 22 Iss 3 pp. 426 – 445
10. Rodney McAdam, Frances Fulton, (2002) "The impact of the ISO 9000:2000 quality standards in small software firms", Managing Service Quality: An International Journal, Vol. 12 Iss: 5, pp.336 - 345.
11. Trevor Cadden Stephen John Downes, (2013),"Developing a business process for product development", Business Process Management Journal, Vol. 19 Iss 4 pp. 715 – 736
12. Berchicci, L. and Bodewes, W. (2005), "Bridging environmental issues with new product development", Business Strategy and the Environment, Vol. 14 No. 5, pp. 272-285.
14. Ki-Hoon Lee, (2009),"Why and how to adopt green management into business organizations?" Management Decision, Vol. 47 Iss 7 pp. 1101 – 1121
15. Ioannis Papadopoulos , Glykeria Karagouni , Marios Trigkas , Zoi Beltsiou , (2014) "Mainstreaming green product strategies: Why and how furniture companies integrate environmental sustainability?", EuroMed Journal of Business, Vol. 9 Iss: 3, pp.293 – 317
16. Stanley Kam-Sing Wong, (2012),"The influence of green product competitiveness on the success of green product innovation", European Journal of Innovation Management, Vol. 15 Iss 4 pp. 468 – 490
18. Allen H. Hu Chia-Wei Hsu, (2010),"Critical factors for implementing green supply chain management practice", Management Research Review, Vol. 33 Iss 6 pp. 586 – 608
19. Sang M. Lee Sung Tae Kim Donghyun Choi, (2012),"Green supply chain managementand organizational performance", Industrial Management & Data Systems, Vol. 112 Iss 8 pp. 1148 – 1180
20. Su-Yol Lee, (2008),"Drivers for the participation of small and medium-sized suppliers in green supply chain initiatives", Supply Chain Management: An International Journal, Vol. 13 Iss 3 pp. 185 – 198
21. Sreejith, B. (2012), "A hierarchical framework of barriers to green supply chain management in the construction sector", Journal of Sustainable Development, Vol. 10, pp. 15-27.
22. Hosseini, A. (2007), "Identification of green management of system's factors – a conceptualized model", International Journal of Management Science and Engineering Management, Vol. 2 No. 3, pp. 221-228.
23. Watchaneeporn Setthasakko, (2009),"Barriers to implementing corporate environmental responsibility in Thailand", International
25. Rao, P. and Holt, D. (2005), "Do green supply chains lead to competitiveness and economic performance", International Journal of Operations & Production Management, Vol. 5 No. 25, pp. 898-916.
26. Jiju Antony, Maneesh Kumar, Christian N. Madu, (2005) "Six sigma in small- and medium-sized UK manufacturing enterprises: Some empirical observations", International Journal of Quality & Reliability Management, Vol. 22 Iss: 8, pp.860 - 874
27. Sharma, B.P., Singh, M.D. and Neha (2012), "Modeling the knowledge sharing barriers using an ISM approach", International Conference on Information and Knowledge Management, Vol. 45, pp. 233-238.
28. Daine, H. (2009), "An analysis of the drivers affecting the implementation of green supply chain management", Journal of Resources, Conservation and Recycling, Vol. 55 No. 6, pp. 659-667.

29. Mudgal, R.K., Shankar, R., Talib, P. and Raj, T. (2010), "Modeling the barriers of green supply chain practices: an Indian perspective", International Journal of Logistics Systems and Management, Vol. 7 No. 1, pp. 81-107.

Using Experimental Design in Assistive Learning Technology Trials: An Experience-Based Review

Gede Pramudya Ananta[1,a], Helmi Adly Mohd Noor[2]Sharifah Nadiyah Razali[3]Asmala Ahmad[1], Norzihani Yusof[1], and Siti Nur Azreen Ruslan[1]

[1] *Faculty of Information and Communication Technology, Universiti Teknikal Malaysia Melaka*
[2] *Malaysian Institute of Industrial Technology, Universiti Kuala Lumpur*
[3] *Unit of Research and Innovation, Kolej Komuniti Selandar*

Abstract. Assistive learning technologies are products that are aimed at assisting people with disabilities in improving their learning with minimum intervention of caregivers. Their efficacy has been assessed by means of experimental research trials. An experiment is characterized by the treatments and experimental units to be used, the way treatments are assigned to units, and the responses that are measured. The treatments and experimental units require representative participants or sample. However, because of the limits of numbers of participants or sample sizes, such kinds of studies have been delicate but challenging experiences. This paper attempts to review such valuable research experiences.

1 Introduction

In general assistive technologies (AT) are defined as items, pieces of equipments, or product systems, whether acquired commercially, modified, or customized, that is used to increase, maintain, or improve functional capabilities of individuals with disabilities [2]. Assistive learning technologies include computer-based products that are aimed at assisting people with disabilities in improving their learning with minimum intervention of caregivers. Further study from the same resource found that about 80 percent of published studies on AT conducted to investigate the efficacy of the use of AT in impacting participants' skills and nearly half of them employed experimental design [1].

Using experimental design for the trials of AT efficacy is unique, delicate and challenging experiences [20] [21] [22]. Participant of the trials are people or children with disabilities, such as with Autism Spectrum Disorder (ASD) or partially strokes. There have been several problems related with gaining access, parent or caregiver concern or permission, legal issues, and reporting the findings [19] [23]. In the following parts of the paper, the reviews on experiences are attempted based on two selected trials, which are a mathematics tutor, called as *MathTutor*, and a serious game diagnosis, called as *Vi-Per Games*, for children with ASD.

2 Trial on *MathTutor*

2.1 Design and Setting

This research aimed mainly to examine the efficacy of a computer assisted instruction or tutor in helping students with autism to learn elementary concepts of addition. For best result from the trial, pretest and post-test design of experiment was planned for the population of Malaysia resident children with ASD. The design can be depicted by the Table 1 as follows.

Table 1. Design of Experiment 1

O_1	P	O_2
O_3	X	O_4

Access to the centers of caregivers or schools was sought from related authorization bodies. However, the access could not be gained for some months until the population changed to the children with those disabilities in Melaka. Besides, while at the process of selecting participants, pre-interviews were conducted with the teachers and basic information about the students had been retrieved. Later, the participants were observed separately to verify if they did portray the required skills for the experiment. At the end of the observation period, 40 (forty) students with ASD who met these skills were selected as participants for the experiment, in which they were randomly divided into two groups; the intervention group that employed the *MathTutor*, and the control group that utilized the teacher-assisted learning approach. There were from several primary schools in Ayer Keroh, Melaka.

The participants were required to have certain prerequisite skills such as knowing how to use computers and mouse, being able to follow instructions orally or written, being able to focus on the activity for at least 10 minutes, and being capable to recognize numbers between 1 and 10. The research was held in an individualized education

[a]Gede Pramudya Ananta

classroom. During the experimental session, there was no other person in the classroom other than the researcher and the student. A laptop was used to present the intervention modules which were developed and presented in the tutor during the experimental session. Minimum prompt and assistance were given to the participants if they failed to follow the written or verbal instructions.

2.2 Results and Discussion

The following reports the performance of participants throughout the experiment. During the experiment, pre-test and post-tests were carried out to measure the effectiveness of the tutor as a learning tool. Before initiating the experiment, all the participants were required to sit for a pre-test in order to determine their initial skill level before treatment. Lastly, at the end of the last session, the participants were required to sit for post-test, where the questions were similar to those found in the pre-test.

As for the experiment, the control group consisted of randomly selected 20 students with ASD and they utilised the normal tutoring system throughout the experiment. The Wilcoxon Signed-Rank Test was used in order to determine whether there is a significant difference between the pre-test and post-test scores of the control group. As a result, the Wilcoxon Signed-Rank Test successfully determined the significant variance between the pre-test and post-test scores for before and after the experiment carried out within the control group ($Z = -3.91$, $p < 0.05$). This result indicated that a change did take place in the score for the control group. In precise, a difference was present between the pre-test and post-test scores in measuring the skill level of the control group.

On top of that, the use of box plots and scatter plots exemplify the distribution of pre-test and post-test scores of the participants from the control group. The plot indicates that 90% of the pre-test scores were distributed below 50% score. Hence, 90% of the participants scored less than or equivalent to 50% score. The right box, on the other hand, which shows the post-test scores, points out that about 35% of the participants attained above 60% score. This suggests that about 60% of the participants had high post-test scores (between 60% and 80%), thus displaying improvement in the control group for before and after the experiment.

In the experiment, the intervention group consisted of randomly selected 20 student diagnosed with autism, in which they experienced the tutor that ascertains the suitability of learning material based on the level of users' skill throughout the experiment. As such, the Wilcoxon Signed-Rank test was used in order to determine whether there is a significant difference between the pre-test and post-test scores of the experiment group. As a result, the test identified a significant difference between the pre-test and post-test scores derived from the experiment group ($Z= -3.92$, $p<0.05$). This result hints that a change did occur in the experiment group. In other words, a significant difference was detected between the pre-test and post-test scores in measuring the skill level for the intervention group.

The experimental case study was carried out to determine the significant difference between the intervention and control groups. As such, the post-test scores obtained from both groups were used to determine the results of the experiment. The Mann-Whitney U test was employed to compare the level of skill among participants in the experiment group, precisely looking into their basic addition concept in mathematics after the experiment, besides ensuring the presence of a significant variance between the post-test scores obtained from the intervention and control groups. Besides, the findings of the Mann-Whitney U test with the aim of determining the significant difference between post-test scores of the intervention and control groups.

It also portrays that the post-test rank average of the intervention group is 29.92, while that for the control group is 11.10. Hence, a vast difference of 18.82 points is recorded between the rank averages of the two groups. Based on the analysis performed, the value of ($U = 11.50$, $p<0.05$) obtained indicates a rather significant difference between the intervention and control groups. In short, the method exposed to the intervention group proved to be more effective, in comparison to the technique applied in the control group.

During the experiment, an informal observation was carried out on both control and experiment groups. The observation was performed when the participants attempted using the tutor. Besides, notes were recorded from the observations pertaining to their behaviour, progress, and if they had required guidance or assistance.

During the first session, all the participants exhibited some interest in using the computer as a learning tool to learn the addition skill although some seemed playful and needed help to stay focused at the initial stage of the session. Besides, some of the participants required guidance and assistance when accomplishing reinforcement exercises, especially those questions that involved numbers larger than five. As for the second session, some participants kept repeating the same mistakes as they did in the previous session. Moreover, they seemed distracted while using the tutor system, thus seeking guidance throughout the session. Meanwhile, in the third session, most of the participants required both guidance and assistance when attempting the sums. Furthermore, similar in the previous session, they were distracted by the surrounding while using the tutor system, hence they needed both guidance and assistance throughout the session. Next, in the fourth session, most of the participants stopped showing interest or were unmotivated during the lesson. Besides, the participants required heavy guidance and assistance at their sums. Additionally, a majority of the participants also seemed to rush in answering the reinforcement exercise. Later, during the fifth session, those distracted participants were unable to focus on the lesson, while more failed to show interest or motivation during the lesson. During this particular session, some participants required heavy guidance throughout the session. As for the sixth and the last sessions, almost all the participants were able to do the exercise with minimal guidance, especially during the reinforcement exercise.

On top of that, during the experiment, some participants did display improvement in behaviour, where they maintained good behaviour throughout the remaining experimental session, in comparison to the first three sessions. Furthermore, some participants showed that they did enjoy the experiment session. Nevertheless, the tutor system failed to attract the attention of these students to keep them engaged throughout the experiment.

In the experiment group session, during the first session, all the participants exhibited interest in using the computer as a learning tool to learn the addition skill, although they seemed playful and needed help to stay focused at the initial stage of the session. In fact, some participants showed that they knew how to calculate using their fingers even though the calculations made by some of them were not in line with verbal counting. However, some had required guidance and assistance during the reinforcement exercise, especially with questions involving numbers larger than five. During the second session, some participants portrayed that they did remember the lesson learned during the last session. Moreover, some counted using fingers, besides using the pictures provided with the questions. Nonetheless, some kept on repeating the same mistakes they did in the last session. In addition, a majority of the participants seemed disturbed by the sound of the answer button. Hence, in this session, a number of participants demanded guidance and assistance, especially when attempting reinforcement exercises. Next, in the third session, most of the participants proved to remember the lesson learned during the last session. Besides, most of the participants had managed to focus on both learning and assessment during the session. Nonetheless, some participants had needed guidance, while some others required both guidance and assistance when doing the calculations. In fact, most of the participants also seemed to enjoy the session and requested for the lesson to be repeated.

Meanwhile, in the fourth session, a majority of the participants displayed interest and motivation during the lesson, although some were distracted by the sound of the answer button. During this session, most of the participants only required guidance when attempting the calculations. In fact, they also seemed to enjoy the session and requested the lesson to be repeated. As for the fifth session, similar to the previous session, most participants exhibited interest and motivation during the lesson, although some were distracted by the sound of the answer button. Hence, those distracted had needed guidance and assistance while at reinforcement exercises. As for the sixth and the last sessions, all participant seemed eager to start the lesson. Furthermore, almost all participants had been able to accomplish the exercise efficiently, while only some required minimal guidance instructions during reinforcement exercise. The participants also seemed to enjoy the session and requested for a repeat. On top of that, during the experiment, almost all the participants displayed improvement in behaviour for they maintained good behaviour throughout the remaining experimental session, in comparison to the first two sessions. The participants also showed that they enjoyed the experiment session as they requested for the lesson to be repeated and regularly requested to use the computer. Thus, the method employed in the tutor, such as images, animations, and sounds, seemed to attract these students with autism, besides retaining their focus when learning took place. As a result, the tutor had been proven to enhance both engagement and interest among students with autism during the learning sessions. Moreover, all the students with autism actively participated during the lessons, where they counted together with the tutor. Hence, the participants were interested to pursue the learning sessions and asked to repeat the lesson that they had covered. Besides, the tutorial and the examples shown to them attracted their attention, thus aided them in memorizing the lesson learned.

3 Trial on *Vi-Per Games*

3.1 Design and Setting

Early identification and diagnosis to children with ASD is required to introduce them with early intervention program in order to overcome, or to minimize at least, the problems related with ASD. The objective of this research is to develop serious games, called as *Vi-Per Games*, as a tool for caregivers or special education teachers to diagnose visual perception problems in students with ASD. The Chalfont diagnosis theory was employed for development of the tool [3]. Then a trial was designed for the population of students or children with ASD in Melaka in order to achieve external validity of the design of experimental research. As above, gaining access was initiated from centers of rehabilitation or special classes of children or students with ASD. Even though, within the extra long time limit, the access was granted, but there are only a small number of students or children was able to participate in the study because small number of returned concern forms from the parents or caregivers. At last, only 10 (ten) students are willing to be the participants of this study.

This study utilized the mixed method, using both the quantitative and the qualitative methods. The usage of quantitative method is to validate the accuracy of the serious games for diagnosing and generating comprehensive and systematic diagnosis reports on visual perception problems in students with autism. On the other hand, the qualitative approach applied in this research is to validate the model and the perception of teachers towards the use of the serious games for diagnosing visual perception problems in students with ASD.

The aspects of research design such as the instruments used in this study are also addressed in this chapter. ADDIE model is used as the research design to develop the visual perception diagnostic tool. This five-phase systematic model, namely Analysis, Design, Development, Implementation and Evaluation, is used to guide through the process of creating multimedia products for a variety of settings. Each phase represents a series of tasks that help to ensure development efforts stay on track, on target and on time. Completing each phase adequately increases the chance that the information presented ultimately remains relevant to the needs of the research. Selection of the participants can be considered as a purposive sampling. According to Chaudary [5], Bogdan and Biklen [6] since the participants were identified and selected in accordance with some purposive principles such as accessibility, hence the selection of the participants can be regarded as purposive sampling instead of random sampling. In addition, specific information obtained from the participants were relevant to the purpose of the study.

Data collection was conducted with the participation of special education school teachers and students with ASD of SMK Bandar Baru Uda and SMK Tun Syed Nasir Ismail, both are secondary schools in Johor Bahru. The details related to the collection methods used in this study is also reviewed and described. Data about each participant level of visual perception are ordinal type. They were coded or assigned an ordinal scale measurement. The Wilcoxon Signed-Rank test, a non- parametric test was employed on every sample in conforming no significant difference between participant conventional and *Vi-Per Games* test results. Meanwhile, data that describe the participant perception, statistic on using the prototype, comment, and general related responses are ordinal and/or nominal type. For the purpose of data analysis, descriptive statistics such as percentage, median, percentiles were employed. In general, this study involves the development of the *Vi-Per Games* system for diagnosing visual perception for autistic students and how these games are utilized by the teachers in the special education class in school. A diagnostic test was taken by the students in the initial stage. In addition, this method selected for this study and describes the research approach, research design, sampling strategy, data collection, as well as the data analysis of the study.

3.2 Research Approach

Dominant concepts of research domain demands details understanding towards research approach. According to Brewer et al. [15], quantitative research is referred to hypothesis testing research in which studies initiate with statements of theory from which the research hypotheses are derived. In addition, a qualitative research approach based on observations, document studies and interviews taken from real world is argued in order to meet the research objectives. However, a few alternative research approaches will also be discussed. From literature and discussions on various traditions and approaches to "good" research [7] [8] [9] [10] [11], four potential research strategies have been considered for the research project. There are logical theoretical research, quantitative experimental research, qualitative observational research and participatory action research.

Along with research strategies, two types of methods were applied in this study, which are quantitative method with experiment and qualitative method with use of questionnaires. The data was collected by employing the survey method and the opinions of the sample population were gathered using a questionnaire. Malhotra [12] and Sekaran [13] stated that face- to-face sessions can aid in collecting data for a survey. Besides that, data can be collected via telephone interviews, observations, e-mails, the Internet or personally administered questionnaires. Then, the Likert scale, a popular method of collecting data for surveys was used. This method is employed to measure attitudes. It requires respondents to provide answers to a number of statements. The answers range from *strongly agree* to *strongly disagree* as recommended by Saunders et al.[14]. Therefore, personally administered techniques and the Internet were considered as the most applicable method to collect the data.

Quantitative and qualitative approaches were chosen to be applied in this study. The quantitative research applied in this research is to validate the accuracy of the serious games for diagnosing visual perception problems in students with autism and generate comprehensive and systematic diagnosis reports on visual perception problems. Meanwhile, the qualitative research applied in this research is to validate the model and the perception of teachers towards the use of the serious games for diagnosing visual perception problems in students with ASD.

In order to describe how the prototype was used, separate-sample pre-test and post- test complemented by a case study design was employed. The participants were divided into size of 3, 3 and 4 separate and independent samples. One group pre-test and post-test pre-experimental design is described by Tuckman [16] and Kerlinger [17] as shown in Table 2.

Table 2. Design of Experiment 2

O_1	X	O_2

An observation or measurement is designated as O and a treatment or the presence of the treatment is designated as X. For ease of identification and referral each O carries an arbitrary subscript. In this study, to provide some information about the sample of participants, observation or measurement O_1 was carried out by applying pre-test or survey. Then, a treatment X was experimented on the subjects. Finally, to access the effect of the treatment X, another observation or measurement O_2, was made on the members of the group. To overcome history bias which was the main inherent deficiency in this design, Tuckman [16] the one group pre-test and post-test was applied three times. The term *history bias* refers to the preconception triggered by events occurring in the environment concurrently when the experimental variable is being tested. The modified design adapted by Pramudya [18], which is called separate-sample pre-test and post-test design is illustrated in Table 3.

Table 3. Design of Experiment 3

O_1	X	O_2		
	O_3	X	O_4	
		O_5	X	O_6

For this study, O_1, O_3, and O_5 consecutively represent the pre-test procedure applied to obtain data on participant levels of visual perception diagnosis by using the conventional method, which is Visual Perception Diagnostic Test Instrument, Ministry of Education Malaysia on three different separate samples. X stands for the instructional treatment, which is use for the prototype *Vi-Per Games*, and O_2, O_4 and O_6 designate the post-test procedure. Questionnaires and data of autistic student statistics at the database were used to gather information about participant level of visual perception, teacher perception of the prototype, and other general related comments from the three samples after the treatment.

3.3 Results and Discussion

The results from the three separate samples indicate that there were no significant differences between the pre-test and the post-test scores of participants in this study. It shows that, in each of the three samples, the participant's pre and post-test scores were the same. As the collection of score marks was performed in three different independent samples on different timelines, the history bias in this study was avoided. In other words, it is unlikely that some other event would have occurred simultaneously to affect the use of the serious game environment in the three samples. This supports the conclusion that the test scores resulting from the use of the serious games are accurate. This suggests that the serious game approach by using *Vi-Per Games* as an alternative method may benefit teachers and autistic students in the diagnosis of the level of visual perception in autistic students.

Since *Vi-Per Games* might improve the diagnosis process, encourage students, and enhance their interest and motivation, these games can be used to assist autistic students to undergo diagnostic testing. Apparently, these findings are coherent with a previous study, which also indicate autistic student compatibility with games. In addition, the findings also highlight that games could help to analyze visual perception performance.

4 Conclusion

This paper is intended to review experiences on using experimental design in two different studies. The main purposes of the studies were to investigate the efficacy of two assistive learning technology named *MathTutor* and *Vi-Per Games* in helping students with ASD to learn elementary mathematics concepts of addition and assisting caregivers to diagnose children with ASD. Representative samples were designed in order to meet the external validity of the design or in order to generalize the findings. However, because of the limited concern from the parents or caregivers of the children as well as minimum access gained, only small samples or participants are employed. This forced the researchers to shift the paradigm, scope, and data analysis of the studies.

References

[1] S. Alper and S. Raharinirina, Journal of Special Education Technology **21**, 47-64 (2006)

[2] A. G. Dell, D. A. Newton, and J. G. Petroff, *Assistive Technology in the Classroom: Enhancing the School Experiences of Students with Disabilities* (Allyn & Bacon, Boston, 2012) 10-15

[3] H. A. Noor, F. Shahbodin, G. Pramudya, N. C. Mohd, S.N. Razali, M.S. Khalid, R. Kassim, F.A. Rahman, and Z. Baharum, International Journal of Advanced and Applied Sciences **4**, 68-72 (2017)

[4] G. Pramudya and S.N. Ruslan, Proceedings of Mechanical Engineering Research Day 2017 , 235-236 (2017)

[5] C.M. Chaudary, *Resesarch Methodology*(S.K. Parnami, Jaipur,1991)

[6] R.C. Bogdan and S.K. Biklen, *Quantitative Research for Education: An Introduction to Theory and Methods* (Allyn & Bacon, Boston, 1992)

[7] N.K. Denzin and Y.S. Lincoln, *The SAGE Handbook of Qualitative Research* (SAGE, California, 2011)

[8] D.H. Jonassen and T.M. Duffy, *Constructivism & Technology of Instruction: A Conversation* (Lawrence Erlbaum, New Jersey, 1992) 137-148

[9] M.B. Miles and A.M. Huberman, *An Expanded Sourcebook: Qualitative Data Analysis* (SAGE, New York, 1994)

[10] J. Corbin and A. Strauss, *Basics of Qualitative Research: Techniques and Procedures for Developing Grounded Theory* (SAGE, Thousand Oaks,CA, 2008)

[11] G. Walsham, European Journal of Information Systems, **15** 220-230, (2006)

[12] N.K. Malhotra, *Marketing Research: An Applied Orientation* (Prentice Hall, Singapore, 1999)

[13] U. Sekaran, *Research Methods for Business: A Skill-Building Approach* (John Wiley & Sons, New York, 2003)

[14] M.N.K. Saunders, P. Lewis and A. Thornhill, *Research Methods for Business Students* (Pearson, Harlow, 2012)

[15] R. Brewer, F. Biotti, C. Catmur, C. Press, F. Happe, R. Cook, and G. Bird, The Authors Autism Research **00**, 1-10, (2015)

[16] B.W. Tuckman, *Conducting Educational Research* (Harcourt Brace & Co., Florida, 1994)

[17] F.N. Kerlinger, *Foundation of Behavioral Research* (Harcourt Brace & Co., Florida, 1992)

[18] G. Pramudya, *Using an Adaptive Web-Based Learning Environment to Develop Conceptual and Procedural Knowledge* (Ed.D Thesis, University of Wollongong, 2004)

[19] W. C. Mann, K. J. Ottenbacher, L. Fraas, M. Tomita, C. V. Granger, ARCH FAM MED **8**, 210-217 (1999)

[20] E. B. Itzchak, D. A. Zachor, Research in Autism Spectrum Disorder **5**, 345-350 (2011)

[21] S. W. White, K. Keonig, L. Scahill, Journal of Autism and Development Disorders **37**, 1858-1868 (2007)

[22] M. Tincani, J. Travers, A. Boutot, Research & Practice for Persons with Severe Disabilities **34**, 82-90 (2009)

[23] B. Robins, K. Dautenhahn, R. Boekhorst, A. Billard, Proceeding of the 2004 IEEE International Workshop on Robot and Human Interactive Communication 277-284 (2004)

"Talking Phonics for Autism": Developing A Multi-Purpose Touch Screen Technology Software Application Which Utilizes Sound Articulation Point to Teach Autistic Children

Noriha Basir[1],*, *Ameer Fuhaili* Mohamad Hashim[2], *Sharmini* Abdullah[1], *Noor Asliza* Abdul Rahim[1], *Masturah* Sabri[1], *Afifah Hanani* Yusuf[1], *Wan Norhaizar* Harun[1] and *Dipima* Buragohain[1]

[1]CIL,UniMAP,Malaysia
[2]Ikom, UniMAP, Malaysia

Abstract. This paper aims to examine the potentials of the multi-purpose touch screen technology application which utilizes a sound articulation point software called "TALKING PHONICS FOR AUTISM"as an alternative method of teaching phonics to autistic children. "TALKING PHONICS FOR AUTISM"is developed via a collaborative effort between the Centre of International Languages (CIL) and the School of Human Development and techno Communication (iKOM), University Malaysia Perlis (UniMAP). The reading skills of autism children are developed intofive (5) levels ofMalay reading comprehension skills. The first level consists of open syllables - vowels and consonants. The following consecutivethree levels comprised of closed syllables. The fourth level introduces vocal sequence while the fifth level constitutes of Malay language loans. The Malay-language dialect phonological theory by Tajul Aripin Kassin (2000) which is based on Clements &Keysher's (1980) Generative Booking Generic Fonology Model (CV) forms the theorectical framework of this paper.

1 Introduction

This paperwork discusses the development of bilingual phonic software for reading skills among autistic students. Autism belongs to a category of special needs children who have problems in terms of learning. The Department of Special Education (1995) defines children with learning disabilities as those who suffer from one or more of the original psychological processes involving the use of language, speech or writing that will result in imperfect ability to listen, speak, think, read, spell and make math calculations. The children with disabilities are those who have been identified and certified by clinicians as a disability that interferes with the learning process. Children with autism are usually detected since they are small. However, in most cases, they are detected while in school. Based on the observations in the classroom, autism children also have a short, short time span and not interested in theoretical learning. They prefer something that is practical and like to play with moving objects. However, these autistic children can be taught to speak and mingle with the community, rather than being marginalized. Consequently, the behaviour and emotion of these children should be seen from a positive point and these elements should be incorporated into the process of learning Malay and English which is specifically catered for them. The Ministry of Education is also constantly working on addressing children's problems in teaching and learning. Certainly, early actions and interventions can be channelled to problematic students which will help autistic children in Malaysia.

2 Phonics studies in Malaysia

Phonics studies in reading comprehension have been widely discussed by researchers in Malaysia. Tatat Hartati (1998: 64) explained that the Phonetic Reading Method has the same meaning as the Reading Method. The difference between the Phonetic Reading Method and the Reading Method is based on the sound of each letter. For example "B - Beh", "Z - zeh". Children are also given early exposure in knowing every letter and sound and sounding syllables. The next process is to combine these syllables into words. This Phonetic Reading method is suitable as one of the methods to teach reading Malay, because the Malay spelling system is phonemic and systematic. The disadvantages of this method are the same as the Reading Method. Children take a long time before being able proficiently read their own sentences and stories. Mazni (2007) in her action research entitled "Reading Skills Improvement Program" also suggested that the Phonetic Reading method be applied as a reading comprehension technique for school children. Techniques used include the merger of CVCV, CVCVC and VCVC syllables. Tajul Aripin (2008) notes

* Corresponding author: sharminie.abdullah@gmail.com

that the Phonetic Reading is a reading method which is hundred per cent based on the international standard phonetic sounds. In this case, he states that pupils are introduced directly with the phonetic sounds of a phoneme without introducing them to the sounds of the letters. He further explained that the reading method that combines the Reading Method and the Phonetic Reading Technique requires the students to understand both the sound strata of sound and the sound of letters and phonemes (the phonemes) which the letter a is pronounced as [a], the letter i is pronounced as [i], the letter e is pronounced as [e] and [ə], the letter u is pronounced as [u] and o is pronounced as [o]. By using these two methods of reading, students are charged with the concept of recognizing and remembering a sound of the letters with their phonetic sounds which are unrelated to each other. The next discussion covers the aspects related to the development of this phonic software which is the Malay Language Phonology Theory by Tajul Aripin Kassin (2000). Similarly, the Strata Approach to Mastering Malay Language by Tajul Aripin Kassin (2012), the Reading Levels of Malay Language Skills, the Background and Animated Music Sounds proposed in this Phonic Software.

2.1 Theory of Malay Phonology by Tajul Aripin Kassin (2000)

2.1.1 The Malay Language Phonological Words

Tajul Aripin (2000) has explained that the Malay language phonology is the basic word (+ ending). That statement means, any phonological process that occurs in the base word and the term + ending is known as the phonological word. Based on the definition of the above phonological terms, the phonological process of Malay language such as the formation of sliding and geminating occurs in the phonological environment. Nevertheless, the process of gemination that occurs at the boundary of the suffix will not be discussed and not involved in this study. Meanwhile, other phonological processes such as insertion of glotis stops, vocal intermediates, assimilation, etc. are phonological processes that are applied beyond the phonological word domain. Therefore, the mastery of reading Malay language will be easier if the phonological word is applied in the teaching and learning of the Malay language of autism children. This phonological word can help teachers and students in determining when the glacial slide and stopping process comes in between two vocal sequences. For example, the word buah and juruukur. Surface sounds of the word are [bu.wah] dan [ju.ru./u.kur] not [bu./ah], [bu.ah] and [ju.ru.wu.kur] (Tajul Aripin, 2005: 11).

2.2 Strata Approach to Mastering Malay Language Reading

The strata approach to mastery of Malay reading was introduced by Tajul Aripin in 2012. He explains the approach of reading comprehension strata in the early stages of children with levels or strata; Input Representation (RI), Mental Representation (RM) and Phonetic Representation (RF). This statement is the same as the phonemic representation and the phonetic representation described by Zaharani (1998: 1094). He explains that the internal representation or input is an abstract representation that contains all the idiosyncratic information about the mention of morphemes or words in the utterance and this information is stored in the mental lexicon of the speaker. While surface or output representation is a concrete representation that manifests the actual mention of the morphemes or words in the utterance, and the information contained therein consists of idiosyncratic information as well as predictable information.According to Tajul Aripin (2012), the goal of mastery of reading the Malay language is to assess and test the extent to which the output of an input (process of spelling) is optimal or true. Output optimization according to him must be synchronous between RM and RF. Figure 2 below is the Early Childhood Reading Proficiency Strength Strata.

Figure 2: Strata of Proficiency in Reading the Malay Language

Source: Modified from the Strata of Mastery of Reading the Malay Language, Tajul Aripin, (2012)

The main purpose of the application of mastering Malay language skills by Tajul Aripin (2012) in this study is to see the syncreteness of the input or output of the word in Malay. This assessment will be used to describe the extent of reading Malay language among children with autism and the effectiveness of autistic children's Phonetic Reading Method.

2.3 Malay language syllable structure (C)V(C)

Malay Language syllable structure (C)V(C) by Tajul Aripin (2000) is based on Generative Phonology by Clements & Keysher in his thesis for Doctor in Phylosophy titled *The Phonological Word in Standard Malay*. This will be the basis in explaining the phonological phenomenon of the Malay language of autism children. Malay syllable structure is classified in language type IV, with these syllable structures: CV, V, CVC,VC. Various aspects of syllables such as syllable structures, combinations of segments that build syllables, syllable and word processing, and other related aspects can be explained through the model. Zaharani (2008) describes many of the phonological phenomena that apply in the language have been well-managed and solved through syllable theory analysis, for example phenomenon of vocal elongation, vocal insertion, consonant abortion, stress and others.

2.4 Articulation of speech sound in Malay Language

Articulation/ of speech in Malay is vowel and consonant which is important in the preparation of Malay language reading skills among kids with austism in this phonic software.

2.4.1 Vowel

Nor Hashimah (2000: 38) states that vowel sounds involve air that blows out when producing this sound. Vowels are articulated in a manner that the articulators are far enough apart to allow the airflow to exit unhindered. John Lyons (1992: 88) explains that the vowels as sound produced with no obstacle for airflow through the oral cavity. Therefore, vowels are articulated in a manner different to that of consonants. He classifies this form of oral cavity as narrow, wide, front, back, round and upright. There are six vowels of Malay language that is [e, ↔, a, i, o, u]. Vowel /i/ known as high vowels and vowels /u/ are known as high back vowels. Whereas, the vowel /e/ is a narrow half-front vowel and the vowel /o/ is a narrow half-back vowel. Vowel /a/ is also known as broad back vowels (Tajul Aripin: 2005). It shows that the vowels play a very important role in the production of language sounds because without vocal a word cannot be formed and promulgated.

2.4.2 Vokal Sequencing in Malay Language

Vocal sequence exists when two or more vowels appear in sequence either in the beginning, in the middle, or at the end of the word. The sequential vowel presence in the Malay language will show two forms of vocal sound production, either by separate syllables [V.V] or diphthong [VV] (Tajul Aripin, 2009). Separate vocal sequences separated by syllables can often be seen in two forms, namely the vowel sequence with consonants and vowel sequences without any consonants (Tajul Aripin, 2009: 2). According to him, the presence of consonants in the vocal sequence is the result of the phonological process, either through the process of insertion of a consonant or the process of high vocal glide. Instead, a vowel sequence that does not allow any consonant to exist between the vocal sequence and allows the vocal sequence to sound as a separate syllables is known as hiatus. The vowel sequence symbolized as a phoneme and sounded in one syllable is also known as diphthongs (VV). M. Yunus Maris (1980: 40) explains the diphthongs sound refers to the pronunciation of two vowels released with only one breath. Nor Hashimah (2000: 41) states that the diphthong is a sequence of two vowels realized as one sound. The diaphragm sound is produced when the vowel sounds are present after the other vowel sound slides toward the first vowel without showing any volume peaks. The absence of peak volume during diphthong sound making makes it sounded like a single syllable.

2.4.3 Consonant and Consonant Cluster

Consonants are all sounds other than the vowels produced by restricting the airflow released through the lungs. The restriction can be made in the form of a full stop, partial stops or intermittent stops (Abdullah Hassan, 2006: 44). There are three different types of consonants in Malay. The consonant is the original consonant, loan, "*gandingan dan rangkap*". Tajul Aripin *et.al* (2005) states that there are 18 original Malay Language consonants. The consonants are /p, t, k, b, d, g, s, c ʃ |, l, r, m, n, N, J, h, j ʃ, w, y/. In phonetics, the symbols of the letters *ng, ny* and *j* are trascribed as [N], [J] and [jʤ] and a combination of two symbols *ng* and *ny* are considered as one phoneme in Malay Language. Apart from the original consonant, Tajul Aripin et.al (2005) also adds that there are sounds of Arabic language and English language consonants that are present in Malay. The loan consonants are /f, z, ⊗, ξ, Σ/ symbolized in the Malay writing system as *f, z, gh, kh* dan *sy*. In phonetics, the combination of two letters *gh* [⊗], *kh* [ξ], and *sy* [Σ] are considered as one phoneme.

Table 1: BMS Consomants Phonemes

Place of articulation/ Manner of articulation	Labial	Alveaolar	Dental	Hard palate	Soft palate	Glottal/larynx
Plosive Voiceless Voiced	P b	t d			k g	ʔ
Plosive Voiceless Voiced				ʧ ʤ		
Nasal Voiced	m	n		ɲ	ŋ	
Fricative Voiceless Voiced	f		s z	ʃ	x ɣ	h
Vibration/Trill Voiced		r				
Liquid Voiced		l				
Half vocal Voiceless	w		j			

Resource: Adapted from Indirawati and Mardian (2006: 58)

2.5 Recommendations for Levels of Malay and English Reading Skills in Phonics Software.

The author presents some suggestions on the reading skills of the Malay Language for Autistic Children.

2.5.1 Vocal and Consonant Skill

The suggested skill and level is to pronounce the vocal sounds based on Malay vocal sound articulation arrangement as in Table 2 below:

Place of articulation		Sounding of letters
Middle		ə
Front	(high)	i
	(low)	a
Back	(high)	u
	(low)	o

The suggested skill and level is to pronounce the consonant sounds based on Malay consonants articulation arrangement as in Table 3 below:

Place of articulation	Sounding of letters
Voiceless Plosive Voiced Plosive	p t k b d
Voiced Plosive Voiceless Fricative Voiced Liquid Voiced Trill	g s c l r
Voiced Nasal Voiceless Fricative	m n ng ny h

2.5.2 Open syllables Skill

Open syllables skill is a combination of consonant sounds and vowel sounds as recommended in phonics software in Table 4 below:

Combination of consonant sounds and vowel sounds				
pe	pa	pi	po	pu
te	ta	ti	to	tu
ke	ka	ki	ko	ku

2.5.3 Close Syllable Skill

Close syllable skills level in the phonic software is as tabled in Table 4 below:

pən	pan	pin	pun	pon
tən	tan	tin	tun	ton
kən	kan	kin	kun	kon
bən	ban	bin	bun	bon
dən	dan	din	dun	don

2.5.4 Vocal Sequence Skill

Vocal sequencing skill of „ai", „au" and „oi" that is recomended in phonic software is in Table 6 below:

pai	pau	poi
tai	tau	toi
kai	kau	koi
bai	bau	boi
dai	dau	doi

2.5.5 Borrowed Words Skill

Borrowed words skills recommended by the phonic software is as below (Table 7):

tarikh
khat
khas

Next, the author recommends English Reading Skill in the phonic software.

2.5.6 English Reading Skill

2.5.6.1 Vocal pronunciation skill

Letter	Word
a	apple
	album
e	egg
	elbow
i	igloo
	ink
o	orange
	octopus
u	umbrella
	up

2.5.6.2 Consonant pronunciation skill

Letter	Word
b	Bird
	Book
h	Hand
	House
J	Jacket
	Jet

Therefore, in the software, simple English words and examples of pictures close to the world of autistic children is used to attract them to learn English.

2.6 Selection of Attractive Drawings and Moving Elements

To attract autistic children to master reading through phonics, selection of interesting pictures and moving image elements are used in this phonic software. Next the author displays and explains the main menus in this phonic software.

2.6.1 Main menu in the phonic software "Talking Phonics for Autisme"

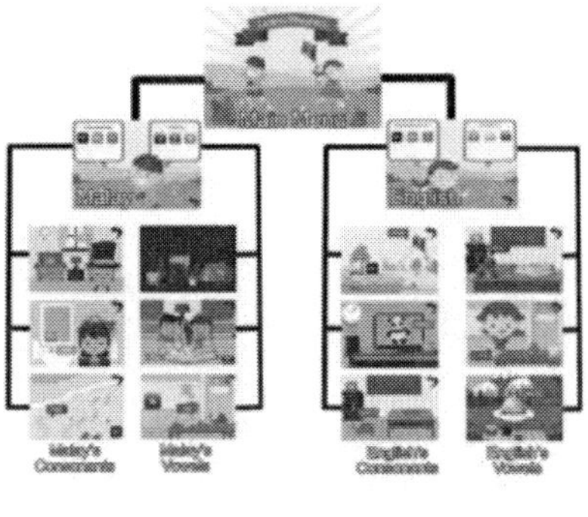

2.6.2 Steps in Phonic Application

(a) Choice of language in Menu : Phonics in Malay Language and Phonics in English

(b) After clicking on the Malay or the English Menu, user may select and choose to go to the Consonant or Vowel section.

(c) Selection of the animated object chosen to accompany sounds and letters

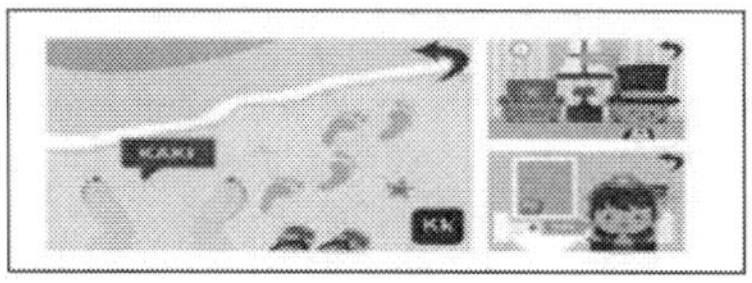

(d) Selection of animated pictures for the letter „U"

(e) English Menu

(f) For consonant „k", the animated pictures given are to illustrate the word „kite" and „key"

(g) Selection of animated pictures for each object

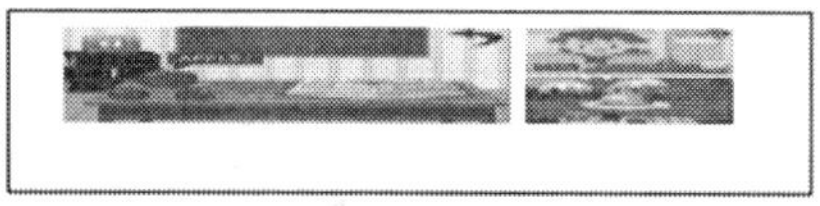

The selection of interesting and animated pictures is to attract autistic children and to foster their interest in browsing through the content of this phonics software, and thus to be able to learn how to read by following the recommended reading skills.

2.7 Songs and Music

Songs and music are an important medium in everyday life, especially in children for a specific message. Therefore, the authors recommend that the song and music medium be applied in the teaching and learning approach of children to overcome the problem of reading Malay.

Abdullah Hassan (1987) stated that the approach was an attitude toward teaching and learning. The term approach in language teaching is a set of principles or assumptions related to language, language teaching and language learning. According to him, the approach is an overview of what the language really is, the way the language is derived, how it is born and understood, and the things that control the language. An approach will also be based on language theory, language teaching theory and specific language learning theories. While teaching according to Kamaruddin Husin (1993) is a complex process filled with various elements, namely student quality, intelligence, talent, interest, motivation, and so on. Therefore, the proper selection of songs and music should be applied in children's teaching and learning sessions. The content of songs that have interesting tones and rhythms are said to be able to deliver messages and positive values in children.

2.8 Child Friendly Technology

This Autism Phonic software has a child friendly interface. It can be installed on tablet and phone, aside from computers, as it is more accessible and convenient to be carried to almost anywhere as compared to software that can only be installed in computers. Touch screen technology used is also more suitable as children nowadays are very accustomed to and familiar with touch screen devices. Through one click, children will be able to surf through the applications that come with this phonic software. If before, only adults are familiar with technology, but now children especially special needs children will be able to enjoy and appreciate the world of technology and internet too.

3 Conclusion

In conclusion, the development of phonics software is very beneficial to the Ministry of Education in general and autism teachers across Malaysia in particular. The findings of the study can be utilized for the phonetic reading and learning process of autism children in Malaysia. The researcher also hopes that this study will be an inspiration to the Malay and English language autism teachers in identifying the appropriate teaching and learning methods for the students to master the reading skills of the Malay language more efficiently. Furthermore, the contribution of this study is to nurture and cultivate the interest of Malay and English language autism teachers to use effective techniques and methods in their teaching and learning process, especially in the application of phonetic reading methods based on certain levels of reading skills.

References

Abdullah Hassan. (1987). Isu-isu Pmbelajaran dan Pengajaran Bahasa Malaysia. Kuala Lumpur: Dewan Bahasa dan Pustaka. Kementerian Pendidikan Malaysia.

Abdullah Hassan. (2006). *Morfologi*. Kuala Lumpur: PTS Professional.

Clements, G.N. and Keyser, S.J. (1983). *CV Phonology* a *Generative Theory of the Syllable*: Cambridge. Massachusetts. London. England: The MIT Press.

Indirawati bt Zahid dan Mardian Shah bt Omar. (2006). *Fonetik dan Fonologi*. PTS Profesional Publishing Sdn. Bhd.

Kamarudin Hj. Husin (1993). Sikap dan Motivasi Pelajar Terhadap Pembelajaran Bahasa Melayu. Kuala Lumpur : Kumpulan Budiman.

Lyons John; Ramli Salleh, Toh Kom Hoi. (1992). *Bahasa dan Linguistik: SuatuPengenalan*. Kuala Lumpur: Dewan Bahasa dan Pustaka

Mazni Othman (2007). *Program Peningkatan Kemahiran Membaca*: Sekolah Kebangsaan Kayan, Sitiawan, Perak.

Nor Hashimah Jalaluddin. (2000). *Asas Fonetik*. Kuala Lumpur:Dewan Bahasa dan Pustaka.

Tajul Aripin Kassin. (2000). *The Phonological Word in Standard Malay*. Tesis Doktor Falsafah. University of Newcastle at Newcastle Upon Tyne: United Kingdom.

Tajul Aripin Kassin. (2009). *Kata di Gelanggang Bahasa, Esei-esei Penghargaan untuk Profesor Mashudi Kader*. Pulau Pinang:Penerbit Universiti Sains Malaysia.

Tajul Aripin Kassin. (2012). *Penguasaan Kemahiran Membaca Bahasa Melayu di Peringkat Awal Kanak-kanak*. Dalam Seminar Jom membaca 1 Malaysia: 31 Mac 2012, 1-14.

Tatat Hartati. (1998). Keberkesanan Program Membaca Penguasaan Membaca Awal Kanak-kanak Pada Peringkat Pra Sekolah: Ijazah Sarjana Pendidikan Universiti Sains Malaysia. Pulau Pinang

Yunus Maris. (1980). *The Malay Sound System*. Kuala Lumpur: Fajar Bakti Sdn. Bhd.

Zaharani Ahmad (1998). Phonology and Morphology Interface in Malay : An Optimality Theoretic Account. Tesis Ph.D. Universiti of Essex.

MATEC Web of Conferences **150**, 05041 (2018) https://doi.org/10.1051/matecconf/201815005041
MUCET 2017

Level of Soft Skill in the Implementation of Work-Based Learning among Community College Students

Azita Binti Ali (Ph.D)[1], *Shuhada* Mahmod[1]

1 Universiti Tun Hussein Onn Malaysia

Abstract. The issue of graduate unemployment often crops up in the mass media; and more often than not, the discussions have centred on the failure of tertiary educational institutions to churn out quality graduates. Thus, the method of work-based learning (WBL) is seen as a way to improve the soft skills of the graduates. The study was conducted using quantitative research survey; the design of the study used an adapted questionnaire as an instrument. Data were analysed using Statistical Package of Social Science (SPSS) version 20. The respondents consisted of 97 students who attended WBL programmes at a community college. Descriptive statistics was used to extract data from the questionnaires for the calculation of mean. The findings reveal that the level of soft skills among community college students was high, and they include these abilities: communication skills, problem-solving skills, learning and information management, professional ethics skills and leadership skills.

1 Introduction

According to Community College Management Sector GCMS (2010), work-based learning (WBL) is an approach to learning and innovative teaching that involves cooperation between Community College Management Sector and the participating companies. WBL is a learning approach that requires work placements as part of the students' learning process in their institutions; this learning style will increase employment-related competencies, real-world work experiences and expertise in industrial applications (Becker, 2013). This may address the issues related to graduate unemployment--often talked about by the academic community and the employers who regard it as a failure of higher educational institutions to produce quality graduates . Students with soft skills could compete in the job market or work places, and they are ready to tackle the challenges of globalisation today (Anuar & Esa (2010); it is evident from various findings that WBL as one of the effective ways to improve the soft skills ommunity college students.

The collaboration between the industry and educational institutions is a major milestone towards the consolidation of higher education. According to Hassan (2007), the cooperation between educational institutions and the industry is not well coordinated and implemented; the former have the duty to include compulsory industrial trainings for the students in the curriculum, and the latter has the responsibility of providing the required training to the students. The cooperation between the educational institutions and the industry is the best way to shape the curriculum in determining the skills required.

*Corresponding authors: azita@uthm.edu.my

Graduate unemployment in Malaysia has become a worrying phenomenon in recent years. This problem occurs not only in Malaysia but also in many other developing countries like the Philippines, Indonesia, India and even in developed countries such as Britain and the United States (Stevens,2014). Employers want employees who have technical or hard skills and soft skills. The employers of any company today are looking for graduates who have a combination of technical and soft skills, graduates in Malaysia have sufficient technical skills, but employers are less satisfied with their soft skills.

Employers and the industry in Malaysia have a limited number of graduates to choose from; many have the knowledge and technical skills, but most of them are not able to communicate properly (Zaharim, 2009). Hassan (2007) states that although the university has put in place various measures to ramp up the quality of teaching and learning, the students still do not have the teamwork skills and experience required by the employers. Syed (2013), states that entrepreneurship is the key transformation factor that can solve the unemployment problem nowadays.

Professional ethics and morals are increasingly taking a prominent role due to the frequent occurrences of white-collar crimes; these are very serious matters and they involve the students (Shahrul Anuar Mohamed, 2011). Leadership skills are important elements in the formation of the nation's cultural identity, but the students are not paying much attention in developing these necessary skills (Esa, 2013).

2 Literature Review

2.1 Communication Skills

According to Zaharim (2009), the employers are satisfied with the technical skills of graduates but are not very happy with their communication skills, which is still at a low level. This statement is supported by Pumphery (2001) which states that a large number of graduates do not have the skills required by the employers. Students will benefit from the implementation of the work-based learning (WBL); their communication skills will improve and indirectly WBL prepares the students to meet the requirements of the industry. This is because students who have undergone WBL have actual experience and are accustomed to working conditions in the industry.

2.2 Critical Skills and Problem-Solving Skills

Critical and problem-solving skills are some of the abilities the tertiary students should acquire as outlined by the Malaysian Higher Education Institutions. Problem solving requires a student to think critically, creatively, innovatively, and analytically; and at the same time, he or she must be capable of applying the knowledge and understanding to new and different problems (Acar & Newman, 2003).

2.3 Teamwork Skills

Teamwork skills, also known as teamwork or group work, are soft skills that can be applied through the implementation of work-based learning (WBL). Hambur, Rowe and Luc (2002) define teamwork skills as a structured collection of individuals who interact with one another to achieve common goals. According to Anderson (2007), teamwork skills are a necessary ingredient for graduates to compete for and complete future work. Work-based learning is an effective approach that can train graduates to have the required teamwork skills for future tasks.

2.4 Learning Skills and Information Management

In this 21st century, the international community is more concerned with increasing knowledge through the concept of continuous learning and information management. Learning skills and information management are processes of democratising education, which include programme acquisition of knowledge, skills and competences, formal or informal experiences, and training in the workplace. Global competitiveness and diversity requirements compel people to constantly update their knowledge and skills. The implementation of work-based learning (WBL) can help improve the soft skills of students through continuous learning and information management, thereby producing all-round employees the employers expect.

2.5 Entrepreneurial skills

According to Harun, (2010), entrepreneurial skills are some of the soft skills which students need to master while they are in their earlier years of schooling or later years of higher education. With entrepreneurial skills, a person has the ability to explore opportunities and develop an awareness of risk, creativeness and innovativeness in business and work-related activities, (Kementerian Pengajian Tinggi , 2006). The statement is supported by Kuratko and Hodgetts (2007), that students require entrepreneurial skills so that they can be shrewd entrepreneurs who are capable of looking for opportunities, taking risks and translating dreams and ideas into reality.

2.6 Professional Ethics and Moral Skills

According to the Kementerian Pengajian Tinggi Malaysia (2006), successful professional practices and fruitful social interactions demand the observance of professional and high moral standards. The Kementerian Pengajian Tinggi Malaysia (2006) highlights three stages in the development of professional ethics moral skills: the ability to understand the impact of economic, environmental and socio-cultural factors on professional practices; the ability to analyse problems related to ethics and make decisions in solving them; and ability to practise ethical behaviours, in addition to having a sense of responsibility to the society.

2.7 Leadership Skills

The employers expect their employees to possess leadership skills. According to the Kementerian Pengajian Tinggi Malaysia (2006), leadership skills involve the ability to practise leadership in a variety of activities through four levels of knowledge about the basic theory of leadership: the ability to lead projects, the ability to understand and interchange roles between the team leader and team members, and the ability to supervise the team members. Students can develop leadership skills by going through work-based learning (WBL).

3 Methodology

Descriptive and inferential surveys were conducted to collect quantitative information. For this study, the population is composed of diploma students of Selayang Community College, Community College Ampang, Community College Segamat 2, Community College Hulu Langat, and Community College Kota Melaka, all of which had carried out the implementation of the work-based learning (WBL). This study used the random sampling technique and the respondents were the students of community colleges in Malaysia. The population in this study consists of 104 people and the selected sample is 97.

3.1 Instrument

The instrument used in this study was the distribution of questionnaires. The questionnaire consists of two parts. Part A includes questions related to the backgrounds of the respondents. Part B aims to identify the level of soft skills of community colleges when implementing WBL.

4.0 Result

This section contains the descriptive analysis for each item of soft skills and its level for respondents undergoing WBL. The overall average mean score for the level of communication skills is (4.1218), which indicates a high level. Overall, the level of soft skills among community colleges which had implemented WBL is high. **Table 4.1** below

https://doi.org/10.1051/matecconf/201815005041

shows the mean score for each soft skill among the community colleges.

Table 4.1: *Mean Score for Soft Skills*

Overall Average Mean Score of Communication Skills	4.1218	High
Overall Average Mean Score of Critical skills and Problem-Solving Skills	4.0946	High
Overall Average Mean Score of Teamwork Skills	4.2297	High
Overall Average Mean Score of Continuous Learning and Information Management	4.1219	High
Overall Average Mean Score of Entrepreneurship Skills	4.0240	High
Overall Average Mean Score of Professional Ethics and Moral	3.9410	High
Overall Average Mean Score of Leadership Skills	4.2104	High

5 Findings and Discussions

The Levels of Soft Skills

The overall findings derived from data analysis show that students who took WBL programmes in the community colleges obtained high mean score for the communication skills. The highest mean score goes to communication skills in English as well. High mean scores are recorded for both critical skills and problem-solving skills. These skills are helpful for students when they are working in the real-world organisations in the near future. This finding contradicts the study of Anderson (2007), which states that all people use critical thinking while solving problems because they lack self-awareness and other features that enable them to interpret and evaluate the problem. Most previous studies show the same trend that students do are not use critical thinking to analyse problems properly. However, the researcher found that most community college students have good critical thinking skills and problem-solving skills.

The analysis of data shows a high score for teamwork skills. This is because when community college students take up industrial training, they develop teamwork skills in order to cooperate with other workers in the companies they are attached to. This proves that WBL is able to improve students' teamwork skills. In fact, students can also cooperate with members of the group in sharing ideas, knowledge and ways of improvement (Rosenshine, 1999).

The results also show that there are high scores for continuous learning and information management among community college students. The finding shows that students who pursue tertiary education have the aim of seeking knowledge, and they are rewarded with what they desire. This is because community college students are open-minded; they are ready to assimilate new ideas or anything that brings positive results.

Other than that, the results also show a high score for entrepreneurial skills; community college students are good at identifying business opportunities. These findings are in line with the studies carried out by Stevens, B. (2014); graduates can identify business opportunities that endeavour to solve the problem of unemployment. This shows that community college students trained under the WBL approach have good entrepreneurial skills to tackle the problem of unemployment today. This study finding is supported by Mitchelmore and Rowley (2010), who says that graduates must think of themselves as an inventor rather than a job seeker.

In the course of investigation in meeting the objectives of this study, the respondents were asked seven questions. The results show that the score for professional ethics and moral is high. As we know, there are lots of immoral or unethical activities happening among students. This problem will adversely affect the process of building up a pool of first-class human capital, which is vital to the future success of the country. However, the community college students have proven that they possess a high level of professional ethics and moral . a Characterised by good attitudes and work ethics, community college students will always behave and conduct business deals with the utmost propriety, wherever they are. The findings state that the community college students can practise ethical behaviours; this is consistent with a study conducted by Zaharim (2009), which reveals that quality employees are not only experts in their work, but they have an admirable character and behaviour.

Based on the results of data analysis, there is a high level of leadership skills among community college students. These findings are supported by the study of Lee Barron (2012), which states that when a person has good leadership skills, he or she can influence and convince people inside and outside organisations to work and help to achieve organisational goals. Based on the findings of this study, the community college students have excellent leadership skills; they can mingle with and judge people well. The implementation of WBL at community colleges is certainly a wise move, which will produce students armed with good leadership skills.

5 Conclusions

Premised on the discussions of the findings, it is concluded that the implementation of WBL programmes in the community colleges has produced students with a high level of soft skills. The findings clearly show that students who have attended community colleges that implemented WBL possess a high level of soft skills. The first question of the study is answered: the community college students have a high level of communication skills, particularly those who speak in languages other than English. This is because most of the students who attend WBL are Malay. The findings also show that the community college students have a poor command of English. The second research question is answered: the community college students have a high level of critical skills and problem-solving skills. This is because students who have attended WBL are exposed to higher-order thinking skills required by the employers. The implementation of WBL at community colleges is effective in training thinking students. In addition, teamwork skills are required in the implementation of the WBL at community colleges. Students can organise group activities together, to acquire and exchange views and ideas. Community college students are also very open-minded, and receptive to the idea that they will benefit from interacting with others. Overall, the community college students have a high level of soft skills. This is in line with the requirements of the Ministry of Higher Education for tertiary institutions to produce students who have strong soft skills. This indicates Community Colleges are successfully implemented WBL to produce students who meet the requirements of the industry

References

1. Acar, B.S and Newman, I.A.(2003). Students as Tutors-Learning Problem-solving Skills by tutoring PBL. *International Journal Engineering Education*, Vol, 19, No.5, 712-716.
2. Anuar, A, & Esa, A (2010). "Penerapan Kemahiran Insaniah di Pusat Giat MARA (PGM): Satu Analisis ATIKAN: Jurnal Kajian Pendidikan, 3(2) Desember 2013 149 Perbandingan" dalam Esa, A & Mustafa, M.Z. Kemahiran Insaniah: Kajian di Institusi-institusiPengajian. BatuPahat, Johor: Penerbit UTHM [UniversitiTun Hussein Onn Malaysia], ms.1-19.
3. Anderson, K. M. (2007). Differentiating instruction to include all students prventing school failure. *International Journal Engineering Education*. Vol 51, Iss 3: .49-54.
4. Becker, G. (2013). Development Education and OtherAcademic Programs. Human Capital, 1-10.
5. Hambur, S., Rowe, K., & Luc, L. T. (2002). Graduate skills assessment. Australian Council For Educational Research. Commonwealth Department of Education Science & Training.
6. Harun, N. (2010). Aspirasi Kerjaya Keusahawanan Dalam Kalangan institusi

Pengajian Awam.Jurnal Pendidikan Malaysia, 11-17.

7. Hassan, A. (2007). Isu dan Strategi ke Arah Pembangunan Modal Insan Sektor Pengajian Tinggi Bertaraf Dunia.Pengurusan Awam, 23-44.
8. Hurley (2008).The Importance of Strong Business Communication Skills. McGraw Hill Inc.
9. Othman,H .(2009.). Pelaksanaan Pendekatan Pembelajaran Berasaskan Pengalaman (PBL dan POBBL) bagi Meningkatkan Kemahiran Insaniah Pelajar. Pemikiran Kritis dan Kemahiran Menyelesaikan Masalah.
10. Kaprawi, N. (2010). Kajian Tinjauan Secara Kuantitatif. dlm. Norainil dris (Pnyt). Penyelidikan dalam Pendidikan, 195-214. Kuala Lumpur. Mc. Graw-Hill (M) Sdn. Bhd.
11. Kementerian PengajianTinggi Malaysia (2006).Modul pembangunan kemahiran insaniah (soft skills) untuk Instutusi Pengajian Tinggi Malaysia.Serdang: Penerbit Universiti Putra Malaysia.
12. Krejcie, R.V, Morgan, D.W. (1970) Education and Psychological Measurement: Determining Sample Size for Research Activities.
13. Kuratko, F., & Hodgetts, M. (2007). Entrepreneurship: Theory, process and practice. Taunton, MA. Thomson & South–Western
14. Lee Barron, J. R. (2012). Martial Arts Training as a Method of Modifying Attitudes and Behaviors in TheClasroom. International Journal of Combat Martial Arts and Sciences ICMAUA.11 (I), ms. 17-22.
15. Mitchelmore, S.& Rowley, J. (2010). Entrepreneurial competencies: a literature review and development agenda. International Journal of Entrepreneurial Behavior & Research, 16(2), 92 -111.
16. Mohd Najib Ghafar. (1997). *Pembinaan dan Analisis Ujian Bilik Darjah*. Skudai: Universiti Teknologi Malaysia (UTM).
17. Pumphery, J.(2001). A Comprehensive Summary of Generic SKilss Requirement. *Paper Research The Council for Administration London*.
18. Rosenshine, B.V. (1999). Effective Teaching In Industrial Education And Training. *Journal Of Industrial Teacher Education*. Vol, 23(2): 5-19.
19. Shahrul Anuar Mohamed, A. A. (2011). Pembangunan Etika dan Moral Dalam Kursus kursus yang Ditawarkan di Universiti Teknikal Malaysia Melaka (UTeM). Journal of Human Capital Development, 141-155.
20. Stevens, B. (2014). Career and Technical Education Work- Based Learning Guide.Richmon, Virgnia: Education, Virginia of Department.
21. Syed, Z.A. (2013). The need for inclusion of entrepreneurship education in Malaysia lower and higher learning institutions.Education + Training. Vol. 55 No. 2, pp. 191-203.
22. Zaharim, A.(2009). Employers "Perceptions and expectstions toward engineering Graduates: A study case. *Proceedings of the 6th WSEAS International Conference on Engineering Education, Rodos Greece,* July 22-24. ISSN: 1790-2796. ISBN: 978-960-474-100-7.23-29

 https://doi.org/10.1051/matecconf/201815005042

Determinant Factors for Managing Innovation in the Malaysian Public Sector

Zety Norhaiza Ab Rahman[1*], *Norain* Ismail[1]

[1]Universiti Teknikal Malaysia Melaka, Hang Tuah Jaya, 76100 Durian Tunggal, Melaka, Malaysia

Abstract. Since 1991, the government has started the forces to encourage public sector to innovate in order to strengthen the capability and solve problems at workplace for the benefit of organization and its people. The focuses of this forces is to create value creation, encourage cost reduction, and provide services that meet aspirations and people's expectations through high impact ideas. However, to achieve these focuses, public sector faces few challenges that affect the management of innovation; which includes lack of leadership skills and poor innovative culture to stimulate innovation. Although there are various initiatives conducted to enhance the management innovation, the innovative performance is still less satisfactory. Therefore, this paper aims to discuss the determinant factors that affect managing innovation in the selected public sectors which consist of two categories; patron and executor agencies. Thematic approach was used to analyse the interview data. Findings from the interview suggested that in the organization, an employee especially a leader should show his full commitment towards innovative culture, he needs to be committed and supportive and communicate well with the employees and he also should provide attractive rewards to encourage the innovative culture for an effective management innovation. These determinant factors need to be considered as high significant for the public sector to develop new strategies or approaches in managing innovation towards sustaining their competitive advantages in a long term for the benefit of public.

1 Introduction

Public nowadays are expecting a better performance of the delivery system provided by the government. Due to this, the government has changed their old paradigm into an improved and effective of the delivery system. This can be seen from the introduction of four concepts including *Cepat, Tepat, Integriti – Produktiviti, Kreativiti dan Inovasi* (CTI–PCI); National Blue Ocean Strategy (NBOS); principle of Value Innovation; and concept of *Merakyatkan Perkhidmatan Awam* (MPA). These four concepts can be achieved through effective management innovation for the improvement in efficiency, effectiveness or quality outcomes especially relevant for the public sector. Thus, Malaysia should redesign and restructure its public sector current systems and processes for a sustainable development and maintaining its competitive advantage [2]. These changes are related to both; management innovation and non-technological innovation, where public sector develops new organizational methods, concepts or other non-technological ways to achieve its organizational objectives [20].

Innovation is related to the radical change in various fields, where it characterizes by the novelty attribute for developing a successful new system. It also describes as a process of transforming new ideas, new knowledge into new products and services, new technological processes and new organizations [19]. The term 'new' is not necessarily originated from new scientific discoveries, but it comes from the combination of existing technology and their application in a new context [10]. Innovation is implemented to overcome some issues related to environmental changes, economic changes and global competitiveness. It also helps organization to minimize other issues related to cost and resources, create value, maintain competitive position in the market and sustain for long term prosperity [7].

Management innovation is defined as the "generation and implementation of a management practice, process, structure, or technique that is new to the state of the art and is intended to further the organizational goals" [4]. Term "new to state of the art" denotes the management as a whole, new to the world or new to the organization that implementing it [4] [15]. It also said that effective management innovation helps organization to enhance its performance [15], through profit margin and survival in marketplace.

In managing innovation, one of the main factor that influences the innovation is leadership skill. Leadership is defined as a process of social influence to execute activities and relationship within organization [23]. The influence is seen from the leader behavior in consulting, delegating and assigning task, where employees are

* Corresponding author: zetyrahman@outlook.com

given freedom to participate in decision making process and enhance the motivation to be more innovate. They will develop innovative solution for problems or challenges faced by the organization [12]. They will also set a direction, developing a vision for the future of organization and built strategies to implement the changes needed to achieve a vision. However, building innovative capabilities are not only one's shoulder, but it should be bear by everyone in the organization [5]. Thus, if the commitment and participation from the top management is poor, the employee is less likely to take risk to innovate [16]. Therefore, to encourage innovative action among employees, particularly in the public sector, leaders should be able to deliver positive message that support innovation, for example through participation in creating new ideas or new ways of doing things [17].

In addition, culture is also important in managing innovation because it enables individual to understand how the organization function [21]. It also gives positive impact on management practices and firm performance in innovation [3] [11]. Culture is also used by the organization to determine the process of daily routines within organization and between organization and its environments. In managing innovation effectively, the public sector should be able to understand the function of culture in the organizational. It is believed through culture, organization will become innovative where the traits of involvement and adaptability to execute and implement innovation are found [18]. The culture of an organization can also be shaped through a leadership. Leadership is said to be a key for organizational culture which influence employees' ethical behavior [9]. Through leadership, public sector able to create a mindset of developing any possible solutions in tackling problems using various mechanism such as training, public services, rewards and recognition [8].

Besides leadership and innovative culture, the reward system offered by an organization could also affect the management of innovation in public sector. Reward orientation consists of emphasizing an equitable reward based on the performance of the employees [6]. This reward orientation provides a guideline to the organization and employees on how to offer rewards, gain rewards and encourage employees to innovate. Thus, organization that used rewards system as one of a way to stimulate innovation able to survive and achieve sustainable profitability [13]. Organizational profitability is sustained through new ideas, methods and programs develop by the employees, whereas organization creates value and makes money from the successful introduction of new products and generation of new ideas [22]. This factor affects management innovation in the organization because most of individual is motivated to the reward system which encourage them to contribute more in the innovative activities

This paper identifies three important determinant factors face by the organization, which include leadership skills, innovative culture and reward system. This paper is divided into four sections, beginning with the introduction. Then followed by the research methodology using semi structured interviews to capture the rich data for data analysis. In third section, findings and discussion are presented and finally, this chapter summarizes and recommends for an improvement to effectively manage the innovation in Malaysian public sector. This study aims to answer these questions:

a) What are the determinant factors affecting management innovation in the Malaysian public sector?
b) What improvement can be made to effectively assist in managing innovation in the Malaysian public sector?

2 Research Methodology

This research used qualitative research method to collect data as it aims to capture in-depth information and respondent experience through the expressions and words [14]. Besides that, a qualitative is referring to an inductive process where data is gathered to build concepts, hypothesis or theories rather than deductive research where hypothesis is test as a positivist research.

Semi structured interview with purposely selected respondent has been adopted in this study. Six respondents from different agencies in public sector were chosen based on these criteria; (a) high involvement in innovation unit and activities, (b) extensive experience in managing innovation, (c) won innovation award locally or internationally and (d) part of coordinator and executor in implementing innovation approaches. For the purpose of anonymity, six interviewees are identified as PA, PB, PC, PD, PE and PF.

This study apply semi structured interview in order to gather in-depth information that derives from face-to-face interactions between interviewer and public sector towards their opinions on experiences in management innovation. The interview questions is structured by covering the research objectives. Each interview lasted for approximately one hour. The researcher asked permission to record the interview session for the purpose of transcription. An axial coding is used for data analysis and from that the information gathered is categorized into three factors which are leadership skills, innovative culture and reward system.

3 Findings and Discussion

Various initiatives are designed by the public sector in implementing innovative activities for the benefit of the society. However, there are certain public sectors that have not successfully implemented the management innovation practiced. This is due to the leadership skills, innovative culture and reward system.

3.1 Leadership Skills

In order to successfully innovate, leader should give full support and commitment towards employees when dealing with innovation activities. This can be made through various forms such as moral support, coaching, training and providing facilities and funding. If the commitment and support from the leader is poorly given, the employees might feel less likely to innovate. All interviewee admitted that commitment and support from the leader is crucial in promoting and managing innovation. Interviewee PA indicates that leader should provide training and give time for the employees to innovate. He said that *"leader should provide training and expertise to provide exposure on innovation. Leader who encourage innovation will give time to innovate".* This was also agreed by interviewee PB, stated that *"the leader should give high commitment when innovate by actively participating in the innovation team and giving new ideas."* Interviewee PF emphasize that *"leader should support any ideas because we (the employees) are still in a learning process; and creative ideas is a good ideas. Leader also should be optimist and try to take all ideas. Employees may feel unmotivated if their voice are not heard. Understanding is important as this team (the employees) may need time to focus on innovation work. Furthermore, leader should propose a mechanism to effectively encourage for innovation. By supporting the employees it could propel the organization to become innovative."*

Therefore, to encourage innovative activities in public sector, leader should deliver positive message that support innovation. He also needs to allow the employees to try new ideas and to be open-minded to suggest new ways for doing things. From this, the employees will feel motivated to involve in innovation activities.

In addition, leaders' behavior could also affect the management innovation in public sector. If a leader fails to perform a good leadership skill, he may not be able to set as example to influence his/her employees to innovate, adopt changes or perform well to achieve institutional goals. Four respondents admitted that leadership skill is crucial in managing innovation in the public sector. Interviewee PA stated that:

"Leader is a mover in an organization [...] therefore, he should prioritize improvement and changes so that, employees will feel motivated to innovate. When employees do not realize the opportunities of commercialisation in their product, the leader should help and let them see it." Interviewee PD agreed and added, *"Leader should help through hand-holding concept and not hands-on only. And every department should also have a leader or someone to push them towards innovation."* Interviewee PE also mentioned that leader should developed strategies to help the organization to implement innovation in a better way. *"If a leader has formed a barrier in supporting employees to be creative and innovate, they will feel disappointed, down and refuse to innovate.*

Leader must understand and applied good leadership skill in order to influence and lead his/her employees to a right direction. The employees will be inspired to involve in innovation and make a change to solve organization problems if the leader manage to set as a role model to his subordinates.

3.2 Innovative Culture

The implementation of organizational culture is important to understand how management innovation is being shaped. Each organization has distinct culture that derive from the core values, assumptions, interpretations and approaches applied in the organization [1]. A culture helps organization to achieve their organizational goals and developed strategy for sustainability. Without a strong innovative culture, an organization will not be able to maintain their performances and competencies. Three interviewees have adopted a strategic system to develop an innovative culture... Interviewee PC mentioned that:

"We need to implement strategic system in every department, where we asked them to join competition and innovation. Each department is required to send at least a team to join the competition. They also have to include discussion about innovation in their meeting." Similarly, interviewee PD also added that: *"We implement a compulsory innovation system in each department. It means that each department need to come out with at least one innovative ideas. It can be in terms of work improvement, social innovation, new ideas or something big such as process engineering. This really helps in encouraging innovative environment in our department."*

Interviewee PC and PD indicates that when an organization actively encourages the employees to join innovation activities, the employees will accept gradually and participation rate is potentially increasing. Besides, the organization should also develop an environment or programs that cultivate innovative culture such as trainings, innovation competitions or innovative programs. Interviewee PC stated that:

"To help organization cultivate and develop understanding about innovation in public sector, we should provide programs on innovation. We also could ask for competition, make it free where everyone can give ideas through write-ups or posters and invite students to join this competition".

Moreover, interviewee PB have different opinion, he pointed out that: *"In every agencies, we encourage innovation element in each stage, which is from the top management to the support level. This is one of the way to cultivate innovation."* A strong culture is developed if all the members in the organization work together. If a leader shows some good ethical behavior as a role model, the employees will get influenced with that behavioral style.

3.3 Reward System

Other factor that affect the management innovation in public sector is the reward system. Reward or recognition could attract and influence people to innovate as they feel appreciated and motivated. Four interviewee's claims that rewards and recognition are crucial in promoting innovation culture and encourage employees to innovate.

Interviewee PA state that: *"organization needs to be creative when giving a reward so that it can attract the employees. Instead of giving money, sometimes the organization may sponsor for an oversea holiday."* Interviewee PC also agreed that: *"we provide gifts, reward and appreciation for those involved for them to be more motivated. This is normal attitude of an individual, they will feel appreciated when they got something from us."* Interviewee PF also identified that the employees are very attracted with competition especially the prize, *"they like the competition because they want the rewards. So, if we want to encourage innovation and new ideas, we need to give high reward for that category"*.

However, two interviewees have different opinions. They avoid from giving high reward to employees as they worry it may cause the employees to lose their focus. Interviewee PE mentioned that: *"when innovate, we did not encourage gifts or reward, we do not want them to be over-confident after winning [...] we should consider the external and internal factors that help you (the organization) to win. I believe a success does not depend on them only but the whole team too."*

Before giving rewards to the employees, it is also important to consider the organization accountability. Interviewee PB stated that: *"winners will receive gift and incentive as a way to motivate them, but the incentive should depends on the accountability of the organization. If they afford, they can provide very high incentives."* With the current condition of limited budget, public sectors have to find their innovative and creative way to stimulate innovative activities at their organization.

4. Conclusion

Public sectors have contributed a lot to the society in providing a better services and supports for the development of nations and society as a whole. They implement innovation in the management and conduct various programs to encourage employees to be innovative and creative. However, due to some restrictions and obligations, there are some drawbacks faced by the public sector in managing innovation effectively. Based on semi-structured interview, this study found three main determinant factors that affecting management innovation; a leader should show his full commitment towards innovative culture, he needs to be committed and supportive and communicate well with the employees, he also should provide attractive rewards to encourage the innovative culture for an effective management innovation.

This study endorses that a program and training development should be built for leaders and employees to sharpen their skills and to be innovative. Moreover, a well-structured and strong culture should be created in the public sector to cultivate innovative capabilities. In order for public sector to sustain its competitive advantage, it should be more adaptive and responsive to employees and environment.

We would like to express our gratitude to the Universiti Teknikal Malaysia Melaka for supporting this research under a grant of FRGS/1/2016/SS01/FPTT/02/F00322

References

1. Alas R., Übius Ü. and Vanhala S. *Connections between Organisational Culture, Leadership and the Innovation Climate in Estonian Enterprises*. E-Leader Vietnam 2011. (2011).
2. Ali K. A. M. and Buang M. *Study on Factors that Influence Innovation in Malaysian Public Sector. Journal of Advanced Research in Business and Management Studies.* **4** (1). P. 60-73. (2016)
3. Baker R. (2014). Innovation and Value Creation: A Cross-Industry Effects Study of Patent Generation. Thesis.
4. Birkinshaw J., Hamel G. And J. Mol M. Management Innovation. Academy of Management Review. **33** (4) P. 825–845. (2008).
5. Borins S. *Encouraging Innovation In The Public Sector*. Journal of Intellectual Capital. **2** (3). P. 310 – 319. (2001).
6. Cheung S.O., Wong P.S.P. and Wu A.W.Y. (2011). Towards An Organizational Culture Framework in Construction. International Journal of Project Management. Volume 29 (1). P. 33-44.
7. Cohn S. *A Firm-Level Innovation Management Framework and Assessment Tool for Increasing Competitiveness.* Technology Innovation Management Review. P.6-15. (2013).
8. Conference Edition *Innovation Framework for the Public Sector Of South Africa*. (2007).
9. Daglio M., Gerson D. and Kitchen H. *Innovating the Public Sector: from Ideas to Impact 12-13.* OECD Conference Centre, Paris. (2014).
10. Hana U. *Competitive Advantage Achievement through Innovation and Knowledge*. Journal of Competitiveness. **5** (1). P. 82-96. (2013).
11. Kalay K. and Lynn G.S. *The Impact Of Strategic Innovation Management Practices On Firm Innovation Performance.* Research Journal of Business and Management. **2** (3). (2015).
12. Kowang T.O., Long C.S. and Rasli A. *Innovation Management and Performance Framework for Research University in Malaysia.* International Education Studies. **8** (6). (2015).
13. Law P.H. (2013). Achieving Sustainable Organizational Profitability: The Management Of Innovation.
14. Merriam S. B. *Qualitative Research: A Guide to Design and Implementation*.P.18. (2009).
15. Mol J. and Birkinshaw J. *The sources of management innovation: When firms introduce new management practices.* Journal of Business Research. P. 1269-1280. (2009).
16. Mulgan G. *Innovation in the Public Sector How Can Public Organisations Better Create, Improve and Adapt.* NESTA. (2014).
17. Public Service Commission. *Creating an Innovative Public Sector Discussion Paper 2013.* (2013).
18. Seen N. *Organizational Culture and Innovation among Malaysian Employees*. The Journal of Human

Resource and Adult Learning. **8** (2). P. 147-158. (2012).

19. Shqipe G., Gadaf R. and Veland R. *Innovation Strategies and Competitive Advantages.* СОВРЕМЕННАЯ ЭКОНОМИКА: ПРОБЛЕМЫ, ТЕНДЕНЦИИ, ПЕРСПЕКТИВЫ. № 8. (2013).
20. Sohail A., Suleman Sabir M. and Zaheer A. *Link between Product Innovation and Non-Technological: Organization Performance.* Asian Journal of Business Management. **3**(4). P. 287-293. (2011).
21. Stresea et al. *Corporate culture and absorptive capacity: The moderating role of national culture dimensions on innovation management.* International Business Review. P. 1149-1168. (2016).
22. Tan C. and Nasurdin A. (2011). Human Resource Management Practices and Organizational Innovation: Assessing the Mediating Role of Knowledge Management Effectiveness. Electronic Journal of Knowledge Management. **9** (2). P. 155-167.
23. Widiartanto and Suhadak. *The Effect of Transformational Leadership on Market Orientation, Learning Orientation, Organization Innovation and Organization Performance.* IOSR Journal of Business and Management. **12** (6). P. 8-18. (2013).

Interactive Online Advertising: The Effectiveness of Marketing Strategy towards Customers Purchase Decision

Nurul Zarirah Nizam[1,*], *Johanna* Abdullah Jaafar[2], and *Siti Hindun* Supaat[3]

[1 2] Faculty of Technology Management & Technopreneurship, Universiti Teknikal Malaysia Melaka, Malaysia
[3] Centre of Graduate Studies, Universiti Teknikal Malaysia Melaka, Malaysia

Abstract. A growing of IT industry made many companies used Internet as their mediators on advertising platform to control over the promotion materials and it possess greater reliability and flexibility. Internet also becoming an important stop point for customers in finding their needs and wants. This study has been designed to measures how significant is the online advertising effectiveness in terms of attitude towards ads, ability to recall ads and frequency of clicking ads in explaining purchase decision in Malacca, Malaysia. A series of survey was conducted from 200 respondents and were analyzed quantatively. The results showed among the attitude towards ads, ability to recall and frequency of clicking ads had significantly gave impact to the purchasing decision by the customers. For the recommendations discussed to the advertisers to have good understanding on how to advertise effectively (using marketing strategy program) and do some innovation of the products in terms of the products association.

1 Introduction

Online advertising is widely used by companies and organization to promote their product and services. The reason is, online advertising able to capture the opportunity to reach their consumers via various online platforms to create an awareness for their products to lead their sales in future [2]. Online advertising as a form of promotion that uses the internet and worldwide web to deliver marketing messages to attract customers. Example of online advertising according to Wikipedia, include contextual ads on search engine results pages, banner ads, blogs, rich media ads, special networking advertising, interstitial ads, online classified advertising, advertising networks and e-mail marketing, including e-mail spam. Many of these types of ads are delivered by an ad server. The view that "rapid growth in online advertising revenues indicates the viability of worldwide web advertising as an alternative to that of traditional media considering internet advertising growth, there is little doubt that the internet is a powerful and viable alternative to traditional media advertising" [1]. On the other hand, online advertising is flexible and makes it easier for the company to update their products, information [4]. Due to the diversity of products and services, increasing customers' desire and demand, producers need to be faster and more efficient method to reach the consumer loyalty [28]. Consumers' attitude is direct human reflection towards the online advertising. The attitude of consumers may be influenced by the factors of online advertising, such as in formativeness, entertainments, creates awareness and annoying. For example, the internet users might found online advertising was informative but less entertaining and it did not encourage them to make purchases even they did not perceive it to increase product prices [12]. In this new medium, consumers have control over the advertising exposure as they can choose when and how much advertisement content they wish to view. Consumers will immediately terminate the visit if they find that the web advertisement is not useful [1]. In addition, consumers behavior on online advertising may evolve from time to time [8] thus there are still lack of understanding and information on how consumers react to online advertising in Malaysia [2].

2.0 Online Advertising

The effectiveness of online advertising in purchase decision and concluded that Malaysian consumers are quite receptive to online advertising with a positive attitude towards online advertising [12]. This study is to target the marketers and advertisers on the consumer preference of the advertising media and to determine whether using online advertising would be effective in reaching and creating awareness among the target audience. Internet has grown in recognition as an advertising platform because it allows 24 hours interactivity between the advertiser and the consumer. As more consumers has turn to the internet as their purchasing platform, it is important for the local

* Corresponding author: zarirah@utem.edu.my

businesses to look into internet advertising. A small enterprise that can offer online purchasing may be gain advantages among the competitors. Thus, it is imperative for the organizations to understand the characteristics of the online customers towards online advertising before adapting this new marketing practices. Other the other hand, the government sector would understand the value of internet advertising and its influence on consumer behaviors from this study. In addition, the government can effectively regulate how internet advertising is delivered among the consumers by acting rationally on laws that could restrict data usage and at the same time protecting the consumers' interest and rights. For scholars and academic researchers, the study will act as a reference for future research on internet advertising. The findings of this study can be resourceful in providing viable information to the researchers and consumers in the future.

2.1 Effectiveness of Online Advertising

Several effectiveness indicators of advertising on the Internet have been developed specifically to measure the persuasive power of this form of communication [15]. Previous academic studies incorporates a vast spectrum of studies on internet advertising effectiveness measured via direct response and branding metrics [22]. Internet advertising is a fairly new medium which has acquired large attention from academics and practitioners. The study of online advertising effectiveness has been carried out through two alternative paradigms. The primary, broadly used in academic research, argues that internet advertisement should be considered as a form of marketing communication used to elevate brand awareness. The second one, predominantly used in empirical studies, contends that the internet is a direct marketing medium, and subsequently a banner ad is likened to a coupon in print media [22]. Numerous studies have examined the effectiveness of Online advertising in the past decade [12]. There are few types of these common measures, attitudes toward the ad, ability to recall ad and clicks on ad, are the focus of the present study. The effectiveness of these three measures was assessed by their ability to predict online purchase decision [12].

2.1.1 Attitudes towards Online Ads

Attitude is a fairly global and enduring evaluation of an object, issue, person or action [6]. In other words, it could be described as an overall evaluation that expresses the extent of favor or disfavor of a person towards an object, issue, person, or action and it tends to be a long lasting emotional feeling. While the multi-attribute attitude model such as the Fishbein model is the most influential model and assumes that a consumer's attitude of an object will rely upon the beliefs she or he has about numerous or many attributes of the object [2]. Malaysia internet users or consumers are aware of advertising and they have a positive and negative perception of web advertising. From the study, they concluded that they still formed positive beliefs on web advertising such as for retrieving up-to-date information about products available in the marketplace and web advertising is a convenient source of good information [1]. On the study of assessing beliefs, attitudes, and behavioral responses toward online advertising in three countries: the five common belief factors, entertainment, information, credibility, economy and value corruption were significant predictors of attitudes towards online advertising [16] [17].

2.1.2 Ability to Recall Online Ads

There are a lot of researchers have studied numerous factors that might have an impact on online advertising recall [12]. These factors such as ad characteristics, internet users' viewing mode, and time spent on viewing and so on. Consumer able to recall the advertisement because of its necessity and usefulness of the ads. Other than that, prices of the product displayed online also influence the customer to recall the advertisement [4]. Furthermore, previous study additionally suggests that there has been a close association between animated advertisement and banner simple text followed by floating advertisement pop up advertisement and embedded video. It is also significant to note that online advertisement is taken into consideration that is important to majority of the respondents are of the opinion that they recall the product through seeing the advertising of the product online [4]. The effects and effectiveness of interactive advertising and found that traditional measures of advertising effectiveness, such as recall, attitude change, and brand choice are only a part of the story of effectiveness of interactive advertising [18]. However, such measures are useful, but they are in the tradition of advertising research that focused on the advertising's influence on the consumer. These measures offer limited insight into what the consumer does to and with advertising. Traditional measures such as awareness, recall, and attitude change and purchase behavior are also relevant in an interactive advertising context [18].

2.1.3 Frequency of Clicking Online Ads

A click-through rate (CTR) is a measure of online advertising effectiveness have examined the relationships between CTR and a series of factors such as web user motives, audience targeting, exposure frequency, copy content and a limited number of design elements of banner ads [22]. Several studies suggest that click-through effectiveness may depend on web user motives. Click-through rates provide the marketer with considerable advantages, as click-through record voluntary behavior in the actual medium environment [22]. On the study of internet advertising effectiveness: the effect of design on click-through rates for banner ads, the largest banner ads are more effective in generating click-through than the smaller-sized banners [22]. The author also suggested that unbranded banners stimulate greater curiosity, hence generating more clicks. On the study of the effectiveness of internet advertising through memorization and click on a banner, found that click-through rate is an operation technically easier to achieve in the sense that we can easily know the number of times the webpage is loaded and the number of times one has clicked on the banner [15].

3. Method

Quantitative method was used in this study and research was conducted through the use of questionnaire. The survey was conducted in Malacca and a total of 200 questionnaires has distributed to the target population. The target population was the consumers who live in Malacca City and their age range is between 19 to 26 years old. In this study, descriptive analysis was used to describe targeted respondent demographic by using frequency and percentage. Another analysis was Pearson Product Moment Correlation Coefficient (PMCC) used to assess the strength of relationship between the two variables as presented in Table 3.

4. Analysis and Result

Table 1. Descriptive Analysis of Respondents Demographic Profile

		Frequency	Percentage
Gender	Male	100	51
	Female	96	49
Age	20 and below	10	5.1
	21 – 30	103	52.6
	31 – 40	72	36.7
	41 and above	11	5.6
Marital Status	Single	118	60.2
	Married	78	39.8
Education Level	SPM	40	20.4
	STPM/DIP/A-levels	38	19.4
	Bachelor's Degree	94	48
	Master	19	9.7
	PhD	5	2.6
Employment Status	Student	65	33.2
	Employment/ Self-employed	112	57.1
	Unemployed/Housewife	19	9.7
Monthly Salary (RM)	1000 and below	81	41.3
	1001 – 2000	29	14.8
	2001 – 3000	35	17.9
	3001 – 4000	30	15.3
	4001 and above	21	10.7

The background of the respondent is important in conducting research. This is because the respondent contributed to the value of the research as well. The general information of the respondents which involved the gender, age, employment status and monthly salary (refer Table 1). The number of male respondents is more than female which represent 100 person (51.02%) and 96 person (48.98%) of each. Specifically, based on Table 1 we can see, gender: Male contribute more with 51%, Age: within 21 – 30 years old represent 52.6%, marital status: single person with 60.2%, educational level: bachelor's degree with 48%, employment status: employment/self-employed contribute 57.1% and monthly salary: 1000 and below with 41.3% which represent dominant in every level of demographic.

Table 2. Frequency click on online advertisement by respondents

		Frequency	Percent	Valid Percent
Valid	Click at least once	48	24.5	24.5
	Occasionally click	88	44.9	44.9
	Frequently click	60	30.6	30.6
	Total	196	100.0	100.0

From the total of 196 respondents, 59 (30.1%) people averagely spend between 16 – 25 hours per week online. The second highest is between 26 – 35 hours with 58 (29.59%) people. There are 38 (19.39%) people average spend between 6 – 15 hours per week online, followed by 25 (12.76%) people average spend 5 hours and below online. Lastly, there are least 16 (8.16%) people send 35 hours and above per week online as showed in Table 2.

Table 3. Pearson's Product Moment Correlation Coefficient (PMCC)

	1	2	3	4
1.Attitude towards ads	1			
2.Ability to recall ads	.700*	1		
3.Frequency of clicking ads	.847**	.763**	1	
4.Purchase decision	.778**	.745**	.821**	1

Note: N=196 **Correlation is significant at the 0.01 level (2-tailed).

Table 3 stated that the *p*-value (sig.) of the correlation between independent variables (attitude towards ads, ability to recall ads and frequency of clicking ads) and the dependent variable (purchase decision) is less than 0.01. This reflects that there is a significant relationship between them. The positive value of correlation coefficient indicated that the relationships between each independent variables and dependent variable are positively related. The strength of the relationships is strong as all the values of correlation coefficient are greater than 0.7.

Among all the independent variables, the strength of association between attitude towards ads and frequency of clicking ads are the strongest (r=0.847), and the correlation coefficient is highly significant (p=0.000). This explained that the attitude towards ads have a positive and strong relationship with the frequency of clicking ads.

Conversely, the strength of association between attitude towards ads and ability to recall ads is the weakest among all the independent variables (r=0.700). This shows that there is a moderate relationship between the two variables.

On the other hand, the *p*-value (sig.) of the correlation between independent variables (attitude towards ads, ability to recall ads and frequency of clicking online ads) and the dependent variable (purchase decision) is 0.000 which is less than 0.01. This reflects that there is a significant relationship between them. The value of correlation coefficient of frequency of clicking ads and purchase decision is 0.821, which indicated that there is a strong positive relationship between this two variable. Followed by attitude towards ads and purchase decision with the value of 0.778 and lastly the ability to recall ads and purchase decision with the value of 0.745.

4.1 Multiple Linear Regression

Table 4. R square

Model	R	R Square	Adjusted R Square	Std. Error of the Estimate
1	.851[a]	.724	.720	.29685

a. Predictors: (Constant), Frequency of Clicking ads, Ability to Recall ads, Attitude toward ads

Table 5. Coefficients

	Beta	t	Sig.
Attitude toward ads	.244	3.387	.001
Ability to Recall ads	.252	4.240	.000
Frequency of Clicking ads	.422	5.296	.000

a. Dependent Variable: Purchase Decision

Regression analysis is used to find the relationship between the two variables. Regression analysis can also be used to predict the values of a dependent variable given the values of one or more independent variables by calculating a regression equation.

Table 5 reported that ANOVA [a] shows that the F-ratio value of 167.873 with a significance level of 0.000[b] (*p*-value = 0.000), which is below 0.01. This indicated that there is a statistically significant difference between the dependent variable (purchase decision) and independent variable (attitude toward ads, ability to recall ads and frequency of clicking ads). Therefore, the overall model is significance.

Standardize coefficients (Beta) among the independents variables ranged from the weakest relationship of 0.244 (between attitude toward ads and purchase decision) to the strongest relationship of 0.422 (between frequency of clicking online ads and purchase decision). Therefore, this shows that the frequency of clicking online ads has the strongest effect on consumer online purchase decision. Whereas, attitude toward ads has the lesser impact on the purchase decision. While ability to recall ads ranked as the second important independent variable with the coefficient of 0.252.

5 Conclusion

5.1 Attitude towards online ads is significantly related to purchase decision.

Referring to Table 6, from the 3 independent variables, attitude towards online advertising has the weakest effect on purchase decision. The beta value of attitude towards online advertising is 0.244 with 0.000 significant value. This indicated that there is a significant relationship between attitude towards online advertising and purchase decision. Attitudes are primarily based on the cognitions or beliefs which mean attitudes can be formed based on thoughts and beliefs we've about the information received [6]. In other words, consumers who believes the information of the product or services will have favorable attitude and purchase intention. When a person wants to purchase a product or services, he or she will make decision based on the available information and also based on the belief of the individual towards the product or services. Attitude towards a product or brand can be anticipated by figuring out these specific beliefs and mixing them to derive a degree of the consumer's overall attitude [1]. Exploring consumers' attitude towards web advertising and its influence on web ad usage in Malaysia [2]. Malaysia internet users or consumers are aware of advertising and they have a positive and negative perception of web advertising [2]. From the study, they concluded that they still formed positive beliefs on web advertising such as for retrieving up-to-date information about products available in the marketplace and web advertising is a convenient source of good information. Besides, individuals can make behavioral decisions via careful consideration of the available information [9]. In other words, purchase decision can be influenced by the available information which based on the attitude of the consumer towards the product or services.

5.2 Ability to recall online ads is significantly related to purchase decision.

Ability to recall online ads has the second strongest effect on purchase decision. The beta value of ability to recall online ads is 0.252 with 0.000 significant values which indicated that there is a significant relationship between ability to recall online ads and purchase decision. Means that, the consumers that have higher ability to recall the ads will most likely purchase the product or services. This is because the consumer spent more time and repeatedly view on the advertisement. There are a lot of researchers have studied numerous factors that might have an impact on online advertising recall [12]. These factors such as ad characteristics, internet users' viewing mode, and time spent on viewing and so on.

5.3 Frequency of clicking online ads is significantly related to purchase decision.

For frequency of clicking online ads has the strongest effect on purchase decision, the beta value of frequency of clicking online ads is 0.422 with 0.000 significant values. This indicated that there is a significant relationship between frequency of clicking online ads and purchase decision. This result shows that consumer with higher frequency of clicking on the advertisement will have chances to purchase the product or services. A click-through rate is a measure of online advertising effectiveness and it record the voluntary behavior in the actual medium environment [21]. The author found that the largest banner ads are more effective in generating click-through than the smaller-sized banners. The author also suggested that unbranded banners stimulate greater curiosity, hence generating more clicks. In other words, sizes of the banner and curiosity will influence the clicks from the consumer. When the consumers click on the advertisement, it will more likely generate the positive impact on the consumer to purchase the product or services. This is supported by the study [11]. Advertisement that appear to be interesting and attractive to create consumers' curiosity and draw their attention, they may click on the advertisement [11]. Consumer will click on the advertisement if they appear to be interesting and stimulating visually when they are doing aimless browsing for something interesting, which may lead to impulse buying [11]. Besides that, a potential barrier to advertising effectiveness is the quantity of ads to which consumers are exposed, which it indicate that online ad position can also play a position in attracting visitors [23]. This causes the interactive rich media ad earned a substantially higher average daily click-through rate.

Nowadays, almost everyone has Internet access in their life due to the technology advancement. Therefore, with the increasing importance of internet in e-commerce, advertisers should fully utilize this online platform for their advertising activities that fulfill their communication purposes with consumers. The purpose of the study was to measure the effectiveness of online advertising on consume behavior in Melaka. According to the literature review, attitude towards ads, ability to recall ads and frequency of clicking ads was the determinants of consumer behavior. The result has indicated that each variable has significant relationship with the purchase decision. Based on the result, frequency of clicking ads with beta value of 0.422 has the strongest impact on purchase decision. In other words, frequency of clicking ads was the most significant determinant that influence purchase decision. This is because when consumer click on advertisement that draws their attention, it will more likely lead them to purchase the product or services. Lastly, from the results of the research, it has provided evidences to support that attitude toward ads, ability to recall ads and frequency of clicking ads are important measurement tools for company or marketer to take into consideration for them to improve their advertising skills. By understanding all the important elements in online advertising, marketer can create a better and effective advertisement to draw attention from consumers.

6 Conclusion and Future Work

For the future research, future researchers are recommended to broaden the research by incorporating a larger sample size of respondents. Sample are recommended to draw from different geographic area and the age of the respondents should also be taken into consideration when drawing the sample. Future researcher should cover the whole Malaysia to obtain a more accurate result. Most of the respondents in this research are young consumer, therefore, the future researchers can target as many as possible different age of consumer. Thus, the result can indicate the effectiveness of online advertising in each level of age. This can also generate a better and more accurate result. Furthermore, the future research questionnaire should be provided in dual languages such as Bahasa Malaysia or Mandarin in order to help the respondents who are poor in understanding English. This method may help the researcher to save time for explanation and enable to reach out to more respondents.

Last but not least, due to lack of information about online advertising causes difficulty for researcher to get information. Therefore, more research should be done in Malaysia. In addition, as technology going fast nowadays, research should not just focus on traditional advertising but also online advertising. This will be easier for future researchers to get information regarding this topic.

Thanks are expressed to Low Yong Jie for his help in collecting the data used in the study while fulfilling his role as an undergraduate student at the time.

References

1. Aziz, N., Yasin, N. & Kadir, B., 2008. Web Advertising Beliefs and Attitude: Internet Users' View. *The Business Review, Cambridge*, 9(**2**), pp.332–339. Available at: http://abufara.com/abufara.net/images/abook_file/Web Advertising Beliefs and Attitude- Internet Users%E2%80%99 View.pdf [Accessed November 25, 2016].
2. Baumeister, R.F. et al., 2008. Free will in consumer behavior : Self-control, ego depletion, and choice. 18, pp.4–13.
3. Busen, S.M.S., Mustaffa, C.S. & Bahtiar, M., 2016. Impacts of online banner advertisement on consumers' " purchase intention : A theoretical framework. , 3(**1**), pp.75–82.
4. Chaubey, D.S., Sharma, L.S. & Pant, M., 2013. Measuring the Effectiveness of Online Advertisement in Recalling a Product : An Empirical Study. , pp.37–47.
5. Hoyer, W. D. & MacInnis, D. J., 2010. Consumer Behavior (5th Ed.). Boston: Cengage Learning.
6. Johnson, E. J., Pham, M. T. & Johar, G. V., 2007. Consumer Behavior and Marketing. In: Social Psychology. Handbook of Basic Principles (2nd Ed.). New York: The Guildford Press, pp. 869-887.
7. Karson, E.J., McCloy, S.D. & Bonner, P.G., 2006. An Examination of Consumers' Attitudes and Beliefs towards Web Site Advertising. *Journal of Current Issues & Research in Advertising*, 28(2), pp.77–91. Available at: http://www.tandfonline.com/doi/abs/10.1080/10641734.2006.10505200.
8. Kim, J.U., Kim, W.J. & Park, S.C., 2010. Consumer perceptions on web advertisements and motivation factors to purchase in the online shopping. *Computers in Human Behavior*, 26(5), pp.1208–1222. Available at: http://dx.doi.org/10.1016/j.chb.2010.03.032.
9. Kotler, P. & Amstrong, G., 2010. Principles of Marketing (13th Ed.). New York: Pearson.
10. Lee, S.M. & Chen, L., 2016. The Impact of Flow on Online Consumer Behavior the Impact of Flow on Online Consumer Behavior University of Nebraska-Lincoln. , 4417(February).
11. Lim, Y., Yap, C. & Lau, T., 2011. The Effectiveness of Online Advertising in Purchase Decision : Liking, Recall and Click. , 5(**9**), pp.1517–1524.
12. Mazaheri, E., Richard, M. & Laroche, M., 2011. Online consumer behavior : Comparing Canadian and Chinese website visitors. *Journal of Business Research*, 64(**9**), pp.958–965. Available at: http://dx.doi.org/10.1016/j.jbusres.2010.11.018.
13. Melissa, H., 2014. CMOs Are Preparing for Digital to Grow to 75% of Marketing Budgets. Available at: http://www.adweek.com/digital/cmos-expect-75-marketing-budgets-be-digital-within-5-years-158481/.
14. Nihel, Z., 2013. The Effectiveness of Internet Advertising through Memorization and Click on a Banner. , 5(**2**), pp.93–101.
15. Nikhashemi, S.R., Paim, L. & Sharifi, S., 2013. The Effectiveness of E-Advertisement towards Customer Purchase Intention : Malaysian Perspective. , 10(**3**), pp.93–104.
16. Nila, S.C.V., 2013. Consumer Attitude towards Internet Advertisement. , (March), pp.264–265.
17. Pavlou, P.A. & Stewart, D.W., 2013. Measuring the Effects and Effectiveness of Interactive Advertising Advertising : A Research Agenda. (September 2014), pp.37–41.
18. Poddar, A., Donthu, N. & Wei, Y., 2009. Web site customer orientations, Web site quality, and purchase intentions: The role of Web site personality. *Journal of Business Research*, 62(**4**), pp.441–450. Available at: http://dx.doi.org/10.1016/j.jbusres.2008.01.036.
19. Rani, P., 2014. Factors influencing consumer behaviour. , pp.52–61.
20. Razzouk, N. & Seitz, V.A., 2008. Banner Advertising and Consumer Recall Banner Advertising and Consumer Recall : An Empirical Study. , 6491(June).
21. Robinson, H. et al., 2015. International Journal of Advertising : Internet advertising effectiveness Internet advertising effectiveness the effect of design on click-through rates. , (January), pp.37–41.
22. Rosenkrans, G., 2013. The Creativeness and Effectiveness of Online Interactive Rich Media Advertising. , (February 2015), pp.37–41.
23. Rosenkrans, G.L., 2006. Online banner ads and metrics challenges. , 3(**3**), pp.193–218.
24. Rani, P., 2014. Factors influencing consumer behaviour. , pp.52–61.
25. Robinson, H. et al., 2015. International Journal of Advertising : Internet advertising effectiveness Internet advertising effectiveness the effect of design on click-through rates. , (January), pp.37–41.
26. Wang, Y. & Sun, S., 2010. Assessing beliefs, attitudes, and behavioral responses toward online advertising in three countries. *International Business Review*, 19(**4**), pp.333–344. Available at: http://dx.doi.org/10.1016/j.ibusrev.2010.01.004.
27. Zourikalatehsamad, Naser Payambarpour, S.A., Alwashali, I. & Abdolkarimi, Z., 2015. The Impact of Online Advertising on Consumer Purchase Behavior Based on Malaysian Organizations. , 9(**10**), pp.3340–3345.

 https://doi.org/10.1051/matecconf/201815005044

A Revised Production Model of Learner-Generated Comic: Validation through Expert Review

Farah Nadia Azman[1*], *Syamsul Bahrin* Zaibon[2], and *Norshuhada* Shiratuddin[2]

[1]Faculty of Information and Communication Technology, Universiti Teknikal Malaysia Melaka, 76100 Malacca, Malaysia
[2]School of Multimedia Technology and Communication, Universiti Utara Malaysia, 06010 Kedah, Malaysia

Abstract. Recent advancement of authoring tools has fostered a widespread interest towards using comics as a Digital Storytelling medium. This technology integrated learning approach is known as learner-generated comic production; where learners constructively produce digital stories in a form of educational comics. However, there were concerns towards the obstacles and challenges of producing learner-generated comics. Hence, a conceptual production model of learner-generated comic was proposed to guide learners in designing and developing digital educational comics. Accordingly, as the decision making stage for validating the proposed model, expert review method was adopted. Results of expert review were coded and classified into flexibility, understandability, completeness, generality, and usability aspects, aligning with dimension of conceptual model characteristics. Consequently, a final appraisal cycle with experts was conducted to approve the revised and redesigned LGC production model based on expert review. In summary, the experts concluded that the proposed model replicates the process of learner-generated comic production very well, visually and descriptively. Suggestion of future research is put forward.

1. Introduction

Recent advancement of authoring tools has fostered a widespread interest towards using comics as a Digital Storytelling medium. This technology integrated learning approach is known as Learner-Generated Comic (LGC) production; where learners produce digital stories in a form of educational comics [1].

According to [2], while learning "from" media and technology concentrates on encoded information in tutor tools, the focus of learning "with" media and technology approach is on the cognitive tools in constructivist environments. Cognitive tools require learners to think mindfully in order to use the application to represent what they know [3] and provide visual representations of tasks and facilitate learners towards their execution [4]. 'Media' on the other hand, is the symbol system that educators and learners use to represent knowledge; while 'technology' is the utilized tool to build and share their knowledge representations with others [2]. This denotes that LGC production belongs to learning "with" media and technology taxonomy; whereas digital educational comic is the symbol system in presenting knowledge.

To date, literature has rapidly reported on the potential, and benefits of LGC production (eg: [5], [6], [7]). However, although learners expressed high interest in LGC production, it was revealed that majority of them were concerned towards the obstacles they would encounter when producing LGC such as presentation of educational content, and utilization of comic elements and storytelling techniques into their LGC [8].

Hence, to undermine these challenges, a conceptual production model of LGC was proposed. As a cognitive tool [3], it was hypothesised that the LGC production model would significantly serve as a guideline for learners to design and develop digital educational comics.

Adopting Design Science Research methodology [9], the proposed model was structurally constructed through content analysis, comparative study of LGC classroom strategies, Digital Storytelling models and frameworks, expert consultation, and user participation. Principally, the core components of LGC production model were classified into phases and supported by task, activity, and flow subcomponents. Accordingly, the proposed production model of LGC was evaluated through user experience testing and expert review. Thus, this paper aims to discuss the selection of experts, findings, and results obtained from the review of LGC production model.

2. Expert Review

Essentially, expert reviews can occur early or late in the design phase [10]. In this study, expert review was chosen as the decision making stage for validating the proposed LGC production model. This was because experts recognize and understand the concept of heuristic [11]. Plus, to compound in trustworthy results, the experts investigated the features of the proposed model [12]. The

* Corresponding author: farah@utem.edu.my

results obtained from expert review were gathered to replicate several key questions of conceptual model evaluation [13] as follows:

- Are the facts described by the model accepted as correct by the domain experts?
- Are the described instances on the desired level of detail?
- Is the model conforming to necessary standards?

2.1 Instrument

The instrument design was based on previous expert review questionnaires [14, 15]. The questions ask about the (i) relevancy of the proposed phases which represent the main stages of LGC production, (ii) necessity of the proposed tasks and the activities within them, (iii) connections and flows of all of the components, (iv) usability of the proposed model, and (v) readability of the proposed model. The rest of the items were adapted from maturity model experts review instrument by [16] because they were specifically designed for model evaluation by domain experts.

2.2 Profile of Selected Experts

Eight experts were involved in this activity where majority of them have either theoretical or industrial contributions in comic field. The numbers of experts were regarded adequate as stated in the conditions set by [17]. Experts involved in the review process were selected based on the criteria described in Table 1.

Table 1. Criteria of Expert Selection

Category	Criteria
Academician	• Have PhD qualifications either in Human Computer Interaction (HCI) or Multimedia or Instructional Design or Art and Design related area OR/AND • Have been studying/researching either in comics, Human Computer Interaction (HCI) or Multimedia or Instructional Design or Art and Design related area for at least five years.
Practitioner	• Have at least five years of professional experience in comic industry.

Demographic information about the experts such as age, gender, position, and experience were collected to support the dependability of the selected experts in evaluating LGC production model (see Table 2).

2.3 Procedure

The review process started with the invitation via email to the identified experts. After the experts agreed and replied, the official appointment letter by the dean, and consent form were sent to them. Next, the experts received an illustration of the proposed model together with expert review form through email. The objective and scope of the proposed model were made clear to the experts where the target users are undergraduate students and the model of LGC production does not involve illustrating comics; but focuses on educational storytelling with the utilization of comic authoring software instead.

The experts were instructed to observe and analyse the supplied illustration and description of LGC production model before carefully filling up the provided spaces in the review form. The experts were required to note problematic features by inspecting the components and items relevancy in the model and predict potential problems when users interact with it. It took approximately four to ten weeks to accomplish the expert review process. The researcher also conducted a face to face meeting with four of the selected experts during the mentioned time frame.

Table 2. Demographic Profile of Experts

Expert	Gender	Age (Year)	Position	Academic Qualification	Affiliation	Experience (Year)
A	Male	55	Professor	PhD	Universiti Teknologi MARA	40
B	Male	44	Associate Professor	PhD	Ramapo College of New Jersey	24
C	Female	39	Associate Professor	PhD	University of North Florida	15
D	Male	46	Academician	PhD	Universiti Teknologi MARA	14
E	Female	42	Academician and Practitioner	PhD	Universiti Teknologi MARA and MOY Sdn Bhd	23
F	Female	33	Academician and Practitioner	Masters	Universiti Teknologi MARA and Gempakstarz	10
G	Male	37	Practitioner	Diploma	Comicore Creative Solution	12
H	Male	35	Practitioner	Bachelor Degree	ABS Holdings	6

3. Findings and Discussion

Initially, results from the expert review demonstrated that; although majority of them approved most components proposed in the model, there were moderate agreement towards "Scriptwriting" phase and "Characters" and "Basic" tasks (refer to Fig 1 and Fig 2). The experts however has come to agreement that the proposed model is usable, has logical flow, and the terminology used is understandable.

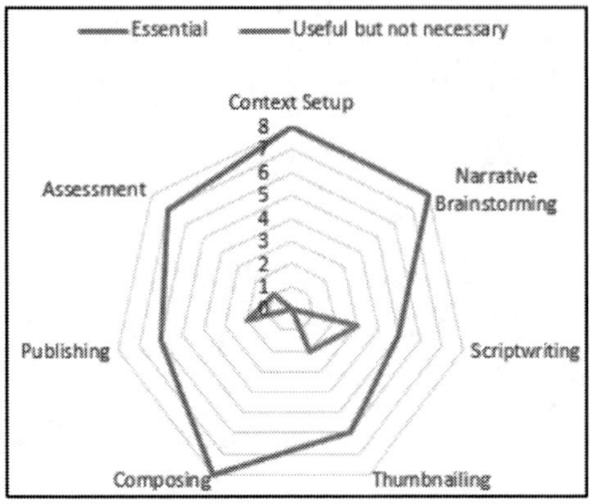

Fig. 1. Frequency of the proposed phases' relevance in LGC production model.

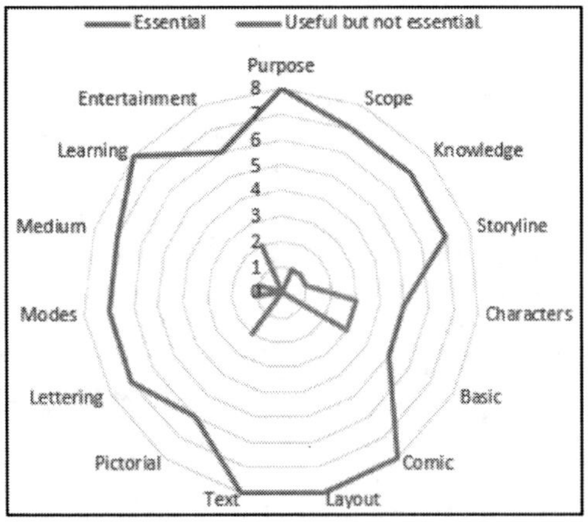

Fig. 2. Frequency of the proposed tasks' relevance in LGC production model.

To further contemplate the feedbacks given by the experts, all of the experts' written comments were also qualitatively analysed. In conveying the clearer meaning, some of the comments were rephrased from the original versions as exhibited in Table 3.

Table 3. Further Comments from the Experts

Expert	Comments
A	(1) Target audience should be included in The Context Setup phase. For example, younger audience. (2) In Narrative Brainstorming phase (narrative type), may include daily scenes and symbols. (3) The description under Character activity may include the aspect of size and physical appearance. Dialogue must be appropriate for target readers. (4) The model is understandable, practical and flexible. (5) Overall, it is very important to identify the target reader, audience or user of the comic from the beginning of the process. This will determine the suitability of the following process.
B	(1) Context Setup phase is very important as it sets up the very purpose of the project. (2) You did a nice job in giving the scope of how students should pull together their ideas for the project. (3) You might add a discussion activity as part of Narrative Brainstorming or after it. It would be good for students to provide feedback for each other as they begin to plan and craft their comics. (4) Scriptwriting or Thumbnailing is certainly something to include, but given the time constraints of a typical classroom, perhaps it is something you would consider making optional. (5) Good job with giving a broad range of Composing phase options in this field. Perhaps consider providing ideas for outlets for publication/exhibition of the comics in Publishing phase. (6) I like how Assessment phase component is included so that there is a clear outcome to work toward. (7) Model description is very clear and understandable. (8) You have done an excellent job in the creation of this model. You have provided a very usable and effective model. I think it will be very useful and effective for educators. As you move forward, you may want to provide actual examples of these steps in action, perhaps through video. You might also consider providing an addendum to the model for classrooms that don't have immediate access to technology (that is, using hand-drawn comics). Overall, I think this is excellent work.
C	(1) How are you defining Basic phase? Basic of what? (2) Yes, the model will have logical flow, usable and readable once you fixed it. Refer to my notes. (3) How are Layout and Pictorial tasks different? Please specify. (4) Text and Lettering tasks seem to be overlapping too much. Are they more similar or different? (5) I can't wait to see how well this goes. Super fantastic project!

D	(1) Typography is essential because it defines the comic appeal. (2) Anthropomorphic should be added to assess character design. (3) Characters could be grouped under Storyline task. (4) Add narrative structure such as Gustar Freytag model in the description. (5) Overall, the model has listed all the necessary components for developing comic. However, it could also consider inserting these elements – content analysis, Freytag analysis, and anthropomorphism.
E	(1) Add symbol and semiotic elements into the model. (2) Character is considered an element in Storyline. (3) The model should be readable in monochrome. Therefore, consider eliminating colours for the phases' blocks. (4) Visually, the model does not require complex shape for each phases to avoid confusion. Stick to consistent shape for every proposed phase. (5) The model should be able to deliver message in a simpler way.
F	(1) The Context Setup is very essential for setting up the foundation for whole production. Narrative Brainstorming is important for group project. Scriptwriting is not essential because this usually can be included in thumbnail process. Thumbnailing is important because the Rough detail sketches fasten the next process. Composing is required for quality control can be made at this time before publish. Publishing save all the documentation and makes the final product run smoothly. (2) For comic publishing stage, I would add up special phase for digital development process especially for Androids and OS platform. Finishing up the comic is just a first part of the game, the other half is to make sure the comic run smoothly in variety platforms in every screen size of gadgets. Tasks for animated and Interactive comics should be considered by adding GIFs, or essential GUI (Graphic User interface) to be put for easy reading in the digital platform. Simple games can also be added to make the digital comics more interactive. (can easily be done in Flash or Construct). (3) The word 'Iconography' it is kind a bit off for visual metaphor representation in comic process. There is no definite term for this stage but I can only explain it to people as the 'exaggerated symbols' that defines movement and emotion in static world in comics. (4) I love the presentation of the chart flow of the production, but need more detail explanation based on different learning scopes. (5) The model it can be visually practical and understandable, the pipeline description in interactive infographics will be nice.
G	(1) I would not recommend Scriptwriting to be an essential phase because this phase process can be done when thumbnailing the comic. (2) Overall, the model is a decent guideline for effective comic production. But it is up to the comic artist to decide which his/her own preferable methods in developing comics.
H	(1) The model is adaptable for developing comic for different lesson/subjects. (2) The model could include examples from the proposed components. (3) The theme of comic should be clear. (4) Since the aim of educational comic is to motivate learning, help understanding content and encourage high level of thinking, this model is helpful for facilitating the process to achieve those aims through comic development.

Given the mixed results towards certain phases and tasks depicted in Fig. 1 and Fig. 2, the feedback and suggestions from experts in Table 3 were analysed and considered in refining the proposed model. Aligning with dimension of conceptual model characteristics [18], the expert comments from the review were coded and classified into flexibility, understandability, completeness, generality, and usability aspects.

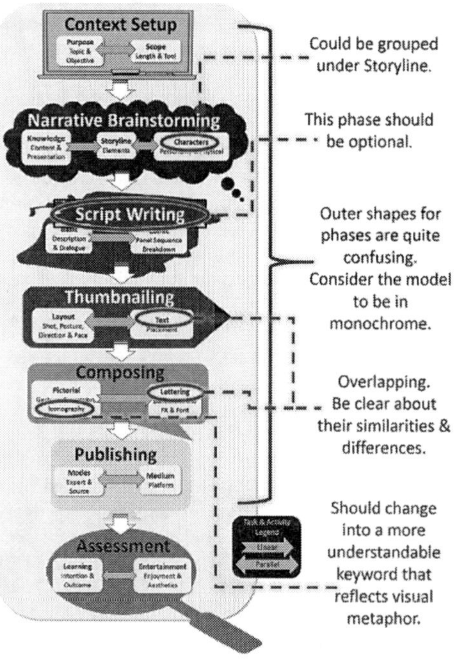

Fig. 3. Several commentary towards LGC production model.

The most apparent critique was towards the flexibility aspect on LGC production model. Primarily, three experts stated "Scriptwriting" phase as useful but not essential. Accordingly, since some practitioners directly draft their script into the sketch thumbnails, Expert F and Expert G recommended that "Scriptwriting" and "Thumbnailing" "phases should be combined. In fact, Expect B prompted that these two phases should be optional due to constraints of a typical classroom. Moreover, Expert C argued that the "Text" task within "Scriptwriting" seems to be intersecting with "Lettering". Hence, to tailor these issues, the "Text" task was replaced by "Script" task and merged under the "Thumbnailing" phase. These criticisms were decisively accepted by the researcher because it had been proven in the user experience testing stage where several participants concentrated straightforwardly on their thumbnails layout rather than the story script. In addition, the revised model also grouped "Character" as an activity within the "Storyline" task as recommended by Expert D and Expert E.

Secondly, the focus was on the understandability aspect of the proposed model. Answering Expert E's comments regarding the visual representation of LGC production model, readability of the diagram could be improved without confusing phase shapes. For example, during the user experience testing activity, participants could not figure out the typewriter shape for "Scriptwriting" phase and computer monitor for "Publishing" phase. Thus, this study visually revised the model by using consistent rounded rectangle shape for each phases. However, Expert E's suggestion on modifying the diagram colours was omitted. This was because the colours matched with the phases' description referred during LGC production. Regarding Expert F's critique about "Iconography" keyword, after a discussion with Expert D and F, this was changed into "Visual Metaphor".

Next, the completeness aspects from the review were scrutinised. First, inclusion of content analysis, narrative models, and anthropomorphism was suggested by Expert D. As a response, the first two elements were embedded in "Knowledge" and "Storyline" tasks with the "Narrative Brainstorming" phase. Depending on the LGC's story, anthropomorphism would be an optional element under the "Character" task. Secondly, addition of symbol and semiotic were proposed by Expert E. These elements were considered inclusive in "Pictorial" and "Text" tasks within "Composing" phase. Thirdly, as addressed by Expert H, the theme element had already included in the "Purpose" task. Fourthly, Expert A highlighted the importance of target audience to be included in the model. This element is certainly associated with the Purpose task. When "Topic" and "Objective" are defined, target audience should also be relatively clarified by the learner. As a whole, most of the suggested elements by experts were included in the revised description section of LGC production model.

The ensuing review focused on generality aspect. Concerning Expert C's argument on the difference between "Layout" and "Pictorial" tasks, the clarification were made in the model description. Despite their resemblance, "Layout" task concentrated on experimenting with the visuals before intensifying them during "Pictorial" task. Although Expert B evoked that discussion should be added into the model, Expert F exclaimed that this embedded activity is more relevant to group projects within "Narrative Brainstorming" phase. In regards to Expert E's recommendation to add a specific process in assuring the comic run smoothly in variety platforms in every screen size of gadgets, this element had been included in "Medium" task within the "Publishing" phase. Thus, responding to the experts' recommendations, these suggested elements were included in the revised description section of LGC production model.

Finally, Expert B, Expert F, and Expert H emphasized on the usability aspect of the proposed model. For instance, Expert B suggested actual examples of LGC production process in action through video. Expert F on the other hand preferred an interactive infographic version of proposed model. Lastly, Expert H advocated to provide examples of the components proposed in LGC production model. These useful feedback were taken into major consideration in future study. This study undoubtedly acknowledged that a multimedia representation of the proposed model is an efficient factor in improving LGC production implementation.

4. Results and Conclusion

Lastly, a final appraisal cycle with experts was conducted to approve the revised and redesigned LGC production model based on expert review as illustrated in Fig. 4. Expert C, Expert F, and Expert H committed to this activity.

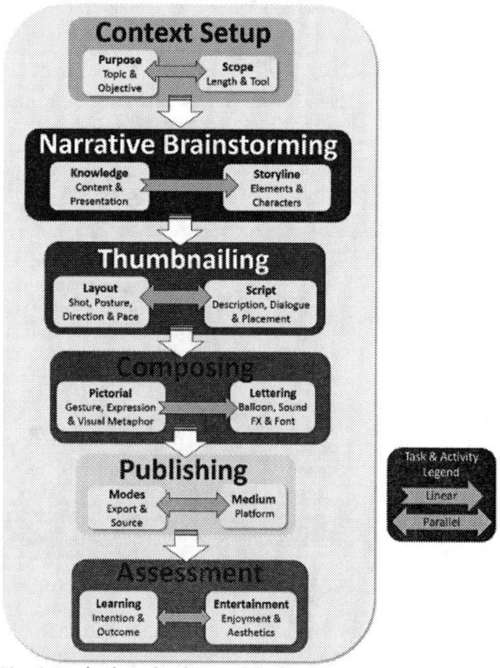

Fig. 4. Revised production model of LGC.

In summary, the experts concluded that the proposed model replicates the process of LGC production very well, visually and descriptively. At this rate, the production model of LGC has been thoroughly validated. In general, the components and description in LGC production model were revised based on the quantitative and qualitative data from expert review. Ultimately, the proposed LGC production model reflects the novelty and practicability of the relevant theories pertaining to comic and multimedia production. This provides new means for educators to adopt solid model of LGC production into their project-based learning classroom practice. Thus, future works should explore on extending the proposed model encompassing production of motion and interactive educational comics.

This work was supported by the Centre for Research and Innovation Management (CRIM), Universiti Teknikal Malaysia Melaka (UTeM).

References

1. F.N. Azman, S.B. Zaibon, & N. Shiratuddin, Journal of Theoretical and Applied Information Technology **89**, 2, 461-469 (2016)
2. D. H. Jonassen, & T. C. Reeves, *Handbook of Research for Educational Communications and Technology* (Macmillan, New York, 1996)
3. D. H. Jonassen, *ITForum Paper*, 67–80 (1994)
4. E. Tambouris, M. Zotou, K. Tarabanis, Education and Information Technologies, **19**, 1 , 61–86 (2014)
5. M. J. Green, Journal of Medical Humanities, **34**, 4, 471–476 (2013)
6. R. D. Watkins, PhD Thesis. Iowa State University (2014)
7. K. Comer, Composition Studies, **43**, 1, 75–104 (2015)
8. F.N. Azman, S.B. Zaibon, N. Shiratuddin, International Review of Management and Marketing **6**, 8S (2016)
9. V. Vaishnavi & B. Kuechler, *Design Science Research Methods and Patterns: Innovating Information and Communication Technology*. (CRC Press, New York, 2007)
10. B. Shneiderman & C. Plaisant, *Designing the User Interface* (Pearson, Addison-Wesley, 2005).
11. S. A. Aziz, S. N. A. Salam, A. A. Mutalib, S. Ismail, Journal of Telecommunication, Electronic and Computer Engineering (JTEC) **8**, 8 (2016)
12. P. Psomos & M. Kordaki, *Procedia - Social and Behavioral Sciences*, 46, 2012, 1213–1218 (2012)
13. F. Wolff & U. Frank, *13th European Conference on Information Systems, Information Systems in a Rapidly Changing Economy, ECIS 2005*, 1283–1294 (2005).
14. S. M. Sarif, PhD Thesis, Universiti Utara Malaysia (2011)
15. N. Aziz, PhD Thesis, Universiti Utara Malaysia (2014)
16. D. Salah, R Paige & P. Cairns, *Product-Focused Software Process Improvement. PROFES 2014. Lecture Notes in Computer Science*, Springer, 8892 (2014)
17. J. Nielsen, *Proceedings of the third international conference on human-computer interaction on Designing and using human-computer interfaces and knowledge based systems*, 394–401 (1989).
18. F.N. Azman, S.B. Zaibon, & N. Shiratuddin, International Journal Of Interactive Digital Media, **4**, 2, 28-32 (2016)

Geographical Information System (GIS) as an Innovative Adoption to Ease Customers in Locating SME Business Premises in Malacca

*Yusri bin Arshad[1], Muhammad Syahir bin Mohd Sani[2], Sharifah Norhafiza binti Syed Ibrahim**

[1, 2] Faculty of Technology Management and Technopreneurship, Universiti Teknikal Malaysia Melaka, Hang Tuah Jaya, 76100 Durian Tunggal, Melaka, Malaysia.
[1] Sustainable IT Economics and Information system research group (SuITE), Centre for Technopreneurship Development (C-TeD) Centre of Excellence, Universiti Teknikal Malaysia Melaka
[*] Faculty of Accountancy, Universiti Teknologi Mara, Jasin Campus, Melaka, Malaysia

Abstract. Geographical Information System (GIS) is one of the recent innovation to ease users in locating places. GIS is a system designed to capture, store, manipulate, analyze, manage, and present all types of spatial or geographical data. It can also be used by customers in locating business premises. Research related to application of GIS in locating SME business premises especially in Malacca as a tourist state is still very limited. This study used a survey method to produce more generalizable findings. Quantitative procedures for data collection and analysis were rigorously applied to ensure reliability and validity of the findings. A total of 160 usable responses was collected from 400 distributed questionnaires. This study found all variables significant and therefore, all hypotheses were accepted. In the future, researchers may extend this study to other areas to understand more about current usage patterns and reasons for its adoption. Studies which cover non-Malaysian respondents and different education levels can further document different perspectives of GIS adoption.

1 Introduction

A geographic information system (GIS) is a system designed to capture, store, manipulate, analyze, manage, and present all types of spatial or geographical data. The acronym GIS is sometimes used for geographic information science (GIScience) to refer to the academic discipline that studies geographic information systems and is a large domain within the broader academic discipline of geoinformatics. The term describes any information system that integrates, stores, edits, analyzes, shares, and displays geographic information. GIS applications are tools that allow users to create interactive queries (user-created searches), analyze spatial information, edit data in maps, and present the results of all these operations. Geographic information science is the science underlying geographic concepts, applications, and systems. One of the Global Positioning System (GPS) functional tool such as Google Map are associated with the small and medium-sized enterprises (SMEs) businesses providing satellite imagery, street maps, 360° panoramic views of streets (Street View), real-time traffic conditions [1], and route planning for traveling by foot, car, bicycle in beta, or public transportation, and an API that allows maps to be embedded on third-party websites, and offers a locator for urban businesses and other organizations in numerous countries around the world that influence the easiness of customer in finding the small and medium-sized enterprises (SMEs) businesses location. The objectives of the study include:

1. To identify the benefits of using GIS software by customer to find the SME business in Malacca.
2. To determine the factor that ease the customer in adopting GIS software.
3. To examine the most significant factor in adopting GIS software.

2 Literature Review

2.1 Geographical Information System

GIS is one of many information technologies that have transformed the ways geographers conduct research and contribute to society. In the past two decades, these information technologies have had tremendous effects on research techniques specific to geography, as well as on the general ways in which scientists and scholars communicate and collaborate.

The acronym GIS is sometimes used for geographic information science (GIScience) to refer to the academic discipline that studies geographic information systems and is a large domain within the broader academic

[*] Corresponding author: ayusri@utem.edu.my

discipline of geoinformatics. The term describes any information system that integrates, stores, edits, analyzes, shares, and displays geographic information. GIS applications are tools that allow users to create interactive queries (user-created searches), analyze spatial information, edit data in maps, and present the results of all these operations. Geographic information science is the science underlying geographic concepts, applications, and systems. The Global Positioning System is the responsibility of the Joint Program Office, a component of the Space and Missile Center at El Segundo, California. In 1973, the Joint Program Office was directed by the US Department of Defense to establish, develop, test, acquire, and deploy a spaceborne positioning system. The present navigation system with timing and ranging Global Positioning System (GPS) is the result of this initial directive. GPS was conceived as a ranging system from known positions of satelites in space to unknown positions on land, at sea, in air and space. Effectively, the satelite signal is continually marked with its own transmission time so that when received the signal transit period can be measured with a synchronized receiver. The original objectives of GPS were the instantaneous determination of position and velocity such as navigation, and the precise coordination of time such as time transfer (GNSS — Global Navigation Satellite Systems, 2008).

2.2 Small and medium-sized enterprises (SMEs)

Small and medium-sized enterprises (SMEs) businesses are largely growing nowadays in a huge amount of business being conducted. They are often described as efficient and prolific job creators, the seeds of big businesses and the fuel of national economic engines. Small and Medium Industries Development Corporation Malaysia (SMIDEC) classified small and medium-sized enterprises (SMEs) based on the yearly sales turnover or quantity of full time worker [2]. Malaysian small and medium-sized enterprises (SMEs) can be defined according to size, turnover and activity but most small and medium-sized enterprises (SMEs) businesses have a quite similar problem that is in term of providing an exact location of their business store to their target customer. Small and medium-sized enterprises (SMEs) can be divided into three sectors which are manufacturing, services and agriculture.

In the broadband economy, SMEs enjoy technology-supported opportunities to interact with customers in ways that were neither possible nor thinkable until recently. However, to effectively sustain SMEs' competitive position, technology must first meet the business needs of small enterprises. In a competitive market like the current one, knowledge and its management have become the best partner to generate competitive advantages in small and medium enterprises. It is important for country economic development that SMEs, as main source of wealth, manage their intangible resources to continue creating value, both business and social, and get to be leader in an international and competitive market, in order to gain advantages to present difficult situation due to financial crisis [3]

2.3 Theoretical Framework

Using the GIS software to relay the location information is all well and good, but how best to impart the relationship reasoning behind it is explained in theoretical framework. A theoretical framework is a type of intermediate concept which has the ability to connect all aspects of inquiries. A framework can help decide and explain the route that should be taken to a certain point [4, 5]. The study framework comprises the dependent variable (Social Influence, Time Management, Effort Expectancy, Performance Expectancy, and Facilitating Condition) and the independent variable (Behavioral Intention to adopt GIS software). Experience acts as the moderator variable.

2.4 Factors Affecting the Adoption of Geographical Information System

From literature, GIS adoption factors can include Social Influence, Time Management, Effort Expectancy, Performance Expectancy, and Facilitating Condition.

2.4.1 Social Influence

Social influence occurs when a person's emotions, opinions, or behaviors are affected by others. Social influence takes many forms and can be seen in conformity, socialization, peer pressure, obedience, leadership, persuasion, sales, and marketing. Social influence is defined as the degree to which a person perceives that important others believe he or she should use the GIS. A study conducted by [6] found the most significant positive impact was of social influence on consumer's intention to adopt mobile banking services in Pakistan. Hence, this study proposed hypothesis → H1: There is positive and significant relationship between social influence and behavioral intention to adopt GIS software.

2.4.2 Time Management

Time management in any organization and in life generally is of essence if the objective of existence of such entity is to be realized. The quest for better and more efficient way of doing things, continuous pursuit of efficiency, increased productivity initially in manufacturing but later expanded to focusing on individual efficiency in offices brought about the study of managing the time at everyone's disposal. Lawson et. al, [7] mentioned that the findings from the existing literature addressing GIS, GPS are limited at this time and the primary issue of GPS as a source of travel data relates to the satellite signal that improve the time taken and advancements in the technology being used for the GPS units appear to offer possible solutions to some of the issues regarding urban canyon effect, tree canopies, and other physical obstructions. The time management of GIS is a major help for the customer on the easiness for them to locate the SMEs premises location.

The efficient management of time is one of the most important aspects of project management; frequently it is the most critical aspect for a client. Time itself is unmanageable, and, if it is critical, must be treated as a diminishing resource from day one. Since time as a resource cannot be increased, it follows that time management really relates to the management of the activities involving any other resource within an allocated period. The effluxion of time is continuous, very easily passing unnoticed, especially in the early stages of a project; and unless a positive attitude to its use is taken from the outset, the project can very quickly be in trouble [14]. Hence, hypothesis two is → H2: There is positive and significant relationship between time management and behavioural intention to adopt GIS software.

2.4.3 Effort Expectancy

Perceived ease-of-use (PEOU) was defined by Davis (1989) as "the degree to which a person believes that using a particular system would be free from effort". Expectancy theory is based on four assumptions [9]. One assumption is that people join organizations with expectations about their needs, motivations, and past experiences. These influence how individuals react to the organization. A second assumption is that an individual's behavior is a result of conscious choice. That is, people are free to choose those behaviors suggested by their own expectancy calculations. A third assumption is that people want different things from the organization (e.g., good salary, job security, advancement, and challenge). A fourth assumption is that people will choose among alternatives so as to optimize outcomes for them personally [2, 13]. Therefore, the study proposed hypothesis three as → H3: There is positive and significant relationship between effort expectancy and behavioral intention to adopt GIS software.

2.4.4 Facilitating Condition

Merriam-Webster defines Awareness as the knowing that something such as a situation, condition, or problem exists, and feeling, experiencing, or noticing something such as a sound, sensation, or emotion knowing and understanding a lot about what is happening in the world or around you. Awareness is the ability to perceive, to feel, or to be conscious of events, objects, thoughts, emotions, or sensory patterns. In this level of consciousness, sense data can be confirmed by an observer without necessarily implying understanding. More broadly, it is the state or quality of being aware of something. In biological psychology, awareness is defined as a human's or an animal's perception and cognitive reaction to a condition or event. Based on the study conducted by [2, 10], effort expectancy is major predictor technology adoption. Many studies incorporated a few variables such as Performance expectancy, effort expectancy, and subjective norm as facilitating conditions to adopt technologies. Thus, hypothesis four is → H4: There is positive and significant relationship between facilitating condition and behavioral intention to adopt GIS software.

2.4.5 Performance Expectancy

Investopedia defines efficiency signifies a level of performance that describes a process that uses the lowest amount of inputs to create the greatest amount of outputs. Efficiency relates to the use of all inputs in producing any given output, including personal time and energy. Efficiency is a measurable concept that can be determined by determining the ratio of useful output to total input. It minimizes the waste of resources such as physical materials, energy and time, while successfully achieving the desired output. Efficiency is often measured as the ratio of useful output to total input, which can be expressed with the mathematical formula $r=P/C$, where P is the amount of useful output ("product") produced per the amount C ("cost") of resources consumed. Hence, hypothesis five is → H5: There is a positive and significant relationship between performance expectancy and behavioral intention to adopt GIS software.

2.4.6 Experience

This study is also interested in evaluating experience as a moderator for GIS adoption. Ramayah et al., [8] study prior experience as moderating variable among students. They found that prior experience helped in solving easier problems only and not harder tasks. Moderating variable function is to test the relationship between independent and dependent variable which is not only surprisingly weak but also inconsistent or non-existence. Thus, the moderating variable is included to reduce or strengthen the relationship.
H1a: The influence of social influence factor on individual intention will be moderated by experience.
H2a: The influence of time management factor on individual intention will be moderated by experience.
H3a: The influence of effort expectancy factor on individual intention will be moderated by experience.
H4a: The influence of performance expectancy factor on individual intention will be moderated by experience.
H5a: The influence of facilitating condition factor on individual intention will be moderated by experience.

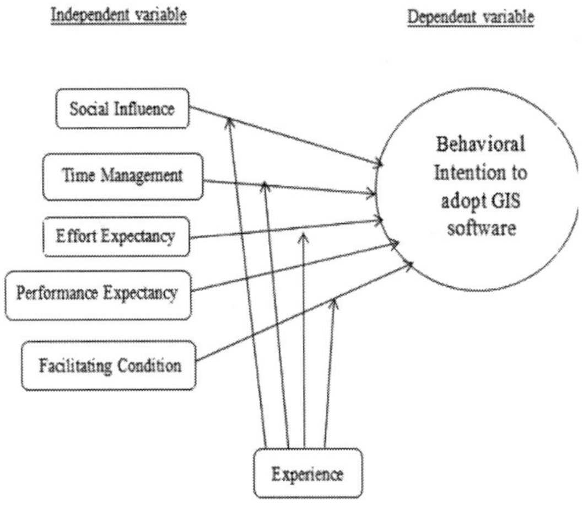

Fig. 1. Research model

3 Methodology

This research employs quantitative survey method. Quantitative method is best at assessing relationship between factors and to test regressions at advanced level [4]. The study managed to get 160 responses or 40% from total 400 surveys distributed to customers on SME premises around Jalan Hang Tuah, Melaka. These 160 respondents were selected randomly. The questionnaire was divided into two parts that are Section A for Demographic Profile and Section B for Variables. A 5-point Likert scale rating was used to assess end user acceptance of a software program. Pre-test was done to get input from experts on the design of the questionnaire. Ten pre-tests feedback was received. A small improvement was made before pilot study was undertaken. From the 160 responses, the researchers used statistic package for the social science (SPSS) to analyse the results. Cronbach Alpha reliability and internal validity checks were carried out [5].

4 Results

4.1 Descriptive Analysis

Pilot test analysis involved 28 responses. Cronbach's alpha for this pilot was 0.824, which is high and accepted as very good level. This section explain the demographic information of respondents including gender, race, GIS usage frequency. The respondents constituted 80% malays and 10% each for indian and chinese. Female dominated as repondents which is around 70% as shown in Table 1 and 2 respectively.

Table 1. Gender of respondents

GENDER

		Frequency	Percent	Valid Percent	Cumulative Percent
Valid	MALE	48	30.0	30.0	30.0
	FEMALE	112	70.0	70.0	100.0
	Total	160	100.0	100.0	

Table 2. Frequencies for using GIS applications

How often do you use GIS application such as Waze or Google Maps whether via mobile apps, web browser, or inside business website itself to locate any desired location?

0		Frequency	Percent	Valid Percent	Cumulative Percent
Valid	Never	40	25.0	25.0	25.0
	Once a week	100	62.5	62.5	87.5
	2 to 4 times a week	10	6.3	6.3	93.8
	5 or more times a week	10	6.3	6.3	100.0
	Total	160	100.0	100.0	

In Table 2, only 6.3% respondents said that they were frequent users of GIS applications to locate SME premises. 6.3% also said they used 2 to 4 times a week whereas a vast majority of 62.5% claimed that the used only once a week. 25% said the have never used GIS applications to locate SME premises.

4.2 Correlation Analysis

A correlation coefficient enables us to quantify the strength of the linear relationship between two ranked or numerical values [5].

Table 3. Correlation analysis

Correlations

		Performance Expectancy	Social Influence	Effort Expectancy	Facilitating Condition	Time Management	Behavioral Intention
Performance Expectancy	Pearson Correlation	1	.462	.371	.441	.667**	.528*
	Sig. (2-tailed)		.072	.158	.088	.005	.035
	N	16	16	16	16	16	16
Social Influence	Pearson Correlation	.462	1	.163	.192	.182	.500*
	Sig. (2-tailed)	.072		.547	.475	.501	.049
	N	16	16	16	16	16	16
Effort Expectancy	Pearson Correlation	.371	.163	1	.388	.445	.560*
	Sig. (2-tailed)	.158	.547		.137	.084	.024
	N	16	16	16	16	16	16
Facilitating Condition	Pearson Correlation	.441	.192	.388	1	.785**	.752**
	Sig. (2-tailed)	.088	.475	.137		.000	.001
	N	16	16	16	16	16	16
Time Management	Pearson Correlation	.667**	.182	.445	.785**	1	.502*
	Sig. (2-tailed)	.005	.501	.084	.000		.047
	N	16	16	16	16	16	16
Behavioral Intention	Pearson Correlation	.528*	.500*	.560*	.752**	.502*	1
	Sig. (2-tailed)	.035	.049	.024	.001	.047	
	N	16	16	16	16	16	16

**. Correlation is significant at the 0.01 level (2-tailed).

*. Correlation is significant at the 0.05 level (2-tailed).

In Table 3 above, all variables show high correlations and p values. For example, the correlation between the Performance Expectancy and Behavioral Intention is 0.528 (p=0.035). This correlation coefficient is therefore a statistically significant. Secondly, the correlation between the Social Influence and Behavioral Intention is

0.500 (p=0.049). Thirdly, the correlation between the Effort Expectancy and Behavioral Intention is 0.560 (p=0.024). Fourthly, the correlation between the Facilitating Condition and Behavioral Intention is 0.752 (p=0.001). Finally, the correlation between the Time Management and Behavioral Intention is 0.502 (p=0.047).

4.3 Hypothesis Testing

There is a significant relationship between independent variable and dependent variable if the significant value was below 0.05 ($p < 0.05$). The multiple correlation coefficients were the linear correlation between the observed and the model predicted values of dependent variable.

Based on Table 4 below, all variables show high p values and therefore, all hypotheses were accepted.

Table 4. Hypothesis testing

Item	Findings	Relationship
Hypothesis 1	Accepted p value = 0.035	Moderate positive relationship r = 0.528
Hypothesis 2	Accepted P value = 0.049	Moderate positive relationship r = 0.500
Hypothesis 3	Accepted P value = 0.024	Moderate positive relationship r = 0.560
Hypothesis 4	Accepted P value = 0.001	High positive relationship r = 0.752
Hypothesis 5	Accepted P value = 0.047	Moderate positive relationship r = 0.502

5 Discussion

Geographically referenced data consist of any type of measurement or observation, whether analog or digital, which have a known distribution across the surface of the ground, and hence can be presented as a map. Computer software programs designed to store, manipulate and present geographical data are known as geographical information systems (GIS) [11].

GIS applications are tools that allow users to create interactive queries (user-created searches), analyze spatial information, edit data in maps, and present the results of all these operations. Geographic information science is the science underlying geographic concepts, applications, and systems. One of the Global Positioning System (GPS) functional tool such as Google Map is associated with the small and medium-sized enterprises (SMEs) businesses providing satellite imagery, street maps, 360° panoramic views of streets (Street View), real-time traffic conditions (Google Traffic), and route planning for traveling by foot, car, bicycle in beta, or public transportation, and an API that allows maps to be embedded on third-party websites, and offers a locator for urban businesses and other organizations in numerous countries around the world that it can search by keyword such as type of business that influence the easiness of customer in finding the small and medium-sized enterprises (SMEs) businesses location [1].

The benefits of using GIS software has also been support by Bing Pan, John C. Crotts, and Brian Muller in their book 'Information and Communication Technologies in Tourism 2007' that saying the handheld tour guide is based on mobile phones with a GPS antenna to communicate with Google Map and deliver real-time and location-sensitive tourist information. Online Charleston trip planner combines destination specific information with Google Map to provide an interactive trip planning tool based on the map metaphor. The handheld tour guide is slow and unstable due to memory limitations and slow Internet connection provided by the mobile phone service. The online trip planner is being adopted by various hotels and conferences. The development effort demonstrated that Google Map API is a flexible tool with reasonable speed for developing destination-specific online services. For mobile tools, the computing power and wireless connections of small devices are the bottlenecks when communicating with online services [12].

Objective 2 is to determine the factors that ease customers in adopting GIS software. Literature mentioned many factors used frequently by past researchers. However, Technology acceptance model (TAM) and UTAUT models 1 and 2 suggests several tested and confirmed acceptance and adoption factors. Hence, the researchers gathered six independent variables that have been used to identify its impact towards behavioral intention. As shown in Figure 1, the theoretical framework was developed to identify the impact of social influence, time management, effort expectancy, performance expectancy, and facilitating condition on behavioral intention.

From the research, it has been found that social influence, effort expectancy, performance expectancy, and facilitating condition factors were adopted from UTAUT model. The theory holds that there are four key constructs: 1) performance expectancy, 2) effort expectancy, 3) social influence, and 4) facilitating conditions. The first three are direct determinants of usage intention and behavior, and the fourth is a direct determinant of use behavior.

Other than that, based from the hypothesis there is a positive and significant relationship between the independent variable of social influence, time management, effort expectancy, performance expectancy, and facilitating condition to the dependent variable of behavioral intention to adopt GIS software by having the 'p' value to be lower than 0.05. This means that the independent variables are significant enough to be considered as the factor in adopting GIS software.

Objective 3 is to examine the most significant factors in adopting GIS software. The regression analysis was done to develop a model of adoption using the behavioral intention as an element. The result of behavioral intention (dependent variable) against all the independent variables can be seen in the table above which $p < 0.05$. In this research, the p value is less than 0.05 thus the model of regression is a significant.

The outcome can be examined from the linear regression that is the Facilitating Condition as the most affecting factor in adopting GIS software because it has the highest significant level and also by increasing 1% of Facilitating Condition will increase the Behavioral Intention for 1.093% and followed by time management as the second highest significant level, and then effort expectancy and experience as the third highest significant level.

5 Conclusion

This research is to study the adoption of Geographical Information System (GIS) to ease customers in locating SME business premises along Jalan Hang Tuah Malacca. Based on previous discussion, this study concludes that the related factors directly impact one another and consequently they influence user's behavioral intention to adopt GIS software. Almost 70% of the respondents were in the category of never use GIS or use GIS less than once a week. Further studies are necessary to understand why there is a slow growth in using GIS for locating business premises or even tourist places. Examples of mostly used GIS or GPS applications are google maps and waze applications to assist in road traveling.

UTAUT and TAM are two theories researchers most referred to in technology adoption and acceptance studies. Based on these models and other related literature, this study developed and tested six variables and it found all variables were significant influencers in the study. Hence, all hypotheses were accepted. It is interesting to note that Facilitating Condition is the most significant factor in adopting GIS software because it has the highest significant level and also by increasing 1% of Facilitating Condition will increase the Behavioural Intention by 1.093%. The next factors ranked by their significant level were time management, effort expectancy and experience.

In the future, researchers may extend this study to other areas to understand more about current usage patterns and reasons for its adoption. Studies which cover non-Malaysian respondents and different education levels can further document different perspectives of GIS adoption.

The authors would like thank to Universiti Teknikal Malaysia Melaka for sponsoring this publication. We would also like to thank Ministry of Higher Education, Sustainable IT Economics research group, Faculty of Technology Management and Technopreneurship (FPTT) and respondents who helped us along the way to completion of this research..

References

1. Wikipedia (2017). Retrieved from https://en.wikipedia.org/wiki/Google_Maps
2. Moorthy, M.K., Tan, A., Choo, C., Wei, C.S., Tan, J., Ping, Y., & Leong, T.K. (2012). A Study on Factors Affecting the Performance of SMEs in Malaysia. Academic Research in Business and Social Sciences, 2(4), 224-239.
3. I. Alvarez, E. Cilleruelo, I. Zamanillo & E. Zarrabeitia (2014) Knowledge Management Practices in SME: Case Study in Basque Country SME, Annals of Industrial Engineering 2012
4. Y. Arshad. (2012). IT Outsourcing Decisions and Implementations In Malaysian Public Healthcare Sector Agencies: Grounding An ITO Relationship Model Using Qualitative Approach. Unpublished thesis. International Islamic University Malaysia.
5. A. R. Ahlan, M. A. Suhaimi, H. Hussin, Y. Arshad (2008). Assessing future needs of IT Education in Malaysia: A preliminary result. Proceedings of 4th WSEAS/IASME International Conference on educational Technology (EDUTE'08), Corfu, Greece
6. J. M. Burger (1992). Social Influence, Desire for Control, Part of the series The Plenum Series in Social / Clinical Psychology pp 63-79
7. C. T. Lawson et al. (2007) GPS Pilot Project Phase One: Literature And Product Review
8. T. Ramayah, I. Joshua and A. Bushra (2005) Personal computer Usage Among Students In A Private Institution of Higher Learning: The Moderating Role Of Prior Experience. The Asia Pacific Journal of Educators and Education (formerly known as Journal of Educators and Education), 20 (1). pp. 1-22. ISSN 2289-9057
9. V. H. Vroom. (1964) Work and motivation. New York: Wiley.
10. F. C. Lunenburg (2011). Expectancy Theory of Motivation: Motivating by Altering Expectations
11. R. W. Marjoribanks (1997). Geographical Information Systems, Geological Methods in Mineral Exploration and Mining
12. B. Pan, J. C. Crotts, B. Muller (2007). Developing Web-Based Tourist Information Tools Using Google Map, Information and Communication Technologies in Tourism 2007
13. M. Renko, K. G. Kroeck & A. Bullough (2012). Expectancy theory and nascent entrepreneurship, Small Business Economics
14. T. Gudehus & H. Kotzab (2012). Time Management, Comprehensive Logistics.

MATEC Web of Conferences **150**, 05046 (2018) https://doi.org/10.1051/matecconf/201815005046
MUCET 2017

Techno-*Da'ie* and Challenges in the 21st. Century

Datuk Su'aidi Dato' Safei[1]

[1]Universiti Teknikal Malaysia Melaka (UTeM), 76100 Melaka, Malaysia

Abstract. Education has always been instrumental in the dissemination of religious ideas, rituals and values. In the Islamic tradition it is the classical method of propagation and, therefore, education as a means of instructing and motivating (prospective) believers is an integral part of *da'wah*. *Da'wah* operates on two fronts: (i) towards Muslim who are perhaps unaware of Islam, maybe because of growing up in the western lifestyle and missing out of a proper Islamic education, and (ii) towards our non-muslim neighbours and colleagues, many of whom may be searching for truth. At the very least, we may be able to counter-act the many incorrect stereotypes of Islam and muslims, no one else will do this for us, we have to speak up and show by our behaviour what Islam is all about. By doing this, we may even bring people into the fold of Islam. Contemporary *du'at* face great challenges to systematize and to apply the practical methodology within the new era of modern context. This paper discusses some suggestions that Muslim *du'at*, individually and collectively can undertake to accomplish *da'wah* goals: i) performing *da'wah* through a worldwide Islamic movement, ii) strengthening *da'wah* with technology, iii) using communication media and conducting seminars and workshops, iv) displaying the real Islamic lifestyle.

1 Introduction

وَمَنْ أَحْسَنُ قَوْلًا مِّمَّن دَعَآ إِلَى ٱللَّهِ وَعَمِلَ صَٰلِحًا
وَقَالَ إِنَّنِى مِنَ ٱلْمُسْلِمِينَ ﴿٣٣﴾

"And who is better in speech than one who invites to Allah, does righteousness and says: Indeed, I am of those who surrender (unto Him)?" (*Fussilat*; 41:33)

Way, technique and method in English, *tariqah*, *manhaj* and *uslub* in Arabic are principles for doing something. *Da'wah* methodology is a discipline which concerns with the question of how *da'wah* is performed systematically and strategically. The deliverer of the islamic *da'wah* ought further bolstering pass on of the individuals those frameworks by which they must take care of their worldly undertakings. The history of the triumph of the islamic *da'wah* is that it is progressive and fit to deal with individual, what's more initiates clinch alongside him a thorough transform. [1, 2]

Da'wah, mission of inviting people to Islam, is a must on every Muslim. It is not honey and dew. It is filled up with complications, problems and difficulties. Many *du'at* invite others, advise them, but at the same time they forget themselves. They think that they are saved from *da'wah*, that it is for others. Those who ignore the duty of reforming and training their own character and ruminate over the reformation and training of others are completely devoid of common sense. Their own house is on fire, yet they are carrying about buckets of water looking for a blaze anywhere else to extinguish. Such people are losers not only in this world but in the hereafter also. The Prophet s.a.w. observed during the night of ascension some people whose lips were being clipped with scissors of fire and Gabriel told him that those are the orators of the *ummah*, they preached piety and fear of God to people, but forgot to follow these teachings in their personal lives. [3, 9]

On one occasion, a person mentioned to Saidina Abdullah ibn Abbas that he wished to instruct people in piety and forbid them to commit evil deeds. He desired to propagate religion among people. Ibn Abbas enquired if he have acquired the proper qualifications for becoming a preacher and the person believed that he is capable of performing the task of a missionary. Ibn Abbas observed: "if you do not apprehend any condemnation in the light of three verses of the Holy *Qur'an*, do preach Islam."

The 1st verse:

*Corresponding author : suaidi@utem.edu.my

https://doi.org/10.1051/matecconf/201815005046

أَتَأْمُرُونَ ٱلنَّاسَ بِٱلْبِرِّ وَتَنسَوْنَ أَنفُسَكُمْ وَأَنتُمْ تَتْلُونَ ٱلْكِتَٰبَ
أَفَلَا تَعْقِلُونَ ﴿٤٤﴾

"Do ye enjoin right conduct on the people, and forget (to practise) it yourselves, and yet ye study the Scripture? Will ye not understand" (*al-Baqarah*; 2:44)

Ibn Abbas then enquired: "Do you follow this verse in your practical life fully?". The man replied: "No". Ibn Abbas then recited the 2nd verse:

يَٰٓأَيُّهَا ٱلَّذِينَ ءَامَنُوا۟ لِمَ تَقُولُونَ مَا لَا تَفْعَلُونَ ﴿٢﴾

"O ye who believe! Why say ye that which ye do not?" (*al-Saf*; 61:2)

Ibn Abbas then asked: "Are you sure you are following this verse?". The man replied "No".

The 3rd verse:

وَمَآ أُرِيدُ أَنْ أُخَالِفَكُمْ إِلَىٰ مَآ أَنْهَىٰكُمْ عَنْهُ

"I desire not to do by your backs that which I ask you not to do" (*Hud*; 11:88)

2 *Da'wah* Methodology In The *Qur'an* And *Prophetic Traditions*

As far as the method of *da'wah* is concerned, we find a large number of verses in the *Qur'an* related to this subject. The basic teaching of *da'wah* methodology has been formerly laid down by the *Qur'an*. This tends to be conceptual guidance which was later applied in reality by the Prophet. It was then adopted by later Muslim missionaries. [4, 7]

Allah commanded the Prophet Muhammad s.a.w. to start his *da'wah* mission with his family members including his relatives:

وَأَنذِرْ عَشِيرَتَكَ ٱلْأَقْرَبِينَ ﴿٢١٤﴾

وَٱخْفِضْ جَنَاحَكَ لِمَنِ ٱتَّبَعَكَ مِنَ ٱلْمُؤْمِنِينَ ﴿٢١٥﴾

"And admonish thy nearest kinsmen, and lower thy wing to the believers who follow thee" (*al-Shu'ara'*; 26:214-5)

As very well-known verse which describes the method of *da'wah* in the *Qur'an* is in *Surah al-Nahl*. It clearly consists of the three ways to perform *da'wah*, namely *hikmah* or wisdom, fair exhortation or goodly counsel and argumentation with a gracious manner:

ٱدْعُ إِلَىٰ سَبِيلِ رَبِّكَ بِٱلْحِكْمَةِ وَٱلْمَوْعِظَةِ ٱلْحَسَنَةِ ۖ وَجَٰدِلْهُم بِٱلَّتِى هِىَ
أَحْسَنُ ۚ إِنَّ رَبَّكَ هُوَ أَعْلَمُ بِمَن ضَلَّ عَن سَبِيلِهِۦ ۖ وَهُوَ أَعْلَمُ بِٱلْمُهْتَدِينَ

"Invite (all) to the Way of thy Lord with wisdom and fair exhortation, and reason with them in ways that are best and most gracious, for thy Lord knoweth best, who have strayed from His Path, and who receive guidance" (*al-Nahl*; 16:125)

Allah has reminded the Prophet to be kind, compassionate, low, humble, loving and friendly to all. It is only through love and mercy that he will be able to bring people together.

فَبِمَا رَحْمَةٍ مِّنَ ٱللَّهِ لِنتَ لَهُمْ ۖ وَلَوْ كُنتَ فَظًّا غَلِيظَ ٱلْقَلْبِ لَٱنفَضُّوا۟ مِنْ حَوْلِكَ ۖ
فَٱعْفُ عَنْهُمْ وَٱسْتَغْفِرْ لَهُمْ وَشَاوِرْهُمْ فِى ٱلْأَمْرِ ۖ فَإِذَا عَزَمْتَ فَتَوَكَّلْ عَلَى ٱللَّهِ ۚ
إِنَّ ٱللَّهَ يُحِبُّ ٱلْمُتَوَكِّلِينَ ﴿١٥٩﴾

"It is part of the Mercy of God that thou deal gently with them. Were thou severe or harsh-hearted, they would have broken away from about thee, so pass over (their faults, and ask for)God's) forgiveness for them, and consult them in affairs (of moment). Then, when thou has taken a decision, put thy trust in God, for God loves those who put their trust (in Him)" (*ali-Imran*; 3:159)

In the process of *da'wah* every Muslim is to say the best or else *satan* may play his dirty tricks. *Du'at* are to use the best approach as they are talking to the *mad'uw* (subject).

وَقُل لِّعِبَادِى يَقُولُوا۟ ٱلَّتِى هِىَ أَحْسَنُ ۚ إِنَّ ٱلشَّيْطَٰنَ
يَنزَغُ بَيْنَهُمْ ۚ إِنَّ ٱلشَّيْطَٰنَ كَانَ لِلْإِنسَٰنِ عَدُوًّا مُّبِينًا ﴿٥٣﴾

"Say to My servants that they should only say those things that are best, for satan does sow dissensions among them, for satan is to man an avowed enemy" (*al-Isra'*; 17:53)

Islamic workers should not address labourers with the language of grave diggers nor the atheists with tongue of emotionalists. They should use appropriate language at the proper place and time. As derived from the Prophet's *hadith*: "We are commanded to speak to people according to their mental capacity." [10, 15]

3 *Da'wah* Methodology In The Modern Context

The deliverer of the Islamic *da'wah* ought to pass on to the individuals the frameworks by which they have to settle their common illicit relationships. The mystery of the achievement of the Islamic *da'wah* is that it is moral force and ready to manage man as a man, and starts in him a far reaching change.

Da'wah operates on two fronts: (i) towards Muslims who are perhaps unaware of Islam, maybe because of growing up in the western lifestyle and missing out on a proper Islamic education, and (ii) towards our non-muslims neighbours and colleagues, many of whom may be searching for truth. At the very least, we may be able to counter-act the many incorrect stereotypes of Islam and muslims, no one else will do this for us, we have to speak up! and show by our behaviour what Islam is all about. By doing this, we may even bring people into the fold of Islam. [13]

Effective du'at perceive that their part is not just in conveyance of the message, but but rather stretches out to the limits of consistent *tarbiyah* (education) of youthful energetic generations from claiming adherents. The *du'at* must plant the seeds of goodness in the hearts of the Muslim youth and then pledge to constantly raise them progressively in accordance with Allah's law. [5, 6]

Contemporary *Du'at* face great challenges to systematize and to apply the practical methodology within the new era of modern context. Provided below are some suggestions that *Muslim Du'at*, individually and collectively can undertake to accomplish *da'wah* goal. [16]

3.1 Performing *Da'wah* through a Worldwide Islamic Movement

The *Qur'an* has obviously directed the *Du'at* to work within group:

وَلْتَكُن مِّنكُمْ أُمَّةٌ يَدْعُونَ إِلَى الْخَيْرِ وَيَأْمُرُونَ بِالْمَعْرُوفِ
وَيَنْهَوْنَ عَنِ الْمُنكَرِ وَأُولَٰئِكَ هُمُ الْمُفْلِحُونَ ﴿١٠٤﴾

"Let there arise out of you a nation inviting to all that is good, enjoining what is right, and forbidding what is wrong, and those will be the successful." (*ali-Imran*; 3: 104)

The technique of *da'wah* through aggregate work or deliberate association was started when *Dar al-Da'wah wal-Irsyad* was established by Shaykh Rashid Rida in the early 19th century. His ambition was to train *Muslim Du'at* within an established *da'wah* college and to send them into Christian lands for propagating the message of Islam. They were also sent into those Muslim countries in which attempts had been made to induce the Muslims to abandon their faith. This approach is quite similar to the idea of al-Maududi and Hassan al-Banna that *da'wah* and politics are inseparable and the mass movement should struggle to create an Islamic order within the society.

3.2 Strengthening *Da'wah* with Technology

There appears to be a consensus on the compelling reason for preachers of Islam to utilise modern instruments to carry out the mission of *da'wah*. Today's culture is a whole lot technocentric. Technology has broken down physical, geographical and political barriers that once may have stood in the way of completing things proficiently and effectively. Consequently there exists a substantial compelling reason for the creation of a new breed of Muslim preachers. This new breed of techno-*da'ie* is a critical perspective for advanced Islam. Without them, it would be troublesome to present the genuine Islam in this post-modern world that we currently live in.

The new era particularly are becoming more techno smart. So what better way to approach the post-modern generation than to apply that which they are most comfortable with? Why not use technology to narrow the gap between the *da'ie* and the community? ICT, for instance, could be utilised to the best by the techno-*da'ie* to share and disseminate info data on Islam. Multimedia could be used to teach kids and adolescent how to read the *Qur'an* correctly. Multimedia is interactive and combines the usage of most senses. This would make learning the *Qur'an* more interesting, especially to today's children.

Information and Communication Technology (ICT) additionally displays a perpetual likelihood of the techno-*da'ie*. Say, for instance, a techno-da'ie were to address a lecture in a mosque. He could make use of web-based presentation tools and online gadgets to make his talk extra charming. The audience would also be capable should recall more adequately if multimedia is utilized. [8, 12]

Likewise, assuming that conventional works of Islam such as the voluminous Imam Syafii's *Kitab al-Umm* or Ibn Khaldun's *Muqaddimah* are digitally transformed and stored in the form of Compact Disc Read-Only Memory (CD-ROM) or Hard Disk Drive (HDD), this would help the techno-*da'ie* carry these references wherever and whenever. The techno-*da'ie* would just need to carry a couple of CD-ROMs as contradicted with several volumes of very thick books around. All the techno-*da'ie* would need is of course a laptop or tablet, which is becoming lighter, much simple and easier to carry as technology engineering advances. [11]

The substances are generally there; the *Qur'an*, prophetic traditions, history of Islam, stories of the prophets, writings on Islamic jurisprudence, scholar works of Muslim *ulama* and many more. These materials need to be inventively repackaged using the latest technology so that they can be used to empower the task of *da'wah*.

Thus, a linking bridge between Muslim preachers and technologies must be established in order for *da'wah* in the 21st century to be successful. Neither of these two assemblies can afford to conduct *da'wah* all alone. One lacks the state-of-the-art technical knowledge of disseminating information but has a great deals of materials for *da'wah*, while the other has the innovation and technology, but fails to offer the content. Reliably these two clusters should facilitate supporting collaborate and work together accordingly, that a win-win circumstances could be created. Over the long haul, Islam will benefit.

It must be noted that the techno-*daie* is embarking to dislodge the customary, traditional and conventional methods and techniques of *da'wah*. This new breed of preachers should be seen as supplementing and fortifying the *da'wah* movement.

Devices like the tablet, laptop, smartphone and the personal digital assistant (PDA) could be used by audience of a religious lecture to take down notes. Listening alone is not sufficient. Research has shown that to fully understand what is being taught, one must listen and jot down what one hears. This, lamentably, is infrequently carried out by Muslims when it comes to learning. [11]

3.3 Using Communication Media and Conducting Seminars/ Workshops

Ismail Faruqi suggested that the method of *mujadalah* should be applied in a kind of academic seminar on a large scale where *Du'at* can freely present and argue with other religions. This approach can achieve its objective because Islam puts in man's rational power to discriminate between the truth and false.

On the other hand, his approach may differ from Ahmad Deedat who eludes a charismatic and saintly appearance and his wit and charm temper what might otherwise be considered an acid tongue. He, moreover, regards Christianity as his chief enemy, and this creates tension between faiths rather than peaceful method required by Islam.

To practise *da'wah* one must be well mannered and conscious. *Da'wah* implies to gently persuade, attract, be polite, be friendly, be caring and understanding, and try to have a sense of humour. Outward appearance has a role to play; i) be clean and neat, ii) don't eat raw onions or garlic before attending mosque or meeting people, and iii) chew gum or something mint for good breath. [14]

3.4 Displaying the Real Islamic Lifestyle

The above discussion is thoroughly concerned with the method of direct *da'wah* or activitistic preaching which is directed to the non-Muslim. In fact, *da'wah* should also be directed toward the Muslim. It is called as indirect/ non-verbal *da'wah* or Islamic lifestyle (*Da'wah bil-Hal*). It is to take Islam to non-Muslim, not by pulpit preaching and mailing Islamic literature but by doing what Muslim ought to do (action); living, drinking, eating, sleeping and behaving as Muslims are enjoined to do. This method is more attractive because people can practically see the beauty of Islam within the *ummah* (muslim community).

Word alone however will not suffice. *Da'wah* can be by writing, speaking, behaviour, attitude, manner, sharing and caring with and for others. The best *da'wah* is to be a good example.

Success in *da'wah* work means first of all knowing Islam and *Qur'an* and being familiar with the traditions and stories of the Prophet Muhammad's life, for only then it is possible in conversation to give answers to followers of other faiths.

Da'wah through humanitarian causes should be pondered thoughtfully in order to prove the fruit of Islam. Without such as this credibility, the whole effort of *da'wah* will be rhetorical and devoid of meaning. Unfortunately, it has been left to Christian missionaries to work on such lines.

Showing the fruit of the teaching is more effective than calling to something abstract and theoretical which is written on papers. For this purpose, Muslims should improve their achievement in every aspect of life. Problems such as poverty and illiteracy within the Muslim *ummah* should be overcome. [14]

4 Conclusion

Educational training has a focal and key part, particularly when it is associated with the push to improve HR quality. It is brought about by the acknowledgment for human standing and glory increment. Socialization of Islamic qualities should be possible just by prominent human asset.

One of the best means for Islamic worker to attain the gentle and effective method of *da'wah* is to know what he wants. This clear conception and analysis of goals dictates the method the work should follow. Knowing what he wants, the Islamic worker will save effort and time and will make well-planned moves. This will enable him to avoid taking arbitrary decisions which are the results of a lack of planning and proper calculations. Al-Hasan al-Basri was quite right in his saying: "The one who acts without knowledge is like the one who travels on no road, and the one who accomplishes what he doesn't want, spoils more than he amends."

Deliverers of the Islamic *da'wah*, are not searching for any common compensate or expecting acclaim from individuals. They request nothing aside from the delight for Allah. May Allah keep us on the straightway, and acknowledge our modest endeavors for His adoration.

References

1. A. Hassan. *Komunikasi untuk Bakal Pendakwah: Panduan Mengubah Sikap dan Tingkah Laku Orang Lain*. (2001).
2. A.G. Don et. al. *Dakwah dan Pengurusan Islam di Malaysia: Konsep dan Pelaksanaan*. (2000).
3. A. Hj Mhd. Zain. *Bimbingan berdakwah: Teori & Penerapan di Alaf Baru*. (2001).
4. A. Ali. *http://www.ilaam.net/pdf/handlingpeople.pdf*. (Accessed: 24 May 2016).
5. C. Schirrmacher. *http://www.islaminstitut.de/uploads/media/IFI Z2003-2-Steps_to_effective_Dawah.pdf*. (Accessed: 24 May 2016).
6. El-Mokhtar, Mohamed. *Islamic Education Between Tradition and Modernity*. (1998).
7. Justdawah.org. *http://www.justdawah.org/resources/training-material*. (Accessed: 24 May 2016).
8. M. A. Dollah. *Islamic Da'wah in the Age of Information and Communication Technology*. (2003).
9. N.J. Kasim. *Islamic Da'wah in Brunei Darussalam Since the Second Half of the 20th. Century: an Analysis*. (2006).
10. P. Larry. *Islamic Dawah in the West: Muslim Missionary Activity and the Dynamics of Conversion to Islam*. (1992).
11. Productive Muslim. *http://productivemuslim.com/dawah-series-2*. (Accessed: 24 May 2016).
12. Shaikh Mohd Saifuddeen bin Shaikh Mohd Salleh. *Memahami Islam Melalui Teknologi Maklumat dan Komunikasi*. (2000).
13 S. Dato' Safei. *http://www.slideshare.net/Dr22s/dakwah-1003394*. (Accessed: 24 May 2016)
14. S S. Dato' Safei. *http://eprints.utem.edu.my/id/eprint/15904*. (Accessed: 25 May 2016
15. Teachislam.com. *http://www.teachislam.com/dmdocuments/96/ Dawah/What is Dawah.pdf*. (Accessed: 24 May 2016).
16. Z. Abdul Ghani. Islam, *Komunikasi dan Teknologi Maklumat*. (2001).

Defining Human Factor and Ergonomic and its related issues in Malaysia Pineapple Plantations

Nur Fazrina Mohamad Salleh[1] and Ezrin Hani Sukadarin[1]

[1]Occupational Safety and Health Program, Faculty of Engineering Technology, Universiti Malaysia Pahang, 26300 Gambang, Pahang, Malaysia

Abstract. Human factor and ergonomic has been a standardized as synonymous for a long time and has great potential to contribute in designing all kinds of systems with people. However, some opinions draw precisely distinctions among these terms. Numerous studies have been conducted in their attempt to understand the concept of human factor and ergonomic. The meaning of each term must be clearly understood before using them in any research to understand how human beings interact with surroundings. So, this paper seeks to review the definition of human factor and ergonomic. English articles and books dated as far 1970 were compiled from Taylor and Francis Online, Google Scholar and Science Direct. The keywords used in the selection of the articles are human factors, human factors engineering, ergonomics, industrial ergonomics, reviews, definition, difference and risk factors. A risk model related to each term is also provided to have more understanding towards it. Based on the finding of the literature review of human factor and ergonomic issue in pineapple plantation were explore and are categorized accordingly.

1 Introduction

Pineapple or *Ananas Cosmosus* is a type of plant in tropical category originated from East area of South Africa. It was commercially introduced in Malaya in the 16th century by the Portuguese. Pineapple plant began to be cultivated in Malaysia at Johor and Selangor area as cash crop in year 1921, along with rubber crop development. Pineapple plantation was then continued to expand in peat soil in Johor [1]. Pineapple plantation that falls under Malaysian Pineapple Industry Board (MPIB), recorded pineapple exports valued at RM 161 million in year 2015 and targets to increase to RM 254 million by year 2020. The plan to increase the country's pineapple production is to meet market demand and high-value agricultural development.

Agricultural sector is one of the primary sources of income in Malaysia especially among the community in the rural areas [2]. Labor works in agricultural industry has also increased from approximately 1.2 million to 1.4 million before year 2010 [3]. Escalation of work force and demand concurrently has increased the risk of occupational related accidents and diseases in agricultural sectors. In 2012, International Labor Organization (ILO) also highlighted that agricultural sector is one of the most hazardous types of work that lead to health issues worldwide. Same trend is observed in the Malaysian agricultural industry where by agriculture sectors are among highest sectors that reported occupational accident and diseases to Department of Occupational Safety and Health, Malaysia [4] .

Occupational related accidents and diseases that happen in agricultural sector especially in pineapple plantations normally are due to human related issue. Manual handling, fatigue due to overload, long duration of exposure sun light, job dissatisfaction, prolonged in an awkward posture of handling work, might be the factors that contribute to the aforementioned issues. Those factors simply used to be addressed as Human Factors and Ergonomics by many researchers. Although the used term is widely accepted, well understanding between those two terms would be beneficial for researchers in conducting more impactful research in this field. Due to that, this paper is about to (1) explore the definition or Human Factors and Ergonomic from published articles, (2) based on the gathered definition, this paper will discuss one published model that represent the factors for the two terms separately, (3) Human Factors and Ergonomics issue that arise in pineapple plantation activities are gathered and then are grouped into Human Factors risk and Ergonomics Risk.

* Corresponding author: ezrin@ump.edu.my

2 Methodology

To identify the definition of human factors and ergonomic, english articles and books dated as far 1970 were compiled from Taylor and Francis Online, Google Scholar and Science Direct. The keywords used in the selection of the articles are human factors, human factors engineering, ergonomics, industrial ergonomics, reviews, definition, difference and risk factors. PEAR Model (People, Environment, Action, Resources) [5] for human factors and a model by Carayon et al in 1999 [6] for ergonomics are also discussed because of those models represent each term very closely so that the understanding of human factors and ergonomics is obtained appropriately. After that, human factor and ergonomic issue in pineapple plantation, are identified and then classified into (1) human factor and (2) ergonomic.

3 Results and Discussion

3.1 Human Factors and Ergonomics in a Glance

Human factors and ergonomic is "the scientific discipline concerned on the perceptive of interactions among humans and other elements of a system, in order to enhance human well-being and overall system performance [5]. The importance of human factors and ergonomics was highlighted by author Carayon [6] as it is an oriented system, design driven to improve performance and well being of population.

The combination of human factors and ergonomic is now well accepted as can be seen, the term is started to be used over the decades. Human Factor and Ergonomic Organization from United Kingdom has been recognized as the oldest professional body since 1946. It is also called "The Chartered Institute of Ergonomics and Human Factors". The Human Factors and Ergonomics Society (HFES) were also founded in 1957 and were the main professional organization in the United States [9]. The International Ergonomics Association (IEA) is the federation of ergonomics and human factors societies around the world entitled as international organization. The IEA mission is to elaborate and enhance ergonomics science and practices. It also has 42 Federated Societies from across the globe.

Many professional considered that the terms ergonomics and human factors are synonymous. Many articles show that the definitions of these two terms are also similar. A number of excellent reviews covering several different topic areas of relevance to human factors and ergonomics, as well as handbooks and textbooks, are published with some constancy. While IEA [10] combines ergonomic and human factor in its definition as "Ergonomic (or human factors is the scientific discipline concerned with the understanding of interactions among humans and other elements of a system. The Ergonomic (or human factors) professional applies theory, principles data and methods to design in order to optimize human well-being and overall system performance. Table 1 shows the publication which combine human factors and ergonomic into one term.

However, according to Cambridge Dictionary, ergonomics is defined as the scientific study of people and their working conditions, especially done in order to improve effectiveness. While human factors is the act of studying how people use systems or equipment in order to design, develop and create technology that is safer and more effective. Human Factor is the discipline that integrates human capabilities, limitations, requirements and expectations in the design of products, workplaces and work systems while Ergonomics is the science of studying people at work and of their working environment [11]. And, in addition, traditionally ergonomic involves the physical aspects of work, while human factor involve perception and cognition [9].

Table 1. Definition of Human Factor/ Ergonomic

Ref	Definition
Brown and Hendrick [12]	"...the relations between man and his working environment..."
Meister [13]	"...the purpose of ergonomics/human factors is to describe, analyze measure, predict, and control the real world of systems functioning operationally..."
Pelsma [14]	"...The application of knowledge about human characteristics and capabilities - physical, psychological, and cognitive - to the design of work system with the goal of improving well-being and optimizing productivity..."
Rubinstein [15]	"....the study of how people and machines interact..."
Thomas [16]	"...focus on the relationship between humans and their work... to design systems and its environment that is more condusive and happier..."

3.2 Human Factor

A field of human factor was motivated by a search for efficient and error-free operations in United States of America (USA) military in the Second World War (WWII). With its origins derived from concerns about piloting aircraft, human factors was originally much more focused on elements of cognition [17]. There are all 10 disciplines of human factor which is shown in Figure 1 [19]. It has long been recognized as a discipline of knowledge for improving people's health, safety, and comfort and both human and system productivity [20]. It involves a study on how humans behave physically and psychologically in relation to particular environments, products, or services. Literature finding also shows that human factor is originated from research in experimental psychology, which the focus was on human performance and systems design [18]. Table 2 shows the definitions of human factor and its references.

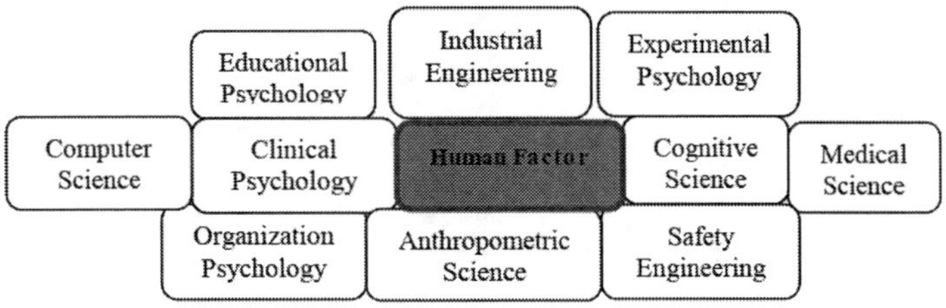

Fig. 1. Human Factor Disciplines

Table 2. The definition of human factors and its references

Ref	Definitions
Bennet, Degan & Spiegel[21]	"...To recognize and understanding of man's characteristics, needs, abilities, and limitations when the procedures and products of technology are being designed..."
Chapanis [18]	"...human abilities, human limitations and other human characteristics that are relevant to design..."
Dhillon [22]	"...Concern on the characteristics of human beings. The term embraces all biomedical and psychosocial considerations..."
Hawkins [23]	"...Human relationship with machines and equipment, with procedures and with the environment..."
Goldenson[24]	"...A branch of psychology concerned with the design of environments and equipment that promote optimum use of human capabilities and optimum efficiency and comfort..."
Kantowitz & Sorkins [25]	"...Optimization the linkage between technology and population..."

3.2.1 A Model related to Human Factor

For better understanding towards human factor term, a model used in describing human factor is discussed. PEAR model has been used as a memory jogger to characterize human factors [5]. PEAR was developed by Dr Michael Maddox and Dr Bill Johnson as a simple way for aviation maintenance personnel to classify human factors that relate to tasks and conditions within the maintenance environment [26]. Human factor is usually associated with the term 'Human Error'. Nearly most accidents and operating upsets involve some kind of human error. For example, over 70% of air-crashes are the result of human error rather than technical failure [26] .In air crashes, errors happen mostly involve failure to comply procedures, taking a short cut or making an (incorrect) assumption about the validity of an instrument reading. Other than that, error may occur during the response phase of an incident [14]. All of these are the examples of human related issues. This issue can be managed if PEAR model as the human factors programs is integrated in the system. There are four important considerations for human factors programs which is people who do the job, their working environment, action that they performed and resources to complete their task (refer Figure 2).

3.2.1.1 People

Focus on a person who execute the work and address the physical, psychological, physiological and psychosocial factors. This also including their mental state, cognitive capacity, physical capabilities and situations that may affect their interaction with others.

3.2.1.2 Environment

There are two types of working environment, physical workplace (eg : ramp or workshop) or organization environment that exists within the company. The physical environment includes ranges of temperature, humidity and workplace design. Oganizational environment are typically related to cooperation, communication and the culture of the company.

3.2.1.3 Action

Job Task Analysis (JTA) is one of the method in human factors approach to identify the knowledge, skills, and attitudes neded to perform each task in a given job. The JTA assist to identify what instructions, tools, and other resources are necessary.

3.2.1.4 Resources

The component of the people, environment, and actions dictate the resources. Many resources are tangible, such as tools, test equipment, computers and technical manuals. However, other resources are less tangible. The examples are the number and qualifications of staff to complete a job, the amount of time allocated, and the level of communication.

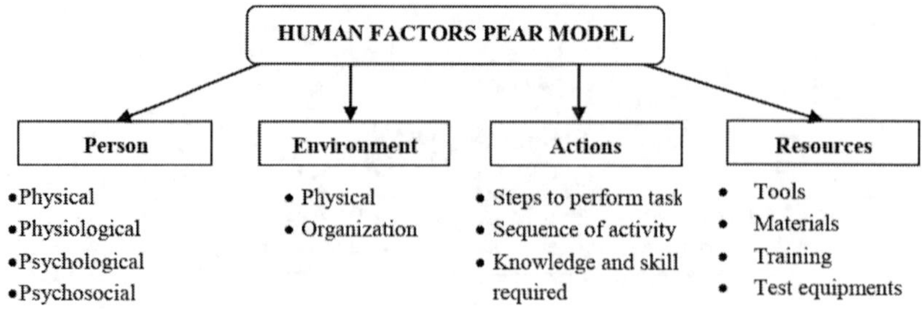

Fig. 2. Johnson and Maddox (2007) PEAR Model

3.3 Ergonomic

The word ergonomic is derived from the Greek words ergo (work) and nomos (laws). It was used for the first time by Wojciech Jastrzebowski in a Polish newspaper in 1857 [27] . Wojciech Jastrzebowski had suggested the term 'ergonomics' to represent the science of human work [29]. The name 'ergonomic' is retained in the title of many major scientific and professional organizations of Europe. Indeed, the name ergonomic is incorporated in many research societies worldwide, including the overall global organization, the International Ergonomics Association (IEA) [17].

In August 2000, the International Ergonomics Association Executive Council adopted the following definition "Ergonomic (or human factors) is the scientific discipline concerned with the understanding of interactions among humans and other elements of a system, and the profession that applies theory, principles, data and methods to design in order to advance human well-being and overall system performance [9]. Table 2 shows the ergonomic fragment and its explanation endorsed by IEA.

Table 2. Ergonomic fragment and its coverage

Ref	Fgt	Coverage
Kroemer [30]	Physical Ergonomic	Anthropometrical, biomechanical, and physiological information to design physical environments
Harris [31]	Cognitive Ergonomic	Scientific understanding of attention, perception, memory, mental models, expertise, and mental workload in designing tools and systems
Chapanis [18]	Organizatio-nal Ergonomic	Optimization of socio-technical work systems including their policies, structures and organizational processes.

Currently, ergonomic practitioners collaborate directly with professionals in industrial engineering, systems safety, occupational medicine and industrial hygiene to reduce or prevent damage and injury on the job. The earlier historical of ergonomic spread largely within the countries that have emphasized on the importance of safe working conditions. Therefore, it is no surprise that ergonomic has initially been considered to be of European origin as in Europe, the focus of ergonomic is to be found in industry and it has been related to an interest in improving worker performance and satisfaction [32]. Table 3 shows the definition of ergonomic as be extracted from published work.

Table 3. Definition of Ergonomics and Its References

Ref	Definition of Ergonomics
NIOSH [33]	Science of fitting workplace conditions and job demands to the capabilities of the working population (1) Effective and successful "fits" assure high productivity (2) Avoidance of illness and injury risks (3) Increased satisfaction among the workforce
ILO [34]	Ergonomics is the systematic study of people at work with the objective of improving the work situation, the working conditions and the task performed
OSHA US [35]	Study of work. It is the science of designing the job to fit the workers, rather than physically forcing the worker's body to fit the job
HSE [36]	Science concerned the "fit" between people and their work. If put people first, taking account of their capabilities and limitation
Humantech [37]	Ergonomics is an approach or solution to deal with a number of problems—among them are work-related musculoskeletal disorders

3.3.1 A Model related to ergonomic

To have better understanding of ergonomic term and its related factor, Job Stress Model by Carayon et al [6] that showing the factors for Work Related Musculoskeletal Disorders (WMSDs) is presented. This model illustrates job stress mechanisms; the work system (e.g. work organization, job design, work environment and technology) had a direct impact on short-term emotional, physiological and behavioural responses. Physical and psychological exposures could give impact to the development of WMSDs and individual factors might moderate the relationships between exposures and outcomes [38,6]. From this model, it is also showing that one of the factors which lead to WMSDs is biomechanical factors. In biomechanical studies, the underlying conceptual model signify that WMSDs are the result of tissue damage and/or pain produced when a worker is exposed to physically demanding work [40]. It is often used in ergonomics for understanding and controlling the risk of WMSDs. Psychosocial profile includes psychosocial risks which can derived from the organisation and design of work are associated with chronic health conditions in workers including depression, anxiety and musculoskeletal disorder [41,42]. Morphological characteristic and generic traits also believed may influence the cause and effect of individual musculoskeletal system which towards the end it will cause strain to human body. However, based on Carayon et al [6], individual biochemical and physiological together with individual body structure may affect individual pain threshold limit due to perception and knowledge that they keep in their mind.

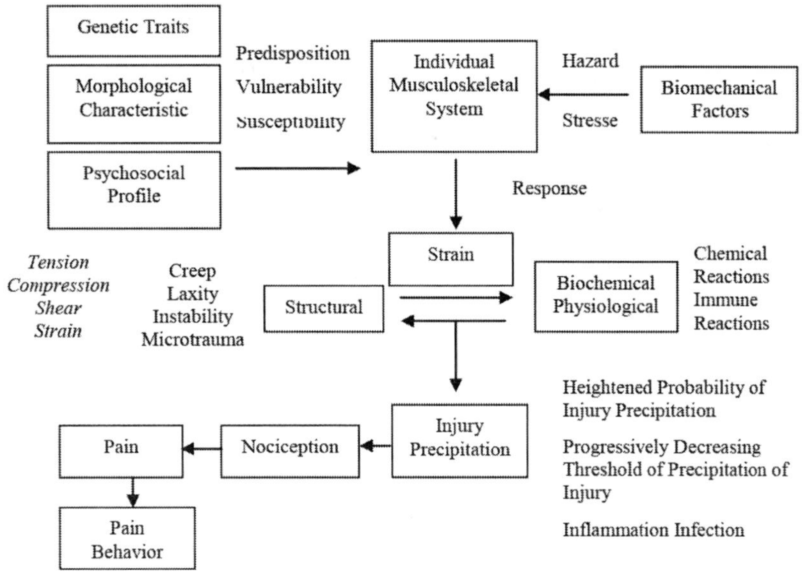

Fig. 3. Carayon et al. (1999) model of job stress

3.4 Human Factor and Ergonomics in Malaysia's Pineapple Plantations

Human factors and ergonomics related terms are considered equivalent by many, although some opinions draw precisely distinctions among these terms. The focus of ergonomic implementation should removes barriers to quality, productivity and most importantly to keep safe human performance by fitting products, tasks, and environments to people [45] . On the other hand, human factor should focus on applying knowledge in designing systems that work, accommodating the limits of human performance and exploiting the advantages of the human operator in the designated process [44]. Salvendy [45] also stated that the word ergonomic implies the study of man at work while word human factor imply the study of man in relation to equipment and environment. From all these opinion, it can be concluded that ergonomic is emphasis on how work affects people while the human factor is often on designs that reduce the potential for human error [43].

After the definition of both (human factor and ergonomic) have revealed, it is easy to capture human factor issue as well as ergonomic issue in pineapple plantation activities. Basically, pineapple plants grow low on the ground. In general, work activities in pineapple plantation requires workers to stoop in a prolonged times. Repetitive task that put a lot of strain on the body, involving the lower limbs such as constant bending over to plant seeds, weed and harvest pineapples are norms [2] . In a study conducted by Guo et al., [46] among 122 pineapple workers in the Southern Taiwan showing that the work activities involved with high prevalence of heavy manual work. It also in line with other studies that focus of manual work in other plantation [47, 48]. Figure 4 shows four main steps in planting pineapples whereby it involves preparing the land, planting, maintaining the crops and lastly, harvesting the pineapple (Figure 4).

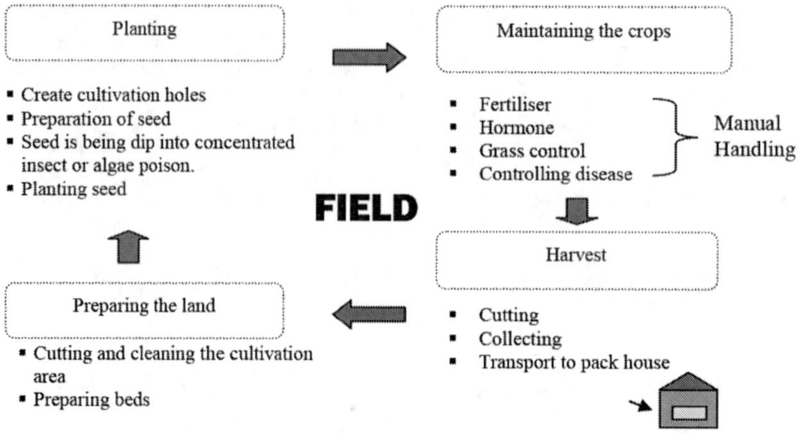

Fig. 4. Main Processes (Steps) in Pineapple Plantation

Table 4. Pineapple plantation activities and the possible issues facing by workers

Activity	Ergonomic	Human Factor	Safety
Preparing the land	Back pain from prolonged and awkward bending	• Poor knowledge to working safely • Low job satisfaction due to environment condition • Low income but physical demanding job • Inadquate facilities such as appropriate place for breaks	Sharp objects (hoe, cleaver) can cause minor or major cut
Planting	Awkward squatting position can cause numbness on knees		Exposure to heat stress
Maintaining Crops	Prolonged standing that can cause muscle strain on back and waist		Chemicals exposure
Harvest	Awkward bending and overexertion of waist and arm and heavy lifting		Manual handling

Ergonomic issues in pineapple plantation such as activities that require workers to bend awkwardly with the use of force most of the time to fulfil the work task are unavoidable. Meanwhile, human factor issue such as satisfaction of workers towards their salary, workers acceptance on the overall of the work system including, human to human, human to tools or machineries, and the relation of human to the surrounding are the things that also need to be taken seriously. Table 4 shows the problems associated with pineapple plantation activities that categorized into human factor, ergonomic and others issue.

4 Conclusions

In Malaysia pineapple plantation, some issues involving Human Factor and Ergonomics is revealed. Ergonomic issue such as constant and prolonged awkward postures is directly associated with Work-related Musculoskeletal Disorders (WMSDs) as this poor postures are one of the common risk factors of developing musculoskeletal problems among workers such as low back pain and knee pain. Human Factor issues such as inadequate facilities and low job satisfaction shows how humans behave physically and psychologically in relation to particular environments. The oucomes from this Human Factor issue is possibly pineapple plantation workers are not willing and happy to do their job, several areas of their work are affected, productivity loss and high employess turnover rates.

Ergonomic approach is aimed to optimise human interactions with systems, on how to make human activities more efficient, safe, comfortable and satisfying. While, human factor is the discipline to do with the design of systems in order to reduce human error and minimize ergonomics problems. So, to attain both goals effectively, the field needs a clear, concise, unambiguous and usable term to describe experts' efforts. As for now, it is clearly can be seen that the term ergonomics will take that role, along with adjectives of physical and cognitive. Regardless of the terms and labels that are used, the aim is always to optimize human efficiency, effectiveness, health, safety, and well-being, within the context of overall system performance operability, reliability, and maintainability. Ergonomic and Human Factor can made environments to be safer and more usable, efficient, and enjoyable for people to interact with. However, in pineapple plantation, it is good to really understand the differences between these two terms, so that planning to inculcate safe and healthy culture is easy to be achieved. Because, the human factor issue needs different approach to be tackled and ergonomic need different technique to be managed in the workplace setting.

This project is financially funded by the grant of Universiti Malaysia Pahang, Malaysia (Grant Number: RDU170308) (www.ump.edu.my).

References

[1] Malaysian Pineapple Industry Board, Retrieve from: http://www.mpib.gov.my/en/;jsessionid2 (2017)

[2] S.B.M. Tamrin & Aumran. A comparison of the hazards, the risks, and the types of control in three selected agricultural industries. *Occupational Safety and Health in Commodity Agriculture :Case Studies from Malaysian Agriculture Perspective*, **1**: 93–146 (2014)

[3] Department of Agriculture, *Agriculture Statistic Handbook*, Websites Retrieved from http://www.doa.gov.my (2008)

[4] Department of Occupational Safety and Health. *Occupational Accidents Statistics by Sector* Website Retrieve from http://www.dosh.gov.my/index.php/en/occupational-accident-statistics/by-sector (2017)

[5] W. Johnson and M. Maddox, A *PEAR Shaped Model For Better Human Factors* Civ. Aviat. Train Mag., 3–4 (2007)

[6] P. Carayon, M. J. Smith, and M. C. Haims, *Work Organization, Job Stress, and Work-Related Musculoskeletal Disorders*. Hum. Factors J. Hum. Factors Ergon. Soc., **41**, 644–663 (1999).

[7] M. Hassall, T. Xiao, P. Sanderson, and A. Neal, International Encyclopedia of the Social & Behavioral Sciences, Second Edi Elsevier, **11** (2015)

[8] J. Dul, R. Bruder, P. Buckle, P. Carayon, P. Falzon, W. S. Marras, J. R. Wilson, and B. Van Der Doelen. *A strategy for human factors / ergonomics : developing the discipline and profession*, 1–27 (2012)

[9] P. G. Dempsey, M. S. Wogalter, and P. A. Hancock. *Defining Ergonomics / Human Factors* Taylor & Francis Group (2006)

[10] International Ergonomic Association, *Definition and Domains of ergonomics*, http://www.iea.cc/whats/index.html (2017).

[11] I. Sutton, Chapter 8 – Human factors and ergonomics, 160–177 (2015)

[12] O. Brown, H. W. Hendrick, (Eds.), *Human factors in organizational design and management* - II. Amsterdam: North-Holland (1986)

[13] D. Meister , *Behavioral analysis and measurement methods*, New York: John Wiley & Sons (1985)

[14] K. H. Pelsma, *Ergonomics sourcebook: A guide to human factors information*. Lawrence, KS: The Report Store (1987).

[15] R. Rubinstein, H. Hersh, The human factor: Designing computer systems for people. *Bedford, MA: Digital Press* (1984)

[16] J. C. Thomas, Organizing for human factors. In Y. Vassilou (Ed.), *Human factors and interactive computer systems* Norwood, NJ: Ablex Publishing, 29-46 (1984)

[17] P. A. Hancock, *Human Factors/Ergonomics*, 2nd ed. Elsevier Inc (2012)

[18] A. Chapanis, *Ergonomics in product development: a personal view*, **38** (1995)

[19] F. Wikipedia, Human factors, **5** 1–13 (2008) Retrieved from https://www.faa.gov/regulations_policies/handbooks_manuals/aircraft/media/AMT_Handbook_Addendum_Human_Factors.pdf

[20] H. W. Hendrick, Good Ergonomics Is Good Economics, *Proc. 40th Annu. Meet. Hum. Factors Ergon. Sciety*, 1–15 (1996)

[21] E. Bennett, J. Degan, J. Spiegel, *Human factors in a technological society. New York: McGraw-Hill*, (1963)

[22] B.S. Dhillon, Human reliability with human factors, *New York: Pergamon Press* (1986)

[23] F. H. Hawkins, Human factors in flight, *England: Gower Technical Press* (1987).

[24] R. M. (Ed.). Goldenson, *Longman Dictionary of Psychology and Psychiatry*, (1984)

[25] B. H. Kantowitz, R.D Sorkin, Human factors: Understanding people-system relationships. New *York: John Wiley & Sons* (1983)

[26] J. Reasons, Human Factors: Resource guide for Engineers. *Civil Aviation Safety Authority Australia* (2001)

[27] P. A. Brennan, D. A. Mitchell, S. Holmes, S. Plint, and D. Parry, Good people who try their best can have problems: Recognition of human factors and how to minimise error,*Br. J. Oral Maxillofac. Surg* **54** 3–7 (2016)

[28] W. Karwowski, *International Encyclopedia of Ergonomics and Human Factors*, 2nd edn.,**1**, Boca Raton, FL: CRC Press (2006)

[29] W. Jastrzebowski, *An Outline of Ergonomics or the Science of Work Based on the Truths Drawn from the Science of Nature* (reprinted by the Central Institute for Labour Protection, Warsaw, Poland) **29** (1857)

[30] K. H. E. Kroemer, E. Grandjean, Fitting the Task to the Human: A Textbook of Occupational Ergonomic, *Taylor & Francis, Bristol, PA.* (2009)

[31] D. (Ed). Harris, Engineering Psychology and Cognitive Ergonomics: Understanding Human Cognition (2013)

[32] G. Salvendy, *Handbook of human factors and*

ergonomics, **9**, 3(1997)

[33] L.C. Alexander, C.G. Christopher, J.F. Lawrence, P.B. Bruce, D.M. James, Elements of Ergonomics Programs - A Primer Based on Workplace Evaluations of Musculoskeletal Disorders (1997)

[34] International Labour Office, **15**, 93 (1998).

[35] OSHA US, *Ergonomics : The Study of Work* (2000)

[36] Health and Safety Executive, *Ergonomics and human factors at work*, 1–10 (2013)

[37] Humantech, Definition of Occupational Ergonomics, 1–4, Website Retrieve from https://www.humantech.com/wpcontent/uploads/2016/06/Definition-of-Occupational-Ergonomics-Position-Paper.pdf (2016)

[38] Sauter, S.L. and Swanson, N.G, An ecological model of musculoskeletal disorders in office work. In Beyond biomechanics: Psychosocial Aspects of musculoskeletal disorders in office work, Bristol, PA: Taylor and Francis, 3–22 (1996)

[39] P. Carayon, M.J. Smith, M.C. Haims, Work organization, job stress, and work- related musculoskeletal disorders, Hum. Factors, **41**, 644–663 (1999)

[40] M.S. Kerr, Workplace Psychosocial Factors and Musculoskeletal Disorders: A Discussion Paper (1998)

[41] E.-M. Backé, A. Seidler, U. Latza, K. Rossnagel, B. Schumann, The role of psychosocial stress at work for the development of cardiovascular diseases: A systematic review, Int. Arch. Occup. Environ. Heal. **85** 67–79 (2012)

[42] H. Sultan-Taïeb, C. Lejeune, A. Drummond, I. Niedhammer, Fractions of cardiovascular diseases, mental disorders, and musculoskeletal disorders attributable to job strain. Int. Arch. Occup. Environ. Health, **84**, 911 – 925 (2011)

[43] N. Jaffar, A.H. Abdul-Tharim, I.F. Mohd-Kamar, N.S. Lop, A literature review of ergonomics risk factors in construction industry,Procedia Eng., **20**, 89–97 (2011)

[44] C.D. Wickens, *Engineering psychology and human performance. Columbus: Charles Merrill Publishing* (1984)

[45] G. Salvendy, Has ergonomics the same meaning in Europe and North America? In I.D. Brown, R. Goldsmith, M.A. Sinclair (Eds.), Ergonomics international London: Taylor & Francis, **85**, 97–98 (1985)

[46] Y. Guo, B. Wang, C. Lee, J. Wang, Prevelance of dermatoses and skin sensitisation associated with use of pesticides in fruit farmers of southern Taiwan, Occup. Environ. Med., **53**, 427–431 (1996)

[47] E. H. Sukadarin, B. M Deros, N. S. M. Syazwani, J. A. Ghani, A. R. Ismail and J. Zakaria, Back Pain and the Observed Factors among Oil Palm Workers, International Journal of Engineering Technology And Sciences (IJETS) **5** (1) (2016)

[48] E. H. Sukadarin, B. M Deros, N. S. M. Syazwani, J. A. Ghani, A. R. Ismail and J. Zakaria, A Preliminary Study of Postural Problems among Oil Palm Workers: Is OWAS method suitable to investigate those problems? International Journal of Engineering Technology And Sciences (IJETS) **5** (1) (2016)

A Descriptive Study towards Green Computing Practice Application for Data Centers in IT Based Industries

Bokolo Anthony Jnr.[1,*], *Mazlina* Abdul Majid[2], and *Awanis* Romli[1]

[1]Faculty of Computer Systems and Software Engineering, Universiti Malaysia Pahang, 26300 Gambang, Pahang, Malaysia

Abstract. The progressive upsurge in demand for processing and computing power has led to a subsequent upsurge in data center carbon emissions, cost incurred, unethical waste management, depletion of natural resources and high energy utilization. This raises the issue of the sustainability attainment in data centers of Information Technology (IT) based industries. Green computing practice can be applied to facilitate sustainability attainment as IT based industries utilizes data centers to provide services to staffs, practitioners and end users. But it is a known fact that enterprise servers utilize huge quantity of energy and incur other expenditures in cooling operations and it is difficult to address the needs of accuracy and efficiency in data centers while yet encouraging a greener application practice alongside cost reduction. Thus this research study focus on the practice application of Green computing in data centers which houses servers and as such presents the Green computing life cycle strategies and best practices to be practiced for better management in data centers in IT based industries. Data was collected through questionnaire from 133 respondents in industries that currently operate their in-house data centers. The analysed data was used to verify the Green computing life cycle strategies presented in this study. Findings from the data shows that each of the life cycles strategies is significant in assisting IT based industries apply Green computing practices in their data centers. This study would be of interest to knowledge and data management practitioners as well as environmental manager and academicians in deploying Green data centers in their organizations.

1 Introduction

Over the years the increasing need for computer related resources has resulted to a major growth in the figures of data center servers deployed along with an increase in the energy consumed by these servers. Data centers provides support to the present day digital economy based on the prevalent application and usage of cloud based services, but since data centers utilizes a substantial amount of power and also releases Carbon dioxide (CO_2) emission to the atmosphere [1], hence there is need for strategies, technics and technologies that can reduce CO_2 emission from data center.

Green computing is one of these technologies that can help lessen CO_2 emissions in data centers. Green computing aims to deploy and utilize technologies for a cleaner and Greener planet. Green computing is mainly the study and practice of utilizing computing resources proficiently [2]. It is a paradigm towards practicing cost-effective and environment friendly use of energy.

Green computing can also be referred to as optimal utilization of Information Technology (IT) for governing the sustainability attainment of an organization. It involves activities that emphases on tactical deployment of IT to vigorously and ethically align organizations aims and objectives with environmental protection in mind during the complete industrial operations. According to Uddin et al. [3] Green computing entails applications of ecological friendly IT based initiatives focused on organizational objectives. Hence Green computing practices can be applied in IT based industries data centers towards CO_2 emissions reduction, cost incurred lessening, proper ethical waste management, depletion of natural resources lessening and high energy utilization decrease.

Data center energy consumption is anticipated to increase IT based industries costs and CO_2 emissions. Hence Green computing practice application in data centers can help lessen these negative effects [4]. In continents such as Europe, the energy consumption of enterprise data centers for cooling and powering server equipment institutes large proportion electricity. Hence organizations such as IT based industries have called on policy makers to resolve data center electricity efficiency in order to certify that the linked impacts, such as societal, economic and environmental issues are mitigated [5].

In IT based industries data centers signify the buildings, rooms and facilities which encompass enterprise servers, networks communication infrastructures, cooling facilities and powering equipment that supports data providing services which may include data handling for intranet, internet, web hosting, IT and telecommunication cables, routers, hubs

* Corresponding author: bkanjr@gmail.com

and switches [6]. Green computing in IT based industry data center involves the deployment of energy efficient IT such as electrical, lighting, computer systems and mechanical equipment with negligible environmental impact on the environment [7-9].

Therefore this research aims to conduct a descriptive study of how IT based industries can apply Green computing practices in their data centers in regards to cost saving decrease, energy efficiency, eco-friendly waste management, natural resource conservation and CO_2 emission reduction towards sustainability attainment. To achieve the aims of this study it is imperative to identify the Green computing life cycle strategies to be applied for Greening data centers in IT based industries

But in order to identify the life cycle strategies to be applied in Greening data centers towards the application of best practices for cost saving decrease, energy efficiency, eco-friendly waste management, natural resource conservation and CO_2 emission reduction, this research presents the life cycle strategies that are drawn from secondary data. The life cycle strategies are confirmed from primary data (questionnaire) collected from more than 133 respondents in IT based industries. More specifically the benefits of this study would have positive implications for IT based industries data center system practitioners and staffs by providing best practice towards the reductions of energy consumption and CO_2 emissions.

The remainder of this article is structured as follows. The next section is the literature review. Then the research methodology is provided. Next, the results and findings from the questionnaire data collection are outlined. We then proceed to the discussion section, after which the research and practical implications are revealed. The article is concludes with conclusion, limitation and future work section.

2 Literature review

A few studies has be published that contributes to support Greening data centers in organizations, among this studies, 5 studies related to data centers Greening are reviewed in this section because they individually contributed to the presentation of Green computing life cycle strategies. Among the 5 studies that contributed to Greening data centers in organizations Karanasios et al. [1] carried out an exploratory case study by identifying the antecedents towards the adoption of techniques, technologies and best practices for Greening data centres. The authors developed a conceptual framework to explain the circumstances that might impact the implementation of the best practices in data centres Greening.

Molla and Cooper [9] investigated the motivation, ability drivers and expectancy for Greening data centres. The researchers resolved the issue relating to the institutional, motivation ability and expectancy concepts based on survey data collected from 96 data centres. Findings form their research shows that effort expectancy, ability and performance are the main drivers that influence practices and technologies implementation towards energy efficiency improvement in data centres.

Pawlish and Varde [6] proposed a decision support system based on decision trees and case based reasoning that collects existing campus data in order to facilitate decision-making for enhances data centers management for Greening campus data centers. Uddin et al. [3] utilized Green IT a developed framework for achieving energy efficient of data centers alongside the application of virtualization technique. The authors deployed Green IT to securely and seamlessly divide data center modules into different resource pools depending on different metrics such as energy utilization ratio, workloads consumption ratio, CO_2 emission etc. The developed framework is grounded on cloud computing and virtualization to reduce the energy utilization ratio of deployed servers.

OGCIS [10] summarized a list of guidelines for implementing Green data center practices. The author utilized Green IT to lessen energy consumption and CO_2 emission of data centres in government organizations. The report presented by the author can be used as a quick reference for Greening data centers environment. Also the practice outlines sets of endorsed practices applicable for attaining energy efficiency and minimising environmental impacts in data centre. The practices are characterized based on 4 lifecycle phases which comprises of design, procurement, operations and disposal.

Findings from the reviewed literatures shows that data centre Greening has been a issues to be addressed by industries such as IT based industries, although researchers such as Pichetpongsa and Campeanu [16] believes that the implementing of Green practices can help reduce the negative effects caused by IT usage in IT based industries.

Each of the review papers aims to contribute to address climatic changes and environmental problems none of the reviewed studies simultaneously addressed sustainability issues in data center in relation to energy efficiency, cost saving, natural resource preservation, waste management, and CO_2 emissions. The reviewed studies are mainly concerned about energy efficiency, CO_2 emission and waste management. Hence lifecycle strategies and metrics related cost saving and natural resource preservation was not fully addressed.

3 Research methodology

This descriptive study reports findings for data collected to verify life cycle strategies for applying Green computing practices in IT based industries data centers. The qualitative data for this study is collected from 133 respondents from IT based industries in Malaysia. The survey instrument was prepared by based on the descriptive nature of the study and following Hair et al. [11] procedures on data collection instrument development. Figure 1 shows the research methods adopted in this study, each of the phase shown in Figure 1 are carried out throughout this research paper.

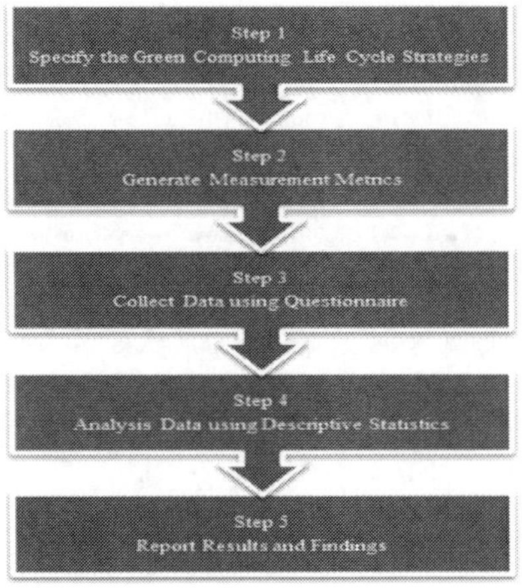

Fig. 1. Research process.

Figure 1 displays the research methods followed in this study. Each of the steps is further explored in sub sections of section 3.

3.1. Green computing life cycle strategies

This section provides a theoretical and practical description of the process lifecycles strategies to be applied in practicing Green computing in IT based industries data centers. In this regards the lifecycle strategies are presented below;

3.1.1 Green design

Green design aims to analyse, create and synthesize ecological friendly products with outstanding efficiency. IT based industries usually disregarded environmental effects during design and as such hazardous wastes were discarded without the ecological issues being considered [12]. Therefore Green design in IT based industries data centers should consider energy proficient and deployment of ecologically sound servers, cooling equipment and computer components such as Light Emitting Diode (LED) monitor, etc. [13, 14].

3.1.2 Green production

This phase in IT based industries data centers involves taken steps toward applying Greener prospect by changing their data centers operations to more competent activities and also increasing the reutilizing and reusability rate of the existing infrastructures deployed in the data center [12]. As suggested by Pichetpongsa and Campeanu [13] in Green production every procedure involved in running the data center electronic components, computers and other related subsystems should indicate a low or no effect on the natural environment. This statement was also supported by Saha [14] who mentioned that in Green production electronic equipment which has negligible impact or no effect on the environment should only be installed.

3.1.3 Green procurement

In Green procurement IT based industries can also apply environmental purchasing practice that involves reduction, recycling and reuse of IT infrastructures in purchasing operations [14]. Moreover Green procurement promotes ecologically considerations and promotes the economically viability of the organization towards selecting and acquiring services and products that reduces environmental pollution [13]. Other Green procurement activities may include the practice of considering the Green track record of IT services and software vendors toward environmental protection [15].

3.1.4 Green operations

This strategy aims to save energy which leads to less emission of CO_2 to the atmosphere when data centers servers are being deployment or implemented to facilitate IT based industries end users, staffs, practitioners and management. Hence, data center administrator should be aware of how Green operation can be applied towards to reducing energy consumption of data centers [12]. Green operation practices also aims to enhance power performance of IT based industries assets towards decreasing energy utilization of cooling and powering enterprise data centers assets, optimizing energy performance of data centers, reducing data centers induced CO_2 emissions [17], practicing low carbon emitting corporate practices and lastly assessing the organizations total environmental footprint [15].

3.1.5 Green disposal

Green disposal provides IT based industries data centers with a flexible and audit policy solution for gathering; re-processing and recycling of end-of-life redundant IT associated equipment [18]. Hence in Green disposal IT base industries can plan to refurbish and reuse old data centers components, while other unwanted electronics components can be prepared for recycling operations [13]. Also OGCIS [10]; Molla [15] recommended that the reuse (extend life) or disposal of obsolete data centers equipment and IT related facilities should be dismantled or refurbished for reuse, while discarding of out-dated infrastructures should adhere to relevant ecological legislations guidelines such as waste disposal ordinance for end of equipment life dumping policy or they can utilize information systems to track the life-cycle of enterprise data center assets and ascertain the cost-benefit of different discarding methods [12].

Based on the identified and discussed life cycle strategies a research model is developed as seen in Figure 2 to be verified by the questionnaire data. The research model is verified to ascertain if each of the strategies are important and should be applied by IT based industries data centers in regards to energy efficiency, cost saving, natural resource preservation, waste management, and CO_2 emissions.

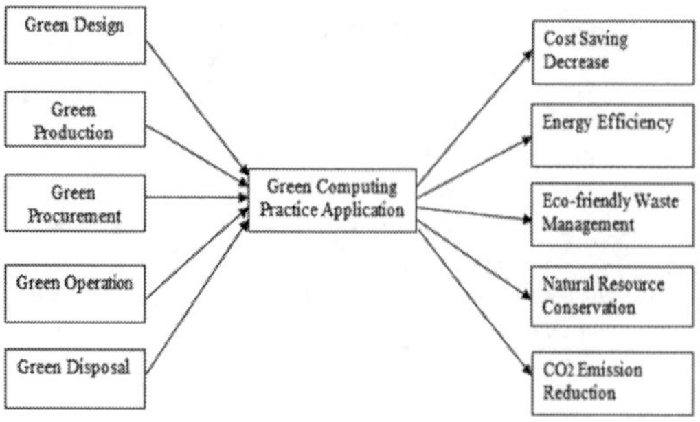

Fig. 2. Research model.

Figure 2 shows the research model developed based on the life cycle strategies that comprises of Green design, Green production, Green procurement, Green operation and Green disposal which influences Green computing practice application in IT based industries data centers towards cost saving decrease, energy efficiency, eco-friendly waste management, natural resource conservation and lastly CO_2 emission reduction.

3.2. Measurement metrics generation

To measure the life cycle strategies presented in section 3.1 and Figure 2, metrics are derived from the literatures to assess each life cycle strategy, hence each of the metrics are shown in Table 1 to measures the degree to which each life cycle strategy relates to the application of Green computing practices in data centers.

Table 1. Operationalization of life cycle strategies and metrics

Life Cycle Strategies	Code	Metrics
Green Design	GDN1	Concerned about the energy consumption of cooling and lighting in our data center.
	GDN2	Concerned about the efficiency of powering our IT infrastructure.
	GDN3	Considers environmental factors in the design of lighting, power delivery, cooling systems and servers, storage and network.
	GDN4	Relocate enterprise data center near clean sources of energy/renewable energy.
	GDN5	Use electricity supplied by Green energy providers in our data center.
	GDN6	Enforces power management.
Green Production	GPN1	Install software to make production more environmentally friendly in our data center.
	GPN2	Retire energy inefficient systems
	GPN3	Analyses IT's energy bill separately from overall corporate bill in our data center.
	GPN4	Engaging the service of a professional service provider to maintain our data center.
	GPN5	Install more energy efficient lightings.
	GPN6	Upgrades to efficient transformers and UPS.
	GPN7	Auditing the power efficiency for lesser power consumption in our data center.
	GPN8	Eliminates and di-commission unused services and systems in our data center.
Green Procurement	GPT1	Install software to make material sourcing more environmentally friendly.
	GPT2	Buys recycled IT equipment.
	GPT3	Makes preference to IT hardware suppliers that have a Green track record.
	GPT4	Gives weight to environmental considerations in equipment procurement.
	GPT5	Deploy environment-friendly IT procurement policy in our data center.
	GPT6	Purchase IT equipment from vendors that offers take back option.
Green Operation	GON1	Environmental consideration in planning IT usage and operations in our data center.
	GON2	Applied power management features of IT equipment regularly used in our data center.
	GON3	Turn off associated data center system when not in use to saves energy.
	GON4	Print data center report on both side of a paper to reduce paper wastage.
	GON5	Utilizes equipment that can monitor workloads and to shut down when unused.
	GON6	Uses free cooling in data centers to reduce incurred energy cost.
Green Disposal	GDL1	Recycle consumable equipment (e.g. batteries, ink cartridges, and paper).
	GDL2	Disposes of IT equipment in an environmentally friendly manner.
	GDL3	Carryout policy on managing electronic waste.
	GDL4	Reuse IT equipment.
	GDL5	Refurbish old, outdated and obsolete IT equipment.

Table 1 show the life cycle strategies and metrics used to verify each life cycle, where "Green design" "Green procurement" and "Green operation" are all measured with 6 different metrics with a 5 point Likert scale ranging from not applied as "1" and fully applied as "5". "Green production" is measured with 8 different items; with a 5 point Likert scale ranging from not applied as "1" and fully applied as "5". Lastly "Green disposal" is measured with 5 different items; with a 5 point Likert scale ranging from not applied as "1" and fully applied as "5".

3.3. Data collection

Data was collected using questionnaire used to verify life cycle strategies, related to Green computing practice application in IT based industries data centers. Before administering the questionnaire to randomly and purposively selected respondents, 3 domain experts in Green computing area assessed the validity of the questionnaire items which lead to the refinement of a few questionnaire metrics or questions based on the feedback from the experts. In this study the reliability was measured using Cronbach's alpha which measure the internal consistency of a test and it is defined as number ranging from 0-9 [11]. Where the value of alpha is given as 0.986 as obtained from SPSS version 22. After which link to the questionnaire survey was sent to the official email of IT practitioners, IT staffs in IT based industries in Malaysia, the email addresses were gotten from their organizations website. The respondents were more than 700 randomly and purposively sampled from IT based industries in Malaysia.

3.4. Data analysis using descriptive statistics

The results were based on survey questionnaire from different IT based industries that possess data center. After sending the survey questionnaire link, data was collected from 133 IT staffs responsible for performing technical and operational functionalities at their respective organizations data centers. The collected data helped to verify whether the lifecycle process presented in this research is feasible to be practically applicable with ease for data centers Greening or not.

Test was carried out to ascertain if all the life cycle strategies; Green design, Green production, Green procurement, Green operation and Green disposal fulfils the desired dimensions of sustainability (social, economic and environmental) and also provides recommendation as best practices on how IT based industries can address energy efficiency related issues, cost saving decrease, eco-friendly waste management, natural resource conservation and ensure that there is a limited amount of CO_2 emission in the atmosphere.

Lastly the collected data also helps to find out if the life cycle strategies presented in this study provides operational solutions in assisting IT based industries data centers to apply more Greener practice.

3.5. Results and findings

The instrument used for data collection was developed into two main sections. Section A collects data regarding respondents' demographic information, whereas section B contained 5 life cycle strategies with associated metrics that measured life cycle strategies. The metrics were derived from review of literature on Greening data center and Green computing practice application. The data were analysed using descriptive statistics, responses to the five life cycle strategies measuring Green computing practice application in IT based industries data centers was analysed and presented as frequency in percentage, mean, standard deviation (SD), maximum (max), minimum (min) and median value.

Table 2. Characteristic of the questionnaire respondents

Demographic Profile	Options	Response
Gender	Male	55.0 %
	Female	45.0 %
Age	< 25	1.01 %
	25-34	41.43 %
	35-44	43.43 %
	45-55	13.13 %
	>55	1.01 %
Education	High School	3.00 %
	Diploma	26.0 %
	Bachelor's Degree	35.0 %
	Master's Degree	23.0 %
	PhD	13.0 %
Industry Country	Malaysia	100 %
	Others	0.00 %
Industry Sector	ICT, Communication and Media	38.00 %
	Education and Research	52.00 %
	Health and Community Services	0.00 %
	Engineering and Construction	2.00 %
	Finance, Banking and Insurance	0.00 %
	Government, Administration and Defence	6.00 %
	Manufacturing	0.00 %
	Business, Wholesale and Retail	0.00 %
	Personal, Professional	2.00 %
	Others Services	0.00 %
Job Title	Business and Systems Analysts and Programmers	24.42 %
	ICT Managers	18.60 %
	ICT Network and Support Professionals	13.95 %
	Non ICT Specialist Managers	2.33 %
	Chief Executives, General Managers and Legislators	0.00 %
	Tertiary Education Lecturers	12.79 %
	Database Systems Administrators, and ICT Security Specialists	6.98 %
	Others	20.93 %
Working Experience	0-5	29.07 %
	6-10	25.58 %
	11-15	24.42 %
	16-20	12.79 %
	>20	8.14 %
Industry Size	Below 50 employees	12.79 %
	51-250 employees	23.26 %
	251- 1000 employees	11.63 %
	Above 1000 employees	52.33 %
Industry Founded Date	Before 1967	5.81 %
	Between 1967-1980	18.60 %
	Between 1991-2000	34.88 %
	Between 2001- 2010	33.72 %
	From 2011- Till date	6.98 %
Industry Annual Revenue	RM 90,000 or below	19.75 %
	RM 90,000 to RM 900, 000	14.81 %
	RM 900,000 to RM 2, 700,000	14.81 %
	RM 2,700,000 to RM 4,500,000	13.58 %
	RM 4, 500,000 to RM 9, 000,000	4.94 %
	RM 9,000,000 or above	32.10 %

Table 2 shows the demographic characteristic of the questionnaire respondents in relation to individual and industries characteristics.

3.5.1 Descriptive statistics

In regards to the life cycle strategies and metrics presented in Table 1, findings for the verification of each life cycle strategy are presented below. The frequency response is represented as percentage distribution and descriptive statistics for Green design is shown in Table 3, where the metrics includes GDN1-GDN6.

Table 3. Descriptive statistic results for Green design metrics

Life Cycle	Items	1 Not Applied	2 Partly Applied	3 Neutral	4 Applied	5 Fully Applied	Mean	SD	Min	Max	Median
Green Design	GDN1	3.23 %	4.84 %	17.74 %	51.61 %	22.58 %	3.85	0.93	1.00	5.00	4.00
	GDN2	6.35 %	4.76 %	20.63 %	42.86 %	24.40 %	3.76	1.08	1.00	5.00	4.00
	GDN3	3.17 %	4.76 %	23.81 %	42.86 %	25.40 %	3.83	0.97	1.00	5.00	4.00
	GDN4	17.46 %	11.11 %	20.63 %	34.92 %	15.87 %	3.21	1.32	1.00	5.00	4.00
	GDN5	15.87 %	12.70 %	20.63 %	36.51 %	14.29 %	3.21	1.29	1.00	5.00	4.00
	GDN6	11.11 %	11.11 %	23.81 %	33.33 %	20.63 %	3.41	1.24	1.00	5.00	4.00

Table 3 shows the findings for Green design metrics items GDN1 to GDN6. It can be seen that all metrics are fully applied. Finding further revealed that GDN1, GDN2 and GDN3 response frequency is higher than the other metrics in terms of application. In relation to the mean value as seen in Table 3 each of the have a mean value greater than 2.5 which is the least required mean value on a scale of 5. Table 3 also shows that the standard deviation value is lower for GDN1 and GDN3. This shows that the responses are quite similar in options selected in relation to Green design application in their data centers. Percentage distribution and descriptive statistics for Green production is shown in Table 4, where the metrics includes GPN1-GPN8.

Table 4. Descriptive statistic results for Green production metrics

Life Cycle	Items	1 Not Applied	2 Partly Applied	3 Neutral	4 Applied	5 Fully Applied	Mean	SD	Min	Max	Median
Green Production	GPN1	14.06 %	10.94 %	14.06 %	46.88 %	14.06 %	3.36	1.25	1.00	5.00	4.00
	GPN2	15.63 %	9.38 %	26.56 %	29.69 %	18.75 %	3.27	1.30	1.00	5.00	3.00
	GPN3	23.44 %	9.38 %	10.94 %	40.63 %	15.63 %	3.16	1.43	1.00	5.00	4.00
	GPN4	17.19 %	9.38 %	17.19 %	40.63 %	15.63 %	3.28	1.32	1.00	5.00	4.00
	GPN5	9.52 %	9.52 %	17.46 %	41.27 %	22.22 %	3.57	1.20	1.00	5.00	4.00
	GPN6	10.94 %	7.81 %	20.31 %	34.38 %	26.56 %	3.58	1.26	1.00	5.00	4.00
	GPN7	14.06 %	9.38 %	25.00 %	26.56 %	25.00 %	3.39	1.33	1.00	5.00	4.00
	GPN8	12.50 %	15.63 %	15.63 %	32.81 %	23.44 %	3.39	1.33	1.00	5.00	4.00

Table 4 shows the findings for Green production metrics GPN1 to GPN8. The results shows that GPN1, GPN3 and GPN4 all have higher relevance priority and are applied in relation to Green production in data centers, although GPN6 and GPN8 are averagely applied, whereas values from GPN2 and GPN7 are perceived by the respondents to neutrally being applied. For the mean all items possess mean value greater than 2.5, with GPN7 and GPN8 having the highest mean of 3.39. in regards to the standard deviation values all metrics values deviates a bit further from "1" meaning that the response from the respondents are further apart from each other. Percentage distribution and descriptive statistics for Green procurement is shown in Table 5, where the metrics includes GPT1-GPT6.

Table 5. Descriptive statistic results for Green procurement metrics

Life Cycle	Items	1 Not Applied	2 Partly Applied	3 Neutral	4 Applied	5 Fully Applied	Mean	SD	Min	Max	Median
Green Procure-ment	GPT1	18.75 %	10.94 %	12.50 %	37.50 %	20.31 %	3.30	1.40	1.00	5.00	4.00
	GPT2	25.00 %	17.19 %	17.19 %	21.88 %	18.75 %	2.92	1.46	1.00	5.00	3.00
	GPT3	18.75 %	17.19 %	20.31 %	29.69 %	14.06 %	3.03	1.33	1.00	5.00	3.00
	GPT4	12.50 %	17.19 %	17.19 %	29.69 %	23.44 %	3.34	1.34	1.00	5.00	4.00
	GPT5	12.70 %	17.46 %	15.87 %	34.93 %	19.05 %	3.30	1.30	1.00	5.00	4.00
	GPT6	14.29 %	12.70 %	22.22 %	38.10 %	12.70 %	3.22	1.24	1.00	5.00	4.00

Table 5 shows the findings for Green procurement metrics GPT1 to GPT1. The results show that all metrics were applied in the respondents data centers except from GPT2 which measure if the respondents organizations "Buys recycled IT equipment for data center use". The percentage was low for this metrics showing that IT based industries do not buy recycled equipment to be deployed and used in their data centers. In regards to the mean, all values where greater than 3.5, MF2 mean value is also given as 2.92 which is still greater than the required 2.5 threshold value. The standard deviation values of each metrics were also greater that "1" showing the response from the respondents are partly spatially deviated in terms of the respondents perception toward Green procurement application in their organizations data centers. Percentage distribution and descriptive statistics for Green operation is shown in Table 6, where the metrics includes GON1-GON6.

Table 6. Descriptive statistic results for Green operation metrics

Life Cycle	Items	1 Not Applied	2 Partly Applied	3 Neutral	4 Applied	5 Fully Applied	Mean	SD	Min	Max	Median
Green Operatio n	GON1	8.20 %	9.84 %	26.23 %	42.62 %	13.11 %	3.43	1.09	1.00	5.00	4.00
	GON2	6.56 %	13.11 %	21.31 %	39.34 %	19.67 %	3.52	1.14	1.00	5.00	4.00
	GON3	3.28 %	6.56 %	14.75 %	39.34 %	36.07 %	3.98	1.03	1.00	5.00	4.00
	GON4	3.28 %	2.28 %	8.20 %	44.26 %	40.98 %	4.16	0.94	1.00	5.00	4.00
	GON5	16.39 %	6.56 %	3.11 %	40.98 %	22.95 %	3.48	1.35	1.00	5.00	4.00
	GON6	16.39 %	8.20 %	21.31 %	37.70 %	16.39 %	3.30	1.30	1.00	5.00	4.00

Table 6 shows the findings for Green operation metrics GON1 to GON6. The results show that all metrics are applied in their industry for Green operation strategy based on the fact that more than 37.70 % respondents as seen from GON6 metrics to 44.26% GON4 metrics agreed that they apply Green operation

* Corresponding author: bkanjr@gmail.com

strategies in their industry data center. The mean were all above 3.00 values, with GON4 possessing a mean of 4.16 outlining that in the respondents individual data centers they print data center reports on both side of a paper to reduce paper wastage thus saving the environment. For the standard deviation values each metric are between 0.94 for GON4 which is very good value based on the value nearness to "1" and the highest standard deviation value is 1.35 for GON5 which is still acceptable. Table 7 shows the findings for Green disposal metrics GDL1 to GDL5. The results show that all metrics are applied in IT based industries data center based on the percentage values on the all the metrics which range from GDL4 as 45.90% to GDL5 which possesses a value of 35.00%. The mean are all above 3.50 and the standard deviation of the five metrics are almost similar ranging from 1.05 for GDL2 to 1.34 for GDL3 Percentage distribution and descriptive statistics for Green disposal is shown in Table 7, where the metrics includes GDL1-GDL5.

Table 7. Descriptive statistic results for Green disposal metrics

Life Cycle	Items	1 Not Applied	2 Partly Applied	3 Neutral	4 Applied	5 Fully Applied	Mean	SD	Min	Max	Median
Green Disposal	GDL1	6.56 %	8.20 %	16.39 %	44.26 %	24.59 %	3.72	1.12	1.00	5.00	4.00
	GDL2	4.92 %	6.56 %	18.03 %	44.26 %	26.23 %	3.80	1.05	1.00	5.00	4.00
	GDL3	13.33 %	11.67 %	13.33 %	36.67 %	25.00 %	3.48	1.34	1.00	5.00	4.00
	GDL4	8.20 %	6.56 %	16.39 %	45.90 %	22.95 %	3.69	1.14	1.00	5.00	4.00
	GDL5	11.67 %	13.33 %	16.67 %	35.00 %	23.33 %	3.45	1.30	1.00	5.00	4.00

4 Discussion

This research study aims to facilitate the Greening of data centers in IT based industries by applying Green computing practices. In the course of this paper 5 different life cycle strategies were derived from literature review as were verified based on associated metrics or item from questionnaire data. Results from this study shows that all the 5 life cycle strategies to be applied for Green computing practice in IT based industries data centers are important and should be considered by data centers administrators, IT practitioners and staffs in IT based industries data centers. Moreover no new life cycle strategy was discovered from the data collected from the questionnaire session.

Finding from this study shows the Green design practice should be applied in IT based industries data center, this finding is consistent with research carried out by Reza et al. [12] where the researcher encourages the application deployment of ecologically sound servers, cooling equipment and computer components such as LED monitor in data centers. Findings from this study regarding Green production is similar with previous study carried out by Pichetpongsa and Campeanu [13] where the author suggested the installing and usage of electronic components, computers and other related subsystems in data centers that possesses low or no effect on the natural environment.

Furthermore findings regarding Green procurement support OGCIS [10] guideline where Green procurement is considered as one of the life cycle phase from Green ICT practice implementation. Our findings also corroborates with case study findings presented by Anthony and Majid [16]; Huang [17] where the authors mentioned that IT industries adheres to sourcing policy and Green purchasing guidelines when purchasing IT facilities and equipment. The findings are regarding Green operation is similar to results presented by Molla [15] in their research where the authors mentioned that IT based industries should deploy power performance management of data centers facilities to decreasing energy utilization thereby decreasing CO_2 emissions from data centers. Additionally the results on Green disposal suggests this life cycle strategy is most adopted Green computing practice applied in IT based industries this is in line with finding presented by Huang [17] in his research on ecological sustainable IS development where the author mentioned that most organization only practice recycling reuse and refurbishing of IT infrastructure as Green computing practice in their industrial operations.

5 Practical and research implication

The practical implication of this study correlates to the life cycle strategies which comprises of Green design, Green production, Green procurement, Green operation and Green disposal that provides an approach to describe how data center managers, IT practitioners and staffs can apply Green computing solutions in their new or existing data centers to make them emits less CO_2, energy efficient and thereby reducing cost incurred. The Green computing life cycle strategies can enhance the efficiency of data center based on 5 steps to be applied to properly deploy IT based industries data centers alongside metrics that can be utilized by data centers administrators, IT practitioner and staffs to measure the sustainability of data center in terms of energy efficiency, cost saving, natural resource preservation, waste management, and CO_2 emissions.

The research implication of this study is associated to the Green computing life cycle strategies and associated Green metrics which can be utilized as a benchmark by data center managers in IT based industries to apply Green practice initiatives in their data center to make the data centers more eco-friendly. The life cycle strategies does not only offer understanding of diverse aspects of Green computing, but can also be used as a guideline tool to assess the activities needed to apply Green computing initiatives into action. As a continuing evaluation tool, the life cycle strategies acts both as a strategic Green computing practice application bridge and as an interim reference manual for policy-making and strategic decision making regarding data centers deployment for enterprise gain.

6 Conclusion, limitation and future work

This research study describes Green computing practice application towards supporting IT based industries data center in regards to cost saving decrease, energy efficiency, eco-friendly waste management, natural resource conservation and CO_2 emission reduction towards sustainability attainment. The presented life cycle strategies also contains individual metrics to categorize data center practice into measureable units contingent on measuring the routine of data center in terms of cost saving decrease, energy efficiency, eco-friendly waste management, natural resource conservation and CO_2 emission reduction.

In this descriptive study secondary data was collected from existing literatures on data center Greening and Green computing, after which the life cycle strategies which comprises of Green design, Green production, Green procurement, Green operation and Green disposal was derived. Primary data was collected based on a questionnaire from more than 133 respondents in IT based industries to verify each of the derived life cycle strategies based on individual metrics. Finding from the analysed questionnaire data shows that all the life cycle strategies are significant and should be considered by IT based industries in Greening their data centers sustainability attainment.

The approach presented in this study provides a formative guide, which could simplify be reflected and applied by novice IT practitioners and staffs who are new to benchmarking and Greening data centers. This study thus provides an effective and clear approach of presenting the general Green computing practice application concept. Hence in summary findings from this study presents a realistic, practical and uncomplicated life cycle strategies, which can easily be deployed in real working environment. The life cycle strategies are feasible for total data center performance and management also meets industry standards for data centers by diminishing overall cost of ownership and lastly provides a simple method for integrating benchmarking, measuring concepts towards data centers Greening. The limitation of this study is related to the data collection which involves only respondents in IT based industries in Malaysia. Hence finding from this study cannot be generalized to other countries. In future work data will be collected from other countries and other domain to verify if the results will be same or different and if any new life cycle strategies will emerge in the cause of data collection.

References

1. S. Karanasios, V. Cooper, H. Deng, A. Molla, S. Pittayachawan, Antecedents to greening data centres: A conceptual framework and exploratory case study (2010).
2. S. L. Hart, Beyond Greening: Strategies for a Sustainable World. *Harvard Bus. Rev.* **75**, 1, 66-76 (1997).
3. M. Uddin, M. Talha, A. A. Rahman, A. Shah, J. A. Khader, J. Memon, Green Information Technology (IT) framework for energy efficient data centers using virtualization. *International Journal of Physical Sciences*, 7, 13, 2052-2065 (2012).
4. S. Murugesan, Going Green with IT: Your Responsibility toward Environmental Sustainability. *Cutter Business-IT Strategies Executive Report*, **10,** 8, 1-25 (2007).
5. A. J. Bokolo, C. P. Noraini, A Framework for Adoption and Implementation of Green IT/IS Practice in IT Governance. *Proceedings of Third International Conference on Green Computing, Technology and Innovation* (ICGCTI2015), SDIWC, Serdang Malaysia, 38-48 (2015).
6. M. J. Pawlish, A. S. Varde, A decision support system for green data centers. *Proceedings of the 3rd workshop on Ph. D. students in information and knowledge management*, 47-56 (2010).
7. R. R. Schmidt, E.E. Cruz, M. K. Iyengar, Challenges of Data Center Thermal Management. *IBM Journal*, **49**, 4/5, 709-723 (2005).
8. G. Schulz, The Green and Virtual Data Center. FL: CRC Press, Taylor & Francis (2009).
9. A. Molla, V. Cooper, Greening data centres: The motivation, expectancy and ability drivers (2014).
10. ogcio.gov.hk, green data centre practice (2013).
11. J. F. Hair, W.C. Black, B.J. Babin, R.E. Anderson, R.L. Tatham, Multivariate Data Analysis, 6th edn, Pearson Prentice Hall, Upper Saddle River, N.J (2006).
12. K. Raza, V. K. Patle, S. Arya, A review on green computing for eco-friendly and sustainable it. *Journal of Computational Intelligence and Electronic Systems*, **1**, 3-16 (2012).
13. N. Pichetpongsa, G. Campeanu. Analysis of Green Information Technology in Dell and Toshiba Companies. *IDT: Malardalen University* (2012).
14. B. Saha, Green computing. *International Journal of Computer Trends and Technology (IJCTT)*, **14**, 2, 46-50 (2014).
15. A. Molla, The reach and richness of green IT: a principal component analysis. *Australian Conference on Information Systems*, 754-764 (2009).
16. B. J. Anthony, M. A. Majid, Development of a Green ICT Model for Sustainable Enterprise Strategy. *Journal of Soft Computing and Decision Support Systems*, **3**, 3, 1-12 (2016).
17. A. H. Huang, A model for environmentally sustainable information systems development. *Journal of Computer Information Systems*, **49**, 4, 114-12 (2009).
18. Environmental Protection Agency (EPA), Report to Congress on Server and Data Center Energy Efficiency – Public Law 109-431, Environmental Protection Agency, Washington, DC (2007).

Significance of mobile learning in learning Mathematics

Mislina Atan[1,*], and *Faaizah* Shahbodin[2]

[1]ISTE, Faculty of Technology Management and Technopreneurship, Universiti Teknikal Malaysia Melaka, Malaysia.

[2]ISTE, Faculty of Information and Communications Technology, Universiti Teknikal Malaysia Melaka, Malaysia.

Abstract. Mobile learning has been expanding due to the significance growth and demand for wireless and mobile computing technologies in recent years. It offers a new paradigme shift in learning dimension that conventional classrooms may fail to offer in the teaching and learning approach. This study explores the integration of mobile learning in mathematics at the higher learning institution as research on the use of mobile learning for the teaching and learning of Mathematics appears to be limited. The study focused to examine student's perceptions and experience of mobile learning in Mathematics classroom and also to examine the significant of mobile learning on attitude towards Mathematics. The study involved a total of 70 students at one of the technical university in Malaysia. The students reported their perceptions and the pre and post experiences through a survey questionnaire. The findings reported that the increase level of students' learning excitement, engaging and positively able to be an independent learner after they experienced mobile learning in formal classroom setting. Hence, mobile learning does significantly increase students' attitude toward significance of Mathematics and reduce Mathematics anxiety.

1 Introduction

The evolution of technology has significance affect in teaching and learning since the past few decades. Mobile devices has been rapidly and significantly evolve to adopt as current and future of primary technology device assistant. Mobile technology are among the tools of the digital age required by students, especially in tertiary level of education. The mobile devices are categorized as 'the must to own' for everyone. The trend has created new perspective and ideas for research especially in educational technology. The educational technology experts termed it as mobile learning or m-learning.

M-learning is acknowledged as the fourth invention of the electronic learning setting [1], where it is an expansion of e-learning [2]. However e-learning is not necessarily m-learning. The delivery of the learning content in m-learning focuses on using mobile computing devices and/or with wireless transmissions as supporting tools, where it can enhance learners' engagement and improve accessibility of course contents as well as increase learners' interaction and collaboration [3].

Referring to [4], m-learning describes as "a process of learning, where learners are constantly on the move and construct a brief, but precise interpretation across space and time in collaboration with peers and teachers". In the same vein, [5] sees m-learning as a novel stage in the growth of technology-enhanced learning using mobile technologies where students can access the information by learning individually or in groups, face-to-face or at a distant place using personal mobile devices as a tool. It is a techno centric, related to e-learning, enriching formal education, student-centred and at the same time must include motivational and affective aspects, such as control, ownership, fun, communicative, learning-in-context and continuity between contexts. These features appear to enhance and make learning more meaningful.

Although there are varying definitions and views of m-learning, the common thread that runs through is that it incorporates the use of mobile technology and mobile devices to facilitate the process of learning or knowledge transfer of user who is always mobile, regardless of within or outside of the formal classrooms, informal learning or lifelong learning [6].

There are subjects cannot just simply learned by only reading and thinking, such as mathematics subject, that involve calculation skills. The subject require students to do mathematics. Learning by doing mathematics is essentially about getting involved in an activity and solving the problem. Through this process, the learners study the theory and practice, as these two components emphasis better understanding and achieve better objectives in the context of learning mathematics. An engagement process between students and the subjects occurs during students' participation in 'learn by doing mathematics', which could leads to positive achievement. A study done by [7] describes positive achievements and successes in mathematics often lead to enhanced attitudes and raised students' confidence, which resulted in increased effort and persistence and further success. Furthermore, through the process of 'discovery' of knowledge and solutions based on learning by doing, it builds students' self-esteem, which is also contributes to the development of positive attitude towards mathematics.

According to [8], it is more meaningful if 'learn by doing mathematics' is performed in a group because it can assist students in identifying their knowledge gap and hence assist successful task comprehension. During the process of learning by doing, learners within the group rely on each other which create positive interdependence to encourage and assistant. This benefits both group and individual learners.

Furthermore, 'learn by doing' method could enhance critical and logical thinking, creative problem solving and visual as well as spatial skills while combating negative attitudes towards higher order thinking questions in mathematics. The engagement level that occurs while learn by doing mathematics in group also increase where more inputs among team members can be assessed. Motivation and encouragement from lecturers are revealed as the main contribution in transforming the cycle from failure to success in mathematics class.

Digital natives, the generation that born in era of digital, spent their daily activities and routine surround gadgets and technological items such as computers, video games, digital music players, smartphones and tablets. Thus, teaching approach need to adapt the digital natives era which mathematics lecturers should integrate technological devices in teaching and learning process.

To facilitate m-learning, mobile devices can be adopted as multimedia access tools, communication tools, capture tools, representation tools, analytical tools, assessment tools, and task managing tools. These tools have been incorporated in various studies.

For instance, [9] adopted mobile devices as multimedia access tools, where the participants in the study study a five-minute video clip on how to measure blood pressure; [10] employed mobile devices as communication tools, where the participants in the study communicated with the supervisors during a teaching practicum; [11] used mobile devices as representational tools, where the participants were required to compose music using software in their mobile devices; [12] used mobile devices as analytical tools, where the participants were required to find mathematical relations in the real world phenomena using Math4Mobile midlets software; other researchers, such as [13], [14] used mobile devices as assessment tools where summative or formative assessments were carried out using students' mobile devices.; and a study by [15], [16] used mobile devices as task management tools, where students utilized the devices as personal information managers.

A student-centered environment requires students to be active learners and participate in their own learning process in order to achieve meaningful learning. Students learn the content by the process of constructing knowledge and interacting with their peers. While the teacher acts as a facilitator who provides support and guidance during students' learning process.

In the context of this study, the learners have the ability to make their own decision regarding what to learn, what navigation paths to take, and how to learn within the context. Students have the control over their learning in terms of content delivery and interactivity using 'learn by example' principle. Learning by worked-out example which consists of a problem formulation and the solution process that lead to final solution of the problem, is used widely in Mathematics and Physics subjects. Learning by examples provides quicker algorithm and solutions of problems, requires less time for learning, needs less time for solving problems and helps to minimize mistakes.

In addition, learning by worked-out examples can effectively and efficiently assist students to get better understanding of the problem solving of that particular subject mainly during the early cognitive skill achievement [17].

Furthermore, [18] stressed that the American Statistical Association recommends using active learning, technology, and formative assessment in teaching mathematics. Thus, incorporating technology, activities, and cooperative learning motivates students and leads to effective learning of mathematics. Apart from that, [19] proposed to include learning environment, learning interaction and learning design in implementing online collaboration among learners and lecturers, especially in mobile learning environment.

Referring to study done by [20], there were two components extracted by factor analysis in determining students attitude towards one of the mathematics courses, which are, the significance of the subject and the students' anxiety. Based on these two factors, this study examines the significance different when students utilized mobile learning in their learning method.

The objectives of this paper are:

i. To examine students' perception on mobile learning in mathematics

ii. To examine significant differences of mobile learning on significant of mathematics and anxiety towards mathematics

2 Research Methodology

2.1. Sample

A group of 70 university students who register a mathematics class at technical based university in Malaysia, aged between 20 to 24 years old and agreed to participate in the study. They filled in a consent form to participate in this study.

2.2 Research Design

Prior to implementation stage, a questionnaire was on assessing students' attitude towards Mathematics was distributed. The mobile learning activities were conducted in line with the subject's course outline. It started in Week 3 based on the academic calendar of the institution, and it took 2 weeks to complete the activities. During the session, students were encouraged to use their own mobile devices since they were familiar with the features of their own mobile devices. After the implementation stage, a questionnaire regarding students' perception on mobile learning in mathematic class and student's attitude towards Mathematics was distributed.

3 Results

Out of the 70 sample, 38 (54%) were males and 32 (46%) were females. Hundred percent of the participants owned a mobile device. The most common activities performed using mobile devices by participants are communication through social media (97%), followed by making calls (91%). More than half of the participants used their mobile devices for leisure activities such as listening to music (75%) and watching video (69%). Checking emails or emailing and sending or receiving SMS/ MMS were the least common activities (31%).

3.1 Students' Perceptions of Learning Using Mobile Learning Platforms

The following result in Table 1 displays a descriptive analysis on the student's perception of mobile learning.

Table 1 Students' Perception Of Mobile Learning

	Mean, M	Standard deviation, SD
Mobile learning is an exciting experience	4.11	0.202
Mobile learning can assist me to be an independent learner	4.30	0.914
Activity in mobile learning helped me learn the course content	4.04	0.850
Mobile learning captures my interest to focus on my study	3.97	0.979
It is helpful to learn using my mobile devices	4.21	0.951
Mobile learning engages me more in my own learning	4.43	0.884
I would like to continue learning using mobile learning platform in any course in future	4.40	0.914
Mobile learning increases my interest in the topic being discussed	3.97	0.985

More than half of the participants agreed that mobile learning is an exciting experience (mean = 4.11), and could assist them to be an independent learner (mean = 4.30). They also agreed that learning through their mobile device was helpful (mean = 4.21), and it could capture their interest to focus on their study and increases interest in the topic being discussed (mean = 3.97). Almost half of them agreed that they would like to continue learning using mobile learning platform in any course in the future (mean = 4.40). Through mobile learning, they agreed that they were engaged in their independent learning (mean = 4.43).

3.2 Significance Differences on Students' Attitude Towards Mathematics After Experiencing Mobile Learning

This section reports on the students' attitude towards Mathematics before and after experience learning through mobile learning. This analysis was obtained based on the paired sample t-test, and divided into two components; the significant of Mathematics and anxiety of taking Mathematics course.

Table 2 shows the summary of descriptive statistics in terms of mean, standard deviation and standard error mean for each condition of the students' attitude towards the significance of learning Mathematics before and after mobile learning.

Table 2 Paired Sample Statistics Of Attitude Toward Significant Of Mathematics

	Mean	Standard Deviation	Std.Error Mean
Before mobile learning	4.0159	.38898	.06575
After mobile learning	4.5111	.52546	.08882

Based on Table 2, the mean of significance of learning Mathematics before mobile learning and after mobile learning slightly differs, in which the mean 'after mobile learning' greater than 'before mobile learning'.

Table 3 shows the Pearson correlation between the two conditions; before and after mobile learning.

Table 3 Pearson Correlation

	Correlation	Sig.
Significance before and after mobile learning	.480	.003

For these data the experimental conditions yields a positive correlation coefficient (r= 0.480) and are significantly correlated since $p < .05$ (Sig= 0.03 < .05).

https://doi.org/10.1051/matecconf/201815005049

This is mainly because the data in each condition come from the same people and there are possibility consistency in their response. Since the correlation coefficient is positive, it will be more useful to conduct the test using a paired sample *t*-test.

Table 4 shows a summary of the paired sample *t*-test for the students' attitude towards significance of learning Mathematics.

Table 4 Paired Samples Test For Attitude Towards Significant Of Mathematics

	Paired Differences		Sig. (2-tailed)
	Mean	Std. Deviation	
Significance before and after mobile learning	.09524	.48067	.02

The significance value shown in the table is 0.02, which is less than the 0.05, therefore, it showed that there was enough evidence to support the claim that students' attitude towards significance of Mathematics was different after experiencing mobile learning.

Table 5 shows the summary of descriptive statistics in terms of mean, standard deviation and standard error mean for each condition of the students' attitude towards Mathematics anxiety, before and after experiencing through mobile learning.

Table 5 Paired Samples Statistics For Mathematics Anxiety

	Mean	Standard Deviation	Std. Error Mean
Before Mobile Learning	3.119	.57957	.09797
After Mobile Learning	2.5476	.74817	.12646

Based on Table 5, the mean for attitude towards Mathematics anxiety before the mobile learning and after the mobile learning slightly differs, where the mean 'after the mobile learning' is less than 'before the mobile learning'.

Table 6 shows the Pearson correlation between the two conditions, i.e. before and after mobile learning.

Table 6 Paired Samples Correlations For Mathematics Anxiety

	Correlation	Sig.
Anxiety before and after mobile learning	.508	.002

For these data the experimental conditions yield a positive correlation coefficient (r= 0.508) and are significantly correlated since $p < 0.05$ (Sig= 0.002 < 0.05). Since the correlation coefficient is positive, it will be more useful to conduct the test using a paired sample *t*-test.

Table 7 shows a summary of the paired sample *t*-test for the students' attitude towards Mathematics anxiety.

Table 7 Paired Samples Test For Mathematics Anxiety

	Paired Differences		Sig. (2-tailed)
	Mean	Std. Deviation	
Significance before and after mobile learning	.07143	.67433	.005

The Table 7 also reports the significance value less than 0.05. Thus, it does support the claim that there was a significant difference on the mean of the students' attitude towards mathematics anxiety, before and after learning through mobile learning. Thus, there was enough evidence to support the claim that there was a significant different on attitude towards significance of Mathematics and attitude towards Mathematics Anxiety, before and after learning through mobile learning.

4 Discussions

A number of participants agreed on the items which having positive feeling towards the mobile learning. They mentioned through mobile learning activity, their understanding of the subject was enhanced. Thus, in the proses their critical thinking were developed. Higher percentage of students agreed that it was fun activities and learning Mathematics through mobile learning. The participants also agreed that they enjoyed and could feel the positive impacts of the mobile learning to their understanding, learning satisfaction and motivation level.

Mobile learning platform able the participants to exchange opinions, ideas, and participate in active discussion through the project's guidelines. The platform also enhance their critical thinking as the participants challenge themselves think creative ways to problem solving and having fun and enjoyable moments at the same time. Thus, mobile learning activity was preferred by the participants.

In terms of learning objectives, the participants agreed that they understood the subject better through discussions and guidance from the lecturer who plays role as facilitator. Apart from the positive feeling reported by students on mobile learning, a positive discussion among students could be observed, as most of the students enjoyed and participate the discussions to increase positive output for the mobile learning activities.

This is a positive sign indicate that students could be actively engaged in their independent learning if a student-centered classroom learning environment created and less formal as the conventional type of teaching and learning.

The findings showed mobile learning brought positive impacts in learning motivation and engagement among students in Mathematics class. Moreover, mobile learning did significantly increase student's attitude toward significance of Mathematics, and also it did significantly reduce students' Mathematics anxiety. Further research should improvise more on framework of effective instructional design for mobile learning in learning mathematics subject.

References

1. G. Salmon, *E-moderating:The Key to Teaching & Learning Online*. London, New York: RoutledgeFalmer, 2004.
2. G. Kambourakis, D.P. Kontoni, and I. Sapounas, Introducing Attribute Certificates to Secure Distributed E-Learning or M-Learning Services. *Proceedings of the IASTED International Conference*, (pp. 436-440). Innsbruck, Australia, 2004.
3. P. Wentzel, R.V. Lammeron, M. Molendijk, S. Bruin, and A. Wagtendonk, *Using Mobile Technology to Enhance Students' Educational Experiences*, 2009.
4. M. Sharples, J. Taylor, and G. Vavoula, Retrieved from www.mlearn.org.za: www.mlearn.org.za/CD/papers/Sharples- Theory of Mobile.pdf
5. T. Chan, J. Roschelle, S. Hsi, Kinshuk, M. Sharple, T. Brown, T, One-to-one technology- enhanced learning: An opportunity for global research collaboration. *Research and Practice in Technology Enhanced Learning*, *1* (1), 3-29.
6. A. Mislina, A.C. Koo., and B.H. Madhubala, (2010). Mobile Learning: Will this innovation provide better support to Malaysian tertiary level students?. *International Conference of Education, Research and Innovation.* Madrid, Spain.
7. P. Ernest (2009). "The Mathematics Attitudes, Beliefs and Ability of Students", University of Exeter
8. E. Soloway and C. Norris (2012). "Devices are ready-at-hand", Communications of the ACM 44(6), 15-20.
9. N. Maniar, (2008). The effect of mobile phone screen size on video based learning. *Journal of Software.* 3(4), 51-61.
10. P. Seppala and H. Alamaki, (2003). Mobile learning in teacher training. *Journal of Computer Assisted Learning.* 19, 330-335.
11. M. Polishook, (2005). Handheld composing: Reconceptualizing artistic practice with PDAs.
12. W. Daher and N. Bayaa' (2009). Learning Mathematics in an authentic mobile environment: The perceptions of students. iJim (3), 6-14.
13. A.C.Ganger and M. Jackson, (2003). Wireless handheld computers in the preclinical undergraduate curiculum. *Medical Education Online,* 8(3).
14. N. Fujimura and M. Doi, (2006). Collecting students' degree of comprehension with mobile phones. In *Proceedings of the 34th Annual* EPL, **84** (2008) *ACM SIGUCCS Conference on Users Services* (pp. 123-127). Edmonton, Alberta: Canada.
15. D. Corlett, M. Sharples, S. Bull and T. Chan, (2005). Evaluation of a mobile learning organizer for university students. Journal of Computer Assisted Learning, 21, 162-170.
16. I. Olney and G. Lefoe, (2007). Introducing mobile technologies: Preparatory staff development issues. In *ICT:Providing choices for learners and learning. Proceedings ascilite Singapore 2007.* http://www.ascilite.org.au/conferences/singapore07/procs/olney.pdf
17. R. Atkinson, S. Derry, A. Renkl, and D. Wortham, (2000). Learning from examples:instructional principles from the worked examples research. *Review of Educational Research*, *70*, 181-214.
18. C. Franklin and J. Garfield, (2006), The guidelines for assessment and instruction in statistics education (GAISE) project: Developing statistics education guidelines for Pre K-12 and college courses. In G. F. Burrill, (Ed.), *Thinking and Reasoning about Data and Chance: Sixty-eighth NCTM Yearbook* (pp. 345-375). Reston, VA: National Council of Teachers of Mathematics.
19. R. Sharifah Nadiyah, S. Faaizah, H. Hanipah, and B. Norasiken, (2015). Online collaborative learning elements to propose an online project based collaborative learning model. *Jurnal Teknologi*, pp 55-60. Universiti Teknologi Malaysia.
20. A. Mislina, A.C. Koo, and B.H. Madhubala, (2009). Factor Analysis on the students' attitude towards Probability. *Simposium Kebangsaan Sains Matematik ke 17. 2*, pp. 622-627. Melaka: Universiti Putra Malaysia.

Modelling Students' Experience Towards the Development of Alumni Involvement and Alumni Loyalty

U. N. Saraih[1*], *Nor Irwani* Abdul Rahman[1], *Norshahrizan* Noordin[1], *Sayang Nurshahrizleen* Ramlan[1], *Razli* Ahmad[2], *Mohd Fo'ad* Sakdan[2] and *M. Harith Amlus*[1]

1School of Business Innovation and Technopreneurship, Universiti Malaysia Perlis (UniMAP), Malaysia
2School of Human Development and Technocommunication, Universiti Malaysia Perlis (UniMAP), Malaysia.

Abstract. The purpose of this study is to investigate the influence of students' experience towards the development of alumni involvement and alumni loyalty in Malaysia. This study utilized the survey (questionnaire) method. Data are gathered from 347 respondents in one of the Public Higher Educational Institution (PHEI) in Malaysia. Results presented that students in this institution have low levels of alumni involvement (mean=1.23) and alumni loyalty (mean=1.76). Results also presented that in this institution, students' experience is significantly related to alumni involvement (β=.42, p<.01) and alumni loyalty (β=.57, p<.01). This study contributed some new knowledge which can be channelled to all PHEIs in Malaysia to improve the level of alumni involvement and alumni loyalty. Good experience of students can be cultivated by improving several aspects on managerial side. Some recommendations to increase the level of students' experience are discussed.

1 Introduction

Alumnus is one of significant party in any university's community. Thus, fostering a lifelong connection with current students, and later the alumni, is a potential source of support for institutions [1]. Based on the existing literature, undergraduate students who have a positive and satisfying experience in the university tend to develop lifelong relationships as alumni with the institution. Positive experiences include enjoying classes, developing friendships and obtaining a useful degree [2]. Meanwhile, being satisfied with the undergraduate experience could mean obtaining a job upon graduation or getting connected to worthwhile efforts that offer challenge for the graduate.

Recognizing the importance of alumni to the university in enhancing networks and connecting community for the local and international level, Ministry of Higher Education Malaysia (MoHE) has taken the initiative by making alumni as one party to realize the goals of internationalization. Therefore, to realize the goals of universities' internationalization, it is expected that alumni need to be involved and loyal to the university. Alumni studies are useful not only to assess institutional effectiveness but also to aid institutional planning and revenue generation. Alumni are important sources of information and support, and alumni studies should occupy a prominent place in the institutional research portfolio [3]. The first objective of this study is to reveal the relationship between students' experience in the university and their intention to involve when they become the alumni for the university. According to [4], an active involvement of alumni enables them to contribute their wealth to the former alma mater. Besides alumni involvement, loyal alumni also play an important role to the institution performance. Hence, understanding on how loyalty is formed is a key factor to shape loyalty while alumni are students. Therefore, the second objective of this study is to reveal the relationship between students' experience in the university and their intention to be loyal to the university after they become the alumni. As stated by [5], the benefit of alumni loyalty to institutions is immeasurable. This include volunteerism of alumni, hiring graduates, recruiting students, and making financial contributions. Moreover, in this current economic situation, alumni loyalty is something which all institutions cannot ignore [5].

Given that alumni loyalty is formed while they are students [5], and alumni who are more satisfied with their educational experience have greater level of alumni involvement [6]; thus, forming an understanding of the experiences which students have with their alma maters would provide insights into the future alumni attitudes (eg. alumni involvement, alumni loyalty). In other word, students' experience in the university is the best platform to reveal the intention of the future alumni involvement and alumni loyalty.

* Corresponding author: ummi@unimap.edu.my

2 Literature Review

2.1 Students' Experience

Students' experience became a popular term among the higher educational managers but is still theoretically under-developed [7]. Alumni who enjoyed their students' experience gave positive feedback to people in their networks, acted as role models for the current students, shared their skills and experiences, and advised the program of the alma mater [8]. This author found a significant correlation between students' experience and future perceptions of the institution. Many studies have stated that the strongest giving predictor is the satisfaction a student had with his or her students' experience [9]. Graduates who are satisfied with their college experience are more likely to contribute to the college or university [10]. As highlighted by [11] studying the undergraduate experience may provide a glimpse into the source of loyalty for individuals who financially contribute to their alma mater. Also, [12] found that there was a relationship between students' involvement and satisfaction with the college experience.

Besides of college experience, satisfactory interactions with faculty and staff were also critical to students' positive academic experience [13]. A research of academic experience by [14] found that the quality of teaching experienced by students is the key determinant of student loyalty. In addition, higher education has a high involvement, experientially rich service, characterized by a relatively lengthy experience [15]. Therefore, educational experience is also salient in shaping the future alumni giving.

In local context, [16] assessed how alumni relations can help the ministry in spreading information and promote the higher learning institutions. The findings indicate that relationship between students and university create a remarkable experience to alumni throughout their studies and formed a communal relationship. Thus, in this study the researchers attempt to investigate the satisfaction of students' experience in one of the PHEI in Malaysia. Based on report as presented by the Quality Department in this PHEI, the level of satisfaction among students towards this institution is still moderate. Therefore, studying students' experiences would provide insights into the existing literature in the domain of alumni
involvement and alumni loyalty.

2.2 Alumni Involvement

Alumni involvement is referred to the relationship which alumni had with their undergraduate institution after graduation. It include of wearing the institution's brand, promoting the institution to future students, and giving of their time, talent or treasure. There are many ways that alumni can remain involved in their alma mater. They can attend alumni events, speak in classes, serve on university or alumni association committees, assist with admissions, offer internships, and support numerous other areas. Several researchers have studied the impact of alumni involvement has on giving and found it has a positive correlation. Existing literature also confirmed that alumni are more likely to be donors, are more likely to be repeated donors, with size of the gift strongly related to the involvement.

According to [5], alumni involvement referred to those who feel included and participated in university activities and engaged in some way with the life of the institution. Examples would be holding volunteer positions, attending alumni events or engaging in some other way with the university. In a local context, [4] stated that the active involvement of the alumni contributed their wealth to their former university. As presented by [6] alumni, who are more satisfied with their educational experience, and who are more involved in students' life during undergraduates, have greater involvement in alumni activities (high alumni involvement). This was supported by [13] who presented that alumni involvement predicted donation behavior; and alumni involvement is predicted by several factors such as MBA satisfaction, university prestige, and current salary.

2.3 Alumni Loyalty

Most of the previous studies tend to examine the factors of students' loyalty instead of studying the factors of alumni loyalty. For example, factors of university's image perceived by students and student satisfaction have been tested as potential predictors of students' loyalty in Portugal [17]. Similarly, other factors such as image of study program, image of the University College and student satisfaction have been examined as the predictors of students' loyalty in a Norwegian University College [18]. Therefore, several factors of students' loyalty have been investigated in the previous studies, however limited evidence is known on a comprehensive explanation of how long-term students' loyalty is generated and sustained [19].

Although [14] presented that students' loyalty only involved in long-term relationships, [19] argued that it seemed unclear on how post-graduation loyalty (alumni loyalty) can be generated unless it is backtracked to the school years in order to understand its sources and antecedents. Loyalty can mean different things to different people and can have many definitions depend on the context in which it is used [20]. As defined by [21] loyalty was unswerving in allegiance, faithful to a cause, ideal, custom, institution, or product. Students' loyalty has been as a deep relational and emotional bond the student develops for the university [14]. As claimed by [22] efforts should be taken to ensure that alumni are connected to their university to enhance their loyalty. Thus, all institutions must strategically identified what was the most important to the graduates and determined what made the greatest long term impact on their educational experience to develop the alumni loyalty [20]. Given all the literature, the specific hypotheses are developed as follows:

H_1 : There is a moderate level of alumni involvement in this institution.
H_2 : There is a moderate level of alumni loyalty in this institution.
H3: There is a significant relationship between students' experience and alumni involvement in this institution.
H4: There is a significant relationship between students' experience and alumni loyalty in this institution.

3 Methodology

The population of this research included all of final year students from all faculties in the institution. This study utilized the quantitative method (eg. questionnaire). Questionnaires are distributed randomly for 500 respondents. The SPSS software (version 22) has been used for the statistical analysis. The instruments used for alumni involvement and alumni loyalty have been adapted from previous measurements. However, the instrument used for students' experience had been developed based on the environment of the institution. Students need to respond their satisfaction towards the four dimensions of experience consist of teaching and learning activities (3 items), dealing with several departments in the institution (6 items), co-curricular activities (2 items), and their intention as the future alumni (2 items). As for the alumni attitudes, 3 items of [23] and 5 items of [20] have been adapted for alumni involvement and alumni loyalty.

Based on the pilot test, the reliability values of students'
experience, alumni involvement and alumni loyalty are reported at .87, .83 and .62 each. In this study, the researchers used the criteria of Cronbach's alpha as established by [24]. Therefore, the reliabilities of variables used in this research are ranged from acceptable to good which mean that all variables can be used for further analysis.

4 Results and Discussions

A quantitative approach has been used in this study. Questionnaires has been utilized as the primary instrument to collect data from the respondents. The total returned questionnaires that could be used was 347, which was 69.4% of the total distributed questionnaires.

4.1 Respondent's Background

Most of the respondents in this study were female, 58.8% (204 participants); and the remaining 41.2% (143 participants) were male. The highest percentage of responses (20.5%) are contributed from School of Business Innovation and Technopreneurship; and the lowest percentage of responses (7.6%) are obtained from School of Material Engineering. Other responses were from School of Electrical Systems Engineering (14.1%), School of Microelectronics Engineering (12.1%), School of Computer and Communication Engineering (9.5%), School of Human Development and TechnoCommunication (9.8%), School of Bioprocess Engineering (8.9%), School of Environmental Engineering (8.9%) and School of Mechatronics Engineering (8.6%).

4.2 Reliability Analysis

The Cronbach's alpha values are reported at .76 for alumni involvement, .79 for alumni loyalty and .82 for students' experience. According to [24], the Cronbach's alpha values showed that all of the variables are good in this study.

4.3 Mean and SD Analysis

In this study, the interpretation of mean scores has been referred to the previous authors [25]. According tho these authors, the mean scores ranged from low (1.00-2.00), moderately low (2.01 -3.00), moderately high (3.01 -4.00) to high (4.01 -5.00). As presented in Table 1, the mean value for students' experience was 1.25. Therefore, low acceptance is observed for this variable. Almost similar means score are recorded for alumni involvement; 1.23 and alumni loyalty; 1.76. Thus, *H1: There is a moderate level of alumni involvement in this institution* and *H2: There is a moderate level of alumni loyalty in this institution* are not supported in this study.

Table 1. Mean and SD Values

Variables	N	Mean	SD
Students' Experience	347	1.25	.21
Alumni Involvement	347	1.23	.20
Alumni Loyalty	347	1.76	.29

As for the SD analysis, it is noted that the SD values for students' experience (.21), alumni involvement (.20) and alumni loyalty (.29) were also low. The low level of SD showed that most respondents were close to the mean for variables in this study.

4.4 Correlation Analysis

Pearson correlation coefficient (r) is computed to examine the correlations and directions among the study variables. This analysis had been conducted to determine the interdependency of the study variables. As presented in Table 2, strong relationship are observed between students' experience and alumni loyalty ($r=.53$, $p<.01$); and between alumni loyalty and alumni involvement ($r=.57$, $p<.01$). Meanwhile, moderate relationship is remarked for correlation between students' experience and alumni involvement ($r=.42$, $p<.01$). In sum, all correlations are reported as significantly and positively between each other.

Table 2. Correlation Analysis

Variables	N	1	2	3
Students' Experience	347	-		
Alumni Involvement	347	.42**	-	
Alumni Loyalty	347	.53**	.57**	-

***Correlation is significant at the 0.01 level (2-tailed)*

4.5 Regression Analysis

As presented in Table 3, students' experience only explained 18% of variance in alumni involvement (R2=.18). To further clarify, only 18% is explained by students' experiences, while the rest of 82% of variance in alumni involvement is explained by the other variables. Meanwhile, the adjusted R2 value also taken into account as .17. Students' experiences is observed had strong influences on alumni involvement (β=.42, p=.00). Thus, H3: There is a significant relationship between students' experience and alumni involvement in this institution is supported in this study.

Next, regression result presented that students' experience explained 27% of variance in alumni loyalty (R2=.27). This clarified that only 27% of variance in alumni loyalty is explained by students' experiences, while the rest of 73% of variance in alumni loyalty is explained by the other variables. Meanwhile, the adjusted R2 value also taken into account with .26. Students' experiences had strong influences on alumni loyalty (β=.57, p=.00). Thus, H4: There is a significant relationship between students' experience and alumni loyalty in this institution is supported in this study.

Table 3. Regression Analysis

Variables	Students' Experience			
	Beta	Sig.	R^2	Adjusted R^2
Alumni Involvement	.42**	.00	.18	.17
Alumni Loyalty	.57**	.00	.27	.26

***Correlation is significant at the 0.01 level (2-tailed)*

5 Conclussion

The results of this study presented that the level of alumni involvement and alumni loyalty among students in this institution were in the low indications with the mean value reported at 1.23 and 1.76 each. Therefore, *H1: There is a moderate level of alumni involvement in this institution*; and *H2: There is a moderate level of alumni loyalty in this institution* are not supported this study.

In addition, this study provides some new knowledge pertaining the influence of students' experiences on alumni involvement and alumni loyalty. As presented in Table 4, students' experience are positively associated with alumni involvement (β =.42, p<.01) and alumni loyalty (β=.57, p<.01).

Therefore, *H3: There is a significant relationship between students' experience and alumni involvement in this institution*, and *H4: There is a significant relationship between students' experience and alumni loyalty in this institution* are supported in this study.

Based on the results, it can be concluded that by increasing the level of students' experiences among the current students in this institution, the future level of alumni involvement and alumni loyalty among them can be enhanced. In the theoretical perspective, these findings had been consistent with the previous findings as reported by several authors.

For instance, [13] found that alumni involvement is predicted by factors such as MBA satisfaction and university prestige. It is presented that alumni, who are more satisfied with their educational experience have greater involvement in alumni activities [6]. Moreover, [18] presented that positive relationships between students' satisfaction and alumni loyalty in Norwegian University College. Previous findings of [11] also found that undergraduate experience are related to alumni loyalty to their alma mater.

In the practical perspective, the top management may allocate several new efforts to increase the level of students' experience such as: 1) To create a platform for further interaction between students and the academic staff even after teaching and learning session; 2) To create a platform for further interaction between students and the non-academic staffs even after office hours; 3) To enhance the implementation of the buddy system; and 4) To provide a more effective interaction medium in serving students pertaining several complaint related to the University's library, Bursary Department, ICT Department, Residential Colleges, Students Affair and Alumni Department, etc.

References

1. Pulley, J. The journey is the reward. *Currents*, **39(1)**, 16-21 (2013)
2. Berquam, L.M. Affinity development in undergraduate students at a large research institution, Doctoral Dissertation, Colorado State University (2013)
3. Volkwein, J.F. Assessing alumni outcomes. New Directions for Institutional Research, 125-139. doi:10.1002/ir.335 (2010)
4. Muhammad, F., Abdul Razak, A., Mohd Hussin, M.Y. & Awang, S.A. Willingness of the alumni to contribute to their university: A study among Muslim teachers in Johor, *Shariah Journal*, **22(2)**, 259-288 (2014)
5. Vanderbout, J.L. Impact the undergraduate student experience has on the development of alumni loyalty, Doctoral Dissertation, University of MissouriColumbia (2010)
6. Hoyt, J.E. Understanding alumni giving: Theory and predictors of donor status. Paper presented at the meeting of the Association of Institutional Research (2004)

7. Koenig-Lewis, N., Asaad, Y., Palmer, A. & Petersone, E. The effects of passage of time on alumni recall of student experience. *Higher Education Quarterly*, **70(1)**, 59-80 (2016)
8. Hansen, L.L. Once students, forever alumni: A study analyzing drivers of alumni loyalty, Master Thesis, Carthage College (2015)
9. Stephenson, A.L. & Bell, N. Motivation for alumni donations: A social identity perspective on the role of branding in higher education. *International Journal of Nonprofit and Voluntary Sector Marketing,* **19**, 176-186 (2014)
10. Sun, X., Hoffman, S. & Grady, M. A multivariate causal model of alumni giving: Implications for alumni fundraisers. *International Journal of Educational Advancement,* **7(4)**, 307-332 (2007)
11. Mercatoris, M.E. Alumni loyalty: Examining the undergraduate college experience and alumni donations, Doctoral Dissertation, University of Texas (2006)
12. McAdoo, A.D. Factors affecting the institutional perception of alumni of the University of Arkansas, Doctoral Dissertation, University of Arkansas (2010)
13. Baruch, Y. & Sang, K.J.C. (2012). Predicting MBA graduates' donation behavior to their alma mater. *Journal of Management Development*, **31(8)**, 808-825 (2012)
14. Hennig-Thurau, T., Langer, M.F. & Hansen, U. Modeling and managing student loyalty: An approach based on the concept of relationship quality. *Journal of Service Research*, **3(4)**, 331 -344 (2001)
15. Woodall, T., Hiller, A. & Resnick, S. Making sense of higher education: Students as consumers and the value of the university experience. *Studies in Higher Education*, 1 -20.doi: 10.1080/03075079.2011648373 (2012)
16. Mohd Rashid, N.A., Kee, C. P. & Mohamed Salleh, S. Internationalization of Malaysia Higher Education through experience of Indonesian alumni with alma mater, *Journal of Social Science and Humanities*, **10(2)**, 261 -273 (2015)
17. Alves, H. & Raposo, M. Conceptual model of student satisfaction in Higher Education, *Total Quality Management*, **18(5)**, 571 -588 (2007)
18. Helgesen, Ø. & Nesset, E. What accounts for students' loyalty? Some field study evidence. *International Journal of Educational Management*, **21(2)**, 126-143 (2007)
19. Rojas-Mendez, J.I., Vasquez-Parraga, A.Z., Kara, A. & Cerda-Urrutia, A. Determinants of student loyalty in higher education: A tested relationship approach in Latin America. *Latin American Business Review*, **10**, 21 -39 (2009)
20. Breland, J.R. Development and validation of the alumni loyalty scale, Doctoral Dissertation, University of Alabama at Birmingham (2012)
21. loyalty. 2017. In *Merriam-Webster.com*. Retrieved June 2017, from https://www.merriamwebster.com/dictionary/loyal
22. Lertputtarak, S. & Supitchayangkool, S. Factors influencing alumni donations, *International Journal of Business and Management*, **9(3)**, 170-178 (2014)
23. Sharkness, J. & DeAngelo, L. Measuring student involvement: A comparison of classical test theory in the construction of scales from students surveys, Research in Higher Education, **52(5)**, 480-507 (2011)
24. George, D. & Mallery, P. *SPSS for Windows step by step: A simple guide and reference* (4thed.). Boston: Allyn & Bacon (2003)
25. Norasmah, H.O. & Salmah, I. Kecenderungan terhadap pemilihan kerjaya keusahawanan mengikut persepsi peserta skim usahawan siswa. *Jurnal Teknologi,* **56**, 47-63 (2011)

The Influence of Self-Efficacy on Entrepreneurial Intention among Engineering Students

U. N. Saraih[1,*], *Ain Zuraini Zin Aris*[1], *Suhana Abdul Mutalib*[2], *Tunku Salha Tunku Ahmad*[1], *Sharmini Abdullah*[3] *and M. Harith Amlus*[1]

1School of Business Innovation and Technopreneurship, Universiti Malaysia Perlis (UniMAP), Malaysia
2School of Human Development and Technocommunication, Universiti Malaysia Perlis (UniMAP), Malaysia.
3Centre of International Language, Universiti Malaysia Perlis (UniMAP), Malaysia.

Abstract. The objective of this study is to examine the influence of self-efficacy on entrepreneurial intention amongst engineering students from Public Higher Educational Institution (PHEI) in Malaysia. This study employs a quantitative method using the questionnaire instrument. Data is obtained from 345 respondents comprising of final year students from various public institutes of higher learning in Malaysia. Findings revealed that the students from these public institutes demonstrated a high interest in entrepreneurial intention (mean=3.67, SD=.54) and a moderate level of sel-efficacy (mean=3.22, SD=.71). Findings also pointed out that self-efficacy is significantly associated with entrepreneurial intention (β=.45, p<.01) as exhibited by the engineering students in these institutes. These findings further reinforces an element of Bandura Social Learning Theory which states that self-efficacy is able to influence the entrepreneurial intention amongst engineering students in public institutions. As a result, public instituitions can emphasize strategies to increase the degree of self-efficacy amongst students to enhance the level of entrepreneurial intention. Theoretical and practical implications of the findings are also discussed along with recommendations for the further improvement of institution management.

1 Introduction

Empirical works reveal that most countries nowadays are facing a surge in unemployment problems. Thus, entrepreneurial knowledge and skills may help the countries to reduce the unemployment problem [1]. According to [2], limited opportunities to gain employment occured mostly in competitive environments. In line with this statement, graduates from universities and colleges also found some difficulties to secure their employment in both public and private sectors due to the current volatile economic environment [3].

In the local context, the Ministry of Higher Education or MoHE tries to switch and focus on entrepreneurship as a gateway to the employment creation in order to stabilize the economies. This initiative is consistent with [4] who stated that the entrepreneurship development could solve unemployment problems.

The discussion on the intentions in entrepreneurship research has attracted the attention of scholars for more than thirty years [5]. As claimed by these authors intentions have recently become a research field which promised to help us better understand the phenomenon of entrepreneurial activity and provided some answers to a basic question in entrepreneurship research such as what makes individuals more entrepreneurial?

Therefore, entrepreneurship education is an important key point to promote entrepreneurship development to the countries' economy. In the Malaysian context, MoHE has taken several initiatives to render entrepreneurship as a compulsory subject to all its educational programs in the public universities. Hence, entrepreneurial disciplines has been introduced as among the most important component to produce students with entrepreneurial minds (National Education Blueprint, Higher Education).

Given that, self-employment as an entrepreneur is a career that must be given attention as an alternative for graduates alongside those employed in facing the volatile limited job market and high retrenchment rate [6]. Hence, an investigation is warranted to reveal the level of entrepreneurial intention among students who are not in the non-economic or in the business programs (eg. engineering students). This is because entrepreneurship can provide an alternative source of income when an economy cannot yet supply enough jobs [7].

* Corresponding author: ummi@unimap.edu.my

Based on the MoHE statistics, the unemployment percentage for the first degree graduates from the sub field of technical and engineering are reported at 33.7% in 2015 [8]; and 33.2% in 2014 [9]. Thus, it revealed that the percentages of unemployment among graduates in these sub fields seemed high. Therefore, this implied that the current Malaysian students in engineering field have a low to moderate level of intention to become future entrepreneurs. As a result, the researchers attempted to investigate the level of entrepreneurial intention among students from various engineering programs in one public university in Malaysia

However, although self-efficacy has been proven as a predictor to individuals' entrepreneurial intention, there is a lack of evidence that demonstrates the influence of this factor on entrepreneurial intention particularly amongst the Malaysian engineering students. Therefore, this study is aimed to examine the influece of self-efficacy on entrepreneurial intention among the engineering students in the Malaysian context. Should the findings of this study reveal that self-efficacy has the ability to create influences on students' entrepreneurial intention; then appropriate initiatives can be well-planned to decrease the unemployment percentage among the future engineering graduates in Malaysia, particulary in the aspect of enhancing the level of students' self-efficac

2 Literature Review

2.1. Entrepreneurial Intention

Entrepreneurial intention has become a popular topic of research. However, only a limited number of studies has focused on the entrepreneurial intention among students [10]. The empirical work of entrepreneurial intentions first began in research of the psychologist Bandura [11, 12] with the introduction to the concept of self-efficacy in the social cognitive theory which is defined as the belief in one's capabilities to organize and execute the courses of action required to manage prospective situations.

Another author [13] define entrepreneurial intention as a commitment to start a new business. Meanwhile, [14] defined entrepreneurial intention as the intention to start a new business. According to [15], the decision to become an entrepreneur and create a new business is a deliberate and conscious decision that requires time, considerable planning and a high degree of cognitive processing.

As stated by [16], entrepreneurial intention describes the degree of commitment directed towards the performance of the entrepreneurial endeavour of putting up a business for self-employment. However, other researchers in China [17] defined entrepreneurial intention as a mental orientation such as desire, wish and hope that can influence their choice of entrepreneurship.

In this study, the researchers attempt to examine the influence of self-efficacy on entrepreneurial intention among the engineering students. According to [11], selfefficacy is a key construct in social learning theory. However, there is a lack of evidence in the local context, particularly regarding the influence of self efficacy on the level of entrepreneurial intention among the Malaysian engineering students. Therefore, this study will contribute some useful insights into this field of study.

2.2 The Influence of Self-Efficacy on Entrepreneurial Intention

Self-efficacy is widely recognized as a key construct in social learning theory [11]. Self-efficacy refers to people's judgments regarding their ability to perform a given activity [11, 12, 18] and is proposed to influence individual choices, goals, emotional reactions, effort, ability to cope, and persistence [19].

As defined by Bandura [12], self-efficacy is the taskspecific consideration of perceived fitness to perform a particular activity. Therefore in the case of entrepreneurship, entrepreneurial self-efficacy may be comprised of deliberation of those tasks that relate to the initiation and development of new ventures. A different approach to clarifying entrepreneurial efficacy is to consider the broader human competencies associated with new venture development.

A study of [10] in Nigeria found that self-efficacy determined entrepreneurial intention of students in tertiary institutions with $R=.679$; $R2=.461$; $F(2, 465)=53.007$; $P<.05$). Thus, this confirms a significant prediction of self-efficacy on the tertiary students' entrepreneurial intention in Nigeria. In the United States, a similar relationship between self-efficacy and entrepreneurial intention is presented. These authors [20] also revealed a positive correlation between self-efficacy and entrepreneurial intention among students who took the introductory to entrepreneurship course. This similar pattern of relationship is supported by [21] who conducted a study on entrepreneurial intention among 215 new undergraduate students in one reputable university in Indonesia. These authors presented that self-efficacy has a positive and significant influence on entrepreneurial intention.

Meanwhile, a study of [5] on entrepreneurial intention in Poland found that the level of entrepreneurial intentions is gender-dependent and men had higher intentions towards venture creation than women. However, self-efficacy is not a major factor in shaping entrepreneurial intentions in this study. This is because the reason for lower entrepreneurial intention among women in this study is not significantly connected with their self-efficacy.

Martinez Campo in [22] investigated the mediating role of gender in the relationship between self-efficacy and the development of individual intentions to become entrepreneurs in a group of people in BarranquillaColombia. They focused on the influence of self-efficacy entrepreneurial intentions by using gender as a mediator. However, the results do not provide evidence to consider gender as a mediator in the relationship between selfefficacy and the development of individual intentions to become entrepreneurs.

Entrepreneurial self-efficacy is viewed as having the capabilities that can modify a person's belief in his or her likelihood of completing the tasks required to successfully initiate and establish a new business [18]. In this study, the views of entrepreneurial intention are obtained from the formulation of self-efficacy among the Malaysian engineering students.

Given the above literature, specific hypotheses are developed as follows:

H_1: There is a moderate level of entrepreneurial intention among the engineering students in this institution.
H_2: There is a moderate level of self-efficacy among the engineering students in this institution.
H_3: There is a significant relationship between selfefficacy and entrepreneurial intention in this institution.

3 Methodology

This study involved a sample of the engineering students from one of PHEI located in the Northern Region of Malaysia. The Questionnaire Set consisted of three (3) sections: Section A (entrepreneurial intention), B (selfefficacy) and C (demographic). Likert scales are commonly used to measure attitude, providing a range of responses to a given question or statement. In this study, the scale for entrepreneurial intention and self-efficacy ranged from (1) strongly disagree to (5) strongly agree.

The adapted instruments used for this study has been presented in Table 1.

Table 1. Instrumentation for Variables

Variables	No of Items	Authors
Entrepreneurial Intention	9	[23]
Self-efficacy	8	[24]

The 9 items of [23] which have been asked for entrepreneurial intention in this study are: I will choose a career as an entrepreneur; I will choose a career as an employee in a companylan organization; I prefer to be an entrepreneur rather than to be an employee in a companylan organization; The idea is appealing of one day starting your own business; I would rather found/form a company than to be a manager of an existing one; I want the freedom to express myself in my own business; I would rather be my own boss than have a secure job; I enjoy the challenge of creating a new business; and You can only make big money if you are self-employed.

Whereas, the 8 items of [24] which have been used to obtain the information on students' self-efficacy are: I will be able to achieve most of the goals that I have set formyself; When facing difficult tasks, I am certain that I will accomplish them; In general, I think that I can obtain outcomes that are important to me; I believe I can succeed at most any endeavor to which I set my mind; I will be able to successfully overcome many challenges; I am confident that I can perform effectively on many different tasks; Compared to other people, I can do most tasks very well; and Even when things are tough, I can perform quite well.

Results of pilot test presented that the Cronbach alpha values for entrepreneurial intention and self-efficacy are reported at 0.68 and 0.88 each. Generally, the acceptable reliability lower limit value of Cronbach's Alpha should be more than .70. However, Cronbach's Alpha of .60 could also be accepted as stated by [25]. Thus, these suggest that the specified indicators are acceptable to be used for further analysis.

4 Results and Discussions

Out of 500 distributed questionnaires, only 345 could be used for further analysis. With reference to gender composition based on the Questionnaire Set, the respondents were made up of 136 males (39.4%) and 209 females (60.6%). 223 participants (64.6%) are Malays, 91 participants (26.4%) are Chinese, 10 participants (2.9%) are Indians and the remaining 21 participants (6.1%) represents other races. Hence, the subjects came from the three main ethnic groups of the Malaysian populations; namely, Malay, Chinese and Indian.

Results represents that over than ¾ of the participants (83.5%) are in the range of 21 to 23 years old. Next, there are 15.9% of the participants come in the range of 24 to 26 years old. Apart from that, .3% of participants are in the group of 27 to 29 years old. The remaining .3% comes from the category of above than 30 years old.

With regards the schools categories, the participants came from the various types of schools/faculties. The majority of respondents came from School of Engineering Technology which contributed about 42.0%. This figure is followed by respondents from School of Computer and Communication Engineering (35.4%), School of Manufacturing Engineering (21.2%) and School of Mechatronics Engineering (1.2%). The small percentage of the respondents which contributed of .3% came from School of Microelectronics Engineering.

In the aspect of family business background, more than half of the participants (65.8%) stated that they came from families that have a business background.

4.1 Mean and SD Analyses

Table 2 presents the means for all variables which are relatively high, with the means for all variables exceeding the scale mid-point of three. In this study, the interpretation of mean scores have been interpreted in the ranged of low (1.00-1.66), moderate (1.67-3.33) and high (3.34-5.00). Based on the results, it is noted that the mean for entrepreneurial intention is high at 3.67. Therefore, H_1 is not supported in this study. In the perspective of predictor, the means is noted as moderate which is 3.22 for self-efficacy. Thus, it can be confirmed that respondents in this study have a moderate level of self efficacy. Therefore, H_2 is supported in this study.

Table 2. Mean and SD Values

Variables	N	Mean	SD
Entrepreneurial Intention	345	3.67	.54
Self-efficacy	345	3.22	.71

The SD values for entrepreneurial intention and selfefficacy are reported at .54 and .71 respectively. These low levels of all SD values demonstrated that most respondents were close to the mean for both variables in this study.

4.2 Correlation Analysis

A correlation analysis was performed to assess the strength and direction of a linear relationship between the variables of the study. The correlation coefficients indicate the strength and the direction between two variables. The closer the coefficient is to either -1.0 or + 1.0, the stronger the correlation between the variables. This analysis was conducted on both variables in this study, which are entrepreneurial intention and selfefficacy. In interpreting the strength of relationships between variables, the correlation coefficient is as follow:
Very high correlation (0.90 to 1.00), high correlation (0.70 to 0.89), moderate correlation (0.40 to 0.69), low correlation (0.20 to 0.39) and very low correlation (0.00 to 0.19). As shown in Table 2, entrepreneurial intention are moderately correlated with self-efficacy (r=.45, p<.01).

Table 2. Correlation Analysis

Variables	N	1	2
Entrepreneurial Intention	345	-	
Self-efficacy	345	.45	-

***Correlation is significant at the 0.01 level (2-tailed)*

4.3 Regression Analysis

As shown in Table 3, 20.0% of the variance in entrepreneurial intention is explained by self-efficacy (R2=.20)

Table 3. Regression Analysis

Variables	Entrepreneurial Intention			
	Beta	Sig.	R^2	Adjusted R^2
Self-efficacy	.45	.00	.20	.20

***Correlation is significant at the 0.01 level (2-tailed)*

The remaining of 80.0% of variance in entrepreneurial intention among these engineering students is explained by the other variables. In addition, the adjusted R^2 value also taken into account as .20. Based on the results, entrepreneurial intention is positively associated with self-efficacy (β=.45, p<.01).

Thus, this implied that self-efficacy is significantly related with entrepreneurial intention amongst the engineering students in this institution. Therefore, H_3 is supported in this study

5 Conclusion

The results of this study provided new knowledge regarding the level of students' entrepreneurial intention as well as self-efficacy amongst the engineering students in one of the PHEI in Malaysia. Based on the findings, it is presented that the level of entrepreneurial intention is high in this institution with the mean score at 3.67. However, the result on self-efficacy revealed that there is a moderate level of this variable among the engineering students in this institution with the mean score of only 3.22. This study also revealed that students' self-efficacy are positively and significantly correlated to entrepreneurial intention (r=.45, p<.01).

Therefore, by identifying students' strengths as well as their weaknesses especially on the factors of self-efficacy towards entrepreneurial intention, it provides a clearer career path for these students. Hence, the institutions are able to identify from what angle they can increase the level of students' self-efficacy that in turn will help shape the students' intention towards determining their own employment career.

According to [18], education serves as a preparatory function in relation to new venture creation, transferring of knowledge and the relevant skills that increase the selfefficacy. Therefore, self-efficacy will predict the effectiveness of the future entrepreneurs. The author also presented that perceived self-efficacy concerns with the judgments of how well a person can execute a specific action to deal with certain situations [12]. In this study, the institution needs to boost the elements of students' self-efficacy. Only then can these students demonstrate specific actions to deal with certain situations as future entrepreneurs.

Previous research indicated that student levels of entrepreneurial self-efficacy are high, thus, the tendency to become an entrepreneur is also high [26]. Although the level of self-efficacy is moderate in this study, however early aspirations towards entrepreneurship can be formed at the first year of students at the tertiary level.

As stated by [27] intention is a function of perceived selfefficacy, which also means the control over behavior. Thus, other authors [28] claimed that self-efficacy and perceived behavioral control are used interchangeably since the concept of perceived behavioral control is closely related to self-efficacy. Meanwhile, [29] claimed that perceived behavioural control is the same construct as self-efficacy.

The results of this study is consistent with the previous research [10, 20, 28] of who presented that self-efficacy and entrepreneurial intention is positively associated. However, the findings of this study is inconsistent with the previous research [30, 31] who argued that selfefficacy does not have significant direct effect towards entrepreneurial intention. For example, as stated by [30] when individuals have a high self-efficacy, this will increase their commitment towards entrepreneurship and lead to greater motivation to start a business.

This research has been conducted in other foreign countries. It may be different with the Malaysian context,as students also possess different types of the sociological factors such as personal characteristics, educational program background and family background. Not only do students in this institution have different cultures but the differing environments also play a contributing factor to form the domain of entrepreneurial intention in themselves

References

1. Iqbal, A., Melhem, Y. & Kokash, H. Readiness of the university students towards entrepreneurship in Saudi Private University: An exploratory study. European Scientific Journal, 8 (15), 109-131 (2012)
2. Keat, O.Y., Selvarajah, C. & Meyer, D. Inclination towards entrepreneurship among university students: An empirical study of Malaysian university students. International Journal of Business and Social Science, 2(4), 206-220 (2011)
3. Teshome, T. Attitudes of private higher education students towards entrepreneurship: A case of distance learners in Wolaita Sodo and Hossana Towns. MiddleEast Journal of Scientific Research, 19(2), 277-285 (2014)
4. Dilanchiev, A.Relationship between entrepreneurship and unemployment: The Case of Georgia. Journal of Social Sciences, 3(2), 5-9, (2014)
5. Kurczeska, A. & J. Bialek, Is the interplay between self-efficacy and entrepreneurial intention genderdependent?, Argumenta Oeconomica, 2(33), 23-38(2014)
6. Mazdan, A. A. Involvement of graduates in the field of entrepreneurship in Kuala Lumpur, Malaysia. International Journal of Liberal Arts and Social Science, 3(9), 46-53 (2015)
7. Kelly, D.J., Bosma, N. & Amoros, J.E. Global Entrepreneurship Monitor (GEM) (2010)
8. Ministry of Higher Education (MoHE). "Laporan kajian pengesanan graduan 2015" Retrieved from: http://graduan.mohe.gov.my/skpgreport/Penerbitan/SKPG%202015/6)%20Bab%203% 20-%20Status%20Pekerjaan%20Graduan%202015.pdf (2015)
9. Ministry of Higher Education (MoHE). "Laporan kajian pengesanan graduan 2014" Retrieved from: http://graduan.mohe.gov.my/skpgreport/Penerbitan/SKPG%202014/6)%20Bab%203% 20-%20Status%20Pekerjaan%20Graduan%202014.pdf (2014)
10. Iro-Idoro, C.B. & Iro-Idoro, E.U. Self-Efficacy as correlates of entrepreneurial intention of tertiary institution students in Ogun State, Nigeria. Self, 5(2) (2015)
11. Bandura, A. Self-efficacy: Toward a unifying theory of behavioral change. Psychological Review, 84, 191- 215 (1977).
12. Bandura, A. Self-efficacy mechanism in human agency. American Psychologist, 37, 122-147 (1982).
13. Krueger, N.F. The impact of prior entrepreneurial exposure on perceptions and new venture feasibility and desirability. Entrepreneurship Theory and Practice, 18, 5-21 (1993)
14. Pillis, E. & Reardon, K.K. The influence of personality traits and persuasive messages on entrepreneurial intention: A cross-cultural comparison. Career Development International, 12(4), (2007)
15. Wilson, F., Kickul, J. & Marlino, D. Gender, entrepreneurial self-efficacy, and entrepreneurial career intentions: Implications for entrepreneurship education. Entrepreneurship Theory & Practice, 31(3), 387-406, (2007)
16. Souitaris, V., Zerbinati, S. & Al-Laham, A. Do entrepreneurship programmes raise entrepreneurial intention of science and engineering students? The effect of learning, inspiration and resources. Journal of Business Venturing, 22(4), 566-591, (2007)
17. Peng, Z., Lu, G. & Kang, H. Entrepreneurial intentions and its influencing factors: A survey of the University Students in Xi'an China. Creative Education, 3, 95- 100 (2012)

18. Bandura, A. Social foundations of thought and action: A social cognitive. Englewood Cliffs, NJ: Prentice Hall. (1986).
19. Gist, M.E., Stevens, C.K. & Bavetta, A.G. Effects of self-efficacy and post-training intervention on the acquisition and maintenance of complex interpersonal skill. Personnel Psychology, 44, 837-861, (1991)
20. Shinnar, R.S., Hsu, D.K. & Powell, B.C. Self-efficacy, entrepreneurial intentions, and gender: Assessing the impact of entrepreneurship education longitudinally, The International Journal of Management Education, 12(3), 561-570, (2014)
21. Rachmawan, A., Lizar, A.A. & Mangundjaya, W.L. The role of parent's influence and self-efficacy on entrepreneurial intention. The Journal of Developing Areas, 49(3), 417-430, (2015)
22. Martínez Campo, J.L. Analysis of the influence of self-efficacy on entrepreneurial intentions. Prospect, 9(2), 14-21, (2011)
23. Indari & Kristiansen. Empirical studies of selfemployment. Journal of Economic Survey, 13, 381- 415 (2003)
24. Chen, G., Gully, S.M. & Eden, D. Validation of a new general self-efficacy scale, Organizational Research Methods 2001, 4(1), 62-83, (2001).
25. Hair, J.F. Jr., Anderson, R.E., Tatham, R.L. & Black, W.C. Multivariate data analysis (5th ed.). New Jersey: Prentice Hall (1998)
26. Wan Nur Azlina, I., Ab Rahim, B., Soaib, A., Shamsiah, M. & Noor Syamilah, Z. Impact of Entrepreneurship Education on the Entrepreneurial Intentions of Students in Technical and Vocational Education and Training Institutions (TVET) In Malaysia, International Education Studies, 8(12), 141- 156 (2015)
27. Liñán, F., Rodríguez-cohard, J.C. & Rueda-cantuche, J.M. Factors affecting entrepreneurial intention levels. 45th Congress of the European Regional Science Association, Amsterdam (2005).
28. Zaidatol, Akmaliah & Lope Pihie. Entrepreneurship as a career choice: An analysis of entrepreneurial selfefficacy and intention of University students. European Journal of Social Sciences, 9(2), 338-349 (2009)
29. Schwarz, E., Wdowiak, M. & Breitenecker, R. The effects of attitudes and perceived environment conditions on students' entrepreneurial intent: An Austrian perspective. Education Training, 51(4), 272-291 (2009)
30. Elfving, J., Brännback, M. & Carsrud, A. Toward a contextual model of entrepreneurial intentions. International Studies in Entrepreneurship, 24(1), 23-33 (2009)
31. Paço, A.M.F., Ferreira J.M., Raposo, M., Rodrigues, R.G. & Dinis, A. Behaviours and entrepreneurial intention: Empirical findings about secondary students. Journal of International Entrepreneurship, 9(1), 20-38 (2011)

AUTHOR INDEX

AUTHOR INDEX

AUTHOR INDEX

AUTHOR INDEX

AUTHOR INDEX

AUTHOR INDEX

Malaysia Technical Universities Conference on Engineering and Technology (MUCET 2017)

MATEC Web of Conferences Volume 150 (2018)

Penang, Malaysia
6 – 7 December 2017

Part 2 of 2

Editors:

S.A. Aljunid
M.A.A. Mohd Salleh
C.B.M. Rashidi
P.J. Soh
K.N.F. Ku Azir

ISBN: 978-1-5108-5872-5

Printed from e-media with permission by:

Curran Associates, Inc.
57 Morehouse Lane
Red Hook, NY 12571

proceedings.com

Some format issues inherent in the e-media version may also appear in this print version.

Printed by Curran Associates, Inc. (2018)

For additional information, please contact EDP Sciences – Web of Conferences at the address below.

EDP Sciences – Web of Conferences
17, Avenue du Hoggar
Parc d'Activité de Courtabœuf
BP 112
F-91944 Les Ulis Cedex A
France

Phone: +33 (0) 1 69 18 75 75
Fax: +33 (0) 1 69 28 84 91

contact-edps@webofconferences.org

Additional copies of this publication are available from:

Curran Associates, Inc.
57 Morehouse Lane
Red Hook, NY 12571 USA
Phone: 845-758-0400
Fax: 845-758-2633
Email: curran@proceedings.com
Web: www.proceedings.com

TABLE OF CONTENTS

PART 1

PART 2

https://doi.org/10.1051/matecconf/201815005052

Conceptual Paper: Digital Preservation Strategies in Archival Institution

Adila Ismail[1,*], and *Habee* Bullah Affandy[1]

[1]School of Human Development and Technocommunication, Universiti Malaysia Perlis

Abstract. In this paper, the overall overview of concept of preservation regardless if it is preservation of physical records/archives or digital records/archives is discussed. The concept of the preservation discussed includes preventive preservation, restorative preservation, and content preservation. Then the concept of digital preservation. Furthermore, this paper highlighted the strategies and methods used in implementing the preservation of digital records/archives. The issues those are involved in the digital preservation implementation are also discussed.

1 Introduction

The advancement of technology has given birth to records that exist in digital forms. With the fast rate of technological advancement, more and more records are produced in digital form. Also, the characteristics of digital form records that are easily access further promotes the transition from typical 'physical' documents to digital forms. According to Cumming and Findlay (2010), to ensure that digital records can be reliable and are of value, digital records need to have certain characteristics. Firstly, digital records need to be meaningful and trustworthy to be relied on for business. The second characteristic is that digital records must be fixed, inviolate representations of business activity, preserved in context and protected from loss and alteration.

Electronic records have been used widely in many organizations and the problem arises when it comes to the preservation of electronic records for the purpose of long term access. On a national level in Malaysia, the responsibility of preserving the digital records falls on the National Archive. An archival institution is the facility responsible for preserving records of enduring value. The preservation of records involves both ensuring that the record contains the same information it did when it was archived and ensuring that the record can be viewed using existing technology. It also involves maintaining security and intellectual property rights [1]. Various strategies, policies, and standards have been established by different organization around the world to ensure that the preservation process is effectively implemented. The problem of preserving digital records for long-term access seems straightforward but requires careful consideration about process and technology. Until today there has been no storage platform that can be trusted to store the last existing copy of a critical electronic record. Preserving digital information is more difficult than preserving records on materials such as paper or film. The sheer volume and the volatility introduced by digital demand new software architecture capable of scaling and of preventing accidental changes to the records [1].

Archives preserves a trivial portion of the totality of records and, in the current environment, the digital record they do preserve is not the most important part of that record nor acquired in a manner that ensures its authenticity. Nevertheless, by transferring digital records to archives, archivists become responsible for its preservation which they are, mostly, ill-equipped to provide [2].

Various organizations from foreign countries have designed various strategies in preserving the digital records either on a national level or on an institutional level. Due to the importance of digital records these strategies are seen as initiatives to protect the longevity of the digital records.

Digital records play an important role to support the government functions and its accountability as well as valuable sources of information [3]. For an organization that are responsible for the preservation of digital records on a national level, it is important for the

National Archive of Malaysia to provide and implement a well-designed strategy to ensure that the preservation procedures protect the records existence itself and also its characteristics that ensures the value of the records.

2 Overview of the Concept of Preservation

According to MS ISO 223-1 (2009), preservation are processes and operations involved in ensuring the technical and intellectual survival of authentic records through time. This refers to preservation is seen as means to protect the records within an organization from any threats that may cause the records to be damaged or destroyed over time. The records within an organization are very important in supporting the ongoing business of an organization. Records are used to solve business problems, evidence to business transactions and also functions as the corporate memory for the organizations. Due to these values, it is important or an organization to protect their records in ensuring that it can be accessed for active use and for long term use (Archives). This is where preservation comes in. The concept of preservation can be categorized in three categories. This includes preventive preservation, restorative preservation and content preservation.

Preventive Preservation: According to the National Library of France (2011), Preventive Preservation refers to the range of direct and indirect actions undertaken on collections and aimed at preventing degradation and prolonging document life spans: carrying out environmental checks, preparing and monitoring an emergency plan, and transferring to alternative media. This statement by the National Library of France, relates to the overall activities in ensuring the physical protection of the materials. The activities involved ensure that the materials are protected from any threats that can put the materials at risk of damages and destructions. Preventive preservation can significantly reduce or even prevent premature deterioration by activities such as environmental control, disaster preparedness, security, storage and handling, reformatting, binding and repair, and conservation treatment.

Restorative Preservation: This concept focuses more on the physical records rather than the digital records. Paper records may not be as sensitive as electronic records, but cautions must be taken in handling and protecting paper records. In doing so, the Public Office of Northern Ireland has developed the Code of Practice for Staff Handling Records to ensure that the records are well handled; such as by not folding and bending the documents as it can crumples the documents; or the staff of the records office must never mark documents with any type of pen or erase anything on them and do not use glue, self-adhesive tape, or correcting fluid anywhere near the documents. Even though the original paper records have stain on it, never used correction fluid to erase as it will affect the originality of the records and might also make it worst. Other than that, the use of self-adhesive tape is prohibited due to the possibility of gluing together unrelated documents; etcetera.

Content Preservation: This concept of preservation regards more with digital records. Digital records are more at risk to content changes rather that paper records. Due to its existence in a digital form, it is at risk of alteration. The content of the digital records must be preserved in order to maintain the authenticity and reliability of electronic records. In content preservation, the need of various hardware and software is never low especially in accumulating the content. The requirements in content preservation are to master the variety of computing technologies to do the routine management such as backup and update [4]. The concept of content preservation described by Han (2004) described the management of digital records based on the usage of various tools. The sensitivity and vulnerability that resides within digital records resulted in the need of detail in the management of the records. The combination of technology such as hardware and software can achieve the main objective of content preservation which is to ensure accessibility of the digital content.

2.1 The Concept of Digital Preservation

Digital preservation is defined very broadly and refers to all of the actions required to maintain access to digital materials beyond the limits of media failure or technological change. Those materials may be records created during the day-to-day business of an organization, "born digital" materials created for a specific purpose or the products of digitization projects [5]. Digital preservation consists of the processes aimed at ensuring the continued accessibility of digital materials. To do this involves finding ways to re-present what was originally presented to users by a combination of software and hardware tools acting on data. [6]. To achieve this requires digital objects to be understood and managed at four levels: as physical phenomena; as logical encodings; as conceptual objects that have meaning to humans; and as sets of essential elements that must be preserved in order to offer future users the essence of the object.
According to UNESCO (2011), the most significant threats to digital continuity concern loss of the means of access.

Digital materials cannot be said to be preserved if the means of access have been lost and access becomes impossible. Only with proper access that a digital records can be fully utilized. The purpose of preserving digital materials is to maintain accessibility which is the ability to access their essential, authentic message or purpose. Digital preservation involves choosing and implementing an evolving range of strategies to achieve the kind of accessibility discussed above, addressing the preservation needs of the different layers of digital objects.

3 The Conceptual Framework

In implementing digital preservation within an organization, certain strategies must be developed to support and ensure that the entire procedure and processes are well designed and managed. According to the National Archive of Malaysia (2003), the strategy should be developed and implemented organization-wide targeted at identifying, managing, preserving and ensuring continued access to digital records. This statement indicates that the design of the strategy must be on an organizational scale consisting of different departments that relates with the preservation process, human resources, financial support and technological infrastructure. In other words, the strategy must include all activities regarding the preservation of electronic records. The strategy must also cover the procedures in ensuring the access towards the digital records after preservation. Without proper access, the records are useless even though they are well preserved.

Based on several literatures, a framework can be developed to describe the digital preservation strategy implementation. The strategy is divided into two stages: i) Initial Stage and ii) Implementation Stage. The framework is shown in Fig. 1.

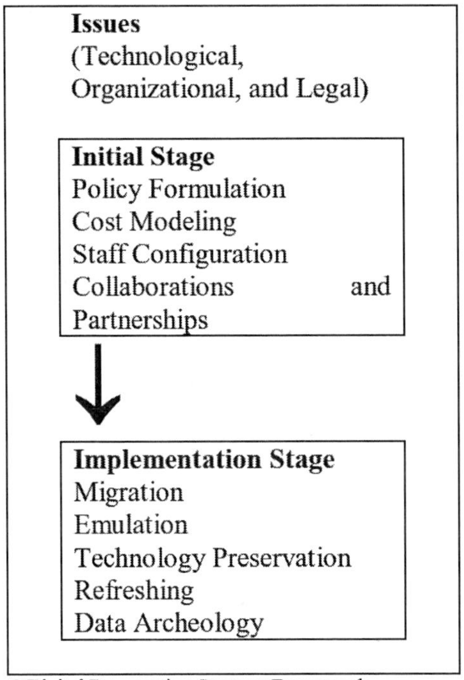

Fig. 1. Digital Preservation Strategy Framework

The framework described 3 major elements in the digital preservation strategy. The first is the initial stage of the strategy. The initial stage refers to works being done before the actual digital preservation processes takes place. The works involved during this stage are policy formulation, cost modeling, staff configuration, and collaborations and partnerships. The next stage in the framework is the Implementation Stage. This refers to the actual digital preservation works. There are five digital preservation techniques that are described in the framework; includes migration, emulation, technology preservation, refreshing and data archeology. The third element in the digital preservation strategy framework is issues. This refers to the issues surrounding the implementation of the digital preservation works. Institutions that are involved in the digital preservation works should address these issues to ensure that the digital preservation works within the institution are well implemented.

3.1 Initial Stage

This is the stage within the strategy that establishes a foundation for the implementation. Every process within this stage ensures that the preservation can proceed and sustained.

Policy Formulation: One vital element in the digital preservation strategy is the policy development. The Oxford English Dictionary (OED) puts it more concisely and defines "preservation" as the art of "keeping safe", "keeping alive", "maintaining" and "retaining", while defining a policy simply as: "a course or general plan of action". Therefore, a preservation policy at its most basic is a plan of action for safe keeping. In some of the literature on preservation, the words policy and strategy are often used interchangeably. According to the OED a strategy is "the art of war"; in our case, war against decay and destruction; it is also "the management of an army in a campaign", in other words: the management of all the tools and skills at our disposal to fight against decay. The difference is that a policy, a plan of action, should address the questions of what needs to be preserved, why, for what purpose and for how long, while a strategy addresses the questions of how this should be done and in what order. An effective digital records preservation strategy should incorporate formal policies and procedures governing the agency approach for the maintenance and preservation of digital records. The development of a good public policy requires the appropriate political climate as well as the cooperation of numerous stakeholders [7]. Cloonan and Sannet suggested that in designing the policy, it must be in line with the current politic environment. This is because policies that go against current political environment may cause a misunderstanding among many parties. It is also important that in a digital preservation policy development, an organization should cooperate with various stakeholders in the design process. This is to ensure that the policy is well communicated to the stakeholders and these policies are well accepted. Cloonan and Sannet (2002) indicated that policy will naturally evolve rapidly at the outset of a program when the practitioners encounter new, possibly unanticipated features of the program that require policy decisions.

There are several reasons why digital preservation policies are needed. With ever-dwindling resources and many more claims on those that are available, the need for accountability and the need to justify expenditure have become more pressing. Organizations and institutions, particularly their governors and funders, but also their users, expect a policy as a sign of accountability and as a tangible expression of intent, as well as something that can be monitored [8]. A policy is often required when bidding for funds and to demonstrate that such funds can and will be used responsibly and to good long term effect. Only with adequate funding can a digital preservation procedures be fully implemented. A policy is also required for developing a preservation strategy and to plan coherent preservation programs [8]. With a well-designed strategy, it will ensure that the digital preservation processes are well managed.

Cost Modeling: Decisions about preserving information should consider the costs [9]. However, there are no proven techniques for estimating the costs of long-term preservation of digital information. Russell and Weinberger (2000) suggest that the ongoing costs of digital preservation span a more extended timeframe than traditional preservation and will therefore require resource commitments of a different nature. According to Cloonan and Sanett (2002), the strategies in the preservation of electronic records must include the estimation of cost; includes the cost for staff, equipment, space, energy, research and development, and other related costs. These elements are important to ensure that the preservation processes and procedures can be ongoing and be completed in time.

The cost modeling should also include the cost for the access of the materials, cost for long term creation and maintenance of digital materials, production of metadata, supplies, contingency and administration [7]. It is important for digital records to be maintained regularly to ensure its accessibility at present and for future use. The cost of the maintenance is highlighted so that electronic records can be sustained as evidential value. Cloonan and Sannet (2002) also indicated that the cost of staff and equipment consistently need large figures because the cost for staff and equipment will vary over time. New upgraded equipment needs larger cost and with the cost modeling, it will help to estimate the cost needed for the preservation equipments.

Staffing Configuration: A well designed digital records preservation strategy must consider the staffing configuration. In Malaysia, the National Archive of Malaysia advises the public offices to assign responsibility for the management of long term digital records to an appropriate area within an organization [3]. This is to ensure that the management of the digital records within an organization is under the supervision and management of a specific department. This department must consist of professionals with adequate knowledge and skills in records management practice.

Collaborations and Partnerships to Preserve Digital Records: According to the National Archive of Georgia (2007), another element in the digital records preservation strategy is through collaborations and partnerships. This is considered more of a program initiative rather than a procedure. This includes:

a) Identify appropriate partners and stakeholders that are able to contribute to the statewide effort
b) Establish agreements on responsibilities and roles
c) Pursue agreements that provide a reliable basis for ongoing accessibility over time
d) Help identify and develop policies, procedures and tools to support such an aim
e) Work with creators, publishers and re-users of digital content to encourage practices that will enable, rather than hinder, preservation
f) Work with government agencies and officials to develop legislative and funding frameworks that will enable cost-effective preservation.

3.2 Implementation Stage

After the foundation works have been done during the initial stage, only then can the preservation work itself (dealing with the materials) can be done. This is where the strategy focuses on the implementation stage. The strategy involved the techniques to be used in preserving digital records or archives. Bear in mind that not all technique must be done in the preservation works. Archival institutions should only used techniques that are suitable for them.

There are various digital preservation techniques implemented by organizations in preserving their digital records. Different technique will give different impact to the organization whether it is bad or good. According to Knight (2010), the approach to digital preservation must be related to the provenance, context, authenticity and integrity. The process of digital preservation must not neglect the characteristics of records which are needed to be maintained and preserved. Records are not considered of any value if its authenticity and integrity is abandoned. There are various techniques that include migration, emulation, technology preservation, refreshing and data archeology. Developed countries may not implement all these techniques but only those techniques that are suitable for their organizations.

Migration: The migration of digital information refers to the periodic transfer of digital materials from one hardware and software configuration to another, or from one generation of computer technology to a subsequent generation [10]. The purpose of migration is to preserve the integrity of digital objects and to retain the ability of users to retrieve, display, and use them in the face of constantly changing technology.
There are several countries that implement migration as one of the procedures in their digital records preservation strategy. According to Lee et al. (2002), National Archives and Records Administration or NARA has undergone a project which has been led by San Diego Supercomputer Center which aims to develop a persistent archive to support the storage of archival materials, search for archival materials as well as the preservation of digital collections. The ultimate goal of the project is to preserve not only original data but also the context that permits the data to be interpreted [11]. The National Archive of United Kingdom is seen as one organization that implemented migration in their digital records preservation strategy. Migration according to the National Archives is overcoming technological obsolescence by transferring digital resources from one hardware or software generation to the next. It is seen that migration is done by transferring the old version of software and hardware to the current version to facilitate better access for the digital records. Technology advances in a rapid rate thus it is important that digital records are migrated from old version to newer ones to

avoid the records being obsolete and unreadable. Also with new technology it will assist in increasing efficiency both in term of process and access.

Another organization that opted for the migration technique is the National Archive of Australia [3]. The National Archive implemented migration when the older formats of documents became obsolete. Based by the definition of the National Archive of Australia, migration is the process of converting a digital object from one data format to another [12]. The authors also indicate that migration is the action done by archivist to preserve digital records. The action involved the change of format of the content of the records from obsolete format to the current format. This method is mainly done to ensure the accessibility of those records before it becomes obsolete and unreadable.

Based on the migration technique implemented by three different organizations, all organization shares the same intent in choosing migration technique in their digital preservation strategy which is to ensure accessibility of the digital records.

Emulation: According to the National Archive of United Kingdom, emulation refers to a technique in overcoming technological obsolescence of hardware and software by developing techniques for imitating obsolete system on future generations of computers. The authors indicated that in order to preserve electronic records, a replica of the old system should be developed. The author view of emulation is reproducing the same systems or technologies for digital records activities by using the current generations of technologies. The author argued that it is important to ensure that the whole content, context and structure of the digital records will never change although emulation has been applied in the digital preservation works. The National Archive recommended that emulation camp must be done for active file. This is mainly because active files have been used for many purposes thus it crucial to emulate those files because it will affect many activities in the National Archive with regards to the digital records. Emulator programs can be designed and run on future computer platforms. The emulator is programmed to mimic the behavior of old hardware platforms and operating system software, such as for games and executable files [13. However, this strategy does not involve preserving ageing hardware and original operating system software. The goal of emulation is to preserve the look and feel of the digital object as well as its functionality. The essence of this strategy is to copy the technical context of the resource allowing the original object or a refreshed copy of the original object to be used in the future [11]. This is in line with the statement by Heslob, Davis and Wilson (2002), where they indicated that the emulation process always maintains the exact look of the records that need to be preserved or in other words the originality of the structure of the records. This is why the National Archive of Australia opted for this technique due to its characteristic that preserved all aspects of the digital records.

Technology Preservation: Technology preservation is based on preserving the technical environment that runs the system, including software, media drives and so on [14]. This solution supposes that complete museums of obsolete equipment could be maintained in order to replicate any old configuration of hardware and software. This strategy involves preserving an original application program, operating system software, and hardware platform [11]. The advocates of this strategy emphasize that the original environment needs to be run to really preserve the behavior as well as the look and feel of the digital object. For some digital objects, this may be the best solution at least in the short-term because it ensures that the material is accessible by preserving the access tools as well as the object itself [11].

However, various issues including space, maintenance, and costs may make this impossible in the longer term. Specifically, equipment ages and breaks, documentation disappears, vendor support vanishes, and the storage medium as well as the equipment deteriorates [11]. This strategy also limits the portability of the resource since it will depend on hardware stored in specific locations [14]. Technology preservation is more of a disaster recovery strategy for use on digital objects that have not been subject to proper digital preservation strategy. It offers the potential of coping with media obsolescence, assuming the media has not decayed beyond readability [14].

Refreshing: The physical media that stores the digital records can deteriorate in time. The condition of the physical media might not be the same as when it is created because of degradation. In order to overcome the eventual degradation of the storage media, there will generally be some point at which the information contained in these carriers will have to be duplicated or reconstructed on another instance of medium, or a new medium all together [15]. According to the Cornell University Library (2007), refreshing involved copying content from one storage medium to another and its targets only media obsolescence and is not a full-service preservation strategy. This refers to the movement of the digital content from an obsolete physical medium to the new medium. The objective of this technique is to cope with technological changes that can cause digital media storage becoming obsolete. Obsolete physical storage might affect the readability of the digital content.

Data Archeology: Digital or Data Archeology includes methods and procedures to rescue content from damaged media or from obsolete or damaged hardware and software environments [14]. Cornell University Library emphasized that digital archeology is related to the preservation of the data when there is possibility of the tools in managing the electronic records become obsolete. The data will be preserved in archive which contains all information about the records. Data archives ensure long term availability of data through standardizing formats, migrating formats as technologies change, and guarding against general degradations in the data [16]. The authors stressed that the availability of data is dependent on technology. Technology which involves the tools in data archeology can affect the accessibility of the data especially for long term preservation. The data that resides in electronic records might achieve long term preservation if the changes in technology can be overcome and solve. On the other hand, data archives can provide a third party system for ensuring the integrity and accessibility of research data for the future [16]. The authors indicated that data archive can help data archeology in terms of maintaining the integrity of the original data which means the original characteristic of the data.

According to the National Archive of Malaysia (2003), there are five factors that any agencies and organizations need to consider before implementing certain digital preservation technique. These include cost implementation, technical complexity, technology compatibility, impact on business operations and overall effectiveness of the approach. Cost implementation is one factor in choosing the suitable methods of digital preservation, as it is quite expensive to implement. Due to this, it is not advisable for archival institutions to implement all methods. The process of choosing the right method is vital for an archival institution.

3.3 Issues in Digital Preservation

Another element in the digital preservation framework is issues. Not just the framework concern with the works involved but also it concern with the issues surrounding the digital preservation works. It is important for archival institutions to analyze the issues that exist within the scope of digital preservation. Digital preservation requires considerations from all aspects on all level of implementation from the initial planning, the implementations of the processes and after the completion of the processes. Several issues are recognized regarding the preservation of digital records, and are categorized into three major categories which are Technological Issues, Organizational Issues and Legal Issues.

a) Technological Issues

Digital Media: Digital materials are especially vulnerable to loss and destruction because they are stored on fragile magnetic and optical media that deteriorate rapidly and that can fail suddenly from exposure to heat, humidity, airborne contaminants, or faulty reading and writing devices [17]. Digital media are subject to destruction and deterioration in new ways, though unintended loss can be avoided if procedures are adapted to the needs of the technology. Precautions can be taken which will help significantly to reduce the danger of loss and include: Storing in a stable, controlled environment, implementing regular refreshment cycles to copy onto newer media, making preservation copies (assuming licensing/copyright permission), implementing appropriate handling procedures and transferring to "standard" storage media [5].

Changes in Technology: Unlike the situation that applies to books, digital archiving requires relatively frequent investments to overcome rapid obsolescence introduced by galloping technological change [18]. Because digital material is machine dependent, it is not possible to access the information unless there is appropriate hardware, and associated software which will make it intelligible. The certainty that there will be frequent technological change poses a major challenge. Precautions can, and should be taken, which will greatly reduce the risk of inadvertently losing access to a resource because of changes in technology [5].

Authenticity and Context: The approach to digital preservation of digital records must be related to the provenance, context, authenticity and integrity [19]. This means that the electronic records must maintain its original condition without alteration after the preservation process. The authenticity and integrity of electronic records are mutual concern [7] because records cannot be considered as records if the records do not maintain its authenticity and integrity.

b) Organizational Issues

Costs: The cost of digital preservation cannot be easily isolated from other organizational expenses, nor should it be. Digital preservation is essentially about preserving access over time and therefore the costs for all parts of the digital life cycle are relevant. In that context even the costs of creating digital materials are integral in so far as they need to include cost elements which will ultimately facilitate their long-term preservation [5]. Nonetheless the exercise of calculating costs, however complex, is a valuable and necessary task to establish cost-effective practices and a reliable business model. The cost of the labor required for digital preservation will be the most significant by far and includes not only dedicated experts but varying proportions of many staff such as administration, management, IT support, legal advisers etc [5].

Expertise: The need for digital preservation expertise is high: asked to rate staff as expert, intermediate, or novice, only 8 of the 54 institutions considered their staff at the expert level. [17]. The dramatic speed of technological change means that few organizations have been able even fully to articulate what their needs are in this area, much less employ or develop staff with appropriate skills. In addition, there is little in the way of appropriate training and "learning by doing" can often be the most practical interim measure [5].

Selection: The enormous quantity of information being produced digitally, its variable quality, and the resource constraints on those taking responsibility to preserve long-term access, makes selectivity inevitable if the objective is to preserve ongoing access. In the digital environment, it is possible to by-pass the traditional distribution channels, as well as filtering and quality control processes. While there are benefits for users in terms of swift access, there are also difficulties in terms of quality control. Selecting quality materials for long-term retention therefore places a burden on organizations in terms of resources and also in terms of the potential impact of selection [5]. In cases where there may be multiple versions, decisions must be made in selecting which version is the best one for preservation, or whether more than one should be selected. Sampling dynamic resources as opposed to attempting to save each change, may be the only practical option but may have severe repercussions if the sampling is not undertaken within a well-defined framework and with due regard to the anticipated contemporary and future needs of the users [5].

c) Legal Issues

Intellectual property rights (IPR) and preservation: Copyright and other intellectual property rights (IPR) such as moral rights have a substantial impact on digital preservation. The preservation of digital materials is dependent on a range of strategies, which has implications for IPR in those materials. The IPR issues in digital materials are arguably more complex and significant than for traditional media and if not addressed can impede or even prevent preservation activities. Consideration may need to be given not only to content but to any associated software. Simply copying (refreshing) digital materials onto another medium, encapsulating content and software for emulation, or migrating content to new hardware and software, all involve activities which can infringe IPR unless statutory exemptions exist or specific permissions have been obtained from rights holders [5].

Privacy and Confidentiality: Information held within the repository may be subject to the Data Protection Act or similar privacy legislation protecting information held on individuals. Information may also be subject to confidentiality agreements. Privacy and confidentiality concerns may impact on how digital materials can be managed within the repository or by third parties, and made accessible for use [5].

4 Conclusion

MS ISO 223 was a standard regarding records management that was established to assist Malaysia in becoming a structured country. According to MS ISO 223-2 (2009), preservation strategies for records especially electronic (digital) records may be selected on the basis of their ability to maintain the accessibility, integrity and authenticity of the records over time, as well as cost effectiveness. The nature of the electronic (digital) records must be maintained even though the records had undergone preservation activities such as copying, conversion and migration.

The National Archive of Malaysia has indicated the importance of the digital records in supporting the government functions but there are no publications or researches that describe in detail the strategies being done by the National Archive of Malaysia (as it is the institution that is responsible in the digital preservation on a National basis). Without detailed research, it is difficult to ensure that the strategies being implemented are effective and suitable or not.

References

1. A. Rodriguez, *Computer Technology Review,* **25**(3), 17-19 (2005)
2. D. Bearman, Addressing selection and digital preservation as systemic problems. *International Conference organized by Netherlands National Commission for UNESCO Koninklijke Bibliotheek (National Library of the Netherlands).* The Hague, 4-5 November 2005.
3. National Archive of Malaysia. (2003). *Electronic records management and archives management policy.* Kuala Lumpur: National Archive of Malaysia.
4. Y. Han, Digital content management: *The research for a content management system.* Library Hi Tech. **22** (4), 355-356 (2004)
5. Digital Preservation Coalition. (2008). *Digital Preservation Definition and Concepts.* Retrieved from www.dponline.org.
6. UNESCO. *Concept of Digital Preservation.* (2011). Retrieved from www.unesco.org.
7. M. V. Cloonan, S. Sanett, The American Archivist. **65**, 70-106 (2002)
8. M. Foot, (2001): Building Blocks for a Preservation Policy. *NPO Preservation Guidance Preservation Management Series.* 1-13. Retrieved from http://www.bl.uk/services/preservation/npo8.pdf
9. T. Hendley, (1998). Comparison of Methods & Costs of Digital Preservation. *British Library Research and Innovation Report.* Retrieved from http://www.ukoln.ac.uk/services/elib/papers/tavistock/hendley/hendley.html
10. J. Garrett, D. Waters, *Preserving Digital Information, Report of the Task Force on Archiving of Digital Information.* The Commission on Preservation and Access and The Research Libraries Group. Washington DC and Mountain View CA (1996)
11. K. Lee, O. Slattery, R. Lu, X. Tang, V. Mc Crary, Journal of Research of the National Institute of Standards and Technology **107** (1), 93-106 (2002)
12. H. Heslob, S. Davis, A. Wilson, (2002). *An approach to the preservation of digital records. National Archives of Australia.* Retrieved from: http://www. Naa.gov.au/images/an-approach-green-paper_tcm2-888.pdf.
13. S. Granger (2000). Emulation as a Digital Preservation Strategy. *D-Lib Mag.* 6(10). Retrieved from http://www.dlib.org/dlib/october00/granger/10granger.html
14. *Cornell University Library.* (2007). Digital preservation. Retrieved from: http://www.library.cornell.edu/preservation/tutorial/preservation/preservation-03.html.
15. N. Pozo, A. S. Long, D. Pearson, D. (2010). *Land of the lost: a discussion of what can be preserved through digital preservation.* Library Hi Tech, **28**(2): 290-300 (2010)

16. J. J. Albright, J. A. Lyle, *Political Science and Politics* **43**(1), 17-21(2010)
17. M. Hedstrom, S. Montgomery, (1998). Digital preservation needs and requirements in RGL member institutions. United States of America: Research Libraries Group. Retrieved from http://www.oclc.org/research/activities/past/rlg/digpresneeds/digpres.pdf.
18. M. Feeney, (1999). *Digital Culture: Maximising the Nation's Investment.*
19. S. Knight, S. (2010). Early learnings from the National Library of New Zealand's National Digital Heritage Archive project. *Electronic Library and Information Systems*. **44** (2): 85-97 (2010)

https://doi.org/10.1051/matecconf/201815005053

The Right of the Accused in Saudi in Criminal Procedure during Investigation Process and Arbitrary Dentation and Prohibits Torture and Protects the Rights of Suspects to Obtain Legal Council

Faten Mohmed Alqahtani[1,*], *Amani* Ali[1], *Abdulrahman* M.A.AlBelihi[1] and *Metwally* Ali[1]

[1]School of Human Development and Technocomunication (iKOM), Universiti Malaysia Perlis, Perlis, Malaysia

Abstract. Saudi Arabia has recent in years pass several important legislations to ensure a fair and balanced justice system, including: 1) The Law of Procedure Before Sharia Courts of September 2001, which grants defendants the right to legal representation and outlines the process by which please, evidence and experts are heard by the courts. 2) The Code of Law Practice of January 2001, which outlines requirements necessary to become an attorney and defines the duties and right of lawyers, including the right of attorney-client privilege. 3) The Law of Criminal Procedure of May 2001, which protects a defendant right with the regard to interrogation, investigation, and incarceration; outlines a series of regulations that justice and law enforcement authorities must follow during all stages of legal process, from arrest and interrogation to trial and sentencing; prohibits torture and protects the right of suspects to obtain legal counsel; and limits the period of arbitrary detention. The main objective of this study is to examine the operation of the criminal justice system in Saudi Arabia in light of the international human rights standards pertaining to the administration of justice. It involves the international human rights treaties and focuses mainly on the instruments in which the right to a fair trail in Saudi Arabia.

1 Background Research

Saudi Arabia has in recent years passed several important legislations to ensure a fair and balanced justice system, including:

1. The Law of Procedure Before Sharia Courts of September 2001, which grants defendants the right to legal representation and outlines the process by which pleas, evidence and experts are heard by the courts.
2. The Code of Law Practice of January 2002, which outlines the requirements necessary to become an attorney and defined the duties and rights of lawyers, including the right of attorney-client privilege.
3. The Law of Criminal Procedure of May 2001, which protects a defendant right with regard to interrogation, investigation, and incarceration; outlines a series of regulations that justice and law enforcement authorities must follow during all stages of the legal process, from arrest and interrogation to trial and sentencing; prohibits torture and protects the rights of suspects to obtain legal counsel; and limits the period of arbitrary detention.

In 2008 the government continued its reform initiative with the allocated sum of approximately £ 1.3 billion for the planned reforms. The new rules, which emphasize the independence of judges, set up a Supreme Court, the main function of which is to oversee the implementation of the Sharia (Islamic Law) as well as the various laws issued by the government. The new laws also set up new specialized courts for issues relating to commerce, labour, personal status, traffic disputes, together with a fund for training old and new judges. Currently, justice in Saudi Arabia is administered by a system of Sharia courts, and judges have wide discretion to issue rulings according to their own interpretation of the Sharia texts.

* Corresponding author: bkaifi@hotmail.com

Saudi Arabia has never promulgated a criminal code. However, a committee was appointed in 2005 to codify penal provisions but is yet to produce a draft.

The Study Significance:
The main argument in this research work may be in the following statement: 'The human rights of suspect during the fair trial procedure have many issues in the Saudi Arabian legal system and is partly compliant with international human rights standard. In order to evaluate the compatibility between the Saudi Arabia fair trial and the International human rights standard, this study examines and evaluates the mechanism of the of the pre-trial and in-trial process within the Saudi Arabian legal systems and compares it with international human right standards, this study examines and evaluates the mechanism of the pre-trial and in-trial processes within the Saudi Arabian legal system and compares it with international human right standards found in international treaties. United Nations, Office of the High Commissioner for Human Rights, Status of Ratifications accessed 15 January 2013

In pursuit of these research, the study attempts to answer the following questions:

1. In Saudi Arabian is the domestic Law of Criminal Procedure consistent with Saudi Arabia's international human rights obligations with respects to the right to a fair trial and real according to Islamic law?
2. Do Saudi Arabia judicial apply the international human right standards without violating Islamic law?
3. What is the difference the system between international human right and Islamic law?

1.1 Significant of the study

The importance of this study lies in its timing. As stated earlier, the Saudi legal system is currently undergoing substantial changes both structurally and procedurally. The government's reform initiative takes place in the context of its desire to better achieve the objectives contained in the Sharia. Human Rights Watch conducted research missions to Saudi Arabia in November and December 2006, and again in May 2007. We found pervasive injustices in the Saudi criminal justice system and systematic and multiple violations of defendants right. Individuals in Saudi Arabia may find themselves detained and arrested for behavior that is not inherently criminal, or for apparently (and unwittingly) offending vague legal prohibitions. They may then find themselves in solitary confinement and subject to form of ill-treatment. The authorities often do not inform individuals of the crime of which they are accused, or the evidence supporting the accusation. An accused person typically does not have access to a lawyer, faces abuse when refusing to incriminate him or herself, and waits excessive periods of time before trial, where he or she is often to examine witness or evidence and present a legal defense, not least because of a presumption of guilt and shifting charges. *Ministry of Justice Saudi Arabia, 'History of justice in Saudi Arabia' accessed 15 September 2012*

2 Methodology

Comparative methods are applied, as the protection of fair trial under Saudi Arabian domestic law is compared to the international human rights standards. The research examines two group of case related to the right to a fair trial: first, the domestic case, that is to say those which have been provided by Saudi Arabian authority in the Ministry of Justice.
With the exception of three group interviews with detainees in al-Ha'ir Correctional Facility and one interview with intellectuals in Riyadh, we conducted all interviews privately and individually. During the group interviews with prisoners in al-Ha'ir Correctional Facility, we asked about their individual criminal cases, cases of abuse and cases of death in custody known to them, and their individual experiences during interrogation and trial. We cross-checked allegation across these groups and with former detained account for consistency.

3 Finding

3.1. Statutory law and Islamic law

The division of respective of influence between the absolute rulers of the House of Sa'ud and the Wahhabi religious establishment has endure through subsequent centuries and periods of temporary demise of the Saudi state. The religious establishment in Saudi Arabian has board influence over everyday life. Its scholars and officials write and vet textbooks used in schools. Officials in the Commission to Promote Virtue and Prevent Vice(CPVPV) lecture at social gatherings, teach the Quran in prisons and social institutions, and keep a watchful eye over the moral behavior of the general public. Religious officials preach in local mosques where prayers attendance in mandatory. The judiciary, too, is almost the exclusive province of the religious establishment. *'King Abdul-Aziz institution announcement' (Al Riyadh newspaper 26 May 2008) 14581 accessed 29 August 2012*
The Saudi government does not publish an official interpretation of Sharia. In the area of criminal law, the government has not published an interpretative text carrying the force of law of the precise definitions of acts that constitute offenses, such as "disobeying the ruler". Unlike Qatar, which also follows the Hanbali School, Saudi Arabia has no written penal code.

"The Courts shall apply rules of the Islamic Sharia in cases that are brought before them, according to the Holy Quran and the Sunna." Sharia, however, is silent on many areas in which modern life requires the application of precise legal norm. To fill this void, Saudi Arabia's prime minister (a post held currently by the king), may issue positive, or stator, law-called regulations to differentiate them from God-given laws of Sharia-as long as they do not conflict with Sharia precepts. Article 48 of the Basic Law, itself one such statutory law, also obliges the courts to apply Sharia rules "according to laws which are decreed by the rules in the agreement with the Holy Quran and the Sunna." Saudi Arabia has published hundreds of such statutory laws to regulate areas where Sharia precedents or interpretations have little bearing, such as traffic and banking laws.

Sheikh Mohammad Al Abdullah, the kingdom's chief prosecutor, told Human Right Watch that "all of our work follows the criminal procedure code all in cases." Most of those interviewed told Human Right Watch that the Saudi government implemented the criminal laws unevenly, and sometimes not at all. Defendants described numerous and specific instances in which prosecutors, arresting officers, and judges did not act in conformity with the law's provisions. Jeddah-based lawyers Aiman told Human Right Watch, "The criminal procedure code is still new for prison officials. A detainee has no insists on his right and know them. Nobody will tell him his rights or facilities, access to them." Hisham, a Dammam-based lawyer with trial experience, confirmed to Human Right Watch that "Judges are not very conversant in the criminal procedure code*." Al-Sulami, 'Terror Cell Plotted to Kill Top Officials' Arab News accessed 18 January 2013*

3.2 Codification of criminal laws

Saudi Arabia does not have a written penal code, but relies on judge's interpretation of the Sharia for determination of which actions constitute crimes and what the attendant punishment should be. The definitions of crimes and severity of punishment may vary far from case to case. In 2005, as Saudi Arabia was negotiating its accession to the World Trade Organization, Saudi officials revived an idea first touted by King Abd al-Aziz in the 1930s, to provide citizen, law enforcement officials, and judges with a clear, written formulation of what constitutes a crime. According to Sheikh Abd-al-Mukhsin al-Ubaikan, the Justice Ministry's judicial adviser and member of the Saudi Shura Council, the country's "highest leadership" approved a plan to compile Islamic jurisprudence (fiqh) in the form of articles of law "to be used by the courts but without being compulsory."

4 Conclusion

The kingdom of Saudi Arabia has sought to address a number of elements inherent within and outcomes associated with human rights decisions and the right to a fair trial. Nevertheless, such involvement has increase dramatically following the domestic reform implemented across Saudi Arabia. However, during the past twenty years, a remarkable shift has been witness within Saudi Arabia domestic law, which reflects positively on the human rights of suspects, initiated at the point at which constitutional reform was adopted through the Consultative Council, and spanning up until the establishment of two human rights entities, the NSHR and the HRC.

The establishment of the LCP was one the practical implementations of the right to a fair trial, after its provision in the BLG. However, these remain notable challenges in relation to the compatibility of Saudi domestic law with international human rights standards. But still have a long way to go because so many cases couldn't find justice, that's because there are no sentencing guidelines, including on the meaning of guilt proven beyond reasonable doubt for discretionary sentences. There is no guidelines for introducing evidence, including guidelines for assessing the credibility of witness. Ensure that witness for the prosecution and defense actually appear in court so defendants may cross-examine them and review of the decision to charge a suspect and to remind him or her in custody.

References

1. A. Abdul-Wahhab, "Human rights in *qasas* crimes between Sharia and international law" (PhD Thesis, Naif Arab University for Security Sciences 2000)
2. Abu Zahrah, Crime and punishment in Islamic jurisprudence (Dar alFikr Al-Arabi 1984)
3. A. Al-Barrak, "The Right of Accused to defend himself in the Saudi Criminal Justice System (Naif Arab University 2000)
4. S. Alghozy, 'The legal status of lawyer in Saudi law' (PhD Thesis, Naif Arab University 2008)
5. N. Al-Joufan, "Public hearing in the judicial domain" (2009) 5 Al-Adl 10
6. Allahim, Abdulkarim, Ta'aeweed a'n an sejn (2002) 12 Al-Adl Journal 1422
7. H. Al-Marsafawi, *The fundamental of criminal procedure* (Dar Al-Marifah, Egypt, Egypt, 1998) 187
8. H. Al-Qahtani, The Specialisation of thr Provinces Between Practice and Criminal Procedure Code in Saudi Arabia (PhD Thesis, Naif Arab University 2007)
9. F. Al-Shihri, "The High Administrative Court in the Board of Grievances" (Phd Thesis Arab University, Saudi Arabia 2009)
10. H. Al-Subaie, "Crime Control officer to perform some of Investigation acts under Saudi Criminal Procedure Act; Analytical and comparative study" (Master Dissertation, Naif Arab University for Security Sciences, 2006)
11. A. Al-Washmi, Fitnat Al-qul be Taleem Al Banaat (Arabic Education Center,Morocco, 2009)
12. K. Badaywe, Ijraat Raf'a Al-Dawa Fi N izam Al-Murafaa Al-Sharia, (Qibba, Cairo 2006)
13. B. Hassuna, The evidence in Hudud crime in Sharia and Law (PhD Thesis, Naif Arab University Saudi Arabia 2002)

14. Ibn Qudamah, Al Mugni, 429/7, and Jawahir Al-Ikleel, Imam Malik (Dar ihiya alturath Al-Arabi, 1985)
15. Ibn Taymiya, Al-Syasa al-Shar'iya, (ed Al-Afaq Al-Jadidah, Beirut 1983)
16. Ishak, 'Mabda al bad fe alkada al shari (2002) 8 Al-Adl Journal
17. Y. Kilzi, Human right in the face of the authority of law enforcement officer (Naif Arab University for Security Sciences 2007)
18. Mohammad al-Ramli, Nihayat al Muhtaj fe Sharh al minhag 5/71
19. A. Shalabe, Falsafat al-tshre al islami (Dar al Nahdah al-Masria 1998)
20. A. Waleed, The Independences of the commission of investigation and public prosecution in Saudi Arabia (Master Thesis, Naif Arab University 2007)

Jawi Writing in Malay Archipelago Manuscript: A General Overview

Makmur Haji Harun[1,*], *Mohd Kasturi Nor* Abd Aziz[2], *Edzham Armin* Abd Rahim[2], *Shuhairimi* A.[2], and *Yasmin* Ahmad[2]

[1]Faculty of Languages and Communication, Universiti Pendidikan Sultan Idris(UPSI), 35900 Tanjong Malim, Perak, Malaysia
[2]School of Human Development and Techno-communication, Universiti Malaysia Perlis, Perlis, Malaysia

Abstract. Jawi writing is the main form of writing in various manuscripts recovered around the Archipelago, especially during the early arrival of Islam. These manuscripts are found to record history, art, culture, language, social, as well as ancient knowledge. . The usage of Jawi writings had covered a number of calligraphy which are high in value with deep philosophical meaning, are full with creative notion as well as countless of patterns that went along with the call to Islam around the Archipelago in a relatively short period of time. The development of Jawi writing had been immortalized in various genres such as books, manuscripts and letters written by individual writers and in groups as a local way of living. This writing was also introduced through a long process using certain methods of writing and reading, and kept as individual as well as institutional collections all around the Archipelago. This article is focused in various manuscripts that not only were exclusively used as learning and teaching aids, but were also garnished as beautiful symbols and philosophical Islamic art that deserves a high degree of honour. This article's objective is to deeply analyze the usage of this writing from manuscripts found around the Archipelago as a proof on the importance of this form of writing. The methodology of this article is library study, through a number of theories and methods. The implication of this article is hoped to form a certain standard as well as its own identity for the local society through their acceptance of this writing form, especially during the early arrival of Islam in the Archipelago thus deemed as a national treasure.

1 Introduction

Through the reading of Malay manuscripts it can be one of the discloser to treasure Jawi writing in the form of historical, cultural, social, science, and knowledge of the past. Jawi cover various types of calligraphy with elements of high value art, has a philosophical, creative laden with meaning, have diverse patterns of spread together with Islam in a short time almost all regions of the country. Raising various manuscripts written in Jawi, it is not just as a teaching and learning, but also can be a beautiful decoration in the form of works of art and heritage of Jawi calligraphy patterns, styles, colors, symbols and philosophy of credible Islamic art. Various collections of Jawi manuksrip can be found as calligraphy using nasakh, but there is also a wide variety of other khat as a step to strengthen the position of writing and mastery of many beautiful and wonderful as a civilized and dignified nation's heritage. This study is conucted to discuss the issue as a joint search for identity in addition to getting to know and understand the use of Jawi writing a paper base to produce manuscripts written in Jawi in Malay, in addition to being one of the discourse of science that can uncover the richness of spoken and written of the Muslim community in Nusantara.

2 Brief view on Malay manuscript

Jawi was instrumental in establishing communication through writing, even a spoken language among the Malay community in the archipelago. Language and Jawi revealing the identity of the Malay, Jawi to relay communication and language of the Malays who come grow together with the science of philology and their manuscripts. Therefore, the Jawi script in various manuscripts in the Malay used mostly as a tool to understand the culture of his people, especially the language used in the review of the manuscript heritage as a nation. The following display is the debate on the study of ancient manuscripts that flourished in the

* Corresponding author: makmur@fbk.upsi.edu.my

archipelago that is the legacy of their ancestors from time to time. Multiple exposure will be displayed as a study in the question of whether art, culture, language, sciences, and other knowledge through various mansukrip in Jawi.

In the 4th century AD where several cities in West Asia have college as a center of research knowledge that comes from the Renaissance, Greek, Byzantine and Western, and Alexandria. All of these places become centers of research manuscripts and manuscript whose influence has spread from time to time. Scientific studies also cover various fields of science, philosophy, art, culture, science and technology come to grow and spread until it reaches areas of Malaya. Since the archipelago nation, especially the Malays literacy, especially after the advent of Islam to the Malay world, the greater part of the society is to cultivate write in their manuscript copies and so it can form a document that gives information about their life in the past. Among the Asian nations that are considered pretty much had the documents they are relics of the past India. It was revealed from various research and studies to document their research mainly in the form of writing such as stone inscriptions, manuscripts and inscriptions of ancient relics of their ancestors when compared with the Malays. While the early development of writing in Indonesia is not yet known, because it is considered as the expertise of its own and only certain people only knew, not even many of the civil society who control except from certain groups or writers castle alone.

There is language in the script used in conveying information, also used in the writing of the manuscript as a tool to know the culture of a nation included in the Malay world. The existence of the manuscripts in the Malay heavily influenced by the languages that have the time, among others derived from Sanskrit, Tamil, Arabic, Persian and regional languages cognate's manuscripts. But the languages most influential manuscripts in the archipelago are Sanskrit and Arabic. Therefore, to understand and study the manuscripts are available in the archipelago in the study of philology, then both of these languages are necessary to learn and even be controlled, such as studies manuscripts Javanese texts and copies of classic Malay.

In ancient manuscripts they are very much influenced by Sanskrit, such as the effect on the absorption of vocabulary, phrases, syntax and excerpts sometimes without translation. As an example of these influences can be seen in Uttarakanda, Kakawin Ramayana and Sanghyang Kamahikan. Although the ancient Malay manuscripts also involve Sanskrit but only the form of a loan word that remains sahaja required. For old Malay manuscripts, knowledge of Arabic is also needed; especially for study of manuscripts rather than the influence of Islam in particular that contains the teachings of Islam and Sufism or Suluk. Because in these manuscripts contained many serpan such as words, phrases, expressions, and excerpts-excerpts in Arabic.

In addition to foreign languages, the regional languages in the archipelago are also indispensable in the cultivation of which produces manuscripts archipelago, that language is a communication substitutions tribal areas of the archipelago which is closely related to the language manuscripts. Knowledge of languages is important because controlled cultivation of manuscripts sometimes disukarkan reading words that it was not coming out of a foreign language, but precisely than the local language and the local language, including the manuscripts beraksara Jawi, Java script pegon and other characters.

3 Expansion Study on Malay Manuscript

The development of the expansion of the manuscript on the next stage of Malaya, estimated in the 12th century AD, the community has not yet used the Jawi script immediately but they use scripts and manuscripts Jawi existing for the purpose of writing, so it appears inscribed stone, which is the initial document evidence of Jawi writing in the Malay already begun. His appearance is directly related to the arrival of Islam to the archipelago. This character is based on the Arabic script used for writing Malay speech. Thus, there is inevitably some point additions or modifications to adjust the sound does not exist in Arabic (eg, speech / o /, / p /, or / n /).

After the earliest evidence of Jawi script is in Malaysia with an inscribed Terengganu Stone is dated 702 H or 14th century AD (the date is quite a problem because the number of years of writing, not with numbers). Here only can read seven hundred and two (702 H). But this can be followed by two other words namely 20 to 29 or - eight -> two eight -> "eight". This word can also be followed by the word "nine". With this, it becomes more likely date being 702, 720-729, or 780-789 H. But since this Inscription Stone also mentioned that this year is the "Year of the Crab", then there are only two possibilities left namely in 1326 M or 1386 M. But according to Muhammad Naquib al-Attas, that the inscribed stone that was written in the year 702 H equal to the year 1303 AD its content emphasizes the rulings of fiqh, also using the Jawi script which is a translation of religious books entitled *'Aqa' id al-Nasafi.*

Jawi manuscript writing and keeping pace with the spread of Islam and managed to build a society with a culture in line with Islam (Ismail Hussein, 1984). Activities continued with writing letters of various materials there is a peak in the 17th century AD when it appears influential figures such as Hamzah Fansuri and Nuruddin ar-Raniri (r.h) and so on. Persian scholars and scholars from many contributions in the fields of politics, trade, social, culture and religion, but also well-known as a center of Pasai Malay language and literature (T. Iskandar, 1995). Developments in the field of literature very rapidly as writing literary works in Malay and Jawi writing is widely used not only in society even among people of the castle (Ismail Hussin, 1984).

4 Jawi Written Manuscript Studies in the Malay World

In this century, hundreds and even thousands of manuscripts written in various areas in the Malay world, especially in the palaces and administrative centers. While writing the Malay manuscripts which often featured in masyarakatny is covering all fields. Malay Jawi manuscripts collection estimated at 265 title handwritten manuscripts written in Jawi Malay and other languages. This has opened up a special research in the field of Malay writing. Among the titles that is often referred to Sulalatus Salatin, Hikayat Amir Hamzah, Hikayat Indera Putera, Panji and others (Digital Library DBP). The following description of some examples of Malay manuscripts which often become the reference in the archipelago. According to the detailed description as follows;

4.1 Manuscript *'Hikayat Raja-Raja Pasai'*

The manuscript is believed to be the oldest book in Malay literature. It is a story about the discovery of the 16th century. The manuscript is believed to be one of the original copies of previous issues. Hikayat Raja-Raja Pasai this is a literary masterpiece that is the oldest history of Islam Nusantara. In the manuscript was told of important events that occurred between the years 1250-1350 AD. This period is the reign of King Merah Silu who later converted to Islam and changed his name as Malik al-Salih.

The manuscript of this saga is the only step the history of the kingdom Pasai ever successful when used in the archipelago. According to estimates by Russel Jones saga was written in the 14th century. This saga also includes the establishment of the Sultanate of Samudera Pasai until the conquest by the kingdom of Majapahit. In addition, the contents of the Hikayat Raja-Raja Pasai is told about the elements of legalization of family members of the ruling, stating the origin of the sacred family. In addition, the manuscript also has a didactic function, such as a tyrannical king will be punished, and the country will be destroyed. Likewise with Sultan Malik al-Mansur who seized his concubine. At the same where Sultan Ahmad who is jealous of his sons, causing their second murder.

Manuscript "*Hikayat Raja-Raja Pasai*" has striking similarities in the composition of the subject and verse with "Sejarah Melayu." "Hikayat Raja-Raja Pasai" is one source of the story of the coming of Islam to the West it also is a work in Malay which tells the story of the first Islamic kingdom in the archipelago, Samudera-Pasai which is now located in Nanggroe Aceh Darussalam.

4.2 Manuscript *'Hikayat Amir Hamzah'*

Hikayat Amir Hamzah was initially written by a Persian writer with Qissa'i title Emir Hamzah and then translated into Arabic with the title Sirat Hamzah. This manuscript was brought by missionaries from India to Melaka in Melaka Malay Sultanate era. This manuscript was copied by hand by the author of the palace of King Melaka when it takes over 15 years to complete the whole text. Besides Malay language, this manuscript is written in Java, Bali, Aceh, Turkish and Hindi.

The main character in this story is Hadrat Hamzah; the Islamic warriors were killed in the battle of Uhud. In accordance with the saga of the story has been exaggerated. Amir Hamzah described so great for example as follows:

> *Due to the strength of Amir Hamzah shouts can be heard up to 18 winds away. The tiger heard the shout of surprise ran. The camel and horse flopped. The arrows in the hands of his enemies fall to the ground.*

4.3 Manuscript *'Hikayat Hang Tuah'*

Hikayat Hang Tuah manuscript is written in Jawi. This manuscript is a heroic epic spanning all time, the rise and fall of the Malacca Empire. Hang Tuah was a major warlord as a Malay warrior who obeys his king, all the more so during the reign of Sultan Mansur Shah of Malacca in the 15th century. This saga is believed to be written around the 1700s (Leiden Manuscript Cod.Or.1762.), But a complete manuscript is dated 1849 AD One excerpt from the manuscript of the Hikayat Hang Tua as follows:

> *At the time of his youth, Hang Tuah and four of his friends, Hang Jebat, Hang Kasturi, Hang Lekir and Hang Lekiu, kill a bunch of pirates and two others who run amok and create chaos in the village. The Treasurer (together with the Prime Minister in modern terms) Melaka saw their talent and take their duty at the palace. Hang Tuah is enshrined as a hero in the Sejarah Melayu and also narrated in the Hikayat Hang Tuah.*

4.4 Manuscript *'Hikayat Merong Mahawangsa'*

Hikayat Merong Maha Wangsa or history is a saga that tells Kedah traceability Merong Maha Wangsa with the founding of the founding of the state, one of which is located in Malaysia. Although there are some historical facts, this story also contains allegations that impressive. The era covered by this text starting from the opening of the state by Merong Maha Wangsa who claimed to be descendants of Alexander the Great of Macedonia until the arrival of Islam in the 12th century AD Merong Maha Wangsa is Hindu and there are nine other Hindu previous government under his control, which called Phra Ong Maha Wangsa who converted to Islam in 1136 AD and then took the name as Mudzafar Sultan Shah. In addition, this tale also illustrates the Chola attack on the territory of Kedah.

5 Discussion and Suggestion

Here is some debate as to the findings for further discussions, to serve as input a discussion on Jawi writing applications in various manuscripts in the Malay world, such as the following description:

5.1 Discussion

The results of our study as a whole, in this case we can say that the contribution and work of scholars in the form of manuscripts and manuscript turned many into a means of development Jawi writing from time to time by touching everything, especially about the history, language, literature, social, economic, art, culture and so on. In addition, we also found that the most widely used for the works of other manuscripts also based on Jawi writing as promotion of the development of this next as an optional extra in the Jawi script reveals treasures.

Although the Jawi script was influenced by calligraphy that has subtle art of West Africa, in addition to competent bentuknnya, pattern, piled aesthetically attractive features, displaying the beauty of the impressive motif. But the provisions of the Jawi still pose difficulties to the Malays who are not familiar with the intricacies of the text. In order to popularize the use of Jawi writing this, it is not surprising that a lot of players this paper seek ways and methods in order to keep the text as written treasures that remain until anytime. In addition, this paper also expected to be used as a kind of writing that the community in a variety of materials and books, as the frequency of use is said to be of much help to its development. In addition to his writings, also commonly seen Muslim Malaysia in particular and the country in general, to increase the variety of writing that can help the reader to read the works of Jawi through various manuscripts in Malay.

Meanwhile, the existence of Jawi writing which has been intensified by the existence Za'aba and Matlob also began to be associated with Spelling Za'aba (EZ) and the enhanced Spelling (SI) when they were asked to get involved in a committee entrusted with reviewing the Glossary Malay: Rumi -Say-Jawi, first edition, in two volumes, three mold for 2001, 2003 and 2004. Among its members is Tuan Hj. Hamdan Abdul Rahman, Prof. Emeritus Dr. Extremely Mario Moain, Matlob, Prof. Dr. Muhammad Bukhari Lubis, and Ismail Dahaman (former employee DBP). They held a workshop last important on 22-24 March 2002 (Volume I) and 8 to 13 April 2002 (Volume II). Finally came the book Glossary Malay: Rumi-mention-Jawi, the second edition of one volume only, in 2008. All these efforts can produce more perfect method to master Jawi writing up the application are available in a variety of Jawi in Malay manuscripts.

Even their role Matlob Za'aba and they both equally have proved it, they came to highlight their works are impressive, even demanding lovers of Jawi writing since the fifties again, because their works are not solely concerned with writing , but he also developed the science of writing and maintaining quality and quantity. In addition, their publications constantly improved over the years, whether in the form of teaching, for the reading of children, adolescents and adults and so on. Even books and his works remain awake writing accuracy and methods that are structured and accuracy of the method used in it. Perhaps to keep this all, it has resulted in works of their authoritative so widespread in the Malay and even can help the readers and researchers to identify more closely the writings of Jawi are available in various manuscripts in Nusantara guided by the rules of Jawi was founded by two great figures in Jawi.

5.2 Suggestion

Here is some debate about the findings of the study should be submitted as a proposal reviewer, to be used as an ingredient in realizing the struggle and Putting the right place Jawi writing methods that have been applied in various manuscripts in Malay so felt almost all over Asia and even the world.

Through this study, it can be stated that almost all states in Peninsular Malaysia including Sabah and Sarawak, according to the luggage museum, institution or library regarding the application of Jawi in various works of the manuscript, it is not impossible to be able to produce generations of leaders and lovers treasure Jawi script for Malays in the Malay world for teaching and learning continues on an ongoing basis. Then applying the method of Jawi in several places throughout the country up to cover all of Southeast Asia, is expected to apply the methods of teaching and learning of Jawi writing correctly and according to the existing rules.

Jawi writing in a variety of applications in the Malay manuscripts can make this paper more precious and valuable to be used as written in various places and fields. And adequate financial assistance shall be extended to further advance the field of Jawi writing of this manuscript, especially in producing new mansukrip required by the Malays. In addition, the Department of Islamic states, is expected to help the development and writing of Jawi writing so it made correctly and interesting, thus giving the impression of external and internal for lovers and researchers. For this, the Islamic Religious Department in the states, or the Writers' Association states, foundations, colleges, institutions, universities and so on are required in order to establish an association or club lovers Jawi, in order to find a talented writer who really can be heir to the greatness of this method ever gemilag in the past.

Also hope the Ministry of Education (MOE), to create a syllabus in secondary schools in teaching writing skills by using the special skills of Jawi writing skills. Then ask also to the lovers and fans of Jawi writing to further enhance the skills to read and write Jawi script and mastery of methods, ranging from the School of Early Childhood (kindergartens and nurseries), Primary School and Sekolah Menengah Kebangsaan (SMK) to the adult stage. In addition, there should be a university or institution which can diversify the program. If the use of Jawi methods necessary to ensure proper use, especially regarding the names, spelling, grammar, uslub and so on. In fact, competition should be held occasionally read Jawi manuscripts all levels in the Malay world to raise the power of the manuscript treasures.

6 Conclusion

In conclusion, should ponder that the application is one of Jawi writing skills in Jawi font with unique diversified in all requirements, in addition to the support from the Arabic, al-Quran and al-Hadith as writing concept sketched. In addition, the use of Jawi writing in various manuscripts in the Malay became one of the fields accounting for the source of Islamic art handicraft as the identity of Malay identity in showing patterns Malay culture in particular and society in general Southeast Asia. It is not independent of meaning, philosophy, shapes, and patterns of work and writing has been described by the authors Jawi cedikiawan ago.

Jawi writing in a variety of applications in the Malay manuscript was complete methods, models, styles and techniques that are useful for Jawi heritage lovers and collectors of old manuscripts in Nusantara. In fact, this makes writing applications in various manuscripts in Malay can make it more valuable and precious to be used as treasures in various places and fields. In addition, the development and activities of the Jawi script should continue to be made with good and interesting, thus giving the impression of external and internal for lovers and researchers in the country. For this, too, must be taken seriously by the Islamic Religious Department in the states, Writers Association states, foundations, colleges, institutions, universities and so on in order to foster the spirit of love for the treasures of the nation in addition to being a remarkable heritage from time to time.

The debate was the focus of discussions among the study is the existence of numerous manuscripts written in Jawi are commonly found in the Malay world with various titles and topics for discussion, such as; *Book 'Aqai'id al-Nasafi, Hikayat Raja-Raja Pasai, Manuscript War Batak, Hikayat Amir Hamzah, Hikayat Bayan Budiman, Hikayat Indera Putera, Hikayat Hang Tuah, Hikayat Merong Maha Wangsa, Mass Manuscripts Malay Manuscripts Kitab Tib, Hukum Kanun or law Melaka, Hikayat Abdullah, Manuscript Sulalatul-Salatin, Singapore Burned* poetry and so on. Hopefully the exposure of all copies of the manuscript and can provide important information on the heritage of the archipelago to all who need it from time to time.

References

1. Abdul Razak Ab. Karim, *Melestari manuskrip Melayu* (Kuala Lumpur: Perpustakaan Negara, 2003)
2. Ahmad Nazuki @ Marzuki Yaakub dan rakan-rakan. (2006). *Penggunaan kaedah transliterasi dalam buku teks bahasa Arab: Satu kajian kes.* Laporan Akhir Penyelidikan. Selangor: Unit Penyelidikan, Pembangunan dan Pengkomersilan Universiti Teknologi Mara.
3. Ahmad Shah Mohd. Noor. (1977). *Hikayat raja-raja Pasai.* Penyelenggaraan A.H. Hill: Suatu Analisa. T.tp.
4. Ali Hj. Ahmad. Ed, *Hikayat Inderaputera* (Kuala Lumpur: Dewan Bahasa dan Pustaka, 1979)
5. Alis Murni, *Ulasan: Hikayat Inderaputera* (Kuala Lumpur: Dewan Bahasa dan Pustaka, 1969)
6. Amat Juhari Moain, *Sejarah askara Jawi* (Kuala Lumpur: Dewan Bahasa dan Pustaka, 1996)
7. Anas Hj. Ahmad. *Sastera Melayu lama dan baru* (Selangor: Syarikat Cahaya, 1984)
8. Anas Hashim Musa, *Sejarah perkembangan tulisan Jawi* (Kuala Lumpur: Dewan Bahasa dan Pustaka, 1999)
9. Arenawati. *Perkembangan kesusasteraan Melayu* (Kuala Lumpur: Pustaka Antara, 1966)
10. A. Samad Ahmad. Ed, *Hikayat Amir Hamzah* (Kuala Lumpur: Dewan Bahasa dan Pustaka, 1987)
11. S. B. Baried, *Pengantar teori filologi* (Yogyakarta: BPPF Seksi Filologi, Fakultas Sastra UGM, 1994)
12. Che Wan Shamsul Bahri bin Che Wan Ahmad, *Transliterasi mesin bahasa Melayu: Jawi lama-rumi berasaskan petua.* Proceeding of the International Conference on Artificial Intelligence in Computer Science and ICT (AICS 2013), 25-26, (2013)
13. Ding Choo Ming, *Kajian manuskrip Melayu: Masalah, kritikan dan cadangan* (Kuala Lumpur: Utusan Publication and Distributions, 2010)
14. Djamaris, Edward. *Hikayat Nabi Mikraj, Hikayat Nur Muhammad dan Hikayat Drama Tasiya* (Jakarta: Departemen Pendidikan & Kebudayaan RI, 1983)
15. Dzulkifli Salleh, *Hikayat Merong Mahawangsa sebagai sumber sejarah* (Kuala Lumpur: Dewan Bahasa dan Pustaka, 1965)
16. E. S. kadjati, *Katalog induk naskah-naskah Nusantara jilid 5A: Jawa Barat koleksi lima lembaga* (Jakarta: Yayasan Obor Indonesia, 1999)
17. A. T. Gallop, The British library journal **17** (2), (1991)
18. A. T. Gallop, *Warisan warkah Melayu* (London: The British Library, 1991)
19. Harun Mat Piah, *Kesusasteraan Melayu tradisional* (Kuala Lumpur: Dewan Bahasa dan Pustaka, 1993)
20. Harun Mat Piah, *Cerita-cerita panji Melayu* (Kuala Lumpur: Dewan Bahasa dan Pustaka, 1977)
21. Hashim Abd Hamid (2004) *Doksologi dalam Manuskrip.* Kertas Kerja dalam Seminar Manuskrip Islam pada 1-2 Mac 2004, Anjuran Jabatan Perdana Menteri Negara Brunei Darussalam.
22. Ibrahim Saad, *Kesusateraan Melayu: Satu pengenalan* (Kuala Lumpur: Utusan Melayu, 1974)

https://doi.org/10.1051/matecconf/201815005054

23. Ismail Hussien, *Sejarah pertumbuhan bahasa kebangsaan kita* (Kuala Lumpur: Dewan Bahasa dan Pustaka, 1984)
24. Jones, Rusell, *Hikayat raja Pasai* (Petaling Jaya: Penerbit Fajar Bakti, 1987)
25. *Kamus Padat Bahasa Malaysia-Inggeris* (Sri Serdang: Penerbit Minda Sdn. Bhd., 2008)
26. *Kamus Dewan edisi keempat*, (Kuala Lumpur: Dewan Bahasa dan Pustaka, 2010)
27. Kassim Ahmad, *Hikayat Hang Tuah (Menurut naskhah Dewan Bahasa dan Pustaka)* (Kuala Lumpur: Dewan Bahasa dan Pustaka, 1964)
28. Khalid Hussein, *Taj us-Slatin* (Kuala Lumpur: Dewan Bahasa dan Pustaka, 1966)
29. Muhammad Bukhari Lubis, Ali Haji Ahmad, *Transliterasi kata Arab-Parsi-Turki* (Kuala Lumpur: Utusan Publication & Distributors Sdn. Bhd., 2006)
30. Muhammad Bukhari Lubis, Makmur Haji Harun dan Alizah binti Lamri, *Jawi: Warisan warga Alam Melayu* (Tanjong Malim: Penerbit Mentari, 2013)
31. H. Muhammad Yusuf, *Persejarahan Melayu Nusantara* (Kuala Lumpur: Teks Publishing Sdn. Bhd., 1988)
32. Raja Ali Haji Raja Haji Ahmad. (1892). *Bustan al-Katibin*. T.tp.
33. S. Siti Hawa, *Hikayat Merong Mahawangsa* (Kuala Lumpur, University of Malaya Press, 1970)
34. Siti Hawa Salleh, *Masalah pengeditan manuskrip Melayu.* (Kuala Lumpur, University of Malaya Press, 1978)
35. H. Soebadio, *Penelitian naskhah lama Indonesia* (Buletin Yaperna, 2010)
36. N. S. Suryani, *Teori filologi.* (Bandung, Fakulti Sastera Unpad, 2006)
37. N. S. Suryani, *Filologi* (Bogor, Ghalia Indonesia, 2012)
38. Sutrisno, Sulastin, *Hikayat Hang Tuah: Kajian struktur dan fungsi* (Yogyakarta, Gajah Mada University Press, 1983)
39. Syed Muhammad Naguib Al-Attas, *Islam dalam sejarah dan kebudayaan Melayu* (Bangi, Universiti Kebangsaan Malaysia, 1972)
40. Wan Ali Wan Mamat, *Pemuliharaan buku dan manuskrip.* (Kuala Lumpur, Dewan Bahasa dan Pustaka, 1988)
41. R. O. Winstedt, *A History of classical Malay literature* (Kuala Lumpur: Oxford University Press, 1977)
42. R.O. Wolters, *History, culture and religion in Southeas Asian perspectives* (Singapura: Institute of Southeas Asian Studies, 1882)
43. Website
http://ms.wikipedia.ogr/filologi_asia, india, nusantara dan moden
http://www.naskah-naskah-lama.blogspot.com/naskahaceh.blogspot.com/
http://www.manuskriplama.blogspot.com.
http://www.utusan.com.my/utusan/Bicara_Agama/20130118/ba_06/Kerja-penulisan manuskrip-Melayu#ixzz2lLMU6ckk ©Utusan Melayu (M) Bhd.

Social Media and the Role in Spreading Social Unification Mutes in the Family

Amani Ali[1,*], *Hassad* Hassan[1], and *Siti Nurayu* Mohd Basir[1]

[1]School of Human Development and Technocomunication (iKOM), Universiti Malaysia Perlis, 02600 Perlis, Malaysia

Abstract. The effect of new media social not only on the children; but on the whole family was and still has Serious impact and dangerous to the community and the next generation and probably a lot of family relationship failed, and deteriorating. Perhaps explained in this paper some of which can be done to avoid those problems that brought us more of the West and traditions. They try to be with the children as a family with communication. Approximately 60% of educational concepts and ethics, behavior and belief sourced media. The Toxins That Threaten Our Brain (The reflection of communication in family relation) a. Understanding the wellbeing of the children and young people. b. Issues for consideration in conceptualizing the wellbeing of the children and young people. c. Identify emerging trends concerning the psychological and emotional wellbeing needs of children and young people. Identify and determine the relationships between parents and young people make them able of dealing with others, not only online and away their social and family mutes. At the same time though, overall Internet use keeping family member apart. Online social networks are more than just a fad among the younger generation, "Lynn Franco, director of The Conference Board Consumer Research Center, in a statement. "They've become an integral part of our personal and professional lives. They're an effective way to keep in touch with people, connect with friends and family, and network with colleagues and all that online if they see each other they don't jnow how to communicate. "

1 Background of Study

Family Life may actively follow and track use of social media through its internal logging and audit procedures. All use of Family Life ICT resources may be monitored. But before engaging in social media on behalf of Family Life individuals must be authorized by family to do so. Individuals may not comment as a representative of Family Life unless authorized to do so.

1.1 Family rules to use of Social Media

Social media as representative of Family Life individuals must: Always disclose they are worker of Family Life, Ensure that their comments/actions are not harassment. Not respond to material that is offensive, obscene, defamatory, threatening, harassing, bullying, discriminatory, hateful, racist or sexiest. Not use or disclose any confidential or secure information Ensure their comments/actions do not cause an damage to Family Life *Devorah Heitner is the author of "Screenwise: Helping Kids Thrive (and Survive) in Their Digital World."2017.*

To be effective, family communication must be an exchange of feelings and information. Doors of communication will swing open in the home if members will realize time and participation on the part of all are necessary ingredients. In family discussions, differences should not be ignored, but should be weighed and evaluated calmly *Creating a family social media policy In: kids, policy, privacy, rules, social media policy May 26, 2011.* One's point or opinion is not as important as a healthy, continuing relationship. Courtesy and respect in listening and responding during discussions are basic in proper dialogue. As we learn to participate together in meaningful associations, we are able to convey out through of love, dependence, and interest. When we are inclined to give up in despair in our efforts to communicate because other family members have failed to respond, perhaps we would do well not to give up, but rather to give and take our conversations. How important it is to know how disagree with another's point of view without being disagreeable. How important it is to have discussion periods ahead of decisions. Jones Stephan wrote, " I have learned that the head does not hear anything until the heart has listened, and that what the heart know today the head will understand *tomorrow March 8, 2016 Family technology rules: What kids expect of parents,Jennifer Langston , UW News. "*

1.2 Social media kill communication skills

With social media consuming so much of our time, conversing with one another face to face has suffered.

* Corresponding author: amaniali@unimap.edu.my

The busyness of our lives and over scheduling also take a toll on the family. The percentage of families that sit down together to share dinner is steadily declining. Deb McMahon, A Better Iowa contributor Published Sept. 30, 2015.www.desmoinesregister.com/...social-media-kill-communication-skills/73067760/

The dinner table used to be a place where those gathered shared their news of the day. Often it was where families discussed current events and had actuals conversations. Fast food meant was ready at the allotted time, not a bag of greasy takeout to be devoured in record time.

Email, Facebook, Instagram, Twitter, Snap chat and a host of other ways (so many web side) to communicate have often taken the place of talking and sharing. Yes they are fast, available at all hours and easy, who take the place of verbal discussions. *A Work-In-Process Literature Review: Incorporating Social Media in Risk and Crisis Communication .Shari R. Veil,Tara Buehner,Michael J. Palenchar First published: 4 April 2011Full publication history DOI: 10.1111/j.1468-5973.2011.00639.x View/save citation Cited by (CrossRef): 109 articles*

Throughout the last decade as social media has progressed into the homes of nearly 2 billion people around the world many lasting relationships have been formed, friendships have bonded and, wait for it, personal social interaction has seemingly declined. What is the point in meeting a friend face to face when we chat with them all night long if we choose to and we never have to leave the comfort of our own home?

2 Family Communication Plan for Parents and Kids

2.1 Simply solve family problem

For many families, eating together is a lost art. According to recent surveys, less than families sit down to a meal together on a regular basis. Typically, everyone grabs something to eat on their way to band or football practice, a soccer game, a play rehearsal or when coming home late from work or the gym. What's dinnertime like in your home? The importance of the event cannot be underestimated. The evening meal is a key time for families to reconnect, share the day's event, and continue deepening their knowledge of each other. Simply put, making time to eat dinner. How Social Media Kills Face-to-Face Conversation by Meg Lenzkes ..https://prezi.com/v6yz-wvpizfc/how-social-media-kills-face-to-face-conversation/Jun 6, 2017.

2.2 Example for that things that are being killed by the internet conversation

Today our thumbs get more exercise than our vocal chords. Like a premise for a dystopian novel, our generation has almost forgotten how to speak because we're too busy typing. We're guilty of texting instead of talking over coffee and Facebook chatting instead of conversing on the phone. It's as if we've sacrificed the art of conversation for the art of convenience. We opt for three-hour text conversations over face- to-face ones. Texting and Facebook chatting allow us to step away and take time to think and respond, while talking in person allows us to cultivate our skill of asking, listening, and responding in the moment.

3 Result and Solution

- If you are too busy for a family meal, you may too busy. Keep things simple. Plan ahead for a quick meal and be flexible if adjustments are needed at the last minute. It's worth it to get everyone together.
- Turn off the TV. A blaring television is distracting and discourages conversation. If you normally watch TV during dinner, decrease the habit slowly by having one or two TV-free meals everyday week. When your kids see you follow through, they'll begin to understand the dinner time is important.
- Don't slave the other stove. Elaborate meals are not a key ingredient for family fellowship.
- Involve the kids. Have your children help plan, prepare, and serve meals. This builds teamwork and cooperation. Give children "joy" jobs as setting the table, reading recipes, mixing and stirring food, and clearing the table.
- Show, don't tell. When you make nurturing family mealtimes a priority, your kids will notice and, someday, understand benefits.

3.1 Encouraging children to communicate

Good communication is important. It can help you and your children do well in work and school. Families who communicate get along better. There are many ways to communicate. One way to better communicate with your children is to spend more time with them. Isa group that helps families do things together. 4-H teaches us how to communicate better. Any children ages 5-9 can be a member of adults can volunteer their time to work with 4-H as leaders has lots of projects that families can do together. 4-H helps youth members learn to speaking public. Members can also learn communicate through writing. In members and adults leaders learns to work well with others. Members and leaders become part of a team. Being part of the team means members communicate with other children and adults in their area.

3.2 Learning to listen each other

Listening to your family shows that you care and that you think the other person is important. Listening takes some time and means you really have to understand what the other person is saying. Maybe you never be taught how to listen to someone else. Learning how to listen is easy. You must look at the person, concentrate or think about what is said, and repeat it. This shows that you heard what the person meant to say. This shows you care enough to listen to what your partner is saying. A good way to remember how to listen is: Look, Think, and Repeat. For example, when your partner says, "Work was really hard today," what is he/she saying? How can you show you are listening? Look, think, and repeat. First, look into your partner's eyes to show you are listening. Next, think about what your partner saying. Work was really hard today might means a lot of different things. It might mean someone at work was making it hard to get work done or someone angry or that your partner was too tired to get work done. Pick one, pick what you think it means and repeat it. Say, "Work was hard today because you were tired." Even if you are not right about your partner said, you tried. Your partner will either agree with what you heard or tell you something like, "I was angry because the supervisor made us re-do what we did last week."

3 Conclusion

Web technology itself hasn't undermined the foundations of the institution of the family. We've been doing that to families without it for centuries. What seems to be happening is that out online activity is actually reversing a worrying trend. So go on, post that video of Grandma on the Xbox to YouTube. She may not thank you for it now, because we trapped in new media more than any time ever. *www.telegraph.co.uk, Technology,Sep 4, 2009 - The internet has wrought huge changes on our lives*

Another way that the world of social media is killing you through the decreasing levels of self-esteem and overall confidence that can quickly occur because of the cyber-bullying and overall negativity that is directed toward people online. For example, you might decide to post a picture of yourself during a past beach vacation and automatically become bombarded by comments from your friends, coworkers and relatives about how unflattering you look in that particular picture *Learning to listen to one another NIcholas Becker, https://a9a1263f9caafb223a0e-rackcdn.com/images/pgmisc/pg-logo-800px-transparentbg.png.*

Even if they are joking, these types of negative comments cause people to become more self-conscious about their appearance to the point where they are skeptical about posting any other pictures of themselves. In more extreme cases, this major deficiency in self-image and overall confidence can force a person to live an introverted lifestyle, feeling that they would probably just be better off not exposing themselves to other people at all instead off just through online pictures. Studies have proven that a key reason why many people would rather post pictures of other people, viral images and other subject instead of pictures of themselves is primarily because they are ashamed of the way that they look and are afraid about how other people may react if they were to see those pictures *https://familylife.com.au/about-family-life/governance...role-of.../social-media-policy/ Family Life's social media policy is based on the Department of Justice Social Media.*

I wonder what would have happened if religious teaching would have just mass texted all his teachings, Facebook posted all his parables, and started a blog with pictures of all his miracles, before and after. "Follow me by putting in your email address!" I'm thinking it wouldn't have had the same impact as his face-to-face ministry.

References

1. "Author Nicholas Carr: The Web Shutters Focus, Rewires Brains | Magazine." Wired.com. 4 July 2010
2. "Boston Globe: Instant gratification is making us perpetually impatient, Feb 2013."
3. "Carr, Pinker, the shallows, and the nature-nurture canard: neutron culture". Scienceblogs.com. 20 September 2011.
4. "Children and Parents: Media Use and Attitudes Report"(PDF). Ofcom. 23 October 2012. February 2013.
5. "CNN: Online viewers ditch slow-loading video after 2seconds, November 2012".
6. "Digital power: exploring the effects of social media on children's spirituality".
7. "NPR Morning Edition: In Video-Streaming Rat Race, Fast is Never Fast Enough, October 2012".
8. "Patience is a Network Effect, by Nicholas Carr, November 2012"
9. "Science Daily: How Online Video Streaming Quality Affect Viewer Behavior, November 2012".
10. "The effects of the Internet: Fast forward." The Economist. July 2010
11. "Video Stream Quality Impact Viewer Behavior, by Krishnan and Sitaraman, ACM Internet Conference" (PDF). November 2012.
12. 2011, Neuron Culture : the shallows and nature-nurture.Sciencebloges.com
13. Cyber bullying: Bullying in the Digital Age.
14. Goldsmith, Belinda (31 march 2007). "Web news readers have greater attention span: study." Reuters. February 2013
15. Helen Briggs (11 January 2012). "Web addict have brain changes, research suggests". BBC News. 17 March 2013
16. Jessie Connell, *"Parent' Use of Technology and the Internet"* (University of Minnesota, 2012)
17. Q. Li, Journal of aggression, Maltreatment and trauma, 372 – 392 (2010)
18. Mary Madden, Sandra Cortesi, Urs Gasser, Amanda Lenhart, Maeve Duggan (20 November 2012. "Parents, Teens, and Online Privacy". Pew Research Center. 10 February 2013
19. G. Mesch, The Journal of Family Communication: 119 – 138 (2007)
20. Pinker, Steven (2010-06-10). "Mind over Mass Media." The New York Times. 4July 2010.

21. Rowan Pelling (22 May 2012)."How technology is taking hold of our children's lives". Telegraph. 26 May 2012
22. G. Sanford, M. H. Stephanie, P. Susie, Y. Amber, Advances in Consumer Research **29**, 66- 70 (2002)
23. Schmidt, Alex (12 October 2011). "Immigrant Parent Rely On Kids For Help Online." NPR. 10 February 2013.
24. M. Sussman, S. Suzanne, *Handbook of Marriage and the Family* (New York: Plenum Press, 1987)
25. The Patience Deficit, by Nicholas Carr, December 2013.
26. K. Yuan, W. Qin, G. Wang, F. Zeng, L. Zhao, "Microstructure Abnormalities in Adolescents with Internet Addiction Disorder". 6(6), 20708 (2013) .doi:10.1371/journal.pone.0020708. PMC 3108989. PMID 21677775.17 March 2013

The Theory of Interpretation in Solving Contemporary Legal Issues: With A Focus on the Instrument of Ijtihad

Abdulrahman M.A.Albelahi[1,*], *A.* Ali[1], *Faten* Mohmed[1], and *Metwally* Ali[1]

[1]School of Human Development and Technocomunication, Universiti Malaysia Perlis, Perlis, Malaysia

Abstract. Since the beginning, legal theory has concerned itself with the establishment of principles and precepts that govern the procedure of legal interpretation, from the initial stages of the judicial reasoning down to the promulgation of ruling and their implementation, Islam is a total way of life. Muslims are obliged to abide by the rules of Allah in every aspect of their lives, always and wherever they live. However, the actual rules of Allah as given in the Qur'an and the sunna are limited. The Qur'an contains only six hundred verses directly related to laws, and there are approximately two thousand hadiths. The function of interpretation is to discover the intention of the Lawmaker of the matter, therefore, interpretat primarily concerned with the discovery of that which is rot self-evident the objective of interpretation is to ascertain the intention c the Lawmaker with regard to what has been left unexpressed as a matter of necessary interference from the surrounding circumstances. Sometimes, the textual sources did not provide detailed guidelines in which to derive the law, and then the role of interpretation is important to determine the law. In Islamic law the role of Ijtihad undoubtedly important in order to meet new problems. But some of the Jurist contended that the role of Ijtihad had ended and we have to follow the rule that has been stated. An explanation given to this trend is that a point had been reached at which all essential question of law had been thoroughly discussed and further deliberation was deemed unnecessary. In Common law, man-made law and legislation are related to one another within a philosophy of law. Parliament makes law and it is the duty of the courts to give effect to them if properly enacted. While courts may rule that a particular statute or section is invalid for various reasons such as unconstitutionality, they cannot say, "We shall change this Act because it is not appropriate". That function belongs to Parliament (Wu Min Aun 1990: 120). So as in Islamic law, the Lawmaker is Allah S.w.t and the sacred text (Quran) is legislated due to His intention whereas Sunnah of the Prophet Muhammad is enacted due to the Prophet's intention. Therefore, Ulama of Usul Fiqh, in making any Ijtihad, they are du y bound to be guided by Quran and Sunnah.

1 Background Search

Islam, the fastest growing religion in the world, with over 1.2 billion followers, has the astounding capability of adapting to ever changing conditions. Recent international events have intensified interest among religious leaders, scholars and educational professionals about the way Shariah (Islamic law) is taught & adopted in Muslim society. The Quran has given the fundamental education and knowledge but the Muslim scholars must interpret this fundamental knowledge wisely in accordance with the spirit of the time in which we live .Shariah (Islamic law) is subject to interpretation in order to meet the ever-changing needs of society. Its guiding principles were designed to protect the individual and the society so that it was not established as a set of fixed rules. To respond to the changing needs of Muslim societies, muslim jurists and scholars have relied on the well-established process of change, ijtihad. This process is based not only on the holy book (Qur'an) and religious tradition (Sunna), but also on reason, deduction, and prioritization. Ijtihad is practiced by the jurisprudent to discover secondary divine legislation (laws). Ijtihad is defined in different ways.

The terms 'mujtahid' and 'ijtihad' are nowadays among those which have acquired great currency, even sanctity, among the Shi'ah. One would be surprised to know that the term ijtihad was formerly, from the times of the Prophet (S) and for several successive centuries, a Sunni term. It became Shia after undergoing a change of meaning, or what would be more precise to say, the term remained specifically Sunni for several centuries and became 'Muslim', in the wider sense, that is, after undergoing a change of meaning and dissociating itself from its earlier particular sense.

'Ijtihad' in the Sunni Tradition Sunni scholars narrate a hadith that the Prophet (S), while sending Mu'adh to Yemen, asked him as to on what he would base his judgment. "In accordance with the Book of Allah", replied Mu'adh, "But what if you don't find it there?" inquired the Prophet (S). "According to the Sunnah of the Apostle of Allah", replied Mu'adh. "But what if you don't find it there too?" asked the Prophet (S) again. 'I will exert my own opinion', replied Mu'adh. The

* Corresponding author: bkaifi@hotmail.com

Prophet (S) put his hand on Mu'adh's chest and said: "Thank God for assisting His Apostle with what he loves." They have narrated other traditions on the subject to the effect that either the Prophet (S) directly commanded his Companions to exercise ijtihad in case they could not

Find a rule in the Book and the Sunnah, or to the effect that he approved of the practice of his Companions that practiced ijtihad. To the Sunnis, this is something definite, confirmed by consensus (*ijma'*). About the Holy Prophet (S) himself, they have said that some of his injunctions were purely based on personal ijtihad not on revelation. Even in their works on jurisprudence (*'ilm al-'usul*) the problem is raised whether or not the Prophet (S) could make errors in his personal ijtihad. They have narrated traditions in this regard and transmitted reports of the Companions as to how they justified their own actions or those of others on the basis of ijtihad. We abstain from quoting any of them here for the sake of brevity.

1.1 Definition of Ijtihad

Ijtihad is defined literally as striving, exerting, self-endeavor or making an effort to deduce a principle and exerting oneself to the utmost. While legally it refers to the striving of the jurists to the point of mental exhaustion to derive principles and rules of law from evidence found in the sacred texts of sources (*Codd 1999: 115*
Mashood A. Baderin - 2017 - the Islamic legal Theory, volume 1.)

However, it has also been described as a rethinking *(Fatawa Qadi Han, (I-VI), Eqypt: Bulaq RAHMAN, Fazlur (1962). "Post-Formative Developments in Islam" in: Islamic Studies, Karachi, 1, 4 RIDA, M. Rashid (1324AH))* or most commonly, as independent reasoning.

It is an activity, a struggle and a process to discover the law from the texts and to apply it to the facts awaiting decisions. The jurist stay as close to the meaning of the texts as is possible. The purpose is to discover the true intent of the Lawmaker.
In jurisprudential sense, the meaning of Ijtihad is exercising of one's opinion or judgment. It means that in a case where the Divine dicta is absent or implicit, one should see what would be more acceptable to one's intelligence and taste, or nearer to truth and justice, or analogous to other Islamic laws and to adopt it for his judgment (Mutahhari 2002: 2).

1.2 Basic of ijtihad

Ijtihad is the most important source of Islamic law next to the Quran and Sunnah. The main difference between ijtihad and the revealed sources of the shariah lies in the fact that ijtihad is a continuous process of development whereas divine revelation and Prophetic legislation have discontinued with the demise of the Prophet.

The role of ijtihad had been recognized in the Quran and Sunnah for example in

Surah Al-Ankabut 69 ; *And those who strive for Us - We will surely guide them to Our ways. And indeed, Allah is with the doers of good.*

Which means, "And those whose strive in Our (cause) We will certainly guide them to Our Paths. For verily God is with those who do right". This ayat clearly allowed the foundation of the principle of Ijtihad to meet the growing of legal ruling in the future. It is because Quran only provides basic principles and few details legal provisions, but mostly it needs elucidation and elaboration of those matters. This process of inference is intended to meet the diverse needs of ever growing life in the developing society in the future. By this inference and logical deduction, then all future problem of human society could be solved by evolving guiding rules

Another surah which encourages to search a knowledge so as to arrive at sounh, decisions, can be seen in Surah An-Nisa' 8**3:** *And when there comes to them information about [public] security or fear, they spread it around. But if they had referred it back to the Messenger or to those of authority among them, then the ones who [can] draw correct conclusions from it would have known about it. And if not for the favor of Allah upon you and His mercy, you would have followed Satan, except for a few.*

Which means, "And when there comes to them tidings of safety or fear, they spread all around, whereas if they had referred it to the Prophet and to such other persons as are in authority those among them who can search out of knowledge of it, would have known it".
The term used in this verse is 'to such other persons as are in authority' or ulul amri in Arabic means that men who are in authority, or who are in a position to search out the knowledge. The Quran, while thus encouraging the process of ijtihad, indicates that the function should be carried out by the Prophet and by ulul amr or men in authority, power, wisdom and understanding.
The famous and well-known hadith, where Muaz bin Jabal has been appointed as the Governor of Yemen; The Prophet asked him:

"0 Muaz, by what rule would you be guided?" He replied, "By the Law of the Quran". "But if you find no direction therein?" asked the Prophet. "Then I will act according to the Sunnah of the Prophet", replied Muaz "But what if that also does not help?" he was again asked. "Then I will exercise my judgment and act accordingly" He replied it. The Prophet raised his hands in
satisfaction and thankfulness to God and uttered, 'Praise be to Allah, who guides the Messenger of His Apostle as He pleases".

This hadis shows the approval of the principle of ijtihad by the Prophet. This is involving the exercise of one's own judgment when there are no specific directions in the Quran or the Sunnah. It further shows that during the Prophet time, the companions were aware of the principle of ijtihad and it was freely resorted to when necessary in order to meet new circumstances. Subsequently, companions followed the principle of ijtihad in order to meet the changes of time and society and it was proved to be successful.

2 Procedure of Ijtihad

Since ijtihad occurs in a various forms such as Quran, Sunnah, qiyas, istihsan, maslahah, custom (urf), etc, each of this is regulated by its own rules. There is, in other words, no uniform procedure for ijtihad as such.

From the viewpoint of the procedure that it employs, ijtihad may occur in any of the following four varieties. First in the form of a juridical analogy, this is found on an effective course. The second variety of ijtihad consists of a probability without the presence of any effective course, such as practicing ijtihad in regards to ascertaining the time of solat or the direction of qiblah. The third type of ijtihad consists of the interpretation of the source materials and the deduction of ahkam from existing evidence. This type of ijtihad is called explanatory ijtihad which takes priority over the analogical ijtihad. The last type of ijtihad referred to as ijtihad istislahi which is based on maslahah and seeks to deduce the ahkam in pursuance of the shariah, which may take the form of istislah, juristic preference, the obstruction of means, etc. Imam Shafii however, accepts only the first type, namely analogical ijtihad but to the majority of ulama, ijtihad is not confined to analogical deduction and may take the form of any of the foregoing varieties *(Kassab 1984: 24 THE CONCEPT OF IJTIHAD IN THE HISTORY OF ISLAMIC JURISPRUDENCE)*

As far as sources of law are concerned, Ouran ranks in priority and followed by the Sunnah. When the matters are silence both under the revealed sources of law then the rule of interpretation plays an important role. Accordingly, ijtihad is also accounted as one of the sources of Islamic legislation, like the Ouran and Sunnah, although not as a source parallel to these two. As long as a rule is to be found in the Ouran and the Sunnah, the need for ijtihad does not arise.

3 The Need and Justification of Ijtihad in Modern Times

The role of ijtihad cannot be denied due to changes in the society and circumstances. A new role has to be formulated in order to meet a modern era in which the need and justification of ijtihad in the modern times is very important.

Mohd Abduh had advocated the need for fresh reinterpretation of the principles found in the divine revelation as the basis of the legal reforms. He also argued the ijtihad was not only the right of modern day jurists but the only way to which Islam could adapt itself to the needs of today's society (Glassse, *The Concise Encyclopaedia of Islam,* 1999: 121).

The role of ijtihad can be seen during the time of Caliph Umar and Ali practiced the ijtihad even in the presence of a clear Ouranic and Sunnah injunctions. So, in the modern times, ijtihad is used in a largely peripheral manner and cannot be used in any other way unless there is a complete overhaul of the legal system. Therefore, to be reused properly within a new Islamic framework to effect modern legal reform, then ijtihad will have to be modified.

In order to implement ijtihad in the modern time, then it must be freed from the limitation imposed on it previously ("Al-Ghazali employed dialectical methods in presenting his theology and also claimed for himself the right of independent *Ijtihad* which had been curtailed by the orthodoxy" (Gatje, *The Qur'an and its exegesis,* 1999: 129). The example of the limitation is that the requirement of the mujtahid itself, which is quite impossible, and hardly be found especially in the modern times.

4 Qualifications of Mujtahid

Interpretation will be reliable only of those who engage themselves in inquiry and investigation, having deep knowledge and insight in legal matters supported by technical expertise in determining the meaning, import and applicability of a particular injunction. The interpreters must have an ability to understand worldly affairs as well as public policy. This can be referred to Surah Aanbyaa Ayat 78-79. *We bestowed the same favour upon David and Solomon. Recall, when they gave judgement regarding a tillage into which the sheep of some people had strayed at night, and We were witnesses to their judgement. We guided Solomon to the right verdict, and We had granted each of them[70] judgement and knowledge. We made the mountains and the birds celebrate the praise of Allah[71] with David. It was We Who did all this.* So, it is clear from the above statement that not everybody is qualified to form a valid ijtihad but only certain people who had fulfilled all the requirements to be a mujtahid.
Weiss in his article said that "if every man was a mujtahid, there would be no law in the ordinary sense; every man would be law unto himself. The implications, which this would have for society, can hardly be comprehended. Therefore, it is in interest of social stability that not all choose to acquire the qualifications which are prerequisites of ijtihad" (AI-Amini IJTIHAD - Interpreting Islamic Principles1986: 207).The mujtahid should have knowledge of the Quran and Hadith, he should be able to distinguish the nasikh from the mansukh, the amm from the khass, the mujmal from the mubayyan and the muhkam from the mutashabih. He should be familiar with the Quranic vocabulary and terminology, know the circumstances in which a particular verse was revealed and have knowledge of the successive generations of narrators and transmitters of hadith. He should also be able to reconcile the apparently conflicting traditions. In addition to all that, he should practice ijtihad and exercise his personal judgment in particular case.

It seems at these qualifications in certain aspects are very good, where they are very experts in Islamic jurisprudence and Islamic laws and they can answer all the questions raised to them. But on the other hand, these qualifications hardly to be achieve at the present times. But, indeed, they should be able to know all the related fields of the Quran and Sunnah.

According to al-Shawkaniyy, one who attempted to conduct ijtihad had to acquire a certain degree of knowledge and other requirements:

- Who master all requirements for general ijtihad.
- Who are self-sufficient to understand the sources.
- Who know the reference of the shariah.
- Who specialized in a certain branch of non-religious sciences yet attempting to understand the shariah. .

Few the actual conditions, some ulama has aptly observed that. Qualifications required of a mujtahid would seem to be extremely moderate and I there can be no warrant for supposing that men of the present day are unfitled to.

He had made relatively easy and possible the requirement for mujtahid since he is the most radical among his contemporary informers towards the reopen "the closing of the door of ijtihad". Some Maliki, Hanbali and Zahiri ulama also held the view that when a person is learned in a particular area of the shariah he may practice ijtihad in that area only. This would be no way violating any of the accepted principles of ijtihad (Badran Can Faith-Based NGOs Advance Interfaith Reconciliation? 1984: 486).

Some observers have suggested that the practice of ijtihad WC\s abandoned partly because the qualifications required for its practice w re made so immaculate and rigorous and were set so high that they were humanly impossible of fulfillment (Kassab 1984: 38 Previous source). This is however a implausible supposition which has been advanced mainly by the proponents of taqlid with a view to discourage the practice of ijtihad. Acquire such qualifications

5 Instruments of Ijtihad

5.1 In the Form of Juridical Analogy (Qiyas)

If new cases occurred which were not provided in the Quran, the Sunnah or the consensus of opinion, the jurists found them compelled, in seeking solutions to have recourse to reason, logic and opinion for example, applying Qiyas. However, they were not entirely free or unrestrained for they had to adhere to scientific rules and premise which they classified in a new category known as analogy. The jurists derived this new source from the fundamental judicial premise that all rules are based upon objectives and interest and those objectives and interests are the causes for the rules. Hence from the rules they deduced their causes. Having done so in reference to any particular problem, they were able to apply the same rule to another problem whenever the cause for both was identical (Ziadeh, A Study of a Revivalist Movement in Islam 1987:79).

A case in point is the prohibition of seeking marriage with a woman already engaged to another person. The prohibition is derived from the hadith of Prophet; "Do not ask for a woman in marriage when another Muslim has already done so". The ratio decided of this prohibition lies in the fact that this would violate the brotherhood and unity among fellow Muslims. This rule can simply be applied to other similar contract such as sale of property, hire purchase, rent, etc.

In The Form of Interpretation of the Source Materials al *Quran*

The Quran is the first and important sources of Islamic law. If the Quran proves something, it is considered binding. The related Quranic verse (And We sent not before you except men to whom We revealed [Our message]. So ask the people of the message if you do not know. Al-Nahl: 43) provides the authority to pursue ijtihad and contemplated of the existence of learned man at all times. As far as Quran is concerned, its authenticity is unquestioned. What is expressed in the Quran is the expression of the absolute authority of God. But it is fairly clear that the Quran, which is essentially a religiomoral scripture, does not provide comprehensive provision for the formation and the development of the shariah. The Quran only supplied elementary rules for few cases (Hidayatullah The Origins of Islamic Law - Constitutional Rights Foundation 1975: 75).

So we may say that Quran only provides general principles and only a few of normative verses, which, of course, need to be interpreted especially for the changes of the society of the modern times. In other words, it must be interpreted in the light of modern ages. Therefore the Quran contains a significant number of verses with legal bearing. It is these verses, which provide Muslims raw material out of which legal elaboration, were made and around which their legal speculation remained focused. For example, the Quranic verse, which needs further elaboration as follows: *"Prohibited to you (for marriage) your mothers, daughters". Prohibited to you [for marriage] are your mothers, your daughters, your sisters, your father's sisters, your mother's sisters, your brother's daughters, your sister's daughters, your [milk] mothers who nursed you, your sisters through nursing, your wives' mothers, and your step-daughters under your guardianship [born] of your wives unto whom you have gone in. But if you have not gone in unto them, there is no sin upon you. And [also prohibited are] the wives of your sons who are from your [own] loins, and that you take [in marriage] two sisters simultaneously, except for what has already occurred. Indeed, Allah is ever Forgiving and Merciful.Al-Nisa': 23.* It is unanimously agreed that marriage with one's mother and daughter is prohibited. However, the word "daughter" could be taken for its literal meaning which would be a female child born to a person either through marriage or zina. Judicially, the word daughter can only mean a legitimate daughter. Jurists had exercised ijtihad to come to the own conclusion. For example, Hanafi stated that this prohibition applies to both legitimate and illegitimate daughters, whereas Shafii opined that this prohibition only applies to legitimate daughters. Even the child born out of wedlock, but the father of the child stills the same man and considers the father of the child then he cannot be permitted to marry his own child.

5.1.1 Sunnah

Traditional theory is that Sunnah confirms the Quran, assists in its interpretation or fill in the lacunae. The Sunnah is the second most important source of Islamic law. Besides getting proof of continuous ijtihad in hadith, the Sunnah itself

According to AI-Faruqi needs ijtihad. For example, where the rulings of Sunnah were changed in order to reflects the changes of circumstances. The Prophet had granted the right to attend the mosque for congregational prayer. Due to change of circumstances, however, Aishah later changed this ruling as she concluded "had the Messenger of God observed what was happening to woman, he would have forbidden them from attending the mosque" (AI-Ghazaliyy: 48).

5.2 In the form of a probability without the presence of any effective course

5.2.1 Custom

Custom is defined by some jurists as recurring practices, which are acceptable to the people of sound nature. Thus, the legal theory of Islam, on the whole, does not recognize custom as an independent or formal source of law. But the ulama accepted the authority of general usage, satisfying the juristic standards, for the purposes of interpretation and application of legal rules dealing with worldly affairs. They admitted a limited contribution of custom in the growth of law. The mainspring of this growth was however, the interpreter (mujtahid) (Santillana 1931: 307).

No custom; whatever is the degree of its utility therefore, became law i any judicial sense unless it was authoritatively recognized. In order to constitute a valid basis for legal decisions, customs must be sound, and reasonable. It has been accepted as a binding authority and been accepted in order to solve contemporary problems. In Malaysia, the most popstar. Custom, which has been recognized and enacted by Islamic law via the ins Rent of ijtihad, is jointly acquired property. It was defined as property acquire during the subsistence of a marriage by a husband and wife out of their joint res6urces or by their joint effort (Mat 1985: 92). A claim over the jointly acquired property often arises from divorce or from the death of either husband or wife. It is clear cut evident that there is no specific provision under Islamic law concerning the said property and yet, the claim of jointly acquired property has been practiced in courts and judicial decisions on the claim are mainly based on custom.

It is to be noted that most of the states in Malaysia have special provisions concerning JAP. For instance, section 58 of the Selangor Islamic Family Law Enactment No. 4 of 1984 provides to order the division between the parties (husband and wife) of any assets acquired by them during the marriage by their joint efforts.

In the case of Hajah Lijah bt. Jamal V. Fatimah bt. Mat Diah, (1956) 16 MLJ 63, Briggs J, says that since there is a clear recognition on jointly acquired property in many states and courts, thus, any regulation concerning jointly acquired property is to be part of the personal law in the state. Tun Salleh Abbas in his judgment in the case of Boto' bt. Taha v. Jaafar bin Muhamed, (1985) 2 MLJ 98 also says that jointly acquired property is a custom practiced by the Malays Muslim. Thus, since custom has been recognized as a basis of law, ijtihad has been made in determining the validity of jointly acquired property under Islamic law in Malaysia.

5.3 In the form deducting hokum in pursuance of the purpose of shariah

5.3.1 Maslahah

According to AI-Ghazali, maslahah consists of consideration which secure a benefit and prevent a harm but harmonious with the objective of shariah, that is to protect five values; Religion, life, lineage, intellect and property. The usefulness of this principle of law in contemporary issues is beyond doubt since both Quran and Sunnah do not textually cover current issues.

The Islamic Religious Council of Johore, issued a fatwa, effective as of 13 November 2001, which requires mandatory HIV/AIDS blood test for prospective Muslim couples prior to registration of marriage. Reports indicate that the fatwa was under consideration over a period of months prior to its eventual approval and implementation. Upon announcing its proposed implementation, the Menteri Besar of Johor made a brief statement in which he said that the fatwa was taken in accordance with the principle of syara' in Islam, which safeguards religion, life, property, intellect and lineage. Fatwa can become the binding law if the Islamic Religious Council submits it to the legislative authority for ratification, either by
Parliament or the state legislature as may be the case (Karnali 2001: 116).

The shariah doctrine of maslahah entitles the political authority and those in charge of the community affairs, the ulul amr to rn, decisions, enact legislation and issue fatwa, which are designed to secure the public welfare. It is the Islamic duty of the government to promote and advance the maslahah of the people. maslahah as such is an important goal and hilosophy of Islamic government.

5.3.2 Istihsan

The jurist who resorts to istihsan may find the law to be either too general or too specific and inflexible. In both cases, istihsan may offer a means of avoiding hardship and solution, which is harmonious with the higher objectives of the shariah. For example, the ruling of

Saydina Umar, where during famine he was not enforce a hadd penalty of the amputation of the hand for theft. Saydina Umar had departed from existing law in order to preserve public interest and justice.

Istihsan enable a choice between alternative solutions, it contemplates the relatives merit, and demerits of each of the alternatives. In this sense, istihsan offers considerable potential for innovation and for imaginative solutions to legal problems. The question is istihsan is not merely to find a solution to a particular problem but to find a better solution, which is already, exists.

Generally speaking, the scientific and medical advancement primarily aim at the preservation of some benefit in Islam especially in the area of biomedical technology. For example, the donation of human body organs in its contemporary state are not textually covered by Quran and Sunnah and is the new issue which the ancient jurists not dealt with. It is agreed among the jurists that the basic rule is the forbidding of the utilization and exploitation of the parts of the human body, either because of its dignity or value or because no legitimate use can be made of the parts. This prohibits the donations of human body organs. This is based on Surah An-Nisa' 29-30, Allah says:

O you who have believed, do not consume one another's wealth unjustly but only [in lawful] business by mutual consent. And do not kill yourselves [or one another]. Indeed, Allah is to you ever Merciful. And whoever does that in aggression and injustice - then We will drive him into a Fire. And that, for Allah, is [always] easy.**Which means, "Nor kill (destroy) yourselves, for verily God had been to you most merciful".**

The Prophet also said :
"The breaking of the dead's person bone is just like the breaking of his bone while alive".
On the other hand, as a result of the advancement in this medical issue, there is need for a new ruling. The recognition of the jurists right to exercise istihsan in donation of human bod organs as a result of scientific advancement requires departure from the principle by allowing the donation to prevail over such prohibition. Therefore, based on the right of the jurists to exercise istihsan, that is to choose the least of the two harms in order to reject the worst of the two, the donation of human organ is permitted.

6 Fatwa Making Procedure in Malaysia

Fatwa means a response given by a qualified person who expounds the ruling 9 shariah on a particular issue that is put to him by a person or a group of persons. Fatwa also can be defined as a formal legal opinion given by a Mufti in an answer to a question submitted to him either by a judge or private individual (Ishak 1981: 9).

Fatwa is often used synonymously with ijtihad, both of which involve intellectual effort of a qualified scholar to provide shariah-based responses to real issues. Fatwa or ijtihad that is based on interpretation and personal opinion of the mufti is normally not binding on anyone unless and until it is published in the States's gazette. The important element in fatwa is that it always involves a question and an answer to a new case, but if a hukum is issued without any query, it is no called fatwa but a mere ijtihad.

In Malaysia, Fatwa Committee of every state is established under the respective States' Administration of Islamic Law enactments. The Committee shall consists of the Mufti, Deputy

Mufti, two members of the Majlis nominated by the Majlis, not less than two fit and proper persons to be appointed by the Majlis and an' officer of the Islamic Religious Department of the particular states to be appointed by the Majlis, who shall be the secretary.
Whenever the mufti proposes to make a fatwa he shall call for a meeting of the Committee for the purpose of discussing the proposed fatwa. Before the mufti makes fatwa, he may cause such studies or research to be conducted as he may direct and a working paper prepared.

In issuing any fatwa or certifying any opinion, the mufti shall ordinarily follow the accepted views of mazhab Shafii. However, if the mufti considers that following the view of the mazhab Shafii will lead to a situation, which is repugnant to public interest, the mufti may follow the accepted views of mazhab Hanafi, Maliki or Hanbali. For example, the use of local currency to replace the use of rice in zakat fitrah payment is in accordance with the opinion of mazhab Hanafi (Shuaib 2002 :273).

The mufti shall make and publish in the Gazette, a fatwa on any unsettled question relating to Isla ic law on the direction of the Yang di Pertuan Agong or maybe he do so on his own initiative or on the request made by letter addressed to a mufti. Upon publication in the Gazette, a fattwa shall be binding on every muslim residents in the particular state as dictate 0 this religion and it shall be the muslims' duty to up little unless he is permittee by Islamic law to depart from the fatwa because of personal observance, belief or opinion.
A fatwa shall also be recognized by all syariah courts im the particular states as authoritative of all matters laid down but not civil courts. However, once the civil courts call it, it should be observed and respected.

7 Problem

There are certain qualifications to be fulfilled before the person can be called as mujtahid. For example, he must possess t e knowledge of the sources of the shariah, knowledge in Arabic. The problem faced at present is that despite the door of ijtihad having been declares widely open, but still we fail to detect any movement towards the regeneration of ijtihad. There are opinions also which stated that qualification of mujtahid is heavy and exacting. According to some scholar, this is another taqlid-oriented assertion by those who wished to bring ijtihad to be closed.

Free development of the law was already existed. Rigidity was greatly increased by the development of the doctrine of the closing of the gate of independent reasoning. The jurists were restricted to the interpretation of the authoritative manuals of the schools, which generally were books written not by ancient masters of the schools but by the comparatively late scholars. In theory, therefore Islamic jurists since then have been mere imitators and copyists. In practice, of course, the development of the law has not stood still although it was, no doubt, slowed (Liebesny 1964: 18).

8 Suggestion

The contention made is not true since the qualifications were not excessive and were fulfilled frequently. Furthermore, the uncertainties surrounding ijtihad have in modern times been exacerbated by the spread of secularism and the fact that the state has become the sole law making authority in its own territorial domain. Assuming that mujtahid has been recognized, and then it should be possible to devise a procedure, which would integrate ijtihad in the legislation processes of Government.

Universities and legal professions in many Muslim countries are engaged in training lawyers and barristers in modern law streams, then these are their duties to institute an effective program of training for the prospective mujtahid. Abdul Wahab Khallaf suggests that the government in every Muslim country should specify certain conditions for attaining to the rank of mujtahid and this contingent upon obtaining a recognized certificate, which would enable the government to identify the mujtahid and to verify their views when the occasion so requires (Khallaf 1978 :49-50).

The view for collective ijtihad is mainly supported by Mohd Iqbal (1982: 174) in his work "Reconstruction of Religion Thought in Islam" that the power to carry out ijtihad and ijma' should be rested in the Muslim legislative assembly. There are suggestion also in support of the institutionalization of both ijma' and ijtihad within the fabric of modern government. Ijtihad in the past has been related to the juristic concept and remain preserving of the mujtahid. So ijtihad in the sense of self-exertion is a method of finding solutions to new issues in light of the guidance of wahyu. Scholars of shariah as well as experts in other disciplines must exercise so wider propositions and they must acquire mastery of the ruler data, especially in the Quran and Sunnah pertaining to his subject According to Coulson (Layish 1987: 225) there can be no doubt that the review of the existing commercial laws imported from the west in the light of the basis principles of the textual sources will eventually lead to a genuine Islamic law of contracts and commerce that will give shariah legitimacy to the principle of contractual freedom.

The enactment of modern codes and statutes is not enough, so this is the task of the court to adjust the new enactment to practical needs. If the courts, as has been the case occasionally, are too literal and strict in their interpretation and allow formalism to stifle and search for truth and the equitable application of legal rules, the law is likely to remain alien and often bewildering to the people at large .In order to serve rapidly changing social, economic and political conditions and new systems will, however also have to preserve flexibility.

References

1. Al-Ghazaliyy, Abu Hamid. Ihya Ulum ai-Din. 2nd Edition. Cairo: Darul Fikr. Khallaf. Abdul Wahhab, *Ilmu Usul al-Fiqh* (Kuwait: Darul Qalam, 1978)
2. F. Al-Attas, Islamic Perspective Journal **1**(2), 7–16 (2009)
3. M. T. Amini, *Fundamentals of Ijtihad* (Delhi, 1986)
4. Badran, Abu Al-Ayrayn Badran, *Usul al-Fiqh al-Islami* (Alexandria: Muassasah Shihab al-Jamiah, 1982)
5. Codd, A Critical Analysis of The Role of ljtihad in Legal Reforms in the Muslim World. Arab Law Quarterly, Kluwer Law International **14** (1999)
6. W. B. Hallaq, Journal of Studies Islamica **24** (63), 129–141, (1986)
7. C. Malik, *God and Man in Contemporary Islamic Thought* (Centennial Publishers, Beirut, 1972)
8. W. B. Hallaq, Journal of Middle East Studies 16(1), 3–41 (1984)
9. M. Hidayatullah, The Role of the Quran in the Development of the Shariah, Islam and the Modern Age (1975)
10. M. Iqbal, *Reconstruction of Religious Thought in Islam* (Ashraf Printing Press, Lahore, Pakistan, 1982)
11. M. Iqbal, *The Reconstruction of Religious Thought in Islam* (Lahore: Sh Muhammad Asyraf, 1982)
12. Ishak, Othman. 1981. Fatwa Dalam Perundangan Islam. Kuala Lumpur: Fajar Bakti Sdn. Bhd.
13. G. N. Jalbani, Teaching of ShahWaliyullah of Delhi (Ashraf Printing Press, Lahore, Pakistan, 1967)
14. H. Kamali, IIUM Law Journal **9**(2), (2001)
15. A. A. L. Kassab, *Adwa Hawl Qadiyyah Al-ljtihad Fi AlShariah Al-Islamiyyah* (Cairo: Dar Al-Tawfiq, 1982)
16. A. Layish, Reforms In Islamic Commercial Law-Towards A Reopening Of The Gate To ljtihad, Asean And African Studies 21 (1987)
17. B. Lewis, The Muslim Discovery of Europe (London Redwood Burn Ltd. Publishers, UK, 1982)
18. J. Herbert Liebesny, *Stability and Challenge in Islamic Law* (Middle East Institute, 1964)
19. M. K. Mas'ud, *Iqbal's reconstruction of Ijtihad* (Iqbal Academy Publishers, Lahore, Pakistan, 2003)
20. I. Mat, Adat and Islam in Malaysia: A Study On Legal Conflict and Resolution, Temple University Press. Santillana. 1931. Law and Society in the Legacy Of Islam (1985)

21. M. M. Mutahhari, The Role of Ijtihad in Legislation. Translated from the Persian by Mahliqa Qara'l, **4**(2), (2002)
22. F. Rahman, *Post Formative Developments in Islam* (Karachi: Islamic Studies, 1962)
23. A. Rahnema, *Pioneers of Islamic Revival* (Zed Books Ltd Publishers, London, UK edition, 1995)
24. B. Weiss, The American Journal of Comparative Law **26**(2), 199–212 (1978)
25. B. Weiss, The American Journal Of Comparative La **26**, (1978)
26. Wu Min Aun. *The Malaysian Legal System* (Longman Malaysia Sdn. Bhd., 1990)
27. R. Zia, R.:Transmission of Values in Muslim Countries: Religious Education and Moral Development in School Curricula. In CERC Studies in Comparative Education, Springer Netherlands **18**(3), 119–134 (2007)
28. J. Ziadeh, The Philosophy of Jurisprudence. H IZBI (1987)

Cross Ethnic Friendship among Multi-ethnic Students and Teacher's Role in Supporting Cultural Diversity in School

Yasmin Ahmad[1,*], *Mohd Kasturi Nor* Abd Aziz[1], *Suzana* Sulaiman[1], *Suhana* Abdul Mutalib[1] and *Nurul Naimah* Rose[1]

[1]School of Human Development and Technocomunication, Universiti Malaysia Perlis, Perlis, Malaysia

Abstract. This article presents an in-depth discussion about cross ethnic friendship among students and teacher's role in supporting the cultural diversity that exists in school. The school which consist of students from various ethnic groups provide space and opportunities for students to interact socially with peers either from the same or other ethnic groups. On the other hand, the school that consists of only one ethnic group limits the opportunity for students to interact with friends from different ethnic groups. Students who have attended the schools that are not diverse in terms of ethnicity were reported having more friends from the same ethnic group. A positive relationship between individuals from different ethnic groups led to the reduction in prejudice, enhance the sense of common identity and closeness among individuals. Teachers as agents of unity should play an important role in assisting students to acquire the necessary social skills that enables them to interact effectively with students from different ethnic, cultural and languages which consequently create a harmony cross ethnic friendships among multi-ethnic students in school.

1 Introduction

Malaysia is a country that consists of a multi-ethnic society, culture, way of life and history [1]. In every society, the education system is closely related to the needs of society due to the symbiotic relationship that exists between education and society. The education system cannot ignore the political, economic and cultural ideologies that shape their environment. In Malaysia, since independence, one of the objectives is national unity. Therefore, all educational policies have outlined the unity as the most important objective [2].

Razak Report in 1956 is the basis of the formation of national policies which related to education such as the Rahman Talib Report. The education system in Malaysia reflects the diversity of cultural and linguistic, especially at the primary level. Under the Education Act 1961 and the Education Act 1996, there are several streams using different medium of instruction across the primary, secondary and higher education level [3]. Students can learn their native language, while the same curriculum and syllabus were used to build national identity and the spirit of patriotism. In addition, Malay Language is compulsory in SJKC (National Chinese School) and SJKT (National Tamil School), while English is compulsory for all. At the secondary school level, most of the students from National Primary School, SJKC and SJKT will proceed to National Secondary School. But, there are also Chinese independent schools and Islamic religious schools that will attract small number of students [3. Multiple streams at the primary school level allow the evacuation of linguistic and cultural from Malaysia's three main communities, the fact is students from different ethnic groups tend to remain separate until the secondary schools and higher education level [3].

The government have implemented a variety of approaches to promote national unity and patriotism among the diverse ethnic groups in Malaysia [1]. However, the issue of racism that threatens the unity and harmony of the country still occurred [1]. Every ethnic group would protect their rights and privileges, thus creating the social boundaries. These boundaries can lead to ethnic tensions and conflicts between ethnic groups [2]. It is important to emphasize the value of ethnic diversity and national unity. Students from various ethnic backgrounds should know and appreciate the attitudes, institutions and traditions shared by all Malaysians [4].

2 Interethnic Groups Relation

The relations which have occurred among students from the same or diverse ethnic groups will influence the development of ethnic identity of an individual [5-6]. Theory of intergroup contact suggested that under a certain circumstances, an individual who have more

* Corresponding author: yasminahmad@unimap.edu.my

intergroup contact, exhibit a positive traits against friend from out group [7].

Study has shown that a positive relationship between individuals from different ethnic groups lead to a reduction of prejudice [8], increase the sense of common identity [9] and closeness between individuals [10]. Pettigrew and Tropp (2006) reported that a relationship between groups has a positive effect in the reduction of prejudice against a whole group of a friends and not just specific to a friend who has a relationship with an individual.

In one study that examined the relationship between intergroup contact and ethnic's identity, it was discovered that Koreans who lives in China and have more social interaction with Chinese society have a low levels of ethnic identity because discrimination often occurs [11].

Based on the context of relations between ethnic groups in Malaysia, ethnic relations are very complex, complicated and sensitive, especially when it comes to the issues of religion, culture and language which plays as an essential features in the competitions of identity based on a phenomenon that was formed by the British in the context of colonialism [12].

The dynamics relationship between group determine the changes in the construction of one's identity and way of life of an individuals who are living in a multicultural society and how they open and close with each other from social and cultural ethnic perspectives [13]. When individuals have an interaction, they cannot ignore in-group members, which is what is specific, distinctive and unique to the culture of their own ethnic group, but also a similar feature to outside groups [13].

3 Cross Ethnic Friendship

A closeness of cross ethnic friendship will reduce prejudice between ethnic groups [14-18]. According to Yusof (2006), cross ethnic friendships are more common in schools with students from various ethnic groups. Most of the research on cross ethnic friendship have explained about an effect of closeness in cross ethnic friendships [17]. Research conducted in four European countries have shown that the majority ethnic who build a cross ethnic friendships with minorities exhibit a reduction in bias against ethnic minorities [17].

Based on the study of the relationship between different ethnic groups, it consistently shows that the relations between ethnic groups can be positively managed to reduce prejudice against friends from different ethnic groups [7]. For example, respondents from ethnic majority in Germany, France, Netherlands and United Kingdom who have a good friends from minority ethnic groups were reported not only have lower prejudice against their friends but also diminish in prejudice against other minority ethnic groups as well [17]. Based on a study conducted on more than 1,000 senior students at the University of Michigan, it was found that seniors who have more contact with students from different ethnic groups, have more close friends from other ethnic groups and those who have more interaction with colleagues from other ethnic groups were reported to have more intellectual engagement, critical thinking and support the diversity exist among them [19].

Cross ethnic friendship is a complex process but certainly it would be benefit for the students in a positive way. Students could develop more positive attitudes toward intergroup interaction such as they would be more comfortable to mix around with friends of different ethnic background. It also develops a strong sense of empathy and caring for others [20]. This would create a harmonious and balance social interaction whereby everyone is included in the group. It will lead to low the level of conflict among students. In terms of psychological aspect, students would feel safe and less victimized by peers [21]. According to a study conducted by Hamm (2000), he found that by having a cross ethnic friendship, it also have been associated with having high scores on the importance and positive perception to any other ethnic group.

Research on friend selection among students in the secondary schools that contained of various ethnic students have shown that friendship among students were from the same ethnic group [22-23]. The diversity of peers based on ethnicity may be less important to the students whether in the minority or majority ethnic group from a residential area consisting of a multi-ethnic society [19]. A friendship pattern among students who have attend a school consisting of many students from the same ethnic group show a tendency to have a friend from the same ethnic group [22].

Theory on intergroup contact suggested that individuals should have the opportunity to interact with someone from other ethnic groups in everyday relationships [24]. Many studies have proved the importance of an opportunity to interact in the context of macro variables such as ethnic diversity [25].

The school consists of students from various ethnic group provide an opportunities for students to interact with friend from the same ethnic and also different ethnic background [24]. On the other hand, by attending schools consisting of student from the same ethnic group, students reported having more friends from the same ethnic group [22]. Schools that have a students which are less diverse in term of ethnicity, however only provide opportunities for students to interact with peers from same or different ethnic but not both in the same time [24].

4 Multi-ethnic School Environment

Multi-ethnic school environment in Malaysia is like in National School and National Secondary School. This type of school usually consist a students from various ethnic groups such as Malay, Chinese, Indian and other minority ethnic groups. The school consists of students from various ethnic groups provide an opportunity for students to build a friendships with peers from different ethnic background, unlike school attended by only one ethnic group where students do not interact with peers from different ethnic backgrounds [26]. Exposure to diversity in the classroom and cross ethnic friendship can lead to social benefits in the future [18]. Theory of intergroup contact suggested that individuals should have an opportunity to interact with their peers from other ethnic groups in their daily interactions. Many scholars have discussed the importance of an opportunity to have a relationship in the context of macro variables such as a diversity [25].

Schools which are less diverse in terms of ethnicity, however, only provide opportunities for students to interact with the same or different ethnicity but not both in the same time [24]. According to a study conducted by Yip, Seaton and Sellers (2010), although there are opportunities to interact with peers from the same ethnic or different ethnic background, students actually does not interact based on an opportunity of the diversity. One of the factors was the barriers from the institution, e.g. class arrangement that do not provide opportunities for students from different levels of academic achievement to interact [24, 27].

Studies of Dutch adolescent without influenced by attending a school that consist of a students from diverse ethnic background whereby at least half of the student population is non-Dutch have discovered that 16% of them were reported having no contact with peers from non-Dutch [28]. Outside of the school, where students are expected to have more opportunities whether to get involved or not in the intergroup contact, it was reported that 45% of Dutch students have no inter-ethnic relation or only happened at minimum levels [28].

A study which examines the ethnic diversity in schools and a choice of friend among students in secondary schools have found that the higher proportion of students based on ethnicity, students will be more inclined to build a friendships with peers from the same ethnic group [22-23].

The study of Black and White students who attending the integration school have shown that most of the students were reported having a close friend from other ethnic groups [29].

5 Cross Ethnic Friendships among Schools' Students in Malaysia

Country such as Malaysia has a diversity of ethnics resulted from period of colonial history. Therefore, early exposure to cultural differences is vital in order to build a positive attitude towards other ethnic group. A positive attitude is a strong foundation to strengthen the relationship among multi ethnic society. The early exposure of cultural difference begins from school particularly in primary school. Students from various ethnic backgrounds such as Malays, Chinese, Indians and others would form cross ethnic friendships through interaction during the learning process. Most common factors to form a friendship are based on similarity and availability [21].

Despite the positive benefits of cross ethnic friendship, school students prefer to be in the same ethnic group [4]. Similarly, the friendship among students from multi-ethnic groups in Malaysia are still inclined with a friend from the same ethnic group. This situation exists because most of students e.g. Malays, Chinese and Indians students at the primary school level have attended schools mostly based on their ethnicities. Ethnic segregation in educational context led the interaction between students from different ethnicities to be limited and rare [30-31]. The majority of students from various ethnic groups in secondary schools are mostly comfortable hanging out with friends from the same ethnic group [32]. Based on the study conducted by Vellymalay (2014), students from various ethnic groups in secondary schools tend to build a friendships with peers from the same ethnic group as compared to build a cross ethnic friendship. According to a study conducted by Ahmad (2011), even though students from a multi-ethnic school environment has a lot of opportunities for social interaction with friends from different ethnic group, however, the social interaction mostly occurred among same ethnic group. Students in a multi-ethnic school environment still have less social interaction with friends from different ethnic groups. Consequently, this led to the construction of the friendship based on the same ethnics group than building a cross ethnic friendships.

According Jamil et al., (2004), students from SJKC and SJKT are having difficulty to build relationships with students from different ethnic background when they were in secondary school. According to a study conducted by Ahmad (2011), Malay students who have attend schools that consist of majority students from Malay ethnic group are not willing to carry out activities together with students from other ethnic groups. This situation makes a cross-ethnic friendship among students are difficult to build especially when there were in secondary school.

6 Teachers' Role in Supporting a Cultural Diversity in Schools

The school environment and teachers play an important role where teachers themselves need to be aware of cultural diversity that exists in the classroom [1]. An anti-bias approaches in teaching and learning focusing on identity formation among students is to provide students with the ability to connect with the real world which contains diversified cultures, languages and ethnicity [1]. Chow (2007) stated that to teach multi-ethnic students effectively, teachers must be willing to bear a difficult responsibility and have a critical thinking. It is important to create an atmosphere where students feel safe to interact across the differences that exist between ethnic groups [18].

Teachers play an important role in promoting cross ethnic friendship among students [33]. Every student who comes to school brings together their culture, way of life and ethnic identity. It is influenced and shaped by the social background of students, culture, families, communities and beliefs or their religion [33-34]. Children should be educated and made them understand from the early stage about life in multicultural society. Understanding, respect, tolerance and the erosion of prejudice must be established among children to create a harmonious of social interaction [35].

According to Banks (2001), teachers and school administrators should carry out the following principles in establishing the relationship between ethnic groups existed among the students;

I. Schools must create or make a key feature of cross ethnic friendship among students as a step to improve relations between students from different ethnic groups.
II. Students should learn about stereotypes and other related bias would negatively affect ethnic relations.
III. Students should be taught about the values shared by almost all ethnic groups such as justice, equality, freedom, peace, humanity and virtue.
IV. Teachers should help students acquire the necessary social skills to interact effectively with students from different ethnic groups, culture and language.
V. Schools should provide opportunities for students from different ethnic group, cultural and language to interact socially in conditions designed to reduce fear and anxiety between ethnic groups.

In order to preserve democracy in terms of social changes and reducing ethnic polarization, teachers should set primary goals as the way to be an effective teacher who teaches students from different cultural and ethnic backgrounds [33]. One of these goals is that the teacher must help students to describe their ethnic identity and to be use it effectively with the members of their ethnic groups. Students must be clear with ethnic and their own identity before they can interact socially in a positive way with people from other ethnic groups. Teachers should also educate students about acceptance of their self and not promoting the racism. Although there are a students who are from a multi-ethnic society backgrounds, students must be taught to accept their own ethnic identity and become comfortable with it. Student must also learn how to use their ethnic identity effectively. They must also learn how to interact with their peers from other ethnic groups and how to practically solve conflicts [33].

Students from minority ethnic groups must work between the two ethnic social environments, which are in their ethnic and community groups and within the dominant ethnic group. In the context of Malaysia, teachers should help students from minority ethnic groups, which are Chinese and Indian students to learn how to interact in their own ethnic group and with the majority ethnic group. With the role of teachers in helping students to learn harmonies ethnic relations, it will help ethnic minority groups to act in good possible way in other culture and ethnic groups [33].

Teachers can also play a role in either to reinforce or reject multiculturalism [36]. Teachers who are sensitive to culture always understand and deal directly with the sensitivities that exist between cultures [37-38]. As agents who are sensitive to culture, education among teachers should promote the construction of knowledge, skill and a required sensitivity [39]. Dukes and Ming (2006), emphasized the importance of teachers' role as an agents of cultural diversity. The teachers' role is very important because they serve as a model for students to achieve goals in understanding the cultural diversity that exist among them [1]. Aspects of teaching that is not in accordance with the diversity of students' cultural backgrounds will lead to a cultural clashed [40-42]. Grootboom (2007), state that student in a class consisting of diverse cultures tend to have a lot of problems based on the culture mixing.

Teachers should help students recognize that not only an ethnic groups have a differences and diverse, but individuals from the same ethnic group are also different from each other [43-44]. At the same time, students should also be encouraged to identify common features between peers from different ethnic groups. Learning about ethnic cultural diversity should involve all students without forming any gap or separation between the ethnic minority and majority students as an effort to encourage students to interact with each other [44-45].

7 Conclusion

A multi-ethnic school environment provides an opportunity for the students from various ethnic group to interact socially with peers from other ethnic groups and build a cross ethnic friendships. However, based on a review, it shows that there are multiple findings in terms of building a cross ethnic friendship even though the students were in the multi-ethnic school environment. Students from various ethnicities reportedly still carry out activities either in classroom or outside the classroom and build a friendship based on their respective ethnic groups. Therefore, teachers should play an important role in promoting ethnic relations among students with a range of programs either in classroom or outside the classroom to ensure the students interact socially with peers from other ethnic groups. Teachers should also be trained with multicultural teaching methods as a basis to teach students from different ethnic groups and cultures.

References

1. R.M. Majzub, S. Hashim, E. H. Johannes, Procedia - Social and Behavioral Sciences, **15**, 1573-1579 (2011)
2. R. Mustapha, N. Azman, F. Karim, A. R. Ahmad, M. A. Lubis, Social Integration among Multiethnic Students at Selected Malaysian Universities in Penisular Malaysia: A Survey of Campus Social Climate. AJTLHE, **1**(1), 35-44 (2009)
3. Government Transformation Programme: The Roadmap. (2010). Prime Minister's Department, Malaysia.
4. A. R Hamdan, M. N. Ghafar, A. A. Ghani, *Fostering inter group contacts among multiracial students in Utm.* Paper presented at International Conference on Learner Diversity: ICELD 2010. Bangi:UKM.
5. R. Alba, V. Nee, *Remaking the American mainstream: Assimilation and contemporary immigration* (Cambridge, MA: Harvard University Press, 2003)

6. A. Portes, R. G. Rumbaut, Legacies: The Story of the Immigrant Second Generation. New York: Russell Sage Foundation.
7. T. F. Pettigrew, Intergroup contact theory. Annual Review Psychology **49**, 65-85 (1998)
8. L. E. Molina, M. A. Wittig, Journal of Social Issues **62** (3), 489-509 (2006)
9. J. A. Nier, S. L. Gaertner, J. F. Dovidio, B. S. Banker, M. W. Christine, C. R. Mary, Changing interracial evaluations and behavior: The effects of a common group identity." Group Processes & Intergroup Relations **4**(4), 299-316 (2001)
10. L. Tropp, Social Psychology Quarterly **70**(1), 70-81(2007)
11. L. Lee, H. Carollee, C. Brandt, Merrill-Palmer Quarterly (53) 3, Article 3 (2007)
12. S. A. Baharuddin, *Many ethnicities, many cultures, one nation: The Malaysian experience. Institut Kajian Etnik* (Universiti Kebangsaan Malaysia, 2008)
13. M. Filpisan, Tomuletiu Adriana-Elena, A. Moraru, M. Stoica, B. Gorea, A. Solovastru, Procedia - Social and Behavioral Sciences **15**, 3263-3267 (2011)
14. A. Eller, D. Abrams, D. (2003). “Gringos” in Mexico: Cross-sectional and longitudinal effects of language school-promoted contact on intergroup bias. *Group Processes & Intergroup Relations, 6*, 55–75(2003)
15. S. Levin, L. Van Lar, W. Foote, Journal of Applied Social Psychology **36**(6), 1471-1501 (2006)
16. S. Paolini, M. Hewstone, A. Voci, J. Harwood, E. Cairns, (2006). Intergroup contact and the promotion of intergroup harmony: The influence of intergroup emotions. In R. Brown & D. Capozza (Eds.), *Social identities: Motivational, emotional and cultural influences* (pp. 209–238). Hove, England: Psychology Press.
17. T. F. Pettigrew, Generalized Intergroup Contact Effects on Prejudice. Personality and Social Psychology Bulletin **23**(2), 173-185 (1997),doi: 10.1177/01461672 9723 2006.
18. T. B. Smith, L. Sylva, Journal of Counseling Psychology **58**, 42-60 (2011)
19. P. Gurin, E. L. Dey, S. Hurtado, G. GurinDiversity and higher education: Theory and impact on educational outcomes. Havard Educational Review **72** (3), 330-366 (2002)
20. X. Chen, S. Graham, Cross-ethnic friendship and intergroup attitudes among Asian American adolescents. Child Development, 749-764 (2015)
21. S. Graham, A. Muniksma, A. Juvonen,Child Development 469-483 (2014)
22. G. Kao, K. Joyner, Social Science Quarterly **87**, 972–992(2006).
23. E. Vaquera, G. Kao, Sociological Quarterly, 47: 375–396(2006)
24. T. Yip, E. Seaton, R. Sellers, Interracial and Intraracial Contact, School-Level Diversity, and Change in Racial Identity Status Among African American Adolescents. Child Development, **81**(5), 1431-1444 (2006)
25. X. De Souza Briggs, City & Community, 6: 263–290 (2007). doi:10.1111/j.1540-6040.2007.00228.x
26. H. McGlothlin, M. Killen, C. Edmonds, British Journal of Developmental Psychology, **23**, 227–250 (2005). doi:10.1348/026151005x26101.
27. W. N. Kubitschek, M. T. Hallinan, Social Psychology Quarterly **61**(1):1–15 (1998)
28. C. N. Masson, CM. Verkuyten, Journal of Applied Social Psychology **23**,156–168 (1993)
29. D. L. DuBois, B. J. Hirsch, School and neighbourhood friendship patterns of blacks and whites in early adolescence. Child Development **61**, 524–536 (1990)
30. S. Osman, *Ikatan etnik dan kelas di Malaysia* (Bangi, Universiti Kebangsaan Malaysia, 1989)
31. N. M. Yusoff, (2006). Pola interaksi sosial dalam kalangan murid multietnik di sekolah kebangsaan. Tesis PhD. USM.
32. S. K. Vellymalay, Journal of Human Capital Development **7**(2), 47-61
33. J. A. Banks, *Race, culture, and education: The selected works of James A. Banks* (Oxon: Routledge, 2006)
34. J. A. Banks, (1989). Multicultural education: Characteristics and goal. In J. A. Banks & C. Banks (Eds.), Multicultural education: Issues and perspective. Boston: Allyn and Bacon.
35. H. Jamil, Pendidikan berbilang budaya: Meningkatkan kefahaman pelajar tentang kepelbagaian budaya masyarakat melalui pendidikan sejarah. Diges Pendidik **2**, 30-40 (2002)
36. J. A. Spinthourakis, E. Karatzia-Stavlioti, Y. Roussakis, Pre-service teacher intercultural sensitivity assessment as a basis for addressing multiculturalism. Intercultural Education, **20**(3), 267-276 (2009)
37. G. M. Chen, W. J. Starosta, Foundations of intercultural communication. (Boston, MA: Allyn & Bacon, 1998)
38. A. E. Fantini, A. E. (2000), A central concern: Developing intercultural competence’. In SIT Occasional Paper Series. Brattleboro (VT), 25-42 (2000)
39. A. H. Hains, E. W. Lynch, P. J. Winton, Moving towards cross-cultural competence in lifelong personnel development: A review of the literature: Early Childhood Research Institute on Culturally and Linguistically Appropriate Services University of Illinois at Urbana-Champaign. Retrieved from; http://clas.uiuc.edu/techreport/tech3.html.
40. J. Necochea, Thrust for Educational Leadership **27**, 20-23 (1997)
41. C. Casteel, The Journal of Educational Research **92**, 115-120 (1998)
42. G. Cartledge, C. Kea, D. Ida, Teaching Exceptional Children **32**, 30-37 (2000)
43. A. D. Pellegrini, Nature, **467**, 27 (2010)
44. L. Pellegrini, European Journal of Teacher Education, 33(2), 113-125 (2010)
45. L. Derman-Sparks, P. G. Ramsey, *What if all the kids are white?* (New York: Teachers College Press, 2006)

The Influence of Social Media Towards Emotions, Brand Relationship Quality, and Word of Mouth (WOM) on Concert's Attendees in Indonesia

Yulia Rahmadini[1] and *Rizal Edy* Halim[1*]
[1]Faculty of Economics and Business, Universitas Indonesia, Indonesia

Abstract. Social media has been developing significantly from year to year. In Indonesia, 72% of the internet users use the internet to access social media. This indicates that social media are effective brand communication and promotion tools. Music industry is one of the business industries which utilizes social media as communication and promotion tools, especially related to music concerts. The purpose of this study is to identify the impact and relationship of social media towards emotional attachment and brand relationship quality, in which both will be then compare towards word of mouth (WOM) on concert's attendees. This research uses descriptive design research conducted in one period (cross sectional design). Respondents of this study are 160 persons who attended concerts last year. The five-hypotheses research model in this study is tested using the Structural Equation Modelling (SEM). This research shows that social media interaction may result in emotional attachment. Emotional attachment felt by concert attendees may influence positively and significantly to brand relationship quality and creates word of mouth among the concert attendees.

1 Introduction

Across the time, social media keeps growing, pioneered by Friendster which dominated the usage of social media at that time. In 2004, Facebook was established, followed by Twitter in 2006. Facebook and Twitter are the most used and visited social media in the world. Right now, Facebook has 1.44 trillion active users, with 13% increment from last year's record. As for Twitter, it has 308 million active users, with 18% increment form last year's record, reaching 500 million tweets per day.

Based on the data from The Indonesian Association of Internet Service Provider (Asosiasi Penyelenggara Jasa Internet Indonesia/APJII), up until the end of 2014, there are 88 million Internet users in Indonesia. The majority of these 72% internet users use internet to access the communication channel. Here, the social media is the main internet activity, where its usage exceeds other activities' like information browsing, chatting, news browsing and accessing the email.

A survey from Global Web Index presented that Indonesia is one of the most

* Corresponding author: rizaledy@gmail.com

active countries in terms of social media usage. Indonesia has 79,7% active social media users, ahead of Filipina with for 78% active users, Malaysia with for 72% active users and China with 67% active users. Facebook still dominates social media market in Indonesia with 62% active users, followed by Twitter, Google Plus and LinkedIn [28]. The social media users in Indonesia are expected to reach 75,4 millions users in 2015 [29].

Social media is an internet-based application that established based on ideology framework and Web 2.0 technology which enables the idea creation and information exchange among its users [1]. According to [2] social media is an online community consists of internet users who have the same interest and have a desire to communicate with each other. Nowadays, some of social media that remain popular in Indonesia are Twitter, Facebook, Instagram, and Path. Additionally, Kaplan and Haenlein stated that social media has been the dominating topic in almost all aspects of business and industries.

Social media evolves to be the dominant means of digital communication [3]. However, as postulated by Pew Researcher Centre [4], among 67% internet users who also use social media, only a few of them know about the role of social media in influencing emotions towards brand and whether social media based relationship could result in favourable result, for example positive word of mouth.

Based on previous illustration, music industry is one of industries that incorporate social media as a means of communication and promotion tools, especially in promoting music concert. The growth of music concert, especially in Jakarta, keeps escalating. Between 2012 and 2014, there are at least 20 music concerts held in each year. Almost all of the performers, who have held their concert, own social media account. One of the examples is Afgan, a young talented singer from Indonesia who just held his own concert titled "Dari Hati Afgan" in Februari 2015. He owns various type of social media like Twitter, Facebook and Instagram. In his Twitter, Afgan has 9.812 tweets that garnered 7.9 millions of followers. In his Facebook fan-page, Afgan garnered 1.5 million likes, while his Instagram account has 1.049 posts and garnered 476 thousands of followers. Besides Afgan, there is Katy Perry, an international musician, who also has held concert in Indonesia. Just like Afgan, Katy Perry owns various type of social media like Twitter, Facebook, Instagram, etc. In twitter, Katy Perry posts for 6.488 tweets and has garnered 69.7 million followers. Meanwhile, her Facebook fan-page is followed by 72 million of fans. As for her Instagram, it has 293 posts and garnered 19 million followers. There was at least 20 concerts held yearly between 2011 – 2015. Those music concerts utilize social media as a means of communication medium and promotion tools.

The previous illustration shows that social media is a right media to build emotions through interactions between the musicians and their fans. For the fans, social media is not only useful in exchanging information, but also in acquiring the most updated news about their favourite musician, such as their daily activities, rehearsal activities, concert schedule, etc. Knowing the fact that such activates could strengthen the relationship with their fans, often musicians will return to social media comment section to give love or likes, to re-tweet, and so forth.

Music concert as a brand has a distinctive characteristic, especially on its social media usage which occurs on certain time frame. As asserted by [5], the hedonic nature of music festivals turns them into an ideal context to empirically test the influence of emotions on brand relationship quality, as the value of the experience depends on the ability of the festival to engage one's subjectivity to arouse feelings or to provoke emotional reactions. Although there are only few studies on brand ownership and brand control related to music concerts, Mossberg and Getz asserted that brand can also be applied to concerts, as it can be applied to products, services, an organization, a person, a team or a symbol [6]. Here, it can be concluded that music concert can also be classified as a brand. According to [5], music concerts in particular are more popular than ever, where now they are attracting millions of fans.

Based on above explanation, this study wants to test and to analyse the effect of interaction between musicians and their fans in the social media on the level of emotional attachment, brand relationship quality, and word of mouth. First of all, this study will examine the effect of media social interaction on music concert. Emotion is the positive and negative feeling of a person [7]. Next, this study will try to discover the effect of media social interaction and emotional attachment on music concert's attendees brand relationship quality, as per stated by [8], emotion is the component of brand relationship, whereas brand relationship quality a significant indicator for the strength and depth of consumers 'relational behaviour towards brands. The last but not least, this study will investigate the effect of emotional attachment and brand relationship quality on the music concert attendee's Word of Mouth. Here, word of mouth refers to a voluntary communication from the consumers where they promote a product or a brand to friends or relatives based on particular motives.

2 Theoretical Background, Conceptual Model & Hypotheses

2.1 Social Media Marketing

Social media consist of channel and content that are spread through the interaction between individuals and organizations, in which social media creates a two-way interaction between the organization and its consumers [11]. The true power of social media is the notion of "we are all connected", which means that the communication occurs in social media is not a one-way communication, but rather a two-way communication whereas it enables a user to interact with not only other users, but also with the organization as well [10]. This is also supported by Kim & Ko [11] who stated that a brand and its consumers interact with each other without time, place and media boundary, that is the dynamic changes from a one-way communication to a two-way communication. The benefit of social media has been proven as the company using social media as a communication platform successfully improved its relationship with community and expanded consumer's prospect, actively and passively.

The transformation from product centric to customer centric emphasize the importance for the company to be more focus on its consumers, in which social media is considered as one of the best media to not only interact with the consumers, but also to build the image of the company. The importance of social media signify with the facilitation of immediate, interactive, and low-cost communications [12]. Thus, it require the company to increase the use of social media in its marketing activity. Moreover, it is also important to conduct the analysis of social media marketing activity and its influence towards the brand and the company as a whole. [11] explained that social media activity is influenced by 5 indicators, namely entertainment, interaction, usefulness, word of mouth and trendiness.

2.2 Social Media Interaction

Social media is the means of electronic communication which can be personalized by its user to share ideas, information, opinion and message [9]. Social media site like Facebook and microblogging site like Twitter have been the main source of communication to seek acquaintances and brands. These kind of social media site can be incorporated by the marketers and consumers to build a community which can attract other potential consumers to join.

It is asserted that people visit social media for 3 major reasons [28]. The first reason is because of the brand or the product promotional campaign conducted by the company. The second reason is because the brand has a relevant content with the social media. The content is considered to be relevant when the delivered message has a meaning, or in the other words, the delivered message creates social connection. The third reason is because the popular and favourable content. The popularity of social media platform and content is important to make the users feel engaged with the brand. The fourth reason is because of the application used to access the social media. As the consumers can not automatically engage with the social media by themselves, the company has to analyse the characteristics of its target consumers then find the most effective social media platform to be utilized to interact with them.

In entertainment industry, Rothschild [19] observed that social media is utilized, organized and sensed by the company. It is discovered that more than a half (57%) number of industries that utilize social media have higher income compared to those who did not. This finding supported previous study which asserted that when the brand initiates social media interaction with the consumers, it would lead to a positive impact for the brand. This positive impact can be in form of cognitive or affective [22]. Basically, cognitive is related to thinking, while affective is related to feeling. The affective system has 5 basic characteristics, which are: 1) Affective system has an immediate response towards the environment, 2) Society has a little control towards affective system of an individual, 3) Affective response can be observed physically from the individual's body motion, 4) Affective system can react to various kinds of stimulus, and 5) Although getting the same stimulus, the affective system of each individual is different, as it is based on each environment. One of the impact of affective system is emotion. Emotion is the feeling that triggers an individual to react towards internal or external stimulus.

2.3 Emotion

Emotion is a positive or negative feeling of an individual [7]. Emotion is part of consumer evaluation towards a stimulus which can be depicted as the consumer affective process [27]. The result of this evaluation will produce emotion towards the brand, and this emotion can be a positive one like happy, satisfied, and delighted or a negative one like angry, irritated, and upset. Human behaviour is easily affected by emotion. Consumers often feel so emotional and intuitive in their behaviour, in which it happens out of their consciousness. The marketers are interested to attract consumers emotionally as the emotionally-engaged consumers is desirable for the marketers.

Peter and Olson [22] asserted that in analysing the consumers, the marketers have to consider 3 aspects, namely consumer affective and cognitive state, consumer behaviour and consumer environment. Peter and Olson [22] also divided affective response into four types of response, which are emotion, special feeling, mood and evaluation. Emotion is the external factor that cannot be disregarded. Extra attention on the brand helps to create positive emotion, while negative emotions are the result of failing to meet a minimum standard [22].

Emotion towards the brand is one of the components of brand relationship. Consumers who have strong emotional relationship towards the brand are indirectly attached to that brand [20]. Emotional attachment with the brand makes individual to develop a delightful feeling towards the brand and to feel sad when the brand is not there [27]. The emotional relationship with the brand can be beyond the cognitive behaviour towards the brand [27], which is indicated by the strong relationships between brand and self-concept.

Emotional attachment can predict the strength of commitment towards the brand.

However, this emotional attachment along with the consumer's personality can change the way consumers behave when they feel incompatible with the brand. Consumers who are emotionally-attached with the brand identify and recognize the brand really well so that they often show a quick dissatisfaction response when having a negative experience. This can endanger the emotional relationship between the consumers and the brand since the negative experience can lead to a strong negative reaction. This also shows that the bigger emotional attachment possessed by the consumers, the larger the tendency for them to lose trust and commitment toward the brand. Emotional attachment is the highest form of relationship as it is resulted from investment in the relationship between the company or brand and its consumers. To deal with this emotional attachment issue, the marketers can create a different product category with other brand, analyse the market better so that the consumers' needs are relevant with the brand, and create a more intimate relationship between the brand and the consumers.

2.4 Brand Relationship Quality

Brand relationship quality is also known as consumer-brand relationship or the relationship between the brand and its consumers. Brand relationship quality is depicted as the indicator for the strength and depth of consumers' relational behaviour towards brands [8]. Phau and Lau [23] illustrated that consumers project their personal characteristics into the brand when the brand itself can represent the consumers. Consumers who project themselves into the brand show a strong brand attachment. Similarly, [24] postulated that when consumers interact with brands, they develop an active relationship with those brands like people did with friends.

Brand relationship quality portrays the depth and strength of the relationship between the brand and its consumers. Brand relationship is considered superior in providing brand knowledge for the consumers and understanding consumers' needs [18]. Brand relationship as an interdependent, voluntary relationship between an individual and a brand, characterized by the unique history of the interaction and the anticipation of future event [18] . It means that the interdependent relationship between the consumers and the brand is a key indicator of the notion that relationship with the brand has the same quality with the relationship with other human.

This study asserted that the relationship quality between the brand and its consumers depicts the strong motivation towards the brand [18], in which this motivation influences the extensive evaluation and also identifies the condition and process of the mediation process for the brand relationship quality effect. The traditional brand researchers examined how the consumers see and evaluate the brand, checked the brand attitude, and evaluated the brand or the quality of the brand. These researcher noted that the consumers not only have the difference in brand perception, but also in the way of how they interact with the brand [18].

Conceptually, brand relationship quality has a richer and deeper meaning compared to the evaluation of a brand as a whole or behavioural intention. Brand relationship quality depicts the emotional attachment and strong motivation with the brand. There are a lot of concept to keep this relationship alive, which are more than just building positive feeling. The concepts are affective and sociomotive attachment, the consists of love/passion and self-connection; behavioural ties, the consists of interdependence and commitment; and supportive cognitive beliefs, the consists of intimacy and brand partner equity. Concepts of brand relationship quality, which are love/passion, self-connection, interdependence, commitment, intimacy, and brand partner quality [18].

2.5 Word of Mouth

Word of Mouth is the process of transferring information from one individual to other and has an important role in consumer decision making [19]. Brown and Reingen argued that WOM plays an important role in consumer behaviour and attitude formation [6]. WOM includes consumer attitude in sharing the brand, opinion or reaction towards business, product and service to others [20]. A positive WOM is a powerful marketing media for the company in influencing the consumers.

Some studies in recent decades found consistent results regarding the importance of personal interaction and WOM, especially in influencing market choice [20] and in distributing information about new product. WOM is usually created by the consumers who do not have personal agenda in recommending particular brand or product. Therefore, the recommendation via WOM is considered to be more credible and can gain more attention from the market [21].

When the consumers are satisfied and have positive attitude towards the brands, it does not automatically make them recommend the brand or involve in WOM. The consumers have to feel motivated first in order to initiate the WOM. There are some factors that can make the consumers initiate the WOM, and even better, initiate the positive WOM. Understanding these factors can make the marketers to be more proactive in finding the best way to organize and influence the positive WOM, thus making the marketing communication work more effectively. The internal factors that can motivate consumers to initiate positive WOM are as following: strength of social ties, altruism, seeking confirmation of own judgement, and expressing sense of achievement [25].

Based on literature background above, we testing five hypothesis using the model in Figure 1 adopted from [26]:

H1: Social media interaction positively influence emotional attachment
H2: Social media interaction positively influence brand relationship quality
H3: Emotional attachment positively influence brand relationship quality
H4: Emotional attachment positively influence word of mouth
H5: Brand relationship quality positively influence word of mouth

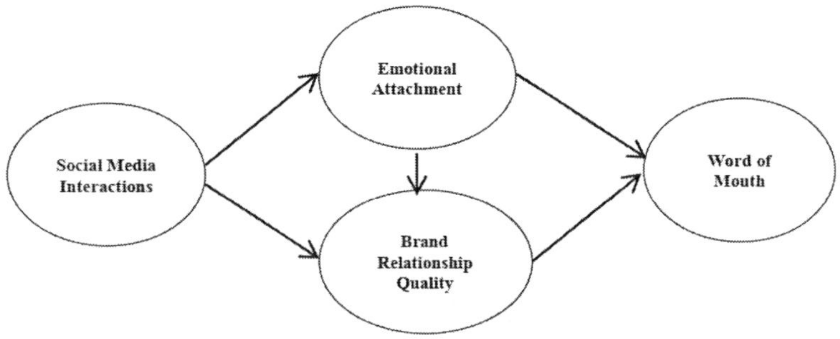

Fig. 1. Research Model

3 Method

This study used descriptive research design with single cross sectional time frame. The data was collected by distributing the questionnaire directly and via online. The online questionnaires was distributed using Google docs. The respondents of this study were the one who have watched live concert at least once in this year, and followed the social media of their favorite musicians. The number of respondents needed in this study were 160 respondents. To acquire the respondents, this study utilized convenience sampling method, in which it gave freedom for the researcher to gather the sample from the places that are accessible and convenient for the researcher. For the places themselves, this study acquired the sample from Plenary Hall Jakarta and Artist Fanbase of Afgan and Taeyang. Here, the researcher came directly to the aforementioned place to meet the fans of Agfan and Taeyang.

The questionnaire of this study was measured by using a six-point Likert type scale. Emotion was measured by adapting items, which are affectionate, friendly, loved, peaceful, passionate, delightful, captivated, connected, bored and attached. As for brand relationship quality, it was measured by using 16 items from [18]. For word of mouth, it is a statement of respondents' willingness to recommend music concert from certain musicians to their friends or relatives.

4 Results

Out of 160 respondents, 84% are female and 16% are male. These 160 respondents are the one who at least have watched 1 concert in this year, and already followed the social media of their favourite musicians.

Next, this study conducted the goodness of fit analysis. The goodness of fit scores of this study are as follow: GFI = 0.71 (marginal fit), RMSEA = 0.10 (mediocre fit), NFI = 0.71 (marginal fit). Also Parsmonious Fit Index shows a good fit score. The results show that the model has a good fit since it already fulfils 4 - 5 criteria of goodness of fit. A model has to meet at least 3 criteria of goodness of fit to be considered as a good fit model.

Figure 2 below depicts the t value scores of each hypotheses used in the study. Four hypotheses in this study are supported as their t-values are > 1.96, meanwhile 1 hypotheses is not supported as its t-value is < 1.96.

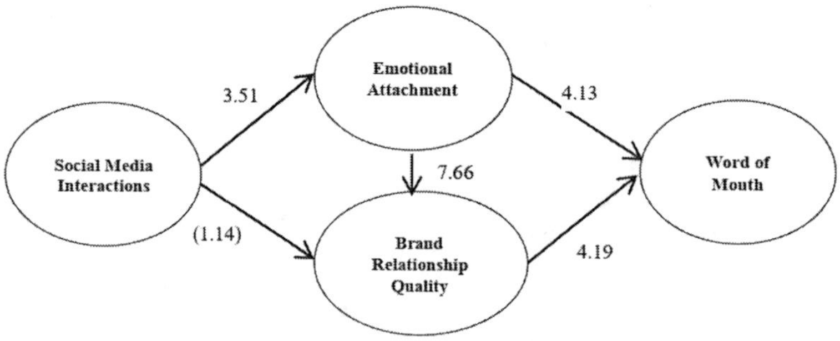

Fig. 2. T-Value Output

5 Conclusion

Based on the results, several conclusions are presented to answer the objectives of the study. First of all, social media interaction has a positive and significant effect on emotional attachment. This significant effect is an indication that should the attendees of music concert experience a high level of interaction in the social media, their emotional attachment will be increased. Then, it is apparent that social media interaction has no significant and positive effect on brand relationship quality. This insignificant result might happen because the brand relationship quality, either related to the musicians or the concerts itself, has to be influenced first by the depth and strength of the relationship between the consumers and the brand [8].

As for emotional attachment, it has a positive and significant effect on brand relationship quality. This significant effect indicates that when the brand relationship of the consumers is high, then their emotional attachment will be high as well. Maintaining the relationship with the consumers will not only lead to positive feelings, but also to affective and sociemotive attachment which consist of love/passion and self-connection; behavioural ties, the consists of interdependence and commitment and supportive cognitive beliefs, the consists of intimacy and brand partner quality.

It is also proven that brand relationship quality has a positive and significant effect on word of mouth. This significant effect shows that when the brand relationship quality of the consumers is high, then their willingness to conduct word of mouth will be high as well. The same case also applied for emotional attachment, where it also has a positive and significant effect on word of mouth. This significant effect is an indication that should the emotional attachment of the consumers is high, then their willingness to conduct word of mouth will be high as well.

5.1 Limitations and Recommendation for Future Research

This study has several limitations that affect the result of the study as a whole. First, this study was only conducted in the Plenary Hall, Jakarta and within the fan base of Afgan and Taeyang. It would be better if it was conducted on several music concert venues instead of only one, and on more varied fan base, like female musicians fan base, for example Raisa, so that it would be more representative in explaining the music concert attendes' characteristics in general. Then, most of the questionnaires were only distributed on Jakarta area, thus most of the respondents were the residents of Jabodetabek and the total respondents acquired were only 160. It would be better to gain a larger number of respondents in order to increase the reliability and validity of the study. Moreover, the distribution of physical questionnaires were only conducted on the fan base of Afgan and Taeyang, in which the majority of the members are female. This could create a bias on the study result as the majority of the respondents were female. For future studies, it is suggested that the distribution of the questionnaire is not only limited on the fan base of male musicians, but also on the fan base of female musicans in order to capture a larger and more balance picture of music attendes characteristics. It is also suggested to improve the range the area of study as to not limit it to Jabodetabek area only. By doing so, the result of study is believe to be richer. Doing the comparison across several areas will also improve the generalibility and external validity. Moreover, future studies can also add purchase intention as one of the variable in the research model.

https://doi.org/10.1051/matecconf/201815005058

References

1. Aaker, David, *Building Strong Brands* (New York: The Free Press, 2006)
2. A. d'Astous, F.Colbert, E. d'Astous, International Journal of Arts Management 8 (2), 14–23 (2006)
3. A.M. Kaplan, M Haenlein, Users of The World, Unite! The Challenges and Opportunities of Social Media. Business Horizon, **53**, 59-68 (2010)
4. A. Mecke, I. Lee, J.R. Baker jr., M.M. Banaszak Holl, B.G. Orr, Eur. Phys. J. E **14**, 7 (2004)
5. B. Chappuis, B. Gaffey, P. Parvizi. Are your customers becoming digital junkies. McKinsey Quarterly, July, 1–4 (2011)
6. J. J. Brown, P. H. Reingen, Journal of Consumer Research **14**, 350-362 (1987)
7. M. Cheung, M. M. Anitsal, I. Anitsal, Journal of Marketing Theory and Practice 15(3), 235-249 (2007)
8. Erdogmus, Irem Eren dan Cicek, Mesut, Procedia – Social and Behavioral Sciences **58**, 1353 – 1360 (2012)
9. E. Smit, F. Bronner, M. Tolboom. Journal of Business Research **60**, 627-633 (2007)
10. S. Fournier, Journal of Consumer Research **24**, (1998)
11. F. De Lillo, F. Cecconi, G. Lacorata, A. Vulpiani, EPL, **84** (2008)
12. F.P William. Social networking sites: How to Stay Safe Sites. Monthly Security tips Newsletter **4**(3) (2009)
13. R. E. Halim, Elszuary Abrar Uzi Zulkarnain, Journal of Distribution Science **15**(4), 15- 23 (2017)
14. R. E. Halim, The Impact of Trust-Distrust toward Quality of Life: The Case of Poor and non-Poor Household in Indonesia. European Research Studies Journal Vol XX (Issue 3A) page 801-819. (2017)
15. J. A. Kim, E. Ko, Journal of Global Fashion Marketing **1**, (2011)
16. L. Mossberg, D. Getz, Scandinavian Journal of Hospitality and Tourism **6** (4), 308-326 (2006)
17. T. D. L. Luigi, Propulsion physics (EDP Sciences, Les Ulis, (2009)
18. M. Ben Rabha, M.F. Boujmil, M. Saadoun, B. Bessaïs, Eur. Phys. J. Appl. Phys. 47(1), (2009)
19. K. D. Miller, F. Fabian, S. J. Lin, Strategic Management Journal **30**, 305-322 (2009)
20. J. P. Peter, J. C. Olson, *Consumer Behavior 6th edition* (McGraw-Hill. New York, 2002)
21. Pew Internet. Social media update (2014)
22. I. Phau, C. K. Lau, The Journal of Brand Management 8(6), 428-444 (2001)
23. R. Hanna, A. Rohm, V. L. Crittenden. We're all connected: The power of the social media ecosystem. Bussiness Horizon 2011, **7**, 881 (2011)
24. M. L. Richins, T. Root-Shaffer, The Role of Involvement and Opinion Leadership in Consumer Word-of-Mouth: An Implicit Model Made Explicit. Advance in Consumer Research, **15**, 32-36. (1998)
25. Social media data, "user of social media" http://gsbipb.com, (accessed on 2014)
26. H. Simon, M. S. Roth, T. J. Madden, R. Hudson. The effects of social media on emotions, brand relationship quality, and word of mouth: An empirical study of music festival attendees. Tourism Management Vol. 47 (2015)
27. T. Suryanto, Investor Reaction and Fraud Prevention International Journal of Economic Perspectives **10**(1), 138-150 (2016)
28. D. J. M. Thomson, C. Whan Park. Journal of Consumer Psychology **15**(1), 77–91 (2005)
29. Y. Lee Y, J. Kim, JH. Back, Prev Med **48** (1), 86-90 (2009)

The Factors Influencing Organizational Citizenship Behaviour of Employees in Private Universities in South Sumatra Province

Desi Ulpa Anggraini[1,*], *Hanes* Riady[2] and *Harapid* Hafid[3]

[1]STIE Rahmaniyah, Sekayu, South Sumatera, Indonesia
[2]Kwik Kian Gie Institute, Jakarta, Indonesia
[3]Halu Oleo University, Kendari, South east Sulawesi, Indonesia

Abstract. The objectives of this study is to analyze the factors affecting Organizational Citizenship Behavior (OCB) Nonacademic Employees in Private Universities in South Sumatra.The research design used in the study is explanatory design which explains the causal relationship among variables and scoring all indicators for each variables. The study location is in South Sumatra Province while the research objects are 13 Private Universities with 200 nonacademic employees as respondents. This dissertation used primary data from questionnaire filled out by the respondents.The sample selection used simple random sampling. The method of research used descriptive statistics using SPSS and Structural Equation Modeling (SEM) using Lisrel.The descriptive analysis shows that the level of OCB, Work Satisfaction, Transformation Leadership, Work Motivation, and Organization Culture are at a positive level.The results of relationship among variables stated that the Transformational Leadership has no significant impact on OCB, Work Motivation has significant impact on OCB, Organization Climate has significant impact on OCB, Work Motivation has significant impact on Work Satisfaction,Organization Climate has significant impact on Work Satisfaction,Transformational Leadership has significant impact on Work Satisfaction, Transformational Leadership has no significant impact on Work Motivation, Organization Climate has significant impact on Work Motivation, and Work Satisfaction has significant impact on OCB.

1 Introduction

One of psychology that plays role in human resources management is *organizational citizenship behavior (OCB).* OCB is a helpful attitude shown by organization' members, which is constructive, appreciated by the company but it isn't directly related to individual productivity [1]. A success organization needs employees who will do more than their formal tasks and want to give performance beyond expectation. In today's dynamic world of work, where the task is always done in a team, flexibility is the most important. Organization wants its employees are ready to the tasks which are not stated in their job descriptions. According to Robbins and Judge, the fact shows that an organization which has employees with good OCB' quality, will have better performance in organization [2].

Based on some research showed that the employees' OCB enhancement can be triggered by job satisfaction, where this satisfaction is affected by some factors such as transformational leadership, motivation and organizational climate.

Transformational leadership is a leader's behavior which is able to motivate employees to work harder and want to work more than what they should do to be able to make changes and achieve goals. Some previous studies conclude that transformational leadership has significant positive effect toward employees job motivation [4-6]. Bass in Purvanova states that transformational leadership can make employees be more involved and caring on their jobs, give more attention and time to their jobs, and less attention to their personal interests [7].

Besides transformational leadership style, job satisfaction is also affected by job motivation factor, in line with the research result of Mehryar Motevalli dan Francoise Chevalier [8]. His research findings showed that job motivation is very significant where the salary is the factor that influences the most important motivation. By increasing the employees' job motivation so they will be happy and have higher job satisfaction.

Many researches have been done related to organizational climate, satisfaction and employees' commitment. Research on business company done by Castro and Martin who has tested the relationship of organizational climate and job satisfaction in an information and technology organization in South Africa [10]. Research findings showed that 9 of 12 organizational climate dimensions showed strong positive relationship toward job satisfaction variable.

University is an open organization in running their activity process involves and is affected by many elements, either the elements inside or the elements outside. In order to get clarity about the problem and its

* Corresponding author: author@e-mail.org

causes, need to be revealed through elements related to the quality of higher education institutions. In the universities environment, there are two important elements in supporting college goals and mission from the human resource side, those are lectures and administration staff. The lectures will help in improving the quality of students' resources, while the administrative staffs play a key role in the students service process. The performance of these two elements is able to affect university competitive advantage level to be able to win the competition.

Private universities in South Sumatera whose its guidance below the Ministry of Research, Technology and Higher Education Kementerian Riset, Teknologi, dan Pendidikan Tinggi Republik Indonesia and in this case under control of "kopertis wilayah 2" which is as an extension hand of the ministry of research, technology and university in doing supervision, controlling, and guiding to the private universities of three provinces, those are:South Sumatera province, Lampung province, Bengkulu province. Research objects will be studied are the employees of the administrative headquarters only at 13 private universities in South Sumatera, those are: PGRI, Baturaja, IBA, Kader Bangsa, Muhammadyah, Bina Darma, Syahyakirti, Tamansiswa, Tridinanti, Islam OKI, IGM, Musi Rawas, Katholik Musi Charitas, dan universitas Palembang.

Based on the problems above, so this research was conducted by using five variables, those are: transformational leadership style, job motivation, and organizational climate as the factors affected job satisfaction and organizational citizenship behavior of the administrative headquarters employees at private universities in South Sumatera Province.

Organ states that Organizational Citizenship Behavior (OCB) as individual behavior that is discretionary, not directly or explicitly recognized by the formal reward system, and that in the aggregate promotes the effective functioning of the organization [9]. But the most commonly used dimension to conceptualize OCB is the dimensions are developed by Organ [9]. According to Organ there are five OCB dimensions, those are as follow:1.Altruism, 2.Civic virtue, 3.Conscientiousness, 4. Courtesy, 5. Sportsmanship [10].

Job satisfaction, according to Mathis and Jackson is employee's emotional condition where there is or there is no interaction between the value repayment of employee's services from the company with the level of the value of the remuneration that is desired by the employees concerned, it can be financial or non financial [10]. The dimension of job satisfaction according to Smith, Kendall & Hulin and Luthan there are some that can be used to reveal the important characteristics of work, where the people are able to respond. These dimensions are: 1) Work It self, 2) Supervision, 3). Co-Workers,4). Promotion, 5). Pay [12-13].

Robbins & Judge states that Transformational leadership inspire followers to transcend their own self interest and who are capable of having a profound and extraordinary effect on follower [2]. The impact is profound and extraordinary on these followers reflected on positive behavior that is shown at work environment so are be able to support an organization or company to reach its goal effectively. Robbins & Judge and Northouse states transformational leadership dimension, those are:1) idealize influence, 2) inspirational motivation, 3) intellectual stimulation, 4) individualize consideration [2 , 14].

Motivation is a psychological condition from the interaction result between employees' needs and external factor that affect an employee's [14]. According to George and Jones states that there are three indicators in job motivation, those are: 1). Direction of behavior, referring to behavior that people choose when they work, 2). Level of effort, referring to how hard a person attempt in working, 3).Level of persistence, focus on someone's mental in facing problem [16].

While the concept of organizational climate according to Richard M. Hodgetts "Human relations is the process by which management and workers interact and attain their objectives" [17]. Stringer states the six of organizational climate dimension, such as:1). structure, 2). Standards, *3).* Reponsibility, *4).* Recognition, *5).* support, 6). Commitment.

2 Research method

This research methodology aimed to get complete, profound, and credible data. Research methodology was used in this research is depended on research goals which have been stated before. To answer the research goals those are to know the influence of leadership (X1), job motivation (X2), organizational climate (X3) toward job satisfaction (Y1)and *Organizational Citizenship Behavior* (OCB) (Y2)of private universities' employees in South Sumatera province by using verificative research methodology that has systematically test goal about some allegations related to relationship between variables and problems are being investigated in the hypothesis. Hypothesis will be tested by path analysis for research hypothesis testing, so the determination steps and operational variables, formulation of hypothetical test, types determination and research model analysis.

Population in this research spreads out in thirteen universities with the total number of it was 517 persons, while sample was taken related to Hair opinion that the number of sample that was taken from population can be decided by determining 5 – 10 times of total variable used in an analysis design, and at least 200 samples [18]. According Hair the number of universities' employees sample that was observed is 200 persons. While for sampling technique used *proportionate simple random sampling* [18].

Technique for collecting the data in this research used communication and questionnaire filling. Questionnaire was distributed to the employees of administrative headquarters and interview was done to the employees of administrative headquarters and universities' leaders and related parties in a university.

Questionnaire instrument testing was done by using validity and reliability test. Validity shows the extent of

https://doi.org/10.1051/matecconf/201815005059

the score/value/size which is obtained really states the measurement/observation result that will be measured its validity generally questionable regarding the result of psychological or non – physical. Related to psychological characteristics, the measurement results obtained are actually expected to describe or to give score/value of another characteristics that will be main concern.

Validity is generally classified into three major categories, those are content validity, criterion-related validityand construct validity. In this research will be discussed it concerns validity to test whether the question has measured the same aspect. So that in this research construct validity was used. Reliability is an index indicates the extent to which a measuring device is reliable. Every measuring devices should has capability to give measurement result are relatively consistent over time.

Validity testing process was done by using minimum data testing of 30 respondents. Because of the sum contains elements of each question, so its correlation result must be corrected first. Then, the correlation result was compared by correlation value at Correlation *Product Moment*table. If correlation value of counting result is more than correlation value table, so the questions are valid, and vice versa.

After questionnaire is declared valid, questionnaire testing are continued to reliability testing. Measuring tool was used is *Cronbach's Alpha test.* If the value of *onbach's Alpha* counting result is higher than 0.7 so the questionnaire is reliable [19].

Analytical tools used consists of two, those are:

1. An analytical tool to describe research variables through indicators, dimensions and variables using descriptive statistical method those are average and percentage. In an average analysis, measurement scale 1, 2, 3, 4, and 5 will be completed in form of scare range so results of analysis data can be set qualitatively at a certain range. The formula was used to determine the scale range is:

Rs = (k-m) / k where: k is the largest scale value, m is the smallest scale value = 4/5 or 0.8.

So that, the measurement scale ranges are:

1 – 1.80 = strongly disagree/worst
> 1.80 – 2.60 = disagree/worse
> 2.60 – 3.40 = doubtful/disagree
> 3.40 – 4.20 = agree/good
> 4.20 – 5.00 = strongly agree/very good

For the company, if each dimensions has no positive average or disagree that is the average range value under 3.40, so it's problematic dimension and need to be followed up.

By using Delphi technique the value of measurement scale 1, 2, and 3 of each indicators, is gained a minimum cumulative percentage 68% so those indicators will be problem for a company so it need to be followed up.

Data processing tool used computer with SPSS software. Besides quantitative analysis with questionnaire data, this descriptive study also comes with qualitative exposure toward problematic result analysis. Qualitative exposure will be supported by primary data of a part of respondents interview result.

2. Analysis tool for testing research hypothesis by using *Structural Equation Modeling* (SEM) method supported by Lisrel software.

SEM structural equation model involves a number of systematic symbols. It can be seen such as in LISREL software package was used in this research. Research structure model used complete indicators, symbols, and these explanations can be seen in Figure 1

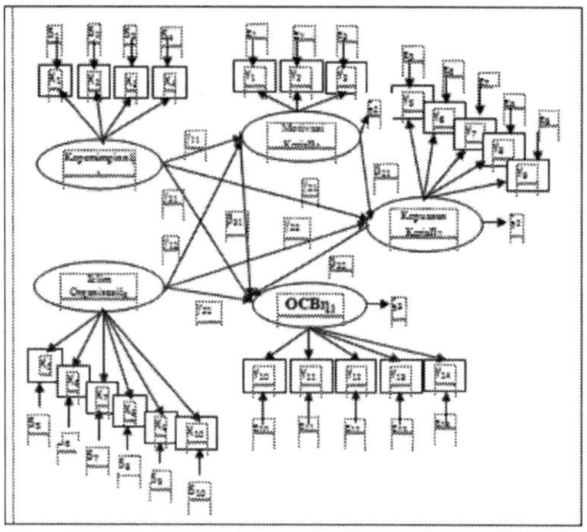

Figure 1. Research model with LISREL notation

Based on this complete research model, then it can be set structural model and measurement model. Structural equation is an equation which states the relationship between indicator variables and latent indicators.

a. The equations of Structural Model:

$$\eta_1 = \gamma_{11}\,\xi_1 + \gamma_{12}\,\xi_2 + \zeta_1 \qquad (1)$$
$$\eta_2 = \gamma_{21}\,\xi_1 + \gamma_{22}\,\xi_2 + \beta_{21}\,\eta_1 + \zeta_2 \qquad (2)$$
$$\eta_3 = \gamma_{32}\,\xi_2 + \gamma_{31}\,\xi_1 + \beta_{31}\,\eta_1 + \beta_{32}\,\eta_2 + \zeta_3 \qquad (3)$$

b. The equations of Measurement Model:

$$X = \lambda_x \xi + \delta \qquad (4)$$

3 Research findings and discussions

3.1 Research findings

Figure 2 is a computer output in *basic modelestimates* that is a complete output (hybrid) that describes great influence between all latent variable and observed variable, and between latent variable and others latent variables.

https://doi.org/10.1051/matecconf/201815005059

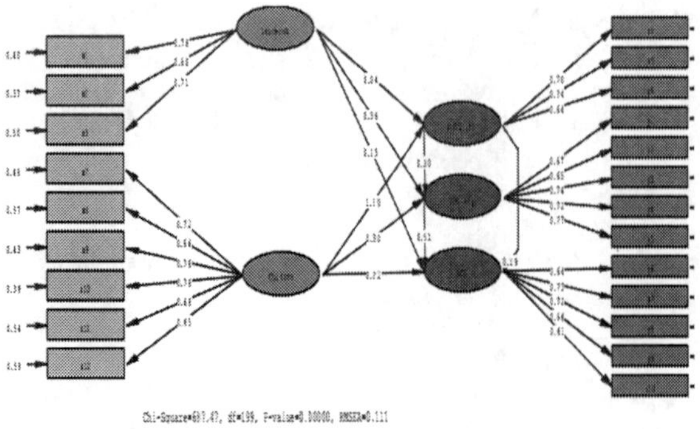

Figure 2. Complete model of research findings

Figure 3 is a computer output in t statistic value (*basic model* T-values) that is a complete output (hybrid) that describes significant level whether great influence between all latent variables toward observed variable, and latent variable's influence toward other latent variables are significant or not.

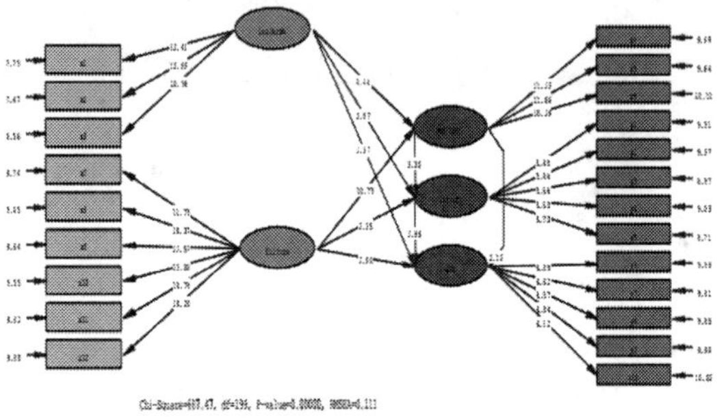

Figure 3. Complete Model (t statistic value)

For analysis needs, two pictures above are simplified for research goals and research model, as in Figure 4. that show great influence among variables and Figure 5 shows t statistic value (T-values).

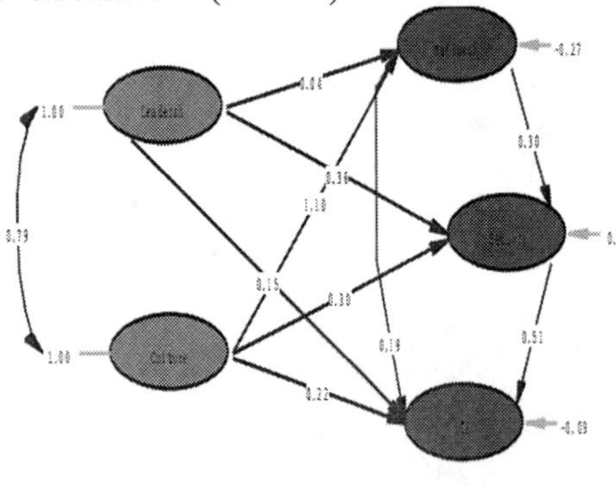

Figure 4. Great influence among variables

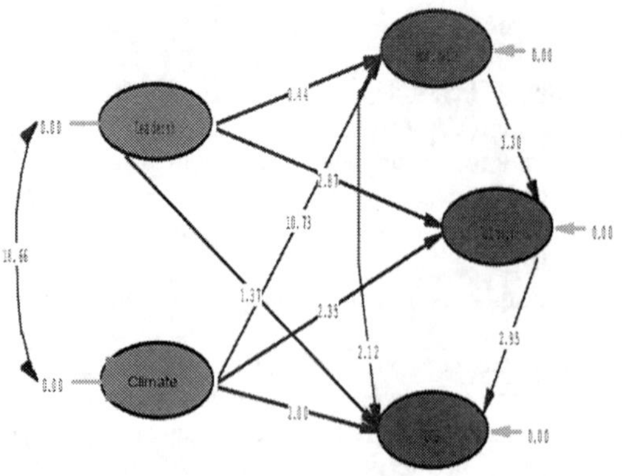

Figure 5. T statistic value

Based on influence value and t - distribution value of one variable to the other variables as in Figure 4 and Figure 5 can be known the influences among variables in Table 2.

Table 2. Great influence among variables

The Influence among Variables	Great Influence	T Value	Influence Value
Transf Leadership. -> OCB	0.15	1.37	No effect
Job Motivation -> OCB	0.19	2.12	Positive
Organizational Climate-> OCB	0.22	2.00	Positive
Job Motivation ->Job Satisfaction	0.30	3.30	Positve
Organizational Climate ->Job Satisfaction	0.30	2.35	Positive
Trans Leadership. ->Job Satisfaction	0.36	2.87	Positive
Trans Leadership ->Job Motivation	0.04	0.44	No effect
Organizational Climate ->Job Motivation	1.10	10.73	Positive
Job Satisfaction -> OCB	0.51	2.95	Positive

Besides to know the influences of each variables, it can be set either direct great influence of one variable with other variables or great influence on this moderating variables in Table 2.

Table 3. Great influence on moderating variable

No.	Plot	*Intervening Variable*	Direct Influence	Indirect Influence	Total Effect
1	Leadership->Satisfaction	Motivation	0.36	0.04 x 0.3 = 0.012	0.37
2	Leadership->OCB	Motivation	0.15	0.04 x 0.19 = 0.01	0.16
3	Leadership ->OCB	Job satisfaction	0.15	0.36 x 0.51 = 0.18	0.18
4	Climate-> OCB	Motivation	0.22	1.1 x 0.19 = 0.21	0.43
5	Climate-> OCB	Job Satisfaction	0.22	0.3 x 0.51 = 0.15	0.37

Besides in table form, influence value and t distribution value in the picture form can be seen on Figure 6.

Picture 4.5. Basic Model of Research Findings

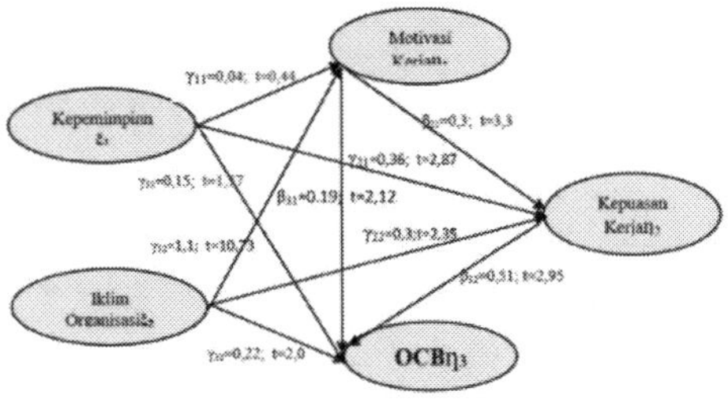

Figure 6. Basic model of research findings

Based on table 48 or Figure 6 can be known the great influence between latent variables both directly and indirectly, and so its t – distribution value.

1. leadership influence toward job satisfaction is 0.36 and significant, but if it's through job motivation, the influence of leadership is unknowable impact on job satisfaction because leadership influence toward job motivation is not significant.

2. leadership influence toward OCB is 0.15 and isn't significant, but if through job satisfaction its influence is significant and as big as 0.36 x 0.51 = 0.1836 and if its influence on job motivation, the leadership influence is unknowable its effect on OCB because the influence of leadership toward job motivation is not significant.

3. the influence of organizational climate toward job satisfaction is 0.30 and significant, if through job motivation its influence is significant with the result 1.1 x 0.3 = 0.33 so job motivation variable is important to increase job satisfaction based on organizational climate.

4. the influence of organization climate toward work satisfaction is 0.22 and significant. If through work satisfaction the effect is significant as amount 0.3 x 0.51 = 0.153.

Based on the descriptive analysis results related to all question items in questionnaire and the result of influence analysis between variables are found some findings need to be discussed, those are:

3.1.2 Descriptive Analysis Result

1. Superior has weakness in motivating his subordinates to be able to do some innovations and creative in working to make work is more effective and efficient. Non – academic employees still weak in theiron time attendance at campus because of many reasons.
2. Non – academic employees stated that financial and non financial reward doesn't accordance with the job performance.
3. Non – academic employees still weak in showing their altruism, courtesy and sportivity.

3.1.2 Causality Result Analysis

1. The quality of transformational leadership has no effect toward employees' job motivation.
2. The quality of transformational leadership has no effect toward OCB.

4 Discussions

Discussions or this research finding will be reviewed of practical and theoretical views. Reviewed from practical view attempts to set solution of findings in a campus organization and employees' attitude and behavior. But in theoretical view will be discussed why there are research findings which reject hypothesis.

4.1 The Discussion of Descriptive analysis Result

4.1.1 The quality of Superior is weak.

From five factors of transformational leadership those are needed by a superior, according to Robbins & Judge and Northouse, P.G. [2 , 13], three factors among them are still weak on the leadership at private universities, those are:

a. based on this research is known that the leaders of private universities observed haven't had to behave for affecting their subordinates with strong emotion but positive.
b. based on research findings are known that the leaders of private universities observed haven't been able to give inspiring motivation to their subordinates yet.
c. based on research findings are known that the managers or the leaders of private universities observed

haven't been able to give inspiring motivation to their subordinates yet.

4.1.2. Campus presence is Weak

The presence of employees on campus, either their arrival, existence on work field or returning time has not shown the presence in accordance with applicable rules. Laziness, feeling bored, saturated with job and traffic jam are two reasons why the employees came late.
The employees are outside because of their personal business be one of the other reasons. And also unconditional home time is the cause of why the presence of employees in campus is still constrained. Management should provide a system which is able to monitor the presence of employees in campus. *hand-key* machine is one of monitoring presence tools, the obligatory to get superior' permission to go out of the campus is also one way to keep employees' existence in campus.

4.1.3 Financial and non financial reward is weak

In this research, reward is given by universities to non academic employees are still weak, it is shown by employees' attitude that still lack of spirit in working because they think that there is no reward if they are success in doing something and in contrast there is also no punishment for them who aren't able to finish their task.

Wage is financial compensation that is so needed at least to fulfill our daily needs. Corresponding to the motivation concept of Maslow, the lowest requirement of an employee are fulfillment of clothes, food, and shelter. Therefore, salary or wage is so needed. It will be different if an employee has been in the highest level of Maslow hierarchy that is self actualization. In this level, an employee willing to spend their wage to do a specific job.

Because of this research used non academic employees' opinion of private universities, so the need of financial wages is still high. College parties should be able to increase the income, not only from tuition fee, but must be able to get funs of other sources. Through more effective and efficient working process and fund income of many sources, so larger opportunities may occur so significant increase in wage for the employees can be fulfilled.

4.1.4. Altruism, politeness and sportsmanship is weak.

OCB is a voluntary attitude of someone in doing tasks beyond his responsibilities and duties to sake the progress or profit of his organization. the weakness of three OCB indicators are caused by many reasons as stated by non academic employees in private universities, such as:

1. This attitude weakness of altruism is indicated by unwillingness to work on the colleagues work who are not present, the reason is they didn't want to interfere with someone's job and other said that they worried to make mistake because it isn't his job and it relate to persons' responsibility. Beside that altruism weakness is shown by giving services to the students, lecturers, and guests with laziness and half-hearted services. Their reasons are they feel disturbed when they are working someone asks for help, because it can disturb their concentration.
2. Weak decency – This attitude is shown by the non – academic employees by unwillingness to be involve in other activities those are not part of their tasks, their reasons are lazy to do some additional tasks because it will increase the time on campus, and there are some of them refused because of the wage is unbalance with their sacrifice.
3. The weak attitude of Sportsmanship – The weak of this attitude is shown by non – academic employees by their complaints, such as too many tasks to do and sometimes their annoyance had an impact to their services to the students. And sometime their complain is caused by theyaren'tinvolved in an activity which will get some wages. Frankly they showed that they don't like and try to find these activities' weakness in order to make as if these activities failed or run unwell.

4.2 The Discussions of Inferential Analysis Result.

4.2.1 The transformational leadership quality of Superiors have no effect to the employees' job motivation

According to descriptive exposure above that the transformational leadership quality at private universities observed is still low, so the capability to increase employees' motivation is low, too.

As we know that transformational leadership seeks to improve relationship that increase motivation and moral. The leader with this type motivates his subordinates to do their job or task better than their subordinates want to and higher than previously thought. Transformational leadership prioritize agreement on what is done by subordinate and promising rewards to be earned when they are able to be achieved, less appreciate the importance of motivating employees. Based on these research findings, where the transformational leadership quality has no effect toward job motivation indicated two alternatives, those are: 1) the leadership implementation is chosen is transactional leadership, and not a transformational leadership as desired by non – academic employees; 2) the implementation of transformational leadership that have been applied by the superiors aren't as desired by non – academic employees expectation.

And the impact is employees' motivation haven't formed well to work better. So that, increasing work ability as a superior needs to be improved by management through the chairman's guidanceor through the participation of leadership trainings.

4.2.2 Superior leadership quality has no effect toward OCB

Research findings showed that universities transformational leadership in South Sumatera province have no effect toward OCB.

The same as the result findings above which states that transformational leadership has no effect toward job motivation, and it didn't affect to OCB. So that, leadership style which is applied by the superior to the non – academic employees is non – transformational style, that is transactional style.

Based on these finding results are known that transformational leadership owned by the private universities' leaders are unable to encourage non – academic employees to show their OCB attitude at the universities where they work.

Therefore, the previous application of OCB by the employees was not the effect of how the leader or superior are able to plan and control OCB practice, but OCB' application in the universities are supported by other factors. The results supported a research was conducted by Yustina.

5 Conclusions

After finishing research formulation, research model, research hypothesis, and continued to collecting data process, statistic analysis and discussions of research findings, so in this part is research conclusion.

1) Conclusion based on descriptive analysis are known that:
    a. Transformational leadership of private universities in South Sumatera province was good. The universities leaders have had and given ideally effect and inspiring motivation to their subordinates, however intellectual stimulation to their subordinate is still weak so the subordinates are less motivated to create creative and innovative ideas in increasing their performance.
    b. Job motivation of non – academic employees of private universities in South Sumatera province is good. The employees feel what have been given by university were enough to give them satisfaction so it can motivate them to work harder. However the direction of employees' behavior is still underestimated, it showed that the employees are lack of motivation to come on time to the office.
    c. Organizational climate of private universities in South Sumatera province is conducive. It proves that university has been an effective organization because employees feel that university' party has involve them into organization. These involvement are in communication, team work, and participation in implementation of organizational policies. However reward forms are depended on their performance are given by the universities considered are still less.
    d. Job satisfaction of private universities non academic employees in South Sumatera province is high. The employees feel that the universities have fulfilled all needs they expected. However, the satisfaction on salary or wage is still weak. It showed that the salary or wage is given by universities still under their expectation.
    e. OCB of private universities non – academic employees in South Sumatera province isn't good, it can be seen from the average value of OCB dimension, which is only two dimension can reach above 4.0. while there are three dimension under this value. The leaders of universities are still weak in motivating their employees to voluntary work beyond their responsibility to reach organization' progress and profit. But, the attitude of civilian employees' virtues is considered strong, where they want to follow organizational changes and depelopments and want to read announcement of management. besides that, careful attitude shown by the employees is high by willing to be on campus in accordance work schedule and speak as necessary in work even phone.

2) Conclusion based on the inferential analysis, are known that:

According to causality analysis, known that there are two rejected hypothesis those are (1) there is no influence of transformational leadership toward OCB **and** (2) there is no influence of transformational leadership toward job motivation.

So that, causality research findings which support hypothesis of this research adding new and reinforcing evidence like the previous researches.

6 Implications

Research process is begun from early stage to the final stage have their strength and weaknesses either correlate to real sector that is for private universities or further researches.

6.1 Suggestion for Management

For private universities management are suggested to do some trainings at middle management level or for the supervisors so will be able to behave as a closer leader with non – academic employees staff, able to control employees' performance for administrative academic task, so the data and learning information are saved well. Universities parties should give rewards or punishments to the employees in a balanced manner according to their performance, with expectation it can motivate the employees to work harder and will be able to increase OCB' quality. Beside that OCB quality also can be increased by creating an activity such as employees gathering to keep good relation among employees.

6.2 Suggestion for advanced research

For research advanced are suggested to do advanced research toward transformational leadership in affecting job motivation and OCB by adding other variables and by expanding the scope of similar object studied.

References

1. M, Bhutto, Job Satisfaction as a Predictor of Organizational Citizenship Behavior A Study of Faculty Members at Business Institutes (2012)
2. S. P. Robbins, T. Judge. *Oganizational Behavior*, 12, 382-387 (New Jersey: Prentice Hall, 2007)
3. K. Chen, *Factors affecting job satisfaction of public sector employees in Taiwan,* [Dissertation] (Taiwan: Nova Southeastern University, 2004)
4. N. P. S. Mariatni, *Pengaruh Kepemimpinan Terhadap Disiplin dan Kinerja*, (Dosen Universitas Warmadewa Denpasar, Tesis.Udayana : Magister Manajemen, 2007)
5. P. Anoop, M. Lokman, International Journal of Hospitality Management, **28**(2), 254-262 (2009)
6. R. K. Purvanova, J. E. Bono, J, Dzieweczynski, Human Performance, **19**(1), 1-22 (2006)
7. M. Mehryar, C. Francoise, Int. J. Mgmt Res. & Bus. Strat (2015)
8. M. L. Castro, N. Martin, SA Journal ofindustrial psychology, **36**,1-9 (2010)
9. D. W. Organ, *Organizational Citizenship Behavior. Ilustration Edition*, (Sage Publication, Inc, 2006)
10. D. W. Organ, P. M. Podsakoff, S. B. MacKenzie, *Organizational Citizenship Behavior* (Its Nature, Antecendents, and Consequences. California: Sage Publications, Inc, 2006)
11. R. L. Mathis, J. H. Jackson. *Human Resource Management,* **10, (**Thomson South Western, United State of America, 2003)
12. P. C. Smith, L. M. Kendall, C. L. Hulin, *The measurement of satisfaction in work and retirement, (*Chicago, IL: Rand McNally, 1969)
13. F. Luthans, *Organization Behavior*, **8**, (New York: The McGraw-Hall Co,INC, 2006)
14. P. G. Northouse, *Leadership : Theory and Practice* **6**, 174-178 (California: Sage, 2013)
15. S. Danim, *Leadership Motivation and Group Effectiveness*, Issue Rineka Cipta, 2004)
16. G., Jones, *Understanding and Managing Organizational Behavior,* **4**, (Pearson Prentice Hall, 2005)
17. H. M. Richard, *Modern Human Relations at work*, **8**, (Ft Worth: Harcourt College publishers, 2002)
18. J. F. Hair, *Multivariate data analysis*, **7**, (New Jersey: Pearson Education Inc, 2010)
19. J. C. Nunnally, *Psychometric Theory,* **2**, (New York: McGraw-Hill Book, 1978)

The Role of Talent Management in the Relationships between Employee Engagement: A Study of GLCs

Afdzal Aizat Ramli[1,*], *Aerni* Isa[1], *Nur Lyana* Baharin[1] , and *Hazril* Izwar Ibrahim[2]

[1]Department of Management and Human Resources Universiti Tenaga Nasional, Kampus Sultan Haji Ahmad Shah, Bandar Muadzam Shah, Pahang
[2]School of Management, Universiti Sains Malaysia, Pulau Pinang, Malaysia

Abstract. Talent management practice is crucial in relating the potential talent. The GLCs need to have a potential talent in order to continue the value creation for business competitiveness. Moreover, talent that possesses high level of commitment and engagement will add value in the organization. This study aims to examine the relationship between talent management practices and employee engagement of employees at GLCs. The sample included 230 employees, which were selected randomly. For this study the method for data collection uses a structured questionnaire and were analyzed using partial least square. The hypothesis in this study is accepted whereby on the empirical evidences, it proves the relationship between talent management practices and employee engagement..

1 Introduction

In the contemporary world of work, to compete effectively, companies not only must recruit the top talent, but must also inspire and enable employee to apply their full capabilities to their work. GLICs are investment arms of the government that allocate government funds to the GLCs. In addition to having ownership in GLCs the Malaysian government also has an influence on the appointment of members of the board of directors and senior management positions. Many GLCs are rapidly growing, improving productivity and showing strong performance in their core businesses. GLCs are leading the way in performance in certain sectors such as banking, with Maybank and BCHB/CIMB showing significant improvement in their fundamentals. In addition to that, some GLCs have made waves in the international market, raising the profile of Malaysian businesses internationally, as well as raising industry standards.

To strive for success in the era of globalization, GLCs need to have a potential talent to continue the value creation for business competitiveness. Talent that possesses high level of commitment and engagement will add value in the organization. Contemporary organizations need employees, who are psychologically connected to their work; who are willing and able to invest themselves fully in their roles; who are proactive and committed to high quality performance standards. They need employees who feel energetic and dedicated, i.e., who are engaged with their work the applicable criteria that follow [1, 2] define an engaged employee as a worker who is fully involved in and enthusiastic about his or her work. Engagement was essentially defined as "an innate human desire to contribute something of value in workplace Employee engagement as a key to the retention of talent (one-of-a kind hire in 100 employees; [3] is an area in which the leads has been taken by practitioners [4-9]. Employee engagement is viewed as an important construct because it is has been linked to increasing the retention of top talent as an outcome of the talent management process for e.g. [10].

Talent management has now been identified as a critical enabler for the Government Transformation Programme (GTP), Economic Transformation Programme (ETP) and the 10th Malaysia Plan (2011-2015) to propel Malaysia towards the competitive status of a high income economy and achieve a vision 2020. It is time to tap the local talents to move forward. Malaysians need to synergize the strength to achieve the Strategic Reform Initiative of ETP in developing quality workforce and reducing dependency on foreign labour. GLCs should be given the opportunity to develop their own capacities and capabilities and consider sourcing talents with a global vision and culture.

Bureaucratic meddling, conflicting objectives, over-centralization of decision making and inadequate capitalization and the lack of managerial skills are also pressing problems for GLCs [11]. Directorship and senior managerial positions in GLCs where political patronage positions for retired military and high level civil servants, or for relatives and friends of powerful political leaders [12-13] According to [14] GLCs are facing issues on internal control, lacking in strategic direction resulting in poorer return on capital and shareholder value, lower productivity, higher gearing ratio, inefficient procurement handling, and ineffective performance management system.

Another issue that raised attention is the weak linkage between employee's performance and the reward scheme in GLCs [15]. The issues on ineffective performance management system thus creating job dissatisfaction. Among the employees which leads to changing of job attitudes, followed in the longer term by responses that are more retaliatory such as quitting [16]. When the evaluation process is not transparent, the outcome will be imbalance and it may give detrimental impact on employee engagement. Based on the finding by [17], it shows the unfairness in the performance management system and top management will always have the right to override the KPI points. Furthermore, the top management have the power to reduce or increase the total KPI points of the employees.

1.1 Talent Management Practices.

According to TalentCorp (Malaysia), talent management practices refer to a process of identifying, developing the potential talent. GLCs need potential talent to strive a success in an organization, as most of the GLCs expanding the operation internationally. Through the implication of talent management practices it helps the organization in retaining the potential talent to stay in the existing organization. Based on the study done by [18], there is insignificant relationship between talent management and employee engagement. Therefore this study will be more beneficial as empirical evidence is provided to prove the relationship between talent management practices and employee engagement. Employees that highly engaged contribute greater outcome, they are aware of business context and work closely with other colleagues at the workplace. Employee engagement is a vital for the organization in retention of top talent as being proposed by existing studies, for example by [10]. Furthermore, the key driver of employee engagement is the strong implementation of career development and this relates with talent management practices. As effective talent management practices can result in more engaged employees and lower turnover, it will bring forward positive outcome in enhancing the performance at GLCs [19].

1.2 Employee Engagement.

Employee engagement as a key to the retention of talent. Employee engagement has a substantial impact on employee productivity and talent retention. [20] see engagement as engaging employee into the strategic process which is different than the attitude and the behavior. Engagement is part of the "fair process" with explanation and clarity of expectations. Based on their findings, employees care about the justice of the process as much as they do the outcome itself. They claim that fair process affects people's attitudes and behaviors. In their model: Engagement is part of the fair process which is followed by trust and commitment (attitude). Then the employee voluntarily cooperates (behavior) and exceeds expectations (strategy execution). The authors define engagement as "involving individuals in the strategic decisions that affect them by asking for their input and allowing them to refute the merits of one another's idea and assumptions" (p.175).

1.3 The Relationship between Talent Management Practice and Employee Engagement

Employee engagement is the act of an employee being involved in, enthusiastic about, and satisfied with his or her work. Talent management practices and policies that demonstrate commitment to human result in more engaged employees and lower turnover [21]. The study done by [22] shown that there is a significant contribution between talent management practice (employee career development and rewards and recognition) and employee engagement. Based on the study done by [18], there is insignificant relationship between talent management and employee engagement. Therefore this study will be more beneficial as empirical evidence is provided to prove the relationship between talent management practices and employee engagement. Employees that highly engaged contribute greater outcome, they are aware of business context and work closely with other colleagues at the workplace. Employee engagement is a vital for the organization in retention of top talent as being proposed by existing studies, for example by [10]. Furthermore, the key driver of employee engagement is the strong implementation of career development and this relates with talent management practices. As effective talent management practices can result in more engaged employees and lower turnover, it will bring forward positive outcome in enhancing the performance at GLCs [23]. If employees feel they have the opportunity to expand their career, they are more likely to have positive feeling about the organization. In this context, career development for employees at GLCs is crucial, if the employees do not have the opportunity for career growth this lead to job dissatisfaction. Therefore, every organization needs to have a strong support on talent management practice which results in more engaged employees and lower turnover. The challenge today is not just retaining talented people, but fully engaging them, capturing their minds and hearts at each stage of their work life. The literature provides considerable support for the significant link between talent management practices and employee engagement. Therefore, the following hypothesis is proposed:

H1: There is a significant positive relationship between talent management practices and employee engagement.

2 Methodology

A questionnaire survey was used to gather primary data for this study. Stratified random sampling technique is use to select the respondent for this study. The respondents for the study are drawn from GLCs employees in Malaysia. Questionnaire were reported using 5 likert scale ranging from (1) strongly agree to (5) strongly disagree. In this study, the questionnaire for talent management practices is adapted from research by [24] and [25] and employee engagement will be measured by 9-items Utrecht Work Engagement Scale (UWES-9) based from [26].

To investigate the research model, the statistical software application SmartPLS 3.0 [27] is used to compute the PLS path model. In evaluating and reporting the results, it is done by following recent guidelines for PLS-SEM given by [28] and [27]. Two step approaches were used, (1) confirmatory factor analysis, to confirm validity and reliability (2) structural equation model, to assess the path model [29]. IBM Statistical Package for Social Science (SPSS) version 22 was also used for importing data and to explain the descriptive statistics.

3 Results and Analysis

Table 1: Demographic

Items	Characteristic	Frequency	Percentage
Age	20 and below	4	1.9
	21-38	146	68.5
	39-49	44	20.7
	50 and above	19	8.9
Gender	Male	90	42.3
	Female	123	57.7
Marital Status	Married	131	61.5
	Single	78	36.6
Education	Secondary	8	3.8
	School	32	15.0
	Diploma	136	63.8
	Degree	35	16.4
	Master	1	.5
	Ph.D.		
Service	Less than 5	70	32.9
	years	62	29.1
	6-10 years	44	20.7
	11-15 years	35	16.4
	Above 16 years		
Salary	RM2,000 and	44	20.7
	below	46	21.6
	RM2,001.00 -	47	22.1
	RM 4,000.00	74	34.7
	RM4,001.00 -		
	RM 6,000.00		
	RM 6,001.00		
	and above		

Table 1 above are the demographic profile of the respondent. The respondents aged between 21 and 38 years constituted the biggest category (at 68.5%). Those aged between 39 to 49 years constituted 20.7% of the sample. The respondents aged between 40 and 49years constituted the remaining 18.4% of the sample. It can be concluded from the above table that majority (57.7%) of the respondents for this study are female. The remaining 14.6% of the respondents are males. At the same time majority (61.5%) of the respondents were married. The remaining 36.6% of the respondents are single. Maximum numbers of respondents (63.8%) are degree holder, 16.4% of the respondents are master holders, and 15.0% of the respondents are diploma holder. Majority of the respondents (32.9%) are having less than 5 years of experience. As for the salary, it could be conclude that, most of the respondent salary are at the range of RM 6001.00 and above (34.7%).

3.1 Measurement Model

3.1.1 Convergent validity

In analyzing the measurement model convergent validity need to be assess. Convergent validity is the degree to which that constructs that are expected to be related are, in fact, related. For convergent validity, it will be verified through average variance extracted (AVE) [30]. AVE measure the amount of variance captured by a construct by calculating the ratio of the amount of the variance captured by the construct and the measurement variance. Convergent validity is adequate when constructs have an AVE of at least 0.5 with factor loading of each item at ≥ 0.6 [31, 32]. Base from table 1 above, all constructs statistically achieved the AVE >0.5, where talent management practice with AVE 0.582 and employee engagement with 0.773. All factor loading are also range from 0.651 to 0.905 which is as recommended by [31]. Comply with the aforementioned scholar's convergent validity of each dimension inspection standards, the various aspects of this study have convergent validity.

3.1.2 Discriminant validity

Discriminant validity is the degree to which item distinguish among the construct [33]. Discriminant validity was measured by evaluating the correlations between constructs and the square root of AVE for that construct [31]. Besides that, the square root of AVE must be higher than the correlations values in the row or column. [34]. Table 2 shows the discriminant validity of this study. Therefore, the measurement model is accepted as it fits all the criteria of convergent and discriminant validity.

3.1.3 Reliability analysis

Reliability analysis was done to determine the reliability of the questionnaire. As for this study reliability will be assess through (1) internal consistency and (2) cronbach's alpha. To measure internal consistency the composite reliability were tested. As recommended [34] for composite reliability, all the items must be more than 0.70 as the minimum value is 0.70. From table 2 we can see that all value for composite reliability for all the items are more than

0.70. The value of composite reliability for items of talent management practice is 0.967, and value for employee engagement items is 0.968. Therefore it could be concluded that high level of internal consistency reliability have been demonstrated among latent variables. As for the cronbach's alpha, the suggested value for cronbach's alpha is above 0.6 [35]. Based on the result, all the cronbach's alpha value for each construct are above 0.6 where the value for talent management practice are 0.964 and employee engagement are 0.963. Align with the composite reliability, all the construct are exceed the suggested value and it's showed all the measurement are reliable.

3.1.4 Structural model

After reliability and validity testing is completed, next is the path for the PLS model coefficient test and predict estimated structural model analysis. Consistent with [36] bootstrapping (500 sample) was used to generate t-statistics. This allows us to access the statistical significance of the path coefficients. The use of the t-value to estimate the p-value, to test the significance of coefficient the carried out to determine the hypothesis whether was established [37]. Examining the path coefficient value, a researcher is able to know the strength of the relationship between variable [38]. Table 4 and Fig. 1 show the result of the structural model from the PLS output.

Based on the result, hypothesis which were proposed were accepted as the t-value is significant at $p<0.01$. In structural model, R^2 of the variables could expresses the proportion of variance in the independent variables (talent management practice) explained by the structural relationships for the dependent variable (employee engagement). Based on the result in table 4, R^2 was found to be 0.614 which indicates that the PLS regression model can explain 61.4% of total variance in employee engagement. It is similar with the assessment suggested by [39] 0.26 substantial, 0.13 moderate and 0.02 weak. Thus, in this study R^2 here is considered substantial. Another criterion to assess the quality of the model is using the blindfolding procedure to generate the cross-validate communality and cross-validated redundancy, where Q^2 values will be calculated. According to [31] the model will have predictive quality if the cross-redundancy value was found to be more than 0, otherwise the predictive relevance of the model cannot be concluded. From the results in table 4, the obtained cross validated redundancy was found to be 0.417. This result supports the claim that the model has an adequate prediction quality.

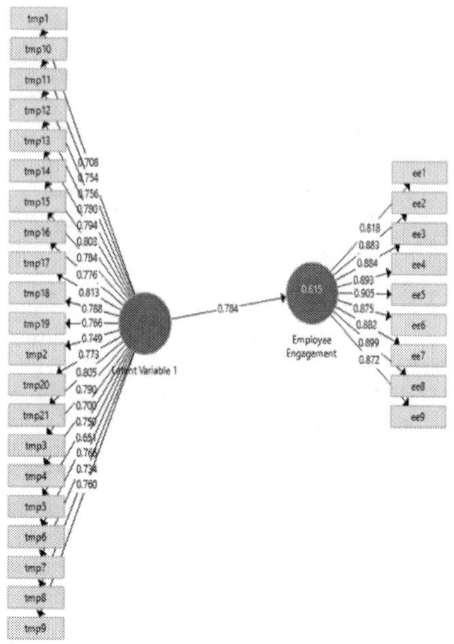

Fig. 1. structural model

Table 2: Result of measurement model

Construct	Measurement items	Loadings	AVE	CR	Cornbach's Alpha
Talent management practice	TMP1	0.708	0.582	0.967	0.964
	TMP2	0.749			
	TMP3	0.79			
	TMP4	0.700			
	TMP5	0.750			
	TMP6	0.651			
	TMP7	0.766			
	TMP8	0.734			
	TMP9	0.760			
	TMP10	0.754			
	TMP11	0.756			
	TMP12	0.780			
	TMP13	0.794			
	TMP14	0.803			
	TMP15	0.784			
	TMP16	0.776			
	TMP17	0.813			
	TMP18	0.788			
	TMP19	0.766			
	TMP20	0.773			
	TMP21	0.805			
Employee engagement	EE1	0.818	0.773	0.968	0.963
	EE2	0.883			
	EE3	0.884			
	EE4	0.893			
	EE5	0.905			
	EE6	0.875			
	EE7	0.882			
	EE8	0.899			
	EE9	0.872			

Note: AVE= Average Variance, CR=Composite reliability, TMP=talent management practice, EE=employee engagement

Table 3: Discriminant validity

Construct	#1	#2
#1 Talent management practice	**0.784**	
#2 Employee engagement	0.763	**0.879**

Note: Values on the diagonal (bolded) are square root of the AVE while the off-diagonals are correlations

Table 4: Summary of structural model

Hypothesis	Standard data	Standard error	t-value	Decision	R^2	Q^2
Talent management practice -> Employee engagement	0.784	0.786	24.883**	Accepted	0.614	0.437

Note: t-value is greater than 2.33 at **p<0.01

4 Conclusion

It has been empirically shown the bigger the R^2, the higher the predictive power of the model. The proposed model shows that 61.4 % of the proportion in variance. This finding concurs with previous study by [40,22]. Therefore this study would be beneficial for GLCs as there are limitations on empirical evidence on the study of talent management. Based on the empirical evidences in this study it proves the positive relationship between talent management practices and employee engagement. As Malaysia aspires to transform into a developed and more competitive economy by 2020, talent will play a crucial role in order to strive for success and to sustain strong economic growth. In order for the GLCs to retain the talent it is important for the organization to focus on nurturing and developing talent of the human capital in GLCs.

The authors also acknowledge that the study is somewhat limited in size and scope in that we only test on few GLCs. The respondents were drawn from only five GLCs, while different types of GLCs may report different results if surveyed.

References

1. G. Bakker, A.B., M. P. Leiter, *Work engagement: A handbook of essential theory and research* (New York, 2010)
2. T. Tritch, Gall. Man. J., (2003)
3. E. K. Glen, C. H. Elaine, L. S. Mathew, ACAD MANAGE J **49**(5), 1031-1057 (2006)
4. A. Parsley, Man. Ser., **50** (1), 10-11 (2006)
5. R. Baumruk, B. Gorman, R. E. Gorman, J. Ingham, "Why Managers are Crucial to Increasing Engagement" (2006)
6. C. Woodruffe, Brit. J. Admin. Man., **50**, 28-9 (2005)
7. Gall. Man. J., http://gmj.gallup.com (2006)
8. M. Bennett, A. Bell, *Leadership & Talent in Asia* (Wiley, Singapore, 2004)
9. Hay Group, Inc. Reward program implementation – missed opportunities. www.haygroup.com (2002)
10. J. Bhatnagar, Talent Management Strategy of Employee Engagement in Indian ITES (2007)
11. M. M. Shirley, P. W. Patrick, *Public versus private ownership*, **2420**. World Bank Publications, Strategic HR Review **5**(2), 24-7 (2000)
12. W. L. Megginson, R. C. Nash, M. V. Randenborgh, J. Fin., **49**(2), 403-452 (1994)
13. A. R. Vining, E. B. Anthony, Pub. Choice, **73**(2), 205-239 (1992)
14. Y. Azman, An Overview of a Nation Building Institution http://tkr.khazanah.com.my/nation-building-institution/ (2004)
15. A. B. Abdullah, *"Culture of High Performance for GLC's", Keynote Address Presented at the Seminar on Culture of High Performance for GLCs"* www.pmo.gov.my/WebNotesApp/ABDULLAH.nsf/ (Theatrette, Ministry of Finance, Putrajaya, 2004)
16. E. Vigoda, J. Voc. Behav., **57**, 326-347 (2000)
17. R. Abu Bakar, *Understanding Factors Influencing Employee Engagement: A Study of the Financial Sector in Malaysia. Doctoral Thesis.* (School of Management, RMIT University, 2013)
18. K. Rooper, *The relationship between employees' perceptions of talent management practices and levels of engagement: A Botswana organisation case study*, (Master Thesis, 2009)
19. A. Isa, H. I. Ibrahim, Int. J. Bus., Eco. L., **4**(1), (2014)
20. W. C. Kim, R. Mauborgne, *Blue Ocean Strategy* (Boston, Massachusetts: Harvard Business School Publishing Corporation, 2005)
21. B. Lockwood, *The political economy of decentralization* (Edward Elgar Publishing, 2006)
22. N. E. Alias, H. Mat Noh, R. Hassan, J. Hum. Res. Man. L. Stud., **2**(2), 227-242 (2014)
23. A. Isa, H. I. Ibrahim, Int. J. Bus., Eco. L., **4**(1), (2014)
24. P. Bernsen, M. Segers, H. Tillema, Int. J. Hum. Res. Dev. Man., **9**(4), 358–373 (2009)
25. H. Baert, J. W. Learn. (2010)
26. B. Schaufeli, W. H. Macey, K. M. Barbera, Edu. Psycho. Mea., **66** (4), 701-716 (2006)

27. Hair, Ringle, Sarstedt, J. Market. T. Pra., **19**(2), 139-152 (2011)
28. W. W. Chin, B. L. Marcolin, P. R. Newsted, Info. Sys. Res, **14**(2), 189-217 (2003)
29. J. C. Anderson, W. G. David,"Structural equation modeling in practice: A review and recommended two-step approach." *Psychological bulletin* 103, no. 3, 411 (1988)
30. Sabri Ahmad, Zulkurnain, N. N. A., Khairushalimi, & Izzati, F., (2016)
31. Fornell, Claes, David F. Larcker, J. Market. Res., 382-388, (1981)
32. J. F. Hair, E. Rolph, B. J. B. Anderson, C. B. Wiiliam, *Multivariate data analysis: A global perspective* (NJ: Pearson, 2010)
33. C.M.K. Cheung, M.K.O. Lee, **4**(1), 161-172 (2010)
34. D. Compeau, C. A. Higgins, S. Huff, MIS quarterly, 145-158 (1999)
35. J. C. Nunnally, I. Bernstein, Psycho. Theo., **3**(1), 248-292 (1994)
36. J. Henseler, C. M. Ringle, R. R. Sinkovics, Adv. Int. Market., **20**(1), 277-319 (2009)
37. L. L. Bove, S. J. Pervan, S. E. Beatty, E. Shiu, J. Bus. Res, **62**(7), 698-705 (2009)
38. A. Martinez-Ruiz, T. Aluja-Banet, Stat. J., **7**(3), 265-290 (2009)
39. Cohen, Stat. Pwr. Ana, (1988)
40. M. Dass, S. Muthaly, J. Annakis, J. Cont. Iss. Bus. Gov., **21**(1), (2015)

Environmental Balance Management In Nonformal Education Learning

Abdul Rahmat[1,*], *Tulus Suryanto*[2], *Supli Effendi Rahim*[3]

[1]Gorontalo State University, Gorontalo, Indonesia
[2]Raden Intan Islamic University, Lampung, Indonesia
[3]Higher Education of Health Sciences BinaHusada, Palembang, South Sumatra, Indonesia

Abstract. This study aims to determine the direct effect of the learning environment on the learning outcomes about environmental balance, interest in learning on the results of learning about environmental balance, and learning environment learning to interest in learning. This research is a survey research using path analysis technique with the sample of research is student package C in SKB BonebolangoGorontalo 2016/2017 Lesson Year, which amounts to 60 people. There are three variables that are used as research objects, namely two exogenous variables that include learning environment and interest in learning and endogenous variable that is the result of learning about environmental balance. The results of the study indicate that: (a) there is a positive direct effect of the learning environment on the learning outcomes of environmental balance, (b) there is a direct positive influence of interest on the learning outcomes of environmental equilibrium, (c) there is a direct positive influence of the learning environment on learning interest, d) there is an indirect influence of the learning environment on learning outcomes through interest in learning.

1 Page layout

Learning as a process is influenced by several factors. As stated by the Syah (2012: 146), that "success in the learning process is influenced by two factors: internal factors (*factors from within students*), namely the state or condition of the students' physical and spiritual, interests and motivation and external factors (*external factors students*), ie environmental conditions around students. From the preliminary survey results it can be argued that the average learning outcomes on Environmental Balance are relatively low when compared with other materials.

This is due to the lack of interest in learning from students and also because of the condition of the learning environment in their poor schools. Uno [2], that learning is a process that produces behavioral changes done deliberately to gain knowledge, skills and new experiences towards the better. According to learning environment includes three components, namely family environment, school environment and community environment. So the learning environment is a condition where one can carry out the learning process.

Shah, learning is a processed activity and is a very phenomenal element in the conduct of each type and level of education [1]. Learning as a process allows one to change his behavior, Djamarah, suggests that learning is a series of activities of body and soul to obtain a change of behavior as a result of individual experience in interaction with the environment [3]. Environmental balance is the ability of the environment to overcome the pressure of nature and from human activities, and the ability of the environment in maintaining the stability of life in it. Environmental balance will be achieved when there is an interaction of organisms with environmental factors and intercomponents interactions in an environment can run proportionally [4].

Soemarwoto, explains that the environment is a space occupied by a living human being along with other living things (plants, animals, microorganisms), and inanimate objects, such as air composed of various gases, water in the form of steam , liquid and solid, soil, minerals in soil, rocks and fossils, which support each other to meet human needs.

Experts divide the learning environment into several kinds. Sartain in Purwanto, divides the environment into three parts, namely (1) external or physical environment, is everything that exists in this world that is not human, like humans, plants, water, climate, animals etc. (2) internal inveronment, is everything that has been included in us, which can affect our physical growth (3) social environment, is all other people or humans that affect us.

To be able to see the success of the process of teaching and learning activities, all factors related to teachers and students must be considered. This interest

* Corresponding author: abdulrahmat@ung.ac.id

in students is one of the signs of interest. According to the Shah, interest (interest) means a high tendency and excitement or a great desire for something [1]. The same thing is also expressed by Slameto, that interest is a sense of preference and a sense of attachment to a thing or activity, without anyone telling [5]. The same opinion by Winkel, says that interest is the tendency of sedentary subjects, to appeal to a particular field of study or subject and to feel good about learning the material.

2 Research Methods

The study was conducted within 2 months from April to May 2017. Type of research used in this study is a survey by using Path Analysis. Target population is all students of package C in SKB BonebolangoGorontalo Lesson Year 2016/2017. Testing the validity of grains of learning results about environmental balance using the formula biserial correlation is the correlation between the test score score with the total test score [6].

3 Results and Discussion

3.1 Learning Outcomes of Environmental Balance (Y)

Average score ($\overline{Y}$) is 9,467; median (Me) of 14,630; mode (Mo) of 14,891; and standard deviation (SD) of 2,456. Calculation results can be seen in the appendix. If the price of mode (Mo), median (Me) and average ($\overline{Y}$), then based on the assessment of the norm norm score of learning result of environmental balance of students tend to be high.

3.2 Learning Environment (X1)

Average score ($\overline{X}_1$)was 84,800; median (Me) of 85,804; mode (Mo) of 86.833; and standard deviation (SD) of 13,097. Calculation results can be seen in the appendix. If the price of mode (Mo), median (Me) and average ($\overline{X}_1$), then based on the assessment of the norm norm score of student learning environment tends to be high.

3.3 Interest to learn

Average score ($\overline{X}_2$) was 89,800; median (Me) of 89.260; mode (Mo) of 88.409; and standard deviation (SD) of 12,291. Calculation results can be seen in the appendix. If the price of the mode (Mo), median (Me) and the average ($\overline{X}_2$), then based on the assessment of the norm reference score of student learning interest tend to be normal.

The coefficient of the learning environment pathway to learning interest is similar to the relationship between learning interest and the learning environment. The value of the path coefficient as output is in Table 1.

Table 1. Correlations

		Learning Environment	**Interest to learn**
Learning Environment	Pearson Correlation	1	.930**
	Sig. (2-tailed)		.000
	N	60	60
Interest to learn	Pearson Correlation	.930**	1
	Sig. (2-tailed)	.000	
	N	60	60
**. Correlation is significant at the 0.01 level (2-tailed).			

From Table 1 obtained correlation coefficient r_{12} = 0,930. The value of correlation coefficient is also the value of the coefficient path. Test significance correlation coefficient obtained Sig value. (2-tailed) 0.000.Karena two-party test then the value of Sig must be divided 2. The division by two obtained 0,000. Because the value of Sig. = 0,000 < 0.050 then the correlation coefficient (path coefficient) interest in learning with learning environment significant (meaningful).

4 Conclusions

The result of learning about the balance of the student environment is directly influenced by the positive learning environment. This means that the more conducive the student learning environment the higher the achievement of their learning results about the material of environmental balance. Results of learning about the environmental balance of students are directly influenced by the positive interest in learning. This means that the higher interest in student learning the higher the achievement of their learning results about the material of environmental balance. Students' interest in learning is directly influenced by the positive learning environment. This means that the more conducive learning environment that each student has, the higher his interest in learning about the material of environmental balance.

References

1. M. Syah, *Learning Psychology*, 63-152 (Jakarta: Raja Grafindo Persada, 2012)
2. H. Uno, *Learning with PAILKEM Approaches*, 138 (Jakarta: Bumi Aksara, 2012)
3. Djamarah, S. Basri, *Learning Psychology*, 13 (Jakarta: Rineka Cipta, 2002)
4. Miarsyah, M. Luvfiati, D. Mutiara, T. Ernawati, *Natural Sciences for SMK and MAK Class XII,* 62 (Jakarta :Erlangga, 2006)
5. Slameto, *Learning and its Affecting Factors*, 57 (Jakarta: RinekaCipta, 2012)
6. Djaali, Muljono, dan Pudji, *Measurement in the education*, 90 (Jakarta: PT Gramedia, 2008)

The Strategy of Welfare Improvement for Salt Farmers in Indonesia

Bambang Sudaryana[1,*], *Puji* Pramesti[1]

[1]Politeknik LP3I Bandung West Java Indonesia

Abstract. Until now, production of salt in Indonesia is done by salt farmers and PT. Garam Indonesia as the only state company that produces salt. As the perpetrators of production which contribute greatly to the national salt production, salt farmer is considered not prosperous. In general, they depend on marine and coastal resources, use requires large investment and rely heavily on season. Climate and weather conditions that are often hostile, the price mechanism and the salt market tend to be contrast to the condition of farmers, this is what makes salt business is risky. The goals of this research are: 1. Improving the Welfare of salt farmers through a focused and effective empowerment. 2. Improving government'spolicies on salt. To conclude, Empowerment of salt farmers in Indonesia should be focused and effective, salt governance and policy revisions significantly affect the welfare of salt farmers.

1 Introduction

Indonesia is in the on-going process to Achieve food security, as mentioned in Act No.7 of 1996. It is supported by the very spacious potential of marine, with the facts that Indonesian coastline is very long, and, Indonesia is a country with Tropical climate. It is very likely that Indonesia becomes the country with largest salt producer in the world. Official Data states that the ratio of domestic salt production needs is not comparable with national results of salt. According to the calculation of demand, the need for salt in the country Reached 3,251,691 tons, while the national salt production only 1,113,118 tons. To meet the short comings of the required commodity, it is crucial to import about 2,6152 million tons. (Speech of the Minister of Maritime Affairs in the Congress of People's Salt, Bangkalan, Madura Indonesia July 2012). It's not unusual why this salt commodity can be interesting and important to be studied. Salt is one of the strategic commodities, that the country Includes it in to nine basic material needs of society. Salt is not only used for domestic consumption, but is also used for the needs of industry (pharmaceutical, mining, fertilizers and others). Problems in national salt production influences Several major issues, namely the production, infrastructure, institutional, marketing and supply demand. When we look at the history of Indonesia during the Dutch colonial period, salt is an important commodity and quite profitable, so it was monopolized private parties. (Handbook of the Netherlands Indies 1930:121).

The solution is to continue to maintain and improve their food prosperity through various government programs that support the achievement of national food security. The government policy to import salt has been controversial (Nurhayat, 2012;Sujatmiko, 2013; Firdaus, 2013; Rismana, 2013). The government policy to import salt, particularly the salt for industry is contrary to the logic of the potential length of the beach Indonesia which correlates with the ability of the national salt production. This shows the error of national government's salt policy, in the relation with the fact that Indonesia is still importing salt. Salt as agricultural commodities must be managed through a synergic policy, from infrastructure, social, economic, and education, so, that there is production continuity of salt to meet the needs of national demand. National salt category can be grouped into three general scopes, namely, the production, marketing, and, supply and demand. Problems of production refer to the difficulty of salt production that is very dependent on the climate. The technology used is still very reliant on weather factors. Indonesian that has relatively short drought climate is strongly related to the production of salt. Salt production is a line of' labour intensive business that needs a deep consideration related to capacity of substantial power. Problem of location is significant to salt works that has varies of scales. The condition of existing manufacturers that is generally covered with weak-economic farmers, and, not to mention, the ownership structure of land by landowners, and sharecroppers with a distribution system that is not fair to the tiller; their limited working capital so that farmers are trapped in debt bondage system; and the low price of salt farmers causing some people in salt works land has

* Corresponding author: Bambangsudaryana58@gmail.com

been converted and this led to the low production of salt (Latitude, 2013).

There are problems of supply and demand on the national salting, According Latitude(2013). It Consist of the supply, a certain moments, excess of supply or shortage of supply, seasonal production, monopolistic goods or no substitution. The offered salt in the form of detached and processed, it is same with the origin production of PT. Garam Indonesia and also Farmers. On the demand side, there are problems related to its sustainability, the increase of demand, in-elastic demand. It is especially for food consumption, the fishing industry, and food industry, the demand is quite jumbled. The last one, problem also covers the scattered and uneven request. Since the production is abundant, the power position of farmers is in weak position, so, price is fluctuated each year. Potential production of domestic salt has a different ability due to technical differences between the salt production of PT. Garam Indonesia and salt farmers. In rare circumstances that caused the decline of the salt supply in the country (due to climate), the government made policies to import salt. The import of salt is done to meet the needs of a salt shortage in the country but on the salt trade system which is apply no, it considered needs some improvement.

This certainly complex problems that can't be solved by only one party, and in a short time. It requires master plan from government as a regulator, as well as the executor who has power in politics and national economy. Synergy between government institutions themselves, for example the Ministry of Maritime Affairs and Fisheries, the Ministry of Industry, Ministry of Health, Ministry of Commerce, Ministry of Finance, Local Government and DRPD district centres of salt, and all parties associated with policy making and execution of the production chain, distribution, and national salt consumption is considered very important. This synergy will be a long-term solution for Indonesia's inability to meet the needs of national salt, both households and industry. It is proper for us as a nation to look far ahead in the settlement of the current problems in the field with one stop solution. So, the expected settlement does not patchy and only a mere momentary solution.

Conference on the Food and Agriculture in 1943 initiated the concept of a "secure, adequate and suitable supply of food for everyone". The definition of food security varies greatly, but generally refers to the definition of the World Bank (1986) and Maxwell and Frankenberger (1992), namely "access by all people at all times to the sufficient food for healthy living (secure access at all times to sufficient food for a healthy life). Food Agriculture Organization (FAO) states that the supply that dependent on food imports to the country of more than 100 millions, will make it difficult to advance the nation and independent. One of the serious problems faced by Indonesia is self-sufficiency, and food security. As a country endowed with fertile agricultural natural resource, our nation also faces the issue of resilience and self-sufficiency. Food politics is supposed to be in favour to the consumer, where the orientation is meeting the needs of consumers, especially for salt commodities, but it becomes pragmatic. Imports can actually be fatal, both in terms of import costs that need special allocations from the state budget, not to mention the risk the salt farmers that certainly would be jeopardized income. Today, the challenges of increasingly complex salt farming sector must be addressed not only about how the government's ability to increase production in order to maintain the resilience and food security, but more than that, also regarding how to improve the welfare of the actors in the agro-food sector, such as jobs and increase the income of farmers.

It should be recognized that the agricultural sector policies on salt is prioritizing the aspects of production to meet the needs of local consumers, while the welfare and needs of farmers it is still marginalized.

PT. Garam Indonesia is a State-owned Enterprise which has the vision of becoming the leading salt company in Southeast Asia and providing value added to the stakeholders. Salt production and marketing of raw materials and processed salt of PT. Garam Indonesia tends to fluctuate from year 2004-2011. And it is in accordance with that criticism by the Ministry of Maritime Affairs and Fisheries to produce salt.

Productivity of PT. Garam Indoensia is considered very low, but this state receiving State Capital Participation in billions of rupiah. PT. Garam Indonesia consumption of salt production in 2012 is far from complete, it is only capable to produce 340,000 tons. Yet, the first report to the Ministry of Industry reaches 385,000 tons.

Displays the translation for the Conference of the Food and Agriculture in 1943 which initiated the concept of a "secure, adequate and suitable supply of food for everyone". The definition of food security varies greatly, but generally refers to the definition of the World Bank (1986) and Maxwell and Frankenberger (1992), namely "access by all people at all times to the sufficiency of food for healthy living (secure access at all times to sufficient food for a healthy life). "According to the food Agriculture Organization (FAO), supply dependence on food imports to the country is more than 100 million. It will make it difficult to advance the nation and independent. One of the serious problems faced by Indonesia is self-sufficiency and food security. As a country endowed with fertile agricultural natural resource, our nation also faced the issue of resilience and self-sufficiency. Food politics should be in favour of the consumer, where the orientation is meeting the needs of consumers for salt commodities salt. It is now still pragmatic. Imports can actually be fatal, both in terms of import costs that need special allocations from the state budget, not to mention the risk the salt farmers that certainly would be jeopardized income. Today, the challenges of increasingly complex salt farming sector must be addressed not only about how the government's ability to increase production in order to maintain the resilience and food security, but more than that, also regarding how to improve the welfare of the actors in the agro-food sector, such as jobs and increase the income of farmers.

During this time, the domestic market requires two types of salt, the salt that is destined for consumption and industry. Salt Consumption with Na Cl of 94.7

https://doi.org/10.1051/matecconf/201815005062

percent is used not only for consumption but also for salting and human consumption of food and livestock. Meanwhile, Industrial salt with Na Cl content of 97 percent or higher levels of salt consumption is widely used for the leather industry, pharmaceuticals and textiles. The need for salt industry must import 100 percent or it reaches 3.3 million tons. While the national salt production1,113,118 tons that only meet the shortfall, and the rest of it must be imported.

Total salt imported during January 2014 reached 278 thousand tons, up 78% in volume. While the value of imports Reached US $ 13.4 million, up to 75% in value of imports. In the same period years ago, in January 2013, the import of salt recorded only $ 156 thousand tons with the import value of US $ 7.7 million. Based on supply-needs calculation, the total demand for salt Indonesia is 3 to 3,2 million. That isused for consumption, fish preservation, and so forth about 1.2 to 1.4 million tons and 1.8 million tons of industrial salt. In 2004 and 2012, the volume of imports is Increased annually.

Here are the data of imported salt to Indonesia:

a) Australia, which is the biggest salt supplier for Indonesia. In June 2013, the import of salt by Indonesia from Australia Reached 111 thousand tons or US $ 5.4 million. Meanwhile, the previous month (May 2013) the entry of imported salt from Australia amounted to 98 thousand tons or US $ 4.8 million. The cumulative (January to June 2013), the import of salt from Australia recorded 733 thousand tons or US $ 34.2 million.

b) India, in June 2013, India did not supply salt to Indonesia. However, for the month of May 2013, the salt import from India reached 47 thousand tons, worth US $ 1.97 million. If accumulated over 6 months in 2013, total imports of salt of India is 189 thousand tons or US $ 7.89 million.

c) Germany, the import volume in June 2013 was 34 tons or US $ 119 thousand. Previous months, imports salt from Germany is not too much different. Within six months, salt imports from Germany Reached 177 tons or $ 445 thousand.

d) New Zealand, salt imports from New Zealand in June 2013 Reached 48 tons or US $ 19 thousand. While in May 2013, salt imports from New Zealand Reached 480 tons or US $ 194 thousand. While in the period January to June 2013, total salt imports from New Zealand Reached 816 tons or US $ 325 thousand. e) Singapore, the amount of salt imported from Singapore in June 2013 Reached 293 kg or US $ 1.012. During January to June 2013, the salt imported from Singapore entered at 7.2 tons or US $ 57 thousand. In addition there is a collection of other countries with total imports of salt during the month of June amounted to 25.3 tons, or US $ 4.370 and for a period of 6 months in 2013 amounted to 663.9 tons or US $ 142 thousand

Those data and facts show that Indonesia in fulfilling the needs of domestic salt by importing salt from other countries. Midgley (1995) describe three important strategies in improving the welfare of the community, namely through an individual approach (individualist or enterprise approach), through a community approach (Communitarian approach), or by the role of the state (statist approach). And when it is viewed from the side of an individual approach, the welfare of the whole society will be achieved when each individual can improve their welfare. The efforts undertaken by this approach is to promote a culture of entrepreneurship, encourage the emergence of small businesses, and to encourage the participation of individuals in order to optimally utilize entrepreneurial culture and business climate. As described earlier, the condition of the salt farmers in the village is very alarming. Hence, their capacity building (empowerment) is important to up grad, looking at their contribution to the fulfilment of the national salt. Observing aspects of income derived from the salt farmers struggle during this time, it does not provide a significant contribution to economic growth. Economic growth tends to be slow, erratic and difficult to read. Delays of economic growth automatically interact with the process of economic equality.

In general, the farmers are in poverty, where they are mostly located in rural areas. To reduce poverty, it is good to outwit the farmers since it is very helpful in the process of independence and establish their identity. Their identity and existence of this form of independence became the driving side. When the policy was first adopted, importing is salt based on the consideration to meet domestic salt need that can not be met by the producers of industrial salt and salt consumption. That rule seems to consider the possibility of the impact of the legality of imported salt. In the preamble of the regulation stated "that in order to meet domestic demand for salt as a raw material in industries, as well as, increase the income and welfare of salt farmers, it is need to regulate the import of salt provisions". So, the government itself seems to also face a dilemma related to the trade of salt: on the one hand requires the supply, but on the other hand could certainly be a boomerang for the existence of salt people. Associated with government policy on import of salt, many people doubt whether the policy of import of salt, especially salt intake really is based on accurate data or research results.

The willingness of governments as reflected in its policies through the relevant ministry can not automatically be implemented in line with expectations. However, both the rules on paper and in practice turn out that the government lacks the ability to control the conditions in the field. Prices of salt on the market in reality only determined and controlled by a particular company, created a cartel in the salt trade system. Therefore, even though the government has tried to protect the people of salt at the time of harvest, in fact imported salt is still flooding which caused the price of salt people is very low. In fact, based on the findings in the field, farmers are generally not aware of any protection from the government, because there is no significant market uptake activity and still outstanding import salt.

2 Material and Methodology

2.1 Research Location

The research location is determined intentionally (purposive) in Cirebon. The choice of location is based on the consideration that the salting Cirebon District could represent the people/ farmers.

Population and Sample

The population in this study are share croppers salt. Method of sampling is done proportional random sampling. Number of samples /respondents are about 50 people. This amount is deemed to have met the numbers as expressed by Singarimbun and Effendi(1995) that samples with greater than 30 are included in a large sample.

2.2. Method of collecting data

The data used in this research are quantitative data and qualitative data that are obtained from both primary and secondary sources. The primary data are obtained by observation, questionnaires, and interviews. While secondary data are obtained from the village offices, district offices, the Department of Marine, Fisheries and Livestock, PT. Garam Indonesia and other related institutions.

This study uses survey, by interviewing the respondents using questionnaire. Sampling method in this research is done by using purposive sampling, the sampling technique with certain consideration. The method can be used if the sources correspondents interviewed are experts or working in a field, for example research on food, then the data source or resource person is the person who is addition (Sugiyono 2010). Purposive sampling method or judgment, in which the determination of sample obtained from consideration of the interviewer, with a note that the respondents who were interviewed are people who are experts or working in the field of research that is being studied or the late respondent meets the criteria specified by the interviewer (Fauzi2001).

$$n = \frac{N}{N(d2)+1} \quad (1)$$

The data used in this research is primary data and secondary data. According to the governor (2011) in Yuliriane (2012), the primary data is data obtained directly from study subjects using a measuring device or appliance makers as a source of information of data, such as interviews, questionnaires, or observation. Secondary data were obtained with a literature study of the relevant agencies. Primary data were obtained from the respondent, while secondary data are obtained from relevant agencies such as the Central Bureau of Statistics(BPS), the Local Government Office of Cirebon District, and the books that are relevant to the research.

2.3 Statistical Analysis

Data obtained from the results of the study were analyzed using quantitative descriptive method. Descriptive method aimed to tell and interpret data with respect to the situation in a systematic, factual and accurate information on the facts and the relationship between variables to get the truth. Whereas quantitative methods aimed at raising the facts, state variables, and the phenomenon that occurs when current and present their study (Sugiono 2003). However, the criteria that should be analyzed in this study are described in the next section. The instrument is analyzed by using Pearson Product formula Moment Correlation(Pearson Product Moment Correlation),as follows:

$$r_{xy} = \frac{N\sum xy - (\sum x)(\sum y)}{\sqrt{\{N\sum x^2 - (\sum x)^2\}\{N\sum y^2 - (\sum y)^2\}}} \quad (2)$$

Structural equation model to be tested take the form of $Y = \rho y^{x_1} X_1 + \rho y^{x_2} X_2 + \rho y\varepsilon$. And The Multiple Linear Regression Analysis as follows: $Y = a + b_1X_1 + b_2X_2 + \varepsilon$.

3 Result and Discussions

1. Correlation between The Strategy of Salt Production Governance(X1), the implementation of policy (X2), with Welfare improvement of the Salt Farmer (Y)

By usingSPSS17, it is known that the multiple coefficient correlation between variables The Strategy (X1), the implementation of Policy (X_2), with welfare

Improvement of the Salt Farmer (Y) are shown in the table as follow

Table 1 Correlations

			Strategy	Policy	Welfare
Spear man's rho	Strategy	Correlation Coefficient	1.000	.732**	.821**
		Sig. (2-tailed)	.	.000	.000
		N	55	55	55
	Policy	Correlation Coefficient	.732**	1.000	.771**
		Sig. (2-tailed)	.000	.	.000
		N	55	55	55
	Welfare	Correlation Coefficient	.821**	.771**	1.000
		Sig. (2-tailed)	.000	.000	.
		N	55	55	55

**. Correlation is significant at the 0.01 level (2-tailed).

Based on Table 1 the results of correlation Kendall's tau b is stated that variable The Strategy of Salt Production Governance (X_1) has a very close relationship with the variable Welfare improvement of the Salt Farmer (Y) of 0.938, or by 93.80% and variable The Implementation of Policy (X_2) also has close links with the variable of Welfare Improvement of the Salt Farmer (Y) of 0.874 or 87.40% and the Strategy (X_1) has a close relationship with the variable The Implementation of Policy (X_2) of 0.817 or 81.70%

Multiple correlation coefficient of 0.989 means that the relationship between The Strategy of Salt Production Governance (X_1), The Implementation of Policy (X_2), and Welfare Improvement of the Salt Farmer(Y) is very close. Value R^2=0, 989 means that changes to Improve the Welfare of the Poor by 98.90% due to changes in the Strategy of Salt Production Governance(X_1) and The Implementation of Policy (X_2).

2. The Strategy of Salt Production Governance **(X_1)** significantly influences on Welfare Improvement of the Salt Farmer **(Y)**

Based on the correlation data that prove the influence of variables The Strategy of Salt Production Governance (X_1) to Improve the Welfare of the Salt Farmer (Y) that is calculated with a correlation coefficient of 0, 989 or(rx_1y) =0, 989 suggests the effect is very close, when compared with the coefficient as follows:

1. 0,00 - 0,20 the relationship is very small and can be ignored
2. 0,20 - 0,40 small relationship (not tight)
3. 0,40 - 0,70 the relationship is moderate (close enough)
4. 0,70 - 0,90 tight relationship
5. 0,90 - 1,00 a very close relationship

Meanwhile, to declare the size of the contribution of variable$X_1$to Y variable or determinant coefficient is stated as=r^2X 100% or $(0{,}766)^2$ x 100% = 58,68 %, while the remaining 41.32% is determined by other variables. Then to find significant levels of correlation coefficients X_1 on Y with one sided methods (one tailed) of output. Measured probability is yields 0.00 figure. Because the probability is much below 0.05, then, the influence of The Strategy of Salt Production Governance to Improve the Welfare of the Salt Farmer is significant.

Table. 2. Summary Model[b]

Model	R	R Square	Adjusted R Square	Std. Error of the Estimation	Durbin-Watson
1	.766[a]	.806	.917	1.54556	.532

a.Predictor (Constant), strategy, policy
b. Dependent Variable: Welfare

Table 3. Coefficients[a]

Model	Unstandardized Coefficients		Standardized Coefficients	T	Sig.	Collinearity Statistics	
	B	Std. Error	Beta			Tolerance	VIF
1 (Constant)	1352	1.496		.509	.707		
Strategy	.611	.319	.733	13.323	.000	.412	2.430
Policy	.427	.323	.723	9.284	.000	.412	2.430

a. Dependent Variable: Welfare

The coefficient table illustrates that simple regression equation is as follows:

$\acute{Y} = a + b_2x_2 = 13{,}52 + 0{,} 723 X_2$

Is:

X_2 = The Implementation of Policy

Y = Improving the Welfare of the Salt Farmer

Constants of 13,52 states that there is no increase in the value of the variable The Implementation of Policy (X_2), then the value of Welfare improvement of the Salt Farmer (Y) is 13,52. The regression coefficient of 0,723states that any additions (for the sign+) of the scores or grades Quality of care would rise to a score of 0,723

T test is used to test the significance of constant and variable Y (Improving the Welfare of the Salt Farmer). Coefficient tests the following variables:

Ha: The Implementation of Policy is significantly affect Welfare Improvement of the Salt Farmer

Ho: The Implementation of Policy does not significantly affect Improving the Welfare of the Salt Farmer

The hypothesis in statistical form:

Ha : $rx_2y \neq 0$

Ho : $rx_2y = 0$

Basis for decision making: by comparing the value of t arithmetic with t table as follows: If thet count >t table, then Ho is rejected it means a significant regression coefficient

If the value oft<t table, then Ho accepted means of regression coefficients were not significant Value=7.230 (taken at coefficient value for the variable (X_2)

*The level of significance(α) =0.05

*df (degrees of freedom) amount n-2=55-2=53

*Test done one side, so that the value t table=1.671

Decision:

Because the value of t count>t table, or7,230>1.671, then Ho is rejected seen in the column sig (significant) coefficient sig 0.000 or less than the probability value 0.05 or 0.05 value >0,000, then Ho is rejected and Ha is accepted, meaning that the regression coefficient is significant. Thus The Implementation of Policy significantly affects Welfare Improvement of the Salt Farmer . Thus the regression equation is:

$\acute{Y} = a + b_1x_1 + b_2x_2 = 13{,}52 + 0,\ 766\ X_1 + 0,\ 723X_2$

is :

X_1 = The Strategy of Salt Production Governance

X_2 = The Implementation of Policy

Y = Improving the Welfare of the Salt Farmer

Constants of 16.44 states that if there is no increase in the value of the variable The Strategy of Salt Production Governance (X_1) and The Implementation of Policy (X_2), then the value of Welfare Improvement of the Salt Farmer (Y)is13,52. The regression coefficient of 0,766 and 0,723 states that any additions(for the sign+) of the scores or grades The Strategy and The Implementation of Policy will give rise to a score of 0, 766 and 0, 723

3. The Strategy of Salt Production Governance (X_1) and The Implementation of Policy (X_2) together – equally influence significantly on Welfare Improvement of the Salt Former (Y)

F test in the ANOVA table is to test the significance of the constants and the dependent variable(Improving the Welfare of the Salt Farmer). Criteria regression coefficient test of The Strategy of Salt Production Governance and The Implementation of Policy to Improve the Welfare of the Salt Farmer is as follows: Hypothesis to three proposals are

Ha : The Strategy of Salt Production Governance (X_1) and The Implementation of Policy (X_2) together –equally influence significantly on Welfare Improvement of the Salt Farmer (Y)

Ho : The Strategy of Salt Production Governance(X_1) and The Implementation of Policy (X_2) together –equally influence significantly on Welfare Improvement of the Salt Farmer (Y)

The hypothesis instatistical form:
Ha : $r\ x_1x_2y \neq 0$
Ho : $r\ x_1x_2y = 0$

Table 4. ANOVA[b]

Model	Sum of Squares	df	Mean Square	F	Sig.
1 Regression	6979.349	2	3489.675	1023.433	.000[a]
Residual	156.936	55	2.092		
Total	7136.286	53			

a. Predictors: (Constant), Strategy, Policy

b. Dependent Variable: Welfare

Taken from the table anova[b] F count=10,23Basic decision-making: by comparing the value of F arithmetic with F table value is as follows:

If F count >F table, then Ho is rejected. It means, this is significant regression coefficient.
If the value of F arithmetic<F table, then Ho is accepted. It means, the regression coefficients were not significant. Finding the value of F calculated using the F table with the formula:

Significance level(α) =0.05 significance level (α) = 0,05
$F_{table} = F(1-\alpha)(dk{=}k),\ (dk = n\text{-}k\text{-}1)$
$= F(1-\alpha)(dk = 3),\ (dk = 55 - 2 - 1)$
$= F(1\text{-}0{,}05)\ (3,\ 52)$
How to find F table =3, as the numerator and the denominator 52

F table=4.08 (interpolation)

Decision :

It turned out that F count>F table, or 10,23>4.08 then Ho is rejected and Ha is accepted. It means that The Strategy of Salt Production Governance(X_1) and The Implementation of Policy (x_2) together-equally influence significantly on Welfare Improvement of the Salt Farmer(Y). It appears that at column Sig(significant) in the table Anova[b] the Sig. 0,000or smaller than the probability value 0.05 or 0.05 value >0.000, then Ho is rejected and Ha accepted, meaning that multiple regression coefficient is significant. So, The Strategy of

Salt Production Governance (X_1) and The Implementation of Policy (X_2) together –equally influence significantly on Improving the Welfare of the Salt Farmer (Y)

4 Conclusion and Discussion

Research findings indicate various conclusions, related to:

1. The relationship / correlation is very strong / strong between The Strategy of Salt Production Governance (X_1) and The Implementation of Policy(X_2) together with variable Improving the Welfare of the Salt Farmer (Y) of 0.766 or 76,60% and the remaining 23.40% are other factors that are not included in the research model (Epsilon).
2. There is significant influence of variables The Strategy of Salt Production Governance (X_1) to Improving the Welfare of the Salt Farmer (Y)
3. There is significant influence of variables The Implementation of Policy (X_2) on Improving the Welfare of the Salt Farmer (Y)
4. There is significant influence of variables The Strategy of Salt Production Governance (X_1) and The Implementation of Policy (X_2) together toward Improving the Welfare of the Salt Farmer (Y).

Acknowledgment

We are gratefully acknowledged to Local Government of Cirebon, West Java. Special thanks to the Regent of the respective Public Administration Institution, Local Government of Cirebon and all subjects who participated in the study are gratefully acknowledged.
Conflict of interest declaration: There is no conflict of interest among the authors.
Funding: This study received no specific grant from any funding agency.
Authors' contributions Bambang Sudaryana, conception of study, data collection, analysis and interpretation of results, drafting of manuscript, review of manuscript and interpretation of results.

References

1. Adiprayoga, Gatot. "Penanggulangan Kemiskinan Melalui Pemberdayaan Masyarakat," dalam Jurnal *ASPIRASI*, Vol. XVIII No. 1, Jember: Fakultas Ilmu Sosial dan Ilmu Politik Universitas Jember, Juli 2008

2. Dillon, HS. "Kemiskinan di Negara Berkembang: Masalah Konseptual dalam Global", dalam Jurnal Prisma No. 3/XIII/1993. Jakarta: LP3ES, 1993

3. Elliot, T. Charles. *Perfect Empowerment*, New York: Unesco. http://www.Dinas Kelautan dan Perikanan Kabupaten Lombok Barat, 1987

4. Iskandar Dinata, Pemberdayaan petani garam dalam mewujudkan kesejahteraan sosial (potret petani garam di desa pandai kecamatan woha kabupaten bima). Volume 9, Nomor2, Juli-Desember 2013

5. Nasirin, Chairun.. "Alternatif Model Kebijakan dalam Upaya Penanggulangan Kemiskinan di Indonesia", dalam*ASPIRASI* Vol. XVIII, No. 1 Juli 2008, Fakultas Ilmu Sosial dan Ilmu Politik Universitas Jember, 2008

Empirical Analysis on The Existence of The Phillips Curve

Mohd Shahidan Shaari*, *Diana Nabila Chau* Abdullah, *Razleena* Razali, *Mohamad Luqman Al-Hakim* Md Saleh

Abstract. The Phillips curve shows the trade-off relationship between the inflation and unemployment rates. A rise in inflation due to the high economic growth, more jobs are available and therefore unemployment will fall. However, the existence of the Phillips curve in high-income countries has not been much discussed. Countries with high income should have low unemployment rate, suggesting a high inflation. However, some high-income countries, the United States in 1970s for example, could not avert stagflation whereby high unemployment rate and inflation occurred in the same time. This situation is contrary to the Phillips curve. Therefore, this study aims to investigate the existence of the Phillips curve in high-income countries for the period 1990-2014 using the panel data analysis. The most interesting finding of this study is the existence of a bidirectional relationship between unemployment rate and inflation rate in both long and short runs. Therefore, the governments should choose to stabilize inflation rate or reduce unemployment rate

1 Introduction

Phillips [12] introduced the Phillips curve, which illustrates the trade-off relationship between the inflation and unemployment rates. As inflation rises because of the high economic growth, better job opportunities are available to people, implying a low unemployment rate. Stable inflation rate and low unemployment rate are the macroeconomic goals of developing and developed countries. The Phillips curve can cast a light on the problem of stabilizing inflation or reducing unemployment rate. A country that decides to stabilize inflation anticipates a higher unemployment rate, and should therefore be poised to confront the economic problem. Policy makers need to this matter carefully. A wrong decision on policies can aggravate the economy.

Numerous studies, such as those of Resurreccion 13] in Philippines, Al-zeaud and Al-hosban [1] in Jordan, and Furuoka and Munir [5] in Malaysia, have investigated the existence of the Phillips curve in various countries. Their findings were consistent, that is, the Phillips curve exists in their respective countries. Meanwhile, a limited number of studies explored this existence in high-income countries, such as Haug and King et. al [6] in the United States, and their findings confirmed the existence of the Philip curve. However, the study focused only on one country and did not include other high-income countries. Countries with high incomes should have low unemployment rate, suggesting a high inflation rate. However, some high-income countries, the United States in 1970s for example, could not avert stagflation, thus high unemployment rate and inflation occurred in tandem. Frideman et. al [3] argued that the existence of the Phillips curve might be valid only in the short run. Inflation can drive workers to demand higher wages to adjust their real income. Hence, it can result in higher production cost. Firms have to incur the rising cost for every additional worker. As a result, the profits of firms fall and ensue layoffs. This situation is contrary to the Phillips curve. Therefore, this study aims to investigate the existence of the Phillips curve in high-income countries in both short and long runs.

The relationship between the unemployment and inflation rates in high-income countries is intriguing and merits attention because their economies are prosperous. Fig. 1, 2, and 3 show trends in inflation and unemployment rates in Switzerland, the United States, and Sweden, respectively. In 2009, the inflation rate was low while the unemployment rate was high in all the countries, suggesting the existence of the Phillips curve. The United States also experienced the same trend, recording the highest inflation rate at 5.4% and unemployment rate at 5.6% in 1990. In Switzerland, the highest inflation rate was recorded at 5.9% in 1991, while the unemployment rate was 1.8%, which was low. In Sweden, the highest inflation rate was recorded at 10.5% in 1990, while the unemployment rate was very low at 1.8%. However, inflation rate hit a negative value of 0.5% in 2009, while the unemployment rate was very high at 8.4%.

* Corresponding author: shahidanshaari@gmail.com

Figure 1: Inflation rate and unemployment rate in Switzerland

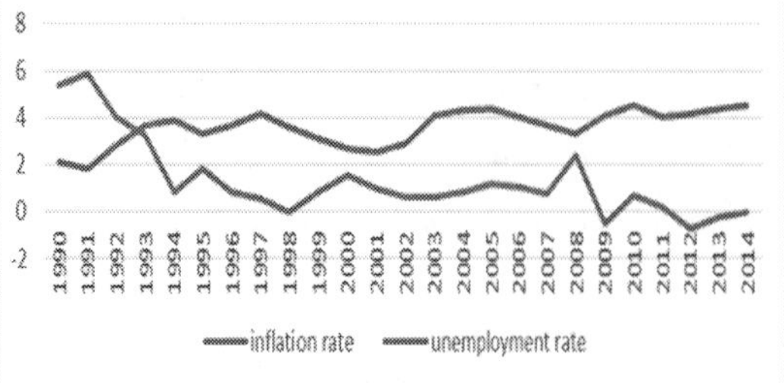

Figure 2: Inflation rate and unemployment rate in the United States

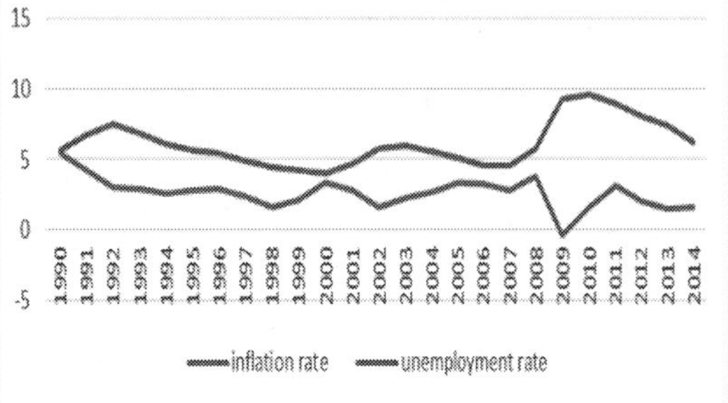

Figure 3: Inflation rate and unemployment rate in Sweden

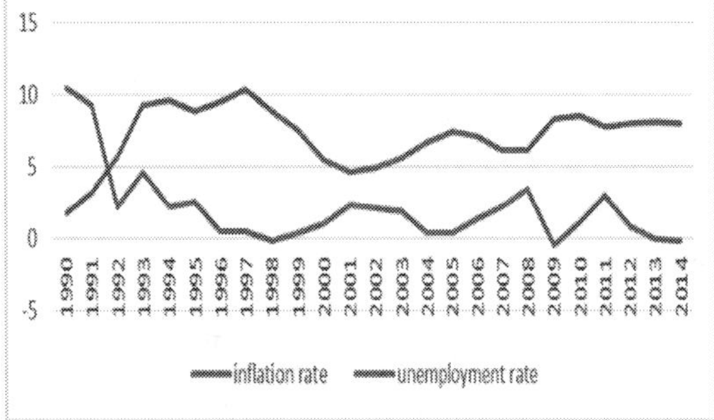

2 Literature Reviews

Numerous studies have been done to identify the existence of the Philip curve in various countries such as Greece by Dritsaki and Dritsaki [2] Jordan by Al-Zeaud & Al-hosban, [1], Malaysia by Furuoka et. al [4] etc. However, their findings were mixed probably due to the economic condition. Most of previous studies discovered the existence of Philip curve. Furuoka and Munir [5] explored the connection between unemployment rate and inflation rate in Malaysia. Malaysia was chosen as it is a developing country and many previous studies have focused on developed countries. The data from 1975 to 2004 were analysed using the Johansen co-integration and error correction model tests. The results showed that there is a connection running from the unemployment rate to inflation rate in both short and long run. Therefore these findings confirmed the existence of the Phillips curve in Malaysia. Shaari et al. [15] also examined the same connection in the same country. The study analysed the data from 1982 to 2010 using the Johansen c-integration and Granger causality tests. The finding revealed the existence of a negative connection between inflation and unemployment. Schreiber and Wolters [14] examined the existence of a connection between unemployment and inflation in Germany for the period 1977:2-2002:3. The study employed vector auto regression model and vector error correction model and the findings showed that there is a negative connection between inflation and unemployment in the long run. Resurreccion et. al [13] extended the previous study by examining the connection between unemployment and inflation and economic growth within 1980-2009 but the study was done in Philippines. The analysis was based on the ordinary least square test was employed and the findings showed that an increase in the inflation rate or economic growth can cause the unemployment rate to decrease. This inverse connection corresponded with the Phillips curve.

Karahan et al. [8] examined the connection between unemployment and inflation in Turkey for the monthly data from 2006 to 2011. The Autoregressive Distributed Lag (ARDL) Bounds test was conducted and the findings showed that unemployment negatively affects inflation in the short run but there is no causal connection in the long run. Ismael and Sadeq et. al [7] explored the connection between unemployment and inflation in Palestine for the quarterly data from 1996:2 to 2015:3. The study conducted the co-integration and Granger-causality tests and the findings revealed that inflation can affect unemployment negatively in the long run but positively in the short run. Haug and King (2011) [6] also investigated the long-run connection between inflation and unemployment in the United States for the period 1952 to 2010. The band-pass filter approach was used and the findings however showed a positive connection between inflation and unemployment. Therefore, a rise in inflation rate can result in higher unemployment rate.

Several studies examined the connection between inflation and unemployment in numerous countries and the existence of the Phillips curve can be confirmed. Katria et al. [9] selected 8 SAARC countries including Afghanistan, Bangladesh, Bhutan, India, Maldives, Nepal, Pakistan and Sri Lanka, and 6 expected countries of SAARC, namely, Republic of China, Russia, Indonesia, Iran, Myanmar and South Africa to analyze the connection between inflation and unemployment for the same period of 1980-2010. The analysis was based on the multiple regression and the results indicated that there is a negative connection between inflation and unemployment rate in the SAARC Countries. Therefore, this finding is similar to the Phillips curve.

Bhattarai et. al [10] chose 28 OECD countries to examine the connection between inflation and unemployment for the period 1990:1-2014:4. The panel co-integration and panel granger causality tests were performed and the study showed varied findings whereby a significance of a negative connection between inflation and unemployment in some countries including Australia, Denmark, France, Italy, Netherlands, Spain, New Zealand, the UK and the US. However, different results of no connection were found in Austria, Germany, Israel and Norway. Surprisingly, a positive connection was ascertained in Korea, Russian and Slovak Republic.

3 Research Methodology

This study used data on unemployment rate and inflation rate from 10 high-income countries. The data from 1990 to 2014 were analyzed to see the existence of trade-off relationship between inflation rate and unemployment rate. The panel data methods were employed in this study as these methods are more powerful than times series methods. The panel unit root tests were conducted to see the stationarity of the data. Then the panel co-integration tests were performed to see the co-integrating relationship. The panel Granger causality tests were carried out to explore the existence of the Phillips curve in the short run. The equation for the simple Phillips curve is as follows:

$$IR_t = \beta_1 + \beta_2 UR_{t-1} + \varepsilon_t \tag{1}$$

Where

IR represents inflation ration rate

β_1 is constant

β_2 represents the unemployment rate

ε is the error correction terms

There are various tests of unit root that can be conducted in the panel data analysis such as Levin, Lin and Chu (LLC), Im, Pesaran and Shin (IPS) and Breitung. However this study employed LLC and this test is enough to see the stationarity of the data. This study is easy in making estimation and widely used in empirical studies.

After the unit root tests were conducted, the panel co-integration tests were performed to see the existence of a co-integration between the two variables (inflation rate and unemployment rate). The panel co-integration proposed by Pedroni et. al [11] is employed to explore the existence of a long-run co-integration between inflation rate and unemployment rates. Predoni introduced three group mean panel statistics for panel co-integration. Two sets of statistics were proposed and the first set of three statistics based on pooling the residuals within the dimension of the panel. Therefore, the equations for the statistics as follows:

$$Z_{\hat{v},N,T} = T^2 N^{3/2} \sum_{i=1}^{N}\sum_{t=1}^{T} \hat{L}_{11i}^{2} \hat{e}_{i,t\ 1}^{2} \tag{2}$$

$$Z_{\hat{\rho}N,T\ 1} = T\sqrt{N} \sum_{i=1}^{N}\sum_{t=1}^{T} \hat{L}_{11i}^{2} \hat{e}_{i,t\ 1}^{2} \quad ^{1} \sum_{i=1}^{N}\sum_{t=1}^{T} \hat{L}_{11i}^{2} \left(\hat{e}_{i,t\ 1} \Delta\hat{e}_{i,t} \quad \hat{\lambda}_i\right) \tag{3}$$

$$Z_{tN,T} = \tilde{\sigma}_{N,T}^{2} \sum_{I=1}^{N}\sum_{T=1}^{T} \hat{L}_{11i}^{2} \hat{e}_{i,t\ 1}^{2} \quad ^{1/2} \sum_{i=1}^{N}\sum_{t=1}^{T} \hat{L}_{11i}^{2} \hat{e}_{i,t\ 1}^{2} \left(\hat{e}_{i,t\ 1} \Delta\hat{e}_{i,t} \quad \hat{\lambda}_i\right) \tag{4}$$

The second set of the statistics are based on pooling residuals between the dimension of the panel and it follows for a heterogeneous autocorrelation parameter across members and the equations are as follows:

$$\tilde{Z}_{\hat{\rho}N,T\ 1} = \sum_{i=1}^{N} \sum_{t=1}^{T} \hat{e}_{i,t\ 1}^{2} \quad ^{1} \sum_{t=1}^{T} \left(\hat{e}_{i,t\ 1} \Delta\hat{e}_{i,t} \quad \hat{\lambda}_i\right) \tag{5}$$

$$\tilde{Z}_{tN,T\ 1} = \sum_{i=1}^{N} \sum_{t=1}^{T} \hat{e}_{i,t\ 1}^{2} \quad ^{1/2} \sum_{t=1}^{T} \left(\hat{e}_{i,t\ 1} \Delta\hat{e}_{i,t} \quad \hat{\lambda}_i\right) \tag{6}$$

This study also conducted the VECM test with a dynamic error correction term based on the analysis in Holtz-Eakin et al. (1989). The empirical models are as follows:

$$\Delta IR_{it} = \alpha_{1t} + \sum_{p=1}^{m} \alpha_{11ip} \Delta IR_{it-p} + \sum_{p=1}^{m} \alpha_{12ip} \Delta UR_{it-p} + \alpha_{1i} ECT_{it-1} + \mu_{1it} \tag{7}$$

$$\Delta UR_{it} = \alpha_{2t} + \sum_{p=1}^{m} \alpha_{21ip} \Delta UR_{it-p} + \sum_{p=1}^{m} \alpha_{22ip} \Delta IR_{it-p} + \alpha_{2i} ECT_{it-1} + \mu_{2it} \tag{8}$$

Where Δ is the lag operator, p represents the lag length. The equations show the directional relationship between inflation rate and unemployment rate. In the short run if H_0: $\alpha_{12ip}=0$ for all i and p, it indicates that unemployment rate does not Granger cause inflation rate and if H_0: $\alpha_{22ip}=0$ for all i and p, it suggests that inflation rate does not Granger cause unemployment rate.

4 Finding

Results of the panel unit root tests based on the LLC, IPS, ADF and PP are presented in Table 1. The results show that at first difference, the null hypothesis is rejected. Therefore the two variables (inflation rate and unemployment rate) are stationary at first difference.

Table 1: Unit Root Results

Variables	Intercept		Intercept + Trend	
	LLC Test		LLC Test	
	Level	First difference	Level	First difference
IR	-6.28694 (0.0000)	-15.4159 (0.0000)	-4.44800 (0.0000)	-14.0441 (0.0000)
UR	-3.43764 (0.0003)	-8.05717 (0.0000)	-1.85266 (0.0320)	-3.69446 (0.0000)
Variables	**Intercept**		**Intercept + Trend**	
	ADF		ADF	
	Level	First difference	Level	First difference
IR	85.6816 (0.0000)	182.276 (0.0000)	59.8629 (0.0000)	165.195 (0.0000)
UR	47.3115 (0.0005)	95.3149 (0.0000)	35.8506 (0.0160)	58.9044 (0.0000)

Note: * and ** indicate the rejection of the null hypothesis of non-stationary at 1% and 5% significance level, respectively. In addition, P-values are in brackets.

Results of the panel Co-integration test are recorded in Table 2. For all the statistics, the null hypothesis of no co-integration can be rejected and thus suggesting inflation rate and unemployment rate are co-integrated in 10 selected high-income countries.

Table 2: Panel Co-integration Results

	Intercept		Intercept + Trend	
Within Dimension	**Statistic / Prob.**	**Statistic / Prob.**	**Statistic / Prob.**	**Statistic / Prob.**
Panel v-Statistics	2.256329 (0.0120)	2.110181 (0.0174)	-0.158291 (0.5629)	-0.454314 (0.6752)
Panel rho-Statistics	-3.858839 (0.0001)	-4.544042 (0.0000)	-2.282684 (0.0112)	-2.897784 (0.0019)
Panel PP-Statistics	-4.397037 (0.0000)	-4.986474 (0.0000)	-4.206640 (0.0000)	-5.042186 (0.0000)
Panel ADF-Statistics	-4.042474 (0.0000)	-4.853926 (0.0000)	-4.045164 (0.0000)	-5.018442 (0.0000)
Between Dimension				
Group rho-Statistics	-3.011402 (0.0013)		-1.637835 (0.0507)	
Group PP-Statistics	-5.740562 (0.00000)		-5.631865 (0.0000)	
Group ADF-Statistics	-5.442847 (0.0000)		-5.032436 (0.0000)	

Results of the FMOLS tests are presented in Table 3. The results show that there is a significant effect of unemployment rate on inflation rate with feedback in the long run. The negative values of the coefficients suggest that there are negative relationships. Therefore an increase in unemployment rate can cause inflation rate to increase simultaneously and vice versa.

Table 3: FMOLS Results

Dependent Variable = UR		
	Coefficient	**Prob.**
IR	-0.200295**	0.0325
Dependent Variable = IR		
	Coefficient	**Prob.**
UR	-0.178032**	0.0275

Note: ** indicates the rejection of the null hypothesis of non-stationary and 5% significance level.

Results of the panel VECM are recorded in Table 4. The results show that the coefficient of the error correction term (ECT) is negative and significant, thus supporting the long-run relationship between inflation rate and unemployment rate in 10 selected high-income countries. Apart from that the results also indicate that unemployment rate can influence inflation rate negatively in the short run. Therefore, a rise in unemployment rate can reduce inflation rate in the short run.

Table 4: Panel Granger Causality based on VECM

Dependent variable: ΔUR			
Variables	**ΔIR**	**Constant**	**ECT_{i-1}**
Coefficients	0.008233* (0.8520)	-0.010768 (0.8410)	-0.100754* (0.0000)
Dependent variable: ΔIR			
Variables	**ΔUR**	**Constant**	**ECT_{i-1}**
Coefficients	-0.440630* (0.0001)	-0.137570 (0.1289)	-0.082795 (0.0914)

Note: * indicates the rejection of the null hypothesis of non-stationary and 5% significance level.

5 Conclusion

This study aims to investigate the existence of the Phillips curve in high-income countries for the period 1990-2014, using the panel data analysis. The panel unit root test was conducted and the results show that all the data becomes stationary at first difference. Then, the panel co-integration test was performed and the results show that there is a co-integrating relationship between inflation rate and unemployment rate. The FMOLS method was employed and the results show that there is a negative relationship between inflation rate and unemployment rate in the long run. Results of the VECM test reveal that the value of ECT is negative and significant; therefore, it can confirm that there is a long run relationship between inflation rate and unemployment rate. Apart from that the VECM test can show the relationship in the short run. Based on the results, it shows that there is a negative relationship between inflation rate and unemployment rate.

Therefore, all the results confirm the existence of the Phillips curve in the both short and long run. These findings are important for policy implications. If policy makers formulate policies to stabilize inflation, the country has to accept high unemployment rate and vice versa. This is because there is a trade-off relationship between inflation rate and unemployment rate.

References

1. Che Mansor bin Che Tom, Al-zeaud, H., & Al-hosban, S. Does Phillips curve really exist? An empirical evidence from Jordan. European Scientific Journal, 11(10), (2015)
2. Dritsaki, C. & Dritsaki, M., Inflation, unemployment and the NAIRU in Greece', Procedia Economics and Finance, 1, 118–127, (2012)
3. Friedman, M. The Role of Monetary Policy, The American Economic Review, 58(1), 1-17, (1968)
4. Furuoka, F, Does the Phillips curve really exist? New empirical evidence from Malaysia, Economics Bulletin, 5 (16), 1-14, (2007)
5. Furuoka, F. & Munir, Q., "Unemployment and Inflation in Malaysia: Evidence from Error Correction Model", Malaysian Journal of Business and Economics, 1 (1),35-45, (2014)
6. lfred A. Haug & Ian P. King., "Empirical Evidence on Inflation and Unemployment in the Long Run". Research Paper, 1128, (2011).
7. Ismael, M. & Sadeq, T. , Does Phillips Exist in Palestine? An Empirical Evidence. MPRA Paper No. 70245, (2016)
8. Karahan, O., O. Colak and O.F. Bolukbaşı, Tradeoff between Inflation and Unemployment in Turkey, The Empirical Economics Letters, 11(9): 973-980, 2012
9. Katria, S., N.A. Bhutto, F. Butt, A.A. Domki, H.A. Khawaja & J. Khalid, Trade off between inflation and unemployment. International Conference on Business Management. 1-18, 2011
10. Bhattarai, K. Unemployment–inflation trade-offs in OECD countries. Economic Modelling, 58, 93–103, (2016)
11. Pedroni, P., Panel Cointegration: Asymptotic and Finite Sample Properties of Pooled Time Series Tests with an Application to the PPP Hypothesis, Working Papers in Economics,no. 95 – 013, Indiana University, Bloomington, IN, (1995)
12. Phillips, A.W. The Relationship between unemployment and the rate of change of money wage rates in the United Kingdom, Economica, 25, 258-299, (1958)
13. Resurreccion, P. F. , Linking unemployment to inflation and economic growth: Toward a better understanding of unemployment in the Phillipines, Asian Journal of Economic Modelling, 2(4), 156-168, (2014)
14. Schreiber, S., and Wolters, J. The Long-Run Phillips Curve Revisited: Is the NAIRU Framework Data-Consistent? Journal of Macroeconomics, 29 (2), 355–367, (2007)
15. Shaari, M.S., Rani, M.J.A., Ismail, M.S., & Hussain, N.E. Inflation and unemployment in Malaysia. Malaysian Technical Universities Conference on Engineering and Technology (MUCET) 2012, (2012)

The Creation of Women from the Perspective of Mufassirun and Muhaddithun Between Past and Present Scholars

Roshimah Shamsudin[1,*], *Aiza Maslan* Baharudin

[1]Universiti Sains Malaysia

Abstract. The creation of women is under constant debate, not only among the Islamic scholars but outspreads to feminists both from the west as well muslims.This debate stems from the differing views of the preceding and contemporary scholars in interpreting the texts from the Quran and the Hadiths on the creation of human particularly Eve as the first woman created by Allah. The majority of the *mufassirun* and hadith scholars in the past pioneered the view that the origin of human; that Eve was created from the rib of Adam a.s. This view was regarded by the feminists as derogatory to women. The contemporary *mufassirun* and *muhaddithun* are however of the opposite opinion whereby Eve's creation was equal to that of Adam's a.s; thereby indicating that creation of Eve did not originate from Adam a.s. Hence this study was undertaken with the aim of reviewing the opinions of the tafsir and hadith figures by comparing the views between the past and present scholars in interpretation of Quranic texts as well as hadiths pertaining to the origin and creation of women. This study also extends to examining the arguments on which the opinions of each group were deduced.In order to achieve the above objectives, this qualitative form of study will employ data collection method, conducted through library research to obtain data relating to the origin of creation of women, in particular *turath* literatures in tafsir for example *Tafsir Ibn Kathir* and tafsir works by the leading tafsir scholars in the like of Syed Qutb; in addition to the past and present books on *syarah hadith*. The findings of this study show that the previous and present scholars of tafsir or hadith took different methodology in understanding both textual authorities.Unlike the previous scholars of tafsir and hadith, some of the contemporary scholars' understanding is that the texts are not to be construed literally; instead the texts need to be understood as parables and allegories thereby indicating that the creation of women is not from the ribs of Adam a.s.

1 Introduction

The issue of creation of women is not spared from intense discussion, particularly in the context of women's position to that of men.The issue stems from the view that women creation was derived from the rib of Adam. Wan Siti Nur Haini argues that the above view which had long been inherited and permeated into the psyche of most Muslims implicates the status of women which generally are regarded as second class citizens. The idea was reinforced by the existence of texts that show women were created from the ribs as evidenced by an authentic hadith narrated by Al-Bukhari and Muslim whose narrations form the two major hadith compilations that are consensually agreed upon by the hadith scholars as authentic, second only to the holy al-Quran. Thus it is not unfathomable on why this women creation idea has lasting influence in the Islamic traditions.

Unfortunately, the notion that women was created from Adam's rib compromised women of their rights from various aspects such as religion, politics and social [25]. The denial of the rights of women according to Halimah Basri is attributable to the concept that women are not created equal to men, and to top it off was created out of men as opined by certain *mufassirun*. In this context, she did not only associate this creation issues with the texts from hadiths alone but also links to the Quranic texts that were interpreted by past scholars

2 Materials and Methods

This qualitative study relies solely on library research. This means data was collected from primary scriptures – from the books of *tafsir* and hadith. From the works of *tafsir*, data was taken from principal books of *tafsir* written by earlier scholars such as *Tafsir Ibn Kathir, Tafsir al-Tabari, Tafsir al-Maraghi,* books written by other contemporary tafsir scholars such as Muhammad Abduh and Rashid Redha and reference books of Tafsir written by a number of scholars from the archipelago such as Hamka, M. Quraish Shihab and Ahjmad Sonhadji. Data are also taken from key books of hadith such as *Sahih al-Bukhari, Sahih Muslim and Jami' al-Tirmidhi* and commentaries of hadith such as

Fath al-Bari, written by Ibn Hajar and *Sharh Sahih Muslim*, authored by al-Nawawi. The data were then analyzed using the inductive, deductive and comparative method.

3 Results

In the context of creation of women, Halimatussa'diyah opines that the sources are mostly from the *Israilliyyat*, Torah, Bibble and Talmud compared to the al-Quran itself, whereby the latter does not even have clear absolute passages as evidence thereof. There are nevertheless three Quranic verses that led to the story of creation of women. For instance, verse 1: Al-Nisa'; verse 189: surah Al-A'raf and verse 6: surah Al-Zumar. However verse 1 of surah Al-Nisa' is more often quoted by *mufassirun* to prove the origin of women.

يَا أَيُّهَا النَّاسُ اتَّقُوا رَبَّكُمُ الَّذِي خَلَقَكُم مِّن نَّفْسٍ وَاحِدَةٍ
وَخَلَقَ مِنْهَا زَوْجَهَا وَبَثَّ مِنْهُمَا رِجَالًا كَثِيرًا وَنِسَاءً ۚ
وَاتَّقُوا اللَّهَ الَّذِي تَسَاءَلُونَ بِهِ وَالْأَرْحَامَ ۚ إِنَّ اللَّهَ كَانَ
عَلَيْكُمْ رَقِيبًا

Meaning: O mankind, reverence your guidence Lord who created you from a single person, created of like nature his mate and from them twain scattered countless men and women.

There are three words that form the basis of the creation of women namely *wahidah, minha,* and *zaujaha* all of which are given different interpretation by the *mufassirun* [6]. The differences had sparked disagreements among scholars of exegesis themselves. The ancient scholars in the likes of at-Tabari literally interprets *min nafs wahidah* to be referring to Adam while the word *zaujaha* refers to his spouse; Eve. In other words the verse, according to him shows that from a single Adam, Eve was created from his left rib to be his spouse. Al-Tabari further brings in the story of Eve's creation from Adam's rib while he was asleep as narrated by al-Suddi [23].

Al-Tabari's view is supported by a prominent *mufassir* such as al-Razi who not only opines that the creation of Eve was from Adam's left rib, but he also brings in a hadith to prove that Eve was created from Adam's bend rib. In spite of the detailed interpretation, al-Razi quotes the views of other scholars on the meaning of the said words. For instance Abu Muslim al-Isfahani exegizes the word *nafs* to refer to the nature of Adam based on verse 72: Surah al-Nahl meaning Allah has made for you mates from your nature.Further to the above verse al-Razi also brings in a view from al-Qadi who rejects the interpretation of Abu Muslim. By alleging that if Eve was created from the nature of Adam, then this could infer that two beings had been created and thereby, the proper grammatical application in the Quranic text should be *makhluqain min nafsain* and not *min nafs wahidah*. The word *min* in the verse refers to the beginning of creation and hence dedicated to Adam.This discussion clearly demonstrates that the view of al-Razi is consistent with al-Tabari in interpreting the word *nafs* to be referring to Adam [17].

Ibn Kathir also concurs with al-Tabari and al-Razi on the origin of Eve from Adam's rib. As a leader exegesist who mastered the science of Hadiths, Ibn Kathir did not neglect to produce arguments from the hadiths to support his interpretations as is the case with al-Razi and asserts that since the hadith was narrated by Muslim thereby it indicates its authenticity in addition to having a sahih status and recorded in the grand hadiths compilation books. In addition, Ibn Kathir does not elaborate on the creation of Eve as does he on the creation of man from clay without a mother and a father, creation of Jesus from only his mother without a father and creation of mankind other than Adam, Eve and Jesus through the process of pregnancy. This discussion signifes that the creation of man can be understood from various dimensions based on its formative elements [10]. The same can be said of Suyuti who seems to have similar views to the scholars preceding him. However though the editor of his *tafsir* book by the name Fakh al-Din opines that the hadith pertaining to the creation of Eve is indeed authentic, however, the texts are in the form of *tamthil or* parables to showcase the stubbornness of women compared to men [22]. Ibn Asyur too is found to have supported the notion that *nafs wahidah* to be referring to Adam from whose rib Eve was created as can be deduced from the word *minha* [8].

However al-Maraghi from amongst the ancient scholars have a dissenting view from the consensus of *mufassirun*. To him, *mufassirun jumhur*'s interpretations are not based on the Quranic passages but instead are made on a natural basis that Adam is the father of humankind as mentioned in Torah. He however does not affirm whether he denies of concurs with this belief. He presents his exegesis by bringing in the opinions of a number of scholars who opine that *nafs wahidah* does not mean Adam but instead it means one single self or one single type. The word *zaujaha* means wife or spouse and the *dhamir "ha"* that is found in the word *minha"* refers to *nafs wahidah* which is defined as one type or from that very one type, that the spouse of Adam was created; and thereby means Adam and Eve were created from the same material, that is clay [13].

The view of al-Maraghi is in line with the views of certain present *mufassir* figures like Muhammad Abduh. The term *min nafs wahidah* according to him does not refer to Adam because preceding the term is the sentence *wa baththa min huma rijalan kathiran wa nisa'* that shows the term is in the form of *nakhirah* i.e not intended or devoted to any particular party. If the sentence were intended to refer to Adam, the correct sentence should be *wa baththa min huma jami' al-rijal wa al-nisa.* The creation of Eve from Adam's rib according to Muhamad Abduh is a story that is found in the Old Testament [1].

Echoing Muhamad Abduh's opinion, Rashid Redha comments that the phrase *nafsin wahidah* purpoted to be referring to Adamis not derived from Quranic texts but instead relied only on belief that Adam is the father of humankind. To him there is no absolute proof that suggests Adam is the forefather of humankind (Ridha 1367H)[18]. According to Halimatussa'diyah, the above mentioned verse that which understood by Rashid Ridha as men originating from the same substance and therefore the word *nafs wahidah* symbolises that human originates from humanly nature which welcomes goodness and rejects evil; and thereby negating the contentions that the verse is about the questions on the beginning of creations [6]. Al-Qasimi, the present *mufassir* figure in his work opines Eve was created from the human kind, and not from rib as per understood by scholars in the past [16]. Nevertheless, studies have found that not all the present exegesis scholars agree with the opinion of al-Maraghi, Muhammad Abduh and Rashid Redha, as evidenced by the interpretation of Said Hawwa which is in line with the consensus of ancient *mufassirun* in their views that the term *nafs wahidah* refers to Adam.

However, Said Hawwa is found to have quoted the view of al-Alusi which says Allah had created 30 Adams prior to the creation of the Adam who is demed as the father of humankind; with the time span of 1000 years for each Adam. Addition to that, Said Hawwa advises that the opinion of al-Alusi needs to be taken with cautions due to the fact that there are no authentic textual evidence to back the absurd version up. Said Hawwa further states that Allah is able to create Adam and Eve from clay but the reason why He created Eve and Adam was to prove his power that which is able to create being out of another living being [19].

A few contemporary scholars from the Malay Archipelago has also interpreted the same phrase. Among them is M. Quraish Shihab who views the old school scholars' opinions of Eve created from Adam has triggered negative thoughts among the muslim community giving rise to belief that women are created from men. M. Quraish Shihab cited the view al-Tabataba'i to give credit to his opinion that creation of Eve was from the same kind and not from Adam's rib [20].

Hamka joins the bandwagon with his own view. Hamka opposes opinions of the ancient scholars because he views that the term does not mention that the one soul as Adam with his partner Eve, created from his rib despite the preceding scholars the likes of Ibn Abas, Ibn Abi Shaibah, Ibn Jarir, Ibn Mundhir and Ibn Abi Hatim all opine to the contarary. Similar to Rashid Redha, Hamka views the idea of Eve being created out of Adam's rib is nothing more than a fairy tale plagiated from the Ibrani race and specifically the Jews from their Torah under the Book of Genesis [7]. Ahmad Sonhadji while interpreting this term had quoted al-Maraghi who stated that the scholars who interpreted *nafs* as Adam are basically not guided by the proofs contained in the Quran but only rely on mere beliefs that Adam is the father of mankind. He also comments that al-Maraghi argues using the historical relics like the discovery of human skeletal remains which predate the history of the existance of Adam [3]. Ahmad Sonhadji's view is in line with the opinion of Rashid Redha.

Besides, the discussion regarding the original creation of women has been in the attention of the *muhaddithun* since there exist in the master books of Hadiths, i.e *al-Kutub al-Sittah* the main and famous reference for Muslims, a considerable number of Hadiths that tell women were created from the rib. This hadith is found compiled in *Sahih al-Bukhari*, *Sahih Muslim*, *Jami' al-Tirmidhi*, *Musnad Imam Ahmad* and *Sunan al-Darimi*, collective efforts by four companions namely Abu Hurayrah, Aishah Abu Dhar and Samurah. Al-Bukhari, Muslim and al-Tirmidhi record only narrations by Abu Hurayrah, while Ahmad bin Hanbal takes narrations from all four *Sahabah*. Al-Darimi on the other hand records narrations by Abu Hurayrah and Abu Dhar. Although the *matan* of the hadith is similarly narrated, however there is one difference in terms of the word (من ضلع) *min dila'in* and *ka al-dila'i* (كالضلع) that triggers a different understanding of the hadith as will be discussed further.

In the *Sahih al-Bukhari*, this hadith narrated by Abu Hurayrah is mentioned in three places. i.e in the *Kitab al-Ahadith al-Anbiya'* found in the chapter of the creation of Adam and his offsprings, and two hadiths in the *kitab of al-Nikah* in the chapter of doing good towards women. The *matan* of the hadith is as follows [2] :

عن أبي هريرة رضي الله عنه قال : قال رسولُ اللهِ ﷺ : استَوصُوا بِالنسَاءِ , فَإنَّ المَرْأة خُلِقَتْ مِنْ ضِلَعٍ, وإنَّ أعْوَجَ شَىْءٍ فِى الضِلَعِ أعلاَهُ, فَإنْ ذَهَبْتَ تُقِيمُهُ كَسَرتَهُ, وَإنْ تَرَكْتَهُ لَمْ يَزَلْ أعْوَجَ, فَاستَوصُوا بالنسَاءِ

Meaning : Narrated Abu Hurayrah, Rasulullah ﷺ said : Treat women nicely, for a women is created from a rib, and the most curved portion of the rib is its upper portion, so, if you should try to straighten it, it will break, but if you leave it as it is, it will remain crooked. So treat women nicely.

عن أبي هريرة أن رسول الله ﷺ قال : الْمَرْأَةُ كَالضِلَعِ، إنْ أَقَمْتَهَا كَسَرْتَهَا، وَإنِ اسْتَمْتَعْتَ بِهَا اسْتَمْتَعْتَ بِهَا وَفِيهَا عِوَجٌ

Meaning: Narrated Abu Hurayrah, Rasulullah ﷺ said, "The woman is like a rib, if you try to straighten her, she will break. So if you want to get benefit from her, do so while she still has some crookedness.

Ibn Hajar in his book, *Kitab al-Ahadith al-Anbiya'* discusses a hadith narrated with a *matan min dila'in* is found to have been using *sighah tamrid* which can be seen in the text قيل and can be understood as not from his absolute opinion. This means Ibn Hajar was aware that there are opinions that claim that Eve was created from Adam's rib and from other opinion from a short bend rib. In addition, Ibn Hajar quotes other narrations by Ibn Ishaq that there are additional attributes that is from the left rib before being sent to the heaven. Unsatisfied, Ibn Hajar further quotes the view of al-Qurtubi who opines that the hadith talks about women being created from the rib end thus it is like a rib. In *Kitab al-Nikah*, Ibn Hajar once again hinted the connection of this hadith with the narration of Ibn Ishaq; from Ibn Abbas that Eve was created from the short left most rib of Adam while he was asleep. Followed was the description on the meaning of the hadith; that is women were created from bend substance. This is not in conflict with the matan *ka al-dila'i* afore mentioned equating women with rib. For Ibn Hajar, there is a common ground in the allegation that women temperament are bend like ribs because they originate from ribs [9].

Al-Kirmani when discussing this hadith in *Sahih al-Bukhari* states that women were created from bend rib. He further elaborate that this hadith is pertaining to the creation of the mere first woman Eve from the rib of Adam. He quotes the view of al-Tibi who says the gist of this hadith is somewhat akin to a will that demands compliance in discharging the rights of women and in furtherance to that encouragement to treat women well and to have patience with their temperaments 12].

Ibn al-Mulaqqin on the other hand in his book brings the view of Ibn al-Tin who views that mentioning the word rib in the hadith does not presuppose towards women but instead is a mere allegory to rib. Regardless, he still views that the rib women were created from is the left rib of Adam [11].

As for one hadith figure of the present day, Ibn ‘Uthaymim when arguing a hadith narrated by al-Bukhari; elaborates that this hadith comes under the category of parables from the prophet s.a.w intended to mean that the prophet equates women with ribs because ribs are closest to a person. He concludes that this hadith is didactic in nature in which it expresses the need to interact with women who come in many forms; as wives, mothers, sisters in the most delicate and prudent way. He adds that the bend nature of women temperament if straightened will break because bend thing is unstraightenable [24].

Another contemporary hadith figure, al-Albani is also inclined to understanding this hadith in parable form. To him, this hadith is not to be taken literally or factually. His arguement is that there are no sahih and hasan hadith that expresses the creation of Eve from Adam’s rib. It is in other categories of hadiths that the stories of Eve coming from Adam’s rib are found expressive. As such, according to his evaluation, this hadith falls under *munkar jiddan* status because the *sanad* bears a narrator by the name Abdul Rahman bin Zayd whom the consensus of scholars judge as dhaif. As for the hadith that speaks about Eve being created while Adam was asleep is having a dhaif *sanad* and it is in fact a *mauquf* narration; whereby it was sourced from the companion and more inclined towards *israilliyat*. In addition, al-Albani brings in an arguement to corroborate his claim that narrations by al-Bukhari and Muslim are allegories in nature by the use of *sigha al-tashbih* found in the narrations i.e *ka al-dila‘i* [4].

The view of al-Albani is consistent with that of Shu‘aib al-Arna’ut, a present hadith scholar from Jordan who views the rib in question as merely parables that exemplify women’s attitude that is bend and imperfect. This view is based on the term *ka al-dila‘i* which he understands as a form of parables [21].

It is clear from the discussion of hadith on the creation of women is understood in two different meanings according to *Muhaddithun*, specifically literal understanding that women were created from the rib of Adam and secondly by allegory that women were created like a rib, refering to women’s attitude that needs to be handled gently.

4 Discussion

The outcome of the discussions above regarding the creation of Eve from the al-Quran and hadith clearly shows that the earlier and the present *mufassirin* and *muhaddithun* held two differing perspectives. The majority of the earlier *mufassirun* and *muhaddithun* uphold the view that Eve was created from Adam's rib based on the following passages from the Quran من نفس واحدة وخلق منها زوجها and hadith إن المرأة خلقت من الضلع. Interpretation of the pronouncement من نفس واحدة وخلق منها زوجها which they understood to be the reference to the creation of Eve was further reinforced by the hadith إن المرأة خلقت من الضلع. Their views were contrary to the thoughts of some contemporary *mufassir* and *muhaddith* who believed that the creation of Eve did not originate from Adam’s rib. The Quranic verses concerned are understood to mean that Adam and Eve were both created from the same substance which is the soil and there is also the view that they were both created equally in terms of humanity. The hadith is also understood to metaphorically describe women as the rib.

Although there are scholars who believe that there is a close link between the Quran and hadith that is seen as mutually supportive, there are also allegations of a difference of opinion between the two. This assertion is written by Qasim Qul Bek of Ibn Qimas who believes that the hadith written on the creation of women from the rib of men contradicts the Quranic verse خلقكم من نفس واحدة which to him meant that the creation of men and women are equal physically and intellectually. It does not mean that Adam was first created and from Adam, Eve was then created. In response to these allegations, Qasim Qul Bek stated that the word *min* in the Quran means to explain. It does not mean showing some parts as can be witnessed in some verses of the Quran, for example خلقلكم من أنفسكم أزواجا and الله الذي خلقكم من ضعف. Consequently, according to Qasim Qul Bek further, the verse carries this meaning that God created Eve from the same substance as that of Adam which is the soil and not from the light or fire from which angels and *jinns* were formed. In the context of understanding this hadith he is seen as emphasizing on this perspective that women are equated to a bent rib.

Thus, he stresses that true understanding of the meaning of the verse shows that there is no contradiction between the two. The researcher agrees with the view that there is no contradiction between the Quran and the hadith concerned as both are revelations and comes from the same source. Furthermore the hadith is a hadith which has been rated as authentic by al-Bukhari, Muslim and other figures, although there are claims that the source is from the *Israiliyyat* traditions, thus the concern for its legitimacy. Issues concerning the *Israiliyyat* and its connection to this hadith as well as *Tafsir* scholars’ understanding of the Quranic verse related to the creation of Eve from Adam must be construed in the context that not all of the *Israiliyyat* tradition is rejected as prescribed by scholars. There is a category of the *Israiliyyat* that is seen as valid through the al-Quran and al-Sunnah. Grounded on this fact, the similarities between the al-Quran and al-Sunnah with the scriptures of the People of The Book are not a means to amaze to their fabricated scriptures as points of argumentation. Instead, this knowledge is grounded in divine revelation which is the al-Quran and al-Sunnah [14]. Therefore, the issue of ruling out the hadith and *Tafsir* of previous scholars of the Quran on the basis of Judaism are not sound.

In addition, not all the hadiths of the Prophet can be interpreted and understood literally. In this context, al-Qaradawi clarified that apart from the type of hadiths that are easily understood and clearly delineated, there are also hadiths which relies extensively on the use of *Majazi* in its sentences. This is due in part to the Arab culture of that era which uses a great deal of figurative speech. According to al-Qaradawi, *Majazi* sentences is made up of several forms such as *majaz lughawi, majaz 'aqli, isti'arah, isti'arah al-tamthiliyyah* and *kinayah* as spoken by the Prophet p.b.u.h. to refer to actions, words or certain conditions. An understanding of the hadiths of the Prophet in this form should be based on *qara'in* or specific instructions to engender to the true meaning 15]. Therefore, in the setting of the hadith concerning the creation of women from the rib, understanding of it should be in its figurative sense. Furthermore, there is also the *qarinah* which shows this fact using the word *ka al-ḍila'i* in other authentic hadith traditions.

5 Conclusion

The interpretation of the majority of earlier *mufassirun* and *muhaddithun* against the Quranic verses and hadiths which became the basis for discussing the origins of women and Eve, were found to conflict with the views of later *muhaddithun* and *mufassirun* scholars. The *min nafs wahidah* pronouncement which was understood by the early *mufassirun* as a mention to Adam and Eve and Eve's creation from Adam contradicts the interpretation of contemporary scholars who read its meaning as Eve being created out of the same substance as Adam. Also, the rendering of the present *muhaddith* is found to disagree with the reading of past *muhaddith* with regards to the creation of woman from the rib. This Hadith is understood to be a parable by the current *muhaddith* and this differs to the understanding of the past *muhaddith* who understands it as a fact. However, the contentions put forward by the present *mufassir* and *muhaddith* scholars is seen as stronger in discussing the origins and the creation of Eve.

Equations should be centred and should be numbered with the number on the right-hand side.

$$T_s\,(l,t) = T_g\,(l,t) \tag{1}$$

$$T_s\,(l,t) = T_g\,(l,t)\; T_b\,(x \rightarrow -\beta,\, t) = 0 \tag{2}$$

Use italics for variables (u) and bold (**u**) for vectors. The order for brackets should be {[()]}, except where brackets have special significance.

6 Acknowledgement

The research was done under Short Term Grant of Universiti Sains Malaysia, No 304/PHUMANITI/6313200

References

1. Abduh, Muhammad and Muhammad 'Ammarah. *Al-'Amal al-Kamilah*. 1st ed. Beirut: Dar al-Syuruq, (1993)
2. Al-Bukhari, Abi 'Abdullah Muhammad b. Ismail. *Sahih al-Bukhari*. 1st ed. Beirut: Dar Ibn Kathir, (2002)
3. Ahmad Sonhadji Mohammad. *Tafsir al-Quran (Tafsir al-Quran Di Radio)*. 2nd ed. Kuala Lumpur: Pustaka al-Mizan, (1992)
4. Al-Albani, Muhammad Nasir al-Din. *Silsilah al-Ahadith al-Da'ifah wa al-Mawdu'ah*. 1st ed. Riyad: Maktabah al-Ma'arif, (2004)
5. Halimah Basri. *Penciptaan Wanita*, Jurnal Studi Gender & Anak, Vol. 5, No. 1. Indonesia: Universiti Islam Negeri (UIN), Makassar, (2010)
6. Halimatussa'diyah. *Tafsir Kebencian : Studi Bias Gender Dalam Tafsir al-Quran Karya Zaitunah Subhan*. Jurnal Ilmu Agama, Vol. 16, No. 1. Indonesia: Universiti Islam Negeri Raden, Palembang, (2015)
7. HAMKA. *Tafsir al-Azhar*. 1st ed. Singapore: Pustaka Nasional PTE LTD, (1990)
8. Ibn Asyur, Muhammad Tahir. *Tafsir al-Tahrir wa al-Tanwir*. Tunisia: Dar al-Tunissiah, (1984).
9. Ibn hajar, Ahmad b. 'Ali b. Hajar al-'Asqalani. *Fath al-Bari*. 1st ed. Riyad: Maktabah al-Malik Fahdi al-Wataniah, (2001)
10. Ibn Kathir, Ismail b. 'Umar b. Kathir al-Qurasyi al-Dimsyiqi and Sami b. Muhammad al-Salamah. *Tafsir al-Quran al-'Azim*. 1st ed. Saudi: Dar Tayyibah, (1997)
11. Ibn al-Mulaqqin, Siraj al-Din Abi Hafs 'Umar b. 'Ali b. Ahmad al-Ansari al-Syafi'.*Al-Tawdih li Syarh al-Jami' al-Sahih*. 1st ed. Qatar: Wizarah al-Awqaf wa Syu'un al-Islamiyah, (2008)
12. Al-Kirmani, Muhammad b. Yusuf al-Kirmani Syamsuddin. *Al-Kawakib al-Darari fi Syarh al-Bukhari (Syarh al-Kirmani 'ala al-Bukhari)*. 1st ed. Beirut: Dar Ihya' al-Turath al-'Arabi, (1981)
13. Al-Maraghi, Ahmad Mustafa. *Tafsir al-Maraghi*. Masr: Maktabah al-Mustafa al-Babi al-Halabi, (1946)
14. Qasim Qul Bek. *Syubhah Wurud Hawla al-Ahadith al-Khassah bi al-Mar'ah*. 1st ed. Jordon: Dar al-Nafa'is, (2014)
15. Al-Qardawi, Yusuf. *Kaifa Nata'ammul Ma'a al-Sunnah al-Nabawiyah*. 2nd ed. Kaherah: Dar al-Syuruq, . (2002)
16. Al-Qismi, Muhammad Jamal al-Din and Muhammad Fu'ad Abd al-Baqi. *Tafsir al-Qasimi Mahasin al-Ta'wil*. 1st ed. Masr: Dar Ihya' al-Kutub al-'Arabiyah, (1957)

17. Al-Razi, Muhammad al-Razi Fakhr al-Din Ibn 'Alamah al-Diya' al-Din Umar. *Tafsir al-Kabir wa Mafatih al-Ghaib.* 1st ed. Beirut: Dar al-Fikr, (1981).
18. Ridha, Muhammad Rasyid. *Tafsir al-Manar.* 3rd ed. Masr: Dar al-Manar, (1367H)
19. Said Hawwa. *Al-Asas fi al-Tafsir.* 1st ed. Kaherah: Dar al-Salam, (1985)
20. Shihab, M. Quraish. *Tafsir al-Misbah : Pesan, Kesan dan Keserasian al-Quran.* 1st ed. Indonesia: Lentera Hati, (2000)
21. Shu'aib al-Arna'ut and Abi Zakariya b. Syaraf al-Nawawi. *Riyad al-Salihin.* 3rd ed. Beirut: Muassasah al-Risalah, (2001)
22. Al-Suyuti, Jalal al-Din and al-Mahalli, Jalal al-Din. *Tafsir al-Jalalain al-Muyassar.* 1st ed. Beirut: Lubnan, (2003)
23. Al-Tabari, Abi Ja'far Muhammad b. Jarir and 'Abdullah b. 'Abd al-Muhsin al-Turki. *Tafsir al-Tabari Jami' al-Bayan 'an Ta'wil ay al-Quran.* 1st ed. Kaherah: Dar Hijr, (2001)
24. Al-'Uthaymin, Muhammad b. Salih and Muhammad Nasir al-Din al-Albani. *Syarh Sahih al-Bukhari.* 1st ed. Kaherah: Maktabah al-Islami, (2008)
25. Wan Siti Nur Haini. *Penciptaan Wanita Dari Tulang Rusuk (Studi Kritis Tentang Hadis Penciptaan Wanita Dari Tulang Rusuk).* Semarang: Institut Agama Islam Negeri Walisongo, (2013)

Patient's Behaviour with Coronary heart desease Viewed from Socio-Cultural aspect of Aceh Society in Zainoel Abidin Hospital

Marniati[1*], *Soekidjo* Notoatmodjo[2], *Sutomo* Kasiman[1] and *R kintoko* Rohadi[1]

1Faculty of Public Health, University of Sumatera Utara
[2]Universitas Indonesia

Abstract. Coronary heart disease (CHD) is a chronic inability of heart to draw an adequate blood on myocardium due to specific problem in coronary artery. When the artery that supplies myocardium is interrupted, thus the heart is unable pumping the blood effectively to fill up an adequate blood perfusion to vital organs and other peripheral tissues. If the perfusion is impaired in time of oxygenation, the patient will threatened by a sudden death. This is a quantitative, analytical and observational research with a design of case study control, in which the subject is CHD patients and the control is non-CHD patients. The research populations were 2318 outpatients with CHD who visited Heart Polyclinic at Zainoel Abidin Hospital Banda Aceh. The sample studied were 92 patients with CHD and 92 patients with non-CHD (control). We used primary and secondary data that processed through editing, coding, entry, cleaning and tabulating. The data was analyzed using univariate, bivariate and multivariate analysis. The results showed that the CHD occurrences are significantly influenced by eating behavior (p = 0,000), smoking habit (p = 0,000) and sporting habit (p = 0,000). Therefore, health workers are expected to increase health promotion in order to improve the quality of life of people, especially in elderly people, by adopting a healthy lifestyle and avoiding CHD risk factors. Moreover, people are also suggested to take regular lipid profile check up and keep off smoking habit in order to do early prevention of CHD.

1 Introduction

World Health Organization (WHO) estimated that about 50% of global populations die from heart and blood vessel diseases each year. According to World Health Statistic Report 2008, 17.3 million of the deaths were caused by Cardiovascular Disease (CVD), representing 30% of all global deaths. Of these deaths, an estimated 7.4 million were due to coronary heart disease and 6.7 million were due to stroke. These deaths are predicted to continue to increase up to 23.4 million people in 2030. Over 80% of CVD deaths take place in low- and middle-income countries and more than a half of the deaths were nearly equal between men and women. Therefore, it has been recognized as the number one killer and causing disability in the world [24].

Heart disease is also the leading cause of the death in America. Each year, 478.000 people die from Coronary Heart Disease (CHD), 1.5 million people have heart attack, 407.000 people experience heart surgery and other 300.000 people undergo an angioplasty. In 2010, the prevalence of CHD was greatest among persons aged ≥65 years (19.8%), followed by those aged 45--64 years (7.1%) and those aged 18--44 years (1.2%). CHD prevalence was greater among men (7.8%) than women (4.6%), however the prevalence will be drastically accelerate in postmenopausal women (CDC, 2011). According to American Heart Association (AHA) in Heart Stroke Statistic 2010, it was indicated that every 25 seconds, someone will have coronary event and every one minute one will die [4].

The deaths caused by CHD occurrences in Indonesia are also high, approximately reach 26%. According to National Household Health Survey, it continues to increase within the last 10 years. In 1991, the death rates from CHD were 16% and significantly increased in 2001 to 26.4%. In 2007, the rate was more than double, growing to 59.5%. According to the Basic Health Research (Riskesdas) 2013 issued by Health Research and Development Agency of Health Ministry of Indonesia, the prevalence of CHD based on interview diagnosed by doctor was 0.5% while based on doctors diagnose or symptom was 1.5%. The mortality rate due to the disease in Indonesia reached 7.6 million people per year, while the number of patients of all ages were also significantly increased, from 10,600 people in 2010 to 12.800 in 2013 [12].

In a research conducted by Ismail *et al.* [10] to men and women 14-45 years of age in the South Asia region stated that the risk of myocardial infarct for active smoking was 3.82 higher compared to the control group (OR=3.82; 95%; CI=1.47 – 9.94), while for those with an increased serum cholesterol was 1.67 higher than the control group (OR=1.67; 95%; CI=1.4 – 2.45). In addition, another research conducted by Smeltzer and Bare [23] found that the CHD risk is substantially increased for 2-4 times in smokers compared to the never smokers. Smoking will reduce the oxygen supply to the heart and body which in turn led to the raise of blood pressure and pulse rate, reduce the cholesterol level (HDL), cause the blood clot to form and damage the endothelial cells. To reduce the risk of CHD, people are recommended to do exercise for about 30 minutes per day. In the research conducted by Harvard Alumni Study in Lipoeto [17] it clearly proved that physical activity/exercise (even in a short term) can significantly reduce CHD risk as it increases the HDL concentration. Women with less physical activity face higher risk of it.

2 Conceptual Framework

Conceptual framework is a relationship framework among concepts or variables to be studied. Variable is a characteristic that attached to population which is varied from one to another and is investigated in a research. The variables in this research comprised of CHD occurrence (as dependent variable) and dietary, smoking and sporting habit (as independent variables). The conceptual framework of this research is presented as follow:

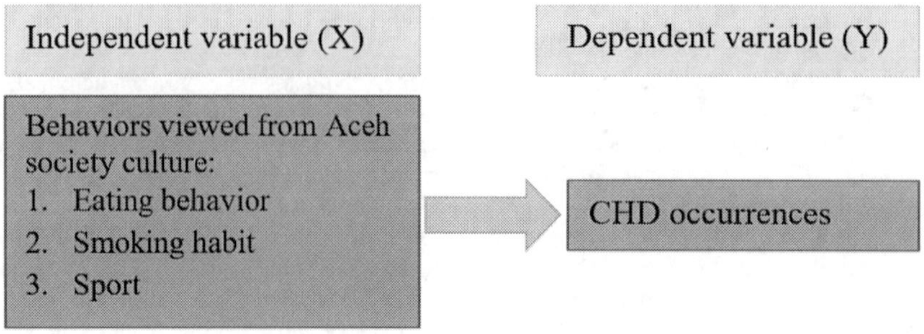

Fig. 1. Conceptual framework of this research

3 Research Methodology

This is a quantitative, analytical and observational research with a design type of case study control, in which the subject is CHD patients and the control is non-CHD patients [15].

This research was conducted in Zainoel Abidin Hospital Banda Aceh from November 2014 to September 2015, began with preliminary research and ended with final oral exam. The populations in this study were 2318 outpatients with CHD who visited Heart Polyclinic at Zainoel Abidin Hospital Banda Aceh. The size of samples needed in this research was determined using the following formula:

$$n = \frac{Z^2_{1-\alpha/2} P(1-P)N}{d^2(N-1) + Z^2_{1-\alpha/2} P(1-P)} \quad (1)$$

where P denotes the subject proportion of previous researches that equals to 0.5, d is the precision level of 0.1, Z is the level of trust of 1.96, n is the number of samples, and N is the number of population of 796.

$$n = \frac{(1.96)^2 0.5(1-0.5)2319}{(0.1)^2(2319-1) + (1.96)^2 0.5(1-0.5)}$$

$$n = 92{,}26 \text{ people} = 92 \text{ people} \quad (2)$$

The samples to be studied are 92 CHD patients and 92 non-CHD patients (control). The sampling method used was simple random sampling.

The inclusive criteria of the samples are: (1) CHD outpatients who hospitalized at Heart Polyclinic of Zainoel Abidin Hospital Banda Aceh; (2) patients aged >30 years; (3) able to write and read; and (4) willing to participate as respondent. While the exclusive criteria are: (1) CHD patients hospitalized at Heart Polyclinic of Zainoel Abidin Hospital Banda Aceh with complication of heart and kidney failure; and (2) A drastically decreased on health status.

3.1 Operational Defination

Below are the operational definitions of the studied variables in this research.

Table 1. Operational definitions of the studied variables

No.	Variable	Operational Definition	Measurement Scale	Measurement Manner	Category
Independent Variable					
1	Behavior	The CHD patients' response or reaction to the CHD occurrence which is viewed from Aceh society culture, encompassing dietary, smoking and sporting habit.			
2	Eating behavior	The eating/drinking habit of meals and beverages which contain saturated fat, like meat, offal, egg, milk, etc., that consumed daily in a period of time.	Ordinal	Questionnairesand interview	1. Frequently, everyday - 3 times a week. 2. Moderate, 1 – 2 times a week. 3. Infrequently, once a week or never.
3	Smoking habit	The habit/behavior of smoking cigarretes or ever smoking which is measured from the number of cigarretes consumed per day.	Ordinal	Questionnairesand interview	1. Heavy smoker (>20 cigarretes a day). 2. Moderate smoker (10-20 cigarretes per day). 3. Light smoker (<10 cigarretes per day)

4	Sporting habit	Ideally sporting habit is doing a certain exercise with an average time of 30 minute		Questionnair esand interview	1. Never exercise, never doing any exercise. 2. Non-ideal exercise, doing a certain exercise less than 3 times a week and 30 minutes a day. 3. Ideal exercise, doing a certain exercise at least 3 times a week and 30 minutes a day.
Dependent Variables					
1.	CHD occurren ce	Cardivascular disorder is characterized by a typical chest pain—felt like a heavy pressure on chest—and the increased of cardiac enzym. It can be screened using electrocardiography.	Nominal	Questionnair esand interview	1. Yes (if have CHD. 2. No (if have no CHD).

Place the figure as close as possible after the point where it is first referenced in the text. If there is a large number of figures and tables it might be necessary to place some before their text citation.

3.2 Data Analysis

Data in this study were obtained from both primary and secondary sources. The primary data were directly obtained from respondents who filled-in the questionnaires, while the secondary data were obtained from available reports and documents owned by the hospital. The data processing was involved a series of following steps, i.e.:

1) *Editing*. Recheck and further investigate all the collected data and fix the failures and the shortcomings one in order to improve the quality of data.
2) *Coding*. Give numerical codes for the data which consisted of several categories.
3) *Entry*. Enter the data into computer to be analyzed in accordance with their respective variables

4) *Cleaning*. Identify and correct errors of the entered data by observing the distribution or deployment frequency of each studied variable.
5) *Tabulating*. Put the data in a frequency distribution table in order to facilitate the analysis, data processing and conclusions to be made.

The collected data were processed using computer software and further analyzed to see the distribution overview of each variable, as well as to test the research hypothesis for observing the relationship between independent and dependent variables and finding a fit model for CHD occurrence detection. A series of analysis carried out in this research are as follow:

1) *Univariate analysis*. Analyze each of research variables in order to find out their distribution [22].
2) *Bivariate analysis*. Examine the relationship between two variables, independent and dependent, using chi-square test [9]. In this research, the analysis was also used in selecting potential independent variables to be included into further analysis, that is variables with $p < 0.25$ (Mickey and Greenland in Hosmer and Lemeshow, 2000). It was also used further in obtaining OR value and 95%CI as well as in making decisions with significance level of 0.05 (α=5%) according to the following provisions:

- If p value is ≤ 0.05, the relationship is significant;
- If p value is > 0.05, the relationship is insignificant.

3) *Multivariate analysis*. Simultaneously observe the relationship between several independent variables and several dependent variables. In this research, the multivariate analysis used in analyzing the data was Multiple Logistic Regression Test. The intention of this test is to obtain the most appropriate model which illustrates the influential determinants on CHD occurrences (Ariawan, 2008). The regression formula is as follow:

$$\gamma = \ln \frac{p}{1-p} = \beta o + \beta_1 x_1 + ... + \beta_i x_i \quad (3)$$

where p denotes CHD occurrences.

$$p = \frac{1}{1+e^{-(\beta_0+\beta_1 x_1 ..... +\beta_i x_i}} \quad (4)$$

where $\beta_i = 0, 1, 2, ..., n$ denotes the parameters of logistic regression model and $x_i = 1, 2, 3, ..., n$ denotes independent variables observed.

4 Results and Discussion

4.1 Univariate Analysis

Characteristic of Respondents

Table 2. Distribution of characteristic frequency of respondents

No	Characteristics	Case		Control	
		N	%	n	%
1	**Age**				
	- 30-45 years old	19	20.7	32	34.8
	- 46-60 years old	60	65.2	43	46.7
	- 61-75 years old	13	14.1	17	18.5
	Total	**92**	**100**	**92**	**100**
2	**Level of Education**				
	- Low (Elementary School)	20	21.7	13	14.1
	- Middle (High School)	50	54.3	65	70.7
	- High (University)	22	23.9	14	15.2
	Total	**92**	**100**	**92**	**100**
3	**Type of Occupation**				
	- Farmer	17	18.5	22	23.9
	- Enterpreneur	52	56.5	58	63.0
	- Civil Servant	23	25.0	12	13.0
	Total	**92**	**100**	**92**	**100**

The underlying characteristics of respondents in this research were classified based on three categories, i.e.: age, level of education and type of occupation (Table 2). According to the table, CHD cases mostly occurred in age 46-60 years (65,2%). Similarly, the control group were also found to be vulnerable for people aged 46-60 tahun (46,7%). With regard to level of education, both CHD and control group were found to be mostly experienced by those who were with middle level of education, respectively 54,3% and 70,7%. It happened the same to type of occupation, where the CHD and the control group were mostly had by those who work as enterpreneur, respectively 56,5% and 63,0%.

Respondents Behaviors and CHD Occurrence

Table 3. Influence of respondents' behaviors on CHD occurrence

No	Respondents Behaviors	Case		Control	
		N	%	N	%
1	**Eating behavior**				
	- Frequently	62	67.4	20	21.7
	- Moderate	18	19.6	37	40.2
	- Infrequently	12	13.0	35	38.0
	Total	**92**	**100**	**92**	**100**
2	**Smoking habit**				
	Heavy smoker	71	77.2	23	25.0
	Moderate smoker	15	16.3	32	34.8
	Light smoker	6	6.5	37	40.2

Total		**92**	**100**	**92**	**100**
3	**Sporting habit**				
	Never exercise	68	73.9	27	29.3
	Non-ideal exercise	17	18.5	36	39.1
	Ideal exercise	7	7.6	29	31.5
Total		**92**	**100**	**92**	**100**

The table 3 above presents the influence of respondents' behaviors on CHD occurrence amongst the outpatients at Heart Polyclinic Zainoel Abidin Hospital using univariate analysis. The behaviors are relevant to eating behavior, smoking habit and sporting habit. In regard to eating behavior, the case group was dominated by people who have frequent meals (67.4%) while the control group was dominated by those with moderate eating habit (40.2%). Regarding the smoking habit, 77.2% persons in the case group were identified as heavy smokers, while in the contrary, 40.2% persons in the control group were identified as light smokers. To the sporting habit, mostly people (73%) in the case group were known to had never exercises, while in the control group mostly of them (39.1%) were had non-ideal exercise.

4.2 Bivariate Analysis

The Influence of Respondents Behavior on CHD Occurrence

Table 4. Influence of respondents behavior on CHD occurrence

No	Respondents Behavior	Case		Control		*p value*
		N	%	N	%	
1	**Eating behavior**					
	- Frequently	62	67.4	20	21.7	0.000
	- Moderate	18	19.6	37	40.2	
	- Infrequently	12	13.0	35	38.0	
Total		**92**	**100**	**92**	**100**	
2	**Smoking habit**					
	Heavy smoker	71	77.2	23	25.0	0.000
	Moderate smoker	15	16.3	32	34.8	
	Light smoker	6	6.5	37	40.2	
Total		**92**	**100**	**92**	**100**	
3	**Sporting habit**					
	Never exercise	68	73.9	27	29.3	0.000
	Non-ideal exercise	17	18.5	36	39.1	
	Ideal exercise	7	7.6	29	31.5	
Total		**92**	**100**	**92**	**100**	

According to the chi-square test, this study concluded that:

1) There is a significant influence between eating behavior and CHD occurrence as mostly CHD patients in this research were dominated by those who have frequent meals (67.4%);

2) There is a significant influence between smoking habit and CHD occurrence as mostly CHD patients in this research were identified as heavy smokers (77.2%);

3) There is also a significant influence between sporting habit and CDH occurrence as mostly CHD patients in this research were known to had never exercise (73.9%).

4.3 Multivariate Analysis

After multivariate analysis for observing the relationship of each variable on CHD occurrence was done, the multivariate analysis was then carried out for determining the most dominant variable that influence the CHD occurrence by using Multiple Logistic Regression Model.

The variables with p value of <0.25 were then included into regression test. According to the chi square test result (Table 4), all the tested variables were had p<0.25, thus they all were able to be included to the multivariate analysis. Detail is shown below (Table 5).

Table 5. Variables selection result

Variable	*p value*	Description
Eating behavior	0.000	P<0.25
Smoking habit	0.000	p<0.25
Sporting habit	0.000	P<0.25

Tabel 6. Multiple Logistic Regression Model and CDH occurrences

Variabel	B	Sig.	OR	95% C.I	
				Lower	Upper
Eating behavior	1.394	0.000	4.031	2.316	7.016
Smoking habit	1.662	0.000	5.269	2.962	9.373
Sporting habit	1.438	0.000	4.214	2.322	7.647
Constant	**7.736**	**0.000**	**0.000**		

According to multivariate analysis, the most dominant variable that influence CHD occurrence is smoking habit (p=0.000; OR=5.3) which means that CHD risk in heavy smoker is 5.3 times higher than in light smokers.

4.4 Discussion

The Influence of Eating Behavior on CHD Occurrence

The research results shows that there is a significant influence between eating behavior and CHD occurrence (p=0.000). The majority patients of CHD (case group) were those who have frequent meals (67.4%), while the non-CHD patients (control group) were those who have moderate eating behavior (40,2%).

 https://doi.org/10.1051/matecconf/201815005065

Ridwan (2009) stated that almost 90% of diseases, both degenerative and cardiovascular, are caused by the poor eating habits in people daily life. Foods recommended to avoid are those that contain high saturated fat and bad cholesterol like meat and fast foods.

According to Ratna et. al [19] eating behavior plays an important role in CHD prevention and treatment. People consuming 35 grams fiber foods per day have lower risks of CHD compared to people who consume less than 15 grams fiber per day. Low calorie foods, like seeds, fruits and vegetables can lowering cholesterol and maintaining healthy weight. Most of cardiovascular patients are overweight and have high blood cholesterol. To lose the weight, the calorie intake must be cut back by doing more physical activities.

The Influence of Smoking Habit on CHD Occurrence

The result of this study shows that there is a significant influence between smoking habit and CHD occurrence (p=0.000). According to the results, 77.2% persons in the case group were identified as heavy smokers, while in the contrary, 40.2% persons in the control group were identified as light smokers. In addition, the multivariate analysis was clearly suggested that smoking habit is the most dominant variable that influences CHD occurrence (p=0.000; OR = 5.3, which means that heavy smokers are 5.3 times have increased risks of CHD). This result is in line with the previous researches conducted by Lewis *et al.* (2007) and Smeltzer and Bare (2002) which suggested that heavy smokers were likely to have 2-4 times higher risks of CHD than the non-smokers.

Cigarrete contains hundreds of harmful chemical substances, like tar, nicotine, carbonmonoxide, etc., that can reduces oxygen supply to the heart, increases blood and pulse pressure, decreases HDL cholesterol level, clots blood and causes endotel dysfunction.

Gray *et al.* (2002) stated that CHD risk rises with the increasing of blood pressure. This study suggested that blood pressure levels ranging from 120-129 mmHg systolic/80-84 mmHg diastolic were associated with two times higher risk of cardiovascular disease compared to levels of less than 120/80 mmHg.

The Influence of Sporting Habit on CHD Occurrence

The result of this study also shows that there is a significant influence between sporting habit and CHD occurrence as mostly CHD patients in this research were known to had never exercise (73.9%), while the non-CHD patients were mostly had non-ideal exercise (39.1). This is in accordance with Bender et al. [2] research results presented in their *Cardiology* book which indicated that there is a strong relationship between physical activity and CHD occurrence. The more exercises people do, the lesser CHD risk they have. Kabo (2008) stated that lack physical activity causes various diseases, including CHD. In regard to the disease, inactive people are 1.9 times more likely to develop CHD than those who are active.

According to Kusmana et. al [13], the frequency of physical activity/exercise determines its quality. The best exercise frequency that gives optimal result is carried out in a regular basis, 3- 6 times a week. The longer the exercise, the more the blood supplied to the heart and the more oxygen get distributed to the whole body. Having a 30-minutes exercise per day can double the effect, not only increases the blood supply but also turn off the body's generation of fat and cholesterol which eventually decreased CHD risk.

 https://doi.org/10.1051/matecconf/201815005065

5 Conclusion and Suggestions

This research was intended to observe CHD patients behaviors which are viewed from socio-cultural aspect of Aceh society in Zainoel Abidin Hospital Banda Aceh with the following conclusions:

1) There is a significant influence between eating behavior and CHD occurrence (p=0.000);
2) There is a significant influence between smoking habit and CHD occurrence (p=0.000);
3) There is also a significant influence between sporting habit and CHD occurrence (p=0.000).

According to the results, we also provide several suggestions as follow:

1) Health workers are expected to increase health promotion in order to improve the quality of life of people, especially in elderly people, by adopting a healthy lifestyle and avoiding CHD risk factors;
2) People are also suggested to take regular lipid profile check up and keep off smoking habit in order to do early prevention of CHD.

References

1. Backer G, et al. Overweight and obessity:a major challenge for coronary heart disease secondary prevention in clinical practice in Europe,European Heart Journal 2002; 21: 808-813, (2002)
2. Bender J. R, Russel K. S, Rosenfeld L. E, and Chaundry S,. Oxford American Handbook of Cardiology. New York: Oxford. p. 256-260,(2011)
3. Dahlan, S.M, Analisis Multivariat Regresi Logistik. PT. Epidemiologi Indonesia. Jakarta, (2012)
4. Dalusung-Angosta, A. Coronary Heart Desease Knowledge and Risk Faktors among Filipino-Americans connected to Primary Care Services. University of Hawai at Manoa. ProQuest Dissrtations and Theses, (2010)
5. Gray, H.H, Dawkins, K.D., Morgan, J. M & Simpson, I.A, Lectures Note on Cardiology (4th ed). Southampton. Blackwell Science Ltd, (2002)
6. Hastono, S.P,; Sabri, Luknis,. Statistik Kesehatan. Penerbit Raja Grafindo, Jakarta, (2006)
7. Hendramartono,. Obesitas sebagai faktor risiko penyakit kardiovaskular. Surabaya. Majalah Kedokteran UDAYANA. Volume 33, (2002)
8. Honigmann, John, J.. The World of Man Dalam Pengantar Ilmu Antropologi. Rineka Cipta. Jakarta, (1959)
9. Hosmer, David .W, and Lemeshow, Stanley,. Applied Logistic Regression 2nd. Ed.John Wiley & Sons.Inc., New York, (2000)
10. Ismail, TH jafar, FH Jafari, F White, AM Faruqui, N Chaturvedi, Risk factors for non-fatal myocardial infarction in young south asian adults, Heart on Line 2004; 90: 259-263

11. Kabo, P.. Mengungkap Pengobatan Penyakit Jantung Koroner : Kesaksian seorang ahli jantung dan ahli obat. Gramedia Pustaka Utama. Jakarta, (2008)
12. Safaat, Bagaimana Menggunakan Obat-Obat Kardiovaskuler Secara Rasional. Jakarta : Balai Penerbit FKUI,(2010)
13. Kusmana D. dan Hanafi M.,. Patofisiologi Penyakit Jantung Koroner In : Buku Ajar Kardiologi Fakultas Kedokteran Universitas Indonesia, 1st ed, Jakarta : 159-165, (1996)
14. Lemeshow, S,. Besar Sampel dalam Penelitian Kesehatan. Penerjemah Pramono, Gadjah Mada University Press, Yogyakarta, (1997)
15. Lapau, Buchari,. Metode Penelitian Kesehatan : Metode Ilmiah Penulisan Skripsi, Tesis dan Disertasi. Buku OBOR. Jakarta, (2013)
16. Lewis, S. L., Heitkemper, M.M., Dirksen, S.R., O'brien, P.G. & Bucher, L, Medical Surgical Nursing : Assesment and Management of Clinical Problems. Servent Edition. Volume 2. Mosby Elsevier, (2007)
17. Lipoeta, I.,. Zat Gizi dan Makanan Pada Penyakit Kardiosvascular, Andalah InsistPress. Yogyakarta, (2006)
18. Notoatmodjo.. Promosi Kesehatan Teori & Aplikasi. PT Rineka Cipta, Jakarta, (2010)
19. Ratna,. Penyakit- Penyakit Mematikan. Nuha Medika. Yogyakarta, (2013)
20. Riskesdes,. Penyajian Pokok-Pokok Hasil Riset Kesehatan Dasar. Badan Penelitian dan Pengembangan Kesehatan Kementerian Kesehatan RI. Jakarta, (2007)
21. Penyajian Pokok-Pokok Hasil Riset Kesehatan Dasar. Badan Penelitian dan Pengembangan Kesehatan Kementerian Kesehatan RI. Jakarta, (2013)
22. Riyanto, A, Pengolahan dan Analisis Data Kesehatan. Nuha Medika. Yogyakarta (2010)
23. Smeltzer, C, & Bare, B.G,. Buku Ajar Keperawatan Medical Bedah : Brunner & Suddarth. Edisi 8.Volume 2. Jakarta. EGC, (2002)
24. World Health Organization (WHO), Cardiovascular Disease. Available from:www.who.int/mediacentre, (2011)

https://doi.org/10.1051/matecconf/201815005066

The Factor Related to Mother's Motivations in Carrying out the Circumcision of the Female Children in a Teuk Jawo Village Baiturrahman Sub District, Banda Aceh in 2014

*Marniati**, *Rahmayani* , and *Ayuwari* Yunita

Abstract. Circumcision is the activity carried out to remove the partial foreskin of penis. The preliminary survey result of 15 mothers that were carried out the circumcision to their female children mention that 7 mothers carried out the circumcision because of the culture/tradition from generation to the generation, 5 mothers brought their children because of the family supports. There are 121 female children that are 0 – 5 years old in Ateuk Jawo Village. The objective of this research is to detect the factors that affect the mothers' motivations in carrying out the circumcision to female children in Ateuk Jawo Village, Baiturrahman Sub District, Banda Aceh. The methodology used in this research is analytical descriptive with 121 population of people. The samples in this research are 55 people. The sampling technique is purposive sampling. The data was collected by distributing the questionnaires. This research was conducted in Ateuk Jawa Village, Baiturrahman Sub Village, Banda Aceh on July 2014. The results of this research mention that Chi –Square tests show that from 28 respondents who chose the culture were not affected to the low motivations of the mothers in carrying the circumcision, there were 16 respondents (57.1%) with the value of p = = 0.003 ($p < 0.05$). From the 32 respondents who chose the family supports were higher affected of the mothers' motivations, there were 26 respondents (81.2%) with the value of p= 0.004 ($p < 0.05$). The last was from the 25 respondents who chose the religion factor were the most affected of the mothers' motivations, there are 21 respondents (84%), with the value of $p = 0.010$ ($p < 0.05$).This research concluded that there are the relations among culture, family supports and religions factor that were affected the mothers' motivations in carrying out the female children circumcision. It is recommended to the pregnant women to improve the knowledge about the circumcision to their female children later.

1 Introduction

Circumcision practice to the female children becomes the tradition in the particular community groups. The main objective is to control the sexual libido of the women. There is an assumption that the dirt on the clitoris can cause the women sexual libido become uncontrollable. whereas the circumcision practice is harmful to physical health and sexual psychic of the women [5].

The update estimation data from UNICEF mention that about 120 millions of the female children and the women had been carried out the female circumcision in 29 countries. Based on these current trends, as many as 30 million female children under 15 years old still have the risk to be carried out the circumcision. The consolidation Programme between the Funds of the United Nations Population (UNFPA) and UNICEF on the female children circumcision has been conducting the progress in preventing female children and female future generations from the circumcision practice. The latest estimation data of the UN General Committee that had been ratified with the unanimous vote on December 2012 was to urge the member countries in intensifyng efforts to eliminate the whole female circumcision practice [13]. The various type of the female circumcision practice, both symbolically and female genital mutilation/cutting practice is a form of discrimination and violence against women that have to be abolished. This activity is an arbitrary action against the female body integrity. When the country fails to effectively oppose this practice, it will encourage the public perceptions that the other person has the right to control the sexuality of a woman or female children [8].

The female children circumcision in Indonesia is carried out to 0-18 years old female depending on the local culture. The female circumcision generally is done to the baby after birth, 70 % is in Java and Madura, 78 % is in North Sumatra and 64% is in West Sumatra, are carried out to under 1 year old and partly to 7 – 9 years old [2].

The provision of female circumcision is arranged in Health Minister Regulation (Permenkes) No.1636 / 2010. This Permenkes mentioned that the definition of female circumcision is an activity that only scratching the skin that covers the front of the clitoris, without injuring the clitoris. Female circumcision can only be carried out by the medical officers such as Doctors, Midwives, and Nurses. In addition, the process of circumcision has to be done in the sterile place, room and equipment. Circumcision also has be carried out with adequate lighting and the availability of the clean flow [8].

The practice of female circumcision is expected that had been started since 4000 years ago, before the organized religion came. Anthropologists revealed that the practice of female circumcision had been popular in the ancient Egyptian community; this is evidenced by the pure discoveries. Some people consider that female circumcision is a tradition related to religion, so rise to the pro and contra. Female circumcision is also carried out in in 28 countries, mostly carried out in the most of African countries, partly in Asian, Latin America, North America and Europe Countries In Indonesia, mentioned that the tradition of female circumcision is a habit that came together of Islamic that continue and common happen since the ignorance period (Tonang, 2006).

Data of female circumcision in Indonesia is still very difficult to obtain. However, it does not mean that female circumcision is not carried out in Indonesia. The problem of female circumcision is a hidden problem because the prevalence is very big but not raised up. No complication reports because of the female circumcision activity in Indonesia, it is because there is lack of knowledge about the female reproductive organs or the shame culture and feeling fear to reveal the incidence of female circumcision (Juliansyah, 2009).

The female circumcision especially in Indonesia is most done as a aprt of the culture or tradition. At the beginning, it was only the community tradition, not religion. But when the

https://doi.org/10.1051/matecconf/201815005066

religion meets the tradition, then it was given the response. The religion response that accommodated the tradition then become the culture (Meiwita, 2006).

The relatives can be the motivator to the parents, because by the supporting of the families/relatives to carry out the circumcision to the female childred, so that the mothers will do so to their daughters [9].

Based on the preliminary study that has been done in Ateuk Jawo Village, it was found that many mothers brought their daughters to have the circumcision practice. Even from the health factor, the female children circumcision is not allowed anymore because it will give the negative effects to the female children. The mothers said that the female children circumcision had been their tradition or habit since before.

From the preliminary study that has been done by the researcher on January 3rd 2014 in Ateuk Jawo Village, Baiturrahman Sub District, Banda Aceh, found that there the circumcision were carried out to 45 baby girls that had just been born by the midwife. According to preliminary survey result to 15 mothers that carried out the circumcision practice to their daughters, it was found that 7 mothers took the baby to have circumcision because of the culture/tradition from generation to generation, 5 mothers told that they took the babies because of the family supports. There are 121 female children that are 0 – 5 years old in Ateuk Jawo Village.

From the above phenomena and background, the writers are interested in assessing **"the** Dari Fenomena dan latar belakang tersebut di atas, maka penulis tertarik untuk melihat " **What The Factors Related To Mothers' Motivations In Carrying Out The Circumcision Of The Female Children In Ateuk Jawo Village, Baiturrahman Sub District, Banda Aceh are"**

Problem Finding

The problem finding in this research is how the The Factors Related To Mothers' Motivations In Carrying Out The Circumcision Of The Female Children In Ateuk Jawo Village, Baiturrahman Sub District, Banda Aceh are.

2 Research Objectives

Main Objective

To detect the factors that affects the mothers' motivations in carrying out the circumcision to female children in Ateuk Jawo Village, Baiturrahman Sub District, Banda Aceh.

Special Objective

a. To detect the relation between the culture and mother motivation in carrying out the circumcision to the female in Ateuk Jawo Village, Baiturrahman Sub District, Banda Aceh.

b. To detect the relation between the family support and mother motivation in carrying out the circumcision to the female in Ateuk Jawo Village, Baiturrahman Sub District, Banda Aceh.

c. To detect the relation between the religion factor and mother motivation in carrying out the circumcision to the female in Ateuk Jawo Village, Baiturrahman Sub District, Banda Aceh.

Research Benefits

1. to *Puskesmas*

Improving the concept and reference about the female children circumcision and improving the health services especially in handling female children circumcision

 https://doi.org/10.1051/matecconf/201815005066

2. to Respondents
The result can be the suggestion and recommendation the pregnant women to improve the knowledge about the circumcision to their female children later.

3. to Educational Institution
To Improve the concept and reference in giving the guideline material of obstetrics and gynecology.

4. to Researchers
To improve the concept and reference about the female children circumcision so that the research can be continued by using the different variables.

Concept Framework
The research concept framework is the framework of the relation among the observed concepts and be measured through the research done. In this research, the researchers only want to detect the factors that affects the mothers' motivations in carrying out the circumcision to female children in Ateuk Jawo Village, Baiturrahman Sub District, Banda Aceh.

According to the MOH [8], factors that affected mothers' motivations in carrying out the circumcision to female children are because of three factors, they are: predisposing factors, enabling factors, attitudes, beliefs, moral/norms / culture, confidence, and capacity Enabling factor consists of the availability of health resources, reachable of health resources, government regulations, priorities and community commitment to health, and health counseling from health professionals, peers, teachers, employer/supervisor, health care providers, community leaders, and policy makers. / keputusan. But the researchers only want to focus on cultural factors, family support, and religion that have the huge relation to the female circumcision.

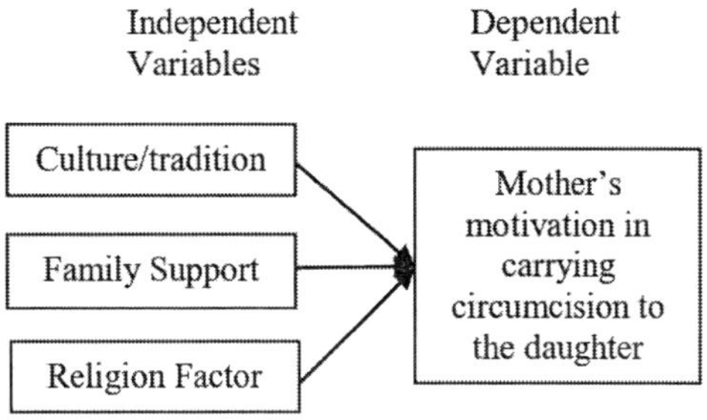

Research Type
This research is analytical descriptive with cross sectional study design that has the objective is to the factors that affects the mothers' motivations in carrying out the circumcision to female children in Ateuk Jawo Village, Baiturrahman Sub District, Banda Aceh.

Population and Sample
The population in the research is all mothers whose daughters are 0 – 5 years old on January – December 2013. The total are 121 who living in Ateuk Jawo Village, Baiturrahman Sub District, Banda Aceh.

The sampling technique is *Purposive Sampling* means the samples are chosen based on the researchers considerations using the characteristics or features of the population known before (Notoatmodjo, 2005). The criteria are shown as below:

https://doi.org/10.1051/matecconf/201815005066

a. Mothers whose daughters are 0-5 years old
b. Mothers that have capable in reading and writing
c. Mothers willing to be respondents

To determine the amount sample of the population, the researchers use the Slovin Formula (Notoatmodjo, 2005).

$$n = \frac{\mathrm{N}}{1 + \mathrm{N}(d^2)}$$

then:

$$n = \frac{121}{1 + 121(0{,}01)}$$
$$n = \frac{121}{1 + 1{,}21}$$
$$n = \frac{121}{2{,}21}$$
$$n = 54{,}75$$
$$n = 55 \text{ sampel} \qquad (1)$$

Description:

N = The amount of population
N = The amount of sample
d =Precision determined of 10 % with the confidence level of 90 % (0,1)

There are 55 samples calculating from the above formula.

Location and Period of the Research

The location of the research is in Ateuk Jawo Village, Baiturrahman Sub District, Banda Aceh. The research was conducted on July 2014.

Data Collection

a. Primary Data

The research is using the structured interview that the interview questions are detail compiled in a check- *(*Arikunto, 2006)

1. Data collection Part A is Demography data that is divided in 3 questions that is used as the opening interview questions. The demography data consists of: No, age of the respondent, and the last education of the respondent.
2. Data collection Part B is to assess the mother's motivation, culture factor, family support, and religion factor, by using the Guttman scale question that have 10 items for each question. For the positive statement, the answer of "yes" gets value 1 and "no" gets value 0. For the negative statement, the answer of "yes" gets value 0 and "no" gets value 1.

Data Processing

According to Notoatmodjo [9] data processing is done by some phases such as: Editing, Coding, Transferring, Tabulating.

Captions should be typed in 9-point Times. They should be centred above the tables and flush left beneath the figures.

3 Result Research

1. Unvaried Analysis

a. Mothers' Motivations

Table 4.1: Distribution of Mothers' Motivations Frequency in Carrying Out the Circumcision to the Female Children Ateuk Jawo Village, Baiturrahman Sub District, Banda Aceh

No	Live Motivation	f	(%)
1.	High	35	63,6
2.	Low	20	36,4
	Total	55	100

Table 4.1 shows that from 55 respondents interviewed, the mother motivation in carrying out the daughter circumcision is high. There are 35 respondent (63,6%) choose it.

b. Culture

Table 4.2: Distribution of Culture Frequency in Carrying Out the Circumcision to the Female Children Ateuk Jawo Village, Baiturrahman Sub District, Banda Aceh

No	Culture	f	(%)
1	Affected	27	49,1
2	Not Affected	28	50,9
	Total	55	100

Table 4.2 shows that from 55 respondents interviewed mentioned that the culture is not affected in carrying out the daughter circumcision. There are 28 respondent (50,9%) choose it.

c. Family Support

Table 4.3: Distribution of Family Support Frequency in Carrying Out the Circumcision to the Female Children in Ateuk Jawo Village, Baiturrahman Sub District, Banda Aceh

No	Family Support	f	(%)
1.	Support	32	58,2
2.	Not Support	23	41,8
	Total	55	100

Table 4.3 shows that from 55 respondents interviewed mentioned that the family gives support in carrying out the daughter circumcision. There are 32 responden (58,2%).

d. Religion Factor

Table 4.4: Distribution of Religion Factor Frequency in Carrying Out the Circumcision to the Female Children in Ateuk Jawo Village, Baiturrahman Sub District, Banda Aceh

No	Religion	f	(%)
1.	Affected	25	45,5
2.	Not Affected	30	54,5
	Total	55	100

Table 4.4 shows that from 55 respondents interviewed mentioned that the religion is not affected in carrying out the daughter circumcision. There are 30 respondent (54,5%) choose it.

2. Bivariate Analysis

a. The relation between culture and mother's motivation in carrying the daughter circumcision

Table 4.5: The Relation between culture and mother's motivation in carrying the daughter circumcision in Ateuk Jawo Village, Baiturrahman Sub District, Banda Aceh

No	Culture	Mother's Motivation				Total		P-value
		High		Low				
		f	%	f	%	f	%	
1	Berpengaruh	23	85,2	4	14,8	27	100	
2	Tidak Berpengaruh	12	42,9	16	57,1	28	100	0,003

Table 4.5 shows that from 28 respondents choosing that the culture is not affected show that the mother's motivation is low in carrying the daughter circumcision, there are 16 respondents (57,1%). Meanwhile from 27 respondents choosing that the culture is affected show that the mother's motivation is high in carrying the daughter circumcision, there are 23 respondents (85,2%).

After doing statistical test using chi-square test with a confidence level of 95% was obtained that p-value of 0.003 is smaller than α-value of 0.05. It can be concluded that there is a relationship between culture with mother's motivation in carrying out the female children circumcision in Ateuk Jawo Village, Baiturrahman Sub District, Banda Aceh.

b. The relation between family support and mother's motivation in carrying the daughter circumcision

Tabel 4.6: The Relation between family support and mother's motivation in carrying the daughter circumcision in Ateuk Jawo Village, Baiturrahman Sub District, Banda Aceh

No	Family Support	Mother's Motivation				Total		P-value
		High		Low				
		f	%	f	%	f	%	
1	Support	26	81,2	6	18,8	32	100	
2	Not Support	9	39,1	14	60,9	23	100	0,004

Table 4.6 shows that from 32 respondents choosing that the family support show that the mother's motivation is high in carrying the daughter circumcision, there are 26 respondents (81,2%). Meanwhile from 23 respondents choosing that the family is not support show that the mother's motivation is low in carrying the daughter circumcision, there are 14 respondents (60,9%).

After doing statistical test using chi-square test with a confidence level of 95% was obtained that p-value of 0.004 is smaller than α-value of 0.05. It can be concluded that there is a relationship between family support with mother's motivation in carrying out the female children circumcision in Ateuk Jawo Village, Baiturrahman Sub District, Banda Aceh.

c. The relation between religion factor and mother's motivation in carrying the daughter circumcision

Tabel 4.7: The Relation between religion factor and mother's motivation in carrying the daughter circumcision in Ateuk Jawo Village, Baiturrahman Sub District, Banda Aceh

No	Religion	Mother's Motivation				Total		P-value
		High		Low				
		f	%	f	%	f	%	
1	Affected	21	84	4	16	25	100	0,010
2	Not affected	14	46,7	16	53,3	30	100	

Table 4.7 shows that from 25 respondents choosing that the religion is affected show that the mother's motivation is high in carrying the daughter circumcision; there are 21 respondents (84%). Meanwhile from 30 respondents choosing that the religion is not affected show that the mother's motivation is low in carrying the daughter circumcision, there are 16 respondents (53,3%).

After doing statistical test using chi-square test with a confidence level of 95% was obtained that p-value of 0.010 is smaller than α-value of 0.05. It can be concluded that there is a relationship between religion factor with mother's motivation in carrying out the female children circumcision in Ateuk Jawo Village, Baiturrahman Sub District, Banda Aceh.

4 Discussion

1. **The relation between the culture and Mother's motivation in carrying the circumcision to the female children.**

Based on the research done result that from 28 respondents choosing that the culture is not affected show that the mother's motivation is low in carrying the daughter circumcision, there are 16 respondents (57,1%). Meanwhile from 27 respondents choosing that the culture is affected show that the mother's motivation is high in carrying the daughter circumcision, there are 23 respondents (85,2%).

After doing statistical test using chi-square test with a confidence level of 95% was obtained that p-value of 0.003 is smaller than α-value of 0.05. It can be concluded that there is a relationship between culture with mother's motivation in carrying out the female children circumcision in Ateuk Jawo Village, Baiturrahman Sub District, Banda Aceh.

According Syarifuddin (2009), culture comes from sangketa language (budhaya) is the plural of budhhi which means "mind" or "sense", all matters relating to the intellect. Culture is all of the complex parts, , in which consists of knowledge, belief, art, morals, habits acquired by the person as a member of community. Culture is a condition that describes the non-physical characteristics, such as moral, beliefs, attitudes, customs that are agreed by the community and is inherited from one generation to the next generation.

The circumcision has become the part of Indonesia Culture. In the rules of Islam, circumcision became one guidance that have to be executed by the followers, both men and women. On the recent development the male become more dominant to be the subject of the circumcision (Rahima, 2009).

Research has been done by Sartika (2011) about the factors affect the circumcision to the female children in Ladang Bambu Village, Medan Tuntungan Sub District. The results from the research mentioned that cultural factors affected in carrying out female children circumcision is 59 people (95.2%). Based chi square test resulted that the value of $p = 0.08$. it was smaller than the value of $\alpha = 0.05$, thus there is a relationship between culture with mother's motivation in carrying the circumcision to the female children.

According to the research assumption, there are relation between the culture and mother's motivation in carrying the circumcision to the female children, because the culture is a community tradition not religion, but when the religion meets the tradition, the religion gives the responds to accommodate the tradition or culture and then become the religion part. When the religion spreads to other area, the community practices the circumcision to the female children.

2. **The relation between the family support and Mother's motivation in carrying the circumcision to the female children.**

Table 4.6 shows that from 32 respondents choosing that the family support show that the mother's motivation is high in carrying the daughter circumcision, there are 26 respondents (81,2%). Meanwhile from 23 respondents choosing that the family is not support show that the mother's motivation is low in carrying the daughter circumcision, there are 14 respondents (60,9%).

After doing statistical test using chi-square test with a confidence level of 95% was obtained that p-value of 0.004 is smaller than α-value of 0.05. It can be concluded that there is a relationship between family support with mother's motivation in carrying out the female children circumcision in Ateuk Jawo Village, Baiturrahman Sub District, Banda Aceh.

Notoatmodjo et. al [9] mentioned that supporting factors from other party, for example, from husband/wife, parents, parents in-laws, friends, neighbors and midwives itself also affects the mother's attitude toward health service. Family can be the motivator for parents, because the most need other instruction very, but the family should also be able to explain about the benefits of circumcision so that motivation and knowledge can walk together. This motivation drives a person to behave and do the activity in order to reach the goals. Because in the fact the motivation is needed by the parents in order to encourage them in carrying out the circumcision to both male and female children [14].

Research conducted by Icha (2012), about "Factors Affecting Mother's Motivation in Carrying Out The Circumcision to the female children in Labuy Village, Baitussalam Sub District, Aceh Besar District indicated that the respondents who received family support proved to have high motivation in carrying out the circumcision of the female children as much as 16 respondents (76.2%) compared to the respondents who have no family support proved to have low motivation in carrying out the circumcision of the female children as much as 5 respondents (29.4%). Based on the chi square test, it was resulted that value of the $p = 0.007$. it was smaller than the value of $\alpha = 0.05$, thus there is a relationship between family support and mother's motivation in carrying out the female children circumcision.

According to the assumption of the researchers, there is the factor affected between the family support and mother's motivation in carrying out the female children circumcision, because by having the family support, the motivation is needed very much to facilitate someone in carrying out the circumcision both to the male and the female children, the circumcision can cause the improvement of the children fertility growth.

3. **The relation between the religion and Mother's motivation in carrying the circumcision to the female children.**

Based on the research conducted to 32 respondents choosing that the religion is affected show that the mother's motivation is high in carrying the daughter circumcision; there are 21 respondents (84%). Meanwhile from 16 respondents choosing that the religion is not affected show that the mother's motivation is low in carrying the daughter circumcision, there are 16 respondents (53,3%).

After doing statistical test using chi-square test with a confidence level of 95% was obtained that p-value of 0.010 is smaller than α-value of 0.05. It can be concluded that there is a relationship between religion factor with mother's motivation in carrying out the female children circumcision in Ateuk Jawo Village, Baiturrahman Sub District, Banda Aceh.

Islam teaches us to act proportionately. One is how we are used to controlling ourselves, including controlling our sexual libido. Female circumcision is expected to be a brake for women to control her sexual libido. Because according to the authentic history, women's desires are multiplied many times comparing to the males. In Islamic female circumcision is from the language khatana which means cut, the meaning in only cut/scratch the foreskin that cover the tip female genital to clean from dirty or called thahur which means cleaning (Umar, 2020).

According to the researchers assuming, there is relation between religion and mother's motivation in carrying out the female children circumcision, the religion is recommended to carry out the circumcision both to male or female, because the practice can control themselves including their sexual libido.

5 Conclusion

Based on the research done to 55 respondents summarize the statistical test as below:

1. There is the relation between the culture and mother's motivation in carrying out the female children circumcision in Ateuk Jawo Village, Baiturrahman Sub District, Kota Banda Aceh, with the p value is 0,003.

2. There is the relation between the family support and mother's motivation in carrying out the female children circumcision in Ateuk Jawo Village, Baiturrahman Sub District, Kota Banda Aceh, with the p value is 0,004.

3. There is the relation between the religion factor and mother's motivation in carrying out the female children circumcision in Ateuk Jawo Village, Baiturrahman Sub District, Kota Banda Aceh, with the p value is value 0,010.

6 Recommendation

1. to *Puskesmas*

It is estimated to improve concept and reference about the female children circumcision and improving the health services especially in handling female children circumcision

2. to respondent

https://doi.org/10.1051/matecconf/201815005066

It is estimated that the research result can be as input and information to mothers whose daughters are 0-5 years old carry out the female children circumcision.
3. to the educational institution
It is estimated to improve the concept and reference in giving generally guideline material in of obstetrics and gynecology.
4. to the next researchers
It is expected to improve the concept and reference about the female children circumcision so that the research can be continued with the different research variable.

References

1. Arikunto, S. *Research Procedure of A Practice Approach*. Jakarta: Rineka Cipta, (2006).
2. Azhari. A, Circumcision. www.arhamazhari.com, (2007) (Viewed on February, 15th 2015)
3. Budiarto, Eko, *Biostatic to Medical and Community Healthy,* Penerbit EGC, Jakarta, (2006)
4. Budiharsana,M, dkk., *"Research Reprot". Female Circumcision in Indonesia: Extend., Implication and Posible Intervention to Uphold Women's Helath Right,* Population, Jakarta, (2009)
5. Khomar, *MOH Policy of the Female Circumcision Medicalization.*www.pdpersi.co.id./banne rbursa.jpg.(Viewed on February, 1st 2014), ((2007)
6. Lutfi, Ahmad Fathullah. *Fiqh Khitan Perempuan*, Al-MughniPress.8, Jakarta, (2006)
7. MOH, *Health Minister Regulation (Permenkes) No 163.MENKES/PERXI/2010 about Female, Law and Organization Bureau, Jakarta,* (2010)
8. MOH, *Health Minister Regulation (Permenkes) No. 1636 of 2010 about Female Circumcision: Ensure Security and Protection of Female Reproductive System.* www. www.kesehatanibu.depkes.go.id/archives/167 (Viewed on January, 12th 2014), (2011)
9. Notoatmodjo, S. *Health Promotion and Behavior Knowledge.* Jakarta : Rineka Cipta, (2005)
10. Setiawati. *Learning Process in Health Education*, Tran Info Media, Jakarta, (2008)
11. Sunaryo. *Psychology*, Medical Manual Publisher. EGC, Jakarta, (2008)
12. WHO, *Psychology Module to the Maternity Women Psikologi Ibu Bersalin*, EGC Publisher, Jakarta, (2008)
13. WHO. *Female Genital Mutilation, An Overview,* Gneva, Jakarta, (2012)
14. Widayatun .T.R *Behavior Knowledge*, Sagung Seto, Jakarta.z (2006)

Traditional Belief and Practice on Postpartum Recovery among Mothers in East Coast of Peninsular Malaysia

Radiah Abdul Ghani[1,*] and *Saezah* Salehudin[1]

[1]Department of Biomedical Science, Kulliyah of Allied Health Sciences International Islamic University Malaysia,25200 Kuantan Pahang

Abstract. Postpartum care is highly important to prevent maternal and infant morbidity. In Malaysia, there are several traditional postpartum care practices by mothers. Different states adopted different ways to perform those practices. In spite of long and established practices in Malaysia, its effect and safety profile is not well understood. This study aims to gain an understanding of the traditional practices that Malay women follow in relation to postpartum care and the rationales underpinning such practices. A cross sectional study (n=100) was conducted using self-administered questionnaire consists of demographic information, practice of postpartum care and knowledge of postpartum care. Data was analyzed using Statistical Package Social Software (2.0). The findings shown that, there was no significant association of socio-demographic data with confinement period ($p>0.05$). For postpartum dietary practice, there were only association between age and encourage more water intake ($p=0.047$, $p<0.05$) and also between number of children and prohibit greasy food ($p=0.032$, $p<0.05$). For association between socio-demographic data and postpartum physical practice, there was only association between age and body scrub ($p=0.046$, $p<0.05$). The most significant factor that influenced postpartum care practice and its knowledge was family tradition with 83% and 97%, respectively. Traditional postpartum care practice is still significant and dominant among mothers in Kuantan, Pahang. This information is important for health care professional to educate women and provide strategies to help them to integrate their beliefs and the practices recommended in contemporary health care practice.

1 Introduction

Pregnancy and birth are seen as socially important transition periods that should be taken seriously. For women that delivering their first baby, that moment is probably the most significant experience and gives impact in their life [16]. The process to become a mother is defined as a process of learning, discovery, appreciation and acceptance of new role of a woman which leads in a worthwhile and valuable experience [7]. While the postpartum period begins immediately after the birth of the baby and extends for about six

* Corresponding author: radiah@iium.edu.my

weeks during which, the mother's body return to the non-pregnant state (WHO,1998). There are also physical changes, strong emotions, altered and new relationships and adaptation to a new role [3].

There are major changes occur during postpartum period in the body that determine mothers' well-being and expected for a healthy future [17]. Complications may come up and if they are not treated immediately and efficaciously can lead to ill-health and death to mothers, babies or both [1]. The risk of morbidity and mortality is highest in the first six weeks after birth but stays high after 6 weeks because of issues such as untreated anaemia or repeat pregnancy [7]. Thus, improvement of delivery of post-partum care has been neglected as a strategy for improving maternal, infant, and child health.

The practices of postpartum care aim to restore the normal function of sexual and reproductive organs, increasing wellbeing and energy of the mothers. The normal feedback mechanism of human body will involve such as wound healing promotion and weight loss. Due to these reasons, postpartum care is a priority programme worldwide which aims to restrain maternal deaths and for aesthetic purposes.

Women in Asian countries are doing well after childbirth compared to those in Western countries [2]. They find that many postpartum complications are nearly non-existent especially postpartum depression. Conversely, 14% of postnatal mothers suffer moderate to severe postpartum depression in developed countries. They believe this is due to the traditional postpartum care. However, it has still not been fully understood whether postpartum care practices give any significant effect on women's health or not. The health care providers are not able to come out with the benefits or negative effects on health due to inadequate study on the postpartum care practice. In spite of long and established practices in Malaysia, its effect and safety profile is not well understood.

Hence, this study was conducted to identify knowledge and practice of postpartum care among mothers in one area in East Coast of Peninsular Malaysia; Kuantan, Pahang. Moreover, this study aims to determine association of socio-demographic data with confinement period, postpartum dietary practice and postpartum physical practice. It is also to identify factors that influence postpartum care practice among mothers in Kuantan and lastly to investigate factors that contribute to knowledge of postpartum care among mothers in this selected population.

2 Material and Methods

Study design and sampling: A cross-sectional study was conducted among 100 respondents among mothers in selected area in Peninsular Malaysia which is Kuantan, Pahang. Subjects were selected by using the method of purposive sampling based on the inclusion criteria which were women who have given live birth from 2013 until 2016, given live birth in Kuantan and the resident of Kuantan, Pahang and exclusion criteria which were women who do not understand Malay or English language and cannot communicate effectively, women who have chronic disease or on medication and also miscarriage and still-birth cases.

The determined sample size for this study was calculated using single proportion sample size calculation formula. The selected subjects were asked to complete a prepared questionnaire which included all appropriate issue of knowledge and practice of postpartum care. The questionnaire was divided into three categories which were personal details, practice of postpartum care and knowledge of postpartum care. The multiple choice and Likert scale format were selected for simplicity in scoring and data management.

Instrument development: The prepared questionnaire included all appropriate issue of knowledge and practice of postpartum care. The questionnaire was divided into three categories which were personal details, practice of postpartum care and knowledge of postpartum care. The questions of personal details were age, ethnicity, and religion, level of education, occupation and number of children. Apart from that, for practice of postpartum care, it included confinement period, types of practices, complications, dietary intakes and supplement consumed. The questions in the category of knowledge of postpartum care were to evaluate the basic information about postpartum care such as the origin of knowledge, reason behind the practice and purposes of the practice. The multiple choice and Likert scale format were selected for simplicity in scoring and data management.

Validity and reliability of questionnaire: To validate the accuracy and validity of questionnaire, the content id reviewed by panel of expert. A pilot study was conducted within selected group of 30 among International Islamic University's staff. Results from pilot study showed the Cronbach alpha 0.67 and 0.73 for section B and C, respectively.

Ethical consideration: The approval to conduct this study was obtained from Dean, Kulliyyah of Allied Health Sciences and International Islamic University Malaysia Ethics Committee. The guideline stated was strictly followed before conducting this study. Prior to taking part in this study, respondents were informed about the objective of the study. All respondents were asked to sign on consent form confirming their willingness to participate. All respondents were assured that confidentiality would be maintained.

Statistical analysis:All the collected data was quantitatively analyzed using Statistical Package for Social Sciences (SPSS) version 12. The test used was Chi-square test (to determine association of socio-demographic data with confinement period, postpartum dietary practice and postpartum physical practice). For Chi- square test, the data must meet the assumption by cell with and expected count less than five must be $< 20\%$. The minimum expected count must be ≥ 1. If the assumption was not met, the Fisher's exact test can be used. The data showed statistical significance relationship with p-value < 0.05.
Figures and tables, as originals of good quality and well contrasted, are to be in their final form, ready for reproduction, pasted in the appropriate place in the text. Try to ensure that the size of the text in your figures is approximately the same size as the main text (10 point). Try to ensure that lines are no thinner than 0.25 point.

3 Results

There were 100 eligible mothers participated in this study. The age of respondents ranged from 20 to 49 years old. Most of them were in the age group of 20-29 followed by 30-39. Most of the respondents were Malays, Muslims, employed and had one child. The highest level of education was up to tertiary school. There was no significant association of demographic data which were age, level of education, occupation and number of children with confinement period ($p>0.05$) (Table 1).

Table 1
Chi-square test for demographic data and confinement period

Variable	≤ 44 days (n=54) n (%)	≥ 45 days (n=46) n (%)	Chi-square (d.f)	*p*- value
Age				
≤29	31 (57.4)	19 (41.3)	2.576	0.108
≥30	23 (42.6)	27 (58.7)	(1)	
Level of education				
No formal education	0 (0.0)	1 (2.2)	-	0.460*
With education	54 (100.0)	45 (97.8)		
Job status				
Unemployed	13 (24.1)	12 (26.1)	0.054	0.817
Employed	41 (75.9)	34 (73.9)	(1)	
Number of children				
≤4	52 (96.3)	44 (95.7)	-	1.000*
≥5	2 (3.7)	2 (4.3)		

Chi-square test for independent (with Fisher's exact test*) indicated that there was statistical significant association between socio-demographic data and confinement period ($p < 0.05$) among mothers in Kuantan.

Table 2
Chi-square test for age and postpartum dietary practice

Variable	Prohibit cold food		Chi-square (d.f)	*p*- value
Age	**Yes (n=99) n (%)**	**No (n=1) n (%)**		
≤29	49 (49.5)	1 (100.0)	-	1.000*
≥30	50 (50.5)	0 (0.0)		
Variable	**Prohibit acidic food**		**Chi-square (d.f)**	***p*- value**
Age	**Yes (n=77) n (%)**	**No (n=23) n (%)**		
≤29	35 (45.5)	15 (65.2)	2.767	0.096
≥30	42 (54.5)	8 (34.8)	(1)	
Variable	**Prohibit greasy food**		**Chi-square (d.f)**	***p*- value**
Age	**Yes (n=78) n (%)**	**No (n=22) n (%)**		
≤29	35 (44.9)	15 (68.2)	3.730	0.053
≥30	43 (55.1)	7 (31.8)	(1)	
Variable	**Limit water intake**		**Chi-square (d.f)**	***p*- value**
Age	**Yes (n=29) n (%)**	**No (n=71) n (%)**		
≤29	13 (44.8)	37 (52.1)	0.437	0.509
≥30	10 (55.2)	34 (47.9)	(1)	
Variable	**Encourage warm food**		**Chi-square (d.f)**	***p*- value**
Age	**Yes (n=57) n (%)**	**No (n=43) n (%)**		
≤29	27 (47.4)	23 (53.5)	0.367	0.545
≥30	30 (52.6)	20 (46.5)	(1)	
Variable	**Encourage more water intake**		**Chi-square (d.f)**	***p*- value**
Age	**Yes (n=29) n (%)**	**No (n=71) n (%)**		
≤29	10 (34.5)	40 (56.3)	3.934	0.047
≥30	19 (65.5)	31 (43.7)	(1)	

Chi-square test for independent (with Fisher's exact test *) indicated that there was statistical significant association between age and postpartum dietary practice ($p < 0.05$) among mothers in Kuantan.

https://doi.org/10.1051/matecconf/201815005067

Table 3
Chi-square test for number of children and postpartum dietary practice

Variable	Prohibit cold food		Chi-square (d.f)	p- value
Number of children	Yes (n=99) n (%)	No (n=1) n (%)		
≤ 4 ≥ 5	95 (96.0) 4 (4.0)	1 (100.0) 0 (0.0)	-	1.000*
Variable	**Prohibit acidic food**		**Chi-square (d.f)**	**p- value**
Number of children	**Yes (n=77) n (%)**	**No (n=23) n (%)**		
≤ 4 ≥ 5	73 (94.8) 4 (5.2)	23 (100.0) 0 (0.0)	-	0.571*
Variable	**Prohibit greasy food**		**Chi-square (d.f)**	**p- value**
Number of children	**Yes (n=78) n (%)**	**No (n=22) n (%)**		
≤ 4 ≥ 5	77 (98.7) 1 (1.3)	19 (86.4) 3 (13.6)	-	0.032*
Variable	**Limit water intake**		**Chi-square (d.f)**	**p- value**
Number of children	**Yes (n=29) n (%)**	**No (n=71) n (%)**		
≤ 4 ≥ 5	27 (93.1) 2 (6.9)	69 (97.2) 2 (2.8)	-	0.577*
Variable	**Encourage warm food**		**Chi-square (d.f)**	**p- value**
Number of children	**Yes (n=54) n (%)**	**No (n=42) n (%)**		
≤ 4 ≥ 5	54 (94.7) 3 (5.3)	42 (97.7) 1 (2.3)	-	0.632*
Variable	**Encourage more water intake**		**Chi-square (d.f)**	**p- value**
Number of children	**Yes (n=29) n (%)**	**No (n=71) n (%)**		
≤ 4 ≥ 5	27 (93.1) 2 (6.9)	69 (97.2) 2 (2.8)	-	0.577*

Chi-square test for independent (with Fisher's exact test*) indicated that there was statistical significant association between number of children and postpartum dietary practice ($p < 0.05$) among mothers in Kuantan.

Table 4
Association of Socio- Demographic Data and Postpartum Physical Practice

Variable	**Body scrub**		**Chi-square (d.f)**	**p- value**
Age	**Yes (n=20) n (%)**	**No (n=80) n (%)**		
≤ 29 **≥ 30**	**6 (30.0)** **14 (70.0)**	**44 (55.0)** **36 (45.0)**	**4.000 (1)**	**0.046**

It has been found that not all demographic data (age, level of education, job status and number of children) affected the postpartum dietary practices which consisted of prohibit cold food, prohibit acidic food, prohibit greasy food, limit water intake, encourage warm food and encourage more water intake. There were only association between age and encourage more water intake (ρ=0.047, p<0.05) (Table 2) and also between number of children and prohibit greasy food (ρ=0.032, p<0.05) for postpartum dietary practice (Table 3). Other than these two mentioned practices were not affected by any respondents' socio-demographic data.

There were ten postpartum physical practices which were hot compress, corset, massage, herbs bath, param, pilis, sauna, body scrub, omam and oil bath associated with socio-demographic data (age, level of education, job status and number of children). However, there was only association between age and body scrub (ρ=0.046, p<0.05) (Table 4).

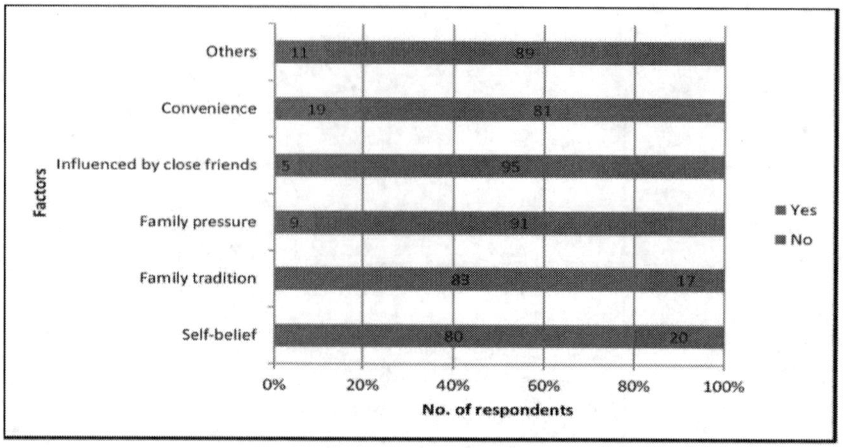

Figure 1
Factors that influence postpartum care practice

The factors that influenced respondents to do postpartum care practice were self-belief, family tradition, family pressure, influenced by close friends, and convenience. It has been found that the family tradition was the most influenced factor among respondents with 83 % of the mothers. Then, followed by self-belief (80 %), convenience (19 %), other factors (11%), and family pressure (9%). The effect of close friends was the least factor that influenced respondents to experience postpartum care practice (5%) (Fig. 1).

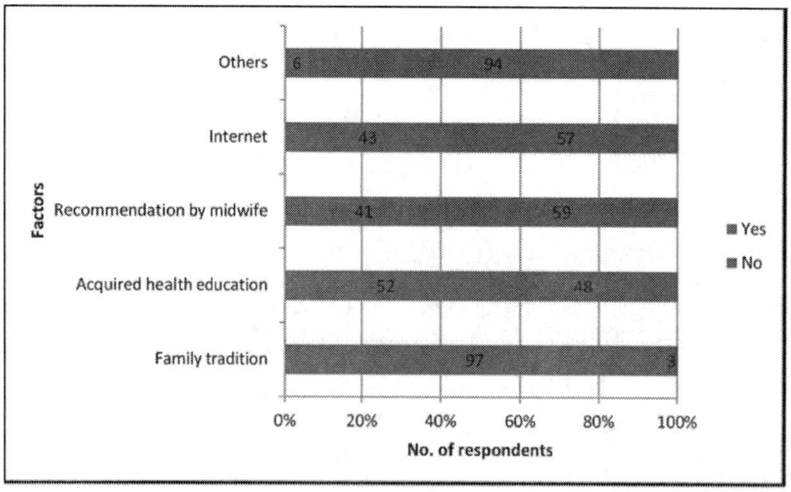

Figure 2
Factors that contribute to knowledge of postpartum care

There were five elements that contribute to knowledge of postpartum care which were family tradition, acquired health education, recommended by midwife, internet and others. Family tradition has contributed the most to knowledge of postpartum care among mothers with 97 %. The least factor that contributed to postpartum care knowledge was other factors (6%). Acquired health education, internet and recommended by midwife have contributed to knowledge of postpartum care with 52 %, 43 % and 41 %, respectively (Fig. 2).

4 Discussion

There was no significant association between socio-demographic data with confinement period, which means socio-demographic data did not affect confinement period of the mothers after delivered a baby. This finding was similar to the study done in Penang which was the association between socio-demographic data and confinement period is not statistically significant [9]. It means that the perceived minimum confinement period is not affected by the different settings of the study but it is more likely to be based on the belief of the mothers. Malay women especially Muslim practice a 40-day period of rest based on Islamic beliefs (Stern, 1983). In contrast, Chinese and Indian women practice their confinement for 30 days [4]. Thus, it shows that the confinement period strongly depends on the belief of the mothers and does not influenced by other factors.

For association of socio-demographic data and postpartum dietary practice, there were only two statistically significant associations between socio-demographic data and postpartum dietary practice. There was association between age and encourage more water intake and association between numbers of children and prohibition of greasy food. The encouragement of water intake is good because it can decrease the risk of dehydration and constipation among mothers. Water is the most crucial nutrient for human and the absence of it solely can be lethal within days. Apart from that, the short and long-term effects of water on human health are very clear [10]. However, it was reported that association between socio-demographic data and postpartum dietary practice shows that the mothers are more likely to restrict their water intake even though the limitation of water and certain foods like vegetable and fruit intake may lead to a risk for constipation, dehydration, and malnutrition [10]. It shows that people are not really aware the importance of water in maintaining their health. Nevertheless, the result of a study in Penang on association of socio-demographic and prohibit greasy food by is consistent with this study [9].

The association between socio-demographic data and postpartum physical practice has been explored in this study as well. There was only a significant association between age and body scrub. Majority of the respondents in this study were 20 to 29 years old. Thus, they are likely to concern more on their skincare as young women are very particular about their skin. This is because they want to ensure there are no skin problems especially after childbirth. Skin changes often occur during pregnancy until after delivery due to physiologic changes which can cause several skin problems such as hyperpigmentation and stretch mark [8]. By practicing body scrub, chances to have those problems can be reduced. The ingredients in the scrub and mask products can give healthy, smooth and radiant skin by exerting their effect through absorbingdirt, removing dead cells, brightening the skin and improve blood circulation [6].

However, the same study done in Penang, Malaysia found an opposite result on association between socio-demographic data and postpartum physical practice [9]. They mentioned that there is a statistically significant result between association of socio-demographic data and consumption of traditional herbs as older women are more likely to use or consume traditional herbs. The difference of finding regarding association of socio-demographic data and postpartum physical practice might be due to the exclusion of women more than 49 years old in this study which leads to no elderly group of the respondents.

Apart from that, family tradition was the most influenced factor of postpartum care practice in this community. The tradition that passes from generation to generation without fail ensures the continuity of postpartum care practice in each family. This finding was supported by the study done in Mangalore, India which claimed that the cultural background of the new mother influences their practice during postpartum period [14]. The studies done in China and Cambodia also have similar result which the new mothers say that they are doing such practices in order to respect the tradition and following the advice of the elders [3,12]. However, the opposite result regarding this issue which the practice stem from self-belief of the new mothers also was reported. The factors that influencing the postpartum practice among mothers depends on whether the mothers are the first child mothers or multiparous mothers [5]. The first child mothers tend to practice diligently due to lack of motherhood experience and knowledge. It seems that factors that influenced practice of postpartum care among mothers are very wide and rely on the standpoint of the mothers.

Furthermore, the finding of this study showed that the family tradition has contributed the most to knowledge of postpartum care among mothers. The continuation of postpartum care practice that originated from old generation enhance the knowledge of the mothers. Majority of the families follow the beliefs and practices of their parents and grandparents and adopt them because their ancestors are more knowledgeable about those practices [12]. Family that have been engaged with certain postpartum care practices for decades will ensure the practices are passed to their next generations [5]. Thus, their descendants will also experience the benefits of the practices.

Note that this study is very important in order to support government's mission and vision as tabulated in United Nations Millennium Development Goals (MDG). One of the MDG's goal is to improve maternal health by 2015 [15]. This goal should not be overstated because one of the most leading causes of death for adolescent girls is complications during pregnancy or childbirth. However, most maternal deaths in developing countries are preventable by receiving a proper health care and adequate nutrition. In addition, good postpartum care is not only can save mother's lives but their babies too Consequently, improving maternal health can be the key to achieve MDG 4 which is to reduce child mortality. It can be seen that this study can help to achieve not just one but two of the MDG's goals which are improving maternal health and reducing child mortality.

Hence, by conducting this study, the beneficial knowledge about postpartum care can be disseminated to the mothers to expand their existing knowledge. Consequently, they would be able to distinguish between good and bad in order to make sure their health are assured. Morever, the good postpartum care practices can be encouraged among mothers particularly those that can reduce the complications during confinement. The reduction of postpartum complications can indirectly reduce the number of deaths among the mothers. Hence, this study can contribute to the achievement of the vision and mission of the government.

5 Conclusion

In conclusion, knowledge of postpartum care among mothers in Kuantan was good and the practice was prevalent among them. This study indicated that socio-demographic data have influenced certain postpartum dietary practice and postpartum physical practice. Moreover, this study found that the most important factor that influenced both postpartum care practice and knowledge was family tradition. For recommendation, mothers should be exposed to the possibility of complications due to extreme practice of postpartum care. Therefore, the harmful practice can be avoided. Apart from that, this study can be used as a foundation to conduct a study among multiracial population. The study would yield interesting results based on racial points of view.

References

1. Ching-Yu, C., Eileen, R.F.,& Lorraine, O.W. Postpartum Maternal Health Care in the United States: A Critical Review. The Journal of Perinatal Education. 15 (2006)
2. Fort, M.A.L., Kothari, T & Abderrahim, N. Postpartum Care: Levels and Determinants in Developing Countries. USA.15. (2006)
3. Koung Lo., Postpartum Practices among Cambodian Mother in Preah Vihear Province: A Qualitative Study of Beliefs and Practices.Thesis (M.Arts).Mahidol University, (2007)
4. Laderman, C., Destructive health and cooling prayer: Malay humoralism in pregnancy, childbirth and the postpartum period. Social Science and Medicine. 25 (4). 357–365, (1987)
5. Le, M.T., Wanawipha, P., & Oratai, R., Traditional postpartum practices among Vietnamese mothers in Anthi district, HungYen province. Thesis. Mahidol University, (2003)
6. Magestuti., Subehan., Aty Widyawaruyanti., Syed Faisal Haider Zaidi., Suresh Awale., & Shigetoshi Kadota., Traditional Medicine of Madura Island in Indonesia. Journal of Traditional Medicine. 24 (3). 90-103, (2007)
7. Martell, L. K., Heading toward the new normal: A contemporary postpartum experience. Journal of Obstetric, Gynecologic, and Neonatal Nursing, 30 (2001). 496–506, (2001)
8. Martin, A.G., & Leal-Khouri, S., Physiologic Skin Changes Associated with Pregnancy. International Journal of Dermatology. 31(6). 375-378, (2007)
9. Munirah Hishamshah., Muhammad Sirri 'Ammar bin Ramzan., Abdul Rashid., Wan Nur Hidayah bt Wan Mustaffa., Raihanah Haroon., & Nur Baizura bt Badaruddin., Belief and Practices of Traditional Post Partum Care Among a Rural Community in Penang Malaysia. The Internet Journal of Third World Medicine, 9 (2), (2011)
10. Poh, B.K., Wong, Y.P., & Norimah A. Karim., Postpartum Dietary Intakes and Food Taboos Among Chinese Women Attending Maternal and Child Health Clinics and Maternity Hospital, Kuala Lumpur. Mal J Nutr, 11(1), 1-21, (2005)
11. Popkin, B.M., D'Anci, K.E., & Rosenberg, I.H., Water, Hydration and Health. Nutrition Reviews. 68(8). 439-458, (2010)
12. Raven, J. H., Chen, Q., Tolhurst, R. J., & Garner, P., Traditional beliefs and practices in the postpartum period in Fujian Province, China: a qualitative study. BMC Pregnancy and Childbirth, 7(8), (2007)
13. Stern, G., & Kruckman, L., Multi-disciplinary perspectives on postpartum depression: An Anthropological Critique.Social Science and Medicine. 17 1027–1041, (1983)
14. Sunanda, B., & Shynee, P., A Study on the Cultural Practices of Postnatal Mothers in Selected Hospitals at Mangalore. Nitte University Journal of Health Science. 3(3), (2013)
15. United Nations. The Millennium Development Goals Report 2013. Retrieved April,29,2014fromhttp://www.undp.org/content/undp/en/home/mdgoverview/mdg_goals/mdg5/,(2013)

16. World Health Organization, Postpartum Care of the Mother and Newborn: A practical Guide. Geneva, . (1998)
17. World Health Organization., WHO Technical Consultation on Postpartum and Postnatal Care. Geneva, (2010)

Perception of Teacher on Cooperative Learning

Nur Salimah Alias*, *Huzili* Hussin, *Junainor* Hassan, *Nor Syamimi* Mohamed Adnan, *Mohamad Hashim* Othman, *Kamarudin* Hussin

Abstract. The participation of students in the process of learning is one of the keys to ensure effective learning outcomes either in terms of knowledge or skill. However, teacher-centred teaching methods which have been practiced in this country after so long avert students to fully engage in the learning process especially in a classroom. Various attempts have been made by the government to emphasize student's development in term of communication, creativity and innovation through the of process learning in class. It requires school to implement student-centred teaching methods such as cooperative learning. Cooperative learning is a teaching method that requires teachers not only interacts with the students but also with other teaching aids such as audio, video and computer. Therefore, this study was conducted to identify teachers' perception on the implementation of cooperative learning in the teaching and learning process in class. 25 teachers were selected as respondents. Mixed methods were used in this study, where the instrument of questionnaires and interviews were used to collect the data. Quantitative data analysis was made using descriptive statistics of frequency, mean and percentage. While, qualitative data analysis was based on the recommendations of the Miles & Huberman (1994). The findings showed teachers have a positive attitude toward cooperative learning and they are ready and feel confident to practice it. However, the constraints are classroom control, time and limited teaching aids

1 Introduction

Education plays an important role in developing countries to achieve the status of a developed country. This is because education can act a catalyst to produce creative, innovative and kwowledgable students. It can be a determinant of civilization for every country. Therefore, our goverment intervenes to improve the quality of education I n our country to ensure that we can accomplish economic development.

Nowadays, education in Malaysia can furnish students with knowledge and it can also help them develop interpersonal skills, soft skill, and generic skills [4]. We should incorporate appropriate teaching and learning methods in school curriculum in order to improve students' academic performance and enhance their soft skills such as communication skills, leadership and generic skills, simultaneously. One of the methods is cooperative learning.

According to [9] the purpose of coorperative learning is to avail students of the opportunity to think on their own. Therefore, teachers should play an imopertant role in ensuring this method can be successfully and effectively implemented in school. Also, according to the study, teachers can encourage curiosity of students by asking them questions. Thus, the students can make connections between the questions and ideas that have been generated in their mind.

In addition, Che Mansor bin Che Tom [1] stated that teachers should encourage students to think effectively in class in order to produce future generation who are able to think creatively and innovatively. Hence, educators do not only deliver their lesson, but emphasize cooperative learning in their class. Students also need to participate in the leaning process so that they can expand their mind.

Nevertheless, this method cannot be successfuly and effectively implemented if teachers have scant knowledge about how to use it. The good quality of the implementation should come from teachers who have positive perception and strong awareness about the benefits of cooperative learning. This is because teachers are the individuals who will perceive the learning outcomes as well as the one who will design the learning process including activities and methods used in class. The teaching plan in using cooperative learning methods is important. A good plan can help students realize about the importance of knowing the group and their respective functions [7] thus ensuring the effectiveness of the learning process.

2 Purpose of Study

The purpose of this study is to identify teachers' perception on cooperative learning.

3 Research Methodology

* Corresponding author: salimah@unimap.edu.my

This is a pilot study in which researchers used mix method where quantitative and qualitative methods were used. Respondents involve 25 teachers of one school in Johor Bahru. It comprises the entire population of teachers in the school. While for qualitative methods, only 6 teachers from the sample were randomly selected for the interviews.

3.1. Quantitative Method Instruments

A set of questionnaires are used as an instrument to collect data in this study. There are 25 items that emphasize issues related to teachers' perceptions of cooperative learning which is divided into five elements, namely 1) cooperative learning attractions, 2) role of teachers, 3) teacher requirements, 4) social benefits and 5) cognitive benefits. Each element consists of 5 items. Researchers have established that respondents expressed their views on the statement of the items in the form of approval. Likert scale is used as a level of approval. The level approval indicates whether respondents have a positive or negative opinion on the statement.

3.2 Qualitative Methods Instruments

In order to collect data using qualitative methods, a semi-structured interview were held. Semi-structured interviews were chosen to answer the research questions about the perception of teachers in cooperative learning as well as to explore the issues of cooperative learning without limitation of opinions and not rely on the limited answer in questionnaire. A set of questions are also provide as a guidance for the researcher to ask related question.

4 Quantitative Findings

The aim of the study is to determine the teacher perception towards cooperative learning by examine their consent to the advantages and disadvantages of cooperative learning through the distributed questionnaires. Five elements used to assess teachers' perceptions of cooperative learning.

4.1 Coorperative Learning attractions

Overall, the items for this element is at the high level of agreement based upon the scale of min. Respondents agree with the items that "cooperative learning increases the level of learning for weak students (mean: 4.18)", "is a very useful teaching approach (mean: 4.12)" and a "efficient learning strategies in the classroom (min: 4:08)". However, a teacher also believes that the implementation of cooperative learning requires a high cost with a mean value of 2.92.

4.1 Role of Teachers

Cooperative learning is seen by teachers as a medium for students to gain knowledge in which the teachers agreed with the statement that "with cooperative learning, students learning will not only depend by teachers alone" with a mean of 3.92. The teachers do not agree that "classes with cooperative learning activities are hard to control by the teacher" and they do not consider cooperative learning leads unfinish syllabus, with mean 3.32 and "cooperative learning give too much responsibility to the students, so the teacher does not play a role " with mean 3.16. However, most respondents agreed that "Cooperative learning takes too much time for teachers to prepare" with the smallest mean value of 2.64

4.2 Teachers Requirements

Items that have the highest mean is the item with the statement "I am confident to perform cooperative learning (CL) although this method requires more control than a normal teaching class methods" with a mean value of 3.92. However, for this elemen only one item has mean value at high levels while other items are at a moderate level, such as items with the statement "I feel pressured by other teachers to use CL", with mean value of 3:48 and the statement "I have a little teaching experience to perform CK " with a mean value of 3.24. Items with the statement "I prefer the normal teaching methods" had the smallest mean value with a mean value of 2.84.

4.3 Social Benefits

Social benefits refer to how teachers see the benefits obtained by students through cooperative learning. Statement "Cooperative learning promotes confidence in face to face interaction" and "Cooperative learnings trained students to accept the opinions of others" share the same mean value and the highest with value 4:28. The other items also recorded as high mean value, "Cooperative learning can foster social interaction among students" and "Engage with the class" with mean value 4.24 and 4.08 respectively. However, there is one item with moderate mean value with the statement "Cooperative learning gives too much emphasis to the development of social skills of students" with a mean value of 2.72.

4.4 Cognitive Benefits

As with the previous element, this element also shows high mean values on most items. "Cooperative Learning allows students' creativity flourish and be able to give their own ideas during the discussion" with a mean value of 4.24. The other item also at the highest level is the item with the statement "cooperative learning helps improve the understanding of students" with a mean value of 3.92, followed by items with the statement "The concept of a subject more easily grasp by the students when using cooperative learning" and " cooperative learning increased student achievement " with each having a mean value of 3.88 and 3.84. However, there is one item with moderate level mean value,3:48, item with the statement "Engaging in cooperative learning disrupt academic progress"

5 Qualitative Findings

Results show interviewed respondents have some different answers on the implementation of cooperative learning. Overall respondents saw cooperative learning as a very effective practice in class but there are also a number of shortcomings in its implementation. Based on the findings in the interview, there are four sub-themes obtained to conclude the perceptions of respondents. The four sub-themes are 1) students involvement in cooperative learning, 2) classroom control, 3) time and 4) teaching aids.

5.1 Student Involvement

Student participation refers to the views of respondents about the involvement of students in the classroom when they implement cooperative learning. The response given is as below.

"..it bring more good, it builds students self esteem, provide opportunities for students to speak and get involved." (Mr C)

"Cooperative learning is good to encourage students to become more active in the classroom"
(Madam E)

"Students become more daring, more satisfied and their ideas flow faster"
(Madam D)

Based on the statements above, respondents have very positive views on how cooperative learning can increase student participation in the learning process. Furthermore, the impact of the interaction happened during the learning process boost the formation of students social skills.

5.2 Class Control

Although most respondents have a positive perception with cooperative learning there are also some constraints or problems that respondents perceived in cooperative learning methods, one of it is class control. Below are some of the statements of respondents about classroom control.
"To train students are difficult to control, for example if there are fast and slow learner in a group, fast learner can follow quickly than slow learner, so it become problematic to make sure all students contribute to the group" (Madam D)
"it require more control over the class, need to always round the class and need to always be creative" (Madam F)

5.3 Time

Apart from the control class, time is also one of the constraints in the implementation of cooperative learning. Here are some of the respondents' statements about it,
"It is good to encourage students to be active, but we have time constraints, for example when there is a syllabus needs to be spent" (Madam E)
"Apart from that, we also lack of time to implement cooperative learning" (Mr B)
Based on the statements of teacher E, time constraints happens because of the need for teachers to finish the subject syllabus and teacher B said about teacher lack of times to implement cooperative learning is in terms of preparation and the time required for the activities in cooperative learning.

5.4 Teaching Aids

The last finding on respondents' views on cooperative learning is about teaching aids. Teaching aids refers to material and equipment that is use by respondents during the process of teaching and learning. Here are the statements of respondents about teaching aids,

"Sometimes we are lack of materials, especially if you want to use audio, because the school is only have one audio device, so if there is clash, one of the teachers need to change their teaching plan" (Mr A)

"We do have shortage of material, but the teachers here are willing to share, students also doesn't have problem to bring the materials from home as ordered" (Mr B)

Based on the given statements, the lack of teaching aids are solved with the exchange of teaching planning, purchase the material using their own cost, share the materials and ask students to bring required material from home. This element shows an implementation of cooperative learning looks like a burden to the teachers.

6 Discussion

Based on the five elements in quantitative data, this study show teachers have positive perceptions on cooperative learning. Positive perceptions of cooperative learning can be seen when respondents give more positive responses in the interviews and the analysis of the questionnaires in the survey.

Teachers agree that cooperative learning is one method that can turn on an attractive atmosphere in the classroom as well as to build a more positive attitude in the learning process. Cooperative learning can also form a student-centered teaching and learning compared to existing methods that are more teacher-centered.

However, the study also found an efficient implementation of cooperative learning requires a strong commitment from teachers. Teachers need to spend more time preparing for a lesson designing their teaching plan especially when they need to use teaching aids such as audio and video equipment. Mimi Haryani et.al., [4] also believes that the implementation of cooperative learning methods require much more time compared to traditional teaching methods, especially for teachers who are inexperienced.

Although respondents expressed the constraints in the implementation of cooperative learning, this study also proves that these constraints do not prevent teachers to implement it. Items in the teacher needs element, most of the respondents are willing to implement cooperative learning even require more control and commitment. Therefore, even though cooperative learning is not easy to implement but the teacher is still willing to do it for the positive impact on the students gained from this method.

At the same time, as a whole, more respondents agreed with the positive effects from the social and cognitive benefits derived from cooperative learning. These findings are consistent with studies Mohaffyza Mimi et. al., [6]] which concluded that cooperative learning makes students more confident to provide ideas and be actively involved in learning. In addition, cooperative learning also improves communication and increase student interaction as defined by Effendi et. al., (2006), Ibrahim et al, (2011), Zahara et. al., [8] and Mimi Haryani et. al., [4]. Each item in the element of cognitive benefit got a positive response from respondents. This proves that the respondents agreed with the cooperative learning positive impact on the cognitive development of students which can be judged from their understanding, presentation ideas, problem solving skills and academic achievement.

7 Summary

Overally, teachers have a positive view towards ooperative learning. Eventhough it requires a higher commitment than traditional teaching methods but all the teachers are confidence to practice cooperative learning in order to improve student achievement and make students to become a better perosn. Cooperative learning is positively viewed as effective learning practice. However, some improvement in terms of equipment availability and school management should be reviewed to make cooperative learning to be more effective.

References

1. Che Mansor bin Che Tom., Jurnal Bestari : Koleksi Guru Johor. Jabatan Pendidikan Negeri Johor. (hlmn 48), (2000)
2. Meor Ibrahim Bin Kamaruddin & Nurul Amira Binti Ahmad , Persepsi Guru Sains Terhadap Kaedah Pembelajaran Koperatif (PK) Berdasarkan Kemahiran Sosial Pelajar Dan Minat Guru, Fakulti Pendidikan, Universiti Teknologi Malaysia, (2010)
3. Miles, M. B. dan Huberman, A. M. Qualitative data analysis. 2nd edition. Thousand Oaks, CA: Sage, (1994)
4. Mimi Haryani Hassim , Mohd. Kamaruddin Abd. Hamid , Mohd. Ariffin Abu Hassan ,
5. Khairiyah Mohd. Yusof , Syed Ahmad Helmi Syed Hassan, Mazlina Esa, Enhancing Learning Through Cooperative Learning: UTM Experience. Conference on Engineering Education, Kuala Lumpur, Dec 14-15,. (2004)
6. Mimi Mohaffyza binti Mohamad, Md.Nazaruddin b. Sarji & Masitah binti Misman, Strategi Pengajaran Pembelajaran Koperatif Dalam Tajuk Rekabentuk Dan PenghasilanProjek (Matapelajaran Kemahiran Hidup). Universiti Tun Hussein Onn Malaysia, (2008)
7. Nor Azizah Salleh, Siti Rahayah Ariffin & Musa Daia, Penerapan Nilai Murni Melalui Pembelajaran-Kooperatif Dalam Sains. Jurnal Pendidikan 27 (2001) 47-57, (2001)
8. Zahara Aziz dan Suzela Bustam, Kesan Strategi Pembelajaran Koperatif Terhadap Pencapaian Geografi Tingkatan Satu – Topik Tumbuhan Semula Jadi dan Hidupan Liar (The Impact of Cooperative Learning Strategy on the Geography Achievement of Form One: Topics Forest and the Wild Life). Jurnal Pendidikan Malaysia. 36(1):1-10, (2011)
9. Zainudin B Abu Bakar & Nor Hidayatul Aini Ismail, Persepsi Dan Sikap Pelajar Terhadap Pendekatan Konstruktivisme Serta Kesan Terhadap Pencapaian Dalam Mata Pelajaran Matematik . Journal Of Science And Mathematics Educational, Vol 3, Sept 2011, Pages 92-107. Universiti Teknologi Malaysia, (2011)

Screening Drug, Alcohol and Substance Abuse the Psychometric Measures

Mohamad Hashim Othman, *Huzili* Hussin, *Kamarudin* Hussin *and N.S.* Alias

Abstract. Urinalysis was used in previous studies among higher institution students (n=16252) in Malaysia to answer the question of whether university students are involved in drug abuse. However, the use of urinalysis had faced some problems. The problems were related to human rights issues and the cost to perform the urinalysis was expensive and quite impossible to be implemented to a large population of university students. To overcome this problem, this study was conducted to examine the effectiveness of psychometric measures in screening drug, alcohol and substance abuse. The Substance Abuse Subtle Screening Inventory A2 (SASSI-A2) was used for this purpose. SASSI-A2 is a brief screening tool designed to identify individuals who have a high probability of having a substance use disorder, including both substance abuse and substance dependence. SASSI-A2 comprises of 72 items that are rated on a two point scale with response; true and false. SASSI-A2 was translated into Malay language and it was refined through a back-translation technique and focus group approach. Psychometric testing was undertaken on a sample of 750 university students from five public universities in Malaysia. All participants were aged between 19 and 20 years. Internal consistency coefficients were calculated for the total scale and its subscales. Chronbach's alpha obtained for SASSI-A2 was 0.72. This relatively high level of Chronbach's alpha showed relatively high level of reliability. The results demonstrated that the whole SASSI-A2 meets the fundamental measurement properties and can discriminate groups of higher institution students from high to low on the substance dependency variable. The accuracy of the test has been found to be unaffected by gender, ethnicity, age and years of education. Although more rigorous validation studies are needed, it is recommended that SASSI-A2 be considered for usage to higher institution students populations when a brief, objective, and accurate screening tool for chemical dependency is needed.

1 Introduction

Drug use and abuse remain as a critical problem in most countries and are often associated with several social and economic consequences. In recent years many adolescent including students in Higher Education Institutions (HEIs) engage in drug abuse. Since 2005, the Ministry of Higher Education (MOHE) has implemented a urinalysis for students of Higher Education Institutions (HEIs). A total of 16252 of students were involved in this study. Urinalysis were carried out to detect the involvement in drug abuse among HEIs students. In urinalysis conducted by Kamaruddin et al., (2013), 0.52% of respondents was detected involved in drug abuse. Despite its widespread use, urinalysis as a measure of unreported drug use is constrained by several limitations:

(i) Cost inefficient and it is impossible to administer to HEIs population.

(ii) Time consumption and requires expertise to conduct the urinalysis.

(iii) Human rights issue in collecting the urine specimens.

(iv) A narrow window of detection (36 to 72 hours), because many drugs are rapidly excreted (Baumgartner, [3]; Baumgartner, et al., [4]; Berka, 1991; Mieczkowski, et al., [15].

(v) Urine specimens require special handling and may pose a potential health risk to those collecting and assaying samples [15].

The above mentioned situations have created problems to conduct urinalysis in the future. Therefore, an alternative to the urinalysis should be identified. Substance use screening instruments can be helpful in determining if an individual has a problem with substance use, is at risk for developing one, or is in need of further assessment. Adger and Werner et. al [1] state that the goal of screening is to determine whether a problem exists and the "need for an in-depth assessment" (p. 124). Currently, a wide array of screening instruments is used to assess Substance Used Disorder (SUD) such as the Addiction Severity Index (ASI; McLellan, Luborsky, Woody, & O'Brien, 1980; McLellan et al., [13], the Addiction Acknowledgement and Addiction Potential Scales (AAS; APS; Weed, Butcher, McKenna, & Ben-Porath, 1992), the Drug Abuse Screening Test (DAST; Skinner, [18], and the Substance Abuse Subtle Screening Inventory (SASSI; Miller, 1985). SASSI is one of the most frequently used measures and more comprehensive instrument. It seems to be "especially effective in identifying early-stage chemically dependent individuals who are either in denial or are deliberately trying to conceal their chemical dependency pattern" [14]. Previous research has established that the SASSI-A2 has good psychometric properties [6,19]. Reliability coefficients are equally impressive and has a good level of validity to be used to identify users / abusers substance / drug of diverse backgrounds and gender [11]. Miller & Lazowski (2001) reported that the percentage of accuracy with African American ethnicity was 90.2%; Asian American 100%; 94.5% Caucasian; Hispanic American 93.8%; 98.2% Native American and other ethnic 93.5%. Accuracy of measurement for men is 93.8% and 95.4% for females. Based on the scores obtained from SASSI-A2, respondents can be classified into four (4) categories that are; Low Probability (LP), Low Probability Frequently Abuse (LPFA), Substance Dependent Abuse (SDA) and Substance Dependent Disorder (SDD). The third (SAD) and fourth (SDD) categories are classified as at risk. The SASSI is an alcohol and drug screening measure and, therefore, was designed to distinguish individuals who may have an alcohol or drug use problem and expose the need for further assessment.

2 Study objectives

The aim of the present study was to examine the psychometric properties of a Malay version of the SASSI-A2 and the effectiveness of psychometric inventories to identify HEIs students involved in drug abuse.

3 Method

For the purpose of this study, the researcher use a cross-sectional survey design and descriptive approach in a representative sample of 750 students from five Malaysian public HEIs. The respondents comprised of students who are at risk, such as; engaged in misconduct at the university, low achievement, health problems and other matters related to the university regulations.

3.1 Instrumentation

The psychometric properties of the Substance Abuse Subtle Screening Inventory-A2, an 72-item self-report questionnaire was evaluated. SASSI-A2 has also been translated in Japanese [9] Persian [20] and Spanish language [12, 16]. In this study the SASSI-A2 was independently translated from English into Malay by expert (Hambleton & Jong, 2003). One of these expert were from Educational Psychology while the remaining two were bilingual expert. First, the researcher translated the items from the English language into Malay; secondly, the expert translated the items from Malay back into the English language. Finally, the researcher discussed the results and developed a common proposal and found that about 20.8% of the items conform to the original item, 8.3% need to revise while the remaining 70.9% correction should be made only on aspects of grammar and sentence structuring. Subsequently, the researchers also performed interrater reliability. Interrater reliability is the extent to which two or more individuals (coders or raters) agree. The number of experts selected was determined following the recommendations made by Crocker, Llabre, and Miller (1988) for obtaining useful estimates to adequately calculate interrater agreement. The recommendation is to select at least three interrater for each item (Lynn, 1986). Cohen's kappa (k) (Landis & Koch, 1977) tests were used to determine the scores between the different raters. A pilot test was then conducted to determine the reliability of the instrument. A total of 152 samples were randomly selected from a public university to respond at SASSI-A2 instrument which has been refined. Analysis performed using a split-half reliability method. Results showed the coefficient obtained is 0.77.

3.2 Analysis

Internal consistency was assessed using split half reliability. In the split-half reliability method, the SASSI-A2 was first divided into two equivalent halves and the correlation coefficient between scores of these half-test was found. In the present study the correlation coefficient was calculated by using Spearman formula and Brown formula. Meanwhile descriptive analysis was performing to determine the effectiveness of the SASSI in classifying respondents into four categories (LP, LPFA, SAD & SDD).

4 Results and Discussion

Investigation into the reliability of the SASSI-A2 was conducted using the reliability tool under scale in the SPSS programme. Data was subjected to internal consistency reliability analyses; first, Cronbach's Alpha and second, split-half reliability. For the split-half reliability, both Spearman Brown Split-Half and Guttman Split-Half were obtained for the scale in general. The results of reliability test presented in Table 1 were obtained from the analysis conducted on the 72 items translated in Malay version of SASSI-A2.

Table 1: Reliability For SASSI-A2

No. of Items	72	p
Cronbach's Alpha	.722	<0.05
Spearman-Brown Coefficient	.729	<0.05
Guttman Split-Half Coefficient	.728	<0.05

Effectiveness of the SASSI-A2

SASSI-A2 effectiveness in identifying respondents involved in drug abuse is based on four categories: LP, LPFA, SDA and SDD as can be seen in Table 2.

Table 2 : Effectiveness of SASSI-A2 in Identifying Drug Abuse Among HEIs in Malaysia

SASSI-A2		Finding
Low Probability (LP)	371	49.5%
Low Probability Frequently Abuse (LPFA)	336	44.8%
Substance Abuse Dependent (SAD)	34	4.5%
Substance Dependent Disorder (SDD)	9	1.2%
Total	750	100%

Discussion

The aim of the present study was to examine the psychometric properties of a Malay version of the SASSI-A2 and the effectiveness of psychometric inventories to identify HEIs students involved in drug abuse. These findings indicate that drug screening can be done by using SASSI-A2. Undoubtedly urinalysis is more accurate, but this method can only be used on the small target group. In addition, urinalysis is not able to identify those who have the potential to be involved in drug abuse. The results of this study show that the SASSI-A2 can categorize respondents into four groups, namely LP, LPFA, SAD and SDD. This information allows the responsible party to plan ahead in order to curb drug abuse among groups that have been identified. This condition can help in providing interventions that focus on groups that have been identified through screening. The SASSI-A2 can also be conducted on the entire population of students of institutions of higher education in Malaysia through the online system. This system can save time, cost and analysis can be done immediately. The accuracy of the test has been found to be unaffected by gender, ethnicity, age and years of education. Although more rigorous validation studies are needed, it is recommended that SASSI-A2 be considered for usage to HEIs students populations. Based on the results of this study, SASSI-A2 has a high potential to be used as a screening test to identify HEIs students who abuse drugs.

References

1. Adger, H., & Werner, M. J., The paediatrician. Alcohol Health and Research World, 18, 121-126, (1994)
2. Baer,J,D., Baumgartner, W.A., Hill, V.A., Blahd, W.H., Hair analysis for the detection of drug in pre-trial probation and parole populations. Fed Probat 55, 3-10, (1991)

3. Baumgartner, W.A., Hill, V.A., Blahd, W.H., Hair analysis for drugs of abuse. Journal Forensic Science 34, 1433-1453, (1989)
4. Baumgartner, A.M., Jones, P.F., Baumgartner, W.A., Black, C.T., Radioimmunoassay of hair for determining opiate-abuse histories. Journal Nuclear Medicine 20, 748-52, (1979)
5. Berka C., New drug test offers significant advantages over traditional methods. Health Link, 4, 1-2, (1991)
6. Clements, R., Psychometric properties of the Substance Abuse Subtle Screening Inventory-3. Journal of Substance Abuse Treatment, 23(4), 419-423, (2002)
7. Cohen J., A coefficient of agreement for nominal scales. Educational Psychological Measurement, 20, 37-46, (1960)
8. DeVellis, R., Scale development: Theory and application. SAGE: Newbury Park, CA, (1991)
9. Kikuchi A, Wada K Nihon, The Japanese version of Substance Abuse Subtle Screening Inventory and its psychometric assessment. Arukoru Yakubutsu Igakkai Zasshi, 35(6), 388-401, (2000)
10. Landis, J.R., Koch, G.G., The measurement of observer agreement for categorical data. Biometrics, 33, 159–174, (1977)
11. Lazowski, L.E.; & Miller, F.G., The adult substance abuse subtle screening inventory (3rd Ed.) manual. Springville, IL: The SASSI Institute, (1999)
12. Lazowski,L.E., Boye, M.W., Miller, G.A., & Miller, F.G., The development and validation of the Spanish SASSI. Springville, IN: The SASSI Institute (2002)
13. McLellan, A. T., Luborsky, L., Woody, G. E. & O'Brien, C. P., An improved diagnostic evaluation instrument for substance abuse patients: The Addiction Severity Index. Journal of Nervous and Mental Disease, 168, 26-33, (1980)
14. Merta, Rod J., Addictions counseling. Counseling and Human Development, 1-3
15. Mieczkowski, T., Landress, H.J., Newel, R., Coletti, S.D. (1993). Testing hair for illicit drug use. Washington, DC: National Institute of Justice, Document NCJ 137057, pp 1-5, (2001)
16. Miller, G. A., Supplement to the substance abuse subtle screening inventory manual. Bloomington, IN: Glenn A. Miller, (1994)
17. Polit, D.F., & Beck, C.T., Nursing research: generating and assessing evidence for nursing practice. (8th. ed.) Philadelphia: Lippincott Williams & Wilkins, (2008)
18. Skinner, H. A, The Drug Abuse Screening Test. Addictive Behaviors, 7, 363-371, (1982)
19. Swartz, J.A., Adapting and Using the Substance Abuse Subtle Screening Inventory-2 with Criminal Justice Offenders. Criminal Justice and Behaviour, 25(3), 344-365, (1998)
20. Vahid Sadeghi Firouzabadi, Reza Rostami, Nima Ghorbani, The reliability and validity of Substance Abuse Subtle Screening Inventory. Journal of Psychology, 13, 206-224, (2009)
21. Weed, N. C., Butcher, J. N., McKenna, T., & Ben-Porath, Y. S., New measures for assessing alcohol and drug abuse with the MMPI-2: The APS and the AAS. Journal of Personality Assessment, 58, 389-404, (1992)

Can Retail Selling Prices of Petrol and Diesel Trigger Inflation in Malaysia?

Mohd Shahidan Shaari, *Afifah Hanani* Yusuf, *Aidanazima* Abashah *and Tan Lee* Pei

Abstract. Previously, many studies explored the effects of oil price on inflation by using the price of crude oil as the proxy of oil price. The novelty of this current study is that it used several types of fuel prices including RON95, RON97 and Diesel. This study employed the ARDL approach and analysed the monthly data from 2010 to 2015 in Malaysia. The result shows that there are significant effects of the price of RON95, RON97 and diesel on inflation in the long run in Malaysia. However, in the short run, the price of RON97 does not affect inflation while the prices of RON95 and diesel do affect inflation in Malaysia.

1 Introduction

The uncertainty in oil prices has attracted a large number of researchers to explore the effects of oil prices on economy such as inflation. Some ascertained that higher oil prices can have negative impacts and some discovered that higher oil prices can have positive impacts especially for oil producing countries.

Oil prices are inextricably related to the cost of production. Therefore, the rising oil price can lead to an increase in the cost of production. Higher oil prices can hike the costs of raw materials up, thus suppliers should adjust their price of goods so that they can maintain their profits [5]. As a result, the cost of increasing oil will be incurred by consumers and thus causing inflation. Sovuthea [22] stated that a rise in oil price results in inflation. Alem and Soderbom, [1] added that countries with higher oil price might face inflation. Therefore, governments should play an important role in controlling the price.

Most previous studies were only concerned about the relationship between oil price and inflation and used the price of crude oil as a proxy of oil prices [4, 23]. However, this current study uses RON95, RON97 and diesel prices as the proxy of oil prices instead of aggregate oil price to examine the effects of the retail selling petrol and diesel prices on inflation in Malaysia. Petrol is largely consumed by households for various activities. Petrol can be split into several types; RON95 and RON97. Households usually use these types of petrol for their vehicles. Diesel is another petroleum product which has the same function as that of petrol. However, its use is limited to only certain transports such as lorry and bus. All these petroleum products are indispensable for not only domestic economy, but also world economy [21]. Due to the importance of the commodity, many economists agreed that subsidizing petroleum price can shield economy from deleterious impacts of higher petroleum prices.

Since many issues arose due to an increase in oil price, some countries such as Malaysia subsidized petroleum prices. Nevertheless, in order to ensure its financial stability and improve its economic efficiency, the Malaysian government removed subsidies from this commodity. The reduction in the subsidy on oil triggered indignation among oil consumers. Most people suffered from the cost of increasing oil price. The rising prices for RON95, RON97 and diesel in Malaysia caused many people including economists to cast aspersions on it.

Fig.1 shows the price of RON95, RON97 and diesel in Malaysia from 2008 to 2015. In June 2008, the price of RON97 went up from RM1.92 to RM2.70 per litre. Therefore, to reduce the government's expenditure, the Malaysian government decontrolled the price of RON95 and RON97 starting from 1st November 2008. The sharp rise in petrol price resulted in inflation for all consumer products (Shaari *et al.*, 2012). The increasing trend was exhibited for RON97 with an increase of 41% over the period of September 2009 to May 2011. In July 2010, the government of Malaysia announced that the price of RON97 was determined by the global oil price (The Central Bank of Malaysia, 2011). This led to an increase of 25 cents per litre in the price of RON97 in Malaysia. Furthermore, the oil subsidy revision in Malaysia was implemented on December 4, 2010, causing a reduction of 5% in the subsidy for RON95. Thus, the price of RON97 increased by 50 cents to RM2.80 per litre from January to July, 2011. However, the price of RON97 grew sharply at the rate of 15.38% from RM2.60 per litre in July 2012 to RM3.00 per litre in September 2012 due to the hike of the world's oil price in the market. Furthermore, the price of RON97 increased from RM2.70 in August, 2013 to RM2.85 in September, 2013. However, the increase in the price of RON97 was lower than the increase in the price of RON95 and diesel in 2013 due to the increasing supply of oil in the world market from the OPEC member countries [11]. The reduction in the subsidy for RON95, RON97 and diesel aggravated the burden for Malaysian especially the low-income group. Hence, to reduce the burden for those in the low-income group, the Malaysian government announced the BR1M (Bantuan Rakyat 1Malaysia), a program that was introduced to curb the increasing cost of living pressures experienced by low-income group in 2010 [10].

Since mid-2014, the global oil price dropped by 50% due to surge in oil supply and a weak demand [15]. To enhance the country's revenue and to take advantage of the lowest oil price, the government of Malaysia has officially ended the subsidy of all fuels including RON95, RON97 and diesel on 1st December 2014. Nonetheless, the elimination of subsidy for RON95, RON97 and diesel has affected most Malaysians financially. The inflation rate in 2015 was affected by the implementation of the new pricing mechanism for petrol prices in which there was an influence of global oil price volatility into domestic prices that had affected the domestic fuel products in Malaysia [13]. Fluctuation of oil prices kept happening starting from January to December 2015.

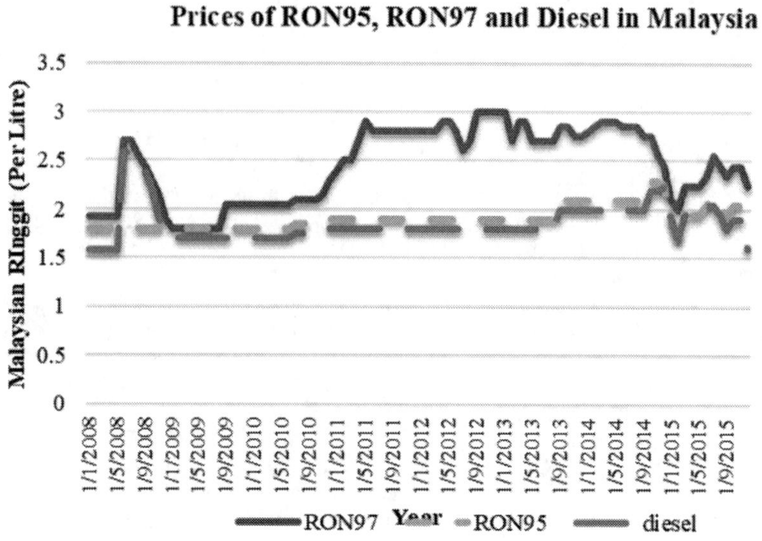

Fig. 1: Prices of RON95, RON97 and diesel in Malaysia.
(Source: Ministry of Domestic Trade, Co-operatives and Consumerism, 2016)

2 Literature Review

Numerous studies on examining the effects of oil price on inflation have produced mixed findings [24, 4,] and Taghizadeh Hesary, 2014. However, their findings were not consistent with some other studies that found a positive effect of oil price on inflation [2, 14, 17]. A limited number of previous studies found that there is no correlation between oil price and inflation [20, 15, 3]. The inconsistency were due to different economic conditions. In China [24] analysed the impact of oil price shocks on inflation from 1990 until 2013. Using the model of New Area-Wide and Dynamic Stochastic General Equilibrium, the study found that oil supply shocks can cause inflation in China, especially in the short run. [16] investigated the impact of global oil prices changes on consumer and producer prices separately from 1990 to 2012 in Turkey. The study found that the impact of oil price shocks on producer prices was two times higher than the impact on consumer prices.

Celik and Akgul [14] also studied the relationship between changes in fuel oil price and consumer price index in Turkey. The study used the Vector Error Correction Model to investigate the linkage between oil price shock and inflation in Turkey from 2005 until 2010. The study found that price fluctuations created by the distribution firms were adjusted to the price in order to suit the developments in the international oil market. Using the Vector Error Correction Model to analyse the data, the result indicated that a 1% increase in fuel oil prices will automatically increase the price of consumer index by 1.26%. Besides, the study also found that an increase in fuel oil prices has increased the consumer price index in Turkey.

Jiranyakul [19] estimated the linkage between oil price shocks (nominal oil price) and domestic inflation (consumer price index and industrial production index) in Thailand for the period of 1993 to 2013. Using the Autoregressive Distributed Lag (ADDL) model, Error Correction estimation, Constant Conditional Correlation (ccc-GARCH) and Granger

 https://doi.org/10.1051/matecconf/201815005070

Causality test, the study found that domestic oil price shocks can cause inflation to rise. Oil price uncertainties do not cause inflation in the long run.

In a study by Chou and Tseng [15] on the effects of oil price on Taiwan's inflation rate (food, clothing, accommodation, transportation, medicine and health, education and leisure, and miscellaneous) for the period of 1982M3 to 1992M2, they have concluded that pass-through effect of oil prices on inflation is not significant in the short run (except for accommodation). The results also indicated that there is a significant effect of global oil prices shocks on Taiwan's consumer price index in the long run.

Alvarez *et al*. [3] analysed the impact of oil price shocks on consumer price inflation in Spain, Germany, France and Italy. Using the Linear and Nonlinear Model, Dynamic Stochastic General Equilibrium and Bank of Spain's Quarterly Model (MTBE) for the period of 1997 to 2007, the study discovered that there is a limited impact of oil price shocks on consumer price inflation in Spain, Germany, France and Italy. Nevertheless, the impact of oil price shocks on inflation in Spain is higher than that in Germany, France and Italy due to the diminishing share of indirect taxes in Spain.

3 Methodology

This study aims to examine the effects of retail selling petrol and diesel prices on inflation in Malaysia. An empirical analysis on the variables is conducted. Monthly data from 2010 to 2015 were collected to determine the effects of petroleum price on inflation. All the data were extracted from the Central Bank and the Ministry of Domestic Trade, Cooperative and Consumerism.

$$\mathbf{CPI_t = \beta_0 + \beta_1 M3_t + \beta_2 RON95_t + \beta_3 RON97_t + \beta_4 DIESEL_t + \varepsilon_t} \tag{1}$$

Whereas CPI is consumer price index, t refer to year, β is coefficient, M3 is money supply, RON5 is the price of petrol for RON95, and RON97 is the price of petrol for RON97. DIESEL is the price of diesel and ε is an error. In this study we include the money supply as a controlled variable. Unit root test was conducted to determine the stationary series in the level and in the first difference test based on the augmented Dickey–Fuller (ADF). The variable series is stationary or does not have a unit root; implying the alternative hypothesis is accepted and the null hypothesis is rejected. If the stationary test is not significant, suggesting that the variable series is non-stationary, the variable has a unit root (null hypothesis is accepted). The hypothesis in this test is presented as follows:

H0: $\delta = 0$ (have a unit root test / not stationary)

H1: $\delta \neq 0$ (no unit root test / stationary)

If the value of the t-statistic is larger than the critical value of the ADF and PP, the null hypothesis is accepted (unit root test exists). If the t-statistic is lower than the critical value of the ADF and PP, I does not have the unit root (the alternative hypothesis is accepted). First, the unit root test was conducted in level (unit root test in level with constant and unit root test in level with constant and trend). Then, the unit root test was performed in the first difference (unit root test in first difference with constant and unit root test in first difference with constant and trend). Equation 3, 4, and 5 express the equations in level without constant and trend, in level with constant only, and with constant and trend, respectively.

$$\Delta Y_t = \alpha_0 + \sum_{i=1}^{a} \emptyset_1 Y_{t-1} + \sum_{j=1}^{b} \theta_j \Delta Y_{t-1} + \varepsilon_t \quad (2)$$

Without constant and trend:

$$\Delta Y_t = \delta Y_{t-1} + \mu_t \quad (3)$$

With constant:

$$\Delta Y_t = \alpha + \delta Y_{t-1} + \mu_t \quad (4)$$

Constant and trend:

$$\Delta Y_t = \alpha + \beta T + \delta Y_{t-1} + \mu_t \quad (5)$$

The co-integration technique of autoregressive distributed lag (ARDL) is employed to determine the order of co-integration of each variable. The advantage of this technique lies in the fact that it obviates the need to classify variables into I(1) or I(0). Each equation must be estimated by employing ordinary least square (OLS) approach. Then Wald test (F-statistic) is calculated to discern the long run relationship among the variables. Applying the assumptions by Pesaran et al. (2001) in case III (intercept and trend), the equation for the relationship between dependent variable and independent variables is as follows:

$$CPI_t = \beta_0 + \beta_1 CPI_{t-1} + \beta_2 M3_{t-1} + \beta_3 RON95_{t-1} + \beta_4 RON97_{t-1} + \beta_5 DIESEL_{t-1} + \beta_{6,i} \sum_{i=1}^{p} \Delta CPI_{t-i} + \beta_{7,i} \sum_{i=1}^{q1} \Delta M3_{t-i} + \beta_{8,i} \sum_{i=1}^{q2} \Delta RON95_{t-i} + \beta_{9,i} \sum_{i=1}^{q3} \Delta RON97_{t-i} + \beta_{10,i} \sum_{i=1}^{q4} \Delta DIESEL_{t-i} + \beta U_t \quad (6)$$

Whereas Δ is the first level of differentiation as described in the logarithm. Other than that, equation (6) can also be considered as ARDL model. For the selection of lag in this study, the Akaike's Information Criterion (AIC) is referred. Wald test is also used to calculate the F-statistic. This is to confirm the existence of the long run relationships among variables. So, the hypothesis of the Wald test is as follows:

$$\mathbf{H_0: \beta_1 = 0 \text{ and } \beta_2 = \beta_3 = \beta_4 = 0}$$

(No relationship in long run)

$$\mathbf{H_1: \beta_1 \neq 0 \text{ and } \beta_2 \neq \beta_3 \neq \beta_4 \neq 0}$$

(There is a relationship in long run)

Then, the boundaries test is conducted by comparing F-statistic with critical value (according to the Pesaran schedule/table). If the F-statistic is lower than the critical value, the hypothesis H_0 will be accepted at a certain significance level. The error correction model can be defined in the ARDL framework as follows:

$$\begin{aligned} &\mathbf{CPI_t = \mu + \textstyle\sum_{i=1}^{p} \emptyset_i\, \Delta CPI_{t-i} +} \\ &\mathbf{\textstyle\sum_{j=1}^{q=1} \varphi_j\, \Delta M3_{t-j} +} \\ &\mathbf{\textstyle\sum_{m=1}^{q=2} \gamma_m\, \Delta RON95_{t-m} +} \\ &\mathbf{\textstyle\sum_{p=1}^{q=3} \eta_p\, \Delta RON97_{t-p} +} \\ &\mathbf{\textstyle\sum_{s=1}^{q=4} \kappa_s \Delta DIESEL_{t-s} + vecm_{t-1} + \varepsilon_t} \end{aligned} \tag{7}$$

Where $\emptyset, \varphi, \gamma, \eta, and\ \kappa$ in equation (7) refers to short-term dynamic coefficient and vecm indicates the speed of adjustment for the model (1) above.

4 Finding

Unit Root Results: Results from the Unit Root test shows that all the variables are not stationary at level for intercept, and intercept and trend. However, the variables become stationary at the first difference for intercept, and intercept and trend. Therefore, ARDL test can be conducted to see the long-run and short-run effects of retail selling prices of petrol and diesel on inflation.

Table 1: Unit Root Results

	Intercept		Intercept and Trend	
Variables	Level	First Difference	Level	First Difference
M3	0.936171 (0.7711)	9.718928* (0.0000)	1.410553 (0.8496)	9.731209* (0.0000)
R97	1.833817 (0.3615)	7.256789* (0.0000)	1.488713 (0.8245)	7.490991* (0.0000)
R95	2.819971	9.718928*	1.410553	9.731209*

	(0.0605)	(0.0000)	(0.8496)	(0.0000)
DS	2.458855 (0.1299)	7.2768* (0.0000)	2.405163 (0.3735)	7.329222* (0.0000)
CPI	0.375964 (0.9805)	-8.024582* (0.0000)	-2.639825 (0.2647)	-5.547181* (0.0001)

Note: * and ** represent significant level at 1% and 5%, respectively.

ARDL Results: The Bound test was conducted to see whether there is a long-run relationship between retail selling prices of petrol and diesel prices, and inflation in Malaysia. Results from the Bound test show that F-statistic is higher that upper bound and therefore, it can prove that there is an existence of the long-run relationship. Therefore, we can proceed to the ARDL approach to see which price can affect inflation.

Table 2: Bounds Test for the Existence of a Long Run Relationship

F-statistic		
6.047937		
Critical value		
	Lower	Upper
1% significant level	3.03	4.06
5% significant level	3.47	4.57
10% significant level	4.40	5.72

Notes: Critical value bounds obtain from Pesaran et al (2001), table CI (V) case V: Unrestricted intercept and unrestricted trend.

The Diagnostic and stability tests were performed to determine the goodness of the ARDL model. Table 3 and Fig. 2 and 3 for CUSUM and CUSUMQ tests are presented. The results suggest that the model pass the Breusch-Godfrey Serial Correlation LM test, Ramsey RESET stability test and ARCH test in the first stage. Therefore, there is no evidence of autocorrelation, no serial correlation and heteroscedasticity. Then, the cumulative sum of recursive residual (CUSUM) and the cumulative sum of squares of recursive residual (CUSUMSQ) are presented. The plots for both the CUSUM and CUSUMQ are within the boundaries. Therefore, it can be concluded that there is a stability in the ARDL models.

Table 3: Diagnostic Tests

Test statistic	F-statistic
A: Breusch-Godfrey Serial Correlation	1.0958
B: Ramsey RESET stability	1.8401
C: ARCH	0.54298

A: Langrange multiplier test of residual serial correlation
B: Functional Form
C: Heteroscedasticity

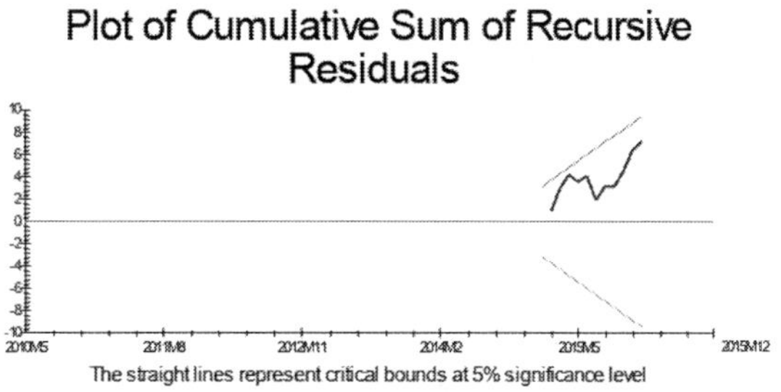

Fig. 2: Cumulative Sum of Recursive Residuals

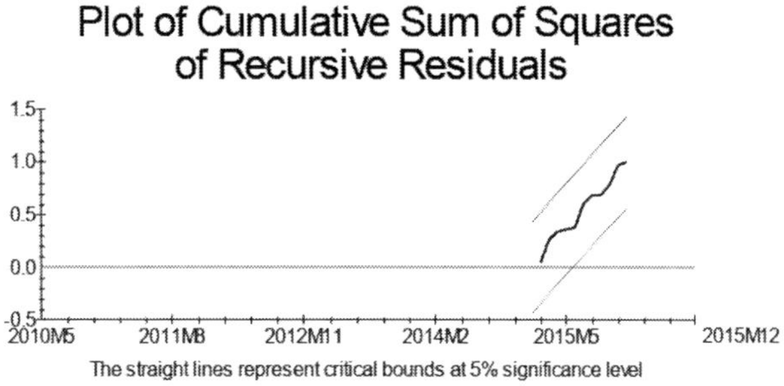

Fig. 3: Cumulative Sum of Squares of Recursive Residuals

The long-run coefficients of the variables under investigations are shown in the Table 4. The result shows that all the independent variables have effects on inflation in the long run. An increase in money supply can cause inflation in the long run. The price of RON97 can affect inflation positively in the long run. Therefore, a rise of 1 unit in the price of RON97 can increase inflation by 0.64261. The results also show that there is a significant effect of the price of RON95 on inflation in the long run. Thus, the price of RON95 changes by 1 unit can cause inflation to change by 7.59. The price of diesel can also influence inflation negatively in the long run. Therefore, an increase of 1 unit of the price of diesel will lead to a reduction of -2.9130 in inflation.

Table 4: Estimated Long run Coefficients Using the ARDL Approach

Regressor	Coefficient	Standard Error	T-ratio
DM3	0.00001	0.000004	2.9619[.005]*
DR97	0.64261	0.29672	2.1657[.035]**
DR95	7.5895	1.2769	5.9436[.000]*
DDS	-2.9130	1.1683	-2.4934[.016]**
C	24.6059	9.0275	2.7257[.009]*

Notes: *, and **indicate significant at 1%, and 5% significant level.

Table 5 shows results from the error correction representation for the selected ARDL Model test. The result shows that both money supply and RON97 does not positively and significantly affect inflation in the short run. However, the price of RON95 and diesel can affect inflation in the short run. An increase of 1 unit of the price of RON95 and diesel will lead to a rise of 5.38 and a decrease of -2.44 in inflation, respectively. Furthermore, the result also shows that the R-Squares of this study is high which is at 92%, whereby the dependent variable can be explained by all independent variables by 92%, and the DW-statistic is 1.7345 which is close to 2.

Table 5: Error Correction Representation for the Selected ARDL Model

Regressor	**Coefficient**	**Standard Error**	**T-Ratio[Prob]**
dDM3	0.0000008	0.000002	0.50802[.613]
dDR97	0.25481	0.18545	1.3740[.175]
dDR95	5.3806	1.0483	5.1328[.000]*
dDDS	-2.4363	0.94998	-2.5646[.013]*
dC	20.5787	6.5188	3.1568[.003]*
R-squared	0.91541		
DW-statistic	1.7345		
Akaike Info. Criterion	37.6316		

5 Conclusion

This paper aims to investigate the effects of retail selling prices of petrol and diesel on inflation in Malaysia. The ADF unit root test was employed. All variables were found to be non-stationary in level and stationary in the first difference. Results from the long run coefficients using the ADRL approach shows that there are significant effects of retail selling prices of petrol and diesel in the long run. In the short run, the higher price of RON95 and Diesel can result in inflation in Malaysia. This study will be instrumental in the formulation of policies to ensure that inflation can be controlled in Malaysia. The Malaysian government should subsidize the price of RON95, RON97 and diesel to cushion its impacts. If the price was influenced by the world oil price, it might be detrimental to the general price in Malaysia.

References

1. Alem, Y., & Soderbom, M., Household-Level Consumption in Urban Ethiopia: The Effects of a Large Food Price Shock. World Development, 40(1), 146–162, (2011)

2. Alom, F., Economic Effects of Oil and Food Price Shocks in Asia and Pacific Countries : An Application of SVAR Model. In New Zealand Agricultural & Resources Economics Society (pp. 1–43). Nelson: NZARES (2011)
3. Alvarez, L. J., Hurtado, S., Sanchez, I., & Thomas, C., The Impact of Oil Price Changes on Spanish and Euro Area Consumer Price Inflation. Economic Modelling, 28(1-2), 422–431, (2010)
4. Belke, A., & Dreger, C., The Transmission of Oil and Food Prices to Consumer Prices-Evidence for the MENA Countries. International Economics and Economic Policy, (2013)
5. Bolaji, B. O., & Bolaji, G. A., Investigating the Effects of Increase in Oil Prices on Manufacturing Companies in Nigeria. Journal of Science and Technology, 11(2), 387–390, (2010)
6. Central Bank of Malaysia., Press Release Central Bank of Malaysia Annual Report 2008. Kuala Lumpur: Central Bank of Malaysia Publishing, (2009)
7. Central Bank of Malaysia., Central Bank of Malaysia Annual Report 2010. Kuala Lumpur: Central Bank of Malaysia Publishing, (2011)
8. Central Bank of Malaysia., Central Bank of Malaysia Annual Report 2011. Kuala Lumpur: Central Bank of Malaysia Publishing, (2012)
9. Central Bank of Malaysia., Central Bank of Malaysia Annual Report 2012. Kuala Lumpur: Central Bank of Malaysia Publishing, (2013)
10. Central Bank of Malaysia., Variations in Household Propensity to Consume Across Income Segments. Kuala Lumpur: Central Bank of Malaysia Publishing (2013)
11. Central Bank of Malaysia., Central Bank of Malaysia Annual Report 2013. Kuala Lumpur: Central Bank of Malaysia Publishing, (2014)
12. Central Bank of Malaysia., Price Reforms: Motivation, Impact and Mitigating Measures. Central Bank of Malaysia (Vol. 53). Kuala Lumpur: Central Bank of Malaysia Publishing (2014)
13. Central Bank of Malaysia., Central Bank of Malaysia Annual Report 2014. Kuala Lumpur: Central Bank of Malaysia Publishing (2015)
14. Celik, T., & Akgül, B., Changes in Fuel Oil Prices in Turkey: An Estimation of the Inflation Effect Using VAR Analysis. Journal of Economics and Business, XIV(2), 11–21, (2011)
15. Chou, K. W., & Tseng, Y. H., Pass-Through of Oil Prices to CPI Inflation in Taiwan. International Research Journal of Finance and Economics, 69, 73–83, (2011)
16. Dedeoglu, D., & Kaya, H., Pass-Through of Oil Prices to Domestic Prices: Evidence from An Oil-Hungry but Oil-Poor Emerging Market. Economic Modelling, 43, 67–74, (2014)
17. Gichana, H. M., Relationship Between Crude Oil Prices and the Consumer Price Index in Kenya (Doctoral Diessertation, University of Nairobi), (2013)
18. Institute For Energy Research., 2014 Record Year for U.S. Oil Production; Largest Increase in Over 100 Years. Institute For Energy Research, (2015)
19. Ibrahim, M. H., & Said, R., Disaggregated Consumer Prices and Oil Price Pass-Through: Evidence from Malaysia. China Agricultural Economic Review, 4(4), 514–529, (2012)
20. Jiranyakul, K., Oil Price Shocks and Domestic Inflation in Thailand. Available at SSRN 2578836, (2015)
21. Noordin, N. S. S., Oil Price Shock and Malaysian Sectoral Stock Market Return. Unpublished Masters in Business Administration Thesis, Faculty Business and Accountancy, University of Malaya, Kuala Lumpur, (2009)

22. Sovuthea, H., Government Response to Inflation Crisis and Global Financial Crisis.CDRI, (2013)
23. Yoshino, N., & Taghizadeh Hesary, F., Economic Impacts of Oil Price Fluctuations in Developed and Emerging Economies. IEEJ Energy Journal, 9(3), 58–75, (2014)
24. Zhao, L., Zhang, X., Wang, S. Y., & Xu, S. Y., The Effects of Oil Price Shocks on Output and Inflation in China. Energy Economics, 1–10, (2014)

Business Lures Employed by Malay Kelantanese Entrepreneurs

Mohd Nizam Sahad, *Suhaila* Abdullah, *Atikullah Haji* Abdullah *and Shuhairimi* Abdullah

Abstract. The foundation of a successful business is essentially the level of consumer response to a product or service offered. Entrepreneurs' abilitities to cater to their consumers' needs are one of their secrets to success. Hence, the objective of this study is to uncover the secrets of business success, particularly the business lures that have helped Kelantan entrepreneurs to popularise products or services they offer. A qualitative approach was used to analyse the interview data according to themes. The study found five factors which determined the business success of the selected Malay Kelantanese entrepreneurs. These factors were: they were persistent in asking for help from Allah, they strived to be approachable entrepreneurs, and they cultivated positive entrepreneurial virtues such as having an honourable personality, being creative, and innovative, as well as being bold in their marketing strategy. Although the findings are not representative of all Malay entrepreneurs in Malaysia, this study can, however, serve as a source of reference to encourage other entrepreneurs to emulate the business success of the selected entrepreneurs. Apart from that, it can encourage awareness among the Muslim entrepreneurs, about the importance of managing a business in accordance with the Islamic laws

1 Introduction

Entrepreneurship has long been acknowledged as a career option and has also been also given priority in various Malaysian government's policies [13]. One of the ministries in Malaysia, which plays a role in the development of *Bumiputera* entrepreneurs is the Ministry of Entrepreneur Development. It has a crucial role in producing entrepreneurs of high quality, potentiality, and ability to form the Commercial and Industrial Community (MPP), as envisaged in Vision 2020 [35].

Meanwhile, in the context of Islam, entrepreneurship is a form of worship to meet the needs of life that are balanced in terms of worldly and hereafter [30]. It is crucial to integrate spiritual elements in business activities because Islam, in essence, operates on the basis of encapsulating every aspect of life in its entirety, regardless of the profession and the way of life led by a person [44]. In fact, Prophet Muhammad SAW was renowned in the field of business, where he had followed his uncle, Abu Thalib in a trade convoy to Damascus, Syria from a young age of 12 [8,10]. Armed with a vast experience in business, Prophet Muhammad SAW had shown that entrepreneurship is a reputable and lucrative career which should be highly valued by the Muslims [1].

However, the process of building and strengthening a business is not a simple affair because today's entrepreneurs are faced with many challenges, such as shortage of capital,

* Corresponding author: nizamsahad@usm.my

https://doi.org/10.1051/matecconf/201815005071

competition, technology, management, communication, changes in demands, and economic uncertainty [45,15,42,36,48,35]. Therefore, many entrepreneurs have suffered losses and were declared as bankrupts. The statistics recorded for the period of three years from 2006-2009 by the Malaysian Insolvency Department showed that a total of 4,067 entrepreneurs of the small and medium enterprises (SMEs) had experienced bankruptcy [14].

Nevertheless, there are also entrepreneurs who have succeeded in their business. Behind the success of their business, there are, of course, practices, measures, and a certain thinking paradigm based on the theories and concepts of conventional entrepreneurship as well as complying with the religion of Islam. Therefore, this study aims to uncover the secrets and lures that have shaped the success of several selected entrepreneurs.

1.1 Business and Entrepreneurship in Islam

Although business and entrepreneurship are recognised in Islam, Allah SWT forbids His followers from earning a living based on *haram* (prohibited) resources, as Allah SWT says in Surah an-Nisa' [4]:29;

Translation: O you who believe! Eat not up your property among yourselves in vanities: but let there be amongst you traffic and trade by mutual good-will: nor kill (or destroy) yourselves: for verily God has been to you Most Merciful! ('Ali, 2005, p. 90).

Every Muslim businessperson or entrepreneur should understand the restrictions of *halal* (permitted) and *haram*, and practice them in their business activities. Based on Osella and Osella's [37] point of view, entrepreneurs who practice the teachings of Islam have inscribed their specific business interests and practices into the rhetoric of the common good. In other words, the combination of material success with moral values is seen as an exemplary contemporary way of being a modern, moral Muslim [37]. According to Adas [3], Islamic entrepreneurs may not be different from any other entrepreneurs in terms of their desire to expand their business and increase their profits, but their strong adherence to an Islamic identity, through which they give meaning to their economic incentives and activities, simultaneously redefine their Islamic identity in relation to their business activities. This is what distinguishes them from others.

Besides that, in Islam, the *ummah* (people) is encouraged to be self-employed and to venture into entrepreneurship (Ab. Mumin & Suhaili, [1] to earn a living, as contained in a *hadith* narrated by Al-Bazzar:

Rifa'ah bin Rafi' said that the Prophet Muhammad SAW was once asked; "What is the best means of livelihood?" And Prophet SAW answered, "A person working with his hands and every sale that is clean."

1.2 Malay Kelantanese Business

Seeking for a source of income through business and entrepreneurship has become common pursuits among the Malays in Kelantan. A review of the Kelantan history has shown that business activity is part of the traditional socio-economic patterns of the state of Kelantan, where the majority of its population is Malay [28]. According to Ilhaamie [18] a majority of the Malay Muslims in Kelantan venture into business, whether in a small or large scale.

The state of Kelantan has shown an outstanding development since 2009 with the presence of the trade industry, construction, consultation, manufacturing, property, and agriculture [4]. Up to 2013, there was a total of 2,075 businesspersons registered at the Malaysian Malay Business Chamber Kelantan (DPMMK) [5]. The rapid development of business and entrepreneurship in Kelantan is due to the government's efforts in encouraging and facilitating the Malays and *Bumiputera*s to venture into the business domain [28].

1.3 Entrepreneurial Ethics in Islam

Ethics can be defined as a theory, a set of values, moral standards, rules, and principles relating to the behaviour of individuals in a particular social group that allows them to determine good or bad, right or wrong [33]. Thus, ethics are guidelines in making judgments and decisions on various matters. Meanwhile, Islamic ethics are referred to as a moral character, which is guided by the values of Islam, such as faith, justice, and universal happiness [19].

Ethics in Islam is divided into two, that is, etiquette to Allah SWT (as a caliph and worshipper) and etiquette to humans (comradeship) [2]. According to Abuznaid [2] these ethics are fundamental to business success, both in this world and for the life hereafter. Abuznaid [2] defines the entrepreneurial ethics of Islam as a code of moral principles in business described in the al-Qur'an and *as-Sunnah* (legal ways of the Prophet SAW), such as righteousness, honesty, justice, patience, keeping promises, generous, responsible, humble, respecting others, and have *taq'wa* (piety) to Allah SWT.

Meanwhile, Kalsom and Mohd Ismail [22] placed five principles of Islamic entrepreneurial ethics, namely honesty (in business communication), trust (in resource management), sincerity (in terms of looking after the needs and safety of the customers), comradeship (among partners and customers), and justice (to all parties, as in pricing).

Islamic entrepreneurial ethics should be chosen by Muslim entrepreneurs as a form of reaction to the era of rapid development, where human greed is clearly evident (Zen, 2014). Hence, humans are at this time, in dire need for guidance to lead them out of digression, as contained in the *hadith* narrated by Abu Huraira RA that the Prophet SAW said;
"A time will come upon the people when one will not care how one gains one's money, legally or illegally." (Al-Bukhari, 1994, p. 465, no. 988).

1.4 Entrepreneurial Ethics in Islam

According to Ahmad Zahiruddin [6], management is a part of the sub-systems in life that cannot be separated from religion. It is because the orientation and the ultimate goal of Muslims in every aspect of their lives are to achieve the contentment of Allah SWT, and this distinguishes the Islamic management from the conventional form of management that is merely sufficient to fulfill human needs and desires [6]. Bearing this in mind, the practices of managing a business should be aligned with the Islamic entrepreneurial ethics. The merging of entrepreneurial or business practices with Islamic values, principles and business ethics has brought forth the Islamic entrepreneurial practices [19].

In managing a business, an entrepreneur should always monitor and plan with care so that every programme can be carried out smoothly. In fact, the quality and excellent performance in accomplishing any work is an emphasis in Islam [21], as contained in the *hadith* narrated by al-Bayhaqi RA in which the Prophet SAW said;
"Allah, the Almighty wants that when one of you does a job, he or she does it well." (Chapter 4, No. 5312)

Moreover, an entrepreneur should offer *halal* products or services to customers. Business activities which are clearly prohibited in Islam include to sell goods deemed to be harmful to others, such as alcohol, drugs, and prostitution; to take usury (interest); to practice unjust competition, and to practice fraud and forgery in business [49]. Hence, Allah SWT has called upon the human beings to earn a living based on *halal* sources, as Allah SWT says in Surah al-Baqarah [2]:172;

Translation: O you who believe! Eat of the good things that We have provided for you, and be grateful to God, if it is Him you worship. ('Ali, 2005, p. 44)

The purpose of such practices is to purify the human souls so that they will be rewarded in this life and in the hereafter. When Muslims do good deeds, they will be rewarded equitably, as Ilhaamie (2014) states that such practices will bring blessings to entrepreneurs. Thus, according to Ibn Khaldun (1995), the foundation of Islamic management is *irasati ad-din wa siyasatun dunya bih* (putting religion at the highest place and lead the world with it).

2 Methodology

The study employs a qualitative approach, in which, 10 successful Malay Kelantanese entrepreneurs were selected for an interview. The Malay Kelantanese entrepreneurs were selected for this study because a majority of the Malay Muslims in Kelantan venture into business, whether in a small-scale or large-scale (Ilhaamie, 2014). The interview transcriptions were translated from *Bahasa Melayu* to the English language, and then, analysed from the perspective of how the Islamic entrepreneurship was carried out. In addition, a literature review was conducted to enable resources and data from documents, such as books, journals, manuscripts, and so forth about the cultures and practices of entrepreneurship to be used.

3 Results and Discussion

According to the Oxford Advanced Learner's Dictionary (2007), an entrepreneur can be defined as a person who makes money by starting or running businesses, especially when this involves taking financial risks (Hornby, 2010, p. 509). Ashmore (1989) defines entrepreneurs as individuals who run their own businesses. Meanwhile, Nor Aini Idris (2003) and Wim Naude (2010) consider entrepreneurs as a group of businesspersons and dealers who are willing to take risks, always strive for innovation, and smart, as well as creative in finding ways to increase wealth, power, and social status. Van Gelderen (2007) defines entrepreneurial skills as the ability that is translated into the behaviour or actions of an entrepreneur. These skills include the capability to start a new business, taking care of business, making decisions, identifying opportunities, managing time, communicating well, solving problems creatively, and managing stress, all of which are regarded as basic skills in business management (Soaib Asimiran & Rosnani Jusoh, 2008).

Meanwhile, in Pihie, Akmaliah, and Elias' (2004) study, 11 main features of entrepreneurs were identified and they are: having high commitment, determination and perseverance, motivated to succeed and thrive, oriented to opportunities and goals, good at taking initiative and responsibility, skilful in problem solving, realistic, smart in searching and using feedbacks, possess internal control locus (highly confident of their capabilities to reach success), willing to take risks, not overly expecting power and high status, and with the integrity of reliability.

From the Western perspective, the main purpose of an entrepreneur setting up a business is to maximise profits, while continuously strive to expand the business (Longenecker et al., 2006; Madura, 2007; Scarborough & Zimmerer, 2006). However, from the perspective of Islam, entrepreneurs who apply the values of Islam and take actions based on the Islamic ethics will be able to overcome the obstacles in business and at the same time, be able to bring themselves closer to the Creator. According Nor'Azzah (2009), among the main factors that lead to entrepreneurial success and gaining a huge profit in business is working hard and implementing Islamic ethics in conducting business.

Previous research have shown that successful entrepreneurs have indeed adopted certain practices, such as conducting *majlis tahlil* (prayer ceremony) and *doa selamat* (safe prayer or supplication), and giving alms (Soaib Asimiran & Rosnani Jusoh, 2008). Meanwhile, there are also a handful of entrepreneurs who use talisman, amulets, *ibu duit* (mystical way to 'breed' money), and *wafak* (a branch of *hikmah* (supernatural) knowledge, in which there are practices of using verses of the Al-Qur'an and *Asmaul Husna* (the best 99 names of Allah SWT)) as mystical business lures to the detriment of the Islamic *aqidah* (faith) and leads to sin by associating Allah SWT with other forms or beings (Mohd Nizam, 2015).

Entrepreneurs who manage their business based on the Islamic ethics will receive considerable amount of goodness and blessings. In other words, their work performance improves as a result of their beliefs that success, prosperity, and security actually depends entirely on Allah SWT, and an improvement in terms of confidence, optimism, perseverance, contentment, gratitude, and patience to face any outcome after performing a task in earnest (Isa Mohammed, Zulkarnain, & AAhad, 2011). Indeed, righteous entrepreneurs will be guided in all matters, and the value of sincerity within them will be radiated (Isa Mohammed, Zulkarnain, & AAhad, 2011), so that they will be rewarded with a high position and rank close to Allah SWT, as contained in the *hadith* narrated by Abu Sa'id RA that the Prophet SAW said;

"The truthful, trustworthy merchant is with the Prophets, the truthful, and the martyrs." (At-Tirmidhi, 2007, p. 24, No. 1209)

After reviewing, researching, and analysing the interviews conducted with 10 selected successful Malay entrepreneurs of Kelantanese descent for this study, the identifiable practices and secrets of their success include instilling the love of being charitable and generous to anyone, whether they are customers or non-customers; perform *sunnah* (supererogatory) prayers, such as *tahajjud* (falls between the prayer times of the *isha'a* (night-time obligatory daily prayer) prayers and the *fajr* (dawn time obligatory daily prayer) prayers), *hajaat* (the prayer of need), and *dhuha* (morning prayer; begins after sunrise and ends at meridian); allocating time and taking the opportunity to supplicate in seeking for help and sustenance from Allah SWT; and readiness to extend help to anyone. Meanwhile, other factors which also contribute to their success in business include producing creative and innovative products in the market and strengthening their promotions and marketing strategies. The following are the examples of the interviews and discussions based on the source of Al-Qur'an, *hadith*, and other contemporary sources.

Generosity and friendliness

Respondent A, who has been running her own food store for nearly seven years in the city of Pasir Puteh, Kelantan, said;

"As businessmen and businesswomen, we have to be friendly, generous, and avoid being stingy. That is the key to our success in business, of which, at the same time, make us feel loved and are remembered by customers.

Generous and hospitable attitude, which are the opposite of parsimonious and arrogance, are good values in Islam and they bring blessings in business, as mentioned by Allah SWT in Surah al-Layl [92]:8-11;

Translation: But he who is a greedy miser and thinks himself self-sufficient, And gives the lie to the Best,- We will indeed make smooth for him the path to Misery; Nor will his wealth profit him when he falls headlong (into the Pit). ('Ali, 2005, p. 653)

Perform supererogatory prayers

According to Respondent B, the success in her business was due to her reliance on Allah SWT, and thus, she sought help only from Allah SWT. She advised other entrepreneurs to perform prayers for the love of Allah SWT, as a sign of their gratefulness, and not simply because they needed something. In other words, every form of worship should be performed sincerely to gain *mardhatillah* (the pleasure of Allah SWT), as mentioned by Allah SWT in Surah al-Bayyinah [98]:5;

Translation: And they have been commanded no more than this: to worship God, offering Him sincere devotion, being true (in faith); to establish regular prayer; and to practise regular charity; and that is the Religion Right and Straight. ('Ali, 2005, p. 662)

Moreover, entrepreneurs are encouraged to read the Qur'an to find inner peace, as well as getting the rewards promised by Allah SWT, as mentioned in a hadith narrated by Imam Abu Dawud Sulayman ibn al-Ash'ath as-Sijistani RA (Book 8, *Hadith* No. 41), which means;

'Uqbah bin 'Amir al-Juhani said: When we were in the Suffah, the Messenger of Allah SAW asked: "Which of you would like to go out every morning to Buthan or Al-'Aqiq and bring two large humped and fat she-camels without being guilty of sin and severing ties of relationship?" They (the people) said: "Messenger of Allah, we would all like that." He said: "If any one of you goes out in the morning to the mosque and learns two verses of the Book of Allah, the Exalted, it is better for him than two she-camels, and three verses are better for him than three she-camels, and so on than their numbers in camels."

One respondent expressed her opinion regarding the practice of reciting the Qur'an:

"Verses of the Qur'an should be frequently read because it is necessary to be *istiqamah* (consistent) and constant in every practice so that they are well instilled within us. Not to just reading the Qur'an, hoping for rewards from Allah or that Allah fulfils our dreams and desires. No such thing! In reading and implementing such practices, they have to be parallel with our intention to gain the pleasure of Allah, and not in hoping for worldly rewards. *InsyaAllah* (God willing), only then will Allah bestow what is best for us. Allah would not simply give us what we ask for."

The rewards from Allah SWT to His servants who perform religious duties and practices are not necessarily in terms of material gains, but they may also be in terms of inner strength, such as happiness and tranquillity. Thus, a Muslim should submit (*tawakkal*) to Allah SWT and trust that indeed, Allah's arrangement is the best, as stated by Allah SWT in Surah al-Baqarah [2]:216;

Translation: But it is possible that you dislike a thing which is good for you, and that you love a thing which is bad for you. But God knows, and you know not. ('Ali, 2005, p. 50)

Respondent C, who owns a mixed rice food store who runs three outlets, believed that by performing supererogatory prayers, the relationship between the slave and his/her Creator will be firm and steady, as mentioned in her statement;

"There is no specific practice, but the thing which I never forget is performing *tahajjud* and *dhuha* supererogatory prayers. I am obliged to perform them. Not only because I want an abundant of sustenance, but that is the way I draw closer to Allah, to connect with Him."

Familiarise with supplication in every affair

Among the *du'a* (supplication) practiced by some respondents are Surah al-Fatihah and al-Waqi'ah as an opening of store *du'a*, whereby, they are steadfast in reading these *surahs*, as excerpted from the statements of Respondent D and E;

"Once I reach the store, I start off by reading Surah al-Fatihah from the beginning until the end. Therefore, I try not to be late otherwise I will be in a hurry till it becomes slipshod and easy to forget, and that my mouth will be locked from reading the *du'a*."

Meanwhile, respondent E reads Surah al-Waqi'ah as a *du'a* before he opens his shop, as mentioned in his statement;

"Previously, there was an *Ustaz* (religious teacher) who taught me to practise reading up Surah al-Waqi'ah till the end before opening the store. It is believed that by reading Surah al-Waqi'ah as an opening of the store's *du'a*, the store will be shielded from treasonous sorcery of others and it is also able to popularise the business."

The following is the answer of Respondent E when asked about the impact of forsaking the recitation of Surah al-Waqi'ah:

"If we do not read it, our store will be easily inflicted with treasonous sorcery, causing customers not to see our store. Since we are running our business in an open space, where there are many others who are engaged in business, there is a possibility of others to be jealous of us. Then, they would go see a shaman to ask for something or seek for help to bring us down, for treason. If you do not have time to read it, you can just turn on the cassette."

In addition, there are also posters of *surahs* (any of the 114 chapters of the Al-Qur'an) found suspended and pasted on the walls of the stores of a number of respondents. According to a respondent, the purpose of the suspension of these *surahs* is not merely to popularise the business, since all sustenance and blessings come from Allah SWT. In other words, a business can only be successful with the help of Allah SWT. Thus, the practice of reading or even suspending *surahs* is a kind of effort and endeavour to seek the help and pleasure of Allah SWT. According to the respondent as well, the suspension of these *surahs* will be able to remind them of Allah SWT in times of negligence.

Produce creative and innovative products or services in the market

According to Shuhairimi (n.d.), innovation and creativity are necessary skills for every entrepreneur. Creative skill is the key factor which distinguishes one entrepreneur after another. In fact, Allah SWT has bestowed a variety of sources to complement human needs. However, on a general scale, the endowment will not be able to accommodate the needs of a man if the optimal results are not manipulated as maximal as possible [27]. Therefore, entrepreneurs will have to bring out the creative and innovative values from within themselves in order to meet the diverse needs and desires of the communities.

Luecke and Katz [25] defined innovation as the introduction of a substance or method in which it is an image, a combination or synthesis of knowledge that exists in the natural state, has an interest and value to a product or process or a new service. According to Muhammad Hisham [31], a company does not necessarily produce the element of innovation only in the context of the creation of goods or services, but innovation can also exist in the form of new business models, such as the concept of a business franchise. Moreover, innovation can also exist in the marketing strategy, as some of the companies which offer services after office hours or on weekends. Meanwhile, businessmen and businesswomen in the Siti Khadijah Market, Kelantan choose to keep running their business on weekends.

Creativity and innovation is described in a *hadith* of Prophet Muhammad SAW. It was narrated from Simak that he heard Musa bin Talhah bin 'Ubaidullah narrating that his father said;

"I passed by some palm trees with the Messenger of Allah SAW, and he saw some people pollinating the trees. He said: 'What are these people doing?' They said: 'They are taking something from the male part (of the plant) and putting it in the female part.' He said: 'I do not think that this will do any good.' News of that reached them, so they stopped doing it, and their yield declined. News of that reached the Prophet SAW and he said: 'That was only my thought. If it will do any good, then do it. I am only a human being like you, and what I think may be right or wrong. But When I tell you: "Allah (SWT) says," I will never tell lies about Allah (SWT).'" (Sunan Ibn Majah, Vol. 3, Book 16, *Hadith* No.2470)

Strengthen the promotion and marketing

The entrepreneurs should strive tenaciously and give emphasis to their marketing strategy. This is because a good product or service will not sell by itself without the effort to market, promote, and advertise it [39].

The respondents were of the view that the strategies and methods for introducing a product or service offered to customers play an important role in the smooth running of their business activities. The strategies and methods include strengthening their promotional activities, testimonials, and product marketing to gain consumers' trusts. For instance, Dato' Sri Dr. Vida expanded her product marketing by sponsoring the Kelantan football team and customising the team name from 'The Red Warriors' to 'Pamoga Qu Puteh The Red Warriors'. The decisions made by Dato' Sri Dr. Vida, as an entrepreneur, have shown her courage in marketing her products. This trait is in line with Nor Aini's (2003) statement that a smart entrepreneur is one who is able to seize economic opportunities, identify potential products, analyse potential markets and customers, and understand the desires and preferences of customers.

4 Conclusion

Based on the secrets of success in business, which have been shared by the respondents, it is apparent that the practices adopted were in accordance with the teachings of Islam and the *sunnah* of Prophet Muhammad SAW. The respondents' aim of adopting these practices is to popularise their business in an captivating way that is permitted in Islam, whilst drawing themselves closer to the Al-Khaliq (Creator). Thus, becoming an entrepreneur is a *fardhu kifayah* (obligatory acts which exist in the Islamic community regardless of who becomes one. However, Allah SWT does not require them to be accomplished by each person or some specific person, since the interests of the community can be realised through the efforts of a few who are morally responsible. Hence, it does not entail everyone to fulfill this obligation) and worship for sustenance. Meanwhile, in the process of popularising their business, there are surely certain techniques and practices which must be adopted by the entrepreneurs in order to attract the customers.

Acknowledgement

This study is funded by the short-term grant of Universiti Sains Malaysia (304/phumaniti/6313196), Penang.

References

1. Ab. Mumin Ab. Ghani & Suhaili Sarif, Penerapan budaya keusahawanan dalam masyarakat Islam. Paper presented at the Seminar Keusahawanan Islam Peringkat Kebangsaan, 6 September, Academy of Islamic Studies, Universiti Malaya, Kuala Lumpur, (2005)
2. Abuznaid, S.A., Business ethics in Islam: The glaring gap in practice. International Journal of Islamic and Middle Eastern Finance and Management, 2(4), 278-288, (2009)
3. Adas, E.B., The making of entrepreneurial Islam and the Islamic spirit of capitalism. Journal for Cultural Research, 10(2), 113-137(2006)
4. Ahmad, M., The influence of culture on Malay Muslim business practice: A case study of Malaysia with special reference to the state of Kelantan Dar al-Naim (Unpublished Ph.D thesis). University of Birmingham, UK, (2009)
5. Ahmad, M., Md. Hussain, M.N., Palil, R., & Dolah, N.H., The Islamic image of market place in Malaysia: A case study presentation. SEA Journal of Contemporary Business, Economic and Law, 2(1), 86-88, (2013)
6. Ahmad Zahiruddin Mohd Zabidi, Hebatnya Rasulullah: Teladani peribadi dan pengurusan kekasih Allah. Shah Alam, Selangor: Karya Bestari @ Grup Buku Karangkraf Sdn. Bhd, (2014)
7. Al-Bukhari, Summarized sahih Al-Bukhari: Arabic-English (Compiled: A-I.Z.A.A.L. Az-Zubaidi, Translated: M.M. Khan). Riyadh: Dar-us-Salam Publications, (1994)
8. Al-Buthy, M.S.R., Sirah Nabawiyah: Kajian mendalam manhaj gerakan Islam pada masa Rasulullah (s.a.w.) (Translated by: Aunur Rafiq Shaleh Tamhid). Shah Alam, Selangor: Pustaka Dini, (2005)
9. 'Ali, A.Y., The meaning of the holy Qur'an. Subang Jaya, Selangor: Secretariat for Asia Assembly of Ulama', (2005)
10. An-Nadwi, A.H.A., Riwayat hidup Rasulullah (Translated by: H. Bey Arifin & Yunus Ali Muhdhar). Surabaya: PT Bina Ilmu, (2006)
11. Ashmore, M.C., The power of the entrepreneurial vision. Vocational Education Journal, 64(8), 28-29, (1989)
12. At-Tirmidhi, A.I.M.I.I., English translation of jami' At-Tirmidhi (Vol. 1) (Compiled: H.A.E.M.I.E. At-Tirmidhi, Translated: A. Khaliyl). Riyadh: Darussalam, (2007)
13. Azmi Abdul Manaf, Nik Hairi Omar & Lee, K.Y, Faktor kritikal kejayaan usahawan dalam perniagaan. Journal of Social Sciences and Humanities, 7(1), 34-45, . (2012)
14. Hansard Parliament, Pemberitahuan pertanyaan Dewan Rakyat Malaysia, Soalan 1, Bertarikh 27/10/2009. Retrieved from www.parlimen.gov.my/files/hindex/pdf/DR-27102009.pdf (accessed 1 November 2011), (2010)
15. Hess, D. W., Relevance of small business courses to management needs. Journal of Small Business Management. 25(1): 26-34, (1987)
16. Hornby, A.S., Oxford Advanced Learner's Dictionary (8th ed.). Oxford: Oxford University Press, (2010)
17. Ibnu Khaldun, Muqaddimah (translated). Kuala Lumpur: Dewan Bahasa dan Pustaka, (1995)
18. Ilhaamie Abdul Ghani Azmi, The implementation of Islamic business practices of women traders: A study in an Islamic state in a developing country. GJAT, 4(2), 17-28, (2014)

19. Ilhaamie Abdul Ghani Azmi & Mohamed Dahlan Ibrahim, Wanita dan pelaksanaan amalan keusahawanan Islam: Satu kerangka konseptual. In the proceeding of International Conference on Islamic Management, 2013, Department of Shariah and Management, Academy of Islamic Studies, Universiti Malaya, Kuala Lumpur, (2009)
20. Isa Mohammed Adamu, Zulkarnain Kedah, & Aahad Osman-Gani, Spirituality in entrepreneurship from Islamic perspective. A conceptual analysis on the effects on entrepreneurial motivation and social responsibility. Proceeding of the 10th International Conference of the Academy of HRD (Asia Chapter), 3-6 December 2011, Kuala Lumpur , (2011)
21. Jabnoun, Naceur, Islam and management. Kuala Lumpur: Institut Kajian Dasar, (1994)
22. Kalsom, A. & Mohd. Ismail, Compliance to Islamic marketing practices among businesses in Malaysia, Journal of Islamic Marketing, 1(3), 286-297, (2010)
23. Kent, C.A., Integrating entrepreneurship in the secondary curriculum: Economics and other courses. In Kent, C.A. (Ed.), Entrepreneurship Education. Connecticut: Quorum Book, (1990)
24. Longenecker, J.G., Moore, C.W., Petty, J.W., & Palich, L.E., Small business management: An entrepreneurial emphasis (13th ed.). Mason, OH: Thompson South-Western, (2006)
25. Luecke, R., & Katz, R., Harvard business essentials: Managing creativity and innovation. Harvard: Harvard Business School Press, (2003)
26. Madura, J., Introduction to business (4th ed.). Mason, OH: Thompson South-Western,k (2007)
27. Maisarah Ahmad & Suhaila Abdul Kadir, Characteristics of entrepreneurs and the practice of Islamic Values in influencing the success of small medium enterprises in Kelantan and Selangor. Journal of Social and Development Sciences, 4(5), 229, (2013)
28. Mohd Fauzi Yaacob, Peniaga dan perniagaan Melayu: Suatu kajian di Kota Baharu, Kelantan. Kuala Lumpur: Dewan Bahasa dan Pustaka, Kementerian Pelajaran, Malaysia, (1981)
29. Mohd Nizam Sahad, Penggunaan azimat pelaris perniagaan menurut perspektif akidah Islam. Melayu: Jurnal Antarabangsa Dunia Melayu, 8(1), 18-44, (2015)
30. Mohd Zain Mubarak, Asyraf Ab. Rahman & Mohd Rafi Yaacob, Elemen kerohanian dalam keusahawanan Islam: Kajian terhadap usahawan berjaya di negeri Kelantan. Journal of Business and Social Development, 3(1), 43-53, (2015)
31. Muhammad Hisyam bin Mohamad, Membudayakan inovasi. Berita Harian, (2013)
32. Nor Aini Idris, Daya saing usahawan wanita Melayu menghadapi cabaran globalisasi. Paper presented at the Persidangan Kebangsaan Pusat Pengajian Sosial, Pembangunan & Persekitaran (Cabaran Pembangunan Dilema Persekitaran), Universiti Kebangsaan Malaysia, (2003)
33. Nor'Azzah Kamri, Nilai dan etika dalam pengurusan Islam (Values and ethics in the management of Islam) (unpublished MA of Shariah dissertation). Department of Shariah and Management, Academy of Islamic Studies, Universiti Malaya, Kuala Lumpur, (2002)
34. Nor'Azzah Kamri, Etika usahawan muslim berpandukan kepada al-Quran dan al-Sunnah, In Keusahawanan Islam: Konsep dan isu-isu perlaksanaan. Universiti Malaya: Jabatan Syariah dan Pengurusan, Akademi Pengajian Islam, (2009)
35. Norasmah Othman, Keberkesanan program keusahawanan remaja di sekolah menengah (Ph.D dissertation). Universiti Putra Malaysia, Serdang, (2002)

36. Norhashim Nordin, Towards strengthening SMIS. The Bank Pembangunan perspective. Paper presented at the 1st National Conference for SMIS, Kuala Lumpur, Malaysia, (1994)
37. Osella, F. & Osella, C., Muslim entrepreneurs in public life between India and the Gulf: Making good and going good. Journal of the Royal Anthropological Institute, 15(1), 202-221, (2009)
38. Pihie, L., Akmaliah, Z., & Elias, H., Keupayaan usahawan Bumiputera melaksanakan kemahiran keusahawanan: Satu kajian kes. Pertanika Journal of Social Sciences & Humanities, 12(1), 61-70, (2004)
39. Sazelin Arif, Keupayaan pemasaran dalam kalangan pengusaha kecil makanan halal di Melaka Tengah: Satu kajian awal. Malim-Jurnal Pengajian Umum, 10 (2009), 183-208, (2009)
40. Scarborough, N.M. & Zimmerer, T.W., Essentials of entrepreneurship and small business management. Upper Saddle River, New Jersey: Pearson/Prentice Hall, (2006)
41. Shuhairimi Abdullah (n.d.). Nilai Inovasi, Kreativiti Perlu Dalam Keusahawanan, Retrieved from http://mukmin.com.my/ ID 1163, Kategori 12.
42. Sieh, L.M.L., Malaysian industrial and entrepreneur profile. Malaysian Management Review, 25(2), 3-10, (1990)
43. Soaib Asimiran, S. & Rosnani Jusoh, R., Membangunkan kemahiran generik usahawan Malaysia (KeGUM): Satu kajian kes ke atas usahawan makanan. Proceeding of the Seminar Kemahiran Insaniah dan Kesejahteraan Sosial (SKIKS) 2008. Penerbit Universiti Teknikal Melaka. ISBN: 978-983-2948-33-9, (2008)
44. Solahuddin Abdul Hamid & Che Zarrina Sa'ari, Reconstructing entrepreneur's development based on al-Qur'an and al-Hadith. International Journal of Business and Social Science, 2(19), 110-116, (2011)
45. Timmons, J.A., Smollen, L.E., & Dingee, A.L.M., New Venture Creation (2nd ed.). Homewood, Illinois: Irwin, (1985)
46. Van Gelderen, M., Research based yet action oriented: Developing individual level enterprising competencies. Paper presented at the 17th Global Conference – Internationalizing Entrepreneurship Education and Training, 8-11 July, 2007, Faculty of Management and Economics, Gdansk University of Technology, Gdansk, Poland, (2007)
47. Wim Naude, Promoting entrepreneurship in developing countries: Policy challenges. World Institute for Development Economics Research of the United Nations University (UNUWIDER) Policy Brief, 42010, 1-8, (2010)
48. Zaidatol Akmaliah, Lope Pihie, & Habibah, Keusahawanan dan Motivasi Diri. Serdang: Penerbit Universiti Putra Malaysia, (1997)
49. Zen Abdul Rahman, Strategi pemasaran cara Rasulullah. Batu Caves, Selangor: Al-Hidayah House of Publishers Sdn. Bhd, (2014)

Promoting Counseling Competence using Silf-Reflection

Norliza Alis[a,*], *Wan Marzuki Wan* Jaafar[b] *and Ahmad Fauzi* Mohd Ayub[c]

[a]School of Applied Psychology, Social Work and Policy, University Utara Malaysia.
[b]Department of Counselor Education and Counselling Pyschology, Faculty of Educational Studies, University Putra Malaysia.
[c]Department of Foundations of Education, Faculty of Educational Studies, University Putra Malaysia.

Abstract. Self-Reflection (SR) has received a high degree of attention in extant Counseling Competence (CC). The present study sought to determine the relation of SR subscale; self awareness, and managing strategies with the three general areas in CC; performing helping skill, managing the counseling process, and handling challenging counseling situations. A total of 100 counselor trainees completed The Self-awareness and Management Strategies (SAMS) Scale to measured the SR, and the Counselor Activity Self-Efficacy Scales (CASES) to measured the CC. Result of the Pearson Product Moment Coefficient indicated there is a significant relationship between SR; Self-awareness with CC;handling challenging counselling, and SR;management strategies with all the three general areas in CC. A disscussion of the findings are presented and several implications and recommendations for further development of CC a proposed.

1 Introduction

Self-reflection in counselling practices is an action that is based on theories, beliefs and assumptions. All three of these components are drivers toward the understanding of a counsellor to his clients, in guiding them when choosing the most appropriate intervention for their clients [4].

The process of self-reflection is also seen as a continuation of the development of supervisory practices. Morrisette et. al [11] emphasized that reflectivity is fundamental in the process of self-supervision. In fact, she also stated that self-monitoring, self-assessment, self-evaluation, self-reporting and self-management are among the elements of reflection. Morrisette also postulated that reflection can improve self-consciousness or self-awareness, and the ability to apply a clinical approach in work from various perspectives, as well as reducing countertransferences issues.

Self-reflection is done by questioning the assumptions made and the expectations held, beliefs, and values that led to the elements of open-mindedness, thoughtfulness, and intellectual responsibility. Regardless of how counsellors choose to do self-reflection, the

* Corresponding author: wanmarzuki@upm.edu.

principles set out above are necessary to support the self-assessment process. Finally, reflection can be done naturally and spontaneously. However, for some counsellors, the presence of a supervisor who acts as mentors can assist them in improving the practice of reflection.

In general, self-reflection is a complex undertaking that includes (a) introspection, (b) self-awareness, and (c) self-assessment. Ultimately, understanding, skills, and the counsellor's function can be enhanced [11].

2 Research Objective

The focus of this study is to:
1. identify the stages of SR among counsellor trainee.
2. identify the relationship between SR and CC; performing helping skill, managing the counseling process, and handling challenging counseling situations Figures and tables, as originals of good quality and well contrasted, are to be in their final form, ready for reproduction, pasted in the appropriate place in the text. Try to ensure that the size of the text in your figures is approximately the same size as the main text (10 point). Try to ensure that lines are no thinner than 0.25 point.

3 Literature Review

3.1 Self-Reflection

The process of reflection is also seen as a continuation in the development of supervisory practices [12]. Morrisette et. al [11] emphasized that reflectivity is the cornerstone in the self supervision process. He also mentioned that self-monitoring, self-assessment, self-evaluation, self-reporting and self-management are the elements of reflection. In addition, Morrisette stated that reflection can enhance self-awareness and the ability to apply the clinical approach in one's work from various perspectives, as well as reducing countertransferences.

The practice of reflection can be a very useful process in the practice of counseling. Reflection can help the counselor trainee remember their counseling practices. The situation allows for the trainees to better understand themselves, are less likely to accentuate personal issues to their clients, and are more likely to treat the issues faced by their clients.

Reflection in applied disciplines is adopted based on the contribution from the results of previous studies, including Dewey (1933), Shcön (1987) and Kolb [7] These three researchers stressed on the reflective thinking aspect as a means to resolve issues arising in the workplace. For example, diagnosis and clinical strategies are crucial in nursing courses whereby both actually derive from reflective thinking [5]. Reflective thinking involves internal assessment and exploration which will provide awareness to individuals who combines changes that eventually trigger the acceptance of a concept from a new perspective (Boyd & Fales, 1983).

3.2 Counselling Competence

Urbani et al. [15] focused their study on 18 counselling skills which are divided into three stages of counselling process, namely; exploring, understanding, and acting. Each level consists of six counselling skills. Six skills in the exploring stage are, eye contact, body language, tone of voice, questioning, paraphrasing, and summarizing. The understanding stage consists of skills to identify content and feelings, self-disclosure, concrete and specific expressions, urgency (of the situation, actions, and feelings), and confrontation. Meanwhile, the acting stage consists of decision-making skills, making selections, creating expectations, gaining approval, making an appointment and reviewing all goals and actions to determine the outcome. These elements encourage the efficacy of counselling and growth of cognitive complexity [15].

In fact, counselling skills include all elements of the skills that can help mobilize and facilitate clients when meeting counsellors, ensuring a comfortable atmosphere. The basic counselling components of warmth, sincerity, empathy, unconditional acceptance and congruence are the basic skills that are important and must be mastered by a counsellor trainer [13]. Therefore, supervisors should emphasize and focus on the supervisory process.

Basic counselling skills such as open and closed questions, paraphrasing, reflection of contents, reflecting feelings, asking for clarification, confrontation, identification and information are the second most important component that should be attained by a counsellor [13]. These components also incorporate all other counselling approach which differentiates counselling session interactions with normal conversations. Therefore, acquiring control and skills over the two components is essential towards being a good counsellor.

Skills development in therapeutic efficiency is affected not only by educational experience, but also by the students' view on their counselling ability to perform counselling sessions effectively (Larson and Daniels, 1998; Bandura, 1991; Kirsch, 1986). A strong belief in the ability to perform a wide range of required skills, allow trainees to blend their skills in a counselling environment [14,15] with monitoring by supervisors

3.3 Experiential Learning Theory

Experiential Learning Theory explains how people learn and acquire knowledge [7]. This theory views knowledge as a result of the transformational process of experience. The combination of understanding and transforming experiences will provide new inputs and is influenced by various factors such as the type of personality, educational background and career endeavor. Kolb Learning Experience theory is an experience-based learning approach that has gained the attention of various fields of study, especially those in need of practical training.

Experiential learning consists of two dimensions, namely; (I) the understanding and acceptance of experience, and (ii) the transformation of experience. The dimension of understanding and acceptance is formed by two processes, namely through "Concrete Experience" and "Conceptual Abstract". Meanwhile, the dimensions of transformation are "Reflective Observation" and "Active Experimentation".

This theory can be explained in the form of a process cycle. This cyclical process starts from concrete experience (what is happening?), followed by observation of reflection (think back, what happened?). Then, this process will be followed by a conceptual abstract, which is a reflection on what sparked the idea. Active experimentation process refers to the actions taken. This reaction stems from the idea that is driven by conceptual abstract [7].

According to Anderson et al. [1] the underpinning of Kolb's Learning Experience theory consist of three basic principles, namely;

i. Learning involves the whole individual person; sensory, cognitive and emotion.
ii. The learning of new experience is affiliated with the individual's past experience.

 https://doi.org/10.1051/matecconf/201815005072

iii. Regular practice of self-reflection allows the experience to be understood in depth.

Kolb et. al [7] stated that the quality of reflection will bring people to learn the important learning outcomes, rather than the experience itself. This indicates that experience without reflection will not bring favourable results. Internship training provides an opportunity for counsellors to learn through practical experience in a real working environment. Learning can only happen through experience, based on its continuity and interactions.

4 Methodology

This study is a correlational study. The focus of this study is to identify the relationship of self-reflection with counseling competance among counselor trainees.

4.1 Measurement

The study used a set of questionnaire consisting of three sections: background of respondent, self-reflection, and competence of counsellor trainee. Two questionnaires were used to obtain the required information.

The instrument SAMS Scale was used to measure self-reflection. The SAMS scale designed by Williams, Hurley, O'Brien, and DeGregoria et. al [16]. The SAMS scale consists of 25-items divided into two dimensions: self-awareness and management strategies. Items within each dimension are rated on a likert scale of 1-5.

Meanwhile, the instrument CASES was used to measure competence of counsellor trainee. The CASES designed by Lent, Hill, and Hoffman [9]. It consists of 41-items divided into three dimensions: performing helping skills, managing the counselling process, and handling challenging counselling situations. Items within each dimension are rated on a likert scale of 0-9.

4.2 Sampling and Recruitment

The sample consists of 100 counselor trainee from four universities who are undergoing internship training. Stratified random sampling was salected to ensure a balance of sample was taken from the population.

First of all elements in the population isolated in groups according to the university. The samples will then be selected at random in each group were segregated. The study population consisted of counselor trainees from four Universities in Malaysia. Samples were selected based on the percentage of revenue derived from the ratio of counselors to the number of trainees in universities. The advantage of using stratified sampling method is more accurate findings obtained in which, through isolation strata of the population to provide an opportunity for each element to be selected.

4.3 Analysis

Data were analyzed using descriptive analysis involving frequencies, percentages, means and standard deviations. At the same time, inferential analyses were tested using Pearson's Product-Moment Correlation Coefficient.

5 Result and Discussion

The following are the findings of this study:

Table1: Distribution of the Respondents according to Demography.

Demography	Frequencies	Percentages
Gender		
Male	22	22%
Female	78	78%
Race		
Melay	88	88%
Chinese	9	9%
Indian	3	3%
Internship Setting		
School	44	44%
University	44	44%
Government organization	8	8%
Non-Government organization	4	4%

Table 1 shows the distribution of the respondents according to demography. Distribution of sample based on gender shows that male respondents are 22 person (22%) and female are78 person (78%). The distribution of the respondents according to race shows that Malay respondents are 88 person (88%), Chinese are 9 person (9%), and Indian are 3 person (3%), meanwhile the distribution of the respondents according to internship setting shows that respondents who are doing in school are 44 person (44%), university are 44 person (44%), government organization are 8 person (8%), and non-government organization Indian are 4 person (4%).

Stages of Self-Reflection and Counselling Competence

The mean for self-reflection is 81.45 (SD = 7.99). Frequency distribution according to level of self-reflection shows that the majority of respondents were at a moderate level, totalling 92 respondents (92%). Only eight respondents (8%) reached a high level of self-reflection and one reached a low level. Table 2 details out the result of the study.

Table 2: Frequency and percentage of self-reflection level

Self-Reflection		Frequency	Percentage
High	(93 to 125)	8	8%
Moderate	(59 to 92)	92	92%
Low	(25 to 58)	1	1%

Mean=81.45 Std. deviation = 7.99

Meanwhile, the mean for ability to apply counselling skills is 263.23 (SD = 40.93). Frequency distribution according to their capacity to apply counselling skills showed that the majority of respondents are at a high level, totalling 65 respondents (65%), while 35 respondents (35%) were at a moderate level. No one scored the lowest level.

Table 3: Frequency and percentage of the ability to apply counselling skills

Counseling Competence		Frequency	Percentage
High	(247 to 369)	65	65%
Moderate	(124 to 246)	35	35%
Low	(0 to 123)	0	0%
Mean = 263.23		std. deviation = 40.93	

Correlation between Of Self-Reflection and Counselling Competence

As indicates in table 4 there is a small positive correlation between self reflection with performing helping skills ($r_{(99)}$ = .255, p=.010) and handling challenging counselling situations ($r_{(99)}$ = .270, p=.007). Meanwhile no significant correlation between selfreflection with managing the counselling process ($r_{(99)}$= .158, p=.116).

Table 4. Correlations Coefficients between self- and Counselling Competence

Variables	**Self reflection**			
	r	**Sig.**	**M**	**Std. Deviation**
Performing helping Skills	.255	.010	100.39	15.535
Managing the counselling process	.158	.116	70.95	10.897
Handling challenging counselling situations	.270	.007	91.89	20.393
	Mean=81.45		Std. Deviation=7.998	

The study had comprehensively discussed and analyzed the self-reflection variable based on the framework of Experience Learning Theory. Self-reflection factor is measured based on the perspectives of counsellor trainees in the context of supervisory internship counselling.

Based on Kolb Learning Experience theory, this reflection process starts from concrete experience (what is happening?), followed by observation of reflection (think back, what happened?). Then, this process will be followed by a conceptual abstract, which is reflection on what sparked the idea. On the other hand, active experimentation process refers to the actions taken. This reaction stems from the idea that is driven by conceptual abstract [7].

Magnuson and Norem [10] found that self-reflection is able to help counsellors' development, and can even enhance the ability of counsellors to improve their efficiency as a counsellor. Counsellors who constantly practice self-reflection on their experience are able to think critically about their clients as well as their interaction sessions. Continuous reflection process allows the efficiency of these counsellors to grow.

The current study's findings support previous studies like Larson and Daniels (1998), whereby they reported self-reflection can improve the self-efficacy of counselling. Bolin [3] also found that the process of self-reflection promotes change and improve the practice of an occupation. The practice of self-reflection should be managed on an ongoing basis which involves the process of planning, taking action, deliberating and making reflections.

The Kolb Learning Experience theory also emphasizes on a regular practice of the reflection process. The practice of self-reflection allows the experience to be understood more in depth. Self-reflection also involves a learning process by involving the entire sensory, cognitive and emotion [1].

6 Conclusion

Research findings can provide a general overview counselling competence among trainee counselor in Malaysia. Accurate counselling competence may be challenging to obtain. Yet, such precision in measuring competence may be necessary for a supervisor to maintain and enhance his or her supervisee skills as well as gaining insight into the more global awareness boundaries of competence. The hope of this current research is to build on prior findings and add a small but important increase in our understanding, by seeing if counselors-in-training can obtain accurate self-assessments of their skills and examining why this might be.

References

1. Anderson, S. A., Schlossberg, M., & Rigazio-DiGilio, S., Family therapist trainees' evaluations of their best and worst supervision experiences. Journal of Marital and Family Therapy, 26, 79–92, (2000)
2. Bandura, A., Social cognitive theory in self-regulation. Organizational Behavior and Human Decision Processes, 50 (2), 248-287.Boyd & Fales, 1983), (1991)
3. Bolin, F. S., Helping student teachers think about teaching. Journal of Teacher Education, 39(2), 48-54, (1988)
4. Griffith, B., & Frieden, G., Facilitating reflective thinking in counselor education. Counselor Education and Supervision, 40, 82-93, (2000)
5. Kessler, P., & Lund, C., Reflective journaling: Developing an online journal for distance education. Nurse Educator, 29(1), 20-24, (2004)
6. Kirsch, I., Early research on self-efficacy: What we already know without knowing we knew. Journal of Social and Clinical Psychology, 4, 339-358, (1986)
7. Kolb, D. A., Boyatzis, R., & Mainemelis, C., Experiential learning theory: Previous research and new directions. Dalam R. Sternberg and L. Zhang (Eds.) Perspectives on cognitive learning, and thinking styles. Mahwah, NJ: Lawrence Erlbaum Associates, (2001)
8. Larson, L. M., & Daniels, J. A., Review of the counselling self-efficacy literature. The Counseling Psychologist, 26(2), 179-218, (1998)
9. Lent, R. W., Hill, C. E., & Hoffman, M. A., Development and validation of the Counsellor Activity Self-Efficacy Scale. Journal of Counseling Psychology, 50(1), 97-108, (2003)
10. Magnuson, S., & Norem, K., Reflective counselor education and supervision: An epistemological declaration. Reflective Practice, 3, 167-173, (2002)

https://doi.org/10.1051/matecconf/201815005072

11. Morrisette, P., Reflectivity: The essence of self-supervision, dalam Self-supervision: A primer for counselors and helping professionals (pp. 77-108). New York: Brunner-Routledge, (2001)
12. Norzaliza Alis, Wan Marzuki Wan Jaafar & Ahmad Fauzi Mohd Ayub., The Influence of Self-Reflection Towards Counselor Trainee Self-Development. Middle-East Journal, 19, 85-88, (2014)
13. Othman Mohamed., Penulisan tesis dalam bidang sains sosial dan terapan. Serdang, Selangor: Penerbit Universiti Putra Malaysia, (2001)
14. Scott, J., Practicum experience. Dalam Urbani, S., Smith, R.M., Maddux, C.D., Smaby, M.H., Torres-Rivera, E., & Crews, J. (2002). Skills-based training and counseling self-efficacy. Counselor Education and Supervision , 42(2) , 92-106, (1995)
15. Urbani, S., Smith, R.M., Maddux, C.D., Smaby, M.H., Torres-Rivera, E., & Crews, J., Skills-based training and counseling self-efficacy. Counselor Education and Supervision, 42(2), 92-106, (2002)
16. Williams, E. N., Hurley, K., O'Brien, K. N., & DeGregoria, A., Development and validation of the Self-Awareness and Management Strategies (SAMS) Scale for therapist. Psychotherapy Theory, research, Practice, Training, 40 (4), 278-288, (2003)

Supervision Outcomes as Predictor to The Supervisory Relationship and Supervision Contextual Factors: Study on The Internship Trainee Counsellors

Nor Mazlina Ghazali[a], *Wan Marzuki Wan* Jaafar[b] *and Azzahrah* Anuar[a]

[a]*Department of Counselling, Faculty of Cognitive Sciences & Human Development, UniversitiMalaysia Sarawak*
[b]*Department of Counselor Education and Counselling Pyschology, Faculty of Educational Studies, Universiti Putra Malaysia*

Abstract. The purpose of this study was to investigate the influence of the supervisory relationship and contextual supervision factors on the supervision outcomes among trainee counsellors. Respondents were 120 trainee counsellors and 18 supervisors from four public universities in Malaysia. Eight instruments were used in measuring the variables. The Supervisory Working Alliance Trainee Inventory (SWAI-T) was administered to measure the supervisory working alliance among trainee counsellors and the Role Conflict Role Ambiguity Inventory (RCRAI) was administered to measure the role conflict among trainee counsellors. Meanwhile, the Supervision Interaction Questionnaire – Supervisee and Supervisor Inventory (SIQ-S) was used to measure the interaction between trainee counsellors and supervisor and the Counsellor Rating Form – Short (CRF-S) was used to measure the characteristics of the supervisors in supervision. The Selective Theory Sorter (STS) inventory was used to measure the counselling orientations among the trainee counsellors and supervisors whereas the Multicultural Counselling Knowledge and Awareness Scale (MCKAS) measures the knowledge and awareness toward multicultural counselling among trainee counsellors. The Supervision Outcomes Survey (SOS) and the Counsellor Performance Inventory (CPI) were utilized to measure the satisfaction and performance among trainee counsellors. Results have revealed that there was a significant correlation between the supervisory relationship (supervisees' working alliance, supervisees' role conflict, supervision interaction, supervisors' attributes) and supervision outcomes, $r(118) = .53$; $p < .05$. Other factors that have contributed to the significant correlations of supervision outcomes were supervisees' working alliance, supervisees' role conflict, and supervisors' attributes, $r(120) = .55$; $p < .05$; $r(120) = .21$; $p < .05$; and $r(116) = .50$; $p < .05$ respectively. However, the result has shown that there was no significant correlation between the supervision contextual factors (supervisees' and supervisors' counselling orientation and supervisees' cultural knowledge and awareness) and supervision outcomes. The Multiple Regression analyses reported that the supervisory relationship had an influence on the supervision outcomes, $R^2 = .28$, F

[1] Corresponding author: wanmarzuki@upm.edu.

https://doi.org/10.1051/matecconf/201815005073

(1,105) = 40.2, $p < .05$. Meanwhile, the supervision contextual factors had no influence on the supervision outcomes. Based on the research findings, the model signified that the supervision process could bring out changes in the supervisees. Practically, the supervisees' working alliance was a significant factor that has influenced the supervisees' development. Therefore, the academic supervisor should consider the supervisees' role conflict, supervision interaction, and supervisors' attributes during supervision. It is recommended that the differences between supervision interaction of the supervisors and the supervisees are to be examined in the future research.

1 Introduction

Counselling supervision is one of the fundamental courses developed to monitor the counselling practices and to ensure that the existing counselling practices meet the standard professional requirement. Much attention has been given by the Board of Counsellors (Malaysia) to the curriculum of counselling education programme particularly the supervision process. It is imperative to provide a practical guideline in the supervision process to facilitate the supervisor and trainee counsellors' personal and professional development, promote their counselling competencies, and preserve the accountability of their practices. Therefore, the counselling educators have placed great emphasis on the supervision component in the counselling training programme especially in the practicum and internship courses.

Through supervision, trainee counsellors could improve self - development, enhance skills, implement therapeutic relationships, and protect the clients [23]. Supervisory relationships can be a positive form of working alliance between the trainee counsellors and supervisors because the supervision sessions have correlations with the satisfaction, performance, and supervisors' multicultural competence [16].

Supervision can be defined as a dynamic relationship in which a supervisor evaluates the trainee counsellors' performance or professional growth and competency in the clinical process [18]. The purpose of supervision focuses more on the therapeutic growth rather than the clients [5]. Evidently, the role of the supervisor in supervision has an influence on the supervisory relationship.

Supervisory relationship and supervision contextual factors determine the supervision outcomes. There is no distinct research that explains variables that emphasize on the supervision process or outcomes such as the pre and post supervision satisfaction [17]. Conceptually, the supervision outcomes are beneficial for researchers in managing and conceptualising the supervision experience holistically.

2 Supervisory Relationship

The supervisory relationship involves the relationship between the supervisor and trainee counsellors. The supervisory relationship comprises of the component of power and authority, the role of a trainee counsellor's and supervisor's values, issues of trust between trainees and their supervisors, and conflicts between supervisor and supervisee (Haynes et al., 2002). Supervisors may train their supervisees to effectively deal with a range of challenges, such as coping with doubts and fears, recognizing personal needs, recognizing countertransference, and understanding diverse value systems of clients. The foundation of supervision is the relationship between supervisor and trainee counsellors. The supervision process is an important indicator of a good professional relationship. Supervision is an educative process whereby the supervisee learns certain specific knowledge and skills.

The supervisory relationship component includes the supervisee's working alliance, role conflict, supervision interaction, and supervisor's attributes. The working alliance is correlated with the therapeutic relationship between the counsellor and the client [3].The supervisory working alliance is a collaborative process achieved based on the mutual agreement concerning the goals and tasks of supervision, as well as the development of a strong emotional bond [3]. Research has found that a strong supervisory working alliance signified a high supervisory satisfaction. For example, if the supervisory working alliance were positive, it is highly likely that the trainee counsellor would be more satisfied with the supervision process. Bordin et. al [3] emphasised on the importance of working alliance for a therapeutic supervisory relationship. The supervisory working alliance is a proposed primary concept in the supervisory relationship which result in the effectiveness of supervision [22].

The study also examined the influence of supervisees' role conflicts on the supervision outcomes. Supervisees at the advanced stage have tendency to involve with disagreement with their supervisors. They might be having conflicts with the supervisor due to the inconsistency evaluation and feedback from the supervisor. Role conflict is related with the negative self-statement and poor performance. Previous research has indicated that a high score in role conflict might be correlated with low satisfaction and poor performance among trainee counsellors. Role conflict is related to the negative self-statements which can lead to anxiety and low performance in supervision [9].

Another related variable in a supervisory relationship is supervision interaction. The supervision interaction revolves around control and conflict which demands power struggle and involvement between supervisor and trainee counsellor. The interaction can be assessed from the supervisor and trainee counsellor's perspectives. Power creates impact on the supervision outcomes such as satisfaction and performance. Green and Dekkers [10] also indicated that power has a direct or indirect influence on the outcomes of supervision. The ineffective management of the components in the supervisory relationship between supervisor and trainee counsellors might lead to negative outcomes. Previous literatures have reported about the link between the use of power in formal or informal supervision and the supervision outcomes. According to Quarto et. al [20] the supervisory interaction can be distinguished by the level of response from each person on behaviours indicative of power or submission and friendliness or hostility in the dyad relationship. Even though the supervisor has the formal power during the supervision process, it still requires the interaction between supervisor and trainee counsellors.

The characteristics of supervisors or supervisors' attributes consist of a sense of humour, empathy, respect, genuineness, personal warmth, supportive attitudes, ability to confront, immediacy, concern for supervisee's growth, and well-being of clients' welfare (Resources from power point slide issues of supervision). In addition, according to Ladany and Bradley, [17], the supervisors' attributes are described as the commitment shown by the supervisor during counselling and supervision. The supervisors' attributes include the alertness of the supervisor on the trainee counsellors' growth or progress as well as their well being [17] The criteria for choosing supervisor includes the level of competency and achievement in the helping activities [17].

3 Supervision Contextual Factors

Supervision contextual factors are categorized as the supervisor's and supervisees' counselling orientations and supervisees' cultural knowledge and awareness. The following are the descriptions of these constructs. Previous literatures have reported that there was a positive supervision outcome when the trainee counsellor has been equipped with the knowledge and awareness of culture of the clients. According to Buchanan (2006), supervisors' lack of multicultural competence has been quite alarming resulting in the lack of multicultural competence among trainee counsellors. The cultural competence awareness is salient because the supervisees typically have direct contacts with the clients. Moreover, the clinical supervision is crucial for the supervisees' professional growth and development in multicultural counselling skills [4]. An effective supervision generates a transformation in the trainee counsellor's in terms of gaining a holistic understanding and capability in working with the clients from various cultural background (Buchanan, 2006). The theoretical orientation could keep the supervisor and supervisees on the right track while managing the supervision and counselling process.

Counselling orientation is referred to as the supervisees' preferred theories or approaches. The supervisees are likely to perform when applying their own preferred theories [13]. A study on the clients' perception of the appropriateness of the orientation of the counsellor and the client has generated a successful outcome of the therapy and became a unique contributor to the supervision outcome [12]. Hence, using similar principle, this study has investigated the relationship between the supervisor's and supervisees' counselling orientation. Contextual variables have to be considered when determining the satisfaction of trainee counsellors during supervision process including the counselling orientation of both parties, supervisors and trainee counsellors. The interaction of counselling orientation also affects the perceptions of supervisors [19].

The supervisees' cultural knowledge and awareness have typically focused on gender, ethnicity, race, sexual orientation, religious beliefs, and personal values. Cultural diversity in the supervision has been linked with the supervisees' satisfaction and supervision's outcomes in the previous research (Gatmon et al., 2001; Inman [16]; Murphy & Wright, 2005 as cited in Green & Dekkers, 2010). When the supervisors are competent in valuing the cultural diversity of the supervisees, it increases the supervisees' satisfaction.

4 Supervision Outcomes

Supervision outcomes are described as the composition of supervisor external events, supervisor characteristics, supervisee characteristics of supervision, and supervisee external events during the supervision process (Bradley & Ladany, 2001). The Supervision Outcomes Survey (SOS) and Counsellor Performance Inventory (CPI) were used to gather the information on the supervision outcomes (supervisees' satisfaction and performance). The SOS measures the supervisee's satisfaction [24] whereas the CPI (Iannelli, 2000) measures the supervisees' performance Supervisee's satisfaction and performance are defined as the trainee counsellors' ideas on the common quality of supervision experiences related to the individual motivational and practical needs (Ladany et al., as cited in Madani et al., 2010). Setting a high standard in the evaluation process, using indirect forms of communication, and emphasizing shortcoming were among reasons that have contributed to the poor supervision experienced by the family therapy trainee counsellor's (Anderson et al., 2000). On the other hand, the presence of a helpful supervisor in the supervision has been perceived as a good supervision experience because it is a supervisory relationship that promotes sensitivity in accommodating the needs of the counsellors-in-training [14]. Based on the findings of this study, it can be inferred that the supervisees' satisfaction and performance were related with the trainee counsellors' perception of the general quality of the supervision process of which contributes to motivation and the importance of the facilitative relationship in the professional development of trainee counsellors.

Performance is an important component in determining the clients' feelings, counsellors' interests, listening skills, and as well as counsellors' competency in counselling technique because these characteristics influence outcomes of the supervision process [11]. Performance might generation useful information such as understanding personal satisfaction of supervisees in the supervision process (Ilgen, et. al, as cited in Fried & Tiegs, 1995).

5 Trainee Counsellors or Supervisees

A trainee counsellor is an individual seeking licensure as a professional counsellor. Trainee counsellors are those who are at the final stage of the study programme who spend considerable time at the designated site(s) (Robert, 2001). The trainee counsellors were the selected undergraduate students who were in their final year of study in the counselling programme and were undergoing an internship either at the agency, school, or hospital setting.

6 Methodology

This research utilizes a quantitative design to measure the supervisory relationship and supervision contextual factors on the supervision outcomes. The instruments that have been administered include the Supervisory Working Alliance Inventory - Trainee version (SWAI-T) with 19 items, Role Conflict Role Ambiguity Inventory (RCRAI) with 12 items, Counsellor Rating Form Short (CRF-S), and Supervision Inventory Questionnaire (SIQ).

The CRF-S is a 12-item instrument consist of three components which include attractiveness, expertness, and trustworthiness whereas the SIQ - Supervisee and SIQ – Supervisor is a 20-item instrument that assesses the supervisees and supervisors' interaction and its influence on the supervision outcomes. Meanwhile, the supervision contextual factors are measured using the Selective Theory Sorter (STS), a 60-item instrument.

Other instruments that have been administered in the study were as follows: The Multicultural Counselling Knowledge and Awareness Scale (MCKAS) with 32 items, the Supervision Outcomes Survey (SOS), and the Counsellor Performance Inventory (CPI).

Section A: Demographic Information

In this section, respondents were required to respond to the demographic information including gender, races, level of education, year of study, and setting of internship or practicum.

Section B: Supervisory Relationship measures

To measure the supervisory relationship, the following instruments were administered: (i) the Supervisory Working Alliance Inventory - Supervisee version (SWAI-T), (ii) the Role Conflict Role Ambiguity Inventory (RCRAI), (iii) the SIQ – supervisor and SIQ – supervisees inventory, and (iv) the Counsellor Rating Form – Short (CRF-S).

Pilot Study
The purpose of a pilot test is to design a new research field in determining the variables that are measureable with different tests and conditions (Dell et al., 2000). There are no specific number of respondents in the pilot test, however, they rely on the researcher's experience or fieldwork, because there is no prior data on the particular research. Data from the pilot tests are seldom used to determine the sufficiency of variables of the research.

Validity and Reliability
The content validation has been conducted by the experts specializing in counselling with a vast experience in the counselling practice. The evaluation and rating by the experts were useful to ensure the accuracy of the content of each instrument. 25 counselling interns and 25 supervisors at a higher education institution in Malaysia were recruited as sample of the study. Seven male and 18 female counselling interns as well as 19 male and six female supervisors participated in the pilot study. They were identified based on the list gathered from the internship coordinator. During the pilot test, they were in the midst of the internship process. The respondents answered all the questions within one month. The Cronbach alpha from $\alpha = 0.61$ to $\alpha = 0.97$ were reported in the study.

6.1 Participants

Respondents of the research were 120 counselling who undergo the internship at organisations. Respondents were required to rate the scale on each measures involved. The sample consisted of 19.2 percent males (N = 23) and 80.8 percent females (N = 97). Majority of the respondents were females (Table 1). Ethnically, there were 75.8 percent Malays (N =91), 8.3 percent Chinese (N =10), followed by 2.5 percent (N = 3) Indians and 13.3 percent were from other ethnic groups (i.e: Dusun, Bajau, Iban, Bidayuh, Melanau). Malay respondents formed the majority (Table 1). As for the supervisors' ethnic background, Malays made up 66.7 percent of the sample (N = 12). 16.6 percent (N = 3) were Chinese, and 16.6 percent (N = 3) were from other ethnic groups (i.e; Dusun, Murut, Iban, and Bidayuh). The highest degree level attained by the supervisors were the M.A/M.Ed/M.Sc (55.6%) and 44.4 percent PhD/Ed.D.

Table 1: Descriptive Information on the Demography of the Trainee Counsellors and Supervisors

Descriptive information	Frequency	Percent
Gender of Trainee Counsellors		
Male	23	19.2
Female	97	80.8
Total	120	100.0
Etnnicity of Trainee Counsellors		
Malay	91	75.8
Chinese	10	8.3
Indian	3	2.5
Others	16	13.3
Total	120	100.0

Supervisors' Ethnicity		
Malay	12	66.7
Chinese	3	16.6
Others	3	16.6
Total	18	100.0
Highest degree level attained by supervisors		
M.A/M.Ed/MSc.	10	55.6
PhD/Ed.D	8	44.4
Total	18	100.0

6.2 Data Collection and analysis

Data of the research were analysed by the IBM Statistical Programme Social Sciences (SPSS) Version 23.0. Descriptive and inferential analyses were used in the data analysis. The descriptive analyses were used to analyse the demographic factors. The inferential analyses were used to analyze the supervisory relationship (supervisees' working alliance, supervisees' role conflict, supervision interaction, and supervisors' attributes) and supervision contextual factors (supervisors' and supervisees' counselling orientation and supervisees' cultural knowledge and awareness) and how these variables might relate and influence the supervision outcomes (supervisees' satisfaction and performance). Correlation coefficient and multiple regression were appropriate for measuring various types of issues in the behavioural research (Cohen & Cohen, 1983, Pedhazur, 1982; Mc Neil et al., 1975 and Ward & Jennings, 1973). Therefore, the Pearson product moment correlation coefficient and Multiple Regression analysis tests were chosen to analyse the data.

7 Finding

In this research, each independent variable is described as the supervisory relationship. Table 2 reports the descriptive statistics of the supervisory relationship. Results on the independent variables are as follows: supervisees' working alliance (M = 99.87, SD = 21.14); rapport (M = 67.48, SD = 14.91) which was higher than the mean score of client focus (M = 32.39, SD = 6.73); supervisees' role conflict (M = 49.00, SD = 8.60;); supervision interaction (M = 137.90, SD = 18.38); supervisees' control (M = 28.90, SD = 9.67) had a higher score than the supervision conflict (M = 14.01, SD = 5.38), and the supervisors' control (M = 24.55, SD = 5.20). Results on the supervisors' perceptions revealed the supervisors' control (M = 37.42, SD = 8.21) had a higher score than the supervision conflict (M = 34.01, SD = 9.11); supervisors' attributes (M = 71.58, SD = 10.89); trustworthiness (M = 24.11 and SD = 3.88) showed a higher score then expertness (M = 24.03, SD = 3.48) and attractiveness (M = 23.08, SD = 5.18) and supervision contextual factors (supervisors' and supervisees' counselling orientation (M = 655.31, SD = 47.51) and supervisees' cultural knowledge and awareness (M = 140.58, SD = 24.11); supervisees' cultural knowledge was (M = 87.34, SD = 15.45), and supervisees' awareness was (M = 43.87, SD = 8.12).

Table 2: Description on means and standard deviations of independent variables

Scales		n	Mean (*M*)	Standard Deviation (*SD*)
Supervisory Relationship				
SWA		120	99.87	21.14
	Rapport	120	67.48	14.91
	Client Focus	120	32.39	6.73
SRC		119	49.00	8.60
SI		116	137.90	18.38
	Supervisee Control	119	28.90	9.67
	Supervision Conflict – Supervisee	117	14.01	5.38
	Supervisor Control – Supervisee	119	24.55	5.20
	Supervisor Control – Supervisor	120	37.42	8.21
	Supervision Conflict - Supervisor	120	34.01	9.11
SA		118	71.58	10.89
	Attractiveness	120	23.08	5.18
	Expertness	120	24.03	3.48
	Trustworthiness	118	24.11	3.88
Supervision Contextual Factors				
SSCO		94	655.31	47.51
SCKA		116	140.58	24.11
	Knowledge	119	87.34	15.45
	Awareness	118	43.87	8.12

*SWA – supervisory working alliance; SRC – supervisees' role conflict; SI – supervision interaction; SA – supervisors' attributes; SSCO – supervisor and supervisees' counselling orientation and SCKA – supervisees' cultural knowledge and awareness.

The correlations between the supervisory relationship (supervisees' working alliance, supervisees' role conflict, supervision interaction, and supervisors' attributes) and supervision outcomes were analysed using the Pearson product moment correlation coefficient. The sample respondents were trainee counsellors who were undergoing the internship programme at the organisations. Based on Cohen (1992) formula, the minimum sample to conduct the analysis in this study was 85. The output showed that the supervisory relationship (supervisees' working alliance, supervisees' role conflict, supervision interaction, and supervisors' attributes) had a coefficient $r = .53$, which was significant at $p < .05$ (Table 3). As a conclusion, the supervisory relationship was related to the supervision outcomes. Based on the result, the null hypothesis was rejected.

Table 3: Pearson product moment correlation coefficient of supervisory relationship and supervision outcomes.

Scales	r	Sig.
SR	.53	.00
SO	1.00	.00

* *SR = Supervisory Relationship; SO = Supervision Outcomes*

Based upon the Pearson product moment correlation coefficient test results, the supervision contextual factors (supervisors' and supervisees' counselling orientation and supervisees' cultural knowledge and awareness) were not significantly correlated with the supervision outcomes, $r = .18$ at $p > .06$. As a conclusion, the independent variable was not related to the dependent variable. Therefore, the null hypothesis was accepted (Table 4).

Table 4: Pearson product moment correlation coefficient of supervision contextual factors (supervisors' and supervisees' counselling orientation and supervisees' cultural knowledge and awareness) and supervision outcomes.

Scales	r	Sig
SCF	.18	.00
SO	1.00	.00

A stepwise Multiple Regression analysis was conducted to investigate whether the supervisory relationship and supervision contextual factors influenced the supervision outcome. The prediction model contained two predictors. The correlations were statistically significant $F (1,105) = 40.2$, $p < .05$, and accounted for approximately 28 percent of the variance of the supervision outcome ($R^2 = .28$, Adjusted $R^2 = .27$). Supervision outcome was primarily influenced by the higher level of supervisory relationship and lower level of Supervision Contextual Factors. Supervisory relationship highly accounted for approximately 53 percent of the variance in supervision outcome but supervision contextual factors accounted for only 7.8 percent of the variance of supervision outcome. The regression equation for predicting the supervision outcome is as follows: Predicted supervision outcome = (0.85) supervision outcomes + (-0.22). Therefore, the null hypothesis was rejected. Therefore, the supervisory relationship predicted the supervision outcomes. However, the supervision contextual factors did not predict the outcomes of supervision. Details on this description are in Table 5.

Table 5: Influence of supervisory relationship and supervision contextual factors on supervision outcomes.

Predictor	b	SE-b	Beta	Pearson r	sr^2	Structure Coefficient
SO (Constant)	-.22	3.33	.00	1.0	.00	.00
SR	.85	.13	.53	.53	.28	.53
SCF	.00	.00	.07	.20	.01	.00

*SO = Supervision Outcomes; SR = Supervisory Relationship; Supervision Contextual Factors

8 Conclusion and Implications

The supervisory relationship had the highest influence on the supervision outcomes in this study. In the supervisory relationship, the supervision working alliance has contributed greatly to the supervision outcomes compared to other variables identified in the study. However, the supervision contextual factors did not significantly influence the supervision outcomes.

Implications from this research can be seen from the contributing factors that influenced the supervision outcomes which are of concern to the supervisors, trainee counsellors, universities, site supervisors, as well as the Counselling Board (Malaysia). Factors that have contributed towards positive supervision outcomes were the supervisees' working alliance, supervisees' role conflict, supervision interaction, and supervisors' attributes. These factors are statistically proven through this research.

Based on the current research, it is recommended that future researchers are to include the variable of cultural knowledge and awareness from the supervisor's perspectives. In addition, the additional explanation on the supervision contextual factors and the difference of supervision interaction between supervisor and trainee counsellor by the researcher are highly encouraged because it will help the respondents to understand the questions.
Place the figure as close as possible after the point where it is first referenced in the text. If there is a large number of figures and tables it might be necessary to place some before their text citation.

References

1. Anderson, S., A., Schlossberg, M., & Rigazio –DIGillio, S., Family therapy trainees' evaluations of their best and worst supervision experiences. Journal of Marital and Family Therapy, 26, 79-91, (2000)
2. Bordin, E.S., The generalizability of the psychoanalytic concept of working alliance. Psychotherapy: theory, research, and practice, 16, 252-260, (1979)
3. Bordin, E., S., A working alliance based model of supervision. The Counselling Psychologist, 11, 35-42, (1983)

4. Buchanan, G., F., The multicultural supervisor competency indicator: a behaviorally anchored rating scale approach. PhD Dissertation. Southern Illinois University Carbondale, (2006)
5. Cheon, H.S., Blumer, M. L.C., Shih, A.T., Murphy, J.M., & Sato, M., The influence of supervisor and supervisee matching, role conflict, and supervisory relationship on supervisee satisfaction. Journal of Family Therapy, 31.52-67.doi 10.1007/s10591-008-9078-y, (2008)
6. Cohen, J., Quantitative methods in psychology: A power primer. Psychological Bulletin, 112(1), 155-159, (1992)
7. Cohen, J., Applied multiple regression/ correlation analysis for the behavioral sciences. Hillsdale NJ:Erlbaum, (1983)
8. Fried, Y. & Tiegs. Supervisors' role conflict and role ambiguity differential relations with performance ratings of subordinates and the moderating effect of screening ability. Journal of Applied Psychology, (2), 282-291, (1995)
9. Friedlander, M., L., Keller, K., E., & Peca-Baker, P., T., A., Effects of role conflict on counsellor trainees' self-statement, anxiety level, and performance. Journal of Counselling Psychology, 33, 74-77, (1986)
10. Green, M., S., & Dekkers, T., D., Attending to power and diversity in supervision: An exploration of supervisee learning outcomes and satisfaction with supervision. Journal of Feminist Family Therapy, vol. 22, 293-312, (2010)
11. Grigg, A. E., & Goodstein, L. D. The use of clients as judges of the counselors performance. Journal of Counseling Psychology, 1957, 4, 31-36.
12. Grunebaum, What if family therapy were a kind of psychotherapy? A reading of handbook of psychotherapy and behaviour change. Journal of Marital and Family Therapy, 14, 195-199, (1988)
13. Hill, C.E., Charles, D., & Reed, K., G., A longitudinal analysis of changes in counselling skills during doctoral training in counselling psychology. Journal of Counselling Psychology, 28, 428 – 438, (1981)
14. Hutt, C. H., Scott, J., & King, M., A phenomenological study of supervisees' positive and negative experiences in supervision. Psychotherapy: Theory, Research, and Practice, 20, 118-123, (1983)
15. Iannelli, R. J., A structural equation modeling examination of the relationship between counseling self-efficacy, counseling outcome expectations, and counselor performance. Unpublished doctoral dissertation, University of Missouri-Kansas City, (2000)
16. Inman, A.G., Supervisor multicultural competent and its relation to supervisory process and outcome. Journal of Marital and Family Therapy, 32, 73-85, (2006)
17. Ladany, N., & Bradley, J., B., Counselor Supervision. Taylor & Francis: New York, (2010)
18. Liddle, H., & Saba, G., Teaching family therapy at the introductory level: A conceptual model emphasizing a pattern, which connects training and therapy. Journal of Marital and family Therapy, 8, 63-72, (1982)
19. Putney, Worthington, Jr and McCullough, Effects of supervisor and supervisee theoretical orientation and supervisor-supervisee matching on interns' perceptions of supervision. Journal of Counselling Psychology, vol. 39 (2), 258-265, (1992)
20. Quarto, C. J., Supervisors' and supervisees' perceptions of control and conflict in counselling supervision. The Clinical Supervisor. 21, 21 – 37. Retrieved on 6 January, 2013 from http://www.gallaudet.edu/documents/academic/cou_supervisionmodels[1].pdf, (2002)

21. Roberts, Walter, B., Jr., Morotti & Allan, A., Professional School Counseling, 4 (3), 208-Retrieved from http://eric.ed.gov/?id=EJ630582 - 10k, (2001)
22. T Thome, S., Effects of working alliance on trainee counseling skills and personal development. PhD Dissertation. University of Nevada, Reno, (2006)
23. Tromski-Klingshrim, D.M. & Davis, T.E., Supervision: Supervisees perceptions of their clinical supervision: A study of the dual role of clinical and administrative supervisor. Counsellor Education and Supervision, 46, 294-304, (2007)
24. Worthen, V., E., & Isakson, R., L., The Supervision Outcomes Survey (SOS). In Falender & Shafranske (2004).

How does the Ambience of Cafe Affect the Revisit Intention Among its Patrons? A S on the Cafes in Ipoh, Perak

Sharmeela-Banu Syed AbuThahir[1,*], *Gengeswari* Krishnapillai[1]

[1]Department of Marketing, Faculty of Business and Finance, University Tunku Abdul Rahman, Jalan Universiti, Bandar Barat, 31900 Kampar, Perak, Malaysia

Abstract.

Food service industry is growing rapidly as a result of the changing consumer lifestyle. The food service industry is highly competitive due to the increasing number of new entrants offering inventive food products and services. In order to be outstanding in such competitive industry, retailers nowadays opt to emphasize on their store environment. Past studies discovered that store environment stimulates emotions that significantly boost customer revisit intention. As a result, retailers attempt to differentiate their store by combining various environmental stimuli to create an attractive ambience that will in turn draw in the customers. Hence, this study attempts to investigate the impact of various café ambience factors on the patrons' revisit intention. The patrons of cafes in Ipoh, Perak were selected using purposive sampling technique to be the respondents of this study. 250 questionnaires were collected and Partial Least Square technique was used to analyse the data collected. Findings show that all the five factors of café ambience namely lighting, music, decoration, cleanliness and layouts were significantly influencing the patrons' revisit intention. Of these five factors, lighting was most influential while music was the least influential in affecting the patrons' revisit intention. Accordingly, this study lists several recommendations for practitioners and academics with regards to the store environment and its impact on the repurchase intention.

1 Introduction

Food service retailers are one of the largest employers' around the world. Food service industry is growing rapidly as a result of the changing consumer lifestyle.Consumer trust in the food industry resulted into the further consumers demandauthenticity, transparency and credibility from all the industry players [57].

* Corresponding author: sharmeelas@utar.edu.my

According to Agriculture and Agri-Food Canada [1], food service industries includes cafes and bar, self-service restaurant, full-service restaurants, home delivery and take away, street kiosks, as well as fast food restaurants. Putiy [49] stated that cafe and bar, full service restaurants and fast food restaurants make up the main segment of the food service industry. It consists of more than 8 million food outlets ranging from small family-owned units to large multinational chains operating thousands of stores worldwide.

The global food service industry is projected to reach $2.2 trillion in revenues by 2015 as reported in the Business Wire, 2009. Global food market growth is driven by rising disposable earnings, increasing demand for outside food in emerging market and demographic pattern transformation which further boosts the needs for food services. Likewise in Malaysia, majority of Malaysian consumers select to dine out during weekdays or even weekends and public holidays due to their hectic schedules in urban cities and the increased household income [19].

Cafes are typically present in most urban and suburban areas. Due to their creative concepts and relaxing ambience, cafes are usually more exclusive and premium priced than a regular restaurant [10]. Cafes are located at convenient locations around the consumers' place to offer affordable products in a comfortable environment [19]. In addition to that, the increasing disposable income and growing young populations are contributing to the expansions of cafes in Malaysia [68] where these young people characteristically enjoy hanging out with friends at cafes and willing to spend more time and money.

According to Agriculture and Agri-Food Canada [1] the cafe industry in Malaysia is expected to continue to develop and grow as more consumers are choosing to visit these outlets more frequently. Based on the Fig. 1.1, the cafe industry is predicted to achieve a market value of US$ 3,707.1 million in coming year 2017.

Fig.1.1: Forecast Market Value and Growth of Malaysian Foodservice CAGR

Forecast Market Value and Growth of Malaysian Foodservice by Subsector, US$ millions

	2013	2014	2015	2016	2017	CAGR % 2013-17
Consumer Foodservice (Total)	10,511.2	11,064.4	11,648.7	12,275.5	12,939.1	5.3
Full-Service Restaurants	3,594.8	3,779.6	3,975.1	4,191.5	4,431.3	5.4
Cafés/Bars	3,027.4	3,186.8	3,353.4	3,527.5	3,707.1	5.2
Street Stalls/Kiosks	2,098.2	2,189.9	2,292.2	2,405.9	2,529.5	4.8
Fast Food	1,513.2	1,614.3	1,718.0	1,823.5	1,929.1	6.3
Self-Service Cafeterias	204.1	211.7	219.4	229.2	237.1	3.8
100% Home Delivery/Takeaway	73.5	82.1	90.6	97.9	105.0	9.3

CAGR= compound annual growth rate.
Source:Agriculture and Agri-Food Canada, Global Analysis Division. (2014). *Foodservice Profile: Malaysia.* Ottawa, ON:Author.

Store atmosphere is defined as the attributes which enhances the ambience of the store with diverse combination of elements such as lighting, music, colour, and scent (Levy & Weitz, 2009). According to Namasivayam & Lin [46] store atmosphere is crucial as it affects customer's behaviour as well as their cognitive, emotional, and physiological state. In addition to that, it is proposed that store environment stimulates emotions that significantly boost customer revisit intention [54].

Various past studies [35, 43] adopted the Mehrabian-Russell model to understand the influence of store atmosphere on customer's response. According to the model, environmental stimuli which denotes the physical and non-physical element of a store affects the emotional state of the customer in terms of pleasure, arousal or dominance which in turn influences their approach or avoidance response. According to Chen, Peng, and Hung (2015), this model hasbeen applied in various contextssuch as retail, service, leisure, and tourism.

Figure 1.2: The Mehrabian-Russell Model

Figure 1
The Mehrabian-Russell Model

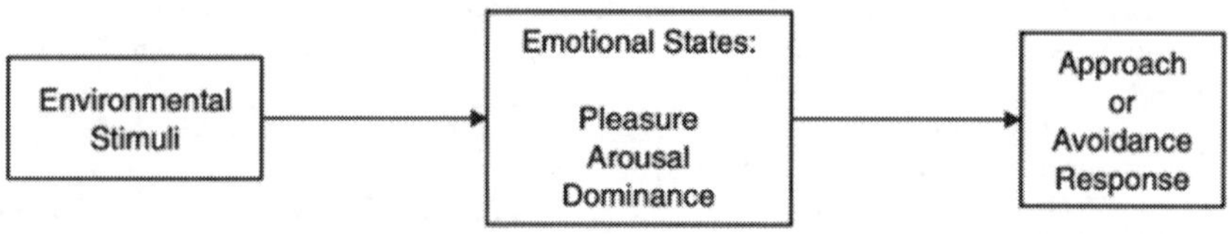

Source:Ryu, K., & Jang, S. C. (2007).The effect of environmental perceptions on behavioral intentions through emotions: The case of upscale restaurants.

According to Voon [65] the food service industry is highly competitive due to the increasing number of new entrants offering inventive food products and services. As a result, retailers attempt to differentiate their store by combining various environmental stimuli to create an attractive servicescape that will draw in the customers [69]. It has been found that most of the previous studies only focused on a particular servicescape dimension instead of evaluating several dimension together at once [53, 28 ,41]. As a result, those studies are not suitable to the retailersnowadays who usually combine several atmospheric stimuli to enhance their store appearances. As for this research, it intended to provide accurate result by using several store environment stimuli to evaluate its impact on customer revisit intention.

Hussain and Ali [34] supported by stating that customer nowadays are requesting for additional elements such as store environment instead of food quality. Cafes with pleasant store environment are more likely to attract walk-in customers which trigger their revisit intention. In accordance to that, retailers are willing to spend a substantial amount of resources to improve their store design and other related activities [37]. Despite, most of the retailers are uncertain of which store attributes are deemed important by customers.

Although there are ample studies conducted on the effect of store environment on customer behaviour in the retail world [30] there are limited empirical research evaluating the effect of store environment on customer satisfaction and revisit intention in the food service industry, cafes in particular. In fact, several researchers who studied on the food service industry particularly at foreign countries such as Hong Kong and Sri Lanka. Therefore, this study will concentrate on the customer behaviour towards different store environment attributes which influences their revisit intention. Additionally, the finding may provide an insightful indication to the café owners around Ipoh on consumer behaviour.

2 Literature Review

Chen et al. [15] described customer revisit intention as customer's intention to revisit the same restaurant and theytend to recommend it around their circle (i.e. family and friends). The revisit intention is further explained as the repurchase intention and action that shows

 https://doi.org/10.1051/matecconf/201815005074

the customer's willingness to provide positive information about the service provider [61]. In the meantime, revisit intention also connected to the customer's decision to build an everlasting relationship with the service provider [33]. According to the past studies, customer satisfaction is the influencing factor for customer's revisit intention [26] hence; customer satisfaction enhances customer loyalty [3] which in turn encourages the customers' revisit intention. Conversely, Chiu, Hsu, Lai, & Chang [16], emphasized that customer satisfaction leads to customer retention whereas retaining current customers are highly cost-effective than attracting new customers. Also, Jiang & Rosenbloom, 2005 added that greater customer retention cause positive escalation on the firms' business profitability.

Customer preferences towards music affect their customer satisfaction (Shashikala & Suresh, 2013). Hence, Han et al. [27] explained that café owners should fine-tune background musicby changing its volume (i.e. loud to soft), tempo (i.e. fast to slow), and genre (i.e. classical or jazz), based on customer's interest to arouse consumers' emotional states. Similarly, Brattico and Jacobsen [12] claimed that music has an impact on emotions, mood, and pleasure. Furthermore, Rea, MacDonald, and Carnes [51] agreed that the mood of the listener can be positively or negatively influenced depending on the type of music played. They added that classical and pop music increases listener's feeling of ease and decreases feeling of worry or tension. Besides, music affects the customer viewpoint of waiting time for service at restaurants and cafes [9, 48]. According to Kim [39], popular music increased the duration customers spend in a café compared to unpopular music which influences dine-in experience.

On the whole, Soriano (2002) insisted that music as one of the elements of ambience which have major effect on the customer revisit intention. Besides, music has been directly stimulates the consumer behaviour to revisit the café (Bonn, Joseph-Mathews, Dai, Hayes, & Cave, 2007).Therefore, the following hypothesis is proposed:

H1: Music has a significant effect on the revisit intention to the cafes.

According to Morin, Dube, & Chebat [45], lighting affects customer's behavior in restaurants. Custers, de Kort, IJsselsteijn, & de Kruiff [17] supported by explaining that lighting influences customer's emotions, mood, and cognition. Dimmed lighting createsan intimate environment for the customer [17] however, Jacquier and Giboreau (2012) indicated that brightness of the lighting is important for customers to read the cafes' menu. Past studies by Houser, Tiller, Bernecker, and Mistrick [32] Vaccaro, Yucetepe, Torres-Baumgarten, and Lee [63] Ariffin, Bibon, & Raja Abdullah, [5] found that adequate lighting influences customer's impression towards the café ambience and enhances customer revisit intention.

Ryu and Jang [56] supported that lighting elements of store have great impact on the customer intention to visit the retailer for repurchase action. Moreover, the customers believe that the pleasant lighting in the café indicates the quality of the service provider [57]. Furthermore, Ramlee & Said [50] asserted that lighting is one of the important characteristics of intangible dimensions which greatly
affecting non-visual sense of consumer perception towards café selection. Therefore, the following hypothesis is proposed:

H2: Lighting has a significant effect on the revisit intention to the cafés.

According to Barber, Goodman, and Goh [8] the cleanliness of a foodservice business includes the dining room and tableswares, the kitchen, as well as the restroom. Hussain & Ali [4] described cleanliness as the store's appearance that improves the atmosphere and enhances customer's reaction towards the store. Cleanliness is the most significant attribute that influences customer's perception of service quality [8] and satisfaction level [58]. According to Kaul (2005), a store with clean physical facilities yields satisfaction and develop revisit intention. Besides that, cleanliness generates positive evaluation among customer and encourages them to stay longer [2, 34]. Long et al. (2011) found that people are typically more comfortable in a clean environment, which in turn affects the amount of time they spend in the cafe. Customers will predict the cleanliness of the kitchen based on the dining room surroundings and the appearance of the employees [56] which also influences their evaluation of the food quality [41].

Banat & Wandebori [6] has argued that the cleanliness of café affects the customer feeling towards the café whether to revisit the café again in the future. Furthermore, Gajanayake, Gajanayake & Surangi [22] supported that the level of cleanliness able to create an image of comfort and luxury among customers' which affect the customer revisit intention. Therefore, the following hypothesis proposed:

H5: Cleanliness has a significant effect on the revisit intention to the cafes.

VanBaren [64] claimed that the store layout is the design ofa store's interior to provide maximum exposure of merchandise. Well-planned layouts are essential to facilitate in-store traffic patterns [66] and increased customers' efficient movement through the store [62]. Store layout in café should emphasize on the seating arrangement, equipment and furnishing which increases customer satisfaction and leads them to spend longer hours in the café [30]. Hence, those items are able to facilitate café performance in increasing customer revisit intention.

Study of Tlapana(2009) ascertained that the layout and display such as furnishing and the arrangements of the equipment in the café would influence the customer revisit intention. Be sides, Hart, Farrell, Stachow, Reed, & Cadogan [29] added that the spatial condition of the store influences the consumer attitude towards café selection. Therefore, the following hypothesis is proposed:

H3: Store layout has a significant effect on the revisit intention to the cafes.

Decoration and artifacts are vital components for developing attraction of consumers towards the pleasant appearance of the café setting which influences the consumer overall satisfaction and loyalty [35, 39]. The decoration of cafe enhances customer satisfaction and thus increases their intention to revisit café [14]. Besides that, decorations and artifacts help customers to develop their first impression of the overall ambience of the café [60]. Long, Ka, Fong, and Lo (2011) added that when the cafe is well decorated with superior quality artifacts, customers are more likely to classify the café as providing high-class and superior service quality. The determination of superior quality artifacts influenced by various aspects of interior designs such as color scheme applied in the café wall or ceiling, paintings, floor coverings, and quality furniture which able to create the overall visual gratification [41].

Decoration and artifacts is the physical environment that induces perceptual and emotional responses among customer that affect the consumer revisit intention (Ali & Amin, 2014). Meanwhile, Wilson, Zeithaml, Bitner, & Gremler [67] ; Ryu & Han [55] stressed that the decorations and artifacts of the café not only create aesthetic impressions, but it also employed as an indicator of differentiation of the chosen dining environment which leads to the customer revisit intention. Therefore, the following hypothesis is proposed:

H4: Decorations and artifacts have a significant effect on the revisit intention to the cafes.

3 Research Methodology

This study was conducted in Ipoh, Perak where regular visitors of 10 chosen cafes were approached to participate. The 10 cafes were chosen due to their popularity and as they were recommended by numerous sources. Purposive sampling technique was used to select the target respondents where there were filtered based on their familiarity and frequency of visit. Hinkin et. al [31] ascertained that an ideal sample size should have an item to response ration of between 1:4 to 1:10. Accordingly, the ideal sample sizeof this study should range from 96 to 240 as there were 24 items in the research instrument. On the other hand, Cattell recommended that a research with three or more independent variables should have a minimum sample size of 250 elements [18]. In view of the typology of Cattell and Hinkin, it was decided to have a sample size of 250 elements in this study. Prior to the actual fieldwork, a pilot study was conducted among 30visitors of three most popular cafes in Ipoh, Perak. Based on the feedbacks received, the word "store" was changed to "café" in order enhancesthe clarity of the items in the instrument. All the 250 sets of responses were retained for further analysis after ticking the requirements of data cleaning process. The data was analysed using Partial Least Square technique.

4 Discussions

As shown in Table 1, there were more female (64%) respondents and majority of the respondents were within the age group 21-25 years old (38.4%) and visited the café at least thrice in a month (42%).In order to further apprehend the impact of these demographic characteristics on the research variables investigated, independent sample t-Test (i.e. as for the Gender) and one-way analysis of variances tests (i.e. as for the Age and Frequency of visits) were done.

Table 1: Summary of Demographic Characteristics

Demographic Characteristics	Frequency	Percentage, %
Gender		
Male	90	36
Female	160	64
Age		
Below 20 years old	69	27.6
21 – 25 years old	96	38.4
26 – 30 years old	50	20
Above 30 years	35	14
Frequency of visits (in a month)		
1	36	14.4
2	74	29.6
3	105	42
4	35	14

Of these 3 demographic characteristics, only "frequency of visit" had contributed a significant difference on all the research variables except the variable of "music" and "light" (refer to Appendix 1). Table 2 summarises the descriptive statistics and an extract of the mean comparison tests conducted. Despite the insignificant overall effect, "frequency of visit" was found to have a significant difference on the "music" between respondents who visit the café 3 times and 4 times in a month. Similarly, these two groups of respondents (i.e. who visit the café 3 times and 4 times in a month) were also found to have significant differences on the "layout", "decoration", "cleanliness" and "revisit intention". Respondents who visit the café 4 times in a month were noted to have higher mean values as compared to respondents who visit the café thrice in a month. This implies that the visitors who frequently visit the café tend to place more importance on the various aspect of café ambience alongside their higher revisit intention ascompared to the respondents who visit the café less frequently.

Table 2: One Way Analysis of Variance by Frequency of Visits

Variables	Frequency of visit	N	Mean	STD	F	Sign.	Post-hoc (LSD) Sign.
Music	1	36	3.7153	0.9452	1.979	.118	3 vs. 4
	2	74	3.6892	0.9895			
	3	105	3.6024	1.0230			
	4	35	4.0571	0.6617			
Light	1	36	3.6667	1.0089	1.173	.321	NA
	2	74	3.5845	1.0479			
	3	105	3.4762	1.1242			
	4	35	3.8500	0.9099			

layout	1	36	3.6458	1.0271	2.978	**.032**	3 vs. 4
	2	74	3.6926	1.0183			
	3	105	3.4214	1.1289			
	4	35	4.0000	0.7525			
decoration	1	36	3.7083	0.9608	3.922	**.009**	2 vs. 4
	2	74	3.6757	1.0621			3 vs. 4
	3	105	3.4262	1.1451			
	4	35	4.1071	0.6510			
cleanliness	1	36	3.8264	0.8718	3.300	**.021**	1 vs. 4
	2	74	3.8784	0.9338			2 vs. 4
	3	105	3.7381	0.9722			3 vs. 4
	4	35	4.2786	0.3364			
revisit	1	36	3.7639	0.9315	2.571	**.045**	3 vs. 4
	2	74	3.7230	0.9387			
	3	105	3.5595	1.0252			
	4	35	4.0571	0.5426			

The proposed research model of this study was analysed using structural equation modelling (SEM) technique which is a famous second-generation technique. There are two variations in SEM technique namely Covariance-based SEM (CB-SEM) and variance-based SEM (PLS-SEM) where the former is primarily confirmatory in nature while the latter is being [25]. As the current study was not developed to underpin and hence confirm any underlying theory, PLS SEM technique was used to analyse the data collected. The lenient requirements of the measurement model and model complexity were also key reasons to utilise PLS SEM technique in this study.

All the variables of this model are reflective constructs where assessment of construct validity is a must for the reflective constructs [24] Construct validity of this study was assessed based on the model's convergent and discriminant validity. Fornell [20] suggested referringto factor loading (FL) value, average variance extracted (AVE) and composite reliability (CR) inassessingthe convergent validity of the model. Table 3 exhibits the FL, AVE and CR of the study's measurement model.

Table 3: Assessment of Convergent Validity

Items		FL	CR	AVE
Revisit Intention				
CR1	I would like to consider the café as one of my choices in the future.	0.8793	0.9021	0.6975
CR2	I would like to invite my friends and family along to the café again.	0.8143		
CR3	I would like to experience the pleasant feeling in the café again.	0.8184		
CR4	I would like to build an everlasting relationship with the service provider.	0.8271		
Decoration & Artifacts				
DA1	The concept of the decorations and artifacts improves my first impression on the cafe.	0.9039	0.9226	0.7488
DA2	The uniqueness of the decorations and artifacts encourages me to spend more time evaluating the cafe.	0.8499		
DA3	The style of the decorations and artifacts encourages me to categorize the cafe as high-class.	0.8308		
DA4	The design of the decorations and artifacts affects my satisfaction level towards the cafe.	0.8752		
Lighting				
L1	The brightness of the lighting allows me to browse through the menu easily.	0.8915	0.9216	0.7462
L2	The dimness of the lighting gave me an intimate / relaxed feeling.	0.8539		

L3	The clarity of the lighting allows me to evaluate the overall ambience of the cafe.	0.827		
L4	The attractiveness of the lighting (e.g.: shape and colour of bulb) enhances my perception of the café's image.	0.8815		
Music				
M1	The right genre of music influences my mood in a positive manner.	0.8731	0.9093	0.7149
M2	The tempo of the music increases my willingness to wait for my food.	0.8326		
M3	The rhythm of the music enhances my enjoyment.	0.844		
M4	The appropriate volume of the music helps to reduce my stress level.	0.8317		
Store Layout				
SL1	The accessibility of the store layout enhances my positive evaluation.	0.8896	0.9178	0.7364
SL2	The design of the store layout encourages me to walk around and select more food and beverages (e.g.: from the fridge or pastries display)	0.832		
SL3	The efficiency of the store layout eases my entry and exit.	0.8489		
SL4	The creativeness of the store layout (e.g.: seating arrangements) encourages me to stay longer.	0.8609		
Cleanliness				
C1	The cleanliness of the café encourages me to have a positive impression towards the cafe.	0.879	0.9019	0.6971
C2	The cleanliness of the café improves my evaluation of the food quality.	0.8204		
C3	The cleanliness of the café increases my desire to stay longer.	0.8009		
C4	The cleanliness of the café enhances my overall satisfaction level.	0.8376		

All the items were found to have more than 0.8 FL value (which ranges from 0.8009 to 0.9039)while each construct had achieved a minimum 0.6 of AVE value (which ranges from 0.6971 to 0.7488) and good CR score(which rangesfrom 0.9019 to0.9226). Such results evidence that the measurement model hadachieved a fair convergent validity.

On the other hand, discriminant validity was tested by assessing the comparison of items' loading and cross-loading value besides comparing the construct'sAVE and otherconstructs' correlation coefficients (as suggested by Gefen & Straub, [23]). Table 4 depicts the loading and cross-loading of items where it is notable that each item hadhigher loading value on its respective construct than on other constructs.

Table 4: Item's Loadings and Cross-Loadings

Construct/Item	Cleanliness	Decoration	Store Layout	Light	Music	Revisit
CRI1	0.7281	0.6821	0.6744	0.676	0.583	**0.8793**
CRI2	0.6269	0.6915	0.6512	0.6538	0.5457	**0.8143**
CRI3	0.703	0.664	0.6479	0.6971	0.6083	**0.8184**
CRI4	0.6426	0.6879	0.6717	0.6838	0.5471	**0.8271**
DA1	0.6242	**0.9039**	0.5757	0.6032	0.4888	0.7341
DA2	0.5876	**0.8499**	0.5419	0.5277	0.4626	0.6736
DA3	0.5535	**0.8308**	0.5693	0.5416	0.4601	0.6951
DA4	0.592	**0.8752**	0.5795	0.5777	0.4539	0.719
L1	0.6645	0.6024	0.583	**0.8915**	0.5021	0.7503
L2	0.5555	0.5524	0.5193	**0.8539**	0.4627	0.6679
L3	0.5546	0.5025	0.4439	**0.827**	0.4859	0.6477
L4	0.6173	0.5861	0.5659	**0.8815**	0.4651	0.732
M1	0.5407	0.4506	0.4855	0.5015	**0.8731**	0.5986
M2	0.4573	0.4569	0.4744	0.4586	**0.8326**	0.5608

M3	0.4406	0.4647	0.4614	0.4466	**0.844**	0.5718
M4	0.4353	0.4512	0.4895	0.4659	**0.8317**	0.582
SL1	0.6152	0.5967	**0.8896**	0.5856	0.5164	0.7136
SL2	0.5187	0.5478	**0.832**	0.4587	0.4887	0.6275
SL3	0.5332	0.522	**0.8489**	0.4927	0.4831	0.6618
SL4	0.5557	0.5791	**0.8609**	0.5618	0.4531	0.7104
C1	**0.879**	0.6228	0.5785	0.6763	0.5002	0.7561
C2	**0.8204**	0.5762	0.5419	0.5549	0.4478	0.6617
C3	**0.8009**	0.5313	0.4905	0.497	0.4524	0.6103
C4	**0.8376**	0.5396	0.5516	0.5757	0.4507	0.6641

In comparing the AVE and correlation coefficients, [21] suggested that theconstruct's AVE square root should be larger than other constructs' correlation coefficients. As shown in Table 5, all the constructsof this studyhad higher square root of AVE than other constructs' coefficients. Based on the two tests conducted, it can be wisely concluded that the modelhad achieved fairdiscriminant validity.

Table 5: Assessment of Discriminant Validity

Construct	Cleanliness	Decoration	Store Layout	Light	Music	Revisit
Cleanliness	**0.8349**					
Decoration	0.6814	**0.8653**				
Store Layout	0.6489	0.655	**0.8581**			
Light	0.6944	0.651	0.614	**0.8638**		
Music	0.5549	0.539	0.565	0.5541	**0.8455**	
Revisit	0.8094	0.8158	0.792	0.8117	0.6842	**0.8352**

***BOLD** value is the square root of AVE.

Upon confirmingthe reliability and validity of the model, the study's structural model was assessed. It is suggested to assess the possible collinearity issues before assessing the key results of the structural model [25]. Table 6 summarises the collinearity statistics as well as the path's coefficient and T-statistics. All the exogenous constructs were found to have the Tolerance value and Variance inflation factor (VIF) value below than the recommended threshold value of 0.2 and 5, respectively. On the other hand, it was noted that all the five aspects of café ambience had significant (T-statistics > 1.96) and positive influence towards customers' revisit intention to the café. Therefore, all the proposed hypotheses were supported. Of these 5 aspects, "Light" was noted to have highest effect (β =0.2695) on the revisit intention and closely followed by the "Decoration" (β =0.2674). In addition, "Music" was found to contribute the least effect (β =0.1419) on the revisit intention. As a whole, all these 5 aspects of café ambience were found to explain about 88.94% variances in the customer revisit intention to café. This implies that this study's model has substantial predictive accuracy of R2 = 0.8894.

Table 6: Collinearity Statistics and Path Analysis

Path	Collinearity Statistics		Beta	T-value	Result
	Tolerance	VIF			
Cleanliness ->Revisit (H1)	.455	2.199	0.2079	3.1364	Supported
Decoration ->Revisit (H2)	.423	2.366	0.2674	4.1444	Supported
Store layout ->Revisit (H3)	.585	1.709	0.2362	3.9283	Supported
Light ->Revisit (H4)	.394	2.536	0.2695	3.9987	Supported
Music->Revisit (H5)	.432	2.316	0.1419	3.1777	Supported

5 Implications

This study empirically shows that the five aspects of the cafe music, lighting, cleanliness, store layout, decorations and artifacts have a positive influence on the intention to revisit among the cafe customers. Among all the contributing factors, lighting was the focus of attraction which was noted to have the greatesteffect, followed by the music with has theleast impact towards customer revisit intention to café. In particular, all the five aspects of café ambience explained about 88.94 percent of variances in the customer revisit intention.

Based on the results of the study, lighting is the most important thing that customers are concerned in the selection to revisit their preferred cafe (Table 6). This shows that the lighting elements being the natural pull factor to change customers' perceptions towards their preferred cafe.

The choice of lighting indeed greatly influences the customer revisit intention [56]. Ramlee & Said [50] affirmed that lighting highly influences the customers in selecting a suitable café for their dine-in activities. In certain conditions, the choice of dim lighting which creates the pleasant and amusing environment demonstrate the quality of the café in providing services [57].

As a guideline, the usage of lighting that is lively with natural elements such as wood, cotton, leather and seagrass to name a few are able to stimulate customer stay in the café. A part from that, the primary concern also centers on outdoor lighting at night using lumens, foot-candles which creates the pleasant and warm environment.

In addition,the decoration and artifactselement of the café werenoted to have second best influence (β = 0.2674) towards customer revisit intention to the cafe. Past studies found that customer impression towards a café is based on the design elements thatportray uniqueness of the café environment [4]. It was further emphasized that theappropriate choice of decorations and artifacts closely related to the differentiation depictedby the café as high-class service provider [67].

This shows that the café management need to develop an aesthetic café environment to stimulate customer perception and selection of the café. The decorations and artifacts such as classical mix western should emphasizesince it almost certainly able to differentiate an exclusivecafé from thecompetitors.Theattractivefacility aestheticsof the café (i.e. ceiling/wall décor, carpeting/flooring, paintings/pictures, plants/flowers, furniture, and color)are able to develop customer satisfaction and revisit intention.

On the contrary, musicwasfound to contribute the least effect (β = 0.1419) on the customer revisit intention (Table 6). This may cause of the choice of music selection in the café which leads to endow a modest effect on customer revisit intention.

Past study of North & Hargreaves [47] found that customers are giving more considerationto the music that they encountered at home rather than music in public situations in which customers felt they have less control over the music heard. In public, considering the device is important, as customers are using mobile listening devices allow for them to exert control.Supported by Krause, North, and Hewitt [40] discovered that theselection ofinstrument in the café affects the customer's response which influences the customer moodto stay longer in the café.

This suggests that the selection of music such as classical, jazz, and slow tempo music should be blended around the time to create mix-mode environment to the customer. Meanwhile, facilitate customers to choose their own preferred music exerts a strong influence over the ways in which a customer perceives it.

6 Limitation and Recommendation for Future Researches

Quantitative research method has been applied to conduct this research. Therefore, it provides modest information on contextual factors to impart deeper understanding which explains the variations in customer's preferences. Moreover, quantitative research focuses on hypothesis testing rather than on hypothesis generation. As a result, the actual store atmosphere affecting customer's revisit intention might be left out from this study.

Future researchers are suggested to pursue either a mixed-method research or qualitative researchmethod in order to evaluatecustomer's opinion or viewpoint, whereby therespondent are able to provide detailed description of their personal experiences for enhanced interpretation.

Another limitation of this study is that it used a cross-sectional study. A cross-sectional study only evaluates a situation at one point in time. It does not consider the effect of store atmosphere on customer revisit intention over a period of time. Customers behavioral emotions affecting the revisit intention is not always consistentbased on the factors reviewed since the cross-sectional study less emphasizedthetime factor and its impact tothe variables as a whole. Prospective scholarscanconsider adoptinglongitudinal study to promote a better understanding of customer revisit intention. Longitudinal approach can be used to predict behavioral intention over the time. As such, the model should be validated at different points in time.

6 Acknowledgement

The authors would like to extend their appreciationto Chow Pooi Yan, Ho Sim Yee, Ho Tze Yikand Prasna Devi for the help in the data collection process.

References

1. Agriculture and Agri-Food Canada, Global Analysis Division., Foodservice Profile: Malaysia. Ottawa, ON:Author. (2014)
2. Akinyele, S. T., Customer satisfaction and service quality: customer's repatronage perspectives. Global Journal of Management and BusinessResearch, 10(6), 83-90, (2010)
3. Alegre, J., & Cladera, M., Analysing the effect of satisfaction and previous visits on tourist intentions to return. European Journal of Marketing, 43, 670-685, (2009)

4. Ali, F., & Amin, M., The influence of physical environment on emotions, customer satisfaction and behavioural intentions in Chinese resort hotel industry.Journal for Global Business Advancement,7(3), 249-266, (2014)
5. Ariffin, H. F., Bibon, M. F., & Raja Abdullah, R. P. S., Restaurant's Atmospheric Elements: What the Customer Wants. Procedia -Social and Behavioral Sciences, 38, 380-387, (2012)
6. Banat, A., & Wandebori, H. S. T., Store Design and Store Atmosphere Effect on Customer Sales per Visit Economics, Management and Behavioral Sciences, (2012)
7. Barber, N., & Scarcelli, J. M., Enhancing the assessment of tangible service quality through the creation of a cleanliness measurement scale. Measuring Service Quality, 20(1), 46-69, (2010)
8. Barber, N., Goodman, R. J., & Goh, B. K., Restaurant consumers repeat patronage: A service quality concern. International Journal of Hospitality Management, 30(2), 329-336, (2011)
9. Berman, B., & Evans, J. R., Retail Management: A Strategic Approach(10thed.). Upper Saddle River, New Jersey: Prentice Hall, (2007)
10. Bernson,A., The Social Space of The Café: How Service and Physical Design Condition Social Performances(Doctoral dissertation, Wesleyan University), (2011)
11. Bonn, M. A., Joseph-Mathews, S. M., Dai, M., Hayes, S., & Cave, J., Heritage/cultural attraction atmospherics: Creating the right environment for the heritage/cultural visitor.Journal of Travel Research,45(3), 345-354, (2007)
12. Brattico, E., & Jacobsen, T., Subjective Appraisal of Music: Neuroimaging Evidence. Ann NY Acad Sci 1169, 308–317, (2009)
13. Campo-Martine, S., Garau-Vadell, J. B., & Martinez-Ruiz, M. P., Factors influencing repeat visits to a destination: The influence of group composition. Tourism Management, 31, 862–870, (2010)
14. Chang, K., The Effect of Facility Aesthetics and Music onBehaviouralIntention Through Emotion in The Upscale Restaurant Environment. Unpublished master's thesis, The University of Guelph, Guelph, Ontario, Canada, (2012)
15. Chen, A., Peng, N., Hung, K. P., The effects of luxury restaurant environments on diners' emotions and loyalty. International Journal of Contemporary Hospitality Management, 27(2), 236-260, (2015)
16. Chiu, C. M., Hsu, M. H., Lai, H., & Chang, C. M., Re-examining the influence of trust on online repeat purchase intention: The moderating role of habit and its antecedents.Decision Support Systems,53(4), 835-845, (2012)
17. Custers, P. J. M., de Kort, Y. A. W., IJsselsteijn, W. A., & de Kruiff, M. E., Lighting in retail environments: Atmosphere perception in the real world. Lighting Res. Technol, 42,331–343, (2010)
18. De Winter, J. C. F., Dodou, D., & Wieringa, P.A., Exploratory Factor Analysis with Small Sample Sizes. Multivariate Behavioral Research, 44, 147-181, (2009)
19. Euromonitor International., Cafés/Bars in Malaysia. Retrieved July 12, 2015, from http://www.euromonitor.com/cafes-bars-in-malaysia/report, (2015)
20. Fornell, C., A second generation of multivariate analysis. 2. Measurement and evaluation(Vol. 2). Praeger Publishers, (1982)
21. Fornell, C., & Larcker, D. F., Evaluating structural equation models with unobservable variables and measurement error. Journal of marketing research, 39-50, (1981)
22. Gajanayake, R., Gajanayake, S., & Surangi, H. A. K. N. S. , The impact of selected visual merchandising techniques on patronage intentions in

supermarkets.Paper presented at the Conference on Business and Economic Research, Kedah, MY, (2011)
23. Gefen, D., & Straub, D., A practical guide to factorial validity using PLS-Graph: Tutorial and annotated example. Communications of the Association for Information systems, 16(1), 5, (2005)
24. Gil-Garcia, J. R., & Helbig, N., Exploring e-government benefits and success factors. In A.-V. Anttiroiko & M. Malkia (Eds.), Encyclopedia of digital government. Hershey, PA: Idea Group Inc, (2006)
25. Hair Jr, J. F., Hult, G. T. M., Ringle, C., & Sarstedt, M., A primer on partial least squares structural equation modeling (PLS-SEM). Sage Publications, (2014)
26. Han, H., & Ryu, K., The roles of the physical environment, price perception, and customer satisfaction in determining customer loyalty in the restaurant industry. Journal of Hospitality & Tourism Research, 33(4), 487-510, (2009)
27. Han, H., Back, K. J., & Barrett, B., Influencing factors on restaurant customers' revisit intention: The roles of emotions and switching barriers. International Journal of Hospitality Management, 28(4), 563-572, (2009)
28. Harris, L. C., & Goode, M. M., Online servicescapes, trust, and purchase intentions.Journal of Services Marketing,24(3), 230-243, (2010)
29. Hart, C. A., Farrell, A. M., Stachow, G., Reed, G., & Cadogan, J. W., Shopping Experience Enjoyment: Impact on Customers' Repatronage Intentions, and Gender Influence.Service Industries Journal,27(5), (2007)
30. Heung, V. C. S., & Gu, T., Influence of restaurant atmospherics on patron satisfaction and behavioral intentions. International Journal of Hospitality Management, 31(4), 1167-1177, (2012)
31. Hinkin, T. R., A review of scale development practices in the study of organizations.Journal of management,21(5), 967-988, (1995)
32. Houser, K. W., Tiller, D. K., Bernecker, C. A., & Mistrick, R. G, The subjective response to linear fluorescent direct/indirect lighting systems. Lighting Research and Technology, 34, 243–264, (2002)
33. Hume, M., Mort, G. S., & Winzar, H., Exploring Repurchase Intention in a Performing Arts Context: Who Comes? and Why Do They Come Back? International Journal of Nonprofit and Voluntary Sector Marketing, 12, 135-148, (2007)
34. Hussain, R., & Ali, M., Effect of Store Atmosphere on Consumer Purchase Intention. International Journal of Marketing Studies, 7(2), 35-43.Jacquier, C., & Giboreau, A. (2012). Perception and Emotions of Colored Atmospheres at the Restaurant. Proceedings of the 3rd International Conference on Appearance, 165-167, (2015)
35. Jang, S. S., & Namkung, Y., Perceived quality, emotions, and behavioral intentions: Application of an extended Mehrabian–Russell model to restaurants.Journal of Business Research,62(4), 451-460, (2009)
36. Jiang, P., & Rosenbloom, B., Customer intention to return online: price perception, attribute-level performance, and satisfaction unfolding over time.European Journal of Marketing,39(1/2), 150-174, (2005)
37. Kaltcheva, V. D., & Weitz, B. A., When Should a Retailer Create an Exciting Store Environment? Journal of Marketing, 70, 107-118, (2006)
38. Kim, J. H., The effect of amount of information and music on consumer shopping behaviors in an online apparel retailing setting(Doctoral dissertation, The Ohio State University), (2006)

39. Kim, W. G., & Moon, Y. J., Customers' cognitive, emotional, and actionable response to the servicescape: A test of the moderating effect of the restaurant type. International Journal of Hospitality Management, 28, 144–156, (2009)
40. Krause, A. E., North, A. C., & Hewitt, L. Y., Music-listening in everyday life: Devices and choice.Psychology of music,43(2), 155-170, (2015)
41. Lam, L. W., Chan, K. W., Fong, D., & Lo, F., Does the look matter? The impact of casino servicescape on gaming customer satisfaction, intention to revisit, and desire to stay.International Journal of Hospitality Management,30(3), 558-567, (2011)
42. Levy, M., & Weitz, B. A., Retailing Management (7th ed.).Boston, MA: McGraw-Hill/Irwin, (2008)
43. Lin, I. Y., & Mattila, A. S., Restaurant servicescape, service encounter, and perceived congruency on customers' emotions and satisfaction.Journal of Hospitality Marketing & Management,19(8), 819-841, (2010)
44. Liu, Y., & Jang, S. S., The effects of dining atmospherics: An extended Mehrabian–Russell model.International Journal of Hospitality Management,28(4), 494-503, (2009)
45. Morin, S., Dube, L., & Chebat, J-C., The role of pleasant music in servicescapes: A test of the dual model of environmental perception. Journal of Retailing, 83(1), 115–130, (2007)
46. Namasivayam, K. & Lin, I., "The servicescape", in Jones, P., The handbook of Hospitality Operation and IT. New York: Elsevier Publishers, (2008)
47. North, A. C., & Hargreaves, D. J., Music and consumer behaviour.Oxford handbook of music psychology, 481-490, (2009)
48. Oakes, S., & North. A. C., The impact of background musical tempo and timbre congruity upon ad content recall and affective response. Applied Cognitive Psychology, 20, 505-520, (2006)
49. Putiy, A., Food Service -Hotel Restaurant Institutional.Moscow, RU: USDA Foreign Agricultural Service, (2011)
50. Ramlee, N., & Said, I., Review on Atmospheric Effects of Commercial Environment.Procedia-Social and Behavioral Sciences,153, 426-435, (2014)
51. Rea, C., MacDonald, P., & Carnes, G., Listening to classical, pop, and metal music: An investigation of mood. Emporia State Research Studies, 46(1), 1-3, (2010)
52. Ribeiro Soriano, D., Customers' expectations factors in restaurants: The situation in Spain.International Journal of Quality & Reliability Management,19(8/9), 1055-1067, (2002)
53. Rosenbaum, M. S., & Massiah, C., An expanded servicescape perspective. Journal of Service Management, 22(4), 471-490, (2011)
54. Ryu, K., DINESCAPE, emotions, and behavioral intentions in upscale restaurants. Unpublished doctoral dissertation, Kansas State University, Manhattan, KS, (2005)
55. Ryu, K., & Han, H. New or repeatcustomers: How does physical environment influence their restaurant experience? International Journal of Hospitality Management, 30, 599-611, (2011)
56. Ryu, K., & Jang, S., DINESCAPE: A scale for customers' perception ofdining environments. Journal of Foodservice Business Research, 11(1), 2-22, (2008)
57. Ryu, K., Lee, H. R., & Kim, W. G., The influence of the quality of the physical environment, food, and service on restaurant image, customer perceived value, customer satisfaction, and behavioral intentions. International Journal of Contemporary Hospitality Management, 24(2), 200-223, (2012)

58. Seung, A. Y., Customer Perceptions of Restaurant Cleanliness: A Cross Cultural Study. Unpublished master's thesis, Virginia Polytechnic Institute and State University, Blacksburg, Virginia, (2012)
59. Shashikala, R., & Suresh, A. M., A Study on Consumer Perception of Servicescape in Shopping Malls. Vishwakarma Business Review, 3(2), 68-75, (2013)
60. Siu, N. Y., Wan, P. Y. K., & Dong, P., The impact of the servicescape on the desire to stay in convention and exhibition centers: The case of Macao. International Journal of Hospitality Management, 31, 236-246, (2012)
61. Teng, C-C., & Kuo, T-H, The combined effects of employee hospitality performance, environment, and entertainment on customer affective response and revisit intention in restaurants. Advances in Hospitality and Tourism Marketing and Management Conference Proceeding, 3-7, (2011)
62. Tlapana, T. P., Store layout and its impact on consumer purchasing behavior at convenience store in Kwa Mashu. Unpublished master's thesis, Durban University of Technology, (2009)
63. Vaccaro, V., Yucetepe, V., Torres-Baumgarten, G., & Lee, M. S., The Relationship of Music-Retail Consistency and Atmospheric Lighting onConsumer Responses. Review of Business Research, 8(5), 214-221. Retrieved July 26, (2008)
64. VanBaren, J. Store Layout Definition. eHow.Retrieved July27, 2016, from http://www.ehow.com/facts_6951533_store-layout-definition.html, (2012)
65. Voon, B. H.Service Environment of Restaurants: Findings from the youth customers. Journal of ASIAN Behavioral Studies, 1(2), 45-56, (2011)
66. Vrechopoulos, A., O'Keefe, R., Doukidis, G., & Siomkos, G., Virtual Store Layout: An Experimental Comparison in the Context of Grocery Retail. Journal of Retailing, 80(1), 13-22, (2004)
67. Wilson, A., Zeithaml, V. A., Bitner, M. J., & Gremler, D. D., Services marketing: Integrating customer focus across the firm. McGraw Hill, (2012)
68. Yee, Y. S., & Yuen, J. L. F., The Recruitment of Migrant Workers in the Food Service Industry in Malaysia. A Study of Old Town White Coffee and Pappa Rich Kopitiam. International Journal of Business, Economics and Management, 1(10), 291-304, (2014)
69. Zeynep, E., & Nilgun, G., Congruence between atmospheric cues and store image in fashion retailing. Turkey: Izmir University of Economics, (2011)

Problem-based Learning (PBL) in Sociolinguistics as a Way of Encouraging Active Learning

Engku Haliza Engku Ibrahim[1]

[1]English Language Division, Centre for Languages and Pre-university Academic Development, International Islamic University Malaysia, Jalan Gombak, 53100 Kuala Lumpur

Abstract. The major concern of this paper is to advocate the integration of PBL strategies in classroom instruction as a way of promoting active learning. It is undoubted that the benefits of problem-based learning (PBL) are numerous. In the sciences, PBL has been well integrated in the curriculum. This research reports of an experience of integrating problem-based learning in an introductory Sociolinguistics course for 60 undergraduates of a Bachelors of English programme through a semester that ran for 14 weeks. A focused group interview and questionnaire were used to find out the perceptions of the students undergoing the hybrid PBL course. The findings of this study reveal that students generally enjoyed the PBL approach and found that they had little choice but to become active learners. Some challenges faced by the learners were also highlighted. These findings have implications for the integration of PBL in the field of social sciences.

1 Introduction

The advent of globalization has both presented many opportunities as well as challenges for the society today. This has forced individuals to attain different sets of skills to ensure they are competitive in the job market. The education sector plays a key role in ensuring students are equipped with a plethora of soft skills such as critical thinking skills, problem solving skills to name a few. Many approaches have been adopted in the education sector to harness these essential but often neglected areas. The problem-based learning is one such approach that has gained popularity in the past decade.

PBL has gained widespread acceptance in all levels of education; from kindergarten right up to the tertiary level. It has also been widely practised in a multitude of disciplines; Art, Mathematics, English and Education [3] even though the history of PBL can be traced to have started in the field of medicine at McMaster University in Hamilton, Ontario, Canada [4]. It was introduced in the field of medicine as a solution to overcome student boredom in classes, student not being able to apply what has been taught to clinical practice and to improve the professional skills of new graduates [12]. Its advantages in promoting a learner-centered education that builds on principles of self-directed and active learning have convinced more and more professionals to practice it as an attempt to better prepare learners for the world of professional practice.

In the PBL environment, the learner is presented with the problem as a stimulus for learning [5]. These problems are ill-structured which allow for free enquiry on the part of the learners. Through the process of problem solving, they have to rely on teamwork where critical thinking skills, self- directed learning abilities and negotiation skills come into play. These are the needed skills when learners are in their workplace [9].

Many studies have indicated positive results in terms of engaging learners in active and meaningful learning which results in deeper understanding and longer retention [2, 1, 7]. These research have also yielded positive feedback from learners. It is also said to be a student-centred approach that is in line with the principles of constructivism where students are focused on solving stimulating and open-ended questions that are carried out through collaborative learning. (Hmelo-Silver, [9]; Savery & Duffy, 1995). According to Freeman, Elly and McDonough [6] PBL leads to active learning which in turn increases the performance of students. Additionally, Hartling , Spooner , Tjosvold & Oswald [8] conclude that students have to take up the inquisitive style of learning in PBL. This leads them to become self-directed learners. Malan and Ndlovu [10] has also shown that the integration of PBL in a university foundation programme increase self-directed learning.

However, a transition from the traditional didactic curricular to PBL is not without challenges. One important change is that in a PBL environment the focus shifts from that of teaching to the learning process which involves active participation from all learners.

Baptiste [3] lists several essential underlying values that group members have to have in order to ensure effective group functioning which could be summarised as:

- Partnership

Group members are encouraged to minimise competition within the group. Infact, they are to capitalise on the notion of partnership and collaboration in the process of learning. Learners and facilitators are part of the same system with both parties gearing towards the same direction.

- Honesty and openness

Learners should be aware of the learning issues from the very beginning so that they can be kept on track. They should also be open to each other so they can learn from the experience of their peers. The general learning climate should be on the basis of open relationship between

learners and facilitators so that learning can happen from both ends of learners and facilitators.

- Respect

Respect on the part of learners can be manifested many ways; punctuality, paying attention, asking of questions and being engaged in the learning process. Facilitators would also contribute in the giving of proper feedback to the learners in the process of inquiry.

- Trust

Trust is the other essential value after the development of partnership values, honesty, openness and respect, as only with trust can risk-taking take place. An indication that the group is functioning well is when risk-taking becomes a natural process of the group dynamics.

In order to foster the above characteristics, the roving facilitator would have to play a major role at the early stages. This can be the role of the coach of a team, always asking and probing from every member of the team in their discussions. The coach has to also be observant of the participation of every group member. If one seems withdrawn, it is the role of the facilitator to encourage more involvement, if one is always late, the coach has to inquire. The facilitator has to also inquire about different learner roles in the group. Peer feedback form is a useful way of ensuring everyone actively participates in group work while allowing others to provide feedback on individual involvement and contribution. It also trains learners to evaluate their peers in the group contribution. In addition, it also highlights their individual contribution towards the group dynamics.

One of the biggest benefits of using the PBL approach is the focus on learner inquiry in the process of problem solving. Learners in general are not used to having to carry out independent inquiry and problem solving and they have to start off "groping in the dark". Many learners are typically unsure of how to progress and which direction to follow. This could be attributed to the fact that a majority of our students are a product of traditional learning methods where independent learner inquiry was never a skill required of them. One of the reservations of our traditional system of education is that we produce graduates who are used to passively acquiring knowledge in isolation. They are more accustomed to being told exactly what to do and memorising and regurgitating facts from textbooks. In these kinds of learning, everything that is needed to be learnt has been given to them by the lecturer. Whatever needs to be learnt in the course has spelled out for them, without requiring much effort and thought form the learners. The knowledge acquired is also frequently seen as inapplicable to the real world when they leave the university. In a PBL environment, it is very normal to see learners become extremely frustrated at the beginning of being presented with an unstructured problem. Many learners cannot cope with 'not knowing' how to proceed. Some find it extremely disturbing not knowing which grounds they should cover and to what extent, while others may be insecure that they have researched enough. Generally speaking, this literature review presented the evidence that there are numerous advantages when PBL is introduced as an instructional method. The challenges experienced were also presented.

Thus the main objective of this research was to find out if the introduction of PBL in the teaching of Sociolinguistics would be able to produce active, self-directed and self-regulated learners.

2 Methodology

Purposive sampling was used in this single case-study. The participants were 81 students from 2 classes of an introductory Sociolinguistics course. They were all Malaysian students in the first and second year of the Bachelors of English Language and Literature Programme. The semester was made up of 14 weeks. To accommodate for the different learning styles of our learners, during the first half of the semester (Week 1 – Week 7), the traditional lecture method of teaching was used. The first half of the semester was mainly used to focus on the key concepts of Sociolinguistics and to build a framework of ideas related to sociolinguistics. During the second half of the semester (Week 8 – Week 14), students were assigned to groups heterogeneously by year of study and by academic ability by the instructor. This was to ensure that learners of mixed ability could pool their talents collaboratively in their process of inquiry. Each group was assigned a different set of problems on learning issues that had not been covered in the lectures during first 7 weeks of the semester. They were given four in-class sessions to function in groups with a roving facilitator. This is consistent with the recommendation of Gallagher [7] that students are to be given academic scaffolding to ensure accessibility of subject matter. During these sessions, the facilitator observed group functionings, asked probing questions to ensure equal understanding of all group members, steered the group in the right direction if needed and mostly ensured that learners were comfortable in the new learning environment.

At the end of the semester, as part of their assessment, learners were expected to produce the following:

- An action plan – proposal of problem solution
- A group PowerPoint presentation

- A handout for their classmates
- A mini research paper on the problem
- Provide peer feedback on the contributions of each group member to the group functioning based on the provided peer feedback form

Prior to the actual PBL assignment, students were given a trial run of the PBL approach in the first half of the semester when the traditional didactic approach was used. This was found to be effective in promoting learner confidence to slowly train learners to deal with real-world problems as a learning task albeit with the key concepts covered in lectures. After a lecture session on one of the learning issues, a problem was designed and assigned to learners. They were then expected discuss in groups for possible solutions to the problem based on topics covered in the lectures. They were then asked to present their solutions with justifications to the class. The various solutions proposed were then discussed in depth and at great length. The main difference with an actual PBL exercise is that learners are assigned problems without prior knowledge.

The author obtained written consent from the students who were involved in the study. The participants' anonymity and confidentiality were maintained throughout the study.
At exit point, learners were given a set of questionnaires to elicit their responses on the PBL experience. Focus group interviews (with 5 to 8 students in a group) were also used to further probe for learner response. Data from observation of group functioning was also taken into account in the data reporting of this study.

SAMPLE OF PROBLEM 1
Source: Reader's Digest; Topic: Politeness across societies

The *Reader's Digest* (R.D.) survey made world headlines on June 21, 2006. The survey concluded that New Yorkers are the most polite and civilized people in the world while Mumbaikars are the rudest on the earth. For Mumbaikars this was an awakening to start behaving well, otherwise they will be tagged as uncivilized folks. Newspaper headlines read "New York the politest city, London half courteous," "N.Y. very polite, Asia quite rude" and "If in Mumbai, learn to be rude."

Do you think this is a true reflection of Asians? Do you agree or disagree with the report?

SAMPLE OF PROBLEM 2
Source: Jokes from the internet; Topic: Style, context and register.

Joke 1
A patient has a sore throat and goes to a doctor to get treatment for it.

Doctor: Your tonsils gotta come out.
Patient: I wanna second opinion.
Doctor: Okay, you're ugly, too.

Joke 2

Tech Support: "I need you to right-click on the Open Desktop."

Customer: "Ok."

Tech Support: "Did you get a pop-up menu?" Customer: "No."

Tech Support: "Ok. Right click again. Do you see a pop-up menu?"

Customer: "No."

Tech Support: "Ok, sir. Can you tell me what you have done up until this point?"

Customer: "Sure, you told me to write 'click' and I wrote click'.

Joke 3
How do you keep a programmer in the shower all day?
Give him a bottle of shampoo which says "lather, rinse and repeat".

Joke 4
One reason the Military Services have trouble operating jointly is that they don't speak the same language. For example, if you told Navy personnel to "secure a building," they would turn off the lights and lock the doors. The Army would occupy the building so no one could enter. Marines would assault the building, capture it, and defend it with suppressive fire and close combat. The Air Force, on the other hand, would take out a three-year lease with an option to buy.

3 Finding

What was apparent was that, in general, learners' responses towards their first exposure to PBL were very positive and encouraging in nature. After the initial stage of struggling with the ambiguity of assigned problems and handling group dynamics, learners found the experience very rewarding in terms of encouraging learner autonomy and self-directed learning. There was positive feedback on deeper understanding of the learning issues assigned which lead to a greater sense of confidence.

As far as this study is concerned, success can be viewed from the summary of positive remarks made by learners on an exit questionnaire, focused group interview as well as observation.

Table1: Summary of positive aspects of PBL

Summary of findings based on triangulation of exit questionnaire (Q), focused group interview (FG) and observation (O)
Active learning • *It forced us to actively participate in groups to ensure successful problem solving (Q)* • *It allowed active participation of students (Q)* • *It required us to think critically, analytically, and creatively (FG)* • *We had to learn to take responsibility for our own*

learning. The more we read on our topic, the more confident we are
Self-directed learning • *We had to carry out extensive research in order to effectively understand and solve the problem (FG)* • *We had to meet regularly with our group mates for discussions (FG)* • *It made us responsible for our own learning (Q)* • *We were not being spoon fed by lecturers (Q)* • *We have to think out of the box to solve the problems assigned to us (FG)* • *Many students took the initiative to research, using multiple resources, for ideas in the attempt to solve problems (O)*
Self-regulated learning • *We had to ensure that we were prepared to participate in our groups when we came to class (Q)* • *I feel so motivated to find the solution to the problem. It is a challenge (Q)* • *It forced us to be systematic in clarifying our ideas so that different people can understand us(Q)* • *We had to listen and evaluate the ideas of our group mates (FG)* • *Our group had to look for solutions to the problems from many sources – the internet, books and also experts in the field (FG)* • *It allowed students themselves to be more sensitive to their own learning needs (Q)* • *We had to be independent in learning to solve the problems and can no longer just wait for the lecturer to tell us what to read. (FG)*
Collaborative skills • *It taught us to learn to cooperate with each other in our groups (FG)* • *Brainstorming in the group helped to generate ideas in problem solving (FG)* • *It forced us to interact with other people in our class – we would not have interacted with them if not for forced grouping by the lecturer (Q)* • *It taught many students the value of interaction and negotiation with others and how more ideas contribute to more options of problem solution (O)*
In depth understanding • *We have to make sure that we have in-depth understanding of the topic rather than just depending on rote memory (Q)* • *Students showed a visible deeper understanding of concepts during class discussions, mini research proposals and presentations(O)*
Real-world application

https://doi.org/10.1051/matecconf/201815005075

• *It developed our problem-solving skills based on real world scenarios (FG)* • *We have to apply what we had learnt in class to the real world (FG)* • *It is very interesting to know that what we learn in class is actually applicable to the real world; most times what we learn in class is only theory and mostly cannot be applied to our real world experience (FG)* • *We have to apply some of the things (concepts) that we had been exposed to since the beginning of the semester (FG)* • *It gave students a sense of achievement to know that what they have learnt in class is relevant to the real world(O)*
Student-centred learning • *We had to look up for the solutions ourselves (FG)* • *The teacher is not giving us all the information as usual, we had to research ourselves (Q)* • *Students had to manage their own learning and were very resourceful (O)*

In the focused group interview learners were asked to list some of the things that they didn't quite like about the PBL approach, if there were any. Their responses could be summarised in the categories listed below:

Table 2: Challenges of PBL

Time consuming • *We spent lot of time trying to solve the problem* • *We have to negotiate with all the group members to agree on one solution* • *Dividing the work can take a lot of time especially if member of the group did not do what they are supposed to do*
Uncooperative group members • *We had problems when some group members were not cooperative eg. Some did not come for meetings* • *Some group members did not want to listen to the ideas of others* • *Some group members were very quiet and just agreed with all suggesstions*
Inquiry process • *It was stressful in the beginning stages of being assigned the problems, we were unsure of how to begin solving problems* • *Difficult to understand certain concepts at the beginning of the PBL session; we had to do a lot of research to be sure we are on the right track*
Change in assessment technique • *We were nervous about the expectations of the lecturer since most of our previous assessments consist of tests*

 https://doi.org/10.1051/matecconf/201815005075

and exams.
• *Nervousness during class presentation*

However, when asked which learning mode students preferred, their response could be summarized in Fig. 1.

Fig. 1: Students' preference of classroom environment

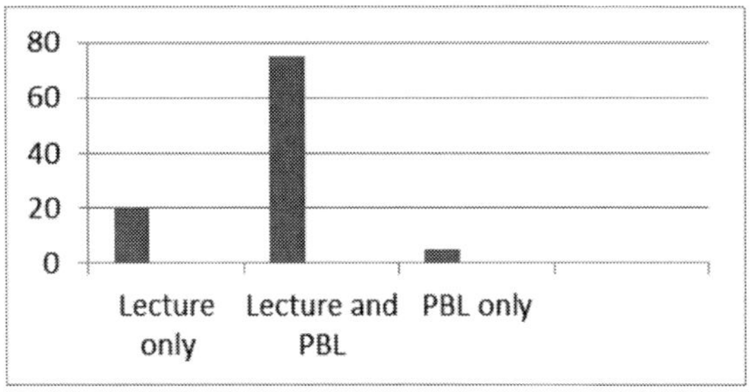

Fig. 1 shows that a majority of the students (75%) preferred the Lecture and PBL mode. 20% preferred Lecture only while only 5% wanted PBL only. In the focus group interviews, students were asked if they would like PBL only in future. A majority mentioned that they needed the lecture to provide them with the basics of the course content. In other words, most of the students felt that they needed the lecture during the early part of the semester even though the lecture was not on the topic areas covered by PBL. Further analysis of the questionnaire revealed that a majority of the students strongly agreed that PBL helped in building interest (85.6%), made them have a deeper understanding of the subject matter (90%) and promote self-directed learning (93.3%). 94% agree that PBL motivated them to be more active learners and 85 % strongly argued that it facilitated cooperative learning. The PBL sessions were highly rated by almost all participants. Overall, 95 % subjects strongly recommended that more PBL sessions should be incorporated in the curriculum.

Despite the challenges faced by students who were mostly not familiar with the PBL environment, the first exposure appears promising as the change in paradigm was quite drastic for the majority of students who have been mainly brought up in a system of traditional didactic learning environment. None of the students had ever participated in a PBL learning environment.

4 Discussion and Conclusion

Clearly, learners themselves recognize the benefits of the PBL when compared to the traditional lecture method of teaching. This can be seen as they realise that PBL enhances active learning while forcing them to take accountability and responsibility for their own learning. They also had to learn to function collaboratively and effectively in groups of differing personalities.

The findings of this study point to an area of future research, also highlighted by Nariman & Chrispeels [11], which is the issue of how to teach students to ask their own questions related to standards and set learning goals, and how to engage them in their search to find answers. Upon reflection of this PBL experience, despite the challenges that accompanies the PBL approach, the positives far outweigh the negatives. The PBL approach has also been an eye opener and a reminder that as an educator, one never ceases to learn new things and this has amplified the researcher's enthusiasm and energy for the teaching of learning. There was active learning on both parts of the facilitator and the learners. Learners were so involved in solving the ill-structured problems that they didn't notice that they were more active and involved in their own learning. The only thing that was more satisfying was the realisation that at the end of the semester both learners and facilitators greatly improved their communication and interpersonal skills. The challenge that remains at this stage is to extend PBL to other courses in the curriculum.

References

1. Ahmed Yaqinuddin,. Problem-Based Learning as an Instructional Method. Journal of the College of Physicians and Surgeons Pakistan, 23, 1: 83-8, (2013)
2. Bajaj Preeti, Ahuja Ashish and Gosavi Shriram, Journal of Clinical and Diagnostic Research, 7, 12: 2896-2897, (2013)
3. Baptiste, S. E., Problem-based learning: A Self Directed Journey. New Jersey: Slack Incorporated. Barrows, H. S., & Tamblyn, R.M. (1980) Problem-Based Learning: An Approach to Medical Education. New York: Springer Publishing Compan, (2003)
4. Barrows, H. and R. Tamblyn, Problem-based Learning: An approach to medical education. New York, Springer, (1980)
5. Boud, D., and G. Feletti, The Challenge of Problem-based Learning. London: Kogan Page, (1998)
6. Freeman, S., Eddy, S. L., McDonough, M., Smith, M. K., Okoroafor, N., Jordt, H., and M.P., Wenderoth, Active learning increases student performance in science, engineering, and mathematics. Proceedings of the National Academy of Sciences of the United States of America, 111, 23: 8410–8414, (2014)
7. Gallagher, S. A., What do you need to know? Becoming an effective PBL teacher. In MacFarlane, B, and Stambaugh, T. (Eds.), Leading Change in Gifted Education: The Festschrift of Dr. Joyce VanTassel-Baska, 337–350, Waco, TX: Prufrock Press, (2009)
8. Hartling L, Spooner C, Tjosvold L, and A. Oswald,. Problem-based learning in pre-clinical medical education: 22 years of outcome research. Med Teach, 32: 28-3,(2010)
9. Hmelo-Silver, C. E., Derry, S. J., Bitterman, A., and N. Hatrak,. Targeting transfer in a STELLAR PBL course for preservice teachers. The Interdisciplinary Journal of Problem-based Learning, 3, 2: 24–4, (2009)
10. Malan, S.B. and M. Ndlovu, Introducing problem-based learning (PBL) into a foundation programme to develop self-directed learning skills. South African Journal of Education, 34, 1, (2014)
11. Nahid Nariman and J. Chrispeels,. PBL in the Era of Reform Standards: Challenges and Benefits Perceived by Teachers in One Elementary School. The Interdisciplinary Journal of Problem-based Learning, 10, 1, (2016)
12. Newman, M.J.,. An Introduction and Overview of the Key Features of the Approach. Journal of Veterinary. 32, 1: 12 – 20, (2005)

English Language Textbooks and Portrayal of Culture: A content analysis

Ruzai Syarilili Aiyu Abd Rashid[1,*], *Engku Haliza* Engku Ibrahim

[1]English Language Division, The Centre for Languages and Pre-University Academic Development (CELPAD), International Islamic University Malaysia, 53100, Jalan Gombak, Kuala Lumpur,Malaysia

Abstract. Research has shown that there is a need to use materials that are cultural sensitive and context-appropriate in the EFL classroom. It is imperative that teaching materials match with the goals, objectives and philosophy of the programme. This is to ensure that the materials are congruent with the learners' attitudes, religious beliefs and preferences. Thus, this study aims to find out if there were materials that are not in harmony with Islam. The sampling of this study was taken from 5 textbooks that were being used in an intensive English course. The findings from this study showed that in general the contents of the ELT textbooks could be categorised into three categories of neutral, positive and negative. Additionally, it was also discovered that these themes fall under community knowledge, beliefs and values of the western culture in Byram's (1993) cultural content checklist. This study has strong implications for the use of ready-made textbooks in the EFL context.

1 Introduction

Many researchers have highlighted that textbooks play an important role in teaching and learning, especially in foreign language classroom. It has been emphasised by Hatoss [26] the importance of culture teaching through language learning that without textbooks, English teachers would find it difficult to teach a cultural content to their students as textbooks supply this need without transporting the students to the native speakers' country. "The presence of cultural content in course books is also important for the development of students' intercultural communicative competence" [53]. Today there is a plethora of commercially available textbooks in the market, thus the growing need to determine whether the teacher or the decision maker has made the right choice concerning the right textbook for the right purpose. As Anjaneyulu [9] affirmed, "the analysis of the textbook would yield insights as to its suitability... It was therefore important to examine whether it corresponded to the learners needs of the particular situation" (p.181).

Wright [55] who studied the factors shaping the attitudes of secondary education learners in Great Britain towards French, found that textbooks are viewed by learners of languages as having a significant influence on their attitudes towards the target culture and

* Corresponding author: ruzai_syarilili@iium.edu.my

that there is a significant positive correlation between this influence and the attitudes of many pupils. Additionally, Yamada [57] stated that a textbook is not merely a commercially available reference book which is used in a language teaching programme, it also serves as an important means to deliver political and ideological orientation. She further pointed out that "textbooks can be understood and examined as curricular materials, which project images of society and culture. They reflect historical, social, cultural and socio-cultural, economic and political contexts" (p. 20). Dellinger [16] too asserted that language is a type of social practice used for representation and signification. Texts, including textbooks, are written by socially situated persons. Thus, textbook descriptions can be examined as a social product. In other words, textbooks determine not only material conditions for teaching and learning but also define what elite or legitimate culture is passed on [10]. Apple and Christian-Smith too advocate that a textbook signifies a "particular constructions of reality, particular ways of selecting and organizing that universe of possible knowledge" (1991, p.3).

The constantly changing landscape of higher education has compelled the teaching and learning of English to be culturally sensitive to the context in which the language is taught and used (Kachru, [30]; Prodromou, [38]; Fawwaz and Oqlah, [19]; Ratnawati, [41]; Phillipson, [37]; Al-Attas, [4]; Pennycook, [36]; Canagarajah, [12]; Cardona, [13]; Yamada, [57] and Ainol and Isarji, [3]) This is especially true after the 9/11 tragedy. The teaching and learning of English began to undergo a robust push for the utilisation of teaching materials that meet the needs of learners, institutions, and industry. It necessitates ELT materials and learning outcomes to be well-matched. Despite the realisation of the importance of meeting the needs of ESL learners, institutions, and industry, it is extremely difficult for teaching materials to be socio-culturally neutral because it has to express a social order and value system, implicitly or explicitly [43] With the absence of proper guidance, the tasks could turn out to be complicated in terms of deciding whose value system or social order needs to be incorporated.

On this account, there is an urgent need to study ELT materials from socio-cultural perspectives and its appropriateness to Muslim learners to ensure that the materials complement learners' needs and the vision and mission of the institution.

Literature review: The literature on second language teaching and learning and ELT materials have addressed the need for both learners and teachers to be aware of the dynamic interplay of language, thought and action and the macro and micro cultures on ELT theory and practices. Many studies have been done on socio-cultural perspectives in second language learning. These include the works of Haja Mohideen [22,23,24], Casewit [14], Shamimah (2008), Mojgan Majdzadeh Tabatabaei [31], Mohamed S. Haded [33], Abdullah [2], Yadullah [56], Ratnawati [41,42], Tariq [52], Hyde [28], Fawwaz and Oqlah [19], Ali Ahmed [6], Dahiru [15], Mohamed Ismail [32], Grace [21], Qiang and Wolf [40], Sohail and Pennycook [36], Mohammad Shafi [34], Farzad [18], Pennycook [36], Sami [44], Ainol and Isarji [3], Isarji and Mohd. Feham (2000). Alghamdi and El-Hassan [5] emphasised specifically that "English-language teaching must be responsive to the local culture and realities and must be driven by local needs" (p. 1).

Casewit [14] for example, stated that in developing appropriate materials for learners in Muslim societies, one must look beyond and beneath such superficial confirmation of the social psychological theory of foreign language acquisition. He continues that multicultural TEFL materials could be prepared in such a way as to promote more favourable attitudes and stronger motivation on the part of learners in Muslim societies.

Similarly, Haja Mohideen [23] articulated that language teaching and learning materials need to include a section on offensive expressions to warn users against their usage because English today is not used exclusively by native speakers of English and there is a vast number of English as second language speakers who have to use the language but need to know that the target language is appropriate to their religious and cultural values.

Washima et al. [54] highlighted the need to actively promote materials related to the home culture of the learners. This is an effort to minimise the negative effect of ELT materials that could include unwelcome and contradictory values. With the incorporation of local cultures, it encourages both students and teachers to be aware of potential conflicts, to discuss and to critically reflect on how they might respond appropriately.

In selecting reading materials, teachers of Muslim learners need to exercise discretion in the choice of reading texts. In those classes where the learners have a reasonable degree of maturity, the teacher will need to develop the skill of critical reading with particular reference to how easily the unsuspecting reader can be misled into accepting values and norms which are incongruous with the Islamic way of life.

More recently, Alshammari [7] argued that "the existing ESL content lack[s] the diverse nature of English of today and conflict[s] with the Islamic discourses and values" (p. 369). Similarly, Ahmar Mahboob and Tariq Elyas [8] argued that historically, English in Saudi Arabia has been nativised to reflect local cultural practices.

Even though Casewit [14], Washima et al. [54], Haja Mohideen [23] Sugirin et al. [48], Abbas [1], Hermawan [27], Dweikat and Shbeitah [17], Syahri and Susanti [49] and many others have provided critical appraisal of ELT materials in L2 language teaching and learning, there is a need to include empirical evidence based on data collected from ELT materials. More importantly, there is a lack of empirical evidence on the analysis of the content of ELT textbooks which are considered culturally inappropriate to Muslim learners. Considering this inadequacy and in view of the fact that learners are the most important clients of an education institution, it is imperative to study this concern systematically to ensure that the needs of the learners and the institution are accurately met.

The main purpose of the study is to analyze the socio-cultural and religious content of ELT textbooks used in the English Language Programme at the International Islamic University Malaysia. Based on a content analysis, this study also seeks to find out how the materials were being portrayed according to an adaptation of Byram's (1993) Cultural Content Checklist with specific reference to neutral, positive and negative contents.

Specifically, this research will address the questions below:

1. Are there contents that relate to Islam or Muslims in the textbooks?
2. How are the contents portrayed?

2 Material and Methods

Data for this study was collected through three major steps; document search, conceptual content analysis and content categorisation based on an adaptation of Byram's (1993) cultural content checklist.

5 textbooks were analysed for the purpose of this study.

1. Interactions 1: Listening and Speaking (T1)
2. Interactions 2: Listening and Speaking (T2)
3. Interactions 2-Reading (T3)
4. Consider the Issue: Listening and Critical Thinking Skills 3rd Ed. (T4)
5. Engaging Writing-Paragraphs and Essays. (T5)

The first step in the data analysis was deciding which data would be analysed throughout the study; which in this case, the chosen textbooks. The concepts covered were the cultural elements of western culture which relate to Islam and the Muslims. At the same time the researcher investigated whether there were western culture elements that were not in harmony with Islam. The selected textbooks were also evaluated to find out how Islam and Muslims were portrayed and depicted.

The content analysis applied in this study was utilised as a method of inquiry to explore the cultural content focusing on the different features of western culture which is not in harmony with Islam. References to cultural content of Islam and the Muslim culture mentioned in the textbooks were also identified throughout the materials and were extracted word for word and then coded according to the four elements of Byram's (1993) cultural content checklist as listed below:

- belief and behaviour
- socialisation and life cycle
- stereotypes and national identity
- social and political institution

3 Results

Content analysis of the 5 textbooks chosen could be summarised as:

- T1:1 neutral cultural content
- T2:2 neutral cultural contents and 1 positive cultural content.
- T3; 11 neutral cultural contents and 6 negative cultural contents.
- T4: no positive, neutral or negative contents were found
- T5: 2 negative cultural contents

In total there were 23 contents identified, with 14 neutral cultural contents, 1 positive cultural content and 8 negative cultural contents.

Neutral Contents: Based on the analysis, there were thirteen contents that had a neutral portrayal of Islam and Muslims. Some examples are listed below:

- Nabil and the pet: cat.
- Hip-hop in Istanbul.
- Water.
- Islamic Wedding – Bedouin and Henna.
- Bedouin Culture- Henna.
- Mehndi-Islamic cultures.
- United Arab Emirate-unmarried Emeriti men to marry Emeriti women.
- Churches, mosques and palace-structures.
- Arabesques- exquisite design.
- Mosques-Islamic religious worship
- Christianity- religion- Islam.
- Spread of Islam to the East.
- The Middle east-Reality TV show-The Big Brother.
- Banking in the United Arab Emirate.

According to Byram's [11] cultural content checklist, 5 relates to belief and behaviour, another 5 to socialisation and life cycles, 3 to stereotypes and national identity and only one relates to social and political institution.

Positive Contents: There was one positive portrayal of Islam and Muslims found in T2 under the category of stereotypes and national identity. The content was regarding the numeral symbol. According to Byram's [11] cultural content checklist, this falls under stereotypes and national identity.

Negative Contents: Based on the analysis, it was found that there are seven contents that had negative portrayals of Islam and Muslims. Three could be categorised under Byram's [11] category of belief and behaviour, four under socialisation and life cycles and the last remaining one under the category of social and political institution. The following examples can be found:

- Islam and Buddhist Art.
- Term: Holy Place.
- Emeriti Wedding.
- Bedouin wedding involves money.
- Emeriti wedding are too lavish.
- Kamal Ataturk.
- Arab's Hospitality.
- The Quran and Hadith

Under Byram's [11] cultural content checklist, these contents relate to belief and behaviour (3), socialisation and life cycles (4) and one to Social and political institution.

4 Discussions

The findings of this study support Risager's [43] contention that it is extremely difficult for teaching materials to be socio-culturally neutral because they have to express a social order and value system, implicitly or explicitly. One of the significant findings which emerged from this study is that the contents of the five textbooks seem to portray the social order and value system of the Judeo-Christian. However, it is interesting to note that while the Judeo-Christian social order and value system were being portrayed positively, contents related to Islam and Muslims were presented in a negative manner.

In view of the fact that these textbooks are intended for an international audience, it can be assumed that the writers not only had very limited knowledge of Islam and the Muslim culture, but also projected a negative representation of Islam and Muslims. As a result of the negative representation, Muslim learners might feel frustrated or demotivated in their effort to acquire English. A prolonged exposure to negative representations of Islam and Muslims in ELT textbooks, without proper guidance of experienced teachers, may corrode their worldview and Aqidah.

The study shows that 23 contents mentioned Islam and Muslims. Of the 23 contents, 69.5% were presented in a neutral way; and 30.4% were presented in a negative manner. It is assumed that this is done unintentionally due to a lack of understanding of different cultural values.

The findings of this research support the assumption that students' motivation may be negatively affected if the materials that they are dealing with are offensive or are not in harmony with their personal worldview. Similarly, teachers may be put in a difficult situation if students strongly voice their objection when these materials are used in the classroom. These findings are in line with Dweikat and Shbeitah's content analysis [17] where they too found that:

the most frequent cultural values were related the American culture which indicated an obvious bias towards the foreign cultural values in general and the American cultural values in particular at the expense of the Arab–Islamic culture as no unit talked about Arabs or Muslims except the mentioning of the word "Arab" which was mentioned only once in the textbook and in a way that even displeases the Arabs. (p.564)

It is also important to realise that if the content relates to universal values like honesty, accountability, social justice, beauty and marriage; the author has to use extreme caution because the comparison that is made in the textbooks is substantial when two cultures are compared with each other. For example, in T5, there are short articles that talk about celebrations in different countries. An article about Fiji presents a celebration as follows;

Awards are presented to young people for the annual essay competition, and there is a singing contest between church choirs….sometimes winners playfully teases the losers, but there is no serious competition. After the games, players stay to chat with their friends.

Finally, at home a big celebration is underway. Friends and, relatives arrive early to help make palusami-a mixture of chicken, onion and tomato, wrapped in a taro leaf and cooked in coconut milk. The head of the household makes the traditional drink, yaqona, in a large bowl and offers it to the guest in coconut shells. After the yaqona drinking, everyone eats palusami, rice and salad. Fiji Day is our most popular celebration because it incorporates all the things Fijians love most: our national heritage, sports, food and drink, and socializing with family and friends. (p.84)

The portrayal ends in a positive manner. However, the following is used to describe a celebration in Islamic culture;

Another important aspect of Arab culture is hospitality. In Arab countries from Morocco to Iraq, visitors are very warmly welcomed. Relatives, friends, and even strangers are always offered coffee or tea and perhaps some bread or sweets, and if they happen to arrive at mealtime, they are urged to stay and eat. Guests cannot refuse these offers without offending their host. This tradition of hospitality dates back thousands of years. To survive in the desert, nomadic tribesman had to depend on one another, so it became a matter of honour to give and receive hospitality. The Quran and the books known as the Hadith have helped preserve this tradition of hospitality by urging people to practice kindness. But the holy books do not say that kindness and hospitality must be given without limit. Guest must know when to leave, and hospitality must be returned. Hospitality can strengthen the bonds between the members of society only when its rules are followed well. (p.98)

Quite often it might seem as if the depiction of a certain culture is neutral and harmless. However, upon scrutiny, we will see that they are often suggestive in the sense that one culture is cast in positive light while another is depicted negatively. Whereas the presentation of some cultures or regions shows both the positive and the negative sides and is therefore seen as neutral, for other cultures or countries only the good side is presented.

In general, the findings of this study suggest that the evaluated textbooks are loaded with features of the western culture, which affirms Risager's [43] observation that it is rather challenging for ELT materials to be socio-culturally neutral. In the context of this study, it is at the expense of Islam and Muslims. To a certain extent, the findings are coherent with those of Shatawani [46] that there is a clear religious and cultural bias in ELT materials. For example, when the authors make reference to the concept of water (in T2), they make an assumption that both Muslims and Christians have the same perception of water. This could stem from the publisher's lack of awareness of the way water is used as prescribed by the teachings of Islam.

The analysis of the textbooks revealed that there is a noticeable partiality towards foreign cultural values in general and this is at the expense of the Islamic culture.

5 Conclusion

There were contents that relate to Muslims and Islam in the textbooks chosen for this research. In term of Byram's [11] cultural content checklist, a majority of the contents that are related to Islam and Muslims are related to the category of socialisation and life cycles.

It is recommended that publishers consult experts on the cultures depicted in their textbooks so that they can be suitable and do not come across as offensive to certain groups of readers. Another option that these particular publishers should consider, and many today have done so, is to publish local editions of their textbooks in order to make the contents more culturally responsive to students' needs.

References

1. Abbas, I., Palestinian and global cultural dimensions in English for Palestine Textbooks for Grades 11 and 12. Journal of Al-Quds Open University for Research and Studies, pp. 69-91, (2011)
2. Abdullah Hamad, The Problem of Culture in Language Teaching. Muslim Education Quarterly. Vol. 16. No.2. The Islamic Academy, Cambridge. UK, (1999)
3. Ainol Madziah, Z and Isarji, S., November. Islamization of English Language Teaching. Proceedings of the 2nd International Conference on Language, Education and Diversity, 2007, University of Waikato, Hamilton, New Zealand, (2007)
4. Al-Attas, S.M.N, Islam and Secularism. Kuala Lumpur, Malaysia. Angkatan Belia Islam Malaysia, (1978)
5. Alghamdi, A. K., and El-Hassan, W.S., Multiliteracies and the Pedagogy of Empowerment: The Perspective of Saudi Female Students. The Journal of Teaching English for Specific and Academic Purposes, pp. 417-434, (2016)
6. Ali Ahmed Hussein., ESP Materials Development: the Case of Islamic ESP. English and Islam: Creative Encounters. p. 4, (1996)
7. Alshammari, A. K., Developing the English Curriculum in the Kingdom of Saudi Arabia: Possibilities and Challenges. Arab World English Journal, p. 369, (2015)
8. Ammar Mahboob and Tariq Elyas,. English in the Kingdom of Saudi Arabia. World Englishes, p. 138, (2014)
9. Anjaneyulu, T. A critical analysis of the English language textbooks in. ELT Research Journal, pp. 181-200, (2014)
10. Apple, M. and Linda Christian-Smith, eds,. The Politics of the Textbook. London. Routledge, (1991)
11. Byram, M,. Language and Culture: The Need for Integration. In M. Byram (ed.), Germany, Its representation in textbook for teaching German in Great Britain Germany, Frankfurt am Main: Diesterweg, pp. 3-16, (1993)
12. Canagarajah, S., Resisting Linguistics Imperialism in English Teaching. Oxford, England: Oxford University Press, (1999)
13. Cardona, S. M., Teaching English the Puerto Rican way: developing ESL materials that mirror the source culture for the Puerto Rican elementary English classroom. Puerto Rico: University of Puerto Rico, (2008)
14. Casewit, S. D,. Teaching English as a foreign language in Muslim Countries. Muslim Education Quarterly 2(2), Cambridge: UK, (1985)
15. Dahiru Muhammad Argungu,. English, Muslims and Islamisation: Between Needs and Deeds. English and Islam: Creative Encounters. p. 9, (1996)
16. Dellinger, B,. Critical Discourse Analysis. Retrieved on April 21, 2008 http://users.utu.fie.bredelli/cda.html, (1995)
17. Dweikat, K. A., and Shbeitah, G,. Investigating the Cultural Values in EFL Textbooks A Case Study of NorthStar Intermediate Textbook. The Islamic University of Educational and Psychological Studies Journal, pp. 564-589, (2013)

18. Farzad Sharifian. The Impact of Macro-cultures and Micro-cultures on ELT Theory and Practice. Retrieved May 3, 2008.http://www.ecu.edu.au/ses/research/CALLR/onlinejournal/1999/Sharifian99.htm, (1999)
19. Fawwaz al-Abed al-Haq and Oqlah Smadi., Spread of English and Westernization in Saudi Arabia. World Englishes. Vol 15 No 3, p. 6, (1996)
20. Fitzpatrick, M,. Engaging Writing-Paragraphs and Essays. Pearson Longman, (2005)
21. Grace Ebunlola Adamo., Globalisation, Terrorism and the English Language in Nigeria. English Today 84, Vol.21, No Cambridge University Press, (2005)
22. Haja Mohideen M.A., Second Language Teaching and Learning from an Islamic Perspective. Muslim Education Quarterly Vol.16.No.2 the Islamic Academy. Cambridge .U.K, (1999)
23. Haja Mohideen M.A., 'Mind Your Language': An Islamic Perspective. Muslim Education Quarterly Vol.17.No.2. The Islamic Academy. Cambridge .U.K, (2000)
24. Haja Mohideen M.A., Islamic Correctness in Language Use. Muslim Education Quarterly Vol.19. No.1. the Islamic Academy. Cambridge .U.K, (2001)
25. Hartmann, P and Kirn, E,. Interactions 2-Reading. McGraw-Hill, (2007)
26. Hatoss, A., A model for evaluating textbooks as mediators of culture. Babel Vol. 39 No. 2 pp. 25-32, retrieved 3rd December.2008.http://eprints.usq.edu.au/archive/00001137/02/Hatoss.pdf, (2004)
27. Hermawan, B., Traces of cultures in English textbooks for primary education. Indonesian Journal of Applied Linguistics, (2012)
28. Hyde, M., The Teaching of English in Morocco: The Place of Culture. ELT Journal Vol 48. No 4. Oxford University Press, (1994)
29. Isarji, S and Mohd. Feham., Islamic English and English Language Teaching (ELT) Materials. Proceedings of the International Conference on Islamization of Human Sciences, the International Islamic University Malaysia, Kuala Lumpur, (2000)
30. Kachru, B.B., The speaking tree: A medium of plural canons. In Georgetown Round Table in Languages and Linguistics (GURT). Washington, D.C.: Georgetown University Press, (1994)
31. Majdzadeh Tabatabaei, M., National identity and civic values in the pre-revolution and post-revolution English as a foreign language textbook in Iran. Ph.D. dissertation, Loyola University Chicago, United States -- Illinois. Retrieved June 17, 2008, from Dissertations and Theses: Full Text database. (Publication No. AAT 3212975) (2006)
32. Mohamed Ismail A.S., An Approach to the Islamization of the Teaching of English: Teaching Islamic Literature in the English Language Classroom. Muslim Education Quarterly Vol.16. No.2. p. 5, (1999)
33. Mohamed S.H., Focus on The Muslim Learner of English: The Communicative Approach and the Teacher's Role. Muslim Education Quarterly. Vol 16.No.1 1998 the Islamic Academy, Cambridge. U.K, (1998)
34. Mohamad Shafi,. Teaching of English as a foreign language: The Islamic Approach. Muslim Education Quarterly Vol. 1. No. 1. p. 5, (1983)
35. Numrich, C., Consider The Issue. Listening and Critical Thinking Skills (3rd ed.) Longman. NY, (2003)
36. Pennycook, A., English and Islam. Extracted from The Cultural Politics of English as an International Language. Retrieved on May 3 2008.http://www.tesolislamia.org/articles/pennycook.pdf, (1994)

37. Phillipson, R.H.L., Linguistic Imperialism. Oxford: Oxford University Press, (1992)
38. Prodromou, L., ELTJ correspondence. ELTJ, Vol. 52 no.3. pp. 266–267, (1998)
39. Prodromou, L., The good language teacher. In Teacher development: Making the right moves, selected articles from the English Teaching Forum 1989-1993, ed. T. Kral. Washington, DC: U.S. Information Agency, (1994)
40. Qiang, N and Wolff, M., Is EFL a Modern Trojan Horse? English Today Vol. 21, Issue 4 Cambridge University Press.UK, (2005)
41. Ratnawati, M.A,. The Cultural Implications of Teaching English as a Second or Foreign Language. Muslim Education Quarterly. Vol 14, No 4. The Islamic Academy, Cambridge, U.K, (1997)
42. Ratnawati M.A., English and Islam: A Clash of Civilization. Journal of Language, Identity and Education Vol.4.No.2, (2005)
43. Risager, K., Cultural references in European textbooks: An evaluation of recent tendencies. In: Buttjes, D.; Byram, M. (eds.), Mediating languages and cultures. Clevedon, UK, Multilingual Matters, (1990)
44. Sami Samra., The Cultural Factor in L2 Acquisition: Between Theory and Application. Proceedings of the Association of Professor of English and Translation at Arab Universities First International Conference, Jordan University. Retrieved May 15 2008. http://www.ndu.edu.lb/rfr.htm?http://www.ndu.edu.lb/academics/palma/20010701/CulturalFactorL2.htm, (2000)
45. Shamimah, M., Introducing Islamic Elements in English Language Teaching. Proceedings of the International Language Conference 2008, International Islamic University Malaysia, (2008)
46. Shatanawi, M., The Cultural Dimension in TEFL: A Case Study of the Cutting Edge Series, Unpublished Ph.D. Dissertation. Amman Arab University for Graduate Studies. Jordan, (2005)
47. Sohail Karmani and Pennycook, A., Islam, English and 9/11. Journal of Language, identity and Education. Vol 4.No.2. pp 157-172, (2005)
48. Sugirin, Sudartini, S., Suciati, and Nurhayati, L,. A study on cultural integration in the English textbooks for junior high schools. LITERA, pp. 235-246, (2011)
49. Syahri, I. and Susanti, R. An Analysis of Local and Target Culture Integration in the English Textbooks for Senior High. Journal of Education and Human Development, pp. 97-102.
50. Tanka, J and Most, P., Interactions 1, Listening and Speaking. McGraw-Hill, (2007)
51. Tanka, J and Baker, L. R., Interactions 2, Listening and Speaking. McGraw-Hill, (2007)
52. Tariq Rahman, The Muslim Response to English in South Asia: With Special Reference to Inequality, Intolerance and Militancy in Pakistan. Journal of language, Identity and Education. Vol 4 No 2 p. 9, (2005)
53. Teo, A. and Kaewsakul, W., Cultural Contents in Fundamental English Coursebooks. NIDA Journal of Language and Communication, pp.1-17, (2016)
54. Washima Che Dan, Harshita Aini Haroon and Naysmith, J., English and Islam in Malaysia: Resolving the Tension. World Englishes. Vol. 15.No.2. p. 6, (1996)
55. Wright. M., Influences on Learner Attitudes towards Foreign Language and Culture. Educational Research Vol. 41. No. 2. p. 12, (1999)
56. Yadullah Kazmi., The Hidden Political Agenda of Teaching English as an International Language. Muslim Education Quarterly. Vol.15. No. 1. The Islamic Academy, Cambridge. UK, (1997)

57. Yamada, M., The representation of race and ethnic relations in Japanese junior high school English language textbooks from 1987 to 2002. Ph.D. dissertation, Western Michigan University, United States -- Michigan. Retrieved June 17, 2008, from Dissertations and Theses: Full Text database. (Publication No. AAT 3221660), (2006)

Tables

Table 1: The portrayal of Islam and Muslims in English language textbooks.

	Neutral	Positive	Negative	Total
Interactions 1: Listening/ Speaking	1	0	0	1
Interactions 2: Listening/Speaking	2	1	0	3
Interactions 2-Reading	11	0	6	17
Engaging Writing: Paragraphs and Essays	0	0	2	2
Consider the Issue: Listening and Critical Thinking Skills 3rd ed.	0	0	0	0
TOTAL	14	1	8	23

Table 2: The category based on Byram's (1993) cultural content checklist.

No.	Category	Contents
1.	Belief and behaviour	8
2.	Socialisation and life cycles	9
3.	Stereotypes and national identity	4
4.	Social and political institution	2
	Total	23

Table 3: The portrayal of Islam and Muslims as Neutral in English language textbooks.

No.	Content	Category	Textbook
1.	Nabil and the pet: cat	Belief and behaviour	Interactions 1: Listening Speaking
2.	Hip-hop in Istanbul	Belief and behaviour	Interactions 2: Listening/Speaking
3.	Water	Belief and behaviour	Interactions 2: Listening/Speaking
4.	Islamic Wedding – Bedouin and Henna	Socialisation and life cycles	Interactions 2-Reading
5.	Bedouin Culture- Henna	Socialisation and life cycles	Interactions 2-Reading
6.	Mehndi-Islamic cultures	Socialisation and life cycles	Interactions 2-Reading

7.	United Arab Emirate-unmarried Emeriti men to marry Emeriti women	Socialisation and life cycles	Interactions 2-Reading
8.	Churches, mosques and palace-structures	Stereotypes and national identity	Interactions 2-Reading
9.	Arabesques-exquisite design	Stereotypes and national identity	Interactions 2-Reading
10.	Mosques-Islamic religious worship	Stereotypes and national identity	Interactions 2-Reading
11.	Christianity- religion- Islam	Belief and behaviour	Interactions 2-Reading
12.	Spread of Islam to the East	Belief and behaviour	Interactions 2-Reading
13.	The Middle east-Reality TV show-The Big Brother	Socialisation and life cycles	Interactions 2-Reading
14.	Banking in The United Arab Emirate	Social and political institution	Interactions 2-Reading

Table 4: The category for Neutral contents based on Byram's (1993) cultural content checklist

No.	Category	Contents
1.	Belief and behaviour	5
2.	Socialisation and life cycles	5
3.	Stereotypes and national identity	3
4.	Social and political institution	1
	Total	14

Table 5: The portrayal of Islam and Muslims as Positive in English language textbooks.

No.	Content	Category	Textbook
1.	Numeral Symbol	Stereotypes and national identity	Interactions 2: Listening/Speaking

Table 6: The category for Positive contents based on Byram's (1993) cultural content checklist

No.	Category	Contents
1.	Belief and behaviour	0
2.	Socialisation and life cycles	0

3.	Stereotypes and national identity	1
4.	Social and political institution	0
	Total	1

Table 7: The portrayal of Islam and Muslims as Negative in English language textbooks.

No.	Content	Category	Textbook
1.	Islam and Buddhist Art	Belief and behaviour	Interactions 2-Reading
2.	Term: Holy Place	Belief and behaviour	Interactions 2-Reading
3.	Emeriti Wedding	Socialisation and life cycles	Interactions 2-Reading
4.	Bedouin wedding involves money	Socialisation and life cycles	Interactions 2-Reading
5.	Emeriti wedding are too lavish	Socialisation and life cycles	Interactions 2-Reading
6.	Kamal Ataturk	Social and political institution	Interactions 2-Reading
7.	Arab's Hospitality	Socialisation and life cycles	Engaging Writing
8.	The Quran and Hadith	Belief and behaviour	Engaging Writing

Table 8: The category for Negative contents based on Byram's (1993) cultural content checklist

No.	Category	Contents
1.	Belief and behaviour	3
2.	Socialisation and life cycles	4
3.	Stereotypes and national identity	0
4.	Social and political institution	1
	Total	8

MATEC Web of Conferences 150, 05077 (2018) https://doi.org/10.1051/matecconf/201815005077
MUCET 2017

Alternative Dispute Resolution for Islamic Finance in Malaysia

N. Khalidah Dahlan [1,*]

[1]Institute of Malaysian and International Studies (IKMAS), Universiti Kebangsaan Malaysia, 43600 Bangi, Selangor.

Abstract. The law lays down rules that govern private disputes, self-help actions or open-conflicts that may occur in the society. Here, courts, tribunals, arbitration and other dispute resolution forums are the main secular institutions in dispensing just resolutions in a society. As important as the court proceedings, Alternative Dispute Resolution (ADR) process is neutral, it is like a judge who served as the finder of fact and decision maker for the parties involved. By adapting the ADR methods in resolving the disputes relating to Islamic financial products, this method will able to fulfil with the demands on Islamic financial markets in Malaysia. As the highest law of the land places Islam, the religion associated with one of the main ethnicity, as the official religion of the nation and allowed the duality of economic system, in particular banking system, in which conventional banking system exists side-by-side with the Sharia-compliant, Islamic banking. This paper will reviewing the pertinent literature on the development of Islamic banking cases, the Alternative dispute resolution methods on Islamic banking and its contributions to Malaysian society regardless of ethnicity and religion. The discussion in this paper therefore will be directed towards highlighting the alternative dispute resolution methods on Islamic banking and its implication to enhance the application of Islamic banking laws to consumers regardless of ethnicity and religion, in order to maintaining fairness and justice in a multi-ethnic society.

1 The Alternative Solutions (Alternative Dispute Resolution) For Any Disputes

The objective of dispute resolution is to solve the dispute between the parties who seek justice [1]. Who has suffered injury or loss and wanted to claim or refunded any damage or loss that they obtained. Alternative Dispute Resolution (ADR) is a quick way to resolve the dispute, based on freedom of choice on the laws and procedures of their dispute. There are other dispute resolution mechanisms available to the parties to resolve their disputes without going to court.

Over 1,400 years ago, ADR methods have been used by Muslims to settle their differences amicably and this method has been encouraged by Islam and considered as a source of legislation. This is because Islam is a religion that emphasizes the concept of harmony instead of hostilities and embrace the principle of compromise rather than confrontation.

Justice without trial court was raised by Roscoe Pounds since 1912, he said that the court was crowded; delay in the trial is endemic; high costs; shortage of lawyers; overall substantive legislation that traditional legal and not current. Even Pound also said there is a difference in the usage and adoption of legislation among wealthy or poor. This is because the practice of law itself is often dominated by the rich for their favour. Clearly that the justice of the law does not guarantee equality of the overall concept [2].

Pound had put the two concepts of 'justice without law' and 'justice according to law'. As dissatisfaction with legal institutions increased in the early decades of the twentieth century, there was renewed interest in alternatives to litigation, particularly Conciliation and Arbitration [3-4]. Both were touted as a quick, inexpensive procedure to waive counsel, the delay in the trial would result in an expensive cost to produce legal action.

'Alternative dispute resolution' or ADR is described as an alternative way other than through a court case to resolve the problems of the parties involved. Some examples of ADR done to solve the Islamic finance disputes are the Arbitration (tahkim), mediation, conciliation and *sulh* methods.

2 Sects' view on Alternative Dispute Resolution Methods

Each view of the four sects have different views in regards to alternative dispute resolution methods, especially methods of arbitration. Aishath Muneeza explains that the method of alternative dispute resolution, particularly arbitration method according to the Hanafi school is a contract in which the parties involved would prefer a more open and achieved mutual consent on that particular matter [5]. An arbitrator will act as the representative of the parties involved and due to that the decision reached by the arbitration method has a lower level authority than a court's judgment.

According to Maliki, an arbitrator may be selected by any of the parties involved and the appointment of

[1] Corresponding author: mr_palil@ukm.my

an arbitrator cannot cancelled the proceedings once it have been commenced. Based on Shafie sect, the arbitration proceedings are different from the process of a court proceedings. An arbitrator can change a decision before they issued a valid arbitration decision. Finally, in Hambali, the alternative dispute resolution methods such as arbitration rules have the same effect as the process of a court proceedings. Therefore, an arbitrator shall have the same qualifications as a judge and any decision made by him will bind all the parties involved.

Conflicts or disputes often occur in a community or in the family institution. Noone described a dispute between individuals resulting from certain conflicts which is inevitable and is considered as a normal scenario in life [6]. Often methods of conflict resolution in the community involving court institution, nonetheless the concept of alternative dispute resolution (ADR) has long been practiced in the community. For example, most societies, nations and religions of the world have been adopting the rules in religion or tradition to resolve their disputes (civil) (each nation and religion has its own terms in implementing this practice). As stated by Syed Khalid Rashid, traditionally the Chinese people using non-litigation settlement [7]. Their method is to negotiate peace (Youhao Xieshang which means friendly consultation). This method is carried out as a first step to resolve the dispute in the field of commercial, industrial, family and neighbourhood communities.

In Malaysia, since the time of the Malacca sultanate, peaceful settlement of disputes has been practiced. This practice has been incorporated in accordance with the teachings of Islam which demands the issue of *musyawarah*. The practice of alternative dispute resolution remain intact from the colonial period until today. Now, one of the alternative dispute resolution method which had been adapted with the court proceeding *(Sulh* method) had been proven and accepted as practice to achieve solution for family cases in the courts of law in several states in Malaysia.

3 Alternative dispute resolution for Islamic finance disputes

The Malaysian regulatory framework provides two mechanisms for settlement of disputes in Islamic banking. The mechanisms are litigation and arbitration in which had been provided in Section 56 of the Central Bank Act 2009. The provision states that any proceedings relating to Islamic banking and finance can be made through the courts and arbitration rules. However the proceedings must be in reference to the Syariah Advisory Council [8]. In fact, with reference to the decision in the case of *Bank Islam Malaysia Bhd vs. Adnan bin Omar (1994*), the parties may choose to settle their disputes on Islamic finance matters thru arbitration method or mediation method provided that such provision should have included in the agreement (section 24 (4))of the Courts of Judicature Act 1964).

During the proceeding of arbitration or mediation, the parties involved have the right to choose an arbitrator or mediator who specializes in Islamic finance to resolve the case. The arbitrator or the mediator's decision is final and there is no need to refer the court [8].

According to Sundra Rajo (director of the Kuala Lumpur Regional Centre for Arbitration KLRCA) [9], stated that the statistics of cases registered for the arbitration rules were 85 cases in 2012, while in 2013, registered a case for using arbitration tara is growing by about 110 cases. This includes 20% of them were international cases. In fact, there are many Malaysian researcher have revealed their finding that Malaysian society is now increasingly exploring other methods to settle their dispute and not only seeking solutions in court [5 , 7 , 10 , 11]. Other researches that also indicated the method of alternative dispute resolution are such as mediation method [12], mediation and arbitration method [13], Sulh method [14-15] and [16-17] and lastly research on the arbitration rules adopted globally [18].

According to Umar Oseni [11], any disputes that resolved through an alternative dispute resolution is considered as a better solution. Especially, a mediation session that lead by a private negotiations in which the dispute is chaired by third parties. For example cases involving improper management of Waqf properties are better resolved through alternative dispute resolution instead of through a declaration by the court.

The proceeding of mediation and sulh method in Sharia law involve actions for compromise, which is a practical demonstration of the give-and-take the initiative. Furthermore in Malaysia, all disputes relating to the issue of Islamic finance will continue to refer to the Shariah Advisory Council of Bank Negara. Therefore, if the parties involved choose to solve their dispute thru alternative dispute resolution in which this case is the Sulh method, the Sulh officer will refer to the Shariah Advisory Council of Bank Negara to help reduce the number of problems that exist in Islamic finance industry in Malaysia.

The initiative to establish a system of Alternative Dispute Resolution to help speed up and reduce the backlog of cases leads to the implementation of Sulh in the Syariah Court. This idea is also driven by the effectiveness implementation of the mediation method practiced in other countries. Despite that, Sulh and mediation methods have different names but the purpose and objectives of their practice are the same, the parties settle the dispute by binding memorandum of understanding.

4 The Evolution On Civil Law And Islamic Law In Dealing With Islamic Banking Transactions.

Clash of Sharia and civil courts jurisdictions has given rise to a number of problems, especially when it comes to litigation procedures. According to Zainal Amin Ayub [19], there should be harmonization of aspects of civil litigation, especially the discovery and appeal procedures, with the procedures currently in use in the Sharia courts, which are based on the precepts of Islamic law. There are also opinions that the presence of Sharia experts in the civil High Court is necessary. In relation to Islamic banking cases, this would create a hybrid legal system whereby there would be two courts – the civil court and the Sharia court - which have parallel jurisdictions in the field of Islamic banking [20].

Previously, Court cases involving Islamic banking issues had often brought to the limelight the issue of clash of jurisdiction of civil court and Sharia court. Among the cases wherein this issue was raised are *Tinta Press Sdn Bhd v Bank Islam (M) Bhd (1987), Bank Islam Malaysia Berhad v. Adnan Omar (1994), Dato 'Nik Mahmud Bin Daud v Bank Islam Malaysia Berhad (1996), Bank Malaysia Bhd v Emcee Corporation Sdn Bhd (2003), Bank Islam Malaysia Bhd v. Pasar raya Peladang Sdn Bhd. (2004), Arab Malaysian Merchant Bank Bhd v Silver Concept Sdn Bhd (2005), Malayan Banking Bhd v Ya'kub Oje & Anor (2007).*

Overall the majority of cases involving Islamic banking often include restrictions on the deep understanding of the legal system and also in terms of Islamic banking practices, namely in the form of technical documentation. Appreciation and analysis of these cases were also formulated by Rusni Hassan [20] that serious obstacles faced in Islamic banking in Malaysia is an obstacle in the legal system and barriers in the documentation.

Barriers in the legal system is basically related to the Islamic Banking Act 1983, Takaful Act and other acts relating to Islamic banking law in Malaysia. Before the start of the Islamic banking in Malaysia, all legislation in the banking business is based on civil law. And this has given rise to barriers to Islamic banking. This can be seen by the cases decided in court. For example, in the case of *Bank Islam Malaysia Bhd v Adnan bin Omar (1994),* there is a dispute of Order 83 Rules of the High Court 1980 where interest or usury should be investigated practice. This is because the practice of usury is contrary to the principles and laws of the Islamic financial system.

Obstacles in the documentation shows that the accuracy in a contract that would reflect the intention of the parties. Where for some other it reflects a contract that involves a complicated procedure. In Islamic banking, content and order of documents also represent an important aspect. An Islamic banking documents shall be valid and lawful under the laws of Islam. As a result, the burden of drafting documentation for Islamic banking business is very hig Obstacles encountered in the documentation of Islamic banking in Malaysia can be seen through the results of the case by the court. For example, in the case of *Malayan Banking Bhd v Ho Marilyn Siok Lin (2006)*, Islamic banking documents submitted to the court has proved that this case does not have the compatibility of Sharia perspective. This is because the main purpose of the document is not explicitly stated. It becomes more difficult when someone who does not understand the amenities in product Bai Bithaman Ajil drafting the document and gave it to clients who are not Muslims. As a result, the documentation that is designed to have failed to reach their objectives and reflect the intention of the parties.

Nonetheless, the issue of homogenization of Islamic banking products offered by all financial institutions in Malaysia is however far from being resolved. Commonly all the legal documents are in English (Civil Procedurals) and common law lawyers who draft those documents could not appear in the Sharia courts. The law is an English law, not a Sharia law. However, the current ongoing harmonization of civil procedural and substantive laws to accommodate the Sharia peculiarities of Islamic banking principles and practices implies that instances of clashes of jurisdictions between the civil and Sharia courts will be a thing of the past in the nearest future. This harmonization process has addressed the concern of Islamic banking consumers that they are being denied the full measure of justice they deserve [22].

After researching a number of court cases in Malaysian legal system on the issue of financial law, especially in the issue of Islamic banking has also undergone a transformation in the dispute resolution process between civil law and Syariah law. However, effective dispute resolution and with a strong foundation is needed to ensure that the Islamic financial system to meet the needs of each nation and religion in order to achieve the demands of the market, justice and equality. It is evident that Islamic banking law and practice are currently undergoing a transformation process which will make them compatible with the civil law. The challenge is to ensure that the transformed Islamic banking law and practice meet the needs of the market demands and the way of life, if it is to achieve equality for all races and religions.

Looking at the diversity of barriers in the legal framework, Islamic finance may be required by a method other than through court proceedings to resolve the dispute by means of alternative dispute resolution (ADR). Alternative dispute resolution can be defined as a procedure that acts as an alternative to legal action to resolve disputes. It can be formal (e.g. arbitration rules) or informal (e.g. methods of negotiation and mediation), and is a private process and its decision will be binding ADR process when done. The dispute has been settled will be given an award which is made by a process which independent tribunal (e.g. by an arbitrator, mediator and more).

Conclusion

In general, the differences between solving dispute thru court proceedings and the Alternative Dispute Resolution are more focused on the operation procedures and its impact towards the parties involved. Not to mention as well as the role and decisions made by a third party (the judge / arbitrator / conciliator /

mediators / Sulh officer). During the process of Alternative Dispute Resolution method, the third party will become as the facilitator, decision makers and problem solver for the parties involved in order to reach a decision [23]. Moreover, the Alternative Dispute Resolution process must be carried out with the consent of both parties and it is binding between the parties. Except when there is a reasonable doubt in which the decision had been made fraudulently and bias. This method is vital to achieve good relationship between both sides [24].

Arbitration and mediation are often practiced in Malaysia when involved with financial disputes. Arbitration is a dispute resolution methods that involve disputants present their cases before an independent tribunal, impartiality and an expert who has knowledge of the law. When inspecting the facts in that case the arbitrator will make a decision. If the parties involved have agreed the decision would be binding on the parties. Unlike the methods of mediation is a process in which a third party would be a neutral party for the parties to a dispute to reach an agreed settlement to their conflict.

Most of the disputes associated with Islamic financial transactions in the court, will refer to specialists such as the Shariah Advisory Council of Bank Negara Malaysia. Therefore, the expertise of the Shariah Advisory Council of Bank Negara Malaysia plays an important role in helping to reduce the problems that exist in the Islamic finance industry in Malaysia. So it is not an impossible matter if their expertise can also be practiced through alternative dispute resolution methods. With further research on the Alternative Dispute Resolution method (namely judgement without a court trial), it can be expected to become not only as the alternative way but it will become one of the main methods in the future to solve the disputes in the Islamic financial system.

References

1. J. Rawls, 'Justice as Fairness: Physical not Metaphysical'. In Thomas Morawetz. *Justice*. (England: Datmouth Publishing Co. Ltd. 1998)
2. P., Roscoe., *Mo. Bar Ass'n Proc.*, 110, 112; Reginald Heber Smith, Justice and the Poor **15** (1919)
3. Pound, Roscoe., The limits of effective legal action, 3 *American Bar Ass'n J.* (Jan 1917), 55-57; (1917)
4. Pound, 1922. An introduction to the philosophy of law (*new haven*, 1922), **54** (1922)
5. A. Muneeza, Malayan Law Journal Articles, *4*, 2. (2010)
6. Noone, M., Australian Dispute Resolution Journal, **17**, 96-100 (2006)
7. Rashid, S.K., *Alternative dispute resolution: The emerging new trend of informal justice.* (Kuala Lumpur: International Islamic University Malaysia, 2002)
8. H. Rusni, M. A. Hussain, A. Yusoff, 25 KANUN (1). June 2013: 83-106. (2013)
9. Anon, Datuk Sundra Rajoo's Speech at the Launch of KLRCA's Revised Rules 2013. http://klrca.org/speeches/datuk-sundra-rajoos-speech-at-the-launch-of-klrcas-revised-rules2013/ (2013)
10. M. Surianom, A. H. Nor 'Adha, The Second Asian Business and Management Conference 2011, 224-234, ISSN: 2186-5914, Proceedings URL: http://iafor.org/abmc_proceedings.html (2011)
11. U.A. Oseni, (2012)
12. P. G. Lim., Malayan Law Journal 4 (cv-cxii.) (1998)
13. Bukhari, K. Zaman, *In 8th general Assembly 2003 by Badan Peguam Malaysia*. (2003)
14. A. Raihanah, *Kertas Kerja Seminar Isu-isu Mahkamah Syariah VII. Anjuran Universiti Islam Antarabangsa Malaysia,* (2005)
15. A. Raihanah,. *Sulh Dalam Kes Kekeluargaan Islam*, (Kuala Lumpur: Penerbit Universiti Malaya 2008)
16. Safei, S. A., Int. Sem. Comp. L., 237. (2008)
17. M. R. Zainudin, *Muzakarah Pegawai-Pegawai Sulh Seluruh Malaysia (Kali Pertama) Anjuran Hotel Royal Penang, Pulau Pinang,* (2009)
18. Saleh, S. *Commercial Arbitration in the Arab Middle East: Jordan, Kuwait, Bahrain, and Saudi Arabia.* (United Kingdom: Aseanlex Ltd., 2011)
19. Ayub, Z. A. Hussain, M. Azam. A. Shariff, Nurretina, Ali, Hassan, **2** CLJ ix (2007)
20. Shariff, M. Ismail, 1 MLJ cxiv (1998)
21. Hassan, R., 5 MLJ ixxx & 6. (2011)
22. Markom, Ruzian. Pitchay, S. Ali. Zainol, Z. Ashiqin. A. Rahim, Anita. A.R. Merican, R. Merican. Eur. J. L. Eco. Spr. (2012)
23. A. Adam., J. Legal Aspets Sports. **12**, 153-182. (2002)
24. Jady @ Z. Hassim, K.H. Hassan, Z. Z. Ashiqin. *Penyelesaian Pertikaian Mekanisme dan Perundangan*, 93-111. (Bangi: Penerbit Universiti Kebangsaan Malaysia, 2014)

Awareness and willingness among the residents in Johor Bahru City Council and Johor Bahru Tengah Municipal Council of Iskandar Malaysia-Im (Wpi@Sjer) towards recycling

Nora Farina Mohd Halim[1,*], *Nur Zaidani* Wati Mohd Darwis[1], and *Juliana* Martin[1]

[1]Universiti Teknologi Mara (UiTM) Cawangan Johor, Kampus Pasir Gudang, Malaysia

Abstract. Solid waste is increasingly becoming an important urban challenge as a result from population growth and also the development of economy and culture which has improved the standard of living and the lifestyle of the people. If left unchecked, solid waste could lead to severe environmental degradation. In order to reduce the environmental impact of waste disposal, recycling has been introduced. About twenty percent of our solid waste can be recyclable, rather than disposing it through landfills or incinerators. Residential area is the best place to practice recycling because most of the waste is generated there every day. This research focus on the awareness level and the recycling knowledge among the residents in Johor Bahru City council and Johor Bahru Tengah Municipal Council of Iskandar Malaysia. A total of 500 residents were surveyed and the questionnaires were focused on the waste management and also the recycling knowledge. The results discovered that more than 50% of the residents' state that they practicing recycle and they also aware of the environmental and economic benefits from recycling. As a conclusion, the residents are prepared for recycling if the recycling amenities are available and easy to reach by them.

1 Introduction

In line with the vision 2020, the development of economic and urban transformation is emerging tremendously. The population in Malaysia has extended up to 30 million people in year 2016 according to the Department of Statistics Malaysia. The growing of the population contributes to the increase of solid waste production in Malaysia.

Besides that, the changes of living standard and a variety of daily activities among Malaysian will also lead to this excessive solid waste production [1]. Even though "Recycling" has been accepted worldwide as a method of waste disposal, Malaysia's domestic recycling rate is still low at about five percent [2]. Iskandar Malaysia has presented a severe waste generation growth from time to time.

Waste reduction and recycling are no longer a preference but a needed step to ensure a healthy life style and at also to protect our surrounding. Conferring to [3], Malaysia is one of the Asian countries that produced the most municipal solid waste

Recycling is the method of changing waste materials into reusable objects to avoid waste of possibly useful materials, lessen the intake of fresh raw materials, energy usage, air pollution and water pollution .Recycling is a key element of modern waste reduction and is the third element of the "Reduce, Reuse and Recycle" waste hierarchy [4].

Recycling is a process of reusing items that would be rejected as waste. There are few types of recycling include upcycling, which involves adding value to an element for reuse, and downcycling, which involves breaking down an element or material into its component to reuse anything that can be reclaimed.

Precycling is a different approach which involves avoiding the acquisition of excessive items that would have eventually have to be recycled or dealt with as waste.

* Corresponding author: norafarina@johor.uitm.edu.my[1] , nurzaidaniwati@johor.uitm.edu.my[2] , julianamartin@johor.uitm.edu.my[3]

2 Study area

The leading southern development corridor in Johor is Iskandar Malaysia (IM) which previously known as South Johor Economic Region (SJER). On 8 November 2006 Iskandar Malaysia was established. The development is managed by Iskandar Regional Development Authority (IRDA) and was named after the late Sultan of Johor, Almarhum Sultan Iskandar ibni Almarhum Sultan Ismail [5]. Figure 1 and Figure 2 show the whereabout and geographical coverage of Iskandar Malaysia.

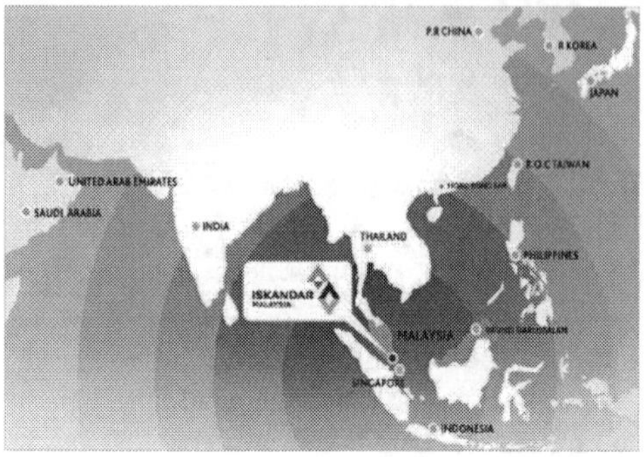

Figure 1: Location of Iskandar Malaysia

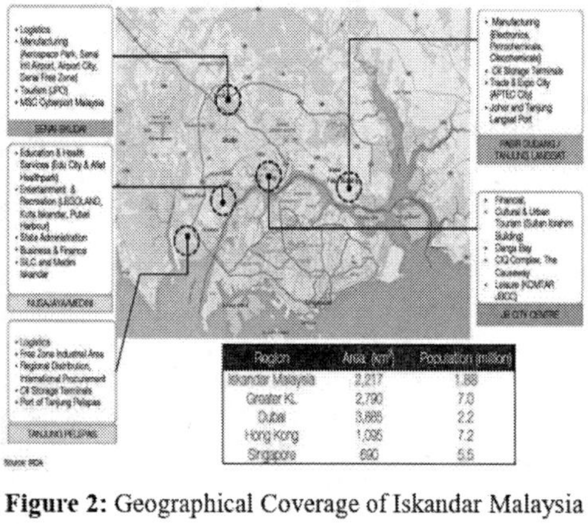

Figure 2: Geographical Coverage of Iskandar Malaysia

The total area of the development region of IM is 2217 square kilometres which cover the city of Johor Bahru, Pontian, Senai, Pasir Gudang and the new administrative investment in Iskandar Puteri. The district of Johor Bahru is included in its whole, as well as the parts of the district of Pontian. Five local government authorities have the jurisdiction over the covered area, including Majlis Bandaraya Johor Bahru, Majlis Perbandaran Johor Bahru Tengah, Majlis Perbandaran Pasir Gudang, Majlis Perbandaran Kulai,and Majlis Daerah Pontian . In this research we just focus on the awareness level and the recycling knowledge of the two local jurisdiction of the local authority in Iskandar Malaysia which is Majlis Bandaraya Johor Bahru and Majlis Perbandaran Johor Bahru Tengah

3 Methodology

3.1 Research design

Questionnaires run as the main data while literature reviews run as minor data. The random sample for this research will be any houses in Iskandar Malaysia selected randomly by accessibility. Two jurisdictions of local authority in Iskandar Malaysia will be involved in this study which is *Majlis Bandaraya Johor Bahru* (Johor Bahru City Council) and *Majlis Perbandaran Johor Bahru Tengah* (Johor Bahru Tengah Municipal Council). For this research there will be a total of about 500 respondents for the two jurisdictions of local authority.

3.2 Analysis mechanism

The analysis mechanism used for data collection is a questionnaire. It will be deliberate as accurate as possible for respondents to be able to answer the question without difficulty. The study was steered using structured questionnaire that covered two basic principles which is about the waste management and the recycling knowledge. In this research, the questionnaire will be distributed to the respondent involved by hand. The final data were statistically analysed using the SPSS.

4 Results and discussion

4.1 Evaluation of Solid Waste Management in Majlis Bandaraya Johor Bahru (Johor Bahru City Council) and Majlis Perbandaran Johor Bahru Tengah (Johor Bahru Tengah Municipal Council)

Table 1: Results questionnaire for Waste Management

	Question	**Answer**	**Number of Respondents**	
			MPJB	**MPJBT**
A1	The fulfilment of the respondent with the waste collection system in the area.	Yes	148	171
		No	77	39
		Not sure	22	41
A2	The efficiency of the collection system.	Yes	118	146
		No	68	54
		Not sure	60	50

https://doi.org/10.1051/matecconf/201815005078

A3	Frequent in a week the trash overflowed.	0-2 times	154	180
		3-5 times	85	65
		5-7 times	6	5
A4	Is odour, waste scattered, and breeding of flies are common in your area?	Rarely	88	113
		Average	122	111
		Often	37	27
A5	Amount of waste generate each day.	<0.6 kg/day	78	90
		0.6 kg/day	90	94
		>0.6 kg/day	74	65

Question A1 was to classify the fulfillment of waste collection systems in the area and the outcomes shows that 59.9% of the residents were pleased with the collection system in MPJB while 68.1% were pleased in MPJBT. The number of residents were not pleased with the current method of waste collection were 31.2% for MPJB and 15.5% for MPJBT.

For question A2, 48% of the residents in MPJB thought positively of the current collection system while 58.4% thought positively in MPJBT. 27.6% for MPJB and 21.6% for MPJBT did not consider the collection system is effective.

Results for question A3 displayed the residents opinions on the excess of trash bins 0-2 times a week were 62.9% for MPJB and 73% for MPJBT, 3-5 times per week were 34.7% for MPJB and 26% for MPJBT and 5-7 timed per week were 2.4% for MPJB and 2% for MPJBT.

Furthermore, the results for question A4 showed 35.6% for MBJB and 45% for MPJBT of the residents thought of trash contamination in their area as exceptional, followed by 15% for MPJB and 10.8% for MBJBT showed contamination rate at trash house are often.

Results indicate that in question A5 32.2% for MPJB and 36.1% for MPJBT residents expected waste generation amount is less than 0.6kg/day while 30.6% for MPJB and 26.1% for MPJBT of residents will generate more than 0.6kg/days.

4.2 Awareness and Recycling knowledge of respondent

Table 2 : Results questionnaire for Recycling Knowledge

	Question	**Answer**	**Number of Respondents**	
			MPJB	**MPJBT**
B1	Knowledge for 3R concept.	Yes	162	164
		No	48	39
		Not sure	35	44
B2	What are 3R stands for?	Reduce; Reuse; Recycle	122	115
		Reduce; Reuse; Recycle	69	65
		Reduce; Reuse; Recycle	51	65
B3	Do you recycle?	Yes	134	140
		No	85	76
		Not sure	27	36
B4	Frequency recycle monthly.	1 times	42	50
		2 times	60	68
		3 times	41	44
		>3times	35	21
B5	Separated the trash or not.	Never	67	67
		Yes	209	197
		No	25	24
		Not sure	12	29
B6	Do you separate garbage for recycling?	Yes	182	187
		No	63	63
B7	Number of recycling bins to be provided.	2 bin	35	53
		3 bin	151	137
		4 bin	44	50
		5 bin	17	12
B8	Participation in recycle if implemented in Malaysia.	Yes	223	223
		No	10	6
		Not sure	14	22

Results for question B1 showed a total of 66.1% for MBJB and 66.4% for MPJBT residents understood the concept of 3R. Meanwhile 19.6% for MPJB and15.8% for MPJBT residents did not know about 3R concept.

For question B2, 50.4% for MBJB and 46.9% for MPJBT residents knew the right order of 3R. Result for question B3 showed that 54.5% for MBJB and 55.5% for MPJBT residents carry out the recycling while 34.6% for MBJB and 30.2% for MPJBT residents did not perform recycling activities.

Results for question B4 showed that a sum of 17.1% residents in MBJB performed at least 1 time the recycling activities each month followed by 24.5% who had 2 times recycling per month,16.7% residents ran 3 times per month,14.3% residents in MBJB exceeded three times and 27.4% did not carried out any recycling activities. For residents in MPJBT, 20% carried out at least 1 time,27.2% carried out two times per month,17.6% carried out three times per month,8.4%

https://doi.org/10.1051/matecconf/201815005078

exceeded three times per month and 26.8% of residents in MPJBT did not recycle at all.

Furthermore, question B5 indicate that 85% for MBJB and 78.8% for MPJBT residents will separate waste at source while 10.2% for MBJB and 9.6% for MPJBT residents will not separate the waste at source. Question B6 showed 74.3% for MBJB and 74.8% for MPJBT will separate waste at source with the facilities available while 25.7% for MBJB and 25.2% for MPJBT residents thought not to separate the waste at source.

Result B7 showed that both residents in MBJB and MPJBT had a higher percentage of suggested the providing of 3 bins recycle bins which were 61.1% and 54.8% respectively. It is due to the people in Malaysia has been aware with the concept of 3 recycling bins. Question B8 showed a positive response of 90.3% for MBJB and 89.2% for MPJBT residents will discrete waste if recycling facilities are applied such as 2-bins or 3-bins recycle bins.

Results for question B1 showed a total of 66.1% for MBJB and 66.4% for MPJBT residents understood the concept of 3R. Meanwhile 19.6% for MPJB and15.8% for MPJBT residents did not know about 3R concept. For question B2, 50.4% for MBJB and 46.9% for MPJBT residents knew the right order of 3R. Result for question B3 showed that 54.5% for MBJB and 55.5% for MPJBT residents carry out the recycling while 34.6% for MBJB and 30.2% for MPJBT residents did not perform recycling activities.

Results for question B4 showed that a sum of 17.1% residents in MBJB performed at least 1 time the recycling activities each month followed by 24.5% who had 2 times recycling per month,16.7% residents ran 3 times per month,14.3% residents in MBJB exceeded three times and 27.4% did not carried out any recycling activities. For residents in MPJBT, 20% carried out at least 1 time,27.2% carried out two times per month,17.6% carried out three times per month,8.4% exceeded three times per month and 26.8% of residents in MPJBT did not recycle at all.

Furthermore, question B5 indicate that 85% for MBJB and 78.8% for MPJBT residents will separate waste at source while 10.2% for MBJB and 9.6% for MPJBT residents will not separate the waste at source. Question B6 showed 74.3% for MBJB and 74.8% for MPJBT will separate waste at source with the facilities available while 25.7% for MBJB and 25.2% for MPJBT residents thought not to separate the waste at source.

Result B7 showed that both residents in MBJB and MPJBT had a higher percentage of suggested the providing of 3 bins recycle bins which were 61.1% and 54.8% respectively. It is due to the people in Malaysia has been aware with the concept of 3 recycling bins. Question B8 showed a positive response of 90.3% for MBJB and 89.2% for MPJBT residents will discrete waste if recycling facilities are applied such as 2-bins or 3-bins recycle bins.

5 Conclusion

Due to the growth in population and economic development, reducing and recycling household waste has become progressively vital because waste generation has been increasing [6]. The balance of landfill areas is deteriorated and that makes recycling not only sensible but essential. Currently the waste disposal practice is that the household are required to place their waste in the bins that usually located in front of their houses and the collectors will gather the wastes twice a week. However, not all waste is disposed in such way.

From this study, most of the respondents have the knowledge of recycle, know the advantages of recycling and aware about the purpose of recycling but amazingly only few of the respondent practice recycle frequently. The reasons of most of the respondents is that there were not enough facility for waste separation and some of them stated that they were busy and did not have time to practice recycling.

There are many effective ways to promote recycle among household and one of the most operative way is television. Television can disseminate information to people with regards to recycle campaign. Nevertheless, the neighborhood themselves which is family and friends can deliver about recycling because people who practicing it frequently stated that they are the most influence in recycling resource. Since the respondents also cited that lack of facilities is one of the main reasons for not practicing recycle, the authority should provide more recycling facilities

The funding provided by the Ministry of Higher Education (MOHE) Malaysia and the management of Universiti Teknologi Mara (UiTM) Malaysia under the RAGS research grant scheme (600-RMI/RAGS 5/3 (178/2014)) are thanked

Reference

1. R. Othman, M. Yuhaniz, Centre of Environment Behaviour Studies, Faculty of Architecture, Planning and Surveying, UiTM Malaysia (2013)
2. S. Chandravathani,. Waste Reduction : No longer An option But a Necessity,Bernama.com, Bernama.com database, Institute of Biologocal Sciences: Kuala Lumpur (2008)
3. J. Othman, Household preferences for Solid Waste Management in Malaysia (2009)
4. A.V. Shekdar, Sustainable solid waste management : An integrated approach for Asian Countries, Waste Mangement **29**(4), 1438-1448 (2009)
5. IRDA, Integrated Solid Waste Management Blueprint For Iskandar Malaysia (2011)
6. M. Akil, F. Johar, H. C. Siong, Household participation in recycling programs : A case study from Malaysia,Tata Loka Vol. **17**, No. 2 (2015)

Colleges Students' Misconception about Type of Bonding

Ardiansah[1,2,*]

[1]Sebelas Maret University
[2]Indonesia Endowment Fund for Education

Abstract. This research is based on writer's previous misconceptions research on chemical bonding. The study aims to calculate the percentage of misconceptions and misconceptions types about chemical bonding. The method used is descriptive. Samples used in this study was 75 students in college. Data collection tool used in this study was an online modified two-tier diagnostic test. The results showed that types of student misconceptions occur in sub concept: determining the type of bonding of a compound based on the difference in electronegativity, determine the type of bonding may occur if the metal and non-metal bind together and determine two examples of ion or covalent bond compounds.

1 Introduction

Chemistry is one of the subjects must be taught in natural science program in Senior High School in Indonesia. Based on the ministerial decree of the ministry of education and culture about the national education system, the purpose of learning chemistry is to make the student understand about chemistry concepts and apply it to solve the problem in real life and in science and technology life. Application of chemistry in real life start with an understanding of chemistry concepts, principles, laws, and theories. But, chemistry is one of the subjects that difficult to understand by students and teachers too [1]. Most of the students hate the abstract concept that requires more attention when they learn chemistry. Furthermore, the difference between chemistry word and reality word also make a relational and interpretation problems for the students.

The knowledge level chemistry subject in senior high school can be categorized into three types, that is sub-microscopic (abstract and cannot detectable with ears, eyes, noses, tongue, and skin), macroscopic (can be detectable with our body's instrument) and symbolic (a symbol represents a chemistry formula or subject) [2-4].

The complexity of this subject also one of cause the misconception happened in chemical bonding [5]. Chemical bonding is one of difficult concept that student should understand clearly. The concept of chemical bonding also related with another concept such as thermochemistry, chemical equilibrium, and others. The lack of concept in chemical bonding will influence the next topics related each other.

Chemical bonding generally gives many conceptions in teaching and learning the process. There are several correct conceptions we cannot find in a textbook. Glynn and Duit said that the importance of teachers helps not only the commonalities among representations but also their limitations. The teachers' misconception also happened in several countries such as Indonesia [6], Turkey [7], and China [8]. They were assessed with a specific instrument to measure the number of conceptions.

A misconception is defined as a different idea, concept and subjects' thought that is different with scientific conception. The literature also called misconception as misunderstandings, naıve conceptions, and alternative conceptions. Ozmen also cited that other terms of misconception is misconceptions [9-25], alternative conceptions [13, 26-31] , preconceptions [32-33], alternative frameworks [13, 14, 30, 34-36], naıve beliefs [37], naive theories [38], naive conceptions [39], children's scientific intuitions [40], conceptual frameworks [41], children's science [42-43], common sense understanding [44], common sense concepts [45], alternative conceptual framework [25], intuitive conceptions [46], intuitive science [47], common alternative science conceptions [14] students' intuitive theories [48], prescientific conceptions [49], alternate perceptions [50], students' descriptive and explanatory systems [51] and spontaneous knowledge [52].

Misconception in a student can be detected using various detection instruments. It includes interview [53-55], open reasoning [56-57], and multiple choice [58-59]. Every instrument has different advantages and disadvantages to assess the misconception. The accuracy and precision of the result are can be represented the real if the instrument can represent the real condition in school or subject class. Multiple choice is one of the instrument can be assessed the misconception. The option should be represented the unique of conception. It must not four or five options

*Corresponding author: ardiansah@student.uns.ac.id

every question. The option depends on the complexity of the question and opportunity of conception in the subject. The tiered instrument is a way to help more accurate result that consists of two or more tier. It mapped the conception and reduce the guess of the student. The student should choose the first tiered reason and choose the possible reason to their answer.

There were many misconceptions are found by researchers about chemical bonding reviewed by Ozmen [61], but in the type of bonding is not find in specific research. The method used were multiple choice, tiered question and interviews with some of the subjects. But, the interview cannot be recommended during the lack of time although it can detect deep insight about students' conception.

An alternative method to detect college students' understanding in a different place in Indonesia is with online with two-tier question inventories. Two tiers question consist of a multiple-choice question (2-4 options) in the first tier. The second tier is certainty of subjects' reason after answering the option that includes "Yes" for certain answer and "No" for the uncertain question. These questions were developed from previous research about the certainty of response index [62] and multiple-tiered [63]. The test used was adapted from author previous research. Distractors derived alternative conceptions may happen in the subject.

2 Materials and Methods

The material used in this research were adapted from author previous research. The reliability, validity, and distractor were tested and no more testing needed for this research.

Seventy-five college student of diversity educational background that got chemical bonding as their subject in senior high school. Actually, there were ninety-seven respondents participated in this survey. But, some of them wrote never got that subject in their school. They consisted of 40% of graduate students (18 to 30 years old), 49.3% master degree students (23 to 41 years old) and 10.7% from doctorate degree students (22 to 41 years old).

There were three type of conception in this research that is a good concept, misconception and lack of knowledge. Based on responses given, possible combinations of answer were defined. If the respondents answer correctly and stated sure, then it is possible to conclude that she/he possesses a scientific knowledge of the examined concept. We can be certain that the respondent possesses a misconception if they answer incorrectly and stated sure. Another possibility is she/he answer correctly but stated not sure then can be classified as lucky guess or lack of confidence. The last combination, incorrect-not sure, indicate a lack of knowledge.

3 Result and Discussion

This research includes five question established based on author previous research. In the first indicator, determining the bonding type of compound based on differences of electronegativity, two question were established. The second indicator, determine the type of bonding may occur if the metal and non-metal bind together, include one question. While the third indicator, determine two examples of ion or covalent bond compounds, consists of two question.

The question represents each indicator. First question about ionic bonding and second question about covalent bonding. The subject should use the electronegativity value to determine molecule's bonding type. The question of this indicator found in the first and second question. The third question is represented the second indicator while fourth and fifth question represents the third indicator. The fourth question about the example of ionic bonding and fifth question about the example of covalent bonding. Below is shown by Table 1 the questions and explanations of degree of conception from the subject that author got in this research.

Table 1: Five question established based on author previous research

Question No	Questions
#1	Salt or sodium chloride is formed from Na^+ and Cl^- ions. Based on this data, the type of NaCl perhaps happen in this molecule after make a bonding is .... (you can consider about ionic character after two ions composed to make a bonding, electronegativity value for Na about 0.93 and Cl about 3.16) a. Ionic bonding* b. Covalent bonding Are you certain with your answer? a. Yes b. No
#2	Wurtzite crystal contains zinc sulphide molecules. Zinc form Zn^{2+} ion while sulphur forms S^{2-} ion. Based on this datum, type of bonding when ZnS form a molecule is .... (considering of ionic character after they make a molecule bonding, Zn: 1.and S 2.58). a. Ionic bonding b. Covalent bonding*

	Are you certain with your answer? a. Yes b. No
#3	Both of $BeCl_2$ and CaF_2 are the example of a compound composed of metal and a non-metal element. Based on this description, type of bond in these molecules is ... a. Ionic bonding b. Covalent bonding * c. Covalent bonding and ionic bonding d. Polar covalent bonding and ionic bonding Are you certain with your answer? a. Yes b. No
#4	This pair of molecules that both of them bind by ionic bonding is ... (Electronegativity value for Na: 0.93; Cl: 3.16; H: 2.20; Ag: 1.93; B:2.04; F: 3.98; Br: 2.96; N:3.04; Al:1.61 O:3.44. You may choose more than one option.) a. NaCl and HCl b. NaCl and AgCl c. BF_3 and KCl d. NaBr and NH_3 e. Al_2O_3 and CsCl * Are you certain with your answer? a. Yes b. No
#5	This pair of molecules that both of them bind by covalent bonding is ... Electronegativity value for Na: 0.93; Cl: 3.16; H: 2.20; Ag: 1.93 B:2.04; F: 3.98; Br: 2.96; N: 3.0; Al:1.61; O:3.44. You may choose more than one option. a. NaH and AgCl* b. HBr and KCl c. NaCl and H_2O d. NH_3 and AlF_3 e. BH_3 and NaBr Are you certain with your answer? a. Yes b. No

There was one misconception type occurred in this question. While two type lack of knowledge. 76% of the subject was chosen ionic bonding when they asked about the type of bonding in sodium chloride molecules. It can be accepted because almost all of literature in chemistry textbook write sodium chloride as their examples for ionic bonding. Furthermore, the subject also remembers this concept all along their life. Previous research said the subject spontaneously answer the question without further thinking when they found this question. They also did not use the electronegativity value written in question ion from metal (Na^+) and ion from non-metal (Cl^-) will compose ionic bonding.

Only 5.3% subject certainly chosen sodium chloride is covalent bonding as their answer. This subject can be categorised as a misconception. They can be simplified as forgot the right concept. It same with another research. This is different with science concept that a compound can be categorised an ionic bonding if they have a difference of electronegativity value more than 1.7 [64]. The electronegativity differences between Na (electronegativity value: 0.93) and Cl (electronegativity value: 3.16) is 2.23. This value shows that sodium chloride is ionic bonding when bond each other.

While 18.6% chosen the answers uncertainty for this question. It consists of 13.3% uncertain chosen ionic bonding and 5.3% uncertain chosen covalent bonding as type bonding of sodium chloride. It means the subject in this research cannot remember they have gotten in previous school degree although more than half of them remember.

The scientific concept of this problem is based on Effendy's book entitled "Ikatan Kimia (Chemical Bonding)". A molecule calls have a covalent bonding if the difference of electronegativity value between atom formed is lower than 1.7 in Pauling Scale. Zinc sulphide consists of Zn which is 1.65 value of electronegativity while S is 2.58 so that the electronegativity value differences is 0.93. It indicates that zinc sulphide is formed by covalent bonding.

This question captured student conception about this problem. Almost a quarter of them knows the right concept about covalent bonding. There was a misconception type occurred in this problem, while two type lack of knowledge. 21.3% subject was chosen ionic bonding when they answered about the type of bonding in zinc sulphide molecules. It means they know the concept of covalent bonding without any uncertainty.

40% of subject indicated as a misconception. It similar to author's previous research [65]. Some of the subjects only decided the type of bonding based on metal and non-metal characteristics. They directly answer the question without further thinking when they found this question. They also did not use the electronegativity value written in question.

While 38.7% chosen the answers uncertainty for this question. It consists of 20% uncertain chosen ionic bonding and 18.7% uncertain chosen covalent bonding as type bonding of zinc sulphide. It means the subject in this research is not enough knowledge for this problem.

Subject conceptions' regarding the determination of the type of bonding that may occur between metallic and non-metallic elements in a molecule. Subjects were asked to specify the type of bonding in compounds $BeCl_2$ and CaF_2. The conception of scientists' topic is based on the ionic characteristic, metallic and non-metallic elements can form compounds that bind covalently-ion and depends on the percentage of ionic character. Binary compound that has more than 1.7 electronegativity difference will form compounds that bind ions whereas the compound having electronegativity difference of less than 1.7 will form polar covalent compounds. $BeCl_2$ compound has the electronegativity difference of 1.5 while CaF_2 compound electronegativity difference of 3.0 so as to have an ionic bond.

In the first option that mentions the type of $BeCl_2$ and CaF_2 is ionic bonding chosen by 36% of the subject. It consists of 28% certain with their answer and 8% uncertain answer this option as their answer. Like previous research, the subject may believe that every element of metal and non-metal will definitely form ionic bonds. They spontaneously answer ionic bond when it finds a molecule concerning the determination of the type of bonding between metallic and non-metallic elements without further thinking. Some of the literature of chemistry textbook in senior high school write this manner to determine the type of bonding.

The scientist conception, B option, only chosen by 17.3%. It consists of 9.3% certain and 8% uncertain when they chose this answer. This means less than 10% is understanding the correct conception about not all the metallic and non-metallic element form ionic bonding. This conception that not all of student understand though it sometimes beryllium chloride wrote in the textbook as covalent bonding. It caused by the example about beryllium chloride never mention as metal and non-metal element. They associated the calcium fluoride formed by ionic bonding too similar as beryllium chloride.

The third and fourth option can be combined because of looks similar which almost a half of subject chosen this answer. As 24% and 22.7% were chosen this option. The "C" option chosen by 10.7% certainty subject and 13.3% uncertainty subject while the "D" option chosen by 10.7% certainty subject and 12% uncertainty subject. This mean subject understood the type of beryllium chloride that sometimes appear in chemistry textbook as covalent bonding, but they did not understand the calcium difluoride bonding type and associated as ionic bonding. It caused by both calcium and fluoride were known as reactive and located in metal and non-metal elements.

In previous research author [65] also found five type of misconception as follow: (1) the subject looked at bonding that occurs when metallic and non-metallic elements bonded by a large electronegativity difference without specify the value of the difference in electronegativity to form ionic bonds, (2) the subject is only looking at ionic bonds based on the properties of metals and non-metals without using the data value of the difference in electronegativity in determining the type of bonding compound, (3) the subject considers the compound $BeCl_2$ two Cl atoms attract electrons from the atoms Be, whereas CaF_2 compound F atom attracts two electrons from an atom Ca. Cl atoms and F will be a negative ion is $2Cl^-$ and $2F^-$, (4) the subject considers $BeCl_2$ formed from two elements Be^{2+} and Cl while $CaCl_2$ is formed of two elements of Ca^{2+} and Cl, (5) the subject assumes that the ionic bonds occur in $BeCl_2$ and $CaCl_2$ compound which is metallic and non-metallic elements causing electrostatic forces.

In the third indicator, determine two examples of ion or covalent bond compounds, two question were established. First question about ionic bonding and second question about covalent bonding. Both of question made with previous author research about chemical bonding topic [65]. But, the subject used here is different with another chemistry teacher instrument and some of the modifications has made. The subject should use the electronegativity value to determine molecule's bonding type. The first question translation show above:

Subject conceptions' regarding the determination of sample molecules that bind to ionic bonding in a molecule consist of 56 subjects chosen a single answer for the question and 19 subjects chosen more than one answer. In this question, the subject asked to specify examples of compounds that bind ionic bonding. The subject was allowed to choose one or more answer in this question. The conception of scientists in this matter is a binary compound that has more than 1.7 electronegativity difference will form compounds that bind ionic compound. The Al_2O_3 has an electronegativity difference 2.0 and the CsCl has an electronegativity difference 2.4 so that the ionic bonds occur in the compounds Al_2O_3 and CsCl [64].

Only a person or 1.8% subject certainly was chosen aluminium oxide and caesium chloride as an example of ionic bonding. This mean very small amount of subject understand and can determine a pair of the compound in ionic bonding. He/she using difference value of atomic compiling based on electronegativity value each. Another two people (3.6%) were chosen this option with the uncertainty that means they are got a lucky guess or uncertain with their choice.

In the other hand, another option is dominated by subject. The first option, sodium chloride and hydrochloride acid were chosen by about a quarter of subject. It can be accepted because both of them is consists of IA and VIIA groups. Some people assume that this group will make an ionic bonding when two atoms bind a molecule. It similar with author previous research that subject did not use the data value of the difference in electronegativity compounds in determining the type of bond despite having been supplied electronegativity value compound constituent elements. They were thought the formation of ionic bonds based on the position of the element in Table of Element Periodic.

Most of the subject were chosen sodium chloride and silver (I) chloride as an example of ionic bonding. They include 37.5% of the subject. Previous research said that subject assumed the properties of metals and non-metals of an element and by anion and cation will form an ionic bond. This is not in accordance with the conception of scientists who claimed that the binary compound that has more than 1.7 electronegativity difference will form compounds that bind ions [64]. Electronegativity value of Na (0.93) and Cl (3.16), then the difference in electronegativity is 2.23. This value indicates that the compound has the sodium chloride ionic bond after binding [64]. However, silver (I) chloride compounds have an electronegativity difference of 1.07, so it has the properties of a polar covalent bond.

Three people were certainly chosen c and d option as their answer. The c option, BF_3 and KCl, chosen by two people and the option d, NaBr, and CH_3 chosen by a person as their answer. Unlike previous option, this option chosen by fewer people. It caused by the molecules is unfamiliar. Also, each one person chose these options as uncertainty answer. As a conclusion, their conception in about this question is rarely found. This categorised as minority chosen by subject.

There were 19 people chosen more than one option as their answer count separately. Most answer chosen by subject is NaCl, HCl and AgCl as molecules bind by ionic bonding about 26.3%. The second choice most chosen is NaCl, AgCl, BF_3 dan KCl about 15.8%. While other option: NaCl, AgCl, Al_2O_3 and NH_4Cl; NaCl, HCl, BF_3 and KCl; NaCl, HCl, AgCl, Al_2O_3, and NH_4Cl; and NaCl, HCl, AgCl, BF_3 and KCl, chosen by a person each. It also consists of uncertainty answer: NaCl, HCl, Al_2O_3, and NH_4Cl; NaCl, AgCl, BF_3, and KCl; NaCl, HCl, Al_2O_3, and NH_4Cl; NaCl, HCl, AgCl, BF_3 KCl, NaBr, and NH_3; was also chosen by a people. They dominant uncertainty chose NaCl, HCl, and AgCl as their answer about 15.8%. This data captured the conception happen in the subject was based on metal and non-metallic element only.

Previous research captured that these conceptions caused by: (i) Subject only looking at compounds that bind to ions formed from metal and non-metal elements to the compound. They will answer ionic bond is formed by metal and non-metal elements that spontaneously when they find a compound concerning the determination of the type of bonding between metallic and non-metallic elements without thinking about the truth and electronegativity differences value. (ii) They view about of the positive ions and negative ions form ionic bonds in compounds. They only looked that the bonds between the elements that form the anion and cation will have an ionic bond. This is different with scientist concept that there are several compounds that are formed by ions can form covalent bonds [64] (iii) Compounds are formed from metal and non-metal elements which have a large electronegativity difference. Subject's explanations that assume that every element of metal and non-metallic elements to form ionic bonds because it has a large electronegativity difference regardless of data values electronegative constituent elements of a compound.

Subject conceptions' regarding the determination of sample molecules that bind to covalent bonding in a molecule consist of 62 subjects chosen a single answer for the question and 13 subjects chosen more than one answer. In this question, the subject asked to specify examples of compounds that bind covalent bonding. The subject was allowed to choose one or more answer in this question. The conception of scientists on this matter binary compounds that have an electronegativity difference of less than 1.7 will form covalent bonds with polar compounds [64]. Both compounds have an electronegativity difference of less than 1.7, ie NaH 1.2 and AgCl of 1.07 so that couples covalently bonded compounds are NaH and AgCl. [64].

This question has several conceptions captured. 6.5% of the subject was certainly given sodium hydride and silver (I) chloride for their answer as an example of covalent bonding. This is mean only small of the subject can understand and determine a pair of a compound bind by covalent bonding. There was 11.3% of subject uncertainly chosen this option as their answer which means they got a lucky guess with their answer.

In the other side, "D" option is more chosen by subject. 37.1% subject certain and 19.4% uncertain chosen NH_3 and AlF_3 bind by covalent bonding. It indicated that result in this research same with previous author research. Their reason this option at least caused by: (1) Subject are only looking at the compound ammonia covalent bonds due to the sharing of a pair of electrons without specifying the type of bond on aluminium fluoride. However, they associate that aluminium fluoride compounds have the same kind of bond with the compound ammonia is a covalent bond. (2) Both of ammonia and AlF_3 are derived from non-metallic elements. Subject associate Al including metallic elements will form covalent bonds with the elements. This is contrary to scientist conception that aluminium fluoride is ionic bonding. The differences electronegativity between Al and F about 2.37 so that make aluminium fluoride has ionic bonding. The "B", and "E" option chosen by a fewer subject while the "C" option was chosen by nobody. The "B" option, HBr and KCl as an example of covalent bonding, was chosen by 6.5% of the subject. This mean they do not understand the potassium chloride bind by covalent bonding they did not consider the electronegativity difference for this problem. Only a person chose the option "E". Like previous option, they have a similar problem occurred.

There were 13 people chosen more than one option as their answer count separately. Three people answer chosen NH_3, AlF_3, BH_3 and NaBr as another person chose HBr, KCl, NaCl and H_2O as their answer. It also consists of uncertainty answer: NaCl, H_2O, NH_3 and AlF_3; HBr, KCl, BH_3 and NaBr; NaCl, H_2O, NH_3 and AlF_3; NaH, AgCl, HBr, KCl, NaCl, H_2O, NH_3, AlF_3, BH_3 and NaBr; NaH, AgCl, NaCl and H_2O; was also chosen by a people (7.7%). They dominant uncertainty chose NaCl, H_2O, NH_3 and AlF_3 was chosen by two people and NH_3, AlF_3, BH_3 and NaBr was chosen by three people. This data captured the conception happen in the subject was based on metal and non-metallic element only.

The conclusion of this research was found many types of misconception happen in the subject about determination the type of bonding using two-tier diagnostic test. The subjects understand the proper concepts that written in chemistry textbook only. They did not use the electronegativity value to determine the bonding type. Thus, metallic characteristic was a popular concept used by subjects when considering the type of bonding in a molecule.

Below is shows the Table 2, the summary of conception found.

Table 2. Subject Conception Summary about Type of Bonding in Molecules.

Item No	Subjects' Conception	%
#1	Sodium chloride bind by covalent bonding while making a molecule	5.3%
#2	Zinc sulphide is binding from ionic bonding	40%
#3	a. The bonding types' of $BeCl_2$ and CaF_2 is ionic bonding	28%
	b. The bonding type of beryllium chloride is covalent bonding, calcium difluoride bonding type bind ionic bonding.	10.7% (10.7%)
#4	a. Sodium chloride and hydrochloride acid were binds by ionic bonding	25%
	b. Sodium chloride and silver (I) chloride as an example of ionic bonding.	37.5%
	c. BF_3 and KCl are the examples of ionic bonding.	1.8%
	d. NaBr and CH_3 are the examples of ionic bonding	1.8%
	e. NaCl, HCl, and AgCl is molecules bind by ionic bonding	26.3%
	f. NaCl, AgCl, BF_3 and KCl is molecules bind by ionic bonding	15.8%
#5	a. The NH_3 and AlF_3 molecule bind by covalent bonding.	37.1%
	b. The HBr and KCl molecule bind by covalent bonding.	6.5%
	c. The NH_3 and NaBr molecule bind by covalent bonding.	1.6%
	d. The NH_3, AlF_3, BH_3 and NaBr are molecules bind by covalent bonding.	23.1%
	e. The HBr, KCl, NaCl and H_2O molecules bind by covalent bonding.	7.7%

This work was supported by Indonesia Endowment Fund for Education from the Indonesia Ministry of Finance. Any opinions, findings, and conclusions or recommendations expressed in this material are those of the authors and do not necessarily reflect the views of the Indonesia Ministry of Finance

References

1. M. B. Nakhleh, J. Chem. Edu. **69**, 191–196 (1992)
2. D. Gabel, J. Chem Edu, *76*(4), 548–554 (1999)
3. J. K. Gilbert, D. Treagust, *Multiple representations in chemical education.* (The Netherlands: Springer, 2009)
4. A. H. Johnstone, Sch. Sc. Rev., *64,* 377–379 (1982)
5. A. K. Griffiths, K. R. Preston, J. Res. Sc. Teach., **29**, 611–628 (1992)
6. Ardiansah, E. Enawaty, I. Lestari, J. Pendidikan dan Pembelajaran Untan, **3**(9), 1-18 (2014)
7. Kolomuç, Tekin, Eura. J. Phys. Chem. Educ., **3**(2), 84-101, (2011)
8. Cheung, Ma, Yang, Int. J. of Sc. Maths. Edu., **7**(6), 1111–1133 (2009)
9. I. O. Abimbola, Sc. Edu., **72**, 175–184 (1988)
10. D. E. Brown, J. of Res. Sc. Teach. **29**, 17–34 (1992)
11. S. K. Chambers, T. Andre, J. Res. Sc. Teach., **34**, 107–123 (1997)
12. Y. Din, J. Bio. Edu. **33**, 21 (1998)
13. R. Driver, J. Easley, St. Sc. Ed., **5**, 61–84 (1978)
14. F. M. Gonzalez, Sch. Sc. Maths. **97**,68 (1997)
15. A. K. Griffiths, *Proceedings of the 1994 International Symposium on Problem Solving and Misconceptions in Chemistry and Physics*, (The International Council of Association for Science Education Publications, 70-99, 1994)
16. A. K. Griffiths, K. Thomey, B. Cooke, G. Normore, J. Res. Sc. Teach., 25, 709–719 (1988)
17. H. Helm, Phys. Edu., **15**, 91–97 (1980)
18. P. W. Hewson, M. G. Hewson, Inst. Sc,, **13**, 1–13 (1984)
19. A. E. Lawson, L. D. Thompson, J. Res. Sc. Teach., **25**, 733–746 (1988)
20. J. Michael, Adv. Physio. Edu., **26**, 5–6 (2002)
21. M. B. Nakhleh, J. Chem. Edu.,**69**,191–196 (1992)
22. J. Nussbaum, Int. J. Sc. Edu., **3**, 159–169 (1981)
23. S. J. Sanger, Sc. Edu., 81, 123–135. (1997)
24. K. S. Taber, J. Sc.Edu., 20,597–608 (1998)
25. L. R. Astudillo, M. Niaz, J. Sc. Edu. Tech., 5, 131–140 (1996)
26. J. Gilbert, D. Swift, Sc. Edu., 69, 681–696 (1985)
27. M. Niaz, J. Sc. Edu. Tech., 10, 205–211 (2001)
28. D. Palmer, Int. J. Sc. Edu., 23, 691–706 (2001)

29. K. S. Taber, Chem. Edu.: Res. Prac. Euro., 2, 43–51. (2001)
30. H. Wandersee, J. J. Mintzes, J. D. Novak, Res. Alt. Con. Sc., *Handbook of Research on Science Teaching and Learning*, (McMillan, New York,177–210, 1994)
31. M. Z. Hashweh, J. Res. Sc. Teach., 25, 121–134 (1988)
32. J. D. Novak, *Theory of Education*, (Cornell University Press, Ithaca, NY, 1977)
33. D. R. Driver, Euro. J. Sc. Edu., **3**, 93–101 (1981)
34. J. Kuiper, Int. J. Sc. Edu., **16**, 279–292 (1994).
35. K. S. Taber, Edu. Chem., **36**, 135–137 (1999)
36. A. Caramazza, M. McCloskey, B. Green, Cog., **9**, 117–123 (1981)
37. L. Resnik, Sc. Edu*.*, **64**, 59–84 (1983)
38. A. Champagne, R. Gunstone, L, Klopfer, Res. Sc. Tech. Edu., **1**, 173–183 (1983)
39. C. R. Sutton, Euro. J. Sc. Edu., **2**, 107–120 (1980)
40. S. A. Southerland, E. Abrams, C. L. Cummins, J. Anzelmo, Sc. Edu., **85**, 328–348 (2001)
41. J. K. Gilbert, R. J. Osborne, P. J. Fensham, Sc. Edu., **66**, 623–633 (1982)
42. R. J. Osborne, B. F. Bell, J. K. Gilbert, J. Sc. Edu., **5**, 1–14 (1983)
43. G. Hills, *Proceedings of the International Seminar on Misconceptions in Science and Mathematics*, (June Cornell University, New York, 245–256, 1983)
44. I. A. Halloun, D. Hestenes, A. J. Phys., **53**, 1056–1065 (1985)
45. Y. Lee, N. Law, Int. J. Sc. Edu., **23**, 111–149 (2001)
46. P. Preece , Euro. J. Sc. Edu., **6**, 7–10 (1984)
47. S. B. Boujaoude, J. Res. Sc. Teach., **29**, 687–699 (1992)
48. R. Good, J. Res. Sc. Teach., **28**, 387 (1991)
49. C. S Carter, N. W. Brickhouse, J. Chem. Edu., **66**, 223–225 (1989)
50. A. B. Champagne, L. E Klopfer, R.F. Gunstone, Edu. Psycho., **17**, 31–53 (1982)
51. L. Pines, L. West, Sc. Edu., **70**, 583–604 (1986)
52. S. Novick, J. Nussbaum, Sc. Edu., **62**(3), 273-281 (1978)
53. B. Butts, R. Smith, Res. Sci. Educ., **17** (1), 192-201 (1987)
54. M. J. S., T. J. Greenbowe, Res. Sci. Teach., **34** (4), 377-398 (1997)
55. T. Pinarbasi, M. Sozbilir, N. Canpolat, Chem. Educ. Res. Pract., **10** (4), 273-280 (2009)
56. C. Tanahoung, R. Chitaree, C. Soankwan, Eura. J. Phys. Chems. Edu., **2**(2), 82-94 (2010)
57. E. Uzuntiryaki, Ö. Geban, Inst. Sc., **33**(4), 311-339 (2005)
58. C. F. Herrmann-Abell, G. E. DeBoer, Chemistry Education Research and Practice, **12**(2), 184-192 (2011)
59. D. D. Milenković, T. N. Hrin, M. D. Segedinac, S. Horvat, J. Sub. Did., **1**(1), 3-12. (2016)
60. H. Özmen, J. Sc. Edu. Tech., **13**(2), 147-159, (2004)
61. S. Hasan, D. Bagayoko, E. L. Kelley, Phys. Edu., **34**(5), 294-299 (1999)
62. D. F. Treagust, Int. J. Sc. Edu., **10**, 159–169 (1988)
63. Effendy, *Ikatan Ionik dan Cacat-Cacat pada Kristal Ionik.* (Malang: Bayumedia, 2008)
64. Ardiansah., *Miskonsepsi Guru dalam Materi Ikatan Kimia pada SMA Negeri di Kabupaten Bengkayang Menggunakan Certainty of Response Index* (*CRI*) (Unpublished undergraduate thesis, Tanjungpura University, Pontianak, Indonesia, 2014)

A Conceptual Model of Folktale Classification as a Visual Guide to a Malaysian Folktale Classification System Development

Harryizman Harun[1] , and *Zulikha* Jamaludin[2]

[1]School of Multimedia Technology and Communication,
[2]School of Computing, UUM College of Arts and Sciences, Universiti Utara Malaysia, 06010 UUM Sintok, Kedah, Malaysia

Abstract. In a study to systematically preserve the Malaysian folktales as one of Malaysia's intangible cultural heritage, a Malaysian Folktale Classification System (MFCS) is proposed to be developed as encouraged by UNESCO. Such a classification system is currently absent in Malaysia. In order to develop a comprehensive classification system, three folktale units are integrated and utilized: function, motif, and type. The use of the three folktale units ensures that the MFCS covers two important facets of folktale: structure and content. The integration of the classification system warrants a complicated classification process. Therefore, a conceptual model, which is central to this article, is constructed as a visual guide to assist the classification process. It illustrates a flow of analysis and all components required to classify namely the three folktale units and their guiding factors, and a primary classification method. A pictorial representation method is utilized to construct the conceptual model. With the conceptual model constructed, it is expected that the analysis of the Malaysian folktales toward the development of the MFCS becomes apparent and guided.

1 Introduction

Cultural heritage is important to society be it in the form of tangible or intangible. Intangible cultural heritage (ICH), which is the focus of this study, is a kind of heritage that is fleeting and ephemeral, unlike the tangible counterpart. According to UNESCO, ICH is the customs, portrayals, manifestations, wisdom, talents and also the tools, entities, relics and cultural areas that societies, crowds and at some points individuals acknowledged as a piece of their cultural heritage [1]. Folklore is a significant ICH, and one of its branches is prose narrative. Prose narrative or in a layman term, tale, is composed of three forms: folktale, legend, and myth [2-3]. The current study chose the folktale as its scope of ICH because there is an urgent need for its preservation considering a very minimal and scarce attempt by far [4-6]. In order to preserve the ICH of the world, UNESCO suggests that, among others, scientific studies to be conducted and a comprehensive register of folklore to be created [1-7].

This study aimed to preserve the Malaysian folktales through a systematic folktale classification coined as the Malaysian Folktale Classification System (MFCS) to meet the suggestion of UNESCO which is found absent in the context of Malaysia. Having a standardized folktale classification system guarantees consistency in classifying, archiving, and retrieval of folklore data and it is vital because the ancient knowledge preserved is not only the thing in the past but also helps to discover ideas and life's philosophy of the past and current generations [8]. At the same time, it is also conceded that the folktale classification process is a lengthy, complex, and meticulous examination [9]. To that end, it became a question in this study whether a conceptual model can be constructed as a visual guide to shed clarity on the development of the MFCS. Therefore, preceding the classification process, this study proposed to construct a conceptual model of folktale classification in order to answer the question imposed.

This article begins with an introduction to the study which eventually triggers the need for the conceptual model. Then, a method to construct the model is explained. Next, the result is presented which is the conceptual model constructed, and a discussion of the model follows. The article ends with a conclusion and references used. That said, the subsequent section elaborates on the method employed to construct the conceptual model.

2 Method

A model aids in channeling internal thought-process to the outer boundary to communicate ideas and solutions on a particular subject matter to other. Considering the mentioned purpose, the magnitude and gravity of forming a standard model representation method are undeniable. A standard representation method would ensure all users comprehend the model equally and this alone makes it fundamental. Fortunately, there happen to be three categories of representation methods that are useful in forming a standardize model specifically textual, pictorial, and multi-faceted representation [10-11].

The first representation method is textual. A text-based representation method is a flexible and the simplest type of model representation. It uses narrative text to briefly explain the content and stimulate visualization of the structure of model components such as objective, output, scope, and flow of the model. This method is practical in a situation that beckons for a quick decision on the suitability of a model in a certain condition. Unquestionably there are benefits and shortcomings of this method. The textual presentation is flexible for it can be narrated in many ways and styles namely sentences, point form or table. Besides, it is also rapid in development compared to the other methods. Nevertheless, this very method also has its drawbacks. The flexibility of the method can also be a bane because it can cause unclear explanation of the model. Thus, it is vital that the structure and the content of the representation be organized well, and the use of standard keywords is practiced. In addition, the validation of the method's correctness is more suitably implemented via a subjective technique such as expert review instead of an objective technique. Not to neglect, since the use of narrative text involved languages, the understandability of a particular language's semantic among users of the model is of paramount.

Differs from the textual representation method, a pictorial representation is a notch upgrade. It is because this method is more effective in a sense that it permits a representation of information via pictures (shape/symbol) associated with links (arrows/lines). The use of pictures in representing the model enables the information structure to flow non-sequentially, which is fitting for a more intricate model. This method is appropriate to be used during the model development as opposed to the early development phase as the previous method. Typically, the diagram is utilized in the pictorial representation method as it signifies information with shapes or symbols linking them with arrows and lines.

The third representation method is a multi-faceted presentation. Even though this method is claimed to be the best in documenting a complete model, it is apt to be used when there is more than a single diagram required in forming the model. The method also becomes the platform for verification for the consistency of model components [12]. Apart from that, the textual representation is also presented as part of the method for a more comprehensive model and to denote its different components. One of the types of the representation method is Unified Modeling Language (UML).

In the context of the conceptual model of the study, the pictorial representations method was borrowed and adapted in modeling the complex classification process. Though the original domain of application is a simulation, it is found that the pictorial representation method is fitting with what the study aims to achieve with the model. Additionally, its concept is known to be applicable in a broader gamut of implementation [13]. The main reason the method is adapted being its aptness to be used during the stage of model development and it clearly maintains the information regarding the configuration and association amid the factors of the issue represented [10 , 14].

Moreover, since the method communicates complex non-sequential information effectively via pictures, its adaptation in this study is astute because the folktale classification is by no means simple and sequential. The classification by function, motif, and type exhibits a complex and entangled configuration of narrative information. Via the use of shapes, symbols, and linkage, the components and the flow required to analyze the folktales for classification were lucidly envisaged. With that said, the next section presents the conceptual model of folktale classification of the study.

3 Results

Based on the pictorial representation method, the conceptual model that the study utilized as the visual guide in the classification process was constructed (depicted in Figure 1 overleaf). The components visualized by the conceptual model are the three folktale units (the function, the motif, and the type), the guiding factors used to identify the units in the Malaysian folktales analyzed (the seven-sphere of actions for the first-level classification, the actor, the item, and the incident for the second-level classification, and the theme for the third-level classification), and the main method employed to classify (a structural-semantic analysis). The shapes and the connecting lines and arrows were drawn manually to depict the components and the flows of the classification process toward the development of the MFCS. The following section discusses the conceptual model in details.

4 Discussion

The conceptual model exemplified in Figure 1 was visualized such a way to represent the unique nature of the classification process needed to be performed on the Malaysian folktales. The conceptual model evidently displays that in order to develop the MFCS (signified by the solid rectangle), each of the Malaysian folktales collected (signified by the dashed rectangle) was analyzed using the primary method elected in the study: the structural-semantic analysis for narrative folklore (situated on the right side of the dashed rectangle) [15-17]. This particular method was created to analyze folktales, among other folklores, to objectively obtain the type of a folktale, particularly by analyzing its structure and semantic (content). Thus, it can be seen from the model that the analysis of the

Malaysian folktales toward the classification was divided into two: the structure and the content (both encased in dashed rounded rectangle). Such analysis ensures that in the end, the comprehensive folktale classification system was obtained because its development was based on the both important aspects of the folktale.

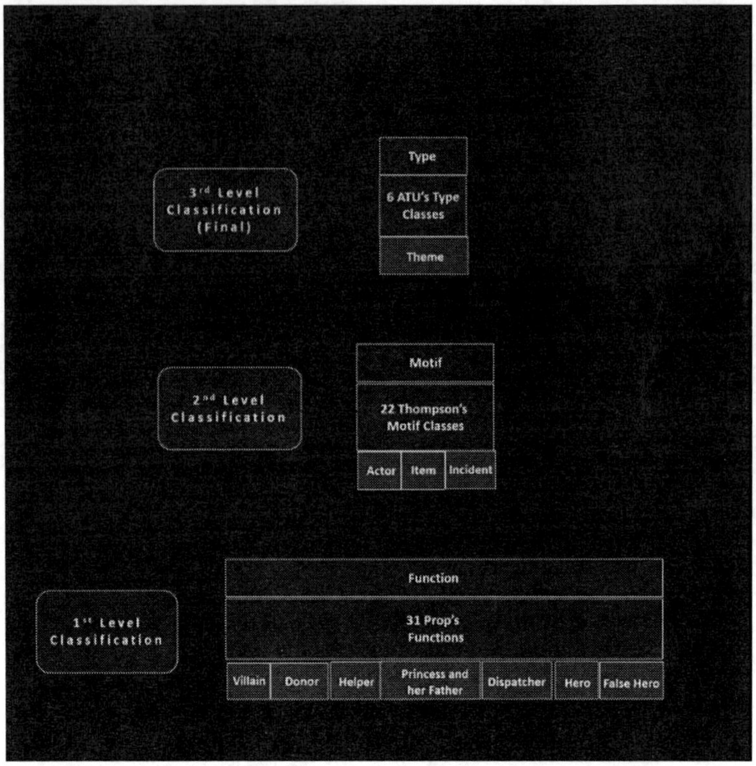

Fig. 1. Conceptual model of folktale classification for the MCFS development

Based on the model, the two aspects of analysis was supported by the three folktale units which gave birth to the three levels of classification (signified by the dark blue rounded rectangles). The horizontal arrows from each dark blue rounded rectangle point to the specific unit of folktale representing the analysis for each level. The first-level classification was to obtain the structure of the folktale through the function unit and the second and the third-level classification the content via the motif and the type units.

The classification begins with the structure because the content is built with the support of its structure and the structure is the frame that forms the content [18]. Once the structure is affirmed, the content is sought because the structure alone does not suffice for a classification system to be comprehensive. A tale is a unique and not a solid entity owing to its specific and in-depth contents instead of being general and abstract like the structure [19]. Therefore, to achieve the comprehensive folktale classification system this study aimed, all the three levels of classification were necessary. The flow of the analysis was represented by the vertical arrows directing the analysis from the bottom to the top of the model. For each level, the guiding factors were available as the guides to identify the function, the motif, and the type in the folktales (signified by the light blue rectangle).

It is evident that from the ground, the first level classification begins with the analysis on the Malaysian folktales based on the 31 functions of folktale's morphology [20] (signified by the dark blue rectangle). The function is the action of the significant actors in the folktales. The sequence of the action of the actor forms the structure of the folktale. At this level, the Malaysian folktales were analyzed thoroughly for the functions in the folktales, and the folktales with a similar sequence of the structure were classed together. In order to identify the functions in the Malaysian folktales, the guiding factors were utilized. For this level, the seven spheres of action (Villain, Donor, Helper, Princess and her Father, Dispatcher, Hero, False Hero) were the guiding factors that aided the researcher in identifying the functions in the folktales analyzed. The spheres of action are general characters that perform the functions in the folktale. The result of this level was the structures of the Malaysian folktales based on the sequences of function.

Next is the second-level classification based on the motif-index of folk literature. In the context of the study, there were 22 motifs utilized to analyze the content of the folktale [21] (signified by the dark blue rectangle). The motifs are Mythological, Animal, Tabu, Magic, The Dead, and Marvels to name a few. The motif is the smallest narrative element in a tale with the ability to endure in tradition. Its endurance is due to the fact that it is an element of extraordinary and striking attributes in the tale. In order to identify the motifs in the Malaysian folktales, the three categories of motifs were used as the guiding factors: the significant actor, the item, and the incident. The actor is the principal characters in the tale that contribute to the meaning of the folktale such as the most handsome prince and the cruel stepmother. The same goes with the item. The item in the folktale is considered as a motif when it adds meaning to the tale, for instance, the magical ring and the talking fish-helper. As for the incident, it is the happening in the tale that provides meaning to the content of the folktale, for example, the princess cursed by the moon and the prince poisoned by the stepmother. Based on the three guiding factors, the motifs in the Malaysian folktales' content were identified, and the folktales that share the same motifs were classed together according to the 22 motifs. Thus, the outcome of this level was the motifs of the Malaysian folktales based on the content.

The last and third-level classification rests on the type. There were six types of ATU (Aarne–Thompson–Uther) type-index employed to analyze the content of the folktale in the context of this study]21-24] (signified by the dark blue rectangle). The types are Animal Tales, Tales of Magic, Realistic Tales (Novelle), Tales of the Stupid Ogre (Giant, Devil), Anecdotes and Jokes, and Formula Tales. The type is the overall theme of a tale built from a single or a combination of motifs in a reasonably fixed order. There is only one guiding factor which is the theme of the folktale analyzed. Considering the theme is a non-concrete entity unlike the motif, each of the Malaysian folktales' content was analyzed as a whole to obtain the right theme which was linked to the best type representing. To make the identification of the theme conceivable, the motifs identified in the second-level classification in a particular folktale worked as an additional assistance to identify the theme of the folktale (hence, the arrow pointing from the motifs to the theme). It is because the spectrum of motifs in a folktale embodies the theme (s) of the folktale which in the end, shapes the tale type [25-26]. As soon as the types were identified, the Malaysian folktales that possess similar theme were classed together based on the six types. Consequently, the outcome of this level was the types of the Malaysian folktales based on the content.

With the completion of the three levels of classification, the classes of the Malaysian folktales based on the structure and the content were obtained. Hence the MFCS was theoretically developed. Using the conceptual model as the visual guide to the classification process, the flow of analysis that needs to be conducted at all the three levels of classification is apparent. Besides the flow of the analysis, the pictorial representations of the folktale units, the guiding factors, the division of structure and content, and the structural-semantic analysis method employed to classify are clearly visualized as an implication of constructing the conceptual model.

After discussing the individually visualized parts of the conceptual model, it is only fair to discuss a limitation of the model. Since the conceptual model was constructed as the visual guide for the classification process, the lack of visualized details regarding the 31 functions, the 22 motifs, and the six types is questionable to onlookers who aspire to undertake the classification of folktale proposed in the study. Nevertheless, in defense of the conceptual model constructed, it is tailored more toward researchers and classifiers who are well versed regarding the specifics of the three units but are unfamiliar with the integration of the three units for the purpose of classification of a folktale.

5 Conclusion

The objective of this study is to develop the MFCS for the systematic folktale preservation by integrating the three important folktale units: the function, the motif, and the type. The integration of the three units ensures the MFCS covers the two significant and connected aspects of a folktale study which are the structure and the content. Such integration leads to the comprehensive yet complex classification system. Therefore, the conceptual model that visualizes the complex classification process was considered necessary as the visual guide in this study and was constructed via the pictorial representation method. Being the emphasis of this article, the construction of the conceptual model of folktale classification marked the achievement of the proposed model. It also answered the question imposed in the study whether the conceptual model can be constructed as the visual guide to the MFCS development for the systematic Malaysian folktale preservation.

As future works, the researcher plans to engage two matters. The first is regarding the limitation stated. The limitation of the conceptual model is a matter that can be justly addressed by upgrading the details of the conceptual model's visualization regarding the particulars of the three folktale units. By doing so, it is hoped that the scope of users of the model and its applicability are more wide-ranging than the current model hence overcoming the limitation indicated. The second matter is a verification of the model. The verification of the model is vital before it is implemented as the visual guide for the classification process. It affirms the model was constructed correctly thus crucial since a flawed model affects the quality of the analysis of the Malaysian folktales toward the development of the MFCS.

References

1. UNESCO., Convention for the safeguarding of the intangible cultural heritage.

http://unesdoc.unesco.org/images/0013/001325/132540e.pdf (2016)

2. W. Bascom, The Journal of American Folklore, **78**: 3-20 (1965)
3. J. Y. Menon, *Malaysian folk tales: A study of archetypal patterns in selected tales.* (Saarbrucken, Germany: LAP Lambert Academic Publishing, 2012)
4. C. M. Desai, Children's Literature in Education, **37**: 163-184 (2006)
5. N. Ismail, T. Masron, A. Ahmad, 4th International Conference on Tourism Research (4ICTR), Sabah, Malaysia (2014)
6. Z. Abidin, M. I., A. Abd. Razak, Information Technology in Childhood Education Annual, 29-44 (2003)
7. UNESCO., Recommendation on the safeguarding of traditional culture and folklore. http://portal.unesco.org/en/ev.php-URL_ID=13141&URL_DO=DO_TOPIC&URL_SECTION=201.html (1989)
8. H.-J. Uther, Folks Art - Croatian Journal Of Ethnology and Folklore Research, **46**, 15-32, (2009)
9. H.-J. Uther, Journal of Folklore Research, **34**, 209-220, (1997)
10. B. S. Onggo, *Conceptual modelling for discrete-event simulation.* (Boca Raton, FL: CRC Press., 337-354, 2010)
11. W. Wang, R. J. Brooks, Simulation Conference, Winter, Washington, DC (2007)
12. B. S. Onggo, Journal of Simulation, **3**, 40-49, (2009)
13. S. Robinson, Journal of the Operational Research Society, 59, 278-290 (2008)
14. J. H. Larkin, H. A. Simon, Cognitive Science, **11**: 65-100 (1987)
15. H. Harun, Z. Jamaludin, *Islamic perspectives relating to business, arts, culture and communication*, (Singapore: Springer Singapore., 127-135, 2015)
16. B. Kerbelyte, Structural-semantic principles of formation of the types of the folk tale. website: http://hdl.handle.net/10400.1/2534 (1995)
17. B. Kerbelyte, *The meaning of Lithuanian folklore works,* (Kaunas, Lithuania: Vytautas Magnus University,2011)
18. C. Levi-Strauss., *Theory and history of folklore (3th ed.),* (Minneapolis, MN: The University of Minnesota Press., 167-188, 1997)
19. R. A. Georges, Journal of Folklore Research, **34**: 203-208 (1997)
20. V. Propp, *Morphology of the folktale (14th ed.)*, (Austin, TX: University of Texas Press, 1998)
21. S. Thompson, *Motif-index of folk-literature,* **1-6**, (Bloomington, IN: Indiana University Press, 1966)
22. H.-J. Uther, The types of international folktales. A classification and bibliography (2nd ed., Vol. **2**). Helsinki, Finland: Academia Scientiarum Fennica (2011a)
23. H.-J. Uther, The types of international folktales. A classification and bibliography (2nd ed., Vol. **3**). Helsinki, Finland: Academia Scientiarum Fennica (2011b)
24. H.-J. Uther, The types of international folktales: A classification and bibliography (2nd ed., Vol. **1**). Helsinki, Finland: Academia Scientiarum Fennica (2011c)
25. J. D'Huy, The Retrospective Methods Network Newsletter, **13**-29 (2014)
26. H. El-Shamy, *Types of the folktale in the Arab world: A demographically oriented tale-type index* (Bloomington, IN: Indiana University Press, 2004)

Islamic English in Islamic Talks

K. Othman[1,*] and *A. I.* Ismail[2]

[1]Centre for Languages and Pre-University Academic Development
[2]Department of English Language and Literature, International Islamic University Malaysia

Abstract. The international nature of the English language has affected many communities across the globe and this has led to the emergence of varieties of English, specifically to meet the needs of non-native speakers of English. For Muslim speakers, Islamic English has been proposed which aims to maintain Arabic terms in the English language when there is an absence of equivalent English words. An attempt to translate would lead to distortion in meanings. This paper aims to highlight the presence of Islamic English employed by a prominent international Islamic speaker. Content analysis method is employed. The findings indicate the functions of Islamic English in the speaker's talks are to provide the accurate meaning of the terms, correct misconceptions and lack of equivalent word in the English language, hence the need for Islamic English for Muslim speakers of English.

1 Introduction

Recent decades have witnessed the spread of the English language across the globe which has become the *lingua franca* of the 21st century. It has empowered and brought mobility to its users, especially non-native speakers of the English language. In fact, as puts it [1], the English language has become localised in its journey to become an international language. He further elaborates that people from different cultures have adopted the language to express themselves based on their cultures which leads to 'nativisation' of language and development of new varieties of English. Examples of these varieties of English include Pakistani English, Saudi English, Korean English, Singaporean English and Indian English [2-6]. For Muslims, they do not only use the language in communication, education, politics and business but also as a means to convey the message of Islam to the global audience, especially to non-Muslim English speaking audience. The tendency for Muslim speakers to maintain some Arabic terms due to the absence of equivalent meaning in the English language has led to the usage of Islamic English, a new variety of English. This paper aims to highlight the presence of Islamic English used by a prominent international Islamic speaker and its functions, hence the importance of maintaining Arabic Islamic terms in discourse particularly among Muslims.

2 Islamic English

The significant rise in the number of English-speaking Muslims around the world has called for more attention on the proper usage of Islamic or divine terms in the English language. The preservation of religious and spiritual values of Arabic words in the English language is essential to avoid change, loss of meaning and obliteration through translation or transliteration process.

According to al-Faruqi who advocates Islamic English, the term refers to "the modified English language to enable it to carry Islamic proper nouns and meanings without distortion" which involves "the infusion of religious, spiritual and cultural terms of Islam to modern English" [7]. The aim is to "serve the needs of the Muslim users of the English language" in an attempt to protect English speaking Muslims from being influenced by the negative values and culture of the west. Abdul Salam concurs that any language that Muslims use, not just English needs to be "Islamised with features that enable it to convey Islamic belief, values, and heritage" [8]. Since majority of Pakistanis and Saudis are Muslims, it is found that Pakistani English and Saudi English reflect Islamic values and beliefs [2 , 3].

Al-Faruqi discusses the nature of current distortion in transliteration of Prophet Muhammad's name [7]. The spelling of Prophet Muhammad's name for instance, has been changed into Maumet, Mawmet and Mahound. Webster International Dictionary (www.webster-dictionary.org) for instance, misspells Prophet Muhammad's name as Mawmet and defines it as "a false god or idol arising from a belief that Mohammedans worshiped images of Mohammed." [7]. This inaccurate and false definition of Prophet Muhammad could lead to confusion not only to non-Muslim English speakers but also to young Muslims. Therefore, the meanings of Arabic names and words must be translated properly to preserve its divine meanings.

The needed rectification in translation has also been discussed by Al-Faruqi [7]. The titles of verses in Al-Quran, Allah's and Prophets' divine names and terms in Islamic laws and concepts are the general rules listed by Al-Faruqi as a guideline for translation in Islamic English. For example, *surah Al-Ikhlas, Al-Azim, Muhammad* and *halal* [7].While the spelling of these Arabic words must be preserved, the meanings are equally important. For instance, *halal* in Islam means

*Corresponding author: khairiah@iium.edu.my

"that which Allah s.w.t. has made legitimate" [7]. Al-Faruqi emphasizes that Arabic words must not be easily translated and defined into English. He is concerned that when translated, users might change the meaning of the Arabic divine words and further cause misconceptions in Islam.

3 Method

The talks of a prominent contemporary international Islamic speaker, Dr Zakir Naik are chosen due to his achievements and recognition in the field of propagation of Islam and inter-religious speeches. Three speeches are selected to ensure consistency in the findings. There are as follows:

Speech 1: "Islam and the 21st Century" was delivered on 11 February 2011 at the Oxford Union Historic Debate in Oxford University, England. https://www.youtube.com/watch?v=rdoCzv28IOw. The audience were educated non-Muslims [9].

Speech 2: "The Solution for Humanity 2015" was delivered on 13 January 2015 at the Dubai Conference Centre, United Arab Emirates. https://www.youtube.com/watch?v=mzPtHsb3fiw. The audience were Muslims [10]

Speech 3: "Islam's View on Terrorism and Jihad" was delivered on 7 October 2015 at the Putra World Trade Centre, Kuala Lumpur, Malaysia. https://www.youtube.com/watch?v=c7Hr8bW5PyU. The audience consisted of Muslims and non-Muslims [11]

The speeches were first transcribed before the Islamic English terms were identified and analysed.

4 Finding and Discussion

The findings indicate the presence of Islamic English in all the three speeches by Zakir Naik. The main function as to why the Arabic terms are used is to correct misconceptions and to give the accurate meanings. In speech one, for instance, when Zakir Naik talked about "Islam and the 21st century", he stated that the most misunderstood word in Islam is the word *jihad*. He said in the 21st century, *jihad* is often mistranslated as holy war that aims for personal gain, fame, honour or land. The truth of what he is saying could be seen for example, in the definition of jihad which is given by the official Collins English Dictionary online which defines *jihad* as a "a holy war which Islam allows Muslims to fight against those who rejects its teachings", besides "a holy war against infidels undertaken by Muslims in defence of the Islamic faith" and "the personal struggle of the individual believer against evil and persecution" [12]. The third edition of Cambridge Advanced Learner's Dictionary has similarly defined *jihad* as "a holy war which is fought by Muslims against people who are a threat to the Islamic religion or who oppose its teachings" [13]. Zakir Naik corrected this information by stating that the word holy war was first used during the Crusades war where the Crusaders forced people to accept the religion of Christianity. In Islam, religion should not be forced, but embraced by one own's free will. Unfortunately, asserted Zakir, the word holy war is used to describe the propagation of Islam by force and war. From the Islamic context, *jihad* means to strive and struggle in order to improve the society, for self-defence, against injustice operation, or one's own selfish intention. He even gave an example of a student who struggles to pass an examination is in fact doing his *jihad*. The extract of his speech is as follows:

"And the most misunderstood word regarding Islam is the word *jihad*. It is not only misunderstood by the Muslims, but it is also misunderstood by the Muslims. Most of the people whether they are Muslims or non-Muslims, they think that any war, by any reasons, whether it is for personal gain, whether it be for fame, honour or land, any war by any Muslim is called *jihad*. *Jihad* is derived from the Arabic word *jahada*, which means to strive, to struggle. In the Islamic context, *jihad* means to strive or to struggle against one own intention. *Jihad* also means to struggle, to strive to make the society better. *Jihad* also means to strive and struggle for self-defence. So *jihad* basically means to strive and struggle. For example, a student is struggling to pass a certain examination, and we will say that he is doing *jihad*. So *jihad* basically means to strive and struggle".

Similarly, in his speech entitled "Islam's View on Terrorism and Jihad", he reiterated the word *jihad* as misunderstood not only by non-Muslims, but also by Muslims. He again elaborated on the accurate meanings of jihad as follows;

"The most misunderstood word in Islam is *jihad*. It is not just been misunderstood by the non-Muslims, but also misunderstood by the Muslims. *Jihad* is derived from the Arabic word, *jahada,* which means to strive or to struggle. In Islam, *jihad* means to struggle against one's own desire. *Jihad* means to strive and struggle to make the society better. *Jihad* means to strive and struggle against operation. *Jihad* also means to strive and struggle in the battle field for self-defence. *Jihad* basically means to strive and struggle. *Jihad* is always misunderstood as the war by Muslims against the non-Muslims, for any reason, whether it be for personal gain, whether it be for fame, for wealth, for religion, for fame for power, it's called as *jihad."*

Zakir Naik also expressed his disappointment when the Muslims themselves translated the word *jihad* incorrectly.

"Most of the orientalists, they translate *jihad* as holy war. Unfortunately, Muslims scholars inverted comas, even they translate *jihad* as a holy war. Holy war if you translate in Arabic word is *harbun muqddasa*. If you read the Quran, never it is mentioned the word *alharbun muqaddasa* in the Quran. There is no hadith where the word *harbun muqaddasa* is mentioned. The word holy war was first coined by the Crusades and by the Christian. If you read history, the Christian crusaders, they forced and killed thousands of human beings in the name of Christianity...Today it is used in the English language to define *jihad*. In short, *jihad* does not mean holy war".

In this speech, he addressed that even the Muslim scholars had carelessly translated the word *jihad* as holy war. Zakir Naik referred to the history of the word holy war which the term was first used by the Christian Crusaders, not by the Muslims. Muslims do not use the term holy war or translated in Arabic as *harbun muqddasa* to mean jihad. Zakir Naik explained the divine term *jihad* in his speeches. It shows his urgency to correct misunderstanding by providing the correct translation of the word *jihad* as to strive and struggle.

Zakir Naik also took the opportunity to explain the concept of *hijab* which has been misunderstood as extremely patronizing and degrading for women in his talk at Oxford University entitled "Islam in the 21st century" when he was asked by one of his audience, the secretary head of Oxford Union. He emphasized that to wear *hijab* is to protect and uplift the status of women as it is a symbol of modesty as can be seen in the following extract of his talk;

"…if women wear revealing clothes, there are more chances of being raped … on March 9, 2009 in Britain, one out of seven British people believe that women who wear revealing and sexy clothes should be hit, I disagree with this… one article in 2005, in the same paper, the Sunday Times, it said that 26% of the British people who wear revealing clothes is partially or totally responsible for the rape. So what I say, the more modest you dress up, the more you are respected. So, Islam has prescribed the modest hijab for the woman, not to degrade her but to uplift her. I do agree maybe there are cultural differences. Islam cannot force anyone to adopt it. There are cultural differences for example, in some societies, even looking at the woman is immodest. Some societies feel that looking is no problem, but touching women is immodest. Some societies feel that shaking hand is no problem, some societies feel kissing is no problem. Some societies feel that doing anything as long as both parties agree, no problem. Different societies have different cultures, different rules and regulations. When I went to America, while I was giving a talk, one of the Americans said, "you Eastern woman, you immodest!" I was shocked. I said, "why do you call the Eastern woman immodest?" He told me Eastern women exposed their belly. So, exposing your belly is immodesty. In India, exposing belly is not immodest, wearing shorts is immodesty. So what you have to realize sister, there are different cultures, there are different systems. Islam cannot force anyone to adopt. It's clearly mentioned in *surah Al-Baqarah,* chapter 2 verse 256; there is no compulsion in religion", but if some women want to adopt the hijab because they feel modest and they feel respected, I feel no other woman should disagree. When I went to U.K., I've seen hundreds and thousands of women who cover their hair and they feel uplifted because of its modesty. Hope that answer the question".

Another function of the usage of Islamic English in Zakir Naik's speeches is due to lack of equivalent word in the English language. In his speech entitled "The Solution for Humanity 2015" for instance, Islamic term *solah* has been introduced as an alternative to the word pray in the English language.

According to the Cambridge Advanced Learner's Dictionary [13], pray is "to speak to god either privately or in religious groups, to express gratitude, love and admiration or to ask for something" The Longman Dictionary of Contemporary English termed pray as "to wish or hope to God very strongly that their wish will come true" These definitions of pray do not capture the meaning as understood in Islam [14].

In this speech, an audience asked Zakir Naik as to why Muslims have to offer their prayers five times a day, which possibly questions the practicality of performing *solah* five times a day when a person could be preoccupied with works and daily problems. Dr. Zakir Naik then answered his question as shown in the extract below:

"Why do Muslims pray 5 times and Christians pray less? The reason is, for Muslims, *solah* is the programming towards righteousness. Normally people, they say pray. It's not the right translation of *solah*. Pray means to ask for help. In Oxford Dictionary, pray means besiege. In *solah*, we don't merely ask for help. Besides asking for help, we are asking guidance from Allah s.w.t. In other religions, they only ask for help. In Islam, besides asking for help, we are asking guidance from Allah s.w.t. For example, if the *Imam,* after *surah Fatihah*, he recites a verse from the Quran, *surah al-Maidah,* chapter number 5, verse number 90; " O you believe, all intoxicants and gambling, dedication of stones, divination of arrows, abstain from this endeavour and you may prosper". Here we are getting guidance from Allah *swt*, in our *solah* that we will not have intoxicants, avoid gambling, avoid fortune telling, avoid idol worshipping, abstain and we as Muslims shall prosper."

In short, Zakir Naik clarified that *solah* in Islam does not only mean to ask God for something, but also to ask for help, forgiveness and seek guidance from Allah s.w.t. For this reason, the word pray cannot be substituted with the term *solat* as *solat* implies more meanings. As discussed by Noor, wrong definitions, spelling and pronunciation of Islamic divine terms could lead to mistaken belief, confusion and misinterpretation [15]. Each Islamic concept has its own meanings and related to Islamic notions of faith. In fact, Zakir Naik proposes the term *solat* to mean programming oneself towards righteousness to ensure one will always be on the right track.

"I prefer the meaning of *solah* as programming towards righteousness. If we analyse it today, the amount of evil around us, you know, bad things happening, evil thing, obscenity, cheating, there are high chances that we must be programmed. Our God, Almighty Allah *Subhanahu wa ta'ala* knows that we have to be programmed 5 times a day, so that we will be on the safe and straight track. For He is our creator, He knows that."

Besides to correct misconception and to provide an alternative to the absence of equivalent English words,

Zakir Naik also use Islamic English to praise the Creator, *Allah Subhanahu wa ta'ala* and to cite the Quranic verses as mentioned earlier like *surah Fatihah, surah al-Baqarah* and *surah al-Maidah*.

In conclusion, Zakir Naik employed Islamic English as an attempt to correct mistranslation of Arabic divine words such as *jihad, hijab* and *solah*. It could also be seen that he utilized Islamic English to maintain the Arabic terms in the English language due to the absence of equivalent English words. Besides, he also used Islamic English to praise the Creator and Quranic verses. These points have been raised by al-Faruqi when he proposed the need for Islamic English to be used in the English language [7]. Similarly, Maher Zain also employed Islamic English in his songs for the purposes of praising Allah and Prophet Muhammad and the absence of equivalent English words. Unlike Maher Zain who also used Islamic English as a form of *du'a* and to evoke special attachment to Islam, Zakir Naik mainly utilized it to correct misconceptions that people have regarding the Arabic words [16].

4 Conclusion

As the English language has become more prominently used by the people around the world, Arabic words on the other hand have suffered changes and loss of meaning through translation process from Arabic to the English language. In order to correct mistranslation and misconception of Arabic divine words, such as *jihad, hijab* and *solah,* Zakir Naik provided clarifications to his audience. The attempt to define Arabic words in the English language may not capture the meaning as intended in Islam. In short, Zakir Naik made use of Islamic English language in his speeches mainly to show the urgency to correct the mistranslation of Arabic words into English words.

This study was funded by a research grant from the International Islamic University Malaysia.

References

1. F. Sharifian, *Language and intercultural Communication* (New York/London: Routledge, Taylor and Francis Group, 2013)
2. A. Mahboob, English as an Islamic language: A case study of Pakistani English, **28**(2), 175-189. (2009)
3. A. Mahboob, T. Elyas, World Englishes, **33**(1), 128-142 (2014)
4. H. Ahn, World Englishes, **33**(2), 195-222 (2014)
5. P.G.L. Chew, World Englishes, **32**(3), 380-394 (2013)
6. J. Lambert, World Englishes, **33**(1),112-127 (2014)
7. I.R. Al-Faruqi, Toward Islamic English. Herndon: International Institute of Islamic Thought, 7-45 (1986)
8. A.S. Abdussalam, Islam And Language, 8 (Kuala Lumpur: Al-Hilal Publishing, 1999)
9. Zakir Naik's *"Islam and 21st century"*. Retrieved March 24, 2015 https://www.youtube.com/watch?v=rdoCzv28IOw (2015)
10. Zakir Naik's *"Ask Dr.Zakir Naik in Dubai International Peace Convention"* .Retrieved March 24, 2015 https://www.youtube.com/watch?v=mzPtHsb3fiw (2015)
11. Zakir Naik's *"Islam's View on Jihad and Terrorism* "Retrieved March 24, 2015. https://www.youtube.com/watch?v=c7Hr8bW5PyU (2015)
12. Collins English Dictionary/ Always Free online. Retrieved December 30,2016 https://www.collinsdictionary.com/dictionary/english (2006)
13. Cambridge Advanced Learner's Dictionary, Cambridge: Cambridge University Press, 774 (2008)
14. Longman Dictionary of Contemporary English, 2003. Oxford: Pearson Education, 1283 (2003)
15. M. H. Noor, *A linguistic analysis of definitions of Islamic terms in the Oxford Advanced Learner's Dictionary (2005), the Macmillan English Dictionary for Advanced Learner's (2007) and the Cambridge Advanced Learner's (2008).* 1112 (Unpublished master's thesis. Gombak, International Islamic University Malaysia, 2011)
16. K. Othman, F. Abdul Malik, Islamic English in Maher Zain's songs: Is there a need?, 1st World International Conference on Islamic Sciences, 5th - 7th December 2016, Kuala Lumpur. (2016)

Teaching Styles, Learning Styles and the ESP Classroom

Lee Mei Ph'ng[1,*]

[1]Department of Languages, Center for Languages and Human Development Universiti Teknikal Malaysia Melaka, Hang Tuah Jaya, 76100 Durian Tunggal, Melaka, Malaysia

Abstract. Learner diversity that exists in the classroom plays a role in influencing the teaching and learning process in the classroom. It should be acknowledged in order for the teaching and learning process to be a meaningful and effective process. Thus, this study examined the learning styles preference of engineering students and the teaching styles preferences of their Technical Communication lecturers. The study also looked at whether the students' learning styles preferences were influenced by their field of study, gender and ethnic backgrounds. Felder and Solomon's Index of Learning Styles was administered to 588 engineering students while Grasha and Riechmann-Hruska's Teaching Style Survey was administered to 10 Technical Communication lecturers. The findings revealed that the students have a marked preference for the visual learning style but balanced preferences for the other learning styles dimensions. The students' field of study, gender and ethnic backgrounds did not seem to influence the students' learning styles preferences. As for their Technical Communication lecturers, they seem to favour the student-centered teaching approach. All the data support the notion of adopting a balanced teaching approach in the Technical Communication classroom.

1 Introduction

Learners and educators are becoming equally diverse. The diversity that occurs among the students means that learners will react differently to their lecturers based on their learning styles preferences [1]. In addition, diversity in teaching and learning styles preferences would result in matches and mismatches between the lecturers' teaching styles and students' learning styles. Learning styles refer to students' preferred learning approaches for all learning situations while teaching styles refer to the lecturers' behaviour, beliefs and selected instructional methods used to present lessons to students [2]. There are researchers who believe that lecturers' teaching styles should match the students' learning styles for effective learning to occur, ensure student motivation and achievement [3-5].

However, the mismatches which are bound to happen may be helpful to students to stretch their learning experiences as this would mean adopting their non-preferred learning styles [6]. In other words, all types of learners still need to be addressed [1]. This study believes that a balanced teaching approach that uses various teaching approaches should be adopted for the classroom.

Variety in instructional approaches can be used to address student diversity [7]. Addressing the preferred learning styles of the students acknowledges the uniqueness and differences of the students [8-9]. The different learning styles of students indicate that they have different strengths and preferences in the ways they absorb and process information [10-12]. In addition, students' learning styles have been identified as one of the factors that influence the quality of learning [13-14].

Emphasis has been placed on learning styles and teaching styles in engineering education [8 , 15]. One reason is lecturers' teaching styles preferences and students' learning styles preferences do influence the teaching and learning process [16-17]. Studies have been conducted to identify the learning styles preferences of engineering students with regards to their fields of study [18-20], gender [21-23] and ethnic backgrounds [24-26]. Researchers have also looked at the teaching styles of language lecturers [27-28].

Shindler stated, five principles for succeeding with student learning diversity can be used as a guide for the teaching and learning process [29]. Two of the principles included the identification of the teaching styles preferences of educators and the learning styles preferences of learners. Thus, this study attempted to profile the student diversity of Universiti Teknikal Malaysia Melaka (UTeM) engineering undergraduates in terms of their learning styles preferences according to their field of study, gender and ethnic background. It also sought to examine the teaching styles of their Technical Communication lecturers.

The Technical Communication classrooms are English for Specific Purposes (ESP) in nature and thus knowledge of the students' learning styles preferences may be useful information for the lecturers when planning their lessons. As highlighted by Sinha and Sadorra, the ability to use a variety of in-class activities

[1] Corresponding author: meiphng@utem.edu.my

is one characteristic central to an ESP teacher for creating a climate for learning [30]. Having an understanding of the learning styles of students and also the tendency the lecturers tend to teach will be able to assist in the creation of the in-class activities in the technical communication classroom.

2 Method

2.1 Sampling

This study involved 588 engineering students from four engineering faculties namely the Faculty of Electrical Engineering (FKE), Faculty of Electronics and Computer Engineering (FKEKK), Faculty of Manufacturing Engineering (FKP) and Faculty of Mechanical Engineering (FKM). They were from 16 out of the 17 degree programs comprising 146 students from FKE; 163 students from FKEKK; 128 students from FKM and 151 students from FKP. The respondents comprised of 300 male students and 288 female students. 487 of them were Bumiputera respondents, 88 of them Chinese and 13 of them Indians. The sample also included ten Technical Communication lecturers.

2.2 Research Instruments

Felder and Solomon's Index of Learning Styles was employed to determine the students' learning styles preferences. This instrument was chosen for this study as it was designed with engineering students in mind and various studies have discussed its reliability and validity [11, 21-32] . It was administered and the data analysed to identify students' learning styles preference according to their field of study, gender and ethnic backgrounds.

There are 44 questions in the Index of Learning Styles. Respondents were required to choose an answer from two options, 'a' or 'b'. For example, one of the questions asks "I understand something better after I"... and the options given are (a) try it out and (b) think about it in detail. Learners are categorised according to eight learning styles preferences (active, reflective, sensing, intuitive, visual, verbal, sequential and global) which are further categorised into four learning styles dimensions namely the active-reflective, sensing-intuitive, visual-verbal and sequential-global dimensions.

Active learners, for example, learn by doing and working with others. Reflective learners, on the other hand, prefer to work alone on the material. Sensing learners are generally oriented towards facts while intuitive learners are generally oriented towards theories. Visual learners learn more from things they see while verbal learners learn more from written and spoken words. As sequential learners, they prefer to learn step by step compared to global learners who prefer to learn in a holistic manner.

The second research instrument was the Teaching Style Survey by Anthony F. Grasha and Sheryl Riechmann-Hruska's, which was used to determine the teaching styles of the Technical Communication lecturers. The Teaching Style Survey was chosen as it was based on a model that was derived from thematic analysis of the universal teaching styles found in college education [33]. His analysis focused on how and why educators taught in particular ways and universal qualities of educators from multiple disciplines and classroom environments. He suggested that every educator has elements of every teaching style in the model but in varying degrees [33]. The Teaching Style Survey classifies one's teaching styles preferences into five categories namely the formal authority (sets standards and defines acceptable ways of doing things), delegator (develops students' ability to function autonomously), facilitator (guides and directs by asking questions, exploring options, suggesting alternatives), personal model (teaches by illustration and direct example) and expert styles (transmitter of information) [34]. There are 40 questions in the Teaching Style Survey and the respondents were required to rank their responses based on a 5-point Likert scale (1 = strongly disagree, 2 = moderately disagree, 3 = undecided, 4= moderately agree, 5 = strongly agree).

2.3 Data Analysis

The first stage involved manual calculation of the scores before they could be input into SPSS for analysis to determine the distribution of the respondents' learning styles preferences. The basis for the manual calculation is based on the calculation by Felder.

The Index of Learning Styles has 44 questions and each question has two options i.e. 'a' or 'b'. Let's take the visual-verbal learning style dimension as an example. The 'a' options for these questions refer to the visual style while the 'b' options refer to the verbal style. In determining a student's learning style preference, the following method was used. For example, if a student chose 8 'a' and 3 'b' options, he/she will be described as inclining towards the visual learning style based on the calculation of (8a-3b = 5a) as there are more 'a' options. Once the scores have been manually calculated, the scores were recorded and entered into SPSS Ver. 19 for statistical analysis to determine the distribution of the students' learning styles preferences for the following dimensions:

- active-reflective dimension,
- sensing-intuitive dimension,
- visual-verbal dimension
- sequential-global dimension.

Taking the visual-verbal dimension for example, students are categorised as either visual or verbal learners in Felder's model. This study however considers the possibility that students may have balanced preferences for the particular learning style dimension.

Since the paper version of the Teaching Style Survey was used in this study, the researcher manually calculated the scores to determine the individual teaching styles of the ten lecturers. The strength of the lecturers' preferences was determined based on the calculation provided by Grasha [33]. Their preferences were categorised as either low, moderate or high.

3 Results

Data analysis revealed that students were generally inclined towards having balanced preferences for the active-reflective, sensing-intuitive and sequential-global learning styles dimensions. However, there was an inclination towards the visual learning styles for the visual-verbal learning styles dimension (Table 1).

3.1 Learning Styles Preferences According to Field of Study

A reasonably evenly distributed sample population was selected from each faculty for the study (Table 2). The findings revealed that the patterns of preference of students from all faculties were fairly similar. All groups showed preference for the visual learning styles and balanced preferences for the active-reflective, sensing-intuitive and sequential-global learning styles dimensions.

3.2 Learning Styles Preferences According to Gender

An approximately equal number of male and female respondents took part in the study (Table 3). The analysis showed the students as having fairly similar patterns of preferences according to their gender. The students had a marked preference for the visual learning styles and balanced preferences for the active-reflective, sensing-intuitive and sequential-global learning styles dimensions. From the findings it appeared that there were no marked differences in learning styles across gender.

Table 1: General Learning Styles Preferences of UTeM Students (% Count)

Act.	Bal.	Ref.	Sns.	Bal.	Int.	Vis.	Bal.	Vrb.	Seq.	Bal.	Glo.
23	68	9	16	71	13	64	34	2	17	71	12

Act. = Active, Ref. = Reflective, Sns. = Sensing, Int. = Intuitive,
Vis. = Visual, Vrb. = verbal, Seq. = Sequential, Glo. = Global, Bal. = Balanced

Table 2: Learning Styles Preferences of UTeM Students: According to Field of Study (% Count)

Fac	Act	Bal	Ref	Sns	Bal	Int	Vis	Bal	Vrb	Seq	Bal	Glo
FKE (N=146)	27	66	7	15	69	16	61	37	2	16	77	7
FKEKK (N=163)	17	71	12	20	64	16	60	38	2	25	67	8
FKM (N=128)	27	64	9	8	77	15	70	28	2	14	68	18
FKP (N=151)	23	69	8	20	74	6	65	33	2	12	74	14

Fac = Faculty, Act. = Active, Ref. = Reflective, Sns. = Sensing, Int. = Intuitive,
Vis. = Visual, Vrb. = verbal, Seq. = Sequential, Glo. = Global, Bal. = Balanced

Table 3: Learning Styles Preferences of UTeM Students: According to Gender (% Count)

Gender	Act	Bal	Ref	Sns	Bal	Int	Vis	Bal	Vrb	Seq	Bal	Glo
Male (N=300)	24	66	10	14	63	13	65	34	1	13	73	14
Female (N=288)	23	69	8	19	69	12	63	35	2	21	69	10

Act. = Active, Ref. = Reflective, Sns. = Sensing, Int. = Intuitive,
Vis. = Visual, Vrb. = verbal, Seq. = Sequential, Glo. = Global, Bal. = Balanced

3.3 Learning Styles Preferences According to Ethnic Background

The Indian respondents were excluded from the analysis due to their relatively small number (n=13).

The analysis highlighted fairly similar learning styles preferences for both groups of respondents (Table 4). The students had a distinct preference for the visual learning styles and were balanced in their preferences for the active-reflective, sensing-intuitive and sequential-global learning styles dimensions. From the findings it appeared that there were no marked differences in learning styles across the students' ethnic background.

3.3 Lecturers' Teaching Styles Preferences

Ten lecturers comprising of six lecturers and four language teachers participated in this part of the study. Their preferences are shown in Table 5. L8 was the only lecturer who had a high preference for the delegator style. Five lecturers (L3, L4, L5, L6 and L7) had preference for the facilitator teaching style followed by the expert style (L1, L2 and L5). Incidentally, L8 was the only lecturer who had a high preference for the delegator teaching style which was also the least prevalent teaching style while L9 was the only lecturer who recorded a low preference for her teaching style. Interestingly, majority of the lecturers (eight lecturers) had moderate teaching style preferences.

Table 4: Learning Styles of UTeM Students: According to Ethnic Background (% Count)

EB	Act	Bal	Ref	Sns	Bal	Int	Vis	Bal	Vrb	Seq	Bal	Glo
B (N=487)	25	67	8	14	72	14	64	34	2	16	71	13
C (N=88)	12	74	14	31	61	8	62	38	0	17	72	11

EB = Ethnic Background, B = Bumiputera, C. = Chinese, Act. = Active, Ref. = Reflective, Sns. = Sensing, Int. = Intuitive, Vis. = Visual, Vrb. = verbal, Seq. = Sequential, Glo. = Global, Bal. = Balanced

Table 5: Lecturers' Teaching Styles Preferences

Lecturer	Teaching Styles Preference	Strength of score
L1	Expert	Moderate
L2	Expert	Moderate
L3	Formal Authority & Facilitator	Moderate
L4	Personal Model & Facilitator	Moderate
L5	Expert & Facilitator	Moderate
L6	Facilitator	Moderate
L7	Facilitator	Moderate
L8	Delegator	High
L9	Personal Model	Low
L10	Personal Model	Moderate

4 Discussion

One of the objectives of this study was to profile UTeM engineering undergraduates in terms of their learning styles preferences. In addition, this study sought to determine whether the students' field of study, gender and ethnic background influenced their learning styles preferences. Analysis revealed the students had a marked preference for the visual learning styles but generally balanced preferences for the active-reflective, sensing-intuitive and sequential-global learning styles dimensions.

These findings matched previous work done on the learning styles of engineering students in the sense that they are similar in their preferences for the visual learning styles [20 , 35-36]. The difference between this study and those of previous researchers was that this study looked at the students' learning styles preferences for every learning styles dimension by determining whether the students indicated any balanced learning styles preferences for a particular learning style dimension. For example, previously reviewed studies usually identified whether students have preference for the active or reflective learning style. This study takes into consideration students may also have balanced preferences for every learning style dimension as well.

This study also looked at their preferences according to gender and revealed similar results to previous studies in the sense that both groups of respondents were similar in their marked preferences for the visual learning styles. Palou found similar findings where the respondents had clear preferences for the visual learning styles. The findings of this study have shown that the students had balanced preferences for the other three learning styles dimensions [21].

This study also focused on the students' learning styles preferences according to their ethnic backgrounds. It only looked at the learning styles preferences of Bumiputera and Chinese students due to the small number of Indian respondents. The students had a marked preference for the visual learning styles and were similar in terms of their balanced preferences for the other three learning styles dimensions. This finding was similar to their preferences for these dimensions according to their field of study and gender. These findings were both a contrast and similarity to Zualkernan et al.'s cross-cultural comparison of students' learning styles preferences which revealed engineering students as having preferences for the active, sensing, visual and sequential learning styles [37].

UTeM engineering students did have a similar marked preference for the visual learning style as other engineering students regardless of their field of study, gender or ethnic background. However, the UTeM students generally had balanced preferences for the other three learning styles dimensions regardless of their field of study, gender or ethnic background. Similarly, studies on engineering students [21-22] and non-engineering students [24 , 26] proved that field of study, gender and ethnic background had minimal effect on the students' learning styles preferences.

This study also sought to examine the teaching styles preferences of the Technical Communication lecturers. This was deemed as necessary as lecturers' teaching styles partly determines engineering students' success in their engineering class [38]. Thus, doing so can indicate whether these Technical Communication lecturers have teaching styles preferences which are inclined towards teacher-centred teaching or student-centred teaching. For example, lectures and demonstrations are examples of teacher-centred approaches while lab activities and learning centres are student-centred approaches [39].

The findings illustrated that the dominant preferred teaching styles was the facilitator teaching styles as it was preferred by five of the lecturers. The least preferred teaching style was the delegator teaching styles as it was preferred by only one of the lecturers. Interestingly, the findings appeared to be partially similar to Juliawati's study which revealed that English lecturers preferred the facilitator, personal model and delegator teaching styles [28]. The findings of the study also appeared to be contradictory to that of Liu et al. who identified language instructors as favouring the teacher-centred style [27].

Based on the discussion above, the findings can be used to enhance the teaching and learning process in the Technical Communication classroom. For example, lecturers should create sufficient opportunities for interaction in the classroom. Another factor lecturers should take note is that some students are shy and they prefer the lecturers to begin interacting with the students. In addition, although preference for classroom interaction is usually associated with active learners, it can encourage reflective learners (who prefer to work alone) to be out of their comfort zones and expand their learning experiences.

Besides that, opportunities for interaction can be created in the classroom by encouraging the students to ask questions or getting the students to report to the lecturer regarding their progress in doing their written assignments, particularly the technical report. This would be in line with the active learners who prefer interaction. As for the reflective learners, they will be encouraged to speak. For example, the students can be encouraged to meet the lecturer group by group. This way, they get to have personal interaction with the lecturer and address issues that may have arose when they were doing their assignments. This also fosters lecturer-student relationship. Besides that, language games on the grammar component can be conducted to encourage class interaction. Besides allowing the students to check their understanding of the topic, it can serve as a break from the routine grammar exercises and add the element of fun into the classroom. Getting students to work together also encourages the development of the teamwork soft skill.

In line with the constructivism view of teaching which calls for the lecturer to play the role as a facilitator, students should be allowed to take charge of their learning at some point in time. For example, the students can be allowed to choose their own topic for their assignments. Doing so allows them to relate their interests to their assignments and gives them room to explore their true potential. This will encourage the sensing learners who have preference learning facts and apply the knowledge they have learnt. The intuitive learners may have preference for the learning of the theoretical aspects of things but doing so encourages them to stretch their learning experiences by making them look for facts and learn to apply the knowledge.

Lecturers can provide students with the overview of the course or assignments before introducing the lessons step by step. This will benefit both students who prefer the global style (course overview) and sequential style (learning step by step). The lecturers can also provide the students with mindmaps of the lessons or concepts being taught.

In addition, lecturers can incorporate the use of examples in their lessons. For example, when conveying the expectations for the coursework, lecturers can show samples of good coursework done by previous students. This sets a benchmark for the kind of work expected from the students and indirectly encourages them to work towards producing a good piece of work. Another way for using examples would be to show the students videos related to the lessons. Examples include video recordings of meetings, presentations and job interview tips for students' reference. With regards to the sensing-intuitive learning style dimension, the learners can pick out the dos and don'ts related to meetings, presentations and job interviews. These videos can be used to attract the students' attention (visual-verbal learning style dimension).

5 Conclusion

The success of the teaching and learning process depends on various factors which include the classroom environment, suitability of the venue and student motivation. The roles played by lecturers and students are also of equal importance towards the creation of a successful and meaningful teaching and learning process. While the lecturers may have their own teaching styles preferences and the students have their own learning styles preferences, they have their own roles to play. From the lecturers' perspective, they should consider the students' learning needs when designing their lessons. Thus, the choice depends on lecturers to understand the learning styles preferences of their students. Lessons which have a combination of all modes of learning are believed to be effective. Such lessons also become better avenues for students who have different learning styles to learn [40].

The findings have indicated that UTeM engineering students have a marked preference for the visual learning styles and balanced preferences for the other three learning styles dimensions. The findings also showed that the students' field of study, gender and ethnic background did not appear to influence the students' learning styles preferences. Secondly, they also indicated the students' learning styles preferences were generally inclined towards the balanced preferences. Their Technical Communication lecturers, on the other hand, generally favoured the student-centred approach.

Thus, the findings seem to point to the possibility of having a balanced teaching approach for the Technical Communication classroom [41]. A balanced teaching approach could improve student learning [42]. Having a balanced teaching approach means there should be variety in the teaching activities in order to maintain student interest. This integration of a variety of teaching styles is also a characteristic of student-centred teaching [43-44]. Doing so meets the needs of certain learning styles favoured by the students. It simultaneously encourages students to use learning styles less favoured by them as a way to stretch their learning abilities [7 , 45]. Such variety widens the students' learning experiences [46].

Thus, it is hoped that the findings and suggestions made in the study will be viewed as supporting information in effort to enhance the teaching and learning in the Technical Communication classrooms and for staff development purposes. When lecturers better understand their students and in the context of this study via their learning styles preferences, it allows lecturers to reach out to their students.

The author would like to thank Universiti Teknikal Malaysia Melaka for supporting this research.

References

1. D. M. Jepsen, M. M. Varhegyi, S. T. T., Edu. Train., **57**(5), 575-587 (2015)
2. M. P. Lee, *Students' preferences and perceptions of teaching styles and learning styles in an ESP context at a Malaysian technical university*, (Doctoral Dissertation, Universiti Kebangsaan Malaysia, Malaysia, 2014)
3. K. S. Mak, WAHANA Akademik, **1**(1), 58-67 (2002)
4. A. Briggs, (PhD Thesis, Walden University, 2003)
5. L. Bostrom, L. M. Lassen, Edu. Train., **48**(2/3), 178-189 (2006)
6. M. Zhou, International Education Studies, **4**(1), 73-77 (2011)
7. P. R. Durben, D. M. Byrd, *Methods for effective teaching: Meeting the needs of all students.* (United States of America: Pearson Education Inc., 2010)
8. M. Ayre, A. Nafalski, *Proceedings of the 30th ASEE/IEEE Frontiers in Education Conference*, 18-21 (2000)
9. B. Samsiah, WAHANA Akademik, **3**(1), 47-53 (2004)
10. N. Hativa, *Teaching for effective learning in higher education,* (Netherlands: Kluwer Academic Publishers, 2000)
11. R. M . Felder, J. Spurlin, Int. J. Eng. Edu., **21**(1), 103-112 (2005)
12. H.S. Fry, S. Ketteridge, S. Marshall, *A handbook for teaching and learning in higher education: enhancing academic performance,* 8-26 (Great Britain: Kogan Page Limited, 2009)
13. B. Uğur, B. Akkoyunlu, S. Kurbanoğlu, Edu. Info. Techno., 1–19. http://doi.org/10.1007/s10639-009-9109-9 (2011)
14. N. Moussa, Inst. Learn. Styles J., **1**, 19–27, http://www.auburn.edu/academic/education/ilsrj/Journal Volumes/Fall 2014 Vol 1 PDFs/Learning Styles Nahla Moussa.pdf (2014)
15. L. F. Capretz, Int. J. Com. Info. Sc., **4**(1), 46-49 (2006)
16. R. M. Felder, L. K. Silverman, Eng. Edu., **78**(7), 674-681 (1988)
17. J. W. Wood, *Teaching students in inclusive settings: adapting and accommodating instructions* (United States of America: Pearson Education, 2006)
18. H. Broberg, P. Lin, K. Griggs, J. Eng. Techno., 25(1), 10-17 (2008)
19. W . Ni, Y.Y. Koh,M. W. Khin, *International Conference on Teaching and Learning: Achieving Excellence in Education Through Quality teaching and Learning.* http://ictl.intimal.edu.my/ictl2007/proceeding/Full_Paper/3C-06-Paper%2090%20(Malaysia).doc (2009)
20. S. L. Franzoni, S. Assar, Edu. Techno. Soct., **12**(4), 15-29 (2009)
21. E . Palou, J. Food Sc. Edu., **5**(7), 51-57 (2006)

22. M. S. Zywno, M. F. Stewart, M. F, European Society for Engineering Education, Portugal, (2007)
23. H. Dermikan, O. O. Demirbas, Proc. Soc. Beha. Sc., 2(2), 1390-1394 (2010)
24. M. E. Roig, *The relationship between learning style preference and achievement in the adult student in a multicultural college,* (Doctor of Education Thesis, Walden University, 2008)
25. P. Prasanthi. *An exploratory study of undergraduates' learning styles*, (Doctor of Education Thesis, Auburn University, 2008)
26. R. A. Boondao, A. J. Hurst, J. I. Sheard, Int. J. Human Soc. Sc., **4**(9), 691-695 (2009)
27. R. Liu, X. Qiao, Y. Liu, The Arizona Working Papers on Second Language Acquisition and Teaching **13**, 77-91 (2006)
28. Juliawati Ibrahim. *Gaya pengajaran guru bahasa daerah Hulu Langat: satu kajian tinjauan.* (Projek penyelidikan Ijazah Sarjana Pendidikan, Fakulti Pendidikan, Universiti Kebangsaan Malaysia, 2006)
29. Shindler, J., Teaching for the success of all learning styles: five principles for promoting greater teacher effectiveness and higher student achievement for all students http://www.calstatela.edu/faculty/jshindl/ls/5principles.htm (2005)
30. A. C. Sinha, L. C. Sadorra, *A primer on ESP for teachers of English*, (Manila: De La Salle University Press, 1991)
31. M. S. Zywno, American Society for Engineering Education Annual Conference and Exposition. Organised by the American Society for Engineering Education Annual Conference and Exposition, Tennessee, 22-25 June, (2003)
32. T. A. Litzinger, S. H. Lee, J. C. Wise, R. M. Felder, J. Eng. Edu., **96**(4), 309-319 (2007)
33. A. F. Grasha, *Teaching with style: a practical guide to enhancing learning by understanding teaching and learning styles*, (United States of America: Alliance Publishers, 2002)
34. A. F. Grasha, *The Professional & Organizational Developmental Network in Higher Education,* **7**(5) http://www.podnetwork.org/publications/teachingexcellence/95-96/V7,%20N5%20Grasha.pdf (1995)
35. A. Kolmos, J. E. Holgaard, J. E., Learning styles of science and engineering students in problem and project based education. *SEFI 2008 Annual Conference.* http://www.sefi.be/wp-content/abstracts/1243.pdf (2008)
36. K. M. Ismail, Puteh, S. Mohamed, A preliminary study on the learning style of civil engineering engineering students in Malaysia. *Proceedings of the International Conference on Engineering Education (ICEED),* 87-91 (2009)
37. I. A. Zualkernan, J. Allert, G. Z. Qadah,., *IEEE Transactions on Education,* **49**(4), 443-450 (2006)
38. J. Larsen, P. R. McCright, G. Weisenborn, *Proceedings of the 2004 American Society for Engineering Education Annual Conference & Exposition, USA.* http://soa.asee.org/paper/conference/paper-view.cfm?id=20229 (2004)
39. P. R. Burden, D. M. Byrd, *Methods for effective teaching: Meeting the needs of all students.* (United States of America: Pearson Education Inc, 2010)
40. A. Ganesh, U. P. Ratnakar, SCMS Journal of Indian Management, **11**(3), 26-37 (2014)
41. M. P. Lee, S. M. Thang, M. K. Radha, Nambiar, The Social Sciences, **11**(12), 2977-2982 (2016)
42. S. Meera, L. Sudak, P. Egberts, Mapping conventional teaching methods and learning styles in engineering dynamics. In *American Society of Engineering Education Annual Meeting in Engineering Dynamics* (2016)
43. R. M. Felder, E. R. Henriques, Foreign Language Annals, **28**(1), 21-31 (1995)
44. B. L. Brown, *Teaching style vs. learning style*, (United States of America: Clearinghouse on Adult, Career and Vocational Education, 2003)
45. B. Joyce, M. Weil, E. Calhoun, *Models of teaching.* (United States of America: Pearson Education Inc, 2011)
46. C. K. Lee, S. S. Manjit, Edu. Techno. Soc., **18** (3), 266-281 (2015)

Influence Of Demographic Factors On Memory And Attention (Cognitive Ability) Among Hearing-Impaired

Ummi Habibah Abd Rani[1,*] and *Najib* Ahmad Marzuki[1]

[1]School of Applied Psychology, Social Work and Policy, University Utara Malaysia, Malaysia

Abstract. The aim of this study was to examine the influence of demographic factors on cognitive ability (memory and attention) among hearing-impaired students. Data from 163 hearing-impaired students were collected using simple random sampling at five polytechnics in Malaysia. The Cambridge Neuropsychological Test Automated Battery (CANTAB) software were utilised to examine cognitive ability (memory and attention). Demographic factors consist of gender, hearing aid usage and parents' hearing status. The results revealed that attention and memory differed according to hearing aid usage and parent's status of hearing. No significant differences were found for memory and attention according to hearing-impaired gender. The findings of this study will enable the education authority to direct these special students in a way that best suits the institution and students goals.

1 Introduction

Hearing-impaired can be related to impediment to access an information in educational experiences. The perception of hearing-impaired students as concrete, literal thinkers has now been replaced by a better understanding of the interactions of language and intellectual development in hearing-impaired students. Hearing-impaired was linked with significant delays in spoken language [1]. The language development seems to be specifically influence learning process which is linked to general cognitive process of association, attention and memory. However, it was identify that the absence of auditory stimulation and language delay appear to affect neurocognitive domains, such as memory, attention, auditory and inhibition [1].

According to Myklebust, individuals with hearing impairment were cognitively different and interpreted as having limited memory capacities and language-related cognitive deficits. In mainstream programs with hearing-impaired students in segregated (residential or separate) programs often assumed that all other things are equal. But all other things are seldom equal when it comes to hearing-impaired. Most investigations of this issue have not taken into account the intellectual abilities, language fluencies, or social skills of hearing-impaired students. Such findings should not be taken to imply that hearing-impaired students will always have learning difficulties. Instead, they should be seen as emphasizing the need for a more complete understanding of hearing-impaired students' cognitive ability such as attention and memory if we are to foster educational success [2].

Individuals with Disabilities Education Act suggest the assessment of cognitive abilities is essential for differential diagnosis in educational settings such as identifying eligibility for special education programs in large part on the basis of cognitive abilities [3].

Based on the literature, memory and attention of hearing-impaired students were influenced by demographic factors. Apparently, the hearing-impaired cognitive ability can influence various aspects of life in some individuals, their families and society. Research focusing on specific aspects of cognitive aspect, such as memory and attention and influence of demographic factors have yielded contradictory results specifically among hearing-impaired in Malaysia. Therefore, accomplishing this study expands the knowledge of memory and attention (cognitive ability) as well as hearing disabilities and special education study in Malaysia.

In this study, cognitive ability was defined based on Horn and Noll, describe cognitive ability as intellectual human ability to accomplish and manage something complex through mental analysis [4]. The attention and memory are the components of cognitive ability. Throughout this paper, the term attention refers to how well students can focus when faced with internal distractions such as mind wandering and boredom; while students' cognitive ability in memory refers to how well students utilized their memory when faced with external distractions. Hearing-impaired mentioned to people with limited hearing and deafness stated to the audiological condition of not hearing [5]. According to World Health Organization [WHO], hearing Impairment is defined as an absolute or partial loss of the capacity to hear out of one or two ears. The degree of impairment can be mild, moderate, severe or profound [6]. Whereas, a hearing aid is a device that amplifies and customizes incoming sound signals to meet specific requirements of each hearing impaired.

Studies done found no significant gender difference in general intelligence [7-9]. Moreover, reports of male and females not differ in reaction time [10]. There are, however males better cognitive ability performance [11].

There also appears to be a gender difference in episodic memory with females performing better in most tasks, although this area has not been the subject of meta-analysis. The finding on the recall test is consistent with previous research showing better female performance on long term memory tasks [12].

Hearing-impaired people, in contrast, have been said to rely heavily on visual-spatial short-term memory codes [2]. The most frequently cited support for the notion that hearing-impaired individuals tend to use visual short-term memory codes comes from a series of studies conducted by Parasnis reports that there were no gender differences in cognitive ability of hearing-impaired people [13].

The origins of these cognitive gender differences are still not fully understood [9]. Although there is no doubt that sex hormones and sexual brain dimorphisms contribute to sex differences in specific cognitive abilities, it is also clear that social priming, and especially gender stereotypes can significantly affect male's and female's cognitive performance [14-15]. In fact, gender stereotypes might be a central variable in environmental influences on sex differences in intelligence [9].

Gatehouse show a significant interactions between hearing impairment and cognitive ability, and the extent to which, when tested unaided, listeners can derive and gain advantage from temporal structure [18]. This is similar with Mulrow study who found changes in cognitive aspect among aided hearing-impaired [16].

According to Marschark, hearing-impaired who's born into a family with hearing-impaired members, for example, is often exposed to sign language early in life, and hearing-impairment is not viewed as negatively as it might be in a family with no such history [17].

Studies show that hearing-impaired born in hearing-impaired families and growing up in an environment where the conversation is naturally dependent on visual have an advantage compared to those who raised by normal hearing parents [18]. Accordingly, hearing-impaired students of hearing-impaired parents may be able to attain higher levels of cognitive activities.

One might expect that hearing-impaired of hearing-impaired parents would perform comparably to hearing peers on most cognitive tasks, but these expectations are not met with any consistencies. More research is needed with larger numbers of students before conclusive findings can be obtained. The lack of any causal link between language delays and cognitive abilities in hearing-impaired indicate that there are undiscovered factors that influence evaluations of cognitive development in hearing-impaired students.

2 Methods

2.1 Research Design

The study followed a prospective cross-sectional design for examine the influence of demographic factors on memory and attention who are hearing-impaired students. The main reason for choosing a quantitative method for the cross-sectional correlational survey design was due to the required degree of generalization of results. The rationale for this research stems from the fact that the design allows researchers to examine various variables.

2.2 Participants

Participants for this study involved the students with hearing-impaired from five polytechnics in Malaysia. The participating students were identified as students with hearing disabilities by their polytechnic and are currently receiving special education services at their polytechnic. The participants consisted of both male and female students. For this study, researcher select the participants through simple random sampling technique. A random size of 163 participants was taken based on Krejcie and Morgan table [19], from a population size of 198.

2.3 Instrument

The cognitive ability (memory and attention) was assessed by the cognitive test battery. The software program called Cambridge Neuropsychological Test Automated Battery (CANTAB) [20], which provide an effective method of cognitive ability assessment, was utilised to test the cognitive ability of hearing-impaired students at Malaysian Polytechnics. CANTAB has been specifically designed for assessing the cognitive abilities from animals to humans. Application of test batteries, including diagnosis, cognitive development and occupational work is a "law" that was highlighted by CANTAB. The CANTAB is a computed nonverbal battery of visually presented neuropsychological tests. The test were administered on a touch-sensitive computer screen and were completed by the participants.

2.4 Data Analysis

The Statistical Package for the Social Sciences (SPSS) version 21 was performed to analyse the data. The independent t-test was conducted to investigate the differences in mean between (attention and memory) cognitive ability and demographic factors.

3 Results and Discussion

3.1 Differences of attention and memory based on gender

The result showed that there were no significant differences of memory with [t= .94, p> .05] and attention [t= .34, p> .05] between both gender. Referring to table 1, there were no significant differences of attention and memory mean between male and female hearing-impaired.

Table 1: Attention and memory between gender

Cognitive ability	Gender	N	Mean	SD	t
Attention	Male	79	3.78	1.09	.94
	Female	84	3.94	1.00	
Memory	Male	79	3.22	1.11	.34
	Female	84	3.15	1.12	

As can be seen, male and female hearing-impaired in polytechnics appeared to demonstrate attention and memory irrespective of the gender status. This discovery supports studies that found that no significance gender differences in cognitive ability [7-8] and attention [10]. Nevertheless, the current findings in contrast with Wai and Lincoln,which found males tend to have higher cognitive ability [21].

3.2 Differences of attention and memory based on hearing aid usage

As demonstrated in table 2, there were significant differences of memory with [t= 4.76, p< .05] and attention [t=5.02, p< .05] between hearing aid usage. Results indicating that there were significant differences of attention and memory between hearing aid usages of respondents.

Table 2: Attention and memory between hearing aid usage

Cognitive ability	Hearing aid usage	N	Mean	SD	t
Attention	Yes	60	4.24	1.55	4.76*
	No	103	3.23	0.79	
Memory	Yes	60	3.67	1.55	5.02*
	No	103	2.54	1.27	

Notes: *p< .05

With reference to the table, it appears that memory and attention of hearing-impaired in polytechnics differed based on hearing aid usage. The result is supported by Gatehouse, who found that cognitive ability of aid and unaided hearing-impaired differed [22]. In this study, the significant difference achieved by students may be due to the ability of hearing-impaired with hearing aid to speak, hear and communicate better than unaided hearing-impaired. Thus, it will contribute to enhance their attention and memory during the learning process. According to Zand and Pierce, hearing aids provide sustained benefits for individuals with hearing impairment [23].

3.3 Differences of attention and memory based on parents' hearing status

The findings showed that there were significant difference of memory with [t= 4.07, p< .05] and attention [t=3.99, p< .05] between parent's hearing status. The outcome in table 1 revealed that there were significant differences of attention and memory between normal hearing parent and hearing-impaired parent.

Table 3: Attention and memory between parent's hearing status

Cognitive ability	Parents' hearing status	N	Mean	SD	t
Attention	Normal	123	3.39	1.22	4.07*
	Hearing impaired	40	4.26	1.03	
Memory	Normal	123	2.67	1.45	3.99*
	Hearing impaired	40	3.70	1.34	

Notes: $^{*}p< .05$

The results showed that parents' hearing status based on attention and memory differed between normal hearing parent and hearing-impaired parent. As reported by Tyrose, parent hearing status are associated with the cognitive ability. In this study, it can be assumed that the cognitive ability are heightened by communication style of hearing-impaired parent such as using sign language, hand and body movement which involve visual movement and memory coding [24]. According to Marschark, sign language contribute to cognitive development and visual imagery [2].

4 Conclusion

In conclusion, attention and memory differed according to hearing aid usage and parent's status of hearing. The study also found that no significant differences for memory and attention according to hearing-impaired gender. Thus, a few suggestions are recommended to increase the memory and attention of hearing-impaired students. The ability to profit from both formal and informal instruction at education system requires that students have skills in areas such as attention, memorizing and have a positive attitude toward learning. More educators are needed to serve as role models and to be fluent communicators with students who are hearing-impaired. It can match accessible teaching strategies and materials to the needs of hearing-impaired learners in order to provide them with fair and appropriate education. The exposure to normal hearing peers and educators to sign language skills program may leads to a supportive environment and may influence attention and memory of hearing-impaired students. Relevant education authorities should take necessary steps in improving the hearing impaired students' attention and memory (cognitive ability).

References

1. D. Koo, K. Crain, C. LaSasso, G.F. Eden, Ann. N.Y. Aca. Sc., 1145, 83–99, doi: 10.1196/annals.1416.025 (2008)
2. M. Marschark, C. Morrison, J. Lukomski, G. Borgna, C. Convertino, Learn. Indv. Diff., **25**, 156–162. doi: http://doi.org/10.1016/j.lindif.2013.02.006 (2013)
3. Individuals with Disabilities Education Act Amendments (1997).http://www.ed.gov/offices/OSERS/IDEA/index.html. (2014)
4. J. L. Horn, J. A. Noll, *Contemporary intellectual assessment: Theories, tests, and issues,* (New York: Guilford, 53-91, 1997)
5. *National association of the deaf., The Guide for Deaf and Hard of Hearing People (National Association of the Deaf)*, http://ebooklibrary.space/?book=1563686449, **6**, (Washington, DC: Gallaudet University Press, 2015)
6. World Health Organization, *Deafness and hearing impairment.* http://www.who.int/mediacentre/factsheets/fs3 00/en/index .html (2006)
7. A. Saggino, M. Balsamo, A. Grieco, M. R. Cerbone, N. N. Raviele, N.N. Perce. Mot. Skills, 98,840–848. doi:10.2466/pms.98.3.840-848 (2006)
8. M. Burgaleta, K. Head, J. Álvarez-Linera, K. Martínez, S. Escorial, R. Haier, et al., Intel, **40**, 60–68. doi.org/10.1016/j.intell.2011.10.006. (2012)
9. D.G. Halpern, M.L. LaMay, Edu. Psycho. Rev, **12**, 229–246 https://www.academia.edu/2758826/The_Smarter_Sex_A_Critical_Review_of_Sex_Differences_in_Intelligence (2000)
10. D. Reilly, D. Neumann, G. Andrews, Intel, **54**, 147–158. doi: 10.1016/j.intell.2015.12.004 (2016)
11. L. H. Summers, *Remarks at NBER Conference on Diversifying the Science & Engineering Workforce*, (http://www.harvard.edu/president/speeches/summers_2005/nber.php (Cambridge, M.A., 2005)
12. C. Flores-Mendoza, K.F. Widaman, H. Rindermann, R. Primi, M. Mansur-Alves, C. C. Pena, Intel, **41**, 70–84, doi: http://dx.doi.org/10.1016/j.intell.2012.11.002 (2013)
13. I. Parasnis, *Cultural and Language Diversity and the Deaf Experience*, 79–98, (Cambridge: Cambridge University Press, 1998)

14. M. Hausmann, D. Schoofs, H. E. S. Rosenthal, K. Jordan, Psychoneuro., **34**, 389–401, doi:10.1016/j.psyneuen.2008.09.019 (2009)
15. M. Hirnstein, N. Freund, M. Hausmann, Zeits. Psycho., **220**,70–77. doi: 10.1027/2151-2604/a000098 (2012)
16. C. D. Mulrow, M. R. Tuley, C. Aguilar, J. Speech Hear. Res., **35**, 1402-1405, doi:10.1044/jshr.3506.1402 (1992)
17. M. Marschark, H. G. Lang, J. A. Albertini, *Educating Deaf Students: From Research to Practice*, (Oxford: Oxford University Press, 2002)
18. E. Jambor, M. Elliott, M., Stud. D. Edu., **10**(1), 63–81. Retrieved 18 March,, 2014, doi:10.1093/deafed/eni004 (2005)
19. R. V. Krejcie, D. W. Morgan, *Determining sample size for research activities. Educational and Psychological Measurement.* https://opa.uprrp.edu/InvInsDocs/KrejcieandMorgan.pdf (1970)
20. CANTAB:*Cognitive Test Neuropsychological Test Automated Batteries*, **5**, (England: Cambridge Cognition Ltd, 2015)
21. J. Wai, D. Lincold, Intel, **54**, 1-32. doi.org/10.1016/j.intell.2015.11.002_(2016)
22. S.J. Zand, K. J. Pierce, K. (Eds.)., *Resilience in deaf children: Adaption through emerging adulthood,* (New York, Springer, 2011)
23. H. Dillon, Inter. J. Audio., **42**, S77–S85. doi:10.3109/14992020309074627. (2003)
24. M. P. Tayrose, *The Developmental Organization of Cognitive Abilities of Deaf and Hard of Hearing (D/HOH) People as Compared to those who are Normal Hearing* http://repository.lib.ncsu.edu/ir/bitstream/1840.16/6766/1/etd.pdf (North Carolina State University Raleigh, North Carolina 2011)

 https://doi.org/10.1051/matecconf/201815005084

Emotional Intelligence and Psychological Well-Being among Hearing-Impaired: Does it Relates?

Najib Ahmad Marzuki[1,*] , *Azlizamani* Zubir@Salim[1] and *Ummi* Habibah Abd Rani[1]

[1]School of Applied Psychology, Social Work and Policy Universiti Utara Malaysia, Kedah, Malaysia

Abstract. This paper explores the relationship between emotional intelligence and psychological well-being among hearing-impaired students. Emotional intelligence consists of five dimensions namely intrapersonal, interpersonal, adaptability, stress management and general mood. A total of 130 hearing-impaired students were chosen as participants via simple random sampling at four polytechnics in Malaysia. The Bar-On Emotional Quotient: Short (EQ-i: S) and Ryff's Psychological Well-Being instruments were utilised to measure emotional intelligence and psychological well-being. It was found that in general, hearing impaired students have moderate emotional intelligence level and psychological well-being. Results also found that intrapersonal, interpersonal and adaptability dimensions correlated significantly with psychological well-being. No correlations were found between stress management and general mood dimensions with psychological well-being. Though limited by their ability to speak and to hear, emotional intelligence among hearing-impaired students does play a role in enhancing their individual capability to learn and to experience positive psychological well-being in life.

1 Introduction

World Health Organisation (WHO) has estimated that there are 328 million hearing-impaired adults in the world [1]. Hearing impairment is defined as an absolute or partial loss of the capacity to hear out of one or two ears and may be considered as either mild, moderate, severe or profound [2]. In Malaysia, 53,357 people with disability (hearing-impaired) have registered with the Department of Social Welfare. However, only 1 percent of the overall total of people registered has acquired education and continues their study in educational institutions that provide special education in Malaysia including polytechnics. Nevertheless, the special education to develop this particular group into individual with emotional and psychological stability could not be delivered effectively. Parents fear and anxiety of permitting their children to enrol in this special program hinders this group from improving their quality of life [3]. According to [4], hearing impairment is significantly associated with cognitive development, social and emotional aspect of an individual.

The hearing-impaired community often faces barrier and this contribute towards negative effects in terms of emotional and psychological development. In [5] as well as [6] stipulated that individuals with hearing impairment faces problems in communication and are not able to make themselves understood by normal person. This situation affects the social, emotional and cognitive development of the hearing-impaired [7] and play a pivotal role in their overall psychological well-being.

According to [8], the most important needs for the hearing-impaired is to enhance their capability of engaging with the larger community, sensitive to their own emotional and psychological needs and aware of effects of their actions to the community. In general, hearing-impaired students are different from normal students in terms of communication with others and this may influence their emotions. In [9] argued that the emotional and psychological issues among hearing-impaired in the education sector were given less attention and are less studied.

It is difficult to define precisely the meaning of emotional intelligence and psychological well-being among hearing-impaired. However, the underlying definitions should not be discriminated between normal individual and the hearing-impaired.

* Corresponding author: author@e-mail.org

In defined emotional intelligence as the ability to perceive accurately, appraise and express emotion [10]; the ability to access and/or generate feelings when they facilitate thought; the ability to understand emotion and emotional knowledge; and the ability to regulate emotions to promote emotional and intellectual growth. In this study, emotional intelligence refers to the definition based on the measurement by [11]. It consists of five dimensions namely intrapersonal, interpersonal, adaptability, stress management and general mood.

In terms of psychological well-being, [12] perceived it as subjective satisfaction of an individual life while [13] defined it as a state that emerges from feeling of satisfaction with one's physical health and oneself as a person and with one's close interpersonal relationships. The psychological well-being definitions in this study refers to [14] dimensions of psychological well-being that combines self-acceptance, positive relations with others, autonomy, environmental mastery, purpose on life and personal growth.

This paper explores the level of emotional intelligence and psychological well-being among hearing impaired students as well as examining the relationship between dimensions of emotional intelligence and psychological well-being.

1.1 Previous Studies

A study by [15] on the psychological well-being of hearing-impaired students demonstrated low scores in dimensions of self-acceptance and positive relationship. On the other hand, [16] stated that no significant relationship was found between social relationship and psychological well-being among hearing-impaired students.

In a study involving psychological well-being and gender of the hearing-impaired done by [17] and [18] differences of psychological well-being between genders were found. All these studies, however, have contradicted the studies done by [19] and [20]. Meanwhile, [21] found that genetic excuse (parents status of hearing) did have a significant association with the students' psychological well-being.

In [22] examined the connection of emotional intelligence with adolescent's psychological well-being (depression and self-esteem). They found significant positive relationship between variables of emotional intelligence and self-esteem but showed negative association between emotional intelligence and depression. They concluded that emotionally intelligent people adopt flexible patterns of life, which allow a person to adopt attitudinal changes, which protect a person to feel failure, neglected and development of depression. Their study found that a stepwise structural chain exists between emotional intelligence and life success.

Another research was done by [23] which involves emotional intelligence and psychological well-being. Their study aimed to identify the relationship between emotional intelligence, personality and array of indicators of psychological well-being among law students. In this study, 43 students had completed all tests. The test instruments comprised of emotional intelligence test, personality test, coping test and social well-being test. The findings showed that emotional intelligence was significantly related to three of the six variables that were chosen to represent psychological health and the Big Five personality factor of neuroticism was found to be the strongest predictor of well-being. However, it seems that emotional intelligence does not relate to the additional variance in well-being over personality.

According to [24] also stressed on the relationship between emotional intelligence and psychological well-being among students. One hundred students were randomly selected for this study. There were 2 scales that had been utilised in this study namely Well-Being Manifestation Measure Scale and Emotional Intelligence Scale. Their findings showed that there was a significant positive correlation between emotional intelligence and psychological well-being indicating that higher emotional intelligence level will lead to higher psychological well-being level. The findings also indicated that girls are more intelligence than boys.

Meanwhile, a study focusing on the same topic but using different setting and sample was conducted by [25]. In this scenario, the researcher had selected 75 males and 75 females from public and private medical college. The scale of emotional intelligence and psychological well-being were used in this study. Result showed that the emotional intelligence and psychological well-being has significant differences on dimensions of psychological well-being (autonomy, environmental mastery and personal growth) and has insignificant differences on three dimensions of psychological well-being (positive relations, purpose in life and self-acceptance) among private and public college.

2 Method

2.1. Research design

The study utilised a quantitative approach via cross-sectional survey by distributing questionnaires to hearing-impaired students from four polytechnics in Peninsular Malaysia.

2.2 Population and sample

The population consists of hearing-impaired students at five polytechnics in Peninsula Malaysia. There are a total of 167 hearing-impaired students in five polytechnics in Malaysia. These polytechnics provide special courses to students with hearing-impaired disability (four are situated in Peninsula Malaysia and one in East Malaysia). However, only four polytechnics in Peninsula Malaysia were chosen. Based on the four polytechnics, the population is 139 students. These consist of male and female hearing-impaired students in all semesters from all courses.

Based on [26] table, a total of 103 respondents are needed for a population of 139 students. For this study, a total of 130 respondents (57 [43.8%] male and 73 females [56.2%]) were chosen through simple random sampling. The breakdowns of samples are as follows: Ungku Omar Polytechnic (17 respondents), Shah Alam Polytechnic (16 respondents), Sultan Ibrahim Polytechnic (68 respondents) and Tuanku Syed Sirajuddin Polytechnic (29 respondents).

2.3 Instruments

The Bar-On Emotional Quotient: Short (EQ-i:S) by Reuven Bar-On [11] and translated by [27] was utilised to measure emotional intelligence. The measurement consists of five dimensions namely intrapersonal, interpersonal, adaptability, stress management and general mood with a five-point Likert scale type response (1=very seldom or not true of me, to 5=very often of me or true of me). The internal reliability Cronbach Alpha value for each dimension in this study are as follows: intrapersonal (0.96), interpersonal (0.94), adaptability (0.94), stress management (0.93), general mood (0.93), and overall EQ (0.82).

To measure psychological well-being, the Ryff's Psychological Well-Being by [14] which consist of 42 items was used. It combines the dimensions of self-acceptance, positive relations with others, autonomy, environmental mastery, purpose in life and personal growth. A seven-point Likert type scale were employed to gather responses (1=Strongly disagree, to 7= strongly agree). The internal reliability for this measurement is 0.95.

2.4 Data analysis

Data has been analysed through the Statistical Packages of the social sciences version 21. Descriptive and inferential statistics were employed for these purposes.

3 Results and Discussion

3.1 Level of emotional intelligence

Table 1 depicts the level of emotional intelligence among the hearing-impaired students. A total of 116 respondents (89.2%) have moderate emotional intelligence while those with low emotional intelligence consist of 7 respondents (5.4%) as well and those with high emotional intelligence which consist of 7 respondents (5.4%).

Table 1: Level of emotional intelligence

	Frequency	Percentage %
High	7	5.4
Moderate	116	89.2
Low	7	5.4
Total	130	100

Based on Table 1, it seems that the majority of hearing-impaired students achieved moderate and high level of emotional intelligence compared to a small number that achieve low level of emotional intelligence. This could be due to lower self-esteem based on communication barrier. Nevertheless, the figure is not alarming since less than 10% have low emotional intelligence.

Being far from home and studying in an environment with normal students is not without difficulties. Hearing-impaired students sometimes have lower self-esteem compared to normal students. Therefore, having moderate emotional intelligence is good enough for sustaining their normal individual life.

3.2 Level of psychological well-being

For psychological well-being, a total of 25 respondents (19.2%) have high psychological well-being compared to 89 respondents (68.5%) that have moderate psychological well-being and 16 respondents (12.3%) with low psychological well-being (Table 2).

Table 2: Level of psychological well-being

	Frequency	Percentage %
High	25	19.2
Moderate	89	68.5
Low	16	12.3
Total	130	100

The well-being of the hearing-impaired students was also at the moderate and high level in terms of psychological well-being. They are neither in the positive or negative state of emotional and social level or self-complacent. Their overall well-being is deemed satisfactorily. Individual with disabilities are those without demanding attitude. They are satisfied with their life and it is somehow subjective in terms of satisfaction if compared to normal individual. Thus, sense of being grateful with their surroundings is one of the attribute of hearing-impaired individual.

3.3 Relationship between emotional intelligence and psychological well-being

Bivariate correlational analysis using Pearson correlations has been utilised to determine the association between dimensions of emotional intelligence and psychological well-being. The matrix is shown in Table 3.

Table 3: Correlations between emotional intelligence dimensions and psychological well-being

	1	2	3	4	5	6
PW						
IN	.65**					
IT	.67**	.94**				
EQ	.59**	.66**	.68**			
AD	.48**	.56**	.60**	.48**		

SM	.18	.07	.07	.13	.05	
GM	-.24	-.22	-.22	-.25	-.05	-.11

$^*p < .05$, $^{**}p < .01$

Notes: PW = psychological well-being, IN = intrapersonal, IT = interpersonal, EQ = overall emotional intelligence, AD = Adaptability, SM = stress management, GM = general mood

Based on Table 3, the association between overall emotional intelligence and psychological well-being was found to be significant ($r = 0.59$, $p < .01$). This indicates that the higher the level of emotional intelligence among hearing-impaired students, the better the individual psychological well-being.

Results also showed that psychological well-being had a significant association with intrapersonal dimension ($r = 0.65$, $p < .01$) as well as with interpersonal dimension ($r = 0.67$, $p < .01$). This support [22] study that found significant association between emotional intelligence and psychological well-being. In addition, the results are also in line with [23] and [24] studies that found significant relationship between elements of emotional intelligence and psychological well-being.

Hearing impaired students who have better emotional self-awareness, independent and possess adequate sense of assertiveness will be able to achieve better psychological functioning in life. In addition, hearing-impaired students with better social responsibility who manage to establish good relationship with others will also achieve better psychological well-being. Being able to perceive accurately their own emotions and the ability to generate feelings that facilitate their own thoughts are the attributes of those with high intrapersonal and interpersonal elements and this will lead to better satisfaction in individual life.

Finally, adaptability dimension also showed a significant relationship with psychological well-being ($r = 0.59$, $p < .01$). The higher the adaptability skills they have, the better they will be in terms of psychological well-being. In other words, hearing-impaired students with flexible and open-minded perspective and have better problem solving skills will be better in terms of having positive attitude, better self-awareness and may develop talents and potential to enhance their quality of life. The result support [25] study where elements of emotional intelligence have been found to be associated with elements of psychological well-being.

Being away from home in a new surrounding requires an individual to be adaptive to the environment. Hearing-impaired students in this study have to be able to generate skills and ability to blend with normal hearing individual. With the help of lecturers and colleagues of both normal and hearing-impaired, their overall life satisfaction is enhanced.

4 Conclusion

Hearing-impaired students need encouragement and support from communities to enable them to function well in everyday life. Higher emotional intelligence will lead to better psychological well-being in order to carry on with whatever they need to perform in daily life. It is suggested that special workshops or self-emotions training be conducted to these students to alleviate their emotional intelligence skills. As people with disabilities, they should be given the chance to enjoy life to the fullest as normal individual. They are the country's human capital assets. In facts, they could be the unsung heroes with their own special abilities.

The authors wish to thank Universiti Utara Malaysia for funding this study under the university research grant.

References

1. World Health Organisation,. WHO global estimates on prevalence of hearing loss: mortality and burden of diseases and prevention of blindness anddeafness. http://www.who.int/mediacentre/factsheets/fs300/en/index.html (2013)

2. World Health Organisation,. Deafness and hearing impairment. http://www.who.int/pbd/deafness/estimates/en/ (2006)

3. Fatimah, Parents Urged To Discard Fear Of Sending Disabled Children To School. The New Straits Times. https://my.news.yahoo.com/parents-urged-discard-fear-sending-disabled-children-school-064208255.html (2012)

4. J. Fellinger, D. Holzinger and R. Pollard, Lancet, **379**, 1037–44 (2012)

5. M. Jacks, D. Marsh and C. Massey,. Welcome to the Silent Zone Multimedia Courseware on Sign Language. ACM SIGBIO Newsletter, **20**, 17-19 (2000)

6. S.D. Antia, K.H. Kreimeyer, (Oxford University Press New York, 164-176, 2003)

7. N. Silvestre, A. Ramspott, I. Pareto, J. D. Stud. D. Edu., **12**, 38–54 (2007)

8. Acuna, J.E., R.Q. Bugas, Int. J. Res. Rev., **5**, 74-91 http://www.researchgate.net/publication/228347395 (2010)

9. H. Fazlinda, *Employability skills, self-esteem and self-efficacy of technical and vocational students with hearing problems in polytechnics and community college, Malaysia.* (Unpublished doctoral dissertation, Universiti Putra Malaysia, 2013)

10. J.D. Mayer, P. Salovey, App. Pre. Psycho., **4**, 197-208 (1995)

11. R. Bar-On, *The Bar-On emotional quotient inventory (EQ-i): A test of emotional intelligence*. (Toronto: Multi-Health Systems, 1997)

12. W. Pavot, E. Diener, *Well- being (inducing life satisfaction). In Rocio, Fernadez-Ballesteror (Eds.) Encyclopedia of Psychological Assessment 2,* 1097-1101, (New Delhi: Sage Publications, 2003)

13. S. O. Salami, J. Int. Soc. Res, **3**, 387-398 (2010)

14. D. Ryff, Int. J. B. Dev., **12**, 35–55 (1989)

15. Keilmann, A. Limberger, W. J. Mann, Int. J. Paed. Otorhino., **71**, 1747-1752 (2007)

16. J. Fellinger, D. Holzinger, U. Dobner, J. Gerich, R. Lehner, G. Lenz, D. Goldberg, Soc. Psyc. Psyc. Epid., **40**, 737–42 (2005)

17. Weisel, A. Kamara, J. D. Stud. D. Edu., **10**, 51–62, (2005)

18. L. Mejstad, K. Heiling, C.G. Svedin,. A. Ann. D., **153**, 504–515 (2009)

19. T. Van Eldik, P. D. Treffers, J. W. Veerman, F.C. Verhulst, A. Ann. D., **148**, 390–395 (2004)

20. N. Wolters, H. Knoors, H. N. Cillessen, L. Verhoeven, J. D. Stud. D. Edu., **17**, 463-482 (2012)

21. G. S. Palmer, P. Boudreault, E. E. Baldwin, M. Fox, J.L. Deignan, Y. Kobayashi, J. S. Sinsheimer, J. Gen. Coun., **22**, 492-507 (2013)

22. T. Mehmood, S. Gulzar, As. J. Soc. Sc. Human., **3**, 178-185 (2014)

23. James, M. Bore, S. Zito, J. Psycho. Ass., **30**, 425-438 (2012)

24. S. Shaheen, H. Shaheen, Int. J. Indi. Psycho., **3**, 206-213 (2016)

25. S. Irshad, Int. J. Inno. Sci. Res., **15**, 50-63 (2015)

26. R.V. Krejcie, D.W. Morgan, Edu. Psycho. Mea., **30**, 607-610 (1970)

27. M. Najib, M. Che Su, M. S. Zarina, Res. J. Org. Psycho. Edu. Stu., **1**, 93-99 (2012)

 https://doi.org/10.1051/matecconf/201815005085

Pedestrian Travel Pattern At Johor Bahru Commercial Centre

Noor Shazreen A.R[1,*], *Doris* Asmani M.Y[1], *Li* Sian Tey[1], *Mohamed* Khatif[1] and *Haron* H.Austo[1]

[1]Department of Civil Engineering, Universiti Teknologi MARA Kampus Pasir Gudang, Malaysia
[2]Faculty of Civil Engineering, Universiti Teknologi MARA Kampus Pulau Pinang, Malaysia

Abstract. Pedestrians are often exposed to complex and dangerous traffic situations. Therefore, planning for pedestrian facilities could reduce some of the problems that are faced by the pedestrians. Thus, this study describes the level of awareness, risky behavior and analyse pedestrian travel pattern in Johor Bahru (JB) commercial centre. Data for the study was collected from questionnaire distribution and video recording. Thirty sets of questionnaire were distributed with the purpose of determining the level of awareness among pedestrian while using the pedestrian facilities provided by the. Meanwhile, the video recording was used to determine the risky behaviour done by the pedestrians at commercial centres. The video was also used to analyse pedestrian travel pattern at commercial centres. The outcome of this study would be beneficial to JB cities and lead to a successful pedestrian travel pattern within a safety, practice, and knowledge of pedestrian facilities provided.

1 Introduction

Johor Bahru (JB) is the capital city of Johor state, stated in southern part of Malaysia. The city has a number of interesting commercial center buildings. However, traffic congestion in Johor Bahru is a daily feature and its impacts on the community are extensive. The commercial centre is typically the most frequent area where pedestrian is exposed to risk since it generates high pedestrian traffic. The purpose of this study is to demonstrate the integration of safety and other important societal goals which are crucial for pedestrians especially in urban areas. Therefore, it should be a high priority for the authorities to cater for the needs and vulnerability of pedestrians. In the study, several objectives have been laid down to address the issues. The objectives of the study are to determine the level of awareness of pedestrians using the facilities provided, identify the risky behaviour that predisposes pedestrians to accidents and analyse pedestrian travel pattern at commercial centres in Johor Bahru.

Road traffic injuries are a major public and development crisis but have been neglected in global public health problem. These situations are predicted to increase if road safety is not properly addressed. According to the statistics provided by World Health Organization (WHO) [1] and World Bank in World Report on Road Traffic Injury Prevention, in 2002, almost 1.2 million people were killed in road traffic accidents, which represents an average of 3242 persons dying each day around the world [2]. Millions of others sustain injuries, with some suffering permanent disabilities. From year 2000 to 2020, the total number of road traffic deaths worldwide is forecasted to rise by some 65% in low-income countries whereas increased as much as 80% in middle-income countries [2]. Vulnerable road users such as pedestrians, cyclists and motorcyclists are the major contributors to the number of death cases On the other hand, road traffic injuries were the 11th leading cause of death worldwide and accounted for 2.1% of all deaths globally. Furthermore, these road traffic deaths are accounted for 23% of all injury deaths worldwide as shown in Figure 1 In Malaysia, 6,811 deaths were reported in 2011 which resulted in a loss of RM 9.0 billion to the country's economy. On average, 19 people are killed every day in Malaysia, which makes road accidents a serious medical challenge to the country. This also indicates the need for a basic action that is effective [3].

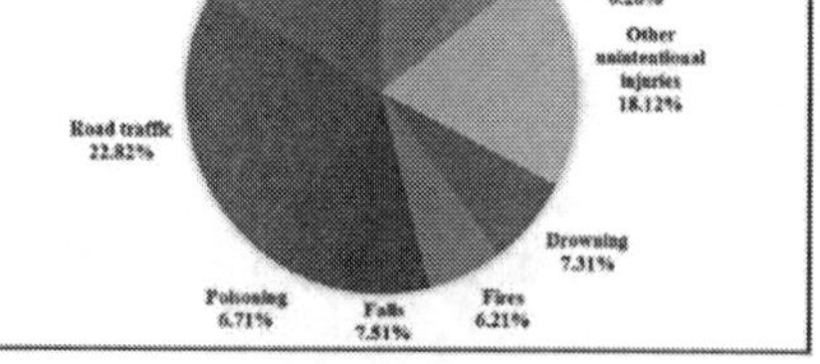

Figure 1: Distribution of global injury mortality by cause (Peden et al., 2004)

* Corresponding author: noorshazreen@johor.uitm.edu.my

 https://doi.org/10.1051/matecconf/201815005085

There are a few factors which influence the pedestrian crashes or fatalities. According to the report prepared by National Highway Traffic Safety Administration (NHTSA) of United States of America, over three-fourth of the pedestrians were killed at non-intersections and less than one-fourth were killed at intersections over the past decade [4]. Apart from that, the report also mentioned that the percentage of pedestrian deaths in crosswalks is less than deaths in roadways. Meanwhile, urban roads accounted for more than two-thirds of pedestrian crash deaths with nearly one-third of those on other principal arterial roads.

Other than that, a studied by Diaz shows that pedestrians fatality can be influenced by pedestrian's characteristics such as gender and age where adult and women is a low-risk group by committing to the rules [5]. According to Hamid pedestrians tend to take their risk by shorten their journey especially on the road that have central refuge or median because of longer waiting time and hastily to continue the other half of their journey [6]. While, Sisiopiku and Akin found that improper design and location of signalized intersections give an impact to the pedestrian behaviour when the travel distance for pedestrians become more distant to desired location [7].

The current study was conducted by pedestrian awareness survey and video recording of pedestrian travel pattern and risky behaviour. These methods were used to indicate the 'pedestrian compliance' parameter in order to measure the pedestrian behaviour in utilizing the facility provided. To provide more understanding, this paper provides Section 2, consisting of a description of study background and methodology of data collection. Section 3 presents the analysis of the result from the field surveys and Section 4 concludes the main findings

2 Methods

The study took place at the most developed state of Malaysia, which is Johor. Johor is located in the southern portion of Peninsular Malaysia. Johor Bahru is the capital city of Johor, having an area of approximately 220 square kilometres and a population of 497 067 as of 2010 [8]. The city is administered by the Johor Bahru City Council (Majlis Bandaraya Johor Bahru – MBJB) and has obtained the city status since January 1st, 1994. The study area is focused at three main site locations at commercial centers area in Johor Bahru, including Galleria @ Kotaraya (Site 1), Johor Bahru City Square (Site 2) and Jalan Dato Onn opposite Johor Bahru City Council (Site 3).

Galleria @ Kotaraya is a five-storey shopping complex and previously known as Plaza Kotaraya. It is located in the heart of Johor Bahru with a gross area of approximately 546, 530 square feet and part of 11-acre Kotaraya Development Project, mooted by Johor Corporation [9]. The building is surrounded by office buildings, financial institutions, hotel, Persada Johor Convention Centre and tourist attraction. Figure 2 shows the layout plan for site 1.

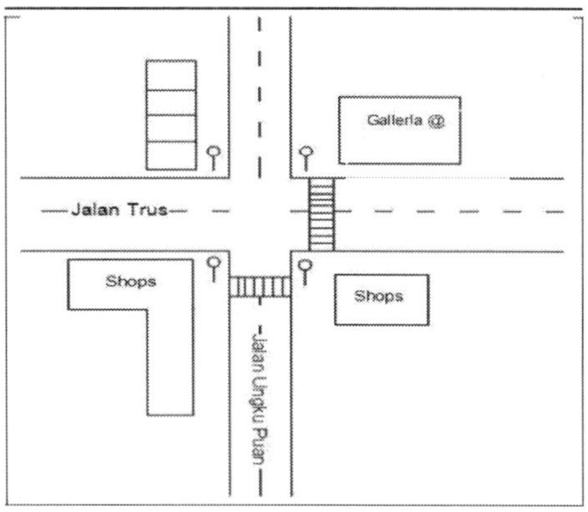

Figure 2: Layout Plan for Galleria @ Kotaraya (Kotaraya)

City Square is strategically located in the central business district of Johor Bahru and it is one of the landmarks in the city. The seven storey mall is located at the busy road of Jalan Wong Ah Fook and connected to Sultan Iskandar Customs, Immigration and Quarantine Complex with pedestrian bridges. Apart from that, the building is comprised of thirty-six storey office block, which is divided into the lower, middle and upper zones and also three floors of basement car park. Figure 3 shows the layout plan for site 2.

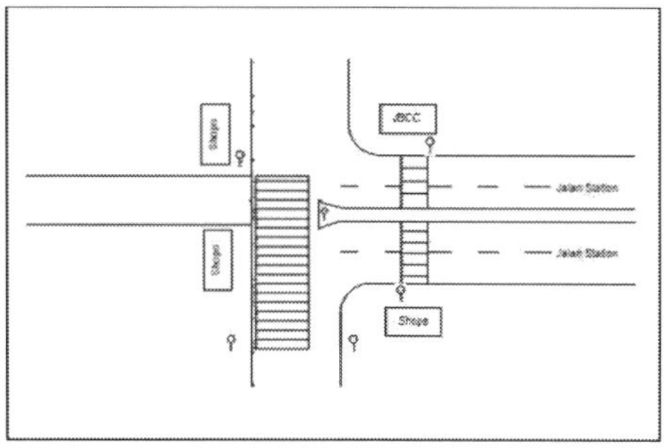

Figure 3: Layout Plan for Johor Bahru City Square (City Square)

Johor Bahru City Council (Majlis Bandaraya Johor Bahru – MBJB) is a local authority that administered Johor Bahru city centre and other areas within the perimeter. Among their responsibilities are to ensure that public health and sanitation, waste removal and management, town planning, environment protection and building control, social and economic development and general maintenance functions of urban infrastructure are well managed and monitored. The headquarters of MBJB is located at Jalan Dato Onn, Johor Bahru. Figure 4 shows the layout plan for site 3.

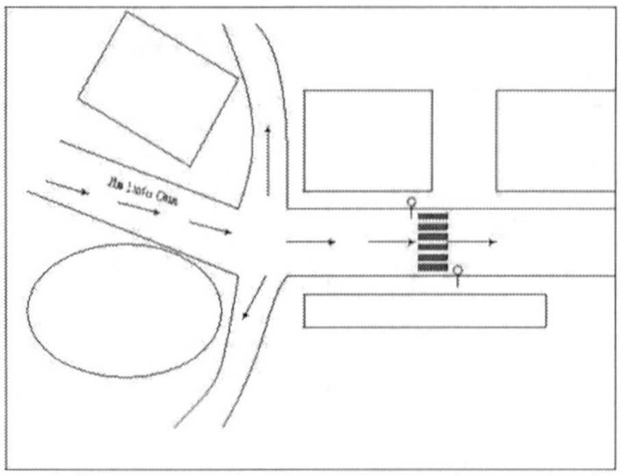

Figure 4: Layout Plan for MBJB

This location has been selected based on high pedestrian activities due to the attractions of the areas. The pedestrian facilities available at the study areas are shown in Table 1.

Table 1: Pedestrian facilities available at the study areas

Facilities	Venue		
	Commercial Centre		
	Johor Bahru		
	Site 1	Site 2	Site 3
Pedestrian signal	√	√	√
Pedestrian crossing (zebra/ladder)	√	√	-
Median	-	√	-
Footwalk/footpath	√	√	√
Pedestrian bridge	-	-	-

There are two methods used in the study to collect the data, namely questionnaire distribution and video recording. The purpose of distributing a set of questionnaire for pedestrians present at the particular area and time are to determine their awareness while using the pedestrian facilities provided by authority and for them to answer it that best reflects their level of agreement or disagreement with the questions. Meanwhile, the video recording is used to determine the risky behaviour done by the pedestrians at commercial centres. Apart from that, the video recording is also used to analyse pedestrian travel pattern at commercial centres.

The level of awareness of pedestrians towards the facilities provided at the study areas were collected by distributing a set of questionnaire. There are three different variables included in the questionnaire, which are knowledge, practice and concern. Questions on knowledge represent the level of knowledge that the respondents have on the pedestrian facilities provided at the study areas while questions on practice represent whether the respondents apply or use the knowledge they have while using the facilities. The question on concern is about whether the respondents are concerned about their safety while dealing with the facilities provided for them to use.

A pilot study was conducted before collecting the actual data by distributing 10 sets of questionnaire at the study areas. The questionnaire were then analysed to check its validity and reliability using Cronbach's Alpha (Table 2) in SPSS. Cronbach's Alpha measured the internal consistency on how closely related a set of items are as a group. Some of the questions unrelated for the study were discarded to ensure that the objectives of the study will be successfully achieved. The score from the analysis were interpreted as follows.

Table 2: Reliability coefficient of Cronbach's Alpha

Score	Reliability	Score	Reliability
> 0.9	Excellent	> 0.6	Questionable
> 0.8	Good	> 0.5	Poor
> 0.7	Acceptable	< 0.5	Unacceptable

Once the questionnaire had been revised and changed, the new sets of questionnaire were then distributed at the study areas, with 30 sets of questionnaire distributed at each location. The responses from the respondents were then analysed and recorded.

From the video recorded at the study areas, it was then analysed by using Statistical Package for Social Science (SPSS) version 23. The pedestrian risky behaviour at the study areas was observed through the video recording. The study area was surveyed to find the suitable place to install the video camera. The location must have a clear view of the pedestrian movement and behaviour as well as the facilities provided within the area of interest.

The first thing that has been observed is the status of compliance of the pedestrians while using the pedestrian facilities. In order to be categorized under compliant category, the pedestrians must use the facilities provided and obey the traffic rules in that particular study areas meanwhile the non-compliant category is vice versa from compliant category. Among the safe actions that can be categorized under compliant categories are using pedestrian crossing or pedestrian bridge to cross the road and obey the pedestrian traffic signal.

Other criteria that have been observed from the video are as follows:

i. Pedestrian gender – male or female

ii. Carrying object while working such as handbag, backpack, sling bag and others

iii. Pedestrian age group – children, teenager, adult or senior citizen

iv. Facing an obstacle while walking such as cars, buses and others

v. Pedestrian characteristics – individual, partner or group

Apart from that, pedestrian travel pattern has also been observed and sorted out based on the compliance behaviour of the pedestrians. The pattern made by pedestrians is either compliant or non-compliant were then marked on the schematic diagram of the facilities provided in the study areas.

3 Result

3.1 Determination of level of awareness

In determining pedestrian's awareness, 30 sets of questionnaire survey was distributed at each site to observe the response of pedestrian towards the pedestrian facilities provided and pedestrian rule adherence when crossing the road. As stated in Section 2, Chronbach's Alpha Coefficient was used to analyze the question. The result, as shown in Table 3 is above 0.5 which reflect that the question is acceptable to be used for the survey.

Table 3: The Chronbach"s Alpha Coefficient for selected study areas

Study Areas	Alpha coefficient (α)
Site 1	0.856
Site 2	0.815
Site 3	0.828

The questionnaire was divided into three variable of interest; 'knowledge', 'practice' and 'safety' in order to justify the level of awareness. Each variable of interest was analyzed using scale factor analysis with scale 1 to 5 (strongly disagree to strongly agree). The average scale data for each variable was calculated using SPSS and represented as means of score value. Table 4 shows the level of awareness of each variable of interest.

Based on the mean score result, the variable of interest for 'knowledge' indicates the highest mean score followed by 'practice' and 'safety' for all the study areas. The level of 'knowledge' records the score of 4.042, 4.219 and 4.039 for location 1, 2 and 3 respectively. Meanwhile, 'practice' records 3.817, 4.056 and 3.839, and as for 'safety' it records 3.647, 3.536 and 3.644 for location 1, 2 and 3 respectively.

Results of the analysis illustrate that pedestrians disregard pedestrian safety by not using the facilities provided even though they know the purpose and function of the facilities. These values reflected that pedestrian do not practice the pedestrian law as much and do not really aware of their own safety even though they have a higher level of knowledge

Table 4: Means of score for each of the study area

Study Area	Variable Interest	Means of score
Site 1	Practice	3.817
	Knowledge	4.042
	Concern	3.647
Site 2	Practice	4.056
	Knowledge	4.219
	Concern	3.536
Site 3	Practice	3.839
	Knowledge	4.039
	Concern	3.644

3.2 Determination of pedestrian risky Behaviour

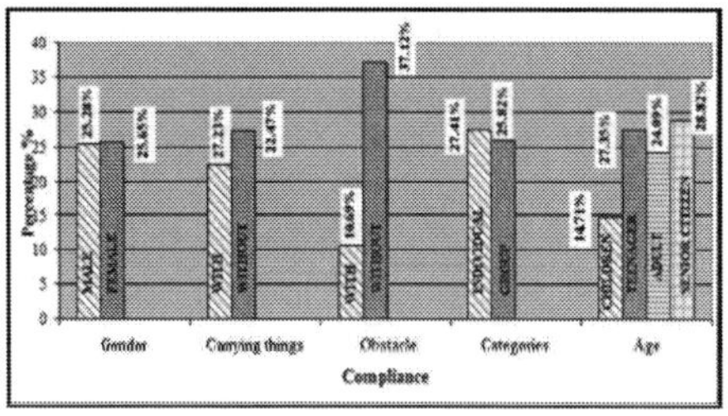

Figure 5: Graph of percentage of compliance according to the pedestrian criteria

Based on the description in section 2, pedestrian movement were classified into five potential variables behavior which were based on gender, carrying things, facing obstacles, categories of group walking and age. Figure 5 shows the result of compliance behavior from video recording observation.

Data from the three study sites was combined for the analysis. Based on the three study areas, 25.65% of females comply the safety behaviour compared to with males, which scored 25.28% of the gender criteria. This proves the statement of the previous studies that women tend to comply the rule and are more careful than men when crossing the roads [10]. For the carrying things compliance criteria, 27.23% of pedestrian are recorded to not carry things and it shows the highest percentage compared to only 22.47% of pedestrian who carry things. Meanwhile, the pedestrian who do not face any obstacle to cross the road tend to comply the criteria with the highest percentage of 37.12% rather than only 10.69% that comply the rule with obstacle. Pedestrians tend to expedite their journey in order to decrease the burden of carrying things and to avoid the obstacles. Walking individually indicate the highest percentage, with 27.41% of the walking categories instead of 25.82% who walk in a group. Lastly, the highest compliance age group criteria were senior citizen, which recorded 28.82% of the total group and the least compliance were children group (14.71%). The finding shows that senior citizen age group has the highest compliance rate, which is coincided with common understanding from previous study which have found that the risk of fatality become higher along with the increasing on the age [4-5 , 11].

3.2 Determination of travel pattern analysis

There are four travel patterns observed based on the observation which are similar in all study area. From Figure 6, the compliance pattern was represented by Pattern 1, whereas pattern 2, 3 and 4 are the non-compliant pattern since the pedestrian cross the road outside of the designated area.

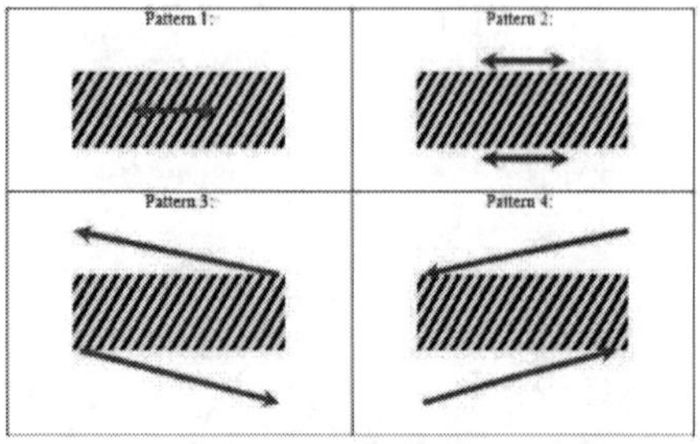

Figure 6: Travel pattern at pedestrian zebra crossing facilities at Site 1, 2 and 3

Figure 7 shows the travel pattern analysis where the highest compliance percentage was in Site 3 with 40.60%, followed by Site 2, 32.71% and Site 1, 17.88%. Unfortunately, overall location shows that more than half of the pedestrian disregard their safety by neglecting the rules and not using the facility provided. The result pattern was contributed by the location itself, where Site 1 and Site 2 were the central point of shopping areas, rather than Site 3 which is surrounded with administrative area. Other than that, the only location facilities provided at Site 3 were pedestrian signal and a foot walk/footpath at straight line crossing that help encourage pedestrians to comply with the rules. As for the junction area as at Site 1 and Site 2, it provides a lot of options to pedestrians to shorten their trip.

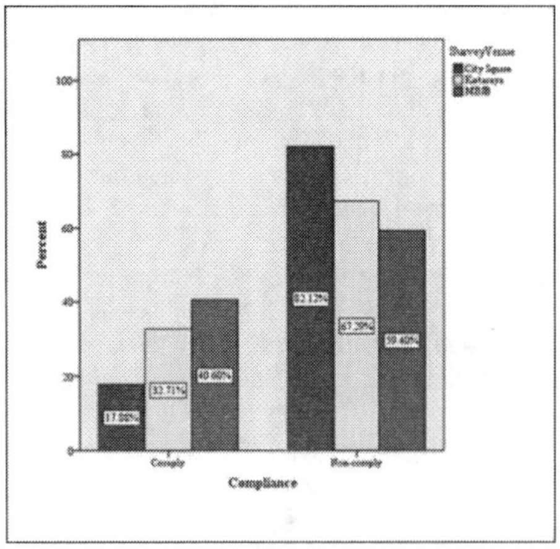

Figure 7: Percentage of compliance according to the survey venue

4. Conclusion

From the overall analysis, it can be concluded that the level of awareness at the three locations indicates that respondents tend to disobey the rules by not practicing the knowledge and neglecting the concern for their safety even though they have high level of knowledge about the benefit of implemented facilities and rules. From the observation, the groups that show the highest percentage of non-compliant behavior were the males, children, pedestrians walking in a group, pedestrian carrying things and facing an obstacle when crossing the road. These groups were exposed to the risk of accident and fatality. In addition, there are variations of pedestrian travel pattern. The results have proven that pedestrian crossing facilities need to be improved in the junction area as pedestrians have many options to shorten the trip even they have to risk their lives by violating the rules.

The authors would like to express sincere gratitude to the Ministry of Higher Education under RAGS 2014 with reference number 600-RMI/RAGS 5/3 (169/2014) and Universiti Teknologi MARA for the opportunity to carry out this research by supporting the financial aid.

References

1. World Health Organization (WHO), World Report on Road Traffic Injury Prevention, pps.who.int/iris/bitstream/10665/42871/ 1/9241562609.pdf (2004)
2. M. Peden, R. Scurfield, D. Sleet, D. Mohan, A. A. Hyder, E.Jarawan, E. Mathers, World Report on Road Traffic Injury Prevention. Geneva. Retrieved August 30, 2016, from injury_prevention/road_safety_status/20 13/en/ (2004)
3. Jabatan Keselamatan Jalan Raya (JKR), Pelan Keselamatan Jalan Raya 2014-2020. (Ministry of Transportation report) http://www.mot.gov.my/SiteCollection, Documents/Darat/Pelan+KJRM+2014+2 020.pdf. (2014)
4. National Highway Traffic Safety Administration (NHTSA), *Traffic Safety Fact 2012 Data*. U.S. Department of Transportation (2014)
5. Diaz, Theory of planned behavior and pedestrians' intentions to violate traffic regulations, Transportation Research Part F (2002)
6. Hamed, Analysis of pedestrian's behavior at pedestrian crossings, 63– 82, (2001)
7. Sisiopiku, Akin, Pedestrian behavior and perceptions towards various pedestrian facilities: an examination based on observation and survey data.Transportation Research Part F (2003)
8. Malaysian Digital Repository, Background of Johor Bahru City Council and History of Johor Bahru http://myrepositori.pnm.gov.my/bitstrea m /123456789/752/3/Majlis%20Bandaraya%20Johor%

20Bahru%20%5BSumber% 20Internet%5D.pdf (2013)

9. Damansara Assets Sdn Bhd. The Latest Retail Signature In Johor Bahru City, http://www.galleriakotaraya.com/usr/deal.aspx?pgid=19 (2016)
10. Tom A & Granie M (2011). *Gender differences in pedestrian rule compliance and visual search at signalized and unsignalized crossroads. Accident Analysis and Prevention*. Elservier
11. Martin A. (2006). *Factor Influencing Pedestrians Safety: A Literature Review*. Retrieved September 2, 2016, fromhttp://content.tfl.gov.uk/factors-influencing-pedestrian-safety-literature-review.pdf

 https://doi.org/10.1051/matecconf/201815005086

Metadiscourse in Academic Writing of Pre-University Arab Students at the International Islamic University Malaysia (IIUM)

Muhamad Khairul Zakaria[1,*] and *Faridah* Abdul Malik[2]

[1]English Language Centre (ELC), Centre of Foundation and Liberal Education (PPAL), Universiti Malaysia Terengganu, Malaysia
[2]English Language Division, Centre for Languages and Pre-University Academic Development (CELPAD), IIUM Malaysia

Abstract. There is lack of studies on the use of metadiscourse markers; especially amongst international students studying in Malaysia and Malaysia are receiving scores of international students particularly from the Middle East annually. This study involves a textual analysis of students' academic writing where the metadiscourse markers in 50 Arab IIUM students' academic texts were identified and analyzed. The findings of this study indicated that Arab writers had a greater inclination for the deployment of the *interactive* markers (Total counts = 919) than *interactional* ones (Total counts = 592) as there was a higher percentage of *interactive* metadiscourse (60.8%) usage than the *interactional* ones (39.2%). It might be useful for English language teachers to integrate cultural considerations within their syllabus with regard to metadiscourse markers in order to prepare relevant materials based on their students' needs as well as to develop the students' awareness of the importance of these linguistic features.

1 Introduction

1.1 Background of study

Writing essays is an important part of a student's life at the university for the reason that they have to write essays for their assignments. The quality of the written assignments would have a significant influence on their academic achievement. Moreover, the capability to compose academic texts competently at institution of higher education is a practical anticipation to be grasped by the tertiary learners [1]. In order to compose a comprehensible writing, the learners need to employ appropriate metadiscourse markers within their written tasks. Tan and Wong (2014) mentioned that metadiscourse is a central semantic means that aids authors to navigate the movement of their written content as well as to address their standpoint to their readers [2].

According to Hyland and Tse (2004), the metadiscourse markers are important resources deployed to designate how the text is organized, to clarify difficult words and jargons for readers, and to put into code what rhetorical acts are being performed [3]. Competent employment of metadiscourse markers will make a text readable.

Thus, the knowledge of metadiscourse is frequently observed to be beneficial not only for students and the anticipated readers of the students' written works, but also for language instructors. Teachers can interpret their students' writing by looking at the students' deployment of metadiscourse. Consequently, the observation on the deployment of metadiscourse might assist the teachers to guide their students to be more thoughtful of the variances between students' national culture and the culture of the discourse or the community to which the text denotes [4].

1.2 Statement of the problem

Effective deployment of metadiscourse markers can be one of the means of enhancing the quality of academic essays written within academic circles [5]. Specifically, the utilization of metadiscourse markers were suggested to improve the writing quality [6], increase the readability [7], and build a relationship with the readers [4, 8]. These features are very important in developing a good quality academic writing [2]. In fact, several studies have shown that the use of metadiscourse can improved the academic writing among students. For instance, a study done by Lu (2011) shows that metadiscourse has helped in improving the academic writing among the Chinese EFL students' in the pedagogical context of composition classrooms in China [9]. Such abundance of studies provide evidence of its prominence within the writing realm, predominantly in academic writing. Moreover, there is an increasing concern with the quality and standard of academic writing among university students especially concerning ESL students [10].

However, there is lack of studies on the use of metadiscourse markers, especially amongst international students studying in Malaysia. This is significant since higher education institutions in Malaysia are receiving scores of international students particularly from the Middle East annually. The study aimed to determine the forms and frequency of metadiscourse markers used in the academic essays written by these students.

1.3 Research objective

The significant of metadiscourse in academic writing prompts this research which attempts to determine the forms and frequency of metadiscourse markers used by Arab Pre-University writers in their opinion-based essays.

1.4 Research Question

In keeping with the aim of the study the following research question was addressed: **RQ1** What are the forms and frequency of metadiscourse used in the opinion-based essays written by Arab Pre-University writers?

1.5 Significance of the study

Primarily, the study will cater to the teaching and learning need of IIUM's multi-national environment where the second biggest group of students is the Arabs who come second after the Malaysians. Besides, metadiscourse markers are important to the targeted groups which include the Arab students of English as ineffective deployment of metadiscourse in their academic writing will affect the quality of their writings. Thus, knowing about metadiscourse markers will assist the language instructors to increase the Arab students' language awareness and improve their writing ability. This is very significant in the case of the Arab students as higher education institutions in Malaysia are receiving scores of international students particularly from the Middle East every year.

2 Materials and Methods

2.1 Research design

This study involves a textual analysis of students' academic writing where the metadiscourse markers in Arab ESL students' academic texts were identified and analyzed.

2.2 Participants

A total of 50 Level 6 Pre-University Arab students from the Centre for Languages and Pre-University Academic Development (CELPAD), IIUM participated in this study.

English and Arabic are the two primary languages used for instructions in IIUM, depending on the specific programme of study. The CELPAD is responsible for ensuring that all IIUM students have a strong command of the languages required for their studies. Applicants wishing to pursue their studies in programmes with English as medium of instruction must present a score of 550 in the Test of English as a Foreign Language (TOEFL) or Band 6.0 in the International English Language Testing Service (IELTS) (Academic) with a minimum Band of 5.5 for each of the Skills - Writing, Reading, Listening/ Speaking (WRLS).

Alternatively, they may sit for the IIUM-administered English Placement Test (EPT) and obtain a minimum Band of 6.0 to fulfill the English language requirement. Applicants who are unable to submit satisfactory results in any of the above tests may take up language proficiency courses at the University to fulfill the admission requirements. ("Centre for Languages and Pre-University Academic Development," n.d.)

2.3 Data collection

Opinion based essays written by the students' in the English Placement Test were analysed to study the deployment of metadiscourse markers by the Pre-University Arab students. The opinion based composition was the kind of essay written and was selected for the reason that the academic written text is opinion centered where the deployment of metadiscourse was anticipated to be apparent in such text category (Silver, 2003). The essays were written 50 Arab ESL learners who were registered in the intensive English course at CELPAD. All these essays were given a score of band 6 which means that the writers were considered to have acquired the minimum band of proficiency in writing and can proceed to their faculty programmes.

2.4 Textual analysis

The essays were analyzed by means of the subsequent stages: categorising, ordering, and inferring. To start with, the metadiscourse markers of ESL students' academic writings were identified. Next, the markers were categorized into either *interactive* or *interactional*. The *interactive* metadiscourse markers were then further categorised into *code glosses*, *transitions*, *endophorics*, *frame markers* or *evidentials*; while the *interactional* metadiscourse markers were then further categorised into *self-mentions*, *booster*, *attitude markers*, *engagement markers*, or *hedges*. This inclusive organisation is grounded on Hyland's taxonomy [5]. Then, the researcher identified the *interactive* and *interactional* metadiscourse markers that were dominantly employed by the learners.

To achieve a higher reliability, the whole corpora were manually analysed word for word because computer-assisted analysis faces the risk of assuming external reference items as metadiscourse and could damage the validity of research [11]. All the procedures were carried out separately by another rater, in order to countercheck as well as to uphold impartiality in the process of detailing of the deployment of metadiscourse markers within the selected corpora. The second rater received sufficient training in doing the task. During the training session, the researcher and the second rater comprehensively

 https://doi.org/10.1051/matecconf/201815005086

discussed any confusion and incongruity with regard to the metadiscourse markers. Series of exercises using samples from the participants' writings were endured by both researchers to make sure that both raters comprehended and perceived the metadiscourse markers in a similar fashion. There was an agreement found between the researcher and the second rater with reference to the frequency count of metadiscourse markers in the students' academic essays.

After the collection of data, the total amount of metadiscourse markers was determined. The total amount of metadiscourse markers in the academic essays written by the Arab Pre-University writers was tallied and checked individually. This presented the answer to RQ1 which aimed to determine the frequency and forms of metadiscourse markers used in the academic essays written by the Arab Pre-University students.

2.5 Theoretical framework

The framework used in this study to investigate metadiscourse usage in academic writing is the International Model of Metadiscourse which was developed Hyland [4]. Even though researchers dealing with metadiscourse have respectively fabricated their personal classification of metadiscourse [12-14], the classification made by Hyland was selected as the main framework over the varieties of models and to be the main outline for this research for the reason that the classification is comprehensive as well as concise.

As illustrated in Table 1, the *interactive* metadiscourse, which is the first dimension of Hyland's model, permits authors to compose a unified as well as a comprehensible writing. The sub-categories of the *interactive* dimension are *endophoric markers*, *code glosses*, *transition markers*, *evidential*, and *frame markers*.

The *interactional* category is another element of metadiscourse. The main purpose of the *interactional* dimension is to enable the authors to deliver their reactions or sentiments in the intended parts of the writings as well as to construct a particular relationship with the person who reads the text. The *interactional* metadiscourse sub-categories are *engagement markers*, *hedges, attitude markers*, *self-mentions* and *boosters*.

Table 1. An *interpersonal* model of metadiscourse [4]

Classification	Function	Instances
Interactive Metadiscourse	Guide and support the reader through the writing	
Transitions	express semantic connection between focal phrases	and / thus / but / in addition
Frame markers	denote to text stages acts, discourse, or sequences	my purpose here is to/ to conclude/ finally
Endophoric markers	indicate statistics and data in other sections of the writing	see Fig / noted above/ in section 2
Evidentials	refer to source of information from other texts	(X, 1990) / Y states / according to Z
Code glosses	help readers grasp functions of ideational material	in other words/ namely/ e.g./ such as
Interactional Metadiscourse	Involve the reader in the writing	
Hedges	withhold writer's full commitment to proposition	possible / about / might / perhaps
Boosters	accentuate writer's certainty or force in proposition	definitely/ it is clear that/ in fact
Attitude markers	express writer's attitude to proposition	I agree/ surprisingly/ unfortunately
Engagement markers	explicitly refer to or build relationship with reader	note that/ you can see that/ consider
Self-mentions	explicit reference to author(s)	I/ my/ our/ we

3 Results

Research question 1: what are the forms and frequency of metadiscourse used in the opinion-based essays written by Arab pre-university writers?

3.1 Interactive metadiscourse

Figure 1 shows the frequency of the different sub-categories within the *interactive* metadiscourse among the Arab pre-university writers in their academic essays. The total number of occurrences of the deployment of *interactive* metadiscourse in the academic texts produced by the Arab students was 919. As can be seen from Fig. 1, the highest usage is for *Transition* (583) while the second and third highest frequency are for *Frame markers* (220) and *Code glosses* (103) correspondingly. Additionally, the fourth highest frequency is for *Evidentials* (10) while the least used *interactive* metadiscourse marker is *Endophoric marker* (3).

https://doi.org/10.1051/matecconf/201815005086

Similar to the Malaysian writers' usage, *Transitions,* which function to mark semantic relation between main clauses in the Arab students' text, were actually the most noticeable *interactive* metadiscourse feature. Furthermore, the lower density of *endophoric markers* and *evidentials* shows that the Arab students did not have sufficient information to make a mention to facts in other sections of the writing. The third least frequently applied *interactive* metadiscourse markers in the students' essays, *Code glosses,* might be owed to the fact that the students habitually undertook that their intentional readers already acquired sufficient contextual knowledge to comprehend and appreciate whatever messages they were conveying in the texts.

Examples of *interactive* markers used by the Arabs students are given in (1) – (5).

(1) Transitions *"...some people do not do well in their studies **but** they have different hobbies **and** skills that can make them to have their own business..."* (Arab student, ID: 72)

(2) Frame markers *"**Secondly,** there is no conclusive doubt that communication is an important factor in education."* (Arab student, ID: 2)

(3) Endophoric markers *"In addition to **the above,** this helps teach ones child to respect elders, while offering a liner emotional support..."* (Arab student, ID: 10)

(4) Evidentials *"For instance, in a **study conducted by** Lawa University, researchers found that almost all youngsters consider their parents as models in their lives."* (Arab student, ID: 6)

(5) Code glosses *"In conclusion, while being a child's best friend is important, yet it is only a part of a greater picture, **i.e.** parenthood."* (Arab student, ID: 10)

The subsequent section will present the findings on the usage of *interactional* metadiscourse markers among the Arab Pre-University writers in their essays.

3.2 Interactional metadiscourse

Figure 2 shows the frequency of the different sub-categories of the *interactional* metadiscourse among the Arab pre-university writers in their academic essays. The total number of occurrence of *interactional* metadiscourse in the academic texts produced by the Arab students was 592.

From the data in Fig. 2, it can be seen that by far the highest usage is for *Self mentions* (276) while the lowest usage is *Engagement markers* (45). The second highest usage is for *Hedges* (105) and followed by *Attitude markers* (87) and *Boosters* (79). Arab students appeared to focused on *self-mentions* more, while showing a comparatively lower reliance on openly constructing relationship with or communicating to their intentional reader since they utilized fewer *boosters* in their texts. The examples of *Interactional* markers used by the Arab students are given in (1) – (5).

(1) Hedges *"Regarding this issue, parents **may** provide advises when their children have emotional problem."* (Arab student, ID: 7)

(2) Boosters *"It is an **obvious fact** that children are born to both father and mother and their role are undeniable."* (Arab student, ID: 7)

(3) Attitude markers *"However, **unfortunately** this group has been showing a bad image about Islam in all over the world."* (Arab student, ID: 14)

(4) Engagement markers *"Lastly, **due to the above** reason I believe that with proper implementation of this approach to parenting, friendship with the children is the best method to raise children with..."* (Arab student, ID: 8)

(5) Self-mentions *"Finally, **we** can conclude this argument by saying that teachers gave **us** a lot, and **we** should not pay back by ignoring their importance, advancement in educational..."* (Arab student, ID: 3)

The results indicate that *interactive* metadiscourse markers were used more by the Arab students than *interactional* resources.

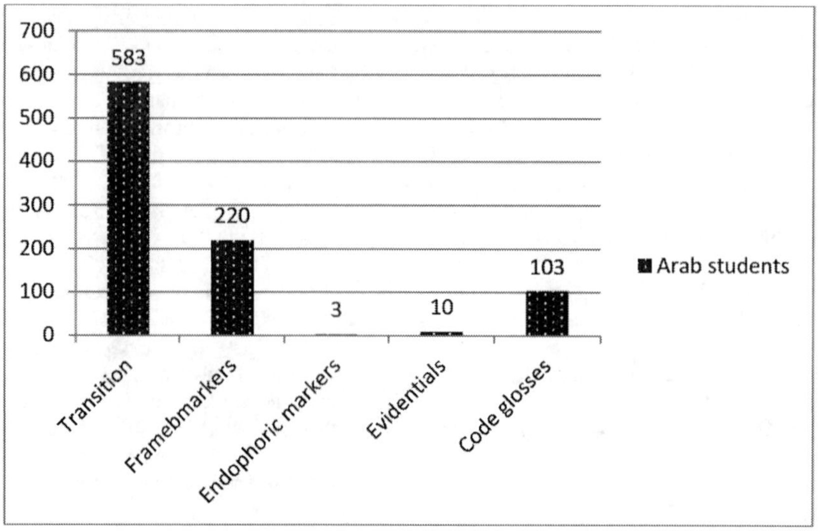

Fig. 1. Frequency of use of the different sub-categories within the *interactive* metadiscourse among the Arab pre-university writers. (***Interactive metadiscourse markers***)

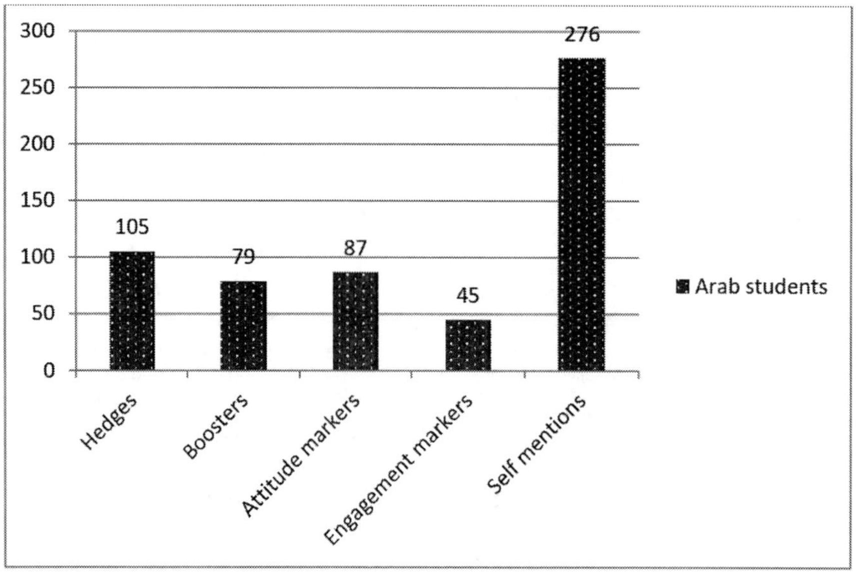

Fig. 2. Frequency of use of the different sub-categories within the *interactional* metadiscourse among the Arab pre-university writers. (***Interactional metadiscourse markers***)

4 Discussion

The total occurrence of metadiscourse in the academic texts produced by the Arabs students was 1511. Another important finding was that Arab pre-university writers had a greater inclination for the deployment of the *interactive* markers than *interactional* ones as there was a higher percentage of *interactive* metadiscourse usage than the *interactional* ones among the students.

These results are in agreement with previous findings [15-16] which showed that undergraduate students have greater tendency in deploying the *interactive* metadiscourse (also known as *textual metadiscourse*) than using the *interactional* metadiscourse (*interpersonal metadiscourse*). One dominant cause might be due the culture differences demonstrated in the students' writing conventions. In the same vein, the result for the Arab students is in agreement with Faghih and Rahimpour's findings which showed that *interactive* metadiscourse (those markers which help to guide the reader through the text) were deployed considerably greater in comparison to *interactional* metadiscourse (those metadiscourse markers which include the person who reads in the text) by Iranians as non-natives of English within their research articles compositions. In addition to the main reason of different socio-culture background amongst the ESL writers, Faghih and Rahimpour provided another possible explanation that the ESL students are more familiar with the usage of certain markers under the *interactive* dimension; for instance, *frame markers, transition markers,* as well as *code glosses* [17].

The result concerning the high rate of deployment of *interactive* metadiscourse, in excess of the *interactional* markers was also found to be in agreement with the result described in a study conducted by Salek [8] on academic synopses written in English by native authors. This consistency evidently advocates the significance of *textual* metadiscourse (also known as *interactive* metadiscourse) components in academic essays in comparison to the *interactional* ones. Furthermore, the different usage of metadiscourse might be due to the authors' linguistic customs and norms attained during their discrete involvements in their respective writing courses. To put it differently, the academic authors from diverse cultural upbringings and backgrounds may have established fundamentally distinctive writing styles well as distinct modes of interaction due to training which they received from their respective instructors. Consequently, the diversity in the individual writing styles among the writers of different backgrounds may be demonstrated by the construction as well as conceptualization of discrete worldviews which are rooted in a specific culture of a specific discourse cluster [17].

The essays used for text analysis in this current study were opinion-based academic essays. The essays required the students to express their opinion towards an issue. In this type of composition, there are usually more *transition* (for instance, but; in addition; thus; and) and *frame markers* (for instance, finally; my purpose is; to conclude) used in directing the target reader from end to end of the writing while more *hedges* (for instance, possible; perhaps; might; about) *boosters* (for instance, in fact; definitely; it is clear that), *self-mentions* (for instance, my; we; I; our; me), *attitude markers* (for instance, unfortunately; surprisingly; I agree), and *engagement markers* (for instance, you can see that; note; consider) will be used to involve the reader in the text. It is expected that in an opinion-based essay, more *interactive* markers should be involved to provide guidance in order to assist the intentional reader to read the written text effortlessly, to manifest semantic relation concerning key passages, to denote discourse items, orders, as well as writing phases, and to provide support for the person who reads to comprehend meanings of ideational material within the written text.

5 Conclusion

The main objective of this research was to determine the deployment of metadiscourse markers in the academic texts produced by Arab Pre-University writers. The examination of the metadiscourse employed by these ESL students has offered interesting findings and thought-provoking insights. One of the significant findings that has emerged from this study is that Arab pre-university writers had a greater inclination for the deployment of the *interactive* markers than *interactional* ones as there was a higher percentage of *interactive* metadiscourse usage than the *interactional* ones among the students. Culture differences demonstrated in the students' writing conventions may be the reason for the differences.

Contribution to body of knowledge. The findings of this research offer further understandings for the use of metadiscourse in the area of text analysis. Moreover, not many academics connected with the field of linguistics and English as a second language, had comprehensive knowledge on the ways metadiscourse markers are realized in English academic written texts. This study extends our knowledge of the use of metadiscourse among the Arab students. This finding is important as it will serve as a source that students as well as the English language educators can refer to and learn about the types of metadiscourse markers which are scarce in the students' writings (such as, *endophoric markers* and *evidentials*) and to be deployed more frequently in their composition so that it will enhance their quality of academic writing.

A key strength of the present study is that it enhances our understanding of the dissimilar deployment of metadiscourse markers among the Arab students' in their academic writing. This is important as higher education institutions in Malaysia are receiving higher intake of international students particularly from the Middle East. Therefore, it is crucial to take care of the education standards especially in the English language in order to sustain the credibility of the Malaysian higher education institutions. By having a text-analysis instrument (i.e. metadiscourse), it will help English language educators to detect any rhetorical variations as well as limitation in written essays quickly and improve the inadequacies in English academic writing among the students especially those from the non-native English countries. Hence, it will address future needs of these students especially in relation to their written proficiency.

Limitations and future studies. A limitation of this study is that it was carried out among the Arab students only. It did not include any other international students. As a result, the generalisability is not conceivable to other ESL students. Also, the present study was restricted to pre-university students; that it did not include the whole array of students' groups; under graduates and post graduates. Also, the scope of the study was limited only to the deployment of metadiscourse markers in academic writing, specifically the opinion based essays.

Although it is possible that proficiency could be accountable for the varieties of metadiscourse patterns among the students as mentioned in some of previous studies [1-2], this cannot be determined since generally most of the studies did not state the level of proficiency of the writers who wrote the texts that they examined. Also, this is not the focus of this study as the participants' proficiency level was presumed to be similar, that is within satisfactory proficiency level.

More research is required to determine the effectiveness of this text-analysis instrument (i.e. metadiscourse) in the future. It is recommended that further research can involve comparisons with other international students who are studying in Malaysia. This will profit Malaysian higher education institutions. In addition, the learners can be more conscious with the deployment of metadiscourse markers in English for academic writing.

References

1. B. Giridharan, *Identifying Gaps in Academic Writing of ESL Students. US- China Education Review*, (Curtin University, Miri, Malaysia, 2012)
2. H. Tan, B. E. Wong, English Language Teaching, **7**(7), (2014)
3. K. Hyland, P. Tse, App. Ling., **25** (2), 156-177, (2004)
4. K. Hyland, *Metadiscourse: Exploring Interaction in Writing* (London, 2005)
5. Continuum, J. App. Res. Eng., **1**(2), 37-44, (2012)
6. P. M. Letsoela, Int. J. Ling., **5**(6), 138-153, (2014)
7. W. Shi, J. Han, Eng. Lang. Teach., Vol. **7**, No. 11.,151-158, (2014)
8. Z. Zarrati, R. M. Nambiar, T. N. R. T. M. Maasum, Effect of Metadiscourse on Reading Comprehension of Iranian EFL Learners. *3L: Language, Linguistics, Literature*, *20*(3), pp 27-38 (2014)
9. M. Salek, Journal of Language Sciences & Linguistics **2** (3), 55-66 (2014)
10. L. Lu, *Metadiscourse and genre learning: English argumentative writing by Chinese undergraduates* (The University of Hong Kong, 2011)
11. H. J. S. Abbas, Acta Ling., Vol. **5**, (2011)
12. G. Mehrnaz, T. Ghaffar, S. Nasrin, Euro. J. Eng. Lang. L. Stud., **2**(2), 1-41 (2014)
13. A. Crismore, *Metadiscourse: What It Is and How It Is Used in School and Non-School Social Science Texts. Technical Report 273*, (University of Illinois: Centre for the Study of Reading, 1983)
14. N. Fairclough, *Discourse and social change* (Cambridge, England: Polity Press,1992)
15. W. J. Vande Kopple, J. Eng. Aca. Pur., **20**, 114–124 (2002)
16. A. Nor Hafizah, E. O. Luyee, J. I. Gabriel R. K. Seyed Ali, Eng. Lang. Teach., **16**(9), (2013)
17. E. Faghih, S. Rahimpour, Ling, **1**, 92-107, (2009)
18. H. Shokouhi, A. Talati Baghsiahi, Poznan Stud. Cont. Ling, **45** (4), 549-568 (2009)
19. M. Silver, J. Eng. A. Pur., (4), 359–374 (2003)

The Impact of Interpersonal Communication toward Customer Satisfaction: The Case of Customer Service of Sari Asih Hospital

Novalia Agung W. A.[1,*]

[1]School of Communication, University of Prof. Dr. Moestopo (Beragama), Jakarta, Indonesia

Abstract. Customer Service has a considerable role. In order to retain the loyal customer, their attitude should be friendly, courteous, patient, and willing to listen to what customer said. Good Customer Service should create customer satisfaction for it is the presence of customer determines the existence of the company. This condition is also true for hospital as it is the case of Sari Asih Hospital in Ciledug, Indonesia. Sari Asih Hospital is a private hospital whose average patients are more than 700 patients monthly. This study assume that the interpersonal communication might be the cause. In constructing the argument, this paper will use the Humanistic Perspective Theory and the Theory of Value Expectancy. Quantitative approach will be the method and the survey will use the accident sampling among customers. It was found that the Impact of Interpersonal Communication Sari Asih Hospital toward Customer Satisfaction is has possitive effect.

1 Introduction

The presence of Public Relations in organization or company is to establish a harmonious relationship and build a positive public opinion of the organization or company. The relationship in this point is mutual relationship between company with internal and external community. Internal community is associated with giving information and motivation to investors and employees as well as to listening their input. While External Public Relations activities is supposed to strengthen the relationship or maintain the loyalty of customers, suppliers, governments, and communities. Communication is considered as an important component in the establishment of business relationships. "However, these variables are often assumed or taken for granted and consequently overlooked as a component of relationship development" [1]. As well as the case of Sari Asih Hospital, eventually every company will seek to maintain their customers. Extra services is provide to make customers feels comfortable and satisfied. This approach is used to mantain customers. Another way is build a good relationship between company and its customers.

Customer Service as a frontline of companies, they are interact directly with the external public. Customer Service is represent to Public Relations in providing services such as: information, product sales and other things according to customer requirements. However, to be Customer Service is not easy because they have tobe able to recognize the characteristics of each customer and their needs. Therefore the most effective approach is interpersonal communication. The role of Customer Service has become one of the determinants of customer satisfaction. Due to customer satisfaction is created not only from facilities and products provided by the company, but also related to the impression of the services provided by Customer Service, including of their communication skillsAs the research results of Hardianti showed that there was the impact of interpersonal communication Customer Service to customer satisfaction Kaltim Bank Syariah Branch Juanda Samarinda [2]. This paper would to examine how the interpersonal communication by Costumer Service has the ability to transform and build customer attitudes towards customer satisfaction. As well as in Sari Asih hospital Ciledug, since it began operating, the number of competitors increases. It is one of the challenges in retaining customers. The results of this research could be an evaluation for hospitals.

Health facility is holds very important role of public welfare in the region. One of the existing health facilities in Tangerang is Hospital. Referring to Law Number 44 Year 2009 regarding Hospital, the definition of a hospital is a health care institution for people with characteristics of its own which is influenced by the development of health science, technological progress and socio-economic life of society who must remain able to improve service more qualified and affordable for the people to manifest the degree of health as high. This means that, the Hospital is a health care institution which organizes personal health services in the plenary that provides inpatient, outpatient, and emergency department. According to data from the Central Bureau of Statistics Tangerang, in 2011 in the city of Tangerang

* Corresponding author: agungnawa@ymail.com

there are 22 hospitals, 51 maternity hospitals, 185 polyclinics, 28 health centers, 11 health centers, 235 doctors and 287 practice of midwifery practice. Overall hospital in the city of Tangerang located in 22 Village and 10 districts. Development in the health sector supposed to make all levels of society can obtain health services easily and evenly. The existence of such efforts are expected to achieve a good degree of public health, which in turn increases productivity.

Table 1. List of Tangerang Hospital

No	The Names of the Hospital	Total Patient
1	RS Usada Insani	20.320
2	RS Mayapada	6.611
3	RS Daan Mogot	1.712
4	RS Aminah	3.317
5	RS Sari Asih – Karawaci	12.804
6	RS An –Nissa	6.408
7	RSK Sitanala	330
8	RSIA Keluarga Ibu	698
9	RS Melati	3.003
10	RS Bhakti Asih	6.715
11	RSIA Mutiara Bunda	1.473
12	RS Sariasih – Ciledug	15.085
13	RS Mulya	3.772
14	RS Medika Lestari	12.888
15	RS Awal Bros	7.033
16	RS Sari Asih – Sangiang	7.442
17	RS Dinda	3.990
18	RSIA Bunda Sejati	906
19	RS Hermina	3.046
20	RS Ariya Medika	7.892
21	RS Aqidah	1.491
22	RS Pratiwi	2.049
23	RSIA Karunia Bunda	5.414
24	RS Karang Tengah Medika	62
	TOTAL	**134,461**

Source:Tangerang City Health Department (2016)

Sari Asih Hospital Ciledug is one of the hospitals in the region Ciledug Tangerang. This hospital is equipped with high-quality medical equipment and adequate facilities such as 155 room number, outpatient facilities, emergency departments and other medical support (http://aldintara.com/overview-rs-sari-asih-ciledug/ acces in 13 Januari 2017, 06.53 am). People take advantage of the presence of hospital. Base on fact, there is an increase from 41.74% to 71.09% on 2006 to 2016. This conditions is mean a very potential market segments for health services. Problems faced by the hospital that stood since 2005 is still able to survive and thrive in the current conditions of the moment. Where competition similar business more competitive in the snatch customers. Customer Service equipped with knowledge of the sale of health products offered, ability to communicate well, dressed accordingly and keep the attitude to dealing with customers. With hope customers will feel comfortable and satisfied with the product and services.

In this section will describe the definition of Interpersonal Communications, Customer Service, the definition of Customer Satisfaction, Perspective Humanistic and The Affective Cognitive Consistency Theory.

1.1 Interpersonal communication

Simple overview of the communication process can be seen According to Hovland [3]; "communication is the process by which an individual (the communicator) transmits stimuli (usually verbal) to modify the behavior of other individuals (the audience)". According to Joseph A. DeVito [4]; "Interpersonal communication is the process of sending and receiving messages between two persons, or among a small group of persons, with some effect and some immediate feedback". Interpersonal communication that is intended in this research is an oral communication in a conversation conducted by Customer Service with customers in Sari Asih Hospital, Ciledug. The ideal process of interpersonal communication is where Customer Service know whether the service that provided could be well received by customers and get the responses that are positive or negative.

This is shown through verbal communication in a conversation feedback directly. Communicators know the response of communication at the moment and know with certainty whether the messages they sent was received or rejected. If not accepted, the communicator will provide greater opportunities to the communicant to ask.

1.2 Customer service

According to Effendy [5]; "customer service is one of public relations activities in carrying persuasive communication process conducted by a person to another person face to face in the work situation and in the organization. The aim is to arouse excitement and activity to work with the spirit of productive cooperation with feeling happy and satisfied". "In practice customer service function is as a receptionist, as deskman, as a salesman, a customer relations officer, and as a communicator" [6]. Customer service in this study is referred to those who conduct persuasive communication in providing information on health care services in Sari Asih Hospital, Ciledug.

1.3 Customer satisfaction

According to Kotler [7]; "customer satisfaction is feeling happy or disappointed someone caused by the performance or the result of a service that is perceived, compared with expectations. If the services earned less than expected, the customer will be dissatisfied, disinterested and disappointed to the provision of services is concerned. Meanwhile, if the services are perceived by customers more than they expected, customers will be satisfied". Customer satisfaction is obtained after the purchase and consumption. Then the researchers can conclude that customer satisfaction will be achieved if a service is considered satisfactory and can complement the needs of the customer according to the desired expectations. As is the customer in this study is, customers who use health care services at Sari Asih Hospital, ciledug.

1.4 Prespective humanistic theory

According to DeVito, that "the Perspective Humanistic approach presents those attributes needed to enhance bonding between individuals [4]. At the risk of being overly repetitious, these two aproaches are not opposites. They do represent two different ways of defining and approaching interpersonal communication effectiveness". Perspective Humanistic theory assumes that individuals interact with one another, feel is linked to one another and do more communication actions. Conversely, individuals who do not have conflicting or interpersonal relationship, then it will only slightly even less in carrying out acts of communication. By using this theory, researchers can measure and determine how the interaction process of interpersonal communication will have cohesion and interactive. The process of interpersonal communication as a form of behavior can change from highly effective to ineffective, and at one time it could be worse communication and at other times could have been better. Five general qualities from this model: openness, empathy, supportiveness, positiveness, and equality [4]. (1) Openess; Referring to our ability to open up where there is a disclosure of information about ourselves that are usually far we cover, and to react honestly to the message that we receive, and referring to the feelings and thoughts that we have. (2) Empathy; is our ability to feel what others feel, try to experience what happened to the others from the viewpoint of the person without losing their own identity. (3) Supportiveness; is an atmosphere that is more directed to a support. (4) Positiveness; is showing positive behavior and praise the people who interact with you. Furthermore, (5) Equality; is felt that two or more people are actually similar in many aspects. The Humanistic models will use as a basis to determine the extent of interpersonal communication is able to create customer satisfaction interaction between the Customer Service Sari Asih Hospital Ciledug with customers.

1.5 The affective cognitive consistency theory

In this theory, examines the relationship between attitudes and beliefs [8]. It has been proposed that attitudes have an affective component (a positive or negative feeling toward the issue) and a cognitive component (beliefs or thoughts about the issue). Rosenberg stated that when the affective and cognitive components of an attitude are mutually consistent, the attitude is in a stable state. In other words, the strength of an attitude is determined by more than just the favourability of the attitude. This means that the strength of an attitude is influenced by the consistency or agreement between the affective and cognitive components of the attitude. Behaviour is generally more influenced by the hope of achieving something desired result (any positive incentives) of the impetus from within. Expectation value based on a balance between trust and evaluation. Value is defined as the quality of service of Customer Service. So the attitude of customers is the result of the evaluation of the services provided by officers. Customer Service with customer expectations will be a quality service. Additionally Customer Service should be able to put itself well to enable the smooth communication with customers, and the most important is the attitude of support and a positive attitude. There are several factors that affect customer satisfaction. (1) Customer expectations (desire customers to served with good and right). Customers very much hope that getting the best quality services. Quality services can be realized if an officer can serve customers by being polite and professional, have alacrity in serving customers, extensive knowledge about the company and have the alertness and speed and reliable in helping customers to solve customer problems, so as to satisfy the customer. And (2) The fact that customer acquired (Person Need or Individual Needs). Customers come to the Customer Service to get their basic service needs (Self Personality).

This expectation can be realized if the workers provide services that show a sense of caring to give individual attention to customers and understand customer needs. If the service provided is able to create a positive belief in customers, it will lead to customer satisfaction. Satisfaction with the quality of service that will reinforce the attitudes and foster customer loyalty to Sari Asih Hospital Ciledug. Customer satisfaction to Sari Asih Hospital Ciledug can be formed if the value of customer expectations can be fulfilled.

2 Materials and methods

2.1 Research framework and hypotheses

Based on the above framework, "the researchers made a temporary answer" [9]. Temporary answer (hypothesis) then must be verifiable. The hypothesis is determined based on the background of the problem, the subject matter, the theoretical basis and framework. Such as:

Ho, rxy = 0 : There is no relationship between the interpersonal communication between Customer Service and customer satisfaction at Sari Asih Hospital, ciledug.

Ha, rxy > 0 : There is a relationship between the interpersonal communication between Customer Service and customer satisfaction at Sari Asih Hospital, ciledug.

Fig. 1. Research Model

Information:
Independent variable (X) = Interpersonal Communication
Dependent variable (Y) = Customer Satisfaction

In this case the independent variable is interpersonal communication Customer Service with customers. Indicators that can be developed from the theory that supports variable inter-personal communication Customer Service are: openess, empathy, supportiveness, positiveness, and equality. For the dependent variable is a variable that is affected or become due, because of the independent variables. In this case is customer satisfaction, using indicators: the desired expectation and personal need.

This paper uses the paradigm of positivism. According to Neuman, "Positivism paradigm is used to show the relationship between the variables to be studied and symptoms of causal relationship" [10]. Positivism is used to reveal the truth of reality and how reality is running. Information on the truth directly to the researchers asked respondents without the intervention of researchers. The information then can be observed, measured accurately, and the researchers tested the hypothesis through the analysis of numbers of measurement results. For this type of research is quantitative eksplanatif. This type of research is able to describe the relationship between research variables and conduct testing hypotheses that have been formulated previously.

2.2 Sample and Data Collection

Data collection and analysis is using quantitative method. The research instrument used questionnaire. The questionnaire aimed to obtain information from respondents. The respondents of this study are customer Sari Asih Hospital Ciledug. The survey was conducted on December, 2016. 985 total of patients into the study is the population. And due to the limitations of time, cost and effort, researchers determined the sample. "Population is represented by the sample. The sample is part of the population" [9].

According to Guilford, "statistics is a way of looking at a population's behavior by taking a sample" [11]. It's usually impossible to survey every member of a population because of money or time. That's where Slovin Formula comes in. The formula used to determine the sample Slovin, with an error rate of 10%.

=

$= 90{,}78 \approx 91$ people

Information :

n = Sample
N = Population
e = 1%, 2%, 3%, 4%,5%, or 10%

Sampling technique used in this research is through accidental sampling, to anyone who happened to meet researchers and seen people who happened suitable as a data source. Data collection techniques in this study using two types of data. First Primary data in the form of a questionnaire. The questionnaire will measure customers. perception of the level of satisfaction of respondents by giving a set of questions or a written statement to answer. The second type of data is from of literature studies. In this technique the researchers conducted a literature study to complete the data. Examples of secondary data are theories and data hospital related to the study. The collection of data through books related to the title of research and other sources (internet). As well as researchers compared with reference materials earlier study with the same problem with what the researchers carefully. In addition, researchers get information about the patient data from the Sari Asih Hospital Ciledug.

2.3 Measurement

To Analysis the questionnaire data is using inferential statistical tests. "Inferential test aims to explain the relationship between two or more variables" [12].

Statistical tests were performed with use of statistical data processing program (SPSS). The early stage of testing is to test the validity and reliability of the questionnaire. If not valid and reliable, the researchers need to revise the research instrument. Afterwards, the instrument spread back up to obtain valid and reliable results. Measurement opinions of respondents used a scale developed by Likert Rensis. This scale is called Likert Scale. "Likert scale used to measure attitudes, opinions, and perceptions of a person or of a phenomenon" [10].

Table 2. Likert Scale

Scale	Values
Strongly Agree	5
Agree	4
Doubtful / No Comment	3
Disagree	2
Very Disagree	1

Source: Sugiyono (2009)

2.4 Data Analysis

This section describes the results of statistical tests. Not all tests performed in the statistics. Relevant tests to the needs of the research purpose only. Among others are a test of Validity and Reliability, Correlation and Regressions. Validity was conducted to determine the extent to which the instrument is able to measure what we want to measure. Is the measuring instrument that can measure the properties of the object that we examine. "While reliability to measure the consistency of the answers to the questionnaire" [10]. According to the table the value of r-Product Moment, the obtained limit r-table is 0,206. Significant level using 10%, with the number of respondents 91 people.

Table 3. Validity Test Result

Interpersonal Communication Customer Service (Variable X)				
No	**Statement**	**r-values**	**r-table**	**Conclusion**
P1	Customer Service provide an explanation to customers openly	0.409	0.206	Valid
P2	Customer Service provides friendly service	0.665	0.206	Valid
P3	Customer Service serves customers not only face-to-face interaction but with other media such as telephone or email	0.531	0.206	Valid
P4	Customer Service is able to greet by name the customer	0.599	0.206	Valid
P5	Customer Service respond each question or customer complaints immediately	0.484	0.206	Valid
P6	Customer Service serve customers with fast, skilled, accurate, and thorough	0.515	0.206	Valid
P7	Customer Service serve all customers without any distinction	0.539	0.206	Valid
P8	Customer Service seriously listen every customer information	0.574	0.206	Valid
P9	Customer Service responds to customer demand with immediately	0.540	0.206	Valid
P10	Customer Service is able to provide solutions	0.600	0.206	Valid
P11	Customer Service serving with friendly	0.533	0.206	Valid
P12	Customer Service concerned with customer complaints	0.682	0.206	Valid
P13	Customer Service serve customers with courtesy	0.589	0.206	Valid
P14	Sari Asih hospital waiting room is comfortable	0.369	0.206	Valid
P15	Customer Service can create a friendly atmosphere with customers	0.442	0.206	Valid
P16	Customer Service is attractive	0.556	0.206	Valid
P17	Customer Service using uniforms or identity	0.460	0.206	Valid
P18	Customer Service respect every customer who comes	0.408	0.206	Valid
P19	Customer Service appreciate every customer who comes	0.512	0.206	Valid
P20	Customer Service seemed close when interacting with customers	0.363	0.206	Valid
P21	Customer Service immediately contacted the relevant sections in the hospital to solve customer problems	0.633	0.206	Valid
P22	Cust Service accompanies the customer until the needs is resolved	0.394	0.206	Valid

Customer Satisfaction (Variable Y)				
No	**Statement**	**r-values**	**r-table**	**Conclusion**
P23	Customer Service accompanies customer until the issue is resolved	0.698	0.206	**Valid**
P24	Customer Service looks reliable in serving customers	0.613	0.206	**Valid**
P25	Customers get clear information from Customer	0.553	0.206	**Valid**

	Service			
P26	Customer Service looked sprightly and quick to responding every complaints	0.600	0.206	**Valid**
P27	The customer wishes can be fulfilled when interacting with Customer Service	0.754	0.206	**Valid**
P28	Customer problems can be resolved with the help of Customer Service	0.737	0.206	**Valid**
P29	Customer Service officer serving with the maximum effort	0.276	0.206	**Valid**
P30	The whole process of health care were answered by the Customer Service	0.512	0.206	**Valid**
P31	Customers feel helped by the presence Customer Service	0.510	0.206	**Valid**
P32	Customer Service is able to ensure the confidentiality of customer information	0.620	0.206	**Valid**
P33	Customer service to provide services according to the customer needs	0.616	0.206	**Valid**
P34	Customer service to provide services according to the customer's	0.742	0.206	**Valid**

Source: Output SPSS

Based on the validity of the test results by using the formula Pearson Product Moment, the whole item questionnaire statement declared invalid because the correlation value is greater (>) in compare to the table r Product Moment. Reliability tests intended to determine whether the respondent answers given credible or reliable (valid), if done twice or more measurements of the research object or to the object and measuring the same. Reliability is the level of accuracy, precision or accuracy of an instrument. Instruments that have been trustworthy or reliable will produce reliable data as well. A construct or variable is said to have a reliable if the Cronbach Alpha > 0.70. From these test results obtained reliability value data from each of the variables as described in the following table 4.

Table 4. Conclusion Test Reliability Questionnaire

No	Variable	Cronbach's Alpha	Realibility Limit	Conclusion
1	X	0.903	0.70	Very Realible
2	Y	0.893	0.70	Very Realible

Source: Output SPSS

Based on the reliability test in Table 4, so that the data can be concluded that the questionnaires used reliable.

3 Results and discussion

3.1 Score test

Furthermore, researchers made the recapitulation of the value of the variable X (Interpersonal Communication Customer Service) and Variable Y (Customer Satisfaction) as follows:

Table 5. Score Value Interpersonal Communication Customer Service (Variabel X)

No Statement	Score Value	Conclusion
P1	388	
P2	387	
P3	376	
P4	361	
P5	384	
P6	382	
P7	392	High
P8	374	
P9	382	
P10	382	
P11	384	
P12	382	
P13	386	
P14	360	Low
P15	360	
P16	378	
P17	388	
P18	389	

P19	388	
P20	380	
P21	383	
P22	375	

Source: Output SPSS

Table 5, shows that the average value of respondents to the overall indicator Interpersonal Communication Customer Service (variable X) are:

Presentage = (**Total Score** : Skor Maximum) x 100%
= (8369 : 10.010) x 100%
= 83.60%

Based on the percentage calculation above, the percentage of votes obtained by the magnitude of respondents to the Interpersonal Communication is 83.60% .When adjusted for interval continuum, then the results are included in the category of Good. That means, respondents said Interpersonal Communication Customer Service runs well enough to be able to grow Customer Satisfaction at Sari Asih Hospital, Ciledug.

Table 6. Score Value Customer Satisfaction (Variabel Y)

No Statement	Score Value	Conclusion
P23	364	
P24	368	
P25	370	
P26	369	
P27	362	
P28	354	
P29	377	High
P30	371	
P31	349	Low
P32	366	
P33	360	
P34	365	

Source: Output SPSS

Based on the results of the development of As for the variable Y (Customer Satisfaction) Table 6, shows that the average value of respondents to the overall indicator is the establishment of Customer Satisfaction:

Presentage = (**Total Skor** : Score Maximum) x 100%
= (4753 : 5915) x 100%
= 80.35%

Percentage of respondents to the Customer Satisfaction ratings of 80.35%. When adjusted for the value continuum interval then the results are included in the category of Good. That means, the respondents said that increasing customer satisfaction is good enough for their interpersonal communication Customer Service interaction with customers.

3.2 Correlation analysis

Acquisition of data that has been given a score and then processed by using SPSS (Statistics Package for Social Science). Further analysis is reinforced by regression analysis to determine the influence that occurred. To provide interpretation of the correlation coefficient is determined large or small, it is guided by the provisions contained in the following table:

Table 7. ntepretation Coefisien Correlation

Coefficients Correlation	Level of Correlation
0.00 – 0.199	Very low
0.20 – 0.399	Low
0.40 – 0.599	Moderate
0.60 – 0.799	Strong
0.80 – 1.00	Very Strong

Source: Sugiyono [13]

"Correlation is a statistical tool that can be used to compare measurements of two different variables, in order to determine the degree of correlation between the variables" [14]. Correlation test the hypothesis that the researcher will answer submitted. The results of correlation test variables X and Y are as follows:

Table 8, Correlations Test Result

Variables		X	Y
X	Pearson Correlation	1	.676**
	Sig. (2-tailed)		.000
	N	91	91
Y	Pearson Correlation	.676**	1
	Sig. (2-tailed)	.000	
	N	91	91

Source: Output SPSS

Based on calculations using SPSS (Statistical Package for Social Science) in the above table, it can be a correlation between these two variables, where the value of the correlation coefficient (r) discovered is 0.676. Means the value of this correlation coefficient correlation between the impact of interpersonal communication Sari Asih Hospital toward customer satisfaction is has possitive effect.

3.3 Regression analysis

Table 9 showed that it can be said the influence of variable X (Interpersonal Communication Customer

Service) with Variable Y (Customer Satisfaction) at Sari Asih Hospital Ciledug are at the level of the Strong correlation coefficients, located at the level of 0.60 to 0.799 with value a coefficient of 0.676 (using Table 7 for the intepretation). The results of R Square = 0457 or in the form of a percentage of 0457 x 100% = 45.7%, while the remaining 100 % - 45.7% = 54.3% influenced by other factors.

Table 9: Regression Test Result

Model	R	R Square	Adjusted R Square	Std. Error of the Estimate
1	.676ᵃ	.457	.451	4.20009

Source: Output SPSS

From the research, there is a strong relationship between interpersonal communication Customer Service Sari Asih Hospital Ciledug to customer satisfaction. This is consistent with the explanation of the Perspective Humanistic theory. That individuals interact with one another will find it has a relation to one another. So that they will more act of communication. Instead, individuals are conflicting (not having interpersonal relationship) will be less in carrying out acts of communication. Interpersonal communication between the Customer Service to customers will be successful if both parties have a mutual openness, empathy, positive attitude, support, and similarity. Furthermore, by using the theory of interpersonal communication effectiveness of humanistic perspektive can be a material evaluation. So that it can be seen the extent to which the communication is going well and is able to improve customer satisfaction Sari Asih Hospital Ciledug.

According to Kotler [7]; customer satisfaction is the customer response to the evaluation of the perceived between prior expectations or other performance expectations and actual performance services perceived after wear. Then the researchers can conclude that customer satisfaction will be achieved if a service is rated to meet customer needs in accordance with the desired expectations. By measuring the level of customer satisfaction then the Sari Asih Hospital Ciledug can provide better service, effectively, and efficiently.

Conclusion

From the analysis concluded that the impact of interpersonal communication toward customer service Sari Asih hospital Ciledug in Best categories. It is based on a statistical calculation of all questionnaires filled out by respondents. This means that the interaction of customer service rated positively to customers. The presence of Customer Service Sari Asih hospital Ciledug is important as a first gate that interact directly with the customer (patient). Cutomer Service could provide a solution of confusion cutomer. Positive impression will be embedded in the customer mind after interacting with Customer Service. This impression will they bring when interacting with other parts of the Sari Asih Hospital Ciledug. Thus the function of public relations in shaping the relationship and strengthen the positive image is reached. It is represented by the presence of Customer Service. Interpersonal communication process can be used to measure customer attitudes after interacting with Customer Service. It becomes necessary for the Customer Service have an attitude of empathy in evaluating their action. Evaluation can be done by looking at an attraction or rejection of customers. Because based on theory, is a response to customers from any activities with Customer Service. With such a response, it can be used in determining the next action or strategy to achieving customer satisfaction. However, simultan evaluation and continuous innovation becomes a solution in the face of competition from other hospitals. By continuing to realize the achievement of the quality of public health. As their vision and mission over the years.

According to [7] means that when the customer has a feel for how the service Customer Service customer will evaluate it. Customers will be satisfied if their expectations are met. And then customers will do more positive behavior. For example, if they are sick, Sari Asih Hospital Ciledug services be an option in his heart and the first choice. Behavioral word of mouth they can do anyway. Customers will recommend others about health care in Sari Asih Hospital Ciledug compassion. Many factors affect satisfaction. In this study focused on the personal form of communication between Customer Service. What is the role of doctors, nurses, facility, speed drug services and so interesting to study. So this into recommendations for the next research.

References

1. P. R. Andersen, Journal of Business & Industrial Marketing, **1**(2), 167-82, (2001)
2. O. Hardianti, Octaviani, ejournal ilmu komunikasi fisip unmul.ac.id, 2016, **4**(3), 237-251, (2016)
3. C. J. Hovland, I. L. Janis, H. H. Kelley, *Communication and Persuasion*, (CT: Yale University Press, New Haven, 1953)
4. J. DeVito, *Komunikasi Antar Manusia, 2nd ed.*, (Professional Book, Jakarta, 2007)
5. O. U. Effendy, *Ilmu Komunikasi Teori Dan Praktek* (PT Citra Aditya Bakti, Bandung, 1999)
6. Kashmir, *Etika Customer Sevice* (Raja Grafindo Persada, Jakarta, 2005)
7. P. Kotler, K. L. Keller, *Manajemen Pemasaran Jilid 1, 13rd ed.*, (Bob Sabran, MM. Erlangga, Jakarta, 2009)
8. M. J. A. Rosenberg, C. I. Hovland, *Attitude organization and change,* (CT: Yale University Press, New Haven, 1960)
9. J. W. Creswell, *Research Design, 3rd ed.*, (Pustaka Pelajar, Yogyakarta, 2010)
10. W. L. Neuman, *Social Research Methods, 5th ed.*, (Pearson Education, Boston, 2003)

11. J. P. Guilford, B. Frucher, *Fundamental Statistics in Psychology and Education,* (MC Graw-Hill, New York, 1973)
12. R. E. Walpole, *Pengantar Statistika, 3rd ed.*, (Gramedia Pustaka Utama, Jakarta, 1992)
13. Sugiyono, *Metode Penelitian Kuantitatif Kualitatif dan R&D,* (Alfabeta, Bandung, 2009)
14. D. Kleinbaum, L. L. Kupper, K. E. Muller, *Applied Regression Analysis and Other Multivariable Methods, 2nd ed.,* (PWS-Kent, Boston, 1988)
15. Al-Hersh, A. Mohammad, Abdelmo'ti Suleiman Aburoub, Abdalelah S. Saaty, Journal of International Journal of Academic Research in Business and Social Sciences, **4**(5) 67-100, (2014)
16. Cangara, H. Hafied, *Pengantar Ilmu Komunikasi* (PT Raja Grafindo Persada, Jakarta, 2014)
17. J. Griffin, *Customer Loyalty: Menumbuhkan dan Mempertahankan Kesetiaan Pelanggan,* (Erlangga, Jakarta, 2000)
18. F. Jefkins, Frank, *Public Relations, 5th ed.* (PT Gelora Aksara Pratama, Jakarta, 2003)
19. Kueh, Shirley, A. H. Bahar Pengiran Bagul, Journal of International Journal of Research in Management & Technology, **3**(4),126-135, (2013)
20. A. Liliweri, *Perspektif Teoritis Komunikasi Antar Pribadi,* (PT Citra Aditya Bakti, Bandung, 2004)
21. S. W. Littlejohn, (Theories of Human Communication, 7th ed. Wadsworth, New Mexico, 2002)
22. K. Miller, *Communication Theories*, (McGraw Hill, New York, 2001)
23. R. West, Richard, *Introducing Communication Theory,* (Salemba Humanika, Jakarta, 2010)
24. V. A. Zeithaml, Bitner, *Service Marketing* (The Mc Graw Hill Companies, New York, 1996)

The Dynamics of Housing Prices in Malaysia: Findings from Focus Group Discussions

H.N. AuYong[1,*] , *C.Y.* Yip[1], *K.H.* Woo[1], and *A.* Senadjki[1]

[1]Faculty of Business and Finance, Universiti Tunku Abdul Rahman, Kampar, Perak, Malaysia

Abstract: Background of the Research: This study examines the factors influencing housing prices in Malaysia. The study explores qualitatively whether there is housing bubble in Malaysia, and whether the housing prices are associated with changes in construction cost, land cost, compliance cost, housing speculation, and mortgage rate. Methodology: The paper is exploratory in nature. The data were collected via focus group discussions among nine property industry players in Malaysia and were analysed using qualitative research methodology. Main Findings: The study reaches the qualitative outcomes that rising housing prices are mainly due to cost factors and housing speculation but may not necessarily be influenced by mortgage rate. The findings suggest that the residential property market is currently not facing housing bubble issue. However the problem is partly due to PTPTN blacklisted borrowers. Conclusion: It is imperative for the Malaysian government to put in further efforts to control housing prices in order to maintain affordability of homeownership.

1 Introduction

Bank Negara Malaysia [1] announced that Malaysia's household debt-to-gross domestic product ratio increased to 89.1% as of 2015, one of the highest household debts in the region. BNM has also put in place various macro-prudential policies such as BNM limiting loan-to-value ratio to 70% for the third home financing as reported in TheEdgeProperty.com [2], and increasing real property gain tax to 30% for sales of real estate within three years to stem real estate market speculation activities since 2014 [3].

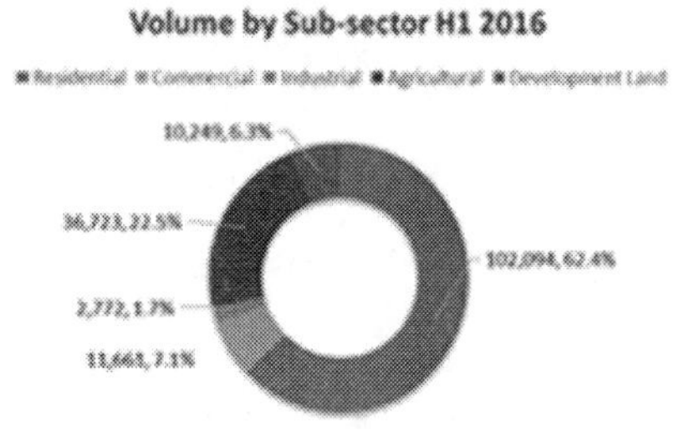

Fig. 1. Volume by Sub-Sector H1 2016
Source: NAPIC (2016b)

According to National Property Information Centre [4], market volume recorded at 362,105 transactions worth RM149.9 billion in 2015, down by a marginal 5.7% in volume and 8.0% in value against 2014. Residential sub-sector continued to lead the overall market, with 65.2% contribution in volume and 49.0% in value. The sub-sector recorded a slight downturn by 4.6% in volume and correspondingly down in value by 10.5%.

There were 102,094 transactions worth RM32.70 billion recorded in the review period, a decline of 14.5% in volume and 10.3% in value. Performance across the states was subdued. The downtrend (volume transaction) in major states namely Kuala Lumpur (-20.2%), Selangor (-14.1%), Johor (-10.9%) and Pulau Pinang (-13.5%) led to the overall drop in the sub-sector. Combined, these states formed 49.7% of the total national residential volume [5].

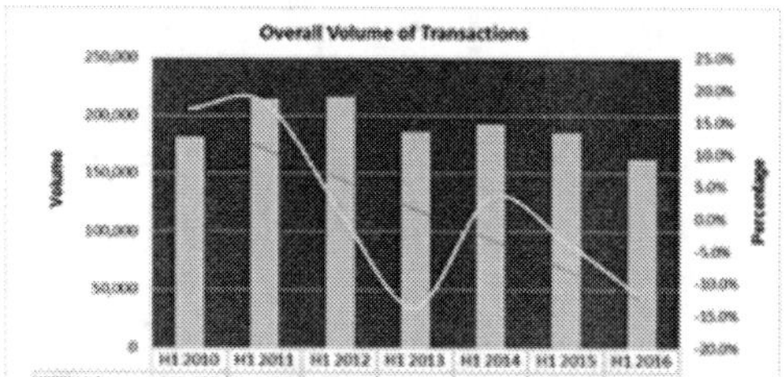

Fig. 2. Overall Volume of Transactions Source: NAPIC (2016b)

According to Khazanah Research Institute (KRI) Making Housing Affordable Report, Malaysian housing market is overpriced as a result of fragmented and inefficient construction business system. KRI has reported that houses in Kuala Lumpur and Penang are "severely unaffordable" and 40% of the middle income groups would need some kind of social housing in the future. The Malaysian all-house price had rose at a compounded annual growth rate (CAGR) of 3.1 per cent from 2000 to 2009. However, between 2009 and 2014, it grew at a CAGR of 10.1 per cent, which was almost three times more than the 2000 to 2009 period [6].

On the other hand, according to Real Estate and Housing Developers" Association (REHDA), properties within the RM200,001 and RM500,000 range are still leading the market [7].

This study explores the current situation of residential property prices in Malaysia based on microeconomic, macroeconomic and financial factors. The continuous upward trend of housing price in Malaysia has raised the attention of home buyers, investors and policy makers. This study is important as it focus on the main factors that affect housing price in Malaysia. These agencies have to treat housing carefully as a significant fluctuation in housing price will affect the wealth of investors and home buyers, as well as the whole country.

2 Research Method

2.1 Description of Focus Groups

This study is exploratory in nature with a qualitative approach. The data is collected via focus group discussion with nine experts from the property and related industries. Focus group questions were developed by the team of researchers. Focus group discussions were conducted with each group being held in a separate session.

There were two focus group discussions (FGDs) conducted. A focus group panel discussion was held on 13 October 2016 with members:

- Director of an architect consultancy SM
- Senior Project Manager of a holdings company SP Setia
- Chief Marketing Officer of a Property Developer MCT

Another focus group panel discussion was held on 15 October 2016 with members:

- Senior Executive Project Management of a Property Developer SW
- Assistant Manager of a construction and engineering firm FB
- Management staff of a Property Developer SS
- Management staff of a Property Developer GL
- Management staff of a building material firm BS
- Management staff of a Property Developer JH

2.2 Data Analysis

Transcriptions of recorded focus-group discussion, lists of facts and views drawn from each group, and individual worksheets are incorporated into the analysis process. This method is applied to create hypotheses that can be tested later by a quantitative survey to be completed by a larger number of participants. For each question, a summary statement will be written to illustrate participants' ideas and viewpoints. A summary report of the key findings will be prepared and reviewed by the researchers as a validity check.

3 Results and Analysis

The study analyses the important factors that influence housing prices from information obtained from the participants of the FGD which focus on the following topics.

3.1 Economic and Housing Market Conditions

3.1.1 What is your opinion on the Malaysian economic conditions?

The responses were mixed. Some said that the condition was not good due to influence of the external economy as well as the internal economic factors. Others have opined that the economic growth was slowed though high inflation made it a good tool to hedge. Yet, some said the condition was reactive, and that market forecast was not accurate while there were also participants with the opinion that the economic condition was positive.

3.1.2 How would you describe the current state of the housing market?

Some participants highlighted the effectiveness of the central bank's cooling off policy and the participation of foreign developers from China. Other participants lamented that while there was no control on developer loan, strict control was imposed on house buyer loan. Nevertheless participants shared the view that the situation of financing for developer/homebuyer depends on the location (state) of the projects as well as the type of project, whether low, medium or high end range. Some participants reflected that different state of the market depends on the state governments; overall low range housing market was better than medium to high range market. Other participants expressed that the housing market was slow with high volume of unsold units in the high-end segment.

3.2 How do the 1990s and 2010s scenarios compare in terms of housing bubble?

The booming property market in the early 1990's collapsed and was followed by the 1997/98 Asian financial crisis. An episode of housing crisis of very much smaller impart recurred in the local market during 2010 following the sub-prime mortgage led housing crisis in the U.S.

Generally, the participants did not think that any housing bubble exists in the domestic market. However it was uncertain of this time with internet marketing.

There was opinion that the housing industry was much localised, i.e. its issue was not housing bubble but rather more of supply and demand and regulatory policies. One participant said that the situations were incomparable with emerge of internet marketing as new driver of economic growth. Another was of the opinion that Developer Interest Bearing Scheme (DIBS) affected the housing price, and it all depended on risk profile of individual, and interest rate level. Another participant was of the opinion that he had more confidence in 1998 period than the 2010 period.

3.3 Housing Price

3.3.1 In what direction is the housing price moving?

The participants were divided for the clear direction as they have different opinions on construction and associated costs, types of properties and government institutional factors. Some participants were still confident that prices would increase in certain areas due to higher land cost, construction cost and compliance cost. Some expressed the importance of the government's actions, i.e. institutional factor and how well the government controls public securities and social unrest elements. Other participants expressed view that price still can increase depending on property types, such as Small Office Virtual Office (SOVO) and Small Office Home Office (SOHO) which still remain promising, and these are purchasers having follower behaviour that causes price change.

3.3.2 Do you expect housing prices to fall rapidly?

Most participants did not think that the housing prices would fall rapidly. Some stressed that bank credit situation that would determine the price movement. Some said that medium range housing would not drop, but high-end housing (RM2 to 3 million per unit) would be affected. Some predicted that the housing price would stay stagnant for some time but would not decline. Some thought positively that developers nowadays were more adaptive to the market and able to produce houses in line with buyers' needs

3.3.3 On what factors are you basing your view?

Some were pessimistic that the price would revise downward 10% to 25%. Others have mixed views that it all depends on supply and demand, i.e. loan availability to developer and end borrower, with impacts from graduates blacklisted due to PTPTN loan. Some stressed the importance of local content policy i.e. construction material cost. Some were optimistic that the current house offerings seen to be competitive with healthy take-up rate.

3.4 Housing Price Changes

3.4.1 What are the factors that could influence housing prices?

In general, it was said that it was due to government policy and construction cost. Some said the housing price would not drop while credit approval depended on the government's lending policy. Furthermore, details of the costs were such as land cost, construction cost and compliance cost; the weightage were different from one country to another, e.g. the ratios in Malaysia were 20%: 70%: 10% vs. Singapore's 50%: 40%: 10%. Lower interest rate would reduce the chances of bubble burst, and encourage FDI.

3.4.2 Has the Malaysian housing market enter into such a volatile situation? What are the significant impacting factors?

Some participants said that in general, housing bubble was not expected in short-term depending on government policies. Some said that there was no bubble and it was not expected to have one in near future. Another participant concurred that there was not bubble in long-term, depending on confidence (increase in FDI that creating jobs and eventually housing demand). Some participants said that it all depends on the government's check and balance policy (relaxing loan vs. non-performing loan). Another participant said government should decentralise public services as 40 departments involved for approval.

3.5 How effective have countervailing policies been in stabilising housing prices?

Over the recent few years of rapid surge in housing prices, the Malaysian government has implemented various policies with the aim to stabilise and check house price hikes. Some said the polices were not effective, while other said that the policies brought the housing prices to fair level, with more people buying for staying rather than for investment, as even PRIMA housing prices were not cheap.

3.6 Characteristic of the Purchasers

3.6.1 Do you see more of first time house buyers during the current period?

Most participants agreed that more of the buyers were first time house buyers as buying property to preserve value and saving rental cost, though some other still maintained that more of investors.

3.6.2 What do you think factors shaping current trend?

Most FGD participants had different opinions on the determinants, some were not sure of these factors, some mentioned availability of bank loan, speculators, some said foreigner price ceiling, though some insisted buyers would appreciate properties with new trends, designs and features and paying premiums for these properties.

3.7 Inflow of Capital

3.7.1 Do government regulatory control measures affect the elasticity of supply of housing?

In the past, one of the measures that the authorities have employed in countering housing market volatility is the restriction on the price level of residential properties for foreign buyers. The participants were divided on their opinions. Some said the policy was good to keep restrictions of house prices for foreigners; however others expressed their doubts of its effectiveness due to small percentage of foreign buyers.

3.7.2 Will liberal inflow of capital have a negative impact on housing prices?

Some participants expressed their doubts of its importance due to insignificant percentage of foreign buyers. Nevertheless, some other participants thought that it was important to obtain FDI moderately. There were participants who highlighted the better way to attract FDI is to speed-up transaction period such as the transaction would be completed in one week in Hong Kong.

3.8 Housing Affordability

3.8.1 Are the current prices of housing comparable to the purchasing capability of the buyers?

It has always been widely concerned among the people and the authorities on the affordability of housing. All the participants were of the opinion that the current level of housing price was healthy based on the ratio of house price/income, furthermore, developers were building affordable houses under the PRIMA and Rumah SelangorKu.

3.8.2 • Are current price levels affordable for respective income groups?

All participants agreed that the current price levels were barely affordable especially newly launched projects

3.9 Bank lending

3.9.1 Does the previously more relaxed bank loan regulation in dispensing mortgage loans contribute to house prices increase?

Recent reports have pointed to the tightening of lending by banks and financial institutions as one of the principal reasons affecting the housing market. All participants were of the view that the dispensing of mortgage loans need to be more relaxed, especially for first time house buyers. However the current problem lie with the substantially large number of blacklisted PTPTN borrowers who though within the home buyers age group do not quality for homeloans. Meanwhile, the growth of housing supply has outpaced that of bank financing.

3.9.2 Do you consider the current housing mortgage rate as too high?

All participants were of opinion that the mortgage rate was reasonable low and therefore not the major factor affecting housing market.

3.10 Are there any factors that you will monitor for signs of market overheating?

Some participants shared the view that overheating signs were such as ads on completed projects and house auctions. Other participants are of the opinion that the Government could control and balance the housing supply through housing industry regulatory measures and financing policies.

4 Discussion

The findings of the FGD indicate that speculators, housing demand, land cost, construction material cost, compliance cost and new house features are positively linked with the housing price. On the other hand, the participants agreed that high mortgage rate was not the factor affecting the industry. Another finding reveals that there is inverse relationship between the financing policies and the housing price. All participants thought that there was no housing bubble, and it was unlikely that housing prices would drop further especially the medium range housing, and most of the house buyers were first time buyers.

It is interesting to find that the industry players consistently do not think that affordability is an issue while government agencies such as Khazanah Research Institute, say otherwise. The mortgage rate is reasonably low. The participants highlighted that the current problems lied at blacklisted PTPTN borrowers. The industry players also thought that there was no issue of overheating as the Government could control housing supply and financing policies.

5 Conclusion

The following Information is for either paper type. Every presentation at the conference must have a submission. The conclusion should provide the thoughtful end of the paper which sums up the authors' view and opinion on the subject matter of the paper. In addition, the future development on the findings and the prophecies espoused on future studies may also be highlighted. There is diverse opinion on the main issues of overheating or affordability of housing prices between the Government and the property industry players; while the Government reports that the housing market is overpriced, the industry says that properties within the RM200,001 and RM500,000 range are leading the market and still available in various locations. In this study, data were collected from focus group discussions involving nine property developers and related industry experts. The discussion shows that cost factors are influencing Malaysian housing prices, while mortgage rate is not. In addition, Real Estate and Housing Developers'' Association (REHDA) said properties within the RM200,001 and RM500,000 range are still leading the market, with residential properties below RM500,000 still available in various locations. As a result, efforts to increase affordability should be led by the Government housing and financing policies with greater emphasis to maintain housing affordability especially for the younger generation.

References

1. Bank Negara Malaysia (BNM). Financial Stability and Payment Systems Report 2015. (2016).
2. Bank Negara imposes maximum loan-to-value (LTV) ratio of 70% for 3rd home loan. TheEdgeProperty.com, 3 November 2010. (2010).
3. Inland Revenue Board (IRB). Real Property Gains Tax. LHDNM/R13/15. (2015)
4. National Property Information Centre (NAPIC). Annual Report 2015. (2016a).
5. National Property Information Centre (NAPIC). First Half Report 2016. (2016b).
6. I. Suraya, I. Nadia, J. P. Marjan, M. M. *Making Housing Affordable*, (Khazanah Research Institute, Kuala Lumpur, 2015)
7. E. Yeong, *REHDA disagrees with Khazanah Institute's Housing Market 'Findings',* The Sun Daily - 14 September 2015 (2015)

The Moderating Role of Risk Perception on The Relationship Between Satisfaction and Repurchase Intention of Bottled Water in Nigeria: A Research Proposition

Abdul Wahid N.[1,*] , *Mohaidin* Z.[1] and *Ojatta* D. O.[1]

[1]Graduate School of Business, Universiti Sains Malaysia (USM), 11800, Pulau, Pinang, Malaysia

Abstract. There has been a surge of bottled water consumption globally to date. This phenomenon requires investigation on why consumers intend to repeat purchase the product. While past literature has applied Oliver's expectation disconfirmation theory (EDT) to explain the link between customer satisfactions (CS) and repurchase intention (RI), other studies have reported conflicting findings. Thus, this study argues that risk perception (RP) should be added as a moderating variable between CS-RI which should improve our understanding of the RI behaviour in the context of EDT, bottled water consumption, policy and marketing strategy research. A new framework is proposed to explain the role of RP in EDT, with the aim of stimulating further enquiry.

1 Introduction

There is a surge of bottled water consumption across the globe to date. China, United States and Mexico were the top three country consumers of BW in a released compounded annual growth rate (CAGR) of bottled water between 2009 and 2014 [1]. China for instance consumed 15% more BW in 2014 than they did in 2009 (from 5,692.30 to 11,459.00 millions of gallons) followed by 5.2% for United States (from 8,453.10 to 10,874.80 millions of gallons) and by 4.6% for Mexico (from 6,891.60 to 8,645.30 millions of gallons) in the same evaluated period. The surge in consumption is also observable for Thailand, India and Indonesia with 20.2%, 13.2% and 12.5% CAGR of BW respectively. In sum, the global CAGR has increased by almost 7% amounting to 74 billion gallons or 2.83 trillion litres between the period [1] ahead of any other in the beverages category. Eighty percent of the growth occurred in only ten countries, namely the US (North America), Mexico, Brazil (South America), Italy, France and Germany (Europe). Others are China, Indonesia, Thailand, and India (Asia).

While the global increase of BW consumption may be due to consumers' perception that BW is safer for drinking than tap water [2], studies have reported violation of BW quality standard and guidelines across the globe[3-6]. The question is why do consumers still repurchase or have intention for repurchase the product whose quality is doubtful? The ability for researchers and practitioners to explain the RI behaviour will provide an insight to consumer behaviour literature, marketing strategy and policy research particularly as these help them understand customer retention (CR). CR has been linked to increased revenue [7], reduced cost of transactions [8], competitive edge [9] profitability [10-11], and long term growth [12].

A review of the literature indicates that the expectation disconfirmation theory (EDT) by Oliver [13] have been popularly adopted or adapted to help explain customer RI behaviours for examples [14-21]. However, research conclusions are emerging that CS with a product does not necessarily mean he will repurchase the product [22-23]. This brings about conflicting findings on the effect of CS on RI. Consequently, prescriptive model needs to be conceptualized. In this study, while all variables in Oliver's EDT are retained as the base of model, risk perception (RP) is proposed to moderate CS-RI relationship. Baron and Kenny recommend adding a variable as a moderator when findings on the relationship between variables are inconsistent [24]. Further, Oliver and Burke had also recommended modification of the EDT model which is discussed in the next section [13].

2 The Expectation Disconfirmation Theory

The EDT has dominated CS/D research over the past couple of decades; flaunted as one of the most versatile model for the study of human behaviours and has been applied in various study domains such as product [25], services [26], online studies [12], [27], governance and policy [28]. Others are education [29], health [30], and family life and marriage [31-32]. Though some earlier studies [33-35] had investigated the disconfirmation as a variable, much of the credit for the evolution of EDT

 https://doi.org/10.1051/matecconf/201815005089

are ascribed to the numerous works of Oliver and his colleagues [13].

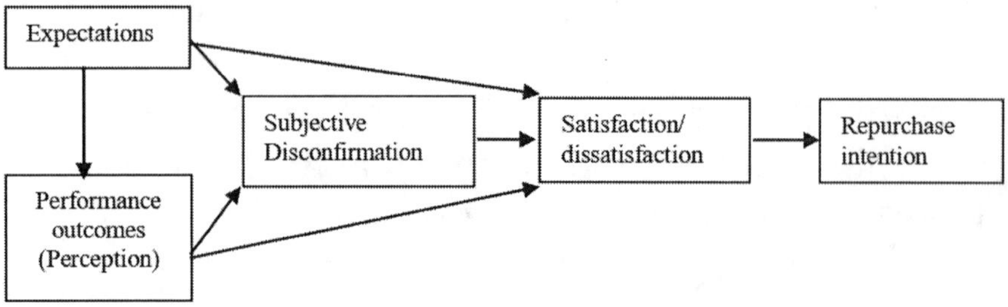

Fig. 1. A model of Expectation Disconfirmation Theory (EDT). Oliver and Burke (1999)

Precisely, the theory has been used or adapted to report assimilation effects for less experienced IT users (16) negative disconfirmation of Employment Assistance Services (EAS) among employees of South African Police Service (SAPS) and relationship between student's level of group satisfaction, engagement and positive performance [36]. Further, EDT has also been applied to investigate the predictors of web customer's repurchase intention in Taiwan [22], users' continuance intention towards e-learning [10], information systems (IS) users' continuance intention [14], as well as the effects of perceived customization and perceived sociability on CS and RI amongst consumers of virtual products [37]. To that effect, p.1190 [38], conclude that research in the field of consumer behaviour "has been grounded theoretically in the expectation-disconfirmation model".

Essentially, the theory is important because it has been used to explain the basis for customer assessment of product or service offerings, what factors deliver higher customer satisfaction and why consumers like or dislike certain offerings. It has also been used in explaining what customers do when they are satisfied and when they are not. In the model,
customer expectation (CE), the customers' pre-trial beliefs about a product [39] is posited as antecedent of performance outcome i.e. perception (CP), subjective disconfirmation (D) and customer satisfaction (CS). Similarly, CP which is a customer's subjective view of how a product performed compared to pre-trial beliefs, is modelled as an antecedent of both D and CS while D, tagged as a subjective feeling of how worse or better a product has performed compared to intial expectation is modelled as an antecedent of satisfaction. Finally, satisfaction is posited to be positively related to RI.

The theory holds that if CE, that is the standard against which CP, D and CS are measured is high, then CP, D and CS will likely be high because customers will assimilate any discrepancy that might have occurred (assimilation theory). This will likely occur under the conditions that the discrepancy is within the zone of or the level of effort expended in the purchase process is high [40]. This is also the position of cognitive dissonance theory [34]. [33] report high CE will lead to high evaluation. Also [35] confirmed that high CE is positively related to customer product evaluation. On the contrary, customers are likely to magnify the discrepancy (contrast theory) when a customer's product evaluation is lower than expectation [33], especially when the discrepancy is outside the zone of tolerance [40]. D and CP equally modelled as antecedents of CS, just as CS is modelled to predict repurchse intention (RI). It is believed that positive quality evaluation and disconfirmation will increase CS and vice versa, which affects a customer's decision to re-buy. Therefore, CS with product quality is regarded as vital because it is one of the major components of customer behaviour [33].

2.1 The gap

Despite the vast application of EDT in investigating consumer RI, the potential threat to the model is the faltering effect of CS on RI. Literature reports diversity of findings in the link between CS and RI. For example whereas some report direct link between CS and RI [13 , 41-43], others differ in their findings [22 , 44-45]. It does not matter whether all these studies adopted EDT or not, because deductive studies underlined by positivist world-views believe there is a "single reality". If the relationship between CS and RI has become inconsistent in some models, this could hold EDT, thereby making it essential to introduce a moderator in a prescriptive model.

Introducing RP as a moderator in the CS/RI link is predicated on the inconsistent findings on the relationship between CS and RI and the importance and relevance of RP as a predictor variable in customers' purchase decisions. The research tradition posits that when findings on the relationship between a predictor and a response variable are weak or inconsistent [23], a third variable called the moderator may be introduced. RP is an important predictor variable in customers' purchase decision [47]. Baron and Kenny state that moderator variables always function as predictor variables [23]. Previous research [48-49], confirms RP as a predictor. Further, related studies suggest that RP is a likely variable that could moderate the CS/RI link, because it is an important antecedent variable in consumption of bottled water

[50]. Specifically, Bauer "proposed that consumer behaviour be seen in the theoretical framework of risk taking" [51]. This study therefore introduces RP as a moderator between CS and RI link.

2.2 Risk perception

RP is the level of perceived, potentially negative consequences attached to a purchase decision. It is a decision maker's evaluation of the risk present in a situation [52]. It is also considered "the negative consequences that can arise from the purchase of a product" [53] RP is considered "fundamental to the understanding of consumer choice behaviour" [47]. that is likely to affect consumer behaviour" in e-commerce [54]. In the context of drinking water, it is a person's subjective evaluation of a drinking water quality in terms of aesthetic and non-aesthetic attributes [55]. A significant number of decisions involve some measures of uncertainty which can be positive or negative. If the uncertain future outcome includes likelihood of negative results, it can be said that the decision is risky. Contextually, the antecedents of RP of BW include organoleptics and external information among [50]. Concede that RP is informed by the level of awareness of drinking water problem [56]. Back to the EDT, since the only portion that is susceptible to threat of inconsistency is the CS/RI link, the remaining part of this paper focuses on RP and the CS/RI relationships.

2.3 The conceptual framework and proposition

Evidently, our prescriptive model is adapted from the EDT discussed in a previous section. From the literature, the predictors of CS include CE, CP and D as shown in Figure 2. The relationship between consumer satisfaction (CS) and RI is well documented in the literature. According to [41 , 23]. CS is a predictor of behavioural intention which includes repurchase intention. In another study it was established that customers' repurchase intention (i.e. intention to continue IS use) was predicated by satisfaction [14]. Additionally, it has been found that RI is related to satisfaction with previous visits [57], and directly affected by CS [42]. Report that satisfaction is associated with online continuance intention.) also associated CS with RI as confirmed by [59]. Recently too, Izogo finds that satisfaction primes repurchase intention [43].

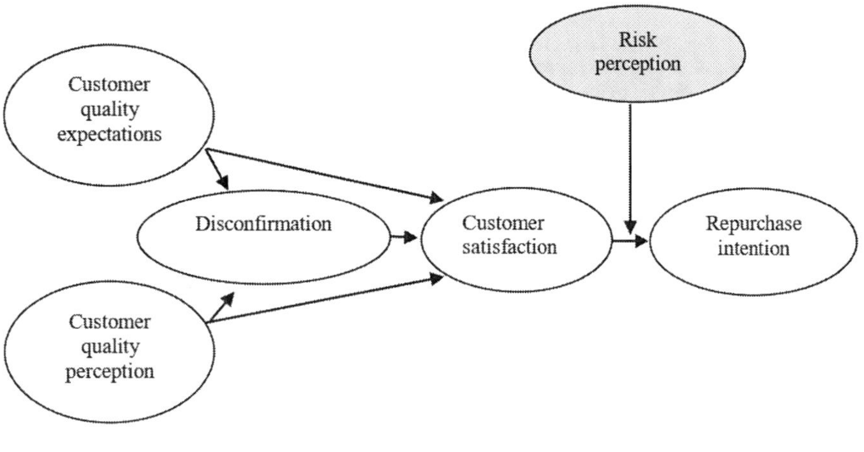

Fig. 2. Customer expectation and satisfactionSource: Adapted from (Oliver & Burke, 1999)

2.4 The moderating role of risk perception on CS/RI relationship

A moderator is a predictor variable which when introduced into a relationship between another predictor and a response variable, weakens, strengthens or changes the direction of the relationship. According to past studies [23] certain conditions must exist to justify introduction of a moderator in a research. One of such is that there should be inconsistency in findings about the relationship from the literature. Furthermore, introduction of a moderator becomes plausible if such relationship is found to be consistently weak.

In the situation under consideration, extant literature presents conflicting findings on the relationship between CS and RI. Whereas the normative model associates CS with RI [13 , 41-43], other [21-22 , 44-46] are divergent in conclusions. The literature indicates that RP can impact the relationships between variables in decision-making. Previously, it has been shown that RP moderate the relationship between incongruity and evaluations as well as being related to use intention. Featherman and Fuller reports that, moderating effect of RP such that higher levels of RP weakens the effect of ease of use on perceived usefulness and adoption intention. Additionally, observe a need for a theoretical basis for the integration of RP as probable direct influences on purchase intention. This necessitates introducing RP to moderate the CS/ RI link in this study. Based on above conditions, we propose as follows. Proposition: The higher the customers' risk perception, the weaker the CS/RI relationship.

3 Summary and Directions for Future Research

Understanding repurchase intention, a predictor of repurchase, is necessary because it is critical to firms' competitive advantage, consumer price tolerance, low operating costs, profitability and long term growth. Though prior research largely point to customer satisfaction as a major antecedent, findings on this relationship remains inconsistent. While some studies find direct association, others find indirect or none at all. To deepen our understanding of this phenomenon, this study develops a research framework and a research proposition that will support future research on consumer repurchase intention. To successfully accomplish this, a theory/industry interface is needed to test the framework. This leads to selection of the expectation disconfirmation theory (EDT) and the bottled water (BW) sub-sector. While the EDT is one of the most appropriate and versatile in the study of consumer behaviours, the BW sub-sector is the fastest growing in the beverage industry, with a market value of US$169.9 bn in 2015 which is expected to reach US$307.2 bn by 2024 [60].

The problem with this industry is that while consumers prefer BW because it is perceived as being free from health risks, empirical evidence points to significant quality standard violation by practitioners. Given this background, it becomes unclear why bottled water consumption is increasing as quality guidelines violation increased instead of the other way round. Based on these, this study advances a framework needed to aid future research on BW consumption, policy and marketing strategy, and extend the EDT. The framework links customer satisfaction and its antecedents to repurchase intention and expands the expectation disconfirmation theory by incorporating risk perception to moderate the link between CS and RI. The proposition that evolve out of the framework present opportunity for future empirical research. The issue in focus is understanding consumers' intention BW that is lacks quality assurance because of the violations of quality standard by practitioners. Researchers' and practitioners' ability to explain the consumers' RI behaviour will provide some understanding to consumer behaviour literature, marketing strategy and policy research, and customer retention (CR) which has been linked to increased revenue, reduced cost of transactions, competitive edge, profit possibilities and long term business growth.

References

1. J. G. Rodwan, International Developments and Statistics, ://www.bottledwater.org/economics/industry-statistics. (New York, 2015)
2. C. Fornell, J. Market., **56**(1), 6–21 (1992)
3. A. Azlan, H. E. Khoo, M. A. Idris, A. Ismail, M. R. Razman, Sc. W. J., 1–10 (2012)
4. Z. Hu, L. W. Morton, R. Mahler, Int. J. Envi. Res. Pub. H., **8**(12), 565–578 (2011)
5. O. A. Igbeneghu, A. Lamikanra, BMC Research Notes, **7**, 859 (2014)
6. W. K. Viscusi, J. Huber, J. Bell, Cont. Eco. Pol., **33**(3), 450–467. (2015)
7. V. Mittal, C. Frennea, R. A. Westbrook, Marketing Review St. Gallen, **31**(3), 46–54. (2014)
8. J. N. Cannon, C.A. Denison, Verifiable Detail as a Valid Source of Disclosure Credibility In Non-Financial Strategy. http://haskayne.ucalgary.ca/files/haskayne/Cannon-and-Denison.pdf (2014)
9. J. J. J. Cronin, M. K. M. K. Brady, G.T. M. Hult, *J. Retail.*, **76**(2), 193–218 (2000)
10. K. C. Murigi, K. Ombui, Int. J. Sc. Res, **3**(3), 192–196 (2014)
11. V. A. Zeithaml, L. L. Berry, J. Parasuraman, Aca. Market. Sc., **21**(1), 1–12 (1993)
12. C. Lin, W. Lekhawipat, Factors, Ind. Man. Data Sys., **114**(4), 597–611 (2014)
13. R. L. Oliver, R. R. Burke, Expectation Processes in Satisfaction Formation: A Field Study, **1**(3), 196–214 (1999)
14. A. Bhattacherjee, Dec. Supp. Sys., **32**(2), 201–214 (2001)
15. E. Ko, J. G. Aca. Market. Sc., **4**(1), 197–211 (1999)
16. N. K. Lankton, H. D. Mcknight, J. Assc. Info. Sys., **13**(2), 88–115 (2012)
17. E. Y. Li, J. C. Hsiao, J. A. Yang, H. R. Yen, The 9th International Conference on Electronic Business, Macau (2009)
18. V. Venkatesh, S. Goyal, MIS Quarterly, **34**(2), 281–303 (2010)
19. W. T. Wang, W. H. Chang, Info. Sys. Frontiers, **16**(5), 887–908 (2014)
20. Y. Yi, S. La, Serv. Ind. J., **23**(5), 20–47 (2003)

21. Chang K.C., M. C. Chen, C. L. Hsu, N. T. Kuo, Industrial Management & Data Systems, **110**, 1420–1443 (2010)
22. H. Saleem, N. S Raja, Middle - East J. Sc. Res., **19**(5), 706–711 (2014)
23. R. M. Baron, D. Kenny, J. P. Soc. Pscho. **51**(6), 1173–1182. (1986)
24. D. Baum, M. Spann, J. Füller, *Proceedings of the 21st European Conference on Information Systems*, 1, (2013)
25. Int, Rev. Retail, Dis. Con. Res., **6**(1), 1–33 (1996)
26. M. Kim, C. A. Vogt, B. J. Knutson, J. Hosp. Tour. Res., **39**(2), 170–197. (2015)
27. T. H. Poister, J. C. Thomas, J. Pub. Admin. Res. Theo., **21**(4), 601-617 (2011)
28. R. Jacobsen, J. W. Snyder J. Pub. Admin. Res. Theo., **25**(3), 831–848 (2015)
29. T. N. Cassidy-Smith, B. M. Baumann, E. D. Boudreaux, J. E. Med., **32**(1), 7–13 (2007)
30. J. Guttmann, A. Lazar, J. Repro. I. Psycho., **22**(3), 147–156 (2004)
31. D. M. Sagiv-Reiss, G. E. Birnbaum, M. P. Safir, M. P., Arch. Sex. Beha., **41**(5), 1241–1251 (2012)
32. R. N. Cardozo, J. Market. Res., **2**, 244–249 (1965)
33. L. Festinger, *A Theory of Cognitive Dissonance*, (Stanford University Press, 1962)
34. R. W. Olshavsky, J. A. Miller, J. Market. Res., **9**(1), 19 (1972)
35. C. Schwarz, Z. Zhu, J. Info. Sys. Edu., **26**(1), 47–59 (2015)
36. W. T. Wang, W. H. Chang, Info. Sys. Frontiers, **16**(5), 887–908 (2014)
37. V. Martínez-Tur, N. Tordera, J. M. Peiró, J. App. Soc. Psycho., **41**(5), 1189–1213 (2011)
38. V. McKinney, K. Yoon, F. Zahedi, F. Info. Sys. Res., **13**(3), 296–315 (2002)
39. V. A. Zeithaml, L. L. Berry, A. Parasuraman, J. Market., **60**(4), 31–46 (1993)
40. M. C. Lee, Com. Edu., **54**(2), 506–516 (2010)
41. E. W. Anderson, C. Fornell, D. R. Lehmann, J. Market., **58**(3), 53–66 (1994)
42. C. Huang, S. Yen, C. Liu, T. Chang, Int. J. Org. Inno., **6**, 106–121 (2014)
43. E. E. Izogo, Int. J. E. Market., **11**(3) (2016)
44. Y. Chen, H. Huaang, Y. Hsu, Com. Info. Sc., **3**(3), 14–22. (2010)
45. S. Jose, N. Khare, F. R. Buchanan, Int. J. Bank Market., **33**(3), 316–329 (2015)
46. L. P. Pleshko, R. A. Heiens, Int, Rev. Retail, Dis. Con. Res., **25**(1), 55–71 (2015)
47. V.-W. Mitchell, P. J. McGoldrick , Int. Rev. Ret., Dist. Con. Res., 6(1), 1–33 (1996)
48. D. T. Griffee, *An Introduction to Second Lanaguage Research Methods: Design and Data. TESL-EJ Publications Berkeley,* (eBook edit). (Berkeley, California, USA: TESL-EJ Publications, 2012)\
49. F. J. Hair, J. M. Sarstedt, L. Hopkins, V. G. Kuppelwieser, V., *A Primer on Partial Least Squares Structural Equation Modeling (PLS-SEM). European Business Review*, **26**, (Washington: Sage Publications, 2014)
50. M. de F. Doria, N. Pidgeon, P. R. Hunter, Sc. Tot. Envi., **407**(21), 5455–5464 (2009)
51. J. N. Sheth, M. Venkatesan, S. J. Market. Res., **5**(3), 307–310 (1968)
52. Sim B. Sitkin and Amy L. Pablo, Aca. Man. Rev., **17**, 1, 9-38 (1992)
53. F. V. Wangenheim, J. Ser. Res., **8**(1), 67–78 (2005)
54. A. Pavlou, Pavlou, Int. J. Elec. Comm., **7**(3), 101–134 (2003)
55. A. K. Harding, E. C. Anadu, J. A. Water Works Assc, **92**(8): 32-41 (2000)
56. M., Holm, Hansen., TOC, **12**(4), 333–352 (2013)
57. A., Camarero, L. García,, 1080/13683500.2014.936834 (2014)
58. N. Mohamed, Ramlah Hussein, Nurul Hidayah Ahmad Zamzuri, Hanif Haghshenas, Ind. Man. Data Systems, https://doi.org/10.1108/IMDS-07-2014-0201, **114**(9), 1453-1476 (2014)
59. M., Faullant, Renzl, Leiter, The Relationship between Personality Traits Extraversion and Neuroticism Emotions and Customer Self Satisfaction (2005)
60. Transparency Market Research., *Global Bottled Water Market : Rising Health Concerns to Drive Demand for Healthy and Hygienic Products , states TMR*, (New York, 2016)
61. M. Morgan, X. D., J. Inter. Res., **13**(2), 44–58 (2008)
62. M. F. Doria, J. W. H., ***27:232.3)***, 2, 271–276. (2006)
63. C. Ferrier, AMBIO: J. Human Envi., **30**(2), 118–119. (2001)

64. F., L., *A Theory of Cognitive Dissonance.* (Stanford University Press, 1962)

65. B. Karoubi, R. Chenavaz, C. Paraschiv, App. Eco., **48**(14), 1317–1329. (2015)

66. K., E., J. G. Aca. Market. Sc., **4**(1), 197–211. (1999).

67. N. Lankton, D. H. McKnight, J. B. J. Thatcher, J. S. Info. Sys., **23**(2), 128–145. (2014)

68. E. D. Olson, *Bottled Water: Pure Drink or Pure Hype? National Resources Defense Council.* https://www.nrdc.org/sites/default/files/bottled-water-pure-drink-or-pure-hype-report.pdf. (1999)

Religiosity and Volunteering Intention Among Undergraduate Malaysian Muslim Students

A.A.A. Sallam[1,*] , *S.* Abdullah[1], *A.J.* Ramli[1], *N.S.* Hussin[1], *Z.* Ahmad[1], and *A.* Bahari[1]

[1]Universiti Malaysia Perlis, Perlis, Malaysia

Abstract. This paper deals with the question: To what extent do religiosity characteristics, contribute to the influence of volunteering intention among Malaysian Muslim students during disasters? To answer this research question, we focused the students in public universities. The finding concerns found that religiosity increases the likelihood of volunteering intention, implying that religious affiliation of youth increases the likelihood of volunteering. This is in line with previous research, that religious attendance is related positively to volunteering. These results confirm the idea that support of the religious attributes community plays quite a large role in volunteering process. . However, it a bear that volunteering is not only dependent on religious community, but also on individual motivation.

1 Introduction

Lately, religiosity has emerged as an attractive field of research [1] and there has been increasing interest in the relationships between various aspects of religion [2-3]. Several studies have systematically tried to unveil the underlying dynamics of religiosity in organizational life [4-5]. The influence of religion and spirituality on individual's behaviour has emerged as an area of interest for researchers [6]. In fact, for NPO (non-profit organization), young people are considered a growing and strong volunteer source and essential for the efforts of non-governmental organization (NGO) marketing [7].

In volunteer work, the participation of Australian younger generation is considered lowest in the age group 25–34 [8]. Although there are about 135,000 full-fledged volunteers in Malaysia, the figure only represents less than 1% of youths aged above 15 [8]. For the explanation and understanding related to youth volunteering, the focus of an individual must be on the factors influencing the motivation or encouraging youth for being a volunteer and positive youth development [9]. A study of youth volunteers of Anti-Drug Program findings shows that most of the participants did not know the membership procedures of this Youth Volunteers for the Anti-Drug Program. It is believed that there are some factors that affect students' inactiveness or lack of interest in voluntary activities and in their intention to volunteering. And they should be given the opportunity to participate in various aspects [10]. Also important to develop a student college volunteer [11].

There are very few studies regarding of volunteering intention among college student [6 , 11]. There is also a lack of research concerning volunteers [12] in developing countries [11], and the relationship between religiosity and volunteers [12]. Recently, the field of testing the influence of religion and spirituality to the psychological behaviour of an individual has been of interest [6 , 13-14]. Despite the researchers scepticism, spirituality and religiosity are considered the influencing factors of emotions, behaviours, cognition, merit of an individual [6 , 15-16]. And volunteering behaviour [17]. Moreover, there are many studies of religion and mental health in the context of Christianity context, but studies in Islamic countries are few [18].

Based on the observations, a gap has been found in the literature related to the understanding of the factors influencing the intention of college or university students to volunteering. This study is an attempt to find the literature gap for the investigation of the influencing of religiosity on volunteering intention among Malaysian Muslim students. It will provide for Malaysian organizations, donors and interested ministries the tools and techniques for development of volunteering behaviour among youth, consistent with previous volunteering researches [19-20].

1.1 Volunteering Intention in Malaysia

Volunteerism is a method to promote active learning, providing experience which helps in the enhancement of commitment to contribute among citizens, the education experience related to expansion, serving the communities more appropriately. Youth volunteerism is very crucial in developing a community.

The majority of people joining volunteering activities are retirees. In a study of volunteerism and the development of Malaysian social care system found 4% had 21-30 years active volunteers and 1% had 31-40 years active volunteers. Time is considered the barrier for greater in the volunteer practices among younger generation. The issue of younger generation being volunteers is of utmost importance because of their higher level of energy and motivation which can result in good outcome.

1.2 Purpose of Study

Although, studies related to volunteering, in previous three decades, have increased [21], little is known about volunteering [22]. Also, about the social and psychological features of volunteers, volunteering motivations, psychological influence [21], and the causal relationship and intentions related to volunteering [19].

The purpose of this study is to examine the relationship between volunteering intention, and how factors of religiosity influence this intention among college students. Several studies have focused on intention or volunteering behaviour of college students [23-24]. A few studies have examined religious preferences and spiritual beliefs among college. However, this study will examine the relationship between volunteering intention and the influence of religiosity specifically among the universities students. There sis notes that, there are very few published studies about religion and health among young persons in Islamic countries [18].

1.3 Religiosity

The term religion in this study means the institution serving as a source of knowledge, practice, and beliefs of an individual. From scholarly perspective, religiosity is the level of individuals being religious, and the importance an individual attached to beliefs and practices of religion. Sociologists of religion have struggled with the definition of religion, but agree that religion is more than simply the belief in the supernatural, god, or gods. Religion consists of practices and institutions that have moral purpose. In other words, religion is the organization that provides a set of beliefs, instructions on appropriate practices, and support for individual members, while religiosity is the acceptance of and commitment to those elements provided by religion. It is a collection of organized practices, norms, and values.

Adam Smith in his Theory of Moral Sentiments analysed religiosity from a rational point of view and noted that religiosity acts as a kind of internal moral enforcement mechanism. Individuals characterized as religious are not only those who hold particular religious beliefs but also practice them in day-to-day life [25-26]. For example, religious people pray, attend a place of worship, in some cases observe fasting and engage in socially beneficial actions (e.g., charity and volunteer work), political and economic activity [27]. As directed or encouraged by their religion [5 ,28]. Many define religiosity as both beliefs and practices relating to an organized religious institution or a specified divine power [29] or to a religious affiliation or to God [30].

1.4 Religiosity and Volunteering Intention

Religion is another important predictor of volunteering [8]. Although, there are differences by types of religion and religious denomination, most religions encourage altruism, helping behaviour [28 , 31] and encourage cooperation between members [32-33] research has shown that religious attendance in particular is related to volunteering behavior [34-36]. It has been shown that the relationship between religiosity and volunteering is mediated by denominational involvement, and is not a direct result of prevailing religious norms and teachings. Crucial are social networks that provide contacts and enhance norm conformity within the group. As well as when investigating influences of individual religious characteristics, it is highly important to control for collective religious characteristics [23] The General Social Survey conducted in 2010 by the Australian Bureau of Statistics shows that volunteering with religious groups was higher among the young (aged 18–24) and older age groups (aged 55–74) and lower in the age group 25–34 [8].

In a study on the role of religiosity for formal and informal volunteering in the Netherlands, found that spirituality increases the likelihood of informal volunteering, implying that openness to other people's needs increases the likelihood of the actual provision of help [23]. They also found in formal volunteering the religious attendance is related positively to formal volunteering, religious as well as secular volunteering [23].

In similarity [23-24 , 36] found current institutional religious involvement emerged as a positive predictor of the likelihood that men would be involved in volunteer work, as well as of the amount of time that men gave to this kind of work. And Men who are more involved in church life or houses of worship may have a higher probability of being exposed to (or invited to participate in) general volunteer activities. Taniguchi & Thomas, argue that religious exclusiveness has a significant positive influence on volunteering only in religious areas [38]. Meanwhile, religious inclusiveness is significantly and positively associated with both religious and secular volunteering.

Volunteering for religious organizations shows an interesting pattern in that there are higher odds of participating for those in the range of 18–24 years, controlling for other factors. None of the other life course variables shows a strong relationship with this type of volunteering. Unsurprisingly, an individual's level of religiosity is strongly associated with this kind of volunteering [8].

2 Research Methodology

2.1 Participants

A total of 356 undergraduate students (246 females and 110 males) from two public universities participated in this study. Participants ranged in age from 18–26 years (mean = 1.2051), and were primarily Malay (97.2%) with a minority of others (2.8%). Most of the respondents came from Kedah which account for (11.2%) and Sarawak with the lower respondent rates of (0.6%) of the total students represented. The results showed that respondents consist of 98 (27.5%) first year students, 113 (31.7%) second year students, 90 (25.3%) third year students and 55 (14.6%) fourth year students. Students were recruited from different colleges from two universities Universiti Malaysia Perlis (UNiMAP) and Universiti Utara Malaysia (UUM) and were asked to complete the following self-report questionnaires. The number of the samples taken from the undergraduate students consisted of 400 respondents which were randomly chosen. 200 sets of questionnaires were distributed, 356 sets were successfully collected back.

2.2 Measures

Computer software – Statistical Package for the Social Sciences (SPSS) Version 19.0.1 was used for data analysis. This software was used in this study for data management and analysis. Religiosity towards volunteering intention Scale was developed for the present study under the guidance and supervision of the Universiti Malaysia Perlis. The scale contained 11 items adapted from [23] studies (Table 3.1). Each item was rated by the respondent on a five-point Likert scale ranging from strongly disagree (1) to strongly agree (5) so that scores on the total scale had a possible range of 18 to 26, with higher scores indicating a more positive religiosity to volunteering intention. Internal reliability was found to be adequate for religiosity (Cronbach's alpha = 0.94) and volunteering intention (Cronbach's alpha = 0.87).
The questionnaire has been developed for measuring religiosity in volunteering intention, a crucial distinction from other measures available which focus on religiosity toward intention to volunteering. Questions include items such as "My religious faith encourages me to volunteer." and "My beliefs about God give me strength to help those who needed assistance." The measure can be found in Table 1.

Table 1. Measurement of Religiosity.

Items
My religious faith encourages me to volunteer.
I look to my religion as a source of inspiration to volunteer in humanity/community services.
My religion provides meaning and purpose towards volunteering activities.
My religion place a high value in helping the needy.
My religion influence my intention to volunteer.
My religious beliefs make me willing to volunteer.
I feel calm and happy if I involve in humanity/community service for the sake of my religion.
I must help the needy in order to please God.
I intent to give money for humanity/community activates because it is highly encouraged by my religious beliefs.
I intend to volunteer to help others because of my religious beliefs.
My beliefs about God give me strength to help those who needed assistance.

Source: Adapted from Freiheit et al. (2006); Hernandez (2011)

3 Results

The mean for volunteering intention variable which was 4.0314 and religiosity was 4.4941 indicates that the respondents on average agreed with the questions in general. The standard deviation value for volunteering intention is .69127, which is large from mean value and this indicates standard deviation for volunteering intention is less concentrate. Meanwhile, standard deviation religiosity is .54825. The standardized coefficient is 0.458 for religiosity. The value of R-Square indicates that .210 of the variance in volunteering intention can be predicted from the variables of religiosity. The significant F value of 0.000 which is significant at $\alpha = 0.05$. Religiosity value $t = 9.695$, $p = 0.000$ is significant and the beta value is .587. As the conclusion the religiosity have significant influence on volunteering intention.

4 Discussion and Conclusion

In this study, religiosity has a positive and significant effect among Malaysian Muslim students. For religion attributes, it is high correlation and has a significant value. This result is also in line with previous studies that suggested that religious attendance is related positively to volunteering. The elements of religiosity across different religions serve beliefs and practices [8 , 23-24 , 36 , 38]. This result is similar to study by Berger, 2006; and [38] where religiosity has positive relationship with volunteering intention. Therefore, viewed as being a follower of a religious group may be the reason for the increase in the number of invitations for volunteering, and increases the knowledge related to volunteering [39]. In addition, religiosity has many interactions with other factors [3 , 23]. However, there are some limitations, the study focused on limited sample for data collection and the sampling of the study that involved the undergraduate Muslim students of the public universities. Because of this reason, the findings may be applicable only to the university students with higher knowledge and better educational and knowledge background. Thus, the study managed to evaluate students' intention from the public universities students' point of view only. It limits the generalizability of the findings. Therefore, it is suggested that future research can be done in larger sample size which include multi-ethnic and other religions that would allow higher response rate for more robust statistical analyses including structural equation modelling.

References

1. R. A. Emmons, R.F Paloutzian, Annual review of psychology, **54**(1), 377-402 (2003)
2. A. F. Shariff, A. K. Willard, T. Andersen, A. Norenzayan, Perso. Soc. Psycho. Rev., 1088868314568811 (2015)
3. S. Vasegh, M. R. Mohammadi, Int. J. Psych. Med., **37**(2), 213-227 (2007)
4. M. Benefiel, L. W. Fry, D. Geigle, Psycho. Rel. Spirit., **6**(3), 175 (2014)
5. F. Ntalianis, W. Darr, Int. J. Org. Ana, **13**(1), 89-102 (2005)
6. M. S. Harcrow, *Relationships between religiosity, spirituality & health behaviors among college students* (The University of Alabama , 2010)
7. E. Briggs, T. Landry, C. Wood, C., J. Non-p.Public Sect. Market, (2007)
8. E. Gray, S. E. Khoo, A. Reimondos, J. Popu. Res., **29**(4), 373-398 (2012)
9. S. Goudeau, *Principal Motives of Positive Youth-Adult Relationships: A Model for Identifying the Motives of Adult Volunteers and Youth-Adult Relationships in Physical Activity-Based Youth Development Programs* (Doctoral dissertation, Emporia State University, 2015)
10. N. A. Suandi, H. A., Suandi, T., Hamzah, S. R. A., Ismail, I. A., Factors Associated with Youth Engagement and Participation in Voluntary Activities in Malaysia (2015)
11. A. Nazilah, I. Rozmi, I. Fauziah, Soc. Sc., **12**(9), 139 (2016)
12. M. E. Abreu, R. Laureano, R. V. Da Silva, P. Dionísio, S.S.A. Faridah, The Drivers of Donations Practices in Religious and Secular Organisations, SSRN 2380085 (2013)
13. A. Agorastos, C. Demiralay, C. G. Huber, Psycho. Res. Behav. Man., **7**, 93-101 (2014)
14. G. Brewer, S. Robinson, A. Sumra, E. Tatsi, N. Gire, J. Rel. Health, **54**(6), 2225-2234 (2015)
15. Khayruzzaman, Int. J. Fin. Bank. Res.. Vol. **2**, No. 1, 2016, pp. 18-23. doi: 10.11648/j.ijfbr.20160201.14 (2016)
16. F. T. Newaz, K. S Fam, R. R. Sharma, J. Fin. Serv. Market., **21**(2), 141-152 (2016)
17. E. A. Minton, L. R. Kahle, T. S. Jiuan, S. K. Tambyah, J. Sci. St. Rel., **55**(2), 365-383 (2016)
18. S. Vasegh, M. R. Mohammadi, Int. J. Psych. Med., **37**(2), 213-227 (2007)
19. J. P. Hill, K. R. den Dulk, J. Sci. St. Rel., **52**(1), 179-197 (2013)
20. R. T. DeAngelis, G. A. Acevedo, X. Xu, Rel., **7**(6), 74 (2016)
21. N. Morrow-Howell, Psycho. Sc. Soc. Sc. , **65**(4), 461-469 (2010)
22. M. E. de Abreu, R. M. S. Laureano, S. F. Syed Alwi, R. V. da Silva, P. Dionísio, Managing Volunteerism Behaviour: The Drivers Of Donations Practices In Religious And Secular Organisations (2015)
23. M. Van Tienen, P. Scheepers, J. Reitsma, H. Schilderman, International J. Volun. Nonp. Org., **22**(3), 365-389 (2011)
24. R. D. Putnam, D. E. Campbell, S. R. Garrett, S. R., Simon and Schuster (2012)
25. E. Olson, P. Hopkins, R. Pain, G. Vincett, Ann. Assc. A. Geo., **103**(6), 1421-1436 (2013)
26. P. Hopkins, E. Olson, M. Baillie Smith, N. Laurie, Transactions of the Institute of British Geographers, **40**(3), 387-398 (2015)
27. P. Cloke, J. May, S. Johnsen, *Swept up lives: re-envisioning the homeless city,* (John Wiley & Sons, 2011)
28. N. Krause, Assessing the Religious Roots of Volunteer Work in Middle and Late Life. Research on aging, **37**(5), 439-463 (2015)
29. K. I. Pargament, H. A. Raiya, Psyke Logos, **28**(2), 25 (2007)
30. E. Dingemans, E. Ingen, Does Religion Breed Trust? A Cross-National (2015)
31. K. M. Loewenthal, Psychology and Charity (2015)
32. K. Coe, C. Keller, J. R. Walker, J. R., J. Rel. Health, **54**(1), 46-60 (2015)
33. H. J. Oh, S. J. Lee, K. S. Moon, K. Hong, S. W. Kim, S. W., H. P. Kang, J. Christ. Edu. Info. Techno., 28 (2015)
34. P. Vermeer, P. Scheepers, M. te Grotenhuis, M, VOLUNTAS: International Journal of Voluntary and Nonprofit Organizations, **27**(3), 1361-1384 (2016)
35. Y. I. Kim, S. J. Jang, Nonprofit and Voluntary Sector Quarterly, 0899764016655619 (2016)
36. L. Hustinx, R. Van Rossem, F. Handy, R. A. Cnaan, Religion and volunteering: complex, contested and ambiguous relationships (2015)
37. J. S. Mattis, W. P. Beckham, B. A. Saunders, J. E. Williams, V. Myers, D. Knight, C. Dixon, J. Adult Dev., **11**(4), 261-272 (2004)
38. H. Taniguchi, L. D. Thomas, VOLUNTAS International Journal of Voluntary and Nonprofit Organizations, (2), 335-355 (2011)
39. I. E. Berger, International Journal of Voluntary and Nonprofit Organizations, **17**(2), 110-127 (2006)

Enhancing Student's Understanding in Entrepreneurship Through Business Plan Simulation

M. Guzairy[1,*] , *N.* Mohamad[2], and *A. R.* Yunus[2]

[1]Faculty of Business Management, Kolej Universiti Agrosains Malaysia, Melaka, Malaysia

[2]Faculty of Technology Management and Technopreneurship, Universiti Teknikal Malaysia, Melaka, Malaysia

Abstract. Business Plan is an important document for entrepreneurs to guide them managing their business. Business Plan also assist the entrepreneur to strategies their business and manage future growth. That is why Malaysian government has foster all Higher Education Provider to set entrepreneurship education as compulsory course. One of the entrepreneurship education learning outcome is the student can write effective business plan. This study focused on enhancing student's understanding in entrepreneurship through business plan simulation. This study also considers which of the factor that most facilitate the business simulation that help the student to prepare effective business plan. The methodology of this study using quantitative approach with pre-and post-research design. 114 students take part as respondent in the business simulation and answer quantitative survey pre-question and post question. The crucial findings of this study are student characteristic factor after playing the simulation contribute much on facilitate business plan learning. The result has shown that the business plan simulation can enhance undergraduate student in understanding entrepreneurship by preparing effective business plan before opening new startup.

1 Introduction

Bank Entrepreneur play a crucial role in economic development on Malaysia targeting to be developed country in year 2020. Various action has been implement by the government to produce more entrepreneur in this country. To expose the citizen with the entrepreneurship, most of the higher learning offer entrepreneurship programs and courses. They also fostering the student to open small business in campus. When we talk about start up a business, the main reason why people afraid to open a business is because of financial sources. Entrepreneur need to have sources of capital for business start-up, however to obtain the financial support is not easy although there are many sources of finance from local bank and government agencies. The requirement or the condition to obtain the funding is to prepare business plan. Business plan are the crucial element that will determine the amount fund that the entrepreneur eligible from the funders.

In entrepreneurship education, student need to learn how to prepare effective business plan. Business plan can be categories as a part of most important factor that can determine the business process growth and sustainability [1-3]. The need of preparing business plan often be the first criteria that will presented by the entrepreneur for get financial assistant from bankers and investors [4-5]. Nowadays entrepreneurship courses are compulsory for the most of Malaysian Higher Education Provider. They must run the Entrepreneurship Program to expose and inculcate students to the entrepreneurial skills and values, which contain of the element of leadership, creativity, innovation, competitiveness, resilience, calculated risk, independence, and able in creating and identifying the opportunities. The objective of this study is to recognize the factor and process in business plan simulation that facilitate business plan learning.

One of the element in studying entrepreneurship course is business plan that remain major objective entrepreneurship program [1]. In certain Higher Education Provider (HEP), they offer business plan as stand-alone subject and it became final year team project. The 4th issue state by the Malaysian Entrepreneurship Policy show that there are still occur lack of effectiveness of entrepreneurship and education and program. In other hand, the third element of Entrepreneurship Development Policy is to Strengthen Entrepreneurship that aim to expose students to the real world of business at the same time assisting them that decide to choose a startup business as a career. Entrepreneurship education in Malaysia Higher Education also is some of the General Subject (MPU) that was established September 2013.

1.1 Entrepreneurial Education

Providing entrepreneurship course as compulsory subject for undergraduate program in most of higher

https://doi.org/10.1051/matecconf/201815005091

education institution will increase the figure of the student that taking entrepreneurship education. Entrepreneurship education play vital role in developing entrepreneurs however the ability to foster entrepreneurship value with traditional teaching method still unclear and the possibility still arguable [6-7]. Entrepreneurship education nowadays still deliver in traditional method in classrooms as teacher centered learning that using didactic approach and the student gain knowledge when the teacher is delivering knowledge [7-8]. In this case the knowledge transfers only occur in a single method and lack of student participant that mean lack of student centered learning. The effectiveness of an entrepreneurship education courses relies mostly on educator's skills and knowledge of different pedagogy especially entrepreneurship teaching method [9]. While Ismail and Ahmad said that in inculcating the entrepreneurial spirit for enterprises education require several innovations in the content, exposure to real work life, assessment method towards more practical and teaching mode [8]. Both discussing on teaching method that will give the effectiveness on delivering entrepreneurship education. With entrepreneurship education student, especially the undergraduates that tend to start business career will help them be prepare on every step they pick to start new enterprise.

1.2 Business Plan Courses

Business plan can offer a roadmap for the future and advantage potential investors to invest the growing business [10]. Business plan is one of the major courses provide in entrepreneurship education define business plan as planning process outcome [1 , 9]. Writing business plan enable to prepare organizations and to predict well for future challenges [2]. In an educational aspect, the business plan plays as a simulation of the process of developing a business, including: (1) confirmation of the need for a given product or service; (2) funding sources, analysis of the financial requirements and potential returns; (3) evidence of a distribution and marketing plan; and (4) evidence of a team with the talent necessary to execute the plan. It is agreed by [11] that business plan as a condition of most potential funders before deciding to invest.

1.3 Experiential Learning

Experiential processes such as reflection, action and conceptualization are used via pedagogy, which include or example, action learning, group work, learning contracts, and self-developme [12]. A study on experiential learning in scope of management, by Kolb has utilized Experiential Learning Theory to make decision and solve problem, strategy formulation and entrepreneurial opportunity finding (see figure 1) [13]. In determining an experiential learning product whether is a true simulation or not is assessed by the degree to which the product has external conformity

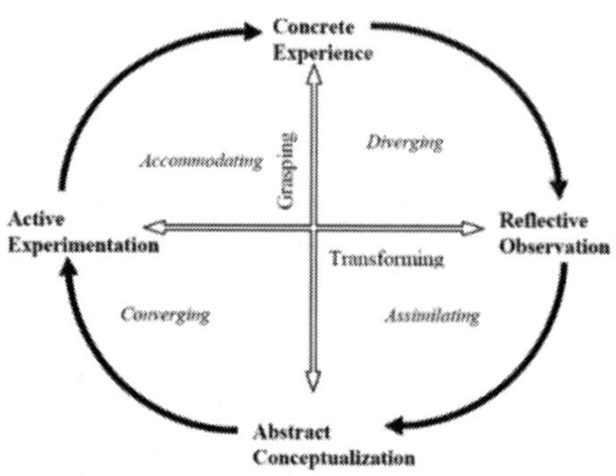

Fig.1 Experiential Learning adapt from Kolb and Kolb, 2008

and thus truly shows a real business environment. Ismail define experiential learning as a learner-centered approach that caters to individual learning styles [8]. Based on study they see that experiential learning provides students a method for adapting and developing their attitudes [14], skill and knowledge. Dellaportas and Hassall, found that to produce long-term learning outcome, experiential learning can be a powerful learning tool that can be apply in education [15]. Learning is categorized as a process, not as an outcome. To enhance learning in higher education, the primary interest should be on attracting students in a process that best improves their learning. Learning is best ease by a process that produce out the students' idea and beliefs about a topic and then they can be tested, examined, and together with new, more refined ideas. All students agreed that experiential learning is important in assisting them to build innovative skills and creative thinking [16]. Ismail and Ahmad mention that the objective of experiential learning is to develop an environment in which students actively engage with the entrepreneurial process rather than just read about it [8]. This process of experiential is representing as an idealized learning spiral or cycle where the learner thinking, reflecting experiencing, doing in a relating process that is reacted to the learning situation and what is being learned. Research showed that experiential learning contributes to enhancing students' employability and workplace readiness [17].

1.4 Business Simulation

Business Simulation courses were important to cope with the compulsion from stakeholders, but are still crucial because they show an innovative way of education. Simulation games are all quite the same in that they need input from the students, process the information and confront the participants with a certain consequence of their decisions, both in right terms of liquidity status, profit, loss, or market share, and about other virtual rivals [18]. Simulation can bring impact to the path of education, it brings the player or the participant to use their current knowledge to face the simulated business world. They tend to enhance their knowledge and skill and the same time learn new thing to stay survive. Considering management area in handling business, simulation enable player to understanding the management function in business environment. Load with experience, most of entrepreneur first knowledge is from their experience, learning by experiencing will allow student to improve their skill especially in business world. Business simulations are effective experiential method that promote business concepts, to appreciate a cross-functional understanding of business and to increase decision-making [19]. Participate in business simulation is an experiential learning that the student can stimulate and feel. Teaching strategies that would facilitate this decision; however, current development efforts, the simulations and games to get more attention from the teaching [20]. Therefore, the team can create a social, a serious game that teaches the concept of identity in an environment of business who has multiple aspects of online communities [21]. Tao found learned that many teachers no longer use business simulation games (BSG), or assistant teacher had just asked them to do a task or a BSG as competition without personal involvement, because of the high complexity management game time and investment necessary to facilitate the process [22]. In addition, by grounding gaming activities supported by project-based learning pedagogy, the characteristics of digital games can be in line with the intended objectives of education and thus facilitate the achievement of their [23]. Kulkarni state that [19] most business simulation enable student to manage a firm in competitive environment; (1) compete with other firm in the same industry, (2) produce, manage resource and market goods to achieve profit, (3) Each team compete each other and require to run the firm by deciding across functional area (marketing, R&D, HR, manufacturing, finance). Simulation create role play by student to act as new character in business area. Business simulations content give a range of learning as an outcome and also developing knowledge, at the same time developed skills after playing the simulation game [24]. In Juan Sebastián and Carlos Mario, view that simulation game is an active learning approach for teaching strategies [25]. According to Tawil, the simulation game provide understanding to decision making that is involved in business life and the effective of a real business [26].

2 Methodology

Quantitative designs adapt from [27-28] were used to find which of the elemen of business simulation that facilitate of business plan preparation and the relationship among them. Students of entrepreneurship courses (114 students) were selected to run business simulation play testing. Students were instructed to play business simulation using computer-based simulation. Before they run the business simulation they need to answer questionnaire of pre-session after playing the business simulation they need to answer a post-session of questionnaire. The objective of the questionnaire was to evaluate the perceptions of the students [22 , 29] towards business simulation games to facilitate business plan learning and its impact between traditional method and after using simulation.

2.1 Sampling

The sample for this study is Entrepreneurship Course students from Bachelor Degree program and Diploma program. The simulation game is Business Plan Game that use Visual Basic for Application (VBA) in Microsoft Excel. This game designed to create simulation business process that follow requirement of Malaysian business entity development. The business simulation game still in prototype mode and will be update after analyze the feedback from the player.

2.2 Measurement

Students were needed to answer a set of pre-test and post-test questionnaire. The measurements were adapted from the works of [30-32]. The questionnaire is divided into 3 sections. Section 1 comprises background and sosio-demographic of student's education and experiences, Section 2 is to assess the traditional method of learning business plan. Section 3 is to test the factor of simulation that facilitate business plan learning. Measurements in section 2 and 3 were based on 5 point Likert scale from highly ineffective strongly disagree to highly effective strongly agree.

2.3 Background of the Business Simulation

The business simulation is design to assist student for make business plan preparation. In other word that this simulation focus on the business setup process, and strategical planning of business function area. The development of this business simulation using Microsoft excel VBA (Visual Basic for Operation) [33]. In this simulation player can play alone or in group [28]. This simulation require player to set up a company, capital, manage resource and manage financial. Player need to carefully determine the allocation of expenses and capital that need to run the business (see Figure 2). Basically, in this simulation it has already put 3 default businesses that the player can choose. The business that set default in the simulation is Small Medium Enterprise (IKS) production like

Starfruit Juice and Catfish Cracker. The Simulation have four-part Administration, Marketing, Operation and Financial. This simulation allows student or participant to understand the process of starting business before they can convert it into strategical business planning.

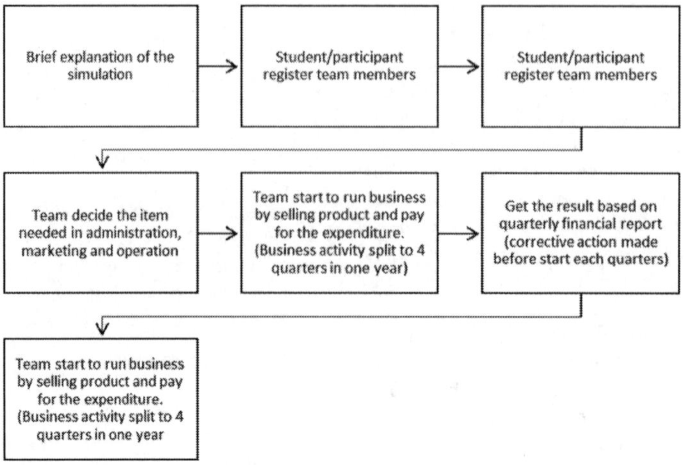

Fig 2. Flow chart of the business plan simulation

3 Results and Analysis

3.1 Respondent Background

The frequencies analysis shows that most of the participant were female 74.5% instead of male only 25.4 % and most of them were 20-25 years old. From the analysis, also there are 43% of management student that involve in preparing business plan of entrepreneurship. 38% of them also are from the matriculation program and about 42% of the participant get 71.9% score CGPA 3.0 and above. 114 participants agree that the business plan is a crucial part in entrepreneurship education. 29% of the participant state that lecture is the traditional way to learn business plan. Before the simulation session begin 80.7% of them decide that the traditional method is effective of learning business plan, however 93% of them interested of using business simulation method in learning business plan.

3.2 Student's Characteristic

Student characteristics is one of the factor that can facilitate business plan learning through playing simulation[34-35]. Its show that how can the characteristic of the student give impact to the learning and how they can shift or upgrade their skill and knowledge from experiencing the business simulation. It is including writing skills, decision making, problem solving etc. Gain new knowledge and ability to recognize the problem show high score of percentage that the participant agreed. Following are the feedback from questionnaire that show student characteristic gain from simulation towards effective business plan.

Table 1. Characteristic of Student (Skills and Knowledge)

Items	Agree %
Help to increase ability to write business plan	84.2
Increase ability to seek for information	82.5
Learn of make decision from various sources	80.7
Gain more knowledge of business plan	80.7
Increase ability to make business decision	82.4
Increase ability to recognize business problem	78.0
Increase skills and knowledge of real business	76.3
Increase potential of planning business competencies	83.3
Increase problem solving skills	78.0
Gain new business knowledge	89.5
Integrating various functional business knowledge	86.9

3.3 Educator's Roles

Educator like teacher, instructor and lecturer are involving in delivering the knowledge [36-37]. The simulation itself might be not complete enough if the educator not involve. The human capital of delivering knowledge still relevant enough this day even though we are surrounding by the technologies. In this result show that student need the assistant from the educator even though they choose to have own group and read the simulation manual. Explanation from the lecturer show high score of percentage level of agree from the student. Following are the feedback from questionnaire that show educator roles that take part of simulation towards effective business plan.

Table 2. Educator's Roles

Items	Agree %
It's hard to run the simulation without clear explanation from the lecturer	71.1
Information given from the lecturer not enough to run the simulation	38.6
The simulation's manual need to be distribute before beginning the simulation	70.1
The student need more times to read the manuals	73.7
The group must be selected by the lecturer	41.2
Student need to have a right to choose their own teams members	76.3
Lecturer need to be always monitoring and give response every single time	81.6
Student need to give their freedom on running the simulation without the lecturer around	43.0

3.4 Simulation Game Characteristics

Previous analysis shows the side factor that effecting the simulation process however the simulation itself is the most important thing that will prove either the simulation working or not to facilitate the business plan learning [38]. The element that contain in the simulation need to be in the right place time and for the right person. As show on the analysis below, student agree that the simulation meet the requirement of business plan preparation and give them the real business experiences. However, most of them not really understand the term and condition of the simulation, they also agree that the time given is not enough and hard to work in group due to that issue. Following are the feedback from questionnaire that show simulation game characteristic towards facilitating effective business plan:

Table 3. Simulation Game's Characteristics

Items	Agree %
It's hard to understand the simulation term and condition	38.6
The information on the manual not help enough	38.5
I don't really familiar with the default business in this simulation	51.8
The turnover time of this simulation take too much longer	41.2
The default business in this simulation is meet the business plan requirement	72.8
The simulation gives us the real business experiences	74.5
The time given to run this simulation is not enough	52.7
This business simulation is design for individual not for the group	46.5
It's hard to work within the group due to lack of time and sources	41.2
Student need to have one-week preparation before run the simulation	50.0
The rating and marks of the simulation distribute properly	62.3

3.5 Infrastructures

The infrastructure provide in the classroom are also one of the element that will affect the process of the simulation. The infrastructures referring on the facilities that provide to ensure the process of the business simulation to run smooth. Almost all the student agree that computer aid simulation make their work in group easy and able to search for extra information before make any decision. Following are the feedback from questionnaire that show the importance of classroom infrastructure towards preparing effective business plan:

Table 4. Infrastructures

Items	Agree %
In class group table layout assist the process of the simulation	61.4
Internet access in classroom assist the process of gathering information	59.7
Computer slide presentation help to deliver simulation information	80.7
Ease of extension socket that enable the use of laptop	75.4
Computerize simulation helped students	87.7
Integrating various functional business knowledge	86.9

3.6 Experiential Learning

Experiential Learning is the core that determine the value of the simulation. Most of the simulations in various industrial types are design to simulate the experiential learning that assist the participant to really understand the process of something using simulative environment before entering the real situation. The result of the analysis show that the business simulation gives real business experience and important to the business process. Following are the feedback from questionnaire that show the experiential learning of student towards preparing effective business plan:

Table 5. The Experiential Learning

Items	Agree %
The business simulation facilitates in preparing business plan	70.2
This business simulation is user-friendly	69.3
This business simulation allows the transfer of knowledge from theory to practical	82.5
This simulation challenge and test knowledge and skills	81.5
This simulation is interesting and impressive	79.8
This simulation is relevant and useful for student	82.5
This simulation is meaningful to the learning process	81.6
This simulation run with organize order	79.8
This business simulation gives the real business experiences	83.3
Simulation is important to the business process	83.3
The simulation provides enjoyable experience	77.2
This simulation clear and easy to understand	75.4
This simulation is difficult and bored	50.9
This simulation brings real business problem in classroom	65.7
The simulation helps to understand difficult business concept	78.1

3.7 Impact of The Simulation

The impact of the simulation show how the simulation facilitate of business plan learning. It is also known as the complete structure of the simulation that contain the element; (1) to test student characteristic, (2) to determine the role of the educator, (3) of the mechanic for game characteristic, (4) that need facilities that provide ease of usage and (5) of experiential learning that can be found during and after playing simulation. The high score of percentage from the analysis show that the simulation facilitates the student to make business decision and they able to integrate all off-business function. Following are the feedback from questionnaire that show the impact of business simulation in preparing effective business plan:

Table 6. The Impact of Business Simulation

Items	Agree %
The simulation facilitates the student about understanding business process	79.8
The simulation enable student to integrate all of business function	85.1
The simulation facilitates the student how to make business forecasting	84.2
The simulation facilitates the student to make properly business decision	86.8
The simulation facilitates the student to distribute resource according to the priorities	83.3
The simulation facilitates the student in preparing financial reporting	81.6
The simulation facilitates the student on how to implement marketing strategies	82.5
The simulation facilitates the student to explore operational aspect in business	85.1
The simulation facilitates the student on how to evaluate the elements of cost saving.	83.4

3.8 The Importance of Business Simulation

This analysis also tries to find the most important aspect in business simulation. From below data shows that the first important thing is Encourage of Cooperation followed by Realistic Content of Business World, Individual Interaction, Joy and Excitement and then the last one is Educational Value. Following are the feedback from questionnaire that show the importance of business simulation in preparing effective business plan:

Table 7. The Importance of Business Simulation

Items	Important %
Individual Interaction	84.2
Realistic content of business world	86.8
Encourage of cooperation	92.1
Joy and excitement	84.2
Educational Value	83.3

3.9 The Simulation's Factor That Facilitate Business Plan Learning

The student also gives responses to factor that facilitate the business plan learning. The four factor of developing business plan are (1) Student acceptance (Student Characteristic on cognitive and affective domain), (2) The Simulation's content (Simulation Game Mechanic and Characteristic) (3) Lecturer role (Delivering and facilitating) (4) Infrastructure and technical aid. From the simulation factor the Lecturer 'roles show 82.5% influence to facilitate business plan learning. Following are the feedback from questionnaire that show the impact of business simulation in preparing effective business plan:

Table 8. The Influence of Simulation factor to facilitate business plan learning

Items	Influence %
Student Acceptance	68.4
The content of Simulation	76.4
Educator's Role	82.5
Infrastructure and technical aid	80.7

3.10 Statistical Result of Regression Analysis

The statistical measure using regression analysis show that the R Square from the model summary is .617 are close to 1. This indicate that the value of the variance to the score of Business Simulation (DV) that is associated with the independence variable (infrastructure, educator's roles, student characteristic, the content of the game and experiential learning) is 61.7% as showed in table 9. The value in table 10 ANOVA show that this model with R square 61.7% is significant, when p < .05.

Table 9. Model Summary

Model	R	R Square	Adjusted R Square	Std. Error of the Estimate
1	.785[a]	.617	.599	3.42185

a. Predictors: (Constant), infrastructure, educator, student, game, experiential

Table 10. ANOVA[a]

Model	Sum of Squares	df	Mean Square	F	Sig.
1 Regression	2034.440	5	406.888	34.750	.000[b]
Residual	1264.577	108	11.709		
Total	3299.018	113			

a. Dependent Variable: Business Simulation
b. Predictors: (Constant), infrastructure, educator, student, game, experiential

As refer to the Table 11, the beta value indicates that small contribution of .150 (15%) from educator. Meanwhile the highest contribution of .340 (34%) si from the score of student characteristic as significantly influence the score of business simulation, after the others contribution in the control model.

Table 11. Coefficients[a]

Model	Unstandardized Coefficients		Standardized Coefficients		
	B	Std. Error	Beta	t	Sig.
1(Constant)	6.780	2.613		2.594	.011
game	.105	.063	.150	1.672	.097
student	.280	.078	.340	3.581	.001
educator	-.208	.095	-.193	-2.184	.031
experiential	.211	.074	.289	2.847	.005
infrastructure	.438	.107	.298	4.092	.000

a. Dependent Variable: Business Simulation

This regression equation carry the meaning of the Business Simulation predict base on following equation: Business Simulation score=.105(game)+.280(student)+.208(educator)+.211(experiential)+.438(infrastructure) + 6.780. Student score (Beta = .340, p,.05 are the best predictor that significant as compare to another score with the overall of R square .617.

4 Finding and Discussion

This study finds that the most facilitate factor that contribute to the development of business simulation is the student characteristic including their knowledge, experience, expectancies, motivation and their affective and cognitive level. As mention from Kadot study the simulation allow student to test their current knowledge and make decision based on the resource that been distribute in the simulation [39]. It shows that the experiential learning by running simulation to facilitate the learning of business plan is helpful to the student while they apply and their knowledge from theory knowledge to be implement in the simulate real business world. Student take decisions across business functional areas when they compete each other to run the firm and defined the decision cycles [19]. They by themselves for the same time will increase their motivation to take care of the business by carefully made decision and motivate them to be the best with proper management. Because every simulation is the same that need input from student to process information and face other participants with certain outcome of decision [18]. This knowledge and skills that gain from playing the simulation make them ready to write the business plan properly with considering the element of integrate business function and understanding the basic business process. Even the element of student characteristic is the most facilitate, this study also agrees that the others element such experiential and educator's role somehow contribute to the successful of business plan simulation. While for the infrastructure that provide by the trainer is known as the element of assisting the progress of the simulation that depend on the present facilities and physical ease. Respond on the element that less contribute to the simulation, it is need some improvement to the content that in other word is try to match the current situation and be more user friendly and put extra mechanic of the simulation. Business plan simulation prove to be effective factor that help student to prepare well before writing the business plan because after playing they have gain to improvise and increase their knowledge, skill and confident level to understanding the way how business will be set up.

5 Conclusion

Preparing effective business plan are the key to gain financial assistant for graduate to start a business as their career. To make it effective student need to know better the process of business, to understand the process they need to run the business. Running business in simulated environment minimizing risk and can restart over to analyze the problem. However, we suggest that in future research to create a business simulation that enable student to run business with real product and real market approach.

1. A. Chwolka, M.G. Raith, J. Bus. Ven., **27**(3), 385–399 (2012)
2. J. Brinckmann, D. Grichnik, Kapsa, J. Bus. Ven., **25**(1), 24–40 (2010)
3. B. Honig, Aca. Man. Learn. Edu., **3**(3), 258–273. (2004)
4. B. Honig, T. Karlsson, Admin. Sc., **3**(4), 266–289 (2013)
5. C. Bianchi, G. W. Winch, C. Grey, *Proceedings of the 16th International Conference of the System Dynamics Society, Quebec '98*, **1**(c), 21 (1998)
6. T. Y. Nian, R. Bakar, A. Islam, A, Students ' Perception on Entrepreneurship Education : The Case of Universiti Malaysia Perlis, **7**(10), 40–49 (2014)
7. E. Auchter, M. Keding, *BRIDGING THE GAP : Transforming Knowledge into Action through Gaming and Simulation. Proceedings of the 35th Conference of the International Simulation and Gaming Association, Munich,* 2004)
8. M. Z. Ismail, S. Z. Ahmad, J. Chinese Entrep., **5**, 144–160 (2013)
9. Z. Arasti, M. Kiani Falavarjani, N. Imanipour, Hi. Edu. Stud., **2**(1), 2–10 (2012)
10. C. Elgood, *How Do You Create Effective Business Simulation Games ? Elgood ' S 10 Step Design Process Tackle Your Issues • Engage Your Staff • Maximise Performance How Do You Create Effective Business Simulation Games ? Elgood Effective Learning*. (2011)
11. C. M. Mason, J. M. Stark, Int. S. Bus. J., 227–248 (2013)
12. L. Pittaway, a, *Simulating Entrepreneurial Learning: Assessing the Utility of Experiential Learning Design,* (2004)
13. A. Y. Kolb, D. H. Kolb, Man. Learn., Edu. Dev., 1–59 (2008)
14. J. A. Poore, D. L. Cullen, G. L. Schaar, Clinic. Sim. Nur., **10**(5), 241–247 (2014)
15. S. Dellaportas, T. Hassall, Brit. Acct. Rev, **45**(1), 24–36 (2013)
16. A. Ayob, A. Hussain, M.M. Mustaf, M.F.A. Shazi Shaarani,., Proc. - Soc. Behav. Sc., **18**, 247–254 (2011)
17. D. Finch, M. Peacock, D. Lazdowski, M. Hwang, Int. J. Man. Edu., **13**(1), 23–36 (2015)
18. D. Keshodarah, Capacity Building in Entrepreneurship Education and Training Using Simulation. In: *International HRD Conference, Mauritius 2013: Excellence in HRD for sustainable growth Capacity*, 1–12 (2013)
19. B. Kulkarni, Using Business Simulations To Introduce Business Concepts, **40**, 393–400 (2013)
20. R.S. Mulcahy, Int. J. D. Learn., **2**, 1–17 (2011)

References

21. M. Critelli, D. I. Schwartz, S. Gold, Serious Social Games: Designing a Business Simulation Game. *4th International IEEE Consumer Electronic Society - Games Innovation Conference, IGiC 2012*, 2–5 (2012)
22. Y. Tao, C. Cheng, S. Sun, Edu. Techno. Soc., **15**, 177–189 (2012)
23. H. Panoutsopoulos, M. A. Lykourentzou, D. G. Sampson, Business Simulation Games as Digital Tools for Supporting School Entrepreneurship Education. *Proceedings of the 2011 11th IEEE International Conference on Advanced Learning Technologies, ICALT 2011*, 155–156 (2011)
24. A F. Barišić, P. Monika, Business Simulation as a Tool for Entrepreneurial Learning The Role of Business Simulation in Entrepreneurship Education. In: *4th International Conference on Entrepreneurial Learning 2014 – ICEL4*, (Cmc) (2014)
25. J. Sebastián, C. Mario, Dev. Bus. Simu. Exp. Learn., **42**(2015), 117–123 (2015)
26. N. M. Tawil, R. Hassan, S. Ramlee, Z. K-Batcha, Journal of Engineering Science and Technology, **10** (Spec. Issue on 4th International Technical Conference (ITC) 2014), 71–80 (2015)
27. L. Vos, R. Brennan, Marketing Simulation Games: Student and Lecturer Perspectives (2010)
28. N.L. Abdullah, M.H. Hanafiah, N. A. Hashim, Int. Edu. Stud., **6**(6), 95–107 (2013)
29. T. Russ, M. Drury-Grogan, Comm. Quart., **61**(5), 584–595 (2013)
30. C.N. Bodea, R. I. Mogo, M. I. Dasc, A. Purnu, N.G. Ciobotar, Amfit. Eco. **17**(38),10–24 (2015)
31. W. J. Wellington, D. Hutchinson, M. Gowing, **38**, 320–326 (2011)
32. S. Shane, E. Locke, C. J. Collins, Human Res. Man. Rev., **13**, 257–279 (2003)
33. Evans, INFORMS Trans. Edu., **1**, 27–37 (2000)
34. B. Baruah, A. Ward, Enhancing Intrapreneurial Skills of Students through Entrepreneurship Education. *ITHET 2014 - 13th International Conference on Information Technology Based Higher Education and Training*. (2015)
35. W. Wellington, and Hutchinson, D., Dev. Bus. Simu. Exp. Learn., **39**, 132–140 (2012)
36. J. Moizer, J. Lean, M. Towler, G. Smith, Int. J. Learn. Techno., **2**(1), 49 (2006)
37. P. H. Anderson, L. Lawton, L, Simu. Gam., **40**, 193–216 (2008)
38. W. M. P. Aalst, Van Der, In: J. vom B. (5) and M. Rosemann, eds., *Handbook on Business Process Management 1*, (Part II., Springer Berlin Heidelberg, 1–34, 2015)

39. E. Cadotte, Ann. Entrep. Edu. Peda, 280–301 (2014)

The Effect of Brief Cognitive Behavioural Therapy Group Intervention on Self Esteem among Drug Users

Huzilil Hussin[1,*], *M.H.* Othman[1], *Kamarudin* Hussin[1] and *N. S.* Alias[1]

[1]School of Human Development and Technocomunication, Universiti Malaysia Perlis, Perlis, Malaysia

Abstract. Self-esteem is related to the individuals understanding and accepting the existing character and attitude to themselves. Failure to understand and accept their own strengths and weaknesses and in the same time do not have the skills to cope with it will caused them to be classified in the group who have low levels of self esteem. The up and down in self esteem will seriously affect the person emotional development and their reasoning process. Thus, reforming self esteem is an important key to help the rehabilition of drug abuser. Therefore, the purpose of this study is to investigate the effects of Brief Group Cognitive Behavioural Therapy on self-esteem among drug users. The quasi experimental design with one treatment group and one control group and four measurements (Pre Test, Post 1, Post 2 and Post 3) were employed in this study. The sample of 108 drugs users was chosen using purposive sampling procedure. Data were collected using standardized psychometric instruments, including Internal Control Index (Patricia Duttweiler, 1984) and The Rosenberg Self-Esteem Scale and were analysed using descriptive analysis. The findings shows the intervention of group Brief Cognitive Behavioural Therapy CBT (B) are succeed in increasing self-esteem mean score compared to control group in different time interval. As a conclusion, self-esteem among the subjects can be enhanced with an intervention of proper and systematic therapy instead of unplanned therapy sessions

1 Introduction

Self-esteem is how one views himself or herself in terms of self-value. It also concerns on how other people sees the values on him or her [1]. Proportionally, it is a self-belief on oneself with self-worthy, self-respect and self-integrity were one of the ingredients in enhancing or lowering one's self esteem [1]. Self-esteem involves on our feelings about ourselves and how we judge these feelings; either to be positive or negatives [2].

Thus, life experience plays important role in building and enhancing these positive and negative emotions [3]. The ups and downs of events that has happens in one's life have significantly contribute to the effect of how individual's sees themselves [1]. Individuals who were high in spirit and always vigilant in promoting positive emotions and thus positive self will always embrace high self-esteem no matter how the life's fluctuations are [3] however, to individuals who has low self-esteem, 'small fluctuations in life event' may mean like 'the whole deal of life' and significantly affect his or her emotions and personal values [1].

1.2 Brief Cognitive Behavioural Therapy (CBT) on Drug Users

Cognitive behavioural therapy is a talk therapy which focuses on current problems (Royal College of Psychiatrists). It gave insights on practical methods in improving the state of mind on daily problem basis without referring to the past events [4]. According to Mahmood [5], CBT is counselling approach based on strategies to recognize, avoid and cope. CBT-based training skills focus on strategies to overcome the wants and the addictions by reinforcing motivation to change, to develop problem-solving skills, to design and to manage high-risk situations that may lead to return and also to develop skills to stay away from drugs.

The aimed of cognitive behavioural therapy is to make clients not to feel apprehensive and able to view current situation not to be stressful and non-threatening as negative stressors will heighten the negative emotions such as stress, anxiety, helplessness and depression [4]. These negative emotions will then lead to ineffective and less positive outcome of the counselling or therapy sessions between both parties [6].

Thus, according to McHugh et al. [7], cognitive brief therapy is the version of short-session that went up to ten or twenty counselling or therapy sessions. Cognitive brief therapy is much focused on specific issue or problem. Meaning that, in cognitive brief therapy the concept of current situation (here and there) was prioritized. The setting of goals and specific objectives in resolving, attending and helping the problem at hand was clearly overlaid and make known to both sides of the therapist and client. Subsequently, it is considered as on the most suitable and effective method in making change of behavioural problem [7]. Truth is the brief counselling has been practiced since 1949. Most counselling sessions in the year therapist in handling the issues of drug addiction [8].

* Corresponding author: huzili@unimap.edu.my

Another supporting study which shows strong correlation on the use of cognitive brief therapy in enhancing self-esteem by looking at the self-efficacy of the drug addict was form [9]. Their study on 90 respondents of drug dependency patients shows significant improvement of coping strategy and significant reduction of addiction symptoms among the 90 drug addicts.

Additionally, study by Milan et al. [10] among 90 drug-dependent patients who were referring to the clinics in Uremia, Iran has also proven positive result in enhancing self-esteem by having the self-confidence (self-efficacy) in implementing and boosting coping strategies, the significant improvement of mental and physical health when using the cognitive brief therapy as a treatment.

1.2 Problem in Self Esteem among Drug Users

There were many studies shows on the causal factor of life's events that serves as environmental stressors for low self-esteem individuals with drug abused [11]. One of the study that was done by Wheeler (2010), involves on girls and drug abused. His study has shown the higher the self-esteem felt by the girls; the less drug abused were being practiced.

Another study that shows the influence of self-esteem and drug abused was from Khajehdalulee et al. [12], whereby their longitudinal study of one year from 2010 to 2011 concerning of 943 high school students from Sarakhs shows high significant correlation between self-esteem and drug abused for 507 of their respondents.

This study is parallel with a study by Stephens et al. [13] to 291 respondent of marijuana addicts. 87% of the respondents stated the reason of taking marijuana is because they have low self-esteem and 79% reSDondents show low internal locus of controls. As the result, they will rely more on surrounding as well as fail to make a better path for their own life. Study by Fauziah et al. [14] shows that factor of low confidence contribute about 64% to repeated drug addicts.

Thus, an intervention that could help to improve their self-confident need to been implement to help drug addicts [14] to regain their self-esteem and self-control simultaneously.

2 Research Methadology

This study tests the effectiveness of Brief Cognitive Behavioural Therapy CBT[B] by using quasi experimental design with one treatment group [CBT (B)] and one control group (KW) and four measurements (pre, post 1, post 2, and post 3. The sample of 108 drug abusers was chosen using purposive sampling procedure. Samples were then distributed randomly into two groups: CBT (B) and control group (KW). Each group consisted of 36 samples. Data were collected with standardized psychometric instruments, including Internal Control Index [15] and Rosenberg Self-Esteem Scale. Data were analyzed using Two-way ANOVA repeated measurement.

3 Analysis on the Study

In this study the effectiveness of the module are measured by the instruments and the results showed based on mean score,mean percentage and the change pattern of self esteem (SE) profil. Therefore, this section will present the results in mean score and standard deviation of self-esteem profiling for each groups CBT (B) and control group (CG) in each time occasion and the different of mean percentage between time occasions (Pre, Post Test 1, Post Test 2 and Post Test 3). Findings analysis for SE profile for group CBT(B) and CG between time occasions show in table below;

Table 1. Analysis findings of SE profile for group CBT (B) and CG between time occasions

	CBT(B)			CG		
	M	*SD*	% Mean Differen ce	*M*	*SD*	% Mean Differenc e
Pre	1.54	.07	-	1.56	.07	-
Post 1	2.63	.35	70.8	2.39	.27	53.2
Post 2	3.00	.39	14.1	2.83	.46	18.4
Post 3	3.27	.61	9.0	2.82	.56	-0.35
Pre-Post			112.3			80.7

Result demonstrated SE mean score increased for group CBT(B) between time occasions from Pre Test M=1.54 (SD=.06), Post Test 1 (M=2.63; SD=.35), Post Test 2 (M=3.00; SD=.39) and Post Test 3 (M=3.27; SD=.61). While the mean score percentage is 70.8% increase for Post Test 1 compare to Pre Test, 14.1% of mean percentage increase for different mean score for Post Test 2 and Post Test 1 and 9.0% mean percentage increase for mean score for Post Test 3 compare to Post Test 2. Overally, 112.3% increase in mean score between Pre Test and Post Test 3.

While for control group (CG), result show the CG mean score for Pra test is M=1.56 (SD=.07), Post Test 1 (M=2.39; SD=.27), Post Test 2 (M=2.83; SD=.46) and Post Test 3 (M=2.82; SD=.56). There are in mean score for group CG between time occasions. In the different of percentage for the mean score the mean score is increase by 53.2% in Post Test 1 compare to the mean in Pra Test, increase of 18.4% between mean in Post Test 2 and Post Test 1 and increase -0.35% for mean score for Post Test 3 compared to Post Test 2. Overall, the increase is 80.7% between mean score percentage of pra test and Post Test 3.

4 Discussion

The increases of SE mean score for treatment group CBT [B] might be because of the input in CBT module are applying skill of identifying characteristic of low self-esteem and how to increase it. In addition, the brief therapy also gives the skills to identify the health of client's emotions (healthy or unhealthy) and how to overcome the negative feelings. Skills to manage these emotions can help clients identifying internal problems, having the ability to think of themselves and more positively, and also achieving self-perfection [16].

Other than that, the improvement of SE is lead by the techniques of 'deserted island' and 'Socratic dialogue' implement to the clients. According to Beck [17], 'Socratic dialogue' help to answers a general questions asked by the clients along the sessions. While according to Curwen, B., Palmer, S. & Ruddel, P. [18] 'deserted island' technique would help to change individual's internal belief to be freer and independence.

Curwen et al. [18] said, instead of changing ones character and personality, the CBT approach fosters individuals to be more responsible for their lives. According to Wilding et al. [19], CBT model is engaged to solve problems in daily life through training skills to increase the level of self-understanding and being responsible for themselves. The important thing is the individual trained to achieve balanced between thinking and behaviour. Thus, they will be able to identify the negative thought and change the negative to the positive behaviour. Moreover, approaches and techniques to control thinking and behaviour are different from detaining or restricting the thinking and the behaviour as they help towards self-understanding to build a better life. This means CBT approach will not only causes the clients to feel good and comfortable but they know more clearly why they feel good and comfortable.

Meanwhile, according to Mahmood [19], a role in rehabilitation counseling can help clients build a new, positive behaviour, develop problem-solving skills and potential, strengthen family relationships, increase self-esteem, self-awareness, positive thinking and help them to develop the independent attitudes. Meanwhile, according to Mahmood et.al [20], a role of counseling in rehabilitation is to help clients have a significant changes in a psychological aspects such as depression, anxiety, self-esteem and to increase self-esteem. Studies summarizes both group and individual counseling intervention can solve the problem of alcohol and drug addiction.

Based on Mior [20], he said that there is significance different between treatment groups which emphasize on therapeutic elements compared to control group which did not apply the therapeutic relationship. This is in line with study done by Schutte et al. which proves the effectiveness of CBT intervention in improving self-esteem of individuals who suffered from prolonged fatigue and also improving self-esteem of individuals who have self-image and self-concept problems [21].

Overally, this study shows that Brief Cognitive Behavioural Therapy Intervention is effective in improving self esteem of drug users compared with the control group that did not receive any treatments. We can conclude that with structured and systematic module and with the influence and cooperation by comitted, dedicated, skilled and experienced counselor can help to ensure the success of the intervention using CBT approach through group counseling methods for drug rehabiliton. It is clear from the results of this study, this module can be used as a reference for the development of experimental research in Malaysi and most important is it can also used as one of the approaches applied in all treatment procedures of drug rehabilitions in Malaysia.

References

1. Alavi, Hamid, Reza, Journal of Addiction and Health **3**(3-4), 119-124 (2011)
2. H. Lotfabadi, *Development Psychology: Adolescence, Young People and Adults* (Tehran: Samt Press, 2009)
3. S. Kermode, D. MacLean, Aust J Adv Nurs **19**(2). 33–40 (2001)
4. K. M. Carroll, S. A. Ball, C. Nich, P. G. O'Connor, D. A. Eagan, T. L. Frankforter, E. G. Triffleman J. Shi, B. J. Rounsaville, Arch Gen Psychiatry **58**, 755–761 (2001)
5. Mahmood Nazar Mohamed (2003), Merawat Penagih Dadah Melalui Kaunseling dan Pendekatan Berperspektif Islam. Prosiding Seminar Kebangsaan Kaunseling Berperspektif Islam. Universiti Utara Malaysia.
6. K. M. Carroll, Therapy Manuals for Drug Addiction. Manual 1: A Cognitive-Behavioral Approach: Treating Cocaine Addiction (1998)
7. R. K. McHugh, B. A. Hearon, M. W. Otto, Psychiatric Clinics of North America **33**(3), 511–525 (2010)
8. A. Molaie, S. Shahidi, S. Vazifeh, S. Bagherian, Journal of Social and Behavioural Sciences, Prodecia **5**, 1180-1184 (2010)
9. H. Kamarzarrin, M. Khaledian, M. Shooshtari, Z. Ahmadimehr, Journal of Engineering and Applied Sciences **3**(13), 1293-1300 (2013)
10. Milan, Hassan Brockie, Kamarzarrin, Hamid, Zare, Hossein, Journal of Research on Addiction Quaterly Journal of Drug Abuse (2014)
11. D.W. Brook, E. Rubenstone, C. Zhang, N.K. Morojele, J.S. Brook, Journal of Soc Sci Med. **72**(9), 1447–1453 (2011)
12. Khajehdaluluee, Mohammad, Zavar, Abbas; Alidoust, Mahbobeh, Pourandi, Razieh, Journal of Iranian Red Crescent Medial Jounal **15**(11), (2013). doi: 10.5812/ircmj.7682
13. R.S. Stephens, R.A. Roffman, L. Curtin, Journal of Consulting & Clinical Psychology **68**, 898-908 (2000)
14. Fauziah Ibrahim, Bahaman Abu Samah, Mansor Abu Talib, Mohd Shattar Sabran, Jurnal Antidadah Malaysia. Jld. 5, 235-251 (2009)

15. P.C. Duttweiler, The Internal Control Index: A Newly Developed Measure of Locus of Control. Educational and Psychological Measurement 1984 **44**, 209 (1984) doi: 10.1177/0013164484442004.
16. K.S. Dobson, Handbook of cognitive-behavioral therapies (New York: The Guilford Press, 2001)
17. J.S. Beck, Cognitive *Behavior Therapy: Basics and beyond* (New York: Guilford Press, 2011)
18. B. Curwen, S. Palmer, P. Ruddel, *Brief cognitive Behaviour therapy* (London: SAGE Publications, 2000)
19. Mahmood Nazar Mohamed, *Mumulihkan Penagih Tegar : Implikasi kepada Modaliti Masa Kini. Prosiding Kerja Sosial* (Universiti Utara Malaysia, 1991)
20. Mahmood Nazar Mohamaed, Md Shuaib Che Din, Lasimon Matokrem, Muhamad Dzahir kasa, Rusli Ahmad, *Penagihan dadah dan residivisme: Aspek-aspek psikososial dan persekitaran* (Sintok: Universiti Utara Malaysia, 1999)
21. S.H. Dworkin, B.A. Kerr, Journal of Counseling Psychology **34**(2), 136-140 (1987)

Model of Islamic Social Entrepreneurship: A Study on Successful Muslim Social Entrepreneur in Malaysia

Mohd Adib Boulven[1], *S.* Abdullah[1] , *Azizan* Bahari[1] , *A. J.* Ramli[1], *N. S.* Hussin[1], *Jamsari* Jamaluddin[1], and *Z.* Ahmad[2]

[1]School of Human Development and Technocomunication, Universiti Malaysia Perlis, Perlis, Malaysia
[2]Institute Pendidikan Guru Malaysia Kampus Perlis, Perlis, Malaysia

Abstract. Since research effort in the area is minimal, there is a clear need to examine the practice of Islamic social entrepreneurship among successful Muslim social entrepreneurs in Malaysia. One such practice is to organize charitable activities to benefit the community through the gains made from entrepreneurial activities that are based on social mission and vision. The research problem is lacking of model on Islamic social entrepreneurship. The main objective of this paper is to develop a Model of Islamic Social Entrepreneurship based on successful Muslim social entrepreneur in Malaysia. The research method used in this study is literature review, content analysis, and interview with 14 participants constituting nine successful Muslim social entrepreneurs and five experts with religious academic backgrounds participated in the study. The research finding shows that model of Islamic social entrepreneurship is the major contribution of the study which could serve as guidelines for successful Muslim social entrepreneurs, particularly young entrepreneurs.

1 Introduction

Social entrepreneurship is an emerging area of research that examines business that is significantly influenced by entrepreneur's society's motives instead of being purely economically driven [1]. Social entrepreneurship is one of the branches of the economy that can alleviation of poverty and able to improve the country's economic status, from other perspectives such as charity and philanthophy as well as social services [2-3]. To seek for the lawful income is obligatory according to the Islamic view and work in the field of business entrepreneurship and make the essentials good for Muslims is obligatory (*fardu kifayah*) (Mohamad Mustafa, 2002). The concept of social entrepreneurship has been rapidly emerging in the private, public and non-profit sectors over the last few years, and interest in social entrepreneurship continues to grow [4]. Social entrepreneurship has become a global phenomenon that impacts the society by employing innovative approaches to solve social problems [5].

Based on current economic conditions, social entrepreneurial activities is very important to reduce social problem especially in developing countries such a Malaysia [6]. Among the initial steps recommended by the Malaysian government is the collaboration between institutions of high potential from various government and private agencies need to adopt the concept of social entrepreneurship to the world of entrepreneurship in the field of research as a whole to help entrepreneurs develop a range of products and services with low prices and quality [6-7].

Social entrepreneurship has become a new phenomenon in a country in order to reduce social problems, eradicate poverty communities and to build up the good relationship between entrepreneurs and society [8-9]. According to figures from the famous corporate entrepreneur Muhammad Ali Hashim (2013), states that in Malaysia it is time for the Muslim entrepreneurs to implement the Islamic social entrepreneurship in the current practice of the concept of jihad business or jihad economy through Islamic social entrepreneurship and Islamic business entrepreneurship. Entrepreneurs are among the individuals and groups that are often discussed in the al-Qur'an, especially in a business context. Social entrepreneurs are entrepreneurs who have entrepreneurial activity in a business and have managed not to get a profit as a result of the activities. The main objective is to seek the pleasure and blessing of Allah SWT in this world and the hereafter [10].

Islamic social entrepreneurship is not foreign to Islam because this is part of business, entrepreneur and economic. Since the advent of Islam, Islamic social entrepreneurship has played a vital role in its spread. Islam has reached many parts of the world. The study based on Islamic social entrepreneurship from the social entrepreneurial activity is still new and lacking, especially in the Islamic perspective [3, 10-13]. As we all know that Islam is a monotheistic religion which is complete and perfect for all Muslims because it covers all aspects of the life of this world and the hereafter. Allah SWT says in the holy Qur'an:

> *"Allah hath permitted exchange and forbidden usury" (Surah Al-Baqarah, 2:275)*

> *"This day, those who disbelieved have given up all hope of your religion; so fear them not, but fear Me. This day, I have perfected your religion for you, completed My favour upon you, and have chosen for you Islam as your religion. But as for him who is forced by severe hunger, with no inclination to sin (such can eat these above mentioned animals),then surely, Allah is Oft-Forgiving, Most Merciful." (Surah Al-Maidah, 5: 3)*

Perfection of Islam not just includes religious aspects, but also the best life procedures outlined by Allah SWT. Al-Qur'an and al-Hadith is the main source of reference and complete guideline to the best human life. In addition, people are supplied with common sense to be used to achieve the purpose of life and closer to their Creator.

Based on the word of Allah which means: -

> *"Who is he that will lend to Allah a goodly loan so that He may multiple it to him many times And it is Allah that decreases or increase (your provisio), and unto Him you shall return." (Surah Al-Baqarah, 2: 245)*

According to Ibnu Kathir, Allah encourages His servants to spend in His cause. Allah mentioned this same Ayah in several other parts of His Glorious Qur'an. The Hadith that mentions that Allah descends (every night down on the nearest heaven to us when the last third of the night remains) states that Allah says: *Who would give a loan to He Who is neither poor nor unjust.*

> *"By no means shall you attain Al-Birr, unless you spend of that which you love; and whatever of good you spend, Allah knows it well." (Surah Ali-Imran, 3:92)*

Therefore, the main objectives of this study are: -

i. To develop model of Islamic social entrepreneurship among successful Muslim social entrepreneur in Malaysia.

2 Research Methodology

A qualitative approach is relevant to this study as it enables researchers to generate thick description of the social actors in their natural setting [14]. The study was conducted based on the methodology epistemology, content analysis and interviews. This study is also based on existing models and theories from previous studies. Comparative analysis approach was used to look at the elements from previous studies. As shown in Fig. 1. This study consists of two phases. The first phase is to analyze the previous models of social entrepreneurship through literature review, content analysis and comparative analysis. The second phase is to develop a model of Islamic social entrepreneurship according to Islamic perspective. The method used for second phase is interview. Interviewing participants was the primary method of data collection in this study. A semi structured interview involving 14 participants constituting nine successful Muslim social entrepreneurs and five experts with religious academic backgrounds participated in the study. The strength and advantage of a semi structured interview is that the interviewer can seek clarification and elaboration of the interviewees' responses and thus encourage further discussion on a particular topic.

Participants were identified through the purposeful sampling technique. We approached nine successful Muslim social entrepreneurs (Inang Suri Enterprise, Former CEO of Johor Corporation, Red Profile Sdn Bhd, Mohd Maliki Bin Endut Enterprise, Colours of Asia Sdn Bhd Group of Companies, Ameqin Sepakat, Prowheels (M) Sdn Bhd, Goncarian Enterprise, and IdSoft Enterprise) and five experts with religious academic backgrounds (Dean of Islamic Business School Universiti Utara Malaysia, Professor in Management and Entrepreneurship Universiti Malaysia Kelantan, two senior lecturers in Islamic Business School, and Religious Officer at Islamic Centre UUM. The data was analysed using thematic analysis within the constructionist paradigm.

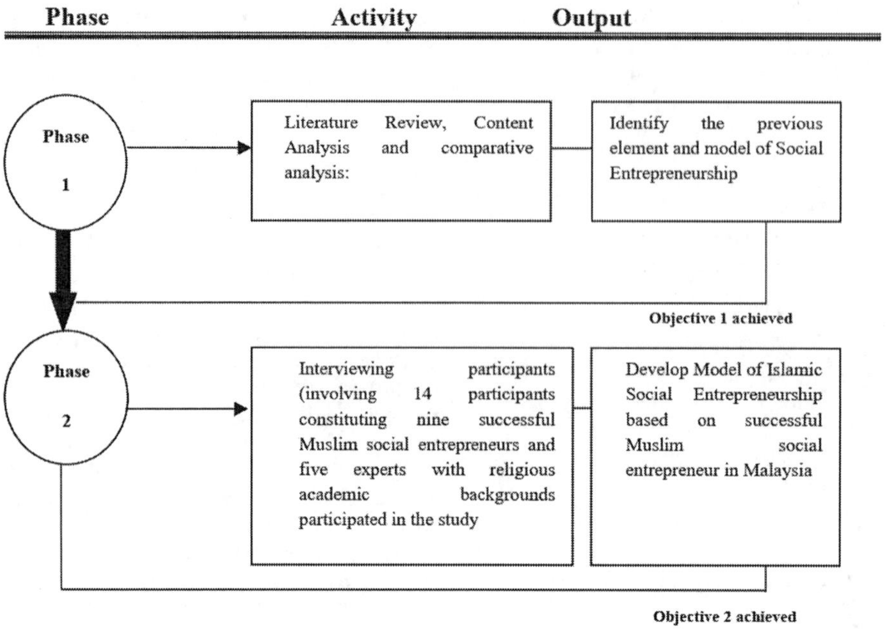

Fig. 1. Phase Analysis and Develop Model of Islamic Social Entrepreneurship

3 Finding and Discussion

All of these elements got from interview analysis, most of participants agree that these elements must have in model of Islamic social entrepreneurship.

Elements	Participants (1, 2, 3, 4, 5, 6, 7, 8, 9, 10, 11, 12, 13, 14)
1. To seek pleasure of Allah s.w.t in the world and hereafter	/ / / / / / / / / / / / / /
2. Concept of Islamic Social Entrepreneurship	/ / / / / / / / / / / / / /
i. Concept of *Amar maa'ruf wa nahi munkar*	/ / / / / / / / / / / / / /
ii. Concept of Welfare	/ / / / / / / / / / / / / /
iii. Concept of Fairness	/ / / / / / / / / / / / / /
iv. Concept of *al-Falah* (success in world and the	/ / / / / / / / / / / / / /
hereafter)	/ / / / / / / / / / / / / /
v. Concept of Developing and improving of Social	/ / / / / / / / / / / / / /
Value providers	/ / / / / / / / / / / / / /
vi. Concept of *Khalifah*	/ / / / / / / / / / / / / /
vii. Concept of Charity	/ / / / / / / / / / / / / /
viii. Concept of *Waqf, hibah* and *zakat*	/ / / / / / / / / / / / / /
ix. Concept of Justice and balance	/ / / / / / / / / / / / / /
x. Concept of Welfare and social security	/ / / / / / / / / / / / / /
xi. Concept of *Jihad* economy	/ / / / / / / / / / / / / /
xii. Concept of Prayer and attending to *Masjid.*	/ / / / / / / / / / / / / /
3. Basic element of Muslim social entrepreneur	/ / / / / / / / / / / / / /
i. *Iman*	/ / / / / / / / / / / / / /
ii. *Taqwa*	/ / / / / / / / / / / / / /
iii. *Ihsan*	/ / / / / / / / / / / / / /
iv. *Sifat Mahmudah*	/ / / / / / / / / / / / / /
i. Taubat	/ / / / / / / / / / / / / /
ii. Ikhlas	/ / / / / / / / / / / / / /
iii. Khauf	/ / / / / / / / / / / / / /
iv. Zuhud	/ / / / / / / / / / / / / /
v. Sabar	/ / / / / / / / / / / / / /
vi. Syukur	/ / / / / / / / / / / / / /
vii. Tawakal	/ / / / / / / / / / / / / /

viii. Mahabbah	/	/	/	/	/	/	/	/	/	/	/	/	/	/
ix. Redha	/	/	/	/	/	/	/	/	/	/	/	/	/	/
x. Zikrul Maut	/	/	/	/	/	/	/	/	/	/	/	/	/	/
4. Element of *al-Maqasid al-Syariah*. These are the preservation of:	/	/	/	/	/	/	/	/	/	/	/	/	/	/
i. Religion (*deen*)	/	/	/	/	/	/	/	/	/	/	/	/	/	/
ii. Life (*nafs*)	/	/	/	/	/	/	/	/	/	/	/	/	/	/
iii. Lineage (*nasl*)	/	/	/	/	/	/	/	/	/	/	/	/	/	/
iv. Intellect (`*aql*)	/	/	/	/	/	/	/	/	/	/	/	/	/	/
v. Property/ wealth (*mal*)	/	/	/	/	/	/	/	/	/	/	/	/	/	/

Source: From Interview Result

3.1 Islamic Social Entrepreneurship based on Islamic perspective

The concept of Islamic social entrepreneurship must refer the primary source in Islamic religion which is al-Qur'an and al-Hadith and must follow objective of the Shariah or *al-Maqasid as-Shari'ah* (preserve religion, life, intellectual, lineage, and property) in order to achieve the concept of *al-Falah* as stated in Fig. 2. Besides, it must fulfil basic requirement to help people in solving the social problem [10]. According to Habibollah et al. [15], religious Waqf has been able in distribution of wealth in society and using finance and property in the line well-meaning humanitarian to develop opportunities and access vulnerable to the main infrastructure development. The tradition of the Prophet Muhammad: *"Allah S.W.T. love the people whom help the poor and needy mankind"*.

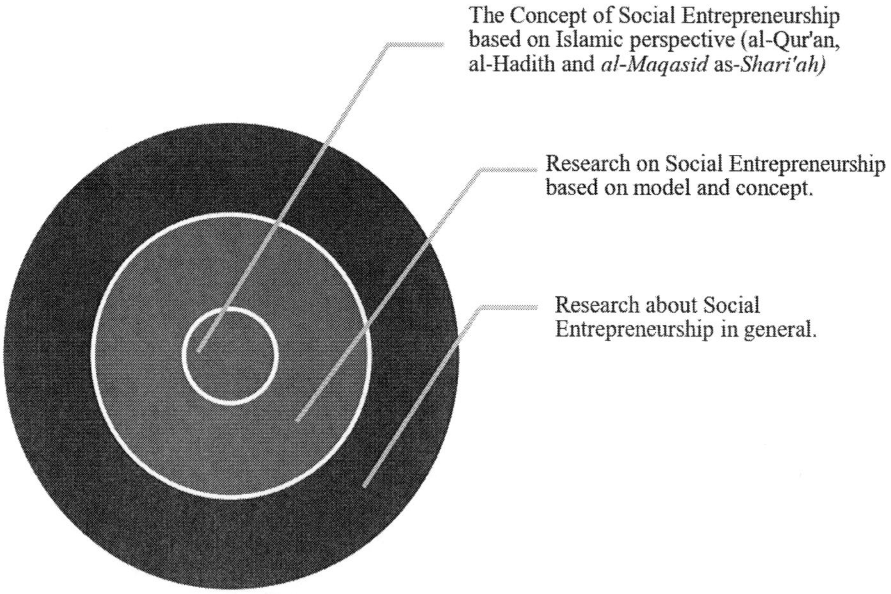

Fig. 1. The Concept of Islamic social entrepreneurship

3.2 Model of Islamic Social Entrepreneurship

Fig. 2 shows the model of Islamic social entrepreneurship based on interview analysis.

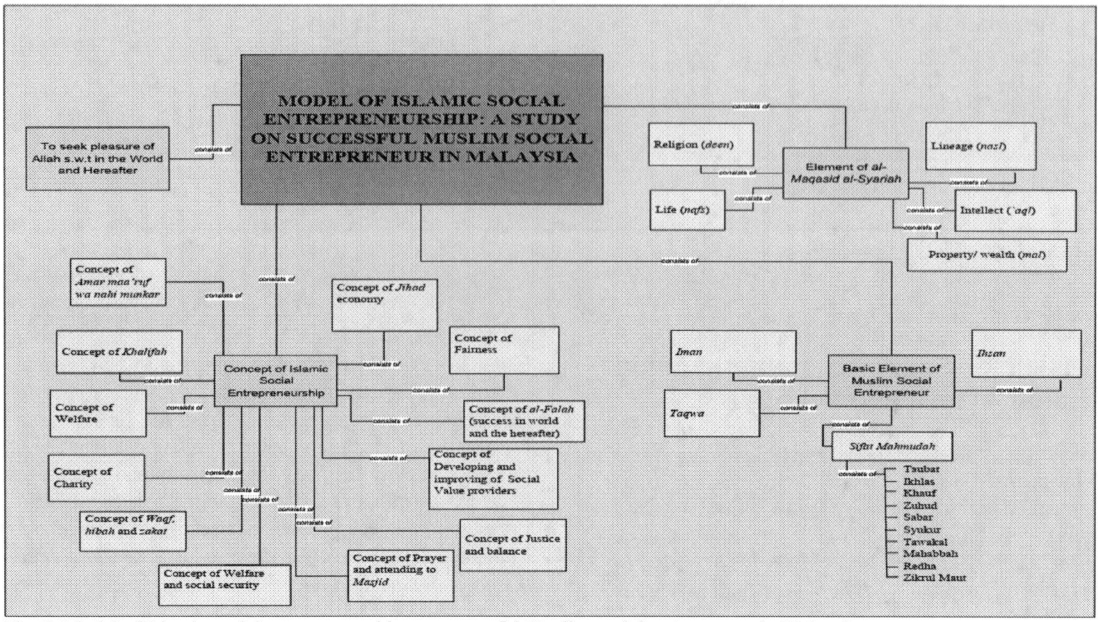

Fig. 2. Model of Islamic social entrepreneurship on successful Muslim social entrepreneur in Malaysia

4 Conclusion

In conclusion, this paper is an attempt to develop a model of Islamic social entrepreneurship on successful Muslim social entrepreneur. This study is based on the Al-Quran and As-Sunnah and need to implement the Maqasid Shari'ah (preserve religion, life, intellectual, lineage, and property) in to the real practices in order to achieve al-Falah (seek the pleasure of Allah s.w.t in this world and hereafter).

By developing this model it will not only helps successful Muslim social entrepreneur to implement the Islamic Social entrepreneurship in Malaysia but it can also be a guide to the new social entrepreneurs to learn and practice. Moreover, findings from the participants has shown that majority of the experts and Muslim social entrepreneurs agreed and perceived the elements in Model of Islamic social entrepreneurship useful and helpful to the Muslim social entrepreneur to implement the Islamic social entrepreneurship; help to seek the pleasure of Allah SWT; help to increase the activity of Islamic social entrepreneurship in Malaysia; help to increase the awareness of Muslim social entrepreneur; and help to reduce the social problems among society in Malaysia.

In addition, this study is very important to increase the economy growth for nation. Therefore, this study hopes that various parties to participate and take initiatives to increase the self-knowledge to help in eradication of social problems through Islamic social entrepreneurship. Besides, it can be additional reading materials in the collection of studies and research on the Islamic perspective. Perhaps, this study also will contribute in a better understanding on the concept of Islamic social entrepreneurship. For future work, other researchers could expand the knowledge on Islamic social entrepreneurship from this study.

5 Acknowledgement

The author would like to thank to Universiti Malaysia Perlis (UniMAP) (http://www.unimap.edu.my) and the Ministry of Education Malaysia (http://www.moe.gov.my) for sponsoring this study.

References

1. J. Almarri, J. Meewella, T. Mainela, Social Entrepreneurship and Islamic Philanthropy. International conference on Excellence in Business. Shariah United Arab Emirates (2012)
2. Saifuddin Abdullah. (2012, 6 Ogos). *"Keusahawanan Sosial Basmi Kemiskinan"*. Akhbar Sinar Harian.
3. Suhaili Sarif. (2016). Social entrepreneurship through an Islamic institution: the Malaysian Zakat experience. Dalam al-Azmi Bakar, Zakaria Mat Yusof, Desa Mat Rabi, Umihani Awang, Rohaizal

Umar, Baharudin Mohamad, Muhammad Yusry Che Dah, Ngeow Yeok Meng, Faezah Kassim, Ponmalar N. Alagappar, Faridah Che Husain, Zamre Abu Hassan dan Wendy Yee Mei Tien (ed.), *Social entrepreneurship and community engagement: Transforming societies.* Selangor: Social Institute of Malaysia (ISM).

4. Johnson, S. (2000), "Literature review on social entrepreneurship", available at:www.bus.ualberta.ca/ccse/Publications/Publications/Lit._Review_SE_November_2000.rtf (accessed February 20, 2011).
5. Robinson, J.A., Mair, J. and Hockerts, K., (Eds) (2009), International Perspectives of Social Entrepreneurship, Palgrave, London.
6. Saifuddin Abdullah. (2009, 29 Julai). *"Keusahawanan Sosial Dalam Penyelidikan".* Akhbar Utusan Malaysia.
7. Mohd Adib Abd Muin, Shuhairimi Abdullah, Azizan Bahar, Journal of Human Development and Communication (JoHDeC) **4**, 21-36 (2015)
8. J.T. Zietlow, Nonprofit Management and Leadership **13** (1), 85-90 (2002)
9. Mohd Adib Abd Muin, Azizi Abu Bakar, Shuhairimi Abdullah, Prosiding: International Conference on Business Innovation, Entrepreneurship and Engineering 2013, 386-399 (2013)
10. Mohd Adib Abd Muin, Shuhairimi Abdullah, Muhammad Fakhirin Che Majid, (2014). Pengurusan Keusahawanan Sosial Islam: Model Amalan Usahawan Berjaya Dalam Amalan Nilai-Nilai Murni. Prosiding: Seminar Penyelidikan Kebangsaan 2014
11. Yazilmiwati Yaacob, Ilhaamie Abdul Ghani Azmi, (2012). Entrepreneurs' Personality from Islamic Perspective: A Study of Successful Muslim Entrepreneurs in Malaysia. Doi: 10.7763/IPEDR. 2012. V46. 16.
12. Shuhairimi Abdullah. (2011). *Aplikasi Nilai- Nilai Murni Dalam Kalangan Usahawan ke Arah Pembangunan Usahawan Muslim Yang Holistik.* Kertas Kerja Seminar Keusahawanan Islam II Peringkat Kebangsaan.
13. Ab. Aziz Yusof, *Pengenalan kepada Usahawan dan Keusahawanan* (Kuala Lumpur: Scholar Mind Publishing, 2010)
14. N.K. Denzin, Y.S. Lincoln, Y. S. (Eds.). (2003). *Collecting and Interpreting Qualitative Materials.* (2nd Ed). Thousand Oaks, CA: Sage.
15. S. Habibollah. A. Hamed, N. Davoud, International Journal of Business and Management **5**(7), (2010)
16. Al-Quran al-Karim Terjemahan (2012). Kajang, Selangor.
17. M.A. Abdullah, A. Hoetoro, International Journal of Management and Business Research **1**(1), 35-46 (2011)
18. Adnan Alias, Mohamed Dahlan Ibrahim, *Keusahawahanan Islam* (Petaling Jaya: Prentice Hall, 2002)
19. Global Entrepreneurship Monitor, *"Report on Social Entreprenurship 2009"* (Report) (London, United Kingdom, 2009)
20. H. Jiao, Social Enterprise Journal **7**(2), 130–149 (2011) doi:10.1108/17508611111156600
21. Mohd. Fadzillah Kamsah, Muhammad Zakaria, *Langkah Bijak Usahawan Terbilang* (Kuala Lumpur: Telaga Biru Sdn Bhd., 2008)
22. Muhammad Ali Hashim, *Khalifah Ganti Superman* (Kuala Lumpur: Utusan Publications & Distributors Sdn Bhd., 2013)
23. Mohamad Mustapha, *Keusahawanan Islam* (Petaling Jaya: Prentice Hall, 2002)
24. P.J. Murphy, S.M. Coombes, Journal of Business Ethics **87**, 325-336 (2009)
25. A. Nicholls, *Social Entrepreneurship: New Models of Sustainable Social* Change (Oxford University Press, Oxford, 2008)
26. Radin Siti Aishah, Zaidatol Akmaliah, "Keusahawanan Sosial Dari Perspektif Islam". *In Proceeding of the International Conference on Research in Ethics Education 2013.* Istanbul Turki (2013)
27. E. Shaw, (2011) Toward A Theory of Entrepreneurship : The Significance and Meaning of Performance and the Emotion Management of Entrepreneurs *A Thesis submitted for the Degree of Doctor of Philosophy By Elizabeth Shaw Brunel School of Business and Management Brunel Unive.* Brunel University.
28. Shuhairimi Abdullah, Journal of Islamic Human Advanced Research **3**, 322-345 (2013)
29. Suhaimi Mhd Sarif, Abdullah Sarwar, Yusof Ismail, Middle East Journal of Scientific Research **14** (11), 1463-1470 (2013)
30. C.S. Teresa, L.N. Patricia, Journal of Research in Marketing and Entrepreneurship 11(1), 49-65 (2009)
31. J. Weerawardena, G.S. Mort, Journal of World Business **41**(1), 21-35 (2006)
32. S.A. Zahra, H.V. Rawhouser, N. Bhawe, D.O. Neubaum, J.C. Hayton, Strategic Entrepreneurship Journal **2**(2), 117-31 (2008)

A Literature Review on Quality Teacher's Working Life

Siti Intan Diyana Ishak[1], *Nordin* Abd Razak[2], *H.* Hussin[1], *Nur Suriaty* Fhiri @ Daud[1] and *Aida Shakila* Ishak[1]

[1]School of Human Development and Technocomunication, Universiti Malaysia Perlis, Perlis, Malaysia
[2]School of Education Studies, Universiti Sains Malaysia, Penang, Malaysia

Abstract. Quality of work life is becoming an imperative issue to achieve the goals of the organization in every sector whether it is education, tourism, service sector, manufacturing, banking sector and other. Quality of work life it about work environment, reward, organizational commitment, recognition, participative management, work life balance, welfare facilities, proper grievances handling, job satisfaction and other. High quality of work life can give a result in better organizational performance, effectiveness and innovativeness. Quality of work life also affects the social responsibility. This is because quality of work life can improves the family life as well as work life of the individual. This paper focuses and analyse the literature review on the quality of teacher work life.justified.

1 Introduction

Today in many organizations, employees are not satisfied with their work. Quality of work life has become one of the organizational issues in the modern era. "Quality of Work Life (QWL)" is a human resource management concept which is used to improve the work life of employees [1].Quality of work life is the quality of relationship between employees and the total working environment, with human dimensions, technical and economic consideration [2]. [12] Believes that most studies done on the quality of work in 60s focused on psychology of individuals and their perception of the industrial environment. Work life quality had been defined as the sum of perceived stress, or lack of stress, and perceived satisfactions, or dissatisfactions [12]. Quality of work life is the quality of relationship between employees and the total working environment, with human dimensions, technical and economic consideration [2].

Quality of work life is a process by which interest groups in the organization will learn how to work better together, how to determine for themselves what activities, changes and improvements are desirable and effective to the organization to be more efficient; as teachers at technical schools have different integration in the organization, work life and social dependence, overall living space from the other theoretical schools, thus, leading to different development of human capability and growth opportunities [3]. [5] Also points out that the concept of "quality of work life" in 80s was a general term to include a set of conditions in different fields such as organization, work environment, and partnership. Efforts to improve the quality of life and changes healthy work environments have produced positive results not only in mental health of organizations, but also in participation rate and economic efficiency.

[13] Defined quality of work life as, "A process by which an organization responds to employee needs by developing mechanisms to allow them to share fully in making the decisions that design their lives at work". [20] Explains quality of work life in terms of eight broad conditions of employee that constitute desirable quality of work life. Fig. 1 shows the component-component of quality of work life.

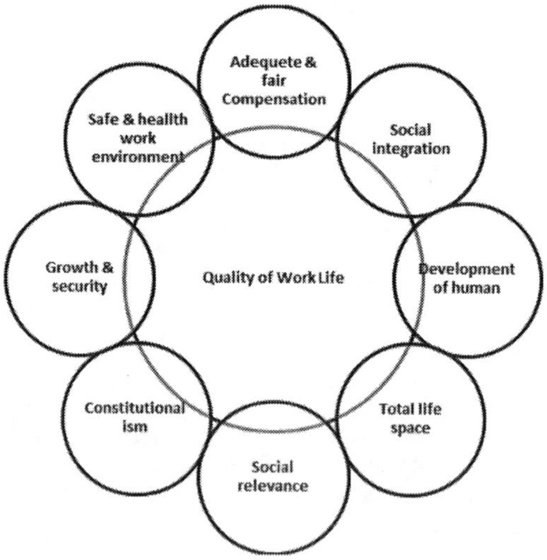

Fig. 1. Component-component in quality of work life.

Besides that, he proposed the following criteria for measuring quality of work life:

i) Adequate and fair salary: The committee on fair salaries defined fair pay as, "the pay which is above the minimum salary below the living age".

ii) Safe and healthy working conditions: Most of the educational organizations provide safe and healthy working conditions due to humanitarian requirements and legal requirements.

iii) Opportunity to use and develop human capacities: The extents that the teacher can exercise more control over his her work, and the degree to which the teaching job embraces an entire meaningful task, but not a part of it. Further quality of work life provides for opportunities like autonomy in work and participation in planning in order to use human capabilities.

iv) Opportunity for career growth: Future opportunity for continued growth and security by expanding one's capabilities, knowledge and qualifications contributes to Quality of work life. Opportunities for promotions are limited in case of teachers either due to educational barriers or due to limited openings at the higher level.

v) Social integration in the work force: Social integration in the work force can be established by creating freedom from prejudice, supporting primary work groups to have a sense of community and inter personal openness, egalitarianism and upward mobility of teachers.

vi) Constitutionalism in the work organization: Quality of work life provides constitutional protection to the teachers only to the level of desirability as it hampers work. It happens because the educational management's action is challenged in every action and bureaucratic procedures that need to be followed at that level and is provided to employees on such matters as privacy, freedom to speak, and equity.

vii) Work and quality of life: Quality of work life provides for the balanced relationship among work, non-work and family aspects of life. In other words family life and social life should not be strained by working hours including overtime work, work during inconvenient hours travel, transfers, and vacations.

viii) Social relevance of work: Quality of work life is concerned about the establishment of social relevance to work in a socially beneficial manner. The self-esteem work life of secondary school teachers. In the changing environment of modern technological advancement the traditional concept of the teaching profession is subjected to rapid changes. A person who enjoys the work and derives satisfaction alone can perform in the best perfect manner. The fulfillment of personal needs and goals leads to satisfaction well-being and happiness. But how far and how long could an individual be satisfied in the profession, if it is full of work related stress and strain.

1.1 Objective of Paper

1. To find out the scope of the quality of teacher working life in the research area.
2. To find out the dimensions used most and least under quality of teacher working life.
3. To examine the relationship between dimensions and quality of teacher working life.

2 Methodology

The study is based on previous research method to review literature on quality of teacher working life. The secondary data is drawn from books, journals and various reports published by the agencies working in this field.

3 Results and Discussion

There are only few researchers' conducted studies to analyze the quality of teachers work across the world. The researches of quality of teacher work life are shown below in chronological order: Results are taken from the research papers which have been done their research on the quality of teacher working life and their dimensions. After doing deep study of the quality of teacher working life which shows the methods of sampling, test used, respondent's rate, method for data collection, research area and findings. List of the important research paper are discussed below:

[7] Investigated effects of teacher quality of work life in secondary schools on commitment and sense of efficacy. The aim for these studied was to model that ties organizational characteristics of the work place to important behaviours, attitudes and psychological characteristics of teachers that effect their teaching is presented. Second, measures of quality of work life in eight schools are examined, along with the relationship between the quality of work life variables and measures of teachers' commitment and sense of efficacy. Finally, the effects of school structure on quality of work life are examined. The results suggested that qualities of work life measures are strongly associated with both dependent variables. The discussion focuses on the implications of the findings and the model for reform strategies.

[18] Investigated of quality of working life amongst teachers. The samples were from 100 secondary schools and 200 primary schools were selected as part of random schools in England. The findings for these studied was teachers have more job satisfaction than other workers, but are neutral about job commitment, teachers feel well supported at work and have positive working relationships with colleagues, teachers feel well supported at work and have positive working relationships with colleagues, secondary teachers would like more responsibility and involvement but primary teachers are undecided about this and Senior staff in primary schools enjoy a higher quality of working life than others in several respects , although primary head teachers feel unsupported. Besides that, teachers experience job security and are satisfied with communication in their schools, teachers are dissatisfied with their salaries and report more stress than other employees, roles and responsibilities impact on quality of working life, as do hours worked and job commitment

is affected by levels of job satisfaction and stress in both sectors, and by levels of support in primary schools.

[8] Studied the examined relationships between quality of work life and work commitment, work stress and work satisfaction. Besides that, this study also investigated the differences in work life quality, work commitment, work stress and work satisfaction based on demographic variables and the relationships amongst the variables. This study was quantitative and conducted with 110 permanent teachers at a government secondary school in Kuching, Sarawak, Malaysia. The findings for this research were found the respondents only had moderate work life quality. Relationships between work life quality and work commitment, stress, and satisfaction also moderate but there were no differences in the quality of work life based on demographic variables but there were differences in work commitment based on gender. The results suggested that work commitment, work stress and satisfaction were not significantly correlated.

[19] Studied the perception of college teachers about quality of work life. The studied aims to help the college teachers to know the level of perception towards QWL and to enhance the same by the educational administrators. The sample for these studies was 239. The findings were the overall quality of work life 59.0% of the respondents have high level of quality of work life and 41.0% of the respondents have low level of quality of work life. Beside that there is no significant difference between sex, type of family, age, various income levels, type of college, native place and their perceived levels of overall quality of work life. There were no significant difference between the department, type of family, the designation, the various income levels of the respondents and their perceived levels of overall quality of work life in teaching environment.

[3] Investigated to compare the rate of quality of work Life among technical and theoretical high school teachers in Kurdistan province in Iran. This studied used cluster sampling with consisted of 410 high school teachers. The data were analyzed by using descriptive and inferential statistics based on the questionnaire of quality of work Life (QWL). Findings for studied show the quality of work Life among technical and theoretical high school teachers in Kurdistan was average and there was no meaningful relationship between high school type (technical and theoretical) and quality of work life. Besides that, there were no differences in types of high schools ' QWL and all technical and theoretical high schools in Kurdistan province have the same quality of work life.

[15] Studied the relationship between quality of work life and teacher motivation among secondary school English as a foreign language (EFL) teachers in Tehran, Iran. The sample for these studied was 160 teachers and addition, 30 of the participants were randomly selected to take part in follow-up interviews which asked why they felt the way they reported. The findings revealed that the participants enjoyed a medium level of quality of work life and experienced a medium-to-low level of motivation and a significant relationship was found between motivation and quality of work life categories. Meanwhile, the subscales of the quality of work life that best predicted teacher motivation were identified.

[9] Investigated the quality of work life for secondary school teachers from Mysore City. This studied used adopting stratified random sampling technique and the sample was 100 teachers. Besides that this studied was adopted the descriptive survey method. The results of this studied were a majority (70.2%) of the secondary school teachers in Mysore was found to possess an average level of quality of work life and (13.9%) were found to possess low level of quality of work life and only (15.9%) of the teachers possessed high level of quality of work life. There was significance difference of Quality of Work life of male and female secondary school teachers comparing teachers mean scores it is found that the female teachers have a high quality of work life than their male counterparts. Meanwhile there was no significance difference between above 10 years' experience and below 10 years of experienced secondary school teachers in their Quality of Work Life and no significant difference is the quality of work life of teachers belonging to government, private aided, and private unaided secondary school teachers.

[10] Investigated the assessing the quality of work life and its components the primary school teachers of the Isfahan city. The population was composed of 862 teachers in Isfahan and 120 Samples were selected based on statistical estimates and simple random sampling. The results for these studied were the quality of life of teachers is less than average. Its means they do not enjoy the good quality of life. However, the desire and motivation to the teachers was high but component of the salary of the teachers had the greatest dissatisfaction. The results indicate that female teachers have the higher quality of life than male teachers. Beside that studied shows the results also suggest that there is no difference between single and married teachers' comments on the quality of work life. Results also show that there is no difference between the views of teachers with less than a bachelor's degree and bachelor's degree or higher in about the quality of work life and its components.

[16] The studied explore various dimensions of quality of work-life (QWL) as it affects the life and attitude at work of teachers of private universities in Lahore, Pakistan. This study was quantitative and conducted with 360 faculty members from private universities in Lahore, in order to find out their perceptions of QWL, and its spill-over effect on employee commitment, engagement, job involvement and reputation of the university. The findings for this research was found that perceived value of work, work climate, work-life balance and satisfaction with relationships in life were the major factors which shaped work attitudes and employee perceptions of overall quality of work-life. The research was carried out in order to discover how teachers in private universities in Pakistan view their work environment and whether this working environment is helping them to become valued human beings or not. Besides that this research makes two contributions, scholarly and practical. The scholarly contribution highlights that the dominant constructs of

QWL play an important role in shaping attitudes towards work, life and relationships of teachers of private universities. On a practical level, the study hints at the possible implications of dissatisfaction and imbalance within employee commitment and engagement, and even the reputation of the university.

[11] Investigated quality of work life of employees in private technical institutions. These studied helps the technical institution employees to know the level of perception towards quality of work life and to enhance the same by the management. The sample consists of 109 employees of a technical institution the results reveals that male employees are more satisfied than female employees. Beside that there are no significance relationship between demographic characteristics of employees and quality of work life. These studied also reveals that Adequacy of Resources are more correlated and Training & Development are less correlated with QWL in teaching staffs and in case of non-teaching staffs compensation & rewards are more correlated and work environment are less correlated with QWL. Furthermore the correlation analysis also reveals that all the dimensions of QWL are positively correlated with QWL of faculties, which indicates that enhancement in the dimensions of QWL, can lead to increase the overall QWL of faculties.

[4] Studied explored about perceived factors that influence teachers' quality of work life in primary schools in one education district in Trinidad and Tobago. Beside that's to develop separate Quality of Work-life (QWL) categories and make comparisons among schools. The sample for these studied consisted of randomly selected 405 teachers from thirty (30) government and assisted schools primary schools from the St. George Education taking into account variables such as size, school type, demography and sex. Furthermore studied also examined the relationship between the demographic and school-level quality of work life factors as well as the predictive power of each independent factor on overall quality of work life. These studied was quantitative method using a cross-sectional design. The findings indicated further that there was a moderate to high correlation among the quality of work life factors and collegial relations among teachers had the greatest influence on their work life in school.

4 Conclusion

After the study of literature review on quality of teacher work life, it is clear out that an education organization cannot get efficiently and effectively outcomes from the employees' without quality of work life. Quality work life is the shared responsibility not only of the management and employees, but also by the society.

To improve quality of teacher work life is first to identify and then try to satisfy employee's important needs through their experience in their working environment. Depending upon the situational requirements, management may select the relevant needs of the employee's to improve them with a short term plan. There is a significant association between quality of work life total and quality of life in teaching environment total.

[6] It can be conclude that most research of QWL has done on the academic section which showed faculty relations and community services is the most positive elements in faculty's work life and in present scenario highly satisfaction in the female regarding QWL dimensions compared to male. The purpose of this paper was to find out the important dimensions of quality of teacher work life from literature review. More than 25 literatures were investigated out of that 11 literatures discussed deeply and the most extracted factors were gauged. The aims of the paper to find out the variable are affect the quality of teacher work life in the researcher life. The finding shows that there are some dimension been used frequently in the literature rather than other dimension. Table 1 is shows the variables are used in most researchers.

Table 1. Variable used in research quality of teacher work life

Extremely Used	Average Used	Least Used
Job satisfaction Demographic factor Pay and benefits Supervision Organization commitment Growth and development Safety and healthy environment	Social integration Attitude and perception Employee participation Work life balance and relationship Rewards Team work Welfare and opportunities Autonomy	Delegation of authority Organization citizenship behavior Training and development Financial ratio Equal job opportunities

5 Suggestions

This paper only focuses in quality of teachers work life in the research area, the dimensions used in most under quality of work life and the relationship between dimensions and quality of teachers work life. The convey further researchers who are interested to do research on quality of teachers work life and having a confusion how to relate a quality of teachers work life with their dimensions because there have been done more research on quality of work life and their variables. Besides that, further researchers also can refine the definition of quality of work life.

There are some dimensions such as such as team work, attitude and perception, pay and benefits, social integration, participative management, organization commitment, growth and development , job satisfaction, , safety and healthy environment and participative management, salary, employee participation, welfare opportunities, rewards are have been used by mostly researchers continuously. There are many dimensions still untouched which affect quality of work life such as growth and profitability of organization, organization citizenship behaviour and personality.

References

1. P. Archana, B.K. Jha, Global Journal of Management and Business Research: A Administration and Management **14** (11), (2014)
2. A.F. Chelte, *Organizational commitment, job satisfaction and quality of work life. U.M.I.* (1983)
3. H. Farideh, M. Bahram, International Journal of Vocational and Technical Education **4**(1), 1-5 (2012)
4. G. George, International Journal of Humanities Social Sciences and Education (IJHSSE) 1(10), (2014)
5. J. Huzzard, The Convergence of the Quality of Working Life and Competitiveness: A Current Swedish Literature Review(National Institute for Working Life, Stockholm, 2003)
6. L.K. Johnsrud, *Studied on Quality of faculty work life* (the University of Hawaii, 2006)
7. L.S. Karen, Effects of Teacher Quality of Work Life in Secondary Schools on Commitment and Sense of Efficacy. School Effectiveness and School Improvement **9**(1), 1-27 (1998)
8. S.H. Kian, W.T. Kok, B. Suriani, Technology Journal 1–15 (2010)
9. N.D. Manju, International Journal of Education and Psychological Research **3**(2), (2014)
10. E. Mohammad Kazem, K. Mahnaz, N. Fateme, Interdisciplinary Journal of Contemporary Research in Business **3** (9), (2012)
11. T.S. Nanjundeswaraswamy, D.R. Swamy, Management Science Letters, **5**(1), 65-78 (2015)
12. D.M. Pelsma, G.V. Richard, R. G. Harrington, J.M. Burry. 1989. Measurement & Evaluation in Counseling & Development **21**(4), 165–176 (1989)
13. S.P. Robbins, *Organizational behaviour Concepts, controversies, applications* (New Delhi: Prentice-Hall of India Private Limited, 1999)
14. G.M. Reyan, Theoretical basic for the QWL concept (University of Siena, 1995)
15. B. Sasan, G. Yahya, Australian Journal of Teacher Education **37**(7), (2012)
16. A. Seema, I. Maryam, Quality Assurance in Education **21** (3), 282 – 298 (2013)
17. A. Shirkovand, Studied the relation between the quality of working life and performance of insurance Company in Tehran (Tehran: Allameh, 2007)
18. L. Sturman, (2003). Survey of Quality of Working life in Teachers. Retrieved from http://www.nfev.ac.uk/research-areas/pims-data/sum
19. B.P. Subburethina, *M. Umaselvi, K.N. Senthil, Quality of work life: Perception of college teachers* (Hallamark Business School, Trichy, Tamilnadu, India, Majan,University College, Muscat, Sultanate of Oman, Anna University Chennai, 2010)
20. R.E. Walton, Improving the quality of work life (Harvard Business Review, 1974)

The Discussion of Social Entrepreneurship: Review of the Literature

Nur Suriaty Daud@Fhiri[1,*], *Siti Intan Diyana* Ishak[1], Suhairimi *Abdullah*[1], A. A. *Azmi*[1] , Aida Shakila Ishak[1,*], and Z. Ahmad[2]

[1]School of Human Development and Technocomunication, Universiti Malaysia Perlis, Perlis, Malaysia
[2]Institute Pendidikan Guru Malaysia Kampus Perlis, Perlis, Malaysia

Abstract. The purpose of this article is to explore the various discussion of social entrepreneurship. Social entrepreneurship provides a unique opportunity and assumptions to question, challenge and rethink from different perspective of management and business research. This article offers a comparative analysis of commercial entrepreneurship and social entrepreneurship using a prevailing analytical model from commercial entrepreneurship. The analysis highlights key differences and similarities between commercial entrepreneurship and social entrepreneurship. This article also presents a framework on how to approach the social entrepreneurial process more effectively and systematically. Social entrepreneurship is a process of creating value by combining resources in new concepts. These process are intended primarily to get the opportunities to create social value by stimulating social change or achieve social needs. When viewed as a process, social entrepreneurship involves the offering of products and services but can also refer to the creation of new organizations. This article focuses and analyses the literature finding of social entrepreneurship.

1 Introduction

Commercial entrepreneurship is important in the growth of the business sector as well as a driving force behind the brisk expansion of the social sector. In addition commercial entrepreneurship is also becoming an important factor in the development and good of societies, this is supported by most academics and economists. Besides that, commercial entrepreneurship is considered to be an important mechanism for economic development through employment, innovation and welfare effects [42,1,53,54]. The dynamics of commercial entrepreneurship can be vastly different depending on institutional context and level of economic development. There are considerable differences across countries in the orientation of entrepreneurial activities [26]. The nature and structure of entrepreneurial activities varies across countries as reflected by, for example, the relative volumes of necessity and opportunity entrepreneurship.

Furthermore, entrepreneurial activities considered to be an important mechanism through employment, innovation and corporate social responsibilities. In stages of depression, due to limited demand which implies limited profits forced private sector of economy to not develop entrepreneurial action. Social economy is a business activity involve government and the private sectors. Social needs and or are related to weak social groups covered by the economic entities that are activated within the context of social economy.

[28] Stressed that commercial entrepreneurship is an attempt to create value through recognition of business opportunities. [49] states that commercial entrepreneurship is a way of thinking, reasoning, and acting that is opportunity obsessed, holistic in approach and leadership balanced. Based on that though of the study, the situation need someone to manage the activities. The important people behind the sense is entrepreneurs but the truth of individuals that think about social vision is social entrepreneurs. [6] states that social entrepreneur is a part breaker with a powerful new idea who combines visionary and real-world problem-solving creativity, has a strong ethical fiber, and is totally possessed by his or her vision for change. This sale entrepreneurs characteristic are mission leader and social value creator. So, social entrepreneurs was the individuals or institution that run or manage the activities known as social entrepreneurship.

Social entrepreneurship as a field that mixes economic and social value creation has a long legacy and a global company. The global effort of Ashoka, founded by Bill Drayton in 1980, to provide seed funding for entrepreneurs with a social vision [3]. Consideration of the role of social vision in the social entrepreneurship movement requires an understanding of the concept. However, there has been some debate about the definition of social entrepreneurship [12,31].

* Corresponding author:nursuriaty88@gmail.com

1.1 Objective of Paper

1. Study of social entrepreneurship in research area.
2. To examine the key differences between commercial entrepreneurship and social entrepreneurship.
3. To examine the similarities between commercial entrepreneurship and social entrepreneurship. Maintaining the Integrity of the Specifications.

2 Review of Literature

There are only few researchers' conducted studies to analyze the social entrepreneurship. This is because social entrepreneur only few in the social sector. The researches of social entrepreneurship are shown below in chronology area:

The concept of commercial entrepreneurship was first defined more than 250 years ago, many researchers have seized it as one of the special forces of human nature. The practice of entrepreneurship as old as trading between villages and tribes. Many different and useful approaches have been used to explain and to analyze entrepreneurship. They have tended to fall within three main streams of research, which contain a focus on the results of entrepreneurship, the causes of entrepreneurship, and entrepreneurial management [28]. In the first stream of research, economists have explored the results and impacts of entrepreneurship. For example, [42], in his seminal article, examined entrepreneurship as a key process done which the economy as a whole is innovative. The second stream of research has intensive on the entrepreneurs themselves. Research in this stream studies entrepreneurship from a sociological and psychological perspective [14,32]. Last but not least, the third stream has focused on the entrepreneurial management process. This various literature includes research on how to substitute innovation within established corporations [9,10], start-ups and venture capital [50], organizational life cycles [38], and analysts of entrepreneurial success [15,23].
Nowadays, social entrepreneurship activities with an inserted social needs has been rises. The growth of the activities (including employment, innovation, and corporate social activities) in the last decade may suggest existed to new forms of organizing for business innovation that is social entrepreneurship. Social entrepreneurship is the field in which entrepreneurs fit their activities to be directly tied with the goal of creating social value. In addition, they often act with little or no intention to gain personal profit. A social entrepreneur "combines the passion of a social mission with an image of business-like discipline, innovation, and determination commonly associated with, for instance, the high-tech pioneers of Silicon Valley" [20]. The term of social entrepreneurship is more popular nowadays. However, confusion and uncertainty are constantly noted about what exactly a social entrepreneur is and does. Besides that, it is not clear how social entrepreneurs create large-scale change. There is should be more define in what is meant by the term of social entrepreneur. Social entrepreneurship and commercial entrepreneurship have their differences and similarity. Social entrepreneurship is still emerging for academic inquiry.

Social entrepreneurship is the innovative activity with a social objective in either the institution profit sector, such as in social purpose commercial ventures [19,25] in corporate social entrepreneurship [4] in the nonprofit sector, across sectors, such as hybrid structural forms which mix profit and nonprofit approaches [20]. Besides that, under the narrow definition, social entrepreneurship usually refers to the situation of applying business expertise and market based skills in the nonprofit sector such as when nonprofit organizations develop innovative approaches to earn profit [39,48]. Moreover, mutual across all definitions of social entrepreneurship is the fact that the primary drive for social entrepreneurship is to create social value rather than personal and shareholder wealth [57], and that the activity is characterized by innovation, or the creation of something new and rare rather than simply the replication of existing enterprises. The dominant driver for social entrepreneurship is the social problems and the particular organizational form a social enterprise takes should be a decision based on which format would most effectively mobilize the resources needed to faces that problems. Therefore, social entrepreneurship is not defined by legal form, as it can be pursued through various activities. Certainly, the examples of social entrepreneurship can be found within nonprofit institution, business, or governmental sectors.

Based on business management literature, entrepreneurship is a unique set of activities carried out by individuals or institution with an exceptional mind-set in order to maximize profit. But it different with social entrepreneurship because social entrepreneurship more on creating social values rather than maximize profit. Therefore, the process is closely tied to success. In addition, the business literature differentiates entrepreneurs from business people by including words such as: entrepreneurs "create needs"; while business people "satisfy needs" (2010 Global Report: Global Entrepreneurship Monitor, 2011).

Social entrepreneurs are conceptualized as individuals who see the innovation to the world and grab the opportunities to create social values. They seize opportunities that other people would not noticed. They will take a risks differently than others. Even the term of social entrepreneur was used by the economies growing rapidly, the field of social entrepreneurship lacks knowledge and is in its infancy compared to the wider field of entrepreneurship. The individual success of solving complex social problems are being used to legitimize the field of social entrepreneurship. For example, Stanford University launched Social E Lab as part of its Entrepreneurial Design for Extreme Affordability course, which promotes the use of entrepreneurship principles to solve social and environmental problems [41].

The absence of consensus on a research topic frequently results in scholars working independently and failing to build upon one another's work, therefore knowledge cannot be accumulated [7].

Social entrepreneurship is a process of combining resources in a new ways to create values. These resource combinations are planned primarily to exploit and explore opportunities to create social value by stimulating social needs. Social entrepreneurship involves the offering of services and products but can also refer when see as a process [31]. Social entrepreneurship or entrepreneurial activity with an embedded social purpose has been on the increased in recent decades. A partial indicator of this field is revealed by the rise in the number of nonprofit organizations, which increased 31% between 1987 and 1997 to 1.2 million, exceeding the 26% rate of new business formation (The New Nonprofit Almanac and Desk Reference, 2002).

3 Results and Discussion

[45] defined social entrepreneurship as "The pursuit of opportunity beyond the tangible resources that you currently control." Thus, with this definition, emphasis is placed upon how opportunity can be recognized, the process of committing to an opportunity, gaining control over the resources, managing the network of resources that may or may not be within a single hierarchy, and the way in which participants are rewarded [45, 47]. The entrepreneurial organization emphases on opportunity, not resources. Entrepreneurs must commit rapidly, but cautiously, to be able to accommodate as new information arises. The process of commitment becomes multistage, limiting the commitment of resources at each stage to an amount sufficient to create new information and success before additional resources are required. Furthermore, the entrepreneurial organization uses the resources that invention within the graded control of others and should achieve the system as well as the order.

[37] Did a study on social entrepreneurship on economic development. He focused on factor, efficiency, or innovation of social entrepreneurship in economic development. Entrepreneurship offers a new competition, and as such promotes improved productivity and healthy economic competitiveness [51]. The term social entrepreneur is ill-defined [5, 52], it is fragmented, and it has no coherent theoretical framework [52]. [42], an entrepreneur is an innovator who implements entrepreneurial change within markets, where entrepreneurial change have five manifestations: 1) The introduction of a new/improved goods; 2) the introduction of a new method of production; 3) the opening of a new market; 4) the exploitation of a new source of supply; and 5) the carrying out of the new organization of any industry. [32], the entrepreneur is a person with a high need for achievement. This need for achievement is directly related to the process of entrepreneurship. Entrepreneur is an energetic moderate risk taker. [29] The entrepreneur recognizes and acts upon market opportunities. The entrepreneur is essentially an arbitrageur. [44] Entrepreneurs take initiative, organize some social and economic mechanisms and accept risks of failure. [11] the entrepreneur is characterized principally by innovative behaviour and will employ strategic management practices in the business.

Kate Barr (2008) studied on the topic of social enterprise. She finds out that social enterprise are a nonprofits, businesspeople interested in nonprofits, and foundations. She keep tripping over the lexicon, though, because she don't think that the commonly used terms are certain, universal, or completely clear. It seems that the "field" encompasses a number of different types of organizations with different definitions and identifiers. Furthermore, looked around the other day to compile definitions for these terms in regular use. Social enterprise is defined by Social Enterprise Alliance as "an organization or venture (within an organization) that advances a social mission through entrepreneurial, earned income strategies." This often reflects a market-based effort to receive earned income in direct exchange for a product or service. Examples of social enterprise include:

- A substance abuse treatment facility operating a moving company
- An organization promoting college enrollment that provides workshops to educators
- A youth services organization opening a pretzel shop or ice cream shop franchise
- Goodwill thrift stores

Social entrepreneurs are defined by Ashoka as those who "act as the change agents for society, seizing opportunities others miss and improving systems, inventing new approaches, and creating solutions to change society for the better. While a business entrepreneur might create entirely new industries, a social entrepreneur comes up with new solutions to social problems and then implements them on a large scale.

A socially responsible business is defined as a venture (generally for-profit) that seeks to "leverage business for a more just and sustainable world" (Social Venture Network) or "help create a better world by building healthy communities, promoting economic equity, and fostering a clean environment" (Social Investment Forum). In addition to generating a profit for shareholders, these businesses have goals in the areas of economic development, employment, environmental practices, or ethical business practices.

The key distinctions between these terms are the following:

- Social enterprises are defined by revenue source
- Social entrepreneurs are defined by innovative vision and strategy
- Socially responsible businesses are defined by the intention and goals of a for-profit business

It may seem to some people that the definitions are just semantics, but Kate Barr think they're important if we want to create resources, find capital, and develop a knowledge base. The needs and demands are probably different if the focus is on revenue sources rather than a game changing strategy.

The environment shaping the economy affects the dynamics of entrepreneurship within any given country. This environment is marked by interdependencies between economic development and institutions, which affect other characteristics, such as quality of governance, access to capital and other resources, and the perceptions of entrepreneurs. Institutions are critical determinants of economic behavior (North 1990) and economic transactions [27] in general, and they can impose direct and indirect effects on both the supply and demand of entrepreneurs.

[36] and [37] define competitiveness according to country economic development, distinguishing three specific stages: (1) factor-driven stage, (2) efficiency-driven stage and (3) innovation-driven stage; and two transitions between these stages. Countries in the factor-driven stage compete through low cost efficiencies in the production of commodities or low value-added products. The first stage is marked with high rates of non-agricultural self-employment. Sole proprietorships is the self-employed probably account for most small manufacturing firms and service firms. Almost all economies experience this stage. These countries neither create knowledge for innovation nor use knowledge for exporting.

[48] was defined social entrepreneur are people who realize where there is an opportunity to satisfy some unmet need that the state welfare system will not and cannot meet and who gather together the necessary resources (generally people, often volunteers, money, and premises) and use this to 'make a difference'.
[20] though that social entrepreneur play the role of change agents in the social sector by adopting a mission to create and sustain social value, recognizing and relentlessly pursuing new opportunities to serve that mission, engaging in a process of continuous innovation, adaptation, and learning: acting boldly without being limited by resources currently in hand: and exhibiting the heightened sense of accountability to the constituencies served for the outcomes created.

This article also presents a framework on how to approach the social entrepreneurial process more effectively and systematically. Definition of entrepreneurship by Stevenson's was used and build on [40] analytical framework from the entrepreneurial management literature. Sahlman's model more captures on the key elements that are critical thoughts for commercial entrepreneurship and thus provides a strong basis for developing a framework for social entrepreneurship. This model shown the creation of a dynamic fit among four interrelated components: the people, the context, the deal, and the opportunity (PCDO) [40]. This is because these elements are dependent and are situationally, the entrepreneur should manage the fit and must adapt continuously to new situations.

The first interrelated components is people. People is defined as those who actively involved in the venture or who bring properties to the venture. They (people) include both those within the organization and external whose should be participate for the venture to succeed. People's contacts, attitudes, skills, knowledge, goals, and values provide the resource mix that contributes centrally to success. The whole person with multiple motivations and capacities creates the energy and controls the environment of the outcome whether in nonprofit or in for-profit organizations.

The second interrelated components is context. Context is defined as those elements outside the control of the entrepreneur that will influence success or failure. Contextual factors include the macro economy, tax and regulatory structure, and sociopolitical environment. However, labor, religion and politics are examples of specific contextual factors that can frame the opportunities and the risks that a new venture faces involve in economic environment, tax policies, employment levels, technological advances, and social movements. With this definition, it is clear that one of the critical elements for success is defining those elements that must be consciously dealt with, and those that can simply play out as they will. In addition, leaving out a single critical element of context can be the cause of failure.

The third interrelated components is deal. Deal is the element of the compact that defines who in a venture gives what, who gets what, and when those deliveries and receipts will take place. Each transaction delivers a bundle of values. They include economic benefits, social recognition, autonomy and decision rights, satisfaction of deep personal needs, social interactions, fulfillment of generative and legacy desires, and delivery on altruistic goals.

Last but not least, the next interrelated components is Opportunity. Opportunity is defined as "any activity requiring the investment of scarce resources in hopes of a future return" [40]. One of the significant problems in the study of entrepreneurship is that the definition of opportunity is not certainly shared by the several populations who should work together to create transformation. Often change affects power relationships, economic interests, personal networks, and even self-image. A critical factor that creates motivation for joint action arises out of the ability to create a common definition of opportunity that can be shared. Even slight perturbation in one of these PCDO domains can have great inference for the others. Changing people often requires a different deal. Furthermore, changing context can render the skill set of one group obsolete and make another group's skills more significant. Thus, different opportunities are perceived in differing situations, and adjusting a deal may attract new players and drive away the old. Entrepreneurs should consciously manage the dynamic fit between these elements (see Fig. 1 [40]).

PCDO framework from commercial entrepreneurship can be applied to social entrepreneurship. First highlights the key of similarities, then identifies significant differences in comparative analysis of the two forms of commercial entrepreneurship for each of the PCDO elements. This research also offers the basis for illustrate out lessons that can be useful for social entrepreneurs and researchers. In the following analysis, the first step is examine opportunity and context, which together determine the scope of the opportunity. Then, turn to a discussion of the people and deal dimensions of the venture.

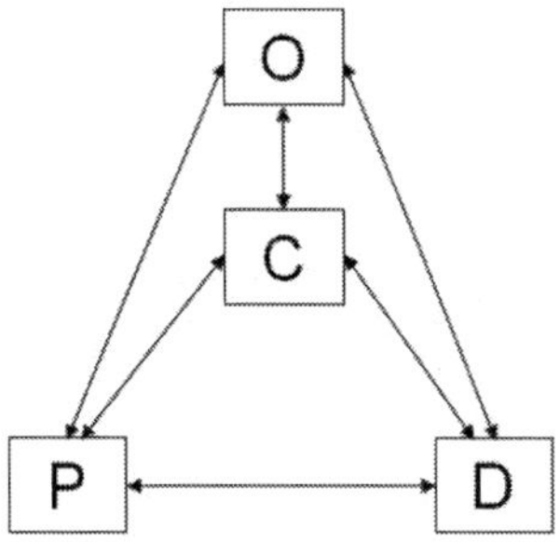

Source: Sahlman (1996)
PCDO: people, context, deal, opportunity.

Fig. 1. PCDO Framework.

Last 2006, Nobel Peace was gave award to Muhammad Yunus. Muhammad Yunus is one of the most renowned social entrepreneur. He got a prize for his endeavors in microfinance. The Grameen Bank has been highly regarded as an exemplar of social enterprises that help address poverty in Bangladesh [55,56]. Although, in the Philippines, Sebastian (2010) illustrates the cases about social enterprises such as ECHOstore, Hapinoy, and Rags2Riches. Gawad Kalinga continues to build on its 'bayanihan economics' (Meloto, 2013), fostering its Center for Social Innovation through the Enchanted Farm at Angat, Bulacan – serving as the 'Silicon Valley for social entrepreneurs' [33].

Nowadays, the field of study and practice is continuously involve social entrepreneurship concepts. However, this article review of literature shows that developing a framework on how to approach the social entrepreneurial process more effectively and systematically. This is supported by Noreen Bautista, General Manager of Consulting and Business Services for Social Enterprises (CBS) under the BCY Foundation. She narrates how neophyte social entrepreneurs often struggle to make sense, understand, and organize their strategies once pitted in the 'real world' of social entrepreneurship. Although the field is in constant development through trial-and-error methods, there is a need for frameworks that are simple, easy-to-understand, yet grounded on theory and easy to apply in the real situations.

In recent times, research on business models are more popular. Many researcher and practitioner involved with social entrepreneurship [16,18] have borrowed the concept of business models – modifying and alteration them to fit the contexts of social enterprises' must balance to money-mission relationships [2,8,25,34,35,56]

Social entrepreneurship has been wrote on the conceptualization by many Scholar. As an academic field, Gregory Dees is one of the pioneers of social entrepreneurship, wrote the seminal article "The Meaning of Social Entrepreneurship" [16] that has been various cited. The other articles and books wrote by Dees focusing on non-profit and for profit social entrepreneurship theory [16,17,18]. Electronic databases such as EBSCOhost and ProQuest feature several peer revised articles related to the topic, signifying the growing interest of scholars all around the world in developing social entrepreneurship concept.

Another framework in doing for-profit social ventures and framing theories for social entrepreneurship, authors have stated the complexity of managing the financial and social bottom lines [16,17,4]. Austin et al. (2006) have tackled the overlaps between the commercial entrepreneurship and social entrepreneur frameworks, building on Sahlman's (as cited in Austin et al, 2006) PCDO framework. Figure 1 describes the components of the framework, namely People, Context, Deal, and Opportunity.

PCDO framework by Austin et al. (2006) proposed a social entrepreneurship framework, putting an enterprise's social value proposition at the center as shown in Fig. 2.

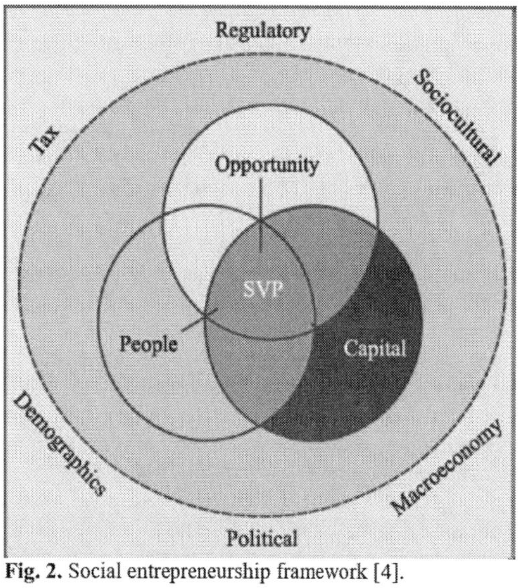

Fig. 2. Social entrepreneurship framework [4].

The social entrepreneurship framework of Austin et al. (2006) shows how the entrepreneurial activities, especially the PCDO components have a connection with the usage of business model frameworks. The business model frameworks guide the researchers used a visual thinking to identifying People, Context, Deals, and Opportunities in organizing value creation. Besides that, in designing a business model, it is important to account for the Context or PESTEL factors and match the strategies based on changing conditions.

The most importance in the foregoing analysis and framework underscore is on the SVP for practitioners. This is because, the fact shows that what drives most social entrepreneurs develop social entrepreneurship and make it more important.

Nevertheless, in practice, social entrepreneurs prefer to achieve an organizational goals as a means to achieve social needs rather than on social impact itself. Although, the practicality of this approach based on a number of factors limit. While people and resources associate the venture's growth are needed, as Austin et al. (2006) analysis has emphasized, handling human and financial resources for social entrepreneurship is really difficult task. The process of procuring resources by social entrepreneurs was very challenges and it becomes an importance of the organization's activities. Thus, social entrepreneurs may become so internally focused on procuring resources to support their organization's growth that the paths to creating social value may become blurred [4]. The resources are often a means of delivering on the SVP, but a broader perspective is needed.

To deliver effectively on the SVP, The social entrepreneur must reach a state of the key components of the framework (opportunity, people, capital, and context). The complicating fact for external alignment is the dynamic nature of the context. Therefore, the SVP that made sense at the time of the venture's founding may in fact evolve dramatically as perturbations in the operating context are continuously occurring [4]. Enduring adjusted to how relative changes can affect the opportunity and the human- and financial-resource environment causing the need for realignment is a critical skill for the social entrepreneur. In addition, researchers should remain conscious of a unique characteristic of the operating context, namely, that the societal demand for social-value creation is huge. Therefore, this will creates a place of opportunities for social entrepreneurs and an associated ever-present extra temptation to address for them. Furthermore, a social entrepreneur's task is then to determine at any given moment how to define the appropriate scope of the opportunity that can be pursued effectively. But, the available people and resources will be dependent on ensuring that the scope is aligned internally. Moreover, the scope can cause a wrong alignment that could effect's the core SVP. Pursuing to address a very broad set of issues, when the organizations resources are spread too little with very limited human and financial resources, may actually result in low social impact. Although, a social entrepreneur may offer significant attention to achieving both external and internal organizational goals, it is also important to remember that social impact can often be more effectively generated from beyond organizational limits.

4 Conclusion

After the study of past research about social entrepreneurship, it is clear out that a social entrepreneurs is a person that important in social entrepreneurship. Then, the social entrepreneurship is important to communities in social needs. Besides that, social entrepreneurship as innovative, social value creating activity that can occur within or across the nonprofit, business, or government sectors. However, most definitions of social entrepreneurship, as well as in the academic literature, focus primarily on social entrepreneurship within and across the nonprofit and business sectors. After the study of literature review on social entrepreneurship, it is clear out that communities needs social entrepreneurs to help them in achieve social needs. An exploration of the history and future trends within the wider social economy movement provides scholars and practitioners with an understanding of the environment or communities and gives them the opportunity to assess various models of social enterprise and their application in practice [43]. Social entrepreneurship focused on social value rather than create profit. Social value is value created by the participants but enjoyed by society as a whole [13]. This article also presents a framework on how to approach the social entrepreneurial process more effectively and systematically. This critical analysis of relevant literature serves as a key step in the understanding of how social enterprise manages and social entrepreneurs can use visual frameworks in management commercial and social value propositions. The researcher encourages scholars and practitioners to enhance and improve the social entrepreneurship through practice. Furthermore, the researcher recommends tough qualitative research in the future, perhaps case study analysis or involved action research in the testing of recommends tough qualitative research in the future, perhaps case study analysis or involved action research in the testing of the proposed framework. Moreover, the social entrepreneurship can be further synthesized with cutting edge management

5 Suggestions

This paper convey further researchers who are interested to do research on social entrepreneurship and having a confusion on the commercial entrepreneurship and social entrepreneurship also with the entrepreneur and social entrepreneur. There are some dimension have been used by mostly researchers continuously such as innovation, risk taker and social entrepreneurship is creating social value.

References

1. Z.j. Acs, B.A. David B, American Economic Review **78** (4), 678-690 (1988)
2. S.K. Alter, Social enterprise models and their mission and money relationships. In A. Nicholls, Social entrepreneurship: New models of sustainable social change Oxford: Oxford University Press, 205-232 (2006)
3. Ashoka Innovators for the Public (2000), *Selecting leading social entrepreneurs*. Washington, DC. Retrieved from http//:www.ashoka.org
4. J.E. Austin, H. Leonard, E. Reficco, J. Wei-Skillern, Harvard Business School Working Paper 05-021, in press.
5. L. Barendsen, H. Gardner, Is the social entrepreneur a new type of leader? *Joshua Venture Newsletter*, 1-7.
6. D. Bornstein, Changing the world on a shoestring. *Atlantic Monthly*, **281**(1), 34-39 (1998)
7. C. Bruyat, P.A. Julien, Journal of Business Venturing **16**(2), 165-178 (2000)
8. M. Bull, Social Enterprise Journal **3**(1), 49-66 (2007)
9. R.A. Burgelman, Management Science **29**, 1349–1364 (1983)
10. R.A. Burgelman, California Management Review **26**, 154–166 (1984)
11. J.W. Carland, F. Hoy, W.R. Boulton, J.C. Carland, Differentiating entrepreneurs from small business owners. *Academy of Management Review*, **9**(2): 354-359 (1984)
12. M. Christie, B. Hong, The Journal of World Business, 1-5 (2006)
13. Clay Shirky. (2008). Here comes everybody. The Power of Organizing without Organizations. Penguin Group. England.
14. O.F. Collins, D.G. Moore, *The enterprising man* (Michigan State University, 1964)
15. A.C. Cooper, A.V. Bruno, *Proceedings of the 35th Annual Meeting*, August 426-428
16. J. Dees, (2001). The meaning of "social entrepreneurship from http://www.caseatduke.org/documents/dees_sedef.pdf.
17. J.G. Dees, B.B. Anderson, (2006) Framing a theory of social entrepreneurship: Building on two schools of practice and thought. Retrieved from http://www.redf.org/from-the community/publications/457.
18. J.G. Dees, B.B. Anderson, International Journal of Entrepreneurship Education **2**, 1-26 (2002).
19. J.G. Dees, B.B. Anderson, International Journal of Entrepreneurship Education (special issue on social entrepreneurship) **2**, 1–26 (2003)
20. J.G. Dees, (1998). *The meaning of "social entrepreneurship."* Comments and suggestions contributed from the Social Entrepreneurship Founders Working Group. Durham, NC: Center for the Advancement of Social Entrepreneurship, Fuqua School of Business, Duke University. Available at http://faculty.fuqua.duke.edu/centers/case/files/dees-SE.pdf
21. J.G Dees, J. Emerson, P. Economy, Enterprising nonprofits: A toolkit for social entrepreneurs (NY: John Wiley and Sons, Inc., 2001)
22. J.G. Dees, J. Emerson, *Economy, Strategic tools for social entrepreneurs: Enhancing the performance of your enterprising nonprofit* (NY: John Wiley and Sons, Inc.,2002)
23. M.J. Dollinger, Academy of Management Journal **27**, 351–368 (1984)
24. J. Emerson, F. Twersky, (Eds). (1996, September). *New social entrepreneurs: The success, challenge and lessons of non-profit enterprise creation*. San Francisco: Roberts Foundation, Homeless Economic Development Fund.
25. J. Emerson, American Economic Review, **78** (4), 678-690 (2003)
26. Erkko Autio (2007), Global Entrepreneurship Monitor. Global Report on High-Growth Entrepreneurship. London Business Scholl.
27. G.W. Jeffrey, The Journal of Economic Perspective **12** (1998)
28. J.K. John H.S. Howard, *Entrepreneurship: "What it is and how to teach it"* (Harvard Business School,1958)
29. I.M. Kirzner, Journal of Economic Literature **35**, 60-85 (1978)
30. C. Letts, A. Grossman, W. Ryan, *High performance nonprofit organizations: Managing upstream for greater impact* (New York, Wiley, 1999)
31. J. Mair, I. Marti, The Journal of World Business, 36-43 (2006)
32. D.C. McClelland, *The achieving society*. Princeton: Van Nostrand (1961)
33. A. Meloto, (2013, October). Welcome remarks. Social business summit. Conducted from Enchanted Farm Village, Angat, Bulacan, Philippines.
34. L. Michelini, D. Fiorentino, Social Responsibility Journal **8**(4), 561-577 (2012)
35. A. Nicholls, Social entrepreneurship: New models of sustainable social change (Oxford: Oxford University Press, 2006)
36. M.E. Porter, *The Competitive Advantage of Nations* (New York, Free Press,1990)
37. M. Porter, J. Sachs, J. McArthur, (2002). Executive summary: Competitiveness and stages of economic development. In M. Porter, J. Sachs, P. K. Cornclius, J.W. McArthur, & K. Schwab (Eds.). *The global competitiveness report 2001-2002* (pp. 16-25). New York: Oxford University Press.
38. R.E. Quinn, K. Cameron, Management Science **29**, 33–51(1983)
39. T. Reis, (1999). *Unleashing the new resources and entrepreneurship for the common good: A scan, synthesis and scenario for action* (Battle Creek, W.K. Kellogg Foundation, 1999)
40. W.A. Sahlman, Some thoughts on business plans. In W.A. Sahlman, H. Stevenson, M.J. Roberts, & A.V. Bhide (Eds). *The entrepreneurial venture* (pp. 138–176). Boston: (Harvard Business School Press, 1996)
41. Samer Abu Saifan, Technology Innovation Management Review (2012)
42. J.A. Schumpeter, *The theory of economic development* (Cambridge, Harvard University Press., 1934)
43. S. Shane, Organization Science **11**, 448-46 (2000).
44. A. Shapero, Small Business Economics **13**, 27-55 (1975)
45. H.H. Stevenson, *A perspective on entrepreneurship*. Harvard Business School Working Paper No. 9-384-131. (Boston, Harvard Business School, 1983)
46. H.H. Stevenson, Harvard Business Review **63**, 85–94 (1985)
47. H.H. Stevenson, J.C. Jarillo, A new entrepreneurial paradigm. Socio-economics: *Toward a new synthesis* (Armonk, NY:,M.E. Sharpe, 1991)
48. J. Thompson, International Journal of Public Sector Management, **15**(5), 412–431 (2002)

49. J.A. Timmons, S. Spinelli, *New Venture Creation: Entrepreneurship for the 21st Century* (8th ed.). (Boston, McGraw-Hill/Irwin, 2008)
50. J.A. Timmons, W.D. Bygrave, Journal of Business Venturing **1**, 161–173 (1986)
51. UNCTAD. (2006). World investment report 2006. FDI from developing and transition economies: implications for development. New York/Geneva: United Nations.
52. J. Weerawardena, G. Mort, The Journal of World Business 21-35 (2006)
53. A.R.M. Wennekers, A.R. Thurik, (1999)
54. J. Baumol William, *The Free Market Innovation Machine: "Analyzing the Growth"* (2002)
55. M. Yunus, *Creating a world without poverty: Social business and the future of capitalism* (New York, Public Affairs, 2007)
56. M. Yunus, *Building social business: The new kind of capitalism that serves humanity's most pressing needs* (New York, PublicAffairs, 2010)
57. S. Zadek, S. Thake, social entrepreneurs. New Statesman **26**, 31 (1997)

The Knowledge about the Dangerous Effect of Opiates among Adolescent

A.S. Ishak[1,*], *Hussin* H.[1], *Alia Asrani* Azmi[1], and *M.H.* Othman[2]

[1]School of Human Development and Technocomunication (iKOM), Universiti Malaysia Perlis, Perlis, Malaysia
[2]School of Education Studies, Universiti Sains Malaysia, Penang, Malaysia

Abstract. Statistics from the National Anti Drug Agency (AADK) showed an increase in adolescent who are in the age range 13 to 18 years involved in drug abuse. Realizing this problem, a study was conducted to examine the knowledge about dangerous effect of drug abuse on opiates among adolescent. There are four type of opiates were tested in this study such as opium, morphine, heroin and codeine. Respondent consisted 759 adolescent aged 16 years old. This quantitative study using the "Soal Selidik Kajian Pengaruh Dadah" as an instrument. The data obtained were analyzed using descriptive statistics to get the frequency and percentage. The result showed that both male and female adolescent know about the dangerous effect of drug abuse. Findings also identify the most popular drugs among adolescent is heroin followed by opium, codein and morphine.

1 Introduction

Drug abuse among adolescent has increased all over the world and the age of initiation of abuse is progressively falling. Encouragement from the peer group, deep desire to be popular, poor parental control and easy availability of the substances make an adolescent to go for drug abuse. Earlier initiation of drug use is found to have more impairment, crime, and difficult to quit and is usually associated with a poor prognosis and a lifelong pattern of deceit and irresponsible behaviour [1].

Knowledge about the age of drug abuse commence is of paramount in importance. There are a few studies has been associated with the knowledge about drug abuse. High knowledge of harmful effects of substance was reported in students [2-7], health professionals [8-9] and general public [10]. Several studies in adolescent population have reported positive association between knowledge about substance abuse and their attitudes toward substances [2-4]. Nahid et al, 2014 conduct the study and gathered data about the type of drug abused, knowledge about short term and long term complications of narcotics and stimulants.

1.1 Objective of Study

The present study was comprehensively assess the knowledge about the dangerous effect of drug abuse on opiates among adolescent. Adolescent is a critical time in one's life, and the habit and behaviour formed during this years can have a long lasting impact on the health of the individual. In Malaysia, opiates are the most commonly abused drug, followed closely by cannabis and amphetamine based stimulant, [11]. The table below shown the statistic of most commonly used addictive drug by Agensi Anti Dadah Kebangsaan, Malaysia (AADK) from 2010- 2015.

Table 1. Statistic of Most Commonly Used Addictive Drug 2010-2015

Year	Opiate	Opium	Cannabis	Metha-phetamin	ATS	Psikotropik Pil	Others	Total
2010	11,664 49.34%	31 0.13%	3,011 12.73%	4,026 17.03%	4,525 19.14%	94 0.40%	291 1.23%	23,642 100%
2011	9,629 49.30%	9 0.05%	2,026 10.37%	7,034 36.01%	604 3.09%	58 0.30%	171 0.88%	19,531 100%
2012	8,472 56.10%	9 0.06%	1,427 9.45%	4,761 31.53%	286 1.89%	66 0.44%	80 0.53%	15,101 100%
2013	16,035 75.07%	0 0.00%	1,885 8.82%	2,901 13.58%	476 2.23%	18 0.08%	46 0.22%	21,361 100%
2014	14,496 64.84%	0 0.00%	1,919 8.58%	4,117 18.42%	1,774 7.94%	6 0.03%	43 0.19	22,355 100%
2015	16,616 60.48%	0 0.00%	1,389 5.05%	8,133 29.60%	1,314 4.78%	1 0.00%	26 0.09%	27,479 100%

Table 1: Type of Drug
Source: Agensi Anti Dadah Kebangsaan, 2015

1.2 Type of Opiates

Drug are psychoactive chemical materials that's affect the central nerve system until a user is in a condition of intoxication, addiction and behavioral problems. Drug are chemical materials dangerous to individuals that partake in them, as it changed how the mind and body function. Drug are a special term referring to substances that harm a user's physical, mental and emotional health as well as behaviour after use. As a result, a drug abuser becomes addicted and is highly dependence on the substance [12]

An opiate is a narcotic analgesic that directly depresses the central nervous system." There are three types of opiates: natural, synthetic, and semi-synthetic. While some of these drugs are used for medicinal purposes, others are illegal and highly dangerous. However, all opiates can become addictive when abused. Listed below are the most commonly used and abused opiate drugs. These substances occur naturally within the poppy plant. While they are often thought to be less harmful than synthetics, they can still become addictive and cause dangerous respiratory depression.

1.2.1 Opium

Opium serves to depress the central nervous system. Opium comes from the poppy seed that turns from a while to brown- blackish in hue. It has a strong odour, is elastic when new and hardens after expose to air. It is most commonly smoked and often combined with other drugs of abuse like marijuana and methamphetamine.

1.2.2 Morphine

Morphine is classified as an opiate drug and is primarily used for short and long-term pain relief. Like other opiate drugs, it is derived from the poppy plant and can also be categorized as a narcotic. Morphine is available in several forms, such as in a syrup, injection, tablet or suppository. Morphine can be injected or taken as a tablet, suppository, or capsule.

1.2.3 Heroine

Heroine is the name for a modified version of morphine that is very addictive and illegal opioid drug. The drug is available on the illicit market in a number of different varieties, with purer product tending to occur in white powder form. Heroin may be smoked, snorted, or injected. Regardless of the type of use, heroin acts quickly in the body to elicit its dramatic results. In short, heroin is very harmful. The speed and intense effects of the substance are main contributors to its harmful nature. They are also factors that lead to the addictive nature of heroin.

1.2.4 Codein

Codein is naturally-occurring alkaloid in opium. However, most codein in the market is produced from morphine. In the medical field, codein is used to alleviate mild pain and as cough medicine. Whether codein is used properly or it is abused, it has many of the same effects as other opiates. Codein may cause nausea, vomiting and drowsiness. Since opiates suppress breathing, the codeine may have slowed breathing or difficulty getting his breath. Codein is swallowed or injected and often abused as a cough syrup and mixed with soda and alcohol.

2 Material and Methods

2.1 Research design

This descriptive study using quantitative methods, namely through the use of set questionnaire. Refers to descriptive statistical methods to describe the characteristics of the the sample [13]. According to Mohd. Majid (1993), a questionnaire more practical and effective use because its use can improve the accuracy and correctness of responses given by the sample. This is because it is not affected by gestures researchers. They are free to express their own opinion to answer each item submitted.

1.2.1 Participant

Female and male adolescent aged 16 years old from three states from northern Peninsular of Malaysia (Perak, Pulau Pinang and Kedah) responded to the questionnaire.

1.2.2 Procedure of Data Collection

In this study, researchers used questionnaires to obtain research data. According to Mohd Majid (1994), a questionnaire was used because it is easier to get cooperation of respondents. The questionnaire can accommodate that many respondents to increase reliability study. The use of the questionnaire is simple and easily administered. Suggestion of the answers will allow respondents to answer the question fast.

The "Soal Selidik Kajian Pengaruh Dadah" which consisted of 3 parts was used. Part A is related to the student demographics; gender, race and current living. Part B is a knowledge / experience regarding drugs. Six type of drug are tested in this study; opium, depresen, marijuana, heroin, codein, morphine, ecstasy, methaphetamin and cocaine.

Analysis was done by SPSS version 22.0. The data was presented in percentages and frequencies, from which the researcher presented it in table, figures and graph.

3 Result

After collecting all the questionnaire, only 759 were used for this study. There were more female (n=425; 56.0%) participant compared to males (n=334; 44%). The majority of the respondents were Malay (n=691; 91.0%). Chinese respondents were (n=6; 0.8%) while the Indians (n=59; 7.8%). There are also respondents from other nations (n = 3; 0.4%). Most of the respondents living at home (n = 671; 88.4%) than in the hostel (n = 88; 11.6%). Some demographic characteristic of the participant are presented in Table 2. Table 2 shows the analysis of the dangerous effects of opiates drugs among adolescents. It was found that both male and female know about the dangerous effects of drugs. Generally its seem that female have a higher level of knowledge about the dangerous effects of opiates. This is because more female respondent than male. The male who know the dangerous effect of opium is (n = 104; 13.70%), morphine (n = 68; 9.0%), heroin (n = 147; 19.4%), and codeine (n = 56; 7.4%). While the female who know the dangerous effects of opium is (n = 116; 15.3%), morphine (n = 84; 11.1%), heroin (n = 185; 24.4%) and codeine (n = 278; 36.6%).

Table 2. Demographic Background

Background		Number	%
Gender	Male	334	44%
	Female	425	56%
Race	Malay	691	91.0%
	Chinese	6	0.8%
	India	59	7.8%
	Others	3	0.4%

Current Living	Home	671	88.4%
	Hostel	88	11.6%

Table 3. Analysis Knowledge of Dangerous Effect of Opiates

Type of Drug		Opium		Morphine		Heroine		Codeine	
		Number	%	Number	%	Number	%	Number	%
Male	Yes	104	13.70%	68	9.00%	147	19.40%	56	7.40%
	No	230	30.30%	266	35.00%	187	24.60%	64	8.40%
Female	Yes	116	15.30%	84	11.10%	185	24.40%	278	36.60%
	No	309	40.90%	341	44.90%	240	31.60%	361	47.60%

4 Discussion

From the finding, we can conclude that knowledge regarding the dangerous effects of drug abuse is considerably low among both female and male adolescent. Indirectly, they can be said to be at higher risk involved with substance abuse problems.

Because the adolescent still do not know about all the type of drug, Prevention-based programs are necessary to decrease demand through informing adolescent about the relative risks of dependence on various illegal drugs based on present knowledge of these risks. As the drug use usually begins after the age of 12 and rises rapidly until 16 years of age [14].

Hence the preventive drug education should begin in early adolescence and should deter or delay drug use through changes in knowledge, attitude, and behavior [15-19]. Adolescent drug education must meet the needs of those naive to drugs as well as those experiencing initial drug exposure. Education provided in the first group is important in modifying young peoples' responses in drug situations and in the latter group it is most meaningful [16].

Besides that, the more information need to provide primarily through the media, principally the television and hoardings. Therefore media and close contact are primarily responsible for imparting knowledge, whether it be a pro or anti-drug use, and this finding are common anywhere, as is seen in others studies. Six hundred and ninety eight (86.5%) student claimed they were aware of drug abuse. Oshikoya & Ali, 2006 founded that the television, radio and billboard advertisement were the students major sources of awareness about drug.

Haddad et al., 2010 reported parents, friends, and teachers also can be a major source of information among adolescents.

5 Conclusion

Using the obtained result of this research, knowledge of adolescent toward drug abuse are identified. The organized continuous decision makings and programs can be implemented to raise knowledge and attitude of the students toward drug abuse and to decrease the rate of dependence to drugs. Considering the importance of adolescent acnt role age in forming personal behaviors in the future and the opportunity provided at schools, paying more intentions in teenagers can have a significant role in decreasing drug abuse in this age group and consequently in the society by increasing their knowledge and correcting their attitude toward drug abuse.

The literature surveyed indicates that knowledge of drugs and perceptions of drug prevalence and the harm associated with drug use tend to be mediated by individual experience of drugs, personal acquaintance with drug users, media coverage of drugs and drug use, and social and demographic variables. Attitudes to drugs and drug use may also depend to some extent on social group membership and the attitudes held by the social group as a whole. An association has also been found between gender and perceptions of drug related harm, with women perceiving drug use as far riskier than men. Societal attitudes towards drug users have similarly complex origins, but research indicates they have a tendency to be negative. These negative attitudes exacerbate the sense of alienation felt by drug addicted individuals and discourage them from seeking help for their problem.

Inspite of being aware of the dangerous effects of substance use, adolescents take up this habit. This requires comprehensive prevention and control programs in schools and the community, targeted toward adolescents and their parents and other family members. Effective measures are required to encourage shaping the attitude of school children toward self-confidence and adequacy, as also to prevent risk behavior among adolescents.

Reference

1. S. Chatterjee, B. Tempalski, E. Pouget, HL. Cooper, CM. Cleland, SR. Friedman, Changes in the Prevalence of Injection Drug Use Among Adolescents and Young Adults in Large U.S. Metropolitan Areas. AIDS and Behavior **15**,1570-1578 (2011)
2. CJ. Schwarz, *Knowledge and attitude of undergraduate and graduate social work students towards addiction and those who use illegal drugs* (California State University, Long Beach, 1997)
3. RA. Gassman, HW. Demone, R. Albilal, Journal of Social Work Education **37**, 137-145 (2001)
4. VJ. Giannetti, JD. Sieppert, MJ. Holosko, Journal of Health and Social Policy **15**, 45-58 (2002)
5. L. Warburton, D. Callfas, SL. Neubauer, Can Pharm J, 140, 366-75 (2007)
6. O. Prakash, OP. Giri, AK. Mishra, S. Kumar, P. Kulhara, J Mental Health & Human Behavior 14, 4-79 (2009)
7. D. Tsering, R. Pal, A. Dasgupta, J Pharmacy and Bioallied Sciences 2, 137-40 (2010)
8. B. Happell, B. Carta, J. Pinikahana, Nursing and Health Sciences 4,193-200 (2002)
9. CJ. Heckman, JL. Dykstra, BN. Collins, Health Education Journal 20, 1-17 (2010)
10. A. Bryan, R. Moran, E. Farrell, M. O'Brien, *Drug-Related Knowledge, Attitudes and Beliefs in Ireland: Report of a Nation-Wide Survey* (Dublin: The Health Research Board, 2000)
11. *Agensi Anti Dadah Kebangsaan* (Kementerian Dalam Negeri, Malaysia, 2015)
12. S. Galea, A. Nandi, D. Viahoc, The social epidemiologi of substance use. Epidemiologic Review 26, 36-52 (2004)
13. Mohd Najib Abd Ghafar. (1999). Penyelidikan Pendidikan. Skudai, Johor. Penerbit Universiti Teknologi Malaysia.
14. D. White, M. Pitts, Addiction **93**,1475-87(1998)
15. HM. Belcher, HE. Shinitzky, Arch Pediatr Adolesc Med **152**, 952-60 (1998)
16. NS. Tobler, T. Lessard, D. Marshall, Effectiveness of school-based drug prevention programs for marijuana use. School PsycholInt **20**, 105-37 (1999)
17. DR. Black, NS. Tobler, JP. Sciacca, J School Health **68**, 87-93 (1998)
18. N. McBride, Health Educ Res **18**, 729-42 (2003)
19. J. Lilja, BU. Wilhelmsen, S. Larsson, D. Hamilton, Subst Use Misuse, 38,1831-63 (2003)

Drug Addicts: Psychosocial Factor Contributing to Relapse

Alia Ashrani Azmi[1,*], *Hussin* H.[1], *Siti Intan Diyana* Ishak.[1] and *Nur Suriaty* Daud@Fhiri[1]

[1]School of Human Development and Technocomunication, Universiti Malaysia Perlis, Perlis, Malaysia

Abstract. This article present about relapse among drug addicts in rehabilitation center. In Malaysia, many drug addicts relapse to drug use after discharged from successful treatment at rehabilitation center. Various effort and the authorities involved in assisting the National Anti-Drug Agency (NADA) efforts to cure and treat drug addicts. This article aims to examine the factors that affect relapse among drug addicts. Therefore, this article will discuss aspects of psychosocial factors including self-efficacy, family support, friends support and community support. Psychosocial factors, such as aspects of self-efficacy, family support, friends support and community support, had significant relationships with relapse. Psychological factors are also vital to design adequate treatment strategies. Thus, this article evaluating psychosocial factor as main factors that influenced the relapse addictions tendency amongst addicts.

1 Introduction

The problem of drug addiction is not a new issue in our country, Malaysia and the world. The problem of drug addiction is a problem that is difficult to recover. Addiction is a chronic disease that will recur in the absence of drug addicts in the body causing pain due to drug withdrawal syndrome. This situation is a cause of drug addicts taking drugs back pain due to drug withdrawal syndrome can be stopped [1].

The drug problem is a global problem. Activities production of various types of drugs is constantly increasing. This is supported by high demand. An international political change with the end of the cold war between the great powers of the world, have a negative impact in the fight against drugs. This is because the world is now considered no longer boundaries and everyone is free and out of a country without restrictions. Communication and transportation systems are increasingly sophisticated help facilitate their movement. For those involved with with drug activity, the development of this kind of operation was expected because they become more smooth and widespread.

Based on the statistics available from the National Anti-Drugs Agency (NADA), from January to June 2010, some 3,095 of 12,079 drug addicts were detected as repeat offenders or relapse.

The number of repeat offenders was found to be increased compared to only 1,136 people repeat offenders were detected from January to June 2009 (NADA 2010). Meanwhile, a total of 2,792 new cases and 1,968 cases of drug addicts has been detected repeatedly during the January to June 2012, in which case repeatedly fell by 24.20%. This situation is alarming because the problem of relapse in Penang and Kuala Lumpur are still exceed his increase.

1.1 Objective of the paper

Objectives of this paper are to find out psychosocial factors of relapse as follows:

a. Self-efficacy
b. Friends support
c. Family Support
d. Community Support

Result are taken from the research papers which have been done their research on the psychological factors of relapse.

2 Methodology

The study is based on previous research method to review psychosocial factor contributing to relapse. The secondary data is drawn from books, journals and various reports published by agencies working in this fields. There are only few researches conducted studies to analyze factors contributing to relapse. Result are taken from the research papers which have been done their research.

3 Result and Discussion

According to Mahmood [2] generally means addicts relapse drug reuse of the original drug or other type after treatment and rehabilitation programs. From the above definitions can be formulated that relapse is the failure of addicts who have been through the process recovery to continue free from drugs. According to Childress [3], relapse is a return to the use the drug after a period free from dependence on the material while they continue constantly trying to recover.

Relapse is a difficult problem and challenge in the process of maintaining the recovery in addicts. Whereas according to Rasmussen [4], relapse defines as the reuse of relapse after stopping taking the drug for a certain period. Dennis [5] measures the definition of relapse of the behavior of the process of returning to a state of unhealthy behaviors that lead to the re-use of drugs and experiencing signs of change in terms of attitudes, thoughts and ways of behaving.

According Fig. 1, [6] Marlatt and Gordon reported that billing factor for relapse started as High Risk Situations. High-risk situations is the state that has a high influence to bring back a former addict takes drugs. High-risk situations for relapse is composed of three situations, namely (i) the negative emotion, (ii) experiencing interpersonal conflict and (iii) to face the social pressure of the surrounding circumstances.

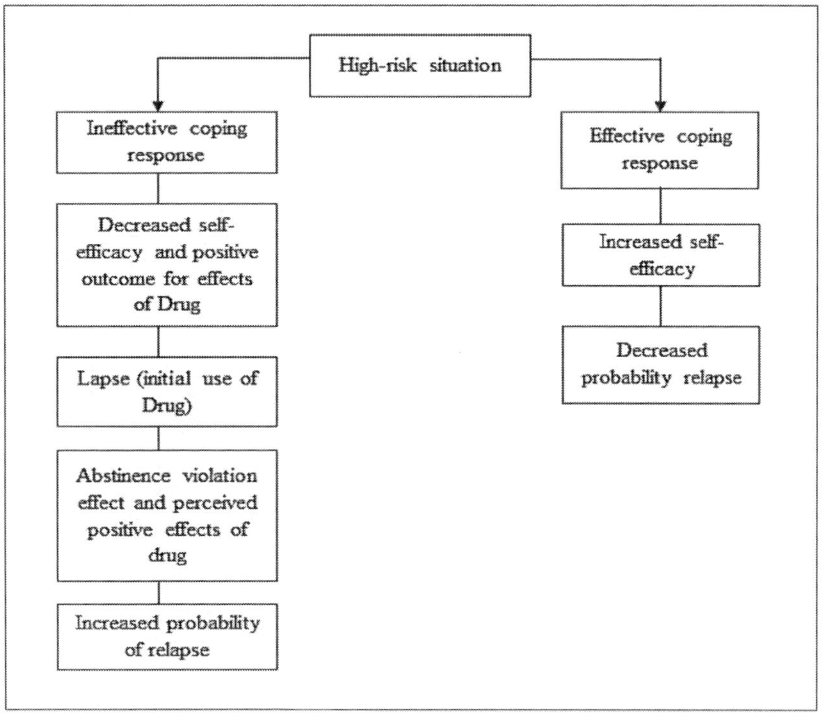

Fig. 1. Cognitive behavior model [6]

3.1 Self-efficiency

Other relapsed promoting factor is self-efficacy, defined as a degree to which an individual feels confident and capable performing a certain behavior in a specific situational context. According to Nurhazlina and Azlinda [7], the study said two from eight respondents fall back using drugs after completion of treatment because they do not have strength and confidence in themselves that they were living without drugs or more clearly faced with self-esteem problem. Reported by Fauziah Ibrahim [8] said most addicts have a low self-esteem to overcome the problems that lie ahead, it is easy to give up and not be able to resolve a problem with a positive and intelligent. Due to these things, so they are easily influenced by things that lead to the occurrence of the issue of relapse. Cheung et al [9], identified that nine people from 21 drug users surveyed managed to refrain from using the drug because it has high self-confidence as a result of satisfaction in their lives who have a job and a steady income.

Recent studies have found the features of relapse for those who have feelings of depression, negative thoughts and think of all the things difficult to resolve, thinking immature, unable to manage feelings and emotions well, often confused, depressed and irritable. In this study, a total of 79.8% of respondents admitted that they often produce feelings of anger by screaming and yelling and some 60% of the respondents admitted often complain about unnecessary things. This situation is giving rise to confusion, disorder and emotional instability among addicts relapse and thus causing them classified as high-risk individuals relapse [4]. Whereas according to Mahmood [10], found that drug addicts are receiving treatment and rehabilitation of drug rehabilitation centers have a poor psychological state. They have a level of self-efficacy moderately low, still use this type of mechanism martial primitive and immature and has a personality that is not normal, such as high levels of anxiety, paranoia, emotional instability, tension and depression are high.

Previous research has shown that characterizes individuals who relapse have a level of depression, stress and anxiety disorders uncontrolled besides no acceptance or family support to free themselves from continuing to use the drug again. For example, some 72% of respondents in the study admitted that they often felt depressed, sad, moody and weepy. This decision gives the impression that an addict relapses have character and personality are less mature and rely on others to get support for maintaining a drug-free lifestyle [5] Dennis. Previous research has established that the study of the question of relapse to 327 subjects were divided into five groups of drunken behavior, smoking, heroin addicts, gamblers and nutrition that cannot be controlled Cummings [11]. Subjects who used to deal with the problem of relapse and the average age ranged from 30 to 45 years. The study found that as (30%) drug addicts involved in the issue of relapse it is because low self-efficacy.

3.2 Friends Support

Amongst other factors that have been associated with relapsed addiction are peer pressures especially from the old friends who still use drugs. Studies on serious drug users found that factors that led to a total of 17 respondents who are drug users and alcohol users who had participated in rehabilitation programs in Queensland, Australia fall back to drug because peer influence Hammer-Bacher [12]. Saodah Siraj [13] in his research found that peer influence has to do with behavior problems. The findings showed that the respondents admitted often invited friends to hang out at the mall and also agreed that the partners better understand their problems compared to their own family members. The findings showed that 50% of respondents agreed that they easily accept a friend invitation them to try drugs.

Kaplan [14], contended that total of 30.6% of respondents agreed that their stand is wobbly when urged by a friend to try the drug. The results showed that most individuals with drug problems is because they are influenced by a reference group which is a social network used by individuals as a reference to his daily behavior. In this case, the reference group can be associated with a group of peers. In adolescence, students spend more time with peers such as within school hours. Pressure from peer group eventually lead students to drug abuse. Insistence that they received led to the establishment of an increasingly shaky to be involved in drug abuse.

Other previous report by Pruitt et al [15] found that among 1,004 adolescents grades 8 and 10 in a small community in Central and Eastern Texas, which again involved the use of materials as having at least involved with the consumption of alcoholic beverages once a week and also have a friend who is involved with the drug marijuana, smell, and other amphetamines. Nik Mohd Zaini found that this research is a field study conducted at the Rehabilitation Centre Besut, Terengganu. The subjects are 190 trainees who have different demographic factor and is undergoing treatment and rehabilitation. The results show the influence of peers is an important factor causing drugs by addicts. Life that cannot be controlled is a major cause of drug treatment participation. The study also found the drug to be influenced by their friends because there is a sense of belonging within the group itself. In addition, some of the conditions specified by members of the group also made them stuck with drugs. Peers have a strong influence on a friend-pawn, especially at the junior level.

Manja Mohd Ludin [16] in his research found that kids in high school are usually form groups according to their respective interests. The extent of the interest of the individual is dependent on the extent to which the strength of attraction peers. When a person begins to engage with peers who use drugs there will be attractions for them also engage in drug abuse. Students will begin to be affected when they often invite their friends to use drugs. A total of 41.7% of respondents agree that they are easy to follow treatment friend that led to drug abuse, such as smoking.

Tec [17] Studies on the influence of the peer group over the behavior of marijuana use. The results of these studies show the use of marijuana as social activities which often occurs in the peer group. This shows that students easily lead to treatment of peers. The study also states that teens who use marijuana are more closely by peer group than the family. The results showed 41.7% of respondents agreed that they do not care if their friends are involved in drug abuse. Apathy or indifference can lead to influence the students to experiment with drugs when many of their peers are involved in drug abuse.

3.3 Family Support

Family support also identified contributing to the relapsed addiction tendency. According to Noor Zalifah, in his case study on cases under the supervision of the National Anti-Drug Agency (NADA), which examines the causes of addiction relapse said the families are factors that most influence them toward drug addiction relapse. Pressure from families who want addicts to recover and turn immediately made them forget the charge for the immediate pressure. In addition, a high desire to drugs to make them re trapped in addiction. Amongst other factors that have been associated with relapse addiction is family.

Mattoo, Chakrabarti and Anjaiah [18] in their study of 30 male addicts to see psychosocial factors associated with relapse. The study looked at the relationship between demographics, medical and psychosocial factors with relapse. The results showed that family history of drug addiction and family have a high rate of relapse that most affect their return to the billing. Fauziah Ibrahim [19] found that a total of 96.3% of drug addicts who engage in the issue of relapse said they received a moderate level of family support to a high level. Of these, 15 (3.8%) of respondents stated that they received a low level of family support. While 300 (75.0%) received the support of a simple family of 85 (21.3%) received a high level of support.

Research by Ibrahim, a total of 36.1% of respondents agreed that their parents often quarrel caused them to get involved in drug abuse. This suggests that the failure of parents to set a good example to the children. Not a good value will result in depressed teens and began to engage in negative things like drugs. Parents certainly have an important role and responsibility in maintaining discipline children. Parents need to demonstrate the value of the moral high enough to guide their children. In addition, children who start moving into adolescence will find peace outside if there was a tense at home. According to a random survey conducted on 60 family units in the north of the peninsula, found 95% of respondents from among the civil society the view that they assume addicts as a group of people who are not useful to the country. Only 18% among them thought that the drug is a disease that must be cured. While 6.0% felt that all addicts should be brought to a place far away from society and not allowed back because of mistakes made [20].

3.4 Community Support

Community support is the effects of psychological factors that contributes towards relapsed addiction tendency amongst addicts. Dittmar, Ratnasinggam [21], Study in Malaysia about the influence of people on drug addicts who look from the point of view of drug addicts and point of view of society in terms of moral and ethical found addicts has shown a positive attitude but the situation is at odds with a society where drug addicts are often considered resist and ignore values community. The negative attitude of the society towards recidivist addicts cause they feel alienated and will repeat the habit relapse drugs for solace and settle the conflict themselves. What is the truth faced by respondents to this study.

Research by Chen [22] related to social support, spiritual and addiction recovery program explained that social support from the local community through meetings of former drug users in association involving integration into society and spirituality programs are important to reduce the negative consequences of the formation pressure encountered during the recovery process. This clearly has pointed out that society should provide support in the form of social and physical as it gives them unconditionally and providing jobs if necessary to drug users so that they have the confidence to continue to change and finally off drugs.

Fauziah Ibrahim and Naresh Kumar [23] study the factors that influence relapse in Malaysia. The study was conducted on 400 people who are repeat offenders in select from eight rehabilitation centers in Peninsular Malaysia. They studied seven variables were considered as the main influence on the relapse of self-esteem, family support, peer support, community support, employer support and the effectiveness of rehabilitation programs. The researchers found that the confidence and support of the most influential collectors of charge. Many drug addicts have self-esteem and low and they do have the power within themselves. 64% also said the society does not accept them back into society and society has always looked askew to those like them. In addition, people also often suspect them when there is an incident such as theft and they continue on the label as a performer.

Mohammad Roslan [24] study states to ensure that no cases of recurrent drug addicts, every member of society must change the perception of former drug addicts. There are a few cases where public support to former drug addicts were able to make this former drug addicts be useful to the society and the state. A former drug addicts involved in drug abuse for fourteen years to recognize the support of family and community acceptance helped her to turn and stop drug abuse. Even former drug addicts just need attention and support from the community, but the attention and support plays a very big role in encouraging a person to go through the recovery process. The criticism of society and less drug resistance makes the bill because there is no place to complain and would not tolerate insults and denigration of the family and the local community.

4 Conclusion

Relapse among drug users in rehabilitation centers occur due to the influence of psychosocial factors. Therefore, as an individual who is directly involved in the helping profession and service delivery in any rehabilitation center in Malaysia should look for these recurring billing problems as a whole, rather than only on the shoulders of drug users involved. A comprehensive commitment to the task and this trust needs to be applied to all levels of personnel to the philosophy of recovery and rehabilitation policy is understood and can be implemented honestly and effectively. There can produce individuals who are advanced, productive and virtuous. The involvement of family, community and voluntary organizations was needed to educate and prevent crime. A safe social environment, free from abuse is certainly the hope of the people and the country.

Reference

1. H. Habil, M. Ali Mohd, *Penyalahgunaan Dadah Hidup Tidak Beerti Maut Menanti* (Dewan Bahasa dan Pustaka: Kuala Lumpur, 1999)
2. N. M. Mahmood, (2006). *Peranan Institusi Keluarga dalam pencegahan gejala sosial dan penagihan dadah*. In Azlin Hilma, H., Fatimah Zaily, A.R., Rusimah, S., dan Yusmarhaini (Ed.), Isu-isu dan Pendidikan Kerja Sosial (Sintok, UUM, 2006)
3. A. R. Childress, *Understanding Addiction*. From http://www.hbo.com/addiction/understanding
4. S. Rasmussen, *Addiction Treatment: Theory and Practice* (Carlifornia. Sage Publication, 2000)
5. C. D. Dennis, C.D. (1989). Relapse Prevention: *Treatment Alternatives and Counseling Aids*. Tab Books Inc, Blue Ridge Summit, PA.
6. G. A. Marlatt, J. R. Gardon, *Relapse Prevention: Maintenance Strategies in the treatment of Addictive Behaviors* (New York: Guilford Press, 1985)
7. M. A. Nurhazlina, A. Azlinda, *Malaysian Journal of Social Policy and Society* **6**, (2009)
8. Fauziah Ibrahim, Bahamah Abu Samah, Mansor Talib, Mohamad Shatar Sabran, Jurnal Antidadah Malaysia Jilid 5 (2009)
9. C. K. Cheung, T. Y. Lee, C. M. Lee, *Factors in successful relapse prevention among Hong Kong drug addicts. Treating Substance Abusers in Correctional Context: New Understanding, New Modalities (2003)* From http://www.HaworthPress.com
10. N. M. Mahmood, *Perubahan Psikologikal dari segi harga diri, profil personality dan penggunaan mekanisme bela diri di kalangan penagih yang mengikuti pemulihan serenti,* (Universiti Kebangsaan Malaysia, Bangi, 1995)
11. C. Cumming, J. R. Gordon, G. A. Marlatt, *Relapse: Prevention and Prediction. The Addictive Behaviours: Drug Abuse, Smoking and Obesity* (Britain: Pergamon Press, 1980)

12. M. Hammer-Bacher, M. Lyvers, Journal Substance Use **11** (6), 387-394 (2006)
13. Saodah Siraj, *Faktor-faktor yang mendorong kepada berlakunya masalah tingkah laku pelajar-pelajar di Sekolah Menengah Kulai* (UTM, Johor, 2000)
14. D. B. Kaplan, Journal of Drugs Issues 26 (1996)
15. B. E. Pruitt, P. M. Kingery, E. Mirzaee, G. Heuberger, R. Hurley, Journal of Drug Education **21** (1), 36-56 (1991)
16. Manja Mohd Ludin, *Psikologi Sosiologi Dan Falsafah Dalam Pendidikan* (Kuala Lumpur: Dewan Bahasa Dan Pustaka, 1990)
17. N. Tec, Journal of the Marriage and the Family **32**, 48-58 (1970)
18. S. K. Matttoo, S. Chakrabarti, M. Anjaiah, *Psychosocial Factors Associated with Relapse in Men with Alcohol or Opiate Dependence* (India: Chandigarh, 2009)
19. Fauziah Ibrahim, Bahaman Abu Samah, Mansor Abu Talib, Mohamad Satar Sabran, *Faktor menyumbang kepada penagihan relaps dalam kalangan penagih dadah PUSPEN di Semenanjung Malaysia* (Bangi: Universiti Kebangsaan Malaysia, 2008)
20. N. M. Mahmood, Jurnal PERKAMA 6 (1996)
21. M. Dittmar, M. Ratnasingam, V. Navaratnam, *A Comparative Analysis of the Psychological Profile of Institusionalised Drug Using Population Minden* (USM: Pusat Penyelidikan Dadah Kebangsaan, 1984)
22. G. Chen, International Journal of Offender Therapy and Comparative Criminology **50**(3), (2006)
23. Fauziah Ibrahim & Naresh Kumar, Asian Social Science Journal **5**(12) (2009)
24. Mohamad Roslan Saad, *Sikap Masyarakat Terhadap Penagih Dadah* (Sabah: Universiti Malaysia Sabah, 2010)

 https://doi.org/10.1051/matecconf/201815005098

Leadership and Followership in Organizational Impact Humanity in Government Sector

Norazilawani Ahmad[1,*], and Hanum Hassan[1]

[1]School of Human Development and Technocomunication, Universiti Malaysia Perlis, Perlis, Malaysia

Abstract. Leadership in organizational behavior can be impact of humanity in Goverrment Sector., tourism, service sector. Communication style in the organization is very important and also key forces in providing employees with job satisfaction, humanity and communication skills. When employees do not get satisfaction from their jobs, morale drops an absences and lateness increases. Any person just follow instruction from a leader behavior and subordinates even the step, instruction good job, take action for respontibilities from communication transaction by interpersonal. Even the staff followers support for leaders is enhanced when their decisions affirm a distinct social identity that is shared with followers. Participants showed less support for a leader who favored in group members who were relatively sympathetic to an out group position than for one who favored in group members who opposed an out group position. A social constructionist view is highlighted. I clarify some of the assumptions of this approach, contrasting them with those of a more leader-centered perspective. In an effort to increase the testability of this approach, In outline a general model, paving the way for generating individual and group-level hypotheses, and discuss implications for practice and for future leadership research

1 Introduction

Literally hundreds of studies have investigated the effects of leader initiating structure and consideration on subordinate satisfaction (Bass, 1999). This research indicates that the subordinates tend to be more satisfied with leaders who exhibit high levels of consideration. The effects of leader initiating structure, on the other hand, are not very consistent, but instead vary as a function of subordinate characteristic like experience, competence and job level and task characteristics like time urgency, physical danger and ambiguity.

The path-goal theory of leadership was developed,in part, to explain the consistent effects of leadership behavior on subordinate satisfaction impact of humanity in Goverrment Sector. Communication style in the organization is very important and also key forces in providing employees with job satisfaction, humanity and communication skills. It is based upon expectancy theory (Vroom, 1998; Porter & Lawler, 2003) proposes that subordinates will be more satisfied when leader behavior is perceived as:

(1) Increasing personal payoffs for work goal attainment.
(2) Clarifying paths to these goals.
(3) Reducing or eliminating barriers to task completion.
(House & Dressler, 2000)

One implication of path-goal theory is that situational characteristics determine the relationship between leader behaviour and subordinate satisfaction impact of humanity in Government Sector. Communication style in the organization is very important and also key forces in providing employees with job satisfaction, humanity and communication skills. For example, when the task is routine, subordinates perceive leader initiating structure negatively because it is not needed for goal attainment. In more complex and ambiguous tasks, on the other hand, leader initiating structure is perceived positively because it is critical for task accomplishment (Szilagyi & Sims, 1990; Vecchio, 1990)

Cranny, Smith, and Stone (1998) estimated that more than 5,000 relevant job satisfaction studies have been published during the twentieth century. Many articles and dissertations credit Hoppock's (1995) study of job satisfaction as pioneering work, but his review of other job satisfaction investigations already included 32 studies. Employees from manufacturing, retailing, and service firms; local, state, and federal government agencies; and schools, colleges, and universities have been participate in job satisfaction research.

1.1 Significant of the study

In generally, the study is significance to prove what dimensions of leadership styles will provide satisfaction on the job among staffs of IPK Kedah (Ibu Pejabat Polis Kontinjen Kedah, Jalan Stadium, 05560 Alor Setar, Kedah (STaRT Department), Ibu Pejabat Polis Daerah Kota Setar, Jalan Raja, 05100 Alor Setar, Kedah (Head

* Corresponding author: Kbc5183@yahoo.com

Department Start- Financial Section, Store, Communication, Transport). By having staff satisfaction will bring benefits to the organization as they are able to provide job well and this act will reflect the good image and reputation of the organization itself. It cannot be deny that in every organization, one of the factors that maintain the reputation and good image rely a lot on the services that provide by the staffs.

We hope that the finding of this study will provide valuable information to the organizations to identify areas of leadership styles (initiating structure, consideration) in order to enhance staffs satisfaction impact of humanity in government Sector. Communication style in the organization is very important and also key forces in providing employees with job satisfaction, humanity and communication skills and to provide the quality staffs.

The result of the study may give better understanding of patterns of leadership styles in relation to job satisfaction and indicated the general influences methods of initiating structure style or consideration style and it is much to provide detailed descriptions of individual staff's satisfied. Future research could be conducted on a bigger population including those employees in others government and private sectors employees throughout Malaysia.

IPK Kedah top management is expected to lead according to the standards police department organization and the organization in charge has to uphold the ethical conduct, in both words and actions. At the same time, they need to create a good environment that makes their staffs feel suitable to work with. So, this research tries to find out whether leadership styles give an impact to the staffs job satisfaction.

The procedural problem of this study was to assess the relationships and the effects of leadership styles (initiating structure and consideration) on job satisfaction among the staff in Logistic Department of IPK Kedah.

1.2 Organization of the study

This study is structured into five chapters. Chapter one provides an introduction to the topic to be studied, statement of problem, significance of the study, objective of study, research questions, and limitation of study.

Chapter two addresses the (a) meaning of leadership, (b) theoretical approach to leadership, (c) meaning of job satisfaction, (d) theories of job satisfaction, (e) leadership behavior and job satisfaction, (f) job satisfaction as criterion variables, (g) conceptual framework, (h) hypothesis and (i) definition terms.

Chapter three discussed regarding the methodology which includes research population, research sampling, technique of sampling, instrumentation, data collection procedures and data analyses procedures for conducting the study.

Chapter four presents the findings of research include survey responses, description of respondents which is selected accordingly (gender, age, marital status, working experiences and educational level), reliability of the instruments, hypothesis test and summary of hypotheses results. Chapter five summarize the conclusion, along with discussion and recommendation for future research. Lastly conclude reference, and appendix.

2 Literature Review

The literature review addressed the (a) meaning of leadership, (b) theoretical approach to leadership, (c) meaning of job satisfaction, (d) theories of job satisfaction, (e) leadership behavior and job satisfaction, (f) job satisfaction as criterion variables, (g) conceptual framework, (h) hypothesis and (i) definition terms.

2.1 Introduction

Leadership is one of the fields of organizational that has interest many scholars. Hundreds of books and articles have been published about leadership. Leadership involves a wide range of aspects such as decision making, traits, effectiveness, motivation, and types
However, everyone has opinions and ideas about leadership. These theories are often founded on research and based on people's own experience or popular literature. This chapter therefore sets out to review the literature on leadership behavior, job satisfaction and the relationship between two concepts of leadership behavior styles (initiating structure, consideration) and job satisfaction.

2.2 Meaning of Leadership

If scholars could agree on one definition of leadership or perhaps settle on a series of kinds of leadership each with its own reliable, the might, through systematic research, arrive at one comprehensive theory. If they could also agree on definitions o the "good or bad outcomes" of leadership they could also assess how effective various kinds of leadership are in various situations. But there is no general agreement here either. This may be due to the "value" nature of judged leadership effectiveness. Hogan, et al. (2002), insists that leaders must be judged by their result on group life.

What would have obtained without the leader? However, while followers can sometimes function without leaders, n leader is effective without followers to carry out plans and we are left with a puzzle about who should get credit (or blame) for with outcome. Unfortunately, total agreement on these issues is not forthcoming and we must do the consensual definitions of what "good" or "bad" leadership outcomes are.

The source of a leader's influence over his/her subordinates is his/her "position power" and "personal power" (Wexley & Yulk, 2003). Position power includes the leader's legitimate authority and his/her control over organizational rewards and punishments. Personal power is derived from the affection loyalty of subordinates and their respect for the leader's expertise. Position power is primarily determined by the organization, while personal power depends upon the personality and behavior of leader.

Leaders serve as true facilitators and coaches (Gardner, 2005). discussed three roles that leaders, whether they are called team leaders, team coordinators, or coaches, served in team-based organizations. The role of the leader appeared to have shifted from one of formal authority in term of decision making, goal setting, directing floor activity, and troubleshooting to one that focused primarily on coaching and training team members, as well as directing special project, ensuring adequate resources, and monitoring overall performance. According to Bornstein and Smith (2001), leadership is now understood by many to imply collective action, orchestrated in such a way as to bring about significant change while raising the competencies and motivation of all those involved that is action where more than individual influences the process.

2.2.1 The Trait Approach

Early research into leadership can be characterized as a search for "the great man". Personal characteristics of leaders were emphasized and the implicit idea was that leaders are born rather than made. All leaders were supposed to have certain stable characteristics that them into leaders. The focus was on identifying and measuring traits that distinguished leaders from non-leaders or effective from ineffective leaders (Hollander & Offermann, 1998). From these distinctions between leaders and non-leaders, a profile of an "ideal" leader could be derived, which could serve as the basis for selection of future leaders.

Some individual traits that appear to be related to leadership success are high energy level, tolerance for stress, emotional maturity, integrity, and self-confidence. High energy level and stress tolerance help people cope with the hectic pace and demands of most leadership position, the frequent role conflicts, and the pressure to make important decisions without adequate information (Yulk, 1998). Leaders with high emotional maturity and integrity are more likely to maintain cooperative relationships with subordinates, peers, and superiors.

Emotional maturity means that a leader is less self-centered, has more self-control, has more stable emotion and is less defensive. Integrity refers to a person's behavior being consistent with expressed values and the person being honest and trustworthy. Self-confidence makes a leader more persistent in the pursuit of difficult objectives, despite initial problems and setback. Judge and Bono (2000) reported that the personality factors of extraversion and agreeableness were also associated with success as a leader.

2.2.1 Charismatic Leadership

Beginning in the 1970s, a number of leadership theories emerged that focused on importance of a leader's charisma to leader effectiveness. Included within this class of theories is House's theory leadership (1999), Bass's transformational leadership theory (2000), and Conger and Kanungo's (2003) charismatic leadership theory. These theories have much in common. They all focus on attempting to explain how leaders can accomplish extraordinary things against the odds, such as turning around a failing company, founding a successful company, or achieving great military success against incredible adds. The theories also emphasize the importance of leaders' inspiring subordinates' admiration, dedication and unquestioned loyalty through articulating a clear and compelling vision.

A first major application of charisma to the study of formal organizations can be found in House (1999), whose theory combines personal traits, leader behavior and situational factors. According ton house, four personal characteristic of leader contribute to charismatic leadership, dominance, self- confidence, need for influence, and a strong conviction of the integrity of one's own beliefs. Charismatic leaders represent their value and beliefs through role modeling. To create a favorable perception with followers they can engage in image building and express ideological goals (a mission). They communicate high expectations of followers and confidence in followers' ability to live up to expectations.

Bass's (2000), transformational leadership theory identifies four aspects of effective leadership, which include charisma, inspiration, intellectual stimulation and consideration. A leader who exhibits these qualities will inspire subordinates to be high achiever and put the long-term interest of the organization ahead of their own short-term interest, according to the theory.

2.2.1 Substitutes for Leadership

Kerr and Kermier (2004) introduced the substitutes for the leadership theory in 1978. The theory's focus is basically concerned with providing an explanation for the lack of stronger empirical support for a relationship between leader traits or leader behaviors and subordinates satisfaction and performance. The substitutes for the leadership theory suggest the characteristic of the organization, the task, and subordinates may substitutes for or negate the effects of leaderships, thus observed relationships between leadership behavior and the importance organizational outcomes.

Substitutes for the leadership make leader behavior such as task-oriented or relationship-oriented unnecessary. Characteristics of the organization that may substitutes for leadership include formalization, group cohesiveness, inflexible ruled and organizational rewards not under the control of the leader. Characteristics of the task that may substitute for the leader include ability, experience, training, and job-related knowledge. The substitutes for the leadership theory have generated a considerable amount of interest because it offers an intuitively appealing explanation for why leader behavior impacts subordinates in some situations but not in others. However, some of its theoretical propositions have not been adequately tested. The theory continues to generate empirical research.

2.3 Meaning of Job Satisfaction impact of humanity in Government Sector

This is probably one of the most researches and constructs in organizational behavior. Literally thousands of articles have been written about its definition and meaning, its antecedents, and its consequences. Job satisfaction impact of humanity in Goverrment Sector. Communication style in the organization is very important and also key forces in providing employees with job satisfaction, humanity and communication skills may be defined as the emotional state resulting from the appraisal of one's job and as such can be negative, positive, or neural. A basic element in this definition is that job satisfaction impact of humanity in Goverrment Sector. Communication style in the organization is very important and also key forces in providing employees with job satisfaction, humanity and communication skills has to do with an effective state or how one "feels" about one's job in contrast to simply describing a job.

Spector (2001) are present three reasons to clarify the importance of job satisfaction. First, organizations can be directed by humanitarian values. Based on these values they will attempt to treat their employees honorably and with employees are dealt with effectively. High levels of job satisfaction could also be a sign of emotional wellness or mental fitness. Second, organizations can take on a utilitarian in whom position in which employees' behavior would be expected to influence organizational operations according to the employee's degree of job satisfaction impact of humanity in Goverrment Sector. Communication style in the organization is very important and also key forces in providing employees with job satisfaction, humanity and communication skills or dissatisfaction.

2.4 Theories of Job Satisfaction impact of humanity in Goverrment Sector. Communication style in the organization is very important and also key forces in providing employees with job satisfaction, humanity and communication skills

2.4.1 Maslow Needs Hierarchy Theory

Maslow believed that human beings have five ascending types of "need" which they seek to satisfy or fulfill within different environments (Maslow, 1998). At the bottom of Maslow's "hierarchy" of human need are the basic physiological needs for food, shelter and clothing. According to the theory, these needs must be "reasonably satisfied" before a person will turn his or her attention to the next higher order need. "Reasonable satisfaction" is culturally defined. One physiological need is met (and there is a certain assurance that will continue to be met), a person turns his or her attention toward the need for the safety from danger.

2.5 Leadership Behaviour and Job Satisfaction

Having discussed the major theories of leadership behavior in the first section and of job satisfaction in the second section reviews the empirical evidence on the relationship between leadership behavior and job satisfaction. As what we known primary responsibility to get things done through others (Fruzi & Patrick, 1998). Effective leader get most of most of their useful information from talking with people (Hill, 1979). In order to communicate with people for information, leader must give sincere effort and commitment. Leader must also work with people for information, leader must give sincere effort and commitment. Good leader must be good coaches who know how to encourage staff to raise their performance at work, improve their knowledge, and realize their full potential (Heller, 1998). Leader must also work with people to solve problems.

"To do this effectively, leader must be able to influence what and how other people do thing" (Hill, 1999). This is a challenge because people are unique and has different personalities, needs goals, desires and background. Each individual's personality is a result from heredity, upbringing, schooling or lack of it, family and friends, work, religion, neighborhoods, and experiences. As a result, managers must always adapt to the employees behavior to take into account the expectations, values, and perceptions of the people they are interacting with (Hill, 1999).

2.5.1 Gender

Gender is another employee characteristic which is often to cause variations in reported levels of job satisfaction Sousa-Poza & Souza-Poza, 2000). There have been many suggested explanations to this phenomenon ranging from genetic differences to an overrepresentation of women with part-time jobs.

Research investigating the relationship between gender and job satisfaction uncovered three possibilities. First, females are more satisfied than males. Second, males are more satisfied than females (Hulin & Smith, 2001; Locke, Fitzpatrick, & White, 2000). Third, no difference exists between males and females with respect to job satisfaction (i.e., D'Arcy, Syrotuik, & Siddique, 2001).

2.5.2 Age

The majority of studies on the relationship of age and job satisfaction have found some association between employee age and job satisfaction. Clark et al. (2004) suggest that job satisfaction is U-shaped in age, with higher levels of morale among young workers but that this declines after the novelty of employment wears off and boredom with the job sets in. Satisfaction raises again later life as workers become accustomed to their role. In contrast, other studies report a linear positive relationship based on age (Lee and Wilber, 2003). In their study of accountants in Singapore, Ang et al. (2001) report a positive linear relationship between age and job satisfaction (except for one respondent group for which the relationship is curvilinear).

2.5.3 Marital Status

The first two employee characteristics dealt with physical off the employees but there have also been reported differences in reported job satisfaction with respect to more mental properties of the employees. Especially educational level has been found to affect the level of job satisfaction and in many job satisfactions has been found to decrease with increasing educational level ,(Robie, et al. 1998, Sousa-Poza & Sousa-Poza, 2000, Wharton, et al, 2000).

By and large, no differences in the satisfaction scores have been found between married and single employees (Hashim, 1998). However, some studies have found married employees to be more satisfied than unmarried (McDonald & Gunderson, 2001), and some others have found single employees to be more satisfied than their married counterparts (Wright, et al. 1998).

2.5.3 Highest Educational Level

3 Methodology

This section will describe the research methodology that was used to test the hypotheses. First we will introduce the study design and discuss the appropriateness and size of the sample. Next, we will describe the actual model to be tested, the data collection process, and then the measure. Finally, we will present the data analysis procedures to establish the measure and test the hypotheses. I am studying in process to defend proposal. All process and step study can be proceed after process defend proposal approval panel and document data. My suggestion will choose as bellow :

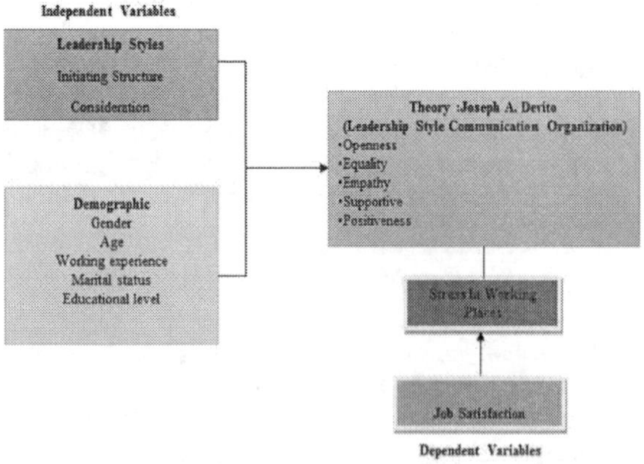

Fig.1. Conceptual Framework

2.5 Leadership Behaviour and Job Satisfaction
3.1 Instrumentation

In this study, we used adapted questionnaire for data collection to elicit the opinions of the staffs about leadership styles of their top management towards their job satisfaction.

The questionnaire used in this study consisted of three sections: Section A concerned the demographic factors including gender, age, race, marital status, highest education qualification, previous working experiences, salary, current position and working experiences of the staffs. However item such as race, salary, and current position are not included in hypotheses test. Section B contained of 27 questionnaires on leadership styles and section C consisted 6 of questionnaires of the job satisfaction. This instruments was selected because it is related with our study and the scale has shown a high level of reliability, good level of validity and high level of consistency.

4 Findings

This purpose of this chapter is to present the findings of the analyses of data which were collected in the study of the relationship between leadership styles and job satisfaction of staffs IPK Kedah. The findings of the study are discussed in this chapter based on the research objectives and hypotheses that have been proposed in the Chapter two. The discussions are divided into four main sections. The first section describes the demographic factors of the respondents. Second section indicates the reliability of the variables (leadership styles [initiating structure and consideration] and job satisfaction). The third section describes the general view of job satisfaction and the last part describes the result of the hypotheses test and summary of the results.

5 Summary of Findings

This chapter presents the analyses and findings for a whole part of this study. The multiple analyses including descriptive statistics, Pearson correlation, regression analysis, and etc. will gives us the multiple view of interpretations of this study.

5.1 Summary, Discussion, Conclusions and Recommendations

This chapter presents the summary, methodology and findings that would be discussed refer to the hypotheses developed for this study. In addition, a brief of conclusion and recommendation of the study for future research are also included.

This study examined the job satisfaction in IPK Kedah. The distribution of demographic factors among the staffs in this organization was analyzed. The relationships and effects between staff's job satisfaction and the independent variables, leadership styles were also determined.

References

1. J. Almarri, J. Meewella, T. Mainela, International conference on Excellence in Business. Shariah United Arab Emirates (2012)
2. Saifuddin Abdullah. (2012, 6 Ogos). "Keusahawanan Sosial Basmi Kemiskinan". Akhbar Sinar Harian.
3. Mohamad Mustapha, *Keusahawanan Islam* (Petaling Jaya, Prentice Hall, 2002)
4. S. Johnson, (2000), "Literature review on social entrepreneurship" available at: www.bus.ualberta.ca/ccse/Publications/Publications/Lit._Review_SE_November_2000.rtf (accessed February 20, 2011).
5. A. Nicholls, Social Entrepreneurship: New Models of Sustainable Social Change, Oxford University Press, Oxford (2008)
6. J.T. Zietlow, Nonprofit Management and Leadership **13** (1), 85-90 (2002)
7. S.A. Zahra, H.N. Rawhouser, N. Bhawe, D.O. Neubaum, J.C. Hayton, Strategic Entrepreneurship Journal **2** (2), 117-31 (2008).
8. Mohd Adib Abd Muin, Azizi Abu Bakar, Shuhairimi Abdullah, Prosiding: International Conference on Business Innovation, Entrepreneurship and Engineering 2013, 386-399 (2013)
9. Muhammad Ali Hashim, *Khalifah Ganti Superman* (Kuala Lumpur: Utusan Publications & Distributors Sdn Bhd., 2013)
10. Mohd Adib Abd Muin, Shuhairimi Abdullah, Muhammad Fakhirin Che Majid, Prosiding: Seminar Penyelidikan Kebangsaan 2014, (2014)
11. Yazilmiwati Yaacob, Ilhaamie Abdul Ghani Azmi, Entrepreneurs' Personality from Islamic Perspective: A Study of Successful Muslim Entrepreneurs in Malaysia, **46**(16), (2012)
12. Shuhairimi Abdullah, Aplikasi Nilai- Nilai Murni Dalam Kalangan Usahawan ke Arah Pembangunan Usahawan Muslim Yang Holistik. Kertas Kerja Seminar Keusahawanan Islam II Peringkat Kebangsaan, (2011)
13. Ab. Aziz Yusof, *Pengenalan kepada Usahawan dan Keusahawanan* (Kuala Lumpur: Scholar Mind Publishing, 2010)
14. *Al-Quran al-Karim Terjemahan* (Kajang, Selangor, 2012)
15. Shuhairimi Abdullah, Journal of Islamic Human Advanced Research **3**, 322-345 (2013)
16. M.A. Abdullah, A. Hoetoro, International Journal of Management and Business Research **1**(1), 35-46 (2011)
17. Suhaimi Mhd Sarif, Abdullah Sarwar, Yusof Ismail, Journal of Scientific Research **14** (11): 1463-1470 (2013)
18. Global Entrepreneurship Monitor. (2009). "Report on Social Entreprenurship 2009". (Report). London. United Kingdom.
19. Saifuddin Abdullah. (2009, 29 Julai). "Keusahawanan Sosial Dalam Penyelidikan". Akhbar Utusan Malaysia.
20. C.S. Teresa, L. N. Patricia Journal of Research in Marketing and Entrepreneurship 11(1), 49-65 (2009)
21. H. Jiao, Social Enterprise Journal **7**(2), 130–149 (2011). doi:10.1108/17508611111156600
22. J. Weerawardena, G. S. Mort, Journal of World Business **41**(1), 21-35 (2006)
23. P. J. Murphy, S.M. Coombes, Journal of Business Ethics **87**, 325-336 (2009)
24. Adnan Alias, Mohamed Dahlan Ibrahim, *Keusahawahanan Islam* (Petaling Jaya: Prentice Hall, 2002)
25. Radin Siti Aishah, Zaidatol Akmaliah, "Keusahawanan Sosial Dari Perspektif Islam". In Proceeding of the International Conference on Research in Ethics Education 2013. Istanbul Turki (2013)
26. S. Habibollah, A. Hamed. N. Davoud, Journal of Business and Management **5**(7), (2010)
27. E. Shaw, (2011). Toward A Theory of Entrepreneurship : The Significance and Meaning of Performance and the Emotion Management of Entrepreneurs A Thesis submitted for the Degree of Doctor of Philosophy By Elizabeth Shaw Brunel School of Business and Management Brunel Unive. Brunel University.
28. J.A. Robinson, J. Mair, K. Hockerts, *International Perspectives of Social Entrepreneurship* (Palgrave, London, 2009)
29. Mohd. Fadzillah Kamsah, Muhammad Zakaria. *Langkah Bijak Usahawan Terbilang* (Kuala Lumpur, Telaga Biru Sdn Bhd., 2008)
30. Mohd Adib Abd Muin, Shuhairimi Abdullah, Azizan Bahari, (2014). Model on Social Entrepreneurship: Identify the Research Gap based on Islamic Perspective. International Conference on Social Entrepreneurship 2014, 7-9 Nov 2014, Genting Highland Pahang.
31. J. Marri, J. Meewella, T. Mainela, Social Entrepreneurship and Islamic Philanthropy. International conference on Excellence in Business. (Shariah United Arab Emirates, 2012)
32. Saifuddin Abdullah. (2012, 6 Ogos). "Keusahawanan Sosial Basmi Kemiskinan". Akhbar Sinar Harian.
33. Mohamad Mustapha, *Keusahawanan Islam* (Petaling Jaya: Prentice Hall, 2002)
34. S. Johnson, (2000), "Literature review on social entrepreneurship", available at: www.bus.ualberta.ca/ccse/Publications/Publications/Lit._Review_SE_November_2000.rtf (accessed February 20, 2011)

Integrated Counselling Strategies for Children in the Substance Abused Family

Abijo Temitayo Serah[1,*], *Huzili* H.[1], and *Hanif* Suhairi[2]

[1]Department of Education & Sport Science, Universiti Malaysia Perlis, Perlis, Malaysia
[2]Department of Information Technology Communication, Universiti Malaysia Perlis, Perlis, Malaysia

Abstract. In spite of the fact that substance abuse remains a public health challenge and social problem of enormous magnitudes, with many adolescents and families involved and affected by substance abuse disarrays the issue of proper counselling still requires more attention. In view of this therefore, this study has explored and appraised conceptually the desirability as well as applicable integrated strategies for counselling of the children in the substance abused family. Based on the appraisal of existing strategies for counseling of substance abused family then this study has been able to establish that if people involved as well as the victims of the abused will be able to learn about concepts such as inner strength, self-praise, and group praise, which may prompt setting up of goals for positive decision on their addiction. Similarly, the strategies will encourage the target family to generate more and more positive self-portraits and, as these may inspire making positive lifestyle changes.

1 Introduction

In families where substance are being abused, behavior becomes recurrently impulsive and communication is uncertain [1]. Family life is characterized by chaos and unpredictability. Behavior can range from loving to withdrawn to crazy. Structure and rules may be either nonexistent or inconsistent. Children, who may not understand that their parent's behavior and mood is determined by the amount of substance in their bloodstream, can feel confused and insecure. They love their parents and worry about them, and yet feel angry and hurt that their parents do not love them enough to stop using. This situation was corroborated by The American Academy of Experts in Traumatic Stress [2] who revealed that in view of the enormous challenges being faced by children in the substance abused family, many families referred to children's social care because of concerns about the children's welfare and evidence of parental substance misuse or domestic violence. More so, that parental drug abuse is a possibility factor for child abandonment and general mistreatment [3].

In spite of the fact that substance abuse remains a public health challenge and social problem of enormous magnitudes, with many adolescents and families involved and affected by substance abuse disarrays [4] the issue of proper counselling still requires more attention [5]. Similarly, Abdu-Raheem [6] reiterated that the issue of substance abuse needs pressing responsiveness due to the startling rate and more so, that the children of these families who are innocently at risk receive less attention.

Many times, children of substance abusers are frightened. They may be the victims of physical violence or incest. They may also witness violence. As noted by NFPI Children of parents who have a drink problem often want to keep it a secret and don't feel they can invite other children home – this affects their ability to make friends and have a social life. They see other families having fun together and notice that their own family is different. Home life is tense and unpredictable. Often children don't know what to expect when they return from school or when their parent comes home after a drinking session. However, in spite of all these challenges, many of the Children of substance abused family are often not aware of how much their parents' drinking is impacting on their lives [1]. Therefore, counselling becomes imperative for children who are in this problem. Correspondingly, Kendler, Ohlsson, Sundquist and Sundquist [7] believed that the widespread prevalence of illicit substance use which is high within the family presents another indication of the need for continued research because it is a risk factor for children. Therefore, this study is set out to explore conceptually the desirability as well as applicable strategies for counselling of the children in the substance abused family.

2 Prevention Strategies for substance abuse impacts on Children

As to substance utilize and abuse, avoidance depends on previous history of conduct, and embarks to foresee and turn away planned negative outcomes from happening. Counteractive action programming as far as substance utilize and abuse is essentially conveyed to the whole school populace and frequently the entire educational system, paying little respect to students' hazard elements. More focused on aversion programs have been utilized to address the issues of students who show the hazard components sketched out in the past segment or who have a background marked by medication utilize [8]. The motivation, skills, and decision-making model of drug abuse prevention. Substance Use & Misuse, 39, 1971–2016. Not with standing, little writing exist with respect to the school guide's part.

At the school level, two methodologies have principally been utilized to address the psychosocial advancement of students. The social impact approach [9] is utilized to improve the familiarity with natural and associate impacts which possibly advance medication utilize, create peer refusal aptitudes, and test sedate desire. The individual and social expertise upgrade approach has additionally been utilized and is viewed as more far reaching in nature [10]. These sorts of counteractive action programs show abatements of 10% to 15 % in school age youngsters starting utilization of substances contrasted with students that are not part of an avoidance program. In particular, these projects both diminish in social impact and increment in individual and social abilities [8].

Prevention strategies targeting youths have evolved over more than twenty years as evaluation research has revealed more about what works [11]. Several strategies are used effectively, especially in combination (Indiana Prevention Resource Center at Indiana University):

• *Information dissemination*. This strategy provides awareness and knowledge of the nature and extent of alcohol, tobacco, and other drug use, abuse, and addiction and their effects on individuals, families, and communities as well as information to increase perceptions of risk. It also provides knowledge and awareness of prevention policies, programs, and services. It helps to set and reinforce norms (e.g., a policy that underage drinking and drug dealers will not be tolerated in this neighborhood).
• *Prevention education*. This strategy aims to affect critical life and social skills, including decision making, refusal skills, critical analysis (e.g., of media messages), and systematic and judgmental abilities.
• *Alternatives*. This strategy provides for the participation of targeted populations in activities that exclude alcohol, tobacco, and other drug use by youths. Constructive and healthy activities offset the attraction to, or otherwise meet the needs that are usually filled by, alcohol, tobacco, and other drug use.
• *Problem identification and referral*. This strategy calls for identification, education, and counseling for youths who have indulged in age-inappropriate use of tobacco products or alcohol or who have indulged in the first use of illicit drugs. Activities under this strategy would include screening for tendencies toward substance abuse and referral for preventive treatment for curbing such tendencies.
• *Community-based process*. This strategy aims to enhance the community's ability to provide prevention and treatment services to alcohol, tobacco, and other drug use disorders more effectively. Activities include organizing, planning, enhancing efficiency and effectiveness of services implementation, interagency collaboration, coalition building, and networking. Building healthy communities encourages healthy lifestyle choices.
• *Environmental approach*. This strategy sets up or changes written and unwritten community standards, codes, and attitudes, influencing the incidence and prevalence of alcohol, tobacco, and other drug use problems in the general population. Included are laws to restrict availability and access, price increases, and communitywide actions.

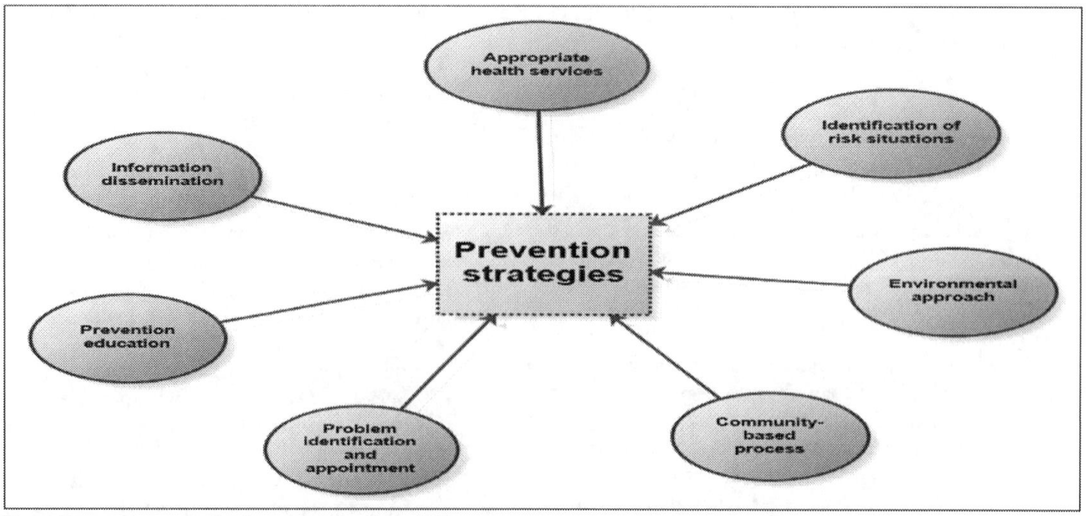

Fig. 1. Prevention strategies for substances abuse [11]

3 Strategies for Substance abusers and Victims Counselors

According to Igidi [12] the spate of alcohol abuse in families in Nigeria has reached alarming point because at present, only few is free of alcohol abuse and unfortunately, nothing is being done about it. Similarly, Chweya and Auya [4] lamented that in Nigeria, the situation shows there is high pervasiveness of substance use especially alcohol among family members and there is high likelihood that the frequency will continue to increase.

Therefore, in view of the fact that most of the children of substance abused family attend school, there is a tendency for them to interact with school instructors on a predictable premise in addition to that, schools are purposively arranged to give decisive counselling to such. The purposeful setup includes the school advocates' capacity to distinguish students' hazard components and hindrances to individual, social, and scholastic improvement [13].

However, in view of the increasing rate of family substance abuse [12] and the challenges and risks resulting from it [6], it thus far indicates that school counselors need to be familiar with students who may be substance abusers; students who might be at risk for becoming substance abusers; and students who have parents who have been or who are substance abusers. According to Lambie and Rokutani [14] counselors for children and students who have issues with substance abuse have three main tasks. These tasks are: identifying the signs of substance abuse; establishing a therapeutic relationship; helping the family system so that change is possible; and to be a liaison between the student, family, school, and community

3.1 Identifying Warning Signs of Substance Abuse

Fulfilling these functions may be challenging due to the lack of education about substance abuse that school counselors receive. School counselors and school counseling students must take it upon themselves to learn more about substance abuse disorders.

Here is a list of concerning behaviors to look for in an adolescent who might be using drugs:

- ✓ Changes in school performance (falling grades, skipping school, tardiness)
- ✓ Changes in peer group (hanging out with drug-using, antisocial, older friends
- ✓ Breaking rules at home, school, in the community
- ✓ Extreme mood swings, depression, irritability, anger, negative attitude
- ✓ Sudden increases or decreases in activity level
- ✓ Withdrawal from the family; keeping secrets
- ✓ Changes in physical appearance (weight loss, lack of cleanliness, strange smells)
- ✓ Red, watery, glassy eyes or runny nose not due to allergies or cold

- ✓ Changes in eating or sleeping habits
- ✓ Lack of motivation or interest in things other teenagers enjoy (hobbies, sports)
- ✓ Lying, stealing, hiding things
- ✓ Using street or drug language or possession of drug paraphernalia/items
- ✓ Cigarette smoking

3.2 Establishing a Therapeutic Relationship and Involving the Family

In substance abuse intervention, treatment, and follow-up, it is important to establish a relationship in order to provide ongoing support [14]. Naturally, a school counselor has other duties to consider as part of the job, but if a school counselor is committed to counseling, then he or she can maintain a relationship with the troubled adolescent. This counseling relationship may be a source of comfort for a student struggling with substance abuse.

Involving the family is not only ethical and probably mandatory in most school districts but can also give the counselor an idea about what the student's.

Involving the family is not only ethical, and probably mandatory in most school districts, but can also give the counselor an idea about what the student's home life is like. The school counselor must consider the family system when thinking about early identification of substance abuse for two reasons: working with the parents can provide useful information about the student's substance abuse, and inviting the family to come into the school to consult with the school counselor can help the school counselor see how the family interacts [14].

3.3 Substance Abuse Prevention and Intervention within the School

In order to try to protect the most students possible, school counselors can advocate for substance-abuse prevention services in schools. School counselors can achieve this goal in a few ways. Counselors can use group counseling sessions to address substance-abuse issues with students who are struggling with anxiety, depression, stress, grief, or social skills. School counselors can also bring in outside agencies to provide services to individual students, targeted groups of students, or prevention programming to all students.

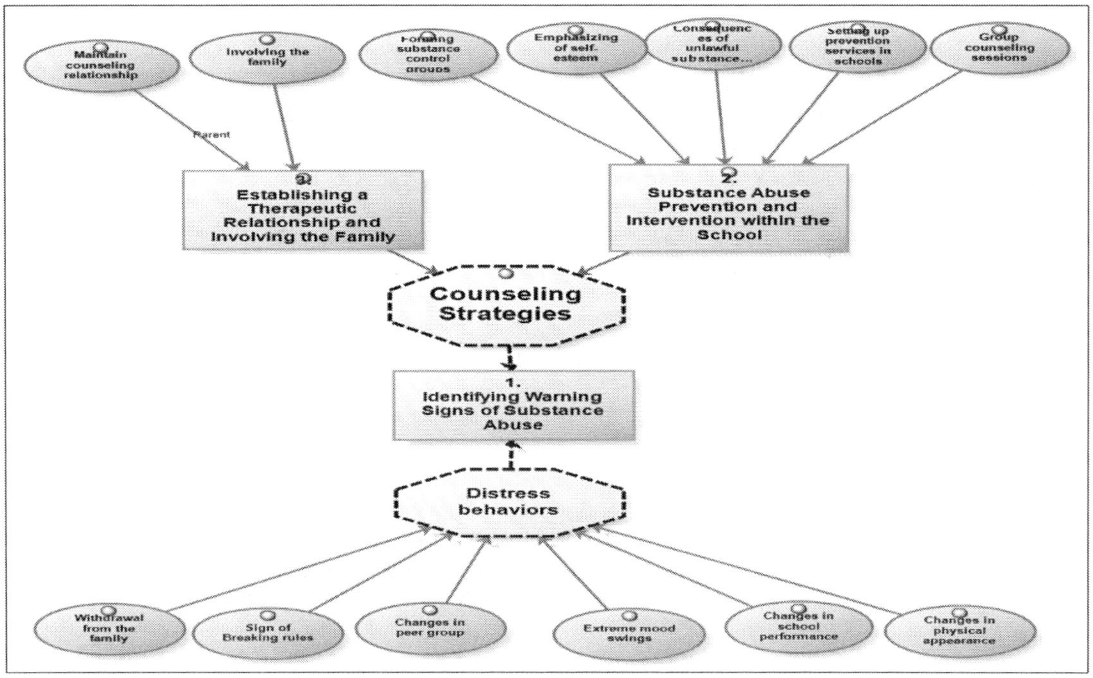

Fig. 2. Counselors Strategies for Substance Abusers and Victims [14]
Source: Lambie and Rokutani (2002).

Arising from the Lambie and Rokutani [14] strategies and in view of the emerging sensitivity to issues of culture in substance abuse as revealed by DHHS [15] the need for an integrated counselling strategies for children of substance abuse families becomes the gaps which this study fills. The integrated strategies which are critical for both long and short-term success by setting clear and achievable goals for counselors will be a main guide for government, counselors, and agencies. Therefore, in line with UNODC, DHHS, and Baird [5, 1-16] the following basic integrated strategies are being recommended for substance counseling for substance abuse family generally.

3.3.1 Confidentiality should be assured

Confidentiality contributes to openness and trust which are essential to any helping relationship. If total confidentiality cannot be guaranteed, counselling staffs must make the victims aware of the possible implications of disclosure right from the beginning of the interaction. Though, due to secrecy involved, children of substance abused family may be unwilling to take up counselling opportunities when confidentiality cannot be guaranteed.

3.3.2 Counselors should know their strengths and limitations

Every counselor need to consider their level of skill and knowledge before intervening, identifying that restrictions which could relate to issues such as available time, confidentiality and victims acceptance.

3.3.3 Counselling should be kept separate from discipline

In some schools it may not be possible to have two separate staff members to provide counselling and discipline. Where possible, however, those roles should be separated in order to increase the effectiveness of intervention in both areas.

3.3.4 De-stigmatizing without condoning drug use

Counselors and teachers should avoid trying to force the student to change, as that may increase their resistance. In the same way, condoning or condemning the student's behaviour is counter-productive. Instead, the student should be treated as a decision-maker who has certain reasons for his or her choices based on how he or she sees the world. Teachers should direct their efforts to helping the student balance the costs and benefits of his or her decisions.

3.3.5 Concentrating on rapport and empathy

Drug use is a sensitive topic and one that young people are often reluctant to talk about. Some of them believe that adults will attempt to convince them to stop, criticize their behaviour or punish them. Establishing a rapport and expressing understanding is vital to a healthy helping relationship.

3.3.6 The role of culture and cultural backgrounds

To develop effective counseling strategies for diverse populations, the counselor must understand the role of culture and cultural backgrounds, recognize the cultural backgrounds of substance abuse family, and know enough about their culture to understand its effect on key counseling issues. This sensitivity is important at every stage of the counseling process. Meanwhile, cultural background can affect attitudes concerning such factors as proper family behavior, family hierarchy, acceptable levels of substance use, and methods of dealing with shame and guilt. Forcing families or individuals to comply with the customs of the dominant culture can create mistrust and reduce the effectiveness of therapy.

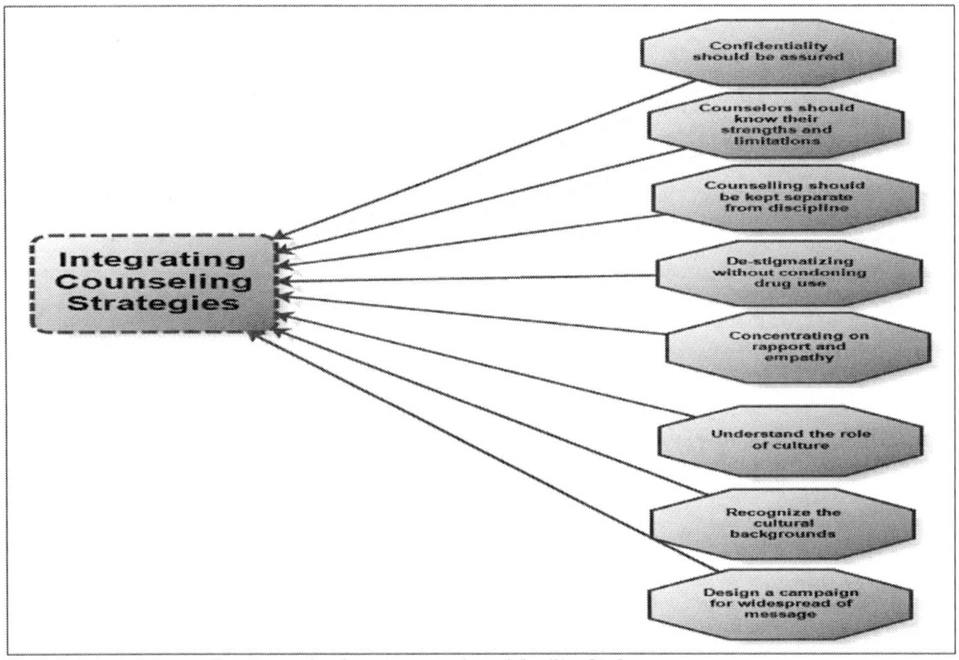

Fig.3. Integrated Counseling Strategies for substance abused families [16]

Reducing the harm to children from parental problem substance use as well as the parent themselves should become a main objective of policy and practice however much attention has not been directed at such. Therefore, this study conceptually synthesize previous strategies with what the handling requires in order to attain an effective counseling which will have major benefits for the innocent victims of family engrossed in substance abuse.

Similarly, as part of the integration, counselor should institute and sustain a campaign that will achieve widespread, frequent, and prolonged exposure to implications of substance abuse as well as the benefits derivable from the acceptance of message.

4 Conclusion

Based on the appraisal of existing strategies for counseling of substance abused family this study has been able to establish that if people involved as well as the victims of the abused will be able to learn about concepts such as inner strength, self-praise, and group praise, which may prompt setting up of goals for positive decision on their addiction. Similarly, the strategies will encourage the target family to generate more and more positive self-portraits and, as these may inspire making positive lifestyle changes.

As new demonstration and innovation projects continue to be implemented, expanded, and evaluated, the field continues to learn more about promising and effective approaches to holistically address the complex needs of families with substance use issues, there is need for a systematic counselling section for all the victims and their families. In particular, there is a continuing call for and movement toward enhanced collaboration among child welfare, substance abuse treatment, courts, and other systems to provide coordinated and comprehensive services of counselling to both children and their parents. Generally according to previous studies, children of parents who engage in substance abuse remained more likely to be having indication of educational difficulties, emotional and behavioral problems as well as other challenges.

References

1. Turning Point, (2016). Bottling it up the effects of alcohol misuse on children, parents and families. Retrieved 27 March, 2016 from http://www.substancesandalcohol.ie/6276/1/3499-3720.pdf.
2. The American Academy of Experts in Traumatic Stress (2014). Effects of Parental Substance Abuse on Children and Families. Retrieved October 6, 2016 from http://www.aaets.org/article230.htm.
3. M. F. Taylor, D. Coall, R. Marquis, R. Batten, Int J Ment Health Addiction (2016) DOI 10.1007/s11469-016-9645-7.
4. M. Chweya, S. Auya, *International* Journal of innovation and scientific research, ***9***(1), 35-39 (2014)
5. Baird, *Educational Specialist* **33** (2011)
6. B. O. Abdu-Raheem, Contemporary Issues in Education Research **6**(2), 233-240
7. K. S. Kendler, H. Ohlsson, K. Sundquist, J. Sundquist, JAMA psychiatry **70**(2), 235-242 (2013)
8. S. Sussman, M. Earleywine, T. Wills, C. Cody, T. Biglan, C. W. DentSubstance Use & Misuse **39**, 1971–2016 (2004)
9. R. I. Evans, Preventive Medicine **5**, 122–127 (1976)
10. G. J. Botvin, E. M. Botvin, Psychological Reports **70**, 171-178 (1992)
11. H. Abadinsky, *Drug use and abuse: A Comprehensive Introduction* (USA: Wadsworth, 2011)
12. T. Igidi, (February 25, 2016). Few families are substance free in Nigeria. Daily Trust Newspaper. Retrieved February 27, 2016 from http://www.dailytrust.com.ng/news/home-front/-few-families-are-substance-free-in-nigeria/135355.html.
13. M. J. Ratts, A. M. Hutchins, Journal of Counseling & Development **87**, 269-275 (2009)
14. G. W. Lambie, L. J. Rokutani, A systems approach to substance abuse Counseling **5**(5), 353-359 (2002)
15. DHHS (2005). Subtance *Substance Abuse Treatment and Family Therapy.* U.S. Department of Health And Human Services, USA: Center for Substance Abuse Treatment.
16. UNODC (2004). Nigerian Country office. Directory of treatment and rehabilitation centers. Nigeria.

Religiosity and Volunteering Intention among Undergraduate Malaysian Muslim Students

Abdullah AbdulElah Ali Sallam[1,*], *Shuhairimi Abdullah*[1], Abdul Jalil *Ramli*[1], *N. S.* Hussin[1] , *Z* Ahmad[2], and *Azizan* Bahari[1]

[1]Universiti Malaysia Perlis, Perlis, Malaysia
[2]Institute Pendidikan Guru Malaysia Kampus Perlis, Perlis, Malaysia

Abstract. This paper deals with the question: To what extent do religiosity characteristics, contribute to the influence of volunteering intention among Malaysian Muslim students during disasters? To answer this research question, we focused the students in public universities. The finding concerns found that religiosity increases the likelihood of volunteering intention, implying that religious affiliation of youth increases the likelihood of volunteering. This is in line with previous research, that religious attendance is related positively to volunteering. These results confirm the idea that support of the religious attributes community plays quite a large role in volunteering process. . However, it a bear that volunteering is not only dependent on religious community, but also on individual motivation.

1 Introduction

Lately, religiosity has emerged as an attractive field of research [1] and there has been increasing interest in the relationships between various aspects of religion [2][3]. Several studies have systematically tried to unveil the underlying dynamics of religiosity in organizational life [4][5]. The influence of religion and spirituality on individual's behaviour has emerged as an area of interest for researchers [6]. In fact, for NPO (non-profit organization), young people are considered a growing and strong volunteer source and essential for the efforts of non-governmental organization (NGO) marketing [7].

In volunteer work, the participation of Australian younger generation is considered lowest in the age group 25–34 [8]. Although there are about 135,000 full-fledged volunteers in Malaysia, the figure only represents less than 1% of youths aged above 15 [9]. For the explanation and understanding related to youth volunteering, the focus of an individual must be on the factors influencing the motivation or encouraging youth for being a volunteer [10] and positive youth development [11]. A study of youth volunteers of Anti-Drug Program findings shows that most of the participants did not know the membership procedures of this Youth Volunteers for the Anti-Drug Program. It is believed that there are some factors that affect students' inactiveness or lack of interest in voluntary activities and in their intention to volunteering [12]. And they should be given the opportunity to participate in various aspects [13]. Also important to develop a student college volunteer [14].

There are very few studies regarding of volunteering intention among college student [6][14]. There is also a lack of research concerning volunteers [15][16]in developing countries [14] and the relationship between religiosity and volunteers [16]. Recently, the field of testing the influence of religion and spirituality to the psychological behavior of an individual has been of interest [17][18][6]. Despite the researchers scepticism, spirituality and religiosity are considered the influencing factors of emotions, behaviors, cognition, merit of an individual [6][19][20][21]. And volunteering behavior [22]. Moreover, there are many studies of religion and mental health in the context of Christianity context, but studies in Islamic countries are few [3]. Based on the observations, a gap has been found in the literature related to the understanding of the factors influencing the intention of college or university students to volunteering [23]. This study is an attempt to find the literature gap for the investigation of the influencing of religiosity on volunteering intention among Malaysian Muslim students. It will provide for Malaysian organizations, donors and interested ministries the tools and techniques for development of volunteering behavior among youth, consistent with previous volunteering researches [24][25][26].

2 Literature Review

2.1 Volunteering Intention in Malaysia

Volunteerism is a method to promote active learning, providing experience which helps in the enhancement of commitment to contribute among citizens, the education experience related to expansion, serving the communities more appropriately [27]. Youth volunteerism is very crucial in developing a community [10].

The majority of people joining volunteering activities are retirees. In a study of volunteerism and the

* Corresponding author: youthcentre97@gmail.com

development of Malaysian social care system found 4% had 21-30 years active volunteers and 1% had 31-40 years active volunteers [28]. Time is considered the barrier for greater in the volunteer practices among younger generation [28]. The issue of younger generation being volunteers is of utmost importance because of their higher level of energy and motivation which can result in good outcome [10].

2.2 Purpose of Study

Although, studies related to volunteering, in previous three decades, have increased [29] little is known about volunteering [30][16]. Also, about the social and psychological features of volunteers, volunteering motivations, psychological influence [29][31] and the causal relationship and intentions related to volunteering [30].

The purpose of this study is to examine the relationship between volunteering intention, and how factors of religiosity influence this intention among college students. Several studies have focused on intention or volunteering behavior of college students [10][32][33][34][35]. A few studies have examined religious preferences and spiritual beliefs among college students [6][36]. However, this study will examine the relationship between volunteering intention and the influence of religiosity specifically among the universities students. There sis notes that, there are very few published studies about religion and health among young persons in Islamic countries [3].

2.3 Religiosity

The term religion in this study means the institution serving as a source of knowledge, practice, and beliefs of an individual [37]. From scholarly perspective, religiosity is the level of individuals being religious, and the importance an individual attached to beliefs and practices of religion [38]. Sociologists of religion have struggled with the definition of religion, but agree that religion is more than simply the belief in the supernatural, god, or gods. Religion consists of practices and institutions that have moral purpose [38]. In other words, religion is the organization that provides a set of beliefs, instructions on appropriate practices, and support for individual members, while religiosity is the acceptance of and commitment to those elements provided by religion [37]. It is a collection of organized practices, norms, and values [38].

Adam Smith (1976, 2010) in his Theory of Moral Sentiments analysed religiosity from a rational point of view and noted that religiosity acts as a kind of internal moral enforcement mechanism [39]. Individuals characterized as religious are not only those who hold particular religious beliefs but also practice them in day-to-day life [40][41]. For example, religious people pray, attend a place of worship, in some cases observe fasting and engage in socially beneficial actions (e.g., charity and volunteer work), political and economic activity [42]. As directed or encouraged by their religion [4][43]. Many define religiosity as both beliefs and practices relating to an organized religious institution or a specified divine power (e.g., [44] or to a religious affiliation or to God [45][46].

2.4 Religiosity and volunteering intention

Religion is another important predictor of volunteering [8]. Although, there are differences by types of religion and religious denomination, most religions encourage altruism, helping behaviour [47][43], and encourage cooperation between members [48]. Previous research has shown that religious attendance in particular is related to volunteering behavior [49][50][51]. It has been shown that the relationship between religiosity and volunteering is mediated by denominational involvement, and is not a direct result of prevailing religious norms and teachings. Crucial are social networks that provide contacts and enhance norm conformity within the group [52]. As well as when investigating influences of individual religious characteristics, it is highly important to control for collective religious characteristics [53]. The General Social Survey conducted in 2010 by the Australian Bureau of Statistics shows that volunteering with religious groups was higher among the young (aged 18–24) and older age groups (aged 55–74) and lower in the age group 25–34 [8].

In a study on the role of religiosity for formal and informal volunteering in the Netherlands, Van Tienen, Scheepers, Reitsma and Schilderman found that spirituality increases the likelihood of informal volunteering, implying that openness to other people's needs increases the likelihood of the actual provision of help. They also found in formal volunteering the religious attendance is related positively to formal volunteering, religious as well as secular volunteering [53].

In similarity [53][53][55]found current institutional religious involvement emerged as a positive predictor of the likelihood that men would be involved in volunteer work, as well as of the amount of time that men gave to this kind of work. And Men who are more involved in church life or houses of worship may have a higher probability of being exposed to (or invited to participate in) general volunteer activities. Taniguchi and Thomas argued that religious exclusiveness has a significant positive influence on volunteering only in religious areas [56]. Meanwhile, religious inclusiveness is significantly and positively associated with both religious and secular volunteering.

Volunteering for religious organizations shows an interesting pattern in that there are higher odds of participating for those in the range of 18–24 years, controlling for other factors. None of the other life course variables shows a strong relationship with this type of volunteering. Unsurprisingly, an individual's level of religiosity is strongly associated with this kind of volunteering [8].

3 Research Methodology

3.1 Participants

A total of 356 undergraduate students (246 females and 110 males) from two public universities participated in this study. Participants ranged in age from 18–26 years (mean = 1.2051), and were primarily Malay (97.2%) with a minority of others (2.8%). Most of the respondents came from Kedah which account for (11.2%) and Sarawak with the lower respondent rates of (0.6%) of the total students represented. The results showed that respondents consist of 98 (27.5%) first year students, 113 (31.7%) second year students, 90 (25.3%) third year students and 55 (14.6%) fourth year students. Students were recruited from different colleges from two universities Universiti Malaysia Perlis (UNiMAP) and Universiti Utara Malaysia (UUM) and were asked to complete the following self-report questionnaires. The number of the samples taken from the undergraduate students consisted of 400 respondents which were randomly chosen. 200 sets of questionnaires were distributed, 356 sets were successfully collected back.

3.2 Measure

Computer software – Statistical Package for the Social Sciences (SPSS) Version 19.0.1 was used for data analysis. This software was used in this study for data management and analysis. Religiosity towards volunteering intention Scale was developed for the present study under the guidance and supervision of the Universiti Malaysia Perlis. The scale contained 11 items adapted from Freiheit, Sonstegard, Schmitt and Vye, (2006) [57]; Hernandez (2011)[45] studies (Table 3.1). Each item was rated by the respondent on a five-point Likert scale ranging from strongly disagree (1) to strongly agree (5) so that scores on the total scale had a possible range of 18 to 26, with higher scores indicating a more positive religiosity to volunteering intention. Internal reliability was found to be adequate for religiosity (Cronbach's alpha = 0.94) and volunteering intention (Cronbach's alpha = 0.87).

The questionnaire has been developed for measuring religiosity in volunteering intention, a crucial distinction from other measures available which focus on religiosity toward intention to volunteering. Questions include items such as "My religious faith encourages me to volunteer." and "My beliefs about God give me strength to help those who needed assistance." The measure can be found in Table 1.

Table 1. Measurement of Religiosity

Items
My religious faith encourages me to volunteer.
I look to my religion as a source of inspiration to volunteer in humanity/community services.
My religion provides meaning and purpose towards volunteering activities.
My religion place a high value in helping the needy.
My religion influence my intention to volunteer.
My religious beliefs make me willing to volunteer.
I feel calm and happy if I involve in humanity/community service for the sake of my religion.
I must help the needy in order to please God.
I intent to give money for humanity/community activates because it is highly encouraged by my religious beliefs.
I intend to volunteer to help others because of my religious beliefs.
My beliefs about God give me strength to help those who needed assistance.

SOURCE: ADAPTED FROM FREIHEIT, SONSTEGARD, SCHMITT AND VYE, (2006) ; HERNANDEZ (2011)

3.3 Results

The mean for volunteering intention variable which was 4.0314 and religiosity was 4.4941 indicates that the respondents on average agreed with the questions in general. The standard deviation value for volunteering intention is 0.69127, which is large from mean value and this indicates standard deviation for volunteering intention is less concentrate. Meanwhile, standard deviation religiosity is 0.54825. The standardized coefficient is 0.458 for religiosity. The value of R-Square indicates that 0.210 of the variance in volunteering intention can be predicted from the variables of religiosity. The significant F value of 0.000 which is significant at $\alpha = 0.05$. Religiosity value $t = 9.695$, $p = 0.000$ is significant and the beta value is .587. As the conclusion the religiosity have significant influence on volunteering intention.

4 Discussion and Conclusion

In this study, religiosity has a positive and significant effect among Malaysian Muslim students. For religion attributes, it is high correlation and has a significant value. This result is also in line with previous studies by Gray, Khoo and Reimondos, (2012); Mattis, Beckham, Saunders, Williams, Myers, Knight, Rencher and Dixon, (2004); Taniguchi and Thomas, (2011); Van Tienen, Scheepers, Reitsma and Schilderman, (2011) Putnam, Campbell and Garrett, 2012 that suggested that religious attendance is related positively to volunteering. The elements of religiosity across different religions serve beliefs and practices. This result is similar to study by Berger, (2006); Taniguchi and Thomas (2011) where religiosity has positive relationship with volunteering intention. Therefore, viewed as being a follower of a religious group may be the reason for the increase in the number of invitations for volunteering, and increases the knowledge related to volunteering [56]. In addition, religiosity has many interactions with other factors [45][3].

However, there are some limitations, the study focused on limited sample for data collection and the sampling of the study that involved the undergraduate Muslim students of the public universities. Because of this reason, the findings may be applicable only to the university students with higher knowledge and better educational and knowledge background. Thus, the study managed to evaluate students' intention from the public universities students' point of view only. It limits the generalizability of the findings. Therefore, it is suggested that future research can be done in larger sample size which include multi-ethnic and other religions that would allow higher response rate for more robust statistical analyses including structural equation modelling.

References

1. RA. Emmons, RF. Paloutzian, "The psychology of religion. Annual review of psychology". 54(1), 377-402 (2003)
2. AF. Shariff, AK Willard, T. Andersen, A. Norenzayan, Religious priming a meta-analysis with a focus on prosociality". Personality and Social Psychology Review (2015)
3. S. Vasegh, MR Mohammadi, The International Journal of Psychiatry in Medicine **37**(2), 213-27 (2007)
4. F. Ntalianis, W. Darr, International Journal of Organizational Analysis **13**(1), 89-102 (2005)
5. M. Benefiel, LW Fry, D Geigle, Psychology of Religion and Spirituality 6(3), 175 (2014)
6. MS. Harcrow, " Relationships between religiosity, spirituality & health behaviors among college students". THE UNIVERSITY OF ALABAMA; 2010.
7. E. Briggs, T. Landry C. Wood, Journal of Nonprofit & Public Sector Marketing **18**(2), 27- 45 (2007)
8. E. Gray, SE. Khoo, A. Reimondos, Journal of Population Research **29**(4), 373-98 (2012)
9. A. Azhar, J. Pereira, "Volunteerism sorely lacking in Malaysia" freemalaysiatoday. December 5, 2012: http://www.freemalaysiatoday.com/category/nation/2012/12/05/volunteerism-sorely-lacking-in-malaysia/
10. Z. Hussin, M. Arshad, "Altruism as Motivational Factors toward Volunteerism among Youth in Petaling Jaya" Selangor" 2012: 46..
11. S. Goudeau, "Principal Motives of Positive Youth-Adult Relationships: A Model for Identifying the Motives of Adult Volunteers and Youth-Adult Relationships in Physical Activity-Based Youth Development Programs" (Doctoral dissertation, Emporia State University).
12. D. Aisyah, M. Ibrahim, M. Sontang, AM. Manam, N. Ab Ghani, Z. Rosufila, Journal of Environmental Research and Development **8**(1),148 (2013)
13. N. A. H. A. Sahimi, T. Suandi , S. R. A. Hamzah, I. A. Ismail, Middle-East Journal of Scientific Research (2014)
14. I. Nazilah, I. Rozmi, Fauziah, Asian Social Science **12**(9),139 (2016)
15. TG. Baum, L. Lockstone, International Journal of Event Management Research **3**(1), 29-41(2007)
16. Eça de Abreu, M., R. M. S. Laureano, S. F. Syed Alwi, Rui Vinhas da Silva, Pedro Dionisio. "Managing volunteerism behaviour: The drivers of donations practices in religious and secular organisations." (2015).
17. C. Agorastos, Demiralay, CG. Huber, "Influence of religious aspects and personal beliefs on psychological behavior: focus on anxiety disorders". Psychol Res Behav Manag, **7**, 93-101(2014)
18. G. Brewer, S. Robinson, A. Sumra, E. Tatsi, N. Gire, Journal of religion and health, **54**(6), 2225-34 (2015)
19. KH. Kim, J. Sobal, Public Health Nutrition **7**(06), 773-81(2004)
20. Khayruzzaman, International Journal of Finance and Banking Research **2**(1), 18-23 (2016) doi: 10.11648/j.ijfbr.20160201.14
21. F.T. Newaz, K.S. Fam, R.R. Sharma, Journal of Financial Services Marketing **21** (2),141-52 (2016)
22. EA. Minton, LR. Kahle, TS. Jiuan S.K. Tambyah, Journal for the Scientific Study of Religion **55**(2),365-83 (2016)
23. MK. Hyde, SR. Knowles, Australian Journal of Psychology **65**(3), 135-45 (2013)
24. JP. Hill, KR. den Dulk, Journal for the Scientific Study of Religion **52**(1),179-97 (2013)
25. H.J. Oh Journal of Christian Education & Information Technology **28**, 7-31(2015)
26. RT. DeAngelis, GA. Acevedo, X. Xu, Religions, **7**(6):74 (2016)
27. M. Sahri, K. Murad, A. Alias, MD. Sirajuddin, Journal of Educational and Social Research (7), 502 (2013)
28. H. Ali, "Volunteerism and the development of Malaysian social care system". International Council on Social Welfare 101-9 (2002)
29. N. Morrow-Howell, The Journals of Gerontology Series B: Psychological Sciences and Social Sciences **65**(4), 461-9 (2010)
30. E. Marta, M.Pozzi, Journal of Adult Development **5**(1), 35 (2008)
31. MK. Johnson, T. Beebe, JT. Mortimer, M.Snyder, Journal of Research on Adolescence **8**(3), 309-32 (1998).
32. KS. Fielding, R. McDonald, WR. Louis, Journal of environmental psychology **28**(4), 318-26 (2008)
33. MH. Lai, MY. Ren, A. Wu, EP.Hung, Journal of Community & Applied Social Psychology **23**(2),128-42 (2013)
34. G. Pauline, International Journal of Event Management Research **6**(1), (2011)
35. MC. Shen, " A study of examining motivation, job satisfaction, and intention to continue volunteering of the volunteers of the 2011 Fubon Taipei Marathon Running Sporting Event in Taiwan" (Doctoral dissertation, UNITED STATES SPORTS ACADEMY).
36. ML. Denton, LD. Pearce and C. Smith, "Religion and spirituality on the path through adolescence". National study of Youth and religion. 2008 Nov;22:2010.
37. B. Meade, "Examining the effects of religiosity and religious environments on inmate misconduct" (Doctoral dissertation, UNIVERSITY OF SOUTH CAROLINA).
38. D. Voas, "Does religion belong in population studies?". Environment and Planning A **39**(5), 1166-80 (2007)
39. B. Torgler. "The importance of faith: Tax morale and religiosity. Journal of Economic Behavior & Organization". 2006 Sep 30;61(1):81-109.
40. E. Olson, P. Hopkins, R. Pain, G. Vincett, "Retheorizing the postsecular present: Embodiment, spatial transcendence, and challenges to authenticity among young Christians in Glasgow, Scotland". Annals of the Association of American Geographers **103**(6), 1421-36 (2013)
41. P. Hopkins, E. Olson, M. Baillie Smith, N. Laurie, "Transitions to religious adulthood: relational geographies of youth, religion and international volunteering". Transactions of the Institute of British Geographers 40(3), 387-98 (2015)

42. P. Cloke, J. May, S. Johnsen, *"Swept up lives?: Re-envisioning the homeless city"* (John Wiley & Sons, 2011)
43. N. Krause, "Assessing the religious roots of volunteer work in middle and late life". Research on aging **37**(5), 439-63 (2015)
44. KI. Pargament, HA. Raiya, "A decade of research on the psychology of religion and coping: Things we assumed and lessons we learned". Psyke & logos. **28**(2), 25 (2007)
45. BC. Hernandez. "The Religiosity and Spirituality Scale for youth: Development and initial validation (Doctoral dissertation, Louisiana State University).
46. E. Dingemans, E.Ingen, Journal for the Scientific Study of Religion (2015)
47. Loewenthal, M. Kate, "Psychology and Charity". (2015)
48. K. Coe, C. Keller, JR. Walker, Journal of religion and health **54**(1), 46-60 (2015)
49. P. Vermeer, P. Scheepers, M. Te Grotenhuis, International Journal of Voluntary and Nonprofit Organizations **27**(3),1361-84 (2016)
50. YI. Kim, SJ. Jang, "Religious Service Attendance and Volunteering A Growth Curve Analysis". Nonprofit and Voluntary Sector Quarterly. 2016 Jun 27:0899764016655619.
51. L. Hustinx, R. Van Rossem, F. Handy, RA. Cnaan, "A cross-national examination of motivation to volunteer: religious context, national value patterns, and nonprofit regimes". InReligion and volunteering: complex, contested and ambiguous relationships 2015. Springer.
52. R. Bekkers, Acta Sociologica **50**(2), 99-114 (2007)
53. M. Van Tienen, P. Scheepers, J. Reitsma, H. Schilderman, International Journal of Voluntary and Nonprofit Organizations **22**(3), 365-89 (2011)
54. JS. Mattis, WP. Beckham, BA. Saunders, JE. Williams, V. Myers, D. Knight, D. Rencher, C. Dixon, Journal of Adult Development **11**(4), 261-72 (2004)
55. RD. Putnam, DE. Campbell, "American grace: How religion divides and unites us". Simon and Schuster; 2012 Feb 21.
56. H. Taniguchi and LD. Thomas, International Journal of Voluntary and Nonprofit Organizations **22**(2), 335-55 (2011)
57. Berger, IE. Berger, International Journal of Voluntary and Nonprofit Organizations **17**(2), 110-27 (2006)
58. SR. Freiheit, K. Sonstegard, A. Schmitt, C. Vye, Pastoral Psychology **55**(1), 27-33 (2006)

MATEC Web of Conferences **150**, 05101 (2018)
MUCET 2017
https://doi.org/10.1051/matecconf/201815005101

Promoting Social Entrepreneurship among Entrepreneur with Disabilities in Contribution for Community

Norhasyikin Rozali[1,*], *Shuhairimi Abdullah, Jamsari* Jamaluddin[1], *Abdul Jalil* Ramil[1] , *Noor Salwani Hussin*[1] and *Aida Zainab* Ahmad[2]

[1]Malaysia Perlis, Perlis, Malaysia
[2]Institusi Pendidikan Guru Malaysia Kampus Perlis, Perlis, Malaysia

Abstract. Looking forward of the uniqueness social entrepreneurship (SE) new phenomena born from entrepreneurships felid. Because of that it is a new step point for the entrepreneur with disabilities (EWD) to be part of social entrepreneurship environments. Where SE, link the positive impact for society creating by entrepreneur that understand that social problem give a big opportunities for profit or nonprofit activity that can be solve creatively. This give a lot of advantage of EWD to contributes to community in the term of "social impact". This conceptual paper maps SE and EWD in contribution to community with the important role play by stakeholder in promoting SE. This conceptualization enhances understanding of EWD in SE and clear picture of role play by stakeholder give chances for EWD contributes to community.

1 Introduction

Social entrepreneurship is an exciting new area to explore, but also to provide a very large impact on society. This uniqueness makes social entrepreneurship actively explored by many researchers. According to [7] social entrepreneurship is more effective than commercial entrepreneurship based on ability to added value through creative efforts to solve social problems either non-profit or profitable approach. Scenario unemployment among the disabled people viewed as social issues that still in conversation and multiple solutions has developed to address this issue. There are various obstacles that drive this group inability to get a job. Based on research [15] still these people with disabilities are not getting the job and face a lot of difficulty to enter job markets opportunity. As the government has introduced a proactive step approach to self-employed or are directly involved in entrepreneurship field. This is the step point from the social entrepreneurship will change the landscape of the people with disabilities and at the same time encouraging entrepreneurs with disabilities to participate in society contributed to achieving the social entrepreneurship context.

2 Promoting Social Entrepreneurship among Entrepreneur with Disabilities

Social entrepreneurship was born from entrepreneurship. Where entrepreneurship generally, to increase the level of economic growth [12] and also the ability to provide jobs [8, 25, 30]. Entrepreneurship and social entrepreneurship is an area with a significant difference when viewed from the angle of the ability of an entrepreneur to see a social problem that can be handled as a potential business opportunity whether it is profitable or disadvantage [7, 28]. Opportunity explored by entrepreneurs from various angles, where entrepreneurs are able to optimize the opportunities and resources to gain an advantage through innovation and creativity [12, 29].

Social entrepreneurship is a business for social purpose through additional motivation to benefit from the disadvantage, which can create social added value [1, 3]. Social entrepreneurs are able to optimize resources and innovative and creative in finding.

By understanding through the definition of social entrepreneurship and the ability to solve problems where the problem is seen as an opportunities [9, 21, 29] to ensure that the solutions to social problems can be solved [11, 29] as well as add value to society [24]. Social problems or issues are matters related to society and the environment such as poverty, hunger, natural disasters, war, drugs, people with disabilities, youth, women, pollution and so on. Social entrepreneurs will see this problem as an opportunity to complete and able to have a positive impact [13] on society and the environment.

First question will be asked might be, why is social entrepreneurship is innovative approaches used to establish the completion for issue unemployment people with disabled? And how this approach will encourage entrepreneurs with disabilities to contribute to society? Above text show the answers for the first question in general explanation.

* Corresponding author: suhairimi@unimap.edu.my

Show that social entrepreneur give big impact to solve social problem in innovative approach [1, 13, 17]. View of the job market that does not guarantee a person with disabilities to get a job and be followed by a variety of obstacles to get the job, and then there is encouragement of stakeholders involved in encouraging people with disabilities to self-employment or entrepreneurship [6]. Among the obstacles are always barriers for people with disabilities to get a job [4, 5, 14, 26] are like discrimination, lack of facilities for people with disabilities as well as employers' perceptions.

Table 1. The Division of Work and the Disabled and Normal

Person	Percentage					
	National Income Low		High-income Countries		All Countries	
	Normal	PWD	Normal	PWD	Normal	PWD
Male	71.2	58.6	53.7	36.4	64.9	52.8
Women	31.5	20.1	28.4	19.6	29.9	19.6
18-49	58.8	42.9	54.7	35.2	57.6	41.2
50-59	62.9	43.5	57.0	32.7	60.9	40.2
60 and over	38.1	15.1	11.2	3.9	26.8	10.4

Sources: World report on disability, 2011

Opportunity to bring in revenue and a positive impact on society. Based on [28] for example, Grameen Bank, founder by Muhammad Yunus, has been successful in implementing micro-credit scheme for poor women out from poverty. Through social entrepreneurship approach has been successfully implemented to help solve the problem of poverty.

Therefore, it is important remedy identify the essence of social entrepreneurship is important in order to make social issues to social fortune to society. Recently, the government has undertaken various initiatives in order to help people with disabilities to overcome barriers to employment such as establish policies and laws encourage joining the government or the private sector [5] [15]. At the same time provide education and training program and activities to encourage the disabled to work or engage in entrepreneurship [6].

The government is still not achieving the desired goals of these groups. Based on study done by [29] show that increased interest in social entrepreneurship potential remedy makes the problem more innovative. To promote social entrepreneurship exposure to entrepreneur with disabilities should be ongoing in order effects or results to society can be maintained not only temporary. According to [23] entrepreneur play important role as a second invisible hand responsible to society by giving innovative solutions for social problem. The role of stakeholders is very important [10, 22] in provide a facility for developing entrepreneurship through social entrepreneurs need to ensure that people with disabilities are not left behind and social goals can be achieved.

The role played by the government [6] in helping to promote and give exposure to the entrepreneur with disabilities is important to sees such a great potential to add value not only to their own group of disabled peoples but also to the community. Through the involvement of the government in promoting social entrepreneurship, other approaches through education and training and under the supervision of program [20, 31] can help to encourage disabled entrepreneurs directly involved.

Response for social problem facing by many country roles play by government is one of important key to provide a solution for social problem. Because of that, an innovative approach by collaboration between government and social entrepreneur will leverage public and private resources and generate transformative and cost-effective solutions to the most challenging social problems facing in the country itself [27]. Based on the Report to the President, 2007, by The Small Business Administration, Office of Advocacy suggest few methods that can be used by government agencies and elected officials to (1) encourage social entrepreneurs to innovate, (2) create enabling environments for their efforts, (3) reward their performance, (4) help scale their successes, and (5) produce knowledge to help them solve social problems[27].

According to [16], also agree that government support for social entrepreneurship as an approach for strengthening communities by social entrepreneur involves private individuals and organizations taking the initiative to address social challenges in their communities.

3 Entrepreneur with Disabilities Contribution to Community in Context of Social Entrepreneurship

Through the encouragement of people with disabilities to engage in entrepreneurial viewed indirect impact on employment opportunities encourage the fate and overcome unemployment [17] [22] and poverty of these people. Table 1 show high of differences between incomes earned from normal people and people with disabilities this is evidence that many of people with disabilities still in poverty because of incomes they earned. Entrepreneur with disabilities will become agents of intermediate action to ensure that people with disabilities are not left out in the same boat, and even contribute back to society [24] in the context of larger social entrepreneurship impact.

The first step is to change a disabled to successful entrepreneurs with disabilities and at the same time makes successful social entrepreneurs with disabilities who can contribute to society, where previous years people with disabilities who are subject to the social mission of the business is not yet come into force [22].

Social entrepreneurship good platform for people with disabilities that will trigger a new world in entrepreneurship or social entrepreneurship. No issue, to question of the ability of entrepreneurs with disabilities if viewed from the angle of equal opportunity which is always highlighted [5, 20, 15] [31]. In addition, the power to motivate and encourage entrepreneurs trials that the disabled have become a powerful starting point to them to put themselves in entrepreneurship [20] should receive due recognition and more opportunities are highlighted.

Social impact that will have a direct impact on disabled people will indirectly resolve an issue faced by disabled people and entrepreneurs can contribute to society. References [23] entrepreneur drive the economy by efficient outcomes trues the systematic identifying social problem with positive externalities and develops approach into economic system. Contribution of entrepreneur are not much differences, entrepreneur also frequently act to social manner but for social entrepreneur they always look for social problem directly in mission social solving add creating social impact [9].

Involvement of entrepreneurs with disabilities in social entrepreneurship will produce social entrepreneurs who will contribute to society through a variety of approaches and channels. Maybe in terms of education or charity or charitable contributions can help people then indirectly add value to society. Contribute to the disabled will be translated more clearly based on previous experience, based on the role played by social entrepreneurs are able to provide more employment opportunities[6, 20, 21], providing better facilities, and education and training directly with more effectively.

4 Summary

This paper provides a conceptual map in social entrepreneurship approach is explained in general about entrepreneurs with disabilities and the role it played by the government to help raise awareness among entrepreneurs about social entrepreneurship. Therefore, it is important to ensure the lack of exploration of entrepreneurship in the context of social entrepreneurship so that they are no longer stuck in unemployment problems or poverty for people with disabilities. At the same time entrepreneurs with disabilities will have the opportunity to contribute to society. In the end of social entrepreneur community will gain the benefit from creative and innovative approach of social entrepreneur.

5 Future Agenda

In future agenda, researchers hope to explore more about experiences face by successful entrepreneurs with disability and build a modal based on social entrepreneur's perspectives for entrepreneur with disabilities. In addition, to explore more in how they can involve with social entrepreneur and at the same time contribute to society.

References

1. S. Abu Saifan, Social entrepreneurship: definition and boundaries. *TechnologyInnovation Management Review*, (February), 22–27. Retrieved from: http://www.timreview.ca/article/523
2. J. A. Ahmad, U. T. Mara, Social Enterpreneurship in Ecotourism: An Opportunity for Fishing Village of Sebuyau , Sarawak Borneo. *Tourism, Leisure and Global Change*, *1*(April 2013), 22–24 (2014)
3. S.H. Alvord, L.D. Brown, C.W. Letts, *Social Enterpreneurship and Social Transformation - An Exploratory Study* (2002)
4. Ang, M. (2012). *Work and Disability: A Malaysian Scenario*. Penerbit Universiti Sains Malaysia.
5. M.C. Ang, Jurnal Pengurusan **41**, 81–89 (2014)
6. W.A.W. Arnidawati, *ECONOMY REPORT APEC FORUM ON HUMAN RESOURCES DEVELEPMENT 2013, Human Resources Development for People with Disabilities "Improvement Employability."*(2013)
7. J. Austin, H. Stevenson, J. Wei-Skillern, Social and commercial entrepreneurship: Same, different, or both? *Entrepreneurship: Theory and Practice* 30(1), 1-22 (2006) http://doi.org/10.1111/j.1540-6520.2006.00107.x
8. P.D. Blanck, L.A. Slander, J.L. Schmeling, H.A. Schartz, *The Emerging workforce of entrepreneurs with Disabilities: Preliminary Study of Entreprenuership in Iowa* (2002)
9. B.J. Boschee, J. Mcclurg, J. (2003). Toward a better understanding of social entrepreneurship: Some important distinctions. *Chief Executive*, 1–5. Retrieved from http://www.se-alliance.org/better_understanding.pdf
10. K. Caldwell, S.P. Harris, M. Renko, Intellectual and Developmental Disabilities, **50**(6), 505–518 (2012).
11. S.T. Certo, T. Miller, Social entrepreneurship: Key issues and concepts. *Business Horizons*, **51**(4),267–271(2008)
12. T. Cooney, T. Irish Business Journal, **4**(1), 119–129 (2008)
13. J.G. Dees, Transaction Social Science and Modern, **44**(3), 145 (2007)
14. S.L. Khoo, T.L. Ta,L.W. Lee, International Jouranl of Sosial, Education, Economics and Management Engeneering, **6**(6), 255–260 (2012)
15. S.L. Khoo, L. Tiun, L.W. Lee, *Kajian Malaysia*, **31**(1), (2013)
16. M.L. Korosec, M. L., & Berman, E. M. (2006). Municipal support for social entrepreneurship. *Public Administration Review*, *66*(3), 448–462.
17. J. Mair, I. Martí, Journal of World Business, 41(1),36-44 (2006)
18. J. Mair, J. Robinson, K. Hockerts, *Social Entrepreneurship. PALGRAVE MACMILLAN* **32** (2006) http://doi.org/10.1057/9780230625655
19. P.J. Murphy, S.M. Coombes, Journal of Business Ethics, **87**(3), 325–336 (2009). http://doi.org/10.1007/s10551-008-9921-y
20. O. Norasmah, *Keusahawanan Orang Kurang Upaya Di Malaysia*. (Dewan Bahasa Dan Pustaka, 2014)

21. Osman, Che Asniza, H. Rahim, Journal of Basic and Applied Sciences, **8**(23), 62–69 (2014)
22. S. Parker Harris, M. Renko, K. Caldwell, Disability & Society, **29**(8), 1275–1290 (2014) http://doi.org/10.1080/09687599.2014.924904
23. F.M. Santos, F. M. *Journal of Business Ethics*, *111*(3), 335–351 (2012) http://doi.org/10.1007/s10551-012-1413-4
24. C. Seelos, J. Mair, Business Horizons, **48**(3), 241–246 (2005) http://doi.org/10.1016/j.bushor.2004.11.006
25. G. Shaheen, W.N. Myhill, Entrepreneurship for Veterans with Disabilities : Lessons Learned from the Field Factors Affecting Employment Among Veterans with Disabilities Entrepreneurship as a Viable Career Option for Veterans with Disabilities Meeting the Needs of Future Veteran Bu. *Brief*, (1), 1–8 (2009)
26. T.L. Ta, K.S. Leng, Challenges Faced by Malaysians with Disabilities in the World of Employment. *Disability, CBR & Inclusive Development*, *24*(1), (2013) http://doi.org/10.5463/dcid.v24i1.142
27. M.A. Wolk, *Social Entrepreneurship & Government* (2007)
28. M. Yunus, B. Moingeon, L. Lehmann-Ortega, Long Range Planning, **43**(2–3), 308–325 (2010). http://doi.org/10.1016/j.lrp.2009.12.005
29. S. Zahra, H. Rowhouser, N. Bhawe, D. Neubaum, J. Hayton, Strategic Entrepreneurship Journal, **2**(2), 117–131 (2008) http://doi.org/10.1002/sej.43
30. A. Zalinawati, M. Norudin, *The Relationship between Environmental Factors and Opportunities to became entrepreneur among visually Impaired* (2007)
31. A. Zinaida, *Kerjaya untuk Orang Kurang Upaya* (PTS Professional, Publishing Shd.Bhd., 2006)

I6-FPS: Automating the ICMPv6 Filtering Rules

Wan Nor Ashiqin Wan Ali[1,*], *Abidah* Mat Taib [2], and *Syed Zulkarnain* Syed Idrus[1]

[1]School of Human Development and Techno-Communication (iKOM), Universiti Malaysia Perlis (UniMAP), Malaysia
[1]Centre of Excellence Geopolymer and Green Technology (CEGeoGTech), Universiti Malaysia Perlis (UniMAP), Malaysia DP Sciences, Editorial Department, 91944 Les Ulis Cedex A, France
[2]Faculty of Computer & Mathematical Sciences, Universiti MARA, Malaysia

Abstract. Enterprises are required to utilize Internet Control Message Protocol version 6 (ICMPv6) when IPv6 is deployed. In IPv4, Internet Control Message Protocol (ICMP) is aggressively filtered by a network administrator while in IPv6, ICMPv6 messages cannot be aggressively filtered due to the function of ICMPv6 message. ICMPv6 security risks increase when ICMPv6 threats and vulnerabilities are exploited. Thus, it is very crucial for enterprises to address the issues. In practice, network researchers must review several resources to identify ICMPv6 related attacks occurring due to the exploitation of ICMPv6 vulnerabilities. Overlooking any of these issues will jeopardize the security of ICMPv6. While conducting the attack scenarios testing, IPv6-Filtering Prototype System (I6-FPS) was developed to overcome the deficiency and limited filtering tools that supported IPv6 filtering rules (ip6table). I6-FPS is used to automate and simplify the writing of ip6table and it was developed using PHP5 and Shell script languages. This research revealed that I6-FPS is significant in the initial phase of securing IPv6 deployment as well as focusing on the ICMPv6 filtering rules. The I6-FPS has the potential to be enhanced and developed over time by including more functions to that system in generating specific filtering ip6table rules.

1 Introduction

In practice, network and security administrators are responsible to strengthen the network security from any intruders. The probability of a network to be intruded is getting higher from the unintended use of IPv6 [1]. In scenarios in which IPv6-enabled devices are deployed on enterprise networks that are intended to be IPv4-only, native IPv6 support and/or IPv6 transition/coexistence technologies could be leveraged by local or remote attackers for a number of (illegitimate) purposes. In general, most of the aforementioned security implications can be mitigated by enforcing security controls on native IPv6 traffic and on IPv4-tunneled IPv6 traffic. Among such controls, is the enforcement of filtering policies to block undesirable traffic is crucial. A prototype is developed to assist the IPv6 beginners and unwell-trained administrators to the IPv6 filtering policies and how to implement those policies into filtering rules.

The IPv6 filtering rules prototype is a part of firewall filtering rules where it helps user to generate and activate the filtering rules. User can study on how IPv6 filtering rules work in Linux Operating System. However, having a threat model that only represent on how to mitigate from the identified threats and vulnerabilities is insufficient due to the possibility that unwell-trained administrator and IPv6 beginners are applying inadequate control measure to mitigate ICMPv6 attacks. The basic control measure that can be applied to mitigate the ICMPv6 attacks is by enforcing filtering rules. Nevertheless, the unwell-trained administrator requires more time to learn the filtering rules. Therefore, in facilitating the generation of filtering rules, I6-FPS is developed using the PHP5 and shell script languages.

2 Related Works

An IPv6 filtering rules prototype is proposed while conducting the testing due to the deficiency for available IPv6 filtering tools. The IPv6 filtering rules prototype is a part of firewall filtering rules where it helps user to generate and activate the filtering rules. User can study on how IPv6 filtering rules work in Linux Operating System.

2.1 Firewall Filtering Rules

Firewall is a hardware-based or software-based network equipment to secure network from any threats and attacks. Firewall implements packet filtering, that uses a set of predefined rules, to filter incoming and outgoing packets [2]. Thus, defining of the filtering rules must be coherent to react with the firewall. Moreover, the firewall is one of the vital tools in securing private network, and it is broadly enforced to secure enterprise network as well [3]. Firewall also acts as a fundamental component based on filtering rules to secure network from illegal traffics, as well as threats and attacks [4].

* Corresponding author: wannorashiqin@gmail.com

Moreover, the firewall is the current use of security measure, which directly becomes crucial in any applications [5].

They also stated that packet filtering firewall is enforced broadly because it is simple, forwards quicker, more efficient and clear to the users. This packet filtering firewall works based on tuples and does not check data segment in a packet. Furthermore, the packet filter rules inside a firewall are only based on the network security's needs, even though it becomes more and more difficult to manage [5]. Hence, there is a high possibility for mistakes to occur in the enterprise network. The firewall filtering rules must be checked, managed and controlled properly in order to avoid it from making mistakes. Rapid growth of technology in firewall filtering rules has contributed to the development of software-based or hardware-based firewall filtering rules. Firestarter [6],[7] and Firewall Builder [8],[9] are the two well-known software-based firewall filtering rules that are being used.

2.2 Firestarter

Firestarter is considered as a present Linux firewall which simplifies the security management inside Linux OS. Firestarter comes with several beneficial features such as it provides GUI, and it is simple. It is also the appropriate software to be used by desktop, gateway or server, and it allows users to define whitelist, blacklist IP addresses as well as inbound and outbound policy. Not only that, Firestarter provides Internet connection for sharing activity. It is also free, and it is open source software which can be downloaded as an individual or host firewall. Firestarter uses Netfilter (iptables/ipchains) system built-in inside Linux kernel, which provides real-time monitoring for network traffics. The Firestarter provides a functional, secured and simple GUI software-based designed for an advanced firewalling technology [7].

Firestarter is an appropriate tool for a user who needs to learn Linux-based firewall. Firestarter works in a faster way since iptables is quite complex for new beginners in Linux filtering rules. The Firestarter solves the problem for average users to enforce proper firewall filtering rules without taking a longer time in designing complex iptables rules [6]. However, Firestarter does not have a command prompt to manipulate the rules, thus limits users to learn iptables to increase their network security. Hence, there is a suggestion for Firestarter to be enhanced and included with a console version to allow users to learn Linux-server rules as well.

2.3 Firewall Builder

Firewall Builder provides three main features which are simplicity, time saving and flexibility [9]. Firewall builder's features such as search-and-replace, shared objects, drag-and-drop GUI have simplified the task of configuring firewall rules. It also supports several firewall platforms such as Private Internet eXchange (PIX), Berkeley Software Distribution (BSD) packet filter and Cisco ASA which makes it more flexible. Firewall Builder has simple GUI and supporting multiple platforms, which allows users to focus on filtering their traffic instead of wasting time to learn and search for filtering rule commands. However, the Firewall Builder designed for large target user such as organization, academic institution and government. A comparison table was constructed to summarize and compare the Firestarter, Firewall Builder and a designed prototype for this research that is known as I6-FPS shown in Table 1:

Criteria	Firestarter	Firewall Builder	Research Design Prototype (I6-FPS)
Download version	i. Open source software ii. Available free of charge	i. Trial open source for 30 days. ii. Need to buy a license to keep using the software.	i. Open source software ii. Available free of charge
Graphical User Interface (GUI)	i. Easy to use Graphical User Interface (GUI)	i. Easy to use Graphical User Interface (GUI) ii. More complex GUI elements compare to Firestarter	i. Easy to use Graphical User Interface (GUI)
Supporting Operating System	i. Linux	i. Linux ii. Windows	i. Linux ii. Can be accessed in Windows, but rules cannot be activated.
Manageme nt tool supports	i. iptables	i. iptables ii. ip6tables iii. Access Control List (ACL)	i. ip6tables
Target Users	i. Linux user	i. Large organization network	i. IPv6 beginner, enterprises and students

Table 1. Comparison Table (Firestarter, Firewall Builder and Research Designed Prototype)

3 Development of I6-FPS

This section discusses the steps involved in I6-FPS development. This I6-FPS is able to be used in the experimentation phase which indirectly eases the researcher in executing the ICMPv6 rules for each node represented in the testing.

3.1 Planning

In the planning phase, some preliminary studies regarding tools in generating the filtering rules have been conducted. A filtering rules prototype was developed to demonstrate on how filtering policies can be converted into filtering rules. The prototype was developed using JAVA programming languages. Figure 1 shows the standalone prototype (rule creator) that is capable to convert the filtering policies into ip6tables rules. The rule creator is developed to proof the concept on how possible the filtering policies can be converted into filtering rules.

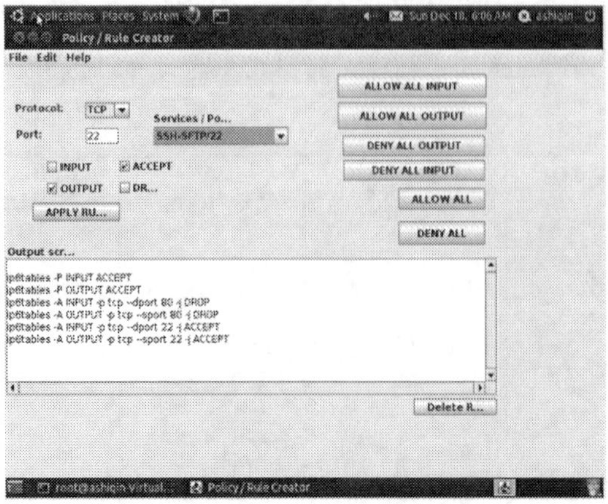

Fig 1. Filtering Rules Prototype (Rule Creator)

The rule creator as shown in Figure 1 is developed to initiate the development of I6-FPS even though the rule creator cannot activate the filtering rules inside the Linux OS. Each of the tasks that are involved in I6-FPS development is described in details in Context Diagram, Sequence Diagram and Level 0 Diagram.

3.2 Analysis

The analysis phase defines the processes involve in gathering and interpreting facts about the filtering rules, diagnosing the problems of the existing filtering system and recommending the development of I6-FPS. Context Diagram and Level 0 Diagram show the flow in the development of I6-FPS.

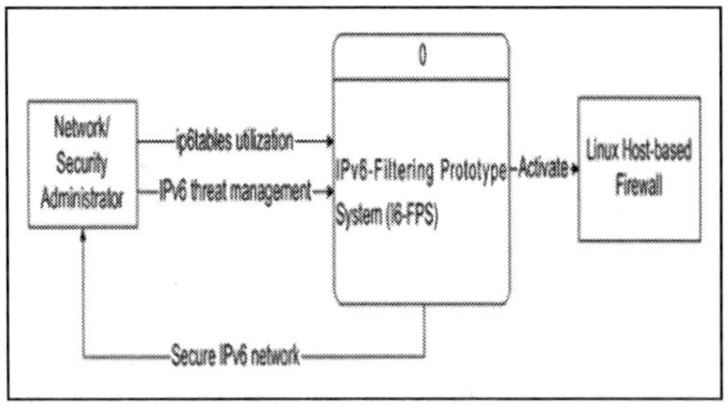

Fig 2. Context Diagram

Referring to Figure 2, there are two entities involved in the system development which are network or security administrator and Linux host-based firewall. The two entities are linked together with the main process known as I6-FPS. The network or security administrator manages the IPv6 threat and utilizes the ip6tables using I6-FPS. Then, I6-FPS activates the Linux host-based firewall for each of the network administrator's nodes. As a result, the network or security administrator can manage their IPv6 network securely. The context diagram has been zoomed in and narrowed down by expanding the main process which is called Level 0.

3.3 Design

The development of I6-FPS is commenced once the required software is installed. A sequence diagram as in Figure 3 was illustrated to show how I6-FPS works.

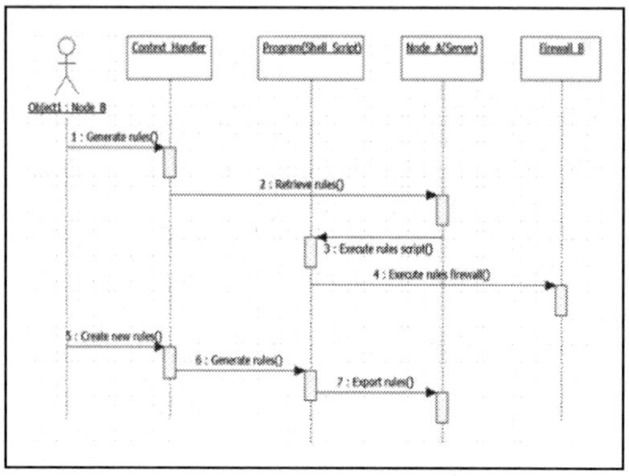

Fig 3. Sequence Diagram of I6-FPS

The object, which is Node_B, will perform two activities such as generating rules and creating new rules. For the generating rules activity, Node_B will generate the rules using the context handler. Then, the context handler will retrieve the rules from the server. The server will execute the rules script to the host firewall. Node_B can also create new rules if there is no existence rule for the current case via the context handler. The rules can be generated by a shell script and then will be exported to the server. The flow of the context handler's functions can be referred to Level 0 diagram.

The Level 0 for I6-FPS development consists of six processes, which have been expanded from the main process as in Figure 4. However, there is still the same number of entity involved. Figure 4 shows the expansion of the main process including; a) Check ip6tables rules, b) Generate ip6tables rules, c) Select ip6tables rules action, d) Check ip6tables status, e) Report an attack, and f) View report.

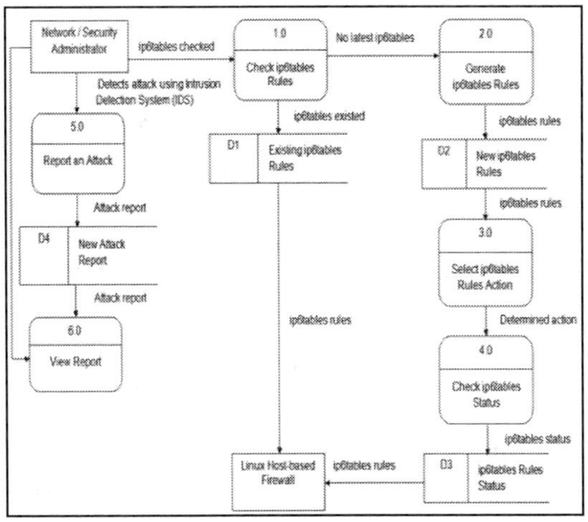

Fig 4. Level 0

3.3.1 Process 1.0: Check ip6tables Rules

The entity "Network/Security Administrator" checks whether the computer already has ip6tables or not. If the required ip6tables already existed, user might proceed to Process 2.0 or Process 3.0, but if the required ip6tables is not yet there, the entity will move to Process 2.0.

3.3.2 Process 2.0: Generate ip6tables Rules

Process 2.0 produces the new required ip6tables rules. Then, the new ip6tables rules will be stored in the database and the entity "Network/Security Administrator" will move to Process 3.0.

3.3.3 Process 3.0: Select ip6tables Rules

After the new required ip6tables rules are stored, then the entity "Network/Security Administrator" will select the action whether to activate or deactivate those ip6tables rules. The selected action will determine whether ip6tables in the entity's "Linux Host-based Firewall" will be activated or deactivated. After that, the entity needs to check the ip6tables rules' status as in Process 4.0.

3.3.4 Process 4.0: Check ip6tables Rules

Process 4.0 happens when the entity "Network/Security Administrator" desires to check the ip6tables' status whether the selected action is successful or not.

3.3.5 Process 5.0: Report an Attack

The entity "Network/Security Administrator" can report the new attack when the entity finds or detects the attack. Report of the attack will be stored in the database once the entity submits the report.

3.3.6 Process 6.0: View Attack

The entity "Network/Security Administrator" can view all the reported attack, and if the entity has any solutions to mitigate the attack, he will generate new ip6tables rules as in Process 2.0. I6-FPS development processes proceed with a flowchart before it moves to the encoding phase.

3.4 Implementation

3.4.1 Identifying System Requirement

I6-FPS was developed in Ubuntu Virtual Machine. The Ubuntu Virtual Machine and software needed in the system are installed. The specification of the computer used for the development is Intel(R) Core(TM) i5-2300 CPU @ 2.80GHz. The software installation is done in the root privilege mode in order to ensure that all the software can be installed successfully. There are three steps in completing the software installation for the system development which are installing MySQL 5, installing Apache2, and installing PHP5 [10].

Figure 5 shows the flowchart of I6-FPS. Identically, every system needs a user to log in to authenticate the user of the system. The user will enter the name and password before can use the system. After the user's information has been authenticated by the system, the user can validate the ip6tables inside their machine, checking whether the ip6tables rules exist or not. If no ip6tables rules exist yet, the user will generate the new ip6tables rules using I6-FPS. Finally, if the user does not have anything to do with I6-FPS such as Process 5.0 and Process 6.0 as in Figure 5, the user can log out from the system.

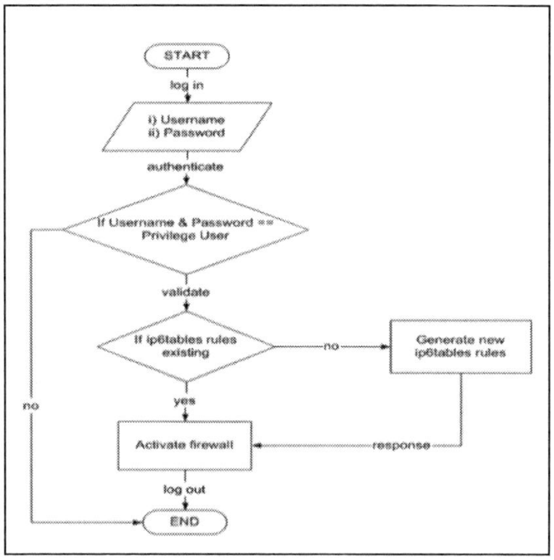

Fig 5. Flowchart of I6-FPS

As has been noted before, I6-FPS is used in the experimentation to ease the user in managing the ip6tables rules efficiently. The system can be accessed through the Local Area Network (LAN). Figure 6(a) and Figure 6(b) show the snapshot of I6-FPS front page.

4 Proposed I6-FPS

Based on Figure 6(a) and Figure 6(b), I6-FPS is used to facilitate the filtering rules generation associated safeguards in the ICMPv6 threat model [11]. The filtering rules used in the testing are translated by adapting the ICMPv6 selective filtering policy [12]. The prototype also facilitates users to write the ip6tables rules with Graphical User Interface (GUI) in the threat experimentation.

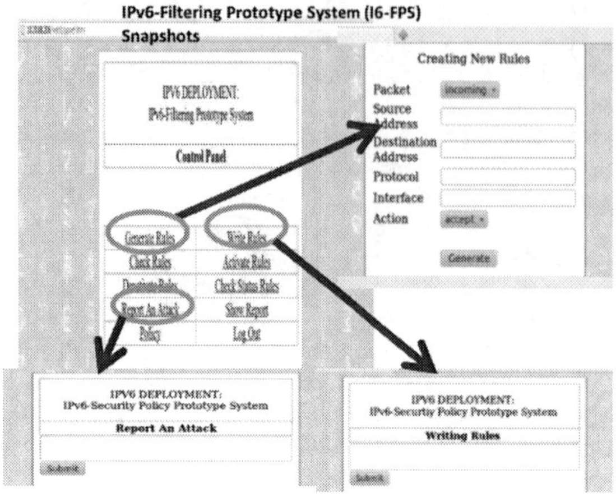

Fig. 6(a) The Snapshots of I6-FPS

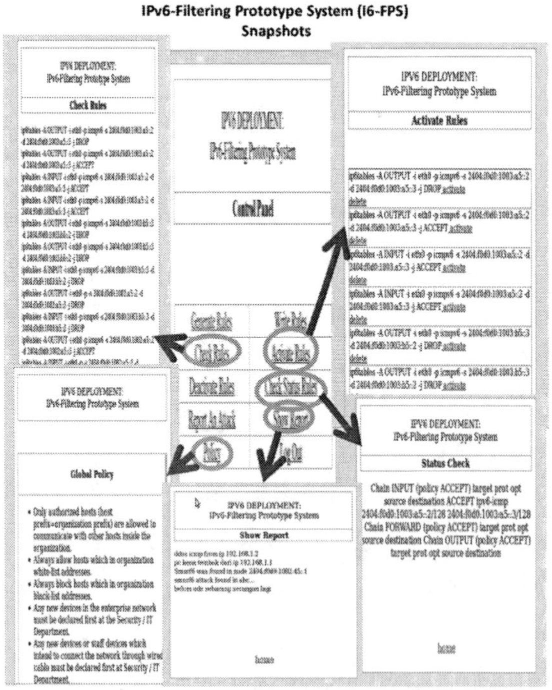

Fig. 6(b) The Snapshots of I6-FPS

5 Conclusions

This research is conducted to automated the writing of the filtering rules (ip6tables) using I6-FPS. The I6-FPS is quite similar to the Firestarter in Linux. Firestarter is used in configuring firewall rules and setting policies for IPv4 only. Thus, this research designed I6-FPS as filtering prototype for IPv6 deployment. Similar to Firestarter, I6-FPS is not a firewall; rather, it is a fronted system to configure rules. Dissimilar to Firestarter, it has been designed to support ip6tables rules. This prototype is also accessible by any users in the network and they do not need to install this prototype in their own machine.

Several attack scenarios testing were conducted to test the associated safeguards for ICMPv6 threats and vulnerabilities using the use of I6-FPS as in [12]. I6-FPS was tested in the attack scenario testing to show that it is capable in generating and activating ip6tables rules. A usability testing was also conducted by distributing questionnaires to assess whether I6-FPS can ease and help the users in writing IPv6 filtering rules (ip6tables).

The I6-FPS is crucial for the enterprise network when deploying IPv6 because it is simple and easy to use for beginners in the enterprise network or individual user to learn firewall scripting (ip6tables rules). The I6-FPS is indirectly contribute in securing the IPv6 deployment. In contrast, without having the I6-FPS, the enterprise networks are exposed and threatened with the potential of ICMPv6 attacks which are probably exploited from ICMPv6 threats and vulnerabilities.

Future works are conducted to enhance the I6-FPS by adding Rule-Based and Case-Based Reasoning approaches. With those approaches, I6-FPS will work more efficient and optimistically will be able to imitate the function of the firewall. Furthermore, the usability testing was conducted if demonstrated that I6-FPS is an easy and simple prototype. Hence, it can be clearly considered that having a GUI prototype system helps users to handle the ip6tables filtering rules.

This research work was supported by Ministry of Education (MOE), Malaysia and Universiti Malaysia Perlis (UniMAP), Malaysia under Research Short Term Grant (STG), Project code: "9001-00527".

References

1. Gont, T., Liu W. (2014). Security Implications of IPv6 on IPv4 Networks. Technical Report: Internet Engineering Task Force (IETF). ISSN: 2070-1721.
2. Benelbahri, M. A., & Bouhoula, A. (2007). Tuple based approach for anomalies detection within firewall filtering rules. Paper presented at the 12th IEEE Symposium on Computers and Communications (ISCC), 2007.
3. Abbes, T., Bouhoula, A., & Rusinowitch, M. (2008). An inference system for detecting firewall filtering rules anomalies. Paper presented at the Proceedings of the 2008 ACM Symposium on Applied Computing.

4. Benelbahri, M. A., Bouhoula, A., & Trabelsi, Z. (2007). XML based open tool for anomalies detection in firewall filtering rules. Paper presented at the 4th International Conference on Innovations in Information Technology (IIT), 2007.
5. Wang, Y.-G., Ge, Y.-m., & Yang, J.-X. (2008, 12-14 Dec. 2008). Research on packet filter rules of the firewall based on visual prolog. Paper presented at the International Conference on Computer Science and Software Engineering, 2008.
6. Jack, W. (2009). Review: Firestarter firewall for Linux. Retrieved May 12, 2012, from http://www.techrepublic.com/blog/products/review-firestarter-firewall-for-linux/667.
7. Tomas, J. (2004). Firestarter manual. Retrieved October 4, 2012, from http://firestarter.sourceforge.net/manual/introduction.php.
8. Kurland, V. (2010). What's new in Firewall Builder 3.0. Linux J., 2010(189), 4.
9. NetCitadel. (2012). Firewall Builder. from http://www.fwbuilder.org.
10. Timme, F., & Schmalfeld, C. (2011). Installing Apache2 with PHP5 and MySQL support on Ubuntu 11.10 (LAMP). Retrieved October 22 2012, from http://www.howtoforge.com/installing-apache2-with-php5-and-mysql-support-on-ubuntu-11.10-lamp.
11. Taib, M. A., Ali, W. N. A. W. & Shaari, N. S. (2013). "ICMPv6 Vulnerability: The Importance of Threat Model and SF-ICMP6", International Journal of Mobile Computing and Multimedia Communications (IJMCMC), 5(2), 78-100, April-June 2013.
12. Ali, W. N. A. W., Taib, A. H. M., Hussin, N. M., & Othman, J. (2012). IPv6 attack scenarios testbed. Paper presented at the IEEE Symposium on Humanities, Science and Engineering Research (SHUSER), 2012.

Engineered nanostructures to carry the biological ligands

Subash C.B. Gopinath[1,2*], *Santheraleka* Ramanathan[1], *Koh* Hann Suk[1], *Mu* Ee Foo[1], *Periasamy* Anbu[3], and *M.N.A.* Uda[1]

[1]School of Bioprocess Engineering, Universiti Malaysia Perlis, 02600 Arau, Perlis, Malaysia.
[2]Institute of Nano Electronic Engineering, Universiti Malaysia Perlis, 01000 Kangar, Perlis, Malaysia.
[3]Department of Biological Engineering, College of Engineering, Inha University, Incheon 402-751, Republic of Korea.

Abstract. Different nanostructures were engineered with the nanoscale dimension lesser than 100 nm. These nanostructures include silver, cellulose nanoparticles and single-walled carbon nanotube (SWCNT). Biological ligands were obtained from the medicinally important herbal plants, such as *Solanum trilobatum* and *Hempedu bumi* and conjugated with the nanostructures silver nanoparticle and SWCNT, respectively. On the other hand, bio-ligands from cow urine were encapsulated in the cellulose nanoparticle. To confirm morphology these nanostructures, they were observed under Field Emission Scanning Electron Microscope and the results displayed the uniformed nanostructures. Further, biological ligand carrying ability of these nanostructures was confirmed by the bacterial inhibition assay on the agar plate. This study provided the evidence on the capability of nanostructures to carry the biological ligands.

1 Introduction

Currently, nanotechnology is a booming discipline implemented effectively in various sectors. The development of high-performance nanomaterials could be a central importance for medical diagnosis, target validation and drug-delivery [1-6]. Careful selection of proper nanomaterials is aiding to interacting biomolecules for various downstream applications. Success in the massive production of nanomaterials is mainly driven by top-down and bottom-up approaches and making sequential alignments among inter-disciplinary sciences [7]. Under laboratory scale, various nanomaterials were fabricated with the aim to bridge the industries. There are many researches have been carried out on engineering the nanoparticles. The obtained particle size plays an importance role in many sectors, due to their nanoscale in dimension. There are many methods can be used to test the toxicity of engineering nanoparticles towards microorganisms which display the antimicrobial activity. Among different nanostructures, insoluble metallic nanoparticle, soluble non-metallic nanoparticle, graphene-based materials are playing a pivotal roles in the biotechnological applications. This study was carried out to attest the bio-carrying carrying capability of silver, cellulose nanoparticles and single-walled carbon nanotubes (SWCNT).

Silver nanoparticles have been widely considered to be a vital nanoparticle, produced by different methods, including physical, chemical and biological methods. Physical methods are less preferred under mass scale preparation due to the mandatory of intensive power and consumption of high energy. On the other hand, chemical methods are considered in several cases using polymers, reducing analytical agents, surfactants and protective agents [8]. However, considering the environmental benefits in most of the cases, it has been advisable to avoid the usage of chemical compounds, especially analytical reagents. Because of that, the wise way to produce the silver nanoparticle is through biological method and it was found to be cost-effective and environmentally friendly method to synthesize a stable silver nanoparticle and dominates over other methods [9-11].

On the other hand, cellulose nanoparticles are the non-metallic have been found to have the advantages such as cheaper and consume lesser energy and easier to prepare in large scale to be beneficial to the food and pharmaceutical industries. Cellulose nanoparticle has been prepared by the emulsification processes and applied for the engineering and biomedical purposes [12]. In the biomedical processes, such as targeted drug delivery and treatment cellulose nanoparticles are ideal to use. In addition, the dispersing properties and mechanical strength of cellulose nanoparticle is higher as a polymer [13].

Carbon is the one the most abundant elements in our life and it plays a very important rule in the world such as life creation. There are lot of allotropes contain carbon like diamond (the hardest material), graphite (the softest material) and buckminsterfullerence (ball-like form). Carbon is in group fourth and has atomic number of six in the periodic table. All elements in atomic size cannot be seen by naked eyes and must be observed by the microscope. In atom, there are three particles which are positive charged proton, negative charged electron and neutral charged neutron. The carbon has six protons as the atomic number is six and it has the electronic ground state configuration of $1s^2 2s^2 2p^2$. The graphene-based

* Corresponding author: subash@unimap.edu.my

material, SWCNTs can be defined as single-walled of carbon nanotubes appear as a hollow cylinder shape. It is just one atom thick in thickness and can be viewed as a roll. Graphene is a sheet of two-dimension of carbon, atomic scale, honey-comb lattice and it is bounded by sp^2 bond (Figure 1). Because of the same element of composition and bond arrangement of SWCNTs and graphene, both having the same properties and characterization. SWCNT has the advantages and their pros in various applications compared to others element [14-15]. Therefore, SWCNTs should be analyzed wisely so that it can give a help in the fields of technology.

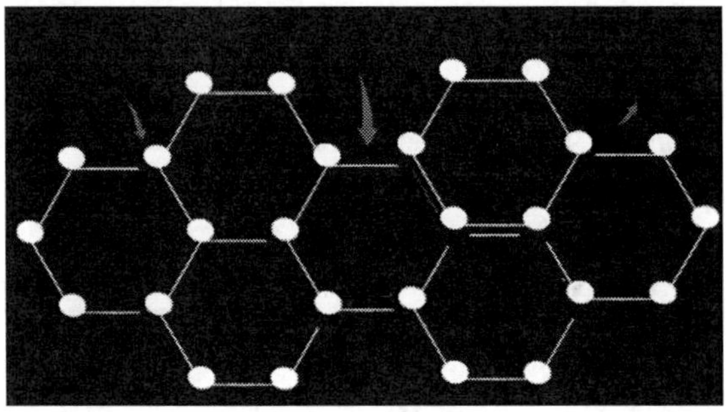

Figure 1. Typical structure of graphene. It has 6 carbon atoms on each ring structure, connected by the molecular bond with the size of ~0.142 nm.

To utilize the above nanomaterials as the base to carry the biological ligands, two herbs were chosen, namely *Solanum trilobatum* and *Hempedu bumi* and extracted the compounds from them. *S. trilobatum* is a commonly used name and also called locally as 'Thoothuvalai', originated from the Asian country, India. This plant is mainly growing in the Southern region of India, which includes the states Tamil Nadu and Kerala and also from some Northern part of India. In addition, this plant is found to grown in Southern part of Asia, including the regions Malaysia, Vietnam and Thailand. The local climate in these parts favour the growth of this plant and appears throughout of the year. *S. trilobatum* is found to have medicinal value for treating the health issues. Mainly to treat respiratory problems, cough, gastric, asthma, constipation and relieve gas in stomach. *S. trilobatum* has been analysed and noticed to have antimicrobial characteristics, make suitable for the current study.

Figure 2. Morphology of *Solanum trilobatum*.

On the other hand, *Hempedu bumi* or its scientific name, *Andrographis paniculata*, is an akin of shrub, belonging to the family of Acanthaceae, also known as Ayurveda Herb (Kalamegha) in India due to its common properties. *A. paniculata* grows abundantly in India, Taiwan and Sri Lanka. However, it is cultivated extensively in China and Thailand. This plant has long been used in Chinese official herbal medicine against broad spectrum of illness. Past a few decades, *A. paniculata* plant is highly valued by local people as one of the medicine that can treat a variety of diseases. It is highly effective for current global illness and health concerns.

Figure 3. Morphology of *Hempedu bumi.*

2 Materials and Methods

2.1 Materials

Silver Nitrate, Sodium Chloride, Tween-20 were purchased from Sigma-Aldrich. Cow urine was collected from the local form in Malaysia. *Solanum trilobatum* and *Hempedu bumi* plants were collected from India and Malaysia, respectively. All analytical grade reagents were from Merck. All chemicals and reagents were stored and used as recommended by the supplier. The bacterium, *Bacillus* sp. was obtained from the culture collection of School of Bioprocess, Universiti Malaysia Perlis, Perlis.

2.2 Extraction of plant extract

To prepare the plant extract 5 gram of plant material was soaked in the 50 ml of water for the period of overnight. Next day, the mixture was boiled at 55°C for 10 min and filtered through the filters. Further, these extracts were filtered again using syringe filter (Millipore) having the pore size of 0.45 μm.

2.3 Preparation of Nanostructures

The plant extract of *Solanum trilobatum* was prepared as mentioned above. Then 1 mM of silver nitrate was mixed with plant extract with the constant stirring. It was continued until the silver nitrate solution become yellowish orange. After the formation of color, the solution was centrifuged at 10,000 xg for 10 min. The obtained pellet was washed with double distilled water twice. For the encapsulation of cow urine, the previous method proposed was followed as outlined [16]. SWCNTs were purchased commercially and conjugated electrostatically with the extract prepared from *Hempedu bumi*.

2.4 Scanning Electron Microscope

All the nanostructures were observed under Field Emission Scanning Electron Microscope (Hitachi, S-4300 SE, Japan) with appropriate scanning speed under the applied voltage of 15 kV. Magnification was done at 5000x.

2.5 Biological inhibition assay

To evaluate the inhibitory activity of the prepared nanostructures conjugated biological ligands, disc diffusion assay was followed. Initially, agar plates using LB media agar were prepared by following the routine procedure. Upon preparation, the culture *Bacillus* sp. was spread on the agar plate. At the same time, the disc made from the filter paper was placed on the agar plate and loaded with 35 μl of the sample to be tested. For the entire test with all samples, the similar volume was used. Ampicillin was used the standard. Then, after continuing the culture at 37°C for overnight, the formation of the clear region was measured.

3 Results and Discussion

The compounds obtained from the herbal plant have gained a considerable amount of importance in the field of medicine and biotechnology. Several herbal plants are found to have medicinal values and used traditionally for a quite long time to treat illnesses and injuries [17]. In several instances herbal plants have been used directly without further purification steps. However, production of engineered nanomaterials [7] will boost the utilization of herbal compounds by conjugating them on the nanomaterials.

Another nanomaterial is chosen for this study is the cellulose nanoparticle, composed of organic compound with the D-glucose units [18]. Cellulose is one of the cheaper mateials available easily all over the word and about 10^{10} to 10^{11} tons of the cellulose is annually synthesized and destroyed. Researchers are taking attempts to recycle the cellulose waste in different ways. Currently, cellulose nanoparticle is considered to be the counter for the metallic nanoparticles and play a vital role [19].

Synthesis and fabrication of nanostructures are the currently main streams in the front of nanotechnology in order to engineer the novel nanostructures. These nanostructures have shown to be nanoparticle, nanofilm, nanotube, nanowire, nanowick and combination of these structures. On the other hand, the downstream application of these nanostructures has been focused mainly in the biomedical utilizations. In the field of biomedicine, nanostructures are mainly used to be the carrier for the purpose of interactive analysis and nanodelivery. In the current study, we have chosen three different types of nanostructures, which include silver nanoparticle, cellulose nanoparticle and SWCNT. To demonstrate the suitability of these nanostructures to carry the biological ligands, both plant extract and cow urine were chosen. Plenty of active compounds can be extracted from the plant; on the other hand, cow urine has the effective bioactive compounds (Figure 4).

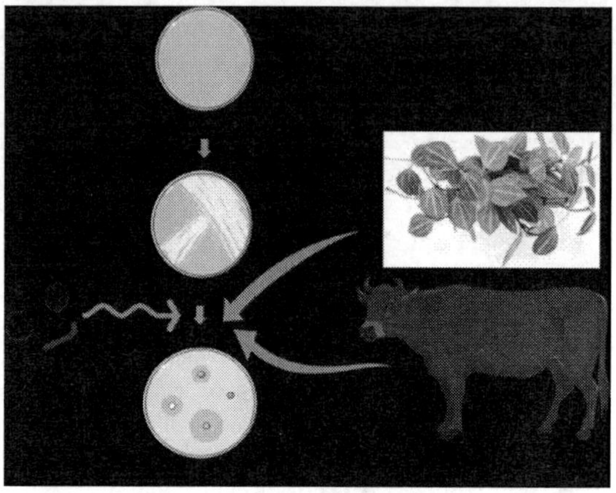

Figure 4. Scheme for the overall study. Either plant extract or cow urine was conjugated on the nanostructures and studied the bacterial inhibition. *Bacillus* sp. was used as the test bacterium.

3.1 Silver nanoparticle as the carrier for biological ligands

Biosynthesis production of silver nanoparticles using medicinal plants is more desirable than physical and chemical methods due to its eco-friendliness. The purpose of the present study is to identify the medicinal properties of silver nanoparticles synthesized by herbal extracts. To demonstrate this purpose the compounds from plant '*Solanum trilobatum*' were chosen as the capping the reducing agents. Upon completion of the production of silver nanoparticle, it can be considered as it is carrying the biological ligands from *S. trilobatum*. After synthesizing the silver nanoparticles, they were observed under Field Emission Scanning Electron Microscope (FESEM) as shown in the Figure 5. As per measurement from the FESEM, the size of the produced nanoparticle is estimated to be about 15 nm. Also confirmed by Field Emission Transmission Electron

 https://doi.org/10.1051/matecconf/201815006002

Microscope (FETEM; Figure 6). It was also noticed that silver nanoparticles are distributed uniformly. Upon preparation of silver nanoparticles, we evaluated the biological inhibitory activity in order to confirm the carrying of biological ligands. The biological inhibition was compared with the ampicillin (1 mg/ml) as the standard.

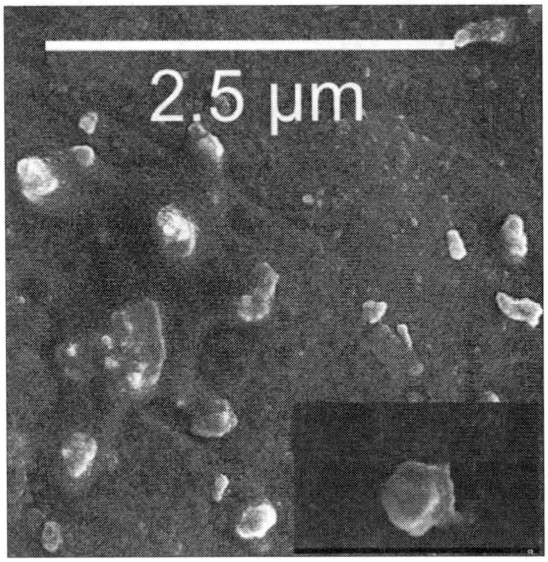

Figure 5. FESEM image of *Solanum trilobatum* plant extract mediated synthesis of silver nanoparticle. Figure inset display the enlarged view. Performed at 15 kV. Scale bar is shown.

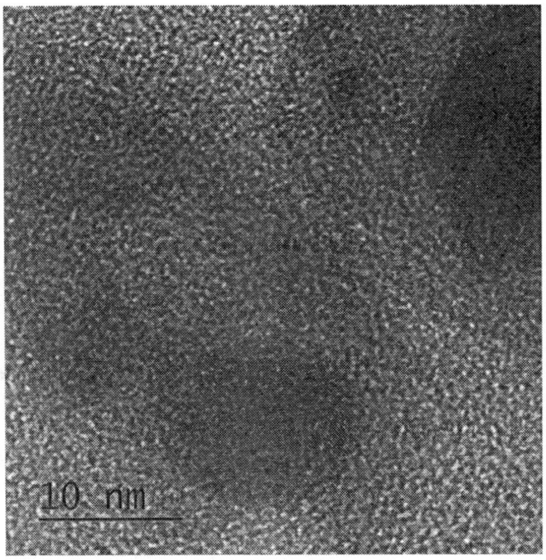

Figure 6. FETEM image of *Solanum trilobatum* plant extract mediated synthesis of silver nanoparticle. Scale bar is shown.

With the ampicillin using 15 µg, we could observe the clear zone against the Gram-positive bacterium *Bacillus* sp. and it shows the clearing region of 15±0.2 mm. Whereas, *S. triblobatum* silver nanoparticle has shown he clearing region of 6.5±0.15 mm (Figure 7).

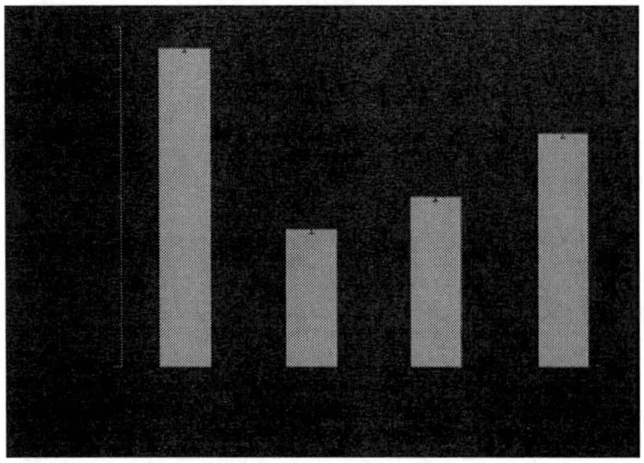

Figure 7. Zone of Inhibition. Measured on the agar plate cultured with *Bacillus* sp. Disc diffusion assay was followed. Ampicillin was used as the standard solution. Silver and SWCNT were conjugated with the plant extract. Cellulose nanoparticle was encapsulated with cow urine.

By comparison, *S. triblobatum* capped silver nanoparticles have shown nearly half of the inhibition region displayed by 15 µg of ampicillin. This analysis confirms the ability of silver nanoparticles to act as the carrier for the biological ligands.

3.2 Cellulose nanoparticle as the carrier for biological ligands

Similar to the above case with the silver nanoparticle, another nanostructure cellulose nanoparticle was also demonstrated for carrying the biological ligands. In this case, cow urine was collected and encapsulated in the cellulose nanoparticle by microemulsion method as stated earlier [16]. The cow urine encapsulated cellulose nanoparticle was observed under FESEM as displayed in the Figure 8.

Figure 8. FESEM image of cow urine encapsulated cellulose nanoparticle. Performed at 15 kV. Scale bar is shown.

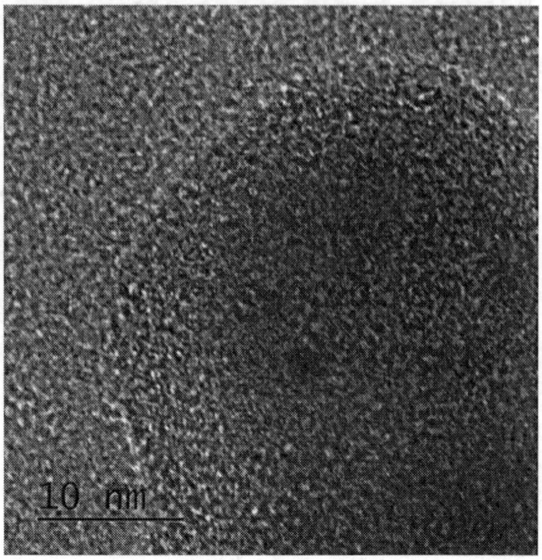

Figure 9. FETEM image of cow urine encapsulated cellulose nanoparticle. Scale bar is shown.

From the FESEM analysis it was noticed that the encapsulated cellulose nanoparticle is agglomerated. This might be due to the soft texture of the cellulose nanoparticle compared to the metallic nanoparticle. This result was further supported by FETEM analysis (Figure 9). As in the case above, we also evaluated the inhibition ability of the cellulose nanoparticle encapsulated cow urine.

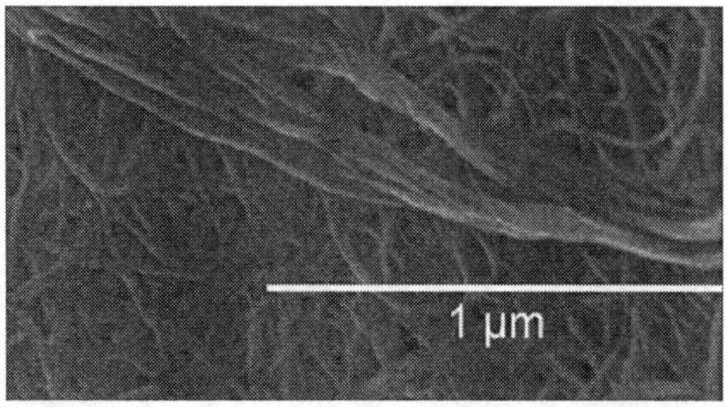

Figure 10. FESEM image of *Hempedu bumi* plant extract conjugated SWCNTs. The scale bar is shown with 1 μm range.

The result was measured to be with the clearing region of 8.0±0.18 mm against the bacterium *Bacillus* sp. (Figure 7). The difference between the clear regions obtained from the above *S. trilobatum* capped silver nanoparticle and cellulose nanoparticle encapsulated cow urine is due to the amount of the biological ligands conjugated and its ability to inhibit the *Bacillus* sp. In the case of cellulose nanoparticle encapsulation is showed more than half level of inhibition compared to 15 μg of ampicillin. This analysis proved as in the case of silver nanoparticle, cellulose nanoparticle can also be the biological carrier.

3.3 SWCNT as the carrier for biological ligands

In the above cases, we used both metallic and non-metallic nanoparticles. As the third nanostructure we preferred different nanostructure fabricated based on graphene material, called SWCNTs. The extracted compounds from the medicinally important herb plant *Hempedu bumi*, was chosen as the biological ligands and conjugated electrostatically on the SWCNTs. SWCNTs were seen under the FESEM and it was obvious with the parallel SWCNTs with the uniform sizes, further SWCNTs were nicely packed (Figure 10). *H. bumi* extract conjugated SWCNTs was also tested for their inhibitory ability against the bacterium *Bacillus* sp. Upon measuring, we could found the clear region of 1.1 ±0.19 mm and this level is close to the level of inhibition obtained with 15 μg of ampicillin (Figure 7). This analysis evidenced that the extracted compounds from *H. bumi* conjugated on SWCNT is active with the antibacterial inhibition.

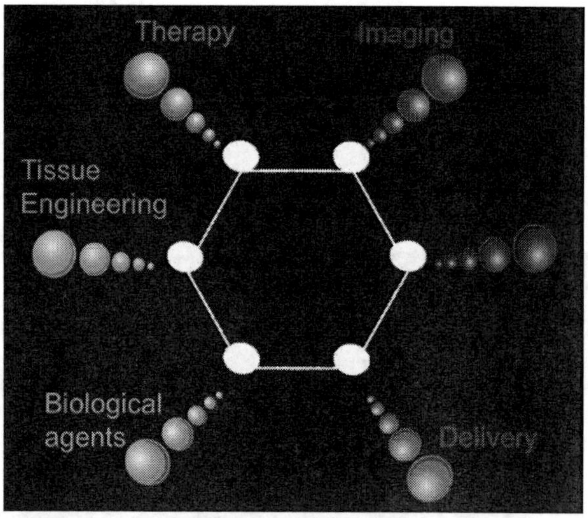

Figure 11. Potential applications of biological ligands conjugated nanomaterials. These applications are highly connected to the field of medicine.

This study clearly demonstrated with different nanomaterials (Silver nanoparticle, Cellulose nanoparticle and SWCNT), for their potential to be active by carrying the biological ligands. Similarly, by mimicking the same strategy for conjugating different biomolecules, it will be highly beneficial for the downstream applications, such as therapy, imaging, biosensor, tissue engineering, targeted delivery (Figure 11). As a proof of concept, in the past several studies have been demonstrated for these applications [20-22].

4 Conclusions

In the present study, three engineered nanostructures have been chosen, namely silver nanoparticle, cellulose nanoparticle and SWCNT. These nanostructures were conjugated with the biological ligands obtained from the herb *S. trilobatum*, cow urine and the herb *H. bumi*, respectively. The engineered nanostructures were well formed and uniformly arranged under FESEM analysis. From the inhibition analyses, it was revealed that all these nanostructures have ability to carry the biological ligands. Based on these, in can be concluded that engineered nanostructures are the potential candidates to carry the biological ligands for biomedical applications, especially for pathogenic inhibition.

References

1. S. Anniebell, S.C.B. Gopinath, Curr Med Chem. **24** (2017) [Epub ahead of print].
2. K.H.Suk, S.C.B. Gopinath, Curr Med Chem. (2017) [Epub ahead of print].
3. S. Ramanathan, S.C.B. Gopinath, Microsyst. Technol. **23** (2017) 4345–4357.
4. K. Nomura, S.C.B. Gopinath, Nat.Commun. **4** (2013) 2855.
5. M.E. Foo, S.C.B. Gopinath. Biomed. Pharmacotherapy **94** (2017) 354-361.
6. N.Z. Jin, S. Anniebell, S.C.B. Gopinath, Y. Chen, Nanoscale Res. Lett. **11** (2016) 399.
7. A. Suominen, Y. Li, J. Youtie, P. Shapira, J Nanoparticle. **18**, (2016) 270.
8. P. Velusamy, C.-H. Su, G. Venkat Kumar, S. Adhikary, K. Pandian, S.C.B. Gopinath, Y.Chen, P. Anbu , PLoS One, **11** (2016) e0157612.
9. S. Ahmed, M. Ahmad, B.L. Swami, S. Ikram, J. Radiation Research and Applied scinces. **9**, (2016) 17–28.
10. D. Nayak, S. Ashe, P.R. Rauta, M. Kumari, B. Nayak, Mater. Sci. Eng. C. **58**, (2016) 44–52.
11. K. Logaranjan, A.J. Raiza, S.C.B. Gopinath, Y. Chen, K. Pandian, S, Nanoscale Res. Lett. **11** (2016) 520.
12. N. Lin, A. Dufresne, Eur. Poly. J. **59** (2014) 302-325.
13. L. Zhou, H. He, M.C. Li, K. Song, H.N. Cheng, Q. Wu, M, Carbohydr. Polym. **153** (2016) 445–454.
14. D. A. Ghuge, A. R. Shirode , V. J. Kadam. Curr. Drug Targets **18** (2017) 724-733.
15. S. Bharec, R. Kumar, J. Mat. Sci. Mechanical Engg. **2**, (2015) 70-73.
16. S.C.B. Gopinath, C.S. Goh, M. Citartan, T. Lakshmipriya, M.K. Arshad, M.F. Fatin, A.R. Ruslinda, U. Hashim, S.V. Chinni, T.H. Tang, Micro Nanosyst. **8**, (2016) 1–6.
17. P.S. Latha, K. Kannabiran, African J. Biotechnol. **5** (2006) 2402–2404.
18. S. Kalia, A. Dufresne, B. M. Cherian, B. S. Kaith, L. Avérous, J. Njuguna, E. Nassiopoulos. Int. J. Polymer Sci. **2011** (2011), 837875.
19. M. A. El-Sheikh, S. M. El-Rafie, E. S. Abdel-Halim, and M. H. El-Rafie. J. Polymers **2013** (2013), 650837.
20. S.C.B. Gopinath, K. Awazu, M. Fujimaki, K. Shimizu, T. Shima, PLoS One **8** (2013) e69121.
21. S.C.B. Gopinath, P.K.R. Kumar, *Acta Biomaterialia* **9** (2013) 8932-8941.
22. S.C.B. Gopinath, D. Balasundaresan, J. Akitomi, H. Mizuno, J. Biochem. **140** (2006) 667-676.

 https://doi.org/10.1051/matecconf/201815006003

Application of Data Mining Techniques for Medical Data Classification: A Review

Saima Anwar Lashari[1,*], *Rosziati* Ibrahim[1], *Norhalina* Senan[1] *and N. S. A. M.* Taujuddin[2]

[1]Faculty of Computer Science and Information Technology
[2]Faculty of Electrical and Electronic Engineering
Universiti Tun Hussein Onn Malaysia, Batu Pahat, Malaysia

Abstract. This paper investigates the existing practices and prospects of medical data classification based on data mining techniques. It highlights major advanced classification approaches used to enhance classification accuracy. Past research has provided literature on medical data classification using data mining techniques. From extensive literature analysis, it is found that data mining techniques are very effective for the task of classification. This paper analysed comparatively the current advancement in the classification of medical data. The findings of the study showed that the existing classification of medical data can be improved further. Nonetheless, there should be more research to ascertain and lessen the ambiguities for classification to gain better precision.

1 Introduction

Data Mining has been used in a variety of applications such as marketing, customer relationship management, engineering, and medicine analysis, expert prediction, web mining and mobile computing [1]. Of late, data mining has been applied successfully in healthcare fraud and detecting abuse cases [2]. Clinical decisions are often made based on doctors' intuition and experience rather than on the knowledge rich data hidden in the database. This practice leads to unwanted biases, errors and excessive medical costs which affects the quality of service provided to patients [3]. This suggestion is promising as data modeling and analysis tools, e.g., data mining, have the potential to generate a knowledge-rich environment which can help to significantly improve the quality of clinical decisions. Successful data mining applications have provided the impetus for the relevant parties to fully utilized them as they have realised that data mining is crucial in the acquisition of valuable information for all sectors involved in healthcare-related industries.

Healthcare insurers are able to identify fraud and abuse cases, health administrators are able to make better decisions especially in managing their customers, and healthcare practitioners are able to deliver better services and treatments. The huge amounts of data generated by healthcare transactions are too complex and voluminous to be processed and analyzed by traditional methods [4]. Data mining can improve decision-making by discovering patterns and trends in large amounts of complex data. Such analysis has become increasingly essential as financial pressures have heightened the need for healthcare organizations to make decisions based on the analysis of clinical and financial data. Insights gained from data mining can influence cost, revenue, and operating efficiency while maintaining a high level of care. Healthcare organizations that perform data mining are better positioned to meet their long-term needs [5].

In recent years, computers and their peripherals have been made cheaper and more readily available and in line with the development of information technology, various kinds of advanced data mining techniques have hit the market. These new age data mining techniques embrace traditional and more recent sophisticated classification algorithms. Both classification techniques are for handling complex datasets such as multidimensionality, user inference and prior knowledge, web data, spurious data points that cause overfitting of models, improvement in human ability, noisy datasets cleaning, mining multimedia datasets and incremental datasets. Interdisciplinary data mining techniques and approaches can be used for all the above mentioned databases for forecasting the impact and discovering meaningful relationships in the data with the purpose of extracting useful information for knowledge generation [6].

To employ data mining algorithms to medical data, researchers' comprehension on the type of data mining algorithms and their functions should be clear. Descriptive (or unsupervised learning) and predictive (or supervised learning) algorithm are the two categories of data mining algorithms [2] [4]. Descriptive data mining group's data by determining the objects' similarity (or records) and detecting patterns that are unknown, or associations in the data whereby users are able to recognize a massive data pool. Descriptive data mining is investigative. It includes clustering, association, summarization, and sequence discovery [5]. Prediction data mining that comprises classification, regression,

* Corresponding author: saima@uthm.edu.my

time series analysis, and prediction [8] implies predicting rules (also known as classification/prediction models) from (training) data and the rules are employed to unpredicted/unclassified data [9].

In assisting researchers to comprehend understand the importance of data mining, and the application of data mining techniques, three of the most widely-used data mining algorithms (classification, clustering, and association) are discussed below, complete with guidelines for their respective use.

Thus, a variety of models have been fitted in order to determine hidden patterns in the data. The approach that is able to produce the most accurate output and relationships pattern in the observed datasets is considered to be the most efficient in the particular model. Such approach fulfils the objective of data mining. Current data mining practices utilizes a range of model functions including classification, regression, clustering, discovering association rules and sequence analysis [8].

However, the challenge increases as the interest in data mining grows rapidly. In order to handle these problems without using the traditional statistical methods, soft computing has emerged to be one of the encouraging data mining solutions in this area [9].

The digital medical data is not only enormous in amount, but also complex in its structure for traditional software and hardware. Some of the contributing factors to the failure of traditional systems in handling these datasets include:

a) The vast variety of structured and unstructured data such as handwritten doctor notes, medical records, medical diagnostic images (magnetic resonance imaging (MRI), computed tomography (CT)), and radiographic films [10].
b) Existence of noisy, heterogeneous, complex, longitudinal, diverse, and large datasets in healthcare informatics [11].
c) Necessity of improving medical issues such as quality of care, sharing, security of patients' data, and the reduction of the healthcare cost, which are not sufficiently addressed in traditional systems [8].

Hence, solutions are needed in order to manage and analyze such complex, diverse, and huge datasets in a reasonable time complexity and storage capacity for enhanced insight and decision-making. Therefore, this paper aims to highlight an assessment of the current and common methods for medical data categorisation. The existing approaches to the categorisation of medical records based on data mining techniques are being revised by highlighting the diverse categorisation algorithms for clinical imaging applications.

This paper is arranged into five sections. Section 2 explains classification in data mining. Comparative analysis of data mining techniques is provided in Section 3, followed by a discussion of the results in Section 4 and the paper ends with a conclusion in Section 5.

2 Classification in Data Mining

In data mining, categorization is formulated to make a forecast of the memberships in a group for data instances. This process utilizes complex analysis of data to determine data connections in huge datasets. Due to its complex features, medical databases provide complications for pattern extortion [7]. There are two approaches to data mining: statistical and machine learning algorithms. The processes in data mining are classified into descriptive and predictive (Fig. 1). Descriptive mining tasks provide the general data properties in the database. For Predictive mining tasks, inference is made on the data for predictions [9] whereby forecast is made on explicit values based on patterns identified by known results. Descriptive data mining, without having any predefined target, provides characteristics and descriptions for the data set.

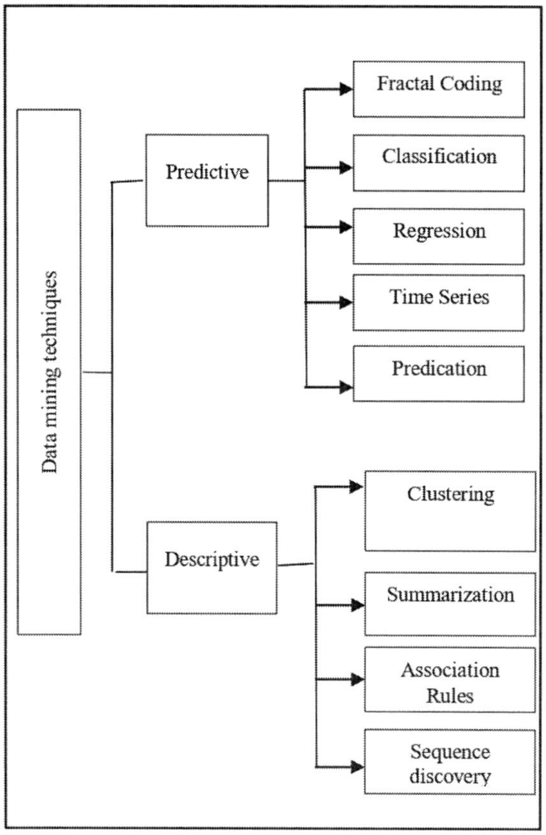

Fig. 1. Data mining techniques

Data mining techniques are effective and predictive for future patterns because: a) it is user friendly and prediction is based on past circumstances b) it operates by learning from past data c) data from numerous resources is managed and only required data is extracted d) models are easily updated by relearning, past information and change in trends. These are what makes it reliable and practical in the medical image categorization [4].

The three learning approaches in data mining algorithms are supervised (the algorithm works with a set of examples with known labels whose values are nominal in classification task, or numerical in regression task), unsupervised (unknown labels in the dataset, and the algorithm typically aims at grouping examples according to the similarity of their attribute values, characterizing a clustering task, or semi-supervised (whereby learning is conducted when there is availability of a small subset of labelled examples, concurrent with a large number of unlabelled examples).The task of categorising is regarded as a supervised technique in which each instance belongs to a class, specified by the value of a special goal attribute or the class attributes [14].

3 Comparative analysis of data mining techniques

Solutions resulted from classification algorithm are commendable but as of now, none is diverse and flexible to be accepted generally in the medical data classification community. Categorical variables in medical data are occasionally useful to arrive at decisions and to generalize information. Categorical data (e.g. classification of disease and non-disease groups) is handy for data mining technique and also easy to extract medical information. Table 1 represents data mining techniques representation briefly.

Table 1. Data Mining Techniques Representation

Approach	Representation	Pattern Recognition Models Recognition Function	Error Estimation
Template Matching	Sample, pixels, curves	Correlation, distance measurement	Classification error
Statitical	Featuers	Discriminatio n function	Classification error
Neural networks	Samples, voxel pixels, featuers	Network function	Mean square error

With the advancement in data mining for diagnosis and prognosis of different diseases, a significant number of attempts have been proposed for a wide variety of medical image classifications. However, the different assumptions and hypothesis have been made in these methods differ considerably. The section reviews the rapidly expanding body of work on the development and application of classification methods to problems of fundamental importance in medical image classification. Meanwhile, in conducting comparative studies, classification researches rely heavily on stored repositories of data (such as UCI repository [15] it allows new algorithm ideas to test its plausibly on known problems. Table 2 represents summary of data mining techniques.

Table 2. Summary of data mining techniques

Data mining technique	Advantage	Disadvantage
SVM	▪ Better Accuracy as compare to other classifier. ▪ Easily handle complex nonlinear data points. ▪ Over fitting problem is not as much as other methods.	▪ Computationally expensive. ▪ The selection of right kernel function. For every dataset different kernel function shows different results. ▪ Training process take more time. ▪ Suitable for the problem of binary class. It solves the problem of multi class by breaking it into pair of two classes such as one against-one and one-against all.
Decision Tree	▪ No requirements of domain prior knowledge in the construction of decision tree. ▪ It minimizes the ambiguity of complicated decisions. ▪ It is easy to interpret and it also handles both numerical and categorical data.	▪ It generates categorical output. ▪ Performance of classifier depends upon the type of dataset.
ANN	▪ Easily identify complex relationships between dependent and independent variables. ▪ Able to handle noisy data	▪ Local minima. ▪ Over-fitting.
Bayesian Belief Network	▪ It makes computations process easier. ▪ Have better speed and accuracy for huge datasets.	It does not give accurate results in some cases where there exists dependency among variables.

K-NN	▪ It is easy to implement. ▪ Training is done in faster manner	▪ It requires large storage space. ▪ Sensitive to noise. ▪ Testing is slow.

Support Vector Machine (SVM)
Artificial Neural Network (ANN)
k- nearest neighbour (K-NN)

Data mining techniques, which are a recent application in the medical domain, are applied in mining medical data, which comprises association rule mining for finding frequent patterns, prediction, classification and clustering. To date, there have been many research on this and intelligent and decision support systems have been developed to make more accurate diagnosis and prediction of diseases especially in predicting heart diseases, lung and breast cancer and remote health monitoring.

Table 3 provides the summary of medical data classification regarding the resolved difficulties that are solved, convenience in medical data mining or implementation of the tools. Therefore, selected researches on classification performance of different classifiers are summarised. Here is shown the effort made for data classification. Nevertheless, it is obvious that benchmarking to determine the best classification algorithm for medical data classification is still lacking.

Table 3. Summary of medical data mining techniques

Author(s), Year	**Medical Dataset**	**Technique (s)**	**Comments**
Khanmohammadi & Chou, 2016	Six datasets (UCI Repository)	GMBD	Discretization process was more concise for representation of continuous variable
Aswal & Ahuja 2016	six datasets (UCI Repository)	K-NN, SVM	Classification of bio medical data
Long, 2015	heart disease	rough sets based attribute reduction and interval type-2 fuzzy logic system (IT2FLS)	heart disease diagnostic system using rough sets based on attribute reduction and IT2FLS
Zuo et al. 2013	Parkinson Disease	Fuzzy K-NN approach	Familiarised an adaptive Fuzzy K-NN approach for diagnosing the disease
Ghofrani et al. 2014	X-Ray dataset	K-NN & SVM	Slow testing, scale depended
Ramana *et al.* 2011	Liver dataset	KNN & SVM	High accuracy
Kharya, 2012	MIAS	Decision tree	Iterative training producer, overtraining sensitive, need pruning
Rajini & Bhavani, 2011	MRI	ANN & k-NN	Sensitive to training parameters, slow training,
Cheng et al. 2009	Dermatology	ANN & Decision Tre	dermatologic diagnosis
Polat et al. 2007	Breast Cancer and Liver Disorders dataset	Fuzzy-AIRS	Modelling and analysis of medical data
Dangare & Apte, 2012	heart disease	Naive Bayes Decision Trees Neural Networks	accuracy for different classification methods with 13 input attributes & 15 input attributes values
Xing et al. 2007	coronary heart disease	Decision Tree Algorithms such as C4.5, C5, and CART	Prediction models
Shim, et al. 2003	Liver diseases	Classification using BYY	-
Tang et al. 2009	Diabetes, Cancer	k-Nearest Neighbour	Classification of Disease using K-NN

Acronym:
Genetic k-Nearest Neighbour (GkNN)
Support Vector Machine (SVM)
k- nearest neighbour (K-NN)
Gaussian mixture model based discretization (GMBD)
Bayesian Ying Yang (BYY)

This part of the paper is to present the necessities in the healthcare industry, and the potential techniques to be utilised. Here are guidelines on the usage of the different data mining techniques: Identification of the unnecessary attributes which impedes the processing task is crucial before the application of the classification technique. Besides acting as noise and disturbing the process, they also affect the classifier performance. To identify these, statistical methods are employed. In contrast, the feature selection methods are engaged to select functional attributes in order to improve the precision and success of the classification model [18]. Thus, the researchers have found that there is no classifier that generates the best result for each dataset. To check the performance of classifiers, each dataset is parted into two division – training and testing. A classifier, which is tested using a testing dataset, is chosen based on its performance in comparison to other available classifiers. Cross validation method is conducted for both training and testing dataset to ensure accuracy.

The main emphasis of classification rules is to discover the class of attributes, but it does not take into account the relationships of attributes. While association is useful for identifying the relationship or association among numerous attributes and generates association rules which in turn helpful for domain experts to remove insignificant association rules and consider only those rules which are useful for making vital decision

4 Discussion

Based on Table 1, each data mining modality has its individual characteristics with which to contend. With all the efforts, there is still no extensively used method to classify medical data. The most explainable reason is that exceptionally precise data and especially very low rate of false negatives are required in the medical domain. There are general methods, however, that can be applied to a variety of data. Specialized methods for particular applications that consider prior knowledge can enable the achievement of better performance. There have been progress in medical data over the past decades in the following three areas:

1) Development and use of advanced classification algorithms
2) Use of multiple features
3) Integration of medical data into classification procedures.

Nevertheless, there are challenges. These include data mining methodology, user interaction, performance and scalability. Other issues are the exploration of data mining application and their social impacts. The selection of an appropriate approach to a classification problem can therefore be a difficult dilemma. Consequently, there is the possibility of further improvement in the current medical data classification tasks. As data mining techniques for medical data classification has not been fully investigated, there is a great potential for further work and interesting directions for future research can be created.

5 Conclusion

In this paper, it is observed that data mining techniques have been employed for medical data classification. There are voluminous records in this medical data domain and because of this, it has become a requisite to use data mining techniques to help in decision support and prediction in the field of healthcare to identify diseases. Therefore, medical data mining contributes to business intelligence which is useful for diagnosing of diseases. This paper throws light into data mining techniques that are used for medical data for various diseases which are identified and diagnosed for human health. For future works, despite many opportunities and approaches for big data analytics in healthcare are presented in this work, there are many other directions to be explored, concerning various aspects of healthcare data, such as the quality, privacy, timeliness, and so forth. This section provides an outlook of big data analytics in healthcare informatics from a broader view, which covers the topics of healthcare data characteristics (e.g., high complexity, large scale, etc.), data analytics tasks (e.g. longitudinal analysis, visualization, etc.), and objectives (e.g. real-time, privacy protection, collaboration with experts, etc.). The future of health informatics will benefit from the exponentially increasing digital health data.

Due to their practicality, data mining applications may benefit the healthcare industry immensely. Nonetheless, it should be cautioned that this benefit depends on how clean the data is. To ensure the success of data mining applications, the capture, storage, preparation and mining of data must be critically considered whereby there should be a standard practice of clinical vocabulary and data-sharing across healthcare establishments. Additionally, the scope and nature of healthcare data can be expanded beyond quantitative data. Text mining can be conducted to harvest data from handwritten medical notes and records by the doctors. Combining data and text mining may enrich the process. Besides, there have also been development in data mining through digital diagnostic images. The researchers highly anticipate that findings from this study can provide advancement to data mining and healthcare resources and enable all relevant parties to be benefitted.

This work is supported by Research, Innovation, Commercialization, and Consultancy (ORICC), with the grant post-doctoral fund, Vote No D001, Universiti Tun Hussein Onn Malaysia.

References

[1] Yoo, Illhoi, Patricia Alafaireet, Miroslav Marinov, Keila Pena-Hernandez, Rajitha Gopidi, Jia-Fu Chang, and Lei Hua. "Data mining in healthcare and biomedicine: a survey of the literature." *Journal of medical systems* 36, no. 4 (2012): 2431-2448.

[2] Witten, Ian H., Eibe Frank, Mark A. Hall, and Christopher J. Pal. *Data Mining: Practical machine learning tools and techniques*. Morgan Kaufmann, 2016.

[3] Chang, Chun-Lang, and Chih-Hao Chen. "Applying decision tree and neural network to increase quality of dermatologic diagnosis." *Expert Systems with Applications* 36, no. 2 (2009): 4035-4041.

[4] U. Fayyad, G. Piatetsky-Shapiro, and P. Smyth, “From data mining to knowledge discovery in databases,” *AI Mag.*, vol. 17, no. 3, p. 37, 1996.

[5] D. Chauhan and V. Jaiswal, "An efficient data mining classification approach for detecting lung cancer disease," in *Communication and Electronics Systems (ICCES), International Conference on*, 2016, pp. 1–8.

[6] M. Cottle, W. Hoover, S. Kanwal, M. Kohn, T. Strome, and N. Treister, "Transforming Health Care Through Big Data Strategies for leveraging big data in the health care industry," *Inst. Heal. Technol. Transform. http//ihealthtran. com/big-data-in-healthcare*, 2013.

[7] Ramana, Bendi Venkata, M. Surendra Prasad Babu, and N. B. Venkateswarlu. "A critical study of selected classification algorithms for liver disease diagnosis." *International Journal of Database Management Systems* 3, no. 2 (2011): 101-114.

[8] Sharmila, K., and S. A. Vethamanickam. "Survey on data mining algorithm and its application in healthcare sector using Hadoop platform." *International Journal of Emerging Technology and Advanced Engineering* 5, no. 1 (2015): 567-71.

[9] Kharya, Shweta. "Using data mining techniques for diagnosis and prognosis of cancer disease." *arXiv preprint arXiv:1205.1923* (2012).

[10] Dangare, Chaitrali S., and Sulabha S. Apte. "A data mining approach for prediction of heart disease using neural networks." (2012).

[11] Shim, Jeong-Yon, and Lei Xu. "Medical data mining model for oriental medicine via BYY binary independent factor analysis." In *Circuits and Systems, 2003. ISCAS'03. Proceedings of the 2003 International Symposium on*, vol. 5, pp. V-V. IEEE, 2003.

[12] Ghofrani, Fatemeh, Mohammad Sadegh Helfroush, Habibollah Danyali, and Kamran Kazemi. "Improving the performance of machine learning algorithms using fuzzy-based features for medical x-ray image classification." *Journal of Intelligent & Fuzzy Systems* 27, no. 6 (2014): 3169-3180.

[13] S. A. Lashari, R. Ibrahim, N. Senan, I. T. R. Yanto, and T. Herawan, "Application of Wavelet De-noising Filters in Mammogram Images Classification Using Fuzzy Soft Set," in *International Conference on Soft Computing and Data Mining*, 2016, pp. 529–537.

[14] S. A. Lashari, R. Ibrahim, and N. Senan, "De-noising analysis of mammogram images in the wavelet domain using hard and soft thresholding," in *Information and Communication Technologies (WICT), 2014 Fourth World Congress on*, 2014, pp. 353–357.

[15] Khanmohammadi, Sina, and Chun-An Chou. "A Gaussian mixture model based discretization algorithm for associative classification of medical data." *Expert Systems with Applications* 58 (2016): 119-129.

[16] Xing, Yanwei, Jie Wang, and Zhihong Zhao. "Combination data mining methods with new medical data to predicting outcome of coronary heart disease." In *Convergence Information Technology, 2007. International Conference on*, pp. 868-872. IEEE, 2007.

[17] Rajini, N. Hema, and R. Bhavani. "Classification of MRI brain images using k-nearest neighbor and artificial neural network." In *Recent Trends in Information Technology (ICRTIT), 2011 International Conference on*, pp. 563-568. IEEE, 2011.

[18] Polat, Kemal, Seral Şahan, Halife Kodaz, and Salih Güneş. "Breast cancer and liver disorders classification using artificial immune recognition system (AIRS) with performance evaluation by fuzzy resource allocation mechanism." *Expert Systems with Applications* 32, no. 1 (2007): 172-183.

[19] Aswal, Shobha, and Neelu Jyothi Ahuja. "Experimental analysis of traditional classification algorithms on bio medical dtatasets." In *Next Generation Computing Technologies (NGCT), 2016 2nd International Conference on*, pp. 566-568. IEEE, 2016.

[20] Tang, Ping-Hung, and Ming-Hseng Tseng. "Medical data mining using BGA and RGA for weighting of features in fuzzy k-NN classification." In *Machine Learning and Cybernetics, 2009 International Conference on*, vol. 5, pp. 3070-3075. IEEE, 2009.

[21] Long, Nguyen Cong, Phayung Meesad, and Herwig Unger. "A highly accurate firefly based algorithm for heart disease prediction." *Expert Systems with Applications* 42, no. 21 (2015): 8221-8231.

[22] Zuo, Wan-Li, Zhi-Yan Wang, Tong Liu, and Hui-Ling Chen. "Effective detection of Parkinson's disease using an adaptive fuzzy k-nearest neighbor approach." *Biomedical Signal Processing and Control* 8, no. 4 (2013): 364-373.

Effects of Land Cover Changes to the Quantity of Water Supply and Hydrologic Cycle using Water Balance Models

CC Caja [1,*], *NL Ibunes* [1], *JA Paril* [1], *AR Reyes* [1], *JP Nazareno* [1], *CE Monjardin* [1] and *FA Uy* [1]

[1] School of Civil, Environmental and Geological Engineering, Mapúa University658 Muralla St., Intramuros, Manila 1002, Philippines

Abstract. The hydrologic cycle is a recurring consequence of different forms of movement of water and changes of its physical state on a given area of the earth. The land cover of a certain area is a significant factor affecting the watershed hydrology. This also affects the quantity of water supply within the watershed. This study assessed the impacts of the changing land cover of the Ipo watershed, a part of the Angat-Ipo-La Mesa water system which is the main source of Metro Manila's water supply. The environmental impacts were assessed using the interaction of vegetation cover changes and the output flow rates in Ipo watershed. Using hydrologic modelling system, the hydrological balance using rainfall, vegetation and terrain data of the watershed was simulated. Over the years, there has been a decreasing land cover within the watershed caused mostly by deforestation and other human activities. This significant change in the land cover resulted to extreme increase in water discharge at all streams and rivers in the watershed and the water balance of the area were affected as saturation and shape of the land terrain changes.

1 Introduction

Water is a very important element in the earth. It comprises about 70% of the earth's surface and it is very easy to find. But then, people can intake only 2.5% of these waters, which is freshwater. Metro Manila gets it share of water supply from the Angat-Ipo-La Mesa system and the Laguna De Bay. But these sources does not only provide water but also is used for irrigation, hydroelectric power supply and for flood control. These purposes show the importance of water to the lives of people in the metro and also for nearby provinces [7].

The Ipo watershed is a mountain reserve that values the inhabitants in the area. It is bounded and secluded by proclamation no. 391 from the total horizon of the Angat watershed. But to sum it all up, it is part of a water system that extends from Angat-Ipo-La Mesa [7]. For the past countless months, there have been major concerns of deforestation. Illegal logging, Slash and Burn Farming or 'Kaingin', and charcoal making were major forest disruptions in the area. This causes an alteration to the vegetation in the area. Vegetation plays an important role in the infiltration and runoff in the soil. Changes in the vegetation cover from woodland to agricultural land lowers the infiltration rate of the soil. This may be one of the causes of the alteration and disruption of the hydrological cycle in the area. Thus, leading to a more major concern of decrease in water level in the Angat-Ipo Watershed.

In recent years, many studies have been conducted to assess the impact of land use changes to the hydrological cycle of an area. However, the effects vary from place to place due to their geographical varieties [1-6].

During the previous decade there have been two situations where the Angat Dam, that it is directly connected to Ipo Dam, dipped below minimum operating levels. First was in 2010 where it reached a low record of 157.56 meter above sea level (masl), which was when the Philippines experienced El Niño. Second one was this summer season of 2014 where it reached 162.74 masl.

The study evaluates the water balance in the Ipo watershed with respect to the scenario of land cover changes in the area. The group will present the study by correlating the water balance and the hydrological response in the Ipo watershed throughout the different scenarios of land cover for the past years. Specifically, it sought answers to the following questions which are

1. What are the environmental impacts of changing land cover to the Ipo watershed?
2. How does changes in the land cover of Ipo Watershed affect the following hydrological cycle and the quantity of the water supply

This study addresses the effects of varying vegetation cover to the water balance of the watershed. The study will be of utmost importance to hydrologists, foresters, civil engineers and water-related practitioners which is of concern to the status of water storage in the Angat-Ipo watershed. The analysis and results to be done from the

* Corresponding author: chennieccaja@gmail.com

study will further enhance the understanding of our experts in the maintenance of the dam and its surrounding watersheds.

This paper may be used as a reference for future studies. This will help the concerned government agencies to analyze the water balance of the Ipo dam and hence, make ways of preventing water crises in the future. This could also educate our people about the diverse effects of deforestation in our everyday living. In general, this study will benefit the entire community.

2 Methodology

This study is conducted based on the framework shown in Figure 1. In preparation for the study, there are basically 3 important data to be gathered. The Digital Elevation Model (DEM) data, Land cover Map, and the Precipitation Gauge data. This will include the conceptual modeling of the domain of the study in ArcGIS. The ArcGIS software is a geographical information system that compiles geographic data and analyzes mapped information that will allow the study to assess and manage the study area.

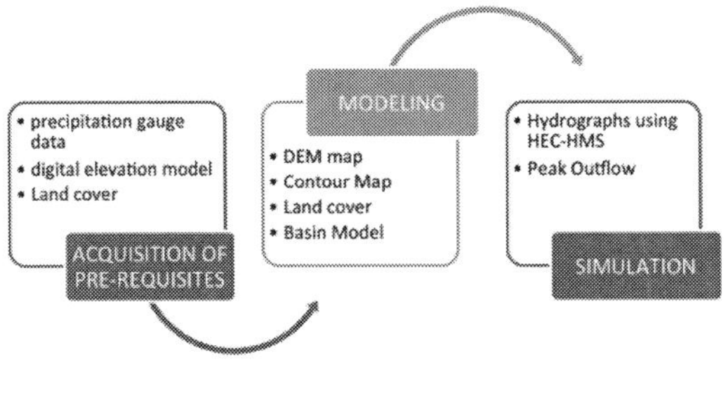

Fig. 1. Conceptual Framework

For the simulation procedure, the study is to use Hydrological Engineering Center – Hydrologic Modeling System (HEC-HMS) software. It is to be modeled in a mathematical condition where there are changing hydrometeorological conditions using set of equations that represent the response of the hydrologic components. It will be a continuous model where in it is simulated at a longer period predicting watershed response with response to the scenarios of precipitation.

This will include different model components namely: State Variables, parameters, boundary conditions and initial conditions. State variables is the condition or the state of the watershed at a certain period. Parameters will be the overall control point of the systems input and output. Boundary conditions are the changing parameter which will affect the state of the watershed and will bring a noticeable change to the system input.

The research pattern of the study starts with the model analysis of the research area. Processing the acquired data into different software in ArcGIS and HEC-HMS. For the research we are to process 3 past scenarios of land cover. We are to analyze the results of the simulation and come up with a conclusion on the relevance of land cover with response to the overall water intake in the watershed

The research setting is at Ipo Watershed declared under proclamation no. 391 at the boundaries of Bulacan, Philippines. The watershed is at a total area of 7,700 ha. According to articles, the current situation at ipo dam is at least 70% of the forest cover of the ipo watershed is gone.

There are 2 basic steps for the study, Modeling and Simulation. Modeling is done with the aid of the ArcGIS software. The DEM data is pre-processed in this software to organize the data's conditions to be used in the simulation. The simulation done in the study will be aided by HEC-HMS software. After organizing the DEM data, it is transferred to the HMS software and added with other supporting data to create a whole simulation of the hydrological cycle in the area. For the study, we are to simulate 3 past years with varying vegetation cover. The process done in the study is to vary the Curve Number of the Model. The curve number represents the vegetation within a specific area to represent its infiltration and other related vegetation parameters.

The land cover maps and the Digital Elevation Model (DEM) of the Ipo watershed were gathered from the National Mapping and Resource Information Authority (NAMRIA). It was then processed using a GIS software. Figure 2 shows the location map of the Ipo watershed [8,9].

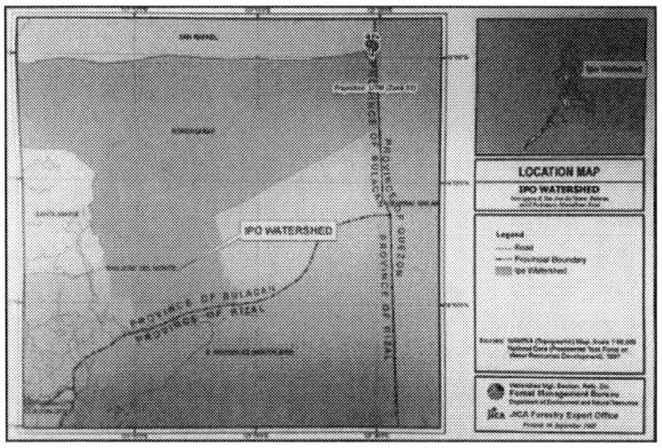

Fig. 2. Location Map (DENR-FMB, 2005)

After the Model analysis of the watershed, the water balance in the watershed is to be simulated in the HEC-HMS software. The different parameters in the watershed is to be input and to be analyzed and processed in the software to create an overall hydrological simulation and water balance in the watershed.

3 Results and Discussion

3.1 Terrain Assessment

The watershed was processed in the ArcGIS and can be seen through different viewpoints. One is by processing its image classification and processing it to view the Landcover. Another is by processing it with different parameters under the GeoHMS toolbar and processing the land area as sub basins to be used in the HMS software. For the image classification, it is classified under each category of land cover and is differentiated by color to identify different types of land cover. The sub watersheds are divided and classed under different vegetation types defined by different shades of color as seen in Figure 3. For the year 2003, most of the area are seen to be closed forest. But, as several years had passed, most then are now classified under wooded grasslands. These shows that there are major disruptions in the land cover in the area.

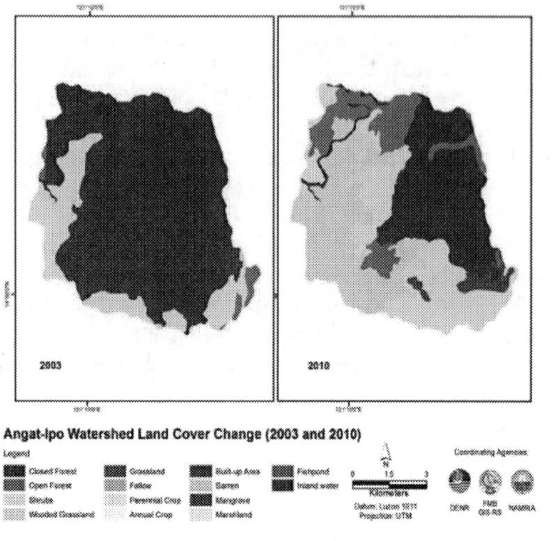

Fig. 3. Land cover change of Angat-Ipo watershed, 2003-2010

As processed, the eastern side of the watershed has been kept solid into a closed forest, but most of the western part of the watershed has been turned into a combination of shrubs and wooden grasslands. This shows an entirely different variation in the perspective of land cover for the watershed.

Each type of vegetation corresponds a certain curve number to indicate flow infiltration and surface runoff. Table II shows the corresponding curve number for each land cover type.

Table 1. Criteria for Curve number.

Land Cover type	Curve number
Closed forest	77
Shrubs	78
Open forest	79
Wooden grassland	82
Grassland	84
Forage	89

3.2 Modelling and Simulation

A hydrologic modelling software was used for the simulation of the model for the different years. The Figure 4 shows the Ipo watershed and its sub basins processed in the software. The different parameters, such as Curve number, were used as input to be analysed and processed in the software to create an overall hydrologic simulation and water balance in the watershed. The hydrographs produced were assessed to study the area's forecast flow, water availability and impacts of sudden changes in the environment.

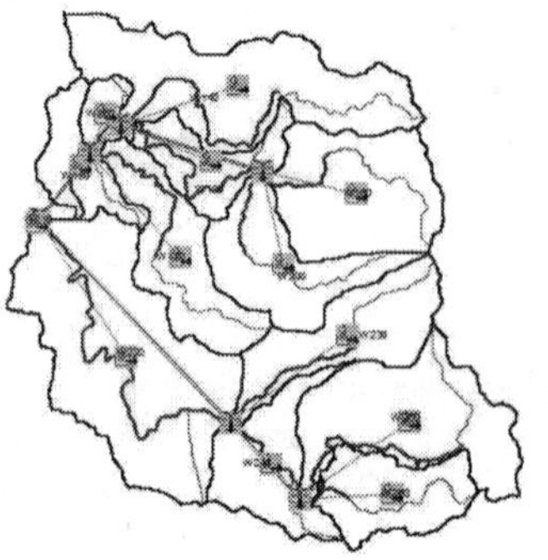

Fig. 4. Ipo watershed and sub basins

After getting the unit hydrograph, a convolution of the unit hydrographs with the precipitation ecess was computed and the runoff for a specific time frame was determined. These hydrographs, once processed into convolution, produced total volume runoffs, and so the flow of the water in the reservoir was determined, predicting how much water is present in the system. Doing the same process for the different year, the overall water balance in the system with respect to the changing years and varying vegetation cover in the area was compared and analysed.

Figure 5 shows the hydrographs for years 2003, 2010 and 2014. With varying inputs of curve number depending on the land cover and with varying precipitation gauge record, the peak outflows computed are 391.1, 400.2 and 779.5 cubic meters for 2003, 2010 and 2014, respectively.

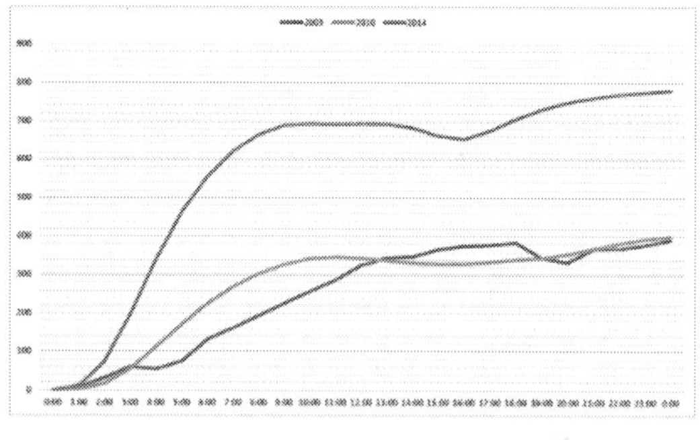

Fig. 5. Time Series Outflow of Ipo Dam

4 Conclusion

This study was conducted to determine the impacts of the varying land cover of the Ipo watershed. It is imminent that with less vegetation in the watershed, rate of infiltration increases. Less forestry and vegetation is a product of deforestation in the area which has been a major problem since the early 2000. With the drastic decrease in number of vegetation, few trees and plants absorb water from the ground, causing the rainwater to directly seep through the soil and find ways to the lower elevations. Thus, the extreme increase in water discharge occurs at nearly all streams and rivers in the watershed including the flow rate measured from the outlet point of the watershed where the Ipo dam is currently situated.
The peak flow in 2014 as determined by the study is almost twice of the peak flow computed in the years 2003 and 2010. With this rapidly decreasing land cover in the watershed and the increase in the peak charges, the water balance of the area will be affected as saturation and shape of the land terrain changes. As per the quantity of water supply, the less rate of infiltration provide an increase in water elevation in the Ipo dam, neglecting the water discharge released by the Angat dam.

References

1. Batra, Yang, Choi, Kumar, Cai, De Fraiture, Understanding Hydrological Cycle Dynamics due to Changing Land Use and Land Cover: A Congo Basin Case Study; University of Illinois, 2008.
2. B.J. Bond, F.C. Meinzer and J.R. Brooks, How Trees Influence the Hydrological Cycle in Forest Ecosystems, John Wiley and Sons, Ltd., 2008.
3. C. D'Almeida, C.J. Vorosmarty, J.A. Marengo, G. Hurtt, S.L. Dingman and B.D. Keim, *A Watershed Balance Model to Study the Hydrological Response to Different Scenarios of Deforestation in Amazonia*, May 2006.
4. J.M. Roberts, The Role of Forests in the Hydrological Cycle Volume 3, Encyclopedia of Life Support Systems.
5. J. Nemec, Engineering Hydrology, Tata McGraw-Hill Publishing Ltd., 1964.
6. K.X. Soulis and J.D. Valiantzas; SCS-CN Parameter Determination using Rainfall-Runoff Data in Heterogeneous Watersheds – the Two-CN Approach; Agricultural University of Athens, Department of Natural Resources Management and Agricultural Engineering, Division of Water Resources Management, Athens, Greece, Oct. 5, 2011.
7. Landcover and Digital Elevation Model of the Ipo Watershed; Photogrammetry Division, Mapping and Geodesy Branch, National Mapping and Resource Information Authority, Fort Bonifacio, Taguig City, Philippines.
8. R.C. Pati, Land Cover Changes and Flood Dynamics in the River Basin of Tanay, Rizal, Philippines, July 15, 2014.
9. Terrain Maps of the Ipo Wateshed, Department of Environment and Natural Resources, Forest Management Bureau, Visayas Ave., Diliman, Quezon City, 1100 Philippines.

Application of Grounded Theory in Determining Required Elements for IPv6 Risk Assessment Equation

Athirah Rosli[1*], *Abidah* Mat Taib[1**], *Wan Nor Ashiqin* Wan Ali[2] and *Ros Syamsul* Hamid[1]

[1]Faculty of Computer and Mathematical Sciences, University Teknologi MARA, 02600 Arau, Perlis.
[2]School of Human Development and Telecommunication (iKOM), Green Advance Computing & Technology (GREAT), Center of Excellence Geopolymer & Green Technology (CEGeoGTech), Universiti Malaysia Perlis, Malaysia.

Abstract. The deployment of Internet Protocol version 6 (IPv6) has raised security concerns among the network administrators. Thus, in strengthening the network security, administrator requires an appropriate method to assess the possible risks that occur in their networks. Aware of the needs to calculate risk in IPv6 network, it is essential to an organization to have an equation that is flexible and consider the requirements of the network. However, the existing risk assessment equations do not consider the requirement of the network. Therefore, this paper presents the adaptation of grounded theory to search for elements that are needed to develop IPv6 risk assessment (IRA6) equation. The attack scenarios' experiments; UDP Flooding, TCP Flooding and Multicast attacks were carried out in different network environment to show how the IPv6 risk assessment equation being used. The result shows that the IRA6 equation is more flexible to be used regardless the network sizes and easier to calculate the risk value compared to the existing risk assessment equations. Hence, network administrators can have a proper decision making and strategic planning for a robust network security.

1 Introduction

Internet dependencies among users and the enthusiasm of online application and social network could result in some of them overlook the security of their network. This may increase risks in the network. Thus, a proper method is required to strengthen the network security. Risk can increase when users are connected to the Internet with uncertain security measure. Moreover, limited control and access by the network administrator over the Internet is also one of the factors of risk increment [1]. With the deployment of IPv6, risk in the network is expected to increase because it is not only inherent IPv4 weaknesses, but it also encounter new vulnerabilities due to the presence of IPv6.

By identifying the risk that could happen, organizations can mitigate the risk from the root. One of the methods to assess risk is by using risk assessment equation. However, most of the existing equations do not take into account security risks for IPv6 alongside with the threats and vulnerabilities of IPv6 deployment. This will bring vulnerability to the network system because attackers can take advantage of the vulnerability to threaten networks' assets. Additionally, the existing risk assessment equations are not specifically focused on evaluating risk for IPv6. The existing risk assessment equations are too subjective because they have been used in many types of field such as in military field and medical field [2].

Dependencies between asset, threat, vulnerability, and security requirement, risk and vulnerabilities inside the organization is essential. However, the existing risk assessment equation unable to relate the dependencies and cause the network to be exposed to attackers [3]. Security goal that emphasis on confidentiality, integrity and availability is crucial as it acts as guidance for the network to achieve the security objectives [4].

The existing risk assessment equation considers assets, threat and vulnerability as the key elements in finding risk value. In existing equation from [5], it stated that asset valuation is vital as organization places their assets on the network. Damage or compromise of the assets may cause loss to the organization and market shares as well as customer confidence towards the organization. However, the value of the asset can be varied from one to another because it comprises different information and different format for every asset [6].

Since network in the organization is evolving through time, the needs of risk assessment that emphasize the aims and objectives of the network is essential in order to calculate risk value. The existing risk assessment equations calculate risk value per asset rather than calculate the risk value for the network entirely. When the equation calculates risk per asset, it

Corresponding author: athirah.rosli@ymail.com*; abidah@perlis.uitm.edu.my** needs to list down all the assets in the organization. This is time consuming and it looks like a process in asset audit rather than risk assessment activity. Thus, this will make the risk value that obtained from the existing equations insufficient because it did not fulfil the security objectives of the network in the organization. Therefore, this paper presents the use of grounded theory method that is capable to identify the element that is needed to the IRA6 equation that can calculate the risk value of the network.

The arrangement of the paper is as follows. Firstly, it presents the current risk assessment equation. Then, it explains the use of grounded theory method to identify the elements that are needed to calculate risk value. After that, it discusses on the development of the IRA6 equation. Subsequently, it presents the result and analysis from the output of the IRA6 equation and discussion from the result. Lastly, it concludes and suggests the future work in the last section.

2 Current Risk Assessment Equation

To determine risk of an attack that happen in the enterprise network, risk value, R is used in risk assessment equation. There is plenty of risk assessment equation exist. Each of the equation has different kinds of functions that may not be compatible with certain conditions. For example, [5] have introduced the following equation.

$$R = (\sum t * v) * a \qquad (1)$$

where,

R = risk value,
t = threat value; $1 \leq t \leq 6$,
v = vulnerability value; $1 \leq v \leq 6$,
a = asset value; $1 \leq a \leq 6$,
$R \geq 0$.

Equation (1) shows that sum of multiplication of threat value and vulnerability value is multiplied with asset value. Asset value in Equation (1) means loss and compromise of assets that happen in the network environment. Unfortunately, this equation encounters issues because there is no standardized technique of assessing the asset's worth [5]. Assessing asset also caused the asset value not to be fully evaluated if there is an issue of missing asset and it will consume time if there is an incidental event occur. While [7] calculate risk value, R by using the following equation:

$$R = a * t * v \qquad (2)$$

where,

R = risk value,
a = asset value; $1 \leq a \leq 5$,
t = threat value; $1 \leq t \leq 6$,
v = vulnerability value; $1 \leq v \leq 6$
$R \geq 0$.

Risk value in Equation (2) is calculated by multiplication of asset value, threat value and vulnerability value. As can be seen in both equations, they emphasis on asset value. Asset value can vary from one to another and it may comprise different information and a different format for every asset [6]. Moreover, both equations do not consider the nature and characteristics of IT software aspects, security requirements and availability of solutions to respond to risks [2]. To develop the IRA6 equation, grounded theory method is used to identify elements that is needed in risk assessment equation.

3 Grounded Theory Method

Grounded theory method has the capability to analyze the gathered data systematically [8]. Furthermore, grounded theory is flexible to be used in any types of data. It also allows researchers to emphasis on theory development as this method is flexible to construct various types of theory [8].

Table 1 shows the gathered data that supports the method to identify the elements need to calculate risk value.

Table 1. Grounded theory key points and codes from literature review.

Id	Key Point	Code
P1	"There is undeniably uncertainty associated with valuation. Often that uncertainty comes from the asset being valued, though the valuation model may add to that uncertainty." [6]	Asset valuation uncertainty
P2	"Analysts also assessed the impacts that could result from a security failure and the possible losses of confidentiality, integrity and availability on information systems, in relation with these seventeen security control families. The risk assessment criteria adapted by the proposed procedure can be replaced with other security criteria depending on varying practice needs of each organization."[9]	Other security criteria needed.

P3	"Identifying security goals is a process of iteration and refinement, leveraging the knowledge and expertise of the analyst to secure not only the core functionality but the security mechanisms as well. Moreover, a comprehensive security plan should include goals for not only preventing a breach, but also for detecting and appropriately responding in case a breach does occur."[10]	Importance of security goal
P4	"If documentation does not exist or such assessments for the organization's IT assets have not been performed, the system and data sensitivity can be determined based on the level of protection required to maintain the system and data's availability, integrity, and confidentiality." [11]	Importance of security goal
P5	"By incorporating security requirements into the design phase of the development life cycle, prevention or protection is easier to implement because it is inherent in the system's or application's design up front." [12]	Importance of security requirement
P6	"This article begins with a discussion of the tradeoffs among seemingly different system security properties such as confidentiality, integrity, and availability. We develop a quantitative framework for understanding these tradeoffs and the issues that arise when those security properties are all in play within an organization." [13]	Importance of security goal
P7	"It views the protection of the integrity, confidentiality, and availability of its information assets and networks as a strategic objective. Although the organization has been performing information security risk assessments for 15 years, the information risk management program has become more robust and formalized in recent years. The fundamental basis of the risk assessment program is to balance the company's security requirements with other factors associated with doing business." [14]	Importance of security goal

As can be seen in Table 1, 'P' indicates the 'key point'. P3 to P7 states the importance of security goal in risk assessment. In P3, it describes that, security goal can handle knowledge and expertise of the analyst. Hence, by having security goal that emphasizes the aims and objectives of the network, organizations can control the security needs as well as the business purpose of the organization.

4 Development of IRA6 Equation

Aims and objectives of the organization relate to security goal of the network. Therefore, base score value is adequate element to be included in calculating risk value because it considers the security goal of the network. Base score has been introduced by Common Vulnerability Scoring System (CVSS) as a metric that represents basic qualities of vulnerability [15]. Base score also has considered the elements of security goal which are confidentiality, availability and integrity. Meanwhile, asset value is inclusive indirectly in the equation as it is part of the vulnerability value [16] because it inherent weaknesses that can be manipulated by the attacker onto the organization's network. The proposed equation is as follows:

$$R = b * t * v \quad (3)$$

where,

R = risk value,

b = base score value; $1 \leq b \leq 10$,

t = threat value; $1 \leq t \leq 6$,

v = vulnerability value; $1 \leq v \leq 6$,

$R \geq 0$

Based on Equation (3), the multiplication of elements represents attributes for the organization's network. Therefore, risk value only becomes zero if base score is equal to zero, or either threat or vulnerability does not exist which means, there is no risk or attack to the network [17]. Base score is added as part of the elements in calculating risk value because it has considered confidentiality, integrity and availability which are security goal for a network.

5 Result and Analysis

To show how the IRA6 equation works, three types of attack have been conducted. The attacks were UDP flooding attack, TCP flooding attack and multicast attack. These three types of attacks are chosen because they are the most common attacks that occur in IPv6 network [18]–[21].

Figure 1 shows that the three attacks were simulated in different network sizes in OMNeT++ simulator to show the capabilities of the equation to be used in different network regardless the size of the network. OMNeT++ is an open source software and it was chosen due to its capabilities of simulating IPv6 network [22]. OMNeT++ also contain protocols especially for IPv6, UDP, TCP and Ethernet [23].

In UDP and TCP flooding attack, attacker sent tremendous amount of UDP and TCP packets to the destination. While for multicast attack, tremendous amount of packets were sent to the victim. At the victim, the packets were captured and analyzed to identify the risk that occur in the network.

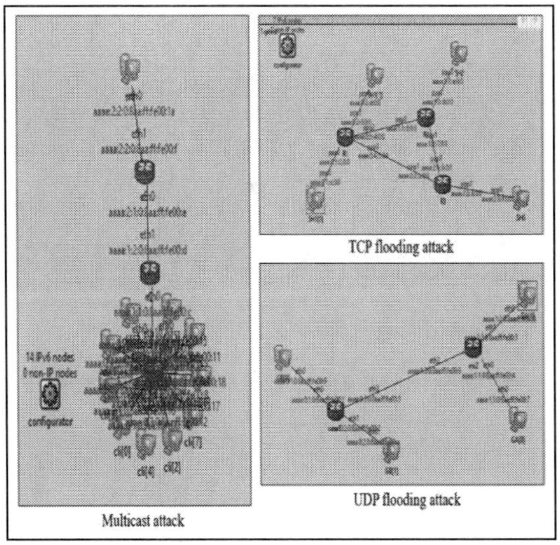

Fig. 1. Attack topologies.

For the scenarios of the attacks, it can be seen in Table 2.

Table 2. Attack scenarios.

Attack Type	Description
UDP flooding attack	Number of host : 4 nodes Parameters: Packet size Packet size used (bps) : 200, 500, 700, 1000, 1500, 2000, 4000, 5000, 9000, 12000, 30000, 50000, 60000.
TCP flooding attack	Number of host : 4 nodes Parameters: Packet size Packet size used (bps): 200, 500, 700, 1000, 1500, 2000, 2500, 3000, 3500, 4000, 4500, 5000.
Multicast attack	Parameters: Number of nodes. Node used: 1, 10, 20, 30, 40, 50

To calculate the risk value, base score values were retrieved from CVSS scoring system. Threat value and vulnerability value were determined based on the attacks information and current condition of the network. Threat value determines the likelihood of the attack while the vulnerability value is based on access authorization to the asset in the network [24].

After identifying the threat value and the vulnerability value, by using the IRA6 equation, the risk value for the three attacks are calculated and presented in Table 3.

Table 3. Base score of IPv6 attacks from experimentation.

	Base Score	Threat Value	Vulnerability Value	Risk Value
UDP Flooding	7.1	4	3	85.2
TCP Flooding	7.8	3	3	70.2
Multicast	4.96	5	3	74.4

The threat value and vulnerability value are sorted according to the range score stated by [15]. As shown in Table III, UDP flooding attack gives the highest value which is 85.2. UDP flooding attack can be exploited both from internal and external network. Some information regarding the network need to be gathered before UDP attack being launched. UDP flooding also can make the network to be unavailable completely. This attack also occurs sometimes and affects multiple assets. It means that the availability of this attack is high and organization should aware and get ready for the attack to occur. This is why the risk value of UDP flooding attack is the highest.

For TCP flooding attack, it also can exploit both local and from the network, which also identified as remotely exploitable. To generate TCP flooding attack, it involves limited skill and requires some information concerning the victim's network. However, this type of attack can cause the network to be unavailable completely, similar to the UDP flooding attack. Moreover, TCP flooding attack is calculated as the lowest risk value due to the likelihood of the attack that happens occasionally and it affects multiple assets. Organization needs to be prepared to face these types of attack.

Multicast attack states risk value of 74.4. Similar to UDP and TCP flooding attack, multicast attack also can be remotely exploitable. To generate multicast attack, an attacker needs to search for additional information concerning the network and they need to generate attack manually. This attack affects multiple assets in the organization. The confidentiality in base score value of this attack is partial which means access to some system files is possible due to the likelihood of the attack that happen frequently. The organization does not have control over entire files in the system. Hence, appropriate actions need to be taken to safeguard the confidential files that are potentially being accessed by the unauthorized party.

6 Discussion

The existing risk assessment equations consist of threat value, vulnerability value and asset value compared to the IRA6 equation. In IRA6 equation, it consists of threat value, vulnerability value and base score value. Threat value of both equations emphasize the likelihood that the attack can happen within a period of time. For vulnerability value, both equations emphasize the accessibility of the asset and asset security.

However, the difference falls on the usage of asset value and base score value. In the existing equations, asset value emphasizes the value of the asset whereas base score value in IRA6 equation focuses on the security goal of the network. In the existing equations, asset value can be determined by using two approaches; qualitative and quantitative assessments. For these assessments, the downside is they need experts, point of view for several decisions regarding asset valuation. Judgment and professional standing might vary. Assets which are not in physical form such as data and information can be difficult to evaluate [25]. It is possible, but it can consume more time and it needs expert judgment along with professional assessment that might be insignificant.

The vulnerability of the asset increase when it is exposed to exploitation and attackers can gain access to it [26]. Most of the physical assets in the organization are connected to the Internet and this will expose them to external and internal traffic that might comprise contaminated packets.

Additionally, every asset has different types of risks and although they are the same types of assets, some particular assets are more efficient in managing certain risks than others [27]. Some of the existing equations valued asset based on the currency. This can cause overvalue or undervalue because it is influenced by the market price. Thus, this situation can affect asset valuation for the organization's network [28].

7 Conclusion

The adaptation of grounded theory method has highlighted that security goal is the important elements to calculate risk value. However, the existing risk assessment equations emphasis on asset value. Compared to the existing equations the IPv6 risk assessment equation takes into account the security goal of the organization due to the adaptation of base score value. The IPv6 risk assessment equation has considered asset value as an element to calculate risk value as it is counted as part of vulnerability. This statement also supported by [16] that stated that when there is asset on a network, the network is exposed to vulnerability. Attackers can manipulate the asset to take over the network.

The IPv6 risk assessment equation also considers security goal as parts of the elements to calculate the risk value. In risk management strategies, one of the steps is risk assessment. Hence, risk assessment must take into account security goal and objectives of the organization's network [29]. Security goal is embedded in base score value which is element to calculate risk value. By identifying the security goal, network administrators can govern organizational needs and requirements for their network.

This paper also has shown how the IPv6 risk assessment equation being used to calculate three types of attacks; UDP flooding attack, TCP flooding attack and multicast attack. The attacks were ranked based on the risk value of the result obtained. Thus, an organization can determine which attacks that they need to prioritize and apply appropriate mitigation strategies. Hereafter, more IPv6 based protocols will be tested to measure the capability of IPv6 risk assessment equation.

Acknowledgements

This research work was supported by Ministry of Education (MOE), Malaysia and Universiti Teknologi MARA, Malaysia under Research Acculturation Grant Scheme (RAGS), Project code: "600-RMI/RAGS 5/3 (9/2014)" and Universiti Malaysia Perlis (UniMAP), Malaysia under Research Short Term Grant (STG), Project code : 9001-00527.

References

[1] S. Roy, C. Ellis, S. Shiva, D. Dasgupta, V. Shandilya, and Q. Wu, "A survey of game theory as applied to network security," *2010 43rd Hawaii Int. Conf. Syst. Sci.*, pp. 1–10, 2010.

[2] J. Lee, "An Enhanced Risk Formula for Software Security Vulnerabilities," *Inf. Syst. Audit Control Assoc.*, vol. Volume 4, 2015.

[3] H. Dawood, "IPv6 Security Vulnerabilities," *Int. J. Inf. Secur. Sci.*, vol. 1, no. 4, pp. 100–105, 2012.

[4] Z. Adelsberger, G. Grubor, and I. Nađ, "Methodological approach to risk assessment in building security," *Coll. Antropol.*, vol. 38, no. 1, pp. 215–227, 2014.

[5] M. Schumacher, E. Fernandez-Buglioni, D. Hybertson, F. Buschmann, and P. Sommerlad, *Security Patterns: Integrating Security and Systems Engineering*, 1st Edition. West Sussex, England: John Wiley & Sons Inc., 2006.

[6] A. Damodaran, *Investment valuation: Tools and techniques for determining the value of any asset*. John Wiley & Sons, 2012.

[7] S. Tanimoto, R. Sato, K. Kato, M. Iwashita, Y. Seki, H. Sato, and A. Kanai, "A Study of Risk Assessment Quantification in Cloud Computing," in *International Conference on Network-Based Information Systems (NBiS)*, 2014, pp. 426–431.

[8] A. Strauss and J. Corbin, "Grounded theory methodology," *Handb. Qual. Res.*, vol. 17, pp. 273–285, 1994.

[9] C.-C. Lo and W.-J. Chen, "A hybrid information security risk assessment procedure considering interdependences between controls," *Expert Syst. Appl.*, vol. 39, no. 1, pp. 247–257, 2012.

[10] M. Riaz, J. Stallings, M. P. Singh, J. Slankas, and L. Williams, "DIGS: A Framework for Discovering Goals for Security Requirements Engineering," in *Proceedings of the 10th ACM/IEEE International Symposium on Empirical Software Engineering and Measurement*, 2016, p. 35.

[11] S. P. NIST, "800-30," *Risk Manag. Guid. Inf. Technol. Syst.*, pp. 800–830, 2002.

[12] M. Gregg and D. Kim, *Inside network security assessment: Guarding your IT infrastructure*, 1st ed. Sams, 2006.

[13] J. Hughes and G. Cybenko, "Quantitative metrics and risk assessment: The three tenets

model of cybersecurity," *Technology Innovation Management Review*, 2013. [Online]. Available: http://timreview.ca/article/712.

[14] A. and I. M. D. United States General Accounting Office, "Information security risk assessment: Practices of leading organizations," 1999.

[15] A. Rosli, A. M. Taib, and W. N. A. W. Ali, "Utilizing the Enhanced Risk Assessment Equation to Determine the Apparent Risk due to User Datagram Protocol (UDP) Flooding Attack," *Sains Humanika*, vol. 9, no. 1–4, 2017.

[16] J. Bhattacharjee, A. Sengupta, and C. Mazumdar, "A formal methodology for enterprise information security risk assessment," in *2013 International Conference on Risks and Security of Internet and Systems (CRiSIS)*, 2013, pp. 1–9.

[17] N. Vavoulas and C. Xenakis, "A quantitative risk analysis approach for deliberate threats," in *International Workshop on Critical Information Infrastructures Security*, 2010, pp. 13–25.

[18] D. Zagar and K. Grgic, "IPv6 security threats and possible solutions," *2006 World Autom. Congr.*, pp. 1–7, 2006.

[19] W. M. Eddy, "TCP SYN flooding attacks and common mitigations," 2007.

[20] X. Yang, T. Ma, and Y. Shi, "Typical dos/ddos threats under ipv6," *Int. Multi-Conference Comput. Glob. Inf. Technol.*, 2007.

[21] C. Caicedo, J. Joshi, and S. Tuladhar, "IPv6 security challenges," *Computer (Long. Beach. Calif).*, 2009.

[22] V. Veselý, P. Matoušek, and M. Švéda, "Multicast simulation and modeling in OMNeT++," in *Proceedings of the 6th International ICST Conference on Simulation Tools and Techniques*, 2013, pp. 142–145.

[23] A. Varga, "OMNeT++," in *Modeling and Tools for Network Simulation*, Springer, 2010, pp. 35–59.

[24] A. Rosli, A. M. Taib, H. Baharin, and W. N. A. Wan, "Enhanced risk assessment equation for IPv6 deployment," *5th Int. Conf. Computer and Informatics (ICOCI 2015)*, 2015.

[25] A. Munteanu, "Information security risk assessment: The qualitative versus quantitative dilemma," in *Managing Information in the Digital Economy: Issues & Solutions-Proceedings of the 6th International Business Information Management Association (IBIMA) Conference*, 2006, pp. 227–232.

[26] V. Visintine, "An introduction to information risk assessment," *SANS Inst.*, vol. 8, 2003.

[27] J. Alwang, P. B. Siegel, and S. L. Jorgensen, "Vulnerability: a view from different disciplines," 2001. [Online]. Available: http://siteresources.worldbank.org/SOCIALPROTECTION/Resources/SP-Discussion-papers/Social-Risk-Management-DP/0115.pdf. [Accessed: 09-Mar-2016].

[28] J. M. Murphy, "Asset value calculators." Google Patents, 20-Jan-1976.

[29] R. Ross, S. Katzke, A. Johnson, M. Swanson, and G. Stoneburner, "NIST SP800-39, managing risk from information systems an organizational perspective." Gaithersberg, MD: NIST, http://csrc. nist. gov/publications/drafts/800-39/SP800-39-spd-sz. pdf, 2008.

A Comparative Study of Feature Selection Techniques for Bat Algorithm in Various Applications

Rozlini, Mohamed[1,*], *Munirah* Mohd Yusof[1], and *Noorhaniza* Wahidi[2]

[1]Department of Software Engineering, Universiti Tun Hussein Onn Malaysia (UTHM), 86400 Batu Pahat, Johor, Malaysia
[2]Department of Multimedia, Universiti Tun Hussein Onn Malaysia (UTHM), 86400 Batu Pahat, Johor, Malaysia

Abstract. Feature selection is a process to select the best feature among huge number of features in dataset, However, the problem in feature selection is to select a subset that give the better performs under some classifier. In producing better classification result, feature selection been applied in many of the classification works as part of preprocessing step; where only a subset of feature been used rather than the whole features from a particular dataset. This procedure not only can reduce the irrelevant features but in some cases able to increase classification performance due to finite sample size. In this study, Chi-Square (CH), Information Gain (IG) and Bat Algorithm (BA) are used to obtain the subset features on fourteen well-known dataset from various applications. To measure the performance of these selected features three benchmark classifier are used; k-Nearest Neighbor (kNN), Naïve Bayes (NB) and Decision Tree (DT). This paper then analyzes the performance of all classifiers with feature selection in term of accuracy, sensitivity, F-Measure and ROC. The objective of these study is to analyse the outperform feature selection techniques among conventional and heuristic techniques in various applications.

1 Introduction

Feature selection is a technique that has an ability to decrease the number of attribute by eliminates the least significant features [1]. However, the problem in feature selection in finding the optimum features. Most of the features in datasets that did not contribute to end result are unknown. Some unimportant or irrelevant features need to be diminished in order to reduce the classification complexity and time processing [1]. Feature selection becomes the important process in order to improve the classification performance. However, not all the feature selection technique reduces the same feature in dataset. For that reason, chosen the feature selection techniques is crucial where subset feature is needed for dimensionality reduction and gives better performance in classification.

Previously, some researcher likely to employ the Chi-Square (CH), Attribute evaluation (AE), Information Gain (IG), Correlation-based Attribute Evaluation (CB) and Symmetrical Uncertainty Attribute evaluation (SU) [2] that knows as conventional techniques. Today, several metaheuristic optimization algorithms seems become the potential technique to become feature selection techniques. The well-known metaheuristic optimization algorithms that widely such as Firefly Algorithm (FA), Genetic Algorithm (GA), Simulated Annealing (SA), Cuckoo Search (CS), Artificial Bee Colony (ABC), Bat Algorithm (BA) and Particle Swarm Optimization (PSO)[3]. Due to the strength of the combination of several metaheuristic algorithms, BA has become more powerful than PSO, GA and Harmony Search (HS) [4]. Thus, BA is seen as one of the possible solutions to resolve problem in data mining problems such as feature selection and classification.

The aim of this study is to find out the outperform feature selection technique by taking into consideration the capability of BA as a feature selection. The main contribution of this study are 1) run the experiment in 14 dataset from various application order to find out the outperform feature selection techniques between CH, IG and BA., 2) analyse the outperform feature selection techniques among conventional and heuristic techniques where kNN, NB and DT were applied in these experiments to evaluate the performance of selected features by using four performance measures accuracy, sensitivity, F-Measure and ROC area. The rest of this paper is organized to provide a brief explanation of BA in Bat Algorithm section. The following section will discuss about Related Work. While, in the next section will present the Methodology, Experimental Result and conclude the finding by Discussion and Conclusion in the last section.

2 Feature selection

Feature selection has been an active and fruitful field of research area in pattern recognition, machine learning, statistics and data mining communities [5]. It is a dimensionally reduction technique that main goal is to reduce irrelevant data and finding a features that increase classification accuracy. The main objective of feature

* Corresponding author: rozlini@uthm.edu.my

selection is to choose a subset of input variables by eliminating features, which are irrelevant or of no predictive information. It has been proven in both theory and practice to be effective in enhancing learning efficiency, increasing predictive accuracy and reducing complexity of learned results [6,7].

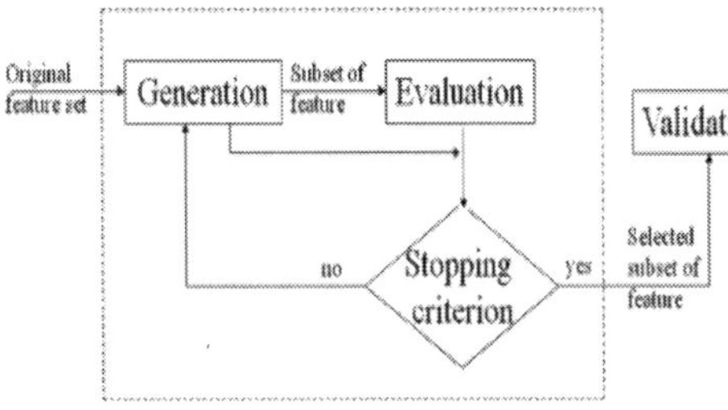

Fig. 1: Feature selection process [8].

There are four basic steps in a typical feature selection process as shown in Fig. 1[8].
The process of feature selection is as below;

- The generation procedure to generate the next candidate subset from original feature set
- The evaluation function to evaluate the subset to determine the relevancy towards the classification task using measure for instances distance, dependency, information and consistency
- Stopping criteria to decide when to stop. This is where it determine the relevant subset or optimal feature subset
- Validation procedure is to check whether the selected feature subset is valid

. There exists a several feature selection method that used by researcher. Some researcher trend to employ conventional method such as information gain and chi-square for instance [3,9,10]. In the other research, heuristic method such as genetic algorithm [11] ACO [12] and [13] in memetic feature selection, noisy data, spam email, binary variables; respectively.

Feature selection also involve as active field of research such as in pattern recognition, machine learning and data mining area [14,15]. Feature selection objective is to reduce irrelevant data and finding the most relevant features that would increase classification accuracy. It has been proven in both theory and practice to be effective in enhancing learning efficiency, increasing predictive accuracy and reducing complexity of learned results [16].

A wrapper feature selection approach based on BA and Optimum Path Forest had been proposed by Nakamura [17]. This approach modeled a problem of feature selection as a binary based optimization technique. Six datasets been used in experiments that demonstrated that the proposed approach provides statistically significant more compact sets and in some cases it indeed improves the classification effectiveness.

Binary Bat Algorithm (BBA) was one of the inspired binary version feature selection that proposed to find the most significant feature in a search space [18]. BBA was proposed to associate each bat a set of binary coordinates that indicate whether that feature belongs to the final set of features or not. It combined the power of bat algorithm and Optimum Path Forest in finding the set of features that maximizes the accuracy of validating sets. It been proved that the proposed techniques can outperformed other well known techniques such as PSO, FFA and GSA.

From [19] proposed bio-inspired method called Bat Algorithm hybridized with a Naive Bayes classifier (BANB). Twelve benchmarks datasets from different domains been used in experiments to compared their performance measures with three well known feature selection techniques; GA, PSO and GPSO in term of the number of selected features from the original datasets. It shows that BANB significantly outperformed other algorithms in selecting significant number of features and lead to maintaining and improving classification accuracy.

Generally these studies found that feature selection techniques are capable to improve the performance of learning algorithms thru increasing the accuracy of the classifier by removing irrelevant attributes. Therefore with high quality features, it makes the classification process accurate, comprehensible and produces better results. For that reason this research will conduct the experiment that focuses to analyze the outperform techniques among conventional and heuristic techniques.

3 Experimental results

In this research, the experiment is setup to analyze the outperform feature selection techniques among conventional and heuristic techniques. The conventional techniques involve in this experiments are Chi-Square (CH) and Information Gain (IG). Meanwhile Bat algorithm (BA) as a heuristic method. This section discusses on the experimental results of 14 datasets from UCI repository in various application with selected attributes using CH, IG and BA. Three benchmark classifier; kNN, NB and DT DT were applied in these experiments to evaluate the performance of selected features by using four performance measures accuracy, sensitivity, F-Measure and ROC area.

Table 1 shows the characteristics of 14 datasets; number of attributes and instances. In general, selected features by CH and IG improve the classification performance for all classifier as shown in Table 2,3,5 until 15. From Table 2 and 3, ROC area for NB is a highest performance measures in Credit Approval and Ecoli with value 0.879 and 0.988, respectively. However the result for Hill Valley dataset is little bit different as shown in Table 4. Selected features by CH and IG only improve the classification performance for NB in all performance measures. But BA improves all the performance measures for KNN. Meanwhile, BA obtains the same result with CH and IG for DT in all

https://doi.org/10.1051/matecconf/201815006006

classification performance. The highest value is ROC area for KNN with value 0.59.

For Image Segmentation dataset, ROC area for NB, 0.946 is the highest value (see Table 5). Meanwhile, the ROC area with BA is 0.926. From Table 6, ROC area for NB, 0.943 is the highest value while ROC area by BA is 0.938. In Table 7, the highest value is ROC area for NB, 0.876. From Table 8 shows, ROC area for NB is the highest value for Plant Species dataset in CH, IG and Bat with value 0.996, 0.996 and 0.993, respectively. In Table 9, the highest value for ROC area for NB in CH, IG and Bat with value 0.965, 0.965 and 0.957, respectively. In Table 10 and 12, the highest value for ROC area in Automobile dataset is 0.806 and in Yeast dataset is 0.817.

In Table 11, the highest performance is ROC area for NB in CH, IG and Bat with value 0.71, 0.71 and 0.704, respectively. For Waveform dataset that shown in Table 13, the highest performance is ROC area for NB in CH and IG is 0.96 and BA is 0.954.From Table 14 and 15, the highest value is ROC area for NB is 0.935 and 0.878, respectively.

4 Discussion and conclusion

This study was used three feature selection techniques and 14 datasets in various applications from UCI repository. The experiment was testing using Weka and Matlab. CH and IG is consider as conventional features selection method while BA as heuristic features selection method. These features selection techniques are used to find the subset features from 14 datasets. Then, analyse the outperform feature selection techniques, it can be determine according to classification performance. There are three benchmark classifier; NB, KNN and DT with four performance measures accuracy, sensitivity, F-Measure and ROC area were used in this experiment

The results shows the outperform feature selection method is determine by ROC area. The outperform feature selection method is conventional techniques, CH and IG. However, BA has a potential to be outstanding feature selection method. The result from Table 4 show BA improves all the performance measures for KNN. And also, results from Table 5, 6, 8, 9, 11 and 13 shows the ROC area for BA is only slightly different with CH and IG.

Table 1. Datasets Characteristics

#	Datasets	Number of Instances	Number of Attributes
1.	Credit Approval	690	15
2.	Ecoli	336	8
3.	Hill_Valley	606	100
4.	Image Segmentation	210	19
5.	Libras Movement	360	90
6.	Steel Plates Faults	1941	27
7.	Plant Species	1600	64
8.	Urban Land	507	147
9.	Automobile	159	25
10.	Abalone	4177	8
11.	Yeast	1484	8
12.	Waveform	5000	21
13.	Ionosphere	351	34
14.	Water Treatment	523	38

Table 2. Results for classification performance for Credit Approval dataset

FS	Irrelevant features	Precision			Recall			Fmeasure			ROC Area		
		NB	KNN	DT	NB	KNN	DT	NB	KNN	DT	NB	KNN	DT
BA	8-11,14,15	0.628	0.609	0.632	0.596	0.61	0.632	0.59	0.61	0.632	0.68	0.604	0.625
CH	1,12	0.783	0.827	0.842	0.765	0.828	0.842	0.756	0.827	0.842	0.879	0.824	0.855
IG		0.783	0.827	0.842	0.765	0.828	0.842	0.756	0.827	0.842	0.879	0.824	0.855

Table 3. Results for classification performance for Ecoli dataset

FS	Irrelevant features	Precision			Recall			Fmeasure			ROC Area		
		NB	KNN	DT	NB	KNN	DT	NB	KNN	DT	NB	KNN	DT
Bat	1,2,4-6	0.694	0.704	0.665	0.735	0.711	0.753	0.712	0.707	0.7	0.936	0.831	0.879
CH	5	0.962	0.932	0.973	0.973	0.932	0.964	0.967	0.931	0.969	0.988	0.96	0.985
IG		0.962	0.932	0.973	0.973	0.932	0.964	0.967	0.931	0.969	0.988	0.96	0.985

Table 4. Results for classification performance for Hill Valley dataset

FS	Irrelevant features	Precision			Recall			Fmeasure			ROC Area		
		NB	KNN	DT	NB	KNN	DT	NB	KNN	DT	NB	KNN	DT
Bat	2,8,20,21,23,27,28, 30-52, 55,57, 59,62,64,66,68,69,73-75 ,77-94,96,98-100	0.515	0.589	0.253	0.505	0.589	0.503	0.432	0.589	0.337	0.491	0.59	0.494
CH	1	0.519	0.584	0.253	0.507	0.584	0.503	0.434	0.584	0.337	0.492	0.585	0.494
IG		0.519	0.584	0.253	0.507	0.584	0.503	0.434	0.584	0.337	0.492	0.585	0.494

Table 5. Results for classification performance for Image Segmentation dataset

FS	Irrelevant features	Precision			Recall			Fmeasure			ROC Area		
		NB	KNN	DT	NB	KNN	DT	NB	KNN	DT	NB	KNN	DT
Bat	1-3,5-9,14,16,19	0.619	0.771	0.781	0.619	0.771	0.776	0.587	0.769	0.777	0.926	0.87	0.902
CH	1,3,4,5	0.781	0.894	0.89	0.786	0.895	0.89	0.774	0.894	0.889	0.946	0.939	0.939
IG		0.781	0.894	0.89	0.786	0.895	0.89	0.774	0.894	0.889	0.946	0.939	0.939

Table 6. Results for classification performance for Libras Movement dataset

FS	Irrelevant features	Precision			Recall			Fmeasure			ROC Area		
		NB	KNN	DT	NB	KNN	DT	NB	KNN	DT	NB	KNN	DT
Bat	1,3,5,7-9,11, 30,41,43,45 ,47-49,51,53-55, 57, 59,61,63,65,67,69,71,73,75,77,79,81,83,85,87 ,89	0.626	0.816	0.701	0.606	0.811	0.694	0.609	0.809	0.69	0.938	0.894	0.867
CH	43,47,51,55,59,57,61,63,65,67,69,71,73,75,77 ,79	0.647	0.865	0.709	0.631	0.861	0.706	0.635	0.859	0.702	0.943	0.921	0.866
IG		0.647	0.865	0.709	0.631	0.861	0.706	0.635	0.859	0.702	0.943	0.921	0.866

Table 7. Results for classification performance for Steel Plates Faults dataset

FS	Irrelevant features	Precision			Recall			Fmeasure			ROC Area		
		NB	KNN	DT	NB	KNN	DT	NB	KNN	DT	NB	KNN	DT
Bat	1-5,8,11,12,15,	0.616	0.642	0.71	0.536	0.641	0.712	0.529	0.641	0.71	0.844	0.767	0.84
CH	10,21,16	0.663	0.707	0.749	0.6	0.708	0.75	0.582	0.707	0.749	0.876	0.809	0.859
IG	21,16	0.667	0.713	0.759	0.604	0.714	0.76	0.589	0.713	0.759	0.876	0.813	0.864

Table 8. Results for classification performance for Plant Species dataset

FS	Irrelevant features	Precision			Recall			Fmeasure			ROC Area		
		NB	KNN	DT	NB	KNN	DT	NB	KNN	DT	NB	KNN	DT
Bat	2,3,5,6,8-10,14,16-19,21-26,30,41-45,48-51,55,57,63,64	0.785	0.644	0.412	0.774	0.639	0.398	0.776	0.636	0.4	0.993	0.818	0.728
CH	16,8,61,60,34	0.849	0.753	0.501	0.842	0.739	0.493	0.843	0.74	0.491	0.996	0.868	0.766
IG		0.849	0.753	0.501	0.842	0.739	0.493	0.843	0.74	0.491	0.996	0.868	0.766

Table 9. Results for classification performance for Urban Land dataset

FS	Irrelevant features	Precision			Recall			Fmeasure			ROC Area		
		NB	KNN	DT	NB	KNN	DT	NB	KNN	DT	NB	KNN	DT
Bat	2,3,18-21,36-42,44,61,62,65,81-84,103,105,107,124,128,145,	0.761	0.751	0.77	0.75	0.738	0.768	0.752	0.738	0.767	0.957	0.845	0.884
CH	13,16-18,34,36,38,39,55,57,59,60,76,78,80,81,87,97-99,101,102,105,118-123,126,139-144,147	0.83	0.816	0.812	0.821	0.804	0.804	0.824	0.806	0.805	0.965	0.88	0.905
IG		0.83	0.816	0.812	0.821	0.804	0.804	0.824	0.806	0.805	0.965	0.88	0.905

 https://doi.org/10.1051/matecconf/201815006006

Table 10. Results for classification performance for Automobile dataset

FS	Irrelevant features	Precision			Recall			Fmeasure			ROC Area		
		NB	KNN	DT	NB	KNN	DT	NB	KNN	DT	NB	KNN	DT
Bat	1,9-13,16,21-25	0.486	0.717	0.641	0.453	0.698	0.648	0.437	0.706	0.64	0.765	0.814	0.858
CH	4,3,15,7,19,20,14,8	0.562	0.855	0.809	0.566	0.836	0.805	0.547	0.843	0.805	0.806	0.905	0.892
IG		0.562	0.855	0.809	0.566	0.836	0.805	0.547	0.843	0.805	0.806	0.905	0.892

Table 11. Results for classification performance for Abalone dataset

FS	Irrelevant features	Precision			Recall			Fmeasure			ROC Area		
		NB	KNN	DT	NB	KNN	DT	NB	KNN	DT	NB	KNN	DT
Bat	3,7,8	0.182	0.19	0.212	0.233	0.192	0.219	0.201	0.191	0.215	0.704	0.549	0.604
CH	1	0.197	0.199	0.2	0.241	0.201	0.206	0.209	0.2	0.202	0.71	0.551	0.585
IG		0.197	0.199	0.2	0.241	0.201	0.206	0.209	0.2	0.202	0.71	0.551	0.585

Table 12. Results for classification performance for Yeast dataset

FS	Irrelevant features	Precision			Recall			Fmeasure			ROC Area		
		NB	KNN	DT	NB	KNN	DT	NB	KNN	DT	NB	KNN	DT
Bat	5-8	0.517	0.487	0.496	0.53	0.486	0.515	0.519	0.486	0.504	0.79	0.659	0.711
CH	7	0.591	0.524	0.564	0.579	0.522	0.575	0.569	0.522	0.567	0.817	0.684	0.741
IG		0.591	0.524	0.564	0.579	0.522	0.575	0.569	0.522	0.567	0.817	0.684	0.741

Table 13. Results for classification performance for Waveform dataset

FS	Irrelevant features	Precision			Recall			Fmeasure			ROC Area		
		NB	KNN	DT	NB	KNN	DT	NB	KNN	DT	NB	KNN	DT
Bat	3,7-9,13,17	0.831	0.745	0.762	0.814	0.746	0.762	0.808	0.745	0.762	0.954	0.809	0.856
CH	1,21	0.841	0.785	0.761	0.81	0.785	0.761	0.798	0.785	0.761	0.96	0.839	0.846
IG		0.841	0.785	0.761	0.81	0.785	0.761	0.798	0.785	0.761	0.96	0.839	0.846

Table 14. Results for classification performance for Ionosphere dataset

FS	Irrelevant features	Precision			Recall			Fmeasure			ROC Area		
		NB	KNN	DT	NB	KNN	DT	NB	KNN	DT	NB	KNN	DT
Bat	3-5,8-11,13,15-17,19-21,23,26,28-31,33	0.788	0.896	0.862	0.789	0.895	0.863	0.788	0.892	0.862	0.881	0.861	0.854
CH	2	0.842	0.871	0.915	0.826	0.863	0.915	0.829	0.857	0.913	0.935	0.825	0.892
IG		0.842	0.871	0.915	0.826	0.863	0.915	0.829	0.857	0.913	0.935	0.825	0.892

Table 15. Results for classification performance for Water Treatment dataset

FS	Irrelevant features	Precision			Recall			Fmeasure			ROC Area		
		NB	KNN	DT	NB	KNN	DT	NB	KNN	DT	NB	KNN	DT
Bat	4,5,7,11-15,17-22,24-38	0.613	0.563	0.596	0.62	0.585	0.61	0.608	0.573	0.599	0.77	0.671	0.712
CH	2, 34, 36, 38	0.739	0.699	0.673	0.734	0.713	0.677	0.732	0.702	0.673	0.878	0.77	0.769
IG		0.739	0.699	0.673	0.734	0.713	0.677	0.732	0.702	0.673	0.878	0.77	0.769

Acknowledgments

This paper has been supported by, Short Term Grant, Universiti Tun Hussein Onn Malaysia (UTHM) (vot U539) for the financial support. This research also supported by GATES IT Solution Sdn. Bhd under its publication scheme.

References

1. R. Amirreza, N. Hossein, A hybrid feature selection approach based on ensemble method for high-dimensional data, 2nd Conference on Swarm Intelligence and Evolutionary Computation (CSIEC), 7-9 March, pp. 16-20, (2017)
2. Y. Zhang, A. Yang, C. Xiong, T. Wang, Z. Zhang, "Feature selection suing data envelopment analysis", Knowledge-Based Systems Journal, 64, pp. 70-80, April 2014, (2014)
3. Y. Sayes, I. Inza, , P. Larranaga, A review of feature selection techniques in bio-informatics, Bioinformatics vol. 23, no. 19, pp. 2507 – 2517, (2007)
4. X.S. Yang, A new metaheuristic Bat-inspired algorithm, in: J.R. Gonzalez, et al.(Eds.), Nature Inspired Cooperative Strategies for Optimization (NISCO 2010). Studies in Computational Intelligence, Springer Berlin, Springer, Berlin, August 2010, pp. 65–74, (2010)
5. J. Han, and M. Kamber, Data Mining; Concepts and Techniques. Morgan Kaufmann Publishers, 2000.
6. H. Almuallim, T. G. Dietterich, T. G. Learning boolean concepts in the presence of many irrelevant features. Artificial Intelligence, vol. 69, no. 1-2, pp. 279– 305, (1994)
7. D. Koller, M. Sahami, M. Toward optimal feature selection. In Proceedings of the Thirteenth International Conference on Machine Learning, pp. 284–292, (1996)
8. M. A. Hall, L. A. Smith, Feature Subset Selection: A Correlation Based Filter Approach, In 1997 International Conference on Neural Information Processing and Intelligent Information Systems, pp. 855- 858, (1997)
9. M. Y. Munirah, M. Rozlini, N. Wahid, A comparative analysis on feature selection techniques for medical datasets, APRN Journal of Engineering and Applied Sciences, vol 11, no 22, November 2016, (2016)
10. C. Shang, M. Li, S. Feng, Q. Jiang, J. Fan, Feature Selection via Maximizing Global Information Gain for Text Classification, *Knowledge-Based Systems*, vol. 54, 298-309, (2013)
11. J. H. Lee, J. R. Anaraki, C. W. Ahn, J. An, Eficient Classification System based on Fuzzy-Rough Feature Selection and Multitree Genetic Programming for intension Pattern Recognition using Brain Signal, Expert Systems with Applications, vol. 42, 1644-1651, (2015)
12. S. Kashef, H. Nezamabadi-pour, An Advanced ACO Algorithm for Feature Subset Selection, Neurocomputing vol. 147, 271279, (2015).
13. Y. Zhang, D. Gong, Y. Hu, W. Zhang,. Feature Selection Algorithm based on Bare Bones Particle Swarm Optimazation, Neurocomputing, vol. 148, pp.150-157, (2015)
14. Y. Shen-Lan, R. Gang, F. Yi-Ping, Multiple kernel learning based feature selection for process monitoring, 2017 IEEE/ACIS 16th International Conference on Computer and Information Science (ICIS), 24-16 May 2017, pp. 809-814, (2017)
15. B. Emel, S. Mustafa, Video classification based on ConvNet collaboration and feature selection, 25th Signal Processing and Communications Applications Conference (SIU), 15-18 May 2017, pp. 1-4, (2017)
16. D. Koller, and M. Sahami, "Toward optimal feature selection". In Proceedings of the Thirteenth International Conference on Machine Learning, pp. 284–292, (1996)
17. Rodrigues, L. Pereira and R.Y.M Nakamura, "Wrapper approach for feature selection based on bat algorithm and optimum-path forest", Expert Systems with Applications, vol.41, pp.2250–2258, (2014)
18. R.Y.M. Nakamura, L. Pereira, M. Acuckoo, K.A Costa, D. Rodrigues, J.P. Papa, X.S. Yang, "BBA: a binary bat algorithm for feature selection'. in 25th, SIBGRAPI Conference on Graphics, Patterns and Images (SIBGRAPI), 22–25 August, IEEE Publication, pp. 291-297, (2012)
19. A.M Taha, A. Mustapha and S.D. Chen, "Naïve Bayes-Guided bat algorithm for feature selection", The Scientific World Journal, vol 2013, http://dx.doi.org/10.1155/2013/325973 ,(2013)

Antioxidant and Total Phenolic Content of Breadfruit *(Artocarpus altilis)* Leaves

Lee Yit Leng[1, a], *Nuramira binti Nadzri*[1,b], *Khor Chu Yee*[1, c], *Norawanis binti Abdul Razak*[1, d] *and Abdul Razak Shaari*[1,e]

[1]Faculty of Engineering Technology, Universiti Malaysia Perlis (UniMAP), Campus of UniCITI Alam Sungai Chuchuh, Padang Besar, 02100 Perlis, Malaysia.

Abstract. This work aims to determine antioxidant, total phenolic content and Fourier transform infrared spectroscopy (FTIR) analysis of breadfruit leaves which are essential in management of diabetes. The methanolic extracts of breadfruit leaves was used to analyze for total phenolic content and antioxidant. Total phenolic content of the extracts was measured using the Folin–Ciocalteu assay while the antioxidant activity of plant extracts was measured by the 1, 1-diphenyl-2-picryhyradrazyl (DPPH) radical scavenging assay. FTIR analysis was used to determine the chemical components in the leaves. Total phenolic content in fresh breadfruit leaves (144.16 mg/g ± 17.98) was comparable to those of green tea. The results showed the breadfruit leaves extracts exhibited potent antioxidant activity. The presence of OH group also suggests antioxidant capacity of breadfruit leaves to deactivate free radicals as glucose itself could react with hydrogen peroxide in the presence of iron and copper ions to form hydroxyl radical.

1 Introduction

Antioxidant is a bioactive compound that protects a biological target against oxidative damage [1]. Antioxidant prevents the formation of reactive species like peroxides or hydrogen peroxide (H_2O_2), deactivate free radicals by capturing free radicals and repair damage at once happened in our bodies. Phenolic nutraceuticals have been suggested as effective antioxidants related to health promotion [2].

Numerous studies have been reported on useful of phenolic compounds in exhibiting potential biological activities such as antioxidant, anti-diabetic, anti-microbial and anticancer [3]. The phenolic compounds from several plant sources have been shown to inhibit a-amylase and a-glucosidase activities and allow for better control of blood glucose levels. Management of diabetes without any side effects is still a challenge to the medical system [4]. Diabetes mellitus is a chronic disease caused by inherited and/or acquired deficiency in the production of insulin by the pancreas, or by the ineffectiveness of the insulin produced. Such a deficiency results in increased concentrations of glucose in the blood, which in turn damage many of the body's systems, in particular the blood vessels and nerves.

Herbal drugs are widely prescribed because of their effectiveness, lack side effects and relatively low cost. Ample bioactive compounds which derived from plants are potential to establish antidiabetic activity. Green, yellow and black tea leaves are generally rich in phenolic compounds [5].

Breadfruit (*Artocarpus altilis*) is a staple food and traditional crop originating in the South Pacific and is now been cultivated widely in tropics. It has been recognized for its potential to prevent hunger in the tropics and developing countries [6]. Some findings suggest that traditional diet based on breadfruit can prevent onset of type II diabetes [7][8] based on field observations, ethnobotanical reports and medical records. All parts of the breadfruit trees are used medicinally, especially the leaves.

Breadfruit leaves are characterized with a pleasing flavor. The leaves of breadfruit have also been consumed by some local people to reduce the glucose level in body. However, there is not much information on the antioxidant capacity and total phenolic content from the breadfruit leaves in regulating blood glucose levels. This study was carried out to determine antioxidant, total phenolic contents and FTIR analysis of breadfruit leaves which are essential in management of diabetes.

2 Materials and Methods

2.1. Samples Collection and Preparation

About 500 g of fresh breadfruit leaves were collected from Perlis, northern part of Malaysia. The leaves were thoroughly cleaned with distilled water prior to extraction. The cleaning process is important to get rid of unwanted substances. The leaves were separated from their petioles.

* Corresponding author: yllee@unimap.edu.my

2.2 Plant Extraction

After the preparation of fresh leaves, the leaves were ground into powder using a blender. About 0.2 g powders were put into conical flask and 20 ml methanol was added. The mixture was placed in water bath at 40°C for 3 hours. Afterwards, the sample was filtered using filter paper. The extract was stored in a cap bottle at 4°C for further analysis.

2.3 Antioxidant Analysis

The 1, 1-diphenyl-2-picryhyradrazyl (DPPH) radical scavenging assay is a standard procedure to determine the antioxidant activity of plant extracts and natural compounds. The free radical scavenging activity of methanolic leaf extract was measured using the method described by Cos et al. (1998) [9].

About 2 ml of methanolic DPPH solution was added to 200 µl of water extract of leaves and then was added with methanol to make final volume of 3 ml. After 60 min, the absorbance was measured against methanol as a blank at 517 nm using spectrophotometer [10]. The free radical scavenging activities of the tested samples were compared with a control. Percentage inhibition of the DPPH radicals was calculated using the formula below:

$$Percent\ of\ inhibition\ (\%) = \left[\frac{Ac\ of\ control - As\ of\ sample}{\text{Ac of control}}\right] x\,100$$

Ac = The absorbance of control
As = The absorbance of sample

2.4 Total Phenolic Content Determination

Total phenolic contents of the extracts were measured using the Folin–Ciocalteu assay developed by Singleton and Rossi (1965) [11]. About 0.5 mL of samples was pipetted into 10 mL volumetric flask containing 0.5 mL of Follin-Ciocalteu's reagent then 5 ml of distilled water and 1.5 ml of sodium bicarbonate (Na2CO3) solution (w/v=20%) were added. The volume was made up with distilled water.

During oxidation of phenolic compounds, phosphormolybdic and phosphortungstic acid, contained in the Folin-Ciocalteu's reagent, were reduced to blue-colored molybdenum and tungsten oxides. After two hours, the absorbance of blue coloration was measured at absorbance 765 nm against a blank sample. The measurements were compared to standard curve of prepared gallic acid solutions (concentration range of 100 to 500 ppm) and expressed as milligram of gallic acid equivalent per 100 g ± S.D. All measurements were performed in triplicate.

2.5 Fourier transform infrared spectroscopy (FTIR) analysis

Perkin Elmer Universal Attenuated Total Reflectance (ATR) sampling accessory was used to obtain the FTIR spectrum of breadfruit leaves. The sample of leaves was in the form of powder. Data collection and processing were tested by Spectrum version 6.2.0.0055 software.

3 Results and Discussion

3.1 Antioxidant capacity and total phenolic content

Antioxidants have the ability to prevent free radical tissue damage by inhibiting free radicals formation, scavenging them and promoting their decomposition. Phenolic components are the primary antioxidants to deliver antioxidant ability and protection against the attack of harmful reactive oxygen species. Total phenolic contents in plants might affect the antioxidant capacity [12].

Total phenol content (TPC) of the fresh breadfruit leaves was quantified from a calibration curve (y = 0.0053x – 0.487) of gallic acid and expressed as mg/g. Total phenolic content in fresh breadfruit leaves (144.16 mg/g ± 17.98) was comparable to those of green tea [13]. This suggests that breadfruit leaves extracts is a potential beverage and its consumption can provide phenolic contents to promote health.

Aging and different chronic diseases including diabetes and cardiovascular diseases could be caused by oxidative stress. The percentage DPPH radical-scavenging inhibition of fresh breadfruit leaves was 63.88% ± 1.87. This finding was consistent with antioxidant activity in fresh ginger reported by [14] Li et al. (2016). The result indicates that the leaf extracts are capable of scavenging the free radicals and prevent the initiation of free radicals by stabilizing them to participate in any deleterious reactions.

3.2 FTIR analysis

The FTIR spectrum of fresh breadfruit leaves is shown in Figure 1. The FTIR Spectrum of breadfruit leaves indicate the presence of primary alcohol CO stretching (1037.74 cm-1 region), alkene C=C (1604.91 cm-1 region), alkyne (2160.24 cm-1 region), alkane CH stretching (2918.13 -2850.09 cm-1 region) and OH bonding group (3264.15 cm-1 region), typical characteristic of tea leaves polysaccharides [15]. The results suggest that the breadfruit leaves preventing the risk of diabetes as tea polysaccharides is a bioactive substance possess antioxidant and immunological activities [16].

High intensity of bands found at 1037.74 cm-1 for indicated the presence of phenolic compound in breadfruit leaves. This indicates that breadfruit leaves are potentially served as a healthy diet for human as phenolic compounds exhibit a wide range of biological and physiological functions, such as anti-allergenic, anti-inflammatory, anti-microbial and antioxidant activities [17][18]. The presence of OH group indicates antioxidant capacity of breadfruit leaves to deactivate free radicals as glucose itself could react with hydrogen peroxide in the presence of iron and copper ions to form hydroxyl radical [19].

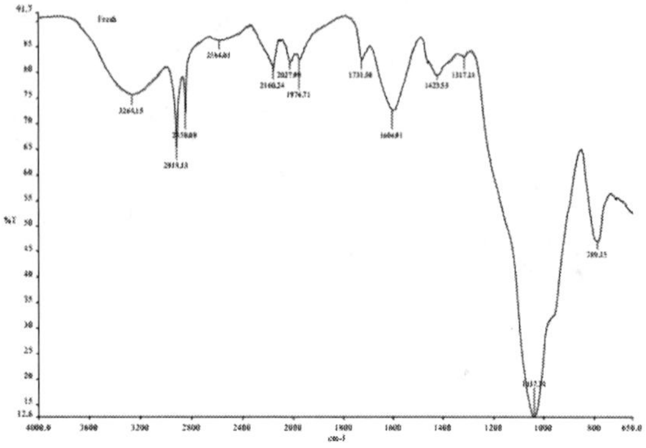

Fig. 1. FTIR spectrum of fresh breadfruit leaves.

4 Summary

Diabetes affects about 5 % of the global population [20]. The outcome of diabetes is determined, to a large extent, by patients'adherence to medication and lifestyle changes, and the maintaining of adherence to oral antidiabetic medications has been one of the important steps in reaching long-term glycemic control. Management of diabetes without any side effects is still a challenge to the medical system. A growing interest of people on the antioxidant research on plants due to their possible uses as natural additives to replace synthetic ones. The results of the present study showed that the breadfruit leaves extracts exhibited potent antioxidant activity. Further investigation of the anti-glucosidase activities of phenolic-rich extracts from Artocarpus altilis leaves in mitigating diabetes is required

The authors acknowledge the financial support received from the Short Term Grant of Universiti Perlis Malaysia, the research facilities of the Institute of Sustainable Agrotechnology (INSAT) and Faculty of Engineering Technology, UniMAP.

References

[1] B. Halliwell, J. Gutteridge, *Free radicals in biology and medicine*, (Fifth ed.)Oxford University Press, Oxford USA (2015)

[2] A. Gioxari, D.A.A. Kogiannou, N. Kalogeropoulos, A.C. Kaliora, *Phenolic compounds: bioavailability and health effects*, B. Caballero, P.M. Finglas, F. Toldrá (Eds.), Encyclopedia of Food and Health, Vol. **1**, Elsevier Science, Kidlington, United Kingdom, UK, 339–345 (2016)

[3] S. Kumar, A.K. Pandey, *Chemistry and biological activities of flavonoids: an overview*, Sci. World J., pp. 162750 (2013)

[4] Kameswara Rao B, Kesavulu MM, Apparao C, *Evaluation of hypoglycemic effect of Momordica cymbalaria fruit in alloxandiabetic rats.* Fitoterapia, Vol. **74**, pp. 7-13 (2003)

[5] M. Kopjar, M. Tadic, *Phenolic content and antioxidant activity of green, yellow, and black tea leaves*, Chem. Biol. Techn. Agric., Vol. **2**, no.1, pp. 1-6 (2015)

[6] A.M.P. Jones, R. Baker, D. Ragone, S.J. Murch, *Identification of pro-vitamin A carotenoid-rich cultivars of breadfruit (Artocarpus, Moraceae)*, J. Food Compost. Anal., Vol. **31**, pp. 51-61 (2013)

[7] C.A. Lans, Enthnomedicines used in Trinidad and Tobago for urinary problems and diabetes mellitus, J. Ethnobiol. Ethnomed., Vol. **2**, 45 (2006)

[8] D.D. Ramdath, C.L.R. Issacs, S. Teelucksingh, S.M.T. Wolever, *Glycaemic index of selected staples commonly eaten in the Caribbean and the effects of boiling v. crushing*, Br. J. Nutr., Vol. **91**, pp. 971–977 (2004)

[9] P. Cos, L.Y. Ying , M. Calomme , J.H. Hu, V.P.K.B. Cimanga, L. Pieters, A.J. Vlietinck, D.V. Berghe, *Structure activity relationships and classification of flavonoids as inhibitors of xanthine oxidase and superoxide scavengers*, J. Nat. Prod., Vol. **61**, pp. 71-76 (1998)

[10] S. Abbas, M.Wink, *Epigallocatechin gallate from green tea (Camellia sinensis) increases lifespan and stress resistance in Caenorhabditis elegans*. Planta Medica, Vol. **75**, pp. 216–221 (2009)

[11] V.L. Singleton, J.A. Rossi, *Colorimetry of Total Phenolics with Phosphomolybdic-Phosphotungstic Acid Reagents*, Am. J. Enol. Vitic., pp. 144-158 (1965)

[12] Duda, S.C. L.A. Mărghitaş, D. Dezmirean, M. Duda, R. Mărgăoan, O. Bobiş, *Changes in major bioactive compounds with antioxidant activity of Agastache foeniculum, Lavandula angustifolia, Melissa officinalis and Nepeta cataria: Effect of harvest time and plant species.* Ind Crops Prod., Vol. **77**, pp. 499–507 (2015)

[13] U.J. Unachukwu, S. Ahmed, A. Kavalier, J.T. Lyles, E.J. Kennelly, *White and green teas (Camellia sinensis var. sinensis): variation in phenolic, methylxan-thine, and antioxidant profiles*. J. Food Sci., Vol. **75**, pp. 541–548 (2010)

[14] Li, Y. Hong, Y. Han, Y. Wan, L. Xia, *Chemical characterization and antioxidant activities comparison infresh, dried, stir-frying and carbonized ginger*, J. Chromatogr. B, Vol. **1011**, pp. 223–232 (2016)

[15] Y.F. Wang, F.F. Mao, X.L. Wei, *Characterization and antioxidant activities of polysaccharides from leaves, flowers and seeds of green tea*, Carbohydr. Polym., Vol. **88**, pp. 146–153 (2012)

[16] J.J. Yang, B. Chen, G. Yan, *Pharmacological evaluation of tea polysaccharides with antioxidant activity in gastric cancer mice*, Carbohydr. Polym., Vol. **90**, no. 2, pp. 943–947 (2012)

[17] N. Balasundram, K. Sundram, S. Samman, *Phenolic compounds in plants and agri-industrial by-products: Antioxidant activity, occurrence, and potential uses*, Food Chem., Vol. **99**, no. 1, pp. 191–203 (2006)

[18] M. Claudine, M. Andrzej, S. Augustin, *Polyphenols and prevention of cardiovascular diseases*, Curr. Opin. Lipidol, Vol. **16**, no. 1,pp. 77–84 (2005)

[19] M. Valko, C.J. Rhodes, J. Moncol, M. Izakovic, M. Mazur, *Free radicals, metals and antioxidants in oxidative stress-induced cancer*, Chem Biol Interact, Vol. **160**, pp. 1-40 (2006)

[20] R. Chakraborty, R. Rajagopalan, *Diabetes and insulin resistance associated disorders: disease and the therapy*. Curr. Sci. India, Vol. **83**: 1533-1538 (2002)

Significant Correlation Study between Cause and Effect of Hand Arm Vibration Exposure among Grass Cutters

N.A. Azmir[1,*], *M.H. Ali*[2], *M.N. Yahya*[1], *I. Zaman*[1], *M.H. Zainulabidin*[1], *M.A. Madlan*[1], *M. Ismon*[1],and M.Z. Kasron[1]

[1]Noise, Vibration & Acoustic – Research Group (NOVIA), Faculty of Mechanical & Manufacturing Engineering, Universiti Tun Hussein Onn Malaysia, Johor, 86400, Malaysia
[2]OSH Management Division, CRD Department, National Institute of Occupational Safety & Health (NIOSH), Bandar Baru Bangi, Selangor, 43650, Malaysia

Abstract. Prolonged exposures to hand-transmitted vibrations from grass-cutting machines have been associated with increasing occurrences of signs of occupational diseases related to the hand-arm vibration syndrome (HAVS). However, there are no specific studies available that cover the health cause-effects of the hand arm vibration risk factors during onsite operations. Therefore, the main objectives of this study are to determine the significant correlation of the objective and subjective measurement variables of the hand arm vibration exposure on hand arm vibration symptoms and disorders. The study was conducted for evaluation of the cause and effect for categorical and continuous variables. Significant sub symptoms element consists of neurological, vascular, musculoskeletal and awareness. Meanwhile, significant sub disorder element consists of loss of hand grip strength and finger blanching that induces to HAVS. The highest correlation significant variable was worked repeatedly done with the forearms and hand with uncomfortable hand position or grip with the value of 0.431. Thus, the cause and effect able to monitor health condition particularly at workers hand due to hand-transmitted vibrations among grass cutters.

1 Introduction

Hand-transmitted vibration (HTV) is a common physical hazard for occupational safety and health which is caused by harmful activities to workers who use vibrating equipment in their routine tasks [1]. The fundamental HTV evaluation is often an initial process for employers to evaluate hand-transmitted assessment due to the hand arm vibration hazard at workplace among their employees [2-4]. With this evaluation, it will assist the employers to measure whether the works or activities are exposed to risk factor that could induce hand arm vibration syndrome (HAVS). By measuring the HTV exposure towards health cause-effect factors; it also assists the employers to decrease the cost of occupational rehabilitations and work-related injuries [5-7]. In European countries, there is a regulatory requirement to perform HTV exposure at workplace in which to improve employees' health quality and this can be associated to the responsibility and accountability of employers [3,4].

It is an important requirement for the risk measurement, evaluation and assessment of hand arm vibration exposure among hand-held vibrating machine workers to be taken in the monitoring process of the safety and health management [1,2,6]. The rationale of this study was established from occupational hand arm vibration knowledge for epidemiological methods used to identify and analyze the health cause-effects induce in a work related with vascular, neurological and musculoskeletal disorders at workplace. Hence, its principle method is to define a categorical and continuous measures linked with the development of hand arm vibration syndrome (HAVS) which has a significant element for HTV in developing the health cause-effect of hand arm vibration exposure in Malaysia [8-12]. The investigation of hand arm vibration exposure to the health cause-effect associated to the HAVS is becoming a major interest for both human factor and occupational health in conducting their research study [10-12].

Hand arm vibration syndromes (HAVS) are a prevalence health difficulty and a main cause of permanent or temporary hand disabilities [13]. Combinations of physiology, psychology and physics vibration risk factors are connected to the development of HAVS. Hand arm vibration risk factors are based on magnitude, frequency, direction and duration of vibration exposure energy to the hand palm while performing the hand-held vibrating equipment [14]. Previous studies have shown that the hand-transmitted vibration effects of HAVS are significant consequences to the vascular disorders, neurological disorders and loss of hand grip strength [15].

Safety and health cause-effect do important procedure for employers and employees to understand how to mitigate the HTV hazard which can benefit grass-cutting operators. Research has been recognized on human response to hand arm vibration measures, but has not investigated categorical and continuous data in

* Corresponding author: azali@uthm.edu.my

correlation with health cause-effect among hand-held grass-cutting workers. Additionally, subjective surveys of HAVS have been proposed by scholars in Malaysia, yet no one has looked at classifying both hand numerical scoring and hand grip strength assessment at the same time. Thus the objective of this research is determining the significant correlation of categorical and continuous measurement variables on hand arm vibration symptoms and disorders.

2 Methodology

The study received written approval from the subcontractors prior to their participation in the study. Information on HAVS symptoms was obtained from interviews and physical observation or explanations to determine the presence of a condition in which workers experience HAVS after working with vibrating equipment. Workers were briefed as to the purpose of the questionnaires and were asked to complete through face-to-face interview. In this process, the interviewees selected the proper answers based on their professional judgements and subjects' feedback in order to prevent misunderstandings. Workers had no knowledge that they would later be interviewed or observed. This will provide true information in answering the HAVS questionnaire. All the information granted was confidential and used only for occupational vibration engineering research activities.

A total of 204 hand-held grass-cutting workers were selected for the study within August 2013 to June 2014. The workers were selected from general cutting operation locations involving flats, sloped and drain areas; the jobs performed at expressway median and verge as well as expressway facility buildings such as rest and service area. Some of the workers were employed full time as grass cutters, while some workers make grass cutting as their part time routine work for maintaining building facilities.

All subjects chosen were interviewed using a modified questionnaire adopted from the Hand-Transmitted Vibration Health Surveillance – Initial Questionnaire and Clinical Assessment published by the Research Network on Detection and Prevention of Injuries due to Occupational Vibration Exposure [16]. Some of the questions in the questionnaire were based on a study of HAVS among a group of construction workers in Malaysia [9], which was found to be reliable and applicable for different work activities. The interviews were conducted based on personal identification, social history, occupational health, occupational history, machine safety inspection and symptoms.

The personal identification information consists of name, work location, age, nationality, height (m), weight (kg), and types of grass cutting machine. Besides that, nominal questions such as workers' dominant hand, marital status and academic level need to be answered by workers. These sorts of questions are able to find information regarding workers' demographic detail. To determine the workers' social history, questions such as "Do you smoke or drink alcoholic beverages", any hobby involving vibration tool and work experience involving neurotoxic chemicals have also been asked. The example of workers' health questions are;

(1) Have you ever sustained an injury to your arm or hand;

(2) Do you suffer from any long-term medical illness;

(3) Are you taking any long-term medication;

(4) Are you exposed to organic solvent chemical (e.g. n-hexane).

Subjects were asked to specifically recall any sustained injury to the body and the healing process to regain normal health conditions.

Occupational history sub questions include working experience with hand-held grass-cutter, exposure to hand-held grass cutter (cutting hours) for 1 day, and the scale questions which require workers to choose either "Yes" or "No":

(1) Is work repeatedly done with the forearms and hands with twisting movement;

(2) Is work repeatedly done with the forearms and hands with forceful movement;

(3) Is work repeatedly done with the forearms and hands with uncomfortable hand position / grips.

This sort of questions can determine the musculoskeletal disorder or ergonomic problem due to repetitive and forceful tasks. Since vibration comes from the hand-held grass-cutting machine itself, the machine's safety inspection questions have been introduced. The questions are;

(1) Do you inspect the entire machine for loss of parts and any damage;

(2) Repair or replace as necessary before using the machine;

(3) Clean and maintain the machine completely after operation.

Similar scale answer required which is "Yes" or "No".

During physical observation, the author looked for colour changes in finger phalanges or finger blanching and any skin abnormalities at hand palm before workers started their work. Details on the health effects information, including colour changes on hands or fingers, tingling, numbness and weakness in handgrip strength for both hands were included in symptoms' identification. Besides that, health effect such as cold fingers, difficulty in picking up, handling or manipulating small objects as well as difficulty in opening tight jars were also asked. The information obtained is important to investigate the disability at hand due to grass-cutting operation. Finally, two occupational safety and health awareness questions have been asked;

(1) Do you agree that regular use of vibratory tools is hazardous to your health;

(2) Are you aware of any devices that can protect you from the vibration hazard.

All the questions in symptoms' identification was based on either "Yes" or "No" answers. By simply responding to "Yes" and "No" scales are claimed to detect positive and negative state of health using questionnaire. According to Paakkonen et al., (2008), the use of questionnaires is appropriate because it is easier to get cooperation from subjects, one of the information gathering tools, economical, practical, as well as effective.

3 Results

In determining the significant variables, the selection of the variables has followed an output result from the data correlation significant test (α=0.05) and supported by extensive review on scientific literature. The collection of epidemiological evidence of the hand arm vibration disease variables was identified. The total mean of age of the participants (N=204) was 32.15 years (SD=9.56) in the range between the ages of 15 to 63 and the total mean working experience was 3.37 years (SD=4.06) in the range between 0.1 to 30 years (Table 1).

Table 1. Demographic of the workers in age-group in grass-cutting operation

Age-Group (year)	Age (year)	Working Experience (year)
	Mean	Mean
15-24 (N=46)	21.96	1.76
25-29 (N=56)	26.96	2.56
30-34 (N=32)	31.56	3.29
35-39 (N=28)	36.89	4.07
40-44 (N=10)	41.30	8.45
45-49 (N=19)	46.37	4.15
50-54 (N=7)	50.57	6.21
55-59 (N=4)	55.50	3.75
60-65 (N=2)	63.00	10.75

3.1. Categorical Variables

The selection of the significant variables based on the categorical data from the questionnaires such as the work location, suffering from any long-term medical illness together with types of illness, any sustained injury at arm or hand, repeated works with forearms and hands with uncomfortable positions or grips, academic level, types of grass cutting machine and maintaining the machine after operation. For the health effect due to grass cutting machine, four symptoms were determined to be associated with the occupational disease: vascular, neurological, musculoskeletal and awareness. All variables were selected by correlation coefficient and significant level with contribution to hand arm vibration syndrome. Fig. 1 shows the significant inclusive sub symptoms associated to categorical cause-effect variables. The outcomes from the significant variables correlate cause effect study had revealed that the sub-symptoms consists of neurological sign due to tingling and numbness, vascular sign due to finger colour change, musculoskeletal sign due to work repeatedly and awareness sign due to less information related to occupational exposure using hand-held machines.

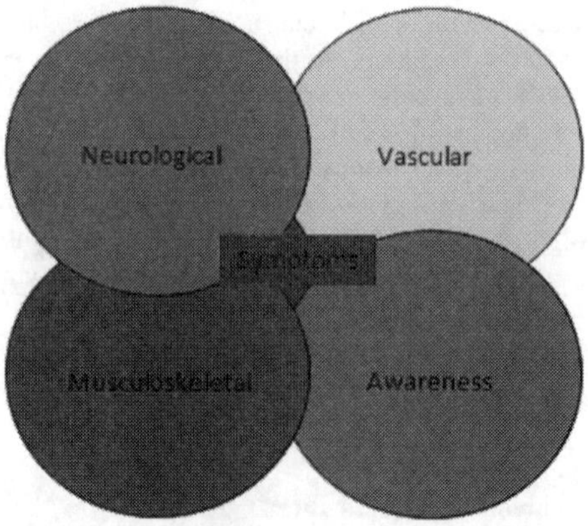

Fig. 1. Significant sub symptoms element that induces to HAVS.

3.1.1 Vascular Symptoms

During the selection of significant cause that affects the finger colour change, the work location or process indicates the weak positive correlation, 0.201 with the significant level of 0.007. In addition, the vascular symptoms were also associated with suffering any long term medical illness such as shoulder pain, itchy, hot body, back pain, foot pain, waist pain, chest pain and knee pain. There is a very weak positive correlation, 0.159 coefficient of these pain associated with finger blanching due to hand-transmitted vibration generated from the back pack of grass cutting machine. Table 2 shows the correlation coefficient and significant level associated with finger colour change of the vascular symptoms.

Table 2. Formatting sections, subsections and subsubsections

Cause / Variables	Correlation coefficient	Significant level
Work location	0.201	0.007
Suffer from any long-term medical illness	0.159	0.032
Type of illness	0.176	0.017

3.1.2 Neurological Symptoms

Repeated works from forearms and hand movement contribute to incapacity of hand movement which is exposed to high risk level. Thus, the prevalence of neurological symptoms was 20% and 18% for finger tingling and finger numbness, respectively. During the selection of significant cause variables due to sensation of numbness and tingling at fingers, the work repeatedly done with the forearms and hand with uncomfortable hand position or grip indicate the medium positive correlation coefficient of 0.431 associated with tingling sensation. Meanwhile, working repeatedly also has medium positive correlation with 0.519 coefficient associated with numbness. Other cause variables had shown weak and very weak positive correlation coefficients for both neurological symptoms as shown in Table 3.

Table 3. Experience a numb and tingling sensation at fingers

Cause / Variables	Tingling sensation		Numbness	
	Correlation coefficient	Significant level	Correlation coefficient	Significant level
Work location	0.179	0.016	0.276	0.000
Academic level	0.078	0.294	0.200	0.007
Sustained any injury at arm or hand	0.235	0.001	0.215	0.004
Suffer from any long-term medical illness	0.207	0.005	0.223	0.002
Type of illness	0.191	0.010	0.194	0.009
Work repeatedly	0.431	0.000	0.519	0.000

3.1.3 Musculoskeletal Symptoms

The feedback from grass-cutting workers in the questionnaire indicated that the cause variables of sustained injury at hand or arm and work repeatedly were significant to cause weakness at hand grip. The left hand correlation coefficient was higher compared to right hand with the value of 0.315 and 0.208, respectively. Both variables show weak positive correlation with high significant level. Thus, the value produced revealed that the left handle was exposed to high vibration value as compared to right handle. The correlation coefficient and significant level for both hands was shown in Table 4.

Table 4. Suffer from weakness at hand grip

Cause / Variables	Left Hand		Right Hand	
	Correlation coefficient	Significant level	Correlation coefficient	Significant level
Sustained any injury at arm or hand	0.236	0.001	0.198	0.007
Work repeatedly	0.315	0.000	0.208	0.000

3.1.4 Awareness Symptoms

The occupational safety and health (OSH) awareness among hand-held grass-cutting workers on the hand arm vibration hazard is still low due to lack of supervision, instruction, training and information. Hence, the questionnaire proved that the awareness of vibratory tools which are hazardous to health significantly weak positive correlation coefficient (0.274) in respect to types of grass cutters. In the significant variables assessed by the regulatory use of vibratory tools hazardous to health, the repeated operation and maintenance of the machine were positive weak correlation (0.298) and positive very weak correlation (0.186) respectively. The results of significant variables associated with awareness symptoms are shown in Table 5.

Table 5. Regular use of vibratory tools which are hazardous to health

Cause / Variables	Correlation coefficient	Significant level
Types of grass cutting machine	0.274	0.000
Work repeatedly	0.298	0.000
Clean and maintain the machine	0.186	0.012

3.2 Continuous Variables

The significance of variables that induces to hand arm vibration exposure disorder measurement system are divided into two sections, namely the musculoskeletal disorder and the vascular and neurological disorder. Both of these disorders were measured using hand grip strength (HGS) and hand numerical scoring (HNS) assessment respectively. Fig. 2 shows the significant inclusive sub disorder associated with continuous cause-effect variables. The outcomes from the significant variables correlate cause effect study had revealed that the sub-disorders consists of loss of hand grip strength due to worker age and finger blanching due to year of exposure operating hand-held grass-cutting machine.

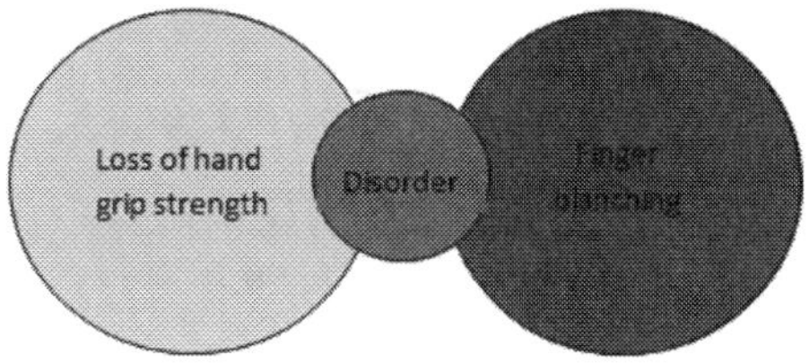

Fig. 2. Significant sub disorder element that induces to HAVS

3.2.1 Loss of Hand Grip Strength Disorder

The significant cause of the process among hand-held grass cutters associated with hand grip strength is categorized into two parts which are right hand and left hand. Both hands have significantly weak positive correlation which contributes to 0.392 (height) and 0.373 (weight) for right hand while 0.349 (height) and 0.353 (weight) for left hand. However, the age variable was weak negative correlation for right hand and differs from left hand. The estimated daily vibration exposure, A(8) has significant weak positive correlation for right hand. The value of correlation coefficient and significant level associated with hand grip strength is shown in Table 6.

Table 6. Measurement of hand grip strength force.

Cause / Variables	Right hand		Left hand	
	Correlation coefficient	Significant level	Correlation coefficient	Significant level
Age	-0.303	0.000	0.282	0.000
Height	0.392	0.000	0.349	0.000
Weight	0.373	0.000	0.353	0.000
Right hand daily vibration exposure, *A(8)*	0.249	0.001	0.105	0.158

 https://doi.org/10.1051/matecconf/201815006008

3.2.2 Finger Blanching Disorder

The significant cause of the variables for hand-held grass-cutting workers due to scoring of finger blanching disorder was separated into two parts which are left hand and right hand. Both hands indicate weak positive correlation for age and working experience with hand-held grass cutter. The values of correlation coefficient and significant level, based on hand numerical scoring are shown in Table 7.

Table 7. Hand numerical scoring due to finger blanching.

Cause / Variables	Right hand		Left hand	
	Correlation coefficient	Significant level	Correlation coefficient	Significant level
Age	0.386	0.000	0.257	0.000
Cigarettes per day	-0.147	0.048	-0.051	0.498
Working experience	0.344	0.000	0.337	0.000

4 Discussion

The determination of the variables has been tested by data screening consisting of data normalization and significant correlation. The main objective of this analysis was to determine the significant correlation level of categorical data symptoms and continuous data disorders. The following sections discuss details of significant variables induced to HAVS categorical symptoms and continuous disorders.

4.1. Categorical Symptoms

An analysis selection revealed that the work repeatedly done with the forearms and hands with twisting movement showed significant medium association for neurological symptoms. According to Soo et al. (2012), the work-related fatigue initiated from repeated exertion had significant decline in muscle contractibility. This is because, the vibration propagated from the hand palm entering other body parts due to vibration intensity increases which may be due to the fact that frequency of machine vibration approaches the natural frequency of vibration at handle. In the categorical questions, the health effect is merely focusing on repeated work which significantly introduces exertion posture when operating vibrating machine. The same finding was revealed by other researchers [2,19], who observed that repeated vibration with sufficiently low magnitude gives rise to sign and symptoms of chronic disorder. Other researcher has confirmed that it is not possible to estimate symptom effects due to repeated vibration exposure at hand [20]. The determination of work repeatedly also found positive association with neurological, musculoskeletal and awareness symptoms. The study by Bovenzi (1998) in the exposure response relationship found that it was complex neurologic symptoms in the upper-limb of vibration which affects exposed workers. Griffin (1994) highlighted the same issues and found that the health symptoms of tingling and numbness as reasonable variables for predicting effects of vibration. So, work repeatedly is the most significant variable that influences hand-held grass-cutting workers for the purpose of neurological symptom.

4.2 Continuous Disorders

According to Griffin (2008), some preventive procedures including health surveillance at hand grip force may reduce the vibration risk. Measurement of hand grip strength is one of the useful indicators to confirm lack of upper limb disorder. In addition, the hand grip strength was best related to the age [23]. In the continuous measurement, the health effect is targeting on hand grip strength before work which is equal to normal grip force condition before exposed to vibrating equipment. The same technique was used by other researchers [24] who measure non-exposed subject to force over the test period. The study done by Kamarul et al. (2006) revealed that there may be potential cause variables which contributed to hand grip strength force based on the workers' population. Only age, weight, height, occupation and gender were correlated with hand grip strength ($p<0.01$). The determination of significant variables was the same with the study done by Kamarul et al. (2006). However, the daily vibration exposure at right hand has been added to the input selection since the study focuses on hand arm vibration disorder. Seri Rahayu et al. (2013) highlighted the same issues and found that the hand grip force is one of the factors that affects workers in handling vibrating tools and could decrease the grip force of workers.

5 Conclusions

In summary, the determination of significant cause-effect health variables to hand arm vibration symptoms and disorders were separated into two categories which are categorical data and continuous data. An analysis was also conducted to test the significant correlation of categorical and continuous cause variables to hand arm symptoms and disorders. For categorical symptoms, an analysis selection revealed that the work repeatedly done with the forearms and hands with twisting movement shown significant medium association (Spearman correlation=0.599; p=0.000) for neurological symptoms. For continuous disorders, the most high significant correlation cause-effect is height for right hand grip strength (Pearson correlation=0.392; p=0.000), weight for left hand grip strength (Pearson correlation=0.353; p=0.000), age for right hand numerical scoring (Pearson correlation=0.386; p=0.000) and working experience for left hand numerical scoring (Pearson correlation=0.337; p=0.000). The continuous cause-effect variables have covered an occupational diseases originating from hand arm vibration exposure such as loss of hand grip strength and finger blanching. In categorical symptoms, the work repeatedly was the most significant cause variables associated to neurological, musculoskeletal and awareness symptoms. In continuous data, the height, weight, estimated right hand daily vibration exposure, age and working experience were the significant cause variables associated to loss of hand grip strength and finger blanching.

References

1. M. Bovenzi, Exposure-response relationship in the hand-arm vibration syndrome: an overview of current epidemiology research. Int Arch Occup Environ Health, **71**, pp. 509–519 (1998)
2. M. J. Griffin, Measurement, evaluation, and assessment of occupational exposure to hand-transmitted vibration. Occupational and Environmental Medicine, **54**, pp. 73–89 (1997)
3. D. J. Edwards & G. D. Holt, Exposure to hand-arm vibration: implications of new statutory requirements. Building Research & Information, **33**(3), pp. 257–266 (2005)
4. R. Cederlund, S. Iwarsson & G. Lundborg, Quality of life in Swedish workers exposed to hand-arm vibration. Occupational Therapy International, **14**(3), pp. 156–169 (2007)
5. J. Friden, Vibration damage to the hand: Clinical presentation, prognosis and length and severity of vibration required. Journal of Hand Surgery, **26**B(5), pp. 471–474 (2001)
6. N. J. Mansfield, *Human Response to Vibration*. United Kingdom: CRC Press LLC (2005)
7. K. F. Chan, C. W. C. Tan, D. S. C. Yeo, H. S. K. Tan, F. L. Tan, E. W. Tan, G. P. Y. Szeto & A. S. K. Cheng, Occupational Rehabilitation in Singapore and Malaysia. Journal Occupational Rehabilitation, **21**, pp. 69–76 (2011)
8. M. Widia & S. Z. M. Dawal, The effect of hand-held vibrating tools on muscle activity and grip strength. Australian Journal of Basic and Applied Sciences, **5**(11), pp. 198–211 (2011)
9. T. A. Su, V. C. W. Hoe, R. Masilamani, & A. B. Awang Mahmud, Hand-arm vibration syndrome among a group of construction workers in Malaysia. Occup. Environ. Med., **68**(1), pp. 58–63 (2011)
10. S. B. M. Tamrin, M. N. Jamalohdin, Y. G. Ng, S. Maeda & N. A. M. Ali, The characteristics of vibrotactile perception threshold among shipyard workers in a tropical environment. Industrial Health, **50**, pp. 156–163 (2012)
11. S. R. Kamat, H. Norhidayah, I. Halim & H. C. Hui, Effect of working posture and hand grip for carpal tunnel syndrome among aerospace workers. International Journal of Mechanical & Mechatronics Engineering, **14**(1), pp. 1–6 (2014)
12. A. T. Su, S. Maeda, J. Fukumoto, N. Miyai, M. Isahak, A. Yoshioka, R. Nakajima, A. Bulgiba & K. Miyashita, A cross sectional study on hand-arm vibration syndrome among a group of tree fellers in a tropical environment. Industrial Health, **52**, pp. 367–376 (2014)
13. R. Sauni, R. Paakkonen, P. Virtema, V. Jantti, M. Kahonen, E. Toppila, I. Pyykko & J. Uitti, Vibration-induced white finger syndrome and carpal tunnel syndrome among Finnish metal workers. Int Arch Occup Environ Health, **82**, pp. 445–453 (2009)
14. Y. H. Ko, L. E. Ooi, & Z. M. Ripin, The design and development of suspended handles for reducing hand-arm vibration in petrol driven grass trimmer. International Journal of Industrial Ergonomics, **41**, pp. 459–470 (2011)
15. A. J. Brammer, W. Taylor & G. Lundborg, Sensorineural stages of the hand-arm vibration syndrome. Scand. J. Work Environ Health, pp. 279–283 (1987)
16. Vibration Injury Network *Guidelines for Hand Transmitted Vibration for Health Surveillance*. Appendix H1A: European Committee. (2011)
17. R. Paakkonen, R. Sauni, J. Uitti & E. Toppila, Evaluation of hand-transmitted vibration exposure on basis of a questionnaire. *Proc. of the Acoustics Conference. Paris. Euronoise*. pp. 2235–2238 (2008)
18. Y. Soo, M. Sugi, H. Yokoi, T. Arai, R. Kato & J. Ota, Quantitative estimation of muscle fatigue on cyclic handgrip tasks. International Journal of Industrial Ergonomics, **42**, pp. 103–112 (2012)
19. J. Jang, S. Kim, S. K. Park, J. Roh, T. Lee & J. T. Youn, Quantitative Exposure Assessment for Shipyard Workers Exposed to Hand-Transmitted Vibration From a Variety of Vibration Tools. America Industrial Hygiene Association (AIHA) Journal, **63**, pp. 305-310 (2002)
20. M. Dado, R. Hnilica, J. Kovac & M. Kucera, Assessment of Hand-transmitted Vibration During Work With Chain Saw. *Proceeding of the 22nd International DAAAM Symposium*, 22(1) (2011)
21. M. J. Griffin, Foundations of hand-transmitted vibration standards. Nagoya Journal Medical Science, **57**, pp. 147–164 (1994)
22. M. J. Griffin, Negligent exposure to hand-transmitted vibration. Int. Arch. Occup. Environ. Health, **81**(5), pp. 645–659 (2008)
23. K. Poole & H. Mason, Relationship between self-reported upper limb disability and quantitative tests in hand-arm vibration syndrome. Disability and Rehabilitation, **29**(5), pp. 359–366 (2007)
24. F. M. Moy, E. W. H. Chang & K. W. Kee, Predictors of handgrip strength among the free living elderly in rural Pahang, Malaysia. Iranian J Publ Health, **40**(4), pp. 44-53 (2011).
25. T. Kamarul, T. S. Ahmad, & W. Y. C. Loh, Hand grip strength in the adult Malaysian population. Journal of Orthopedic Surgery, **12**(2), pp. 172-177 (2006)
26. K. Seri Rahayu, A. Nurulhuda & A. Rohana, Vibration of power hand tool and discomfort experience among malaysian industrial workers. *Proc. of 2th International Conference on Ergonomics*, Universiti Malaya (2013)

High Performance Systolic Array Core Architecture Design for DNA Sequencer

Dayana Saiful Nurdin[1], *Mohd. Nazrin* Md. Isa [1,*] *Rizalafande* Che Ismail[1] and *Muhammad* Imran Ahmad[2]

[1]The Integrated Circuits and Systems Design Group (ICASe),School of Microelectronic Engineering, Universiti Malaysia Perlis, Pauh Putra Campus, 02600, Arau, Perlis, Malaysia.
[2]School of Computer and Communication Engineering, Universiti Malaysia Perlis, Pauh Putra Campus, 02600, Arau, Perlis, Malaysia.

Abstract. This paper presents a high performance systolic array (SA) core architecture design for Deoxyribonucleic Acid (DNA) sequencer. The core implements the affine gap penalty score Smith-Waterman (SW) algorithm. This time-consuming local alignment algorithm guarantees optimal alignment between DNA sequences, but it requires quadratic computation time when performed on standard desktop computers. The use of linear SA decreases the time complexity from quadratic to linear. In addition, with the exponential growth of DNA databases, the SA architecture is used to overcome the timing issue. In this work, the SW algorithm has been captured using Verilog Hardware Description Language (HDL) and simulated using Xilinx ISIM simulator. The proposed design has been implemented in Xilinx Virtex -6 Field Programmable Gate Array (FPGA) and improved in the core area by 90% reduction.

1 Introduction

DNA is made up of thymine (T), guanine (G), cytosine (C) and adenine (A) nucleotides (NTs)[1]. Biological processes such as mutation produce unknown DNA sequences as it alters the natural DNA sequences [2]. Therefore, sequence alignment is important as it facilitates the mutated DNA sequences detection by finding the similarities NT region.

Aligning DNA sequence is an operation that computationally intensive. The task time execution cannot be achieved realistically by desktop computer systems as sequence databases growing exponentially as shown in Fig. 1. Hence, a faster computation platform is needed to overcome the aforementioned problem. Currently, reconfigurable hardware FPGAs has been suggested to enhance the DNA sequence alignment performance [3]. Certainly, FPGAs are attractive platforms to speed up DNA sequence alignment computation compared to General Purpose Processor (GPP).

SA architecture has been introduced by researchers to further improve the DNA sequence alignment process. SA is a pipeline network arrangement where the process task is divided among several processors. It is a build-up of row data processing units called cell or Processing Element (PE). One of the advantages of using SA architecture is that the data is streamed across the array due to the presence of local connection between the cells [4]. The sequence alignment algorithms can take advantage of SA to realize the parallelism on FPGAs.

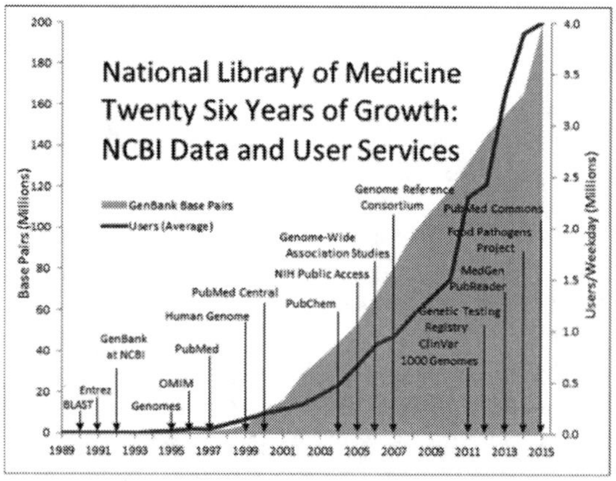

Fig. 1. Database growth from 1989 to 2015 [5].

SW algorithm, a type of Dynamic Programming (DP) algorithm is used in aligning DNA sequence where the common regions between two or more subsequence of DNA sequences represent the optimal sequence alignment [6-7]. DP-based algorithms usually exhaustive due to the accurate analysis. Unlike heuristic sequence alignment such as FASTA, give sub-optimal alignments that help to reduce the computational burden in DP algorithm analysis with uncertain accuracy of the result [8]. Thus, the sequence alignment algorithm can be implemented in SA to speed up its computation time and uses FPGA to improve the DNA sequencer computation process.

* Corresponding author: nazrin@unimap.edu.my

The implementation of FPGA related to DNA sequence alignment architectures are extensively reported in [9], [10], [11], [12], [13] and [14]. All of these architectures are designed based on SA with realization of the SW algorithm with linear gap penalty. The architectures are differentiated based on penalty gap used and also additional features that are included in their designs such additional algorithm for trace back (TB) step. For further information of the related works, please refer to [13].

The remainder of this paper will present the general information of SW algorithm with affine gap, SA and FPGA. Next, comparative timing performance evaluation of the proposed design against other FPGA platforms. Finally, conclusion of this work.

2 Sequence Alignment Algorithms

Pairwise Sequence Alignment (PSA) is one of the alignment analyses to align DNA sequence. It investigates the relationship between a newly discover query sequence and subject sequences that are taken from databases. T. F. Smith and M. S. Waterman have introduce an algorithm in 1981 known as SW algorithm to find the best local alignment between the aforementioned sequences [15]. There are two ways in penalizing insertions and deletions gaps; linear and affine.

A more complex SW algorithm was introduced by Gotoh which is suitable for the affine gaps due to mutation in DNA sequence [16] as shown in (1). Implementation of this algorithm in hardware requires more logic resources and computation time. The time complexity of this algorithm is $O(mxn)$ where m and n are the length of subject and query sequences respectively. Based on (1), d is penalty for open gap, e is penalty for extended gap and $\gamma(r_i,s_j)$ is a score of substitution matrix for subject sequence r_i $(r_1,r_2,r_3 \ldots r_m)$ and query sequence s_j. $(s_1,s_2,s_3 \ldots s_n)$.

$$M(i,j) = \max \begin{Bmatrix} I_x(i-1,j-1) + \gamma(r_i,s_j) \\ I_y(i-1,j-1) + \gamma(r_i,s_j) \\ M(i-1,j-1) + \gamma(r_i,s_j) \end{Bmatrix}$$

$$I_x(i,j) = \max \begin{Bmatrix} I_x(i-1,j) - e \\ M(i-1,j) - d - e \end{Bmatrix} \qquad (1)$$

$$I_y(i,j) = \max \begin{Bmatrix} I_y(i,j-1) - e \\ M(i,j-1) - d - e \end{Bmatrix}$$

$$F(i,j) = \max \left\{ M(i,j), I_x(i,j), I_y(i,j), 0 \right\}$$

The alignment between each NT pair of r, indexed by i and s, indexed by j is computed in the matrix form $F(i,j)$. This matrix computes the three neighbouring cells; the diagonal cell, $M(i,j)$, the top cell $I_x(i-1,j)$, and the left cell, $I_y(i,j)$, for the maximum score as shown in Fig. 2.

The boundary values $F(0,0)$, $F(i,0)$, $F(0,j)$ are set to zero as there is no alignment in either r or s as shown in Fig. 4. TB starts after the scores were completely filled the matrix $F(i,j)$ by locating the maximum score in as the starting point. The next highest score will be the highest value among its neighbouring cells (top, left, diagonal). This process continues until it reached the origin as shown in Fig 4. TB is used to obtain the optimal alignment between r and s as shown in Fig. 3.

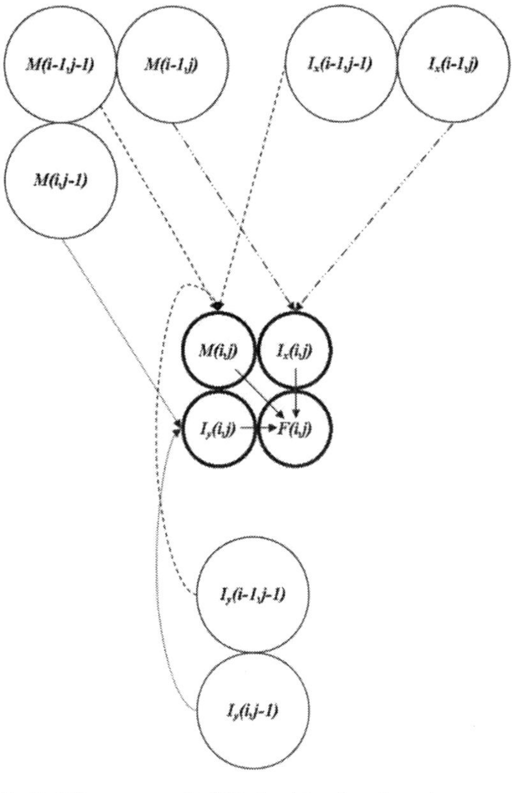

Fig. 2. Affine gap penalty SW algorithm data dependency.

r	–	*T*	*C*	*G*	*T*	*A*
	|	|	|	⋮		|
s	*T*	*T*	*C*	*A*	*T*	*A*
score :	−2	+5	+5	−2	+5	+5 = 16

Fig. 3. Optimal sequence alignment [17].

		T	G	C	T	C	G	T	A
		0	0	0	0	0	0	0	0
T	0	5 I_x:-2 I_y:-2 M:5	0 I_x:-2 I_y:-4 M:-2	0 I_x:-2 I_y:-6 M:-2	5 I_x:-2 I_y:-8 H:5	0 I_x:-2 I_y:-9 H:-2	0 I_x:-2 I_y:-11 H:-2	5 I_x:-2 I_y:-13 H:5	0 I_x:-2 I_y:-9 H:-2
T	0	5 I_x:-4 I_y:-2 M:5	3 I_x:-4 I_y:-4 M:3	0 I_x:-4 I_y:-6 M:-4	3 I_x:-4 I_y:-8 H:3	3 I_x:-4 I_y:-10 H:3	0 I_x:-4 I_y:-11 H:-4	3 I_x:-4 I_y:-13 H:3	3 I_x:-4 I_y:-11 H:3
C	0	0 I_x:-6 I_y:-2 M:-2	3 I_x:-6 I_y:-4 M:3	8 I_x:-6 I_y:-6 M:8	0 I_x:-6 I_y:-6 H:-6	8 I_x:-6 I_y:-8 H:8	1 I_x:-6 I_y:-6 H:1	0 I_x:-6 I_y:-8 H:-6	1 I_x:-6 I_y:-10 H:1
A	0	0 I_x:-8 I_y:-2 M:-2	0 I_x:-8 I_y:-4 M:-4	1 I_x:-6 I_y:-6 M:1	6 I_x:-8 I_y:-8 H:6	0 I_x:-6 I_y:-8 H:-8	6 I_x:-8 I_y:-10 H:6	0 I_x:-8 I_y:-8 H:-1	0 I_x:-8 I_y:-10 H:-1
T	0	5 I_x:-10 I_y:-2 M:5	0 I_x:-10 I_y:-4 M:-4	0 I_x:-8 I_y:-6 M:-6	6 I_x:-8 I_y:-8 M:6	4 I_x:-8 I_y:-8 H:4	0 I_x:-8 I_y:-10 H:-10	11 I_x:-10 I_y:-12 H:11	0 I_x:-10 I_y:-3 H:-3
A	0	0 I_x:-9 I_y:-2 M:-2	3 I_x:-12 I_y:-4 M:3	0 I_x:-10 I_y:-6 M:-6	0 I_x:-8 I_y:-8 M:-8	4 I_x:-10 I_y:-10 H:4	2 I_x:-10 I_y:-10 H:2	0 I_x:-3 I_y:-12 H:-12	16 I_x:-12 I_y:-14 H:16

Fig. 4. Alignment matrix of local alignment with affine gap penalty with its optimal sequence alignment, match=+5, mismatch=-2, d=-12, e=-2 [17]

3 Systolic Arrays

Parallel architectures such as SA have been used in computing technologies to accelerate the algorithm in FPGA [18]. A linear SA consists of an array of PEs with each PE holds one DNA character of a query DNA sequence as shown in Fig. 5. The highest score will be stored in SA after the subject sequence is streamed in. If r is related to s, the score will be high.

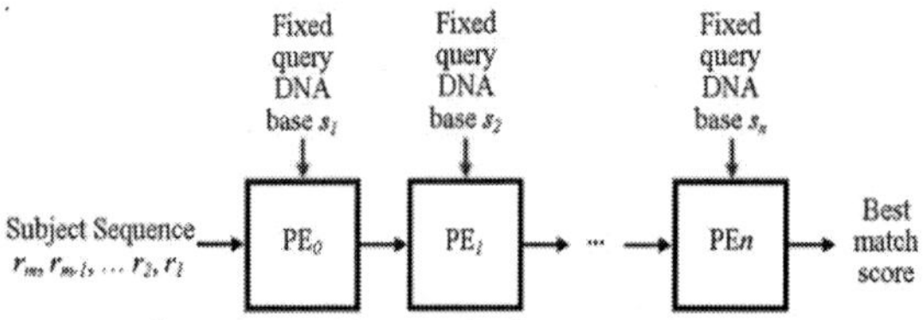

Fig. 5. A linear SA architecture [19].

The internal architecture of PE implement the affine gap penalty SW algorithm as shown in Fig. 6. The computation of the alignment algorithm ise main task of a PE comparing the score in the top, left and diagonal elements which have been described in previous section based on (1).

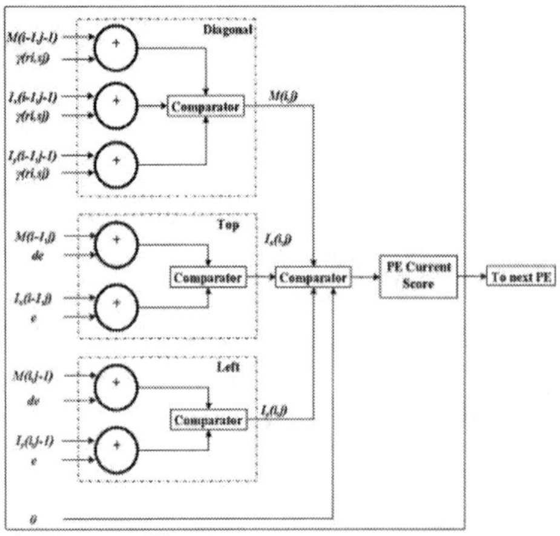

Fig. 6. PE core architecture.

Each PE performs one calculation due to the comparison of a DNA character between the query and subject sequence respectively. In each clock cycle, the PE generates one alignment matrix element. After one clock cycle, a column is generated in each PE where the column size is based on m. The execution of full alignment between the two sequences using SW algorithm in $O(m+n-1)$ time complexity is illustrated in Fig. 7.

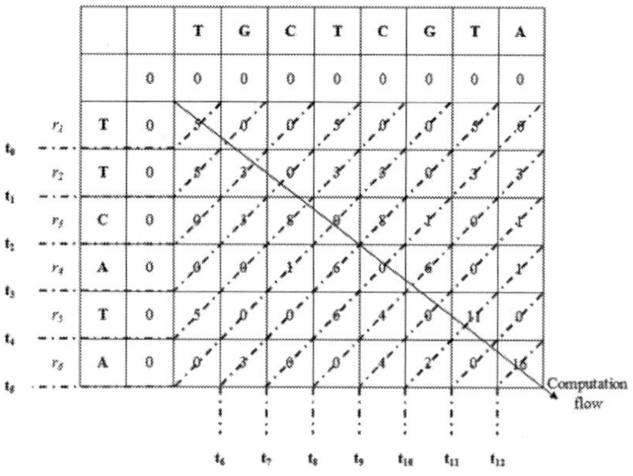

Fig. 7. Diagonal wavefront.

4 FPGA, the accelerator

The affine gap penalty SW algorithm is used as the backbone SA to speed up the computation process in aligning DNA sequence where the FPGA acts as an accelerator. Different FPGA vendors have different FPGA architecture. Altera uses ALMs (Adaptive Logic Module) while Xilinx uses Slices to represent their logic resources [20]. Slices are the combination of Flip Flops (FFs) and Look Up Tables (LUTs). The LUTs can either be 4-input or 6-input depending on devices. Virtex – 4 uses 4-input LUTs whereas Virtex- 6 uses 6-input LUTs. Altera's FPGA devices also uses 4-input LUTs or 6-input LUTs as the building block of ALM. Thus, normalization is required to analyze speed performance

* Corresponding author: nazrin@unimap.edu.my

equally between different logic resources which will be discussed in the next section.

The results were normalized in terms of Slices as the proposed core architecture is implemented in Virtex-6 FPGA. The performance of the core architecture will be evaluated in terms of (Slice / PE), total number of PEs that can be occupied by the device (#PE), peak frequency (Peak Freq), and Peak Cell Update Per Second (Peak CUPS). Slice / PE are used to find the total number of slices that can reside in a PE. #PE can be calculated by dividing the total slices of the device with Slice/PE. Peak Freq is the time in Hertz (Hz) of the longest path between 2 FFs in a PE which can be taken in the timing report generated by Xilinx software. Peak CUPS gives the total time based on #PE. It can be calculated by the multiplying the peak frequency with the #PE.

5 Result and discussion

The Verilog HDL which has been used is targetable for a variety of FPGAs platforms. The score obtained from the matrix elementary operation can be compared with the waveform from the proposed design simulated in Xilinx ISIM simulator as shown in Fig. 8. This is to ensure the best-matched score from the core architecture is the same as the matrix alignment in Fig. 4. The parameters used for the core architecture were the same as the parameters for matrix alignment operation.

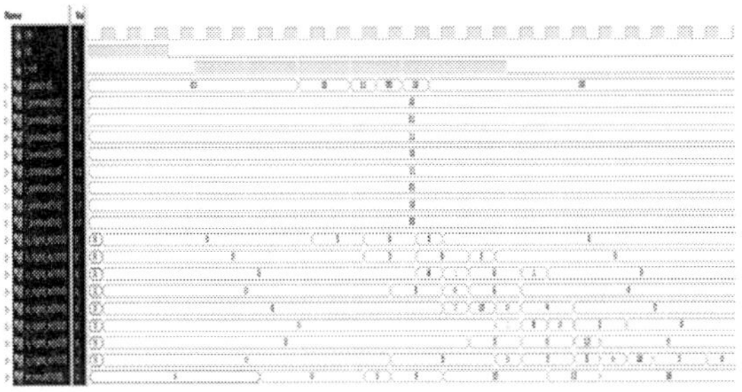

Fig. 8. Affine gap penalty SW algorithm computation output waveform.

In this work, the proposed core architecture is implemented in Virtex-6 FPGA whereby 6-input LUT is the building block of the device. For a fair performance comparison, the equivalent 6-input LUT is determined based on (2) for Stratix IV [20] and Virtex-4 and Spartan III [21] (3) first before determining the equivalent Slices using (4) [22]. The Logic Cell value (LC), Slices and ALM value can be retrieve from User Guide and Device Handbook respectively. Then, the equivalent slices are compared to reference slices as shown in Table 1 for normalized Slice/PE calculation.

$$6 - \text{Input LUT (Equivalent)} = \frac{ALM}{1.2} \quad (2)$$

$$6 - \text{Input LUT (Equivalent)} = \frac{LC}{0.5625} \quad (3)$$

$$\text{Slices (Equivalent)} = \frac{6\text{-Input LUT (Equivalent)}}{4} \quad (4)$$

Table 1. Logic blocks normalization.

FPGA Devices	Slice equivalent ratio with respect to Virtex - 6 XC6VLX75T
Virtex - 6 XC6VLX75T (Reference)	1
Virtex – 7 XC7VX485T	0.1533
Virtex – 5 XC5VLX330T	0.2245
Virtex – 4 XC4VLX100	0.2368
Spartan III XC3S250 XC3S1600E	4.754 0.7890
Stratix IV EP4SGX230KF40C2	0.1531

Virtex-6 FPGA has 11,640 slices where 776 PEs can be implemented in the aforementioned device with a 15 slices/PE. The proposed design was operated 600 MHz for the sequence alignment computation that is shown in Fig. 4. The comparison against other FPGAs platforms is reviewed in Table 2. Usually, the bottleneck for FPGA designs for area is the LUTs number [23] and timing is Peak CUPS [24]. Thus, based on Table 2, the proposed designed is the second slowest. However, the proposed design has the smallest slice as compared to other previous work. Since smaller core area can implement more # PEs, the speed performance is flopped to second place.

6 Conclusions

This paper reviewed on the current FPGA platform on implementing SA architecture for DNA sequence alignment. The proposed design is implemented on Xilinx Virtex - 6 FPGA using the affine gap penalty SW algorithm. The result shows that the proposed design has the smallest area core architecture and gained 465 GCUPS as compared to other architecture. Although it has the second highest in terms of speed, the core area is the smallest. This architecture design is suitable for high speed algorithm computation specifically for early disease detection such as cancer using its genetic profile.

Table 2. Performance comparison among devices.

Ref	Year	FPGA Device	Slice/PE	#PE	Peak Freq (MHz)	Peak GCups
[9]	2011	XC5VLX330T	75	1536	500	768
[10]	2012	XC4VLX100	90	128	60	7.61
[11]	2013	XC3S250E	332	35	50	1.75
[12]	2014	EP4SGX230KF40C2	1187	64	125	8
[13]	2016	XC3S1600E	96	241	98.7	23.79
[14]	2017	XC7VX485T	22	512	200	102.4
		XC6VLX75T	**15**	**776**	**600**	**465**

This work was supported by the FRGS under Grant 9003-00480. Modelling of an Efficient DNA/Protein Sequencer using Hardware Description Language (HDL) for Biomedical Applications.

References

1. F. H. C. Crick, "The Structure of DNA." Dept. of Biological Syst., The Cavendish Laboratory, Cambridge, England, Tech. Rep.
2. D. T. Hoang. "FPGA Implementation of Systolic Sequence Alignment," in *International Workshop on Field Programmable Logic and Applications,* Vienna, Austria, 1992.
3. D. T. Hoang, "Searching genetic databases on Splash 2", in *Proceedings of the IEEE Workshop on FPGAs for Custom Computing Machines,* 1993, pp. 185-191.
4. L. Hasan, Y. M. Khawaja, A. Bais, "A Systolic Array Architecture for The Smith-Waterman Algorithm With High Performance Cell Design," in *Proceedings of IADIS European Conference on Data Mining, Amsterdam, The Netherlands,* Jul., 2008.
5. (2015, Feb. 5). *Congressional Justification FY 2015.* Retrieved September 24, 2016 from https://www.nlm.nih.gov/about/2015CJ.html.
6. T. F. Smith and M. S. Waterman. "Identification of Common Molecular Subsequences," *Journal of Molecular Biology,* vol. 147, no. 1, pp. 195-197, Mar. 25, 1981.
7. R. Giegerich, "A Systematic Approach to Dynamic Programming in Bioinformatics.," *Bioinformatics,* vol. 16, pp. 665-677, 2000.
8. W. R. Pearson, "Using the FASTA Program to Search Protein and DNA Sequence Databases, Computer Analysis of Sequence Data: Part I, Methods in Molecular Biology," vol. 24, pp. 307-331, 1994.
9. A. Pułka and A. Milik, "Considerations on Incremental Approach to Hardware Implementation of Smith-Waterman Algorithm," in *Mixed Design of Integrated Circuits and Systems (MIXDES), 2011, IEEE 18th International Conference,* pp. 283-288, June 16-18, 2011.
10. N. Sebastião, N. Roma and P. Flores, "Integrated Hardware Architecture for Efficient Computation of the n-Best Bio-Sequence Local Alignments in Embedded Platforms," *IEEE Transactions on Very Large Scale Integration (VLSI) Systems,* vol. 20, no. 7, pp. 1262 – 1275, July,2012.
11. H. A. Shah, L. Hasan and N. Ahmad, "An Optimized and Low-cost FPGA-based DNA Sequence Alignment – A Step towards Personal Genomics," in *2013 35th Annual International Conference of the IEEE Engineering in Medicine and Biology Society (EMBS),* pp. 2696 – 2699, July 3 – 7, 2013.
12. J. Marmolejo-Tejada, M. V. Trujillo-Olaya, C. P. Rentería-Mejía and J. Velasco-Medina, "Hardware Implementation of the Smith-Waterman Algorithm using a Systolic Architecture," in *Circuits and Systems (LASCAS), 2014 IEEE 5th Latin America Symposium,* pp. 1 – 4, February, 2014.
13. D. S. Nurdin, M. N. Isa, and S. H Goh, "DNA sequence alignment: A review of hardware accelerators and a new core architecture," in *2016 IEEE 3rd International Conference on Electronic Design (ICED)*, pp. 264-268, August 11 -12, 2016.
14. X. Fei, Z. Dan, L. Lina, M. Xin and Z. Chunlei, "FPGASW: Accelerating Large-Scale Smith-Waterman Sequence Alignment Application with Backtracking pn FPGA Linear Systolic Array," *Interdisciplinary Sciences Computational Life Sciences,* pp. 1-13, 2017.
15. T. F. Smith and M. S. Waterman. "Identification of Common Molecular Subsequences," *Journal of Molecular Biology,* vol. 147 no. 1, pp. 195-197, Mar. 25, 1981.

16. O. Gotoh. "An improved algorithm for matching biological sequences," *Journal of Molecular Biology,* vol. 162, no. 3, pp. 705, 1982.
17. R. Durbin, S. Eddy, A. Krogh and G. Mitchison, *Biological Sequence Analysis:Probabilistic Models for Proteins and Nucleic Acids.* United Kingdom: Cambridge University Press, 1998.
18. K. Benkrid, A. Akoglu, C. Ling, Y. Song, Y. Liu and X. Tian, "High Performance Biological Pairwise Sequence Alignment: FPGA vs. GPU vs. Cell BE vs. GPP," *International Journal of Reconfigurable Computing,* Apr., 2012.
19. C. Fenton, (Aug. 2009,). *Final Project Report.*
20. "Advantages of the Virtex-5 FPGA 6-Input LUT Architecture,"Xilinx Inc., 2007.
21. "FPGA Logic Cell Conversion Ratios," Galorath Incorporated, 2014.
22. "Virtex-6 Family Overview," Xilinx, Inc., 2015.
23. "FPGA Logic Cells Comparison," Core technologies.
24. M. N. Isa, M. I. Ahmad, S. A. Z. Murad, R. C. Ismail, K. Benkrid, "Biological Sequence Alignments: A Review of Hardware Accelerators and a New PE Computing Strategy," *IEEE Region Symposium,* pp. 1-6, 2014.

The study of particle filter for satellite angular rate estimation without rate sensor measurement

Nor Hazadura Hamzah[1,*], *Sazali Yaacob*[2], *Ahmad Kadri Junoh*[1], and *Mohd Zamri Hasan*[3]

[1]Institute of Engineering Mathematics, Universiti Malaysia Perlis, Malaysia
[2]Malaysia Spanish Institute, Universiti Kuala Lumpur, Malaysia
[3]Faculty of Engineering Technology, Universiti Malaysia Perlis, Malaysia

Abstract. This paper studies particle filter algorithm to estimate the angular rate of a satellite without the rate sensor measurements. In this work, the performance of the algorithm is studied in terms of capability to estimate the angular rate by using the Euler angles attitude information only. The effects of the number of particles on the algorithm performance are also investigated in terms of accuracy and computational aspects. The performance of the particle filter algorithm is verified using real flight data of Malaysian satellite.

1 Introduction

Satellite attitude determination is one of the important aspects in Attitude Determination and Control System (ADCS) of a satellite. Satellite attitude is important to be determined in a satellite to be fed back to controller in accomplishing a specific satellite mission such as Earth observation, communication, scientific research and many other missions. However not all states are directly available may be due to malfunction sensor or as a way to obtain a substantial reduction of sensors which represents a cost and hardware complexity reduction.

In most practical implementations of ADCS, the angular rate and attitude information of a spacecraft are obtained respectively from measurement of rate sensor such as gyroscopes and also attitude sensor such as sun sensor, star sensor, or magnetometer. However, gyroscopes are generally expensive and are often prone to degradation or failure [1]. Therefore, as an alternative or backup system to circumvent the problem of rate sensor measurement absence, an estimation to provide the information of angular rate by using only the Euler angles attitude information is designed.

Since decades, a great number of research works have been devoted to the problem of estimating the attitude of a spacecraft based on a sequence of noisy vector observations such as [2][3][4][5]. Different algorithms have been designed and implemented in satellite attitude estimation problem. Early applications relied mostly on the Kalman filter for attitude estimation. Kalman filter was the first applied algorithm for attitude estimation for the Apollo space program in 1960s. Due to limitation of Kalman filter which work optimal for linear system only, several famous new approaches have been implemented to deal with the nonlinearity in satellite attitude system including extended Kalman filter (EKF) [6][7][4], unscented Kalman filter (UKF) [8][9][10], particle filter [11][12][13], and predictive filtering [14][15]. EKF is an extended version of Kalman filter for nonlinear system whereby the nonlinear equation is approximated by linearized equation through Taylor series expansion. UKF, an alternative to the EKF uses a deterministic sampling technique known as the unscented transform to pick a minimal set of sample points called sigma points to propagate the non-linear functions. EKF and UKF approaches is restricted assume the noise in the system is Gaussian white noise process. While, particle filter is a nonlinear estimation algorithm that approximates the nonlinear function using a set of random samples without restricted to a specific noise distribution as EKF and UKF.

In the open literature, spacecraft attitude estimation use different attitude representation either Euler angles, Rodrigues parameter, or quarternion parameter as their kinematic model [16]. Each kinematic model of different parameter is governed by different differential equation [17]. The researchers also studied the performance of estimated states by varying different type of sensor measurement such as gyroscope, magnetometer, sun sensor or star sensor.

To the best of authors' knowledge, there is no study that estimates the satellite's angular rate by implementing particle filter algorithm using nonlinear Euler angles attitude kinematics model. Particle filter is opted in this study because it is still infancy estimation algorithm in gyroless satellite attitude application and is believed or claimed by some researchers that beyond the EKF is the particle filter [21]. The Euler angles attitude representation is preferred in this work rather than quaternion because it can provide direct physical interpretation for analysis, contrary to quaternion. Hence, in this work the particle filter algorithm is studied in this application and is verified using real flight data of Malaysian satellite, RazakSAT.

* Corresponding author: hazadura@unimap.edu.my

The organization of this paper proceeds as follows. Section 2 presents the nonlinear satellite attitude dynamics model. Section 3 describes briefly the particle filter algorithm used in this work. Section 4 presents and discusses the results of the estimation system and Section 5 provides the paper's conclusions.

2 Nonlinear satellite attitude dynamics model

Satellite attitude dynamics model is described by both the dynamics equation of motion and kinematics equation of motion [18].

Dynamic equation of motion relates the satellite's angular rate to the exerted torque as defined by Euler's Moment Equation [18][17]

$$I\dot{\omega} + \omega \times I\,\omega = T \qquad (1)$$

or similarly is written in component-wise

$$\dot{\omega}_x = -\left(\frac{I_z - I_y}{I_x}\right)\omega_y\omega_z + \frac{T_x}{I_x} \qquad (2)$$

$$\dot{\omega}_y = -\left(\frac{I_x - I_z}{I_y}\right)\omega_x\omega_z + \frac{T_y}{I_y} \qquad (3)$$

$$\dot{\omega}_z = -\left(\frac{I_y - I_x}{I_z}\right)\omega_x\omega_y + \frac{T_z}{I_z} \qquad (4)$$

with

$I = diag[I_x, I_y, I_z]$ is satellite moment of inertia matrix.

$\dot{\omega} = [\dot{\omega}_x, \dot{\omega}_y, \dot{\omega}_z]$ is angular acceleration vector.

$\omega = [\omega_x, \omega_y, \omega_z]$ is angular rate vector.

$T = [T_x, T_y, T_z]$ is space environmental disturbances torque vector.

For low Earth orbit satellite, gravity gradient torque must be taken into consideration as part of external torque since it is continuously acting on the spacecraft body and influence the satellite's attitude motion. The external torque dominated by gravity gradient torque is written as [17][19]

$$T = \begin{bmatrix} T_x \\ T_y \\ T_z \end{bmatrix} = \begin{bmatrix} 3\omega_0^2(I_z - I_y)s\emptyset c\emptyset c^2\theta \\ 3\omega_0^2(I_z - I_x)s\theta c\theta c\emptyset \\ 3\omega_0^2(I_x - I_y)s\emptyset c\theta s\theta \end{bmatrix} \qquad (5)$$

While, kinematic equation of motion relates the attitude parameter to the satellite's angular rate. In this work, Euler angles parameter is preferred to represent the satellite's attitude as its straightforward physical interpretation for analysis. Euler angles are defined as the rotational angles about the body axis as follows: $\emptyset$ is rotational angle about X-axis (roll); θ is rotational angle about Y-axis (pitch); and φ is rotational angle about Z-axis (yaw). The kinematic equation of Euler angles parameter using $\varphi - \theta - \emptyset$ (or some literature use notation 3-2-1) sequence rotation is

$$\dot{p} = \begin{bmatrix} \dot{\emptyset} \\ \dot{\theta} \\ \dot{\varphi} \end{bmatrix} \qquad (6)$$

with

$$\begin{aligned}\dot{\emptyset} = {} & [\omega_x + \omega_0 c\theta s\varphi\,] \\ & + s\emptyset t\theta[\omega_y + \omega_0(c\emptyset c\varphi + s\emptyset s\theta s\varphi)] \\ & + c\emptyset t\theta[\omega_z + \omega_0(-s\emptyset c\varphi + c\emptyset s\theta s\varphi)]\end{aligned} \qquad (7)$$

$$\dot{\theta} = c\emptyset[\omega_y + \omega_0(c\emptyset c\varphi + s\emptyset s\theta s\varphi)] - s\emptyset[\omega_z + \omega_0(-s\emptyset c\varphi + c\emptyset s\theta s\varphi)] \qquad (8)$$

$$\dot{\varphi} = \frac{s\emptyset}{c\theta}[\omega_y + \omega_0(c\emptyset c\varphi + s\emptyset s\theta s\varphi)] + \frac{c\emptyset}{c\theta}[\omega_z + \omega_0(-s\emptyset c\varphi + c\emptyset s\theta s\varphi)] \qquad (9)$$

where c, s and t denote cosine, sine, and tangent functions, respectively. While, ω_0 is the orbital rate of the spacecraft.

A complete formulation of the satellite attitude dynamics for low Earth orbit satellite is obtained by combining both the dynamics equation of motion under influence of gravity gradient torque with the kinematics equation of motion such that

$$\begin{bmatrix} \dot{\omega} \\ \dot{p} \end{bmatrix} = \begin{bmatrix} \dot{\omega}_x \\ \dot{\omega}_y \\ \dot{\omega}_z \\ \dot{\emptyset} \\ \dot{\theta} \\ \dot{\varphi} \end{bmatrix} \qquad (10)$$

with

$$\dot{\omega}_x = -\left(\frac{I_z - I_y}{I_x}\right)\omega_y\omega_z + 3\omega_0^2\frac{(I_z - I_y)}{I_x}s\emptyset c\emptyset c^2\theta \qquad (11)$$

$$\dot{\omega}_y = -\left(\frac{I_x - I_z}{I_y}\right)\omega_x\omega_z + 3\omega_0^2\frac{(I_z - I_x)}{I_y}s\theta c\theta c\emptyset \qquad (12)$$

$$\dot{\omega}_z = -\left(\frac{I_y - I_x}{I_z}\right)\omega_x\omega_y + 3\omega_0^2\frac{(I_x - I_y)}{I_z}s\emptyset c\theta s\theta \qquad (13)$$

$$\begin{aligned}\dot{\emptyset} = {} & [\omega_x + \omega_0 c\theta s\varphi\,] \\ & + s\emptyset t\theta[\omega_y + \omega_0(c\emptyset c\varphi + s\emptyset s\theta s\varphi)] \\ & + c\emptyset t\theta[\omega_z + \omega_0(-s\emptyset c\varphi + c\emptyset s\theta s\varphi)]\end{aligned} \qquad (14)$$

$$\dot{\theta} = c\emptyset[\omega_y + \omega_0(c\emptyset c\varphi + s\emptyset s\theta s\varphi)] - s\emptyset[\omega_z + \omega_0(-s\emptyset c\varphi + c\emptyset s\theta s\varphi)] \qquad (15)$$

$$\dot{\varphi} = \frac{s\emptyset}{c\theta}[\omega_y + \omega_0(c\emptyset c\varphi + s\emptyset s\theta s\varphi)] + \frac{c\emptyset}{c\theta}[\omega_z + \omega_0(-s\emptyset c\varphi + c\emptyset s\theta s\varphi)] \qquad (16)$$

3 Particle filter algorithm

Particle filter or also known as Sequential Monte Carlo (SMC) is an on-line, recursive algorithm trying to estimate the true state of a nonlinear system where only some noisy measurements are available. Particle filter does not require any assumption about the state-space or the noise of the system to be Gaussian as restricted in conventional methods EKF and UKF. The underlying idea behind the approach is to approximate the posterior density using a set of particles. Particle filter was developed in 1993 by Gordon et al [20] under the name of the 'bootstrap filter algorithm' by including the resampling technique to reduce the degeneracy problem of the existing Monte Carlo approach.

Particle filter algorithm is described as follows [18]. Given the nonlinear system dynamics and measurement described by continuous-time model [19]

$$\dot{x} = f(x) + w \qquad (17)$$

$$y = h(x) + v \qquad (18)$$

with $x \in R^n$ is the state vector and $y \in R^p$ is the output vector, w and v denote the noise or uncertainty vector in the state and measurement respectively. Then the continuous-time model in Equation (17) and (18) is transformed into the discrete-time model such that

https://doi.org/10.1051/matecconf/201815006010

$$x_k = f(x_{k-1}) + w_{k-1} \quad (19)$$
$$y_k = h(x_k) + v_k \quad (20)$$

Here the subscript of the variables denotes the time step, while w_{k-1} and v_k are state and measurement noises respectively with variance R_w and R_v. Hence, the estimated state is obtained through the following step [21][22]:

Step 1: Set the number of particles N_s and set the initialization

- Initial state estimate: $\hat{x}_0 = \hat{x}_{0|0}$ (21)
- Initial particles: $\chi_0^i = \hat{x}_{0|0}$ for $i = 1,2,\cdots,N_s$ (22)
- Initial weight: $w_0^i = \frac{1}{N_s}$ for $i = 1,2,\cdots,N_s$ (23)

Step 2: Repeat

(i) Sequential importance sampling

- Draw particles:

$$\chi_k^i \sim N\big(x_k; f_{k-1}(\chi_{k-1}^i), R_w\big) \text{ for } i = 1,2,\cdots,N_s \quad (24)$$

- Compute the weight for each particle:

$$w_k^i = w_{k-1}^i N\big(z_k; h(\chi_k^i), R_v\big) \text{ for } i = 1,2,\cdots,N_s \quad (25)$$

- Calculate the total weight:

$$T = \sum_{i=1}^{N_s} w_k^i \quad (26)$$

- Normalize the weight:

$$w_k^i = \frac{w_k^i}{T} \text{ for } i = 1,2,\cdots,N_s \quad (27)$$

(ii) Resampling (To eliminate samples with low importance weights)

- Initialize cumulative sum of weight (CSW):

$$c_1 = 0 \quad (28)$$

- Construct CSW:

$$c_i = c_{i-1} + w_k^i \text{ for } i = 2,3,\cdots,N_s \quad (29)$$

- Start at the bottom of the CSW: $i = 1$ (30)
- Draw a starting point: $u_1 \sim U\left[0, \frac{1}{N_s}\right]$ (31)
- For $j = 1,2,\cdots,N_s$
  - o Move along the CSW:

$$u_j = u_1 + \frac{1}{N_s}(j-1) \quad (32)$$

  - o Set if $u_j > c_i$, then update $i = i + 1$ (33)
  - o Assign particles: $\chi_k^j = \chi_k^i$ (34)
  - o Assign weight: $w_k^j = \frac{1}{N_s}$ (35)

(iii) State estimation

- Compute estimated state:

$$\hat{x}_{k|k} = \sum_{i=1}^{N_s} \chi_k^i w_k^i \quad (36)$$

4 Result and discussion

In this section, the performance of particle filter as an estimation algorithm to estimate the angular rate without the rate sensor measurements is investigated.

The performance of the estimation using particle filter algorithm is validated using real sensors data from the RazakSAT satellite. RazakSAT is a Malaysian satellite which was launched into low Earth orbit near Equatorial in 2009. In the mission, the quaternion attitude was provided by using sun sensor, one of the attitude sensor, while the angular rate was provided by gyroscope sensor. The available measurements are for about six orbits sequentially, available to the ground system at a sampling rate of about 1 minute. The sample of RazakSAT's telemetry data is shown in Fig. 1, while the satellite's characteristics of RazakSAT are given in Table 1, which was provided by Astronautic Technology Sdn Bhd (ATSB), the Malaysian company that responsible for RazakSAT's mission.

TIME	ATTITUDE (QUATERNION)				GYRO OUTPUT (deg/s)		
	Q1	Q2	Q3	Q4	X	Y	Z
10:58:51	-0.775506	0.451796	-0.431151	0.092625	-0.008350	0.031246	0.048060
10:59:51	-0.218972	0.800993	-0.551323	0.080649	0.000562	-0.000396	0.002956
11:00:51	-0.209406	0.802330	-0.549578	0.101878	0.000110	-0.000117	0.000048
11:01:54	-0.209518	0.804907	-0.545130	0.105169	0.000717	0.000156	-0.000647
11:02:54	-0.210605	0.801599	-0.550194	0.101838	0.000130	-0.000080	0.000033
11:03:56	-0.211824	0.801596	-0.549643	0.102305	0.000011	0.000216	-0.000190
11:04:56	-0.210696	0.800860	-0.551182	0.102125	0.000087	0.000042	-0.000018
11:05:56	-0.210882	0.800777	-0.551268	0.101926	0.000031	0.000016	-0.000050
11:06:56	-0.211182	0.800579	-0.551430	0.101980	0.000149	0.000080	0.000033
11:07:56	-0.211328	0.800571	-0.551364	0.102101	0.000147	-0.000235	0.000191
11:08:56	-0.211150	0.800743	-0.551232	0.101833	-0.000098	0.000155	0.000025
11:09:56	-0.211046	0.800647	-0.551360	0.102107	0.000025	-0.000089	0.000095
11:10:56	-0.211360	0.800523	-0.551463	0.101875	0.000015	0.000042	0.000050
11:11:56	-0.211435	0.800452	-0.551498	0.102084	-0.000019	-0.000171	0.000104
11:12:56	-0.211487	0.800628	-0.551240	0.101992	0.000015	0.000056	0.000081

Fig. 1. Sample of RazakSAT telemetry data.

Table 1. RazakSAT's characteristics.

Parameter	Values
Moment of inertia, I_x	25.4 kg.m^2
Moment of inertia, I_y	26.2 kg.m^2
Moment of inertia, I_z	21.0 kg.m^2
Orbital rate, ω_0	0.001063 rad/s or 0.0609 deg/s

The first study is to validate the capability of particle filter to estimate the angular rate by using measurement Euler angles attitude roll, pitch and yaw only. The accuracy of the estimated states is assessed by comparing the estimated states with the real states provided by sensor data of RazakSAT. Figs. 2, 3, and 4 show comparisons between the estimated angular rates using particle filter algorithm and the real angular rates measurement provided by gyroscope sensor in RazakSAT mission, respectively around X-axis, Y-axis, and Z-axis. From all the three figures, it is observed that the behaviour of estimated states does not scattered at random, otherwise they seem deviate from the approximately zero values of the real states as measured by gyroscope. This deviation is due to mis-modeling and the influence of other unmodeled space environmental disturbance torques including aerodynamic, solar radiation pressure and Earth magnetic torques. However as overall, the trendlines of the estimated states are still within 0.1 deg/s ranges which are suitable for moderate accuracy attitude determination mode such as during housekeeping and detumbling task, but unsuitable for pointing mode that requires more precise accuracy such as during imaging task.

https://doi.org/10.1051/matecconf/201815006010

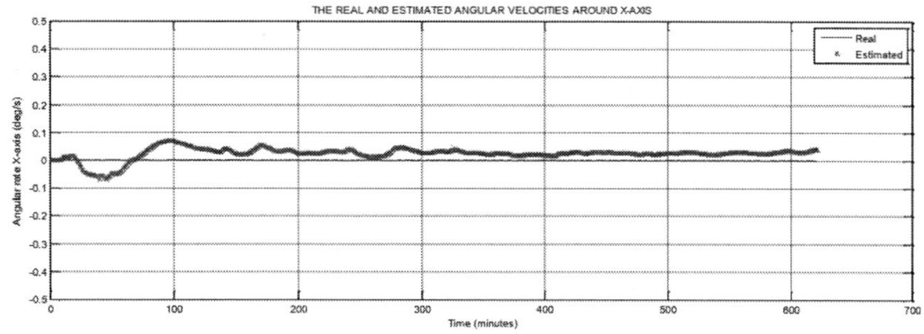

Fig. 2. Comparison between estimated and real angular rate around X-axis.

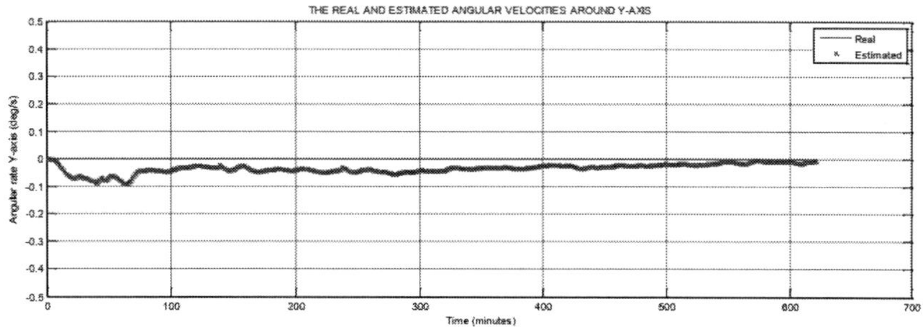

Fig. 3. Comparison between estimated and real angular rate around Y-axis.

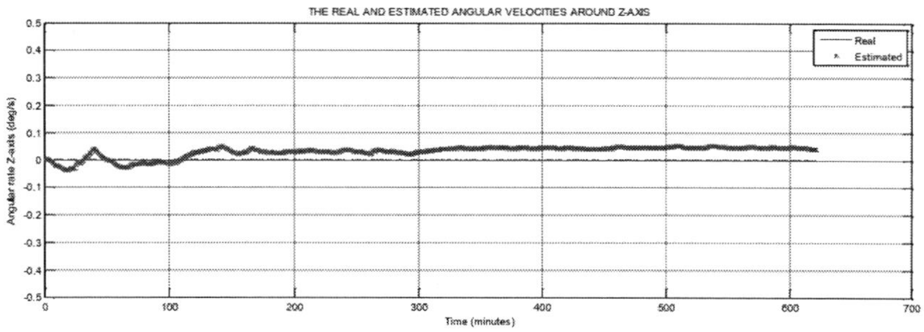

Fig. 4. Comparison between estimated and real angular rate around Z-axis.

The accuracy of the estimated states is validated using Root Mean Squared Error (RMSE) with

$$RMSE = \sqrt{\frac{\sum_{i=1}^{n}\left(x_{i,ESTIMATE} - x_{i,REAL}\right)^2}{n}} \qquad (37)$$

The RMSE is a frequently used measure of the differences between values estimated by an estimator and the values actually observed. It is a representative of the size of an average error. The RMSE value of the estimated angular velocity is tabulated in Table 2. From the table, the average error between the estimated and the real states are 0.0339 deg/s, 0.0371 deg/s, 0.0387 deg/s respectively for angular rate around X-axis, Y-axis, and Z-axis. As overall it can be deduced that the particle filter algorithm is able to provide the information of angular rate in less than 0.1 deg/s average error which is a quite good performance filter.

Table 2. RMSE of estimated angular rate.

States	RMSE value
Angular rate around X-axis, ω_x	0.0339 deg/s
Angular rate around Y-axis, ω_y	0.0371 deg/s
Angular rate around Z-axis, ω_z	0.0387 deg/s

The second study is to investigate the effects of number of particles, N_s on the particle filter algorithm performance in terms of accuracy and computational aspects. The accuracy and computational aspects are assessed respectively using norm of error and CPU time. Norm of error and CPU time generated for different values of the number of particles are tabulated in Table 3. The divergence tracks are resulted for the number of particles $N_s \leq 5$. This is because very small number of particles is not enough to represent the density hence gives divergence result. Fig. 5 plots the norm of error as a function of the number of particles for $6 \leq N_s \leq 1000$. It is observed that the norm of error is drastically decreased until values about 0.07 deg/s at the number of particles $N_s \leq 10$. However, note that as N_s is increased beyond 100, there are insignificant changes of the norm of error values. This is because if the particle numbers are fewer, the importance distribution is not accurate enough to represent the real distribution hence affect the estimation accuracy. While the plot of the CPU time as a function of the number of particles for $6 \leq N_s \leq 1000$ is shown in Fig. 6. It is observed that CPU time increase

* Corresponding author: hazadura@unimap.edu.my

linearly as the number of particles increased. This is because the intensive memory and computational effort is required when the number of particles is increased. Hence it is concluded that the optimal value of number of particles is $N_s = 100$ for efficient implementation in this problem.

All the studies was performed by using Matlab R2013(a) in Windows 7 operation system with the specification of Intel Core i7, 2.0GHz CPU and 4GB RAM.

Table 3. Norm of error and CPU time with different values of the number of particles.

Number of particles, N_s	Norm of error, $\|\|e\|\|$ (deg/s)	Computation/ CPU time, t (s)
1	NaN	-
2	NaN	-
3	NaN	-
4	NaN	-
5	NaN	-
6	0.1199	0.645
7	0.1053	0.710
8	0.0763	0.814
9	0.0743	0.910
10	0.0717	0.998
100	0.0634	9.327
200	0.0643	18.652
300	0.0632	29.314
400	0.0638	38.541
500	0.0658	48.019
600	0.0649	58.673
700	0.0650	67.377
800	0.0646	75.940
900	0.0650	85.399
1000	0.0658	95.131

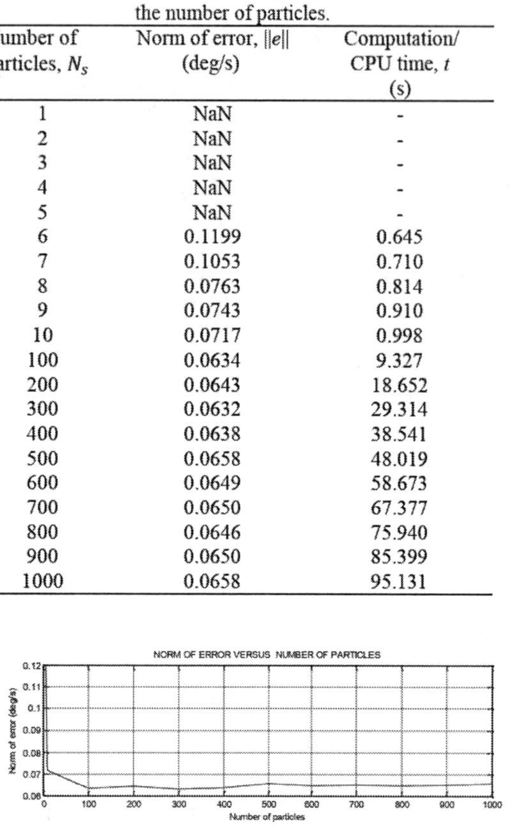

Fig. 5. Norm of error versus number of particles.

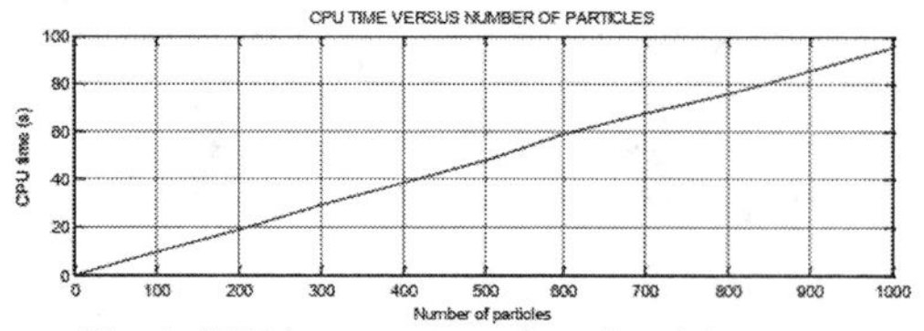

Fig. 6. CPU time versus number of particles.

5 Conclusion

In this paper, particle filter estimation algorithm to provide the angular rate information during the absence of rate sensor circumstances was studied. The capability of the algorithm to provide the angular rate information using Euler angles attitude only is validated using real flight data of RazakSAT, the Malaysian satellite. The results show that the algorithm is able to provide the information of satellite angular rate within 0.1 deg/s accuracy, which is suitable for moderate accuracy attitude determination such as during housekeeping and detumbling mode. Besides, the effects of number of particles on the particle filter algorithm performance are also investigated in terms of accuracy and computational aspects through assessment of the error norm and CPU time. The results show that it is important to determine the optimal value of number of particles for efficient implementation. This work can be used as an alternative or backup attitude determination system of a low Earth orbit satellite during the absence of angular rate measurement due to unexpected sensor failure or for cost reduction. Future work will compare the PF algorithm with the extended Kalman filter (EKF), the most widely employed algorithm in spacecraft community both in real practice and theoretical study.

References

1. J.L. Crassidis, F.L. Markley: An MME-based Attitude Estimator using Vector Observations. *NASA Conf. Publ.* pp. 137 (1995)
2. J.A. Christian, E.G. Lightsey: The Sequential Optimal Attitude Recursion Filter. J. Guid. Control. Dyn. pp. 1787–1800 (2010)
3. H.E. Emara-Shabaik: Spacecraft Spin Axis Attitude. IEEE Trans. Aerosp. Electron. Syst. **vol. 28**, no. 2, pp. 529–534 (1992)
4. P. Appel: Attitude Estimation from Magnetometer and Earth-Albedo-Corrected Coarse Sun Sensor Measurements. Acta Astronaut. **vol. 56**, no. 2, pp. 115–126 (2005)
5. B. Liu, Z. Chen, X. Liu, F. Yang: An Efficient Nonlinear Filter for Spacecraft Attitude Estimation. Int. J. Aerosp. Eng. **vol. 2014**, pp. 1–11 (2014)
6. M.N. Filipski & R. Varatharajoo: Evaluation of a Spacecraft Attitude and Rate Estimation Algorithm. Aircr. Eng. Aerosp. Technol. **vol. 82**, no. 3, pp. 184–193 (2010)
7. M. Fadly, O. Sidek, A. Said, H. Djojodihardjo, A. Ain: Deterministic and Recursive Approach in Attitude Determination for InnoSAT. TELKOMNIKA, **vol. 9**, no. 3, pp. 583–594 (2011)
8. Y.J. Cheon, J.H. Kim: Unscented Filtering in a Unit Quaternion Space for Spacecraft Attitude Estimation. *In 2007 IEEE Int. Symp. Ind. Electron.* pp. 66–71 (2007)
9. M.C. Vandyke, J.L. Schwartz, C.D. Hall: Unscented Kalman Filtering for Spacecraft Attitude State and Parameter Estimation. Department of Aerospace & Ocean Engineering, Virginia Polytechnic Institute & State University, pp. 1–13 (2004)
10. J.H. Bae, Y.D. Kim: Attitude Estimation for Satellite Fault Tolerant System using Federated Unscented Kalman Filter. Int. J. Aeronaut. Sp. Sci., **vol. 11**, no. 2, pp. 80–86 (2010)
11. Y.Liu, X. Jiang, G. Ma: Marginalized Particle Filter for Spacecraft Attitude Estimation from Vector Measurements. J. Control Theory Appl., **vol. 5**, no. 1, pp. 60–66 (2007)
12. Y. Yafei, L. Jianguo: Particle Filtering for Gyroless Attitude/Angular Rate Estimation Algorithm. *In 2010 First Int. Conf. Pervasive Comput. Signal Process*. pp. 1188–1191 (2010)
13. Y. Cheng, J.L. Crassidis: Particle Filtering for Sequential Spacecraft Attitude Estimation. *In Proc. of the 2004 AIAA Guidance, Navigation, and Control Conference*, p. 5337 (2004)
14. J.L. Crassidis, F.L. Markley, E.G. Lightsey, E. Ketchum: Predictive Attitude Estimation using Global Positioning

System Signals. *In NASA Conference Publication*, pp. 1–14 (1997)
15. J.L. Crassidis, F.L. Markley: Predictive Filtering for Attitude Estimation Without Rate Sensors. J. Guid. Control. Dyn., **vol. 20**, no. 3, pp. 522–527 (1997)
16. M.D. Shuster: A survey of Attitude Representation. J. Astronaut. Sciense, **vol. 41**, no. 4, pp. 493–517 (1993)
17. M.J. Sidi: *Spacecraft Dynamics and Control: A Practical Engineering Approach.* Cambridge University Press, Cambridge (1997)
18. J.L. Crassidis, J.L. Junkins: *Optimal Estimation of Dynamic Systems*. CRC Press, Boca Raton (2004)
19. M.M. Aly, H.A. Abdel Fatah, A. Bahgat: Nonlinear Observers for Spacecraft Attitude Estimation in Case of Yaw Angle Measurement Absence. Int. J. Control. Autom. Syst., **vol. 8**, no. 5, pp. 1018–1028 (2010)
20. N.J. Gordon, D.J. Salmond, A.F.M. Smith: Novel Approach to Nonlinear/Non-Gaussian Bayesian State Estimation. IEE Proc. F Radar Signal Process., **vol. 140**, no. 2, pp. 107–113 (1993)
21. B. Ristic, M.S. Arulampalam, N. Gordon: *Beyond the Kalman filter: Particle filters for Tracking Applications*. Artech House, Boston (2004)
22. M.S. Arulampalam, S. Maskell, N. Gordon, T. Clapp: A Tutorial on Particle Filters for Online Nonlinear/Non-Gaussian Bayesian Tracking. IEEE Trans. Signal Process, **vol. 50**, no. 2, pp. 174–188 (2002)

Effect of torrefaction pre-treatment on physical and combustion characteristics of biomass composite briquette from rice husk and banana residue

Nor Amira Atan[1,*], *Munira* Mohamed Nazari[1], and *Fathin* Ayuni Azizan[1]

[1]School of Bioprocess Engineering, Universiti Malaysia Perlis, 02600 Arau, Perlis, Malaysia

Abstract. Biomass is an alternative renewable energy sources that can generates energy almost same as fossil fuel. The depletion sources of fossil fuel had increase the potential use of biomass energy. In Malaysia, rice husk and banana residues are abundantly left and not treated with proper disposal method which later may contribute to environment and health problems. Thus the development of biomass composite briquette made from rice husk and banana residue is one of the potential ways to reduce the problems and hence may contribute the better way to treat the waste by recycling the waste into a form of biomass product. The biomass briquettes are used for thermal applications because it can produce a complete combustion as it has a consistent quality and high burning efficiency. However, the quality of the biomass briquette can be added by application of torrefaction pre-treatment method. Torrefaction is a thermal method that can produce more high quality of the briquette with high calorific value, high fixed carbon content, low volatile matter, and low ash content. This study was conducted to assess the physical and combustion characteristic of the biomass briquette from rice husk and banana residue which was produced through torrefaction process. The biomass briquette, were densified by using hot press machine with temperature of 180°C for about 30 minutes. The briquette produce are 150 μm in particle size with varies in mixing ratio of rice husk to banana residue which are 100:0, 80:20 and 60:40. After the briquetting process, the biomass fuel briquettes have been undergoes parameter testing and the data have been analysed. Result showed the best biomass briquette is developed from torrefied rice husk and banana residue mixed at ratio of 60:40. Moreover, SEM image reveal that torrefaction pre-treatment has shrinkage the fibres size which confirming the thermal stability of the briquette.

1 Introduction

Biomass briquette is a biofuel replacement to lignite-coal and charcoal which usually made from agricultural waste and other organic materials. Briquette is an alternative to fossil fuel as they are renewable, non-polluting and an economic of energy source. Biomass briquette produce through briquetting process where the waste material undergo densification to produce compact compounds with the use of pressure. Briquette is ideally sized for complete combustion as it has a consistent quality and high burning efficiency. However, the best quality of biomass briquette depends on some factors such as the crushing strength, moisture content, density, flame propagation and ash content value. A part from that, to get a good quality of biomass composite briquette it also needs low moisture content, ash content, and slow flame propagation in order to have uniform and complete combustion process.[1]

A research on combustion performance of biomass composite briquette from rice husk and banana residue had been done before and the result shows that the rice husk and banana residue with particle size of 150 μm and mixing ratio of 60:40 have a huge potential to become a commercialized product by briquetting as it meets the properties for a good fuel briquette [2]. Table 1 shows the summary result obtained from the research.

Table 1: The physical and combustion performance of rice husk and banana residue briquette.

Parameter Testing	Value
Density (g/cm$^{3)}$	0.5675
Ignition time (s)	26.32
Burning rate (cm/s)	0.2193
Calorific value (kJ/kg)	16305
Ash content (%)	12.91
Volatile matter (%)	75.43
Fixed carbon content (%)	11.66

* Corresponding author: noramiraatan93@gmail.com

https://doi.org/10.1051/matecconf/201815006011

In that research, the raw material was not treated with any pretreatment and the best combustion performance in term of calorific value is 16205 kJ/kg. However, the performance of the briquette can be better through addition of pre-treatment process. So that, the main purpose of this paper was to enhance the current production method of biomass composite briquette from rice husk and banana residue through torrefaction pre-treatment method.

Torrefaction process is a thermal pretreatment for biomass taking place in temperature range of 200°C to 300°C at up to 30 minutes with the nonexistence of oxygen. According to J. S. Tumuluru *et al.* the torrefaction pretreatment enhance the biomass physical properties and increase from 15 percent to 25 percent of heating value of the briquette [3]. The torrefaction pre-treatment towards the raw material improves the energy density and the overall fuel characteristic of the torrefied briquette. Figure 1 illustrates the flow diagram for basic production of torrefied biomass composite briquette.

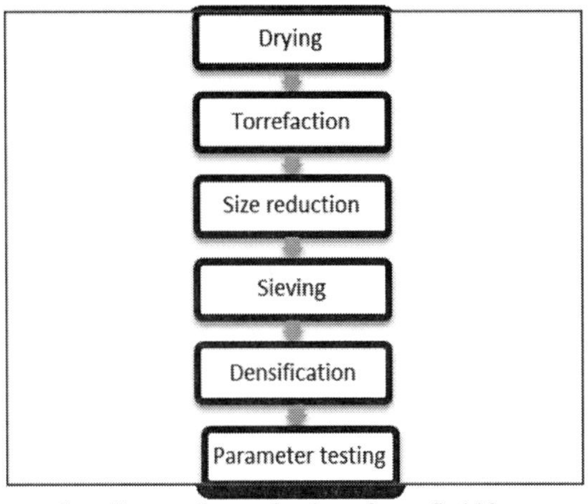

Fig. 1: Flow diagram for production of torrefied biomass composite briquette.

2 Material and method

The production of biomass composite briquette were starts from the collection and preparation of raw material, the pre-treatment process, size reduction, sieving and densification process. Then the biomass briquette produced were analysed of its physical, proximate and morphology.

2.1. Collection and preparation of raw materials

The raw materials which are rice husk and banana residue were collected from KBB Simpang Empat, Perlis rice mill and the nearby fried banana stalls. Then, sun dried method followed by oven dry method with temperature of 105 °C was applied in order to reduce the moisture content of the raw materials until it reached constant moisture content. The moisture content of the raw materials was then calculated as follow:

$$\% \text{ Moisture content} = (W_i - W_f)/ W_i \times 100\% \quad (1)$$

Where W_i and W_f are weight of test raw materials before drying and when oven-dried, respectively in gram unit (g).

2.2 Pre-treatment process

The dried raw materials were then undergoing torrefaction treatment process. The rice husk and banana residue were placed in a furnace and torrefied at 300°C for 30 minutes. Figure 2 and Figure 3 shows the rice husk and banana residue after undergo the torrefaction pretreatment. Both of the raw materials were turns brown to black, which can be mainly because to the change in the chemical composition of the torrefied biomass.

Fig. 2: Torrefied rice husk

Fig. 3: Torrefied banana residue

2.3 Production of biomass composite briquette.

The torrefied samples were ground and sieved into small particle size (150 μm) in order to increase the surface area of the material which improves on the binding and strength of the briquette samples before they were compacted. Then, the briquette was densified by using hot press machine as shown in figure 4 and mould size of 3.0 cm × 3.0 cm with mixing ratio of rice husk and banana residue at 100:0, 80:20 and 60:40. Figure 5 shows the example of briquette produced after the densification process was done at temperature of 180°C for 30 minutes.

Fig. 4: Hot press machine

Fig. 5: Briquette produced made from mixing of torrefied rice husk and banana residue

2.3 Physical and proximate analysis

In this analysis, there were seven tests that have been done in order to complete the proximate analysis of the briquette and each analysis was replicated 3 times to ensure the accuracy of the results obtained. The analysis that were carried out were density, ignition time, combustion rate, ash content, volatile matter, fixed carbon content and calorific value. The methodologies summarized of proximate analysis were discussed below.

2.3.1 Density

Density was determined from the ratio of the mass over the volume of the briquette. Mass was taken by weighing the briquette on the digital weighing scale and volume using the formula of cylinder. Equation 2 shows the formula to calculate the density.

$$\text{Density, } \rho = \text{mass/volume,(kg/m}^3) \quad (2)$$

2.3.2 Ignition time

Each of the samples of the biomass composite briquette was ignited at a free corner. By using stop watch, the time taken for the flame to ignite was recorded as the ignition time.

2.3.3 Combustion rate

Combustion rate can be characterized as the portion of the mass of the fuel burn to the total time taken. The samples were placed on the wire gauze and the burner starts to ignite. Combustion rate was determined as follows:

$$\text{Combustion rate} = \text{distance (cm)/ afterglow time (s)} \quad (3)$$

2.3.4 Ash content

The ash content can be characterizes as approximate measure of the mineral substance and other inorganic matter in biomass. The biomass composite briquette was weighted before the burning process. At that point, the briquette was undergoing burning process by following the ASTM E1755-01 (2007) where the briquette were burned at 575 °C for 4 hours inside the furnace. Then, the weight of ash was determined by using analytical balance. Ash content was determined as follows:

$$\text{\% Ash content} = (\text{Weight}_{\text{ash}})/(\text{Weight}_{\text{sample}}) \times 100 \quad (4)$$

2.3.5 Volatile matter

2g of the briquette sample were filled in the crucible and covered with a lid before it was placed in furnace at temperature of 550 °C for 10 minutes according to the ASTM 872-82 (2006). The crucible was cooled first in air inside desiccators and then the remaining samples were weighed again. Loss in weight was reported as volatile matter on percentage basis. The percentage of the volatile matter was calculated as follows:

$$\text{\% Volatile Matter} = (W_i - W_f)/ W_i \times 100\% \quad (5)$$

Where w_i and w_f is the weight initial and final of the samples briquette.

2.3.6 Fixed carbon content

The percentage fixed carbon was calculated using the equation of the sum of the percentages of ash content and volatile matter was subtracted from 100% in accordance with applying the mass balance for the biomass sample.

2.3.7 Calorific value

The calorific value of material was obtained by using Bomb Calorimeter (C 200 Basic), where the environment of the combustion was 25 atmospheric pressure of oxygen in a closed vessel which will be surrounded by water under controlled surrounding to ensure complete combustion according to test method of DIN 51900. After all briquettes samples were fully burned, the result of calorific value was show in kJ/kg.

2.4 Morphology analysis

The treated rice husk and banana residue were then analyzed and observed by using Scanning Electron Micrograph (SEM).

2.4.1 Scanning Electron Micrograph (SEM) analysis

SEM analysis was done in order to observe the surface morphology of the raw materials before and after the pretreatment process as effect from the torrefaction pretreatment to the rice husk and banana residue.

3 Result and discussion

3.1 Physical and proximate analysis

The result for this proximate analysis was discussed based on the effect of different mixing ration of torrefied rice husk and banana residue and the effect of torrefaction pre-treatment towards the briquette produced by comparing with the data stated in table 1.

3.1.1 Density

Density is one of the important parameter to be tested in order to produce high quality of briquette. Density shows the compactness of briquette and the strength of the briquette is high as the density increases. Figure 6 shows the effect of mixing ratio of torrefied rice husk and banana residue on the density of tested briquettes.

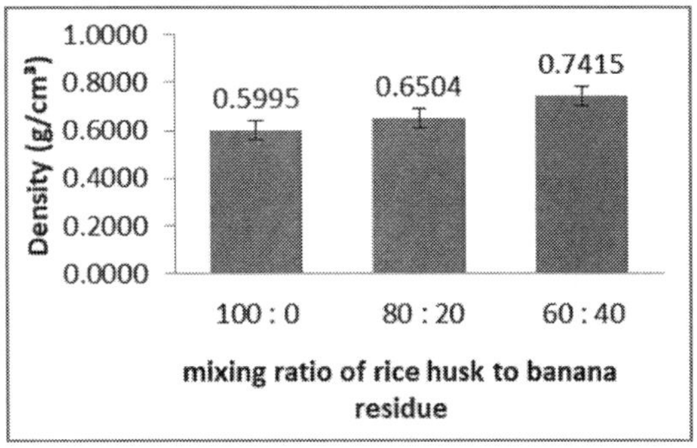

Fig. 6: Effect of different mixing ratio of torrefied rice husk and banana residue on density.

Based on the data obtain, the density for briquette made from mixing ratio of rice husk and banana residue at 100:0, 80:20, and 60:40 were 0.5995 g/cm³, 0.6504 g/cm³ and 0.7415 g/cm³ respectively. The density of the briquettes was gradually increases with the increase ratio of torrefied banana residue. The addition of banana residue to the rice husk at 20% and 40 % had influenced the density of the briquette produced as the density were increases up to 7% to 19% compared to the briquette comprised of 100% of torrefied rice husk. Moreover, the density of the briquette with torrefaction pretreatment also shows a good result with 30% density greater than briquette with untreated rice husk and banana residue. The torrefied raw materials can produce more compacted briquette than untreated because at the higher temperature of torrefaction process, depolymerization and devolatilization was occur where the lignin of the raw material were softens and make the raw materials more suitable for densification. Softened lignin of the raw materials helps in the compaction of the briquettes as one of the basic binding agent in densification process is the lignin content in the materials [3].

3.1.2 Ignition time

Ignition time is the time required for the briquette to start burning. The ignition time was tested in the control condition under fume hood with the movement of air velocity of 4.0m³/s. The data obtained for the effect of mixing ratio of torrefied rice husk and banana residue on ignition time is plotted in Figure 7.

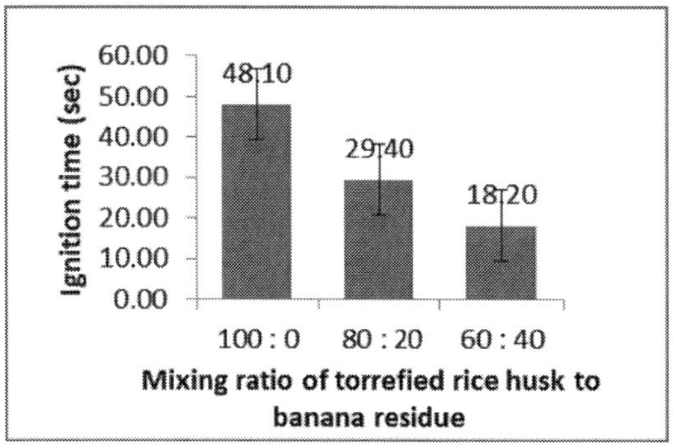

Fig. 7: Effect of different mixing ratio of torrefied rice husk and banana residue on ignition time

From the result it showed that the time required to ignite the flame is increase with decrease of mixing ratio of banana residue. At 100:0 mixing ratio the ignition time was 48.10 seconds whereas there was taken 29.40 seconds for mixing ratio 80:20. Meanwhile, the shortest ignition time of 18.20 seconds was recorded in briquette produced from mixing ratio of 60:40. The upswing in the ignition time with increase mixing ratio of rice husk indicates that rice husk has flame retarding and self-extinguishing as when the mixing ratio was 100 percent of rice husk it took the longest time and difficult to ignite. However, when comparing with the untreated biomass briquette, the torrefaction pretreatment process had contributed to the low ignition time. The ignition time obtain for torrefied biomass briquette was reduced approximately 30% from the untreated biomass briquette made from rice husk and banana residue. The higher ignition time produced from the untreated biomass briquette is due to the silica content inside the rice husk after the burning process. The torrefaction pretreatment process had reduced the silica formation during the combustion at once had reduced the time taken for the briquette to ignite.

3.1.3 Combustion rate

Combustion rate is the burning rate during the burning process of the briquettes. It is measured by measuring the distance of burning briquette over time taken to burn the briquette. Figure 8 illustrated the effect of different mixing ratio of torrefied rice husk and banana residue on combustion rate.

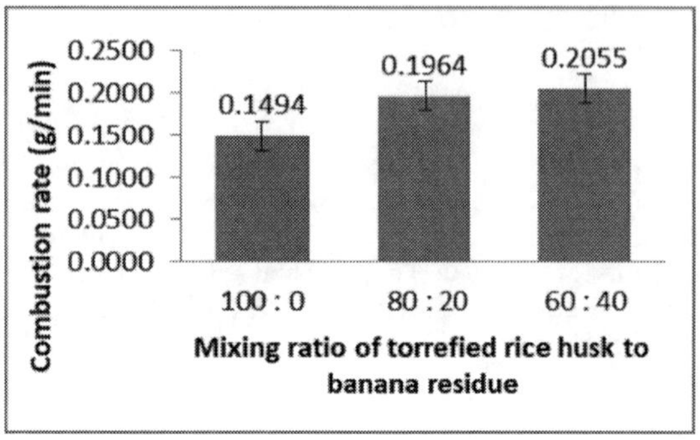

Fig. 8: Effect of different mixing ratio of torrefied rice husk and banana residue on combustion rate

Based on the result, the combustion rate increased with the increased ratio of banana residue. It was recorded that when the mixing ratio is 100:0, 80:20 and 60:20 the combustion rate was 0.1494 g/min, 0.1964 g/min and 0.2055 g/ min respectively. Rice husk can be burnt at longer duration for the combustion process due to its flame-retarding condition. The reason for this situation was same as the reason which stated at ignition time. It was not easily to ignite the briquette with 100 percent of rice husk as these situations occur because of at certain temperature reached, the molecular bond between the silica and carbon inside the rice husk was strengthened and this compound formed a hamper to prevent oxygen reaching to the carbon inside [4]. Thus, it was not easily to burn with open flame. The difficulty of burning cause the burning rate slower and mean that required longer time for combustion.

3.1.4 Ash content

Ash is the non-burnable component of biomass and it impact on the heat transfer to the surface of the briquette as well as the diffusion of oxygen to the briquette surface during the briquette's combustion. The bar chart in Figure 9 shows the effect of different mixing ratio of torrefied rice husk and banana residue on ash content.

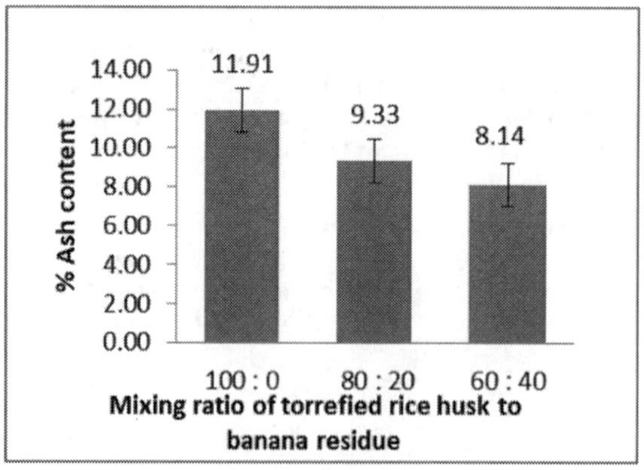

Fig. 9: Effect of different mixing ratio of torrefied rice husk and banana residue on ash content.

The biomass composite briquette with 100% of rice husk burned with highest percentage of ash content followed by composite briquette with mixing ratio of 80:20 and 60:20 of torrefied rusk husk and banana residue. This result indicates that, the higher amount of banana residue, the lower the percentage of ash content. Thus, it proves that, banana residue helps in reducing the non-combustibles component in the briquette. In addition, when comparing with the untreated briquette of rice husk and banana residue, the torrefaction pretreatment were able to reduce the percentage of ash content from 8% to 36 %. The low ash recorded is a reflection of the high heating value obtained in the briquette made with the torrefied raw materials. High ash content produced by the briquette regularly cause to the lower heating value during combustion process .High ash content results into dust emissions which lead to air pollution and affects the combustion volume and efficiency. The higher the fuel's ash content, thereby lowering the briquette's calorific value since it bring the impacts to the burning rate due to minimization of the heat transfer to fuel's interior parts and diffusion of oxygen to the briquette surface during briquette combustion [5]. Hence, briquettes made from torrefaction pre-treatment of raw materials are most suitable for cooking compare to the briquette without treatment as it has lower ash content.

3.1.5 Volatile matter

Volatile matter is the percentage component of oxygen, carbon and hydrogen that turn into vapor when heated [6]. The effect of different mixing ratio of torrefied rice husk and banana residue on volatile matter is shown by bar chart in Figure 10.

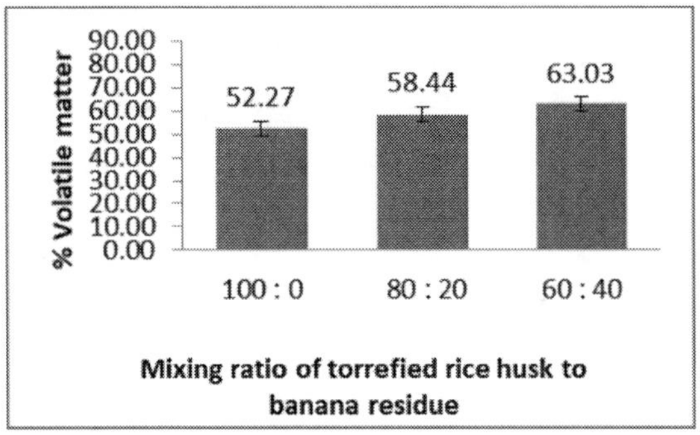

Fig. 10: Effect of different mixing ratio of torrefied rice husk and banana residue on volatile matter.

From the results, the higher percentages of volatile matter release obtain by briquette made from mixing ratio of 60:40 which is 63.03% compare to the briquette made from mixing ratio of 80:20 and 100:0 with 58.44% and 52.27% volatile matter respectively. The additional amount of banana residue had increase the percentage of volatile matter of the briquette. Besides, by comparing to the untreated and torrefied composite briquette of rice husk and banana residue, the torrefied briquette had a better result where the briquette with torrefaction pretreatment had decrease the percentage of volatile matter at 16% compare to the untreated briquette. This result because of during the torrefaction process, the hydrogen and oxygen component in the raw materials were decrease due to the water (H_2O) formation during the thermal process.

3.1.6 Fixed carbon content

Fixed carbon content represents the percentage of carbon available for char combustion after the percentage of volatile matter and ash content is refined off. The fixed carbon content also gives an indication for the percentage of char that remains after the devolatilization phase. Graph in Figure 11 shows the effect of different mixing ratio of torrefied rice husk and banana residue on the briquette's fixed carbon content.

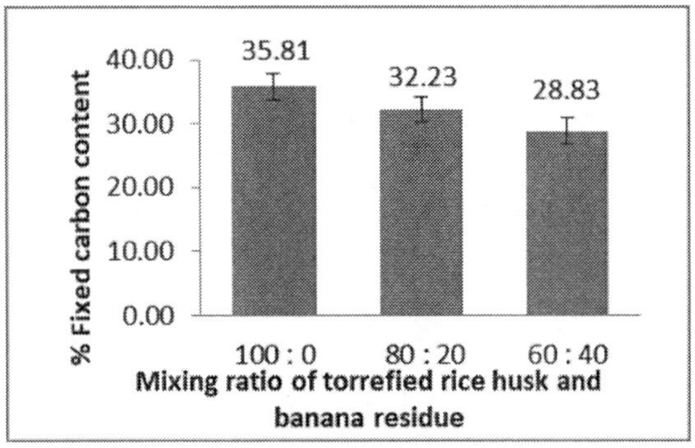

Fig. 11: Effect of different mixing ratio of torrefied rice husk and banana residue on fixed carbon content.

According to the graph above, the briquette comprise of 100 percent of rice husk have the highest percentage of fixed carbon compare to the briquette mixed with mixing ratio of 80:20 and 60:40. This result was obtained because of the influence by the result from percentage of ash content and volatile matter. The percentage of fixed carbon is inversely proportional to the percentage of ash content and volatile matter. The lower the percentage of ash content and volatile matter of the briquette produce, the higher the percentage of fixed carbon. A better briquette is produce when the percentage of fixed carbon content is increase as particularly the corresponding gross calorific value produce from the briquette will increase too. As the result, the higher the fixed carbon content, the higher heat release and making it tends to reduce the cooking time.

3.1.7 Calorific value

The calorific value or heating value is the amount of heat released during the combustion of a specified amount of it. Figure 12 shows the relationship of different mixing ratio of torrefied rice husk and banana residue on the gross calorific value of the tested briquette.

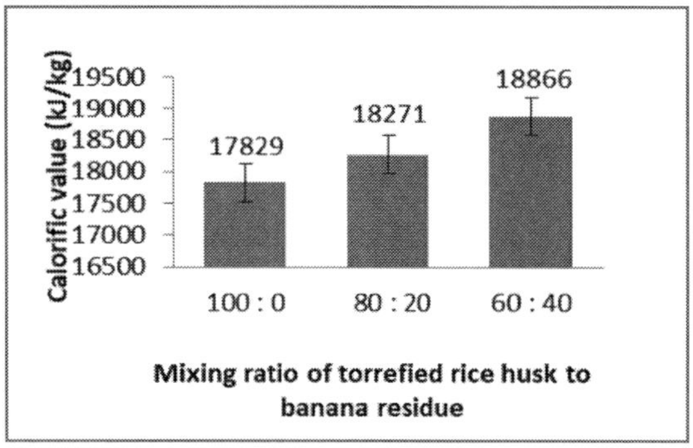

Fig. 12: Effect of different mixing ratio of torrefied rice husk and banana residue on calorific value

Based on the graph in Figure 12 the gross calorific value were increases with the increase ratio of banana residue in the mixture of the briquette. The value recorded were 17829 KJ/Kg, 18271 KJ/Kg followed by 18866 KJ/Kg for the briquette made from 100% of rice husk, 80:20 and 60:40 mixing ratio of rice husk and banana residue respectively. The higher calorific value obtained from the briquette indicated that the briquette produced is better to be burned. In addition, it can be said that the torrefied rice husk and banana residue briquettes were successfully produced. According to the result obtain, by torrefaction pretreatment it had increase 16% of calorific value compared to the untreated rice husk and banana residue briquette that were tested before. The rice husk and banana residue were loses generally more moisture content, oxygen and hydrogen than carbon during torrefaction, which increases the result in calorific value of the briquette

3.2 Morphology analysis

3.2.1 SEM

The micro and macro structure of the samples were observed with the purpose of gaining a more profound into the impact of torrefaction on the structural change of the torrefied material and compare this change with the untreated raw material. The SEM images of the torrefied and the raw samples of rice husk and banana residue were taken under the same analysis condition (magnification was ×500, ×1000, ×3000 and ×5000 at 10kV for the comparison results. The images are presented in Figure 13.

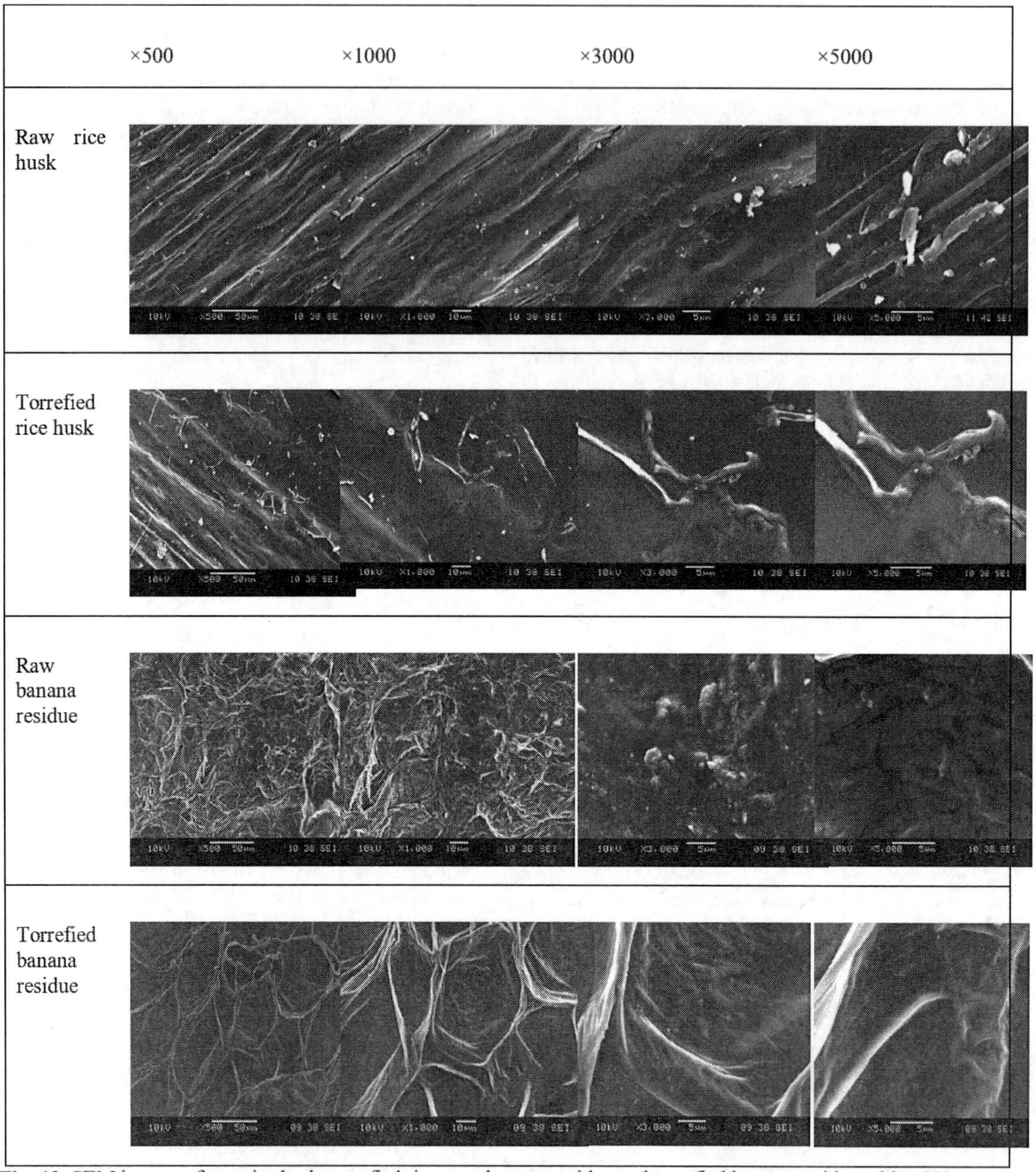

Fig. 13: SEM images of raw rice husk, torrefied rice, raw banana residue and torrefied banana residue with ×500, ×1000, ×3000 and ×5000 magnification.

Based on Figure 13 it is evidence that the untreated and torrefied rice husk and banana residue exhibit different surface morphology under the analytical scanning electron microscope (JOEL JSM 6460 LA). Based on the observation, both untreated raw rice husk and raw banana residue appears lighter in colour with rough surface compare to the rice husk and banana residue that have been undergo the torrefaction pretreatment. It cannot be explained well the effect of pretreatment to the morphology of the rice husk as there are not much different that are noticeable on the surface structure of the both untreated and torrefied samples. However, for both samples of untreated and torrefied banana residue it is clearly observed that the torrefaction pretreatment had changed the raw materials surface as the structure of torrefied banana residue shows a porous structure like honey comb. This situation obviously can be attributed to the thermal pretreatment that may have decomposed the hemicellulose and some weak part of lignin component in the raw material.

4 Conclusions

Based on the physical and combustion performance, the biomass briquette from rice husk and banana residue which undergo torrefaction pre-treatment process exhibit more excellent potential as an alternative energy source compare to biomass briquette without torrefaction process. Torrefaction pre-treatment methods enhance the quality of briquette produce as torrefaction has shrinkage the fibers size which confirming the thermal stability of the briquette. In this project the torrefaction was done at 300°C for 30 minutes. From morphology analysis, it showed torrefaction will shrinkage fiber size on biomass sample which confirming the thermal stability of biomass briquette produce later. This project showed that the torrefaction pre-treatment and the mixing ratio of the raw materials used will affect the parameters such as density, ignition time, combustion rate, ash content and the calorific value. The study of this project shows that there are relatively relationships between the pretreatment and mixing ratio with the parameter tested. From all of the parameter testing, it was found that the best briquette was the briquette made from mixing ratio of torrefied rice husk and banana residue at 60:40 as it meets the properties for a good fuel briquette. The energy values and combustion qualities of the briquettes produced in this study are also sufficient enough to produce the required heat for domestic cooking and also for industrial application especially the energy requirement of the small-scale industries.

The authors would like to acknowledge the support from the Research Acculturation Grant Scheme (RAGS) under a grant number of RAGS/1/2015/UNIMAP/9018-00087 from the Ministry of Higher Education Malaysia.

References

1. Oladeji, J. (2012). Comparative Study of Briquetting of Few Selected Agro-Residues Commonly Found in Nigeria. *Pacific Journal of Science and Technology*, **13**(2), 80–86.
2. M. M. Nazari, C. P. San, N. A. Atan. (2016) Performance of Biomass Composite Briquette from Rice Husk and Banana Residue. The National Seminar of Indonesian Society of Agricultural Engineers (PERTETA).
3. J. S. Tumuluru, S. Sokhansanj, J. R. Hess, C. T. Wright, and R. D. Boardman. (2011) A Review on Classification and Composition, Co-firing Issues and Pretreatment Methods. American Society of Agricultural and Biological Engineers (ASABE).
4. Olivier, P. A. (2002). The Rice Hull House, 1–13.
5. Nasrin, A.B., Ma, A. N., Mohamad, S., Choo, Y. M., Rohaya, & Azali, A. (2006). Production of palm-based biomass briquettes Retrieved on Nov 25, 2016 from MPOB. At: http://palmoilis.mpob.gov.my/publications.
6. T. Nottingham and N. E. User. (2010) "Combustion Characteristics of Biomass Briquettes by Joel Chaney, MSci Physics with French,"

* Corresponding author: noramiraatan93@gmail.com

A Network Coding-Aware Directional Broadcast Routing Scheme for VANET

Normaliza Omar, *Naimah* Yaakob, *Rohani* S.M. Farook, *Zulkifli* Husin, *Aini Syuhada* Md Zain, *Mohamed* Elshaikh, and *Aznor Hanah* Abdul Halim

School of Computer & Communication Engineering, Universiti Malaysia Perlis, Malaysia

Abstract. Recent technological advancements in wireless communications are fundamentally changing the manner by which devices communicates with one another. Modern wireless devices build networks on their own and aid each other in passing information to any device in the network. The aim of this research is to design and develop a robust routing protocol using network coding technique for Vehicular ad hoc networks (VANET). The purpose of the protocol is to disseminate vehicular traffic to the vehicles approaching the problem area, rather than those leaving the problem area. The designed protocol during the research will be more robust, reliable, and efficient with topology change. In this research, network coding and geographical routing models will be used to develop network coding based VANET routing protocol for alerting vehicles approaching the accident area and emergency message dissemination. The network coding model assumed will serve to increase throughput, in this sense reducing packet loss which will provide a robust broadcast routing protocol.

1 Introduction

Nowadays, the rapid evolution and cost reduction experienced by wireless communication technologies during the last decade have made them suitable for a wide spectrum of applications. There are a huge number of vehicles travelling along highways and streets around the world and millions of car accidents occur in the world every year. Millions of people injuries and death incidences via automobile accidents each year. Therefore, there is a growing demand for real-time collision avoidance and warning technology to improve driving safety and traffic condition. Due to rapid advancement of technology, vehicles will be comprised with embedded computers, route planner, GSM/GPRS/WCDMA cellular networks, Global Positioning System (GPS), access to road-side wireless sensors devices and WiFi access point [1]. VANETs consist of smart vehicles on the road and provide useful communication services among arbitrary-formed collections of vehicles that are geo-located and with road side infrastructure. Envisioned applications are ranged from critical safety services and advanced driver assistance systems to traffic management and infotainment applications for intelligent transport applications, and comfort applications [2]. For safety applications to work, it is significant that safety message be reliably delivered to the destination vehicle in a timely manner. Thus a number of studies have been performed in designing and developing a reliable routing protocols in VANET [3][4][5]. Whereas, in VANET, a single hop transmission is usually unable to cover all the intended receivers due to limited radio range across a larger distance. Therefore, multi-hop broadcast protocol is necessary to establish a communication between vehicles. As a result, it is necessary to develop reliable and robust routing protocol in VANETs.

The implementation of message dissemination is one of the main challenges of VANET. In VANET network topology changes more rapidly, and then causes frequent network fragmentation (topology hole between platoon vehicles), frequent route breakage, and packet loss. Thus, the routing protocols have been designed for Mobile Ad Hoc Network (MANET) is not applicable on VANET. There are several simulation based studies have been done to investigate the performance of routing protocols between VANET and MANET, in variety traffic scenarios [6][7][8][9][10].

The broadcast routing protocol is used frequently in VANET, such as data sharing and traffic dissemination, thus many researchers have designed and developed suitable broadcast routing protocols for VANET [4][5][11][12][13]. There are several researches have been executed to study data sharing and establishing a distributed view of the VANET [14][15]. Recent researches have investigated the design of realistic mobility model [16]. Security problems in VANETs are discussed in [17][18][19][20][21][22]. Whereas, this research will fully concentrate on the broadcast routing using network coding. Network coding has recently emerged as a crucial solution of broadcast scenarios in ad hoc networks, and network coding was shown to reliably handle such type of networks and to profoundly enhance the performance. Researchers have investigated network coding in many areas of wireless networking [23][24].

1.1 High Mobility and Large Obstacle Environment in Cities

Analyses of traditional routing protocols for mobile ad hoc networks (MANETs) demonstrated that their performance is poor in VANETs [13][25]. The main problem with these protocols (e.g.., AODV [26], DSR [27], etc) in VANETs environments is their route instability in high mobility and large obstacle environments, like cities. The traditional node-centric view of the routes (i.e., an established route is a fixed succession of nodes between the source and destination)

* Corresponding author: normaliza@unimap.edu.my

leads to frequent broken routes in the presence of VANETs' high mobility, as illustrated in Figure 1. Consequently, many packets are dropped and the overhead due to route repairs or failure notifications increases significantly, leading to low delivery ratios and high transmission delays.

An alternative approach is offered by geographical routing protocols (e.g., GFG [28], GOAFR [29], GSR [25], GPSR [30]) which decouple forwarding from the nodes identity; they do not establish routes, but use the position of the destination and the position of the neighbor nodes to forward data. Unlike node-centric routing, geographical routing has the advantage that any node ensuring progress toward the destination can be used for forwarding. For instance, in Figure 1, geographical forwarding could use node N2 instead of N1 to forward data to D. Route (S, N1, D) established at time t breaks at time t + Δt when N1 moves out of the transmission range of S. Despite better path stability, geographical forwarding does not perform well in city-based VANETs either [25][31]. Its problem is that many times it cannot find a next hop (i.e., a node closer to the destination than the current node).

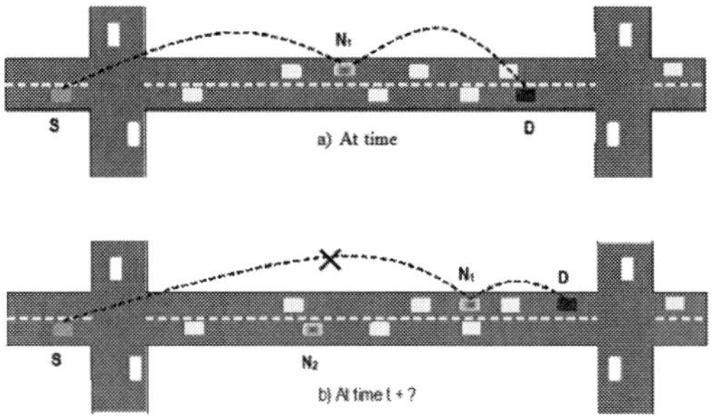

Fig. 1: Routes established as fixed successions of nodes break frequently in highly mobile VANETs.

For example, as shown in Figure 2, it could take road paths that do not lead to destination. Instead of forwarding data on the dotted path, geographical routing sends data to N1 and N2, following the shortest geographical path from S to D on a dead end road. The recovery strategies proposed in literature are often based on planar graph traversals, which were shown not to be as effective in VANETs due to radio obstacles, high node mobility, and the fact that vehicle positions are constrained on roads rather than being uniformly distributed across a region [25].

A number of road-based routing protocols [10][13][25][30][32] have been designed to address this issue. For example, Geographical Source Routing (GSR) [25] used static street map and location information about each node to avoid GPSR [30] problems. A static street map provides the global knowledge of the city topology to GSR algorithm. The Connectivity-aware Routing Protocol (CAR) [13] is another position-based routing protocol for inter-vehicle communication in a city and/or highway environment. In CAR, the routes are found to destination by maintaining a cache of successful routes between various source and destination pairs. In another attempt, a position-based routing scheme called A-STAR [32] has been proposed to handle disconnection problem due to high rise buildings and trees in city environment. A-STAR [32] also used the static street map to compute the sequence of junctions (anchors) to route messages around radio obstacles. Similarly, a reactive location based routing algorithm is proposed [10] that uses cluster-based flooding technique called LORA_CBF. In LORA_CBF, each node either in any place of cluster can be a cluster, a gateway or an intermediate node. Each cluster must have exactly one cluster-head and if a node is connected to more than one cluster is called a gateway. However, many of these approaches fail to factor in the vehicular traffic flow by using the shortest road path between source and destination.

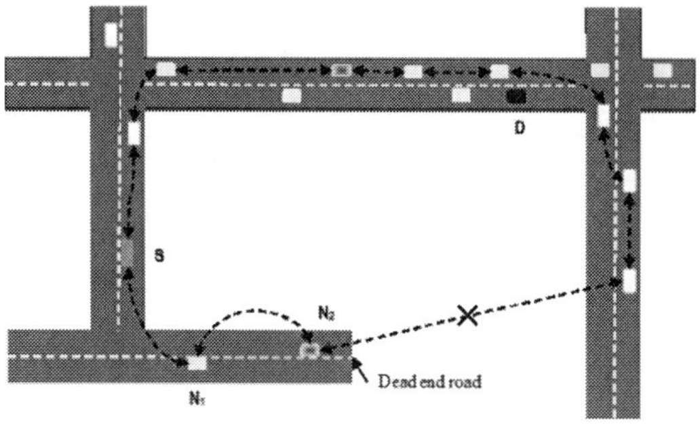

Fig. 2: Geographical routing can route packets toward dead ends causing unnecessary traffic overhead in the network and longer delays for packets

1.2 High Mobility Environment on Highways/Rural Areas

Every year traffic accidents are increasing on highways due to automobile crashes. It causes billions of cost in damaged property, nonfatal injuries, and deaths. While different factors contribute to vehicle crashes, such as vehicle mechanical problems, bad weather, and driver behavior. Recently, due to dramatic increase in wireless communication technologies, inter-vehicle networks provide potential solution to control high accident rates on highways.

The efficient VANET routing protocols can play an important role to take in consideration the specifics of highways, relative car movements and application constraints. The main requirement of routing protocols is to achieve minimal communication time with minimum consumption of network resources. Various routing protocols have been developed for Ad-hoc wireless networks, but they cannot be applied directly to inter-vehicle communication since the characteristics of vehicles movement, information movement and relative speed of mobile nodes are different from that of an Ad-hoc wireless network. The inability of drivers to react in time to emergency situations often creates a potential for chain collisions, in which an initial collision between two vehicles is followed by a series of collisions involving the following vehicles. The existing MANET routing protocols cannot be used in VANET because of highly dynamic topology, frequently disconnection problems, less visibility on slopes, big curves and sharp edges etc (as shown in Figure 3).

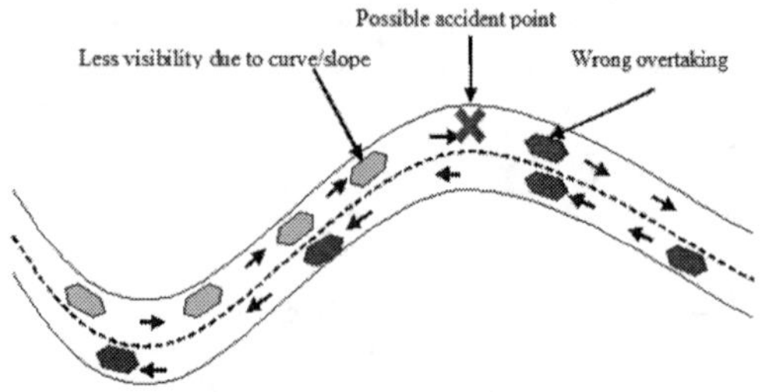

Fig. 3: Rural area scenario on less visibility due to curve/slope

Recently, various VANET routing protocols [11][33] [34][35] have been proposed to handle uncertain situations on highways. An emergency broadcast protocol called BROADCOMM [11] is proposed which is based on a hierarchical structure for a highway network. In BROADCOMM, the highway is divided into virtual cells, which moves as the vehicles move. This protocol outperforms similar flooding based routing protocols in the message broadcasting delay and routing overhead. However, it is very simple and only works with simple highway networks. In [33], authors proposed a cache and distance aware neighborhood selection schemes to deal with the situation of high velocities as compared to regular geocast protocols. A small cache is added within the Zone of Relevance (ZOR) to the routing layer that holds those packets that a node cannot forward instantly due to a local minimum. In [36] also proposed a special geocast routing called abiding geocast. It is a time stable geocast delivered to all nodes that are inside a destination region within a certain period of time. However, the delivery success ratio of these approaches with local message storage is lower in some cases.

2.1 Vehicular Ad hoc Network (VANET)

VANET is a special class of mobile ad hoc network (MANET) with their own unique characteristics. Like MANETs [26][27][37], VANETs also do not require any fix infrastructure. However, permanent network nodes can be used in the form of roadside sensor devices. These roadside devices open up a wide variety of services for vehicular networks, such as acting as a drop point for messages on sparsely populated roads, serving up geographically-relevant data, or serving as a gateway to the Internet. VANET has a distinct high speed controlled mobility patterns than other kinds of ad hoc networks (MANETs). It consists of fast moving mobile nodes and large obstacles (e.g., buildings, trees, etc) on highways and urban areas. Speeds of vehicles moving in the same direction are generally similar, and they can therefore remain in radio contact with one another for much longer periods of time than with vehicles moving in the opposite direction.

Because of the unique characteristics of VANETs, the MANETs routing protocols cannot be directly used in VANET efficiently. Several researchers are trying to develop efficient routing protocols that are suitable for VANETs. There are 3 potential areas of research are under investigation in VANETs which includes efficient routing protocols, security, and fairness of bandwidth. However, the routing protocols have only been discussed to study the effect of traffic patterns and congestion in VANETs.

2.2 Broadcast Routing Protocols

In this section, broadcasting routing methods in VANETs have been discussed. Broadcasting is a frequently used routing method for sharing traffic information, weather, emergency, road condition, advertisements etc in VANETs. The primary objective of this routing method is to distribute information from a source vehicle to many unknown destinations vehicles. In unicast routing protocols broadcasting method is used to find an efficient route to the destination, Figure 4 and Figure 5 present the difference between unicast and broadcast routing protocols in VANETs. Because of the multi-hop nature of vehicular networks, flooding is a fundamental mechanism to implement the multi-hop broadcasting. Flooding guarantees the message will eventually reach to all nodes in the network. Flooding works well for small number of nodes in the network. However, if number of nodes increase, the bandwidth requested for one broadcast message transmission can also increase exponentially. As each node receives and broadcasts the message almost at the same time, this causes contentions and collisions, broadcast storms and high bandwidth consumption. Flooding may have a very significant overhead and selective forwarding can be used to avoid network congestion.

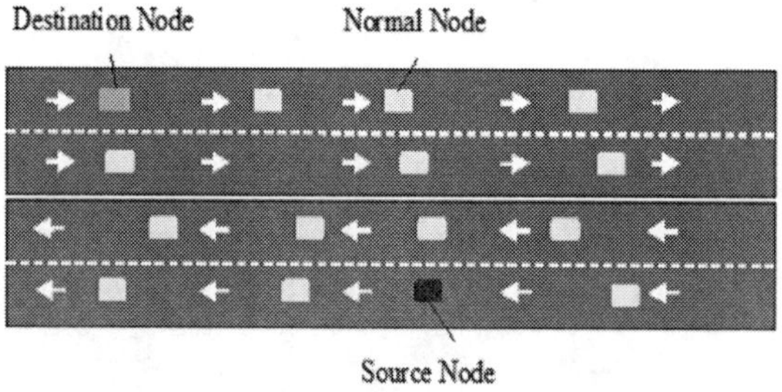

Fig. 4: Unicast routing mechanism in VANETs

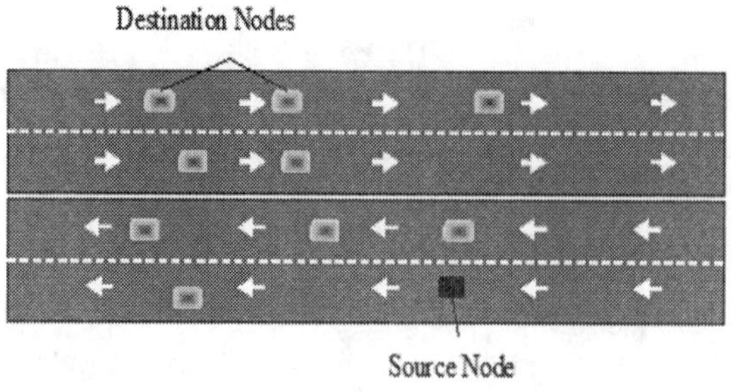

Fig. 5: Broadcast routing mechanism in VANETs

3.1 Simulation Setup

In simulation setup, first, different areas (e.g., 1000m x 1000m, 1500m x 1500m, and 2000m x 2000m) extracted from the TIGER/Line database of the US Census Bureau [38] will be used to form a grid layout of the road segments. Second, simulation of Urban Mobility (SUMO) [39] will be used to generate the movements of the vehicle nodes. SUMO is an open-source, microscopic, space-continuous and time discrete vehicular traffic generator package. SUMO uses a collision-free car-following model to determine the speeds and positions of the vehicles. The map extracted from the Tiger/Line database as well as specifications about the speed limits and number of lanes of each road segment on the map will be entered into SUMO. The traffic light operated intersections as well as priority intersections (less that one-fifth of the intersections are regulated using traffic lights) will be specified. The first 2000 seconds of SUMO output will be discarded to obtain more accurate node movements. The output from SUMO is converted into input files for the movement of nodes in the NS-2 simulator. The experiments in networks will run with different node densities: the 350 nodes scenario represents relatively dense networks, 250 nodes scenario for medium density networks and 150 nodes for sparse networks.

3.2 Assumption & Limitations

In simulation scenarios, different areas extracted from TIGER/Line database of the US Census Bureau [38] will be used to forms a grid layout with different road segments. Bidirectional traffic will be set on each side of road with two lanes in each direction. In this scenario obstacles are not included in order to evaluate the performance of protocols under increased network congestion. This way, a small increase in data sending rate will provide a noticeable increase in the level of contention in the network. The following points are assumed in this research:

Some data in simulation (e.g., obstacles) is not strictly real-time, there are no real-time constraints on the data, thus in some sense, the data is delay-tolerant.

The number of lanes is not an important and effective parameter in analysis. In fact, the existence of several lanes intensifies the assumption of independent mobility patterns in sparse situations, because the vehicles can overtake each other by simple lane transitions without any hindering effect and any need to speed change. However, the most important factors in analysis are the distribution of the speeds for the vehicles at the arrival instants and their mobility patterns along the street. These factors strongly affect the number of vehicles at the street in the steady state.

- Storage of data is not constrained: VANETs are typified by no power constraints and no data storage limitation.
- Communication between vehicles is over a low data rate connection: While this constraint depends on the radio technology used. Currently, 802.11x devices will offer good-put of the order of a few hundred Kbps.
- Push model: Data is being continually "pushed" by the access points to the nodes in the transmission range.
- Multi-hop delivery: It is infeasible to transmit data to more than 2-3 hops.

3.3 Expected Results

Improving the safety of the drivers and passengers by wirelessly exchanging information between vehicles represents a major driving force for the design of vehicular networks. This research proposes the mobility model and a dynamic, cooperative and opportunistic road-based VANET routing mechanism that can cope with heterogeneous vehicles connections. In this research, we are expecting the following results to handle distinguish characteristics of VANETs:

- VANET Routing Protocols for Cities and Highways:
  This research is expecting a class of VANET routing protocols for city-based environments that take advantage of the roads layouts to improve the performance of routing in VANETs. The protocols will use real-time vehicular traffic information to create road-based paths between end-points. To improve the end-to-end performance under high contention, it will also be proposed a distributed next hop self-election mechanism for geographical forwarding.

- Realistic Vehicular Environments:
  The existing VANET routing protocols are instable due to high mobility and large obstacle environments that leads to frequent broken routes. So, many data packets are dropped and the overhead due to route repairs or failure notifications increases significantly, leading to low delivery ratios and high transmission delays. Therefore, it is expected that proposed technique will outperform the existing approaches in terms of average delivery ratio, average delay, average path length, and overhead. The propose protocols forward data along the streets, not across the streets, and take into account the real traffic on the roads, they perform well in realistic vehicular environments in which buildings and other road characteristics such as dead end streets are present. Such results will show that distributed applications that generate a moderate amount of traffic can be successfully implemented in VANETs.

- Heterogeneous Vehicle Connections:
  Heterogenity is important because there may be many vehicles are trying to communicate with each other and exchanging useful information. Therefore, it is expected that proposed vehicular network system will be able to handle such situations.

4 CONCLUSION

It is a big challenge to handle the distinguish characteristics of VANETs such as highly dynamic topology, frequently disconnected networks, various communications environments, and hard delay constraints. The aim of this research is to focus on these VANETs special characteristics. In particular, first mobility model based on car-following and lane-changing Gipps model will be developed to design network protocols. Then, reactive and proactive routing algorithm will be designed to discover routes on-demand basis and generate periodical connectivity to store route information in terms of graphs respectively. Finally, a protocol will be designed to cope with the network delay, loss, and mobility while meeting the application constraints.

REFERENCES

[1] Chao, H. C., and Zeadally, S. (Eds.). Mobility protocols for ITS/VANET. Elsevier, Computer Communications 31 (2008) 2765–2766(2008).

[2] Sichitiu, M., and Kihl, M., Inter-vehicle communication systems: a survey, IEEE Communications Surveys & Tutorials, Vol. 10, Issue 2, pp. 88–105, (2008).

[3] Ros, F., Ruiz, P., and Stojmenovic, I., Reliable and Efficient Broadcasting in Vehicular Ad Hoc Networks. Vehicular Technology Conference IEEE, Barcelona, 26-29 April 2009, pp. 1 – 5 (2009).

[4] Lai, P., Wang, X., Lu, N., and Liu, F., A Reliable Broadcast Routing Scheme Based on Mobility Prediction for VANET. Intelligent Vehicles Symposium IEEE, Xi'an, China, 3-5 June 2009, pp. 1083 – 1087 (2009).

[5] Yadumurthy, R., Chimalakonda, A., Sadashivaaiah, M.,Makanaboyina, R., Reliable MAC Broadcast Protocol in Directional and omni-Directional Transmissions for Vehicular Ad hoc Networks. In Proceedings of the 2nd ACM international workshop on Vehicular ad hoc network, Germany, 2005, pp. 10 - 19 (2005).

[6] Namboodiri, V., Agarwal, M., and Gao, L., ''A study on the feasibility of mobile gateways for vehicular ad-hoc networks.,'' in Proceedings of the First International Workshop on Vehicular Ad Hoc Networks, pp. 66–75, (2004).

[7] Liu, G., Lee, B.-S., Seet, B.-C., Foh, C.H., Wong, K.J., and Lee, K.-K., ''A routing strategy for metropolis vehicular communications,'' in International Conference on Information Networking (ICOIN), pp. 134–143, (2004).

[8] Fu¨ßler H., Mauve M., Hartenstein H., Kasemann M., and Vollmer, D., ''Location-based routing for vehicular ad-hoc networks,'' ACM SIGMOBILE Mobile Computing and Communications Review (MC2R), vol. 7, no. 1, pp. 47–49, January 2003, (2003).

[9] Wang S.Y., Lin C.C., Hwang Y.W., Tao K.C., and Chou C.L., ''A practical routing protocol for vehicle-formed mobile ad hoc networks on the roads,'' in Proceedings of the 8th IEEE International Conference on Intelligent Transportation Systems, 2005, pp. 161–165, (2005).

[10] Santos, R. A., Edwards A., Edwards R., and Seed L., ''Performance evaluation of routing protocols in vehicular ad hoc networks,'' The International Journal of Ad Hoc and Ubiquitous Computing, vol. 1, no. 1/2, pp. 80–91, (2005).

[11] Durresi, M., Durresi, A., and Barolli, L.,. Emergency broadcast protocol for inter-vehicle communications. In ICPADS '05: Proceedings of the 11th International Conference on Parallel and Distributed Systems—Workshops (ICPADS'05), (2005).

[12] Korkmaz, G., Ekici, E., Özgüner, F., and Özgüner, Ü.,. Urban multi-hop broadcast protocol for inter-vehicle communication systems. In ACM International Workshop on Vehicular Ad Hoc Networks, pp. 76–85, (2004).

[13] V. Naumov and T. Gross, "Connectivity-Aware Routing (CAR) in Vehicular Ad Hoc Networks," In Proc. of the IEEE International Conference on Computer Communications (INFOCOM'07), May, (2007).

[14] M.Y.Sanadidi ,A. Nandan, S. Das, G. Pau, and M. Gerla, , ''Co-operative downloading in vehicular ad-hoc wireless networks,'' in Second Annual Conference on Wireless Ondemand Network Systems and Services (WONS), pp. 32–41, (2005).

[15] Nandan A., Das S., Tewari S., Gerla M., and Klienrock L., ''Adtorrent: Delivering location cognizant advertisements to car networks,'' in The Third International Conference on Wireless On Demand Network Systems and Services(WONS 2006), (2006).

[16] Saha, A.K., and Johnson, D.B.,. Modeling mobility for vehicular ad-hoc networks. In Proceedings of the 1st ACM International Workshop on Vehicular Ad Hoc Networks, VANET'04, ACM, New York, NY, USA, (2004).

[17] Choi, Jaeduck and Jung, Souhwan "A security framework with strong non-repudiation and privacy in VANETs" in 6th IEEE Consumer Communications and Networking Conference, (2009).

[18] Broustis, I., and Faloutsos, M.,. Routing in vehicular networks: feasibility, security, and modeling issues. Tech. Rep. UCR-CS-2006-05219, Department of Computer Science and Engineering, University of California, Riverside, Oct 2006, (2006).

[19] Eichler, S., Dötzer, F., Schwingenschlögl, C.,

Caro, F.J.F., and Eberspächer, J.,. Secure routing in a vehicular ad hoc network. In: Proceedings of the IEEE 60th Vehicular Technology Conference, VTC2004-Fall, IEEE, (2004).

[20] Fonseca, E., and Festag, A.,. A survey of existing approaches for secure ad hoc routing and their applicability to VANETS. Tech. Rep. NLE-PR-2006-19, NEC Network Laboratories, Mar 2006, (2006).

[21] Golle, P., Greene, D., and Staddon, J.,. Detecting and correcting malicious data in VANETs. In: Proceedings of the 1st ACM International Workshop on Vehicular Ad Hoc Networks, VANET'04, ACM, New York, NY, USA, (2004).

[22] Raya, M., and Hubaux, J.-P.,. The security of vehicular ad hoc networks. In: Proceedings of the 3rd ACM Workshop on Security of Ad Hoc and Sensor Networks, SASN'05, ACM, New York, NY, USA, (2005).

[23] Fragouli, C., WIdmer, J., and LeBoudec, J.,"Efficient broadcasting using network coding" IEEE/ACM Transactions on Networking, (2007).

[24] Katti, S., Rahul, H., Katabi, D., Hu, W., M´edard, M., and Crowcroft. J., "XORs in the Air: Practical Wireless Network Coding" In ACM SIGCOMM, (2006).

[25] Lochert, C., Hartenstein, H., Tian, J., Herrmann, D., Füßler, H., and Mauve, M.,A routing strategy for vehicular ad hoc networks in city environments. In Proceedings of IEEE Intelligent Vehicles Symposium (IV2003), pp. 156–161, June 2003, (2003).

[26] Perkins, C.E., and Royer, E.M., (1999). Ad-hoc on demand distance vector routing. In Proceedings of the 2nd IEEE Workshop on Mobile Computing Systems and Applications, pp. 90–100, Feb. 1999, (1999).

[27] Johnson, D., and Maltz, D.,. Dynamic Source Routing in Ad Hoc Wireless Networks. Mobile Computing, T. Imielinski and H. Korth, Ed., pp. 153-181. Kluwer, (1996).

[28] Bose, P., Morin, P., Stojmenovic, I., and J. Urrutia,. Routing with guaranteed delivery in ad hoc wireless networks. ACM Wireless Networks, vol. 7, no. 6, pp. 609–616, November 2001, (2001).

[29] Kuhn, F., Wattenhofer, R., Zhang, Y., and Zollinger, A., (2003). Geometric ad-hoc routing: Of theory and practice. In Proceedings of the twenty-second annual symposium on Principles of distributed computing, Boston, MA, USA, July 2003, pp. 63–72, (2003).

[30] Karp, B., and Kung, H.T.,. GPSR: Greedy perimeter stateless routing for wireless networks. In Proceedings of the ACM/IEEE International Conference on Mobile Computing and Networking (MobiCom), (2000).

[31] Li, T., Hazra, S. K., and Seah, W.,. A position-based routing protocol for metropolitan bus networks. In Proceedings IEEE 61st Vehicular Technology Conference VTC-Spring, Stockholm, Sweden, June 2005, pp. 2315–2319, (2005).

[32] Seet, B.-C., Liu, G., Lee, B.-S., Foh, C.-H., Wong, K.-J., and Lee, K.-K.,. A-STAR: A mobile ad hoc routing strategy for metropolis vehicular communications. In: Proceedings of the Third International IFIP-TC6 Networking Conference, Networking Technologies, Services, and Protocols; Performance of Computer and Communication Networks; Mobile and Wireless Communications, NETWORKING 2004, in: Lecture Notes in Computer Science, vol. 3042/2004, Springer, Berlin, Heidelberg, 2004, pp. 989–999, (2004).

[33] Maihöfer C., and Eberhardt, R.,. Geocast in vehicular environments: Caching and transmission range control for improved efficiency. In Proceedings of IEEE Intelligent Vehicles Symposium (IV), pp. 951–956, (2004).

[34] Sun, M., Feng, W., Lai, T.-H., Yamada, K., Okada, H., and Fujimura, K., GPS-based message broadcasting for inter-vehicle communication. In ICPP '00: Proceedings of the 2000 International Conference on Parallel Processing, (2000).

[35] Ko, Y.-B., and Vaidya, N.H.,. Location-aided routing (LAR) in mobile ad hoc networks. Wireless Networks, vol. 6, no. 4, pp. 307–321, (2000).

[36] Maihöfer, C., Leinmüller, T., and Schoch, E., (2005) Abiding geocast: time-stable geocast for ad hoc networks. In Proceedings of the 2nd ACM international workshop on Vehicular ad hoc networks (VANET '05), pp. 20–29, (2005).

[37] Perkins, C., Belding-Royer, E., and Das, S., (2003). Ad Hoc On demand Distance Vector (AODV) Routing. IETF RFC 3561, July 2003, (2003).

[38] TIGER, U.S.Census Bureau - TIGER/Line 2006 Second edition, http://www.census.gov/geo/www/tiger/ (accessed June 2017), (2008).

[39] SUMO, Centre for Applied Informatics (ZAIK) and the Institute of Transport Research German Aerospace Centre. Sumo - simulation of urban mobility. http://sumo.sourceforge.net/, (2009).

The Relationship between Time Based Management towards Manufacturing Capabilities: Investigating in Malaysian Manufacturing Industries

Mohammad Harith Bin Amlus[1,*], *Amlus Ibrahim*[2], *Ahmad Zaidi Abdullah*[1], *U. Naeimah Saraeh*[1], *Adi Anuar Azmin*[1], *Wan Sallha Yusuf*[1] and Nur *Afifah Shuib*[1]

[1]Universiti Malaysia Perlis, Malaysia
[2] Universiti Utara Malaysia, Malaysia

Abstract. Towards Industry 4.0, Malaysian Manufacturing Firms need to identify back their manufacturing capabilities especially to increase the performance and profit. This research is conducted to identify the role of manufacturing practices towards manufacturing capabilities. From this information it can provide valuable ingredients for companies in sustainable their operation. The identified factor of manufacturing practices is time based management and the dependent variables is manufacturing capabilities(focusing on agility. Through the survey, a total of 119 firms representing various industries in Northern Region of Malaysia had responded. The hypotheses involved were tested using correlation and regression techniques. The results of the study support the hypotheses. The multiple regression analysis indicates that there are significant relationships among this role on manufacturing capabilities. It is believed that results of this study will be beneficial to the top management of Malaysian manufacturing industries to focus on tim e based management and agility to increase the manufacturing capabilities.

1 Introduction

Manufacturing industries change dramatically and Malaysian government now set to move on the next stage after focus on agriculture industries and conventional manufacturing industries. Today, it has segued into yet another monumental era – the fourth industrial revolution, Industry 4.0, the name given to the latest evolution in the digitisation and automation of manufacturing processes. Industry 4.0 covers the entire value chain, including suppliers, procurement, design, logistics and even sales, resulting in higher productivity and flexibility. There is less wastage or storage, better monitoring and maintenance of machinery, and improved security and safety. The first industrial revolution started in the late 18th century with the shift from human or animal power to machines run by water or steam.

The second occurred between 1870 to 1914 with the introduction of electricity, and the rise of the steel and oil industries, triggering the era of mass production of goods and vehicles.

The third significant shift began in the 1960s with the entry of the first programmable logistic controllers and early versions of computers, boosting automation and control of production lines.

This spurred the extensive use of computer networks, and the eventual birth of the Internet changed the world in ways that no one could have imagined.

It is aimed at revolutionising the manufacturing industry, by switching from centralised to decentralised networks under which connected equipment and devices communicate with each other to analyse and respond to information received.

Manufacturing capabilities are grounded in the firm's people, skills, knowledge, processes, systems, and equipment. These capabilities can be assembled from different internal and external sources [17] and then deployed to create products and introduce them to the market in a timely manner. In this study, manufacturing capability involves four key elements that need to be addressed: the integration, acuity, control, and agility. Pursuing the idea that multiple capabilities are desirable, a few researchers have addressed the question of which capabilities a company should developed first [13]. Recently, researchers have related that manufacturing capability can be performed with good manufacturing practices, knowledge transfer and training [14]. With the combination of all this important factors, manufacturers would be able to develop their company capability to attain more profit and produce best product for customers.

* Corresponding author: harithamlus@gmail.com

1.1 Manufacturing Capabilities

The concept of strategic capabilities that determine a manufacturing's contribution to the success of a firm is closely related to the notions of strategic resources, competences and priorities. In contrast to capabilities, resources are something a firm possesses or has access to, not what a firm is able to do. Resources can be tangible, e.g. specialized production systems, and intangible, e.g. level of training of workers [9]. Based on such resources, capabilities are developed. For instance, flexible production systems in combination with highly skilled workers or resources, allow to produce in a flexible way, capability. Capabilities allow an enterprise to develop and to exploit resources in order to generate profit through its products and services [1].

1.2 Agility

Agility is the ability to easily move from one manufacturing state to another. Agility divided to two criteria which are volume flexibility and mix flexibility. Volume flexibility refer to the ability of efficiently produce wide ranges in the demanded volumes of products. Mix flexibility is the ability to manufacture a variety of products, over a short time span, without modifying facilities [16]. The new concept agility in manufacturing, these drivers are competition, fragmentation of mass market, cooperative production relationship, evolving customer expectations, and increasing social pressures. The main issue in this new area of manufacturing management is the ability to cope with unexpected changes, to survive unprecedented threats of business environment, and to take advantage of changes as opportunities. This ability is called agility or agile manufacturing.

Agile manufacturing that was sometimes mixed up and confused with previous thought schools of manufacturing management such as flexibility and lean manufacturing has been backed for having novel concepts beyond the former remedies. This has happened thanks to the wide concern it received during the past few years, though in place this has been a natural result of the increasing need to resolve problems with the so-called remedies and increasing pressures on manufacturing companies in competing for success. Agility in concept comprises two main factors. They are responding to change (anticipated or unexpected) in proper ways and due time and exploiting changes and taking advantage of them as opportunities. These, indeed necessitate a basic ability that is sensing, perceiving and anticipating changes in the business environment of the company. An agile manufacturer, in this way is an organisation with a broad vision on the new order of the business world, and with a handful of capabilities and abilities to deal with turbulence and capture the advantageous side of the business.

Agility is an ability and a character that every manufacturing company must have to be able to survive and prosper in the new order of the world business environment. Different organisations are different in the way they should respond to changing business environment, so they need different levels of agility. Agility is a response to the changes that a company faces, and shall become a characteristic of the company. Therefore, the way that the company should act in turbulent circumstances of the business environment, which is called agility, is a direct function of changes in the business environment, and also the business environment itself and the company's situation. Agility in concept is a strategic response to the new criteria of the business world, and in practice, is a strategic utilization of business methods, manufacturing and management processes, practices and tools, most of which are already developed and used by industries for certain purposes, and some are under development to facilitate the capabilities that are required for being agile. Information system/technology in its utmost level of timeliness, coverage, communication ability, data banking and interchange, etc., is a major differentiator of an agile manufacturing company compared to traditional systems.

1.3 Time Based Management

The lead time is the time a customer must wait between order placement and receipt. Reliability or dependability (on-time delivery, OTD) reflects on the company's reliability in delivering a customer's order on or before the quoted delivery date. There is growing recognition of the importance of delivery reliability as a criterion in most markets. Its change toward being a qualifier is part of that competitive perspective. Delivery lead time or speed (Order Fulfilment lead time, OFLT) is about short delivery lead times and involves decisions in production stability, investments in capacity and/or inventory and the control of workflow. Another measure could be Time-to-Market for new products which manufacturing effects. Therefore, it is essential to consider these factors in studying manufacturing capabilities. This study focuses on manufacturing practices namely *time based management* The investigation on the effects of organizational capabilities on the manufacturing capabilities addressed by the following hypotheses;

H_1 Time Based Management has significant influence on Agility

2 Methodology

2.1. The Sample and Data

A survey was conducted to collect cross sectional data in Northern Region of Malaysia which is one of the state introduce as the Silicon Valley of Malaysia. The population of this study was the Top Management and Managers. These grades consist of management and professional employees which identified a group with high potential to engage manufacturing capabilities. The selection of the survey location and the sample was due to the Federation of Manufacturers of Malaysia (FMM).

A set of questionnaire was formulated and designed based on the previous literature in the subject area. The questionnaires were emailed to 160 respondents. 119 sets questionnaires received within two weeks giving the response rate of 74.4%. The sample profile of the survey is shown in Table 1.

Table 1. Sample profile of the respondent

Variables	Item	Frequency	Percentage (%)
Companies's Type	Malaysian owned	75	63.0
	Foreign owned	28	23.5
	Joint Venture	16	13.5
Size of Company(In term of Full Time Employees	Less than 5	-	-
	5 to 50	-	-
	51 to 150	36	30.0
	More than 150	83	69.7
Respondents' Job Designation	General Manager	3	2.5
	Manager	89	74.8
	Executive	27	22.7
Year of designation	Less than 1 years	8	6.7
	1-5 years	31	26.1
	6-10 years	17	14.3
	More than 10 years	63	52.9
Education level	Degree	98	82.3
	Master Degree	20	16.8
	PhD	1	0.9

From Table 1 it is found that more than 63% respondents are Malaysian owned companies and 83 companies had more than 150 workers in their operation. The respondents are 3 General manager, 89 Manager and 27 Executive and 63 from the respondent work more than 10 years. 98 from this respondent are from Degree level.

2.2 Reliability Analysis

An internal consistency analysis was performed separately for the items of each managerial commitment and system perspectives by using the SPSS version 20, reliability procedure. Based on the suggestion by [8], the author highlighted an acceptable alpha value is greater than 0.6. As show in Table 2, the alpha values of reliability analysis for this study ranges from 0.70 to 0.85. From the results obtained, all the alpha values are greater than 0.7. Thus it can be concluded that this instrument has good internal consistency and is therefore reliable.

Table 2: Reliability analysis result

Variables	Number of items	Alpha
Time Based Management	4	0.812
Agility	4	0.864

Descriptive analyses for the organizational capabilities and manufacturing capabilities are shown in Table 4 and Table 5. The levels of organizational capabilities and manufacturing capabilities are based on the levels of mean score range provided in Table 3.

Table 3: Level of Mean Score Range

Range of mean score	Level
1.00 – 2.33	Low
2.34 – 3.67	Medium
3.68 – 5.00	High

3 Findings

3.1. The Level of Manufacturing Capabilities

Manufacturing capabilities among the manufacturing companies are measured based on the values of means and standard deviations. Table 4 shows the ranking of knowledge transfer and the value of mean and standard deviation for each activity.

https://doi.org/10.1051/matecconf/201815006013

Table 4: Mean Values of Manufacturing Capabilities

Activities	Mean	Standard Deviation
Able to efficiently produce wide ranges in the demanded volumes of products	3.64	0.687
Able to manufacture a variety of products, over a short time span, without modifying facilities	3.67	0.634
Able to accelerate or decelerate the rate of production quickly to handle large fluctuations in demand	3.68	0.643
Manufacturing Capabilities	3.66	0.654

Mean values explained the propensity of the respondent to involve with manufacturing capabilities (Li, 2000). The mean values in range of 2.34 to 3.67 is categorized as medium stage.

The mean score for the overall manufacturing capabilities is 3.66. This indicated that the level of manufacturing capabilities among manufacturing companies is at the medium stage.

3.2. The Level of Time Based Management

Table 5 gives the mean values for the eight organizational capabilities. It is shown that one of the organizational capabilities has more than 3.67 mean values which is 3.72. Base on Table 5, the levels of these organizational capabilities are considered as medium stage among the manufacturing companies.

Table 5: Mean Values of Time Based Management

Activities	Mean	Standard deviation
Indicate The Importance Given To Delivery Time	3.61	0.861
Indicate The Importance Given To Engineering Time	3.72	0.713
Indicate The Importance Given To Procurement	3.45	0.762
Indicate The Importance Given To Set-Up Time	3.23	0.731
Indicate The Importance Given To Throughput Time	3.46	0.712
Indicate The Importance Given To Time To Market	3.12	0.720
Indicate The Importance Given To Bottleneck Identification	3.01	0.716
Time Based Management	3.29	0.743

The lowest mean of time based management is *Indicate The Importance Given To Time To Market* and the average mean value of organizational capabilities is at medium levels.

3.3. Correlation Analysis

From the results of correlation analysis it is found that organizational capabilities are significantly correlated to manufacturing capabilities. As shown in Table 6, the correlation coefficients are found to be positive and statistically significant at 0.01 level ($p<0.01$).

Table 6: Pearson Correlation Coefficients between Time Based Management and Manufacturing Capabilities

	Correlations	
Variable	F1	Manufacturing Capabilities
F1	1.00	
Time Based Management	0.736**	1.00

** Correlation is significant at the 0.01 level (1-Tailed)
* Correlation is significant at the 0.05 level (1-Tailed)

Therefore, Hypothesis 1 is supported. There is a significant relationship between time based management with the other variables that is agility (manufacturing capabilities). Hence, Hypothesis 1 is supported.

4 Discussions

This study is to identify empirically the core manufacturing capabilities in manufacturing industries in Malaysia. The rationale of the study stems from the major consideration, that is, the emerging concern of top management and shareholders of the manufacturing companies and the directors of the companies in this industry in particular to develop core manufacturing capabilities.

The analysis based on the respondents' perception of core manufacturing capabilities in manufacturing companies showed that manufacturers took a lot of effort in time based management in achieved high degree of core manufacturing capabilities

The findings show that time based management is the important variable that must be consider by the manufacturers in their operation. These findings are supported by previous researchers who have studied in developing manufacturing capabilities (Swink & Hegarty, 1998; Haifeng, 2006; Li, 2000)

The correlation analysis show that time based management is the priority variables that must be highlight which show the correlations coefficients 0.736 at the 0.01 level (1-Tailed).

5 The Limitations

This survey has been conducted on manufacturing companies in Northern Region of Malaysia. Nevertheless, the result may not be able to generalise the influence of time based management on manufacturing capabilities for the whole country as there are different focus on their operation. This study focuses solely on the manufacturing capabilities in term of agility. Future research should consider other factors that influence the manufacturing capabilities quality, lean and six sigma.

6 Conclusion

This paper has applied empirical analysis on the influence of time based management on manufacturing capabilities of local manufacturers. Survey on manufacturing capabilities on the local manufacturers reveals that the level of manufacturing capabilities is high level. The results show that the time based management contribute to the manufacturing capabilities. The results suggest that, all manufacturers must consider time based management as their focusing aspect to survive in the business.

Companies or manufacturers must increase their time based management in term of agility to gain significantly in the core manufacturing capabilities. The time based management absolutely are very important in develops manufacturing capabilities.

References

[1] Amit, R. & Schoemaker, P. (1993). Strategic assets and organizational rent. *Strategic Management Journal*; 14, (1), pp. 33–46.

[2] Barnes, D. (2001). Research Methods for the Empirical Investigation of the Process of Formation of Operations strategy. *International Journal of Operations & Production Management, 21*(8), 1076-1095.

[3] Beach, R. Muhlemann, A.P. Price, D. Paterson, A. & Sharp, J. (2000). Manufacturing Operations and Strategic Flexibility: Survey and Cases. *International Journal of Operations & Production Management, 20*(1), 7-30

[4] Bhatt, G. & Grover, V. (2005). Types of information technology capabilities and their role in competitive advantage: An empirical study. Journal of Management Information Systems, 22(2), 253–277.

[5] Bolden, R. Waterson, P. Warr, P. Clegg, C. & Wall, T. (1997). A new taxonomy of modern manufacturing practices. *International Journal of Operations & Production Management, 17*(11), 1112-1130.

[6] Coates, T. (2002). An Exploratory Analysis of New Competences: a Resource-Based View. *Journal of Operations Management 20*(5), 435-450.

[7] Haifeng. G. Yezhuang, T. & Zhandong, L. (2006). An empirical analysis on relationships of manufacturing practices and manufacturing capabilities. *Paper presented at the International Conference on Management of Innovation and Technology.*

[8] Hair, J. Anderson, R. Tatham, R. & Black, W. (2007). Multivariate data analysis with readings

[9] Hall, R. (1992). The strategic analysis of intangible resources. *Strategic Management Journal.* Vol 13. 135- 144

[10] Katz, R. Rebentisch, E. & Allen, T. (1996) A Study of Technology Transfer in a Multinational Cooperative Joint Venture. *IEEE Transactions on Engineering Management.*

[11] Li, X. (2000). Manufacturing Capability development in a changing ´ business environment. *Industrial Management & Data Systems, 100*(6), 261-270.

[12] Narasimhan, T. Swink, S. & Kim, D. (2005). An exploratory study of manufacturing practice and performance interrelationships, Implications for capability progression. *International Journal of Operations & Production Management, 25*(10), 1013-1033.

[13] Grobler, A. (2007). A dynamic view on strategic resources and capabilities applied to an manufacturing strategy literature *Journal Of Manufacturing Technology Management,, 18*(3).

[14] Sohal, M. Gordon, M. Fuller, T. & Simon, K. (1999). Manufacturing practices and compatibility : An Australian study. *Technovation, 19*(5), 295 – 304

[15] Narasimhan, T. (2005). An exploratory study of manufacturing practice and performance interrelationships, Implications for capability progression. *International Journal of Operations & Production Management, 25*(10), 1013-1033.

[16] Swink, M. (1998). Core manufacturing capabilities and their links to product differentiation. *International Journal of Operations & Production Management, 18*(4), 374- 396.

[17] Teece D. & Shuen, A. (1997). Dynamic capabilities and strategic management. *Strategic Management Journal, 18*(7), 509-533.

[18] Yang, B., Burns, N.D. & Backhouse, C.J. (2004). Postponement: A Review and Integrated Framework. *International Journal of Operations & Production Management, 24*(5), 468-487.

[19] Zeleny, M. (1986). High technology management, *Human Systems Management*, 6(2), pp. 109-120.

[20] Zikmund, W. G. (2000). Business Research Methods. *Ford Worth:The dryden Press.*

Electromagnetic Properties Performance of MWCNTs/Polyester Composites in X-band

Y. S. Lee[1,2], *F. H. Wee*[2], *K. Y. You*[3], *C. Y. Lee*[3], *Y. Y. Lee*[2], *L. Zahid*[1],
Nur Hidayah Ramli[1], *M. S. Shakhirul*[1], *Nurulbariah Idris*[1], *E. M. Cheng*[4]

[1]Department of Electronic Engineering Technology, Faculty of Engineering Technology, Universiti Malaysia Perlis (UniMAP), Perlis, Malaysia.
[2]Bioelectromagnetics Research Group (BioEM), School of Computer and Communication Engineering, Universiti Malaysia Perlis (UniMAP), Perlis, Malaysia.
[3]Department of Communication Engineering, Faculty of Electrical Engineering, Universiti Teknologi Malaysia (UTM), 81310 UTM Johor Bahru, Johor Darul Ta'zim, Malaysia.
[4]School of Mechatronic Engineering, Universiti Malaysia Perlis (UniMAP), Pauh Putra Campus, 02600 Arau, Perlis, Malaysia.

Abstract. In this paper presents nano-composites based on polyester (PE) matrix with multi-walled carbon nanotubes (MWCNTs) as fillers have been developed for microwave absorbing materials. The MWCNTs/PE composite samples were fabricated with different weight ratio of MWCNTs (3 wt%, 5 wt%, and 10 wt%). The electromagnetic properties of different MWCNTs/PE composite have been determined by using rectangular waveguide technique and Agilent material measurement software. Moreover, the reflection loss (microwave absorption) of MWCNTs/PE composite have been calculated based on the basis of transmission line theory. The performance of electromagnetic properties and microwave absorption of MWCNTs/PE composite were analyzed in X-band frequency. The dielectric loss properties of the composite are increse with increasing in MWCNTs weight ratio. The microwave absorption results show that such 3 wt% MWCNTs/PE composites sample with 4 mm thickness has achieved less than -10 dB values (< 90 % microwave absorption) of reflection loss.

1 Introduction

The rapid growth in science and wireless technology towards gigahertz frequency applications in modern communication has increased the electromagnetic interference (EMI) [1], [2]. EMI is cause by uncontrolled interference of electronic devices or systems. Therefore, the efficiency of EMI shielding or microwave absorbing materials (MAMs) were increasing to control the EMI pollution [3], [4]. For MAMs, it should can absorb microwave energy incident on it and have minimum reflection of the microwave energy. MAMs can use to protect electronic device and also human body from harmful effect of these microwaves energy.

Carbon nanotubes (CNTs) is a high carbon dielectric material which exhibit excellent electronic, mechanical, and material sciences [5], [6]. A composite material can defined as combining two or more materials to give a unique combination of properties. Recent years, composites based on CNTs with polymer matrix are widely used in many application due to the unique and advantageous properties offered by the new composites materials. Multi-walled carbon nanotube (MWCNT) is a tube with many CNTs inside which could be concentric or spirally spread [7]. The real part of dielectric properties is known as dielectric constant which indicate the energy store. Dielectric loss factor is the imginary part of the dielectric properties show the ability of material convert the stored energy to heat and dissipate. Therefore, the dielectric properties ($\varepsilon_r = \varepsilon'_r - j\varepsilon''_r$) of the composites is important for microwave absorber [8], [9]. The interaction of microwave energy with MAMs is microwave energy stored (absorption) in materials and convert the energy to heat.

A good MAMs are require high microwave absorption performance, thin, and light weight. The electromagnetic properties of samples were investigated using network analyzer, and measurements were carried out using the transmission line technique to determine the dielectric properties of MWCNTs/PE samples. Furthermore, the reflection loss (microwave absorption) of the samples were calculated and analyzed.

2 Reflection Loss

In the present study, the reflection loss of the samples were calculated based on the basis of transmission line threory to analyze the best performance in microwave absorption over 8.2- 12.4 GHz frequency. The result can be explained base on the impendance matching as equation (1) and (2). In general, researchers use a perfect electric conductor (PEC) as a metal back plate to measure reflection loss of the MAMs [10], [11]. Figure 1 shows the illustration of the MAMs with metal backed plate.

* Corresponding author: yslee@unimap.edu.my

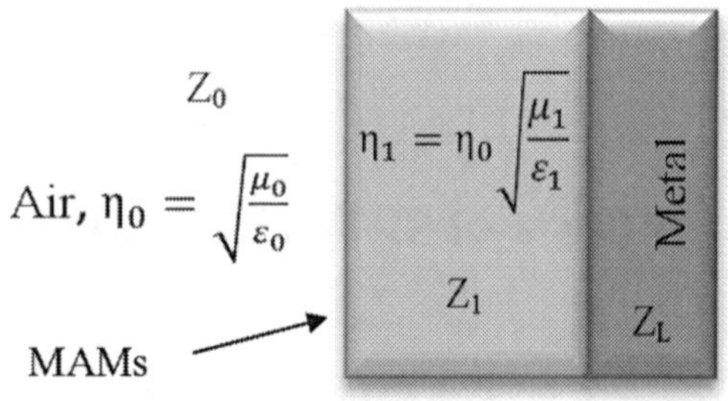

Fig. 1. Illustration of the MAMs with metal backed plate.

MAMs are functional materials that absorb microwave energy with minimum reflection and maximum attenuation/absorption of the microwave energy. The reflection of microwave energy occurs when incident energy reflected from the MAMs interface and from the metal backing of the MAMs. Input impedance (Z_{in}) of MAMs close to air (377 *ohm*), the reflection at the interface can be minimized. When an electromagnetic wave is incident normally to a single layer microwave absorber material is backed a metal back plate, the impedance is given by [12];

$$Z_{in} = Z_0 \tanh\left(j\frac{2\pi}{c}\sqrt{\mu_r \varepsilon_r}\, fd \right) \qquad (1)$$

where, Z_0 is the free space impedance given by, $Z_0 = \sqrt{\mu_0/\varepsilon_0} = 377\ ohm$; c, f ,*and* d are the propagation velocity of the wave in free space, frequency of the incident wave, and thickness of the MAMs, respectively. The material properties namely the dielectric properties, $\varepsilon_r = \varepsilon_r' - j\varepsilon_r''$ and magnetic properties, $\mu_r = \mu_r' - j\mu_r''$; d is the thickness of MAMs; c and f are the velocity of light in vacuum and the frequency of microwaves in free space, respectively. For dielectric MAMs, the permeability, $\mu_r \approx 1 - j0$.

In case of a metal backed single layered absorber, the normalized input impedance with respect to impedance in free space and the reflection loss with respect to the normal incident plane wave are given by [13];

$$R_L(\text{dB}) = -20\log_{10}\left[\frac{(Zin-1)}{(Zin+1)}\right] \qquad (2)$$

There have two reflection phenomenon occurs when the microwave energy interact with MAMs. The first reflection is at the air-absorber interface while the another reflection is the energy propagated through the materials and bounce back from the metal backing of the MAMs. Reflection at the interface can be minimized by making input impedance of MAMs close to that of free space as known as impedance matching.

3 Experimental

The multi-walled carbon nanotubes (MWCNTs) used in this experiment was purity 99 wt% purity. MWCNTs is a nanomaterial which very hard to compacted due to the very large surface area. Hence, MWCNTs in this experiement was composites with polymer (polyester resins). In this experiment, the PE resins and Methyl ethyl ketone peroxide (MEKP) harden agent weight ratio are constants. The different weight ratio of MWCNTs as filler is the main parameter in this study. The MWCNTs/PE composites are prepared using mechanical mixing and MEKP is used as hardent agent for MWCNTs/PE samples. The samples are prepared in different weight ratio of MWCNTs (2 wt%, 3 wt%, and 5 wt%) with dimension 22.860 mm x 10.160 mm x 5 mm. The MWCNTs/PE composites preparation process was mixed the MWCNTs with PE resins and stir for 1 hour shown in Figure 2. Then, the composites of the MWCNTs/PE was mixed with MEKP and poured into the rectangular sample holder mould flange and cured in room temperature 25 °C for 24 hours. The different rectangular samples of MWCNTs/PE was fabricated. Figure 3 shows the fabricated rectangular samples of MWCNTs/PE.

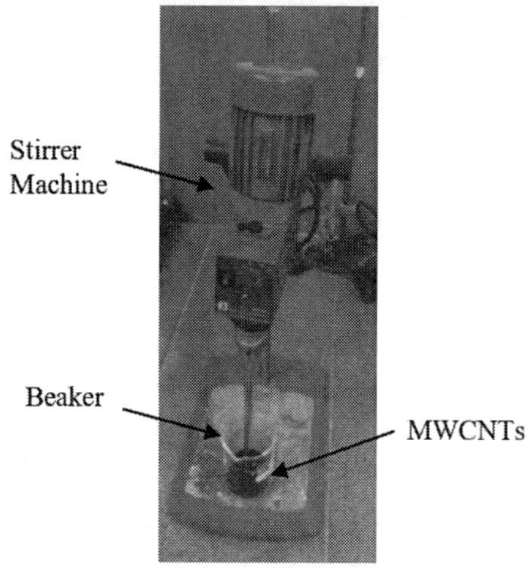

Fig. 2. The MWCNTs composites is stired with PE resins.

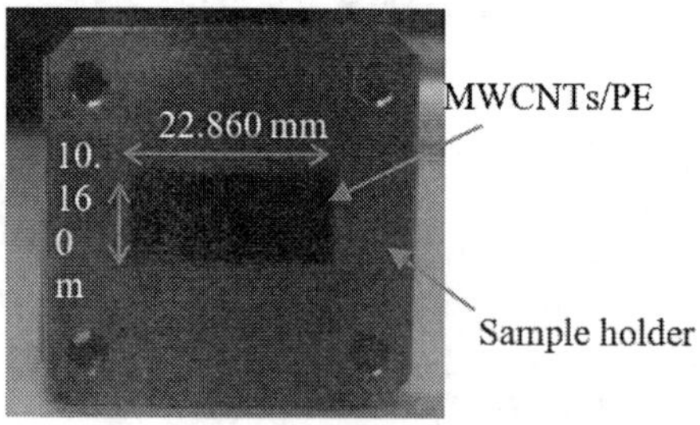

Fig. 3. MWCNTs/PE rectangular samples placed inside the sample holder.

3.1 Material Properties Measurement

The electromagnetic properties of MWCNT/PE nanocomposites with various weight percentage MWCNTs were first measured in the X-band to get an indication of dielectric properties dependence on MWCNT percentage. There have two methods common use in the measure electromagnetic properties of materials; free space method and transmission line method. In free space method, the materials propertis characterization much more flexible to measure under dfferent condition. But the material under test (MUT) required larger than the horn antennas. In a transmission line method, the MUT is placed in a a piece of transmission line or sample holder. The MUT are placed in samples holder between two waveguide adaptors. This method is widely use in measure electromagnetic properties of small samples. The electromagnetic properties of material are characterize based on the basis of the reflection from the MUT and transmission through the MUT [14]. In this study, transmission line method is used to measure the dielectric properties of materials which suitable for small samples.

The rectangular waveguide transmission line technique is used to measure the dielectric properties of the MWCNTs/PE sample. A pair of coaxial cables were connected between Agilent network analyzer model E8362B and the rectangular waveguide adaptors (WR-90). The fabricated MWCNTs/PE samples are inserted into the rectangular sample holder as shown in Figure 4. The sample holder is well connected between two rectangular waveguide adaptors with screws as shown in Figure 4. Figure 5 shows the sample holder is custom manufacture with size a = 22.860 mm, b = 10.160 mm, and d (thickness) = 4 mm.

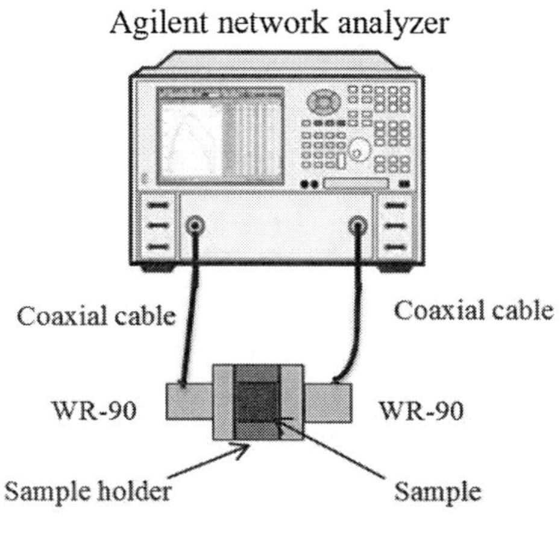

Fig. 4. Rectangular waveguide transmission line technique measurement setup.

Figure 5 shows the Thru-Reflect-Line (TRL) calibration kits of the rectangular waveguide transmission line. The TRL calibration of the measurement is to eliminate the testing error induced by the gap between sample and flange. The transmission line is enclosured with conductor material. Therefore, the microwave just can propagated inside the rectangular waveguide. Then, The dielectric properties of MWCNTs/PE samples are computed by using the Agilent software 85071E [8].

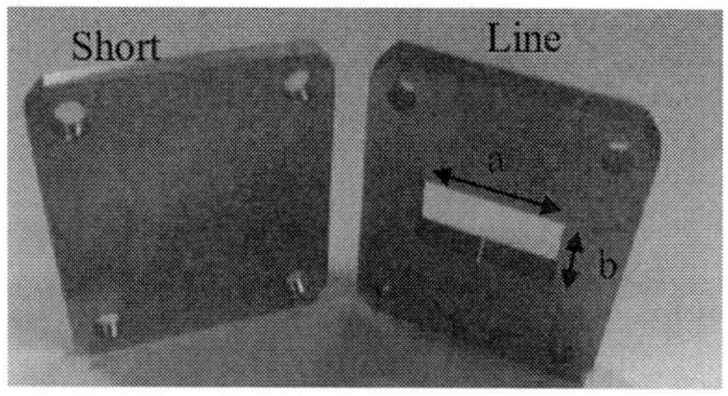

Fig. 5. TRL calibration Line and short for rectangular waveguide.

4 Results and Discussion

Figure 6 (a) and (b) represent the dielectric constant, loss factor, and loss tangent of different MWCNTs filler 3 wt%, 5 wt%, and 10 wt% respectively. The dielectric properties of MWCNTs/PE samples are measured using rectangular waveguide mehod in X-band frequency band. From the dielectric properties results, it can be observed that the weight percentage of MWCNTs filler effect the dielectric constant, loss factor, and loss tangent of dielectric properties. The dielectric properties increase with increasing of wt. % MWCNTs filler over 8.2 GHz to 12.4 GHz. The dielectric constant values increase from 4.08 to 15.52 with increasing the MWCNTs filler from 3 wt% to 10 wt%. Similar to the loss factor values increase from 1.63 to 11.27. For the lower MWCNT weight percentages, the loss factor was relatively flat with frequency. Table 1 shows the average values of dielectric properties of MWCNTs/PE samples in X-band.

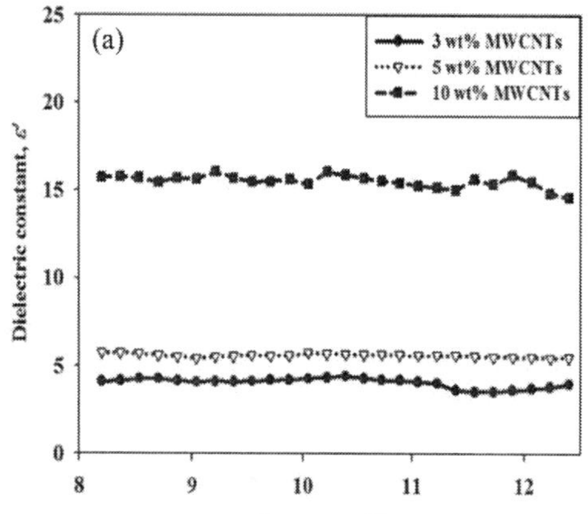

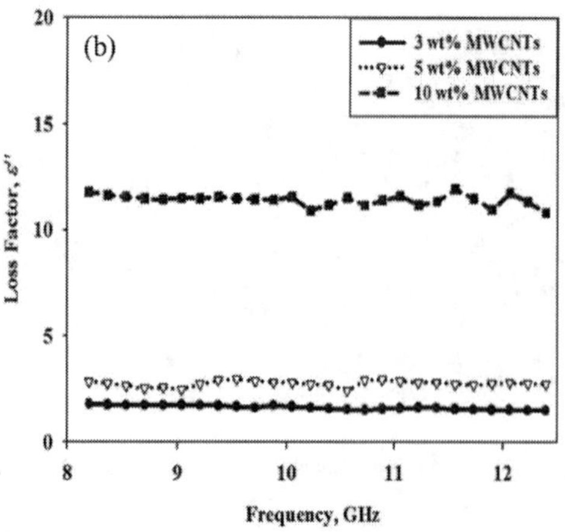

Fig. 6. (a) Dielectric constant and (b) loss factor of 3 wt%, 5 wt%, and 10 wt% MWCNTs fillers (wt%).

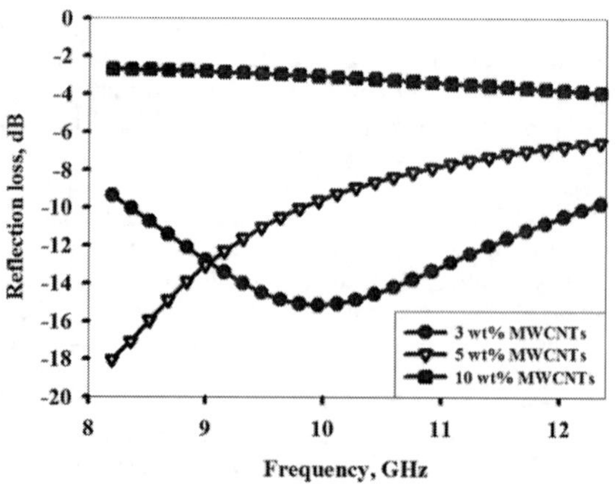

Fig. 7. The reflection loss of 3 wt%, 5 wt%, and 10 wt% MWCNTs fillers (wt.%).

Table 1. The dielectric constant and loss factor of MWCNTs/PE samples in X-band.

Sample	Average values	
	Dielectric constant	Loss factor
3% MWCNTs	4.08	1.63
5% MWCNTs	5.61	2.72
10% MWCNTs	15.52	11.27

From the calculated reflection loss in Figure 7, a broadband MAMs would have a MWCNTs weight percentage 3 wt % of the composite. The results for 5 wt% MWCNT samples cover the X-band with less than -10 dB (< 90 % microwave absorption) shown in Figure 7. Nevertheless, the 3 wt % MWCNTs weight percentage in MWCNTs/PE composites shows better broadband absorption than 5 wt % and 10 wt% of MWCNTs in X-band. The sample with 3 wt% MWCNTs shows average lower values of reflection loss in the frequency of X-band compare to the samples with 5 wt% and 10 wt% MWCNTs. At the frequency of 10 GHz the reflection loss value is close to -15.8 dB which corresponds to 97% of microwave absorption. From Figure 7, observed that sample 5 wt% MWCNTs reflection loss increase with increasing the frequency due to the mismatch of the impedance matching (Z_{in}) at higher freqeuncy. Similar problem for sample with 10wt% MWCNTs, MWCNTs/PE sample with 10 wt% MWCNTs has the highest values of reflection loss (lowest absorption) in X-band. This shows that the sample with high dielectric properties in range, $\varepsilon_r' = 15$ to 15.5 and $\varepsilon_r'' = 11.5$ to12.5 with 4 mm thickness is not suitable use as microwave absorber in X-band. All this mismatch of impedance matching (Z_{in}) over frequency range in X-band can be explained using the equation (1) and (2). The dielectric properties of materials and frequency range are the important factor to determine the absorption of MAMs.

5 Conclusions

This study shows that the different application in microwave absorber with flexibility of the dielectric properties of materials to use as MAMs. MWCNTs/PE samples with different weight ratio (3, 5, 10 wt%) of the same MWCNTs were fabricated. The dielectric properties of MWCNTs/PE samples were measured with rectangulare transmission line technique and an Agilent 85071E software in the X-band frequency. The microwave absorption properties of the MWCNTs/PE samples were design and calculated based on the basis of transmission line theory.

This research was funded by the grant of Fundamental Research Grant Scheme (No. 9003-00591) under Ministry of Education, Malaysia.

References

[1] B. Archambeault, C. Brench, and S. Connor, "Review of Printed-Circuit-Board Level EMI/EMC Issues and Tools," *IEEE Trans. Electromagn. Compat.*, vol. 52, no. 2, pp. 455–461, 2010.

[2] M. H. Al-Saleh and U. Sundararaj, "Electromagnetic interference (EMI) shielding effectiveness of PP/PS polymer blends containing high structure carbon black," *Macromol. Mater. Eng.*, vol. 293, no. 7, pp. 621–630, 2008.

[3] I. B. Darney, *Circuit Modeling for Electromagnetic Compatibility*. Institution of Engineering and Technology, 2013.

[4] P. Saini and M. Arora, "Microwave Absorption and EMI Shielding Behavior of Nanocomposites Based on 2 Intrinsically Conducting Polymers, 3 Graphene and Carbon Nanotubes 4," 2012.

[5] T. Zhao, C. Hou, H. Zhang, R. Zhu, S. She, J. Wang, T. Li, Z. Liu, and B. Wei, "Electromagnetic wave absorbing properties of amorphous carbon nanotubes.," *Sci. Rep.*, vol. 4, p. 5619, 2014.

[6] D. L. Zhao, X. Li, and Z. M. Shen, "Microwave absorbing property and complex permittivity and permeability of epoxy composites containing Ni-coated and Ag filled carbon nanotubes," *Compos. Sci. Technol.*, vol. 68, no. 14, pp. 2902–2908, 2008.

[7] C. M. Chang, J. C. Chiu, W. S. Jou, T. L. Wu, and W. H. Cheng, "New package scheme of a 2.5-Gb/s plastic transceiver module employing multiwall nanotubes for low electromagnetic interference," *IEEE J. Sel. Top. Quantum Electron.*, vol. 12, no. 5, pp. 1025–1030, 2006.

[8] Y. S. Y. Lee, M. M. F. B. A. Malek, E. M. E. E. M. Cheng, W. W. Liu, K. Y. You, M. N. Iqbal, F. H. Wee, S. F. Khor, L. Zahid, and M. F. bin Haji Abd Malek, "Experimental determination of the performance of rice husk-carbon nanotube composites for absorbing microwave signals in the frequency range of 12.4-18 GHz," *Prog. Electromagn. Res.*, vol. 140, no. April, pp. 795–812, 2013.

[9] D. A. Makeiff and T. Huber, "Microwave absorption by polyaniline-carbon nanotube composites," *Synth. Met.*, vol. 156, no. 7–8, pp. 497–505, 2006.

[10] S. Das, G. C. Nayak, S. K. Sahu, P. C. Routray, A. K. Roy, and H. Baskey, "Microwave absorption properties of double-layer composites using CoZn/NiZn/MnZn-ferrite and titanium dioxide," *J. Magn. Magn. Mater.*, vol. 377, pp. 111–116, 2015.

[11] J. W. Kim and S. S. Kim, "Microwave absorbers of two-layer composites laminate for wide oblique incidence angles," *Mater. Des.*, vol. 31, no. 3, pp. 1547–1552, 2010.

[12] A. Ghasemi, A. Paesano, and C. F. C. MacHado, "Magnetic and reflection loss characteristics of terbium substituted cobalt ferrite nanoparticles/functionalized multi-walled carbon nanotube," in *IEEE Transactions on Magnetics*, 2012, vol. 48, no. 4, pp. 1528–1531.

[13] S. Ozah and N. Bhattacharyya, "Nanosized barium hexaferrite in novolac phenolic resin as microwave absorber for X-band application," *J. Magn. Magn. Mater.*, vol. 342, pp. 92–99, Sep. 2013.

[14] Y. S. Lee, F. Malek, E. M. Cheng, W. W. Liu, K. Y. You, K. Y. Wee, L. Zahid, and H. A. Rahim, "Single layer microwave absorber based on rice husk-mwcnts composites," *ARPN J. Eng. Appl. Sci.*, vol. 11, no. 14, pp. 8932–8937, 2016.

https://doi.org/10.1051/matecconf/201815006015

Extraction of Spear Grass (*Imperata Cylindrica*) As Pro-Oxidant In Polymer Blends

M.A. Nuradibah[1,*], *N. A. Abdul Halim* [1], *K.F.* Kasim[1], *S.T.* Sam[1], *H.I.* Zulhaimi[1] *and Muhammad Nur Aiman Uda*[1]

[1]School of Bioprocess Engineering, Kompleks Pusat Pengajian Jejawi 3, Universiti Malaysia Perlis, 02600 Arau, Perlis.

Abstract. Packaging material such as plastic bags is one of the main factors that contribute to the environmental pollution due to slow degradation. The usage of metal oxide as pro-oxidant has been proven to accelerate the degradation of these materials, but the excessive usage of this pro-oxidant will be harmful to the human body. Therefore, in this research, spear grass is investigated to be used as natural based pro-oxidant that can increase the degradation rate of the polymers. In terms of that, spear grass is extracted by using pressurized hot water extraction (PHWE) to obtain the metal element such as zinc (Zn) and ferum (Fe). PHWE is using water as a solvent which is highly favourable due to non-toxicity and non-flammable characteristics that make it easy to handle. Box-Behnken design is used to optimize the temperature, extraction time, and sample-to-solvent ratio to get the maximum amount of Zn and Fe concentration from the extracted spear grass. As a conclusion, the leaf of spear grass contributed the highest amount of Zn and Fe concentration. The highest amount of Zn and Fe concentration is achieved at 150 °C, 20 minutes, and 3 g of sample to 45 ml of water.

1 Introduction

From the last decades, the search for solutions to minimize environmental problem due to polymeric materials such as polyethylene has increase necessity when it disposed in an appropriate environment [1]. Polyethylene represents 64% of plastic materials produced as packaging and bottles, which usually discarded after only brief use due to easily contaminated, recycling these materials is not effortless and recovery cost is higher than cost producing. As the effects, several thousand tonnes of goods made from plastic materials are sent to landfill, and increasing the amount of municipal waste [2]. Due to their low degradability, generating pollution and taking space in landfills, plastics accumulate in environment.

Polyethylene is considered as an inert bulk polymer which is slowly degradable. Polyethylene in its pure form is extremely resistant to environmental degradation. Khabbaz et al. (2001) has estimated that polyethylene would degrade less than 0.5% in 100 years, and 1% if exposed to sunlight for 2 years before biodegradation [3]. Roy et al. (2007) stated that to produce degradable polyethylene is to mix it with pro-oxidant additives that can effectively improve the degradability of these materials [4]. Metallic salt is comprised in the commercial polyethylene as the most additives being used which is harmful to human body if used in excessive quantity. The usage of metal oxides as chemical based pro-oxidant such as is cobalt stearate that led to toxicity that can cause damage to major constituents of biological system.

Imperata cylindrica also known as spear grass in Nigeria, alang-alang in Asia, and cogongrass in America is a perennial rhizomatous grass which has the high up to 1.2 m. Spear grass is the most dominant, competitive, and difficult weed to control in Asia, West Africa, and Latin America [5]. Spear grass can act as natural pro-oxidant that can increase the degradation of the polymer by reducing their molecular weight. In order to do that, spear grass need to be extracted to obtain essential ingredients that can make it act as pro-oxidant. Ferum (Fe) and zinc (Zn) are metal elements that an important content in spear grass that can accelerate the degradation.

In this project, spear grass as pro-oxidant will be extracted by using pressurized hot water extraction (PHWE). PHWE is feasible green solvent extraction that used water as extraction solvent as it utilized pressurized water at elevated temperature and controlled pressure conditions. Water as solvent is highly favourable because water is non-toxicity and non-flammable characteristics that make it easy to handle. In the various report, the polarity of water can be varied close to those of alcohols at certain temperature and applied pressure. Thus water can dissolve a large range of medium and categorized as low polarity analytes [6-7].

2 Materials and Procedures

1.1 Materials

The spear grass was collected from Kuala Perlis. The spear grass was grounded to fine powder. Distilled water was used as solvent for extraction process.

*Corresponding author: nuradibah@unimap.edu.my

1.2 Procedures

1.2.1 Preparation of sample

The plant which is spear grass (*Imperata cylindrica*) was collected from Kuala Perlis, Perlis. The spear grass was separated to different parts which were leaf, root, and straw. It was washed thoroughly under running tap water to get rid dirt, dust, and soil. Then, the plant was undergoing drying process by using oven at 50 °C for 24 hours. After that, the spear grass was crushed by using grinder and sieved to 63 μm particle sized to get powder form.

1.2.2 Extraction of spear grass

To identify the best extraction method, three type of extraction which is cold extraction, hot extraction, and pressurized hot water extraction (PHWE) was conducted. For cold extraction, 2 g of spear grass was mixed with 20 ml of water and leaved at room temperature for about 3 hours. For hot extraction, 2 g of spear grass was mixed with 20 ml water and was stirred using hot plate stirrer for about 3 hours at 100 °C. For PHWE, 2 g of spear grass was mixed with 20 ml water and the sample was extracted at 150 °C for 20 minutes. After that, the extracted sample was analyzed using atomic absorption spectroscopy (AAS) and the best extraction method was proceeding for screening process.

1.2.3 Screening different part of spear grass

After the extraction process, 2 g of different part of spear grass (leaf, root, and straw) were undergone screening process by using pressurized hot water extraction (PHWE) at constant temperature, extraction time, and sample-to-solvent ratio.

After the extraction process has done, the amount of metal element contains in spear grass was determined by using Atomic Absorption Spectroscopy (AAS). The screening process was repeated for three times.

2.2.4 Extraction of selected part using PHWE

The selected part of spear grass with the highest amount of Fe and Zn that obtain from the screening process was extracted using pressurized hot water extractor (PHWE) at different temperature, extraction time, and sample-to-solvent ratio. The extracted spear grass was centrifuge at 4000 rpm at 4 °C for 10 minutes to separate the residue and the liquid. After that, the sample was filtered using syringe filter 0.2 μm to ensure that no residue in the sample. After the extraction process, Atomic Absorption Spectroscopy (AAS) was used to measure and analyze the concentration of Zn and Fe contains in spear grass.

2.2.5 Determining optimum conditions by using Response Surface Methodology (RSM)

The choice of design of experiment can have a large influence on the accuracy of the approximately and the cost of constructing the response process [8]. In this study, Response Surface Methodology (RSM) was used to determine the effect of different parameters that generate maximum response which is the amount of Fe and Zn. The Box-Behnken design was used to optimize the temperature, extraction time, and solvent-to-sample ratio for extraction of selected part of spear grass. Table 1 shows experimental design for optimization under Box-Behnken design.

Table 1. Experimental Design for Optimization under Box-Benhken Design

Standard run	Parameters					
	Temperature		Extraction time		Sample:solvent	
	Coded	Actual (celcius)	Coded	Actual (minute)	Coded	Actual (g/ml)
1	+1	200.00	+1	30.00	0	15.00
2	0	150.00	-1	10.00	+1	20.00
3	+1	200.00	0	20.00	-1	10.00
4	0	150.00	-1	10.00	-1	10.00
5	0	150.00	0	20.00	0	15.00
6	0	150.00	+1	30.00	+1	20.00
7	+1	200.00	-1	10.00	0	15.00
8	0	150.00	0	20.00	0	15.00
9	0	150.00	0	20.00	0	15.00
10	-1	100.00	+1	30.00	0	15.00
11	0	150.00	+1	30.00	-1	10.00
12	0	150.00	0	20.00	0	15.00
13	0	150.00	0	20.00	0	15.00
14	-1	100.00	0	20.00	-1	10.00
15	+1	200.00	0	20.00	+1	20.00
16	-1	100.00	-1	10.00	0	15.00
17	-1	100.00	0	20.00	+1	20.00

3 Results and Discussions

3.1 Comparison of extraction method

Pressurized hot water extraction (PHWE), hot extraction, and cold extraction for the extraction of spear grass, are compared in terms of temperature, extraction time, and sample-to-solvents ratio. Table 2 represents the concentration of zinc (Zn) and ferum (Fe) of spear grass by using different extraction methods.

Based on the Table 2, extraction of spear grass using PHWE is contributed the highest concentration of Zn and Fe with 0.7260 g/mL and 3.4827 g/mL respectively. It indicated that extraction by using PHWE was more efficient compared to cold and hot extraction. One of the greatest advantages of the PHWE is rapidity. An extraction time of 10 minutes provides comparable yield to those 3 hours of cold and hot extraction. The ultimate yield of concentration of Zn and Fe by PHWE was the highest which means PHWE method are 2 times more efficient than cold and hot extraction. This result means a substantial saving time, energy and plant material by PHWE.

3.2 Screening different part of spear grass

The screening process was performed using the pressurized hot water extraction (PHWE). The constant parameters which are 150 °C of temperature, 20 minutes of extraction time, and 15 mL of sample-to-solvents ratio was used to investigate which part of spear grass contributed the highest amount of zinc(Zn) and ferum (Fe) concentration. Table 3 and table 4 represent the average concentration of Zn and Fe after three set of replication.

Table 3. Concentration of zinc, Zn

Sample part	Concentration of zinc, Zn (g/mL)			
	First set	Second set	Third set	Average
Leaf	0.9639	0.8092	0.5627	0.7786
Straw	1.0472	0.6370	0.4092	0.4495
Root	0.1305	0.1408	0.1606	0.1439

Table 2. Concentration of Zn and Fe using different extraction methods.

Extraction method	Parameters			Concentration (g/mL)	
	Temperature (°C)	Extraction time (min)	Sample:solvent (mL)	Zn	Fe
PHWE	150	10	10.00	0.7260	3.4827
Cold	37	180	10.00	0.2111	1.2211
Hot	100	180	10.00	0.3423	0.5404

Table 4. Concentration of ferum, Fe

Sample part	Concentration of ferum, Fe (g/mL)			
	First set	Second set	Third set	Average
Leaf	2.9646	3.8157	4.9571	3.9125
Straw	1.2746	2.2779	1.2762	1.6096
Root	3.8091	2.8215	3.8157	3.4821

Based on the result represents in Table 3 and Table 4, it indicates that the leaf part from the spear grass contributed the highest amount of zinc and ferum with 0.7786 g/mL and 3.9125 g/mL, respectively. Hence, the leaf part of spear grass was chosen to proceed for optimization of spear grass.

3.3 Optimization of spear grass using pressurized hot water extraction (PHWE)

3.31 Regression analysis in Box-Behnken Design

Response Surface Methodology (RSM) is a model which comprises of statistical and mathematical techniques to study the effect of variables and to determine the optimum conditions that generates the maximum response. In the present study, by using the Design Expert 7.1.5 software, the experimental design and the data analysis were easier to observe and interpreted. . From the screening different part of spear grass, the leaf part was chosen as a result of highest concentration of zinc (Zn) and ferum (Fe). A Box-Behnken Design which is one of the components of RSM was applied to analyze the interaction between three parameters with Zn and Fe content in spear grass as a response. Table 5 represents the result based on Box-Behnken design for Zn and Fe concentration.

Table 5. Zinc, Zn concentration (g/ml)

Run no.	Factor			Response	
	A Temperature (Celsius)	B Extraction time (minute)	C Sample:solvent (mL)	Zn Concentration (g/mL)	Fe concentration (g/mL)
1	200.00	30.00	15.00	1.2338	2.9361
2	150.00	10.00	20.00	1.0158	1.3595
3	200.00	20.00	10.00	0.9591	2.2742
4	150.00	10.00	10.00	0.6593	3.1827
5	150.00	20.00	15.00	1.7732	3.0423
6	150.00	30.00	20.00	0.9401	2.4251
7	200.00	10.00	15.00	0.4402	1.2554
8	150.00	20.00	15.00	1.5369	2.9811
9	150.00	20.00	15.00	1.5543	3.0323
10	100.00	30.00	15.00	0.6552	1.6453
11	150.00	30.00	10.00	0.8358	2.3407
12	150.00	20.00	15.00	1.5482	3.0353
13	150.00	20.00	15.00	1.5537	3.0158
14	100.00	20.00	10.00	0.7426	2.2684
15	200.00	20.00	20.00	0.7377	2.0684
16	100.00	10.00	15.00	0.6772	1.7396
17	100.00	20.00	20.00	0.6708	1.2367

3.32 Regression analysis of zinc (Zn) concentration

A regression analysis was performed to fit the empirical model based on the second order equation. The regression equation model is given as equation 1.

$$Y = 1.59 + 0.078A + 0.011B + 0.021C - 0.037AC - 0.063BC - 0.46A^2 - 0.038B^2 - 0.035C^2 \qquad (1)$$

Y was referred as the response of zinc (Zn) concentration (g/mL), A represent temperature, B represent extraction time, and C represent sample-to-solvent ratio. The variable of AB is indicated the interaction effect between temperature and extraction time, AC is indicated interaction effect between temperature and sample-to-solvent ratio, while BC is indicated the interaction effect between extraction time and sample-to-solvent ratio. A^2, B^2, and C^2 were quadratic effects to show the present of curvature in the model.
Table 6 illustrates the statistical data to test the fit of the model.

Table 6. Test of significance for regression square

Standard deviation	0.16
R^2	0.9391
Adjusted R^2	0.8609
Predicted R^2	0.7385
Adequate precision	9.050

For testing the goodness of fit of regression equation, the determination coefficient (R^2) was evaluated. Hence, the value of determination coefficient, $R^2 = 0.9391$ showed that 93.91% of the data obtained were distributed evenly in this quadratic model. The R^2 value ≥ 0.8 was acceptable. The R^2 value which more than 80% represent the good agreement between predicted data and experimental data. Adequate precision measures the signal to noise ratio. A ratio greater than 4 is desirable. Ratio of 9.050 indicates an adequate signal.

3.33 Regression analysis of ferum (Fe) concentration

A regression analysis was performed to fit the empirical model based on the second order equation. The regression equation model is given as equation 2.

 https://doi.org/10.1051/matecconf/201815006015

$$Y = 3.02 + 0.21A + 0.23B - 0.37C + 0.44\,AB + 0.21AC + 0.48BC - 0.75A^2 - 0.38B^2 - 0.31C^2 \quad (2)$$

Y was referred as the response of ferum (Fe) concentration (g/mL), A represent temperature, B represent extraction time, and C represent sample-to-solvent ratio. The variable of AB is indicated the interaction effect between temperature and extraction time, AC is indicated interaction effect between temperature and sample-to-solvent ratio, while BC is indicated the interaction effect between extraction time and sample-to-solvent ratio. A^2, B^2, and C^2 were quadratic effects to show the present of curvature in the model. Table 7 illustrates the statistical data to test the fit of the model.

Table 7. Test of significance for regression square

Standard deviation	0.19
Mean	2.34
R^2	0.9653
Adjusted R^2	0.9208
Predicted R^2	0.4500
Adequate precision	12.328

For testing the goodness of fit of regression equation, the determination coefficient (R^2) was evaluated. Hence, the value of determination coefficient, $R^2 = 0.9653$ showed that 96.53% of the data obtained were distributed evenly in this quadratic model. The R^2 value ≥ 0.8 was acceptable. The R^2 value which more than 80% represent the good agreement between predicted data and experimental data. Adequate precision measures the signal to noise ratio. A ratio greater than 4 is desirable. Ratio of 12.328 of indicates an adequate signal and good agreement between the signals to the noise ratio.

3.34 Summary

Based on the three parameter involved in the research, there were many factors that identified as the significant factor. The concentration of zinc and ferum on PHWE was successfully demonstrated by the RSM. Figure 1 and Figure 2 represents the model analysis of Zn and Fe concentration. The model obtained was good as the R^2 value is 0.9391 for Zn concentration and 0.9653 for Fe concentration. It implies the good agreement. From the model analysis, it showed that each parameter affected the concentration of Zn and Fe concentration. Based on Figure, the red dot in the elliptical indicates the highest concentration of Zn concentration which is 1.55433 g/mL and the highest concentration of Fe which is 3.03237 g/mL.

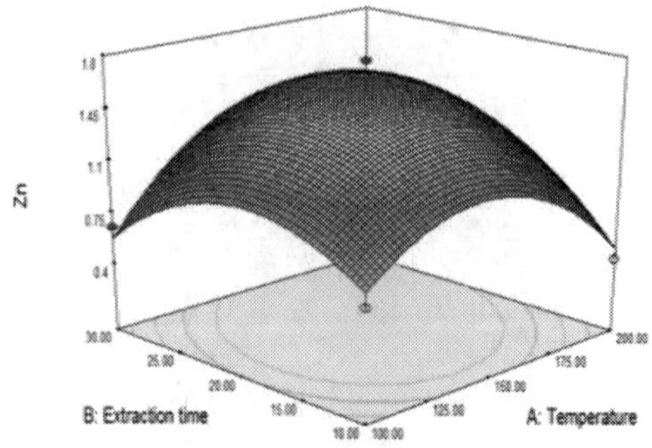

Fig. 1. Highest concentration of zinc, Zn

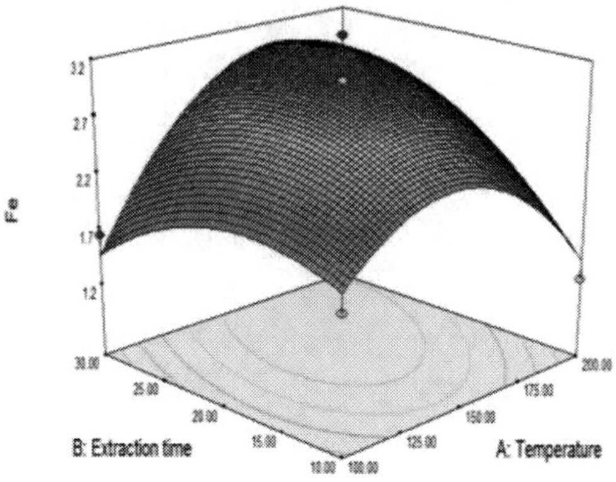

Fig. 2. Highest concentration of ferum, Fe

The sign of regression coefficients indicates that all three parameters have an effect on Zn and Fe concentration. The surface response curves in the both Figure show that the concentration of Zn and Fe increase with temperature, extraction time, and sample-to-solvents ratio, however after achieve the maximum level, the concentration of Zn and Fe will be decrease. At 150 °C of temperature, 20 minutes of extraction time, and 15 mL of sample-to-solvent ratio, the optimum conditions has achieved.

4 Conclusion

Spear grass sample were extracted to obtain metal elements such as zinc and ferum. These type of metal elements can be categorized as natural based pro-oxidant. Pressurized hot water extraction required a shorter time than cold and hot extraction method, provides the quality extracted spear grass. Therefore, in this research, the spear grass was extracted by using pressurized hot water extraction.

After the screening process, the leaf part from the spear grass contributed the highest amount of Zn and Fe concentration. The box-behnken design method was used to optimize the operating conditions parameters. The optimum conditions of temperature, extraction time, and sample-to-solvent ratio were adjusted at 150 °C, 20 min, and 3 g of sample to 45 mL of water, respectively, to achieve the highest amount of Zn and Fe concentration.

5 Acknowledgement

We are grateful to Short Term Research Grant (STG) 9001-00551 from Universiti Malaysia Perlis for funding this research.

6 References

1. I. Kyrikou, B. Demetres, M. Hiskakis, M. E. Babou, Polym. Degrad. Stab. **96(12),** 2237–2252 (2011)
2. M.J. Kirwan, J.W. Strawbridge, Food. Packag.Technol. CRC Press: New York, 174–240.
3. F. Khabbaz, A.C. Albertsson, J. Appl. Polym. Sci. **79(12),** 2309–2316 (2001)
4. P.K. Roy, P. Surekha, C. Rajagopa, V. Choudhary, Express. Polym. Lett. **1(4),** 208–216 (2007)
5. D.P. Garrity, M. Soekardi, M.V. Noordwijk, R. Cruz, P.S. Pathak, H.P.M. Gunasena, N.V. So, G. Huijun, N.M. Majid, Agrofor. Syst. **36**, 3–29 (1997)
6. C.C. Teo, S.N. Tan, J.W.H. Yong, C.S. Hew, E.S. Ong, J. Chromatogr. A. **1217(16),** 2484–2494 (2010)
7. C. Deng, N. Yao, A. Wang, X. Zhang, Anal. Chim. Acta. **536(1–2),** 237–244 (2005)
8. D.B. Hibbert, J. Chromatogr. B. **910**, 2–13 (2012)

Design of Binary Coded Pulse Trains with Good Autocorrelation Properties for Radar Communications

Siti Julia Rosli[1,*], *Hasliza* Rahim[2], *Ruzelita* Ngadiran[3], *K. N.* Abdul Rani[4], *Muhammad Imran* Ahmad[5] and *Wee* Fwen Hoon[6]

[12356] School of Computer and Communication Engineering, Universiti Malaysia Perlis, Perlis
[4] Department of Electronics Engineering Technology, Faculty of Engineering Technology, Universiti Malaysia Perlis, Perlis

Abstract. Finite length of sequences that are modulated both in phase and amplitude and have an ideal autocorrelation function (ACF) consisting of merely a pulse have many applications in control and communication systems. They are widely applied in control and communication systems, such as in pulse compression systems for radar and deep-space ranging problems [1-5]. In radar design, the important part is to choose a waveform, which is suitable to be transmitted because the waveform controls resolution in clutter performance. In addition, it can solve a general signal problem particularly related to the digital processing. Energy ratio (ER), total side lobe energy (SLE), and peak sidelobe level (PSL) are three properties of such sequences interest. This paper presents a method using the Complementation, Cyclic Shift and Bit Addition for synthesizing and optimizing a binary sequence implemented to improve the sequences of a similar quality with the Barker sequence, particularly for lengths greater than 13. All of these methods are guided by the specific parameter with good characteristics in ACF (ER, SLE, and PSL) [6,7,8]. Such sequences can then be effectively used to improve the range and Doppler resolution of radars.

1 Introduction

Sequences with low autocorrelation side lobe levels are useful for channel estimation, radar and spread spectrum communication applications. Sequences achieving the minimum peak autocorrelation side lobe level one are called Barker Sequences. A Barker sequence is a finite length binary sequence with the minimum possible autocorrelation. Currently, there only eight known Barker sequences and it has been conjectured that these are the only Barker sequences that exist, shown in Table 1.

Table 1. Barker Sequences

Code length	Code elements	PSL	ER	SLE
2	+1 −1	0.5	2	0.5
3	+1 +1 −1	0.3333	3	0.2222
4	+1 +1 −1 +1	0.2500	4	0.2500
5	+1 +1 +1 −1 +1	0.2000	5	0.1600
7	+1 +1 +1 −1 −1 +1 −1	0.1429	7	0.1224
11	+1 +1 +1 −1 −1 −1 +1 −1 −1 +1 −1	0.0909	11	0.0826
13	+1 +1 +1 +1 +1 −1 −1 +1 +1 −1 +1 −1 +1	0.0769	13	0.0710

Barker sequences are then used as the best sequences in the highest ER. Binary sequences were initially investigated for the purpose of pulse compression in radar systems. This method results in better range and Doppler resolution without the need to shorten a radar pulse nor increase the power [11,12].

2 The Optimization Problem

Binary sequences with a good periodic autocorrelation functions (ACFs) with low range side lobes are required for many communication applications. However, some good binary sequences have a large sequence length. The efficiency of these codes depends upon the energy content in the range side lobes of their ACF. Peak side lobe level (PSL) and integrated side lobe level (ISL) are the two performance measures for pulse compression codes. Barker sequences are known as the binary sequences that have low PSL and it is only available in length 13.

Many applications require longer codes to achieve a higher signal to noise ratio (SNR). By exhaustive computer search program [13], found all binary sequences up to length of 48 and they had developed an analytical method for generating good binary pulse compression codes.

2.1 Synthesis Using Element Complementation

This scheme is proposed for representing the complementation of (signed and unsigned) binary numbers in which there are sets of sequences that possibly have a good property in the ACF. In mathematics and computing, the method of complementation is a method used to change the sign of

* Corresponding author: sitijulia@unimap.edu.my

one number from another making that value as a representation for a symmetric range (i.e. from 2^k which k is the bit integers). The method which is also known as the simplest test to distinguish between the two possibilities is the complementation test.

Figure 1 shows the flowchart to determine the variation of the different sign of that sequence. In this case, it shows the binary number system and shorthand methods in representing binary codes. With k binary digits, it can represent the 2^k unique patterns from *111....1* until *-1-1-1....-1*. So this method will be run as itteration until 2^k unique patterns to find a good characteristics in ACF . When the signed quantities are represented in the same k digits, it still has only 2^k patterns to work with. Unless it increases the number of digits available (i.e. make k larger), the representation of signed numbers will involve the distribution of 2^k patterns into positive and negative portions. Sign/Magnitude is the simplest way to represent signed numbers, but the most common is the modification of a signed or magnitude called the element complementation.

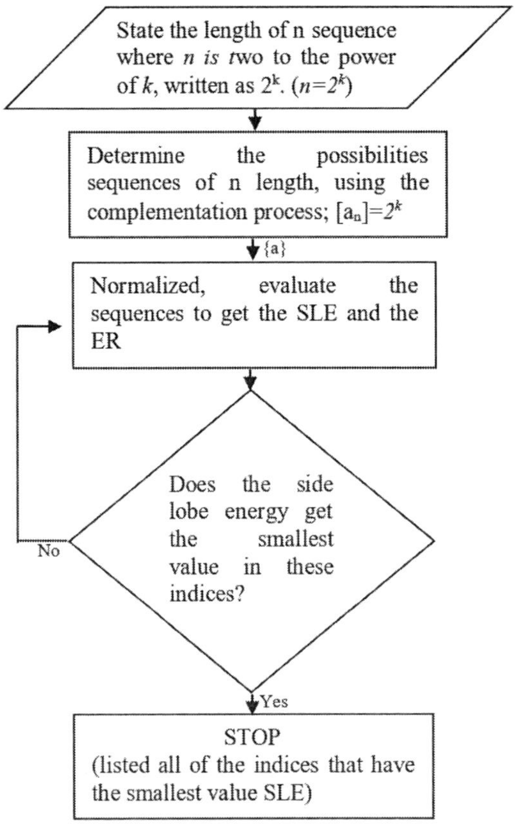

Fig. 1. Complementation Method flowchart

2.1.1 Sign or Magnitude Notation

Sign or magnitude notation is the simplest and one of the most obvious methods of encoding positive and negative bit sequences. This situation is generated as below in example. Example using 4 bits-lengths with 16th of sequences:

Binary=[1111]; [111-1]; [11-11]; [11-1-1]; [1-111]; [1-11-1]; [1-1-11]; [1-1-1-1]; [-1111]; [-111-1]; [-11-11]; [-11-1-1]; [-1-111]; [-1-11-1]; [-1-1-11]; [-1-1-1-1]

2.2 Synthesis Using Cyclic and Bit Addition

2.2.1 Cyclic Shift (Mutation Method)

Mutation is one of the operators that randomly changes by flip method. The purpose of the mutation operator is to check whether the element changes better than itself. The mutation is carried out according to the mutation probability. Here is the mutation methods are implemented for binary genes only:

Flip bit

This mutation method simply changes (flips) a randomly selected bit one until the end of their tap of length:

Before mutation: [-1 1 -1 1 1 1 1 1 -1 -1 1 1 -1]

After mutation: [-1 -1 1 -1 1 1 1 1 1 -1 -1 1 1]

The effectiveness of this method is capable to solve this problem. Combining both of them is aimed to seek out the new multilevel sequences and will be used as an experimental. This method must generate iteratively by repeating the process of inversion more than 10 times to get the samples. From that observation, the inverse filter method must be clipped and normalized to obtain SLE, ER and PSL which would utilize for review. These parameters are very important to fulfil the characteristics of good ACF of a sequence. Responding from that selection, the sequences should be getting through with a mutation procedure to mutate all the elements one by one to scan the best sequences. The flow chart of new methods for multilevel sequence set design is shown in Figure 4.

Bit Addition

The binary addition algorithm operates on two bit patterns and resulted in a bit pattern. Usually all two patterns are the different size or exceed from the first operand whereby all three represent unsigned integers or signed integers. The leftmost bit of the one-bit result is used to carry into the next column. For example, here is the sum of two thirteen-bit integers, shown in Figure 3:

-1	1	-1	1	1	1	1	1	-1	-1	1	1	-1
							the carry into each column					
0	0	0	0	0	0	0	0	0	0	0	0	0
1	1									first operand		
-1	1	-1	1	1	1	1	1	-1	-1	1	1	-1
1	1											

Fig. 3. Example of the sum of the bit integers

From Figure 4 using cyclic shift and bit addition, shows the visually presented flow of data through information of these methods.

The operations performed within the procedure and the starting sequence was considered as the binary sequences with good correlation properties (ex: Barker Code).

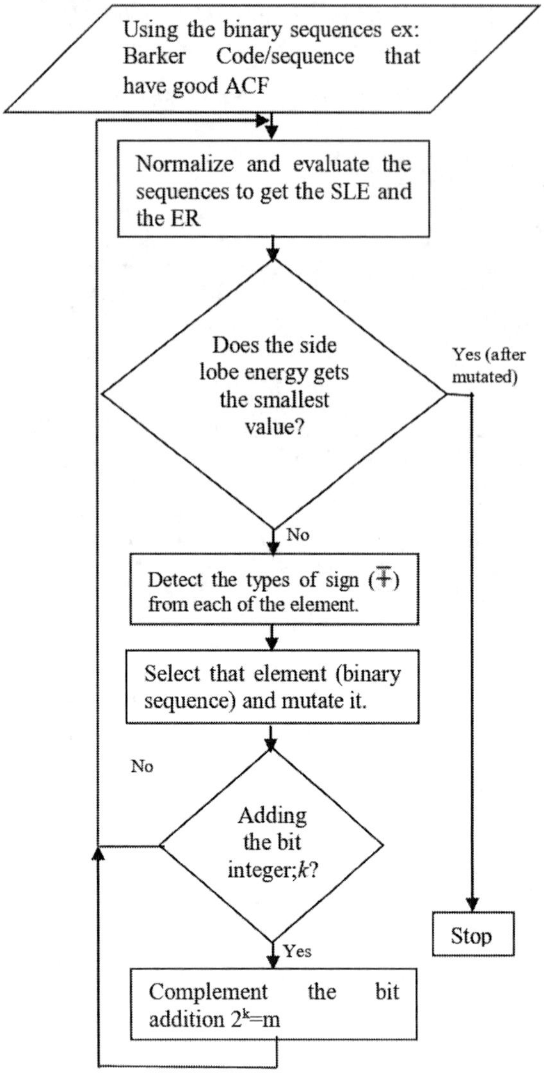

Fig. 4. Cyclic Shift and Bit Addition flowchart

Computers (usually) add two N-bit integers together to produce a N-bit result and a carry-out of the leftmost column. Every bit of the result must have a value. The following shows a 15-bit addition from the carry column that has 13-length.

3 Result and Discussion

Based on Figure 1 using the element complementation, the lengths of the 13 until 21 sequences are generated to acquire all of the possibility sequences that are the lowest SLE. The possibility sequences of length thirteen are listed in Table 2 and here it is able to detect that indices 203 are the Barker sequence. Additional information for the indices of 2657, 5536 and 7990 are the same contributions in SLE, ER and PSL parameter.

Table 2. Listing of the possibility sequences of length=13

Length-13; 2^{13}=8192				
Indices	**Sequences**	**SLE**	**ER**	**PSL**
203	[1 1 1 1 1 -1 -1 1 1 -1 1 -1 1] Barker Sequences	0.0710	13	0.0769
2657	[1 -1 1 -1 1 1 -1 -1 1 1 1 1 1]	0.0710	13	0.0769
5536	[-1 1 -1 1 -1 -1 1 1 -1 -1 -1 -1 -1]	0.0710	13	0.0769
7990	[-1 -1 -1 -1 -1 1 1 -1 - 1 1 -1 1 -1]	0.0710	13	0.0769

From the total 32768 indices only eight sequences are detected with the lowest SLE. As shown in Table 3, 0.1333 is the lowest value in SLE for each indices below.

Table 3. Listing of the possibility sequences of length=15

Length-15;2^{15}=32768				
Indices	**Sequences**	**SLE**	**ER**	**PSL**
822	[1 1 1 1 1 -1 -1 1 1 -1 - 1 1 -1 1 -1]	0.1333	15	0.2000
3822	[1 1 1 -1 -1 -1 1 -1 -1 - 1 1 -1 -1 1 -1]	0.1333	15	0.2000
9288	[1 -1 1 1 -1 1 1 1 -1 1 1 1 -1 -1 -1]	0.1333	15	0.2000
10656	[1 -1 1 -1 1 1 -1 -1 1 1 - 1 -1 -1 -1 -1]	0.1333	15	0.2000
22113	[-1 1 -1 1 -1 -1 1 1 -1 - 1 1 1 1 1 1]	0.1333	15	0.2000
23481	[-1 1 -1 -1 1 -1 -1 -1 1 - 1 -1 -1 1 1 1]	0.1333	15	0.2000
28947	[-1 -1 -1 1 1 1 -1 1 1 1 - 1 1 1 -1 1]	0.1333	15	0.2000
31947	[-1 -1 -1 -1 -1 1 1 -1 -1 1 1 -1 1 -1 1]	0.1333	15	0.2000

The result of the next length are shown in Table 4. It shows that this is the best sequence (binary sequence) respectively and has the distinctive characteristic. Nevertheless, there are risks when the lengths are increase and it caused the stack of data overflow. It has become the out-of-memory error as a warning while the program is running.

Table 4. The possibility sequences of length-17, 18 and 19

Length	**Sequences**	**SLE**	**ER**	**PSL**
12^{17} (131072)	[1 1 1 1 -1 -1 1 1 -1 1 1 -1 1 -1 1 -1 -1]	0.2215	17	0.1176
12^{19} (524288)	[1 1 1 1 -1 1 -1 1 -1 -1 -1 -1 1 1 -1 - 1 1 -1 -1]	0.1607	19	0.1579
12^{21} (2097152)	[1 1 -1 -1 -1 -1 -1 -1 -1 1 1 -1 -1 1 -1 1 -1 1 -1 -1 1]	0.1179	21	0.1429

Figure 2 shows the relation between SLE and length of sequences using the collected results from Table 2, Table 3 and Table 4. The figure clearly shows that for the lowest value in binary sequences is in their length. Through this synthesizing process by using the element of complementation method, is able to generate the best binary sequences and can be created for any desired length. In sum, the results from this method commensurate with any other methods applied.

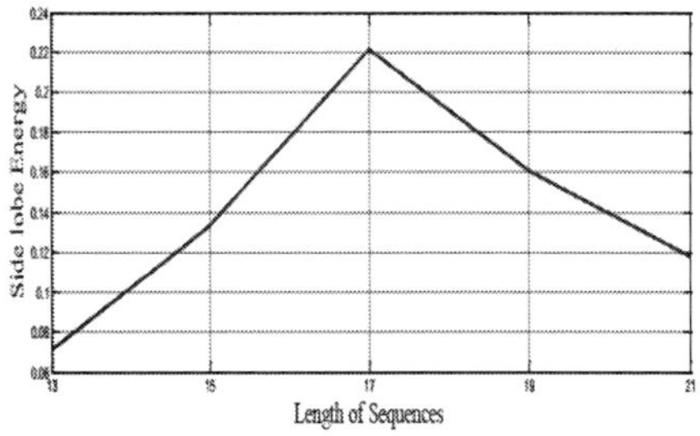

Fig. 2. Side lobe Energy versus Length of Sequences for 13 until 21 lengths using the Synthesis of Element Complementation.

Table 6 is then determined by the cyclic shift by simply changing the sign and unsign numbers which result in flipping are particularly prized. A dimensional cyclic shift of a matrix column is obtained from mutating the sequences by shifting it to the left. The n-1 samples are available where n is the length of the starting sequence when calculated by counting the number shifted from this procedure.

To avoid the same sequences from repeating, the types of sign or unsigned element detection of the sequences are placed in flow procedure. In this context, the gain time is highly emphasized and that particular sequence will be stacked. The addition of bit integer k is for complementing the 2^k to create the new sequences by adding the starting sequence or the shifted sequences. Table 5 shows the result of the bit addition. The Barker sequence of length-13 added with 2^k where k is equal to 2. Therefore, the total element of the sequence is length-15. Here, all of sequences are normalized before the SLE, ER and PSL are calculated.

In the final algorithm, the results from the sequences in Table 7 uses the cyclic shifted to track the lowest SLE. Figure 5 shows the results. Referring from that figure, the lowest value traces is 0.2044 in sm2 and sm3, and has three sets of sequences as shown in Table 7.

sm1=[0.3111, 0.3822, 0.5956, 0.5600, 0.7022, 0.4533, 0.5244, 0.5956, 0.7022, 0.8800, 0.5244, 0.3111, 0.3822, 0.4533, 0.3822;]

sm2=[0.2756, 0.3822, 0.5600, 0.7733, 0.5956, 0.3111, 0.3467, 0.3111, 0.5956, 0.4533, 0.6311, 0.5956, 0.6311, 0.4533, 0.2044;]

sm3=[0.2756, 0.2044, 0.3111, 0.3111, 0.2044, 0.2756, 0.5244, 0.5600, 0.6311, 0.4178, 0.3822, 0.2756, 0.5244, 0.2756, 0.3822;]

sm4=[0.3550, 0.3550, 0.3787, 0.3550, 0.3314, 0.4970, 0.8047, 0.4497, 0.3550, 0.2367, 0.4970, 0.5207, 0.3550, 0.2840, 0.3822;]

Table 5. The Generated Sequences for Bit Addition Method with Complement Number of Two

Sequences	SLE	ER	PSL
Barker=[1 1 1 1 1 -1 -1 1 1 -1 1 -1 1]+2^k	0.0710	13	0.0769
sm15(1)=[1 1 1 1 1 -1 -1 1 1 -1 1 -1 1 1 1]	0.3822	15	0.2000
sm15(2)=[1 1 1 1 1 -1 -1 1 1 -1 1 -1 1 1 -1]	0.2044	15	0.1333
sm15(3)=[1 1 1 1 1 -1 -1 1 1 -1 1 -1 1 -1 1]	0.3822	15	0.2000
sm15 (4)=[1 1 1 1 1 -1 -1 1 1 -1 1 -1 1 - 1 -1]	0.3822	15	0.2000

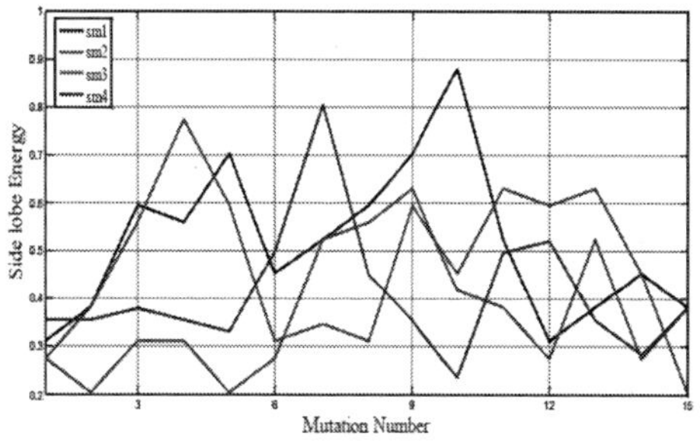

Fig. 5. Side Lobe Energy of the Generated Sequences (Table 5) when applying the Mutation Method

Table 6. The Generated Sequences for Cyclic Shift Method

Sequences	SLE	ER	PSL
Barker=[1 1 1 1 1 -1 -1 1 1 -1 1 -1 1]	0.0710	13	0.0769
Fm(1)	0.1657	13	0.1538
Fm(2)	0.4024	13	0.2308
Fm(3)	0.2308	13	0.4024
Fm(4)	0.1657	13	0.1538
Fm(5)	0.2604	13	0.2308
Fm(6)	0.2308	13	0.3550
Fm(7)	0.4024	13	0.2308
Fm(8)	0.3550	13	0.2308
Fm(9	0.2604	13	0.2308
Fm(10)	0.1657	13	0.1538
Fm(11)	0.4497	13	0.2308
Fm(12)	0.3550	13	0.2308

Table 7. THREE Set of Sequences Obtained using Mutated Process

	Sequences	SLE	ER	PSL
sm2(15)	[11111-1-111-11-111-1]	0.2044	15	0.1333
sm3(2)	[-1111111-1-111-11-11]	0.2044	15	0.2000
sm3(5)	[1-11-1111111-1-111-1]	0.2044	15	0.2000

4 Conclusion

In this chapter, the pulsed compression theory has been introduced to get a high range resolution as well as a good detection probability. One of the basic types of pulse compression is a binary phase coding which encodes the transmitted pulse with the information that is compressed in the receiver radar.

The desired properties and general types of methods for finding and generating such waveform are defined in this chapter. Variety of methods are applied to keep an overview for introducing the existing methods to be done as a routine. The table and the histogram of complete results are presented to elaborate them visually and to predict the future. It has been proved by [14,15] to compare with this method designed.

In conclusion, the binary code pulse trains design can be as the effective way to improve the overall performance of desired properties. It can enhance the radar performance by using the element complementation, cyclic shift and bit addition methods. It has the flexibility to dispose the noise ratio to improve the ER in that pulse trains. This performance result was evaluated via computer simulations and mathematical equation.

References

1. J. R. van der Merwe, W. P. du Plessis, F. D. V. Maasdorp and J. E. Cilliers.: "Introduction of low probability of recognition to radar system classification." IEEE Radar Conference (2016)
2. L. Qian.: "Radar Clutter Suppression Solution Based on ICA. " Fourth International Conference on Intelligent Systems Design and Engineering Applications (2013)
3. A. Khalid., S. A. Sheikh.., I. U. H. Shah and Q. U. Khan.:" Synthesis of linear antenna array using genetic algorithm to reduce peak sidelobe level" 9th International Conference on Electrical and Electronics Engineering ELECO (2015)
4. H. Khan and S. J. Yoo.: "Active interference restriction in OFDM-based cognitive radio network using genetic algorithm" International Conference on Information and Communication Technology Convergence ICTC (2015)
5. P. Pal and D. Mondal.: "Adaptive antenna array pattern synthesis for suppressed sidelobe level and controlled null using genetic algorithm" International Conference for Convergence for Technology (2014)
6. Ghani, F., Rosli, S.J., Ghani, A.H.A., Ahmad, B.R.: "Waveform Design for Improved Range and Doppler Resolution in Radar". In: Proceedings Third International Conference on Intelligent Systems, Modeling and Simulation ISMSI 2011, held at Kota Kina Balu, Malaysia, 8-10 February, 2012, pp 247-251.
7. Rosli, S.J., Ghani, F., Ghani, A.H.A., and Ahmad, B.R.: "Synthesis of Finite Length Multi Level Sequences for Clutter Rejection in Radar". 2012 Fourth International Conference on Computational Intelligence, Communication Systems and Networks.
8. Rosli, S.J., Ghani, F., Ghani, A.H.A., and Ahmad, B.R.: "An Itterative Technique of Amplitude and Phase Modulated Pulse Trains for Radars". Fourth International Conference on Control, Communication and Power Engineering, CCPE 2013, Bangalore.
9. O. Riznyk and B. Balych and I. Yurchak.: "A synthesis of barker sequences is by means of numerical bundles" 14th International Conference The Experience of Designing and Application of CAD Systems in Microelectronics CADSM (2017)
10. A. G. Holubnychyi and G. F. Konakhovych.: "Spread-spectrum control channels for UAV based on the generalized binary Barker sequences" IEEE 2nd International Conference Actual Problems of Unmanned Air Vehicles Developments Proceedings APUAVD (2013)
11. A. Holubnychyi.: "Generalized binary barker sequences and their application to radar technology" Signal Processing Symposium SPS (2013)
12. Pilsu Kim and Eunji Jung and Sua Bae and Kangsik Kim and Tai-kyong Song.: "Barker-sequence-modulated golay coded excitation technique for ultrasound imaging" IEEE International Ultrasonics Symposium IUS (2016)
13. P. Tirumala Rao, P. Siva Kumar, C. Ramesh and Y. Madhu Babu.: "A novel VLSI architecture for generation of Six Phase pulse compression sequences," International Conference on Devices, Circuits and Systems (ICDCS), Coimbatore (2012)
14. Robert C. D. & Vilhelm Gregers-Hansen.: "Code Inverse Filtering for Complete Sidelobe Removal in Binary Phase Coded Pulse Compression Systems" Radar Division, Naval Research Laboratory (2005)
15. Fan, P. & Darnell, M.: "Sequence Design for Communication Applications" John Wiley, Research Studies Press, Taunton (1996)

Determining The Optimal Mix of Institutional Geopolitical Power And ASEAN Corporate Governance on the Firm Value of Malaysia's Multinational Corporations (MNCs)

Yusoff Wan Sallha[1,*], *Handayani* Wuri[1], *Saad* Mohd. Sazli[2] and *Md. Salleh* Mohd. Fairuz[3]
[1]School of Business Innovation and Technopreneurship, Universiti Malaysia Perlis, 01000 Kangar, Perlis, Malaysia.
[2]School of Manufacturing Engineering,Universiti Malaysia Perlis, Ulu Pauh Main Campus, 02600 Arau, Perlis.
[3]School of Accounting, Faculty of Economics and Management, Universiti Kebangsaan Malaysia, 43600 Bangi, Selangor, Malaysia.

Abstract. The purpose of this paper is to examine the relationship between institutional geopolitics, ASEAN corporate governance quality and the firm value of Malaysia's multinational corporation (MNC). We used the data of MNCs in Malaysia that were active from 2009 to 2013 as an evidence of MNCs from emerging market economies. Descriptive analysis, factor analysis and panel data analysis have been utilized to test the equation model. We also propose optimization analysis by using differential evolution method to capture the optimal mix of institutional geopolitics and ASEAN_CG on the firm value of MNC. Results reveal that the geopolitics of G7(Canada, France, German, Italy, Japan, Europe, and the United States), BRICS (Brazil, Russia, India, China, and South Africa), and ASEAN (Brunei Darussalam, Cambodia, Indonesia, Lao PDR, Myanmar, Philippines, Singapore, Thailand, Vietnam, and Malaysia) are highly correlated with the firm value of Malaysia's MNC. The power of institutional geopolitics, namely, military, material, and social power, influences firm value negatively and ASEAN_CG moderate the negative influence of institutional geopolitics on the firm value of MNC. Thus, it is importance for corporate management to understand the geopolitical changes of host countries' and increase the compliance of ASEAN_CG in formulating their market value and segmentation strategies.

1 Introduction

ASEAN Investment Report (AIR) 2015 highlighted that multinational companies (MNCs) remained a key aspect of the ASEAN investment landscape in 2014 and 2015. Foreign and ASEAN MNCs continue to expand their activities in the ASEAN region in a range of industries. The strong regional expansion by MNCs also played an important role in pushing foreign direct investment (FDI) to a higher level in 2014. In addition, some ASEAN MNCs such as Malaysia's MNC are not just expanding in the same host country but also in other ASEAN states members.

Despite a growing number of ASEAN MNCs investing abroad, very little empirical research has looked deeply into the dynamic aspects of the internationalisation and emergence of Malaysia-based MNCs [1]. In fact, little is known about the systemic risk like geopolitical risk and global governance on the value of MNCs originating in Malaysia. Notwithstanding there is a few recent undertaken studies on the international geopolitical risk in financial literature (e.g.[2],[3]) but their focus mainly on countries financial performance. However, lesson from British exit European Union (Bexit) in the age of non polar world has also forced business strategists to make geopolitical uncertainty a component in corporate decision making. Understanding the geopolitical risk and good governance can help MNCs to avoid costly investment mistakes and allow them to take advantage of opportunities that they might not otherwise [4].

In addition, Malaysia's MNCs play an important role to gear up Southeast Asia Nations (ASEAN) as a single market and production base [5]. The relative benefits and costs of systemic risk in the emerging markets may not necessarily at the same magnitude with the developed countries, which could offer a new way of measuring firm value. This study attempts to fill that gap. The study aims to; first, empirically examine the value impact of institutional geopolitics and global governance on the firm value of MNCs originating from Malaysia. Second, formulate the optimal mathematical model of firm value for Malaysia-base MNCs. The study will be using a quantitative method for the first phase and metaheuristic optimisation method for the second phase. The result of the study will among others, underline policies and recommendations to decision maker to manage risk and resources effectively.

* Corresponding author: wansallha@unimap.edu.my

2 Institutional geopolitics, ASEAN corporate governance and firm value

Institutional geopolitics is the term most used by scholars to explain the major power of multilateral institutions (e.g., G7 countries, BRICS (Brazil, Russia, India, China and South Africa) and South East Asian Corporation (ASEAN)) in influencing world political economy [6]. Despite the hard power strategies such as military intervention, economic sanctions and coercive diplomacy; multilateral institutions may use the soft power approach to persuade others to do what they do not want to do [7]; [8];[6]. The soft power strategies are usually associated with ideology, cultural attraction, natural resources and bilateral relations. Both strategies are considered as an important geopolitical mechanism to corporate strategies because the changes of political decision among institutions member's may affect the activities of foreign companies and diminish the investment process.

The case of Brexit for example, provides a powerful motivation to Malaysia-base MNCs on how geopolitical risk affects firm valuation. After the announcement of Brexit most of the MNCs operating in UK and the rest of Europe experience a job cut and profit discrepancies [9]. MNCs need to have at least a basic plan to secure their value chain [4].

As suggested in network governance theory, the interdependent relationship between public, semi-public and private actors such as G7 countries, BRICS, ASEAN and Malaysia-based MNC, that involve in self-organizing negotiation systems need to have a good governance practices to optimize both; private and common goal [10]. Since Malaysia public listed companies (including MNC companies) fully implemented ASEAN Corporate Governance standard and practices (ASEAN_CG) as a new standard that comply with regional and international standard, there is possible that ASEAN_CG able to play as a moderating role to control the geopolitical power of multilateral institution and the value of Malaysia-based MNC. If there is a significant relationship, what are the optimum score of ASEAN_CG that Malaysia's MNC need to comply to maximize the firm value?

To achieve these questions, the quantitative and metaheuristic optimisation methodology will be followed, and hopefully our conclusions may contribute to the construction of ASEAN field of study.

3 Methodological considerations

3.1 Sample selection

The sample of the study consists of MNC companies listed in Bursa Malaysia that are active from 2009 to 2013. The sampling procedure is identified based on the definition of MNC. Based on previous literature, the classification of the MNC companies is as follows:

Table 1. Classification of MNC companies.

No	Classification procedure	Literature Review
1	International segmentation: Companies that operate in at least two countries	[11], [12]
2	Control of assets /sales activities: At least 10 percent of holding assets /sales come from its international subsidiaires	[13], [14]
3	Control of equity: Holding company holds at least 20% equity in its international subsidiaries	[11, 15]

We excluded companies from the financial and banking sectors because these companies are subject to different regulations. We also excluded MNC companies that are not fairly distributed across industries such as from the hotel, mining, and IPC industries. The procedure generated 176 companies with total pooled observations of 880 company years.

3.2 Measures of Firm value

This study used excess value as proxy for firm value. The excess value is calculated as follows:

$$\text{Excess Value (EV)} = \frac{\ln\text{ (the actual capital to sales ratio)}}{\ln\text{(the imputed capital to sales ratio)}} \quad (1)$$

Due to data availability, we employ capital-sales ratio to calculate the actual value of each firm. The formula for capital-sales ratio is as follows :

$$\text{The actual capital to sale} = \frac{MVE + PS + DEBT}{\text{holdings total sales}} \quad (2)$$

Where;

MVE : The market value of equity computed as price per share multiplied by the number of outstanding common shares.

PS : The liquidating value of preferred stock.

DEBT : The value of short-term liabilities net of short-term assets plus the book value of long-term debt.

TA : The book value of total assets.

Next, the imputed value is calculated as the median capital-sales ratio. The median value is obtained from the weighted average of all imputed values computed for each of the firm's segments that operate in the same industry(s). To calculate the imputed value, we group the firm's international segmentation based on G7, BRICS, and ASEAN countries to represent the multilateral geopolitical influence on the firm's international segmentation. Hence, for example, if an plantation MNC has 20% of its sales in the UK, 20% in Canada, 30% in India and 30% in China, then the imputed value is calculated as follows :

https://doi.org/10.1051/matecconf/201815006017

Imputed value = (0.4) (the value of the median pure-play G7 plantation MNCs) + (0.6) (the value of the median pure-play BRICS plantation MNCs).

We assume that the MNC has the same product mix throughout its various international segments because we do not have the breakdown among product market segments and international regions.

3.3 Measures of institutional geopolitical influences

This study used the smart power of G7, BRICS and ASEAN as a proxy to measure the institutional geopolitical influence [16-19]. By using principle component analysis, we measure the smart power of G7, BRICS and ASEAN as follows:

(1) Military power (mp):
1. [ME] Military expenditure (% of GDP)
2. [AF] - ln[Armed forces personnel, total]

(2) Material power (mr):
1. [OIL] - ln[Oil: total proved reserves (thousand million barrels)]
2. [GAS] - ln[Natural gas: total proved reserves (trillion cubic metres)]
3. [NCLR] - ln[Nuclear energy: operable reactors (Mwe)]
4. [P] - ln[Population density (people per sq. km of land area)]
5. [S&T] - Science and technology capability: research and development expenditure (% of GDP)

(3) Social power (sp):
1. [PV_HS] - ln[A personal visit/conference/seminar attended by the heads of state of G7, BRICS, or ASEAN countries to Malaysia**.]
2. [PV_M'PM] - ln[A personal visit/conference/ seminar attended by the prime minister of Malaysia to G7, BRICS, or ASEAN countries**.]

**The score one (1) is given if the above items match and zero (0) for otherwise.

The source of data is taken from The World Bank: World Development Indicator Database, Oil & Gas Journal, BP Journal, Nuclear Energetic Agency and multimedia photo gallery, Prime Minister's Office, Putrajaya Malaysia.

To analyze the influence of institutional geopolitical on the firm value of MNCs, we multiply the above geopolitical scores with firm's segmentation sales. We assume that holding firms that have their segmentation on G7, BRICS, or ASEAN countries will obtain a stronger geopolitical effect compared with firms that have no segmentation on G7, BRICS, or ASEAN countries.

3.4 Measures of ASEAN_CG Score

This study used ASEAN_CG scoring index to measure the moderating effects of ASEAN_CG on the relationship between institutional geopolitics and firm value. We followed ASEAN Corporate Governance Scorecard prepared by [20] as a basic measurement to measure the ASEAN_CG score. However some exemption has been made due to the mandatory disclosure and the availability of data. Mandatory disclosure is excluded because we assume that all firm have comply the requirement.

3.5 Control Variables

Based on several variables that have been widely used by earlier studies ([21-23]), this study used control variables as follows:

$$f(\text{Control}) = \beta 0 + \beta 1\text{TOA} + \beta 2\ \text{ROA} + \beta 3\text{LOA} + \varepsilon \quad (3)$$

Where:
Firm size (TOA)= logarithm of total assets
Profitability (ROA) = EBIT/total assets
Leverage (LOA)= total debt/total assets

3.6 Data Analysis

Descriptive and panel data regression were performed on the gathered data. The equation for the panel data regression is as follows:

$$\begin{aligned} EV_{i,t} = {} & \beta_0 + \beta_1 mpG7_{i,t} + \beta_2 MRBRICS_{i,t} + \beta_3 SPASEAN_{i,t} \\ & + \beta_4 kmG7_{i,t} + \beta_5 mrBRICS_{i,t} + \beta_6 mrASEAN_{i,t} + \beta_7 spG7_{i,t} \\ & + \beta_8 spBRICS_{i,t} + \beta_9 spASEAN_{i,t} + \beta_{10} CGQ_{i,t} + \\ & \beta_{11} mpG7_{i,t}{*}CGQ_{i,t} + \beta_{12} mpBRICS_{i,t}{*}CGQ_{i,t} + \\ & \beta_{13} mpASEAN_{i,t}{*}CGQ_{i,t} + \beta_{14} mrG7_{i,t}{*}CGQ_{i,t} + \\ & \beta_{15} mrBRICS_{i,t}{*}CGQ_{i,t} + \beta_{16} mrASEAN_{i,t}{*}CGQ_{i,t} + \\ & \beta_{17} spG7_{i,t}{*}CGQ_{i,t} + \beta_{18} spBRICS_{i,t}{*}CGQ_{i,t} + \\ & \beta_{19} spASEAN_{i,t}{*}CGQ_{i,t} + \boldsymbol{f}(\text{Control})_{i,t + \psi t + ui + \varepsilon it} \end{aligned} \quad (4)$$

Where;

EV_{it}	= firm value
mp_{it}	= military power
mr_{it}	= material resources
sp_{it}	= social power
G7	= G7 countries – [Canada, France, German, Italy, Japan, Europe and the United States]
BRICS	= BRICS countries – [Brazil, Russia, India, China, and South Africa]
ASEAN	= ASEAN countries – [Brunei Darussalam, Cambodia, Indonesia, Lao PDR, Myanmar, Philippines, Singapore, Thailand, Vietnam, and Malaysia]
CGQ	= Corporate governance quality
f(Control)	= Control variables
ε_{it}	= error term

The i subscript denotes the cross-section dimension, and t denotes the time-series dimension.

3.7 Measures of firm value optimization

By using the EV equation results, this study proceeds with firm value optimization. To measure firm value optimization, we employed one of the evolutionary algorithms (EA) methods, which is differential evolution (DE) by using MatLab Versi 8. The benefit of using EA method is that first, EA search a population of points in parallel, not just a single point. Second, EA do not require derivative information or other auxiliary knowledge; only the objective function and corresponding fitness levels influence the directions of search and third EA are generally more straightforward to apply, because no restrictions for the definition of the objective function exist [24].

Generally, some DE operator must be considered by DE algorithm as shown in Table 2 below:

Table 2. Value of DE operator.

Dimension of problem	D	10
Population size	NP	100
Differentiation constant	F	0.6
Crossover rate	CR	0.9
Generations/Stopping condition	GEN	500

4 Results

4.1 Descriptive analysis

Table 3 shows that the majority of MNC segmentation was from ASEAN countries (69.92%), followed by BRICS (52.10%) and G7 (42.57%) countries. Industrial product shows the highest percentage with 41.54%, and plantation industries obtained the lowest percentage of industry involvement with 3.97%.

Table 3. Descriptive analysis.

	Total years of observation	
Sample/variables	frequency	percentage
Total segmentation of G7	375	42.57
Total segmentation of BRICS	459	52.10
Total segmentation of ASEAN	616	69.92
Total sample by industry:		
1. Construction	45	5.11
2. Consumer	195	22.13
3. Industrial product	366	41.54
4. Plantation	35	3.97
5. Technology	60	6.81
6. Trading and services	180	20.43
Total	881	100

4.2 Regression analysis

Based on panel data analysis procedure, we begins regression analysis with comparing the pooled OLS regression and fixed effects model. The result of the fixed effects model shows that all αi are zero, which means that the OLS estimator is biased and inconsistent. Thus, the null hypothesis is rejected.

We also compared the random effects and fixed effects models to determine the best statistical estimation and conducted the Hausman test to verify the presence of correlations between the unobservable heterogeneity and explanatory variables. Again, the probability is less than 0.05. Thus, we reject the null hypothesis and continued with the fixed effects regression model. To ensure the validity of the statistical results, we conducted a modified Wald test and serial correlation test. Both statistical tests show a heteroskedasticity and serial correlation problem in the fixed effects (FE) model.

Thus, to rectify these problems, we performed the fixed effects (within) regression with Driscoll and Kraay standard errors by using the xtscc command implemented by Daniel Hoechle. The error structure is assumed to be heteroskedastic and autocorrelated up to some lag, and possibly correlated between groups. Figure 4 shows the results of fixed effects (within) regression with Driscoll and Kraay standard errors.

Table 4. The relationship of institutional geopolitics, ASEAN corporate governance and firm value.

Variables	Model (a)		Model (b)	
Pemalar	-0.149		-0.211	
mpG7	-0.212	***	-0.308	
mrG7	-0.125	**	0.067	
spG7	-0.013		-0.135	*
mpBRICS	0.015		0.960	**
mrBRICS	-0.124	***	-0.563	**
spBRICS	-0.103	***	0.143	
mpASEAN	-0.073	**	0.368	**
mrASEAN	-0.124	**	-0.342	
spASEAN	-0.076	**	-0.022	
CGQ			0.521	*
mpG7*CGQ			0.173	
mrG7*CGQ			0.362	
spG7*CGQ			0.211	
mpBRICS*CGQ			1.536	**
mrBRICS*CGQ			0.690	*
spBRICS*CGQ			0.409	
mpASEAN*CGQ			0.776	***
mrASEAN*CGQ			0.370	
spASEAN*CGQ			-0.101	
TOA	-0.362	*	-0.375	*
ROA	0.130		0.089	
LOA	0.253	*	0.234	*
Prob > F	0.0001	***	0.0004	***
R-squared	36.34%		40.02%	
Max lag / (N)	2/(880)		2/(880)	

Model (a) presents the direct effects of institutional geopolitics on firm value. The result supports the idea that the powers of institutional geopolitics are negatively correlated with an MNC's firm value. Specifically, all geopolitical factors, except for military power of BRICS and social power of G7, are statistically significant to firm value. These results also suggest that the military and material power of G7 countries and social power of BRICS are the most important determinant in international location choices for MNCs in Malaysia, with the highest coefficient values of –0.212, –0.125, and –0.103 respectively.

Next, model (b) presents the interaction effects of corporate governance quality on the relationship of institutional geopolitics and firm value. Comparing the results between modal (a) and model (b); the R^2 score increase by 3.68%. This shows that there is an interaction between corporate governance quality and the relationship of institutional geopolitics and firm value. Specifically, the quality of ASEAN_CG weaken the relationship between military power of BRICS, material power of BRICS and social power of ASEAN on the firm value of Malaysia's MNC. Thus on the basis of these results, we derive an econometric model of the institutional geopolitics, corporate governance and firm value as follows:

$$EV = 0.960*mpBRICS + 0.368*mpASEAN - 0.563*mrBRICS - 0.135*spG7 + 0.521*CGQ + 1.536*mpBRICS\_CGQ + 0.776*mpASEAN\_CGQ + 0.690*mrBRICS\_CGQ - 0.375*TOA + 0.234*LOA \quad (5)$$

4.3 firm value optimization

Based on equation 5, we derived firm value optimization. The results in Figure 1 (a) and (b) show that the maximum firm value of Malaysia's MNC can be achieved at 9.2512.

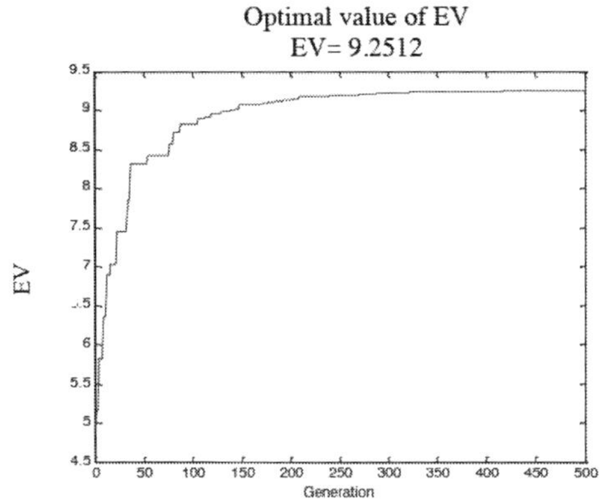

Figure 1 (a). Convergence profile - EV

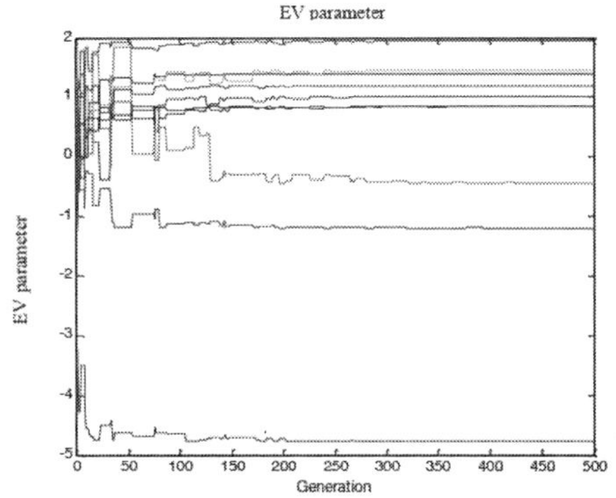

Figure 1 (b). Convergence profile- X parameter

Specifically, we compare the mean results of independent variables between optimization value and regression value. Based on result in table 5; it shows that to maximize firm value to 9.2512 Malaysia's MNC need to comply at least 84.1% of corporate governance score. The current score is 59.5%. Thus MNC company need to increase their corporate governance quality at least 24.6% from the current score.

Table 5. The differences between independent variables optimization mean and regression mean.

No	Independent variables	Optimum mean	Regression mean	Differences
1	mpBRICS	1.404	1.07E-08	1.404
2	mpASEAN	1.016	1.15E-08	1.016
3	mrBRICS	-4.77	-1.04E-08	-4.770
4	spG7	-0.444	-4.42E-09	-0.444
5	TOA	-1.198	0.040	-1.238
6	LOA	1.459	0.371	1.088
7	CGQ	0.841	0.595	0.246
8	mpBRICSCGQ	0.003	0.003	0.000
9	mpASEANCGQ	1.196	-0.008	1.204
10	mrBRICSCGQ	1.945	-0.011	1.956

5 Discussions and Conclusion

In this study, we developed a new conceptual framework to explain how and why institutional geopolitics can be present and have some influence on the firm value of Malaysia MNCs. By introducing a new concept and methodology that are yet unused in the literature, we intend to obtain several evidences. First, institutional geopolitics play an important role in fitting international segmentation decisions. Second, ASEAN_CG is a good moderating function to weaken the negative relationship between institutional geopolitics and the firm value of Malaysia's MNC and third, to maximize the firm value of Malaysia's MNC the top management should comply at least 84.1% of ASEAN-CG score.

This study also contributes to the results of Minority Shareholder Watchdog Group [25] by providing a new evidence that MNCs from emerging market economies are also exposed to international geopolitical challenge. In addition, we complement the report of ASEAN Capital Market Forum (ACMF) [20] by proving that ASEAN_CG able to promote ASEAN as an asset class. This is an important issue for future research.

Reference

1. Ahmad, S.Z., *The internationalization process of third world multinational firms: A study of Malaysia government linked corporation.* International Journal of Business Research, 2008. **8**(5): p. 35-47.
2. Salehi, H., et al., *Foreign-Currency and Monetary Geopolitics of United States and Its Effects on the Future of the International System.* Journal of Economics and Sustainable Development, 2014. **5**(8): p. 96-104.
3. Kapur, D. and M. Suri, *Geoeconomics Versus Geopolitics: Implications for Asia.* The Oxford Handbook of the Economics of the Pacific Rim, 2013: p. 290.
4. Frontier Strategy Group, *Business After Brexit: What MNCs Need to Know*, in *Frontier Strategy Group*. 2016, WordPress.

5. BERNAMA, *Malaysia And ASEAN Gear Up For AEC*, in *Malaysian Digest* 2015.
6. Yusoff, W., et al., *The Geopolitics of ASEAN Cooperation and Firm Value: Evidence from Multinational Corporation in Malaysia.* Journal of Global Business and Social Entreprenuership (GBSE) 1 (2), 2016.
7. Nye, J.S., *Soft power.* Foreign policy, 1990: p. 153-171.
8. Virmani, A., *World Economy, Geopolitics and Global Strategy: Indo-US Relations in the 21st Century.* Economic and Political Weekly, 2006: p. 4601-4612.
9. Fels, J., *What is Brexit's Impact on Global Economy? A Pimco analyst assesses the economic fallout over the coming three to five years.*, in *Wall Street's Best Minds, Barron's Asia*. 2016.
10. Jones, C., W.S. Hesterly, and S.P. Borgatti, *A General Theory of Network Governance: Exchange Conditions and Social Mechanisms.* The Academy of Management Review, 1997. **22**(4): p. 911-945.
11. Ramasamy, B., *Third world multinationals: The case of Malaysia.* Analisis, 1999. **6**(1&2): p. 123-137.
12. Martinez, Z.L. and D.A. Ricks, *Multinational parent companies' influence over human resource decisions of affiliates: US firms in Mexico.* Journal of International Business Studies, 1989: p. 465-487.
13. Hashim, F. and N. Mohd Saleh, *Voluntary annual report disclosures by Malaysian multinational corporations.* 2007.
14. Michel, A. and I. Shaked, *Multinational corporations vs. domestic corporations: Financial performance and characteristics.* Journal of International Business Studies, 1986: p. 89-100.
15. Mustapha, M., *Monitoring Costs of Multinational Companies: An Agency Theory Perspective.* Asian Journal of Business and Accounting, 2014. **7**(2): p. 23.
16. Reynaud, J. and J. Vauday, *Geopolitics and international organizations: An empirical study on IMF facilities.* Journal of Development Economics, 2009. **89**(1): p. 139-162.
17. Armijo, L.E., L. Mühlich, and D.C. Tirone, *The systemic financial importance of emerging powers.* Journal of Policy Modeling, 2014. **36, Supplement 1**(0): p. S67-S88.
18. Teixeira, A.A.C. and M. Dias, *The importance of geopolitics in firms' international location decisions: The Polish case.* Communist and Post-Communist Studies, 2013. **46**(1): p. 79-93.
19. Md Salleh, M.F., Y. Wan Sallha, and N. Basnan, *Does Smart Power of ASEAN Cooperation Influence Firm Value? Evidence from Geopolitical Perspective.* Acta Universitatis Danubius. Œconomica, 2016. **12**(3).
20. ACMF, *ASEAN corporate governance scorecard: Country reports and assessments 2012–2013.* 2013, Asian Development Bank.
21. Ang, J.S., D.K. Ding, and T.Y. Thong, *Political Connection and Firm Value.* Asian Development Review, 2013. **30**(2): p. 131-166.
22. Berger, P.G. and E. Ofek, *Diversification's effect on firm value.* Journal of Financial Economics, 1995. **37**(1): p. 39-65.
23. Brick, I.E. and N.K. Chidambaran, *Board meetings, committee structure, and firm value.* Journal of Corporate Finance, 2010. **16**(4): p. 533-553.
24. Saad, M.S., H. Jamaluddin, and I.Z.M. Darus, *Active Vibration Control of Flexible Beam using Differential Evolution Optimisation.* World Academy of Science, Engineering and Technology, International Journal of Mechanical, Aerospace, Industrial, Mechatronic and Manufacturing Engineering, 2012. **6**(2): p. 446-453.
25. MWSG, *MALAYSIA-ASEAN Corporate Governance Report 2012.* 2013, Minority Shareholder Watchdog Group.

Influence of Code Size Variation on the Performance of 2D Hybrid ZCC/MD in OCDMA System

Rima. Matem[1,*], *S. A.* Aljunid[1], *M. N* Junita[1], *C. B. M* Rashidi[1], *Israa.* Shihab Ahmed[1]

Advanced Communication Engineering Centre of Excellence School of Computer and Communication Engineering (ACE CoE -SCCE), Universiti Malaysia Perlis (UniMAP),Perlis, Malaysia,

Abstract. Several two dimensional OCDMA have been developed in order to overcome many problems in optical network, enhancing cardinality, suppress Multiple Access Interference (MAI) and mitigate Phase Induced Intensity Noise (PIIN). This paper propose a new 2D hybrid ZCC/MD code combining between 1D ZCC spectral encoding where M is its code length and 1D MD spatial spreading where N is its code length. The spatial spreading (N) code length offers a good cardinality so it represents the main effect to enhance the performance of the system compared to the spectral (M) code length according to the numerical results.

1 Introduction

In OCDMA system, each user has its own code word rather of the specific wavelength or time slot which is provided by Wavelength Division Multiple Access (WDMA) and Time Division Multiple Access (TDMA) respectively [1]. This property provides that to users can access the same channel simultaneously and asynchronously [2]. Many 2D coding have been proposed and investigated as [3], [4], [5], [6]. Several classifications has been investigated in OCDMA system such as temporal OCDMA encoding and spectrum-amplitude coding (SAC) OCDMA, spectral-phase encoding (SPE), frequency –hopping (FH) OCDMA, and 2D hybrid coding approach [7]. The 2D OCDMA depicts an enhanced performance compared to 1D OCDMA system in term of low power lost, high cardinality, minimum cross-correlation, high data rate and high auto-correlation with side[8] Several challenges affect the performance of OCDMA system in the implementation level [9-10]. The main challenges are due to the Multiple Access Interference (MAI) of the overlapping chips and Phase Induced Intensity Noise (PIIN) where PIIN is due to the desired signal mixed with the unwanted signal simultaneously at the same wavelength incident on the same photo-detector [11-12]. In the other side, an important factor can be impact the performance of OCDMA system is the code length where it represents one of the essential characteristic to design the system [13-14]. The main obstacle of increasing the code length is that growing complexity and cost, reducing the data transmission and bandwidth efficiency [15-16]. In this paper, combining between one dimensional Zero Cross Correlation code (1D ZCC) [17] and one dimensional Multi Diagonal code (1D MD) [18] where the both represent zero cross correlation property. The new 2D hybrid ZCC/MD code used to eliminate MAI and mitigate PIIN. The increment of spatial code length N significantly influences the system performance compared to the spectral code length M. this paper is organized as follow: section II represents code construction, then section III shows the system performance, section IV depicts on the analytical results and discussion, finally the conclusion.

2 2D hybrid ZCC/MD code construction

The derivation of the 2D hybrid ZCC/MD is based on the Zero Cross Correlation (1D ZCC) code and Multi Diagonal (1D MD) code. This new code is characterized by $(M, N, w, \lambda_a, \lambda_c)$ where M is spectral code length, N is spatial code length,w is the code weight, $\lambda_a$represents auto-correlation and $\lambda_c$is the cross-correlation. The code size of the 2D hybrid code represented by $M \times N$.

2.1 1D Zero Cross Correlation (ZCC) code

The Zero Cross Correlation (ZCC) code for SAC OCDMA was developed by [17] .ZCC is designed with zero cross correlation property between its code words to improve the performance of SAC OCDMA system. ZCC is characterized by a matrix of KxL where K represents number of active users and L represents the minimum code length. The number of active user K and code length L are given as:

$$K = w + 1 \quad (1)$$

$$L = w(w + 1) \quad (2)$$

The construction of the code is designed from the modified double weight (MDW) code [22] where the matrix is generated as follow with the basic ZCC code (w=1):

$$ZCC(w = 1) = \begin{bmatrix} 0 & 1 \\ 1 & 0 \end{bmatrix}$$

* Corresponding author: rymann90@hotmail.com

According to the following maps, the increase of the number of users is occurred:

$$ZCC(w=2)=\begin{bmatrix}0 & Z_1\\ Z_1 & 0\end{bmatrix}$$
$$ZCC(w=3)=\begin{bmatrix}0 & Z_2\\ Z_2 & 0\end{bmatrix}$$

The number of code weight can be increased by the general transformation below:

$$ZCC_w=\begin{bmatrix}A & B\\ C & D\end{bmatrix}$$

Where:
A: is $[1, w(w-1)]$ matrix of zeros.
B: represents w replication of the matrix$\sum_{j=1}^{w} j[0,1]$.
C: contain of the duplication of the matrix from $w-1$.
D: is the diagonal pattern of $[m\times n]$ with exchange column of zeros matrix $[m\times n]$.
An example for the transformation code from $w=1\rightarrow w=2\rightarrow w=3$ as follow:

$$ZCC_{w=1}=\begin{array}{c|c}0 & 1\\ \hline 1 & 0\end{array}$$

A B

$$ZCC_{w=2}=\left[\begin{array}{cc|cccc}0 & 0 & 0 & 1 & 0 & 1\\ \hline 0 & 1 & 0 & 0 & 1 & 0\\ 1 & 0 & 1 & 0 & 0 & 0\end{array}\right]$$

C D

A B

$$ZCC_{w=3}=\left[\begin{array}{cccccc|cccccc}0 & 0 & 0 & 0 & 0 & 0 & 0 & 1 & 0 & 1 & 0 & 1\\ \hline 0 & 0 & 0 & 1 & 0 & 1 & 0 & 0 & 0 & 0 & 1 & 0\\ 0 & 1 & 0 & 0 & 1 & 0 & 0 & 0 & 1 & 0 & 0 & 0\\ 1 & 0 & 1 & 0 & 0 & 0 & 1 & 0 & 0 & 0 & 0 & 0\end{array}\right]$$

C D

2.2 1D Multi Diagonal (MD) code

One-dimensional Multi Diagonal (1D MD) was been developed by [19]. It is characterized by N(code length), w(code weight) and λ_c(in phase cross correlation)
For the code sequence $X=\{x_1,x_2,\dots,x_N\}$, $Y=\{y_1,y_2,\dots,y_N\}$. The cross correlation expression can be expressed by $\lambda_c=\sum_i^N x_i y_i$
1D MD (Multi Diagonal) possesses zero cross correlation $\lambda_c=0$.
The matrix of 1D MD code consists of KxN where K is the number of user, $N$is the code length. The choice of the weight value is free and $N=Kw$.
The 1D MD code can be designed as below:

Step1:
A sequence of diagonal matrix can be constructed using the value of the weight (w) and number of user (K), according to these value the $i(i=1,2,3,\dots.K)$ is the index of rows in each matrix and $j(j=1,2,3,\dots\dots,w)$where j is the number of diagonal matrix.

Step2:
The MD sequences can be computed for each diagonal matrix using the equation below:

$$S_{i,j_w}=\begin{cases}(i_n+1-i), j_w=even\\ i \qquad for\ j_w=odd\end{cases} \quad (3)$$

$$S_{i,1}=\begin{bmatrix}1\\2\\3\\.\\.\\.\\K\end{bmatrix},\ S_{i,1}=\begin{bmatrix}K\\.\\.\\.\\3\\2\\1\end{bmatrix},\ S_{i,1}=\begin{bmatrix}1\\2\\3\\.\\.\\.\\K\end{bmatrix},\dots$$

Each element of the $S_{i,j}$ matrices represents the position of one in $T_{i,j}$matrices with KxKdimensions. Where $T_{i,1}=[S_{i,1}]_{K\times K}, T_{i,2}=[S_{i,1}]_{K\times K},\dots, T_{i,w}=[S_{i,1}]_{K\times K}$

$$T_{i,1}=\begin{bmatrix}1 & 0 & \cdots & 0\\ 0 & 1 & \cdots & 0\\ \vdots & \vdots & \ddots & \vdots\\ 0 & \cdots & \cdots & 1\end{bmatrix}_{K\times K}, T_{i,2}=\begin{bmatrix}0 & \cdots & 0 & 1\\ 0 & \cdots & 1 & 0\\ \cdots & \cdots & \ddots & \vdots\\ 1 & \cdots & \cdots & 0\end{bmatrix}_{K\times K},\dots,$$
$$T_{i,w}=\begin{bmatrix}1 & 0 & \cdots & 0\\ 0 & 1 & \cdots & 0\\ \vdots & \vdots & \ddots & \vdots\\ 0 & \cdots & \cdots & 1\end{bmatrix}_{KxK}$$

Step3:
The construction of the matrix of MD code of power KxN is based on the combination of diagonals matrices. In a matrix each row is a single code sequence

$$MD=[T_{i,1} \vdots T_{i,2} \vdots \cdots\cdots \vdots T_{i,w}]_{KxN}$$

$$MD=\begin{bmatrix}a_{1,1} & a_{1,2} & \cdots & a_{1,N}\\ a_{2,1} & a_{2,2} & \cdots & a_{2,1}\\ a_{3,1} & a_{3,2} & \cdots & a_{3,N}\\ \vdots & \vdots & \cdots & \vdots\\ a_{i_n,1} & a_{i_n,2} & \cdots & a_{i_n,N}\end{bmatrix}$$

The following table give an example of MD code with $K=4,\ w=3$

Table I: The MD code with $K=4,\ w=3$

i	Code sequence
1	100000011000
2	010000100100
3	001001000010
4	000110000001

TABLE II: 2D Hybrid ZCC/MD Code for $k_1 = 3$ and $k_2 = 2$

$A_{g,h}$	[101000100000]	[010010001000]	[000000010101]
$\begin{bmatrix}1\\0\\0\\0\\0\\1\end{bmatrix}$	$\begin{bmatrix}101000100000\\000000000000\\000000000000\\000000000000\\000000000000\\101000100000\end{bmatrix}$	$\begin{bmatrix}010010001000\\000000000000\\000000000000\\000000000000\\000000000000\\010010001000\end{bmatrix}$	$\begin{bmatrix}000000010101\\000000000000\\000000000000\\000000000000\\000000000000\\000000010101\end{bmatrix}$

The cross correlation of newly code can be obtained using four characteristics are given as follow:

$$A^0 = Y^T X \qquad (4)$$
$$A^1 = Y^T \bar{X} \qquad (5)$$
$$A^2 = \overline{Y^T} X \qquad (6)$$
$$A^3 = \overline{Y^T X} \qquad (7)$$

Where A(d) , d∈(0,1,2,3) has been given in [19] represents the characteristic matrices. Where parameters $\bar{X}$ and $\bar{Y}$ are the complementary of X and Y respectively. The cross correlation of hybrid 2D ZCC/MD code $A^{(d)}$ and $A_{g,h}$ is formulated as:

$$R^{(d)}(g,h) =) \sum_{M=1}^{M-1} \sum_{N=1}^{N-1} a_{ij}^{(d)} a_{(i+g)(j+h)} \qquad (8)$$

Where $a_{ij}^{(d)}$ depicts the $(i,j)_{th}$ of $A^{(d)}$ and $a_{(i+g)(j+h)}$ is the $(i,j)_{th}$ of $A_{g,h}$. Table III shows the cross correlation between any two codes $A^{(d)}$and $A_{g,h}$ of 2D Hybrid ZCC/MD code produced from the equation (8).

Table III: Cross Correlation of 2D ZCC/MD hybrid code

	$R^{(0)}$	$R^{(1)}$	$R^{(2)}$	$R^{(3)}$
$g = 0, h = 0$	$k_1 k_2$	0	0	0
$g = 0, h \neq 0$	0	$k_1 k_2$	0	0
$g \neq 0, h = 0$	0	0	$k_1 k_2$	0
$g \neq 0, h \neq 0$	0	0	0	$k_1 k_2$

The cross correlation of $A_{0,0}^0$ and $A_{g,h}$ can be calculated as follows:

$$R^{(0)}(g,h) = \sum_{i=1}^{M-1} \sum_{j=1}^{N-1} a_{i,j}^{(0)} a_{i,j}(g,h)$$
$$= \begin{cases} k_1 k_2 & for\ g = 0, h = 0 \\ 0 & otherwise \end{cases} \qquad (9)$$

In performance analysis there types of noise are considered Phase Induced Intensity Noise (PIIN), Shot Noise and Thermal Noise. In addition the BER is calculated using Gaussian approximation. The general form of the photocurrent noise from the photodiodes can be written as below:

$$\langle i_{noise}^2 \rangle = \langle i_{PIIN}^2 \rangle + \langle i_{shot}^2 \rangle + \langle i_{thermal}^2 \rangle$$
$$= BI^2 \tau + 2eBI + \frac{4K_b T_n B}{R_L} \qquad (10)$$

Where B is the electrical bandwidth, I is the average photocurrent, e is the electron charge, K_b is Boltzmann's constant, T_n is the absolute temperature, $R_L$is the load resistance and τ is the coherence time of the light. By applying the same method as [19], the PIIN, shot noise and thermal noise can be expressed as follow:

$$\langle i_{PIIN}^2 \rangle = B \frac{R^2 P_{sr}^2 w_2 k_1}{\Delta v W} \qquad (11)$$
$$\langle i_{shot}^2 \rangle = 2eB \frac{R P_{sr} w_2}{W} \qquad (12)$$
$$\langle i_{thermal}^2 \rangle = \frac{4K_b T_n B}{R_L} \qquad (13)$$

Note that the probability of sending bit "1" at any time for each user is ½ so:

$$\langle i_{noise}^2 \rangle = \frac{R^2 P_{sr}^2 w_2 k_1}{2\Delta v W} + \frac{eBRP_{sr} w_2}{W} + \frac{4K_b T_n B}{R_L} \qquad (14)$$

where R is the responsivity of the photo-diode given by $R = \eta e / h v_0$, η is the quantum efficiency of the photo-diode, and h is Plank's constant. P_{sr} is the effective source power at the receiver, k_1 is the code weight of spectral encoding, W is the number of active users, N and M are the code lengths of spatial and spectral code sequences respectively.

$$SNR = \frac{I^2}{\langle i_{noise}^2 \rangle} = \frac{\left[\frac{R P_{sr} w_2}{W}\right]^2}{\frac{R^2 P_{sr}^2 w_2 k_1}{2\Delta v W} + \frac{eBRP_{sr} w_2}{W} + \frac{4K_b T_n B}{R_L}} \qquad (15)$$

The BER can then be calculated from SNR as follows [2]:

$$BER = \frac{erfc\left(\sqrt{SNR/8}\right)}{2} \qquad (16)$$

Where

$$erfc = \frac{2}{\sqrt{\pi}} \int_0^\infty \exp(-y^2) dy \qquad (17)$$

3 Results and Discussion

Table IV: Parameters used in analytical calculations.

PD quantum efficiency	$R = 0.75$
Spectral width of broadband light source	$\Delta\lambda = 30nm\ (\Delta\lambda = 3.75\ THZ)$
Operating wavelength	$\lambda_0 = 1.55\mu m$
Electrical bandwidth	$B = 320MHz$
Data transmission rate	$R_b = 622Mbps$
Receiver noise temperature	$T_n = 300K$
Receiver load resistor	$R_L = 1030\Omega$
Boltzmann's constant	$K_b = 1.38 \times 10^{-23}\ W/K/Hz$
Electron charge	$e = 1.60217646 \times 10^{-19}\ coulombs$
Light velocity	$C = 3 \times 10^8\ m/s$

The analysis of the code is based on the parameters shown in the Table IV where the numerical results are exhibited in following figures.

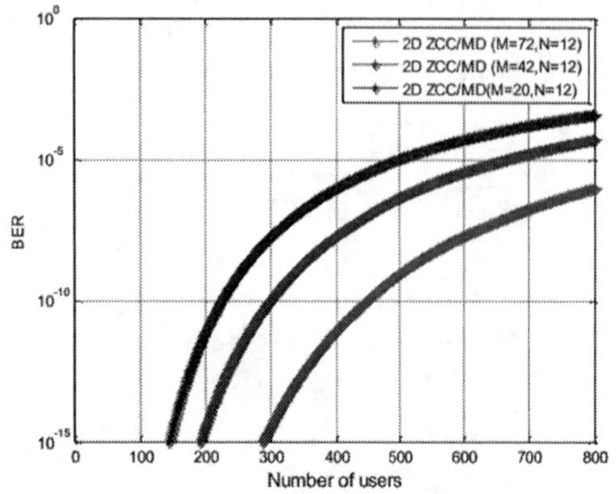

Figure 1: BER versus the number of simultaneous users for 2D ZCC/MD with the fixed value number of time spreading N

Figure1 illustrates the relation between the BER and the number of simultaneous users for 2D ZCC/MD with the fixed number of time spreading N to 12 and the effective source power, Psr is fixed at -10dBm with variation of wavelength M (M=20, M=42, M=72). It can be seen that this variation of M makes a small difference in the number of simultaneous users. The respective number of simultaneous users for groups of (M=20, N=12), (M=42, N=12) and (M=72, N=12) are 280, 345 and 510 respectively when the BER at 10^{-9}.

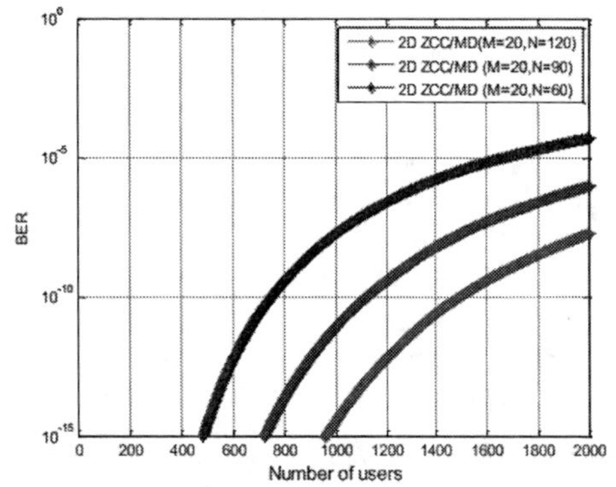

Figure2: BER versus the number of simultaneous users for 2D ZCC/MD with the fixed number of wavelength M.

The figure2 represents the plot of BER versus the number of users for 2D ZCC with the fixed number of wavelength M to 20 and the time spreading N varies at 60, 90 and 120 at Psr=-10 dBm. According to the figure, the curves are separately each from other comparing with the Figure1. The changing of the values of N results the variation of cardinality. From the figure, at BER equal to 10^{-9} the number of simultaneous users for N=60, 900, 90 is 1350, 120 and 1750 respectively. Note that the increases of N causes that the code can accommodate a higher number of simultaneous users but for the lowest number of N offer the lower cardinality in the system.

4 Conclusion

This study is focused to analyze 2D hybrid ZCC/MD code in term of impact size including 1D ZCC a spectral encoding and 1D MD a spatial spreading. From the numerical results it can be seen that the improvement of the cardinality is related to the spatial spreading N where is present the principal role to ameliorate the performance of the system. The large number of N offers the capability to provide a high cardinality in the system at standard acceptable BER when the number of spectral encoding is fixe. The variation of M and fixed value of N makes a small difference in the number of simultaneous users can be obtained. This paper clarifies and defines the relationship between the code size and the system performance.

References

1. Paul R. Prucnal, CRC Press, Florida, (2006).

2. Keraf, N. Din, et al. *Electronic Design (ICED), 2nd International Conference on*. IEEE, (2014).

3. A.R Arief, S.A Aljunid, M.S Anuar, M.N Junita, R.B Ahmad and F. Ghani, IEEE International Conference on Control System, Computing and Engineering, pp. 488-494, (2011)

4. Hongxi Y, Le Ma, Hongbin Li and Lixin Zhu, Photon New Communication, (2010)

5. Cheing-H L, Jingshown Wu and Chun-L Y. Journal of Lightwave Technology, vol. **23**, no.12, pp. 3966-3980, December (2005)

6. Preeti M and Shalini D. International Journal of Computer Science & Engineering Technology IJCSET, pp. 1211-1215, (2012)

7. Arief, A. R., Aljunid, S. A., Anuar, M. S., Junita, M. N., & Ahmad, R. B. Optik-International Journal for Light and Electron Optics, **124**(19), 3786-3793. (2013)

8. Kadhim, R. A., Fadhil, H. A., Aljunid, S. A., & Razalli, M. S. Journal of Theoretical and Applied Information Technology, ***81***(3), 589. (2015)

9. Abdullah, A. R. A. J., Aljunid, S. A., Safar, A. M., Nordin, J. M., & Ahmad, R. B. *Optical Engineering*, ***51***(6), 065007-1. (2012)

10. Keraf, N. Din, et al. *Photonics (ICP), 2014 IEEE 5th International Conference on*. IEEE, 2014.

11. Chang, Yao-Tang, et al. *Optical Fiber Technology* **16.2**: 114-123. (2010)

12. Keraf, N. Din, et al. *Advanced Computer and Communication Engineering Technology*. Springer, Cham. 129-140. (2015)

13. Keraf, N. Din. Malaysia University Conference Engineering Technology. (2014)

14. Jellali, Nabiha, et al. *Optical Fiber Technology* **36**: 26-32. (2017)

15. Dexter, Karl J., Douglas Alexander Reid, and Liam P. Barry. IEEE photonics technology letters 21.22: 1662-1664. (2009)

16. Keraf, N. Din, et al .*Electronic Design (ICED), 2016 3rd International Conference on*. IEEE, (2016).

17. M. S. Anuar, S. A. Aljunid, N. M. Saad & S. M. Hamzah. Optics Communications, **282**(14), 2659-2664. (2009).

18. A .Thanaa Hussein, S.A. Aljunid, H. Adnan, R.A. Ahmad, 2N.M. Saad. Australian Journal of Basic and Applied Sciences, **6**(3): 112-119, (2012)

19. R.A. Kadhim, H. A. Fadhil, S. A. Aljunid & M. S .Razalli. Optics Communications, **329**, pp. 28-33. (2014).

Design and Implementation of an Embedded Smart Intruder Surveillance System

Naseer Sabri[1*], *M. S.* Salim[2], *S.* Fouad[3], *S. Alwee* Aljunid[1], *F.T* AL-Dhief[4], *C.B.M* Rashidi[1]

[1]Computer and Communication Engineering School, University Malaysia Perlis, Malaysia
[2,3]Laser and Optoelectronics Engineering Department, [1*]Computer Engineering Department, AlNahrain University, Iraq
[3]Microelectronic Engineering School, University Malaysia Perlis, Malaysia
[4]Faculty of Electrical Engineering, Department of Communication Engineering, Universiti Teknologi Malaysia, Malaysia

Abstract. Remote and scattered valuable and sensitive locations such as labs and offices inside university campus need efficient monitoring and warning system. As well as scattered area and belonging. This research presents a Real-Time intruder Surveillance System based on a single board computer (SBC). Thus the design and development of a cost effective surveillance management system based SBC that can be deployed efficiently in remote and scattered locations such as universities belonging. The fusion of embedded Python codes with SBC that attached to cameras, Long distance sensors, alerting circuitry and wireless module presents a novel integration based effective cost solution and enhances SBC of much flexibility of improvement and development for pervasive remote locations. The system proves the high integrity of smooth working with web application, it's cost effective and thus can be deployed as many of units to seize and concisely covered remote and scattered area as well as university belonging and departments. The system can be administrated by a remote user sparsely or geographically away from any networked workstation. The proposed solution offers efficient stand alone, flexibility to upgrade and cheap development and installation as well as cost effective ubiquitous surveillance solution. In conclusion, the system acceptable boundaries of successful intruder recognition and warning alert are computed between 1m and 3m distance of intruder from system camera. Recognition rate of 95% and 83% are achieved and the successful warning alert were in the range of 86-97%.

1 Introduction

Nowadays, the demands on video surveillance system are rapidly increasing in the security applications. Video Surveillance is of paramount essence as far as security in general is of high importance. Commercial and public spaces, universities and hospitals, warehouses require sufficient video capturing's which are thus introduced to keep high checks to activities in their domesticity and vicinities for security reasons. The current available technologies require RFIDs which are largely considered costly and hence the security domain and installations becomes highly expensive. For the benefits of this study in curtailing the high expenses, this project aims at using an embedded video surveillance system of single–board computer Raspberry Pi. This justification is as such that embed systems becoming more and more a part of our daily dealings, endeavours and device driven. Embedded system usage has cut across all kinds of devices used in transportation services, communication devices, entertainments, industrial activities, military formation among others [1, 2]. Therefore, the need for befitting solutions using computational prowess is of the essence. This has yearned for solutions and the availability of increased computational power.to manage our daily activities. Currently, the implementations of security surveillance have to aid users for a wide variety of orthogonal designed thereby retaining some semantics based on real time operating system i.e. Raspbian [2, 3]. This project describes the potential use of low cost and single–board computer Raspberry Pi with wireless internet. This is in adherence to the face detection algorithm programmed in Python-based platform as the default programming environment. In comparison to the former, this new approach is less expensive than the latter. The project aims at proposing an efficient system which captures images real time with the ability to display them on a browser using TCP/IP [1]. This study is motivated by; remote and scattered valuable and sensitive locations such as labs and offices at university campus need efficient monitoring and warning system. Beside, a semi-autonomous surveillance system based on human monitoring assistance is not optimum. And finally, a costly solution is emerging for multi-location surveillance system deployment.

The purpose of this research is to design and implement a cost effective solution that would be capable of handling in an efficient manner an intruder detection system that will determine its identity and

* Corresponding author: naseersabri@yahoo.com

location. As well as investigating the existing features and limitations of an embedded real-time surveillance system based on SBC using the raspberry pi platform. To accomplish this goal, developing of a cost effective surveillance management system based SBC that can be deployed efficiently in remote and scattered locations such as universities and detect intrusion is conducted. As well as to implement fast and smart intruder warning system for a university campus.

According to our survey, the surveillance system does not in cooperate immediate alert in case of any unauthorized entry into the universities. This paper is focused on developing a surveillance system that detects stranger with the ability to respond speedily by capturing and relaying it to admin office based wireless module and thus activate the alert system both at intruder location and office admin.

2 Related Work

Extensive related literatures to the research topic have been reviewed in the last few years to examine the gaps and the contribution of various researchers in the field. Studies of many systems for remote monitoring and control are designed for commercial products and experimental research platforms.

A study of a home embedded surveillance system for evaluating the development of a Low-cost security system by using small PIR is presented by Shake and Sumedha S [4]. Developing of a surveillance system which with low cost without losing the semantics of using low power, and less memory and more efficiency using a set of various sensors. Their system helped in monitoring the household activities in real time from anywhere based on microcontroller which are considered nowadays as a limited resources and an open source solution compared to SBC is presented in [5]. Khandavalli Michael Angelo and Katakala Karthik, introduced the use of embedded system video capturing and motion detection system. The embedded web monitoring system uses the powerful ARM9 chip as its processing unit (PU). It provides the system with high embedded processor performance to process the main control, thereby compressing and web processing the captures in a simple form, inexpensive, stable and widespread as compared to others. However, their work did not consider the reaction process of any unusual events [6]. D. Jeevanand et al. in (2014), works on the design of a network video capturing system using Raspberry pi as presented. Their application system captures the videos, shares them among networked systems which also allows the alerts the controlling person with short messages as service alarm as demanded by the client. They designed the system to works in a real time environment and it was supported by Raspberry [7]. In [8], The researchers provided in their study a system that works with a low cost and high effective intelligent monitoring system which are power full to be used in elevators, home security systems and much other application built with low power consumptions. As compared to other embedded systems, this real time system provides client with video monitoring as help [8]. Sneha Singhd et al, described IP Camera Video Surveillance system using Raspberry Pi technology .Their study used this system of low cost to single – board Raspberry Pi technology. The system used a standalone platform for the hosting of image processing. The Researchers aim at developing a system with the capability of capturing real time images and displaying them on browsers using TCP/IP. It further used the algorithm to face detect and implements the design on RaspberryPI which enables live video streaming along the human face detection. Part of the limitation of the study was the lack of including any of surveillance reactions [9]. Mahima F. Chauhan and Prof. Gharge Anuradha P submitted a design and development of a Real Time Video Surveillance System Based on Embedded Web Server Raspberry PI B+ Board. Their system also incurred a low cost, with good openness and portability and it is easy to maintain and upgrade. These are considered as the targets for researchers of this application system to provide better security solutions. This system can be used for security purposes in banking halls, industrial environments, military institutions etc [10].

3 System Design Architecture

Figure 2 depicts the entirety of the system which composed of five parts, namely the Raspberry Pi embedded Linux controller, the PIR sensor, camera, the Wi-Fi adapter and a power supply.

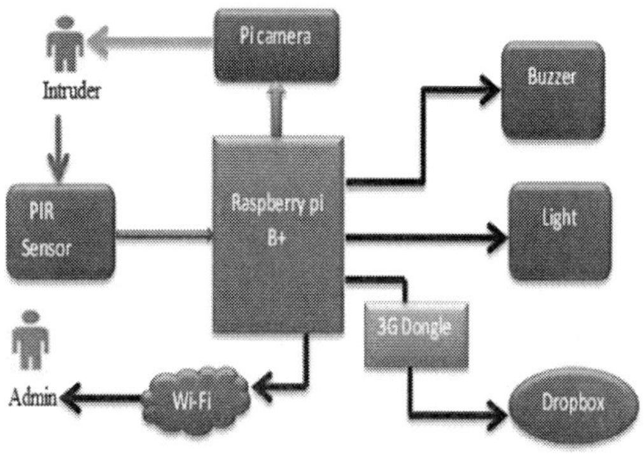

Fig. 1. Surveillance System Architecture

For an embedded real-time surveillance system to be used as an effective monitoring and alert system, it has to have at least three functions, which are detection, picture identifying and alert mechanism. The network of an embedded real-time surveillance system mainly consists of two parts as presented in Figure. 1.

3.1 Hardware Design

In size description, Raspberry Pi is comparable to a Credit card, which is fully featured computer deployed on the Linux operating system. Raspberry Pi has all

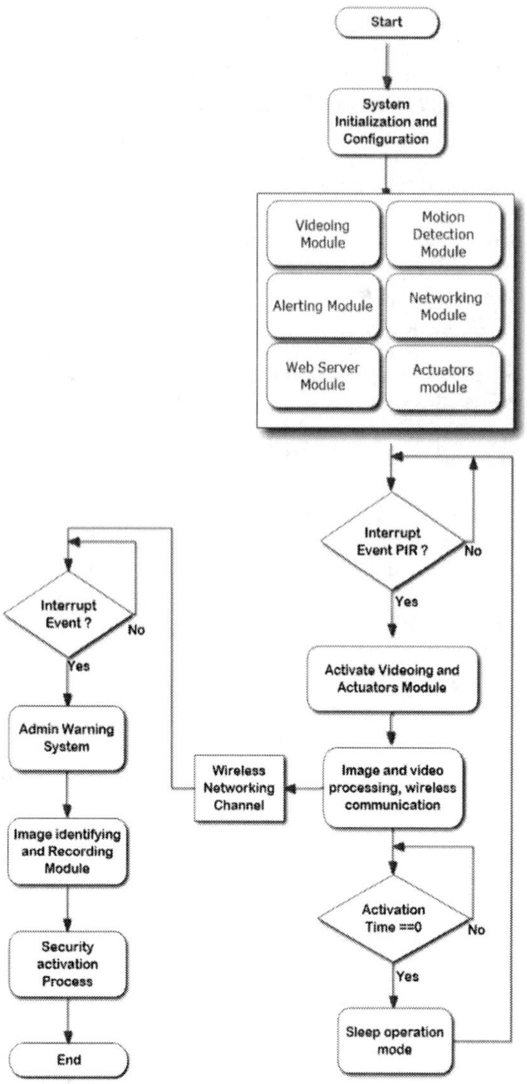

Fig. 2. Research Flowchart processes

incorporated components that aid its connection ports where peripheral devices are plugged by the users. With its HDMI compartments monitor for displays can be plugged through, for pointing and navigation, mouse and keyboard are inserted into the Universal Serial Bus (USB) ports while for the speakers, it provides a reliable 3.5 mm audio jack. In another model B+, there exists an Ethernet socket point for internet connectivity. In general it is as efficient and reliable in terms of functionality as compared to a normal desktop computer with its size compatibility. This is thus an advantage as instance messages, emailing, web surfing, and word-processing, watching high-definition videos etc. could be done on it. Also, its flexibility cuts across diverse kind of electronics project handling and as a vital tool for acquiring programming skills. Raspberry Pi foundation is relatively cheap.

The raspberry pi camera is of a normal resolution of 5 mega pixel, it also has a fixed focus lens on-board. The camera capability in pixel is 2592 x 1944 for static images, still images, while it maintained 1080p30, 720p60 and 640x480p60/90 for videoing. The camera objective is to capture images, videos based on the event of detection.

The PIR sensor, a Pyroelectric device for detecting motion by measurement the changes acquired in infrared levels in its field of view that is emitted by the surrounding objects. Its size is small, inexpensive; its power is low and easy to use for this reason it is found in appliances and gadgets that are used in homes or business outfits. The PIR sensor also referred to as PIR, "Passive Infrared", "Pyroelectric", or "IR motion" sensors. Holds is a basic building block of a Pyroelectric sensor which can detect several levels of infrared radiation. The emission is sometimes on low level radiation, it then tends hotter and the more radiation is emitted. The PIR sensors consist of three pins ground, digital out and 3-5V DC. It is connected to the Pi using female/male jumper connectors. In the event that no motion detected, the digital output will remain low, however, when some motion is detected, the digital out will deliver as high (3.3V) and thus we use our Raspberry Pi in sensing this. It also has the ability of sensing in the area within a ranges less than 20 feet (6 meters) at an angle value of 110°. The Fig. 3 below explains how the connection of the Pi's 5V pin to the sensor's power pin and the Pi's ground pin to the sensor's ground pin. Finally, it connects the sensor's signal pin to pin GPIO16 on the Pi.

The PIR sensor comprises of two slots as depicted on Fig. 4, each of the slots is made from special material which is sensitive to IR.

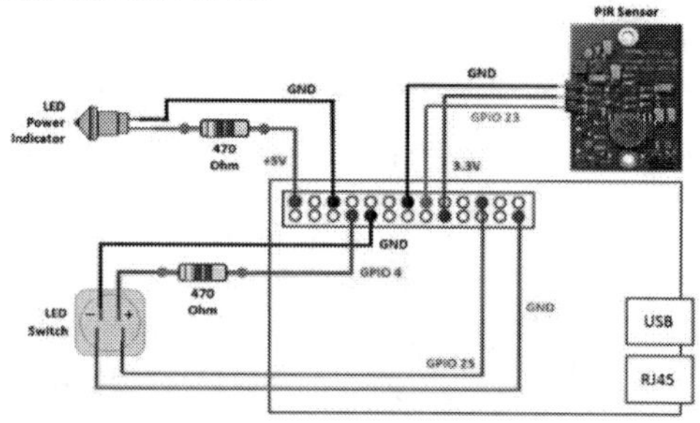

Fig. 3. The Schematic Diagram of PIR Sensor Connect to Raspberry Pi

The lenses used are not really active much and so are observed that the two slots can 'see' out past some distances (basically the sensitivity of the sensor). When the sensor is inactive, both slots detect a replica amount of IR, the ambient amount radiated in the room, walls or outdoors. In an event of human of animal move detection or sensing, the first operation is to intercept a half of PIR sensor, which causes a positive differential alteration between the halves. Consequently, when the warm body departs the sensing area, the reverse of the previous operation occurs, resulting to the generation of negative deferential change. The resulting change pulses are thus what are detected.

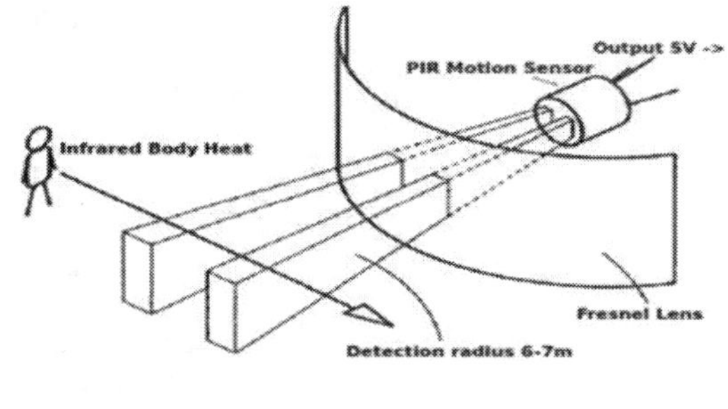

Fig. 4. PIR Sensor Operation

The opto-isolator is a small device that allows to transference of a signal between portions of a circuit while maintaining those two portions electrically isolated. This circuit is used in a wide classification of communications, control and monitoring systems that use light to block electrical high voltage from influence a lower power system receiving a signal. The circuit used in this system to control the voltage source so that can to run lamp or buzzer with main power supply voltage. The opto isolator established to provide solution of very economically, technically distinguished for many applications and it has some general characteristics and prominent features. Such as high input to output voltage isolation, suitable to AC and DC usage, complete solid state construction and low drive current. Figure 5 show circuit diagram to opto isolator connects to Raspberry Pi with lamp 220 V and buzzer.

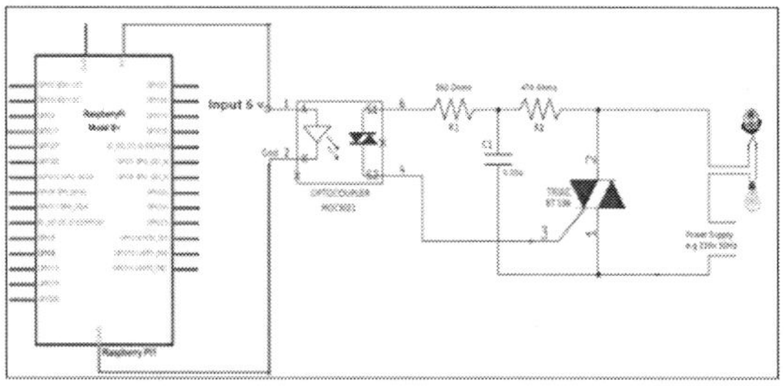

Fig. 5. Opto-isolator interface

Mainly Wi-Fi is using the 2.4GHz and 5GHz ISM radio bands. This technology allows various devices, such as personal computers, smart phones, tablet and digital camera to communicate with network resources such as the Internet via a wireless network access point. Wi-Fi dongle is connected over USB port as an extension. It has wireless speeds that can reach up to 300 Mbps. The dongle is characterized easily to work with IEEE 802.11b/g networks, and some dongles appear as security keys, whereas others serve as adapters as the case may be. Its function in this project is to transfer the captured images of the intruder which are saved on the SD card from raspberry pi to drop box where it is stored into. This provides the easy access to the admin and any staff to see the pictures at ubiquitous positions or anywhere. The following block diagram depicted in the Figure shows the process of transferring captured pictures to the drop box, Figure 6.

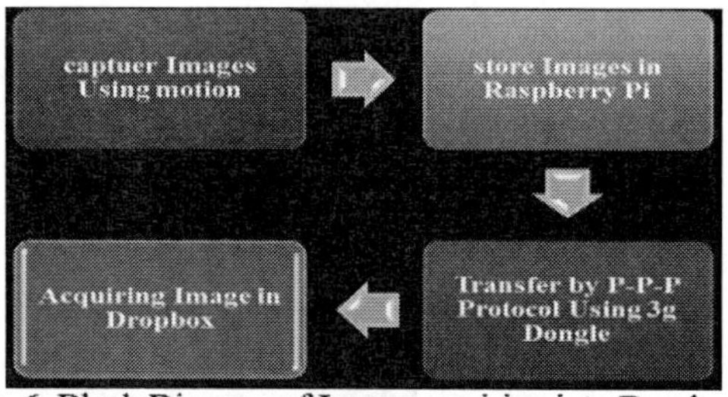

Fig. 6. Block Diagram of Image acquiring into Dropbox

3.2 Software Requirements and Design

Raspberry pi is a single board computer (SBC) base Linux Operating System (OS) which offers high flexibility to modify. Hence its functionality offers a base for open source software additions. The designed system software package is based on PIR module for movement detection activities which yields good activity with light and the capturing of picture by camera module and thus activating the alarm circuitry present in buzzer. The block diagram of system software in Figure 7 shows the main package tasks.

The main procedures in the software contain two parts, the first part being the motion detection phase and secondly, are the signal transmission. The connections between the two parts are further presented in the flow chart. The first part this program will be continuously moderating any activity and movements by the PIR sensor to detect motion. Meanwhile, if the PIR sensor detects an intruder, it will prompt the camera to capture the image of the intruder and in turn allows the raspberry to run lightening. This means that when motion is detected, it will continue taking the culprit images and recording the videos by making sure that the intruder do not escape capture. The second part is the signal transmission that saves the image. The signal transmission relates the raspberry pi to administrator with the main purpose of this is to alert the admin when an intruder is detected. Then the program sends the captured images to the admin for the purpose of storage. The system alarming continuously for one (1) minute then its goes to sleep mode and the whole procedure is repeated again in a continuous fashion.

4 Results and Discussion

This paper discusses on the result gained from experiments and it also focuses on the problems confronted during the progression of this paper. The results obtained in this paper contain the setup and configuration of the designed Surveillance System. The results are partitioned into three parts which are the capturing of an intruder images, the communication between monitoring area and the admin office and finally download of the captured images to the Dropbox.

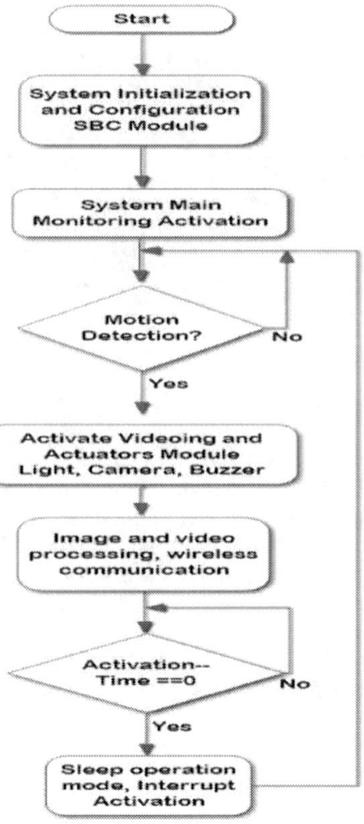

Fig. 7. System Software design flow chart

This system has the ability to capture images continuously in situations where the intruder enters into the surveillance zone. The captured images differ from each other according the PIR sensor sensitivity range. In other words, the camera will start taking pictures of the strange (intruder) when he enters into a lesser range, from 6 meters and whenever the range falls less, the captured picture will be clearer. Figures 7, 8 and 9 represent the screen shots pictures which were photographed by the system as the intruder encroaches into the monitoring zone in different ranges. Fig. 7 shows the picture that was taken at the beginning ingress of the stranger in the monitoring area, for this reason the image was less obvious.

Fig. 7. Intruder Event Detection and System Response

Demonstrated by the result obtained, there was increase in the clarity of the image whenever the stranger comes

nearest to the camera as the result presents in Fig. 8

Fig. 8. System Response for Intruder Detection Image

The final image in the Fig. 9 was more evident because the intruder was closely in front of the camera and through acquaintance on the stranger person who entered the surveillance zone. After testing the system, we found the system to be able to achieve the required results. Here are some pictures that were captured of the intruder of this system which were gotten after the successful execution of the entire system.

Fig. 9. Intruder Image capturing Nearby Detection Zone

The designed system is empirically validated based captured images, 30 samples, recognition rate for various distance with the assumption of good lighting environment. Highest recognition rate achieved for separation distance of 1m with 0.95, since more clear and focus image captured raised the matching ratio. As the distance increased the rate of recognition is decreased especially beyond 3m distance where ratio of recognition is less than 0.78. Figure 10 exploit the distance effects the system recognition rate where 0.95 and 0.4 recognition rate is resulted for apart distance of 1m and 6m respectively.

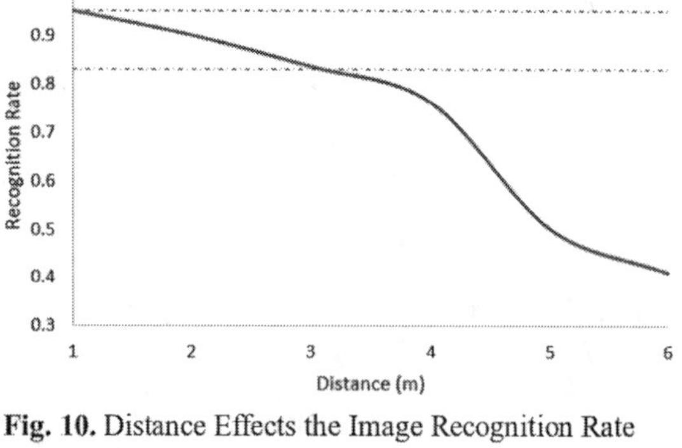

Fig. 10. Distance Effects the Image Recognition Rate

Figure 11 shows the successful intruder alert based distance variation of the intruder. Samples are collected for 100 image and incremental distance step of 1m is adopted. Obviously seen, that the intruder distance is highly effects the successful warning alert, this is true as the recognition rate is higher than apart distance. This rate of warning alert shows that system is working in acceptable region of 3m while as the distance increased system performance is degraded due to imperfect recognition rate.

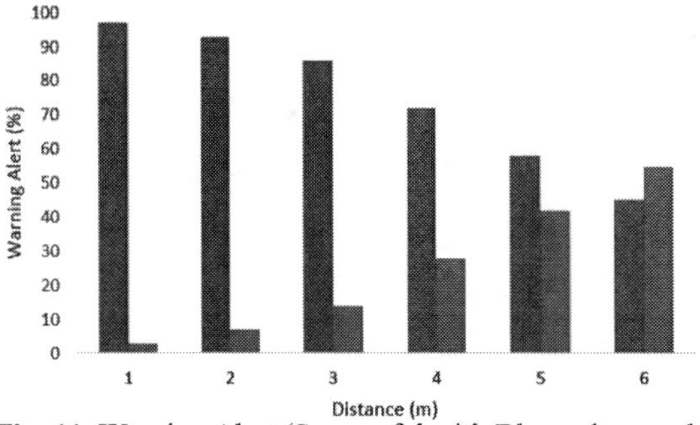

Fig. 11. Warning Alert (Successful with Blue colour and false with red colour) with various intruder distance

In conclusion, the system acceptable boundaries are computed between 1m and 3m distance of intruder from system camera. Recognition rate of 95% and 83% are achieved and the successful warning alert were in the range of 86-97%.

In reality, one of the challenges of the designed system is the spatial positioning in the field and specifically the camera that captured intruder image. The optimum positioning depend on various parameters and can vary from monitoring place to another. The latency time of the system was computed to be in average range of less than 1sec since the capture and alert process. Link availability is of high impact on latency time of the system alert task. This system works efficiently with the lowest time delay in addition to high response speed by capturing and relaying images to admin office based wireless module to activate the alert system both at intruder location sites and the admin office. As shown in Fig. 12, the captured images of the stranger were saved and kept in the dropbox by the 3G dongle.

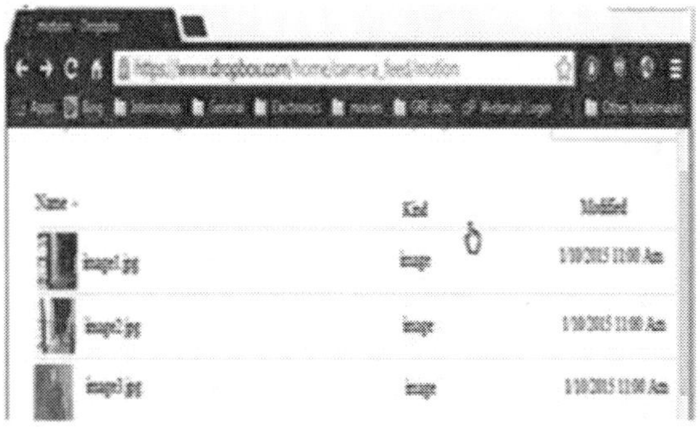

Fig. 12. Dropbox Captured Images

5 Conclusion

A successful system has been applied with low cost ubiquitous surveillance system for intruder detection and identification is presented according to its objectives and desired specifications. The system response was in real time efficient. Also, this project was focused on the design and implementation of cost effective solution that was capable of handling in an efficient way the intruder detection and determining its identity and location. The system performance has been elucidated in relationship to the response speed of captured video presentation at the intruder and transmitted video display to the admin officer The system has the capability to observe multi-targets and has been successfully built with low cost as an achievement.

References

1. Zi-LI Xie and Zong-Han Li, "Design and implementation of a home embedded surveillance system with ultra-low alert power," IEEE Transactions on Consumer Electronics, Feb. 2011, pp. 153-159
2. Moftah Hmaid and Nasser Sabri, (2015). "Design Approach of an intruder Real-Time Ubiquitous Embedded Surveillance System". JATIT,Volume.79. No.02, Issue 1992-8645, 2015
3. Mahima F. Chauhan and Prof. Gharge Anuradha P (2015).Design and Develop Real Time Video Surveillance System Based on Embedded Web Server Raspberry PI B+ Board. International Journal of Advance Engineering and Research Development (Ijaerd), NCRRET-2015, pp: 1-4.
4. Shake, Padmashree S., and Sumedha S. Borde. "Embedded Surveillance System Using PIR Sensor." .Volume No.02, Issue No. 03, March 2014.
5. Maiti, Anindya, and S. Sivanesan. "Cloud controlled intrusion detection and burglary prevention stratagems in home automation systems." Future Internet Communications (BCFIC), 2012 2nd Baltic Congress on. IEEE, 2012.
6. Kandavalli Michael Angelo and Katakala Karthik. Arm Based Remote Surveillance and Motion Detection System by Using MJPEG Algorithm, April 2014, pp.13-16.
7. D.Jeevanand,K.Keerthivasan, .MohamedRilwan and P.Murugan." Real Time Embedded Network Video Capture And SMS Alerting system" , June 2014.
8. Raja, S., Viswanathan, C., Sivakumar, D., & Vivekanandan, M. (2014). Secured Smart Home Energy Monitoring System (SSHEMS) Using Raspberry Pi. Journal of Theoretical & Applied Information Technology, 66(1).
9. Sneha Singh, PradnyaAnap, YogeshBhaigade," IP Camera Video Surveillance using Raspberry Pi", International Journal of Advanced Research in Computer and Communication Engineering Vol. 4, Issue 2, February 2015.
10. Mahima F. Chauhan and Prof. Gharge Anuradha P (2015). Design and Develop Real Time Video Surveillance System Based on Embedded Web Server Raspberry PI B+ Board. International Journal of Advance Engineering and Research Development (Ijaerd), NCRRET-2015, pp: 1-4

Investigation of Empirical Wave Propagation Models in Precision Agriculture

N. Sabri[1*], *Mohammed S.* S[2], *Sarah* Fouad[3], *Syed A.* A[1], *Fahad Taha* AL-Dhief[5], *Auda* Raheemah[4]

[1]Computer and Communication Engineering School, University Malaysia Perlis, Malaysia
[2,3]Laser and Optoelectronics Engineering Dept., [1*]Computer Eng. Dept., AlNahrain University, Iraq
[3]Microelectronic Engineering School, University Malaysia Perlis, Malaysia
[5]Center of Software Technology and Management, University Kebangsaan Malaysia, Malaysia
[4]Computer Engineering Dept., College of Engineering, ALMuthana University, Iraq

Abstract. Wireless Sensor Networks have highly proved its contribution in precision agriculture. Communication channel modelling investigation is a highly demanded to achieve a successful communication system where a good wave propagation model is crucially needed. Precision agriculture degrades the traveling waves in various forms in addition to the effects of the large-scale path losses models. This paper reviews the most known theoretical large-scale path losses models such as free space (FSPL) and Plane earth (PE) models as well as the vegetation models represented by Weissberger, ITU-R, FITU-R and COS235 models. Indeed, this work illustrates the effects of many factors on the total path losses such as separation distance between transceivers, antenna heights and the depth of vegetation that presence in the path propagation. In conclusion, the total path loss is computed based on large-scale path losses and the vegetation losses in protected vegetation environment.

1 Introduction

Wireless sensor networks (WSNs) technologies have dominate a various crucial applications starting from home, precision agriculture, industrial and chemical to space applications. All are aiming monitor and tracking and emergency relief. WSNs mostly, are composed of spatial distributed routers and end nodes which can be in large or small number where network end nodes are assumed to function autonomously in a very cruel situations [1]. Precision agriculture offers the ways for the agricultural practices to be monitored, evaluated and controlled. It covers abroad diversity of agricultural tendencies from diurnal herd management over gardening to yields' cultivation and production. WSNs offer a means for all time monitoring of entity crop and its provisions that will assist farmers to potentially identify a multiplicity of irrigation, climate automation, fertilizers, ventilation, and other necessities which offer the crop with perfect environment and conditions for growing [2]. While end nodes of WSN are spatially located with low elevation from earth surface and thus absences of main ray between network nodes transceivers, even though those nodes have spatially short distance distribution. Hence, WSNs propagation signals may obstacle by dense foliage, fence, building, trees [3]. Thus, the current propagation models of wireless communication might not be accurate defining the propagation channel behaviour of WSN. Therefore, for WSNs a precise model of propagation must be adopted that well describe the signal and its environment [3]. Simulation tools will highly assist with characterization and modelling of WSN propagation channel which is consider as a primarily indication of successful implementation of WSNs [1].

This research adopted four vegetation models of well-known empirical based modelling. Those vegetation models, COST235, ITU- Recommendation (ITU-R), Weissberger and FITU-R, are simulated to find the vegetation losses with WSNs communication nodes.

2 Empirical RF models in simulation environment

WSNs propagation channel assessment is predicted with respect to propagation models within the aids of simulation tools. However, most of the well-known simulators adopt free of sight propagation channel, no obstacles, thus line of sight environment is simulated (LOS), but the results are consider very optimistic and inadequately reflect the empirical real time scenario. Like as, part of plantation area will produce addition attenuation factor over propagation signal and thus affects channel performance of the WSNs nodes. Hence, the most reliable describing of channel propagation must

* Corresponding author: naseersabri@yahoo.com

combine the vegetation losses with other signal attenuation factors like reflection and transceivers distance. The foliage imposed affects in addition to multiple instants of reflected signal from the ground and perchance arbour canopy will formulate the total propagation path loss of small distance near ground agricultural environment, as per recommended in [4][5]. In wireless communications, the average received energy of transmitted wave is computed based on signal path loss models whereas the received power are inversely related to transceivers distance. Such models are characteristically shaped according to environmental characteristics.

Propagation models are mainly categorized as a two well-known channel modelling approaches, theoretical and empirical propagation models [6]. Although theoretical-model approach based on the facts of the physical laws of the communication path such as electrical properties of ground, empirical propagation attenuation models are depend on real radio frequency measurands of communication channels. The simplicity of implementation and its ability to cover all environment-related factors that affect the propagation of the radio wave throughout actual measurements [6] introduces empirical path losses models over theoretical path loss models. The accurate wave propagation model for wireless sensor network are expected to support in achieving suitable valuation with improving of wireless sensors networks performance all over the deployment planning process, minimizing the power consumption of network end nodes and to enhance the localization and target sensing tasks [7].

3 Large Scale Propagation Models

These models are used to estimate the mean received signal strength (RSS) of the propagate wave between transmitter and receiver since they characterize the RSS over large separation distance between transceivers. Mainly large scale models are categorized into well-known two models; free space path loss model (FSPL) and plane earth model (PE) were used in most researches [8].

3.1. Obstacle clear Path Loss propagation model

Obstacle clear propagation model is well known as free space path loss (*FSPL*) model valuated by the loss in the received signal strength (RSS) with respect to transmission signal energy, when there is obstacle free wave traveling path between the transceivers [8].

$$PL_{FSPL}\,(dB) = -\,27.56 + 20\,Log(d) + 20\,log(f) \quad (1)$$

where d is the distance in meter, and f is the signal frequency in MHz;

3.2. Plane Earth Path loss (PE) model

The *PE* model computes the path loss of a traveling wave based on the antenna heights of transmitter and receiver that may present the plane earth reflection of the transmitted signal into the receiver antenna [9]; Path loss can be calculated by equation 2.

$$PL_{PE}(dB) = 40\,Log\,(d) - 20\,Log\,(h_t) - 20\,Log\,(h_r) \quad (2)$$

The separation distance (d) in meter, and the transmitter and receiver antennas height represented by (h_t) and (hr) respectively. The large scale broadcast models is not completely appropriate where the detraction, reflection, and scattering properly disturb the communication channel and affects widely of the received signal as within practical tasks. However, *PE* formula in (2) can be adopted to provide early assessment of what to be predicted [9].

4 Vegetation models

WSN in agricultural application such as greenhouse where the nodes are deployed near ground and there vegetation between these nodes caused many effects on the RSS. Path losses can be modeled experimentally based on the equation (3) [10], where the depth of vegetation (d_f) and frequency are the parameter of PL_{veg}, A, B, and C are constant empirically determined.

$$PL_{veg}(dB) = A.\,f^{B}.\,d_f^{C} \quad (3)$$

4.1. Weissberger model [6]

It is applicable in situations where propagation is occurred through the grove of trees:

$$PL_{weiss}(dB) = \begin{cases} 1.33 \times f^{0.284} \times d_f^{0.588} & 14m < d_f \leq 400m \\ 0.45 \times f^{0.284} \times d_f & 0m \leq d_f < 14m \end{cases} \quad (4)$$

where PL_{Weiss} is the vegetation loss in dB, f is the frequency in GHz, and d_f is the depth of the trees in meter.

4.2. ITU-R model [11]

The Weissberger vegetation model shows fairly matching with ITU-R model. The ITU-R model is computed based on equation (5), where the frequency used (f) in MHz and the transceivers apart distance (d_f) in meters. Equation (5) shows ITU-R vegetation loss;

$$PL_{ITU-R}(dB) = 0.2 \times f^{0.3} \times d_f^{0.6} \quad (5)$$

4.3 Fitted ITU-R (FITU-R) Model [12]

carried out using measurement data at 11.2 and 20 GHz (d < 120 m).

$$PL_{FITU-R}(dB) = \begin{cases} 0.37 \times f^{0.18} \times d_f^{0.59} & out\ of\ leaf \\ 0.39 \times f^{0.39} \times d_f^{0.25} & in\ leaf \end{cases} \quad (6)$$

 https://doi.org/10.1051/matecconf/201815006020

Where the vegetation loss of fitted ITU-R (*PLFTITU-R*) in dB and the propagation frequency (f) in MHz while the vegetation depth of trees (d_f) in meters.

4.4. COST 235 vegetation model

This vegetation model computes the path loss in (dB) with propagation wave frequency in the range of 9.6 GHz to 57.6 GHz throughout a short vegetation distance (d_f) less than 200m of fluting of trees [13]:

$$PL_{COST235}(dB) = \begin{cases} 26.6 \times f^{-0.2} \times d_f^{0.5} & out\ of\ leaf \\ 15.6 \times f^{-0.009} \times d_f^{0.26} & in\ leaf \end{cases} \quad (7)$$

5 Path Losses (P_{Ltot}) in vegetation environment

When WSN nodes deployed in vegetation environments, the receive power effects by vegetation that caused additional losses, then the total path losses can be formulated by combines the losses due PE model with the vegetation losses PL_{veg} that prediction by different vegetation models [3][14].

$$PL_{tot} = PL_{PE} + PE_{veg} \quad (8)$$

Where PL_{tot} is the total path losses, PL_{PE} is the path losses in free-space and PL_{veg} the path losses due to vegetation.

6 Simulation Setup

The simulation assumes a greenhouse of 100 m × 100 m area where wireless sensor nodes are deterministically deployment over this area with 10 m distance between each adjacent nodes. The total number of nodes are 100. The terrain of greenhouse assumes to be flat. The setup of greenhouse wireless sensor network nodes is listed in Table .1.

Table 1. Parameter setting for simulation Setup

Topology	Star topology
Transmit power	0dBm
Frequency operation	2.245 GHz
Data rate	250 kbps
receiver sensitivity	- 95dBm
Antenna Type	Omni-directional
Pattern of deployment	Square grid

7 Results and discussions

WSN nodes when deployment in agricultural environments, propagation wave facing obstruction by the vegetation that existence in these environments, this obstruction caused excess losses added to losses that predicted by large scale propagation models alone. The excess losses must be determined by the abovementioned vegetation models. Investigation of large scale path loss models based on antenna heights and separation distance is depicted in Figure.1. The effects of antenna heights on the path losses that predicted by large scale propagation models (FSPL and PE) when antenna are located at the same level (line of sight). The simulation results explain that the PL_{tot} are decreased with increasing the heights of antenna while it increases with increasing the separation distance between transceivers.

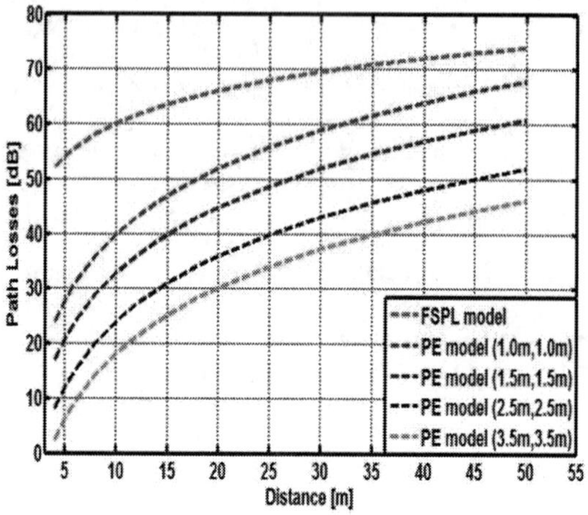

Fig. 1. Path loss versus separation distance at (LOS situation)

Figure.2 show the relation between PL_{tot} that predicted by combined of *PE* model and vegetation models with vegetation depths at different heights of antenna. For comparing between PL_{tot} that predicted by large scale propagation models (*FSPL* and *PE*) alone with PL_{tot} predicated by combined large scale model with vegetation models, explain that, the large scale models (*FSPL* and *PE*) are very optimistic models and don't take into account the effects of vegetation that existence in the path of propagation. Hence, predication by PL_{tot} at such environments required more accurate model.
The total path losses (PL_{tot}) simulated values demonstrate that PL_{tot} computed based vegetation model of equal heights of antenna of line of sight environment is less than PL_{tot} based different antenna heights. Although that,, like this formation is demanded in practical field. The simulation results show that the PL_{tot} that calculates by vegetation models at same heights of antenna (LOS) found be less than PL_{tot} that calculates at different antenna heights but this configuration is required in real field. Our simulation assumes the receive antenna are hanged in roof of greenhouse at height 3.5m. In the simulated field, the relation of PL_{tot} and the vegetation depth is presented in Figure 3. The antennas heights are varied with a range of 0.5m to 2.5m within incremental step of 0.5m.

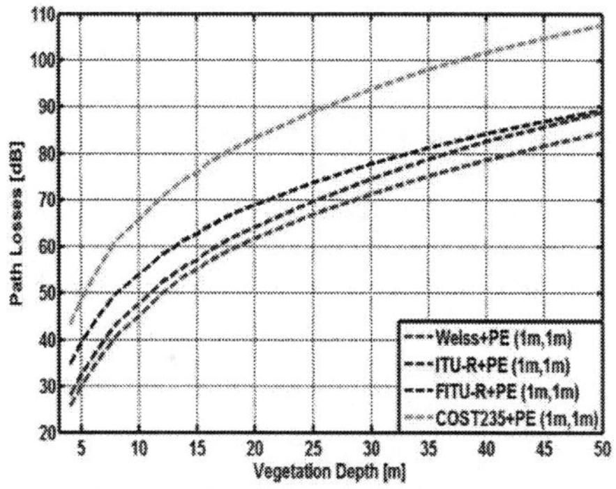

(a) Hr=1m, Ht=1m

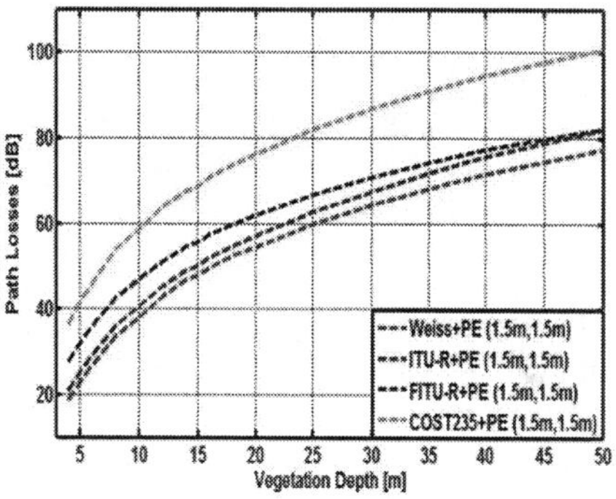

(b) Hr=1.5m, Ht=1.5m

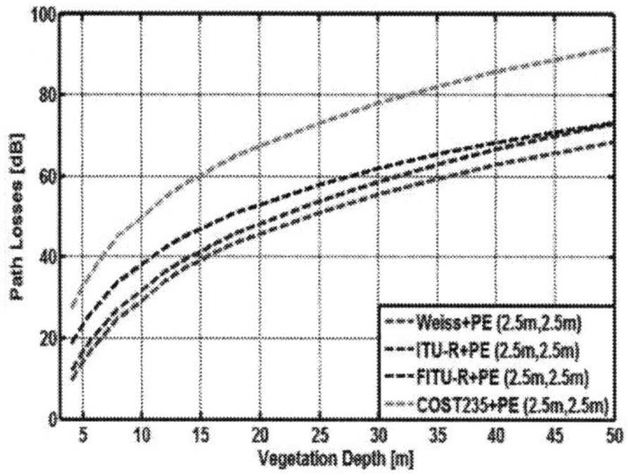

(c) Hr=2.5m, Ht=2.5m

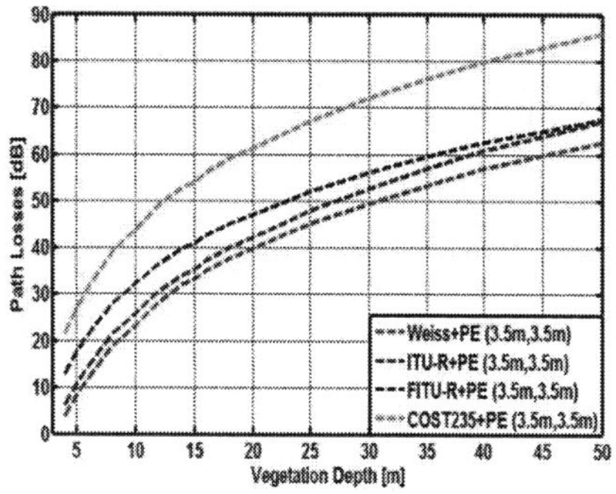

(d) Hr=3.5m, Ht=3.5m

Fig. 2. Total path loss versus separation distance at (LOS situation)

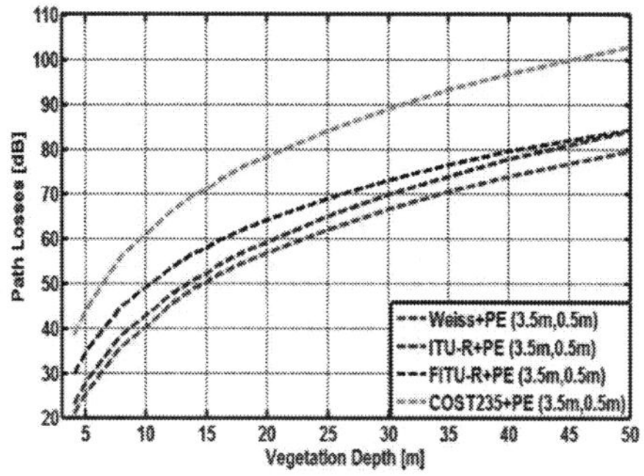

(a): Hr=3.5m, Ht=0.5m

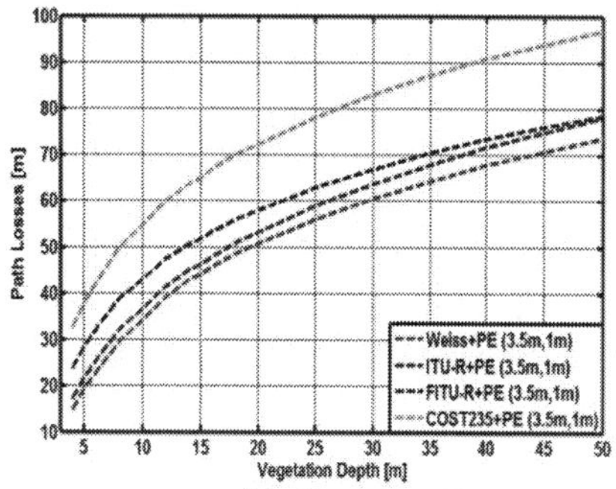

(b): Hr=3.5m, Ht=1m

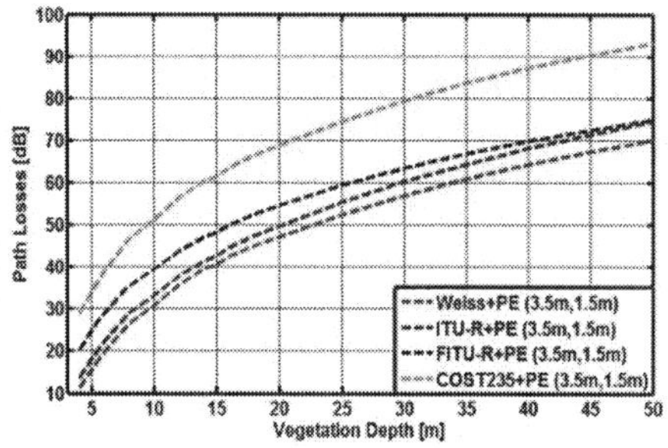

(c): Hr=3.5m, Ht=1.5m

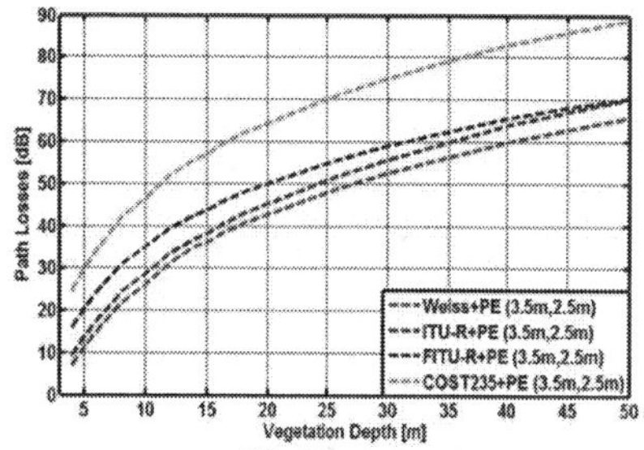

(d): Hr=3.5m, Ht=2.5m

Fig. 3. Total path loss versus separation distance at different antenna heights

* Corresponding author: naseersabri@yahoo.com

The results show that the models have the best PL_{tot} as the location of antenna is raised from the ground. For example the models have the following PL_{tot} (62.14, 65.01, 69.09 and 84.29)dB for (Weiss+PE, ITU-R+PE, FITU-R+PE and COST235+PE) respectively at (d_f=25m) and antenna heights are (3.5m,0.5m) of Figure 3a, whereas, the same models have the following PL_{tot} (52.6, 55.46, 59.55 and 74.74)dB for (Weiss+PE, ITU-R+PE, FITU-R+PE and COST235+PE) respectively at same d_f but when the antenna heights are (3.5m , 1.5) of Figure 3c.

8 Conclusions

WSN nodes when deployment in agricultural environments, propagation wave facing obstruction by the vegetation that existence in these environments. An optimistic results was shown for PL_{tot} , in the simulation, based large scale propagation models and thus these models produce PL_{tot} values far from realistic models. These results do not reflected the effects of objects that existence in real environments. Therefore, cannot depends on large scale propagation models to predicted by PL_{tot} in such environments, an accurate models must be used for reliable communication. The simulation results show the relation between the antenna heights and PL_{tot}, losses will be decrease with increasing the heights of antennas. Also the effect of vegetation existence in the path of propagation was investigated, the results show the relation between PL_{tot} and the depth of vegetation where PL_{tot} will be increase with increasing the depth of vegetation.

References

1. Sabri, N., Aljunid, S. A., Salim, M. S., Fouad, S., & Kamaruddin, R. (2015). Wireless Sensor Network Wave Propagation in Vegetation. In Recent Trends in Physics of Material Science and Technology (pp. 283-298). Springer Singapore
2. Serodio, C., Cunha, J. B., Morais, R., Couto, C., & Monteiro, J. (2001). A networked platform for agricultural management systems. Computers and electronics in agriculture, 31(1), 75-90.
3. Auda Raheemah, Naseer Sabri, Phaklen Ehkan, R. Badlishah and M. S. Salim (2015). Influences of Parts of Tree on Propagation Path Losses for WSN Deployment In Greenhouse Environments. Journal of Theoretical and Applied Information Technology December 2015. Vol. 82. No.2
4. Meng, Y. S., Lee, Y. H., & Ng, B. C. (2010). Path loss modeling for near-ground vhf radio-wave propagation through forests with tree-canopy reflection effect. Progress In Electromagnetics Research M, 12, 131-141.
5. Sabri N, Aljunid SA, Ahmad RB, Malek MF, Yahya A, Kamaruddin R, Salim MS (2012). Performance Evaluation of Wireless Sensor Network Channel in Agricultural Application. American Journal of Applied Sciences. 9 (1), pp: 141-151, 2012.
6. Rappaport, Theodore S. Wireless communications: principles and practice. 2nd Eddition. New Jersey: Prentice Hall PTR, 2002.
7. AlSayyari, A., Kostanic, I., Otero, C., Almeer, M., & Rukieh, K. (2014, March). An empirical path loss model for wireless sensor network deployment in a sand terrain environment. In Internet of Things (WF-IoT), 2014 IEEE World Forum on (pp. 218-223). IEEE.
8. Zhou, G., He, T., Krishnamurthy, S., & Stankovic, J. A. (2004, June). Impact of radio irregularity on wireless sensor networks. In Proceedings of the 2nd international conference on Mobile systems, applications, and services (pp. 125-138). ACM.
9. Kirubanand, V. B., & Palaniammal, S. (2011). Study of performance analysis in wired and wireless network. American Journal of Applied Sciences, 8(8), 826.
10. Sklar, Bernard. Digital Communications: Fundamentals and Applications. 2nd Eddition. New Jersey: Prentice Hall PTR, 2001.
11. Hebel, M. A., Tate, R. F., & Watson, D. G. (2007). Results of wireless sensor network transceiver testing for agricultural applications. In 2007 ASAE Annual Meeting (p. 1). American Society of Agricultural and Biological Engineers.
12. Thelen, J., Goense, D., & Langendoen, K. (2005, April). Radio wave propagation in potato fields. In 1st Workshop on Wireless Network Measurements (Vol. 2, pp. 331-338).
13. Janek, J. F., & Evans, J. J. (2010). Predicting ground effects of omnidirectional antennas in wireless sensor networks. Wireless Sensor Network, 2(12), 879.
14. Auda Raheemah, Naseer Sabri, M.S. Salim, Phaklen Ehkan, R. Badlishah Ahmad (2016). New empirical path loss model for wireless sensor networks in mango greenhouses. Computers and Electronics in Agriculture 127 (2016) 553–560.

Conceptual Framework: Development of Interactive Reading Malay Language Learning System (I-ReaMaLLS)

Nurulisma Ismail[1,*], *Rasimah* Che Mohd Yusoff[2], and *Mohamad Umar* Jiman[1]

[1]School of Computer & Communication Engineering (SCCE), Universiti Malaysia Perlis (UniMAP), c/o School of Manufacturing Engineering Complex, First Floor, Pauh Putra Campus, 02600 Arau, Perlis, Malaysia
[2]Advanced Informatics School, Universiti Teknologi Malaysia, Jalan Sultan Yahya Petra, 54100 Kuala Lumpur, Malaysia

Abstract. Reading is very important to access knowledge. Reading skills starts during preschool level no matter of the types of languages. At present, there are many preschool children who are still unable to recognize letters or even words. This leads to the difficulties in reading. Therefore, there is a need of intervention in reading to overcome such problems. Thus, technologies were adapted in enhancing learning skills, especially in learning to read among the preschool children. Phonological is one of the factors to be considered to ensure a smooth of transition into reading. Phonological concept enables the first learner to easily learn reading such to learn reading Malay language. The medium of learning to read Malay language can be assisted via the supportive of multimedia technology to enhance the preschool children learning. Thus, an interactive system is proposed via a development of interactive reading Malay language learning system, which is called as I-ReaMaLLS. As a part of the development of I-ReaMaLLS, this paper focus on the development of conceptual framework in developing interactive reading Malay language learning system (I-ReaMaLLS). I-ReaMaLLS is voice based system that facilitates the preschool learner in learning reading Malay language. The conceptual framework of developing I-ReaMaLLS is conceptualized based on the initial study conducted via methods of literature review and observation with the preschool children, aged 5 – 6 years. As the result of the initial study, research objectives have been affirmed that finally contributes to the design of conceptual framework for the development of I-ReaMaLLS.

1 Introduction

Education is one of the biggest concerns of a government in developing human resource that can contribute to the nation. The education at preliminary level is a very important phase for children because via preschool education, children will be exposed to a wider learning environment compared to the children that are home-schooled. In the process of teaching and learning Malay language, there are three (3) skills that have to be mastered by the children which are reading, writing and oral skills. Reading is a key to all knowledge [1]. Therefore, it is very important for children to be able to read, especially in Malay language as the first language in Malaysia. At the preliminary education level in most of public preschools in Malaysia, Malay language plays an important role in delivery knowledge to the preschool children, followed by English language. However, despite the importance of reading skills, a lot of research shows that the first preschool learner who is new to learn reading found difficulties to learn reading via only recognizing the alphabet letters. The method to learn reading is easily for the preschool learner to learn reading Malay language is via recognizing the sound of the alphabet letters. This learning method is known as phonological [2], [3], [4], [5]. Learning to read involves the children's capabilities such as attention, memory, knowledge of alphabet letters and knows how to relate the word in a given context. However, during learning a learner tends to lost focus, feeling bored, and difficult to conduct tasks without assistance. Therefore, an interactive learning system can be developed since multimedia technology can be applied to improve reading skills of the preschool learner [6], [7], [8]. Thus, this research aims to propose the conceptual framework in developing interactive reading Malay language learning system (I-ReaMaLLS).

2 Related works

2.1 Phonic method

One of the methods that are commonly used in teaching preschool children to read is by using phonic method. Phonic method teaches the children to recognize letter with its associated sound and not the name of the letter. This method has been introduced in Brunei a few years ago to the preschool children. It had a significant difference between the old methods of teaching the children. The implementation of phonic method at preschool level in Brunei enabled the children learns to read faster than the traditional way which was teaching

* Corresponding author: nurul@unimap.edu.my

children by using syllable. This method enabled the children read newspaper without the help of teachers [9].

2.2 Multimedia System

In the millennium era, it is very important to turn the traditional way of learning into computer-assisted learning mechanism. The Theory of Cognitive states that learning will be easier to be conducted through recognition and observation of objects via a medium such as visual and sound, which is referring to the multimedia elements compared to the old traditional way that depends only on printed materials such as books. Multimedia is the combination of various media: text, graphics, animation, images and video. It is a new technology that benefits to the education field: to ease learning, to give motivation and to give a more effective explanation. Therefore, almost all levels of learning institution in Malaysia implements the interactive learning system that involved almost all the human sensory [10]: visual, audio, and tactile senses.

In order to develop a reading learning system for preschool children, the usages of multimedia have to be fully utilized. A developed system for the children should contain a story element, drawing, designing something, writing a story with the help of images, the understanding of a computer, to develop thinking and to enhance their vocabulary. In addition, the developed system enables the children learning on themselves or in a group. Due to this, a sufficient scaffolding of the developed system should be embedded as the assistive tools for the children's learning. For example, children who have difficulties in understanding a text relies on the articulation tool in a form of audio. Further, the concept of reflection is another significant element in developing an interactive learning system for the preschool children. Every single interaction of the children with the system should have feedbacks to enable the children understand on the learning activities [11].

3 Method

The development of I-ReaMaLLS complies with the initial study in order to understand the problem occurred in the learning reading process among the preschool children. The initial study has been conducted in two (2) methods: literature review and observation. Literature review has been conducted with reviews on few of journals articles that related to the method of teaching and learning reading for the preschool children. The observational method was conducted to identify the teaching and learning process in reading Malay language among the preschool children.

3.1 Review of Literature

A few reviews have been conducted in order to understand the method of teaching and learning reading for the preschool children. Reading process is a difficult phase but an important task for children [12]. This is because reading involves the children's capabilities: attention, memory, knowledge (of letters) and knows how to relate the word in a given context. The main problem for children to learn is they tend to lost focus, become bored, and they cannot begin a new task without assistance. Therefore, there is a requirement to develop an interactive learning system in order to provide a fun method of learning for the children. The following sub-sections highlight on the results of the conducted literature review based on few of journals articles.

3.1.1 Results of literature review

- *Phonic method*

Phonic method is used to assist the children recognize letter by matching the letter with its associated sound. For example, the letter "a" is pronounced with 'aa…', the letter "s" is pronounced with 'ss…', and the letter "z" is pronounced with 'zz…'. Once the children knew the relations between the letter and its pronunciation, they were taught to form a syllable such as 'ba', 'ta', 'cu', and 'ka'. The syllables then were combined to form a word. The word was not formed by spelling the letter, but via combining the syllable that they have learnt. For example, to read 'bata', the children directly combined the syllable 'ba' + 'ta' without spelling each of the letters.

- *Development of application system for children: functionality and usability*

Developing an application system for the children has to consider on two (2) factors: usability and functionality of the developed system. Functionality means the system must function for the specific requirements [13] meanwhile usability means the needs of the system to be friendly, that is easy to be used [14]. Based on the literature review conducted in [13-14], the factors of functionality and usability were very important in the development of an application system to assure that the developed system is well functioned to confirm the usability of the developed system meets the requirements of learning for the children. Lacking in functionality elements leads to the problems in usability of the developed system. For example: the children cannot interact very well with the system because some features of the system were not fully facilitated the requirements of learning environment for the children. Therefore, the issues of functionality and usability are very important elements to be considered in the development of I-ReaMaLLS.

- *Advantages of multimedia in the development of I-ReaMaLLS*

Based on the research conducted [11], claimed that multimedia can stimulate almost all human sensory. Thus, it is crucial to apply multimedia elements in the reading learning system for the preschool children since it is able to expand on the imagination, creativity, fantasy as well as the emotion of the children.

Further, the previous research conducted [13] relied on the observational of the emergent literacy among the preschool children. There were two (2) mechanism used: printed book and electronic book. The results showed that the reading accuracy rate of using electronic book was higher compared to the printed book.

Therefore, it concludes that the use of multimedia learning system can facilitate more effectively in children's reading learning compared to the traditional method: spelling syllable and printed book.

3.2 Observation

Observational method is the action or process of observing something or someone carefully in order to gain information. In this research, the observational method was applied in the initial study to understand the process of the preschool children in learning to read in Malay language.

3.2.1 Results of observation

Observational method was conducted in one of preschool in northern state of Malaysia, Perlis. The observation involved participation of one (1) preschool child aged 5 years who is capable to understand instructions, with the help of one (1) teacher. The preschool child read fluently a printed story book that was written in Malay language. During the reading process, it has been noticed that the child applied the phonic method. She finished reading the story book within a short period.

Parallel to the observation process, the teacher claimed that phonic method is more efficient in teaching the children to read. The children can easily recognize the letters together with the sound of each letters. Once the children clearly understand the pronunciation of each letters, they can easily pronounce the combination letters that finally make up a complete word then a sentence.

In addition, based on the teacher's feedback regarding to learning to read can be implemented in both languages: Malay and English language. There are requirements for two-way interactive of multimedia learning system to motivate the preschool children to learn reading in both languages. At the moment, most of the preschool in northern state of Malaysia used the one-way interactive of multimedia learning system that based on the synchronous method of learning. Sometimes, the children feel bored when the playing multimedia learning system took a long time to finish the syllabus of reading learning.

Therefore, it concludes that the use of phonic method is easy to help the children to learn reading. The reading learning system should integrate the multimedia elements to motivate the children in learning to read. This research focus on the development process of reading learning system in Malay language that applies the phonic method with the integration of voice-based medium. The learning method used would be the asynchronous learning method with approach of the two-way interaction medium: interaction between the preschool children and the reading learning system. This method will be applied in order to achieve the objective of the research to develop an assistive tool to help the preschool children in learning to read in Malay language.

4 Proposed Conceptual Framework for Development of I-ReaMaLLS

As the problem statement has been identified through the literature review and observation conducted, the following research objectives has been formed: (i) to develop a voice based application system as an assistive tool that facilitates the preschool children in learning to read in Malay language (ii) to perform the usability testing of the prototype of I-ReaMaLLS.

Based on the stated objectives, this research will be conducted as a qualitative research. It is to develop a reading Malay language learning system for the preschool learners aged 5 – 6 years. It can be an assistive tool for the preschool children learning reading in Malay language.

The development of I-ReaMaLLS highlights on the usability principles. The research methods considers the research methods in human-computer interaction field, where as the system development focus on software engineering methods that relies on the process of iteration-evolution life cycle based on user-oriented approach. In order to develop I-ReaMaLLS, also focus on the needs of the preschool children in learning to read Malay language via voice-based medium that considered on the Theory of Cognitive [15]. This theory considered a few cognition processes of the preschool children during learning process via voice-based medium of interaction with I-ReaMaLLS. The cognition processes that are taken into consideration for this research include: attention; perception and recognition; memory; and learning. This is according to the usage of voice-based medium of interaction that involved the interaction process between the preschool children with I-ReaMaLLS. The interaction process relied on the voice command spoken by I-ReaMaLLS to the children during learning process to read Malay language. The children give feedback to the I-ReaMaLLS via using the tactile keyboard.

The research involves participatory of the preschool children aged 5 – 6 years who are capable to understand instructions. The participatory of users contributes in two (2) parts of the development process for I-ReaMaLLS. There are:

i. First Part – the initial stage of the research in determining the requirements needed for the development of I-ReaMaLLS.
ii. Second Part – the final stage of the research, which is the testing phase of I-ReaMaLLS. This is to determine the functionality and usability of the system fulfilled the children's requirement in learning to read Malay language. The testing are categorized into two

(2) stages: formative and summative evaluation.

The formative evaluation involves the experts in design and development phase in three (3) iteration cycles. At this stage, the user interfaces and interaction activities of I-ReaMaLLS are based on the cognitive processes of the preschool children to easily and efficiently learning to read Malay language.

The whole development process of I-ReaMaLLS has been conceptualized via the conceptual framework as illustrated in Figure 1. The framework is divided into three (3) phases for the research to be conducted systematically. (i) Phase 1 System Requirements and Specifications, (ii) Phase 2 Development of Interactive Reading Malay Language Learning System (I-ReaMaLLS), and (iii) Phase 3 Evaluation of Interactive Reading Malay Language Learning System (I-ReaMaLLS).

Phase 1 involves the initial study to identify the research problem and requirement specification of the I-ReaMaLLS. This process consists of two (2) parts: (i) Preliminary Study 1, and (ii) Preliminary Study 2. The Preliminary Study 1 involves the process of literature study in order to understand concepts, theories, and principles that related and supported to the development of I-ReaMaLLS for the children to easily and efficiently learning to read Malay language.

Phase 2 involves the process of development for I-ReaMaLLS. The process consists of the process of analysis the requirements and specifications of the realated technology that to be applied in the development of I-ReaMaLLS. User-oriented approach is applied in order to identify the needs of the preschool children in learning to read Malay language. The development of I-ReaMaLLS is based on the software engineering methodology that relies on the process of iteration-evolution. In addition, a development framework is illustrated by a model named as the Iterative-Evolution System Development Life Cycle for Interactive Reading Malay Language Learning System (IE-SDLC-I-ReaMaLLS). This model is based on the three (3) cycles of iteration process of developing prototype: the Throw-away Prototype Model, Prototype Model and Evolutionary Prototype Model. These models are supported by the formative evaluation with the expert. The formative evaluation is based on the functionality testing of I-ReaMaLLS.

Phase 3 is the evaluation phase for the I-ReaMaLLS. It involves the summative evaluation which relies on the usability testing. The usability testing is based on four (4) constructs: Learnability, Efficiency, Usability Error, and satisfaction [16]. The summative evaluation is categorized into two (2) stages: Pilot Study and Field Study. It involves with the end user of I-ReaMaLLS: the preshool children aged 5 - 6 years.

5 Conclusions

This research objective is to develop and evaluate the I-ReaMaLLS as an assistive tool for the preschool children to learn reading Malay language. The development of I-ReaMaLLS is based on user-oriented approach. In order to achieve the research objectives, the development of I-ReaMaLLS is illustrated via Figure 1, the conceptual framework that based on a model: the Iterative-Evolution System Development Life Cycle of I-ReaMaLLS (IE-SDLC-ReaMaLLS. The evaluation of I-ReaMaLLS has two (2) categories: Formative Evaluation (Expert) on the functionality testing of I-ReaMaLLS, and Summative Evaluation on the usability testing (Pilot and Field Study with the preschool children). The usability testing relies on four (4) constructs: Learnability, Efficiency, Usability Error and Satisfaction. Based on the illustrated conceptual framework can lead to a clear of development process for I-ReaMaLLS as the assistive tool for the preschool learner to learn reading Malay language.

References

[1] Z. Tasir and C. Likun, "Perisian Multimedia bagi Pembelajaran Mengenal Huruf dan Nombor untuk Pelajar Prasekolah", pp. 1-18, 2005.

[2] A. Castles and M. Coltheart, "Is there a causal link from phonological awareness to success in learning to read?," *Cognition*, vol. 91, no. 1, pp. 77–111, Feb. 2004.

[3] C. M. Fletcher-Flinn and G. B. Thompson, "Learning to read with underdeveloped phonemic awareness but lexicalized phonological recoding: A case study of a 3-year-old," *Cognition*, vol. 74, no. 2, pp. 177–208, 2000.

[4] M. Melby-Lervåg, S.-A. H. Lyster, and C. Hulme, "Phonological skills and their role in learning to read: a meta-analytic review.," 2012.

[5] C. Hulme, C. Bowyer-Crane, J. M. Carroll, F. J. Duff, and M. J. Snowling, "The causal role of phoneme awareness and letter-sound knowledge in learning to read: Combining intervention studies with mediation analyses," *Psychol. Sci.*, vol. 23, no. 6, pp. 572–577, 2012.

[6] M. Neo, T.-K. Neo, and X.-L. Tai, "A constructivist approach to learning an interactive multimedia course: Malaysian students' perspectives," *Australas. J. Educ. Technol.*, vol. 23, no. 4, 2007.

[7] N. A. Mukti and S. P. Hwa, "Malaysian perspective: Designing interactive multimedia learning environment for moral values education," *J. Educ. Technol. Soc.*, vol. 7, no. 4, 2004.

[8] S. Nusir, I. Alsmadi, M. Al-Kabi, and F. Sharadgah, "Studying the impact of using multimedia interactive programs at children ability to learn basic math skills," *Acta Didact. Napoc.*, vol. 5, no. 2, p. 17, 2012.

[9] Y. Othman, A. M. Daud, A. Othman, D. S. A. P. Mohiddin, and M. M. Sulaiman, "Pelaksanaan pengajaran membaca menggunakan kaedah fonik peringkat prasekolah di Brunei Darussalam", pp. 93-109.

[10] A. Esa, B. Mohamad, and S. N. Mukhtar, "Peranan Multimedia di dalam Pembelajaran Kanak-kanak", November, pp. 18-20, 2007.

[11] Munir and Halimah Badioze Zaman. "Aplikasi Multimedia dalam Pendidikan". Jurnal Bahagian Teknologi Pendidikan. Bil 1. Kementerian Pendidikan, Kuala Lumpur, Malaysia. 1999.

[12] D. M. Popovici, C. Buche, R. Querrec, and F. Harrouet, "An Interactive Agent-based Learning Environment for Children", Cyberworlds, 2004 Int. Conf., pp. 233-240, 2004.

[13] Y. Huang, T. Liang, Y. Su, and N. Chen, "e-Book Learning System for Elementary School Students", no. 60, pp. 703-722, 2012.

[14] E. H. Sibley, "Functionality and Usability".

[15] Sharp, H., Rogers, Y. & Preece, J. Interaction Design: Beyond Human Computer Interaction. John Wiley & Sons Ltd. 2006.

[16] Nielsen, J. Usability Engineering. Morgan Kaufmann. 1993.

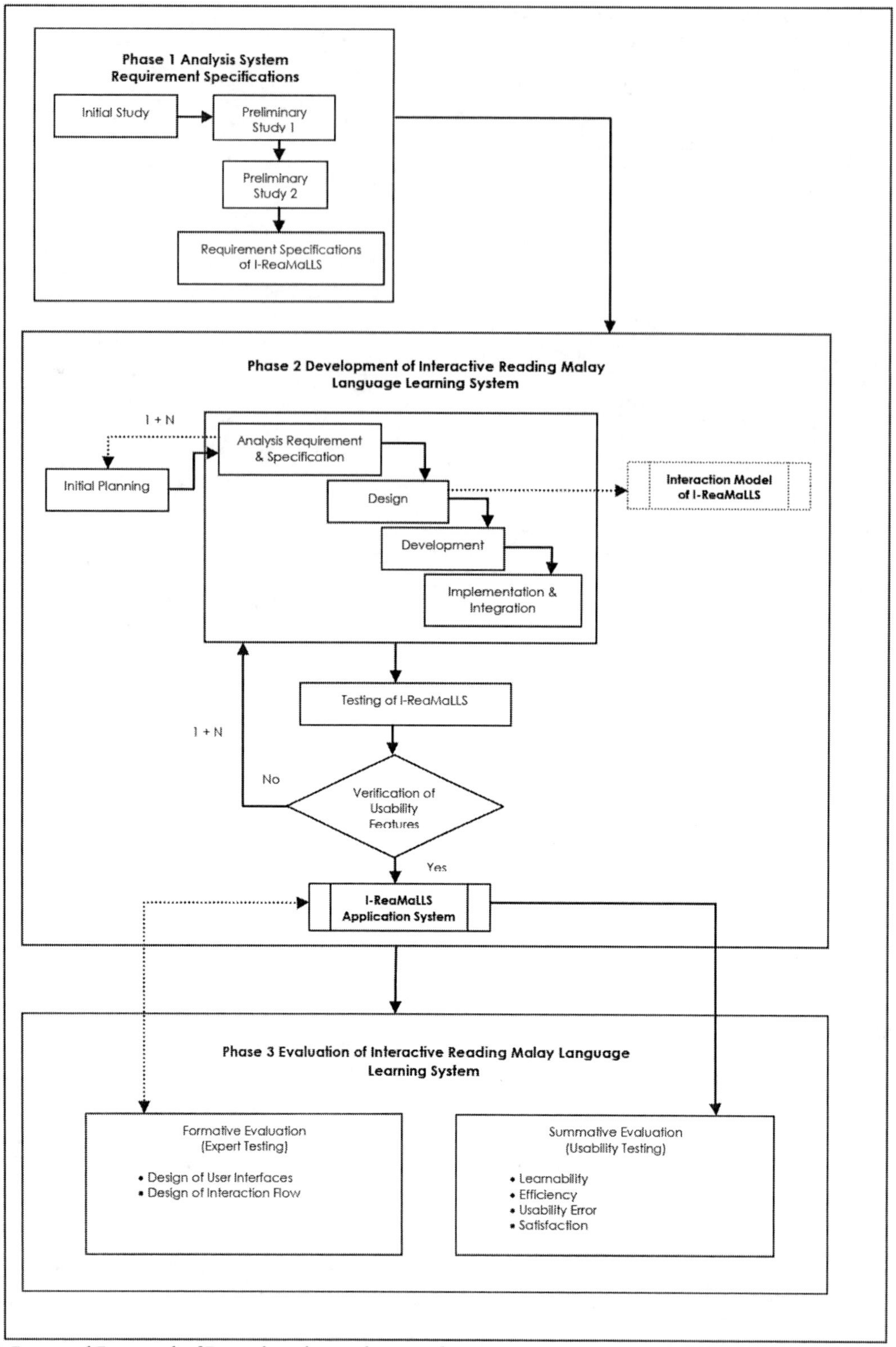

Figure.1 Conceptual Framework of Research on the Development of I-ReaMaLLS

* Corresponding author: nurul@unimap.edu.my

Evaluation of pruning waste of *Mangifera indica* var Harumanis cultivated in greenhouse

Rosfatihah Roslim[1], *Ahmad* Azudin Nordin[1], *Mohd* Hishamudin Che Mat[1], *Ammar* Zakaria[1], and *Mahmad* Nor Jaafar[1,*]

[1]Insititute Sustainable of Agrotechnology (INSAT), Universiti Malaysia Perlis, Malaysia

Abstract. *Mangifera indica* var Harumanis is one of the exotic mango cultivated in Perlis, Malaysia. It economic importance increased in recent years due to the popularity attributed to it excellent taste, sweet and aromatic fragrance. Supply of Harumanis cannot meet market demand, mainly due to the difficulty of raising the crops and the short and uncertain harvesting seasons. The cultivation of Harumanis in greenhouse was found to be feasible to mitigate the short coming. However, the cultural practices of high density planting in greenhouse calls for regular pruning to meet the desired canopy. Regular strategic pruning produced pruning waste with potential economic value. This research was done to quantify the by-products produced by pruning from three ½ acre greenhouses, under the ecological and climatic of Perlis. The fresh biomass of waste collected during primary, secondary and tertiary pruning were determined. The pruning waste produced was 0.676kg per plant equivalent to 146.4kg per greenhouse during primary and secondary pruning. Estimation of fresh biomass waste produced during floral induction determine by regression was 816 kg per greenhouse with the expected total of 40.8 ton from 50 greenhouses. In view of the fact that the mango plant management practice was from the clean agricultural, the waste can be potentially utilized for green manufacturing. This was the first attempted to quantify pruning waste in mango cultivation before further utilization of pruning waste.

1 Introduction

Mango in general is a region based fruit origin, located only in the tropical and subtropical region [1]. It is a woody plant that can have lifespan of more than 100 years. There were more than a thousand mango varieties existed, mostly located in the Pacific region, namely India and Southeast Asian and substantial amount from Central America including Florida. Generally, mango is cultivated in the traditional open field production system. A few selected mango varieties, for instant Irwin, Alphonso, Tommy atkins, Haden, Kensington, Fiji, Keitt and Fairchild were cultivated in greenhouse to ensure quality and to meet market demand. Annually, mango production occupy the fifth position among tropical fruits , it amount to 23.4 million MT globally [2].

Mango cultivation in greenhouse was pioneered by Japanese during 1980s, in Okinawa [3]. Since then, Japan had 200ha greenhouse mango cultivation, followed by Canary Island, Israel and Egypt [4,5]. Currently, there are four types of successful mango cultivated in greenhouse in Japan, for instant Irwin, Sensation, Keitt and Haden. The mango production from greenhouse had increased annually. The reasons were greenhouse production system (GHPS) provide protection from undesired weather condition, insects and disease. The technique to promote pollination either by honeybee or regular fruit flies was sufficient for yield target. Other routine mango cultural practices include pruning, flower peduncle hanging, flower peduncle thinning, fruit thinning and hanging, netting and reflector setting, and harvesting. GHPS has successfully produced high quality mango that promote seasonal gift during summer and festival seasons.

The mango phenological development of plant growth consists of vegetative shoot development, reproductive (flowering) shoot development, and fruit growth and development [6]. It is acknowledged that each branch required at least two times of full cycles of vegetative shoot development (these episodes are called flushes, before the branch is ready for reproductive shoot development. Ramirez proposed that neither vegetative, generative nor fruit growth stages of mango are similar throughout mango varieties [6]. Each full cycle consists of several stages either

* Corresponding author: mahmad@unimap.edu.my

leaf maturity, flower development or fruit growth stages. The vegetative shoot development in believed to be influenced by unknown substance, called vegetative promoter, while for reproductive stage it is called florigenic promoter. There were 8 vegetative stages existed, which are resting, vegetative bud break, vegetative bud elongation, early leaf elongation, elongating green leaf, limp red leaf, immature green leaf, and mature green leaf stages. This cycle of vegetative growth can be repeated at least twice per branch, with 8 weeks for complete stages development.

In high density planting of mango tree, pruning strategies throughout vegetative growth cycles become essential to ensure a well balance canopy with high productive terminal shoots. Conventionally, pruning in mango tree do not have specific rule. It is design based on current purpose of farmer, for example pruning in young mango tree is to shape initial canopy structure of the tree [7]. Davenport classified pruning into three types for commercialize mango tree practice, which are tip, formation and severe pruning [8]. Tip pruning is pruning at the terminal stem to promote new branching in mango tree. The nature of new branches very depended on position of pruning location at the terminal stem. The purposes of tip pruning in mango are to stimulate branching flushes into dense canopy structure, to promote synchronizing flush of vegetative growth of several new branches, and to restore productivity of pruned branches. Over the time, branches in mango tree canopy become overcrowded and unproductive. Thus formation pruning is necessary to restore the original number of stem at lower canopy. Once the tree canopy is beyond reach and unmanageable, severe pruning help to rejuvenate these large tree into smaller size with reasonable height and canopy management.

With several cycles of pruning process, enormous pruning waste will be generated. For instant, in Madrid alone pruning waste from 250000 trees were found to be more than 20,000 ton annually [9,10]. Disposal of pruning waste is an issue in modern farming and merit special attention. Various research for value added product and further application have been propose to utilize pruning waste, for example pruning waste as source of biomass for bio-fuel, compost and fertilizer, application in bio-sorption of toxic heavy metals such as lead (II), cadmium (II) and copper(II), and also as value added in animal feedstuff [10,11].

Since pruning waste of fruit farming mostly consisted of leaves, branches, and young shoots, thus high in fiber, active compound ingredients, and nutrients that are high in value should be capitalized. In pruning of young mango tree the main waste in pruning process are leaf at different maturity stage, young shoot, terminal shoot with early bud, and branches. Each of the wasted plant part has different potential value added revenue that can be generated into by-product as income for the mango farming community.

The aims of this research were to evaluate quantitatively the pruning waste at various stage of strategic pruning for Harumanis mango grown in the greenhouse and to estimate pruning waste that will be generated during pruning for floral induction.

2 Methodology

2.1 Study site

Evaluations were conducted at *Harumanis(HM)* commercial trials farm in the Institute Sustainable of Agrotechnology (InSAT), Universiti Malaysia Perlis (6°39'12.54"N 100°15'51.02"E) during January to December 2016. Three greenhouses planted with high density HM mango crop, were selected (out of total 50 mango cultivation greenhouses) for evaluation. The half-acre (GH), made of galvanized structure, with dimensions of 84x24.2m and 5m gutter height, and designed to withstand 80km/h wind load. The sides are covered with nylon net of 32 meshes; the roof is of 200micron plastic sheet with ultra-violet blocker. It is equipped with automatic micro-irrigation system designed to apply water as required by individual plants. The greenhouse is a tropical natural ventilated modular type, produce by CMF of France. Each GH can accommodate 216 HM plants and were planted in October to December of 2015, with the standard agronomic practice applied to ensure fertility and uniformity of the plants.

2.2 Pruning Waste Determination.

In high density HM mango planting system, strategic pruning is mandatory in order to achieve adequate number productive tertiary branches and later quaternary branches, while controlling the symmetric of canopy. Strategic pruning is a series of pruning at designated position at specific time and done simultaneously for all HM plants in the same greenhouse during the same day. For our in house strategic pruning routine, there were three main reasons involved, which were promoting productive branch at intended position, synchronizing plant/branch/vegetative growth and controlling size of canopy per plant. For example, primary pruning waste was by-product of pruning at main branch to promote primary branch. While secondary pruning waste was pruning at primary branch to promote secondary branch. For complete HM plant vegetative growth, minimum three stages of pruning routine is required, in order to produce tertiary branch. Pruning waste per pruning routine can be determined and quantified. Primary pruning waste is waste of pruning or pruning by-product collected during primary pruning, whilst secondary pruning waste at secondary pruning and tertiary pruning waste is at tertiary pruning routine, as described in Table 1.

Table 1. Illustration of the strategic pruning routine and the timing and waste collected for each respective strategic pruning routine.

Strategic Pruning	Primary	Secondary	Tertiary
Pruning Waste Evaluation	Collection of by-product of primary pruning	Collection of by-product of secondary pruning	Collection of by-product of tertiary pruning
Position of pruning	Pruning at main branch	Pruning at primary branch	Pruning at secondary branch
Illustration	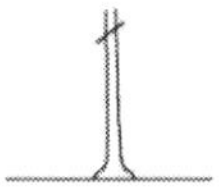		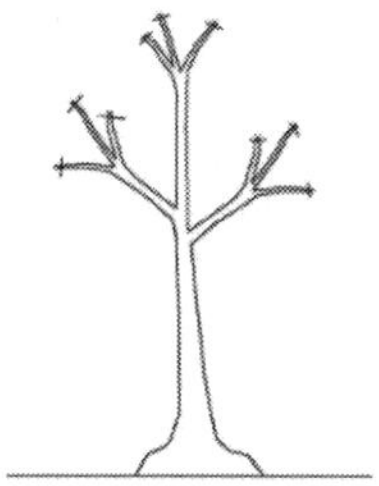
Height	45-80cm	80-120cm	100-150cm
Duration	3-4 months after planting	2-3 months after primary pruning	2-3 months after secondary pruning

All pruning waste were collected right after pruning session and quantified. There were two approached to evaluate the quantification of pruning waste of respective strategic pruning routine. At first, pruning waste of individual waste of HM mango plant were quantified from five replications per greenhouses. It was randomly sampled and applied for all three stages of strategic pruning routine. Secondly, all pruning waste collected once in the greenhouse were sampled. However, since pruning routine employed will include the end of branch, thus pruning waste was classified as consisted of young shoot, the different stages of HM leaf and the branches. Each vegetative growth stages of plant part were determined as BCCH scales [6]. Then biomass of each clustered of fresh plant part was determined on site by mass balance.

2.3 Biomass Determination

Biomass of pruning waste was determined at the same day of pruning waste collection [12]. Fresh sample of pruning waste were placed in oven for overnight at 70°C. Daily measurement of dried weight was done until the weight of dried samples was constant. The procedures were done in triplicate for each of greenhouse. It was done for all primary, secondary and tertiary pruning waste collection respectively.

2.4 Regression Analysis

The pruning waste collections were carried out from strategic pruning during vegetative growth of HM mango plant in the greenhouse. It involved three stages, which were primary, secondary and tertiary stages of pruning waste collection respectively. While, quaternary pruning waste referred as pruning waste from strategic pruning routine for floral induction, which will be practiced annually, after harvest. The estimation of quaternary pruning waste was determined by extrapolated biomass of fresh weight of pruning waste collected respectively. The plot was developed by Microsoft excel, with three possible regression namely linear, quadratic and exponential regression equation. All waste was determined by average of triplicate and with standard deviation. Variable of the graph was determined by r^2 value.

3 Result & Discussion

Complete cycle of strategic pruning for HM plant vegetative growth required at least three pruning routine, thus produced primary, secondary and tertiary pruning wastes. The fresh biomass collections of respective pruning waste of different stages were quantified by sampling from plants replicates and greenhouse (GH) as shown as Table 2. While for Table 3, showed all pruning waste were clustered by stages of vegetative shoot stages [6]. HM Early Bud was young branch that possessed vegetative bud break (VBB) that has the potential use as scion wood for vegetative propagation. HM young shoot was young branch that consisted of vegetative bud at elongation stage (VBE), early leaf elongation stage (ELE), elongating green leaf stage (EGL) and limp red leaf stages (LRL). HM green leaf consisted of only immature green and matured dark green leaves. HM Young Branch was young green branch that was leftover from the remaining waste.

Table 2. The amount of fresh pruning waste (kg) from respective strategic pruning routine.

Strategic Pruning Routine	kg Fresh Biomass/plant *(n=15)*	Estimated Biomass per GH		
		kg Fresh Biomass/GH	kg Dry matter/GH	STDev
Primary	0.073±0.003	15.55	5.57	0.19
Secondary	0.200±0.004	44.64	15.9	0.52
Tertiary	0.403±0.008	86.17	30.8	0.61
Total	**0.676**	**146.36**	**52.27**	

Table 3. The distribution of different parts of pruning waste (kg) from three Greenhouses.

Pruning Waste	HM Early Bud (number)	kg Fresh Biomass/GH			
		HM Young Shoot	HM Green Leaf	HM Young Branch	Total
Primary	29±1.9	2.06±0.18	8.57±0.21	4.93±0.22	15.55
Secondary	108±2.9	3.79±0.26	28.75±1.20	12.11±0.10	44.64
Tertiary	167±3.7	10.54±0.35	52.91±0.87	22.73±0.62	86.17
Total (kg)	**304**				**146.36**

It was observed, pruning waste generated at the end of a branch produced less than one feet of young branch with different vegetative growth stages. Generally, the primary waste consisted of only one young branch waste per plant. Whilst secondary waste produced three less than one feet young branch waste. While third waste produced nine young branch waste. The stages of young branch of waste collected very dependent on plant growth. Total fresh biomass of pruning waste per plant quantified was generally less than one kilogram of waste. However, total waste collected per greenhouse amount to 146kg, which is a very significant amount of waste. Base on the regression correlation the waste produced per GH, can be for casted to be consisted for of approximately less than one feet of 216 young branches from primary waste, while secondary consisted of 648 young branches waste, and 1944 young branches from tertiary wastes. The waste biomass consisted of 15.55kg (10%) from primary waste, 44.64kg (30.5%) from secondary waste, and 86.17kg (58.9%) from tertiary waste.

The estimated waste generated from 50 GH was expected be 7.3 tons; which consisted of varied stages of vegetative shoot. From the clustered sampling, biomass of the major waste produced from HM green leaf was 90.23kg (61.65%) per GH from complete cycle of strategic pruning routine. Whilst, HM young branch produced was 39.77kg (27.2%) and HM young shoot produced was16.39kg (11.2%). While, the number of HM early bud waste produced was 29 (13.4%), 108 (16.7%), and 167 (8.59%) respectively from the three stages of pruning.

Despite of the large amount of waste produced during vegetative growth of HM mango plant, it happened only at the very first year of cultivation. However, strategic quaternary pruning will be carried out to design the total number of productive shoot prior to floral induction to promote uniform flowering or during post harvest period. The yearly quaternary pruning will produce quaternary pruning waste. Several regression equations were developed in order to predict yearly quaternary pruning waste, namely linear, quadratic and exponential equations, as shown in Table 4 and Table 5. The value of biomass obtained was determine base on regression equation with time of pruning in weeks (w) was after planting the x-axis. FBs indicated fresh biomass produced from pruning routine of respective plants and GH, as presented in Table 4. While FB indicated fresh biomass produced from pruning routine of clustered waste respectively, as shown in Table 5.

Table 4. Regression equations to predict fresh biomass (FB) obtained from pruning waste of respective strategic pruning routine weeks after planting.

Regression Analysis		R^2
per plant	$FB = 0.011x - 0.063$	0.987
	$FB = 0.001w^2 + 0.003w + 0.015$	0.999
	$FB = 0.04e^{0.058w}$	0.984
per GH	$FB = 2.439w - 13$	0.993
	$FB = 0.023w^2 + 1.22w - 0.749$	0.999
	$FB = 8.805e^{0.058w}$	0.976

Table 5. Regression equation to predict fresh biomass of clustered plant part from pruning waste of strategic pruning routine.

Regression Analysis		R^2
HM Young shoot	$FB = 0.294w - 1.995$	0.907
	$FB = 0.011w^2 - 0.281w + 3.79$	0.999
	$FB = 1.038e^{0.056w}$	0.984
HM Green Leaf	$FB = 1.529w - 8.681$	0.999
	$FB = 0.005w^2 + 1.23w - 5.676$	0.999
	$FB = 4.83e^{0.062w}$	0.957
HM Young Branch	$FB = 0.614w - 2.324$	0.991
	$FB = 0.006w^2 + 0.27w + 1.136$	0.999
	$FB = 2.92e^{0.052w}$	0.985
HM Early bud	$FB = 4.748w - 18.97$	0.989
	$FB = -0.058w^2 + 7.765w - 49.28$	0.999
	$FB = 17.62e^{0.06w}$	0.911

Based on regression equation, projected quaternary pruning if performed at 107 weeks after planting, approximately at end of 2017, then the pruning waste generated was expected to be 1.1kg per plant, 0.4kg per plant, and 19.8kg per plant estimate linear, quadratic, exponential equations, respectively. Similarly, pruning waste of GH generated will be 248kg per GH, 393kg and 43.65tons per GH. Even thought the regression equations are statically, rational, however, the growth of young age of HM mango plant would high likely follow exponential equation, later slow down to the expected sigmoid curve. The projected quaternary pruning waste was best estimated by assuming the average value of pruning waste generated based on all three regression equations. Thus quaternary waste can be expected to be about 7kg per plant, while equivalent estimated waste per GH was 1.7 tons. In addition, the average of estimated clustered fresh biomass waste produced consisted of 190kg of HM young shoot, 1337kg of HM green leaf, 308kg of HM young branch and 3808 count of HM early bud. It can be expected that total waste of fresh biomass from 50 greenhouses, can be predicted to be in the region of approximately 85 tons per year. This is significant large of amount of waste produced compared to waste of other commercial fruit production from open farming. The waste from HM farming by GHPS amount to 7kg/tree of fresh biomass compared to other commercial fruit farming, such as 2.34kg/tree of apple, 2.45kg/tree of pear and 7.23kg/tree of peach [11].

In order to produce high quality of HM mango all year round, the waste produced is inevitable. Nevertheless, it can be utilized and formed into new discovery. For example, in the case of mango leaf, there were extensive studies of it benefits and practices. Mango leaf can be extracted and had varied of medicinal applications [13]. It can also be used as supplement of animal feedstuff, compost, one of material in silver-nanoparticle synthesis, and bio-sorbent of toxic elements [9,14–16].

4 Conclusion

Although there were many studies of biomass quantification of pruning waste, this was the first attempted to evaluate quantitatively the by-product of pruning of HM mango cultivated in greenhouse. To ensure high productivity and quality, pruning is mandatory, and the waste produce can be scheduled for processing into value added products. The quantification of the value of pruning waste will help in the design of green manufacturing to increase the efficiency of product utilization. The mango leaves as pruning by-product of mango cultivation, are known to have lots of benefits, it can be exploited and thus expected to contribute to the income of HM mango cultivation in greenhouse, and concomitantly, benefited by mango community as whole.

References

1. Bally, I. S. E. *Mangifera indica (mango). Species Profiles for Pacific Island Agroforestry* (2006).
2. Tharanathan, R. N., Yashoda, H. M. & Prabha, T. N. Mango (Mangifera indica L .), ' The King of Fruits '— An Overview. *Food Rev. Int.* **22,** 95–123 (2007).
3. Akinaga, T. & Hasbullah, R. Mango Production Using Plastic Greenhouse in Okinawa. *Acta Hortic.* **575,** 745–749 (2002).
4. Saúco, V. G. Greenhouse Cultivation of Tropical Fruits. *Acta Hortic.* **575,** 727–735 (2002).
5. Medany, M. ., Abdrabbo, M. A. ., Farag, A. ., Hassanien, M. . & Abou-Hadid, A. . Growth and Productivity of Mango Grown under Greenhouse Conditions. *Egypt J. Hortic.* **36,** 373–382 (2009).
6. Ramírez, F. *et al.* Mango trees have no distinct phenology : The case of mangoes in the tropics. *Sci. Hortic. (Amsterdam).* **168,** 258–266 (2014).
7. Poffley, M. & Owens, G. *Mango Pruning in the Top End.* (2006).
8. Davenport, T. L. Pruning Strategies to Maximize Tropical Mango Production From the Time of Planting to Restoration of Old Orchards. *HortScience* **41,** 544–548 (2006).
9. Benito, M., Masaguer, A., Antonio, R. De & Moliner, A. Use of pruning waste compost as a component in soilless growing media. *Bioresour. Technol.* **96,** 597–603 (2005).
10. Benito, M., Masaguer, A., Moliner, A. & Antonio, R. De. Chemical and physical properties of pruning waste compost and their seasonal variability. *Bioresour. Technol.* (2005). doi:10.1016/j.biortech.2005.09.011
11. Bilandzija, N., Voca, N., Kricka, T., Matin, A. & Jurisic, V. Energy potential of fruit tree pruned biomass in Croatia. Spanish J. Agric. Res. 10, 292–298 (2012).
12. Velázquez-Martí, B., Fernández-González, E., López-Cortés, I. & Salazar-Hernández, D. M. Quantification of the residual biomass obtained from pruning of vineyards in Mediterranean area. *Biomass and Bioenergy* **35,** 3453–3464 (2011).
13. Rymbai, H., Srivastav, M., Sharma, R. R., Patel, C. R. & Singh, A. K. Bio-active compounds in mango (Mangifera indica L.) and their roles in human health and plant defence – a review. *J. Hortic. Sci. Biotechnol.* **88,** 369–379 (2016).
14. Rodríguez, J. *et al.* Effects of a natural extract from Mangifera indica L , and its active compound , mangiferin , on energy state and lipid peroxidation of red blood cells. *Biochim. Biophys. Acta* **1760,** 1333–1342 (2006).
15. Philip, D. Mangifera Indica leaf-assisted biosynthesis of well-dispersed silver nanoparticles. *Spectrochim. Acta Part A Mol. Biomol. Spectrosc.* **78,** 327–331 (2011).
16. Saha, R. & Saha, B. Removal of hexavalent chromium from contaminated water by adsorption using mango leaves (Mangifera indica). *Desalin. Water Treat.* **52,** 1–9 (2014).

Comparison of glucose yield from rubberwood sawdust (RSD), growth medium (GM), and mushroom spent medium (MSM) under different sodium hydroxide pretreatment techniques

Siti Aminah Mohd Hassan[1,*], *Muhammad Syarhabil* Ahmad[2], *Amira Farzana* Samat[1], *Nur Zatul 'Iffah* Zakaria[1], *Khairunissa Syairah* Ahmad Sohaimi[1], *Nurfatirah* Nordin[1]

[1]Department of Chemical Engineering Technology, Universiti of Malaysia Perlis, 02100 Padang Besar, Perlis, Malaysia.
[2]School of Bioprocess Engineering, Universiti of Malaysia Perlis, 02600 Arau, Perlis, Malaysia

Abstract. Large quantities of oyster mushroom (*Pleurotus sajor caju.*)spent medium (MSM) were disposed of as agricultural waste to the environment and thereby constituting environmental pollution. Therefore the study was conducted to examine the feasibility of using MSM as glucose source in comparison to RSD and GM under different NaOH pretreatment techniques and at the same enzymatic saccharification condition. The composition analysis proved that decreases in lignocellulosic contents occurred after cultivation of The amount of lignin, hemicelluloses, and cellulose in MSM showed lower values than those in RSD and GM which were 22.40, 27.93, and 27.97% respectively. The surface morphology of MSM appeared to be rough and broken and traces of hyphen on the fibre surface were observed as a result from the mushroom cultivation process. RSD and FM had an even and smooth flat surface, indicating a rigid and highly ordered surface structure. The highest glucose yield were obtained from MSM under all three NaOH pretreatment techniques (autoclaving at 121°C, heating in water bath, soaking at room temperature), that were 31.07, 30.13, and 21.97, g/ dry substrate respectively. The amount of glucose obtained from RSD and GM under the same treatments were comparable.

1 Introduction

Glucose from sugar crops (sugarcane, sugar beets) and starchy food (potato, cassava, corn) can be converted into ethanol via fermentation process with the assist of microbes or enzymes. However, concerns about its production and use related to the increased food prices due to the large amount of arable land required for crops, as well as the energy and pollution balance of the whole cycle of ethanol production has caused a new source of glucose being introduced such as lignocellulosic biomass that may allay these concerns. Concomitantly, over the last decades, research efforts have been devoted to converting the promising feedstock of lignocellulosic biomass into biofuel, especially residues from agricultural and forestry operations due to its great availability, sustainability, and low cost compared to other energy feedstock [1].

The rubber tree or scientific name "*Hevea brasiliensis*" is one of the major agricultural crops grown in Malaysia besides oil palm, cocoa, rice, and coconut. The trees are logged off after 25 to 30 years and utilized mainly for making furnitures. As a result of logging and lumber processes, large amount of residual biomass were generated, which have no significant value except its usage in the making of products like briquetted fuel and compressed powder boards [2]. However, due to its high cellulose content, rubber wood waste represents a potential raw material for bioethanol production [3]. Apart from its potential as bioethanol feedstock, rubberwood sawdust maybe used to cultivate oyster mushroom and then, further used the remaining medium wastes to produce glucose for bioethanol production. Extra income could be generated and necessary pretreatment steps could be reduced prior to the enzymatic hydrolysis process with high reducing sugars yield via the aforementioned processes. According to previous study, lignocellulosic biomass, which has been biologically treated by fungi have been proven to increase the sugar produced during enzymatic saccharafication and lower lignin contents was detected as a result from the fully/partially digested lignocellulosic materials. Wan and Li has reported that, with the result of selective delignification, enzymatic digestibility as high as 60-80% has been obtained from fungal-pretreated corn stover, which was comparable to that obtained from chemical pretreatment [4]. Unlike biologically pretreated biomass, the raw lignocellulosic biomass could not be saccharified by enzymes for high glucose yield without first undergoing several pretreatment including, physical, mechanical, and chemical pretreatment. The main cause for the recalcitrance of lignocellulosic biomass is the presence of lignin and hemicelluloses on the surface of cellulose. They formed a barrier and prevented cellulase from accessing the cellulose in the substrate [5]. The

* Corresponding author: aminahhassanr@unimap.edu.my

https://doi.org/10.1051/matecconf/201815006023

pretreatment of lignocellulosics was primarily employed to increase the accessible surface area of cellulose to enhance the conversion of cellulose to glucose in enzymatic saccharification.

Pretreatment of lignocellulosic materials was considered as the rate-limiting step in an economically feasible process for enzymatic hydrolysis of cellulose. Combination of biological and chemical pretreatment was expected to increase the cellulose amount, which could be feasibly accessed by cellulase in subsequent enzymatic saccharification, after removing the lignin content. Apparently, to produce ethanol from cellulosic feedstock was more costly than to use other starchy source such as corn due to the present of lignin structure in lignocellulosic biomass that form barrier from enzymatic susceptibility. Punsuvan mentioned that, main contributors to the overall costs of bioethanol refinery were the raw materials, pretreatment, and enzyme production [6]. Therefore, it would be worthwhile to explore the potential of mushroom spent medium with partially/ fully delignified structure, which might acquired less intermediate pretreatment steps and low amount of enzyme loading for high glucose yield, as a new second generation energy feedstock. Therefore this study is conducted to explore the potential of MSM in producing high glucose yoeld compared to RSD and GM under different sodium hydroxide pretreatment techniques.

2 Materials and Methods

2.1 Preparation of biomass samples

RSD, GM, and MSM were oven-dried at 100°C for 48 hours and then stored in an air tight container at room temperature until further use. The substrates with size of 254-510 μm were used in this study.

2.2 Analysis of lignocellulosic contents

The lignocellulosic contents analysis of the RSD, GM, and MSM were carried out in five stages. First of all, the ash content was analyzed according to American Standard Test Method (ASTM). Then, the extractives were removed from biomass samples followed by removal of hemicelluloses via dissolution using dilute NaOH solution [7]. The next stage involved the dissolution of lignin using H_2SO_4 dilution method. Finally, the celluloses were calculated as the difference of ash, extractives, hemicelluloses and lignin from 100% of total lignocellulosic content in biomass samples. Extractives, hemicelluloses, lignin, and cellulose content were determined by aforementioned methods as proposed by Li *et al.* [8]. All analytical experiments were performed in triplicates and the average values were presented.

2.3 Analysis of morphology surface

Examinations of the microstructure surface morphologies of RSD, GM, and MSM were carried out at School of Material Engineering, UniMAP. A thin layer of gold was coated on the samples prior to measurement to prevent charging on the surface. SEM micrographs with 1000x magnifications were employed to observe the morphologies of RSD, GM and MSM samples.

2.3 NaOH pretreatment techniques

2.3.1 Autoclaving at 121°C for 20 min

Sample concentrations of 5% (w/v) with particle size in the range of 254-500 μm were treated in an autoclave at 121°C for 20 minutes. As a control, extra samples with the same concentration were immersed in distilled water. After autoclaving process completed, the treated samples were washed extensively over running water using 45 μm sieve tray until the pH reached 7 followed by final rinse with distilled water to remove excess NaOH and by-products that might inhibited the cellulase activity in subsequent enzymatic saccharification. The hydrolysates were, then, dried in an oven at 60°C for 24 hours before further use.

2.3.2 Heating in water bath (30, 60, 90°C) for 2 h

In the second pretreatment method, the same sample concentrations and particle size as Section 3.5.1.1 were heated in different concentration of NaOH (0.5, 1.0, 1.5, and 2.0 M) at different temperature of 30, 60, 90°C for 2 hours with sample immersed in distilled water as control. The samples recovery and drying process were similar to the previous autoclaving pretreatment method.

2.3.3Soaking at room temperature for 2 h

As for soaking method, the same sample concentrations and particle size as Section 3.5.1.1 were immersed in 0.5, 1.0, 1.5, and 2.0 M of NaOH at room temperature for 2 hours with sample in distilled water as control. The samples recovery and drying process were performed as previous autoclaving pretreatment method.

2.4 Enzymatic saccharafication

The efficiency and feasibility of the aforementioned pretreatment methods were determined by total glucose yield obtained from the pretreated samples in enzymatic saccharification. One gram of hydrolysates from different pretreated RSD, GM, and MSM was immersed in 100 ml of 50 mM sodium citrate buffer (pH 4.8) (Leustean et al., 2010) in a 250 ml Erlenmeyer flask, to give a final sample concentration of 1% (w/v). One mililiter of cellulase from Trichoderma reesei, which its activity (67 FPU/ml) has been previously determined, was added and the samples mixtures were incubated at 50°C with rate of agitation at 150 rpm for 48 h. After the saccharification time had elapsed, the flasks were heated at 90°C for 10 minutes to deactivate the enzyme [9]. Then, a total volume of 2.0 ml sample mixture was taken out for centrifugation at 8000 rpm for 10 min. The supernatant was filtered through 0.45 μm nylon membrane filter before analyzing the glucose contents using Biochemistry Analyzer, YSI 2900. The glucose analyzed was reported in gram per 100 grams dry substrate and was calculated according to the Equation 3.1 as proposed by [10]. It should be noted that dry substrate referred to the state of the hydrolysates sample after NaOH pretreatment prior to enzymatic saccharification.

Glucose yield (g/100 g dry substrate)

$$= (C*V)/(W) \times 100 \text{ g} \quad (3.1)$$

where, C was the concentration of glucose (g/l), V was the total volume of the liquid phase (l), and W was the dry weight of the corresponding lignocellulosic waste (g).

3 Results and Discussion

3.1 Lignocellulosic contents

Table 3.1 shows the chemical components; in particular of ash, extractives, hemicelluloses, lignin, and cellulose contents in rubberwood sawdust (RSD), growth medium (GM), and mushroom spent medium (MSM). Among these three sample types, RSD contained the highest amount of lignin (29.30% ± 1.31), which was slightly higher than in GM (29.07% ± 0.72), regardless the previous performed steaming pretreatment prior to P. sajor-caju spawn inoculation. In contrast, the amount of lignin was decreased from GM to MSM, which was 22.40% ± 0.34. This result showed a total decrease of 22.46 % of lignin amount after 5th harvest of *P. sajor-caju* and thus, indicated that the lignin in GM was partially degraded by *P.sajor-caju* during the period of cultivation.

The amount of hemicelluloses and cellulose in MSM was lower than GM and RSD. It was considered that this mushroom species also degraded hemicelluloses and cellulose as well as lignin simultaneously. In addition, it had been found that the amount of ash and extractives contents in MSM gave the highest value of 8.10% ± 0.21 and 12.67% ± 0.18 respectively, implying that carbohydrate and lignin degradation products of low molecular weight were produced by the extracellular enzymes of *P. sajor-caju* during the cultivation. These results of lignocellulosic composition were consistent with previously reported works in the utilization of different type of mushroom waste medium to yield sugars for ethanol production. As for example, Hiyama et al. obtained an amount of lignin decreased from 24.3% to 17.8% after 85 days of shiitake mushroom cultivation on a medium consisted of several woods. The usage of several wood powders as shiitake medium instead of wood log and period of cultivation might contribute to the difference in amount of lignin removal found in both works [11]. From another previous work performed by Irawati et al, using 77 days-spent culture medium of *Auricularia polytricha* (black jelly mushroom) cultivated in a medium mixture containing hardwood meal of *Alnus japonica*, they observed a decrease in lignin, holocellulose and cellulose content from 23.2 to 11.6%, 81.1 to 62.7%, and 44.3 to 31.7% respectively. In addition, the lignin, hemicellulose, and cellulose content in spent culture medium of softwood (*Cryptomeria japonica*) gave a decrease from 32.9 to 22.9%, 78.2 to 60.3%, and 51.6 to 36.5% correspondingly. These results indicated that lignin in the hard- and softwood was partially degraded by A. polytricha during the cultivation period. The amount of lignin and holocellulose in both woodmeal were lower compared to spent culture medium. Thus, indicated that apart of lignin, this mushroom type also degraded cellulose and hemicelluloses during the cultivation [12].

In the utilization of lignocellulosics materials to produce sugars for ethanol fermentation by enzymatic saccharification, lignin would interfered with hydrolysis by blocking the accessibility of cellulase to cellulose [13]. Hence, the removal of lignin from the materials during aforementioned mushroom cultivation could dramatically increase the hydrolysis rate [13], [14], [15]. One of the reasons for high hydrolysis rate was due to the high value of carbohydrate/lignin ratio [16]. According to the previous table of sample composition analysis, the ratio of carbohydrate/ lignin of MSM showed the highest. The cellulose/lignin ratio of RSD, GM, and MSM were 1.15, 1.17, and 1.24 respectively. Collectively, the lignocellulosic components characterization of MSM indicated that such medium have better potential for enzymatic saccharification than RSD and GM due to its low lignin content that would resulted in easy access of the cellulase towards the available amount of cellulose. However, additional pretreatment method such as chemical and thermo-chemical could be employed on MSM to remove the remaining lignin content to obtain high glucose yield in subsequent enzymatic saccharification.

Table 3.1: Chemical components of RSD, GM, and MSM in percentage (dry weight)

	Ash	Extracti ves	Hemicellul ose	Lignin	Cellulose
RSD	1.87 ± 0.14	3.45 ± 0.11	31.52 ± 0.69	29.30 ± 1.31	33.86 ± 1.17
GM	1.90 ± 0.18	3.78 ± 0.17	31.23 ± 1.09	29.07 ± 0.72	34.02 ± 0.52
MSM	8.10 ± 0.21	12.67 ± 0.18	27.93 ± 0.28	22.40 ±0.34	27.97 ± 0.52

3.2 Surface morphology of RSD, GM and SRSD

The morphological changes that were induced by biological pretreatment during the period of cultivation of *P. sajor-caju* were examined by Scanning Electron Microscope (SEM) to obtain insight into the structural modification of the raw rubberwood sawdust. The SEM images of MSM were compared with RSD and GM to observe the relationship between the changes in the structural properties at different stages of oyster mushroom cultivation. The morphology surface of untreated RSD, GM, and MSM are shown in Figure 4.1. The biologically untreated RSD and GM showed an even and smooth, flat surface, indicating a rigid and highly ordered surface structure (Figure 3.1 (a) (b)). Regardless the steaming process performed on the GM, its surface structure was not differed much compared to RSD. Additionally, in both sample types, there were traces of small holes that might originated from the previous furnishing and lumbering process of the rubber tree. On the other hand, the MSM had a rugged, rough, and broken surface (Figure 3.1 (c)). The fragments were separated from the initial connected structure and fully exposed, thus increasing the external surface area and porosity resulting in loose structure of the fibre bundles. This would obviously favour the enzymes contacting the inner linkage, hence accelerating the saccharification process. The SEM figures of the MSM also clearly showed traces of hyphae on the fibe surface resulting from the oyster mushroom cultivation. Taniguchi et al. also discovered a loosening structural of cells with a simultaneous increase in porosity of rice straw resulted from cultivation of *P. ostreatus* for a period of 36 and 60 days. The phenomenon was due to the partial degradation of the lignin that was responsible for preventing the penetration of cellulase in the rice straw in enzymatic saccharification [17].

a)RSD

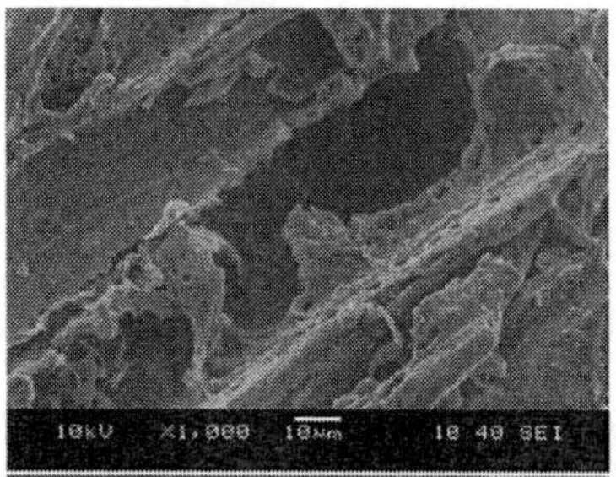

b)GM

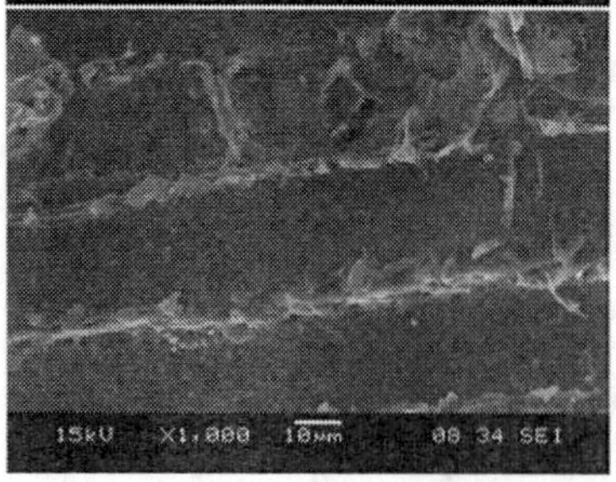

c)RSD

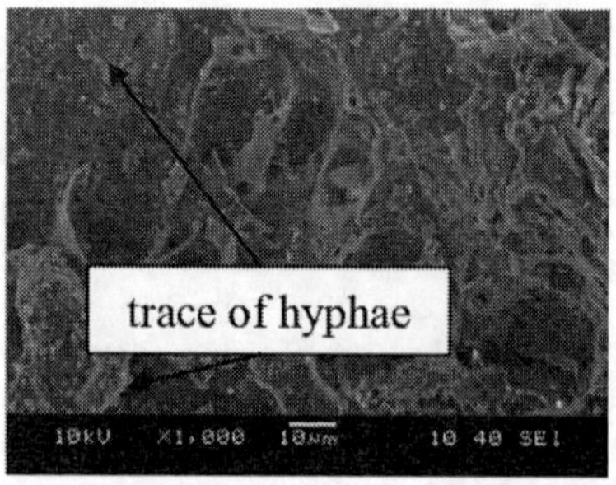

Figure 3.1: Images of morphology structure of (a) RSD, (b) GM, and (c) MSM at magnification of 1000x

3.3 Glucose yield

Figure 3.2 shows the amount of glucose obtained from RSD, GM, and MSM after NaOH pretreatment via autoclaving method at 121°C for 20 min. Based on the results of the work, it showed that, the NaOH treatment improved most of the fiber properties as the glucose yield obtained from the three different samples increased with pretreatment severity. The highest glucose yield was resulted from NaOH pretreatment at 2.0 M for all three samples. The yields of glucose of RSD, GM, and MSM under the employed pretreatment condition were 21.97, 19.81, and 31.07 g/100 g dry substrate respectively. The applied pressure at elevated temperature and high concentration of 2.0 M of NaOH during autoclaving process had loosened the structure of RSD, GM, and MSM, and increased the degree of delignification. The effect could be clearly seen from the increment of glucose yield towards the increasing of the applied NaOH concentration. The yield of glucose based on NaOH-treated MSM was markedly better compared to the original untreated MSM. Compared to untreated MSM, the 2.0 M NaOH treated-MSM gave an increase of 3.32 fold of glucose yield. This result indicated that, the combination of biological and thermo-chemical pretreatment on MSM could improve the glucose yield. In a study performed by McIntosh and Vancov in NaOH (0.5 M) pretreatment of sorghum straw (Sorghum bicolour var. MR Buster) via autoclaving at 121°C with temperature of 60°C, a 5.6 fold increase in total of reducing sugars was obtained in comparison to samples treated at the same temperature in the absence of NaOH. The result confirmed the importance of alkali inclusion to adequately remove a certain amount of lignin for high production of reducing sugars [18]. Meanwhile, Hamzah et al. has treated the oil palm (*Elaeis*) empty fruit bunches fiber (EFB) with 2.5 mol/L NaOH for 15 min. They obtained a significant reduction in hemicelluloses and lignin amount from 35 to 24.55 and 16.45 to 7.25 g/100g EFB fiber in comparison to the untreated EFB, resulting in high amount of cellulose portion of 66.77 g/100g EFB fiber. They suggested that, the combination of the NaOH swelling effect and high temperature disintegrated the EFB fibre into a pliable fiber, cleaned up the fiber surface, and thus exposed more cellulose component in the EFB fibre [19]. Additionally, according to Astimar et al., a part of effectively removing hemicelluloses and lignin, the NaOH also penetrated and swelled both the accessible amorphous and crystalline regions of cellulose [20].

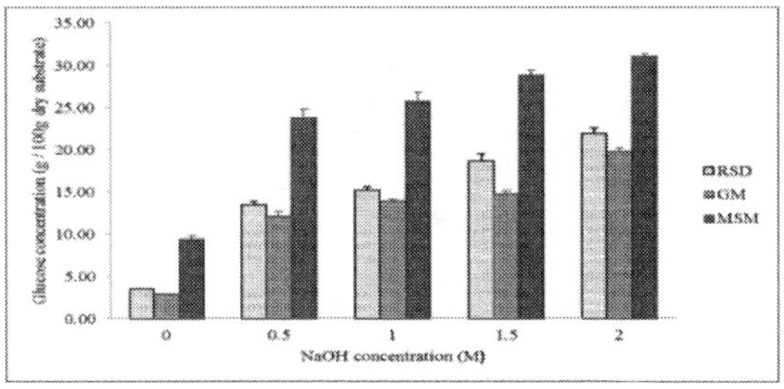

Figure 3.2: Amount of glucose (g /100 g dry substrate) of different concentration NaOH-treated RSD, GM, and MSM (autoclaving method)

Figure 3.3 shows that increasing temperature and NaOH concentration could improve the production of glucose after enzymatic saccharification of RSD, GM, and MSM. As shown in the corresponding figures, treating MSM in 2.0 M NaOH at 30°C, 60°C, and 90°C had enhanced the glucose yield by 8.45, 3.63, and 3.77 fold compared to untreated MSM. In comparison to glucose yield obtained after autoclaving method using 2.0 M NaOH, the glucose yield resulted from heating in a water bath at 90°C with similar NaOH concentration was 30.13 g/100 g dry substrate, which was slightly higher. Treating the MSM sample in water bath at 30 and 60°C gave an amount of glucose yield of 23.63 and 26.53 g/100 g dry substrate after 48 h of enzymatic saccharification. Under all pretreatment conditions of different temperature and NaOH concentration, the glucose yield obtained from RSD and GM was comparable. This indicated that the steaming process on GM before spawn inoculation did not improve and affected much the fiber properties for the feasibility of cellulase susception in enzymatic saccharification. The direct enzymatic saccharification of RSD, GM, and MSM without inclusion of NaOH treatment was ineffective because lignin surrounding cellulose prevented the enzymes from accessing cellulose. However, the glucose yields resulted from MSM at treatment temperature of 30, 60, and 90°C were the highest between the three samples type, which were 2.79, 7.30, and 7.99 g/100g dry substrate, approximately 3.58, 3.08, 2.14 fold of increment, respectively compared to RSD. This might be due to the feasibility of cellulase to access the cellulose as the amount of lignin removed increased with pretreatment temperature and NaOH concentration. In addition, the amount of glucose resulted from untreated MSM was comparable to those obtained from 1.0 M of NaOH treated-RSD at 60°C.

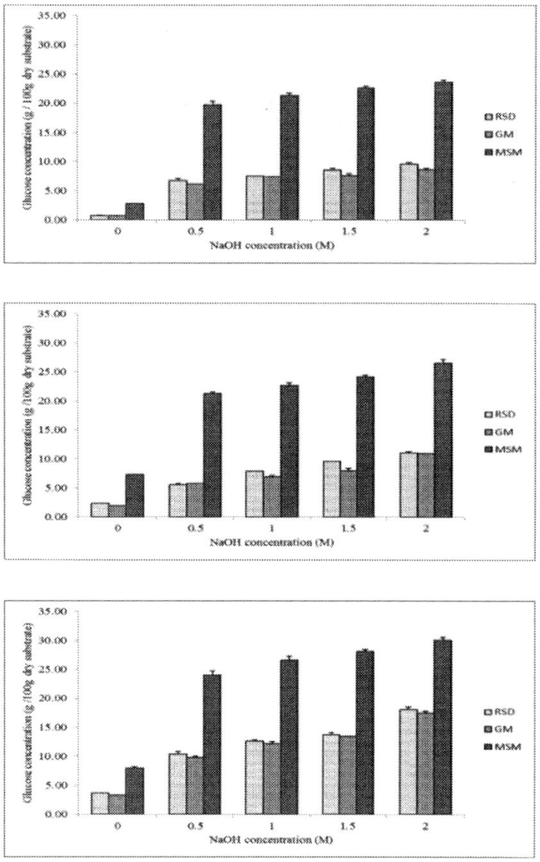

Figure 3.3: Amount of glucose (g /100 g dry substrate) of different concentration NaOH-treated RSD, GM, and MSM (heating method at a)30, b) 60, c) 90°C)

The amount of glucose obtained from different pretreated RSD, GM, and MSM soaked in various concentrations of NaOH pretreament at room temperature are depicted in Figure 3.4. From the figure, the pretreatment condition of soaking MSM in 2.0 M NaOH gave the highest yield of glucose with approximately 2.93 fold compared to RSD. In comparison to the treated MSM in distilled water, the 2.0 M NaOH-treated MSM showed an increase of 10.12 fold of glucose yield. This indicated that the amount of lignin removal after oyster mushroom cultivation increased with the increase in NaOH concentration employed at ambient temperature. The enzymatic saccharification of treated RSD and GM without NaOH inclusion was ineffective because lignin and hemicelluloses surrounding cellulose prevented the enzymes from accessing the cellulose. As for MSM, soaking in distilled water at room temperature was not enough to remove the remaining hyphen and lignin that formed a barrier between cellulase and cellulose. However, after NaOH inclusion, the glucose yield obtained increased with increase in NaOH concentration used for all types of samples, implying that the amount of lignin and hemicelluloses removed also increased. The amount of glucose obtained from GM under all concentration of NaOH pretreatment showed only a slightly different from RSD regardless the previous performed steaming process. It could be concluded that, the combination of biological and thermo-chemical pretreatment on MSM could markedly improve the glucose yield in enzymatic saccharification rather to employ only one separate pretreatment. Therefore, the pretreatment at room temperature only resulted in low yields of glucose.

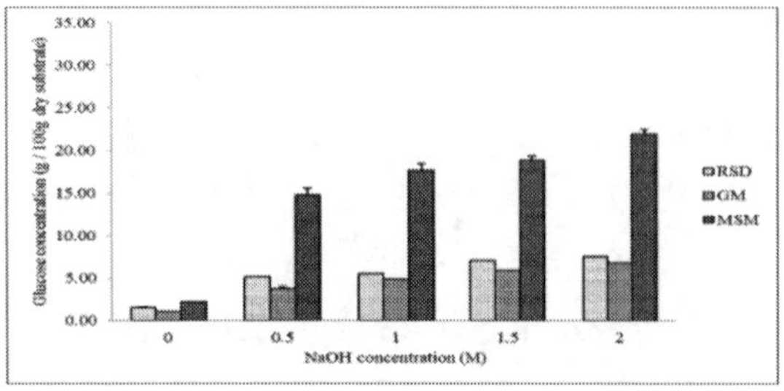

Figure 3.4: Amount of glucose (g /100g dry substrate) of different concentration NaOH-treated RSD, GM, and MSM (soaking method)

4 Conclusion

From the results of lignocellulosic composition and morphology surface analysis of RSD, GM, and MSM, it can be concluded that the yield of glucose produced in enzymatic saccharification can be attributed to three factors: (1) the crystallinity of cellulose, (2) the lignin-hemicellulose sheath that surrounds the cellulose, and (3) the availability of surface area for the enzyme attack. The study showed that MSM was a promising biomass material for enzymatic saccharification to produce bioethanol or other bioproducts as well considering its high glucose yield in comparison to RSD and GM. Under all employed pretreatment techniques, the highest yield of glucose and hydrolysis weight decreases were obtained from MSM. The high glucose production obtained via heating in a water bath at 90°C for 2 h with 2.0 M NaOH proved that the combination of biological and thermo-chemcial pretreatment was effective to reduce the lignin and increase the cellulose content in the MSM. However, increase in thermo-chemical pretreatment severity until certain limit would increase the amount of lignin removed as well as the glucose produced. Under more severerer pretreatment conditions, which lead to greater lignin reductions, did not favour the glucose production due to substantial amount of solid losses.

References

1. Pan, X., Arato, C., Gilkes, N., D.Gregg, & Mabee, W. *Biotechnol. Eng.*, 90, 473-481, (2005).
2. Srinivasakannan, C., & Bakar, M. Z. A. *Biomass Bioenergy*, 27, 89-96. (2004).
3. Alhassan, A. M. University of Putra Malaysia, Serdang, Selangor. (2010).
4. Wan, C., & Li, Y. *Enzym and Microbial Technology*, 47, 31-36. (2010).
5. Koshy, J., & Nambisan, P. *International Journal of Applied Biology and Pharmaceutical Technology*, 3(1), 280-286. (2012).
6. Punsuvon, V. [Electronic Version], 157-178. (2013).
7. Ramsden, M. J., & Blake, F. S. R. *Wood Science and Technology*, 31, 45-50. (1997).
8. Li, S., Xu, S., Liu, S., Yang, C., & Lu, Q. *Fuel Processing Technology*, 85, 1208-1211. (2004).
9. William E. Kaar, M. T. H. *Biomass and Bioenergy*, 18, 198-199. (2000).
10. Neureiter G. L., H. Danner, C. Thomasser, B. Saidi, & Braun., R. *Appli. Biochem. Biotechnol*, 98, 49-58. . (2002)
11. Hiyama, R., Gisusi, S., & Harada, A. *J Wood Sci*, 57, 429-435. (2011)
12. Irawati, D., Yokota, S., Niwa, T., Takashima, Y., Ueda, C., Ishiguri, F. *J Wood Sci*, 58, 180-183.(2012)
13. Cheng, C.-L., Lo, Y.-C., Lee, K.-S., Lee, D.-J., Lin, C.-Y., & Chang, J.-S. *Bioresource Technology*, 102(18), 8514-8523. (2011).
14. Lee, J.-W., Koo, B.-W., Choi, J.-W., Choi, D.-H., & Choi, I.-G. *Bioresource Techn*, 99, 2736-2741. (2008)
15. Yokota, S., Nakajima, R., Suzuki, D., Ishiguri, F., Iizuka, K., & Yoshizawa, N. *Cellulose Chem Technol*, 41, 575-582. (2007).
16. Xu, J., Chen, Y., Cheng, J. J., Sharma-Shivappa, R. R., & Burns, J. C. *Bioresources*, 6(1). (2011).
17. Taniguchi, M., Suzuki, H., Watanabe, D., Sakai, K., Hoshino, K., & Tanaka, T. *Journal of Bioscience and Bioengineering*, 100(6), 637-643. (2005).
18. McIntosh, S., & Vancov, T. *Bioresource Technology*, 101(17), 6718-6727. (2010).
19. Hamzah, F., Idris, A., & Shuan, T. K. *Biomass and Bioenergy*, 35(3), 1055-1059. (2011).
20. Astimar, A.A., Das. K., Husin. M. & Anis. M. *J. Oil Palm Research*. Vol. 14, No. 2, 10-17. (2002).

MANET Routing Protocols Evaluation: AODV, DSR and DSDV Perspective

Fahad Taha AL-Dhief[1], *Naseer* Sabri[2*], *M.S.* Salim[3], *S.* Fouad[4], *S. A.* Aljunid[2]

[1]Faculty of Electrical Engineering, Department of Communication Engineering, Universiti Teknologi Malaysia, Malaysia
[2]School of Computer and Communication Engineering, University Malaysia Perlis, Malaysia
[3,4]Laser and Optoelectronics Engineering Department, [2*]Computer Engineering Department, AlNahrain University, Iraq
[4]Microelectronic Engineering School, University Malaysia Perlis, Malaysia

Abstract. Mobile Ad Hoc Network (MANET) is a set of mobile nodes that communicate with each other to configure an immediate network without utilizing any of infrastructure, the centralized access point or centralized administration in multiple hop manner. There are a lot of routing protocols have been proposed in MANET which are different from each other in the performance and the mechanism. Therefore, the performance study of those protocols in different scenarios is needed. This paper presents the performance comparison between Dynamic Source Routing (DSR), Ad hoc on demand distance Vector Routing (AODV) as reactive routing protocols and Destination Sequenced Distance Vector (DSDV) as a proactive routing protocol to precisely determine which protocol is more effective. Network Simulator (NS) version 2.35 has used to simulate and evaluate the performance of these protocols in terms of the packet delivery ratio, average throughput, average end-to-end delay, and packet loss ratio with respect to the variable number of nodes.

1 Introduction

Mobile Ad Hoc Network (MANET) is a set of mobile nodes that communicate with each other to configure an immediate network without utilizing any of infrastructure, the centralized access point or centralized administration in multiple hop manner, so the network in MANET is infrastructure-less [1]. MANETs are a dynamic topology, where the wireless nodes in MANET work as both router and host to maintain the activation of the communications network. The nodes in MANET are arbitrary can move out or join the network freely and in a random fashion, they can organize themselves [2-3]. In MANET, the nodes collaborate with each other in the job of routing packets from the source node S to the destination node D as every node in the network is capable of communicating just with those nodes that positioned within its transmission radius R, simultaneously, S node and D node can be located at a range very higher than radius those nodes. So, the nodes in MANET need to reroute the packets to other nodes to enable the communication among the nodes that are located outside the transmission range. Thus, the node in MANET changes its link case orderly with other mobile nodes [4].

In recent studies, there are a lot of routing protocol have been proposed in MANET which witnessed a huge interest by researchers. Where the study in [5] has presented a comparison of LAR and OLSR routing protocols in the detection of the forest fire based on MANET. It has used MATLAB simulator to evaluate the performance of these protocols, where the results showed that LAR is better and more efficient than OLSR in detecting the fires and LAR is consumed energy lesser than OLSR. This study [6] presents an evaluation between DSR and AODV routing protocols under mobility model with TCP and UDP traffic sources. It has been shown that AODV is better than DSR in the high mobility conditions. The authors in [7] have presented a comparison in MANET between TCP and UDP for temporarily ordered routing algorithm (TORA) and optimized link state routing (OLSR) with various mobility models. They used NS2 simulation to analyze the performance depending on the different performance metrics like end-to-end delay, packet delivery ratio, and throughput. The results show that TCP is better than UDP in the throughput. Also, TORA in UDP protocol is better than TCP in terms of PDF. This study [8] provides a performance analysis of three different protocols and they are FSR, ZRP and LAR in different pause times. The authors have evaluated these protocols using Qualnet 5.0.2 in terms of the performance measures, where they concluded that LAR is outperformed ZRP and FSR routing protocols. In [9], the authors have evaluated the performance of DSR, LAR and DREAM routing protocols using NS 2 simulator in terms of the performance metrics. The simulation results showed that DREAM has better performance than DSR and the LAR protocol gave the best performance as compared with DSR and DREAM. This study [10] compares the performance of DSDV with OLSR and WRP in terms of packet delivery ratio, throughput, overhead and MAC

*Corresponding author: naseersabri@yahoo.com

collision using MATLAB. The results showed that DSDV is better than WRP and the performance of OLSR protocol showed the best as compared to other protocols.

2 The Routing Protocols in MANET

Routing is the manner that includes the exchange of data from one node to another node in the network. The wireless radios routing in MANET [11] is usually implemented through multi-hop, due to the short range of nodes. Where the message is normally routed by the intermediate mobile nodes.

The routing protocols are available in MANET which defines the routes and transports the packets in the nodes, that is from S node to D node [12]. Various routing protocols have been proposed in MANET. This is important to improve and use the performance of the network properly. There are three main types of routing protocols in MANET namely Proactive, Reactive and Hybrid routing protocols. Figure 1 shows the classification of routing protocols. In proactive protocols, the nodes have a table of routing information. Where the node creates the routes before they are needed. Therefore, the route discovery operation is implemented faster than reactive protocols. Whilst, the nodes in reactive protocols creates the routes just when it is needed. However, the nodes in the hybrid protocols combine the strategies of both reactive and proactive protocols.

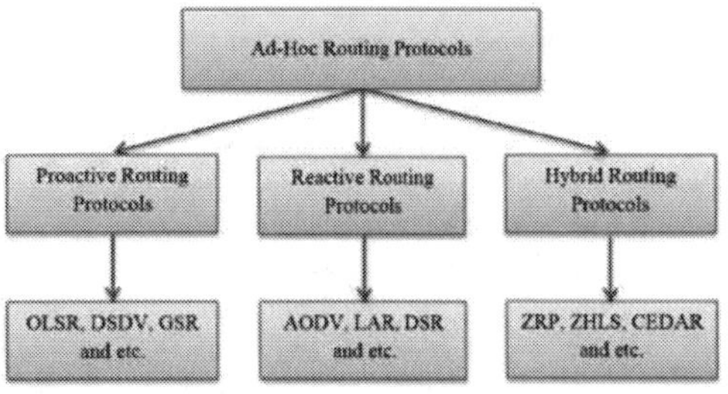

Fig. 1. The Classification of Routing Protocols

2.1 Ad-hoc On-demand Distance Vector (AODV)

AODV is a kind of reactive protocols. Its methodology is hop-to-hop routing. The node establishes the Route Request (RREQ) if it wants to know the route to a particular destination. Then the intermediate nodes forward the route request and at the same time, these intermediate nodes create a reverse route to the destination [13]. When the node receives the request that has the route to the destination, it establishes a Route Reply (RREP) which includes numeral of hops which are required to arrive the destination. Each node that cooperates in sending this reply to the source node, it creates a forward route to the destination [14]. Figure 2 shows the routing of RREQ and RREP in AODV protocol. This route that has been established from source to destination is a hop-by-hop case. Figure 3 shows the advantages and disadvantages of AODV routing protocol.

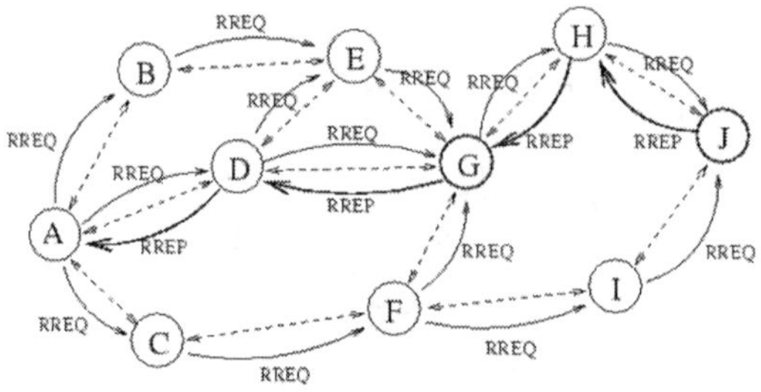

Fig. 2. RREQ and RREP in AODV

2.2 Dynamic Source Routing (DSR)

DSR is a reactive or on-demand routing protocol. This protocol has been designed to reduce the bandwidth wasted via the control packets in wireless networks and that via deleting the periodic table-update messages required in the table-driven approach [15]. In DSR protocol, there is no need for network infrastructure or administration, due to these networks fully self-configured and organized. The source routing is a method which the source packet defines the complete sequence of nodes through which to forward the data packets. The source routing does not need to keep the routing information via the intermediate hops. Figure 3 shows the advantages and disadvantages of DSR protocol [16].

2.3 Destination Sequenced Distance Vector (DSDV)

DSDV is one of the most widely known proactive or table-driven routing protocols for MANETs [17]. The routing algorithm of DSDV is depended on the numeral of hops to arrive at the destination node. To transmit the data packets among the nodes in the network, DSDV protocol utilizing routing tables which are stored in every node. DSDV protocol has three major characteristics which are: decreasing the high routing overhead, solve the "count to infinity" problem and avert the loops. Each mobile node contains a table of routing information which includes all the routes to the destinations and another information [18]. Figure 3 shows the advantages and disadvantages of DSDV protocol.

Protocols	Advantages	Disadvantages
AODV	In AODV, route discovery process is in on demand, which is more efficient in dynamic nature of mobile ad-hoc network.	Due to on demand manner, it won't check route in periodic interval so transmission of data after discover the rote is taking some more delay.
DSR	The route is created only when it is required and the nodes utilize the route cache information efficiently to reduce the overhead and collision.	The route maintenance mechanism does not locally repair a broken link. The delay is higher than in table-driven protocols.
DSDV	DSDV was one of the early algorithms available. It is quite suitable for creating ad hoc networks with small number of nodes.	DSDV requires a regular update of its routing tables, which uses up battery power and a small amount of bandwidth.

Fig. 3. The Advantages and Disadvantages of AODV, DSR and DSDV Routing Protocols.

3 Simulation Scenarios

In this study, we have used ns-2.35 network simulator to compare and evaluate the performance of DSR, AODV and DSDV routing protocols in MANET. This simulation has been used a different number of nodes to deeply verify the performance of these protocols in terms of the performance measures. The number of nodes is 10, 20, 30, 40 and 50. Where the nodes have deployed in the network with speed 5ms and they move randomly. Figure 4 illustrates the simulation environment. The transmission range in all nodes is set to be 250m in the network. The simulation area is 600m x 600m and the simulation time is 160 sec. The packet size in this simulation is 512 bytes. Table 1 shows other simulation parameters.

Table 1. Simulation Parameters

Parameters	Values
Simulator	NS 2.35
Number of Nodes	10, 20, 30, 40, 50
Node Speed	5ms
Transmission Range of Node	250m
Simulation Area	600m x 600m
Simulation Time	160 Sec
Routing Protocols	DSR, AODV and DSDV
Packet Size	512
MAC Type	802.11
Antenna Model	Omni Antenna
Traffic Model	TCP
Mobility Model	RWP
Performance Measures	PDR, Average e2e delay, Average TP and PL

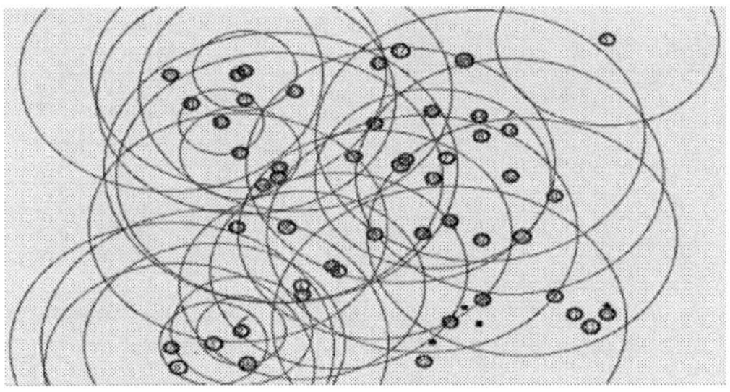

Fig. 4. The Simulation Environment

4 Performance Metrics

We have used different network performance metrics through our simulation between TCP and UDP. These performance metrics are used to evaluate the performance of the protocols.

4.1 Packet Delivery Ratio (PDR)

PDR is the ratio of data packets delivered to the destination to those generated by the sources. It is calculated as follow:

$$\text{PDR}(\%) = \frac{\sum \text{No of packet received}}{\sum \text{No of packet sent}} * 100 \quad (1)$$

4.2 Average Throughput (TP)

Average TP is the number of bytes received successfully and it is calculated as follow:

$$\text{TP} = \text{Number of Bytes Received} * 8 * \text{Simulation Time} * 1000 \text{ kbps} \quad (2)$$

4.3 Average End-to-End Delay (e2e delay)

Average e2e delay is the average time of the data packet to be successfully transmitted across the network from the source to the destination. It is computed as follow:

$$\text{e2e delay} = \frac{\sum \text{arrive time} - \text{send time}}{\sum \text{number of connection}} \quad (3)$$

4.4 Packet Loss (PL)

PL is the different between the number of data packets sent and the number of data packets received. It is calculated as follow:

$$\text{PL} = \text{Number of Data Packets Sent} - \text{Number of Data Packets Received} \quad (4)$$

5 Simulation Results and Discussion

In our simulation, we have compared and evaluated the performance of three different types of routing protocols, they are DSR, AODV which are considered reactive routing protocols and DSDV which is a proactive routing protocol. The simulation has been used ns-2.35 in MANET. Ns-2.35 is a software which is able to handle huge amounts of data efficiently. Also, it is very commonly used software, because of its high capacity to carry out complex mathematical calculations, data analysis and simulations. In this study, the simulation scenario has a different number of nodes. According to the simulation results obtained, Figure 5 shows the performance results of the routing protocols in the throughput. Where AODV has the highest value of throughput, it has achieved 10096.82 of throughput and the performance of DSR protocol is slightly higher than DSDV protocol where DSR has achieved 5004.56 throughput and DSDV protocol has obtained 4479.68. Table 2 shows the results values of the average throughput for the routing protocols in a varied number of nodes.

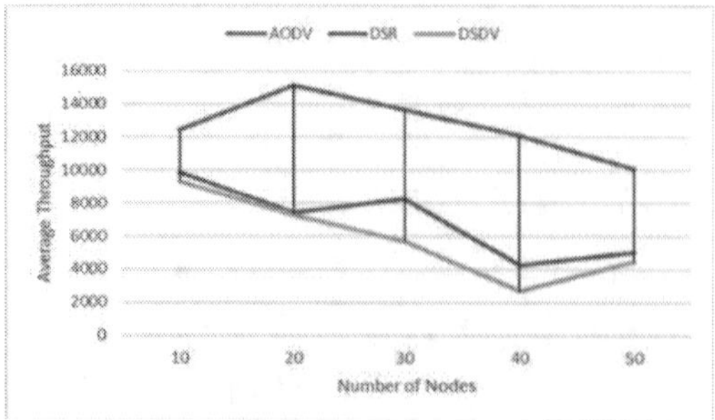

Fig. 5. TP Versus Number of Nodes

Table 2. The Results Values of AODV, DSR and DSDV in Average Throughput

Number of Nodes	AODV	DSR	DSDV
10	12424.28	9882.86	9342.71
20	15123.97	7416.21	7276.56
30	13684.28	8235.16	5672.96
40	12105.33	4251.59	2671.6
50	10096.82	5004.56	4479.68

In PDR as shown in Figure 6, the performance of DSDV is much better than DSR. Where DSDV has obtained 98.87% of PDR at node 10 and 96.64% at node 50. However, AODV is better than DSR in terms of PDR, where AODV has achieved 90.4% of PDR and DSR has achieved 66.67% of PDR. Therefore, it can be seen that the DSDV protocol received a higher amount of data as compared to the AODV and DSR routing protocols. Table 3 shows the results values of PDR for the routing protocols in a varied number of nodes.

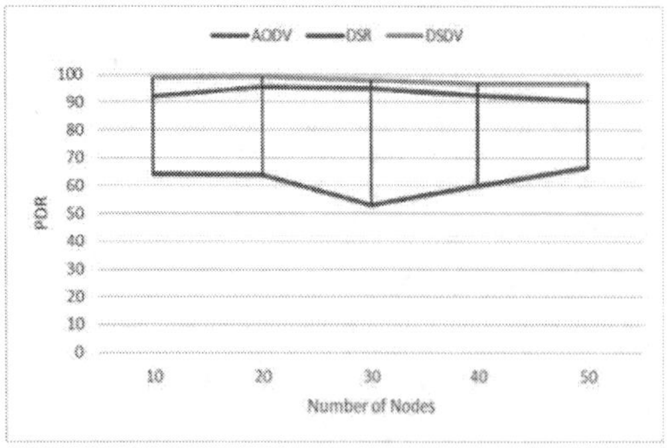

Fig. 6. PDR Versus Number of Nodes

Table 3. The Results Values of AODV, DSR and DSDV in PDR

Number of Nodes	AODV	DSR	DSDV
10	92.11	64.29	98.87
20	95.54	63.64	99.27
30	94.78	52.94	97.92
40	92.56	60	96.67
50	90.4	66.67	96.64

Depending on PDR, the DSDV protocol has the lowest value of PL and DSR has the highest value of PL, where DSDV has 3.36% of PL and DSR has 33.33% at the node 50 as shown in Figure 7. However, AODV protocol has 9.6% of PL. Table 4 shows the results values of PL for the routing protocols in a varied number of nodes.

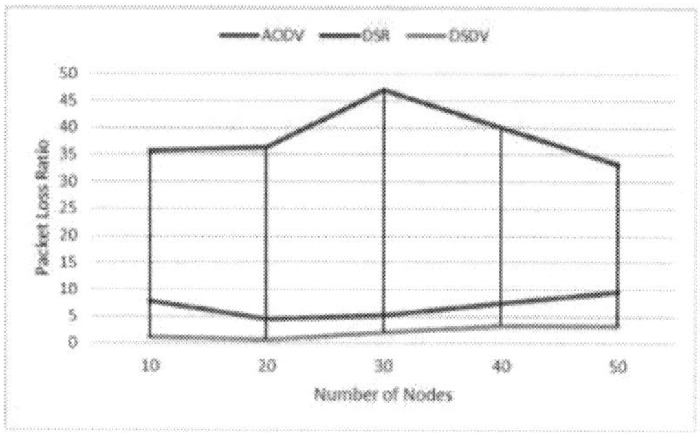

Fig. 7. PL Versus Number of Nodes

Table 4. The Results Values of AODV, DSR and DSDV in PL

Number of Nodes	AODV	DSR	DSDV
10	7.89	35.71	1.13
20	4.46	36.36	0.73
30	5.22	47.06	2.08
40	7.44	40	3.33
50	9.6	33.33	3.36

The routing protocols have different performance in e2e delay, wherein DSR protocol has the lowest value than DSDV and AODV routing protocols. DSR protocol has achieved e2e delay value equal to 0.111149 sec, DSDV protocol has achieved e2e delay value equal to 0.210932 sec and AODV protocol has achieved e2e delay value equal to 0.165754 sec. Therefore, the results showed that the DSR protocol has the best performance in e2e delay as shown in Figure 8. However, AODV protocol is better than DSDV protocol in e2e delay. Table 5 shows the results values of e2e delay for the routing protocols in a varied number of nodes.

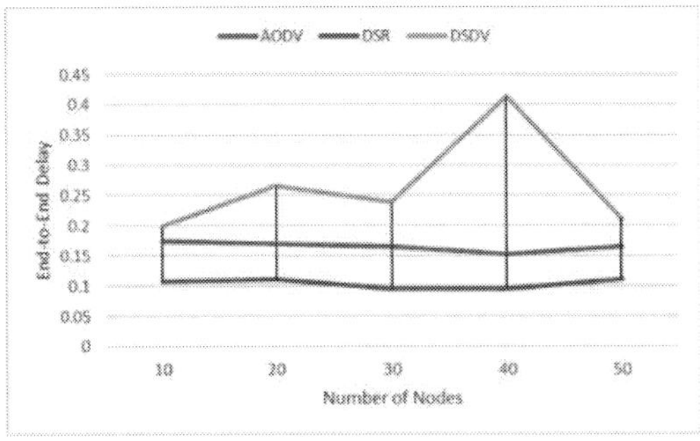

Fig. 8. E2E Delay Versus Number of Nodes

Table 5. The Results Values of AODV, DSR and DSDV in Average E2E Delay

Number of Nodes	AODV	DSR	DSDV
10	0.175002	0.106881	0.199203
20	0.170159	0.109682	0.265643
30	0.165601	0.0945285	0.238772
40	0.153284	0.095455	0.413572
50	0.165754	0.111149	0.210932

6 Conclusion

MANET is a set of mobile nodes that communicate with each other to configure an immediate network without utilizing any of infrastructure, the centralized access point or centralized administration in multiple hop manner. In MANET, there are a lot of routing protocols have been proposed which are different from each other in the performance. This paper has presented the performance comparison between Dynamic Source Routing (DSR), Ad hoc on demand distance Vector Routing (AODV) as reactive routing protocols and Destination Sequenced Distance Vector (DSDV) as a proactive routing protocol to precisely determine which protocol is more effective. Network Simulator (NS) version 2.35 has used to evaluate the performance of these protocols in terms of the performance measures with respect to the variable number of nodes. The results have shown that DSDV is better than DSR and AODV routing protocols in terms of PDR and PL, while AODV is better than DSDV and DSR in TP. DSR is better than other protocols in e2e delay. Due to the importance of PDR, this paper concluded that DSDV is the best protocol as compared to AODV and DSR protocols

References

1. Abdulleh, M.N., S. Yussof, and H.S. Jassim, Comparative Study of Proactive, Reactive and Geographical MANET Routing Protocols. Communications and Network, **7**(02): p. 125 (2015)
2. Singh, G. and A. Singh, Performance Evaluation of Aodv and Dsr Routing Protocols for Vbr Traffic for 150 Nodes in Manets. International Journal of Computational Engineering Research. **2**(5). (2012)
3. Kaur, R. and M.K. Rai, A novel review on routing protocols in MANETs. Undergraduate Academic Research Journal (UARJ), ISSN, 2278: p. 1129. (2012)
4. Chitkara, M. and M.W. Ahmad, Review on manet: characteristics, challenges, imperatives and routing protocols. International Journal of Computer Science and Mobile Computing. **3**(2): p. 432-7 (2014)
5. AL-Dhief, F.T., R.C. Muniyandi, and N. Sabri, Performance Evaluation of LAR and OLSR Routing Protocols in Forest Fire Detection using Mobile Ad-Hoc Network. Indian Journal of Science and Technology. **9**(48). (2016)
6. Kumar, S., R. Rathy, and D. Pandey. Traffic pattern based performance comparison of two reactive routing protocols for ad hoc networks using NS2. in Computer Science and Information Technology, 2009. ICCSIT 2009. 2nd IEEE International Conference on. (2009) IEEE
7. Singh, S.K., R. Duvvuru, and J.P. Singh, TCP and UDP-based performance evaluation of proactive and reactive routing protocols using mobility models in MANETS. International Journal of Information and Communication Technology. **7**(6): p. 632-644 (2015)
8. Thaseen, I.S. and K. Santhi, Performance analysis of FSR, LAR and ZRP routing protocols in MANET. International Journal of Computer Applications, **41**(4). (2012)
9. Kumar, P. and A. Kumar, Simulation Based Analysis of DSR, LAR and DREAM Routing Protocol for Mobile Ad hoc Networks. MIT International Journal of Computer Science and Information Technology. **3**(2): p. 58-62 (2013)
10. Gupta, S., B. Dhaliwal, and R. Malhotra, Performance Comparison of Proactive Routing Protocols: OLSR, DSDV, WRP. International Journal of Advanced Research in Computer Science, **6**(8). (2015)

11. Chennikara-Varghese, J., et al. Survey of routing protocols for inter-vehicle communications. in Mobile and Ubiquitous Systems: Networking & Services, 2006 Third Annual International Conference on. (2006). IEEE
12. Aggarwal, R. and A. Kaur, Energy Efficient Zone Based Location Aided Routing Protocol for MANET. International Journal of Computer Science and Information Technologies, **5**(4): p. 4990-4994 (2014)
13. Perkins, C., E. Belding-Royer, and S. Das, Ad hoc on-demand distance vector (AODV) routing, (2003)
14. Das, S.R., E.M. Belding-Royer, and C.E. Perkins, Ad hoc on-demand distance vector (AODV) routing. (2003)
15. Johnson, D.B., D.A. Maltz, and J. Broch, DSR: The dynamic source routing protocol for multi-hop wireless ad hoc networks. Ad hoc networking, 5: p. 139-172 (2001)
16. Murthy, V., R.P. Kumar, and K. Velala, AODV, DSDV Performance Analysis with TCP Reno, TCP New Reno, TCP Vegas and TCP Tahoe on Mobile Ad-hoc Networks using NS2. IJRCCT, **3**(8): p. 878-883 (2014)
17. He, G., Destination-sequenced distance vector (DSDV) protocol. Networking Laboratory, Helsinki University of Technology, p. 1-9 (2002)
18. JONSON, D., The dynamic source routing protocol for mobile ad hoc networks (DSR). Internet Draft, draft-ietf-manet-dsr-08. txt, (2003)

Brain Tumour Detection using Fine-Tuning Mechanism for Magnetic Resonance Imaging

Mohammed S. H. Al-Tamimi[1], *Nur Hafizah* Ghazali[2,*], *Norfadila* Mahrom[2], *Nurulhuda* Ghazali[3], and *Ghazali* Sulong[4,5]

[1]University of Baghdad, College of Science, Department of Computer Science, Baghdad, Iraq
[2]Universiti Malaysia Perlis, School of Computer and Communication Engineering, Perlis, Malaysia
[3]Universiti Teknologi Mara, Faculty of Computer and Mathematic Sciences, Melaka, Malaysia
[4]Universiti Teknologi Malaysia, Faculty of Computing, Johor, Malaysia
[5]Universiti Malaysia Terengganu, School of Informatics and Applied Mathematics, Terengganu, Malaysia

Abstract. In this paper, new brain tumour detection method is discovered whereby the normal slices are disassembled from the abnormal ones. Three main phases are deployed including the extraction of the cerebral tissue, the detection of abnormal block and the mechanism of fine-tuning and finally the detection of abnormal slice according to the detected abnormal blocks. Through experimental tests, progress made by the suggested means is assessed and verified. As a result, in terms of qualitative assessment, it is found that the performance of proposed method is satisfactory and may contribute to the development of reliable MRI brain tumour diagnosis and treatments.

1 Introduction

In recent years, tumor incidence has become rapidly increasing and this gives cause for concern. A tumor exist when new cells in an organism growth abnormally. These new cells grow very quickly compared to normal cells. Upon growing, these cells can cause damage to adjacent structures or cells and might be fatal. Brain is the most vital organ in human being and therefore brain tumor has always been taken serious by medical practice around the world in order to detect and treat the disease as early as possible. To ensure an accurate diagnosis of the brain tumor, an imaging technique is used to measure the volume of the tumor. A common imaging technique that has been widely used and accepted for the detection of brain tumor image is Magnetic Resonance Imaging (MRI). MRI is a type of scan that is used in radiology to gather pictures of the anatomy and the physiological processes in an organism. The process involved the usage of strong magnetic fields, radio waves and field gradients to yield organ images. From the image, doctors will evaluate and diagnose the condition of the tumor and come out with possible treatments. Most of the times, distinguishing between normal slice and abnormal slice of MRI images are quite challenging. Thus, a proper method is crucial to identify and extract the true location of the tumor. Experiments were conducted on a set of digitized MRI images from three types of dataset that were obtained from Internet Brain Segmentation Repository (IBSR) and multimodal Brain Tumor Segmentation (BRATS) challenge from 1st International Conference on Medical Image Computing and Computer Assisted Intervention (MICCAI 2012).

* Corresponding author: hafizahghazali@unimap.edu.my

The process of detecting abnormality in MRI brain slices is somewhat a thorough, extensive and time-consuming clinical procedure (Singh and Kaur 2012; Kumari 2013; Salankar and Bora 2014). This circumstance is explained by two underpinning factors : (1) homogeneity that occurs among healthy tissues and cancerous cells which are unable to be figured out by naked eyes and some machines, and (2) the presence of humongous amount of slices depending on the type and degree of illnesses (Selvaraj *et al.* 2007; Abdullah *et al.* 2011; Salankar and Bora 2014). Due to these reasons, several trials have been conducted in order to automate the procedure but they were all below the expectations thus entail future improvements. In addition, another glaring matter lies on locating a tumour or cancerous cell present in the abnormal slice where no research has been conducted on this at present. Hence, there is a need for a solution to the problem which will not only equip doctors with a cutting-edge but also to provide a fruitful performance of the following processes which are segmentation, visualization and volume estimation of tumours.

2 Methodology

This part practices the latest procedures to separate the normal slices from the abnormal ones in the brain MR images. There are four steps involved for the proposed method: (1) Extraction of cerebral tissues, (2) Determination of the maximum value for separating the blocks in normal and abnormal kinds, (3) Detection of abnormal blocks via threshold gained in step 2, (4)

Output—identified abnormal slices together with the tumour region.

The threshold value is needed in the process of differentiation of the abnormal MRI slices from the normal ones. Previous works of threshold method used watershed and histogram analysis (Mustaqeem et al. 2012), Inter-Hemisphere Fissure (I-HF) and symmetrical nature threshold (Somasundaram and Kalaiselvi 2010) , and Otsu's threshold (Amrutal et al. 2010). However, in this work, the maximum threshold value is gained from statistical calculation which takes into account the feature parameters including mean, energy and entropy. Once the abnormal blocks inside the slices are detected, the abnormal MRI slices can be gained hence it is fundamental to ascertain the abnormal blocks and refine them in getting the accurate abnormal slice.

There are three subsections present in this section. In detail, Subsection 2.1 explains the method of extracting cerebral tissues. While Subsection 2.2 presents the determining method in gaining the optimal threshold value and subsection 2.3 explains the detection process of tumor block abnormality in the MRI.

2.1. Extraction of Cerebral Tissues

The extraction of the cerebral tissues has four main steps. (1) Image binarization where the transformation of gray scale images (0-255) into binary image through two values of zero and one (Somasundaram and Kalavathi 2011; Jain 2013). This is conducted in order to expel the non-cerebral where in binarization, the threshold level is identified as being either white or black. In this case, Otsu's automatic thresholding algorithm is regarded as a benchmark (Kotte *et.al* 2016; Hamamci *et al.* 2012; Zhang *et al.* 2012).

After that, the determination of the Largest Connected Component or LCC is conducted by locating the component of the largest non-cerebral tissue region through several numbers of connected components inclusive of skull, bone, eyeballs and dura within that slice (Park and Lee 2009; Somasundaram and Kalaiselvi 2011). The proposed technique then separates MRI slice into joined components which combine all the connected white pixels according to its connectivity. In each of these components, 8-side connectivity is used to measure the adjacencies among pixels and they are later labelled orderly using numbers. Similarly labelled pixels represents a component that is extracted through the image to segment each component individually. After all joined components are identified, overlapping might occur where the extension of a component region into another constituent area. As a result, the proposed technique will again locate the connected parts and consider only constituent that possesses the larger area based on pixels. Once all joined components in the MRI slice image are located, the biggest constituents with pixels in every labelled area are found by one loop.

Once the LCC is detected through the previous stage, a binary mask which is based on the LCC pixels is established in the same size as the binary MRI slice. They are then labelled as black and the others as white.

The binary mask for a brain is located through earlier steps where the cerebral tissues are extracted through simple bitwise operation "AND" between the actual MRI slice and the binary mask. It protected the pixels of cerebral tissues corresponding to translucent areas and concurrently restructured the pixels of the masked overlapping image bits with zeros. Fig. 1 displays the binary mask operation of this MRI slice.

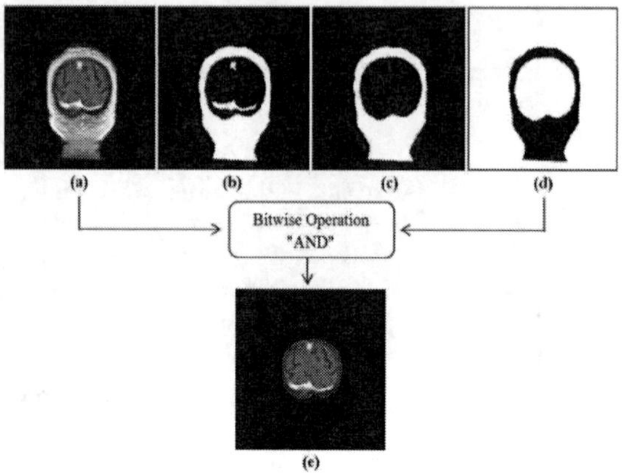

Fig. 1. The cerebral tissue extraction of MR image, (a) Original slice, (b) Binary image with three joined components (white regions), (c) The LCC (the largest white region), (d) The LCC mask (black region), and (e) Extracted cerebral tissue

2.2 Determination of Thresholds

Before the detection of abnormality took place, optimal threshold values are identified statistically through three parameters which are mean, energy and entropy. Then, each MR image that has a set of slices are grouped into non overlapping of 8x8 pixels that is regarded as optimal due to its empirical state that is determined by experiments through several block sizes ranging from 4x4 to 16x16 pixels. After that, for each block, the stated parameters are extracted to identify three threshold values which are the Mean-threshold (T_1), Energy-threshold (T_2), and Entropy-threshold (T_3). Statistically, the characteristics are calculated using equation (1), (2) and (3) respectively.

$$Mean\ (\mu) = \frac{\sum_{x=1}^{w}\sum_{y=1}^{w} I(x,y)}{w \times w} \quad (1)$$

$$Energy\ (e) = \frac{1}{w \times w}\sum_{x=1}^{w}\sum_{y=1}^{N} I^2(x,y) \quad (2)$$

$$Entropy = \frac{1}{w \times w}\sum_{x=1}^{w}\sum_{y=1}^{w} I(x,y)(-lnI(x,y)) \quad (3)$$

where (w=8) is the block size. A correlation between these characteristics and their distribution in the MRI slice is identified in order to gain the optimal threshold values.

2.3 Detection of Abnormal Blocks

The identification of abnormal blocks within each MRI slice is the conducted after the threshold values have been successfully determined. It is clear that if abnormal blocks are present in a particular slice, it will be regarded as abnormal where T_1, T_2 and T_3, are used to locate the abnormal blocks. A block is identified as abnormal if and only if all the rules are satisfied where the rule is shown in Fig. 2.

```
if Mean > T1 then
    if Energy > T2 then
        if Entropy > T3 then
            Block is Tumour
        else
            Block is Non-Tumour
```

Fig. 2. Abnormal block detection rules

Nonetheless, the achieved result might not be 100% accurate as some blocks of tumours might be captured as the non-tumour part. This somehow influences the results of abnormality detection of the slice hence a refined tuning procedure is suggested in order to check, validate and enhance the initial normal block. The tuning procedure is done through three types of 3x3 block masks which are preceding, current, and succeeding block mask as shown in Fig. 3.

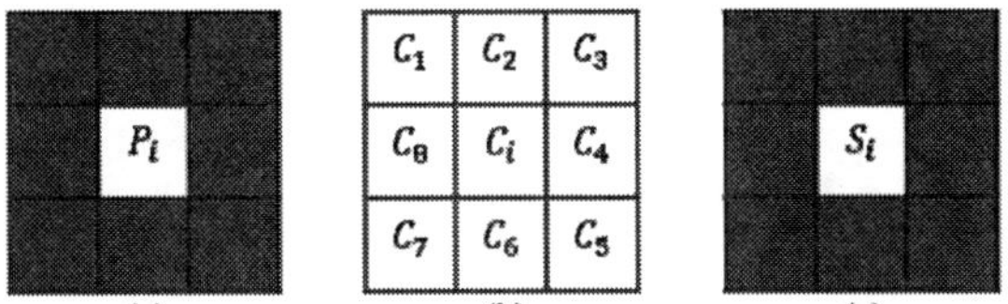

Fig. 3. Three types of block masks with size (3x3) called (a) Preceding mask centred at P_i, (b) Current mask centred at C_i, and (c) Succeeding mask centred at S_i. Blue colour indicates ignored blocks.

This procedure is done by scanning MRI slice block-by-block starting from the top-left to the right-bottom corners through the Current Block Mask starting from the top-left corner and moves a column to another till the last one. After that. it moves down one row and starts again from the first column until it reaches the last column and so on so forth until it reaches the right-bottom corner. The scanned slice is then labelled as tumour slice if it has at least a tumour block. This mechanism does not only identify or locate the tumour slice, it also shows location and size (in terms of block) of tumour regions which is also known as Region of Interest (ROI).

3 Experimental Results and Discussion

Using publicly available IBSR (10Normals_T1), IBSR (536_T1) and challenge MICCAI (BRATS2012-BRATS-1) datasets comprised of 4567 MRI slices of 35 patients out of which 25 are cancerous and 10 non-cancerous, a series of experimentations are handled where qualitative measurements are based on visual human inspections.

Abnormality detection of MRI slices is conducted by doing a sequence of experiments to assess the performance of the proposed abnormal MRI slice detection inclusive of (1) Cerebral tissue extraction, (2) Abnormal block detection (before and after the fine-tuning mechanism), and (3) The abnormal slice detection.

3.1 Results of Cerebral Tissue Extraction

In this stage, qualitative evaluation on various random MR slice images gained from the selected dataset is conducted. Fig. 4 depicts an original MRI slice containing both cerebral and non-cerebral tissues.

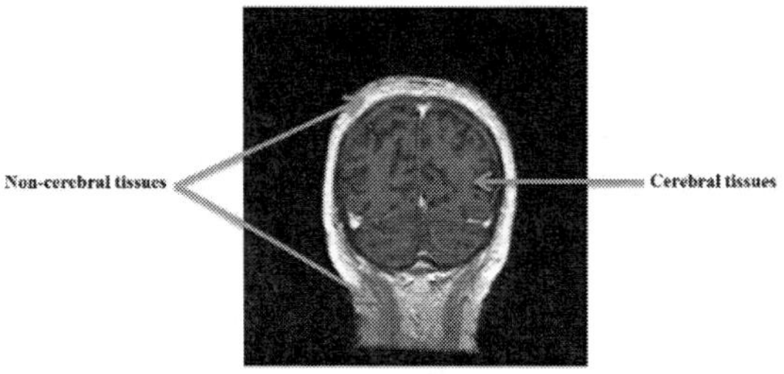

Fig. 4. The original slice showing cerebral and non-cerebral tissues

In order to extract the cerebral tissues via bitwise operations between the original MRI slice with the binary mask of the LLC, subsection 2.1 is conducted. Hence, only the cerebral tissues are maintained in the MRI slice when this mask is done, as illustrated in Fig. 5.

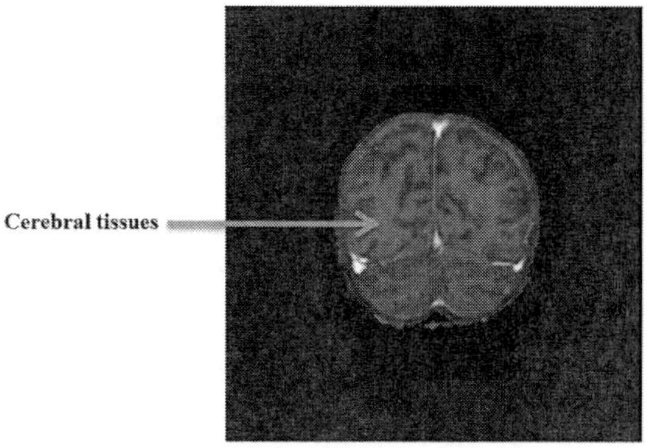

Fig. 5. Result on cerebral tissue extraction

https://doi.org/10.1051/matecconf/201815006025

Fig. 6 shows the experimental results for cerebral tissues of MRI brain slices obtained from IBSR (536_T1) dataset using the proposed method. The left column is the original MRI slice image while the right column is the cerebral tissues gained from the MRI slice.

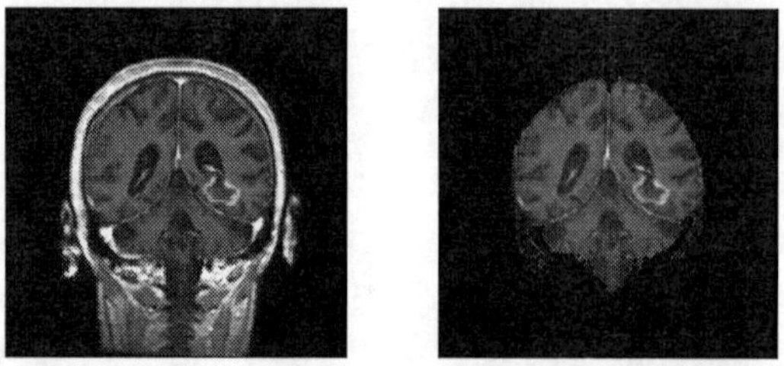

Slice 22

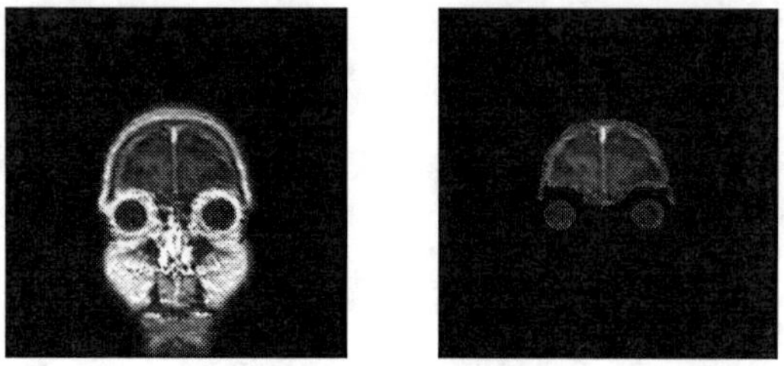

Slice 55

Fig. 6 Results of cerebral tissue extraction from IBSR (536_T1) dataset

The relevance of the proposed method is clearly disclosed by the thorough visual inspection of these extracted cerebral tissue MRI slices. The extraction mechanisms are able of removing all parts of non-cerebral tissues with high accuracy, including skull, bone, eyeballs and the dura.

3.2 Results of the Abnormal Block Detection Before Fine-Tuning

The extracted cerebral tissues are very significant for ROI tumour recognition through non-overlapping block division. Three characteristics such as mean, energy and entropy are obtained from each block where relationship between these features and their distribution in the MRI slice are evaluated using three dimensional (3D) graphs. By implementing Otsu's automatic thresholding algorithm, three distinctive thresholds are developed to represent these extracted features. The Otsu's algorithm is considered as the benchmark in this field. The performance of the current method is assessed using visual inspection mechanism. Fig. 7, 8, and 9 illustrate the experimental results of the block classification using three datasets, where the left column shows the original MRI slices image, the centre is the (8×8) block division using the proposed method and the right column is block detection result where the blue colour represents the normal region (non-tumour area) and the other blocks signifies the abnormal area (tumour area).

Fig. 7 Abnormal block detection result of MRI slice of IBSR (10Normals_T1) dataset

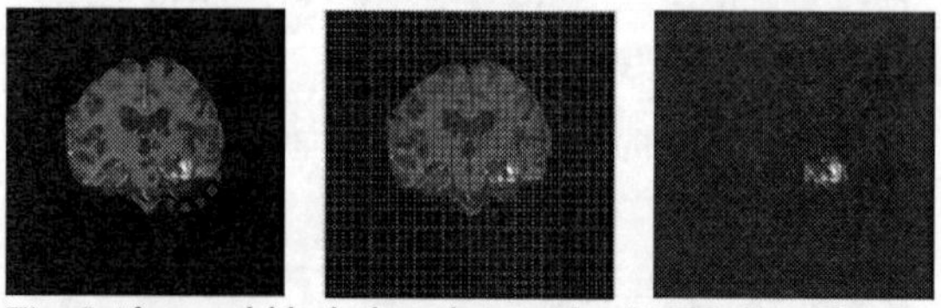

Fig. 8 Abnormal block detection result of MRI slice of IBSR (536_T1) dataset where the tumour initiated by a red circle

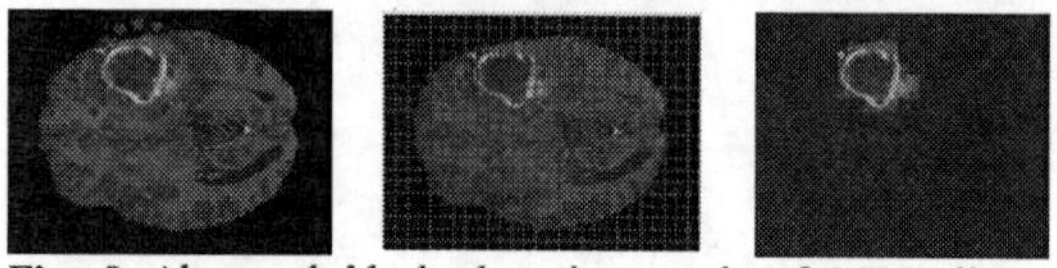

Fig. 9 Abnormal block detection result of MRI slice of MICCAI (BRATS2012-BRAST-1) dataset where the tumour initiated by a red circle

Based on the visual inspection of these slices, it is clearly discovered that the proposed abnormal block detection method, in terms of the brain tumour region, resulted in reasonably excellent result. Nevertheless, several misclassified results were found using the proposed method. Fig. 10, 11, and 12 depict the experimental results for the three datasets, where the proposed method misclassified the block. The term misclassified is defined as "a questioned block is supposed to be tumour but the method labelled it as non-tumour, or vice versa". Due to this, some blocks of tumour region are identified as non-tumour (yellow circle) and numerous normal blocks happened to be the part of the tumour region which is represented by red circle.

Fig. 10 Misclassified block of MRI slice of IBSR (10Normals_T1) dataset without any tumour

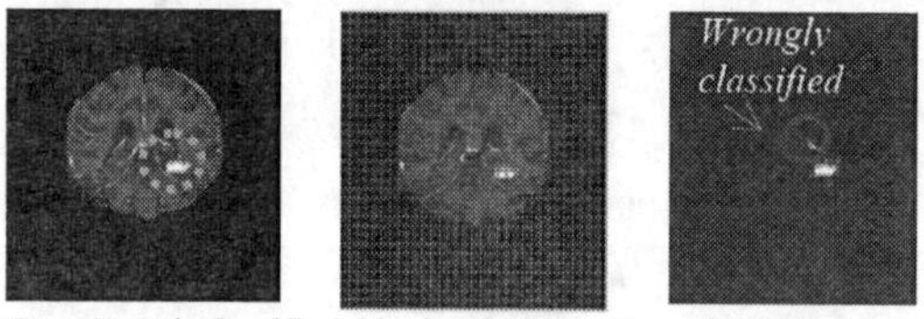

Fig. 11 Misclassified block of MRI slice of IBSR (536_T1) dataset with tumour marked via green circle

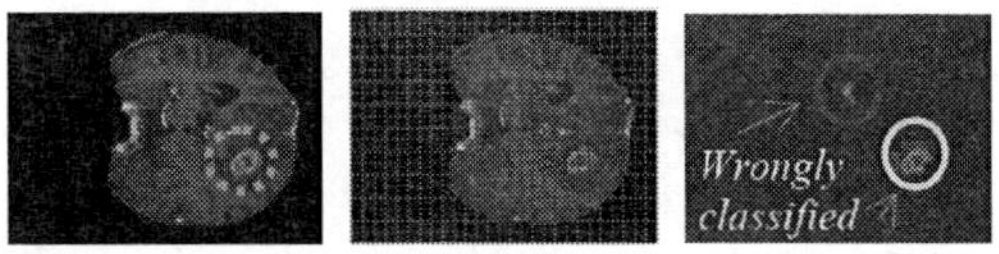

Fig. 12 Misclassified block of MRI slice of MICCAI (BRATS2012-BRAST-1) dataset with tumour initiated by circle

3.3 Results of the Abnormal Block Detection with Fine-Tuning

It is very clear that the inaccurate identification of the tumour region greatly influences the results of slice abnormality detection. This matter is resolved by applying sets of rules based on the neighbouring blocks. First, three types of (3×3) blocks mask namely "Preceding, Current, and Succeeding Blocks Mask" block-by-block are adopted in order to perform scannings of the MR slice image from top to bottom and from left to right direction. Next, all the neighbouring blocks surrounding a centre block in the "Current Block Mask" as well as the centred block of "Preceding, and Succeeding Blocks Mask" are chosen in the final decision of the block classification method. Accurate result is obtained by applying all previously mentioned rules (Subsection 2.3), where the occurrence of inexact classification blocks is then modified through the proposed method. This process is vital as the reliability of abnormality detection is primarily dependent on the accuracy of this mechanism.

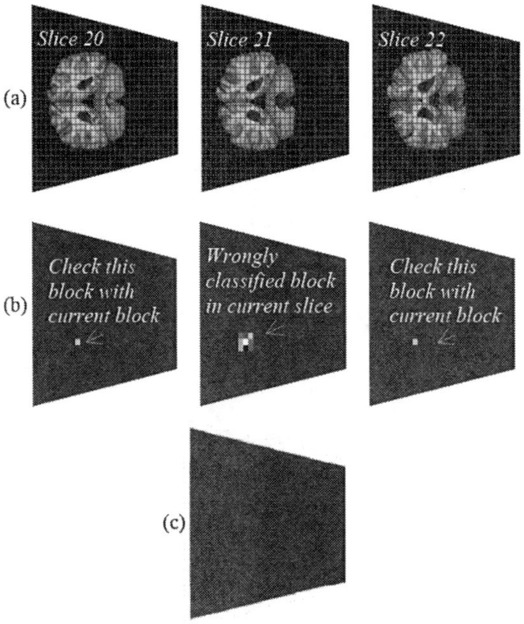

Fig. 13 A fine-tuning result of slice 21 of IBSR (10Normal_T1) dataset: (a) The original non-tumour slice 21, (b) Fine-tuning of the misclassified block of current slice, and (c) Final result of the questioned block

Now the rules are deployed to correct the wrong classifications. Fig. 13, 14, and 15 depict the corrected slices after taking out the inaccuracy. The first, second and third columns show the preceding slice, the current slice, and the succeeding slice, respectively.

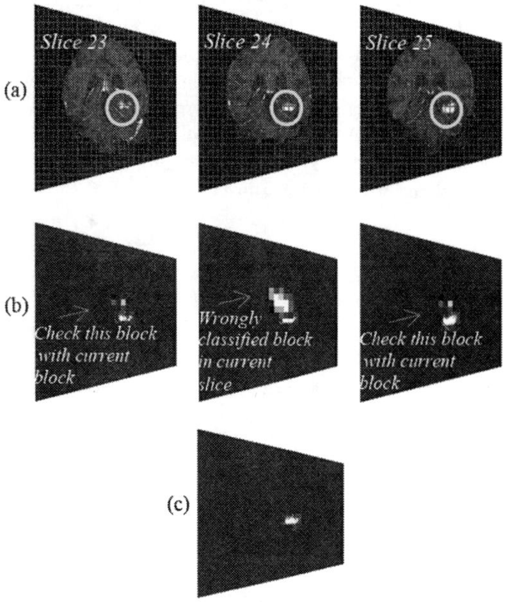

Fig. 14 A fine-tuning result of five tumour blocks of slice 24 from IBSR (536_T1) dataset: (a) The original tumour slice 21, (b) Fine-tuning of the misclassified block of current slice, and (c) Final result of the questioned blocks

It is prominent that from the above findings, the proposed fine-tuning procedure has successfully solved the misclassification issues and enhanced the accuracy of the tumour block detection. In general, the abnormal region posseses all the important characteristics of a tumour and any incorrectness in the identifications of this region has severe consequences with regards to anomaly detection and treatment. Through the proposed abnormal block detection method coupled with the fine-tuning procedure, the tumour region is perfectly selected (i.e. ROI) for better tumour slice detection and segmentation mechanism. As a whole, in terms of the qualitative assessment, the performance of all parts of the current methods, including cerebral tissue extraction, abnormality block classification and fine-tuning procedure are of a satisfactory level.

4 Conclusion

Through a new detection procedure for MRI brain abnormality presented in this paper, a set of pre-processing measures is executed with the MR images containing non-cerebral tissue extraction. These measures included the binarization of MRI slices, followed by their LCC detection. Then, a binary mask is gained from the extracted LCC and bitwise operations are applied between the original slice and the binary mask to extract the cerebral tissue from the MR image. As a result, three distinctive thresholds, which are based on texture features: mean, energy and entropy, are obtained automatically thus permits accurate separation of the MRI slice into normal and abnormal ones. Nevertheless, the abnormality detection showed some normal blocks which are placed incorrectly as abnormal and vice versa. This issue is surmounted by applying the fine-tuning mechanism of which the MRI slice abnormality detection is finally achieved by selecting the abnormal slices along its tumour region. The presented mechanism is suitable to be clinically practiced in the procedures of treating cancers including surgery, cancer drugs, chemotherapy and radiotherapy.

Acknowledgement

The authors would like to thank Universiti Malaysia Perlis for providing the facilities and financial support (Short Term Grant No. 9001-00567) to conduct this research.

References

1. D. Singh, K. Kaur, (2012). International Journal of Engineering and Advanced Technology (IJEAT), *Classification of Abnormalities in Brain MRI Images Using GLCM , PCA and SVM.*, vol.1, no.6, pp.243–248.
2. R. Kumari, (2013). SVM Classification an Approach on Detecting Abnormality in Brain MRI Images. *International Journal of Engineering Research and Applications (IJERA)*, vol.3, no.4, pp.1686–1690.
3. S. S. Salankar, V. R. Bora, (2014). MRI Brain Cancer Classification Using Support Vector Machine. *Electrical, Electronics and Computer Science (SCEECS), IEEE Students' Conference*. pp. 1–6.
4. A. Mustaqeem, A. Javed, T. Fatima, (2012). An Efficient Brain Tumor Detection Algorithm Using Watershed and Thresholding Based Segmentation. *International Journal of Image, Graphics and Signal Processing*, vol.4, no.10, pp.34–39.
5. K. Somasundaram, T. Kalaiselvi, (2010). Automatic Detection of Brain Tumor from MRI Scans Using Maxima Transform. *UGC Sponsored National Conference on Image Processing-NCIMP*. pp. 136–141.
6. A. Amrutal, A. Gole, Y. Karunakar, (2010). A Systematic Algorithm for 3-D Reconstruction of MRI based Brain Tumors using Morphological Operators and Bicubic Interpolation. *Computer Technology and Development (ICCTD), 2nd International Conference on. IEEE*. pp.305–309.
7. H. Selvaraj, S. T. Selvi, D. Selvathi, L. Gewali, (2007). Brain MRI Slices Classification Using Least Squares Support Vector Machine. *International Journal of Intelligent Computing in Medical Sciences and Image Processing*, vol.1, no.1, pp.21–33.
8. N. Abdullah, U. K. Ngah, S. A. Aziz, (2011). Image Classification of Brain MRI Using Support Vector Machine. *Imaging Systems and Techniques (IST), IEEE International Conference*, pp. 242–247.
9. K. Somasundaram, P. Kalavathi, (2011). *Medical Image Binarization Using Square Wave Representation*. First Edition. Springer-Verlag Berlin Heidelberg.
10. S. Jain, (2013). Brain Cancer Classification Using GLCM Based Feature Extraction in Artificial Neural Network. *International Journal of Computer Science and Engineering Technology*, vol.4, no.7, pp.966–970.
11. S. Kotte, P. R. Kumar, S. K. Injeti, (2016). An Efficience Approach for Optimal Multilevel Thresholding Selection for Gray Scale Images Based on Improved Differential Search Algorithm. *Ain Shams Engineering Journal.*
12. A. Hamamci, G. Unal, (2012). Multimodal Brain Tumor Segmentation Using The "Tumor-Cut" Method on The BraTS Dataset. *Process MICCAI-BRATS (Multimodal Brain Tumor Segmentation Challenge)*, pp.19–23.
13. T. Zhang, Y. Xia, D. D. Feng, (2012). Clonal Selection Algorithm for Gaussian Mixture Model Based Segmentation of 3D Brain MR Images. *Intelligent Science and Intelligent Data Engineering. Pringer Berlin Heidelberg.*, pp.295–302.
14. J. G. Park, C. Lee, (2009). Skull Stripping Based on Region Growing for Magnetic Resonance Brain Images. *NeuroImage*, vol.47, no.4, pp.1394–1407.
15. K. Somasundaram, T. Kalaiselvi, (2011). Automatic Brain Extraction Methods for T1 Magnetic Resonance Images Using Region Labeling and Morphological Operations. *Computers in Biology and Medicine*, vol.41, no.8, pp.716–725.

Tourism Led-Inflation: A case of Malaysia

M. S. Shaari[1*], *T. Salha Tunku Ahmad*[1], *R.* Razali[1.]

[1]School of Business Innovation and Technopreneurship, Universiti Malaysia Perlis (UniMAP), Malaysia

Abstract. The importance of the tourism industry has prevailed among developed and even developing countries. It has been perceived to be an important contribution to economic growth. However, in the proliferation of studies on inflation, information on the extent to which tourism industry able to influence inflation, is still sparse. Therefore, this study embarks on investigating tourism as a potential factor towards inflation. Data on consumer price index and the number of tourist arrival from 1986 until 2014 are used in the analysis. This study adopts a quantitative approach employing the Autoregressive Distributed Lag (ARDL) approach. Several controlled variables such as money supply, economic growth, government expenditure, and interest rate are also included. The results suggest that the tourism industry plays an important role in determining inflation in both short- and long-runs. Therefore, governments should take proactive measures in ascertaining that any expansion of the tourism industry can avert inflation.
Keywords: tourism, interest rate, government expenditure, money supply, economic growth

1 Introduction

Recent efforts to understand the possible impacts of the tourism industry on economic growth has led researchers and policy makers to focus on understanding this industry. Throughout the decades, the literature on tourism sector has fuelled debate on the inextricable association between tourism and the two perspectives of the economy; namely, revenue and employment. Studies in the past decades have shown that countries with a large number of tourist arrival can earn higher revenues, hence boost their economic growth [1-7]. This situation is evidenced in both developed and developing countries. Thus any policy on expansions of the tourism industry can yield good results to the country.

Another aspect of tourism that also attract scholars is the relationship between tourism and employment rate. An increase in the number of tourist arrival can cause the demand for domestic goods and services to rise. Therefore, the supply will simultaneously increase to fulfil the high demand. Hence, people can avail themselves of various job opportunities owing to the higher supply. Thus, a rise in the demand for tourism in that country can prompt an increase in employment [8]. High economic growth caused by tourism implies a lower unemployment rate as evidenced in a study by [9]. This can contribute to a larger consumption of goods and service, causing a higher aggregate demand. As a result, prices of goods will increase, ensuing inflation.

Based on the above review, the tourism sector is perceived to be favourable to the economy [3, 4, 7]. Since many studies show that economic growth is one of the macroeconomic goals, policy makers will formulate policies to ensure positive economic growth. However, high economic growth can be detrimental to the economy as it causes inflation [10]. In this regards, the deleterious effects of the tourism sector on the economy is intriguing. Presumably, the inexorable expansion of the tourism industry can be disadvantageous as it may cause inflation. Tourists will spend their money on hotels, transportations, foods, clothes, etc. This may cause the aggregate demand to escalate, hence, ensuing inflation.

To date, extensive studies have been conducted to investigate the inflation factors. Government policies, money supply, lower interest rate and higher government expenditures are among the factors that potentially impacted economic of one's country [11].

Yet, none has treated tourism as a potential variable with regards to inflation. This factor should be taken into account as many studies also highlighted that tourism can boost economic growth and they proposed various policies on the expansion of the sector. If these policies work in the economy, inflation will surely crop up. Because of this concerns, this study embarks on examining the tourism led-inflation.

2 Literature Review

To illuminate the issues on inflation, review on the relevant literature is classified into two perspectives. One is on the predictors of inflation used in the previous studies. Second, is on the findings across the world albeit the similar type of predictors or methods used by researchers. Thus far, the tourism sector is undoubtedly perceived to be favourable to the economy [2, 4]. Parallel with the interest in the economic growth, inflation has become one of the debatable issues in such literature. In studies concerning inflation, predictors such as government policies, money supply, interest rate and government expenditures have been shown to have influenced inflation in one's country.

Most of the studies that use money supplies as inflation predictor, such as those by [12-18], report a positive relationship between money supply and inflation. For instance, [12] found that higher money supply can prompt inflation. This is attributed to the Federal Reserve's policy that attempts to stabilize inflation or increase economic growth by adjusting money supply. Money supply will be increased in order

* Corresponding author: shahidanshaari@gmail.com

to ensure positive economic growth. However, a higher money supply can push prices of goods up, thus leading to inflation. This is due to more money in the market.

In the same demeanor, studies on the interest rate [see 13, 14], show that lower rate can increase the number of borrowers, which can cause people to have more money. The higher amount of money people possess, the more money people will spend, thus triggering undesirable inflation. On the contrary, if the interest rate is lower, the inflation can be decreased.

Apart from both factors, government spending can also be a determinant of inflation [19]. Governments need to spend to stimulate the economy and encourage private investments. However, when more money is spent, inflation may be triggered, as there are more investments from the higher government spending. The number of the employed increases and then consumption will hike.

It is noteworthy that, these studies, regardless of the different methods used, have produced consistent findings. In particular, [18] employed the ARDL approach to examine the linkage between money supply and inflation in two groups of countries: high inflation group and low inflation group. Their study found mixed findings between high and low inflations group whereby there is no significant relationship between money supply and inflation in the low inflation group and there is a significant relationship between money supply and inflation in the high inflation group. This finding strongly suggests that inflation is obviously affected by money supply. Another example of studies using similar predictor but different data samples and approach is [17, 15]. They employed ARDL approach and Johansen cointegration approach respectively. Like their counterpart, they also investigated the relationship between money supply and inflation, by applying the methodology to the Malaysia dataset. Although with different samples of data and methods, their findings are consistent with each other, by which relationship between money supply and inflation was evidenced.

Besides studies that use similar predictor with different methods, it is interesting to review the findings on inflation predictors done by researchers in different countries. Using economic growth as the predictor, numerous studies have been conducted to study its' effects on inflation [8, 20, 21]. [20] employed Johansen cointegration to examine the effects of economic growth in Nigeria. The results showed that there is a significant effect of economic growth on inflation in Nigeria. [8] employed Autoregressive Distributed lagged approach to explore the effects of economic growth on inflation in Bangladesh. Data from 1972 to 2012 were analyzed and the results were split into two dimensions; long run and short run. The results consistently showed that economic growth plays an important role in determining inflation in both short and long runs. [21] studied the effects of economic growth on inflation in Nigeria. The study used Multiple Regression Analysis and Variance Inflation Factor (VIF) model to analyze the data from 1970 until 2008. The results showed the existence of a relationship between economic growth and inflation in Nigeria.

Interest rate can also influence inflation across the world. A significant relationship between interest rate and inflation is exemplified in works by [14] in Asia, [22] in Kenya, and [23] in Nigeria. [14] employed the Pooled Mean Group (PMG) and the panel differenced GMM (General Method of Moment) to analyze Asian countries' data from 1985 to 2012. The results showed that interest rate can significantly affect inflation. [22], using various methods, including Granger causality, correlation and regression, which are different from [14], suggests that there is an existence of a relationship running from interest rate to inflation in Kenya, from 1961 to 2011. While [23] argued that there is no relationship between interest rate and inflation. The study employed the ARDL approach and examined the effects of interest rate on inflation from 1971 to 2010. The findings proved that there is no relationship between interest rate and inflation in Nigeria.

Apart from above factors, the effects of government spending on inflation have also been investigated by previous studies [14, 19, 24]. Some found that government spending is one of the significant determinants of inflation and some found it is not. [19] used Vector Autoregressive Model (VAR) to examine the effects of government spending on inflation in Nigeria from 1970 to 2010. The study found that government spending is an important factor of inflation in Nigeria. [14] agreed that government spending does really affect inflation. [14] examined the effects in India, Vietnam, and Indonesia using Vector Error Correction Model (VECM). Data from 1970 to 2010 were analyzed and the findings consistently proved that government spending is a significant factor of inflation across the three countries. However, [24] argued that government spending does not have significant influence in Iran. They used Smooth Transition Regression (STR) models transition and quarterly data from 1990 to 2014 were collected. Their findings reveal that government spending does not significantly affect inflation.

3 Methodology

This study aims to examine the tourism led-inflation in Malaysia. Data from 1986 to 2014 were collected and analyzed. This study used consumer price index as the proxy for inflation and several controlled variables: consumer price index, money supply (M2) real Gross Domestic Product (GDP), interest rate, and government expenditure. Several tests were conducted in this study such as unit root test, Bound test, long-run and short-run estimations tests using ARDL approach and diagnostic tests. The model specification is as follows:

$$\ln CPI = \alpha + \beta_1 \ln MS + \beta_{21} \ln GDP + \beta_3 \ln IR + \beta_4 \ln TA + \beta_5 \ln GE + \varepsilon$$

where:
lnCPI is Log of Consumer Price Index
lnMS is Log of Money Supply
GDP is Log of Real Gross Domestic Product
IR is Log of Interest Rate

TA is Log of Tourists Arrival
GE is Log of Government Expenditure

3.1 Stationary test

In the analysis of times series data, the unit root test must be conducted to see the stationary of data. Hence, the unit root test based on Augmented Dickey-Fuller (ADF) is performed. Each variable must be estimated at level and first difference. The equation for the unit root is as follows:

$$\Delta Y_t = \beta_1 + \beta_2 t + \delta X_{t-1} + \alpha \sum_{i=1}^{p} \Delta X_{t-1} + u_t \quad (1)$$

X indicates the variable while t is the time trend. p represents the number of lagged term and u represents the white noise. (α_1,α_2,β_1,…β_m) is a set of parameters. If the results of the unit root test show that it is not significant, it means that the data have unit root or are not stationary. If the results show that it is significant, it suggests that the data have no unit root or stationary. Hence, the hypothesis can written as follows:

H_0:δ=0 (unit root/ non stationary)
H_1:δ≠0 (no unit root/ stationary)

3.2 Autoregressive Distributed Lad (ARDL) Model

This study employs the Error Correction Model based on the ARDL modelling to estimate the relationship between inflation and tourist arrival. This study uses the ARDL model as proposed by Pesaran et al. (2001). Since this study is to examine the effect of tourist arrival on inflation, some control variables such as interest rate, money supply, and government expenditure are also included. In order to estimate the ARDL model, there are three approaches need to be completed. The first step is to estimate a cointegrated relationship among the variables using the following equation:

$$\Delta \mathrm{lnCPI}_t = \theta_1 + \sum_{t=1}^{p} \lambda_1 \Delta \mathrm{lnCPI}_{t-1} + \sum_{t=0}^{q} \lambda_2 \Delta \mathrm{lnMS}_{t-1} + \sum_{t=0}^{r} \lambda_3 \Delta \mathrm{lnGDP}_{t-1} + \sum_{t=0}^{s} \lambda_4 \Delta \mathrm{lnGE}_{t-1} + \sum_{t=0}^{t} \lambda_5 \Delta \mathrm{lnIR}_{t-1} + \sum_{t=0}^{u} \lambda_6 \Delta \mathrm{lnTA}_{t-1} + \pi_1 \mathrm{lnCPI}_{t-1} + \pi_2 \mathrm{lnMS}_{t-1} + \pi_3 \mathrm{lnGDP}_{t-1} + \pi_4 \mathrm{lnGE}_{t-1} + \pi_{5l} \mathit{ln}\mathrm{IR}_{t-1} + \pi_6 \mathrm{lnTA}_{t-1} + \mu_{1t}$$

(2)

where, Δ is the first differential operator, (p,q,r,s,t,u) is the optimum lag and μ refer to designation. To determine the existence of long run relationship between the variables in the equation (1), the hypothesis is as follows:

H_0: There is no cointegration:
$(\pi_1 = \pi_2 = \pi_3 = \pi_4 = \pi_5 = \pi_6 = 0)$
H_1: There is cointegration:
$(\pi_1 \neq \pi_2 \neq \pi_3 \neq \pi_4 \neq \pi_5 \neq \pi_6 \neq 0)$

If the F-statistic value is larger than upper bound critical value, it can be concluded that there is an existence in cointegration between the variables. If the F-statistic value is less than the lower bound critical value, then null hypothesis cannot be rejected. If the F-statistic value is between the upper bound and lower bound critical value, so the existence of cointegration cannot be determined. If there is an existence of cointegration, next step is to estimate the long-run relationship using the following equation:

$$lnCPI_t = \theta_{11} + \sum_{i=1}^{p} \pi_{11}\, lnCPI_{t-i} + \sum_{i=0}^{q} \pi_{22}\, lnM3_{t-i} + \sum_{i=0}^{r} \pi_{33}\, lnGDP_{t-i} + \sum_{i=0}^{s} \pi_{44}\, lnGE_{t-i} + \sum_{i=0}^{t} \pi_{55}\, lnIR_{t-i} + \sum_{i=0}^{u} \pi_{66}\, lnTA_{t-i} + \mu_{11t} \quad (3)$$

The final step, the short run ARDL model is estimated by including the error correction terms (ECT). The short-run estimation is depicted via an expression as follow:

$$\Delta \mathrm{lnCPI}_t = \theta_{111} + \sum_{t=1}^{p} \lambda_{111} \Delta \mathrm{lnCPI}_{t-1} + \sum_{t=1}^{q} \lambda_{222} \Delta \mathrm{lnMS}_{t-1} + \sum_{t=1}^{r} \lambda_{333} \Delta \mathrm{lnGDP}_{t-1} + \sum_{t=1}^{s} \lambda_{444} \Delta \mathrm{lnGE}_{t-1} + \sum_{t=1}^{t} \lambda_{555} \Delta \mathrm{lnIR}_{t-1} + \mu_{111t} \quad (4)$$

4 Results

The unit root test was conducted to determine the stationary properties of the time series data. This test was based on the Augmented Dickey-fuller (ADF) approach. The results were divided into two levels: level with trends and level without trends. Table 4.1 shows the results of the unit root test. The results show that all the variables are significant at first difference. It indicates that the variables are stationary at first difference. Based on the results, the ARDL test can be conducted.

Table 4.1: Unit Root Test Results.

	Intercept		Intercept and trend	
	Level	1st difference	Level	1st difference
CPI	-4.19*	-5.93*	-4.21**	-5.86*
MS	-2.36	-2.89***	-1.95	-3.68**
GDP	-1.04	-5.03*	-2.23	-5.00*
GE	-0.38	-5.31*	-2.51	-4.66*
IR	-6.27*	-8.00*	-7.43*	-7.836*
TA	0.01	-4.81*	-2.04	-4.82*

Note: *, **, and *** significant at 1%, 5% and 10% significant respectively.

The Bound test was conducted and the results were recorded in Table 4.2. The results show that the value of F-statistics is 5.167 which is larger than the value of upper bound at a significant level of 1%. Therefore, it can be concluded that an estimation of a long-run relationship can be performed.

Table 4.2 Bound Test for the Existence of a Long Run Relationship

F-statistic		
5.167021		
Critical value		
Significance	**I0 Bound**	**I1 Bound**
1%	3.41	4.68
2.5%	2.96	4.18
5%	2.62	3.79
10%	2.26	3.35

Table 4.3 shows the results of the estimated long run coefficient. The results suggest that money supply can negatively affect inflation at a significant level of 1%. Therefore, a 1% increase in money supply can cause inflation to increase by 17% in the long run. Based on the results, it may suggest that GDP does not affect inflation in the long run as the value of probability is 0.3182 and it is not significant at any level. Apart from that, government expenditure can positively affect inflation as it is significant at 1%. Hence, a rise of 1% in government expenditure can increase by 25.25% in the long run. The results also indicate interest rate can trigger inflation as it is significant at 1%. This implies that a 1% rise in interest rate can prompt inflation to rise by 0.88% in the long run. The results also suggest that tourist arrival can significantly affect inflation as it is significant at 1%. Therefore, an increase in 1% tourist arrival will lead to an increase of 14.87% in inflation. The constant value in this result is also significant at 5% with a coefficient value of 18.1780.

Table 4.3 Estimated Long Run Coefficient using the ARDL Approach

Variable	Coefficient	Standard Error	Probability
MS	17.16	4.93	(0.00)*
GDP	5.49	5.16	(0.31)
GE	23.25	6.02	(0.00)*
IR	0.88	0.22	(0.00)*
TA	14.87	3.91	(0.00)*
C	18.18	6.76	(0.02)**

Notes *, and ** indicate significant at 1% and 5% significant level.

Table 4.4 shows the result of the estimated short-run coefficients. The results indicate that money supply significantly affect inflation and it is significant at 1%. Therefore, a 1% increase in money supply can cause inflation to increase by 14.9% in the short run. The results also suggest that GDP and government expenditure do not have any influence on inflation in the short run. The interest rate is found to be connected with inflation as the results show that it is significant at 1%. It implies that an increase of 1% in interest rate can soar by 0.59% in the short run. Other than that, tourist arrival can have an effect on inflation as it is significant at 5%. This suggests that if tourist arrival goes up by 1%, inflation will surge by 3.5%.

Table 4.4 Estimated Short-run Coefficients Using the ARDL approach

Variable	Coefficient	Standard Error	Probability
dMS	14.90	2.70	0.00*
dGDP	2.73	7.09	0.71
DGE	7.17	6.37	0.29
dIR	0.59	0.13	0.00*
dTA	3.56	1.28	0.02**

Notes * and ** indicates significant at 1% and 5% level respectively.

Table 4.5 shows the results of the diagnostic tests. These tests are to check the goodness of the ARDL model. The Breusch-Godfrey Serial Correlation test, the Ramsey RESET stability test and the ARCH test were conducted. The results show that they are all not significant at any level. Therefore, it can be concluded that no evidence of autocorrelation, serial correlation and heteroscedasticity.

Table 4.5 Diagnostic Tests

Test statistic	F-statistic (Probability)
A: Breusch-Godfrey Serial Correlation	0.49 (0.63)
B: Ramsey RESET stability	0.12 (0.73)
C: ARCH	0.01 (0.91)

In order to ensure the stability of the variables, the cumulative sum (CUSUM) and cumulative sum of squares (CUSUMQ) tests were conducted. Figure 4.6 shows the results of CUSUM and CUSUMQ. The results show that the plots are both located within the margins where both straight lines denote the critical bound at 5% significance level, meaning that this model is good. Therefore, it can be observed that there is a long run coefficient stability in the variables.

Figure 4.6 CUSUM and CUSUMQ

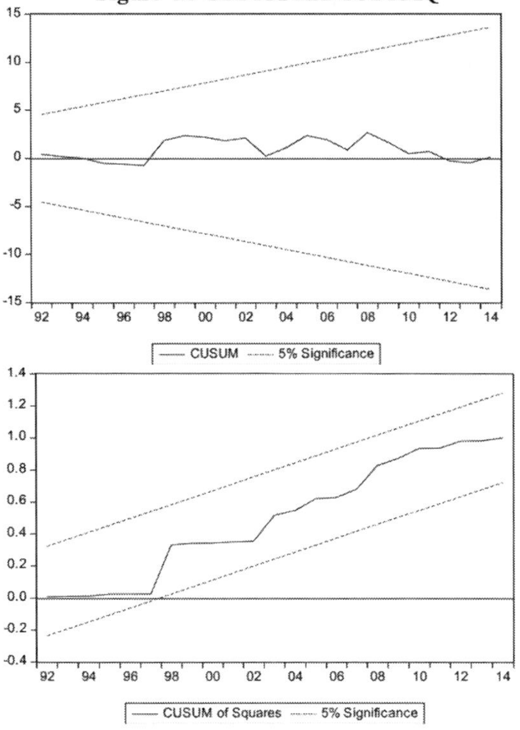

5 Conclusions

This study aims to investigate the tourism led-inflation in Malaysia from 1986 to 2014. The unit root tests were conducted and the results show that most of the variables are non-stationary at level and stationary at first difference. Then, the Bound test was conducted and the results show that there is a cointegration. The long-run estimation using the ARDL approach was carried out and the results show that money supply, interest rate, movement expenditure, and tourist arrival can positively affect inflation in the long run. However, economic growth does not have any effect on inflation. Other than that, the short-run estimation using the ARDL approach was also performed and the results indicate that money supply, interest rate, and tourist arrival can influence inflation in the short run. Nevertheless, government expenditure and economic growth do not have any significant effect on inflation.

This study can be instrumental in the formulation of policies on tourist arrival in order to avert inflation. Inflation is one of the macroeconomic problems that should be controlled by all countries. Since an expansion of the tourism industry can trigger inflation in both short- and long-runs, the government should take possible careful measures to prevent inexorable expansion of the industry, while elevating its potential as country's economic booster.

References

1. Lee, C. C. & Chang, C. P. Tourism development and economic growth: A closer look at panels, Tourism Management, 29, 180–192. (2008).
2. Katircioglu, S. T. Revisiting the tourism-led-growth hypothesis for Turkey using the bounds test and Johansen approach for co-integration. Tourism Management, 30, 17–20. (2009).
3. Payne, J.E., & Mervar, A. The tourism growth nexus in Croatia. Tourism Economics, 16(4), 1089–1094. (2010).
4. Seetanah, B. Assessing the dynamic economic impact of tourism for island economies. Annals of Tourism Research, 38(1), 291–308. (2011).
5. Lee, J.W. and Brahmasrene, T. Investigating the Influence of Tourism on Economic Growth and Carbon Emissions: Evidence from Panel Analysis of the European Union. Tourism Management, 38, 69-76. (2013).
6. Pablo-Romero, M.P., & Molina, J.A. Tourism and economic growth: A review of empirical literature. Tourism Management Perspectives, 8, 28–41. (2013).
7. Salmani, B., Panahi, H., & Razzaghi, S. Assessing the dynamic economic impact of tourism for OIC members. World Applied Sciences Journal, 32(6), 1098–1105. (2014).
8. Inchausti-Sintes F. Tourism: Economic growth, employment and Dutch disease. Annals of Tourism Research, 54, 172-189. (2015).
9. Jebabli, I., Arouri, M., & Teulon, F. On the effects of world stock market and oil price shocks on food prices: An empirical investigation based on TVP-VAR models with stochastic volatility. Energy Economics, 45, 66-98. (2014).
10. Uddin, S., Chowdhury, M.N.M., & Hossain, M.A. Determinants of inflation in Bangladesh: An econometric investigation. Journal of World Economic Research, 3(6), 83-94. (2014).
11. Khan, R.E.A. & Gill, A.R. Determinants of Inflation: A Case of Pakistan (1970-2007). Journal of Economics, 1 (1). 45-51. (2010).
12. Akinbobola T.O. The dynamics of money supply, exchange rate and inflation in Nigeria. Journal of Applied Finance & Banking, 2(4), 117-141. (2012).
13. Anari A. and Kolari J. Dynamics of interest and inflation rates. Journal of Empirical Finance, 39(December 2016), 129-144. (2016)

14. Nguyen V.B. Effects of Fiscal and Money M2 Supply on inflation: Evidence from selected economies of Asia. Journal of Economics, Finance and Administrative Science, 20(38), 49-53. (2015)
15. Khai, T.M. Determinants of inflation in Malaysia 1981-2010. MBA management project, Universiti Sains Malaysia. (2011).
16. Basher, S. A., & Elsamadisy, E. M., (2012). Country heterogeneity and long-run determinants of inflation in the gulf Arab states. OPEC Energy Review, 36(2) 170-203.
17. Hashim, M.J., Osman, I., & Elias, N.L. The determinants of inflation in Malaysia. 3rd International Conference on Accounting, Business and Economics (ICABEC2014) (2014).
18. Lim, Y.C., & Sek, S.K. An examination on the determinants of inflation. Journal of Economics, Business and Management, 3(7). (2015).
19. Olayungbo, D.O. Government spending and inflation in Nigeria: An Asymmetry Causality Test, International Journal of Humanities and Management Sciences, 1(4). (2013)
20. Sictus, A. Agricultural share of the gross domestic product and it implication to rural areas. International Journal of Social Economics, 226-238. (2010).
21. Odusanya, I. A., & Atanda, A. A. Analysis of inflation and its determinants in Nigeria. Pakistan Journal of Social Sciences, 7(2), 97-100. (2010).
22. Mmasi, B.S. An investigation of the relationship between interest rate and inflation in Kenya. MBA thesis, College of Humanities and Social Sciences (CHSS). (2013).
23. Ebiringa, O.T., & Anyaogu, N.B. Exchange rate, inflation and interest rates relationships: An Autoregressive Distributed Lag Analysis. Journal of Economics and Development Studies, 2(2), 263-279. (2014).
24. Mehrara, M., Soufiani, M.B., & Rezaei, S. The Impact of Government Spending on Inflation through the Inflationary Environment, STR approach. World Scientific News. 37, 153-167. (2016).

Examining the relationship between learning capabilities and organizational performance: The mediating role of organizational innovativeness

Khawar Hussain[1*], *Eta Wahab*[1], *Ali Zeb*[1], *Muhammad Asad Khan*[1], *Mudaser Javaid*[1], *Muhammad Ayaz Khan*[1],

[1]Faculty of Technology Management and Business, Universiti Tun Hussein Onn Malaysia, Johor Malaysia

Abstract. This paper investigates the Organizational Learning Capabilities (OLC) affect Organizational innovativeness (OI) on organizational performance (OP). The mediating role of IO on relationship between OLC and OP was also been studied. Analysis is based on data collected from 200 respondents drawn from Pakistan's Telecom Sectors. Multiple regression analysis had been used to test the hypotheses. This study found that both OLC and OI have a positive impact towards OP. OI also mediates the relationship between OLC and OI. Key word: Organizational Learning Capabilities (OLC), Organizational Performance (OP), Organization Innovativeness (OI).

1 INTRODUCTION

1.1 Background of study

OLC has been believed by academics and practitioners as vital for associations generally due to the fast-changing environment. Firms that are able to discover stand a larger chance of detecting, replacing, and adapting on a vibrant nature [1]. Learning capability (LC) is an element of Organizational capabilities (OC) and an essential factor for OI [2]. The ancestor of the LC of the OC extends to use of how organization can grow new items with success of the OI (strategic, market, process, product and behavioral innovativeness) [3]. Innovativeness is believed as an individual and collective discovering procedure that inclines to find new methods of resolving problems [4]. Prior scrutiny suggests that OLC affects product innovation [5]. With the growing rate of rivalry, knowledge exploration and information support to develop the OLC for enhancing innovation and performance [6]. Akgun et al., [7] noted that OLC build up a hierarchical learning capacity, which shows the capacity of an organization to execute the suitable administration practices, method and structures that encourage and empower learning process, in addition, the OLC is essential subject for research scholar and managers [7]. Particularly, the role of the OLC is important for enhancing OI [4]. Further, Developing the LC supports to resolve the business issues, to face the competitor's challenges and to grant the survival and achievement of the organization into the future.

Moreover, the OP is enhanced by OLC and IO. While, the origins of discovering at the organizational level, OLC are mainly established on macro level constructs, such as supportive constructions, and routines that foster collective learning and a smaller amount of research theories have delineated organizational innovativeness as growing across sequential procedures that transpire at disparate levels of scrutiny [5]. They argued OLC as the adoption of a believed whether pertaining to a mechanism, arrangement, procedure, strategy, plan, product or ability that is new to the adopting association [7]. They think Innovativeness as transpiring in three different groups i.e. managerial skill development, new procedure and product [8]. The OLC have been subjected to the attention of scrutiny as it contributes to the survival and sustainability of an association and the OP, temporarily, has not been often times described and has been utilized differently according to the context, as well as being tough to delineate and compute [9].

Finally, this paper is concerned with the organizational learning capabilities toward organizational performance (financial and non-financial) in the service sector. Five facets of

* Corresponding author : khawarpm786@gmail.com

Organizational Learning Capabilities have considered for research purpose, namely: Leadership Commitment & Empowerment, shared vision and mission, dialogue, participative decision making, transfer of knowledge with the intervening impact of Organizational Innovativeness's three dimensions: Behavioral Innovativeness, Process Innovativeness and Strategic Innovativeness onto the association amid Organizational Learning Capabilities and Organizational Performance (financial and non-financial). **Problem Statement**: is pointed out Organizations are striving to sustain themselves in globalized and competitive environment. They are facing ambiguity, complexity and challengeable situations in business processes [10]. Due to intensive economic pressure and competitive environment organizations are facing problems and they are probing different ways to enhance their core competencies with a view to gain competitive advantage. Camison, and Lopez [8] explained that intensive competitive environment force organizations to keep themselves updated and learned about organizational environment in which they operate. This is an issue that how organizations can increase their intellectual assets and learning capabilities with a view to retain in competitive zone.

In addition to gap Analysis**:** The research explorations undertaken pertaining to Organizational Learning capabilities and innovativeness so far, largely encompass the manufacturing sectors while with very little focus onto the service-oriented sector and that also in the Asian and South Asian contexts. The meticulous review of the literature available onto the said spheres together with [9, 11, 12, 13] explorations who recommended that organizational innovativeness may further be assessed with Organizational learning capabilities (OLC) in relation to organizational performance, while Khalib et al., [11] recommended that five facets of OLC namely: (MC) Managerial Commitment, (EAO) experimentation and openness, (DG) Dialogue, (PDM) Participative Decision Making, and (KT) Knowledge Transfer may be tested in the context of service oriented firms, and behavioral, process and strategic innovativeness with regard to innovation in the artificial bodies. This research study will **focus** onto the service-oriented firms on account of them being an eschewed and overlooked area pertaining to learning and innovation research. The five afore-mentioned facets of Organizational Learning Capabilities as Independent Variable, Innovativeness with its three factors (Process, behavioral and strategic Innovation) as intervening variable and organizational performance as the outcome variable have been brought under this research exploration.

First, **research objective**, is to investigate the relationship between the OLC and OP, second is to assess the link between the OLC and OI, third is to examine the relationship between OI and OP and fourth is to analyze the intervening impact of OI on the relationship between OLC and OP. The current paper seeks to answer of **its questions**: Do OLC positively relate to OI? Do OLC relate to OP? Do OLC positively impact OP? Does OI mediate the relationship between OLC and OP?

Finally, the past studies that OLC influence on OI [2] and they determined that OLC affect OP. Rubera and Kirca, [4] explored that OI impact on OP. **The contribution** of present study is examined that OI mediates the relationship between OLC and OP. **Significance of the study**: according to dynamic rapid change **in knowledge,** OLC is a strategic tool in business management to enhance OI and OP for gaining competitive advantage; Learning practices are considered keystone to mitigate the uncertainties and to adopt the external changes in effective manners to gain & sustain the competitive edge [6] [14]. These practices can be organized, analyzed, managed and improved with efficiency & effectively to polish the management learning skills. Continues & effective learning leads to create an innovative culture which move the organizational performance curve in upward trend. Innovation can lead an organization towards growth and improvement in business processes which ultimately leads to get core competencies and competitive edge. Under study variables have significance importance to grow organizational performance [13].

2 Literature Review

2.1 Organizational Learning Capability

OLC can be delineated as the characteristics, traits, habits, dexterities and the agents pertaining management, association and the establishment as a whole that enable the discovering procedures of the association (e.g., producing, buying, disseminating and integrating information/ knowledge) or permits an association to discover [20]. Organizational discovering skill is concerning the skill of one association to apply the precise and appropriate association habits, its constructions, as well as the procedures that enhance, enable and enthuse discovering [15]. According to Argote [16], organizational discovering skill is the organizational and managerial agents that enable the organizational discovering procedure or permitting the association to learn. Salim, & Sulaiman [17] added that the development of this exercise should consequence stronger discovering skill across the organization. The believed of organizational discovering skill gives significance of the enabling factors for organizational discovering or the organizational propensity to discover [18]. On the supplementary hand, Tohidi et al. [19] uttered that OLC are the organizational and managerial factors that give to the enhancement of organizational discovering procedure in organization. As per myriads of scrutiny explorations, OLC is believed as a multi-faceted concept. After going across the works, such conspicuous facets of OLC have been ascertained [8].

2.2 Organizational Learning Capabilities and Performance

This knowledge and vision have connection alongside non-financial and commercial presentation of organization. Benefits of discovering and vision association are alike for the area as well as confidential organizations. Furthermore, Camison and Lopez [8] explained that learning directions and innovative culture within the organization have positive and strong association with organization performance. These factors are behaving like cornerstone to improve the organizational performance. Organizational learning and innovativeness have positive link so there should be learning culture within the organization to enhance the innovative capabilities & performance.

Learning practices within the organization can be considered as antecedent factor for organizational commitment, effectiveness and efficient performance. These learning practices have moderate positive relation with effectiveness and strong positive relation with organizational commitment [12]. Similarly Onag et al. [13] summed up that organizations having learning mechanism and showing progress in their learning processes can improve their performance but financial measures alone are not enough to judge the organizational performance. Employees' measures are effectively & strongly related with learning capabilities. Khalib et al. [11] described that organizational learning has well-built, positive & significant relationship with financial and non-financial measures and performance. As well organizations can build long term relations with employees, customers and suppliers with handsome financial position. Organization learning is considered a vital concept in strategic management due to its effectiveness. In present situation organizations should put organized efforts to gain the strategic learning. Organizations that will manage and develop the higher level learning will gain in terms of profitability and employees value addition relative to their rivals.

Organizational learning has vital role to enhance the organizational performance. Discovering has a momentous correlation alongside performance. Human resource hobbies alter the Discovering of organization. Due to globalization, new discovering necessities are believed vital for the improvement. Discovering habits are the path to accomplish the accomplishment, larger presentation and competitive advantage [9]. Thus H1: there is a relationship between OLC and OP.

2.3 The Association between Organizational Learning Capability and Organizational Innovativeness

OLC supports OI by rising employees' creativity and enhancing their knowledge. Vision allocating within organizations across contact and contact establishes appropriate nature for the collective power of organizational innovativeness [19]. In fact, an outstanding deal of scrutiny has believed new vision as the basis for change [21] understanding innovativeness as an individual and collective discovering procedure that aims to pursue new methods of resolving problems. Innovativeness seems to depend on the company's skill to discover, across that new vision is industrialized, distributed and used. Though, innovativeness can additionally be believed as a catalyst of new vision, as the extremely procedure of and the feedback from prosperous or ineffective aftermath can lead to a new vision of the marketplace or the product [10].

The innovativeness procedure implies enhancing and asking concerning the procedure itself and the possible innovativeness output, that way adaptive and generative discovering or single- and double-loop discovering [8]. One of the early periods in the innovativeness procedure is the creation of an believed or believed that needs hunting for opportunities, vision and brainstorming, that can be believed as a method of communal discovering [13], whereas disparate actors give, interrelate and in the end learn. In sum, change can enhance discovering inside associations [9]. Thus H2: There is a relationship between OLC and OI.

2.4 Innovativeness and Performance

Slater et al. [10] clarified that discovering orders and innovative sophistication inside the association have affirmative and forceful association alongside association performance. These factors are behaving

https://doi.org/10.1051/matecconf/201815006027

like cornerstone to enhance the organizational performance. OI is considerably & affirmatively associated alongside organizational performance. Innovativeness provides a competitive supremacy in the marketplace. In globalization scenario, competent and effectual use of discovering skills are believed vital basis to maximize the organizations' change presentation alongside a think to gain competitive advantage [11]. Thus H3: There is a relationship between OI and OP.

2.5 The Intervening Role of organizational Innovativeness between the Relationship of Organizational Learning Capability and Performance

The works indicate that company innovativeness enhances a firm's commercial performance. Specifically, company innovativeness supports the progress of a firm's core competencies and skills [5]. This method, the competencies a stable possesses and its skills, expertise, experience and the skill to deed enhance the firm's efficiency in noticing and using obtainable resources; these competencies additionally foster the progress of resolutions to company setbacks and arrest market-driven gains that undeviatingly elucidate into higher commercial presentation [7]. Furthermore, comprehensive facts in the works indicates that company innovativeness facilitates the capitalization of changes across lowered creation prices, reductions in transport period and increased operational flexibility [21]. This suggests that the connection amid OLC and commercial presentation can be clarified by incorporating the arbitrating act of innovativeness [9, 11]. Thus H4: OI mediate the relationship between OLC and OP.

2.6 Theoretical Framework

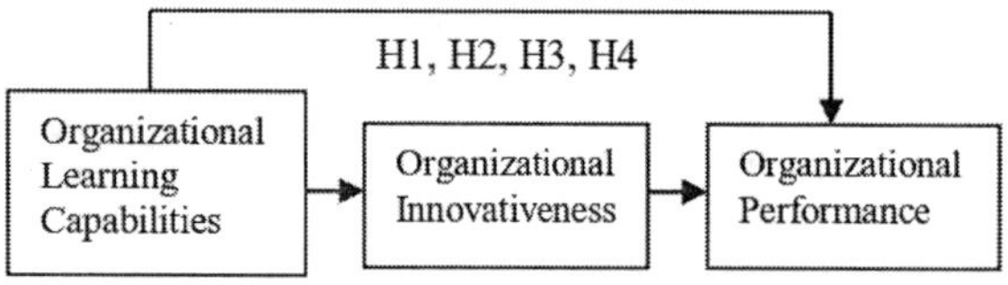

Fig.1. Theoretical Framework

2.7 Research Hypotheses

H1: There is a positive association between Organizational Learning Capability (OLC) and Organization Innovativeness (IO). H2: There is a positive relationship between Organization Innovativeness (IO) and Organization Performance (OP). H3: There is a positive association between Organization Learning Capability (OLC) and Organizational Performance (OP). H4: Organization Innovativeness (IO) positively mediates the association amid Organization Learning Capability (OLC) and Organization Performance (OP).

3 Research Methodology

An preliminary analysis of the literature discovered a lack of suitable items for measuring the defining constructs of OLC that are listed in Figure 1, The researchers chose to adopt and develop appropriate operationalizations to design an instrument that specifically and directly measures these constructs in order to ensure the reliability and validity of the measurement instrument process that we followed to develop and validate the instrument and summarizes the key techniques that we employed. To examine the validity and reliability of our measurement instrument for OLC and OI, we have used exploratory factor analysis.

3.1 Research Design

This is a causal, cross-sectional and quantitative study. In quantitative research design, an analytical and systematic approach has been used to analyze the data for information and to test the proposed hypotheses. Sample Selection and Strategy: 200 employees responded from 2000 employees Population In this study, the employees (Managers and Non-managers) of telecom industries were selected in Rawalpindi and Islamabad to take as population. Non probability convenience sampling method is the selected method for this research that helps gathering data from the people who are easily available. The questionnaire was filled by employees of all levels and, of different age groups and gender.

Unit of Analysis: The unit of analysis in this study is the employees functioning in telecom sector, and other small corporations). Unit of analysis is, individuals including all service providing managers and process owners, top management and administrative staff of telecom industries in Islamabad.

3.2 Construct development and instrument validation

OLC includes five dimensions; (i.e. managerial commitment (04 items) and experimentation and openness (06 items), dialogue (06 items), participative decisions making (05 items) and knowledge transfer (05 items) [13], secondly, OI comprises three factors (i.e. process innovativeness (04 items) and behavioral innovativeness (04 items), and strategic innovativeness (04 items) [3] and thirdly, organization performance (i.e. financial (04 items) & nonfinancial (12 items) [27] respectively, The questionnaire has total items 51 they were used five points Likert to examine the responses. The measurement scale ranged from one (strongly disagree) to five (strongly agree).

3.3 Data Collection Methods for Pilot Testing

It is self-administered survey. Questionnaires were circulated to employees functioning in different banking and telecom organization in Rawalpindi and Islamabad. 80 questionnaires distributed to different organization, 70 were returned and processed for pilot testing.

3.4 Validity of the Instrument

After getting the questionnaire approved by the subject experts it was administered to the sample of participants for pilot testing. The value of Cronbach Alpha should be greater than 0.70 [22, 23]. The internal consistency i.e. the reliability of the items which was greater than point six (0.70) assured that the instrument was reliable and researcher could proceed with it. All of the utilized constructs are well supported in the prior literature. Specifically, the face validity was conducted by professional experts. The pilot testing is conducted on software of SPSS 21 using Cronbach alpha [26].

Table 1. Reliability of items of pilot study

Variable(s)	No. of items	Cronbach's Alpha
OLC	23	.865
OI	12	.727
OP	16	.762

The instrument was pilot tested employing a example of 70 workers from the target population. The result of pilot study conferred seized adequacy [22]. The Cornbach's alpha of all variables ranged from .727 to 0.865 [26].

3.5 Demographics analysis

The summary of the demographics is provided in the details of the respondents. The age of the participants are divided into seven groups in which 14.5% participants were of age 20-25 while 35.5% lies between the ages of 26-30, 21.0% were of 31-35, 17.5% were from 36-40, 9% were from 41-45, 2.5% were in range of 46-45 and only 0.0% were of age 51 and above. Gender group is divided in two parts Male and Female, Male respondents were 70% while only 30% were females. Most of the respondent 80% were belonging to non-manager grade while only 20% respondents were belonging to manager grade. 35% respondents had bachelors' degree, 50.5% were masters, and only 14.5 % was having MS/M.Phil degree.

3.6 Descriptive analysis

The results indicate general agreement of the respondent to the study variables which are organizational learning capabilities, organizational innovation and organizational performance. The mean values ranged from low 3.2609 to highest 3.2811. It shows the results of OLC (M= 3.2811, S.D= .67165, Skewness = -.606, Kurtosis= -.628), OI (M= 3.2988, S.D= .66914, Skewness = -.666, Kurtosis= -.2324), OP (M= 3.2609, S.D= .67827, Skewness = -.482, Kurtosis= -.723). The result of descriptive statistics i.e. Means, S.D, Skewness, further, Skewness results demonstrate OLC to OP to have negative values those are very flat and have a greater number of larger values and Kurtosis show the conformity of the participant's thoughts towards all variables. Kurtosis determined flatness or peaked of a distribution, here, kurtosis results expose negative values of all variables those are vary peaked than normal and data is good for using psychometric. To evaluate the normality of the data, Skewness and Kurtosis were utilized. Literature illustrates that the value of Skewness should be between +1 to -1 to prove the normality of the data [25]. Descriptive values are within the range which proves the normality of the data. Kurtosis is also a measure of the normality of data. The ideal value of kurtosis is 3 [22].

3.7 Reliability Analysis

First, the instrument was adapted, and to check the internal consistency of the items, the following table 2 shows the results.

Table 2. Internal consistency (reliability) of important variables (N=200)

Variables	No of items	Cronbach's Alphas
OLC	23	.834
OI	12	.734
OP	16	.838
Total	51	

The value of Cronbach Alpha should be greater than 0.70 [23]. The results demonstrate that the value of Cronbach Alpha of all the variables is greater than 0.70 [24].

3.8 Correlation Analysis

Table 3. Correlations

	OLC	OI	OP
OLC	1		
OI	.744**	1	
OP	.877**	.475**	1

**. Correlation is significant at the 0.01 level (2-tailed).

The Table 3 explains the correlation among all the constructs that are used in theoretical model. From the table 3, it has been concluded that OLC have the highest positive correlation (0.877) with the OP. The value of correlation (0.744 and .475) indicates that OI has positive and statistically significant association with OLC, and OP. Furthermore, all other variables' correlations show a positive and statistically significant relationship between the variables. Moreover, all the variables results of correlation are significant at the 0.01 level [25] [26].

4 Measurement and Model evaluation

4.1 Multiple Regression Analysis

Hayes [24] explained that the mediation models can be explained in model templates. Therefore, researcher has tested mediation on Process Model 4. PROCESS Models of Hayes [24] has been used in SPSS 21 to test the direct and indirect effect of predictor variables on outcome variables. Regression analysis also shows the fitness of model with the value of R square depicts the explanatory power of the model. The study used one independent variable, one dependent and mediation variables.

Table 4. Direct and indirect relationship with Mediation Analysis using Hayes [24] (N=200)

Steps	Test	B	T	R square
Direct	OLC → IO	.95	3.08	.82
Mediation	IO → OP OLC → OP	.57 .46	5.85 4.45	.804 .80
Direct	OLC → OP	.46	4.45	.80

In this case, while the independent variable (OLC) was a significant predictor for the dependent (OP) and the mediator variables (IO), it is no longer significant in the presence of the mediator variable; confirming the mediation effect. A measure for the indirect effect of X on Y is also presented after the regression models. In this case the effect size was .5740, with a 95% confidence interval which did not include zero; that is to say the effect was significantly greater that zero at $\alpha = .05$.

4.2 Findings

The above table 4 shows that value of R square is 0.82. This shows that 82% variation in the IO is being explained by OLC. P-value is (.000) which means this relationship is also highly significant. OLC is regressed on the mediating variable which is IO was also significant ($\beta = .9569$, t-value= 3.080, p = <.000). It shows that OLC have a significant effect on the IO.

First, in step 2 of the mediation process showed that the mediator (IO (MV)), controlling for OLC, thus, beta value $\beta = .574$, t-value= 5.859 is significant at p-value = .000. The R square result ($R^2 = 0.804$) divulge that the IO has accounted for 80.4 % variation in OP. Thus IO relationship with OP is accepted and positive. Second, controlling for MV (IO), OLC was significant predictor of OP, $\beta = .4669$, t-value= 4.452, p-value =.000. The R square result 0.80 confirms the models fitness. Hence H1, H2, H3, and H4 are both supported. From the above results it can be safely concluded that IO partially mediates the relation between OLC and OP.

OLC as a predictor has positive impact on organizational performance as outcome, therefore, while using SPSS 21 [24] Model 21, we regressed OLC on OP, findings shows (see table 8) that value of R square is 0.80, this indicate that there is 80% variation in the OP is explained by the OLC and beta coefficient was $\beta = .4669$, t-value = 4.452, p-value = <.000. It shows that direct relationship is positive and significant at ($p<.01$) to indicate acceptance of hypothesis. So the result explains that the hypotheses are accepted and indicates positive impact on improving organizational performance.

5 Implications, Recommendations and Conclusion

5.1 Theoretical implication

The current paper is found from past studies that OLC influence on OI [2] and they determined that OLC affect OP. Rubera and Kirca [4] explored that OI impact on OP. The present study is examined that OI mediates the relationship between OLC and OP. Argote [16] counseled that novel methods of coordinating work, adjustments in association actions or company strategies, enhanced forms of association or new configurations of organizational constructions are central in enabling company accomplishment and increased financial performance. OLC affect OP positively. OLC impact on OI positively. According to Tuksinnimit et al. [9] whole research work is determined that OLC influence on OP by mediating role of IO that enhance performance financially and nonfinancially.

5.2 Managerial Implication

Managers should pay attention on OLC for developing skill, knowledge, increasing experience for decision making and performance [7]. Organizations can build long term relation with their employees, customers, suppliers and handsome position for enhancing innovation [2]. In present situation organizations should organized efforts to gain the strategic learning. Further, Organizations manage and develop the higher level learning, to gain innovative ways, employees' value and profitability [19].

5.3 Research Limitations

In Pakistan's organizational context in which examined the effect of organizational learning capabilities (i.e. covered managerial commitment, experimentation and openness, dialogue, participative decision making and knowledge transfer) on organizational innovativeness (i.e. covered behavioral innovativeness, process innovativeness and strategic innovativeness and organizational performance (i.e. covered financial and non financial). Further, positive relationship was tested between OLC, OI and OP. Moreover, it was conducted in Islamabad's Telecom organizations. Small sample size was taken and convenient sampling was used.

5.4 Future Recommendations

Further, demission of OLC, IO and OP may be examined empirically. In addition, sample size can be increased base on demographically and population. Moreover, innovative culture might be used as moderator. Transformational leadership and authentic leaderships are tested as independent variables. Those are most important for enhancing the OLC, OI and OP. Finally, developing the innovative environment in an organization, it might be conducted in different organization e.g. manufacturing, services, food organizations.

5.5 Conclusion

This research proves that organizational learning capabilities have strong effect on organizational innovativeness and organizational performance [27]. Further, in the domain of organizational learning capabilities (Experimentation and openness, participative decision making, Knowledge transferring and managerial commitment) within organizations through behavioral, process and strategic interaction are established appropriate environment for organizational innovativeness. Finally, firms can implement on organizational learning capabilities approach make use for its developing new management technique, process, innovative products and enhancing performance.

References

1. Cheng, C., Lai, M. & Wu, W. (2010), "Exploring the impact of innovation strategy on R&D employees' job satisfaction: a mathematical model and empirical research", Technovation, Vol. 30 Nos 7/8, pp. 459-70.
2. Alegre, J., & Chiva, R. (2013). Linking entrepreneurial orientation and firm performance: the role of organizational learning capability and innovation performance. *Journal of Small Business Management*, *51*(4), 491-507.

3. Wang, C. L., & Ahmed, P. K. (2004). The development and validation of the organizational innovativeness construct using confirmatory factor analysis. *European journal of innovation management*, *7*(4), 303-313.
4. Rubera, G., & Kirca, A. H. (2012). Firm innovativeness and its performance outcomes: A meta-analytic review and theoretical integration. *Journal of Marketing*, *76*(3), 130-147.
5. Cropley, D. H., & Cropley, A. J. (2012). A psychological taxonomy of organizational innovation: Resolving the paradoxes. Creativity Research Journal, 24, 29–40.
6. DiBella, A. J., Nevis, E. C., & Gould, J. M. (1996). Understanding organizational learning capability. *Journal of management studies*, *33*(3), 361-379.
7. Akgun, A.E., Ince, H., İmamoglu, S.Z., Keskin, H. and Kocoglu, İ. (2013), The mediator role of learning and business innovativeness between total quality management and financial performance. International Journal of Production Research, Vol.52, No. 3, 888-901.
8. Camison, C., & Villar-Lopez, A. (2014). Organizational innovation as an enabler of technological innovation capabilities and firm performance. *Journal of Business Research*, *67*(1), 2891-2902.
9. Tuksinnimit, N., Durongwatana, S., & Vadhanasindhu, P. (2015). The Factors Affecting Organizational Innovativeness. *Journal of Management Research*, *7*(2), 561-573.
10. Slater, S. F., Mohr, J. J., & Sengupta, S. (2014). Radical product innovation capability: Literature review, synthesis, and illustrative research propositions. *Journal of Product Innovation Management*, *31*(3), 552-566.
11. Khalib, L. H., Kassim, N. A., Ghazali, F. I., Jaafar, N., & Idris, A. (2015). Organizational Learning Capabilities (OLC) toward Job Satisfaction: A Conceptual Framework.
12. Hogan, S. J., & Coote, L. V. (2014). Organizational culture, innovation, and performance: A test of Schein's model. *Journal of Business Research*, *67*(8), 1609-1621.
13. Onag, A. O., Tepeci, M., & Başalp, A. A. (2014). Organizational Learning Capability and its Impact on Firm Innovativeness. *Procedia-Social and Behavioral Sciences*, *150*, 708-717.
14. Menold, M. J., Jablokow, K., Purzer, S., Ferguson, D. M., & Ohland, M. W. (2014). A critical review of measures of innovativeness. *age*, *24*, 1.
15. Gunsel A., Siachou E. and Acar A. Z., (2011), Knowledge management and learning capability to enhance organizational innovativeness, Procedia Social and Behavioral Sciences, 24, pp.880-888.
16. Argote, L. (2011). 'Organizational learning research: past, present and future', Management Learning, 42, pp. 439–446. Armstrong, J. S. and T. S. Overton (1977). 'Estimating nonresponse bias in mail surveys', Journal of Marketing Research, 14, pp. 396–402.
17. Salim, I. M. and M. Sulaiman (2011). 'Organizational learning, innovation and performance: a study of Malaysian small and medium sized enterprises', International Journal of Business and Management, 6, pp. 118–126.
18. Bess, K.D.,Perkins, D.D., & McCown, D.L. (2011), Testing a mesure of organizational learning capacity and readiness for transformational change in human services. Journal of Prevention and Intervention in the Community, Vol.39, pp.35-49.
19. Tohidi H., Seyedaliakbar S. M., Mandegari M., (2012), Organizational learning measurement and the effect on firm innovation, Journal of Enterprise Information Management, Vol. 25 Iss: 3 pp. 219 – 245.
20. Campbell, C., Silver, I., Sherbino, J., Ten Cate, O. and Holmboe, E. (2010), "Competency-based continuing professional development", Medical Teacher, Vol. 32 No. 8, pp. 657-62.
21. Real, J. C., Roldan, J. L., & Leal, A. (2014). From entrepreneurial orientation and learning orientation to business performance: analysing the mediating role of organizational learning and the moderating effects of organizational size. *British Journal of Management*, *25*(2), 186-208.
22. Sekaran, U. (2003). Research methods for business. Hoboken.
23. Nunnally, J. C. (1979). Citation classic-psychometric theory. *Current Contents/Social & Behavioral Sciences*, (22), 12-12.
24. Hayes, A. F. (2013). *Introduction to mediation, moderation, and conditional process analysis: A regression-based approach*. Guilford Press.
25. Cooper, D. R., & Schindler, P. S. (2008). International Edition: Business Research Methods.
26. Straub, D. W. (1989). Validating instruments in MIS research. *MIS quarterly*, 147-169.
27. Damanpour, F.; Walker, R. M.; Combinative, C. N. A. 2009. Effects of innovation types and organizational performance: a longitudinal study of service organizations, Journal of Management Studies 46(4): 650–675.

The Effect of Aluminium Content on the Bioactivity Behaviour of Magnesium Alloys for Biomedical Applications

Nur Maizatul Shima Adzali[1,2*], *Nur Hidayah* Ahmad Zaidi[1,2], and *Mohd Ashraf* Jamaludin[1]

[1]Schools of Materials Engineering, Universiti Malaysia Perlis, 01000, Kangar, Perlis, Malaysia.
[2]CoE Frontier Material Research Centre, Schools of Materials Engineering, Universiti Malaysia Perlis, 0100, Kangar, Perlis, Malaysia

Abstract. AZ series alloy which consist of aluminium (Al) and zinc (Zn) with magnesium (Mg) as their base are the most investigated for biomedical applications among the Mg alloy. Al content plays a crucial part in differentiating the properties of Mg-based alloy for biomedical applications. Thus, this project picks up pure Mg and AZ31 (with 3% of Al) alloy and the effect of Al content has been investigated. In this study, Mg alloy has been fabricated via powder metallurgy method in order to investigate its bioactivity behaviour. Bioactivity test of pure Mg and AZ31 alloys were conducted by immersing the samples in simulated body fluid (SBF) solution up to 6 hours to observe the formation of apatite layer on the sample surface. The bioactivity behaviour of the samples has been observed using scanning electron microscope (SEM) and X-Ray Diffraction (XRD) analysis. After bioactivity test, sample AZ31 showed the formation of more apatite layers compared to sample of pure Mg with the existence of new phases which were $Mg(OH)_2$ and MgO. This apatite layers might influence the ability of the Mg-based alloy to withstand the corrosive agent that will attack the alloy while being immersed in SBF.

1 Introduction

The implants in biomedical applications, most of them are metallic, are used in the human body mainly for orthopaedic purposes. Metallic materials play a very predominant role in fulfilling almost every difficult factor that arises in implant applications [1]. For example, stainless steels, Co-Cr-Mo alloys, Ti and Ti alloys and Mg and Mg alloys have been used in many biomedical applications. Mg and its alloys have tremendous mechanical properties that have made this metal and its alloy gained a wide popularity in implants applications. Mg and its alloy is one of the various metallic materials that are commonly used as biomedical devices. The main reason magnesium and its alloy were used because, their mechanical properties such as Young's modulus of elasticity, E = 4-45 GPa and density = 1.74-1.84 g/cm^3 are known to be similar that of bone (E = 15-25 GPa and density = 1.8-2.1 g/cm^3). These properties were lower than other biodegradable materials such as iron-manganese (Fe-Mn) and zinc (Zn) based alloys [2].

Aluminium considered as the most influencing element in aluminium and zinc (AZ) series magnesium alloys [3]. AZ series is as a series of aluminium and zinc content in the magnesium based composite. These series of alloy has been study widely to investigate its performance in the physiological condition. Two types of magnesium alloys, which were AZ31 and AZ91, with AZ91 marks the highest aluminium content with 9%. Meanwhile, the lowest aluminium content goes to AZ31 which aluminium only contributed 3%. All of these alloys contain 1% of zinc. Aluminium content affects the degradation rate of magnesium alloy implants. Lunder et al. (1989) witnessed that when aluminium content reaches 8% of mass fraction, the corrosion resistance of magnesium alloy improve in a visible level [4]. Warner et al. (1992) reported that even 5% addition of Al in magnesium alloys is useful in improving their corrosion resistance [5], while Hehmann et al. (1987) revealed that 9% and above Al is helpful in improving of corrosion resistance of magnesium alloy [6].

Formation of apatite layer is crucial and was given a top priority govern with the biomaterials implant. Apatite layer acts as a connector between the implant materials and living tissues. According to British Standards (BS), apatite means group of calcium-phosphates including bone mineral and the main inorganic constituent of bones and teeth similar to hydroxyapatite, which has the composition $Ca_{10}(PO_4)_6(OH)_2$ [7]. Production of apatite layer can be done in the simulated body fluid (SBF) solution, with the solution must has an equal composition in term of blood plasma, and composition as well as structure of mineral bone. When implanted devices placed in human body, a thin layer of calcium and phosphorus forms on its surface before the material connects to the tissue through this apatite layer [7]. Formation of apatite layer is proportional to bioactivity, whereas it is increasing with increment in the rate of bioactivity behaviour.

In this study, pure Mg and AZ31 alloys have been fabricated via powder metallurgy method. Bioactivity

* Corresponding author: shima@unimap.edu.my

test has been conducted to observe the formation of apatite layer on both sample surfaces. The bioactivity behaviour of the sample observed using SEM and XRD.

2 Materials and Methods

2.1 Materials

In this research, commercially available Mg, Al and Zn powders were used as starting materials. For fabrication of AZ31 alloys powder, the mixing of 96% magnesium powders, 3% aluminium and 1% zinc powders were conducted by mixing and milling using roll mill. All these powders were supplied by the Merck Company. Figure 1 shows the SEM micrographs of Mg powders and AZ31 alloy powders.

(a)

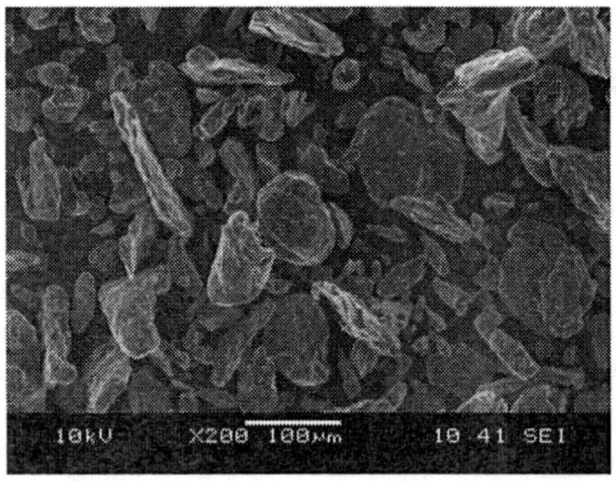

(b)

Figure 1. SEM micrographs of (a) Mg powders (b) AZ31 alloy powders.

2.2 Methods

2.2.1 Powder Metallurgy

The pure Mg and Mg based alloy samples (AZ31) were fabricated by blending, pressing and sintering, followed powder metallurgy route. The AZ31 alloy powder with 3 wt. % of Al particles was mixed in a rotary milling machine for 30 minutes at 130 RPM. The cylindrical green compacts of 13 mm in diameter and 14 mm in height were cold compacted under pressure of 500 MPa. The green samples were then sintered at 400°C for 2 hours in a tube furnace under argon atmosphere. Microstructures of the compacted samples were observed using a scanning electron microscope (SEM) (model JEOL, JSM-6420LA), equipped with energy dispersive spectrum (EDS). The sample was prepared by the standard metallographic methods of wet rotary grinding on a series of SiC papers, followed by polishing on the soft napped cloth (BUEHLER, USA). X-ray diffraction (XRD) was carried out on a Shimadzu (model XRD 6000), Japan radiation for phase analysis. The recorded spectrums were matched in accordance to the data from the JCPDF.

2.2.2 Bioactivity Test

To evaluate the bioactivity of sintered Mg and AZ31 alloy, the samples were immersed in a SBF according to the standard BS ISO 23317:2007 [7]. Table 1 shows the order of the reagents and weight of the reagents needed to prepare the SBF solution. The samples of Mg and AZ31 then were put in container which contain SBF before put in water bath for 6 hours with the starting pH and temperature of SBF were 7.4 and 36.5°C respectively. After 6 hours, the samples were gently cleaned with deionized water and dried at 37°C prior to analysis. After bioactivity test, SEM and XRD analyses were carried out in order to investigate the bioactivity behaviour of the samples.

Table 1. Reagent and amount to prepare 1L SBF solution.

order	reagent	Amount (g)
1	NaCl	8.035
2	$NaHCO_3$	0.355
3	KCl	0.225
4	$K_2HPO_43H_2O$	0.231
5	$MgCl_2.6H_2O$	0.311
6	c(HCl) = 1 mol/l	39
7	$CaCl_2$	0.292
8	Na_2SO_4	0.072
9	TRIS	118
10	c(HCl) = 1 mol/l	0 to 5

3 Results and discussions

3.1. SEM

Figure 2 shows micrograph of SEM for pure Mg and AZ31 alloys after sintering at 400°C in tube furnace (before bioactivity test). In Figure 2 (a), it shows the microstructures of pure Mg with no addition of Al or Zn. The EDS spectrum only shows the presence of Mg element in the samples. In Figure 2 (b), the micrographs revealed the microstructure of AZ31 (magnesium based alloys with 3% of Al and 1 % of Zn) with EDS spectrums shows the presence of Mg, Al and Zn elements. Figure 2 (b) shows there is a presence of pores on the sample surface compared with Figure 2 (a) which have flat surface with no pores.

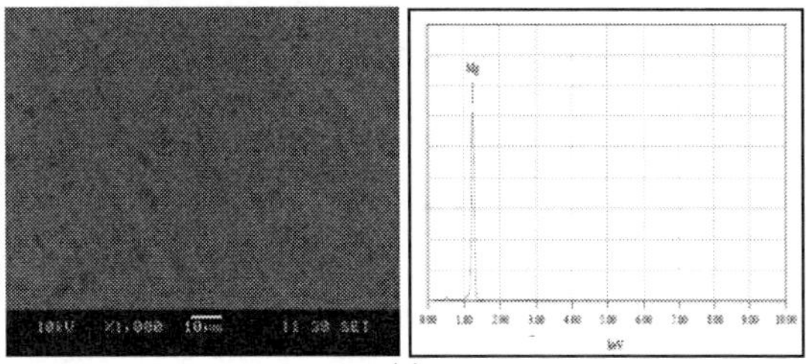

(a)

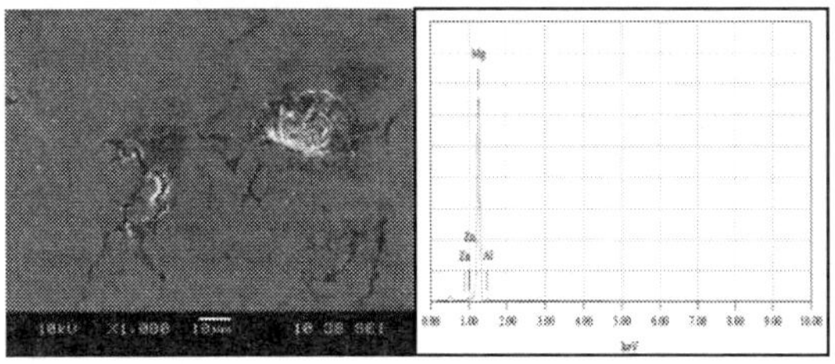

(b)

Figure 2. SEM and EDS results of (a) Mg (b) AZ31 after sintering at 400°C (before immersion test).

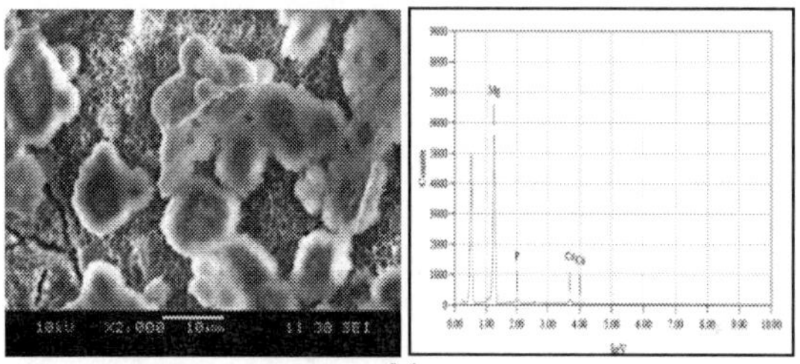

(a)

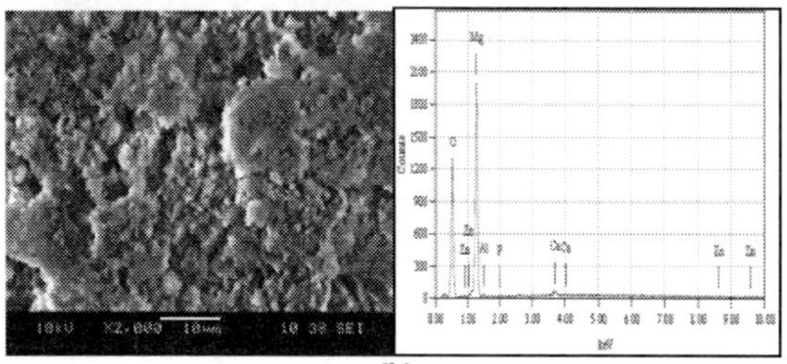

(b)

Figure 3. SEM and EDS results of (a) Mg (b) AZ31 after bioactivity test.

Figure 3 shows the microstructures of the sintered samples surface after immersion in the SBF up to 6 hours. Figure 3 (a) shows that the sample without addition of Al (pure Mg) exhibited the formation of apatite layer with white particles/granular on the corroded surfaces and grain boundaries. A small amount of Ca and P element (from SBF) is detected from EDS results. For samples with addition of Al and Zn (AZ31), Figure 3 (b) shows, the formation of many white particles. These white particles covered the entire sample surface with no grain boundary can be seen. It's revealed that more white particles (spherical shape) precipitated on the surface of the samples as more alloying elements are added; most pores are invisible due to the coverage of the newly formed deposition. The EDS result (Figure 3b) indicates that the precipitate layer composed of small amount of calcium (Ca), and phosphorus (P), Mg, Al, Zn and O. This result indicates that as alloying elements are added to pure Mg, the presence of apatite layer is obvious compared to pure Mg. This result is in agreement with Shamsul et al. (2014) which studied about the addition of filler (HAP) to metal (Co-Cr-Mo). They found that as more HAP is added to Co-Cr-Mo, the formation of apatite layer on the sample surface was increased and covered the entire sample surfaces [8].

From the previous findings, the formation of apatite particles and transformation into the apatite layer as more alloying elements is added can be divided into two stages: nucleus formation and nucleus growth [9-10]. Certainly, the nature and crystallinity of apatite phase phases depend on various parameters, including concentrations of phosphate/carbonate sources, ionic strength and pH of the soaked solution, and the kinetics of the nucleation and growth processes [11]. In general, the formation mechanism of apatite is dissolution-precipitation reaction.

3.2 XRD

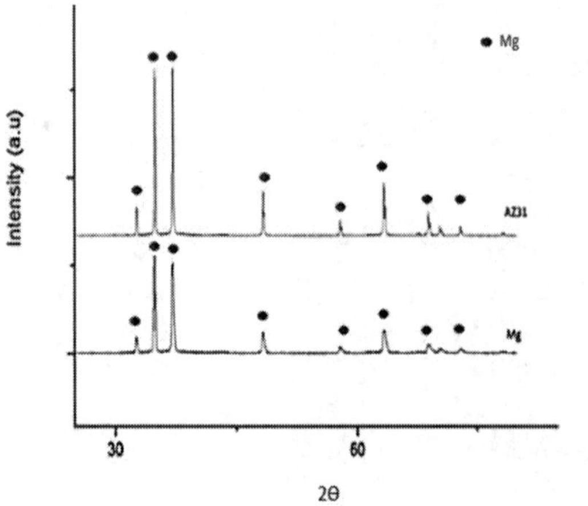

Figure 4. XRD result of Mg and AZ31 before bioactivity.

Figure 4 shows the XRD patterns of the surface of the sintered samples Mg and AZ31before bioactivity. αMg gives the most peaks for both sample. αMg (PDF No.: 03-065-4596) peak located at 35°, 37°, 48°, 53° and 69°, 71° and 74°.

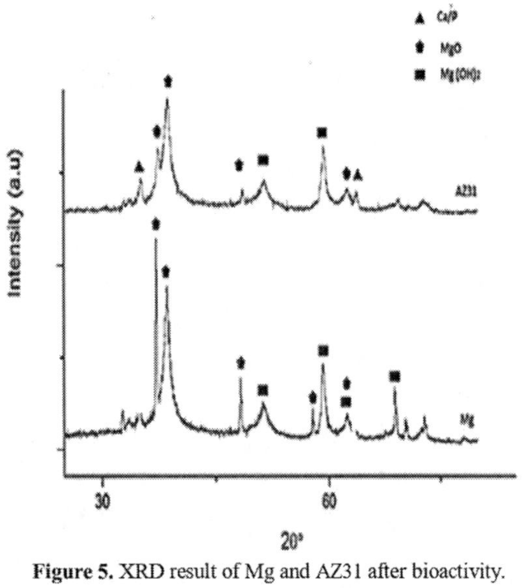

Figure 5. XRD result of Mg and AZ31 after bioactivity.

Figure 5 shows the XRD patterns of Mg and AZ31 after undergo bioactivity test. The test was conducted by immersion samples in SBF for six hours. As shown from the patterns, new phases exist after bioactivity test which were Ca, P, MgO and $Mg(OH)_2$. Ca P phase has its peak existed at the AZ31 sample. MgO (PDF No.: 00-002-1207) phase which may be found in both samples was the result of corrosion products. MgO peak was observed at 37°, 42° and 63°. Reaction between the OH- ions in SBF with both sample resulting in formation of $Mg(OH)_2$ (PDF No.: 01-086-0441) at 51°, 59° and 69°. There is a literature reports that $Mg(OH)_2$ peak intensity will increase if the immersion time in the SBF longer [12].

The present result is also in agreement with Razavi et al. (2010) who conducted the research about Mg/Fluoroapatite (FA). They found that immersion of Mg/FA composite in SBF for 72h shows the formation of white particles of cauliflower shape on the entire surface of Mg/FA sample compared to samples with no FA. XRD presents after bioactivity, a new phase which detected as $Mg_{17}Al_{12}$ [13].

4 Conclusions

This research is conducted to shows the effects of Al contents to the bioactivity behaviour of Mg based alloys. From the results, it shows that it is possible to bioactivate the surface of pure Mg with the addition of Al and Zn to produce apatite layer on the Mg alloy surface. This apatite layer formation considered as the mark of bioactivity. Based on SEM microstructure, its shows the numerous growth of apatite cluster by the additions of Al in Mg based alloys. The result showed that formation of apatite layer increases as the alloying elements is added with the formation of new phases which were Ca, P, MgO and $Mg(OH)_2$. Thus, we conclude that this AZ31 alloys can be good alternative materials for biomedical application.

This work is financially supported from FRGS grant (9003-00367) by MOHE Malaysia.

References

1. U. Kamachimudali, T. M. Sridhar, Baldev Raj, Sadhana **28**, 601-637 (2003).
2. S. Agarwal, J. Curtin, J., B. Duffy, & S. Jaiswal, Materials Science and Engineering: C **68**, 948-963 (2016).
3. A. R. Bharat, B. S. S. Kumar, R. P. Sarath, P. Akhil, G. P. K. Reddy, & B. R. Sunil, Journal of Advances in Biomedical Engineering and Technology **2**, 17-23 (2015).
4. O. Lunder, J. E. Lein, T. K. Aune, & K. Nisancioglu, Corrosion **45(9)** 741-748 (1989).
5. T.J. Warner, N. Thorne, G. Anussbaum, & W.M. Stobbs, J. Surf. Int. Ana. **19** 386-392 (1992).
6. F. Hehmann, F. H. Froes, & W. Young, Journal of Metals **39(8)** 14-21 (1987).

7. British Standard. Implants for surgery —In vitro evaluation for apatite-forming ability of implant materials, BS ISO 23317:2007 (2009).
8. Shamsul Baharin Jamaludin, Nur Maizatul Shima Adzali, & Mohd Nazree Derman, Acta Metallurgica Slovaca **20**, 82-88 (2014).
9. D. Cortes, A. Medina, S. Escobedo, & M. A. Lopez, Journal of Materials Science **40** 3509-3515 (2005).
10. Y. P. Guo, Y. Zhou, D. C. Jia, C. Q. Ning, & Y. J. Guo, Materials Science and Engineering C **30** 472-479 (2010).
11. C. Wu, & J. Chang, Materials Letters **61** 2502-2505 (2007)
12. D. Gopi, P. R. Bhalaji, S. Ramya, & L. Kavitha, Journal of Industrial and Engineering Chemistry **23** 218-227 (2015).
13. M. Razavi, M.H. Fathi, & M. Meratian, Materials Characterization **61** 1363-1370 (2010).

Comparison between Edge Detection Methods on UTeM Unmanned Arial Vehicles Images

Zuraini Othman*, *Asmala* Ahmad, *Fauziah* Kasmin, *Sharifah Sakinah* Syed Ahmad, *Mohd Yazid* Abu Sari and *Muhammad Amin* Mustapha.

Department of Intelligent Computing and Analytics, Faculty of Information & Communication Technology, Universiti Teknikal Malaysia Melaka, Hang Tuah Jaya, 76100 Durian Tunggal,Melaka, Malaysia.

Abstract. Machine vision calls for the use of detectors to ascertain the features and type of object portrayed in the image. The employment of unmanned aerial vehicles (UAVs), which can function freely in active and precarious settings, is currently gaining momentum. These vehicles are mainly used for the detecting, classifying and tracking of an object. However, the achievement of these objectives necessitates the involvement of an effective edge detection procedure. Sobel, Canny, Prewitt and LoG are among the many edge detection procedures presently available. In this endeavour, we opted for the utilization of UTeM UAVs images for an evaluation of these edge detection procedures. During our investigations, the ground truth edge images were corroborated by a specialist in this field. The results obtained from these investigations revealed that in terms of accuracy, precision, sensitivity and f-measure, the Prewitt procedure outperforms the other methods mentioned.

1 Introduction

When it comes to machine vision, the analysis of an image is significantly determined by the particulars related to its edges. Essentially, edge detection is a technique employed to identify the alterations that transpire at the pixel of an image. The mathematical procedures related to edge detection are aimed at distinguishing the points in a digital image at which the luminosity of the image varies abruptly (or put differently, has discontinuities). These points are characteristically structured in an array of arched line segments known as edges. The term step detection denotes the search for discontinuities in one-dimensional signals, while the term change detection denotes the search for signal discontinuities over time. Edge detection is an indispensable instrument for image processing, machine vision and computer vision. This is especially so in matters related to the detection and extraction of features [1].

The acquiring of precise images is especially important in the areas of biometrics, iridology, manufacturing, medical procedures and UAVs [2]–[6]. Precise images serve to ensure that the procedures employed in these areas are appropriate from start to finish.

Edge detection techniques, including those mentioned in [7]–[11], have been the subject of many investigations. Among the techniques available, the most frequently employed are Sobel, Prewitt, Canny, and Laplacian of Gaussian (LoG). Studies conducted on the Canny technique [12], [13] have served to improve its performance in the context of iris, scenery, retina and handwritten images.

This endeavour forwards an innovative UAVs image dataset for the site of Universiti Teknikal Malaysia Melaka (UTeM). This dataset is exceptional as it holds a wide range of features that include buildings, highways and oil farms and nature around UTeM. The edge image ground truth is provided for every image present in this dataset. We also discussed and evaluated the Sobel, LoG, Canny and Prewitt edge detection procedures.

2 Edge Detection

An edge, in the context of an image, is an arc that tracks a route of quick alterations in the intensity of the image. More often than not, edges concern the borders of objects in a setting. The task of edge detection is to make out the edges of an image. In the case of a digital image, the edge is a compilation of pixels with grey values that come with a step or roof change. Here, the edge also denotes the segment where the luminosity of the local area of the image varies considerably. Generally, this is attributed to the swift alteration of a grey value into a substantially different grey value within a small buffer area. Edges are frequently detectable between objects and backgrounds, objects and objects, as well as between primitives and primitives. The edge of an object is revealed by the lack of continuity in its grey profile. This circumstance provides a route towards edge detection. Edge detection involves a scrutiny of the alterations of a sole image pixel in a grey area, and the subsequent employment of the edge neighbouring first order or second order to discern the edge. This process is known as the local operator edge detection procedure. Basically, edge detection has to do with the gauging, distinguishing and locating of alterations in the grey profile of an image. The most fundamental and noticeable feature of any image is its edge and line. The composition of the edge and line in an image reveals the structure of an object. This elevates the importance of edge detection in the realms of graphics processing and feature extraction [14].

* Corresponding author: zuraini@utem.edu.my

https://doi.org/10.1051/matecconf/201815006029

The employment of four edge detection techniques is described below.

2.1 Canny Method

The Canny method was put forward by [15].]. Listed below are the steps for the tradition Canny algorithm [16], [17].

- The Gaussian filter is utilized to smooth the image in order to restrain noise.
- Finite difference on the first order partial derivative is used to work out the gradient magnitude $M(x,y)$ as well as the gradient direction $H(x,y)$ of the image. $M(x,y)$ is computed through the equation

$$M(x,y)=\sqrt{E_x(x,y)^2+E_y(x,y)^2}\ , \quad (1)$$

while H(x, y) is computed through the equation

$$H(x,y)=\arctan(E_x(x,y)/E_y(x,y)) \quad (2)$$

in which E_x and E_y portray the image subsequent to the influence of the filter alongside the row-column route.

- Non-maximum suppression is applied for the gradient magnitude.
- The dual-threshold algorithm is utilized to distinguish and link the edges.
- In order to ascertain the final edges, hysteresis is employed to hold back any edge not linked to an exceptionally durable edge.

2.2 The Sobel method

As displayed in Figure 1, the operator comprises two 3×3 convolution kernels. Essentially, one kernel is a depiction of the other turned around by 90°.

Gx

-1	0	+1
-2	0	+2
-1	0	+1

Gy

+1	+2	+1
0	0	0
-1	-2	-1

Fig. 1. Sobel convolution kernels

These kernels are conceived for maximum reaction to the edges that stretch in a perpendicular and parallel line in relation to the pixel grid. Each of the two perpendicular orientations is represented by one kernel. The application of the kernels to the input image can be carried out independently to realize distinct gradient component dimensions for each orientation (represented as *Gx* and *Gy*). *Gx* and *Gy* can then be merged to ascertain both the gradient's absolute magnitude at each point, as well as its orientation. The gradient magnitude is calculated by way of the equation

$$|G|=\sqrt{Gx^2+Gy^2} \quad (3)$$

More often than not, an estimated magnitude is derived through the equation

$$|G|=|Gx|+|Gy| \quad (4)$$

which facilitates a quicker computation. The angle of orientation of the edge (in relation to the pixel grid) which yields the spatial gradient is calculated through the equation

$$\theta=\arctan(Gy/Gx) \quad (5)$$

2.3 The Prewitt method

The Prewitt operator, which is comparable to that of the Sobel operator, is applied for the detection of vertical and horizontal edges in an image.

$$h_1=\begin{bmatrix}1 & 1 & 1\\ 0 & 0 & 0\\ -1 & -1 & -1\end{bmatrix} \qquad h_3=\begin{bmatrix}-1 & 0 & 1\\ -1 & 0 & 1\\ -1 & 0 & 1\end{bmatrix}$$

Fig. 2. Prewitt convolution kernels

2.4 The LoG method

The Laplacian can be described as a 2-D isotropic calculation of an image's 2nd spatial derivative. The Laplacian of an image draws attention to areas of quick intensity alterations. Frequently employed for edge detection, the Laplacian is regularly used for an image that has undergone smoothing by way of a device resembling a Gaussian smoothing filter. This smoothing process is for the purpose of decreasing the sensitivity of the image to noise. Generally, the operator harnesses one grey level image as input, and generates a separate grey level image as output.

The Laplacian *L(x,y)* of an image with pixel intensity values *I(x,y)* is depicted as:

$$L(x,y)=\frac{\partial^2 I}{\partial x^2}+\frac{\partial^2 I}{\partial y^2} \quad (6)$$

An estimation of the second derivatives in the definition of the Laplacian calls for the involvement of a separate convolution kernel. This is due to the fact that the input image is portrayed as a collection of distinct pixels. Displayed in Figure 3 below are three frequently employed small kernels.

0	1	0
1	−4	1
0	1	0

1	1	1
1	−8	1
1	1	1

−1	2	−1
2	−4	2
−1	2	−1

Fig. 3. LoG kernel

The fact that these kernels are estimating a second derivative measurement on the image renders them exceedingly susceptible to noise. To overcome this setback, the image is usually Gaussian smoothed prior to the introduction of the Laplacian filter. This pre-processing intervention serves to lessen the high frequency noise elements before the onset of the differentiation stage.

As the convolution operation is deemed associative, it is possible to initially coil the Gaussian smoothing filter and the Laplacian filter together. The resulting hybrid filter is then coiled with the image to acquire the targeted result. This move comes with two benefits:

- As both the Gaussian and Laplacian kernels are typically much slighter than the image, this process entails significantly less mathematical activities.
- The LoG kernel can be recomputed beforehand so that the performance of only a single run-time convolution is required on the image.

The focus of the 2-D LoG function is on zero with the Gaussian standard deviation σ depicted as:

$$LoG(x,y) = -\frac{1}{\pi\sigma^4}\left[1-\frac{x^2+y^2}{2\sigma^2}\right]e^{-\frac{x^2+y^2}{2\sigma^2}} \quad (7)$$

During this investigation, the Canny, Sobel, Prewitt and LoG edge detection methods was applied for each UTeM UAVs image. Subsequently, the results attained were weighed against the related ground truth images to determine the value of the measurement.

3 Methodology

A Walkera quad copter UAVs was employed to gather data for this study [18]. This UAVs has a dimension of 620x620x460 mm and weighs 3900 g. Powered by a LiPo 6S 10000 mAh battery, it comes with GPS mechanisms to facilitate controlling and tracking functions. It is also furnished with a RGB digital camera for the gathering of basic data in the form of aerial images. During its foray for data, the UAVs flew at a velocity of roughly 3 ms^{-1} and a height close to 100 m. The location for this study is the main UTeM campus in Durian Tunggal, Melaka, Malaysia. To be more precise, it is sited at a latitude and longitude of 2.3139° N and 102.3212° E respectively. Mission Planner software was employed to plot the navigation route for the gathering of images featuring buildings of interest.

Fig. 4. Study location (Google Maps, 2017).

At the outset, for every building image captured, a reference edge image of the building was drawn by hand. This was achieved by way of a visual inspection of the edges existing in the initial image.

3.1 Comparison between Edge Detection Methods on UTeM Unmanned Arial Vehicle Images

Every image acquired was transformed into a grey scale image. For each image, edge ground images are generated by hand and subsequently validated by a specialist in the geospatial field. The workflow for our recommended procedure is exhibited in Figure 5.

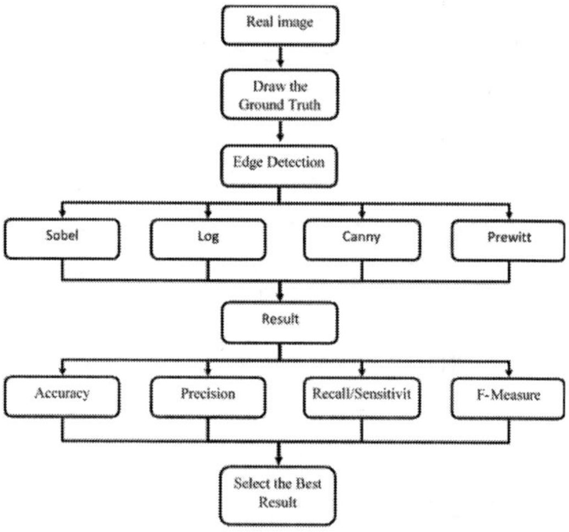

Fig. 5. Recommended procedure

In order to facilitate the computed result display, a graphical user interface (GUI) was developed for comparison. This is illustrated in Figure 6.

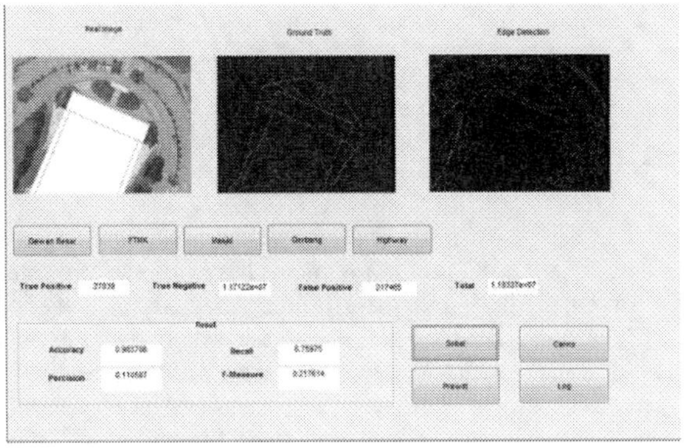

Fig. 6. Utilization of GUI

4 Results and Discussion

The measurement procedures for the results breakdown are as follows:

$$\text{Accuracy} = \frac{\sum \text{True Positive} + \sum \text{True Negative}}{\sum \text{Total Population}} \quad (8)$$

$$\text{Precision} = \frac{\sum \text{True Positive}}{\sum \text{Prediction Positive}} \quad (9)$$

$$\text{Recall / Sensitivity} = \frac{\sum \text{True Positive}}{\sum \text{Condition Positive}} \quad (10)$$

$$\text{F-Measure} = 2 \times \left(\frac{(\text{Precision} \times \text{Recall})}{(\text{Precision} + \text{Recall})} \right) \quad (11)$$

Figure 7 provides examples of the UTeM UAVs images dataset. As can be observed, original images of the main hall, mosque and highway are displayed with their corresponding edge images as a ground truth.

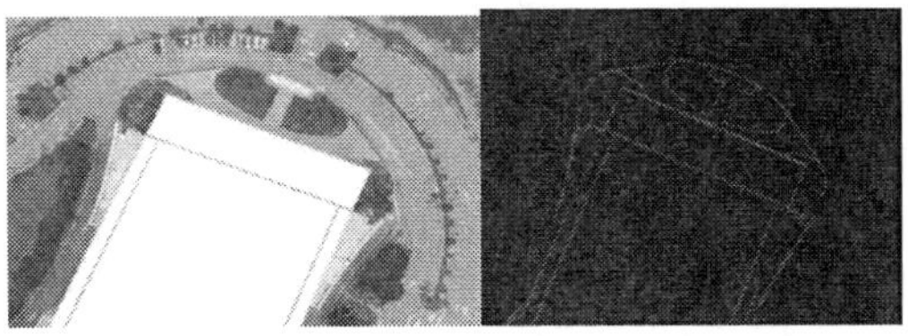

(a)

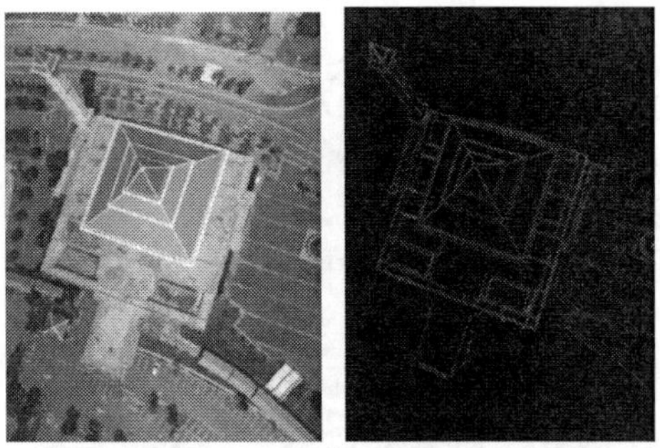

(b)

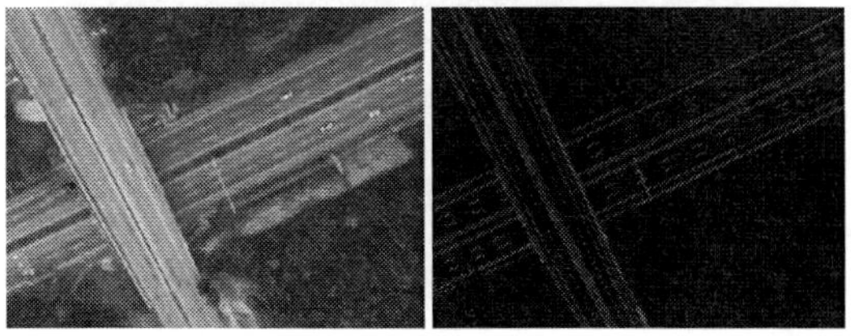

(c)

Fig. 7. Examples of UTeM images dataset with their corresponding ground truth images. (a) Main hall image, (b) mosque image and (c) highway image.

Table 1 displays the results in terms of values regarding the main hall image. In the context of f-measure values, the Prewitt method provided a superior result in comparison to those attained by the Sobel, LoG and Canny methods. The average f-measure results acquired from all images in the UTeM UAVs dataset revealed that the performance of the Prewitt method surpasses those of the other methods. This is portrayed in Table 2.

As for the edge image, the results pointed to the fact that the amount of unnecessary edges generated is less in the Prewitt method than in the other methods. The performance of the Sobel method turned out second in this area, followed by the LoG method, while the Canny method was observed to generate the most unnecessary edges. These results are exhibited in Figure 5.

Table 1. Results derived from the main hall image

	Sobel	Log	Canny	Prewitt
Accuracy	0.983708	0.952329	0.902309	0.983931
Precision	0.110587	0.0376273	0.0238827	0.111776
Recall	6.75975	5.74625	7.28475	6.7575
F-measure	0.217614	0.074765	0.0476093	0.219921

Table 2. Average results from the UTeM UAVs images dataset

	Sobel	Log	Canny	Prewitt
Accuracy	0.986759	0.954141	0.912615	0.986947
Precision	0.256592	0.110509	0.086257	0.257869
Recall	18.06275	15.95981	21.54747	17.99098
F-measure	0.505928	0.219475	0.171797	0.508382

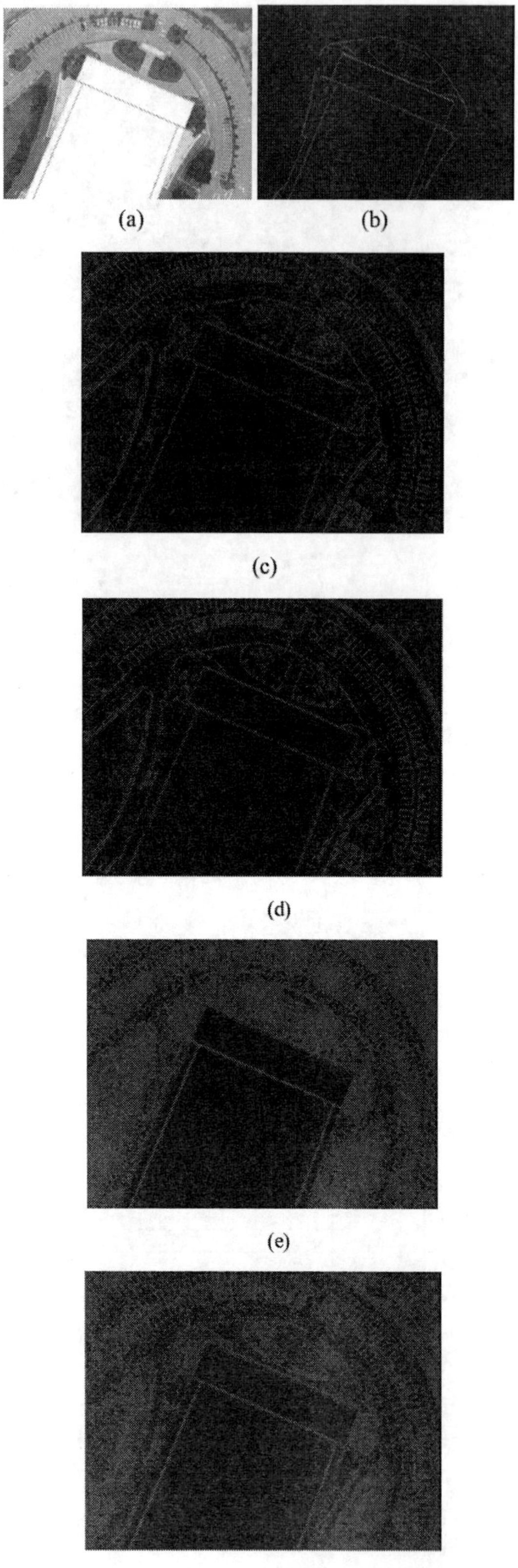

Fig. 5. The acquired image results. (a) Original main hall image, (b) ground truth image, (c) image acquired from Sobel method, (d) image acquired from Prewitt method, (e) image acquired from Canny method and (f) image acquired from LoG method.

5 Conclusions

The results of the experiment have shown that the Prewitt methods outperform other methods in detecting edges for UTeM UAVs image dataset. In fact, Prewitt and Sobel methods have produced approximately the same accuracy. Both of these methods are generally good at detecting edges and their orientations. Canny method gave the most unwanted edges compared to other methods. This is because one of the drawbacks of Canny method, it is sensitive towards weak edges. The UTeM UAVs image dataset is open for comparison on any other techniques and can be used by any researchers. It is hope that the UTeM UAVs image dataset can serve as a reference benchmark in edge detection area and thus will lead to a valid objective evaluation in edge detection. This can eventually support the progression of research work in edge detection for future ahead.

The deepest gratitude and thanks to Universiti Teknikal Malaysia Melaka (UTeM) in supporting this research.

References

1. S. E. Umbaugh, *Digital image processing and analysis : human and computer vision applications with CVIPtools*, Second Edi. CRC Press, 2010.
2. L. Mulrane, E. Rexhepaj, S. Penney, J. J. Callanan, and W. M. Gallagher, "Automated image analysis in histopathology: a valuable tool in medical diagnostics," *Expert Rev. Mol. Diagn.*, vol. 8, no. 6, pp. 707–725, Nov. 2008.
3. H. Zhou, H. Kong, L. Wei, D. Creighton, and S. Nahavandi, "On Detecting Road Regions in a Single UAV Image," *IEEE Trans. Intell. Transp. Syst.*, vol. 18, no. 7, pp. 1713–1722, 2016.
4. M. A. M. Abdullah, S. S. Dlay, W. L. Woo, and J. A. Chambers, "Robust Iris Segmentation Method Based on a New Active Contour Force With a Noncircular Normalization," *IEEE Trans. Syst. Man, Cybern. Syst.*, pp. 1–14, 2016.
5. [5] P. Smith, D. B. Reid, C. Environment, L. Palo, P. Alto, and P. L. Smith, "Otsu1975," vol. 20, no. 1, pp. 62–66, 1979.
6. A. Lucieer, D. Turner, D. H. King, and S. A. Robinson, "Using an Unmanned Aerial Vehicle (UAV) to capture micro-topography of Antarctic moss beds," *Int. J. Appl. Earth Obs. Geoinf.*, vol. 27, pp. 53–62, Apr. 2014.

7. T. Y. Lim and M. M. Ratnam, "Edge detection and measurement of nose radii of cutting tool inserts from scanned 2-D images," *Opt. Lasers Eng.*, vol. 50, no. 11, pp. 1628–1642, Nov. 2012.
8. J. Huang, X. You, Y. Y. Tang, L. Du, and Y. Yuan, "A novel iris segmentation using radial-suppression edge detection," *Signal Processing*, vol. 89, no. 12, pp. 2630–2643, 2009.
9. S. a. Coleman, B. W. Scotney, and S. Suganthan, "Multi-scale edge detection on range and intensity images," *Pattern Recognit.*, vol. 44, no. 4, pp. 821–838, Apr. 2011.
10. W. H. W. He and K. Y. K. Yuan, *An improved Canny edge detector and its realization on FPGA*. Ieee, 2008, pp. 6561–6564.
11. S. B. Kutty, S. Saaidin, P. N. A. Megat Yunus, and S. Abu Hassan, "Evaluation of canny and sobel operator for logo edge detection," *ISTMET 2014 - 1st Int. Symp. Technol. Manag. Emerg. Technol. Proc.*, vol. 2, no. Istmet, pp. 153–156, 2014.
12. Z. Othman, A. Abdullah, and A. S. Prabuwono, "A statistical Approach of Multiple Resolution Levels for Canny Edge Detection," in *Intelligent Systems Design and Applications (ISDA), 2012*, 2012, pp. 837–841.
13. Z. Othman and A. Abdullah, "An Adaptive Threshold Based On Multiple Resolution Levels for Canny Edge Detection," in *IRICT 2017: Recent Trends in Information and Communication Technology*, 2017, pp. 316–323.
14. W. Gao, L. Yang, X. Zhang, and H. Liu, "An improved Sobel edge detection," *Proc. - 2010 3rd IEEE Int. Conf. Comput. Sci. Inf. Technol. ICCSIT 2010*, vol. 5, pp. 67–71, 2010.
15. J. Canny, "A computational approach to edge detection.," *IEEE Trans. Pattern Anal. Mach. Intell.*, vol. 8, no. 6, pp. 679–98, Jun. 1986.
16. B. Wang and S. Fan, "An improved CANNY edge detection algorithm Bing," *2009 Second Int. Work. Comput. Sci. Eng.*, 2009.
17. G. Jie and L. Ning, "An Improved Adaptive Threshold Canny Edge Detection Algorithm," *2012 Int. Conf. Comput. Sci. Electron. Eng.*, pp. 164–168, Mar. 2012.
18. Nasruddin Abu Sari, Asmala Ahmad, MY Abu Sari, S Sahib, AW Rasib (2015) Development Of Rapid Low-Cost LARS Platform For Oil Palm Plantation. Jurnal Teknologi, 77 (20), 99 – 105 (Scopus). EISSN 2180-3722.

Curve Reconstruction In Different Cubic Functions Using Differential Evolution

Zabidi Abu Hasan[1,*], *Zainor Ridzuan* Yahya[1], *Nur Afifah* Rusdi[1] and *Nurshazneem* Roslan[1]

[1]Institute of Engineering Mathematics, University Malaysia Perlis, UniMAP, 02600 Arau, Perlis Malaysia

Abstract. This paper discusses the comparison on using two types of curves for curve reconstruction. Differential Evolution (DE) is used to optimize the parameter in the related spline function. DE minimized the Sum Square Error (SSE) to find the best curve that fit the data. The two curves namely cubic Bézier and cubic Ball is used for comparison purposes. For the curve reconstruction, the cubic Ball consumes less calculation time compare to cubic Bézier and gives better curve approximation based on the errors result. Visualization and numerical comparison shall be given.

1 Introduction

Curve reconstruction is a process of reconstructing a curve or a mathematical function, which has the best fit to a set of data points, possibly subject to certain constraints. It becomes one of the main issue highlighted by recent researchers especially in image processing as well as computer graphics pattern recognition and computer aided geometric design (CAGD) in solving reverse engineering problems [1]. According to [2], the purpose of this curve fitting is to find the set of data points that can accurately represent the given curve. Variety splines have been discussed in the literatures to represent the curve like cubic Bézier curve [1], [2], [3], [4] quartic Bézier curve [5] and B-splines [6], [7]. Other than that, [8] used ball cubic curve to represent the curve.

Soft computing is a recent keyword in information technologies. Soft computing based optimization methods has been used as an attempt to find the best solution in solving various types of problem especially in complex engineering problems. This is mainly due to inefficiency of classical optimization algorithms in solving complex problems. These methods do not guarantee optimal solution but require a relatively short computational time. There are several techniques in soft computing that grabs the attention of researchers such as Genetic Algorithm (GA) [7], [9], Differential Evolution (DE) [5], [10], [11] ,[8] and Artificial Bee Colony algorithm (ABC) [12], [13], [14].

In this study, the comparison of two different curves which are cubic Bézier and cubic Ball with Differential Evolution has been proposed in order to find the best curve in solving curve fitting problem. Differential Evolution (DE) has been chosen as a propose technique in solving curve fitting problem because this algorithm offers great convergence characteristics and require few control parameters. According to [15], the use of mutation operation, crossover operation and greedy selection makes DE a fast converging evolutionary algorithm. Not only that, the minimum number of control parameters of DE makes it easy to be tuned in order to find best solution [16] and [17]. Besides that, DE able to find the acceptable solution with reasonable time demand, but efficiency of the search is sensitive to the setting of control parameters [18].

2 Curve

2.1. Bézier

As been mention in the introduction, Bézier curve has been proposed by many researchers in solving curve fitting problems because of their being computationally efficient and having high level of accuracy to approximate character shapes [19], [20]. Bézier curve is defined as a parametric curve $R(t)$ that use the Bersnstein polynomials as a basis. In general, the equation of Bézier curve is given by:

$$R(t)=\sum_{i=0}^{n} B_i^n(t)P_i \tag{1}$$

where $B_i^n(t)$ is a basis function for Bézier curve and P_i refers to the control points of the curve. The basis function of the curve can be defined as the following:

$$B_i^n(t)=\frac{n!}{(n-i)!(i)!}(1-t)^{n-i}t^i \tag{2}$$

The curve can be represented in any degree n with

* Corresponding author: zabidi@unimap.edu.my

$n+1$ control points [5], [1], [14], [21]. For this study, cubic Bézier curve with has been chosen as a proposed curve. Note that, for cubic Bézier curve, it consists of four control points which are P_0, P_1, P_2, and P_3. This curve interpolates the two end control points P_0 and P_3 and approximate the two intermediate points P_1 and P_2. Figure 1 shows the example of Cubic Bézier curve of degree three.

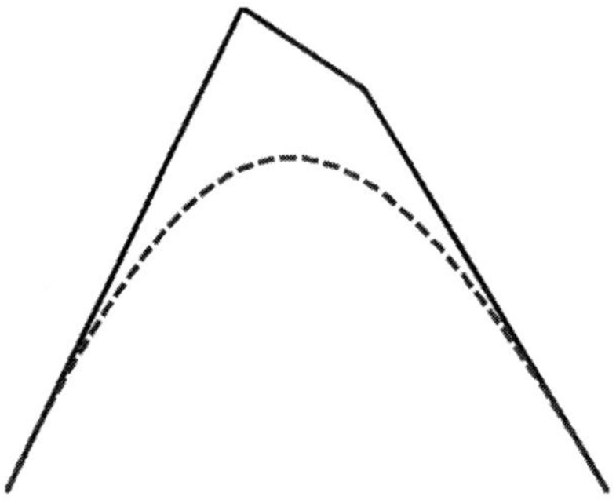

Fig. 1. Example of cubic Bézier curve.

2.2 Cubic ball

Cubic ball basis was introduced by [22]. As blending function, Ball curve have several properties that used in various aspects such as coordinate system independence, convex hull property, symmetry, end point interpolation and etc. The basic functions are:

$$B_o(t)=(1-t)^2, B_1(t)=2(1-t)^2 t, B_2(t)=2(1-t)t^2, B_3(t)=t^2 \quad (3)$$

where $0 \le t \le 1$.

The equation for cubic Ball function similar to cubic Bezier in terms of the number of points needed. However, this function gives a great potential when recently a few research demonstrated the advantages of this function. The most often mentioned from researchers is the ability to generate curves in a quicker time. In other hand, the property of end point interpolation makes this function as a suitable interpolant to connect point by point. Therefore, in curve reconstruction, we need a function that can interpolate every point. Hence, an appropriate interpolant can created by using cubic Ball basis.

Let F_i and F_{i+1} as a two-point edge while $P_{1,i}$ and $P_{2,i}$ are two control points on the cubic Ball. Then, by using the basic functions in (3) Ball cubic parametric curve defined by:

$$P_i(t)=F_i(1-t)^2+2P_{1,i}(1-t)^2 t+2P_{2,i}(1-t)t^2+F_{i+1}t^2 \quad (4)$$

Figure 2 shows the example of cubic Ball curve.

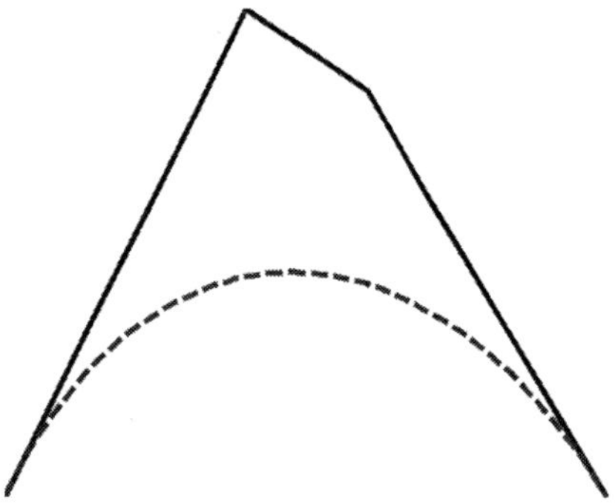

Fig. 2. Example of cubic Ball curve.

3 Differential evolution

Differential Evolution (DE) was proposed by Rainer Storn [23]. DE is an extremely powerful yet simple evolutionary algorithm that improves a population of individuals over several generations through the operators of mutation, crossover and selection for global optimization. It has successfully been applied to diverse domains of science and engineering, such as mechanical engineering design, signal processing, chemical engineering, machine intelligence, and pattern recognition [24]. Besides that, DE also has been used to solve optimal power flow problems [16], [17] and [25]. The process involved in DE algorithm are Initialization, Mutation, Recombination and Selection. These process are described as below and being summarized in Figure 3.

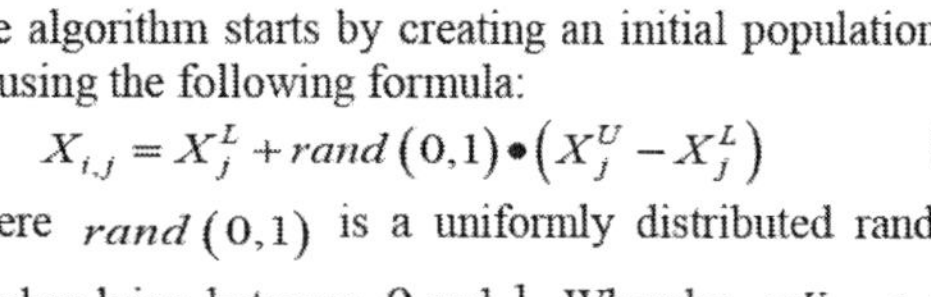

Fig. 3. Process on DE algorithm.

3.1 Initialization

The algorithm starts by creating an initial population of by using the following formula:

$$X_{i,j} = X_j^L + rand(0,1) \bullet (X_j^U - X_j^L) \quad (5)$$

where $rand(0,1)$ is a uniformly distributed random number lying between 0 and 1. Whereby X_j^U and X_j^L is an upper boundary and lower boundary of the population respectively. The population size is an algorithm control parameter selected by the user which remain unchanged throughout the optimization process.

3.2 Mutation

DE process continued with the process of mutation strategy. In this process, each generation (or one iteration of the algorithm) will change the population member known as target vector, $X_{i,G}$, by creating a donor vector $V_{i,G}$. The mutation is based on the difference between two individuals randomly chosen from the current population [26]. There are various method of creating this donor vector that differentiates one DE scheme from another. In this paper, we only focus to the scheme of DE/rand/1 in producing the donor vectors. Several variance of DE mutation strategies are described in the following, [23],[27],[28]. In these scheme, for each vector $X_{i,G}$, a donor vector $V_{i,G}$ is generated according to the following:

$$V_{i,G} = X_{r1,G} + F\left(X_{r2,G} - X_{r3,G}\right) \tag{6}$$

The notation DE/rand/1 specifies that the vector to be perturbed is randomly chosen, and that the perturbation consists of one weighted difference vector. In this process, vector $X_{r1,G}, X_{r2,G}, X_{r3,G}$ are selected randomly such that they are mutually different indices from the current generation. The mutation factor, F is normally used is $0.5 \le F \le 1$, [23] and also chosen $0 < F \le 2$, [16] and [17].

3.3 Recombination

The recombination process incorporates successful solution from previous generation where the trial vector $U_{j,i,G}$ is produced from the target vector $X_{i,G}$ and elements of the donor vector $V_{i,G}$ with the probability of CR. This process is illustrated as follows:

$$U_{j,i,G} = \begin{cases} V_{j,i,G} & if \ rand_{j,i}(0,1) \le CR, \\ & or \ j = j_{rand} \\ X_{j,i,G} & else \end{cases} \tag{7}$$
$$i = 1,2,\ldots,N; \quad j = 1,2,\ldots,D$$

From equation (7), CR become a control parameter of DE just like F. The value of the CR is between the range of $[0,1]$.

3.4 Selection

The next step is called selection. This step will determine whether the target or the trial vector survives to the next generation. The selection operation is described as

$$X_{i,G+1} = \begin{cases} U_{i,G}, & if \ f\left(U_{i,G}\right) \le f\left(X_{i,G}\right) \\ X_{i,G}, & else \end{cases} \tag{8}$$

where $f(X)$ is the function to be minimized. So, the best (lowest) fitness value will be included in the next generation. Mutation process, recombination, and the selection process will be repeated until some stopping criterion is reached. Commonly, stopping criteria are satisfied when a predefined small number of fitness value or a fixed number of generation or iteration have been reached.

4 Boundary extraction and corner detection

The proposed method starts with finding the boundary of the generic shape and then using the output to detect the corner points of the image. The image of the generic shape can be acquired by scanning or by some other method. The purpose of the boundary extraction is to produce an object's shape in graphical representation in a number of boundary points. There are a lots of algorithm that has been used in order to extract the boundary of the generic shape such as [29], [30], and Chain Codes [19]. But in this paper, we used the MATLAB function that called Boundaries in order to extract the boundary of the font images.

After finding out the boundary points, the next step is to find the corner detection of the generic shape. The corner points can also be described as the point that separates the outline into various segments. Some of the possible methods to detect the corner points is Davis algorithm. This algorithm has been applied in the research of [19] and [30]. For the detection of corners, in this paper, we used SAM06 algorithm [31].

According to [31], SAM06 algorithm is robust, simple to implement, efficient and performs well on noisy shapes as well. Detected corners by SAM06 were precise and without detecting any wrong ones. Besides that, we also used Douglas-Peucker algorithm [32] in order to reduce the data in a particular curve. The purpose of this algorithm is to create same curve by using a small number of data. We hope that this algorithm can reduce the computational time.

We have decided to use different types of image such as the Japanese (Kanji) character, and image of heart and fork. Details for the boundary extraction and corner detection of the images is shown in Table 1.

5 Chord length parameterization

In this paper, chord length parameterization is used to estimate the parametric value t associated with each point. Chord length parameterization also has been used by [1], [5], [6], [10], [14], [21], and [27]. Given parametric curve $Q(t)$ over a certain domain $t \in [0,1]$, its chord length at a given point $Q(t)$ is define as:

 https://doi.org/10.1051/matecconf/201815006030

$$t_i = \begin{cases} 0 & if \quad i=1 \\ \dfrac{|p_1p_2|+|p_2p_3|+\ldots+|p_ip_{i+1}|}{|p_1p_2|+|p_2p_3|+\ldots+|p_{n-1}p_n|} & if \quad i<n \\ 1 & if \quad i=n \end{cases} \quad (9)$$

Table 1. Details of the boundary extraction and corner detection

Original image	**Name**	**Boundary extraction and corner detection**
	At(@)	
	Friend (Yuū)	
	Fork	
	Strength (Chikara)	

6 Curve reconstruction

In this section, we apply DE to solve curve reconstruction. The process of the forming curve requires several steps related as we can see in the Algorithm 1 for curve reconstruction. Steps 1, 2 and 3 already described in section 4, while step 4 we will use the DE to solve curve reconstruction by using three different curves as mention in section 2.

However, to fulfil the different result, the method of sum square error (SSE) used in order to calculate the different or errors between the parametric points of generated curve ($P(t_i)$) and given points ($F_{i,j}$). Let $F_{i,j}=(x_{i,j},y_{i,j})$ the data points are given or extracted where $i=1,2,\ldots,n$ and $j=1,2,\ldots,m$. Hence, by using the formula of SSE, we define $S_i=\sum_{j=1}^{m_i}\left[P(t_i)-F_{i,j}\right]^2$, $i=1,2,\ldots,n$ where t_i refer to parameterization as explained in section 5.

Algorithm 1: Curve reconstruction:

Input: Data of image $F_{i,j}=(x_{i,j},y_{i,j})$

Output: Curve fitting, $P(t_i)$

Step 1: Get appropriate image to reconstruct
Step 2: Boundary extraction from the image to acquire data points.
Step 3: Corner points detected
Step 4: The DE for curve reconstruction using:
a) Cubic Ball curve.
b) Cubic Bezier

Step 5: Calculate the number of errors and total time taken.

7 Result and discussions

Generally, DE could be used for curve fitting. In this paper, two type of splines explained. Before using the splines, a few pre-processing steps needed in order to interpolate the data. However, to test which curve have a good result, the curve fitting generated needed to compare with original image to measure the error or known as SSE. As the result demonstrated, the reconstruction have similar image visually, while the error as shown in Table 2 give different result where the cubic Ball lead all from the four examples. Table 3 reflect the result in terms of time where the cubic Ball give impressive result compare to cubic Bezier curve. This is because of the degree in cubic Ball less cubic Bezier. Figure 4 illustrate the curve reconstruction for the image selected.

Table 2. Average error obtained by each image.

Image	**Cubic Ball**	**Cubic Bezier**
AT	199.9212	199.9878
Fork	8.9260	9.2857
Friends	384.2103	411.1150
Strength	572.9575	573.6694

Table 3. Average time taken for each image

Image	Cubic Ball	Cubic Bezier
AT	11.7240	12.8224
Fork	5.1706	5.2744
Friends	11.3632	12.1389
Strength	7.5557	8.1168

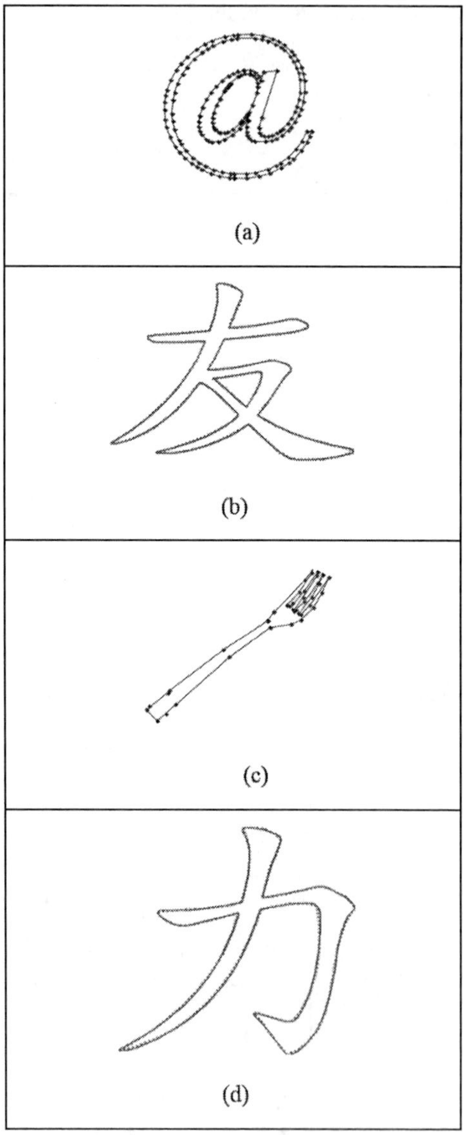

Fig. 4. Curve reconstruction

Acknowledgement

The authors highly acknowledge the referee's valuable comments on the improvement of this paper. The authors also grateful for the financial support provided by UniMAP, under Short Term Grants-(STG) (9001-00531).

References

1. N.A. Rusdi and Z.R. Yahya, Reconstruction of generic shape with cubic Bézier using least square method, International Conference on Mathematics, Engineering & Industrial Applications 2014 (ICoMEIA 2014), AIP Publishing 1660, 050004 (2015)
2. M. Sarfraz and A. Masood, Capturing outlines of planar images using Bézier cubics, Computers & Graphics **31**, 719-729(2007)
3. M. Sarfraz and M.A. Khan, An automatic algorithm for approximating boundary of bitmap characters, Future Generation Computer Systems **20**, 1327-1336 (2004)
4. L. Shao and H. Zhou, Curve fitting with Bezier cubics, Graphical models and image processing **58**, 719-729 (2007)
5. N.A. Rusdi and Z.R. Yahya, Reconstruction of Arabic Font with Quartic Bézier Curve, Sains Malaysiana **44**, 1209-1216 (2015)
6. N. Roslan and Z.R. Yahya, Reconstruction of egg shape using B-spline, International Conference on Mathematics, Engineering & Industrial Applications 2014 (ICoMEIA 2014), AIP Publishing 1660, 050016 (2015)
7. M. Sarfraz, M. Irshad and M.Z. Hussain, Reverse Engineering of Planar Objects Using Gas, Sains Malaysiana **42**, 1167-1179 (2013)
8. Z.A. Hasan, Z.R. Yahya, A.R.M. Piah, Differential Evolution for Reverse Engineering Using Cubic Ball Curve, EDUCATUM Journal of Science, Mathematics and Technology **2**, 58-68 (2015)
9. K. Barkeshli, M. Mokhtari and M. Amiri, Image reconstruction of impenetrable cylinders using cubic B-splines and genetic algorithms, In Antennas and Propagation Society International Symposium, 2001. IEEE **2**, 686-689 (2001)
10. N. Roslan and Z.R. Yahya, Reconstruction of Font with Cubic Bézier Using Differential Evolution Sains Malaysiana **44**, 1203-1208 (2015)
11. P. Pandunata and S.M.H. Shamsuddin, Differential evolution optimization for Bézier curve fitting, In Computer Graphics, Imaging and Visualization (CGIV), 2010 Seventh International Conference on IEEE 68-72 (2010, August)
12. A. Singh, An artificial bee colony algorithm for the leaf-constrained minimum spanning tree problem, Applied Soft Computing **9**, 625–631 (2009)
13. G. Zhu and S. Kwong, Gbest-guided artificial bee colony algorithm for numerical function

optimization, Applied Mathematics and Computation **217,** 3166-3173(2010)

14. N.A Rusdi and Z.R. Yahya, Reconstruction of Arabic Font using Artificial Bee Colony Algorithm, Acceptance for Malaysian Technical Universities Conference on Engineering and Technology 2015 (MUCET 2015), Johor Bharu, 11-13 October 2015.
15. D. Karaboga, and B. Basturk, On the performance of artificial bee colony (ABC) algorithm, Applied soft computing **8,** 687-697 (2008)
16. N. Sinsuphun, U. Leeton, U. Kwannetr, D. Uthitsunthorn and T. Kulworawanichpong, Loss minimization using optimal power flow based on swarm intelligences, Transactions on Electrical Engineering, Electronics, and Communications **9**, 212-222 (2011)
17. S.P. Karthikeyan, K. Palanisamy, L.J. Varghese, L.J. Raglend and D.P. Kothari, Comparison of intelligent techniques to solve economic load dispatch problem with line flow constraints, In Advance Computing Conference, 2009. IACC 2009. IEEE International 446-452 (2009, March)
18. J. Tvrdık, Adaptive differential evolution and exponential crossover. Proceedings of IMCSIT 2008 927-931 (2008)
19. M. Sarfraz and M.A. Khan, Automatic outline capture of Arabic fonts, Information Sciences **140**, 269-281 (2002)
20. M. Sarfraz and M.F.A. Razzak, An algorithm for automatic capturing of the font outlines, Computers and Graphics (Pergamon) **26**, 795-804 (2002)
21. N. Roslan and Z.R. Yahya, Different Mutation Strategies For Reconstruction of Japanese Character, Acceptance for Malaysian Technical Universities Conference on Engineering and Technology 2015 (MUCET 2015), Johor Bharu, 11-13 October 2015.
22. A. A. Ball. CONSURF. Part one: Introduction of the conic lofting tile **6**, 243-249 (1974)
23. R. Storn, On the usage of differential evolution for function optimization, In Fuzzy Information Processing Society, 1996. NAFIPS., 1996 Biennial Conference of the North American. IEEE 519-523 (1996, June)
24. S. Das, A. Abraham, U.K. Chakraborty and A. Konar, Differential evolution using a neighborhood-based mutation operator, Evolutionary Computation, IEEE Transactions on **13**, 526-553 (2009)
25. M. Varadarajan and K.S. Swarup, Solving multi-objective optimal power flow using differential evolution, Generation, Transmission & Distribution, IET **2**, 720-730 (2008)
26. R.S. Prado, R.C. Silva, F.G. Guimarães and O.M. Neto, Using differential evolution for combinatorial optimization: A general approach, In Systems Man and Cybernetics (SMC), 2010 IEEE International Conference on. IEEE 11-18 (2010, October)
27. N. Roslan, and Z.R. Yahya, Different Mutation Strategies For Reconstruction Of Japanese Character, Acceptance For Malaysian Technical Universities Conference On Engineering And Technology 2015 (MUCET 2015), Johor Bharu, 11-13 October 2015.
28. S. Das, A. Abraham and A. Konar, Particle swarm optimization and differential evolution algorithms: technical analysis, applications and hybridization perspectives, In Advances of Computational Intelligence in Industrial Systems. Springer Berlin Heidelberg 1-38 (2008)
29. G. Avrahami and V. Pratt, Sub-pixel edge detection in character digitization, Raster Imaging and Digital Typography II 54-64 (1991)
30. K. Itoh and Y. Ohno, A curve fitting algorithm for character fonts, Electronic publishing **6**, 195-205 (1993)
31. M. Sarfraz, Corner Detection for Curve Segmentation, Interactive Curve Modeling: With Applications to Computer Graphics, Vision and Image Processing 209-240 (2008)
32. The MathWorks Inc. (MATLAB), Ramer-Douglas Peucker algorithm demo by Ligong Han, http://www.mathworks.com/matlabcentral/fileexchange/- 41986-ramer-douglas-peucker-algorithm (2014)

The Reduction of Directed Cyclic Graph for Task Assignment Problem

W.N.M. Ariffin[1]

[1]Institute of Engineering Mathematics, Universiti Malaysia Perlis.

Abstract. In this paper, a directed cyclic graph (DCG) is proposed as the task graph. It is undesirable and impossible to complete the task according to the constraints if the cycle exists. Therefore, an effort should be done in order to eliminate the cycle to obtain a directed acyclic graph (DAG), so that the minimum amount of time required for the entire task can be found. The technique of reducing the complexity of the directed cyclic graph to a directed acyclic graph by reversing the orientation of the path is the main contribution of this study. The algorithm was coded using Java programming and consistently produced good assignment and task schedule.

1 Introduction

Task assignment is one of the most challenging problems in distributed computing environment. An optimal task assignment guarantees minimum turnaround time for a given architecture. Several approaches and techniques of optimal task assignment have been proposed by various researchers ranging from graph partitioning based tools to heuristic graph matching. A good task assignment algorithm and mapping strategy ensure turnaround time minimization. Thus, many researchers have attempted to produce good algorithm to solve the task assignment problem. Previous work put efforts on solving a directed acyclic task graph. But, how about directed cyclic graph ?

Cordella et al. [1] proposed a new algorithm which can handle subgraph isomorphism test efficiently in large graphs. Tong et al. [2,3] proposed an algorithm to find best effort matching in a large graph based on random walk. Another type of work that similar to Linhong et al. which use index to find matches in a large data graph efficiently, Tian and Patel [4] utilize degree information and neighborhood connectivity to filter unmatched node pairs. Zhang et al. [5] proposed another indexing approach based on graph distance. Their methods are not efficient enough to handle graphs with up to millions or billions of nodes and edges. In the preprocessing stage of [5], given a data graph, they generate a set of intersecting subgraphs may increase significantly when the size of data graph grows. Tian et al. [4] used a maximum weighted bipartite graph matching algorithm during the stage of matching important nodes, which can be costly if the bipartite graph is large.

Gutman and Wagner [6] defined the matching energy of a graph and gave some properties and asymptotic results of the matching energy. Shuli and Weigen [7] characterized the connected graph *G* with connectivity k has the maximum matching energy by introducing n-matching n-partite graph. Mohan and Gupta [8] established a methodology for heuristic graph matching, but restricted to a small number of test cases.

As the number and complexities of graph based applications increase, rendering the graphs more compact, easier to understand navigate through are becoming crucial tasks. Another one approach to graph simplification is to partition the graph into smaller parts, so that instead of the whole graph, the partitions and their inter-connections need to be considered. Common approaches to graph partitioning involve identifying sets of edges (edge-cuts) or vertices (vertex-cuts) whose removal partitions the graph into the target number of disconnected components. While edge-cuts result in partitions that are vertex disjoint,in vertex-cuts the data vertices can serve as bridges between the resulting data partitions; consequently, vertex-cut based approaches are especially suitable when the vetices on the vertex-cut will be replicated on all relevant partitions. A significant challenge in vertex-cut based partitioning, however, is ensuring the balance of the resulting partitions while simultaneously minimizing the number of vertices that are cut (and thus replicated).

Ariffin and Salleh [9] put an effort to reduce the graph onto directed acyclic graph. A Kernighan-Lin algorithm is applied to obtain the partition of tasks. Combining the technique of reduction and partitioning lead to an efficient graph-mapping concept. In [10], considered a directed-weighted cyclic task graph. Combining the technique of reduction and bi-partitioning led to an efficient graph-mapping concept.

* Corresponding author: munirah@unimap.edu.my

In this study, a directed cyclic graph (DCG) is proposed as the task graph. It is undesirable and impossible to complete the task according to the constraints if the cycle exists. Therefore, an effort should be done in order to eliminate the cycle to obtain a directed acyclic graph (DAG), so that the minimum amount of time required for the entire task can be found. The main contribution in this study is the problem itself. The previous work did not solve for directed cyclic graph for task assignment problem. The technique of reducing the complexity of the directed cyclic graph to a directed acyclic graph is also the contribution of this study.

2 Problem Statement

An arbitrary task graph, with eight number of nodes is proposed as in Figure 1. A cycle exists for {4,5,7,4} .

How the cycle can be eliminated from the graph to form an acyclic graph?

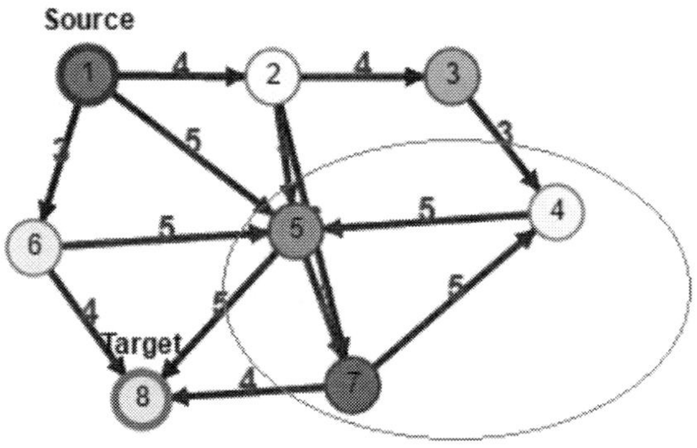

Fig. 1: A DCG with eight number of nodes.

From Figure 1 above, let consider the three nodes that form the cycle {4,5,7,4}.

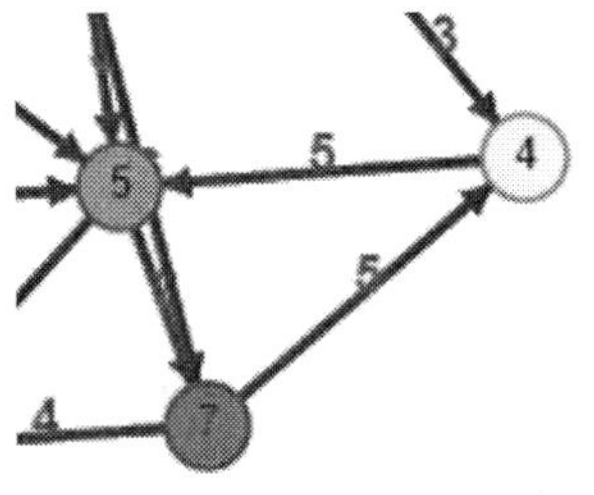

Fig. 2: A cycle in {4,5,7,4}.

The main objective is to eliminate the cycle in the task graph, so that it becomes acyclic. One way to do this is simply drop edges from the task graph to break the cycle. A feedback arc set or feedback edge set is a set of edges which when removed from the graph will leave a DAG. Put another way, it is a set containing at least one edge of every cycle in the graph. Closely related are the feedback vertex set, which is a set of vertices containing at least one vertex from every cycle in the directed graph and the minimum spanning tree, which is the undirected variant feedback arc set problem.

A minimal feedback arc set where one that cannot be reduced in size by eliminating or removing any edges has the additional property that, if the edges in it are reversed rather than removed, then the graph remains acyclic.

Sometimes, it is desirable to drop as few edges as possible, obtaining a minimum feedback arc set or dually a maximum acyclic subgraph. This is a hard computational problem, for which several approximate solutions have been devised.

3 The Decomposition of Strongly Connected Component

A graph $G=(V,E)$ has set of nodes $V=\{1,2,3,\dots n\}$ and an edge set E is a collection of unordered pairs of nodes called the edges of the graph. An orientation or directed graph $\vec{G}$ is obtained by assigning an orientation (ordering) to each edge of G. Each edge has a tail and a head. It is oriented from tail to head in the directed graph. A directed graph might have multiple directed edges between the same two nodes, y and z . This is called directed multi-graphs.

A cycle of a graph G is a subset of the sedge set of G that forms a path such that the start node and the end node are the same. A directed cycle is an oriented cycle such that all directed edges are oriented in the same direction along the cycle. An acyclic directed graph does not contain any directed cycle. The degree of a node y is the number of edges hat have y as an end node. The out-degree of a given node in a directed graph is the number of edges directed out of the node. A node with out-degree 0 is called a sink (however, in this paper the term *sink* is replaced by *target*).

A directed graph is called strongly connected if and only if any two nodes i and j in $\vec{G}$, there exists a directed path from i to j to and from j to i . The strongly connected components of a graph are its maximal strongly connected sub graphs. It is maximal in the sense that a sub graph cannot be enlarged to another strongly connected sub graph by including additional nodes and its associated edges.

4 Equivalent of the Orientation

In this part, choices in orientations of edges which conserve a fixed out-degree of each of the vertices do not alter the decompositions. This is a useful observation for the proofs in the next section and will assist the verification of the algorithm to generate the directed graphs and related decompositions by A. Sljoka [11].

Definition 4.1.

Given a multi-graph G and two orientations $\vec{G}_1$ and $\vec{G}_2$ are equivalent orientations on G if the corresponding nodes have the same out-degree.

Lemma 4.1.

Given two equivalent orientations $\vec{G_1}$ and $\vec{G_2}$ on G, then the two orientations differ by reversals on a set of directed cycles.

Proof 4.1.

Choose an edge $e=(y,z)$ in $\vec{G_1}$ that is oppositely directed in $\vec{G_2}$. Thus, in $\vec{G_1}$ edge e is incoming at node z. Assume that there are k outgoing edges at z in $\vec{G_1}$. Because z has same out-degree k in $\vec{G_2}$, and this edge is reversed to be outgoing in $\vec{G_2}$, there must exist an out-going edge from z in $\vec{G_1}$, say $g=(z,x)$, that is oppositely directed in $\vec{G_2}$. This situation is illustrated in Figure 3 (a) and (b). Walk out of z along g in $\vec{G_1}$ is shown in Figure 3 (c).

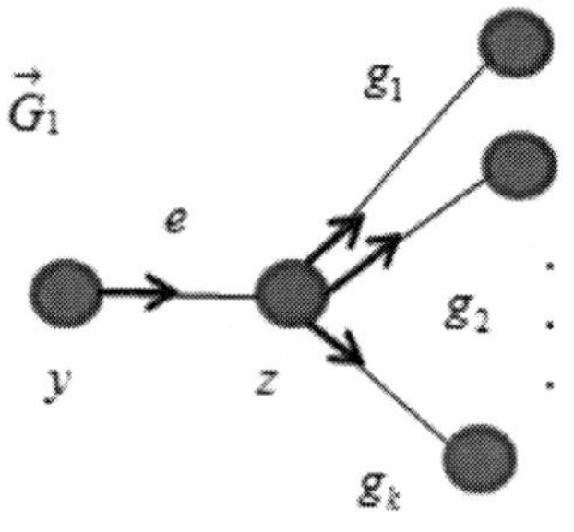

(a)

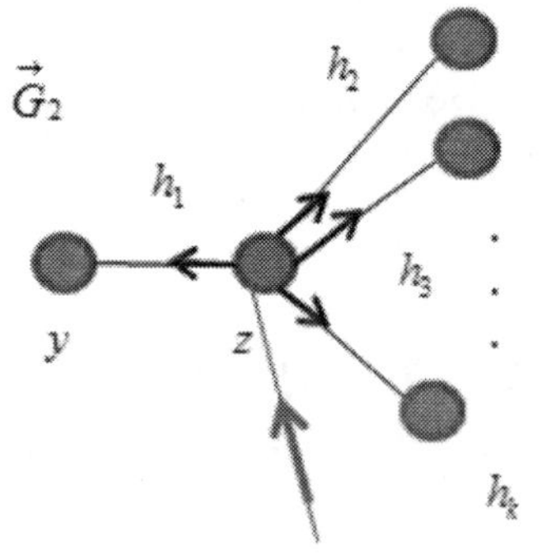

(b)

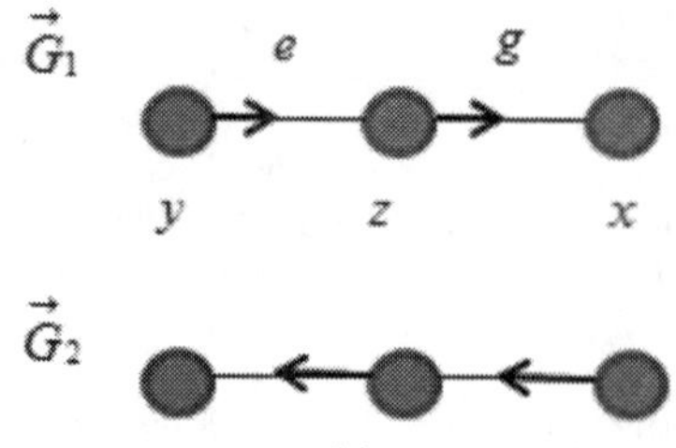

(c)

Fig. 3: Node y has same out-degree in G_1 and G_2. Thus, there exists an edge in G_1 (a) that is oppositely oriented in G_2 (b). The selection of f is shown in (c).

As a new node is inserted, the same argument is applied. This identies a directed path in $\vec{G_1}$ that is oppositely directed in $\vec{G_2}$. As there is only a finite collection of edges that have opposite direction, walking along this directed path until coming back to the same node on this path, identifying a directed cycle in $\vec{G_1}$ that has an opposite orientation in $\vec{G_2}$. Reversing the orientation of such a cycle from the orientation in $\vec{G_1}$ towards the orientation in $\vec{G_2}$. This decreases the number of edges in $\vec{G_1}$ that are oppositely directed from $\vec{G_2}$. The illustration can be seen in Figure 4. Reversing identified cycles is continued until all edges are directed following $\vec{G_2}$.

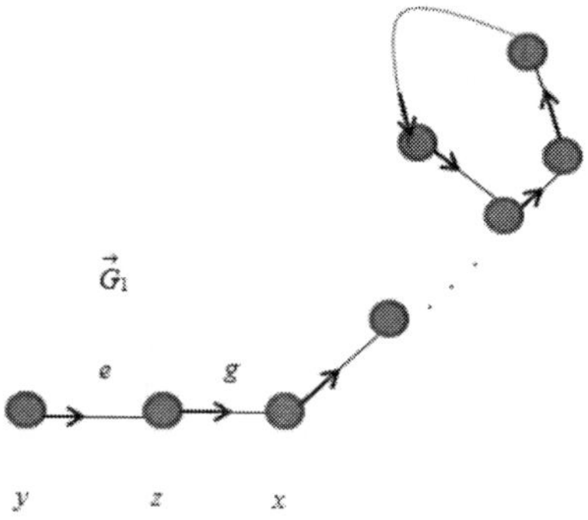

(a)

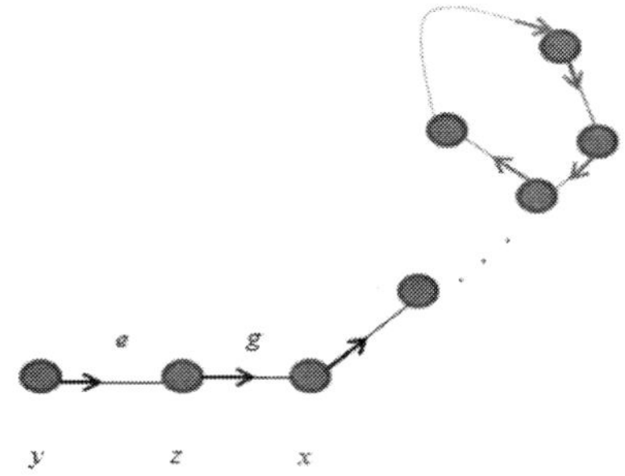

(b)

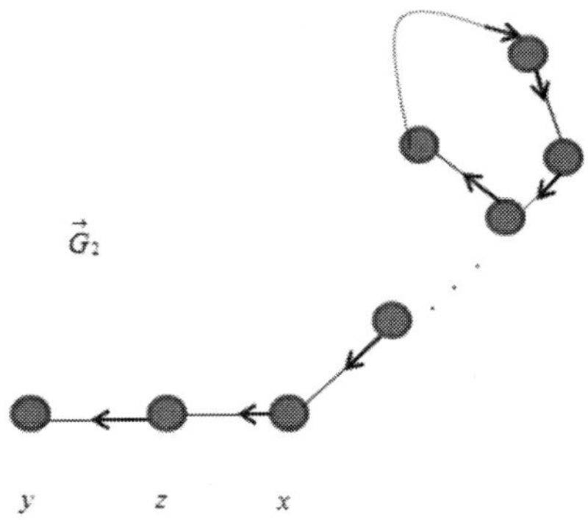

(c)

Fig. 4: Locating a directed path in G_1 that is oppositely oriented in G_2 (a), reversing a first cycle in G_1 (b), which now has the same orientation as in G_2 (c).

A valuable corollary to the lemma:

Corollary 4.1.

Given two equivalent orientations $\vec{G_1}$ and $\vec{G_2}$, the strongly connected components of the decompositions are the same.

Proof.

From Lemma 4.1, $\vec{G_1}$ and $\vec{G_2}$ differ by reversal of set of directed cycles. As cycle reversals do not change the strongly connected components, the decompositions are the same.

5 Proposed Technique to the Directed Cyclic Graph

The discussion in Section 4 is applied to solve the problem in Figure 2. The nodes in Section 4 (v_y, v_z, v_x) are renamed as v_4, v_5, v_7 which indicate node 4, node 5 and node 7 respectively. Note that node v_7 is inserted to form $\vec{G_2}$.

In this study, the technique of removing cycle is presented by reversing the direction of at least for one directed edge. The algorithm is explained as follows:

Step 1: Fix the priority of the tasks (nodes), label them from 1 to n.

Step 2: The task graph contains edges (y, z), where $y < z$.

Step 3: Choose one node as the starting point (parent) with the property that having the least computational time from the previous pointed edge(s).

Step 4: Decompose the nodes with two different edges and share the same child.

Step 5: Recombine the two decomposition in Step 4, and change the direction of the edge.

Step 6: Check again the task graph from node 1 to node n. If cycle exists, repeat Step 1 to Step 5. Else, stop the iteration.

As can be seen in Figure 2, the node {4,5,7,4} consists of a cycle. One can also say that [5,7,4,5] and {7,4,5,7} create the cycle. But, according to the algorithm in Step 3, one node with the least amount of computational time from the previous node is chosen as the starting point, called parent. This is important because a minimum amount of time to complete the task is required. Therefore, from Figure 1, node 3 is directed to node 4 with computational time of three units. Node 5 is directed from node 1, 2, 4, and 6 with bigger amount of computational time, compared to the computational time to node 4. The last one is node 7 is directed from node 2, which is obviously with bigger computational time as it is directed to node 5 previously. Therefore, node 4 is chosen as the parent node, so that the cycle is read as {4,5,7,4}. Then, the nodes are decomposed as shown in Figure 5.

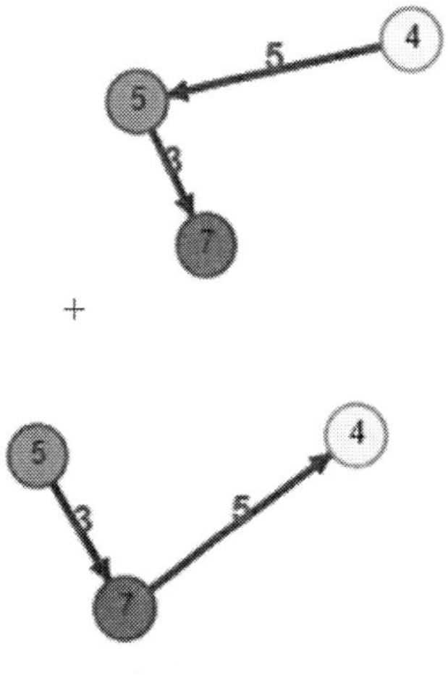

Fig. 5: Decomposition of the three nodes to two set of directed graphs.

The parent node is node 4 is directed to node 5. Therefore, node 5 becomes it child. Next, node 5 is directed out to the node 7. As stated Step 4, with the same child node, it is directed to node 7 and finally back to node 4. Since node 5 consists of directed edge from node 4 and to node 7, therefore the direction remains unchange. Thus, the direction from node 7 to node 4 is changed or reversed, so that it becomes from node 4 to node 7. This can be illustrated as in Figure 6.

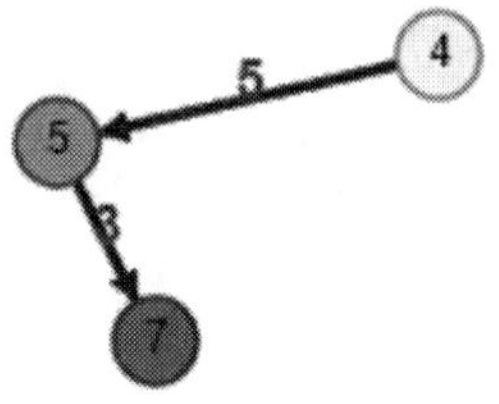

(a)

+

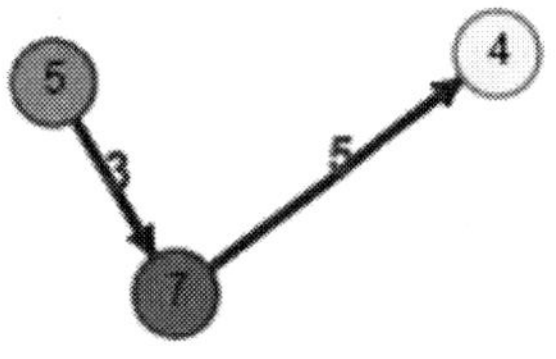

(b)

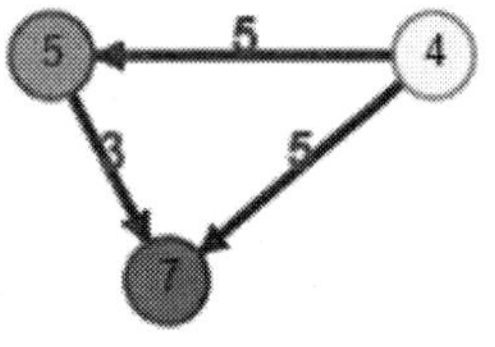

=

(c)

Fig. 6: Decomposition of the three nodes to two set of directed graphs, (a) and (b). Combination of (a) and (b), (c).

6 Formation of Directed Acyclic Graph

The proposed task graph in Figure 1 is now becomes a directed acyclic graph with the reverse direction from node 4 to node 7. Back to the task assignment problem, the nodes represent the beginning and the end of the tasks. Each task takes a certain estimated time to complete, and this is represented by assigning each edge uv a weight $WT(uv)$, being the amount of time required for that particular task.

6.1 Topological Order

A topological ordering of an acyclic digraph G is a permutation of $V(G)=\{1,2,3,\ldots,n\}$ such that $o(u) < o(v)$ whenever $u \rightarrow v$. Thus all edges are directed from smaller to higher node numbers. Notice that only acyclic digraph have topological orderings, since a directed cycle cannot be ordered in this way. Topological orderings are easy to find.

The algorithm is easy to be implemented. It is called a *DCGSimplify*. The topological order is built on the queue. The algorithm begins by placing all vertices with in-degree 0 on the queue. All nodes are arranging on two columns like wrapped-butterfly, inspired by Liao [16]. A source-destination (target) nodes are chosen from the graph. $InDegree[v]$ is then adjusted so that it counts the in-degree of v only from vertices not yet on the queue. This is done by decrementing $InDegree[v]$ according to its in-edges from the queue. When all n nodes are on the queue, the nodes are in the topological order. Note that in order to find the path from the source to the target node, the algorithm does not allow sharing common intermediate nodes and less amount of time on the edge is chosen.

Figure 7 (a) shows the arrangement of the task graph in two columns, while Figure 7 (b) shows its directed acyclic graph based on the proposed algorithm as discussed in Section 5.

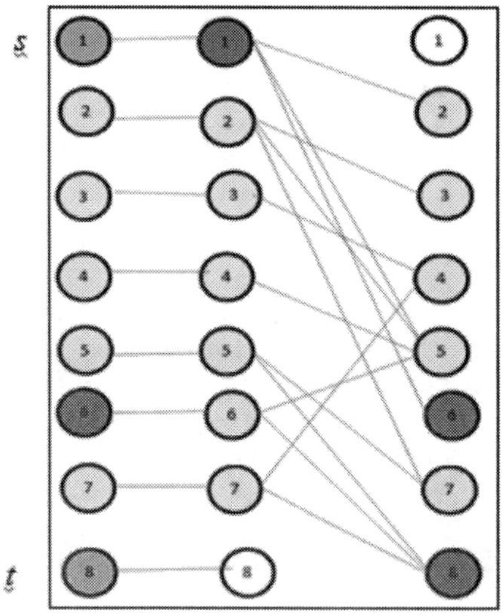

(a)

https://doi.org/10.1051/matecconf/201815006031

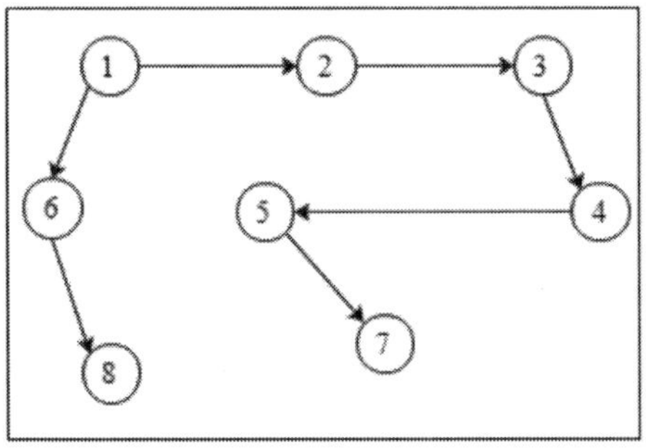

(b)

Fig. 7: All n nodes are arranged in two-column (a) and the directed acyclic graph.

In this problem, node 1 is set as the source node and node 8 as the destination node. The target is to obtain the minimum amount of time to complete the task. This can be refered to find the shortest path from the source node 1, denote as s to the destination node 8, denote as t. So, the path from s to t must be found such that all nodes must be chosen and they must not share common intermediate nodes.

As can be seen from Figure 7 (a), there are three nodes can be chosen from the source node 1. Let say, node 2 is chosen from node 1. Next, from node 2, node 3, 5 and 7 can be reached.

Next, node 3 is chosen. From node 3, only node 4 can be chosen. Next, node 5 is chosen from node 4. After that, node 7 is chosen from node 5. Note that there are two nodes that are not chosed yet. So, again from the source node, node 6 is chosen. From node 6, either node 5 or node 8 can be chosen. Notice that, same intermediate nodes are not allowed. Thus, node 5 is strictly not available to be chosed. Next, from node 6, node 8 can be reached, which is the destination node. Now, the directed acyclic graph is obtained. It is illustrated as in Figure 7 (b).

7 Results

To evaluate our proposed algorithm, we have implemented it using Intel(R) Core(TM) i5 (2.3 GHz) using Javascript programming language with D3 library. The algorithm is applied to test for several number of nodes.

The task graph with small number of nodes (six nodes) is shown in Figure 8 below.

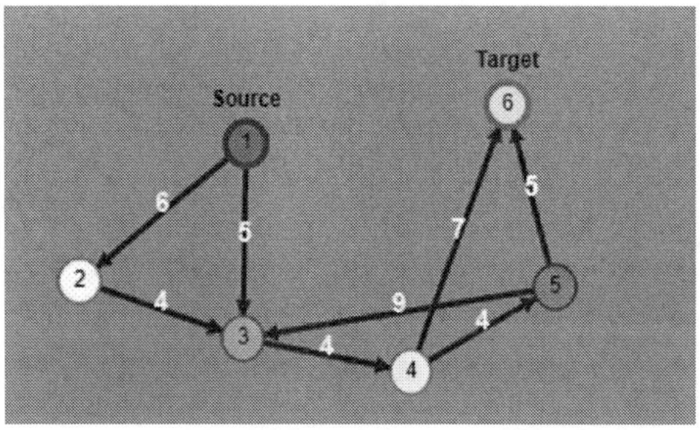

(a)

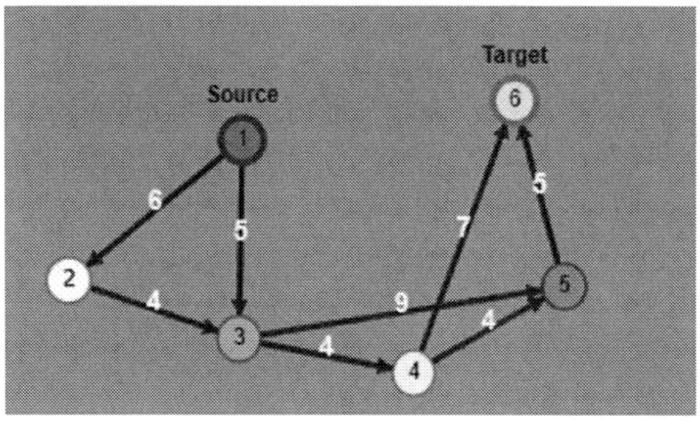

(b)

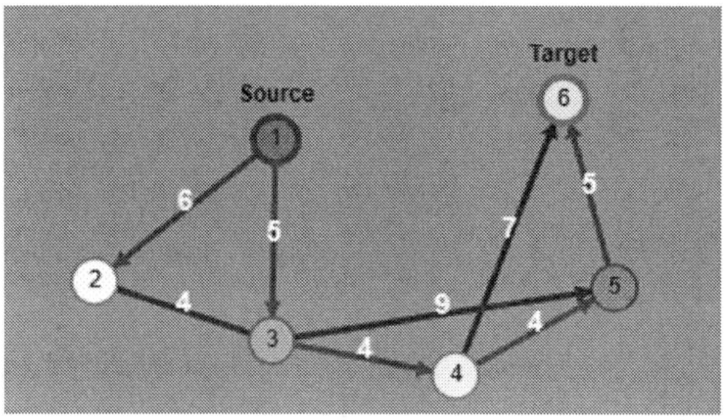

(c)

Fig. 8: Figure 8 (a) The directed cyclic graph, 8 (b) removing the cycle and 8 (c) the directed acyclic graph to be mapped on the processors.

The task graph in Figure 8 (a) is the directed cyclic graph while in (b) after applying the proposed algorithm. Figure 8 (c) shows the final nodes (tasks) that need to be mapped onto processors. This is shown in Figure 9.

P1	2	1		
P2	3	4	5	6

Fig. 9: Mapping the nodes onto two processors.

From Figure 9 above, node 1 and node 2 are located in processor one, while node 3, node 4, node 5 and node 6 are located in processor two. The total schedule length is 18 units.

The other two testing have been done on ten and twenty number of nodes. The task graphs are shown in Figure 10 (a) and (b).

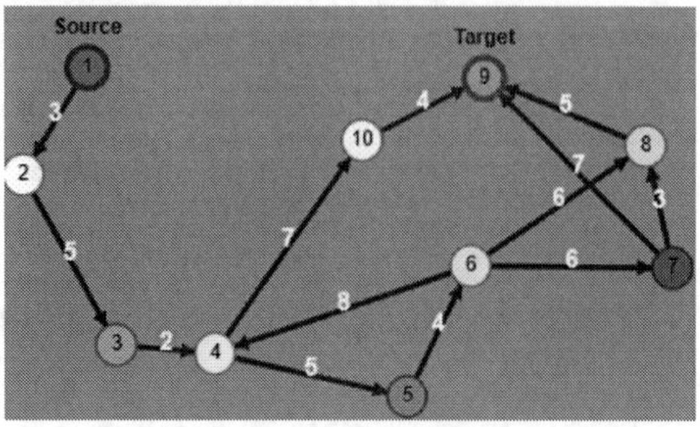

(a)

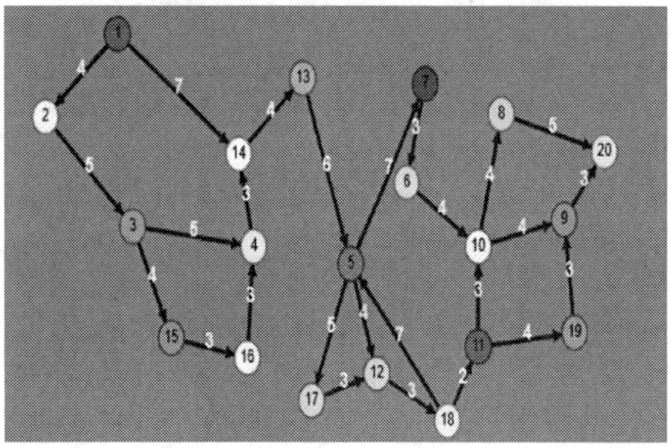

(b)

Fig. 10: The task graph of directed cyclic graph for (a) ten and (b) twenty number of nodes.

The results are scheduled in Figure 11 below.

P1	10	1	4	7	8
P2	9	2	3	5	6

(a)

P1	12	19	6	8	5	2	3	4	13	15	18
P2	17	9	10	20	7	1	11	14	16		

(b)

Fig. 11: Task schedule of (a) ten nodes and (b) twenty nodes.

The algorithm is found easy to be implemented and consistently mapped the task onto prcessors and produce good schedule length.

In the next paper, we will extend the research on load balancing problem.

This work was supported by the Universiti Malaysia Perlis for Short Term Grant under project code 9001-00529.

References

[1] L.P. Cordella, P. Foggia, C. Samsone, M. Vento, *A (Sub)Graph Isomorphism Algorithm for Matching Large Graphs*, IEEE Transaction on Pattern Analysis and Machine Intelligence 26 (10) (2004), 1367-1372.

[2] H. Tong, C. Faloutsos, B. Callagher, T. Eliassi-Rad, *Fast Best-Effort Pattern Matching in Large Attributed Graphs,* Proceedings of the 13th ACM SIGKDD International Conference on Knowledge Discovery and Data Mining, ACM, New York, NY, USA, (2007), 737-746.

[3] H. Tong, C. Faloutsos, *Center-piece Subgraphs: Problem Definition and Fast Solutions*, Proceedings of the 12th ACM SIGKDD International Conference on Knowledge Discovery and data Mining, ACM, New York, NY, USA, (2006), 404-413.

[4] Y. Tian, J.M. Patel, Tale: *A Tool for Approximate Large Graph Matching*. Proceedings of the 24th International Conference on Data Engineering, IEEE Computer Society, Washington, DC, USA, (2008), 963-972.

[5] S.Zhang, S. Li, J. Yang, Gaddi, *Distance Index Based Subgraph Matching in Biological Net- works* Proceedings of the 12th International Conference on Extending Database Technology ACM, New York, NY, USA, (2009), 192-203.

[6] I. Gutman, S. Wegner, *The Matching Energy of a Graph*, Discrete Appl. Math. 160 (2012), 2177-2187.

[7] L. Shuli, Y. Weigen, *The Matching Energy of Graphs with Given Parameters*, Discrete Applied Mathematics 162 (2013), 415-420.

[8] R. Mohan, A. Gupta, *Graph Matching Algorithm for Task assignment Problem*, International Journal of Computer Science, Engineering and Applications (IJCSEA) (2011), Vol. 1, No. 6.

[9] W.N.M. Ariffin and S. Salleh, *The Partitioning Technique Of Directed Cyclic Graph For Task Assignment Problem*, AIP Conference Proceedings 1750, 020010 (2016); doi: 10.1063/1.4954523

[10] W. N. M. Ariffin and S. Salleh, *Bi-Partition Approach Of Directed Cyclic Task Graph Onto Multicolumn Processors For Total Completion Time Minimization Task Assignment Problem*, AIP Conference Proceedings 1775, 030072 (2016); doi: 10.1063/1.4965192.

[11] A. Sljoka, O. Shai, W. Whiteley, *Checking mobility and Decomposition of Linkages Via Pebble Game Algorithm*, in; ASME 2011 International Design Engineering Technical Conferences and Computers and Information in Engineering Conference, Washington, USA, August 28-31, 2011.

The impact of monitoring and business assistance intensity on Malaysian ICT incubatees' performance

Fararishah Abdul Khalid[1], *Aminuddin Ahmad Kayani*[2], *David Gilbert*[3]

[1]Faculty of Technology Management and Technopreneurship, Universiti Teknikal Malaysia Melaka
[2]Faculty of Engineering Technology, Multimedia University, Malaysia
[3]RMIT University, Australia

Abstract. Business incubation has been known in the literature as an economic development tool. Around the world, and in the developing countries particularly, business incubation is deployed to stimulate the growth of small-to-medium sized enterprises or SMEs, which are the lifeblood for many countries. Malaysia's business incubation system has been established since the 1980s and in line with the country's aspirations to become a developed nation by year 2020, much has been done by the government to catalyse the growth of SMEs, particularly, ICT SMEs. Despite establishing numerous ICT incubators over the two decades, the process involved in assisting new entrepreneurs in the incubators remains fragmented. This paper examines a component critical to the business incubation process: monitoring and business assistance intensity and its impact on the performance of incubates. Quantitative method was deployed with a total of 118 incubatees from ICT incubators in Malaysia responding to an online survey questionnaire. Multinomial logistic regression analysis revealed that monitoring and business assistance intensity is statistically significant in predicting incubatee performance. The findings will provide valuable information for entrepreneurs, business incubator managers, and policy-makers on best practices of incubation management and benchmarking towards fourth-generation incubators. This paper fills the gap in the current incubation literature, contributing in several aspects including empirical data, methodology, and noteworthy findings regarding the Malaysian incubation phenomenon.

1 Introduction

The increasingly important role of business incubation as a useful strategy and effective method to accelerate growth and development of technology-based small and medium sized enterprises (SMEs) has been widely acknowledged in the economic and entrepreneurship literature [40] [28] [1] [44]. [3] state that incubators are known for their role to accelerate the growth of new businesses and to create vast employment opportunities through the generation of new businesses. Further, international benchmarking studies, such as the Global Entrepreneurship Monitor (GEM) agrees that new businesses play an important role in enhancing the nation's competitiveness through enhanced degrees of innovativeness and the exploitation of new knowledge and technology. Additionally, incubators have also been suggested to reduce failure of new businesses. These are among the main agendas of business incubators that have been highlighted in extant entrepreneurship literature.

According to [37], despite the growing body of business incubation research, literature on business incubation effectiveness suffers from several deficiencies, including definitional incongruence, descriptive accounts, fragmentation and lack of strong conceptual grounding.

Notwithstanding the growth of research on this domain, understanding of how entrepreneurs and their businesses develop within the business incubator environment remains limited. Given the importance of relational, intangible factors in business incubation and the critical role of business incubation management in orchestrating and optimising such factors, it is suggested that theorising efforts would benefit from a situated perspective.

Hence, notwithstanding the growth of research in this domain since the early efforts to provide frameworks that link business incubation with the incubatee development process [33] [13] [46], there is still a need to understand "how" and "why" incubatee firms grow in a business incubator environment, in processual and longitudinal mode. This paper fills the gap in the current incubation literature, contributing in several aspects including empirical data, methodology, and noteworthy findings regarding the Malaysian incubation phenomenon.

2 Literature review

The NBIA defines business incubator as 'a business assistance program targeted to start-ups and early stage firms with the goal of improving their chances to grow into healthy, sustainable companies' [4]. Alternatively, the [10] defines business incubator as 'a new hybrid type of economic development facility that combines features of entrepreneurship, business facilitation and real estate development'.

The development of business incubation practices has been a subject of significant interest because of its proven ability in stimulating economic growth through job and wealth creation as seen in the United States and the United Kingdom [30] [52] [12]. The reported impacts of business incubation have largely been in the increased number of SMEs as well as increased competitiveness in new venture creations. Subsequently, business incubators are also known to create employment opportunities [8] and have impacted gross domestic product (GDP) of countries such as the US and China [38]. Various agencies from the public and private sectors as well as research institutes and universities have taken deep interest in business incubation, leading to the accumulation of literature on the subject matter [17] [51] [27].

2.1 Monitoring and Business Assistance Intensity

Monitoring of incubatees and providing business assistance to the incubatees have been part of incubator services for quite some time. Literature on incubation acknowledged the need for incubatee monitoring to ensure that businesses progress smoothly at the incubators [49] [41] [19] [13] [20] confirm in their studies that monitoring of incubatees is a source of value that incubators can offer to their incubatees. [7] views it as a critical success factor for incubators. [47] confirmed in their study that business assistance is associated with business incubation performance.

[43] highlighted the significance of monitoring, or coaching, which is referred to as training and educational workshops offered, seminars, programs, either for a fee or free of charge to the incubatees as factors associated with increasing incubatee graduation rates. [19] stated that the incubation process needs to include monitoring and evaluation of incubatee progress to commercialise business ideas, but warned that an overly formal system has the potential to inhibit entrepreneurial flair and thus may fail to take account of the bespoke nature of business development.

2.2 Time Intensity

Studies show that frequent interaction with incubator management creates a better relationship and ultimately contributes to the incubatees' and incubators' success [36] [21]. From a social-capital perspective, more frequent counselling interactions enable the creation of stronger ties that facilitate transfer of knowledge and learning between the incubator management and the venture. This includes venture learning from the incubator management, and for incubator management to learn about the needs of the venture, thus allowing them to offer relevant assistance [21]. The authors hence postulated more frequent counselling interactions can lead to both better business and technical assistance.

Furthermore, [34] suggested that the relationship between the incubator manager and the incubatee is of some importance to the development of the business proposal. A study by [15] found in the context of university incubators that incubator management must form closer ties with incubatees to ensure incubator success. A model proposed by the authors suggests an integrative framework encompassing the involvement of incubator management and the sharing of duties with each incubatee. Rice [34] postulated incubator manager-incubatee dyads co-produce the incubation process, implying that the time intensity of business assistance interventions must be strategically allocated by the incubator manager to the incubatees, and that incubatees must be properly prepared to utilise the advice and insights resulting from such intervention.

[21] acknowledged that prior research supports the notion that counselling interactions are a valuable form of business assistance. They further suggested that more frequent counselling interactions will allow the incubator management to learn better about the needs of the venture, and thus offer more relevant business assistance [48] and the transfer of related knowledge, either directly or by support to the venture to utilise the incubator network successfully [34]. Alternatively, [9] found that advice and frequency of interaction between incubator managers and incubatees do not have a positive influence on economic performance, particularly on job creation.

2.3 Comprehensiveness and Quality

The types of business assistance that incubators claim to provide include administrative-related assistance and services, production-related advice, and operations-related advice [23] [26]. Several studies revealed that the level of business assistance provided at the incubators has a positive influence on the incubation process outcome [34] [48] [5]. The [38] study revealed that while there is no strong correlation between business assistance practices of the incubators and outcomes such as incubatee sales and revenue growth, positive correlations were found between assistance practices and equity investment, patents, research grants, and copyright and licensed intellectual property. Despite that, studies have shown that the range of business assistance provided by a business incubator is instrumental in business incubation success [25].

Literature suggests that incubators ensure the quality of their services by regularly reviewing and obtaining feedback on them [22] [32] . The literature also reveals that incubator managers actively and continuously seek ways to improve the level of customer service satisfaction inside the incubator [45]. Consistent with the findings from a recent study comparing technology incubators and non-technology based incubators in North European Union countries [11], [25], [6] and [22] confirmed that the quality of business assistance provided is essential for successful business incubation.

[44] acknowledged incubator development as one of the main prongs of business incubator-incubation research, alongside research done at incubatee level, entrepreneur level, and system level. Research suggests that incubator level research involves issues that generally relate to the institutional aspects of the incubator; for example, profile of incubators, examination of the physical constitution of incubators, benefits of co-locating within incubators, types of services at the incubators, best practices of business incubators and critical elements of success of the incubators. Incubator level research has been undertaken quite extensively with the purpose of profiling the incubator types according to their objectives, services and facilities offered and their role in enhancing the economic development. Among the studies that have considered the issues related in the incubator level include [2] view on positive environment for entrepreneurs provided by the incubators and [16] study on the role played by incubator organizations in promoting growth-oriented firms. Both studies discussed incubator characteristics and the relationship between incubators and small firms. Similar-themed studies were also found in [14] and [31] where topics discussed include business incubator life cycle, types of funding available for incubators, benefits of incubation, and how incubators play a role in developing new enterprises. Another key research that was done in this area was by [50] where they suggested that business incubation is an effective development tool and requires modest investment while providing excellent return on investment to regional economies.

3 Research Methodology

The objective of this research is to empirically examine the impact of monitoring and business intensity on incubatee performance. To meet this objective, we have addressed the following research question:

> *To what extent does Monitoring and Business Assistance Intensity impact on the incubatee performance?*

3.1 Research design

The study adopts the quantitative approach using survey questionnaire to solicit response from incubates. The survey questionnaire link was distributed via email and in person to 180 ICT incubatees from ICT incubators in Malaysia. The survey yielded a response rate of 65% where 118 valid responses were considered for analysis. The survey questionnaire was developed by the researcher incorporating previously tested and validated scales by [25] and current incubation literature.

'Monitoring and Business Assistance Intensity' refers to "the degree to which the incubator monitors and helps incubatees with the development of their ventures, including helping them to learn about risks involving the resources invested in a business, and about containing the cost of potential (terminal) failure" [47]. [47] state that the time intensity of assistance provided, comprehensiveness of assistance provided, and the quality of the assistance provided all characterised this component of business incubation process. 'Time intensity of assistance provided' refers to "the percentage of working hours devoted to monitoring and assisting incubatees" [47], while 'comprehensiveness of assistance provided' is a measure [47] adapted from [26], and it refers to "the degree to which strategic, operational, and administrative-related assistance are provided by the incubator to the incubatees" [47]. Table 1 presents the items used to measure the Monitoring and Business Assistance Intensity construct.

Table 1: Items for Monitoring and Business Assistance Intensity

Monitoring and Business Assistance Intensity (1=*Strongly Disagree*, 5=*Strongly Agree*)
Time intensity
Q1. On average, our company receives appropriate time in assistance
Q2. On average, our company spends appropriate time interacting with other incubatees in the incubator
Q3. On average, our company receives sufficient time working directly with the incubator manager
Q4. Our company reduces the likelihood of making expensive business mistakes through the interactions with incubator manager and other incubatees.
Comprehensiveness and quality
Q5. Our company receives business planning assistance from the incubator
Q6. Our company receives business feasibility analysis assistance from the incubator
Q7. Our company receives administrative assistance and services from the incubator
Q8. Our company receives production-related advice from the incubator
Q9. Our company receives operations-related advice

from the incubator
Q10. The incubator regularly validates quality of potential new strategic service providers
Q11. Our incubator ensures the quality of its services by regularly reviewing them
Q12. The incubator manager actively seeks ways to continuously improve the level of customer service satisfaction inside the incubator
Q13. The other incubatees teach alternate or new strategies for achieving business success

3.2 Data collection procedures

Participants for the survey questionnaire were initially identified through the websites of their respective incubators. Through contacts with the incubator managers, government agencies such as MDeC and SIRIM, as well we privately formed association such as NINA, basic information regarding the name of the incubatees, email addresses and phone numbers were then obtained. These agencies have shown immense support for this research by providing the researcher with the list of incubatees and expediting their responses for the quantitative part of this research. The participants were made up of companies that are tenants of incubators that have been chosen for the qualitative part of the research. These companies are mostly ICT-based companies with diverse business natures ranging from mobile and wireless communication to internet-based business applications in the financial sector. A letter of invitation was first extended to incubatees through email to obtain their consent to be part of the study.

3.3 Data analysis procedures

Data analyses were undertaken in three principal stages (data screening, exploratory factor analysis, and multinomial logistic regression) using PASW Version 18.0. As part of the preparation and screening process, data were tested for violations of statistical assumptions (e.g., multicollinearity, outliers, and normality) as well as identifying missing data. Data screening revealed that there were no missing data. The statistical procedures involved two main processes: exploratory factor analysis (EFA) and multinomial logistic regression. Factor analysis was conducted to assess the unidimensionality of the four constructs developed in examining relationships with incubation performance including 'Selection Performance', 'Monitoring and Business Assistance Intensity', 'Resource Allocation', and 'Professional Management Services'. This paper discusses the results of the multinomial logistic regression in recognizing the impact of monitoring and business assistance intensity on incubatee performance.

4 Results and Discussion

The individual model analysis examines monitoring and business assistance intensity items and their relationship with incubatee performance. Results of the logistic regression analyses show that the monitoring and business assistance intensity construct was statistically significant Performance ($p < .05$) as shown in Table 2, indicating its strength in predicting incubatee's success. The chi-square values also suggest that similar relationships with high values for F1, F2, and F4, and a lower value for F3. The interaction of all four constructs reveals the strongest effect as a predictor ($p = .003$, $\chi(3) = 14.024$). This suggests that business incubation management will be at its optimum with the inclusion of all factors including proper selection performance, adequate monitoring and business assistance intensity, allocation of resources, and is provision of professional management services

Table 2: Full model evaluation

Predictors	**Chi-Square**	***df***	**Sig.**
Intercept	18.43	3	.000
Selection Performance (F1)	13.02	3	.005
Monitoring and Business Assistance Intensity (F2)	9.50	3	.023
Resources Allocation (F3)	2.75	3	.431
Professional Management Services (F4)	13.39	3	.004
F1Total * F2Total * F3Total * F4Total	14.02	3	.003

The dependent variable in this study is incubate entrepreneur performance which is measured by four categorical outcomes. Logistic regression enables independent variables to predict group memberships, and as this study uses four outcomes, one of the outcomes (our company is barely surviving) has been used as a reference category. Hence, there are three models generating from this data: Model 1, Model 2 and Model 3. Based on data presented in Table 3, the first model shows no significant relationship between the constructs and incubatee performance with all values of p greater than .05. However, Model 2 and 3 show significant relationships with incubatee performance with some constructs having p-values of less than .05.

Table 3: Parameter estimates for the full model

Model 1	β	**SE** β	**Wald's** χ^2	p	e^β **(odds ratio)**
Predictor					
Constant	.176	2.943	.004	.952	
Selection Performance (F1)	-.001	.024	.002	.963	.999
Monitoring and Business Assistance Intensity (F2)	.000	.061	.000	.998	1.000
Resources Allocation (F3)	-.057	.043	1.762	.184	.944
Professional Management Services (F4)	.020	.026	.585	.444	1.020
Model 2					
Constant	-35.271	14.425	5.979	.014	
Selection Performance (F1)	.195	.075	6.731	.009	1.215
Monitoring and Business Assistance Intensity (F2)	.258	.107	5.828	.016	1.295
Resources Allocation (F3)	.026	.084	.098	.754	1.027
Professional Management Services (F4)	.122	.047	6.618	.010	1.130
Model 3					
Constant	-41.092	22.568	3.315	.069	-41.092
Selection Performance (F1)	.265	.134	3.914	.048	1.304
Monitoring and Business Assistance Intensity (F2)	.145	.182	.640	.424	1.156
Resources Allocation (F3)	.083	.138	.356	.551	1.086
Professional Management Services (F4)	.202	.099	4.162	.041	1.223

The Wald statistic is equal to the ratio of β divided by SE squared; it has a chi-square distribution. For each Wald statistic, $df = 1$ and $p = .0000$.

4.1 Goodness-of-fit statistics. Goodness-of-fit statistics assess the fit of a logistic model against actual outcomes. Two descriptive measures are presented in Table 4, which are the R^2 indices, defined by Cox and [18] and [39], respectively. These indices are variations of the R^2 concept defined for the OLS regression model. Due to the limited interpretation of the R^2 in logistic regression [42], the R^2 indices can be treated as supplementary to each other, more useful evaluative indices, such as the overall evaluation model, tests of individual regression coefficients, and the goodness-of-fit test statistic [42]. The Cox and Snell R^2 measure indicates a greater model fit with higher values, but with a limit of less than 1 (<1) [24]. The Nagelkerke R^2 is an adjusted version of the Cox and Snell R^2 and covers the full range from 0 to 1 [24], and therefore it is often preferred. The R^2 values indicate how useful the explanatory variables are in predicting the response variable and can be referred to as measures of effect size.

Table 4: Goodness-of-fit tests of the full model

	Chi-Square	***df***	**Sig.**
Pearson	257.923	297	.951
Deviance	217.464	297	1.000
R^2			
Cox and Snell			.297
Nagelkerke			.326

The research focuses on the *performance outcomes* of the incubatees which is the dependent variable with four categories: *our company is barely surviving; our company has met its break-even and moving on a path*

https://doi.org/10.1051/matecconf/201815006032

toward profitability; our company is making profit; and our company is highly profitable. We obtained a ubiquitous outcome variable for all 118 firms, and found that 32 firms (27.1%) were barely surviving, 44 firms (37.3%) had met their break-even, 38 firms (32.2%) were making profit and 4 firms (3.4%) were highly profitable.

The Model Fitting Information in Table 5 suggests the overall fit of the model. Firstly, the chi-square statistics for this model show that *Comprehensiveness and Quality* of the business assistance contributes significantly to the model, ($p < .05$) while *Time Intensity* of the interaction is not a significant predictor to the model ($p > .05$).

Table 5: Model-fitting information table for Monitoring and Business Assistance Intensity Construct

Predictors	**Chi-Square**	***df***	**Sig.**
Intercept	4.387	3	.223
Comprehensiveness and Quality	10.598	3	.014
Time Intensity	.665	3	.881

The parameter estimates in Table 6 shows that *Comprehensiveness and Quality* of the business services appear to be a significant predictor to the outcome '*our company is making profit*', (p = .003; Wald's χ^2 = 8.925). The odds ratio also suggests that the more comprehensive and better quality the business assistance provided, the more incubatees are making profit.

Table 6: Parameter estimates for Monitoring and Business Assistance Intensity construct

Model 1	β	**SE** β	**Wald's** χ^2	p	e^{β} **(odds ratio)**
Constant	-.882	1.8	.229	.632	
Comprehensiven	.051	.02	3.255	.071	1.052
Time Intensity	-.012	.09	.016	.898	.988
Model 2					
Constant	-	2.2	4.098	.043	
Comprehensiven	.117	.03	8.925	.003	1.124
Time Intensity	.064	.10	.345	.557	1.066
Model 3					
Constant	-.931	3.5	.070	.791	
Comprehensiven	.004	.05	.005	.944	1.004
Time Intensity	-.060	.18	.104	.747	.942

The classification table (Table 7) below for analysis of Monitoring and Business Assistance Intensity elements suggests a 38.0% correct prediction, indicating a reduced ability to predict the model compared to Selection Performance.

Table 7: Classification table predicting membership of outcome categories by Monitoring and Business Assistance Intensity

	Predicted				
Observed	Our company is barely surviving	Our company has met its break-even and is moving on a path toward profitability	Our company is making profit	Our company is highly profitable	Percent Correct
Our company is barely surviving	10	11	8	0	34.5%
Our company has met its break-even and is moving on a path toward profitability	7	13	19	0	33.3%
Our company is making profit	1	17	18	0	50.0%
Our company is highly profitable	1	3	0	0	.0%
Overall Percentage	17.6%	40.7%	41.7%	.0%	38.0%

5 Conclusions

This paper investigates the relationship between monitoring and business assistance intensity and incubatee performance among Malaysian ICT incubators. The results of the multinomial logistic regression showed monitoring and business assistance intensity was statistically significant in predicting incubatee performance categories. This indicates that incubators that provide monitoring and comprehensive business assistance along with adequate interaction with incubator management are related to having incubatees that are making profit. Specifically, the component 'comprehensiveness and quality' appears to be a stronger predictor within this construct than the component 'time intensity'. The significance of the *'Comprehensiveness and Quality'* component suggests that incubators with a range of business assistance and those that seek feedback regarding their services tend to perform better than those without. The second component of the Monitoring and Business Assistance Intensity construct, *'Time Intensity'* revealed non-significance to predicting business incubation performance. This suggests that the amount of interaction between incubatees and incubator managers could not predict the incubatees' outcomes. However, this should not be interpreted as insignificant as a lack of monitoring and business assistance intensity and lower frequency in interaction between incubatees could lead to problems including lack of confidence in incubatees, lack of product sophistication, and limited understanding of market environment leading to delayed graduation of incubatees. Emphasis could be placed on the range of business assistance that fit to the demands of the incubatees. Incubators would only know what fits the demands of the incubatees if they implemented a feedback system to gauge the quality of their current services.

The impact of Monitoring and Business Assistance Intensity on incubatee performance is evident in producing profit-making incubatees and higher number of incubatee graduates. This supports ***the** study's proposition: Incubatees are more likely to perform when monitoring and business assistance are provided.* This finding is consistent with [29] who highlighted the significance of monitoring, or coaching as factors associated with increasing incubatee graduation rates, and [35] who stated that frequent interaction with incubator management results in better relationship and ultimately contributes to the incubatees' and incubators' success. This suggests a positive relationship between providing monitoring and business assistance intensity and incubatee performance.

References

1. Aernoudt, *Incubators: Tool for entrepreneurship?*, Small Business Economics, **23**, 127-135 (2004)
2. Allen and Rahman, *Small business incubators: a positive environment for entrepreneurship*, Journal of Small Business Management, **23**(1) (1985)
3. Atherton and Hannon, *Localised strategies for supporting incubation,* Journal of Small Business and Enterprise Development, **13**(1), 48-61 (2006)
4. Atkins, *Identifying obstacles to the success of rural business incubators. IN NBIA (Ed.)* (2001)
5. A. Bergek, and C. Norrman, *Incubator best practice: A framework.* Technovation, **28**(1-2), 20-28 (2008)
6. A. Chandra, W. He, and T. Fealy, *Business Incubators in China: A Financial Services Perspective.* Asia Pacific Business Review, **13**(1), 79-94 (2007)
7. B. Merrifield, *New business incubators.* Journal of Business Venturing 2, 277- 284 (1987)
8. Besser, *Growing jobs in business incubators,* Iowa State University Extension (1996)
9. B. Vedel, E. Stephany, and I. Gabarret, Selection Strategy, *Assistance and Performance of Incubation Structures: A French Study*, paper presented to ICSB (2011)
10. Biia, *About us* (2008)
11. B.F. Crabtree, and W.L. Miller, (eds), *Doing qualitative research, 2nd edn, Sage, Thousand Oaks* (1999)
12. B. Doyle, and S. Hammond, Technology Incubators Have Significant, Positive Impact on Md. Economy, says TEDCO, viewed 12 September 2008 <http://somd.com/news/headlines/2008/7117.shtml>.
13. C. Campbell, R. Kendrick, and D. Samuelson, *Stalking the latent entrepreneur: business incubators and economic development.* Economic Development Review, **3**(2), 43-48 (1985).
14. Carroll, *The small business incubator as a regional economic development tool: concept and practice*, The Northeast Journal of Business and Economics, **12**(2), 14-43 (1986)
15. C.F. Kuang, H. Hsiao, L. Luong, G. Lin, C,I, and N. Wu, *Development of a New Self-sufficient Model for University Incubator*, International Journal of Innovation and Incubation, **1**(1), 33-50 (2003)
16. Cooper, *The role of incubator organizations in the founding of growth-oriented firms*, Journal of Business Venturing, 175-86 (1985)
17. Cses, *Benchmarking of Business Incubators. European Commission Enterprise Directorate General* (2002)
18. D.R. Cox, and E.J. Snell, *The Analysis of Binary Data, 2nd edn,* Chapman and Hall, London (1989)
19. D. Patton, L. Warren, and D. Bream, *Elements that underpin high-tech business incubation processes*, Journal of Technology Transfer, **34**(6), 621-636 (2009)
20. E. Autio, and M. Klofsten, *A comparative study of two European business incubators,* Journal of Small Business Management, **36**(1), 30-43 (1998)

21. J.L. Scillitoe, and A. Chakrabarti, *The role of incubator interactions in assisting new ventures. Technovation*, **30**(1), 155-167 (2010)
22. J. Costa-David, J. Malan, and R. Lalkaka, *Improving business incubator performance through benchmarking and evaluation: lessons learned from Europe*, paper presented to 16th International Conference on Business Incubation, Toronto, Canada, April 28-May, 1 (2002).
23. H.I. Ansoff, *Corporate strategy: An analytical approach to business policy for growth and expansion*, McGraw-Hill, New York (1965)
24. Hair, Black, Babin and Anderson, *Multivariate Data Analysis 7th Edition*, Pearson (2010)
25. Hackett and Dilts, *Inside the black box of business incubation: Study B - scale assessment, model refinement, and incubation outcomes*, Journal of Technology Transfer, **33**(5) (2008)
26. J.J. Chrisman, *Strategic, administrative, and operating assistance: The value of outside consulting to pre-venture entrepreneurs.* Journal of Business Venturing, **4**(6), 401-418 (1989)
27. Kae-Kuen and Hung-Shun, *The relationship among the resource of incubators, firms' type and performance: An empirical study of academic incubators in Taiwan*, IAMOT (2006)
28. Lee and Yang, *The cradle of Taiwan high technology industry development - Hsinchu Science Park (HSP)*, Technovation, **20**, 55-59 (2000)
29. L. Peters, M. Rice, and M. Sundararajan, *The role of incubators in the entrepreneurial process*. Journal of Technology Transfer, **29**(1), 83-91 (2004)
30. M.G. Colombo and M. Delmastro, *How effective are technology incubators? Evidence from Italy*, Research Policy, **31**(7), 1103-1123 (2002)
31. Martin, *Business incubators and enterprise development: neither tried or tested?*, Small Business and Enterprise Development, 43-11 (1997)
32. M. Abduh, C. D'Souza, A. Quazi, and H.T. Burley, *Investigating and classifying clients' satisfaction with business incubator services*, Managing Service Quality, **17**(1), 74-91 (2007)
33. M. Temali, and C. Campbell, *Business incubator profiles: A national survey*, University of Minnesota, Hubert H. Humphrey Institute of Public Affairs, Minneapolis (1984)
34. M.P. Rice, *Co-production of business assistance in business incubators: An exploratory study*, Journal of Business Venturing, **17**(1), 163-187 (2002)
35. M. McAdam, and S. Marlow, *A preliminary investigation into networking activities within the university incubator*. International Journal of Entrepreneurial Behaviour & Research Policy, **14**(4), 219-241 (2007
36. M. McAdam, and S. Marlow, *The business incubator and the female high technology entrepreneur: a perfect match?*, (2008), viewed 25 March 2009, <http://www.icsb.org/documents/NWBC_ICSB_Best_WOB_Paper_2008.pdf>.
37. N. Theodorakopoulos, N.K. Kakabadse, C. McGowan, *What matters in business incubation? A literature review and a suggestion for situated theorisin*, Journal of Small Business and Enterprise Development, **21**(4), 602 - 622 (2014)
38. Nbia, *Business Incubator Facts, Business Incubation Works* (1997)
39. N.J.D. Nagelkerke, *A note on a general definition of the coefficient of determination*. Biometrika, **78**(1), 691-692 (1991)
40. Oecd, *The Knowledge-based Economy. In National Innovation Systems, P. (Ed.)* (1996)
41. T. O'Neal, *Evolving a Successful University-Based Incubator: Lessons learned from the UCF Technology Incubator*, Engineering Management Journal, **17**(3), 11-25 (2005)
42. Peng, Lee and Ingersoll, *An Introduction to Logistic Regression Analysis and Reporting*, The Journal of Education Research (2002)
43. Peters, Rice and Sundararajan, *The role of incubators in the entrepreneurial process*, Journal of Technology Transfer, **29**, 83-91 (2004)
44. Phan, Siegel & Wright, *Science parks and incubators: observations, synthesis and future research*, Journal of Business Venturing, **20**, 165-182 (2005)
45. R. Lalkaka, *Best Practices' in Business Incubation: Lessons (yet to be) Learned*, paper presented to International Conference on Business Centers: Actors for Economic & Social Development, Brussels, 14-15 November (2001).
46. R.W. Smilor, *Commercializing Technology Through New Business Incubators*, Research Management, **30**(5), 36-41 (1987)
47. S.M. Hackett, and D.M. Dilts, *A real options-driven theory of business incubation*, Journal of Technology Transfer, **29**(1), 41-54 (2004a)
48. S.M. Hackett, and D.M. Dilts, *A systematic review of business incubation research*, Journal of Technology Transfer, **29**(1), 55-82 (2004b)

49. S. Linder, *State of the business incubation industry*, NBIA Publications (2003)
50. Tornatzky, Batts, Mccrea, Lewis and Quittman, *The art and craft of technology business incubation: Best practices, strategies and tools from 50 programs,* In Southern Technology Council, N. B. I. A. (Ed.). Athens (1995)
51. Ukbi, *Business incubation in the East Midlands: A review A summary report for the East Midlands*
52. Voisey, Gornall, Jones and Thomas, *The measurement of success in a business incubation project*, Journal of Small Business and Enterprise Development, **13**(3), 454-468 (2006)

Development of Handheld Haptics Device for Driving System of Unmanned Underwater Vehicles

Syed Mohamad Shazali[1,*]

[1]CeRIA, Faculty of Electrical Engineering, Universiti Teknikal Malaysia Melaka, 76100 Durian Tunggal, Malaysia

Abstract. This paper presents a research aimed at illustrating hydrodynamic force impact on the orientation of a Remotely Operated Underwater Vehicle (ROV) operating underwater by providing kinaesthetic haptic feedback to its handheld steering device. To get more understanding on how this aim can be achieved, a literature review had been done on the haptic feedback which are available to ROV pilots and how it could be delivered through a handheld device. While some achievement were made in providing different cues to pilots on drag force and its influence on its speed, non-have been made to offer insight on how it had affected ROVs orientation through haptic feedback. This study found that currently available handheld haptic device, while successfully delivering tactile feedback, are not capable of providing kinaesthetic feedback at par with the grounded haptic device. To address this, a series of thrusters has been introduced as a new actuation technique in providing kinaesthetic feedback on a handheld device in all three axes. This would allow total illustration of ROV orientation through haptic feedback. This paper has summarized and discussed our findings in our literature review, followed by some details of the proposed method.

1 Introduction

There is increasing demand for an unmanned underwater robot as the marine scientists and related industries are at high speed to grab the best the ocean has to offer and Remotely Operated Underwater Vehicle (ROV) seems to be the best options they have to do it. While undertaking tasks underwater, the vehicles do not enjoy the same luxury of sensors and communication capability like its counterpart on ground or in the sky, which usually presented as a visual cue to an operator. For these reasons, experienced operator had to rely only on visual information as prime feedback for most underwater robots operations [1], displaying information gathered from typically available sensor; depth sensors, compasses, side-scan and other sonars, thermistors, magnetometers and conductivity probes [2].

The operator's ability to comprehend ROV harsh underwater environment and driving the vehicle can badly affected by limited and fluctuating quality of visual feedback [3]. Research had been done to improve the visual feedback by embedding the vision feedback fed to the pilot with additional information on the environment underwater [4] [5]. Virtual Reality (VR) technology also had been used to increase the pilot capability to understand the current state and orientation of the ROV [6] [7]. However, the solution provided by this system had increases pilots' cognitive workload because of high reliance on the visual channel of feedback [8] while maintaining certain orientation in order to enable the vehicle to work. A report [9] indicate two main human factors issues regarding unmanned underwater vehicle; (a) vehicle control and monitoring display (b) perception of the underwater environment. This paper would highlight how our research addressing these issues by having haptic feedback in the system.

This research is about providing kinaesthetic haptic feedback on the orientation on ROV, illustrating hydrodynamic force impact (mainly drag and wave force) on the orientation, for a joystick-like handheld ROV steering device. This feedback is important to the pilot for at least two reasons; (1) it would be a great complement, if not as alternative, to visual feedback currently available to ROV pilots and (2) it would help the pilots to comprehend the impact of hydrodynamic force on the ROV orientation immediately and, in case of wave force, avoid it by diving deeper [10] or at least, in case of drag force, reduced it by reducing the speed or rate of motion of the vehicle.

Many works on haptic had been done, providing haptic feedback for handheld devices but most of them only produce the illusion of force vector [11] [12] [13] through tactile feedback and not a kinaesthetic force feedback which had been enjoyed by grounded haptic devices. This research would introduce the use of a series of thrusters to provide the force feedback in three axes and potential 6 degree of freedom (DOF) with a relatively low computational requirement.

The outlined of this paper is as follows: an overview of recent research on haptic feedback, with more attention

* Corresponding author: syedmohamad@utem.edu.my

https://doi.org/10.1051/matecconf/201815006033

given to the handheld device, is presented in Section II. Section III covers our new proposed technique for illustrating ROV orientation and motion as well as the impact of hydrodynamic force on its orientation. Finally, we conclude our work and some hint of our future work in Section IV.

2 Literature Review

The motivation of this research is to provide haptic feedback on the orientation of an ROV, illustrating the impact of waves, water current and drag force on its orientation and motion. For this, a literature review exercise had been done to gain the understanding of how this forces affected the ROV orientation, what types of haptic feedback would be good for this purpose and currently available haptic feedback to ROV pilot. We also had explored on currently available haptic devices in order to find the best way to provide the required haptic feedback. In this section, the discussion on the impact of hydrodynamic force on ROV's orientation will be followed by the review of haptic feedback and how it could help addressing this issues.

2.1 Wave and Drag Force on ROV orientation

Although the magnitude of force exerted against an ROV, whether from the effects of water flow or water ripples caused by the vehicle or thruster nearby, is not a critical parameter to the operator, the *resulting motion* of the vehicle, which usually lead to changes in the position and orientation or speed of the vehicle, is very important in helping operators to react against it.

Many ROV operation need the operators to be able to tilt or pitch at certain roll or pitch angle frequently and steady enough to work, such as welding, or while driving the ROV at certain diving or pitching angle. This would be difficult as restoring moment, caused by buoyancy force and weight force would work against the motion that pitching or tilting the vehicle. While this phenomenon is very helpful as a passive stabilizing system to the ROV against the effect of wave and water current, some ROV designed to have a small restoring moment which allow the ROV to move and hovering at any tilting or pivoting angle with less effort needed.

A study [10] had shown that while the wave impact on ROVs on its roll and yaw moment would be less than 1Nm, the impact on the vehicle pitch moment is very significant when fully submerge less than 15 metres from the free surface. This finding had resonant with previous research made on ROV motion in uniform current [14] where pitch angle significantly affected by the water current. Drag force, F_D, caused by fluid with P density of the fluid and C_D drag coefficient, on ROV with A projected cross-sectional area moving at u_r absolute speed, can be described as [3];

$$F_D = \frac{1}{2}\rho u_r^2 C_D A \tag{1}$$

As the drag coefficient and cross-sectional area of the ROV would be constant and density of the fluid will be constant at a fixed temperature and pressure, drag force can be described with hydrodynamic coefficient, K_H, as

$$F_D = u_r^2 K_H \tag{2}$$

with

$$K_H = \frac{1}{2}\rho C_D A \tag{3}$$

This shows that drag force would vary with the speed of the vehicle which depend on the motion caused by its thrusters and water current. Drag force does not directly affect the orientation of an ROV but rather oppose the motion involve while the vehicle changing its orientation, either cause by its own propeller or by its surrounding, including any external force acted on it.

2.2 Haptic for ROV

When we interact directly with objects in real life, the *haptic sense* is heavily involved in the perception of that interaction. It comprises two complementary senses; the *tactile sense* operates at the skin level, and allows feeling surfaces and textures and second, the proprioceptive or *kinaesthetic sense*, which is related to the perception of balance and posture of the different parts of the body, relative to each other [15]. Haptic technology recreates the sense of touch, benefitting from a combination of a somatosensory pattern of skin and proprioception of hand position, by applying forces, vibrations, or motions to the user [16].

The function of a haptic system is to provide reliable force feedback of the contact force at the slave side to the user manipulating a haptic device [17] while interacting with remote environments and allow a bi-directional flow of information [16]. While vision feedback dominantly used over the haptic feedback, it is very important when visual feedback is lost, unreliable or inadequate [18], although its limitations need to be tackled [19].

2.3 Handheld Haptic Devices

Haptic devices have a wide variety in workspace sizes, with most of these interfaces are grounded with limited workspaces [15]. Ungrounded haptic feedback devices offer a lot bigger workspace than grounded kinaesthetic devices, which use kinematic chains of rigid links and joints to transmit ground-based reaction forces to the user [20]. Handheld devices, which can take the form of gloves or exoskeleton device, can provide either kinaesthetic or tactile feedback to hands, especially the fingers, while *not hindering the motions of the user* like grounded devices. The leading approaches of providing haptic feedback on a handheld device had been summarized in **Table 1**. As many studies had been done on *handheld haptic devices*, simulating a lot of features of the surface texture, hardness and shape of an object, no research had been done to simulate the perception of net force applied onto an object.

While grounded haptic device enjoy the capability to produce kinaesthetic feedback through series of joints and linkages with reference to ground, handheld devices could only exploit the torque feedback using *gyroscopic effect* (GE) [21] and vibration as a tactile feedback [22] produced by various actuator, such as motors and thin film creating the haptic sensation of being twisted towards one direction the tablet. This approaches however limited to 1 DOF as the device has only one flywheel. Recently, visual feedback had been introduced to improve haptic feedback

Table 1. Leading Approach for Handheld Haptic Device

Criteria	Force Feedback	Torque	Tactile Feedback	Bimanual Operation	Accuracy
Linear Oscillatory Actuator (LOA)	Linear Oscillatory Actuator – pull, push, and vertical force/weight. [26]	No	Simple vibration on horizontal plain	Possible but no specially design device made for bimanual operation	Great for the sense of direction of the force, but not for an accurate and sharp perception of force direction. 2 DOF, 2 axes.
Gyro Effect (GE)	No	Directional torque with more than 360° workspace. [21]	No	Possible but require some modification on the controller for bimanual operation.	Great for directional torque with an anchoring effect.
Electrical Muscle Stimulation (EMS)	Provide the illusion of motion through electrical stimulation without really moving. [32]	Not explored.	Possible	Possible but not tested for bimanual operation yet.	No consistency and required calibration for different users, especially for force feedback [31]. Application without vision feedback had been reported to be annoying/irritating. [13] Better result achieved with vision feedback. [28]
Skin Stretch Feedback (SSF)	Varying speed and magnitude of the tactor motions to superimpose force/ torque movements could create different tactile effects.	Illusion of torque	Yes	Yes.	Great for a sense of direction of the torque and inertia, but not for an accurate and sharp perception of horizontal force direction. 2 DOF

[23].

All these feedbacks, however, did offer none or limited illustration of a force vector direction and *cannot generate net forces on the limb*, as required in providing kinaesthetic feedback, thus requiring the feedback, which usually *only an illusion of force*, to be processed and interpreted by the human before affecting physical behaviour change [20]. Handheld haptic device mostly approaches this issues by stimulating the skin, providing tactile or cutaneous feedback, using vibration [24] [11] or *differential skin stretch feedback* (SSF) providing both translating and angular force illusion to haptic user [12] to enable small, lightweight, and wearable or handheld devices. Although these approaches manage to provide a perception of force in a horizontal direction [25], this approach, however, failed to produce any perception of force in antigravity direction [24].

Linear oscillatory actuator (LOA) using an asymmetrical drive had been proposed as a haptic feedback [11], creating the illusion of being pulled and push as well as weight, achieved by a device based on 2 DOF oscillatory actuator which was significantly smaller and handy [26]. GE had been used to produce torque feedback. iTorqU, a handheld torque feedback device for haptic application [21] built to generate ungrounded torque feedback using a gimbal-controlled gyroscope. This research had been pursuit further with a gyroscopic actuator for TorqueScreen [27], an actuated flywheel imposing angular momentum for a handheld tablet, provided through *electrical muscle stimulation* (EMS) [28], motivated by the unsatisfactory haptic sense induced by EMS in the full spectrum of a haptic event.

GE did not display any translating force feedback and limited to only torque feedback in 2 axes, which is not available from LOA. SSF technique, however, manages to provide both torque and translating force sensation. EMS, on the other hand, had provided the illusion of motion through electrical stimulation but limited to the force caused by the motion of hand rather than the force from outside which apply onto an object. All of this approach, although had managed to provide good force and torque feedback, being ungrounded, none of this approach had provided any *visual sense of orientation* to the user which had been enjoyed by the grounded haptic devices' users.

Both SSF and LOA had the capability to provide tactile feedback with LOA produce a simple vibration on a horizontal plain. Beside GE devices, none of the above, as well as any of the approaches for handheld haptic device describe or mentioned before, had offered kinaesthetic force feedback and rely on illusion created with tactile feedback, which cannot be comparable enough to a grounded multi-DOF haptic device. It is also established that prolong vibration could cause perceptual fatigue [29]. All these lead to another option left to create sustained directional force for kinaesthetic feedback system, ejecting material. This has to be done safely and carefully controlled without annoying the pilot of such vehicle.

3 The Handler

A new handheld haptic device had been design to compliment a steering device which controlling orientation, instead of velocity, of an ROV. This design had, for the first time on a haptic device, a series of thrusters as its actuators. Each thruster comprised of a DC motor with propeller controlled through Pulse Width Modulation (PWM) signal. Acting as a steering device, orientation of the device will be captured through gyro sensor and feed as a *reference orientation angle position* for a driving system of an ROV. This device, *The Handler* (TH), had two systems; (1) ROV steering system and (2) haptic feedback system.

The former would have user desired ROV orientation as the input and ROV controller reference input as the output, while the later would have the ROV and the device orientation as the input and haptic feedback, both tactile and kinaesthetic feedback, as the output, as detailed in **Table 2.**

Table 2. The Handler's Input and Output Requirement

Items	Requirements	Parameters	Resolution
Inputs	Workspace		
	Roll	-90°<θ<90°	
	Pitch	-90°<θ<90°	
	Yaw	-90°<θ<90°	
	Force		
	Rotation	<120N.m, [10]	
Outputs	Workspace		
	Roll	-90°<θ<90°	
	Pitch	-90°<θ<90°	
	Yaw	-90°<θ<90°	
	Force		
	Rotation	0N. m<M<11N.m, [33]	2.25N 1N.m

This research had explored the opportunity to provide both tactile and kinaesthetic feedback. At its current stage, this device had been able to provide 20.6 N.mm for angular force rotating about X axis (roll) with a maximum vibration of 200Hz, as shows in **Figure 1**, and would have 5 thrusters for 3 axes haptic feedback for its final design. The device would control the direction and magnitude of its force feedback (FFB) in each axis with aim to provide *resistance to the motion* of the steering device. It would simulate different size or shape of ROV by varying the drag coefficient which then controls the amount of resistance provided by the device to the user. The device should be able to *vary its magnitude and frequency* of oscillating FFB according to the direction of the wave in respect of the ROV orientation.

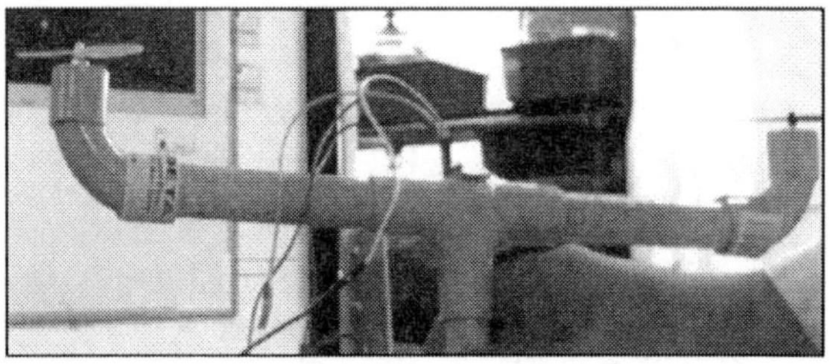

Figure 1. The Handler

3.1 Drag force impact displays algorithm

This research had developed an algorithm to **display drag force** on **rotating ROV** through a haptic device. While previous research had managed to capture the drag force in term of the ROV speed delay caused by the force, the approach was only applied to the drag force caused by a surge motion. As drag force exist when there is relative speed between the fluid and the surface of the vehicle, drag force will be captured by the difference between the **angular orientation** of the vehicle and TH. The velocity vector, V, of an ROV is generally represented [30]as:

$$V = [V_1\, V_2]^T = [u \;\; v \;\; w \;\; p \;\; q \;\; r]^T \qquad (4)$$

with u, v and w, are the component of the linear velocity vector, V_1, in the surge, sway and heave direction, while p, q, and r, are component of angular velocity, V_2, in the roll, pitch, and yaw. On the other hand, position vector, η, of an ROV is generally represented as:

$$\eta = [\eta_1\, \eta_2]^T = [x\, y\, z\, \emptyset\, \theta\, \psi\,]^T \qquad (5)$$

with x, y and z, are the component of position vector, η_1, while ϕ, θ, and ψ, are orientation vector, η_2, for roll, pitch, and yaw. Utilizing simplified equation of drag force, the drag force of rotating ROV can be represented as:

$$F_D = K_H\, V_2^2 \qquad (6)$$

As hydrodynamic coefficient, K_H, would have different value for each side of the vehicle, K_H could be represented as;

$$K_H [K_{Hx} \quad K_{Hy} \quad K_{Hz}]^T \qquad (7)$$

with K_{Hx}, K_{Hy} and K_{Hz} are the hydrodynamic coefficients for each side facing x, y and z-axis. As haptic force, F_H, would have F_D as the reference to provide the force feedback in each axis, with different gain (K_{TH}), pre-set by user, for each axis, the haptic force could be described as

$$F_H = K_{TH} F_D = K_{TH} K_H V_2^2 = \begin{bmatrix} K_{THx} K_{Hx} p^2 \\ K_{THy} K_{Hy} q^2 \\ K_z(\psi_{TH} - \psi_{ROV})^2 \end{bmatrix} \qquad (8)$$

This equation had angular velocity, which is the difference of orientation in a certain of time. In providing the haptic feedback for The Handler, this equation had been modified to have differences between the device orientation, η_{TH}, and the ROV orientation, η_{ROV}, replacing the angular velocity in the equation. With

$$q = \frac{\eta_f - \eta_i}{t_f - t_i},\; t_f = t_i + 1, \text{and } K_n = K_{THn} K_{Hn} \qquad (9)$$

The haptic force can be described with the different between the device orientation, η_{TH}, and the ROV orientation, η_{ROV}, as

$$F_H = \begin{bmatrix} K_x(\emptyset_{TH} - \emptyset_{ROV})^2 \\ K_y(\theta_{THy} - \theta_{ROVy})^2 \\ K_z(\psi_{TH} - \psi_{ROV})^2 \end{bmatrix} \quad (10)$$

As ROV, usually, do not have the uniform or cubic shape, each side would have different size of the surface area which then gives different drag coefficients for rotational motion about each axis. The algorithm should be able to display drag force impact on each surface with independently set drag coefficient. To do this, this device would capture the delay in the motion of the rotational motion and provide force feedback proportional to the amount of the delay, on the *opposite direction of the motion*.

To illustrate the outcome and behaviour of the haptic feedback which will experienced by the user, a simulation had been done by imposing a series of orientation input. Graph shown in Figure 2 shows the input of the driving system in the form of intended orientation in one axis. The user is expected to cause a little delay while moving the device, as shown in Graph B, which is represent the actual orientation of the device. Graph C represent the orientation of ROV which experienced significant delay caused by the drag force against it motion. The different between the orientation angle of ROV and TH had been captured in Graph D which will be the input for the Haptic System.

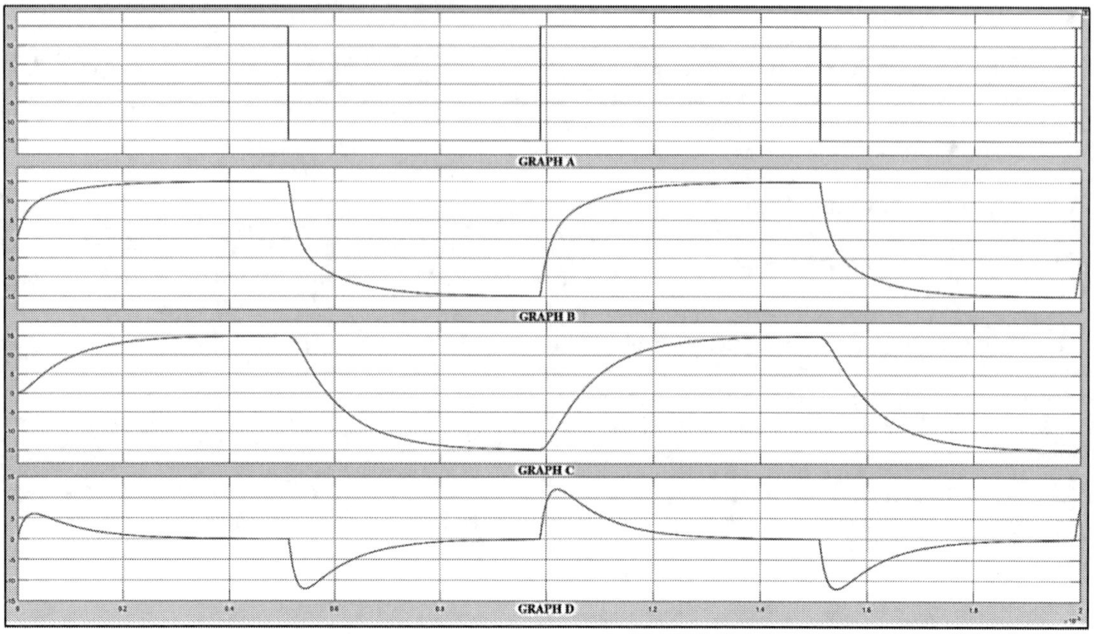

Figure 2. Simulation result illustrating the relation between the intended orientation angle and the haptic feedback produce by TH

3.2 Actuator Control

While the actual angular velocity is available, from an inertial measurement unit (IMU) that provides the gyro sensor reading, this haptic device designed with an objective to provide information regarding the *orientation of the vehicle*, not drag force, to the user through the illusion of force similar to drag force. Since the force of the thrusters on the device would be controlled by varying voltage supply of its motors, the voltage supply, *V*, will be varied between its minimum and maximum to reflect the different between the device orientation, η_{TH}, and the ROV orientation, η_{ROV}

$$V = k_V\,(\eta_{TH} - \eta_{ROV}) \quad (11)$$

$$k_V = (V_H - V_L)/90 \quad (12)$$

where V_H and V_L are the maximum and minimum input voltage for the thrusters, which can be pre-set by the user, to control the magnitude of the **kinaesthetic** feedback.

4 Conclusions and Future Directions

This paper addressed force feedback for the haptic interface of ROV. Because of a direct measurement of applied force underwater is difficult, the haptic force vector had been derived from a proprioceptor sensor, the gyro sensor. The device had also offered both kinaesthetic and tactile feedback using its thrusters. However, the effectiveness of having both types of haptic feedback is yet to be investigated.

On the later part of this research, the study will be made on the impact of kinaesthetic force feedback on handheld device output. Without a based, a handheld device would have the user's hand as the base of its relative motions. While the earth coordinate system would be the reference for the orientation of the device, a handheld device would not have the stiffness enjoyed by a grounded device. This would allow, depends on how the user reacts to the kinaesthetic feedback, significant influence by the actuator of the device on its orientation.

Unlike vibration, which can easily manage by most users, kinaesthetic feedback would be harder to handle. As this research aim at providing haptic feedback for a driving apparatus or device, this study would investigate how the haptic force vector pattern would be tolerated by the users and identify the limitation of the device without seriously affect the handheld device output, especially on the reference point for the ROV controller.

This work was performed under financial support from Universiti Teknikal Malaysia Melaka (UTeM) and The University of Auckland, New Zealand.

References

[1] J. Choi, Y. Lee, H.-T. Choi, J.-J. Kang and J.-H. Ryu, "A preliminary study on development of haptic interface for underwater vehicles," 2015.

[2] X. Lin and S. Guo, "Development of a spherical underwater robot equipped with multiple vectored water-jet-based thrusters," *Journal of Intelligent & Robotic Systems* , vol. 67, no. 3, pp. 307-321, 2012.

[3] K. Le, H. Nguyen, D. Ranthumugala and A. Forrest, "Haptic driving system for surge motion control of underwater remotely operated vehicles," in *Proceedings of the 6th International Conference on In Modelling, Identification & Control (ICMIC)*, 2014.

[4] J. G. Sanchez, B. Patrao, L. Almeida, J. Perez, P. Menezes, J. Dias and P. Sanz, "Design and Evaluation of a Natural Interface for Remote Operation of Underwater Robots," *IEEE computer graphics and applications,* 2015.

[5] C. Domingues, M. Essabbah, N. Cheaib, S. Otmane and A. Dinis, "Human-Robot-Interfaces based on Mixed Reality for Underwater Robot Teleoperation," in *9th IFAC Conference on Manoeuvring and Control of Marine Craft (MCMC)*, 2012.

[6] M. Candeloro, E. Valle, M. R. Miyazaki, R. Skjetne and M. Ludvigsen, "HMD as a new tool for telepresence in underwater operations and closed-loop control of ROVs," in *MTS/IEEE OCEANS 2015*, Washington, 2015.

[7] A. Khadhraoui, L. Beji, S. Otmane and A. Abichou, "Stabilizing control and human scale simulation of a submarine ROV navigation," *Ocean Engineering* , vol. 114 , pp. 66-78, 2016.

[8] A. Vasilijevic, N. Miskovic and Z. Vukic, "Comparative assessment of human machine interfaces for ROV guidance with different levels of secondary visual workload," in *21st Mediterranean Conference on Control & Automation (MED)*, 2013 .

[9] G. Ho, N. J. Pavlovic, R. Arrabito and R. Abdalla, "Human Factors Issues When Operating Underwater Remotely Operated Vehicles and Autonomous Underwater Vehicles (No. DRDC TORONTO-TM-2011-100).," Defence Research and Development Toronto (Canada), Toronto, 2011.

[10] S. A. Malik, P. Guang and L. Yanan, "Numerical Simulations for the Prediction of Wave Forces on Underwater Vehicle using 3D Panel Method Code," *Research Journal of Applied Sciences, Engineering and Technology* , vol. 5, no. 21, pp. 5012-5021, 2013.

[11] T. Amemiya and T. Maeda, "Directional force sensation by asymmetric oscillation from a double-layer slider-crank mechanism," *Journal of Computing and Information Science in Engineering,* vol. 9 (1), pp. 1-8, 2009 .

[12] A. L. Guinan, M. N. Montandon, A. J. Doxon and W. Provancher, "Discrimination thresholds for communicating rotational inertia and torque using differential skin stretch feedback in virtual environments," in *Haptics Symposium (HAPTICS) IEEE*, 2014.

[13] F. Farbiz, Z. H. Yu, C. Manders and W. Ahmad, "An electrical muscle stimulation haptic feedback for mixed reality tennis game," 2007.

[14] M.-C. Fang, J.-H. Chen, J.-H. Luo and C.-S. Hou, "On the behavior of an underwater remotely operated vehicle in a uniform current," *Marine Technology,* vol. 45, no. 4, pp. 241-249, 2008.

[15] A. Talvas, "Bimanual haptic interaction with virtual environments," Doctoral dissertation, INSA de Rennes, 2014.

[16] S. Kapoor, P. Arora, V. Kapoor, M. Jayachandran and M. Tiwari, "Haptics – Touchfeedback Technology Widening the Horizon of Medicine," *J Clin Diagn Res.,* 2014.

[17] T. Opitz and O. Meckel, "Control of Haptic Systems," in *Engineering Haptic Devices*, London, Springer , 2014, pp. 181-225.

[18] R. Balakrishnan and K. Hinckley, "The role of kinesthetic reference frames in two-handed input performance," *Proc. ACM Symp. User Interface Softw. Technol.,* 1999.

[19] N. Gurari and G. Baud-Bovy, "Customization, control, and characterization of a commercial haptic device for high-fidelity rendering of weak forces," *Journal of Neuroscience Methods,* 2014.

[20] J. M. Walker, M. Raitor, A. Mallery, H. Culbertson, P. Stolka and A. M. Okamura, "A dual-flywheel ungrounded haptic feedback system provides single-axis moment pulses for clear direction signals," Philadelphia, USA, 2016.

[21] K. N. Winfree, J. Gewirtz, T. Mather, J. Fiene and K. J. Kuchenbecker, "A high fidelity ungrounded torque feedback device: The iTorqU 2.0.," in *EuroHaptics conference, 2009 and Symposium on Haptic Interfaces for Virtual Environment and Teleoperator Systems. World Haptics 2009.*, 2009.

[22] G. Kato, Y. Kuroda, I. Nisky, K. Kiyokawa and H. Takemura, "HapSticks: A novel method to present vertical forces in tool-mediated interactions by a non-grounded rotation mechanism," 2015.

[23] S.-Y. Kim, J. Kim and K.-B. K. K. Kim, "A Film-type Vibrotactile Actuator for Hand-held Devices," *The Journal of Korea Robotics Society,* 2013.

[24] T. Amemiya and T. Maeda, "Asymmetric Oscillation Distorts the Perceived Heaviness of Handheld Objects," *IEEE Transactions on Haptics,* vol. 1, no. 1, 2008.

[25] T. Amemiya, H. Ando and T. Maeda, "Lead-me interface for a pulling sensation from hand-held devices," *ACM Transactions on Applied Perception,* vol. 5, no. 3, 2008.

[26] M. Kato, Y. Kono, K. Hirata and T. Yoshimoto, "Development of a Haptic Device Using a 2-DOF Linear Oscillatory Actuator," *IEEE Transactions on Magnetics,* vol. 50, no. 11, pp. 1-4, 2014.

[27] M. Murer, B. Maurer, H. Huber, I. Aslan and M. Tscheligi, "TorqueScreen: Actuated flywheels for ungrounded kinesthetic feedback in handheld devices," in *TEI 2015 - Proceedings of the 9th International Conference on Tangible, Embedded, and Embodied Interaction*, 2015 .

[28] T. Ishikawa, T. Toshio and K. Yuichi, "Wearable Pseudo-Haptic Interaction by Using Electrical Muscle Stimulation," *Haptic Interaction,* pp. 135-140., 2015.

[29] S. Coren, Sensation and perception, John Wiley & Sons, Inc., 2003.

[30] L. G. García-Valdovinos, T. Salgado-Jiménez, M. Bandala-Sánchez, L. Nava-Balanzar, R. Hernández-Alvarado and J. A. Cruz-Ledesma, "Modelling, Design and Robust Control of a Remotely Operated Underwater Vehicle," *International Journal of Advanced Robotic Systems* , vol. 11, 2014.

[31] M. Pfeiffer, S. Schneegaß and F. Alt, "Supporting interaction in public space with electrical muscle stimulation," 2013.

[32] H. Kajimoto, "Illusion of motion induced by tendon electrical stimulation," 2013.

[33] V. M. Ciriello, R. V. Maikala and N. V. O'Brien., "Maximal acceptable torques of six highly repetitive hand-wrist motions for male industrial workers," *Human Factors: The Journal of the Human Factors and Ergonomics Society,* vol. 55, no. 2, pp. 309-322, 2013.

Oral Hygiene Improvement using Combined Mouthwash with Plant extracts

Essam A. Makky[1,3,*], *Muna J. Ali*[2], *and Mashitah M. Yusoff*[1]

[1]Faculty of Industrial Sciences & Technology, Universiti Malaysia Pahang, 26300 Gambang, Kuantan, Pahang, Malaysia
[2]Department of pathological analyses, Al-Haweeja Technical Institute, Foundation of Technical Education, Northern Technical University, Kirkuk, Iraq
[3]Center of Excellence for Advanced Research in Fluid Flow (CARIFF), Universiti Malaysia Pahang, 26300 Gambang, Kuantan, Pahang, Malaysia

Abstract: Dental caries considered as one of the most common infectious diseases affecting mankind today, the dissolution of tooth structure by acid produced as a result of the fermentation of dietary carbohydrates by oral bacteria. The antimicrobial susceptibility of medicinal plants including black pepper, black cumin, cinnamon, cardamom on toothpaste and mouthwash against oral isolates for healthcare improvement was studied. Different oral isolates from 50 selected individuals aged 3 to 60 years were obtained from both genders, and a total of 59 bacterial and yeast isolates were collected, purified, and tested against four different commercial medicinal plants extracts for antimicrobial susceptibility profile. A total of 10 mouthwashes was purchased, and the combined action of the medicinal plants with mouthwash was studied against oral isolates. We found a higher bacterial population was in the age group of 3–40 years than other two groups, with approximately 44%. In addition, the combined action of acetone plant extracts (alone) against oral isolates showed increasing effect profile up to 61.02% when combination A (Ci/N) was added. Also, While, the combined action of medicinal plants with mouthwash was improved 100% with combination C. We conclude that the combination C (P/N/Ca/Ci) with mouthwash showed high susceptibility against oral isolates.

1 Introduction

Oral health is closely related to other health and quality of life issues. Oral hygiene is often a neglected susceptibility of daily living that has a major impact on general physical health. As soon as a baby is born, bacterial colonization of the mouth begins. In addition to frequent contact and sufficient inoculation size, colonization by oral bacteria requires a suitable attachment site and nutrients for growth. For example, children whose mouths are colonized by *Streptococcus mutans* are more likely to have been exposed to behaviour that allows for contact with adults' saliva and frequent sugar exposure than children without *S. mutans* [1]. Instead of performing the regular oral hygiene and dietary modifications to avoid dental caries, one of the effective methods by which oral hygiene can be achieved is through the uses of mouthwash. A mouthwash may be recommended to treat infection, reduce inflammation, relieve pain and reduce halitosis or to deliver fluoride locally for caries prevention. There are many brands of mouthwash marketed in Malaysia, each comes with its own deodorant, antiseptic, disinfectant, analgesic and astringent properties. Mouthwashes containing four phenol-related essential oils (thymol, eucalyptol, menthol and methyl salicylate in up to 26% alcohol) claim to penetrate the plaque biofilm and thus kill microorganisms that cause gingivitis. These mouthwashes display broad spectrum antimicrobial susceptibility, prevent bacterial aggregation, slow bacterial multiplication, retard plaque maturation and decrease plaque mass and pathogenicity. Their mechanism of action is thought to involve bacterial cell destruction, bacterial enzyme inhibition, and extraction of endotoxin from Gram-negative bacteria. Mouthwash improvement and development had been done mostly by the manufacturers and the little work that has been done relating to the individual ingredients as their main target to give freshness and eliminate oral bacterial, however, there are many doubts arise whether this mouthwash is really effective or not [2]. Medicinal plants have attracted increasing interest because of their antimicrobial susceptibility against pathogenic oral microorganisms. Plants that are used for traditional medicine contain a wide range of substances that can be used to treat chronic and infectious diseases [3]. The World Health Organization (WHO) estimates that 80% of the population of certain Asian and African countries presently uses medicinal plants in several aspects of primary health care. Studies in the United States and Europe have shown that the use of these plants is less common in clinical settings but has increased in recent

* Corresponding author: essam22001@gmail.com

years, as scientific evidence regarding the effectiveness of herbal medicine has become more widely available [4]. Medicinal plants are important sources for pharmaceutical manufacturing and account for a significant percentage of the pharmaceutical market. For example, in Malaysia, the market for traditional medicine is estimated to be at 1 billion Malaysian ringgit annually [5]. However, an effective combination of medicinal plants in mouthwash against oral isolates has not been reported. The present study aims to obtain extracts combination of four commercial medicinal plants (black pepper, black cumin, cinnamon, cardamom) and screen its antibacterial susceptibility against oral isolates from different 50 selected individuals aged 3 to 60 years. In addition, the antimicrobial susceptibility of different commercial kinds of mouthwash were also evaluated against oral isolates.

2 2 Methodology

2.1 Sample collection and Inoculum Preparation

All oral isolates were collected by dentists from 50 selected individuals at dental clinic Gambang in Gambang, Pahang. Sterile swabs were used for selected individuals of both genders with ages ranging from 3 to 60 years. The collected samples were transferred to the laboratory at Universiti Malaysia Pahang. The oral samples were cultured onto sterile nutrient agar plates and incubated at 37°C for 24 h, and then purified, cultured on agar slants, partial identified against Gram reaction and kept at 4°C in a chiller until use. After initial propagation, all test microbes were grown for 24 hours on Nutrient broth media at 37°C, and inoculums for the assays were prepared by diluting. The density of each microbial suspension was adjusted using a spectrophotometer wavelength 600 nm (GENESYS 10S UV-Vis) to equal that of 0.5 McFarland standards (10^8 CFU/ml).

2.2 Antimicrobial susceptibility test using disc diffusion method

2.2.1 Preparation of plants extracts

The four plant parts were crushed using a blender and then sieved to obtain fine powders. Approximately 25 g of the powdered plants were soaked in 200 ml of four different solvents (95% ethanol, hot water, water, and acetone), kept at room temperature for 72 h, and the suspension was then filtered through a Whatman No.1 filter paper. The filtrates were evaporated to 5 ml the final concentration was 12.5% (g/v), using a rotary evaporator (BUCHI, Rotavapor R-3 Vacuum pump V-700), according to the modified method of [6].

2.2.2 Combination of plant extracts

The combination of plant extracts obtained using the four different solvents was studied at a ratio of 1:1 (v:v). Three different combination groups, A, B, and C, were prepared using the combination of two plant extracts together (A) (P/Ci, P/N, P/Ca, Ci/N, Ci/Ca, N/Ca), three extracts (B) (P/Ci/N, P/Ci/Ca, P/N/Ca, Ci/N/Ca), and all four extracts (C)(P/Ci/N/Ca), respectively. The extracts were stored in sterile bottles and stored at 4 °C until further use. Filter paper discs of 6 mm diameter were prepared using Whatman No. 1, UV sterilized, and loaded with 25 µl plant extracts, left to dry, and then used an antibacterial screening test. All combinations were tested for antimicrobial susceptibility against oral isolates as previously mentioned.

2.2.3 Plants extract against oral isolates

Muller–Hinton agar medium was used for the antimicrobial susceptibility. The plants extract discs from previously prepared solvents as described in previous section were allowed to set onto the inoculated agar surface. Inoculum from primary culture plates were prepared, inoculating into a replicated plate, and incubation at 37 °C for 24 h. After the incubation period, each plate was observed, and the inhibition zone of all isolates were recorded in millimetre (mm) [7] and calculated and represented as previously mentioned. The density of each bacterial suspension was adjusted using 0.5 McFarland standard. Control discs for different solvents were used.

2.3 Selection of commercial Mouthwashes

Ten different commercial mouthwashes (different brands), which were denoted by codes (Bac), (LTC), (FC), (SAC), (OB1), (Wat1), (CP), (Gu), (LCM), and (OX)

2.3.1 Antimicrobial susceptibility test of mixture

2.3.2 Plant extract–mouthwash combination

Ten different commercial types of mouthwash were mixed with the acetone plant extracts of combinations A, B, and C at a ratio of 1:1 (v/v) and antimicrobial susceptibility test performed as previously mentioned as described in pervious section and extract free mouthwash disc was used as a control.

2.4 Statistical Analysis

Statistical analyses were performed using ANOVA for data on disc agar diffusion assays, to test the antimicrobial susceptibility profile of oral isolates within medicinal plant extracts on the zone of inhibition using software Minitab 17. All antimicrobial susceptibility data were determined in diameter (mm), calculated and represented in this study as antimicrobial susceptibility percentage against all 59 oral isolates under study.

3 Results and discussion

3.1 Oral Isolates

Oral isolates were obtained from 50 different selected individuals of different ages and genders, and a total of 59 bacterial and yeast isolates were collected and purified as revealed in Table 1.

Table 1. The oral isolates in Gram reaction

Isolate type	Gram reaction	Number of isolates	Isolate shape and number
Bacteria	Gram negative	2	Bacilli (1) Cocci (1)
	Gram positive	51	Bacilli (8) Cocci (43)
Yeast	-	6	Oval
Total oral isolates		59	

The impact of age group on the infection rates of tooth caries showed that age groups of 3–20 and 20–40 years were more susceptible to infection than the older group of 40–60 years, with the incidence of 32 and 44%, respectively as shown in Table 2.

Table 2. The oral isolates in age group and percentage

No.	Age group	Selected individual number	Percentage (%)
1	3-20(Y)	16	32
2	20-40(Y)	22	44
3	40-60(Y)	12	24
Total		50	

Y: year

This finding may be related primarily to immune shortages of infected people, as well as more microbes were prevalent in younger person. Moreover, the emphasis has been placed on the function of the mother as a source of disease transmission from her infected teeth to the baby, as shown by the similar levels of bacteria in mothers to those found at their children [8].

3.2 Combination of plant extracts using different solvents

The results of antimicrobial susceptibility profile of 11 combinations as shown in Table 3 prepared, with the use of four different solvents to obtain medicinal plants extracts. Among acetone, ethanol, hot, and water extracts, the acetone extract of combination A (Ci/N) showed antimicrobial susceptibility against 59 oral isolates with a higher synergistic effect up to 61.02% than other combinations, followed by acetone extract of combination B (P/N/Ca), with 30.51%. The antimicrobial susceptibility against oral isolate 30.51% was observed in synergistic of group C of acetone extracts (P/Ci/N/Ca). The combination of aqueous extracts of medicinal plants did not exhibit an inhibitory effect against the tested oral microorganisms.

Table 3. Summary of antimicrobial susceptibility profile of plant extracts combination against oral isolates

Group Code	Combination group†	Increasing Combination effects of plant extracts using different solvents (%*)			
		Ethanol 95%	Hot water	Normal water	Acetone
A	P/Ci	22.03	3.38	1.69	2.03
	P/N	15.25	3.38	1.69	50.85
	P/Ca	15.25	1.69	1.69	15.25
	Ci/N	23.72	3.38	3.38	61.02
	Ci/Ca	32.20	6.77	5.08	20.34
	N/Ca	11.86	3.38	3.38	13.56
B	P/Ci/N	28.81	8.64	18.64	25.42
	P/Ci/Ca	30.50	13.55	13.55	15.25
	P/N/Ca	7.11	20.33	20.33	30.51
	Ci/N/Ca	25.42	6.94	20.33	22.03
C	P/Ci/N/Ca	25.42	10.16	5.08	30.51

† Combination group: P: black pepper; Ci: cinnamon; N: black cumin: Ca: cardamom; *All percentage represented in the table explain the calculation of antimicrobial susceptibility percentage of all 59oral isolate.

These results were similar to the finding of Seth (2011), who investigated that the antimicrobial susceptibility of eight spices as an alternative to antibiotics in order to tackle these dangers. The combination of antibacterial susceptibility of common Indian spices namely black pepper, cardamom, cinnamon, clove, coriander seeds, garlic, ginger and turmeric against four potent food borne pathogens namely *E. coli*, *Streptococcus ratti*, *Staphylococcus aureus* and *Bacillus cereus*, he found that the combination of clove with black pepper, clove with turmeric, turmeric with black pepper, garlic with clove, garlic with turmeric, and garlic with black pepper, showed positive results against *E. coli*, *S. ratti*, and *S. aureus* [9].

Our study agrees with Ababutain (2011) that investigated the combination of clove and cinnamon have a broad spectrum antimicrobial susceptibility against *B. subtilis* and *C. albicans* at very low concentration MIC, which can be used as an alternative for antibiotics [10]. Despite the development of antimicrobial susceptibility in products, oral bacterial infections remain a major health problem because of environmental changes, as well as microorganism behaviors. Therefore, developing new antimicrobial compounds with high susceptibility, low toxicity and side effects is urgently needed. Our study showed that all of the studied medicinal plant combinations exhibited high antimicrobial susceptibility against oral isolates, with different synergistic or antagonistic effects.

3.3 Mouthwash combination with plant extract (2 in 1)

Data represented in Fig. 1 (a-c) show the antimicrobial susceptibility percentages against 59 oral isolates of 10 commercial mouthwashes combined with the highest plant extract combinations that were previously obtained. With 100% total mouthwash combined with plant extract combination C (P/Ci/ N/Ca) revealed higher antimicrobial susceptibility against oral isolates. Moreover, the similarity between the combination of mouthwash with plant extract combination A (Ci/N) and B (P/N/Ca) exhibited the highest effect, with 80% antimicrobial susceptibility.

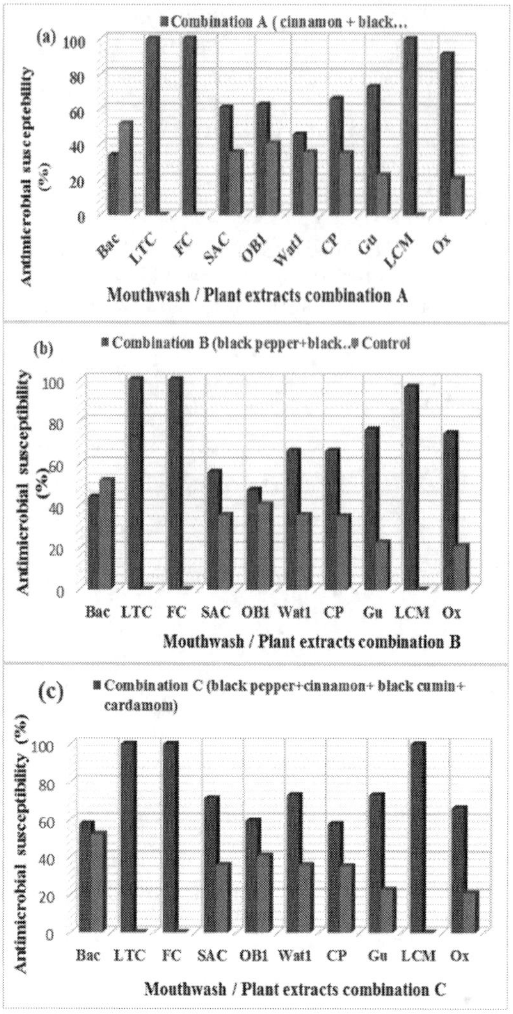

Fig. 1: Antimicrobial susceptibility percentages of combined action between 10 mouthwash (MW) , Bac; LTC; FC; SAC; OB1; Wat1; CP; Gu; LCM; Ox, with plant extracts, (a) combination A(Ci/N); (b) combination B (P/N/Ca); (c) combination C(P/Ci/N/Ca) against 59 oral isolates, control (MW plain).

 https://doi.org/10.1051/matecconf/201815006034

Plants have the major advantage of being the most effective and cheaper alternative source of drugs, medicinal plants extract have the potential to be used as therapeutic agents for chronic gingivitis and periodontitis conditions that have both bacterial and inflammatory components. These results were similar to the finding of Kothiwale et al. (2014), who reported that the mouth rinse containing medicinal plants have antimicrobial susceptibility against oral isolates, which may be useful as an adjunctive to mechanical therapy in the prevention and treatment of periodontal diseases [11]. Chentouf et al. (2012) who revealed a significant reduction in bacteria from children's oral cavity using miswak mouthwash extracts against microbial strains [12]. Mouthwash is used to complete the process of mechanical plaque removal, our results agreement with Mansour et al 2012 who reported that the garlic was a strong antimicrobial agent and acts as an inhibitor against both Gram-positive and Gram-negative bacteria, the efficacy of medicinal plant was higher than chlorhexidine against target bacteria and could be used as an effective mouthwash [13]. Finally, the combinations of medicinal plants A (Ci/N), B (P/N/Ca), and C (P/N/Ca/Ci) can be added to the mouthwash since it is inexpensive and available.

4 Conclusion

The acetonic plant extracts combinations of group A (Ci/N), B (P/N/Ca), and C (P/N/Ca/Ci) with 10 different commercial mouthwash showed high antibacterial susceptibility against oral isolates.

The authors gratefully acknowledge Universiti Malaysia Pahang (UMP), Malaysia for the financial supported by grant GRS 140318 that enables the authors to accomplish this work.

References

1. A. Wan, ., J of Dent R., . **82**,7(2003) p. 504-508.
2. B. Oluremi .,TJ of Pharmaceutical Research,. **9**,6(2010).
3. H.Koo., Archives of oral biology, **45**,2(2000): p. 141-148.
4. Organization, W.H., *WHO traditional medicine strategy 2002-2005*. 2002.
5. Z. Aziz,and N. Tey,. Complementary Therapies in Medicine,. **17**,1(2009): p. 44-50.
6. A. Gahlaut, and A.K. Chhillar,. Inter J of Pharmacy and Pharmaceutical Scie,. **5**,2(2013): p. 372-376.
7. S Mukhtar,. and I. Ghori, . J. Appl. Biol. Pharm. Technol,. **3**,2(2012): p. 131-136.
8. G Rao, G.,. Drugs,. **55**,3(1998): p. 323-330.
9. A. Seth, A., *Anti and phytochemical analysis*,. **11**,5(2011): p. 22-27.
10. I.M. Ababutain, ., Aust J Basic & Appl Sci, . **5**(2011): p. 678-683.
11. S.V. Kothiwale, S.V.,. J of Indian Society of Periodontology,. **18**,3(2014): p. 316.
12. N. Chelli-Chentouf, N., .J of ethnoph,. **144**,1(2012): p. 57-66.
13. A. Mansour, A., K. Maryam, and R. Neda, Jundishapur Journal of Micro,. **2**(2012): p. 398-400.

Optimization parameters for *Mycobacteria confluentis* biodegradation of PAHs

Muna Ibrahim[1], *Essam* A. Makky[1,2*], *Nina* S. Azmi[1], and *Jamil* Ismail[1]

[1]Faculty of Industrial Sciences & Technology, Universiti Malaysia Pahang, 26300 Gambang, Kuantan, Pahang, Malaysia
[2]Center of Excellence for Advanced Research in Fluid Flow (CARIFF), Universiti Malaysia Pahang, 26300 Gambang, Kuantan, Pahang, Malaysia

Abstract - In this study, the effect of process parameters (incubation period, temperature, initial pH-value, concentration of PAHs in the medium, and bacterial inoculum size) on the biodegradation of PAHs using *Mycobacteria confluentis* was studied using One-Factor-At-a-Time (OFAT) method. From the results of the study, it was observed that the studied parameters had significant effects on the degradation process. The capability of *Mycobacteria confluentis* on the degradation of PAHs was found to be maximum when the initial pH of the PAH was 7, temperature 40 °C, PAH concentration of 50 µL, bacterial concentration of 7.5 mL, and incubated for 3 to 5 days. This condition gave a PAH degradation percentage of 40 to 50 %, showing a similarity to most previous studies where the maximum percentage of degradation has been around 40 to 50 %. This study, therefore, concludes that *Mycobacteria confluentis* is a good PAHs degrader which can be of significant important in the management of oil polluted fields and environments.

Keywords: PAHs, Biodegradation, *Mycobacteria confluent, Optimization*

1 Introduction

The polycyclic aromatic hydrocarbons (PAHs) are a complex group of chemicals which are of great concern owing to their abundant environmental distribution, resistance to biodegradation, as well as their harmful effects. They are mainly implicated in the deterioration of water and soil quality. There have been series of treatment options to curb the economic dangers and ecological deterioration caused by the PAHs. Bioremediation is one of such recommended treatment options that holds a promise of delivering a lasting solution for the degradation of PAHs. Biodegradation is the process of degrading the high molecular chains of hydrocarbons using bacteria [1, 2] and fungi [3, 4]. Other microorganisms such as algae have also been reported for their capability of degrading hydrocarbons [5, 6]. Bacteria have been reported as the major and most active degraders of petroleum pollutant [7, 8].

The process of bioremediation involves microorganisms mitigating, degrading and reducing dangerous organic pollutants to the less harmful ones such as CO_2, CH_4, H_2O, without any adverse environment effect [9]. Biodegradation is a primary bioremediation technique where oleophilic microbes are employed for the removal of hydrocarbon pollutants from the environment [5, 10, 11]. The petroleum refinery effluents (PRE) are wastes which originate primarily from oil refining industries during the production of crude oil, lubricants, and petrochemical intermediates [12]. The PREs are complex matrices of organic pollutants; it has been established that oily and hydrocarbon-rich waste waters can be completely mineralized through photocatalytic bioremediation. Interestingly, all the organic substrate found in the PREs are mineralized; therefore, the PREs can be effectively remediated [13].

The ability of microorganisms to degrade hydrocarbons vary; some have the capability to degrade aliphatics, while others can degrade the monoaromatics, polyaromatics or resins. Some of the reported hydrocarbon pollutants degrading microorganisms in the literature include the genera *Achromobacter, Acinetobacter, Azoarcus, Brevibacterium, Cellulomonas, Corynebacterium, Flavobacterium, Marinobacter, Micrococcus, Nocardia, Ochrobactrum, Pseudomonas, Stenotrophomaonas* and *Vibrio* [10, 14]. The bacterial degradation of PAHs and other hydrocarbon pollutants depend on several environmental conditions such as the type of the microorganisms, nature and chemical structure of the compound to be degraded, the temperature of the process, the initial pH of the environment, as well as the number of organisms in the environment. The PAHs are biotransformed into less complex metabolites through mineralization into inorganic minerals such as H_2O, CO_2 under aerobic condition, and CH_4 under anaerobic condition.

* Corresponding author: essammakky@ump.edu.my

The biodegradation of PAHs has been studied using various microbes such as *Pseudomonas aeruginosa, Pseudomonas fluorescence, Mycobacterium spp., Haemophilus spp., Rhodococcus spp., Paenibacillus spp* isolated from the contaminated soil [15]. Although many studies have reported the use of various microbes for the degradation of PAHs, there has been no report on the use of *Mycobacteria confluentis* for the biodegradation od PAHs. This study was, therefore, initiated to study the effect of various parameters such as the period of incubation, the temperature of the process, the initial pH of PAHs, the concentration of PAHs in the medium as well as the concentration of the microbe in the medium using One-Factor-At-a-Time (OFAT) method. The outcome of the study will contribute to the pool of already identified microbes for the bioremediation of hydrocarbon polluted environments.

2 Materials and Methods

2.1 Sample collection and preparation

The refinery wastewater samples were collected in glass bottles, covered with screw caps, and sent to the laboratory for analysis and Liquid-Liquid extraction as described in [16] using Filters Fioroni paper (90 mm diameter), under vacuum pump. The extracted samples were stored at 4 °C for further use.

2.2 Media preparation

An enrichment medium was prepared in demineralized water containing the following constituents (g/ L): 0.2 $MgSO_4$; 0.02 $CaCl_2$; 1.0 NH_4NO_3; 1.0 KH_2PO_4; 0.05 K_2HPO_4; 0.05 $FeCl_3$ [17]. Nutrient agar for the bacterial culture was prepared by adding 23.0 g of nutrient agar (Hardy Diagnostics, USA) to 1000 mL of demineralized water. The nutrient broth was also prepared by adding 20.0 g of nutrient broth (Merck, Germany) to 1000 mL of demineralized water. All the media were autoclaved for 15 minutes at 121 °C and allowed to cool before use.

2.3 Biodegradation experiments

The bacterial suspension was prepared in nutrient broth contained in Falcon tubes. Into the tubes, 25 mL of the already prepared nutrient broth was added, followed by 5 mL of the bacterial broth culture containing about 3.0 x 10^8 CFU/mL. The tubes were incubated at 37 °C for 24 h. For the PAHs biodegradation assay, the prepared bacterial suspension (5 mL) was added into the enriched medium (45 mL), followed by 50 μL of PAHs.

2.4 Effect of each parameter

2.4.1 Effect of incubation period

To study the effect of the incubation period, the initial pH of the PAHs was adjusted to 7 with NaOH; 50 μL of the adjusted PAHs was added into 45 mL of the enriched medium, followed by 5 mL of the bacterial suspension. The medium containing the organism and PAHs was incubated at 37 °C for a total period of 7 days. At the end of every 24 h, an aliquot of the medium was sampled for turbidity check and PAHs degradation percentage.

2.4.2 Effect of temperature

To study the effect of temperature, the initial pH of the PAHs was adjusted to 7 with NaOH; 50 μL of the adjusted PAHs was added into 5 set tubes containing 45 mL of the enriched medium, followed by 5 mL of the bacterial suspension. One set of the tubes was incubated at 5oC, the second, third, fourth, and fifth tubes were incubated at room temperature (25 ± 2 °C), 30, 40, and 55 °C, respectively for 5 days. The samples were analyzed for turbidity and percentage PAHs degradation only after 5 days of incubation.

2.4.3 Effect of initial PAHs pH

To study the effect of pH, the initial pH of the PAHs was adjusted to pH 5, 7, and 9 with NaOH; 50 μL of each adjusted PAHs was added into 3 sets of tubes containing 45 mL of the enriched medium, followed by 5 mL of the bacterial suspension. The media containing the organism and pH-adjusted-PAHs were incubated at 37° C for 5 days. At the end of 5 days, an aliquot of each medium from the three sets was sampled for turbidity check and PAHs degradation percentage.

2.4.4 Effect of bacterial inoculation concentration

To study the effect of bacterial concentration, the initial pH of the PAHs was adjusted to 7 with NaOH; 50 μL of the adjusted PAHs was added into 4 sets of tubes containing 45 mL of the enriched medium, followed by 1, 3, 5, and 7.5 mL of the bacterial suspension into the 1st, 2nd, 3rd, and 4th tubes, respectively. All the media containing the organism and PAHs were incubated at 37° C for 5 days. At the end of 5 days, an aliquot of each of the medium was sampled for turbidity check and PAHs degradation percentage.

2.4.5 Effect of PAHs concentration

To study the effect of PAHs concentration, the initial pH of the PAHs was adjusted to 7 with NaOH; 50, 275, and 500 μL of the adjusted PAHs was added into 3 sets of the tube containing 45 mL of the enriched medium, followed by 5 mL of the bacterial suspension. The media containing the organism and PAHs were incubated at 37 °C for 5 days. At the end of 5 days, an aliquot of each set sampled for turbidity check and PAHs degradation percentage.

2.4.6 Cell growth estimation

The percentage of cell growth in each of the respective assays was turbidimetrically analysed at 600 nm using a microplate reader (Infinite M200 PRO). The percentage growth was calculated using Equation 1. A high cell growth indicates a high utilization of the PAHs as the source of carbon by the organism, indicating a high degradation of the hydrocarbon.

$$Cell\ growth = Abs\ of\ test - Abs\ of\ control \quad (1)$$

where Abs of the test is the absorbance of the tubes containing the media constituent, the PAHS, and the organism; while Abs of the control is the absorbance of the tubes containing only the media and PAHS without the organism.

2.5 PAHs degradation rate

The rate of PAHs degradation at the end of each incubation period was estimated from the turbidimetric analysis of the tubes. The degradation percentage was calculated using Equation 2.

$$PAHs\ \deg radation(\%)(\frac{Abs\ of\ test - Abs\ of\ control}{x100\ Abs\ of\ test}) \quad (2)$$

where Abs of test = Absorbance of the tubes containing all the media supplements, PAHs, and the organism, while Abs of control = the absorbance of the tubes containing all the media supplements and PAHs.

3 Results and Discussions

3.1 Effect of incubation period

The effect of the incubation period on the biodegradation of PAHs using *Mycobacteria confluentis* was studied for a period of 7 continuous days of incubation. The outcome of the study is shown in Table 1. It could be observed from the table that there was a steady utilization of the PAHs as the sole source of carbon by the *Mycobacteria confluentis*, indicated by the steady increase in the absorbance of the media from day 1 to day 3. However, from day 4, there was a progressive decrease in the absorbance of the media, indicating a decrease in the number of active bacteria in the media. The decrease in the number of bacteria could be because of the complete utilization of the carbon source (PAHs), and due to the accumulation of bacterial metabolic waste products which interfered with the growth of the microbes. Kachienga and Momba [18] have reported the biodegradation of hydrocarbon chains of crude oil by-product by selected protozoan isolated from waste water. From the results of that study, the organisms were reported to achieve up to 61 % of hydrocarbon degradation within 5 days of incubation at 30 °C. other studies have similarly reported a complete biodegradation of hydrocarbons within 3 to 5 days of incubation.

Table 1: Effect of incubation period on the biodegradation of PAHs using *Mycobacteria confluentis*

Incubation period (days)	Bacterial growth (mean ± SD)	PAHs degradation (%)
1	0.069 ± 0.001	5.3
2	0.076 ± 0.000	29.7
3	0.105 ± 0.003	51.61
4	0.092 ± 0.003	36.41
5	0.089 ± 0.004	36.0
6	0.076 ± 0.001	28.83
7	0.069 ± 0.000	25.92

n = 3

3.2 Effect of temperature

The effect of the different incubation temperatures on the biodegradation of PAHs using *Mycobacteria confluentis* was studied for a period of 5 days. The results of the study are presented in Table 2. From the table, it could be observed that the bacterial cell growth was affected by the temperature of incubation. A temperature range of less than 30 °C did not support the growth of the bacteria as witnessed in the corresponding rate of PAHs degradation. However, a temperature of 40 °C was better for the growth of the organism as indicated by the high degradation percentage observed. A temperature of more than 40 °C did not support the growth of the organism; the organisms were likely to be inactivated, resulting in a reduced degradation of the hydrocarbon. Although other studies have reported a lower temperature for the biodegradation of hydrocarbons, it should be noted that different organisms, even different species of a genus, have different levels of tolerance to physical factors such as temperature and pH. Shafiee, Shojaosadati, and Charkhabi [19] reported an ideal temperature of 30 °C for the biodegradation of polycyclic aromatic hydrocarbons by aerobic mixed bacterial culture. They reported that the rate of degradation of the PAHs-fluoranthene culture was increased from 20 to 44 % after 10 days of incubation at 30 °C and pH 7. These findings indicate that the right temperature for the effective growth of any organism intended to be used for bioremediation must be determined to achieve the best experience of using the organism.

Table 2: Effect of temperature on the biodegradation of PAHs using *Mycobacteria confluentis*

Temperature (°C)	Bacterial growth (mean ± SD)	PAHs degradation (%)
5	0.078 ± 0.002	27.05
RT	0.077 ± 0.001	26.1
30	0.076 ± 0.001	25.1
40	0.089 ± 0.004	36.0
55	0.070 ± 0.003	18.7

n = 3

3.3 Effect of initial PAHs pH

The effect of the initial PAHs pH on its biodegradation using *Mycobacteria confluentis* was studied for 5 days. The ionic concentration of a medium can affect the growth and metabolic activities of the microorganism. Been that most microbes living in the soil thrive around a Ph range of 6 to 8, changes in the pH of the medium can also affect the bacterial activities of the inherent microbes
[20]. For the studied microorganism, a pH range of 5 to 9 was studied and the results shown in Table 3. It could be observed a neutral Ph was ideal for the bacterial growth and PAHs degradation even though a Ph range of 5 to 9 had a significant support on the growth of the organism. This maximum result achieved at Ph 7 indicated that this organism is a neutrophilic organism. Krishnaswamy and Namasivayam [20] when reporting the effect of Ph, nitrogen source and salts on the degradation of phenol by the bacterial consortium under saline conditions stated that a maximum degradation of phenol was observed at pH 7 using a consortium of bacteria. Similarly, Shafiee, Shojaosadati, and Charkhabi [19] reported an optimal pH of 7 for the biodegradation of polycyclic aromatic hydrocarbons by aerobic mixed bacterial culture. Kauselya and Narendiran [21] also reported that the degradation of benzene was fastest at a neutral pH. All these evidences supported the outcome of this study on the effect of pH on the biodegradation of PAHs using *Mycobacteria confluentis*.

Table 3: Effect of PAHs initial pH on the biodegradation of PAHs using *Mycobacteria confluentis*

Initial PAHs pH	Bacterial growth (mean ± SD)	PAHs degradation (%)
5	0.079 ± 0.001	16.4
7	0.118 ± 0.001	44.0
9	0.097 ± 0.001	31.9

n = 3

3.4 Effect of bacterial inoculation size

The result of the study on the effect of bacterial inoculation size on the biodegradation of PAHs using *Mycobacteria confluentis* is presented in Table 4. It could be observed from the table that the rate of PAHs degradation increases with an increase in the concentration of the microorganism. The result showed that the degradation of PAHs is directly related to the number of active microorganisms in the medium. The maximum observed degradation rate of 100 % was achieved with the maximum studied microbe concentration of 7.5 mL in a 50-mL reaction volume. The study reported by Kauselya and Narendiran [21] on the effect of pH and inoculation size on benzene biodegradation using mixed culture showed that the degradation of benzene increased with inoculation size. With this report, the outcome of this study is, therefore, validated.

Table 4: Effect of bacterial inoculation size on the biodegradation of PAHs using *Mycobacteria confluentis*

Bacterial concentration (mL)	Bacterial growth (mean ± SD)	PAHs degradation (%)
1	0.06 ± 0.001	5.1
3	0.095 ± 0.004	40.1
5	0.101 ± 0.005	43.6
7.5	0.106 ± 0.002	46.3

n = 3

3.5 Effect of PAHs concentration

The effect of the initial concentration of PAHs on its biodegradation using *Mycobacteria confluentis* is shown in Table 5. It could be observed from the table that the initial concentration of PAHs has an impact on the cell growth and rate of PAHs degradation. A moderate concentration of 50 μL favored the optimum growth of the organism and facilitated the degradation of the hydrocarbon; while a higher concentration of 500 μL had a negative inhibiting effect on the growth of the organism and its degradation of the hydrocarbon. This implies that the initial concentration of the hydrocarbon should be kept low for an efficient cell growth and biodegradation. A study on the optimization of growth condition for diesel oil-degrading bacterial strains reported by Mahalingam and Nithya [22] found that diesel at 1 % supported the growth of the studied bacteria better than when supplemented at 5 %.

Table 5: Effect of PAHs concentration on the biodegradation of PAHs using *Mycobacteria confluentis*

PAHs concentration (μL)	Bacterial growth (mean ± SD)	PAHs degradation (%)
50	0.118 ± 0.001	44.0
275	0.114 ± 0.002	42.1
500	0.078 ± 0.001	15.3

n= 3

4 Conclusions

The effect of five process parameters (PAHs concentration, pH, temperature, a period of incubation, and inoculation size) on the biodegradation of PAHs using *Mycobacteria confluentis* was studied using One-Factor-At-a-Time method. From the screening process, it was observed that these factors have a significant impact on the rate of the bacterial growth and its significance on the rate of PAHs degradation. The optimum condition for an effective growth of the cell and degradation of the hydrocarbon was found as follows: PAHs concentration 50 μL, pH 7, temperature 40 ^{o}C, inoculation size 7.5 mL, and an incubation period of 3 - 5 days, giving a degradation of about 40 to 50 %. This study concludes that a combination of parameters as stated will encourage the effective growth of *Mycobacteria confluentis* and enhance its capability in the biodegradation of PAHs.

The authors gratefully acknowledge Universiti Malaysia Pahang (UMP), Malaysia for the financial supported by grant PGRS160393 that enables the authors to accomplish this work.

References

1. N. Hamamura, FEMS Microbiology Ecology. **85**(1), 168 -178 (2013)
2. L. Ferreira, Journal of Chemical Technology and Biotechnology, **90**(3), 500 506 (2015)
3. P. Li, Bulletin of Environmental Contamination & Toxicology, **75**(3), 67-72 (2005)
4. D. Young, PloS one, **10**(6), https://doi.org/10.1371/journal.pone.0130381.
5. S.J. Varjani, V.N. Upasani, International journal of innovative research in science, Engineering and Technology, **2**(10), 5377-5383 (2013)
6. H. Wilkes, Journal of molecular microbiology and biotechnology, **26**(1-3), 138-151 (2016)
7. F. Abbasian et al., Applied biochemistry and biotechnology **176**(3), 670-699 (2015)
8. R.U. Meckenstock et al., Journal of molecular microbiology and biotechnology, **26**(1-3), 92-118 (2016)
9. E.Z Ron, E. Rosenberg, Current Opinion in biotechnology, **27,** 191-194 (2014)
10. S.J. Varjani, V.N. Upasani, Bioresource technology, **222,** 195-201 (2016)
11. B.M. Macaulay, D. Rees, Annals of Environmental Science **8,** 9-37 (2014)
12. M. Harry, New York: McGraw Hill. Inc (1995)
13. B.H. Diya'uddeen, Process Safety and Environmental Protection, **89**(2), 95-105 (2011)
14. S. Varjani, V. Srivastava, Renewable Res. J. **3**(1), 244-249 (2015)
15. A. Haritash, C. Kaushik, Journal of hazardous materials, **169**(1), 1-15 (2009)
16. D. Andrew. L.S.C. Eaton, E. Arnold, A.H. Mary, Water Environment Federation., ed. American Public Health Association (APHA), **20**. (1998)
17. P. Prabhakaran, International Journal of Innovative Research in Science, Engineering and Technology, **3**, 9236-9242 (2014)
18. L.M.M. Kachienga, Xuzhou Sui, Sun & Wang (eds) China University of Mining and Technology Press (2014)
19. P.S. Shafie, S.S. Charkhabi, Iran. J. Chem. Chem. Eng, **25**(3), 73-78 (2006)
20. V.N. Krishnaswamy, International Journal of Biotechnology and Biochemistry, **6** (5), 783-791 (2010)
21. K.R. Kauselya, Journal of Advanced Chemical Sciences, **1**(2015), 20-21 (2015)
22. P.U. Mahalingam, Advances in Applied Science Research, **5**(6), 91-96 (2014)

Thermal Radiation Effects on MHD with Flow Heat and Mass Transfer in Casson Nanofluid over A Stretching Sheet

Yap Bing Kho[1], Abid Hussanan[1], Norhafizah Mohd Sarif[1], Zulkhibri Ismail[1] and Mohd Zuki Salleh[1]*

[1] *Applied & Industrial Mathematics Research Group, Faculty of Industrial Science & Technology, Universiti Malaysia Pahang, 26300 UMP Kuantan, Pahang, Malaysia*

Abstract. The boundary layer heat and mass transfer flow of Casson nanofluid over a stretching sheet with constant wall temperature (CWT) under the magnetic field and thermal radiation effects is investigated numerically. Using similarity transformations, the governing equations are reduced to a set of nonlinear ordinary differential equations (ODEs). These equations are solved numerically by Shooting method. The effects of Casson parameter, magnetic parameter, porosity parameter, radiation parameter, Prandtl number, Brownian parameter and thermophoresis parameter on velocity, temperature and concentration fields are shown graphically and discussed. The results show that increase in Casson parameter causes the wall temperature increase well in the nanofluid.

1 Introduction

Generally, the convectional fluids such as water, ethylene glycol and engine oil used in many industrial processes are poor heat conductors because these fluids have low thermal conductivities. To overcome this situation, Choi [1] was the first who introduced the idea of nanofluids. Nanofluids are a new class of fluids created by scattering nano/micro sized materials (nanoparticles, nanofibers, nanotubes, nanowires, nanorods and nanosheets) in conventional heat transfer fluids. Nanofluids are nanometer-sized particles with large surface area. Since the size of nanoparticles is in nanometer-sized, they have the ability to flow smoothly through the microchannel easily. After the first experimental work of Choi [1] on nanofluids, several researchers have used this concept in their convective heat transfer problems. Amongst them, Buongiorno [2] published a survey article on the convective transport in nanofluids. Ahmed and Pop [3] examined mixed convection flow of nanofluid which the model proposed by Tiwari and Das [4] with the condition that past over a vertical plate embedded in the porous medium. Yacob et al. [5] studied the boundary layer flow of a nanofluid over a stretching/shrinking sheet with convective boundary condition. Chamkha and Aly [6] used implicit finite difference method and analyzed the effect of heat generation or absorption on the steady free convection flow of a nanofluid over permeable vertical plate with suction or injection. Noghrehabadi et al. [7] studied partial slip effects on the boundary layer flow of nanofluid over a stretching sheet with constant wall temperature. Mahdy [8] studied mixed convection flow together with heat transfer in nanofluids due to stretching sheet. He found that the rate of heat transfer at the vertical stretching surface and thermal boundary layer of nanofluids increased as compared to pure water. Qasim et al. [9] studied heat and mass transfer phenomenon in nanofluids with convective boundary conditions whereas Loganathan and Vimala [10] obtained numerical solution of unsteady nanofluids flow lies over vertical flat plate with accretion or ablation at leading edge. Recently, Ganga et al. [11] studied the thermal radiation effects on hydromagnetic flow of a nanofluids and found the analytic solutions by using homotopy analysis method and numerical by the fourth order Runge-Kutta method.

MHD study on nanofluids with thermal effect over stretching sheet have play important rule in causes of high strength collision in plasma, resistivity and kinetic effect. The effects of magnetic field on heat transfer flow of a nanofluids lie over a vertical semi-infinite flat plate were investigated by Hamad [12]. Ellahi [13] had studied the analytical solution of non-Newtonian nanofluids inside a pipe with temperature dependent viscosity with the presence of MHD. Qasim et al. [14] used the conventional base fluid (water) with magnetite (Fe_3O_4) nanoparticles and studied the ferrofluid flow of MHD and heat transfer analaysis over a stretched cylinder with given heat flux and slip condition. Hussanan et al. [15] considered unsteady MHD flow and heat transfer in some nanofluids over an accelerated vertical plate in a porous medium. They indicated that the effect of radiation parameter caused the wall temperature increased but temperature decreases significantly when Prandtl number is increased. Few other attempts made in this direction can be found in Krishna et al. [16] and Haq et al. [17].

All the above mentioned research however, focuses on Newtonian fluid (as water, ethylene glycol and engine oil) containing nanoparticles. Casson nanofluid is one of the fluids that do not obey Newton Laws and it is potentially useful in some applications, viz blood flow. Keeping this in view, Malik et al. [18] studied the Casson nanofluid with boundary layer over an

* Corresponding author: zuki@ump.edu.my

exponentially stretching cylinder vertically and concluded that velocity profile drops as increase the Casson parameter.

As a continuation, our team propose here to study radiation and magnetic field effects on Casson nanofluid over a stretching sheet with constant wall temperature. The resulting governing equations are reduced to ordinary differential equations through similarity transformation and then solved numerically using the Shooting method.

2 Mathematical Formulations

The steady two dimensional boundary layer flow of non-Newtonian Casson nanofluid over a stretching plate embedded in a porous medium. The surface is stretching along x-axis with the velocity $U_w = ax$ at $y = 0$, where $a > 0$ is a constant. A constant magnetic field is applied normal to the sheet, whereas induced magnetic field is neglected by assuming low magnetic Reynolds number (see Fig. 1). Further, we have considered the constant temperature T_w near the sheet and ambient temperature T_∞.

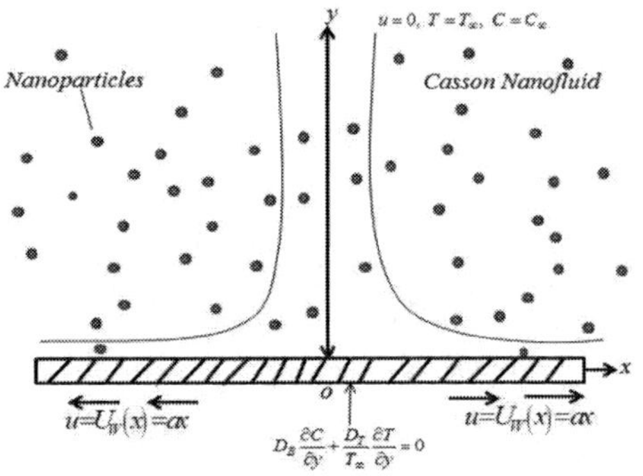

Fig. 1. Physical configuration and coordinate system

The boundary layer flow of a Casson nanofluid are govern as follows

$$u\frac{\partial u}{\partial x}+v\frac{\partial u}{\partial y}=\nu\left(1+\frac{1}{\beta}\right)\frac{\partial^2 u}{\partial y^2}-\frac{\sigma B_o^2}{\rho}u-\frac{\nu}{k_1}u \qquad (1)$$

$$u\frac{\partial T}{\partial x}+v\frac{\partial T}{\partial y}=\alpha_m\frac{\partial^2 T}{\partial y^2}+\tau\left[D_B\frac{\partial C}{\partial y}\frac{\partial T}{\partial y}+\frac{D_T}{T_\infty}\left(\frac{\partial T}{\partial y}\right)^2\right]-\frac{1}{(\rho C_p)_f}\frac{\partial q_r}{\partial y} \qquad (2)$$

$$u\frac{\partial C}{\partial x}+v\frac{\partial C}{\partial y}=D_B\frac{\partial^2 C}{\partial y^2}+\frac{D_T}{T_\infty}\frac{\partial^2 T}{\partial y^2} \qquad (3)$$

subject to boundary conditions as

$$u=U_W(x)=ax,\ v=0,\ T=T_w,\ D_B\frac{\partial C}{\partial y}+\frac{D_T}{T_\infty}\frac{\partial T}{\partial y}=0 \text{ at } y=0 \qquad (4)$$
$$u\to 0,\ v\to 0,\ T\to T_\infty,\ C\to C_\infty \text{ as } y\to\infty$$

According to Rosseland's approximation (Hussanan et al. [19]), the radiative heat flux is simplified as

$$q_r=-\frac{4\sigma^*}{3k^*}\frac{\partial T^4}{\partial y} \qquad (5)$$

where σ^* and k^* are the Stefan-Boltzmann and mean absorption coefficient, respectively. By expanding T^4 in the Taylor series about T_∞ and ignoring the higher order terms, we get

$$T^4 \square \qquad 3T_\infty^4 \qquad (6)$$

By substituting equations (5) and (6) into equation (2), we get

$$u\frac{\partial T}{\partial x}+v\frac{\partial T}{\partial y}=\alpha_m\frac{\partial^2 T}{\partial y^2}+\tau\left[D_B\frac{\partial C}{\partial y}\frac{\partial T}{\partial y}+\frac{D_T}{T_\infty}\left(\frac{\partial T}{\partial y}\right)^2\right] + \frac{16\sigma^* T_\infty^3}{3k^*(\rho C_p)_f}\frac{\partial^2 T}{\partial y^2} \qquad (7)$$

Follow Yasin et al. [20], we introduced the similarity transformations

$$\psi=(a\nu)^{1/2}xf(\eta),\ \eta=\left(\frac{a}{\nu}\right)^{\frac{1}{2}}y,$$
$$\theta(\eta)=\frac{T-T_\infty}{T_w-T_\infty},\ \phi(\eta)=\frac{C-C_\infty}{C_w-C_\infty}. \qquad (8)$$

The stream function ψ is defined as such that $u=\frac{\partial\psi}{\partial y}$ and $v=-\frac{\partial\psi}{\partial x}$. After the substitution of above transformations and equations (1), (3) and (7) take the following forms

$$\left(1+\frac{1}{\beta}\right)f'''(\eta)+f'(\eta)f''(\eta)-f'^2(\eta) -(Q+k_p)f'(\eta)=0 \qquad (9)$$

$$\frac{1}{\Pr}\left(1+\frac{4}{3}R\right)\theta''(\eta)+Nb\theta'(\eta)\phi'(\eta)+Nt\theta'^2 +f(\eta)\theta'(\eta)=0 \qquad (10)$$

$$\phi''(\eta)+Sc.f(\eta)\phi'(\eta)+\frac{Nt}{Nb}\theta''(\eta)=0 \qquad (11)$$

Where f, θ and ϕ are functions of η and prime denotes derivatives with respect to η. The corresponding boundary conditions will follow the following form:

$$f(0)=0,\ f'(0)=1,\ \theta(0)=1,\ \phi(0)=1,$$
$$f'(\infty)=0,\ \theta(\infty)=0,\ \phi(\infty)=0. \qquad (12)$$

The interests of physical quantities are skin friction coefficient C_f, local Nusselt number Nu_x and Sherwood number Sh_x which are defined as

$$C_f=\frac{\tau_w}{\rho U_W^2},\ Nu_x=\frac{xq_w}{kT_\infty} \text{ and } Sh_x=\frac{xq_m}{D_B(C_w-C_\infty)}. \qquad (13)$$

where the shear stress τ_w, surface heat flux q_w and surface mass flux q_m are given by

$$\tau_w=\mu\left(\frac{\partial u}{\partial y}\right),\ q_w=-k\frac{\partial T}{\partial y} \text{ and } q_m=-D_B\frac{\partial C}{\partial y} \text{ at } y=0. \qquad (14)$$

Using the similarity transformations given in equation (8), we obtain

$$C_f\left(\mathrm{Re}_x\right)^{\frac{1}{2}}=f''(0),\ \frac{Nu_x}{\left(\mathrm{Re}_x\right)^{\frac{1}{2}}}=-\theta'(0)\text{ and }\frac{Sh_x}{\left(\mathrm{Re}_x\right)^{\frac{1}{2}}}=-\phi'(0). \tag{15}$$

where $\mathrm{Re}_x=\dfrac{xU_w(x)}{v}$ defined as local Reynold's number.

3 RESULT AND DISCUSSIONS

The graphical results for temperature $\theta(\eta)$, velocity $f'(\eta)$ and concentration $\phi(\eta)$ profiles for different involved parameters such as Casson parameter β, magnetic parameter M, porosity parameter k_p, radiation parameter R, Prandlt number Pr, Brownian motion parameter Nb and thermophoresis parameter Nt. The non-linear ordinary differential equations are derived manually together with boundary conditions. Then, these equations are solved by using Shooting method. Those non-linear ordinary differential equations are of second order in θ and ϕ, third order in f, are then reduced into simultaneous ordinary equations. With the initial conditions of $\eta\to\infty$, we can get the value of $f'(\eta)$, $\theta(\eta)$ and $\phi(\eta)$. Thus, we may obtain unknown initial conditions $\eta=0$ by using Shooting method and assumed the initial values for boundary value problem.

Table 1. Comparison of results for the local Nusselt number $-\theta'(0)$ when $\beta=\infty$, Le=1, $R=K=M=Nb=Nt=0$.

Pr	Khan and Pop [25]	Gorla and Sidawi [26]	Wang [27]	Present study
0.20	0.1691	0.1691	0.1691	0.1698
0.70	0.4539	0.5349	0.4539	0.4539
2.00	0.9113	0.9114	0.9114	0.9113
7.00	1.8954	1.8905	1.8954	1.8954
20.00	3.3539	3.3539	3.3539	3.3539

The calculated boundary values should be similar with the real boundary values, hence, the value should as close to the boundary values are appropriate to consider. In this study, there are no consideration of variation in temperature, velocity and concentration where we took large infinity condition but finite value for η. We chose the value of $\eta_{\max}=8$, which is appropriate and sufficient to achieve asymptotic for all examined parameter with the boundary conditions. Table 1 show the comparison values of $-\theta'(0)$ with previous results. It concluded that this method work efficiently and the results presented here are accurate.

The effects of Casson parameter β on velocity and temperature profiles are shown in Fig. 2 and 3. It is found that an increase in Casson parameter β leads to velocity profiles decrease but temperature profiles increase. When increase the values of Casson parameter β means to reduce yield stress, thus Casson fluid behaves like Newtonian fluid $(\beta\to\infty)$. It is also concluded that Casson fluid velocity is greater than that of Newtonian fluid.

Fig. 4 and 5 show that the magnetic parameter M effects on velocity and temperature profiles, respectively. The velocity profiles drops when magnetic parameter increases. This is because when magnetic parameter increases, the Lorentz force also will increase and caused to oppose the fluid motion at the same time. Fig. 5 shows that an increase in the magnetic parameter M leads to an increase in temperature. Further, this enhancement is very significant near the sheet, however this effect is almost negligible away from the sheet.

Fig. 6 and 7 illustrated the temperature and concentration profiles for different values of Brownian parameter Nb. It is observed that increasing value of Nb cause the temperature field increases but the concentration profiles decreases. Since Brownian motion refers to the particle movement, hence, the active the movement of particles, the more heat produced and temperature will rise. Fig. 8 and 9 demonstrated the temperature and concentration profiles of thermophoresis parameter respectively. It is observed that increasing the value of Nt resulted the temperature and concentration profiles increase.

Fig. 10 reveals the temperature profile of radiation parameter. Based on the graph, the result shown that increase the value of R then wall temperature increase as well. Fig. 11 and 12 is plotted to explain the effect of Casson nanofluid and Magnetic parameter on local Nusselt number and skin friction coefficient, respectively. The greater the value of Casson parameter will cause the higher temperature gradient. But, with the larger value of magnetic force and it will reduce the thickness of boundary layer. Fig. 12 shows that the skin friction coefficient decreases when the value of Casson parameter and magnetic parameter increase. Influence of Brownian motion parameter and thermophoresis parameter on local Nusselt number is illustrated in Fig. 13. It can be seen that the dimensionless temperature rate decreases at all time when increases the value of Nb and Nt. Figure 14 shows the effect of Prandlt number and radiation parameter on temperature gradient. With increasing the Pr value, it is also increase the local Nusselt number. However, the change of heat transfer decreases once the radiation parameter increases.

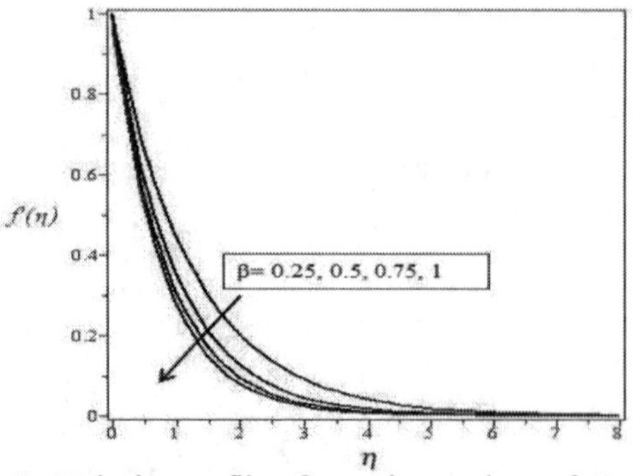

Fig. 2. Velocity profiles for various values of Casson parameter β, when $M=0.2, K=2, R=5, Nb=0.5, Nt=0.5$, $Sc=2$ and $\Pr=7$.

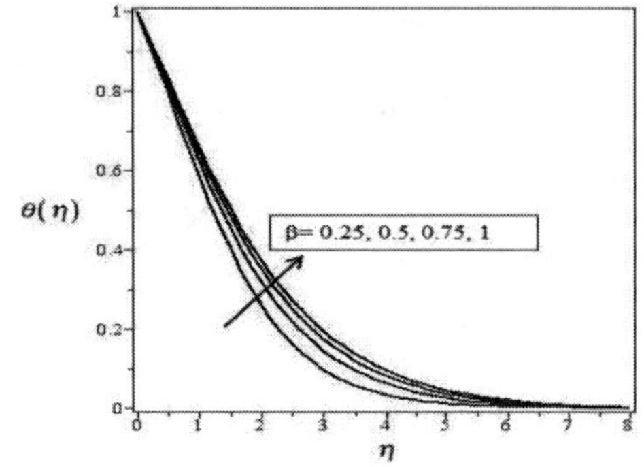

Fig. 3. Temperature profiles for various values of Casson parameter β, when $M=0.2, K=2, R=5, Nb=0.5, Nt=0.5$, $Sc=2$ and $\Pr=7$.

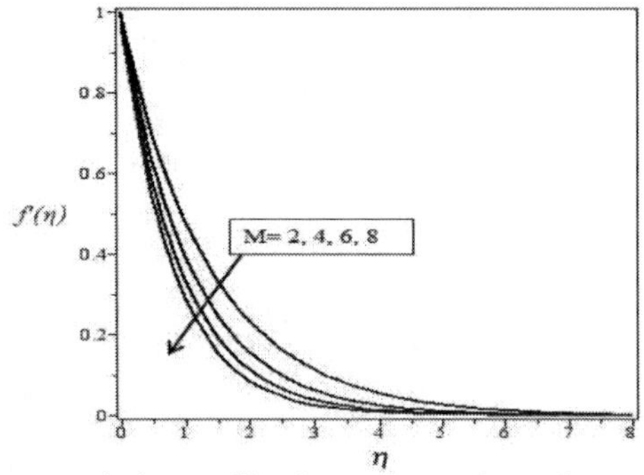

Fig. 4. Velocity profiles for various values of magnetic parameter M, when $\beta=0.2, K=2$, $R=5$, $Nb=0.5, Nt=0.5$, $Sc=2$ and $\Pr=7$.

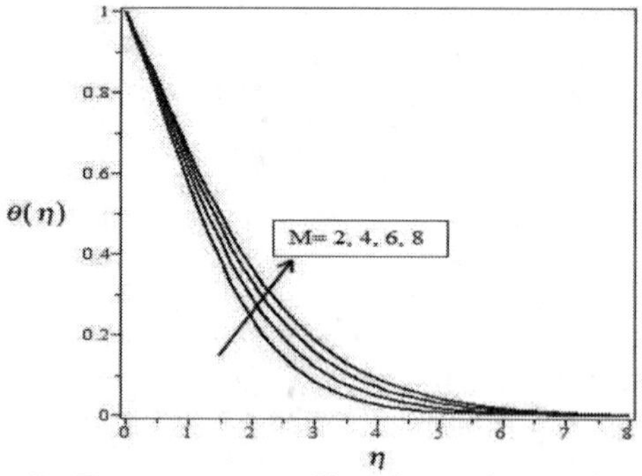

Fig. 5. Temperature profiles for various values of magnetic parameter M, when $\beta=0.2, K=2, R=5$, $Nb=0.5, Nt=0.5$, $Sc=2$ and $\Pr=7$.

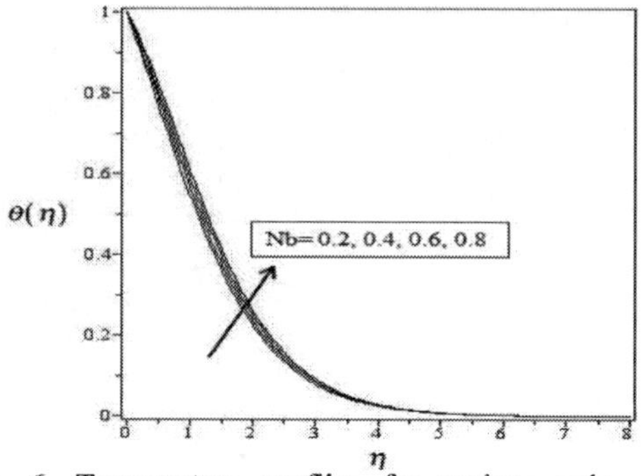

Fig. 6. Temperature profiles for various values of Brownian motion parameter Nb, when $M=0.2, K=2$, $R=5, \beta=0.2, Nt=0.5$, $Sc=2$ and $\Pr=7$.

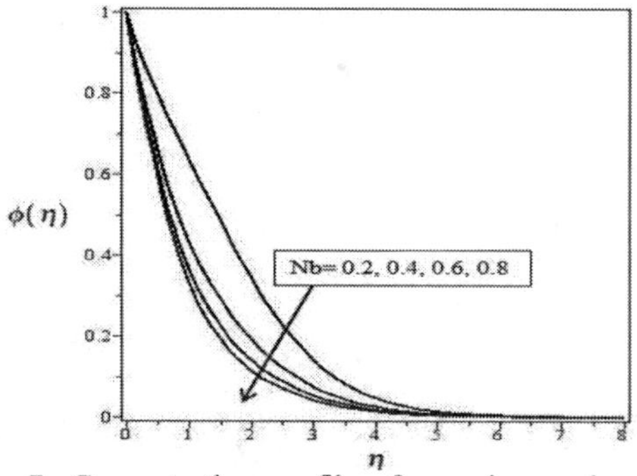

Fig. 7. Concentration profiles for various values of Brownian motion parameter Nb, when $M=0.2, K=2$, $R=5, \beta=0.2, Nt=0.5$, $Sc=2$ and $\Pr=7$.

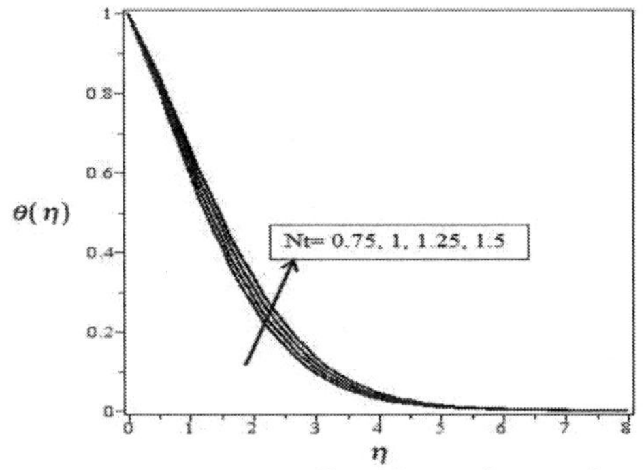

Fig. 8. Temperature profiles for various values of thermophoresis parameter Nt, when $M = 0.2, K = 2$, $R = 5, \beta = 0.2, Nb = 0.5$, $Sc = 2$ and $\Pr = 7$.

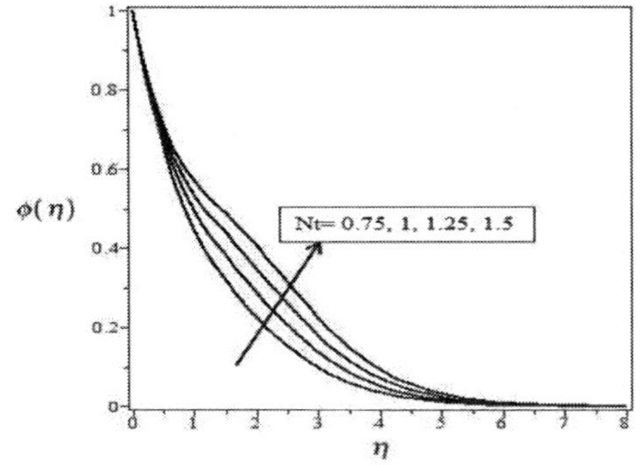

Fig. 9. Concentration profiles for various values of thermophoresis parameter Nt, when $M = 0.2, K = 2$, $R = 5, \beta = 0.2, Nb = 0.5$, $Sc = 2$ and $\Pr = 7$.

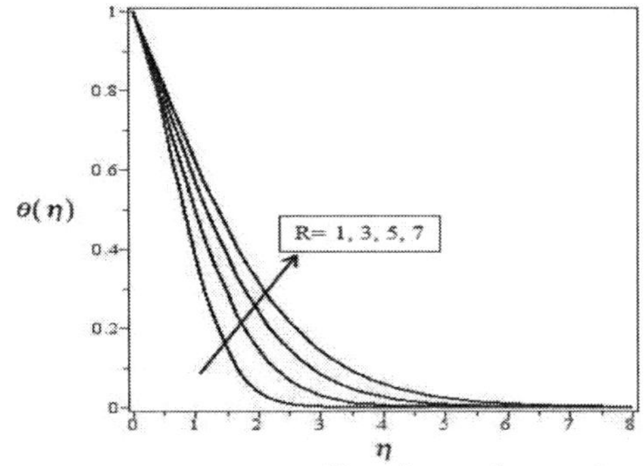

Fig. 10. Temperature profiles for various values of radiation parameter Pr, when $M = 0.2, K = 2$, $\Pr = 7, Nt = 0.5, \beta = 0.2, Nb = 0.5$ and $Sc = 2$.

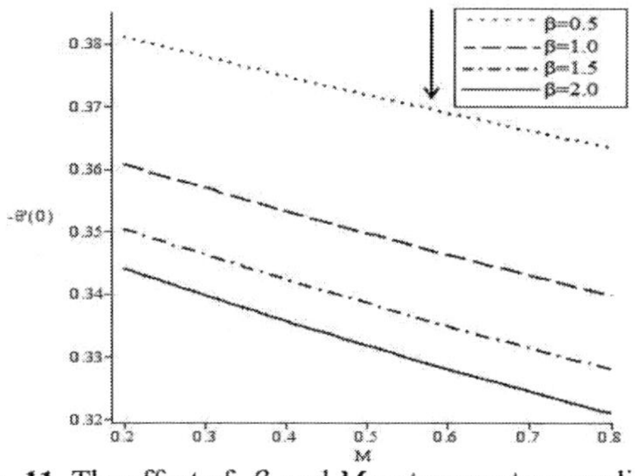

Fig. 11. The effect of β and M on temperature gradient.

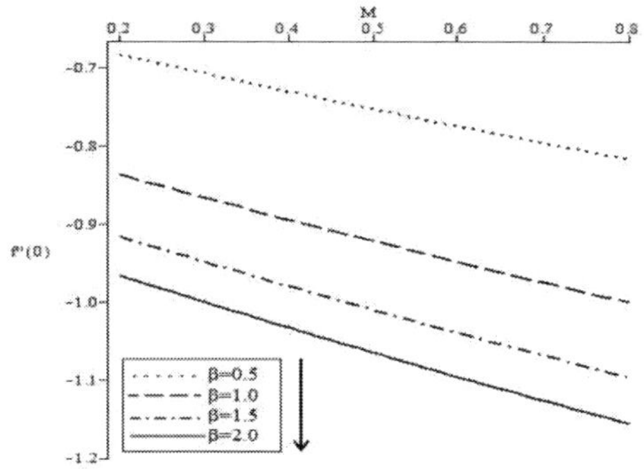

Fig. 12. The effect of β and M on skin friction coefficient.

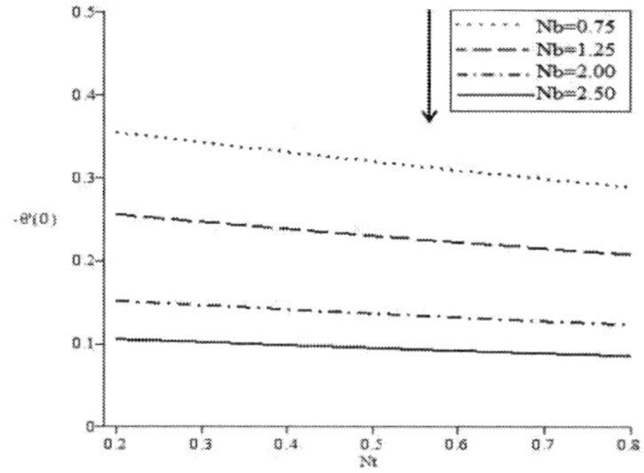

Fig 13. The effect of Nb and Nt on temperature gradient.

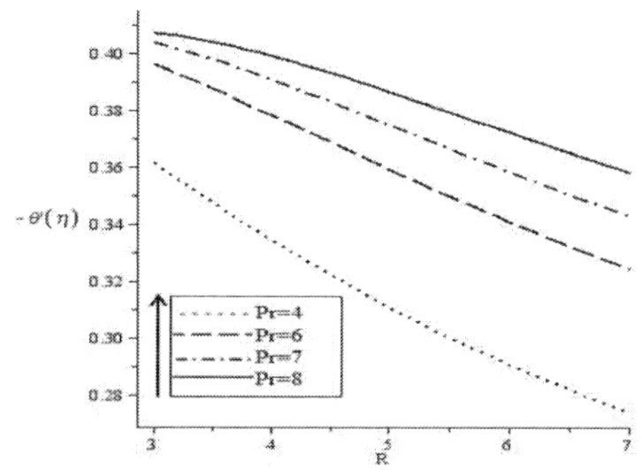

Fig. 14. The effect of Pr and R on temperature gradient.

4 Conclusions

Thermal radiation effects on MHD flow and heat mass transfer over stretching sheet in Casson nanofluid is studied numerically. The nonlinear ordinary differential equations are solved numerically using the Shooting method. The result of velocity, temperature and concentration profiles are then presented graphically for different values of the pertinent parameters. Also, the skin friction coefficient, local Nusselt number and Sherwood number are examined with various parameters. The results are summarized as below:

- Boundary layer thickness increase in temperature and concentration profiles but velocity profiles decreases when Casson nanofluid parameter, magnetic parameter and porosity parameter increase.
- Increase in Brownian motion parameter causes the concentration profiles to drop but the wall temperature increases.
- Increases thermophoresis parameter also causes temperature and concentration profiles increase.
- Temperature profile increases as the radiation parameter increases.
- Skin friction coefficient and local Nusselt number decreases at all the time as β and M increase in the fluid.
- Local Nusselt number decreases and Sherwood number increases as Nb and Nt increase in the fluid.
- The higher the value of $\Pr$ and R, the lower the value of local Nusselt number.

We would like to thank Universiti Malaysia Pahang for the financial support from Ministry of Higher Education Malaysia (RDU170358 and RDU150101).

References

1. S. Choi, ASME. **66,** 7 (1995)
2. J. Buongiorno, J. Heat Trans-t ASME **128,** 11 (2006)
3. S. Ahmed, I. Pop, Int. Commun. Heat Mass **37**, 5 (2010)
4. R.K. Tiwari, M.K. Das, Int. J. Heat Mass Trans. **50,** 9 (2007)
5. N.A. Yacob, A. Ishak, I. Pop, K. Vajravelu, Nanoscale Research Letters. **6**, 8 (2011)
6. A.J. Chamkha, A.M. Aly, Chem. Eng. Commun. **198**, 17 (2011)
7. A. Noghrehabadi, R. Pourrajab, M. Ghalambaz, Int. J. Therm. Sci. **54**, 28 (2012)
8. A. Mahdy, Nucl. Eng. Des. **249**, 8 (2012)
9. M. Qasim, I. Khan, S. Shafie, Math. Probl. Eng. 7 (2013)
10. P. Loganathan, C. Vimala, Indian J. Phy. **88**, 5 (2014)
11. B. Ganga, S.M.Y. Ansari, N.V Ganesh, A.K.A. Hakeem, J. Taibah Uni. Sci. (2016)
12. M.A.A. Hamad, I. Pop, A.I.M. Ismail, Nonlinear Analysis: Real World Applications. **12**, 19 (2011)
13. R. Ellahi, Appl. Math. Model. **37**, 17 (2013)
14. M. Qasim, Z.H. Khan, W.A. Khan, I.A. Shah, Plos One. **9**, (2014)
15. A. Hussanan, I. Khan, H. Hashim, M.K.A Anuar, N. Ishak, N.M. Sarif, M.Z. Salleh, J. Teknologi. **78**, 6 (2016)
16. P.M. Krishna, V. Sugunamma, N. Sandeep, Chem. Process Eng. Res. **25**, 14 (2014)
17. R.U. Haq, Z.H. Khan, W.A. Khan, Phys. Eng. **63**, 8 (2014)
18. M.Y. Malik, M. Naseer, S. Nadeem, A. Rehman, Appl. Nanoscience. **4**, 5 (2014)
19. A. Hussanan, M.Z. Salleh, I. Khan, R.M. Tahar, Z. Ismail, Maejo Int. J. Sci. Tech. **9**, 22 (2015)
20. M.H.M. Yasin, A. Ishak, I. Pop, J. of Magnetism and Magnetic Materials. **407**, 6 (2016)
21. W.A. Khan, I. Pop, Int. J. Heat Mass Trans. **53**, 7 (2010)
22. R.S.R. Gorla, I. Sidawi, Appl. Scientific Res. **52**, 11 (1994)
23. C.Y. Wang, ZAMM J. of Appl. Math. and Mechanics, **69**, 3 (1989)

The Cascaded Enhanced *k*-Means and Fuzzy *c*-Means Clustering Algorithms for Automated Segmentation of Malaria Parasites

Aimi Salihah Abdul Nasir[1,*], *Haryati* Jaafar[1], *Wan Azani* Wan Mustafa[1] and *Zeehaida* Mohamed[2]

[1]Faculty of Engineering Technology, Universiti Malaysia Perlis, UniCITI Alam Campus, Sungai Chuchuh, 02100 Padang Besar, Perlis, Malaysia
[2]Department of Microbiology & Parasitology, School of Medical Sciences, Health Campus, Universiti Sains Malaysia, 16150 Kubang Kerian, Kelantan, Malaysia

Abstract. Malaria continues to be one of the leading causes of death in the world, despite the massive efforts put forth by World Health Organization (WHO) in eradicating it, worldwide. Efficient control and proper treatment of this disease requires early detection and accurate diagnosis due to the large number of cases reported yearly. To achieve this aim, this paper proposes a malaria parasite segmentation approach via cascaded clustering algorithms to automate the malaria diagnosis process. The comparisons among the cascaded clustering algorithms have been made by considering the accuracy, sensitivity and specificity of the segmented malaria images. Based on the qualitative and quantitative findings, the results show that by using the final centres that have been generated by enhanced *k*-means (EKM) clustering as the initial centres for fuzzy *c*-means (FCM) clustering, has led to the production of good segmented malaria image. The proposed cascaded EKM and FCM clustering has successfully segmented 100 malaria images of *Plasmodium Vivax* species with average segmentation accuracy, sensitivity and specificity values of 99.22%, 88.84% and 99.56%, respectively. Therefore, the EKM algorithm has given the best performance compared to *k*-means (KM) and moving *k*-means (MKM) algorithms when all the three clustering algorithms are cascaded with FCM algorithm.

1 Introduction

Malaria is an infectious disease caused by the Plasmodium blood parasite, with high prevalence in tropical and subtropical regions. World Health Organization (WHO) has given the analysis report for the year 2015 as around 214 million peoples on earth suffers from malaria, with 90% of the victims were from African Region amassing a total of 438,000 death cases [1]. Across the globe, it was recorded that 306,000 children under the age of five lost their lives due to malaria, with 292,000 of them are from Africa [1].

Malaria develops to become life-threatening without immediate action. Currently, microscopy-based diagnosis is the most widely used approach to diagnose malaria. The manual microscopic diagnosis is often characterized by its high sensitivity and accuracy. However, its time-consuming and labour intensive characteristics seem to outweigh benefits as it consumes approximately 20 to 60 minutes to examine each slide [2]. The final diagnosis ultimately depends on the ability and experience of the experts, and may be prone to human error [3].

Recognition of malaria parasite requires the segmentation of the malaria parasite or infected red blood cells (RBCs) in malaria image. However, the segmentation process is difficult since most of the parasites are often a part of the RBCs that they have infected; presented in a complex and irregular visual aspect. As a result, various applications of image processing techniques for segmenting the malaria parasites have been reported to necessitate the malaria diagnosis process. These include the use of thresholding [4-5], marker controlled watershed with morphological approach [6-7], fuzzy rule base segmentation [8], active contour base segmentation [9] and clustering approach [10-11] as the latest trend.

Clustering is an unsupervised procedure that aims to group a set of data into meaningful groups, based on the distances between data points. *K*-means, fuzzy *c*-means and mean shift are among the popular clustering algorithms that have been utilized to segment the region of interest in various types of medical images. *K*-means (KM) clustering [12] employs the concept of data allocation to the nearest cluster centre based on the shortest Euclidean distance measurements.

Though KM algorithm is very popular for general clustering, this algorithm is sensitive to cluster centres initialization and has the dead centre problem [13]. The limitations of KM algorithm could lead to poor representation of data. As a result, several clustering algorithms such as fuzzy *c*-means and enhanced *k*-means algorithms have been proposed to overcome the limitations of KM algorithm.

* Corresponding author: aimisalihah@unimap.edu.my

In fuzzy *c*-means (FCM) clustering [14], each data is assigned a membership grade indicating the degree of belonging of the data to each centre rather than assigning each data to one centre as in KM algorithm. The centre is an average of all data points in the cluster, weighted by their degree of belonging of the data to the centre. Substantially, these attributes could minimize the dead centre problem in FCM algorithm. However, this algorithm is still sensitive to cluster centres initialization [15]. The limitations of FCM algorithm could lead to poor segmentation result regardless of the resultant cell images being under-segmented or over-segmented as shown in [16-17].

Another type of clustering algorithm that has been proven to produce good segmentation performance on medical image is enhanced *k*-means (EKM) clustering [18] algorithm. In EKM clustering, the concept of variance and a new version of transferring process for clustered members are used to assist the assignation of data to the proper centre during the process of clustering, so that good segmented image can be generated. In addition, comparisons of different image segmentation techniques show that EKM algorithm has successfully segmented the malaria parasite from the overlapping RBCs compared to the results of KM and FCM clustering [18].

Based on the advantages of EKM algorithm in image segmentation, the current study will utilize its potential by cascading the EKM with FCM algorithm. In addition, the other two popular clustering algorithms which are KM and moving *k*-means (MKM) [19] will also be cascaded with FCM algorithm to compare the performance of segmentation among these three cascaded algorithms.

2 Methodology

2.1 Image Acquisition

The malaria thin blood smear samples were obtained from Department of Microbiology and Parasitology, Hospital Universiti Sains Malaysia (HUSM), Malaysia. The malaria images were acquired by using a Luminera Infinity-2 digital camera mounted to the *Leica DLMA 1200* digital microscope with 100X magnifications. Images were captured in the bitmap (*.bmp) format at the resolution of 800×600 pixels. This study have analysed a total of 100 images obtained from eight malarial *Plasmodium Vivax* slides. The images consist of malaria parasites in ring, trophozoite, schizont and gametocyte stages.

2.2 The Proposed Cascaded Enhanced *k*-Means and Fuzzy *c*-Means Clustering Algorithms

As previously mentioned, FCM clustering algorithm was introduced as a fuzzy version of KM algorithm for reducing the weaknesses of KM. However, its performance is depending on the choice of the initial cluster centre. If a good initial cluster centre that is close to the actual final cluster centre is used, the FCM algorithm will converge faster and the processing time can be deducted. Without a proper initialization, a set of poor final cluster centres would be generated by this algorithm hence producing poor clustering results. Even though FCM is sensitive to cluster centre initialization, the concepts of fuzziness makes this algorithm becomes more flexible in producing the best possible centre.

Thus, to take advantage of EKM and FCM clustering algorithms, the cascaded EKM and FCM clustering algorithm is proposed. Technically, the cascaded clustering algorithm combines the fundamental theory of EKM algorithm which employs the variance concept for data allocation to its cluster centre, with the concept of FCM algorithm that allows each data to be simultaneously assigned to multiple clusters by different degrees of membership. Thus, the final centres generated by EKM clustering will become the initial centres for FCM clustering. Besides, fuzzy membership function is used along with Euclidean distance and variance calculations, in order to assist the assignation of the members to the proper centre during the clustering process.

In order to describe the cascaded EKM and FCM clustering algorithms, consider an image with resolution of $X \times Y$ pixels to be clustered into n_c regions. Let $p(x,y)$ as an input pixel to be clustered and c_j is the j-th centre (cluster) ($x = 1, 2, \ldots, X$, $y = 1, 2, \ldots, Y$ and $j = 1, 2, \ldots, n_c$). The cascaded EKM and FCM clustering algorithms for image segmentation is described as follows:

1. Initialize the centres using:

$$c_j = min_{p(x,y)} + (2j-1)\left(\frac{max_{p(x,y)} - min_{p(x,y)}}{2n_c}\right) \quad (1)$$

   where $min_{p(x,y)}$ and $max_{p(x,y)}$ are the minimum and maximum pixel intensity in the image.
2. Set the values of $\alpha_0 = 0.05$, $\alpha_b = 0.05$ and $\alpha_a = \alpha_0$, where α_0 and α_b are small constant values.
3. For each pixel of an image, calculate the Euclidean distance, d using:

$$d = \left\| p(x,y) - c_j \right\| \quad (2)$$

4. Assign all pixels to the nearest centre based on d and calculate the centre positions using:

$$c_j = \frac{1}{n_j}\sum_{y \in c_j}\sum_{x \in c_j} p(x,y) \quad (3)$$

5. Check the variance between the members with their centre using:

$$\sigma^2(c_j) = \frac{1}{n_j}\sum_{y \in c_j}\sum_{x \in c_j}\left(p(x,y) - c_j\right)^2 \quad (4)$$

   Note: The variance calculation is applied for measuring the relationship between the centre and its members.

6. Find c_s and c_l, the centre that has the smallest and the largest value of variance between the members with their centre.
7. If $\sigma^2(c_s) < \alpha_a \sigma^2(c_l)$,
   a. Find c_n, the nearest centre to c_l by calculating the Euclidean distance between the c_l and other centres.
   b. If $c_n < c_l$, assign the members of c_l to the nearest centre if $p(x,y) < c_l$, where $x,y \in c_l$, and leave the rest of the members to c_l.
   c. Else if $c_n > c_l$, assign the members of c_l to the nearest centre if $p(x,y) > c_l$, where $x,y \in c_l$, and leave the rest of the members to c_l.

   Note: The transferring process of c_l members in (b) and (c) is performed in order to reduce the variance of c_l by giving up members for $p(x,y) < c_l$ and $p(x,y) > c_l$, respectively to the appropriate cluster centre so that the variance of c_l will be kept at a reasonable value.

   d. Recalculate the positions of c_l and c_n according to:

$$c_l = \frac{1}{n_l} \sum_{y \in c_l} \sum_{x \in c_l} p(x, y) \quad (5)$$

$$c_n = \frac{1}{n_n} \sum_{y \in c_n} \sum_{x \in c_n} p(x, y) \quad (6)$$

   Note: n_l and n_n in (5) and (6) are the number of the new members of c_l and c_n respectively, after the reassigning process in step 7d.

8. Update α_a according to $\alpha_a = \alpha_a - \alpha_a / n_c$ and repeat steps 6 and 7 until $\sigma^2(c_s) \geq \alpha_a \sigma^2(c_l)$.
9. Reassign all pixels to the nearest centre and recalculate the centre positions using Eq. 3.
10. Update α_a and α_b according to $\alpha_a = \alpha_0$ and $\alpha_b = \alpha_b - \alpha_b / n_c$ respectively, and repeat steps 5 to 9 until $\sigma^2(c_s) \geq \alpha_b \sigma^2(c_l)$.
11. Calculate membership function, $M_{jp(x,y)}$ according to Eq. 7 to 9.

$$M_{jp(x,y)} = 1 \Bigg/ \sum_{k=1}^{n_c} \left(d_{jp(x,y)} / d_{kp(x,y)} \right)^2 \quad (7)$$

where,

$$d_{jp(x,y)} = \left\| p(x,y) - c_j \right\| \text{ and } d_{kp(x,y)} = \left\| p(x,y) - c_k \right\| \quad (8)$$

$$\left. \begin{array}{l} M_{kp(x,y)} = 1 \\ M_{jp(x,y)} = 0; \text{for } p(x,y) \neq k \end{array} \right\} \text{if } d_{kp(x,y)} = 0 \quad (9)$$

12. Calculate the centre positions using:

$$c_j = \sum_{y \in c_j} \sum_{x \in c_j} M^2_{jp(x,y)} p(x,y) \Bigg/ \sum_{y \in c_j} \sum_{x \in c_j} M^2_{jp(x,y)} \quad (10)$$

In order to ensure a better clustering process, the basic concept of EKM algorithm suggests that each cluster must have strong relationship strength between the members with their centre so that the differences between them could be minimized. After the final centres produced by EKM algorithm are obtained, these centres will be fed as initial values to FCM algorithm for further clustering process. In FCM algorithm, the fuzziness concept is used for allowing each data to assign a membership grade indicating the degree of belonging of the data to each centre simultaneously. However, in the proposed cascaded EKM and FCM clustering algorithms, steps 11 and 12 are used once and not be repeated as in the conventional FCM algorithm. This is done in order to test the capability of each single clustering algorithm in producing good initial cluster centre to FCM algorithm. If a good initial cluster centre is used, hence a set of good final cluster centres would be generated by FCM algorithm after performing steps 11 and 12. Therefore, the proposed cascaded EKM and FCM clustering algorithms combines the characteristics of both clustering algorithms to produce the best clustering process.

2.3 Segmentation of Malaria Parasites

In order to obtain the segmented malaria parasites, the image segmentation process has been conducted based on the procedures as proposed in [20]. In [20], these procedures were used for comparing the performance of malaria parasite segmentation using different colour components. Meanwhile, this study utilizes the similar segmentation procedures for comparing the performance of three cascaded clustering algorithms in segmenting the malaria image. The three cascaded clustering algorithms are obtained by cascading KM, MKM and EKM with FCM clustering algorithm. The procedures for segmentation of malaria parasite are illustrated in Fig. 1.

Initially, the pre-processing step is conducted by applying the modified global contrast stretching (MGCS) technique to generate high contrast image. Then, segmentation process using the three cascaded clustering algorithms is performed on green component image to isolate the RBCs and background regions, followed by 7×7 pixels median filter for noise reduction. Finally, seeded region growing area extraction (SRGAE) algorithm is used to remove the large unwanted regions such as artefacts, platelets and RBCs in the image.

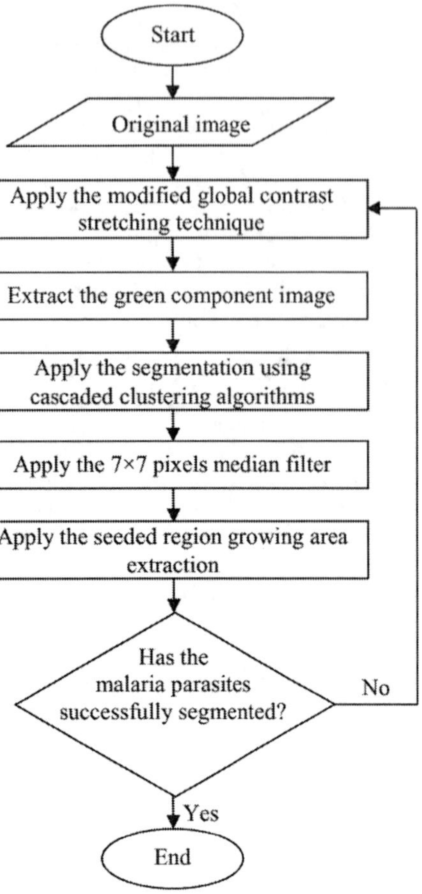

Fig. 1. The procedures for segmentation of malaria parasite.

3 Results and Discussions

The three cascaded clustering algorithms have been proposed in order to combine the advantages of KM, MKM and EKM with the FCM clustering. Therefore, comparisons among these three cascaded clustering algorithms have been made in order to measure the performance of each cascaded clustering algorithm for image segmentation. Figs. 2(a) and (b) show the original malaria images of Schizont and Gametocyte, respectively. The results obtained after applying the MGCS technique with min_P=0.5% and max_P=10% on the two malaria images are shown in images (c) and (d).

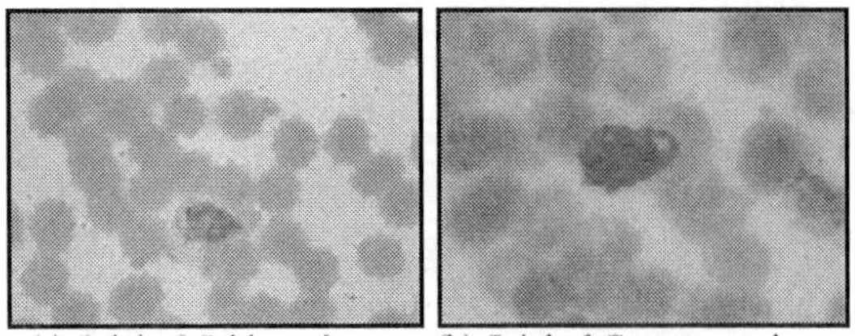

(a) Original Schizont image (b) Original Gametocyte image

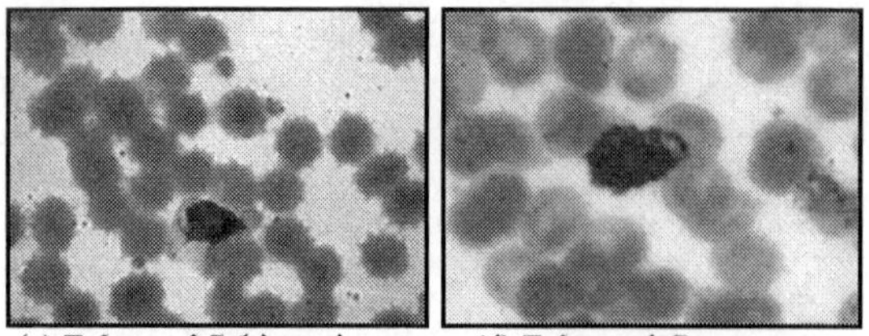

(c) Enhanced Schizont image (d) Enhanced Gametocyte image

Fig. 2. The original and enhanced malaria images.

In order to easily isolate the malaria parasite from the RBCs and background regions, the green component information has been extracted from the enhanced RGB image. Then, FCM and the three cascaded clustering algorithms have been applied for segmenting the malaria image. Fig. 3 shows the results of images after applying the clustering algorithms on the green component image.

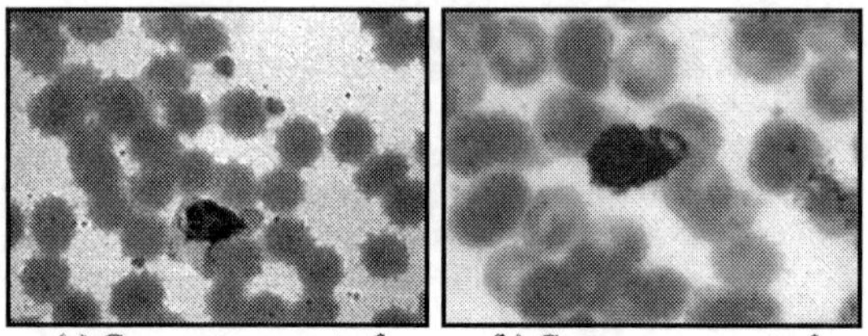

(a) Green component of Schizont image (b) Green component of Gametocyte image

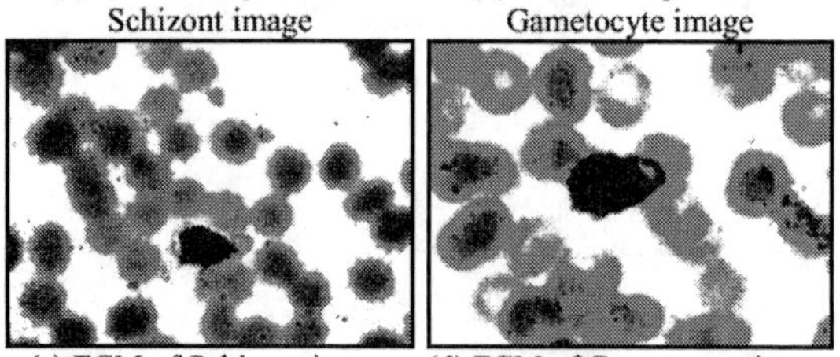

(c) FCM of Schizont image (d) FCM of Gametocyte image

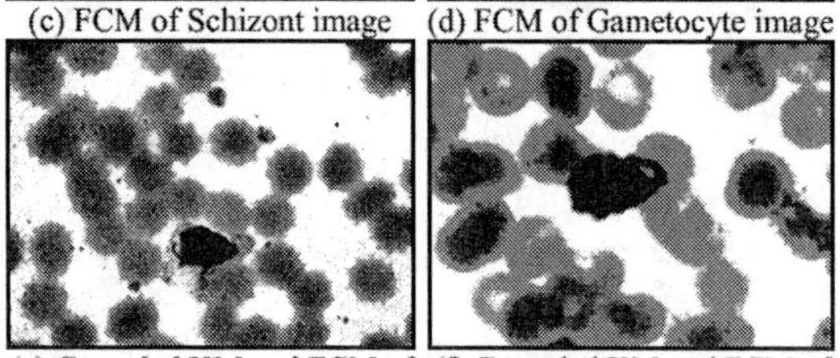

(e) Cascaded KM and FCM of Schizont image (f) Cascaded KM and FCM of Gametocyte image

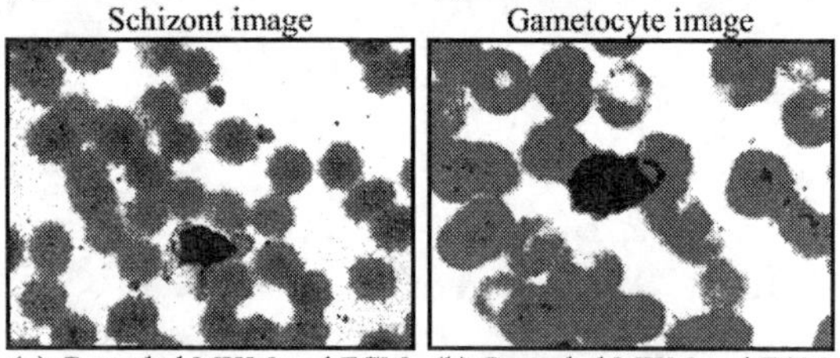

(g) Cascaded MKM and FCM of Schizont image (h) Cascaded MKM and FCM of Gametocyte image

https://doi.org/10.1051/matecconf/201815006037

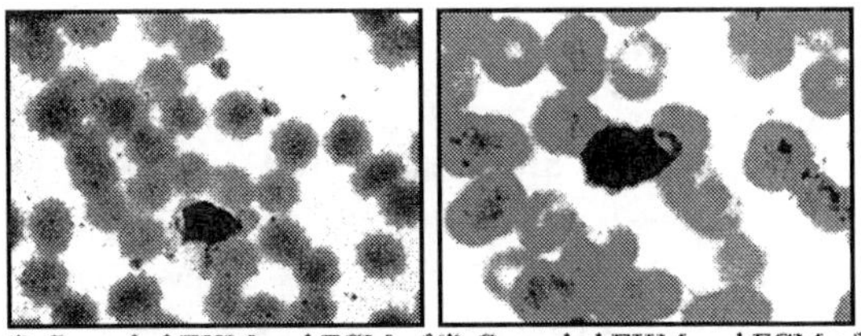

(i) Cascaded EKM and FCM of Schizont image (j) Cascaded EKM and FCM of Gametocyte image

Fig. 3. Results of images for the four clustering algorithms.

Later, the resultant images will undergo filtering process using 7×7 pixels median filter for noise reduction and image smoothing. Finally, SRGAE algorithm has been applied to remove any unwanted regions which are outside the ranges of 3500 to 30000 pixels from the image and the results are shown in Fig. 4. Based on the final segmented images, segmentation using the cascaded KM and FCM clustering is unable to produce a clean segmented malaria image, as the malaria parasite and some of the RBCs component have been clustered into a single region. This over-segmentation is due to pixel similarity of green component between the malaria parasite and RBCs regions especially in the case of overlapping RBCs. In contrast, the cascaded MKM and FCM, as well as the cascaded EKM and FCM clustering algorithms are capable to produce a clean segmented malaria image without or less appearance of RBCs region in which can be seen in the resultant images of cascaded KM and FCM clustering.

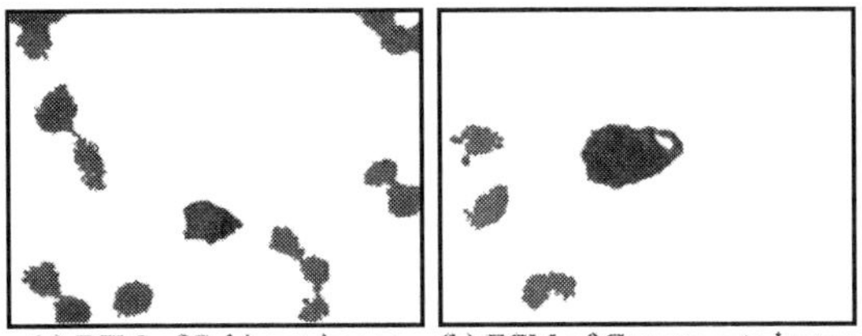

(a) FCM of Schizont image (b) FCM of Gametocyte image

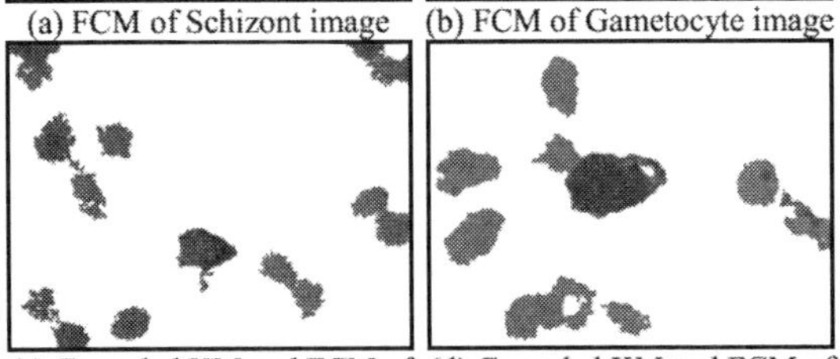

(c) Cascaded KM and FCM of Schizont image (d) Cascaded KM and FCM of Gametocyte image

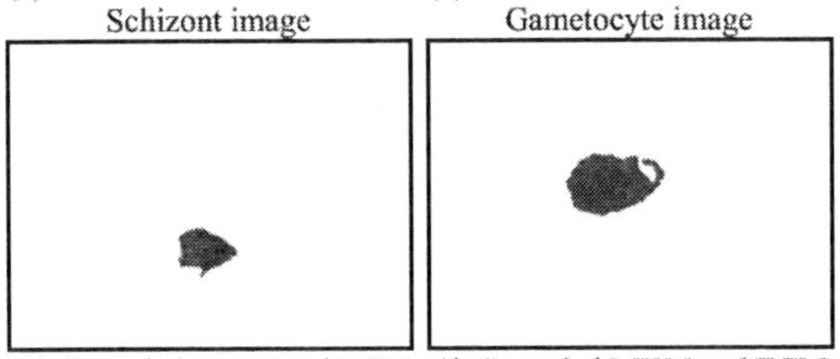

(e) Cascaded MKM and FCM of Schizont image (f) Cascaded MKM and FCM of Gametocyte image

(g) Cascaded EKM and FCM of Schizont image (h) Cascaded EKM and FCM of Gametocyte image

Fig. 4. Results of final segmented images for Schizont and Gametocyte.

In order to ascertain segmentation performances provided in the previous results, the quantitative readings of 100 malaria images are collected. Table 1 summarizes the average segmentation performances based on sensitivity, specificity and accuracy that have been collected from 100 malaria images. The final segmented images are compared to manually segmented images to quantify the segmentation performance for clustering algorithm. The accuracy of segmentation can be obtained by computing the percentage of pixels that are correctly segmented as malaria parasite or background region in the image. The sensitivity is obtained by computing the percentage of pixels that are correctly segmented as positive region, while the specificity is obtained by calculating the percentage of pixels that are correctly segmented as negative region. Higher pixel similarity indicates better segmentation performance.

Based on Table 1, segmentation using the cascaded KM and FCM clustering has proven to be the best in obtaining the fully segmented malaria parasite with sensitivity of 89.9816%. Even though the sensitivity result that has been achieved is high, segmentation using this algorithm is unable to produce a clean segmented malaria image. In contrast, segmentation using the cascaded MKM and FCM clustering has proven to be the best in producing a clean segmented malaria image with specificity of 99.5603%, which is slightly higher than cascaded EKM and FCM clustering (99.5560%).

As for segmentation accuracy, segmentation using the cascaded EKM and FCM clustering has proved to be slightly better than the cascaded MKM and FCM clustering in segmenting the entire area in malaria image with accuracy of 99.2196%. In this study, the main criterion for evaluating the performance of each clustering algorithm is based on its ability to produce fully segmented malaria parasite region as well as separating the malaria parasite from the overlapping RBCs. Thus, the cascaded EKM and FCM clustering algorithm has given the best performance by providing good sensitivity as well as high specificity and accuracy values compared with the results provided by other two cascaded clustering algorithms. This result also shows that by using the final centres that are generated by EKM clustering as the initial centres for FCM clustering, has resulted in good segmentation of malaria images.

Table 1. Segmentation performances for the segmented malaria images.

Analysis	Segmentation Technique	Schizont	Gametocyte	Average of 100 Images
Sensitivity (%)	FCM	95.6705	92.1226	89.4300
	Cascaded KM and FCM	**98.6198**	**95.5749**	**89.9816**
	Cascaded MKM and FCM	96.9926	87.3575	87.6128
	Cascaded EKM and FCM	97.4575	88.6024	88.8437
Specificity (%)	FCM	90.4158	97.1253	98.2454
	Cascaded KM and FCM	89.8885	88.1643	97.3182
	Cascaded MKM and FCM	**99.9242**	**99.9835**	**99.5603**
	Cascaded EKM and FCM	99.9172	99.9750	99.5560
Accuracy (%)	FCM	90.4894	96.9306	97.9489
	Cascaded KM and FCM	90.0109	88.4526	97.0682
	Cascaded MKM and FCM	99.8812	99.4921	99.1770
	Cascaded EKM and FCM	**99.8828**	**99.5324**	**99.2196**

4 Conclusions

This paper has proposed an image segmentation technique for malaria images, by cascading two clustering algorithms which are EKM and FCM algorithms. Overall, the EKM algorithm has given the best performance compared to KM and MKM algorithms when all the three algorithms are cascaded with FCM algorithm. In addition, a better segmentation result has been achieved by cascading two clustering algorithms as compared of using a single clustering algorithm for image segmentation. Here, the segmentation performances using 3 clusters for the cascaded EKM and FCM algorithms is found to be slightly effective than the EKM algorithm with average segmentation accuracy, sensitivity and specificity values of 99.22%, 88.84% and 99.56%, respectively. Overall, these findings prove that the proposed cascaded EKM and FCM clustering algorithms have good potential in the application of malaria image segmentation as compared to the other segmentation techniques.

The authors gratefully acknowledge and thank Hospital Universiti Sains Malaysia (HUSM) for providing the malaria blood samples and validating the results. We also would like to acknowledge Universiti Malaysia Perlis (UniMAP) for providing the financial support under Short Term Grant Scheme (Grant No. 9001-00514).

References

1. *World malaria report 2015* (World Health Organization, Switzerland, 2016)
2. B.S. Kakkilaya, Lab Medicine **34**, 602 (2003)
3. K. Mitiku, G. Mengistu, B. Gelaw, Ethiopian Journal of Health Development **17**, 197 (2003)
4. V.V. Panchbhai, L.B. Damahe, A.V. Nagpure, P.N. Chopkar, International Journal of Image, Graphics and Signal Processing **4**, 54 (2012)
5. J.E. Arco, J.M. Górriz, J. Ramírez, I. Álvarez, C.G. Puntonet, Expert Systems with Applications **42**, 3041 (2015)
6. J. Soni, Int J Eng Sci Technol **3**, 5260 (2011)
7. D.K. Das, M. Ghosh, M. Pal, A.K. Maiti, C. Chakraborty, Micron **45**, 97 (2013)
8. C. Dallet, S. Kareem, I. Kale, *2014 IEEE International Symposium on Circuits and Systems (ISCAS)* (2014)
9. X. Gual-Arnau, S. Herold-García, A. Simó, Med Biol Eng Comput **53**, 623 (2015)
10. J. Somasekar, B.E. Reddy, Computers & Electrical Engineering **45**, 336 (2015)
11. S.S. Savkare, A.S. Narote, S.P. Narote, *Proceedings of the Third International Symposium on Computer Vision and the Internet* (2016)
12. J. MacQueen, *Proceedings of 5th Berkeley Symposium on Mathematical Statistics and Probability* (1967)
13. S.S. Khan, A. Ahmad, Pattern Recognition Letters, **25**, 1293 (2004)
14. J.C. Bezdek, *Pattern recognition with fuzzy objective function algorithms* (Plenum Press, New York, 1981)
15. M.C. Hung, D.L. Yang, *Proceedings 2001 IEEE International Conference on Data Mining* (2001)
16. N.A.M. Isa, M.Y. Mashor, N.H. Othman, *Proceeding of International Conference on Robotics, Vision, Information and Signal Processing* (2003)
17. N.H. Harun, M.Y. Mashor, H. Rosline, IFMBE Proceedings (2011)
18. A.S. Abdul Nasir, M.Y. Mashor, Z. Mohamed, Journal of Telecommunication, Electronic and Computer Engineering (JTEC) (to be published)
19. M.Y. Mashor, International Journal of the Computer, The Internet and Management **8**, 50 (2000)
20. A.S. Abdul-Nasir, M.Y. Mashor, Z. Mohamed, Journal of Biomimetics, Biomaterials and Biomedical Engineering **24**, 43 (2015)

Classification of Security Attacks in VANET: A Review of Requirements and Perspectives

Mohammed Ali Hezam Al Junaid[1], Syed A. A[1], Mohd Nazri Mohd Warip[1], Ku Nurul Fazira Ku Azir[1], Nurul Hidayah Romli[1]

[1] School of Computer and Communication Engineering University Malaysia Perlis, Malaysia.

Abstract. Vehicular Ad Hoc Network (VANET) is a pillar of the envisioned Intelligent Transport System (ITS) and a subset of Mobile Ad Hoc Network that grants the communication in between the vehicles alongside with the absence of established communication infrastructure. Exposure to vulnerabilities of Vehicular Ad-Hoc Network (VANET) has been shown to be related to its nature of the environmental. For this reason, VANET security becomes a critical challenge that need to be resolved. In this paper, we assess the VANET security issues and discuss the challenges in VANET. Equally important, we comparatively review the security requirements, the type of attacks and capabilities of attackers present in VANET.

1 Introduction

Recently, Vehicular Ad Hoc Network (VANET) is fast becoming a key catalyst in the Internet of Things (IoT) area. In general, Vehicular Ad Hoc Network (VANET) is a portion of Mobile Ad Hoc Network (MANET) which receiving numerous attention from researchers and automotive industry [1]. In addition, VANET is capable in improving road safety by allowing each of vehicles on the road communicates to each other with inadequate fixed infrastructures [2]. VANET applications are graded into safety applications and non-safety applications [3]. The former relates to the safety-comfort application in which linked to the safety of users. It aids in providing an alert and warning information to the users regarding to any incidents occurs on the road such as accidents. The latter consists of the non-safety applications responsible in providing comfort to the users and acts as traffic-enhancer. Additionally, VANET provides legitimate information to the users on the road in order to increase the road and users safety. However, it is not guaranteed that vehicular network environment is get off from any jeopardized since VANET is exposed to the vulnerabilities.

It One of the key challenges in the implementation of VANET in relation to security is providing secure vehicular communication. The authors in [4] examined that there are lots of attacks and threat that can compromise the network and communication. The most potential attacks that VANET faces are classified into data threat and VANET system threat. Denial of service attacks is one of the malicious attacks that can deny the On-board units (OBU) or Road Side Units (RSU) from entering the network as well as interruption to the radio channels.

This paper is organized as follows; In Section 2, we discuss the architecture of VANET. Section 3 presents the characteristics of VANET. Section 4 reviews the taxonomy of security concept in VANET and divided into three subsections. The first sub-section reviews the type of the attackers, the second sub-section discuss the capabilities of the attackers while the last sub-section presents the series of attacks that compromised in VANET. In Section 5, we present the network challenges in VANET. Section 6 explains the security requirements in VANET. Conclusion is provided in Section 7.

2 The VANET Architecture

VANET can be categorized into three which are (1) Pure IV. TAXONOMY OF SECURITY CONCEPT IN VANET Cellular/WLAN architecture, (2) Pure ad hoc architecture and (3) Hybrid architecture. Figure 1 exhibits the network In VANET, vulnerabilities are presented due to the architecture in VANET wireless medium used. As a result, VANET are exposed to various kinds of attacks that could disrupt the communication operations between vehicles. Thus, the security concept in VANET as shown in Figure 2 can be classified into three which are (1) Attackers, (2) Attackers Capabilities and (3) Type of the Attacks. In sub-section 4.1, six types of the attackers are discussed. Sub-section 4.2 reviews four potentials that can.

* Corresponding author: aljunaid200@gmail.com

https://doi.org/10.1051/matecconf/201815006038

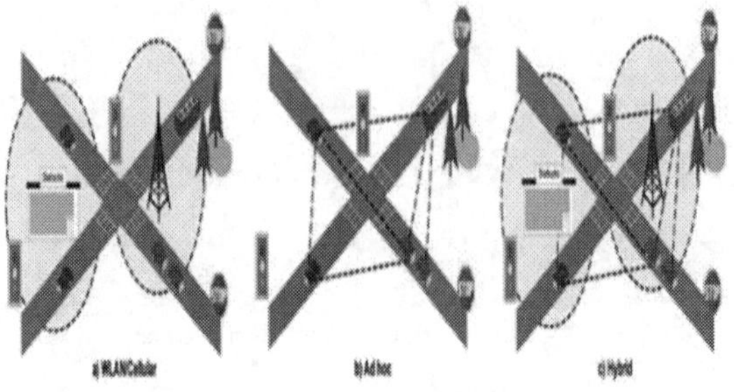

Fig. 1. The Network Architecture of VANET [3].

Generally, the communication and information exchanged in VANET independents from any fixed infrastructure. This possibly because of VANET environment is highly dynamic. According to authors in [3] proposed, the architecture of Fig. 1. The Network Architecture of VANET [3]

The communications patterns in VANET can be classified into.

- Vehicle-to-Vehicle (V2V) communications
- Vehicle-to-Infrastructure (V21) communications
- Infrastructure-to-Infrastructure (I2I) communications

3 VANET Characteristics

VANET is well-known as the subspace of MANET. However there are few characteristics of VANET that makes different from MANET, whereas VANET exhibited complexity in designing it as well as more challenges compared to MANET [3].

3.1. Frequent Disconnected Network

Vehicles are moving while exchanging information. Due to the rapid topology changes, the connections between two vehicles are easily disconnected. Usually the disconnections occur in infrequent networks.

3.2. Rapid Topology Changes

Due to the fast moving of vehicles, VANET topology changes quickly.

3.3. Battery Power and Storage Capacity

The communications in MANET consumes battery power conversely in VANET, the power and storage is boundless.

3.4. Communication Environment.

Obstacles in VANET are presented in dense network as well as sparse network. Trees, buildings, and other objects could obstruct the communications in VANET especially in dense network. For this reason, routing protocol for sparse and dense networks should be considered.

3.5. Mobility Modelling

The pattern of mobility in VANET is dependent on the traffic environment, vehicle's speed and driving behaviour. facilitate the attackers to launch the attacks. Sub-section 4.3 defines the type of attacks based on (1) Attacks on Confidentiality, (2) Attacks on Integrity, (3) Attacks on Availability, (4) Attacks on Authentication/Identification, and (5) Attacks on Privacy.

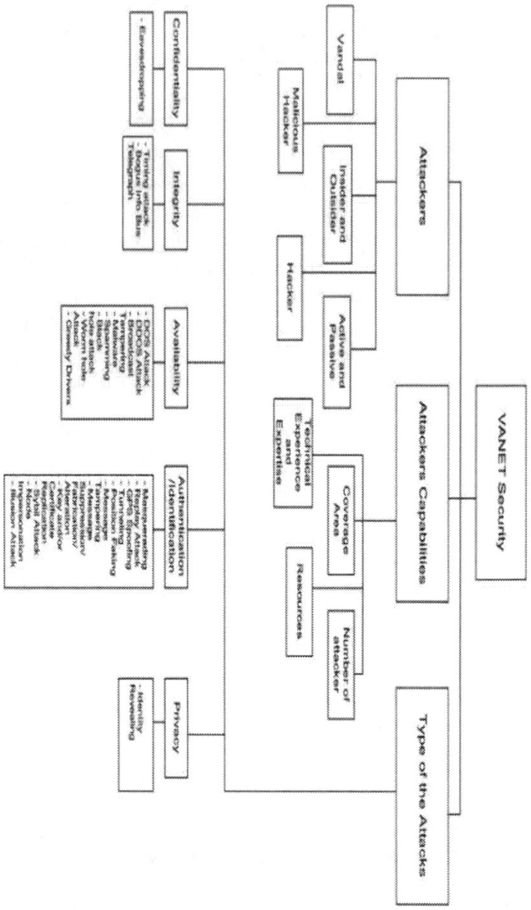

Fig. 2. Taxonomy of Security Concept in VANET

3.5.1. Type of the Attackers

According to the [4], attackers can be categorized based on the following category:

- Vandal: This kind of attacker is ill-motivated. The just want to show their abilities to attacks.
- Hacker: The hacker is motivated by the enthusiasm and interest without getting back any benefit from the attacking.
- Malicious hacker: The malicious hacker is driven by the monetary purposes of organizational or for personal/political gain.
- Insider vs. Outsider: The former relates to the attacker that is authenticated in the network and

known best about the network. The latter uses the attacker from the outside having a limited knowledge about the network that they want to penetrate.

3.5.2. Capabilities of Attackers

- Technical: Experience and Expertise With adequate experiences and skills, attackers are able to generate attacks against the network for example like extracting the program code and secret keys in order to launch the attacks.
- Resources: Attackers are dependent on three main key resources like tools, budget and manpower. Without these key resources, it is hard for the attackers to achieve their goals.
- Coverage Area: The coverage area of the attackers is depending on the attackers 'capabilities and nature of attacks. The rookie attackers could be controlled one of the Dedicated Short-Range Communication (DSCR) channel within the range of 1000 meters whilst the expertise could covered more area and DSCR channels compared to the rookie attackers.

3.5.3. Type of Attacks

The communication medium used in VANET is through the open air which make possible for attackers to penetrate and invade the networks. The real intentions of the attackers are to create problems with the legitimate vehicles in the network [5]. The attacks are classified as the following:

- Attacks on Confidentiality: Eavesdropping in VANET is targeting against the confidentiality and occurs in network layer. It operates by sniffing the conversation in between two nodes. As a result, this attack enables the attacker to intercept the communication, steal the password and important data. The attacker can masquerade itself as one of the node, or located itself as false RSU with the aim to catch up valuable information [6][7].
- Attacks on Integrity: Timing Attack, without altering any contents in the message, the attacker carries on this attack by adding delay which causes the user to receive the message behind the time. Consequently, user may face traffic congestion or even worse, accidents. It is important to realize that the information and messages deliveries in VANET should be received by the users at the appropriate time [3][8]. In addition, timing attack is divided into two levels which are (1) Basic Level and (2) Extended Level. Both levels are targeting users in the vehicular network, where the basic level only pointed to the user in peer-to-peer (P2P) communication while extended level is rigorous compared to basic level since it focuses on group of users [9]. Figure 1 and 2 shows the basic level of timing attack occurs in VANET.

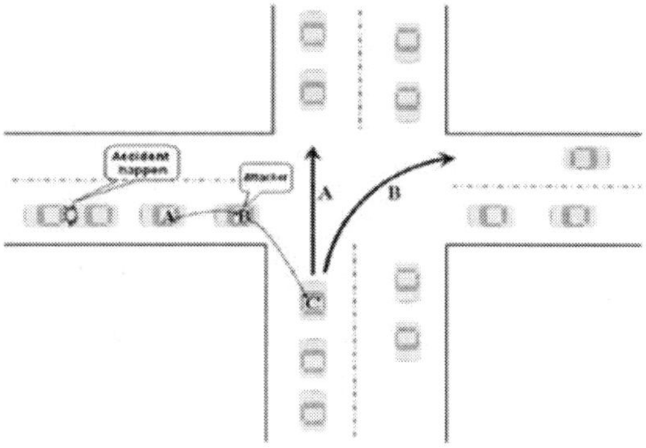

Fig. 3. Before Basic Level Timing Attack Occurs [9]

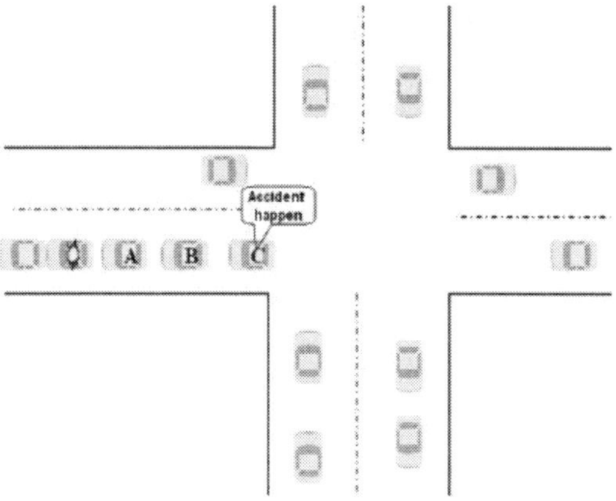

Fig.4. After Basic Level Timing Attack Occurs [9]

In Figure 3 presents the extended level of timing attack occurs in
VANET.

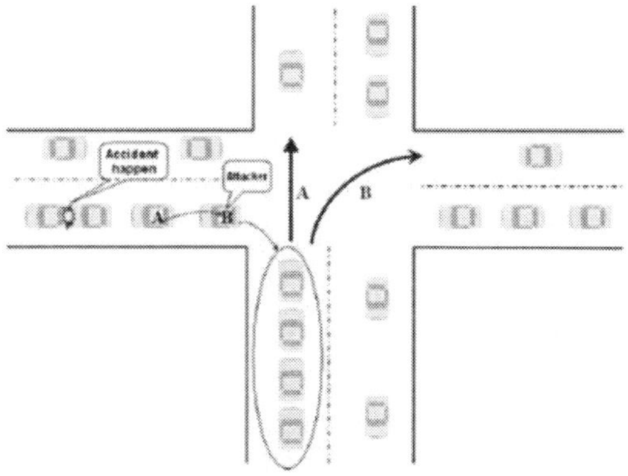

Fig. 5. Extended Level of Timing Attack [9]

Bogus Information and Bush Telegraph Bogus information attacks involving forwarding the counterfeit or false information throughout the network, intended to ignite disarray. In return, the attacker will gain the benefit from it [10].

- Attacks on Availability: Denial of Service Attack (DOS) Denial of Service attack is the most common intrusive attack against the availability. In fact, it can take place in each layer across the network. The purposes of the attacker could be denying the legal vehicles from accessing the network, control the vehicle resources and jamming the communication channels [11]. DOS can be classified into three levels of attacks which are (1) Basic Level, (2) Extended Level and (3) Distributed Denial of Service (DDOS) Attack [5].
- Distributed Denial of Service Attack This kind of attack is dangerous because it is launched from different locations. Consequently, the impact of this attack is dispersed in the network.
- Broadcast Tampering Through broadcast tampering attack, attacker practically injecting erroneous messages which lead to disturbance to the network. Likewise, this attack is akin to bogus information attacks. But in contrary, broadcast tampering only involving internal attackers [12] [8].
- Malware Malware is abbreviation of malicious software which created in order to contaminate the nodes or the network system in VANET. Malware such like worms or virus could take an action during software exchange or software update likely initiated by the insider [12]
- Spamming This attack is very hard to control due to the missing of fixed infrastructure and centralised administration in VANET. Spam messages could increase the latency in communication; thus, will causing the nodes not to receive the messages on time.
- Black Hole Attack In this type of attack, the malicious node will attracts the victims by advertising itself with a fresh route altogether with low hop count. The attacker is free to reply Route Reply (RREP) packet to victim without concerning its routing table. A false route will be created and sent to the victim during the flooding of Route Request (RREQ) packet in flooding-based routing protocol [13].
- Worm Hole Attack Entangle two attackers; a band wormhole is created when these two attackers are next to each other and advertising themselves to the other nodes that they have the shortest path to the destination. Both of the attackers will placed themselves in the most vital position in the network and forming an overlay tunnel over the wireless medium in order to intercept the communications throughout the network. This is the most preferable attack by the attackers because they can strengthens themselves when throughout the establishment at strong strategic location in VANET [14].
- Greedy Drivers The purpose of greedy drivers is to conquer all the network resources for its own used. They can create havoc by set up forged traffic jammed information so that other nodes will deviated from the attacker path. The attacker wisely will be modified the MAC layer parameters in order to expedite this attack [15].

3.5.4. Attacks on Authentication / Identification

Masquerading To perform masquerading attack, an attacker needed to enter the network and has functioning onboard unit. Attacker can disguised itself as legitimate node and carry out any of attacks. The easiest attack to be launched in a network is masquerading attack [12].

- Replay Attack: By injecting back the received packet into the network, the location table of the node is being poisoned by replaying beacons. This kind of attack is called as replay attack. On the other hand, in order to continuously defended from replay attack, VANET must maintained accurate source of time that used to keep cache of the received messages [12].
- Position Faking It is easier for an attacker to alter its position in a network with unsecured communications. Under that circumstance, accurate authentication and frequent reporting of node positions are required. Through location reporting, masquerading or impersonation is futile. Possibly through position faking in unsecured communication will lead the attacker to falsify their own position, creating additional vehicle identifiers which known as Sybil attack or blocking other nodes from receiving important messages [12].
- GPS Spoofing By making nodes thinking that they are in the different location, nodes are easily fooled by the attackers. This kind of attack can be carried out by producing false reading in the GPS devices. GPS spoofing permits the attackers to generate stronger signal than signal that generated by genuine satellite by using GPS satellite simulator [12]
- Message Tampering In this attack, attackers could modify the messages that being exchanged in the V2V and V2I communications. Message tampering attacks are arising from the vulnerabilities in authentications [12].
- Message Suppression/Fabrication/Alteration There are two ways those attackers can disable itself from responding to any beacons; either physically disabling the inter-vehicle communication or altering the application [12].
- Key and/or Certificate Replication In this case, attackers could erode the system by making several nodes with the same identity. This attack can be carried out by replicating key management and/or certificate. The purpose of the attacker is to bemuse the authorities and forbid the identification of attacker [12].
- Sybil Attack The attack is launched by sending wrong numerous messages to the other node but with different forgery identity of the sender. As a result, for example, other nodes will leave the road in order to ease the attacker to pass through the road freely. In this case, the real identities of the sender are hidden and the attackers created a delusion to the other nodes.
- Node Impersonation During the communication between the sender and receiver, the attacker may modified the message from the originator and send it to the receiver. Therefore, the malicious message may seems like originated from the sender. This problem could be curbing by assigning the unique identifier to each vehicle nodes. Henceforth, the real identity of the message originator could be detected.

3.5.5. Privacy

Identity Revealing The identity and the location of the target nodes are divulging in this attack. The attack is carried by monitoring the target node and sends "virus" to the nearby vehicle node. When the nearby vehicles are overwhelmed with virus will take the ID and the location of the target node.

4. Network Challenges in VANET

The key fundamental challenges in deploying vehicular system in real-life are not an easy task as it needs to address many issues and identified the required solutions. According to [16], the main challenges that encountered by VANET are as following aspects;

4.1. Wireless Access Technology

There are limited technologies available to be used as core base in VANET. Those technologies are included cellular technology; IEEE 802.11p based technology and combined wireless access.

4.2. Spectrum Issues

The FCC in US has allocated 75MHz of spectrum at 5.9GHz for car-to-car and car-to-infrastructure communications. Unfortunately, a continuous spectrum is not available in Europe. After that Car2Car CC has proposed allocations of 2x10 MHz for primary use which is for the safety applications at 5.9 GHz range. Conversely, this band is utilized as control channel in US and it allocation in Europe would grants as world-wide harmonization.

4.3. C. Routing Issues

Three main routing algorithms that available in VANET can be combined with concept 'carry and forward'. Those algorithms are (1) opportunistic forwarding, (2) trajectory based forwarding and (3) geographic forwarding. Furthermore, mixing two or multiple approaches also could be done as hybrid solutions.

4.4. D. Security and Privacy

Security and privacy are two issues that need to be concentrated in developing the solutions. This is because several threats are available that potentially could disrupted the traffic as well as compromising the private information like driver's information.

However, authors in [7] classified the challenges in VANET which sub-divided to (1) Technical Challenges and (2) Social and Economic Challenges. The first aspect is with regard to the technical challenges; there are five key issues that have been discussed as follows;

Article I. A. Network Management

Since one of characteristic of VANET is high mobility, the network topology and channel face rapid changes. Because of that, structures that able to maintain while rapid topology changes occur, should be deployed.

Article II. B. Congestion and Collision Control

The limitless network creates congestion and collision during peak hours since the traffic load is high.

4.5. C. Environmental Impact

In VANET, electromagnetic waves are used in communication. Unfortunately, electromagnetic waves can affect the environment. Thus, the environment should be taken into account before deploying it.

4.6. D. MAC Design

Since VANET are using the shared medium, MAC design become big challenges in VANET.
Article III. E. Security

Security of VANET must be fulfilled since VANET provides road safety applications.

Equally important, the social and economic challenges are being discussed based on the issues on how to convince the manufacturers to build up an application that can reveal the traffic violation. This application additionally will help the user to get information on the road such as police trap. While in [17] and [18], VANET challenges are listed out into (1) Mobility, (2) Volatility, (3) Network Scale and (4) Bootstrap. In addition, authors in [8] added up (5) Privacy vs. Authentication and (6) Privacy vs. Liability as additional challenges. The challenges issued are explained as follow;

A. Mobility

VANET has high mobility due to rapid changes of the topology.

B. Volatility

The connection between vehicles will not last long due to fast movements and changes in the directions of vehicles.

C. Network scale

Since the number of vehicles around the world increasing around to 800 millions, hence, network scalability becomes difficult. Throughout the time, another issue arises due to the absence of global authority that used to manage the standard for the network.

D. Bootstrap

Up until now, only few vehicles are equipped with DSCR. Thus, the communications only limited to a few vehicles and developers must take into account about this issue.

5. Security Challenges in VANET

Security issues in VANET are critical due to vulnerabilities exist during information transmission which causing VANET exposed to the attacks. In order to maintain a secure vehicular communication and networks, VANET security system should satisfy with the requirements. Some of the requirements are essential for all networks, but some are definite for VANET only [17]. Those requirements are;

A. Authentication

In order to allow the communication between vehicles which sending and receiving information, VANET should authenticate each of them. This process may comprise the identification of the sender identity and the legitimacy of the sender to use the network.

B. Availability

Availability is defined as the degree of the VANET system that must be operable and available when needed. A fast response time also must applicable for some applications.

C. Privacy

Privacy is one of the most important requirements in VANET. Privacy must ensure that the identity of the drivers and the location of the vehicles are not being exposed.

D. Integrity

The information exchange in between the sender and the receiver should be free from the alteration attacks. Thus, information can be trusted.

E. Non-repudiation

It ensures that the origin of the information cannot be denying that it has sending the information.

6 Conclusion

In summary, safety is the main attention for the road users. VANET has an ability to provide safety requirements by providing the information on the roads to the users. However, VANET is not immune from any vulnerabilities and threats. Because of that, secure solutions in order to enhance the security of the information in VANET must be deployed. Therefore, it is important to maintain the network availability and develop trust to the information in VANET.

References

[1] A. Agrawal, A. Garg, N. Chaudhiri, S. Gupta, D. Pandey, and T.
Roy, "Security on Vehicular Ad Hoc Networks (VANET): A Review Paper," vol. 3, no. 1, 2013.

[2] C. Kruegel and T. Toth, "A survey on intrusion detection systems," TU Vienna, Austria, vol. 109, no. 1, pp. 6–15, 2000.

[3] O. Pattnaik and B. K. Pattanayak, "SECURITY IN VEHICULAR AD HOC NETWORK," vol. 11, no. 2, pp. 337–346, 2014.

[4] C. Y. Shim, "A taxonomy for DOS attacks in VANET," 2014 14th Int. Symp. Commun. Inf. Technol., pp. 26–27, 2014.

[5] H. Hasbullah, I. A. Soomro, and J. A. Manan, "Denial of Service (
DOS) Attack and Its Possible Solutions in VANET," World Acad. Sci. Eng. Technol., vol. 4, no. 5, pp. 411–415, 2010.

[6] J. M. De Fuentes and A. I. González-tablas, "Overview of security issues in Vehicular Ad-hoc Networks," Science (80-.)., 2010.

[7] R. S. Raw, M. Kumar, and N. Singh, "Security Challenges, Issues and Their Solutions for VANET," Int. J. Netw. Secur. Its Appl., vol. 5, no. 5, pp. 95–105, 2013.

[8] M. Bharat, K. S. Sree, and T. M. Kumar, "Authentication Solution for Security Attacks in VANETs," vol. 3, no. 8, pp. 7661–7664, 2014.

[9] I. A. Sumra, J. A. B. Manan, H. Hasbullah, and B. S. Iskandar,
"Timing Attack in Vehicular Network 2 VANET Applications and Time," pp. 151–155.

[10] J. T. Isaac, S. Zeadally, and J. S. Cámara, "Security attacks and solutions for vehicular ad hoc networks," IET Commun., vol. 4, no. 7, p. 894, 2010.

[11] K. M. A. Alheeti, A. Gruebler, and K. D. Mcdonald-maier, "An Intrusion Detection System Against Malicious Attacks on the
Communication Network of Driverless Cars," IEEE Consum. Commun. Netw. Conf., no. 12, pp. 916–921, 2015.

[12] S. Zeadally, R. Hunt, Y.-S. Chen, A. Irwin, and A. Hassan,
"Vehicular ad hoc networks (VANETS): status, results, and challenges," Telecommun. Syst., vol. 50, no. 4, pp. 217–241, 2012.

[13] V. Bibhu, "Performance Analysis of Black Hole Attack in Vanet," no. October, pp. 47–54, 2012.

[14] S. M. Safi, A. Movaghar, and M. Mohammadizadeh, "A novel approach for avoiding wormhole attacks in VANET," 2nd Int. Work. Comput. Sci. Eng. WCSE 2009, vol. 2, pp. 160–165, 2009.

[15] M. N. Mejri and J. Ben-Othman, "Detecting greedy behavior by linear regression and watchdog in vehicular ad hoc networks,"

Glob. Commun. Conf. (GLOBECOM), 2014 IEEE, no. ii, pp. 5032–5037, 2014.

[16] J. Jakubiak and Y. Koucheryavy, “State of the Art and Research Challenges for VANETs,” 2008 5th IEEE Consum. Commun. Netw. Conf., pp. 912–916, 2008.

[17] S. S. Shinde and S. P. Patil, “Various Issues in Vehicular Ad hoc Networks : A Survey,” vol. 1, no. 2, pp. 399–403, 2010.

[18] G. Samara, W. a. H. Al-Salihy, and R. Sures, “Security Analysis of Vehicular Ad Hoc Nerworks (VANET),” Netw. Appl. Protoc. Serv. (NETAPPS), 2010 Second Int. Conf., pp. 55–60, 2010.

Implementation of Estrus Synchronization and Artificial Insemination Program (GBIB) in West Waringin Kota District, Central Kalimantan Province of Indonesia

Tresia Kristiana[1]*and *Harapin* Hafid[2]

[1]Program Study of Government Science, Faculty of Social and Political Sciences, Kristen University [2]Palangka Raya, Central Kalimantan, Indonesia
[2]Program Study of Animal Science, Faculty of Animal Science, University of Halu Oleo Kendari South East Sulawesi, Indonesia

Abstract. The government has launched a program of self-sufficiency in meat as an effort to overcome the shortage of beef supply to meet the needs of the community, especially when the day of religious festivities. GBIB program Implementation is expected to meet the needs of the community will be beef, as demand for beef is always soaring demand every year, especially ahead of religious festivals. To discuss and review the implementation of the program stretcher author uses the theory EDWARDS III to determine the success or failure of the implementation of the program in the district of Kota Waringin Barat. Issues to be examined are 1) the quality and quantity of resources, 2) communication and dissemination program, 3) Attitude 4) Bureaucracy. This will be the focus to see the successful implementation of the program of snapping / sync Lust and optimize of GBIB in the district of Kota Waringin. This study uses the concept of a qualitative descriptive approach with the aim of providing an overview of how the implementation of the field program and then analyzed using the theory of Edwards III. The results of this study show that: 1) Aspects of Resource (HR, Information, Privileges and amenities), some support as the support means the production of drugs / vitamins and breeding stock (straw) superior adequate and well as the authority and the information available is quite good, while some aspects that do not support such as: HR, execution time, geographical conditions, transport facilities support, 2) aspect Communications (Transmission, Clarity, Consistency and Coordination) strongly supports the implementation of the program GBIB is 3) aspects of attitude (attitude implementer, Support Leader , Incentives for implementing and Transparency), relatively supportive in this program, and 4) Structural Aspects of Bureaucracy (SOP and Fragmentation) still do not support the implementation of the Program snapping Sexual arousal and this Artificial Insemination..

1 Introduction

General policy of agricultural development is to create added value and competitiveness of agriculture by enhancing the role and participation of the agricultural community and the implementation of good organizational management based on the application of science and technology. Of public policy are translated into a technical policy where implementation is organized on the principle of economic unity, so that all elements of the powers that be in synergy with one another.

Law No. 7 of 1996 concerning food security, the fulfillment of food for households which is reflected in the availability of adequate food, both in number and quality, safe, equitable and affordable. Food security can be simplified by focusing on three dimensions of different but interrelated namely food availability (production), access to food by households (distribution), and utilization of food by individuals (consumption).

The situation of national consumption of animal food is still dominated by a relatively high consumption of meat, where meat is a basic requirement of the most important and is considered by the government, either the stock / reserves and its market price. Then to secure food situation should be supported by the population of livestock enough, one of the steps taken by the government, namely the implementation of artificial insemination programs in order to stimulate the birth rate of cattle in order to meet the food needs of the community.

Directorate General of Livestock and Animal Health is a government agency that has the task of formulating and implementing policies and technical standardization in the field of animal husbandry and animal health functions include breeding, feed, livestock farming, animal health and veterinary public health. Challenges in implementing the tasks and functions is not easy in the midst of challenges faced include:
(1) The population growth and the increasing need of the productive age group of animal-based food supply that is

Corresponding Author: tresiakristiana@yahoo.co.id

safe, healthy, whole and kosher (FOSTER) becomes larger;
(2) The transformation of the use of agricultural land to non-agricultural become a limiting factor in increasing production and productivity of the agricultural sector, including livestock;
(3) Competition not only for the use of food commodities (food), feed (feed) and energy (fuel), for example maize;
(4) Sentra producer and consumer centers are located far apart in need of transportation facilities and adequate logistics .;
(5) Synchronization central policy areas as the impact of regional autonomy;
(6) The liberalization of world trade pose a threat of imported products; and
(7) The global climate change.

In the midst of the challenges faced, the Directorate General of Livestock and Animal Health has been and will carry out various activities in order to support the duties and functions of which are: Development of Livestock Breeding, Aquaculture Development Cattle Artificial Insemination (AI). In an effort to increase the population and productivity of cattle and buffaloes that lead to self-sufficiency in beef and buffalo.

Government through the Directorate General of Livestock and Animal Health Ministry of Agriculture in 2015 launched a program through bullying Lust Optimization Birth of Artificial Insemination (GBIB). Kalimantan Tengah into one of the working area of the program with technical companion Livestock Breeding Center for Excellence (BPTU) Pleihari as Technical Implementation Unit (UPT) of the Directorate General of Livestock and Animal Health. Benefits of Artificial Insemination efficiency is more time, where to marry the cattle ranchers no longer need to look for bulls (bull), simply contact inseminator breeders in their area and determine the desired seed.

Acceleration of Population Increased activity by snapping Lust and Artificial Insemination (GBIB) in Central Kalimantan Province Regional TA.2015 will be made with a target of 5000 for snapping tail oestrus in 14 Districts / Municipalities. For West Kotawaringin target for GBIB many as 800 animals or 16% of the target provinces, (Evaluation Report Activity deploy GBIB, BPTU-HMT Pelaihari, 2015)

Sexual arousal bullying programs and Insminasi Artificial (GBIB) is a program for increasing the population increase through estrus synchronization and optimization of Artificial Inseminisasi. GBIB program which is one of the government programs in order to fulfill the needs of meat adequacy society. Sexual arousal bullying program and Artificial Insemination (GBIB) very well be applied in order to help farmers / ranchers in an attempt to create a pregnancy cows simultaneously. GBIB is one right way to further increase the lust of cows and buffaloes to ease the process of mating. That way, potential pregnancy cows to be faster to produce new offspring.

At first in Kotawaringin Barat livestock population improvement program carried out through a process of natural mating. Animal mating process is done by cows are married males / natural. However, it is felt less than optimal because it was feared happened inbreeding (inbreeding), so the program is necessary to prevent GBIB descended from the male parent and incest (in breeding) because it can produce offspring that are less good his performance.

Artificial Insemination Program (IB) alone in West Kotawaringin have dilaksanan since the period of 1990 but lack of government financial support. Artificial Insemination (AI) has aggressively implemented around the period of 2010 until now. The problems studied were: 1) How to program implementation snapping Sexual arousal and Artificial Insemination (GBIB) in Kalimantan Tengah Kotawaringin Barat?. 2) How is the target and the realization of snapping Sexual arousal and Artificial Insemination (GBIB) in Kotawaringin Barat Provinsi Kalimantan Tengah Indonesia?. The benefits of this research are expected to provide feedback / input to the Regional Government of Kota waringin Barat especially Agriculture and Livestock Kota waringin Barat towards the implementation of the Program snapping Sexual arousal and Artificial Insemination and can make a positive contribution in order to achieve food needs in Kotawaringin Barat and follow-up of GBIB program in the future.

2 Methodology

This study used a qualitative approach, where data shown generally shaped descriptions and sentences is a description of the factual and accurate, as well as the relationships among the issues examined in a qualitative approach, the researchers went to the field to examine the object of its studies and hold a direct interaction with society aims to get in-depth information about the professionalism including obstacles faced and the efforts made to overcome them. This is based on qualitative research purposes to generate descriptive data in the form of words written or spoken of people and behaviours that can be observed [5].

The data collection is done by observation, interview (Interview) and documentation. The analysis began to formulate and explain the problem, before plunging into the field, and continue until the writing of the research results. The data have been collected later clarified to be analyzed using inductive analysis approach. Furthermore, using data analysis developed by Miles and Huberman, with three types of activities, namely; data reduction, data presentation, and conclusion / verification.

3 Research Result

The number of cattle population in Kalimantan Tengah is still far from sufficient to meet the needs of people of Kalimantan Tengah meat. Overview cattle population in Kalimantan Tengah can be seen from Table 1 below.

Table 1. Cattle Population development and buffalo. Fixed figures (ATAP) in 2015 and a Preliminary figures (ASEM) in 2016 in the province of Central Kalimantan.

No	Kabupaten/ Kota		Populasi					
ATAP 2015	Jumlah 2015	Populasi						
ASEM 2015	Jumlah 2016							
		Sapi	Kerbau		Sapi	Kerbau		
1	Kotawaringin Barat		14.397	0	14.397	16.933	0	16.933
2	Kotawaringin Timur		6.948	192	7.140	7.203	211	7.414
3	Kapuas	3.076	14	3.090	3.200	20	3.220	
4	Barito Selatan	1.125	9.773	10.898	1.150	10.202	11.352	
5	Barito Utara	3.080	135	3.215	3.125	146	3.271	
6	Sukamara	2.794	0	2.794	2.934	0	2.923	
7	Lamandau	3.816	0	3.816	3.925	0	3.925	
8	Seruyan	7.767	6	7.773	8.155	0	8.155	
9	Katingan	7.894	331	8.225	8.829	368	8.657	
10	Pulang Pisau	8.825	15	8.840	9.293	15	9.308	
11	Gunung Mas	4.034	141	4.175	4.100	150	4.250	
12	Barito Timur	1.261	789	2.050	1.324	1.093	2.417	
13	Murung Raya		1.721	8	1.729	1.763	10	1.773
14	Kota Palangka Raya		1.793	13	1.806	1.999	15	2.014
	Kalteng	68.531	11.417	79.948	73.393	12.230	85.623	

Kota Waringin Barat is the region with the highest number of cattle population throughout Kalimantan Tengah. But the production and productivity of livestock still cannot meet local demand. Various efforts and programs have been undertaken to stimulate the growth of birth rate among other Cows Integration Oil, Utilization ex. mine for cattle, cows integration with agricultural crops and Artificial Insemination program / GBIB, [13].

Program snapping Sexual arousal and Artificial Insemination (GBIB) initiated by the central government through the Ministry of Agriculture in this case the Directorate General of Livestock and Animal Health intends to accelerate the increase in population through a snapping / synchronization of oestrus and optimization of Artificial Insemination (GBIB) as well as the prevention of reproductive disorders in cattle / buffalo. Sexual arousal bullying program and Artificial Insemination (GBIB) aims to: 1) Optimizing the implementation of the IB; 2). lust in an effort to accelerate the birth; 3). Increased acceptor IB; 4). expand the range of services IB; 5). Settlement of the case of reproductive disorders 6) establishment of livestock, especially cattle population increase in the target areas of activity, 7) ensuring adequate support in the provision of livestock in the framework of the national meat self-sufficiency and 4) increasing prosperity and income of farmers / ranchers with additional births superior cattle. To support the implementation of the program GBIB necessary target data as in the Table 2.

Table 2. Program targets bullying Sexual arousal and Artificial Insemination by decree of the Head of Agriculture and Livestock of Central Kalimantan Number: 313 / Kpts / Nak-400 / IV / 2015 dated April 16, 2015.

No	Kabupaten Kota	Populasi (ekor)	Betina Produktif (ekor)		Target Akseptor (ekor)		Gertak Birahi IB (ekor)	Regular IB (ekor)
1	Kotawaringin Barat		12.554	6.193	4.440	800	600	
2	Barito Utara	2.592	1.189	1.009	350	300		
3	Kapuas	3.655	947	900	400	350		
4	Pulang Pisau	8.350	4.500	3.400	1.000	700		
5	Barito Selatan	868	363	203	50	25		
6	Katingan	8.081	2.257	1.750	750	500		
7	Kotawaringin Timur		6.872	2.787	2.000	750	800	
8	Palangka Raya	1.725	842	550	200	800		
9	Sukamara	2.687	1.901	300	100	250		
10	Seruyan	7.314	3.400	1.500	400	100		
11	Lamandau	2.917	817	700	200	600		
12	Gunung Mas	3.865	950	350	150	150		
13	Murung Raya	1.039	411	350	100	250		
14	Barito Timur	1.787	859	350	200	300		
Jumlah	64.046	27.217	17.773	5.700	5.325			

From the above table it is known that out of the fourteen districts in the Province of Kalimantan Tengah, Kota waringin Barat received an allocation of snapping Lust IB 800 and IB Regular tail 600 tail. The allocation is distributed according to ability and absorption area per district. Six districts targeted GBIB program is Kumai, South Arut, North Arut, Pepper Base, Base Bull, and Kota waringin. Regular IB Program is a program implemented regularly served by the clerk inseminator throughout the year ranging from the beginning of the year until the end of the year.

Sexual arousal snarled IB GBIB done for their program. GBIB program for 2015 has just been implemented planning beginning of September 2015 until the end of 2015 (4 months). GBIB program allocated to target farmers who are not served by regular IB activities. GBIB Program Results can be seen in the following Table 3.

Table 3. The realization results of implementation activities GBIB 2015 in Kalimantan.

Realization Results of Implementation Activities GBIB and Gangrep 2015 in Kalimantan

No	Kabupaten/Kota		Target GBIB (ekor)		Jenis Kegiatan							
			Selakai (ekor) Penyuntikan hormon (ekor) IB (ekor)									
				I	II	III	Jumlah	I	II	III	Jumlah	
1	Kotawaringin Barat		800	560	560	74		634	560			560
2	Barito Utara	350	179	179	37		216	176			176	
3	Kapuas	400	456	456	164		610	456			456	
4	Pulang Pisau	700	364	364			364	364			364	
5	Barito Selatan50		50	50	12		62	35	13		48	
6	Katingan	750										
7	Kotawaringin Timur		750	283	283	176		459	284	7		291
8	Palangka Raya200		52	52	32		84	52			52	
9	Sukamara	100	46	46			46	46			46	
10	Seruyan	300	48	48	17		65	48	2		50	
11	Lamandau	200	273				273	271			271	
12	Gunung Mas	150	66	66			66	66			66	
13	Murung Raya	100	102	102	16		118	102			102	
14	Barito Timur	200	121	124			124	124			124	
Jumlah	5.000	2.808	2.811	518	0	3.392	2.710	22	0	2.732		

GBIB target achievement in West Kotawaringin highest throughout Kalimantan Tengah although it cannot reach 100% of the specified target. GBIB of the province are set to Kota waringin Barat is 800 breeding. Target achievement during the period of 4 (four) months produce a pregnancy rate of 560 head or about 80%. If the program is implemented for a period of 1 (one) year results will be three times that of the achievements of current or tail 560 X 3 = 1680 tail pregnancy. If the tail is calculated from 1680, mean performance target of 800 animals of about 200%. So if the government program was implemented at the beginning of the year until the end of the outcomes are 1.680 tail mean exceeded the target set at 800 animals.

Economically if in view the realization of the achievement of the program as much as 560 tails GBIB calf's birth, if the selling price is calculated derivative / calf superior to average around Rp. 6,000,000 / tail then the impact of the economic value created from GBIB program in the community is about 560 tails X Rp. 6,000,000 = Rp. 3.36 billion (three billion three hundred and sixty million). Comparing the results of IB derivatives with natural mating process, the selling price higher derivatives IB results because of a genetic child produced superior results than children. The average difference obtained difference Rp.2.000.000 field, - s / d 2.500.000, - / tail, higher selling prices calf IB results.

Sexual arousal bullying program and Artificial Insemination (GBIB) through decision of the Directorate General of Animal Husbandry and Health No. 209 / KPTS / OT.160 / F / 03/2015 ON Steering And Implementing Activity Team to Accelerate Improved snapping population through Birahi And Optimization of Artificial Insemination (GBIB) 2015 revised budget funds are expected to increase the population of hearts Frame Capable of supporting an increase in the Regional Food Security. In the West Kotawaringin, cultivation Cattle Enough Operating long cultivated traditional, Livestock Just released dikebun or dilahan and NOT require expensive Investment and labor, while cattle require intensive maintenance Patterns a More Attention from peternaknya. Thus requiring additional cost between others includes power for feed, cages Management and Others.

- Analysis Model Program by Edwards III

The benefits of a program will be felt if the program can be implemented properly. Implementation of a program is a dynamic process that includes the interaction of many factors.

- Theoretically, according to Edwards III

There are four variables that influence the implementation of the program, namely communication, resources, attitudes and bureaucratic structure.

3.1 Communications (Communications)

According to Edwards III, there are three important things that can be discussed in the communication process program, namely the transmission, clarity (clarity) and consistency.

3.1.1 Transmission

In connection with the Program snapping Birahi Artificial Insemination (GBIB) in West Kota waringin and based on interviews with the Head of Development of Livestock Production Department of Agriculture and Animal Husbandry West Kota waringin note that the implementation of the program in West Kota waringin already know and understand what to do in connection with the implementation Sexual arousal snarled Artificial Insemination program (GBIB) this.

3.1.2 Clarity (Clarity)

Clarity of information about the aims and objectives snapping lust artificial insemination (GBIB), based on the observation of rules and regulations regarding the implementation of this program has been described in detail through the General Guidelines for the

https://doi.org/10.1051/matecconf/201815006039

Implementation of bullying Sexual arousal and Artificial Insemination (GBIB), Technical Guidelines Program snapping Sexual arousal and Insemination artificial (GBIB) as well as the decision of the Directorate General of Animal Husbandry and Health No. 209 / KPTS / OT.160 / F / 03/2015 on the Steering Committee and Project Implementation Acceleration Population Increased by snapping Sexual arousal and Optimization of artificial Insemination (GBIB) 2015 revised budget funds. Agriculture and Livestock of Central Kalimantan Province, Permentan, GBIB General Guidelines and Technical Guidance has been submitted to the Department of Agriculture and Livestock

Support Regulations / Legal Foundation Recipient Regulation

1. General Guidelines for the Implementation of bullying Sexual arousal and Optimization of Artificial Insemination (GBIB) 2015
2. Technical Guide Budget Management Tasks Directorate General of Livestock and Animal Health had been
3. Technical Guidelines snapping Activity Sexual arousal and Optimization of Artificial Insemination (GBIB) 2015
4. Already decision of the Directorate General of Animal Husbandry and Health No. 209 / KPTS / OT.160 / F / 03/2015 on the Steering Committee and Project Implementation Acceleration Population Increased by snapping Sexual arousal and Optimization of Artificial Insemination (GBIB) funds APBNP 2015

Source: Dinas Pertanian dan Pertenakan Kabupaten Kota Waringin Barat.

3.2 Resource

3.2.1 Human Resources

The resources in the implementation of programs in the district of Kota Waringin GBIB West still less support is the Human Resources (HR) in the Livestock Production Development Division amounted to 15 (fifteen) people and has included Head of Division and Head of Section. There should be additional staff, especially staff parts of the field as Agricultural Extension Officers (PPL). For inseminator officers have formed an independent and spontaneous inseminator no division but due to spontaneous inseminator officer status, so as to coordinate official considerably less flexible. PPL's position is under the Office of Agricultural Extension and Food Security, when PPL entered into the Organization Department of Agriculture and Animal Husbandry then expected to add more strength to the Department of Agriculture and Animal Husbandry field.

3.2.2 Facilities

Facilities provided include the provision of seeds / straw superior cows of different races / nations veal obtained from the national superior male sperm to adjust to the circumstances and demands of the farmers on the location of activities. It is absolutely necessary given the circumstances and the situation of livestock breeders who are expecting are married injection / IB with superior seedlings. The facilities provided in the form of the hormone prostaglandin 2 alpha, for the implementation of estrus synchronization in order to obtain some cows were estrus IB simultaneously and in unison. Provision of Medicines and vitamins are the initial stimulus for farmers who have become acceptors GBIB and applications performed by veterinariansadan paramedic's trained animal health.

3.2.3 Attitude

The attitude of the public in this case farmers strongly supports the program GBIB, seen from the success made by the inseminator which has been designated by the Department of Agriculture and Animal Husbandry Kota waringin Barat, although implementation in just four (4) months showed a pretty good result, 800 tail which is targeted to achieve the realization of as many as 560 cows pregnant. The involvement of various parties, especially from the Agriculture and Livestock Kota waringin Barat show a good attitude in the implementation of bullying / lust Insemination (GBIB).

3.2.4 The bureaucratic structure

Bureaucratic structure which is owned not enough to support the implementation of the Program snapping Sexual arousal and Artificial Insemination (GBIB) is in terms of the SOP and fragmentation factor. In the conduct of the Program snapping Sexual arousal and Artificial Insemination (GBIB) The SOP and the section that deals specifically with this program yet, which resulted in the duties and authorities of each section is less clear and this program is less able to run with the maximum because the workload is greater submitted the Head of Development of Livestock Production.

4 Conclusion

1. Implementation Program GBIB implemented in the district of West Kota Waringin has been executed, communication goes well, sosialisasi implemented so understanding of the technical and operational guidelines to understand the program's implementation GBIB inseminator who have by the Agriculture and Livestock District Municipality, Resources limited human does not hinder the implementation of the program, support the availability of adequate facilities with medicines and other facilities. Bureaucracy is limited because the implementing specialized field of structural deal is not yet available.
2. Despite the achievements did not meet the target of 800 (eight hundred) cows for the Fiscal Year 2015, but when seen from the results 560 (five hundred

sixty) tail which eventuated in a period of four (4) months showed remarkable success, because it will bring social and economic impact for the community, and will have an impact on government policy in fulfilling the needs / stock of meat, especially the need for the feast.

5 Suggestion

1. The successful implementation of the program GBIB already carried out by the City District Government of Kota Waringin Barat should proceed, the local government should support the policies of the central government, to create an advanced form of Perbub policy or legislation as a follow-up of Policy Forms that have been made by the central government.
2. The City District Government of Kota Waringin Barat through the Agriculture and Livestock should improve services to farmers, especially in the implementation of the program implementation unit GBIB there should be special handling in order to avoid overlapping execution of tasks. Power inseminator should be increased in order to reach the waitress to farmers in the district of Kota Waringin Barat can be evenly distributed, not only in certain regions.

References

1. S. Abdul Wahab, Analysis of Public Policy. Publishing Unit University of Malang. UNM Press Malang (1997)
2. L. Agustino, *Fundamentals public Alfabeta Policy* (Bandung, 2012)
3. S. Arikunto, *Research Procedure a Practice Approach*. (Rineka Reserved. Jakarta, 2002)
4. Badjuri, Abdulkahar, Yuwono, Teguh, *Public Policy: Concepts and Strategy* (Semarang, MPS UNDIP, 2002)
5. Bogdan, Taylor, 1975 in J. Moleong, Lexy. 1989. Metodologi Qualitative Research. Bandung: Remadja work.
6. BPS, Profile of Districts: Figures Kobar in Pangkalan Bun (2014)
7. N.W. Dunn, Introduction to Public Policy Analysis. (Translation) Yogyakarta, GadjahMada University Press (1999)
8. K. Dwiyanto, Agricultural Innovation Development **1** (3): 173-188 (2008)
9. D. Hastuti, S. Nurtini, R. Widiati, Socio-Economic Assessment Implementation Cattle 8. 10. Artificial Insemination in Kebumen. Media Agro **4**: 1-12 (2008)
11. T. Herath, A. Anggraeni, L. Praharani, D. Utami, A. Argiris. Informatics Agriculture **21**: 81-88 (2012)
12. Howlett, Michael, M. Ramesh, "Studying Public Policy; Policy Cycles and Policy subsystems "Oxford, Oxford University Press (1995)
13. Reports of Agriculture and Livestock Livestock Statistics 2014 Kota waringin Barat.
GBIB Implementation Report 2015, BBPT-HMT Pelaihari, 2015
14. H. Mintzberg, The Manger, s Job. Academy, ed. NY. (1990)
15. L. J. Moleong, Metodologi Qualitative Research. Rosdakarya. Duo (2012)
16 A.R Mustopadidjaja, Policy Process Management, Jakarta, Institute of Public Administration (1984)
17. H. Nawawi, Methods of Social Affairs, Yogyakarta: GadjahMadaPress (1993)
18. M. Nazir, Methods (Jakarta: Ghalia Indonesia, 1988)
19. D. Conscience, Journal Fisip UI (2009)
20. Regulation of 72 of 2012 on Public Service Standard Service Field Agriculture.
21. PRSP. 2011. Final Results Release Livestock Census 2011. Kementan - BPS, Jakarta.
22. W.J.S. Poerwadarminta, *The Indonesian General Dictionary* (Jakarta: Balai Pustaka, 1985)
23. Quade, S. Steward, *Analysis for Public Decisions* (New York elsevior Science Publishing Company, 1982)
24. unggono, Bambang, "Based Dynamic wisdom to the Public Policy Analisys. Yogyakarta: Gava Media (1994)
25. Suwitri, Anailisis of Public Policy Implementation. Jakarta: Student Library (2010)
26. H. Sutardjo, Artificial Insemination Service Quality on the DVO Malang. Graduate Program, University of GadjahMada MAP (2003)
27. R. M. Toelihere, Reproduction in Animal Physiology. Space. Bandung.1981b. Artificial insemination in cattle. Bandung: Angkasa (1981)
28. Law of the Republic of Indonesia Number 7 Year 1996 on Food.
29. Winarno, Budi, *Theory and Public Policy Process* (Yogyakarta, Media Pressindo 2002)

AUTHOR INDEX

AUTHOR INDEX

AUTHOR INDEX

AUTHOR INDEX

AUTHOR INDEX

AUTHOR INDEX